Power System Dynamics

Power System Dynamics

Stability and Control

Third Edition

Jan Machowski
Faculty of Electrical Engineering
Warsaw University of Technology
Poland

Zbigniew Lubosny
Faculty of Electrical and Control Engineering
Gdansk University of Technology
Poland

Janusz W. Bialek
formerly Durham University and Edinburgh University
UK

James R. Bumby
formerly Durham University
UK

WILEY

This edition first published 2020

Edition History
Wiley-Blackwell (1e, 2008)
John Wiley & Sons (2e, 2012)

Registered Offices
John Wiley & Sons, Inc., 111 River Street, Hoboken, NJ 07030, USA
John Wiley & Sons Ltd, The Atrium, Southern Gate, Chichester, West Sussex, PO19 8SQ, UK

Editorial Office
The Atrium, Southern Gate, Chichester, West Sussex, PO19 8SQ, UK

For details of our global editorial offices, customer services, and more information about Wiley products visit us at www.wiley.com.

Wiley also publishes its books in a variety of electronic formats and by print-on-demand. Some content that appears in standard print versions of this book may not be available in other formats.

Library of Congress Cataloging-in-Publication Data

Names: Machowski, Jan, author. | Lubosny, Zbigniew, author. |
Bialek, Janusz W., author. | Bumby, James R., author.
Title: Power system dynamics : stability and control /
Jan Machowski, Faculty of Electrical Engineering, Warsaw University of Technology, Poland,
Zbigniew Lubosny, Faculty of Electrical and Control Engineering, Gdansk University of Technology, Poland,
Janusz W. Bialek formerly Durham University and Edinburgh University, UK,
James R. Bumby formerly Durham University, UK.
Description: Third edition. | Hoboken, NJ, USA : John Wiley, 2020. |
Includes bibliographical references and index.
Identifiers: LCCN 2019032080 (print) | LCCN 2019032081 (ebook) |
ISBN 9781119526346 (hardback) | ISBN 9781119526384 (adobe pdf) |
ISBN 9781119526360 (epub)
Subjects: LCSH: Electric power system stability. | Electric power systems–Control.
Classification: LCC TK1010 .M33 2020 (print) | LCC TK1010 (ebook) |
DDC 621.319/1–dc23
LC record available at https://lccn.loc.gov/2019032080
LC ebook record available at https://lccn.loc.gov/2019032081

Cover Design: Wiley
Cover Images: (Power system diagram) courtesy of Jan Machowski, (abstract wave) © oxygen/Getty Images

Set in 9.5/12.5pt STIXTwoText by SPi Global, Pondicherry, India

Printed and bound by CPI Group (UK) Ltd, Croydon, CR0 4YY

10 9 8 7 6 5 4 3 2 1

Contents

Part III
Advanced Topics in Power System Dynamics 467

About the Authors

Professor Jan Machowski received his MSc and PhD degrees in Electrical Engineering from Warsaw University of Technology in 1974 and 1979, respectively. After obtaining field experience in the Dispatching Center and several power plants, he joined the Electrical Faculty of Warsaw University of Technology, where presently he is employed as a Professor. His areas of interest are electrical power systems, power system protection, and control.

In 1989–1993 Professor Machowski was a Visiting Professor at Kaiserslautern University in Germany, where he carried out two research projects on power swing blocking algorithms for distance protection and the optimal control of FACTS devices.

In 2005–2016 Professor Machowski was Director of Power Engineering Institute at Warsaw University of Technology.

Professor Machowski is the co-author of three books published in Polish: *Power System Stability* (WNT, 1989), *Short Circuits in Power Systems* (WNT, 2008), and *Control of Electric Power Systems* (WPW, 2017). He is also a co-author of *Power System Dynamics and Stability* (John Wiley & Sons, 1997) and *Power System Dynamics: Stability and Control* (John Wiley & Sons, 2008).

Professor Machowski is the author and co-author of over 60 papers published in international fora. He has carried out many projects on electrical power systems, power system stability, and power system protection commissioned by the Polish Power Grid Company, Electric Power Research Institute in the United States, Electroinštitut Milan Vidmar in Slovenia and the Ministry of Science and Education of Poland.

Professor Zbigniew Lubosny received his MSc and PhD degrees in Electrical Engineering from Gdansk University of Technology in 1985 and 1991, respectively, and an MSc in Management also from Gdansk University of Technology in 1987. From 1985 to 2004 he was a lecturer with Gdansk University of Technology. In 2004–2006 he worked at the University of Edinburgh in the UK. In 2006 he again joined Gdansk University of Technology, Faculty of Electrical and Control Engineering, where presently he is employed as a Professor. His areas of interest are electrical power systems, renewable energy systems, and power systems' protection and control.

In 2001 he was EU Marie Curie Individual Fellow in Otto-von-Guericke University Magdeburg in Germany, where he carried out a research project on the influence of renewable energy sources on electric power systems.

Professor Lubosny is the author of a monograph published in English: *Self-organising Controllers of Generating Unit in Electric Power System* (GUT, 1999), the author of a book published in English: *Wind Turbine Operation in Electric Power Systems: Advanced Modeling* (Springer-Verlag, 2003), the author of three books published in Polish: *Wind Turbine Operation in Electric Power Systems* (WNT, 2006), *Wind Farms in the Electric Power System* (WNT, 2009), *Electrical Protection of Wind Farms* (WNT, 2014), and the co-author of a book published in Polish: *Power System Stability* (PWN, 2018).

Professor Lubosny is the author and co-author of over 100 research papers. He has carried out many projects on the protection and control of power systems commissioned by the Polish Power Grid Company, Distribution System Operator Energa SA (EU Horizon 2020 Project), Psymetrix Ltd, UK, the Ministry of Science and Higher Education and other power companies in Poland.

Professor Lubosny is a member of the presidium of the Committee on Electrical Engineering of the Polish Academy of Science, a member of IFAC Power Plants and Power Control Committee, and an editor of the scientific journal *Acta Energetica*.

Professor Janusz W. Bialek received his MEng and PhD degrees in Electrical Engineering from Warsaw University of Technology in 1977 and 1981, respectively. From 1981 to 1989 he was a lecturer at Warsaw University of Technology. After moving to the UK in 1989, he held Chair Professor positions at the University of Edinburgh (2003–2009), Durham University (2009–2014), and since 2019 at Newcastle University. In 2014–2019 he was the Director of Center for Energy Systems at the newly formed Skolkovo Institute of Science and Technology (Skoltech) in Russia. He is a Fellow of the Institute of Electrical and Electronics Engineers (IEEE).

Professor Bialek's main research interest is in the application of advanced mathematical methods to power system analysis. He has published widely on power system dynamics, smart grids, preventing electricity blackouts, power markets, and the technical and economic integration of renewable generation in power systems. He has been the Principal Investigator of a number of major research grants funded by the UK's Engineering and Physical Sciences Research Council (EPSRC) and a consultant to the UK government, International Energy Agency, European Commission, Scottish government, and a number of power companies. Since 2015, he has also been the non-executive independent director of KEGOC (Kazakhstan Electricity Grid Operating Company).

Dr. James R. Bumby received his BSc and PhD degrees in Engineering from Durham University in 1970 and 1974, respectively. From 1973 to 1978 he worked for the International Research and Development Company, Newcastle-upon-Tyne, on superconducting machines, hybrid vehicles, and sea-wave energy. He joined the Engineering Science Department (now School of Engineering) at Durham University in 1978, where he worked as a Lecturer, Senior Lecturer, and Reader in Electrical Engineering until his retirement in 2009. He worked in the general area of electrical machines and systems for over 35 years, first in industry and then in academia.

Dr. Bumby is the author or co-author of over 100 technical papers and three books in the general area of electrical machines and power systems and control. He has also written numerous technical reports for industrial clients. He published his first book, a research monograph, *Superconducting Rotating Electrical Machines* in 1983 (Oxford University Press). He is a co-author of *Power System Dynamics and Stability* (John Wiley & Sons, 1997) and *Power System Dynamics: Stability and Control* (John Wiley & Sons, 2008). These papers and books have led to the award of a number of national and international prizes, including the Institute of Measurement and Control prize for the best transactions paper in 1988 for work on hybrid electric vehicle drive systems and the IEE Power Division Premium in 1997 for work on direct drive permanent magnet generators for wind turbine applications.

List of Symbols & Abbreviations

Notation

Italic type denotes scalar physical quantity (e.g. R, L, C) or numerical variable (e.g. x, y).
Phasor or complex quantity or numerical variable is underlined (e.g. $\underline{I}$, $\underline{V}$, $\underline{S}$).
Italic with arrow on top of a symbol denotes a spatial vector (e.g. $\vec{F}$).
Italic boldface denotes a matrix or a vector (e.g. $\boldsymbol{A}$, $\boldsymbol{B}$, $\boldsymbol{x}$, $\boldsymbol{y}$).
Unit symbols are written using roman type (e.g. Hz, A, kV).
Standard mathematical function are written using roman type (e.g. sin, cos).
Numbers are written using roman type (e.g. 5, 6).
Symbols representing numbers are written using roman type (e.g. π, e = 2.718282 – John Napier number, j – angular shift by 90°, a – angular shift by 120°).
Matrix transposition T is written using roman type.
Differential and partial differential coefficients are written using roman type (e.g. $\frac{\mathrm{d}f}{\mathrm{d}x}, \frac{\partial f}{\partial x}$).
Symbols describing objects are written using roman type (e.g. TRAFO, LINE).
Subscript relating to objects is written using roman type (e.g. $\underline{I}_{\mathrm{TRAFO}}$, $\underline{I}_{\mathrm{LINE}}$).
Subscript relating to physical quantity or numerical variable is written using italic type (e.g. A_{ij}, x_k).
Subscripts A, B, C refer to three-phase axes of a generator.
Subscripts d, q refer to the direct- (d-axis) and quadrature-axis (q-axis) components.
Lower case symbols normally denote instantaneous values (e.g. v, i).
Upper case symbols normally denote RMS or peak values (e.g. V, I).

Symbols

a, a^2	operators shifting the angle by 120° and 240°, respectively.
B_{μ}	magnetizing susceptance of a transformer.
B_{sh}	susceptance of a shunt element.
D	damping coefficient.
E_{k}	kinetic energy of the rotor relative to the synchronous speed.
E_{p}	potential energy of the rotor with respect to the equilibrium point.
e_{f}	field voltage referred to the fictitious q-axis armature coil.
e_{q}	steady-state emf induced in the fictitious q-axis armature coil proportional to the field winding self-flux linkages.
e'_{d}	transient emf induced in the fictitious d-axis armature coil proportional to the flux linkages of the q-axis coil representing the solid steel rotor body (round-rotor generators only).

e'_q	transient emf induced in the fictitious q-axis armature coil proportional to the field winding flux linkages.
e''_d	subtransient emf induced in the fictitious d-axis armature coil proportional to the total q-axis rotor flux linkages (q-axis damper winding and q-axis solid steel rotor body).
e''_q	subtransient emf induced in the fictitious q-axis armature coil proportional to the total d-axis rotor flux linkages (d-axis damper winding and field winding).
$\underline{E}$	steady-state internal emf.
$\underline{E}_f$	excitation emf proportional to the excitation voltage V_f.
E_{fm}	peak value of the excitation emf.
E_d	d-axis component of the steady-state internal emf proportional to the rotor self-linkages due to currents induced in the q-axis solid steel rotor body (round-rotor generators only).
E_q	q-axis component of the steady-state internal emf proportional to the field winding self-flux linkages (i.e. proportional to the field current itself).
$\underline{E}'$	transient internal emf proportional to the flux linkages of the field winding and solid steel rotor body (includes armature reaction).
E'_d	d-axis component of the transient internal emf proportional to flux linkages in the q-axis solid steel rotor body (round-rotor generators only).
E'_q	q-axis component of the transient internal emf proportional to the field winding flux linkages.
$\underline{E}''$	subtransient internal emf proportional to the total rotor flux linkages (includes armature reaction).
E''_d	d-axis component of the subtransient internal emf proportional to the total flux linkages in the q-axis damper winding and q-axis solid steel rotor body.
E''_q	q-axis component of the subtransient internal emf proportional to the total flux linkages in the d-axis damper winding and the field winding.
$\underline{E}_r$	resultant air-gap emf.
E_{rm}	amplitude of the resultant air-gap emf.
$\boldsymbol{E}_G$	vector of the generator emf's.
f	mains frequency.
f_n	rated frequency.
$\vec{F}$	magnetomotive force (mmf) due to the field winding.
$\vec{F}_a$	armature reaction mmf.
$F_{a\ ac}$	ac armature reaction mmf (rotating).
$F_{a\ dc}$	dc armature reaction mmf (stationary).
$\vec{F}_{ad}$, $\vec{F}_{aq}$	d- and q-axis components of the armature reaction mmf.
$\vec{F}_f$	resultant mmf.
G_{Fe}	core loss conductance of a transformer.
G_{sh}	conductance of a shunt element.
H_{ii}, H_{ij}	self and mutual synchronizing power.
i_A, i_B, i_C	instantaneous currents in phases A, B, and C.
$i_{A\ dc}$, $i_{B\ dc}$, $i_{C\ dc}$	DC component of the current in phases A, B, and C.
$i_{A\ ac}$, $i_{B\ ac}$, $i_{C\ ac}$	ac component of the current in phases A, B, and C.
i_d, i_q	currents flowing in the fictitious d- and q-axis armature coils.
i_D, i_Q	instantaneous d- and q-axis damper winding current.
i_f	instantaneous field current of a generator.

$\boldsymbol{i}_{\mathrm{ABC}}$	vector of instantaneous phase currents.
$\boldsymbol{i}_{\mathrm{fDQ}}$	vector of instantaneous currents in the field winding and the d- and q-axis damper windings.
$\boldsymbol{i}_{\mathrm{0dq}}$	vector of armature currents in the rotor reference frame.
$\underline{I}$	armature current.
$\underline{I}_{\mathrm{d}}$, $\underline{I}_{\mathrm{q}}$	d- and q-axis component of the armature current.
$\underline{I}_{\mathrm{S}}$, $\underline{I}_{\mathrm{R}}$	currents at the sending and receiving end of a transmission line.
$\underline{\boldsymbol{I}}_{\mathrm{R}}$, $\underline{\boldsymbol{I}}_{\mathrm{E}}$	vector of complex current injections to the retained and eliminated nodes.
$\underline{\boldsymbol{I}}_{\mathrm{G}}$, $\underline{\boldsymbol{I}}_{\mathrm{L}}$	vector of complex generator and load currents.
$\Delta\underline{\boldsymbol{I}}_{\mathrm{L}}$	vector of load corrective complex currents.
J	moment of inertia.
j	operator shifting the angle by 90°.
k_{PV}, k_{QV}	voltage sensitivities of the load (the slopes of the real and reactive power demand characteristics as a function of voltage).
k_{Pf}, k_{Qf}	frequency sensitivities of the load (the slopes of the real and reactive power demand characteristics as a function of frequency).
$K_{E_{\mathrm{q}}}$	steady-state synchronizing power coefficient (the slope of the steady-state power angle curve $P_{E_{\mathrm{q}}}(\delta)$).
$K_{E'_{\mathrm{q}}}$	transient synchronizing power coefficient (the slope of the transient power angle curve $P_{E'_{\mathrm{q}}}(\delta')$).
$K_{E'}$	transient synchronizing power coefficient (the slope of the transient power angle curve $P_{E'}(\delta')$).
K_i	reciprocal of droop for the i-th generating unit.
K_{L}	frequency sensitivity coefficient of the system real power demand.
K_{T}	reciprocal of droop for the total system generation characteristic.
l	length of a transmission line.
L_{AA}, L_{BB}, L_{CC}, L_{ff}, L_{DD}, L_{QQ}	self-inductances of the windings of the phase windings A, B, and C; the field winding; and the d-and the q-axis damper winding.
L_{d}, L_{q}	inductances of the fictitious d- and q-axis armature windings.
L'_{d}, L'_{q}, L''_{d}, L''_{q}	d- and q-axis transient and subtransient inductances.
L_{xy}	where x, y∈{A, B, C, D, Q, f} and x ≠ y are the mutual inductances between the windings denoted by the indices as described above.
L_{S}	minimum value of the self-inductance of a phase winding.
ΔL_{S}	amplitude of the variable part of the self-inductance of a phase winding.
$\boldsymbol{L}_{\mathrm{R}}$	submatrix of the rotor self- and mutual inductances.
$\boldsymbol{L}_{\mathrm{S}}$	submatrix of the stator self- and mutual inductances.
$\boldsymbol{L}_{\mathrm{SR}}$, $\boldsymbol{L}_{\mathrm{RS}}$	submatrices of the stator-to-rotor and rotor-to-stator mutual inductances.
M	coefficient of inertia.
M_{f}, M_{D}, M_{Q}	amplitude of the mutual inductance between a phase winding and, respectively, the field winding and the d- and q-axis damper winding.
N	generally, number of turns of a winding.
p	number of poles.
P, Q	active (real) and reactive power, respectively.
P_{acc}	accelerating power.
P_{D}	damping power.
P_{e}	electromagnetic air-gap power.

$P_{E_q\ cr}$	critical (pull-out) air-gap power developed by a generator.
$P_{E_q}(\delta)$, $P_{E'}(\delta')$, $P_{E'_q}(\delta')$	air-gap power curves assuming E_q = constant, E' = constant, and E'_q = constant.
P_g	in induction machine (Chapter 7), real power supplied from the grid (motoring mode) or supplied to the grid (generating mode).
P_L	real power absorbed by a load or total system load.
P_m	mechanical power supplied by a prime mover to a generator; also mechanical power supplied by a motor to a load (induction machine in motoring mode, Chapter 7).
P_n	real power demand at rated voltage.
P_R	real power at the receiving end of a transmission line.
P_{rI}, P_{rII}, P_{rIII}, P_{rIV}	contribution of the generating units remaining in operation toward covering the real power imbalance during the first, second, third, and fourth stage of load frequency control.
P_{sI}, P_{sII}, P_{sIII}, P_{sIV}	contribution of the system toward covering the real power imbalance during the first, second, third, and fourth stage of load frequency control.
P_s	stator power of induction machine (Chapter 7).
P_S	real power at the sending end of a transmission line or real power supplied by a source to a load or real power supplied to the infinite busbar.
P_{SIL}	surge impedance (natural) load.
$P_{sE_q}(\delta)$	curve of real power supplied to the infinite busbar assuming E_q = constant.
P_T	total power generated in a system.
P_{tie}	net tie-line interchange power.
$P_{V_g}(\delta)$	air-gap power curve assuming V_g = constant.
$P_{V_g\ cr}$	critical value of $P_{V_g}(\delta)$.
Q_L	reactive power absorbed by a load.
Q_G	reactive power generated by a source (the sum of Q_L and the reactive power loss in the network).
Q_n	reactive power demand at rated voltage.
Q_R	reactive power at the receiving end of a transmission line.
Q_S	reactive power at the sending end of a transmission line or reactive power supplied by a source to a load.
R	resistance.
r	total resistance between (and including) the generator and the infinite busbar.
R_A, R_B, R_C, R_D, R_Q, R_f	resistances of the phase windings A, B, and C; the d- and q-axis damper winding; and the field winding.
$\boldsymbol{R}_{ABC}$	diagonal matrix of phase winding resistances.
$\boldsymbol{R}_{fDQ}$	diagonal matrix of resistances of the field winding and the d- and q-axis damper windings.
s	Laplace operator.
s	slip of induction motor.
s_{cr}	critical slip of induction motor.
S_n	rated apparent power.
S_{SHC}	short-circuit power.
t	time.
T'_d, T''_d	short-circuit d-axis transient and subtransient time constants.
T'_{do}, T''_{do}	open-circuit d-axis transient and subtransient time constants.
T'_q, T''_q	short-circuit q-axis transient and subtransient time constants.

T'_{qo}, T''_{qo}	open-circuit q-axis transient and subtransient time constants.
T_a	armature winding time constant.
$\boldsymbol{T}$	transformation matrix between network (a, b) and generator (d, q) coordinates.
v_A, v_B, v_C, v_f	instantaneous voltages across phases A, B, and C and the field winding.
v_d, v_q	voltages across the fictitious d- and q-axis armature coils.
v_w	wind speed.
$\boldsymbol{v}_{ABC}$	vector of instantaneous voltages across phases A, B, and C.
$\boldsymbol{v}_{fDQ}$	vector of instantaneous voltages across the field winding and the d- and q-axis damper windings.
V	Lyapunov function.
V_{cr}	critical value of the voltage.
$\underline{V}_d$, $\underline{V}_q$	d- and q-axis component of the generator terminal voltage.
V_f	voltage applied to the field winding.
$\underline{V}_g$	voltage at the generator terminals.
$\underline{V}_s$	infinite busbar voltage.
$\underline{V}_{sd}$, $\underline{V}_{sq}$	d- and q-axis component of the infinite busbar voltage.
$\underline{V}_S$, $\underline{V}_R$	voltage at the sending and receiving end of a transmission line.
V_{sh}	local voltage at the point of installation of a shunt element.
$\underline{V}_i = V_i\angle\delta_i$	complex voltage at node i.
$\underline{\boldsymbol{V}}_R$, $\underline{\boldsymbol{V}}_E$	vector of complex voltages at the retained and eliminated nodes.
W	work.
$\boldsymbol{W}$	Park's modified transformation matrix.
$\boldsymbol{W}$, $\boldsymbol{U}$	modal matrices of right and left eigenvectors.
X_a	armature reaction reactance (round-rotor generator).
X_C	reactance of a series compensator.
X_D	reactance corresponding to the flux path around the damper winding.
X_d, X'_d, X''_d	d-axis synchronous, transient, and subtransient reactance.
x_d, x'_d, x''_d	total d-axis synchronous, transient, and subtransient reactance between (and including) the generator and the infinite busbar.
$x'_{d\,PRE}$, $x'_{d\,F}$, $x'_{d\,POST}$	pre-fault, fault, and post-fault value of x'_d.
X_f	reactance corresponding to the flux path around the field winding.
X_l	armature leakage reactance of a generator.
X_q, X'_q, X''_q	q-axis synchronous, transient and subtransient reactance.
x_q, x'_q, x''_q	total q-axis synchronous, transient, and subtransient reactance between (and including) the generator and the infinite busbar.
X_{SHC}	short-circuit reactance of a system as seen from a node.
Y_T	admittance of a transformer.
$\underline{\boldsymbol{Y}}$	admittance matrix.
$\underline{\boldsymbol{Y}}_{GG}$, $\underline{\boldsymbol{Y}}_{LL}$, $\underline{\boldsymbol{Y}}_{LG}$, $\underline{\boldsymbol{Y}}_{LG}$	admittance submatrices, where subscript G corresponds to fictitious generator nodes and subscript L corresponds to all the other nodes (including generator terminal nodes).
$\underline{Y}_{ij} = G_{ij} + jB_{ij}$	element of the admittance matrix.
$\underline{\boldsymbol{Y}}_{RR}$, $\underline{\boldsymbol{Y}}_{EE}$, $\underline{\boldsymbol{Y}}_{RE}$, $\underline{\boldsymbol{Y}}_{ER}$	complex admittance submatrices, where subscript E refers to eliminated and subscript R to retained nodes. $Z = \sqrt{r^2 + x_d x_q}$.
Z_c	characteristic impedance of a transmission line.

$\underline{Z}_s = R_s + jX_s$ internal impedance of the infinite busbar.
$\underline{Z}_T = R_T + jX_T$ series impedance of the transformer.
$\Delta\omega$ rotor speed deviation equal to $(\omega - \omega_s)$.
γ instantaneous position of the generator d-axis relative to phase A.
γ_0 position of the generator d-axis at the instant of a fault.
Φ_a armature reaction flux.
Φ_{ad}, Φ_{aq} d- and q-axis component of the armature reaction flux.
$\Phi_{a\ ac}$ ac armature reaction flux (rotating).
$\Phi_{a\ dc}$ dc armature reaction flux (stationary).
Φ_f excitation (field) flux.
Ψ_A, Ψ_B, Ψ_C total flux linkage of phases A, B, and C.
Ψ_{AA}, Ψ_{BB}, Ψ_{CC} self-flux linkage of phases A, B, and C.
$\Psi_{a\ ac\ r}$ rotor flux linkages produced by $\Phi_{a\ ac}$.
$\Psi_{a\ dc\ r}$ rotor flux linkages produced by $\Phi_{a\ dc}$.
$\Psi_{a\ r}$ rotor flux linkages produced by the total armature reaction flux.
Ψ_D, Ψ_Q total flux linkage of damper windings in axes d- and q.
Ψ_d, Ψ_q total d- and q-axis flux linkages.
Ψ_f total flux linkage of the field winding.
Ψ_{fa} excitation flux linkage with armature winding.
Ψ_{fA}, Ψ_{fB}, Ψ_{fC} excitation flux linkage with phases A, B, and C.
$\boldsymbol{\Psi}_{ABC}$ vector of phase flux linkages.
$\boldsymbol{\Psi}_{fDQ}$ vector of flux linkages of the field winding and the d- and q-axis damper windings.
$\boldsymbol{\Psi}_{0dq}$ vector of armature flux linkages in the rotor reference frame.
τ_e electromagnetic torque.
τ_m mechanical torque.
τ_ω fundamental-frequency subtransient electromagnetic torque.
$\tau_{2\omega}$ double-frequency subtransient electromagnetic torque.
τ_d, τ_q d- and q-axis component of the electromagnetic torque.
τ_R, τ_r subtransient electromagnetic torque due to stator and rotor resistances.
ε rotor acceleration.
φ_g power factor angle at the generator terminals.
δ power (or rotor) angle with respect to the infinite busbar.
δ_g power (or rotor) angle with respect to the voltage at the generator terminals.
$\hat{\delta}_s$ stable equilibrium value of the rotor angle.
δ' transient power (or rotor) angle between E' and V_s.
δ_{fr} angle between the resultant and field mmf's.
λ_R frequency bias factor.
$\underline{\lambda}_i = \alpha_i + j\Omega_i$ eigenvalue.
ω angular velocity of the generator (in electrical radians).
ω_s synchronous angular velocity in electrical radians (equal to $2\pi f$).
ω_T rotor speed of wind turbine (in rad/s).
Ω frequency of rotor swings (in rad/s).
$\boldsymbol{\Omega}$ rotation matrix.
ρ static droop of the turbine-governor characteristic.
ρ_T droop of the total system generation characteristic.
ϑ transformation ratio.

γ	propagation constant of a transmission line.
β	phase constant of a transmission line.
$\mathfrak{R}$	reluctance.
$\mathfrak{R}_d$, $\mathfrak{R}_q$	reluctance along the d- and q-axis.
ζ	damping ratio.

Abbreviations

AC	alternating current
ACE	area control error
AGC	automatic generation control
AGSS	automatic generation shedding scheme
ANN	artificial neural network
AR	autoregressive
ARMAX	autoregressive moving average with exogenous input model
ARX	autoregressive with exogenous input model
AVR	automatic voltage regulator
BESS	battery energy storage system
BFP	breaker failure protection
BIMP	Bureau international des poids et mesures
BTA	balanced truncation approximation
CCGT	combined cycle gas turbine
CCPP	combined-cycle power plant
CCT	critical clearing time
CESA	continental Europe synchronous area
CIM	current injection model
COI	center of inertia
CSC	controlled series capacitor
CT	current transformer
d-axis	direct-axis of a generator
DC	direct current
DFIG	doubly-fed induction generator
DFIM	doubly-fed induction machine
DSA	dynamic security assessment
DSO	distribution system operator
EHV	extra high voltage
emf	electromotive force
EMS	energy management system
ENTSO-E	European Network of Transmission System Operators for Electricity
EPS	electric power system
FACTS	flexible AC transmission system
FC	fixed capacitor
FRC	fully rated converter
FRT	fault ride through

GEP	generator-exciter-power system
GV	governor valves
GPS	global positioning system
GTO	gate turn-off thyristor
HAWT	horizontal-axis wind turbine
HP	high pressure
HRB	heat-recovery boiler
HRSG	heat-recovery steam generator
HV	high voltage
HVDC	high voltage direct current
HVRT	high-voltage ride through
IGBT	insulated gate bipolar transistor
IGCT	integrated gate-commutated thyristor
IGV	inlet guide vanes
Im	imaginary axis
IP	intermediate pressure
IV	intercept valves
K3, K2E, K2, K1	short circuits: three-phase, two-phase-ground, two-phase, single-phase
LFC	load and frequency control
LLFT	lower linear fractional transformation
LP	low pressure
LQE	linear-quadratic estimator
LQI	linear-quadratic-Gaussian regulator with tracking loop based on integral element
LQR	linear-quadratic regulator
LTC	load tap changer
LVRT	low-voltage ride through
MAWS	mean annual wind speed
MIMO	multiple inputs and multiple outputs
mmf	magnetomotive force
MSV	main emergency stop valves
OEL	overexcitation limiter
OLEC	overload emergency state control
OLTC	on-load tap changer
OPF	optimal power flow
OST	out-of-step tripping
PCC	point of common coupling
PCS	power conditioning system
PDC	phasor data concentrator
PMU	phasor measurement unit
POD	power oscillation damper
PSB	power swing blocking
PSD	power swing detection
PSP	pole-slip protection
PSS	power system stabilizer
pu	per unit

PV	photovoltaic
PWM	pulse-width modulation
q-axis	quadrature-axis of a generator
rad	radian
rad/s	radians per second
Re	real axis
RES	renewable energy sources
RMS	root-mean-square
RoCoF	rate of change of frequency
RoMF	range of modal frequencies
ROW	right of way
rpm	revolutions per minute
RSV	reheat emergency stop valves
SCADA	supervisory control and data acquisition
SCL	stator current limiter
SIL	surge impedance load
SISO	single-input single-output system
SMES	superconducting magnetic energy storage
SMIB	single machine infinite bus
SOSP	special out-of-step protection
SPS	special protection system
SSC	switched series capacitor
SSSC	static synchronous series compensator
STATCOM	static compensator
SVC	static VAR compensator
SVG	static VAR generator
TCBR	thyristor-controlled braking resistor
TCPAR	thyristor-controlled phase angle regulator
TCR	thyristor-controlled reactor
TGR	transient gain reduction
TSBR	thyristor-switched braking resistor
TSC	thyristor-switched capacitor
TSO	transmission system operator
TSR	thyristor-switched reactor
UCTE	Union for the Coordination of Transmission of Electricity (Europe)
UEL	underexcitation limiter
ULTC	under-load tap changer
UPFC	unified power flow controller
UTC	coordinated universal time
VAWT	vertical-axis wind turbine
VHP	very high pressure
VI	virtual inertia
VIGV	variable inlet guide vanes
VSC	voltage source converter
VT	voltage transformer

WAC	wide area control
WAM	wide area monitoring
WAMPAC	wide area measurement, protection, and control
WAMS	wide area measurement system
WAP	wide area protection
WAPOD	wide-area power oscillation damper
WTGS	wind turbine generator system

Part I

Introduction to Power Systems

1

Introduction

1.1 Stability and Control of a Dynamic System

In engineering, a *system* is understood to be a set of physical elements acting together and realizing a common goal. An important role in the analysis of the system is played by its *mathematical model*. It is created using the system structure and fundamental physical laws governing the system elements. In the case of complicated systems, mathematical models usually do not have a universal character but rather reflect some characteristic phenomena which are of interest. Because of mathematical complications, practically used system models are usually a compromise between a required accuracy of modeling and a degree of complication.

When formulating a system model, important terms are the *system state* and the *state variables*. The system state describes the system's operating conditions. The state variables are the minimum set of variables $x_1, x_2, ..., x_n$ uniquely defining the system state. State variables written as a vector $\boldsymbol{x} = [x_1, x_2, ..., x_n]^{\mathrm{T}}$ are referred to as the *state vector*. A normalized space of coordinates corresponding to the state variables is referred to as the *state space*. In the state space, each system state corresponds to a point defined by the state vector. Hence, a term "system state" often refers also to a point in the state space.

A system may be *static* when its state variables $x_1, x_2, ..., x_n$ are time invariant or *dynamic* when they are functions of time, that is $x_1(t), x_2(t), ..., x_n(t)$.

This book is devoted to the analysis of dynamic systems modeled by ordinary differential equations of the form

$$\dot{\boldsymbol{x}} = \boldsymbol{F}(\boldsymbol{x}) \quad \text{or} \quad \dot{\boldsymbol{x}} = \boldsymbol{A}\boldsymbol{x} \tag{1.1}$$

where the first of the equations above describes a *nonlinear system* and the second describes a *linear system*. $\boldsymbol{F}(\boldsymbol{x})$ is a vector of nonlinear functions and $\boldsymbol{A}$ is a square matrix.

A curve $\boldsymbol{x}(t)$ in the state space containing system states (points) in consecutive time instants is referred to as the *system trajectory*. A trivial one-point trajectory $\boldsymbol{x}(t) = \hat{\boldsymbol{x}} =$ const is referred to as the *equilibrium point (state)*, if in that point all the partial derivatives are zero (no movement), that is $\dot{\boldsymbol{x}} = \boldsymbol{0}$. According to Eq. (1.1), the coordinates of the point satisfy the following equations

$$\boldsymbol{F}(\hat{\boldsymbol{x}}) = \boldsymbol{0} \quad \text{or} \quad \boldsymbol{A}\hat{\boldsymbol{x}} = \boldsymbol{0} \tag{1.2}$$

A nonlinear system may have more than one equilibrium point because nonlinear equations may have generally more than one solution. In the case of linear systems, according to Cramér's theorem concerning linear equations, there exists only one uniquely specified equilibrium point $\hat{\boldsymbol{x}} = \boldsymbol{0}$ if and only if the matrix $\boldsymbol{A}$ is nonsingular ($\det \boldsymbol{A} \neq 0$).

All the states of a dynamic system, apart from equilibrium states, are dynamic states because the derivatives $\dot{\boldsymbol{x}} \neq \boldsymbol{0}$ for those states are nonzero, which means a movement. *Disturbance* means a random (usually unintentional) event affecting the system. Disturbances affecting dynamic systems are modeled by changes in their coefficients (parameters) or by nonzero initial conditions of differential equations.

Power System Dynamics: Stability and Control, Third Edition. Jan Machowski, Zbigniew Lubosny, Janusz W. Bialek and James R. Bumby.

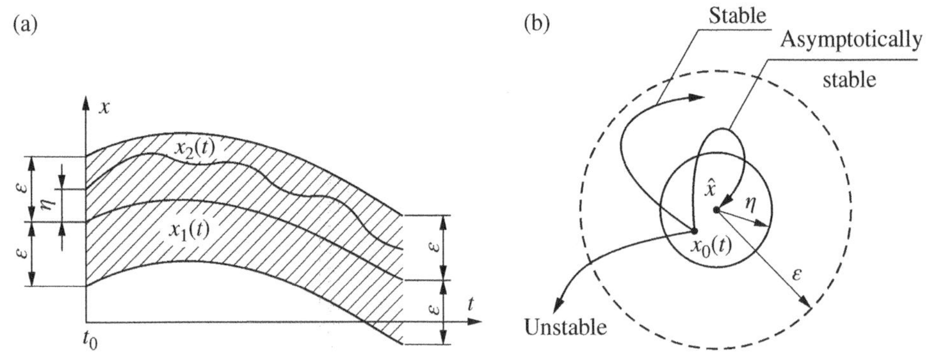

Figure 1.1 Illustration of the definition of stability: (a) when the initial conditions are different but close; (b) in a vicinity of the equilibrium point.

Let $\boldsymbol{x}_1(t)$ be a trajectory of a dynamic system (Figure 1.1a) corresponding to some initial conditions. The system is considered *stable in the Lyapunov sense* if for any t_0 it is possible to choose a number η such that for all the other initial conditions satisfying the constraint $\|\boldsymbol{x}_2(t_0) - \boldsymbol{x}_1(t_0)\| < \eta$ there holds $\|\boldsymbol{x}_2(t) - \boldsymbol{x}_1(t)\| < \varepsilon$ for $t_0 \leq t < \infty$. In other words stability means that if the trajectory $\boldsymbol{x}_2(t)$ starts close enough (as defined by η) to the trajectory $\boldsymbol{x}_1(t)$ then it remains close to it (number ε). Moreover, if the trajectory $\boldsymbol{x}_2(t)$ tends with time toward the trajectory $\boldsymbol{x}_1(t)$, i.e. $\lim_{t\to\infty} \|\boldsymbol{x}_2(t) - \boldsymbol{x}_1(t)\| = 0$, then the dynamic system is *asymptotically stable.*

The above definition concerns any trajectory of a dynamic system. Hence it must also be valid for a trivial trajectory such as the equilibrium point $\hat{\boldsymbol{x}}$. In this particular case (Figure 1.1b), the trajectory $\boldsymbol{x}_1(t)$ is a point $\hat{\boldsymbol{x}}$ and the initial condition $\boldsymbol{x}_2(t_0)$ of trajectory $\boldsymbol{x}_2(t)$ lies in the vicinity of the point defined by η. The dynamic system is stable in the equilibrium point $\hat{\boldsymbol{x}}$ if for $t_0 \leq t < \infty$ the trajectory $\boldsymbol{x}_2(t)$ does not leave an area defined by the number ε. Moreover, if the trajectory $\boldsymbol{x}_2(t)$ tends with time toward the equilibrium point $\hat{\boldsymbol{x}}$, i.e. $\lim_{t\to\infty} \|\boldsymbol{x}_2(t) - \hat{\boldsymbol{x}}\| = 0$, then the system is said to be *asymptotically stable* at the equilibrium point $\hat{\boldsymbol{x}}$. On the other hand, if the trajectory $\boldsymbol{x}_2(t)$ tends with time to leave the area defined by ε, then the dynamic system is said to be *unstable* at the equilibrium point $\hat{\boldsymbol{x}}$.

It can be shown that the stability of a linear system does not depend on the size of a disturbance. Hence if a linear system is stable for a small disturbance then it is also globally stable for any large disturbance.

The situation is different with nonlinear systems as their stability generally depends on the size of a disturbance. A nonlinear system may be stable for a small disturbance but unstable for a large disturbance. The largest disturbance for which a nonlinear system is still stable is referred to as a *critical disturbance.*

Dynamic systems are designed and constructed with a particular task in mind and assuming that they will behave in a particular way following a disturbance. A purposeful action affecting a dynamic system which aims to achieve a particular behavior is referred to as a *control.* The definition of control is illustrated in Figure 1.2. The following signals have been defined:

$\boldsymbol{u}(t)$ – a control signal which affects the system to achieve a desired behavior
$\boldsymbol{y}(t)$ – an output signal which serves to assess whether or not the control achieved the desired goal
$\boldsymbol{x}(t)$ – system state variables
$\boldsymbol{z}(t)$ – disturbances.

Control can be open loop or closed loop. In the case of open-loop control (Figure 1.2a), control signals are created by a control device which tries to achieve a desired system behavior without obtaining any information about the output signals. Such control makes sense only when it is possible to predict the shape of output signals from the control signals. However, if there are additional disturbances which are not a part of the control, then their action may lead to the control objective not being achieved.

In the case of closed-loop control (Figure 1.2b), control signals are chosen based on the control task and knowledge of the system output signals describing whether the control task has been achieved. Hence the control is a function of its effects and acts until the control task has been achieved.

Closed-loop control is referred to as *feedback control* or *regulation*. The control device is then called a *regulator* and the path connecting the output signals with the control device (regulator) is called the *feedback loop*.

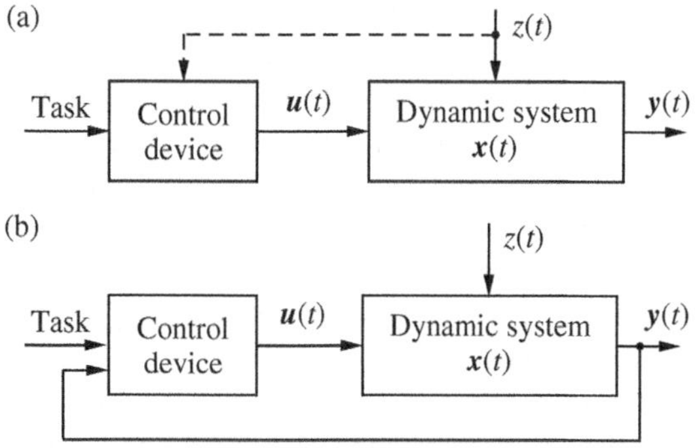

Figure 1.2 Illustration of the definition of: (a) open-loop control; (b) closed-loop control.

A nonlinear dynamic system with its control can be generally described by the following set of algebraic and differential equations

$$\dot{\boldsymbol{x}} = \boldsymbol{F}(\boldsymbol{x}, \boldsymbol{u}) \quad \text{and} \quad \boldsymbol{y} = \boldsymbol{G}(\boldsymbol{x}, \boldsymbol{u}) \tag{1.3}$$

while a linear dynamic system model is

$$\dot{\boldsymbol{x}} = \boldsymbol{A}\boldsymbol{x} + \boldsymbol{B}\boldsymbol{u} \quad \text{and} \quad \boldsymbol{y} = \boldsymbol{C}\boldsymbol{x} + \boldsymbol{D}\boldsymbol{u} \tag{1.4}$$

It is easy to show that, for small changes in state variables and output and control signals, Eqs. (1.4) are linear approximations of nonlinear Eqs. (1.3). In other words linearization of (1.3) leads to equations

$$\Delta\dot{\boldsymbol{x}} = \boldsymbol{A}\Delta\boldsymbol{x} + \boldsymbol{B}\Delta\boldsymbol{u} \quad \text{and} \quad \Delta\boldsymbol{y} = \boldsymbol{C}\Delta\boldsymbol{x} + \boldsymbol{D}\Delta\boldsymbol{u} \tag{1.5}$$

where $\boldsymbol{A}$, $\boldsymbol{B}$, $\boldsymbol{C}$, and $\boldsymbol{D}$ are the matrices of derivatives of functions $\boldsymbol{F}$ and $\boldsymbol{G}$ with respect to $\boldsymbol{x}$ and $\boldsymbol{u}$.

1.2 Classification of Power System Dynamics

An electrical power system (EPS) consists of many individual elements connected together to form a large complex and dynamic system capable of generating, transmitting, and distributing electrical energy over a large geographical area. Because of this interconnection of elements, a large variety of dynamic interactions are possible, some of which will only affect some of the elements, others fragments of the system, while others may affect the system as a whole. As each dynamic effect displays certain unique features, power system dynamics can be conveniently divided into groups characterized by their cause, consequence, timeframe, physical character, or the place in the system where they occur.

Of prime concern is the way the EPS will respond to both a changing power demand and to various types of disturbance, the two main causes of power system dynamics. A changing power demand introduces a wide spectrum of dynamic changes into the system, each of which occurs on a different timescale. In this context the fastest dynamics are due to sudden changes in demand and are associated with the transfer of energy between the rotating masses in the generators and the loads. Slightly slower are the voltage and frequency control actions needed to maintain system operating conditions until finally the very slow dynamics corresponding to the way in which the generation is adjusted to meet the slow daily demand variations take effect. Similarly, the way in which the system responds to disturbances also covers a wide spectrum of dynamics and associated timeframes. In this case the fastest dynamics are those associated with the very fast wave phenomena that occur in high-voltage transmission lines. These are followed by fast electromagnetic changes in the electrical machines themselves before the relatively slow electromechanical rotor oscillations occur. Finally, the very slow prime mover and automatic generation control actions take effect.

Based on their physical character, the different power system dynamics may be divided into four groups defined as: *wave*, *electromagnetic*, *electromechanical*, and *thermodynamic*. This classification also corresponds to the timeframe involved and is shown in Figure 1.3. Although this broad classification is convenient, it is by no means absolute, with some of the dynamics belonging to two or more groups while others lie on the boundary between groups. Figure 1.3 shows the fastest dynamics to be the *wave* effects, or surges, in high-voltage transmission lines and correspond to the propagation of electromagnetic waves caused by lightning strikes or switching operations. The timeframe of these dynamics is from microseconds to milliseconds. Much slower are the *electromagnetic dynamics* that take place in the machine windings following a disturbance, operation of the protection system, or the interaction between the electrical machines and the network. Their timeframe is from milliseconds to a second. Slower still are the *electromechanical dynamics* due to the oscillation of the rotating masses of the generators and motors that occur following a disturbance, operation of the protection system, and voltage and prime mover control. The timeframe of these dynamics is from seconds to several seconds. The slowest dynamics are the *thermodynamic changes* which result from boiler control action in steam power plants as the demands of the automatic generation control are implemented.

Careful inspection of Figure 1.3 shows the classification of power system dynamics with respect to timeframe to be closely related to where the dynamics occur within the system. For example, moving from the left to right along the timescale in Figure 1.1 corresponds to moving through the EPS from the electrical RLC circuits of the transmission network, through the generator armature windings to the field and damper winding, then along the generator rotor to the turbine until finally the boiler is reached.

The fast wave phenomena, caused by lightning and switching overvoltages, occur almost exclusively in the network and basically do not propagate beyond the transformer windings. The electromagnetic phenomena mainly involve the generator armature and damper windings and partly the network. These electromechanical phenomena, namely the rotor oscillations and accompanying network power swings, mainly involve the rotor field and damper windings and the rotor inertia. As the power system network connects the generators together this enables interactions between swinging generator rotors to take place. An important role is played here by the automatic voltage control and the prime mover control. Slightly slower than the electromechanical phenomena are the frequency oscillations, in which the rotor dynamics still play an important part, but are influenced to a much greater extent by the action of the turbine governing systems and the automatic generation control. Automatic generation control also influences the thermodynamic changes through the boiler control action in steam power plants.

The fact that the timeframe of the dynamic phenomena is closely related to where it occurs within the EPS has important consequences for the modeling of the system elements. In particular, moving from left to right along

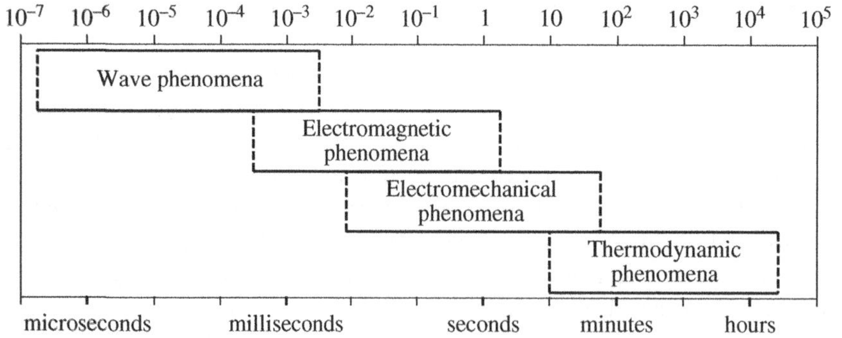

Figure 1.3 Timeframe of the basic power system dynamic phenomena.

Figure 1.3 corresponds to a reduction in the accuracy required in the models used to represent the network elements, but an increase in the accuracy in the models used first to represent the electrical components of the generating unit and then, further to the right, the mechanical and thermal parts of the unit. This important fact is taken into account in the general structure of this book when later chapters describe the different power system dynamic phenomena.

1.3 Two Pairs of Important Quantities

This book is devoted to the analysis of electromechanical phenomena and control processes in power systems. The main elements of electrical power networks are transmission lines and transformers which are usually modeled by four-terminal (two-port) RLC elements. Those models are connected together according to the network configuration to form a network diagram.

For further use in this book, some general relationships will be derived below for a two-port π-equivalent circuit in which the series branch consists of only an inductance and the shunt branch is completely neglected. The equivalent circuit and the phasor diagram of such an element are shown in Figure 1.4a. The voltages V and E are phase voltages, while P and Q are single-phase powers. The phasor $\underline{E}$ has been obtained by adding voltage drop $j X \underline{I}$, perpendicular to $\underline{I}$, to the voltage $\underline{V}$. The triangles OAD and BAC are similar. Analyzing triangles BAC and OBC gives

$$|\mathrm{BC}| = XI\cos\varphi = E\sin\delta \quad \text{hence} \quad I\cos\varphi = \frac{E}{X}\sin\delta \tag{1.6}$$

$$|\mathrm{AC}| = XI\sin\varphi = E\cos\delta - V \quad \text{hence} \quad I\sin\varphi = \frac{E}{X}\cos\delta - \frac{V}{X} \tag{1.7}$$

Active (real) power leaving the element is expressed as $P = VI\cos\varphi$. Substituting (1.6) into that equation gives

$$P = \frac{EV}{X}\sin\delta \tag{1.8}$$

This equation shows that active power P depends on the product of phase voltages and the sine of the angle δ between their phasors. In power networks, node voltages must be within a small percentage of their nominal values. Hence such small variations cannot influence the value of active power. The conclusion is that large changes of active power, from negative to positive values, correspond to changes in the sine of the angle δ. The characteristic $P(\delta)$ is referred to as the *power-angle characteristic*, while the angle δ is referred to as the *power angle* or the *load angle*. Because of stability considerations discussed in Chapter 5, the system can operate only in that part of the characteristic which is shown by a solid line in Figure 1.4b. The smaller the reactance X, the higher the amplitude of the characteristic.

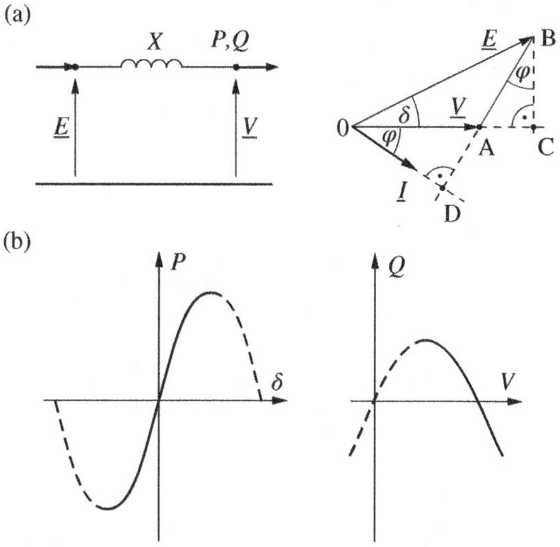

Figure 1.4 A simplified model of a network element: (a) equivalent diagram and phasor diagram; (b) real power and reactive power characteristics.

The per-phase reactive power leaving the element is expressed as $Q = VI \sin \varphi$. Substituting (1.7) into that equation gives

$$Q = \frac{EV}{X} \cos \delta - \frac{V^2}{X} \tag{1.9}$$

The term $\cos\delta$ is determined by the value of active power because the relationship between the sine and cosine is $\cos \delta = \sqrt{1 - \sin^2 \delta}$. Using that equation and (1.8) gives

$$Q = \sqrt{\left(\frac{EV}{X}\right)^2 - P^2} - \frac{V^2}{X} \tag{1.10}$$

The characteristic $Q(V)$ corresponds to an inverted parabola (Figure 1.4b). Because of the stability considerations discussed in Chapter 8, the system can operate only in that part of the characteristic which is shown by a solid line.

The smaller the reactance X, the steeper the parabola, and even small changes in V cause large changes in reactive power. Obviously the inverse relationship also takes place: a change in reactive power causes a change in voltage.

The above analysis points out that Q, V and P, δ form two pairs of strongly connected variables. Hence one should always remember that voltage control strongly influences reactive power flows and vice versa. Similarly, when talking about active power P one should remember that it is connected with angle δ. That angle is also strongly connected with system frequency f, as discussed later in the book. Hence the pair P, f is also strongly connected and important for understanding power system operation.

1.4 Stability of a Power System

Power system stability is understood as the ability to regain an equilibrium state after being subjected to a physical disturbance. Section 1.3 showed that three quantities are important for power system operation: (i) angles of nodal voltages δ also called power or load angles; (ii) frequency f; and (iii) nodal voltage magnitudes V. Those quantities are especially important from the point of view of defining and classifying power system stability. Hence power system stability can be divided (Figure 1.5) into: (i) rotor (or power) angle stability; (ii) frequency stability; and (iii) voltage stability.

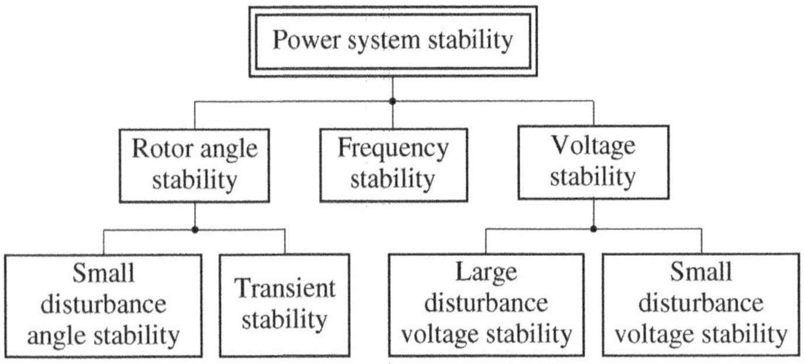

Figure 1.5 Classification of power system stability (based on CIGRE Report No. 325). *Source:* Reproduced by permission of CIGRE.

As power systems are nonlinear, their stability depends on both the initial conditions and the size of a disturbance. Consequently, angle and voltage stability can be divided into small- and large-disturbance stability. Small disturbance angle stability is also referred to as the *steady-state stability*.

Power system stability is mainly connected with electromechanical phenomena (Figure 1.3). However, it is also affected by fast electromagnetic phenomena and slow thermodynamic phenomena. Hence, depending on the type of phenomena, one can refer to *short-term stability* and *long-term stability*. All of them are discussed in detail in this book.

1.5 Security of a Power System

A set of imminent disturbances is referred to as *contingencies*. *Power system security* is understood as the ability of the EPS to survive plausible contingencies without interruption to customer service. Power system security and power system stability are related terms. Stability is an important factor of power system security, but security is a wider term than stability. Security not only includes stability but also encompasses the integrity of an EPS and assessment of the equilibrium state from the point of view of overloads, under- or overvoltages and underfrequency.

From the point of view of power system security, the operating states may be classified as in Figure 1.6. Most authors credit Dy Liacco (1968) for defining and classifying these states.

In the *normal state*, an EPS satisfies the power demand of all the customers, all the quantities important for power system operation assume values within their technical constraints, and the system is able to withstand any plausible contingencies.

The *alert state* arises when some quantities that are important for power system operation (e.g. line currents or nodal voltages) exceed their technical constraints due to an unexpected rise in demand or a severe contingency, but the EPS is still intact and supplies its customers. In that state a further increase in demand or another contingency may threaten power system operation and preventive actions must be undertaken to restore the system to its normal state.

In the *emergency state* the EPS is still intact and supplies its customers but the violation of constraints is more severe. The emergency state usually follows the alert state when preventive actions have not been undertaken or have not been successful. A power system may assume the emergency state directly from the normal state following unusually severe contingencies like multiple faults. When a system is in the emergency state, it is necessary to undertake effective corrective actions leading first to the alert state and then to the normal state.

A power system can transpose to the *in extremis state* from the emergency state if no corrective actions have been undertaken and the system is already not intact due to a reduction of power supply following load shedding or when generators were tripped because of a lack of synchronism. The extreme variant of that state is a partial or complete blackout.

To return an EPS from an *in extremis* state to an alert or normal state, a *restorative state* is necessary in which power system operators perform control actions in order to reconnect all the facilities and restore all system loads.

Assessment of power system security can be divided into static and dynamic security. Static security assessment (SSA) includes the following computational methods:

- for the pre-contingency states, determine the available transfer capability of transmission links and identify network congestion;
- for the post-contingency states, verify the bus voltages and line power flow limits.

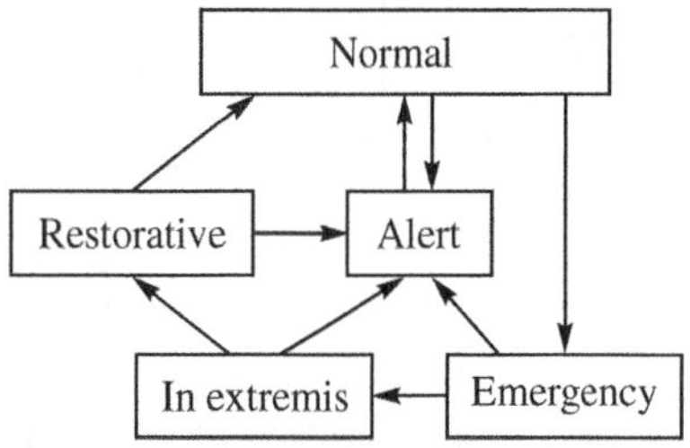

Figure 1.6 Classification of power system operating states (based on CIGRE Report No. 325). *Source:* Reproduced by permission of CIGRE.

Those tasks of SSA have always been the subject of great interest for power dispatching centers. However, when the industry was still vertically integrated (see Chapter 2), security management was relatively easy to execute because any decisions affecting the outputs or control settings of power plants could be implemented internally within a utility controlling both generation and transmission. Security management is not that easy to execute in the unbundled industry structure when the system operator has no direct control of generation. Any decisions affecting outputs or control settings of power plants have to be implemented using commercial agreements with power plants or enforced through the Grid Code. Especially, the analysis of available transfer capacity and congestion management have important implications for power plants as they directly affect their outputs, and therefore revenues.

SSA methods assume that every transition from the pre- to post-contingency state takes place without suffering any instability phenomena. Dynamic security assessment (DSA) includes methods to evaluate stability and quality of the transition from the pre- to post-contingency state. Typical criteria for DSA include:

1) rotor (power) angle stability, voltage stability, frequency stability;
2) frequency excursion during the dynamic state (dip or rise) beyond specified threshold levels;
3) voltage excursion during the dynamic state (dip or rise) beyond specified threshold levels;
4) damping of power swings inside subsystems and between subsystems on an interconnected network.

Criteria (1) and (2) are assessed using computer programs executing *transient security assessment* (TSA). Criterion (3) is assessed by programs executing *voltage security assessment* (VSA), and criterion (4) is assessed using programs executing *small-signal stability assessment* (SSSA).

Recent years have seen a number of total and partial blackouts in many countries of the world. These events have spurred a renewed interest among system operators in the tools for SSA and DSA. There are a variety of online DSA architectures. Figure 1.7 shows an example of the DSA architecture. The main components are denoted by boxes drawn with dashed lines.

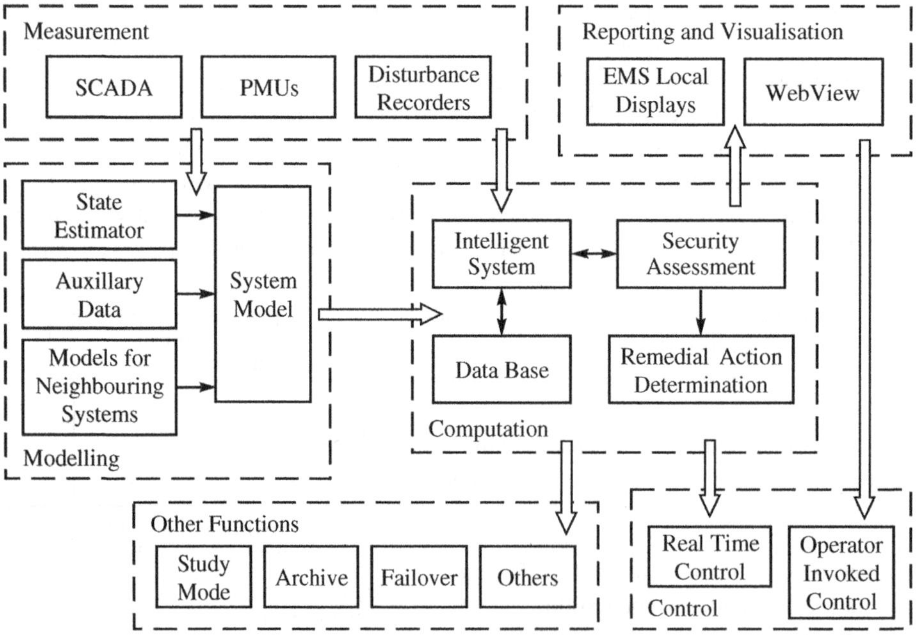

Figure 1.7 Components of DSA according to CIGRE Report No. 325. *Source:* Reproduced by permission of CIGRE.

The task of the component "measurement" is online data acquisition and taking a snapshot of power system conditions. Supervisory control and data acquisition (SCADA) systems usually collect measurements of real and reactive power in network branches, busbar voltages, frequency at a few locations in the system, status of switchgear, and the position of tap changers of transformers. As is shown in Section 2.7, new SCADA systems are often augmented by phasor measurement units (PMUs) collecting synchronized voltage phasor measurements.

The "modeling" component uses online data from the "measurement" component and augments them with offline data, obtained from a database, describing the parameters of power system elements and contingencies to be analyzed. The task of the "modeling" component is to create an online power system model using the identification of the power system configuration and state estimation. That component may also contain computer programs for the creation of equivalent models of neighboring systems. Contingencies vary according to the type of security being examined and in general need to be able to cater for a variety of events like short-circuits at any location, opening any line or transformer, loss of the largest generating unit or largest load in a region, multiple faults (when considered to be credible), and so on.

The next important component is "computation." Its task is system model validation and security assessment. The accuracy of the security assessment depends on the quality of the system model. Offline data delivered to the "modeling" component are validated through field testing of devices. Online data of the network configuration and system state obtained from the "measuring" component are validated using bad measurement data identification and removal which is made possible by redundancy of measurements. The best methodology for power system model validation is via a comparison of simulation results of the dynamic response of the EPS with recorded responses following some disturbances. To achieve this, the "measurement" component sends data from disturbance recorders to the "computation" component. The tools for the security assessment consist of a number of computer programs executing voltage stability analysis, small-signal stability analysis, transient stability analysis by hybrid methods combining system simulation, and the direct Lyapunov method described in the textbook by Pavella et al. (2000). Intelligent systems are also used employing learning from the situations previously seen.

The "reporting and visualization" component is very important for a system operator employing the described architecture. Computer programs of the "computation" component process a huge amount of data and analyze a large number of variants. On the other hand, the operator must receive a minimum number of results displayed in the most synthetic, preferably graphic, way. Some DSA displays have been shown in CIGRE Report No. 325. If the EPS is in a normal state, the synthetic results should report how close the system is to an insecure state to give the operator an idea of what might happen. If the system moves to an alert state or to an emergency state, the displayed result should also contain information about preventive or corrective action. This information is passed on to the "control" component.

The "control" component assists the operator in preventive and corrective actions executed to improve the power system operation. Some information produced by security assessment programs may be used to produce remedial control actions, which can be automatically executed by real-time control. The description of the current state of the art in DSA can be found in CIGRE Report No. 325.

2

Power System Components

2.1 Introduction

Modern-day society requires a large amount of energy for use in industry, commerce, agriculture, transportation, communications, domestic households, and so on. The total energy required for one year is called the *annual energy demand* and is satisfied using naturally occurring primary energy resources, principally fossil fuels such as coal, oil, natural gas, and uranium. In the current world energy scene these fossil fuels are also the main fuels used in the generation of electrical energy with the renewable energy resources such as hydro, biogas, solar, wind, geothermal, wave, and tidal energy being used to a lesser extent. In the future it is likely that the share of the energy market taken by renewables will increase as environmental issues play a more dominant role on the political agenda.

Perhaps the most important, and unique, feature of an electric power system (EPS) is that *electrical energy cannot easily and conveniently be stored in large quantities.* This means that at any instant in time the energy demand has to be met by corresponding generation. Fortunately, the combined load pattern of an EPS normally changes in a relatively predictable manner, even though individual consumer loads may vary quite rapidly and unpredictably. Such a predictable system demand pattern goes some way in allowing the daily generation schedule to be planned and controlled in a predetermined manner.

If a power utility is to provide an acceptable supply of electrical energy to its consumers, it must address the following issues.

2.1.1 Reliability of Supply

High reliability of supply is of fundamental importance as any major interruption of supply causes, at the very least, major inconvenience to the consumer, can lead to life-threatening situations, and, for the industrial consumer, may pose severe technical and production problems. Invariably in such situations the electrical supply utility also incurs a large loss in financial revenue. High reliability of supply can be ensured by:

- high quality of installed elements;
- the provision of reserve generation;
- employing large interconnected power systems capable of supplying each consumer via alternative routes;
- a high level of system security.

Power System Dynamics: Stability and Control, Third Edition. Jan Machowski, Zbigniew Lubosny, Janusz W. Bialek and James R. Bumby.

2.1.2 Supplying Electrical Energy of Good Quality

Electrical energy of good quality is provided by:

- regulated and defined voltage levels with low fluctuations;
- a regulated and defined value of frequency with low fluctuations;
- low harmonic content.

Two basic methods can be used to ensure a high quality of electrical supply. Firstly, the proper use of automatic voltage and frequency control methods and, secondly, by employing large, interconnected, power systems which, by their very nature, are less susceptible to load variations and other disturbances.

2.1.3 Economic Generation and Transmission

The majority of electricity is generated by first converting the thermal energy stored in the fossil fuel into mechanical energy and then converting this mechanical energy into electrical energy for transmission through the EPS to the consumer. Unfortunately the efficiency of this overall process is relatively low, particularly the first-stage conversion of thermal energy into mechanical energy. It is therefore vital that the operation of the overall system is optimized by minimizing the generation and transmission costs. Once again some saving can be achieved by connecting, and operating, a number of smaller systems as one larger, interconnected, system.

2.1.4 Environmental Issues

Modern society demands careful planning of generation and transmission to ensure as little effect as possible on the natural environment while meeting society's expectations for a secure electrical supply. Consequently, air and water pollution produced by power generation plants is limited to prescribed quantities, while the pathways for transmission lines are planned so as to cause minimal disturbance to the environment. In addition new plans for power stations and transmission lines are subject to close public scrutiny.

Environmental issues are now playing an ever-increasingly important role on the political agenda. Power generation has always been a major sources of air pollution and much effort has been devoted to developing cleaner generation technologies. However, the relatively recent concerns about global warming and sustainability have started to change the way power systems operate and expand. It is estimated that power generation contributes about one-third of the global CO_2 emissions, and so many countries in the world have set a target for renewable generation to contribute 20% or more of their total energy production by about 2020. The consequences of this for the power industry are discussed later in this chapter.

Another consequence of the environmental pressure is that power utilities must continually seek ways of making better use of their existing system. Obtaining planning permission for new transmission lines and generation sites has become more difficult and stringent.

It is within this political and operational framework that an electrical power utility generates, transmits, and distributes electrical energy to its consumers. Consequently, the purpose of this chapter is to describe how the different elements of a power system function and the effect they have on both power system operation and control.

2.2 Structure of the Electric Power System

The basic structure of a contemporary EPS is illustrated schematically in Figure 2.1 and shows the EPS to be divided into three parts: generation, transmission, and distribution. Historically, the power supply industry tended to be vertically integrated with each utility responsible for generation and transmission, and, in many cases, also

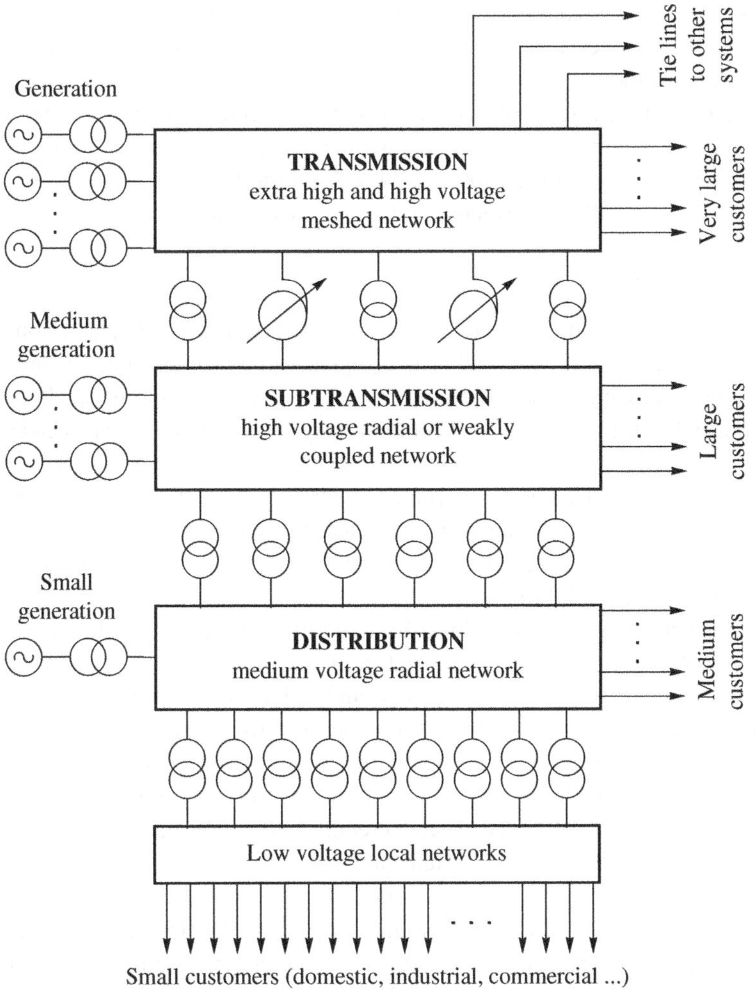

Figure 2.1 Structure of an EPS.

distribution in its own service (or control) area. The main justification for this was economies of scale and scope. It was also thought that, in order to optimize the overall power system planning and operation, a utility should be able to have full control of both transmission and generation, and sometimes also distribution. This situation has changed since the 1990s. In order to improve the overall efficiency of the industry, many countries have decided to introduce a liberalized competitive market for the industry. That has required *unbundling*, that is splitting, the vertically integrated utilities. In a typical liberalized model, the generation sector is divided into an number of private companies each owning individual power stations and competing with each other. The transmission tends to be operated by one monopoly company, referred to as the *system operator*, which is independent of the generation and regulated by an industry regulator. The distribution is also often split into separate distribution companies (*wires businesses*) which own and manage the distribution network in a given area, while retail, that is buying power on the wholesale markets and selling it to final customers, is handled by a number of competing *supply companies*. Customers are free to choose their suppliers, although in many countries that is restricted to industrial and commercial, but not domestic, customers.

That reorganization of the industry has created many challenges to the way power systems are being planned and operated. This book, however, focuses on the technical aspects of power system operation and does not discuss in detail the challenges brought about by liberalization.

Different parts of the EPS operate at different voltages. Generally, voltages can be considered *low* voltages if they are below the 1 kV mark, while *medium voltages*, used in distribution systems, are typically between 1 and 100 kV. The *high voltages* used in sub-transmission networks are between 100 and 300 kV and the *extra-high voltages* used in transmission networks are above 300 kV. This classification is loose and by no means strict.

2.2.1 Generation

Traditionally, power system operation has been based around a relatively small number of large power plants connected to the transmission system. Those plants are usually thermal or hydro plants in which electricity is produced by converting the mechanical energy appearing on the output shaft of an engine, or more usually a turbine, into electrical energy. The main thermal energy resources used commercially are coal, natural gas, nuclear fuel, and oil.

The conversion of mechanical to electrical energy in traditional thermal or hydro plants is almost universally achieved by the use of a synchronous generator. The synchronous generator feeds its electrical power into the transmission system via a step-up transformer (Figure 2.1) in order to increase the voltage from the generation level, (10–20) kV, to the transmission level (hundreds of kilovolts).

As mentioned earlier, concerns about global warming and sustainability have recently spurred interest in renewable generation. Generally, there are three main ways the industry can reduce its CO_2 emissions: (i) by moving from the traditional coal/gas/oil-based generation to renewable generation (wind, solar, marine); (ii) by moving toward increased nuclear generation which is largely CO_2 free; (iii) by removing CO_2 from exhaust gases of traditional thermal generation using, for example, carbon capture and storage technology. Discussing the relative merits of those three options is not the subject of this book. However, it is important to appreciate that the last two options retain the traditional structure of the EPS, as that based around a relatively few large generating units, and would therefore not require major changes to the way power systems are designed and operated. The first option, however, would require a major shift to the current practices, as generation would be increasingly based around a large number of small renewable plants. This is because renewable energy has a low energy density so that renewable power stations tend to be small with capacities of individual plants being between hundreds of kilowatts to a few megawatts. Such small plants are often connected at the distribution, rather than transmission, network due to the lower cost of connection. Such plants are referred to as *distributed*, or *embedded*, generation. Wind plants usually use induction generators, fixed speed or double fed, in order to transform wind energy into electricity, although sometimes inverter-fed synchronous generators may be used. Solar plants can be either thermal or photovoltaic with an inverter feeding a synchronous generator. Renewable generation is treated in more detail in Chapter 7.

2.2.2 Transmission

One significant advantage of electrical energy is that large traditional plants can be constructed near the primary fossil fuel energy resource or water reservoirs and the electrical energy produced can be transmitted over long distances to the load centers. Since the energy lost in a transmission line is proportional to the current squared, the transmission lines operate at high or very high voltages. The electrical network connects all the power stations into one system, and transmits and distributes power to the load centers in an optimal way. Usually, the transmission network has a mesh structure in order to provide many possible routes for electrical power to flow from individual generators to individual consumers, thereby improving the flexibility and reliability of the system.

One cannot overemphasize the importance of transmission for overall power system integrity. The transmission network makes the EPS a highly interacting, complicated mechanism, in which an action of any individual component (a power plant or a load) influences all the other components in the system. This is the main reason why transmission remains a monopoly business, even under the liberalized market structure, and is managed by a single system operator. The system operator is responsible for maintaining power system security and for optimizing power system operation.

As the electrical energy gets closer to the load center, it is directed from the transmission network into a sub-transmission network. When an EPS expands with the addition of new, high-voltage transmission lines some of the older, lower-voltage lines may become part of the sub-transmission network. There is no strict division of the network into transmission and sub-transmission networks and smaller power generation plants may feed directly into the sub-transmission network, while bulk power consumers may be fed directly from the transmission or sub-transmission network (Figure 2.1).

2.2.3 Distribution

Most of the electrical energy is transferred from the transmission, or sub-transmission, network to distribution high-voltage and medium-voltage networks in order to bring it directly to the consumer. The distribution network is generally connected in a radial structure as opposed to the mesh structure used in the transmission system. Large consumers may be supplied from a weakly coupled, meshed, distribution network, or, alternatively, they may be supplied from two radial feeders with a possibility of automatic switching between feeders in case of a power cut. Some industrial consumers may have their own on-site generation as a reserve or as a by-product of a technological process (e.g. steam generation). Ultimately, power is transformed to a low voltage and distributed directly to consumers.

Traditionally, distribution networks have been passive, that is there was little generation connected to them. Recently, the rapid growth in distributed and renewable generation has changed that picture. Power flows in distribution networks may no longer be unidirectional, that is from the point of connection with the transmission network down to customers. In many cases the flows may reverse direction, when the wind is strong and wind generation high, with distribution networks even becoming net exporters of power. That situation has created many technical problems with respect to settings of protection systems, voltage drops, congestion management, and so on.

Typically, about (8–10)% of the electrical energy appearing at the generator terminals will be lost on its way to the consumers in the transmission and distribution level.

2.2.4 Demand

The demand for electrical power is never constant and changes continuously throughout the day and night. The changes in demand of individual consumers may be fast and frequent, but as one moves up the power system structure (Figure 2.1) from individual consumers, through the distribution network, to the transmission level, the changes in demand become smaller and smoother as individual demands are aggregated. Consequently, the total power demand at the transmission level changes in a more or less predictable way that depends on the season, weather conditions, way of life of a particular society, and so on. Fast global power demand changes on the generation level are usually small and are referred to as *load fluctuations*.

2.3 Generating Units

The block diagram of a generating unit is shown in Figure 2.2. Electrical energy is produced by a synchronous generator driven by a prime mover, usually a turbine or a diesel engine. The turbine is equipped with a *turbine governor*, which controls either the speed or the output power according to a preset power-frequency characteristic.

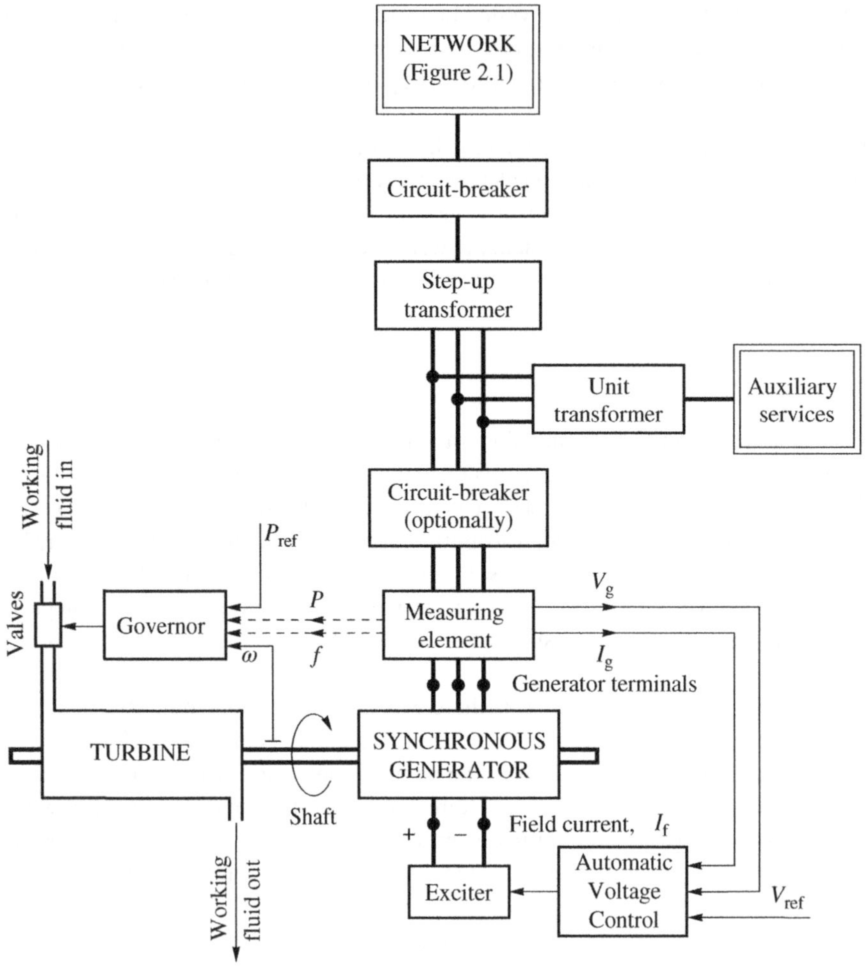

Figure 2.2 Block diagram of a power generation unit.

The DC excitation (or field) current, which is required to produce the magnetic field inside the generator, is provided by the *exciter*. The excitation current, and consequently the generator's terminal voltage, is controlled by an *automatic voltage regulator* (AVR). Generally, a synchronous generator is connected to the transmission network via a step-up transformer. In the case of a small unit the generator and the transformer are connected by cables, while a large, high-power generator may be connected to its transformer by a number of single-phase screened busbars. The generator transformer is usually located outdoors and is of the tank type. Power from the transformer is fed to the substation busbars via high-voltage cables or a short overhead line. An additional *unit transformer* may be connected to the busbar between the generator and the step-up transformer in order to supply the power stations auxiliary services comprising motors, pumps, the exciter, and so on.

The generating unit is equipped with a *main circuit-breaker* on the high-voltage side and sometimes also with a *generator circuit-breaker* on the generator side. Typical configurations are shown in Figure 2.3. In the first configuration (Figure 2.3a) there is only one circuit breaker and the generating unit operates as a common block, which must be switched off in the case of maintenance or a fault in any of the generating unit elements (generator, step-up transformer, unit transformer). In the second configuration (Figure 2.3b) there are two circuit breakers and, in case

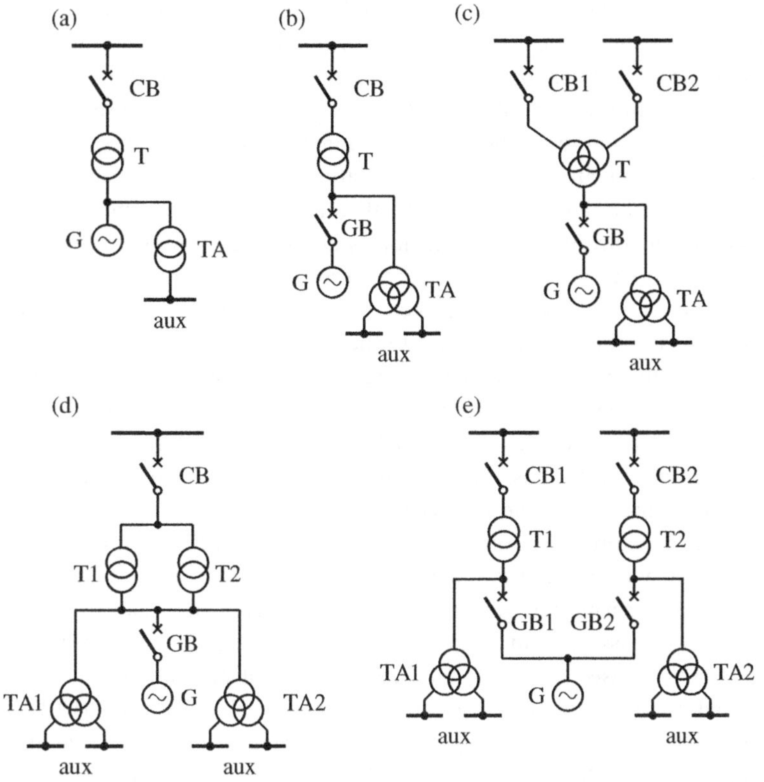

Figure 2.3 Examples of generating unit structures: G – synchronous generator; T – step-up transformer; CB – main circuit breaker; GB – generator circuit breaker; TA – auxiliary transformer (unit transformer); aux – busbars of auxiliary service.

of a fault in the generator or its maintenance, the generator circuit-breaker may be opened while the auxiliary services can be fed from the grid. On the other hand, with the main circuit-breaker open, the generator may supply its own auxiliary services. In the third configuration (Figure 2.3c) a three-winding transformer is used to connect a generator to two separate busbars. Such a configuration is used when the short-circuit power on the busbars must be reduced. In the fourth and fifth configurations two step-up transformers are used to connect a large-power generator with the network.

2.3.1 Synchronous Generators

Synchronous generators can be loosely classified as either high-speed generators, driven by steam or gas turbines (and often called *turbogenerators*), or low-speed generators, driven by water turbines. To reduce centrifugal forces high-speed turbogenerators have relatively low diameter but large axial length and are mounted horizontally. Typically, they will have two or four electrical poles so that in a 50 Hz system a generator would be driven at 3000 or 1500 rpm, respectively. In contrast, low-speed generators operate at typically 500 rpm and below, have a large number of electrical poles, large diameter, and shorter axial length. The actual number of magnetic poles depends on the required speed and nominal frequency of the EPS.

All generators have two main magnetic parts termed the *stator* and the *rotor*, both of which are manufactured from magnetic steel. The *armature winding*, which carries the load current and supplies power to the system, is placed in equidistant slots on the inner surface of the stator and consists of three identical phase windings.

The rotor of a high-speed generator also contains slots for the DC excitation winding, while the excitation winding for low-speed generators is wound on the salient poles of the rotor. The rotor also has additional short-circuited *damper*, or *amortisseur, windings*, to help damp mechanical oscillations of the rotor. In high-speed, nonsalient pole, generators the damper windings are usually in the form of conductive wedges mounted in the same slots as the excitation winding. In low-speed generators the damper windings are mounted in axial slots in the pole face.

The rotor excitation winding is supplied with a direct current to produce a rotating magnetic flux the strength of which is proportional to the excitation current. This rotating magnetic flux then induces an electromotive force (emf) in each phase of the three-phase stator armature winding which forces alternating currents to flow out to the EPS. The combined effect of these AC armature currents is to produce their own *armature reaction* magnetic flux which is of constant magnitude but rotates at the same speed as the rotor. The excitation flux and the armature reaction flux then produce a resultant flux that is stationary with respect to the rotor but rotates at synchronous speed with respect to the stator. As the resultant flux rotates relative to the stator it is necessary to laminate the stator iron core axially in the shaft direction to limit the iron losses due to eddy currents. However, as the magnetic flux is stationary with respect to the rotor, the rotor is normally constructed from a solid steel forging.

If, for some reason, the rotor speed deviates from synchronous, the flux will not be stationary with respect to the rotor and currents will be induced in the damper windings. According to Lenz's law, these currents will oppose the flux change that has produced them and so help restore synchronous speed and damp the rotor oscillations.

Historically, there has been a universal tendency to increase the rated power of new power stations and individual generators as capital cost and operating cost (per-unit megawatt) decrease with increased megawatt rating. This economy of scale results in lower generator mass per-unit megawatt, smaller buildings and power station area, and lower auxiliary equipment and staffing costs. However, the increased use of natural gas since the 1990s has halted the trend of increasing rated power of power stations with combined cycle gas turbine (CCGT) plant utilizing air-cooled generators up to typically 250 MW becoming the norm. Consequently, modern synchronous generators have ratings ranging from about 100 MW to more than 1300 MW and operate at voltages of between 10 and 32 kV.

2.3.2 Exciters and Automatic Voltage Regulators

The generator excitation system consists of an exciter and an AVR and is necessary to supply the generator with DC field current, as shown in Figure 2.2. The power rating of the exciter is usually in the range (0.2–0.8)% of the generator's megawatt rating. In the case of a large generator this power is quite high, in the range of several megawatts. The voltage rating of the exciter will not normally exceed 1000 V, as any higher voltage would require additional insulation of the field winding.

2.3.2.1 Excitation Systems

Generally exciters can be classified as either rotating or static. Figure 2.4 shows some typical systems. In the rotating exciters of Figure 2.4a–c, the excitation current is supplied either by a DC generator or by an AC generator with rectifiers. As DC generators usually have relatively low power ratings, they are cascaded to obtain the necessary output, Figure 2.4a. Because of commutation problems with DC generators, this type of exciter cannot be used for large generators which require large excitation currents.

As the number of cascaded DC generators increases, the dynamic properties of the exciter deteriorate, resulting in an increase in the equivalent time constant. Nowadays, DC generators have been almost entirely replaced by alternators, which are simpler and more reliable. This change to alternators has been possible because of advances in power electronics which allow cheap, high-power rectifiers to be used in conjunction with the AC exciter.

The exciter shown in Figure 2.4b is a reluctance machine (inductor generator) operating at about (500–600) Hz so that the rectified current requires little smoothing. With this exciter both windings (AC and DC) are on the stator side. One disadvantage of this system is that slip rings are required to feed the rectified excitation current to the

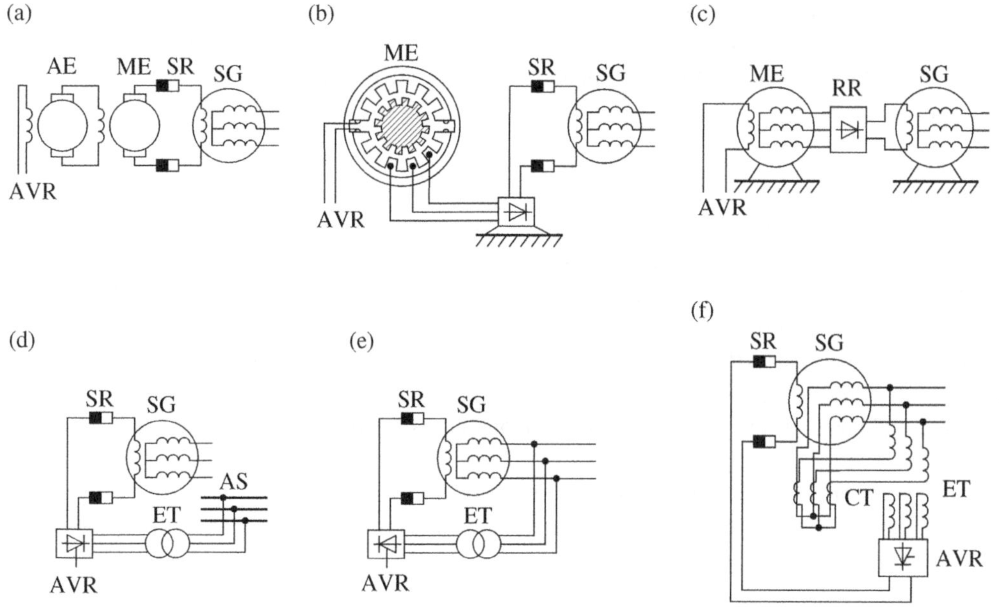

Figure 2.4 Typical exciter systems: (a) cascaded DC generators; (b) reluctance machine with rectifier; (c) inside-out synchronous generator with rotating rectifier; (d) controlled rectifier fed from the auxiliary supply; (e) controlled rectifier fed from the generator terminals; (f) controlled rectifier fed by the generator's voltage and current. SG – synchronous generator; SR – slip rings; ME – main exciter; AE – auxiliary exciter; RR – rotating rectifier; ET – excitation transformer; AS – auxiliary service busbars; CT – current transformer; AVR – automatic voltage regulator.

rotating field winding of the main generator. A further disadvantage is that the exciter itself tends to be quite large. This is a direct result of the way in which the sinusoidal flux changes, necessary to induce the alternating emf in the armature, are produced solely by the changes in reluctance due to the rotation of the salient rotor teeth.

The exciter shown in Figure 2.4c has neither commutator nor slip rings. The principal excitation source is an inside-out synchronous machine with the field winding on the stator and armature winding on the rotor. The induced current is rectified by diodes, which are also mounted on the rotor, and fed directly to the excitation winding of the main generator. One limitation of this type of exciter is that the current supplied to the main generator can only be controlled indirectly via field control of the exciter. This tends to introduce a time constant of about (0.5–1) s into the exciter control system. One solution to this problem is to use rotating thyristors, rather than diodes, and control the exciter output via the firing angle of the thyristors. Unfortunately, controlling the firing angle of a rotating thyristor is not easy and the reliability of such systems tends to be compromised by stray fields causing unscheduled thyristor firing.

Some alternative exciter systems using static thyristor converters are shown in Figure 2.4d–f. In these exciters the thyristor rectifiers are controlled directly by a voltage regulator. The main difference between the systems is in the type of supply used. Figure 2.4d shows an exciter supplied by an additional auxiliary service transformer. Figure 2.4e shows an alternative, and simpler, solution in which the exciter is fed from the generator output via a transformer. However, should a short-circuit occur, particularly one close to the generator terminals, the decrease in the generator terminal voltage will result in a possible loss of excitation. With careful design the exciter can operate when the short-circuit is further away from the generator terminals, for example at the high-voltage terminals of the step-up transformer. More flexibility can be obtained by modifying the supply to the rectifier, as shown in the exciter design of Figure 2.4f. In this system the generator does not lose excitation because its supply voltage is augmented, or compounded, by a component derived from the generator load current.

The main disadvantage of all static exciters is the necessity of using slip rings to feed current to the rotor of the main generator. This is offset to a large extent by the rapid speed with which they can react to control signals. As the cost of high-power rectifiers decreases, and reliability increases, static exciters are becoming the main source of excitation for high-power generators.

2.3.2.2 Automatic Voltage Regulators

The AVR regulates the generator terminal voltage by controlling the amount of current supplied to the generator field winding by the exciter. The general block diagram of the AVR subsystem is shown in Figure 2.5. The measuring element senses the current, power, terminal voltage, and frequency of the generator. The measured generator terminal voltage V_g is compensated for the load current I_g and compared with the desired reference voltage V_{ref} to produce the voltage error ΔV. This error is then amplified and used to alter the exciter output, and consequently the generator field current, so that the voltage error is eliminated. This represents a typical closed-loop control system. The regulation process is stabilized using a negative feedback loop taken directly from either the amplifier or the exciter.

The load compensation element, together with the comparator, is shown in Figure 2.6. The voltage drop across the compensation impedance $\underline{Z}_C = R_C + jX_C$ due to the generator current $\underline{I}_g$ is added to the generator voltage $\underline{V}_g$ to produce the compensated voltage V_C according to the function

$$V_C = |\underline{V}_C| = \left|\underline{V}_g + (R_C + jX_C)\underline{I}_g\right| \tag{2.1}$$

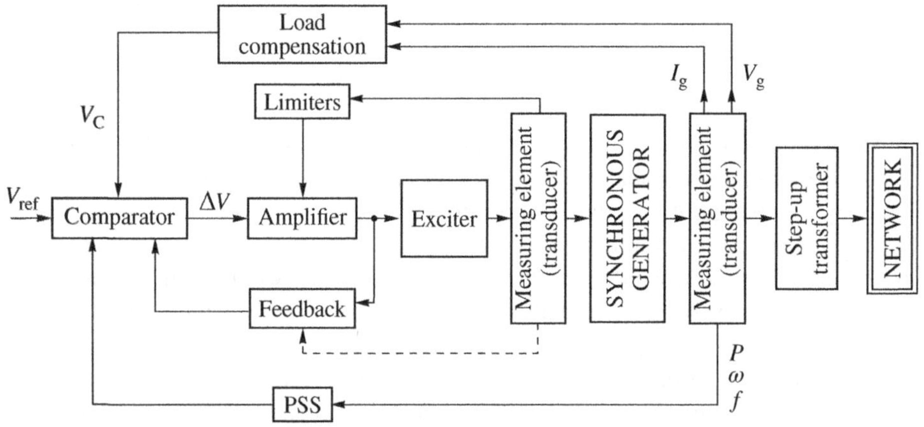

Figure 2.5 Block diagram of the excitation and AVR system. PSS – power system stabilizer.

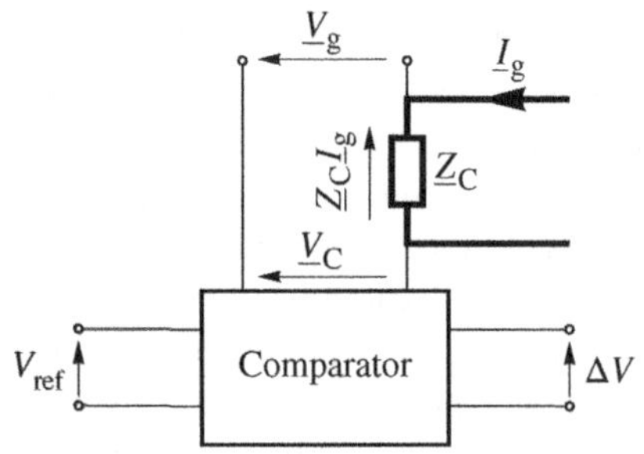

Figure 2.6 Load compensation element together with the comparator.

If the load compensation is not employed $\underline{Z}_C = \underline{0}$ then $V_C = V_g$ and the AVR subsystem maintains constant generator terminal voltage. The use of load compensation ($\underline{Z}_C \neq \underline{0}$) effectively means that the point at which constant voltage is maintained is "pushed" into the network by a distance that is electrically equal to the compensation impedance. The assumed direction of the phasors in Figure 2.6 means that moving the voltage regulation point toward the grid corresponds to a negative compensation impedance.

In the case of parallel generators supplying a common busbar the compensation impedance must be smaller than the impedance of the step-up transformer in order to maintain stable reactive power dispatch between the parallel generators. Usually, $X_C \approx -0.85X_T$, where X_T is the reactance of the step-up transformer. In this case the regulator maintains a constant voltage value at a distance of $0.85X_T$ from the generator terminals toward the network or at a distance of $0.15X_T$ from the high-voltage terminal toward the generator.

The AVR subsystem also includes a number of limiters whose function is to protect the AVR, exciter, and generator from excessive voltages and currents. They do this by maintaining the AVR signals between preset limits. Thus the amplifier is protected against excessively high input signals, the exciter and the generator against too high a field current (*overexcitation limiter* [OEL]), and the generator against too high armature current (*stator current limiter* [SCL]), and too high a power angle (*underexcitation limiter* [UEL]). The last three limiters have built-in time delays to reflect the thermal time constant associated with the temperature rise in the winding.

A power system stabilizer (PSS) is sometimes added to the AVR subsystem to help damp power swings in the system. PSS is typically a differentiating element with phase shifting corrective elements. Its input signals may be proportional to rotor speed, generator output frequency, or the electrical active power output of the generator.

The AVR parameters have to be chosen in such a way that an appropriate quality of voltage regulation is maintained. For small disturbances, that quality can be assessed by observing the dynamic voltage response of a generator to a step change in the reference value. This is illustrated in Figure 2.7 for a step change of reference value by $\Delta V = V_{ref+} - V_{ref-}$. Three indices assess the quality of regulation: (i) settling time t_ε; (ii) overshoot ε_p; and (iii) rise time t_r. These indices are defined as follows:

- Settling time t_ε is the time necessary for the signal to reach its steady-state value with a tolerance of ε.
- Overshoot ε_p is the difference between the peak value of the voltage and a reference value, usually expressed as a percentage of the reference value.

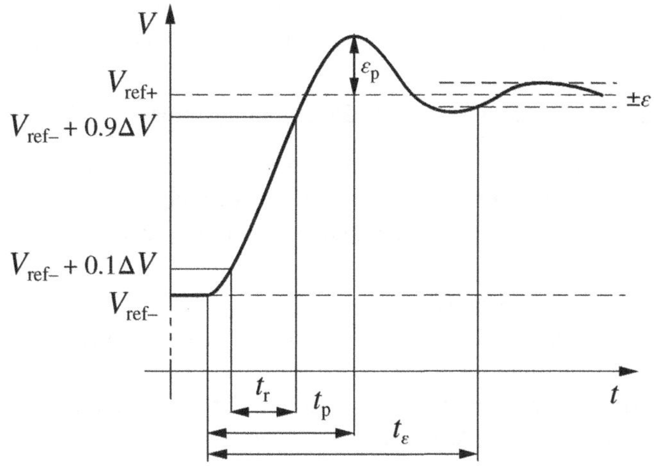

Figure 2.7 Dynamic voltage response to the step change in reference value.

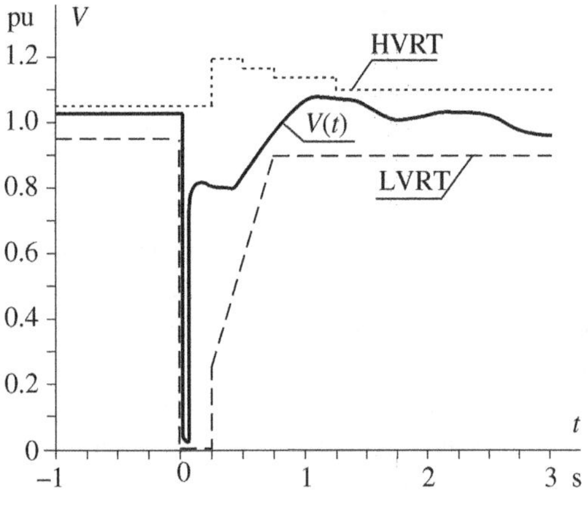

Figure 2.8 Example of fault ride through profile of a synchronous generator driven by steam turbine.

- The time to reach the peak value is denoted as t_p.
- Rise time t_r is the time taken for the voltage to rise from 10 to 90% of $\Delta V = V_{\text{ref}+} - V_{\text{ref}-}$. In this interval the speed at which the voltage increases is about $0.8 \cdot \Delta V/t_\text{r}$.

Usually, it is assumed that with an accuracy of regulation $\varepsilon \leq 0.5\%$ and with 10% step change of the voltage reference value, the settling time is $t_\varepsilon \leq 0.3$ s for static exciters and $t_\varepsilon \leq 1.0$ s for rotating exciters. The overshoot is usually required to be $\varepsilon_\text{p} \leq 10\%$ for step changes of the reference value when the generator is off-load. The speed of voltage increase should not be less than 1.5 U_ref per second.

All generating units operating on extra high voltage (EHV) and high voltage (HV) networks must have the capability to stay connected with the network during and after the fault occurs if the time profile of voltage at the connection point of the generating unit passes between the HVRT and LVRT boundaries as shown in Figure 2.8. The lower LVRT boundary, referred to as *low voltage ride through*, is a survival curve for electric drives and devices operating in the auxiliary service of the power station. If the voltage falls under this curve, some motors may stall. Parameters of the LVRT curves for various types of generating units may be found in documents describing requirements for generating units and grid connections such as ENTSO-E (2011). The upper HVRT boundary, referred to as *high voltage ride through*, results from the permissible operation of induction motors with supply voltage higher than the rated value. Details can be found in standards relating to motors such as NEMA Publications (2016).

The proper time profile of voltage $V(t)$ can be assured by the use of fast protections of busbars and network elements (line and transformers) and fast AVR of generating units with small overshot and small settling time.

2.3.3 Turbines and Their Governing Systems

In an EPS, the synchronous generators are normally driven by either steam turbines, gas turbines, or hydro turbines, as shown in Figure 2.2. Each turbine is equipped with a governing system to provide a means by which the turbine can be started, run up to the operating speed, and operated on load with the required power output.

2.3.3.1 Steam Turbines

In coal-burn, oil-burn, and nuclear power plants the energy contained in the fuel is used to produce high-pressure (HP), high-temperature steam in the boiler. The energy in the steam is then converted to mechanical energy in axial flow steam turbines. Each turbine consists of a number of stationary and rotating blades concentrated into groups, or stages. As the HP steam enters the fixed set of stationary blades it is accelerated and acquires increased kinetic energy as it expands to a lower pressure. The stream of fluid is then guided onto the rotating blades, where it experiences a change in momentum and direction thereby exerting a tangential force on the turbine blade and output torque on the turbine shaft. As the steam passes axially along the turbine shaft its pressure reduces so

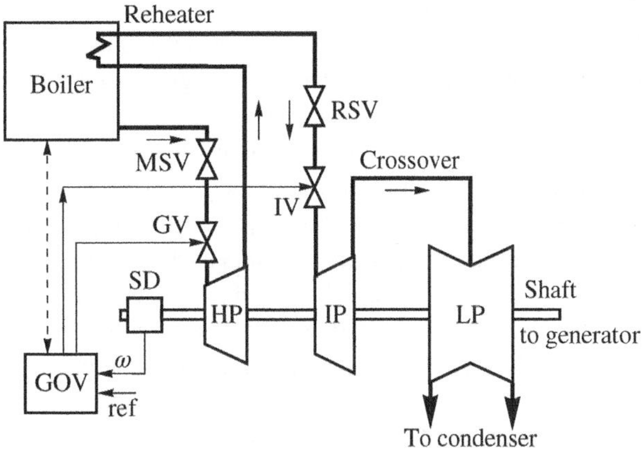

Figure 2.9 Steam configuration of a tandem compound single-reheat turbine.

its volume increases and the length of the blades must increase from the steam entrance to the exhaust to accommodate this change. Typically, a complete steam turbine will be divided into three or more stages, with each turbine stage being connected in *tandem* on a common shaft. Dividing the turbine into stages in this way allows the steam to be reheated between stages to increase its enthalpy and consequently increase the overall efficiency of the steam cycle. Modern coal-fired steam turbines have thermal efficiency reaching 45%.

Steam turbines can be classified as either *non-reheat, single-reheat*, or *double-reheat* systems. Non-reheat turbines have one turbine stage and are usually built for use in units of below 100 MW. The most common turbine configuration used for large steam turbines is the single tandem reheat arrangement shown diagrammatically in Figure 2.9. In this arrangement the turbine has three sections the *high-pressure* (HP), *intermediate-pressure* (IP), and *low-pressure* (LP) stages. Steam leaving the boiler enters the steam chest and flows through the *main emergency stop valve* (MSV) and the *governor control valve* (GV) to the HP turbine.[1] After partial expansion the steam is directed back to the boiler to be reheated in the *heat-exchanger* to increase its enthalpy. The steam then flows through the *reheat emergency stop valve* (RSV) and the *intercept control valve* (IV) to the IP turbine, where it is again expanded and made to do work. On leaving the IP stage the steam flows through the crossover piping for final expansion in the LP turbine. Finally, the steam flows to the *condenser* to complete the cycle. Typically, the individual turbine stages contribute to the total turbine torque in the ratio 30% (HP): 40% (IP): 30% (LP).

The steam flow in the turbine is controlled by the *governing system* (GOV). When the generator is synchronized the emergency stop valves are kept fully open and the turbine speed and power regulated by controlling the position of the GV and the IV. The speed signal to the governor is provided by the *speed measuring device* (SD). The main amplifier of the governing system and the valve mover is an *oil servomotor* controlled by the *pilot valve*. When the generator is synchronized, the *emergency stop valves* are only used to stop the generator under emergency conditions, although they are often used to control the initial start-up of the turbine.

Besides the tandem compound single reheat turbine shown in Figure 2.9, other turbine arrangements are also used. Double-reheat turbines have their first HP section divided into the very-high-pressure (VHP) turbine and the HP turbine with reheat between them. In this arrangement the individual turbines typically contribute to the total torque in the ratio: 20% (VHP): 20% (HP): 30% (IP): 30% (LP). Control valves are mounted after each of the reheaters and before the VHP section. In contrast to the single-shaft arrangements just described, *cross-compound, two-shaft turbines* are sometimes used where one of the shafts rotates at half the speed of the other. These turbines may have single- or double-reheat steam cycles.

---

1 The governor valves are also referred to as the main control valves or the HP control valves, while the intercept valves are also referred to as the IP intercept valves or simply the IP control valves.

2.3.3.2 Gas Turbines

Unlike steam turbines, gas turbines do not require an intermediate working fluid, and instead the fuel thermal energy is converted into mechanical energy using the hot turbine exhaust gases. Air is normally used as the working fluid with the fuel being natural gas or heavy/medium fuel oil. The most popular system for gas turbines is the open regenerative cycle shown in Figure 2.10, which consists of a *compressor* C, *combustion chamber* CH, and *turbine* T. The fuel is supplied through the governor valve to the *combustion chamber* to be burnt in the presence of air supplied by the *compressor*. The hot, compressed air, mixed with the combustion products, is then directed into the *turbine* where it expands and transfers its energy to the moving blades in much the same way as in the steam turbine. The exhaust gases are then used to heat the air delivered by the compressor. There are also other, more complicated cycles that use either compressor intercooling and reheating, or intercooling with regeneration and reheating. The typical efficiency of a gas turbine plant is about 35%.

2.3.3.3 Combined Cycle Gas Turbines

A significant technological step forward in the use of gas turbines came with the introduction of the CCGT, illustrated in Figure 2.11. In this system the exhaust heat from the gas turbine is directed into a *heat-recovery boiler* (HRB) to raise steam, which is then used to generate more electricity in a steam-driven generating unit. Generally, the temperature of the gas turbine exhaust gases is quite high, typically around 535 °C, so by adding a steam turbine cycle at the bottom end of the gas cycle the otherwise wasted heat can be utilized and the overall cycle efficiency significantly increased. Modern CCGT plant can have an efficiency approaching, or even exceeding, 60%. Usually, CCGT power

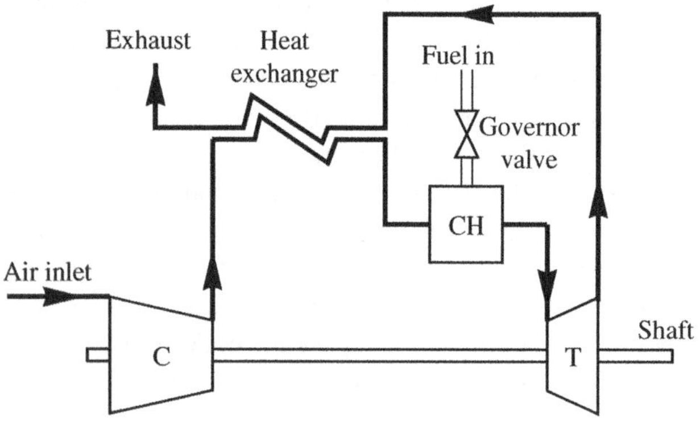

Figure 2.10 Open regenerative cycle of the gas turbine.

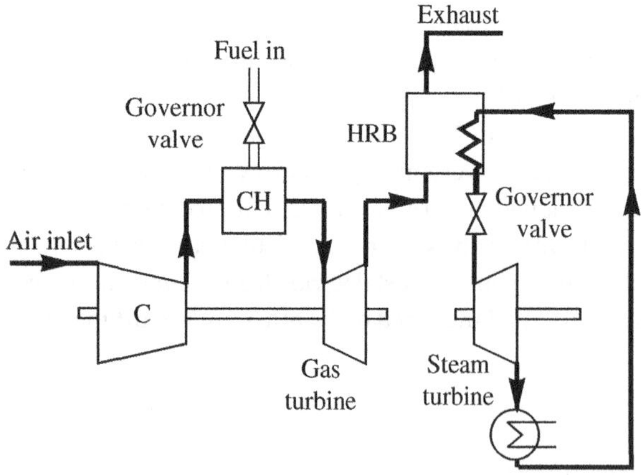

Figure 2.11 Example of a combined cycle gas turbine.

stations utilize the exhaust gases from two or three gas turbines to raise steam for one steam turbine with both types of turbines driving separate generators. More recently, *single-shaft modes* have become popular where both the gas and the steam turbines are mounted on the same shaft and drive the same generator. In some CCGT designs the HRB may be equipped with supplementary firing to increase the temperature of the HP steam. In addition, some combined cycle plants are designed to produce steam for district heating or for use in the process industry.

CCGT plants, apart from higher thermal efficiency, also have other important advantages over more traditional coal-fired plants. They have a short construction time and low capital construction cost, both about half that of the equivalent coal-fired plant, they are relatively clean with almost no SO_2 emission, they require little staffing, and the materials handling problem of gas versus coal and ash is much simpler.

2.3.3.4 Hydro Turbines

The oldest form of power generation is by the use of waterpower. Hydraulic turbines derive power from the force exerted by water as it falls from an upper to a lower reservoir. The vertical distance between the upper reservoir and the level of the turbine is called the *head*. The size of the head is used to classify hydroelectric power plants as *high-head, medium-head*, and *low-head (run-of-river) plants*, although there is no strict demarcation line.

Low- and medium-head hydroelectric plant is built using *reaction turbines* such as the *Francis turbine* shown in Figure 2.12a. Because of the relatively LP head reaction, turbines typically use a large volume of water, require large water passages, and operate at low speed. Because of the low rotational speed, the generators have a large diameter. In operation, water enters the turbine from the intake passage or *penstock* through a spiral case and passes through the stay ring and the movable *wicket gates* onto the *runner*. On leaving the runner, the water flows through the *draft tube* into the *tail-water reservoir*. The movable wicket gates, with their axes parallel to the main shaft, control the power output of the turbine. Francis turbine runners have the upper ends of their blades attached to a crown and the lower ends attached to a band. For low-head operation the runner has no crown or band so that the blades are unshrouded. The blades themselves may be either fixed or adjustable. For adjustable-blade runners

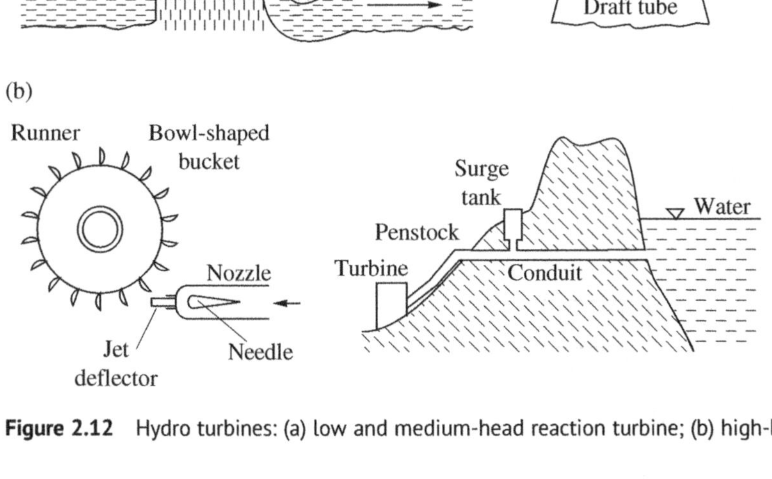

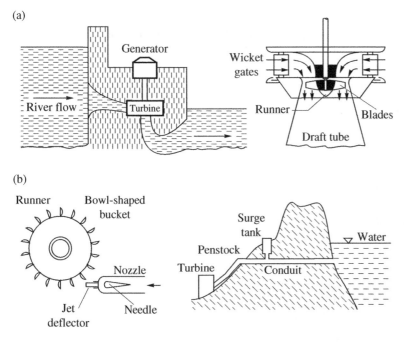

Figure 2.12 Hydro turbines: (a) low and medium-head reaction turbine; (b) high-head Pelton wheel.

the governor can change both the blade angle and the wicket gate opening (*Kaplan-type turbine*). The blades are adjusted by means of an oil-operated piston located within the main shaft.

In high-head hydroelectric power plants, *Pelton wheel* impulse turbines, shown in Figure 2.12b, are used. In these turbines the HP water is converted into high-velocity jets of water by a set of fixed nozzles. The high-velocity water jets impact on a set of bowl-shaped buckets attached around the periphery of the runner which turn back the water, thus impacting the full effect of the water jet to the runner. The size of the jet, and thus the power output of the turbine, is controlled by a needle in the center of the nozzle. The movement of the needle is controlled by the governor. A jet deflector is located just outside the nozzle tip to deflect the jet from the buckets in the event of sudden load reduction.

2.3.3.5 Turbine Governing Systems

For many years turbine governing systems were of a *mechanical-hydraulic* type and used the *Watt centrifugal mechanism* as the speed governor. The original Watt mechanism used two *flyballs* as the speed-responsive device, but on new machines the Watt governor has been replaced by an *electrohydraulic* governor. However, it is useful to understand the operation of the traditional mechanical hydraulic system, shown in Figure 2.13, as it is still in use in various forms on older machines and it is a good way to illustrate the general principle of turbine control.

The pair of spring-loaded weights in the centrifugal governor is driven by a motor that receives its power from the turbine shaft such that the height of the weights depends on the speed. When the turbine mechanical torque is equal to the counteracting generator electromagnetic torque, the rotational speed of the turbine-generator is constant and the position of the weights does not change. If the electrical torque increases, due to a change in load, so that it is greater than the mechanical driving torque, the rotational speed ω decreases and the weights move radially inwards under centrifugal action. This causes point A on the governor floating lever to rise and the floating lever A-B-C rotates around point C. This rotation results in point B and the pilot valve moving upwards so allowing HP oil to flow into the upper chamber of the main servomotor. The differential pressure across the piston now forces the piston to move downwards so partially opening the turbine valve and increasing the turbine power. The

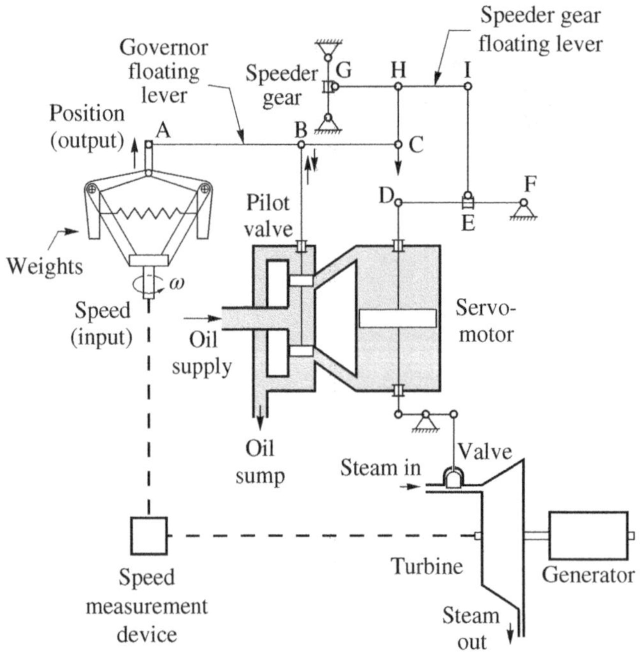

Figure 2.13 Mechanical-hydraulic governing system of the steam turbine.

displacement of the main servomotor piston downwards causes points D, E, I, and H to lower and the speeder gear floating lever to rotate downwards about point G. This lowers point C, around which the lever A-B-C rotates, and partially closes the pilot valve to reduce the oil flow into the upper chamber.

This governing system has two negative feedback loops: the main speed feedback loop through the turbine speed measuring device and the centrifugal governor and the second valve position feedback loop through the steam valve, piston, and points D, E, I, H, and C. This latter feedback loop ensures that the static speed-power characteristic of the turbine has a negative slope. As will be explained later in this section, such a characteristic is fundamental to turbine control as it ensures that any speed increase will be met by a corresponding reduction in turbine torque and vice versa. The slope, or gain, of the characteristic may be changed by moving point E horizontally on the lever D-E-F.

The purpose of the *speeder gear* is twofold. Firstly, it controls the speed of the unsynchronized generator and, secondly, it controls the power output of the synchronized generator. To see how the speeder gear works, assume that the generator is synchronized and that it is required to increase the power output. As the generator is synchronized its speed will be constant and equal to synchronous speed. If the speeder gear is used to raise point G then points C and B and the pilot valve will also rise. HP oil will then enter the upper chamber of the main servomotor, the piston will fall, and the steam valve will be opened, thereby increasing the steam flow through the turbine and the power output. As the servomotor piston falls, so too do points D, E, I, and H. This movement lowers point C and returns the pilot valve to its equilibrium position. The schematic diagram of the mechanical hydraulic governor is shown in Figure 2.14a with the position of the speeder gear setting the *load reference set point*.

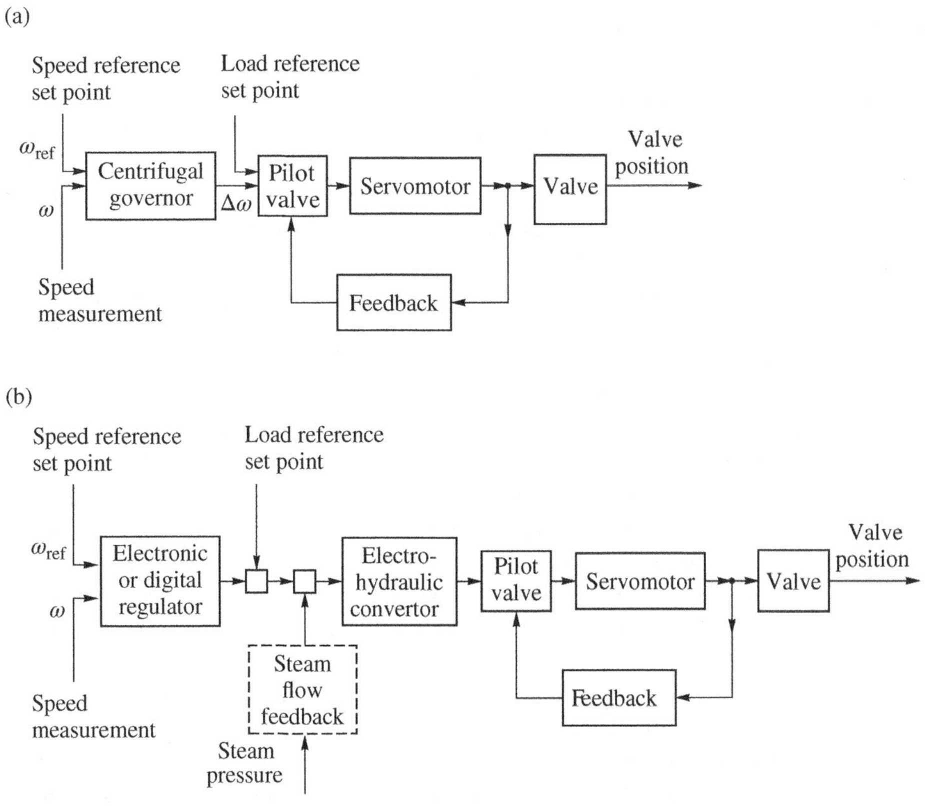

Figure 2.14 Schematic diagram of the governing system: (a) mechanical hydraulic; (b) electrohydraulic.

The main disadvantages of the Watt centrifugal governor are the presence of deadbands and a relatively low accuracy. The size of the deadbands also tends to increase with time due to wear in the moving mechanical elements. For this reason the Watt centrifugal mechanism was first replaced with an electronic regulator and later by a digital regulator. In such systems the turbine rotor speed is measured electronically, with high accuracy, using a toothed wheel and a probe. The resulting electrical signal is amplified and acts on the pilot valve via an electrohydraulic converter. The schematic diagram of the electrohydraulic system in Figure 2.14b shows that its operation does not differ much from that of the mechanical hydraulic system shown in Figure 2.14a, but the flexibility of electronic or digital regulators enables additional control loops to be introduced that link the boiler and the turbine control systems. The dashed line in Figure 2.14b symbolizes the steam flow feedback, and its function is to prevent the valves being opened by the speed regulator when the steam inlet pressure is too low. The reference speed is set electronically in the *speed reference set point*. It is also possible to change the turbine power using an additional signal that is added to the control circuit at the *load reference set point*.

Higher forces are required to move the control gates in hydro turbines than the valves in steam turbines and, as a result, hydro turbine governing systems usually employ two cascaded servomotors. The first, low-power, *pilot servomotor* operates the *distributor* or *relay valve* of the second, high-power, *main gate servomotor*. Just as in the steam turbine, the pilot servomotor has a pilot valve which is controlled either by a mechanical Watt-type governor or by an electronic regulator via an electrohydraulic converter. The turbine governing system is similar to that used in steam turbines (Figure 2.14), but the number of servomotors is higher and the feedback loop has an additional dashpot for reasons described in Chapter 11 (Section 11.3).

2.3.3.6 Turbine Characteristics

For stable operation the turbine must have a power-speed characteristic such that as the speed increases the mechanical input power reduces. Similarly, a decrease in speed should result in an increase in the mechanical power. This will restore the balance between the electrical output power and mechanical input power.

To examine how such a characteristic can be achieved, Figure 2.15 shows the idealized power-speed characteristics for an unregulated and a regulated turbine. Point A is the rated point which corresponds to the optimal steam flow through the turbine, as determined by the turbine designers. Consider first the unregulated characteristic and assume that the turbine is initially operating at point A with the turbine control valve fully open. The generator is assumed to be synchronized with the system, and its speed can only change if the system frequency changes. If, for some reason, the system frequency rises then so too does the speed of the rotor. As the main valve is fully open, the speed increase causes additional losses in the turbine and the efficiency of the steam flow drops (with respect to the optimal point A) with a corresponding reduction in power, as shown by the dashed curve 1. Similarly, a decrease in the system frequency causes the rotor speed to drop with a corresponding drop in power as shown by curve 2. The rapid reduction in turbine power with reduction in system frequency can be explained as follows. The steam flow through the turbine depends on the performance of the boiler and the boiler feed pumps. As the performance of these pumps is strongly dependent on frequency, a reduction in system frequency (and rotor speed) reduces their performance. This causes a decrease in the amount of steam flowing through the turbine and a further drop in the turbine torque.

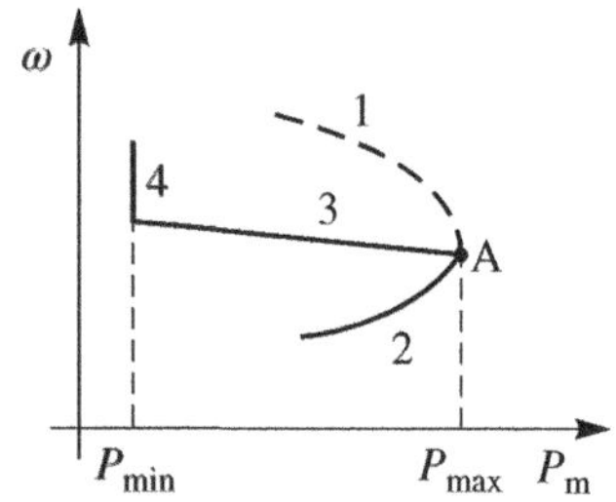

Figure 2.15 Turbine power-speed characteristic for the unregulated turbine (lines 1 and 2) and the regulated turbine (lines 3, 2, and 4).

The task of the turbine governor is to set a characteristic corresponding to line 3 which has a small droop. As explained below, such a characteristic is necessary to achieve stable operation of the turbine.

Let us consider the governor functional diagrams in Figure 2.14. If the steam flow feedback in the electrohydraulic governing system is neglected and the governor response assumed to be dominated by the time constant of the servomotor, both the mechanical hydraulic and the electrohydraulic governors shown in Figure 2.14 may be represented by the simplified block diagram shown in

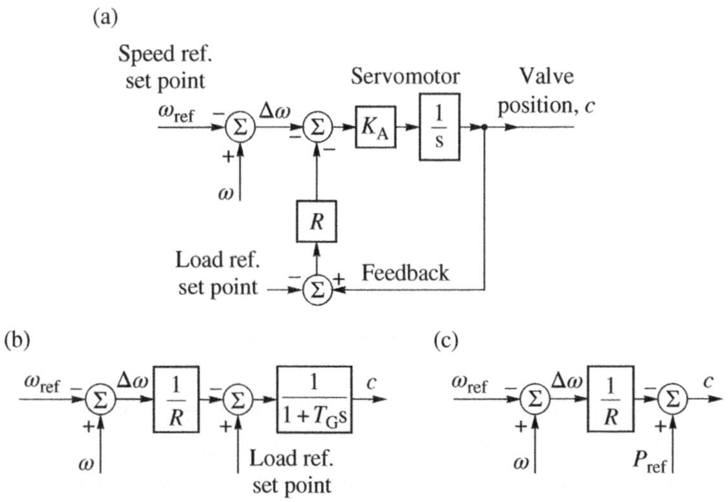

Figure 2.16 Simplified model of the steam turbine governing system: (a) block diagram with negative feedback; (b) equivalent block diagram; (c) equivalent block diagram for the steady state.

Figure 2.16. The coefficient K_A in Figure 2.16a corresponds to the amplification gain of the servomotor, while coefficient R corresponds to the gain of the feedback loop. Transformation of the block diagram allows R to be eliminated from the feedback loop by moving it into the main loop to obtain the block diagram shown in Figure 2.16b, where $T_G = 1/(K_A R)$ and is the effective governor time constant.

The block diagram of Figure 2.16b allows an approximate analysis of the static and dynamic properties of the turbine-governor system. In the steady-state $t \to \infty$, $s \to 0$, and the turbine block diagram can be simplified to that shown in Figure 2.16c, where P_{ref} is the load reference set point expressed as a fraction of the nominal or rated power, P_n. If the valve position c is assumed to vary between 0 (fully closed) and 1 (fully open) then a small change in turbine speed $\Delta\omega = \omega - \omega_{ref}$ will produce a corresponding change in valve position $\Delta c = -\Delta\omega/R$. Normally, $\Delta\omega$ is expressed as a fraction of rated speed ω_n so that

$$\Delta c = -\frac{1}{\rho}\frac{\Delta\omega}{\omega_n} \quad \text{or} \quad \frac{\Delta\omega}{\omega_n} = -\rho\Delta c \tag{2.2}$$

where $\rho = R/\omega_n$ is referred to as the *speed-droop coefficient* or simply the *droop*. The reciprocal of droop $K = 1/\rho$ is the *effective gain* of the governing system. The definition of ρ is illustrated in Figure 2.17.

Physically, droop can be interpreted as the percentage change in speed required to move the valves from fully open to fully closed. If a linear relationship is assumed between the valve position and mechanical power then the turbine power output ΔP_m expressed as a fraction of the nominal or rated power output P_n is given by $\Delta P_m/P_n = \Delta c$ and

$$\frac{\Delta\omega}{\omega_n} = -\rho\frac{\Delta P_m}{P_n} \quad \text{or} \quad \frac{\Delta P_m}{P_n} = -K\frac{\Delta\omega}{\omega_n} \tag{2.3}$$

Equation 2.3 describes an idealized turbine power-speed characteristic. In (P_m, ω) co-ordinates this gives a straight line of gradient ρ shown in Figure 2.15 by line 2. However, it is important to realize that once the steam valves are fully open no more control can be exerted over the turbine so that, should the speed drop, the turbine would follow characteristic 4 in the same way as the unregulated turbine.

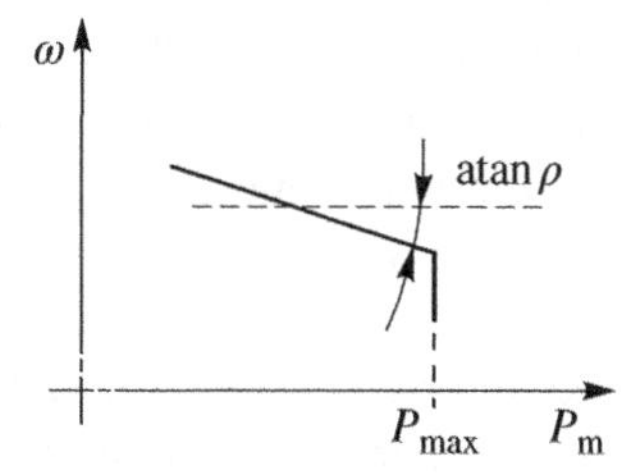

Figure 2.17 Illustration of the definition of the speed-droop coefficient.

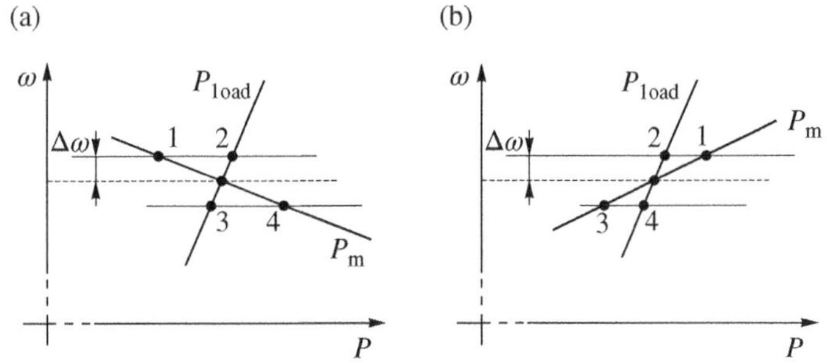

Figure 2.18 The equilibrium point between the turbine power and the load power: (a) stable point; (b) unstable point.

A good control system should ensure that any load fluctuation ΔP_m would only produce a small speed change $\Delta\omega$. This is achieved by making the droop ρ small. However, it should be emphasized that the droop cannot be zero or negative. Figure 2.18 illustrates this point. The system demand is dominated by electrical loads for which an increment in the active power ΔP_{load} is weakly dependent on the change in the system frequency and therefore on the change in the rotational speed of the synchronous generators. As a result the static load characteristic $\omega(P_{load})$ is almost a vertical line in the (P,ω) plane but with a slight positive slope that reflects the frequency dependence. The point of intersection of the load characteristic $\omega(P_{load})$ and the turbine characteristic $\omega(P_m)$ is the equilibrium point where the mechanical torque and the opposing electromagnetic torque acting on the shaft are equal in magnitude and the rotational speed is constant.

Figure 2.18a shows the case where the droop is positive ($\rho > 0$) so that, according to Eq. (2.3), the turbine power increases when its rotational speed decreases. In this case a small disturbance in frequency (turbine speed) causes the system to return automatically to the equilibrium point. For example, if there is a momentary increase in frequency then the disturbed load power (point 2) is greater than the disturbed turbine power (point 1). This means that the opposing electromagnetic torque is greater than the driving torque, and so the system is forced to return to the equilibrium point. Similarly, if the frequency decreases then the load power (point 3) is less than the turbine power (point 4), and this excess mechanical power causes the turbine speed to increase and the system to return to the equilibrium point.

On the other hand, Figure 2.18b shows a case where the droop is assumed to be negative ($\rho < 0$). The situation is then reversed as any increase in frequency will cause the turbine power (point 1) to be greater than the load power (point 2). This would result in an increase in turbine speed and a further increase in the power imbalance. Similarly, a decrease in frequency would result in the turbine power (point 3) being less than the load power (point 4) with further movement away from the equilibrium point. Such a droop characteristic is not resistant to disturbances and the system is unstable.

The case when $\rho = 0$ is the marginal stability case and corresponds to the absence of the negative feedback loop on the valve position. For the Watt regulator of Figure 2.13 this corresponds to the absence of the lever H-E, which realizes the position feedback between the servomotor piston and the pilot valve. Neglecting the steady-state error, the governing system without this negative feedback would be of the constant speed type. Such a governor cannot be used if two or more generators are electrically connected to the same system, since each generator would be required to have precisely the same speed setting (which is technically impossible) or they would "fight" each other, each trying to pull the system frequency to its own setting. If generators connected to the system have speed-droop characteristics with $\rho > 0$ then there will always be a unique frequency at which they will share load. This is described in Chapter 9.

Typical values of the speed-droop coefficient (per unit) are between 0.04 and 0.09 with the lower value corresponding to turbogenerators and the higher to hydrogenerators. This is connected with the relative ease, and speed, with which a hydro plant can accept a change in load compared with a thermal plant.

2.3.3.7 Governor Control Functions

Having established the basic workings of the governor, it is now prudent to look at the overall control functions required of a practical turbine governor. These control functions can be sub-divided into *run-up control*, *load/frequency control*, *overspeed control*, and *emergency overspeed trip*. *Run-up control* of the unsynchronized generator is not considered further other than to comment that this is one area where control may be carried out using the main stop valves with both sets of control valves and the interceptor stop valves fully open. Primary *load/speed control* and secondary *frequency/tie-line power control* are achieved via the GVs with the IVs fully open. This control action is fundamental to turbine operation and is discussed in more detail in Chapter 9. If a severe disturbance occurs then the turbine speed may increase quickly and the aim of the *overspeed control* is to limit the maximum overspeed to about 110%. If overspeed control were possible only via the governor control valves then, depending on the droop setting, the generator speed would increase to, say, 105% (5% droop) before the main valves were shut. Although this would quickly reduce the HP torque, the entrapped steam in the reheater would reduce only slowly, typically with a time constant of 5 s or more, resulting in a slow decay of both the IP and the LP torque. As these typically contribute 70% of the torque, the turbine speed would continue to increase until the steam flow had time to reduce. Consequently, the purpose of the overspeed control is to shut the IVs and, as these are at the inlet to the IP turbine, they have an immediate effect on reducing the IP and LP torque, thus limiting the overspeed. Typically, the IVs will be kept fully open until the generator speed has reached an overspeed of, say, 104%, when the IVs will be closed.

In addition to IV closure, electrohydraulic governors may also be equipped with additional *fast-valving* control logic, which uses auxiliary control signals such as acceleration, electrical power, generator current, and so on, to fast-close the control valves when a large disturbance close to the generator is sensed. These fast-valving control functions are discussed in more detail in Section 10.2.2. The final stage of protection, the *emergency overspeed trip*, is independent of the overspeed control. If this trip function is activated, both sets of control valves and the emergency stop valves are shut and the boiler tripped to ensure that the turbine is quickly stopped.

2.4 Substations

A substation can be regarded as a point of electrical connection where the transmission lines, transformers, generating units, system monitoring, and control equipment are connected together. Consequently, it is at substations that the flow of electrical power is controlled, voltages are transformed from one level to another, and system security is provided by automatic protective devices.

All substations consist of a number of incoming and outgoing circuits located in *bays*. These incoming and outgoing circuits are connected to a common *busbar* system and are equipped with apparatus to switch electrical currents, conduct measurements, and protect against lightning. Each electrical circuit can be divided into a *primary* circuit and a *secondary* circuit. The primary circuit includes the transmission line, power transformer, busbars, and so on, and the high-voltage side of voltage and current transformers. The secondary circuit consists of the measurement circuits on the low-voltage side of the voltage and current transformers and the control circuits for the circuit-breakers and isolators, protection circuits.

The busbar constitutes a point of electrical contact for individual lines and transformers. In indoor substations the busbar consists of flat conductors made of aluminum or copper and supported by insulators, while in outdoor substations the busbars are stranded conductors made of steel and aluminum and suspended on insulators. A number of different busbar arrangements are possible, each of which differs in the flexibility of possible electrical connections and the ease with which maintenance can be carried out without disturbing either the operation of the substation or system security. The type of busbar system used will depend on the role and importance of the substation in the EPS, the voltage level, the installed capacity, and the expected reliability of network operation. Bigger substations tend to use more elaborate busbar systems requiring higher capital investment and operating costs. A description of different types of substation layout can be found in Giles (1970) and McDonald (2003).

2.5 Transmission and Distribution Network

The transmission and distribution network connects all the power stations into one supplying system and transmits and distributes power to individual consumers. The basic elements of the network are the overhead power lines, underground cables, transformers, and substations. Auxiliary elements are the series reactors, shunt reactors, and compensators, switching elements, metering elements, and protection equipment.

2.5.1 Overhead Lines and Underground Cables

Overhead lines are universally used to transmit electrical energy in high-voltage transmission systems. while underground cables are normally only used in low- and medium-voltage urban distribution networks. Because of their high cost, and the technical problems associated with the capacitive charging current, high-voltage underground cables can only be used under special circumstances such as in densely populated urban areas, wide river crossings, or areas of major environmental concern. For example, short-distance cables are sometimes used to connect a power station to a substation.

Whenever current flows through any network element, active power is lost. As this power loss is proportional to the square of the current, transmission lines operate at high voltage and low current. Generally, the more power that is sent over a transmission line, the higher will be its voltage. For practical reasons there is a standardization of voltage levels within different regions of the world. Unfortunately, these standard voltages tend to vary slightly between regions but are not too dissimilar. Typical transmission voltage levels are 110, 220, 400, and 750 kV for Continental Europe; 132, 275, and 400 kV for the United Kingdom; and 115, 230, 345, 500, and 765 kV for the United States.

The maximum theoretical voltage value at which an overhead transmission line can be built is limited by the electrical strength of air and is estimated to be about 2400 kV. Currently, the maximum voltage at which commercial lines have been built is 765 kV (Canada) and 750 kV (former Soviet Union). Experimental lines have been built to operate at 1100 kV in Japan and 1200 kV in the former Soviet Union (CIGRE 1994).

Because of the high right-of-way costs associated with overhead lines, multi-circuit lines are usually built where more than one three-phase circuit is supported on the same tower. If a large increase in transmitted power is predicted for the future, space may be left on the transmission towers for extra circuits to be added later.

Distribution networks generally operate at lower voltages than the transmission network. Here the voltage standards used, both by different countries and by different areas in one country, can be quite varied, partly because of the way the system has developed. Historically, different parts of a network may have belonged to different private companies, each of which would have followed its own standardization procedures. For example, there are 12 different standard distribution voltages in the United States, in the range between 2.4 and 69 kV. In the United Kingdom, the distribution voltages are 6.6, 11, 33, and 66 kV.

2.5.2 Transformers

Transformers are necessary to link parts of the power systems that operate at different voltage levels. In addition to changing voltage levels, transformers are also used to control voltage and are almost invariably equipped with taps on one or more windings to allow the turns ratio to be changed. Power system transformers can be classified by their function into three general categories:

- *generator step-up transformers*, which connect the generator to the transmission network, and *unit transformers* which supply the auxiliary service (Figure 2.2);
- *transmission transformers* which are used to connect different parts of the transmission network, usually at different voltage levels, or connect the transmission and distribution networks;
- *distribution transformers*, which reduce the voltage at load centers to a low voltage level required by the consumer.

Generator and transmission transformers have ratings from several tens of megavolt-amperes to more than 1000 MVA and are usually oil cooled. The transformer core is placed inside a tank filled with oil which acts as both a coolant and insulator for the transformer windings. Heat, due to core loss and ohmic loss in the transformer windings themselves, is removed from the oil through external radiators. The circulation of oil inside the transformer is either natural or forced. The air circulation outside the transformer is usually forced using fans. Because of transportation problems, large, high-power transformers are usually constructed as three separate single-phase transformers. Smaller power transformers are usually of an integrated three-phase design.

Generator transformers step up the voltage from the generator level of typically (10–20) kV to the transmission or sub-transmission voltage. In a power station employing large generators of typically (200–500) MW and above, each generator may have its own transformer consisting of three interconnected *two-winding transformers*. In contrast to this the generators in a smaller power station may operate with two generators connected to one *three-winding*, three-phase transformer.

Generator step-up transformers are usually Δ-Y connected with the neutral grounded. The delta low-voltage winding closes the path for the circulating current resulting from asymmetrical loading and the undesirable third-harmonic magnetizing current, caused by the nonlinear B-H characteristic of the transformer core, so that these currents remain trapped inside it. In a large power station with many generating units some of the transformer neutrals may not be grounded to limit the single-phase short-circuit currents in the transmission network.

Transmission transformers connect different parts of the transmission and sub-transmission networks operating at different voltage levels, supply distribution networks, and connect large industrial consumers directly to the transmission network, as shown in Figure 2.1. The windings of the transformers tying transmission and sub-transmission networks are normally Y-Y connected with the neutral grounded. These transformers often also have a low-power, medium-voltage, Δ-connected tertiary winding to provide a path for the circulating current when the high-voltage winding is asymmetrically loaded. This additional winding can also be used to supply local loads inside a substation or to connect a reactive power compensator.

If the required transformation ratio is not too high, the two-winding transformer shown in Figure 2.19a can be replaced by the one-winding *autotransformer* shown in Figure 2.19b. In the autotransformer, parts of the primary winding, w_1, and the secondary winding, w_2, are common, giving an obvious economy. Autotransformers are normally used to connect networks at consecutive voltage levels, for example 132/275, 275/400 kV in the United Kingdom; 138/230, 230/345, 345/500 kV in the United States; and 110/220, 220/400 kV in Continental Europe.

Distribution networks are normally supplied from transmission and sub-transmission networks by transformers with the high-voltage side connected in star and the medium-voltage side connected in delta to help minimize any possible load asymmetry. Autotransformers linking parts of distribution networks operating at different, but close, voltage levels are usually star connected with the neutral grounded.

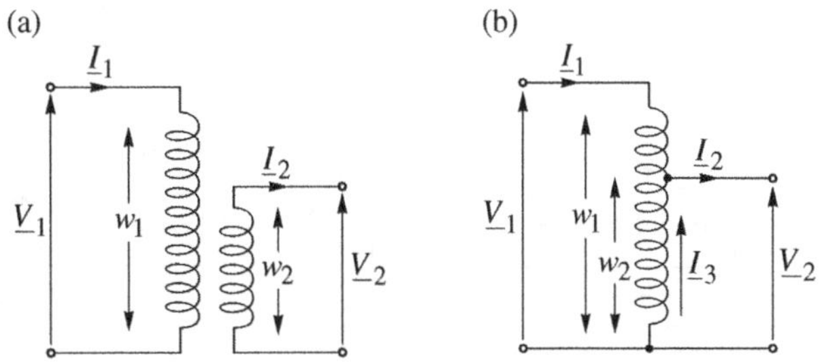

Figure 2.19 Transmission transformers: (a) two-winding transformer; (b) autotransformer.

Each of the above transformers can be made with a controllable voltage transformation ratio and with, or without, phase shift control. The former is used for voltage or reactive power flow control, while the latter controls the flow of active power.

2.5.2.1 Tap-changing Transformers

Controlling the voltage transformation ratio without phase shift control is used for generator step-up transformers as well as for transmission and distribution transformers. The easiest way to achieve this task is by using tap changers to change the transformation ratio.

Control of the transformation ratio without phase shift control is usually achieved by using taps on one of the windings. In this way the transformation ratio is changed step-wise, rather than continuously. Tap-changing facilities can be made to operate either off load or on load.

The off-load tap changer requires the transformer to be de-energized while tap changing takes place. A typical range of regulation is $\pm 5\%$. This method is used for low-rating transformers operating in medium- and low-voltage distribution networks. Change of the transformer ratio is usually done manually according to season, typically twice a year.

The *under-load tap changer* (ULTC), also called the *on-load tap changer* (OLTC) or *load tap changer* (LTC), allows the taps to be changed while the transformer is energized. A typical range of regulation is $\pm 20\%$.

A simplified block diagram of the control system of a regulating transformer is shown in Figure 2.20. The transformer is subject to disturbances $z(t)$ which could be network loading changes or network configuration changes. The regulator acts on the transformer via a tap changer. The regulator receives signals of measurements of the voltage $\underline{V}_T$ and current $\underline{I}_T$ on a chosen side of the transformer. By comparing these with a reference value, it forms a control signal and executes a required control task. The regulator may additionally obtain external control signals V_X from, for example, a supervisory controller.

Depending on the point of installation of the transformer and its function in the system, the controlled variables may be the voltage at a certain point in the network or reactive power flowing through the transformer. When controlling the voltage at a desired location, the control signal is obtained using current compensation, that is by adding the voltage drop on the assumed compensation impedance to the transformer voltage, as in Figure 2.6.

Transformer taps may be situated in the same tank as the main winding. The taps are usually installed on the high-voltage side of the transformer (because of the lower current) and near the neutral end of the winding (where the voltage with respect to ground is smallest). In autotransformers taps are also on the high-voltage side, but near the common part of the winding. The regulator of the tap changer usually tries to minimize the number, or frequency, of tap changes per day in order to prolong the life of the tap changer.

The principle of operation of the OLTC is shown in Figure 2.21. For simplicity, only five taps and a part of the winding have been shown. The choice of taps selected for operation is done by two tap selectors, S1 and S2.

In the first solution, Figure 2.21a, both tap selectors are set on the same tap during normal operation. The load current of the transformer flows through both parallel chokes X. This causes an increase in the reactance by the value $X/2$, which is a disadvantage of the solution. When the tap is to be changed, first selector S1 is moved while selector S2 remains at the initial position. During that time, the part of the windings between the taps is

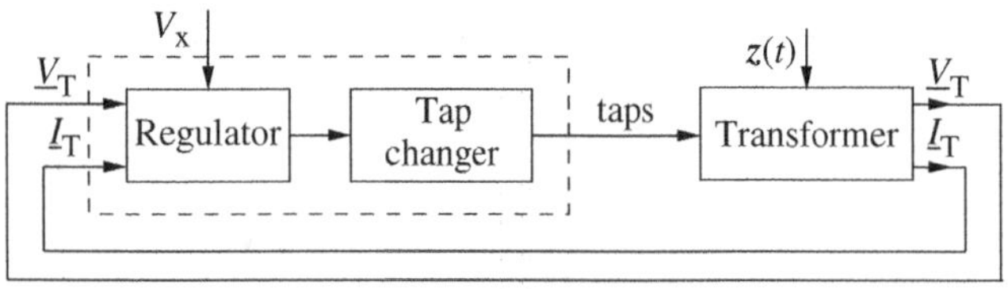

Figure 2.20 Block diagram of the transformation ratio control system.

short-circuited by reactance 2*X*. This reactance reduces the short-circuit current and it is an advantage of the solution. Then selector S2 is moved so that both selectors are in a new position on the chosen tap. An appropriate tap drive system is necessary to ensure that the selectors are moved without creating a gap in the circuit.

In the second solution, Figure 2.21b, two resistors *R* and a diverter switch D are used. During normal operation on a chosen tap, the diverter switch D is at the extreme position, that is the left position in the diagram. The load current flows through the conductor short-circuiting the resistor. Just before the switching sequence is started, the diverter switch D is moved to the middle position. The current then flows through two parallel resistors and the circuit resistance is increased by *R*/2. The selector S1 is moved to the new position and the part of the winding between the taps is momentarily short-circuited by resistance 2*R*. This resistance limits the short-circuit current which is an advantage of the solution. Then selector S2 is moved so that both selectors are in a new position on the chosen tap. Finally, the diverter switch D returns to its extreme left position.

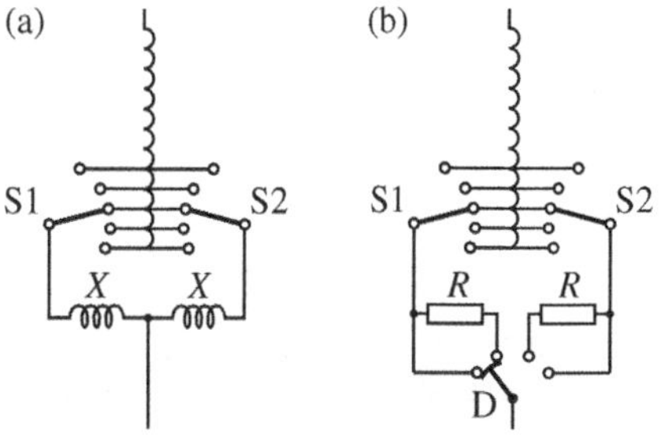

Figure 2.21 Principle of operation of the on-load tap changer: (a) with reactors; (b) with resistors.

Both elements (diverter switch and the selectors) may be parts of one mechanism but they operate in two separate compartments. Both selectors and resistors are located in the lower compartment, which is in the transformer's tank. The diverter switch is in the upper compartment with its own oil, outside the transformer's tank. Thanks to this separation, oil used during tap changes (i.e. during breaking the circuit) does not contaminate oil in the transformer's tank. Oil in the small diverter compartment is replaced more often than oil in the big transformer's tank. The resistors are used only momentarily and if the switching mechanism blocks when the resistors are operating, the transformer must be disconnected.

Sometimes the tap selectors and diverter switch are combined into one switch, as shown in Figure 2.22. That switch is made up of several fixed contacts spread in a circle and one triple moving contact. For simplification, only the left side of the fixed contacts has been shown together with corresponding taps. An important role for the switching sequence is played by the empty space between the fixed contacts. The moving contact consists of the main (middle) contact and two side contacts. There are resistors in the circuit of those contacts. Movement of this triple contact is executed in the following way. The width of the contacts and of the empty space is selected such that, before the main moving contact leaves a given fixed contact, side contacts move from the neighboring empty spaces into neighboring fixed contacts. This causes a momentary short-circuit, through resistors, of the neighboring fixed contacts. Further movement of the triple contact causes a connection of the main contact with the fixed contact and movement of the side contacts into empty space. The short circuit is interrupted and normal operation through a new fixed contact, and on a new tap, is restored.

2.5.2.2 Phase Shifting Transformers

Phase shifting transformers control the voltage transformation ratio together with the voltage phase angle in order to control active power flows in transmission networks. The regulation is executed using a *series transformer* referred to as a *booster transformer* which is fed by an excitation transformer. Examples of connecting booster and excitation transformers are shown in Figure 2.23.

With the transformer connections shown in Figure 2.23, the voltage can be regulated in both magnitude and phase with the degree of regulation depending on the connection made between the tertiary winding of the main transformer MT (or the excitation transformer ET) and the series transformer ST. A number of possible schemes are illustrated Figure 2.24.

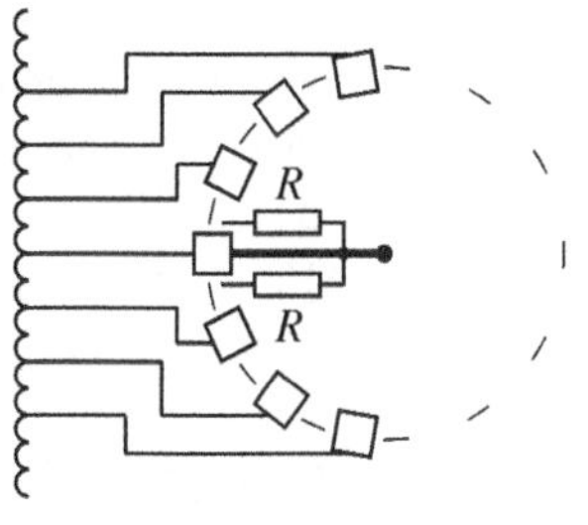

Figure 2.22 Principle of operation of the selector-type tap changer.

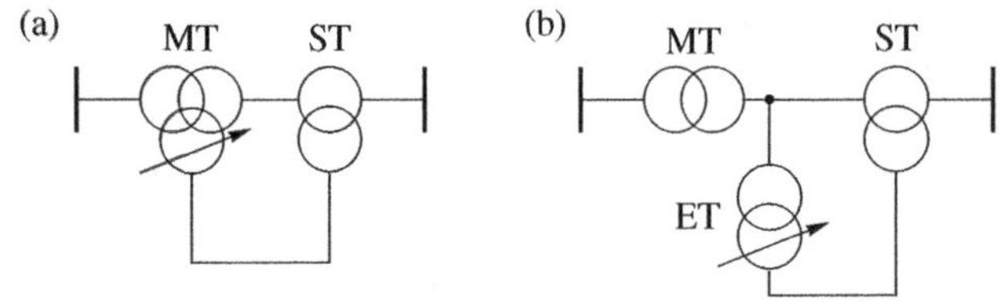

Figure 2.23 Two ways of supplying the series transformer: (a) from the tertiary winding of the main transformer; (b) from a separate excitation transformer. MT – main transformer; ST – series (booster) transformer; ET – excitation transformer.

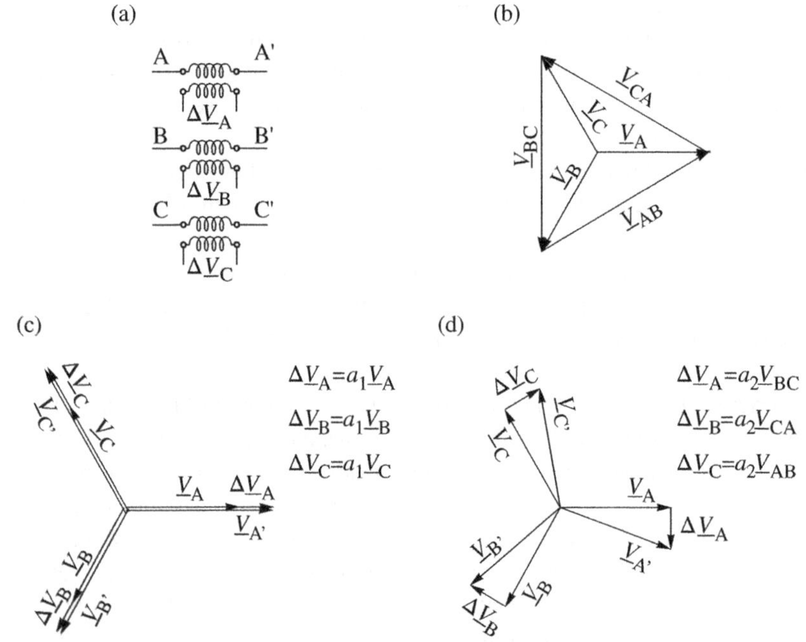

Figure 2.24 Complex transformation ratio: (a) windings of the series transformer; (b) the triangle of phase and line voltages; (c) in-phase booster voltages; (d) quadrature booster voltages.

Figure 2.24a shows the series transformer windings where $\Delta\underline{V}_A$, $\Delta\underline{V}_B$, and $\Delta\underline{V}_C$ are the voltages supplied from the excitation transformer (not shown) to be injected into each phase of the main circuit. Figure 2.24b shows the phase relationship of the phase and phase-to-phase voltages used to supply the excitation transformer. The same voltages are also on the primary side of the series transformer. If the excitation transformer is constructed so that the voltages $\Delta\underline{V}_A$, $\Delta\underline{V}_B$, and $\Delta\underline{V}_C$ are proportional to, and in phase with, the primary-side phase voltages $\Delta\underline{V}_A$, $\Delta\underline{V}_B$, and $\Delta\underline{V}_C$ then the series transformer will produce a change in the voltage magnitude as shown in Figure 2.24c. The secondary voltages of the series transformer are $\underline{V}_{A'}$, $\underline{V}_{B'}$, and $\underline{V}_{C'}$.

Alternatively, as shown in Figure 2.24d, the excitation transformer can be constructed so that the voltages $\Delta\underline{V}_A$, $\Delta\underline{V}_B$, and $\Delta\underline{V}_C$ are proportional to the phase-to-phase values $\Delta\underline{V}_{BC}$, $\Delta\underline{V}_{CA}$, and $\Delta\underline{V}_{AB}$ of the primary-side voltages. As the line voltage between two phases in a three-phase system is always in quadrature with the voltage of the third phase, the series transformer will introduce a change in the voltage angle and a small change in voltage magnitude. This type of booster transformer is referred to as a *quadrature booster transformer*.

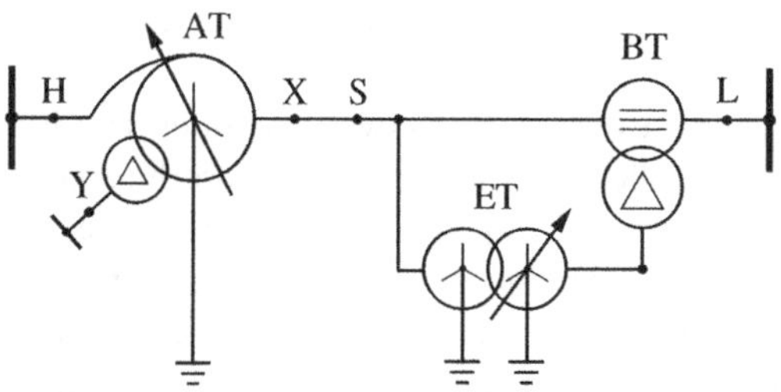

Figure 2.25 Transformer combination for independent in-phase and quadrature regulation.

Generally the excitation transformer can be constructed so that it supplies the series transformer with a net voltage made up of an in-phase component and a quadrature component. The voltage change in all three phases can be then expressed as

$$\Delta \underline{V}_{A} = a_1 \underline{V}_{A} + a_2 \underline{V}_{BC}, \quad \Delta \underline{V}_{B} = a_1 \underline{V}_{B} + a_2 \underline{V}_{CA}, \quad \Delta \underline{V}_{C} = a_1 \underline{V}_{C} + a_2 \underline{V}_{AB} \tag{2.4}$$

where a_1 and a_2 are the voltage transformation ratios associated with the in-phase component and quadrature component, respectively. Both these voltage ratios can be adjusted to allow control of both the voltage magnitude and angle. In this case the transformation ratio is a complex number, $\underline{\vartheta} = \underline{V}_{A'}/\underline{V}_{A} = \underline{V}_{B'}/\underline{V}_{B} = \underline{V}_{C'}/\underline{V}_{C} = \vartheta e^{j\theta}$, where θ is a phase shift angle.

It is possible to construct a single transformer in which both in-phase and quadrature regulation are done in the same tank. In practice, the two modes of regulation tend to be executed in separate transformers, as shown in Figure 2.25. Quadrature regulation is executed using the booster transformer BT and the excitation transformer ET situated in a common tank. In-phase regulation is executed independently in the main autotransformer AT which has its own tank. The main advantage of this solution is operational flexibility. In the case of a maintenance outage or failure of the tap changer of the excitation transformer ET, the autotransformer AT may operate after a bypass is inserted between terminals S and L (see Figure 2.25).

2.5.3 Shunt and Series Elements

Shunt and series elements, such as series capacitors and shunt compensators (static and rotating), are used in transmission networks for a number of purposes. From the point of view of this book their use will be considered for the purposes of reactive power compensation and stability improvement.

2.5.3.1 Shunt Elements

Section 3.1.2 shows how the ever-changing real and reactive power demand may cause large variations in the network voltage profile. Generally, reactive power cannot be transmitted over long distances and should be compensated for close to the point of consumption. The simplest, and cheapest, way of achieving this is by providing *shunt compensation*, that is by installing capacitors and/or inductors connected either directly to a busbar or to the tertiary winding of a transformer. Shunt elements may also be located along the transmission route to minimize losses and voltage drops. Traditionally, static shunt elements are breaker switched either manually or automatically by a voltage relay. Modern solutions, involving the use of thyristors, are described in the next subsection.

When the system power demand is low, reactive power produced by the transmission line capacitance may prevail over the reactive power consumed by the inductance and the transmission line may be a net source of reactive power (Section 3.1.2). The effect of this may be to boost the network voltages to unacceptably high values. In such circumstances *shunt reactors* can be used to consume the surplus reactive power and depress the voltages. Usually, shunt reactors are required for transmission lines longer than about 200 km. During heavy loading conditions some of the reactors may have to be disconnected and *shunt capacitors* used to supply reactive power and boost local voltages.

Another traditional means of providing shunt compensation is by the use of a *synchronous compensator*. This is a salient-pole synchronous motor running at no load whose field is controlled so as to generate or absorb reactive power. When overexcited, the synchronous compensator is a source of reactive power; when underexcited, it absorbs reactive power. Although relatively expensive, synchronous compensators play an important role in the control of voltage and reactive power at the transmission, and especially sub-transmission, voltage levels. They are used in order to improve stability and maintain voltages within desired limits under varying load conditions and contingency situations. When used in new substations synchronous compensators are often supplemented by switched shunt capacitor banks and reactors so as to reduce installation and operating costs. The majority of synchronous compensator installations are designed for outdoor operation and operate unattended with automatic control of start-up and shutdown. Small synchronous compensators of several megavolt-amperes are usually connected to the tertiary winding of the transmission transformer while larger units of up to a few hundred megavolt-amperes are connected by individual step-up transformers to the busbars of a high-voltage substation. The small units are generally air cooled, while the bigger units are either water cooled or hydrogen cooled.

2.5.3.2 Series Elements

Series capacitors are connected in series with transmission line conductors in order to offset the inductive reactance of the line. This tends to improve electromechanical and voltage stability, limit voltage dips at network nodes, and minimize the real and reactive power loss. Typically, the inductive reactance of a transmission line is compensated to between 25 and 70%. A full 100% compensation is never considered as it would make the line flows extremely sensitive to changes in angle between the voltages at the line terminals and the circuit would be series resonant at the fundamental frequency. Moreover, high compensation increases the complexity of protection equipment and increases the probability of subsynchronous resonance as discussed in Section 4.5.

Normally, series capacitors are located either at the line terminals or at the middle of the line. Although fault currents are lower, and line protection easier, when the capacitors are located at the mid-point, the access necessary for maintenance, control, and monitoring is significantly eased if the capacitor banks are positioned at the line terminals. For this reason, the compensating capacitors and associated shunt reactors are usually split into two equal banks positioned at each end of the line. Typically, each bank will be capable of compensating the line to a maximum of 30%. A detailed discussion of the benefits and problems of locating series capacitors at different points along the line can be found in Ashok Kumar et al. (1970) and Iliceto and Cinieri (1977).

Sometimes during power swings or heavy power transfers the line reactive current is high and the voltage may rise excessively on one side of the series capacitor. In this case the system must be designed to limit the voltage to acceptable levels or capacitors of appropriately high-voltage rating must be used. Normally, the voltage drop across a series capacitor is only a small percentage of the rated line voltage. A short-circuit on one side of the capacitor may, however, produce a temporary voltage across the element approximately equal to the line rated voltage. As such a fault is rare, it is uneconomical to design the element to withstand such a high voltage. So normally provision is made for the capacitor to be bypassed during such a fault and re-inserted after fault clearing.

The traditional way of bypassing the capacitor is to include a spark gap across either the capacitor bank itself or each module of the bank. A better solution is to use nonlinear, zinc oxide resistors which provide almost instantaneous re-insertion. Figure 2.26 shows some alternative bypass schemes (ABB 1991). The single-gap protective scheme in Figure 2.26a uses a single spark gap G which bypasses the capacitor if the voltage exceeds a preset value, normally equal to about three to four times the rated voltage of the capacitor. The short-circuit current flowing through the capacitor is damped in the damper D. When the gap current is detected, the bypass breaker S is closed, diverting the current from the gap. When the line current returns to normal the bypass breaker opens within (200–400) ms and the capacitor is re-inserted.

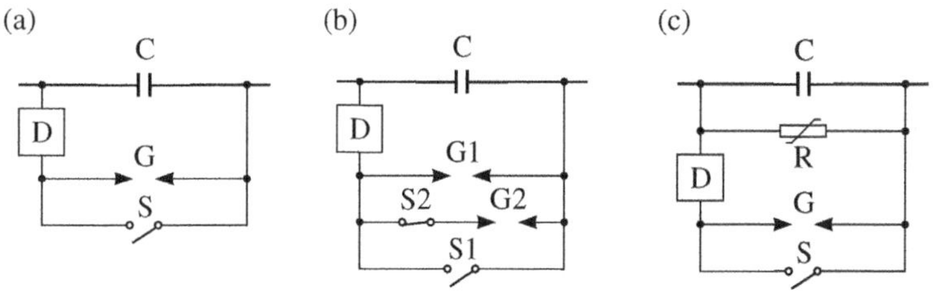

Figure 2.26 Series capacitor protective schemes: (a) single-gap scheme; (b) dual-gap scheme; (c) zinc oxide scheme.

A faster reinsertion time, around 80 ms, is provided by the dual-gap scheme in Figure 2.26b. When the fault occurs, the spark gap G2, which is set lower than G1, sparks over first bypassing the capacitor. Breaker S2, which is normally closed, opens immediately upon sensing the normal line current and re-inserts the capacitor. In this way capacitor re-insertion is not delayed by de-ionization time. The other gap G1 and the bypass breaker S1 serve as back-up protection.

Due to its nonlinear properties, the zinc oxide resistor shown in Figure 2.26c limits the voltage across the capacitor bank during a fault and re-inserts the bank immediately when the current returns to normal. The spark gap G does not normally operate and is provided only as back-up overvoltage protection for the resistor.

2.5.4 FACTS Devices

Traditionally the main control actions in an EPS, such as transformer tap changers, have been achieved using mechanical devices and were therefore rather slow. However, continuing progress in the development of power electronics has enabled a number of devices to be developed which provide the same functions but with much faster operation and with fewer technical problems. Transmission networks equipped with such devices are referred to as *FACTS* (flexible AC transmission systems), while the electronic devices themselves are referred to as *FACTS devices*. At the heart of FACTS devices is a controlled semiconductor, the *thyristor*.

The first thyristor developed in the early 1970s was the *silicon-controlled rectifier* (SCR), which had turn-on but no turn-off capability. Such a thyristor is now referred to as a *conventional thyristor*. It was at the heart of a rapid expansion of power electronics. It was also used to construct the first FACTS devices using thyristor valves. A thyristor valve is constructed using conventional thyristors and may be a circuit-breaker or a current controller (Figure 2.27).

Figure 2.27a shows a thyristor valve consisting of two thyristors that allow regulation of the current flowing through a shunt reactor. Regulation of alternating current is executed by cutting out a part of the sine waveform. The resulting alternating current contains harmonics. Hence any FACTS device using this type of regulation must be equipped with additional harmonic filters to help smooth the current waveform. Such filters are quite expensive and constitute a substantial part of the overall cost.

Figure 2.27 Two applications of thyristor valves: (a) thyristor-controlled reactor; (b) thyristor-switched capacitor.

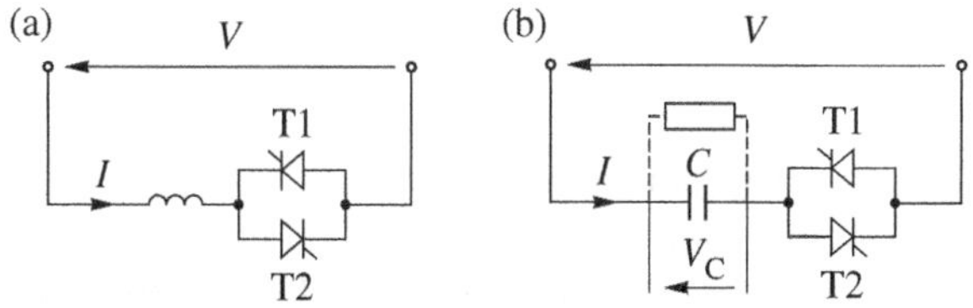

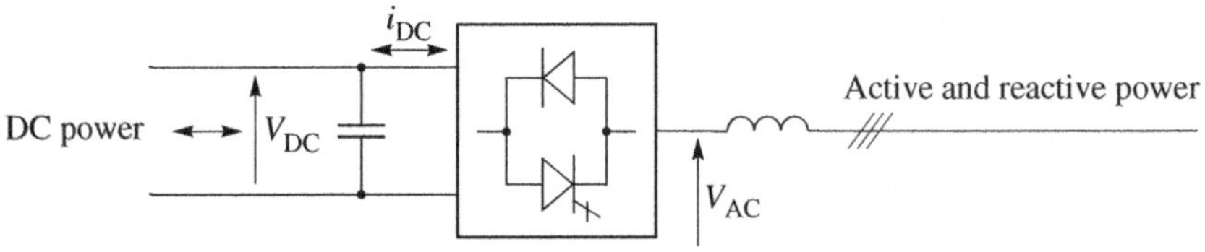

Figure 2.28 Basic principle of voltage source converter.

In the case of the capacitor it is not possible to obtain smooth control of the current due to the long time constant associated with the capacitor charge/discharge cycle so that the thyristor valve can only switch the capacitor on or off as shown in Figure 2.27b. When the flow of the current is blocked, the capacitor is discharged via the discharge resistor.

The next stage in the development of power electronics was the invention of the *gate turn-off thyristor* (GTO), which has both turn-on and turn-off capability. GTOs have found application in a number of more advanced FACTS devices based on voltage source converters and current source converters. The basic principle of these converters is shown in Figure 2.28. To differentiate it from the conventional thyristor, the GTO is denoted by an additional slanted line.

The voltage source converter (Figure 2.28) connects a DC system with an AC three-phase system (the three slanted lines on the right denote a three-phase system). Generally, power can flow in either direction, that is the DC system can either send or receive power. The DC voltage always has one polarity and power reversal takes place through reversal of the polarity of the direct current. Therefore, the converter valve has to be bidirectional and is made up of an asymmetric turn-off GTO device with a parallel diode connected in reverse. The capacitor on the DC side must be large enough to handle a sustained charge/discharge current that accompanies the switching sequence of the converter valve. On the AC side, the voltage source converter is connected with the AC system through a small reactance (usually a transformer) in order to ensure that the DC capacitor is not short-circuited and discharged rapidly into a capacitive load such as transmission lines of the AC system. In a particular case the DC side may consist of only a capacitor and then the active power of the DC system is equal to zero. In that case there is only reactive power on the AC side.

Current source converters used in low-voltage power electronic devices tend not to be used in high-voltage power electronics as they would require AC filters on the AC side, which is expensive. Hence they are not be discussed here.

The main disadvantages of GTOs are their bulky gate drivers, slow turn-off, and costly snubbers. Research continues to overcome those problems. It is likely that in coming years GTOs will be replaced in FACTS devices by new, more advanced thyristors such as the *integrated gate-commutated thyristor* (IGCT) or *MOS-controlled thyristor* (MCT).

Detailed description of thyristor-based FACTS devices can be found in Hingorani and Gyugyi (2000) and Akagi et al. (2007). Below only a short description will be given, necessary for understanding the rest of this book.

Depending on the way FACTS devices are connected to an EPS, they can be divided into shunt and series devices. Main *shunt FACTS* devices are reactive power compensators, energy storage (e.g. superconducting or battery-based), and braking resistors. Among various *series FACTS devices* are series compensators, phase angle regulators, and power controllers.

2.5.4.1 Static VAR Compensator

Static VAR compensators (SVCs) based on conventional thyristors have been used in power systems since the 1970s, long before the concept of FACTS was formulated. The role of the SVC is to adjust the amount of reactive

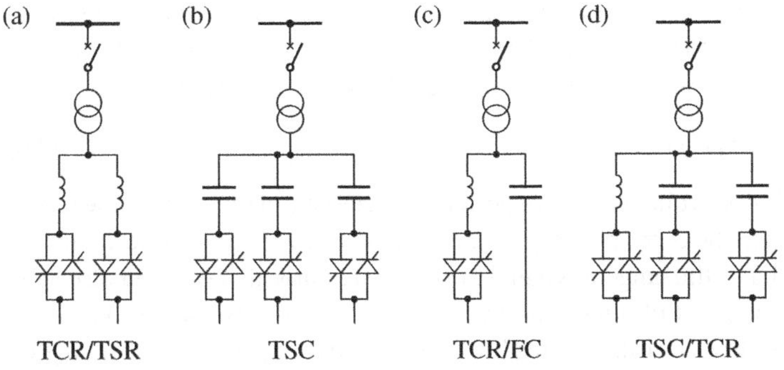

Figure 2.29 Types of SVCs. TCR – thyristor-controlled reactor; TSR –thyristor-switched reactor; TSC – thyristor-switched capacitor; FC – fixed capacitor.

power compensation to the actual system needs. A flexible and continuous reactive power compensation scheme that operates in both the capacitive and inductive regions can be constructed using the thyristor-switched and/or thyristor-controlled shunt elements shown in Figure 2.27. Using these elements, it is possible to design a variety of SVC systems. Some typical configurations are shown in Figure 2.29.

In Figure 2.29a one reactor is thyristor controlled and the other is thyristor switched. When the inductive VAR demand is low only the thyristor-controlled reactor operates. When demand increases, both reactors are switched on with the thyristor-controlled reactor being used to control the actual amount of reactive power needed.

Figure 2.29b shows a thyristor-switched bank of capacitors. The reactive power control (in the capacitive region only) can be accomplished in steps by switching consecutive capacitors in or out.

The SVC shown in Figure 2.29c consists of a bank of shunt capacitors connected in parallel with a thyristor-controlled shunt reactor. The thyristor valve enables smooth control of the lagging VARs produced by the reactor. With the reactors switched fully on, the parallel reactor-capacitor bank appears to be net inductive, but with the reactors fully off, the bank is net capacitive. By controlling the reactor current it is possible to achieve a full control range between these two extremes. A similar principle is used in the system shown in Figure 2.29d, which additionally contains a bank of thyristor-switched capacitors.

Each of the above systems can be associated with a static voltage-reactive power characteristic $V(Q)$. This will be discussed using the TSC/TCR compensator as an example.

The thyristor firing circuits used in SVCs are usually controlled by a voltage regulator (Figure 2.30), which attempts to keep the busbar voltage constant by controlling the amount and polarity of the reactive power injected into the busbar. The TSCs and the TCRs are equipped with a controller, shown on the right-hand side of the diagram, enforcing a required total value of the equivalent compensator susceptance B. This susceptance controller executes the overall control strategy and is very important for the operation of the whole system. The regulator, shown on the left-hand side of the diagram, creates a signal dependent on the controller transfer function and a voltage error in the node where the compensator is connected. Obviously, the value of the total susceptance is between the total susceptance of the capacitor bank and the reactor susceptance when the capacitors are switched off. In the steady state, the regulator's transfer function $G(s)$ is such that for $t \to \infty$, that is $s \to 0$, it is equal to

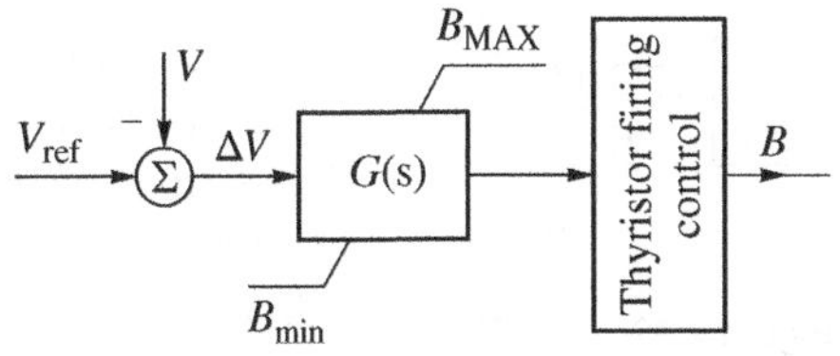

Figure 2.30 Simplified block diagram of SVC.

$$G(s)|_{s=0} = K \tag{2.5}$$

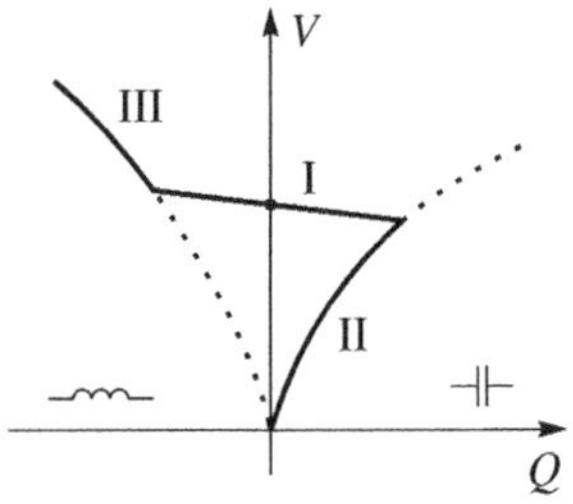

Figure 2.31 Static characteristic of SVCs equipped with voltage regulator.

namely a gain. Hence in the steady state, $\Delta B = K \cdot \Delta U$, that is the change of susceptance is proportional to the voltage change. For voltages close to the rated voltage it may be assumed that $V \cong V_{ref}$, that is $\Delta Q = \Delta B \cdot V^2 \cong \Delta B \cdot V_{ref}^2$. Hence

$$\Delta Q \cong (K \cdot V_{ref}^2) \cdot \Delta V \tag{2.6}$$

Figure 2.31 shows a voltage-reactive power characteristic of the device. The part of the $V(Q)$ characteristic corresponding to (2.6) is denoted by I. This characteristic has a small *droop*, that is the tangent is equal to $1/K$, the reciprocal of the regulator gain. The voltage at the point of intersection with the vertical axis is equal to V_{ref}.

The part of the characteristic denoted by II corresponds to a parabola $Q = B_{MAX} \cdot V^2$, that is the maximum value of the capacitive susceptance when all the capacitors are switched on and the reactors switched off. The part of the characteristic denoted by III corresponds to a parabola $Q = B_{min} \cdot V^2$, that is the minimum value of the susceptance when all the reactors are switched on and all the capacitors switched off.

SVC can operate in transmission networks as a voltage regulator with an additional PSS regulator added to damp power oscillations. This is discussed in Chapter 11.

Nowadays, SVCs based on conventional thyristors are regarded as old technology. Particularly troublesome and expensive is the necessity to smooth the current deformed by TCR. The cost of such an SVC is typically several times that of an uncontrolled bank of shunt reactors or fixed capacitors and a considerable part of that cost is due to the filters. A modern solution to the same problem of thyristor-based reactive power compensation is the static compensator (STATCOM) based on a voltage source converter.

2.5.4.2 Static Compensator

The STATCOM, also called the *static VAR generator* (SVG), provides shunt compensation in a similar way as the SVCs but utilizes the voltage source converter. Consequently, it incorporates a very high content of power electronics, but its conventional components are reduced to only a transformer and a capacitor.

The operating principle of the STATCOM is illustrated in Figure 2.32. On the DC side of the voltage source converter, there is only a capacitor. Compared with the block diagram in Figure 2.28, there is no source or demand of active power. The voltage source converter is equipped with a *pulse-width modulation* (PWM) controller operating with two control parameters m and ψ. The AC voltage produced by this converter is given by

$$\underline{V}_{ac} = mkV_{dc}(\cos\psi + \mathrm{j}\cos\psi) \tag{2.7}$$

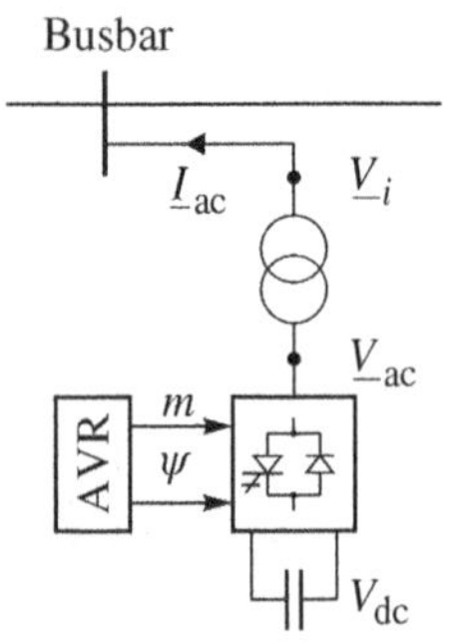

Figure 2.32 STATCOM based on voltage source converter.

A change in m enables the converter to change the magnitude of the AC voltage and therefore it influences a change of alternating current flowing through the transformer reactance X:

$$\underline{I}_{ac} = (\underline{V}_i - \underline{V}_{ac})/\mathrm{j}X \tag{2.8}$$

If $V_{ac} > V_i$ then $\underline{I}_{ac}$ leads $\underline{V}_i$ and reactive power is delivered to the busbar. The compensator acts like a capacitor. Conversely, if $V_{ac} < V_i$ then $\underline{I}_{ac}$ lags $\underline{V}_i$ and reactive power is drawn from the busbar. The compensator acts like a reactor. For a transformer reactance of 0.1 pu, a $\pm$ 10% change in V_{ac} produces a $\pm$ 1 pu change in the inserted reactive power.

Changing ψ, responsible for the phase of AC voltage – see Eq. (2.7) – makes it possible to control the active power fed to the capacitor, which is necessary to keep a constant value of the DC voltage.

To compensate reactive power in an EPS, the STATCOM must be equipped with an AVR. Its function is to enforce appropriate reactive power changes by affecting the regulation parameters m and ψ of the converter controller.

The transfer function of the voltage controller, as for the SVC, enforces a required small droop around the reference voltage V_{ref} in the static voltage-reactive power characteristic. The regulator has a stabilizing feedback loop fed from the compensator current. The regulator also has voltage and current limiters. Current limiters stop regulation after a maximum value of the current is reached. They correspond to vertical lines $I_{\min}$ and I_{MAX} in Figure 2.33a, and diagonal lines $Q = VI_{\min}$ and $Q = VI_{\text{MAX}}$ in Figure 2.33b. Voltage limiters switch off the device when allowed values have been exceeded. The limiters correspond to a break in the characteristics at values $V_{\min}$ and V_{MAX}.

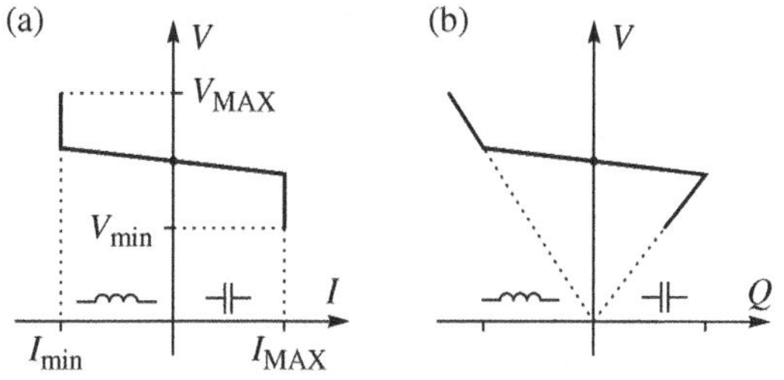

Figure 2.33 Static characteristics of STATCOM equipped with voltage regulator: (a) voltage vs current; (b) voltage vs power.

The STATCOM can operate in transmission networks as a voltage regulator with an additional PSS regulator added to damp power oscillations. This is discussed in Section 10.5.

2.5.4.3 Energy Storage System

Figure 2.28 shows that a voltage source converter may operate with a DC device sending or receiving power. That power will appear on the AC side as active power sent to, or received from, an EPS.

The *battery energy storage system* (BESS) shown in Figure 2.34 has a chemical battery connected to its DC side. The battery voltage V_{dc} may be considered constant so that, according to (2.7), changes in parameters m and ψ allow the magnitude and phase of the AC voltage $\underline{V}_{\text{ac}}$ to be regulated and therefore a regulation of real and reactive power flows. Regulation of m and ψ is executed by a *power conditioning system* (PCS). That regulator enforces a required active power flow discharging or charging the battery at an acceptable rate. Depending on needs, BESS can also control reactive power flows just like the STATCOM.

BESSs connected at the distribution level range in size from less than 1 MW to over 20 MW and find many applications. One of them is smoothing of power flows due to intermittent loads or renewable generators.

In transmission networks, BESSs with ratings of several tens of megawatts or higher may potentially find an application for spinning reserve or frequency control in the first instance after a large power unit is lost (see Chapter 9). Active power of BESS could also be used for the damping of power swings and stability enhancement (Section 10.5).

Superconducting magnetic energy storage (SMES) is functionally similar to BESS but with a superconducting coil used to store energy in the magnetic field of the coil. The active and reactive power available from SMES depends on the direct current stored in the coil. For a given direct current, the SMES power can be regulated in four quadrants of the complex power domain within a circular range limited by

$$[P_s(t)]^2 + [Q_s(t)]^2 \leq |S_{\text{MAX}}|^2 \tag{2.9}$$

where S_{MAX} is the maximum available apparent power. Applications of SMES are similar to those of BESS.

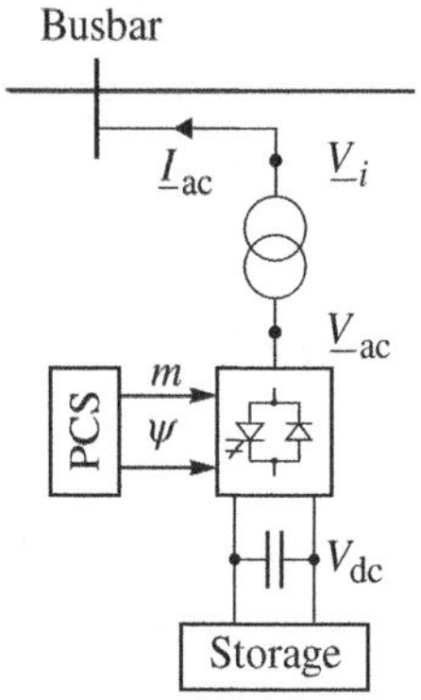

Figure 2.34 Battery energy storage system.

2.5.4.4 Thyristor-Controlled Braking Resistor

The braking resistor is used exclusively for transient stability enhancement. It acts as an additional resistive load capable of absorbing some of the surplus generation in case a severe fault occurs near a generator, thus preventing loss of synchronism.

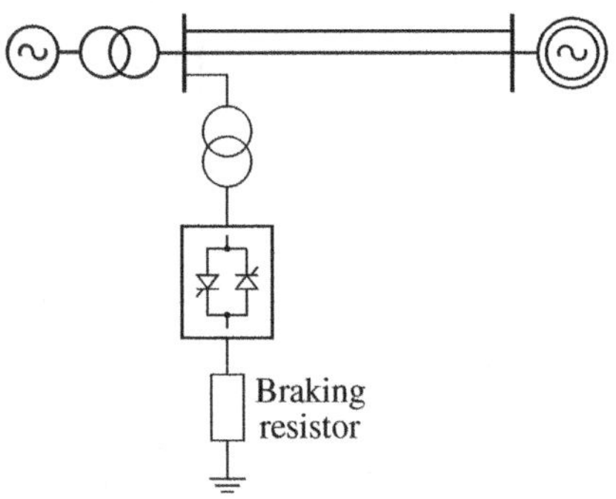

Figure 2.35 Thyristor-switched braking resistor.

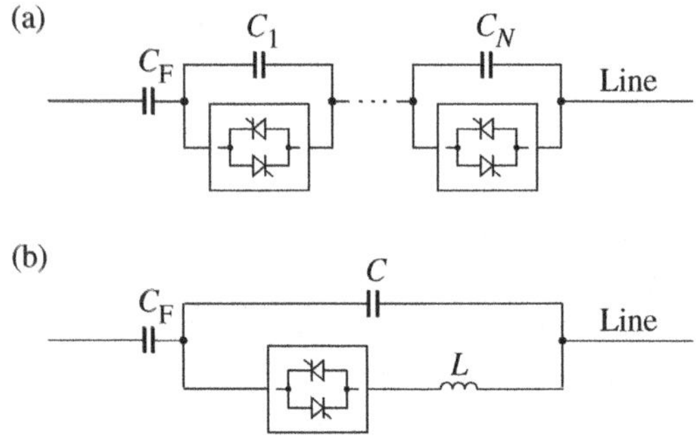

Figure 2.36 Series compensators based on conventional thyristors: (a) thyristor-switched series capacitor; (b) controlled series capacitor.

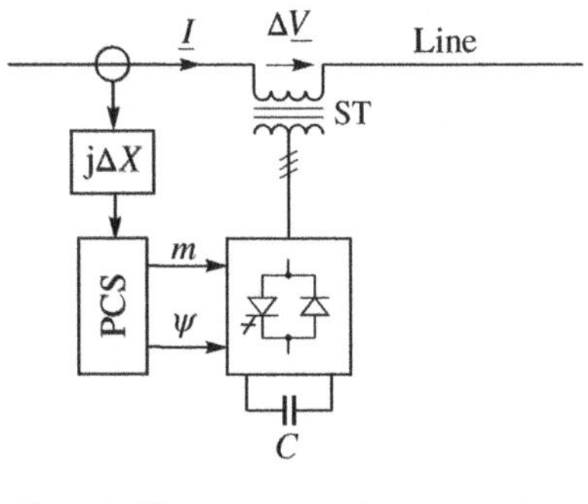

Figure 2.37 Static synchronous series compensator.

Traditionally, braking resistors were made as cast-iron resistors switched on by a mechanical circuit-breaker for a short time after clearing the fault. Because of the restricted lifetime of the mechanical circuit-breaker, the resistor would be switched on and off only once after a fault occurs.

In newer solutions, the mechanical circuit-breaker can be replaced by an electronic switch made of conventional thyristors connected back to back (Figure 2.35). Such a device is referred to as a *thyristor-switched braking resistor* (TSBR). The number of switches allowed is no longer restricted and it is possible to apply bang-bang control with a number of switches on and off after a fault occurs.

Theoretically, it is possible to use voltage source converters to control the braking resistor. In that case the resistor may be connected on the DC side, similar to BESS. The voltage source converter may smoothly control active power absorbed by the resistor from the AC system. Such a device could be referred to as a *thyristor-controlled braking resistor* (TCBR). Like BESS, reactive power could also be controlled within the capacity of the converter.

2.5.4.5 Series Compensators

Traditional series condensers are switched by mechanical circuit-breakers. Consequently, their control properties are limited and they usually operate with a constant capacitance.

Modern FACTS series compensators, apart from traditional compensation of the series line reactance, can also be used to regulate the total reactance of the transmission system and therefore they can provide regulation of active power flows. Such compensators can be made using conventional thyristors or voltage source converters.

Figure 2.36 shows two examples of series compensators based on conventional thyristors. In the first device shown in Figure 2.36a, referred to as the *switched series capacitor* (SSC), the series condensers consist of fixed capacitor C_F and a bank of series capacitors $C_1, C_2, \ldots, C_N$. The thyristor control system can short-circuit or open-circuit a number of capacitors. Thus the total series capacitance inserted into the network can change stepwise with the step equal to the capacitance of one series compensator. The thyristors are protected against overvoltages by nonlinear zinc oxide resistors, as shown in Figure 2.26.

In the second device shown in Figure 2.36b, referred to as the *controlled series capacitor* (CSC), a condenser of capacitance C is bypassed using a TCR. The condenser current is compensated by the reactor current. Consequently, the reactor current control is equivalent to controlling the resultant reactance of a parallel-connected condenser and reactor. The control is smooth but a disadvantage is the thyristor control of the reactor current as it deforms the sine waveform by cutting out a part of it. This leads to harmonics and the need to use smoothing filters.

Modern series compensators use thyristor converters. Figure 2.37 illustrates the *static synchronous series compensator* (SSSC). The source of the AC voltage is a voltage source converter loaded by a condenser on the DC side.

The capacitance of the condenser is such that it maintains a constant DC voltage. The converter operates as a voltage source synchronous with the AC network. The AC voltage $\Delta\underline{V}$ produced by the converter is inserted in the transmission link by the series (booster) transformer ST. Construction of the SSSC is similar to the STATCOM. Hence the SSSC is sometimes referred to as the *series STATCOM*.

As in the STATCOM, the voltage source converter is controlled using two parameters m and ψ responsible for the magnitude and phase of the AC voltage. Those parameters are controlled using PCS.

The SSSC compensates the reactance of a transmission line if its regulator ensures that the series booster voltage is always proportional to the current flowing in that line. This can be proved using Figure 2.38. Let us assume that the booster voltage is given by

$$\Delta\underline{V} \equiv -\mathrm{j}\Delta X\underline{I} \tag{2.10}$$

Using the equivalent diagram of the transmission link including the booster transformer (Figure 2.38a) one gives

$$\underline{V}_k - \underline{V}_j = \mathrm{j}X_{\mathrm{L}}\underline{I} \tag{2.11}$$

$$\underline{V}_k = \underline{V}_i - \Delta\underline{V} \tag{2.12}$$

or $\underline{V}_i - \underline{V}_j = \mathrm{j}X_{\mathrm{L}}\underline{I} + \Delta\underline{V} = \mathrm{j}X_{\mathrm{L}}\underline{I} - \mathrm{j}\Delta X\underline{I} = \mathrm{j}(X_{\mathrm{L}} - \Delta X)\underline{I}$. Hence finally

$$\underline{V}_i - \underline{V}_j = \mathrm{j}(X_{\mathrm{L}} - \Delta X)\underline{I} \tag{2.13}$$

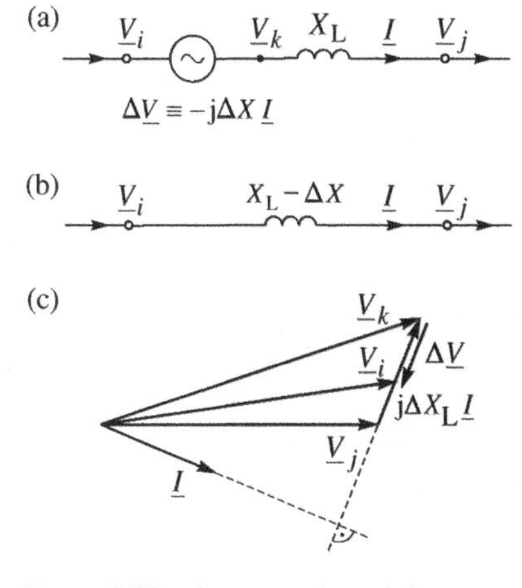

Figure 2.38 Compensation of the transmission line reactance using a current-controlled voltage source: (a) transmission line with a series-connected voltage source; (b) equivalent reactance; (c) phasor diagram.

The last equation corresponds to the equivalent diagram shown in Figure 2.38b in which the equivalent line reactance is equal to ($X_{\mathrm{L}} - \Delta X$). This means that adding the voltage expressed by Eq. (2.10) in parallel with the line is equivalent to compensating the line reactance X_{L} by ΔX. In the phasor diagram shown in Figure 2.38c, the voltage inserted in series with the line compensates the voltage drop on the line reactance.

It should be remembered that the SSSC compensates the reactance only if the voltage source is controlled using Eq. (2.10). That condition must be enforced by the regulator of the device.

Reactive power losses on the equivalent reactance ($X_{\mathrm{L}} - \Delta X$) are smaller than the reactive power losses on the transmission line reactance X_{L}. It follows then that the source $\Delta\underline{V}$ must deliver the capacitive reactive power necessary to compensate the difference between reactive power losses in the transmission line reactance and the equivalent reactance. That power must come from the condenser loading the converter on the DC side.

Series condensers may be used to regulate power flows in the steady-state. Their speed of operation may also be used for the damping of power swings and that application is discussed in Section 10.6.

2.5.4.6 Thyristor-controlled Phase Angle Regulator

Mechanical tap changers for the quadrature voltage regulation previously shown in Figure 2.24 can be replaced by FACTS technology. Figure 2.39 shows the *thyristor-controlled phase angle regulator* (*TCPAR*) where thyristor valves are used as electronic switches to replace the traditional mechanical tap changers. The number of electronic switches is restricted to three because the supply (excitation) transformer ET has three sets of windings on the secondary with a turns ratio of 1 : 3 : 9. The thyristor-switched system allows any one of these three sets of windings to be switched in series with the series transformer ST in either a positive or negative direction. Consequently, each of these sets of windings can therefore be in one of three states, giving $3^3 = 27$ possible values of the voltage supplying the series transformer. This corresponds to 27 taps in a traditional quadrature booster. For example, the booster value 1 is obtained by inserting only winding 1 in the supply circuit. The booster value $2 = 3 - 1$ is obtained if winding 3 is inserted in the opposite direction to winding 1. The booster value $4 = 3 + 1$ is obtained if windings 3 and 1 are connected in the same direction. The other ± 13 values are obtained in a similar way.

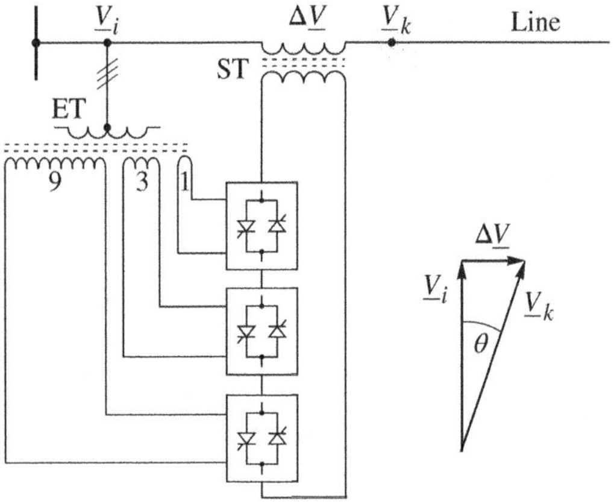

Figure 2.39 Thyristor-controlled phase angle regulator.

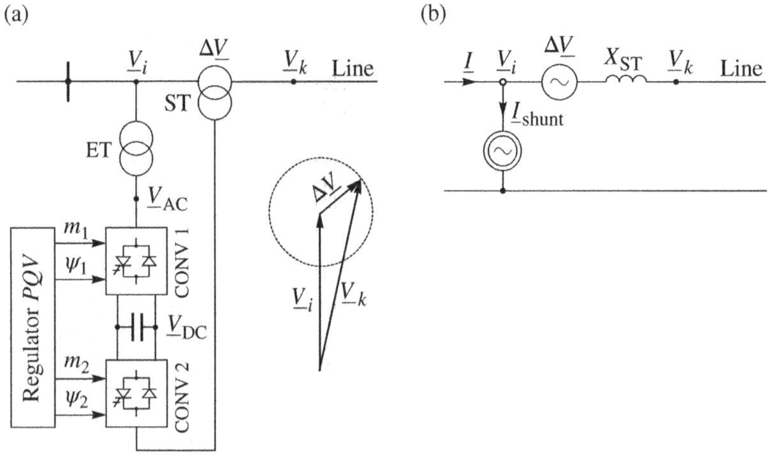

Figure 2.40 Unified power flow controller: (a) functional diagram and the phasor diagram; (b) equivalent circuit.

2.5.4.7 Unified Power Flow Controller

The unified power flow controller (UPFC), shown in Figure 2.40, consists of a shunt and series part. The shunt part consists of an ET and a voltage source converter CONV 1. The series part consists of a voltage source converter CONV 2 and an ST. Both voltage source converters CONV 1 and CONV 2 are connected back to back through the common DC link with a capacitor. Each converter has its own PWM controller, which uses two control parameters, respectively m_1,ψ_1 and m_2,ψ_2, as shown in Figure 2.40.

The shunt part of the UPFC works similarly to the STATCOM (Figure 2.32). Converter CONV 1 regulates voltage $\underline{V}_{\mathrm{ac}}$ and thereby also the current received by the UPFC from the network. The voltage is expressed by

$$\underline{V}_{\mathrm{ac}} = m_1 k V_{\mathrm{dc}}(\cos\psi_1 + \mathrm{j}\,\sin\psi_1); \quad \underline{I}_{\mathrm{shc}} = \frac{\underline{V}_i - \underline{V}_{\mathrm{ac}}}{\mathrm{j}X_{\mathrm{ET}}} \tag{2.14}$$

The controller enforces a required value $\underline{V}_{\mathrm{ac}}$ by choosing appropriate values of m_1 and ψ_1.

The series part of the UPFC works similarly to the SSSC (Figure 2.37). Converter CONV 2 regulates both the magnitude and the phase of the AC voltage $\Delta\underline{V}$ supplying the booster (series) transformer. That voltage is expressed by

$$\Delta \underline{V} = m_2 k V_{dc}(\cos \psi_2 + j \sin \psi_2) \tag{2.15}$$

The controller enforces the required value of $\Delta \underline{V}$ by choosing appropriate values of m_2 and ψ_2. Thanks to the control of both the magnitude and phase of the booster (series) voltage, the voltage $\underline{V}_k$ at the beginning of the transmission line may assume any values within the circle created by the phasor $\underline{V}_i$. This is illustrated on the phasor diagram in Figure 2.40.

Regulation of the magnitude and phase of the booster voltage corresponds to the operation of the phase shifting transformer shown in Figure 2.25. The main constraints of the regulation are the allowed voltage $\Delta \underline{V}$ and the allowed current flowing through the BT.

The simplified steady-state equivalent circuit of the UPFC shown in Figure 2.40 contains a series voltage source $\Delta \underline{V}$, reactance of the booster transformer X_{ST}, and shunt current source $\underline{I}_{shunt}$ responsible for the reactive power consumption by the shunt part necessary to maintain a constant value of the DC voltage. Obviously, the model must also include the limiters described above for both the series and shunt parts.

The UPFC can execute the following control functions:

1) Control of active power flows P by controlling the quadrature component Im $(\Delta \underline{V})$ of the booster voltage in the series part.
2) Control of reactive power flows Q by controlling the direct component Re $(\Delta \underline{V})$ of the booster voltage in the series part.
3) Control of the voltage V_i in the connection node by controlling the reactive current Im$(\underline{I}_{shunt})$ supplied by the network to the shunt part.

The UPFC can also work similarly as the series compensator SSSC (Figure 2.37). In this case, the direct and quadrature components of the booster voltage must be chosen by the regulator in such a way that the booster voltage phasor is perpendicular to the current phasor so that the condition Eq. (2.10) is satisfied and therefore the reactance of the transmission element is compensated.

As with other FACTS devices, the fast-acting UPFC can also be used for the damping of power swings, discussed in Section 9.7 and Section 10.7.

2.6 Protection

No system element is completely reliable and can be damaged by some internal or external fault. If the damaged element is not immediately disconnected, it may suffer further damage and be completely destroyed. A damaged element may also disturb operation of the neighboring elements so as to threaten the operation of the whole power system and the continuity of energy supply to the consumer. Protective equipment is therefore needed to detect a fault and disconnect the faulty element. Typically, power system protective equipment consists of current and/or voltage transformers, relays, secondary circuits supplying the relays and controlling the circuit-breakers, and auxiliary power supplies for the relays.

Operation of the protection must be fast, reliable, and selective. A fast *speed of response* and *high reliability* are vital to limit the damage that could be caused by a fault. In addition the protection must be *selective* so that only the faulty element is switched off. Reliability is achieved by using high-quality equipment and by using two different protection schemes for each element called the *main protection* and the *back-up protection*. The main protection should operate according to different physical principles than its back-up. If back-up protection is placed in the same substation bay as the main protection then it is termed *local back-up*. If the main protection of a neighboring element is used as the back-up protection of the given element then it is called *remote back-up*.

There are a number of books available which deal in detail with power system protection (Phadke and Thorap 1988; Wright and Christopoulos 1993; Ungrad et al. 1995). This section contains only a brief overview of some of the main protection schemes and introduces terms necessary for the remainder of the book.

2.6.1 Protection of Transmission Lines

The main faults on transmission lines are short-circuits. Overhead transmission lines are shielded from lightning strikes by ground wires, hung above the phase conductors, and surge diverters connected to the conductors themselves. Nevertheless, lightning is still the most common cause of faults on overhead transmission lines, with single-phase faults contributing (75–90)% of all faults. In contrast, multiple phase-to-ground faults constitute (5–15)% of faults, while multi-phase faults with no ground connection are the rarest at (5–10)%. Other rare causes of faults are insulator breakages, swinging of wires caused by strong winds, and temporary contact with other objects.

The majority, (80–90)%, of the faults on overhead lines are of a temporary nature and are caused by flashovers between phase conductors or between one or more of the phase conductors and earthed metal or the ground caused by, for example, lightning stroke. The remaining (10–20)% of faults are either semi-temporary or permanent. Temporary faults can be dealt with by switching off the line until the arc or arcs are extinguished, and then switching it on again after a certain period, termed the *dead time*. The whole procedure is referred to as *auto-reclosing* and significantly improves the continuity of energy supply. Obviously, in the case of a permanent fault, the re-energized line will be tripped again by its protection. There may be two or three such attempts, but usually only a single-shot reclosure is used in high-voltage transmission networks.

The oldest type of protection is the *differential-current protection* based on Kirchhoff's current law stating that the sum of currents flowing into and out of a circuit is equal to zero. Figure 2.41 shows the basic differential-current protection scheme where the current transformers installed at both terminals of a line are connected via resistors and pilot wires so as to oppose each other. Under healthy conditions or external faults (like F_2 in Figure 2.41), no current flows in the interconnecting cables. A fault occurring within the protected zone (fault F_1) creates a potential difference between the resistors, and a small circulating current flows through the pilot wires energizing the overvoltage relays and operating the circuit-breakers.

Because of the need to transmit signals continuously with fairly high accuracy between the ends of the protected line, the differential scheme with pilot wires is only used to protect short transmission lines with a maximum length of about (20–30) km. For longer lines the *interlock scheme* is used which employs directional relays sited at each end of the protected zone. These relays initiate the opening of the circuit-breakers at each end of a line if both sets of relays indicate that currents are flowing into the line. The information which must be transmitted between the relays is a logical yes or no, as opposed to the analogue signals used in the differential scheme. To avoid the expense of the pilot cables, the logic signal is usually transmitted over the conductors of the protected line using high-frequency signals and is called the *power line carrier*.

Another, more popular, protection scheme utilizing the power line carrier is the phase-comparison scheme in which the phase of the current at the two ends of the protected circuit is compared. Under normal operating conditions, or in case of an external fault, the currents at the opposite ends of the line are almost in phase with each other, whereas they are displaced by large angles when internal faults are present.

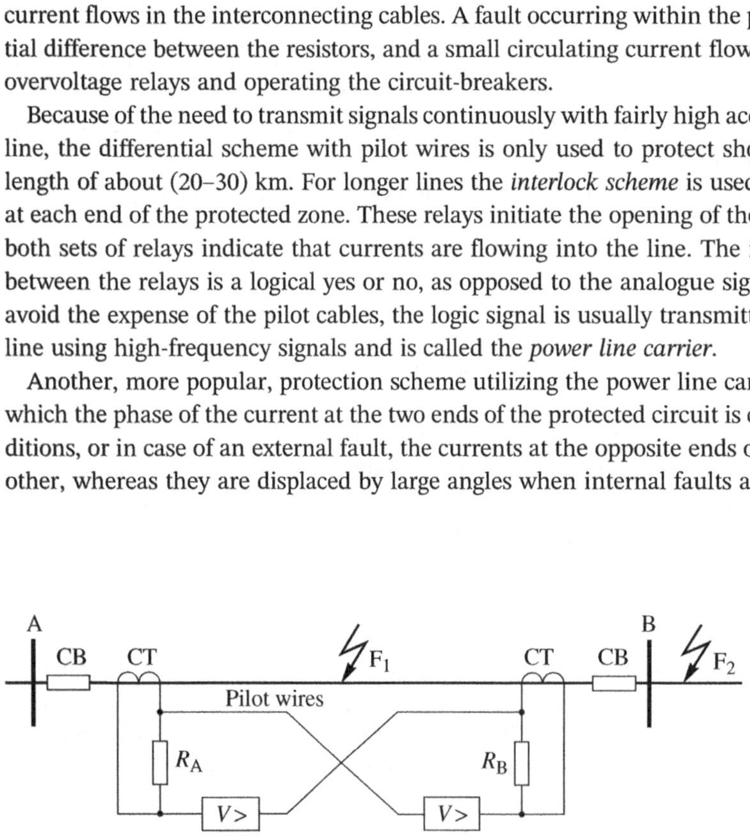

Figure 2.41 Differential-current protection. A, B – substation busbars; CB – circuit-breaker; CT – current transformer; V> – overvoltage relay; F_1, F_2 – internal and external faults.

The frequency used in the power line carriers is typically in the range (20–200) kHz. Lower frequencies increase the cost of the carrier, while higher frequencies cause too high an attenuation of the signal. The lines must be equipped with *line traps*, which are tuned circuits to block the high-frequency signals so that they do not enter other lines.

The *directional comparison* protection scheme utilizes the traveling waves which accompany the short-circuit. Any short-circuit produces voltage and current traveling waves which travel in both directions from the point of fault. The fault is detected by comparing the direction of the traveling waves at both ends of the line. An internal fault will cause the waves to travel in opposite directions, while an external fault will result in waves traveling in the same direction through the protected zone.

The development of optical-fiber technology has opened new possibilities for transmitting information over long distances. Optical-fiber links placed inside ground wires of transmission lines are currently used in directional comparison and phase comparison schemes, replacing pilot wires or power line carriers.

The most popular scheme for protecting transmission lines is *distance protection*. Its main advantage is that it does not require pilot wires or power line carriers. An additional advantage is that it may provide a back-up protection for neighboring network elements (lines and transformers). The principle of operation recognizes that the impedance of a high-voltage transmission line is approximately proportional to its length. This means that the apparent impedance measured during a fault by the relay is proportional to the distance between the point of fault and the relay. If this impedance is less than the series impedance of the line then it may be concluded that a fault has occurred inside the protected zone and the line should be tripped. Unfortunately, measurement of the apparent impedance has low accuracy due to the errors introduced by the current and voltage transformers, the relay itself and other factors such as the impedance of the fault. Consequently, the relay must have a time characteristic with a few zones corresponding to different impedance settings and tripping times, as shown in Figure 2.42. Distance protection of line AB at busbar A (marked by the small solid rectangle) has three zones, Z_{A1}, Z_{A2}, and Z_{A3}, with tripping times t_{A1}, t_{A2}, and t_{A3}, respectively. To make sure that a distance relay will not over-reach the protected zone, that is unnecessarily trip for faults outside the zone, the first protection zone is usually set between 85 and 90% of the line length. As this first zone, Z_{A1}, cannot protect the entire line, the distance relay is equipped with a second zone, Z_{A2}, which deliberately reaches beyond the remote terminal of the transmission line. The second zone is slowed down in order that, for faults in the next line BC, the protection of the next line (located at B) will operate before the second zone of the distance relay at A. The second zone of A also partially backs up the distance relay of the neighboring line BC. In order to extend this back-up as much as possible into neighboring lines, it is customary to provide yet another zone, Z_{A3}. The third protection zone is obviously the slowest.

Choosing the maximum reach of the third zone requires great care and analysis of the relay operation under heavy load, especially during an emergency state with depressed voltages. With high line currents and depressed voltages, the apparent impedance measured by the relay may approach the characteristic of the distance relay and even encroach into the third tripping zone. This may cause unnecessary tripping of the transmission line and further stressing of an already weakened system, possibly leading to a blackout. Examples of such blackouts are discussed in Chapter 8.

A distance relay may also cause unnecessary tripping of a transmission line during power swings. To prevent this, the relays must be equipped with power swing blocking relays or power swing blocking functions. This is further discussed in Section 6.6.

2.6.2 Protection of Transformers

Power transformers are an important link in high-voltage transmission networks and must be protected against both external and internal faults. Internal faults may be due to earth faults on windings, inter-phase faults, inter-turn faults, and inter-winding faults. This classification is important as it also corresponds to the different types of protection used.

The main form of transformer protection is differential-current protection. Its principle of operation is similar to the current differential protection used to protect a transmission line, Figure 2.41. The transformer protection scheme has, however, some special features because the magnitude and phase of the current are different on the primary and the secondary sides. The transformer protection scheme must also cope with the possible presence of a large *magnetizing inrush current*. Depending on the instant when an unloaded transformer is energized, a magnetizing current several times the rated current of the transformer may flow and, because the losses in the transformer are small, it will decay slowly with a time constant of several seconds to its normal small value. Because of the nonlinearity of the magnetizing characteristic, this inrush current has a high second harmonic, and this distortion can be used to distinguish it from a normal fault current.

Differential protection, despite its sensitivity, is not able to detect inter-turn faults. The transformer is protected against these faults by the *Buchholz protection* installed in the pipe between the main tank of the oil-filled transformer and the conservator. Localized heating or arcing associated with inter-turn or inter-winding faults decomposes the oil and produces gases such as hydrogen and carbon monoxide. The gases rise from the transformer and pass up the sloping pipe through the Buchholz relay, filled with oil, toward the conservator. On their way they rise inside the Buchholz relay and become trapped in the top of the casing, so displacing oil. As a result a pivoted float or bucket falls and, depending on the amount of gas released, causes an alarm or trips the relay.

Large transformers are also equipped with two distance protection devices, one on each side of the transformer. The first and third zones of both relays are directed toward the impedance of the transformer and their tripping signals are passed to the trip-coil circuits of both circuit-breakers. This constitutes a local back-up of the transformer's differential protection. The second zone is directed in the opposite direction toward the network. This constitutes the main protection from external faults and the local back-up for busbar protection.

In addition, transformers may also be equipped with earth fault protection (supplied with zero-sequence currents), combined differential, and restricted earth fault protection, overload protection in the form of an overcurrent relay (supplied from the current transformers), and a thermal relay reacting to the temperature inside the transformer tank.

2.6.3 Protection of Busbars

Faults on substation busbars are relatively rare compared with those on overhead transmission lines. The most common cause of busbar faults is flashover on the insulators, power apparatus failure, and, quite often, human error due, for example, to opening or earthing the isolator when on load. The consequences of a fault on a substation busbar may be far more severe than for a fault on a transmission line. The busbars of less important substations, in which all the outgoing circuits are protected with distance protection, are themselves protected by the second zone of the distance protection at neighboring stations. This is illustrated in Figure 2.42, where the busbars

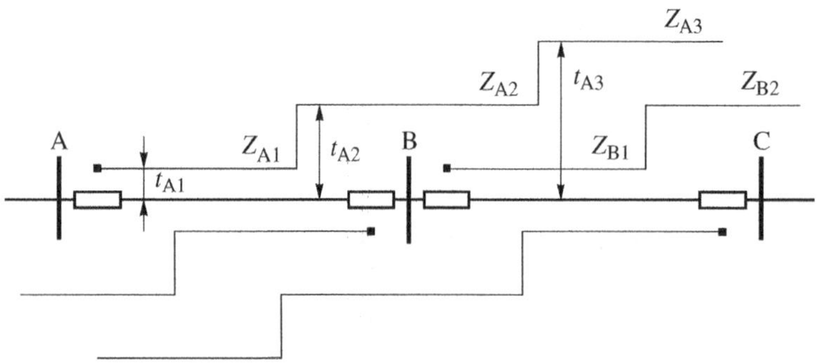

Figure 2.42 Distance protection zones of two neighboring lines.

at substation B are protected by the distance protection at substations A and C. An obvious drawback of this scheme is the long fault-clearing time associated with the second zone of the distance protection. This cannot be tolerated for more important substations, so they are normally equipped with differential-current protection for each of their circuits, or with modern phase comparison protection schemes. In this case the distance protection of outgoing circuits constitutes a remote back-up for the busbar protection.

2.6.4 Protection of Generating Units

As shown in Figure 2.2, a generating unit is a complex system comprising the generator, exciter, prime mover, step-up transformer, and often a unit transformer supplying the auxiliary services. As the generating unit may be subject to a variety of faults or disturbances, it is normally protected by a number of protection systems. The most important is the differential protection against faults inside the generator and transformer. This normally consists of three differential protection systems: one for the generator, a second for the unit transformer, and a third for the generator with its step-up transformer (or for the generator with both step-up and unit transformers). Large generating units are protected from external faults by the distance protection directed toward the network, in a similar manner as for transformers. The unit transformer is usually protected from external faults by an overcurrent relay. Similar overcurrent protection is used to protect the stator winding from both overloading and load asymmetry, and the rotor winding from overloading. Additional protection systems are used to protect the generator from loss of excitation, loss of synchronism (pole-slip protection), faults in stator windings (under-impedance protection), earth faults in the rotor windings, and from failure of the prime mover (motoring protection).

The generator is also equipped with protection from nonelectrical disturbances due to low vacuum, lubrication oil failure, loss of boiler fire, overspeeding, rotor distortion, excessive vibration, and difference in expansion between rotating and stationary parts.

2.7 Wide Area Measurement Systems

Wide area measurement systems (WAMS) are a measurement systems based on the transmission of analogue and/or digital information using telecommunication systems and allowing synchronization (time stamping) of the measurements using a common time reference.

Measuring devices used by WAMS have their own clocks synchronized with the common time reference using synchronizing devices. This concept is not new and for many years radio signals sent from ground stations have been used. The reference UTC (coordinated universal time) has been used as specified by the BIMP (*Bureau international des poids et mesures*) located in France. UTC is specified using time obtained from about 200 atomic clocks located in various places around the globe. A number of ground radio stations have been constructed to transmit the UTC signals. In Europe the DCF77 transmitter is used. It is located in Mainflingen near Frankfurt in Germany. Many supervisory control and data acquisition (SCADA) systems, which are used for monitoring and control of power system operation, utilize the DCF77 signals for synchronization. The accuracy of the time reference obtained using DCF77 is (1–10) ms. Such accuracy is good enough from the point of view of SCADA systems which measure magnitudes of current and voltages and corresponding real and reactive powers. Currently, much better accuracy, at least 1 μs, is obtained using satellite GPS (global positioning system).

The possibility of measuring voltage and current phasors in an EPS has created new control possibilities:

- Monitoring the operation of a large power system from the point of view of voltage angles and magnitudes and frequency. This is referred to as *wide area monitoring* (WAM).
- Application of special power system protection based on measuring phasors in large parts of an EPS. Such protection is referred to as *wide area protection* (WAP).

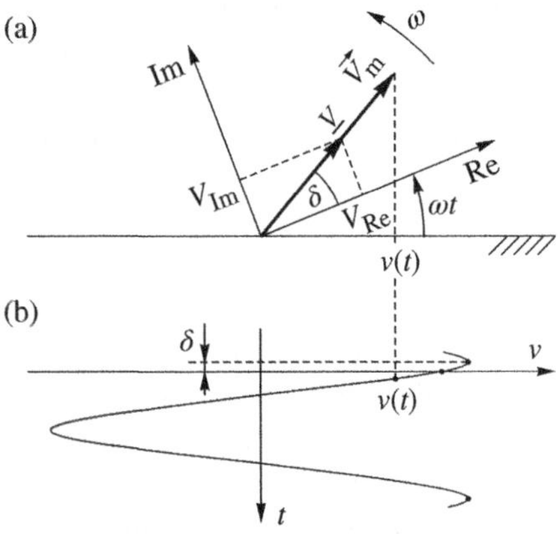

Figure 2.43 Illustration of the definition of phasors: (a) a rotating vector; (b) a corresponding time-domain signal.

- Application of control systems based on measuring phasors in large parts of an EPS. Such control is referred to as *wide area control* (WAC).

WAMS integrated with WAM, WAP, and WAC is referred to as *wide area measurement, protection, and control* (WAMPAC).

Recent years have seen a dynamic expansion of WAMPAC systems. Measurement techniques and telecommunication techniques have made rapid progress, but the main barrier for the expansion of WAMPAC systems is a lack of WAP and WAC control algorithms based on the use of phasors. There has been a lot of research devoted to that problem, but the state of knowledge cannot be regarded as satisfactory.

The definition of a phasor is closely connected to the representation of a periodic waveform as a rotating vector. This is illustrated in Figure 2.43. A vector $\vec{V}_m$ is rotating with angular velocity ω with respect to a stationary reference axis. Its position at any instant of time is given by

$$\vec{V}(t) = V_m e^{j(\omega t + \delta)}, \tag{2.16}$$

where V_m is the amplitude and δ is the phase shift with respect to the reference frame Re. This reference frame together with an orthogonal axis Im constitute the rotating complex plane Re–Im.

Projection of vector $\vec{V}(t)$ on the horizontal axis is periodically time-varying (Figure 2.43b), and is expressed as $v(t) = V_m \cos(\omega t + \delta)$. Frequency f of the periodic changes is related to the angular velocity by $\omega = 2\pi f = 2\pi/T$, where T is the period of rotation. The effective value of the sine waveform is given by $V = V_m/\sqrt{2}$. Equation (2.16) can be transformed in the following way

$$\vec{V}(t) = V_m e^{j(\omega t + \delta)} = \sqrt{2}\frac{V_m}{\sqrt{2}} e^{j\delta} e^{j\omega t} = \sqrt{2}\, V e^{j\delta} e^{j\omega t} = \sqrt{2}\, \underline{V}\, e^{j\omega t} \tag{2.17}$$

Vector $\underline{V} = V\, e^{j\delta}$ is referred to as the *phasor*. Its length (magnitude) is V and it is equal to the effective value of the periodic waveform $v(t)$. Its angle δ is defined by the location of the rotating vector with respect to the axis Re. The phasor components in the complex plane Re–Im can be determined from

$$\underline{V} = V_{Re} + jV_{Im} = Ve^{j\varphi} = V(\cos\varphi + j\sin\varphi) \tag{2.18}$$

The phasor contains information about both the effective value and the phase shift with respect to the reference frame. Knowing the components V_{Re} and V_{Im} of the phasor, it is easy to calculate its length V and the phase shift δ.

The above definition assumes that the reference frame Re and the complex plane Re–Im rotate with the same velocity ω as the vector $\vec{V}_m$. Generally, those two velocities may be different, i.e. vector $\vec{V}_m$ may rotate with velocity ω while the reference frame may rotate with velocity $\omega_{ref} \neq \omega$. In that case the phase shift δ is not constant but changes with a velocity equal to the difference between the two velocities $d\delta/dt = \Delta\omega$, where $\Delta\omega = \omega - \omega_{ref}$. In a special case when ω oscillates around ω_{ref}, the movement of the phasor on the complex plane is referred to as *swinging*.

An electrical network has generally $i = 1, 2, \ldots, n$ nodes. The phasors of all the nodal voltages can be placed in common complex coordinates Re–Im as shown in Figure 2.44. The voltage at node i can be then expressed as

$$\underline{V}_i = V_i\, e^{j\delta_i} = V_i(\cos\delta_i + j\sin\delta_i) \tag{2.19}$$

where V_i and δ_i are the effective value (magnitude) and the phase angle of the voltage, respectively. Section 3.5 shows that the electrical state of a network is determined by the voltage magnitudes and differences between the voltage angles. This means that the common coordinates can be changed by rotating them by an angle, because adding the same value to all the phase angles does not change the differences $(\delta_i - \delta_j)$. This leads to an important conclusion that in order to measure voltage phasors one can assume any, but common, coordinates (any common reference frame).

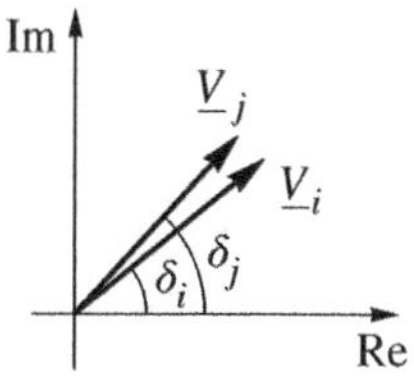

Figure 2.44 Two phasors in the complex plane.

The freedom to choose any common coordinates for an EPS is important for the methodology of phasor measurement using WAMS. The common coordinates for the whole WAMS are obtained by synchronizing the measurements using the 1 PPS signal obtained from the GPS. The GPS signal is received at any point on the earth, and therefore at any measurement system. Thus assuming the same 1 PPS signal as the time base ensures common coordinates for all the measurements in WAMS.

A measurement system allowing measurement of the phasors of voltages and currents in an EPS is referred to as the *phasor measurement unit* (PMU) and is shown schematically in Figure 2.45.

Voltages and currents for which phasors are to be determined are measured, using current and voltage transformers, as three-phase analogue signals and delivered to the PMU. Each analogue signal is filtered using an anti-aliasing filter and sent to an analogue-to-digital converter A/D. Here the signal is sampled, that is converted into digital samples. The sampler impulses are generated by an oscillator operating with the GPS receiver in the phase-locked loop system. Consequent data samples are sent to a microprocessor together with their time stamps.

The positive-sequence phasor of each measured quantity is stored, together with its time stamp, e.g. every 40 ms (25 measurements per second) or every 100 ms (10 measurements per second) depending on the needs. For storing, a data format is used appropriate for the telecommunication port of the PMU. Then the telecommunication port transmits data to other WAMS devices.

Some PMU devices may also calculate the frequency of the measured voltage, perform harmonic analysis, and send data to SCADA systems.

Increasingly, manufacturers of other microprocessor devices, such as disturbance recorders, equip their devices with software functions executing PMU tasks.

WAMS, and constructed on their basis WAMPAC, may have different structures depending on the telecommunication media used. With point-to-point connections, the structure may be multi-layer when PMU data are sent to phasor data concentrators (PDCs). One concentrator may service (20–30) PMUs. Data from the concentrators are then sent to computers executing SCADA/EMS functions or WAP/WAC phasor-based functions. An example of a three-layer structure is shown in Figure 2.46.

In each stage of data transmission, delays are incurred. Concentrators in the lowest layer service PMUs. As the delays are smallest at that stage, the concentrators may supply data not only for monitoring (WAM) but also for protection (WAP) and control (WAC).

The middle-layer concentrators combine data from individual areas of an EPS. The data may be used for monitoring and for some WAP or WAC functions.

The top, central concentrator services the area concentrators. Since at that stage the delays are longest, the central layer may be used mainly for monitoring and for those SCADA/EMS functions that do not require a high speed of data transmission.

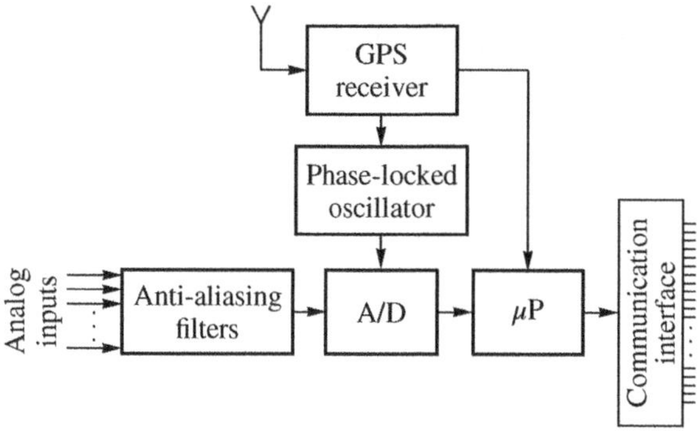

Figure 2.45 Functional diagram of PMU.

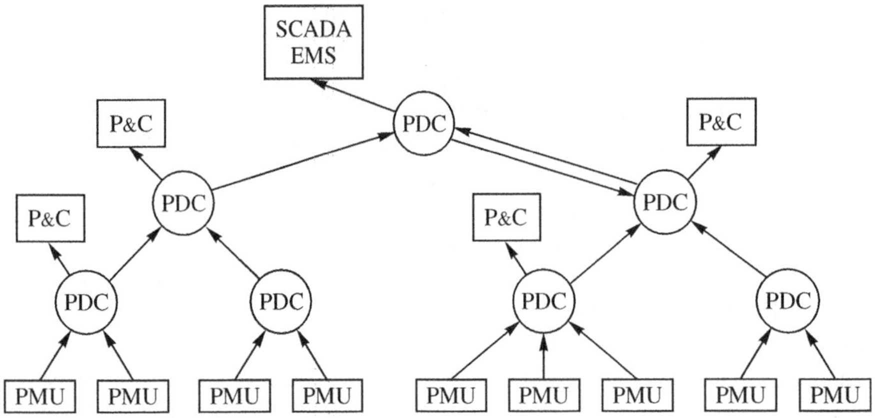

Figure 2.46 An example of a three-layer structure of WAMPAC. PMU – phasor measurement unit; PDC – phasor data concentrator; P&C – protection and control based on phasors.

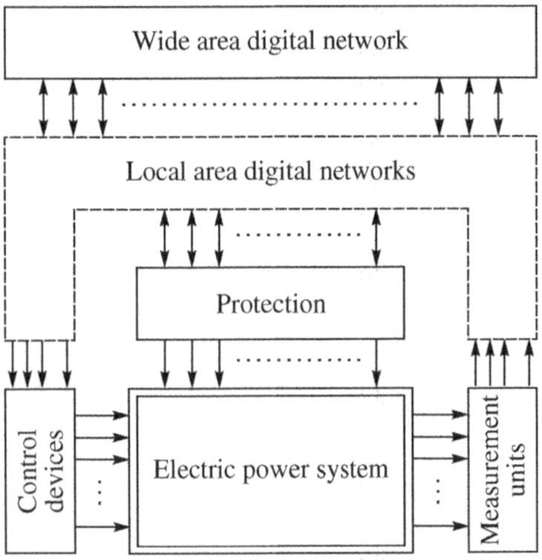

Figure 2.47 WAMPAC structure based on a flexible communication platform.

The main advantage of the layered structure is the lack of direct connections between area concentrators. Such connections may make it difficult, or even impossible, to execute those WAP or WAC functions that require data from a number of areas. The only way to get access to data from another area is via the central concentrator, which incurs additional delays. That problem may be solved by adding additional communication between area concentrators. This leads to more complicated communication structures as more links are introduced.

Computer networks consisting of many local area networks (LANs) and one wide area network (WAN) offer the best possibilities of further WAMPAC development and application. Such a structure is illustrated in Figure 2.47. The digital LAN services all measurement units and protection and control devices in individual substations. The connecting digital WAN creates a flexible communication platform. Individual devices can communicate with each other directly. Such a flexible platform may be used to create *special protection* and control systems locally, for each area and centrally. The platform could also be used to provide data for local and central SCADA/EMS systems.

3

The Power System in the Steady State

One of the characteristic features of power system operation is the continual need to adjust system operating conditions in order to meet the ever-changing load demand. Although the demand from any particular load can vary quite significantly, the total demand, consisting of a very large number of individual loads, changes rather more slowly and in a predictable manner. This characteristic is important as it means that within any small time period the transmission and subtransmission systems (shown in Figure 2.1) can be regarded as being in the steady state and, as time progresses, considered to move slowly from one steady-state condition to another.

To help describe and quantify the system behavior under these steady-state conditions this chapter develops steady-state mathematical models of all the main system elements such as transmission lines, transmission cables, transformers, generators, and so on. The behavior of the main types of system loads is also considered so that the way in which the demand may change with both voltage and frequency is understood. Having established such component models, the way they can be linked together to form the full set of network equations is described.

3.1 Transmission Lines

Figure 3.1 shows the single-phase equivalent circuit of a transmission line suitable for analyzing its symmetrical three-phase operation. In this circuit the voltages and the currents at the line terminals are the variables, while the line parameters are assumed to be uniformly distributed over the line length.

The parameters describing this circuit are:

r series resistance per unit length per phase (Ω/km);
$x = \omega L$ series reactance per unit length per phase (Ω/km) and L is the series inductance per phase (H/km);
g shunt conductance per unit length per phase (S/km);
$b = \omega C$ shunt susceptance per unit length per phase (S/km) and C is the shunt capacitance per phase (F/km);
l line length (km),

where $\omega = 2\pi f$ and f is the frequency. The series impedance and shunt admittance per phase per unit length are defined as $\underline{z} = r + jx$ and $\underline{y} = g + jb$. Each of the parameters in the equivalent circuit has a physical meaning and is linked with a particular aspect of the transmission line behavior. Thus the resistance r represents the joule (heating) loss that occurs due to current flow and depends on the type, construction, and diameter of the conductor used. The series inductance L depends on the partial flux linkages within the conductor cross-section itself and the external flux linkages with the other conductors. The shunt conductance g represents the corona loss and the leakage current on the insulators. This value is not usually constant as the corona loss depends on air humidity, while the leakage current depends on dirt and salt contamination on the surface of insulators. In power lines g is small and is usually neglected. The shunt capacitance C is due to the potential difference between the conductors. As the

Power System Dynamics: Stability and Control, Third Edition. Jan Machowski, Zbigniew Lubosny, Janusz W. Bialek and James R. Bumby.

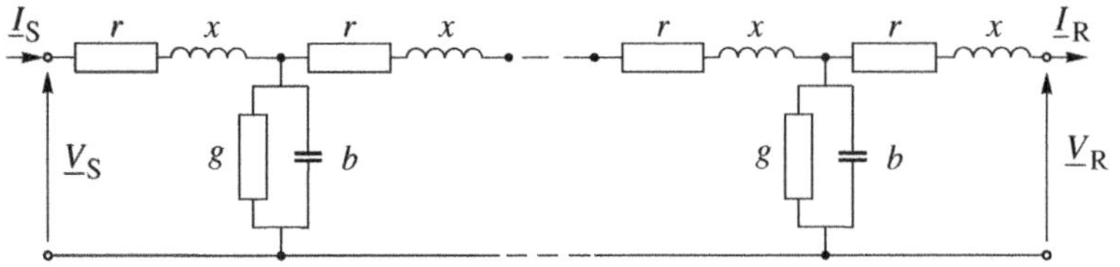

Figure 3.1 Single-phase equivalent circuit of a transmission line with distributed parameters.

voltages are AC voltages this shunt capacitance is alternately charged and discharged resulting in the flow of the *line charging current.*

3.1.1 Line Equations and the π-equivalent Circuit

For steady-state analysis the variables of interest are the voltages and currents at the line terminals $\underline{V}_R$, $\underline{I}_R$, $\underline{V}_S$, and $\underline{I}_S$ where the subscripts R and S signify the receiving end and sending end of the line, respectively. The voltages and currents are linked by the *long-line equation*

$$\begin{bmatrix} \underline{V}_S \\ \underline{I}_S \end{bmatrix} = \begin{bmatrix} \cosh \underline{\gamma} l & \underline{Z}_C \sinh \underline{\gamma} l \\ \sinh \underline{\gamma} l / \underline{Z}_C & \cosh \underline{\gamma} l \end{bmatrix} \begin{bmatrix} \underline{V}_R \\ \underline{I}_R \end{bmatrix} \tag{3.1}$$

where $\underline{Z}_C = \sqrt{\underline{z}/\underline{y}}$ is the *characteristic* (or *surge*) *impedance* of the line and $\underline{\gamma} = \sqrt{\underline{z}\underline{y}}$ is the *propagation constant.* As both the constants $\underline{\gamma}$ and $\underline{Z}_C$ are complex quantities, the propagation constant can be expressed as $\underline{\gamma} = \alpha + j\beta$ where α is termed the *attenuation constant* and β the *phase constant.* The four elements of the matrix linking the sending-end and the receiving-end voltages and currents are often referred to as the *ABCD* constants where $\underline{A} = \underline{D} = \cosh \underline{\gamma} l$, $\underline{B} = \underline{Z}_C \sinh \underline{\gamma} l$ and $\underline{C} = \sinh \underline{\gamma} l / \underline{Z}_C$. Interested readers are referred to Grainger and Stevenson (1994) or Gross (1986) for the detailed derivation of Eq. (3.1).

As power system networks consist of many lines, the use of Eq. (3.1) for analysis purposes is very inconvenient and a simpler way is to replace each line by its π-*equivalent* shown in Figure 3.2. A little circuit analysis shows that the π-equivalent parameters $\underline{Z}_L$ and $\underline{Y}_L$ are given by

$$\underline{Z}_L = \underline{Z}\frac{\sinh \underline{\gamma} l}{\underline{\gamma} l}; \qquad \underline{Y}_L = \underline{Y}\frac{\tanh\left(\underline{\gamma} l/2\right)}{\underline{\gamma} l/2} \tag{3.2}$$

where $\underline{Z} = \underline{z}l$ is the line total series impedance per phase and $\underline{Y} = \underline{y}l$ the total shunt admittance per phase.

It should be emphasized that the π-equivalent parameters $\underline{Z}_L$ and $\underline{Y}_L$ are generally not equal to the total impedance and admittance of the line, i.e. $\underline{Z}_L \neq \underline{Z} = \underline{z}l$ and $\underline{Y}_L \neq \underline{Y} = \underline{y}l$. However, for a typical power line $\underline{\gamma} l$ is small and the hyperbolic functions can be approximated as $\sinh\left(\underline{\gamma} l\right) \cong \underline{\gamma} l$ and $\tanh\left(\underline{\gamma} l/2\right) \cong \underline{\gamma} l/2$. Substituting these values into Eq. (3.2) gives the *medium-length line* (l between 80 and about 200 km) parameters

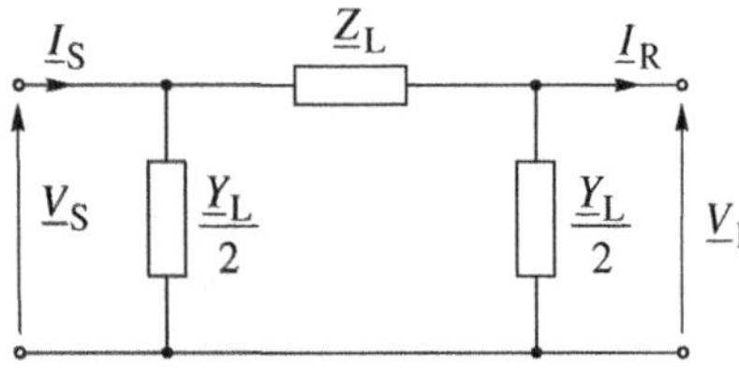

Figure 3.2 The π-equivalent circuit of a transmission line.

$$\underline{Z}_L = \underline{Z}; \qquad \underline{Y}_L = \underline{Y} \tag{3.3}$$

For a *short-length line* ($l < 80$ km) the charging current (and the capacitance *C*) may be neglected when the parameters are

$$\underline{Z}_L = \underline{Z}; \qquad \underline{Y}_L = 0 \tag{3.4}$$

Nowadays, with the wide use of computers, such approximations have limited practical value as the parameters can be easily calculated using Eq. (3.2). For manual calculations the most convenient, and practical, formulae to use are those of the medium-length line, Eq. (3.3), when the parameters in the nominal π-equivalent circuit represent the total series impedance Z and the total shunt admittance Y.

3.1.2 Performance of the Transmission Line

For a typical high-voltage transmission line g can be neglected while $r \ll x$. An additional insight into line performance may be obtained by considering the *lossless line*, that is by neglecting r altogether. With r and g neglected, the characteristic impedance is purely resistive

$$\underline{Z}_C = Z_C = \sqrt{\underline{z}/\underline{y}} = \sqrt{L/C} \tag{3.5}$$

and the propagation constant $\underline{\gamma}$ is purely imaginary

$$\underline{\gamma} = \sqrt{\underline{z}\underline{y}} = j\omega\sqrt{LC} \tag{3.6}$$

so that $\alpha = 0$ and $\beta = \omega\sqrt{LC}$. With these values of Z_C and γ the hyperbolic functions become $\sinh \underline{\gamma} l = j \sin \beta l$ and $\cosh \underline{\gamma} l = \cos \beta l$ so that Eq. (3.1) simplifies to

$$\begin{aligned} \underline{V}_S &= \underline{V}_R \cos \beta l + jZ_C \underline{I}_R \sin \beta l \\ \underline{I}_S &= \underline{I}_R \cos \beta l + j(\underline{V}_R/Z_C) \sin \beta l \end{aligned} \tag{3.7}$$

and the voltage and current are seen to vary harmonically along the line length. The *wavelength* of the full cycle can be calculated as $\lambda = 2\pi/\beta$. For 50 Hz lines the wavelength is nearly 6000 km, while for 60 Hz lines it is $60/50 = 1.2$ times shorter and is equal to about 5000 km.

Power engineers often find it convenient to compare the actual loading on the line with the *natural load*, or *surge impedance load* (SIL), where SIL is defined as the power delivered at rated voltage to a load impedance equal to Z_C, that is

$$P_{SIL} = \frac{V_n^2}{Z_C} \tag{3.8}$$

Substituting $\underline{I}_R = \underline{V}_R/Z_C$ into the first, and $\underline{V}_R = Z_C\underline{I}_R$ into the second, of the equations in (3.7) shows that when a lossless transmission line is loaded at SIL

$$\underline{V}_S = \underline{V}_R e^{j\beta l} \quad \text{and} \quad \underline{I}_S = \underline{I}_R e^{j\beta l} \tag{3.9}$$

indicating that:

- the voltage and current profiles are flat, $V_S = V_R$ and $I_S = I_R$;
- the voltage and current at both ends (and at any point along the line) are in phase so that the reactive power loss in the line is zero (in other words, the reactive power generated by C is compensated by the reactive power absorbed by L).

As the reactive power loss of the line is zero, the natural impedance loading is an optimum condition with respect to voltage and reactive power control.

Table 3.1 shows typical values of SIL and other characteristic parameters of overhead lines.

Unfortunately, the loading on the power line is rarely equal to the natural load. At night the loading may be a small fraction of SIL, whereas during peak load periods the loading may be significantly greater than SIL. To examine the effect this has on the line voltage, Figure 3.3 shows three plots of the sending-end voltage V_S required to

Table 3.1 Examples of overhead lines parameters (Kundur 1994).

f_n	V_n	r	$x = \omega L$	$b = \omega C$	β	Z_C	P_{SIL}
Hz	kV	Ω/km	Ω/km	μS/km	rad/km	Ω	MW
50	275	0.067	0.304	4.14	0.00112	271	279
(British)	400	0.018	0.265	5.36	0.00119	222	720
	230	0.05	0.488	3.371	0.00128	380	140
60	345	0.037	0.367	4.518	0.00129	285	420
(USA)	500	0.028	0.325	5.2	0.0013	250	1000
	765	0.012	0.329	4.978	0.00128	257	2280
	1100	0.005	0.292	5.544	0.00127	230	5260

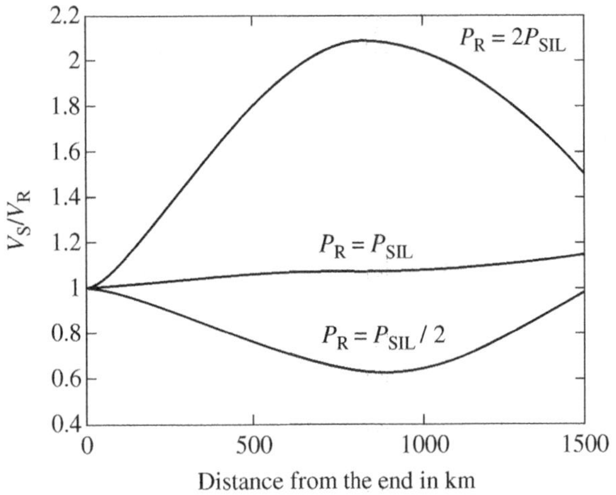

Figure 3.3 Sending-end voltage required to supply a given active power load at rated voltage and at unity power.

deliver power P_R at rated voltage and unity power factor at the receiving-end, each plotted as a function of line length. These calculations have been done using the full line model described by Eq. (3.1) (with resistance included) assuming the parameters of the 400 kV line given in Table 3.1. When the delivered power P_R is less than SIL then V_S is smaller than V_R, while $P_R > P_{SIL}$ requires V_S to be larger than V_R. Delivering $P_R = P_{SIL}$ does not actually result in the voltage along the line being constant but it is seen to increase slightly toward the sending end. This is due to including the line resistance in the calculation. Note the harmonic variation of the voltage with length.

3.1.2.1 Real Power Transmission

In Chapter 1, approximate Eqs. (1.8) and (1.9) are derived for real and reactive power transmission in a line represented by just its series reactance $X = xl$. Now the equation for real and reactive power transmission will be derived using the full π-equivalent model of the line.

Let $\underline{V}_R$ be the reference phasor and assume that $\underline{V}_S$ leads $\underline{V}_R$ by angle δ_{SR}, that is $\underline{V}_R = V_R$ and $\underline{V}_S = V_S e^{j\delta_{SR}}$. The angle δ_{SR} is referred to as the *load angle* or *transmission angle*. Assuming a lossless line, the receiving-end current can be calculated from Eqs. (3.7) as

$$\underline{I}_R = \frac{\underline{V}_S - \underline{V}_R \cos\beta l}{jZ_C \sin\beta l} = \frac{V_S}{Z_C \sin\beta l} e^{j(\delta_{SR} - \pi/2)} - \frac{V_R \cos\beta l}{Z_C \sin\beta l} e^{-j\pi/2} \tag{3.10}$$

allowing the complex apparent power at the receiving end to be calculated as

$$\underline{S}_R = \underline{V}_R \underline{I}_R^* = \frac{V_R V_S}{Z_C \sin\beta l} e^{j(\pi/2 - \delta_{SR})} - \frac{V_R^2 \cos\beta l}{Z_C \sin\beta l} e^{j\pi/2} \tag{3.11}$$

The active power at the receiving end P_R can be now obtained as

$$P_R = \text{Re}\,[\underline{S}] = \frac{V_S V_R}{Z_C \sin\beta l} \sin\delta_{SR} \tag{3.12}$$

showing that for sending-end and receiving-end voltages of specified magnitude, *an increase in the receiving-end power P_R leads to an increase in the load angle δ_{SR}*, while the maximum power transfer occurs when $\delta_{SR} = \pi/2$ and is equal to

$$P_{R,\max} = \frac{V_S V_R}{Z_C \sin\beta l} \approx \frac{P_{SIL}}{\sin\beta l} \tag{3.13}$$

Table 3.1 shows that βl is small for medium-length lines (l less than about 200 km). Thus, it can be assumed that the following simplifications hold

$$\sin\beta l \cong \beta l, \qquad \cos\beta l \approx 1, \qquad Z_C \sin\beta l \cong \sqrt{\frac{L}{C}} \omega\sqrt{LC}\, l = \omega L l = X \tag{3.14}$$

where X is the total inductive reactance of the line. In this case Eq. (3.12) simplifies to

$$P_R \cong \frac{V_S V_R}{X} \sin\delta_{SR} \tag{3.15}$$

showing that the power transfer limit is approximately inversely proportional to the line series inductance and the length. Any attempt to increase power transfer above $P_{R,\max}$ will result in instability and is connected with *steady-state stability* discussed in detail in Chapter 5. Note that Eq. (3.15) corresponds to Eq. (1.8) derived for a simplified line model.

It is worth noting that a transmission line constitutes only part of the transmission link that connects a generator with the system or connects two parts of a system together. This means that it is the equivalent voltages of the generator and the system which should be considered constant and used in Eq. (3.12), not the terminal voltages of the line itself. Obviously, the equivalent impedance of the generator and the system will then have to be added to the line impedance. As a result δ_{SR} is smaller than the angle between the equivalent voltages and, in practice, the power transfer limit occurs at $\delta_{SR} < \pi/2$. In addition when transient stability effects, discussed in Chapter 6, are taken into account the practical limit on δ_{SR} is significantly less than $\pi/2$.

Although Eq. (3.13) suggests that the maximum line loading is due to the power transfer limit set by the load angle δ_{SR}, this is not always the case, and other effects such as the thermal rating of the line and the maximum allowable voltage drop between the ends of the line must also be considered. For short lines, less than about 100 km, the maximum line loading is usually limited by the thermal rating rather than the maximum power transfer limit. On the other hand, for lines longer than about 300 km, the maximum line loading is determined by the power transfer limit, which may be substantially below the limit set by thermal considerations. Typically utilities will give the maximum power loading of a line as a fraction (or multiple) of P_{SIL}.

3.1.2.2 Reactive Power Considerations

If for this discussion the sending-end voltage is assumed to be constant, the question then arises as to what happens to the receiving-end voltage if (a) the reactive power Q_R at the receiving end changes and (b) if the active power P_R at the receiving end changes.

Changes in Q_R will be considered first. The reactive power at the receiving end may be found from the imaginary part of Eq. (3.11) as

$$Q_R = \text{Im}\left[\underline{V}_R \underline{I}_R^*\right] = \frac{V_S V_R}{Z_C \sin \beta l} \cos \delta_{SR} - \frac{V_R^2 \cos \beta l}{Z_C \sin \beta l} = \frac{V_R}{Z_C \sin \beta l}(V_S \cos \delta_{SR} - V_R \cos \beta l) \tag{3.16}$$

The simplifications of Eq. (3.14) allow Eq. (3.16) to be approximated as

$$Q_R \cong \frac{V_R}{X}(V_S \cos \delta_{SR} - V_R) \tag{3.17}$$

This equation corresponds to Eq. (1.8) derived for a simplified line model. Normally, for the stability reasons considered earlier, the angle δ_{SR} is kept small so that $\cos\delta_{SR} \approx 1$. Strictly speaking, this assumption corresponds to considering the reactive power transfer only so that $P_R = 0$ and, according to Eq. (3.12), $\delta_{SR} = 0$. This simplification gives

$$Q_R \cong \frac{V_R(V_S - V_R)}{X} \tag{3.18}$$

and shows very clearly that the reactive power, Q_R, is strongly dependent on the magnitude of the voltage at both the sending end and the receiving end of the line and that it flows from the higher voltage to the lower voltage. Assuming that V_S = constant and plotting the variation of Q_R against the receiving-end voltage V_R produces the parabolic relationship shown in Figure 3.4a with a maximum at $V_R = V_S/2$. As system voltages must be kept close to V_n, the operating point is always to the right of this peak with the condition $V_R > V_S/2$ always being satisfied. Thus, an increase in Q_R leads to a decrease in V_R, while a decrease in Q_R leads to an increase in V_R. This observation has important repercussions regarding the introduction of reactive power compensation.

The sending-end reactive power Q_S can be calculated using a similar derivation as in Eq. (3.16) to give

$$Q_S = \frac{V_S}{Z_C \sin \beta l}(V_S \cos \beta l - V_R \cos \delta_{SR}) \tag{3.19}$$

which, assuming that βl is small, gives

$$Q_S \cong \frac{V_S}{X}(V_S - V_R \cos \delta_{SR}) \tag{3.20}$$

Assuming δ_{SR} is also small gives

$$Q_S \cong \frac{V_S}{X}(V_S - V_R) \tag{3.21}$$

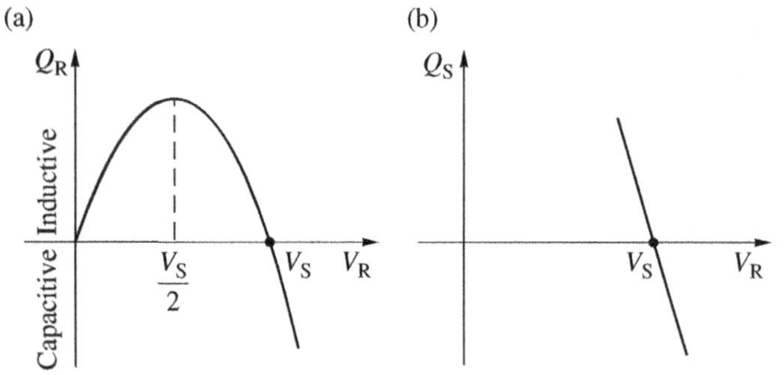

Figure 3.4 The approximate effect on receiving-end voltage of: (a) changes in Q_R and (b) changes in Q_S, when the influence of the active power flow is neglected.

Again the strong dependence of reactive power with voltage magnitude is apparent and, as V_S is assumed constant, the approximate variation of Q_S with V_R is linear, as shown in Figure 3.4b.

Now consider how the inclusion of the active power transfer ($P_R \neq 0$ and hence $\delta_{SR} \neq 0$) modifies the above considerations. To investigate this, consider the reactive power loss in the line itself due to the flow of P_R. Generally, the reactive line loss can be positive or negative, so that the line can be a net source or sink of reactive power. The reactive power is generated in the line capacitance and depends on the voltage which is normally kept close to the rated value. This relatively constant reactive power generation is offset by the consumption in the line inductance. This consumption varies as the square function of the line current ($Q = I^2X$) and strongly depends on the actual loading of the line. To quantify this effect the reactive power loss $\Delta Q = Q_S - Q_R$ can be calculated from Eqs. (3.16) and (3.19) as

$$\Delta Q = Q_S - Q_R = \frac{V_S^2 \cos\beta l - 2V_S V_R \cos\delta_{SR} + V_R^2 \cos\beta l}{Z_C \sin\beta l} \tag{3.22}$$

Assuming $V_S \cong V_R \cong V_n$ and eliminating cos δ_{SR} using Eq. (3.12) gives

$$\Delta Q(P_R) \approx \frac{2P_{SIL}}{\sin\beta l}\left(\cos\beta l - \sqrt{1 - \left(\frac{P_R \sin\beta l}{P_{SIL}}\right)^2}\right) \tag{3.23}$$

The characteristics produced by Eq. (3.23) are shown in Figure 3.5 for transmission lines that operate at different voltage levels. The reactive power loss is zero only when $P_R = P_{SIL}$. If $P_R < P_{SIL}$ then the line is a net source of reactive power and if $P_R > P_{SIL}$ the situation is reversed and the line is a net sink of reactive power.

The consequence of Figures 3.4 and 3.5 is profound. According to Figure 3.5, as the active power increases the reactive power loss in the line also increases. This loss must be supplied from the sending end giving an increase in Q_S. Figure 3.4b then shows that for this to occur with V_S constant then V_R must reduce. If the resulting reduction in V_R is unacceptable then, according to Figure 3.4a, this can be compensated by reducing Q_R in some way, perhaps by the introduction of some form of reactive power compensation (Section 2.5.3).

3.1.3 Underground Cables

From a mathematical point of view an underground cable can be modeled exactly the same way as an overhead transmission line. The only difference is in the values of the characteristic parameters. Because there are many methods of cable construction the parameters of these different cables can also be quite different. In particular the shunt capacitance of the cable depends strongly on whether the three-phase conductors are screened or constitute separate single-phase cables. Typically, the per-unit-length series reactance of a cable is about half that of a

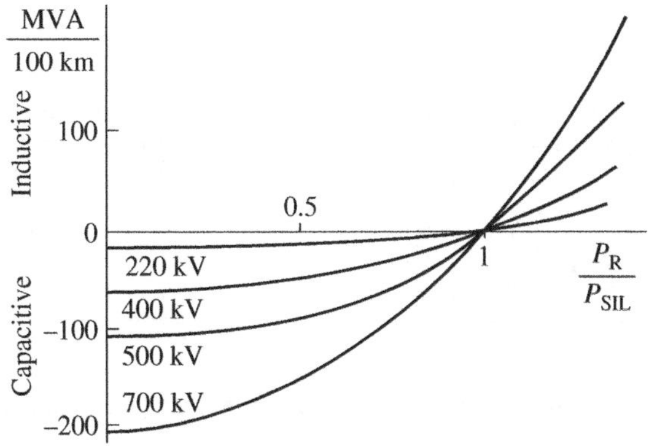

Figure 3.5 Examples of reactive power absorbed by a lossless line as a function of its real load for various voltage ratings.

similarly rated overhead line. On the other hand, the per-unit-length charging current is about 30 times more. This means that even for a short cable run of several tens of kilometers the charging current in the cable constitutes a substantial portion of its thermally admissible maximum current and severely limits its transmission capacity. In the extreme case of a critically long cable the charging current would be equal to the maximum current and there would be no capacity left for any power transmission. The charging current barrier is the main obstacle for the practical application of AC cables in power transmission networks.

3.2 Transformers

The main types of transformers are discussed in Section 2.5.2. In this section the equivalent circuit of the transformer are derived, together with a method of dealing with off-nominal transformation ratios.

3.2.1 Equivalent Circuit

The equivalent circuit of a transformer is shown in Figure 3.6a. The main element of the circuit is the *ideal transformer* with transformation ratio $\vartheta = N_1/N_2$, where N_1 and N_2 are respectively the number of turns on the primary and secondary winding. The resistance R_1 and R_2 account for the I^2R loss in the primary and secondary winding, while the reactance's X_1 and X_2 account for the *leakage flux*, that is the flux component that does not link the two windings but only the primary or the secondary winding. The sinusoidal changes in the supply voltage cause cyclic remagnetization of the transformer core as determined by the hysteresis curve of the core steel. This results in the flow of a *magnetization current* I_μ even at no load. I_μ is in phase with the flux and is therefore delayed with respect to the induced electromotive force (emf) E_1 by $\pi/2$. This current is modeled by the shunt susceptance B_μ. The cyclic changes of the core flux also dissipate a certain amount of energy in the form of heat in the transformer core. This component of the core loss is referred to as the *hysteresis loss*. As the core itself is an electrical conductor the flux changes induce emf's which result in the flow of circulating *eddy currents*. These eddy currents produce a thermal energy loss called the *eddy-current loss*. The sum of the eddy-current loss and the hysteresis loss is the *core loss* P_{Fe}. This energy loss in the core is provided for by an additional current I_{Fe} flowing in the primary but this time in phase with the induced emf. This is modeled by inserting a shunt conductance G_{Fe} in the equivalent circuit of Figure 3.6a. At no load the primary-side current is equal to the phasor sum of the magnetization current and the core loss current and is referred to as the *excitation current* I_E. When the transformer is loaded, the excitation current is superimposed on the load current.

The transformation ratio ϑ can be eliminated from the diagram if all the quantities in the equivalent circuit are referred to either the primary or secondary side of the transformer. For example, Figure 3.6b shows the case when all voltages, currents and impedances of the circuit of Figure 3.6a are referred to the primary. To produce this

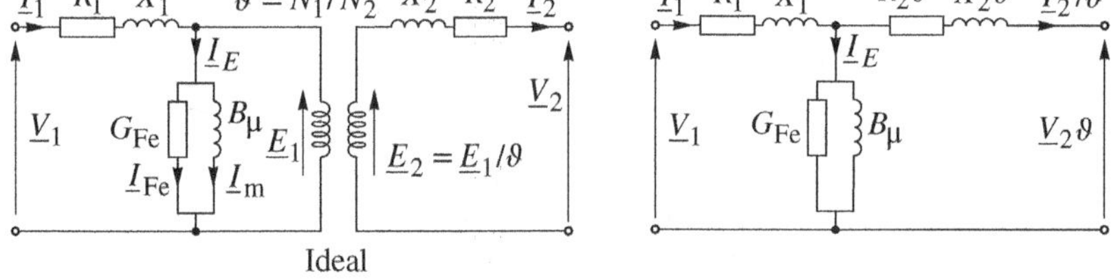

Figure 3.6 Single-phase, two-winding transformer: (a) equivalent circuit; (b) equivalent circuit with secondary referred to the primary.

model of the transformer, the secondary voltage is multiplied by ϑ while the secondary current is divided by ϑ. Consequently the secondary impedance must be multiplied by ϑ^2.

The equivalent circuit shown is of the T type but it is often more convenient in network analysis to deal with π-type circuits. If the shunt element in the transformer equivalent circuit cannot be neglected then the circuit of Figure 3.6b can be transformed into the π form using the standard star-delta transformation. However, the π circuit will not be symmetrical if the series impedances of the primary and the secondary branches in Figure 3.6b are not equal. To avoid this the following approximations are usually made:

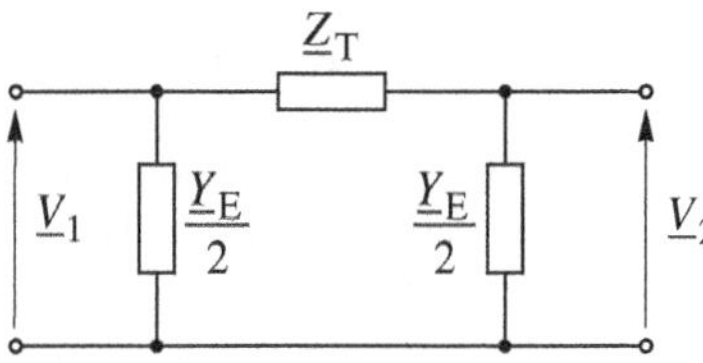

Figure 3.7 The approximate, symmetrical, equivalent π-circuit of a two-winding transformer.

- the secondary series impedance referred to the primary ($\underline{Z}_2 = R_2\vartheta^2 + jX_2\vartheta^2$) is equal to the primary series impedance ($\underline{Z}_1 = R_1 + jX_1$);
- the shunt impedance $1/(G_{Fe} + jB_\mu)$ is much greater than the total series impedance $\underline{Z}_T = \underline{Z}_1 + \underline{Z}_2$.

With these assumptions the parameters of the π-equivalent circuit, shown in Figure 3.7, are given by

$$\underline{Z}_T = \underline{Z}_1 + \underline{Z}_2 = R + jX; \qquad \underline{Y}_E = G_{Fe} + jB_\mu \tag{3.24}$$

The value of the parameters in the transformer equivalent circuit can be determined from the *no-load* test and the *short-circuit* test. In both of these tests the supply voltage, current, and active power are measured. Table 3.2 shows some examples of test data.

On no load the secondary winding of the transformer is open-circuited and the primary supplied with rated voltage $V_1 = V_n$. In this condition the series winding impedance is small compared with the shunt admittance and can be neglected so that the measured current and power correspond entirely to the shunt branch and relate to the excitation current I_E and iron loss P_{Fe}. Parameters of the shunt branch can then be computed, per unit (pu) to the unit rating, from

$$G_{Fe} = P_{Fe(pu)}; \qquad Y_E = I_{E(pu)}; \qquad B_\mu = \sqrt{Y_E^2 - G_{Fe}^2} \tag{3.25}$$

where $P_{Fe(pu)}$ and $I_{E(pu)}$ are also expressed pu.

During the short-circuit test the secondary winding is short-circuited. The primary winding is supplied with a voltage that is of sufficient magnitude to circulate a current in the short-circuited secondary that is equal to the rated load current ($I_2 = I_n$). This primary voltage is usually referred to as the *short-circuit voltage* V_{SHC} and is much smaller than the rated voltage $V_{SHC} \ll V_n$. Small primary voltage means that the excitation current I_E is much smaller than $I_2 = I_n$ and it can be neglected. Hence, the parameters of the series branch, in pu to the unit rating, can be found from

Table 3.2 Typical values of transformer parameters in pu on unit ratings.

S_n	V_{SHC}	P_{Cu}	I_E	P_{Fe}
MVA	pu	pu	pu	pu
150	0.11	0.0031	0.003	0.001
240	0.15	0.0030	0.0025	0.0006
426	0.145	0.0029	0.002	0.0006
630	0.143	0.0028	0.004	0.0007

$$Z_T = V_{SHC(pu)}; \qquad R_T = P_{Cu(pu)}; \qquad X_T = \sqrt{Z_T^2 - R_T^2} \tag{3.26}$$

where $V_{SHC(pu)}$ and $P_{Cu(pu)}$ are given in pu's.

The equivalent circuit for a multi-winding transformer is formed in a similar way. For example, the equivalent circuit of a three-winding transformer consists of three winding impedances connected in star with the shunt admittances connected to the star point.

3.2.2 Off-nominal Transformation Ratio

A power system usually operates at a number of voltage levels. The equivalent circuit shown in Figure 3.7 is not very convenient as the secondary voltage must be referred to the primary, resulting in the equivalent circuit including the transformer turns ratio. If the parameters of the equivalent circuit are expressed in pu then, as explained in Appendix A.1.5, the transformation ratio ϑ must be related to the nominal network voltages on both sides of the transformer. If the transformer rated voltages are equal to the nominal network voltages then the pu nominal transformation ratio is equal to unity, $\vartheta = 1$, and may be neglected in the transformer equivalent circuit shown in Figure 3.6a. However, the pu transformation ratio may not be unity for two reasons: (i) the rated transformer voltages are slightly different to the nominal network voltages and (ii) the tap changer adjusts the turns ratio away from the nominal setting, thereby changing the transformation ratio.

A convenient way to account for the off-nominal turns ratio is to replace the actual turns ratio by some fictitious reactive shunt elements in such a way that these elements change the voltage up or down, as required. As an example consider the transformer from Figure 3.6a with the shunt branch neglected and the secondary series impedance referred to the primary, as shown in Figure 3.8a. The series impedance $\underline{Z}_T$ has been replaced by its reciprocal $\underline{Y}_T$ in order to use the node voltage analysis. The transformer turns ratio will be taken as a complex number as this allows the general case of the phase shifting transformer to be considered.

For the ideal part of the transformer equivalent circuit shown in Figure 3.8a the apparent power on both sides is the same and the current in the primary is equal to the secondary current divided by the conjugate of the turns ratio, that is $\underline{I}_2/\underline{\vartheta}^*$. The two-port network corresponding to the branch with admittance $\underline{Y}_T$ can be described by the following nodal equation

$$\begin{bmatrix} \underline{I}_1 \\ -\underline{I}_2/\underline{\vartheta}^* \end{bmatrix} = \begin{bmatrix} \underline{Y}_T & -\underline{Y}_T \\ -\underline{Y}_T & \underline{Y}_T \end{bmatrix} \begin{bmatrix} \underline{V}_1 \\ \underline{\vartheta}\underline{V}_2 \end{bmatrix} \tag{3.27}$$

Eliminating the turns ratio $\underline{\vartheta}$ from the voltage and current vectors gives

$$\begin{bmatrix} \underline{I}_1 \\ -\underline{I}_2 \end{bmatrix} = \begin{bmatrix} \underline{Y}_T & -\underline{\vartheta}\underline{Y}_T \\ -\underline{\vartheta}^*\underline{Y}_T & \underline{\vartheta}^*\underline{\vartheta}\underline{Y}_T \end{bmatrix} \begin{bmatrix} \underline{V}_1 \\ \underline{V}_2 \end{bmatrix} \tag{3.28}$$

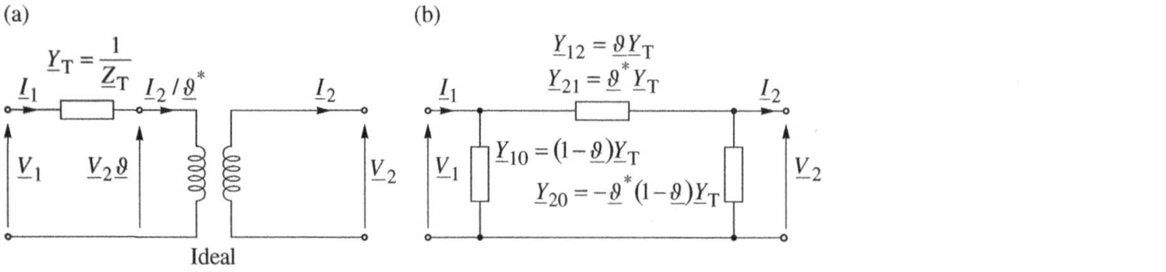

Figure 3.8 Equivalent circuit of a two-winding transformer: (a) using the ideal transformer; (b) using fictitious shunt elements.

The above equation can be interpreted as a nodal voltage equation $\underline{I} = \underline{Y}\underline{V}$ describing a π-equivalent network. The admittance of the series branch of this network is equal to the off-diagonal element (with the sign inverted) in the nodal admittance matrix, that is

$$\underline{Y}_{12} = \underline{\vartheta}\underline{Y}_{\mathrm{T}}; \qquad \underline{Y}_{21} = \underline{\vartheta}^{*}\underline{Y}_{\mathrm{T}} \tag{3.29}$$

while the admittances of the shunt branches are equal to the sum of elements in corresponding rows of the nodal admittance matrix, that is

$$\underline{Y}_{10} = (1-\underline{\vartheta})\underline{Y}_{\mathrm{T}}; \qquad \underline{Y}_{20} = -\underline{\vartheta}^{*}(1-\underline{\vartheta})\underline{Y}_{\mathrm{T}} \tag{3.30}$$

The equivalent network corresponding to this nodal admittance matrix is shown in Figure 3.8b. It may not be symmetric because generally $\underline{\vartheta} \neq 1$. In the case of a phase shifting transformer, the transformer turns ratio is a complex number $\underline{\vartheta} = \gamma + \mathrm{j}\beta$ and the series branch of the equivalent network is *anisotropic*, that is as $\underline{Y}_{12} \neq \underline{Y}_{21}$ it has different admittance values depending on which side the transformer is viewed from.

The transformer ratio in the circuit diagram of Figure 3.8b is modeled by the shunt elements. The flow of current in these elements causes a change in the voltages in the series branch that corresponds to the voltage transformation in a real transformer.

For some computation methods and theoretical analyses the anisotropic branch of the above model may constitute an additional barrier. In such cases another transformer model may be used. It is easy to transform Eq. (3.28) in the following way

$$\begin{bmatrix} \underline{I}_1 \\ -\underline{I}_2 \end{bmatrix} = \begin{bmatrix} \underline{Y}_{\mathrm{T}} & -\underline{Y}_{\mathrm{T}} \\ -\underline{Y}_{\mathrm{T}} & \underline{Y}_{\mathrm{T}} \end{bmatrix} \begin{bmatrix} \underline{V}_1 \\ \underline{V}_2 \end{bmatrix} - \begin{bmatrix} 0 & (\underline{\vartheta}-1)\underline{Y}_{\mathrm{T}} \\ (\underline{\vartheta}^{*}-1)\underline{Y}_{\mathrm{T}} & (1-\underline{\vartheta}^{*}\underline{\vartheta})\underline{Y}_{\mathrm{T}} \end{bmatrix} \begin{bmatrix} \underline{V}_1 \\ \underline{V}_2 \end{bmatrix}$$

or

$$\begin{bmatrix} \underline{I}_1 \\ -\underline{I}_2 \end{bmatrix} + \begin{bmatrix} \Delta\underline{I}_1 \\ \Delta\underline{I}_2 \end{bmatrix} = \begin{bmatrix} \underline{Y}_{\mathrm{T}} & -\underline{Y}_{\mathrm{T}} \\ -\underline{Y}_{\mathrm{T}} & \underline{Y}_{\mathrm{T}} \end{bmatrix} \begin{bmatrix} \underline{V}_1 \\ \underline{V}_2 \end{bmatrix} \tag{3.31}$$

where

$$\begin{bmatrix} \Delta\underline{I}_1 \\ \Delta\underline{I}_2 \end{bmatrix} = \begin{bmatrix} 0 & (\underline{\vartheta}-1)\underline{Y}_{\mathrm{T}} \\ (\underline{\vartheta}^{*}-1)\underline{Y}_{\mathrm{T}} & (1-\underline{\vartheta}^{*}\underline{\vartheta})\underline{Y}_{\mathrm{T}} \end{bmatrix} \begin{bmatrix} \underline{V}_1 \\ \underline{V}_2 \end{bmatrix} \tag{3.32}$$

The admittance matrix in Eq. (3.31) is symmetrical and describes a two-port network shown in Figure 3.9. The series branch in this network is isotropic and has an admittance value equal to $\underline{Y}_{\mathrm{T}}$ independent from which side the transformer is viewed from. Additional nodal currents $\Delta\underline{I}_1$, $\Delta\underline{I}_2$ depend on nodal voltages as described by Eq. (3.32). Such a transformer model is referred to as a *current-injection model*, or CIM.

If all phase-shifting transformers are modeled by CIMs then the admittance matrix describing the whole network is symmetrical, what makes some computations and analyses simpler.

If it is necessary to include the shunt branch corresponding to the exciting current of the transformer then the equivalent circuits in both models (admittance model and CIM) must be modified slightly by the addition of two shunt admittances to the circuit (as in Figure 3.7): shunt admittance $\underline{Y}_{\mathrm{E}}/2$ to the left-hand side of the circuit and shunt admittance $\vartheta^{2}\underline{Y}_{\mathrm{E}}/2$ to the right-hand side.

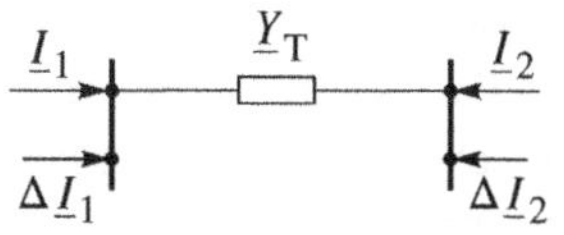

Figure 3.9 Current-injection model of a transformer.

3.3 Synchronous Generators

Chapter 2 describes how a synchronous generator consists of a stator, on which the three-phase armature winding is normally wound, and a rotor, on which the DC field winding is wound. The rotor damper windings do not affect the steady-state operation of the generator and are not considered in this chapter. The rotor is on the same shaft as the prime mover and may have salient or nonsalient magnetic poles.

The steady-state performance of a synchronous machine with a nonsalient round rotor will be analyzed first before generalizing the analysis to include the effects of rotor saliency. Of particular importance in this analysis is the way the rotor magnetic field interacts with the stator magnetic field to produce electromagnetic torque. An understanding of this interaction is more important than the detailed equations themselves. Having established the mechanisms by which torque and emf are created, the role of the generator in the EPS as a source of both active and reactive power will be analyzed.

In the steady state the magnetic field of synchronous generator (Section 2.3.1) is produced by the rotor and stator windings. The rotor (excitation) winding is supplied with DC current from the exciter. Its flux rotates with the rotor and in each phase of the stator (armature) winding induces the emf. If the generator is loaded, its three-phase stator (armature) winding also produces a rotating flux. In steady state both stator and rotor fluxes rotate with the same synchronous speed and are shifted by the *load angle* also called the *power angle*. For further analysis it is important to take assumptions regarding the direction of the rotor rotation and the mutual position of the axes, which define the position of rotor flux and the emf. In this book the following assumptions are taken:

a) The rotor rotates counter-clockwise, i.e. in the direction usually assumed to be positive in electrical engineering.
b) The *direct-axis* (d-axis) is defined as the rotor axis and has the same direction as the rotor-winding flux.
c) The *quadrature-axis* (q-axis) lags the d-axis by 90° and has the same direction as the synchronous emf, which lags the rotor-winding flux by 90°.

Some authors assume the clockwise rotor rotation and consequently an opposite mutual position of the d- and q-axes. Those assumptions are important for the sign of generator electric quantities describing the generator in the steady-state operation.

3.3.1 Round-rotor Machines

The schematic diagram of a two-pole generator on no load is shown in Figure 3.10. For simplicity, only the center conductor of each of the distributed windings is shown. The beginning and end of the field winding are denoted by f_1 and f_2, while the beginning and end of each of the phase windings are denoted by a_1 and a_2 (phase A), b_1 and b_2 (phase B), and c_1 and c_2 (phase C). The stator has three axes, A, B, and C, each corresponding to one of the phase windings. The rotor has two axes: the direct-axis (d-axis), which is the main magnetic axis of the field winding, and the *quadrature-axis* (q-axis), $\pi/2$ electrical radians behind the d-axis. The dashed lines show the path taken by the rotating *field* (or *excitation*) *flux*, Φ_f, produced by the field winding, and the field leakage flux, Φ_{fl}. $\vec{F}_f$ shows the direction (or the peak value) of the *magnetomotive force* (mmf) wave produced by the field current. The angle $\gamma = \omega_m t$, where ω_m is the rotor angular velocity, defines the instantaneous position of the rotor d-axis with respect to the stationary reference assumed here to be along the A-axis. It is assumed that the rotor rotates anti-clockwise.

For a two-pole machine one complete mechanical revolution corresponds to one electrical cycle and one electrical radian is equal to one mechanical radian. However, if the generator has p poles then one mechanical revolution corresponds to $p/2$ electrical cycles. In this general case one mechanical radian is equal to $p/2$ electrical radians and

$$\gamma_e = \frac{p}{2}\gamma_m \tag{3.33}$$

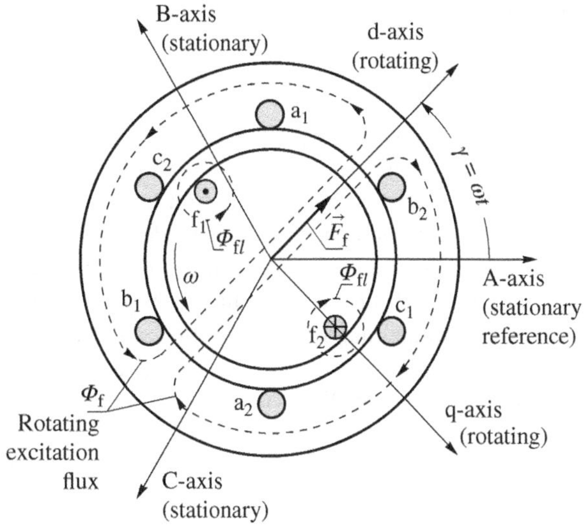

Figure 3.10 Symbolic representation of the generator and its fluxes at no load (counter-clockwise rotation is assumed).

where γ_m is the angle γ expressed in mechanical radians (or degrees) and γ_e is the same angle expressed in electrical radians (or degrees). Similarly, the rotor speed ω_e expressed in electrical radians per second can be related to the rotor speed in mechanical radians per second ω_m by

$$\omega_e = 2\pi f = \frac{p}{2}\omega_m \tag{3.34}$$

where f is the system nominal frequency (50 Hz in Europe, 60 Hz in United States). Often, the rotor speed is expressed in revolutions per minute (rpm) when the relation between the rotor speed n and the system frequency is $n = 120\, f/p$.

For simplicity the generator equations will be developed by reference to the two-pole machine shown in Figure 3.10 when both the speed and the angle are the same in electrical and mechanical units and the subscripts "e" and "m" can be dropped. The generator will be analyzed using only the fundamental component of the stator and rotor spatial mmf waves. Although the equations will be developed for a two-pole generator, they are equally applicable to a p-pole machine when all angles and speeds are expressed in electrical units. All the equations are valid whether expressed in SI or pu notation. If SI units are used it must be remembered that any power expression refers to the power per phase. In pu notation they can be interpreted as generator power (see Appendix A.1).

3.3.1.1 The Generator on No Load

To begin the analysis, assume that the generator is on no load, that is it is not generating any power and the current in the armature is zero. The DC field current i_f produces an mmf wave which is approximately sinusoidally distributed around the circumference of the stator. The peak value of the mmf

$$F_f = N_f i_f \tag{3.35}$$

lies along the d-axis and is shown in Figure 3.10 by the vector $\vec{F}_f$. In this equation N_f is the effective number of field winding turns per pole and is smaller than the actual number of turns N_F in order to account for the winding geometry and the actual trapezoidal mmf distribution of the field winding. It can be shown (McPherson and Laramore 1990) that $N_f = (1/p)(4/\pi)N_F k_{wF}$ where the winding geometry is accounted for by the winding factor k_{wF} and the mmf distribution by the factor $(4/\pi)$.

The field winding mmf drives the *excitation* (or *field*) *flux* Φ_f around the magnetic circuit. The flux per pole is

$$\Phi_f = \frac{F_f}{\mathfrak{R}} = \frac{N_f i_f}{\mathfrak{R}} \tag{3.36}$$

where $\mathfrak{R}$ is the *reluctance* of the path per pole. As the reluctance of the path in iron is negligibly small compared with that in air, $\mathfrak{R}$ is approximately directly proportional to the width of the air-gap.

The flux density produced by the field mmf is sinusoidally distributed around the circumference of the stator and, for a round-rotor machine, its peak coincides with the peak of the mmf wave. As the rotor rotates at synchronous speed, the excitation flux rotates with it and produces a time-varying flux linkage with each phase of the armature winding. Each of the phase flux linkages Ψ_{fA}, Ψ_{fB}, and Ψ_{fC} reaches a maximum when the d-axis of the rotor aligns with the magnetic axis of the respective phase winding. Taking phase A as the reference gives

$$\begin{aligned} &\Psi_{fA}(t) = \Psi_{fa}\cos\omega t = N_\phi \Phi_f \cos\omega t = N_\phi \frac{N_f i_f}{\mathfrak{R}}\cos\omega t = M_f i_f \cos\omega t \\ &\Psi_{fB}(t) = M_f i_f \cos\left(\omega t - \frac{2\pi}{3}\right), \qquad \Psi_{fC}(t) = M_f i_f \cos\left(\omega t - \frac{4\pi}{3}\right) \end{aligned} \tag{3.37}$$

In these equations $\Psi_{fa} = N_\phi \Phi_f$ is the amplitude of the excitation flux linkage of an armature phase winding, $M_f = N_\phi N_f/\mathfrak{R}$ is the *mutual inductance* between the field and the armature winding and $N_\phi = k_w N$, where N is the number of turns in series in each phase winding and k_w is the armature winding factor.

The time-varying flux linkages induce an *excitation emf* (also called the *internal voltage*) in each of the phase windings. Faraday's law gives

$$\begin{aligned} &e_{fA} = -\frac{d\Psi_{fA}(t)}{dt} = \omega M_f i_f \sin\omega t, \qquad e_{fB} = -\frac{d\Psi_{fB}(t)}{dt} = \omega M_f i_f \sin\left(\omega t - \frac{2\pi}{3}\right) \\ &e_{fC} = -\frac{d\Psi_{fC}(t)}{dt} = \omega M_f i_f \sin\left(\omega t - \frac{4\pi}{3}\right) \end{aligned} \tag{3.38}$$

In the absence of any armature current these emf's appear at the generator terminals as the *no-load terminal voltage*. Figure 3.11a shows the time variation of the phase flux linkages and the reference emf e_{fA} as a function of $\gamma = \omega t$, while the phasor representation of the flux linkages $\underline{\Psi}_{fA}$, $\underline{\Psi}_{fB}$, and $\underline{\Psi}_{fC}$ and the emf's $\underline{E}_{fA}$, $\underline{E}_{fB}$, and $\underline{E}_{fC}$ induced by them are shown in Figure 3.11b. The root-mean-square (RMS) value of each of these emf's (or the length of the phasors $\underline{E}_{fA}$, $\underline{E}_{fB}$, and $\underline{E}_{fC}$) is

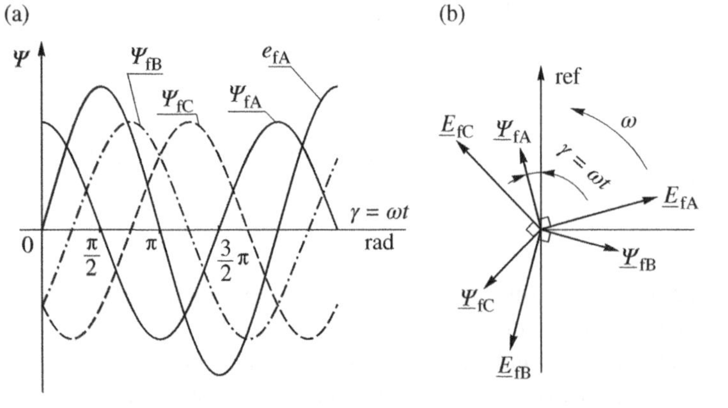

Figure 3.11 The phase excitation flux linkages and the emf's induced by them shown as: (a) time-varying waveforms; (b) rotating phasors.

$$E_f = \frac{1}{\sqrt{2}}\omega\Psi_{fa} = \frac{1}{\sqrt{2}}\omega N_\phi \Phi_f = \frac{1}{\sqrt{2}}\omega M_f i_f \cong 4.44\, fM_f i_f \tag{3.39}$$

This equation is the well-known transformer equation and illustrates how the primary (field) winding current induces an emf in the secondary (armature) winding. The emf is proportional to both the frequency and the field current i_f. The mutual inductance M_f is in practice not constant but depends on the saturation of the magnetic circuit.

3.3.1.2 Armature Reaction and the Air-gap Flux

Now consider the effect of loading the generator when an *armature current* flows in each phase of the stator winding. As the three phase windings are magnetically coupled, Section 11.1 shows how the equivalent circuit can be developed by considering their self- and mutual inductances. In this chapter another, simpler, approach is followed when the combined magnetic effect of all three phase windings is considered. This is referred to as the *armature reaction*.

Generally, the stator phase currents will be delayed by an angle λ with respect to the reference flux linkages $\Psi_{fA}(t)$ and may be written as

$$i_A = I_m \cos(\omega t - \lambda), \quad i_B = I_m \cos\left(\omega t - \lambda - \frac{2\pi}{3}\right), \quad i_C = I_m \cos\left(\omega t - \lambda - \frac{4\pi}{3}\right) \tag{3.40}$$

where I_m is the maximum (peek) value of the armature current. Each of the phase currents produces a pulsating phase mmf per pole

$$\begin{aligned} F_A(t) &= N_a I_m \cos(\omega t - \lambda); \qquad F_B(t) = N_a I_m \cos\left(\omega t - \lambda - \frac{2\pi}{3}\right); \\ F_C(t) &= N_a I_m \cos\left(\omega t - \lambda - \frac{4\pi}{3}\right) \end{aligned} \tag{3.41}$$

where $N_a = (1/p)(4/\pi)N_\phi$ is the effective number of turns per phase, per pole, and, as before $N_\phi = k_w N$.

Because the magnetic axes of the phase windings are shifted in space from each other by $2\pi/3$ electrical radians, the phase mmf's are shifted in both space and time. A convenient way to account for the shift in space is to represent the mmf's as *space vectors* directed along their respective phase axes with instantaneous values given by Eqs. (3.41). A space vector will be denoted by an arrow on top of the symbol. The vector of the resultant *armature reaction mmf* $\vec{F}_a$ can then be obtained by adding the component phase vectors.

A neat way to analyze the space position of the phase mmf's is to introduce a complex plane which has its real axis directed along the A-axis and its imaginary axis $\pi/2$ ahead (counter-clockwise). The space operator $e^{j\theta}$ then introduces a phase shift in the complex plane which, when multiplied by the value of the phase mmf, will direct it in space along the magnetic axes of the winding. Axis B is at an angle $2\pi/3$ with respect to A, while axis C is at an angle $4\pi/3$. The value of F_B must be multiplied by $e^{j2\pi/3}$ to obtain a vector directed along the B-axis, while the value of F_C must be multiplied by $e^{j4\pi/3}$ to obtain a vector directed along the C-axis. The vector of the resulting armature mmf per pole $\vec{F}_a$ is then

$$\begin{aligned} \vec{F}_a &= \vec{F}_A + \vec{F}_B + \vec{F}_C = N_a i_A e^{j0} + N_a i_B e^{j2\pi/3} + N_a i_C e^{j4\pi/3} \\ &= N_a I_m \left[\cos(\omega t - \lambda) + \cos\left(\omega t - \lambda - \frac{2\pi}{3}\right)e^{j2\pi/3} + \cos\left(\omega t - \lambda - \frac{4\pi}{3}\right)e^{j4\pi/3}\right] \end{aligned} \tag{3.42}$$

Using the identity

$$\cos(\alpha - \beta) = \cos\alpha\cos\beta + \sin\alpha\sin\beta \tag{3.43}$$

gives

$$\begin{aligned}\vec{F}_{\mathrm{a}} &= N_{\mathrm{a}}I_{\mathrm{m}}\{\cos(\omega t - \lambda) + [-0.5\cos(\omega t - \lambda) + 0.866\sin(\omega t - \beta)](-0.5 + \mathrm{j}0.866)\\ &\quad + [-0.5\cos(\omega t - \lambda) - 0.866\sin(\omega t - \lambda)](-0.5 - \mathrm{j}0.866)\}\\ &= N_{\mathrm{a}}I_{\mathrm{m}}[1.5\cos(\omega t - \lambda) + 1.5\mathrm{j}\sin(\omega t - \lambda)] = 1.5N_{\mathrm{a}}I_{\mathrm{m}}\mathrm{e}^{\mathrm{j}(\omega t - \lambda)}\end{aligned} \tag{3.44}$$

This equation shows that $\vec{F}_{\mathrm{a}}$ is a vector of constant magnitude $F_{\mathrm{a}} = 1.5N_{\mathrm{a}}I_{\mathrm{m}}$ which rotates in the complex plane with an angular velocity ω. As this is also the rotational speed of the generator, the rotor and stator mmf's are stationary with respect to each other. To find the relative spatial position of the mmf's recall that the angle λ was defined as a time delay with respect to the linkages $\Psi_{\mathrm{fA}}(t)$. As $\vec{F}_{\mathrm{f}}$ is in phase with $\Psi_{\mathrm{fA}}(t)$ (that is Ψ_{fA} reaches maximum when $\vec{F}_{\mathrm{f}}$ aligns with A-axis), Eq. (3.44) shows that $\vec{F}_{\mathrm{a}}$ must lag $\vec{F}_{\mathrm{f}}$ by the spatial angle λ.

As the two rotating mmf's $\vec{F}_{\mathrm{a}}$ and $\vec{F}_{\mathrm{f}}$ are stationary with respect to each other, they can be combined to give a *resultant mmf* $\vec{F}_{\mathrm{r}} = \vec{F}_{\mathrm{f}} + \vec{F}_{\mathrm{a}}$ which drives the resultant *air-gap flux* Φ_{r}. This is shown in Figure 3.12. It is important to realize that the magnetic circuit does not "see" $\vec{F}_{\mathrm{a}}$ or $\vec{F}_{\mathrm{f}}$ alone but $\vec{F}_{\mathrm{r}} = \vec{F}_{\mathrm{f}} + \vec{F}_{\mathrm{a}}$. The peak flux density in the air gap of a round-rotor generator coincides with the peak of $\vec{F}_{\mathrm{r}}$ and it is in the same direction. Figure 3.12 shows the typical relative position of the mmf's with $\pi/2 < \lambda < \pi$. It can be seen that the armature reaction field demagnetizes the generator and the resultant mmf is smaller than the excitation mmf alone.

As the resultant mmf is the sum of two rotating, sinusoidally distributed mmf's, both the resultant mmf per pole and the density of the air-gap flux driven by it are sinusoidally distributed around the air gap. In practice, magnetic saturation will cause a slight flattening of the flux density wave, which results in third-harmonic voltages being induced in the armature winding. Connecting the generator windings in Δ or ϒ with the star point not earthed prevents the third-harmonic currents appearing at the generator terminals.

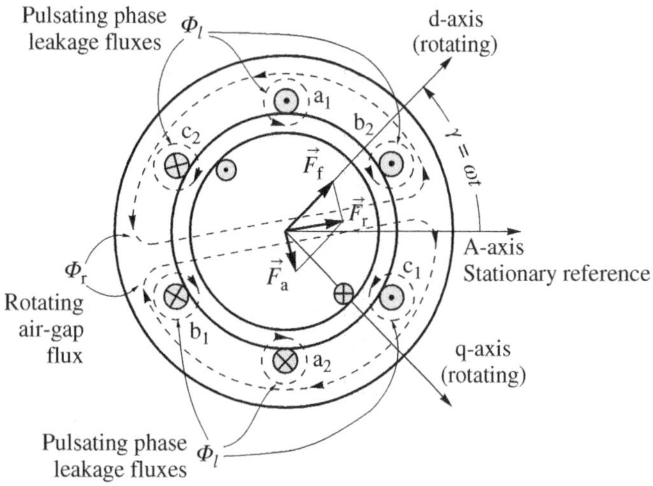

Figure 3.12 Resultant flux in the round-rotor generator operating at lagging power factor (counter-clockwise rotation is assumed).

3.3.1.3 Equivalent Circuit and the Phasor Diagram

The rotating air-gap flux produces sinusoidally changing flux linkages with each of the stator phase windings. The linkages with phase A, F_{rA}, can be calculated by projecting the component rotor and stator mmf waves onto the A-axis in order to obtain the resultant mmf along this axis. Recall that F_f is assumed to align with the A-axis at $t = 0$ and that F_a is delayed with respect to F_f by λ. This gives

$$F_{rA}(t) = F_f \cos \omega t + F_a \cos(\omega t - \lambda) = N_f i_f \cos \omega t + 1.5 N_a I_m \cos(\omega t - \lambda) \tag{3.45}$$

This resultant mmf must be divided by the reluctance, $\mathfrak{R}$, and multiplied by the number of armature turns N_ϕ to obtain the resultant flux linkages with phase A

$$\Psi_{rA}(t) = N_\phi \frac{F_{rA}(t)}{\mathfrak{R}} = M_f i_f \cos \omega t + L_a I_m \cos(\omega t - \lambda) \tag{3.46}$$

where $M_f = N_\phi N_f / \mathfrak{R}$ is the mutual inductance between the field and the armature winding and $L_a = 1.5 N_a N_\phi / \mathfrak{R}$ is the *armature reaction inductance* or *magnetizing inductance*. In the round-rotor machines with a uniform air gap the reluctance $\mathfrak{R}$ does not depend on the flux position.

The linkages $\Psi_{rA}(t)$ induce the *air-gap emf* in phase A equal to

$$e_{rA} = -\frac{d\Psi_{rA}}{dt} = \omega M_f i_f \sin \omega t + \omega L_a I_m \sin(\omega t - \lambda) = e_{fA}(t) + e_{aA}(t) \tag{3.47}$$

where $e_{fA}(t) = \omega M_f i_f \sin \omega t$ and $e_{aA}(t) = \omega L_a I_m \sin(\omega t - \lambda)$. Equation (3.47) shows that the trick in modeling the synchronous generator is to represent the resultant air-gap emf by a sum of two fictitious emf's. The first is the *excitation* (or *internal*) *emf* $e_{fA}(t)$ due to the rotor field and is equal to the no-load terminal voltage, Eq. (3.38). The second is the *armature reaction emf* $e_{aA}(t)$, due to the armature reaction field, which lags the A-phase current by $\pi/2$; compare Eq. (3.47) with Eq. (3.40). Using phasor notation, this corresponds to multiplying the current phasor by $-j$ so that the armature reaction emf phasor is $\underline{E}_a = -jX_a\underline{I}$ where $X_a = \omega L_a$ is the *armature reaction reactance* or *magnetizing reactance* and $\underline{I}$ is the current phasor of magnitude $I = I_m/\sqrt{2}$. The phasor of the air-gap emf can be therefore expressed as

$$\underline{E}_r = \underline{E}_f + \underline{E}_a = \underline{E}_f - jX_a\underline{I} \tag{3.48}$$

where $\underline{E}_f$ is the phasor of the internal emf with magnitude E_f given by Eq. (3.39). Thus $-\underline{E}_a = jX_a\underline{I}$ has the properties of a reactance drop resulting from the effect of the armature current. When $-\underline{E}_a$ is replaced by the reactance voltage drop $jX_a\underline{I}$, the circuit model becomes that to the left of the phasor $\underline{E}_r$ in Figure 3.13.

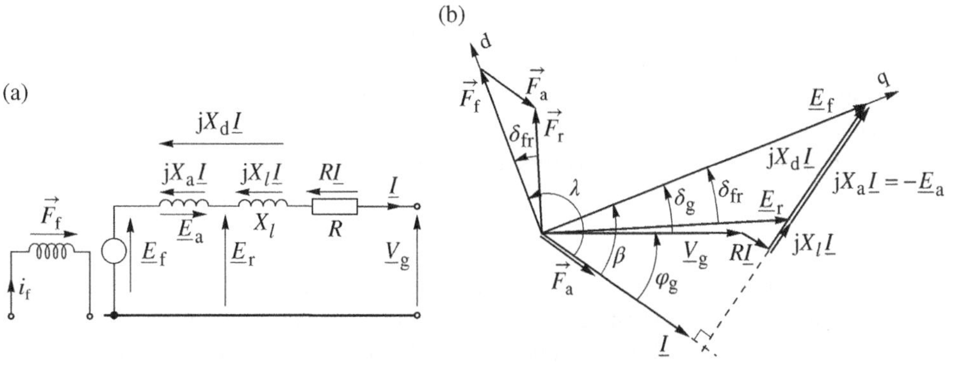

Figure 3.13 Round-rotor synchronous generator: (a) equivalent circuit diagram; (b) phasor/vector diagram for operation at lagging power factor.

Figure 3.13a shows how the full equivalent circuit can be obtained by accounting for electrical and magnetic imperfections of the machine. Firstly, each armature winding has some resistance R, which produces the voltage drop $R\underline{I}$. In the case of the synchronous generator this armature resistance is very small and can often be neglected. Secondly, although most of the magnetic flux produced by the armature crosses the air gap to link with the rotor winding, a part of it, called the *leakage flux*, does not. This pulsating leakage flux is shown as Φ_l in Figure 3.12 and only links with the armature winding. The leakage flux can be accounted for by inserting a *leakage reactance* X_l into the circuit. As the leakage flux closes its path largely through air, $X_l \ll X_a$.

Figure 3.13a shows the total equivalent circuit of the round-rotor synchronous machine with all the imperfections included. The terminal voltage $\underline{V}_g$ is

$$\underline{V}_g = \underline{E}_f - jX_a\underline{I} - jX_l\underline{I} - R\underline{I} = \underline{E}_f - jX_d\underline{I} - R\underline{I} \tag{3.49}$$

where $X_d = X_a + X_l$ is the *synchronous reactance*, or more precisely the *direct-axis synchronous reactance*. As $L_l \ll L_a$, $X_l \ll X_a$ and to all practical purposes $X_d \approx X_a$. The internal emf E_f is sometimes referred to as the *voltage behind the synchronous reactance*. The phasor diagram resulting from Eq. (3.49) is shown in Figure 3.13b for a generator operating at lagging power factor.

The phasor diagram is normally constructed knowing only the generator voltage V_g and its per-phase active and reactive load powers P and Q. From these the current and the power factor angle are calculated:

$$I = \frac{\sqrt{P^2 + Q^2}}{V_g}; \qquad \varphi_g = \arctan\frac{Q}{P} \tag{3.50}$$

Knowing the length and the direction of the current phasor relative to the reference voltage V_g the length and direction of the voltage drops $\underline{I}R$, $j\underline{I}X_l$, and $j\underline{I}X_a$ can be found and the phasor diagram constructed.

Apart from the usual voltage and current phasors, Figure 3.13b also shows the space vectors of the mmf's and the position of the d- and q-axes. Combining both space vectors and phasors on the same diagram is very useful and is accomplished by drawing the voltage (and current) phasors and the mmf vectors on the same complex plane at $t = 0$, taking phase A as the reference.

To find the relative displacement between the phasors and the vectors on the diagram consider the relative position of the armature current I and the armature mmf $\vec{F}_a$. At $t = 0$ the space angle of $\vec{F}_a$ is $(-\lambda)$ with respect to the A-axis, Eq. (3.44). At the same time the phase angle of the A-axis current is also $(-\lambda)$, Eq. (3.40), indicating that the current phasor $\underline{I}$ and the mmf vector $\vec{F}_a$ are in phase on the phasor/vector diagram. As $\underline{E}_a = -jX_a\underline{I}$, it follows that $\underline{E}_a$ lags $\vec{F}_a$ by $\pi/2$. A similar argument shows that all the emf's must lag their mmf's by $\pi/2$.

The phasor diagram of Figure 3.13b provides additional insight into the meaning of all the angles on it. As all the emf's are perpendicular to their mmf's the mmf triangle (F_f,F_a,F_r) is similar to the voltage triangle (E_f,E_a,E_r) and all the angles shown in Figure 3.13 have a dual meaning: they are, at the same time, the spatial angles between the rotating mmf's and the phase shifts between the ac voltages. For example, the angle δ_{fr}, which is the space angle between the field and the resultant air-gap mmf, is at the same time the phase shift between $\underline{E}_f$ and $\underline{E}_r$.

3.3.1.4 The Torque Creation Mechanism

The rotor of the synchronous generator is driven by a prime mover which exerts a mechanical torque τ_m on it. For the speed of the rotor to remain constant, the machine must develop an equal, but opposing, electromagnetic torque τ_e. Resolution of the air-gap mmf into the stator and rotor component mmf's provides a means of understanding how the electromagnetic torque is developed. The component stator and rotor mmf's can be compared with two magnets rotating at the same speed and trying to co-align, with the north magnetic pole of one attracting the south pole of the other, and vice versa. The torque produced by these attractive forces can be calculated from the three-phase air-gap power P_{ag}, which in its electrical form is equal to

$$P_{ag} = 3E_f I \cos\beta \tag{3.51}$$

where β is the angle between $\underline{E}_f$ and $\underline{I}$ and is referred to as the *internal power factor angle* (Figure 3.13b). Neglecting mechanical losses, P_{ag} must be equal to the mechanical power $\tau_m\omega_m$ supplied by the prime mover so that for a p-pole machine

$$\tau = \frac{1}{\omega_m} P_{ag} = \frac{p}{2}\frac{1}{\omega} P_{ag} \tag{3.52}$$

Substituting for E_f from Eq. (3.39) and noting that $I = I_m/\sqrt{2}$ gives

$$\tau = \frac{3}{4} p \Phi_f N_\phi I_m \cos\beta \tag{3.53}$$

This equation can be written in terms of the angle λ and the armature mmf F_a by noting that $N_a = \frac{1}{p}\frac{4}{\pi}N_\phi$ when, from Eq. (3.44)

$$F_a = \frac{3}{2} N_a I_m = \frac{3}{2}\left(\frac{1}{p}\frac{4}{\pi} N_\phi\right) I_m \tag{3.54}$$

Figure 3.13b shows that $\lambda = \pi/2 + \beta$ so that $\sin\lambda = \cos\beta$ and

$$\tau = \frac{\pi}{8} p^2 \Phi_f F_a \sin\lambda = \frac{\pi}{8} p^2 \frac{F_f F_a}{\mathfrak{R}} \sin\lambda \tag{3.55}$$

where λ is the angle between F_f and F_a. Inspection of Figure 3.13b shows that $F_a \sin\lambda = F_r \sin\delta_{fr}$ and Eq. (3.55) can be rewritten as

$$\tau = \frac{\pi}{8} p^2 F_r \frac{F_f}{\mathfrak{R}} \sin\delta_{fr} = \frac{\pi}{8} p^2 F_r \Phi_f \sin\delta_{fr} \tag{3.56}$$

where the angle δ_{fr} is referred to as the *torque angle*. For a two-pole machine this equation simplifies to

$$\tau = \frac{\pi}{2} F_r \Phi_f \sin\delta_{fr} \tag{3.57}$$

If the rotor field leads the air-gap field, as in Figure 3.12, then the electromagnetic torque acts in the opposite direction to the rotation and opposes the mechanical driving torque so the machine acts as a generator. On the other hand, if the rotor field lags behind the air-gap field, then the electromagnetic torque acts in the direction of rotation and the machine acts as a motor.

3.3.2 Salient-pole Machines

Because of the relative ease of balancing round rotors, and their ability to withstand high centrifugal forces, round-rotor generators are normally used for turbo units driven by high-speed steam or gas turbines. Generators operating at a lower speed, such as those driven by hydro turbines, need many magnetic poles in order to operate at 50 or 60 Hz. As the centrifugal forces experienced by the rotors of these low-speed machines are lower than in the corresponding turbo-generators, salient poles can be used and the rotor diameter increased. Normally, salient-pole rotors have more than two poles and the angles (and speed) expressed in mechanical units are related to those expressed in electrical units by Eqs. (3.33) and (3.34). To simplify considerations a two-pole salient generator will be considered here so that the angles expressed in electrical and mechanical radians are the same. The simplified cross-section of such a generator is shown in Figure 3.14.

The main problem with modeling a salient-pole machine is that the width of the air gap varies circumferentially around the generator with the narrowest gap being along the d-axis and the widest along the q-axis. Consequently,

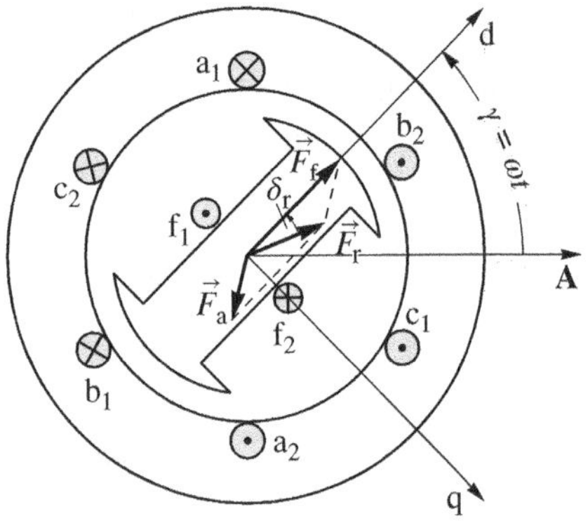

Figure 3.14 A simplified salient-pole generator.

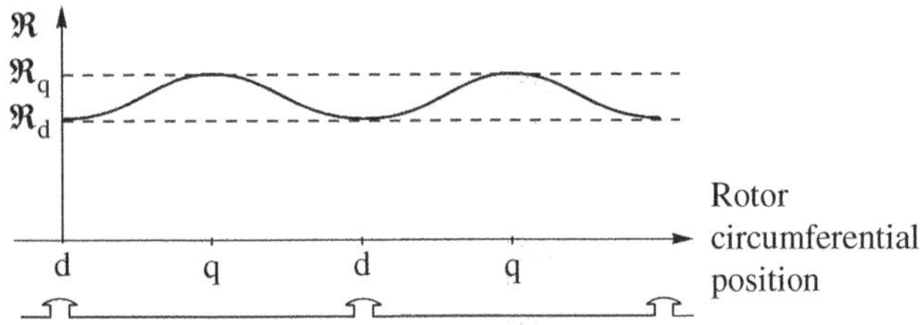

Figure 3.15 Approximate variation of reluctance with circumferential position.

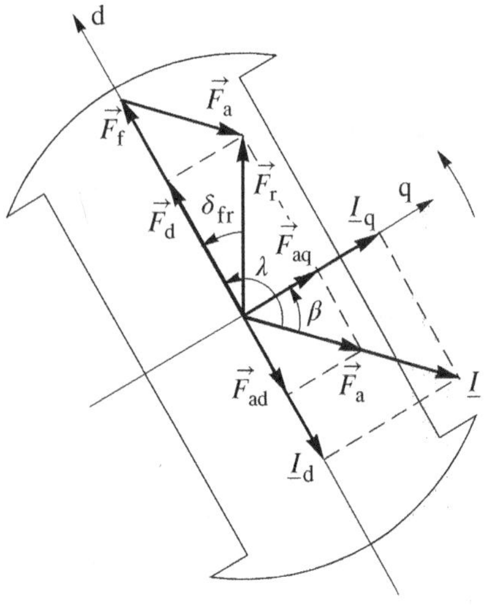

Figure 3.16 Resolution of the mmf's and the current into d- and q-axis components.

the reluctance of the air-gap flux is not uniform but varies between its minimum value $\mathfrak{R}_d$ and its maximum value $\mathfrak{R}_q$, as shown by the approximated function in Figure 3.15. This creates a problem as one of the fundamental assumptions for modeling the uniform-gap, round-rotor generator was that the peak density of the flux wave coincides with the peak density of the mmf wave or, in other words, the mmf and the flux vectors are in phase. It is important to realize that this assumption is no longer valid for salient-rotor machines. As flux tends to take the path of least reluctance, the flux vector is shifted toward the d-axis (where reluctance is least) when compared with the mmf vector. The only instance when the two vectors are in phase is when the mmf vector lies along either the d-axis (which is the position of minimum reluctance) or the q-axis (when the attractive forces that tend to shift the flux toward the d-axis balance each other). At any other position the flux and mmf are out of phase and the simple analysis used for the round-rotor machine is no longer valid. To overcome this problem, A. Blondel developed his *two-reaction theory*, which resolves the mmf's acting in the machine along the d- and q-axes. The emf's due to these component mmf's are then considered separately, assigning different, but constant, values of reactances to the fluxes acting along these axes.

This concept is illustrated in Figure 3.16. The armature mmf $\vec{F}_a$ and the armature current $\underline{I}$ are resolved into two components, one acting along the d-axis ($\vec{F}_{ad}$, $\underline{I}_d$) and the other acting along the q-axis ($\vec{F}_{aq}$, $\underline{I}_q$). There is no need to resolve the excitation mmf $\vec{F}_f$ as this always acts only along the d-axis. The resultant mmf $\vec{F}_r$ may then be expressed as

$$\vec{F}_{\rm r} = \vec{F}_{\rm d} + \vec{F}_{\rm q} \tag{3.58}$$

where $\vec{F}_{\rm d} = \vec{F}_{\rm f} + \vec{F}_{\rm ad}$ and $\vec{F}_{\rm q} = \vec{F}_{\rm aq}$. Similarly the current $\underline{I}$ may be expressed as

$$\underline{I} = \underline{I}_{\rm d} + \underline{I}_{\rm q} \tag{3.59}$$

Similarly, as for the round-rotor generator, the resultant air-gap emf is equal to the sum of the component emf's, each of which is due to a corresponding mmf. In this case the resultant emf is assumed to be equal to the sum of three components due to $\vec{F}_{\rm f}$, $\vec{F}_{\rm ad}$, and $\vec{F}_{\rm aq}$ respectively. Because the excitation mmf $\vec{F}_{\rm f}$ always acts along the d-axis, the internal emf $E_{\rm f}$ due to $\vec{F}_{\rm f}$ depends only on the d-axis reluctance $\mathfrak{R}_{\rm d}$ and is constant for a given value of field current. As the emf's lag their mmf's by $\pi/2$, $\underline{E}_{\rm f}$ is directed along the q-axis.

The emf due to the d-axis armature mmf $\vec{F}_{\rm ad}$ is proportional to $\underline{I}_{\rm d}$ and delayed by $\pi/2$ with respect to it. This emf is therefore directed along the q-axis and may be expressed as

$$\underline{E}_{\rm aq} = -{\rm j}X_{\rm ad}\underline{I}_{\rm d} \tag{3.60}$$

where $X_{\rm ad}$ is the *direct-axis armature reaction reactance*. As $\vec{F}_{\rm ad}$ acts across the shortest gap, $X_{\rm ad}$ is inversely proportional to $\mathfrak{R}_{\rm d}$.

The emf due to the q-axis armature mmf $\vec{F}_{\rm aq}$ is proportional to $\underline{I}_{\rm q}$ and delayed by $\pi/2$. This emf is therefore directed along the d-axis and may be expressed as

$$\underline{E}_{\rm ad} = -{\rm j}X_{\rm aq}\underline{I}_{\rm q} \tag{3.61}$$

where $X_{\rm aq}$ is the *quadrature-axis armature reaction reactance*. As $\vec{F}_{\rm aq}$ acts across the widest gap, $X_{\rm aq}$ is inversely proportional to $\mathfrak{R}_{\rm q}$. The resultant air-gap emf is then

$$\underline{E}_{\rm r} = \underline{E}_{\rm f} + \underline{E}_{\rm aq} + \underline{E}_{\rm ad} = \underline{E}_{\rm f} - {\rm j}X_{\rm ad}\underline{I}_{\rm d} - {\rm j}X_{\rm aq}\underline{I}_{\rm q} \tag{3.62}$$

The terminal voltage $\underline{V}_{\rm g}$ is obtained by subtracting from this emf the voltage drops due to the armature leakage reactance and resistance to give

$$\begin{aligned}\underline{V}_{\rm g} &= \underline{E}_{\rm r} - {\rm j}X_l\underline{I} - R\underline{I} = \underline{E}_{\rm f} - {\rm j}X_{\rm ad}\underline{I}_{\rm d} - {\rm j}X_{\rm aq}\underline{I}_{\rm q} - {\rm j}X_l\left(\underline{I}_{\rm d} + \underline{I}_{\rm q}\right) - R\underline{I} \\ &= \underline{E}_{\rm f} - {\rm j}(X_{\rm ad} + X_l)\underline{I}_{\rm d} - {\rm j}(X_{\rm aq} + X_l)\underline{I}_{\rm q} - R\underline{I}\end{aligned} \tag{3.63}$$

or

$$\underline{E}_{\rm f} = \underline{V}_{\rm g} + {\rm j}X_{\rm d}\underline{I}_{\rm d} + {\rm j}X_{\rm q}\underline{I}_{\rm q} + R\underline{I} \tag{3.64}$$

where $X_{\rm d} = X_{\rm ad} + X_l$ is the *direct-axis synchronous reactance* and $X_{\rm q} = X_{\rm aq} + X_l$ is the *quadrature-axis synchronous reactance*. As the reluctance along the q-axis is the highest (because the gap is the widest) $X_{\rm d} > X_{\rm q}$.

3.3.2.1 Phasor Diagram and the Equivalent Circuit

Figure 3.17 shows the phasor diagram resulting from Eq. (3.64). Its construction is more complicated than that for the round-rotor generator shown in Figure 3.13. In order to determine $\underline{I}_{\rm d}$ and $\underline{I}_{\rm q}$, it is necessary first to determine the angle $\delta_{\rm g}$ so as to locate the q-axis relative to $\underline{V}_{\rm g}$. To solve this problem recall that the emf $\underline{E}_{\rm f}$ lies along the q-axis. Rearranging Eq. (3.64) gives

$$\underline{E}_{\rm f} = \underline{V}_{\rm g} + R\underline{I} + {\rm j}X_{\rm q}\underline{I} + {\rm j}(X_{\rm d} - X_{\rm q})\underline{I}_{\rm d} = \underline{E}_{\rm Q} + {\rm j}(X_{\rm d} - X_{\rm q})\underline{I}_{\rm d} \tag{3.65}$$

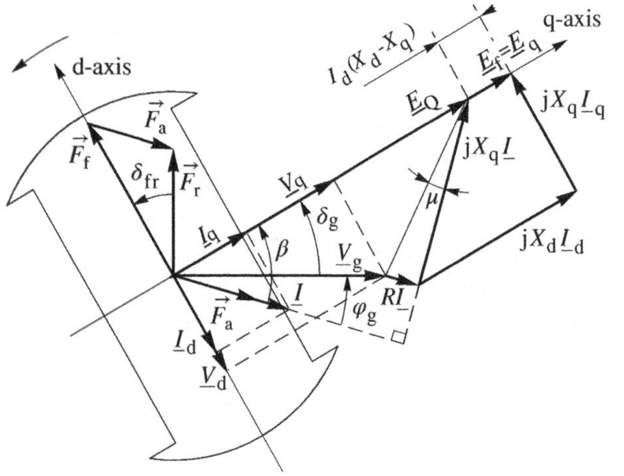

Figure 3.17 Phasor diagram for a salient-pole generator.

where

$$\underline{E}_Q = \underline{V}_g + (R + jX_q)\underline{I} \tag{3.66}$$

Because multiplication of $\underline{I}_d$ by j shifts the resultant phasor $\pi/2$ ahead, the second term in Eq. (3.65), equal to $j(X_d - X_q)\underline{I}_d$, is directed along the q-axis. As $\underline{E}_f$ itself lies on the q-axis, $\underline{E}_Q$ must also lie on the q-axis. Knowing $\underline{V}_g$ and $\underline{I}$, the phasor $(R + jX_q)\underline{I}$ can be added to $\underline{V}_g$ to obtain $\underline{E}_Q$. This determines the direction of the q-axis and the angle δ_g. Once the location of the q-axis is known, $\underline{I}$ can be resolved into its components $\underline{I}_d$ and $\underline{I}_q$ and the phasor diagram completed. For the round-rotor generator $X_d = X_q$ and $\underline{E}_Q$ becomes equal to $\underline{E}_f$ (and $\underline{E}_q$).

Resolving Eq. (3.64) into its d- and q-axis components allows the generator equivalent circuit to be constructed. As multiplication of a phasor by j gives a phasor shifted $\pi/2$ counter-clockwise, the phasor ($jX_q\underline{I}_q$) is directed along the d-axis. Similarly, the phasor ($jX_d\underline{I}_d$) is $\pi/2$ ahead of the d-axis, that is it is directed in the opposite direction to the q-axis so that its q-axis component is negative.[1] Consequently, the d- and q-axis components of Eq. (3.64) are

$$\begin{aligned} &\text{d-axis}: \quad E_d = V_d + RI_d + X_q I_q = 0 \\ &\text{q-axis}: \quad E_q = V_q + RI_q - X_d I_d = E_f \end{aligned} \tag{3.67}$$

These equations can be written in matrix form as

$$\begin{bmatrix} E_d \\ E_q \end{bmatrix} = \begin{bmatrix} 0 \\ E_f \end{bmatrix} = \begin{bmatrix} V_{gd} \\ V_{gq} \end{bmatrix} + \begin{bmatrix} R & +X_q \\ -X_d & R \end{bmatrix} \begin{bmatrix} I_d \\ I_q \end{bmatrix} \tag{3.68}$$

It is important to realize that all the variables in each of these equations are in phase with each other and are real (not complex) numbers. The d- and q-axis components of the terminal voltage and current are

$$\begin{aligned} V_d &= -V_g \sin\delta_g; \qquad & V_q &= V_g \cos\delta_g \\ I_d &= -I \sin\beta; \qquad & I_q &= I \cos\beta \end{aligned} \tag{3.69}$$

1 Figure 3.17 shows the phasor of ($jX_d\underline{I}_d$) to be along the q-axis. This is due to I_d being negative.

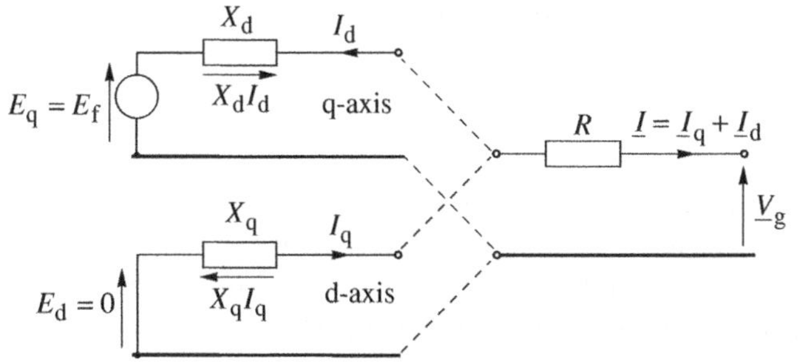

Figure 3.18 Equivalent d- and q-axis circuit diagrams of the salient-pole generator.

and $\beta = \delta_g + \varphi_g$. The minus sign in the d-axis components of the voltage and the current indicates the fact that they are directed in the opposite direction to the d-axis. The equivalent circuit for the salient-pole generator can now be drawn in two parts, one corresponding to the d-axis and the other to the q-axis, as shown in Figure 3.18. As all the variables in each of the Eq. (3.67) are in phase, the reactances X_d and X_q are shown using the symbol of resistance, rather than reactance, to give a voltage drop that is in phase with the current.

A slightly misleading feature of the equivalent circuit shown in Figure 3.18 is that the d-axis current flows into, rather than out of, the generator. This is a consequence of assuming that the d-axis leads the q-axis. As δ_g is assumed to be positive for generator action, this assumption gives negative values for the d-axis components of the terminal voltage and current; see Eq. (3.69). The selection of the d-axis leading the q-axis is purely arbitrary and is recommended by IEEE (1969) on the grounds that the d-axis current usually causes demagnetization of the machine and should be negative in the generator reference frame. Many authors, however, assume the d-axis to lag the q-axis, which removes the minus sign from Eq. (3.69) and changes the sign of the reactive terms in Eq. (3.67).

3.3.2.2 The d- and q-axis Armature Coils

As will be seen in later chapters it is often helpful to consider the effect of the three-phase armature winding to be produced by two equivalent windings phase displaced by 90°. If one of these equivalent armature coils is located along the d-axis and the other along the q-axis then the equivalent circuit of Figure 3.18 can be seen to take on a more direct physical meaning. These two equivalent armature coils are known as the *d- and q-axis armature coils* and are assumed to have the same number of turns as an actual armature phase winding.[2] To account for the DC nature of the equations in (3.67) these two equivalent armature coils are also assumed to rotate with the generator rotor, with the voltages $E_q = E_f$ and E_d being injected into the respective armature coil. As E_f and E_d are proportional to the rotational speed ω, these voltages are known as *rotational voltages*. This concept of d- and q-axis armature coils is examined in detail in Section 11.1.1 (Figure 11.1).

3.3.2.3 Torque in the Salient-pole Machine

The mechanism by which electromagnetic torque is produced was explained earlier with regard to the round-rotor generator, when it was shown that torque is proportional to the product of the stator mmf and rotor flux and the sine of the angle between them, Eq. (3.55). In the case of the salient-pole generator a similar torque creation mechanism applies, but now d- and q-axis components of the mmf must be considered. Resolution of the armature mmf gives

$$F_{ad} = F_a \cos\lambda; \quad F_{aq} = F_a \sin\lambda \tag{3.70}$$

2 This assumption is not strictly necessary but is the assumption made in this book. A fuller explanation of this is given in Chapter 11.

As the angle between Φ_f and F_{aq} is $\pi/2$, application of Eq. (3.55) leads to the electromagnetic torque in a two-pole generator being equal to

$$\tau_q = \frac{\pi}{2}\Phi_f F_{aq} \tag{3.71}$$

This equation represents only part of the tangential force on the armature as there will be other forces due to the interaction of other components of the d-axis flux with the q-axis mmf. The effect of these other d-axis flux components is included by algebraically adding all the d-axis fluxes before multiplying by the q-axis mmf. In the steady state the only other flux along the d-axis is the flux produced by the d-axis mmf itself so that the torque expression becomes

$$\tau_q = \frac{\pi}{2}(\Phi_f + \Phi_{ad})F_{aq} \tag{3.72}$$

where $\Phi_{ad} = F_{ad}/\Re_d$. There will also be a similar interaction between any q-axis flux and the d-axis mmf that will produce an additional torque component

$$\tau_d = \frac{\pi}{2}(\Phi_{aq})F_{ad} \tag{3.73}$$

where $\Phi_{aq} = F_{aq}/\Re_q$. As F_{ad} is produced by the d-axis current (Figure 3.18) which flows into, rather than out of, the generator the torque τ_d acts in the opposite direction to τ_q. The total torque is equal to the difference between the two components giving

$$\tau = \tau_q - \tau_d = \frac{\pi}{2}(\Phi_f + \Phi_{ad})F_{aq} - \frac{\pi}{2}(\Phi_{aq})F_{ad} \tag{3.74}$$

Regrouping the terms and substituting $\Phi_{ad} = F_{ad}/\Re_d$; $\Phi_{aq} = F_{aq}/\Re_q$; $F_{aq} = F_a \sin\lambda = F_r \sin\delta_{fr}$ and $F_{ad} = F_a\cos\lambda = F_r\cos\delta_{fr}$ finally gives

$$\tau = \frac{\pi}{2}\Phi_f F_r \sin\delta_{fr} + \frac{\pi}{4}F_r^2\frac{\Re_q - \Re_d}{\Re_q\Re_d}\sin 2\delta_{fr} \tag{3.75}$$

This equation shows that the torque developed in a salient-pole generator consists of two components. The first component, proportional to $\sin\delta_{fr}$, is identical to the torque expressed by Eq. (3.57) for a round-rotor generator and is termed the *synchronous torque*. The second component, termed the *reluctance torque*, arises as the rotor tries to assume a position of minimum magnetic reluctance by moving toward the air-gap mmf. This additional torque is due to the nonuniform air gap and is a direct consequence of the air-gap mmf and flux not being in phase. The reluctance torque is present even without any field excitation and is proportional to $\sin 2\delta_{fr}$. It vanishes for both $\delta_{fr} = 0$ and $\delta_{fr} = \pi/2$. In both these positions the air-gap mmf and flux are in phase and the tangential forces acting on both poles balance each other. In the round-rotor generator, $\Re_d = \Re_q$ and the torque expression reverts to that given in Eq. (3.57).

3.3.3 Synchronous Generator as a Power Source

The synchronous generator is connected to the high-voltage transmission network via a step-up transformer (Figure 2.2), and together they form a *generator-transformer unit*. From the EPS point of view, the unit is a source of real and reactive power. This subsection contains a description of the mathematical model of the unit together with its main characteristics.

3.3.3.1 Equivalent Circuit of the Generator-transformer Unit

The equivalent steady-state circuit diagram and phasor diagram of a unit consisting of a round-rotor generator and a step-up transformer is shown in Figure 3.19. The transformer is modeled using the impedance $R_T + jX_T$. The

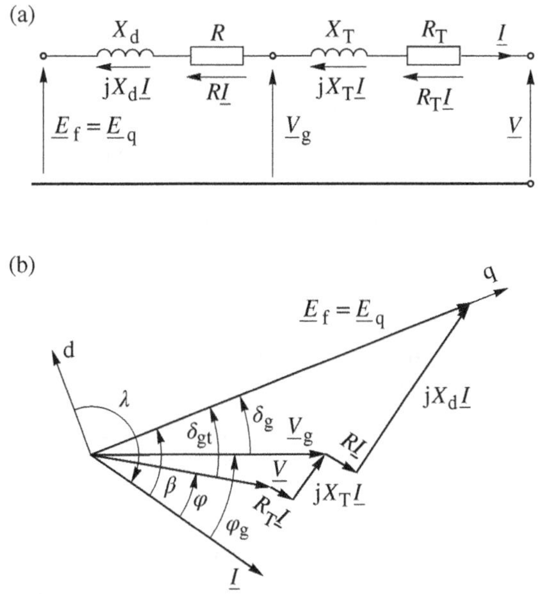

Figure 3.19 Equivalent steady-state circuit diagram (a) and phasor diagram (b) of the round-rotor generator with a step-up transformer.

phasor diagram of the generator alone is the same as that shown in Figure 3.13 but with a voltage drop on the transformer impedance added to the voltage $\underline{V}_g$ in order to get the terminal voltage $\underline{V}$. The following equation can be formed similar to Eq. (3.49)

$$\underline{V} = \underline{E}_f - jX_d\underline{I} - R\underline{I} - jX_T\underline{I} - R_T\underline{I} = \underline{E}_f - j(X_d + X_T)\underline{I} - (R + R_T)\underline{I} \tag{3.76}$$

or simply

$$\underline{V} = \underline{E}_f - jx_d\underline{I} - r\underline{I} \tag{3.77}$$

where $x_d = X_d + X_T$ is the generator reactance increased by the transformer reactance and $r = R + R_T$ is the generator resistor increased by the transformer resistor.

In the case of a salient-pole generator, the block diagram for the generator-transformer unit is similar to that for the generator alone (Figure 3.17) but, similar to Figure 3.19, it is necessary to subtract a voltage drop on the transformer impedance from the generator voltage $\underline{V}_g$ in order to get the voltage $\underline{V}$ on the transformer terminals. For the salient-pole generator, an equation similar to Eq. (3.65) can be derived

$$\begin{aligned}\underline{V}_g &= \underline{E}_f - jX_d\underline{I}_d - jX_q\underline{I}_q - R\underline{I} - X_T\underline{I} - R_T\underline{I} \\ &= \underline{E}_f - jX_d\underline{I}_d - jX_q\underline{I}_q - R\underline{I} - X_T\left(\underline{I}_d + \underline{I}_q\right) - R_T\underline{I}\end{aligned} \tag{3.78}$$

Or in the compact form

$$\underline{V}_g = \underline{E}_f - jx_d\underline{I}_d - jx_q\underline{I}_q - r\underline{I} \tag{3.79}$$

where $x_d = X_d + X_T$; $x_q = X_q + X_T$; and $r = R + R_T$. Hence, as in (3.68),

$$E_q = V_q + rI_q - x_dI_d = E_f \tag{3.80a}$$

$$E_d = V_d + rI_d + x_qI_q = 0 \tag{3.80b}$$

where

$$I_d = -I\sin\beta \text{ and } I_q = I\cos\beta \tag{3.81a}$$

$$V_d = -V\sin\delta_{gt} \text{ and } V_q = V\cos\delta_{gt} \tag{3.81b}$$

where δ_{gt} is the angle of the generator emf with respect to the voltage at the terminals of the generator-transformer unit.

Similarly to Eq. (3.68), Eqs. (3.80) may be written in matrix form as

$$\begin{bmatrix} E_d \\ E_q \end{bmatrix} = \begin{bmatrix} V_d \\ V_q \end{bmatrix} + \begin{bmatrix} r & +x_q \\ -x_d & r \end{bmatrix} \begin{bmatrix} I_d \\ I_q \end{bmatrix} = \begin{bmatrix} 0 \\ E_f \end{bmatrix} \tag{3.82}$$

In the rest of the book, the total reactances and resistances of the generator-transformer unit are denoted by lower case letters. It should be remembered that the phase angle between the voltage and the current is different for the generator terminal voltage $\underline{V}_g$ and the high-voltage side of the transformer $\underline{V}$. Consequently, powers measured on both sides of the transformer differ by the value of power losses on the transformer.

3.3.3.2 Real and Reactive Power of the Generator-transformer Unit

From the EPS point of view, the values of interest are the voltage and powers on the high-voltage side of the generator-transformer. Single-phase power measured on the generator terminals can be calculated from the general expression $P_s = V_s I\cos\varphi$. Figure 3.19 shows that $\varphi = \beta - \delta_{gt}$. Hence

$$P = VI\cos\varphi = VI\cos(\beta - \delta_{gt}) = VI\sin\beta\sin\delta_{gt} + VI\cos\beta\cos\delta_{gt} \tag{3.83}$$

Substituting into Eqs. (3.81) gives

$$P = V_d I_d + V_q I_q \tag{3.84}$$

The current components I_d and I_q can be calculated by solving Eqs. (3.80)

$$I_d = \frac{1}{z^2}\left[r(E_d - V_d) - x_q(E_q - V_q)\right];\; I_q = \frac{1}{z^2}\left[r(E_q - V_q) + x_d(E_d - V_d)\right] \tag{3.85}$$

where $z^2 = r^2 + x_d x_q$.

Equation (3.85) can also be written in matrix form. Solving the matrix Eq. (3.82) yields

$$\begin{bmatrix} I_d \\ I_q \end{bmatrix} = \frac{1}{z^2}\begin{bmatrix} r & -x_q \\ +x_d & r \end{bmatrix} \begin{bmatrix} E_d - V_d \\ E_q - V_q \end{bmatrix} \tag{3.86}$$

Taking into account that in the steady state $E_d = 0$, and substituting Eqs. (3.85) into (3.84), gives

$$P = \frac{1}{z^2}\left[-r\left(V_q^2 + V_d^2\right) - (x_d - x_q)\, V_d V_q + E_q\left(rV_q - x_q V_d\right)\right] \tag{3.87}$$

Substituting Eqs. (3.81b) into (3.85) yields

$$P = \frac{E_q V}{z}\frac{x_q}{z}\sin\delta_{gt} + \frac{1}{2}\frac{V^2}{z}\frac{x_d - x_q}{z}\sin 2\delta_{gt} + \frac{E_q V}{z}\frac{r}{z}\cos\delta_{gt} - \frac{V^2}{z}\frac{r}{z} \tag{3.88}$$

The resistance of the stator and transformer windings is quite small so for approximate calculations the last two components may be neglected. Assuming $r \cong 0$ and $z^2 \cong x_d x_q$, the above equation simplifies to

$$P = \frac{E_q V}{x_d}\sin\delta_{gt} + \frac{V^2}{2}\frac{x_d - x_q}{x_d x_q}\sin 2\delta_{gt} \tag{3.89}$$

The first component is dominant and depends on the sine of the angle δ_{gt} between the voltage and the generator emf. The second component, referred to as the *reluctance power*, exists only in salient-pole generators ($x_d > x_q$). For round-rotor generators ($x_d = x_q$), the reluctance power vanishes and the above equation further simplifies to

$$P = \frac{E_q V}{x_d} \sin \delta_{gt} \tag{3.90}$$

Equations for the reactive power can be derived in a similar way. The expression $\varphi = \beta - \delta_{gt}$ is substituted to the general equation $Q = VI \sin \varphi$. Taking into account Eqs. (3.81) gives

$$Q = -V_q I_d + V_d I_q \tag{3.91}$$

Using Eqs. (3.85) and (3.81b) gives

$$Q = \frac{E_q V}{z} \frac{x_q}{z} \cos \delta_{gt} - \frac{V^2}{z} \frac{x_d \sin^2 \delta_{gt} + x_q \cos^2 \delta_{gt}}{z} - \frac{E_q V}{z} \frac{r}{z} \sin \delta_{gt} \tag{3.92}$$

For the round-rotor generators and when neglecting the resistance ($r = 0$), the above equation simplifies to

$$Q_s = \frac{E_q V_s}{x_d} \cos \delta_{gt} - \frac{V_s^2}{x_d} \tag{3.93}$$

All the equations in this chapter are valid for pu calculations. If phase voltages are used in Eqs. (3.90) and (3.93) then the resulting real and reactive powers are per-phase. If line voltages ($\sqrt{3}$ higher than the phase voltages) are used in Eqs. (3.90) and (3.93) then the resulting real and reactive powers are totals for all three phases.

3.3.4 Reactive Power Capability Curve of a Round-rotor Generator

The synchronous generator is a source of real and reactive power which can be conveniently regulated over a wide range of values. This can be shown using the equations derived above for real and reactive power. Equations (3.90) and (3.93) show that, assuming a given value of the generator reactance, the real and reactive power produced by the generator depends on:

- the emf $E_q = E_f$ which is proportional to the generator field current I_f;
- voltage V on the terminals of the step-up transformer;
- power angle δ_{gt}.

The voltage V of a generating unit on the network side cannot be changed in a wide range, typically less than ±10% of the network rated voltage. Consequently, a wide range of active power changes P – see Eq. (3.90) – corresponds to changes in δ_{gt}. Worth noting is that the mechanical power produced by the turbine must appear as the electrical power P of the generator (minus small losses). Hence, in the steady state, any change in the turbine power corresponds to an almost equal change in P and therefore an almost proportional change in sin δ_{gt}. When the angle is δ_{gt} and the required power is P, the field current I_f can be used to control the emf $E_q = E_f$ and the reactive power Q that the generator produces; Eq. (3.93).

Limits in the real and reactive power control of the generating unit are the result of the following constructional and operating regimes:

i) Stator (armature) current I_g must not cause overheating of the armature winding. Therefore, it must be smaller than a certain maximum value $I_g \leq I_{g\,max}$, where (as mentioned in Section 2.3.2) $I_{g\,max}$ is a parameter of the stator current limiter (SCL).
ii) Rotor (field) current I_f must not cause overheating of the field winding. Hence, it must be smaller than a certain maximum value $I_f \leq I_{f\,max}$ or $E_q \leq E_{q\,max}$, where (as mentioned in Section 2.3.2) $I_{f\,max}$ is a parameter of the overexcitation limiter (OEL).

iii) The power angle must not be higher than a maximum value due to stable generator operation $\delta_{gt} \leq \delta_{max}$, where (as mentioned in Section 2.3.2) δ_{max} is a parameter of the underexcitation limiter (UEL). Steady-state stability is explained in Chapter 5.
iv) The temperature in the end region of the stator magnetic circuit must not exceed the maximum value.
v) The generator active power must be within the limits set by the turbine power, that is $P_{min} \leq P \leq P_{max}$.
vi) Voltage V_g on the generator terminals must be maintained close to the rated value within a small tolerance level, typically less than ±5%.

On the plane with coordinates corresponding to the real and reactive power of the generating unit all above conditions can be expressed as a family of curves and lines, as shown in Figure 3.20. Points satisfying all conditions constitute an area of allowed loading of the generating unit, referred to as the *power capability area*. In Figure 3.20 this area is surrounded by bolded sections of the curves and the lines connecting points H, I, E, D, K, and J. The characteristic created by these sections is referred to as the *capability curve* of the generating unit.

Equations describing curves and lines that constitute the capability curve will be derived from a mathematical description of the above conditions and equivalent circuit diagrams of the generating unit shown in Figure 3.20. The analysis described below is based on a publication of Machowski and Kacejko (2016).

It is assumed that the capability curve relates to the active power P and the reactive power Q of the generating unit on the high-voltage side of the step-up transformer, i.e. the power supplied to a transmission network (Figure 3.21a). The synchronous generator is modeled (Figure 3.21b) by synchronous emf $\underline{E}_q$ behind a synchronous reactance, i.e. $\underline{Z}_g \cong jX_d$. The step-up transformer is modeled by a series branch $\underline{Z}_T \cong jX_T$ and an ideal transformation ratio ϑ inserted on the secondary side of the transformer model. For the simplicity of further analysis the resistances of the generator and the step-up transformer are neglected. Analysis is conducted on pu values.The voltage on the secondary side of the step-up transformer (network side) is divided by the rated voltage of the transmission network. Voltage on the primary side of the step-up transformer (generator side) is divided by the rated voltage of the generator. The transformation ratio ϑ is defined as the ratio of the network-side voltage to the generator-side voltage. When the series impedance $\underline{Z}_T$ of the step-up transformer is replaced by its reciprocal $\underline{Y}_T = 1/\underline{Z}_T$, the circuit diagram shown in Figure 3.21a can be described by nodal admittance equations similar to Eq. (3.27) and may be replaced by the π-equivalent network. The circuit diagram obtained in the considered case is shown in Figure 3.21c. Using the well-known formulas of the star-delta transformation, the circuit diagram

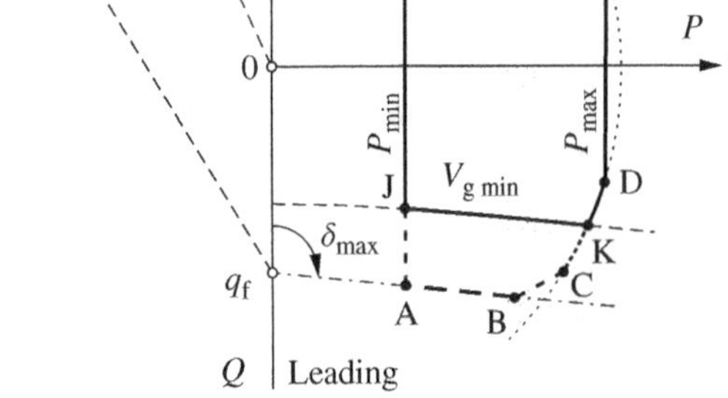

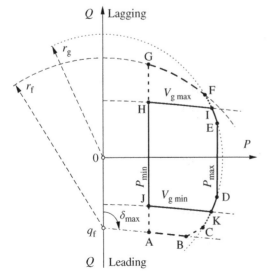

Figure 3.20 Power capability area for a given voltage at the busbar.

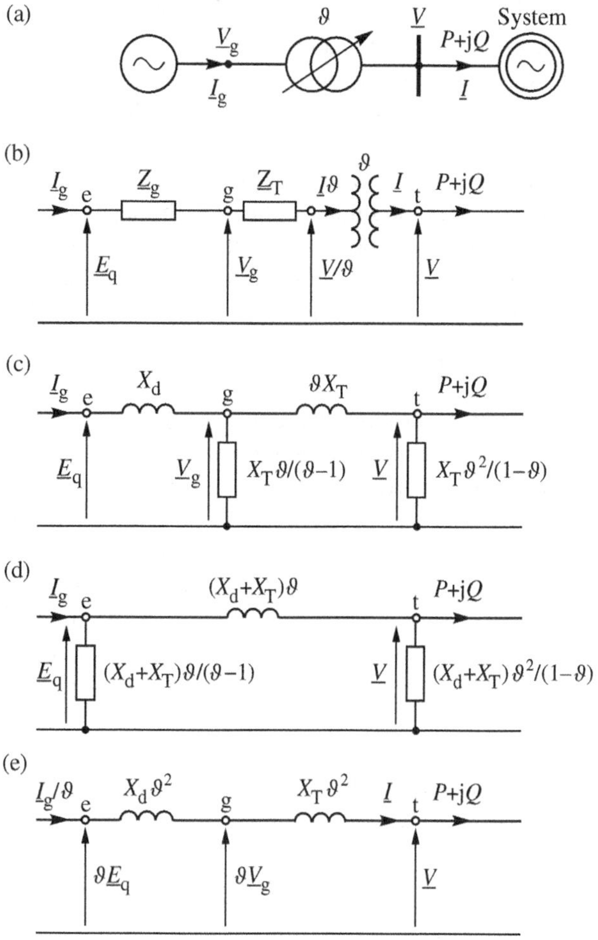

Figure 3.21 Equivalent circuit diagrams of generating unit.

in Figure 3.21c may be transformed into the circuit diagram in Figure 3.21d. It is noteworthy that in the obtained equivalent two-port network, all of the branches depend on the sum $x_d = (X_d + X_T)$, which is multiplied by factors that depend on the ratio ϑ in the same manner as in the π-equivalent network that replaces the step-up transformer (Figure 3.21c). The circuit diagram presented in Figure 3.21e is obtained from the initial circuit diagram (Figure 3.21b) by bringing the voltage and impedance to the secondary side of the transformer, i.e. the network side.

All of the circuit diagrams presented in Figure 3.21 are equivalent and may be used depending on the values that are calculated and the relationships that are searched among them.

For a two-port network consisting of a series reactance X the real and reactive output powers are given by Eqs. (1.8) and (1.9) described in Section 1.3. Application of these formulas to the circuit diagram from Figure 3.21e gives:

$$P = \frac{E_q V}{\vartheta\,(X_d + X_T)} \sin \delta_{gt} \tag{3.94}$$

$$Q = \frac{E_q V}{\vartheta\,(X_d + X_T)} \cos \delta_{gt} - \frac{V^2}{\vartheta^2 (X_d + X_T)} \tag{3.95}$$

where δ_{gt} (as in Figure 3.19) is the difference between arguments of $\underline{E}_q$ and $\underline{V}_s$. In both formulas the transformation ratio ϑ appears in the denominator.

Condition (i) assumes that the load current cannot be higher than the permissible current of the stator winding (armature heating limit). On the plane with coordinates P, Q this condition corresponds to a circle whose radius and location of center result from the following reasoning. Real and reactive power in pu values are determined by the following formulas: $P = VI\cos\varphi$ and $Q = VI\sin\varphi$. Considering that $I_g = I\vartheta$ and $I = I_g/\vartheta$, and after squaring both formulas on both sides and summing them up, the dependence $P^2 + Q^2 = (VI_g/\vartheta)^2$ is obtained because $\sin^2\varphi + \cos^2\varphi = 1$. On the plane (P, Q), the above equation corresponds to the circle of radius VI_g/ϑ and the center at the beginning of the coordinate system. For the given network voltage V and the load of the generating unit with the maximum current $I_g = I_{g\,max}$, the following equation is obtained

$$P^2 + Q^2 = \left(VI_{g\,max}/\vartheta\right)^2 \tag{3.96}$$

which corresponds to the circle of radius $r_g = VI_{g\,max}/\vartheta$. In Figure 3.20 the fragment of this circle has been dotted. Points corresponding to the powers, for which the condition $I_g < I_{g\,max}$ is fulfilled, lie inside the circle. The circle radius is inversely proportional to the ratio ϑ, which means that if a step-up transformer controller increases the ratio, the radius of the circle gets smaller. If the controller of a step-up transformer reduces the ratio, the radius of the circle gets larger. The maximum value of the stator current is to be adopted according to the manufacturer's recommendations, e.g. $I_{g\,max} = I_{rG}$ or $I_{g\,max} = 1.05\,I_{rG}$, where I_{rG} is the rated current.

Condition (ii) assumes that the excitation current cannot be higher than the permissible current of the rotor winding (field heating limit). On the plane with coordinates P, Q this condition corresponds to a circle whose radius and location of the center result from the following reasoning. Equations (3.94) and (3.95) can be rewritten as follows

$$\frac{E_q V}{\vartheta(X_d + X_T)}\sin\delta_{gt} = P \tag{3.97}$$

$$\frac{E_q V}{\vartheta(X_d + X_T)}\cos\delta_{gt} = Q + \frac{V^2}{\vartheta^2(X_d + X_T)} \tag{3.98}$$

After squaring both the formulas on both sides and summing them up, the following equation is obtained

$$P^2 + \left[Q + \frac{V^2}{\vartheta^2(X_d + X_T)}\right]^2 = \left[\frac{E_q V}{\vartheta(X_d + X_T)}\right]^2 \tag{3.99}$$

For $E_q = E_{q\,max}$, Eq. (3.99) takes the following form

$$P^2 + \left[Q + \frac{V^2}{\vartheta^2(X_d + X_T)}\right]^2 = \left[\frac{E_{q\,max} V}{\vartheta(X_d + X_T)}\right]^2 \tag{3.100}$$

which corresponds to the circle of the radius and the center shift determined by the following formulas

$$r_f = \frac{E_{q\,max} V}{\vartheta(X_d + X_T)};\ q_f = -\frac{V^2}{\vartheta^2(X_d + X_T)} \tag{3.101}$$

Hence, the circle center shift is inversely proportional to the square of the transformation ratio, while the radius is inversely proportional to the transformation ratio. The circle is crossed by a vertical axis (reactive power for active power equals zero) at the point

$$r_f + q_f = \frac{V}{\vartheta(X_d + X_T)}\left(E_{q\,max} - \frac{V}{\vartheta}\right) \tag{3.102}$$

For the given values of real power, bus voltage, and transformation ratio, the reactive power determined by condition (ii) can be computed from the following formula resulting from Eq. (3.100)

$$Q = \sqrt{\left[\frac{E_{q\,max}\,V}{\vartheta(X_d + X_T)}\right]^2 - P^2} - \frac{V^2}{\vartheta^2(X_d + X_T)} \tag{3.103}$$

The synchronous emf of a generator is approximately twice and more as high as the rated voltage – see Eq. (3.108) and Example 3.1 below – and it can be assumed that $E_{q\,max} \gg V$. In this case, according to (3.102), the tap changing control after increasing the ratio ϑ moves down the arc GF resulting from the excitation current limiter (Figure 3.20), and therefore reduces the area of allowed loading of the generating unit.

The synchronous emf E_q that occurs in the above equations depends on the excitation current. The maximum value of $E_{q\,max}$ corresponds to the permissible current of a rotor winding. In general, manufacturers do not allow excitation current values above the rated values. Assuming that the generator rotor winding is not overloaded, i.e. permissible current corresponds to rated current $I_{f\,max} = I_{rf}$ and $E_{q\,max} = E_{rq}$, the rated value of emf E_{rq} can be calculated as the multiplicity of the rated voltage of a generator with the multiplication factor depending on synchronous reactance and the power factor. This assumption results from the reasoning presented below. The following formulas can be written for the equivalent circuit diagrams of the generator (the left side in Figure 3.21b)

$$P_g = \frac{E_q\,V_g}{X_d}\sin\delta_{ge};\; Q_g + \frac{V_g^2}{X_d} = \frac{E_q\,V_g}{X_d}\cos\delta_{ge} \tag{3.104}$$

where δ_{ge} is the difference between the arguments of the generator emf and the generator terminal voltage. By squaring the equations on both sides and summing them up, the following is obtained

$$P_g^2 + \left[Q_g + \frac{V_g^2}{X_d}\right]^2 = \left[\frac{E_q\,V_g}{X_d}\right]^2 \tag{3.105}$$

$$E_q^2 = V_g^2\left[\left(1 + X_d\frac{Q_g}{V_g^2}\right)^2 + \left(X_d\frac{P_g}{V_g^2}\right)^2\right] \tag{3.106}$$

Equation (3.106) is correct for any power and, in particular, for rated powers $P_{rG} = S_{rG}\cos\varphi_{rG}$, $Q_{rG} = S_{rG}\sin\varphi_{rG}$, and rated voltage V_{rG}, where $\cos\varphi_{rG}$ is a rated power factor. It is known that synchronous reactance $X_d = X_{d\,pu}\,V_{rG}^2/S_{rG}$, where $X_{d\,pu}$ is the reactance in pu values. After substituting these dependences in Eq. (3.106), the following is obtained

$$E_{rq}^2 = V_{rG}^2\left[\left(1 + X_{d\,pu}\sin\varphi_{rG}\right)^2 + \left(X_{d\,pu}\cos\varphi_{rG}\right)^2\right] \tag{3.107}$$

Therefore, after simple transformations, the following formulas are obtained for the synchronous emf of a generator with a rated load

$$E_{rq} = k_{rf}V_{rG},\qquad k_{rf} = \sqrt{1 + X_{d\,pu}\left(X_{d\,pu} + 2\cdot\sin\varphi_{rG}\right)} \tag{3.108}$$

The rated emf of a generator is a multiplication of the rated voltage.

Example 3.1 A synchronous generator with rated power $S_{rG} = 1\,042$ MVA and rated voltage $V_{rG} = 27$ kV has the synchronous reactance $X_d = 223\,\% = 2.23$ pu, and rated power factor $\cos\varphi_{rG} = 0.85$. From Eq. (3.108) it results

$$k_{rf} = \sqrt{1 + X_{d\,pu}\left(X_{d\,pu} + 2\cdot\sin\varphi_{rG}\right)} = \sqrt{1 + 2.23\,(2.23 + 2\cdot 0.85)} = \sqrt{9.764} \cong 3.125$$

and $E_{rq} = k_{rf}V_{rG} = 3.125\cdot V_{rG} = 84.37$ kV

******

Condition (iii) concerns the limitation of a load angle (underexcitation limit also referred to as *stability limit*). On the plane with coordinates P, Q this condition corresponds to a straight line whose location and inclination result from the reasoning presented below. Dividing both sides of Eqs. (3.97) and (3.98) by each other gives

$$Q = \tan^{-1}\delta_{gt} \cdot P - \frac{V^2}{\vartheta^2(X_d + X_T)} \tag{3.109}$$

Therefore, after substituting $\delta_{gt} = \delta_{max}$, the following equation is obtained

$$Q = m\,P + c; \qquad m = \tan^{-1}\delta_{max}; \qquad c = -\frac{V^2}{\vartheta^2(X_d + X_T)} \tag{3.110}$$

On the plane with coordinates P, Q Eq. (3.110) delineates a straight line that for $P = 0$ crosses the reactive power axis Q at an angle δ_{max} at point

$$q_f = c = -\frac{V^2}{\vartheta^2(X_d + X_T)} \tag{3.111}$$

It is noteworthy that this point corresponds also to the circle center shift determined by Eq. (3.101). In Figure 3.20 the straight line corresponding to Eq. (3.110) is a dash-dotted line crossing points A,B. The shift (3.111) of the straight line Eq. (3.110) is inversely proportional to the squared ratio. Therefore, it can be stated that, with the transformation ratio reduced by the tap changing control, the straight line Eq. (3.110) gets lower, and thus the limitation of reactive power (section AB in Figure 3.20) moves down, i.e. the capability area gets bigger in the lower part.

Condition (iv), concerning the limitation of temperature in the end region of the stator magnetic circuit, cannot be expressed in simple mathematical dependencies (end region heating limit). A relevant curve must be determined by the generator manufacturer. In Figure 3.20 this curve is BC.

Condition (v), concerning the mechanical power of a prime mover (turbine), depends on the turbine type and its adaptation to the operation across a broad range of active power changes (prime mover limit). In Figure 3.20 the relevant limitations have been marked with straight vertical lines corresponding to P_{max} and P_{min}.

Condition (vi) concerns the following limitation of the generator terminal voltage (generator voltage limit)

$$V_{g\ max} \le V_g \le V_{g\ min} \tag{3.112}$$

For currently used generators, permissible changes in terminal voltage and frequency are small. For a rated frequency, the permissible range of the generator terminal voltage changes is ±5% of the rated value V_{rG}.

The formula of characteristic $Q(P)$ for the given value V_g can be derived in the simplest way by using the circuit diagram presented in Figure 3.21a, or, strictly speaking, with the part of the circuit diagram that concerns the step-up transformer. Likewise, for Eq. (3.94) for the middle part of the circuit diagram from Figure 3.21b, the following can be written

$$P = \frac{V_g}{X_T}\frac{V}{\vartheta}\sin\delta_T; \qquad Q + \frac{V^2}{\vartheta^2 X_T} = \frac{V_g}{X_T}\frac{V}{\vartheta}\cos\delta_T \tag{3.113}$$

where δ_T is the difference between voltage arguments on both sides of the transformer impedance. After squaring Eq. (3.113) on both sides and summing them up, the following equation is obtained

$$P^2 + \left(Q + \frac{V^2}{\vartheta^2 X_T}\right)^2 = \left(\frac{V\,V_g}{\vartheta X_T}\right)^2 \tag{3.114}$$

The equation corresponds to a circle whose radius and center shift are determined by the following formulas

$$r_T = \frac{V_g\,V}{\vartheta X_T}; \qquad q_T = -\frac{V^2}{\vartheta^2 X_T} \tag{3.115}$$

By solving Eq. (3.114) with regard to reactive power, the following formula is obtained

$$Q = \sqrt{\left(\frac{V V_g}{\vartheta X_T}\right)^2 - P^2} - \frac{V^2}{\vartheta^2 X_T} \tag{3.116}$$

This formula determines reactive power as a function of active power and both voltages.

The characteristics $Q(P)$ given in formula (3.116) are arcs of a circle determined by Eq. (3.114). As $X_T \ll X_d$, both the shift and radius of this circle are many times larger than the shift and radius of the circle determined by Eq. (3.100), i.e. $r_T \gg r_f$ and $|q_T| \gg |q_f|$. Therefore, arcs $V_{g\,max}$ and $V_{g\,min}$ corresponding to the circles (3.114) shown in Figure 3.20 look almost like straight lines.

Equation (3.114) can be also transformed to the following bi-quadratic equation:

$$a\vartheta^4 + b\vartheta^2 + c = 0 \tag{3.117}$$

where:

$$a = P^2 + Q^2; \quad b = \frac{V^2}{X_T}\left(2Q_s - \frac{V_g^2}{X_T}\right); \quad c = -\frac{V^2}{X_T} \tag{3.118}$$

For given values of voltages on both sides of the step-up transformer V_g, V and real and reactive power P, Q injected to transmission network Eq. (3.117) allows us to compute the value of the transformation ratio ϑ.

If a step-up transformer has a constant ratio $\vartheta = \vartheta_0 = \text{const}$, for the given values of network voltages, the limitations $V_{g\,max}$ and $V_{g\,min}$ resulting from condition (3.112) may lie within the area ABCDEFG (Figure 3.20). In this case, the capability area is limited only to the area HIEDKJ. As a result, the decisive limitation in reactive power generation may be the limitation of generator voltage rather than the limitation of exciting current. A similar observation is mentioned in Reimert (2006).

The location of the above-mentioned arcs $V_{g\,max}$ and $V_{g\,min}$ depends on network voltage V and is not constant. Fortunately, the impact of network voltage changes is favorable and compliant with the needs of the EPS. The arcs cross a vertical axis (reactive power for zero real power) for the following values

$$Q\left(V_{g\,max}\right) = r_T + q_T = \frac{V}{\vartheta X_T}\left(V_{g\,max} - \frac{V}{\vartheta}\right) \tag{3.119}$$

$$Q\left(V_{g\,min}\right) = r_T + q_T = \frac{V}{\vartheta X_T}\left(V_{g\,min} - \frac{V}{\vartheta}\right) \tag{3.120}$$

For positive reactive power (lagging side in Figure 3.20), the difference between voltages on the right side of formula (3.119) is positive and increases when network voltage V drops owing to the shortage of reactive power in the system. As a consequence, arc $V_{g\,max}$ that limits reactive power rises, making it possible to return the growing volume of reactive power. The increase in ratio ϑ has a similar influence, i.e. results in raising the arc $V_{g\,max}$. For negative reactive power (leading side in Figure 3.20), the difference between voltages on the right side of (3.119) is negative, and its absolute value increases when network voltage V rises owing to the excess of reactive power in the system. As a result, arc $V_{g\,min}$ that limits reactive power falls, making it possible to absorb the growing volume of reactive power. The reduction in ratio ϑ has a similar impact, i.e. it causes the lowering of arc $V_{g\,min}$.

It is noteworthy that the above-mentioned electrical conditions (i), (ii), and (iii) depend on voltage V at busbars to which a generating unit is connected. The value of this voltage affects radius r_g of circle (3.95), radius r_f, and shift q_f of circle (3.100), as well as shift q_f of the straight line (3.110). The higher the value of voltage V, the wider the area ABCDEFG. Therefore, characteristic $Q(P)$ presented in Figure 3.20 is always constructed for a given value of voltage V.

Example 3.2 A generating unit is connected to a transmission network of 400 kV and has the following data: $S_{rG} = 1\ 042$ MVA, $V_{rG} = 27$ kV, $X_d = 223\%$, $\cos\varphi_{rG} = 0.85$, permissible voltage range $V_g = V_{rG} \pm 5\%$; maximum stator current $I_{g\ max} = 1.05\ I_{nG}$, $S_{rT} = 1\ 050$ MVA, $X_T = 17.71\%$, $V_{rT\ primary} = 27$ kV, $V_{rT\ secondary} = 407\ \text{kV} \pm 7 \times 1.25\%$. For comparison, this analysis has also been conducted for an option that assumes an installed step-up transformer of constant transformation ratio and rated voltage $V_{rT\ secondary} = 420$ kV.

The following base units have been assumed: $[S] = 1\ 000$ MVA, $[V] = 400$ kV, $[V_g] = 27$ kV. After calculating the rated data for pu values, the following is obtained: $S_{rG} = 1.042$, $V_{rG} = 1$, $S_{rT} = 1.050$, $V_{rT\ g} = 1$, and $V_{rT\ s} = 1.0175$. The reactance in pu values was calculated according to the formula $X_{pu} = X{\cdot}V^2_{r\ pu}/S_{r\ pu}$, where $V_{r\ pu}$ and $S_{r\ pu}$ are rated voltage and rated apparent power in pu values. The relative transformation ratio for the central tap is $\vartheta_0 = (407/400)/(27/27) = 1.0175$. The ratio control range is $\vartheta = 1.0174 \pm 7 \times 0.0125$, i.e. $0.930 \le \vartheta \le 1.105$. It was assumed for the load angle limiter that $\delta_{max} = 90^{\circ}$. For the option with a transformer that has a constant transformation ratio, the relative transformation ratio is obtained as $\vartheta_0 = (420/400)/(27/27) = 1.05$.

Figure 3.22a concerns the option where the transformer has a constant transformation ratio. It is evident that a large upper part of the capability area cannot be used without exceeding the permissible voltage of the generator. Here, the limitation of the terminal generator voltage significantly reduces the capability area. In the lower part, such a reduction of the power capability area is much smaller. If condition (3.112) is fulfilled, only the area delineated with the bolded continuous line is available.

Figure 3.22b concerns the option with a tap changing transformer and automatic control of the transformation ratio. In the upper part of the power capability area, when the reactive power generation increases, the tap changer controller increases the transformation ratio. As a consequence, arc $V_{g\ max}$ rises, thus enlarging the available area

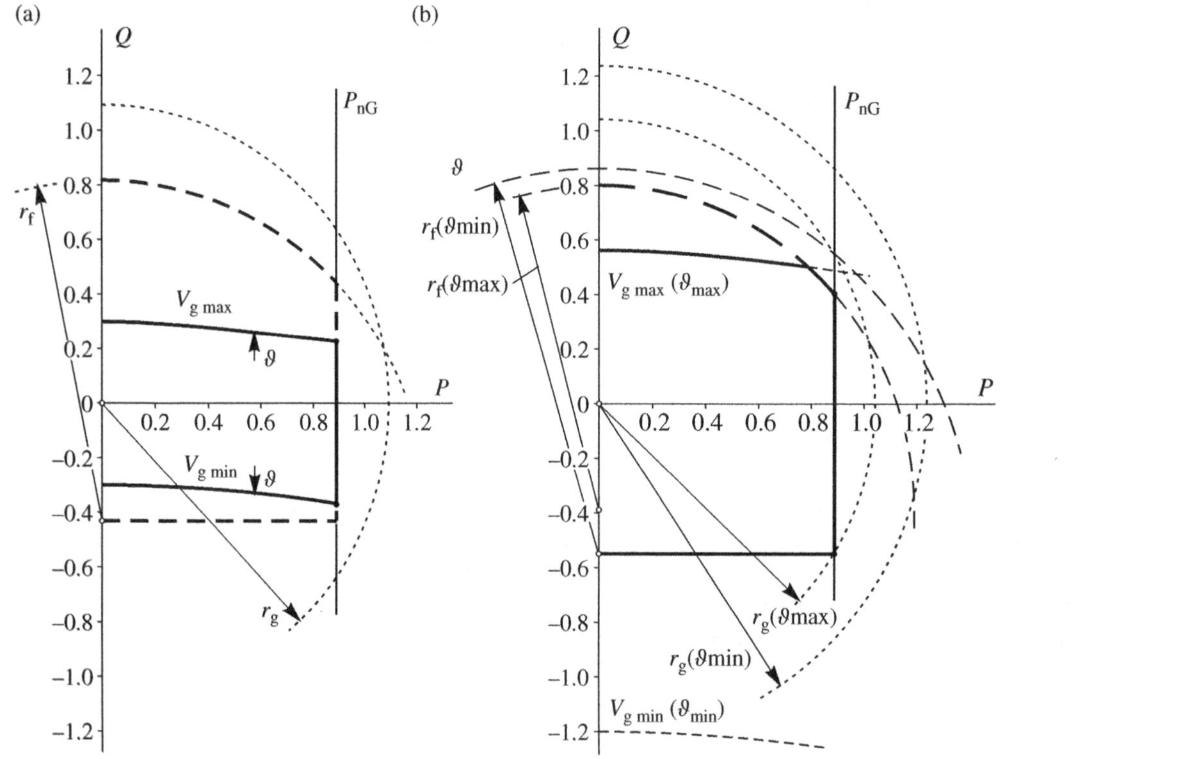

Figure 3.22 The power capability area for $V_s = 1.05$ in case of: (a) step-up transformer with a constant transformation ratio $\vartheta = 1.05$ = const; (b) tap changing step-up transformer with $0.930 \le \vartheta \le 1.105$.

within the reactive power. In the lower part of the capability area, after increasing the absorption of the reactive power by a generator, the tap changer controller decreases the transformation ratio. As a consequence, arc $V_{g\,min}$ decreases significantly (outside the underexcitation limit). Thus, the limitation (3.112) becomes irrelevant. Owing to the control of the ratio, the entire area bounded by bolded continuous lines and curves is available.

As shown by the comparison of Figure 3.22a and b, the area obtained using the ratio control is much larger than the area with no control. Similar figures for $V = 1.0$ and $V = 0.95$ can be found in a publication of Machowski and Kacejko (2016).

******

From the above consideration it can be concluded that optimal range of changes of the step-up transformer ratio should satisfy two conditions:

a) For lagging current (upper part of power capability area) maximal transformation ratio ϑ_{max} should guarantee that the generator voltage limit is shifted behind the field heating limit.
b) For leading current (lower part of power capability area) minimal transformation ratio ϑ_{min} should guarantee that the generator voltage limit is shifted below the underexcitation limit.

Condition (a) is satisfied when for $V = V_{max}$ and $V_g = V_{g\,max}$ the value given by (3.119) is larger than the value given by (3.102). Solving such inequality leads to the following value:

$$\vartheta_{max} \geq \frac{x \cdot V_{max}}{(x+1) \cdot V_{g\,max} - E_{q\,max}} \tag{3.121}$$

where $x = X_d/X_T$.

Condition (b) is satisfied when for $V = V_{min}$ and $V_g = V_{g\,min}$ the value given by (3.120) is smaller than the value given by (3.111). Solving such inequality leads to the following value

$$\vartheta_{min} \leq \frac{V_{min}}{V_{g\,min}} \frac{x}{x+1} \tag{3.122}$$

For a generating unit with data described in Example 3.1 $\vartheta_{min} \leq 0.927$ and $\vartheta_{max} \geq 1.159$ are obtained. It means that the range $0.930 \leq \vartheta \leq 1.105$ assumed in the data described in Example 3.1 is too small (especially for ϑ_{max}). This fact is also evident from Figure 3.22a, where the generator voltage limit is still a constrain for reactive power in the upper part of the power capability area.

For step-up transformers, which are not equipped with the on-load tap changer, usually such a value of ϑ is selected that the arc HI shown in Figure 3.19 lies beyond point F or even point G.

When the step-up transformer is equipped with the on-load tap changer, the control system of the generating unit includes both a generator and a step-up transformer. Such control concerns two facilities that differ in terms of speed and nature of control. Control of a synchronous generator is continuous and fast. Control of a step-up transformer is discrete (stepwise) and much slower than in the case of a generator. In order to provide the proper control process for these two different devices (a generator and a transformer), their controllers should not operate according to the same criterion, i.e. maintain the same quantity (e.g. terminal voltage of a generator or busbar voltage of a substation). In order to provide proper process of control, a transformer controller should maintain the reference value of the voltage on the primary side of the step-up transformer (i.e. voltage at the generator side), while a generator controller should maintain the reference value on the secondary side of the step-up transformer (i.e. voltage at the network side).

A functional diagram of the considered control system is shown in Figure 3.23. When circuit breakers are closed, $V = V_b$. Obviously, control of the voltage V at the substation busbar must be done (as described in Section 3.3.5) with the use of current compensation to obtain a small droop (bias) in the voltage characteristic. Current compensation must be carried out toward a step-up transformer. As a result, the generating unit is seen from the side of the transmission network as a source of voltage behind a small reactance that corresponds to the small impedance of the current compensation applied in the automatic voltage regulator (AVR).

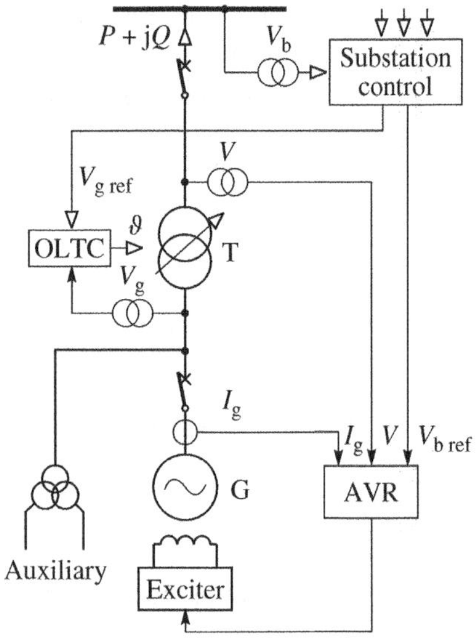

Figure 3.23 Functional diagram of a generation unit control. AVR – automatic voltage regulator, OLTC – on-load tap changer.

The on-load tap changing control of the step-up transformer (Figure 3.23) must be realized as the voltage regulation operating according to the following rule

$$\begin{cases} V_g > (V_{g\,ref} + \Delta V) & \text{increase } \vartheta \\ V_g < (V_{g\,ref} - \Delta V) & \text{decrease } \vartheta \end{cases} \tag{3.123}$$

where V_g is the generator terminal voltage, $V_{g\,ref}$ is the reference value provided by the substation control system, and ΔV is the dead zone of a controller selected according to the following formula

$$\Delta V = (0.6 - 0.7) \cdot \Delta V_{tap} \tag{3.124}$$

where ΔV_{tap} is the voltage on the tap of the transformer winding. This selection of a dead zone guarantees stable control, because the width of the dead zone $\pm\Delta V$ is greater than the tap voltage, i.e. $2\Delta V \geq (1.2 - 1.4)\Delta V_{tap}$. The reference value of the tap changing controller can be set by the substation controller (Figure 3.23), depending on the reactive power of a generating unit according to the following rule

$$V_{g\,ref} = \left(1 + \kappa \frac{Q_s}{S_{rG}}\right) \cdot V_{rG} \tag{3.125}$$

where V_{rG} and S_{rG} are, respectively, the rated voltage and the rated apparent power of a generator, and κ is a small positive number (gain). A recommended value of this gain is equal to

$$\kappa = \frac{V_{g\,max} - V_{rG}}{V_{rG}} \cdot \frac{1}{\sin \varphi_{rG}} \tag{3.126}$$

For $V_{g\,max} = 1.05 \cdot V_{rG}$ this gain is equal to $\kappa = 0.05/\sin \varphi_{rG}$.

For $Q_s > 0$ (lagging current) (3.125) provides $V_{g\,ref} > V_{rG}$. Particularly for $Q = Q_{rG} = S_{rG} \sin \varphi_{rG}$ it is obtained that $V_{g\,ref} = V_{g\,max}$ and generator operates with maximal permitted voltage. For $Q = 0$, (3.125) provides $V_{g\,ref} = V_{rG}$. For $Q < 0$ (leading current), it is obtained that $V_{g\,ref} < V_{rG}$. Obviously, value $V_{g\,ref}$ computed from (3.125) must be passed through a limiter $[V_{g\,min}; V_{g\,max}]$ determining permitted generator voltage.

In order to minimize the number of tap changes made during a day, a delay function can be used depending on the deviation in the controlled voltage or on the distance between the current operating point and the capability curve.

It is noteworthy that in rule (3.123), if voltage V_g is low, the transformation ratio ϑ is reduced, while in the case of high voltage it is increased. At first glance, it may seem that the opposite should be the case. However, this dependence is correct. Given the above definition of a transformation ratio, the primary voltage is obtained by dividing the secondary voltage by the transformation ratio. Therefore, in order to increase the primary voltage (generator side), the transformation ratio must be reduced, and vice versa: to reduce the primary voltage (generator side), the transformation ratio must be increased.

Inside the control area, both controllers (AVR and OLTC) shown in Figure 3.23 cooperate in the following way. When network voltage falls, in the upper side of the power capability area (owing to the shortage of reactive power in the system), generator controller AVR increases the excitation current, which increases the generator terminal voltage and, according to rule (3.123), controller of the OLTC increases the ratio. As a result, arc $V_{g\,max}$ rises as well. This enlarges the actual power capability area of a generating unit (Figure 3.23). When the network voltage rises, at the lower side of power capability area (owing to excessive reactive power in the system), generator controller AVR lowers the excitation current, which lowers the generator terminal voltage and, according to rule (3.123), controller of the OLTC lowers the ratio. As a result, arc $V_{g\ min}$ decreases, which results in the enlargement of the actual power capability area of a generating unit (Figure 3.23).

3.3.5 Voltage-reactive Power Capability Characteristic *V*(*Q*)

The capability curve $Q(P)$ shown in Figure 3.20 characterizes the generator-transformer unit as a source of real and reactive power while taking into account the constraints and assuming that the transformer terminal voltage is a parameter. When assessing the generator as a voltage and reactive power source, it is important to consider a capability characteristic $V(Q)$ of the unit equipped with the AVR discussed in Section 2.3.2 and assuming that active power P is a parameter. The task of the AVR is to maintain a required value of the voltage at a given point in the network while observing the constraints.

According to Figure 2.5, the AVR maintains a measured voltage

$$\underline{V}_C = \underline{V}_g + \underline{Z}_C\underline{I} \tag{3.127}$$

which is the generator terminal voltage with added voltage drop on the compensator impedance $\underline{Z}_C = (R_C + jX_C)$. It is equal to the voltage at a fictitious point inside the transmission network displaced from the generator terminals by the impedance $\underline{Z}_C$. The task of the AVR is to maintain the voltage $\underline{V}_C$ at that fictitious measurement point.

Figure 3.24 shows a simplified equivalent circuit and a phasor diagram to illustrate the principle of voltage regulation. The resistance of the generator and transformer has been neglected. The compensation resistance has also been neglected, $R_C \cong 0$, so that $\underline{Z}_C \cong -jX_C$, where $X_C = (1-\kappa)X_T$. Hence the fictitious measurement point is assumed to be located at a distance κX_T from the transformer terminals and at a distance $X_k = (1-\kappa)X_T$ from the generator terminals; see Figure 3.24. $\underline{V}_g$ is the generator terminal voltage and $\underline{V}$ is the transformer terminal voltage.

The characteristic $V(Q)$ of the generator as a reactive power source, with active power P as a parameter, will be determined by considering four characteristic operating regimes:

- The field current is less that its limit, $I_f < I_{f\ max}$, and the generator controls voltage at a given point in the transmission network.
- The field current I_f is at maximum, $I_f = I_{f\ max}$, so that the generator operates with a constant electromotive force (emf) $E_q = E_{q\ max}$.
- The power angle δ_{gt} is at its maximum, $\delta_{gt} = \delta_{max}$, and the reactive power Q is controlled within the limits.
- The current I of the generating unit is at its maximum, $I = I_{max}$, and the reactive power Q is controlled within the limits.

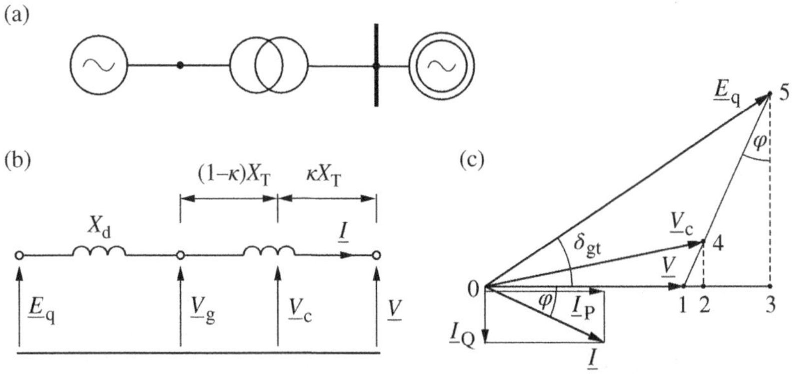

Figure 3.24 The choice of the voltage regulation point for a generator-transformer unit: (a) block diagram; (b) simplified equivalent circuit; (c) phasor diagram.

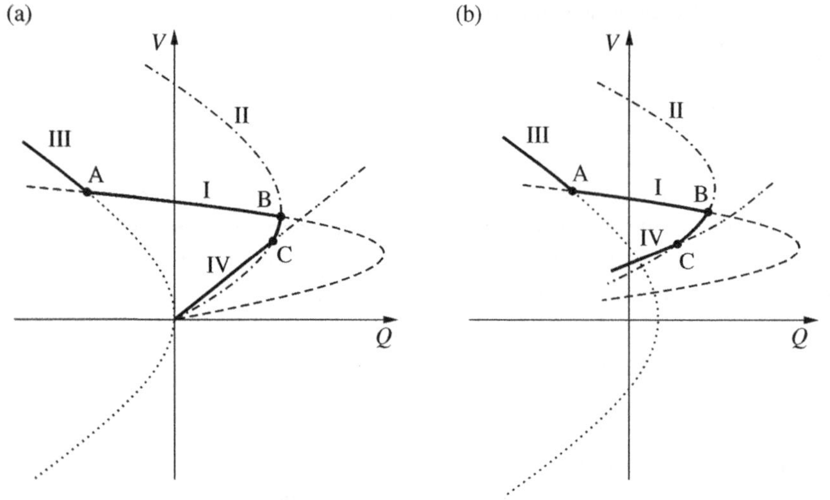

Figure 3.25 Characteristic $V(Q)$ and its elements when: (a) $P = 0$; (b) $P \neq 0$.

Each of the four operating conditions can be described by a separate characteristic $V(Q)$ which will be combined to form the overall $V(Q)$ characteristic of the generator-transformer unit shown in Figure 3.25.

3.3.5.1 $I_f < I_{f\,max}$ and AVR Controls the Voltage

For this operating regime, the characteristic $V(Q)$ is determined by the position of the voltage regulation point. When $\kappa = 1$, the AVR controls the voltage at the generator terminals, and when $\kappa = 0$, the AVR controls the voltage at the transformer terminals (Figure 3.24). In practice, a small positive value of κ is chosen so that the voltage regulation point is located inside the transformer, close to its high-voltage terminals.

The voltages on both sides of the transformer are $\underline{V}_C$ and $\underline{V}$; see Figure 3.24. For a reactance κX_T between the generator transformer and the voltage regulation point, a relationship similar to (3.99) can be derived

$$P^2 + \left(Q + \frac{V^2}{\kappa X_T}\right)^2 = \left(\frac{V_P V}{\kappa X_T}\right)^2 \tag{3.128}$$

When the field current limit is not exceeded, the regulator maintains a reference voltage value $V_C = V_{C\,ref}$, where $V_{C\,ref}$ is the set value of the voltage at the regulation point. Simple algebra transforms the above equation to

$$Q = \sqrt{\left(\frac{V_{C\,ref} V}{\kappa X_T}\right)^2 - P^2} - \frac{V^2}{\kappa X_T} \tag{3.129}$$

When $P = 0$, the equation simplifies to

$$Q = \frac{V}{\kappa X_T}(V_{C\,ref} - V) \tag{3.130}$$

The resulting characteristic $Q(V)$ is denoted by Roman numeral I in Figure 3.25 and shown using a dashed line. It is an inverted parabola crossing the V axis at $V = 0$ and $V = V_{C\,ref}$. The peak of the parabola reaches the value $Q_{max} = V_{C\,ref}^2/4\kappa X_T$ when $V = V_{C\,ref}/2$. The smaller the value of κ, the higher the value of Q_{max} and the parabola becomes slimmer. When $P \neq 0$, the parabola moves along the V-axis toward positive values and its peak Q_{max} becomes smaller.

3.3.5.2 Maximum Field Current

When the field current is maximum, $I_f = I_{f\,max}$, the generator becomes a voltage source with constant emf $E_q = E_{q\,max}$ behind the reactance $x_d = (X_d + X_T)$. After simple algebra, an equation similar to (1.10) is obtained

$$Q = \sqrt{\left(\frac{E_{q\,max} V}{x_d}\right)^2 - P^2} - \frac{V^2}{x_d} \tag{3.131}$$

The resulting characteristic $Q(V)$ is denoted by Roman numeral II in Figure 3.25 and shown using a dot-dashed line. It is also an inverted parabola. For $P = 0$ the parabola crosses the V axis at $V = 0$ and $V = E_{q\,max}$, while its peak reaches the value $Q_{max} = E^2_{q\,max}/4x_d$ when $V = E_{q\,max}/2$. For $P \neq 0$, the parabola moves along the V-axis toward positive values and its peak Q_{max} becomes smaller.

3.3.5.3 Maximum Power Angle

When $\delta_{gt} = \delta_{gt\,max}$ and $\vartheta = 1$, Eq. (3.109) can be transformed to

$$Q = P \cdot \tan^{-1} \delta_{gt\,max} - \frac{V^2}{x_d} \tag{3.132}$$

where $x_d = X_d + X_T$. The resulting characteristic $Q(V)$ is denoted by Roman numeral III in Figure 3.25 and shown using a dotted line. It is an inverted parabola symmetrical with respect to the V-axis. Its peak is at $Q_{MAX} = P \cdot \tan^{-1} \delta_{gt\,max}$ while its roots are at $V = \pm \sqrt{P \cdot x_d \cdot \tan^{-1} \delta_{gt\,max}}$.

3.3.5.4 Maximum Stator Current

Now Eq. 3.96 can be transformed to

$$V = \frac{\sqrt{P^2 + Q^2}}{I_{max}} \tag{3.133}$$

The resulting characteristic $Q(V)$ is denoted by Roman numeral IV in Figure 3.25 and shown using a double-dot-dashed line. When $P = 0$ the equation simplifies to $V = Q/I_{max}$ and the characteristic $V(Q)$ becomes a straight line crossing the origin ($V = 0$, $Q = 0$). When $P \neq 0$ a curved line crosses the V-axis at $V = P/I_{max}$.

3.3.5.5 Combined Characteristic

The combined characteristic taking into account all the constraints consists of the four segments I, II, III, and IV and is shown in Figure 3.25 using a bold line. The left-hand side of the characteristic corresponds to a capacitive loading of the generator. In the area limited by segment III, regulator operation is constrained by the power angle and the voltage depends strongly on reactive power. Increasing capacitive loading causes a strong voltage rise. The middle part of the characteristic between points A and B corresponds to segment I when the regulator can change the field current while controlling the voltage. Changing reactive loading hardly influences the transformer terminal voltage. The slope of the characteristic depends on how slim the characteristic is, as explained below.

When point B is reached, the generator starts to operate with a maximum field current and cannot produce more reactive power. When the voltage drops, the reactive power produced decreases according to segment II until it reaches point C. Further reduction in the voltage results in the stator current becoming the limiting factor, as shown by segment IV.

These considerations show that a synchronous generator operating as the reactive power source has a quite complicated nonlinear characteristic due to the limits imposed by the AVR. It is important to appreciate the following:

- Along segment I, when the voltage drops, the generator produces more reactive power. When the constraints are reached, the generator behaves the other way around, that is when the voltage drops, the generator produces less reactive power moving along segments II and IV. This may cause worsening of the reactive power balance leading to voltage instability (Chapter 8).

- The influence of active power on the shape of the characteristic is very strong, as revealed by comparing both diagrams in Figure 3.25. When active power production is small, the distance between points A and B is large. When active power production increases, points A and B get closer to each other so that the stabilizing action of the AVR is possible for only a small range of reactive power changes. Moreover, when active power output is high, the curvature of the characteristic is stronger, which means that falling voltage causes a faster drop in reactive power output and therefore is more dangerous from the voltage stability point of view.

The slope of parabola (3.129) or (3.130) in the segment AB depends on the value of $\kappa X_{\rm T}$, that is on the choice of the voltage regulation point; see Figure 3.24. This can be proved deriving the derivative $\mathrm{d}V/\mathrm{d}Q$ under a simplifying assumption that $P = 0$ when Eq. (3.130) holds. For the upper part of the parabola, the voltage can be calculated from Eq. (3.130) as

$$V = \frac{1}{2}V_{\rm C\ ref} + \frac{1}{2}\sqrt{V_{\rm C\ ref}^2 - 4\kappa X_{\rm T} Q} \tag{3.134}$$

After differentiation

$$\frac{\mathrm{d}V}{\mathrm{d}Q} = -\frac{\kappa X_{\rm T}}{\sqrt{V_{\rm C\ ref}^2 - 4\kappa X_{\rm T} Q}} \tag{3.135}$$

The square root in the denominator can be eliminated using (3.134) leading to

$$\frac{\mathrm{d}V}{\mathrm{d}Q} = -\frac{\kappa X_{\rm T}}{2V - V_{\rm C\ ref}} \tag{3.136}$$

If $V = V_{\rm C\ ref}$ then

$$\left.\frac{\mathrm{d}V}{\mathrm{d}Q}\right|_{V = V_{\rm C\ ref}} = -\frac{\kappa X_{\rm T}}{V_{\rm C\ ref}} \tag{3.137}$$

This equation shows that the smaller is $\kappa X_{\rm T}$, the smaller is the slope of the voltage capability characteristic shown in Figure 3.24, segment A–B. If $\kappa X_{\rm T}$ is small then the parabola denoted by I is slim and with a peak far away from the vertical axis, that is $Q_{\max}$ is large. When the slope is small, the changes in reactive power hardly influence the transformer terminal voltage V. This is understandable because, for small values of $\kappa X_{\rm T}$, the voltage regulation point is close to the transformer terminals.

The minus sign in (3.137) means that increased reactive power causes reduction in the voltage; see Figure 3.25. A general equation for $\mathrm{d}V/\mathrm{d}Q$ when $P \neq 0$ could also be derived using(3.129). This is, however, more complicated and would not illustrate the underlying mechanisms as clearly.

3.3.5.6 Voltage–current Regulation Characteristic $V(I_{\rm Q})$

In practice, engineers use much simpler equations to describe a voltage–current regulation characteristic $V(I_{\rm Q})$, where $I_{\rm Q}$ denotes the reactive component of the current. The relevant equations can be derived using the phasor diagram shown in Figure 3.24. The distance between points 1 and 4 corresponds to the voltage drop on the reactance $\kappa X_{\rm T}$, so it is equal to $|1-4| = \kappa X_{\rm T} I$. Hence the distance between point 1 and 2 is $|1-2| = \kappa X_{\rm T} I \sin\varphi = \kappa X_{\rm T} I_{\rm Q}$ and the distance between points 0 and 2 is $|0-2| = V + \kappa X_{\rm T} I_{\rm Q}$. It can be assumed that the distance between points 0 and 4 is approximately the same as the distance between points 0 and 2. Hence $V_{\rm C} = |0-4| \cong |0-2| = V + \kappa X_{\rm T} I_{\rm Q}$ so that

$$V \cong V_{\rm C} - \kappa X_{\rm T} I_{\rm Q} \tag{3.138}$$

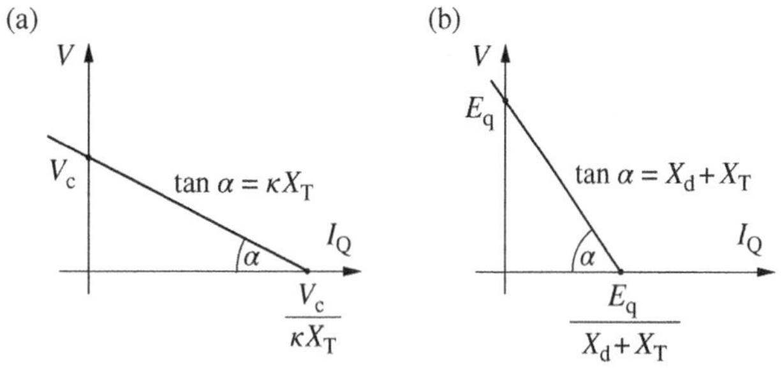

Figure 3.26 Voltage–current characteristics with: (a) acting AVR controlling V_C; (b) constant generator synchronous emf E_q.

A similar analysis can be applied to the distances between points 1, 5, 3, and 0. Taking into account that the distance between points 1 and 4 corresponds to the voltage drop on the reactance $x_d = X_d + X_T$, one gets $E_q = |0-5| \cong |0-3| = V + (X_d + X_T)I_Q$. Hence

$$V \cong E_q - (X_d + X_T)I_Q \tag{3.139}$$

Equations (3.138) and (3.139) correspond to two characteristics shown in Figure 3.26. The characteristic shown in Figure 3.26a is valid when the AVR is active, and it is a rough approximation of the linear characteristic I from Figure 3.25. The slope of that characteristic corresponds to an angle α where $\tan\alpha = \kappa X_T$. This matches Eq. (3.137) but with an addition that $V_{C\ ref}$ is used, as (3.137) was concerned with reactive power, while Eq. (3.138) is concerned with the reactive component of the current.

The tangent of the angle α is referred to as the droop of that characteristic. It depends on the coefficient κ defining the value κX_T and the position of the voltage regulation point.

The characteristic shown in Figure 3.26b is valid when the AVR is not active and the generator synchronous emf is constant. It is a rough linear approximation of the characteristic II from Figure 3.25. The slope of the characteristic corresponds to angle α for which

$$\tan\alpha = x_d = X_d + X_T \tag{3.140}$$

The slope is steeper as $(X_d + X_T) >> \kappa X_T$. A steeper slope means that, when the AVR is not acting, changes in the generator reactive power loading cause large changes in the voltage at the transformer terminal.

Generally, the compensation impedance does not have to match the position of the measurement point corresponding to $\underline{V}_C$; see Figure 3.24. When a generator is connected to the transmission via a radial transmission line, $\underline{Z}_C$ may encroach into the line.

The main aim of using current compensation in the AVR is to make the power station terminal voltage insensitive to reactive power loading. The droop of the regulator characteristic depends on the compensation reactance. Equations. (3.137) and (3.138) and Figure 3.26 show that the smaller the droop of the characteristic, the smaller the sensitivity of the terminal voltage to the reactive power.

3.3.6 Including the Equivalent Network Impedance

Synchronous generators are rarely used to supply individual loads but are connected to an EPS that consists of a large number of other synchronous generators and loads linked by the transmission network (Figure 2.1). The power rating of an individual generator is usually many times smaller than the sum of the power ratings of all the remaining generators in the system. Therefore, for a simplified analysis, these remaining generators in the system can be treated as one equivalent, very large, generating unit with an infinite power rating. This equivalent generating unit is referred to as the *infinite busbar* and can be represented on a circuit diagram as an ideal voltage

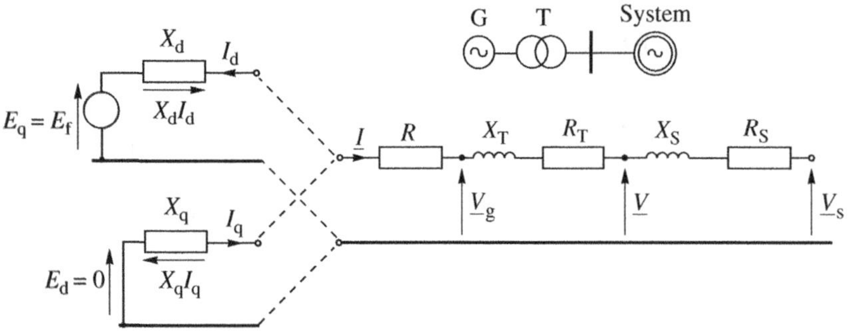

Figure 3.27 Equivalent circuit of the SMIB model in the steady-state.

source behind an equivalent system impedance. The infinite busbar maintains a constant terminal voltage and is capable of absorbing all the active and reactive power output of the generator in question. This simplified model of the EPS is referred to as the *single machine infinite bus* (SMIB) model.

This concept is illustrated in Figure 3.27, where the generator is connected to the system via a step-up transformer, represented by the series impedance $\underline{Z}_T = R_T + jX_T$ (the shunt admittance is neglected). It is assumed that the ideal transformer has been eliminated from the circuit diagram by either using pu's or recalculating all the quantities using a common voltage level. The rest of the EPS is represented by the infinite busbar, that is by the ideal voltage source $\underline{V}_s$ behind the equivalent system impedance $\underline{Z}_s = R_s + jX_s$ (again the shunt admittance is neglected). The impedance $\underline{Z}_s$ combines the transmission network and remaining generators in the system.

The infinite busbar is assumed to have constant voltage and frequency, neither of which is influenced by the action of an individual generator. This means that the voltage $\underline{V}_s$ can be used as a reference and the phase angles of all the other voltages and currents in the circuit measured with respect to it. Of particular importance is the *power angle* δ, which defines the *phase shift* between $\underline{E}_f$ and $\underline{V}_s$. As all the angles in the phasor diagram have a dual time/space meaning, δ is also the *spatial* angle between the two synchronously rotating rotors: that of the generator under consideration and that of the fictitious generator replacing the EPS. This spatial angle is referred to as the *rotor angle* and has the same numerical value (in electrical radians) as the power angle. As the "rotor" of the infinite busbar is not affected by an individual generator, this "rotor" also provides a synchronously rotating reference axis with respect to which the *space position* of all the rotors may be defined. This dual meaning of rotor/power angle is very important and will be used extensively throughout this book.

The elements of the equivalent circuit may be combined to give the total parameters

$$x_d = X_d + X_T + X_s;\ x_q = X_q + X_T + X_s;\ r = R + R_T + R_s \tag{3.141}$$

All the equations derived earlier in this chapter and the phasor diagram shown in Figure 3.17 can still be used by replacing X_d by x_d, X_q by x_q, R by r, and $\underline{V}_g$ by $\underline{V}_s$. Also the angle δ_g must be replaced by the phase shift δ between E_f and V_s. With these modifications the round-rotor generator can be described by the following equation (corresponding to Eq. (3.49))

$$\underline{E}_f = \underline{V}_s + r\underline{I} + jx_d\underline{I} \tag{3.142}$$

while the salient-pole generator can be described by an equation similar to Eq. (3.64)

$$\underline{E}_f = \underline{V}_s + r\underline{I} + jx_d\underline{I}_d + jx_q\underline{I}_q \tag{3.143}$$

This equation can be broken down into its d- and q-axis components to obtain equations similar to Eq. (3.67)

$$\begin{aligned} E_{\mathrm{d}} &= V_{\mathrm{sd}} + rI_{\mathrm{d}} + x_{\mathrm{q}}I_{\mathrm{q}} = 0 \\ E_{\mathrm{q}} &= E_{\mathrm{f}} = V_{\mathrm{sq}} + rI_{\mathrm{q}} - x_{\mathrm{d}}I_{\mathrm{d}} \end{aligned} \tag{3.144}$$

which can be expressed in matrix form as

$$\begin{bmatrix} E_{\mathrm{d}} \\ E_{\mathrm{q}} \end{bmatrix} = \begin{bmatrix} 0 \\ E_{\mathrm{f}} \end{bmatrix} = \begin{bmatrix} V_{\mathrm{sd}} \\ V_{\mathrm{sq}} \end{bmatrix} + \begin{bmatrix} r & x_{\mathrm{q}} \\ -x_{\mathrm{d}} & r \end{bmatrix} \begin{bmatrix} I_{\mathrm{d}} \\ I_{\mathrm{q}} \end{bmatrix} \quad \text{or} \quad \boldsymbol{E}_{\mathrm{dq}} = \boldsymbol{V}_{\mathrm{sdq}} + \boldsymbol{Z}_{\mathrm{dq}}\boldsymbol{I}_{\mathrm{dq}} \tag{3.145}$$

where

$$\begin{aligned} V_{\mathrm{sd}} &= -V_{\mathrm{s}} \sin\delta; \qquad & V_{\mathrm{sq}} &= V_{\mathrm{s}} \cos\delta \\ I_{\mathrm{d}} &= -I \sin\beta; \qquad & I_{\mathrm{q}} &= I \cos\beta \end{aligned} \tag{3.146}$$

and $\beta = \delta + \varphi$.

Equations (3.145) and (3.146) allow formulae for the active and reactive power supplied by the generator to the system to be derived.

3.3.6.1 Real Power

Solving Eq. (3.145) with respect to the currents gives

$$\begin{bmatrix} I_{\mathrm{d}} \\ I_{\mathrm{q}} \end{bmatrix} = \frac{1}{Z^2} \begin{bmatrix} r & -x_{\mathrm{q}} \\ x_{\mathrm{d}} & r \end{bmatrix} \begin{bmatrix} 0 - V_{\mathrm{sd}} \\ E_{\mathrm{q}} - V_{\mathrm{sq}} \end{bmatrix} \tag{3.147}$$

where $Z^2 = \det \boldsymbol{Z}_{\mathrm{dq}} = r^2 + x_{\mathrm{d}}x_{\mathrm{q}}$.

The active power supplied by each phase of the generator to the network (the infinite busbar in Figure 3.27) is given by

$$\begin{aligned} P_{\mathrm{s}} &= V_{\mathrm{s}}I \cos\varphi = V_{\mathrm{s}}I \cos(\beta - \delta) = V_{\mathrm{s}}I \sin\beta \sin\delta + V_{\mathrm{s}}I \cos\beta \cos\delta \\ &= V_{\mathrm{sd}}I_{\mathrm{d}} + V_{\mathrm{sq}}I_{\mathrm{q}} = \frac{1}{Z^2}\left\{V_{\mathrm{sd}}\left[-rV_{\mathrm{sd}} - x_{\mathrm{q}}\left(E_{\mathrm{q}} - V_{\mathrm{sq}}\right)\right] + V_{\mathrm{sq}}\left[-x_{\mathrm{d}}V_{\mathrm{sd}} + r\left(E_{\mathrm{q}} - V_{\mathrm{sq}}\right)\right]\right\} \\ &= \frac{1}{Z^2}\left[-r\left(V_{\mathrm{sq}}^2 + V_{\mathrm{sd}}^2\right) - \left(x_{\mathrm{d}} - x_{\mathrm{q}}\right)V_{\mathrm{sd}}V_{\mathrm{sq}} + E_{\mathrm{q}}\left(rV_{\mathrm{sq}} - x_{\mathrm{q}}V_{\mathrm{sd}}\right)\right] \end{aligned} \tag{3.148}$$

which, after substituting for V_{sq} and V_{sd} from Eq. (3.146), gives

$$P_{\mathrm{s}} = \frac{E_{\mathrm{q}}V_{\mathrm{s}}}{Z}\frac{x_{\mathrm{q}}}{Z} \sin\delta + \frac{V_{\mathrm{s}}^2}{2}\frac{x_{\mathrm{d}} - x_{\mathrm{q}}}{Z^2} \sin 2\delta + \frac{E_{\mathrm{q}}V_{\mathrm{s}}}{Z}\frac{r}{Z} \cos\delta - \frac{V_{\mathrm{s}}^2}{Z}\frac{r}{Z} \tag{3.149}$$

The second component in Eq. (3.149) is referred to as the *reluctance power* and corresponds to the reluctance torque in the salient-pole generator. It depends on $\sin 2\delta$, vanishes for $\delta = 0^0$ and $\delta = 90^0$, and is proportional to $x_{\mathrm{d}} - x_{\mathrm{q}} = X_{\mathrm{d}} - X_{\mathrm{q}}$ and inversely proportional to Z^2. As $x_{\mathrm{d}} - x_{\mathrm{q}} = X_{\mathrm{d}} - X_{\mathrm{q}}$ is independent of X_{s}, the Z^2 term tends to dominate this expression so that the reluctance power of a generator connected to the system through a long transmission link (large X_{s} and Z) can often be neglected.

For round-rotor generators $X_{\mathrm{d}} = X_{\mathrm{q}}$ and the reluctance term vanishes irrespective of the value of the power angle δ. For such generators $Z^2 = r^2 + x_{\mathrm{d}}^2$, and Eq. (3.149) can be re-arranged as

$$\begin{aligned} P_{\mathrm{s}} &= \frac{E_{\mathrm{q}}V_{\mathrm{s}}}{Z} \cos\mu \sin\delta + \frac{E_{\mathrm{q}}V_{\mathrm{s}}}{Z} \sin\mu \cos\delta - \frac{V_{\mathrm{s}}^2}{Z} \sin\mu \\ &= \frac{E_{\mathrm{q}}V_{\mathrm{s}}}{Z} \sin(\delta + \mu) - \frac{V_{\mathrm{s}}^2}{Z} \sin\mu \end{aligned} \tag{3.150}$$

where $\sin\mu = r/Z$ and $\cos\mu = x_{\mathrm{d}}/Z$. The active power load on the generator is larger than the active power supplied to the system because of the I^2r loss in the resistance. If the resistance of the generator and the network is neglected ($r \ll Z$) and P_{s} is equal to the active power load on the generator. In this case, for the salient-pole generator

$$P_s = \frac{E_q V_s}{x_d} \sin\delta + \frac{V_s^2}{2} \frac{x_d - x_q}{x_d x_q} \sin 2\delta \tag{3.151}$$

while for the round-rotor generator the active power is

$$P_s = \frac{E_q V_s}{x_d} \sin\delta \tag{3.152}$$

In pu's these equations can also be used to express the generator air-gap torque.

3.3.6.2 Reactive Power

The reactive power supplied to the network is given by

$$\begin{aligned} Q_s &= V_s I \sin\varphi = V_s I \sin(\beta - \delta) = V_s I \sin\beta\cos\delta - V_s I \cos\beta \sin\delta \\ &= -V_{sq} I_d + V_{sd} I_q = \frac{1}{Z^2}\left\{V_{sq}\left[r V_{sd} + x_q\left(E_q - V_{sq}\right)\right] + V_{sd}\left[-x_d V_{sd} + r\left(E_q - V_{sq}\right)\right]\right\} \\ &= \frac{1}{Z^2}\left(E_q V_{sq} x_q - V_{sq}^2 x_q - V_{sd}^2 x_d + E_q V_{sd} r\right) \end{aligned} \tag{3.153}$$

which, after substituting for V_{sq} and V_{sd} from Eq. (3.146), gives

$$Q_s = \frac{E_q V_s}{Z} \frac{x_q}{Z} \cos\delta - \frac{V_s^2}{Z} \frac{x_d \sin^2\delta + x_q \cos^2\delta}{Z} - \frac{E_q V_s}{Z} \frac{r}{Z} \sin\delta \tag{3.154}$$

For the round-rotor generator $x_d = x_q$ and the second term is independent of δ. Equation (3.154) can then be written as

$$Q_s = \frac{E_q V_s}{Z} \cos(\delta - \mu) - \frac{V_s^2}{Z} \cos\mu \tag{3.155}$$

If the resistance r is neglected then $Z = x_d$, $\mu = 0$, and $\cos\mu = 1$ and this equation simplifies to

$$Q_s = \frac{E_q V_s}{x_d} \cos\delta - \frac{V_s^2}{x_d} \tag{3.156}$$

A generator supplies reactive power to the system, and operates at a lagging power factor, when the field current, and consequently E_q, is high enough to make the first component in Eq. (3.156) larger than the second. This state of operation is referred to as *overexcitation*.

On the other hand a low value of the field current, and E_q, will make the second component of Eq. (3.156) larger than the first; the generator will then supply negative reactive power to the system and will operate at a leading power factor. This state of operation is referred to as *underexcitation*.

As is explained in Section 3.5, a typical power system load contains a high proportion of induction motors. Such a load operates at a lagging power factor and consumes positive reactive power. For this reason generators are usually overexcited, operate at a lagging power factor, and supply positive reactive power to the system. Overexcited generators operate at a high value of E_q which is also important for stability reasons, as explained in Chapters 5 and 6.

3.3.6.3 Steady-state Power-angle Characteristic

Figure 3.17 shows the phasor diagram of a generator operating under a particular load as defined by the length and direction of the current phasor $\underline{I}$. Obviously, the diagram will change if the load is changed. However, the way in which the diagram will change depends on whether the AVR is active or not. When the AVR is active, it will try to maintain a constant voltage at some point after the generator terminals by changing the excitation (Section 2.3.2).

This mode of operation is closely related to steady-state stability and is considered in Chapter 5. The present chapter only considers the case when the AVR is inactive. The excitation is then constant, $E_f = E_q =$ constant, and any change in load will change the generator terminal voltage V_g, the angle δ_g, and the power angle δ. If the resistance r is neglected then the active power supplied by the generator to the system is given by Eq. (3.151), which is repeated here for convenience

$$P_{sE_q} = \frac{E_q V_s}{x_d} \sin\delta + \frac{V_s^2}{2} \frac{x_d - x_q}{x_d x_q} \sin 2\delta \tag{3.157}$$

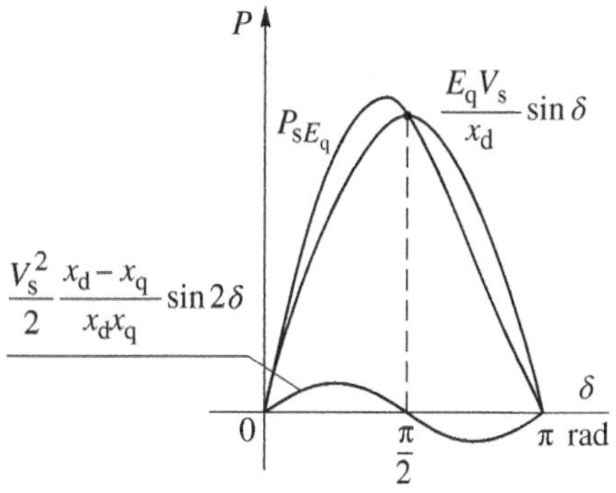

Figure 3.28 Power-angle characteristic $P_{sE_q}(\delta)$ for $E_q =$ const.

In this equation the subscript E_q in P_{sE_q} has been added to emphasize that this formula is valid for the case $E_q =$ constant. Note that P_{sE_q} is a function of the power angle δ only. Function $P_{sE_q}(\delta)$ is referred to as the *power-angle characteristic* of the generator operating on the infinite busbar and is shown in Figure 3.28.

The reluctance power term deforms the sinusoidal characteristic so that the maximum of $P_{sE_q}(\delta)$ occurs at $\delta < \pi/2$. For the round-rotor generators the maximum of $P_{sE_q}(\delta)$ occurs at $\delta = \pi/2$.

3.4 Power System Loads

The word "load" can have several meanings in power system engineering, including:

- a device connected to the EPS that consumes power;
- the total active and/or reactive power consumed by all devices connected to the EPS;
- the power output of a particular generator or plant;
- a portion of the system that is not explicitly represented in the system model, but treated as if it were a single power-consuming device.

This section deals with loads that conform to the last definition. Chapter 2 explains that EPSs are large, complex, structures consisting of power sources, transmission and subtransmission networks, distribution networks, and a variety of energy consumers. As the transmission and subtransmission networks connect the main generation and load centers, they are quite sparse, but as the distribution networks must reach every consumer in their service area they are very dense. This means that a typical power system may consist of several hundred nodes at the transmission and subtransmission levels, but there could be a hundred thousand nodes at the distribution level. Consequently, when power systems are analyzed, only the transmission and subtransmission levels are considered and the distribution networks are not usually modeled as such, but replaced by equivalent loads, sometimes referred to as *composite loads*. Usually, each composite load represents a relatively large fragment of the system, typically comprising low- and medium-voltage distribution networks, small power sources operating at distribution levels, reactive power compensators, distribution voltage regulators, and so on, and includes a large number of different component loads such as motors and lighting and electrical appliances. Determining a simple and valid composite load model is therefore not an easy problem and is still the subject of intensive research (IEEE Task Force 1995). This chapter only describes a simple static composite load model; Section 11.8 describes dynamic load models.

In the steady state the demand of the composite load depends on the busbar voltage V and the system frequency f. The functions describing the dependency of the active and reactive load demand on the voltage and frequency $P(V, f)$ and $Q(V, f)$ are called the *static load characteristics*. The characteristics $P(V)$ and $Q(V)$, taken at constant

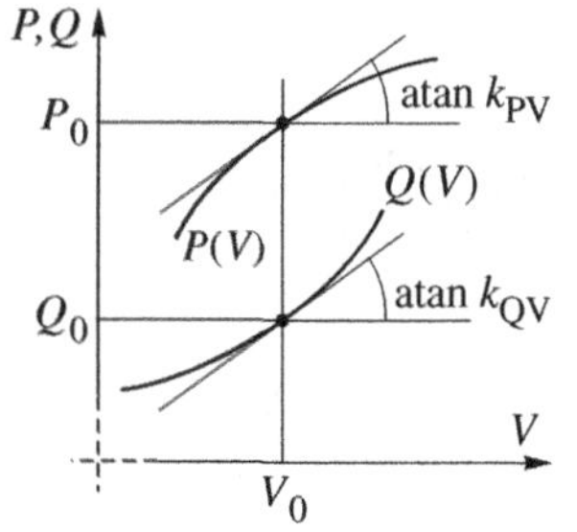

Figure 3.29 Illustration of the definition of voltage sensitivities.

frequency, are called the *voltage characteristics*, while the characteristics $P(f)$ and $Q(f)$, taken at constant voltage, are called the *frequency characteristics*.

The slope of the voltage or frequency characteristic is referred to as the *voltage (or frequency) sensitivity* of the load. Figure 3.29 illustrates this concept with respect to voltage sensitivities. Frequency sensitivities are defined in a similar way.

The voltage sensitivities k_{PV} and k_{QV} and the frequency sensitivities k_{Pf} and k_{Qf} are usually expressed in pu with respect to the given operating point

$$k_{PV} = \frac{\frac{\Delta P}{P_0}}{\frac{\Delta V}{V_0}}; \quad k_{QV} = \frac{\frac{\Delta Q}{Q_0}}{\frac{\Delta V}{V_0}}; \quad k_{Pf} = \frac{\frac{\Delta P}{P_0}}{\frac{\Delta f}{f_0}}; \quad k_{Qf} = \frac{\frac{\Delta Q}{Q_0}}{\frac{\Delta f}{f_0}} \tag{3.158}$$

where P_0, Q_0, V_0, and f_0 are the real power, reactive power, voltage, and frequency at a given operating point.

A load is considered *stiff* if, at a given operating point, its voltage sensitivities are small. If the voltage sensitivities are equal to zero then the load is *ideally stiff* and the power demand of that load does not depend on the voltage. A load is voltage sensitive if the voltage sensitivities are high and small changes in the voltage cause high changes in the demand. Usually, the voltage sensitivity of active power demand is less than the voltage sensitivity of reactive power demand.

As the characteristics of the composite load depend on the characteristics of its individual components, it is first of all necessary to examine the characteristics of some of the more important individual loads. This will then be developed into a more general composite load model.

3.4.1 Lighting and Heating

About one-third of electricity consumption is on lighting and heating. Traditional bulb lighting prevails in residential areas, while discharge lights (fluorescent, mercury vapor, sodium vapor) dominate in commercial and industrial premises. Traditional electric bulbs consume no reactive power and their power demand is frequency independent but, as the temperature of the filament depends on the voltage, the bulb cannot be treated as a constant impedance. Figure 3.30a shows the relevant characteristic.

Fluorescent and discharge lighting depends heavily on the supply voltage. When the voltage drops to below (65–80)% of the rated value the discharge lights extinguish and will restart with a (1–2) s delay only when the voltage recovers to a value above the extinguish level. When the voltage is above the extinguish level, the real and reactive power vary nonlinearly with voltage, as shown in Figure 3.30b.

Heating loads basically constitute a constant resistance. If a heater is equipped with a thermostat, the thermostat will maintain constant temperature and power output despite any variations in voltage. In such cases the load can be modeled as a constant power rather than a constant resistance.

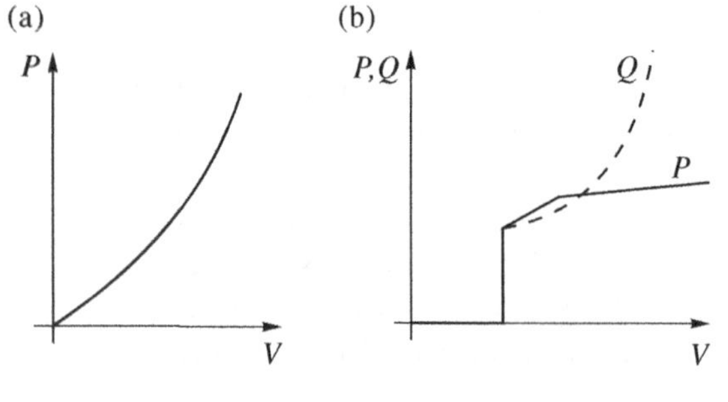

Figure 3.30 Voltage characteristics of: (a) electric bulbs; (b) discharge lighting.

3.4.2 Induction Motors

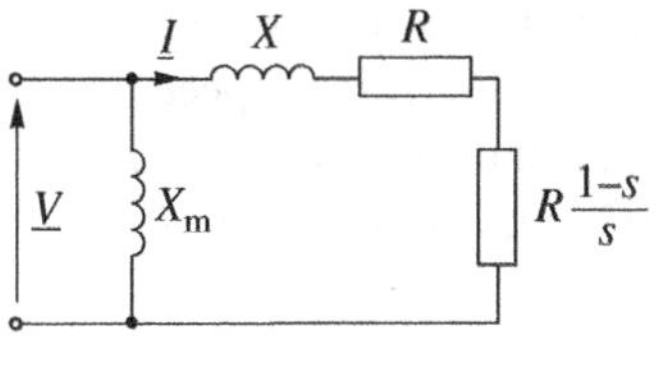

Figure 3.31 Simplified equivalent circuit of induction motor.

About (50–70)% of all electricity is consumed by electric motors with about 90% of this being used by induction motors. Generally, such motors dominate industrial loads to a far greater degree than they do commercial or residential loads.

Figure 3.31 shows the well-known simplified equivalent circuit of the induction motor where X is the equivalent reactance of the stator and rotor windings, R is the rotor resistance, X_m is the magnetizing reactance and s is the *motor slip* defined as $s = (\omega_s - \omega)/\omega_s$. This equivalent circuit is discussed in detail in Chapter 7. Here it is used to derive the reactive power-voltage characteristic of an induction motor as the load in power system.

3.4.2.1 Power-slip Characteristic

The square of the current flowing through the series branch is $I^2 = V^2/[X^2 + (R/s)^2]$ where V is the supply voltage. The active power demand due to this current is then

$$P_e = I^2 \frac{R}{s} = V^2 \frac{Rs}{R^2 + (Xs)^2} \tag{3.159}$$

The solid lines in Figure 3.32a show the variation of P_e with slip for different values of V. By differentiating the power expression it can be shown that the maximum loading $P_{max} = V^2/2X$ is reached at the *critical slip* $s = s_{cr} = R/X$. Note that s_{cr} does not depend on the voltage.

The stable operating part of the induction motor characteristic is to the left of the peak, where $s < s_{cr}$. To prove this consider point 1′ lying to the right of the peak on the highest characteristic. Ignoring for the moment any losses, if point 1′ is an equilibrium point then the electrical power supplied P_e is equal to the mechanical power P_m shown on Figure 3.32a as P_0. Now assume that the motor experiences a momentary disturbance causing a decrease in slip (i.e. an increase in speed). This will cause the supplied electrical power to increase. As P_m can be assumed to remain unchanged, the motor draws more power than the load can absorb ($P_e > P_m$) and the excess power accelerates the rotor so that the speed increases further. This increase in speed results in the operating point moving further to the left away from point 1′. Now consider another momentary disturbance causing an increase in slip (drop in speed). This will cause P_e to drop resulting in a deficit of power and further slowing down of the rotor. The slip increases further and the motor finally stops at $s = 1$.

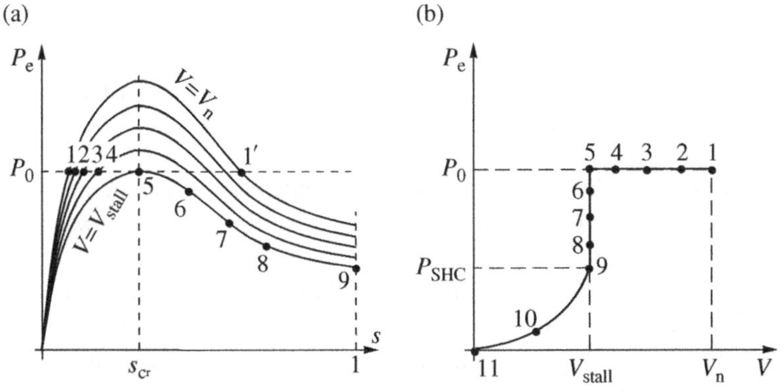

Figure 3.32 Induction motor characteristics: (a) family of $P_e(s)$ characteristics for different values of V; (b) $P_e(V)$ characteristic of the motor loaded with a constant torque.

The situation is reversed at the stable equilibrium point 1. At this point any disturbance causing a decrease in slip (i.e. an increase in speed) will result in a decrease in the power supplied to the motor. This deficit of power ($P_e < P_m$) will result in the motor slowing down (increasing the slip) until it returns to point 1. In a similar way the motor will return to point 1 when subjected to a disturbance causing an increase in slip.

3.4.2.2 Real Power-voltage Characteristic

The motor's operating point corresponds to the intersection of the motor characteristic $P_e(s)$ with that of the mechanical load $P_m(s)$. Mechanical loads can be categorized as either *easy starting* (zero torque at starting) or *heavy starting* (a nonzero torque at starting). For further analysis the special case of a heavy starting load with a constant mechanical torque τ_m will be considered. This type of load will allow the voltage characteristics of the motor to be derived. These characteristics will then be generalized to include other types of mechanical load.

If τ_m is constant then the mechanical power developed by the motor is $P_m = \tau_m\omega = \tau_m\omega_s(1-s) = P_0(1-s)$, where $P_0 = \tau_m\omega_s$, and is represented in the equivalent circuit by the power loss in the resistance $R(1-s)/s$. The total motor power demand P_d is obtained by adding P_m to the power loss in the rotor resistance where $P_{loss} = I^2R$, that is

$$P_d = P_m + P_{loss} = P_0(1-s) + I^2R = P_0(1-s) + \frac{V^2Rs^2}{(Xs)^2 + R^2} \tag{3.160}$$

The value of the operating slip is given by the intersection of the $P_e(s)$ and the $P_d(s)$ characteristics. Equation $P_e(s) = P_d(s)$ gives

$$\frac{V^2Rs}{(Xs)^2 + R^2} = P_0\,(1-s) + \frac{V^2Rs^2}{(Xs)^2 + R^2} \tag{3.161}$$

which, assuming that the motor is not stalled ($s \neq 1$), gives the following equation, after some simple algebra

$$s^2 - 2as_{cr}s + s_{cr}^2 = 0 \tag{3.162}$$

where $a = V^2/(2P_0X)$ and $s_{cr} = R/X$. The roots of this equation are

$$s_{1,2} = s_{cr}\left[a \pm \sqrt{a^2-1}\right] \tag{3.163}$$

The roots are real only if $|a| \geq 1$, that is $V \geq \sqrt{2P_0X}$. Hence the minimum supply voltage for which the motor can still operate, referred to as the *stalling voltage*, is $V_{stall} = \sqrt{2P_0X}$. For $V > V_{stall}$ there are two different roots, s_1 and s_2, with the smaller root corresponding to the stable operating point (i.e. that part of the characteristic to the left of the peak). Substituting this root into Eq. (3.159) gives

$$P_e|_{s=s_{1,2}} = \frac{V^2}{X}\frac{Rs_{cr}\left(a \pm \sqrt{a^2-1}\right)}{R^2 + X^2s_{cr}^2\left(a \pm \sqrt{a^2-1}\right)^2} = \frac{V^2}{X}\frac{a \pm \sqrt{a^2-1}}{1 + \left(a \pm \sqrt{a^2-1}\right)^2} = \frac{V^2}{2aX} = P_0 \tag{3.164}$$

This shows that for a heavy-starting load with τ_m = constant the active power demand is independent of both the supply voltage and the slip. This is shown in Figure 3.32a by the dashed horizontal line $P = P_0$. Figure 3.32b shows the resulting $P_e(V)$ characteristic obtained in the following way.

Assume that the motor initially operates at rated voltage ($V = V_n$) so that the equilibrium point corresponds to point 1 on the top characteristic. A decrease in the supply voltage will lower the power-slip characteristic and shift the operating point to the right, as shown by points 2, 3, and 4 in Figure 3.32a. As active power demand is constant, the $P_e(V)$ characteristic corresponding to points 1, 2, 3, and 4 is horizontal, as shown in Figure 3.32b. The lowest characteristic in Figure 3.32a corresponds to the stalling voltage $V = V_{stall}$. The operating point is at point 5 and the slip is critical $s = s_{cr}$. Any further slight decrease in voltage will result in the motor slip increasing along the lowest, unstable, characteristic (points 6, 7, and 8) until the motor finally stalls ($s = 1$) at point 9. As the motor moves along this stalling characteristic, the active power decreases, while the voltage is constant and slip increases. Consequently, the $P_e(V)$ characteristic for points 6, 7, and 8 is vertical, as shown in Figure 3.32b. At point 9 the slip is $s = 1$ and the motor consumes its short-circuit power, $P_{SHC} = V^2R/(X^2 + R^2)$. Any further decrease in voltage will cause a decrease in the active power demand, which can be calculated from Eq. (3.160) as

$$P_e(s)|_{s=1} = \frac{V^2 R}{X^2 + R^2} \tag{3.165}$$

This is shown by the parabola 9, 10, 11 in Figure 3.32b.

3.4.2.3 Reactive Power-voltage Characteristic

The reactive power consumed by each phase of the motor is made up of two components corresponding to each of the two parallel branches shown in Figure 3.31b

$$Q_m = \frac{V^2}{X_m} \text{ and } Q_s = I^2 X = \frac{P_e s}{R} X = P_e \frac{s}{s_{cr}} \tag{3.166}$$

The component Q_m is associated with the motor magnetizing reactance while the component Q_s depends on the motor load. For a heavy starting-load, with τ_m = constant, the active power demand is $P_e = P_0$ and the slip corresponding to the stable operating point is $s = s_{cr}\left[a - \sqrt{a^2 - 1}\right]$ (Eq. [3.163]). Substituting these values into the expression for Q_s gives

$$Q_s = \frac{V^2}{2X} - \sqrt{\left(\frac{V^2}{2X}\right)^2 - P_0^2} \quad \text{for } V > V_{stall} \tag{3.167}$$

Figure 3.33 shows Q_m, Q_s, and $Q = Q_m + Q_s$ as a function of the voltage. The component $Q_m(V)$ is a parabola starting from the origin and increasing to infinity. The component $Q_s(V)$ tends to zero as $V \rightarrow \infty$. As V decreases, $Q_s(V)$ increases until at $V = V_{stall}$ it reaches a value corresponding to point 5. Any decrease in voltage will cause $Q_s(V)$ to move along the unstable part defined by the vertical line 5, 6, 7, 8, 9. At point 9 the motor stalls ($s = 1$). Any further decrease in voltage causes $Q_s(V)$ to decrease along a parabola 9, 10, 11 describing the short-circuit characteristic.

The resulting characteristic, $Q(V) = Q_s(V) + Q_m(V)$, is shown by the continuous line in Figure 3.33. Inspection of this characteristic leads to the following conclusions:

- for voltages close to the rated voltage V_n the slope of the characteristic is positive;
- as voltage decreases, the characteristic first becomes flatter (indicating a reduced voltage sensitivity) and then, with a further reduction in voltage, the reactive power increases rapidly to a large value when the motor stalls.

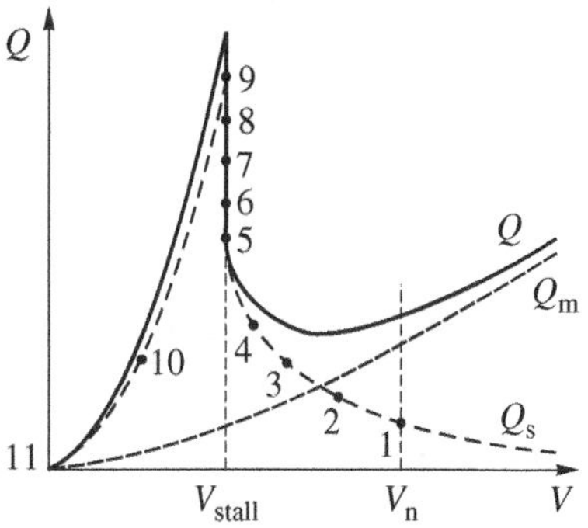

Figure 3.33 Reactive power-voltage characteristics of an induction motor.

Although a heavy-starting load has been considered here, the actual stalling voltage depends on the type of mechanical load that is being driven by the motor. For heavily loaded motors, driving heavy-starting loads, it can be quite close to the rated voltage. For lightly loaded motors, especially those driving easy-starting mechanical loads, the stalling voltage can be quite small.

3.4.2.4 Influence of Motor Protection and Starter Control

The above considerations do not take into account motor protection or starter control. Many industrial motors have starter controls with electromechanically held AC contactors. If the voltage is too low, such contactors immediately drop out, tripping the motor supply. The dropout voltage ranges from about (0.3–0.7) pu. The effect of this can be modeled by assuming zero real and reactive power demand for voltages lower than the dropout voltage, as shown in Figure 3.34. Only heavy-loaded motors may stall above the dropout voltage and give a large increase in the reactive power demand, Figure 3.34a. When the stalling voltage is smaller than the dropout voltage, the motor will not stall or exhibit a rapid increase in reactive power demand; see Figure 3.34b.

When operating at a low voltage the motor may also be tripped by the overcurrent protection. This usually involves some time delay.

Motors operating under residential and commercial loads are usually oversized, operate at less than 60% of rated power, and therefore have characteristics similar to those shown in Figure 3.34b. On the other hand, large industrial motors are usually properly sized and, when driving heavy-starting mechanical loads, may have characteristics similar to those shown in Figure 3.34a.

3.4.3 Static Characteristics of the Load

The aggregate characteristic of the load depends on the characteristics of its individual components. A rough estimate of the aggregate characteristic, viewed from the medium-voltage side (the secondary of the feeder transformer), can be obtained by summing the individual load characteristics. Figure 3.35 shows two examples of load characteristics obtained by this technique. Figure 3.35a shows an industrial load characteristic with a predominance of heavily loaded induction motors and discharge lighting. Near the nominal operating point (voltage V_n),

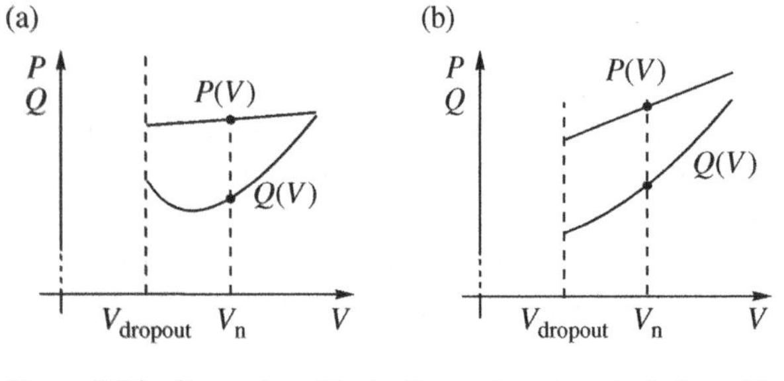

Figure 3.34 Examples of induction motor characteristics with starter control: (a) heavy-starting load; (b) easy-starting load.

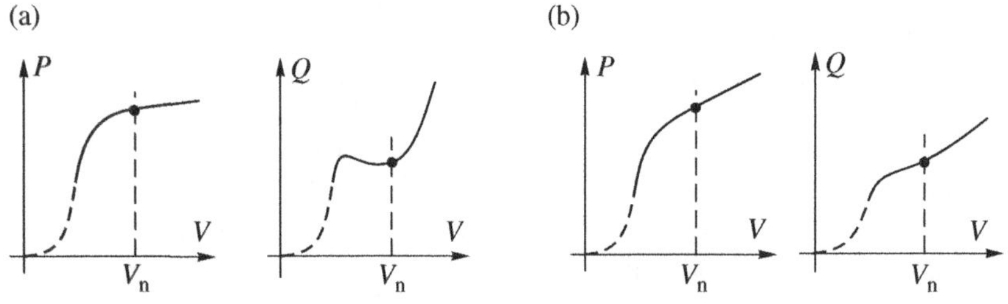

Figure 3.35 Examples of the voltage characteristics of a load: (a) dominated by large, heavily loaded induction motors; (b) dominated by lighting and heating.

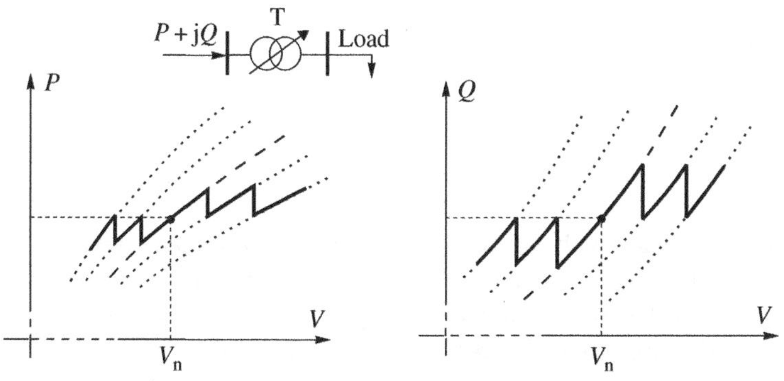

Figure 3.36 Influence of a tap-changing transformer on the voltage characteristic of a composite load.

the $P(V)$ curve is flat while the $Q(V)$ curve is steeper with a positive slope. As the voltage decreases, the $Q(V)$ curve becomes flatter and even rises, owing to the increased reactive power demand of the stalled motors. When the voltage drops below about 0.7 pu, the $P(V)$ and $Q(V)$ curves rapidly decrease, owing to the tripping of the induction motors and extinguishing of the discharge lighting.

Figure 3.35b shows an example of a residential/commercial load that is dominated by traditional bulb lighting and heating. Near the nominal voltage both the $P(V)$ and $Q(V)$ curves are quite steep. Again, the real and reactive power demand drops rapidly at about 0.7 pu. As the induction motors stall voltage is now below the dropout voltage, dropout is not preceded by an increase in the reactive power demand.

The curves shown in Figure 3.35 can only give an indication of the kind of shape a load voltage characteristic may have. They cannot be treated in a general manner, because the characteristic of a particular load may be quite different. For example, reactive power compensation can cause the $Q(V)$ curve to be flatter near the nominal voltage. Also relatively small, nonutility generation embedded in the load area will significantly affect the load characteristic.

There is also a difference in the characteristic as seen from the primary and secondary side of the feeder transformer. First of all, the real and reactive power loss in the transformer must be added to the load demand. Secondly, the feeder transformer is usually equipped with an OLTC to help control the voltage in the distribution network, and this also affects the characteristic, as illustrated in Figure 3.36.

In Figure 3.36 the middle dashed bold line represents the load voltage characteristic at the nominal transformation ratio. Tap changing is controlled in discrete steps so that, if the transformer tap setting is changed, the voltage characteristic moves to the left or right in discrete steps, as shown by the dotted lines. The extreme left and right characteristics represent the tap-changer limits. A dead zone is also present in the regulator in order to prevent any tap changes if the voltage variations are within limits. The resulting voltage characteristic is shown by the bold line and is quite flat within the regulation range, as can be seen by sketching an average line through the resulting characteristic.

3.4.4 Load Models

The previous subsection describes how the real and reactive power of particular types of load depends on the load voltage but does not explain how these could be represented by a mathematical model. Since all power system analysis programs, for example power flow or dynamic simulation, require such a load model, this subsection describes some of the most popular models currently in use.

3.4.4.1 Constant Power/Current/Impedance

The simplest load models assume one of the following features:

- a constant power demand (P)
- a constant current demand (I)
- a constant impedance (Z).

A constant power model is voltage invariant and allows loads with stiff voltage characteristics $k_{PV} \cong k_{QV} \cong 0$ to be represented. This model is often used in load-flow calculations (Section 3.6) but is generally unsatisfactory for other types of analysis, like transient stability analysis, in the presence of large voltage variations. The constant current model gives a load demand that changes linearly with voltage $k_{PV} \cong 1$ and is a reasonable representation of the active power demand of a mix of resistive and motor devices. When modeling the load by a constant impedance the load power changes proportionally to the voltage squared $k_{PV} \cong k_{QV} \cong 2$ and represents some lighting loads well but does not model stiff loads at all well. To obtain a more general voltage characteristic the benefits of each of these characteristics can be combined by using the so-called *polynomial* or *ZIP model* consisting of the sum of the constant impedance (Z), constant current (I), and constant power (P) terms

$$
\begin{aligned}
P &= P_0\left[a_1\left(\frac{V}{V_0}\right)^2 + a_2\left(\frac{V}{V_0}\right) + a_3\right] \\
Q &= Q_0\left[a_4\left(\frac{V}{V_0}\right)^2 + a_5\left(\frac{V}{V_0}\right) + a_6\right]
\end{aligned}
\tag{3.168}
$$

where V_0, P_0, and Q_0 are normally taken as the values at the initial operating conditions. The parameters of this polynomial model are the coefficients (a_1 to a_6) and the power factor of the load.

In the absence of any detailed information on load composition, the active power is usually represented by the constant current model, while the reactive power is represented by a constant impedance.

3.4.4.2 Exponential Load Model

In this model the power is related to the voltage by

$$
P = P_0\left(\frac{V}{V_0}\right)^{n_p} \quad \text{and} \quad Q = Q_0\left(\frac{V}{V_0}\right)^{n_q} \tag{3.169}
$$

where n_p and n_q are the parameters of the model. Note that by setting the parameters to 0, 1, and 2, the load can be represented by constant power, constant current, or constant impedance, respectively.

The slope of the characteristics given by Eq. (3.169) depends on the parameters n_p and n_q. By linearizing these characteristics it can be shown that n_p and n_q are equal to the voltage sensitivities given by Eq. (3.158), that is $n_p = k_{PV}$ and $n_q = k_{QV}$.

3.4.4.3 Piecewise Approximation

None of the models described so far will correctly model the rapid drop in load that occurs when the voltage drops below about 0.7 pu. This can be remedied by using a two-tier representation with the exponential, or polynomial, model being used for voltages close to rated and the constant impedance model being used at voltages below (0.3–0.7) pu. Figure 3.37 shows an example of such an approximation which gives similar characteristics to those shown in Figure 3.35.

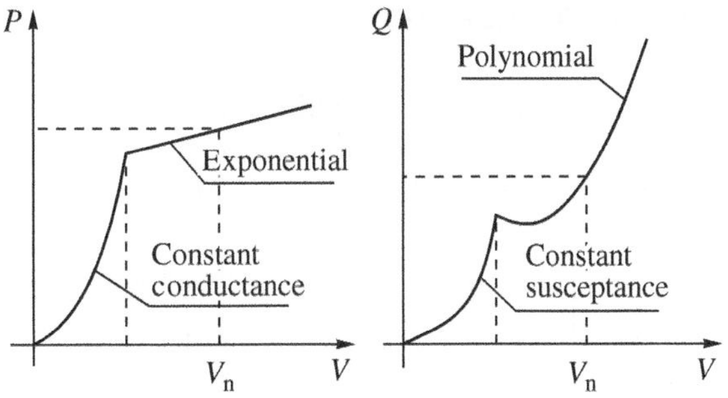

Figure 3.37 Example of a two-tier approximation of the voltage characteristics.

Table 3.3 Typical load model parameters (IEEE 1993).

Type of load	Power factor	k_{PV}	k_{QV}	k_{Pf}	k_{Qf}
Residential	0.87–0.99	0.9–1.7	2.4–3.1	0.7–1	−1.3 to −2.3
Commercial	0.85–0.9	0.5–0.8	2.4–2.5	1.2–1.7	−0.9 to −1.6
Industrial	0.8–0.9	0.1–1.8	0.6–2.2	−0.3–2.9	0.6–1.8

3.4.4.4 Frequency-dependent Load Model

Frequency dependence is usually represented by multiplying either a polynomial or an exponential load model by a factor $[1 + a_f(f - f_0)]$, where f is the actual frequency, f_0 is the rated frequency, and a_f is the model frequency sensitivity parameter. Using the exponential model, this gives

$$P = P(V)\left[1 + k_{Pf}\frac{\Delta f}{f_0}\right]; \qquad Q = Q(V)\left[1 + k_{Qf}\frac{\Delta f}{f_0})\right] \tag{3.170}$$

where $P(V)$ and $Q(V)$ represent any type of the voltage characteristic and k_{Pf} and k_{Qf} are the frequency sensitivity parameters, $\Delta f = f - f_0$.

3.4.4.5 Test Results

A number of papers have been published that describe the results of field tests used to identify the parameter values for different load models; a survey of such papers can be found in Concordia and Ihara (1982) and Vaahedi et al. (1987). An alternative to having to conduct field tests is to use a *component-based approach* (IEEE 1993) where the composite load model is constructed by aggregating individual components. With this technique the component characteristics are determined for particular classes of loads, for example residential, commercial, or industrial, either by theoretical analysis or laboratory experiment. The composite load model is then constructed by aggregating the fraction of the load consumed by each particular class of load. This approach has the advantage of not requiring field measurements and of being adaptable to different systems and conditions. Table 3.3 shows some typical voltage and frequency sensitivity coefficients obtained in this way.

3.5 Network Equations

All electrical networks consist of interlinked transmission lines and transformers, each of which can be modeled by the π-equivalent circuits described in Sections 3.1 and 3.2. These individual models are combined to model the whole network by forming the *nodal network equation*

$$\begin{bmatrix} \underline{I}_1 \\ \vdots \\ \underline{I}_i \\ \vdots \\ \underline{I}_N \end{bmatrix} = \begin{bmatrix} \underline{Y}_{11} & \cdots & \underline{Y}_{1i} & \cdots & \underline{Y}_{1N} \\ \vdots & \ddots & \vdots & & \vdots \\ \underline{Y}_{i1} & \cdots & \underline{Y}_{ii} & \cdots & \underline{Y}_{iN} \\ \vdots & & \vdots & & \vdots \\ \underline{Y}_{N1} & \cdots & \underline{Y}_{Ni} & \cdots & \underline{Y}_{NN} \end{bmatrix} \begin{bmatrix} \underline{V}_1 \\ \vdots \\ \underline{V}_i \\ \vdots \\ \underline{V}_N \end{bmatrix} \quad \text{or} \quad \underline{\boldsymbol{I}} = \underline{\boldsymbol{Y}}\,\underline{\boldsymbol{V}} \tag{3.171}$$

The suffixes i and j represent node numbers so that $\underline{V}_i$ is the voltage at node i and $\underline{I}_i$ is the *current injection* at node i and is equal to the algebraic sum of the currents in all the branches that terminate on node i, $\underline{Y}_{ij}$ is the *mutual admittance* between nodes i and j and is equal to the negative of the branch series admittance $\underline{Y}_{ij}$ that links nodes i and j, $\underline{Y}_{ii} = \sum_{i=1}^{N} \underline{Y}_{ij}$ is the *self-admittance* of node i and is equal to the sum of all the admittances terminating on node i (including any shunt admittance $\underline{Y}_{i0}$), and N is the number of nodes in the network.

The matrix $\underline{\boldsymbol{Y}}$ is called *the nodal admittance matrix.* Within this matrix, the algebraic sum of all the elements in any row i is equal to the shunt admittance Y_{i0} connecting node i to the reference node (ground), that is $\underline{Y}_{i0} = \sum_{j=1}^{N} \underline{Y}_{ij}$. The nodal admittance matrix is singular if it does not have any shunt branches. In this case the sum of the elements in all rows is zero and det $\underline{\boldsymbol{Y}} = 0$. If the matrix is not singular then its inverse $\underline{\boldsymbol{Z}} = \underline{\boldsymbol{Y}}^{-1}$ exists and Eq. (3.171) can be rewritten as

$$\begin{bmatrix} \underline{V}_1 \\ \vdots \\ \underline{V}_i \\ \vdots \\ \underline{V}_N \end{bmatrix} = \begin{bmatrix} \underline{Z}_{11} & \cdots & \underline{Z}_{1i} & \cdots & \underline{Z}_{1N} \\ \vdots & \ddots & \vdots & & \vdots \\ \underline{Z}_{i1} & \cdots & \underline{Z}_{ii} & \cdots & \underline{Z}_{iN} \\ \vdots & & \vdots & & \vdots \\ \underline{Z}_{N1} & \cdots & \underline{Z}_{Ni} & \cdots & \underline{Z}_{NN} \end{bmatrix} \begin{bmatrix} \underline{I}_1 \\ \vdots \\ \underline{I}_i \\ \vdots \\ \underline{I}_N \end{bmatrix} \quad \text{or} \quad \underline{\boldsymbol{V}} = \underline{\boldsymbol{Z}}\,\underline{\boldsymbol{I}} \tag{3.172}$$

where $\underline{\boldsymbol{Z}}$ is *the nodal impedance matrix.*

Any off-diagonal element $\underline{Y}_{ij}$ is nonzero only if there is a branch linking nodes i and j. If there are L branches in the system then $k = L/N$ is the ratio of the number of branches to the number of nodes and $\underline{\boldsymbol{Y}}$ has N^2 elements, $(2k + 1)N$ of which are nonzero. The ratio of the number of nonzero elements to the total number of elements is $\alpha = (2k + 1)/N$. High-voltage transmission networks are rather sparse so that typically k has a value of between 1 and 3. Thus, for example, if a network has 100 nodes, then only (3–7)% of the elements in the admittance may be nonzero. Such a matrix is referred to as a *sparse matrix.* Power network matrices are usually sparse and, to save computer memory, only nonzero elements are stored with additional indices being used to define their position in the matrix. All matrix manipulations are performed only on the nonzero elements using *sparse matrix techniques.* The description of such techniques is beyond the scope of this book but can be found in Tewerson (1973) or Brameller, Allan, and Hamam (1976). As the principal objective of the present text is to describe power system dynamics and stability from an engineering point of view, full, formal matrix notation will be used.

For any node i, the current injection at the node can be extracted from Eq. (3.171) as

$$\underline{I}_i = \underline{Y}_{ii}\underline{V}_i + \sum_{j=1;j\neq i}^{N} \underline{Y}_{ij}\underline{V}_j \tag{3.173}$$

where the complex voltage and admittance can be generally written as $\underline{V}_i = V_i\angle\delta_i$ and $\underline{Y}_{ij} = Y_{ij}\angle\theta_{ij}$. Using polar notation, the apparent power injected at any node i is expressed as

$$\begin{aligned}\underline{S}_i = P_i + \mathrm{j}Q_i = \underline{V}_i\underline{I}_i^* &= V_i\mathrm{e}^{\mathrm{j}\delta_i}\left[Y_{ii}V_i\mathrm{e}^{-\mathrm{j}(\delta_i+\theta_{ii})} + \sum_{j=1;j\neq i}^{N} V_jY_{ij}\mathrm{e}^{-\mathrm{j}\left(\delta_j+\theta_{ij}\right)}\right]\\ &= V_i^2Y_{ii}\mathrm{e}^{-\mathrm{j}\theta_{ii}} + V_i\sum_{j=1;j\neq i}^{N} V_jY_{ij}\mathrm{e}^{\mathrm{j}\left(\delta_i-\delta_j-\theta_{ij}\right)}\end{aligned} \tag{3.174}$$

Separating the real and imaginary parts gives

$$\begin{aligned}P_i &= V_i^2Y_{ii}\cos\theta_{ii} + \sum_{j=1;j\neq i}^{N} V_iV_jY_{ij}\cos\left(\delta_i-\delta_j-\theta_{ij}\right)\\ Q_i &= -V_i^2Y_{ii}\sin\theta_{ii} + \sum_{j=1;j\neq i}^{N} V_iV_jY_{ij}\sin\left(\delta_i-\delta_j-\theta_{ij}\right)\end{aligned} \tag{3.175}$$

Alternatively the rectangular coordinate system (a, b) shown in Figure 3.38 can be used when the apparent power injected at a node is written as

$$\underline{S}_i = P_i + \mathrm{j}Q_i = \underline{V}_i\underline{I}_i^* = (V_{\mathrm{a}i} + \mathrm{j}V_{\mathrm{b}i})(I_{\mathrm{a}i} - \mathrm{j}I_{\mathrm{b}i})$$

and

$$P_i = V_{\mathrm{a}i}I_{\mathrm{a}i} + V_{\mathrm{b}i}I_{\mathrm{b}i};\ Q_i = V_{\mathrm{b}i}I_{\mathrm{a}i} - V_{\mathrm{a}i}I_{\mathrm{b}i} \tag{3.176}$$

Expressing the admittance in rectangular coordinates as $\underline{Y}_{ij} = G_{ij} + \mathrm{j}B_{ij}$, the current injection at each node i, given by Eq. (3.171), can be written as

$$\underline{I}_i = I_{\mathrm{a}i} + \mathrm{j}I_{\mathrm{b}i} = \sum_{j=1}^{N}\underline{Y}_{ij}\underline{V}_j = \sum_{j=1}^{N}\left(G_{ij} + \mathrm{j}B_{ij}\right)\left(V_{\mathrm{a}j} + \mathrm{j}V_{\mathrm{b}j}\right) \tag{3.177}$$

Separating real and imaginary parts gives

$$I_{\mathrm{a}i} = \sum_{j=1}^{N}\left(G_{ij}V_{\mathrm{a}j} - B_{ij}V_{\mathrm{b}j}\right);\ I_{\mathrm{b}i} = \sum_{j=1}^{N}\left(B_{ij}V_{\mathrm{a}j} + G_{ij}V_{\mathrm{b}j}\right) \tag{3.178}$$

With this notation the complex network Eq. (3.171) can be transformed to the real number domain to give

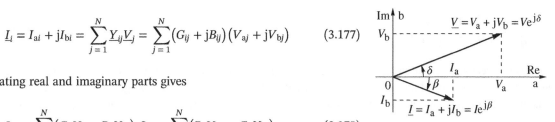

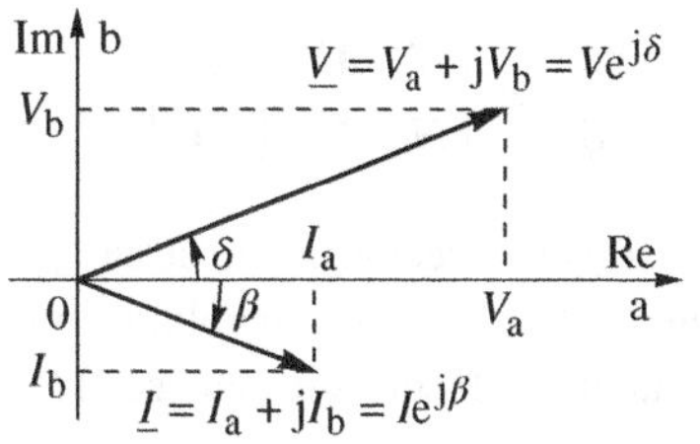

Figure 3.38 Voltage and current in the complex plane; a, b – rectangular coordinates (counter-clockwise angles are positive and clockwise angles are negative).

$$\begin{bmatrix} \boldsymbol{I}_1 \\ \vdots \\ \boldsymbol{I}_i \\ \vdots \\ \boldsymbol{I}_N \end{bmatrix} = \begin{bmatrix} \boldsymbol{Y}_{11} & \cdots & \boldsymbol{Y}_{1i} & \cdots & \boldsymbol{Y}_{1N} \\ \vdots & & \vdots & & \vdots \\ \boldsymbol{Y}_{i1} & \cdots & \boldsymbol{Y}_{ii} & \cdots & \boldsymbol{Y}_{iN} \\ \vdots & & \vdots & & \vdots \\ \boldsymbol{Y}_{N1} & \cdots & \boldsymbol{Y}_{Ni} & \cdots & \boldsymbol{Y}_{NN} \end{bmatrix} \begin{bmatrix} \boldsymbol{V}_1 \\ \vdots \\ \boldsymbol{V}_i \\ \vdots \\ \boldsymbol{V}_N \end{bmatrix} \quad \text{or} \quad \boldsymbol{I} = \boldsymbol{Y}\,\boldsymbol{V} \tag{3.179}$$

where all the elements are now real submatrices of the form

$$\boldsymbol{I}_i = \begin{bmatrix} I_{ai} \\ I_{bi} \end{bmatrix};\ \boldsymbol{V}_i = \begin{bmatrix} V_{ai} \\ V_{bi} \end{bmatrix};\ \boldsymbol{Y}_{ij} = \begin{bmatrix} G_{ij} & -B_{ij} \\ B_{ij} & G_{ij} \end{bmatrix} \tag{3.180}$$

with dimension (2 × 1) and (2 × 2), respectively.

Quite often it is convenient to "mix" the coordinate systems so that the voltages are expressed in polar coordinates as $\underline{V}_i = V_i \angle \delta_i$ while the admittances are expressed in rectangular coordinates as $\underline{Y}_{ij} = G_{ij} + \mathrm{j}B_{ij}$. Equation (3.175) then takes the form

$$\begin{aligned} P_i &= V_i^2 G_{ii} + \sum_{j=1;j\neq i}^{N} V_i V_j \left[B_{ij}\sin\left(\delta_i - \delta_j\right) + G_{ij}\cos\left(\delta_i - \delta_j\right)\right] \\ Q_i &= -V_i^2 B_{ii} + \sum_{j=1;j\neq i}^{N} V_i V_j \left[G_{ij}\sin\left(\delta_i - \delta_j\right) - B_{ij}\cos\left(\delta_i - \delta_j\right)\right] \end{aligned} \tag{3.181}$$

Because of the mixing of complex variables in both the polar and rectangular coordinate systems, the above equations are called *hybrid network equations.*

3.5.1 Linearization of Power Network Equations

The real and reactive power injection at each node is a nonlinear function of the system voltages, that is $P = P(V, \delta)$ and $Q = Q(V, \delta)$, and it is often both convenient and necessary to linearize these functions in the vicinity of the operating point. This linearization is carried out using a first-order Taylor expansion and neglecting the higher-order terms. The change in the real and reactive power injection at all the system nodes can then be written, using matrix algebra and the hybrid network equations, as

$$\begin{bmatrix} \Delta\boldsymbol{P} \\ \Delta\boldsymbol{Q} \end{bmatrix} = \begin{bmatrix} \boldsymbol{H} & \boldsymbol{M} \\ \boldsymbol{N} & \boldsymbol{K} \end{bmatrix} \begin{bmatrix} \Delta\boldsymbol{\delta} \\ \Delta\boldsymbol{V} \end{bmatrix} \tag{3.182}$$

where $\Delta\boldsymbol{P}$ is the vector of the active power changes at all the system nodes, $\Delta\boldsymbol{Q}$ is the vector of reactive power changes, $\Delta\boldsymbol{V}$ is the vector of voltage magnitude increments, and $\Delta\boldsymbol{\delta}$ is the vector of voltage angle increments. The elements of the Jacobi submatrices $\boldsymbol{H}$, $\boldsymbol{M}$, $\boldsymbol{N}$, and $\boldsymbol{K}$ are the partial derivatives of the functions in (3.181), that is

$$H_{ij} = \frac{\partial P_i}{\partial \delta_j};\ M_{ij} = \frac{\partial P_i}{\partial V_j};\ N_{ij} = \frac{\partial Q_i}{\partial \delta_j};\ K_{ij} = \frac{\partial Q_i}{\partial V_j} \tag{3.183}$$

In order to obtain simpler and more symmetrical Jacobi submatrices, Eq. (3.182) is often modified by multiplying the submatrices $\boldsymbol{M}$ and $\boldsymbol{K}$ by the nodal voltage magnitude. This then requires the voltage increments $\Delta\boldsymbol{V}$ to be divided by the voltage magnitude and Eq. (3.182) takes the form

$$\begin{bmatrix} \Delta P_1 \\ \vdots \\ \Delta P_N \\ \hdashline \Delta Q_1 \\ \vdots \\ \Delta Q_N \end{bmatrix} = \left[\begin{array}{ccc:ccc} \frac{\partial P_1}{\partial \delta_1} & \cdots & \frac{\partial P_1}{\partial \delta_N} & V_1\frac{\partial P_1}{\partial V_1} & \cdots & V_N\frac{\partial P_1}{\partial V_N} \\ \vdots & \ddots & \vdots & \vdots & \ddots & \vdots \\ \frac{\partial P_N}{\partial \delta_1} & \cdots & \frac{\partial P_N}{\partial \delta_N} & V_1\frac{\partial P_N}{\partial V_1} & \cdots & V_N\frac{\partial P_N}{\partial V_N} \\ \hdashline \frac{\partial Q_1}{\partial \delta_1} & \cdots & \frac{\partial Q_1}{\partial \delta_N} & V_1\frac{\partial Q_1}{\partial V_1} & \cdots & V_N\frac{\partial Q_1}{\partial V_N} \\ \vdots & \ddots & \vdots & \vdots & \ddots & \vdots \\ \frac{\partial Q_N}{\partial \delta_1} & \cdots & \frac{\partial Q_N}{\partial \delta_N} & V_1\frac{\partial Q_N}{\partial V_1} & \cdots & V_N\frac{\partial Q_N}{\partial V_N} \end{array}\right] \begin{bmatrix} \Delta \delta_1 \\ \vdots \\ \Delta \delta_N \\ \hdashline \Delta V_1/V_1 \\ \vdots \\ \Delta V_N/V_N \end{bmatrix} \tag{3.184}$$

or

$$\begin{bmatrix} \Delta \boldsymbol{P} \\ \Delta \boldsymbol{Q} \end{bmatrix} = \begin{bmatrix} \boldsymbol{H} & \boldsymbol{M'} \\ \boldsymbol{N} & \boldsymbol{K'} \end{bmatrix} \begin{bmatrix} \Delta \boldsymbol{\delta} \\ \Delta \boldsymbol{V}/\boldsymbol{V} \end{bmatrix}$$

with the elements of the Jacobi matrix being

$$\begin{aligned} H_{ij} &= \frac{\partial P_i}{\partial \delta_j} = -V_iV_j\left[B_{ij}\cos(\delta_i-\delta_j) - G_{ij}\sin(\delta_i-\delta_j)\right] \qquad \text{for } i \neq j \\ H_{ii} &= \frac{\partial P_i}{\partial \delta_i} = \sum_{\substack{j=1 \\ j\neq i}}^{N} V_iV_j\left[B_{ij}\cos(\delta_i-\delta_j) - G_{ij}\sin(\delta_i-\delta_j)\right] = -Q_i - V_i^2B_{ii} \end{aligned} \tag{3.185}$$

$$\begin{aligned} N_{ij} &= \frac{\partial Q_i}{\partial \delta_j} = -V_iV_j\left[G_{ij}\cos(\delta_i-\delta_j) + B_{ij}\sin(\delta_i-\delta_j)\right] \qquad \text{for } i \neq j \\ N_{ii} &= \frac{\partial Q_i}{\partial \delta_i} = \sum_{\substack{j=1 \\ j\neq i}}^{N} V_iV_j\left[G_{ij}\cos(\delta_i-\delta_j) + B_{ij}\sin(\delta_i-\delta_j)\right] = P_i - V_i^2G_{ii} \end{aligned} \tag{3.186}$$

$$\begin{aligned} M'_{ij} &= V_j\frac{\partial P_i}{\partial V_j} = V_iV_j\left[G_{ij}\cos(\delta_i-\delta_j) + B_{ij}\sin(\delta_i-\delta_j)\right] \qquad \text{for } i \neq j \\ M'_{ii} &= V_i\frac{\partial P_i}{\partial V_i} = 2V_i^2G_{ii} + \sum_{\substack{j=1 \\ j\neq i}}^{N} V_iV_j\left[G_{ij}\cos(\delta_i-\delta_j) + B_{ij}\sin(\delta_i-\delta_j)\right] = P_i + V_i^2G_{ii} \end{aligned} \tag{3.187}$$

$$\begin{aligned} K'_{ij} &= V_j\frac{\partial Q_i}{\partial V_j} = V_iV_j\left[G_{ij}\sin(\delta_i-\delta_j) - B_{ij}\cos(\delta_i-\delta_j)\right] \qquad \text{for } i \neq j \\ K'_{ii} &= V_i\frac{\partial Q_i}{\partial V_i} = -2V_i^2B_{ii} + \sum_{\substack{j=1 \\ j\neq i}}^{N} V_iV_j\left[G_{ij}\sin(\delta_i-\delta_j) - B_{ij}\cos(\delta_i-\delta_j)\right] = Q_i - V_i^2B_{ii} \end{aligned} \tag{3.188}$$

Note that Eqs. (3.185) and (3.188) show an important property of matrices $\boldsymbol{H}$ and $\boldsymbol{N}$. The diagonal elements of those matrices are equal to the negative sum of all the off-diagonal elements. Consequently, the sum of all the elements in a row is equal to zero

$$H_{ii} = \sum_{j\neq i}^{N} H_{ij} \quad \text{and} \quad \sum_{j=1}^{N} H_{ij} = 0 \tag{3.189}$$

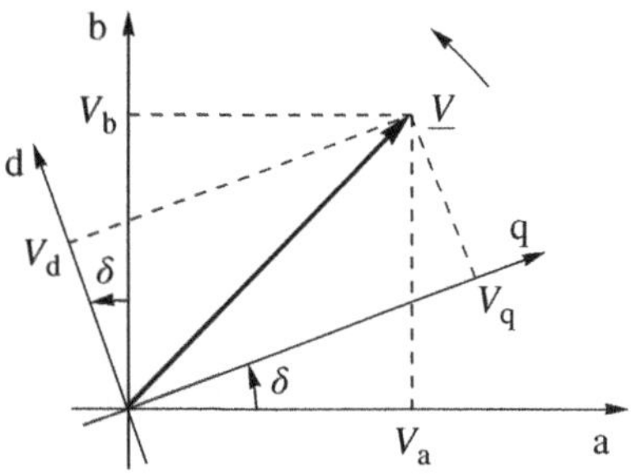

Figure 3.39 Relative position of the generator's rectangular (d, q) coordinates with respect to the network's complex (a, b) coordinates. Counter-clockwise rotation is assumed.

$$N_{ii} = \sum_{j \neq i}^{N} N_{ij} \quad \text{and} \quad \sum_{j=1}^{N} N_{ij} = 0 \tag{3.190}$$

That property generally does not hold for the submatrices $\boldsymbol{M}'$ and $\boldsymbol{K}'$, owing to the presence of shunt admittances in the network. These admittances are added to the diagonal, but not off-diagonal, elements of the matrices.

3.5.2 Change of Reference Frame

The network equations, namely Eqs. (3.171) and (3.179), are in the network's complex coordinates (a, b) while the generator equations, namely Eqs. (3.145), are in the generator's (d, q) orthogonal coordinates. Figure 3.39 shows the relative position of the two systems of coordinates. The q-axis of a given generator is shifted with respect to the network's real axis by the rotor angle δ. The relationship between the two systems of coordinates is

$$\begin{bmatrix} V_a \\ V_b \end{bmatrix} = \begin{bmatrix} -\sin\delta & \cos\delta \\ \cos\delta & \sin\delta \end{bmatrix} \begin{bmatrix} V_d \\ V_q \end{bmatrix} \quad \text{or} \quad \boldsymbol{V}_{ab} = \boldsymbol{T}\ \boldsymbol{V}_{dq} \tag{3.191}$$

The transformation matrix $\boldsymbol{T}$ is unitary as $\boldsymbol{T}^{-1} = \boldsymbol{T}$. The inverse transformation is governed by the same matrix

$$\begin{bmatrix} V_d \\ V_q \end{bmatrix} = \begin{bmatrix} -\sin\delta & \cos\delta \\ \cos\delta & \sin\delta \end{bmatrix} \begin{bmatrix} V_a \\ V_b \end{bmatrix} \quad \text{or} \quad \boldsymbol{V}_{dq} = \boldsymbol{T}\ \boldsymbol{V}_{ab} \tag{3.192}$$

All the other current and voltage phasors can be transformed in a similar way. Obviously, each generator may operate at a different rotor angle δ which gives as many (d, q) systems of coordinates as there are generators in the network.

3.6 Power Flows in Transmission Networks

Transmission networks have a meshed structure, and there are a very large number of possible parallel routes between generators and loads. The actual flow of power through each network element is determined by Kirchhoff's and Ohm's laws. Generally, the flow in each line and transformer cannot be directly controlled, as it is a function of generations and demands in all network nodes. There are only limited possibilities of directly controlling the flow of power in transmission lines, as discussed in Section 3.6.3.

Predicting future power flows plays a major role in planning the future expansion of an EPS as well as in helping to run the existing system in the best possible way. In this book two problems are discussed: (i) the power (or load) flow problem, that is calculation of power flows determining the steady state of power system; and (ii) control of power flow in the network in both the steady state and dynamic state.

Basically, a power flow solution predicts what the electrical state of the network will be when it is subject to a specified loading condition. The result of the power flow is the voltage magnitude and angle at each of the system nodes. These bus voltage magnitudes and angles are defined as the system *state variables* (or *independent variables*) as they allow all the other system quantities such as the real and reactive power flows, current flows, voltage drops, power losses, and so on to be computed.

The starting point for any power flow study is the set of generation and load data with the electrical network being described by the nodal admittance matrix $\underline{\boldsymbol{Y}}$ of Eq. (3.171). Usually, the generation and load data are given in terms of the *scheduled* real and reactive power generation at the generator nodes and the *predicted* real and reactive power demand at the load nodes rather than in terms of current injections. This means that the relationship between the data inputs (real and reactive power nodal injections) and the state variables (nodal voltage magnitudes and angles) is nonlinear, as indicated in Eqs. (3.175) or (3.181). Generator nodes have a different specification to load nodes.

3.6.1 Input Data

For the load nodes, the real and reactive power demand can be predicted and the voltage magnitude and angle must be calculated when solving network equations. In the power flow program the real and reactive power at load nodes can be kept as constant or modeled by voltage characteristics. For the generation nodes the active power and voltage magnitudes can be specified, while the voltage angle and reactive power must be calculated. To understand this, refer to Figure 2.2. Generating units have two main controllers: the turbine governor and the AVR. The turbine governor allows the reference value of the active power output to be set. In the power flow program the active power at the generation node can be kept constant or changed when the turbine control is taken into account. On the other hand, the AVR maintains the generator terminal voltage at a constant value by adjusting the generator excitation, subject to any operating limits. This means that the reactive power generation of a power station is controlled indirectly by specifying the voltage demanded at the station terminals. As the voltage magnitude is specified, there is no need for the power flow program to calculate it, and only the voltage angle needs to be found.

3.6.1.1 Power Mismatch Balanced by One Slack Node

In this variant, turbine control is not taken into account and for generation nodes the active power is kept as constant and equal to the set reference value. The input data for all the nodes in the system are summarized in Table 3.4.

For the generation nodes, referred to as *PV nodes*, the required inputs are the net active power injection (scheduled generation minus predicted local demand) and the magnitude of the bus voltage, while the voltage angle is the unknown state variable. The unknown net reactive power injection will be determined once all the system state variables are calculated. For the load nodes, referred to as *PQ-nodes*, the inputs are in the form of the predicted real and reactive power demand (negative injections) with the unknown state variables being the bus voltage magnitude and angle. In advanced power flow programs the real and reactive power at load node can be represented by voltage characteristics (Section 3.4.4).

Usually one of the generator nodes is selected as the *slack node*, also referred to as the *slack bus* or the *swing bus*. At this node the voltage angle and magnitude are specified and the unknown values of the real and reactive power injection (generation or demand) determined once the state variables at all the other nodes are calculated. Any power system imbalance will then appear as the required generation from the slack generator.

Table 3.4 Node types in the power-flow problem.

Node type	Number of nodes	Quantities Specified	Unknown state variables	Other unknown variables
Slack	Usually 1	$\delta = 0, V$	—	P, Q
PV (source)	N_G	P, V	δ	Q
PQ (load)	$N - N_G - 1$	P, Q	δ, V	—
Total	N	$2N$	$2N - N_G - 2$	$N_G + 2$

3.6.1.2 Power Mismatch Balanced by a Number of Generation Nodes

Single slack node can be used only in the cases when active power assumed for generation nodes well matches the sum of the active power at load nodes and network losses. When a power flow program is used to analyze the effect of the outage of generating units then any mismatch of the active power (caused by the analyzed outage) is located in the slack node giving an unrealistic power flow in the network. In such cases the power flow program must distribute the active power imbalance among all generating units participating in automatic frequency and generation control (Section 9.1.5). This can be simulated in the power flow program by updating the active power at generation nodes in the following way

$$P_i = P_{0i} + \alpha_i \Delta P_\Sigma \quad \text{for} \quad i = 1, \dots, N_G \tag{3.193}$$

and

$$P_i = P_{\max} \text{ if } P_i \geq P_{\max} \quad \text{or} \quad P_i = P_{\min} \text{ if } P_i = P_{\min} \tag{3.194}$$

where P_{0i} are the initial values specified for generation nodes or reference values set at the governors of the generating units (Figure 2.14 and Figure 11.25), α_i are participation factors of individual generation nodes such that $\alpha_1 + \alpha_2 + \dots + \alpha_{N_G} = 1$ and

$$\Delta P_\Sigma = \sum_{i=1}^{N_G} \Delta P_i \tag{3.195}$$

is the total power mismatch in the EPS. If generating units connected to a given generation node do not participate in covering of power imbalance then for such nodes $\alpha_i = 0$.

The power flow program must also check constraints resulting from the reactive power capability curve described in Section 3.3.4. It is obvious from Figure 3.20 that for each value of active power satisfying conditions (3.194) there are limits for reactive power. If the value of reactive power at any generation node obtained by solving a network equation exceeds those limits then the power flow program repeats calculations to find a new and closest value of the voltage magnitude V_i at such nodes for which

$$Q_{\min} \leq Q_i \geq Q_{\max} \tag{3.196}$$

Such modification of the voltage at the generation node made by the power flow program simulates reaction of the AVR of the generating unit (Figure 2.2).

The above approach is described by Lotfalian et al. (IEEE 1985). It was later adapted by Castro et al. (2012) to studies of power systems with wind farms with frequency-dependent characteristics. Section 9.8 describes how this approach can be used to find snapshots of the power flow at characteristic phases of the long-term power system dynamic response to a sudden tripping of a generating unit.

3.6.2 Calculation of Power Flows

The power flow problem described by Eqs. (3.175) or (3.181) is nonlinear and therefore must be solved iteratively. The first power flow computer programs used the *Gauss–Seidel method* because this required little computer memory. Nowadays, with increased computer speed and on-chip memory, the *Newton–Raphson method* is used almost exclusively.

These two methods are briefly described below. A detailed description of these methods is beyond the scope of this book but can be found in a number of textbooks on power system analysis, for example Gross (1986) and Grainger and Stevenson (1994).

3.6.2.1 Gauss–Seidel Method

The Gauss–Seidel method can be used to solve nonlinear algebraic equations $\boldsymbol{F}(\boldsymbol{x}) = \boldsymbol{0}$ that can be rearranged to the form of $\boldsymbol{F}(\boldsymbol{x}) = \boldsymbol{f}(\boldsymbol{x}) - \boldsymbol{x} = \boldsymbol{0}$, i.e. $\boldsymbol{x} = \boldsymbol{f}(\boldsymbol{x})$. The following is an iterative formula

$$\boldsymbol{x}_{l+1} = \boldsymbol{f}(\boldsymbol{x}_l) \tag{3.197}$$

where l is the number of the next approximation of the solution of equations. Beginning from the start value $\boldsymbol{x}_0$ subsequent approximations are calculated $\boldsymbol{x}_1, \boldsymbol{x}_2,...\boldsymbol{x}_l, \boldsymbol{x}_{l+1}$ until the desired accuracy is obtained

$$|\boldsymbol{x}_{l+1} - \boldsymbol{x}_l| < \varepsilon \tag{3.198}$$

where ε is an acceptable tolerance.

In the case of equations describing the EPS network, the equation

$$\underline{S}_i^* = \underline{U}_i^* \underline{I}_i = \underline{U}_i^* \underline{U}_i \underline{Y}_{ii} + \underline{U}_i^* \sum_{j \neq i} \underline{Y}_{ij} \underline{U}_j \tag{3.199}$$

is used, where $\underline{S}_i = P_i + jQ_i$ is the nodal apparent power. After dividing both sides of this equation by $\underline{U}_i^* \underline{Y}_{ii}$ the following iterative formula is obtained

$$\underline{U}_{i(l+1)} = \frac{\underline{S}_{i(l)}^*}{\underline{U}_{i(l)}^* \underline{Y}_{ii}} - \sum_{j \neq i} \frac{\underline{Y}_{ij}}{\underline{Y}_{ii}} \underline{U}_{j(l)} \tag{3.200}$$

The second component on the right side of this equation can be separated into two components

$$\underline{U}_{i(l+1)} = \frac{\underline{S}_{i(l)}^*}{\underline{U}_{i(l)}^* \underline{Y}_{ii}} - \sum_{j<i} \frac{\underline{Y}_{ij}}{\underline{Y}_{ii}} \underline{U}_{j(l+1)} - \sum_{j>i} \frac{\underline{Y}_{ij}}{\underline{Y}_{ii}} \underline{U}_{j(l)} \tag{3.201}$$

Thanks to this separation of components for nodes with numbers $j > i$, the values $\underline{U}_{j(l+1)}$ calculated in a given iteration are used instead of the values $\underline{U}_{j(l)}$ from the previous iteration. This speeds up the calculations.

The calculation algorithm based on the iterative formula (3.201) is as follows. The calculation begins with loading data and an accepted starting point. If there are no start values in the data, the nominal values of voltages and zero values of their arguments are assumed for load nodes. By using (3.201) successively for all nodes, complex voltage values are calculated and, on their basis, modules and arguments, respectively. These values are assumed as new approximations for load nodes, and only voltage arguments are modified for generating nodes. After determining the new voltage values, the reactive powers of the generating nodes are calculated. If the reactive power of the generating node is not exceeded, the recently calculated voltage values are accepted. Then, using the inequality (3.198), the accuracy is checked. If the inequality is not met, the next iteration is resumed. If the inequality is met, the currents and powers are calculated in network branches and losses of real and reactive power.

The advantage of Gauss–Seidel's method is its simplicity. The disadvantage is the slow convergence around the solution. Errors $|\boldsymbol{x}_{l+1} - \boldsymbol{x}_l|$ quickly decrease with each step at the beginning of the iterative process, when approximations are still far from solution. When the successive approximation approaches the solution, a small improvement in the accuracy is obtained in each consecutive step. As a result, a large number of iterations are required to obtain a solution with high accuracy.

3.6.2.2 Newton–Raphson Method

The Newton method assumes that $\boldsymbol{x}_{l+1}$ is the solution of the equation $\boldsymbol{F}(\boldsymbol{x}) = \boldsymbol{0}$, and $\boldsymbol{x}_l$ is its approximation. Expanding function $\boldsymbol{F}(\boldsymbol{x})$ in the Taylor series, such a $\Delta\boldsymbol{x}_{l+1} = (\boldsymbol{x}_{l+1} - \boldsymbol{x}_l)$, is looked for, at which

$$\boldsymbol{F}(\boldsymbol{x}_l) + \left(\frac{\partial \boldsymbol{F}}{\partial \boldsymbol{x}}\right)_l \Delta \boldsymbol{x}_{l+1} + \frac{1}{2}\left(\frac{\partial^2 \boldsymbol{F}}{\partial \boldsymbol{x}^2}\right)_l \Delta \boldsymbol{x}_{l+1}^2 + \ldots = \boldsymbol{0} \tag{3.202}$$

After omitting the higher-order components, one gets

$$\left(\frac{\partial \boldsymbol{F}}{\partial \boldsymbol{x}}\right)_l \Delta \boldsymbol{x}_{l+1} = -\boldsymbol{F}(\boldsymbol{x}_l) \quad \text{or} \quad \Delta \boldsymbol{x}_{l+1} = -\left(\frac{\partial \boldsymbol{F}}{\partial \boldsymbol{x}}\right)_l^{-1} \boldsymbol{F}(\boldsymbol{x}_l) \tag{3.203}$$

Therefore,

$$\boldsymbol{x}_{l+1} = \boldsymbol{x}_l + \Delta \boldsymbol{x}_{l+1} \quad \text{or} \quad \boldsymbol{x}_{l+1} = \boldsymbol{x}_l - \left(\frac{\partial \boldsymbol{F}}{\partial \boldsymbol{x}}\right)_l^{-1} \boldsymbol{F}(\boldsymbol{x}_l) \tag{3.204}$$

where $[\partial \boldsymbol{F}/\partial \boldsymbol{x}]$ is the Jacobi matrix. As can be seen, in a given step to calculate a new approximation the linear Eq. (3.204) must be solved. In the case where the number of equations is very large (as in real large-scale power systems), these equations are not solved by explicit calculation of inversion of the Jacobi matrix, but by its factorization and forward and backward substitutions using sparse matrix techniques (Duff et al. 1986; Pissanetzky 1984).

In the case of power systems the Newton method is used for solving Eqs. (3.175) or (3.181) assuming (as in Table 3.4) that for the load nodes both real and reactive powers are known, and for generation nodes only the real power. With this assumption Eq. (3.182) can be written as

$$\begin{bmatrix} \Delta \boldsymbol{P}_{\mathrm{L}} \\ \Delta \boldsymbol{P}_{\mathrm{G}} \\ \Delta \boldsymbol{Q}_{\mathrm{L}} \\ \Delta \boldsymbol{Q}_{\mathrm{G}} \end{bmatrix} = \begin{bmatrix} \boldsymbol{H}_{\mathrm{LL}} & \boldsymbol{H}_{\mathrm{LG}} & \boldsymbol{M}_{\mathrm{LL}} & \boldsymbol{M}_{\mathrm{LG}} \\ \boldsymbol{H}_{\mathrm{GL}} & \boldsymbol{H}_{\mathrm{GG}} & \boldsymbol{M}_{\mathrm{GL}} & \boldsymbol{M}_{\mathrm{GG}} \\ \boldsymbol{N}_{\mathrm{LL}} & \boldsymbol{N}_{\mathrm{LG}} & \boldsymbol{K}_{\mathrm{LL}} & \boldsymbol{K}_{\mathrm{LG}} \\ \boldsymbol{N}_{\mathrm{GL}} & \boldsymbol{N}_{\mathrm{GG}} & \boldsymbol{K}_{\mathrm{GL}} & \boldsymbol{K}_{\mathrm{GG}} \end{bmatrix} \begin{bmatrix} \Delta \boldsymbol{\delta}_{\mathrm{L}} \\ \Delta \boldsymbol{\delta}_{\mathrm{G}} \\ \Delta \boldsymbol{V}_{\mathrm{L}} \\ \Delta \boldsymbol{V}_{\mathrm{G}} \end{bmatrix} \tag{3.205}$$

where: $\Delta \boldsymbol{P}$ and $\Delta \boldsymbol{Q}$ are vectors of power increments (real and reactive) of all nodes; $\Delta \boldsymbol{V}$ and $\Delta \boldsymbol{\delta}$ are vectors of increments of nodal voltages and their arguments; and $\boldsymbol{H}$, $\boldsymbol{M}$, $\boldsymbol{N}$, and $\boldsymbol{K}$ are matrices of partial derivatives calculated according to Eqs. (3.185)–(3.188). For generation nodes, the voltage values are specified, and therefore $\Delta \boldsymbol{V}_{\mathrm{G}} = \boldsymbol{0}$. This allows us to reduce the Eq. (3.205) to the form

$$\begin{bmatrix} \Delta \boldsymbol{P}_{\mathrm{L}(l)} \\ \Delta \boldsymbol{P}_{\mathrm{G}(l)} \\ \Delta \boldsymbol{Q}_{\mathrm{L}(l)} \end{bmatrix} = \begin{bmatrix} \boldsymbol{H}_{\mathrm{LL}} & \boldsymbol{H}_{\mathrm{LG}} & \boldsymbol{M}_{\mathrm{LL}} \\ \boldsymbol{H}_{\mathrm{GL}} & \boldsymbol{H}_{\mathrm{GG}} & \boldsymbol{M}_{\mathrm{GL}} \\ \boldsymbol{N}_{\mathrm{LL}} & \boldsymbol{N}_{\mathrm{LG}} & \boldsymbol{K}_{\mathrm{LL}} \end{bmatrix} \begin{bmatrix} \Delta \boldsymbol{\delta}_{\mathrm{L}(l+1)} \\ \Delta \boldsymbol{\delta}_{\mathrm{G}(l+1)} \\ \Delta \boldsymbol{U}_{\mathrm{L}(l+1)} \end{bmatrix} \tag{3.206}$$

where suffixes (l) and $(l+1)$ have been used to obtain an iterative formula. The use of this formula is as follows. For the given values $\boldsymbol{\delta}_{\mathrm{L}(l)}$, $\boldsymbol{\delta}_{\mathrm{G}(l)}$, and $\boldsymbol{U}_{\mathrm{L}(l)}$ the following vectors $\boldsymbol{P}_{\mathrm{L}(l)}$, $\boldsymbol{P}_{\mathrm{G}(l)}$, and $\boldsymbol{Q}_{\mathrm{L}(l)}$ are calculated and the mismatches $\Delta \boldsymbol{P}_{\mathrm{L}(l)}$, $\Delta \boldsymbol{P}_{\mathrm{G}(l)}$, and $\Delta \boldsymbol{Q}_{\mathrm{L}(l)}$ relative to the given power values as well. Then, Eq. (3.202) is solved resulting in the required corrections of voltage angles and magnitudes $\Delta \boldsymbol{\delta}_{\mathrm{L}(l+1)}$, $\Delta \boldsymbol{\delta}_{\mathrm{G}(l+1)}$, and $\Delta \boldsymbol{V}_{\mathrm{L}(l+1)}$.

In the Newton method, the Jacobi matrix is modified in each step, i.e. the values of its elements are calculated for each iteration. There are several variants of the simplified versions of this method. The most widespread is the Newton–Raphson method, in which the Jacobi matrix is calculated once at the beginning of the iterative process or is modified every few steps. A decoupled method is also used, in which the Eq. (3.202) is replaced by two form equations

$$\begin{bmatrix} \Delta \boldsymbol{P}_{\mathrm{L}(l)} \\ \Delta \boldsymbol{P}_{\mathrm{G}(l)} \end{bmatrix} \cong \begin{bmatrix} \boldsymbol{H}_{\mathrm{LL}} & \boldsymbol{H}_{\mathrm{LG}} \\ \boldsymbol{H}_{\mathrm{GL}} & \boldsymbol{H}_{\mathrm{GG}} \end{bmatrix} \begin{bmatrix} \Delta \boldsymbol{\delta}_{\mathrm{L}(l+1)} \\ \Delta \boldsymbol{\delta}_{\mathrm{G}(l+1)} \end{bmatrix} \quad \text{and} \quad [\Delta \boldsymbol{Q}_{\mathrm{L}(l)}] = [\boldsymbol{K}_{\mathrm{LL}}]\,[\Delta \boldsymbol{U}_{\mathrm{L}(l+1)}] \tag{3.207}$$

The use of the decoupled method is justified by the fact that the active power is influenced most by the voltage arguments, and the reactive power of the voltage modules.

3.6.3 Control of Power Flows

Real and reactive power flows in a network can be modified to some extent by using controllable network elements without changing the overall generation and demand pattern. Generation control is discussed in Chapter 9.

Figure 1.4 and the simplified power flow (1.8) and (1.9) show that the flow of real and reactive power through a network element (i.e. a line or a transformer) is mainly a function of:

- voltage magnitudes at both ends of the element;
- the load (or power) angle, that is the difference between the terminal voltage angles;
- the series reactance of the element.

Real power and reactive power are two strongly connected quantities in AC transmission networks. However, as discussed in Section 1.3, active power flow is mainly affected by the load angle, while reactive power flow is mainly affected by the voltage magnitudes. Hence reactive power flow control is executed by changing voltage magnitudes through: (i) changing generator voltages; (ii) changing transformation ratios; and (iii) changing reactive power consumed/generated by reactive power compensation elements, as discussed in Section 2.5.3. As reactive power cannot be transmitted over long distances (Section 3.1.2), reactive power control is a local problem.

Control of active power flows is executed using the remaining two quantities, for example the load angle and the series reactance. Such regulation can be implemented using: (i) phase shifting transformers (Section 2.5.2); (ii) series compensators such as SSSCs; and (iii) series FACTS devices such as UPFCs or TCPARs (Section 2.5.4). The control of the load angle using a phase shifting transformers or a FACTS device, such as UPFC or TCPAR, is discussed below.

The dependence of active power flow on the load angle is illustrated in Figure 1.4. It was shown that the load angle can change an active power flow over a wide range from negative to positive values. The way that a change in the load angle can be enforced is illustrated in Figure 3.40. To simplify considerations, a case is considered of two parallel transmission lines I and II with identical parameters. The terminal voltages are $\underline{V}_i$ and $\underline{V}_j$ and the load angle (i.e. the difference between the voltage angles) is δ.

The phasor diagram for line I is shown in Figure 3.40b. The line current is $\underline{I}_{\mathrm{I}}$, while the active power flow is given by $P_{\mathrm{I}} = (V_i V_j / X) \sin \delta$. The phasor diagram for line II is shown in Figure 3.40c. The line has an installed quadrature booster, and the terminal voltages, the same as for line I and II, are equal to $\underline{V}_i$ and $\underline{V}_j$. The diagram shows that booster voltage $\Delta\underline{V}_k$, in quadrature with $\underline{V}_i$, is added to $\underline{V}_i$. Consequently, the voltage at the beginning of line II but after the quadrature booster is $\underline{V}_k = \underline{V}_i + \Delta\underline{V}_k$. The load angle for line II is $(\delta + \Delta\delta)$, while the line current is $\underline{I}_{\mathrm{II}}$ and the active power flow is given by $P_{\mathrm{II}} = (V_k V_j / X) \sin(\delta + \Delta\delta)$. As $(\delta + \Delta\delta) > \delta$, power P_{II} is greater than P_{I}. The total power entering both lines is $P = P_{\mathrm{I}} + P_{\mathrm{II}}$.

Changing the booster voltage $\Delta\underline{V}_k$ causes a change in the load angle $\Delta\delta$ and therefore also a change in the line flow P_{II}. The booster voltage $\Delta\underline{V}_k$ can be controlled from negative to positive values. When $\Delta\underline{V}_k$ is negative, the load angle is decreased and therefore the flow P_{II} is reduced. When $\Delta\underline{V}_k = 0$, that is the quadrature booster is not acting, the power entering both lines is divided equally between them, that is $P_{\mathrm{I}} = P_{\mathrm{II}} = P/2$. The conclusion is that changing the booster voltage

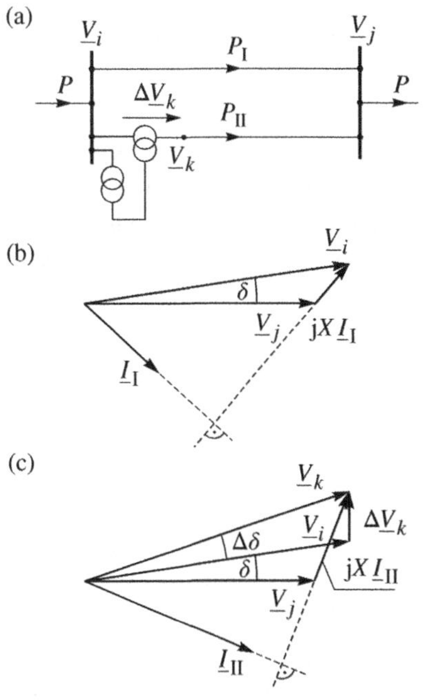

Figure 3.40 Controlling the active power flow by changing the load angle: (a) circuit diagram; (b) phasor diagram for line I; (c) phasor diagram for line II.

$\Delta \underline{V}_k$ changes proportions between power flows P_{I} and P_{II} in parallel lines. Obviously, the booster transformer cannot change the total power flow in the network.

The discussed principle of power flow control in parallel lines can also be used for flow control in more complicated network configurations, and in particular in parallel transmission corridors in meshed networks. Usually, quadrature boosters are used for:

- increasing loading on under loaded, or decreasing loading on overloaded, parallel transmission corridors;
- elimination of circulating power in meshed networks;
- changing import (or export) direction in interconnected power systems;
- prevention of unwanted loop flows entering subsystems in interconnected power systems.

Each of these options will now be briefly discussed.

A transmission network is usually meshed, and parallel transmission corridors are loaded in inverse proportion to the reactances of the corridors, assuming no quadrature boosters are present. Hence it may occur that short transmission links are overloaded, while long ones are under loaded. If that happens, a desired loading of lines can be enforced using quadrature boosters. Reducing the flow on overloaded lines and increasing the flow on underloaded lines has the effect of increasing a transfer capacity between areas.

Large interconnected networks may suffer from *circulation of power* between subsystems when power enters a subsystem through one transmission corridor and returns through another. This means that there may be large power transfers between regions while net power exchanges, when circulating power is taken away, are small. Quadrature boosters may eliminate or significantly reduce circulation of power. This usually results in reduced load on some transmission corridors at the cost of a small increase in transmission losses.

Figure 3.41 illustrates the application of a quadrature booster to a change of import (or export) direction in an interconnected system. It is assumed that system A has a surplus of power and exports it to systems B and C. The operator of system A would like to enforce an increased export to system B at the cost of reducing export to system C. This can be achieved by installing a quadrature booster or a FACTS device in one of the tie-lines, say line A-B. The total export from an area is equal to the difference between generation and demand, that is the total export from system A is $(P_{\mathrm{AB}} + P_{\mathrm{AC}}) = \left(P_{\mathrm{T}}^{\mathrm{A}} - P_{\mathrm{L}}^{\mathrm{A}}\right)$. Obviously, no quadrature booster or FACTS device can change the total value of export from an area. The device only changes the distribution among imported areas, that is it changes the proportions between P_{AB} and P_{AC} while keeping the total $(P_{\mathrm{AB}} + P_{\mathrm{AC}})$ constant. It should be appreciated that installation of a quadrature booster has to be agreed between all partners. Unilateral action of one system operator against the will of other system operators may be futile since the other system operators might install

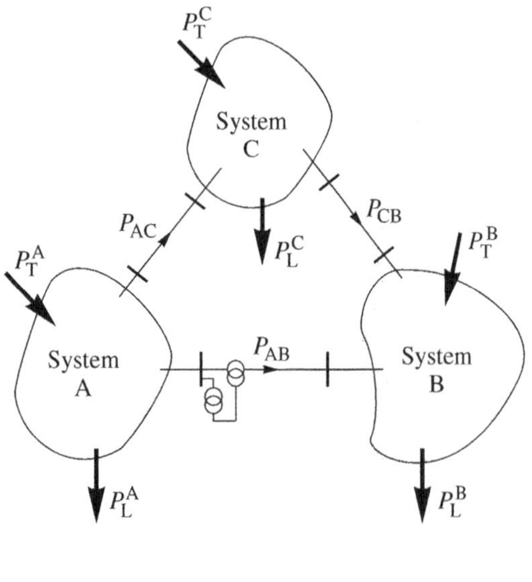

Figure 3.41 Power exchanges between three control areas; P_{T} – power generation, P_{L} – power demand.

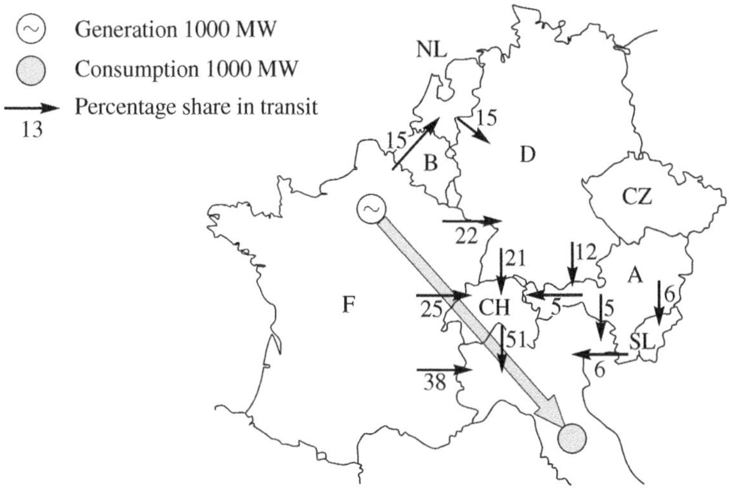

Figure 3.42 Percentage shares through different transit routes for a trade from northern France to Italy (Haubrich and Fritz 1999). *Source:* Reproduced by permission of H.J. Haubrich.

a similar quadrature booster acting in the opposite direction with the net effect of no change of flows and at a significant cost of installation of the devices.

Section 2.1 briefly discusses the recent liberalization of electricity markets. One of the effects of liberalization was a significant increase in cross-border (or interarea) trades in interconnected networks. The problem with cross-border trades is that a trade does not travel along an agreed "contract path" between a seller and a buyer but flows over many parallel routes, as discussed earlier in this section. The flows outside the agreed contract path are referred to as *parallel flows*, or *loop flows*. For example, Figure 3.42 shows the different routes through which an assumed 1000 MW trade between northern France and Italy would flow (Haubrich and Fritz 1999). Only 38% of the power would flow directly from France to Italy; the remaining 62% would flow through different parallel routes loading the transit networks. Note that 15% of the power would even flow in a round-about way via Belgium and the Netherlands.

Parallel flows did not cause major problems before 1990, as interarea exchanges were usually agreed well in advance by system operators and were relatively small. Since 1990, interarea trades have not only increased significantly in volume but also started to be arranged by independent agents, rather than system operators. Consequently, system operators often find their networks loaded with power transfers they have little idea about as they were not notified about the trades, causing loop flows. Such a situation endangers secure power system operation and actually led to a few blackouts that were just avoided in Belgium in the late 1990s. In recent years the situation was made worse by increased penetration of renewable generation, mostly wind. Wind is an intermittent energy source and the actual wind generation may be different from the one predicted a day ahead. A changed wind generation may strongly affect power flows in a network, endangering secure operation. For example, a significant change in actual wind generation was one of the contributing factors to a widespread disturbance leading to the shedding of 17 GW of load in a UCTE network in November 2006 (UCTE 2007a).

System operators can prevent loop flows from entering their systems by installing phase shifting transformers in tie-lines. This solution has been implemented by European countries strongly affected by loop flows, such as Belgium, Switzerland, Poland, the Czech Republic, and Slovenia. Obviously, such a hardware solution is quite expensive and a more cost-effective solution would be to notify all system operators in the interconnected network about all trades in the system and paying compensation for the utilization of transit networks. However, such a harmonization of arrangements is quite difficult to achieve in a multinational network, owing to political and institutional obstacles (Bialek 2007).

Part II

Introduction to Power System Dynamics

4

Electromagnetic Phenomena

Chapter 1 explains how the different types of power system dynamics can be categorized according to their timescale. It also identifies the fastest dynamics of interest to this book to be those associated with the electromagnetic interactions that occur within the generator immediately after a disturbance appears on the system. These dynamics lead to the generation of high currents and torques inside the generator and typically have a timescale of several milliseconds. Over this time period the inertia of the turbine and generator is sufficient to prevent any significant change in rotor speed so that the speed of the rotor can be assumed constant. Later chapters consider longer timescale electromechanical dynamics when the effect of rotor speed changes must be included.

To help understand how the fault currents and torques are produced this chapter uses basic physical laws to explain the electromagnetic interactions taking place within the generator. Although this approach requires some simplifications to be made, it does allow a physical, qualitative approach to be adopted. These explanations are then used to produce equations that quantify the currents and torques in a generator with a uniform air gap. Unfortunately complications arise if the air gap is nonuniform, for example in the salient-pole generator. However, the currents, and torques, in such generators are produced by exactly the same mechanism as in a generator with a uniform air gap, only their effects are a little more difficult to quantify, the details of which are best left to the more advanced analysis techniques presented in Chapter 11. Nevertheless, the way in which the current and torque expressions are modified in the salient-pole generator is discussed at the end of each section.

The chapter opens by considering the basic principles with regard to flux linkage and its application to short circuits in single-phase circuits. These principles are then used to study short-circuit effects in the synchronous generator where the interactions between the stator and rotor circuits must be included. Two types of fault are considered: first of all, a three-phase short circuit on the generator terminals and, secondly, a phase-to-phase short circuit. Fortunately, terminal three-phase short circuits are rare but, should one occur, the generator must be able to withstand the high currents and forces produced and remain intact. From a theoretical point of view, the analysis of such a fault allows a number of important terms and parameters to be introduced and quantified. A more common type of fault is the phase-to-phase fault. This type of fault introduces asymmetry into the problem and is considered later in the chapter. Finally the chapter analyses the currents and torques developed when a generator is synchronized to the grid.

4.1 Fundamentals

4.1.1 Swing Equation

Section 2.3 describes the constructional features of a turbine and explains how a multi-stage turbine drives the generator rotor through a common drive shaft. A diagram of a multi-stage turbine consisting of high-pressure, intermediate-pressure, and low-pressure stages is shown in Figure 4.1, where each turbine stage contributes a

Power System Dynamics: Stability and Control, Third Edition. Jan Machowski, Zbigniew Lubosny, Janusz W. Bialek and James R. Bumby.

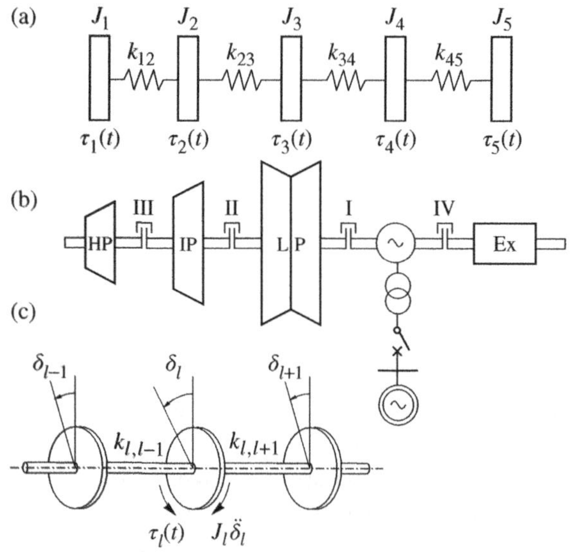

Figure 4.1 Generating unit as an oscillating system: (a) division of the rotor mass into individual sections; (b) schematic diagram; (c) torsional displacement. HP, IP, LP – high-pressure, intermediate-pressure, and low-pressure section of the turbine; G – generator; Ex – rotating exciter; J – moment of inertia of individual sections; τ – external torque acting on a mass; k – stiffness of a shaft section; δ – angular displacement of a mass; I, II, III, IV – shaft couplers.

proportion of the total mechanical driving torque. The drive system can be modeled by a series of rotational masses, to represent the inertia of each turbine stage, connected together by springs, to represent the torsional stiffness of the drive shaft and coupling between the stages. Such a model can be used to compute the torsional natural frequencies of the drive system and can also be used in a detailed computer simulation to obtain information on the actual shaft torques that occur following a major fault or disturbance. One of the natural frequencies of the turbine/generator drive system will be at 0 Hz and represents free-body rotation where the turbine and generator inertias move together with no relative displacement of the individual rotor masses. When connected to the EPS this free-body rotation will appear as a low-frequency oscillation of typically (1–2) Hz. It is this free-body rotation that is addressed in this section.

When considering free-body rotation, the shaft can be assumed to be rigid when the total inertia of the rotor J is simply the sum of the individual inertias. Any unbalanced torque acting on the rotor will result in the acceleration or deceleration of the rotor as a complete unit, according to Newton's second law

$$J\frac{\mathrm{d}\omega_{\mathrm{m}}}{\mathrm{d}t} + D_{\mathrm{d}}\,\omega_{\mathrm{m}} = \tau_{\mathrm{t}} - \tau_{\mathrm{e}} \tag{4.1}$$

where J is the total moment of inertia of the turbine and generator rotor ($\mathrm{kg \cdot m^2}$), ω_{m} is the rotor shaft velocity (mechanical rad/s), τ_{t} is the torque produced by the turbine ($\mathrm{N \cdot m}$), τ_{e} is the counteracting electromagnetic torque, and D_{d} is the damping-torque coefficient ($\mathrm{N \cdot m \cdot s}$) and accounts for the mechanical rotational loss due to windage and friction.

Although the turbine torque τ_{t} changes relatively slowly, owing to the long thermal time constants associated with the boiler and turbine, the electromagnetic torque τ_{e} may change its value almost instantaneously. In the steady state the rotor angular speed is synchronous speed ω_{sm} while the turbine torque τ_{t} is equal to the sum of the electromagnetic torque τ_{e} and the damping (or rotational loss) torque $D_{\mathrm{d}}\omega_{\mathrm{sm}}$

$$\tau_{\mathrm{t}} = \tau_{\mathrm{e}} + D_{\mathrm{d}}\omega_{\mathrm{sm}} \quad \text{or} \quad \tau_{\mathrm{m}} = \tau_{\mathrm{t}} - D_{\mathrm{d}}\omega_{\mathrm{sm}} = \tau_{\mathrm{e}} \tag{4.2}$$

where τ_m is the net mechanical shaft torque, that is the turbine torque less the rotational losses at $\omega_m = \omega_{sm}$. It is this torque that is converted into electromagnetic torque. If, owing to some disturbance, $\tau_m > \tau_e$ then the rotor accelerates; if $\tau_m < \tau_e$ then it decelerates.

In Section 3.3 the rotor position with respect to a synchronously rotating reference axis is defined by the rotor, or power, angle δ. The rotor velocity can therefore be expressed as

$$\omega_m = \omega_{sm} + \Delta\omega_m = \omega_{sm} + \frac{d\delta_m}{dt} \tag{4.3}$$

where δ_m is the rotor angle expressed in mechanical radians and $\Delta\omega_m = d\delta_m/dt$ is the *speed deviation* in mechanical radians per second.

Substituting Eq. (4.3) into (4.1) gives

$$J\frac{d^2\delta_m}{dt^2} + D_d\left(\omega_{sm} + \frac{d\delta_m}{dt}\right) = \tau_t - \tau_e \quad \text{or} \quad J\frac{d^2\delta_m}{dt^2} + D_d\frac{d\delta_m}{dt} = \tau_m - \tau_e \tag{4.4}$$

Multiplying through by the rotor synchronous speed ω_{sm} gives

$$J\omega_{sm}\frac{d^2\delta_m}{dt^2} + \omega_{sm}D_d\frac{d\delta_m}{dt} = \omega_{sm}\tau_m - \omega_{sm}\tau_e \tag{4.5}$$

As power is the product of angular velocity and torque, the terms on the right-hand side of this equation can be expressed in power to give

$$J\omega_{sm}\frac{d^2\delta_m}{dt^2} + \omega_{sm}D_d\frac{d\delta_m}{dt} = \frac{\omega_{sm}}{\omega_m}P_m - \frac{\omega_{sm}}{\omega_m}P_e \tag{4.6}$$

where P_m is the net shaft power input to the generator and P_e is the electrical air-gap power, both expressed in watts. During a disturbance, the speed of a synchronous machine is normally quite close to synchronous speed so that $\omega_m \cong \omega_{sm}$ and Eq. (4.6) become

$$J\omega_{sm}\frac{d^2\delta_m}{dt^2} + \omega_{sm}D_d\frac{d\delta_m}{dt} = P_m - P_e \tag{4.7}$$

The coefficient $J\omega_{sm}$ is the *angular momentum* of the rotor at synchronous speed and, when given the symbol $M_m = J\omega_{sm}$, allows Eq. (4.7) to be written as

$$M_m\frac{d^2\delta_m}{dt^2} = P_m - P_e - D_m\frac{d\delta_m}{dt} \tag{4.8}$$

where $D_m = \omega_{sm}D_d$ is the damping coefficient. Equation (4.8) is called the *swing equation* and is the fundamental equation governing the rotor dynamics.

It is common practice to express the angular momentum of the rotor in terms of a normalized *inertia constant* when all generators of a particular type will have similar "inertia" values regardless of their rating. The inertia constant is given the symbol H, defined as the stored kinetic energy in megajoules at synchronous speed divided by the machine rating S_n in MVA so that

$$H = \frac{0.5J\omega_{sm}^2}{S_n} \quad \text{and} \quad M_m = \frac{2HS_n}{\omega_{sm}} = J\omega_{sm} \tag{4.9}$$

The units of H are seconds. In effect H simply quantifies the kinetic energy of the rotor at synchronous speed in terms of the number of seconds it would take the generator to provide an equivalent amount of electrical energy when operating at a power output equal to its MVA rating.

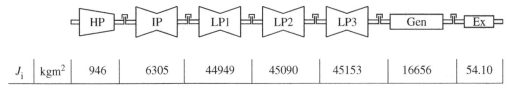

J_i	kgm^2	946	6305	44949	45090	45153	16656	54.10

Figure 4.2 Rotor masses of and their inertias for a large generating unit (abbreviations as in Figure 4.1).

Example 4.1 All rotor masses of a large generating unit with rated power $S_n = 1325$ MVA and individual inertias are shown in Figure 4.2. Total inertia of the rotor, as the sum of individual inertias, amounts to $J = 159153.1\ \text{kgm}^2$. Substituting these values into Eq. (4.9) it is obtained that:

$$H = \left(314^2 \cdot 159153.1\right) / \left(2 \cdot 1325 \cdot 10^6\right) = 5.92\ \text{s}$$

******

In Continental Europe the symbol T_m is used for *mechanical starting time* (also referred to as *mechanical time constant*), where

$$T_m = \frac{J\omega_{sm}^2}{S_n} = 2H \quad \text{and} \quad M_m = \frac{T_m S_n}{\omega_{sm}} \tag{4.10}$$

Again, the units are seconds but the physical interpretation is different. In this case, if the generator is at rest and a mechanical torque equal to S_n/ω_{sm} is suddenly applied to the turbine shaft, then the rotor will accelerate, its velocity will increase linearly, and it will take T_m seconds to reach synchronous speed ω_{sm}.

Section 3.3 shows that the power angle and angular speed can be expressed in electrical radians and electrical radians per second, respectively, rather than their mechanical equivalent, by substituting

$$\delta = \frac{\delta_m}{p/2} \quad \text{and} \quad \omega_s = \frac{\omega_{sm}}{p/2} \tag{4.11}$$

where p is the number of poles. Introducing the inertia constant and substituting Eq. (4.11) into Eq. (4.8) allows the swing equation to be written as

$$\frac{2HS_n}{\omega_s}\frac{d^2\delta}{dt^2} + D\frac{d\delta}{dt} = P_m - P_e \quad \text{or} \quad \frac{T_m S_n}{\omega_s}\frac{d^2\delta}{dt^2} + D\frac{d\delta}{dt} = P_m - P_e \tag{4.12}$$

where D, the damping coefficient, is $D = pD_m/2$. The equations in (4.12) can be rationalized by defining an *inertia coefficient* M and *damping power* P_D such that

$$M = \frac{2HS_n}{\omega_s} = \frac{T_m S_n}{\omega_s}, \quad P_D = D\frac{d\delta}{dt} \tag{4.13}$$

when the swing equation takes the common form

$$M\frac{d^2\delta}{dt^2} = P_m - P_e - P_D = P_{acc} \tag{4.14}$$

where P_{acc} is the net accelerating power. The time derivative of the rotor angle $d\delta/dt = \Delta\omega = \omega - \omega_s$ is the *rotor speed deviation* in electrical radians per second. Often it is more convenient to replace the second-order differential Eq. (4.14) by two first-order equations:

$$\begin{aligned} M\frac{d\Delta\omega}{dt} &= P_m - P_e - P_D = P_{acc} \\ \frac{d\delta}{dt} &= \Delta\omega \end{aligned} \tag{4.15}$$

It is also common power system practice to express the swing equation in a *per-unit (pu)* form. This simply means normalizing Eq. (4.14) to a common MVA base. Assuming that this base is the MVA rating of the generator, then dividing both sides of Eq. (4.14) by S_n does not modify the equation structure but all parameters are now normalized to the three-phase MVA base.

4.1.2 Law of Constant Flux Linkages

The currents and fluxes inside the generator after a fault can be analyzed using the *law of constant flux linkages*. This law, based on the principle of *conservation of energy*, states that the magnetic flux linking a closed winding cannot change instantaneously. Since the energy stored in the magnetic field is proportional to the flux linkage, this simply says that an instantaneous change in the stored energy is not possible.

The application of the law of constant flux linkages to a simple coil fed from a DC source is illustrated in Figure 4.3a. At time $t = 0$ the switch disconnects the coil from the source and, at the same time, short-circuits the coil. Before the switching occurs the coil flux linkages are $\Psi_0 = N\Phi_0 = Li_0$, where L is the coil inductance, N is the number of turns, and i_0 is the current. The law of constant flux linkages requires the coil flux linkage just before, and just after, the short circuit to be the same, that is $\Psi(t = 0^+) = \Psi(t = 0^-)$. If the coil resistance is zero, the circuit is purely inductive and the current will remain constant at a value $i(t) = i_0$, as shown by the horizontal dashed line in Figure 4.3b. Normally, the coil resistance is nonzero so that the magnetic energy stored in the circuit dissipates over a period of time and the current decays exponentially to zero with a time constant $T = L/R$. This is shown by the solid line in Figure 4.3b.

The law of constant flux linkage can now be used to analyze the way in which the generator responds to a short circuit. A natural first step is to consider the response of the simple RL circuit shown in Figure 4.4a as this circuit contains a number of features that are similar to the generator equivalent circuit described in Section 3.3. In this

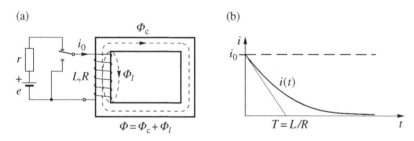

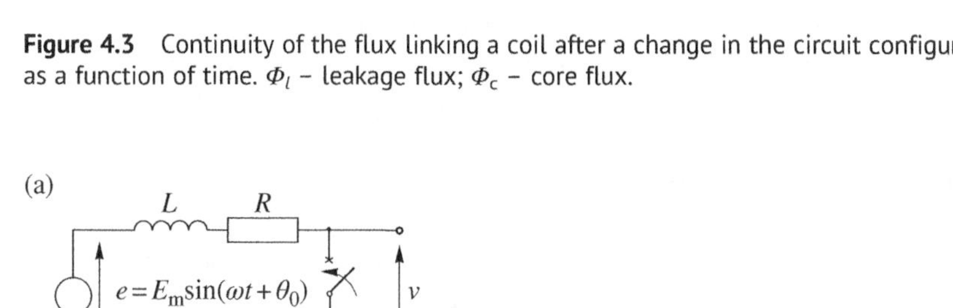

Figure 4.3 Continuity of the flux linking a coil after a change in the circuit configuration: (a) circuit diagram; (b) current as a function of time. Φ_l – leakage flux; Φ_c – core flux.

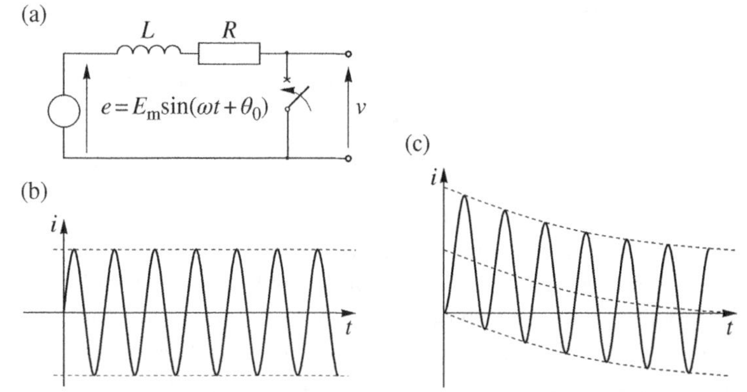

Figure 4.4 Single phase RL circuit: (a) equivalent circuit; (b) short-circuit currents with $(\theta_0 - \phi) = 0$; (c) short-circuit currents with $(\theta_0 - \phi) = -\pi/2$.

circuit the AC driving voltage is $e = E_m \sin(\omega t + \theta_0)$ and, as the circuit is assumed to be initially open-circuit, $i(0^+) = i(0^-) = 0$. If the switch is now suddenly closed, a short-circuit current $i(t)$ will flow which can be determined by solving the differential equation

$$E_m \sin(\omega t + \theta_0) = L\frac{di}{dt} + Ri \tag{4.16}$$

with the initial condition $i(0) = 0$. Solving this equation gives the current as

$$i(t) = \frac{E_m}{Z}\sin(\omega t + \theta_0 - \phi) - \frac{E_m}{Z}\sin(\theta_0 - \phi)e^{-\frac{R}{L}t} \tag{4.17}$$

where θ_0 defines the point in the AC cycle where the fault occurs, $\phi = \arctan(\omega L/R)$ is the phase angle and $Z = \sqrt{\omega^2 L^2 + R^2}$. The instant of the short-circuit is taken as the time origin.

The first term in Eq. (4.17) gives the forced response of the circuit. This alternating current is driven by the sinusoidal electromotive force (emf) and will be the current remaining when the circuit settles down into its final steady-state condition. The second term constitutes the natural response of the circuit and is referred to as the *DC offset*. The AC component has a constant magnitude equal to E_m/Z, while the initial magnitude of the DC offset depends on the point in the AC cycle where the fault occurs and decays with the time constant $T = L/R$. Figure 4.4b shows the current waveform when $(\theta_0 - \phi) = 0$ and there is no DC offset. Figure 4.4c shows the current waveform when $(\theta_0 - \phi) = -\pi/2$ and the initial DC offset is at its highest value.

The response of a synchronous generator to a short circuit is similar to that described above in the sense that it also consists of an AC component (the forced response) and a DC component (the natural response). However, Eq. (4.17) must be modified to account for the three-phase nature of the generator and the effect of a varying impedance in the emf source.

4.2 Three-phase Short Circuit on a Synchronous Generator

4.2.1 Three-phase Short Circuit with the Generator on No Load and Winding Resistance Neglected

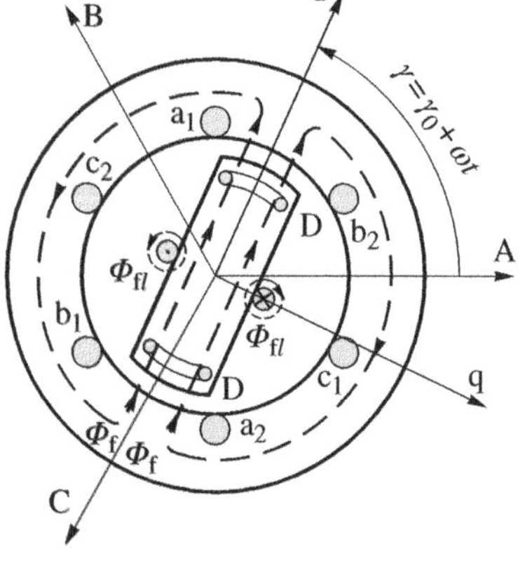

Figure 4.5 The generator and its windings.

The cross-section of a basic generator in Figure 4.5 shows the relative position of all the windings. This diagram symbolizes both the round-rotor and salient-pole generator and, compared with the cross-section in Figure 3.14, shows the rotor to have one additional winding: the rotor direct-axis (d-axis) damper winding denoted as D. At any point in time the position of the rotor is defined with reference to the axis of phase A by the angle γ. However, as a fault may occur when the rotor is at any position $\gamma = \gamma_0$ the time of the fault is taken as the time origin so that the rotor position at some time instant t after the fault is given by

$$\gamma = \gamma_0 + \omega t \tag{4.18}$$

Before the fault occurs the generator is assumed to be running at no load with the three armature phase windings open-circuited when the only flux present in the generator is the excitation flux produced by the rotor field winding. This flux links with each phase of the armature winding producing the armature flux linkages shown in Figure 3.11a and given by Eq. (3.37). When the fault occurs $\gamma = \gamma_0$ and the net flux linking each of the armature phase windings is

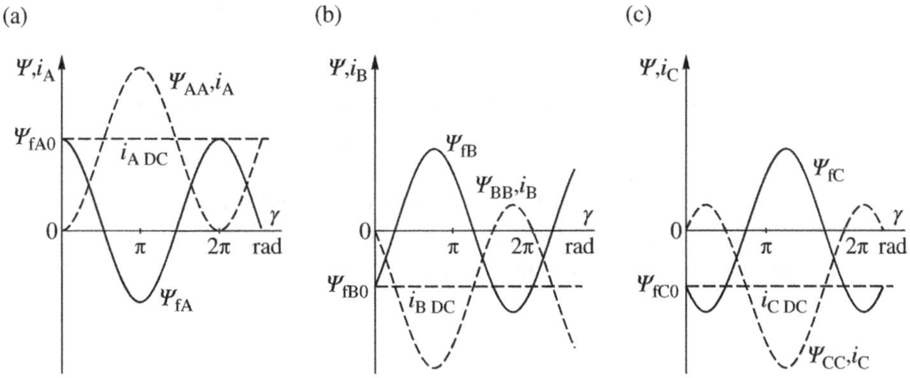

Figure 4.6 Application of the law of constant flux linkages to determine the fault currents due to a three-phase short circuit occurring at the instant $\gamma_0 = 0$: (a) phase A; (b) phase B; (c) phase C.

$$\Psi_{\text{fA0}} = \Psi_{\text{fa}} \cos \gamma_0; \quad \Psi_{\text{fB0}} = \Psi_{\text{fa}} \cos (\gamma_0 - 2\pi/3); \quad \Psi_{\text{fC0}} = \Psi_{\text{fa}} \cos (\gamma_0 - 4\pi/3) \tag{4.19}$$

where $\Psi_{\text{fa}} = N_\phi \Phi_{\text{f0}}$ is the amplitude of the excitation flux linkage of an armature phase winding before the fault occurs, Φ_{f0} is the pre-fault excitation flux per pole, and $N_\phi = k_{\text{w}}N$. In this last equation N is the number of turns on each of the armature phase windings and k_{w} is the armature winding factor. An example of these phase flux linkages is shown by the horizontal dashed lines in Figure 4.6 which is drawn assuming that the fault occurs at $\gamma_0 = 0$.

If the resistance of the armature winding is neglected then, according to the law of constant flux linkage, the value of the flux linking each phase must remain constant after the fault at the value Ψ_{fA0}, Ψ_{fB0}, and Ψ_{fC0} defined by the equations in (4.19). After the fault the rotor continues to rotate and the linkages Ψ_{fA}, Ψ_{fB}, and Ψ_{fC} continue to change sinusoidally, as shown by the solid lines in Figure 4.6 and have an equivalent effect to the emf in Figure 4.4a. To keep the total flux linkages of each of the phase windings, constant additional currents i_{A}, i_{B}, and i_{C} must be induced in the short-circuited phase windings to produce the flux linkages Ψ_{AA}, Ψ_{BB}, and Ψ_{CC} shown by the sinusoidal dashed lines in Figure 4.6. The total flux linkage of each phase winding can now be obtained by adding the two components of flux linkage together to give

$$\begin{aligned} \Psi_{\text{A}}(t) &= \Psi_{\text{AA}} + \Psi_{\text{fA}} = \Psi_{\text{fA0}} = \text{constant} \\ \Psi_{\text{B}}(t) &= \Psi_{\text{BB}} + \Psi_{\text{fB}} = \Psi_{\text{fB0}} = \text{constant} \\ \Psi_{\text{C}}(t) &= \Psi_{\text{CC}} + \Psi_{\text{fC}} = \Psi_{\text{fC0}} = \text{constant} \end{aligned} \tag{4.20}$$

which on rearranging gives the flux linkages Ψ_{AA}, Ψ_{BB}, and Ψ_{CC} as

$$\Psi_{\text{AA}} = \Psi_{\text{fA0}} - \Psi_{\text{fA}}; \quad \Psi_{\text{BB}} = \Psi_{\text{fB0}} - \Psi_{\text{fB}}; \quad \Psi_{\text{CC}} = \Psi_{\text{fC0}} - \Psi_{\text{fC}} \tag{4.21}$$

As the flux linkage is equal to the product of inductance and current, these flux linkages can now be used to obtain expressions for the phase current provided that the inductance value is known.[1] In the case of a generator with a uniform air gap the equivalent inductance of each of the phase windings L_{eq} is the same and does not depend on rotor position (Section 3.3). In this case the short-circuit phase currents $i_{\text{A}} = \Psi_{\text{AA}}/L_{\text{eq}}$, $i_{\text{B}} = \Psi_{\text{BB}}/L_{\text{eq}}$, and $i_{\text{C}} = \Psi_{\text{CC}}/L_{\text{eq}}$ have the same shape as the flux linkages Ψ_{AA}, Ψ_{BB}, and Ψ_{CC} and consist of both an AC and a DC component. The AC component currents can be expressed as

1 The inductance value required is not simply the winding self-inductance but an equivalent inductance that takes into account the way the winding is coupled with all the other windings in the generator. The importance of this coupling is examined in Sections 4.2.3 and 4.2.4 when actual values for L_{eq} are discussed.

$$i_{\mathrm{A\,AC}} = -i_{\mathrm{m}}(t)\cos(\omega t + \gamma_0); \qquad i_{\mathrm{B\,AC}} = -i_{\mathrm{m}}(t)\cos(\omega t + \gamma_0 - 2\pi/3)$$
$$i_{\mathrm{C\,AC}} = -i_{\mathrm{m}}(t)\cos(\omega t + \gamma_0 - 4\pi/3) \tag{4.22}$$

and are proportional to the flux linkages Ψ_{fA}, Ψ_{fB}, and Ψ_{fC}, while the DC component currents can be expressed as

$$i_{\mathrm{A\,DC}} = i_{\mathrm{m}}(0)\cos\gamma_0; \qquad i_{\mathrm{B\,DC}} = i_{\mathrm{m}}(0)\cos(\gamma_0 - 2\pi/3); \qquad i_{\mathrm{C\,DC}} = i_{\mathrm{m}}(0)\cos(\gamma_0 - 4\pi/3) \tag{4.23}$$

and are proportional to Ψ_{fA0}, Ψ_{fB0}, and Ψ_{fC0}. In order to generalize these two sets of equations the maximum value of the current in Eq. (4.22) has been defined as $i_{\mathrm{m}}(t)$, while in Eq. (4.23) it is $i_{\mathrm{m}}(0)$, the value of $i_{\mathrm{m}}(t)$ at the instant of the fault. When the winding resistance is neglected, $i_{\mathrm{m}}(t)$ is constant and equal to its initial value $i_{\mathrm{m}}(0)$. The net current is obtained by adding the component currents to give

$$\begin{aligned} i_{\mathrm{A}} &= -i_{\mathrm{m}}(t)\cos(\omega t + \gamma_0) + i_{\mathrm{m}}(0)\cos\gamma_0 \\ i_{\mathrm{B}} &= -i_{\mathrm{m}}(t)\cos(\omega t + \gamma_0 - 2\pi/3) + i_{\mathrm{m}}(0)\cos(\gamma_0 - 2\pi/3) \\ i_{\mathrm{C}} &= -i_{\mathrm{m}}(t)\cos(\omega t + \gamma_0 - 4\pi/3) + i_{\mathrm{m}}(0)\cos(\gamma_0 - 4\pi/3) \end{aligned} \tag{4.24}$$

where the DC offset depends on γ_0, the instant in the AC cycle when the fault occurs, and is different for each phase of the generator.

The combined effect of the AC and the DC components of the three armature phase currents is to produce an armature reaction magnetomotive force (mmf) that drives an armature reaction flux across the air gap to link with the rotor windings and induce currents in them. Consider first the effect of the AC phase currents $i_{\mathrm{A\,AC}}$, $i_{\mathrm{B\,AC}}$, and $i_{\mathrm{C\,AC}}$, as shown in the upper row of diagrams in Figure 4.7. These AC phase currents, Figure 4.7a, produce an AC armature reaction mmf $F_{\mathrm{a\,AC}}$ that behaves in a similar way to the steady-state armature reaction mmf F_{a} discussed in Section 3.3. Recall that in the steady state F_{a} rotates at the same speed as the rotor (and the excitation mmf F_{f}) demagnetizes the machine and produces a torque proportional to the sine of the angle between F_{a} and F_{f}. During the short circuit, the developed electrical torque and power are both zero so that the angle between $F_{\mathrm{a\,AC}}$ and F_{f} is

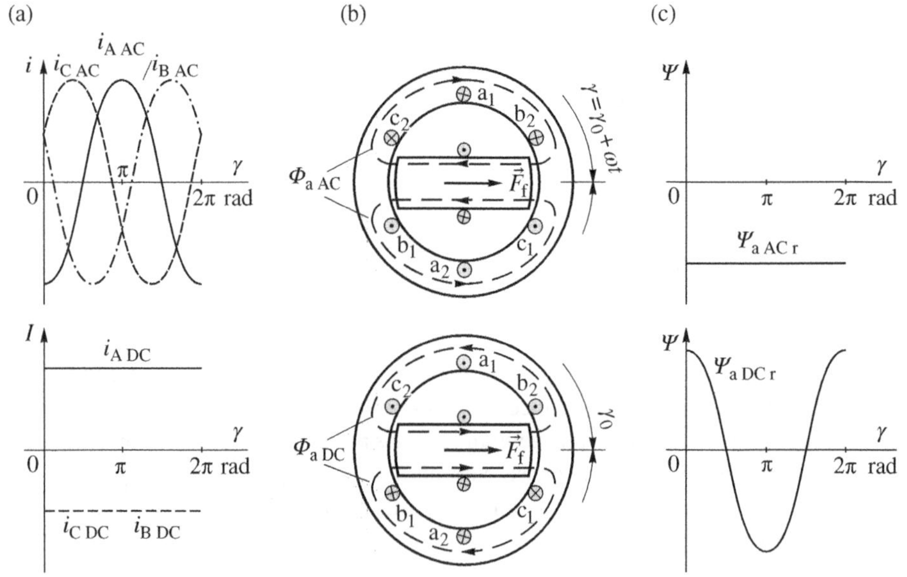

Figure 4.7 Effect of the AC component (upper row) and the DC component (lower row) of the armature currents: (a) AC and DC component of the currents; (b) path of the armature reaction flux (rotating for the AC component, stationary for the DC component); (c) linkages with the rotor windings. Position shown at the instant of fault $\gamma_0 = 0$.

180°; $F_{a\,AC}$ directly opposes F_f, and the flux $\Phi_{a\,AC}$, produced by $F_{a\,AC}$, takes the path shown in Figure 4.7b. Consequently, the rotor flux linkages $\Psi_{a\,AC\,r}$ produced by $\Phi_{a\,AC}$ are constant and negative (with respect to the excitation linkages), as shown by the solid line in Figure 4.7c. Here the subscript *r* has been added to emphasize that $\Psi_{a\,AC\,r}$ is the rotor flux linkage produced by the armature flux $\Phi_{a\,AC}$.

The lower row of diagrams in Figure 4.7 shows the combined effect of the DC phase currents $i_{A\,DC}$, $i_{B\,DC}$, and $i_{C\,DC}$. These currents, Figure 4.7a, produce a stationary DC mmf $F_{a\,DC}$ which drives the stationary armature flux $\Phi_{a\,DC}$ shown in Figure 4.7b. The space direction of $F_{a\,DC}$ can be obtained by adding the component mmf's forced by currents $i_{A\,DC}$, $i_{B\,DC}$, and $i_{C\,DC}$ in a similar way as in Eq. (3.40) when the resulting mmf $F_{a\,DC}$ is found to be always directed at an angle γ_0 with respect to the A-axis. As the rotor is also at angle γ_0 at the instant of fault, $F_{a\,DC}$ is always aligned initially with the rotor d-axis and then counter-rotates with respect to it. This means that the rotor flux linkage $\Psi_{a\,DC\,r}$, produced by $\Phi_{a\,DC}$, is initially positive (magnetizes the machine) and then, as the rotor rotates, $\Psi_{a\,DC\,r}$ changes cosinusoidally, as shown in Figure 4.7c.

As the rotor field and damper windings are closed, their total flux linkage must remain unchanged immediately after the fault. Consequently, additional currents must flow in these windings to compensate the armature flux $\Psi_{a\,r} = \Psi_{a\,AC\,r} + \Psi_{a\,DC\,r}$ that links with the rotor. Figure 4.8 shows the rotor linkages equal to $-\Psi_{a\,r}$ necessary for this and the field and damper currents which must flow to set up such linkages. Both currents contain DC and AC components. Consequently, the resulting rotor flux, as seen by the armature, can be assumed to be unchanged by the fault. These linkages are shown as Ψ_{fA}, Ψ_{fB}, and Ψ_{fC} in Figure 4.6.

To summarize, a three-phase fault causes short-circuit armature currents to flow which have both an AC and a DC component. At the instant of fault both the AC and the DC armature mmf's are directed along the rotor d-axis but then, as the rotor rotates, the AC armature mmf rotates with it inducing additional DC currents in the rotor, while the DC armature mmf is stationary inducing additional AC currents in the rotor. There is always a complementary pair of stator to rotor currents of the form AC → DC *and* DC → AC. The value of the DC offset may be different for each stator phase and depends on the instant of the fault.

4.2.2 Including the Effect of Winding Resistance

A winding dissipates energy in its resistance at a rate proportional to the current squared. Consequently, the stored magnetic energy decays with time and the induced direct currents maintaining the flux linkages decay exponentially to zero at a rate determined by the circuit time constant $T = L/R$.

The DC component of the armature phase current decays with the *armature time constant* T_a determined by the equivalent inductance and resistance of the phase winding. To account for this decay the direct currents in Eq. (4.23) must be modified to

$$i_{A\,DC} = i_m(0)e^{-t/T_a}\cos\gamma_0; \qquad i_{B\,DC} = i_m(0)e^{-t/T_a}\cos(\gamma_0 - 2\pi/3); \\ i_{C\,DC} = i_m(0)e^{-t/T_a}\cos(\gamma_0 - 4\pi/3) \tag{4.25}$$

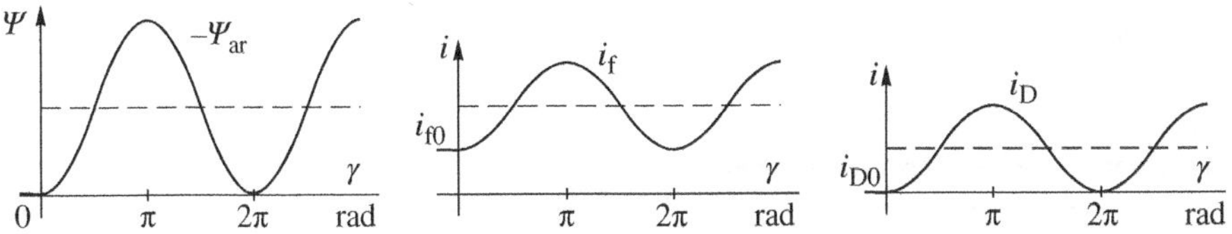

Figure 4.8 Application of the law of constant flux linkage to determine the currents flowing in the rotor windings: Ψ_r – induced flux linkages necessary in the rotor windings in order to compensate the armature reaction flux linkages; i_f – current in the field winding; i_D – current in the damper winding.

when the total phase currents become

$$\begin{aligned} i_{\mathrm{A}} &= -i_{\mathrm{m}}(t)\cos(\omega t+\gamma_0) + i_{\mathrm{m}}(0)\mathrm{e}^{-t/T_{\mathrm{a}}}\cos\gamma_0 \\ i_{\mathrm{B}} &= -i_{\mathrm{m}}(t)\cos(\omega t+\gamma_0-2\pi/3) + i_{\mathrm{m}}(0)\mathrm{e}^{-t/T_{\mathrm{a}}}\cos(\gamma_0-2\pi/3) \\ i_{\mathrm{C}} &= -i_{\mathrm{m}}(t)\cos(\omega t+\gamma_0-4\pi/3) + i_{\mathrm{m}}(0)\mathrm{e}^{-t/T_{\mathrm{a}}}\cos(\gamma_0-4\pi/3) \end{aligned} \tag{4.26}$$

These equations are identical in form to that obtained for the single-phase circuit in Eq. (4.17) with $\gamma_0=\theta_0$ and a phase angle $\phi=\pi/2$. However, in Eq. (4.17) $i_{\mathrm{m}}(t)=E_m/Z$ is constant, whereas for the synchronous generator it varies with time for reasons that will now be explained.

The currents $i_{\mathrm{A\ AC}}$, $i_{\mathrm{B\ AC}}$, and $i_{\mathrm{C\ AC}}$ induce DC currents in the rotor windings. These DC currents decay with a time constant determined by the particular rotor circuit in which they flow. Consequently, the DC component of the damper winding current decays with the damper winding time constant, called the *d-axis subtransient short-circuit time constant* T''_{d}, while the DC component of the field current decays with the field winding time constant, called the *d-axis transient short-circuit time constant* T'_{d}. Normally, the resistance of the damper winding is much higher than the resistance of the field winding so that $T''_{\mathrm{d}} < < T'_{\mathrm{d}}$ and the DC component of the damper winding current decays much faster than the DC component of the field current. As the DC rotor currents induce AC stator currents, the magnitude of the AC component of the stator current $i_{\mathrm{m}}(t)$ will decay to its steady-state value with these two time constants. Conversely, as the DC phase currents $i_{\mathrm{A\ DC}}$, $i_{\mathrm{B\ DC}}$, and $i_{\mathrm{C\ DC}}$ induce AC currents in the field and the damper windings, these AC rotor currents will also decay with the time constant T_{a}.

Table 4.1 summarizes the relationship between the relevant components of the stator and the rotor currents and their decay time constants.

Figure 4.9 shows a typical time variation of the generator currents resulting from a three-phase terminal short circuit. The diagrams constituting Figure 4.9 have been organized so that the first row shows the currents in phase A of the stator armature winding, the second row the field winding currents, and the third row the damper winding currents. Only phase A of the armature winding is shown because the currents in phases B and C are similar but shifted by $\pm 2\pi/3$. The first column in Figure 4.9 shows the DC component of the armature current and the corresponding induced rotor currents, the second column the AC component of the armature current and the corresponding induced rotor currents, while the third column shows the resultant currents in the armature, the field and the damper winding made up from their AC and DC components. The dashed lines indicate the exponential envelopes that correspond to the different time constants.

Note that the AC component of the armature current does not decay completely to zero but to a steady-state value. The field current decays to its pre-fault value i_{f0}, while the damper winding current decays completely to zero, its pre-fault value.

Now it is possible to compare the generator short-circuit armature current shown in the top diagram of Figure 4.9c with the response of a short-circuited RL circuit shown in Figure 4.4c. Both currents have a DC component that depends on the instant the short circuit is applied and that decays with a time constant determined by the resistance and inductance of the winding. However, while the AC component shown in Figure 4.4c has a

Table 4.1 Pairs of coupled stator-rotor currents.

Three-phase stator winding	Rotor windings (field and damper)	Means of energy dissipation	Time constant
DC	AC	Resistance of the armature winding	T_{a}
AC	DC	Resistance of the damper winding Resistance of the field winding	T''_{d} T'_{d}

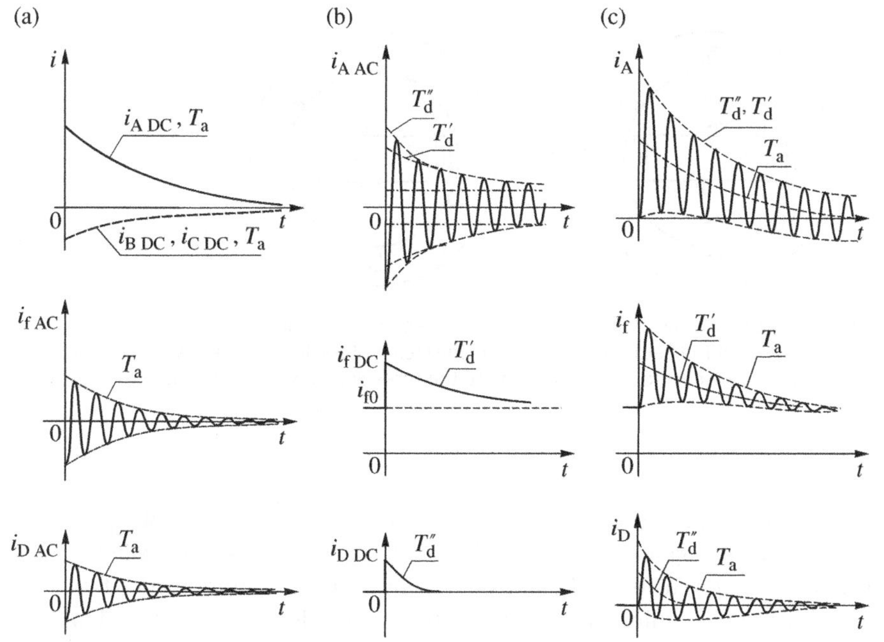

Figure 4.9 Short-circuit currents in the generator: (a) DC component of the phase current and the corresponding AC component of the field and damper winding current; (b) AC component of the current in phase A and the corresponding DC component of the field and damper winding current; (c) the resulting current in phase A, the field and the damper winding as the sum of the currents shown in (a) and (b).

constant amplitude, the amplitude of the AC component of the armature current shown in Figure 4.9c is decaying with two time constants depending on the resistance and inductance of the damper and field windings. The changing amplitude of the alternating current suggests that the internal impedance of the generator is not constant, as in Figure 4.4, but is changing with time because of interactions between the armature, field, and damper windings. This is discussed further in the next section.

If $T_a > T''_d$, the fault current in phase A for $\gamma_0 = 0$ is positive for the first few cycles and does not cross zero until the DC offset decays. This is typical for large generators. If $T'_d > T_a$ the field current i_f will always oscillate above its steady-state value i_{f0}. If $T''_d < T_a$, the damper current i_D may go negative.

4.2.3 Armature Flux Paths and the Equivalent Reactances

Section 3.3 explains how the effect of the steady-state AC armature flux on the generator performance could be accounted for in the equivalent circuit model by the voltage drop across the synchronous reactances X_d and X_q. A similar approach can be adopted during the fault period, but now the value of the armature reactance will be different as the additional currents induced in the rotor windings force the armature flux to take a different path to that in the steady state. As the additional rotor currents prevent the armature flux from entering the rotor windings, they have the effect of *screening the rotor* from these changes in armature flux.

Figure 4.10 shows three characteristic states that correspond to three different stages of rotor screening. Immediately after the fault, the current induced in both the rotor field and damper windings forces the armature reaction flux completely out of the rotor to keep the rotor flux linkages constant, Figure 4.10a, and the generator is said to be in the *subtransient state*. As energy is dissipated in the resistance of the rotor windings, the currents maintaining constant rotor flux linkages decay with time, allowing flux to enter the windings. As the rotor damper winding

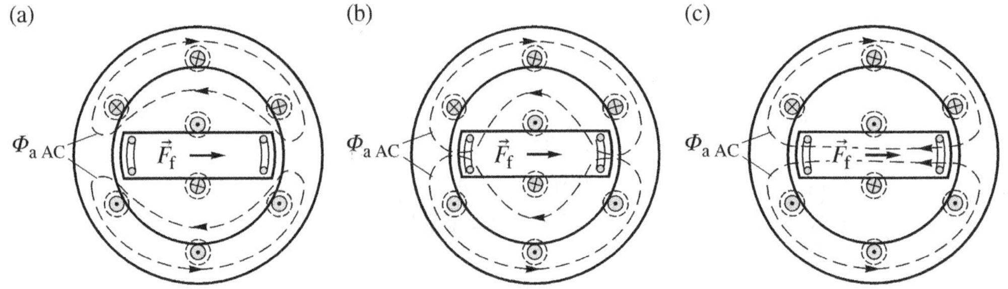

Figure 4.10 The path of the armature flux in: (a) the subtransient state (screening effect of the damper winding and the field winding); (b) the transient-state (screening effect of the field winding only); (c) the steady state. In all three cases the rotor is shown to be in the same position but the actual rotor position corresponding to the three states will be separated by a number of rotations.

resistance is the largest, the damper current is the first to decay, allowing the armature flux to enter the rotor pole face. However, it is still forced out of the field winding itself (Figure 4.10b) and the generator is said to be in the *transient state.* The field current then decays with time to its steady-state value, allowing the armature reaction flux eventually to enter the whole rotor and assume the minimum reluctance path. This steady state is illustrated in Figure 4.10c and corresponds to the flux path shown in the top diagram of Figure 4.7b.

It is convenient to analyze the dynamics of the generator separately when it is in the subtransient, transient, and steady state. This is accomplished by assigning a different equivalent circuit to the generator when it is in each of the above states, but in order to do this it is first necessary to consider the generator reactances in each of the characteristic states.

The inductance of a winding is defined as the ratio of the flux linkages to the current producing the flux. Thus, a low-reluctance path results in a large flux and a large inductance (or reactance) and vice versa. Normally, a flux path will consist of a number of parts each of which has a different reluctance. In such circumstances it is convenient to assign a reactance to each part of the flux path when the equivalent reactance is made up of the individual path reactances. In combining the individual reactances it must be remembered that parallel flux paths correspond to a series connection of reactances, while series paths correspond to a parallel connection of reactances. This is illustrated in Figure 4.11 for a simple iron-cored coil with an air gap in the iron circuit. Here the total coil flux Φ consists of the leakage flux Φ_l and the core flux Φ_c. The reactance of the leakage flux path is X_l, while the reactance of the core flux path has two components. The first component corresponds to the flux path across the air gap X_{ag} and the second component corresponds to the flux path through the iron core X_{Fe}. In the equivalent circuit X_l is connected in series with the parallel combination X_{ag} and X_{Fe}. As the reluctance of a flux

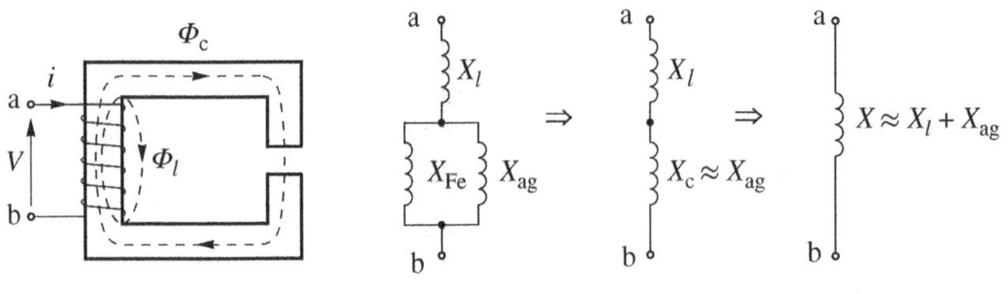

Figure 4.11 A coil with an air gap and its equivalent circuit diagram.

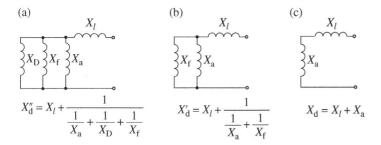

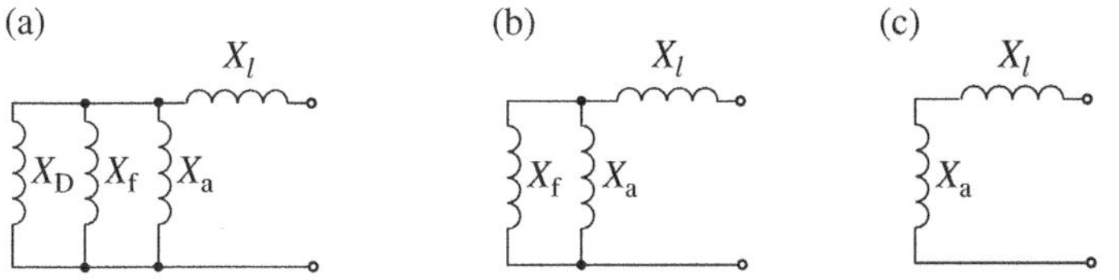

(a) $X''_d = X_l + \dfrac{1}{\dfrac{1}{X_a} + \dfrac{1}{X_D} + \dfrac{1}{X_f}}$ (b) $X'_d = X_l + \dfrac{1}{\dfrac{1}{X_a} + \dfrac{1}{X_f}}$ (c) $X_d = X_l + X_a$

Figure 4.12 Equivalent reactance of a synchronous generator in the: (a) subtransient state; (b) transient state; (c) steady state.

path in iron is very small compared to that in air $X_{Fe} \gg X_{ag}$ and the total reactance of the winding is dominated by those parts of the path that are in air: the air gap and the flux leakage path. Consequently, $X \cong X_{ag} + X_l$.

These principles are applied to the synchronous generator in Figure 4.12 for each of the three characteristic states. It also introduces a number of different reactances, each of which corresponds to a particular flux path:

X_l corresponds to the path that the armature leakage flux takes around the stator windings and is referred to as the *armature leakage reactance*;
X_a corresponds to the flux path across the air gap and is referred to as the *armature reaction reactance*;
X_D corresponds to the flux path around the damper winding;
X_f corresponds to the flux path around the field winding.

The reactances of the damper winding and the field winding are also proportional to the flux path around the winding. Thus X_D and X_f are proportional to the actual damper and field winding reactance respectively.

In Figure 4.12 the armature reactances in each of the characteristic states are combined to give the following equivalent reactances:

X''_d – d-axis subtransient reactance
X'_d – d-axis transient reactance
X_d – d-axis synchronous (steady-state) reactance.

In the subtransient state the flux path is almost entirely in air and so the reluctance of this path is very high. In contrast, the flux path in the steady state is mainly through iron with the reluctance being dominated by the length of the air gap. Consequently, the reactances, in increasing magnitude, are $X''_d < X'_d < X_d$. In the case of large synchronous generators X'_d is about twice as large as X''_d, while X_d is about 10 times as large as X''_d.

As previously discussed, following a fault the generator becomes a dynamic source that has both a time-changing synchronous reactance $X(t)$ and an internal voltage $E(t)$. Dividing the generator response into the three characteristic states with associated constant reactances makes it easier to analyze the generator dynamics. Rather than considering one generator model with time-changing reactances and internal voltages, it is convenient to consider the three states separately using conventional AC circuit analysis. This is illustrated in Figure 4.13. The RMS value of the AC component of the armature current $I_{AC}(t)$ is shown in Figure 4.13a. This AC component was previously shown in the top diagram of Figure 4.9b. The continuously changing synchronous reactance X, shown in Figure 4.13b, can be calculated by dividing the open-circuit emf E by the armature current $I_{AC}(t)$. In each of the three characteristic states, the generator will be represented by a constant emf behind a constant reactance X''_d, X'_d, and X_d, respectively. Dividing the emf by the appropriate reactance will give the subtransient, transient, and steady-state currents.

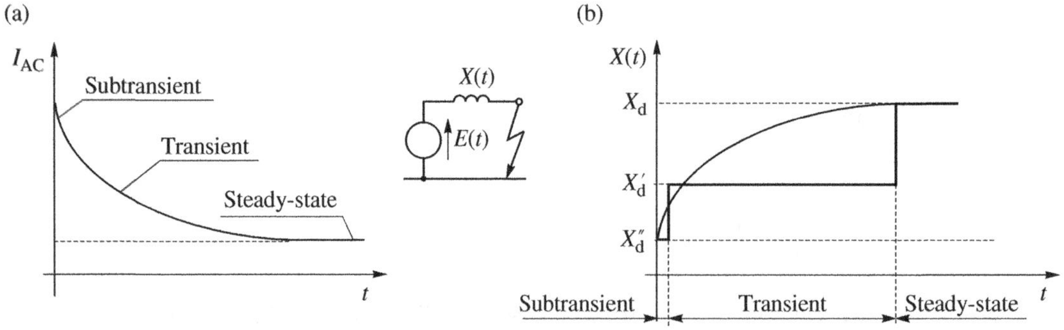

Figure 4.13 Three-step approximation of the generator model: (a) RMS value of the AC component of the armature current; (b) generator reactances.

4.2.3.1 Quadrature-axis Reactances

Figure 4.10 shows the path taken by the AC armature flux in the three characteristic states and is drawn assuming that the armature mmf is directed along the d-axis. This is true when the generator is on no load prior to the three-phase short circuit, but generally, for other types of disturbances or for a generator operating on-load prior to a disturbance, the armature mmf will have both a d- and a q-component. In this more general case it is necessary to analyze the influence of the two armature mmf components separately using the two-reaction method introduced in Section 3.3.2.

If a generator is in the subtransient state, and the armature mmf is directed along the rotor d-axis, then the armature reaction flux will be forced out of the rotor by the currents induced in the field winding, the damper winding, and the rotor core. This flux path corresponds to the d-axis subtransient reactance X''_d. On the other hand, if the armature mmf is directed along the rotor q-axis, then the only currents forcing the armature reaction flux out of the rotor are the rotor core eddy currents and the currents in the q-axis damper winding. If a generator only has a d-axis damper winding, the q-axis screening effect is much weaker than that for the d-axis, and the corresponding *quadrature-axis (q-axis) subtransient reactance* X''_q is greater than X''_d. This difference between X''_q and X''_d is called *subtransient saliency*. For a generator with a damper winding in both the d-axis and the q-axis, the screening effect in both axes is similar, subtransient saliency is negligible, and $X''_q \cong X''_d$.

When the generator is in the transient state, screening is provided by the field winding, which is only in the d-axis. However, in the round-rotor generator some q-axis screening will be produced by eddy currents in the rotor iron with the effect that $X'_q > X'_d$. The actual value of X'_q is somewhere between X'_d and X_q with typically $X'_q \cong 2X'_d$. In the salient-pole generator the laminated rotor construction prevents eddy currents flowing in the rotor body, there is no screening in the q-axis, and $X'_q = X_q$. Because of the absence of a field winding in the q-axis, there is some degree of *transient saliency* in all types of generator.

Table 4.2 summarizes the different types of saliency and the reasons for it, while typical values for all the generator parameters are tabulated in Table 4.3.

In Table 4.3 the q-axis subtransient short-circuit time constant T''_q and the transient short-circuit time constant T'_q are defined in the same way as for the d-axis time constants T''_d and T'_d in Table 4.1. For the round-rotor generator, since there is no field winding in the q-axis, it is assumed that rotor body eddy currents contribute to the q-axis parameters.

Values of the short-circuit time constants T'_d, T'_q, T''_d, and T''_q correspond to the armature winding when short-circuited. Some manufacturers may quote the time constants for the armature on open circuit as T'_{do}, T'_{qo}, T''_{do}, T''_{qo}.

An approximate relationship between the open- and short-circuit time constants may be found from the equivalent circuits shown in Figure 4.12. If X_l is neglected then

$$X'_d \cong \frac{X_a X_f}{X_a + X_f}, \quad \frac{1}{X'_d} \cong \frac{1}{X_a} + \frac{1}{X_f} \tag{4.27}$$

$$X''_d \cong \frac{1}{\frac{1}{X_D} + \frac{1}{X'_d}}; \quad X_D \cong \frac{X'_d X''_d}{X'_d - X''_d} \tag{4.28}$$

Table 4.2 Saliency in the three characteristic states of the synchronous generator.

State	Generator type	Reactance	Saliency: Yes/No	Saliency: Reason
Subtransient	Any type but with a d-axis damper only	$X''_q > X''_d$	Yes	Weaker screening in the q-axis because of the lack of damper winding
	Any type but with both d- and q-axis damper windings	$X''_q \cong X''_d$	No	Similar screening in both axes
Transient	Round-rotor	$X'_q > X'_d$	Yes	Strong screening in the d-axis, owing to the field winding, but weak screening in the q-axis, owing to the rotor body currents
	Salient-pole	$X'_q = X_q$	Yes	No screening in the q-axis because of the laminated rotor core
Steady-state	Round-rotor	$X_q \cong X_d$	No	Symmetrical air-gap in both axes
	Salient-pole	$X_q < X_d$	Yes	Larger air-gap on the q-axis

Table 4.3 Typical parameter values for large generators. Reactances are in pu to the rated MVA and time constants are in seconds.

Parameter	Round-rotor: 200 MVA	Round-rotor: 600 MVA	Round-rotor: 1500 MVA	Salient-pole rotor: 150 MVA	Salient-pole rotor: 230 MVA
X_d	1.65	2.00	2.20	0.91	0.93
X_q	1.59	1.85	2.10	0.66	0.69
X'_d	0.23	0.39	0.44	0.3	0.3
X'_q	0.38	0.52	0.64	—	—
X''_d	0.17	0.28	0.28	0.24	0.25
X''_q	0.17	0.32	0.32	0.27	0.27
T'_d	0.83	0.85	1.21	1.10	3.30
T'_q	0.42	0.58	0.47	—	—
T''_d	0.023	0.028	0.030	0.05	0.02
T''_q	0.023	0.058	0.049	0.06	0.02

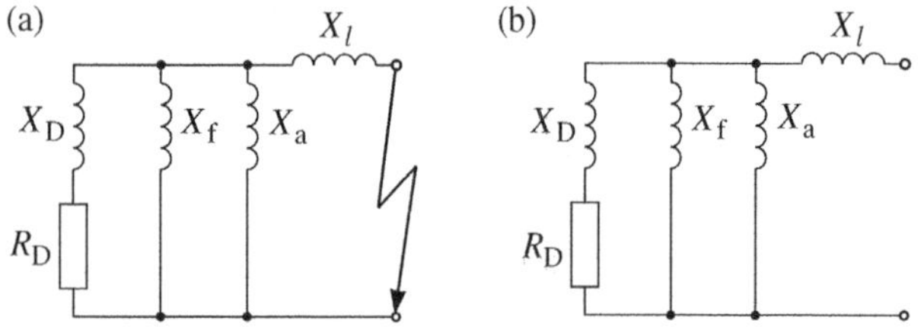

Figure 4.14 Equivalent circuits for determining the subtransient time constants: (a) short circuit; (b) open circuit.

In order to determine the time constants T''_d and T''_{do} it is necessary to insert the resistance R_D into the branch with X_D in Figure 4.12a. This is illustrated in Figure 4.14. If X_l is neglected then the short circuit in Figure 4.14a bypasses all the shunt branches and the time constant is

$$T''_d = \frac{X_D}{\omega R_D} \tag{4.29}$$

When the circuit is an open circuit, as in Figure 4.14b, the time constant is

$$T''_{do} = \frac{X_D + \dfrac{X_a X_f}{X_a + X_f}}{\omega R_D} \cong \frac{X_D + X'_d}{\omega R_D} \tag{4.30}$$

Dividing Eq. (4.29) by Eq. (4.30), and substituting for X_D from Eq. (4.28), gives $T''_d/T''_{do} \cong X''_d/X'_d$. Following a similar procedure for the remaining time constants in the d- and q-axes gives the relationships

$$T''_d \cong T''_{do}\frac{X''_d}{X'_d}, \quad T''_q \cong T''_{qo}\frac{X''_q}{X'_q}, \quad T'_d \cong T'_{do}\frac{X'_d}{X_d}, \quad T'_q \cong T'_{qo}\frac{X'_q}{X_q} \tag{4.31}$$

The open-circuit time constants are larger than the short-circuit ones.

4.2.4 Generator Electromotive Forces and Equivalent Circuits

Equivalent circuits were first used in Section 3.3 to analyze the steady-state behavior of the generator and in this subsection their use is extended to cover all three characteristic states. As these equivalent circuits are valid only for AC circuit analysis purposes, they can only be used to analyze the AC component of the armature current, and the DC component must be neglected. What follows is a physical justification for using such equivalent circuits to model the synchronous generator, with a full mathematical derivation being left until Chapter 11.

In the steady state the generator can be modeled by the equivalent circuit shown in Figure 3.17 where the generator is represented in both the d- and q-axes by constant internal emf's E_q and E_d acting behind the synchronous reactances X_d and X_q. The emf E_q is proportional to the excitation flux and, as the excitation flux is proportional to the field current, is also given the symbol E_f. As there is no excitation in the q-axis, the corresponding emf E_d is zero. Consequently, the total internal emf is $\underline{E} = \underline{E}_q + \underline{E}_d = \underline{E}_q = \underline{E}_f$.

A similar representation can be used when the generator is in the subtransient and transient states. However, as each of the three generator characteristic states is characterized by a different pair of reactances, there are now three different equivalent circuits, each valid at the beginning of a corresponding state, in which the generator is represented by a pair of constant internal emf's behind the appropriate reactances. It is important to realize that, in each of the states, the internal emf's will be different and equal to that part of the rotor flux linkages that are assumed to remain constant in that particular characteristic state.

In the steady state the three emf's E, E_f, and E_q are all equal but during a disturbance they have a different interpretation and it is important to differentiate between them. E_q is proportional to the field current (which produces a d-axis mmf) and will vary in proportion to the changes in the field current. During a disturbance, there are also currents induced in the q-axis rotor body so that E_d will no longer be zero but will vary in proportion to these rotor body currents. Consequently, neither E_q nor E_d is constant during either the subtransient or transient state.

The emf E_f is defined as being proportional to the voltage V_f applied to the field winding and it is only in the steady state when the field current $i_f = i_{f0} = V_f/R_f$ that $E_f = E_q$ (R_f is the field winding resistance). In all other characteristic states $i_f \neq V_f/R_f$ and $E_f \neq E_q$. The importance of E_f is that it reflects the effect of excitation control.

4.2.4.1 Subtransient-state

During the subtransient period, the armature flux is forced into high-reluctance paths outside the rotor circuits by currents induced in the field and damper winding. This is shown in Figure 4.10a for the flux acting along the d-axis only, but in general the armature flux will have both d- and q-components. In this more general case the flux path is distorted not only by the d-axis field and damper currents but also by the q-axis damper current (if there is a q-axis damper winding) and q-axis rotor body currents. The reactances associated with the flux path are X''_d and X''_q. As the rotor flux linkages in both axes are assumed to remain constant during the subtransient state, the internal emf corresponding to these linkages may also be assumed to remain constant and the generator may be represented by a constant *subtransient internal emf* $\underline{E}'' = \underline{E}''_q + \underline{E}''_d$ acting behind the subtransient reactances X''_d and X''_q. The generator circuit equation is then

$$\underline{E}'' = \underline{V}_g + R\underline{I} + j\underline{I}_d X''_d + j\underline{I}_q X''_q \tag{4.32}$$

where $\underline{I} = \underline{I}_d + \underline{I}_q$ is the armature current immediately after the fault. Its value can be calculated once $\underline{E}''$ is known. The emf E''_q is proportional to the rotor flux linkages along the d-axis, that is the sum of the field winding and the d-axis damper winding flux linkages. Similarly, E''_d is proportional to q-axis rotor flux linkages, that is the sum of the q-axis rotor body and the q-axis damper winding flux linkages.

It is important to understand the difference between the steady-state and subtransient emf's. Consider first E_q and E''_q. E_q is proportional to the field current and therefore to the field winding self-flux linkages. In order to screen the field winding from the change in armature flux, the field current varies following the fault and E_q varies accordingly. The emf E''_q is proportional to the total rotor d-axis flux linkages (field and damper windings) and includes the linkages due to the armature flux. This total flux linkage must remain constant and equal to its pre-fault value following the fault. If the generator is on no load prior to the fault then the pre-fault armature current and flux are zero. In this case the total pre-fault rotor flux linkages are equal to the linkages resulting from the excitation flux only and the pre-fault values of E_q and E''_q are the same and equal to the terminal voltage. If, however, the generator is on load prior to the fault then the armature current is not zero and the total rotor flux linkages will include the effect of the pre-fault linkages resulting from the armature flux. In this case the pre-fault values of E_q and E''_q will not be the same. A similar argument is valid for the q-axis flux linkages and emf's. Thus the pre-fault value of E''_d is equal to zero only when the generator is on no load prior to the fault.

The emf $\underline{E} = \underline{E}_q + \underline{E}_d$ can be found by observing that

$$\begin{aligned}\underline{E} &= \underline{V}_g + R\underline{I} + j\underline{I}_d X_d + j\underline{I}_q X_q \\ &= \underline{V}_g + R\underline{I} + j\underline{I}_d\left(X_d - X''_d\right) + j\underline{I}_d X''_d + j\underline{I}_q\left(X_q - X''_q\right) + j\underline{I}_q X''_q \\ &= \underline{E}'' + j\underline{I}_d\left(X_d - X''_d\right) + j\underline{I}_q\left(X_q - X''_q\right)\end{aligned} \tag{4.33}$$

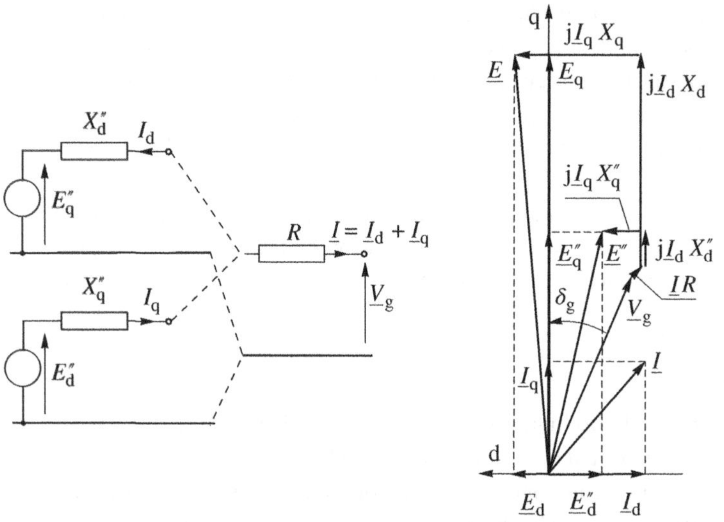

Figure 4.15 Equivalent circuit and phasor diagram of the generator in the subtransient state.

Figure 4.15 shows the equivalent circuit and the phasor diagram of the generator in the subtransient state. $\underline{E}$ is no longer along the q-axis and has both E_d and E_q components. Usually, R is small so the length of IR is small too. The length of IR has been exaggerated in Figure 4.15 to clearly show its influence.

4.2.4.2 Transient-state

During the transient period, the armature flux is forced into high-reluctance paths outside the field winding by currents induced in the field winding. This is shown in Figure 4.10b for the flux acting along the d-axis only, but in general the armature flux will have both d- and q-components. In this more general case the flux path is distorted not only by the d-axis field current but also by the q-axis rotor body currents. The reactances associated with the flux path are X'_d and X'_q. As the rotor flux linkages in both axes are assumed to remain constant during the transient state the internal emf corresponding to these linkages may also be assumed to remain constant and equal to their pre-fault values. Thus the generator may be represented in the transient state by constant *transient internal emf's* E'_q and E'_d acting behind the transient reactances X'_d and X'_q. The circuit equation of the generator is then

$$\underline{E}' = \underline{E}'_q + \underline{E}'_d = \underline{V}_g + R\underline{I} + \mathrm{j}\underline{I}_d X'_d + \mathrm{j}\underline{I}_q X'_q \tag{4.34}$$

where $\underline{I} = \underline{I}_d + \underline{I}_q$ is the armature current at the beginning of the transient period. Its value is different from that given by Eq. (4.32) and can be calculated once $\underline{E}'$ is known. E'_q is proportional to the field winding flux linkages Ψ_f, whilst E'_d is proportional to the flux linkage of the q-axis rotor body. Both components include the effect of the pre-fault armature current and are assumed to remain constant during the transient state. Following a similar argument as for the subtransient state, the pre-fault values of the emf's are $E'_{q0} = E_{q0} = V_g$ and $E'_{d0} = E_{d0} = 0$ only if prior to the fault the generator is on no load (zero armature reaction flux). If prior to the fault the generator is on load then E'_q and E'_d include the effect of the load current and $\underline{E}'_0 \neq \underline{E}''_0 \neq \underline{E}_0$.

Similar to the subtransient state the emf $\underline{E}$ can be found from

$$\underline{E} = \underline{V}_g + R\underline{I} + \mathrm{j}\underline{I}_d X_d + \mathrm{j}\underline{I}_q X_q = \underline{E}' + \mathrm{j}\underline{I}_d (X_d - X'_d) + \mathrm{j}\underline{I}_q \left(X_q - X'_q\right) \tag{4.35}$$

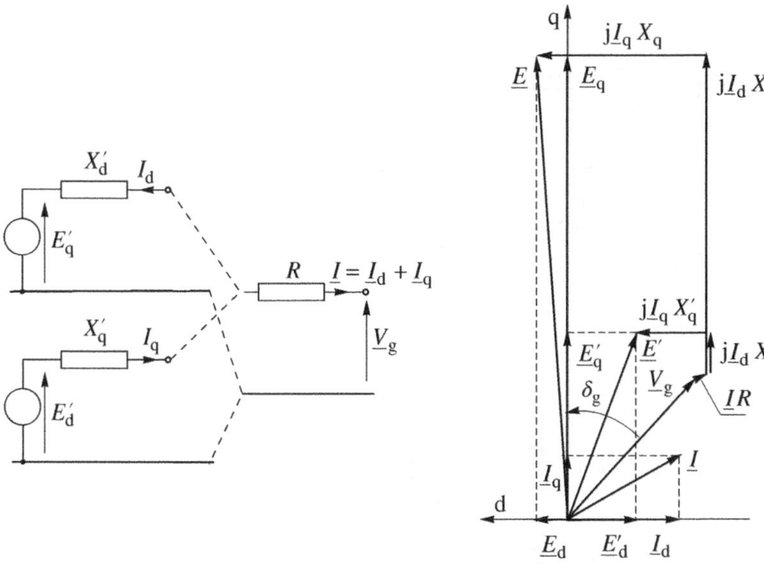

Figure 4.16 Equivalent circuit and phasor diagram of the round-rotor generator in the transient state. The length of IR has been exaggerated for clarity.

Figure 4.16 shows the equivalent circuit and the phasor diagram of the round-rotor generator in the transient state. The transient emf of such a generator has both d- and q-components. For the salient-pole generator with laminated rotor there is no screening in the q-axis and $X'_q = X_q$, so that $E'_d = 0$ and $E' = E'_q$.

4.2.4.3 Establishing Initial emf Values

The previous discussion explained why the subtransient and transient emf's do not change during their respective periods (neglecting the influence of resistances) and are equal to their pre-fault values.

The simplest situation occurs when the generator is no load prior to the fault when the initial values of the transient and subtransient emf's are equal to the pre-fault steady-state emf E (and the generator pre-fault terminal voltage). When the generator is on load the influence of the pre-fault armature current $\underline{I}_0$ on the internal emf's must be taken into account. Note that the three equivalent circuits shown in Figures 3.18, 4.15, and 4.16 are valid for different states of the generator and therefore for different values of current $\underline{I}$. However, as the subtransient and transient emf's are equal to their pre-fault values, the three equivalent circuits must also be valid for the pre-fault current $\underline{I}_0$. Thus, according to Kirchhoff's law, the three equivalent circuits can be combined for the same current $\underline{I}_0$, as shown in Figure 4.17a.

By following a similar procedure to that described in Section 3.3 the initial values of $\underline{E}_0$, $\underline{E}'_0$, and $\underline{E}''_0$ can then be found from the phasor diagram, Figure 4.17b, assuming that $\underline{I}_0$ and its phase angle φ_{g0} are known. As excitation is only in the d-axis, E_0 acts along the q-axis, while $\underline{E}''_0$ and $\underline{E}'_0$ usually have nonzero direct and quadrature components. For the salient pole-generator, Figure 4.17c, $E_{d0} = 0$ and $\underline{E}'_0 = \underline{E}'_{q0}$. In the phasor diagram the value of the voltage drop across the generator resistance has been exaggerated for clarity, but, in practice, the resistance is very small and can often be neglected. Once the initial values of the transient and subtransient emf's have been found, Eqs. (4.32) and (4.34) can be used to calculate the magnitude of the alternating current component at the start of the subtransient and transient periods, respectively.

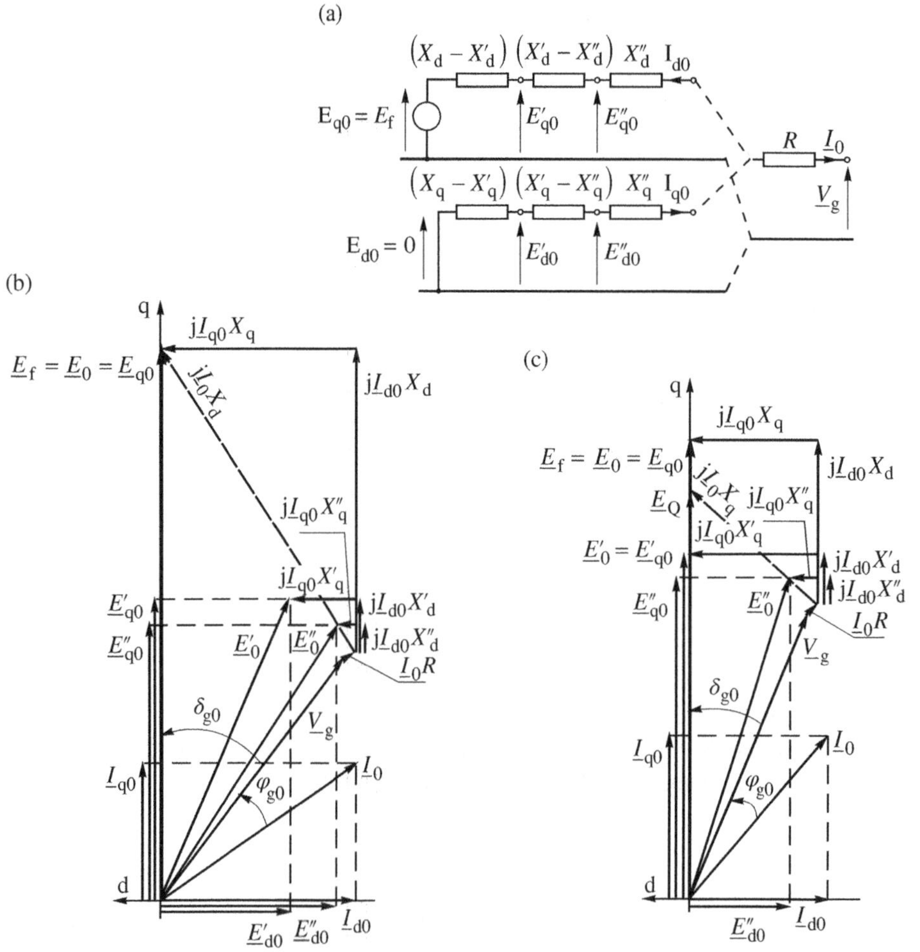

Figure 4.17 Finding the initial values of the emf's: (a) equivalent circuits; (b) phasor diagram of the round-rotor generator; (c) phasor diagram of the salient-pole generator. The length of IR has been exaggerated for clarity.

Example 4.2 A 200 MVA round-rotor generator with the parameters given in Table 4.3 is loaded with 1 pu of active power and 0.5 pu of reactive power (lagging). The voltage at the generator terminals is 1.1 pu. Find the pre-fault values of the steady-state, transient, and subtransient emf's. It is assumed that the armature resistance is neglected and $X_d = X_q = 1.6$.

Assuming the generator voltage to be the reference, the load current is

$$\underline{I}_0 = \left(\frac{\underline{S}}{\underline{V}_g}\right)^* = \frac{P - jQ}{V_g} = \frac{1 - j0.5}{1.1} = 1.016\angle -26.6^\circ$$

so that $\varphi_{g0} = 26.6^\circ$. The steady-state internal emf is

$$E_{q0} = \underline{V}_g + jX_d\underline{I}_0 = 1.1 + j1.6 \times 1.016\angle -26.6^\circ = 2.336\angle 38.5^\circ$$

Thus $E_{q0} = 2.336$ and $\delta_{g0} = 38.5^\circ$. The d- and q-components of the current and voltage are

$$I_{d0} = -I_0 \sin\left(\varphi_{g0} + \delta_{g0}\right) = -1.016 \sin\left(26.6^\circ + 38.5^\circ\right) = -0.922$$

$$I_{q0} = I_0 \cos\left(\varphi_{g0} + \delta_{g0}\right) = 0.428$$

$$V_{gd} = -V_g \sin\delta_{g0} = -1.1\sin 38.5^{\circ} = -0.685, V_{gq} = V_g \cos\delta_{g0} = 0.861$$

Now the d- and q-components of the transient and subtransient emf's can be calculated from the phasor diagram in Figure 4.17 as

$$E'_{d0} = V_{gd} + X'_q I_{q0} = -0.685 + 0.38 \cdot 0.428 = -0.522$$

$$E'_{q0} = V_{gq} - X'_d I_{d0} = 0.861 - 0.23 \cdot (-0.922) = 1.073$$

$$E''_{d0} = V_{gd} + X''_q I_{q0} = -0.612$$

$$E''_{q0} = V_{gq} - X''_d I_{d0} = 1.018$$

******

Example 4.3 Solve a similar problem to that in Example 4.2, but for the 230 MVA salient-pole generator in Table 4.3.

The main problem with the salient-pole generator is in finding the direction of the q-axis. Equation (3.66) gives $\underline{E}_Q = \underline{V}_g + jX_q\underline{I}_0 = 1.1 + j0.69 \times 1.016\angle -26.6^{\circ} = 1.546\angle 23.9^{\circ}$. Thus $\delta_{g0} = 23.9^{\circ}$ and

$$I_{d0} = -1.016\sin(26.6^{\circ} + 23.9^{\circ}) = -0.784, I_{q0} = -1.016\cos(26.6^{\circ} + 23.9^{\circ}) = 0.647$$

$$V_{gd} = -1.1\sin 23.9^{\circ} = -0.446, V_{gq} = 1.1\cos 23.9^{\circ} = 1.006$$

$$E_{q0} = V_{gq} - I_{d0}X_d = 1.006 - (-0.784)0.93 = 1.735$$

$$E'_{d0} = -0.446 + 0.69 \cdot 0.647 = 0, E'_{q0} = 1.006 - 0.3 \cdot (-0.784) = 1.241$$

$$E''_{d0} = -0.446 + 0.27 \cdot 0.647 = -0.271, E''_{q0} = 1.006 - 0.25 \cdot (-0.784) = 1.202$$

******

4.2.4.4 Flux Decrement Effects

Although the subtransient and transient emf's remain constant immediately after a fault, their values change with time as the armature flux penetrates the rotor circuits. The timescale of these changes is such that changes during the subtransient period affect the short-circuit currents and torques, as is discussed later in this chapter, while changes during the transient period can affect generator stability and are discussed at length in Chapters 5 and 6. Although a full mathematical treatment of these *flux decrement* effects must be left until Chapter 11, a basic understanding of the mechanisms behind them can be obtained by considering the simple d-axis coupled circuit shown in Figure 4.18. This circuit models the field winding and the d-axis armature coil, and the coupling that exists between them, and gives the field flux linkage as

$$\Psi_f = L_f i_f + M_{fd} i_d \tag{4.36}$$

where L_f is the self-inductance of the field winding and M_{fd} is the mutual inductance between the two windings. From Eq. 11.18 derived in Section 11.1.3 it results that $M_{fd} = kM_f$, where $k = \sqrt{3/2}$. According to the theorem of constant flux linkage, Ψ_f will remain constant immediately following any change in i_f or i_d and, as $E'_q \propto \Psi_f$, E'_q will

Figure 4.18 Coupling between the field winding and the d-axis armature coil.

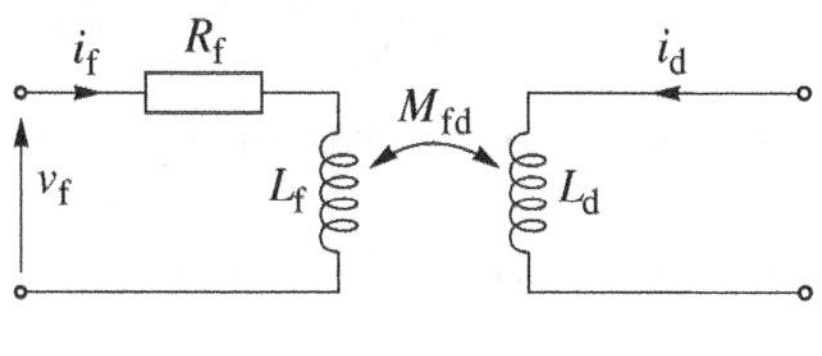

also remain constant. However, as the field winding has a resistance greater than zero, this will allow some of the magnetic stored energy to be dissipated and the winding flux linkage to change. This change in flux linkage is governed by the differential equation

$$\frac{d\Psi_f}{dt} = v_f - R_f i_f \tag{4.37}$$

which, when rearranging and substituting Eq. (4.36), gives

$$v_f = L_f \frac{di_f}{dt} + R_f i_f + M_{fd} \frac{di_d}{dt} \tag{4.38}$$

At this point it is convenient to consider the changes Δi_f, Δi_d, and Δv_f rather than absolute values when introducing the Laplace operator, and writing Eq. (4.38) in terms of these changes gives

$$\Delta v_f(s) = L_f s \Delta i_f(s) + R_f \Delta i_f(s) + M_{fd} s \Delta i_d(s) \tag{4.39}$$

This equation can then be written in transfer function form as

$$\Delta i_f(s) = \frac{1/R_f}{(1 + sT'_{do})} \Delta v_f(s) - \frac{M_{fd}/R_f}{(1 + sT'_{do})} s \Delta i_d(s) \tag{4.40}$$

where $T'_{do} = L_f/R_f$. Substituting Eq. (4.40) into Eq. (4.36) gives

$$\Delta \Psi_f(s) = \frac{L_f/R_f}{(1 + sT'_{do})} \Delta v_f(s) + \frac{M_{fd}}{(1 + sT'_{do})} \Delta i_d(s) \tag{4.41}$$

As $\Delta E'_q \propto \Delta \Psi_f$ and $\Delta E_f \propto \Delta v_f$, this equation can be written in terms of $\Delta E'_q$ and ΔE_f. Figure 4.16 shows that for a generator I_d flows into the winding while I flows out of the winding. Thus, assuming no change in the rotor angle, Δi_d can be replaced by a term proportional to $(-\Delta I)$ when a more rigorous analysis similar to the above, but using the relationships established in Chapter 11 to evaluate the constants of proportionality, would show that

$$\Delta E'_q(s) = \frac{1}{(1 + sT'_{do})} \Delta E_f(s) - \frac{K}{(1 + sT'_{do})} \Delta I(s) \tag{4.42}$$

where K is a constant. As can be seen, a change in the field winding flux linkage, and hence a change in E'_q, can be produced by either a change in the excitation voltage or a change in the armature current. The rate at which these changes occur is determined by the transient time constant T'_{do} with the field winding filtering out high-frequency changes in ΔI and ΔE_f. Assuming for the moment that $\Delta I = 0$ and that a step change is made on E_f then E'_q will change exponentially as shown in Figure 4.19a. Similarly, a step increase in the armature current I, such as occurs during a short circuit, would lead to an exponential reduction in E'_q, as shown in Figure 4.19b.

In practice the armature will be connected to the system (a voltage source) by a transmission line of finite reactance when the actual time constants involved will depend on the impedance in the armature circuit. For a generator connected to an infinite busbar through a transmission link, Anderson and Fouad (1977) show that this modifies Eq. (4.42) to

$$\Delta E'_q(s) = \frac{B}{(1 + sBT'_{do})} \Delta E_f(s) - \frac{AB}{(1 + sBT'_{do})} \Delta \delta(s) \tag{4.43}$$

where a change in the generator loading is now reflected by a change in the power angle $\Delta\delta$ (measured with respect to the system) and B is a constant that takes into account the effect of the impedance in the armature circuit. If both the generator armature resistance and the resistance of the transmission link are neglected then

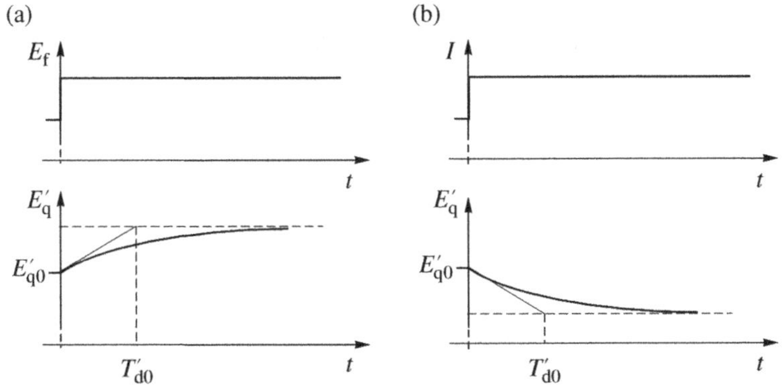

Figure 4.19 Changes in E'_q (a) to a step change in excitation ΔE_f and (b) to a step change in load ΔI.

$B = (X'_d + X_s)/(X'_d + X_s) = x'_d/x_d$ and $A = [(1 - B)/B]\ V_s \sin \delta_0$. The importance of Eqs. (4.42) and (4.43) cannot be overemphasized as they show how changes in the excitation and generator loading alter E'_q.

4.2.5 Short-circuit Currents with the Generator Initially on No Load

To consider the time variation of the fault current, assume the generator to be on no load prior to the fault. In this case the pre-fault values of the three internal emf's E, E', and E'' are all equal to the terminal voltage, that is $E'' = E' = E = E_f = V_g$, their d-components are all equal to zero, and only the q-axis variables need to be considered. This means that the separate d- and q-axis diagrams shown in Figures 3.18, 4.15, and 4.16 can be replaced by the single d-axis equivalent circuit shown in Figure 4.20. The amplitude of the AC fault current component in each of the three characteristic states can be obtained directly from these equivalent circuits by short-circuiting the generator terminals, when

$$\text{for the subtransient state}: \quad i''_m = \frac{E_{fm}}{X''_d} \tag{4.44}$$

$$\text{for the transient state}: \quad i'_m = \frac{E_{fm}}{X'_d} \tag{4.45}$$

$$\text{for the final steady-state}: \quad i^{\infty}_m = \frac{E_{fm}}{X_d} \tag{4.46}$$

where $E_{fm} = \sqrt{2}E_f$.

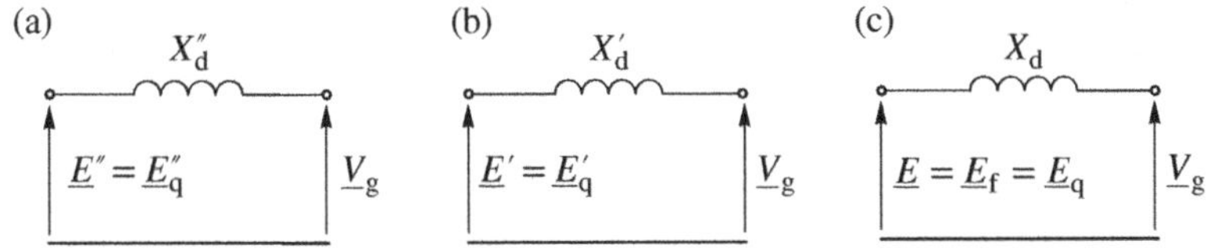

Figure 4.20 Equivalent circuit for a synchronous generator in (a) the subtransient state; (b) the transient state; (c) the steady state. The generator is assumed to be on no load prior to the fault.

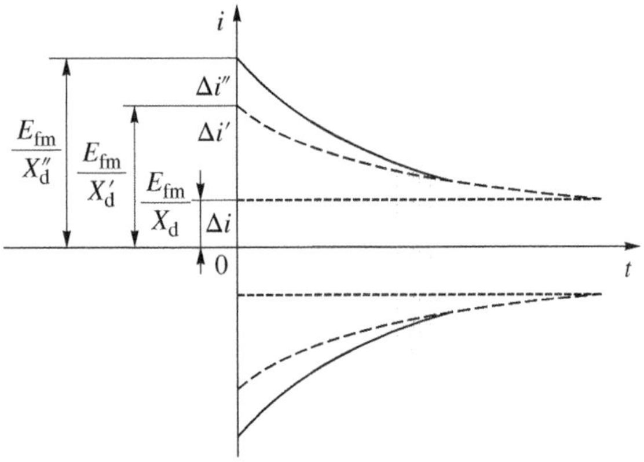

Figure 4.21 Envelopes of the three characteristic AC components of the short-circuit current.

As the AC component of the phase current, and its decay time constant, in all three characteristic states is known, it is now possible to derive formulae for the short-circuit currents as a function of time. Figure 4.21 shows the envelopes of the maximum value of the AC component of the short-circuit current corresponding to those shown in the top diagram of Figure 4.9b. The resultant envelope is defined by $i_{\rm m}(t)$ and is obtained by adding the three components, each of which decays with a different time constant, to give

$$i_{\rm m}(t) = \Delta i''_{\rm m} {\rm e}^{-t/T''_{\rm d}} + \Delta i' {\rm e}^{-t/T'_{\rm d}} + \Delta i \tag{4.47}$$

where

- $\Delta i = i^{\infty}_{\rm m} = E_{\rm fm}/X_{\rm d}$ is the maximum value of the AC component neglecting the screening effect of all the rotor windings; Eq. (4.46);
- $\Delta i + \Delta i' = E_{\rm fm}/X'_{\rm d}$ is the maximum value of the AC component including the screening effect of the field winding but neglecting the screening effect of the damper winding; Eq. (4.45);
- $\Delta i + \Delta i' + \Delta i''_{\rm m} = E_{\rm fm}/X''_{\rm d}$ is the maximum value of the AC component when the screening effect of both the damper winding and the field winding are included; Eq. (4.44).

Some simple algebra applied to Figure 4.21 then gives the equations for the three components of the short-circuit current as

$$\Delta i = E_{\rm fm}\frac{1}{X_{\rm d}}, \quad \Delta i' = E_{\rm fm}\left(\frac{1}{X'_{\rm d}} - \frac{1}{X_{\rm d}}\right), \quad \Delta i''_{\rm m} = E_{\rm fm}\left(\frac{1}{X''_{\rm d}} - \frac{1}{X'_{\rm d}}\right) \tag{4.48}$$

Substituting (4.48) into (4.47) gives the following expression for the envelope of the AC component of the fault current

$$i_{\rm m}(t) = E_{\rm fm}\left[\left(\frac{1}{X''_{\rm d}} - \frac{1}{X'_{\rm d}}\right){\rm e}^{-t/T''_{\rm d}} + \left(\frac{1}{X'_{\rm d}} - \frac{1}{X_{\rm d}}\right){\rm e}^{-t/T'_{\rm d}} + \frac{1}{X_{\rm d}}\right] \tag{4.49}$$

The initial value of Eq. (4.49) at $t = 0$ is

$$i_{\rm m}(0) = \frac{E_{\rm fm}}{X''_{\rm d}} \tag{4.50}$$

Substituting Eqs. (4.49) and (4.50) into (4.26) gives the expression for the short-circuit currents as

$$\begin{aligned} i_A &= -\frac{E_{fm}}{X''_d}\left[g_3(t)\cos(\omega t + \gamma_0) - e^{-t/T_a}\cos\gamma_0\right] \\ i_B &= -\frac{E_{fm}}{X''_d}\left[g_3(t)\cos(\omega t + \gamma_0 - 2\pi/3) - e^{-t/T_a}\cos(\gamma_0 - 2\pi/3)\right] \\ i_C &= -\frac{E_{fm}}{X''_d}\left[g_3(t)\cos(\omega t + \gamma_0 - 4\pi/3) - e^{-t/T_a}\cos(\gamma_0 - 4\pi/3)\right] \end{aligned} \quad (4.51)$$

The function $g_3(t)$ is defined as

$$g_3(t) = X''_d\left[\left(\frac{1}{X''_d} - \frac{1}{X'_d}\right)e^{-t/T''_d} + \left(\frac{1}{X'_d} - \frac{1}{X_d}\right)e^{-t/T'_d} + \frac{1}{X_d}\right] \quad (4.52)$$

and accounts for the decay of the AC component from the subtransient state to the transient state and then to the steady state. The subscript "3" relates to the three-phase short circuit.

The maximum instantaneous value of the current in each phase depends on the instant in the AC cycle when the fault occurs. For example, the current in phase A reaches its maximum value when the fault is applied at $\gamma_0 = 0$, that is when the excitation flux linking phase A, Ψ_{fA}, reaches its maximum value and the voltage E_{fA} is zero. Figure 4.9 shows currents flowing in a large generator after a three-phase fault assuming the armature time constant T_a to be much greater than the AC cycle time. With such a long armature time constant the current i_A may reach a value almost equal to $2E_{fm}/X''_d$. On the other hand, when the fault occurs at $\gamma_0 = \pi/2$, the excitation flux linkage Ψ_{fA} is zero and the current i_A is minimal as it does not contain a DC component. However, the DC component will be evident in currents i_B and i_C.

4.2.5.1 Influence of the Rotor Subtransient Saliency

When saliency effects are included, the different reluctance on the d- and q-axes requires the effect of the armature mmf on the short-circuit current to be analyzed using the two-reaction theory (Jones 1967). However, an intuitive understanding of the effects of saliency can be obtained by considering the AC and DC armature mmf components separately.

Section 4.2.1 explains that the armature mmf due to the stator fault currents consists of a rotating component $F_{a\,ac}$ directed along the d-axis and a stationary component $F_{a\,dc}$ which is also initially directed along the d-axis but then counter-rotates with respect to the rotor. As $F_{a\,ac}$ is always directed along the d-axis, it is associated with a reactance equal to X''_d, and the AC term in Eq. (4.23) remains unaffected by rotor saliency.

On the other hand, the stationary mmf component $F_{a\,dc}$ drives a flux across the air gap the width of which appears to be continuously changing. This has two effects. First of all, the maximum value of the DC component of the fault current, given by the second component in Eq. (4.51), has to be modified by making it dependent on the mean of the subtransient reactances and equal to $E_{fm}\left(1/X''_d + 1/X''_q\right)/2$. Secondly, an additional double-frequency component of magnitude $E_{fm}\left(1/X''_d - 1/X''_q\right)/2$ is introduced into the fault current, owing to $F_{a\,dc}$ lying twice every cycle along the d-axis (q-axis). Both the DC and the double-frequency components decay with the same time constant equal to the mean of the d- and q-axis time constants $T_a = \left(X''_d + X''_q\right)/(2\omega R)$ while their values depend on the instant of the fault.

The short-circuit current in phase A described by Eq. (4.51) can now be adapted to include the modified DC component and the double-frequency AC component

$$i_A = -\frac{E_{fm}}{X''_d}\left[g_3(t)\cos(\omega t + \gamma_0)\right] + \frac{E_{fm}}{2}e^{-t/T_a}\left[\left(\frac{1}{X''_d} + \frac{1}{X''_q}\right)\cos\gamma_0 + \left(\frac{1}{X''_d} - \frac{1}{X''_q}\right)\cos(2\omega t + \gamma_0)\right] \quad (4.53)$$

At time $t = 0$ the second component in this equation is equal to $E_{fm} \cos \gamma_0 / X''_d$ and corresponds to the DC component of the armature mmf always being in phase with the rotor d-axis when the fault is applied. In the case of a generator with damper windings in both axes $X''_q \cong X''_d$ and Eq. (4.53) reduces to (4.51).

4.2.6 Short-circuit Currents in the Loaded Generator

When the generator is on load prior to the fault, the initial values of the emf's E''_{q0}, E'_{q0}, E''_{d0}, and E'_{d0} are functions of the load current and, knowing the initial loading conditions, their value may be determined from the phasor diagram of Figure 4.17. These emf's now further modify Eq. (4.53) to account for the pre-fault load current. Both d- and q-axis AC emf's and currents have now to be considered. The q-axis emf's will force the flow of d-axis alternating currents (changing as cosine with time) while d-axis emf's will force the flow of q-axis alternating currents (changing as sine with time). Denoting the angle between the q-axis and the terminal voltage V_g by δ_g, the following expression for the short-circuit current flowing in phase A is obtained

$$\begin{aligned} i_A = &-\left[\left(\frac{E''_{qm0}}{X''_d} - \frac{E'_{qm0}}{X'_d}\right) e^{-t/T''_d} + \left(\frac{E'_{qm0}}{X'_d} - \frac{E_{qm0}}{X_d}\right) e^{-t/T'_d} + \frac{E_{qm0}}{X_d}\right] \cos(\omega t + \gamma_0) + \\ &+ \left[\left(\frac{E''_{dm0}}{X''_q} - \frac{E'_{dm0}}{X'_q}\right) e^{-t/T''_q} + \frac{E'_{dm0}}{X'_q} e^{-t/T'_q}\right] \sin(\omega t + \gamma_0) + \\ &+ \frac{V_{gm0}}{2} e^{-t/T_a} \left[\left(\frac{1}{X''_d} + \frac{1}{X''_q}\right) \cos(\gamma_0 + \delta_g) + \left(\frac{1}{X''_d} - \frac{1}{X''_q}\right) \cos(2\omega t + \gamma_0 + \delta_g)\right] \end{aligned} \tag{4.54}$$

All the emf's, and the terminal voltage, include an additional subscript "0" to emphasize the fact that they are pre-fault quantities. The first two components represent the fundamental-frequency alternating currents due to the direct and quadrature components of the emf's. The last component consists of a DC term and a double-frequency term resulting from the subtransient saliency.

4.2.6.1 Influence of the AVR

Depending on the type of exciter used, the action of the AVR can have a considerable effect on the shape of the short-circuit current. In the case of a rotating exciter, Figure 2.3a, the fault does not influence the generating capabilities of the excitation unit and the large voltage error, equal to the difference between the reference voltage and the actual value of the terminal voltage, quickly drives up the excitation voltage. This increases the value of the short-circuit current compared with its value without the AVR being active, as shown in Figure 4.22a. As E_q follows the field current, E_q will not now decay to E_{qo} but to a higher value corresponding to the new field current. Consequently, E'_q will not decay to the same extent as when the AVR was inactive. If a static exciter is used, fed solely from the generator terminal voltage as in Figure 2.3e, the three-phase short-circuit reduces the excitation

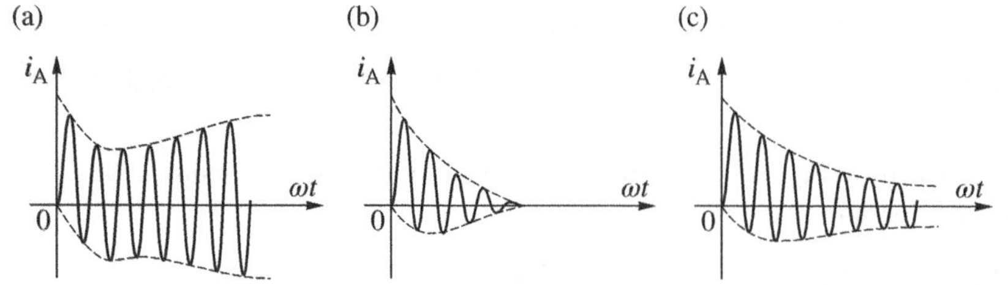

Figure 4.22 Influence of the AVR on the short-circuit current: (a) rotating exciter; (b) static exciter fed from the generator terminal voltage only; (c) static exciter fed from the generator terminal voltage and current.

voltage to zero, the unit completely loses its excitation capabilities and all the emf's will decay to zero (Figure 4.22b). If the static exciter is fed by compounding both the generator voltage and current, as shown in Figure 2.3f, then excitation will not be lost during the short circuit as the exciter input voltage is augmented by the fault currents. The short-circuit current will not now vanish to zero but its shape will depend on the strength of the current component in the compounded excitation signal. An example of this is shown in Figure 4.22c.

4.2.7 Subtransient Torque

The torque creation mechanism is discussed in Section 3.3.2, Eq. (3.73), with regards to a two-pole generator in the steady state. This mechanism is equally applicable here except that now the d- and q-axis damper flux Φ_D and Φ_Q must be added into Eq. (3.74) to give

$$\tau = \tau_d - \tau_q = \frac{\pi}{2}(\Phi_f + \Phi_D + \Phi_{ad})F_{aq} - \frac{\pi}{2}(\Phi_Q + \Phi_{aq})F_{ad} \tag{4.55}$$

In this equation the field flux, Φ_f, is made up from two components Φ_{f0} and $\Delta\Phi_f$ where Φ_{f0} is the flux due to the initial field current i_{f0} and $\Delta\Phi_f$ is the change in the flux due to the change in the field current. If resistance effects are neglected, the rotor flux produced by the damper windings Φ_D and Φ_Q and the change in the field flux $\Delta\Phi_f$ completely cancels the armature reaction flux so that $\Phi_D + \Delta\Phi_f = -\Phi_{ad}$ and $\Phi_Q = -\Phi_{aq}$. Substituting these values into Eq. (4.55) shows that the electromagnetic torque is produced solely by the interaction of the armature q-axis mmf with the original field flux Φ_{f0}. Denoting this torque as τ_ω gives

$$\tau_\omega(t) = \frac{\pi}{2}\Phi_{f0}F_{aq} \tag{4.56}$$

As explained earlier, the AC component of the short-circuit armature mmf acts along the d-axis and does not contribute to the q-axis armature mmf. In contrast the DC component of armature flux is directed along the d-axis at the moment when the fault occurs, but then rotates relative to the rotor with the angle λ between the field and the armature mmf's changing as $\lambda = \omega t$. The amplitude of the DC armature mmf can be evaluated as $F_{a\,DC} = 1.5N_a i_m(0)$ by using a similar technique to that used to obtain Eq. (3.44). The q-axis armature mmf with respect to the rotor is then given by

$$F_{aq} = F_{a\,DC}\sin\lambda = \frac{3}{2}N_a i_m(0)\sin\omega t \tag{4.57}$$

As F_{aq} is solely dependent on the DC component of armature fault current, the torque is created by the initial field flux interacting with the DC component of the armature reaction mmf. This can be compared with two magnets, one rotating and the other stationary, as shown diagrammatically in Figure 4.23a. It is important to note that, for the three-phase fault, the angle λ is independent of the time that the fault is applied and simply varies as ωt. The torque variation is obtained by substituting (4.57) into Eq. (4.56) to give

$$\tau_\omega(t) = \frac{\pi}{2}\Phi_{f0}N_a\frac{3}{2}i_m(0)\sin\omega t \tag{4.58}$$

For the two-pole generator considered here $N_\phi = \pi N_a/2$, Eq. (3.41), when substituting for $E_f = \omega N_\varphi \Phi_{f0}/\sqrt{2}$ from Eq. (3.39) and for $i_m(0)$ from Eq. (4.50), finally gives the short circuit torque as

$$\tau_\omega(t) = \frac{3}{\omega}\frac{E_f^2}{X_d''}\sin\omega t \quad \text{Nm} \tag{4.59}$$

This equation shows that the short-circuit electromagnetic torque is independent of the fault instant but varies as $\sin\omega t$. During the first half-cycle, the electromagnetic torque opposes the mechanical driving torque, while during the second half-cycle it assists the driving torque. The average value of the torque is zero. Equation (4.59) expresses

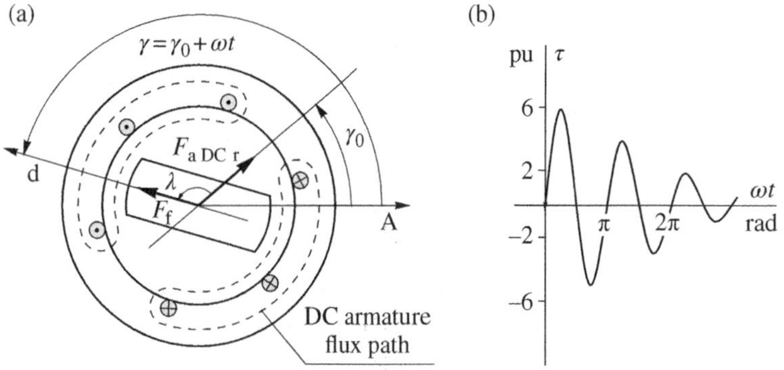

Figure 4.23 Subtransient electromagnetic torque following a three-phase fault: (a) torque creation mechanism; (b) example of torque variation with time.

the torque in SI units but, as explained in Appendix A.1.1, this can easily be changed to pu by multiplying by $\omega/3$ to give the pu form $\tau_\omega(t) = E_f^2 \sin \omega t / X_d''$ and is valid for any generator regardless of pole number.

The effect of including winding resistance is twofold. First of all, the DC phase currents that sustain the constant stator flux linkages $\Psi_{a\,dc}$ decay with a time constant T_a. Secondly, the rotor currents sustaining the constant rotor linkages decay with time constants T_d'' and T_d' allowing the armature reaction flux to penetrate the rotor. The net effect is that the torque τ_ω will decay to zero with the armature time constant T_a and the subtransient and transient time constants T_d'' and T_d' which modify Eq. (4.59) to give

$$\tau_\omega(t) = \frac{3}{\omega}\frac{E_f^2}{X_d''} g_3(t) e^{-t/T_a} \sin \omega t \quad \text{Nm} \tag{4.60}$$

where $g_3(t)$ is defined by Eq. (4.52). As the electromagnetic torque is equal to the product of two decaying functions, it vanishes very quickly (Figure 4.23b).

Strictly, the torque expression in Eq. (4.60) should be modified to take into account the torque developed due to the power loss in the armature resistance. As this power loss is due to the flow of AC armature currents, which decay with the subtransient and transient time constants defined by the function $g_3(t)$ in Eq. (4.52), the torque corresponding to the power loss in all three phases $\tau_R(t)$ may be expressed as

$$\tau_R(t) = \frac{3}{\omega}\left[\frac{E_f}{X_d''} g_3(t)\right]^2 R \quad \text{Nm} \tag{4.61}$$

Similarly, the alternating currents induced in the rotor windings (damper and field) also cause a power loss. These currents decay with time constant T_a and the torque corresponding to the rotor losses may be approximately expressed as

$$\tau_r(t) = \frac{3}{\omega}\left(\frac{i_m(0)}{\sqrt{2}} e^{-\frac{t}{T_a}}\right)^2 r = \frac{3}{\omega}\left(\frac{E_f}{X_d''}\right)^2 r e^{-\frac{2t}{T_a}} \quad \text{Nm} \tag{4.62}$$

where r is an equivalent resistance of all the rotor windings, referred to the stator (similarly as in a transformer). The torque τ_r is initially high but rapidly decays to zero.

As the armature resistance of a large generator is usually very small, the torque τ_R, corresponding to the stator losses, is several times smaller than the torque τ_r corresponding to the rotor losses. However, the former decays at a much slower rate.

The resultant torque acting on the rotor during the fault is equal to the sum of the three components τ_ω, τ_R, and τ_r. None of these components depends on the fault instant γ_0. The function $g_3(t)$, given by Eq. (4.52), does not decay to zero, so that the component, τ_R, decays to a steady-state value corresponding to the losses that would be incurred by the steady-state short-circuit currents. If the generator is on load prior to the fault then the initial value of the torque would correspond to the pre-fault load torque rather than zero, as shown in Figure 4.23.

Equation (4.53) suggests that there should also be an additional periodic torque component corresponding to the double-frequency currents caused by the rotor subtransient saliency. Assuming that the resistance of the armature windings is small, the double-frequency torque $\tau_{2\omega}$ may be expressed as

$$\tau_{2\omega}(t) = -\frac{3\,E_f^2}{2\,\omega}\left(\frac{1}{X_d''} - \frac{1}{X_q''}\right)e^{-\frac{2t}{T_a}}\sin 2\omega t \qquad \text{Nm} \tag{4.63}$$

Typically $X_d'' \cong X_q''$ and the torque $\tau_{2\omega}$ only slightly distorts the single-frequency torque τ_ω. In some rare cases of high subtransient saliency the double-frequency component may produce a substantial increase in the maximum value of the AC torque.

4.3 Phase-to-phase Short Circuit

The phase-to-phase fault will be analyzed in a similar way as the three-phase fault with the generator initially considered on no load and with both winding resistance and subtransient saliency neglected. The effect of saliency and resistance on the short-circuit current will then be considered before finally developing expressions for the short-circuit torque. The short-circuit will be assumed to occur across phases B and C of the stator winding.

4.3.1 Short-circuit Current and Flux with Winding Resistance Neglected

Figure 4.24 shows the armature coils with a short circuit across phases B and C. The short circuit connects the armature coils such that the current flows in opposite directions in the two coils and $i_B = -i_C$. In the case of the three-phase fault there were three closed armature circuits, one for each of the phase windings, and the flux linkage of each of these windings could be considered separately. In the case of the phase-to-phase short circuit there is only one closed circuit, the series connection of the two phase windings. When the fault occurs, the total energy stored in this closed circuit cannot change instantaneously and it is the net flux linkage $\Psi_B - \Psi_C$ of this closed circuit that must remain constant.

The flux linkages Ψ_B and Ψ_C of the two short-circuited windings consist of two components: the winding self-flux linkages Ψ_{BB} and Ψ_{CC} produced by the fault currents and the flux linkages produced by the excitation flux Ψ_{fB} and Ψ_{fC}. As the fault current in the two short-circuited phases are equal and opposite, $\Psi_{BB} = -\Psi_{CC}$ and the net flux linkage $\Psi_B - \Psi_C$ can be expressed as

$$\Psi_B - \Psi_C = (\Psi_{BB} + \Psi_{fB}) - (\Psi_{CC} + \Psi_{fC}) = \Psi_{fB} - \Psi_{fC} - 2\Psi_{CC} \tag{4.64}$$

With the generator on no load prior to the fault $\Psi_{CC}(0^-) \propto i_C(0^-) = i_B(0^-) = 0$ and the only flux linking the windings is the excitation flux. Thus

$$\Psi_B(0^-) - \Psi_C(0^-) = \Psi_{fB0} - \Psi_{fC0} \tag{4.65}$$

The law of constant flux linkage requires the net flux linkage $(\Psi_B - \Psi_C)$ just before and just after the fault to be equal so that

$$\Psi_{fB} - \Psi_{fC} - 2\Psi_{CC} = \Psi_{fB0} - \Psi_{fC0} \tag{4.66}$$

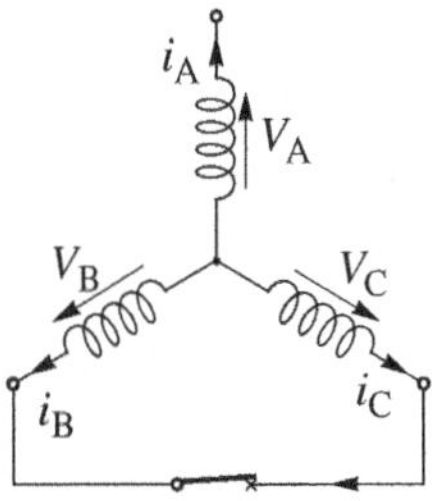

Figure 4.24 Phase-to-phase short circuit: $i_B = -i_C$; $V_{BC} = 0$; $i_A = 0$; $V_A \neq 0$.

Rearranging gives

$$\Psi_{CC} = \frac{1}{2}[(\Psi_{fB} - \Psi_{fC}) - (\Psi_{fB0} - \Psi_{fC0})] \tag{4.67}$$

The variation of the flux linkages Ψ_{fB} and Ψ_{fC} are shown in Figure 4.6, with the initial values defined by Eqs. (4.19), so that

$$\begin{aligned} &\Psi_{fB} = \Psi_{fa}\cos(\gamma - 2\pi/3); \qquad &&\Psi_{fC} = \Psi_{fa}\cos(\gamma - 4\pi/3) \\ &\Psi_{fB0} = \Psi_{fa}\cos(\gamma_0 - 2\pi/3); \qquad &&\Psi_{fC0} = \Psi_{fa}\cos(\gamma_0 - 4\pi/3) \end{aligned} \tag{4.68}$$

which, when substituted into Eq. (4.67), give

$$\Psi_{CC} = \frac{\sqrt{3}}{2}\Psi_{fa}(\sin\gamma - \sin\gamma_0) \tag{4.69}$$

Recall from Eq. (3.39) that $E_{fm} = \omega\Psi_{fa}$ while immediately after the fault the subtransient current $i_C = \Psi_{CC}/L''_d$. This allows a formula for the phase-to-phase short-circuit current to be written as

$$i_C = -i_B = \frac{\sqrt{3}}{2}\frac{E_{fm}}{X''_d}(\sin\gamma - \sin\gamma_0) \tag{4.70}$$

This short-circuit current consists of two components, an AC component

$$i_{C\,AC} = -i_{B\,AC} = \frac{\sqrt{3}E_{fm}}{2X''_d}\sin\gamma \tag{4.71}$$

whose magnitude is independent of the instant of the fault and a DC component

$$i_{C\,DC} = -i_{B\,DC} = -\frac{\sqrt{3}E_{fm}}{2X''_d}\sin\gamma_0 \tag{4.72}$$

the value of which depends on the instant of the fault. If the fault occurs at $\gamma_0 = 0$, the voltage of phase A is zero, and the fault current is purely sinusoidal with no DC component present. If, on the other hand, the fault occurs at $\gamma_0 = -\pi/2$, the voltage in phase A is at its negative peak, and the DC component of the fault current will be at its maximum value.

The flux linkages for the case when $\gamma_0 = 0$ and for $\gamma_0 = -\pi/2$ are shown in Figure 4.25a and b, respectively. When $\gamma_0 = 0$ the flux linkages Ψ_{fB0} and Ψ_{fC0} are identical, the net flux linkage $(\Psi_{fB0} - \Psi_{fC0})$ is zero and no DC component is required to maintain the flux linkages. The flux linkages Ψ_{BB} and Ψ_{CC}, and hence i_B and i_C, simply vary sinusoidally with opposite sign (Figure 4.25a). However, when the fault occurs at $\gamma_0 = -\pi/2$ the voltage in phase A is a maximum and the flux linkages Ψ_{fB0} and Ψ_{fC0} now have the same magnitude but opposite sign. The net flux linkage is $\sqrt{3}\Psi_{fa}$ and a large DC circulating current $i_{C\,DC} = -i_{B\,DC}$ is necessary in order to maintain constant flux linkages (Figure 4.25b).

The resultant armature mmf $\vec{F}_a$ can be found by the vector addition of the individual mmf's $\vec{F}_B$ and $\vec{F}_C$ produced by the stator fault currents i_B and i_C. Taking phase A as a reference (real axis in the complex plane) gives

$$\begin{aligned} \vec{F}_a &= \vec{F}_B + \vec{F}_C = F_B e^{j2\pi/3} + F_C e^{j4\pi/3} = F_C\left(-e^{j2\pi/3} + e^{j4\pi/3}\right) \\ &= -j\sqrt{3}F_C = -j\sqrt{3}N_a i_C = j\sqrt{3}N_a\frac{\sqrt{3}}{2}\frac{E_{fm}}{X''_d}(\sin\gamma - \sin\gamma_0) \end{aligned} \tag{4.73}$$

The result of this phasor addition is shown in Figure 4.26 where the resultant armature mmf $\vec{F}_a$ is at all times perpendicular to the axis of phase A and proportional to the short-circuit current i_C. Consequently, $\vec{F}_a$ is stationary

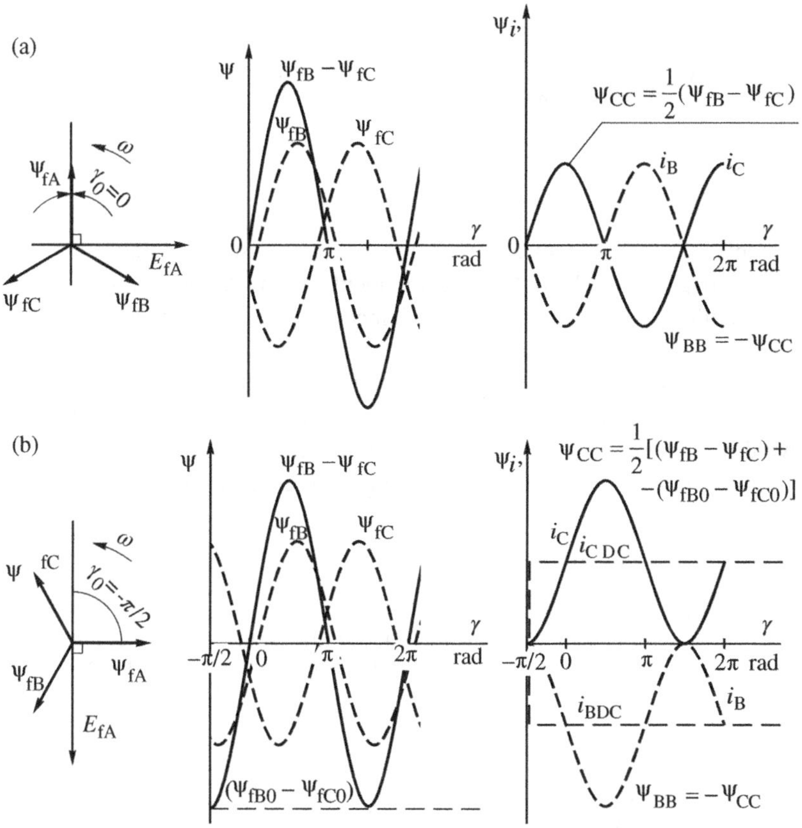

Figure 4.25 Application of the law of constant flux linkages to determine the phase-to-phase fault currents when the fault instant is: (a) $\gamma_0 = 0$; (b) $\gamma_0 = -\pi/2$.

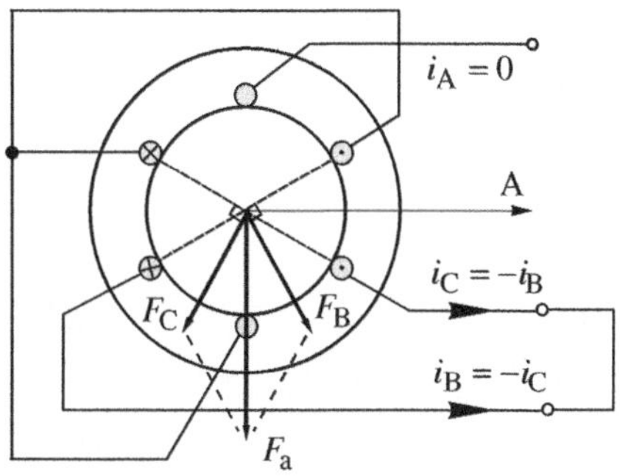

Figure 4.26 Phase-to-phase fault currents creating a stationary armature mmf F_a perpendicular to the axis of phase A. Generator windings are star connected.

in space but pulsates at frequency ω. As the rotor position at any instant in time is determined by the angle γ measured with respect to the A-axis, this stationary armature mmf counter-rotates with respect to the rotor with the angle between the armature and the field mmf changing as $\lambda = \gamma + \pi/2$. This produces mmf components along the d- and q-axes

$$F_{ad} = F_a \cos\lambda = -\sqrt{3}\,F_C \sin\gamma; \qquad F_{aq} = F_a \sin\lambda = \sqrt{3}\,F_C \cos\gamma \tag{4.74}$$

which, when substituting for i_C from Eq. (4.70), give

$$\begin{aligned} F_{ad} &= -\frac{3}{2}N_a\frac{E_{fm}}{X''_d}(\sin\gamma - \sin\gamma_0)\sin\gamma = -\frac{3}{4}N_a\frac{E_{fm}}{X''_d}(1 - 2\sin\gamma_0\sin\gamma - \cos 2\gamma) \\ F_{aq} &= \frac{3}{2}N_a\frac{E_{fm}}{X''_d}(\sin\gamma - \sin\gamma_0)\cos\gamma \end{aligned} \tag{4.75}$$

The d-axis armature mmf has three components: a DC component, an AC component of fundamental frequency, and an AC component of double frequency. The magnitude of both the DC component and the double-frequency component is independent of the instant of the fault, while the value of the fundamental AC component depends on γ_0.

This variation in the d-axis armature mmf also explains the changes that occur in the rotor field current and the d-axis damper current following the fault. The d-axis armature mmf forces an armature flux across the air gap that produces flux linkages Ψ_{ar} with the rotor windings. These flux linkages are proportional to F_{ad} and are shown in Figure 4.27. As the flux linkage of the rotor circuits must remain constant following the fault, additional currents

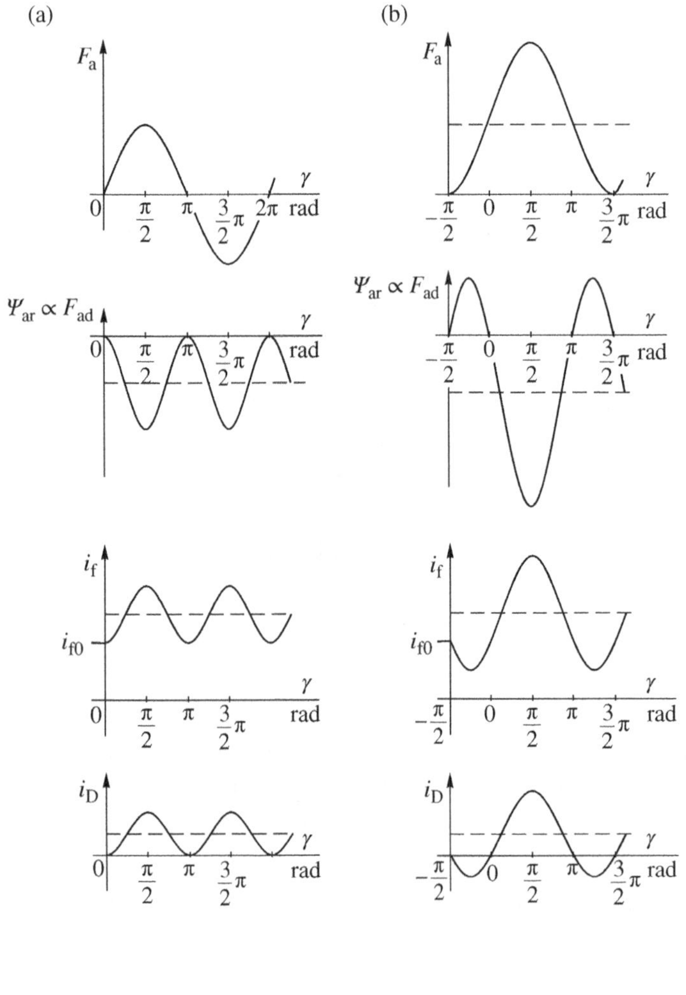

Figure 4.27 Illustration of the influence of the armature flux on the rotor currents when the fault occurs at: (a) $\gamma_0 = 0$; (b) $\gamma_0 = -\pi/2$.

flow in both the field winding and the d-axis damper winding to produce a flux linkage that is equal and opposite to that produced by the armature flux. These additional field and damper currents therefore have the same form as F_{ad} but are of opposite sign. The actual magnitude of the field current will depend on how well the field winding is screened by the d-axis damper winding. If perfect screening is obtained then the field current will remain constant with the armature flux being totally compensated by the d-axis damper current. If the fault occurs at $\gamma_0 = 0$ then the armature current and the resultant armature mmf do not contain a DC component (Figure 4.27a) and the flux linkages in the field and damper windings produced by the armature mmf simply pulsate at twice the system frequency. The induced rotor currents are then similar to those flowing during a three-phase fault, except they are now at twice the system frequency. If, on the other hand, the fault occurs at $\gamma_0 = -\pi/2$ (Figure 4.27b) then the armature mmf pulsates with a large DC component. The rotor flux linkages now contain both fundamental and double-frequency components and the induced rotor currents have a different shape than before in that both the field and the damper currents initially reduce their values, as shown in Figure 4.27b.

The q-axis mmf F_{aq} will be compensated by damper currents in a similar manner as F_{ad}, and its influence will be discussed when the torque is considered.

4.3.2 Influence of the Subtransient Saliency

The armature current Eq. (4.70) is valid for a magnetically symmetrical rotor where the magnetic reluctance encountered by the armature flux is constant and does not depend on the rotor position. However, if the machine has substantial subtransient saliency then the reluctance of the d- and q-axes will be different and the solution must be obtained using two-reaction theory (Ching and Adkins 1954; Kundur 1994). The result of such an analysis shows that in Eq. (4.70) X''_d is replaced by $\left(X''_d \sin^2\gamma + X''_q \cos^2\gamma\right)$ to give

$$i_C = -i_B = \frac{\sqrt{3}E_{fm}(\sin\gamma - \sin\gamma_0)}{2\left(X''_d \sin^2\gamma + X''_q \cos^2\gamma\right)} = \frac{\sqrt{3}E_{fm}(\sin\gamma - \sin\gamma_0)}{X''_d + X''_q - \left(X''_d - X''_q\right)\cos 2\gamma} \tag{4.76}$$

Generators without a q-axis damper winding may have a large value of subtransient saliency which will distinctly distort the fault currents. Figure 4.28 shows these distorted currents for the same γ_0 as the sinusoidal waveforms shown previously in Figure 4.25 when X''_d was equal to X''_q.

The harmonic content of the distorted fault current can be quantified by expanding Eq. (4.76) into a Fourier series[2]

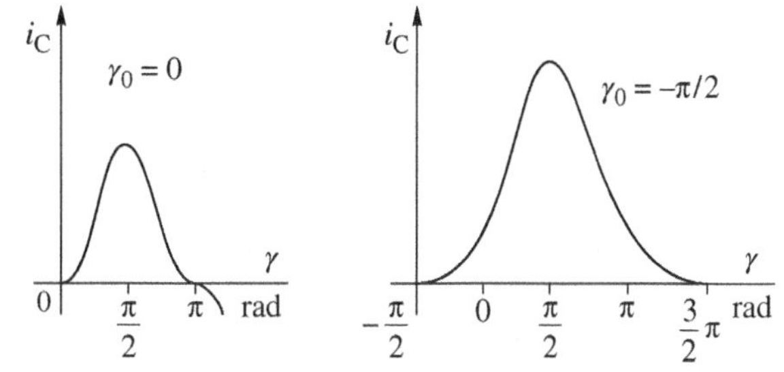

Figure 4.28 Phase-to-phase fault current waveforms strongly distorted because of subtransient saliency.

2 Fourier series expansion of the two components in Eq. (4.76) gives

$$\frac{\sin\gamma}{A + B - (A - B)\cos 2\gamma} = \frac{1}{A + \sqrt{AB}}\left(\sin\gamma - b\sin 3\gamma + b^2\sin 5\gamma - b^3\sin 7\gamma + \dots\right)$$

$$\frac{1}{A + B - (A - B)\cos 2\gamma} = \frac{1}{\sqrt{AB}}\left(\frac{1}{2} - b\sin 2\gamma + b^2\cos 4\gamma - b^3\cos 6\gamma + \dots\right) \quad \text{where} \quad b = \frac{\sqrt{B} - \sqrt{A}}{\sqrt{B} + \sqrt{A}}$$

$$i_C = -i_B = \frac{\sqrt{3}E_{fm}}{X''_d + \sqrt{X''_d X''_q}}(\sin\gamma - b\sin 3\gamma + b^2\sin 5\gamma - b^3\sin 7\gamma + \ldots) - \frac{\sqrt{3}E_{fm}\sin\gamma_0}{\sqrt{X''_d X''_q}}\left(\frac{1}{2} - b\cos 2\gamma + b^2\cos 4\gamma - b^3\cos 6\gamma + \ldots\right) \tag{4.77}$$

where the asymmetry coefficient, b, is

$$b = \frac{\sqrt{X''_q} - \sqrt{X''_d}}{\sqrt{X''_q} + \sqrt{X''_d}} = \frac{\sqrt{X''_d X''_q} - X''_d}{\sqrt{X''_d X''_q} + X''_d} \tag{4.78}$$

In the example shown in Figure 4.28 the generator possesses a high degree of subtransient saliency with $X''_q = 2X''_d$, giving an asymmetry coefficient $b \cong 0.17$, while the magnitude of the third, fifth, and seventh harmonics are respectively 17, 3, and 0.5% of the fundamental.

Even-numbered harmonics appear only when the fault current contains a DC component, that is when $\sin\gamma_0 \neq 0$ and subtransient saliency is present. However, the odd-numbered harmonics appear whenever the generator possesses subtransient saliency, $b \neq 0$, regardless of the instant of the fault. The source of the odd-numbered harmonics is the negative-sequence component of the fault current, as explained below.

4.3.2.1 Symmetrical Component Analysis of the Phase-to-phase Fault

Symmetrical components allow any unsymmetrical set of three-phase currents to be expressed as the phasor sum of three symmetrical AC components: the positive-sequence component $\underline{i}_1$, the negative-sequence component $\underline{i}_2$, and the zero-sequence component $\underline{i}_0$, the detailed derivation of which can be found in any standard textbook on power system analysis, such as Grainger and Stevenson (1994).

The phase-to-phase fault currents shown in Figure 4.25a are given by $i_A = 0$ and $i_B = -i_C = i$. Applying the symmetrical component transformation gives

$$\begin{bmatrix} \underline{i}_0 \\ \underline{i}_1 \\ \underline{i}_2 \end{bmatrix} = \frac{1}{3}\begin{bmatrix} 1 & 1 & 1 \\ 1 & a & a^2 \\ 1 & a^2 & a \end{bmatrix}\begin{bmatrix} 0 \\ \underline{i} \\ -\underline{i} \end{bmatrix} = \frac{i}{3}\begin{bmatrix} 0 \\ a - a^2 \\ a^2 - a \end{bmatrix} = \frac{i}{\sqrt{3}}\begin{bmatrix} 0 \\ j \\ -j \end{bmatrix} \tag{4.79}$$

where $a = e^{j2\pi/3}$. The positive-sequence component, $\underline{i}_1$, leads the current in the B phase by $\pi/2$, while the negative-sequence component, $\underline{i}_2$, lags the current in the B phase by $\pi/2$. The zero-sequence component is equal to zero. The reverse transformation from the symmetric components to phase components gives the actual phase currents, which can be conveniently written as the sum of two three-phase systems representing the positive-sequence and negative-sequence phase currents

$$\begin{bmatrix} \underline{i}_A \\ \underline{i}_B \\ \underline{i}_C \end{bmatrix} = \begin{bmatrix} 1 & 1 & 1 \\ 1 & a^2 & a \\ 1 & a & a^2 \end{bmatrix}\begin{bmatrix} 0 \\ \underline{i}_1 \\ \underline{i}_2 \end{bmatrix} = \begin{bmatrix} \underline{i}_{A1} \\ \underline{i}_{B1} \\ \underline{i}_{C1} \end{bmatrix} + \begin{bmatrix} \underline{i}_{A2} \\ \underline{i}_{B2} \\ \underline{i}_{C2} \end{bmatrix} \tag{4.80}$$

where $\underline{i}_{A1} = \underline{i}_1$, $\underline{i}_{B1} = a^2\underline{i}_1$, $\underline{i}_{C1} = a\underline{i}_1$, $\underline{i}_{A2} = \underline{i}_2$, $\underline{i}_{B2} = a\underline{i}_2$, and $\underline{i}_{C2} = a^2\underline{i}_2$.

Figure 4.29 shows the phase and symmetrical components of the phase-to-phase fault currents.

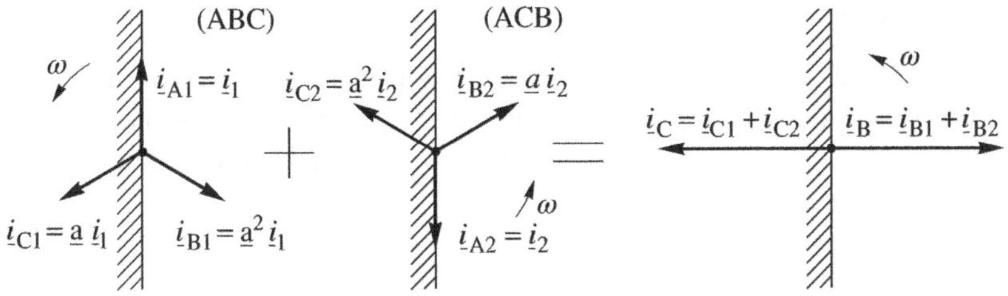

Figure 4.29 The phase-to-phase fault current transformed into its positive- and negative-sequence components rotating in opposite directions.

4.3.2.2 Phase-to-phase Fault Current Harmonics Explained by Symmetrical Components

The phase-to-phase fault current defined by Eq. (4.76) can be decomposed into two fundamental parts: an AC part containing terms in $\sin\gamma$ and a DC part containing terms in $\sin\gamma_0$, each part of which may be analyzed separately using symmetrical components.

The AC part of the fault current produces an armature mmf $F_{a\ AC}$ which is stationary with respect to the stator but pulsates with time. By representing the AC component of the fault current by a positive-sequence and negative-sequence current, the effect of this pulsating mmf can be replicated by two counter-rotating mmf waves produced by the sequence currents. The positive-sequence currents (i_{A1}, i_{B1}, and i_{C1}) produce an mmf $F_{a\ AC1}$ which rotates with the rotor and is stationary with respect to it. Changes in the flux linkages produced by this mmf are opposed by direct currents flowing in the rotor windings, just as in the generator with no subtransient saliency shown in Figure 4.27. The negative-sequence currents (i_{A2}, i_{B2}, and i_{C2}) produce mmf $F_{a\ AC2}$, which rotates in the opposite direction to the rotor and induces double-frequency currents in the rotor windings, again as shown in Figure 4.27 for a generator with no subtransient saliency. In turn, these double-frequency rotor currents produce a pulsating mmf which is stationary with respect to the rotor. This pulsating mmf can be represented by two mmf waves, one rotating with respect to the rotor with velocity (-2ω) and the other with velocity 2ω, with the relative magnitudes of these two mmf waves depending on the degree of subtransient saliency. As the rotor itself rotates with angular velocity ω, these two rotor mmf components rotate with respect to the stator with velocities $(-\omega)$ and 3ω and introduce a third harmonic component into the fault current. Analyzing the effect of this third harmonic on the rotor currents, and then the effect of the rotor currents on the armature, leads to an explanation for the presence of the fifth harmonic in the fault current. This analysis may be continuously repeated to account for all the odd-numbered harmonics in the armature fault current. For a generator with no subtransient saliency the magnitude of the rotor mmf wave that rotates at 3ω relative to the armature is zero and so no third- or higher-order harmonics are present.

When $\sin\gamma_0 \neq 0$, a DC component of the fault current appears and produces a stationary armature mmf $F_{a\ DC}$ which counter-rotates with respect to the rotor. The rotor windings oppose this mmf by inducing fundamental-frequency alternating currents. These rotor currents produce a sinusoidally pulsating mmf which is stationary with respect to the rotor. As before, this mmf can be represented as two counter-rotating mmf's, but in this case with angular velocities ω and $(-\omega)$ with respect to the rotor. Again, the relative magnitudes of these two mmf's depend on the degree of subtransient saliency present. As the rotational velocity of the rotor is ω, one of these mmf's is stationary with respect to the armature while the other rotates with velocity 2ω and induces a second-harmonic component into the fault current. Analyzing the influence of this harmonic explains the presence of the fourth and the other even-numbered harmonics in the fault current when subtransient saliency is present. The phase-to-phase fault current may then be expressed as a sum of the harmonics

$$i_C = -i_B = i_{(\omega)} + \underbrace{i_{(3\omega)} + i_{(5\omega)} + \dots}_{\text{Induced by the negative sequence currents}} + \underbrace{i_{DC} + i_{(2\omega)} + i_{(4\omega)} + i_{(6\omega)} + \dots}_{\text{Induced by the DC currents depending on the fault instant}} \tag{4.81}$$

4.3.3 Positive- and Negative-sequence Reactances

Equation (4.77) can now be used to determine the reactance with which the generator opposes the flow of the negative-sequence currents. Figure 4.30a shows the representation of the generator for the positive- and negative-sequence currents, while Figure 4.30b shows the connection of the sequence equivalent circuits required to represent the phase-to-phase fault (Grainger and Stevenson 1994).

Applying Ohm's law to the circuit of Figure 4.30b gives

$$\underline{I}_1'' = -\underline{I}_2'' = \frac{\underline{E}''}{\mathrm{j}(X_1'' + X_2'')} \tag{4.82}$$

where X_1'' is the positive-sequence subtransient reactance and X_2'' the negative-sequence subtransient reactance. Using the inverse symmetrical component transformation gives the phase currents

$$-\underline{I}_\mathrm{B} = \underline{I}_\mathrm{C} = \frac{\sqrt{3}\,\underline{E}''}{X_1'' + X_2''} \tag{4.83}$$

The positive-sequence component of the fault current produces a rotating flux whose interaction with the rotor is similar to that produced by the AC component of the three-phase short-circuit current. As the generator opposed the flow of currents produced by this flux with the reactance X_d'', the reactance with which the generator opposes the flow of the positive-sequence currents is also equal to X_d'' and

$$X_1'' = X_\mathrm{d}'' \tag{4.84}$$

Comparing Eq. (4.83) with the first part of Eq. (4.77) allows the negative-sequence reactance X_2'' to be evaluated as

$$X_2'' = \sqrt{X_\mathrm{d}'' X_\mathrm{q}''} \tag{4.85}$$

and the generator opposes the flow of the negative-sequence currents with a reactance equal to the geometric average of the two subtransient reactances. Finally, the second part of Eq. (4.77) shows that the generator opposes the flow of the DC component of the fault current with a reactance that has the same value as the negative sequence reactance, that is $\sqrt{X_\mathrm{d}'' X_\mathrm{q}''}$.

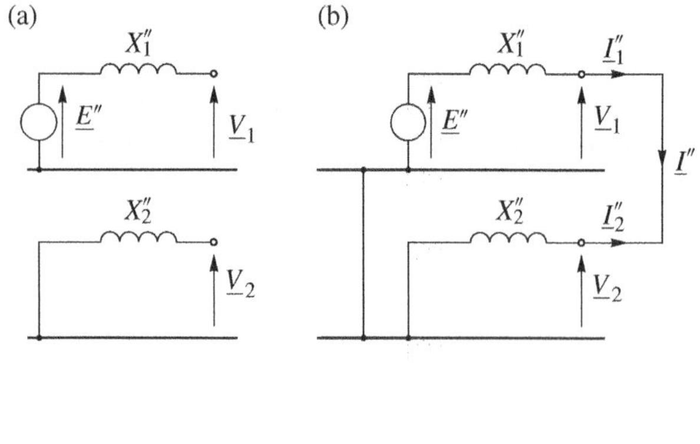

Figure 4.30 Equivalent circuits of the generator: (a) positive- and negative-sequence circuits; (b) connection of the circuits to account for the phase-to-phase fault.

4.3.4 Influence of Winding Resistance

The winding resistance dissipates the magnetic energy stored in the winding and causes a decay in the DC component of the current flowing in that winding. As explained above, the generator opposes the flow of the DC component of the phase-to-phase fault current with the reactance $X''_2 = \sqrt{X''_d X''_q}$. The resistance of the phase winding will then produce a decay in this DC component of the fault current with a time constant

$$T_a = \frac{X''_2}{\omega R} = \frac{\sqrt{L''_d L''_q}}{R} \tag{4.86}$$

Equation (4.76) can now be modified by multiplying the $\sin\gamma_0$ term by the exponential function $\exp(-t/T_a)$.

The decay of the AC component of the fault current can best be explained by considering its positive- and negative-sequence components as derived in the previous section. The positive-sequence component produces a rotating flux whose penetration into the rotor is opposed by DC rotor currents. The resistance of the rotor windings causes this component of the current to decay, first in the damper windings and then in the field winding, so that the armature reaction flux penetrates deeper and deeper into the rotor. As each of these flux conditions corresponds to the subtransient, transient, and steady-state conditions, the generator effectively opposes the flow of the positive-sequence component of the fault current with the same reactances as for the three-phase fault:

$X''_1 = X''_d$ in the subtransient state
$X'_1 = X'_d$ in the transient state
$X_1 = X_d$ in the steady state

The armature mmf produced by the negative-sequence current rotates in the opposite direction to the rotor. The penetration of the flux produced by this mmf into the rotor is prevented by the flow of double-frequency rotor currents so that the same reactance opposes the flow of the negative- sequence fault currents in all three characteristic generator states. The value of this negative- sequence reactance is

$$X''_2 = X'_2 = X_2 = \sqrt{X''_d X''_q} \tag{4.87}$$

As the positive- and negative-sequence reactances of the generator are connected in the same way for the transient state and the steady state as for the subtransient state, Figure 4.30, the AC phase-to-phase transient and steady-state fault currents are, respectively, $(X''_d + X_2)/(X'_d + X_2)$ and $(X''_d + X_2)/(X_d + X_2)$ times smaller than the subtransient fault currents. Equation (4.76) can be used to express the AC component of the fault current in the three characteristic states as

$$\begin{aligned} i''_{C\,AC}(t) &= \frac{\sqrt{3}E_{fm}\sin\gamma}{2\left(X''_d \sin^2\gamma + X''_q \cos^2\gamma\right)} \\ i'_{C\,AC}(t) &= i''_{C\,AC}(t)\frac{X''_d + X_2}{X'_d + X_2}; \qquad i_{C\,AC}(t) = i''_{C\,AC}(t)\frac{X''_d + X_2}{X+_d + X_2} \end{aligned} \tag{4.88}$$

Similarly, as during the three-phase fault, the difference $\left(i''_{C\,AC} - i'_{C\,AC}\right)$ decays at a rate determined by the subtransient time constant, while the difference $\left(i'_{C\,AC} - i_{C\,AC}\right)$ decays with the transient time constant. Following similar arguments to those used to produce Eqs. (4.48) and (4.51), the following expression is obtained for the phase-to-phase fault current with the effect of winding resistance included

$$i_C(t) = \frac{\sqrt{3}E_{fm}}{2\left(X''_d \sin^2\gamma + X''_q \cos^2\gamma\right)}\left[g_2(t)\sin\gamma - e^{-t/T_a}\sin\gamma_0\right] \tag{4.89}$$

where the function

$$
\begin{aligned}
g_2(t) = (X''_\mathrm{d} + X_2) & \left[\left(\frac{1}{X''_\mathrm{d} + X_2} - \frac{1}{X'_\mathrm{d} + X_2}\right)\mathrm{e}^{-t/T''_\beta}\right. \\
& \left. + \left(\frac{1}{X'_\mathrm{d} + X_2} - \frac{1}{X_\mathrm{d} + X_2}\right)\mathrm{e}^{-t/T'_\beta} + \frac{1}{X_\mathrm{d} + X_2}\right]
\end{aligned}
\tag{4.90}
$$

is responsible for describing the decay of the AC component of the fault current (subscript "2" denotes the phase-to-phase fault).

As the reactances with which the generator opposes the flow of the phase-to-phase fault current are different from those opposing the flow of the three-phase fault current, the subtransient and transient time constants must be changed in proportion to the ratio of the reactances

$$
T''_\beta = T''_\mathrm{d}\left(\frac{X'_\mathrm{d}}{X''_\mathrm{d}}\right)\left(\frac{X''_\mathrm{d} + X_2}{X'_\mathrm{d} + X_2}\right); \quad T'_\beta = T'_\mathrm{d}\left(\frac{X_\mathrm{d}}{X'_\mathrm{d}}\right)\left(\frac{X'_\mathrm{d} + X_2}{X_\mathrm{d} + X_2}\right)
\tag{4.91}
$$

4.3.5 Subtransient Torque

Similarly as in the three-phase fault, the flux produced by the damper windings during the phase-to-phase fault is equal and opposite to the d- and q-axis armature flux, and the electromagnetic torque is due solely to the interaction between the field flux and the quadrature component of the armature reaction mmf. Neglecting subtransient saliency and substituting Eq. (4.75) into (4.56), similar to the three-phase fault, gives

$$
\tau_\mathrm{ac} = \frac{\pi}{2}\Phi_{\mathrm{f}0}\frac{3}{2}N_\mathrm{a}\frac{E_\mathrm{fm}}{X''_\mathrm{d}}(\sin\gamma - \sin\gamma_0)\cos\gamma
\tag{4.92}
$$

but, as $E_\mathrm{f} = \frac{1}{\sqrt{2}}\omega\left(\frac{\pi}{2}N_\mathrm{a}\right)\Phi_{\mathrm{f}0}$, the subtransient torque can be written as

$$
\tau_\mathrm{ac} = \frac{3}{\omega}\frac{E_\mathrm{f}^2}{X''_\mathrm{d}}(\sin\gamma - \sin\gamma_0)\cos\gamma \quad \mathrm{Nm}
\tag{4.93}
$$

This equation neglects winding resistance, the effect of which is to cause the rotor currents to decay with time as determined by the function $g_2(t)$, while the decay of the stator currents is determined by the functions $g_2(t)$ and $\exp(-t/T_\beta)$. Multiplying the corresponding components of Eq. (4.93) by these functions gives

$$
\begin{aligned}
\tau_\mathrm{ac} &= \frac{3}{\omega}\frac{E_\mathrm{f}^2}{X''_\mathrm{d}}\left[g_2(t)\sin\gamma - \mathrm{e}^{-t/T_\beta}\sin\gamma_0\right]g_2(t)\cos\gamma \\
&= \frac{3}{\omega}\frac{E_\mathrm{f}^2}{X''_\mathrm{d}}\left[\frac{1}{2}g_2^2(t)\sin 2\gamma - \sin\gamma_0\mathrm{e}^{-t/T_\beta}g_2(t)\cos\gamma\right] \quad \mathrm{Nm}
\end{aligned}
\tag{4.94}
$$

Unlike the torque due to the three-phase fault, the electromagnetic torque produced during a phase-to-phase fault depends on the instant of the fault. The component of the torque that depends on γ_0 has a periodic variation at fundamental frequency and a maximum value when the fault occurs when the voltage in phase A reaches its negative maximum, that is when $\gamma_0 = -\pi/2$. In this situation the fundamental-frequency component is initially the highest and has a value twice that of the double-frequency component. As time passes, the magnitude of the fundamental- frequency torque component reduces until, in the steady state, it vanishes completely (see Figure 4.31) and the torque varies at double frequency.

Including the effect of subtransient saliency introduces considerable complications in the formulae. First of all, the application of two-reaction theory leads to a term $X''_\mathrm{d}\sin^2\gamma + X''_\mathrm{q}\cos^2\gamma$ appearing in the denominator, similar to the expression for the current. Secondly, as in the case of the three-phase fault, an AC component at double

Figure 4.31 Example of the changes in the electromagnetic torque during phase-to-phase fault when $\gamma_0 = -\pi/2$.

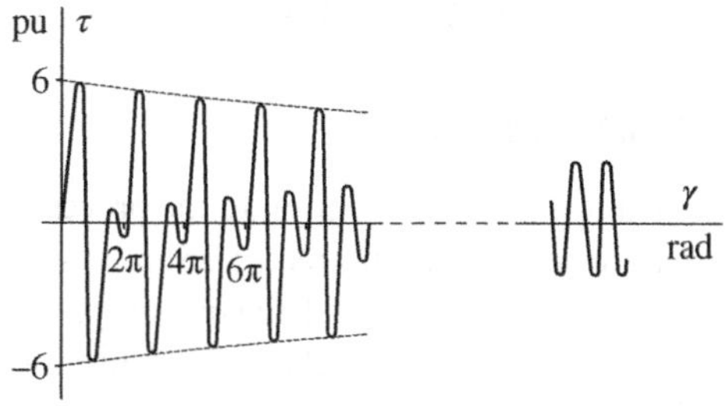

frequency appears that is dependent on $\left(X''_q - X''_d\right)$. The torque is then expressed as

$$\tau_{ac} = \frac{3}{\omega} \frac{E_f^2}{X''_d \sin^2\gamma + X''_q \cos^2\gamma} \left\{ \left[g_2(t) \sin\gamma - e^{-t/T_\beta} \sin\gamma_0 \right] g_2(t) \cos\omega t + \frac{1}{2} \frac{\left(X''_q - X''_d\right) \left[g_2(t) \sin\gamma - e^{-t/T_\beta} \sin\gamma_0 \right]^2 \sin 2\gamma}{X''_d \sin^2\gamma + X''_q \cos^2\gamma} \right\} \ \text{Nm} \tag{4.95}$$

When the difference $\left(X''_q - X''_d\right)$ is small, the torque given by Eq. (4.95) is similar in both shape and maximum value to that obtained from Eq. (4.94). If subtransient saliency is high then the second part of Eq. (4.95) will distort the torque profile and may also increase the instantaneous value of the torque. Figure 4.32 shows the variation of the torque in a case of high subtransient saliency when $X''_q \cong 2X''_d$.

Equation (4.95) defines one part of the electromagnetic torque occurring during a phase-to-phase short circuit and, just as in the case of the three-phase fault, there will be other torque components resulting from the power losses in the armature phase windings and the rotor field and damper windings. However, these are usually small compared with the torque defined in Eq. (4.95).

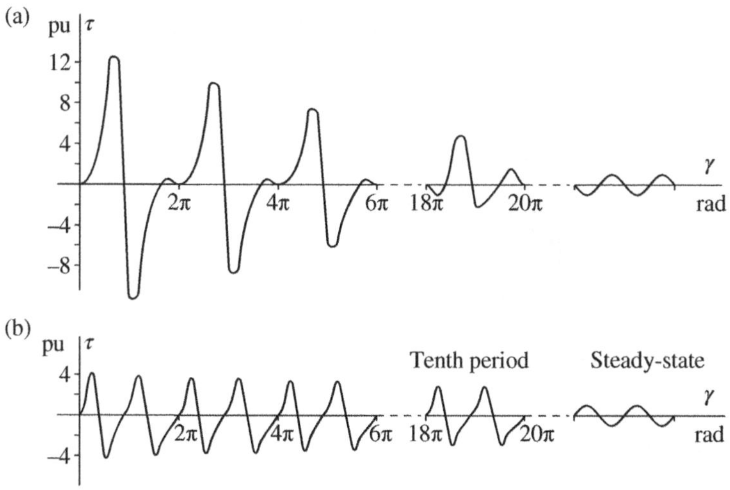

Figure 4.32 Example of torque variation during phase-to-phase fault in a case of high subtransient saliency of $X''_q \cong 2X''_d$ when the fault instant is: (a) $\gamma_0 = -\pi/2$; (b) $\gamma_0 = 0$.

4.4 Switching Operations

4.4.1 Synchronization

Synchronization is the name given to the process of connecting a generator, with its field winding excited, to the EPS. Such a synchronization process is shown schematically in Figure 4.33a where the EPS is replaced by an infinite busbar of voltage $\underline{V}_s$ behind an equivalent reactance X_s. It is assumed that just before the switch is closed the generator rotates at a speed ω close to synchronous speed ω_s and that the excitation produces a no-load terminal voltage E_f close to the system voltage V_s.

Ideal synchronization occurs when $\omega = \omega_s$, $\underline{E}_f = \underline{V}_s$, and $\delta = 0$ so that when the synchronizing switch is closed no circulating current will flow. Normally, a generator is connected to an EPS when these conditions are almost, but not exactly, satisfied, when closing the synchronizing switch will result in the flow of a circulating current that contains both an AC and a DC component. If the generator is assumed to be no local load prior to synchronization then it can be represented in the subtransient state by the excitation electromotive force (emf) E_f behind the subtransient reactance X''_d. The circulating current can then be evaluated using Thévenin's theorem based on the voltage $\Delta\underline{V}$ across the synchronizing switch and the impedance as seen from the switch terminals. For simplicity armature resistance and rotor subtransient saliency will be neglected by assuming $R = 0$ and $X''_q = X''_d$.

The subtransient armature current can be found by resolving the voltage across the switch $\Delta\underline{V}$ into two orthogonal components directed along the a- and b-axes, as shown in Figure 4.33c. It is important to stress that although the a- and b-axes are directed along the d- and q-axes, the resulting (a, b) voltage and current components have a different meaning to the (d, q) components. The (a, b) components are simply the components of the phase quantities resolved along two axes in the complex plane while the (d, q) voltage and current components are associated with fictitious, rotating, orthogonal armature windings, as explained in Section 3.3. As the circuit is assumed to be purely reactive the current must lag the forcing voltage by $\pi/2$. This means that the a-component voltage ΔV_a will force the flow of the b-component of current I_b, and vice versa.

Applying Thévenin's theorem gives the maximum value of the b-axis current, i_{bm}, owing to ΔV_a as

$$i_{bm} = \frac{\Delta V_{am}}{x''_d} = \frac{V_{sm}\cos\delta - E_{fm}}{x''_d} \tag{4.96}$$

where $x''_d = X''_d + X_T + X_s$, $E_{fm} = \sqrt{2}E_f$, and $V_{sm} = \sqrt{2}V_s$. As the voltage component ΔV_a acts along the q-axis its effect in the Thévenin equivalent circuit, when looking from the terminals of the switch, is similar to that of the emf

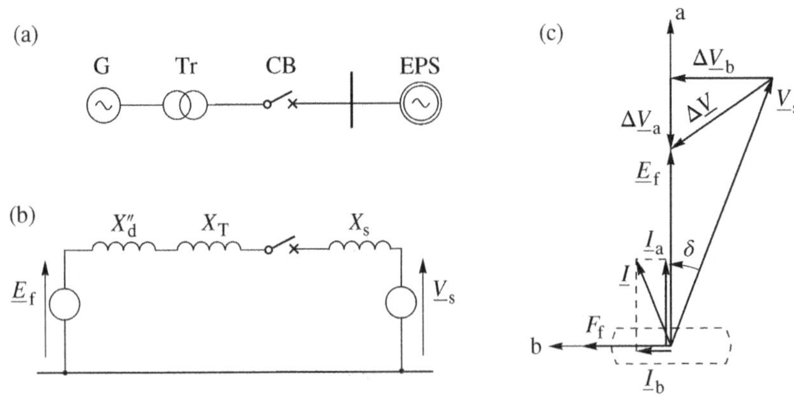

Figure 4.33 Synchronization: (a) schematic diagram; (b) equivalent circuit in the subtransient state; (c) phasor diagram. G – generator; Tr – transformer; EPS – electric power system; CB – circuit-breaker (synchronizing switch).

E_f during the three-phase short circuit on an unloaded generator. As a result, alternating and direct currents will be induced in the armature windings that are similar to those expressed by Eqs. (4.22) and (4.23) but with i_m replaced by i_{bm}.

As in the case of the three-phase fault, the AC component of the phase currents due to I_b produce an armature magnetomotive force (mmf) which rotates with the rotor and is in phase with the excitation mmf F_f (Figure 4.33c). As there is no angular displacement between these two mmf's, they cannot produce an electromagnetic torque. On the other hand, the DC component of the phase currents produces a stationary mmf which counter-rotates with respect to the rotor. Consequently, the angular displacement between the field flux and the armature mmf varies with time, producing a periodic torque given by Eq. (4.58) as

$$\tau_I = -\frac{3}{2}\frac{1}{\omega}E_{fm}i_{bm}\sin\omega t = -\frac{3}{\omega}\frac{E_f}{x''_d}(V_s\cos\delta - E_f)\sin\omega t \quad \text{Nm} \tag{4.97}$$

As compared with Eq. (4.58), the minus sign is due to the voltage $\Delta\underline{V}_a$ directly opposing E_f.

The b-axis voltage $\Delta\underline{V}_b$ produces the a-axis circulating current $\underline{I}_a$. The amplitude of this current is

$$i_{am} = \frac{\Delta V_{bm}}{x''_d} = \frac{V_{sm}\sin\delta}{x''_d} \tag{4.98}$$

$\underline{I}_a$ produces an mmf which rotates with the rotor and is directed along its q-axis (Figure 4.33c). The interaction between this component of armature mmf and the excitation flux produces a constant driving torque given by

$$\tau_{II} = \frac{3}{2}\frac{1}{\omega}E_{fm}i_{am} = \frac{3}{\omega}\frac{E_fV_s}{x''_d}\sin\delta \quad \text{Nm} \tag{4.99}$$

The DC component of the phase current driven by $\Delta\underline{V}_b$ produces a stationary flux shifted by $\pi/2$ with respect to the stationary flux driven by ΔV_a, so that the resulting periodic torque is equal to

$$\tau_{III} = \frac{3}{2}\frac{1}{\omega}E_{fm}i_{am}\sin\left(\omega t - \frac{\pi}{2}\right) = -\frac{3}{\omega}\frac{E_fV_s}{x''_d}\sin\delta\cos\omega t \quad \text{Nm} \tag{4.100}$$

At this point it is worth noting that although the circulating current resulting from synchronization has a similar shape to the three-phase fault current, the direct component can be larger, or smaller, than the three-phase fault current depending on the actual values of X_T, X_s, and δ. In addition, the synchronization torque may be quite different from the three-phase fault torque, owing to the influence of the voltage component $\Delta\underline{V}_b$. Adding the three torque components in Eqs. (4.97), (4.99), and (4.100), gives

$$\tau = \tau_I + \tau_{II} + \tau_{III} = \frac{3}{\omega}\left[\frac{E_fV_s}{x''_d}\sin\delta(1-\cos\omega t) - \frac{E_f}{x''_d}(V_s\cos\delta - E_f)\sin\omega t\right] \quad \text{Nm} \tag{4.101}$$

which can be rewritten by substituting $\cos\delta = 1 - \sin\delta\cdot\tan(\delta/2)$ as

$$\tau = \frac{3}{\omega}\left[\frac{E_fV_s}{x''_d}\sin\delta\left(1-\cos\omega t + \tan\frac{\delta}{2}\sin\omega t\right) + \frac{E_f(E_f - V_s)}{x''_d}\sin\omega t\right] \quad \text{Nm} \tag{4.102}$$

The maximum value of the torque produced on synchronization is highly dependent on the angle δ at which the switch is closed. When $E_f \cong V_s$, the second component in the above equation is zero and the torque is determined by the trigonometric expression

$$T_\delta(t) = \sin\delta\left(1-\cos\omega t + \tan\frac{\delta}{2}\sin\omega t\right) \tag{4.103}$$

The variation of this expression with time for three different values of δ is shown in Figure 4.34a, while Figure 4.34b shows the maximum value of $T_\delta(t)$ as a function of δ. From these diagrams it can be seen that when

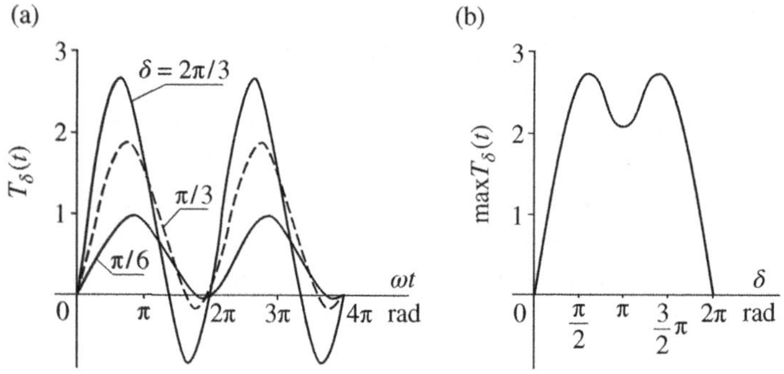

Figure 4.34 Expression (4.103): (a) as a function of time for various values of the synchronization angle δ; (b) its maximum value as a function of the synchronization angle δ.

δ is large the maximum torque is large and occurs early in the cycle with the highest torque occurring when $\delta = 2\pi/3$.

If $E_f \neq V_s$ then according to Eq. (4.102) the torque is increased by a periodic component dependent on the difference $(E_f - V_s)$.

Example 4.4 The pu network data referred to in the generator units are: $X''_d = X''_q = 0.18$; $X_T = 0.11$; $X_s = 0.01$; $E_f = V_s = 1.1$. The pu subtransient torque produced on synchronization is obtained from Eq. (4.102) by multiplying by $\omega/3$ to obtain $\tau = T_\delta(1.1)^2/(0.18 + 0.11 + 0.01) = 4T_\delta$. The torque may be evaluated by taking $T_\delta(t)$ from Figure 4.34b for various values of the synchronization angle δ. The results in the table show that the torque may exceed the rated torque:

Synchronization angle	$\pi/6$	$\pi/3$	$\pi/2$	$2\pi/3$	$5\pi/6$	π
Torque in pu	4.08	7.48	9.64	10.4	9.64	8

******

The equations derived in this section are valid immediately after synchronization, that is during the subtransient period. As time progresses, magnetic energy will be dissipated in the winding resistances, the circulating current will decay, and the generator will move from the subtransient state, through the transient state, until it eventually reaches the steady state. As the circulating current decays, so too do the corresponding torque components.

4.4.2 Short Circuit in the Network and Its Clearing

Section 4.2 discusses in some detail the effect of a three-phase short circuit across the generator terminals. Fortunately, this type of fault does not occur very often, and much more common are faults some distance from the generator elsewhere in the EPS. Figure 4.35 shows a general case where the generator is connected to the EPS via a transformer and two parallel lines.

When a three-phase-to-earth fault occurs at the beginning of one of the lines, point F2, then the fault can be treated as a generator short circuit but with some of the reactances and time constants modified to include the effect of the transformer. First of all, the transformer reactance X_T must be added in series to the reactances X''_d, X'_d, and X_d to form the resultant reactances

$$x''_d = X''_d + X_T; \quad x'_d = X'_d + X_T; \quad x_d = X_d + X_T \tag{4.104}$$

This has the effect of suppressing the magnitude of the currents given by Eqs. (4.44)–(4.46) and the electromagnetic torque containing both periodic and aperiodic components as given by Eqs. (4.60)–(4.63). Secondly, the

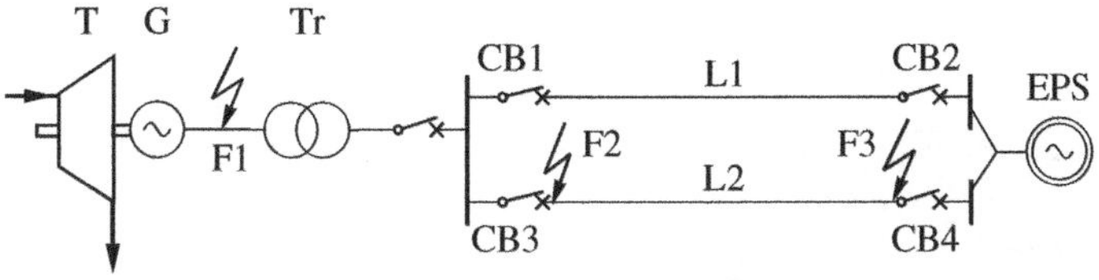

Figure 4.35 Schematic diagram of a fragment of an EPS with examples of the short-circuit points F1, F2, F3. G – generator; Tr – transformer; CB – circuit breaker; EPS – electric power system (infinite busbar).

transformer resistance R_T increases the rate at which the stored magnetic energy can be dissipated so that the DC component of the short-circuit current decays more rapidly. The time constants have to be modified, as in Eq. (4.91), to

$$T''_{d(network)} = T''_d \left(\frac{X'}{X''_d}\right)\left(\frac{X''_d + X_T}{X' + X_T}\right); \quad T'_{d(network)} = T'_d \left(\frac{X_d}{X'_d}\right)\left(\frac{X'_d + X_T}{X_d + X_T}\right) \tag{4.105}$$

As a consequence of the increase in the time constants, the instantaneous value of the short-circuit current may pass through zero even on the first cycle.

The shapes of the currents and torques during the short circuit and following fault clearing are shown in Figure 4.36. During the short circuit, high currents flow in all three phases and the torque oscillates around near-zero average values. As the clearing time is usually small, the rotor angle δ shown by the dashed line on the torque diagram (Figure 4.36), increases only slightly. When the fault is cleared, the changes in the torque and the currents are similar to those occurring during synchronization when the rotor angle first increases and then decreases, but the torque oscillates around an average value.

In the case of an unsymmetrical fault in the network, the generator currents are distorted by the step-up transformer. This effect is illustrated in Figure 4.37, assuming the transformer to be star-delta connected when a phase-to-phase fault on the secondary of the transformer, Figure 4.37a, is seen as a three-phase unsymmetrical current on

Figure 4.36 Three-phase fault and its clearing in the line connecting a generator to the system: i_A, i_B, i_C – phase currents; τ_e – electromagnetic torque; i_f – field current; δ – rotor angle. *Source:* Adapted from Kulicke and Webs (1975). Reproduced by permission of VDE Verlag Gmbh.

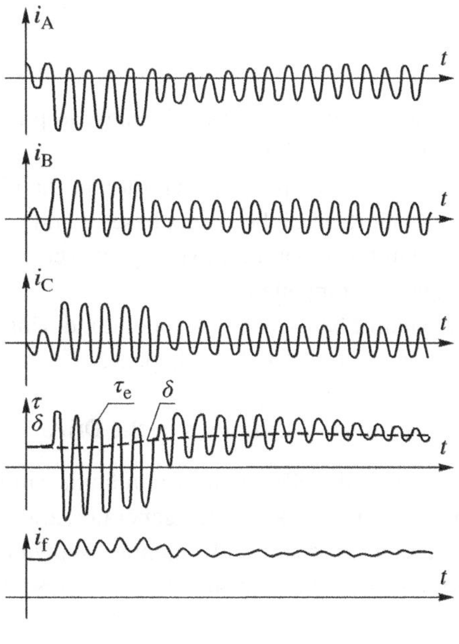

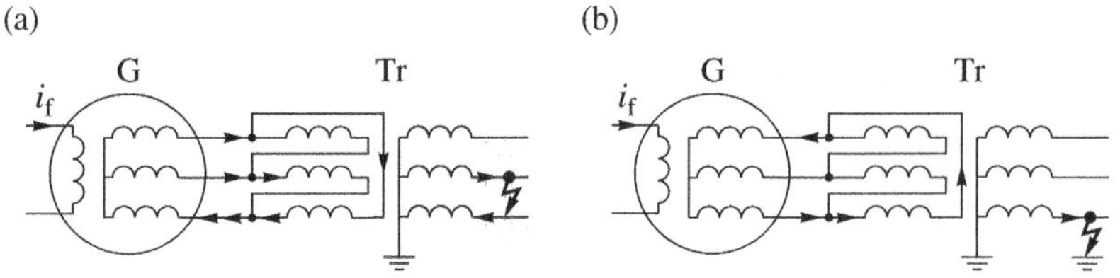

Figure 4.37 Transformation of the short-circuit current in the star-delta connection of the transformer: (a) phase-to-phase fault; (b) single-phase fault.

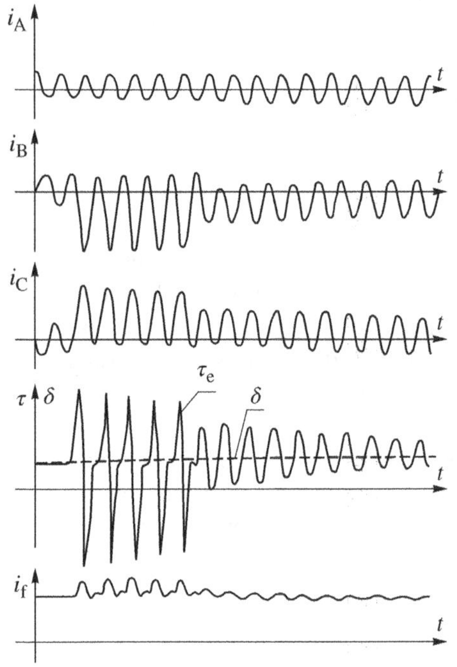

Figure 4.38 Phase-to-phase fault and its clearing in the line connecting a generator to the system, i_A, i_B, and i_C – phase currents; τ_e – electromagnetic torque; i_f – field current; δ – rotor angle. *Source:* Adapted from Kulicke and Webs (1975). Reproduced by permission of VDE Verlag Gmbh.

the primary. In the case of a single-phase-to-phase fault on the secondary, Figure 4.37b, this is seen on the primary side as a two-phase fault.

Figure 4.38 shows the time variation of the phase currents, air-gap torque, and field current during, and after, clearing a short circuit between phases B and C in the line connecting the generator to a system. Such an unsymmetrical fault produces negative sequence currents, and consequently, the electromagnetic torque contains a double-frequency component.

The dynamic behavior of the generator during the short-circuit period and the importance of the clearing time is considered in more detail in Chapter 5.

4.4.3 Torsional Oscillations in the Drive Shaft

In this section the effect that system faults can have on the torques in the shafts and couplings connecting each turbine stage, the turbine to the generator, and the generator to the exciter is discussed. In the previous sections of this chapter these shafts and couplings, shown in Figure 4.1, are assumed to be very stiff when the total drive system moves as a rigid body and can be represented by a single inertia. In practice, the drive shafts and couplings have a finite stiffness so that each turbine mass will be slightly displaced relative to the others. Consequently, any change in the external torque on any of the rotor components initiates a movement of the equivalent mass. The shaft connecting this element to its neighbors twists slightly and the resulting torsion transmits torque to the neighboring element(s). This process is repeated down the shaft creating torsional oscillations in the shaft. These torsional oscillations are superimposed on the external torque which may either increase or reduce the torque in the shaft sections, depending on the direction of the twist and the phase and frequency of the changes in the external torque, so that under certain conditions the shaft torques can become very large. Such large shaft torques can result in a reduction in the shaft fatigue life and possibly shaft failure. It is therefore important to understand the mechanisms leading to these high shaft torques and in which situations they may arise.

4.4.3.1 The Torsional Natural Frequencies of the Turbine-generator Rotor

As the turbine-generator rotor can be considered as a number of discrete masses connected together by springs, a lumped-mass model of the rotor can be developed in order to calculate the natural frequencies of the rotor system. At each of these natural frequencies the rotor system will vibrate in a particular way, defined by the *mode shape*.

Figure 4.39 shows the torques acting on each rotor mass. Newton's second law gives the equation of motion for the general mass l as

$$J_l \frac{\mathrm{d}^2\delta_{\mathrm{m}l}}{\mathrm{d}t^2} = \tau_l + \tau_{l,l+1} - \tau_{l,l-1} \tag{4.106}$$

where J_l is its moment of inertia, τ_l is the external applied torque to mass l, and $\tau_{l,\,l+1}$ and $\tau_{l,\,l-1}$ are the torques in the two shafts (l, $l+1$) and (l, $l-1$). These shaft torques are, respectively,

$$\tau_{l,l+1}(t) = k_{l,l+1}(\delta_{\mathrm{m}l+1} - \delta_{\mathrm{m}l}) + D_{l,l+1}\left(\frac{\mathrm{d}\delta_{\mathrm{m}l+1}}{\mathrm{d}t} - \frac{\mathrm{d}\delta_{\mathrm{m}l}}{\mathrm{d}t}\right) \tag{4.107}$$

$$\tau_{l,l-1}(t) = k_{l,l-1}(\delta_{\mathrm{m}l} - \delta_{\mathrm{m}l-1}) + D_{l,l-1}\left(\frac{\mathrm{d}\delta_{\mathrm{m}l}}{\mathrm{d}t} - \frac{\mathrm{d}\delta_{\mathrm{m}l-1}}{\mathrm{d}t}\right) \tag{4.108}$$

where $k_{l,\,l-1}$ is the stiffness of the shaft section between masses l and $l-1$, $\delta_{\mathrm{m}l}$ is the angle of the l-th mass in mechanical radians and $D_{l,\,l-1}$ is the shaft damping coefficient. The shaft damping is caused by the energy lost in the shaft as it oscillates and goes through a stress/strain hysteresis cycle. The above two equations can now be substituted into Eq. (4.106) to give the equation of motion

$$\begin{aligned} J_l \frac{\mathrm{d}^2\delta_{\mathrm{m}l}}{\mathrm{d}t^2} &= \tau_l(t) - k_{l,l-1}(\delta_{\mathrm{m}l} - \delta_{\mathrm{m}l-1}) - k_{l,l+1}(\delta_{\mathrm{m}l} - \delta_{\mathrm{m}l+1}) \\ &\quad - D_{l,l+1}\left(\frac{\mathrm{d}\delta_{\mathrm{m}l}}{\mathrm{d}t} - \frac{\mathrm{d}\delta_{\mathrm{m}l+1}}{\mathrm{d}t}\right) - D_{l,l-1}\left(\frac{\mathrm{d}\delta_{\mathrm{m}l}}{\mathrm{d}t} - \frac{\mathrm{d}\delta_{\mathrm{m}l-1}}{\mathrm{d}t}\right) - D_{l,l}\frac{\mathrm{d}\delta_{\mathrm{m}l}}{\mathrm{d}t} \end{aligned} \tag{4.109}$$

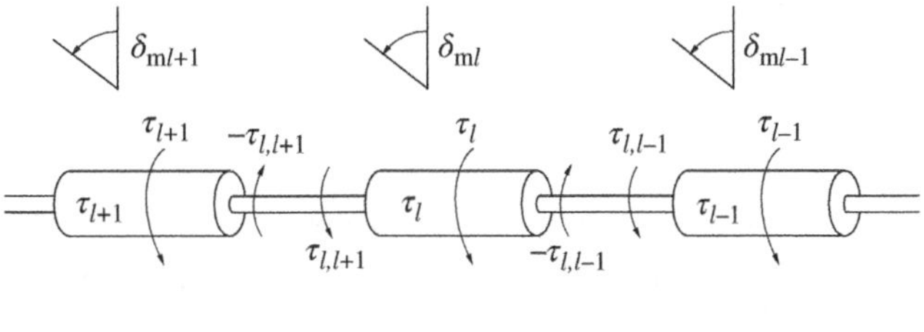

Figure 4.39 Torques acting on mass l; τ_l is the applied torque, $\tau_{l,\,l-1}$ and $\tau_{l,\,l+1}$ are the shaft torques, $\delta_{\mathrm{m}l}$ the displacement of the mass, and J_l the inertia.

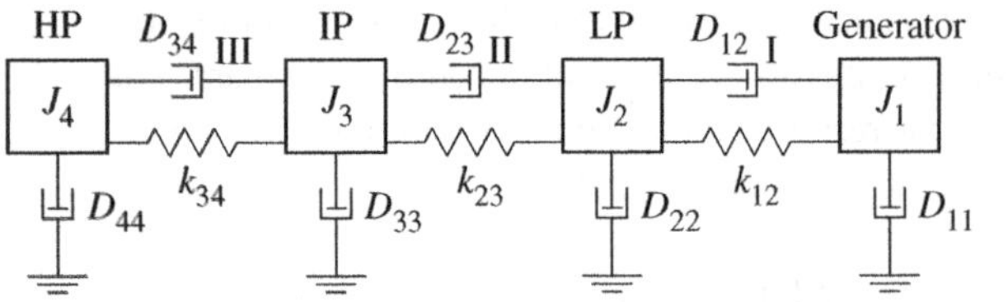

Figure 4.40 Four-mass turbine-generator rotor model.

In this equation $D_{l,\,l}$ is an additional damping term used to represent the damping effect arising at each turbine stage due to the flow of steam through the turbine. Generally, both the shaft damping and the steam damping mechanisms are small. Equation (4.109) is in SI units but a pu version can be obtained by following a similar procedure as in Section 5.1.[3]

Now consider a simple turbine-generator rotor consisting of three turbine stages and a generator, as shown in Figure 4.40. A static exciter is assumed to act with this system so that no rotational mass is required to represent this component. Applying Eq. (4.109) to each rotor mass in turn gives a set of four second-order differential equations

$$
\begin{aligned}
J_1\frac{d\delta_{m1}^2}{dt^2} &= \tau_1 - k_{12}(\delta_{m1}-\delta_{m2}) - D_{12}\left(\frac{d\delta_{m1}}{dt}-\frac{d\delta_{m2}}{dt}\right) - D_{11}\frac{d\delta_{m1}}{dt}\\
J_2\frac{d\delta_{m2}^2}{dt^2} &= \tau_2 - k_{12}(\delta_{m2}-\delta_{m1}) - k_{23}(\delta_{m2}-\delta_{m3}) - D_{12}\left(\frac{d\delta_{m2}}{dt}-\frac{d\delta_{m1}}{dt}\right) - D_{23}\left(\frac{d\delta_{m2}}{dt}-\frac{d\delta_{m3}}{dt}\right) - D_{22}\frac{d\delta_{m2}}{dt}\\
J_3\frac{d\delta_{m3}^2}{dt^2} &= \tau_3 - k_{23}(\delta_{m3}-\delta_{m2}) - k_{34}(\delta_{m3}-\delta_{m4}) - D_{23}\left(\frac{d\delta_{m3}}{dt}-\frac{d\delta_{m2}}{dt}\right) - D_{34}\left(\frac{d\delta_{m3}}{dt}-\frac{d\delta_{m4}}{dt}\right) - D_{33}\frac{d\delta_{m3}}{dt}\\
J_4\frac{d\delta_{m4}^2}{dt^2} &= \tau_4 - k_{34}(\delta_{m4}-\delta_{m3}) - D_{34}\left(\frac{d\delta_{m4}}{dt}-\frac{d\delta_{m3}}{dt}\right) - D_{44}\frac{d\delta_{m4}}{dt}
\end{aligned}
\tag{4.110}
$$

This set of second-order equations is then written as a series of eight first-order equations by noting that for mass i

$$
\frac{d\delta_{mi}}{dt} = \omega_{mi} - \omega_{sm} = \Delta\omega_{mi}; \qquad \frac{d^2\delta_{mi}}{dt^2} = \frac{d\Delta\omega_{mi}}{dt} \tag{4.111}
$$

Using this substitution, and taking mass l as an example, gives

$$
J_1\frac{d\Delta\omega_{m1}}{dt} = \tau_1 - k_{12}(\delta_{m1}-\delta_{m2}) - D_{11}\Delta\omega_{m1} - D_{12}\Delta\omega_{m1} - \Delta\omega_{m2}\frac{d\delta_{m1}}{dt} = \Delta\omega_{m1} \tag{4.112}
$$

By considering small perturbations and defining a set of state variables $\boldsymbol{x}$ so that

$$
\begin{aligned}
&x_1 = \Delta\delta_{m1}; \quad x_2 = \Delta\delta_{m2}; \quad x_3 = \Delta\delta_{m3}; \quad x_4 = \Delta\delta_{m4}\\
&x_5 = \Delta\omega_{m1}; \quad x_6 = \Delta\omega_{m2}; \quad x_7 = \Delta\omega_{m3}; \quad x_8 = \Delta\omega_{m4}
\end{aligned}
\tag{4.113}
$$

and substituting into the Eqs. (4.112) gives a linearized matrix equation of the form

$$
\dot{\boldsymbol{x}} = \boldsymbol{A}\boldsymbol{x} + \boldsymbol{B}\boldsymbol{u} \tag{4.114}
$$

---

3 In pu form J becomes $2H/\omega_s$, whilst τ and k are expressed in pu obtained by dividing the MKSA values by $T_{base} = \frac{S_{base}}{\omega_{sm}} = \frac{S_{base}}{\omega_s}\frac{p^2}{4}$ and $k_{base} = T_{base}\frac{p}{2} = \frac{S_{base}}{\omega_s}\frac{p^2}{4}$, respectively. δ_l is now expressed in electrical radians.

where $\boldsymbol{A}$ is the plant or state matrix, $\boldsymbol{B}$ the driving matrix, $\boldsymbol{x}$ a vector of state variables, and $\boldsymbol{u}$ a vector of inputs, in this case the change in the external torque $\Delta\tau$ applied to each mass. A general solution of the state Eq. (4.11) is discussed in Section 12.1.5. The plant matrix is given by

$$\boldsymbol{A} = \begin{bmatrix} \mathbf{0} & \mathbf{1} \\ \boldsymbol{K} & \boldsymbol{D} \end{bmatrix} \tag{4.115}$$

where $\mathbf{0}$ is a null matrix, $\mathbf{1}$ the identity matrix, $\boldsymbol{K}$ the stiffness matrix, and $\boldsymbol{D}$ the damping matrix. For the four-mass problem being considered these matrices are

$$\boldsymbol{K} = \begin{bmatrix} \frac{-k_{12}}{J_1} & \frac{k_{12}}{J_1} & & \\ \frac{k_{12}}{J_2} & \frac{-k_{12}-k_{23}}{J_2} & \frac{k_{23}}{J_2} & \\ & \frac{k_{23}}{J_3} & \frac{-k_{23}-k_{34}}{J_3} & \frac{k_{34}}{J_3} \\ & & \frac{k_{34}}{J_4} & \frac{-k_{34}}{J_4} \end{bmatrix} \tag{4.116}$$

$$\boldsymbol{D} = \begin{bmatrix} \frac{-D_{12}-D_{11}}{J_1} & \frac{D_{12}}{J_1} & & \\ \frac{D_{12}}{J_2} & \frac{-D_{12}-D_{23}-D_{22}}{J_2} & \frac{D_{23}}{J_2} & \\ & \frac{D_{23}}{J_3} & \frac{-D_{23}-D_{34}-D_{33}}{J_3} & \frac{D_{34}}{J_3} \\ & & \frac{D_{34}}{J_4} & \frac{-D_{34}-D_{44}}{J_4} \end{bmatrix} \tag{4.117}$$

while the driving matrix $\boldsymbol{B} = [\mathbf{0}\ \ \boldsymbol{J}^{-1}]^{\mathrm{T}}$, where $\boldsymbol{J}^{-1}$ is simply a (4×4) diagonal matrix with entries $1/J_1$, $1/J_2$, $1/J_3$, and $1/J_4$. The structure of all these matrices is very clear, allowing more inertia elements to be easily added if required.

The rotor natural frequencies are found from Eq. (4.11) by assuming that there is no change in the applied torque when the input vector $\boldsymbol{u} = \mathbf{0}$ gives

$$\dot{\boldsymbol{x}} = \boldsymbol{A}\boldsymbol{x} \tag{4.118}$$

Section 12.1 shows that the solution of this equation is of the form

$$x_k(t) = \sum_{i=1}^{n} w_{ki} e^{\lambda_i t} \sum_{j=1}^{n} u_{ij} x_{j0} \tag{4.119}$$

where λ_i is the eigenvalue of matrix $\boldsymbol{A}$, w_{ki}, and u_{ij} are the elements of a matrix made up of eigenvectors of matrix $\boldsymbol{A}$, and x_{j0} is the initial condition of state variable $x_j(t)$.

The eigenvalues λ_i correspond to the rotor natural frequencies. As the damping is small the system will be underdamped and complex conjugate pairs of eigenvalues will be returned in the form

$$\lambda_{i,i+1} = -\zeta_i \Omega_{\text{nat}\ i} \pm \mathrm{j}\Omega_i \tag{4.120}$$

where ζ_i is the damping ratio associated with the particular oscillation mode, Ω_i is its damped natural frequency of oscillations (in rad/s), and $\Omega_{\text{nat}\ i}$ its undamped natural frequency (in rad/s). The damped, and the undamped, natural frequencies are related by the standard expression $\Omega = \Omega_{\text{nat}}\sqrt{1-\zeta^2}$.

To obtain the mode shapes (Section 12.1), all the damping coefficients are set to zero when the eigenvalues are purely imaginary and give the undamped natural frequencies. The eigenvector corresponding to each of these undamped natural frequencies is now real and defines how the different rotor masses will be displaced relative to each other if excited at this particular frequency. This is the mode shape, and either the elements of the eigenvector associated with the speed deviation or the angle deviation can be used. It is also normal practice to normalize the eigenvectors so that the maximum displacement is set to unity.

The sensitivity of any particular shaft to a harmonic forcing torque on the generator can also be found via Eq. (6.113) by defining an output equation

$$\boldsymbol{y} = \boldsymbol{C}\boldsymbol{x} \tag{4.121}$$

where $\boldsymbol{y}$ is the output required, in this case the torque in a particular shaft or shafts, with the elements of the output matrix $\boldsymbol{C}$ following directly from Eqs. (4.107) or (4.108). If all the forcing torques are set to zero, except the generator torque which is assumed to be sinusoidal, the transfer function relating the shaft torque $\boldsymbol{y}$ to the input $\boldsymbol{u}$ can be found by substituting Eq. (4.121) into (4.11) to give the standard transfer function matrix equation

$$\boldsymbol{G}(s) = \boldsymbol{C}(s\mathbf{1} - \boldsymbol{A})^{-1}\boldsymbol{B} \tag{4.122}$$

where $\boldsymbol{B}$ now simply becomes $[0\ \ 0\ \ 0\ \ 0\ \ 0\ \ 1/J_1\ \ 0\ \ 0\ \ 0]$ and $G_{11}(s) = y_1/u$, $G_{12}(s) = y_2/u$, and so on. The inertia entry in $\boldsymbol{B}$ is that of the generator so that its position in the matrix will depend on which mass represents the generator.

The evaluation of the frequency response associated with Eq. (4.122) is best achieved using standard software package, such as MATLAB (Hicklin and Grace 1992) when the matrices $\boldsymbol{A}$, $\boldsymbol{B}$, $\boldsymbol{C}$, and $\boldsymbol{D}$ need only be defined and standard subroutines used to do the rest of the work.

Example 4.5 The shaft configuration for a 577 MVA, 3000 rpm thermal power plant is shown in Figure 4.41 with the parameters detailed in Table 4.4.

The plant matrix, Eq. (4.11), is formed using the submatrices defined in Eqs. (4.116) and (4.117) and the eigenvectors and eigenvalues found using standard MATLAB routines. Figure 4.41 shows the undamped natural frequencies and mode shapes. As is standard practice, all the mode shapes have been normalized with the

Table 4.4 The 577 MVA, 3000 rpm, turbine-generator parameters (Ahlgren et al. 1978).

Node	H_i	k_{ij}	D_{ii}	D_{ij}
—	s	pu torque/rad	pu torque/rad/s	pu torque/rad/s
1	0.09	—	0.0	—
1–2	—	0.7	—	0.00007
2	0.74	—	0.0	—
2–3	—	110	—	0.001
3	1.63	—	0.001	—
3–4	—	95	—	0.001
4	1.63	—	0.001	—
4–5	—	87	0	0.001
5	1.63	—	0.001	—
5–6	—	40	—	0.001
6	0.208	—	0.001	—

maximum deflection set to 1.0. Mode 0 represents free-body rotation and, when the generator is connected to the system, the frequency of this mode would increase and become the rotor frequency of oscillations. This can easily be shown by connecting the generator to the infinite system through an "electromagnetic spring" of stiffness $K_{E'}$, where $K_{E'}$ is the transient synchronizing power coefficient. This concept is introduced in Section 5.4. As the value of $K_{E'}$ depends on the system loading condition, a figure of $K_{E'} = 2.0$ pu has been assumed. This "electromagnetic spring" will replace the change in generator torque. As $\Delta\tau_2 = -K_{E'}\Delta\delta_2$, $K_{E'}$ must now be included in the stiffness matrix so that element $\boldsymbol{K}[2,2]$ becomes $\boldsymbol{K}[2,2] = -(k_{12} + k_{23} + K_{E'})/J_2$. The eigenvalues now give the rotor frequency of oscillations as 1.14 Hz but the rotor system continues to rotate as a rigid body, as shown by the mode shape at the bottom of Figure 4.41.

Following the mode shapes through for each individual shaft indicates some degree of movement in the generator-exciter shaft at all the natural frequencies. In mode 1, at 5.6 Hz, the turbine and the generator components remain stationary and the exciter moves relative to them. Mode 4 shows a large amount of movement in the HP-LP_1 shaft, while the LP_3-generator shaft displays movement in all modes except mode 1.

To study further the frequency sensitivity of the generator-exciter shaft and the LP_3-generator shaft the frequency response of these two shafts to a sinusoidal generator torque is shown in Figure 4.42. To obtain this frequency response the $\boldsymbol{B}$ and $\boldsymbol{C}$ matrices are defined as

$$\boldsymbol{B} = \left[0,0,0,0,00,0,\frac{1}{J_2},0,0,0,0\right]$$

$$\boldsymbol{C} = \begin{bmatrix} -k_{12} & k_{12} & 0 & 0 & 0 & 0 & -D_{12} & D_{12} & 0 & 0 & 0 & 0 \\ 0 & -\mathrm{k}_{23} & \mathrm{k}_{23} & 0 & 0 & 0 & 0 & -D_{23} & D_{23} & 0 & 0 & 0 \end{bmatrix}$$

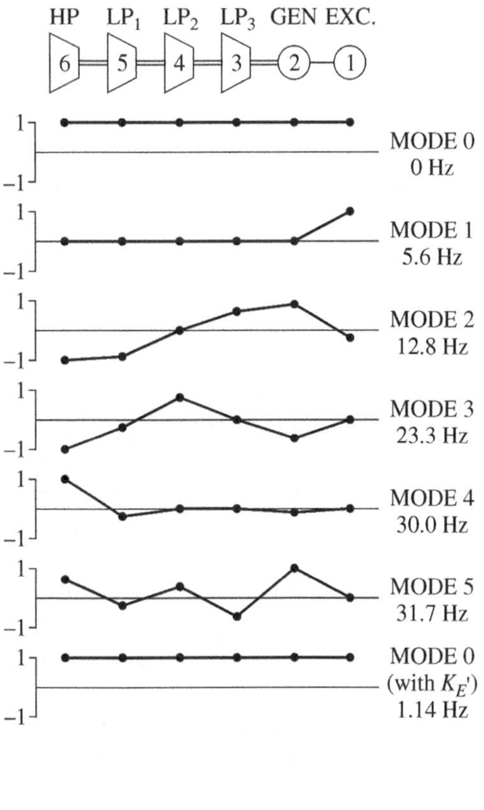

Figure 4.41 The 577 MVA, 3000 rpm, turbine-generator torsional natural frequencies and associated mode shapes.

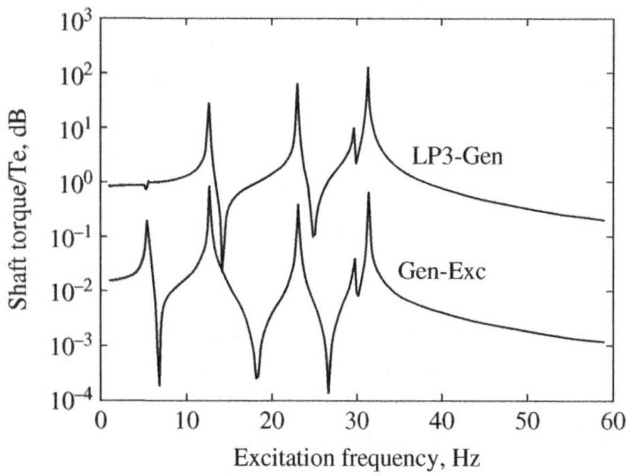

Figure 4.42 Frequency sensitivity of the LP_3-generator and the generator-exciter drive shafts for the 577 MVA, 3000 rpm generator.

As expected, the generator-exciter shaft is sensitive to all five modes but the torque is somewhat less than that in the LP_3-generator shaft, which is particularly sensitive to frequencies around 31.7 Hz but is insensitive to mode 1 type of oscillation. The distinct nature of these curves is due to the low degree of damping present in the system. Additional damping would reduce the peak values of these curves but unfortunately is not usually available, although a small amount of additional damping may be introduced by the electrical system.

Note that the natural frequencies are substantially removed from 50 and 100 Hz so that the periodic changes in external torque at fundamental or double frequency which occurs under fault conditions induce a relatively low value of shaft torque.

******

Such eigenvalue techniques can readily be extended to include details of the EPS and are invaluable when conducting full system stability studies or when assessing the feasibility of new generator designs (Bumby 1982; Bumby and Wilson 1983; Westlake et al. 1996).

4.4.3.2 Effect of System Faults

When a system fault occurs, the highest shaft torque is usually in the main shaft connecting the generator and the turbine because the applied torque on these two rotor components is in opposing directions. The shaft between the last two turbine stages, the LP and the IP sections shown in Figure 4.1, can also experience high torques. The three charts in Figure 4.43 show typical changes in the shaft torque which may occur in the LP-generator shaft (denoted as I in Figure 5.1) when a generator is subjected to three different electromagnetic torques: a step change, a 50 Hz variation, and a 100 Hz variation. The highest momentary values of shaft torque occur when the rotor is subject to an aperiodic external torque and are due to the oscillatory shaft torque being reinforced by the external torque on each backswing of the rotor masses. As a result the shaft torque oscillates around the average value of the external torque at a frequency determined by the different torsional natural frequencies. In this example a frequency response similar to that in Figure 4.42 showed the LP-generator shaft to be particularly sensitive to 12 and 24 Hz oscillations, and these two frequencies can be observed to dominate the time response.

The above considerations lead to the conclusion that the highest shaft torque corresponds to an electromagnetic torque with a high aperiodic component which, as is shown in the previous sections, appears either after synchronization of a generator when the synchronizing angle is high or after clearing a fault in the network. Figure 4.44

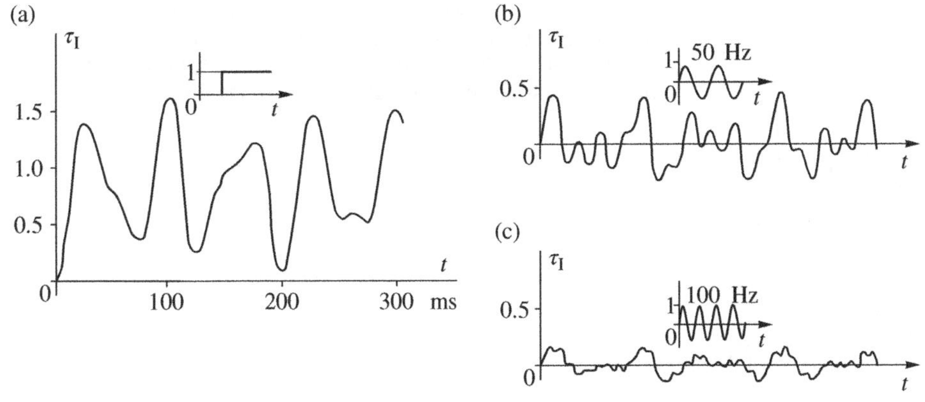

Figure 4.43 Examples of the changes in the torque in the main coupler due to: (a) aperiodic component of the electromagnetic torque; (b) fundamental-frequency component; (c) double-frequency component. *Source:* Based on Bölder et al. (1975). VDE Verlag Gmbh.

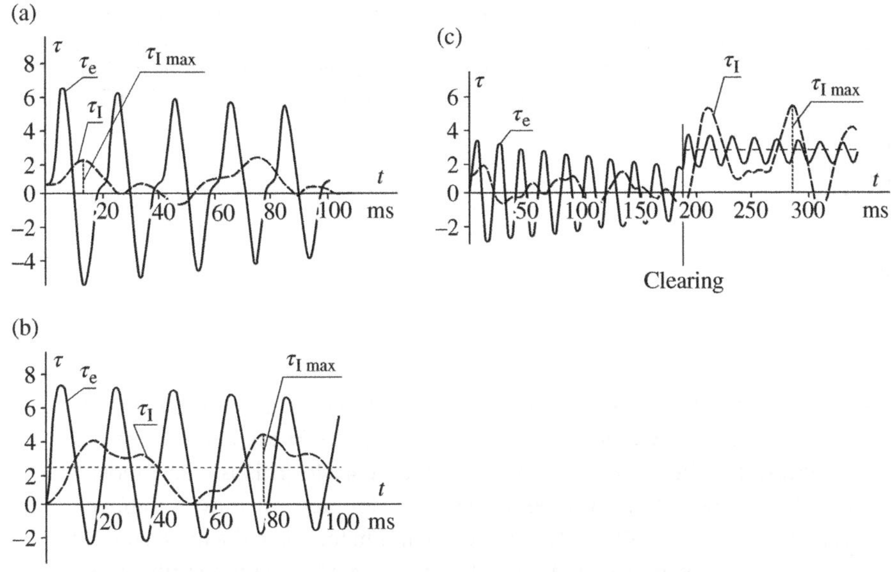

Figure 4.44 Torque acting on the main coupler for the case of: (a) phase-to-phase short circuit on the generator terminals; (b) synchronization when $\delta = 2\pi/3$; (c) fault in the network cleared after 0.187s. *Source:* Based on Läge and Lambrecht (1974). VDE Verlag Gmbh.

shows examples of the torque in the main coupler *I* during a phase-to-phase short circuit on the generator terminals, after synchronization at a large angle and after clearing a three-phase short circuit in the network.

It is interesting to compare the last two cases. The average value of the shaft torque shown by the dashed line is the same in both cases, but the maximum value is much higher in case (c). This is a direct result of the shaft already being twisted when the fault is cleared, this twist being such that the aperiodic electromagnetic torque appearing on fault clearance reinforces the twist. The influence of the degree of shaft twist on the shaft torque, after the fault is cleared, is shown in Figure 4.45, where it can be seen that, when the shaft torque has a negative value at the instant of clearing, the resulting shaft torque is substantially increased. This effect may be compounded by unsuccessful automatic reclosing.

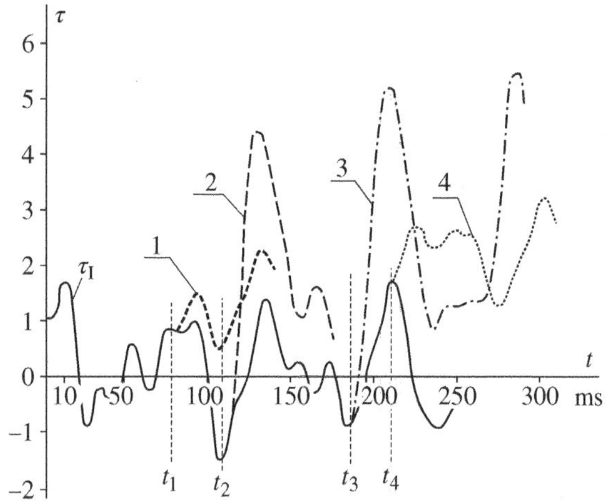

Figure 4.45 Influence of the initial shaft twist on the shaft torque when the fault is cleared: 1 – after $t_2 = 80$ ms; 2 – after $t_2 = 110$ ms; 3 – after $t_2 = 187$ ms; 4 – after $t_2 = 213$ ms. *Source:* Based on Kulicke and Webs (1975). VDE Verlag Gmbh.

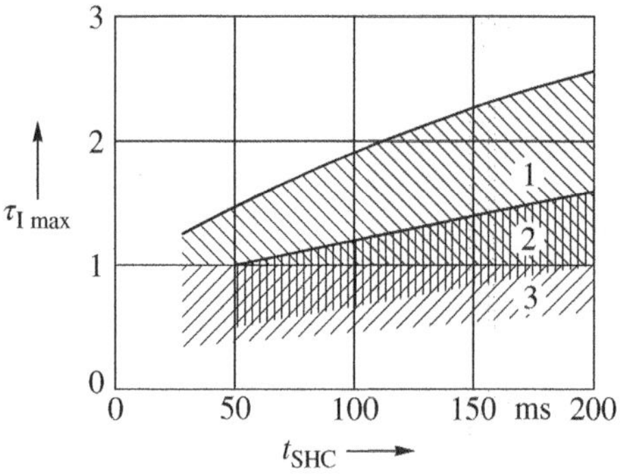

Figure 4.46 Approximate relationship between the maximum torque in the main coupler and the clearing time when the rating of the generator is: 1 : S >> 500 MVA; 2 : S < 500 MVA; 3 : S < 300 MVA. The value of τ_I is relative with respect to the value of torque accompanying a phase-to-phase fault on the generator terminals. *Source:* Based on Bölder et al. (1975). VDE Verlag Gmbh.

The example shown in Figure 4.44a corresponds to a phase-to-phase short circuit on the generator terminals applied when the generator torque loading is high. However, the corresponding torque in the main coupler is much smaller than that occurring after clearing a three-phase system fault, Figure 4.44c. This is characteristic of big machines, whereas in medium- and low-power machines the highest shaft torques generally accompany faults on the generator terminals. An approximate relationship between the maximum torque in the main coupler and the clearing time is shown in Figure 4.46. The greater susceptibility of large machines to short circuits in the network is primarily due to the lower pu impedance of the generator transformer associated with these large machines so that its influence in the transfer reactance is less than for smaller rated generators. Also the pu rotor inertia decreases as unit rating increases so that with large units even short clearing times may result in a relatively large angle increase and a high value of aperiodic torque when the fault is cleared.

4.4.4 Switching Operations in Transmission Networks

Switching operations in a transmission network can force an electromagnetic transient state with duration of approximately 1 s and characterized by sudden changes in electric currents in the network elements and sudden changes in active power of generating units and thereby a sudden change in the electromagnetic torque of synchronous generators. In order to reduce possible disadvantageous effects of such sudden changes the switching operations in modern power systems are controlled by protection devices referred to as a *synchro-check*. Optimal setting of such devices requires a method for calculating the initial values of the changes in electric currents in synchronous generators and network elements (lines and transformers). An efficient method proposed by Machowski et al. (2014) is described below. The method is based on the nodal impedance matrix very commonly used in the short-circuit computer programs.

4.4.4.1 Impedance Network Model

A network model to be used for calculating initial switching currents is shown in Figure 4.47. Set {B} includes generation buses, while set {G} comprises fictitious nodes behind the generator impedances and their step-up transformers. Set {L} is a set of load nodes. Loads are replaced by constant admittances. Nodes a and b are poles of a circuit breaker that gets switched on. Voltages across the circuit breaker poles are denoted by $\underline{V}_{a}$, $\underline{V}_{b}$, and $\underline{V}_{ab} = (\underline{V}_{a} - \underline{V}_{b})$, respectively. Difference of the voltage phasor angles is denoted by θ_{ab}, as in Figure 4.47b.

Similarly, as in the case of calculating initial short-circuit currents, for the initial switching current calculations, each synchronous generator should be represented as for the subtransient state, that is with the application of subtransient emf $\underline{E}''$ behind subtransient reactance at the assumption that $X''_{q} \cong X''_{d}$. Subtransient emf $\underline{E}''$ should be calculated for the preset loading conditions in the EPS. They make up the voltage sources in the transmission network model (Figure 4.47).

When a circuit breaker is closed, there is no voltage difference between its terminals a and b. The zero voltage value across the poles can be replaced by two voltage sources $\underline{V}_{ab}$ of the opposite orientation (Figure 4.48a). The value $\underline{V}_{ab}$ is selected so that it corresponds to the voltage difference across the circuit-breaker poles before its closing, that is for the instant $t = 0_{-}$. As the discussed network is linear, according to the superposition principle, it can be divided into two networks presented in Figure 4.48b and Figure 4.48c, respectively.

The network shown in Figure 4.48b corresponds to the condition before the circuit breaker gets closed, i.e. when it is still open. In this state the generators are loaded with currents $\underline{\boldsymbol{I}}_{G}$ and there are voltages $\underline{V}_{i}$ and $\underline{V}_{j}$ at the arbitrary nodes i and j, respectively.

Supplementary network shown in Figure 4.48c is a fictitious network that corresponds to the difference between the closed and open states of a circuit breaker. That fictitious network is a passive network supplied from one voltage source $\underline{V}_{ab}$. This network is very useful for the discussed calculations as it includes all of the searched quantities:

1) current $\underline{I}_{ab} = \underline{I}_{0+}$ that corresponds to the initial switching current flows through the $\underline{V}_{ab}$ source (Figure 4.48c),
2) currents $\Delta\underline{\boldsymbol{I}}_{G} = \Delta\underline{\boldsymbol{I}}_{G0+}$ that correspond to the changes in generator currents caused by closing of a circuit breaker flow through the generator branches (Figure 4.48c).

The discussed network (Figure 4.48c) is a fictitious network, where nodal voltages correspond to the difference of the voltages of the closed-breaker state and the voltages of the open-breaker state. In arbitrary i and j nodes there

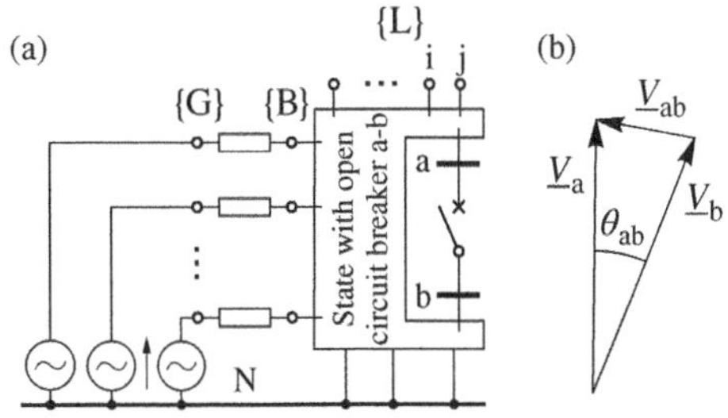

Figure 4.47 Illustration of the mathematical network model elaboration.

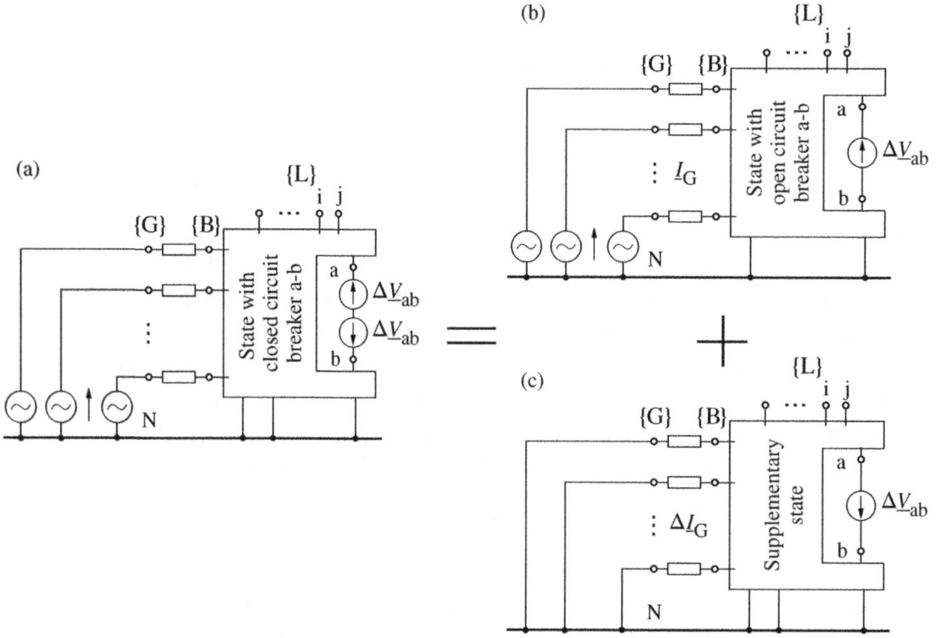

Figure 4.48 Application of the superposition method for the closed circuit breaker state.

are voltages $(\underline{V}_i^+ - \underline{V}_i)$ and $\left(\underline{V}_j^+ - \underline{V}_j\right)$, respectively. The superscript + corresponds to the state after the circuit breaker is closed. Voltages $(\underline{\boldsymbol{V}}_G^+ - \underline{\boldsymbol{V}}_G)$ are present at the {B} nodes, where generating units are connected to.

It follows from the above considerations that the network shown in Figure 4.48c can be used to determine changes in generator currents $(\Delta\underline{\boldsymbol{I}}_G)$ caused by the circuit breaker closing. For that purpose, the voltage source $\underline{V}_{ab}$ (Figure 4.48c) is replaced by two nodal currents $+\underline{I}_{ab}$ and $-\underline{I}_{ab}$, as shown in Figure 4.49.

The network shown in Figure 4.49 can be described by the following nodal matrix equation

$$\begin{bmatrix} \underline{\boldsymbol{V}}_B^+ - \underline{\boldsymbol{V}}_B \\ \underline{V}_a^+ - \underline{V}_a \\ \underline{V}_b^+ - \underline{V}_b \\ \underline{\boldsymbol{V}}_L^+ - \underline{\boldsymbol{V}}_L \end{bmatrix} = \begin{bmatrix} \underline{\boldsymbol{Z}}_{BB} & \underline{\boldsymbol{Z}}_{Ba} & \underline{\boldsymbol{Z}}_{Bb} & \underline{\boldsymbol{Z}}_{BL} \\ \underline{\boldsymbol{Z}}_{aB} & \underline{Z}_{aa} & \underline{Z}_{ab} & \underline{\boldsymbol{Z}}_{aL} \\ \underline{\boldsymbol{Z}}_{bB} & \underline{Z}_{ba} & \underline{Z}_{bb} & \underline{\boldsymbol{Z}}_{bL} \\ \underline{\boldsymbol{Z}}_{LB} & \underline{\boldsymbol{Z}}_{La} & \underline{\boldsymbol{Z}}_{Lb} & \underline{\boldsymbol{Z}}_{LL} \end{bmatrix} \cdot \begin{bmatrix} \boldsymbol{0} \\ -\underline{I}_{ab} \\ +\underline{I}_{ab} \\ \boldsymbol{0} \end{bmatrix} \tag{4.123}$$

where the nodal impedance matrix is an inverted nodal admittance matrix. Subscripts B, a, b, and L correspond to the nodes {B}, a, b, and {L}, respectively. It needs to be kept in mind that all the voltage and current values are complex numbers and have to be given within a common frame of reference. Nodal currents occur only in nodes a and b that correspond to the circuit breaker poles.

As nodal currents of the {B} and {L} nodes are zero valued, in Eq. (4.123) only a part of the impedance matrix is important. For the a and b nodes, the following equation can be written

$$\begin{bmatrix} \underline{V}_a^+ - \underline{V}_a \\ \underline{V}_b^+ - \underline{V}_b \end{bmatrix} = \begin{bmatrix} \underline{Z}_{aa} & \underline{Z}_{ab} \\ \underline{Z}_{ba} & \underline{Z}_{bb} \end{bmatrix} \cdot \begin{bmatrix} -\underline{I}_{ab} \\ +\underline{I}_{ab} \end{bmatrix} \tag{4.124}$$

This matrix equation corresponds to the following two scalar equations

$$\underline{V}_a^+ - \underline{V}_a = -\underline{Z}_{aa}\underline{I}_{ab} + \underline{Z}_{ab}\underline{I}_{ab} \tag{4.125}$$

$$\underline{V}_b^+ - \underline{V}_b = -\underline{Z}_{ba}\underline{I}_{ab} + \underline{Z}_{bb}\underline{I}_{ab} \tag{4.126}$$

For the closed-breaker state it appears that $\underline{V}_b^+ = \underline{V}_a^+$, because there is no voltage difference at the closed circuit breaker. Considering and assuming that $\underline{z}_{ab} = \underline{z}_{ba}$, after having performed both-sides subtraction of Eq. (4.125) from Eq. (4.126), the following can be obtained

$$\underline{V}_{ab} = \underline{V}_a - \underline{V}_b = (\underline{z}_{aa} + \underline{z}_{bb} - 2\cdot\underline{z}_{ab})\cdot\underline{I}_{ab} \tag{4.127}$$

Hence, the initial switching current is given by the following equation

$$\underline{I}_{ab} = \frac{\underline{V}_{ab}}{\underline{z}_{aa} + \underline{z}_{bb} - 2\cdot\underline{z}_{ab}} \tag{4.128}$$

where: $\underline{V}_{ab}$ is the voltage across the circuit breaker poles, before the breaker gets closed, $\underline{z}_{aa}$, $\underline{z}_{bb}$, and $\underline{z}_{ab}$ are the elements of the nodal impedance matrix.

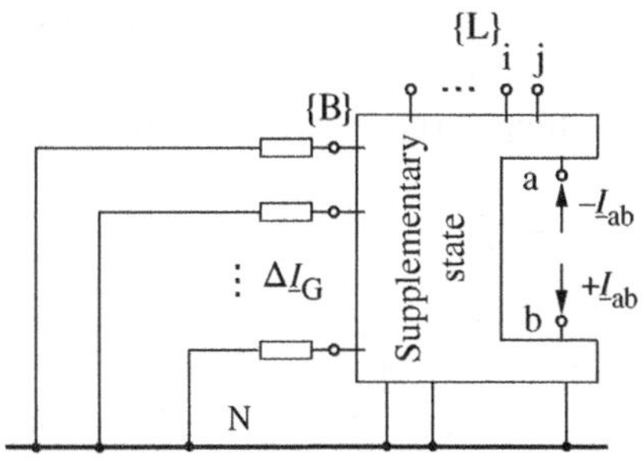

Figure 4.49 Illustration of the impedance method.

Equation (4.124) fully describes the whole network seen looking into the nodes a and b that correspond to the circuit breaker poles. Thus, it can be stated that the network model reduced to the nodes a and b can be described by the following nodal admittance matrix

$$\underline{\mathbf{Y}}_\pi = \begin{bmatrix} \underline{y}_{aa} & \underline{y}_{ab} \\ \underline{y}_{ba} & \underline{y}_{bb} \end{bmatrix} = \begin{bmatrix} \underline{z}_{aa} & \underline{z}_{ab} \\ \underline{z}_{ba} & \underline{z}_{bb} \end{bmatrix}^{-1} \tag{4.129}$$

Matrix (4.129) corresponds to the π-equivalent model shown in Figure 4.50. It follows from the definition of a nodal admittance matrix and the π-equivalent model that

$$\underline{y}_{ab} = -\underline{Y}_{ab} = -\frac{1}{\underline{Z}_{ab}}; \quad \underline{y}_{aa} = \frac{1}{\underline{Z}_a} + \frac{1}{\underline{Z}_{ab}}; \quad \underline{y}_{bb} = \frac{1}{\underline{Z}_b} + \frac{1}{\underline{Z}_{ab}} \tag{4.130}$$

where: $\underline{Z}_a$, $\underline{Z}_b$, and $\underline{Z}_{ab}$ are impedances of the π-equivalent model branches (Figure 4.50). From Eq. (4.129) it follows that

$$\begin{bmatrix} \underline{z}_{aa} & \underline{z}_{ab} \\ \underline{z}_{ba} & \underline{z}_{bb} \end{bmatrix} = \underline{\mathbf{Y}}_\pi^{-1} = \frac{1}{\det\underline{\mathbf{Y}}_\pi} \begin{bmatrix} \underline{y}_{bb} & -\underline{y}_{ab} \\ -\underline{y}_{ba} & \underline{y}_{aa} \end{bmatrix} \tag{4.131}$$

where: $\det\underline{\mathbf{Y}}_\pi = \underline{y}_{aa}\underline{y}_{bb} - \underline{y}_{ab}\underline{y}_{ba}$. After having substituted the values resulting from Eq. (4.130) to Eq. (4.131) the following formulas can be obtained

$$\underline{z}_{aa} = \frac{\underline{Z}_a(\underline{Z}_b + \underline{Z}_{ab})}{\underline{Z}_a + \underline{Z}_b + \underline{Z}_{ab}}; \quad \underline{z}_{bb} = \frac{\underline{Z}_b(\underline{Z}_a + \underline{Z}_{ab})}{\underline{Z}_a + \underline{Z}_b + \underline{Z}_{ab}}; \quad \underline{z}_{ab} = \underline{z}_{ba} = \frac{\underline{Z}_a\underline{Z}_b}{\underline{Z}_a + \underline{Z}_b + \underline{Z}_{ab}} \tag{4.132}$$

Elements $\underline{z}_{aa}$ and $\underline{z}_{bb}$ correspond to impedances seen by the nodes a and b, respectively. This can be easily verified by calculating impedances seen looking into those nodes within the π-equivalent model shown in Figure 4.50.

The equivalent branch connecting nodes a and b and impedance $\underline{Z}_{ab}$ are important for the π-equivalent model, because they represent the transmission network seen by the circuit breaker poles and considerably influence the initial switching current value.

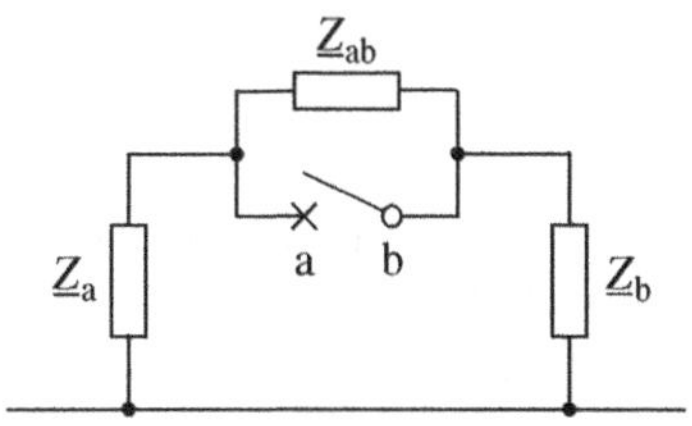

Figure 4.50 π-equivalent model seen looking into nodes a and b.

4.4.4.2 Thévenin's Theorem and the Nodal Impedance Method

Equation (4.128) determines the switching current $\underline{I}_{ab}$ as a function of $\underline{z}_{aa}$, $\underline{z}_{bb}$, and $\underline{z}_{ab}$, i.e. as a function of the nodal impedance matrix elements. This current can also be calculated using Thévenin's theorem for the original network shown in Figure 4.47 or for the π-equivalent model shown in Figure 4.50. According to Thévenin's theorem, when a circuit breaker is closed the switching current can be expressed by the following formula

$$\underline{I}_{ab} = \frac{\underline{V}_{ab}}{\underline{Z}_{Th}} \quad \text{and} \quad I_{ab}^2 = \underline{I}_{ab}\underline{I}_{ab}^* = \frac{V_{ab}^2}{Z_{Th}^2} \tag{4.133}$$

where: $\underline{V}_{ab}$ is the voltage across the poles of a circuit breaker before it is closed, $\underline{Z}_{Th}$ is the Thévenin impedance seen looking into the nodes a and b.

Comparative analysis of Eqs. (4.133) and (4.128) indicates that the Thévenin impedance can be expressed in the following way, with the use of nodal impedance matrix elements

$$\underline{Z}_{Th} = \underline{z}_{aa} + \underline{z}_{bb} - 2\cdot\underline{z}_{ab} \tag{4.134}$$

where $\underline{z}_{aa}$, $\underline{z}_{bb}$, and $\underline{z}_{ab}$ are nodal impedance matrix elements. By substituting the values resulting from Eq. (4.132) to Eq. (4.134) the following can be obtained

$$\underline{Z}_{Th} = \frac{\underline{Z}_{ab}(\underline{Z}_a + \underline{Z}_b)}{\underline{Z}_a + \underline{Z}_b + \underline{Z}_{ab}} = \frac{\underline{Z}_a + \underline{Z}_b}{\underline{\xi}}; \qquad \underline{\xi} = 1 + \frac{\underline{Z}_a + \underline{Z}_b}{\underline{Z}_{ab}} \tag{4.135}$$

is a coefficient (generally a complex one) that represents the dependence of the Thévenin impedance on the impedance $\underline{Z}_{ab}$ of the equivalent branch.

Equation (4.135) is consistent with the π-equivalent model of Figure 4.50, because when looking at nodes a and b, the parallel connection of the impedance $\underline{Z}_{ab}$ with in-series connected impedances $\underline{Z}_a$ and $\underline{Z}_b$ can be seen. This precisely gives the impedance given by Eq. (4.135).

Concluding, it should be noted that the matrix considerations using a nodal impedance matrix lead to the same results as the Thévenin's theorem for the π-equivalent model (Figure 4.50).

4.4.4.3 Initial Switching Current as a Function of Switching Angle θ_{ab}

For an optimal setting of the protection device (such as a synchro-check) it is important to know how the absolute value of the initial switching current $I_{ab} = |\underline{I}_{ab}|$ depends on switching angle θ_{ab}. Function $I_{ab}(\theta_{ab})$ can be found on the basis of the phasor diagram shown in Figure 4.47b and the law of cosines

$$V_{ab}^2 = V_a^2 + V_b^2 - 2V_aV_b\cos\theta_{ab} \tag{4.136}$$

By dividing both sides of Eq. (4.136) by V_b^2 and introducing the coefficient $\nu = V_a/V_b$ the following can be obtained

$$\frac{V_{ab}^2}{V_b^2} = \nu^2 - 2\nu\cos\theta_{ab} + 1 \quad \text{or} \quad V_{ab}^2 = V_b^2\left(\nu^2 - 2\nu\cos\theta_{ab} + 1\right) \tag{4.137}$$

By substituting V_{ab}^2 in Eq. (4.133) with the value resulting from Eq. (4.136), the following expression can be obtained

$$I_{ab} = \frac{V_b}{Z_{Th}}\sqrt{\nu^2 - 2\nu\cos\theta_{ab} + 1} = \frac{V_b}{Z_{Th}}\sqrt{(\nu - \cos\theta_{ab})^2 + \sin^2\theta_{ab}} \tag{4.138}$$

In the particular case when there is no voltage difference $V_a = V_b = V$ and $\nu = 1$, the following can be obtained from Eq. (4.138)

$$I_{ab} = \frac{2V}{Z_{Th}}\sin\frac{\theta_{ab}}{2} \quad \text{for} \quad \nu = V_a/V_b = 1 \tag{4.139}$$

because $(1 - \cos\theta_{ab}) = 2\sin^2(\theta_{ab}/2)$.

It should be kept in mind that Eqs. (4.138) and (4.139) define only the AC component of the switching current. That component is complemented by the DC component (as in the short circuit case). It is also worth noting that, according to Eq. (4.135), in the discussed equations the impedance Z_{Th} depends on the coefficient ξ.

4.4.4.4 Changes in the Active Power of Synchronous Generators

A sudden change in the active power of the synchronous generator causes a sudden change in its electromagnetic torque and can be dangerous for the shaft of the generating unit.

A change in currents of the generators caused by the circuit breaker closing can be calculated with the use of Eq. (4.123), which yields the following result

$$\underline{\boldsymbol{V}}_{\text{B}} - \underline{\boldsymbol{V}}_{\text{B}}^{+} = (\underline{\boldsymbol{Z}}_{\text{Ba}} - \underline{\boldsymbol{Z}}_{\text{Bb}})\, \underline{I}_{\text{ab}} \tag{4.140}$$

and for an arbitrary i-th generator

$$\underline{V}_i - \underline{V}_i^{+} = \left(\underline{z}_{i\text{a}} - \underline{z}_{i\text{b}}\right) \underline{I}_{\text{ab}} \tag{4.141}$$

Equation (4.141) defines the change in voltage at a generating bus caused by the circuit breaker closing. It can be assumed that during the subtransient state the emf $\underline{E}''$ is constant. Then the change in currents of a generating unit can be calculated using Ohm's law

$$\Delta I_{\text{G}i} = \underline{Y}_{\text{G}i}\left(\underline{V}_i - \underline{V}_i^{+}\right) = \frac{\left(\underline{z}_{i\text{a}} - \underline{z}_{i\text{b}}\right)}{\underline{Z}_{\text{G}i}} \underline{I}_{\text{ab}} \tag{4.142}$$

where $\underline{Z}_{\text{G}i} = 1/\underline{Y}_{\text{G}i}$ is the impedance of a generator and its step-up transformer, while $\underline{z}_{i\text{a}}$ and $\underline{z}_{i\text{b}}$ are elements of the nodal impedance matrix. Hence, the change of generator apparent power can be calculated from the following formula

$$\Delta S_{\text{G}i} = \underline{E}''_{\text{G}i} \cdot \Delta \underline{I}^{*}_{\text{G}i} = \underline{E}''_{\text{G}i} \cdot \frac{\left(\underline{z}^{*}_{i\text{a}} - \underline{z}^{*}_{i\text{b}}\right)}{\underline{Z}^{*}_{\text{G}i}} \underline{I}^{*}_{\text{ab}} \tag{4.143}$$

When using the above formula it should be kept in mind that the underlined symbols denote complex numbers within a common reference frame. By substituting Eq. (4.128)–(4.143) the following can be obtained

$$\Delta S_{\text{G}i} = \frac{\underline{E}''_{\text{G}i}}{\underline{Z}^{*}_{\text{G}i}} \cdot \frac{\left(\underline{z}^{*}_{i\text{a}} - \underline{z}^{*}_{i\text{b}}\right)}{\underline{z}^{*}_{\text{aa}} + \underline{z}^{*}_{\text{bb}} - 2 \cdot \underline{z}^{*}_{\text{ab}}} \cdot \underline{U}^{*}_{\text{ab}} \quad \text{and} \quad \Delta P_{\text{G}i} = \text{Re}\, \Delta S_{\text{G}i} \tag{4.144}$$

where the lower case symbols $\underline{z}$ with relevant subscripts denote elements of the nodal impedance matrix (4.131).

4.4.4.5 Discussion of Methods Used in Practice

According to Eq. (4.135), the Thévenin impedance seen looking into the breaker poles depends on the coefficient $\underline{\xi}$ that is given by Eq. (4.135). With the use of that coefficient, Eq. (4.133) can be rewritten the following way

$$\underline{I}_{\text{ab}} = \frac{\underline{V}_{\text{ab}}}{\underline{Z}_{\text{Th}}} = \frac{\underline{V}_{\text{ab}}}{\underline{Z}_{\text{a}} + \underline{Z}_{\text{b}}} \underline{\xi} \tag{4.145}$$

In the particular case when $|\underline{Z}_{\text{ab}}| \gg |\underline{Z}_{\text{a}} + \underline{Z}_{\text{b}}|$, it results from Eq. (4.135) that $\underline{\xi} \cong 1$ and the switching current can be expressed by the following equation

$$\underline{I}_{\text{ab}} \cong \frac{\underline{V}_{\text{ab}}}{\underline{Z}_{\text{a}} + \underline{Z}_{\text{b}}} \quad \text{for} \quad \underline{\xi} \cong 1 \tag{4.146}$$

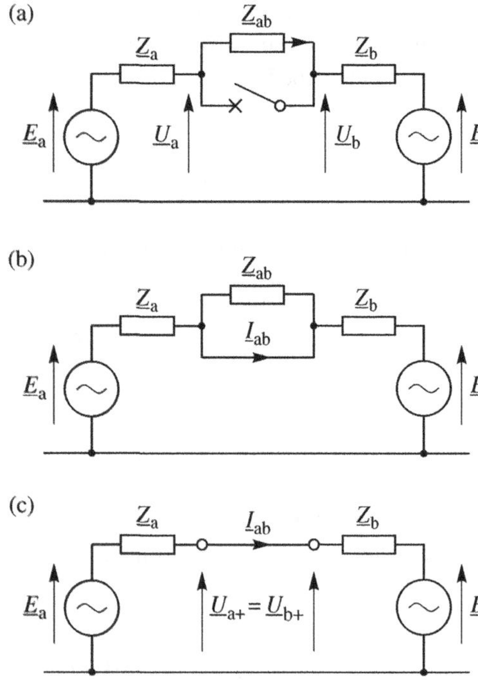

Figure 4.51 Illustration for the initial switching current calculations using $\underline{E}_a$ and $\underline{E}_b$.

That simplification is equivalent to neglecting of the equivalent branch connecting nodes a and b (impedance $\underline{Z}_{ab}$) in the π-equivalent model (Figure 4.50). Alas, in most cases of the real transmission networks, the equivalent branch connecting nodes a and b cannot be neglected, as the coefficient ξ is much higher than unity. In practice, the ξ values can considerably influence the value of the initial switching current $\underline{I}_{ab}$ that is given by Eq. (4.145). Generally, the simplified Eq. (4.146) should not be used. However, in many publications this simplified equation is used and regarding the above considerations it can be deemed incorrect.

Obviously, the branch connecting nodes a and b (impedance $\underline{Z}_{ab}$) that is present in the π-equivalent model (Figure 4.50) could be neglected, if in the equation for the initial switching current calculation the voltage $\underline{V}_{ab}$ were replaced by the difference of emfs $\underline{E}_a$, $\underline{E}_b$ of the equivalent voltage sources connected to branches $\underline{Z}_a$ and $\underline{Z}_b$, respectively. This is illustrated by Figure 4.51.

It follows from the diagram shown in Figure 4.51a that if before closing the circuit breaker there is voltage $\underline{V}_{ab}$ between nodes a and b, then the $\underline{V}_{ab}/\underline{Z}_{ab}$ current must flow through the $\underline{Z}_{ab}$ branch of the equivalent network. On the basis of Kirchhoff's voltage the following values of the source voltages can be calculated

$$\underline{E}_a = \underline{V}_a + \frac{\underline{V}_{ab}}{\underline{Z}_{ab}}\underline{Z}_a \quad \text{and} \quad \underline{E}_b = \underline{V}_b - \frac{\underline{V}_{ab}}{\underline{Z}_{ab}}\underline{Z}_b \tag{4.147}$$

The plus and minus signs in Eq. (4.147) result from the current flow direction (Figure 4.51a). By subtracting both these equations it is obtained that

$$\underline{E}_a - \underline{E}_b = \underline{V}_a + \frac{\underline{V}_{ab}}{\underline{Z}_{ab}}\underline{Z}_a - \underline{V}_b + \frac{\underline{V}_{ab}}{\underline{Z}_{ab}}\underline{Z}_b = \underline{V}_{ab} + \frac{\underline{V}_{ab}}{\underline{Z}_{ab}}(\underline{Z}_a + \underline{Z}_b)$$

where $\underline{V}_{ab} = \underline{V}_a - \underline{V}_b$. Hence

$$\underline{E}_a - \underline{E}_b = \underline{V}_{ab}\left(1 + \frac{\underline{Z}_a + \underline{Z}_b}{\underline{Z}_{ab}}\right) \quad \text{or} \quad \underline{E}_a - \underline{E}_b = \underline{\xi}\cdot\underline{V}_{ab} \tag{4.148}$$

where $\underline{\xi}$ is a coefficient given by Eq. (4.135). Figure 4.52 presents a phasor diagram of voltages and emf's and shows that the voltage difference can be much lower than the difference of emf's.

Taking Eq. (4.148) into account, the switching current $\underline{I}_{ab}$ can be calculated on the basis of the diagram of Figure 4.51c, in the following way

$$\underline{I}_{ab} = \frac{\underline{E}_a - \underline{E}_b}{\underline{Z}_a + \underline{Z}_b} = \frac{\underline{V}_{ab}}{\underline{Z}_a + \underline{Z}_b}\underline{\xi} \tag{4.149}$$

Equation (4.149) is consistent with Eq. (4.145) that has been earlier obtained with the use of Thévenin's theorem. Obviously, for the case when $\underline{Z}_{ab} = \infty$, there is $\underline{\xi} = 1$, and $\underline{V}_a = \underline{E}_a$ and $\underline{V}_b = \underline{E}_b$ can be obtained. However, it is a very particular case. In practice, when $\xi > 1$ Eqs. (4.145) or (4.149) should be applied. The importance of the equivalent branch connecting nodes a and b (Figure 4.50) and the coefficient $\underline{\xi}$ is illustrated below by example for a real large-scale network.

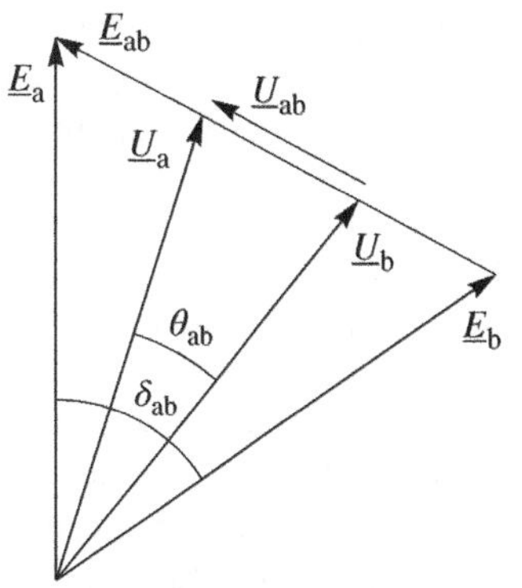

Figure 4.52 Phasor diagram of voltages and emf's.

Example 4.6 An analysis has been performed for an active power system with 664 lines and transformers operating at 400 kV and 220 kV and 3933 lines and transformers operating at 110 kV. Coefficient $\underline{\xi}$ has been calculated for all these network elements. Statistical results are illustrated by the curve shown in Figure 4.53.

For a given value ξ the diagram shown in Figure 4.53 determines a percent number of network elements, for which the coefficients ξ are higher than the given value. For instance, $\xi \geq 1.5$ has been obtained for 45% of the network elements; $\xi \geq 2.0$ for 25% of the elements; and $\xi \geq 3.0$ for 10% of them. This means that, when the branch connecting nodes a and b is neglected in the π-equivalent model and Eq. (32) is used instead of Eq. (31), for 45% of the network elements the calculated value of the switching current $\underline{I}_{ab}$ is 1.5 times smaller than the proper value, for 25% of the network elements it is two times smaller, and for 10% of the network elements it is three times smaller. These are obviously absolutely unacceptable errors.

******

A description of a computer program based on the above method can be found in Machowski et al. (2014).

4.4.5 Synchro-check Device

As described in Section 4.4.4, the switching current and changes in active power of synchronous generators depend on the voltages at the poles of the open circuit breaker. In order to reduce possible harmful effects of the switching

Figure 4.53 Statistical distribution of the ξ value for an example transmission network.

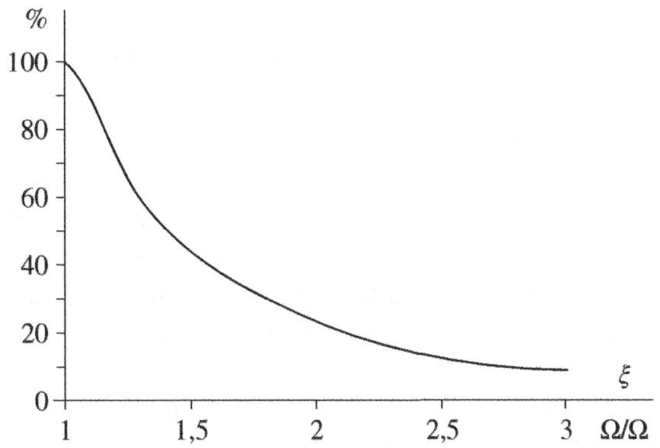

operation on power system elements a protection device (referred to as a *synchro-check device*) can be used to control the circuit breaker closing. Such a protection device measures voltages at poles of the circuit breaker and allows to close it if the following quantities

- differences in the voltage magnitudes
- differences in phase angles (switching angle)
- differences in frequency (voltage slip)

are smaller than relevant set threshold values.

As far as the threshold values are concerned, pertinent publications and guides recommend setting values that considerably differ among one another. The biggest differences concern the threshold value of the closing angle. For example, MAAC (Mid Atlantic Area Council) and Bartylak (2011) recommend $\theta = 60^{\circ}$ for lines that are electrically far from power stations. For lines that are electrically close to power stations, Bartylak (2011) recommends $\theta = 20^{\circ}$, while MAAC recommends to determine an allowed threshold value θ of the switching angle on the basis of the detailed shaft torque analysis. In recommendations of Wunderlich et al. (1994) and Hazarika and Sinha (1998) the threshold value of the closing angle depends on voltage rating of the network element in the following way: (20–30)° for (400–500) kV, (30–40)° for 230 kV, and (50–60)° for 132 kV. Oprea et al. (2007) propose (20–30)° for all network elements.

In power engineering practice the settings of the switching angle are chosen between 20° and 60°. The power system protection engineers usually conservatively tend to set the lowest values (20–30)°. The power system dispatchers demand values as large as possible, for instance (50–60)°, because low settings can cause difficulties in performing switching operations, when they should be performed from the viewpoint of the power system security. When in hard-loading condition a synchro-check device with low setting blocks an attempt to close the circuit breaker the dispatcher must undertake the reduction of the standing phase angle. This usually requires rescheduling of active power outputs in a number of power plants and takes valuable time.

A good example is the blackout described in the UCTE report (2003). The whole north–south corridor, from northern Europe to Italy, was overloaded, which caused cascade tripping of many transmission lines and finally a blackout over a large part of the European power system. During the initial emergency state, a network operator at one of the subsystems tried to reduce the overload by switching on a transmission line. However, the switching operation was blocked by a synchro-check device, because its setting was at 30°, while the switching angle on that line in the emergency state was 42°. Higher setting of this synchro-check device (for example 45°) could possibly change the situation.

Too small or too large a setting of the synchro-check devices may be harmful for an EPS or its elements. For this reason the maxim allowed threshold values should be determined by analysis taking into account all possible harmful effects such as:

- Dynamic damage of a circuit breaker that can occur when its switching capability is exceeded (CIGRE Report 2006).
- Unwanted operation of distance protection and automatic switching-off of a given network element right after its switching-on (Cook 1985).
- Deformation of transformer windings and/or generator end-windings that occur because of the action of forces caused by high inrush current (Reimert 2006).
- Torsional oscillations and fatigue of generating unit shafts that can cause considerable shortening of their lifetime (Rotating Machinery Committee 1982).

Mathematical formulation of above harmful effects is used below to determine four conditions, which must be satisfied by settings of the synchro-check device.

Condition C1 Short-time withstand current of circuit breaker

Short-time withstand of circuit breakers is determined by the rated peak withstand current i_B, which includes both AC and DC components of short-circuit or switching current. In the case of switching operation the peak value of the switching current must be smaller than the rated peak withstand current of the circuit breaker. The peak value of the switching current contains an AC and a DC component and is equal approximately to $i_p = \sqrt{2}\,k_p I_{ab}$, where I_{ab} is the RMS value of the AC current at the closing instant $t = 0_+$, $k_p \leq 2$ is the peak factor taking into account the effect of the DC component

$$\sqrt{2}k_m k_p I_{ab} < i_B \quad \text{or} \quad I_{ab} < \frac{i_B}{\sqrt{2}k_m k_p} \tag{4.150}$$

where $k_m > 1$ is the safety margin. For the EHV and HV networks it can be assumed that $k_p \cong 2$. Substituting I_{ab} in (4.150) by the right-hand side of Eq. (4.139), it is easy to obtain

$$\sin\frac{\theta_{ab\ max}}{2} < \frac{i_B}{\sqrt{2}k_m k_p}\frac{Z_{Th}}{2V} \quad \text{or} \quad \theta_{ab\ max} < 2\arcsin\frac{i_B}{\sqrt{2}k_m k_p}\frac{Z_{Th}}{2V} \tag{4.151}$$

Inequality (4.151) limits the set value of the switching angle $\theta_{ab\ max}$ to the value for which there is no threat to damage the circuit breaker.

Condition C2 Unwanted operation of distance protection

It is assumed (Figure 4.54) that a transmission line (which must be switched on) is protected by the distance protection. When the difference in the voltages at the circuit breaker poles is large then the initial switching current is large too and the impedance measured by distance protection at the switching instant can be small enough to encroach into the distance protection zones.

For further consideration it is important to remember that $\underline{V}_a$ and $\underline{V}_b$ are voltages for the state with open circuit breaker (Figure 4.51a) and must be treated as voltages for instant $t = 0_-$. For instant $t = 0_+$ (when circuit breaker is closed) both voltages are equal one to each other $\underline{V}_{a+} = \underline{V}_{b+}$, and generally $\underline{V}_{a+} \neq \underline{V}_a$ and $\underline{V}_{b+} \neq \underline{V}_b$. On the basis of Kirchhoff's voltage law and the circuit diagram from Figure 4.51c it results that

$$\underline{V}_{b+} = \underline{E}_b + \underline{I}_{ab}\underline{Z}_b \tag{4.152}$$

where the initial switching current $\underline{I}_{ab}$ is given by Eq. (4.145), while the equivalent source voltage $\underline{E}_b$ is given by Eq. (4.147). Substituting $\underline{E}_b$ in Eq. (4.152) by the right-hand side of Eq. (4.147) it is obtained that

$$\underline{V}_{b+} = \underline{V}_b - \underline{V}_{ab}\frac{\underline{Z}_b}{\underline{Z}_{ab}} + \underline{I}_{ab}\underline{Z}_b \tag{4.153}$$

The impedance measured by the distance protection in the transmission line bay (Figure 4.54) can be expressed by the following equation

$$\underline{Z}_{dist} = \frac{\underline{V}_{b+}}{\underline{I}_{ab}} = \frac{\underline{V}_b}{\underline{I}_{ab}} - \frac{\underline{V}_{ab}}{\underline{I}_{ab}}\frac{\underline{Z}_b}{\underline{Z}_{ab}} + \underline{Z}_b \tag{4.154}$$

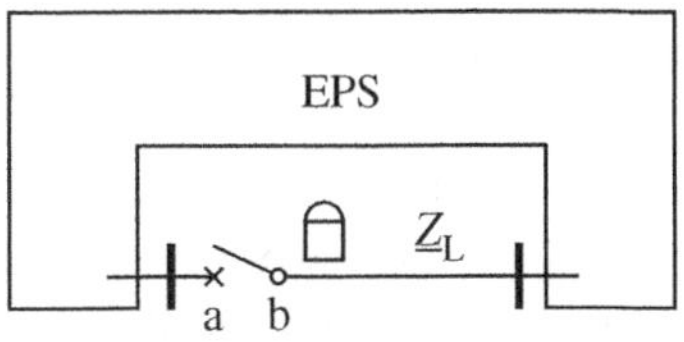

Figure 4.54 Illustration to switching-on a transmission line.

Now in accordance with Eq. (4.133) the initial switching current $\underline{I}_{ab}$ can be replaced by $\underline{V}_{ab}/\underline{Z}_{Th}$ and Eq. (4.154) takes the following shape

$$\underline{Z}_{dist} = \frac{\underline{V}_b}{\underline{V}_{ab}}\underline{Z}_{Th} + \left(1 - \frac{\underline{Z}_{Th}}{\underline{Z}_{ab}}\right)\underline{Z}_b \tag{4.155}$$

However, from Eq. (4.135) it results that

$$1 - \frac{\underline{Z}_{Th}}{\underline{Z}_{ab}} = \frac{\underline{Z}_{ab}}{\underline{Z}_a + \underline{Z}_b + \underline{Z}_{ab}} = \frac{1}{\underline{\xi}} \tag{4.156}$$

Taking into account Eq. (4.156) and Eq. (4.135), it is easy to transform Eq. (4.155) to the following shape

$$\underline{Z}_{\text{dist}} = \frac{\underline{V}_{\text{b}}}{\underline{V}_{\text{a}} - \underline{V}_{\text{b}}} \frac{\underline{Z}_{\text{a}} + \underline{Z}_{\text{b}}}{\underline{\xi}} + \frac{\underline{Z}_{\text{b}}}{\underline{\xi}} \tag{4.157}$$

For further consideration it is convenient to assume that phasor $\underline{V}_{\text{b}}$ lies on the real axis and therefore $\underline{V}_{\text{b}} = V_{\text{b}}\text{e}^{\text{j}0} = V_{\text{b}}$ and $\underline{V}_{\text{a}} = V_{\text{a}}\text{e}^{\text{j}\theta_{\text{ab}}}$. For such assumption the following equation can be written

$$\frac{\underline{V}_{\text{b}}}{\underline{V}_{\text{a}} - \underline{V}_{\text{b}}} = \frac{1}{\dfrac{\underline{V}}{\underline{V}_{\text{b}}} - 1} = \frac{1}{\nu \cdot \text{e}^{\text{j}\theta_{\text{ab}}} - 1} \tag{4.158}$$

where $\nu = V_{\text{a}}/V_{\text{b}}$. Now Eq. (4.158) enables to transform Eq. (4.157) in the following way

$$\underline{Z}_{\text{dist}} = \frac{1}{\nu \cdot \text{e}^{\text{j}\theta_{\text{ab}}} - 1} \left(\underline{Z}'_{\text{a}} + \underline{Z}'_{\text{b}}\right) + \underline{Z}'_{\text{b}} \tag{4.159}$$

where

$$\underline{Z}'_{\text{a}} = R'_{\text{a}} + \text{j}X'_{\text{a}} = -\underline{Z}_{\text{a}}/\underline{\xi} \quad \text{and} \quad \underline{Z}'_{\text{b}} = R'_{\text{b}} + \text{j}X'_{\text{b}} = \underline{Z}_{\text{b}}/\underline{\xi} \tag{4.160}$$

After simple rearrangements of Eq. (4.159) it is obtained

$$\nu \cdot \text{e}^{\text{j}\theta_{\text{ab}}} = \frac{\underline{Z}_{\text{dist}} - \left(-\underline{Z}'_{\text{a}}\right)}{\underline{Z}_{\text{dist}} - \left(+\underline{Z}'_{\text{b}}\right)} \tag{4.161}$$

From the mathematical point of view the right-hand side of Eq. (4.161) is the homographic function of the complex variable $\underline{Z}_{\text{dist}}$. Taking into account that $\left|\text{e}^{\text{j}\theta_{\text{ab}}}\right| = 1$ and Eq. (4.161), it can be written that

$$\nu = \frac{\left|\underline{Z}_{\text{dist}} - \left(-\underline{Z}'_{\text{a}}\right)\right|}{\left|\underline{Z}_{\text{dist}} - \underline{Z}'_{\text{b}}\right|} \tag{4.162}$$

On the plane with coordinates $R = \text{Re}\,\underline{Z}$ and $X = \text{Im}\underline{Z}$ each value $\underline{Z}$ determines a point. It is assumed (Figure 4.55) that points A and B conform with $\underline{Z}'_{\text{b}} = \underline{Z}_{\text{b}}/\underline{\xi}$ and $\underline{Z}'_{\text{a}} = \underline{Z}_{\text{a}}/\underline{\xi}$, respectively. For $\nu = \text{const}$ Eq. (4.162) determines a set of points for which the ratio of the distance to point A and the distance to point B is constant and equal to ν. For a given value of ν such a set of points creates the circle of Apollonius. When ν is treated as a parameter there is a family of circles. The center of each circle lies on the line passing through points A and B. For particular value $\nu = 1$ the circle of Apollonius is open and constitutes a straight line orthogonal to the section AB (dashed line in Figure 4.55). For $\nu > 1$ the circle center lies above point B and for $\nu < 1$ below point A.

The argument of the left-hand side of Eq. (4.161) is equal to θ_{ab} and the argument of the right-hand side of Eq. (4.161) is equal to the difference of the arguments of $\left[\underline{Z}_{\text{dist}} - \left(-\underline{Z}'_{\text{a}}\right)\right]$ and $\left[\underline{Z}_{\text{dist}} - \underline{Z}'_{\text{b}}\right]$. Therefore

$$\theta_{\text{ab}} = \arg\left[\underline{Z}_{\text{dist}} - \left(-\underline{Z}'_{\text{a}}\right)\right] - \arg\left[\underline{Z}_{\text{dist}} - \underline{Z}'_{\text{b}}\right] \tag{4.163}$$

This means (Figure 4.55) that for a given value of $\underline{Z}_{\text{dist}}$ the section AB is seen under angle θ_{ab}.

In order to avoid unwanted operation of the distance protection it is sufficient to limit the differences in the voltage magnitudes V_{ab} and phase angles θ_{ab} to such values for which the apparent impedance $\underline{Z}_{\text{dist}}$ measured by distance protection at the switching instant lies outside the characteristic of the fault detector characteristic of the distance protection. The shapes of the fault detector characteristics of various types of distance relays differ very much. Therefore, in order to simplify the further analysis, a rectangular characteristic (Figure 4.55) is considered assuming that values R_{r} and X_{r} are so large that the fault detector characteristic of the distance relay places inside that rectangle.

On Figure 4.55 point D is the intersection point of line AB and the orthogonal straight line corresponding with $\nu = V_a/V_b = 1$. Its coordinates can be found from Eq. (4.161). For $\nu = 1$ and $\theta_{ab} = 180^\circ$ it is obtained from this equation that $[\underline{Z}_{dist} - (-\underline{Z}'_a)] = -[\underline{Z}_{dist} - (+\underline{Z}'_b)]$ and hence

$$\underline{Z}_{dist\,D} = \frac{\underline{Z}'_b - \underline{Z}'_a}{2} = \frac{R'_b - R'_a}{2} + j\frac{X'_b - X'_a}{2} \tag{4.164}$$

and therefore

$$R_D = \frac{R'_b - R'_a}{2}; \quad X_D = \frac{X'_b - X'_a}{2} \tag{4.165}$$

From Eq. (4.165) and Figure 4.55 it can be concluded that point D is the middle point of the section AB. If $X_D > X_r$ then point D lies outside the rectangle representing the fault detector characteristic, and the considered condition is inessential for any switching angle θ_{ab}. However, if

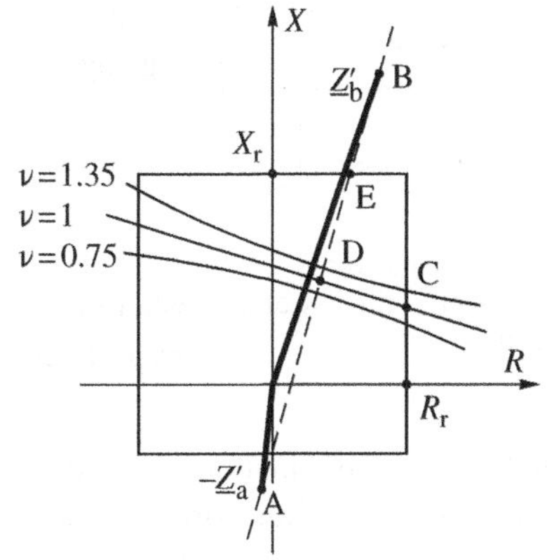

Figure 4.55 Simplified characteristic of fault detector and parts of the circles of Apollonius.

$$X_D = \frac{X'_b - X'_a}{2} < X_r \tag{4.166}$$

then depending on the value of the switching angle θ_{ab} the impedance $\underline{Z}_{dist}$ may encroach the distance protection zone and may cause unwanted transmission line tripping. In such a case it is necessary to calculate maximal value of the switching angle θ_{ab} for which the impedance $\underline{Z}_{dist}$ lies outside the rectangle representing the fault detector characteristic. For such analysis it is useful to realize the fact that for values of $\nu = V_a/V_b$ in the technically reasonable range $0.75 \le \nu \le 1.35$ the arcs (being the parts of the circles of Apollonius) lie (Figure 4.55) very close to the straight line corresponding with $\nu = V_a/V_b = 1$. Therefore, the maximal value of the switching angle θ_{ab} for $0.75 \le \nu \le 1.35$ can be approximately determined for $\nu = 1$. For this purpose the coordinates of point C must be found and then the impedance $\underline{Z}_{dist}$ at point C must be expressed by the switching angle θ_{ab}.

The line passing through points A and B is described by the following equation

$$X - X'_b = \frac{X'_a + X'_b}{R'_a + R'_b}(R - R'_b) \tag{4.167}$$

Point D lies on the line AB and has coordinates determined by Eqs. (4.165). The line passing through points D and C is orthogonal to line AB and is described by the following equation

$$X - \frac{X'_b - X'_a}{2} = \frac{R'_a + R'_b}{X'_a + X'_b}\left(R - \frac{R'_b - R'_a}{2}\right) \tag{4.168}$$

Substituting in Eq. (4.168) values R, X_C with $R = R_C = R_r$, and $X = X_C$ it is obtained

$$X_C = \frac{X'_b - X'_a}{2} + \frac{R'_a + R'_b}{X'_a + X'_b}\left(R_C - \frac{R'_b - R'_a}{2}\right) \tag{4.169}$$

At point C the impedance $\underline{Z}_{dist}$ measured by distance protection is equal $\underline{Z}_{dist} = R_C + jX_C$. For such value and $\nu = 1$ it is obtained from Eq. (4.161) that

$$e^{j\theta_{ab}} = \cos\theta_{ab} + j\sin\theta_{ab} = \frac{\underline{Z}_{zab} + \underline{Z}'_a}{\underline{Z}_{zab} - \underline{Z}'_b} = \frac{(R_C + R'_a) + j(X_C + X'_a)}{(R_C - R'_b) + j(X_C - X'_b)} \tag{4.170}$$

and hence

$$\cos\theta_{AB} = \frac{(R_C + R'_a)(R_C - R'_b) + (X_C + X'_a)(X_C - X'_b)}{(R_C - R'_b)^2 + (X_C - X'_b)^2} \tag{4.171}$$

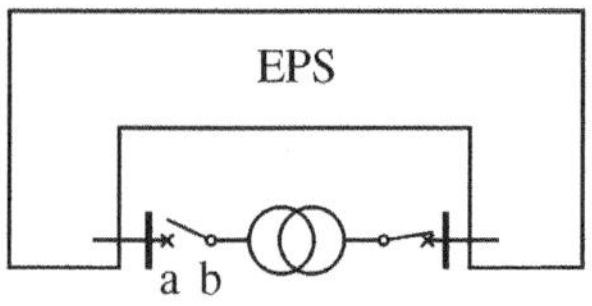

Figure 4.56 Closing the circuit breaker in the bay of a transformer.

Equation (4.171) enables us to calculate the maximal allowed value of the switching angle from the point of view of the unwanted operation of distance protection

$$\theta_{\text{ab max}} = \arccos x \quad \text{where} \quad x = \frac{\left(R_C + R_a'\right)\left(R_C - R_b'\right) + \left(X_C + X_a'\right)\left(X_C - X_b'\right)}{\left(R_C - R_b'\right)^2 + \left(X_C - X_b'\right)^2} \tag{4.172}$$

where R_a', X_a', R_b', and X_b' are given by Eqs. (4.160), and R_C and X_C by Eq. (4.160), respectively.

Condition C3 Deformation of transformer windings

It is assumed that a transformer already connected to the network from one side must be connected to the network from the other side by closing the circuit breaker with poles denoted in Figure 4.56 as a and b. When the difference in the voltages at the circuit breaker poles is large, the dynamic forces caused by the inrush switching current can damage the transformer windings.

Standards relating to transformers do not formulate requirements for the switching currents. The capability of transformers to withstand the dynamic effect of the inrush currents is designed for the short circuits at the transformer terminals. Therefore, from the standpoint of the threat of the windings damage, the synchro-check device should block the switching operation in all cases when the switching current could be larger than the three-phase short-circuit current. Such a situation may happen for particular network parameters and a large switching angle.

For the given parameters of a network equivalent model (Figure 4.50) the maximal allowed switching angle can be found from the following inequality: $I_{ab} \le I_{K3}$, where I_{K3} is the current flowing through the transformer during the three-phase short-circuit at transformer terminals (node b in Figure 4.56). For approximate analysis the network resistances can be neglected. Then the short-circuit current is expressed by equation $I_{K3} \cong V/X_b$ and from Eqs. (4.139) and (4.135) it can be easily concluded that

$$I_{ab} \cong \frac{\xi 2V}{X_a + X_b} \sin\frac{\theta_{ab}}{2} \tag{4.173}$$

For such values of the switching and short-circuit currents above inequality $I_{ab} \le I_{K3}$ is satisfied if

$$\sin\frac{\theta_{\text{ab max}}}{2} \le \frac{1}{2\xi}\left(\frac{X_a}{X_b} + 1\right) \tag{4.174}$$

As results from this inequality the maximal allowed switching angle $\theta_{\text{ab max}}$ depends on ratio X_a/X_b and coefficient ξ given by Eq. (4.135). Small values of $\theta_{\text{ab max}}$ are obtained when $X_a \ll X_b$ and $\xi > 1$ (Figure 4.53). For example, for $X_a = 0.5X_b$ and $\xi = 1.5$ it is obtained that $\theta_{\text{ab max}} = 60^\circ$. For $X_a = 0.5X_b$ and $\xi = 2.0$ a smaller value $\theta_{\text{ab max}} = 44^\circ$ is obtained.

Condition C4 Fatigue of generating unit shafts

As described in previous sections of this chapter, all electrical disturbances in the network such as line switching, faults, automatic reclosing, and generator synchronizing out-of-phase cause a sudden change in the electrical torque imposed across the air gap onto the generator rotor. Peak shaft torque values during a transient state give an indication of the severity of the event. However, torque magnitude alone does not quantify damage. Damage is cumulative and may be measured as a percentage of fatigue life loss during the event. The fatigue life of a shaft section is expressed by a curve, like in Figure 4.57. This plot relates to the number of stress cycles a material can withstand at each values of stress magnitude. The number of tolerable stress cycles increases as stress magnitude decreases. The value of the stress magnitude for which the curve becomes asymptotic (dashed line) is referred to as the *fatigue limit*. It is assumed that stress below this value does not produce loss of life.

Figure 4.57 Example of lifecycle curve. *Source:* Adapted from Reimert (2006).

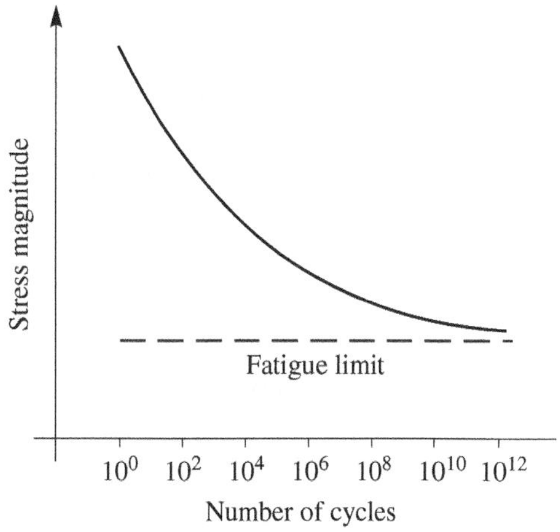

The shaft fatigue in percent per incident for various electrical disturbances is shown in Figure 4.58. These data are the result of studies on two large steam-turbine generating units done by Joyce et al. (1978). A loss of life equal to 100% indicates that shaft crack is possible. The dark parts of the indicating bars concern those two studied generating units. The white extensions to the bar are estimates for generating units with a wide range of parameters. Obviously, data shown in Figure 4.58 should be viewed as qualitative ranking of event severity, not as a basis for loss-of-life evaluation.

From Figure 4.58 it can be concluded that the most severe events (to which a shaft can be exposed) are the multi-phase faults in the network close to the power plants cleared with unsuccessful three-phase automatic reclosing of the faulted line. This is the reason why many power system operators for the lines close to the power plants use at the side of the power plant the three-phase automatic reclosing with the check of the voltage restoration from the side of the EPS. For such a solution the circuit breaker at the side of power plant is reclosed only in the case when automatic reclosing from the side of the system is successful.

In Figure 4.58 also notable is the relative loss of life for the faulty generator synchronization with the angle $90^\circ \leq \theta \leq 120^\circ$, what is a rather rare incidental case.

The shaft fatigue per incident caused by a switching operation in the transmission network depends on the change in the active power of the generating unit. Usually, the switching operations are more frequent than the multi-phase faults and therefore it is reasonable to assume that each switching in transmission network cannot cause the shaft fatigue larger than the three-phase short circuit at the high-voltage terminals of the generating unit. To satisfy such a requirement a report prepared by the IEEE Working Group on the effects of switching on turbine-generators (1980) recommends to avoid switchings which cause the changes in active power of generators larger than 50% of the power rating. From this point of view the synchro-check device of a transmission line should permit closing of the circuit breaker for all values of the switching angle θ_{ab} for which the following condition is satisfied for all generators in the system

$$\Delta P_G \leq 0.5 S_{Gr} \tag{4.175}$$

where ΔP_G is the step change in active power of a given generator caused by the switching and S_{Gr} is the rated power of a given generator. In the calculation method described in Section 4.4.4 the change in active power of generators is given by Eqs. (4.143) and (4.144). It is worth emphasizing here that following a switching operation in the transmission network the active power of a synchronous generator alters with fundamental frequency, as shown in Figure 4.59. For the setting of a synchro-check device only the value ΔP_G is important.

Fault conditions and switing operations		Fatigue per incident (%)					
		negligible	moderate		severe		
		0.001	0.01	0.1	1	10	100
Normal line switching	$\Delta P_G \leq 0.5$ pu	■□					
	$\Delta P_G > 0.5$ pu		■□				
Synchronizing	Automatic	■□					
	Manual ($\theta \leq 10°$ slip = 0.7%)	■□					
	Faulty ($90° \leq \theta \leq 120°$)				□■		
Full-load rejection		■					
Three-phase fault	Generator terminals			■			
	High voltage terminals		□■□				
System faults with multiple transmission lines	Line to graund	□■					
Unsuccessful recloasing	Line-line fault		■□				
	Three phase fault		■				

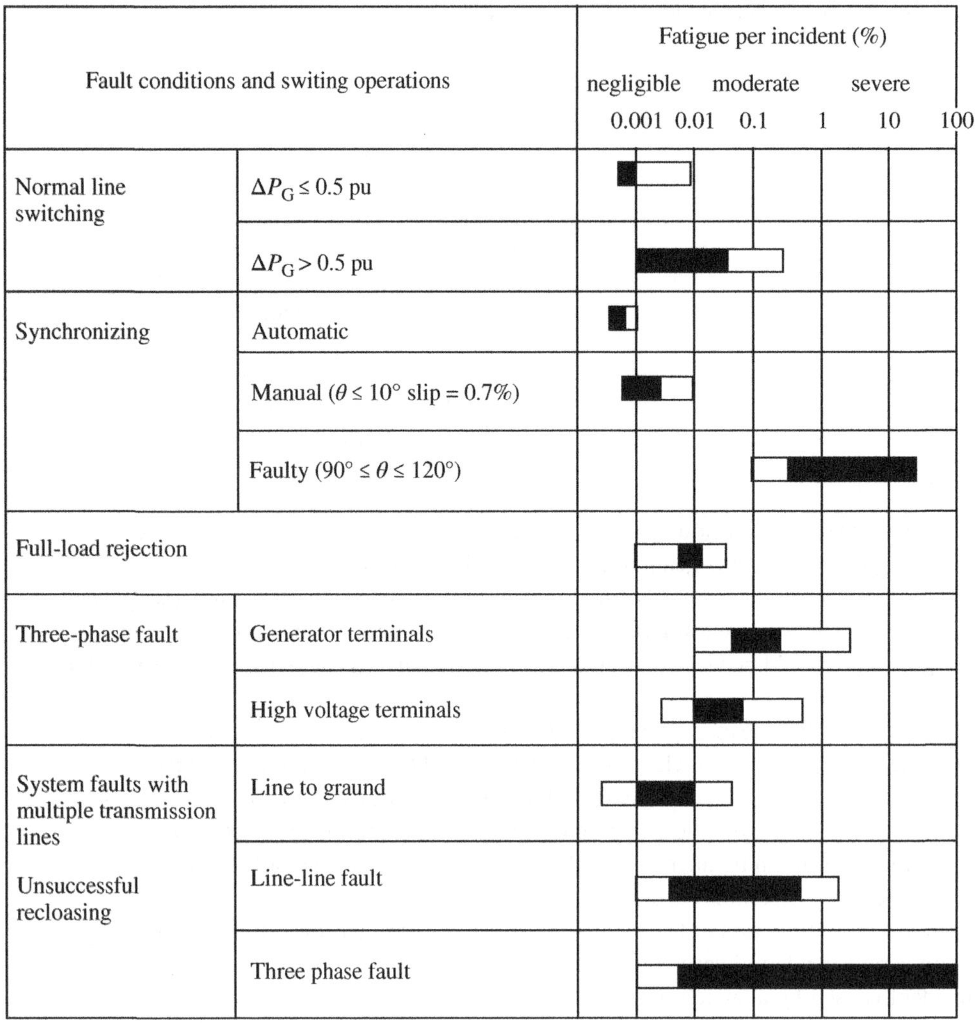

Figure 4.58 Shaft fatigue for switching and fault conditions. *Source:* Reproduced by permission of IEEE.

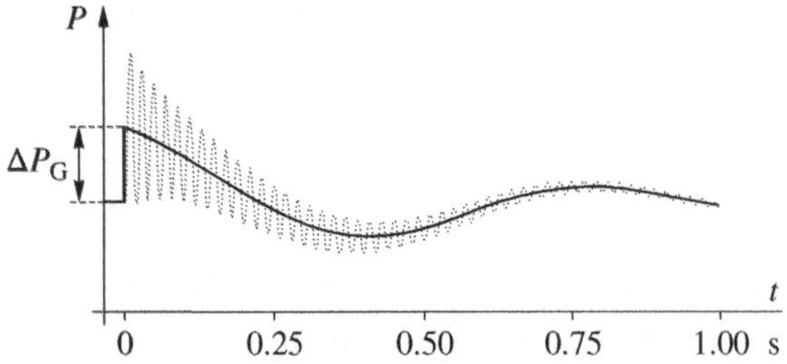

Figure 4.59 Example of the changes in active power of a synchronous generator caused by a switching operation in the network.

When condition (4.175) is satisfied then the effect of the switching operation on the shaft fatigue is negligible (Figure 4.58). However, for emergency states of an EPS such a condition can be too restrictive and can hinder the dispatcher from switching on a line important for power system security. Therefore, the power system operator should have a remote control activated in emergency states enabling them to block the synchro-check devices and switching on important transmission lines even for switching angles that result from condition (4.175). As shown in Figure 4.58, incidental switching operation causing $\Delta P_{G} > 0.5 S_{Gr}$ only slightly contributes to the cumulative shaft fatigue.

Above conditions C1–C4 can be checked using the calculation method described in Section 4.4.4. A relevant computer program is described in the publication by Machowski et al. (2014). Test results for a large-scale power system have shown that for typical data of power system above conditions C1 and C3 are not decisive. For short transmission lines (especially the ones located in the vicinity of large generating units driven by steam turbines), above condition C4 (shaft fatigue) is decisive, while for very long transmission lines it can be condition C2 (distance protection).

4.5 Subsynchronous Resonance

Section 4.2 explains how, when a fault occurs in an EPS, large currents are produced that comprise a DC offset and an AC term. The AC term depends on the type of fault and, in the case of a symmetrical fault, will consist of a positive-sequence component, while with unsymmetrical faults both positive-sequence and negative-sequence components will be present. Owing to the relative speed of the rotor with respect to the stator, and the space distribution of the stator armature windings, these different current components (DC, positive, negative) produce rotor torques at f_s, 0, and $2f_s$ Hz, respectively. Such observations assume that the transmission system is purely inductive and no capacitance compensation is present in the system. Such a system is often referred to as *uncompensated.*

If now the transmission line reactance is compensated by a series capacitor SC, as shown in Figure 4.60a, then, depending on the relative magnitude of the total circuit resistance R, inductance L, and capacitance C, the transient component of any current may now be an oscillation at a frequency f_d that decays slowly with time. This oscillation frequency is closely related to the undamped natural frequency of the line, $f_n = 1/2\pi\sqrt{LC}$ Hz. Such oscillations produce rotor currents and torques at a frequency $(f_s - f_n)$ which, if close to one of the rotor torsional natural frequencies, can produce very high shaft torques and ultimate shaft failure. The frequency $(f_s - f_n)$ is known as the *frequency complement* of the line natural frequency f_n as both these frequencies sum to f_s, the system frequency.

The way that such a damped oscillatory current is produced can be understood by considering the equivalent circuit in Figure 4.60b and calculating the current when the switch S is suddenly closed. The differential equation governing the flow of current is

$$L\frac{di}{dt} + Ri + \frac{1}{C}\int i dt = E_m \sin(\omega_s t + \theta_0) \tag{4.176}$$

Differentiating gives

$$\frac{d^2 i}{dt^2} + \frac{R}{L}\frac{di}{dt} + \frac{1}{LC} i = \omega E_m \cos(\omega_s t + \theta_0) \tag{4.177}$$

the solution of which is of the form

$$i(t) = i_{trans}(t) + \frac{E_m}{Z}\sin(\omega_s t + \theta_0 + \psi) \tag{4.178}$$

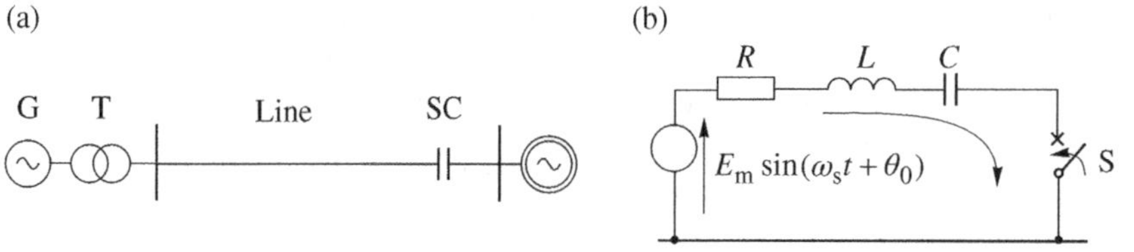

Figure 4.60 Series-compensated transmission line: (a) schematic diagram; (b) simple equivalent circuit to calculate the form of the current when shorting the line.

where $Z = \sqrt{R^2 + (X_L - X_C)^2}$, $X_L = \omega_s L$, $X_C = 1/(\omega_s C)$, and the phase angle $\psi = \text{atan}\ [(X_C - X_L)/R]$. The left-hand side of Eq. (4.177) is of the standard second-order form (see Appendix A.3)

$$\frac{d^2x}{dt^2} + 2\zeta\Omega_n\frac{dx}{dt} + \Omega_n^2 = 0 \tag{4.179}$$

the solution of which determines the form of transient current. Equation (4.179) will have a different type of solution depending on the value of the damping ratio ζ. Of particular interest is the underdamped solution obtained when $0 < \zeta < 1$, which is of the form

$$i_{trans}(t) = Ae^{-\zeta\Omega_n}\sin(\Omega t + \psi_2) \tag{4.180}$$

where A and ψ_2 are constants and the damped natural frequency $\Omega = \Omega_n\sqrt{1-\zeta^2}$ (in rad/s). The undamped natural frequency Ω_n (in rad/s) and the damping ratio, ζ, can be found in terms of the circuit parameters by comparing Eq. (6.179) with the left-hand side of Eq. (4.177) to give

$$\Omega_n = \sqrt{\frac{1}{LC}} = \omega_s\sqrt{\frac{X_C}{X_L}} \quad \text{and} \quad \zeta = \frac{R}{2}\sqrt{\frac{C}{L}} \tag{4.181}$$

Substituting Eq. (4.180) into (4.178) gives the total current for the underdamped condition as

$$i(t) = Ae^{-\zeta\Omega_n}\sin(\Omega t + \psi_2) + \frac{E_m}{Z}\sin(\omega_s t + \theta_0 + \psi) \tag{4.182}$$

With currents flowing in each armature phase at frequencies Ω and ω_s, torques are induced in the rotor at the sum and difference of these frequencies to produce rotor currents and torques at frequencies below (subsynchronous) and above (supersynchronous) system frequency given by $(\omega_s - \Omega)$ and $(\omega_s + \Omega)$, respectively. Of particular importance are the subsynchronous torques and the way in which they interact with the turbine and generator rotor. If any subsynchronous torques appear on the rotor at, or near to, any of the rotor torsional natural frequencies, additional energy can be fed into the mechanical vibrations producing a resonance effect in the mechanical system and large shaft torques. Such large shaft torques produce high shaft stresses, a reduction in shaft fatigue life, and, possibly, shaft failure. To avoid such *subsynchronous resonance* effects it is important to ensure that no subsynchronous torques are present at, or near to, any of the shaft torsional natural frequencies. Typically, these torsional natural frequencies will be in the range (10–40) Hz for a 3000 rpm, 50 Hz generator.

For the simple system shown in Figure 4.60a the subsynchronous frequencies can be estimated using $\Omega_n = \omega_s\sqrt{X_C/X_L}$, Eq. (4.181), assuming that the degree of compensation employed is known. The results of such a calculation are shown in Table 4.5 for a 50 Hz system. Assuming the generator operating in a simple system, such as the one shown in Figure 4.60 with ~30% compensation, the results of Table 4.5 would suggest that subsynchronous resonance with the 23.3 Hz, mode 3, torsional natural frequency might be a problem.

Table 4.5 Line natural frequency as a function of the degree of compensation.

Degree of compensation X_C/X_L	**Line natural frequency** in Hz $f_n = \Omega_n/2\pi$	**Frequency complement** in Hz $(50 - f_n)$
0.1	15.8	34.2
0.2	22.4	27.6
0.3	27.4	22.6
0.4	31.6	18.4
0.5	35.4	14.6

A fuller analysis of subsynchronous resonance problems that includes all the system damping mechanisms would require the use of a full system eigenvalue study such as that conducted by Ahlgren et al. (1978). This and other techniques can be used to study more practical, interconnected, power systems where a number of subsynchronous resonant frequencies may be present (Anderson et al. 1990).

5

Electromechanical Dynamics – Small Disturbances

In the previous chapter the currents and torques produced in a synchronous generator as the result of a system disturbance are discussed and, as the duration of the disturbance is very short, the generator rotational speed could be considered constant. In this chapter a longer timescale is considered during which the rotor speed will vary and interact with the electromagnetic changes to produce *electromechanical dynamic* effects. The timescale associated with these dynamics is sufficiently long for them to be influenced by the turbine and the generator control systems.

The aim of this chapter is both to provide an explanation of how, and why, mechanical movement of the generator rotor is influenced by electromagnetic effects and to examine how this movement varies depending on the operating state of the generator. During the course of this discussion, some important stability concepts will be introduced together with their basic mathematical description and an explanation of the physical implications.

5.1 Swing Equation

For the analysis of electromechanical dynamics enumerated in Figure 1.3 (Section 1.2), the generator rotor swing following a fault in the network plays a fundamental role for the analysis of electromechanical dynamics. For the analysis of torsional oscillations (Section 4.4.3), generating units with a long drive shaft must be treated as a system of oscillating masses (Figures 4.39 and 4.40). For the analysis of electromechanical dynamics, the shaft of the generating unit can be assumed to be rigid and the rotor swing can be described by Eq. (4.15). For the reader's convenience, this swing equation is rewritten here in the following way

$$M\frac{\mathrm{d}\Delta\omega}{\mathrm{d}t} = P_{\mathrm{m}} - P_{\mathrm{e}} - D\Delta\omega \qquad \frac{\mathrm{d}\delta}{\mathrm{d}t} = \Delta\omega \tag{5.1}$$

where the time derivative of the rotor angle $\mathrm{d}\delta/\mathrm{d}t = \Delta\omega = \omega - \omega_{\mathrm{s}}$ is the *rotor speed deviation* in electrical radians per second (rad/s), D is the damping coefficient, M is an *inertia coefficient*, and P_{m} and P_{e} are the mechanical and electrical power, respectively.

5.2 Damping Power

Damping of the rotor motion by mechanical loss is small and can be neglected for all practical considerations. The main source of damping in the synchronous generator is provided by the *damper*, or *amortisseur*, windings described in Section 2.3.1. The damper windings have a high resistance/reactance ratio and act in a similar

Power System Dynamics: Stability and Control, Third Edition. Jan Machowski, Zbigniew Lubosny, Janusz W. Bialek and James R. Bumby.

way to the short-circuited squirrel-cage rotor windings in an induction motor. In the subtransient state these windings act as a perfect screen and the changes in the armature flux cannot penetrate them. In the transient state the air-gap flux, which rotates at the synchronous speed, penetrates the damper windings, and induces an electromotive force (emf) and current in them whenever the rotor speed ω is different from the synchronous speed ω_s. This induced current produces a damping torque which, according to Lenz's law, tries to restore the synchronous speed of the rotor. As this additional torque only appears when $\omega \neq \omega_s$, it is proportional to $\Delta\omega = d\delta/dt$ and is referred to as the *asynchronous torque*.

Damper windings can be on both rotor axes, or on the direct-axis (d-axis) only. In the round-rotor generator the solid-steel rotor body provides paths for eddy currents which have the same effect as damper windings. Machines with laminated salient poles require explicit damper windings for effective damping.

Rigorous derivation of an expression for damping power is long and complicated but an approximate equation for the generator-infinite busbar system can be derived if the following assumptions are made:

- The resistances of both the armature and the field winding are neglected.
- Damping is produced only by the damper windings.
- The leakage reactance of the armature winding can be neglected.
- Excitation does not affect the damping torque.

With these assumptions an equation for damping power can be derived using the induction motor equivalent circuit shown in Figure 3.31.

Figure 4.12a showed that in the subtransient state the equivalent reactance of the generator, as seen from the network, consists of the armature winding leakage reactance X_l connected in series with the parallel-connected damper winding leakage reactance X_D, the field winding leakage reactance X_f and the armature reaction reactance X_a. Figure 5.1a shows a similar equivalent circuit of the generator-infinite busbar system with the generator operating as an induction machine. The reactance X represents the combined reactance of the step-up transformer and the network while V_s is the infinite busbar voltage. Owing to assumption (iv) the field winding is closed but not excited. To take the speed deviation into account the equivalent resistance in the damper branch is divided by $s = \Delta\omega/\omega_s$ in the same way as for the induction machine. Initially rotor saliency is neglected.

Assumption (iii) allows the reactance X_l to be neglected when the subtransient and transient reactance's are given approximately by the expressions in Eqs. (5.12) and (5.13)

$$X'_d \cong \frac{1}{\frac{1}{X_f} + \frac{1}{X_a}}, \quad X''_d \cong \frac{1}{\frac{1}{X_f} + \frac{1}{X_a} + \frac{1}{X_D}} \tag{5.2}$$

The first of these equations allows the parallel connection of X_f and X_a in Figure 5.1a to be replaced by X'_d, as shown in Figure 5.1b. Equation (5.2) also allows the damper winding leakage reactance to be approximately expressed as a function of the subtransient and transient reactances as

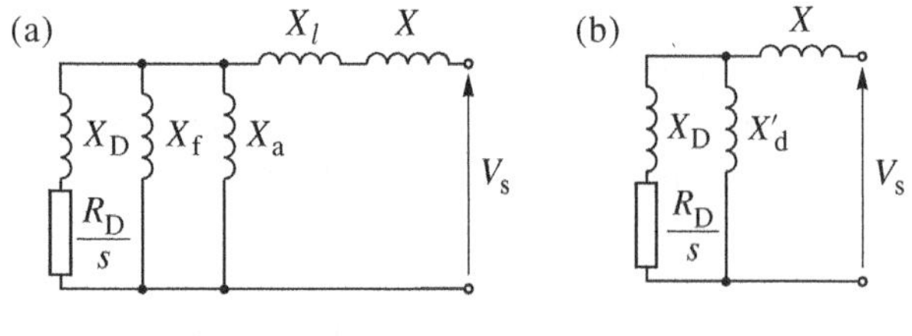

Figure 5.1 The equivalent circuit of the synchronous generator operating as an induction machine: (a) with leakage reactances included; (b) with leakage reactances neglected.

$$X_{\mathrm{D}} \cong \frac{X'_{\mathrm{d}} X''_{\mathrm{d}}}{X'_{\mathrm{d}} - X''_{\mathrm{d}}} \tag{5.3}$$

The subtransient short-circuit time constant $T''_{\mathrm{d}} = X_{\mathrm{D}}/\omega_{\mathrm{s}} R_{\mathrm{D}}$ can now be calculated as

$$T''_{\mathrm{d}} = \frac{X_{\mathrm{D}}}{\omega_{\mathrm{s}} R_{\mathrm{D}}} \cong \frac{X'_{\mathrm{d}} X''_{\mathrm{d}}}{\omega_{\mathrm{s}} R_{\mathrm{D}} \left(X'_{\mathrm{d}} - X''_{\mathrm{d}}\right)} \tag{5.4}$$

allowing the damper winding resistance (R_{D}/s) as seen from the armature, to be written in the form

$$\frac{R_{\mathrm{D}}}{s} = \frac{X'_{\mathrm{d}} X''_{\mathrm{d}}}{X'_{\mathrm{d}} - X''_{\mathrm{d}}} \frac{1}{T''_{\mathrm{d}} \omega_{\mathrm{s}} s} = \frac{X'_{\mathrm{d}} X''_{\mathrm{d}}}{X'_{\mathrm{d}} - X''_{\mathrm{d}}} \frac{1}{T''_{\mathrm{d}} \Delta\omega} \tag{5.5}$$

For small values of speed deviation the term R_{D}/s is large and the current flows mostly through X'_{d}, allowing the series connection of X and X'_{d} to be treated as a voltage divider. With this assumption the voltage drop across X'_{d} is equal to $V_{\mathrm{s}} X'_{\mathrm{d}}/\left(X + X'_{\mathrm{d}}\right)$. Ohm's law applied to the damper equivalent branch then gives

$$I_{\mathrm{D}}^2 \cong V_{\mathrm{s}}^2 \left(\frac{X'_{\mathrm{d}}}{X + X'_{\mathrm{d}}}\right)^2 \frac{1}{\left(\dfrac{R_{\mathrm{D}}}{s}\right)^2 + X_{\mathrm{D}}^2} \tag{5.6}$$

and the damping power as

$$P_{\mathrm{D}} = I_{\mathrm{D}}^2 \frac{R_{\mathrm{D}}}{s} \cong V_{\mathrm{s}}^2 \frac{\left(X'_{\mathrm{d}}\right)^2}{\left(X + X'_{\mathrm{d}}\right)^2} \frac{\dfrac{R_{\mathrm{D}}}{s}}{\left(\dfrac{R_{\mathrm{D}}}{s}\right)^2 + \dfrac{\left(X'_{\mathrm{d}} X''_{\mathrm{d}}\right)^2}{\left(X'_{\mathrm{d}} - X''_{\mathrm{d}}\right)^2}} \tag{5.7}$$

Substituting (R_{D}/s) from Eq. (5.5) gives

$$P_{\mathrm{D}} \cong V_{\mathrm{s}}^2 \frac{X'_{\mathrm{d}} - X''_{\mathrm{d}}}{\left(X + X'_{\mathrm{d}}\right)^2} \frac{X'_{\mathrm{d}}}{X''_{\mathrm{d}}} \frac{T''_{\mathrm{d}}\ \Delta\omega}{1 + \left(T''_{\mathrm{d}}\ \ \Delta\omega\right)^2} \tag{5.8}$$

When rotor saliency is taken into account a similar formula can be derived for the quadrature-axis (q-axis). The resultant damping power can be found by replacing the driving voltage V_{s} in both the d- and q-axis equivalent circuits by the respective voltage components, $V_{\mathrm{d}} = -V_{\mathrm{s}} \sin\delta$ and $V_{\mathrm{q}} = V_{\mathrm{s}} \cos\delta$ to give

$$P_{\mathrm{D}} = V_{\mathrm{s}}^2 \left[\frac{X'_{\mathrm{d}} - X''_{\mathrm{d}}}{\left(X + X'_{\mathrm{d}}\right)^2} \frac{X'_{\mathrm{d}}}{X''_{\mathrm{d}}} \frac{T''_{\mathrm{d}}\ \Delta\omega}{1 + \left(T''_{\mathrm{d}}\ \ \Delta\omega\right)^2} \sin^2\delta + \frac{X'_{\mathrm{q}} - X''_{\mathrm{q}}}{\left(X + X'_{\mathrm{q}}\right)^2} \frac{X'_{\mathrm{q}}}{X''_{\mathrm{q}}} \frac{T''_{\mathrm{q}}\ \Delta\omega}{1 + \left(T''_{\mathrm{q}}\Delta\omega\right)^2} \cos^2\delta\right] \tag{5.9}$$

The damper power depends on the rotor angle δ and fluctuates with the rotor speed deviation $\Delta\omega = \mathrm{d}\delta/\mathrm{d}t$. For small speed deviations the damping power is proportional to the speed deviation, while for larger speed deviations it is a nonlinear function of speed deviation and resembles the power-slip characteristic of the induction motor (Figure 3.32).

For a small speed deviation $s = \Delta\omega/\omega_{\mathrm{s}} \ll 1$ and the $\left(T''_{\mathrm{d}}\ \Delta\omega\right)^2$ term in the denominator of Eq. (5.9) can be neglected when Eq. (5.9) simplifies to

$$P_{\mathrm{D}} = V_{\mathrm{s}}^2 \left[\frac{X'_{\mathrm{d}} - X''_{\mathrm{d}}}{\left(X + X'_{\mathrm{d}}\right)^2} \frac{X'_{\mathrm{d}}}{X''_{\mathrm{d}}} T''_{\mathrm{d}}\ \sin^2\delta + \frac{X'_{\mathrm{q}} - X''_{\mathrm{q}}}{\left(X + X'_{\mathrm{q}}\right)^2} \frac{X'_{\mathrm{q}}}{X''_{\mathrm{q}}} T''_{\mathrm{q}}\ \cos^2\delta\right] \Delta\omega \tag{5.10}$$

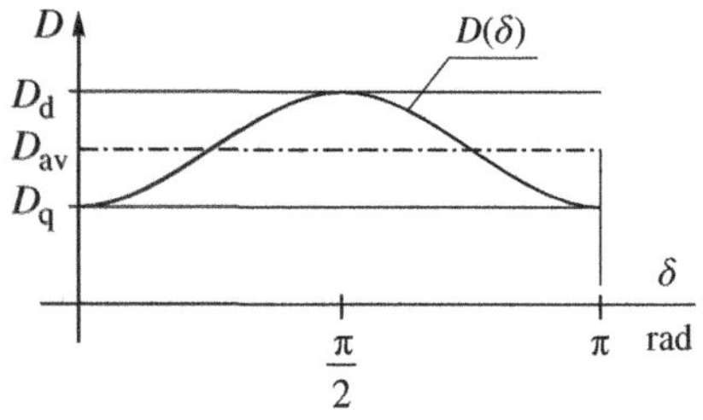

Figure 5.2 Damping coefficient as a function of rotor angle.

This equation is identical to the expression derived by Dahl (1938) and quoted by Kimbark (1995). In both equations the time constants T''_d and T''_q are the subtransient short-circuit time constants.

When δ is large, damping is strongest in the d-axis, and when δ is small, the q-axis damper winding produces the stronger damping. Equation (5.10) can usefully be rewritten as

$$P_D = \left[D_d \sin^2\delta + D_q \cos^2\delta\right] \Delta\omega = D(\delta)\Delta\omega \tag{5.11}$$

where $D(\delta) = D_d\sin^2\delta + D_q\cos^2\delta$ and D_d and D_q are damping coefficients in both axes. Figure 5.2 shows the variation of the damping coefficient D with δ as expressed by Eq. (5.11). The function reaches extrema for $\delta = 0, \pi$ or $\delta = \pi/2,\ 3\pi/2$ with the corresponding extremal values of damping coefficient being equal to D_d and D_q, respectively. The average value of the damping coefficient is $D_{av} = (D_d + D_q)/2$.

The network equivalent reactance X has a significant influence on the damping (asynchronous) power because its squared value appears in the denominator of Eqs. (5.9) and (5.10). In comparison, its influence on the synchronous power, Eq. (3.90), is much less where it appears in the denominator and is not squared.

When analyzing the damping power over a wide range of speed deviation values it is convenient to rewrite Eq. (5.9) as

$$P_D = P_{D(d)} \sin^2\delta + P_{D(q)} \cos^2\delta \tag{5.12}$$

where $P_{D(d)}$ and $P_{D(q)}$ both depend on the speed deviation. Both components are proportional to nonlinear functions of the speed deviation of the form $\alpha/(1+\alpha^2)$, where $\alpha = T''_d\ \Delta\omega$ for the first component and $\alpha = T''_q\ \Delta\omega$ for the second component. This function reaches a maximum for $\alpha = 1$ so that each component of the damping power will normally reach a maximum critical value at a different *critical speed deviation* given by

$$s_{cr(d)} = \frac{\Delta\omega_{cr(d)}}{\omega_s} = \frac{1}{T''_d\ \omega_s}; \quad s_{cr(q)} = \frac{\Delta\omega_{cr(q)}}{\omega_s} = \frac{1}{T''_q\ \omega_s} \tag{5.13}$$

with

$$P_{D(d)cr} = \frac{V_s^2}{2}\frac{X'_d - X''_d}{(X + X'_d)^2}\frac{X'_d}{X''_d}; \quad P_{D(q)cr} = \frac{V_s^2}{2}\frac{X'_q - X''_q}{\left(X + X'_q\right)^2}\frac{X'_q}{X''_q} \tag{5.14}$$

Figure 5.3 shows the variation of both $P_{D(d)}$ and $P_{D(q)}$ as a function of the speed deviation. Both factors are equal to zero when $\Delta\omega = 0$ and increase their value as the speed deviation increases until the critical value is reached,

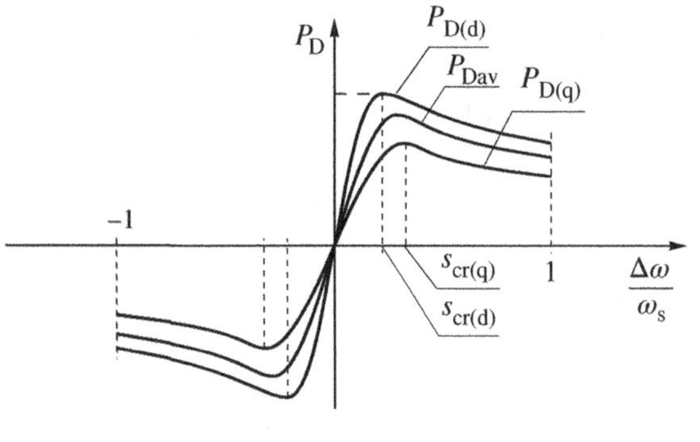

Figure 5.3 Average value of the damping power as a function of speed deviation.

after which they decrease. With δ changing, the damping power P_D will assume values between $P_{D(d)}$ and $P_{D(q)}$ with the average value lying between the axes characteristics, as shown by the bold line in Figure 5.4.

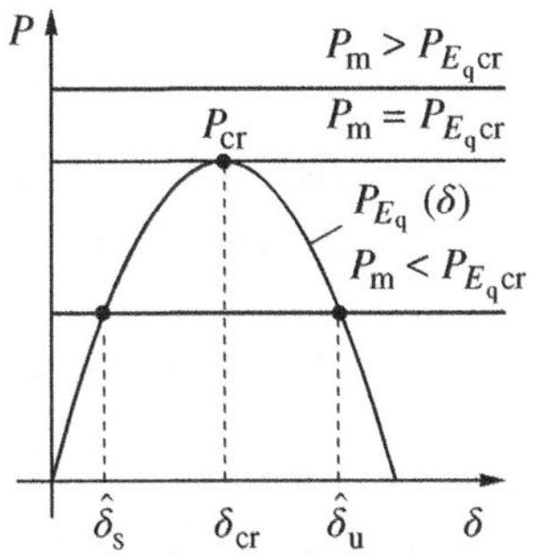

Figure 5.4 Equilibrium points for various values of mechanical power.

5.3 Equilibrium Points

Section 5.1 shows how the accelerating power is dependent on the difference between the turbine power P_m and the electrical air-gap power P_e (minus the damping power P_D). The mechanical power is supplied by the turbine and its value is controlled by the turbine governor. The electrical air-gap power depends on the generator loading and varies depending on the generator parameters and the power angle. It also depends on the operating state of the generator, but in this section only the steady-state model of the generator-infinite busbar system is considered. The infinite busbar voltage V_s, introduced in Section 3.3.6 and shown in Figure 3.27, will be assumed as the datum so that the rotor of the fictitious system's equivalent generator will provide a reference axis rotating at a constant speed. Expressions will be derived for P_e assuming that the combined generator and system resistance r is small and can be neglected. However, it is important to realize that the air-gap power is the power supplied to the system P_s plus the power loss in the equivalent resistance I^2r.

Section 3.3 showed that in the steady state the generator can be represented by a constant emf E_q behind the synchronous reactances X_d and X_q. If all the resistances and shunt admittances of the single machine infinite bus (SMIB) system shown in Figure 3.27 are neglected then the air-gap power P_e is equal to the power delivered to the system P_{sEq} and is given by Eq. (3.151)

$$P_e = P_{E_q} = \frac{E_q V_s}{x_d}\sin\delta + \frac{V_s^2}{2}\frac{x_d - x_q}{x_q x_d}\sin 2\delta \tag{5.15}$$

where $x_d = X_d + X$, $x_q = X_q + X$, and $X = X_T + X_s$ is the combined reactance of the step-up transformer and the equivalent network. The suffix E_q indicates that the air-gap power is calculated with E_q assumed to be constant.

Section 3.3 shows that the angle δ is the angle between the $\underline{E}_q$ and $\underline{V}_s$ phasors (referred to as the *power angle*) and, at the same time, it is the spatial angle between the generator rotor and the fictitious system generator (referred to as the *rotor angle*). This is of vital importance as it allows the swing equation, Eq. (5.1), describing the dynamics of the rotor, to be linked with Eq. (5.15), describing the electrical state of the generator.

Equation (5.15) describes the steady-state, or static, power-angle characteristic of the generator. For constant E_q and V_s the characteristic becomes a function of the power/rotor angle δ only, that is $P_e = P_e(\delta)$, and Eq. (5.1) can be rewritten as

$$M\frac{d^2\delta}{dt^2} = P_m - P_e(\delta) - D\frac{d\delta}{dt} \tag{5.16}$$

in order to emphasize that P_e is a function of δ and that P_D is proportional to $d\delta/dt$.

When in equilibrium the generator operates at synchronous speed $\omega = \omega_s$ so that

$$\left.\frac{d\delta}{dt}\right|_{\delta=\hat{\delta}} = 0 \quad \text{and} \quad \left.\frac{d^2\delta}{dt^2}\right|_{\delta=\hat{\delta}} = 0 \tag{5.17}$$

where $\hat{\delta}$ is the rotor angle at the equilibrium point. Substituting the above conditions into Eq. (5.16) shows that at equilibrium $P_m = P_e(\hat{\delta})$. To simplify considerations the round-rotor generator with $x_d = x_q$ is assumed when the expression for the air-gap power simplifies to

$$P_e(\delta) = P_{E_q}(\delta) = \frac{E_q V_s}{x_d} \sin\delta \tag{5.18}$$

This characteristic is drawn in Figure 5.4. The maximum value of $P_{E_q}(\delta)$ is referred to as the *critical power* $P_{E_q\mathrm{cr}}$, while the corresponding value of the rotor angle is referred to as the *critical angle* δ_{cr}. For the round-rotor generator described by Eq. (5.18) $P_{E_q\mathrm{cr}} = E_q V_s / x_d$ and $\delta_{\mathrm{cr}} = \pi/2$.

As the mechanical power depends only on the flow of the working fluid through the turbine, and not on δ, the turbine mechanical power characteristic can be treated as a horizontal line P_m = constant on the (δ, P) plane. The intersection between the horizontal P_m characteristic and the sine-like $P_e(\delta)$ characteristic gives the equilibrium points of the generator. Three situations, shown in Figure 5.4, are now possible:

- $P_m > P_{E_q\mathrm{cr}}$. Clearly no equilibrium points exist and the generator cannot operate at such a condition.
- $P_m = P_{E_q\mathrm{cr}}$. There is only one equilibrium point at δ_{cr}.
- $P_m < P_{E_q\mathrm{cr}}$. There are two equilibrium points at $\hat{\delta}_s$ and $\hat{\delta}_u$. This condition corresponds to normal operation and is discussed in the next section.

It should be emphasized that the single critical point (with coordinates δ_{cr} and $P_{E_q\mathrm{cr}}$) exists only for the SMIB system. For a multi-machine power system there is an infinite number of critical points creating a surface surrounding the steady-state stability area in the state space. Some points on such a surface may have coordinates $\delta > \pi/2$ as well $\delta < \pi/2$. This is explained in Section 12.2.2 and illustrated by Figure 12.6.

5.4 Steady-state Stability of Unregulated System

Classically a system is said to be steady-state stable for a particular operating condition if, following any small disturbance, it reaches a steady-state operating point which is identical, or close to, the predisturbance condition. This is also known as *small-disturbance*, or *small-signal, stability*. A small disturbance is a disturbance for which the equations that describe the dynamics of the electric power system (EPS) may be linearized for analytical purposes.

The dynamics of the generator, and its stability, are generally affected by automatic control of the generator and the turbine (Figure 2.2). To simplify considerations, analysis of the generator dynamics will be considered in two sections. In this section the *unregulated system* is considered when the mechanical power and the excitation voltage are assumed to be constant. This corresponds to analyzing the system *steady-state inherent* or *natural stability*. The influence of an automatic voltage regulator (AVR) is considered in Section 5.5.

5.4.1 Pull-out Power

Section 4.4 shows that when a generator is about to be synchronized to the system it must rotate at synchronous speed and its terminal voltage must be equal to, and in phase with, the busbar voltage. When the synchronizing switch is closed, the steady-state equilibrium point is reached at $\delta = 0$ and $P_m = 0$ and corresponds to the origin of the power-angle characteristic shown in Figure 5.4. If now the mechanical power P_m is slowly increased by a small amount, the electrical power P_e must follow the changes so that a new equilibrium point of $P_m = P_e$ is reached. In other words, the system is steady-state stable if an increase (decrease) in mechanical power causes a corresponding increase (decrease) in electrical power.

If the system reaction is opposite to this, that is an increase in mechanical power is accompanied by a decrease in electrical power, then no equilibrium point can be reached. These stability considerations are illustrated in Figure 5.5.

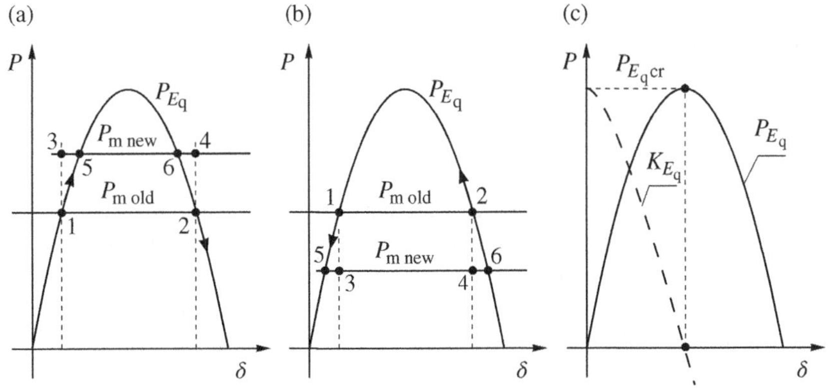

Figure 5.5 Illustration of the conditions for steady-state stability: (a) increase in mechanical power; (b) decrease in mechanical power; (c) the generator steady-state power and synchronizing power coefficient.

For a certain value of mechanical power, marked as "old," there are two equilibrium points: 1 and 2. If mechanical power is increased to a "new" value (Figure 5.5a) then this gives rise to a surplus of power at point 1. This surplus acceleration power is equal to segment 1–3 and will, according to Eq. (5.1), accelerate the rotor and so increase the power angle and the electrical power. The resulting motion is shown by the arrow and is toward the new equilibrium point 5.

The opposite situation occurs at equilibrium point 2. Here the acceleration power, equal to segment 2–4, again accelerates the rotor and leads to an increase in the power angle, but this now results in a reduction in the electrical power. The motion, as shown by the arrow, is away from the new equilibrium point 6.

A similar response is obtained if the mechanical power is reduced (Figure 5.5b). For equilibrium points on the left-hand side of the power-angle characteristic the rotor motion is from point 1 toward the new equilibrium point 5. On the other hand, when starting from equilibrium point 2, on the right-hand side of the characteristic, it is not possible to reach the new equilibrium point 6 as the rotor motion is in the opposite direction. Obviously, an increase in the mechanical power to a value exceeding $P_{E_q\mathrm{cr}}$ results in a loss of synchronism because of the lack of any equilibrium point.

From this discussion it is apparent that the generator-infinite busbar system with constant excitation emf E_f is steady-state stable only on the left-hand side of the power-angle characteristic, that is, when the slope K_{E_q} of the characteristic is positive

$$K_{E_q} = \left.\frac{\partial P_{E_q}}{\partial \delta}\right|_{\delta = \hat{\delta}_s} > 0 \tag{5.19}$$

K_{E_q} is referred to as the *steady-state synchronizing power coefficient* and the critical power $P_{E_q\mathrm{cr}}$ is often referred to as the *pull-out power* to emphasize the fact that a larger mechanical power will result in the unregulated generator losing synchronism with the rest of the system. Figure 5.5c shows the plot of $K_{E_q}(\delta)$ and $P_{E_q\mathrm{cr}}$. The value of $P_{E_q\mathrm{cr}}$ is also referred to as the *steady-state stability limit* and can be used to determine the *steady-state stability margin* as

$$c_{E_q} = \frac{P_{E_q\mathrm{cr}} - P_m}{P_{E_q\mathrm{cr}}} \tag{5.20}$$

where P_m is the actual loading of the generator. The stability margin varies between $c_{E_q} = 1$ (when the generator is unloaded) and $c_{E_q} = 0$ (when the generator is critically loaded).

It should be emphasized that the pull-out power is determined by the steady-state characteristic $P_{E_q}(\delta)$ and the dynamic response of the generator to a disturbance is determined by the transient power-angle characteristic described in the next subsection.

5.4.2 Transient Power-angle Characteristics

Chapter 4 explains how any disturbance acting on a generator will produce a sudden change in the armature currents and flux. This flux change induces additional currents in the rotor windings (field and damper) that expel the armature flux into high-reluctance paths around the rotor so as to screen the rotor and keep the rotor flux linkage constant. As the emf E_q is proportional to the field current the additional induced field current will cause changes in E_q so that the assumption of constant E_q used to derive the static power-angle characteristic (5.18) is invalid in the analysis of post-disturbance rotor dynamics.

Chapter 4 also explains how the induced rotor currents decay with time as the armature flux penetrates first the damper windings (subtransient period) and then the field winding (transient period). Usually, the frequency of rotor oscillations is about (1–2) Hz which corresponds to an electromechanical swing period of about (1–0.5) s. This period can be usefully compared with the generator subtransient open-circuit time constants T''_{do} and T''_{qo}, which are in the region of few hundredths of a second. On the other hand, the d-axis transient time constant T'_{do} is in the region of a few seconds, while the q-axis time constant T'_{qo} is about a second; see Table 4.3 and Eq. (4.31). Consequently, on the timescale associated with rotor oscillations, it may be assumed that changes in the armature flux can penetrate the damper windings but that the field winding and the round-rotor body act as perfect screens maintaining constant flux linkages. This corresponds to assuming that the transient emf's E'_d and E'_q are constant. This assumption will modify the generator power-angle characteristic, as described below.

5.4.2.1 Constant Flux Linkage Model

Assume that the generator is connected to the infinite busbar as shown in Figure 3.27 and that all the resistances and shunt impedances associated with the transformer and network can be neglected. The corresponding equivalent circuit and phasor diagram of the round-rotor and salient-pole generator in the transient state are then as shown in Figure 5.6. The fictitious rotor of the infinite busbar serves as the synchronously rotating reference axis. The reactances of the step-up transformer and the connecting network can be combined with that of the generator to give

$$x'_d = X'_d + X; \quad x'_q = X'_q + X \tag{5.21}$$

where X'_d and X'_q are the d- and q-axis transient reactances of the generator and $X = X_T + X_s$. The voltage equations can now be constructed using Figure 5.6b to give

$$E'_d = V_{sd} + x'_q I_q, \quad E'_q = V_{sq} - x'_d I_d \tag{5.22}$$

where V_{sd} and V_{sq} are the d- and q-axis components of the infinite busbar voltage V_s, given by $V_{sd} = -V_s \sin\delta$ and $V_{sq} = V_s \cos\delta$. As all resistances are neglected the air-gap power is $P_e = V_{sd}I_d + V_{sq}I_q$. Substituting the values for V_{sd} and V_{sq} and the currents I_d and I_q calculated from Eq. (5.22) gives

$$\begin{aligned} P_e = P_s = V_{sd}I_d + V_{sq}I_q &= -\frac{E'_q V_{sd}}{x'_d} + \frac{V_{sd}V_{sq}}{x'_d} + \frac{E'_d V_{sq}}{x'_q} - \frac{V_{sd}V_{sq}}{x'_q} \\ P_e = P_{E'}(\delta) &= \frac{E'_q V_s}{x'_d}\sin\delta + \frac{E'_d V_s}{x'_q}\cos\delta - \frac{V_s^2}{2}\frac{x'_q - x'_d}{x'_q x'_d}\sin 2\delta \end{aligned} \tag{5.23}$$

Equation (5.23) defines the air-gap power as a function of δ and the d- and q-axis components of the transient emf and is valid for any generator (with or without transient saliency). The phasor diagram in Figure 5.6b shows that

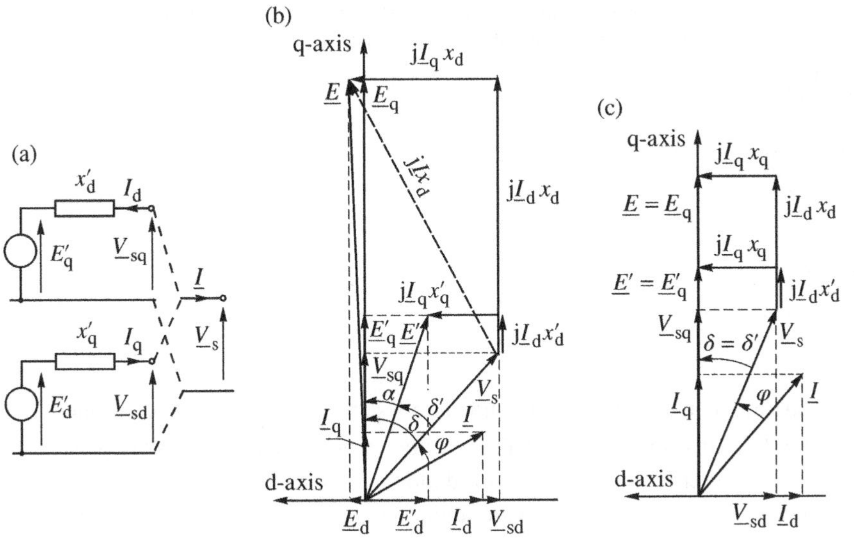

Figure 5.6 SMIB system in the transient state: (a) circuit diagram; (b) phasor diagram of the round-rotor generator; (c) phasor diagram of the salient-pole generator ($x'_q = x_q$).

$E'_d = -E' \sin\alpha$, $E'_q = E' \cos\alpha$, and $\delta = \delta' + \alpha$. Substituting these relationships into Eq. (5.23) gives, after some simple but tedious algebra,

$$
\begin{aligned}
P_e = P_{E'}(\delta') = {} & \frac{E' V_s}{x'_d}\left[\sin\delta'\left(\cos^2\alpha + \frac{x'_d}{x'_q}\sin^2\alpha\right) + \frac{1}{2}\left(\frac{x'_q - x'_d}{x'_q}\right)\cos\delta' \sin 2\alpha\right] \\
& - \frac{V_s^2}{2}\frac{x'_q - x'_d}{x'_d x'_q}\sin 2(\delta' + \alpha).
\end{aligned}
\tag{5.24}
$$

Assuming constant rotor flux linkages, the values of the emf's E'_d and E'_q are also constant, implying that both E' = constant and α = constant. Equation (5.24) describes the generator power-angle characteristic $P_e(E', \delta')$ in terms of the transient emf and the transient power angle and is valid for any type of generator.

A generator with a laminated salient-pole rotor cannot produce effective screening in the q-axis with the effect that $x'_q = x_q$. Inspection of the phasor diagram in Figure 5.6c shows that in this case E' lies along the q-axis so that $\alpha = 0$ and $\delta' = \delta$. Consequently, in this special case, Eq. (5.24) simplifies to

$$
P_e = P_{E'_q}(\delta')\Big|_{x'_q = x_q} = \frac{E'_q V_s}{x'_d}\sin\delta' - \frac{V_s^2}{2}\frac{x_q - x'_d}{x_q x'_d}\sin 2\delta'
\tag{5.25}
$$

5.4.2.2 Classical Model

The constant flux linkage model expressed by Eq. (5.24) can be simplified by ignoring transient saliency, that is by assuming $x'_d \cong x'_q$. With this assumption, Eq. (5.24) simplifies to

$$
P_e = P_{E'}(\delta')\big|_{x'_d \cong x'_q} \cong \frac{E' V_s}{x'_d}\sin\delta'
\tag{5.26}
$$

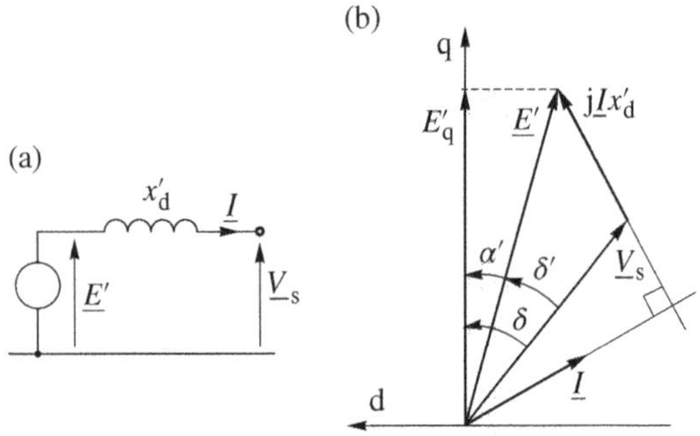

Figure 5.7 Classical model of the generator in the transient state: (a) circuit diagram; (b) phasor diagram.

The assumption of $x'_d = x'_q$ allows the separate d- and q-axis circuits shown in Figure 5.6b to be replaced by one simple equivalent circuit shown in Figure 5.7. In this *classical model* all the voltages, emf's, and currents are phasors in the network reference frame rather than their components resolved along the d- and q-axes.

Generally, as Table 4.3 shows, there is always some degree of transient saliency and $x'_d \neq x'_q$. However, it should be noted that the generator is connected to the infinite busbar through the reactance $X = X_T + X_s$, so that $x'_d = X'_d + X$ and $x'_q = X'_q + X$. The influence of X is such that, as its magnitude increases, the closer the term x'_d / x'_q approaches unity, the closer the term $\left(x'_q - x'_d\right) / x'_d x'_q$ approaches zero. Consequently, when the network reactance is large, the classical model and the constant flux linkage model defined by Eq. (5.24) give very similar results even for a generator with a laminated salient-pole rotor.

It is important to note that δ' is the angle between V_s and E' and not the angle between V_s and the q-axis. However, during the transient period the emf's E'_d and E'_q are assumed to be constant (with respect to the rotor axes) and α is also constant with

$$\delta = \delta' + \alpha; \quad \frac{d\delta}{dt} = \frac{d\delta'}{dt} \quad \text{and} \quad \frac{d^2\delta}{dt^2} = \frac{d^2\delta'}{dt^2} \tag{5.27}$$

This allows δ' to be used in the swing equation instead of δ when Eq. (5.1) becomes

$$M\frac{d^2\delta'}{dt^2} = P_m - \frac{E'V_s}{x'_d}\sin\delta' - D\frac{d\delta'}{dt} \tag{5.28}$$

An important advantage of the classical model is that the generator reactance may be treated in a similar way to the reactance of the transmission lines and other network elements. This has particular importance for multi-machine systems when combining the algebraic equations describing the generator, and the network is not as easy as for the SMIB system. Owing to its simplicity, the classical model is used extensively throughout this book to analyze and explain rotor dynamics.

5.4.2.3 Steady-state and Transient Characteristics

It is now important to understand how the dynamic characteristic of a generator is located with respect to the static characteristic on the power-angle diagram. For a given stable equilibrium point, when $P_e = P_m$, the balance of power must be held whichever characteristic is considered so that both the static and dynamic characteristics must intersect at the stable equilibrium point. Generally, the angle α between δ_0 and δ'_0 is not equal to zero and the transient characteristic is shifted to the right. For salient-pole generators $\alpha = 0$ both characteristics originate at the same point, and the intersection between them at the equilibrium point is due to a distortion in the sine shape of the transient characteristic.

Example 5.1 The round-rotor generator considered in Example 4.2 is connected to the power system (infinite busbar) via a transformer with series reactance $X_T = 0.13$ pu and a transmission line with series reactance $X_L = 0.17$ pu. Find, and plot, the steady-state and the transient characteristics using both the constant flux linkage and the classical generator model. As in Example 4.1 the generator active power output is 1 pu, the reactive power output is 0.5 pu, and the terminal voltage is 1.1 pu.

From Example 4.2, $I_0 = 1.016$, $\varphi_{g0} = 26.6^\circ$, $E_{q0} = 2.336$, $\delta_{g0} = 38.5^\circ$, $I_{d0} = -0.922$, $I_{q0} = 0.428$, $E'_q = E'_{q0} = 1.073$, and $E'_d = E'_{d0} = -0.522$. Angle α can be found from $\alpha = \arctan\left(E'_d/E'_q\right) = 26^\circ$. The total reactances are $x_d = x_q = X_d + X_T + X_L = 1.9$, $x'_d = X'_d + X_T + X_L = 0.53$, and $x'_q = X'_q + X_T + X_L = 0.68$.

Taking $\underline{V}_g$ as the reference, the phasor of the transient emf is $\underline{E}' = 1.193\angle 12.5^\circ$. It is now necessary to calculate the system voltage $\underline{V}_s$ and calculate the position of $\underline{E}_q$ and $\underline{E}'$ with respect to $\underline{V}_s$. The system voltage can be calculated from

$$\underline{V}_s = \underline{V}_g - \mathrm{j}(X_T + X_L)\underline{I} = 1.1 - \mathrm{j}0.3{\cdot}1.016\angle 26.6^\circ = 1.0\angle -15.8^\circ$$

The angle δ_0 is therefore equal to $38.5 + 15.8 = 54.3^\circ$, δ'_0 is $12.5 + 15.8 = 28.3^\circ$, and φ_0 is $26.6 - 15.8 = 10.8^\circ$. The d- and q-axis components of the system voltage are: $V_{sd} = -1 \cdot \sin 54.3^\circ = -0.814$, $V_{sq} = 1 \cdot \cos 54.3^\circ = 0.584$.

The steady-state power-angle characteristic is

$$P_{E_q}(\delta) = \frac{E_q V_s}{x_d}\sin\delta = \frac{2.336{\cdot}1}{1.9}\sin\delta = 1.23\sin\delta$$

The transient characteristic (constant flux linkage model) can be calculated from Eq. (5.23) as

$$\begin{aligned} P_{E'}(\delta) &= \frac{1.07{\cdot}1}{0.53}\sin\delta + \frac{-0.5224{\cdot}1}{0.68}\cos\delta - \frac{1^2}{2}\frac{0.68 - 0.53}{0.68 \times 0.53}\sin 2\delta \\ &= 2.02\sin\delta - 0.768\cos\delta - 0.208\sin 2\delta \end{aligned}$$

The approximated transient characteristic for the classical model can be calculated assuming $x'_d = x'_q$. The transient emf can be calculated with respect to $\underline{V}_s$ as $\underline{E}' = V_s + \mathrm{j}x'_d\underline{I}_0 = 1 + \mathrm{j}\,0.53 \times 1.016\angle -10.8^\circ = 1.223\angle 25.7^\circ$. Thus $E' = 1.223$, $\delta'_0 = 25.7^\circ$, and $\alpha = \delta - \delta' = 54.3 - 25.7 = 28.6^\circ$. The approximated transient characteristic can now be calculated from Eq. (5.26) as

$$P_{E'}(\delta') \cong \frac{1.223 \times 1}{0.53}\sin\delta' = 2.31\sin\delta'$$

This characteristic is shifted with respect to $P_{E_q}(\delta)$ by 28.6°. Figure 5.8a shows all three characteristics. It can be seen that the classical model gives a good approximation of the constant flux linkage model.

******

Example 5.2 Recalculate all the characteristics from Example 5.1 for the salient-pole generator considered previously in Example 4.3.

From Example 4.3, $I_0 = 1.016$, $\varphi_{g0} = 26.6^\circ$, $E_q = E_{q0} = 1.735$, $I_{d0} = -0.784$, $I_{q0} = 0.647$, $E'_q = E'_{q0} = 1.241$, $E'_d = 0$, $\delta_{g0} = 23.9^\circ$, and $\alpha = 0$.

The total reactances are $x_d = X_d + X_T + X_L = 1.23$, $x_q = x'_q = X_q + X_T + X_L = 0.99$, and $x'_d = X'_d + X_T + X_L = 0.6$.

Taking $\underline{V}_g$ as the reference, the phasor of the transient emf is $\underline{E}' = 1.241\angle 23.9^\circ$. The system voltage was calculated in Example 5.1 and is $\underline{V}_s = 1.0\angle 15.8^\circ$. Thus $\delta_0 = \delta'_0 = 23.9 + 15.8 = 39.7^\circ$ and $\delta_0 + \varphi_0 = 39.7 + 10.8 = 50.5^\circ$. The d- and q-axis components of the system voltage are $V_{sd} = -1 \cdot \sin 39.7^\circ = -0.64$ and $V_{sq} = 1 \cdot \cos 39.7^\circ = 0.77$.

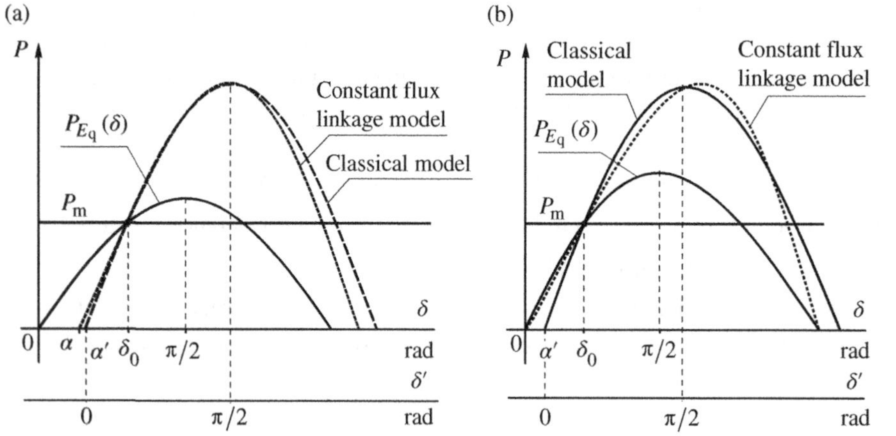

Figure 5.8 Steady-state and transient characteristics: (a) round-rotor generator ($\alpha > 0$); (b) laminated salient-pole machine ($\alpha = 0$).

The steady-state power-angle characteristic is

$$P_{E_q}(\delta) = \frac{E_q V_s}{x_d} \sin\delta + \frac{V_s^2}{2}\frac{x_d - x_q}{x_d x_q} \sin 2\delta = \frac{1.735\cdot 1}{1.23} \sin\delta + \frac{1}{2}\frac{1.23 - 0.99}{1.23 \times 0.99} \sin 2\delta =$$
$$= 1.41 \sin\delta + 0.099 \sin 2\delta \cong 1.41 \sin\delta$$

The transient characteristic (constant flux linkage model) can be calculated from Eq. (5.25) as

$$P_{E'}(\delta) = \frac{1.241\cdot 1}{0.6} \sin\delta - \frac{1^2}{2}\frac{0.99 - 0.6}{0.99 \times 0.6} \sin 2\delta = 2.07 \sin\delta - 0.322 \sin 2\delta$$

The approximated transient characteristic for the classical model can be calculated assuming $x'_d = x'_q$. The transient emf, calculated with respect to $\underline{V}_s$, is then $\underline{E}' = V_s + jx'_d\underline{I} = 1 + j\,0.6 \times 1.016\angle -10.8^\circ = 1.265\angle 28.3^\circ$. Thus $E' = 1.265$, $\delta' = 28.3^\circ$, and $\alpha = \delta - \delta' = 39.7 - 28.3 = 11.4^\circ$. The approximated transient characteristic can now be calculated from Eq. (5.26) as

$$P_{E'}(\delta') \cong \frac{1.265\cdot 1}{0.6} \sin\delta' = 2.108 \sin\delta'$$

This characteristic is shifted with respect to $P_{E_q}(\delta)$ by $\alpha = 11.4^\circ$. Figure 5.8b shows all three characteristics. It can be seen that the classical model gives a good approximation of the constant flux linkage model.

******

Figure 5.8 shows that by neglecting transient saliency the classical model does not generally significantly distort the transient characteristics. As $x_d > x'_d$, the amplitude of the transient characteristic is greater than the amplitude of the steady-state characteristic. Consequently, the slope of the transient characteristic at the stable equilibrium point, referred to as the *transient synchronizing power coefficient*

$$K_{E'} = \left.\frac{\partial P_{E'}}{\partial \delta'}\right|_{\delta' = \hat{\delta}'_s} \tag{5.29}$$

is steeper than the slope of the steady-state characteristic, K_{E_q}, defined in Eq. (5.19). For the classical model $K_{E'} = E' V_s \cos\hat{\delta}'_s / x'_d$.

When the generator loading is changed, P_m changes and the stable equilibrium point is shifted to a new position on the steady-state characteristic $P_{E_q}(\delta)$, providing that P_m does not exceed the pull-out power. This increased load modifies the transient characteristic, as illustrated in Figure 5.9a, which shows three transient characteristics

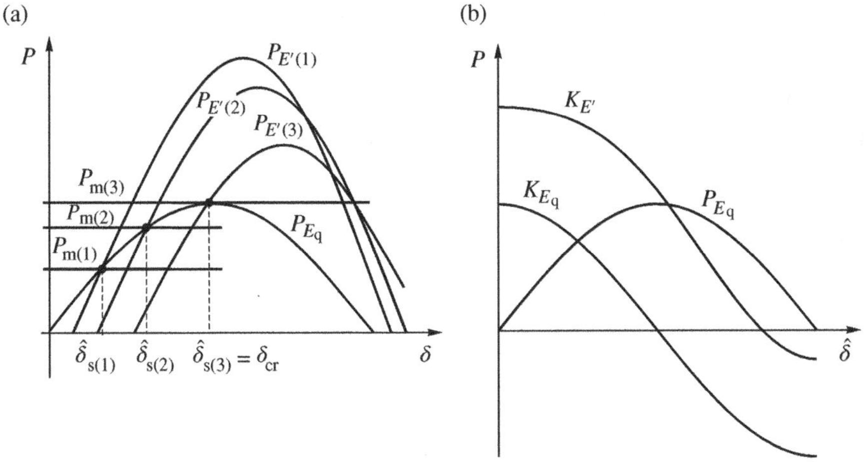

Figure 5.9 Effect of increased loading on the round-rotor generator: (a) the steady-state and transient characteristics for three different loads; (b) the steady-state electrical power and the steady-state and transient synchronizing power coefficients as a function of the steady-state equilibrium rotor angle.

corresponding to three different mechanical powers P_{m1}, P_{m2}, and P_{m3}. The generator excitation (and therefore the amplitude of the steady-state characteristic) is assumed to be unchanged. Each new steady-state equilibrium point corresponds to a different value of E' and a different transient characteristic $P_{E'}(\delta)$ crossing the equilibrium point. Note that increased loading results in a smaller transient emf E' so that the amplitude of the transient characteristic is reduced. This can be verified by inspecting the predisturbance d-axis phasor diagram (Figure 4.17), assuming E_f is constant. This shows that an increase in the armature current I_0 results in a larger voltage drop across the reactances $\left(x_d - x_d'\right)$ and $\left(x_q - x_q'\right)$ and therefore a smaller $E_0' = E'$ value. Figure 5.9b illustrates the same effect shown in Figure 5.9a but in a slightly different way as it shows the steady-state electrical power P_{E_q}, the steady-state synchronizing power coefficient K_{E_q}, and the transient synchronizing power coefficient $K_{E'}$ as a function of the generation loading as expressed by the steady-state equilibrium angle $\hat{\delta}_s$. The case of the round-rotor generator is shown. An increase in the generator loading results in smaller steady-state and transient synchronizing power coefficients K_{E_q} and $K_{E'}$, but nevertheless $K_{E'}$ is always larger than K_{E_q}.

5.4.3 Rotor Swings and Equal Area Criterion

The generator models derived in the previous subsection can now be used to describe and analyze the rotor dynamics assuming that the generator is subjected to a sudden disturbance. Such a disturbance will result in additional currents being induced in the rotor windings so as to maintain constant rotor flux linkages and therefore constant E'. As the synchronous emf's E_d and E_q follow the changes in the field winding current and the rotor body current respectively, they cannot be assumed to be constant and any rotor swings must therefore follow the transient power-angle curve $P_e = P_{E'}(\delta')$.

Disturbances in the generator-infinite busbar system may arise following a change in the turbine mechanical power or a change in the equivalent system reactance. The effect of such practical disturbances is considered in detail later, but here it suffices to consider the effect of disturbing the rotor angle δ from its equilibrium value $\hat{\delta}_s$ to a new value $\left(\hat{\delta}_s + \Delta\delta_0\right)$. Although such a disturbance is unlikely from a technical point of view, it does allow a number of important concepts to be introduced that are fundamental to understanding the effects of other, more practical types of disturbance. The initial disturbed conditions for the solution of the system differential equations are

$$\Delta\delta(t = 0^+) = \Delta\delta_0 \neq 0; \quad \Delta\omega(t = 0^+) = \Delta\omega_0 = 0 \tag{5.30}$$

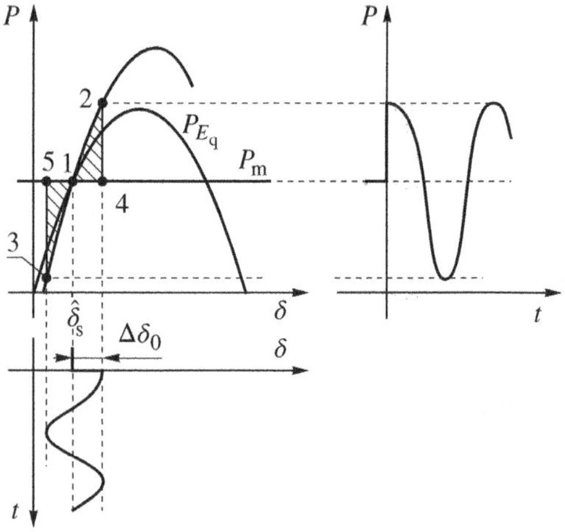

Figure 5.10 Rotor and power swings following a disturbance.

The reaction of the system to this disturbance is illustrated in Figure 5.10, which shows both the steady-state and the transient power characteristics as a function of the rotor angle δ. As the disturbance cannot change the rotor flux linkage, the initial (disturbed) generator operating point will be at point 2 on the transient characteristic $P_{E'}(\delta)$ which crosses the predisturbance stable equilibrium point 1.

Any movement requires work to be done so that by increasing the rotor angle from $\hat{\delta}_s$ to $\left(\hat{\delta}_s + \Delta\delta_0\right)$ the disturbance performs work on the rotor. In rotational motion, work is equal to the integral of the torque acting over the angular displacement and in this case the net torque is equal to the difference between the electrical (transient) torque and the mechanical torque. As power is equal to the product of torque and angular velocity, and assuming $\omega \cong \omega_s$, the work done by the disturbance is proportional to the integral of the net power acting over the angular displacement, that is

$$W_{1-2-4} = \int_{\hat{\delta}_s}^{\hat{\delta}_s + \Delta\delta_0} [P_{E'}(\delta) - P_m]\, d\delta = \text{area (1-2-4)} \tag{5.31}$$

Throughout this book the quantity W is treated as energy (or work), although, strictly speaking, it should be divided by the synchronous speed ω_s. As the rotor speed deviation $\Delta\omega$ at point 2 is assumed to be zero (the speed is synchronous), the kinetic energy of the rotor is the same as that at equilibrium point 1. This means that the work W_{1-2} done by the disturbance increases the system potential energy (with respect to equilibrium point 1) by

$$E_p = W_{1-2-4} = \text{area (1-2-4)} \tag{5.32}$$

This initial potential energy provides the impetus necessary to move the rotor back toward its equilibrium point 1. At the disturbed rotor position 2 the mechanical driving torque (and power) is less than the opposing electrical torque (and power), the resulting net deceleration power (equal to segment 4–2) starts to reduce the rotor speed (with respect to synchronous speed) and the rotor angle will decrease. At the equilibrium point 1 all the potential energy in Eq. (5.32) will be converted into kinetic energy (relative to synchronous speed) and the deceleration work done is equal to

$$E_k = W_{1-2-4} = \text{area (1-2-4)} = \frac{1}{2}M\Delta\omega^2 \tag{5.33}$$

The kinetic energy will now push the rotor past the equilibrium point $\hat{\delta}_s$ so that it continues to move along curve 1-3. On this part of the characteristic the mechanical driving torque is greater than the opposing electrical torque and the rotor begins to accelerate. Acceleration will continue until the work performed by the acceleration torque (proportional to the integral of the accelerating power) becomes equal to the work performed previously by the deceleration torque. This happens at point 3 when

$$\text{area (1-3-5)} = \text{area (1-2-4)} \tag{5.34}$$

At this point the generator speed is again equal to the synchronous speed but, as $P_m > P_{E'}$, the rotor will continue to accelerate, increasing its speed above synchronous, and will swing back toward $\hat{\delta}_s$. In the absence of any damping the rotor will continually oscillate between points 2 and 3, as described by the swing equation (5.16). The resulting swing curve is shown in the lower part of Figure 5.10 with the corresponding power swings being shown to the right.

Equation (5.34) defines the maximum deflection of the rotor in either direction on the basis of equalizing the work done during deceleration and acceleration. This concept is used in Chapter 6 to define the equal area criterion of stability.

5.4.4 Effect of Damper Windings

Equation (5.10) shows that for small deviations in rotor speed the damper windings produce a damping power $P_D = D\Delta\omega$ that is proportional to the rotor speed deviation. To help explain the effect of the damper windings on the system behavior it is convenient to rewrite the swing equation (5.16) as

$$M\frac{d^2\delta}{dt^2} = P_m - [P_e(\delta) + P_D] \tag{5.35}$$

when the damping power is seen either to add to or to subtract from the electrical air-gap power $P_e(\delta)$ depending on the sign of the speed deviation. If $\Delta\omega < 0$ then P_D is negative effectively opposing the air-gap power and shifting the resulting $(P_{E'} + P_D)$ characteristic downward. If $\Delta\omega > 0$ then P_D is positive effectively assisting the air-gap power and shifting the resultant characteristic upwards. The rotor will therefore move along a modified power-angle trajectory, such as that shown in Figure 5.11. To help increase clarity, this diagram shows an enlarged part of the power-angle diagram in the vicinity of the equilibrium point.

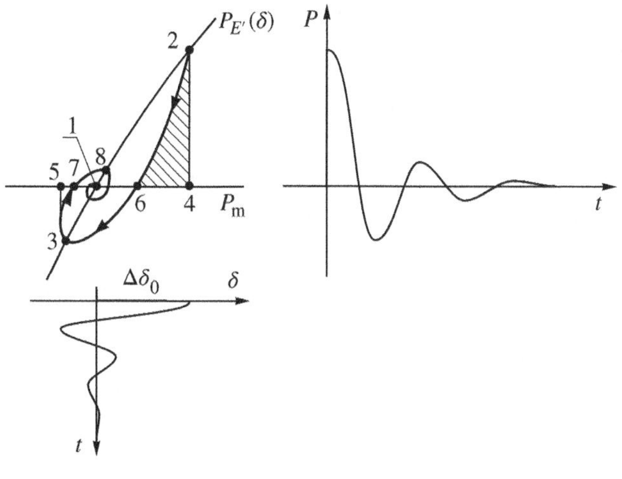

Figure 5.11 Rotor and power oscillations with damping included.

As before, the rotor is initially disturbed from equilibrium point 1 to point 2. At point 2 the driving mechanical power is less than the opposing electrical power and the decelerating torque will force the rotor back toward the equilibrium point. On deceleration the rotor speed drops and P_D becomes negative, decreasing the resulting decelerating torque. The rotor therefore moves along the line 2–6 when the work done by the decelerating torque is equal to the area 2-4-6. This is less than the area 2-4-1 in Figure 5.10 which represents the work that would have been done if no damping were present. At point 6 the rotor speed reaches a minimum and, as it continues to move along the curve 6–3, the accelerating torque counteracts further movement of the rotor and is assisted by the negative damping term. The rotor again reaches synchronous speed when the area 6-3-5 is equal to the area 2-4-6 which is achieved earlier than in the case without damping. The rotor then starts to swing back, still accelerating, so that the speed increases above synchronous speed. The damping term changes sign, becoming positive, and decreases the resulting accelerating torque. The rotor moves along the curve 3–7 and the work performed during the acceleration is equal to the small area 3-5-7. As a result the rotor reaches synchronous speed at point 8, much earlier than in the case without damping. The rotor oscillations are damped and the system quickly reaches equilibrium point 1.

5.4.5 Effect of Rotor Flux Linkage Variation

The discussion so far has assumed that the total flux linking the field winding and rotor body remains constant during rotor oscillations so that E' = constant. However, as the armature flux enters the rotor windings (Chapter 4) the rotor flux linkages change with time and the effect this has on the rotor swings must now be considered.

5.4.5.1 Linearized Form of the Generator Equations

For small-disturbance analysis, suitable for steady-state stability purposes, the equations describing the generator's behavior can be linearized in the vicinity of the predisturbance operating point. Assume that this operating point is defined by the transient rotor angle $\hat{\delta}'_s$ and the transient emf E'_0. An approximate expression for the change in power $\Delta P = P_e(\delta') - P_m$ with respect to the steady-state equilibrium point can be obtained as a function of the change in the transient rotor angle $\Delta\delta = \delta'(t) - \hat{\delta}'_s$ and a change in the transient emf $\Delta E' = E'(t) - E'_0$ as

$$\Delta P_e = \left.\frac{\partial P_e(\delta', E')}{\partial \delta'}\right|_{E'=E'_0} \Delta\delta' + \left.\frac{\partial P_e(\delta', E')}{\partial E'}\right|_{\delta'=\hat{\delta}'_s} \Delta E' = K_{E'}\Delta\delta' + D_{\delta'}\Delta E' \tag{5.36}$$

where $P_e(\delta', E')$ is the air-gap power given by Eqs. (5.24), (5.25), or (5.26), $K_{E'} = \partial P_e/\partial\delta'$ is the transient synchronizing power coefficient and $D_{\delta'} = \partial P_e/\partial E'$ is a constant coefficient, all of which depends on the generator load. As the angle $\alpha = \delta - \delta'$ is assumed to be constant, $\Delta\delta' = \Delta\delta$. This allows the swing equation (5.16) to be linearized in the vicinity of the operating point as

$$M\frac{d^2\Delta\delta}{dt^2} + D\frac{d\Delta\delta}{dt} + K_{E'}\,\Delta\delta + D_{\delta'}\Delta E' = 0 \tag{5.37}$$

The transient emf can be resolved into two components E'_d and E'_q along the d- and q-rotor axes. Analyzing the time changes of both components is complicated as the variations take place with different time constants. To simplify the discussion, the case of the salient-pole machine will be considered when $\delta' = \delta$, $E' = E'_q$, and $E'_d = 0$, and only the flux linkages of the field winding need be considered.

Now assume that the salient-pole generator operates at a steady-state equilibrium point defined by $\delta = \hat{\delta}_s$ and $E'_q = E'_{q0}$. The incremental swing equation (5.37) is

$$M\frac{d^2\Delta\delta}{dt^2} + D\Delta\omega + K_{E'_q}\,\Delta\delta + D_{\delta'}\Delta E'_q = 0 \tag{5.38}$$

where $\Delta\omega = \mathrm{d}\Delta\delta/\mathrm{d}t$ and the values of the coefficients $K_{E'_q} = \partial P_e/\partial\delta'$ and $D_{\delta'} = \partial P_e/\partial E'_q$ can be obtained by differentiating the expression for the air-gap power of the salient-pole generator, Eq. (5.25). Analysis of Eq. (5.38) suggests that if the changes $\Delta E'_q$ are in phase with the rotor speed deviation $\Delta\omega$, then, just as with the damper winding, additional positive damping power will be introduced into the system.

Chapter 4 shows that the rate at which flux can penetrate the field winding is primarily determined by the field winding transient time constant T'_{do}, although the generator and system reactances and the operating point will also have some influence. Assuming that the excitation emf E_f and the infinite busbar voltage V_s are constant, Eq. (4.28) can be rewritten as

$$\Delta E'_q = -\frac{AB}{1 + sBT'_{do}}\Delta\delta \tag{5.39}$$

where A and B are constants that depend on the operating conditions (angle δ_0) and the generator and network reactances. If both the generator armature resistance and the network resistance are neglected then

$$A = \left(\frac{1-B}{B}\right) V_s \sin\delta_0 \quad \text{and} \quad B = (X'_d + X)/(X_d + X) = x'_d/x_d$$

The time constant BT'_{do} is often referred to as the *effective field winding time constant.*

If sinusoidal variations of $\Delta\delta$ are considered then the frequency response of Eq. (5.39) can be obtained by setting $s = \mathrm{j}\Omega$, where Ω is the frequency of the rotor swings (in rad/s) and $2\pi/\Omega$ is the swing period, as discussed later in this chapter; Eq. (5.53). This allows the phase of $\Delta E'_q$ to be compared with that of $\Delta\delta$. As T'_{do} is typically much longer than the swing period and $T'_{do} \gg 2\pi/\Omega$ it can be assumed that $BT'_{do}\Omega > 1$ and

$$\Delta E'_q(\mathrm{j}\Delta\omega) \cong -\frac{AB}{\mathrm{j}\Omega BT'_{do}}\Delta\delta(\mathrm{j}\Omega) = \mathrm{j}\frac{A}{T'_{do}\Omega}\Delta\delta(\mathrm{j}\Omega) \tag{5.40}$$

Thus, the changes $\Delta E'_q$ are seen to lead the changes $\Delta\delta$ by $\pi/2$, that is they are in phase with the rotor speed deviation $\Delta\omega = \mathrm{d}\Delta\delta/\mathrm{d}t$ thereby providing some additional positive damping torque.

Equation (5.40) also shows that the magnitude of the variation in $\Delta E'_q$ depends, via the coefficient $A = \left(\frac{1-B}{B}\right) V_s \sin\delta_0$, on the power angle δ_0 at the linearization (operating) point. Thus the same variation in δ will result in higher variations in $\Delta E'_q$ if the system is heavily loaded (higher value of δ_0) and the assumption of constant flux linkage ($E' =$ constant) is more accurate when the generator is lightly loaded (small δ_0) and the variations in $\Delta E'_q$ are small.

The effect of variations in the rotor flux linkages will be examined for two cases using the equal area method. In the first case the generator mechanical power $P_m < P_{E_q\mathrm{cr}}$, while in the second case $P_m = P_{E_q\mathrm{cr}}$. In both cases the effect of the damper winding will be neglected in the discussion in order to see more clearly the effect that the variation in rotor flux linkage has on the damping.

5.4.5.2 Equilibrium Point for $P_m < P_{E_q\,\mathrm{cr}}$

Figure 5.12a shows how Figure 5.10 has to be modified to include the effect of the variation in rotor flux linkage. Again, only that part of the power-angle characteristic in the vicinity of the operating point is shown. As before, the disturbance causes the electrical power to move from point 1 to point 2 on the predisturbance transient $P_{E'_{q0}}(\delta)$ curve as the field winding acts initially like a perfect screen maintaining constant rotor flux linkages. As the rotor starts to decelerate, its speed drops so that the speed deviation $\Delta\omega$ becomes negative. The resistance of the field winding dissipates magnetic energy and the rotor linkages start to decay, reducing the value of transient emf E'_q as shown in Figure 5.12b. Consequently, the electrical power is less than would be the case with $E'_q =$ constant and the rotor motion is along the curve 2–6 rather than 2–1. The resulting deceleration area 2-4-6 is smaller than

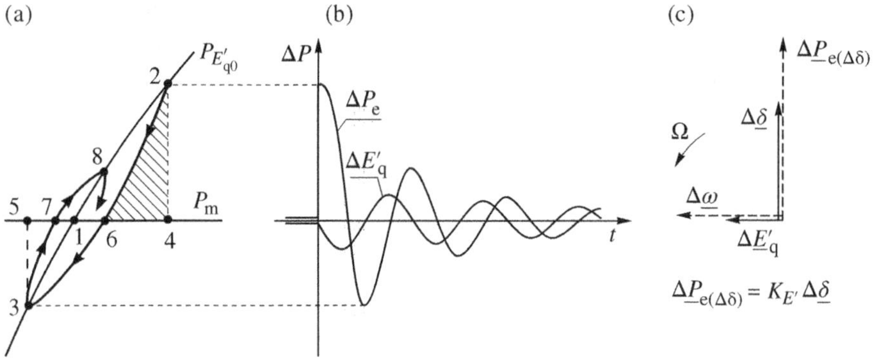

Figure 5.12 Including the effect of variation in rotor flux linkage: (a) trajectory of the operating point; (b) time variation of electrical power and transient emf oscillating with the swing frequency Ω; (c) rotating phasors of increments $\Delta\delta$, $\Delta E'_q$, and $\Delta\omega = d\delta/dt$.

area 2-4-1 thereby reducing the kinetic energy that is responsible for providing the impetus for the backswing. As the rotor passes point 6 it starts to accelerate, the emf E'_q starts to recover, and it reaches its predisturbance value when $\Delta\omega$ is zero at point 3 such that area 6-3-5 is equal to area 2-4-6. The rotor then continues to accelerate and swings back toward the equilibrium point. The speed deviation $\Delta\omega$ increases and E'_q continues to increase so modifying the dynamic characteristic which now lies above $P_{E'_{q0}}(\delta)$. This reduces the accelerating area and the kinetic energy, providing the impetus for the forward swing. The accelerating area is 3-5-7, the forward swing ends at point 8, and the whole cycle is repeated with a reduced amplitude of rotor swings.

Figure 5.12b shows that the E'_q variations are in phase with $\Delta\omega$ and lead the $\Delta\delta$ variations by $\pi/2$. Figure 5.12c illustrates the same effect using a phasor representation. All the increments of quantities shown in Figure 5.12c oscillate with the swing frequency Ω discussed later in this chapter; Eq. (5.53). Hence they can be shown on a phasor diagram in the same way as any sinusoidally changing quantities. Obviously, the phasors shown rotate with the swing frequency Ω rather than 50 or 60 Hz as is the case for AC phasors. Strictly speaking, the phasors shown represent the phase and initial values of the quantities as their root-mean-square (RMS) values decay with time. The synchronous power $\Delta\underline{P}_e$ is in phase with the rotor angle $\Delta\underline{\delta}$, while $\Delta\underline{E}'_q$ is in phase with the rotor speed deviation $\Delta\underline{\omega}$. Obviously, $\Delta\omega$ is a derivative of $\Delta\delta$ and leads it by $\pi/2$. It should be emphasized that the changes in E'_q are very small, in the range of a small percentage.

Figure 5.12 shows that the field winding provides a similar damping action to the damper windings, but much weaker. The combined effect of the two damping mechanisms helps to return the rotor quickly to the equilibrium point. If the damping effect is large then the rotor motion may be aperiodic (without oscillations).

The constants A and B in Eqs. (5.39) and (5.40) were stated assuming that the effect of the network resistance could be neglected. A fuller analysis would show that A and B depend on the network resistance (Anderson and Fouad 1977) and that at large values of this resistance A can change sign when, according to Eq. (5.40), the term with $\Delta E'_q$ changes sign to negative (with respect to $\Delta\omega$) and negative damping is introduced to the system. This may happen when a generator operates in a medium-voltage distribution network (Venikov 1978) and such a case is illustrated in Figure 5.13. If $\Delta E'_q$ lags $\Delta\delta$ then during the backswing the rotor will move from point 2 along a characteristic, that is higher than the initial one, thereby performing deceleration work equal to the area 2-3-2′. This area is larger than if the rotor followed the initial characteristic 2–1. To balance this deceleration work the rotor must swing back to point 4 such that area 3-4-4′ is equal to area 2-3-2′. Consequently, the amplitude of the rotor swings increases and, if this negative damping is larger than the positive damping introduced by the damper windings, the generator may lose stability.

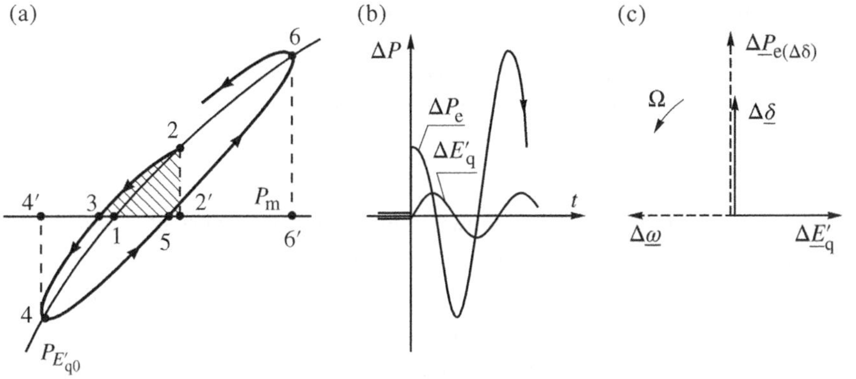

Figure 5.13 Negative damping: (a) trajectory of the operating point; (b) time variation of electrical power and transient emf; (c) relative position of phasors of oscillating increments.

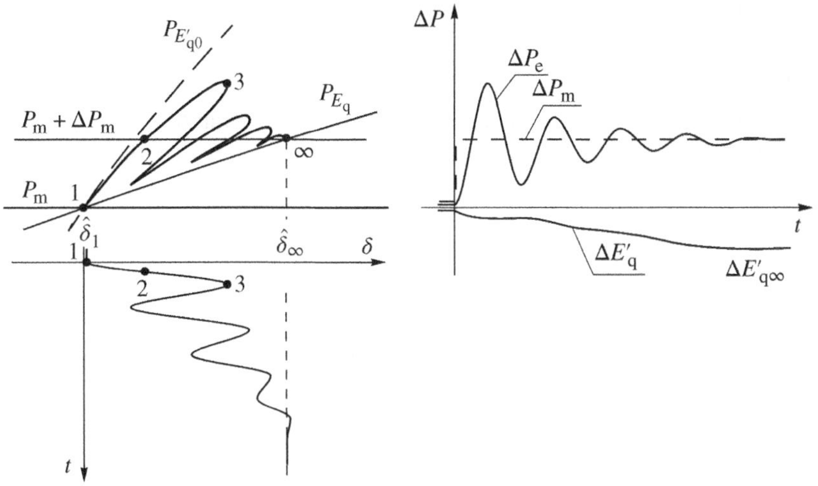

Figure 5.14 Rotor and power swings due to a disturbance in mechanical power.

By analyzing rotor swings similar to those shown in Figures 5.12 and 5.13 it can be shown that the positive damping due to the flux linkage variation described above can only take place when the slope of the transient characteristic is steeper than the slope of the steady-state characteristic, $K_{E'} > K_{E_q}$ (Machowski and Bernas 1989). Consequently, the steady-state stability condition defined in Eq. (5.19) must be modified to take into account the rotor dynamics as

$$K_{E_q} = \frac{\partial P_{E_q}}{\partial \delta} > 0 \quad \text{and} \quad K_{E'} = \frac{\partial P_{E'}}{\partial \delta} > K_{E_q} \tag{5.41}$$

Another type of disturbance consisting of a small increase in the mechanical power to $(P_m + \Delta P_m)$ is shown in Figure 5.14. Again, the characteristics are shown only in the vicinity of the equilibrium point. Initially, the system operates at point 1, the intersection between the characteristics P_m, P_{E_q}, and $P_{E'_{q0}}$. The increased mechanical loading produces a new, final equilibrium point ∞. The final value of the transient electromotive force (emf) $E'_{q\infty}$ is smaller than the initial value E'_{q0} as indicated in Figure 5.9. Consequently, E'_q must reduce as the rotor moves from point 1 toward point ∞. This means that the rotor oscillations will be similar to those shown in Figure 5.12, but occurring along a series of dynamic characteristics of decreasing amplitude, all of them lying below the initial characteristic $P_{E'_{q0}}$. Small oscillations of E'_q around its average decaying value cause some damping similar to that shown in Figure 5.12. This damping again adds to the main damping coming from the damper windings.

5.4.5.3 Equilibrium Point at $P_m = P_{E_q\,cr}$

If $P_m = P_{E_q cr}$ then there is only one steady-state equilibrium point. As shown in Figure 5.9a the transient characteristic $P_{E'_q}(\delta)$ crossing point $(\delta_{cr}, P_{E_q cr})$ has a positive slope, and a momentary disturbance in the angle δ will produce oscillations around the equilibrium point that are damped by the damper and field windings. Although this would suggest stable operation, the system is in practice unstable, as is explained below.

Any generator will be subjected to minor disturbances resulting from vibrations, slight changes in supplied power, switching in the network, and so on. When the system operates at $P_m = P_{E_q cr}$, any such disturbance may cause a relative shift between the P_m and $P_{E_q}(\delta)$ characteristics. This is shown in Figure 5.15, where it is assumed that P_m is increased by ΔP_m such that the new mechanical $(P_m + \Delta P_m)$ characteristic lies above $P_{E_q}(\delta)$. Initially, as the transient $P_{E'_q}(\delta)$ characteristic has a positive slope, the rotor starts to oscillate around $(P_m + \Delta P_m)$ in a similar way to that shown in Figure 5.14. Gradually, the armature flux penetrates the field winding, the emf E'_q decays, and the rotor follows $P_{E'_q}(\delta)$ characteristics of declining amplitude. Eventually, the decay in E'_q is such that the $P_{E'_q}(\delta)$ curve lies below the new mechanical $(P_m + \Delta P_m)$ characteristic and the generator loses synchronism. The behavior of the generator during asynchronous operation is described in Section 6.5.

The oscillatory loss of stability shown in Figure 5.15 takes place when the influence of the damper windings is neglected. In practice, damping at high loading can be significant when the loss of stability may occur in an aperiodic way for the reasons discussed in the next subsection.

5.4.6 Analysis of Rotor Swings Around the Equilibrium Point

In this subsection a quantitative analysis of rotor dynamics around the equilibrium point is attempted. Assuming the constant flux linkage generator model with constant E', the incremental swing equation (5.37) becomes

$$M\frac{d^2\Delta\delta}{dt^2} + D\frac{d\Delta\delta}{dt} + K_{E'}\,\Delta\delta = 0 \tag{5.42}$$

with the initial disturbed conditions being

$$\Delta\delta(t = 0^+) = \Delta\delta_0 \neq 0 \quad \text{and} \quad \Delta\omega = \Delta\dot{\delta}(t = 0^+) = 0 \tag{5.43}$$

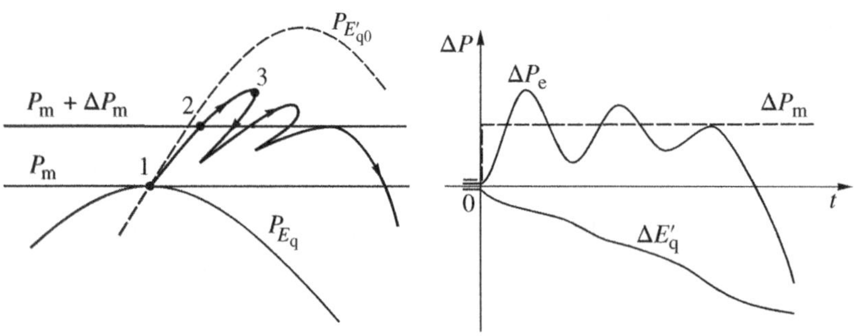

Figure 5.15 Rotor, power, and transient emf oscillations after a small increase in P_m at the critical equilibrium point.

Equation (5.42) is a second-order linear differential equation. As shown in Appendix A.3, the solution of a linear differential equation of any order is of the form: $\Delta\delta(t) = \mathrm{e}^{\lambda t}$. For the solution of that form one gets

$$\Delta\delta = \mathrm{e}^{\lambda t}; \quad \frac{\mathrm{d}\Delta\delta}{\mathrm{d}t} = \lambda\,\mathrm{e}^{\lambda t}; \quad \frac{\mathrm{d}^2\Delta\delta}{\mathrm{d}t^2} = \lambda^2\,\mathrm{e}^{\lambda t} \tag{5.44}$$

Substituting this equation into Eq. (5.42) and dividing by the nonzero term $\mathrm{e}^{\lambda t}$ gives the following algebraic equation

$$\lambda^2 + \frac{D}{M}\lambda + \frac{K_{E'}}{M} = 0 \tag{5.45}$$

That *characteristic equation* describes values of λ for which the assumed function $\Delta\delta = \mathrm{e}^{\lambda t}$ constitutes the solution of the differential Eq. (5.42). The values of λ that solve the characteristic equation are referred to as the *roots* of the characteristic equation.

The characteristic Eq. (5.45) has two roots λ_1 and λ_2 given by

$$\lambda_{1,2} = -\frac{D}{2M} \pm \sqrt{\left(\frac{D}{2M}\right)^2 - \frac{K_{E'}}{M}} \tag{5.46}$$

Three cases are now possible (Appendix A.3).

- The roots are real and distinct and the solution is of the form $\Delta\delta(t) = A_1\mathrm{e}^{\lambda_1 t} + A_2\mathrm{e}^{\lambda_2 t}$, where A_1 and A_2 are the integration constants. This case is described in detail in Example A.3.2, in Appendix A.3. Substitution of the initial conditions defined in Eq. (5.43) gives the aperiodic response

$$\Delta\delta(t) = \frac{\Delta\delta_0}{\lambda_2 - \lambda_1}\left[\lambda_2\mathrm{e}^{\lambda_1 t} - \lambda_1\mathrm{e}^{\lambda_2 t}\right] \tag{5.47}$$

where $-1/\lambda_1$ and $-1/\lambda_2$ are time constants.

- The roots are real and equal $\lambda_1 = \lambda_2 = \lambda$ and the solution is of the form $\Delta\delta(t) = \mathrm{e}^{\lambda t}(A_1 + A_2 t)$. This case is described in detail in Example A.3.3 discussed in Appendix A.3. Substitution of the initial conditions in Eq. (5.43) gives the aperiodic response

$$\Delta\delta(t) = \Delta\delta_0\mathrm{e}^{\lambda t}(1 - \lambda t) \tag{5.48}$$

where $-1/\lambda$ is the time constant.

- The roots form a complex conjugate pair

$$\lambda_{1,2} = -\frac{D}{2M} \pm \mathrm{j}\sqrt{\frac{K_{E'}}{M} - \left(\frac{D}{2M}\right)^2} \tag{5.49}$$

Denoting

$$\alpha = -\frac{D}{2M}; \quad \Omega = \sqrt{\frac{K_{E'}}{M} - \left(\frac{D}{2M}\right)^2} \tag{5.50}$$

one gets $\lambda_{1,2} = \alpha \pm \mathrm{j}\Omega$ where Ω is the frequency of oscillations (in rad/s) while α is the damping coefficient. The coefficient

$$\zeta = \frac{-\alpha}{\sqrt{\alpha^2 + \Omega^2}} \tag{5.51}$$

is referred to as the *damping ratio*. This notation allows Eq. (5.42) to be rewritten as the standard second-order differential equation

$$\frac{d^2\Delta\delta}{dt^2} + 2\zeta\,\Omega_{nat}\frac{d\Delta\delta}{dt} + \Omega_{nat}^2\Delta\delta = 0 \tag{5.52}$$

with the roots

$$\lambda_{1,2} = -\zeta\,\Omega_{nat} \pm j\Omega \tag{5.53}$$

where Ω_{nat} is the *undamped natural frequency* (in rad/s) of rotor swings for small oscillations, ζ is the *damping ratio* and $\Omega = \Omega_{nat}\sqrt{1-\zeta^2}$ where Ω is now referred to as the *damped natural frequency* (in rad/s) of rotor swings. This case is described in detail in Example A.3.4 and Example A.3.5, in Appendix A.3. Comparing Eq. (5.52) with Eq. (5.42) gives $\Omega_{nat} = \sqrt{K_{E'}/M}$ and $\zeta = D/2\sqrt{K_{E'}M}$. The solution for $\Delta\delta(t)$ is now given by

$$\Delta\delta(t) = \frac{\Delta\delta_0}{\sqrt{1-\zeta^2}}e^{-\zeta\,\Omega_{nat}t}\cos\left[\Omega t - \varphi\right] \tag{5.54}$$

where $\varphi = \arcsin\zeta$. The damping ratio ζ determines the amount of damping present in the system response expressing how quickly the amplitude of rotor swings decreases during subsequent periods. Let us express time as the multiplier of periods $t = 2\pi N/\omega_{nat}$ where N is the number of oscillation periods. Then Eq. (5.54) gives

$$k_N = \frac{\Delta\delta(N)}{\Delta\delta_0} = \frac{e^{-2\pi N\zeta}}{\sqrt{1-\zeta^2}} \tag{5.55}$$

where $\Delta\delta(N)$ denotes the amplitude of oscillations after N periods. For a practical assessment of damping, it is convenient to analyze $N = 5$ periods. A plot of the function given by (5.55) for $N = 5$ is shown in Figure 5.16. For example, the amplitude decreases after $N = 5$ periods to 39% for $\zeta = 0.03$, to 21% for $\zeta = 0.05$, and to 4% for $\zeta = 0.10$.

In practice, damping of rotor swings is considered to be satisfactory if the damping ratio $\zeta \geq 0.03$.

As the values of the roots $\lambda_{1,2}$ depends on the actual values of $K_{E'}$, D, and M so too does the type of response. The inertia coefficient M is constant while both D and $K_{E'}$ depend on the generator loading. Figure 5.2 shows that the damping coefficient D increases with load and Figure 5.9b shows that the transient synchronizing power coefficient $K_{E'}$ decreases with load.

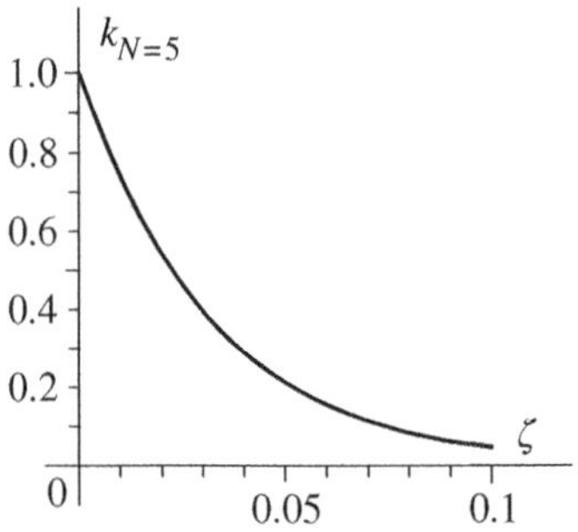

Figure 5.16 Amplitude $\Delta\delta/\Delta\delta_0$ as a function of the damping ratio ζ after $N = 5$ periods.

Equation (5.46) shows that if $K_{E'} > 0$ then, depending on the actual values of $K_{E'}$ and D, the roots of the characteristic equation can be either real or complex. For small initial values of power angle $\hat{\delta}_s$ the damping coefficient is small while the transient synchronizing power coefficient $K_{E'}$ is large so that $(K_{E'}/M) > (D/2M)^2$ and the two roots form a complex conjugate pair. In this case the solution of the differential equation is given by Eq. (5.54). The system response is oscillatory with the amplitude of rotor oscillations decaying with time, as shown schematically in the inset near point A in Figure 5.17. The frequency of oscillations is Ω and is slightly smaller than the undamped natural frequency Ω_{nat}.

As the initial value of the power angle $\hat{\delta}_s$ increases the synchronizing power coefficient $K_{E'}$ decreases while damping increases. Consequently, Ω_{nat} decreases but the damping ratio ζ increases with the result that the rotor oscillations become slower and more heavily damped, as shown in the inset near

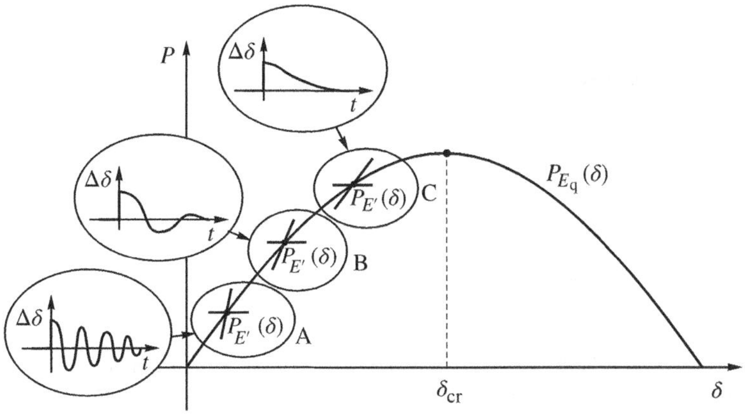

Figure 5.17 Examples of rotor swings for different stable equilibrium points.

point B in Figure 5.17. At some point, when $(K_{E'}/M) = (D/2M)^2$, the damping ratio ζ is unity and the oscillations are said to be *critically damped*. In this case the roots are real and equal $\lambda_1 = \lambda_2 = -\Omega_{nat}$ giving the aperiodic response expressed by Eq. (5.48) and shown in the inset near point C in Figure 5.17.

As the initial power angle $\hat{\delta}_s$ is further increased, at some point it holds that $(K_{E'}/M) < (D/2M)^2$, the two roots are real and negative, and the rotor swings are expressed by Eq. (5.47). In this case the damping ratio $\zeta > 1$, the rotor swings are overdamped, and the response is sluggish and vanishes aperiodically with time. Typically, this condition may occur when the initial operating point is near the peak of the steady-state characteristic in Figure 5.17.

When the power angle $\hat{\delta}_s$ is equal to its critical value δ_{cr}, it is not possible to analyze the system dynamics using the constant flux linkage model (E' = constant), because the effect of flux decrement must be included. This is explained in Figure 5.15.

Chapter 12 considers a similar problem but in a multi-machine system. The differential equations are there represented in the matrix form $\dot{\boldsymbol{x}} = \boldsymbol{A}\boldsymbol{x}$, where $\boldsymbol{x}$ is the state vector and $\boldsymbol{A}$ is the state matrix. Using the matrix form, Eq. (5.42) would become

$$\begin{bmatrix} \Delta\dot{\delta} \\ \Delta\dot{\omega} \end{bmatrix} = \begin{bmatrix} 0 & 1 \\ -\dfrac{K_{E'}}{M} & -\dfrac{D}{M} \end{bmatrix} \begin{bmatrix} \Delta\delta \\ \Delta\omega \end{bmatrix} \tag{5.56}$$

The eigenvalues of the above state matrix can be determined from solving

$$\det \begin{bmatrix} -\lambda & 1 \\ -\dfrac{K_{E'}}{M} & -\lambda - \dfrac{D}{M} \end{bmatrix} = \lambda^2 + \frac{D}{M}\lambda + \frac{K_{E'}}{M} = 0 \tag{5.57}$$

Clearly, this equation is identical to Eq. (5.45). Hence the roots of the characteristic Eq. (5.45) of the differential Eq. (5.42) are equal to the eigenvalues of the state matrix of the state Eq. (5.56).

As results from Eq. (5.50) the frequency of oscillations depends on inertia coefficient and synchronizing power, which in turn depends on loading conditions. In real multi-machine power systems the frequency of oscillations (rotor swings) varies in a wide range (0.2–2.5) Hz depending on loading conditions and location and type of fault exciting these oscillations. Typical values for an EPS with nominal frequency 50 Hz are given Table 5.1. The fastest are *internal oscillations* between generating units belonging to the same power plant excited by a fault appearing within the substation of the given power plant. Slower are *local oscillations* of generating units versus the rest of power system excited by a fault in the transmission network. The slowest are *inter-area oscillations* between subsystems belonging to the large interconnected system.

Table 5.1 Exemplary values of frequency of oscillations.

Type of oscillations	Frequency of oscillations	
	Hz	**rad/s**
Internal	1.8–2.5	11.3–15.7
Local	0.8–1.2	5.0–7.5
Inter-area	0.2–0.4	1.2–2.5

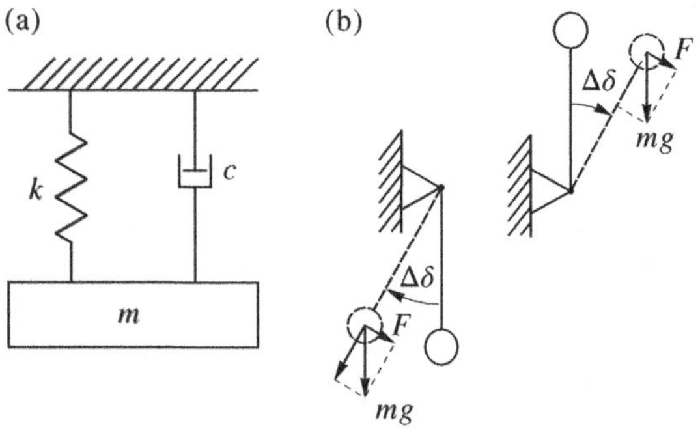

Figure 5.18 Mechanical analogues of the generator-infinite busbar system: (a) mass/spring/damper system; (b) a pendulum at stable and unstable equilibrium points.

5.4.7 Mechanical Analogues of the Generator-infinite Busbar System

Further insight into the generator response to a small disturbance can be obtained by comparing the generator-infinite busbar system with the standard mass/spring/damper system shown in Figure 5.18a described by the equation

$$m\frac{\mathrm{d}^2\Delta x}{\mathrm{d}t^2} + c\frac{\mathrm{d}\Delta x}{\mathrm{d}t} + k\Delta x = 0 \tag{5.58}$$

Comparing this equation with Eq. (5.42) suggests that the generator-infinite busbar system can be treated as an "electromagnetic spring", where any increase in the spring extension Δx is equivalent to an increase in the rotor angle $\Delta\delta$. In this analogous system the mass m is equivalent to the inertia coefficient M, the spring damping coefficient c is equivalent to the generator damping coefficient D, and the spring stiffness k is equivalent to the synchronizing power coefficient $K_{E'}$. This analogy allows the generator response, as expressed by Eqs. (5.47)–(5.54), to be directly related to the response of the standard mass/spring/damper system. Unlike the mechanical spring, the "electromagnetic spring" is nonlinear because $K_{E'}$ depends strongly on the initial value of the power angle $\hat{\delta}_s$ (Figure 5.9b).

Another useful mechanical analogue is that of the pendulum of mass m and length l shown in Figure 5.18b. The pendulum has an upper and lower equilibrium point as shown where the weight of the ball is counterbalanced by the arm. When the pendulum is disturbed from its equilibrium position, a force equal to $F = -mg\sin\delta$ is developed and, neglecting damping, the motion of the pendulum is described by the differential equation

$$m\frac{\mathrm{d}^2\delta}{\mathrm{d}t^2} = -\frac{mg}{l}\sin\delta \tag{5.59}$$

Near the lower (stable) equilibrium point the force F always acts toward the equilibrium point so that the pendulum oscillates around the equilibrium point. Near the upper (unstable) equilibrium point the force F pushes the pendulum away from the equilibrium point causing instability. The behavior of the pendulum at these two equilibrium points can be directly compared with the behavior of a generator at the stable and unstable equilibrium points shown in Figure 5.5.

The pendulum can also provide a useful analogue when analyzing the generator dynamics. In this case Eq. (5.59) has to be compared with the classical generator model in Eq. (5.28). Again, both equations are of identical form.

5.5 Steady-state Stability of the Regulated System

The previous section considers the power-angle characteristics of a simple generator-infinite busbar system and the resulting steady-state, or small-signal, stability conditions when the excitation voltage (and therefore the excitation emf E_f) are assumed to be constant. This section considers steady-state stability when the action of an AVR is included. The influence of the AVR will be considered in three stages. First of all, the modified steady-state power-angle characteristic will be derived. Secondly, the possibility of operation beyond the critical point (as defined by the pull-out power) will be discussed, and finally the influence of the AVR on rotor swings will be analyzed.

5.5.1 Steady-state Power-angle Characteristic of Regulated Generator

The static $P_{E_q}(\delta)$ power-angle characteristic, Eq. (5.15), was derived assuming that in steady state the excitation emf $E_f = E_q =$ constant. In practice every generator is equipped with an AVR which tries to maintain the voltage at the generator terminals constant (or at some point behind the terminals) by adjusting the value of the excitation voltage and, consequently, E_f. As the resulting formulae for the active and reactive power are more complicated than when $E_f =$ constant the following discussion will be restricted to the case of a round-rotor generator ($x_d = x_q$) with resistance neglected ($r = 0$). For this case the steady-state equivalent circuit and phasor diagram are shown in Figure 5.19. The formulae for the active and reactive power will be derived by resolving the voltages and currents along the a- and b-axes, where the a-axis is located along the system voltage V_s.

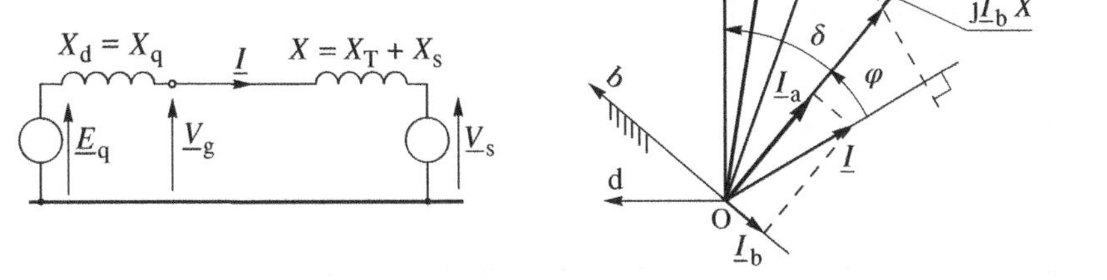

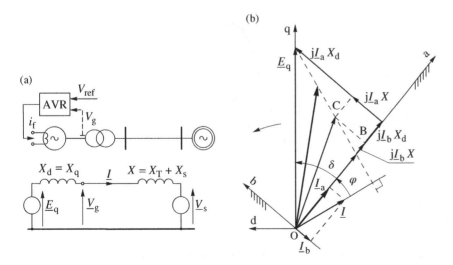

Figure 5.19 Generator operating on the infinite busbars: (a) schematic and equivalent circuit; (b) phasor diagram in the (d, q) and (a, b) reference frames.

The coordinates of $\underline{E}_\mathrm{q}$ in the (a, b) reference frame are

$$E_\mathrm{qa} = E_\mathrm{q}\cos\delta; \quad E_\mathrm{qb} = E_\mathrm{q}\sin\delta \tag{5.60}$$

Inspection of the phasor diagram gives

$$I_\mathrm{a} = \frac{E_\mathrm{qb}}{X_\mathrm{d} + X}; \qquad I_\mathrm{b} = \frac{E_\mathrm{qa} - V_\mathrm{s}}{X_\mathrm{d} + X} \tag{5.61}$$

while the coordinates of the current are $I_\mathrm{a} = I\cos\varphi$ and $I_\mathrm{b} = -I\sin\varphi$. Pythagoras's theorem applied to the triangle OBC yields $(V_\mathrm{s} + I_\mathrm{b}X)^2 + (I_\mathrm{a}X)^2 = V_\mathrm{g}^2$, which, after taking into account Eq. (5.61), gives

$$\left(E_\mathrm{qa} + \frac{X_\mathrm{d}}{X}V_\mathrm{s}\right)^2 + E_\mathrm{qb}^2 = \left[\frac{X_\mathrm{d} + X}{X}V_\mathrm{g}\right]^2 \tag{5.62}$$

This equation describes a circle of radius $\rho = (X_\mathrm{d}/X + 1)V_\mathrm{g}$ with the center lying on the a-axis at a distance $A = -X_\mathrm{d}V_\mathrm{s}/X$ from the origin. This means that with V_g = constant and V_s = constant, the tip of $\underline{E}_\mathrm{q}$ moves on this circle. Figure 5.20 shows the circular locus centered on the origin made by the phasor V_g = constant, and another circular locus (shifted to the left) made by phasor $\underline{E}_\mathrm{q}$.

The circle defined by Eq. (5.62) can be transformed into polar coordinates by substituting Eq. (5.60) to give

$$E_\mathrm{q}^2 + 2\frac{X_\mathrm{d}}{X}E_\mathrm{q}V_\mathrm{s}\cos\delta + \left(\frac{X_\mathrm{d}}{X}V_\mathrm{s}\right)^2 = \left[\frac{X_\mathrm{d} + X}{X}V_\mathrm{g}\right]^2 \tag{5.63}$$

One of the roots of this equation is

$$E_\mathrm{q} = \sqrt{\left(\frac{X_\mathrm{d} + X}{X}V_\mathrm{g}\right)^2 - \left(\frac{X_\mathrm{d}}{X}V_\mathrm{s}\sin\delta\right)^2} - \frac{X_\mathrm{d}}{X}V_\mathrm{s}\cos\delta \tag{5.64}$$

which corresponds to the $E_\mathrm{f} = E_\mathrm{q}$ points that lie on the upper part of the circle. Substituting Eq. (5.64) into the round-rotor power-angle equation, Eq. (5.18), gives the generated power as

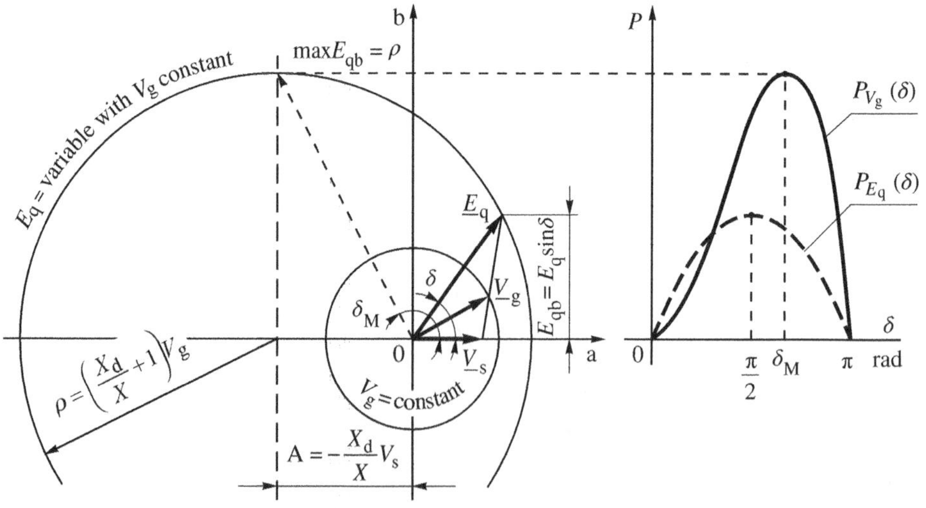

Figure 5.20 The circle diagrams and the power-angle characteristics for the round-rotor generator operating on the infinite busbars.

$$P_{V_g}(\delta) = \frac{V_s}{X_d + X}\sin\delta\sqrt{\left(\frac{X_d + X}{X}V_g\right)^2 - \left(\frac{X_d}{X}V_s\sin\delta\right)^2} - \frac{1}{2}\frac{X_d}{X}\frac{V_s^2}{X_d + X}\sin 2\delta \tag{5.65}$$

Equation (5.65) describes the power-angle characteristic $P_{V_g}(\delta)$ with $V_g = \text{constant}$ and is shown, together with $P_{E_q}(\delta)$, in Figure 5.20. Comparing the two shows that the AVR can significantly increase the amplitude of the steady-state power-angle characteristic.

The maximum value of the power given by Eq. (5.65) can be easily found by examining Figure 5.20. Substituting the second of the equations in Eq. (5.60) into Eq. (5.18) gives

$$P_{V_g}(\delta) = \frac{V_s}{X_d + X}E_{qb} \tag{5.66}$$

indicating that the generator power is proportional to the projection of E_q on the b-axis. The function defined by Eq. (5.66) reaches its maximum value when E_{qb} is a maximum. As can be seen from Figure 5.20, this occurs at the point on the E_q locus that corresponds to the center of the circle. At this point E_q has the following coordinates

$$E_{qb} = \rho = \left(\frac{X_d}{X} + 1\right)V_g \quad \text{at} \quad \delta_M = \tan^{-1}\left(\frac{\rho}{A}\right) = \tan^{-1}\left(-\frac{X_d + X}{X_d}\frac{V_g}{V_s}\right) \tag{5.67}$$

The angle δ_M at which $P_{V_g}(\delta)$ reaches maximum is always greater than $\pi/2$ irrespective of the voltages V_g and V_s. This is typical of systems with active AVRs. Substituting the first of the equations in (5.67) into (5.66) gives

$$P_{V_gM} = P_{V_g}(\delta)\big|_{\delta = \delta_M} = \frac{V_g V_s}{X} \tag{5.68}$$

showing that the amplitude of the power-angle characteristic of the regulated system is independent of the generator reactance. It does, however, depend on the equivalent reactance of the transmission system. The steady-state synchronizing power coefficient of the regulated system is $K_{V_g} = \partial P_{V_g}(\delta)/\partial\delta$ and $K_{V_g} > 0$ when $\delta < \delta_M$.

5.5.1.1 Physical Interpretation

The sin2δ component in Eq. (5.65) has a negative sign, making the maximum of the $P_{V_g}(\delta)$ characteristic shown in Figure 5.20 occur at $\delta_M > \pi/2$. For small rotor angles $\delta \ll \pi/2$ and the characteristic is concave, while for $\delta > \pi/2$ the characteristic is very steep. The sin2δ component has nothing to do with the reluctance power (as was the case with $P_{E_q}(\delta)$) as Eq. (5.65) has been derived assuming $x_d = x_q$. The distortion of the characteristic is entirely due to the influence of the AVR.

Physically, the shape of the $P_{V_g}(\delta)$ characteristic can be explained using Figure 5.21. Assume that, initially, the generator operates at point 1 corresponding to the characteristic $P_{E_{q1}} = P_{E_q}(\delta)\big|_{E_q = E_{q1}}$ shown by the dashed curve 1. An increase in the generator load causes an increase in the armature current, an increased voltage drop in the equivalent network reactance X, Figure 5.19, and therefore a decrease in the generator voltage V_g. The resulting voltage error forces the AVR to increase the excitation voltage so that E_q is increased to a value $E_{q2} > E_{q1}$ and a new operating point is established on a higher characteristic $P_{E_{q2}} = P_{E_q}(\delta)\big|_{E_q = E_{q2}}$, denoted by 2. Subsequent increases in load will cause the resulting $P_{V_g}(\delta)$ characteristic to cross at points 2, 3, 4, 5, and 6 lying on consecutive $P_{E_q}(\delta)$ characteristics of increased amplitude. Note that starting from point 5 (for $\delta < \pi/2$) the synchronizing power coefficient $K_{E_q} = \partial P_{E_q}(\delta)/\partial\delta$ is negative while $K_{V_g} = \partial P_{V_g}(\delta)/\partial\delta$ is still positive.

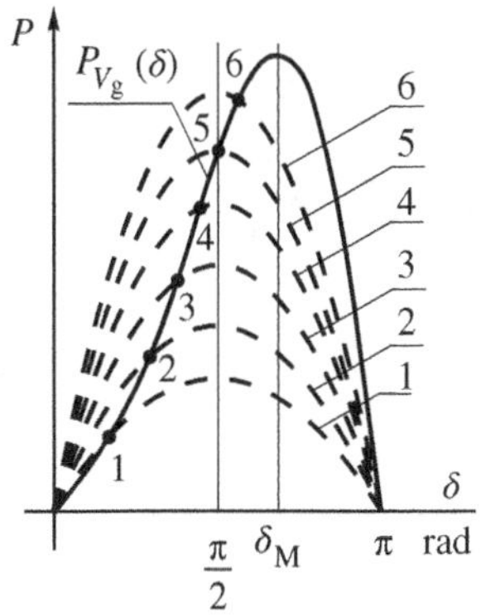

Figure 5.21 Creation of the $P_{V_g}(\delta)$ characteristic from a family of $P_{E_q}(\delta)$ characteristics.

5.5.1.2 Stability

If the AVR is very slow acting (i.e. it has a large time constant) then it may be assumed that following a small disturbance the AVR will not react during the transient state and the regulated and unregulated systems will behave in a similar manner. The stability limit then corresponds to point 5 when $\delta = \pi/2$ (for a round-rotor generator) and the stability condition is given by Eq. (5.19). If the AVR is fast acting so that it is able to react during the transient state, the stability limit can be moved beyond $\delta = \pi/2$ to a point lying below the top of the $P_{V_g}(\delta)$ curve. In this case stability depends on the parameters of the system and the AVR, and the system stability is referred to as *conditional stability*.

A fast-acting AVR may also reverse the situation when the stability limit is lowered (with respect to the unregulated system) to a point $\delta < \pi/2$, for example to point 4, or even 3, in Figure 5.21. In this situation the system may lose stability in an oscillatory manner because of the detrimental effect of the AVR. Such a situation, and the conditional stability condition, are discussed later in this section.

5.5.1.3 Effect of the Field Current Limiter

Equation (5.65) was derived, and Figure 5.20 drawn, under the assumption that the AVR may change $E_q = E_f$ in order to keep the terminal voltage constant without any limit being placed on the maximum value of $E_q = E_f$. In practice the AVR is equipped with various limiters, described in Chapter 2, one of which limits the field current and hence E_f. This limiter operates with a long time delay. If the exciter reaches the maximum field current value during slow changes in operating conditions then any further increase in the load will not increase the field current despite a drop in the terminal voltage V_g. Any further changes will take place at $E_q = E_{fMAX}$ = constant and the operating point will follow a $P_{E_q}(\delta)|_{E_q = E_{qMAX}}$ characteristic.

Whether or not the field current limiter will act before the top of the $P_{V_g}(\delta)$ characteristic is reached depends not only on the field current limit set but also on the equivalent network reactance X. Equation (5.68) shows that the amplitude of the $P_{V_g}(\delta)$ characteristic depends on the reactance X. If X is large then the amplitude is small and the field current limit may not be reached. If X is small then the amplitude is very large and the field current limit is encountered before the peak of the characteristic is reached. Such a situation is illustrated in Figure 5.22. The limit is reached at point Lim. Below this point the steady-state characteristic is $P_{V_g}(\delta)$ while above this point the generator follows the $P_{E_q}(\delta)|_{E_q = E_{qMAX}}$ curve. The resulting characteristic is shown in bold.

From the stability point of view a more interesting case is when the field current limit is not reached. However, it is important to remember that the thermal field current limit may be reached before the stability limit.

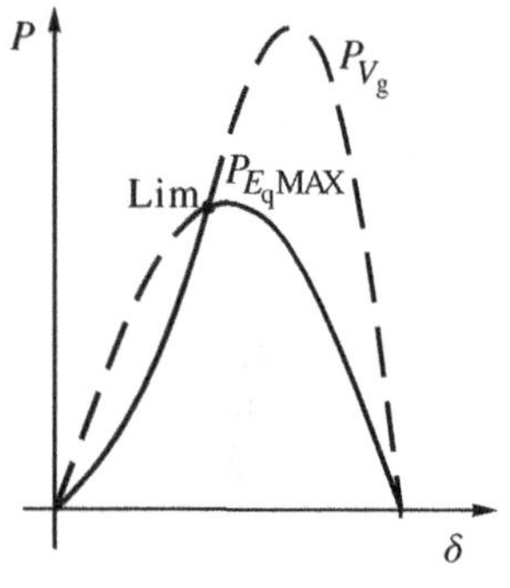

Figure 5.22 Example of the influence of the field current limiter on the steady-state power-angle characteristic.

5.5.2 Transient Power-angle Characteristic of the Regulated Generator

If the AVR or the exciter has a large time constant then the regulation process is slow and the rotor swings follow the transient power-angle characteristic, as discussed in Section 5.4. The transient characteristics of both the regulated and unregulated system are the same, the only difference being that the increased loading in the regulated system will cause an increase in the steady-state field current and therefore a higher value of E'_q and amplitude of the $P_{E'}(\delta')$ characteristic. Moreover, the angle δ' will reach the critical value $\pi/2$ before δ reaches its critical value δ_M. This is illustrated in Figure 5.23, which is a repetition of a fragment of the phasor diagram in Figure 5.20.

For the critical angle $\delta = \delta_M$ the phasor of V_g lies on the vertical axis, Figure 5.23a, and $\delta' > \pi/2$ because the emf E' leads V_g. This means that at the critical point δ_M the transient synchronizing power coefficient

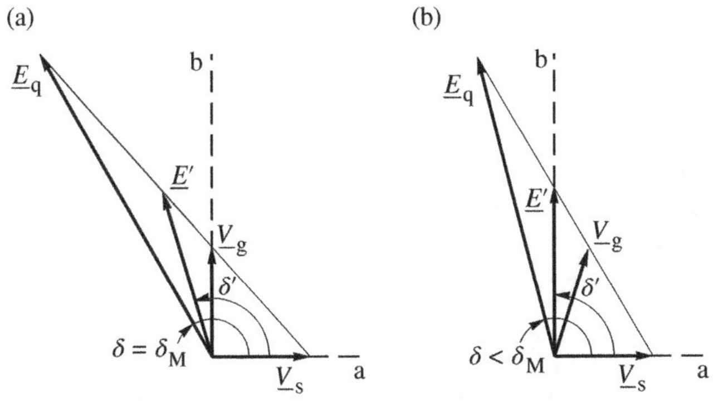

Figure 5.23 Phasor diagram of the regulated system with (a) $\delta = \delta_M$ and (b) $\delta' = \pi/2$.

$K_{E'} = \partial P_{E'}/\partial\delta$ is negative. When $\delta' = \pi/2$, Figure 5.23b, the emf E' lies on the vertical axis and $\delta < \delta_M$. The question now arises: at which point on the $P_{V_g}(\delta)$ curve does the transient synchronizing power coefficient become zero?

This question is most easily answered if the classical generator model is used with transient saliency neglected. The following equation, similar to Eq. (5.62), can be obtained for the transient emf when the phasor diagram of Figure 5.19b is used

$$\left(E'_a + \frac{X'_d}{X}V_s\right)^2 + E'^2_b = \left[\frac{X'_d + X}{X}V_g\right]^2 \tag{5.69}$$

This equation describes a circle on which the tip of $\underline{E}'$ moves when the power angle and excitation are increased. This circle is similar to the circle shown in Figure 5.20 but has a radius and horizontal shift dependent on X'_d. Substituting $E'_a = E'\cos\delta'$ and $E'_b = E'\sin\delta'$ into Eq. (5.69) gives

$$E'^2 + 2\frac{X'_d}{X}E'V_s\cos\delta' + \left(\frac{X'_d}{X}V_s\right)^2 = \left[\frac{X'_d + X}{X}V_g\right]^2 \tag{5.70}$$

Solving this equation with respect to E' yields

$$E' = \sqrt{\left(\frac{X'_d + X}{X}V_g\right)^2 - \left(\frac{X'_d}{X}V_s\sin\delta\right)^2} - \frac{X'_d}{X}V_s\cos\delta \tag{5.71}$$

For $\delta' = \pi/2$ the second component in this equation is zero and the transient emf is

$$E'|_{\delta'=\pi/2} = \frac{V_g}{X}\sqrt{(X'_d + X)^2 - \left(X'_d\frac{V_s}{V_g}\right)^2} \tag{5.72}$$

Substituting this value to Eq. (5.26) and noting that $x'_d = X'_d + X$ gives

$$P_{V_{gcr}} = P_{V_g}(\delta')|_{\delta'=\pi/2} = \frac{V_sV_g}{X}\sqrt{1 - \left(\frac{X'_d}{X'_d + X}\right)^2\left(\frac{V_s}{V_g}\right)^2} \tag{5.73}$$

According to Eq. (5.68), the factor V_sV_g/X in this equation corresponds to the amplitude of the $P_{V_g}(\delta)$ curve. This means that the ratio of power at which $K_{E'} = 0$ to the power at which $K_{V_g} = 0$ corresponds to the square-root expression in Eq. (5.73) and is equal to

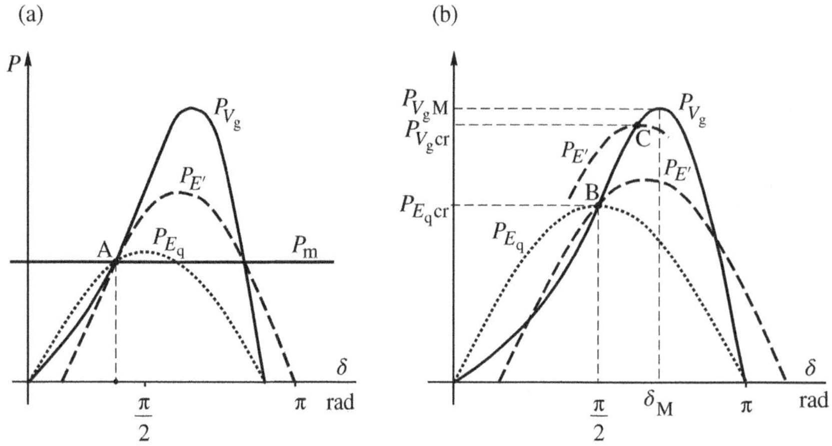

Figure 5.24 Power-angle characteristics of the regulated generator: (a) all synchronizing power coefficients are positive; (b) points above the limit of natural stability ($\delta = \pi/2$).

$$\alpha = \frac{P_{V_g}(\delta' = \pi/2)}{P_{V_g}(\delta = \delta_M)} = \frac{P_{V_g\,cr}}{P_{V_g\,M}} = \sqrt{1 - \left(\frac{X'_d}{X'_d + X}\right)^2 \left(\frac{V_s}{V_g}\right)^2} \tag{5.74}$$

This coefficient depends strongly on the equivalent network reactance X. If X is large then α is close to unity and the transient synchronizing power coefficient $K_{E'} = 0$. This takes place at a point close to the peak of $P_{V_g}(\delta)$ at which point $K_{V_g} = 0$. On the other hand, if X is small then $K_{E'}$ reaches zero at a point lying well below the peak of $P_{V_g}(\delta)$. This is illustrated in Figure 5.24.

Figure 5.24a shows the case when the power angle is small, all the characteristics have positive slope, and all the synchronizing power coefficients are positive. Figure 5.24b shows two operating points. The lower point corresponds to the natural stability limit ($\delta = \pi/2$) when $K_{E_q} = 0$ but $K_{E'} > 0$. Further increasing the load (and the power angle) causes $K_{E_q} < 0$, while $K_{E'} > 0$ until the top point is reached where the $P_{E'}(\delta')$ curve, shown as a dashed line, intersects at its peak with $P_{V_g}(\delta)$. Above this point $K_{E'}$ becomes negative despite $K_{V_g} > 0$. The ratio of both peaks is given by the coefficient α defined in Eq. (5.74).

5.5.2.1 Stability

After a disturbance the rotor swings follow the transient power-angle characteristic $P_{E'}(\delta')$. The system is unstable above the point $P_{V_g\,cr}$ where $K_{E'} = \partial P_{E'}/\partial\delta < 0$ and no deceleration area is available. Therefore, the necessary stability condition is

$$K_{E'} = \frac{\partial P_{E'}}{\partial \delta} > 0 \tag{5.75}$$

At the point $P_{V_g\,cr}$ any small increase in the mechanical power will cause asynchronous operation because the mechanical power is greater than the electrical power. Whether or not the generator can operate on the $P_{V_g}(\delta)$ characteristic below this point, that is when $P_m < P_{V_g\,cr}$, depends on the generator, network, and AVR parameters. Two factors are decisive: (i) the influence that the regulator has on the variations in E'_q because of changes in the excitation flux linkage during rotor swings and (ii) the influence that the regulator has on damping torques caused by additional currents induced in the damper windings.

5.5.3 Effect of Rotor Flux Linkage Variation

The influence of changes in the excitation emf E_f on the transient emf E' is given by Eq. (4.28). For the salient-pole machine when $E' = E'_q$ this equation can be rewritten as

$$\Delta E'_q = \Delta E'_{q(\Delta\delta)} + \Delta E'_{q(\Delta E_f)} \tag{5.76}$$

where

$$\Delta E'_{q(\Delta\delta)} = -\frac{AB}{1+sBT'_{d0}}\Delta\delta, \quad \text{and} \quad \Delta E'_{q(\Delta E_f)} = +\frac{B}{1+sBT'_{d0}}\Delta E_f \tag{5.77}$$

and $\Delta\delta = \Delta\delta'$. These two components are due to the rotor swings and voltage regulation, respectively. The influence of $\Delta E'_{q(\Delta\delta)}$ is described in Section 5.4.5 following Eq. (5.39) where it is shown in Figure 5.12 that this component is in phase with the speed deviation $\Delta\omega$ and introduces an additional damping torque in Eq. (5.38). A question arises as to what influence the voltage control component $\Delta E'_{q(\Delta E_f)}$ has on the problem. To answer this question it is necessary to determine the phase shift of $\Delta E'_{q(\Delta E_f)}$ with respect to $\Delta\omega$ (or $\Delta\delta$). This problem can be better understood with the help of Figure 5.25, which shows how a change in the rotor angle influences $\Delta E'_{q(\Delta E_f)}$.

The first block reflects the fact that (assuming constant infinite busbar voltage) a change in $\Delta\delta$ causes a voltage regulation error ΔV. The second block is the transfer function of the AVR and the exciter. Its effect is to convert the regulation error ΔV into a change in the excitation emf ΔE_f. The third block reflects the changes in $\Delta E'_q$ due to excitation changes and corresponds to Eq. (5.77).

The first block in Figure 5.25 constitutes a proportional element, as an increase in the rotor angle by $\Delta\delta$ causes a decrease in the generator voltage by $\Delta V_g \cong (\partial V_g/\partial\delta)\Delta\delta$ so that the following voltage error is produced

$$\Delta V = V_{ref} - V_g = -\frac{\partial V_g}{\partial\delta}\Delta\delta = K_{\Delta V/\Delta\delta}\Delta\delta \tag{5.78}$$

An expression for the proportionality coefficient $K_{\Delta V/\Delta\delta}$ can be obtained from Eq. (5.70). Solving this equation with respect to V_g gives

$$V_g = \frac{\sqrt{E'^2 + 2\frac{X'_d}{X}E'V_s\cos\delta' + \left(\frac{X'_d}{X}V_s\right)^2}}{\frac{X'_d}{X}+1} \tag{5.79}$$

Differentiation of this expression at the linearization point defined by δ'_0, E'_0, and V_{g0} gives

$$K_{\Delta V/\Delta\delta} = -\frac{\partial V_g}{\partial\delta'} = \frac{X'_d X}{(X'_d + X)^2}\frac{E'_0}{V_{g0}}V_s\sin\delta'_0 \tag{5.80}$$

This coefficient is positive over a wide range of angle changes, which means that the voltage regulation error given by Eq. (5.78) is always in phase with the angle changes $\Delta\delta$. The amplitude of ΔV depends on the generator load. For a small load (and δ'_0) the coefficient $K_{\Delta V/\Delta\delta}$ is small and the resulting voltage error is small. As the load is increased, changes in ΔV caused by the changes in $\Delta\delta$ become bigger.

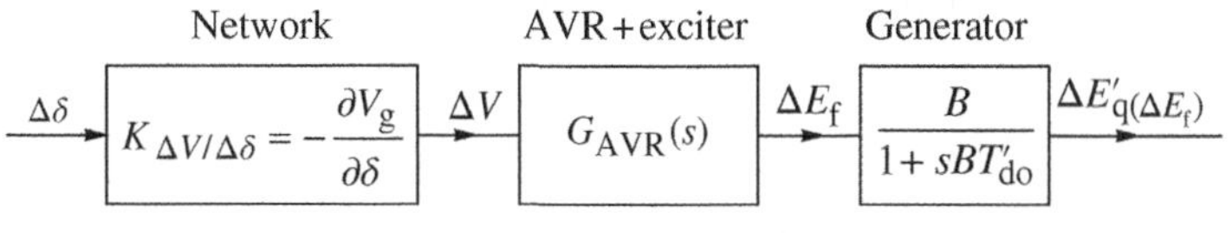

Figure 5.25 Components determining the phase shift between $\Delta\delta$ and $\Delta E'_{q(\Delta E_f)}$.

The second block in Figure 5.25 introduces a phase shift between ΔE_f and ΔV dependent on the transfer functions of the AVR and exciter. In the case of a static exciter, Figure 2.4d–f with a proportional regulator, the phase shift is small and it may be assumed that ΔE_f is in phase with ΔV. In comparison a DC cascade exciter, or AC exciter with rectifier, Figure 2.3a–c, behaves like an inertia element which introduces a phase shift of a few tens of degrees for a frequency of oscillation of about 1 Hz.

The generator block in Figure 5.25 introduces a phase shift which, as in Eq. (5.40), is equal to $\pi/2$. However, the minus sign in Eqs. (5.39) and, (5.40) and the first of the equations in (5.77) will cause $\Delta E'_{q(\Delta\delta)}$ to lead the variations in $\Delta\delta$ by $\pi/2$, while the plus sign in the second of the equations in (5.77) will cause $\Delta E'_{q(\Delta E_f)}$ to lag the changes in $\Delta\delta$ by $\pi/2$.

A knowledge of all these phase shifts allows a phasor diagram similar to that shown in Figure 5.12c to be drawn but with both of the components of Eq. (5.76) taken into account. This is shown in Figure 5.26, which contains two phasor diagrams of increments rotating with the swing frequency Ω (rad/s) drawn for two general types of AVR systems. In both diagrams the phasors of increments $\Delta\underline{\delta}$ and $\Delta\underline{V}$ are in phase; Eq. (5.80). The component $\Delta\underline{E}'_{q(\Delta\delta)}$ leads $\Delta\underline{\delta}$ in the same was as in Figure 5.12c.

The phasor diagram shown in Figure 5.26a is valid for a proportional AVR system when $\Delta\underline{E}_f$ and $\Delta\underline{V}$ are almost in phase. The component $\Delta\underline{E}'_{q(\Delta E_f)}$ lags $\Delta\underline{E}_f$ by $\pi/2$ and directly opposes $\Delta\underline{E}'_{q(\Delta\delta)}$. This diagram clearly shows that voltage regulation, represented by $\Delta\underline{E}'_{q(\Delta E_f)}$, weakens the damping introduced by the field winding and represented by $\Delta\underline{E}'_{q(\Delta\delta)}$. If the magnitude of $\Delta\underline{E}'_{q(\Delta E_f)}$ is greater than that of $\Delta\underline{E}'_{q(\Delta\delta)}$ then the voltage regulation will introduce a net negative damping into the system. This negative damping is enhanced by:

- large generator load (large value of δ'_0) resulting in a large value of the coefficient $K_{\Delta V/\Delta\delta}$, Eq. (5.80);
- large gain $|G_{AVR}(s)|$ in the AVR controller determining the magnitude of $\Delta\underline{E}_f$;
- large network reactance X determining the value of the coefficient $K_{\Delta V/\Delta\delta}$.[1]

Figure 5.26b shows an AVR system with cascaded DC exciter or an AC exciter with a rectifier when $\Delta\underline{E}_f$ lags $\Delta\underline{V}$ typically by a few tens of degrees. The phasor $\Delta\underline{E}'_{q(\Delta E_f)}$ lags $\Delta\underline{E}_f$ by $\pi/2$ giving rise to two components with respect to the direction of $\Delta\underline{\delta}$: (i) the quadrature component which introduces negative damping, as in Figure 5.26a, and (ii) the in-phase component which is in phase with $\Delta\underline{P}_e$ and does not influence damping. The effect of the latter

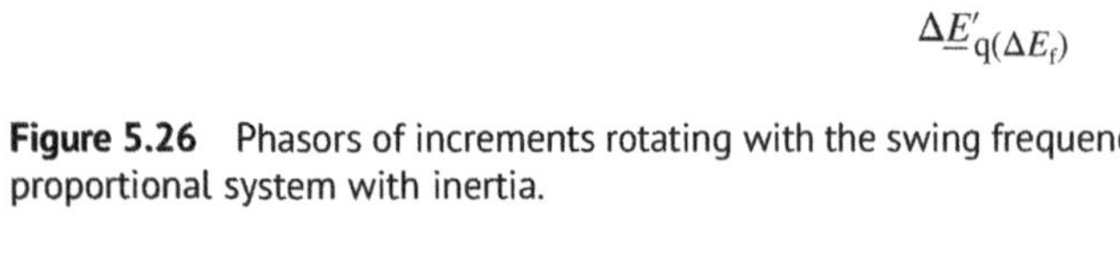

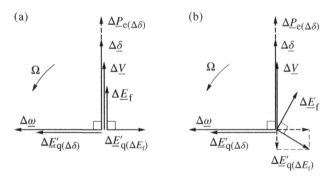

Figure 5.26 Phasors of increments rotating with the swing frequency Ω for: (a) the AVR proportional system; and (b) the AVR proportional system with inertia.

1 The maximum value of $K_{\Delta V/\Delta\delta}$ is obtained for $X = X'_d$. Usually in practice $X \ll X'_d$. Under that assumption the higher network reactance X, the higher the coefficient $K_{\Delta V/\Delta\delta}$.

component is to reduce the synchronizing power coefficient $K_{E'_q}$, Eq. (5.38), and therefore change the frequency of oscillations.

The above analysis of the influence of an AVR system on the generator damping is of a qualitative nature only, with the aim of helping the understanding of these complicated phenomena. Detailed quantitative analysis can be found in Demello and Concordia (1969), later enhanced in Anderson and Fouad (1977), Yu (1983), and Kundur (1994).

5.5.4 Effect of AVR Action on the Damper Windings

Section 5.5.3 describes how the AVR system can influence the damping torque due to the field winding, that is the last component in the swing equation (5.38). The second component in this equation, $P_D = D\Delta\omega$, corresponds to the damping power introduced by the damper windings. Assuming constant excitation voltage, that is E_f = constant, the damping power is given by Eq. (5.10). Recall that the mechanism by which this power is developed is similar to that on which operation of the induction machine is based. A change in the rotor angle δ results in the speed deviation $\Delta\omega$. According to Faraday's law, an emf is induced which is proportional to the speed deviation. The current driven by this emf interacts with the air-gap flux to produce a torque referred to as the *natural damping torque*. To simplify considerations, only the d-axis damper winding will be analyzed.

Figure 5.27a shows a phasor diagram for the d-axis damper winding, similar to that shown in Figure 5.26. The emf induced in the winding $\underline{e}_{D(\Delta\omega)}$ is shown to be in phase with $\Delta\underline{\omega}$. The damper winding has a large resistance which means that the current due to speed deviation, $\underline{i}_{D(\Delta\omega)}$, lags $\underline{e}_{D(\Delta\omega)}$ by an angle less than $\pi/2$. The component of this current which is in-phase with $\Delta\underline{\omega}$gives rise to the natural damping torque. The quadrature component, which is in phase with $\Delta\underline{\delta}$, enhances the synchronizing power coefficient.

Now consider the influence of the AVR on the damper windings. The d-axis damper winding lies along the path of the excitation flux produced by the field winding, Figure 4.3. This means that the two windings are magnetically coupled and may be treated as a transformer (Figure 5.27b), supplied by $\Delta\underline{E}_f$ and loaded with the resistance R_D of the damper winding. Consequently, the additional current $\underline{i}_{D(\Delta E_f)}$ induced in the damper winding must lag $\Delta\underline{E}_f$. Figure 5.27c shows the position of phasors. The horizontal component of $\underline{i}_{D(\Delta E_f)}$ directly opposes the horizontal component of $\underline{i}_{D(\Delta\omega)}$. As the former is due to the AVR while the latter is due to speed deviation and is responsible for the natural damping, it may be concluded that voltage regulation weakens the natural damping. This weakening effect is referred to as *artificial damping*.

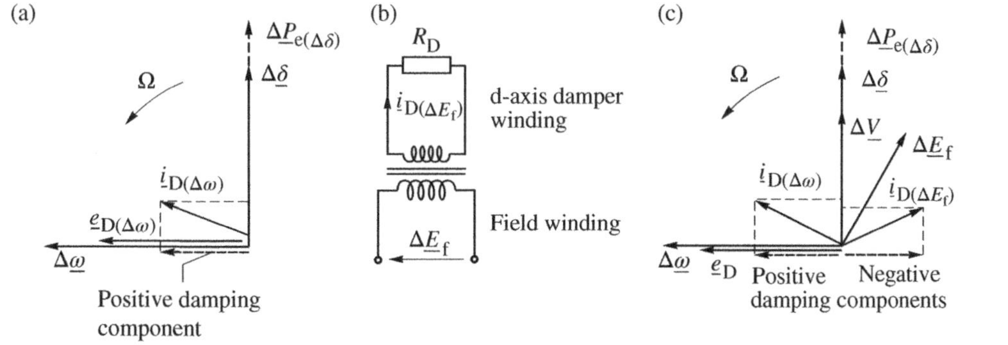

Figure 5.27 Phasor diagram of increments oscillating with the swing frequency Ω (in rad/s) for the damper windings: (a) natural damping only; (b) field and damper windings as a transformer; (c) natural and artificial damping.

Artificial damping is stronger for larger $\underline{i}_{D(\Delta E_f)}$ currents. This current is, in turn, proportional to the variations in $\Delta\underline{E}_f$ and $\Delta\underline{V}$ caused by $\Delta\underline{\delta}$. Some of the factors influencing this effect are described in the previous subsection and are: generator load, reactance of the transmission network, and gain of the voltage controller.

5.5.5 Compensating the Negative Damping Components

The main conclusion from the previous two subsections is that a voltage controller, which reacts only to the voltage error, weakens the damping introduced by the damper and field windings. In the extreme case of a heavily loaded generator operating on a long transmission link, a large gain in the voltage controller gain may result in net negative damping leading to an oscillatory loss of stability. This detrimental effect of the AVR can be compensated for using supplementary control loops referred to as *power system stabilizer* (PSS), which are discussed in more detail in Section 10.1. PSS is widely used in the United States, Canada, and Europe.

6

Electromechanical Dynamics – Large Disturbances

The previous chapter explains how an electric power system (EPS) responds to a small disturbance and determines the conditions necessary for the system to remain stable when subjected to such a disturbance. Much more dramatic from a stability point of view is the way in which a system responds to a large disturbance such as a short circuit or line tripping. When such a fault occurs, large currents and torques are produced and often action must be taken quickly if system stability is to be maintained. It is this problem of large-disturbance stability, and the effect such a disturbance has on the system behavior, that is addressed in this chapter.

6.1 Transient Stability

Assume that, before the fault occurs, the EPS is operating at some stable steady-state condition. The power system transient stability problem is then defined as that of assessing whether or not the system will reach an acceptable steady-state operating point following the fault.

As the subtransient period is normally very short compared with the period of the rotor swings, the effect of the subtransient phenomena on the electromechanical dynamics can be neglected. This allows the classical model of the generator to be used to study the transient stability problem when the swing equation is expressed by Eq. (5.1) and the air-gap power by Eq. (5.26). During a major fault, such as a short circuit, the equivalent reactance x'_d appearing in Eq. (5.26) will be subject to change so that the air-gap power $P_e = P_{E'}$ will also change and the power balance within the system will be disturbed. This will result in energy transfers between the generators producing corresponding rotor oscillations. Usually, there are three states accompanying a disturbance with three, generally different, values of x'_d: (i) the pre-fault state when the reactance $x'_d = x'_{d\,PRE}$; (ii) the fault state when $x'_d = x'_{d\,F}$; and (iii) the post-fault state when $x'_d = x'_{d\,POST}$. This section starts by considering a fault that is cleared without any change in the network configuration being required. In this case $x'_{d\,POST} = x'_{d\,PRE}$.

6.1.1 Fault Cleared Without a Change in the Equivalent Network Impedance

Figure 6.1a shows single machine infinite bus (SMIB) model in which a fault is cleared by tripping the faulty element but without changing the equivalent network impedance. It is assumed that only line L1 of a double- circuit connection is in use with line L2 energized but not connected at the system end. If a fault occurs on the unconnected line L2, and is then cleared by opening the circuit-breaker at the generator end of the line, the pre-fault and the post-fault impedance between the generator and the system are the same.

Power System Dynamics: Stability and Control, Third Edition. Jan Machowski, Zbigniew Lubosny, Janusz W. Bialek and James R. Bumby.

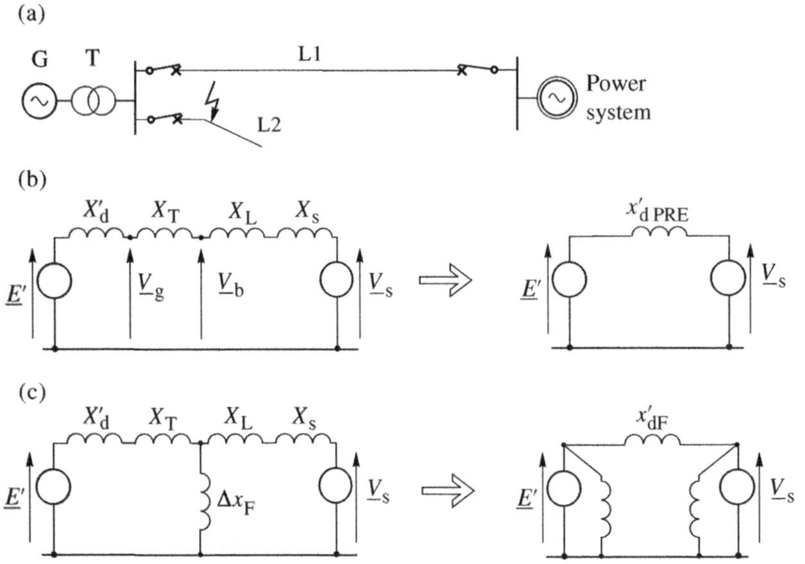

Figure 6.1 Example of a fault with the same pre- and post-fault impedance: (a) schematic diagram; (b) equivalent circuit for the pre- and post-fault state; (c) equivalent circuit during the fault.

6.1.1.1 Reactance Values During the Pre-fault, Fault, and Post-fault State

The equivalent circuit for the system is shown in Figure 6.1b. The generator is represented by the classical model with a constant transient electromotive force (emf) E' behind the transient reactance X'_d, while the system is represented by a constant voltage V_s behind the equivalent reactance X_s. The reactance of the transformer and line L1 are X_T and X_L, respectively. The pre-fault equivalent reactance $x'_{d\,PRE}$ of the whole transmission link is

$$x'_{d\,PRE} = X'_d + X_T + X_L + X_s \tag{6.1}$$

The use of symmetrical components allows any type of fault to be represented in the positive-sequence network by a fault shunt reactance Δx_F connected between the point of the fault and the neutral, as shown in Figure 6.1c. The value of the fault shunt depends on the type of fault and is given in Table 6.1, where X_1, X_2, and X_0 are, respectively, the positive-, negative-, and zero-sequence Thévenin equivalent reactances as seen from the fault terminals.

Using the star-delta transformation, the fault network can be transformed, as shown in Figure 6.1c, so that the voltages E' and V_s are directly connected by the equivalent fault reactance

$$x'_{d\,F} = X'_d + X_T + X_L + X_s + \frac{(X'_d + X_T)(X_L + X_s)}{\Delta x_F} \tag{6.2}$$

The value of this reactance is heavily dependent on the value of the fault shunt Δx_F given in Table 6.1. When the fault is cleared, by opening the circuit-breaker in line L2, the equivalent circuit is the same as that in the pre-fault period so that $x'_{d\,POST} = x'_{d\,PRE}$.

Table 6.1 Shunt reactances representing different types of fault.

Fault type:	Three-phase K3	Double-phase-to-ground K2E	Phase-to-phase K2	Single phase K1
Δx_F:	0	$\frac{X_2 X_0}{X_2 + X_0}$	X_2	$X_1 + X_2$

The circuit diagram of Figure 6.1c corresponds to the positive-sequence network so that when the reactance given in Eq. (6.2) is used in the power-angle characteristic, Eq. (5.40), only the torque and power due to the flow of positive-sequence currents is accounted for. The influence of negative- and zero-sequence fault currents and torques is neglected from further considerations in this chapter.

6.1.1.2 Three-phase Fault

Figure 6.2 shows how the equal area criterion, described in Section 5.4.3, can be used to analyze the effect of a three-phase fault on the system stability. To simplify the discussion, damping will be neglected ($P_D = 0$) and the changes in the rotor speed will be assumed to be too small to activate the turbine governor system. In this case the mechanical power input P_m from the turbine can be assumed to be constant.

For a three-phase fault $\Delta x_F = 0$ and, according to Eq. (6.2), $x'_{d\,F} = \infty$. Thus power transfer from the generator to the system is completely blocked by the fault with the fault current being purely inductive. During the fault, the electrical power drops from its pre-fault value to zero as illustrated by line 1–2 in Figure 6.2 and remains at zero until the fault is cleared by opening the circuit-breaker. During this time the rotor acceleration ε can be obtained from the swing equation, Eq. (5.1), by dividing both sides by M, substituting $P_e = 0$, $P_D = 0$, and writing in terms of δ' to give

$$\varepsilon = \frac{d^2\delta'}{dt^2} = \frac{P_m}{M} = \text{constant} \tag{6.3}$$

Integrating Eq. (6.3) twice with the initial conditions $\delta'(t = 0) = \delta'_0$ and $\Delta\omega(t = 0) = 0$ gives the power angle trajectory as

$$\delta' = \delta'_0 + \frac{\varepsilon t^2}{2} \quad \text{or} \quad \Delta\delta' = \delta' - \delta'_0 = \frac{\varepsilon t^2}{2} \tag{6.4}$$

This corresponds to the parabola a-b-d in Figure 6.2a. Before the fault is cleared, the rotor moves from point 2 to point 3 on the power-angle diagram and acquires a kinetic energy proportional to the shaded area 1-2-3-4.

When the fault is cleared at $t = t_1$ by opening the circuit-breaker, the rotor again follows the power angle characteristic $P_{E'}(\delta')$ corresponding to the reactance given by Eq. (6.1) so that the operating point jumps from point 3 to

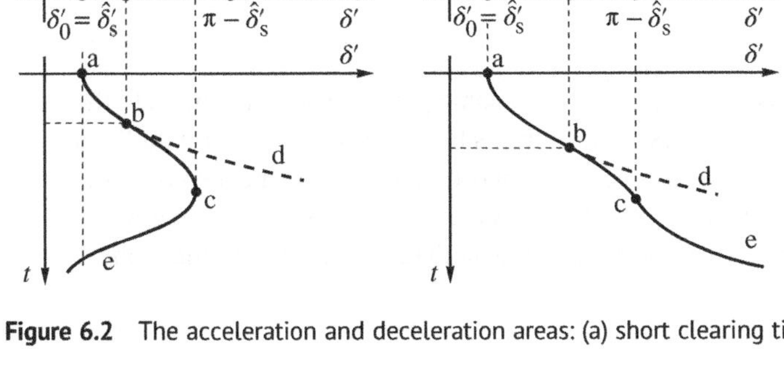

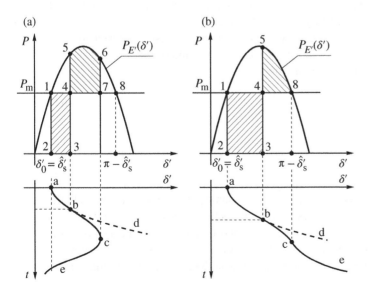

Figure 6.2 The acceleration and deceleration areas: (a) short clearing time; (b) long clearing time.

point 5. The rotor now experiences a deceleration torque, with magnitude proportional to the length of the line 4–5, and starts to decelerate. However, owing to its momentum, the rotor continues to increase its angle until the work done during deceleration, area 4-5-6-7, equals the kinetic energy acquired during acceleration, area 1-2-3-4. The rotor again reaches synchronous speed at point 6 when

$$\text{area (4-5-6-7)} = \text{area (1-2-3-4)} \tag{6.5}$$

In the absence of damping the cycle repeats and the rotor swings back and forth around point 1 performing *synchronous swings*. The generator does not lose synchronism and the system is stable.

Figure 6.2b shows a similar situation but with a substantially longer fault clearing time $t = t_2$ when the kinetic energy acquired during acceleration, proportional to the area 1-2-3-4, is much larger than in Figure 6.2a. As a result the work performed during deceleration, proportional to the area 4-5-8, cannot absorb the kinetic energy acquired during acceleration and the speed deviation does not become equal to zero before the rotor reaches point 8. After passing point 8 the electrical power $P_{E'}(\delta')$ is less than the mechanical power P_m and the rotor experiences a net acceleration torque which further increases its angle. The rotor makes an *asynchronous rotation* and loses synchronism with the system. Further asynchronous operation is analyzed in more detail in Section 6.5.

Two important points arise from this discussion. The first is that the generator loses stability if, during one of the swings, the operating point passes point 8 on the characteristic. This point corresponds to the transient rotor angle being equal to $\left(\pi - \hat{\delta}'_s\right)$, where $\hat{\delta}'_s$ is the stable equilibrium value of the transient rotor angle. Area 4-5-8 is therefore the *available deceleration area* with which to stop the swinging generator rotor. The corresponding transient stability condition states that the available deceleration area must be larger than the acceleration area forced by the fault. For the case shown in Figure 6.2a this criterion is

$$\text{area 1-2-3-4} < \text{area 4-5-8} \tag{6.6}$$

As the generator did not use the whole available decelerating area, the remaining area 6-7-8, divided by the available deceleration area, can be used to define the *transient stability margin*

$$K_{\text{area}} = \frac{\text{area 6-7-8}}{\text{area 4-5-8}} \tag{6.7}$$

The second important observation is that fault clearing time is a major factor in determining the stability of the generator. This is borne out by Eq. (6.4), where the accelerating area 1-2-3-4 is seen to be proportional to the clearing time squared. The longest clearing time for which the generator will remain in synchronism is referred to as the *critical clearing time* (CCT). The relative difference between the critical clearing time and the actual clearing time can be used to give another measure of the transient stability margin

$$c_t = \frac{t_{cr} - t_f}{t_{cr}} \tag{6.8}$$

where t_{cr} and t_f are the critical and actual clearing times.

6.1.1.3 Unbalanced Faults

During an unbalanced fault at least one of the phases is unaffected, allowing some power to be transmitted to the system. The equivalent fault reactance $x'_{d\,F}$ does not now increase to infinity, as for the three-phase fault, but to a finite value defined by Eq. (6.2). The increase in the reactance is inversely proportional to Δx_F and depends on the type of fault, as shown in Table 6.1. This allows faults to be listed in order of decreasing severity as (i) a three-phase fault (3 ph), (ii) a phase-to-phase-to-ground fault (2 ph-g), (iii) a phase-to-phase fault (2 ph), and (iv) a single-phase fault (1 ph).

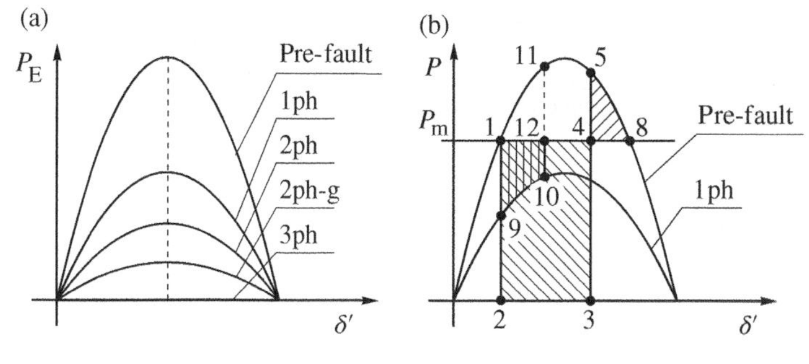

Figure 6.3 Effect of unbalanced faults: (a) comparison of power-angle characteristics; (b) accelerating and decelerating area during a three-phase fault and a single-phase fault.

The corresponding power-angle characteristics during the fault are illustrated in Figure 6.3a.

The effect of an unbalanced fault on system stability is examined by considering the least severe single-phase fault. A fault clearing time is assumed that is slightly longer than the critical clearing time for the three-phase fault. The acceleration and the deceleration areas are shown in Figure 6.3b. In the case of the three-phase fault the acceleration area 1-2-3-4 is larger than the deceleration area 4-5-8 and the system is unstable, as in Figure 6.2a.

For the single-phase fault the power transfer is not completely blocked and the air-gap power drops from point 1 on the pre-fault characteristic to point 9 on the fault characteristic. The accelerating torque, corresponding to line 1–9, is smaller than that for the three-phase fault (line 1–2), the rotor accelerates less rapidly and, by the time the fault is cleared, the rotor has reached point 10. At this point the rotor angle is smaller than in the case of the three-phase fault. The acceleration area 1-9-10-12 is now much smaller than the maximum available deceleration area 11-8-12 and the system is stable with a large stability margin. Obviously, a longer clearing time would result in the generator losing stability, but the critical clearing time for the single-phase fault is significantly longer than that for the three-phase fault. Critical clearing times for other types of faults are of the same order as the Δx_F value given in Table 6.1.

6.1.1.4 Effect of the Pre-fault Load

Figure 6.4 shows a generator operating at load P_{m1} prior to a three-phase fault. The fault is cleared when the acceleration area 1-2-3-4 is smaller than the available deceleration area 4-5-8. The system is stable with stability margin 6-7-8. Increasing the pre-fault load by 50% to $P_{m2} = 1.5P_{m1}$ increases the acceleration power $P_{acc} = P_m - P_{E'}(\delta') = P_m$ by one and a half times so that, according to Eqs. (6.3) and (6.4), the change in the power angle $\Delta\delta'$ also increases by a factor of 1.5. Consequently, as each side of the accelerating area rectangle 1-2-3-4 has increased 1.5 times, the acceleration area 1-2-3-4 is now much larger than the available deceleration area 4-5-8 and the system is unstable.

The pre-fault load is an important factor with regard to determining the critical clearing time and generator stability. The higher the load, the lower the critical clearing time.

6.1.1.5 Influence of Fault Distance

So far it has been assumed that in Figure 6.1 the fault occurs close to the busbar. If the point of the fault is further along the line, as shown in Figure 6.5a, then the impedance of the faulted line Δx_L is proportional to the fault distance and the per-unit (pu) length reactance of the line. The resulting equivalent circuit during the fault period is shown in Figure 6.5b. The equivalent series reactance $x'_{d\,F}$ can again be obtained from Eq. (6.2) but with Δx_F replaced by $\Delta x = \Delta x_F + \Delta x_L$.

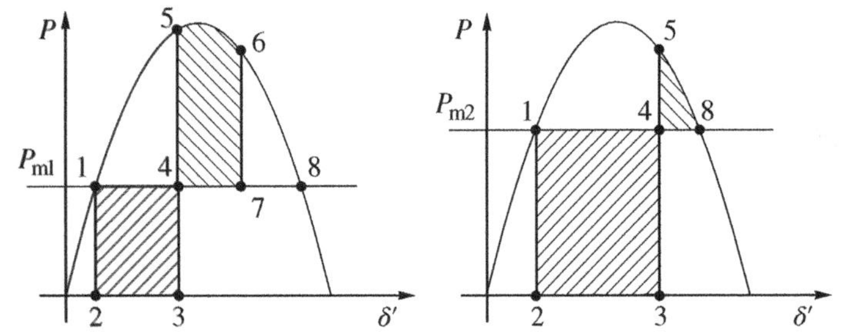

Figure 6.4 Acceleration and deceleration areas for two different pre-fault loads P_{m1} and $P_{m2} = 1.5P_{m1}$. The fault clearing time is the same for both cases.

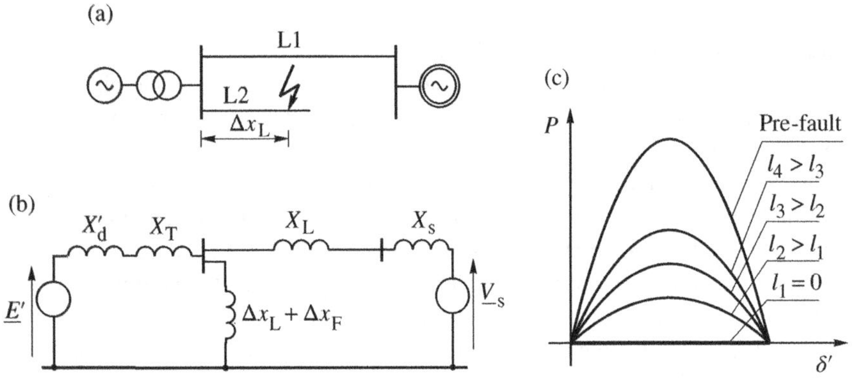

Figure 6.5 Influence of the fault distance: (a) schematic diagram; (b) equivalent circuit diagram; (c) power-angle characteristics before the fault and for various fault distances.

Figure 6.5c shows a family of power-angle characteristics for a three-phase fault ($\Delta x_F = 0$) occurring at increasing distances along the line. In comparison with the discussion on unbalanced faults it can be seen that the further the distance to the fault, the less severe the fault and the longer the critical clearing time.

In the case of unbalanced faults, $\Delta x_F \neq 0$ and the magnitude of the power-angle characteristic during the fault is further increased compared with the three-phase fault case. As a result the effect of the fault is less severe. In the case of a remote single-phase fault the disturbance to the generator may be very small.

6.1.2 Short-circuit Cleared with/without Auto-reclosing

The previous section described a particular situation where the network equivalent reactance does not change when the fault is cleared. In most situations the events surrounding the fault are more complex. First of all, the fault itself would normally be on a loaded element, for example on line L2 when it was connected at both ends. Secondly, the fault will not usually clear itself and the faulted element must be switched out of circuit.

The majority of faults on transmission lines are intermittent so that, after clearing the fault by opening the necessary circuit-breakers, the faulty line can be switched on again after allowing sufficient time for the arc across the breaker points to extinguish. This process is known as *automatic reclosing*. The sequence events in a *successful automatic reclosing cycle*, shown in Figure 6.6, would be:

- both lines operate (before the fault) – Figure 6.6b;
- a short circuit – Figure 6.6c;
- the faulted line is tripped and only one line operates – Figure 6.6d;
- the faulted line is automatically reclosed and both lines again operate – Figure 6.6b.

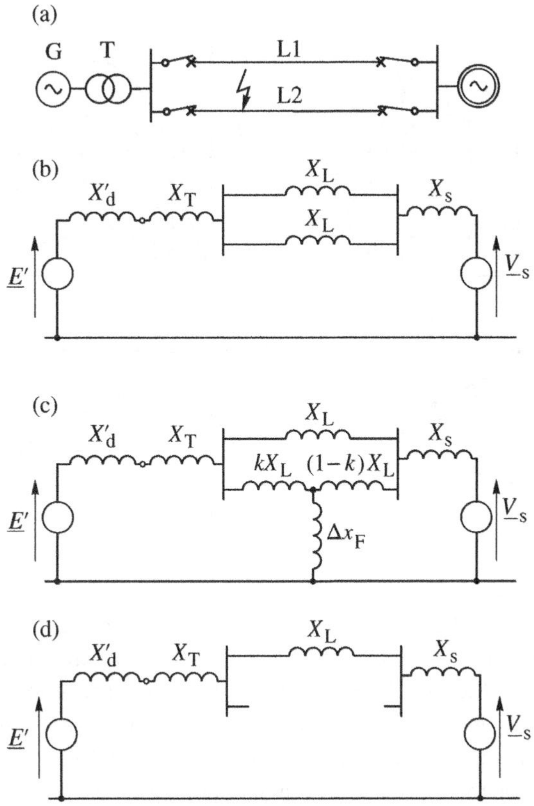

Figure 6.6 Automatic reclosing cycle: (a) schematic diagram; (b) equivalent circuit with both lines operating; (c) short circuit on one of the lines; (d) one line operating.

In Figure 6.6 the fault is assumed to occur on line L2 at some distance k from the circuit-breaker. Each state is characterized by a different equivalent reactance x'_d in Eq. (5.40) and a different power-angle characteristic corresponding to that reactance.

Figure 6.7 shows the effect of a three-phase fault with two fault clearing times, one producing a stable response and the other an unstable one. In both cases the accelerating power 1–2 accelerates the rotor from point 2 to point 3 during the fault. When line L2 is tripped, the operating point moves to point 5 and, because of the acquired kinetic energy, moves further along the characteristic c. After the certain auto-reclose time required to extinguish the arc, the auto-recloser reconnects line L2 and the system moves from point 6 to point 7. The power angle moves further along the characteristic a to point 8 when, in the stable case, the decelerating area 4-5-6-7-8-10 is equal to the accelerating area 1-2-3-4. The system is stable with a stability margin corresponding to the area 8-9-10. In the unstable case, Figure 6.7b, the increased clearing time enlarges the accelerating area 1-2-3-4 and the available decelerating area is too small to absorb this energy and stop the rotor. The generator rotor makes an asynchronous rotation and loses stability with the system.

In the case of a solid fault the reclosed line is again tripped and the cycle is referred to as *unsuccessful automatic reclosing.* Such a cycle poses a much greater threat to system stability than successful automatic reclosing. In the case of unsuccessful automatic reclosing the following sequence of events occurs:

- both lines operate (before the fault);
- a short circuit;
- the faulted line is tripped and one line operates;
- a short circuit (an attempt to auto-reclose onto the solidly faulted line);
- the faulted line is permanently tripped so that only one line remains in operation.

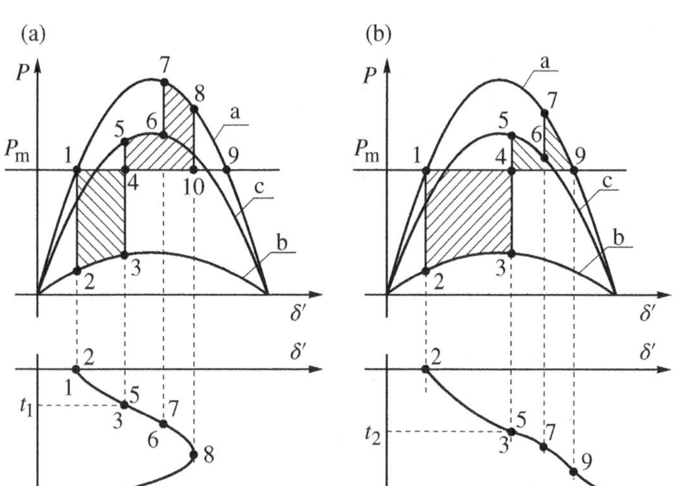

Figure 6.7 Accelerating and decelerating areas for successful automatic reclosing: (a) stable case; (b) unstable case.

An illustration of unstable and stable unsuccessful automatic reclosing is shown in Figure 6.8. In the first case the rotor acquires kinetic energy during the short circuit proportional to the area 1-2-3-4. Then, during the break in the cycle, the rotor is decelerated and loses energy proportional to the area 4-5-6-7. An attempt to switch onto the solidly faulted line results in an increase of kinetic energy proportional to the area 7-8-9-11. When the faulted line is permanently tripped, the decelerating area left is 10-13-11. As the sum of the accelerating areas 1-2-3-4 and 7-8-9-11 is greater than the sum of the decelerating areas 4-5-6-7 and 10-13-11, the rotor passes point 13 and makes an asynchronous rotation.

If now the clearing time and the automatic reclosing time are reduced and, in addition, the pre-fault load on the generator is reduced from P_{m1} to P_{m2}, the system may remain stable, as shown in Figure 6.8b. The sum of the accelerating areas 1-2-3-4 and 7-8-9-11 is now equal to the sum of the decelerating areas 4-5-6-7 and 10-11-12-14

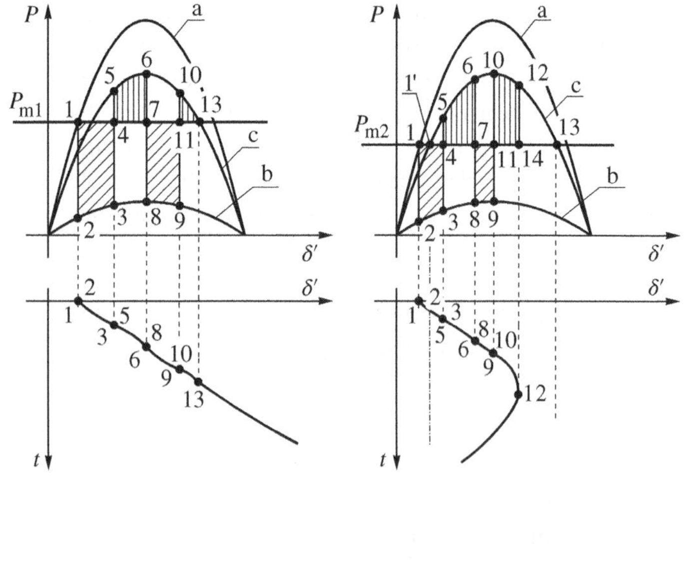

Figure 6.8 Accelerating and decelerating areas for unsuccessful automatic reclosing: (a) unstable case; (b) stable case.

and the system is stable with a stability margin corresponding to the area 12-14-13. Assuming that the oscillations are damped, the new equilibrium point 1′ corresponds to the power characteristic with one line switched off.

6.1.3 Power Swings

The rotor oscillations accompanying a fault also produce oscillations in the generated power. The shape of power variations can be a source of useful, although approximate, information on the transient stability margin. Again consider the system shown in Figure 6.5a and assume that the fault on line L2 is cleared by tripping the circuit-breakers at each end of the line without automatic reclosing. If the stability margin 6-7-8, shown in Figure 6.9a, is small, the power angle oscillations will be large and may exceed a value of $\pi/2$. The corresponding power oscillations will increase until δ' passes over the peak of the power-angle characteristic, when they will start to decrease. During the rotor back-swing the power will initially increase, as the rotor angle passes over the peak of the characteristic, before again decreasing. As a result the power waveform $P_e(t)$ exhibits characteristic "humps" which disappear as the oscillations are damped out. If the transient stability margin 6-7-8 is large, as in Figure 6.9b, then the humps do not appear because the maximum value of the power angle oscillation is less than $\pi/2$ and the oscillations are only on one side of the power-angle characteristic.

It should be emphasized that power swings follow the transient power-angle characteristic $P_{E'}(\delta')$ rather than the static characteristic $P_{E_q}(\delta)$. This means that the value of power at which humps appear is usually much higher than the critical steady-state power $P_{E_q\text{cr}}$ (Figure 5.9).

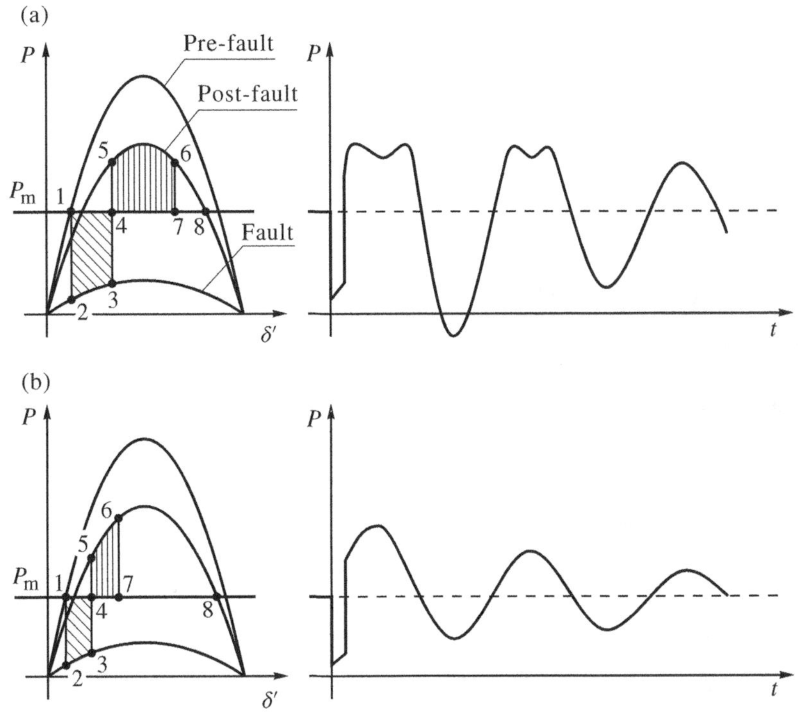

Figure 6.9 Power oscillations in the case of (a) a low stability margin and (b) a large stability margin.

6.1.4 Effect of Flux Decrement

The transient characteristic $P_{E'}(\delta')$ used to analyze transient stability is valid assuming that the flux linkage with the field winding is constant so that $E' = \text{constant}$. In fact, as magnetic energy is dissipated in the field winding resistance, flux decrement effects will cause E' to decrease with time. If the fault clearing time is short then flux decrement effects can be neglected for transient stability considerations, but if the clearing time is long then the decay in E' may have a considerable effect. To understand this refer to Figure 6.2, which illustrates the case of $E' = \text{constant}$. If now flux decrement effects are included, then, following the fault, the amplitude of the $P_{E'}(\delta')$ characteristic will reduce, leading to a decrease in the available deceleration area 4-5-8 and a deterioration in the transient stability. Consequently, the use of the classical model may lead to an optimistic assessment of the critical clearing time.

6.1.5 Effect of the AVR

Section 5.5 explains how AVR (automatic voltage regulator) action may reduce the damping of rotor swings following a small disturbance. In the case of large disturbances the influence of the AVR is similar. However, immediately after a fault occurs and is cleared, a strong-acting AVR may prevent a loss of synchronism. This can be explained as follows.

When a fault occurs, the generator terminal voltage drops and the large regulation error ΔV forces the AVR to increase the generator field current. However, the field current will not change immediately due to a delay depending on the gain and time constants of the AVR, and on the time constant of the generator field winding. To examine the effect of AVR action on transient stability the system shown in Figure 6.5a will be considered, assuming that a three-phase short circuit occurs some distance along line L2 so that $\Delta x_L \neq 0$ and $\Delta x_F = 0$.

When no AVR is present this system may lose stability, as shown in Figure 6.10a. The effect of the AVR, shown in Figure 6.10b, is to increase the field current leading to an increase in the transient emf E', as explained in

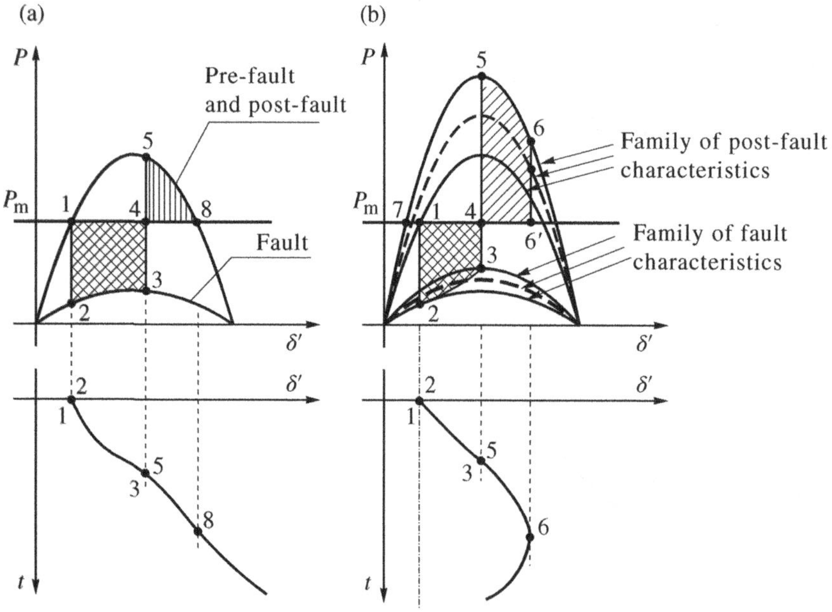

Figure 6.10 The acceleration area and the deceleration area when the influence of the voltage regulator is: (a) neglected; (b) included.

Section 4.2.4 and illustrated in Figure 4.17a. This increase in E' can be accounted for by drawing a family of power-angle characteristics $P_{E'}(\delta')$ for different values of E'. A fast-acting AVR and exciter can increase the excitation voltage up to its ceiling before the fault is cleared, although the change in the field current, and E', will lag behind this, owing to the time constant of the generator field winding. This increase in field current, and hence E', has two positive effects. First of all, as E' increases, the accelerating power decreases and the accelerating area 1-2-3-4 is slightly reduced. Secondly, when the fault is cleared, the system will follow a higher power-angle characteristic resulting from the new E' so that a larger decelerating area is available. In this example the rotor reaches a maximum power angle at point 6, when the decelerating area 4-5-6-6′ equals the accelerating area 1-2-3-4, before starting to move back toward the equilibrium point.

Although a fast-acting AVR reduces the first rotor swing, it can increase the second and following swings depending on the system parameters, the dynamic properties of the AVR, and the time constant of the field winding. Consider the voltage regulation error $\Delta V = V_{ref} - V_g$, created when the fault is cleared. Equation (5.93) is now important as it shows how V_g depends on δ' and the ratio X'_d/X. An example of such dependence is shown in Figure 6.11.

The terminal voltage V_g reaches a minimum when $\delta' = \pi$. The actual value of this minimum depends on the ratio X'_d/X:

- $V_g = 0$ for $\delta' = \pi$ and $X'_d/X = E'/V_s$ (curve 1);
- $V_g = (E' - V_s)/2$ for $\delta' = \pi$ and $X'_d/X = 1$ (curve 2).

As the generator reactance usually dominates other reactances in a transmission link, the case $X'_d/X = 1$ corresponds to a long transmission link. For a short transmission link $X'_d/X > 1$ and the minimum value of V_g is higher (curve 3).

Consider first the case of a long transmission link and assume that when the fault is cleared the angle δ' is large. On clearing the fault, the terminal voltage will recover from a small fault value to a somewhat small post-fault value (curve 1 in Figure 6.11). Consequently, the AVR will continue to increase the excitation current in order to try to recover the terminal voltage. In this case the rotor backward swing will occur on the highest possible transient characteristic $P_{E'}(\delta')$. This situation is illustrated in Figure 6.12a, which shows the system trajectory following fault clearance. The AVR continues to increase the field current during the backward swing, increasing the deceleration area 6-8-7-6′. This results in an increase in the amplitude of consecutive rotor swings such that, in this case, the AVR may have a detrimental effect on the generator transient stability.

Now consider the case of a short transmission link illustrated in Figure 6.12b. In this case the terminal voltage V_g recovers well when the fault is cleared, despite the large value of δ'. As a small increase in the transient emf E' will be forced by the increase in the excitation current during the fault period, the terminal voltage may recover to a value that is slightly higher than the reference value. Subsequently, this high terminal voltage will force the AVR to reduce the field current during the rotor backward swing, the amplitude of the transient power-angle characteristic will decrease and, as a result, the deceleration area 6-8-7-6′ will be reduced. This reduction in the deceleration area will lead to a reduction in the amplitude of subsequent rotor swings. In this case the AVR enhances the transient stability in both forward and backward swing.

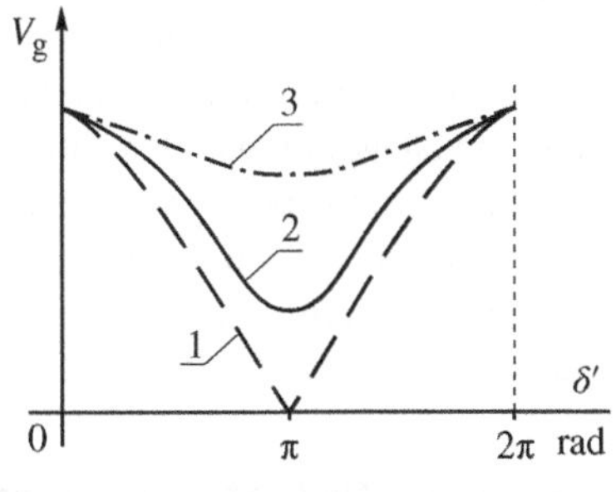

Figure 6.11 Generator terminal voltage as a function of δ': 1 – for $X'_d/X = E'/V_s$; 2 – for $X'_d/X = 1$; 3 – for $X'_d/X > 1$.

Figure 6.12 is also closely related to the damping produced by the AVR when a small disturbance occurs. Section 5.5 concludes that the amount of negative damping produced by the AVR increases with the length of the line as this increases the proportionality coefficient $K_{\Delta V/\Delta\delta}$ between

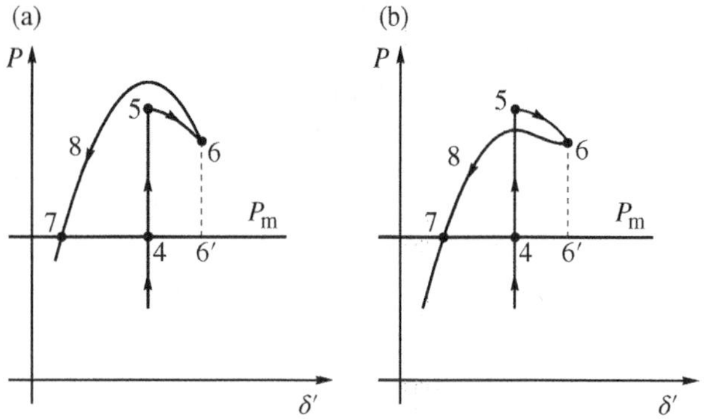

Figure 6.12 Rotor swing after the fault clearance in the case of: (a) a long transmission link and (b) a short transmission link.

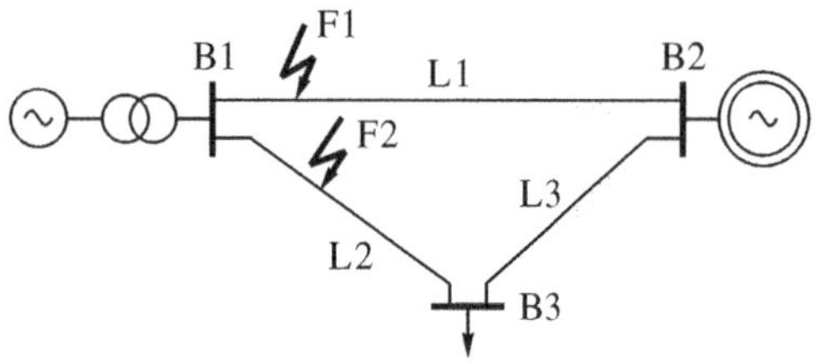

Figure 6.13 Example of a simple power system.

$\Delta\delta'$ and ΔV (Figure 5.26). The example shown in Figure 6.12a corresponds to the situation when this negative damping is greater than the system positive damping, while Figure 6.12b illustrates the reverse case.

As the influence of the AVR strongly depends on the post-fault network reactance, the dynamic system response depends on the fault location and its clearance. This is illustrated in Figure 6.13. If fault F1 appears on line L1 and is cleared by tripping the faulted line, the generator will operate in the post-fault state via a long transmission link that consists of lines L2 and L3. Simulation results for this example are presented in Figure 6.14. Figure 6.14a shows the trajectory of the operating point, while Figure 6.14b shows that the terminal voltage drops during the fault and does not recover very well. This low value of terminal voltage forces the AVR to continue to increase the excitation current so that the rotor backward swing occurs along the upper characteristic in Figure 6.14a. This repeats from cycle to cycle with the effect that the rotor swings are poorly damped (Figure 6.14c).

If the fault occurs on line L2 and is again cleared by tripping the faulted line, the generator operates in the post-fault period via a short transmission link that consists solely of line L1. Simulation results for this case are shown in Figure 6.15. Now, when the fault is cleared, the generator voltage recovers well and stays at a high level (Figure 6.15b). Consequently, the AVR decreases the excitation current and the rotor swings back along the lower power-angle characteristic in Figure 6.15a and the deceleration area is small. This repeats during each cycle of the rotor swings with the result that they are well damped (Figure 6.15c).

These examples show how the rotor oscillations may increase if the transmission link is long and the AVR keeps the excitation voltage at too high a value during the rotor backward swing. Ideally, the regulator should increase the excitation when the power angle δ' increases and lower it when δ' decreases, no matter what the value of the regulation error ΔV is. Section 10.1 shows how the AVR can be provided with supplementary control loops to provide a regulation error that depends on the rotor speed deviation $\Delta\omega$ or the rate of change of real power. Such supplementary loops coordinate the regulation process with the rotor swings in order to ensure correct damping.

6.1.6 Simplified Angle Stability Criteria

Machowski et al. (2015) show that, for a simple power system model represented by a SMIB model, the simple criteria for assessing the steady-state and transient angle stability can be derived based on the short-circuit power of the EPS calculated at the busbar where the generating units are connected.

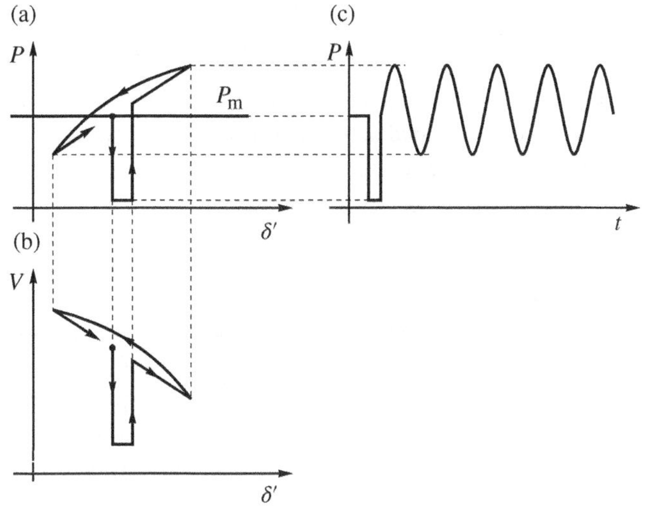

Figure 6.14 Simulation results for a fault on line L1 cleared by tripping the faulted line: (a) equal area method; (b) variation in the generator voltage; (c) power swings.

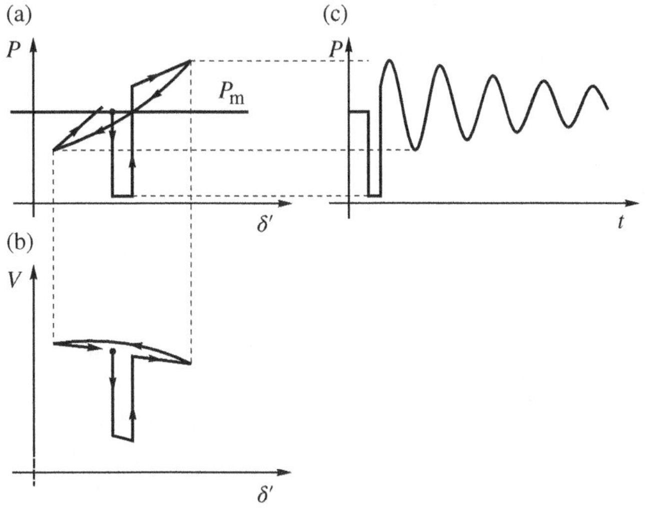

Figure 6.15 Simulation results for a fault on line L2 cleared by tripping the faulted line: (a) equal area method; (b) variation in the generator voltage; (c) power swings.

It is assumed (Figure 6.16a) that the EPS is treated as a voltage source $\underline{V}_s$ behind the short-circuit reactance $X_s = V_n^2/S_K''$, where S_K'' is the short-circuit power of the system without the contribution of the considered generating unit. V_n is nominal voltage of the network. In a steady state the generator is represented by emf $\underline{E}_q$ behind synchronous reactance X_d and in transient state by transient emf $\underline{E}'$ behind transient reactance X_d', respectively. X_T is the step-up transformer reactance. As explained in Section 5.4, the simple model considered here can be described by two power-angle characteristics (Figure 5.8), which for the reader's convenience are shown in Figure 6.16b. Power angles δ and δ' are the arguments of electromotive forces $\underline{E}_q$ and $\underline{E}'$ measured with reference to the voltage $\underline{V}_s$ of the infinite bus. $P_{Eq}(\delta)$ and $P_{E'}(\delta)$ are static and dynamic power angle characteristics.

It is convenient to perform further analysis in pu data with the base voltage equal to the nominal voltage of the network V_n, the base power equal to the rated power of the generating unit S_r, and the base reactance $X_n = V_n^2/S_r$. For such base values the short-circuit reactance in pu is given by

$$X_{s\,pu} = \frac{X_s}{X_n} = \frac{V_n^2}{S_K''}\cdot\frac{S_r}{V_n^2} = \frac{S_r}{S_K''} = \frac{P_r}{S_K''}\cdot\frac{1}{\cos\varphi_r} = \frac{1}{\alpha\cos\varphi_r} \qquad (6.9)$$

where $\cos\varphi_r = P_r/S_r$ is the rated power factor of the generating unit and $\alpha = S_K''/P_r$ is the ratio of the short-circuit power of the system and rated active power of the generating unit. For a typical value of the rated power factor $\cos\varphi_r = 0.85$, the following reactance is obtained

$$X_{s\,pu} \cong 0.3 \quad \text{for} \quad \alpha = S_K''/P_r = 4 \qquad (6.10)$$

$$X_{s\,pu} \cong 0.2 \quad \text{for} \quad \alpha = S_K''/P_r = 6 \qquad (6.11)$$

The steady-state stability margin c_{E_q} is defined by Eq. (5.20), where P_m is the actual loading of the generator and $P_{E_q cr}$ is the critical load referred to as the *pull-out power*. The pull-out power $P_{E_q cr} = E_q V/x_d$ is proportional to the product of the infinite bus voltage V_s and synchronous emf E_q and inversely proportional to reactance $x_d = (X_d + X_T + X_s)$. Hence the pull-out power $P_{E_q cr} = E_q V_s/x_d$ and stability margin c_{E_q} depend on the reactance X_s and at the same time on the short-circuit power $S_K'' = V_n^2/X_s$.

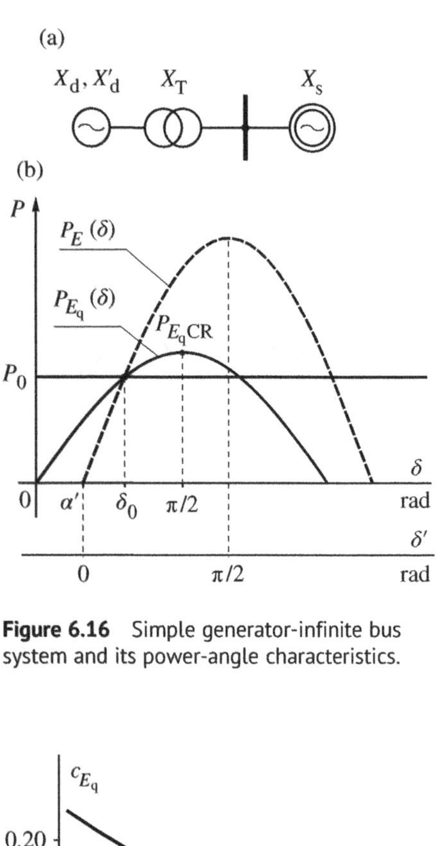

Figure 6.16 Simple generator-infinite bus system and its power-angle characteristics.

Figure 6.17 shows the dependence of the steady-state stability margin c_{E_q} on the generating unit reactance $(X_d + X_T)$ for two values of coefficient α defined in Eq. (6.9). Using this figure, for typical values of the generating unit reactance $(X_d + X_T)$ the following inequalities can be written

$$c_{E_q} > 10\% \quad \text{if} \quad \alpha \ge 6 \quad \text{or} \quad S_K'' \ge 6\cdot P_n \qquad (6.12)$$

$$c_{E_q} > 5\% \quad \text{if} \quad \alpha \ge 4 \quad \text{or} \quad S_K'' \ge 4\cdot P_n \qquad (6.13)$$

The transient state stability margin c_t is defied by Eq. (6.8), where t_{cr} and t_f are the critical and actual clearing times. The magnitude of the dynamic power angle characteristic $P_{E'}(\delta)$ (dashed curve on Figure 6.16) is proportional to the product of the infinite bus voltage V_s and transient emf E' and inversely proportional to reactance $x'_d = (X'_d + X_T + X_s)$. Hence, for a given clearing time of the fault and resulting acceleration area (Figure 6.2), the available deceleration area depends on the transient reactance X_s and at the same time on the short-circuit power $S_K'' = V_n^2/X_s$ and coefficient α.

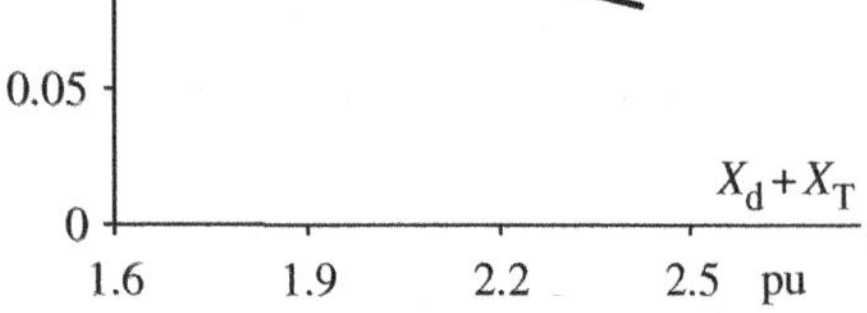

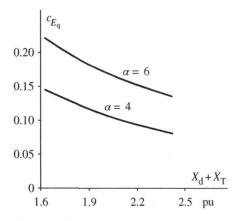

Figure 6.17 Steady-state stability margin as a function of generating unit reactance.

Figure 6.18 shows a relationship between the transient stability margin c_t and the generating unit reactance $(X'_d + X_T)$ assuming that the generating unit is loaded with the rated power $P_r = S_r\cos\varphi_r$ and the short-circuit is cleared with time $t_f = 150$ ms. A mechanical time constant T_m of the generating unit is a parameter for the diagrams. Figure 6.18a shows that for typical values of $(X'_d + X_T)$ and $T_m \ge 6.5$ seconds

$$c_t > 10\% \quad \text{if} \quad S_K'' \ge 6\cdot P_n \qquad (6.14)$$

Figure 6.18b shows that at short-circuit power $S_K'' = 4P_n$ and $T_m = 6.5$ seconds at $(X'_d + X_T) = 0.5$ the coefficient of transient stability margin is close to zero. For greater values of $(X'_d + X_T)$ or smaller values of T_m the system may not be stable if the fault is cleared with time $t_f = 150$ ms.

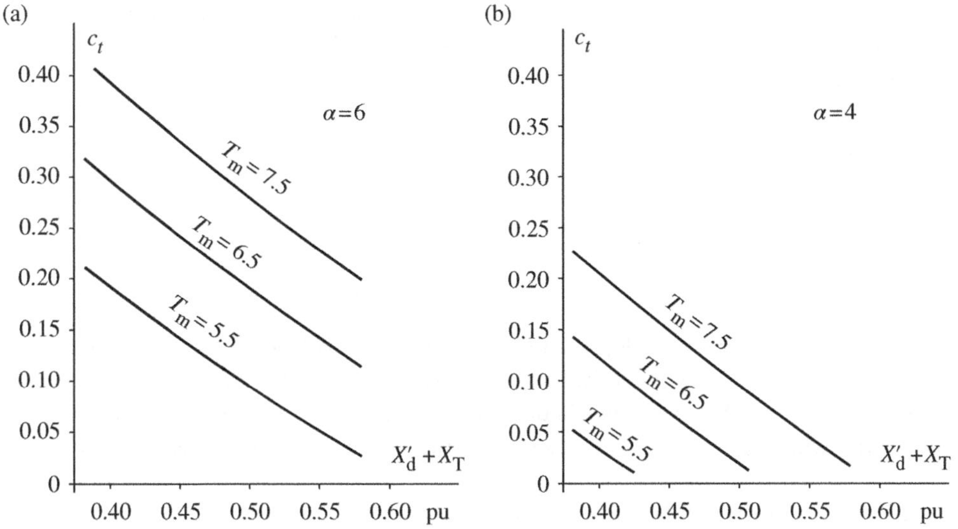

Figure 6.18 Transient stability margin for various mechanical time constants and (a) α = 6; (b) α = 4.

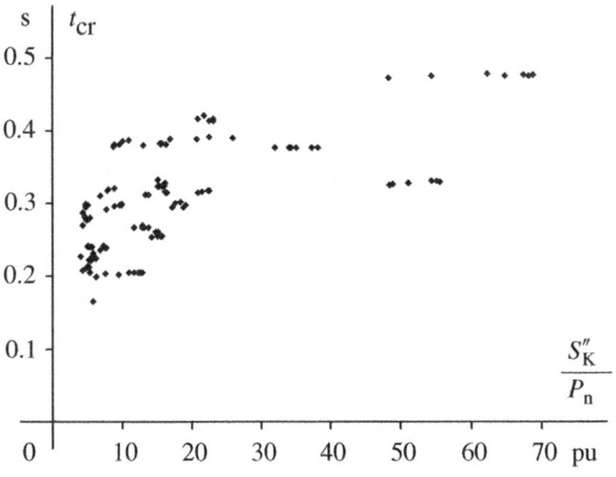

Figure 6.19 Set of points obtained for various locations of balanced short circuit in a large power system.

Equations (6.12) and (6.14) show that, for typical parameters of the EPS and at short-circuit clearing time $t_f \leq 150$ ms, condition $S''_K \geq 6P_r$ must be met after switching off the short-circuited element to ensure both the steady-state and transient stability with stability margin ≥10%.

In order to confirm the above conclusion a multi-case stability analysis has been performed for a real large power system model. Transient angle stability has been analyzed by computer simulation of the transient states for balanced (three-phase) faults using detailed models of power system elements (Chapter 11). Results are described by Machowski et al. (2015) and are presented in Figure 6.19. This figure shows a set of 150 points corresponding to critical clearing times t_{cr} calculated with a computer simulation and relationship between the short-circuit power to the rated power of generating units (installed in a given node) S''_K/P_r. Individual points are spread along the vertical axis of the diagram. Nevertheless, it can be clearly seen that the critical clearing time increases as the value of S''_K/P_r increases. For all the cases the critical clearing time is larger than 150 ms.

The above criterion (6.12) or (6.14) properly identifies the generation nodes for which there are no problems with transient angle stability and can be used as a screening criterion to limit the multi-case analysis of steady-state and transient stability performed for network expansion planning (Chapter 14).

6.2 Swings in Multi-machine Systems

Although the simplification of a multi-machine power system to a single generator-infinite busbar model enables many important conclusions to be made concerning the electromechanical dynamics and stability of the system, this simplification is only possible when the fault affects one generator and has little effect on other generators in

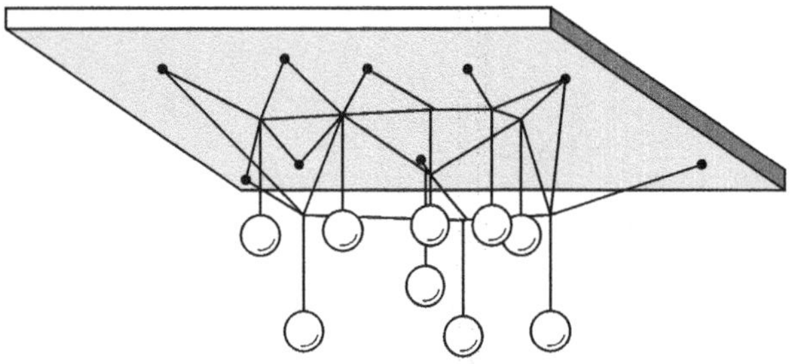

Figure 6.20 Mechanical analogue of swings in a multi-machine system. *Source:* Based on Elgerd (1982).

the system. Contemporary power systems have a well-developed transmission network with power stations located relatively close to each other so that these conditions will not always be met. In these circumstances a fault near one of the power stations will also distort the power balance at neighboring stations. The resulting electromechanical swings in the EPS can be compared with the way the masses swing in the mechanical system, as shown in Figure 6.20.

Section 5.2 shows that a single swinging rotor can be compared with a mass/spring/damper system. Hence a multi-machine system can be compared with a number of masses (representing the generators) suspended from a "network" consisting of elastic strings (representing the transmission lines). In the steady state each of the strings is loaded below its breaking point (steady-state stability limit). If one of the strings is suddenly cut (representing a line tripping), the masses will experience coupled transient motion (swinging of the rotors) with fluctuations in the forces in the strings (line powers).

Such a sudden disturbance may result in the system reaching a new equilibrium state characterized by a new set of string forces (line powers) and string extensions (rotor angles) or, owing to the transient forces involved, one string may break, thus weakening the network and producing a chain reaction of broken strings and eventual total system collapse.

Obviously, this mechanical analogue of the EPS has a number of limitations. First of all, the string stiffnesses should be nonlinear so as to model correctly the nonlinear synchronizing power coefficients. Secondly, the string stiffness should be different in the steady state to the transient state to correctly model the different steady-state and transient state models of the generators.

In multi-machine systems with N_G generating units there are N_G rotor angles δ'_i, which in transient state must be measured with respect to a common reference frame. In order to present the variations of these angles over time, one of the following approaches can be used:

a) All angles δ'_i are measured with respect to a common synchronously rotating axis determined in the pre-fault state. Usually, it is the angle of the slack bus voltage used in the power flow computer program (Section 3.6.2) to solve network equations in the steady-state.
b) One generating unit n is treated as the reference machine and then $(N_G - 1)$ angles δ'_i are recalculated $\delta'_{in} = (\delta'_i - \delta'_n)$ with respect to the angle δ'_n of this reference machine. As a reference machine usually a large generating unit remote from the fault is chosen.
c) All angles δ'_i are recalculated $\delta'_{i,\mathrm{COI}} = (\delta'_i - \delta'_{\mathrm{COI}})$ with respect to the angle δ'_{COI} which is defined as the following weighted average

$$\delta'_{\mathrm{COI}} = \frac{\sum_{k=1}^{N_G} M_i \delta'_i}{\sum_{i=1}^{N_G} M_i} \quad \text{and} \quad \Delta\omega_{\mathrm{COI}} = \frac{\sum_{k=1}^{N_G} M_i \Delta\omega_i}{\sum_{i=1}^{N_G} M_i} \tag{6.15}$$

where M_i is the angular momentum (inertia coefficient), as defined in Section 4.1. A common axis defined by Eq. (6.15) is referred to as the *center of inertia* (COI).

The choice of approach depends on the studied issue related to power system dynamics and stability.

In practice a disturbance may affect the stability of an EPS in one of four ways:

1) The generator (or generators) nearest to the fault may lose synchronism without exhibiting any synchronous swings; other generators affected by the fault undergo a period of synchronous oscillations until they eventually return to synchronous operation.
2) The generator (or generators) nearest to the fault lose synchronism after exhibiting synchronous oscillations.
3) The generator (or generators) nearest to the fault are the first to lose synchronism and are then followed by other generators in the system.
4) The generator (or generators) nearest to the fault exhibit synchronous swings without losing stability, but one, or more, of the other generators remote from the fault loses synchronism with the system.

So far only the first case has been described because it may be represented by the generator-infinite busbar system. In the other three cases instability arises following interaction with other generators located further from the fault. In the second case the generator initially has a chance to retain stability but, as the rotors of the other generators start to oscillate, conditions deteriorate and the generator loses synchronism. In the third case the generator nearest the fault is the first to lose synchronism; this has a major effect on the other generators in the system so that they may also lose synchronism. The fourth case is typical of the situation where some generators, located far from the point of fault, are weakly connected to the system. As the oscillations spread, the operating conditions of the weakly connected generators deteriorate so that they may lose stability. Another example of this fourth instability case is when the network configuration is changed following fault clearance. Although the tripped line may be just one of the lines that connects the nearest generator to the system, it could also be the main connection for a neighboring power station. Example 6.1 illustrates all the above cases (Omahen 1994).

Example 6.1 Figure 6.21 shows a schematic diagram of an EPS that comprises three subsystems connected together at node 2. The capacity of generator 5 is large and can be considered an infinite busbar. Figure 6.22a–d shows the power angles of all the generators, measured with respect to generator 5, for faults located at points a, b, c, and d, respectively. Each of these points corresponds to the four cases of instability. The faults are assumed to be cleared without automatic reclosing.

When line 1–2 is faulted (case a), generator 1 quickly loses synchronism and the rotors of the other generators undergo a period of synchronous oscillation. When the fault is on line 3–13 (case b), then, for a given clearing time,

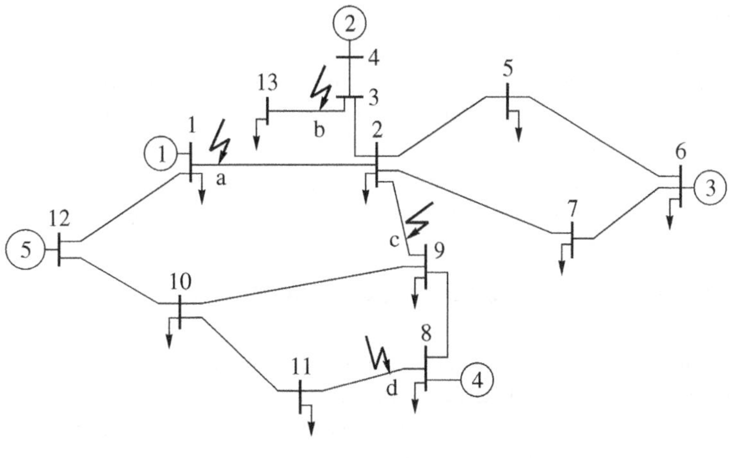

Figure 6.21 Schematic diagram of the test system.

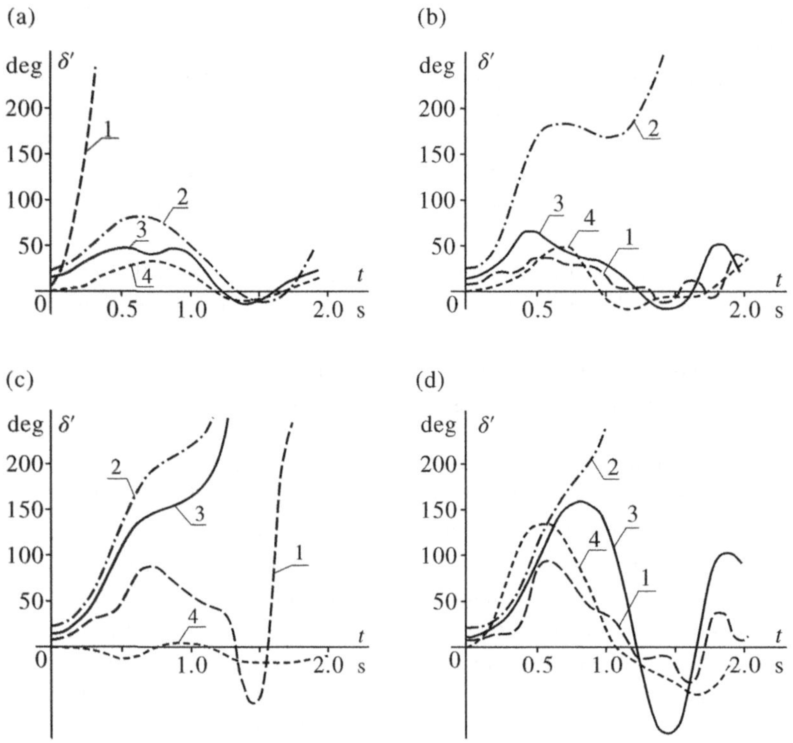

Figure 6.22 Rotor angle variations (measured with respect to generator rotor 5) for different fault locations. Based on Omahen (1994). *Source:* Reproduced by permission of P. Omahen.

generator 2 could remain in synchronism just as if it were operating on the infinite busbars. However, in this particular case the oscillations of the other generator rotors increase the relative difference in the power angles and the situation regarding generator 2 deteriorates and it loses stability on its second swing. When line 9–2 is faulted (case c), the weakly connected generators, 2 and 3, immediately lose stability, while generator 1, which initially remains in synchronism with generator 5, loses synchronism some time later. When line 8–11 is faulted (case d), the power angles of all the generators considerably increase. After the fault is cleared, all the system loads take power from their neighboring generators 1, 3, 4, and 5 but as generator 2 is far from these loads it loses stability.

******

The test system used in Example 6.1 is small in comparison to real power systems. It is, however, a good illustrative example, because in real (large) power systems the generating units may lose synchronism in the same or similar ways. In large power systems (depending on loading conditions and location and type of fault) the frequency of the post-fault oscillations (rotor swings) varies in a wide range of (0.2–2.5) Hz. As shown in Table 5.1 (Section 5.4.6), the fastest are internal oscillations between generating units belonging to the same power plant. The slower are local oscillations of generating units versus the rest of the power system. The slowest are inter-area oscillations between subsystems belonging to the large interconnected system.

6.3 Direct Method for Stability Assessment

Throughout this chapter use is made of the equal area criterion to explain and assess system stability. This method will now be formalized using the *Lyapunov direct method* and an *energy-type Lyapunov function*. In this section the basic concepts on which the Lyapunov direct method is based are described and applied to the simple generator-infinite busbar system. The Lyapunov direct method is also known as *Lyapunov's second method*.

Owing to its potential in assessing power system stability, without the need to solve the system differential equations, the Lyapunov direct method has been the subject of much intensive research. However, the practical application of the direct method for real-time security assessment is still some time off, owing to modeling limitations and the unreliability of computational techniques. When applied to a multi-machine system, especially one operating close to its stability limits, the direct method is vulnerable to numerical problems and may give unreliable results. Interested readers looking for a detailed treatment of this topic are referred to Pai (1981, 1989), Fouad and Vittal (1992), and Pavella and Murthy (1994).

6.3.1 Mathematical Background

Dynamic systems are generally described by a set of nonlinear differential equations of the form

$$\dot{\boldsymbol{x}} = \boldsymbol{F}(\boldsymbol{x}) \tag{6.16}$$

where $\boldsymbol{x}$ is the vector of *state variables*. The Euclidean space determined by $\boldsymbol{x}$ is referred to as the *state space*, whilst the point $\hat{\boldsymbol{x}}$, for which $\boldsymbol{F}(\hat{\boldsymbol{x}}) = \boldsymbol{0}$, is referred to as the *equilibrium point*. For an initial point $\boldsymbol{x}_0 \neq \hat{\boldsymbol{x}}$, $\dot{\boldsymbol{x}}(t = 0) = \boldsymbol{0}$, Eq. (6.16) has a solution $\mathbf{x}(t)$ in the state space and is referred to as the *system trajectory*. The system is said to be *asymptotically stable* if the trajectory returns to the equilibrium point as $t \to \infty$. If the trajectory remains in a vicinity of the equilibrium point as $t \to \infty$ then the system is said to be *stable*.

Lyapunov's stability theory, used in the direct method of stability assessment, is based on a scalar function $V(\boldsymbol{x})$ defined in the state space of the dynamic system. At a given point, the direction of the largest increase in the value of $V(\boldsymbol{x})$ is given by the gradient of the function $\text{grad}V(\boldsymbol{x}) = [\partial V/\partial x_i]$. The points $\tilde{\boldsymbol{x}}$ for which $\text{grad}V(\tilde{\boldsymbol{x}}) = \boldsymbol{0}$ are referred to as the *stationary points*. Each stationary point may correspond to a minimum, a maximum, or a saddle point, as illustrated in Figure 6.23. The stationary point $\tilde{\boldsymbol{x}}$ corresponds to a minimum (Figure 6.23a), if any small disturbance $\Delta\boldsymbol{x} \neq \boldsymbol{0}$ causes an increase in the function, that is $V(\tilde{\boldsymbol{x}} + \Delta\boldsymbol{x}) > V(\tilde{\boldsymbol{x}})$. Similarly, a given stationary point $\tilde{\boldsymbol{x}}$ corresponds to a maximum (Figure 6.23b) if any small disturbance $\Delta\boldsymbol{x} \neq \boldsymbol{0}$ causes a decrease in the function, that is $V(\tilde{\boldsymbol{x}} + \Delta\boldsymbol{x}) < V(\tilde{\boldsymbol{x}})$. The mathematical condition for checking whether a function has a maximum or minimum at a stationary point can be derived by expanding $V(\boldsymbol{x})$ by a Taylor series to give

$$V(\tilde{\boldsymbol{x}} + \Delta\boldsymbol{x}) \cong V(\tilde{\boldsymbol{x}}) + \Delta\boldsymbol{x}^{\text{T}}[\text{grad}V] + \frac{1}{2}\Delta\boldsymbol{x}^{\text{T}}\boldsymbol{H}\Delta\boldsymbol{x} + \ldots \tag{6.17}$$

where $\boldsymbol{H} = [\partial^2 V/\partial x_i \partial x_j]$ is the *Hesse matrix*. At the stationary points, $\text{grad}V(\tilde{\boldsymbol{x}}) = \boldsymbol{0}$ and the increment in $V(\boldsymbol{x})$ caused by the disturbance can be found as

$$\Delta V = V(\tilde{\boldsymbol{x}} + \Delta\boldsymbol{x}) - V(\tilde{\boldsymbol{x}}) \cong \frac{1}{2}\Delta\boldsymbol{x}^{\text{T}}\boldsymbol{H}\Delta\boldsymbol{x} = \sum_{i=1}^{N}\sum_{j=1}^{N} h_{ij}\Delta x_i \Delta x_j \tag{6.18}$$

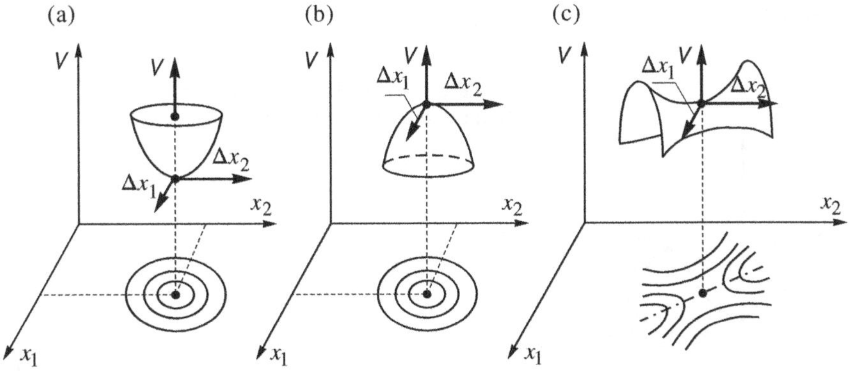

Figure 6.23 A scalar function of two variables with three types of stationary points: (a) minimum; (b) maximum; (c) saddle point.

where h_{ij} is the (i, j) element of matrix $\boldsymbol{H}$. This equation shows that the increment in V is equal to the quadratic form of the state variables constructed using the Hesse matrix. Sylvester's theorem (Bellman 1970) states that such a quadratic form has a minimum at a given stationary point if, and only if, all leading principal minors of the matrix $\boldsymbol{H}$ are positive (Figure 6.23a). The matrix is then referred to as *positive definite*. If all leading principal minors of $\boldsymbol{H}$ multiplied by $(-1)^k$, where k is the column number, are positive then the matrix is *negative definite* and the quadratic form has a maximum at a given stationary point (Figure 6.23b). If some leading principal minors are positive and some negative then the matrix is *nondefinite* and the quadratic form has a saddle point at the stationary point (Figure 6.23c). The dot-dashed line at the bottom of Figure 6.23c shows a ridge line of the saddle going through the saddle point. As the gradient perpendicular to the ridge line is zero for each point on the ridge, function $V(\boldsymbol{x})$ reaches a local maximum in that direction for each point on the ridge. The function reaches a local minimum, in the direction along the ridge line, exactly at the saddle point.

As the scalar function $V(\boldsymbol{x})$ is defined in the state space, each point on the system trajectory $\boldsymbol{x}(t)$ corresponds to a value of $V(\boldsymbol{x}(t))$. The rate of change of $V(\boldsymbol{x})$ along the system trajectory (i.e. the derivative $\mathrm{d}V/\mathrm{d}t$) can be expressed as

$$\dot{V} = \frac{\mathrm{d}V}{\mathrm{d}t} = \frac{\partial V}{\partial x_1}\frac{\mathrm{d}x_1}{\mathrm{d}t} + \frac{\partial V}{\partial x_2}\frac{\mathrm{d}x_2}{\mathrm{d}t} + \ldots + \frac{\partial V}{\partial x_n}\frac{\mathrm{d}x_n}{\mathrm{d}t} = [\mathrm{grad}V(\boldsymbol{x})]^{\mathrm{T}}\,\dot{\boldsymbol{x}} = [\mathrm{grad}V(\boldsymbol{x})]^{\mathrm{T}}\,\boldsymbol{F}(\boldsymbol{x}) \tag{6.19}$$

To introduce the *direct* (or *second*) *Lyapunov method*, assume that a positive definite scalar function $V(\boldsymbol{x})$ has a stationary point (minimum) at the equilibrium point $\tilde{\boldsymbol{x}} = \hat{\boldsymbol{x}}$. Any disturbance $\Delta\boldsymbol{x} \neq \boldsymbol{0}$ will move the system trajectory to an initial point $\boldsymbol{x}_0 \neq \hat{\boldsymbol{x}}$, as illustrated in Figure 6.24. If the system is asymptotically stable, as in Figure 6.24a, then the trajectory $\boldsymbol{x}(t)$ will tend toward the equilibrium point and $V(\boldsymbol{x})$ will decrease along the trajectory until $\boldsymbol{x}(t)$ settles at the minimum point $\tilde{\boldsymbol{x}} = \hat{\boldsymbol{x}}$. If the system is unstable (Figure 6.24b) then the trajectory will move away from the equilibrium point and $V(\boldsymbol{x})$ will increase along the trajectory. The essence of these considerations is summarized in Lyapunov's stability theorem:

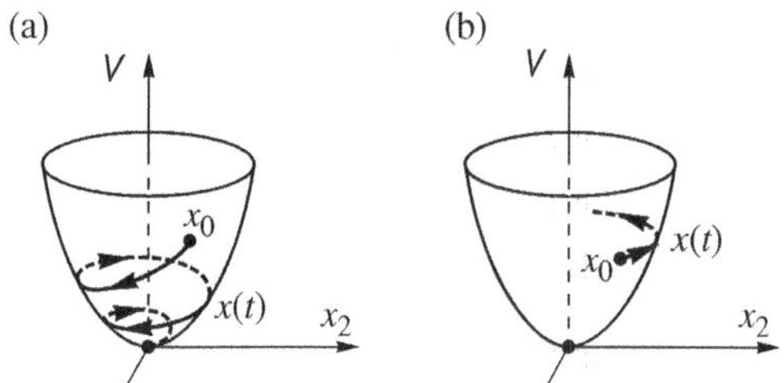

Figure 6.24 Illustration of Lyapunov's theorems on stability: (a) asymptotic stability; (b) instability.

> Let $\hat{\boldsymbol{x}}$ be an equilibrium point of a dynamic system $\dot{\boldsymbol{x}} = \boldsymbol{F}(\boldsymbol{x})$. Point $\hat{\boldsymbol{x}}$ is stable if there is a continuously differentiable positive definite function $V(\boldsymbol{x})$ such that $\dot{V}(\boldsymbol{x}) \leq 0$. Point $\hat{\boldsymbol{x}}$ is asymptotically stable if $\dot{V}(\boldsymbol{x}) < 0$.

One of the main attractions of Lyapunov's theorem is that it can be used to obtain an assessment of system stability without the need to solve the differential Eq. (6.16). The main problem is how to find a suitable positive definite Lyapunov function $V(\boldsymbol{x})$ for which the sign of the derivative $\dot{V}(\boldsymbol{x})$ can be determined without actually determining the system trajectory.

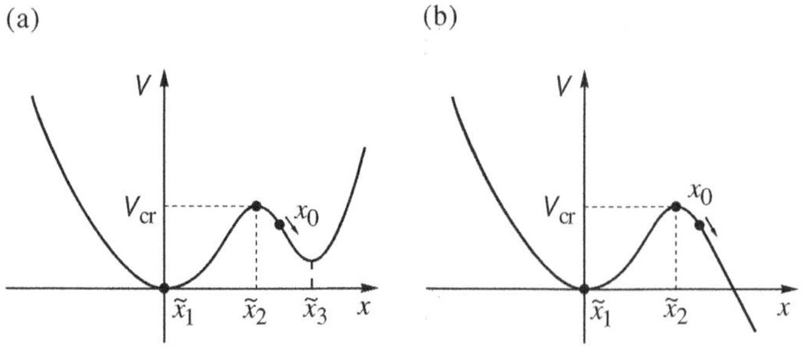

Figure 6.25 Examples of nonlinear functions with more than one stationary point.

Generally, the Lyapunov function $V(\boldsymbol{x})$ will be nonlinear and may have more than one stationary point. Figure 6.25 illustrates this using two functions of one variable: one which has three stationary points (Figure 6.25a) and one which has two stationary points (Figure 6.25b). Assume that in both cases the first of the stationary points corresponds to the equilibrium point $\tilde{\boldsymbol{x}}_1 = \hat{\boldsymbol{x}}$. The Lyapunov function determines the stability only if there is no other stationary point $\tilde{\boldsymbol{x}}_2$ between the initial point $\boldsymbol{x}_0 = \boldsymbol{x}(t = 0)$ and the equilibrium point $\tilde{\boldsymbol{x}}_1 = \hat{\boldsymbol{x}}$. If the initial point $\boldsymbol{x}_0$ is beyond the second stationary point $\tilde{\boldsymbol{x}}_2$ then the condition $\dot{V}(\boldsymbol{x}) < 0$ may mean that the system trajectory will tend toward another stationary point (Figure 6.25a), or run away, as in Figure 6.25b. The value of $V(\boldsymbol{x})$ at the nearest stationary point $\tilde{\boldsymbol{x}}_2 \neq \hat{\boldsymbol{x}}$ is referred to as the *critical value* of the Lyapunov function. These considerations lead to the following theorem:

If there is a positive definite scalar function $V(\boldsymbol{x})$ in the vicinity of the equilibrium point $\hat{\boldsymbol{x}}$ of a dynamic system $\dot{\boldsymbol{x}} = \boldsymbol{F}(\boldsymbol{x})$ and the time derivative of this function is negative ($\dot{V}(\boldsymbol{x}) < 0$) then the system is asymptotically stable at $\hat{\boldsymbol{x}}$ for any initial conditions satisfying

$$V(\boldsymbol{x}_0) < V_{\rm cr} \tag{6.20}$$

where $V_{\rm cr} = V(\tilde{\boldsymbol{x}} \neq \hat{\boldsymbol{x}})$ is the value of Lyapunov function at the nearest stationary point.

It is important to realize that the stability conditions due to Lyapunov's theorems are only sufficient. Failure of the candidate Lyapunov function to satisfy the stability conditions does not mean that the equilibrium point is not stable. Moreover, for any given dynamic system, there are usually many possible Lyapunov functions each of which gives a larger, or smaller, area of initial conditions that satisfies the stability theorems. This means that a particular Lyapunov function usually gives a pessimistic assessment of stability as it covers only a portion of the actual stability area. The function giving the largest area, and at the same time being closest to the actual stability area, is referred to as the *good Lyapunov function*. Usually, good Lyapunov functions have a physical meaning.

6.3.2 Energy-type Lyapunov Function

In Section 5.4, Eq. (5.45) showed that the integral of the acceleration power is proportional to the work done by the acceleration torque. The proportionality constant $\omega_{\rm s}$ was neglected and the integral of power was treated as "work" or "energy". A similar approach is adopted here.

6.3.2.1 Energy Function

Assume that the generator can be represented by the classical model defined by Eqs. (5.1) and (5.26) when the equation describing the generator-infinite busbar system is

$$M\frac{\mathrm{d}\Delta\omega}{\mathrm{d}t} = P_{\rm m} - b\sin\delta' - D\frac{\mathrm{d}\delta'}{\mathrm{d}t} \tag{6.21}$$

where $b = E'V_{\rm s}/x'_{\rm d}$ is the amplitude of the transient power-angle curve $P_{E'}(\delta')$ and $\Delta\omega = \mathrm{d}\delta'/\mathrm{d}t = \mathrm{d}\delta/\mathrm{d}t$ is the speed deviation. This equation has two equilibrium points

$$\left(\hat{\delta}'_{\rm s}\ ;\ \Delta\hat{\omega} = 0\right) \quad \text{and} \quad \left(\hat{\delta}'_{\rm u} = \pi - \hat{\delta}'_{\rm s};\ \Delta\hat{\omega} = 0\right) \tag{6.22}$$

Multiplying Eq. (6.21) by $\Delta\omega$, neglecting the damping term, and moving the right-hand side to the left gives

$$M\Delta\omega\frac{\mathrm{d}\Delta\omega}{\mathrm{d}t} - (P_{\rm m} - b\sin\delta')\frac{\mathrm{d}\delta'}{\mathrm{d}t} = 0 \tag{6.23}$$

As the function on the left-hand side of this equation is equal to zero, its integral must be constant. Integrating the function from the first of the equilibrium points defined in Eq. (6.22) to any point on the system transient trajectory gives

$$V = \int_{0}^{\Delta\omega} (M\Delta\omega)\, \mathrm{d}\Delta\omega - \int_{\hat{\delta}'_s}^{\delta'} (P_m - b\sin\delta')\mathrm{d}\delta' = \text{constant} \tag{6.24}$$

Evaluating the integrals gives the following form of the function

$$V = \frac{1}{2}M\Delta\omega^2 - \left[P_m\left(\delta' - \hat{\delta}'_s\right) + b\left(\cos\delta' - \cos\hat{\delta}'_s\right)\right] = E_k + E_p = E \tag{6.25}$$

where

$$E_k = \frac{1}{2}M\Delta\omega^2;\ E_p = -\left[P_m\left(\delta' - \hat{\delta}'_s\right) + b\left(\cos\delta' - \cos\hat{\delta}'_s\right)\right] \tag{6.26}$$

E_k is a measure of the system kinetic energy, while E_p is a measure of potential energy, both taken with respect to the first equilibrium point $\left(\hat{\delta}'_s;\ \Delta\hat{\omega} = 0\right)$ and, for the remainder of this chapter, is treated as energy. By neglecting damping, Eq. (6.24) shows the sum of potential and kinetic energy $V = E_k + E_p$ to be constant.

It is now necessary to check whether the function V, defined by Eq. (6.25), satisfies the definition of a Lyapunov function, that is whether: (i) it has stationary points at the equilibrium points defined in Eq. (6.22); (ii) it is positive definite in the vicinity of one of the equilibrium points; and (iii) its derivative is not positive ($\dot{V} \le 0$).

The first of the conditions can be checked by calculating the gradient of V. Differentiating Eq. (6.25) gives

$$\text{grad}V = \begin{bmatrix} \dfrac{\partial V}{\partial\Delta\omega} \\ \dfrac{\partial V}{\partial\delta'} \end{bmatrix} = \begin{bmatrix} \dfrac{\partial E_k}{\partial\Delta\omega} \\ \dfrac{\partial E_p}{\partial\delta'} \end{bmatrix} = \begin{bmatrix} M\Delta\omega \\ -(P_m - b\sin\delta') \end{bmatrix} \tag{6.27}$$

This gradient is equal to zero at the stationary points where $\Delta\tilde{\omega} = 0$ and the electrical power is equal to mechanical power given by

$$\tilde{\delta}'_1 = \hat{\delta}'_s;\ \tilde{\delta}'_2 = \pi - \hat{\delta}'_s \tag{6.28}$$

Both these points are the equilibrium points of Eq. (6.21).

The second condition can be checked by determining the Hesse matrix given by

$$\boldsymbol{H} = \begin{bmatrix} \dfrac{\partial^2 V}{\partial\Delta\omega^2} & \dfrac{\partial^2 V}{\partial\Delta\omega\ \partial\delta'} \\ \dfrac{\partial^2 V}{\partial\delta'\ \partial\Delta\omega} & \dfrac{\partial^2 V}{\partial\delta'^2} \end{bmatrix} = \begin{bmatrix} M & 0 \\ 0 & b\cos\delta' \end{bmatrix} \tag{6.29}$$

Sylvester's theorem says that this matrix is positive definite when $M > 0$ (which is always true) and when $b\cos\delta' > 0$, which is true for $|\delta'| < \pi/2$ and holds for the first stationary point $\tilde{\delta}'_1 = \hat{\delta}'_s$. Thus the function V is positive definite at the first equilibrium point $\hat{\delta}'_s$.

The third condition can be checked by determining $\dot{V} = \mathrm{d}V/\mathrm{d}t$ along the trajectory of Eq. (6.21). As V represents the total system energy, its derivative $\dot{V} = \mathrm{d}V/\mathrm{d}t$ corresponds to the rate at which energy is dissipated by the damping. This can be proved by expressing the derivative of V as

$$\dot{V} = \frac{\mathrm{d}V}{\mathrm{d}t} = \frac{\mathrm{d}E_k}{\mathrm{d}t} + \frac{\mathrm{d}E_p}{\mathrm{d}t} \tag{6.30}$$

The derivative of the kinetic energy can be calculated from Eq. (6.25) as

$$\frac{\mathrm{d}E_k}{\mathrm{d}t} = \frac{\partial E_k}{\partial\Delta\omega}\frac{\mathrm{d}\Delta\omega}{\mathrm{d}t} = M\Delta\omega\frac{\mathrm{d}\Delta\omega}{\mathrm{d}t} = \left[M\frac{\mathrm{d}\Delta\omega}{\mathrm{d}t}\right]\Delta\omega \tag{6.31}$$

The factor in the square brackets corresponds to the left-hand side of Eq. (6.21). Replacing it by the right-hand side gives

$$\frac{dE_k}{dt} = \frac{\partial E_k}{\partial \Delta\omega}\frac{d\Delta\omega}{dt} = +[P_m - b\sin\delta']\Delta\omega - D\Delta\omega^2 \tag{6.32}$$

Differentiating the potential energy defined by Eq. (6.25) gives

$$\frac{dE_p}{dt} = \frac{\partial E_p}{\partial \delta'}\frac{d\delta'}{dt} = -[P_m - b\sin\delta']\Delta\omega \tag{6.33}$$

Substituting Eqs. (6.32) and (6.33) into Eq. (6.21) yields

$$\dot{V} = \frac{dV}{dt} = -D\Delta\omega^2 \tag{6.34}$$

which illustrates how the total system energy decays at a rate proportional to the damping coefficient ($D > 0$) and the square of the speed deviation ($\Delta\omega^2$). As $\dot{V} < 0$, the function $V(\delta', \Delta\omega)$ is a Lyapunov function and the first equilibrium point $\hat{\delta}'_s$ is asymptotically stable.

The second equilibrium point $\hat{\delta}'_u = \pi - \hat{\delta}'_s$ is unstable because the matrix $\boldsymbol{H}$ (Eq. [6.29]) calculated at this point is not positive definite.

6.3.3 Transient Stability Area

The critical value of the Lyapunov function V introduced in Eq. (6.20) corresponds to the value of V at the nearest stationary point which, for the system considered here, is equal to the second equilibrium point $\left(\pi - \hat{\delta}'_s;\ \Delta\hat{\omega} = 0\right)$. Substituting these values into Eq. (6.25) gives

$$V_{cr} = 2b\cos\hat{\delta}'_s - P_m\left(\pi - 2\hat{\delta}'_s\right) \tag{6.35}$$

According to Eq. (6.20), the generator-infinite busbar system is stable for all initial conditions $\left(\delta'_0, \Delta\omega_0\right)$ satisfying the condition

$$V\left(\delta'_0, \Delta\omega_0\right) < V_{cr} \tag{6.36}$$

In this context the initial conditions are the values of the transient rotor angle and the rotor speed deviation at the instant of fault clearance when the generator starts to swing freely.

Careful examination of Eq. (6.35) shows that the critical value of the Lyapunov function (total system energy) depends on the stable equilibrium point $\hat{\delta}'_s$ which, in turn, depends on the generator load P_m. At no load, when $\hat{\delta}'_s = 0$, the value of V_{cr} is greatest. Increasing the load on the generator, and hence $\hat{\delta}'_s$, reduces V_{cr} until, for $\hat{\delta}'_s = \pi/2$, its value is $V_{cr} = 0$. Figure 6.26 shows equiscalar contours $V(\delta', \Delta\omega)$ = constant on the phase plane $\Delta\omega$ versus δ'.

Figure 6.26a shows the case $\hat{\delta}'_s = 0$. For small values of $V(\delta', \Delta\omega) < V_{cr}$ the equiscalar contours $V(\delta', \Delta\omega)$ = constant are closed around the equilibrium point $\hat{\delta}'_s = 0$. The critical value $V(\delta', \Delta\omega) = V_{cr}$ corresponds to the closed contour shown in bold crossing the stationary points $\pm\pi$ equal to the unstable equilibrium points. As the equiscalar energy increases, the contours start to open up. The equilibrium point $\hat{\delta}'_s = 0$ is the point of minimum total energy, while the stationary points $\pm\pi$ are the saddle points. Now assume that at the instant of fault clearance (i.e. at the initial point $\delta'_0, \Delta\omega_0$) the system trajectory is inside the contour $V(\delta', \Delta\omega) < V_{cr}$. If damping is neglected then, according to Eq. (6.34), the value of $V(\delta', \Delta\omega)$ remains constant and the trajectory will be a closed curve with

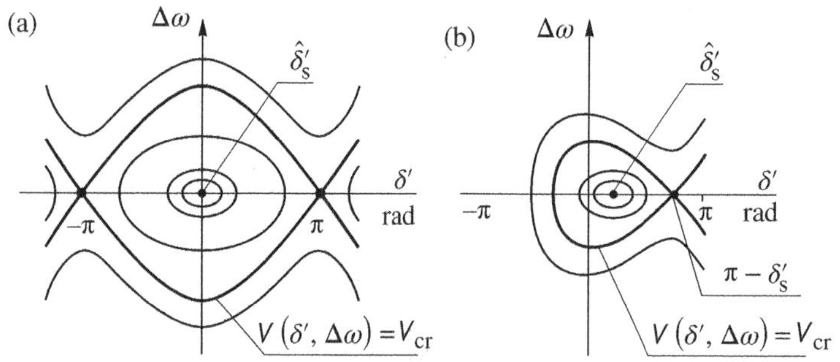

Figure 6.26 Equiscalar contours of the total system energy for the two equilibrium points: (a) $\hat{\delta}'_s = 0$; (b) $\hat{\delta}'_s < \pi/2$.

$V(\delta',\Delta\omega) = V(\delta'_0,\Delta\omega_0) =$ constant. If damping is included, $D > 0$, then $V(\delta', \Delta\omega)$ will decrease with time and the trajectory will tend to spiral in toward the equilibrium point $\hat{\delta}'_s = 0$.

Figure 6.26b shows the case of $\hat{\delta}'_s > 0$ but less than $\pi/2$. In this case the curve $V(\delta', \Delta\omega) = V_{cr} =$ constant corresponds to a contour crossing the stationary point $(\pi - \hat{\delta}'_s)$ equal to the unstable equilibrium point. The area enclosed by this boundary contour is now much smaller than before.

If, at the instant of fault clearance, the transient stability condition (6.36) is satisfied then the calculated value of the total energy $V_0 = V(\delta'_0, \Delta\omega_0)$ can be used to calculate a *transient stability margin*

$$K_{energy} = \frac{V_{cr} - V_0}{V_{cr}} \tag{6.37}$$

which describes the relative difference between the critical value of the total energy (transient stability boundary) and the energy released by the disturbance. K_{energy} determines the distance from the actual contour to the boundary contour of the transient stability area in terms of energy.

6.3.4 Equal Area Criterion

The equal area criterion was introduced in Eq. (6.6) as a means of assessing transient stability and, like the Lyapunov stability condition in Eq. (6.36), is also based on energy considerations. Both approaches can be shown to be equivalent by considering Figure 6.27, which shows the same three-phase fault as that illustrated in Figures 6.1 and 6.2a.

The first term of the energy-based Lyapunov function defined in Eq. (6.25) corresponds to the kinetic energy

$$\frac{1}{2}M\Delta\omega^2 = \text{area 1-2-3-4} = \text{area A} \tag{6.38}$$

The second component, equal to potential energy, can be expressed as

$$-\int_{\hat{\delta}'_s}^{\delta'} (P_m - b\sin\delta')\,d\delta' = -[\text{area 1-7-8-4} - \text{area 1-7-8-5} = \text{area B}] \tag{6.39}$$

The Lyapunov function is equal to the sum of the two components

$$V(\delta', \Delta\omega) = \text{area A} + \text{area B} \tag{6.40}$$

The critical value of the Lyapunov function is equal to the value of the potential energy at the unstable equilibrium point, that is

$$V_{cr} = -\int_{\hat{\delta}'_s}^{\pi-\hat{\delta}'_s} (P_m - b\sin\delta')\mathrm{d}\delta' = \text{area 1-7-9-6-5} - \text{area 1-7-9-6} = \text{area B} + \text{area C} \tag{6.41}$$

Thus, the stability condition $V\left(\delta'_0, \Delta\omega_0\right) < V_{cr}$ can be expressed as

$$\text{area A} + \text{area B} < \text{area B} + \text{area C} \quad \text{or} \quad \text{area A} < \text{area C} \tag{6.42}$$

which is equivalent to the equal area criterion.

The equiscalar contour $V(\delta', \Delta\omega) = V_{cr}$ in the $(\delta', \Delta\omega)$ plane determines the stability area (Figure 6.27c). If, during the short circuit, the trajectory $\boldsymbol{x}(t)$ is contained within the area then, after clearing the fault, the trajectory will remain within the area and the system is stable. When $D > 0$ the trajectory tends toward the equilibrium point and the system is asymptotically stable. As the clearing time increases, the initial point approaches the critical contour and reaches it when the fault time is equal to the critical clearing time. If the clearing time is longer than this then area A > area C, the trajectory leaves the stability area and the system is unstable.

For the case considered here the transient stability margin defined in Eq. (6.37) is equal to

$$K_{\text{energy}} = \frac{V_{cr} - V_0}{V_{cr}} = \frac{\text{area B} + \text{area C} - \text{area A}}{\text{area B} + \text{area C}} = \frac{\text{area 10-6-11}}{\text{area 1-5-6}} = K_{\text{area}} \tag{6.43}$$

which is exactly the same as the transient stability margin defined by the equal area criterion in Eq. (6.7). Consequently, for the generator-infinite busbar system, the equal area criterion is equivalent to the Lyapunov direct method based on an energy-type Lyapunov function.

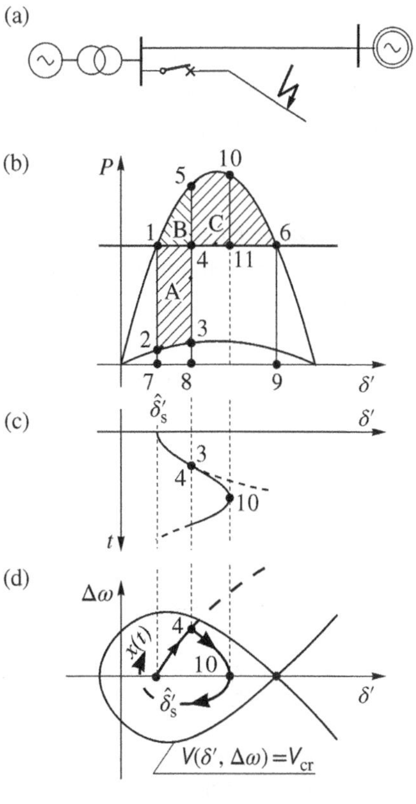

Figure 6.27 Equivalence between the Lyapunov direct method and the equal area criterion: (a) schematic diagram of the system; (b) transient power-angle characteristics with acceleration and deceleration areas; (c) angle variation; (d) stability area and trajectory of rotor motion.

6.3.5 Lyapunov Direct Method for a Multi-machine System

In a simplified stability analysis each generator is modeled by employing the classical model, that is by using the swing equation and constant emf behind the transient reactance. The swing equations for all generators can be written as

$$\frac{\mathrm{d}\delta'_i}{\mathrm{d}t} = \Delta\omega_i \tag{6.44a}$$

$$M_i\frac{\mathrm{d}\Delta\omega_i}{\mathrm{d}t} = P_{mi} - P_i(\boldsymbol{\delta}') - D_i\Delta\omega_i \tag{6.44b}$$

where δ'_i and $\Delta\omega_i$ are the transient power angle and rotor speed deviation respectively, P_{mi} and $P_i(\boldsymbol{\delta}')$ are the mechanical and electrical power, $\boldsymbol{\delta}'$ is the vector of all transient power angles in the system, D_i is the damping coefficient, and M_i is the inertia coefficient (Sections 4.1 and 5.1), $i = 1, \ldots, N$.

Transient reactances of generators are included in the network model. All the network elements (lines and transformers) are modeled by their π equivalents. For the simplified power system transient stability analysis, the transmission network is modeled explicitly, while power injected from/to the distribution network is modeled as a load. Each such load is replaced by a constant nodal shunt admittance. The load nodes are denoted as {L}. Fictitious generator nodes behind generator transient reactances are denoted as {G}.

The resulting network model of a multi-machine system is described by the nodal admittance equation (Section 3.5). That model is then reduced by eliminating the {L} nodes using the elimination method described in Section 18.2. Once the load nodes {L} have been eliminated, the reduced network contains only the fictitious generator nodes {G}. The reduced network is described by the following nodal admittance equation

$$\underline{\boldsymbol{I}}_{\mathrm{G}} = \underline{\boldsymbol{Y}}_{\mathrm{G}}\underline{\boldsymbol{E}}_{\mathrm{G}} \tag{6.45}$$

Expanding this equation for the i-th generator gives

$$\underline{I}_i = \sum_{j=1}^{N} \underline{Y}_{ij}\underline{E}_j \tag{6.46}$$

where $\underline{E}_j = E_j e^{j\delta'_j}$ and $\underline{Y}_{ij} = G_{ij} + jB_{ij}$ are the elements of the reduced admittance matrix. It should be remembered that the off-diagonal elements of the nodal admittance matrix are taken with a minus sign and correspond to admittances of branches linking the nodes. Hence $\underline{y}_{ij} = -\underline{Y}_{ij}$ is the admittance of the equivalent branch linking nodes i and j. Such a branch linking fictitious generator nodes i and j is referred to as the *transfer branch*, while the corresponding admittance $\underline{y}_{ij}$ is referred to as the *transfer admittance*.

6.3.5.1 Conservative Model

Equations expressing nodal injections (here also power generated) corresponding to the currents expressed by Eq. (6.46) are analogous to Eq. (3.181), described in Section 3.5, that is

$$P_i = E_i^2 G_{ii} + \sum_{j \neq i}^{N} E_i E_j G_{ij} \cos\left(\delta'_i - \delta'_j\right) + \sum_{j \neq i}^{N} E_i E_j B_{ij} \sin\left(\delta'_i - \delta'_j\right) \tag{6.47}$$

or

$$P_i = P_{0i} + \sum_{j=1}^{N} b_{ij} \sin \delta'_{ij} \tag{6.48}$$

where $b_{ij} = E_i E_j B_{ij}$ is the magnitude of the power-angle characteristic for the transfer equivalent branch and

$$P_{0i} = E_i^2 G_{ii} + \sum_{j \neq i}^{N} E_i E_j G_{ij} \cos\left(\delta'_i - \delta'_j\right) \tag{6.49}$$

Generally, P_{0i} depends on power angles δ'_i and is not constant during the transient period when the rotors are swinging. Stability analysis of the multi-machine system by the Lyapunov direct method is straightforward, assuming that $P_{0i} \cong P_{0i}\left(\hat{\boldsymbol{\delta}}'\right) = \text{constant}$, that is the power is constant and equal to its value at the stable equilibrium point, where

$$P_{0i}\left(\hat{\boldsymbol{\delta}}'\right) = E_i^2 G_{ii} + \sum_{j \neq i}^{N} E_i E_j G_{ij} \cos\left(\hat{\delta}'_i - \hat{\delta}'_j\right) \tag{6.50}$$

That assumption in practice means that transmission losses in transfer branches are assumed to be constant and added to the equivalent loads connected at the fictitious generator nodes.

The simplified power system model resulting from Eqs. (6.44) and (6.48) can be summarized in the following set of state-space equations

$$\frac{\mathrm{d}\delta'_i}{\mathrm{d}t} = \Delta\omega_i \tag{6.51a}$$

$$M_i \frac{\mathrm{d}\Delta\omega_i}{\mathrm{d}t} = (P_{\mathrm{m}i} - P_{0i}) - \sum_{j=1}^{n} b_{ij} \sin\delta'_{ij} - D_i \Delta\omega_i \tag{6.51b}$$

The third component on the right-hand side of Eq. (6.51b) dissipates energy. No energy is dissipated in the transmission network because power losses on transfer conductance have been added to the equivalent loads as constant values. Such a model is referred to as the *conservative model.*

6.3.5.2 Energy Function Proposed by Gless

Neglecting rotor damping, Eq. (6.51b) can be written as

$$M_i \frac{\mathrm{d}\Delta\omega_i}{\mathrm{d}t} - (P_{\mathrm{m}i} - P_{0i}) + \sum_{j=1}^{N} b_{ij} \sin\delta'_{ij} = 0 \tag{6.52}$$

As the righ- hand side of this equation is equal to zero, integrating gives a constant value $V(\delta', \Delta\omega) = E_{\mathrm{k}} + E_{\mathrm{p}}$ equal to the system energy, where

$$E_{\mathrm{k}} = \sum_{i=1}^{N} \int_{0}^{\omega_i} M_i \Delta\omega_i \mathrm{d}\omega_i = \frac{1}{2} \sum_{i=1}^{N} M_i \Delta\omega_i^2 \tag{6.53}$$

$$E_{\mathrm{p}} = -\sum_{i=1}^{N} \int_{\hat{\delta}'_i}^{\delta'_i} (P_{\mathrm{m}i} - P_{0i})\, \mathrm{d}\delta'_i + \sum_{i=1}^{N} \int_{\hat{\delta}'_i}^{\delta'_i} \left(\sum_{j=1}^{N} b_{ij} \sin\delta'_{ij} \right) \mathrm{d}\delta'_i \tag{6.54}$$

where $\hat{\delta}'_i$ is the power angle at the post-fault stable equilibrium point, while E_{k} and E_{p} are the kinetic and potential energy of the power system conservative model. As damping has been neglected, the model does not dissipate energy and the total energy, equal to the sum of the kinetic and potential energy, is constant (the conservative model).

There is a double summation in Eq. (6.54) that corresponds to a summation of elements of the square matrix

$$
\begin{array}{cc}
 & \begin{array}{cccc} & \quad i \quad & \quad j \quad & \end{array} \\
\begin{array}{c} \\ \\ i \\ j \\ \\ \end{array} &
\left[\begin{array}{c|c|c|c}
\ddots & \vdots & \vdots & \\ \hline
\cdots & 0 & b_{ij}\sin\delta'_{ij} & \cdots \\ \hline
\cdots & b_{ji}\sin\delta'_{ij} & 0 & \cdots \\ \hline
 & \vdots & \vdots & \ddots
\end{array}\right]
\end{array}
\tag{6.55}
$$

The diagonal elements of the matrix are equal to zero because $\sin\delta'_{ii} = \sin\left(\delta'_i - \delta'_i\right) = \sin 0 = 0$. The lower off-diagonal elements have the same value as the upper elements but with an inverted sign because $\sin\delta'_{ij} = -\sin\delta'_{ji}$. Hence

$$\int \sin \delta'_{ij} \mathrm{d}\delta'_i + \int \sin \delta'_{ij} \mathrm{d}\delta'_j = \int \sin \delta'_{ij} \mathrm{d}\delta'_i - \int \sin \delta'_{ij} \mathrm{d}\delta'_j = \int \sin \delta'_{ij} \mathrm{d}\delta'_{ij} = -\cos \delta'_{ij} \tag{6.56}$$

Thus instead of integrating both the upper and lower off-diagonal elements of the matrix with respect to $\mathrm{d}\delta'_i$, it is enough to integrate only the upper elements with respect to $\mathrm{d}\delta'_{ij}$. Consequently, the second component on the right-hand side of Eq. (6.54) can be written as

$$\sum_{i=1}^{N} \int_{\hat{\delta}'_i}^{\delta'_i} \left(\sum_{j=1}^{n} b_{ij} \sin \delta'_{ij} \right) \mathrm{d}\delta'_i = -\sum_{i=1}^{N-1} \sum_{j=i+1}^{N} b_{ij} \left[\cos \delta'_{ij} - \cos \hat{\delta}'_{ij} \right] \tag{6.57}$$

It is now worth considering indices in the sums on the right-hand side of Eq. (6.57). As only the upper off-diagonal elements are summed, the summation is for $j > i$. Moreover, the last row may be neglected since it contains no upper off-diagonal elements. Hence the summation is for $i \leq (N-1)$.

Finally, Eq. (6.54) gives

$$E_{\mathrm{p}} = -\sum_{i=1}^{N} (P_{\mathrm{m}i} - P_{0i}) \left(\delta'_i - \hat{\delta}'_i \right) - \sum_{i=1}^{N-1} \sum_{j=i+1}^{N} b_{ij} \left[\cos \delta'_{ij} - \cos \hat{\delta}'_{ij} \right] \tag{6.58}$$

The total system energy is equal to the sum of the kinetic and potential energy

$$V(\boldsymbol{\delta}', \Delta\boldsymbol{\omega}) = E_{\mathrm{k}} + E_{\mathrm{p}} \tag{6.59}$$

The necessary condition for this function to be a Lyapunov function is that it is positive definite and with a minimum (Figure 6.24) at the stationary point $\hat{\boldsymbol{\delta}}'$; $\Delta\hat{\boldsymbol{\omega}} = \mathbf{0}$. That point is also the equilibrium point of differential Eqs. (6.51a) and (6.44b). Following considerations similar to those related to Eqs. (6.17) and (6.18), the condition is satisfied if the Hesse matrix of function (6.59) is positive definite. Hence it is necessary to investigate the Hesse matrices of the kinetic and potential energy.

The Hesse matrix of the kinetic energy $\boldsymbol{H}_{\mathrm{k}} = \left[\partial E_{\mathrm{k}}^2 / \partial \Delta\omega_i \partial \Delta\omega_j\right] = \mathrm{diag}\,[M_i]$ is a positive definite diagonal matrix. Therefore, function V_{k} given by Eq. (6.53) is positive definite.

The Hesse matrix of the potential energy $\boldsymbol{H}_{\mathrm{p}} = \left[\partial E_{\mathrm{p}}^2 / \partial \delta'_{iN} \partial \delta'_{jN}\right]$ is a square matrix that consists of the generator's self- and mutual synchronizing powers

$$H_{ii} = \sum_{j=1}^{N-1} b_{ij} \cos \delta'_{ij}; \quad H_{ij} = -b_{ij} \cos \delta'_{ij} \tag{6.60}$$

It is in Section 12.2.2 that this matrix is positive definite for any stable equilibrium point such that for all generator pairs $\left|\delta'_{ij}\right| < \pi/2$. Therefore, in the vicinity of the post-fault equilibrium point, the potential energy V_{p} given by Eq. (6.58) is positive definite.

Since the total system energy $V(\boldsymbol{\delta}', \Delta\boldsymbol{\omega})$ is the sum of two positive definite functions, it is also positive definite. It can be treated as a Lyapunov function for the system model defined in Eqs. (6.51a) and (6.44b) providing the time derivative $\mathrm{d}V/\mathrm{d}t$ along the system trajectory is negative. The time derivative $\mathrm{d}V/\mathrm{d}t$ of the function in Eq. (6.59) along any system trajectory can be expressed as

$$\dot{V} = \frac{\mathrm{d}V}{\mathrm{d}t} = \frac{\mathrm{d}E_{\mathrm{k}}}{\mathrm{d}t} + \frac{\mathrm{d}E_{\mathrm{p}}}{\mathrm{d}t} \tag{6.61}$$

where

$$\frac{\mathrm{d}E_{\mathrm{k}}}{\mathrm{d}t} = \sum_{i=1}^{N} \frac{\partial E_{\mathrm{k}}}{\partial \Delta\omega_i} \frac{\mathrm{d}\Delta\omega_i}{\mathrm{d}t} \tag{6.62}$$

$$\frac{\mathrm{d}E_\mathrm{p}}{\mathrm{d}t} = \sum_{i=1}^{N} \frac{\partial E_\mathrm{p}}{\partial \delta'_i} \frac{\mathrm{d}\delta'_i}{\mathrm{d}t} = \sum_{i=1}^{N} \frac{\partial E_\mathrm{p}}{\partial \delta'_i} \Delta\omega_i \tag{6.63}$$

Differentiating Eqs. (6.53) and (6.54) gives

$$\frac{\partial E_\mathrm{k}}{\partial \Delta\omega_i} = M_i \Delta\omega_i \tag{6.64}$$

$$\frac{\partial E_\mathrm{p}}{\partial \delta'_i} = -(P_{\mathrm{m}i} - P_{0i}) - \sum_{j \neq i}^{N} b_{ij} \sin \delta'_{ij} \tag{6.65}$$

Substituting Eqs. (6.64) into (6.62) gives

$$\frac{\mathrm{d}E_\mathrm{k}}{\mathrm{d}t} = \sum_{i=1}^{N} \frac{\partial E_\mathrm{k}}{\partial \Delta\omega_i} \frac{\mathrm{d}\Delta\omega_i}{\mathrm{d}t} = \sum_{i=1}^{N} \Delta\omega_i M_i \frac{\mathrm{d}\Delta\omega_i}{\mathrm{d}t} \tag{6.66}$$

Finally, substituting Eqs. (6.51b) into (6.66) gives the time derivative of V_k as

$$\frac{\mathrm{d}E_\mathrm{k}}{\mathrm{d}t} = \sum_{i=1}^{N} \Delta\omega_i (P_{\mathrm{m}i} - P_{0i}) - \sum_{i=1}^{N} \Delta\omega_i \sum_{j=1}^{N} b_{ij} \sin \delta'_{ij} - \sum_{i=1}^{N} D_i \Delta\omega_i^2 \tag{6.67}$$

Similarly, the time derivative of the potential energy is obtained by substituting the relevant components in Eq. (6.63) by (6.65) to give

$$\frac{\mathrm{d}E_\mathrm{p}}{\mathrm{d}t} = -\sum_{i=1}^{N} \Delta\omega_i (P_{\mathrm{m}i} - P_{0i}) + \sum_{i=1}^{N} \Delta\omega_i \sum_{j \neq i}^{N} b_{ij} \sin \delta'_{ij} \tag{6.68}$$

The derivative of the potential energy expressed by this equation and the first two components of Eq. (6.67) are the same but with opposite sign. This indicates that there is a continuous exchange of energy between the potential and kinetic energy terms. Adding the two equations as required by Eq. (6.61) gives

$$\dot{V} = \frac{\mathrm{d}V}{\mathrm{d}t} = \frac{\mathrm{d}E_\mathrm{k}}{\mathrm{d}t} + \frac{\mathrm{d}E_\mathrm{p}}{\mathrm{d}t} = -\sum_{i=1}^{N} D_i \Delta\omega_i^2 \tag{6.69}$$

That concludes the proof that function (6.59) is a Lyapunov function if the components of this functions (kinetic and potential energy) are given by (6.53) and (6.58), respectively. Such function may be used to investigate the transient stability of an EPS. The function was originally proposed by Gless (1966).

Similar to the generator-infinite busbar system discussed in Section 6.3.3, the stability condition for a given post-fault state in a multi-machine power system is given by

$$V\left(\delta'_0, \Delta\omega_0\right) < V_\mathrm{cr} \tag{6.70}$$

where V_cr is the critical value of a Lyapunov function and $\boldsymbol{\delta}'_0$ and $\Delta\boldsymbol{\omega}_0$ are the initial conditions at the post-fault state.

6.3.5.3 Energy Function Proposed by Lüders

The kinetic energy expressed by Eq. (6.53) is calculated using the speed deviations $\Delta\omega_i$ taken with respect to the synchronously rotating reference axis. It is be explained in Chapter 9 that the fault clearance may cause a power imbalance leading to a change in frequency. In such a situation the rotors of all generating units have a common speed deviation (Figure 9.16) equal to the speed deviation of the COI defined by Eq. (6.15)

$$\Delta\omega_c = \frac{\sum_{i=1}^{n} M_i \Delta\omega_i}{\sum_{i=1}^{n} M_i} = \frac{1}{M_c}\sum_{i=1}^{n} M_i \Delta\omega_i \tag{6.71}$$

where M_i is inertia coefficient occurring in Eq. (6.51b) and

$$M_c = \sum_{i=1}^{n} M_i \tag{6.72}$$

is inertia coefficient (angular momentum) of all generating units operating in the system. Hence, the kinetic energy resulting from the common motion of all generator rotors may be expressed by the following equation

$$E_c = \frac{1}{2} M_c \Delta\omega_c^2 = \frac{1}{2M_c}\left[\sum_{i=1}^{n} M_i \Delta\omega_i\right]^2 \tag{6.73}$$

This energy does not contribute to the loss of synchronism in the system, because the loss of synchronism is determined by the relative angles and relative speed deviations $\Delta\omega_{ij}$ calculated for all possible pairs of generators. This means that the value of Lyapunov function (6.59) contains energy, which is irrelevant to power system stability. Lüders (1971) noticed this fact and proposed to use Lyapunov function based on the kinetic energy resulting only from the relative speed deviations $\Delta\omega_{ij}$. This can be done by subtracting energy E_c determined by Eq. (6.73) from energy E_k determined by Eq. (6.53)

$$E_{kc} = E_k - E_c = \frac{1}{2}\sum_{i=1}^{n} M_i \Delta\omega_i^2 - \frac{1}{2M_c}\left[\sum_{i=1}^{n} M_i \Delta\omega_i\right]^2 \tag{6.74}$$

This equation can be rearranged in the following way

$$E_{kc} = \frac{1}{2}\frac{M_c \sum_{i=1}^{n} M_i \Delta\,\omega_i^2 - \left[\sum_{i=1}^{n} M_i \Delta\omega_i\right]^2}{M_c} = \frac{1}{2}\frac{A - B}{M_c} \tag{6.75}$$

where

$$A = M_c \sum_{i=1}^{n} M_i \Delta\omega_i^2 = \left[\sum_{i=1}^{n} M_i\right]\left[\sum_{i=1}^{n} M_i \Delta\omega_i^2\right] \tag{6.76}$$

$$B = \left[\sum_{i=1}^{n} M_i \Delta\omega_i\right]^2 \tag{6.77}$$

Component (6.76) may be expressed as the multiplication of two sums

$$A = [M_1 + M_2 + \dots + M_N]\left[M_1\Delta\omega_1^2 + M_2\Delta\omega_2^2 + \dots + M_N\Delta\omega_N^2\right]$$

This is the multiplication of two sums and can be treated as the sum of all elements of the following table

$$A = \sum \left\{ \begin{array}{c:c:c:c:c} M_1M_1\Delta\omega_1^2 & M_1M_2\Delta\omega_2^2 & M_1M_3\Delta\omega_3^2 & \dots & M_1M_n\Delta\omega_n^2 \\ \hdashline M_2M_1\Delta\omega_1^2 & M_2M_2\Delta\omega_2^2 & M_2M_3\Delta\omega_3^2 & \cdots & M_2M_n\Delta\omega_n^2 \\ \hdashline \vdots & \vdots & \vdots & \ddots & \vdots \\ \hdashline M_nM_1\Delta\omega_1^2 & M_nM_2\Delta\omega_2^2 & M_nM_3\Delta\omega_3^2 & \cdots & M_nM_n\Delta\omega_n^2 \end{array} \right\} \tag{6.78}$$

Components of the second sum multiplied by M_1 are in the first row of this table. The second row presents components of the second sum multiplied by M_2 and so on. Summation of all elements in table (6.78) can be divided into three stages: (i) summation of all diagonal elements, (ii) summation column by column of all upper elements, (iii) summation column by column of all lower elements

$$A = \sum_{i=1}^{n} M_i \Delta\omega_i^2 + \sum_{i=1}^{n-1} \sum_{j=i+1}^{n} M_i M_j \Delta\omega_i^2 + \sum_{i=1}^{n-1} \sum_{j=i+1}^{n} M_i M_j \Delta\omega_j^2 \tag{6.79}$$

In the same way, the above component B given by Eq. (6.77) can be expressed as the multiplication of two identical sums

$$B = [M_1\Delta\omega_1 + M_2\Delta\omega_2 + \ldots + M_n\Delta\omega_n]\,[M_1\Delta\omega_1 + M_2\Delta\omega_2 + \ldots + M_n\Delta\omega_n]$$

and hence

$$B = \sum \left\{ \begin{array}{c|c|c|c|c} M_1M_1\Delta\omega_1^2 & M_1M_2\Delta\omega_1\Delta\omega_2 & M_1M_3\Delta\omega_1\Delta\omega_3 & \ldots & M_1M_n\Delta\omega_1\Delta\omega_n \\ \hline M_2M_1\Delta\omega_2\Delta\omega_1 & M_2M_2\Delta\omega_2^2 & M_2M_3\Delta\omega_2\Delta\omega_3 & \cdots & M_2M_n\Delta\omega_2\Delta\omega_n \\ \hline \vdots & \vdots & \vdots & \ddots & \vdots \\ \hline M_nM_1\Delta\omega_n\Delta\omega_1 & M_nM_2\Delta\omega_n\Delta\omega_2 & M_nM_3\Delta\omega_n\Delta\omega_3 & \cdots & M_nM_n\Delta\omega_n^2 \end{array} \right\} \tag{6.80}$$

Table (6.78) is symmetrical and summation of its elements can be divided into two stages: (i) summation of all diagonal elements, (ii) summation column by column of all lower elements and multiplication by 2

$$B = \sum_{i=1}^{n} M_i \Delta\omega_i^2 + 2\sum_{i=1}^{n-1} \sum_{j=i+1}^{n} M_i M_j \Delta\omega_i \Delta\omega_j \tag{6.81}$$

Taking into account Eqs. (6.79) and (6.81) it is obtained

$$A - B = \sum_{i=1}^{n-1} \sum_{j=i+1}^{n} M_i M_j \Delta\omega_i^2 - 2\sum_{i=1}^{n-1} \sum_{j=i+1}^{n} M_i M_j \Delta\omega_i \Delta\omega_j + \sum_{i=1}^{n-1} \sum_{j=i+1}^{n} M_i M_j \Delta\omega_j^2$$

$$A - B = \sum_{i=1}^{n-1} \sum_{j=i+1}^{n} M_i M_j \left[\Delta\omega_i^2 - 2\Delta\omega_i\Delta\omega_j + \Delta\omega_j^2\right] = \sum_{i=1}^{n-1} \sum_{j=i+1}^{n} M_i M_j \left[\Delta\omega_i - \Delta\omega_j\right]^2$$

and finally

$$A - B = \sum_{i=1}^{n-1} \sum_{j=i+1}^{n} M_i M_j \Delta\omega_{ij}^2 \tag{6.82}$$

Substituting (6.82) and (6.72) into (6.75) gives

$$E_{\mathrm{kc}} = \frac{1}{2} \frac{\sum\limits_{i=1}^{n-1} \sum\limits_{j=i+1}^{n} M_i M_j \Delta\omega_{ij}^2}{\sum\limits_{i=1}^{n} M_i} \tag{6.83}$$

This equation accounts for the kinetic energy of the relative motion of rotors of all generating units in given power system. The sum of (6.83) and (6.58) determines the following function

$$V = \frac{1}{2} \frac{\sum\limits_{i=1}^{n-1} \sum\limits_{j=i+1}^{n} M_i M_j \Delta\omega_{ij}^2}{\sum\limits_{i=1}^{n} M_i} + \sum_{i=1}^{n} (P_{\mathrm{m}i} - P_{0i})\left(\delta'_i - \hat{\delta}'_i\right) - \sum_{i=1}^{n-1} \sum_{j=i+1}^{n} b_{ij}\left[\cos\delta'_{ij} - \cos\hat{\delta}'_{ij}\right] \tag{6.84}$$

which can be used as the Lyapunov function.

It is worth to emphasize here that function (6.84) proposed by Lüders gives a better assessment of stability area than function (6.59) proposed by Gless (1966). This results from inequality (6.70). Critical value V_{cr} on the right side of this inequality depends on the potential energy in an unstable equilibrium point and is the same for both functions (Gless and Lüders). For given values of $(\delta'_0, \Delta\omega_0)$, the value of the Lüders function is smaller then the value of the Gless function and inequality (6.70) is satisfied for larger values of $(\delta'_0, \Delta\omega_0)$. Therefore, the Lüders function determines an area closer to the real stability area than the Gless function. Nevertheless, both functions provide a rather pessimistic assessment of transient stability of the EPS.

6.3.5.4 Critical Value

The critical value of a given Lyapunov function for the generator-infinite busbar system is calculated at the unstable equilibrium point (6.35). The main difficulty with the multi-machine system is that the number of unstable equilibrium points is very large. If a system has N generators then there is (N-1) relative power angles and there may be even 2^{N-1} equilibrium points, that is one stable equilibrium point and $(2^{N-1} - 1)$ unstable equilibrium points. For example, if $N = 11$ then $2^{10} = 1024$, that is over a thousand equilibrium points. If $N = 21$ then $2^{20} = 1\ 048\ 576$, that is over a million equilibrium points.

Although 2^{N-1} is the maximum number of equilibrium points that may exist, the actual number of equilibrium points depends on the system load. Figure 5.5 shows that, for the generator-infinite busbar system (i.e. for $N = 2$), there are $2^{2-1} = 2^1 = 2$ equilibrium points when the load is small. As the load increases, the equilibrium points tend toward each other and, for the pull-out power (steady-state stability limit), they become a single unstable equilibrium point. As the load increases further, there are no equilibrium points. Similarly, in a multi-machine power system, there may be a maximum number of 2^{N-1} equilibrium points when the load is small. As the load increases, the pairs of neighboring stable and unstable equilibrium points tend toward each other until they become single points at the steady-state stability limit. When the load increases even further, the system loses stability and the equilibrium points vanish.

The choice of a proper unstable equilibrium point determining V_{cr} is difficult. Choosing the lowest unstable equilibrium point in which $V(\delta')$ is smallest requires searching a large number of points and usually leads to a very pessimistic transient stability assessment. A more realistic transient stability assessment is obtained if V_{cr} is calculated assuming an unstable equilibrium point lying on the trajectory $\delta'(t)$ enforced by a given disturbance. There are many methods of selecting such a point, each of them having their own advantages and disadvantages. Figure 6.28 illustrates the *potential boundary surface method* proposed by Athay et al. (1979).

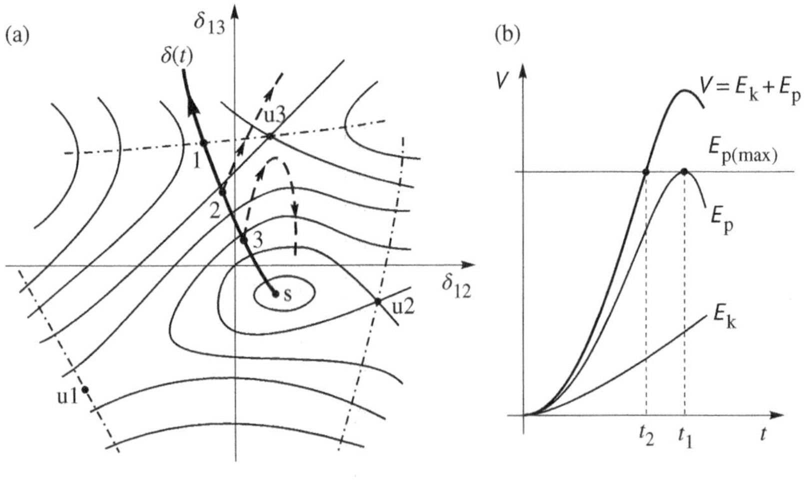

Figure 6.28 Illustration of potential boundary surface method.

The solid lines in Figure 6.28a correspond to equiscalar lines of the potential energy for a three-machine system under certain loading conditions. The stable equilibrium point is denoted by the letter "s". There are three unstable equilibrium (saddle) points: u1, u2, and u3. Dot-dashed lines show the ridge lines of the saddles (see also Figure 6.23c). The ridge line of each saddle point corresponds to a different mode of losing the stability. The bold line shows the trajectory $\boldsymbol{\delta}'(t)$ following a sustained fault. The ridge line of saddle u3 is exceeded and that corresponds to generator 3 losing synchronism with respect to generators 1 and 2. If the trajectory exceeded the ridge line of saddle u2, it would mean that generator 2 had lost synchronism with respect to generators 1 and 3. If the trajectory exceeded the ridge line of saddle u1, it would mean that generator 1 had lost synchronism with respect to generators 2 and 3.

The potential energy increases on the way from the stable equilibrium point toward the ridge line of the saddle, owing to increased rotor angles. After the ridge line has been exceeded, the potential energy decreases. Figure 6.28b shows changes in the potential, kinetic, and total energy corresponding to the trajectory $\boldsymbol{\delta}'(t)$ exceeding the ridge line of saddle u3 shown in Figure 6.28a. As the trajectory $\boldsymbol{\delta}'(t)$ corresponds to a sustained short circuit, the kinetic energy continues to increase during the fault because of rotor acceleration. The potential energy reaches its maximum at point 1 which corresponds to $E_{\mathrm{p(max)}}$ in Figure 6.28b. The ridge line is usually flat in the vicinity of a saddle point. Hence it may be assumed that $E_{\mathrm{p}}(\mathrm{u3}) \cong E_{\mathrm{p(max)}}$, that is the potential energy at the saddle point u3 is equal to the maximum potential energy along the trajectory $\boldsymbol{\delta}'(t)$. Hence u3 is the sought unstable equilibrium point and it may be assumed that $V_{\mathrm{cr}} = E_{\mathrm{p}}(\mathrm{u3}) \cong E_{\mathrm{p(max)}}$. The system is stable for $V\left(\boldsymbol{\delta}'_0, \Delta\boldsymbol{\omega}_0\right) < E_{\mathrm{p(max)}}$ which corresponds to the clearing time $t < t_2$ in Figure 6.28b. For clearing time $t = t_2$ the total energy (kinetic and potential) is slightly higher than the potential energy in u3 and the trajectory exceeds the ridge line of the saddle point u3. This is shown in Figure 6.28a by the dashed line from clearing point 2. For $t < t_2$ the total energy is too small to exceed the ridge line of the saddle point u3. The trajectory then changes direction and returns to the stable equilibrium point. This is shown in Figure 6.28a by the dashed line from clearing point 3. The conclusion is that the critical clearing time is slightly smaller than t_2.

The potential boundary surface method is used to find the critical value of a Lyapunov function as the maximum value of potential energy along a trajectory following a sustained short circuit. The disadvantage of assuming a sustained fault is that the transient stability assessment may be misleading when the trajectory dramatically changes its direction after the fault is cleared. Consider, for example, a sustained fault in line L1 near generator G1 in Figure 6.29. The trajectory $\boldsymbol{\delta}'(t)$ tends toward the saddle point that corresponds to the loss of synchronism by generator G1. The rotor of G2 does not accelerate rapidly during the fault because the fault is remote and the generator is loaded by neighboring loads. Generator G1 is strongly connected with the system so that the potential energy in the corresponding saddle point is high. Therefore, the value of V_{cr} determined using the trajectory during the sustained fault is high and results in a large critical clearing time. Unfortunately, this is not a correct transient stability assessment in the considered case. Right after the faulted line L1 is tripped, the swings of generator G1 are quickly damped. However, generator G2 loses synchronism because the tripped line L1 played a crucial role in the system. Once the line is tripped, G2 is connected to the system via a long transmission link L2, L3, L4, and L5 and starts to lose synchronism. Trajectory $\boldsymbol{\delta}'(t)$ dramatically changes its direction and tends toward the saddle point corresponding to the loss of synchronism of generator G2. The potential energy at that point is small and the trajectory $\boldsymbol{\delta}'(t)$ will easily exceed the ridge line. This would correspond to the situation in Figure 6.28a when the trajectory starting at point 3 moves toward the ridge of saddle point u2 where the potential energy is small.

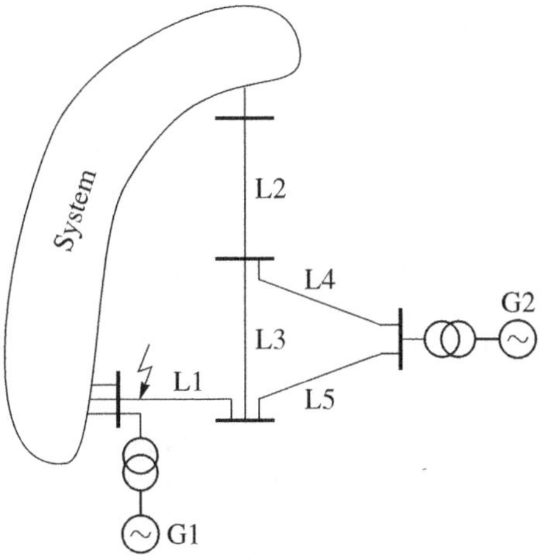

Figure 6.29 Example of an EPS.

The critical value of a Lyapunov function that takes into account a dramatic change of trajectory direction can be determined using methods based on *coherency recognition*, which is discussed in more detail in Chapter 18. A group of generators that have a similar dynamic response and rotor swings after a given disturbance are referred to as a *coherent group*. Finding groups of generators that are approximately coherent limits the number of unstable equilibrium points for which the ridge lines could be exceeded following a given disturbance. This could be explained using the example shown in Figure 6.29. Coherency recognition would show that generators G1 and G2 are not coherent with any other generator for the disturbance considered. Hence three possibilities of losing synchronism can be investigated: (i) generator G1 loses synchronism while G2 remains in synchronism with the rest of the system; (ii) generator G2 loses synchronism while G1 remains in synchronism with the rest of the system; (iii) generators G1 and G2 lose synchronism together. These three ways of losing synchronism correspond to three saddle points and their three ridge lines can be exceeded by the trajectory $\boldsymbol{\delta}'(t)$. The value of V_{cr} is determined from those three saddle points by choosing a point at which E_p is smallest. Such a choice, proposed by Machowski et al. (1986a), gives a prudent transient stability assessment.

Application of the Lyapunov direct method for transient stability assessment is discussed in a number of books, for example Pai (1981, 1989) and Pavella and Murthy (1994).

6.4 Synchronization

Section 4.4 describes the electromagnetic dynamics occurring when a generator is synchronized to the system. This section expands this discussion to include the effect of the electromechanical dynamics that accompany synchronization and in determining those conditions necessary for resynchronization.

Consider the circuit shown in Figure 4.33a and b. When the synchronizing switch is closed the resulting electromagnetic torque will attempt to pull the rotor into synchronization with the EPS by either slowing down or accelerating the rotor until eventually the generator reaches its final equilibrium position defined by the steady-state rotor angle $\hat{\delta}_s = 0$ and the rotor speed $\omega = \omega_s$. To analyze these rotor dynamics, only the transient aperiodic component of the torque will be considered because the subtransient interval is normally too short to have any significant effect on the rotor swings. However, the subtransient torques are vital when determining shaft torque rating and fatigue strength as described in Section 4.2.7.

The generator is represented by the classical model, Eqs. (5.1) and (5.26). As the generator is on no load prior to synchronization, both $E' = E_f$ and $\delta = \delta'$ hold. Now consider the case illustrated in Figure 6.30 when all the synchronization conditions are satisfied apart from the phasors $\underline{V}_s$ and $\underline{E}_f$ being slightly out of phase, that is $P_m \approx 0$, $\omega_0 \approx \omega_s$, and $\delta'(t = 0^+) = \delta'_0 > 0$. When the synchronizing switch is closed, the initial operating point lies at point 1 in Figure 6.30. The equilibrium point is defined by $P_m = P_e$ and $\hat{\delta}'_s = 0$, that is point 3. At the instant of synchronization $P_e(\delta') > 0$, the right-hand side of the swing equation is negative, the rotor decelerates, and the rotor angle δ' decreases. This causes the average value of the electrical torque and power to decrease with the energy used to decelerate the rotor being proportional to the shaded area 1-2-3. At the equilibrium point 3 rotor inertia effects ensure that the rotor angle continues to decrease toward negative δ' until it reaches point 4, where area 3-4-5 equals area 1-3-2. At point 4 the accelerating power is positive and the rotor swings back toward δ'_0. The oscillations continue until the damping torques cause the rotor to settle at its equilibrium angle $\hat{\delta}'_s = 0$.

Changes in δ' due to the swinging rotor produce corresponding oscillations in the torque and current, as shown in Figure 6.30b. Initially, as δ' is large, the average value of the torque, shown by the dashed line, and the amplitude of the current are both large. As energy is dissipated in the windings, the generator changes from the subtransient state to the transient state and then to the steady state, and both the periodic components of the torque and the amplitude of the current vanish with time.

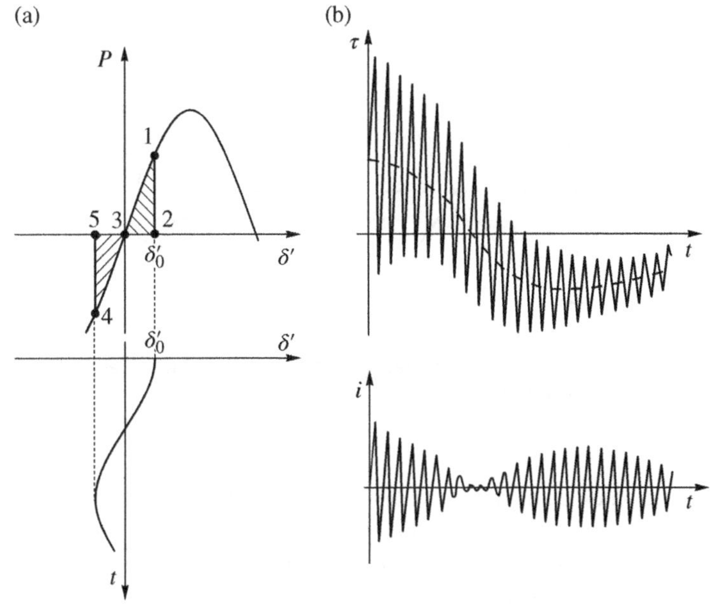

Figure 6.30 Synchronization when $\delta'_0 > 0$, $\omega_0 \cong \omega_s$: (a) power-angle characteristic and the swing curve; (b) time variation of the torque and current.

If at the instant of synchronization $\delta'_0 \neq 0$ and $\omega_0 \neq \omega_s$, the changes in δ' may be much greater than those discussed above. Figure 6.31 shows a case when $\omega_0 > \omega_s$ and the rotor has an excess of kinetic energy. On closing the synchronizing switch, this excess kinetic energy will drive the rotor toward an increasing δ' along line 1–2 until the shaded area is equal to the excess kinetic energy. The rotor then swings back toward the equilibrium point $\hat{\delta}'_s$. The conditions necessary to achieve successful synchronization can be determined using the equal area criterion. At the instant of synchronization, the rotor has an excess kinetic energy relative to the rotating reference frame (infinite busbar). This energy is equal to

$$E_k = \frac{1}{2} M \Delta\omega_0^2 \tag{6.85}$$

where M is the inertia coefficient and $\Delta\omega_0 = \dot{\delta}'(t = 0^+) = \omega_0 - \omega_s$.

The maximum deceleration work that can be done by the generator W_{max} is proportional to the area 4-1-2-3 in Figure 6.31

$$W_{max} = \int_{\delta'_0}^{\pi - \hat{\delta}'_s} \left[\frac{E'V_s}{x'_d} \sin\delta' - P_m \right] d\delta' = \frac{E'V_s}{x'_d} \left(\cos\hat{\delta}'_s + \cos\delta'_0 \right) - P_m \left[\left(\pi - \hat{\delta}'_s \right) - \delta'_0 \right] \tag{6.86}$$

The equal area criterion stipulates that the generator can be synchronized only if $E_k < W_{max}$. Using Eqs. (6.85) and (6.86), this gives

$$\Delta\omega_0^2 < \frac{2}{M} \left[\frac{E'V_s}{x'_d} \left(\cos\hat{\delta}'_s + \cos\delta'_0 \right) - P_m \left(\pi - \hat{\delta}'_s - \delta'_0 \right) \right] \tag{6.87}$$

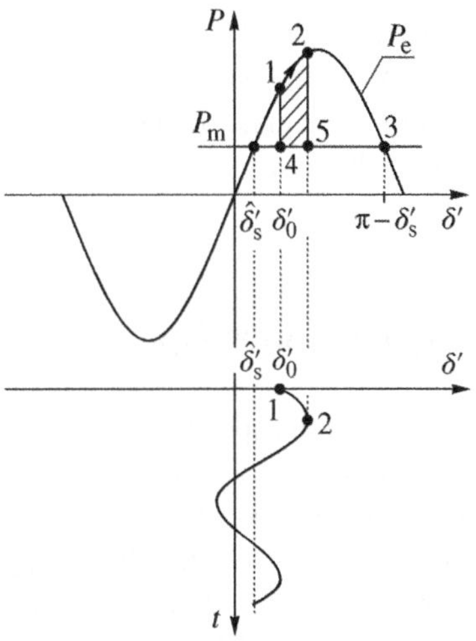

Figure 6.31 Equal area criterion applied to synchronization with $\omega_0 \neq \omega_s$ and $\delta'_0 > \hat{\delta}'_s$.

The maximum permissible value of $\Delta\omega_0$ depends on the synchronization angle δ'_0, the turbine power P_m, and the parameters of the generator and transmission system as expressed by x'_d. Figure 6.32 shows two example

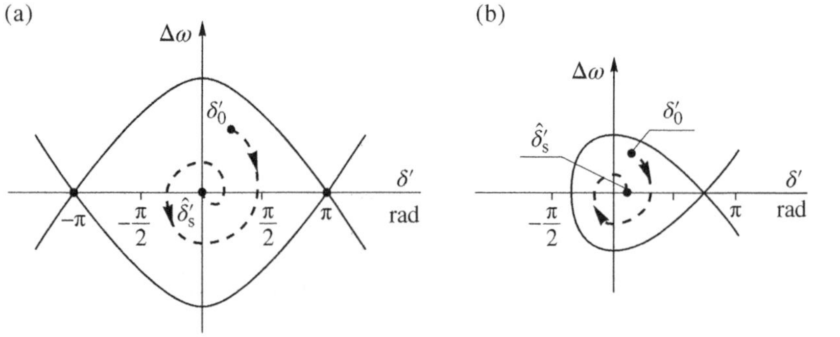

Figure 6.32 The trajectory of the operating point and the synchronization areas on the phase plane for: (a) $P_m = 0$; (b) $P_m = 0.5E'V_s/x'_d$.

areas on the phase plane diagram ($\Delta\omega$, δ) that satisfy the condition for successful synchronization as defined by Eq. (6.87): one when $P_m = 0$ and the other when $P_m = 0.5E'V_s/x'_d$. The dashed lines show the trajectory of the operating point and the solid lines show the critical energy contours obtained from Eq. (6.87) for the limiting conditions on δ'_0 and ω_0. Figure 6.32 shows that an increase in P_m has a detrimental effect on synchronization because the available decelerating area, defined by the area between P_m and P_e for $\delta'_0 < \delta' < \pi - \hat{\delta}'_s$, reduces as P_m increases. For $P_m > E'V_s/x'_d$ the system has no equilibrium point. Although it is unusual to synchronize a generator with the turbine developing large values of torque, similar conditions must be considered when synchronizing certain types of wind turbine (Westlake et al. 1996).

The condition defined by Eq. (6.87) contains useful information regarding possible re-synchronization after a fault is cleared by automatic reclosing action (Figures 6.7 and 6.8). In this case the value of the rotor angle and the speed deviation at the moment when the last switching occurs can be treated as the initial values δ_0 and $\Delta\omega_0$ in the synchronization process.

Section 6.3 shows that the equal area criterion is equivalent to the Lyapunov direct method when an energy-type Lyapunov function is used. The synchronization condition in Eq. (6.87) can therefore be derived from the stability condition in Eq. (6.36). The stability area shown in Figure 6.26 is the same as the synchronization area shown in Figure 6.32. When the synchronization condition in Eq. (6.87) is not satisfied, and the point (δ_0, $\Delta\omega_0$) lies outside the synchronization area, then the generator will start to show *asynchronous operation*. This type of operation is considered in the next section.

6.5 Asynchronous Operation and Resynchronization

The previous section explains how a generator may lose synchronism with the rest of the system when it makes asynchronous rotations at a slip frequency a few hertz above synchronous speed. With such high-speed changes the field winding time constant ensures that the net field winding flux linkage remains approximately constant so that the classical, constant flux linkage, generator model is valid with the synchronous power of the generator being given by $P_{E'}$, Eq. (5.40). However, as $P_{E'}$ changes as $\sin\delta'$ and over one asynchronous rotation δ' goes through 360°, the average value of $P_{E'}$ over one rotation is zero. Furthermore, as the speed increases above synchronous the action of the turbine governing system comes into play, while the asynchronous damping torques, proportional to $\Delta\omega$, are also significant. Indeed, it is the interaction between the turbine-governor characteristic and the asynchronous damping torques that will determine the generator operating point.

As explained in Section 2.3.3, the turbine static characteristic $P_m(\Delta\omega)$ is a straight line of droop ρ which crosses the vertical power axis in Figure 6.33 at point P_{m0} and the horizontal speed deviation axis at point $\rho\omega_s$. The form of the average asynchronous, damping, power characteristic $P_{Dav}(\Delta\omega)$ is discussed in Section 5.2 and is also shown in Figure 6.33. As the average value of synchronous power $P_{E'}$ over each asynchronous rotation is zero, the intersection between the static turbine characteristic and the damping power characteristic will define the generator operating point during asynchronous operation.

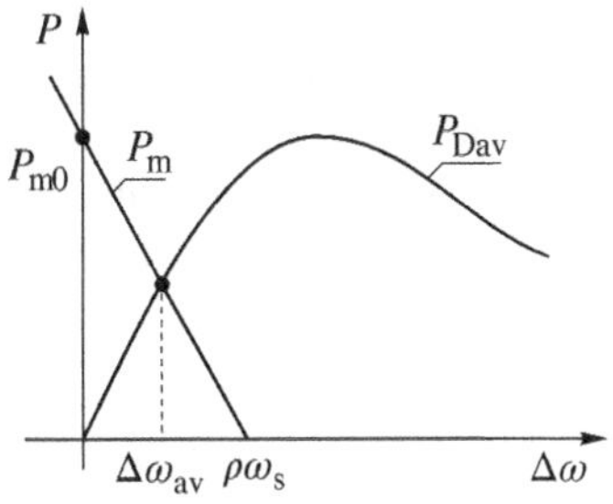

Figure 6.33 Static characteristics of the turbine power and damping power.

6.5.1 Transition to Asynchronous Operation

In order to explain what happens when a generator loses synchronism with a system, consider the system shown in Figure 6.6. Now assume that a three-phase fault occurs on line L2 and is then cleared by tripping the faulted line (without automatic reclosing) after a time such that the acceleration area 1-2-3-4 shown in Figure 6.34a is larger than the available deceleration area 4-5-6. The dashed $P_{E'}$ curve is valid for the pre-fault state, while the solid $P(\delta')$ curve is valid for the post-fault state. During the fault, the rotor speed deviation increases, but when the fault is cleared it starts to fall. Synchronism is lost when the rotor passes point 6 and follows the lower part of the power-angle characteristic. The speed deviation quickly increases because of the large difference between the turbine power and the synchronous power, shown by

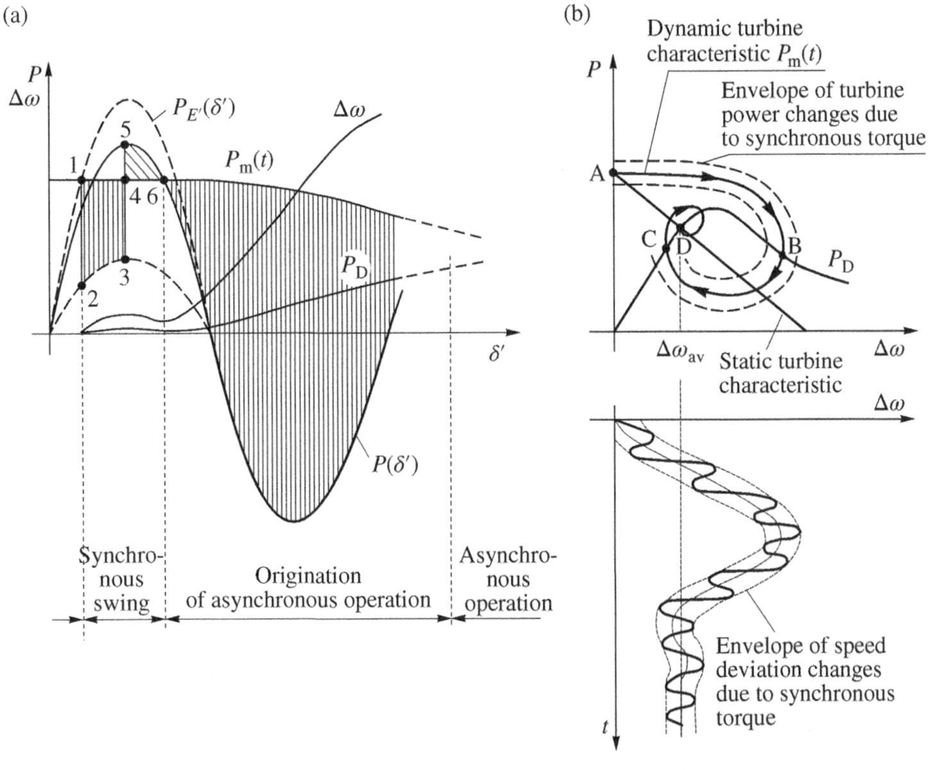

Figure 6.34 Transition to asynchronous operation: (a) development of high-speed deviation and high asynchronous power; (b) settling down of the asynchronous operating point. *Source:* Based on Venikov (1978).

the shaded area in Figure 6.34a. As the speed deviation increases, the turbine governing system starts to close the valves and the turbine mechanical power starts to decrease. Owing to the long time constant of the turbine governing system, the dynamic turbine power $P_m(t)$ does not follow the static characteristic in Figure 6.34b but moves above it. With increased speed deviation the average damping power P_{Dav} increases, according to the characteristic shown in Figure 6.33. As P_{Dav} increases and P_m decreases these two powers become equal at point B. Because of the delay introduced by the turbine time constants the mechanical power continues to decrease after passing point B and is now less than P_{Dav}. Consequently, the speed deviation starts to decrease and the turbine governor opens the valves to increase the turbine power until, at point C, the mechanical power and the asynchronous power are again equal. Again, because of the time delay in the turbine, the mechanical power continues to increase and is now greater than the asynchronous power. The speed deviation begins to increase, the governor valves close, and the cycle repeats until eventually the system operates at point D corresponding to the intersection of the static turbine characteristic and the asynchronous power characteristic. This point defines the average turbine speed deviation $\Delta\omega_{av}$ and power during asynchronous operation and is shown in Figure 6.33.

The effect of the synchronous power $P_{E'}$ is to produce speed deviation oscillations around the mean value as shown in the lower part of Figure 6.34b. The dashed lines in Figure 6.34b depict the envelopes of the power changes and the speed deviation changes, because of the periodic acceleration and deceleration of the rotor caused by the synchronous power.

6.5.2 Asynchronous Operation

Assuming that the generator is allowed to settle at the asynchronous operating point, then typical variations in some of the electrical quantities are as shown in Figure 6.35. The voltage drop ΔV across the equivalent reactance x'_d can be calculated from the phasor diagram shown in Figure 5.7b. The cosine theorem gives

$$\Delta V^2 = \left(Ix'_d\right)^2 = (E')^2 + (V_s)^2 - 2E'V_s\cos\delta' \tag{6.88}$$

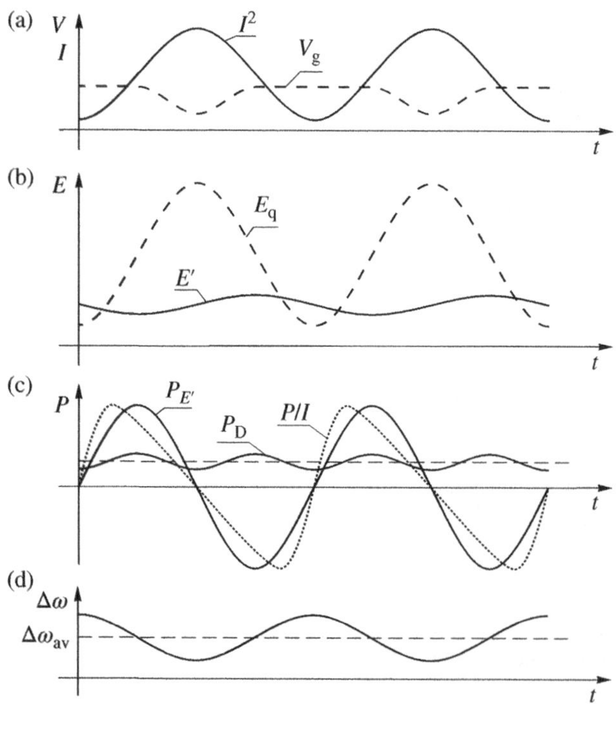

Figure 6.35 Changes in the electrical quantities of a generator operating at the asynchronous operating point: (a) current and terminal voltage; (b) internal emf and transient emf; (c) synchronous and asynchronous power; (d) speed deviation.

Accordingly, the stator current is seen to change as the rotor angle changes, with its maximum value being when $\delta' = \pi$ and its minimum when $\delta' = 0$, while the generator terminal voltage (Eq. [5.93]) is obtained by adding the voltage drop across the system equivalent reactance to the infinite busbar voltage. Accounting for the variations in angle δ' shown in Figure 6.11 allows typical time variations of the current and voltage to be obtained as shown in Figure 6.35a.

As the armature current changes, so too does the armature reaction flux linking the closed rotor circuits and, according to the law of constant flux linkage, currents must be induced in the rotor windings to maintain this flux linkage constant. Consequently, the field current will follow the armature current fluctuations thereby inducing large fluctuations in the internal electromotive force (emf) E_q, and smaller variations in E', as shown in Figure 6.35b. As the transient emf E' is almost constant, the synchronous power $P_{E'}$ is given by Eq. (5.26) and varies as $\sin \delta'$ (Figure 6.35c). Changes in signal $P/I = V \cos \varphi$ are also periodic, but its peak values are shifted with comparison to the peak values of the active power (Figure 6.35c).

The average value of the asynchronous damping power P_{Dav} is shown in Figure 6.35c and can be added to the synchronous power to obtain the resulting air-gap electrical power $P_e = P_{E'} + P_D$. This has the shape of a distorted sinusoid.

Figure 6.33 shows that the average speed deviation $\Delta\omega_{\mathrm{av}}$ is determined by the intersection between the turbine characteristic $P_m(\Delta\omega)$ and the average asynchronous damping power characteristic $P_{\mathrm{Dav}}(\Delta\omega)$. However, owing to the sinusoidal changes in the synchronous power $P_{E'}$, the actual speed deviation will oscillate around $\Delta\omega_{\mathrm{av}}$ in the way shown in Figure 6.35d.

6.5.3 Possibility of Resynchronization

In order to resynchronize with the system, the generator must fulfill the normal requirements for synchronization, that is its speed deviation and power angle must be small; condition (6.87). This subsection investigates under what conditions these two requirements may be met.

Figure 6.34 was constructed assuming that the changes in the speed deviation produced by the synchronous power are small when compared with the average value $\Delta\omega_{\mathrm{av}}$, while $\Delta\omega_{\mathrm{av}}$ itself depends on the asynchronous power characteristic $P_{\mathrm{Dav}}(\Delta\omega)$ shown in Figure 6.33. As the maximum value of $P_{\mathrm{Dav}}(\Delta\omega)$ is inversely proportional to the square of the reactances $x'_d = X'_d + X$ and $x'_q = X'_q + X$ (Eq. [5.10]) then, if the equivalent reactance X is small, the intersection between the $P_{\mathrm{Dav}}(\Delta\omega)$ characteristic and the $P_m(\Delta\omega)$ characteristic may occur at a small value of speed deviation. Moreover, a small value of X produces a large amplitude in the synchronous power $P_{E'}$ characteristic (Eq. [5.26]), so that large variations of $P_{E'}$ during asynchronous operation produce large variations of speed deviation around the average value $\Delta\omega_{\mathrm{av}}$. Large speed deviation variations, combined with a small average value, may at some point in time result in the speed deviation approaching zero, as shown in Figure 6.36a. As the speed deviation approaches zero, the rotor loses its excess kinetic energy, the asynchronous power is zero, and the rotor behavior is determined by the acceleration and deceleration work performed by the synchronous power $P_{E'}$. If the speed deviation reaches zero when $P_{E'}$ is sufficiently small then the generator may resynchronize with the system.

Figure 6.36b shows two cases. In the first case, the speed deviation becomes zero at point 1 when the synchronous power is large and negative. The synchronous power performs work proportional to area 1′-2-4, which is larger than the available deceleration area 4-5-6. Resynchronization is not possible and the rotor makes another asynchronous rotation. In the second case, the speed deviation reaches zero when the synchronous power is much smaller. The acceleration area 1′-2-4 is now much smaller than the available deceleration area 4-5-6 and, after reaching point 5, the decelerating power prevails and the rotor starts to swing back toward smaller values of the rotor angle. After a number of oscillations the rotor will settle at the synchronous equilibrium point 4. Resynchronization has been achieved but whether or not the generator remains stable depends on the subsequent action of the turbine governing system and the voltage regulator.

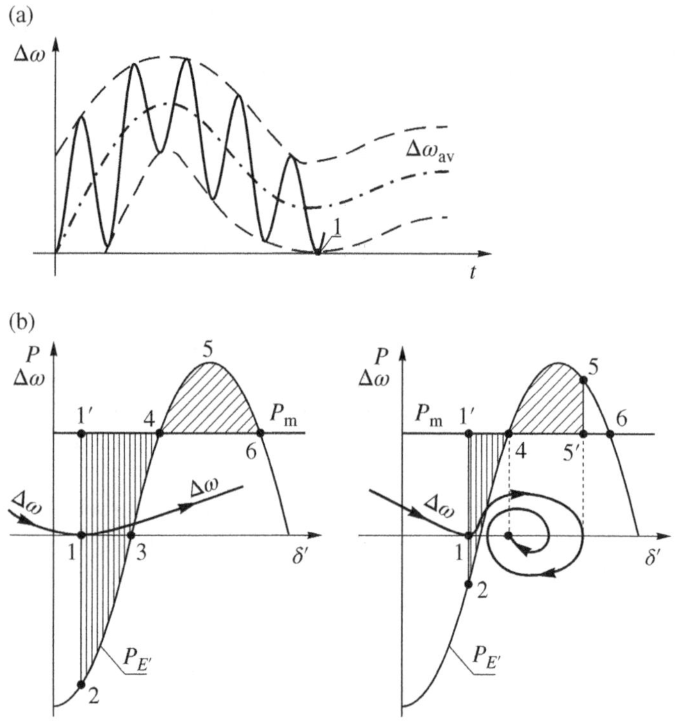

Figure 6.36 Resynchronization of a generator before the speed deviation reaches the average value $\Delta\omega_{av}$: (a) instantaneous value of the speed deviation approaches zero; (b) equal area criterion applied to check the condition for resynchronization. *Source:* Based on Venikov (1978).

6.5.3.1 Influence of the Voltage Regulator

Figure 6.35 shows that during each asynchronous rotation the voltage at the generator terminals drops quite significantly before recovering to a higher value. This means that the regulator error will oscillate at the frequency of the speed deviation oscillations. Normally, this frequency is too high for even a fast AVR to follow. Consequently, the AVR reacts, more or less, to the average value of the error signal, so maintaining a high value of excitation voltage, as shown in Figure 6.37a. As the average voltage value is lower than the pre-fault value, the AVR increases the excitation voltage V_f almost to its ceiling value so that the amplitude of the $P_{E'}(\delta')$ characteristic increases.

It has already been explained that the larger the magnitude of the speed deviation oscillations around the average value, the greater the possibility that the speed deviation will reach zero value. By increasing the excitation voltage, the AVR increases the amplitude of the $P_{E'}(\delta')$ characteristic and so produces larger changes in the speed deviation. This increases the chance of the speed deviation reaching a zero value, as shown in Figure 6.37b. Moreover, as the amplitude of the synchronous power increases, the available deceleration area in Figure 6.36b increases and the range of rotor angles which are acceptable for resynchronization also increases.

Obviously, resynchronization will be followed by a period of large rotor swings which may be damped or aggravated by the action of the AVR for the reasons described in Section 6.1.5.

6.5.3.2 Other Possibilities of Resynchronization

Successful resynchronization depends on the action of the AVR and the turbine governing system. Turbine governors can be equipped with supplementary control loops which allow the turbine power to be quickly reduced after a fault in order to protect the generator from loss of synchronism and/or assist in resynchronization. This is referred to as *fast valving* and is discussed in detail in Section 10.2.2. If a generator is not equipped with such a governor the operating staff may try to resynchronize the asynchronously rotating generator by manually reducing the turbine power to reduce the average speed deviation value. However, in order to avoid large power swings in

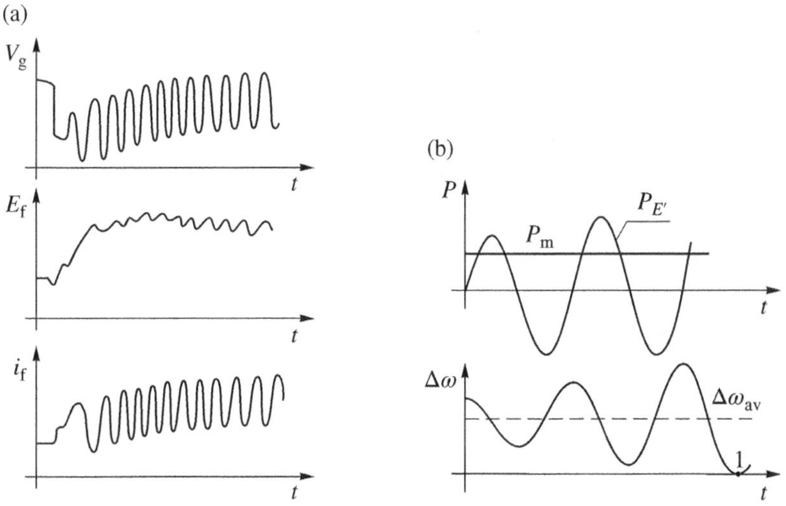

Figure 6.37 Effect of the AVR on resynchronization: (a) examples of the changes in the generator terminal voltage, excitation voltage, and the field current during asynchronous operation following a fault; (b) increase in the amplitude of the synchronous power oscillations and the speed deviation oscillations as a result of voltage regulation.

the system, and possible damage to the generator, generators are usually tripped after a few asynchronous rotations.

6.6 Out-of-step Protection System

Deep synchronous or asynchronous power swings are accompanied by large changes in voltages and currents which pose a serious threat to power system operation. They may cause mal-operation of some protection systems, especially distance protection and underimpedance protection (Section 2.6), which in turn may lead to cascaded outages and blackouts.

6.6.1 Concepts of Out-of-step Protection System

To avoid consequences of deep synchronous or asynchronous power swings, power system protection is usually augmented by additional devices and functions making up the *out-of-step protection system*, as shown in Figure 6.38.

The main elements of out-of-step protection are:

- special out-of-step protection (SOSP) and supplementary control;
- power swing blocking (PSB) of distance and underimpedance protection;
- pole-slip protection (PSP) of synchronous generators;
- out-of-step tripping (OST) in transmission network.

The task of *special protection* is to prevent loss of synchronism and the onset of asynchronous operation. The task of *supplementary control* is to force a rapid damping of power swings. These problems are treated in Chapter 10, which deals with stability enhancement methods. If, despite the special protection and supplementary control systems, there is an onset of asynchronous operation or deep synchronous power swings, then the apparent impedance measured by the distance or underimpedance protection can move away from the normal load area into the

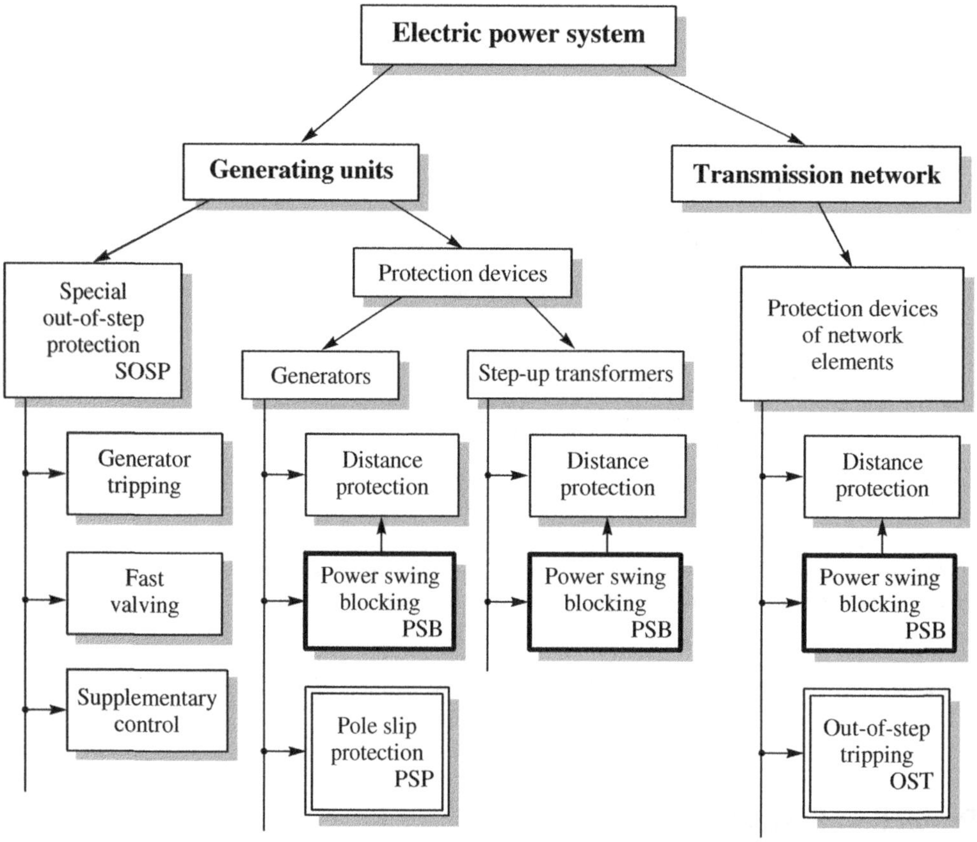

Figure 6.38 Out-of-step protection system. *Source:* Based on Machowski (2012).

distance protection zone causing unnecessary tripping. The task of the PSB relays (or functions) is to detect whether the changes in the measured impedance are due to power swings rather than short circuits. The PSB should be used for distance protection of transmission lines and transformers and also for distance and underimpedance protection of synchronous generators and their step-up transformers. The PSP relays (or functions) are installed as a part of the generator protection system. They isolate the asynchronously operating generators from the system after a set number of asynchronous cycles. The OST relays (or functions) are installed inside the transmission as a part of the network protection. They split the system at predetermined points of the transmission network when asynchronous power swings are observed.

It is important for a power system operator to have a logical strategy of using PSB, PSP, and OST. Examples are discussed in the IEEE Report to the Power System Relaying Committee (1977). Generally speaking, protection of an EPS against the consequences of power swings should consist of efficient blocking of impedance relays during synchronous and asynchronous swings using PSB relays (or functions). Efficient protection against long-lasting asynchronous operation should be achieved by using PSP relays (or functions) for generators and OST for the network.

6.6.2 Impedance Loci During Power Swings

The effect of power swings on the impedance loci can be studied using the simple equivalent circuit shown in Figure 6.39 in which two equivalent synchronous generators are linked via impedances $\underline{Z}'_{\mathrm{a}}$ and $\underline{Z}'_{\mathrm{b}}$. The voltage and current are measured at the relay point between these impedances to obtain the apparent impedance $\underline{Z}(t)$. The current and voltage at the relay point are given by

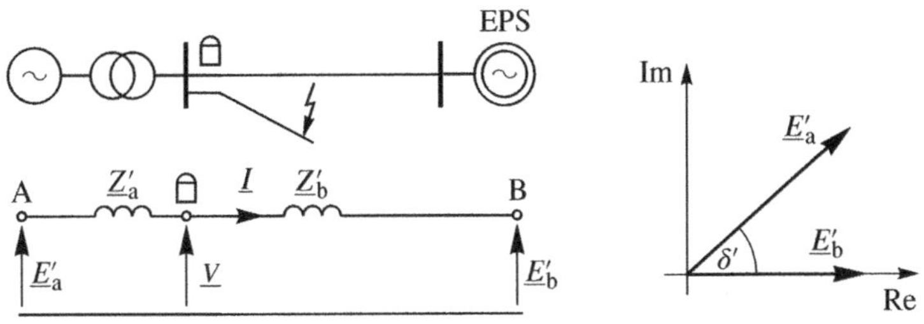

Figure 6.39 Equivalent circuit of the system with a relay point.

$$\underline{I} = \frac{\underline{E}'_a - \underline{E}'_b}{\underline{Z}'_a + \underline{Z}'_b}; \quad \underline{V} = \underline{E}'_a - \underline{I}\,\underline{Z}'_a \tag{6.89}$$

where $\underline{E}'_a$ and $\underline{E}'_b$ are the equivalent generator transient emf's and $\underline{Z}'_a$ and $\underline{Z}'_b$ are the equivalent impedances that include the impedance of the transmission network and the generator transient reactances. If the magnitudes of the equivalent transient emf's are assumed to be constant then

$$\frac{\underline{E}'_a}{\underline{E}'_b} = \frac{|\underline{E}'_a|}{|\underline{E}'_b|} e^{j\delta'} = k e^{j\delta'} \tag{6.90}$$

where $k = |\underline{E}'_a| / |\underline{E}'_b| =$ constant and the transient power angle δ' is the difference between the arguments of the emf's.

The apparent impedance measured by the relay is $\underline{Z}(t) = \underline{V}/\underline{I}$, which, when substituting for $\underline{V}$ and $\underline{I}$ from the equations in (6.89) and including (6.90), gives

$$\underline{Z}(t) = \frac{\underline{V}}{\underline{I}} = \frac{\underline{Z}'_a + \underline{Z}'_b k e^{j\delta'(t)}}{k e^{j\delta'(t)} - 1} \tag{6.91}$$

Note that during power swings the power angle and the apparent impedance are both time dependent. Solving Eq. (6.91) with respect to the power angle gives

$$k e^{j\delta'(t)} = \frac{\underline{Z}(t) - (-\underline{Z}'_a)}{\underline{Z}(t) - \underline{Z}'_b} \tag{6.92}$$

With the exception of the coefficient k, all the variables in this equation are complex. Taking the absolute value of both sides of (6.92) and recognizing that $|e^{j\delta'}| = 1$ gives

$$\left| \frac{\underline{Z}(t) - (-\underline{Z}'_a)}{\underline{Z}(t) - \underline{Z}'_b} \right| = k = \text{constant} = \frac{|\underline{E}'_a|}{|\underline{E}'_b|} \tag{6.93}$$

Equation (6.93) determines the locus in the complex plane of all points with the same value of k. This is illustrated in Figure 6.40. The impedances $(-\underline{Z}'_a)$ and $\underline{Z}'_b$ determine the points A and B, respectively, while the apparent impedance $\underline{Z}(t)$ determines point C (Figure 6.40a). As $|\underline{Z}(t) - (-\underline{Z}'_a)| =$ AC and $|\underline{Z}(t) - \underline{Z}'_b| =$ BC, the ratio $k =$ AC/BC = constant. As $\underline{Z}(t)$ varies, a locus of constant value k will be traced out by point C. The locus for a given value of k is the circle of Apollonius that surrounds either point A or point B.

The diameter of each circle of Apollonius, and the location of its center, depends on the value of k and the impedances $\underline{Z}'_a$ and $\underline{Z}'_b$. The center of each circle lies on the straight line that passes through points A and B and is shown dashed in Figure 6.40b. For $k = 1$ the diameter of the circle tends to infinity and the circle becomes a straight line

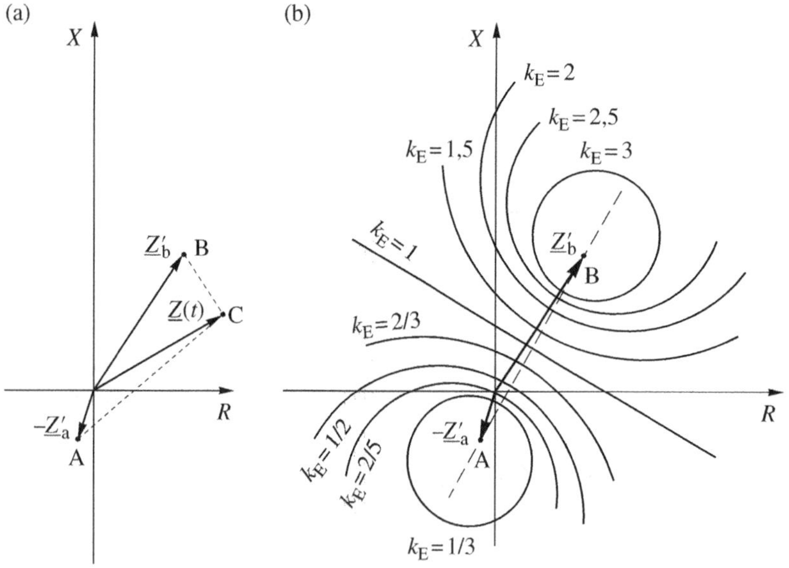

Figure 6.40 Graphical interpretation of Eq. (6.93): (a) impedances on the complex plane; (b) circles of Apollonius determining the impedance loci.

that bisects AB. For $k < 1$, $|\underline{E}'_a| < |\underline{E}'_b|$, the circle lies in the lower part of the complex plane and encircles point A. For $k > 1$, $|\underline{E}'_a| > |\underline{E}'_b|$, the circle lies in the upper part of the complex plane and encircles point B.

6.6.3 PSP of Synchronous Generator

The purpose of this protection is to detect the loss of synchronism of a synchronous generator. The measurement points for the relay are the generator terminals, as shown in Figure 6.41. The impedance $\underline{Z}_a$ to the left of the measurement point corresponds to the generator transient reactance $\underline{Z}_a \cong jX'_d$. The impedance $\underline{Z}_b$ to the right of the measurement point consists of the step-up transformer impedance $\underline{Z}_T$ and the system equivalent impedance $\underline{Z}_S$. The whole transmission system can be divided into two zones. Zone 1 consists of the generator and the step-up transformer, while zone 2 is the rest of the system. At completion of the first half of an asynchronous rotation, that is when $\delta' = 180°$, the equivalent voltages oppose each other so that the voltage at a certain point of the transmission link must be equal to zero. That point is referred to as the *center of power swing* or simply the *electrical center.* Depending on how big the equivalent system impedance $\underline{Z}_S$ is for a given value of $\underline{Z}_a$ and $\underline{Z}_T$, the center of power swing may be inside zone 1 or 2. In Figure 6.41, the center is inside zone 1 and inside the impedance of the step-up transformer. Equation (6.91) shows that when $\delta' = 180°$ the apparent impedance seen by the relay is proportional to the difference between the equivalent impedances, that is

$$\underline{Z}(\delta' = 180°) = \frac{\underline{Z}_a - k\underline{Z}_b}{-k-1} = \frac{k\underline{Z}_b - \underline{Z}_a}{k+1} \tag{6.94}$$

Because k is real, this equation shows that at $\delta' = 180°$ the trajectory $\underline{Z}(t)$ intersects the line crossing points A and B in Figure 6.40. This means that an asynchronous rotation can be identified by a relay with an impedance characteristic that surrounds line AB based on $(-\underline{Z}_a)$ and $\underline{Z}_b$, respectively. Three such characteristics are shown in Figure 6.42. The first type (Figure 6.42a) has an offset mho characteristic 3 with the left and right parts cut off by two directional relays 1 and 2. The second type (Figure 6.42b) has a symmetrical lenticular characteristic

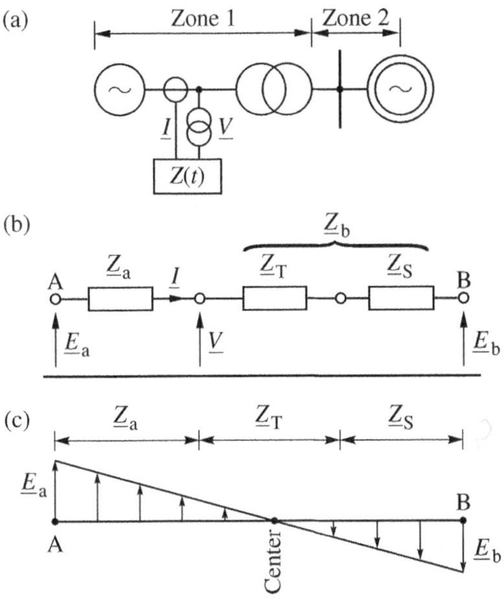

Figure 6.41 Pole-slip protection of synchronous generator: (a) block diagram; (b) equivalent circuit; (c) illustration of the center of power swing.

and is obtained by using two impedance relays with offset mho characteristics shown by the dashed circles 1 and 2. The third type (Figure 6.42c) has an asymmetrical lenticular characteristic.

PSP usually has an additional impedance characteristic that divides $\underline{Z}(t)$ into two zones as shown by line 4 in Figure 6.42. The relay identifies an asynchronous rotation if the impedance locus passes completely through the relay characteristic. If the impedance locus $\underline{Z}(t)$ passes through the impedance characteristic below line 4, it means that the center of power swing (Figure 6.41) is inside zone 1. In that case the generator should be disconnected right after the impedance $\underline{Z}(t)$ leaves the impedance characteristic, that is during the first asynchronous rotation. On the other hand, if the impedance locus $\underline{Z}(t)$ passes through the impedance characteristic above line 4, it means that the center of power swings (Figure 6.41) is inside zone 2, that is inside the transmission network. In that case the generator may be disconnected after 2–4 asynchronous rotations. Such a delay is introduced in order to give a chance of operation for the OST relays installed in the transmission network and splitting the network into islands. The delay is acceptable only if the generator can withhold a thermal and dynamic overload caused by a set number of asynchronous rotations. If there are no OST relays in the network then there is no reason for the delay and the generator may be tripped after one asynchronous rotation independently of where the center of power swing is. Recommended numbers of asynchronous cycles set in pole slip protections are given in Table 6.2.

The above recommendations relate to power plants without SOSP or with preventive SOSP. If there is restitutive SOSP, which in its operating principle allows one asynchronous rotation, then before generator tripping by pole slip protections an opportunity should be given for the SOSP to activate before the pole slip protection activation. In this case, the number of cycles should be set at one cycle more than in Table 6.2.

The way that a generator is tripped depends on the configuration of the unit described in Section 2.3. For generating units containing a step-up transformer, the generator is tripped by opening its main circuit-breaker on the

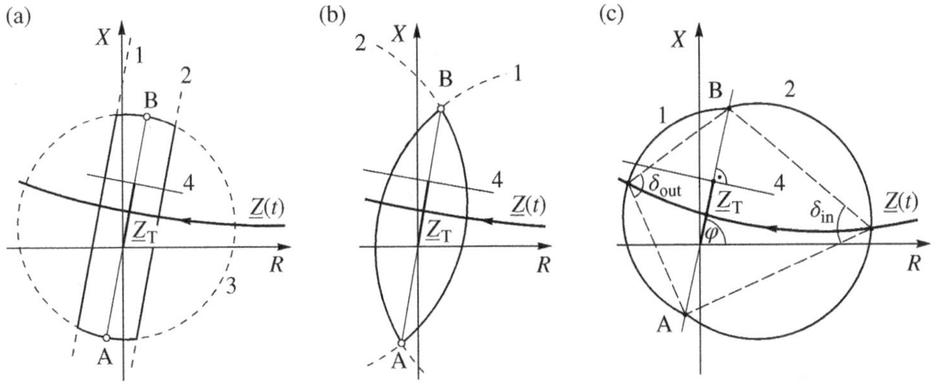

Figure 6.42 Three types of pole-slip protection characteristics: (a) offset mho type; (b) symmetrical lenticular type; (c) asymmetrical lenticular type.

Table 6.2 Recommended number of asynchronous cycles before generator tripping.

–	Power plant without SOSP or with preventive SOSP		
Power swing center location	within generating unit impedance, i.e. zone 1 (inner)	inside transmission network, i.e. zone 2 (outer)	
		with out-of-step tripping protections in the network	with no out-of-step tripping protections in the network
Number of cycles	1	2–4	1–2

high-voltage side (Figure 2.3). The turbine is not disconnected but reduces its power output to the level necessary to supply the unit's auxiliary services. This makes it possible to resynchronize the generator quickly with the system.

6.6.4 Power Swings Blocking of Distance Protection

The effect of a power swing on the trajectory of $\underline{Z}(t)$ can be easily considered for the equivalent circuit shown in Figure 6.39. For the pre-fault operating conditions the voltage V is close to the rated voltage and the current I is small when compared with the fault current. Consequently, $\underline{Z}(t)$ is high and predominantly resistive (since normally the power factor is close to unity). A fault results in a significant drop in voltage and an increase in current so that $\underline{Z}(t)$ drops in value. Assuming that $k = |\underline{E}'_{a}|/|\underline{E}'_{b}| =$ constant then, when the fault is cleared, the value of $\underline{Z}(t)$ rises again. Further changes in $\underline{Z}(t)$ are solely due to the variation of the power angle δ' and must follow one of the circles shown in Figure 6.40.

Figure 6.43a illustrates the trajectory of $\underline{Z}(t)$ in the case of an asynchronous power swing, and Figure 6.43b in the case of a synchronous power swing, following a fault outside the relay protection zone. In both cases the trajectory starts from point O in the normal load area and, when the fault occurs, jumps to point F lying outside protection zones 1 and 2. During the short circuit, the impedance trajectory moves to point C, where the fault is cleared. On fault clearance the trajectory jumps to point P, which does not quite lie on the same circle as point O, because the emf's change their values slightly during the fault. Further changes in δ' will cause $\underline{Z}(t)$ to follow a circle of a constant ratio k or, more precisely, a family of closed circles corresponding to slightly varying emf's and hence ratios k.

When $\delta'(t)$ increases, $\underline{Z}(t)$ decreases and may encroach into the distance relay tripping zone. For the asynchronous power swing, shown in Figure 6.43a, the trajectory will trace a full circle as $\delta'(t)$ completes a full 360° cycle. For a synchronous power swing, shown in Figure 6.43b, the trajectory will reach point B1 on an impedance circle and then move back toward point B2 as $\delta'(t)$ decreases on the backward swing. For constant $|\underline{E}'_{a}|/|\underline{E}'_{b}|$ the trajectory always lies on one circle. A change in the transient emf's due to AVR control will move the trajectory from circle to circle so that when power swings are damped the trajectory tends toward a post-fault equilibrium point in the load area.

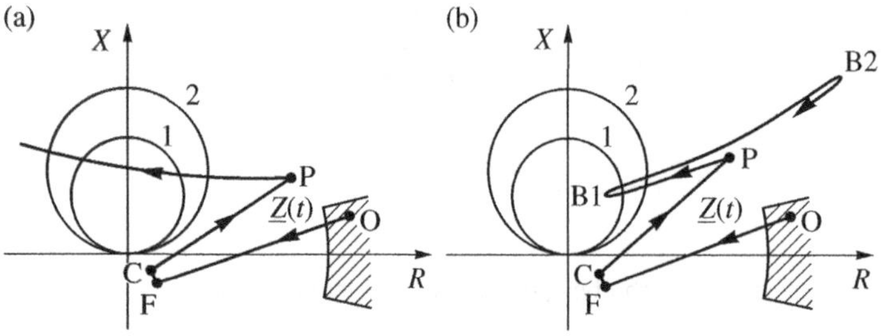

Figure 6.43 Example of the impedance trajectory for (a) asynchronous power swing; (b) synchronous power swing: 1, 2 – tripping zones of the distance relay; load area is shaded.

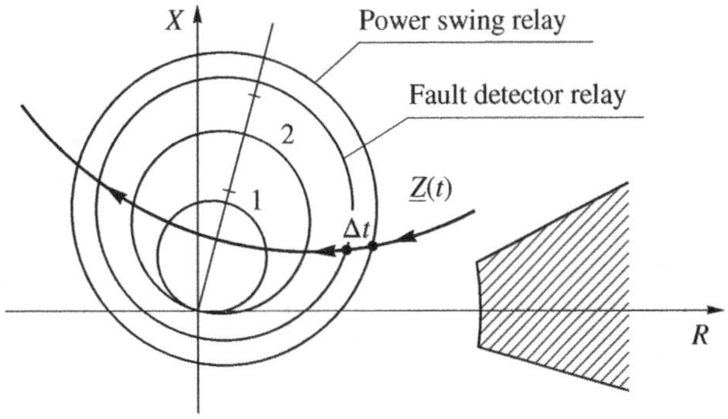

Figure 6.44 Power swing blocking and distance relay characteristics of the offset mho type.

A power swing can be detected by monitoring the speed at which the impedance locus $\underline{Z}(t)$ approaches the characteristic of the distance relay. If the fault is applied so that point F lies outside the relay protection zone (as in Figure 6.43) then the protection zones are approached by $\underline{Z}(t)$ with a finite speed. On the other hand, if the fault is applied so that point F lies inside one of the protection zones, as shown in Figure 6.43, then the impedance locus jumps from point O to a point inside the protection zone almost instantaneously.

Figure 6.44 illustrates the operation of a distance relay with an offset mho-type characteristic. The PSB relay (or function) has the same type of characteristic as the distance relay and surrounds the distance relay characteristic. A timer measures the time Δt taken for the trajectory to pass between the two characteristics. If the fault is inside the protection zone then the power swing relay and the fault detection relay will operate practically simultaneously and a blocking command is not generated. If the fault is outside the protection zone then, during a resulting power swing, the trajectory will take a finite time to move between the two relay characteristics and a blocking command is generated to hold up operation of the circuit-breaker for the time when the impedance locus stays inside the fault detection zone. To avoid blocking of relays during remote unbalanced faults and the dead time of single-phase automatic reclosing, the zero-sequence component of the current is monitored. If the zero-sequence component is present, the blocking command is not generated.

The above blocking principle is also used in distance relays with other types of characteristic such as quadrilateral, rhombus, and oval. Obviously, the power swing relay must have the same type of characteristic as the distance relay.

This method of detecting power swings was widely used in electromechanical relays. Modern digital protection is augmented by additional decision criteria. These criteria are checked by the power swing detection (PSD) function. Additional decision criteria in PSB functions of digital protection consist of monitoring other signals. When trajectories of the signals are smooth and have an expected shape, the blocking command is not generated. Decision criteria in PSD are used because measuring only the time when the measured impedance passes between the two zones is not a reliable method. There are instances when the relay does not operate when it should, and vice versa.

One example of unnecessary activation of the blocking relay based on measuring the time of passing between zones $\underline{Z}(t)$ alone is a high-impedance developing fault. The changes of impedance value during the fault may be so slow that the relay activates and unnecessarily blocks the distance protection. That would be dangerous for both the short-circuited network element and the whole power system.

An example of an out-of-step relay, based on measuring the time of passing $\underline{Z}(t)$ alone, which does not operate, is shown in Figure 6.45. The trajectory of the impedance is shown for a case when the relay is placed at the beginning of line L2 while the short circuit occurs in line L1. At the instant of the short circuit, the trajectory jumps to point F_1 in the fourth quadrant and moves during the short circuit toward point F_2. When the short circuit is cleared, the trajectory jumps to point P which lies inside the second tripping zone. Then the trajectory moves toward point M inside the first tripping zone. The protection activates with the tripping time of the first zone and the healthy line L2 is unnecessarily tripped. The out-of-step relay did not operate because, following the fault clearing, the

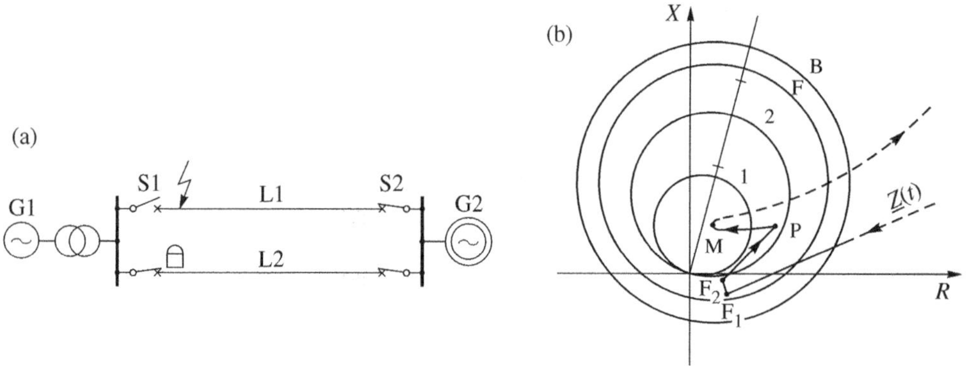

Figure 6.45 Example of the out-of-step relay not operating: (a) block diagram of a simple system; (b) example of impedance changes during synchronous swings.

trajectory did not move outside characteristic B. Hence there was no passing between characteristics B and F which is essential for operation of the relay. Thus augmenting the principle of operation by other decision criteria is clearly necessary.

Description of a number of reasons for unnecessary blocking of distance protection and its unwanted operation is given, for example, by Troskie and de Villiers (2004).

The time after which the blocking command of the distance protection is removed regardless of other criteria is referred to as the *unblocking time*. The choice of the set value of the unblocking time is very important for the operation of the whole out-of-step protection system. The unblocking time must be longer than the time for which impedance trajectory $\underline{Z}(t)$ stays within the distance protection zones during synchronous or asynchronous power swings. When the unblocking time is too short the unblocked distance protection can cause needless tripping in the network. The shortest value of the unblocking time cannot be estimated on the basis of the period of the power swings, because the speed of impedance trajectory on the complex plane is not constant. This can be justified as follows. From Eq. (6.91) it is easy to obtain

$$\underline{Z}(t) = \frac{R_a + jX_a + k\cdot(R_b + jX_b)\cdot(\cos\delta' + j\sin\delta')}{k(\cos\delta' + j\sin\delta') - 1} = R(t) + jX(t) \tag{6.95}$$

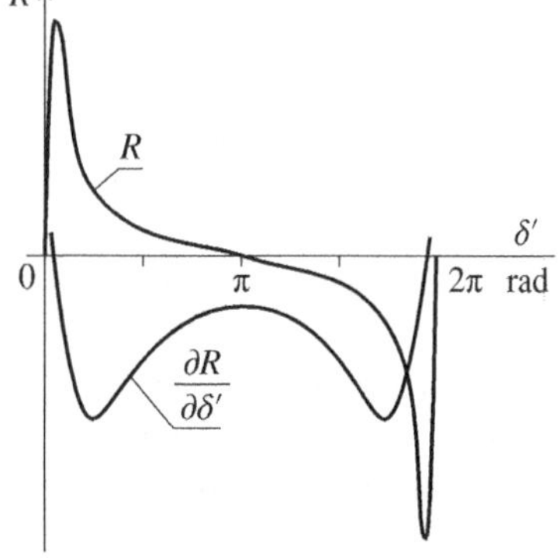

Figure 6.46 Resistance measured at relay point, and its derivative, as power angle functions.

As results from Eq. (6.95) resistance and $R(\delta')$ and reactance $X(\delta')$ measured by distance protection are functions of power angle δ', which during power swings is a function of time t. Therefore, the speed of impedance trajectory along the axes of the complex plane can be measured by time derivatives

$$\frac{dR(t)}{dt} = \frac{\partial R}{\partial \delta'}\cdot\frac{d\delta'}{dt}; \quad \frac{dX(t)}{dt} = \frac{\partial X}{\partial \delta'}\cdot\frac{d\delta'}{dt} \tag{6.96}$$

where: $d\delta'/dt$ is the slip, and derivatives $\partial R/\partial\delta'$ and $\partial X/\partial\delta'$ can be found from Eq. (6.95). It can be concluded from Figure 6.43 that for the estimation of the shortest unblocking time the speed along the horizontal axis (i.e. the rate of change in resistance) is decisive. Figure 6.46 shows an example of $R(\delta')$ and $\partial R/\partial\delta'$ as the function of δ'. In this figure the plot of $R(\delta')$ reaches zero at the power angles equal to a multiplicity of π. Characteristic $R(\delta')$ is the steepest near small angles, and quite flat near the angles close to π. Therefore, near angle π, the derivative $\partial R/\partial \delta'$ assumes small values. It means that, according to formula (6.96), in the vicinity of large power angles close to π (at the same slip $d\delta'/dt$) the time derivative of the resistance dR/dt is lower than at low (but not zero) power angles. The rate of change in $R(t)$ along trajectory $\underline{Z}(t)$ is not constant and depends on the power angle.

In asynchronous swings in the half of asynchronous rotation (i.e. near the imaginary axis) the slip $d\delta'/dt$ is always large and, despite small $\partial R/\partial \delta'$, the time derivative dR/dt reaches quite high values there. This means that in asynchronous swings the trajectory $\underline{Z}(t)$ quickly passes through the distance protection characteristic. In synchronous swings it can be expected that the trajectory $\underline{Z}(t)$ will stay quite long within the distance protection characteristics. This is due to the fact that in such swings both factors in formula (6.96), which determine the rate dR/dt, are small:

- all along trajectory $\underline{Z}(t)$ within the protection characteristics the slip is small, since at the turning point (Figure 6.45) slip $d\delta'/dt$ must change its sign, i.e. first it decreases from positive values to zero and then gradually increases its negative value;
- within distance protection characteristic the power angles are large and close to π, therefore derivatives $\partial R/\partial \delta'$ are small (Figure 6.46).

In synchronous swings the impedance trajectory $\underline{Z}(t)$ can stay within distance protection characteristics much longer than in asynchronous swings. Based on the results of computer simulations it can be assumed that the unblocking time should not be shorter than (2–5) s for transmission lines inside an EPS and (5–10) s for tie-lines connecting the subsystems of an interconnected large-scale power system.

Most publications on distance protections of lines and transformers refer to the application of PSB only for the first and second zones, assuming that the third zone response time is too long to react during power swings. The above presented analysis and operational experience show, however, that this point of view is wrong. Lack of the PSB feature in the third distance protection zones can cause power system blackouts. One example is described in Section 6.6.7.

In some new digital distance relays a user cannot adjust the unblocking time which by default is set to infinity. In such solutions the blocking signal is removed only when impedance trajectory $\underline{Z}(t)$ leaves protection characteristic or when supplementary algorithms identify a short circuit appearing during power swings. Obviously, the reliability of the protection system based on such relays depends on the quality of these supplementary algorithms.

Distance protections augmented with good supplementary fault detectors and infinite unblocking are attractive from the point of view of avoiding needless protection operation during heavy overloads in a transmission network. Many blackouts in power systems in the world began with the needless operation of distance protections during strong overloads of network elements. If a transmission line is heavily overloaded, and the EPS operates at lowered voltages then distance protections (especially of very long lines) can be activated, and can needlessly switch off the overloaded line (i.e. a line important for the power system operation). This may lead to cascading outages (successive loss of system elements) and finally to a system blackout. For years scientists and engineers have contemplated how to avoid unnecessary distance relay interventions during heavy operating overload. It turns out that PSB systems with infinite unblocking time may help here. Such a system recognizes the slow load increase and the accompanying slow encroachment of impedance trajectory $\underline{Z}(t)$ to protection zones as a slow power swing and blocks the distance protection. If during overload the impedance trajectory does not leave the measurement zones, the protection is blocked, and there is no needless tripping unless supplementary fault detectors identify a fault in the overloaded line. Modern multi-criteria distance protections with PSB systems with infinitely long unblocking time may be insensitive to overloads of the protected network elements.

6.6.5 Protections Sensitive to Power Swings

In transmission networks distance protections are used to protect transmission lines and network transformers. Distance protection consists of two relays with the time-stepped zones (Figure 6.47) sited at each end of the protected element. For a transmission line (Figure 6.47a) the second and third zones constitute the backup protection to the adjacent busbars and transmission lines. For a transformer (Figure 6.47b) the second zones are backward directed and constitute the backup protection to the adjacent busbars.

In the simplest solution for transmission lines both relays operate independently and then parts of the line close to the busbars are protected from one side with the time of the second zone. In such a case the opening of the circuit breakers at both ends is nonsimultaneous and the dead time during the auto-reclose cycle must be suitably long. This may be unacceptable from the point of view of power system transient stability. In order to speed up the fault clearing inside the protected line a very common practice is to interconnect both relays by a communications channel. The purpose of the communications channel is to transmit information about the system conditions from one end of the protected line to the other, including requests to initiate or prevent tripping of the remote circuit breaker. There are several schemes of such cooperation of distance relays, which can be divided into two categories: transfer tripping scheme and blocking scheme. The most desirable schemes are obviously such that enable fast instantaneous tripping over the whole length of the protected line and provide the backup protection to adjacent busbars and transmission lines. In the case of transformers both distance relays are sited in the same substation and their interconnection is made directly through the secondary circuits. Both circuit breakers open instantaneously when at least one relay detects the fault in its first zone.

The first (high-speed) zones of two interconnected relays are not sensitive to power swings, because during a power swing the currents measured by relays on both sides of a protected network element have the same directions, while during the internal fault the directions of both currents are opposite. However, the needless tripping of the protected network element (line or transformer) during asynchronous or deep synchronous power swings can be caused by the operation of one of the relays with the time of the second or third zone.

From the power swing and PSB perspective it does not matter whether the protected network element is a line or transformer. In order to illustrate this fact, Figure 6.48 shows a two-machine equivalent power system model, in which two parallel transmission links are identified. The first link consists of two lines, L1 and L2. The second link is made up of two lines, L3 and L4, and two auto-transformers, AT1 and AT2. If power swings develop between the two parts of the EPS, they will have an impact on the distance protections of the lines and auto-transformers.

Distance and underimpedance protections are also used as elements of generating unit protection sets.

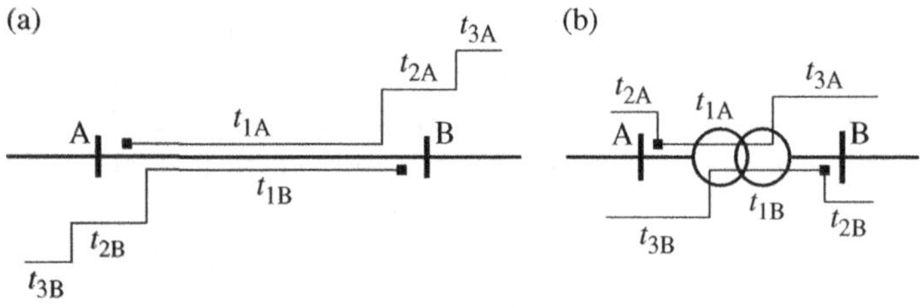

Figure 6.47 Zones of distance protections (a) line; (b) transformer.

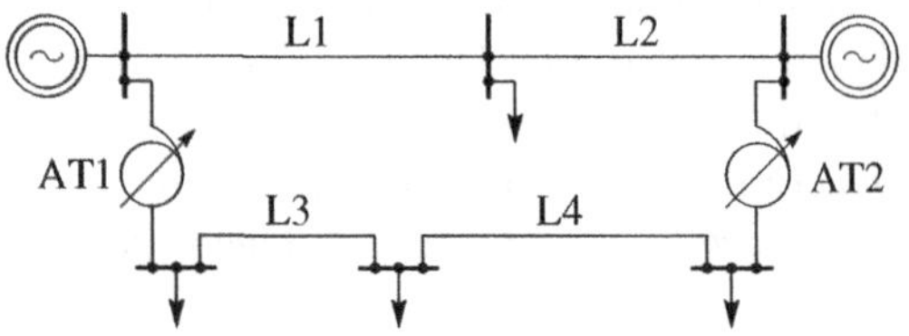

Figure 6.48 Two transmission links consisting of lines and auto-transformers.

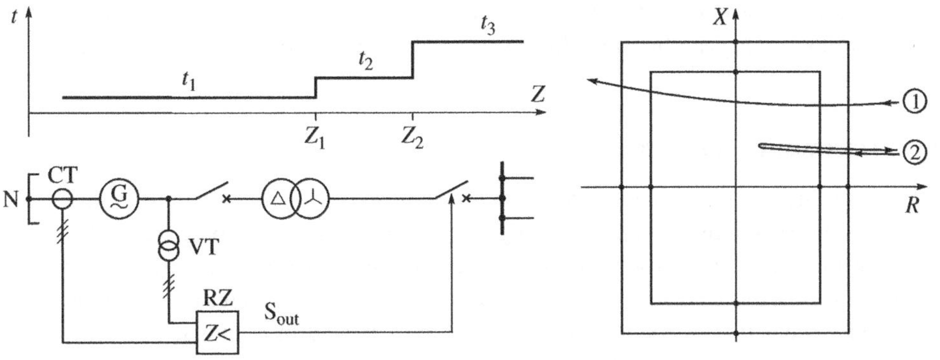

Figure 6.49 Underimpedance generator protection.

Underimpedance generator protection (Figure 6.49) is supplied from current transformers of the generator at its star point and has Mho-type circular characteristics, or rectangular characteristics, symmetrical with respect to the coordinate system. The time-stepped zones are as shown in the discussed figure. This protection is a backup for differential protections, and its first zone has a small delay in relation to the differential protection response time. From the power swing's perspective this protection is fast enough to trip the generator needlessly during synchronous and asynchronous power swings alike. If the swing center (Figure 6.41) is within the generating unit impedance, then the impedance trajectory $\underline{Z}(t)$ may encroach the impedance protection zone (curves 1 and 2 in Figure 6.49). Reaction of the underimpedance protection will be the same as for an internal fault and the generating unit will be completely shut down. In the case of synchronous swings such a reaction is completely needless. Also, in the case of asynchronous swings the full shut-down of the generating unit is needless. A asynchronously operating generator can be tripped to operate its auxiliaries, which enables its fast resynchronization. The task of generator tripping at loss of synchronism should be entrusted to the pole slip protection (Figure 6.42). The generator underimpedance protection should be provided with PSB that effectively blocks it at synchronous and asynchronous swings.

Another protection sensitive to power swings is the distance protection of the generating unit. Such protection is supplied from current transformers at the high voltage side of the step-up transformer (Figure 6.50). The first zone covers the entire step-up transformer and a part of the generator. It is a backup protection for differential protections. Normally no additional time delay is set for this zone. The second zone is directed toward the network and covers the busbars and parts of the power output lines. This is the basic protection of generating unit against external short-circuits. This zone has a time delay adjusted to the offset from the first zones of the distance protections of the power output lines. The distance protection zones facing the step-up transformer and the generator are fast enough to trip needlessly the generating unit during synchronous and asynchronous power swings alike if the power swing center falls within the generating unit impedance.

If the swing center falls in a transmission network but close to the busbars of the power plant, then the second zone of considered distance protection may also cause the unwanted generator tripping, in particular in the case of deep synchronous swings, for which the impedance trajectory returns within the distance protection zone and remains there for a longer time than the second zone tripping time.

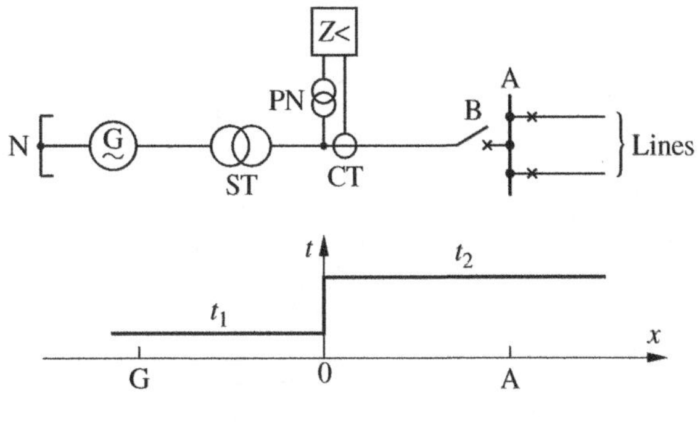

Figure 6.50 Generating unit distance protection.

The final conclusion is that the distance protection of the generating unit should have the PSB that effectively blocks it at synchronous and asynchronous swings in the zones directed toward the generator and toward the network alike.

6.6.6 Out-of-step Tripping in Transmission Network

It has already been mentioned that, during an asynchronous operation, the impedance $\underline{Z}(t)$ measured by the distance protection of transmission lines close to the center of power swing may encroach into the tripping zone of the distance protection. If PSB relays (or functions) are not used, a number of transmission lines may be tripped, leading to a random network splitting into unbalanced zones. Generators in islands with a large surplus of generation will be disconnected by overfrequency protection, while load shedding will be activated in islands with a large deficit of generation. These problems are discussed in Chapter 9.

To prevent a random network splitting during asynchronous operation, it is necessary to use the earlier-described PSB of distance protection in the network and PSP of generators. Additionally, OST relays (or functions) may be installed at locations selected for controlled network splitting in order to trip selected transmission lines once asynchronous operation has been detected. The lines are selected according to the following criteria:

1) The lines must be so close to the center of power swings that the impedance loci encroach into the tripping zone of distance protection of those lines. Otherwise, the relays will not be able to issue tripping commands.
2) After the network has been split, the islands must have roughly balanced generation and demand. Otherwise, generators may be tripped or load shedding activated.
3) Islands have to be internally stable, that is further splitting should not occur.

Obviously, it is not easy to find places in a meshed network that would satisfy all these conditions and at the same time ensure that the controlled network splitting would be satisfactory for all possible locations of short circuits leading to asynchronous operation. Hence controlled splitting in real systems is often restricted to tie-lines which naturally divide the network into strongly connected islands. An example of such structures is shown in Figure 6.51.

The network shown in Figure 6.51a has a longitudinal structure and consists of two weakly connected subsystems. Any disturbance leading to the loss of synchronism will lead to a natural split of the network into two subsystems operating asynchronously. Clearly, the best place for the placement of OST is in the distance protection of the tie-lines. Once an asynchronous rotation has been identified, the relays will open circuit-breakers in the lines, splitting the network into two subsystems.

The second network shown in Figure 6.51b consists of a number of internally well-meshed subsystems connected by relatively weak tie-lines. This situation is typical of an interconnected system. Weakness of connections

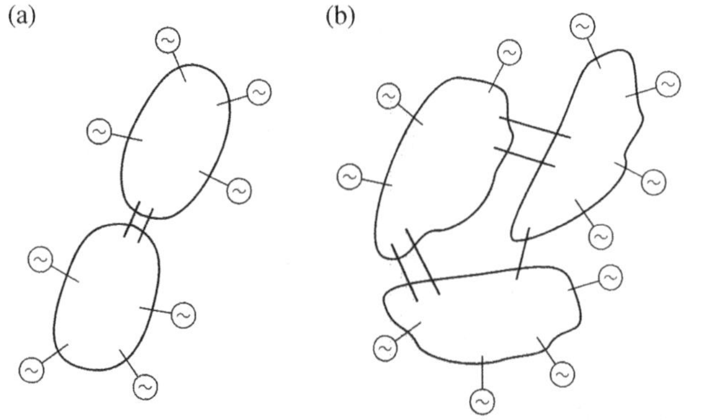

Figure 6.51 Network structures naturally amenable to splitting: (a) longitudinal system; (b) weakly connected subsystems.

may lead to asynchronous rotations in tie-lines following a disturbance. Relays installed in the tie-lines will separate the subsystems after a set number of asynchronous rotations. In both cases shown in Figure 6.51, automatic generation control (Chapter 9) supported by automatic load shedding should lead to balancing of generation and demand in each subsystem. Resynchronization must be preceded by matching frequencies in the subsystems and a check on synchronization.

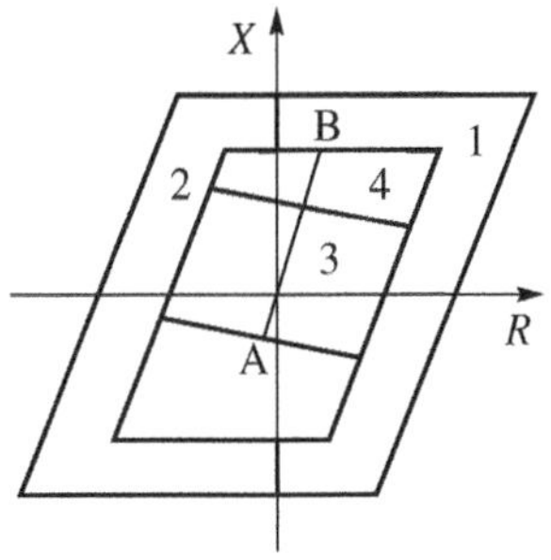

Figure 6.52 Characteristic of an out-of-step function.

It is important to appreciate that there is no natural way of splitting a tightly meshed network. Splitting such a network into islands usually leads to worsening stability conditions of generators operating inside separate islands and further splitting which may lead eventually to a total blackout. Hence it is not recommended to install OST relays (or functions) in tightly meshed transmission networks. Unfortunately some authors suggest that OST may be more widely used and that it is better to split the network into islands than trip generators. This view is valid only with respect to the weakly connected structures shown in Figure 6.51. In tightly meshed networks, keeping the network connected is most important and worth the sacrifice of tripping some generators. When the network is still connected, generators can be easily resynchronized and the power supply restored. On the other hand, when a tightly meshed large network has been split, reconnection is difficult and may take hours, requiring complicated action from system operators. Often some lines may be difficult to connect due to very large differences in voltages or angles in neighboring subsystems.

Splitting a network into islands in predetermined places is achieved by the OST relay or, in newer digital solutions, the OST function added to distance relays. Figure 6.52 shows the impedance characteristic of such a function. To differentiate between power swings and short circuits, two polygon-shaped characteristics are used, the external and internal ones, denoted by 1 and 2, respectively, in Figure 6.52. A short circuit is detected when the trajectory crosses one of the polygon areas in a time shorter than a set one. If the crossing time is longer than the set one, the impedance change is deemed to be due to power swings.

Detected power swings are deemed to be asynchronous if the impedance trajectory enters one side of the polygon, crosses line AB (corresponding to the angle $\delta' = 180^{\circ}$) and then leaves on the other side of the polygon. The protection divides the area of reach into close and remote ones, just as in PSP. This is achieved by two lines dividing the internal area into two areas, denoted by 3 and 4, respectively. If the trajectory crosses area 3, the power swing is deemed to be close. If the trajectory crosses line 4, the swing is deemed to be remote. Each of the areas has its own counter of asynchronous rotations. Tripping of the lines is activated only when the number of identified cycles in a given area reaches a set value.

6.6.7 Example of a Blackout

There have been many papers and reports published describing actual power system disturbances when the lack, or mal-operation, of PSB of distance protection has caused unwanted line tripping. Most of the cases ended up with an alert state and a return to the normal state (Section 1.5). There were cases, however, when an EPS moved from a normal state to an emergency state or even in extremis ended up with a partial or total blackout. Troskie and de Villiers (2004) describe an interesting case of a blackout caused by out-of-step relaying that happened in ESCOM (South Africa).

On 14 September 2001, heavy snowfall in the Drakensberg area resulted in overhead line failures followed by a trip command from zone 3 of the distance protection. This caused a deep power swing in the network. There was no unified concept of using an out-of-step protection system. Some relays had blockades only for tripping zone 1 and others only for tripping zone 2 or 3. The lack of a unified concept and inefficient PSB functionality resulted in more lines tripping and finally a loss of the entire network. Detailed post-mortem analysis and simulation have shown the need to modify the out-of-step protection system and for more careful selection of the settings.

7

Wind Power

The previous two chapters discuss the problems of steady-state stability (Chapter 5) and transient stability (Chapter 6) and are concerned mostly with the system operation of synchronous generators driven by steam or hydro turbines. However, environmental pressures discussed in Chapter 2 have caused many countries to set ambitious targets for renewable generation, often exceeding 20% of energy production. Currently, wind energy is the dominant renewable energy source and wind generators usually use induction, rather than synchronous, machines. As a significant penetration of such generation will change the system dynamics, this chapter is devoted to a discussion of the induction generator and its influence on power system operation.

7.1 Wind Turbines

The power in the wind can be extracted using a wind turbine. Wind turbines can either rotate about a horizontal axis, *horizontal axis wind turbines* (HAWTs), or a vertical axis, *vertical axis wind turbines* (VAWTs). General practice is to use HAWTs with three blades. Although any number of blades can be used, if too many are used they tend to interfere with each other aerodynamically, while using only two blades tends to lead to large power pulsations as the blades pass by the tower; three blades reduce these power pulsations and are also generally deemed aesthetically more pleasing. Three-bladed HAWTs are therefore generally favored. Modern wind turbines extract energy from the wind by using aerodynamic blades that produce a lift force along the length of the blade. This aerodynamic force integrated along the length of the blade produces the torque on the turbine shaft. The tip speed of the turbine is limited to typically (80–100) m/s so that as turbines get bigger their rotational speed reduces such that large multi-megawatt turbines rotate slowly at about (15–20) rpm.

Although a number of generator drive arrangements can be used, described later in this section, generally the wind turbine drives a generator through a gearbox that steps up the speed from about 20 rpm at the turbine shaft to 1500 rpm at the generator. The generator is then connected through a transformer to the main electricity supply. The generator and gearbox, along with other associated equipment, are placed at the top of the turbine tower in a *nacelle*. The layout of a typical turbine is shown in Figure 7.1. Besides the generation system other subsystems are required that will turn the turbine into the wind, the *yaw system*, and provide braking. On the turbine side of the gearbox, power is delivered at low speed and very high torque so that a large-diameter drive shaft is necessary. On the generator side the power is delivered at relatively high speed and low torque so that a thinner shaft can be used. Typically, the generator will operate at 690 V and the transformer will be placed at the bottom of the tower or in a separate building close to the turbine. It is only in offshore wind turbines that the transformer will be located in the nacelle.

Power System Dynamics: Stability and Control, Third Edition. Jan Machowski, Zbigniew Lubosny, Janusz W. Bialek and James R. Bumby.

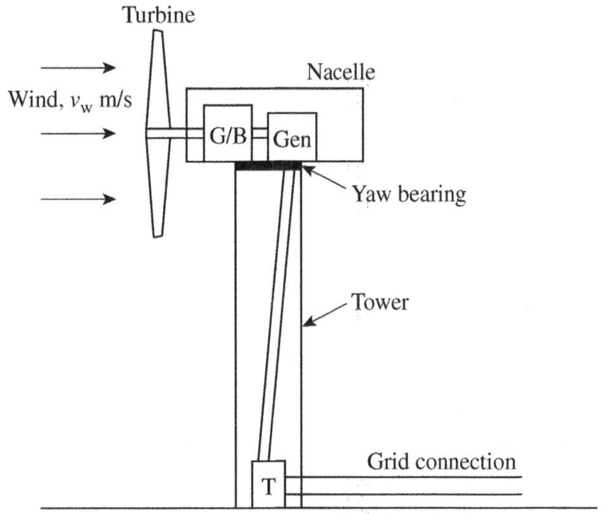

Figure 7.1 Typical arrangement for a wind turbine: G/B – gearbox; Gen – generator; T – transformer.

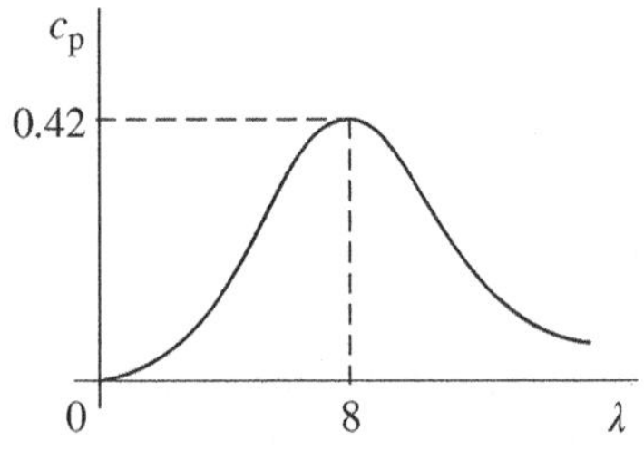

Figure 7.2 Typical c_p/λ curve for a wind turbine.

The power in the wind varies as the cube of wind speed, but unfortunately the wind turbine can only extract a fraction of this given by

$$P = \frac{1}{2}\rho A c_p v_w^3 \quad \text{W} \tag{7.1}$$

where ρ is the density of air, $A = \pi r^2$ the swept area, r the rotor radius, v_w the wind speed, and c_p the coefficient of performance of the turbine. The *swept area* is the area of the circle created by the blades as they sweep through the air. If the turbine were to extract all the kinetic energy from the wind this would mean that the wind velocity behind the turbine was zero. This is not possible as the air flow must be continuous so that the theoretical maximum energy that can be extracted is with $c_p = 16/27 \cong 0.593$ and is termed the *Betz limit*. In practice the coefficient of performance is less than this and also varies with the *tip speed ratio* λ, as shown in Figure 7.2. The tip speed ratio λ is a nondimensionless quantity defined as the ratio of the rotor tip speed to the wind speed v_w, that is

$$\lambda = \frac{\omega_T r}{v_w} \tag{7.2}$$

where ω_T is the rotor angular speed in rad/s.

The c_p/λ curve is unique to a particular design of wind turbine, and to extract maximum power the turbine must be operated at the peak of this curve, referred to as *peak power tracking*. For any given wind speed, Eqs. (7.1) and (7.2) and the c_p/λ characteristic can be used to calculate the turbine power as a function of shaft speed, as shown in Figure 7.3a.

For a given c_p the turbine power as a function of wind speed can be calculated from Eq. (7.1) and is shown schematically in Figure 7.3b. At very low wind speeds the power is very small and the turbine will not operate until the wind speed is above the *cut-in speed*, typically about (3–4) m/s. Above this speed the turbine will produce increasing levels of power until rated wind speed is reached when the power is limited to its rated value. The power now remains constant until the wind speed reaches the *shutdown or furling wind speed*, typically 25 m/s, when the turbine is shut down and turned out of the wind to prevent damage. Turbines are designed to survive winds of about 50 m/s, which is referred to as the *survival wind speed*.

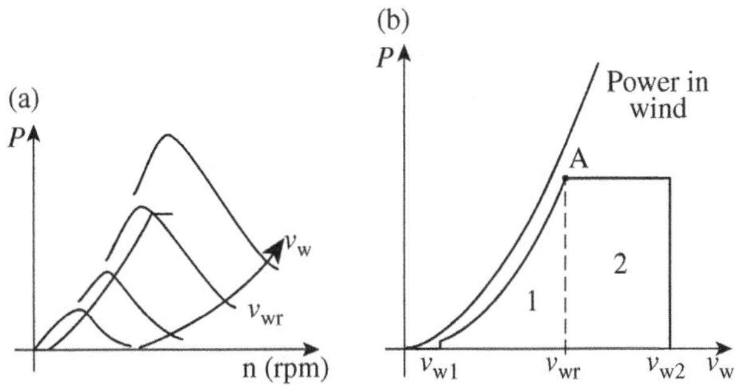

Figure 7.3 Turbine power (a) as a function of shaft speed (b) as a function of wind speed: v_{w1} – cut-in wind speed; v_{wr} – rated wind speed; v_{w2} – shutdown wind speed.

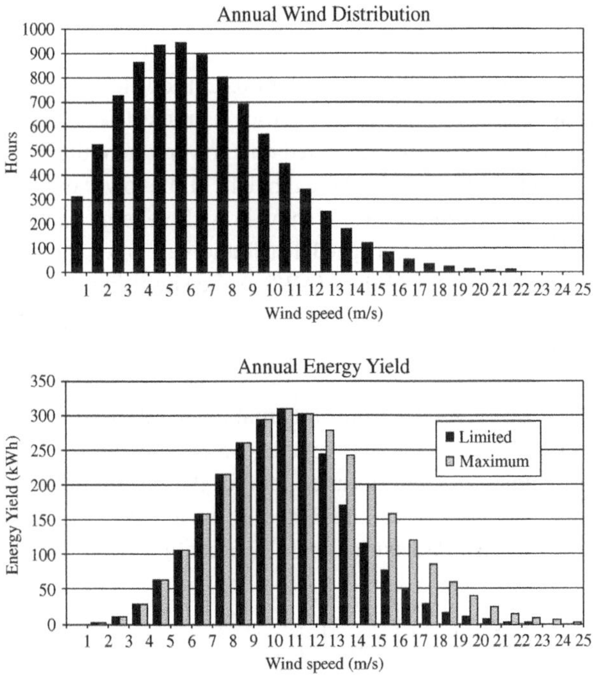

Figure 7.4 Annual energy production by a turbine of 60 m diameter in MAWS of 7 m/s.

As the power increases as the cube of wind speed, at high wind speeds the power can be very large and must be curtailed in some way to prevent damage to the turbine and power conversion equipment. Turbines are generally rated for a wind speed of 12.5 m/s and above this the power is limited in some way to the rated power to give the typical power output curve of Figure 7.3. At these high wind speeds the power content may be large but the number of hours that this occurs per year is small so the energy content is small compared with the energy available in the mid-speed range. This effect is well represented in Figure 7.4 for a site with a *mean annual wind speed* (MAWS) of 7 m/s. This figure shows the number of hours per year that winds of a certain speed occur; it also shows the energy content associated with these winds in 1 m/s "bins." The data are for a turbine of 60 m diameter and show the total energy available at each wind speed and the energy extracted when the power output is limited to the rated value above rated wind speed.

The diameter and rotational speed of a wind turbine are readily calculated for a range of turbine ratings using Eqs. (7.1) and (7.2), as shown in Table 7.1. The rated wind speed is assumed to be 12.5 m/s. Although a typical c_p

Table 7.1 Diameter and operating speed of wind turbines in a 12.5 m/s wind.

Power P (kW)	Area A (m^2)	c_p -	λ -	Diameter $2r$ (m)	Rotational speed ω_T (rpm)
1	2.5	0.3	4.5	1.8	629
10	24.8	0.3	4.5	5.6	199
100	247.7	0.3	5	17.8	70
500	1238.5	0.3	6	39.7	38
1000	2477.1	0.3	6	56.2	27
2000	4954.2	0.3	7	79.4	22
5000	12 385.5	0.3	8	125.6	16

value for a large wind turbine is about 0.42, the values used here have been reduced to account for losses in the electric system so that the power output is an electrical power output.

These turbines must be supported on a tower where the height of the nacelle is typically about the same as the blade diameter, but often the actual height of the tower used depends on location and is a manufacturing option.

The energy data in Figure 7.4 show how the energy capture is distributed through the range of wind speeds. Although the data on wind speed distribution are often measured, a good estimate of the distribution can be generated from knowledge of the MAWS at the site using the *Weibull distribution*

$$F(v) = \exp\left[-\left(\frac{v}{c}\right)^m\right] \tag{7.3}$$

This equation gives the probability that the wind speed v_w is greater than v. In this equation m is the shape factor and c is the scaling factor and depends on the MAWS.

Shape factor varies with location, but for flat terrain in Western Europe a shape factor of two is generally used when the scaling factor can be linked to the MAWS. This special case of the Weibull distribution is sometimes called the *Rayleigh distribution* and the probability that the wind speed is greater than v becomes

$$F(v) = \exp\left[-\frac{\pi}{4}\left(\frac{v}{v_{\text{mean}}}\right)^2\right] \tag{7.4}$$

This equation can be converted to the number of hours per year that the wind speed is greater than v as

$$\text{hrs} = 8760 \times \exp\left[-\frac{\pi}{4}\left(\frac{v}{v_{\text{mean}}}\right)^2\right] = \frac{8760}{\exp\left[\frac{\pi}{4}\left(\frac{v}{v_{\text{mean}}}\right)^2\right]} \tag{7.5}$$

The number of hours that the wind speed is in a (say) 1 m/s bin at v m/s is now easily found by calculating the number of hours at $v - 0.5$ and $v + 0.5$ and subtracting. Such an approach generates a curve similar to that in Figure 7.4. Of particular importance is the *capacity factor* c_f which is defined as the ratio of the actual energy produced *over a designated time* to the energy that would have been produced if the plant had operated continuously at maximum rating. The time period often used is one year (8760 hours)

$$c_f = \frac{\text{actual annual energy production}}{\text{maximum plant rating} \cdot 8760} \tag{7.6}$$

For a site with a MAWS of about 7 m/s the capacity factor is about 30%, but if situated on a site where the MAWS is 5 m/s the capacity factor will drop to about 12%.

The energy carried in the wind is transferred at discrete frequencies with three major energy peaks occurring at approximately 100 hours, 12 hours, and 1 minute, caused by the passing of large-scale weather systems, diurnal variations, and atmospheric turbulence, respectively (Van der Hoven 1957). The longer time variations are relatively predictable, while the short-term fluctuations can have a significant effect on the turbine's aerodynamic performance, and the power can, and does, fluctuate quite widely because of the turbulent nature of the wind. Although the power output of an individual turbine can vary on a second-by-second basis, the aggregation of the output of a number of turbines soon results in a smooth energy flow. Nevertheless, these fluctuations must be accommodated by the electric power system (EPS).

7.2 Generator Systems

A number of generator configurations are used in the conversion of wind energy and these are summarized in Figures 7.5–7.11. In all those diagrams $1:n$ denotes the gear ratio. Most drive systems for large wind turbines use a gearbox to step up the speed from 15 to 30 rpm at the turbine itself to typically (1200–1500) rpm at the generator for a 50 Hz system. Of the generator systems shown, the synchronous generator of Figure 7.5 runs at constant speed and the induction generator system of Figure 7.6 operates at a speed that is very nearly constant and may vary by (2–4)% from no load to full load. Because the change of speed of the induction generator is small, it is often referred to as a *fixed-speed generator*.

The other generator configurations in Figures 7.7–7.11 have different degrees of speed variation. As described in Section 2.2.1, synchronous generators are normally used for power generation with conventional gas, steam, or water turbines. For completeness Figure 7.5 shows the use of such a generator with a wind turbine, but synchronous generators are not normally used for grid-connected wind turbines. The main reason for this is that, as the generator runs at a constant speed, the coupling between the generator and the grid for this system is very stiff with the result that all the transient torques produced in the turbine drive shaft due to wind turbulence produce significant mechanical stress on the gears, reducing the system's reliability. Generator systems with more compliance

1:n
G
T
Distribution system
400 V, 11 kV, 33 kV, 66 kV

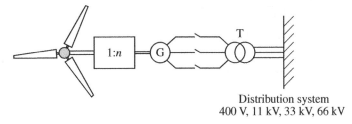

Figure 7.5 Synchronous generator.

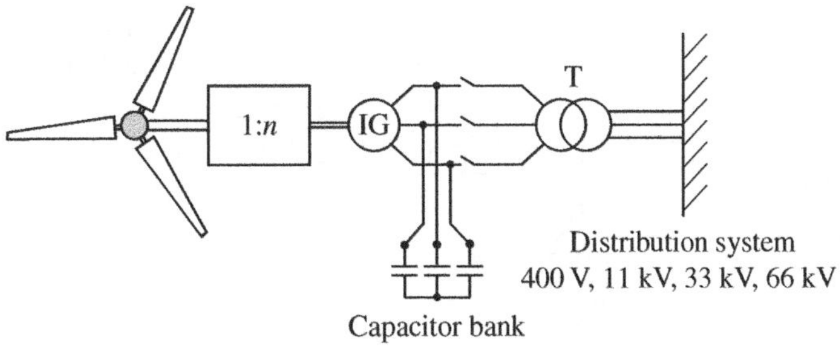
1:n
IG
T
Distribution system
400 V, 11 kV, 33 kV, 66 kV
Capacitor bank

Figure 7.6 Squirrel-cage induction generator.

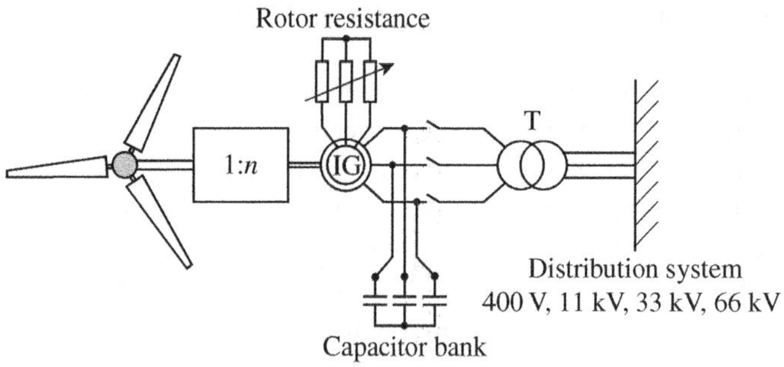

Figure 7.7 Wound-rotor induction generator.

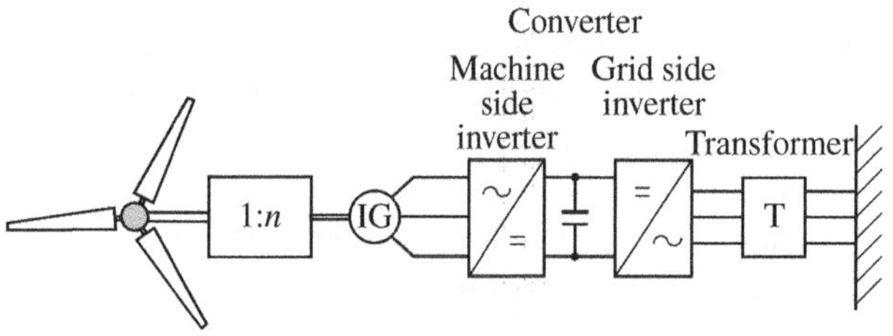

Figure 7.8 Squirrel-cage induction generator with fully rated converter.

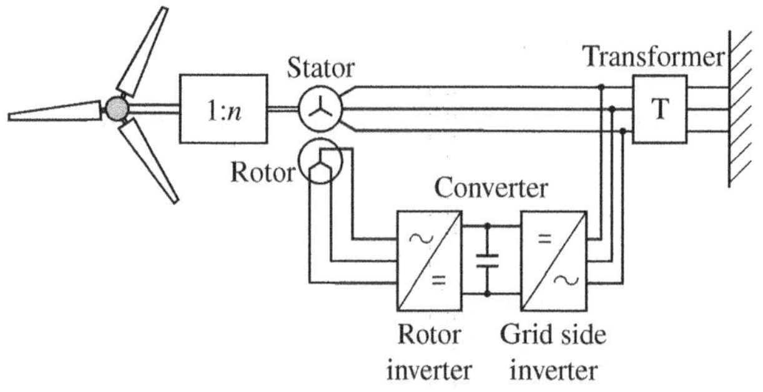

Figure 7.9 Double-fed induction generator.

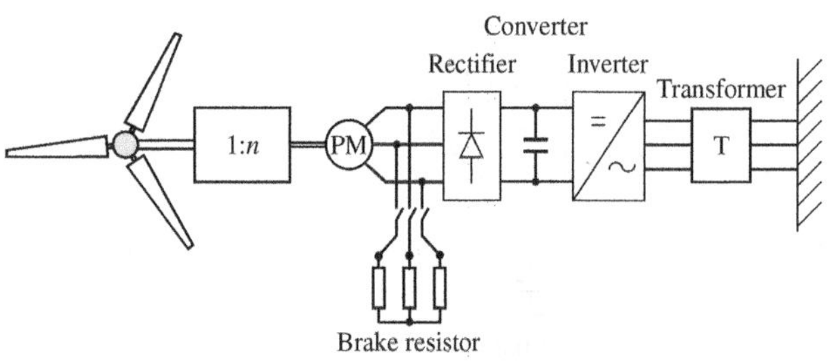

Figure 7.10 Permanent magnet generator with fully rated converter (gearbox optional).

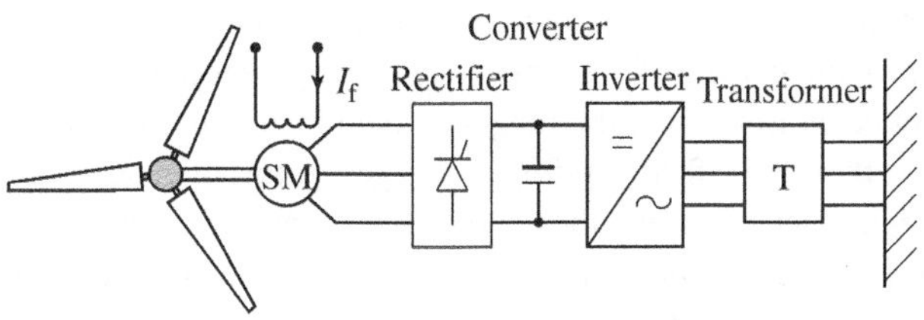

Figure 7.11 Wound field generator with fully rated converter.

are therefore favored. All the other systems shown have this compliance to varying degrees as the speed of the drive system can change when subject to transient torques. However, it is important to realize that synchronous generators may be used to advantage in some stand-alone turbine systems where the frequency of the system can change.

Traditionally, the fixed-speed induction generator of Figure 7.6 has been used with the speed geared up to 1500 rpm in land-based turbines of up to about 750 kW. This arrangement is often referred to as the *Danish concept* with the power generated naturally changing as the wind speed changes (Section 7.2). When the turbine is subjected to a wind gust its speed can change slightly and transient torques are reduced compared with the stiff synchronous system. As this generator arrangement operates at a nominally fixed speed, energy capture cannot be maximized but improvements can be made by using an induction generator with a four-pole and six-pole stator winding. This allows the generator to be operated at two speeds, 1000 and 1500 rpm, for a 50 Hz system, with a corresponding increase in energy captured. Figure 7.7 shows a modification of the fixed-speed induction machine where the squirrel-cage rotor of the fixed-speed machine has been replaced with a wound rotor. By controlling the resistance of the wound rotor, the speed range over which the turbine operates can be increased slightly, providing yet more compliance in the system (Section 7.4). All induction machines must be supplied with reactive power in order to produce the magnetizing flux in the machine. This reactive power must be supplied from the system and to try to reduce this demand, power factor correction capacitors are often fitted at the generator terminals. In operation the wind turbine system generates whatever power it can and has therefore traditionally been regarded as *negative load*. As turbine technology develops, network operators are requiring more control of the turbine so that the generator can contribute to a greater, or lesser, extent to voltage control (control of reactive power) and frequency control (control of real power).

By introducing power electronic control into the generator systems as in Figures 7.8–7.11 the speed range over which the generator operates can be increased, so increasing energy capture. As the speed may now change quite significantly as the turbine power changes, the system has a significant degree of compliance and transient torques are further reduced. In addition, the power electronic control allows the power factor at the generator terminals to be varied as required by the network operator. The induction generator with a *fully rated converter* (Figure 7.8) allows a great degree of control but the power electronic converter must be rated at the full MVA output of the turbine and carry the full output of the generator. Such fully rated converters are expensive and one way of reducing the cost is to use a *partially rated converter* and a *double-fed induction generator* (DFIG). This generator arrangement is shown in Figure 7.9 and is currently the system favored by many manufacturers for multi-megawatt turbine systems. The DFIG is an induction machine with a wound rotor and with the stator connected directly into the system at system frequency. The rotor is fed from a power converter at slip frequency and, as such, is usually rated at (25–30)% of the generator rating. A converter of this rating allows the speed to vary by a similar amount, that is ±(25–30)%, with correct control of the converter allowing both the speed and output power factor of the generator to be controlled (Section 7.5). However, slip rings are necessary with this system to allow the power converter to feed the rotor circuit.

The final two systems in Figures 7.10 and 7.11 have some similar features in that both use fully rated converters and both use synchronous machines. As these generators are decoupled from the network by the converter, they do not suffer the problems of the directly connected synchronous generator of Figure 7.5. Both systems can be used with, or without, a gearbox and both can operate at variable speed. Because of the fully rated power converter, both have full control of the real and reactive power they generate. Figure 7.10 shows a scheme using a permanent magnet generator where the magnetic field inside the generator is produced by permanent magnets on the rotor. Because there is no field winding in this generator, there is no associated I^2R loss and so this type of generator has a very high efficiency, well above 90%. The output of the generator is first rectified before being inverted and connected into the network. Although passive rectifiers could be used, normally on large machines an *active rectifier* using IGBT (insulated gate bipolar transistor) technology is used to give full control over the power delivered to the DC link and also to improve the generator form factor and reduce generator losses and harmonic forces. One special feature of this arrangement is that the permanent magnets always ensure the magnetic field is active so that if the generator is turning it will always induce an electromotive force (emf) in the armature windings. This feature can be taken advantage of because, if the generator windings are short-circuited (perhaps through a small resistance to limit the current), a large electromagnetic torque will be produced that will prevent the turbine rotating, that is it can be used as a braking system.

The arrangement of Figure 7.11 shows a direct-drive synchronous machine without a gearbox. Removing the gearbox is seen to remove one source of failure in the system but requires a low-speed generator. Although a permanent magnet generator is sometimes used without a gearbox, an alternative is to use a wound field machine, as shown in Figure 7.11. By using a wound field machine the strength of the magnetic field inside the generator can be controlled by adjusting the current in the field winding. This allows the magnitude of the induced emf to be controlled as the speed of the generator changes. A thyristor converter or an active IGBT rectifier can be used to rectify the variable frequency-generated AC power to DC power before being finally inverted to fixed-frequency AC power for connection to the network. The fully rated converter allows full control of the real and reactive power at the inverter terminals.

Figure 7.3 shows the need for some form of power control at high wind speeds that limits the turbine power output to the rated value so that the rotational speed of the turbine does not exceed its rated speed. In general two forms of power control are possible. The first is to design the turbine blades so that as the wind speed increases the lift force on the turbine blade reduces and the turbine blade progressively goes into stall along its length. This is termed *passive stall control*. As stall is determined by the relative angle at which the wind attacks the blade, the *angle of attack*, such power control is normally used only on fixed-speed turbines. An alternative form of control is *active pitch control* where the pitch of the blades is changed to reduce the power output. This type of control requires an active control system that changes the pitch of the blades based on some feedback parameter such as rotational speed or output power and is similar in function to the conventional governing systems of Figure 2.14. A third possible control strategy is to pitch the blades in the opposite direction to the conventional pitch-controlled machine when the blades go into stall: this is sometimes termed *active stall*. Table 7.2 classifies the different generation systems of Figures 7.5–7.11 along with the type of power control in a similar way to Hansen, in Ackermann (2005).

The typical variation of power output for a fixed-speed wind turbine with stall control and for a variable speed turbine with pitch control are compared in Figure 7.12. As the rotational speed of a fixed-speed wind turbine does

Table 7.2 Generator options.

Speed Range	Passive Stall Control	Pitch Control
Fixed	Figure 7.6	Figure 7.6
Small	—	Figure 7.7
Limited (±30%) – partially rated converter	—	Figure 7.9
Large – fully rated converter	—	Figures 7.8, 7.10, and 7.11

not change significantly, the tip speed ratio cannot be maintained constant as the wind speed varies (Eq. [7.2]), and the c_p value changes (Figure 7.2). Above rated conditions the blades are designed to stall progressively so that the change in c_p keeps the generated power sensibly constant. However, at lower wind speeds this means that the energy captured is reduced compared with operating at a fixed tip speed ratio and maximum value of c_p. To achieve maximum power output the tip speed ratio must be allowed to vary as wind speed varies so that c_p can be maintained at its peak value; this is called *peak power tracking*. Consequently, allowing the speed of the turbine to vary has a number of advantages over fixed-speed operation including allowing the turbine to operate at the peak of its c_p/λ curve and thereby maximizing energy capture. Also, above rated wind speed the actively controlled machines can maintain a more constant power output, as shown in Figure 7.12. To try and improve energy capture, some fixed-speed turbines can operate at two speeds by using an induction generator with both a four-pole and six-pole winding.

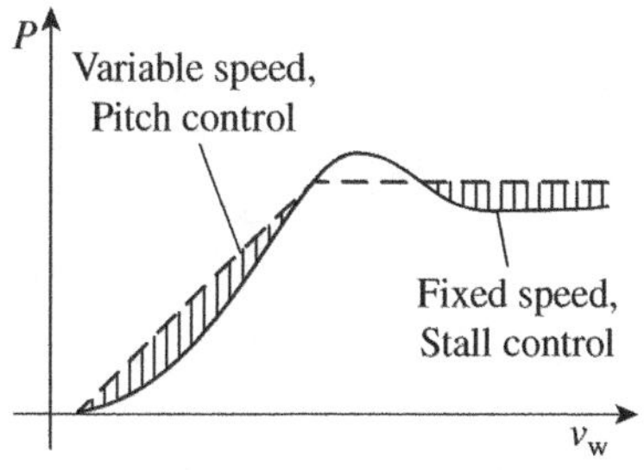

Figure 7.12 Typical power-out curves for a stall-controlled and active pitch-controlled turbine.

Besides increasing the energy capture, allowing the speed of the wind turbine to vary as the wind speed changes also ensures a much softer coupling to the grid than for a fixed-speed turbine, particularly a directly coupled synchronous machine. This has important consequences because introducing compliance into the system by allowing the speed to vary reduces the stress loading on the drive shaft and in the gearbox, thereby increasing its reliability.

7.3 Induction Machine Equivalent Circuit

Induction machines are widely used in many applications, owing to their simple construction and ease of operation. They are mainly used as motors, so, from the power system point of view, they constitute loads. This is discussed in Section 3.4. However, induction machines are also often used as generators in wind farms. To consider how the induction machine can operate as a generator, it is necessary to consider its equivalent circuit in some detail. The induction machine consists of a three-phase stator winding and a rotor. This rotor can either have a squirrel-cage rotor, where the rotor windings are simply connected to large shorting rings at either end of the rotor, or a fully wound rotor with the end of the windings brought out to slip rings. In the case of the wound-rotor machine, the winding is a three-phase winding connected in star with the end of each phase connected to one of three slip rings. These slip rings can then be shorted to form effectively a squirrel-cage winding or connected into some other external circuit to help form the required machine characteristic. Some of the different connection options are described in this section.

The induction machine is essentially a transformer with a rotating secondary winding and can be represented by the equivalent circuit in Figure 7.12. In this circuit $\underline{I}_1$ and $\underline{I}_r$ are the stator and rotor currents, respectively, and $\underline{V}$ is the terminal voltage. Current $\underline{I}_2$ is a fictitious current obtained after subtracting the magnetizing current $\underline{I}_m$ from the stator current $\underline{I}_1$. Current $\underline{I}_2$ flows through the equivalent primary winding of the transformer and is the rotor current $\underline{I}_r$ referred to the stator, or primary winding of the machine. Current $\underline{I}_1$ is seen to flow from the terminal voltage $\underline{V}$ toward the rotor, which is the convention used for motoring, rather than generating. Although the discussion in this section is concerned mainly with induction generators, rather than motors, the motoring convention of signs has been retained mainly because it is the convention that the reader is likely to be more used to.

In Figure 7.13a R_1 and X_1 are the resistance and leakage reactance of a stator phase winding, while R_2 and L_2 are the rotor resistance and leakage inductance per phase. X_m is the *magnetizing reactance* and it is the current $\underline{I}_m$ flowing through this reactance that sets up the rotating magnetic field. Sometimes a resistance is connected in parallel with the magnetizing reactance to represent the stator *iron loss*, that is the hysteresis loss and eddy-current

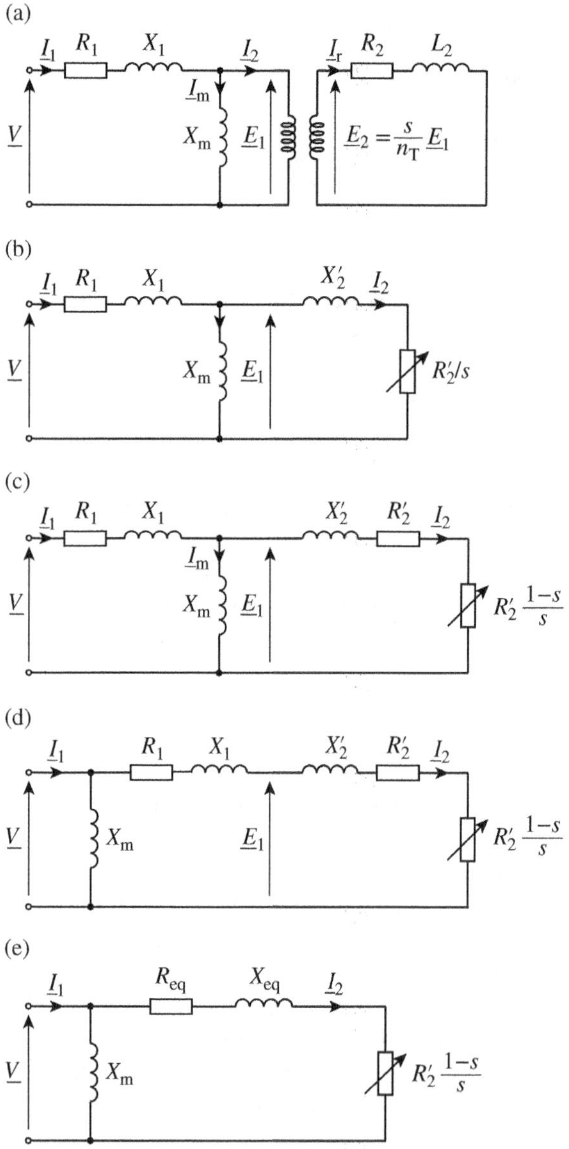

Figure 7.13 Development of the induction machine equivalent circuit (motoring positive).

loss in the stator core. However, the iron loss is generally small so that the equivalent iron loss resistance is very large compared with the magnetizing reactance and is therefore often omitted from the equivalent circuit. The three-phase stator winding produces a magnetic field that rotates at synchronous angular speed ω_{sm} and if the rotor is rotating at an angular speed ω_{rm} slightly different from this an emf will be induced in the rotor at a frequency proportional to the difference between these speeds and inversely proportional to the pole number, the *slip frequency* f_{slip}. The slip speed is defined as the difference between these two speeds, while the per-unit slip, or *slip*, is normalized to the synchronous speed, that is

$$s = \frac{(\omega_{sm} - \omega_{rm})}{\omega_{sm}} = \frac{(\omega_s - \omega_r)}{\omega_s} \tag{7.7}$$

where $\omega_s = 2\pi f = \omega_{sm}p$ is the angular electrical synchronous speed, f is the grid frequency, $\omega_r = \omega_{rm}p$ is the rotor electrical angular speed, and p is the number of pole pairs.

In the equivalent circuit of Figure 7.13a the frequency of the currents in the primary winding is the grid frequency f while the current induced in the rotor is at slip frequency f_{slip}. The rotor current is given by

$$I_r = \frac{\frac{s}{n_T}E_1}{R_2 + j\omega_{slip}L_2} \tag{7.8}$$

where E_1 is the air-gap emf, n_T the turns ratio between the rotor winding and the stator winding, and $\omega_{slip} = 2\pi f_{slip}$ is the slip angular electrical frequency. As the current in the stator and rotor are at different frequencies, simplifying the equivalent circuit as it stands is not possible. However, if the top and bottom of Eq. (7.8) are divided by the slip s (noting that $\omega_{slip} = \omega_s - \omega_r = s\omega$), then referring to the primary stator winding by dividing by the turns ratio gives the rotor current I_2 referred to the stator as

$$\underline{I}_2 = \frac{\underline{E}_1}{\frac{R_2'}{s} + j\omega L_2'} \tag{7.9}$$

In this equation R_2' and L_2' are the rotor resistance and leakage inductance referred to the stator. Equation (7.9) expresses the rotor current referred to the stator at the grid frequency so allowing the equivalent circuit to be simplified to that of Figure 7.13b. In Figure 7.13c the slip-dependent resistance element is conveniently separated into a fixed resistance R_2' and a variable resistance $R_2'(1-s)/s$ that respectively represent the rotor resistance and the mechanical power. This equivalent circuit (Figure 7.13c) is referred to as the *accurate equivalent circuit*. The final simplification is to recognize that the magnetizing current is small compared with the main rotor current and moves the stator resistance and leakage reactance to the rotor side of the magnetizing reactance, as in Figure 7.13d, to produce the *approximate equivalent circuit*. R_1 and R_2' along with X_1 and X_2' can be usefully combined into an equivalent resistance R_{eq} and equivalent reactance X_{eq}, respectively, to give the final approximate equivalent circuit of Figure 7.13e. Both the accurate and approximate equivalent circuits are necessary to fully understand the operation of the induction machine as a generator.

Because of its power system implications, it is important to realize that the rotating magnetic field in the induction machine is always produced by drawing a magnetizing current $\underline{I}_m$ from the supply regardless of whether the machine acts a motor or generator. This magnetizing current is shown in Figure 7.13a and is represented by the magnetizing reactance X_m. As the magnetizing current must be drawn from the supply, the induction machine always absorbs reactive power and must be connected to an EPS that can supply this reactive power if it is to function; only under exceptional circumstances can an induction machine be made to self-excite, and this is not relevant to the subject of this book.

Analysis of the simplified circuit of Figure 7.13d gives

$$I_2 = \frac{V}{\sqrt{\left(R_1 + \frac{R_2'}{s}\right)^2 + \left(X_1 + X_2'\right)^2}} \tag{7.10}$$

This equivalent circuit also defines the power flow through the machine and, if the losses in the stator resistance and the iron core are neglected, the power supplied to the machine from the grid is the same as the power supplied to the rotor and is given by

$$P_s \cong P_{rot} = 3I_2^2\frac{R_2'}{s} = \tau_m\omega_{sm} \tag{7.11a}$$

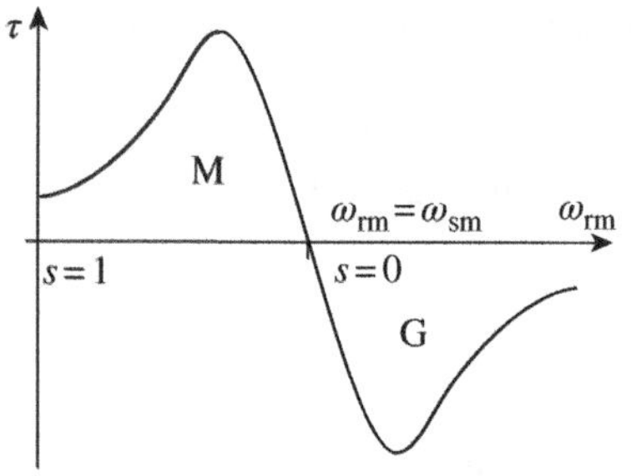

Figure 7.14 Torque-speed characteristics of an induction machine showing both motor and generator action (motoring positive).

while the power loss in the rotor resistance and the mechanical power delivered are

$$\Delta P_{\text{rot}} = 3I_2^2 R_2' = sP_s \tag{7.11b}$$

$$P_m = 3I_2^2 \frac{R_2'(1-s)}{s} = P_s(1-s) = \tau_m \omega_{rm} \tag{7.11c}$$

As the mechanical power output is equal to the product of the torque and the rotor angular speed ω_{rm}, and noting that slip s is given by Eq. (7.7), the power supplied to the rotor can be written in terms of the torque and angular synchronous speed ω_{sm} as $P_s \cong P_{\text{rot}} = \tau_m \omega_{sm}$. This is noted in Eq. (7.11a).

Equation (7.11a) allow the efficiency of the machine to be expressed as

$$\eta_{\text{motor}} = \frac{P_m}{P_s} = (1-s), \quad \eta_{\text{gen}} = \frac{P_s}{P_m} = \frac{1}{(1-s)} \tag{7.12}$$

The shaft torque produced by the machine is obtained from Eq. (7.11c) as

$$\tau_m = \frac{P_m}{\omega_{rm}} = \frac{P_m}{\omega_{sm}(1-s)} = \frac{3}{\omega_{sm}} \frac{V^2}{\left[\left(R_1 + \frac{R_2'}{s}\right)^2 + (X_1 + X_2')^2\right]} \frac{R_2'}{s} \tag{7.13}$$

The variation of torque with slip is shown in Figure 7.14 for both positive and negative slip. Positive slip is when the rotor speed is less than synchronous speed and corresponds to motor action, and is discussed in Section 3.4, while negative slip is when the rotor speed is greater than synchronous speed and corresponds to generator action. Similarly, positive torque corresponds to motor action and negative torque corresponds to generator action. If the machine is driven at a speed greater than synchronous speed, it will naturally generate electrical power into the grid. If the speed drops below synchronous speed, the machine will naturally motor.

The power flow described by Eq. (7.11a) is shown diagrammatically in Figure 7.15, where P_g is the power from, or supplied to, the electricity grid. In this diagram the directions are shown as positive for motor action. If power is now supplied to the grid, that is generator action, P_m and P_s reverse direction and slip s becomes negative. As expected, the direction of the rotor loss ΔP_{rot} remains unchanged regardless of motor or generator action.

7.4 Induction Generator Coupled to the Grid

Section 5.4.7 explains how, when a synchronous generator is connected to the grid, it behaves in a similar way to a mechanical mass/spring/damper system, where the effective spring stiffness is equivalent to the synchronizing power coefficient $K_{E'}$. This creates a very stiff coupling to the grid (Figure 7.16a), which, for a wind turbine system, can result in large stresses in the drive shaft and gearbox because of the way the system responds through the dynamic torques produced by wind turbulence. A coupling to the system that is much "softer" and allows a degree

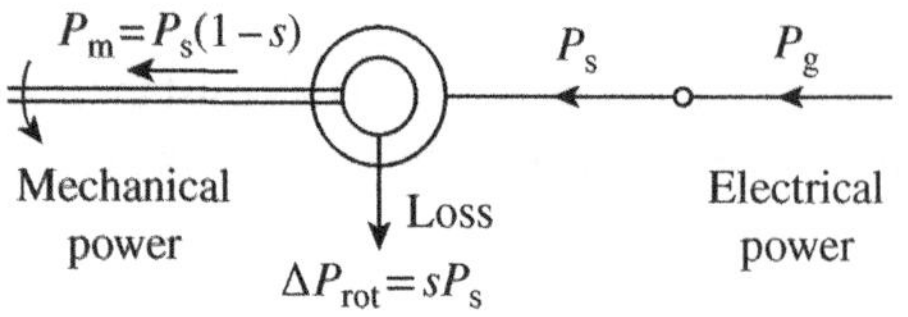

Figure 7.15 Power flow in an induction machine (motoring positive).

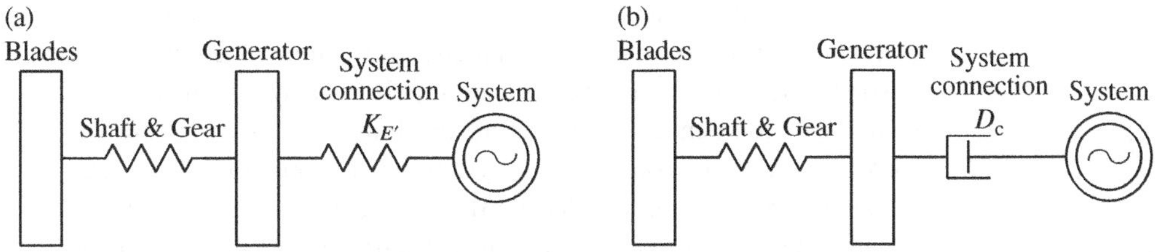

Figure 7.16 Effective system coupling of (a) synchronous and (b) induction generators.

of movement at the generator would help reduce these shock torques. Such a coupling is provided by an induction generator, as described below.

If the slip is small the induction machine torque, given by Eq. (7.13), can be approximated as

$$\tau_m = \frac{3}{\omega_{sm}} \frac{V^2}{\left[\left(R_1 + \frac{R_2'}{s}\right)^2 + \left(X_1 + X_2'\right)^2\right]} \frac{R_2'}{s} \approx \frac{3V^2}{\omega_{sm}} \frac{1}{R_2'} s = D_c \Delta\omega \tag{7.14}$$

where $\Delta\omega = \omega_s - \omega_r$ is the rotor speed deviation with respect to the synchronous speed and D_c is an equivalent "damper constant." Eq. (7.14) shows that the induction machine torque is proportional to speed deviation, implying that the coupling of the induction machine to the grid is analogous to a mechanical damper as shown in Figure 7.16b. Such a coupling introduces a substantial degree of compliance, or "give," into the system and is much softer than the stiff coupling associated with the synchronous generator. This has important advantages for some energy conversion systems, such as large wind turbines, because the additional compliance helps reduce the stress in the drive shaft due to the dynamic torques produced by wind gusting and wind turbulence. Equation (7.14) also shows that the "damper constant" D_c determines the effective compliance and that this can be controlled by changing the rotor resistance.

Generally, when an induction machine is connected to the grid, and used as a generator, it will be within the distribution network rather than the main transmission network. Used in this way, such generation is termed *embedded generation* and is typical of wind generators and other forms of renewable energy generation.

When assessing the performance of an induction generator embedded within the system, the system reactance X_s and resistance R_s impact on the operation of the induction generator and modify the equivalent circuit, as shown in Figure 7.17 (Holdsworth et al. 2001). The system impedance, as seen by the induction generator, is affected by a number of factors:

- The "strength" of the network. If the network is *strong* the reactance between the generator and the system will be small, leading to a large short-circuit level. Short-circuit level being defined at the *point of common connection* as V_s/X_s. On the other hand, a *weak* system will have a large reactance and low short-circuit level.
- For distribution networks, resistance affects are more apparent than at the transmission or subtransmission level with the X/R ratio changing from typically 10 for transmission networks to 2 for distribution networks.

Figure 7.17 Induction machine equivalent circuit including system reactance and resistance. V_{PCC} – voltage at the point of common connection; X_{PFC} – power factor correction capacitance.

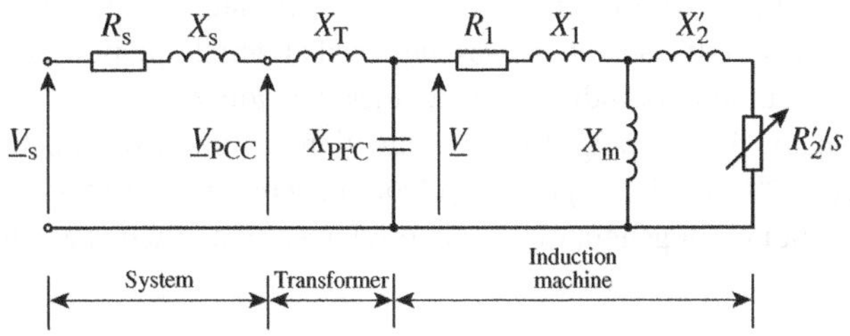

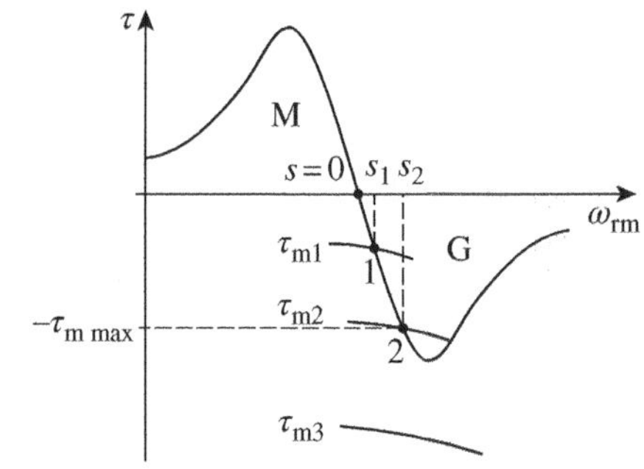

Figure 7.18 Steady-state stability of an induction generator (motoring positive).

Also impacting on the system behavior is the need for power factor correction at the terminals of the induction generator. Whether motoring or generating, the EPS must always supply the magnetizing current to the induction machine, therefore the induction machine always consumes reactive power and the system must supply this. The magnetizing current is represented by the magnetizing reactance in the equivalent circuits of Figure 7.16. The amount of reactive power varies depending on the loading condition, being a maximum at no load and reducing at full load.

The induction machine torque slip curve determines the steady-state stability of the generator. Figure 7.18 shows what happens if the applied mechanical driving torque increases. Initially, the induction generator is operating at point 1 with a mechanical applied torque τ_{m1}. The mechanical torque is now increased to τ_{m2}. As the applied mechanical torque is now greater than the electrical torque the generator speeds up (slip increases) and operates at a slightly higher slip s_2 at point 2. If the applied torque is now increased to τ_{m3} the applied torque and the electrical torque curves do not intersect and there is no steady-state operating point so the system will be unstable. The peak of the torque/slip curve determines the *pull-out torque* and the system steady-state stability limit.

Equation (7.13) can be used to see the way in which system reactance modifies the torque/slip characteristic if the equivalent circuit is modified to a very approximate one, as in Figure 7.16d, with the network resistance R_s incorporated into R_1, the network reactance X_s and transformer reactance X_T incorporated into X_1, and the stator voltage becoming the system supply voltage V_s. However, accurate calculation must use the accurate equivalent circuit of Figure 7.16. Peak torque occurs when $\frac{d\tau_m}{ds}$ is a maximum and occurs when

$$s_{\max} = \frac{R'_2}{\sqrt{R_1^2 + (X_1 + X'_2)^2}} \cong \frac{R'_2}{X_1 + X'_2} \tag{7.15}$$

giving the pull-out torque

$$\tau_{\mathrm{mmax}} = \frac{3}{2\omega_{\mathrm{sm}}} \frac{V_s^2}{\left[R_1 + \sqrt{R_1^2 + (X_1 + X'_2)^2}\right]} \tag{7.16}$$

This equation shows that the pull-out torque is independent of the rotor resistance, but as the system reactance increases, X_1 increases and the pull-out torque reduces, so reducing the generator steady-state stability. This will be most apparent on a weak system where system reactance is greatest (Figure 7.19a). Also the pull-out torque will reduce if the system voltage is reduced for any reason (Figure 7.19b), but the slip at which the pull-out torque occurs is not affected. In contrast Eq. (7.15) shows that the slip at which the maximum torque occurs is determined by the rotor resistance; increasing the rotor resistance increases the maximum slip. This effect is shown in Figure 7.19c. Not only will increased rotor resistance make the system connection more compliant, as described by Eq. (7.14), but also it increases the speed at which the pull-out torque occurs; this can have implications on the transient stability of induction machines (Section 7.9).

Typically, the slip range that the induction generator will operate over will be 0.02 indicating a maximum speed change of 2%, that is 30 rpm for a 1500 rpm generator. As this is a relatively small speed range, turbines using this type of induction generator are often referred to as *fixed-speed machines*.

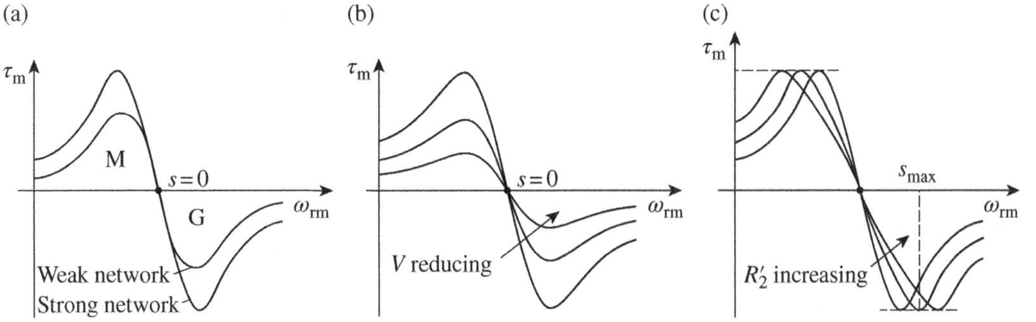

Figure 7.19 Effect of system reactance and voltage on pull-out torque: (a) system reactance; (b) system voltage; (c) rotor resistance (motoring positive).

7.5 Induction Generators with Slightly Increased Speed Range via External Rotor Resistance

In some instances it is required to increase the speed range over which the induction generator operates. For example, in the case of a wind turbine, allowing the speed to vary may increase the energy capture, while allowing the speed to vary will also reduce the shock torques on the turbine and gearbox, owing to wind turbulence. Equation (7.15) shows that the maximum slip increases as the rotor resistance increases but the actual pull-out torque is not affected (Figure 7.19c). To achieve this a wound-rotor induction is normally used with the three-phase rotor winding connected to a variable resistor bank through slip rings (Figure 7.20), although the use of resistors rotating on the shaft is possible in order to avoid the use of slip rings. Such arrangements tend to increase the speed range by about (5–10)%.

The equivalent circuit must now be modified, as shown in Figure 7.21, with R'_{ext} being the external rotor resistance R_{ext} referred to the stator. Using the approximate equivalent circuit of Figure 7.21b, the current and torque expressions of Eqs. (7.10) and (7.13) now become

$$I_2 = \frac{V}{\sqrt{\left(R_1 + \frac{(R'_2 + R'_{ext})}{s}\right)^2 + (X_1 + X'_2)^2}} \tag{7.17}$$

Figure 7.20 Induction generator with variable rotor resistance.

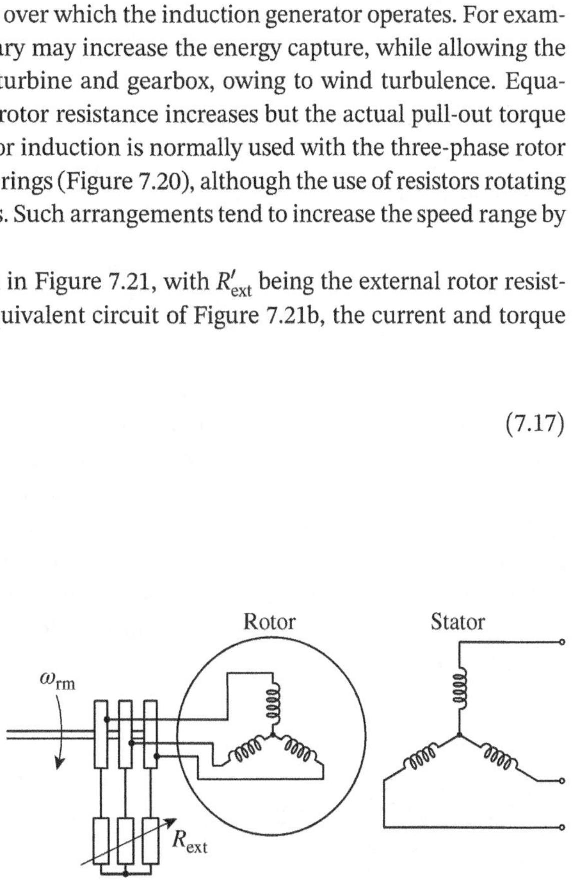

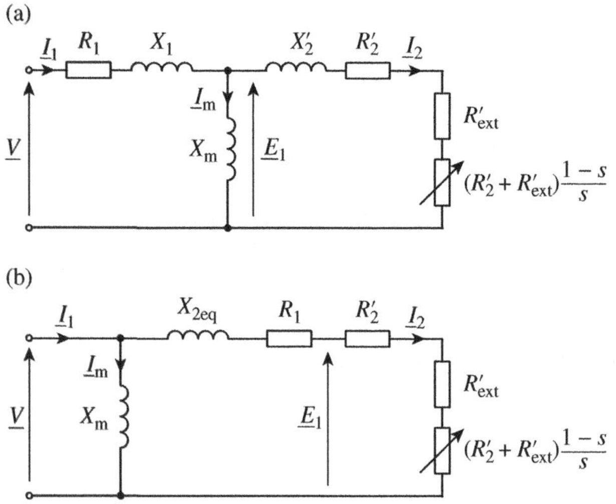

Figure 7.21 Equivalent circuit for induction machine with external rotor resistance: (a) accurate and (b) approximate equivalent circuits (motoring positive).

$$\tau_m = \frac{P_m}{\omega_{rm}} = \frac{P_m}{\omega_{sm}(1-s)} = \frac{3}{\omega_{sm}} \frac{V^2}{\left[\left(R_1 + \frac{(R_2' + R'_{ext})}{s}\right)^2 + (X_1 + X_2')^2\right]} \frac{(R_2' + R'_{ext})}{s} \tag{7.18}$$

and Eqs. (7.11) change to

$$\text{Power supplied} \quad P_s \cong P_{rot} = 3I_2^2 \frac{(R_2' + R'_{ext})}{s} = \tau_m \omega_{sm} \tag{7.19a}$$

$$\text{Power loss in rotor} \quad \Delta P_{rot} = 3I_2^2 (R_2' + R'_{ext}) = sP_s \tag{7.19b}$$

$$\text{Mechanical power} \quad P_m = 3I_2^2 \frac{(R_2' + R'_{ext})(1-s)}{s} = P_s(1-s) = \tau_{sm}\omega_{rm} \tag{7.19c}$$

Equations (7.17) and (7.19) show that if the ratio $(R_2' + R'_{ext})/s$ is held constant then the current and the torque do not change. That is, if the effective total rotor resistance is doubled then the same torque and current will occur at twice the slip. Thus rated torque will always occur at rated current and only the slip (speed) will be different, the slip being scaled by the factor $(R_2' + R'_{ext})/R_2'$ (O'Kelly 1991).

The efficiency is again given by Eq. (7.12) but as the slip is now greater than for a generator with no external rotor resistance the efficiency is less because of the additional loss in the external rotor resistors. However, in the case of a wind turbine generator the gain in energy capture achieved by allowing speed variation coupled with the more compliant grid coupling must be balanced against the loss in efficiency in the generator itself. The power flow is shown in Figure 7.22.

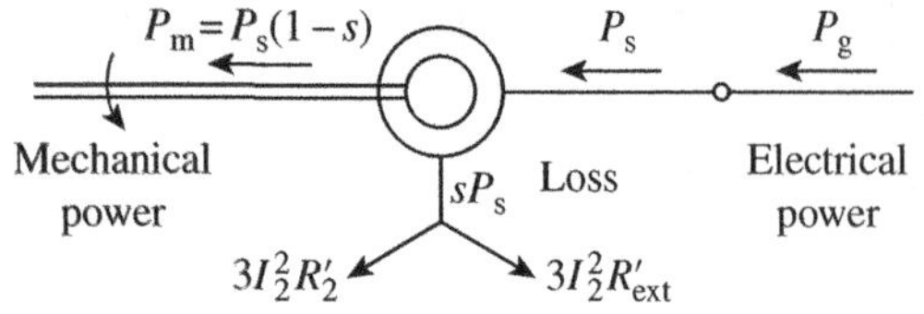

Figure 7.22 Power flow in an induction machine with additional external resistance (motoring positive).

7.6 Induction Generators with Significantly Increased Speed Range

Adding external resistance into the rotor circuit has been shown to allow a small increase in the speed range over which the induction machine can operate but at the cost of reduced efficiency, owing to the losses in the external rotor resistance. The beneficial features of an increased operating speed range can be retained (and expanded) if, rather than dissipating the energy into external resistors, it is fed back into the EPS using a power electronic converter such as that shown in Figure 7.23. This converter consists of two fully controlled IGBT bridge circuits: one, the machine-side inverter, connected to rotor slip rings, and the other, the grid-side inverter, connected to the grid. Together these two inverters produce a four-quadrant converter that can feed power at any frequency or voltage to or from the rotor. The machine-side inverter injects a voltage into the slip rings $\underline{V}_s$ at a slip frequency that is controlled in both magnitude and phase and allows both the torque and the power factor of the machine to be controlled over a large speed range. The grid-side inverter is typically controlled to maintain a constant DC link voltage. As the machine now has power "feeds" to both the stator and rotor from the grid, this type of system is commonly referred to as a *double-fed induction machine* (DFIM). This arrangement is commonly used as a generator with large wind turbines, when it is known as a DFIG, and tends to increase the speed range by about 30%.

The DFIM system shown in Figure 7.23 is very similar to the static Kramer and Scherbius schemes used in the past to control the speed of induction motors (O'Kelly 1991). It differs from the static Kramer scheme in that it uses an IGBT inverter as the machine-side converter instead of a passive rectifier, thereby allowing the injected voltage to be fully controlled in both phase and magnitude. It also allows power flow in both directions to the rotor. In concept it is the same as the static Scherbius system but with the three-phase to three-phase cycloconverter of the Scherbius system replaced by the two fully controlled IGBT bridges.

As the voltage and current injected into the rotor are at slip frequency, the DFIM can be thought of as either an induction machine or a synchronous machine. When the machine operates at synchronous speed the slip frequency is zero, injected rotor current is at DC level, and the machine behaves exactly as a synchronous machine. At other speeds the synchronous machine analogy can still be used but the injected rotor current is now at slip frequency. Analyzing the machine as both an induction machine and a synchronous machine gives a valuable insight into its operation.

The equivalent circuit for the DFIM is shown in Figure 7.24a with $\underline{V}_s$ being the rotor injected voltage. Following the same procedure as in Section 7.2 for the standard squirrel-cage induction motor allows the rotor circuit, operating at slip frequency, to be referred to the stator at grid frequency to give the equivalent circuit of Figure 7.24b. The effect of the transfer is to have, as before, a rotor resistance that varies with slip but now also with an injected voltage that also varies with slip. These two components can be divided into a fixed value and a variable value as shown in Figure 7.24c. The fixed values reflect the actual rotor resistance R_2 and the actual injected voltage $\underline{V}_s$, while the variable terms represent the mechanical power. If required, the accurate equivalent circuit can again be modified to an approximate one by moving the magnetizing reactance in front of the stator components R_1 and X_1 (Figure 7.24d).

Figure 7.23 DFIM system.

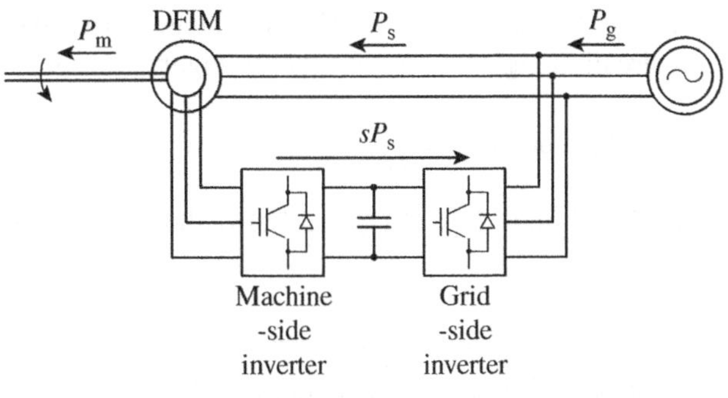

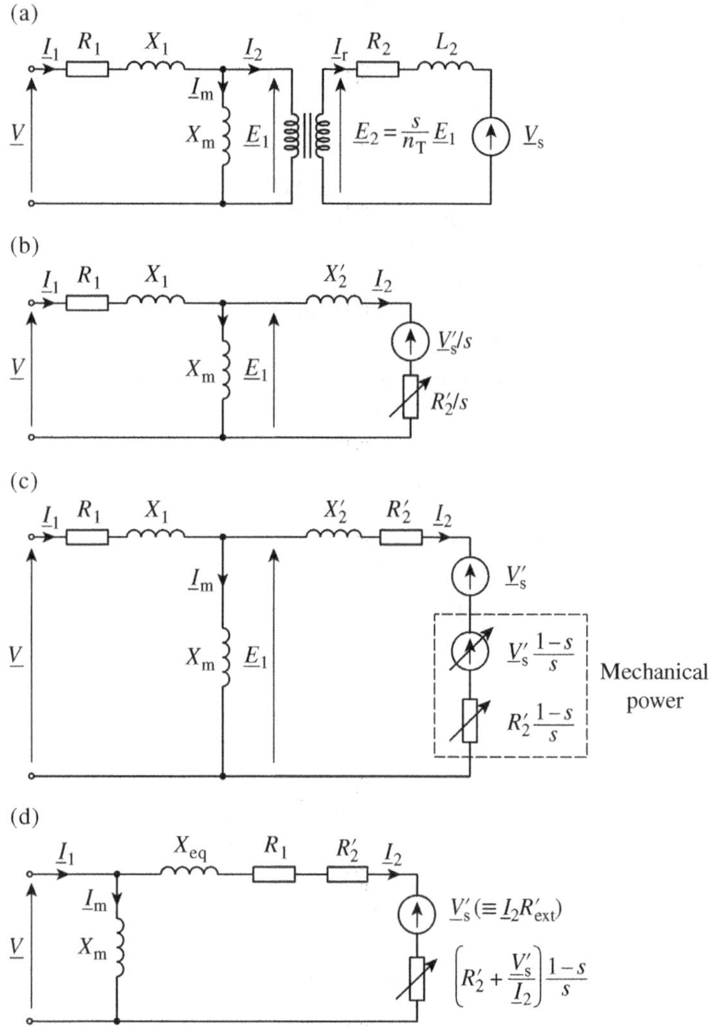

Figure 7.24 DFIM equivalent circuit: (a) stator and rotor; (b) and (c) rotor referred to stator – accurate equivalent circuit; (d) rotor referred to stator – approximate equivalent circuit (motoring positive).

7.6.1 Operation with the Injected Voltage in Phase with the Rotor Current

Assume that the injected voltage $\underline{V}_s$ is in phase with the rotor current. This is equivalent to adding an external resistance, equal to the ratio of the injected voltage to the rotor current, into the rotor circuit. Comparing the equivalent circuit in Figure 7.24c with that in Figure 7.21a they are clearly the same if

$$V'_s = I_2R'_{ext} \quad \text{or} \quad R'_{ext} = \frac{V'_s}{I_2} \tag{7.20}$$

Substituting for R'_{ext} into Eqs. (7.17) and (7.18) allows the current and the torque to be written in terms of the injected voltage as

$$I_2 = \frac{V}{\sqrt{\left[R_1 + \left(R'_2 + \frac{V'_s}{I_2}\right)\frac{1}{s}\right]^2 + (X_1 + X'_2)^2}} \tag{7.21}$$

and

$$\tau_m = \frac{P_m}{\omega_{rm}} = \frac{P_m}{\omega_{sm}(1-s)} = \frac{3}{\omega_{sm}} \frac{V^2}{\left[\left(R_1 + \left(R_2' + \frac{V_s'}{I_2}\right)\frac{1}{s}\right)^2 + (X_1 + X_2')^2\right]} \left(R_2' + \frac{V_s'}{I_2}\right)\frac{1}{s} \tag{7.22}$$

Rearranging Eq. (7.21) with $X_{eq} = X_1 + X_2'$ gives slip as a function of current and injected voltage as

$$s = \frac{I_2 R_2' + V_s'}{\sqrt{V^2 - [I_2 X_{eq}]^2} - I_2 R_1} \tag{7.23}$$

The torque and current for a given slip can be calculated from Eqs. (7.21) and (7.22), but the equations are cumbersome and it easier to consider a range of values for I_2 and calculate the resulting torque and slip from Eqs. (7.22) and (7.23), respectively (O'Kelly 1991). At no load $I_2 = 0$ and Eq. (7.23) reduces to

$$s_0 = \frac{V_s'}{V} \tag{7.24}$$

and the machine operates with a slip that depends on the magnitude and polarity of the injected voltage V_s:

- V_s positive, slip increases and the speed reduces: subsynchronous operation;
- V_s negative, slip becomes negative and the speed increases: supersynchronous operation.

Operation over a wide speed range both above and below synchronous speed is now possible by controlling the magnitude and polarity of the injected voltage.

The slip range over which the machine can operate depends on the magnitude of the injected voltage and that is controlled by the inverter. The greater the no-load slip, the greater the injected voltage. For example, if a speed range of ±30% is required, the injected voltage must be 30% of the nominal supply value. It is this speed range that also determines the rating of the inverter system. The volt-amperes passing through the converter system is

$$VA_{inv} = 3V_s' I_2 = 3s_0 V I_2 \cong s_0 S_{rat} \tag{7.25}$$

where S_{rat} is the volt-ampere rating of the machine.

So for a ±30% speed range the rating of the converter must be 30% of the machine rating. It is this range of speed control with a *partially rated converter* that makes this type of machine economically attractive as opposed to an induction generator or permanent magnet generator with all the machine output passing through an expensive, fully rated converter (Section 7.1).

Substituting for R_{ext}' from Eq. (7.20) into the power flow Eq. (7.19) allows the power supplied by the machine to be written (neglecting the loss in armature resistance) as

$$P_s \cong P_{rot} = 3I_2^2 \frac{R_2'}{s} + 3\frac{V_s'}{s} I_2 = \tau_m \omega_{sm} \tag{7.26}$$

Expanding and rewriting this expression gives

$$\begin{aligned} P_s \cong P_{rot} &= 3I_2^2 R_2 + 3V_s I_2 + 3I_2(V_s + I_2 R_2)\frac{(1-s)}{s} = \tau_m \omega_{sm} \\ &\equiv \text{rotor loss} + \text{injected power} + \text{mechanical power} \end{aligned} \tag{7.27}$$

The first term in this expression is the power lost in the rotor resistance, the second the power that is extracted or injected into the rotor by the converter, and the third the mechanical power produced. Depending on the polarity of the injected voltage, power can be either:

- extracted from the rotor and fed back to the supply; or
- injected into the rotor from the supply.

In Eq. (7.27) the first two terms are associated with the power lost or transferred in the rotor circuit, while the third represents the mechanical power, that is

$$\text{Power transfer}\quad \Delta P_{\text{rot}} = 3I_2^2R_2' + 3V_s'I_2 = sP_s \tag{7.28a}$$

$$\text{Mechanical power}\quad P_m = 3I_2(V_s' + I_2R_2')\frac{(1-s)}{s} = P_s(1-s) = \tau_m\omega_{rm} \tag{7.28b}$$

with the stator and supply powers, respectively, being

$$\text{Stator power}\quad P_s \cong \tau_m\omega_{sm} \tag{7.28c}$$

$$\text{Supply power}\quad P_g = P_s - \Delta P_{\text{rot}} = P_s(1-s) \tag{7.28d}$$

This power flow is shown in Figure 7.25 and, if the rotor resistance loss is assumed to be negligible, the regenerated power sP_s is added algebraically to the stator power to obtain the supply power. As loss in both the armature resistance and rotor resistance has been neglected, the power from or to the supply is the same as the mechanical power and the efficiency is a nominal 100%, that is it is not slip dependent as occurs with external rotor resistance control, because now any additional "loss" in the rotor circuit is fed back to the supply.

Assuming a constant stator power P_s (constant mechanical torque), Eqs. (7.28) explain how this type of machine can be controlled. Figure 7.26 sketches the power flow through the rotor for both motoring (a) and generating (b) over a slip range of about ±30%.

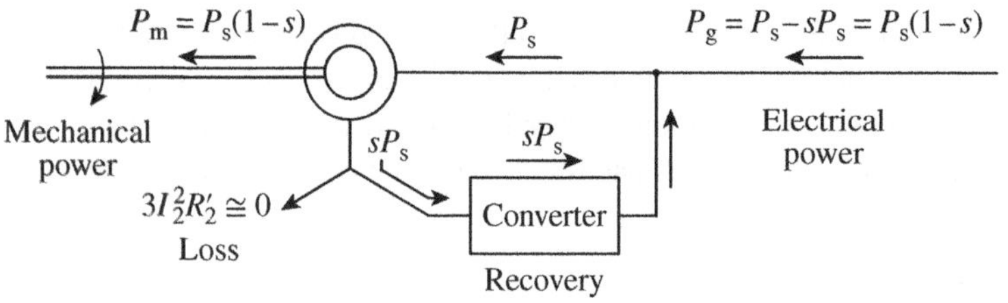

Figure 7.25 Power flow in a DFIM (motoring positive).

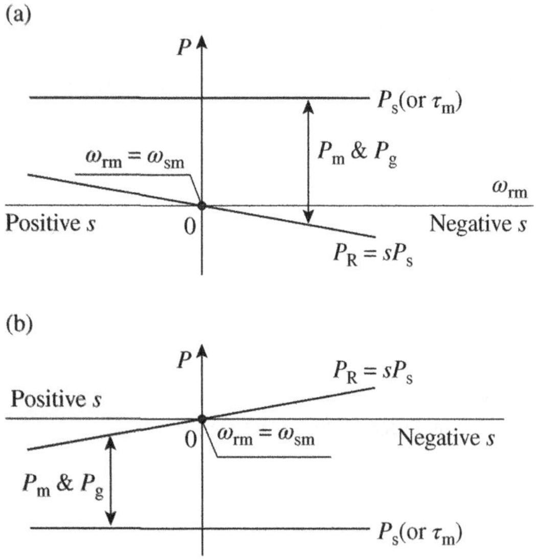

Figure 7.26 Power flow in a DFIM: (a) motoring; (b) generating.

The following observations can be made:

- Rotor speed above synchronous (supersynchronous operation) – slip negative:
  - acts as a motor if power is injected into the rotor from the supply
  - acts as a generator if power is extracted from the rotor and fed back to the supply.
- Rotor speed below synchronous (subsynchronous operation) – slip positive:
  - acts as a motor if power is extracted from the rotor and fed back to the supply
  - acts as a generator if power is injected into the rotor from the supply.
- For a constant slip the amount of power injected or extracted from the rotor determines the stator power and machine torque; increasing the amount of injected or extracted power increases the torque.

This is the basis for a control strategy that can be implemented in the two inverters of Figure 7.23 that supply the rotor circuit and are examined further later in this section.

7.6.2 Operation with the Injected Voltage Out of Phase with the Rotor Current

Equation (7.20) assumes that the voltage $\underline{V}_s$ was injected at slip frequency in phase with the rotor current. If, however, the voltage was injected at some arbitrary controlled phase angle to the rotor current then Eq. (7.20) becomes

$$\frac{\underline{V}_s'}{\underline{I}_2} = R_{ext}' + jX_{ext}' \tag{7.29}$$

and it appears as though the injected voltage is introducing a fictitious impedance $R_{ext}' + jX_{ext}'$. The effect of R_{ext}' has already been discussed, while the influence of X_{ext}' is to increase or decrease the effective rotor impedance depending on the sign and value of X_{ext}' (and will be determined by the relative phase of the injected voltage). This will control the rotor and hence the machine power factor. So, by controlling the magnitude and phase of the injected voltage, both the speed or torque and the power factor of the machine can potentially be controlled.

7.6.3 The DFIG as a Synchronous Generator

The previous discussion considers the DFIM as an induction machine, but it can usefully be considered as a synchronous machine. A synchronous machine rotates at synchronous speed and the current is fed into the rotor field circuit at DC level to produce the synchronously rotating field. In contrast, in the DFIG the speed can vary and the current is fed into the rotor at slip frequency in order to produce the synchronously rotating field.

Figure 7.23 shows that the rotor power converter comprises a grid-side and a machine-side inverter. A common control scheme is to operate both inverters using *pulse width modulation* (PWM) in a *current control mode* with the grid-side inverter being controlled to keep the DC link voltage constant. The machine-side inverter is then controlled to achieve the required torque loading and the required reactive power at the machine terminals. This is achieved by controlling the magnitude of the real and imaginary parts of the rotor current. To understand this control scheme the accurate equivalent circuit of Figure 7.24c is redrawn as in Figure 7.27 by noting that the current through the magnetizing reactance can be written in terms of the rotor $\underline{I}_2$ and stator $\underline{I}_1$ current as

$$\underline{I}_m = \underline{I}_2 - \underline{I}_1 \tag{7.30}$$

when the voltage across the magnetizing reactance $\underline{E}_m$ can be written as

$$\underline{E}_m = j\underline{I}_m X_m = j\underline{I}_2 X_m - j\underline{I}_1 X_m = \underline{E} - j\underline{I}_1 X_m \tag{7.31}$$

This allows the magnetizing reactance to be split between the rotor and stator currents and the equivalent circuit modified to that in Figure 7.27b with the synchronous reactance being the sum of the stator leakage reactance X_1 and the magnetizing reactance X_m (Section 4.2.3). If stator resistance is assumed to be negligible then the

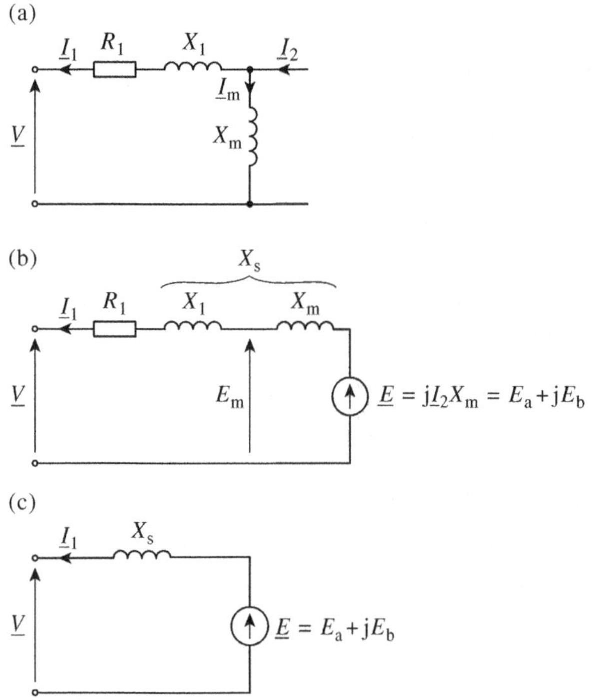

Figure 7.27 A DFIG equivalent circuit as a synchronous machine: (a) stator and rotor current as induction machine; (b) rotor current as a voltage source; (c) equivalent circuit as a synchronous machine (generating positive).

equivalent circuit simplifies to that of the standard synchronous generator shown in Figure 7.27c. Importantly now both the magnitude and phase of the induced electromotive force (emf) relative to the stator voltage can be fully controlled by controlling the real and imaginary components of the induced emf $\underline{E}$. To conform with previous discussions on synchronous machines elsewhere in this book, generator action is taken as positive.

From Eq. (7.31) the emf $\underline{E}$ induced by the rotor current is

$$\underline{E} = \mathrm{j}\underline{I}_2 X_\mathrm{m} \tag{7.32}$$

The rotor current can be expressed in its real and imaginary components

$$\underline{I}_2 = I_{2\mathrm{a}} + \mathrm{j}I_{2\mathrm{b}} \tag{7.33}$$

so that

$$\underline{E} = E_\mathrm{a} + \mathrm{j}E_\mathrm{b} = -X_\mathrm{m}I_{2\mathrm{b}} + \mathrm{j}X_\mathrm{m}I_{2\mathrm{a}} \tag{7.34}$$

and the induced emf $\underline{E}$ is controllable in both magnitude and phase by controlling the in-phase and out-of-phase components of the rotor current $I_{2\mathrm{a}}$ and $I_{2\mathrm{b}}$, respectively.

This is shown in the phasor diagrams of Figure 7.28. The first phasor diagram in Figure 7.28a shows how the real and imaginary rotor current components contribute to the emf $\underline{E}$. If the system voltage $\underline{V}$ is assumed to act along the real axis then Figure 7.28b shows a phasor diagram very similar to that for the standard synchronous generator, only now the current $\underline{I}_2$ will fully control the magnitude and phase of $\underline{E}$. In contrast, in the standard synchronous machine, the emf is induced by a direct current in the rotor so that only the magnitude of the induced emf can be controlled; this allows the control of one output variable only, namely reactive power.

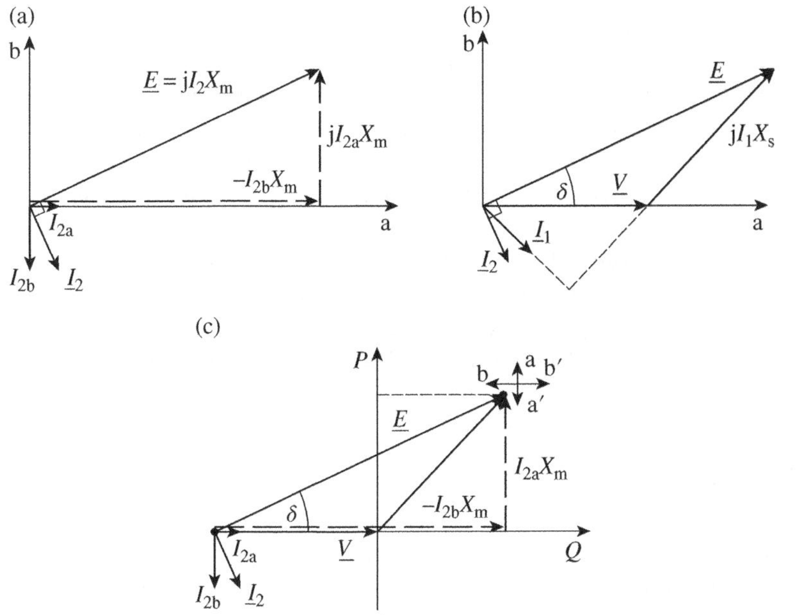

Figure 7.28 Phasor diagrams of a DFIG as a synchronous machine: (a) showing how rotor current controls the induced emf; (b) phasor diagram; (c) operating chart.

With the stator voltage acting along the real axis and defined as

$$\underline{V} = V_a + \mathrm{j}0 \tag{7.35}$$

and neglecting stator resistance

$$\underline{I}_s = \frac{\underline{E} - \underline{V}_s}{\mathrm{j}X_s} = \frac{E_a + \mathrm{j}E_b - V_a}{\mathrm{j}X_s} = \frac{E_b}{X_s} - \mathrm{j}\frac{(E_a - V_a)}{X_s} \tag{7.36}$$

The stator power is then given by

$$\underline{S}_s = 3\underline{V}\,\underline{I}_s^* = P_s + \mathrm{j}Q_s = 3\frac{V_a E_b}{X_s} + \mathrm{j}\,3\frac{V_a(E_a - V_a)}{X_s} \tag{7.37}$$

Substituting for E_a and E_b from Eq. (7.34) gives the important expression

$$S_s = \left[3V_a\frac{X_m}{X_s}\right]I_{2a} + \mathrm{j}\left[\frac{3V_a}{X_s}\right][-X_m I_{2b} - V_a] \tag{7.38}$$

This equation allows the phasor diagram of Figure 7.28b to be redrawn, as in Figure 7.28c, to produce an operating chart similar to that for a synchronous machine (Section 3.3.4) but note that the direction of the P- and Q-axes are now swapped compared with Figure 3.19. This operating chart clearly shows the control options with this type of generator:

- controlling the magnitude of I_{2a}, the component of rotor current in phase with the stator voltage, controls the active power (along aa');
- controlling the magnitude of I_{2b}, the component of rotor current out of phase with the stator voltage, controls the reactive power (along bb').

7.6.4 Control Strategy for a DFIG

The normal control strategy with this type of generator is to control both I_{2a} and I_{2b} independently in order to control both the generator torque (I_{2a}) and the reactive power (I_{2b}). Normally, all calculations are done in a synchronous reference frame that uses the stator voltage space vector as the reference, as in Eq. (7.35), and is termed a *vector controller*. In comparison many motor drive systems use the machine flux vector as the reference and are termed *field-orientated controllers* (Muller et al. 2002). In this stator voltage space vector reference frame I_a would become I_q, and I_b would become I_d, and so on.

When used with a wind turbine, the power output required from the DFIG is defined as a function of rotational speed in order to maximize power output from the wind turbine system, as described in Section 7.1. If the required power P_d (or torque) is known, and the required reactive power output Q_d is defined, the real and imaginary parts of Eq. (7.38) define the demand values of the rotor current components I_{2ad} and I_{2bd}, respectively, that is

$$I_{2ad} = \frac{P_d}{3V_a}\frac{X_s}{X_m} \quad \text{and} \quad I_{2bd} = -\left[\frac{Q_d}{3V_a} + \frac{V_a}{X_s}\right]\frac{X_s}{X_m} \tag{7.39}$$

This control is implemented in the machine-side converter of Figure 7.23, while the grid-side converter is controlled to maintain the DC link voltage constant. Such a control scheme is shown in the schematic of Figure 7.29. This is a complex control structure using fast conversion algorithms to convert from phase values to d and q space vector quantities. *Proportional plus integral* (PI) controllers determine the PWM variables that control the injected rotor currents to obtain the required reactive power and torque. Interested readers are referred to articles by Ekanayake et al. (2003a, 2003b), Holdsworth et al. (2003), Muller et al. (2002), Slootweg et al. (2001), and Xiang et al. (2006).

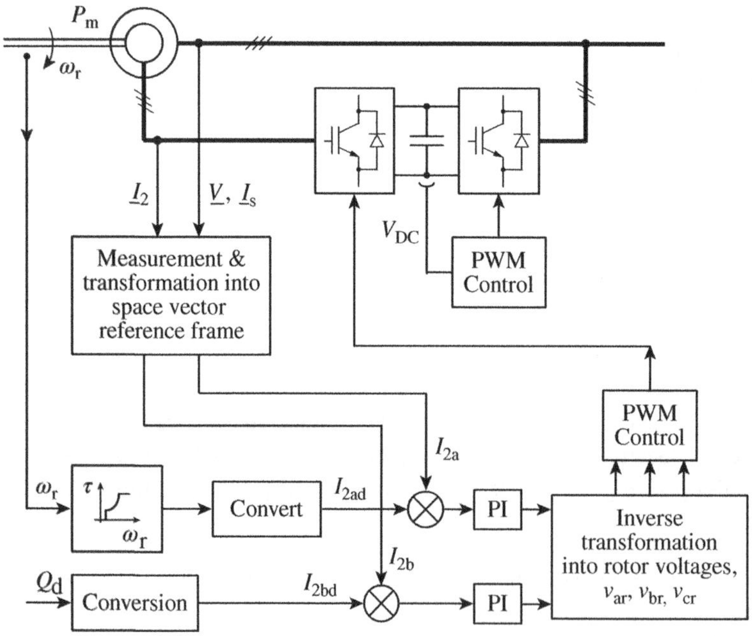

Figure 7.29 A DFIG control implementation.

7.7 Fully Rated Converter Systems (Wide Speed Control)

Instead of using a partially rated converter, fully rated converters can be used to control the real and reactive power injected into the system from either an induction generator or a synchronous generator, as described in Section 7.1. The synchronous generator can have a wound field or can use permanent magnets to provide the rotating magnetic field. Such schemes are used with renewable energy systems and give the widest range of speed control but at the expense of a converter that must be rated to cope with the full power output of the generator. Such a power conversion scheme is shown schematically in Figure 7.30 and differs only in detail on the machine-side inverter for use with induction or synchronous machines. One attraction of the permanent magnet machine is its high efficiency since no magnetizing or field current is necessary to provide the magnetic field and it also allows new generator topologies to be devised to suit a specific application. As developments in power electronics progress, it is likely that the use of fully rated converter systems will replace the partially rated converters and DFIGs.

The fully rated converter system allows full control of real and reactive power. Although a number of control schemes are possible, the machine-side converter is normally operated to control the generator torque loading while the grid-side inverter is controlled to maintain constant voltage on the DC link and, at the same time, controlling the reactive power output in a similar way to a static compensator (Section 2.5.4).

Both converters will normally be current controlled to achieve their control objectives. However, it must be realized that the PWM converters can only adjust the phase and magnitude of the injected phase voltage. So, for example, the grid-side converter can only adjust the phase and magnitude of the injected voltage $\underline{E}$ relative to the system voltage $\underline{V}$. It is these two quantities of phase and magnitude that are adjusted via PI controllers to control the current to its set value; see also Figure 7.29.

7.7.1 Machine-side Inverter

The machine-side inverter is normally controlled to give the required torque loading at any particular speed so as to maximize energy capture. This is achieved by controlling the quadrature-axis (q-axis) current I_q. To show this, consider a permanent magnet machine when the voltage equations are given by Eqs. (3.67) but note that the induced emf E_f is related to the flux linkage per pole produced by the permanent magnets, that is

$$E_f = \omega\psi_{pm} \quad (7.40)$$

Figure 7.30 Fully rated converter system.

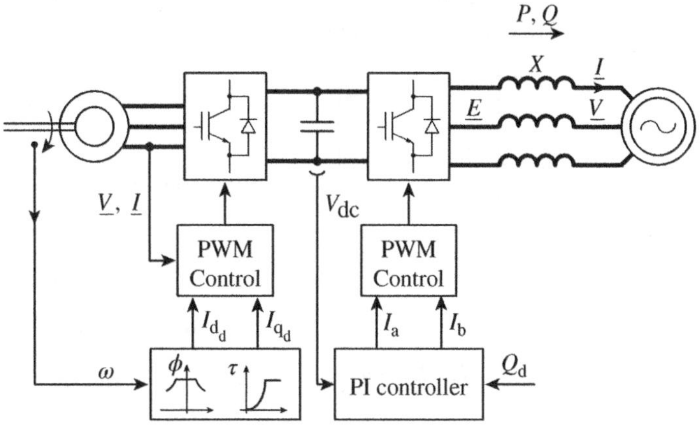

With this change the voltage Eq. (3.67) becomes

$$V_{\rm d} = -RI_{\rm d} - X_{\rm q}I_{\rm q} \quad \text{and} \quad V_{\rm q} = \omega\psi_{\rm pm} - RI_{\rm q} + X_{\rm d}I_{\rm d} \tag{7.41}$$

The terminal power is then given by Eq. (3.84) as

$$P = V_{\rm d}I_{\rm d} + V_{\rm q}I_{\rm q} = I_{\rm d}I_{\rm q}(X_{\rm d} - X_{\rm q}) + \omega\psi_{\rm pm}I_{\rm q} - R\left(I_{\rm d}^2 + I_{\rm q}^2\right) \tag{7.42}$$

The first two terms in this expression define the shaft power and the third term the power lost in the armature resistance so that the machine torque is given by

$$\tau_{\rm m} = \frac{I_{\rm d}I_{\rm q}(X_{\rm d} - X_{\rm q})}{\omega_{\rm rm}} + p\psi_{\rm pm}I_{\rm q} \tag{7.43}$$

With permanent magnet machines the reactance's $X_{\rm d}$ and $X_{\rm q}$ depend on how the magnets are mounted on the rotor; for surface-mounted magnets $X_{\rm d} \cong X_{\rm q}$ as the relative permeability of rare earth permanent magnet material is approximately 1. In this case Eq. (7.43) reduces to

$$\tau_{\rm m} = p\psi_{\rm pm}I_{\rm q} \tag{7.44}$$

That is, the torque loading can be adjusted by controlling the q-axis current $I_{\rm q}$.

In the induction machine, torque can be controlled in a very similar way but now the magnetizing flux must also be produced from the armature and it is important that the flux level in the machine is maintained at the correct value all the time. Equation (3.39) shows how flux, emf, and frequency are related and that to operate at speeds below the rated voltage must be reduced in proportion if the machine is not to be overfluxed (saturated). This means operating at constant volts per hertz at speeds below rated speed and at constant rated voltage, reduced flux, above rated speed. This is achieved by controlling the direct-axis current $I_{\rm d}$.

7.7.2 Grid-side Inverter

The grid-side inverter is normally controlled to transfer power so as to maintain a constant voltage on the DC link capacitor. If the charge on the capacitor increases, the control loop will increase the power transfer to reduce the voltage, and vice versa. The grid-side converter also controls the reactive power delivered to the system. A vital part of this converter is the line reactor X shown in Figure 7.30.

Assume that the system voltage $\underline{V} = V_{\rm a} + {\rm j}0$ acts along the real axis, that the voltage injected by the converter is $\underline{E} = E_{\rm a} + {\rm j}E_{\rm b}$ and that the current injected into the system is $\underline{I} = I_{\rm a} + {\rm j}I_{\rm b}$. The phasor diagram for this system is shown in Figure 7.31. The current through the line reactance is given by

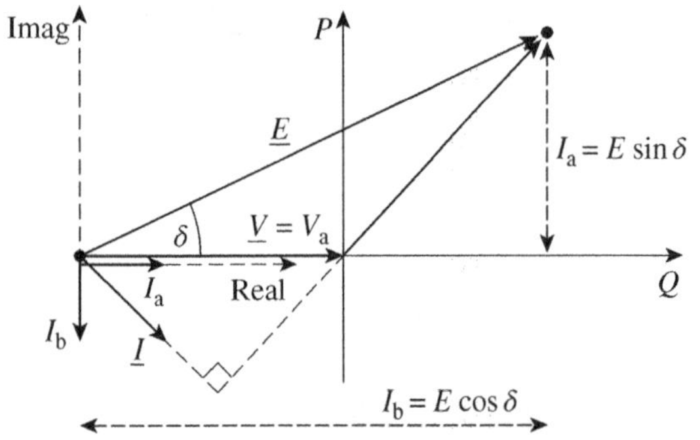

Figure 7.31 Phasor diagram describing the operation of the grid-side inverter.

$$\underline{I} = I_{\rm a} + {\rm j}I_{\rm b} = \frac{\underline{E} - \underline{V}}{{\rm j}X} = \frac{E_{\rm b}}{X} - {\rm j}\frac{(E_{\rm a} - V_{\rm a})}{X} \tag{7.45}$$

that is

$$I_{\rm a} = \frac{E_{\rm b}}{X} \quad \text{and} \quad I_{\rm b} = -\frac{(E_{\rm a} - V_{\rm a})}{X} \tag{7.46}$$

while the apparent power injected into the system is

$$\begin{aligned} \underline{S} &= 3V_{\rm a}\underline{I}* = 3V_{\rm a}I_{\rm a} + 3{\rm j}V_{\rm a}I_{\rm b} \\ &= 3\frac{V_{\rm a}E_{\rm b}}{X} + 3{\rm j}\frac{V_{\rm a}(E_{\rm a} - V_{\rm a})}{X} \end{aligned} \tag{7.47}$$

From the phasor diagram the emf components E_a and E_b can be written in terms of the magnitude of the emf E and its phase δ as

$$E_a = E\cos\delta \quad \text{and} \quad E_b = E\sin\delta \tag{7.48}$$

Substituting into Eq. (7.47) gives

$$\underline{S} = 3\frac{V_a E\sin\delta}{X} + 3j\frac{V_a(E\cos\delta - V_a)}{X} \tag{7.49}$$

Generally, δ is small when Eq. (7.49) becomes

$$\underline{S} \approx 3\frac{V_a E}{X}\delta + 3j\frac{V_a(E - V_a)}{X} \tag{7.50}$$

demonstrating how active power can be controlled by the phase angle δ and reactive power by the magnitude of E. Although there is some cross-coupling between the terms, this can be taken into account in the control structure.

7.8 Peak Power Tracking of Variable Speed Wind Turbines

The discussion in the previous sections explains how an induction machine can be used to generate electrical power and how, with the aid of power electronics, it can be operated over a large speed range. To optimize energy capture from the wind turbine, the electrical torque is generally controlled as a function of rotational speed, while the speed of the generator will increase, or decrease, depending on the mechanical torque produced by the turbine. This change in speed is determined by the equation of motion, Eq. (5.1). The rate at which the speed will change depends on the magnitude of the torque imbalance and the moment of inertia of the turbine and generator system, including the effect of the gearbox.

For example, consider the situation shown in Figure 7.32 for a wind turbine where the generator is operating at point p_1 at rotational speed ω_1 producing power P_1. The wind speed is v_{w1}. At this point the mechanical power (and torque) produced by the turbine is balanced by the electrical power. The line ab is the operating line that the generator is controlled to follow for optimum loading; see Figure 7.3. If now the wind speed increases to v_{w2}, the new optimum steady-state operating point is at p_2. Initially, the speed and electrical power loading of the turbine do not change but the increase in wind speed leads to an increase in the mechanical shaft power ΔP and hence mechanical torque (point c). This increase in mechanical power accelerates the turbine system and the speed changes from ω_1 toward ω_2. As the speed changes the electrical power also changes until the speed settles at the new equilibrium point p_2 when the electrical power again balances the mechanical power. It is important to realize that in practice the wind is turbulent and speed and power changes such as this will be continually taking place (Stannard and Bumby 2007).

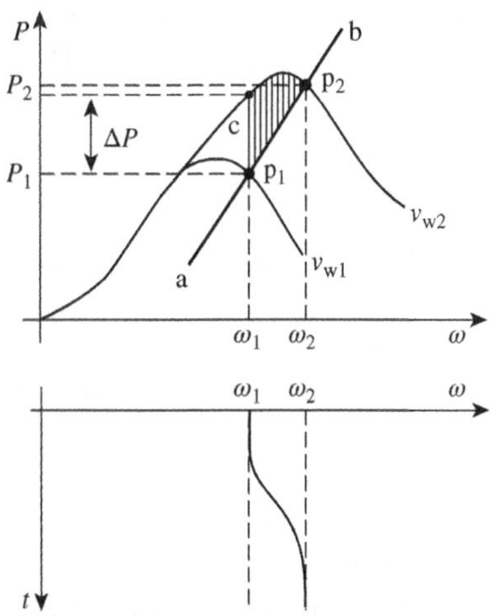

Figure 7.32 Speed changes in a wind turbine system (generating positive).

7.9 Connections of Wind Farms

Although the majority of wind turbines are situated on land, there is a growing demand for wind turbines to be placed offshore with some large wind farms now operational (Christiansen 2003). The reasons for developing offshore sites are both technical and political, with offshore turbines seen to pose fewer planning issues and offshore winds tending to be stronger and more consistent than those onshore. This does not mean that offshore sites are always better than those onshore, as some onshore sites have better wind regimes than sites offshore.

A common problem to all offshore energy conversion systems is the electric cable connection to the onshore substation. This must be by a buried undersea cable and this then raises distance issues because all AC cables have high capacitance and the line charging current for long cable runs can be very high (Section 3.1). Some form of reactive power compensation may be required at the offshore end to help combat the large cable capacitance, while a number of independent cable runs may be necessary in order to transmit the required power from an offshore wind farm. Because of the large cable capacitance, AC cables are currently limited to a distance under the sea of about (100–150) km with the maximum rating of three-core submarine cables currently being about 200 MW at 145 kV (Kirby et al. 2002), although larger ratings are under development. Generally, the outputs of a number of turbines are collected together at an offshore substation for onward transmission to shore. Once the output of a number of turbines has been collected, an alternative to AC transmission to shore is to use DC transmission. New DC transmission technology uses IGBT voltage source converters at the sending end (and possibly also at the receiving end) allowing total control at the sending end. For higher powers, conventional DC technology using gate turn-off thyristors can be used. Interested readers are referred to Kirby et al. (2002).

Currently, offshore wind farms are sufficiently close to shore that AC cables can be used, although a number of cables may be necessary to transmit the required power. One practical point to note is that the distance to shore also includes the shore-based cable run to the shore substation. In some situations this can be substantial.

The problems associated with transferring electrical power to shore from offshore wind farms is also faced by tidal stream generators and wave generators. Tidal stream generators tend to be relatively close to shore, although laying cables in the strong currents where these turbines are situated is not straightforward. Wave energy is in its infancy with the large amounts of resource available some way offshore. Harnessing this energy and transferring it to shore poses a significant challenge.

7.10 Fault Behavior of Induction Generators

Section 7.7 describes the steady-state behavior of induction generators and how their power output might change when used with wind turbines and subjected to changes in wind speed. However, as with any generator, the wind turbine generator connected to the EPS will be subjected to system faults and its behavior during and after these faults is important with regard to system stability. This is examined here.

7.10.1 Fixed-speed Induction Generators

Fixed-speed induction generators are normally regarded as "negative load." That is, they generate power whenever they can and do not contribute to system voltage or frequency support. Section 7.2 explains how these generators always consume reactive power and how this increases as slip increases so that when a fault does occur and is subsequently cleared these generators can have a detrimental effect on the recovery of system voltage. To avoid this it is normal practice to disconnect these generators from the system as soon as a drop in voltage is detected. The generators are then reconnected once the system is restored to normal operation.

Notwithstanding this normal operation, it is instructive to examine the stability of an induction generator following a fault assuming that it remains connected to the system. Consider the system shown in Figure 7.33, where the turbine and induction generator are connected to the system through a transformer and a short line. The effective impedance of the line and transformer will impact on the generator torque-slip curve, as described in Section 7.3. The sequence of events following a three-phase-to-earth fault is shown in Figure 7.34 by reference

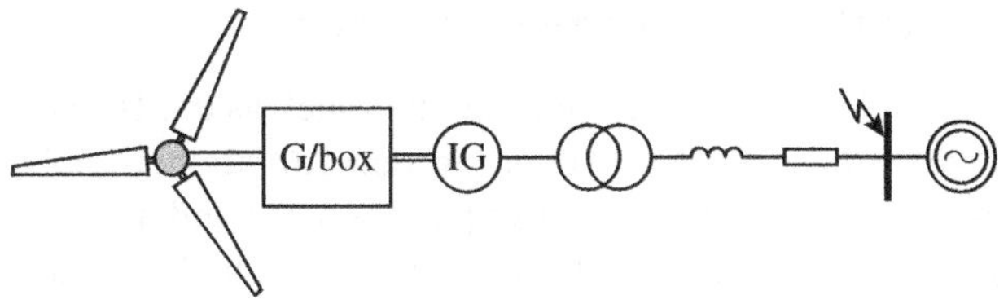

Figure 7.33 Example turbine and fixed-speed induction generator system: G/box – gearbox; IG – induction generator.

to the torque-slip curve of the generator. In Figure 7.34 the torque-slip curve has been inverted so that generator torque is now positive. The torque produced by the turbine as the rotor speed changes is shown by line τ_m.

Initially, the electrical generator torque and the mechanical turbine torque are balanced and the generator and turbine operate at a rotational speed ω_{r1} with a negative slip s_1. When the fault occurs at $t = t_0$ the electrical torque drops to zero (point 2) and the mechanical torque accelerates the rotor and the speed increases. The fault is cleared at time $t = t_1$ when the rotor speed has increased to ω_{r3} and the slip to s_3. At this point the system voltage is restored and the electrical torque increases from zero to τ_5. The electrical torque is now greater than the mechanical torque τ_4 and this acts like a braking torque reducing the speed of the rotor until once again the steady operating point 1 is reached. This system is stable.

If the fault were on for a longer period and not cleared until time $t = t_2$ the rotor speed would have increased to ω_{r6} so that when the electrical torque is restored the electrical torque and the mechanical torque are the same; this is the transient stability limit. If the fault were cleared at time $t < t_2$ the system would be stable, at $t > t_2$ the system would be unstable. This is shown in Figure 7.34.

Note that although a reasoning somewhat similar to that of the equal area criterion (Chapter 5) is applied above, the equal area criterion cannot be applied to Figure 7.34 because the integral of torque with speed is not energy but power. Moreover, re-establishing the flux requires time, so what the diagram does is to show the main effect with regard to stability and that is the relative magnitudes of the electrical and mechanical torques. Anything else would require detailed simulation.

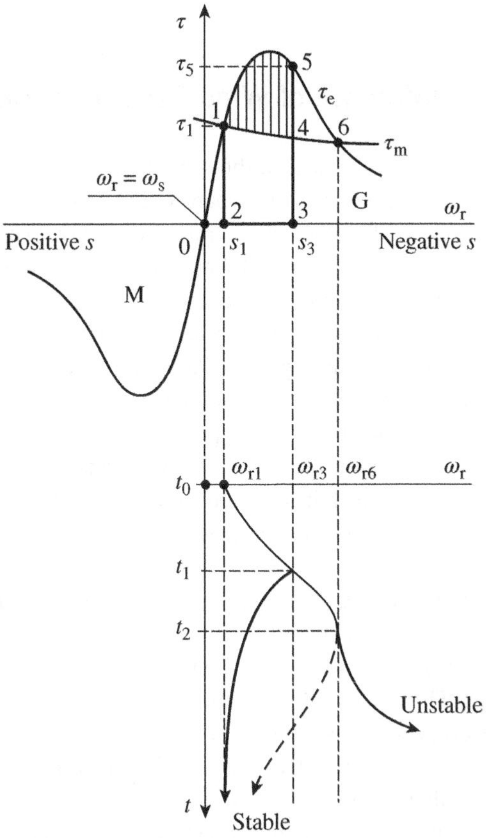

Figure 7.34 Transient stability of a fixed-speed induction generator turbine system (generating positive).

A number of other factors described in Section 7.3 influence the above discussion. First of all, when the fault is removed, the torque-slip curve will take a finite time to re-establish itself because the flux inside the machine must be restored. This in turn will draw significant reactive power from the network. This reactive power draw can depress the system voltage, so reducing the peak of the torque-slip curve. It is for these reasons that detailed modeling of the induction machine, as described in Section 11.4.3, is required. Finally, the transient stability limit could be increased by changing the rotor resistance as this has the effect of moving the peak of the torque-slip curve further to the left (Section 7.3).

7.10.2 Variable-speed Induction Generators

Large, variable-speed induction generators such as the DFIG, described in Section 7.5, and the induction generator with a fully rated converter (Section 7.6) are both able to contribute to system voltage and frequency control by controlling their reactive power output and, to some extent, their active power output, respectively. Such generators are required to "ride through" a fault so that they can contribute to system stability once the fault is removed. To achieve this requires complex control of the power converters, and interested readers are referred to Muller et al. (2002), Slootweg et al. (2001), Ekanayake et al. (2003a), Holdsworth et al. (2003), and Xiang et al. (2006).

7.11 Influence of Wind Generators on Power System Stability

As discussed in Chapter 5, the synchronous generator is stiffly connected to the EPS and exhibits an inherently oscillatory response to a disturbance because its power output is approximately proportional to the sine of the rotor angle. For small values of the rotor angle, power is proportional to the angle itself which produces spring-like oscillations – see Section 5.4.7 and also Figure 7.16a. On the other hand, Section 7.3 explains that squirrel-cage (fixed-speed) induction generators are coupled to the grid less stiffly than synchronous generators. Figure 7.16b shows that the torque of a fixed-speed induction generator is proportional to the speed deviation (slip), hence providing inherent damping of oscillations. This positive influence is counteracted by the vulnerability of fixed-speed induction generators to system faults (Section 7.10.1).

Damping due to variable speed DFIGs depends very much on the particular control strategy employed. Section 7.5 explains that DFIGs have good control capabilities due to the possibility of controlling both the magnitude and phase of the injected voltage. This makes it possible to design a power system stabilizer that improves the damping of power swings without degrading the quality of voltage control provided (Hughes et al. 2006). Fully rated converter systems effectively decouple the generator from the grid, so they offer a very good possibility of improving the damping of power swings. Hence the general conclusion is that a partial replacement of traditional thermal plants employing synchronous generators, which exhibit a relatively poor natural damping, by renewable generators, which exhibit a better damping, will improve the damping of electromechanical swings. This effect will be counterbalanced to some extent by the highly variable nature of renewable sources themselves, such as wind, marine, or solar, but their variability may be effectively managed by either using energy storage or part loading one of the turbines in a farm and using its spare capacity to smooth power oscillations (Lubosny and Bialek 2007).

The network effect of replacing large traditional generators by renewable ones will largely depend on the system in question. Recall that the stability of synchronous generators deteriorates if they are highly loaded and remote and operate with a low, or even leading, power factor. If renewable plants are connected closer to the loads, the transmission networks will be less loaded, which will reduce reactive power consumption by the system and the voltages will rise. This effect can be compensated by reactive power devices, such as reactors or static VAR compensators, but this would require additional investment. If that is deemed uneconomical and the remaining synchronous generators are used for reactive power compensation, their operating points would move toward capacitive loading (leading power factor) so their dynamic properties might deteriorate. As the number of synchronous generators remaining in operation is reduced due to increased penetration of renewables, their overall compensation capabilities will also be reduced. Hence the overall effect might be a deterioration of the dynamic properties of the system (Wilson et al. 2006).

On the other hand, if the renewable sources are located further away from the main load centers, as is the case, for example, in the United Kingdom, then power transfers over the transmission network will increase. Higher transfers will mean larger voltage angle differences between network nodes and deteriorated system dynamic properties (smaller stability margins).

Increased penetration of renewables might also affect frequency stability. Owing to its construction, a wind plant has smaller inertia and speed so that kinetic energy stored in it is reduced by a factor of approximately 1.5 when compared with a traditional plant of the same rating. The reduction in stored kinetic energy will have an effect on system operation and security because the amplitude of frequency variations, discussed in Chapter 9, will increase.

8

Voltage Stability

Chapter 2 explains how an electrical power network is made up from many different elements; is divided into transmission, subtransmission, and distribution levels; and is often organized in such a way that each level belongs to a different company. Because of its immense size, analyzing a complete power system is not possible, even using supercomputers, and normally the system is divided into sensible parts, some of which are modeled in detail and others more superficially. Obviously, the system model used must represent the problem being studied so that when analyzing distribution networks the transmission or subtransmission networks are treated as *sources* operating at a given voltage. On the other hand, when analyzing the transmission or subtransmission networks, the distribution networks are treated (Section 3.4) as *sinks* of real and reactive power and referred to as *composite loads* or simply *loads*.

A distribution network is connected to the transmission network at the *grid supply point* where a change in voltage may cause complicated dynamic interactions inside the distribution network itself, owing to:

- voltage control action arising from transformer tap changing;
- control action associated with reactive power compensation and/or small embedded generators;
- a low supply voltage causing changes in the power demand as a result of induction motors stalling and/or the extinguishing of discharge lighting;
- operation of protective equipment by overcurrent or undervoltage relays, electromechanically held contactors, and so on;
- reignition of discharge lighting and self-start induction motors when the supply voltage recovers.

In this chapter the effect that actions such as these can have on voltage stability is examined using the static characteristics of the composite loads introduced in Section 3.4. This simplified analysis will help to give an understanding of both the different mechanisms that may ultimately lead to voltage collapse and the techniques that may be used to assess the voltage stability of a particular system. These techniques can then be extended to the analysis of large systems using computer simulation methods (Taylor 1994; Kundur 1994; Van Cutsem and Vournas 1998).

8.1 Network Feasibility

The system shown in Figure 8.1 is representative of the general power supply problem in that it shows a generator supplying some composite load. Generally, there will be some limits on the power that can be supplied to the load and this will determine how stable the supply is. To analyze this stability problem the composite load will be represented by its static voltage characteristics when the problem becomes that of determining the solutions to the power network equations, if they exist, and, if they do, determining what limits are placed on the solutions. This process is often referred to as determining the *network feasibility* or determining the *network loadability*.

Power System Dynamics: Stability and Control, Third Edition. Jan Machowski, Zbigniew Lubosny, Janusz W. Bialek and James R. Bumby.

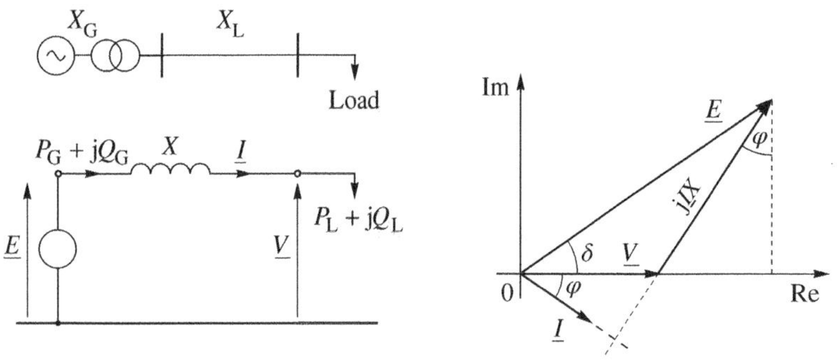

Figure 8.1 Equivalent circuit of the transmission link and its phasor diagram.

The network feasibility problem can be explained by using the simple equivalent circuit shown in Figure 8.1. This circuit consists of an equivalent transmission link and an equivalent generating unit. In the steady state both these elements can be replaced by an equivalent voltage source E behind an equivalent reactance X. Equivalent reactance $X = X_G + X_L$ is the sum of the generating unit reactance X_G and the transmission link reactance X_L. Values of E and X depend on the fact whether the AVR of the equivalent generating unit is operative or not. Under normal conditions the AVR is operative and keeps the generator terminal voltage constant $V_G = \text{const.}$ Then the voltage of the equivalent source is equal to $E = V_G$ and the reactance of the equivalent generating unit is zero $X_G = 0$. In such a case $X = X_L$. If the AVR is not operative, or the equivalent generating unit is operating near its excitation limit then the field voltage remains constant and the equivalent generating unit must be modeled by its synchronous emf E_f acting behind its synchronous reactance $x_d = X_d + X_T$ defined in Section 3.3.3. Then $E = E_f$ and $X = x_d + X_L$, because $X_G = x_d$.

In general the resistance of the generator and transmission link is small and can be neglected, while the equivalent reactance X must combine the source reactance with that of the transformer and the transmission line. The real and reactive power absorbed by the load, $P_L(V)$ and $Q_L(V)$, can be calculated from the phasor diagram in Figure 8.1 by noting that $IX\cos\varphi = E\sin\delta$ and $IX\sin\varphi = E\cos\delta - V$. This gives

$$\begin{aligned} P_L(V) &= VI\cos\varphi = V\frac{IX\cos\varphi}{X} = \frac{EV}{X}\sin\delta \\ Q_L(V) &= VI\sin\varphi = V\frac{IX\sin\varphi}{X} = \frac{EV}{X}\cos\delta - \frac{V^2}{X} \end{aligned} \tag{8.1}$$

The angle δ between the $\underline{E}$ and $\underline{V}$ phasors can be eliminated using the identity $\sin^2\delta + \cos^2\delta = 1$ to give

$$\left(\frac{EV}{X}\right)^2 = [P_L(V)]^2 + \left[Q_L(V) + \frac{V^2}{X}\right]^2 \tag{8.2}$$

This static power-voltage equation determines all the possible network solutions when the voltage characteristics $P_L(V)$ and $Q_L(V)$ are taken into account.

8.1.1 Ideally Stiff Load

For an ideally stiff load (Section 3.4) the power demand of the load is independent of voltage and is constant

$$P_L(V) = P_n \text{ and } Q_L(V) = Q_n \tag{8.3}$$

where P_n and Q_n are the real and reactive power demand of the load at the rated voltage V_n. Equation (8.2) can now be rewritten as

$$\left(\frac{EV}{X}\right)^2 = P_n^2 + \left[Q_n + \frac{V^2}{X}\right]^2 \tag{8.4}$$

Substituting $Q_n = P_n \tan\varphi$ into Eq. (8.4) gives

$$P_n^2 + P_n^2 \tan^2\varphi + 2P_n \frac{V^2}{X} \tan\varphi = \left(\frac{EV}{X}\right)^2 - \left(\frac{V^2}{X}\right)^2 \tag{8.5}$$

After taking into account that $\tan\varphi = \sin\varphi / \cos\varphi$ and $\sin^2\varphi + \cos^2\varphi = 1$ and after some simple math one gets

$$P_n^2 + 2P_n \frac{V^2}{X} \sin\varphi \cos\varphi = \frac{V^2}{X^2}(E^2 - V^2)\cos^2\varphi \tag{8.6}$$

The left-hand side of this equation is an incomplete square of a sum. Hence the equation can be transformed to

$$\left(P_n + \frac{V^2}{X}\sin\varphi\cos\varphi\right)^2 - \left(\frac{V^2}{X}\right)^2 \sin^2\varphi\cos^2\varphi = \frac{V^2}{X^2}(E^2 - V^2)\cos^2\varphi$$

or

$$P_n + \frac{V^2}{X}\sin\varphi\cos\varphi = \frac{V}{X}\cos\varphi\sqrt{E^2 - V^2\cos^2\varphi} \tag{8.7}$$

The voltage at the load bus can be expressed pu as V/E. The above equation can be expressed as

$$P_n = -\frac{E^2}{X}\left(\frac{V}{E}\right)^2 \sin\varphi\cos\varphi + \frac{E^2}{X}\frac{V}{E}\cos\varphi\sqrt{1 - \left(\frac{V}{E}\right)^2 \cos^2\varphi}$$

or

$$p + v^2 \sin\varphi\cos\varphi - v\cos\varphi\sqrt{1 - v^2\cos^2\varphi} = 0 \tag{8.8}$$

where

$$v = \frac{V}{E},\ p = \frac{P_n}{S''_K},\ S''_K = \frac{E^2}{X} \tag{8.9}$$

It is worth emphasizing here that for the equivalent circuit shown in Figure 8.1 the short-circuit current is equal to $I''_K = E/X$ and S''_K is the short-circuit power of the equivalent source. For further considerations it is convenient to express real and reactive power in per-unit (pu) values, where S''_K is the base value.

Equation (8.8) describes a family of curves with φ as a parameter. Figure 8.2 shows such a family of curves for four values of φ. Because of their characteristic shape, the curves are referred to as *nose curves*. For a lagging power factor (curves 1 and 2) the voltage decreases as the real load increases. For a leading power factor (curve 4) the voltage initially increases and then decreases. For $Q_n = 0$, that is for $\varphi = 0$, the peak of the nose curve occurs at $p = 0.5$, that is for $P_n = 0.5E^2/X = E^2/2X = S''_K/2$.

Generally, active power $p = P_n/S''_K$ cannot be treated as a function of $v = V/E$, because (as seen in Figure 8.2) for a leading power factor and for given value of $v > 1$ there exist two values of real power. For this reason, using Eq. (8.8), the nose curve cannot be simply plotted as a function $p(v)$ by substituting consecutive values of v into Eq. (8.8).

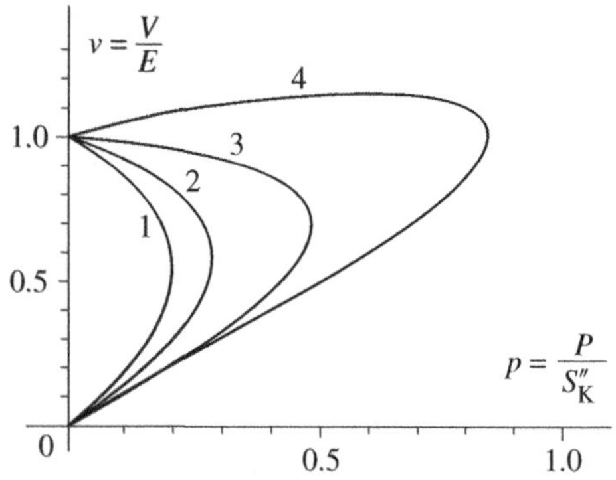

Figure 8.2 A family of nose curves with φ as a parameter: (1) $\varphi = 45°$ lag; (2) $\varphi = 30°$ lag; (3) $\varphi = 0$; (4) $\varphi = 30°$ lead.

It will be shown later in this chapter that in the considered generator-load system shown in Figure 8.1 the upper part of the nose curve (i.e. when the voltages are higher) is stable. The system is unstable in the lower part of the characteristic. It should be emphasized that a condition of the negative derivative $dv/dp < 0$ cannot be a criterion of stability as the curve $v(p)$ has a positive derivative in the upper part of the characteristic for a leading power factor (curve 4 in Figure 8.2). Nevertheless, the usefulness of the nose curve is high in practice as the difference between a current load and the maximum load determined by the peak of the characteristic is equal to the stability margin for a given power factor.

Nose curves $V(P)$ illustrate the dependency of the voltage on active power of a composite load assuming that the power factor is a parameter. The curves $Q(P)$ discussed below are derived assuming that the voltage is a parameter.

For a given value of V, Eq. (8.4) describes a circle in the (P_n, Q_n) plane as shown in Figure 8.3a. The center of the circle lies on the Q_n axis and is shifted vertically down from the origin by V^2/X. Increasing the voltage, V produces a family of circles of increasing radius and downward shift, bounded by an envelope, as shown in Figure 8.3b. For each point inside the envelope, for example point A, there are two possible solutions to Eq. (8.4) at voltage values V_1 and V_2, as defined by the two circles, whereas for any point B on the envelope there is only one value of V for which Eq. (8.4) is satisfied. An equation for this envelope can therefore be obtained by determining those values of P_n and Q_n for which there is only one solution of Eq. (8.4) with respect to V. Rearranging Eq. (8.4) gives

$$\left(\frac{V^2}{X}\right)^2 - \left(\frac{E^2}{X} - 2Q_n\right)\left(\frac{V^2}{X}\right) + \left(P_n^2 + Q_n^2\right) = 0 \tag{8.10}$$

This is a quadratic equation in (V^2/X) and has only one solution when

$$\Delta = \left(\frac{E^2}{X} - 2Q_n\right)^2 - 4\left(P_n^2 + Q_n^2\right) = 0 \tag{8.11}$$

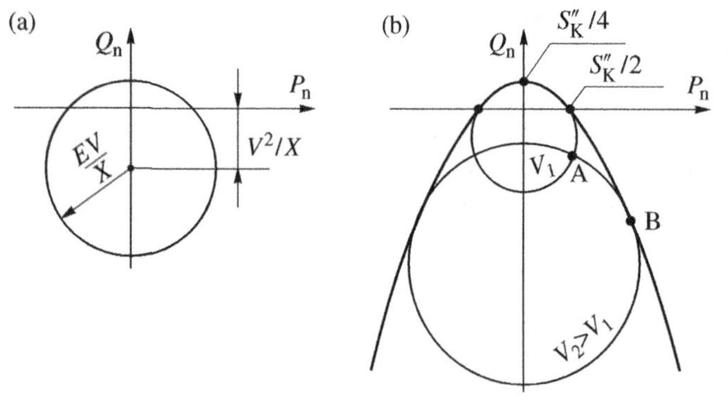

Figure 8.3 Circles determining the power that can be delivered to a ideally stiff load: (a) one circle for a given voltage V; (b) a family of circles and their envelope.

Solving for Q_n gives

$$Q_n = \frac{E^2}{4X} - \frac{P_n^2}{\frac{E^2}{X}} = \frac{S''_K}{4} - \frac{P_n^2}{S''_K} \tag{8.12}$$

which is the equation of an inverted parabola that crosses the P_n-axis at $P_n = E^2/2X = S''_K/2$ and has its maximum at

$$P_n = 0 \text{ and } Q_{n\,MAX} = \frac{E^2}{4X} = \frac{S''_K}{4} \tag{8.13}$$

A point with coordinates $P_n = E^2/2X = S''_K/2$and $Q_n = 0$ shown in Figure 8.3 corresponds to the peak of the nose curve in Figure 8.2 when $\varphi = 0$, that is when $Q_n = 0$.

The parabola described by Eq. (8.12) is important as it defines the shape of the envelope in Figure 8.3b that encloses all the possible solutions to the network Eq. (8.4). Each point (P_n, Q_n) inside the parabola satisfies two network solutions corresponding to two different values of the load voltage V, while each point on the parabola satisfies one network solution corresponding to only one value of voltage. There are no network solutions outside the parabola. In other words, it is not possible to deliver power equal to P_n and Q_n corresponding to any point outside the parabola.

8.1.2 Influence of the Load Characteristics

For the more general case the power demand will depend on the voltage as described by the voltage characteristics $P_L(V)$ and $Q_L(V)$. The possible solutions to Eq. (8.2) will not now be bounded by a simple parabola, as for $P_L(V) = P_n$ and $Q_L(V) = Q_n$, but the shape of the solution area will vary depending on the actual voltage characteristics, as shown in Figure 8.4. In general the less stiff the load, the more open the solution area. For the constant load discussed above the solution area corresponds to a parabola (Figure 8.4a). If the reactive power characteristic is a square function of the voltage, $Q_L(V) = (V/V_n)^2 Q_n$, then the solution area opens up from the top (Figure 8.4b), so that for $P_n = 0$ there is no limit on Q_n. If the active power characteristic is linear $P_L(V) = (V/V_n)P_n$, as in Figure 8.4c, then the solution area is bounded by two parallel, vertical lines. If both real and reactive power characteristics are square functions of the voltage, $P_L(V) = (V/V_n)^2 P_n$ and $Q_L(V) = (V/V_n)^2 Q_n$, then there are no limits on the values of P_n and Q_n, as shown in Figure 8.4d.

Consider again the characteristics of Figure 8.4d where there are no limits on the real and reactive power. This can be proved by expressing

$$P_L(V) = P_n\left(\frac{V}{V_n}\right)^2 = \frac{P_n}{V_n^2}V^2 = G_n V^2; \quad Q_L(V) = Q_n\left(\frac{V}{V_n}\right)^2 = \frac{Q_n}{V_n^2}V^2 = B_n V^2 \tag{8.14}$$

which shows that the load is represented by an equivalent admittance $\underline{Y}_n = G_n + jB_n$ shown in Figure 8.5. Varying the value of P_n and Q_n from zero to infinity corresponds to changing the equivalent admittance from zero (open circuit) to infinity (short circuit). As current will flow in the circuit of Figure 8.5 for any value of $\underline{Y}_n$, so a solution of Eq. (8.2) exists for any P_n and Q_n. This can be proved mathematically by substituting Eq. (8.14) into Eq. (8.2) when the following formula is obtained

$$V = \frac{E}{\sqrt{(G_n X)^2 + (B_n X + 1)^2}} \tag{8.15}$$

confirming that for any G_n and B_n a network solution for V always exists.

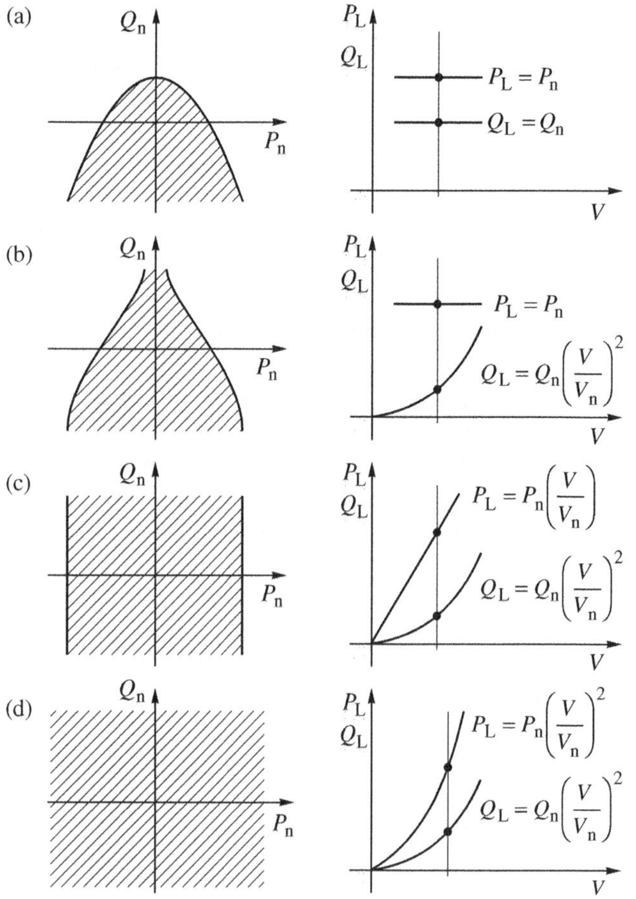

Figure 8.4 Dependence of the network solution area on the shape of the load characteristics.

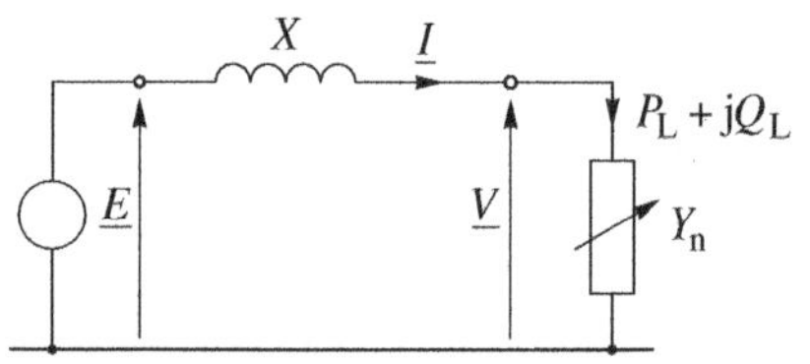

Figure 8.5 Transmission link loaded with an equivalent variable admittance.

8.2 Stability Criteria

For each point inside the envelope of network solutions, for example point A in Figure 8.3b, there are two solutions with respect to the voltage V: one with a higher and one with a lower value of voltage. It is now necessary to examine which of these solutions corresponds to a stable equilibrium point. This problem of *voltage stability* is considered in this section when different, but equivalent, voltage stability criteria are derived.

8.2.1 The dQ/dV Criterion

This classic voltage stability criterion (Venikov 1978; Weedy 1987) is based on the capability of the system to supply the load with reactive power for a given active power demand. To explain this criterion it is convenient to separate notionally the reactive power demand from the active power demand as shown in Figure 8.6. To distinguish between the power supplied by the source at the load node and the load demand itself, let $P_L(V)$ and $Q_L(V)$ be the load demand and $P_S(V)$ and $Q_S(V)$ be the powers supplied by the source to the load.

As the active power is always connected to the transmission link, it holds that $P_L(V) = P_S(V)$. Similarly, during normal operation $Q_L(V) = Q_S(V)$ but, for the purposes of stability analysis, the link between $Q_L(V)$ and $Q_S(V)$ is notionally separated. $Q_S(V)$ is treated as the reactive power supplied by the source and is assumed not to be determined by the reactive power demand of the load. The real and reactive load powers are given by expressions similar to those in Eq. (8.1)

$$P_L(V) = P_S(V) = \frac{EV}{X}\sin\delta \quad \text{and} \quad Q_S(V) = \frac{EV}{X}\cos\delta - \frac{V^2}{X} \tag{8.16}$$

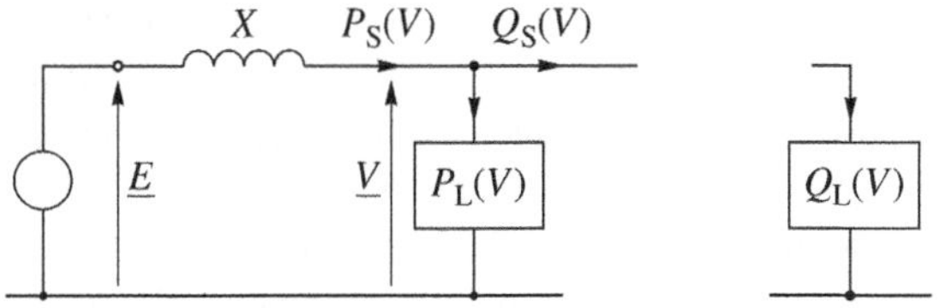

Figure 8.6 Equivalent circuit for determining the reactive power characteristic of the system.

Eliminating the trigonometric functions using the identity $\sin^2\delta + \cos^2\delta = 1$ gives

$$\left(\frac{EV}{X}\right)^2 = P_L^2(V) + \left[Q_S(V) + \frac{V^2}{X}\right]^2 \tag{8.17}$$

and solving for $Q_S(V)$ gives

$$Q_S(V) = \sqrt{\left[\frac{EV}{X}\right]^2 - [P_L(V)]^2} - \frac{V^2}{X} \tag{8.18}$$

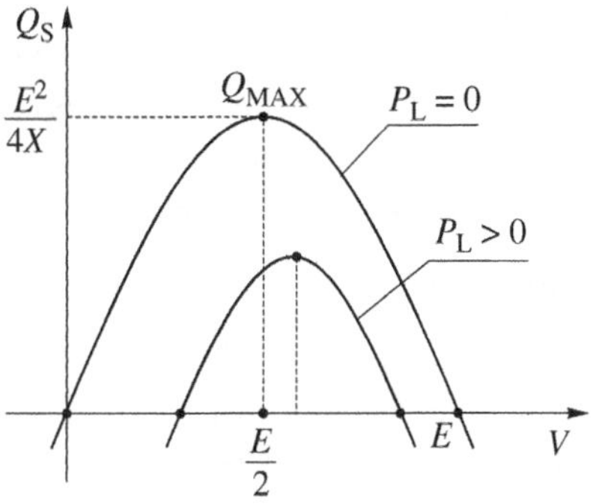

Figure 8.7 $Q_S(V)$ characteristic for $P_L = 0$ and $P_L > 0$.

This equation determines the reactive power-voltage characteristic and shows how much reactive power will be supplied by the source if the system is loaded only with the active power $P_L(V)$ and the load voltage is treated as a variable. For an ideally stiff active power load $P_L(V) = P_L = \text{constant}$ and Eq. (8.18) takes the form of an inverted parabola, as shown in Figure 8.7. The first term in Eq. (8.18) depends on the equivalent system reactance X and the load active power P_L and has the effect of shifting the parabola downward and toward the right, as illustrated in Figure 8.7. For $P_L = 0$ the parabola crosses the horizontal axis at $V = E$ and $V = 0$ and has a maximum at $Q_{MAX} = E^2/4X$ and $V = E/2$. For $P_L > 0$ the maximum value of Q_S occurs at a voltage

$$V = \sqrt{[E/2]^2 + [P_L(V)X/E]^2}$$

which is greater than $E/2$.

If the reactive power of the notionally separated load is now reconnected to the system then both the $Q_S(V)$ and $Q_L(V)$ characteristics can be drawn on the same diagram, as in Figure 8.8a. At equilibrium the supply must be equal to the demand, that is $Q_S(V) = Q_L(V)$, and is satisfied by the two equilibrium points V^s and V^u. This corresponds to the situation shown in Figure 8.3b where for one value of power demand (point A), there are two possible, but different, values of voltage $V_1 \neq V_2$.

Figure 8.8 $Q_S(V)$ and $Q_L(V)$ characteristics: (a) two equilibrium points; (b) illustration of the classic stability criterion.

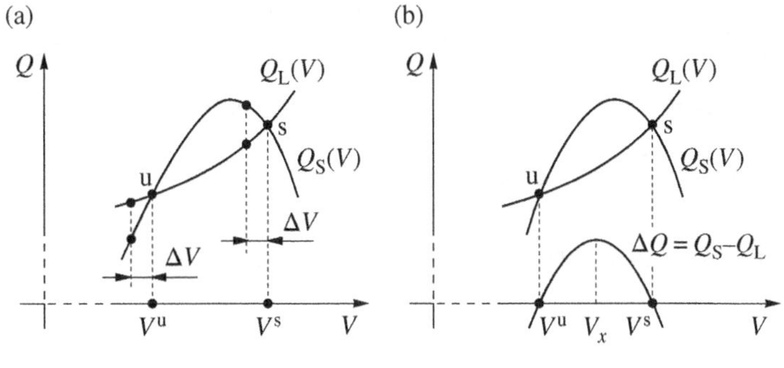

The stability of both equilibrium points can be tested using the small-disturbance method. Recall from Figure 3.4 (Section 3.1.2) that an excess of reactive power was shown to produce an increase in voltage, while a deficit of reactive power resulted in a decrease in voltage. Now consider equilibrium point s in Figure 8.8a and assume that there is a small negative voltage disturbance ΔV. This will result in the supplied reactive power $Q_S(V)$ being greater than the reactive power demand $Q_L(V)$. This excess of reactive power will tend to increase the voltage and therefore force the voltage to return to point s. If the disturbance produces an increase in voltage, the resulting deficit in reactive power will force the voltage to decrease and again return to point s. The conclusion is that equilibrium point s is stable.

On the other hand, a disturbance in the vicinity of the second equilibrium point u which decreases the voltage will produce a deficit of reactive power with $Q_S(V) < Q_L(V)$, which will force a further decrease in voltage. As the disturbed system does not return to the equilibrium point, the equilibrium point u is unstable.

The classic voltage stability criterion is obtained by noting from Figure 8.8b that the derivative of the surplus of reactive power $\mathrm{d}(Q_S - Q_L)/\mathrm{d}V$ is of opposite sign at the two equilibrium points: it is negative at the stable point s and positive at the unstable point u. This is the essence of the classic $\mathrm{d}\Delta Q/\mathrm{d}V$ stability criterion

$$\frac{\mathrm{d}(Q_S - Q_L)}{\mathrm{d}V} < 0 \quad \text{or} \quad \frac{\mathrm{d}Q_S}{\mathrm{d}V} < \frac{\mathrm{d}Q_L}{\mathrm{d}V} \tag{8.19}$$

In the simple system shown in Figure 8.6 the supplied real and reactive powers, expressed by Eq. (8.16), are functions of the two variables V and δ with increments given by

$$\begin{aligned} \Delta Q_S &= \frac{\partial Q_S}{\partial V}\Delta V + \frac{\partial Q_S}{\partial \delta}\Delta\delta \\ \Delta P_L = \Delta P_S &= \frac{\partial P_S}{\partial V}\Delta V + \frac{\partial P_S}{\partial \delta}\Delta\delta \end{aligned} \tag{8.20}$$

Eliminating $\Delta\delta$ from these two equations, and dividing the result by ΔV, gives

$$\frac{\Delta Q_S}{\Delta V} = \frac{\partial Q_S}{\partial V} + \frac{\partial Q_S}{\partial \delta}\left(\frac{\partial P_S}{\partial \delta}\right)^{-1}\left[\frac{\Delta P_L}{\Delta V} - \frac{\partial P_S}{\partial V}\right] \tag{8.21}$$

or

$$\frac{\mathrm{d}Q_S}{\mathrm{d}V} \cong \frac{\partial Q_S}{\partial V} + \frac{\partial Q_S}{\partial \delta}\left(\frac{\partial P_S}{\partial \delta}\right)^{-1}\left[\frac{\mathrm{d}P_L}{\mathrm{d}V} - \frac{\partial P_S}{\partial V}\right] \tag{8.22}$$

where the partial derivatives are obtained from the equations in (8.16) as

$$\frac{\partial P_S}{\partial \delta} = \frac{EV}{X}\cos\delta, \quad \frac{\partial P_S}{\partial V} = \frac{E}{X}\sin\delta, \quad \frac{\partial Q_S}{\partial \delta} = -\frac{EV}{X}\sin\delta, \quad \frac{\partial Q_S}{\partial V} = \frac{E}{X}\cos\delta - 2\frac{V}{X} \tag{8.23}$$

Substituting these partial derivatives into Eq. (8.22) gives

$$\frac{\mathrm{d}Q_S}{\mathrm{d}V} \cong \frac{E}{X}\cos\delta - \frac{2V}{X} - \frac{EV}{X}\sin\delta\frac{X}{EV\cos\delta}\left[\frac{\mathrm{d}P_L}{\mathrm{d}V} - \frac{E}{X}\sin\delta\right] = \frac{E}{X\cos\delta} - \left(\frac{2V}{X} + \frac{\mathrm{d}P_L}{\mathrm{d}V}\tan\delta\right) \tag{8.24}$$

This allows the stability condition defined in Eq. (8.19) to be expressed as

$$\frac{\mathrm{d}Q_L}{\mathrm{d}V} > \frac{E}{X\cos\delta} - \left(\frac{2V}{X} + \frac{\mathrm{d}P_L}{\mathrm{d}V}\tan\delta\right) \tag{8.25}$$

where the derivatives $\mathrm{d}Q_L/\mathrm{d}V$ and $\mathrm{d}P_L/\mathrm{d}V$ are calculated from the functions used to approximate the load characteristics.

Generally, for a multi-machine system, it is not possible to derive an analytical formula for the stability criterion. However, by using a load flow program it is possible to obtain the system supply characteristic $Q_S(V)$ by defining the load node under investigation as a PV node and executing the program several times for different values of the node voltage V. The resulting $Q_S(V)$ characteristic can then be compared with the load characteristic $Q_L(V)$ to check the stability condition.

8.2.2 The dE/dV Criterion

The system equivalent emf E can be expressed as a function of the load voltage by solving Eq. (8.2) for E to give

$$E(V) = \sqrt{\left(V + \frac{Q_L(V)X}{V}\right)^2 + \left(\frac{P_L(V)X}{V}\right)^2} \tag{8.26}$$

where $Q_L(V)X/V$ is the in-phase and $P_L(V)X/V$ the quadrature component of the voltage drop $\underline{I}X$ shown in Figure 8.1.

An example of an $E(V)$ characteristic is shown in Figure 8.9 with the load normally operating at a high voltage corresponding to the right-hand side of the characteristic. As V is large, and much greater than both the in-phase and quadrature components of the voltage drop, then, according to Eq. (8.26), a decrease in voltage will cause the emf $E(V)$ to fall. As V continues to decrease, the $Q_L(V)X/V$ and $P_L(V)X/V$ components become more important and below a certain value of V, they will cause $E(V)$ to rise. Consequently, each value of $E(V)$ may correspond to two possible solutions of the network equations with respect to V. As before, the stability of these solutions can be examined using the small-disturbance method.

First consider the system behavior at point s lying on the right-hand side of the characteristic $E(V)$ in Figure 8.9 and assume that the source emf E is maintained at a constant value. A decrease in the load voltage by ΔV will produce a reduction in the emf $E(V)$ that is less than the source emf E. As E is too large to maintain the lowered load, voltage V is forced to return to the initial equilibrium value. Similarly, a voltage increase by ΔV results in the source emf E being smaller than the emf required to maintain the increased load voltage V so that the source emf E again forces the voltage to return to its initial value.

Now consider a disturbance that produces a voltage decrease by ΔV from the equilibrium point u on the left-hand side of the characteristic. This disturbance results in the source emf E being less than the emf $E(V)$ required to maintain the lowered voltage. As E is too small, the load voltage further declines and does not return to its initial value. The conclusion is that point u is unstable.

From these discussions it is apparent that the system is stable if the equilibrium point lies on the right-hand side of the characteristic, that is when

$$\frac{dE}{dV} > 0 \tag{8.27}$$

Venikov (1978) and Abe and Isono (1983) have both shown this condition to be equivalent to the classic stability condition defined in Eq. (8.19).

Using the stability condition defined in Eq. (8.27) to study the stability of a multi-node system is rather inconvenient when a load flow program is used. The derivative dE/dV is based on the equivalent circuit of Figure 8.1. In the load flow program, the load node should be of the PQ type (specified P_L and Q_L, unknown V and δ) which does not allow the voltage to be set. Consequently, the voltage increase necessary to obtain the derivative dE/dV has to be enforced from the generator nodes (PV type) for which the voltage can be set. As there is normally a large number of generator nodes in the system, a large number of load flows would have to be run to account for all the possible combinations of changes in the voltage settings.

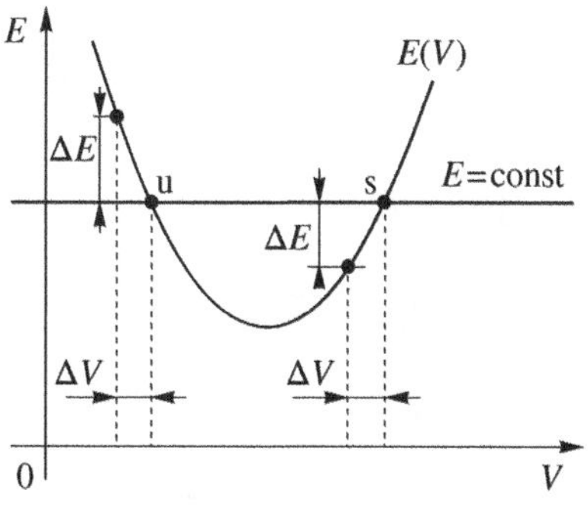

Figure 8.9 Illustration to the stability criterion dE/dV > 0.

8.2.3 The dQ_G/dQ_L Criterion

To understand this criterion it is necessary to analyze the behavior of the reactive power generation $Q_G(V)$ as the load reactive demand $Q_L(V)$ varies. This approach is somewhat different to that taken in the two previous subsections in that $Q_G(V)$ now includes the reactive power demand of both the load, $Q_L(V)$, and the network, I^2X, whereas previously only the reactive power $Q_S(V)$ supplied by the source at the load node was considered.

The equation determining $Q_G(V)$ is similar to Eq. (8.1) but with E and V interchanged, that is

$$Q_G(V) = \frac{E^2}{X} - \frac{EV}{X}\cos\delta \tag{8.28}$$

where both V and δ depend on the demand $P_L(V)$ and $Q_L(V)$. Equation (8.1) allows the second component in this equation to be substituted to give

$$Q_G(V) = \frac{E^2}{X} - \frac{V^2}{X} - Q_L(V) \quad \text{or} \quad \frac{V^2}{X} = \frac{E^2}{X} - Q_L(V) - Q_G(V) \tag{8.29}$$

Substituting this expression into Eq. (8.2) and performing some simple algebra gives

$$Q_G^2(V) - \frac{E^2}{X}Q_G(V) + P_L^2(V) + \frac{E^2}{X}Q_L(V) = 0 \tag{8.30}$$

or

$$Q_L(V) = -\frac{Q_G^2(V)}{S''_K} + Q_G(V) - \frac{P_L^2(V)}{S''_K} \tag{8.31}$$

where, as in Eq. (8.12), $S''_K = E^2/X$. For the case of the ideally stiff active power load with $P_L(V) = P_L =$ constant this equation describes a horizontal parabola in the (Q_G, Q_L) plane, as shown in Figure 8.10a. The vertex of the parabola is at a constant Q_G value equal to $E^2/2X = S''_K/2$, while the maximum value of Q_L depends on P_L and for $P_L = 0$ the maximum is at $E^2/4X = S''_K/4$. Increasing P_L shifts the parabola to the left along the Q_L-axis but without any corresponding shift with respect to the Q_G-axis.

It is worth noting that the vertex of the parabola (the maximum value of Q_L for a given P_L) corresponds to a point on the envelope of the $Q_L(P_L)$ characteristics shown in Figure 8.3b and to the vertex of the $Q_S(V)$ characteristic in Figure 8.7. Obviously, for $P_L = 0$ all three characteristics give the same maximum value $Q_{nMAX} = E^2/4X = S''_K/4$ corresponding to Eq. (8.13).

Figure 8.10b shows how the $Q_G(Q_L)$ characteristic can be used to analyze the system stability. Assuming that the reactive load demand Q_L is lower than its maximum value, there are always two equilibrium points, that is two

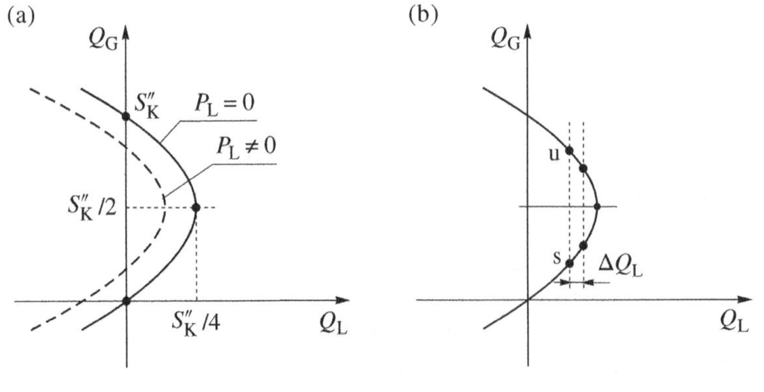

Figure 8.10 Generation and load characteristics: (a) $Q_G(Q_L)$ with P_L as a parameter; (b) the small-disturbance method applied to the $Q_G(Q_L)$ characteristic.

values of reactive generation corresponding to a given demand. At the lower point s, a momentary disturbance ΔQ_L that increases the load reactive power demand results in an increase in the generated reactive power while a disturbance that reduces the demand results in a corresponding drop in the reactive power generation. As the generation follows the demand, the lower equilibrium point s is stable. The situation is reversed at the upper equilibrium point u. Here an increase in Q_L produces a reduction in Q_G, while a reduction in Q_L produces an increase in Q_G. As the changes in reactive generation are now in the opposite direction to the changes in demand, the upper equilibrium point u is unstable.

Consequently, the system is stable if a small change in reactive load demand produces a change in the generation which has the same sign or, in other words, the derivative dQ_G/dQ_L is positive

$$\frac{dQ_G}{dQ_L} > 0 \tag{8.32}$$

It is worth noting that at the maximum loading point at the nose of the $Q_G(Q_L)$ characteristic the derivative dQ_G/dQ_L tends to infinity.

The characteristic defined by Eq. (8.31), and shown in Figure 8.10, is a parabola only for the ideally stiff active power load, $P_L(V) = P_L =$ constant. For a voltage-dependent load characteristic $P_L(V)$ will vary with voltage when it is not possible to obtain an explicit expression for $Q_L(Q_G)$ with $P_L(V)$ as a parameter. However, the $Q_L(Q_G)$ characteristic can be obtained iteratively by solving the network equations for given values of load demand, P_L and Q_L, and emf E at the generator node.

The main advantage of the dQ_G/dQ_L criterion is the ease with which it can be used with a load flow program to analyze a multi-node system (Carpentier et al. 1984; Taylor 1994). The generated reactive power Q_G is replaced by the sum of all the generated reactive powers at all the generator nodes, while the derivative is replaced by the quotient of the sum of all the generated reactive power increments over the increment of the reactive load demand at the examined load node, that is $\sum \Delta Q_{Gi}/\Delta Q_L$.

8.3 Critical Load Demand and Voltage Collapse

Figure 8.4 shows how the network solution area depends on the shape of the load characteristic and Figure 8.8 illustrates the classical $d\Delta Q/dV$ stability criterion. Figure 8.11 extends this discussion and shows how the network equations may have two solutions, one solution or no solution at all depending on the relative position and shape of the $Q_L(V)$ and $Q_S(V)$ characteristics.

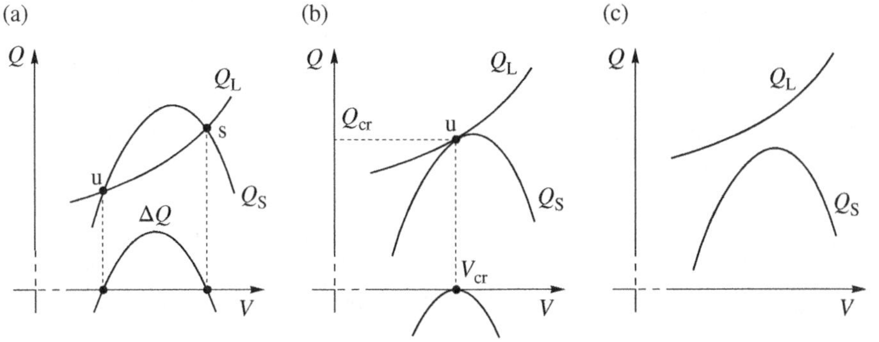

Figure 8.11 Relative position of the generation and load characteristics: (a) two equilibrium points; (b) a single critical equilibrium point; (c) no equilibrium points.

In Figure 8.11a there are two equilibrium points, corresponding to the intersection of the $Q_L(V)$ and $Q_S(V)$ characteristics, of which only point s is stable. For the special case of $Q_L(V)$ = constant and $P_L(V)$ = constant these two voltage values correspond to the voltages V_1 and V_2 designated by point A in Figure 8.3b, only one of which will be stable. For the more general case of $Q_L(V) \neq$ constant and $P_L(V) \neq$ constant shown in Figure 8.4b–d, then a point equivalent to A in Figure 8.3b will exist somewhere inside the envelope with two voltage solutions depicted by the general points u and s in Figure 8.11a. If the point lies on the envelope then there is only one equilibrium point and this corresponds to the one intersection point shown in Figure 8.11b. The power system is then in a *critical* state, the point is referred to as the *critical point*, and its coordinates are the *critical power* and the *critical voltage*. Outside the network solution area in Figure 8.4 there are no equilibrium points and this corresponds to the $Q_L(V)$ and $Q_S(V)$ characteristics having no point of intersection, as in Figure 8.11c, where the $Q_L(V)$ characteristic lies above the $Q_S(V)$ characteristic. In general the area of network solution shown in Figure 8.4 is known as the *steady-state voltage stability area*.

Remember that Figure 8.11 has been drawn for the special case of the ideally stiff load with respect to real power, $P_L(V) = P_L$ = constant, so that the critical power Q_{cr} and P_{cr} is defined by the coordinates (Q_{cr} and P_L) and only Q_{cr} and V_{cr} are marked on Figure 8.11b.

8.3.1 Effects of Increasing Demand

A slow increase in the system demand, such as that due to the normal daily load variations, can have two detrimental effects on the voltage stability. According to Eq. (8.18), an increase in the active power lowers the $Q_S(V)$ characteristic, as shown in Figure 8.7, while an increase in the reactive power raises the $Q_L(V)$ characteristic. As a consequence the stable equilibrium point s moves toward smaller values of voltage and the unstable equilibrium point u moves toward larger values of voltage. As the demand further increases, the equilibrium points move closer together until they finally merge at the critical equilibrium point shown in Figure 8.11b.

When a load operates at this critical point, any small increase in the reactive power demand will produce a deficit in the reactive power, the reactive power demand will be greater than supply, and the voltage will reduce. As the voltage reduces, the deficit in reactive power increases and the voltage falls even further until it eventually falls to a very small value. This phenomenon is generally known as *voltage collapse*, although in some countries the more graphic term of *voltage avalanche* is used. Two forms of voltage collapse are identified in the literature. When the voltage collapse is permanent, some authors refer to it as a *total voltage collapse* (Taylor 1994). On the other hand, the term *partial voltage collapse* is used when a large increase in demand causes the voltage to fall below some technically acceptable limit. As a partial voltage collapse does not correspond to system instability, it is perhaps better to consider the system as being in an emergency state since the system still operates, albeit at a reduced voltage.

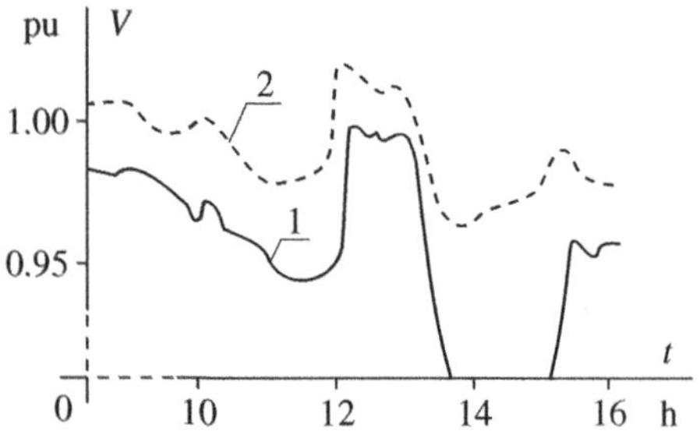

Figure 8.12 An example of the voltage collapse (Nagao 1975): (1) voltage variations during the day of the voltage collapse; (2) voltage variations during the previous day.

An example of an actual voltage collapse is shown in Figure 8.12 (Nagao 1975). In this figure curve 2 represents a typical morning period when the voltage drops slightly as the power demand increases but then recovers by a small percentage during the lunch break as the power demand reduces. At about 13:00, after the lunch break, the power demand again builds up and the voltage drops. However, on one particular day (curve 1) the overall system load is greater, and the local voltage smaller, than normal so that as the load starts to increase after the lunch break the voltage reaches its critical value and then collapses. The system operators then intervene and, after a long interruption, manually restore the normal operating conditions.

From the power system security point of view, knowledge of the critical power and voltage is very important as the operating voltage, and power, at the system nodes should be kept as far as possible from their critical values. Unfortunately, the nonlinearity of the voltage characteristics makes it impossible to derive a general formula for the critical voltage that is valid for any type of active and reactive power variations, even assuming a simple power system model. However, some simple iterative formula can be developed if the following three assumptions are made:

- That when the load demand increases, the power factor is maintained constant by reactive power compensation in the consumer load so that

$$\frac{P_{\mathrm{n}}(t)}{P_0} = \frac{Q_{\mathrm{n}}(t)}{Q_0} = \xi \tag{8.33}$$

where $P_{\mathrm{n}}(t)$ and $Q_{\mathrm{n}}(t)$ are the nominal values of the load demand at the time instant t, P_0, and Q_0 are the initial values of the demand and ξ is the coefficient of demand increase.

- That the power characteristic of an industrial composite load, with a large number of induction motors, can be approximated by the polynomials (Section 3.4.4)

$$\frac{Q_{\mathrm{L}}}{Q_{\mathrm{n}}} = a_2\left(\frac{V}{V_{\mathrm{n}}}\right)^2 - a_1\left(\frac{V}{V_{\mathrm{n}}}\right) + a_0; \quad \frac{P_{\mathrm{L}}}{P_{\mathrm{n}}} = b_1\left(\frac{V}{V_{\mathrm{n}}}\right) \tag{8.34}$$

that is by a parabola and a straight line, respectively.

- That the load composition is constant thereby making the coefficients a_0, a_1, a_2, and b_1 constant.

Equation (8.33) allows the characteristics defined in Eq. (8.34) to be converted to the following form

$$Q_{\mathrm{L}} = \xi\left[\alpha_2 V^2 - \alpha_1 V + \alpha_0\right]; \quad P_{\mathrm{L}} = \xi\beta_1 V \tag{8.35}$$

where

$$\alpha_2 = \frac{Q_0}{V_{\mathrm{n}}^2}a_2; \quad \alpha_1 = \frac{Q_0}{V_{\mathrm{n}}}a_1; \quad \alpha_0 = Q_0 a_0; \quad \beta_1 = \frac{P_0}{V_{\mathrm{n}}}b_1$$

Substituting the second of the equations in (8.35) into Eq. (8.18) gives

$$Q_S(V) = V\sqrt{\left(\frac{E}{X}\right)^2 - \xi^2\beta_1^2} - \frac{V^2}{X} \tag{8.36}$$

Figure 8.11b shows how, at the critical point, the supply characteristic and the demand power characteristic are tangential to each other, which allows the value of the critical voltage V_{cr} to be found by solving the following two equations

$$Q_S(V_{\mathrm{cr}}) = Q_{\mathrm{L}}(V_{\mathrm{cr}}) \tag{8.37}$$

$$\left.\frac{\mathrm{d}Q_S}{\mathrm{d}V}\right|_{V=V_{\mathrm{cr}}} = \left.\frac{\mathrm{d}Q_L}{\mathrm{d}V}\right|_{V=V_{\mathrm{cr}}} \tag{8.38}$$

Substituting for Q_S and Q_{L} from Eqs. (8.35) and (8.36) gives

$$V_{\mathrm{cr}}\sqrt{\left(\frac{E}{X}\right)^2 - \xi_{\mathrm{cr}}^2\beta_1^2} - \frac{V_{\mathrm{cr}}^2}{X} = \xi_{\mathrm{cr}}\left(\alpha_2 V_{\mathrm{cr}}^2 - \alpha_1 V_{\mathrm{cr}} + \alpha_0\right) \tag{8.39}$$

$$\sqrt{\left(\frac{E}{X}\right)^2 - \xi_{cr}^2\beta_1^2} - 2\frac{V_{cr}}{X} = \xi_{cr}\left(2\alpha_2 V_{cr} - \alpha_1\right) \tag{8.40}$$

Equation (8.40) determines the critical voltage as a function of the system parameters and an unknown coefficient ξ_{cr} of the demand increase as

$$V_{cr} = \frac{\sqrt{\left(\frac{E}{\beta_1 X}\right)^2 - \xi_{cr}^2} + \frac{\alpha_1}{\beta_1}\xi_{cr}}{2\frac{\alpha_2}{\beta_1}\xi_{cr} + \frac{2}{\beta_1 X}} \tag{8.41}$$

Multiplying Eq. (8.40) by $(-V_{cr})$ and adding the result to Eq. (8.39) gives

$$\frac{V_{cr}^2}{X} = \xi_{cr}\left[\alpha_0 - \alpha_2 V_{cr}^2\right] \tag{8.42}$$

from which finally

$$\xi_{cr} = \frac{1}{\frac{\alpha_0 X}{V_{cr}^2} - \alpha_2 X} \tag{8.43}$$

Equations (8.41) and (8.43) can now be used in an iterative calculation to find the critical voltage V_{cr} and the critical demand increase ξ_{cr}.

Example 8.1 A composite load of rating $S_n = (150 + j100)$ MVA and $V_n = 110$ kV is supplied from the network by a 130 km, 220 kV transmission line with a reactance of 0.4 Ω/km via two parallel 160 MVA transformers, each with a reactance of 0.132 pu. Assume that the system can be replaced by a 251 kV voltage source with a reactance of 13 Ω and that the load characteristics in kV and MVA are $P_L = \xi\, 0.682\, V$ and $Q_L = \xi\,(0.0122\, V^2 - 4.318\, V + 460)$. Find the critical voltage and the critical coefficient of the load demand increase. Simple calculation leads to the parameter values shown in Figure 8.13. Equations (8.41) and (8.43) then give

$$V_{cr} = \frac{\sqrt{18.747 - \xi_{cr}^2} + 6.331\,\xi_{cr}}{0.03578\,\xi_{cr} + 0.0345}; \quad \xi_{cr} = \frac{1}{\frac{39100}{V_{cr}^2} - 1.037} \tag{8.44}$$

The initial substitution of $\xi_{cr} = 1$ into the first of these equations gives $V_{cr} = 151.69$. Substituting this value into the second equation gives $\xi_{cr} = 1.51$. After five such iterations, the final solution of $\xi_{cr} = 1.66$ and $V_{cr} = 154.5$ kV is obtained indicating that the system would lose stability if the demand increased 1.66 times. The primary voltage $V_{cr} = 154.5$ kV referred to the transformer secondary winding gives 77.25 kV, that is 70% of the rated voltage of 110 kV. Figure 8.13b and c show the system and the load characteristic in the initial and critical states. Initially, the system operated at point s at a voltage of 208 or 104 kV at the transformer secondary, that is at 94.5% of the rated voltage.

******

8.3.2 Effect of Network Outages

The relative position of the supply characteristic defined in Eq. (8.18) depends on the equivalent system reactance. Large changes in this reactance, such as those caused by network outages, may lower the generation characteristic and cause voltage stability problems.

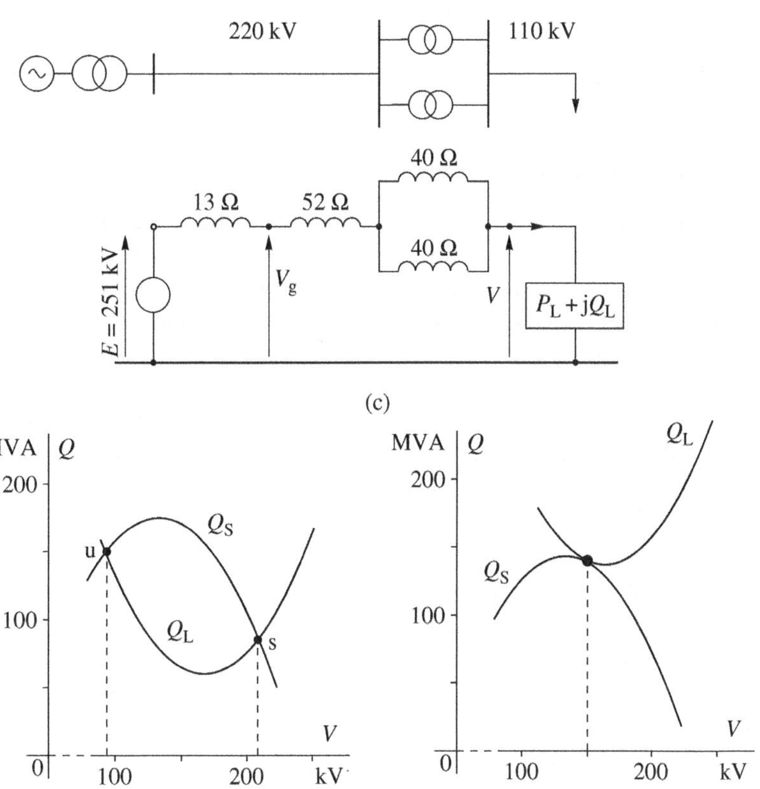

Figure 8.13 Illustration to Example 8.1: (a) equivalent circuit with parameters calculated at 220 kV; (b) initial characteristics; (c) characteristics at the critical state.

Example 8.2 A composite load of $S_n = (240 + j160)$MVA and $V_n = 110$kV is supplied from the network by a double-circuit transmission line of length 130 km and reactance 0.4 Ω/km via two parallel 160 MVA transformers each with a reactance of 0.132 pu. The power system may be represented by an emf of 251 kV acting behind an equivalent reactance of 13 Ω. The load characteristics are: $P_L = 1.09\ V$ and $Q_L = 0.0195\ V^2 - 6.9\ V + 736$. Check the system stability after one of the lines is tripped.

Figure 8.14 shows the load characteristic and the generation characteristic before (Q_S') and after (Q_S'') the line is tripped. Tripping the line causes a reduction in the load voltage to about 170 or 85 kV on the secondaries of the

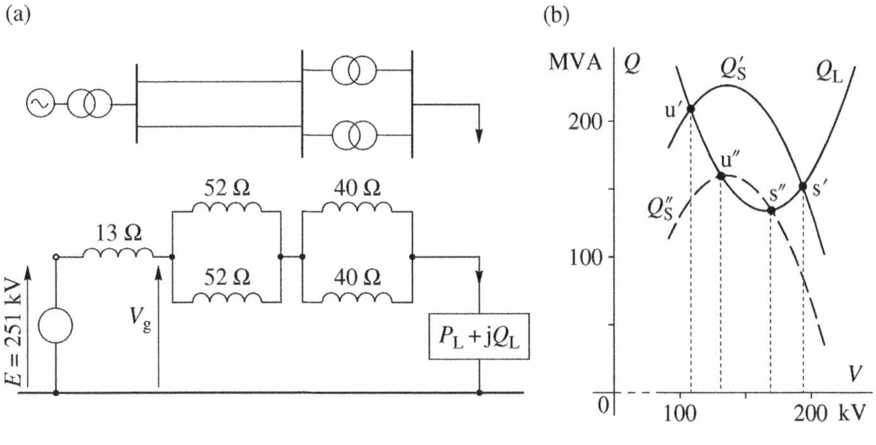

Figure 8.14 Illustration for Example 8.2: (a) equivalent circuit with parameters calculated at 220 kV; (b) voltage characteristics.

transformers, that is to about 77% of the rated voltage. The system is on the verge of losing stability. Calculations similar to those in Example 8.1 show that a demand increase by 3.8% will result in the system losing stability. Owing to the unacceptably low voltage, the system is now in an emergency state with a high risk of instability.

******

8.3.3 Influence of the Shape of the Load Characteristics

Figure 8.7 and Eq. (8.18) demonstrate how an increase in the active power demand shifts the vertex of the reactive power supply characteristic downward. This observation has important implications with regard to the voltage stability of systems where the active power varies with voltage. For example, if the active power demand reduces with reducing voltage then the stability of the system will be improved compared with when the active power characteristics is independent of voltage. This situation is examined further in Example 8.3.

Example 8.3 For the system in Example 8.2 determine the supply characteristics after one of the transmission lines is tripped assuming that the active power characteristics of the load (in MW) are as follows:

$$(1)\, P_L = 240 = \text{constant},\;\; (2)\, P_L = 16.18\sqrt{V},\;\; (3)\, P_L = 1.09V,\;\; (4)\, P_L = 0.004859V^2.$$

The four active power characteristics are shown graphically in Figure 8.15a with curve 1 representing an ideally stiff load. Curves 2, 3, and 4 represent voltage-sensitive loads. The corresponding reactive power supply characteristics can be easily determined using Eq. (8.18) and are shown in Figure 8.15b. For the ideally stiff load (curve 1), the system has no equilibrium points and line tripping results in immediate voltage collapse. The slightly less stiff load 2 is in the critical state and will also lose stability should the line trip. Load 3 is close to the critical state and will lose stability when the demand increases by 3.8% (see Example 8.2), while load 4 is more voltage sensitive and operates with a greater safety margin. In this case stability will be lost if the demand increases by about 15%.

******

Figure 8.4 shows how the network solution area depends on the shape of the load characteristic with Figure 8.4d corresponding to a constant admittance load. The equivalent circuit for such a load is shown in Figure 8.5 and Section 8.1 explains how for a load with such a characteristic every equilibrium point for $V > 0$ is stable and the system cannot suffer voltage collapse. This can also be proved using the classical dQ/dV criterion.

For a constant admittance load the $P_L(V)$ characteristic is given by Eq. (8.14), which, when substituted into Eq. (8.18), gives the reactive power supply characteristic as

$$Q_S = V\left[\sqrt{\left(\frac{E}{X}\right)^2 - (G_n V)^2} - \frac{V}{X}\right] \tag{8.45}$$

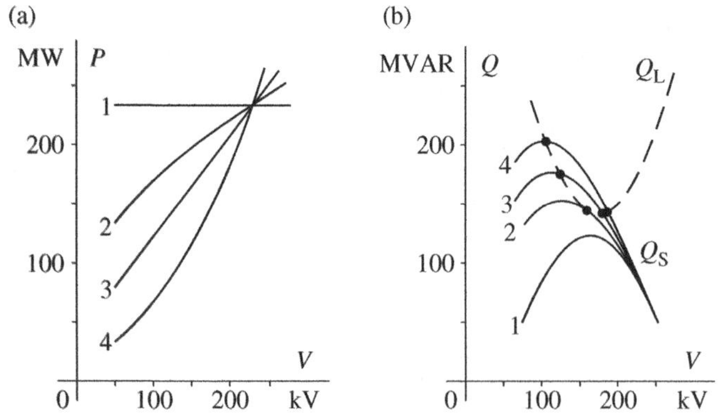

Figure 8.15 Real and reactive power characteristics of Example 8.3.

The resulting characteristic is drawn in Figure 8.16 and shows that for any value of the load conductance G_n the supply characteristic always crosses the origin at $V = 0$. Again from Eq. (8.14) the reactive power demand is given as $Q_L = B_n V^2$ so that the origin is always one of the equilibrium points for any value of the load susceptance B_n. As the load susceptance increases, the parabola defined by $Q_L = B_n V^2$, becomes steeper but the stability condition (8.19) is satisfied for all equilibrium points when $V > 0$. This is true even for the equilibrium point $S_{(4)}$ on the left-hand side of the supply parabola. The constant admittance load is therefore absolutely stable because no value of Y_L can cause the voltage to collapse, although a large value of the load admittance Y_L will result in an inadmissible low operating voltage.

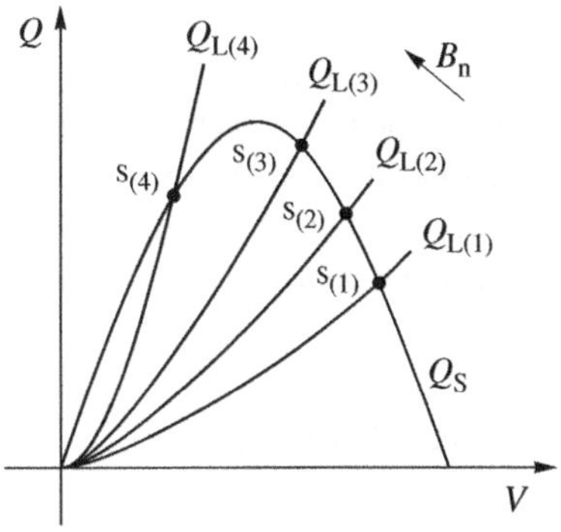

Figure 8.16 A family of reactive demand characteristics for a fixed-load conductance G_n and increased load susceptance B_n.

8.3.4 Influence of the Voltage Control

Section 8.1 explains that the reactance of the equivalent circuit X shown in Figure 8.1 depends on the voltage control capability of the equivalent generating unit. Provided that neither the excitation limits nor the stator current limit is exceeded then the AVR keeps the terminal voltage constant and then voltage and reactance of the equivalent source are $E = V_G$ and $X = X_L$. However, if any of the generator current limits are met then the field voltage will be maintained constant and then the voltage of the equivalent source is $E = E_f$ and the reactance of the equivalent source must be increased to $X = x_d + X_L$. Consequently, as voltage and reactance of the equivalent source depend significantly on the voltage control capability of the generating unit, so too will the voltage stability. Both the area of network feasibility (Section 8.1) and the characteristics used to analyze the voltage stability (Section 8.2) depend on the equivalent reactance of the system. This reactance occurs in two terms in Eq. (8.18), the $-V^2/X$ term and the EV/X term, both of which determine the steepness of the parabola and the position of its vertex. If X is large then the supply characteristic parabola is shallow and its vertex is lowered, as shown in Figure 8.7. This results in a reduction in the maximum reactive power and hence the critical power.

Example 8.4 For the system in Example 8.2 determine the supply characteristic when the system is operating with only one transmission line. Assume that the equivalent system generator is capable of keeping the voltage at the sending end of the line constant.

Before the line is tripped the load operated at 208 kV (referred to the primaries of the transformers). Figure 8.14 shows the voltage drop across the source reactance gives the terminal voltage $V_g = 245$ kV. Assuming that the AVR keeps this voltage constant, the system can now be represented by a voltage $E = V_g = 245$ kV = constant with zero internal reactance. The reactance of one line, and two parallel transformers, give a total $X = 52 + 40/2 = 72\ \Omega$. Figure 8.17 shows that the supply characteristic is now much higher compared with when voltage control was neglected. Calculations similar to those in Example 8.1 show that the system would lose stability when the demand increases by about 12%. Neglecting voltage control gave the critical demand increase as 3.8%.

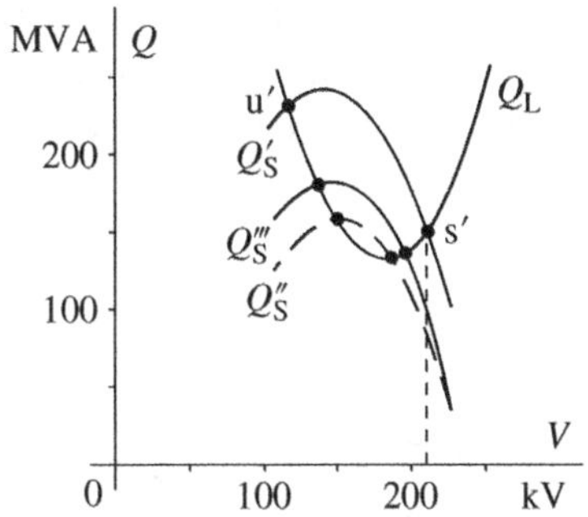

Figure 8.17 Generation and load characteristics analyzed in Example 8.4. Q'_S – before tripping the line; Q''_S – after tripping with E = constant; Q'''_S – after tripping with V_g = constant.

******

It is assumed in the above example that the voltage regulator of the source supplying the load is able to keep the terminal voltage constant. This is an idealized situation. In practice, every source has a limited regulation range. For the synchronous generator, limitations of voltage regulators are discussed in Section 3.3.5. The characteristic of the generator as the source of reactive power is shown in Figure 3.25. That diagram shows that, after the limiting value of the field (rotor) current or the armature (stator) current is reached, the reactive power capability is quickly reduced as the voltage declines. The generator can no longer be treated as a constant voltage source but rather as a source with a large internal reactance equal to the synchronous reactance. For such a source, the generation characteristic Q_S is significantly lowered (curve Q_S'' in Figure 8.17) and voltage stability may be lost via voltage collapse.

8.4 Static Analyses

The analysis so far has concerned the simple generator-load. Voltage stability analysis of a more realistic representation of a multi-machine system supplying many composite loads via a meshed network is much more complicated.

For network planning purposes the critical load power can be calculated offline using a modified load flow computer program and the dQ/dV or dQ_G/dQ_L stability criterion (Van Cutsem 1991; Ajjarapu and Christy 1992; Taylor 1994; Van Cutsem and Vournas 1998). The procedure is similar to that described in Section 8.3 for the simple generator-line-load system, and the idea is to bring the system to a critical state by increasing the power demand of a chosen load or a group of loads. The use of a load flow program allows a variety of effects such as generator voltage control, reactive power compensation, the distribution of active power among the generation units with active governors, the voltage characteristics of the loads, and so on, to be included.

Unfortunately, the method is computationally time consuming and cannot be used for online applications, such as steady-state security assessment, which require a fast assessment of the voltage stability conditions and an estimation of how far a given operating point is from the critical state. The distance from the critical state is usually quantified by one of the so-called *voltage stability indices* (Kessel and Glavitsch 1986; Tiranuchit and Thomas 1988; Löf et al. 1992, among others). A number of different indices are discussed by Taylor (1994).

8.4.1 Simplified Voltage Stability Criterion

Simplified angle stability criteria based on the short-circuit power are described in Section 6.1.6. A similar criterion for voltage stability (also based on the short-circuit power) was derived by Machowski et al. (2015) on the basis of the equivalent circuit shown in Figure 8.1.

It is assumed here, as in Eq. (8.3), that a composite load ($P_n + jQ_n$) is ideally stiff. In such a case all possible solutions of Eq. (8.4) determining the stability area are enclosed by a parabola described by Eq. (8.12) and shown in Figure 8.3b. For the convenience of further considerations, this parabola is shown again in Figure 8.18, and Eq. (8.12) is rewritten here in the following way

$$Q_n = \frac{S_K''}{4} - \frac{P_n^2}{S_K''} \tag{8.46}$$

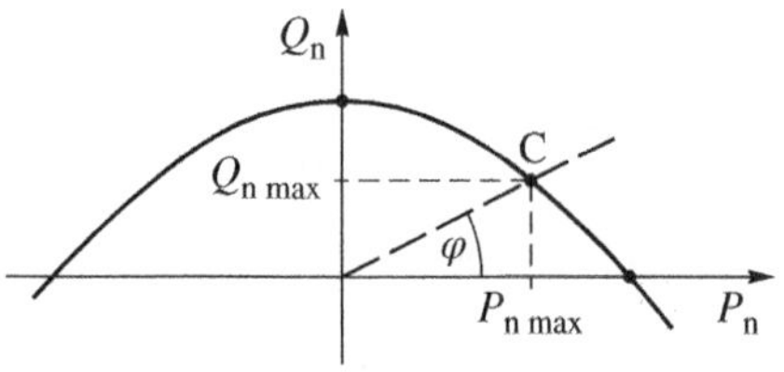

Figure 8.18 Inverted parabola bounding stability area.

For a given value of the power factor $\cos\varphi$ the following relationship occurs between the real and reactive power

$$Q_n = \tan\varphi \cdot P_n \tag{8.47}$$

On the plane with coordinates P_n and Q_n Eq. (8.47) describes a straight line inclined at angle φ. The straight line cuts the parabola (Figure 8.18) bounding the stability area at point C determining the maximum power of the load ($P_{n\,max} + jQ_{n\,max}$) for the given power factor $\cos\varphi$, referred to as *the critical load* or *critical power demand*.

Substituting $P_n = P_{n\,max}$ and $Q_n = Q_{n\,max} = \tan\varphi \cdot P_{n\,max}$ into Eq. (8.46) after simple transformations gives

$$\left(S''_K - 2\tan\varphi \cdot P_{n\,max}\right)^2 - 4P^2_{n\,max}\left(1 + \tan^2\varphi\right) = 0 \tag{8.48}$$

Hence, considering that $1 + tg^2\varphi = 1/(\cos^2\varphi)$, it is obtained

$$\left(S''_K - 2\tan\varphi \cdot P_{n\,max}\right)^2 - \left(\frac{2P_{n\,max}}{\cos\varphi}\right)^2 = 0$$

or

$$\left[\left(S''_K - 2\tan\varphi \cdot P_{n\,max}\right) - \frac{2P_{n\,max}}{\cos\varphi}\right]\left[\left(S''_K - 2\tan\varphi \cdot P_{n\,max}\right) + \frac{2P_{n\,max}}{\cos\varphi}\right] = 0 \tag{8.49}$$

In Eq. (8.49) both factors in the square brackets provide solutions for positive and negative values of the active power respectively. In the considered equivalent circuit (Figure 8.1) the positive value of the active power means the power demand and the negative value of the active power means the power generation. From the point of view of the voltage stability, only the solution for the positive value is important, i.e.

$$\left(S''_K - 2\tan\varphi \cdot P_{Lmax}\right) - \frac{2P_{n\,max}}{\cos\varphi} = 0 \tag{8.50}$$

Hence

$$S''_K = 2P_{n\,max} \cdot \left(\tan\varphi + \frac{1}{\cos\varphi}\right) = 2P_{n\,max} \cdot \frac{1 + \sin\varphi}{\cos\varphi}$$

and finally

$$P_{n\,max} = \frac{S''_K}{2} \cdot \frac{\cos\varphi}{1 + \sin\varphi} \quad \text{and} \quad Q_{n\,max} = \frac{S''_K}{2} \cdot \frac{\sin\varphi}{1 + \sin\varphi} \tag{8.51}$$

Formulas (8.51) determine the maximum values of real and reactive power of a load (with a given power factor $\cos\varphi$) that can be taken from a source with the short-circuit power S''_K. The maximum powers expressed with formulas (8.51) correspond to the maximum power on the nose curve (Figure 8.2).

The maximum (or critical) power of the load is expressed with the following formula

$$\underline{S}_{n\,max} = \frac{S''_K}{2} \cdot \frac{1}{1 + \sin\varphi}(\cos\varphi + \mathrm{j}\sin\varphi) \tag{8.52}$$

and

$$S_{n\,max} = |\underline{S}_{n\,max}| = \frac{S''_K}{2} \cdot \frac{1}{1 + \sin\varphi} \tag{8.53}$$

Sample values calculated from formulas (8.51) and (8.53) are given in Table 8.1. It is assumed for transmission networks that the reactive power in distribution networks is compensated to $\tan\varphi \le 0.4$.

To ensure secure operation, it is necessary that during the power system operation the actual power demand at a given load node is lower than the maximum (critical) value, taking into account a required voltage stability margin. It is then necessary that the apparent power S_L is k_V times smaller than the maximum (critical) power $S_{n\,max}$, i.e. $S_L \le S_{n\,max}/k_V$. Substituting in this inequality $S_{n\,max}$ by (8.53) gives

$$S_L \le \frac{S''_K}{2 \cdot k_V \cdot (1 + \sin\varphi)} \quad \text{or} \quad S''_K \ge 2 \cdot k_V \cdot (1 + \sin\varphi) \cdot S_L \tag{8.54}$$

Table 8.1 Sample values of maximum load power.

φ[deg]	tan φ	sin φ	cos φ	$\frac{P_{max}}{S''_K}$	$\frac{Q_{max}}{S''_K}$	$\frac{S_{max}}{S''_K}$
0	0.0000	0.0000	1.0000	0.5000	0.0000	0.5000
21.8	**0.4000**	**0.3714**	**0.9285**	**0.3385**	**0.1354**	**0.3646**
30	0.5774	0.5000	0.8660	0.2887	0.1667	0.3333
45	1.0000	0.7071	0.7071	0.2071	0.2071	0.2929
60	1.7320	0.8660	0.5000	0.1340	0.2321	0.2679

It means that the short-circuit power S''_K of the source supplying the load should be $2k_V(1 + \sin\varphi)$ times higher than the load power S_L. The value of the margin should be a compromise between the cost of expenditure and the power system security. For $k_V = 1.1$ and a pessimistic assumption that the compensation devices in the distribution network reduce the reactive power demand at a considered load node to $Q_n \leq 0.577 \cdot P_n$ (i.e. $\varphi \leq 30^o$), inequality (8.54) gives

$$S_L \leq S''_K/3.3 \tag{8.55}$$

It means that, for the assumed conditions, the short-circuit power of the transmission network at the load node should be at least 3.3 times higher than the expected power demand.

Usually, a maximum power demand at the load node is also limited by thermal capability of the feeders. Figure 8.19 illustrates a situation where the supply of a load node is limited by the loading capability of the step-down transformer. A distribution network modeled by a composite load is supplied from the transmission network modeled by an equivalent source with short-circuit power S''_K.

Generally, overloading of transformers beyond nameplate rating depends on many factors. From the point of view of voltage stability it is important how much a transformer can be overloaded in a short-time emergency state. Suppose that S_{rT} is the nameplate rating of the transformer supplying a given composite load and that the short-time emergency rating is $1.5 \cdot S_{rT}$. Therefore, the voltage stability must be guaranteed for $S_L = 1.5 \cdot S_{rT}$ with stability margin $k_V \geq 1.1$. Substituting these values into inequality (8.54) gives

$$S''_K \geq 3.3 \cdot (1 + \sin\varphi) \cdot S_{rT} \tag{8.56}$$

It is assumed, for example, that in the considered emergency state the compensation devices were able to reduce reactive power demand to $Q_n = 0.577 \cdot P_n$ (i.e. $\varphi = 30^o$). In such a situation, the following condition is obtained from (8.56)

$$S''_K \geq 5 \cdot S_{rT} \tag{8.57}$$

It means that for the assumed conditions the short-circuit power of the transmission network at the load node should be at least five times higher than the nameplate rating of the step-down transformer supplying the distribution network.

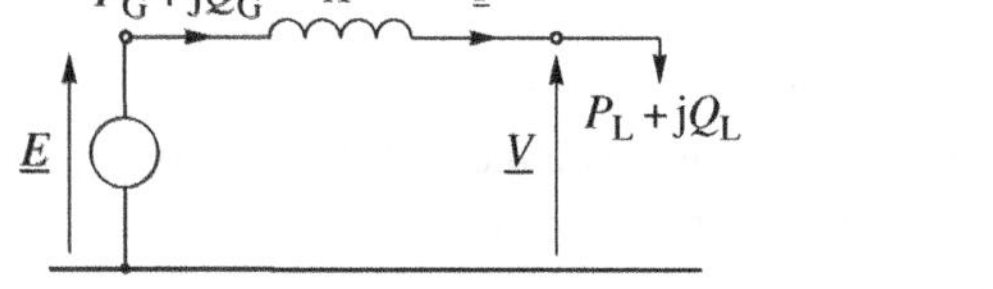

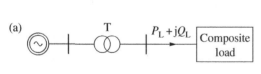

Figure 8.19 Distribution network as a composite load supplied from transmission network via the step-down transformer.

8.4.2 Voltage Stability and Load Flow

Section 8.3 shows that when power demand increases to the critical value, there is only one solution of the network equation and it corresponds to the intersection of the demand characteristic $Q_L(V)$ with the system characteristic $Q_S(V)$. If the demand depends on voltage then its characteristic $Q_L(V)$ is bent at an angle corresponding to the coefficient k_{QV} referred to as the voltage sensitivity (Figure 3.29). In that case (Figure 8.11 and Figure 8.13) the critical point is near the peak of the system characteristic $Q_S(V)$. On the other hand, the voltage characteristic of an ideally stiff load is a horizontal line – $Q_L(V) = Q_n = \text{const}$ – and the critical point corresponds exactly to the peak of the system characteristic $Q_S(V)$, that is the point at which $dQ_S/dV = 0$.

The real and reactive power of an ideally stiff load are correlated by the equation $Q_n = \tan\varphi \cdot P_n$. Hence, for a given $\tan\varphi$, active power reaches a maximum when reactive power reaches a maximum. This means that at the critical state when $dQ_S/dV = 0$ then also $dP_n/dV = 0$. This corresponds to the peak of the nose on the nose curve (Figure 8.2). Hence, when the demand approaches the critical value of the ideally stiff load, the following holds

$$dP/dV \to 0 \text{ and } dQ/dV \to 0 \tag{8.58}$$

That is the derivative of the real and reactive power calculated from the network equations tends to zero. This observation is the basis for determining the critical power demand using load flow programs.

8.4.2.1 Nose Curve and Critical Power Demand

For the simple model of the feeder (Figure 8.1) consisting of an equivalent source and a composite load, the nose curve (Figure 8.2) is determined by Eq. (8.8), where the phase angle of the load φ constitutes a parameter for the family of curves. For a multi-node network model, nose curves can also be determined, but this requires an appropriate method for solving network equations.

The methods for solving Eq. (3.181) described in Section 3.5 are applied to typical operating conditions of an electric power system (EPS) distant from the stability boundary. In these methods, nodes are divided into load nodes (PQ type) and generation nodes (PV type). For load nodes real and reactive power values are set, and the magnitudes and the voltage arguments are calculated. For generation nodes the active power and voltage magnitudes are set and the voltage arguments and reactive power are calculated. When determining the nose curve, the convergence of iterative methods deteriorates as the stability limit is approached and it may not be possible to determine the entire nose curve using the methods discussed in Section 3.6.2. For this reason, to determine the nose curve for multi-node networks, a stepwise method with parameter referred to as the *continuation power flow* is used (Ajjarapu and Christy 1992).

The idea of the stepwise method is illustrated in Figure 8.20a. Each step consists of the prediction and correction of values at the new point. The analysis starts in point 1 corresponding to the operating point of EPS. At this point,

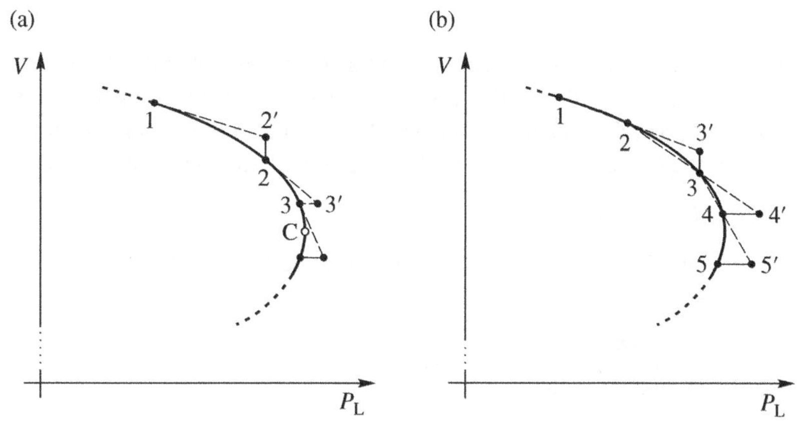

Figure 8.20 Predictor–corrector method used in the continuation power flow: (a) use of tangents; (b) use of secants.

the tangent vector (dashed line) is determined and a new point 2′ is determined for a certain step length. This is the stage of prediction. Point 2′ is treated as a starter for the iterative solution of network equations. It is a correction stage in which the voltage values are determined at a given power demand at the load nodes. If the iterative process is convergent, a new point 2 is obtained. At this point, the tangent vector is determined and for it a new point 3′ is determined under the prediction. This point is treated as a starter for the iterative solution of network equations and determination of voltages at given powers. However, the iterative process here is divergent, because for the given power the voltages that meet the network equations cannot be found (the solution does not exist). After a divergence is detected, the iteration type is changed and for the given 3′ starting point the network equations are resolved again, but this time the voltage is applied to the load nodes and the powers are calculated. The iterative process is convergent and point 3 is obtained. By doing this, the lower part of the nose curve can be determined. However, usually the calculation is interrupted after obtaining the critical point C.

Generally, for any network node, the power balance is expressed by the following equations

$$P_{Gi} - P_{Li} - P_{Ti} = 0; \quad Q_{Gi} - Q_{Li} - Q_{Ti} = 0 \tag{8.59}$$

where P_{Gi}, P_{Li}, Q_{Gi}, and Q_{Li} are generation and load power at a given node, respectively, P_{Ti}, Q_{Ti} – powers in network elements connected to a given node. According to Eq. (3.181), powers P_{Ti} and Q_{Ti} depend on the voltage at the given node and the voltages at the neighboring nodes as well as the sine and cosine functions of the difference between the arguments of these voltages. Powers in the network branches are measured from the node to the network and therefore have the opposite sign than the powers in Eq. (3.181).

The determination of the tangent vector (dashed line in Figure 8.20a) used to predict the new point is based on the following considerations. It is assumed that in a given step the powers in the load nodes (type PQ) are increased according to the following formulas

$$P_{Li} = P_{Li0} + \lambda\,(k_{Li} S_{\Delta} \cos\varphi_i); \quad Q_{Li} = Q_{Li0} + \lambda\,(k_{Li} S_{\Delta} \sin\varphi_i) \tag{8.60}$$

where λ is called the *load increase parameter*, S_{Δ} is the basic increase of apparent power, and k_{Li} is the share coefficient of a given node in the load increase. The active powers in the generation nodes (PV type) are increased according to the formula

$$P_{Gi} = P_{Gi0}\,(1 + \lambda k_{Gi}) \tag{8.61}$$

where k_{Gi} is the growth rate of the generation.

By substituting formulas (8.60) into Eq. (8.59) for load nodes (PQ type) and formulas (8.61) for generation nodes (PV type), a system of nonlinear equations is obtained with the following form

$$\mathbf{F}(\boldsymbol{\delta}, \boldsymbol{V}, \lambda) = \mathbf{0} \tag{8.62}$$

where $\boldsymbol{F}$ is a vector function dependent on vectors of arguments $\boldsymbol{\delta}$ and voltage magnitudes $\boldsymbol{V}$ and the length of the step λ. The dimension of function $\mathbf{F}$ is $(2N_L + N_G)$, where N_L is the number of load nodes and N_G is the number of generation nodes. The dependence of function (8.62) on the arguments and voltage magnitudes results from power P_{Ti} and Q_{Ti} in the branches of the network. By differentiating both sides of Eq. (8.62) one gets

$$\mathbf{F}_{\delta}\Delta\boldsymbol{\delta} + \mathbf{F}_{V}\Delta\boldsymbol{V} + \mathbf{F}_{\lambda}\Delta\lambda = \mathbf{0} \tag{8.63}$$

where $\mathbf{F}_{\delta}$, $\mathbf{F}_{V}$, and $\mathbf{F}_{\lambda}$ are matrices of relevant partial derivatives. The above equation can be written in a matrix form as follows

$$\left[\mathbf{F}_{\delta} \mid \mathbf{F}_{V} \mid \mathbf{F}_{\lambda}\right] \begin{bmatrix} \Delta\boldsymbol{\delta} \\ \hline \Delta\boldsymbol{V} \\ \hline \Delta\lambda \end{bmatrix} = [\mathbf{0}] \tag{8.64}$$

or

$$[\mathbf{F}_\delta \,\vdots\, \mathbf{F}_V \,\vdots\, \mathbf{F}_\lambda]\ [\mathbf{w}] = [\mathbf{0}] \quad \text{where} \quad [\mathbf{w}] = \begin{bmatrix} \Delta\boldsymbol{\delta} \\ \hdashline \Delta\boldsymbol{V} \\ \hdashline \Delta\lambda \end{bmatrix} \tag{8.65}$$

The column matrix $\boldsymbol{w}$ has a dimension of $M = (2N_L + N_G + 1)$. Submatrices $\Delta\boldsymbol{\delta}$, $\Delta\boldsymbol{V}$ included in it, define the tangent vector in which direction the coordinates of the new point should be predicted. The partial derivative matrix on the left-hand side of Eq. (8.65) is rectangular, because it has $(2N_L + N_G + 1)$ columns and $(2N_L + N_G)$ rows. On the right-hand side of Eq. (8.65) is a null column matrix. For these reasons, Eq. (8.65) cannot be solved because to get around this problem Eq. (8.65) must be expanded by one equation as follows. The partial derivative matrix is expanded by one row $\boldsymbol{e}$. To the null column matrix on the right-hand side of Eq. (8.65) one element equal ± 1 is added. In this way, the following equation is created

$$\begin{bmatrix} \mathbf{F}_\delta & \mathbf{F}_V & \mathbf{F}_\lambda \\ \hdashline & \mathbf{e} & \end{bmatrix} [\mathbf{w}] = \begin{bmatrix} \mathbf{0} \\ \pm 1 \end{bmatrix} \tag{8.66}$$

With the appropriate selection of the row matrix $\boldsymbol{e}$, the expanded square matrix of partial derivatives becomes nonsingular and Eq. (8.66) can be solved. In the described method, it is assumed that the row matrix $\boldsymbol{e}$ has zero elements except for one element lying in k-th position. For this element $e_k = 1$ is substituted. The index k for this nonzero element is selected based on the elements of the row matrix $\boldsymbol{w}$ taken from the previous step. It corresponds to the number of the matrix $\boldsymbol{w}$ element $|w_k| = \max(|w_1|, |w_2|, \ldots, |w_M|)$ whose absolute value is the largest of all elements of this matrix. The sign at ± 1 in the column matrix on the right-hand side of Eq. (8.66) is selected based on the changes in the previous steps. The plus sign is assumed when k-th variable increases, and the minus sign when the k-th variable decreases.

It is worth emphasizing here that the acceptance in Eq. (8.66) of the value ± 1 and $e_k = 1$ causes a specific normalization of the value of the column matrix $\boldsymbol{w}$, because for such assumed values in the next step one obtains $|w_k| = 1$. The computed elements of vector $\boldsymbol{w}$ decide on which position value $e_k = 1$ must be placed in the next step (some of them may be greater than one, which will change the position of a nonzero element).

When the vector $\boldsymbol{w}$ is already calculated in the l-th step, the coordinates of the new point $(l + 1)$ are calculated from the formula

$$\begin{bmatrix} \boldsymbol{\delta} \\ \hdashline \boldsymbol{V} \\ \hdashline \lambda \end{bmatrix}_{l+1} = \begin{bmatrix} \boldsymbol{\delta} \\ \hdashline \boldsymbol{V} \\ \hdashline \lambda \end{bmatrix}_{l} + \sigma \begin{bmatrix} \Delta\boldsymbol{\delta} \\ \hdashline \Delta\boldsymbol{V} \\ \hdashline \Delta\lambda \end{bmatrix} \tag{8.67}$$

where σ is the ratio determining the length of the step along the tangent vector. The new point obtained in this way is treated as a starting point for the correction stage, in which the network equations are solved by Newton's method. The critical point (point C in Figure 8.20a) is obtained by tracking the value and the change of the sign of the load parameter λ, respectively.

In Figure 8.20b a modified version of the continuation power flow method is illustrated, but it is also based on the stepwise method with a parameter. In this method, secants of the nose curve are used instead of the tangential vectors to determine the direction. For loads distant from the stability limit, two EPS operation points are determined, for example points 1 and 2. Through these points, a secant 1–2 is defined, on which a point 3′ is established. Using Newton's method, the coordinates of the new point 3 are calculated. Then, the secant 2–3 is determined, etc.

In Figure 8.20 the coordinates V and P_L have been marked, not specifying for which node it is the voltage and active power of the load. Of course, in the multi-node model of EPS there is generally N_L load nodes and the analysis is carried out for voltage vectors $\boldsymbol{V}$ and $\boldsymbol{P}_L$. During the calculations, nose curves are created for each node.

When a critical point (the top of the nose curve) appears for one or several load nodes at the same time, this condition is considered the critical state of EPS.

It is worth noting that in formulas (8.60) one can set any phase angle φ_i of a given load for each load node. This allows to obtain (as in Figure 8.2) critical powers (top of the nose curve) at various power factors. In addition, in formulas (8.60) and (8.61) coefficients k_{Li} and k_{Gi} of the share of nodes in increasing power can be set individually for each node. This allows analyses to performed in various scenarios of power demand increase, for example

- Study of the possibility of supplying the load nodes of the selected EPS area in the event of loss of generation in this area post-fault outages of generating units).
- Study of the possibility of supplying the load nodes of the selected EPS area in the event of a post-fault or operational outage of a part of the transmission lines.
- Study of the possibility of supplying new large customers with the lack or delay of the transmission network expansion.

Of course, every scenario accepted for study must have a technical meaning. The analysis can be performed without taking into account constraints such as the power capability curves of generating units (Section 3.3.4) and thermal capability of network elements (short-time emergency ratings). In this case, a theoretical load limit is obtained, which can be used to calculate voltage stability indices Eq. (8.80). It is also worth repeating the analysis taking into account technical limitations and then one will get information whether the loss of voltage stability under a given power demand increase scenario is technically possible. For example, a composite load that draws power from the transmission network through a transformer with power S_{rT} cannot draw more than the short-time emergency rating of this transformer (e.g. $1.5 \cdot S_{rT}$). The assumption that the power of this load increases several times does not make technical sense.

The range of voltage stability analyzes can be significantly reduced by the simplified method described in Section 8.4.1 based on short-circuit power in the load node (Figure 8.18).

8.4.2.2 Sensitivity and Modal Analyses

For the simple model of the feeder (Figure 8.1) consisting of an equivalent source and composite load with a stiff voltage characteristics ($P_L(V) = \text{const}$, $Q_L(V) = \text{const}$) there is $dQ_L/dV = 0$. From inequalities (8.19) it follows that the condition of stability is $dQ_S/dV < 0$. All stable points are on the right side of the voltage characteristics $Q_S(V)$ of the equivalent source of EPS (Figure 8.1). The voltage stability limit occurs at the top of this characteristic, where $dQ_S/dV = 0$. In the vicinity of stable points occurs $\Delta Q_S/\Delta V < 0$. Thus, $\Delta V/\Delta Q_S < 0$ for stable points and $\Delta V/\Delta Q_S = \infty$ for a critical point (stability limit). In the case of the nose curve (Figure 8.2 the active power and reactive power of the load are in a fixed ratio (depending on the phase angle of load) and for this reason it can be said that the critical point (the top of the nose curve) occurs $\Delta V/\Delta Q_S = \infty$ as well as $\Delta V/\Delta P_S = \infty$.

In the equivalent circuit of the feeder shown in Figure 8.6, by $P_S + jQ_S$ the power supplied by the equivalent source to the load node is indicated. The sign of this power is opposite than in the nodal method. In the nodal method (Section 3.5) for any i-th node, the power injected into the node ($P_i + jQ_i$) is considered positive. The power delivered to the load node thus has a sign opposite to the nodal power sign $(P_{Si} + jQ_{Si}) = -(P_i + jQ_i)$. Taking this into account, it can be said that in a stable point for each load node there is $\Delta V_i/\Delta Q_i > 0$, i.e. the sensitivity of the node to the change in reactive power is positive. In critical condition (stability limit) the factor tends to infinity $\Delta V_i/\Delta Q_i = \infty$.

The above facts indicate that the sensitivity of the load node voltage

$$s_i = \Delta V_i/\Delta Q_i \tag{8.68}$$

to changes in the reactive power of a given node can be used to assess voltage stability. The lower the sensitivity of a given node, the better its situation in terms of voltage stability. Conversely, the greater the sensibility of a given node, the worse its situation in terms of voltage stability. When EPS is overloaded and the critical state (stability limit) is approached, the sensitivities increase and for the worst situation nodes they tend to infinity, i.e. $s_i = \Delta U_i/\Delta Q_i \to \infty$.

If Newton's method is used to determine the EPS operation point, the sensitivity (8.68) of the voltage in the load nodes to changes in reactive power can be calculated using the Jacobi matrix of the network equations.

The linearized network (3.182) described in Section 3.5.1 for convenience of further considerations can be written as follows

$$\begin{bmatrix} \Delta \boldsymbol{Q} \\ \Delta \boldsymbol{P} \end{bmatrix} = \begin{bmatrix} \boldsymbol{J}_{\mathrm{QV}} & \boldsymbol{J}_{\mathrm{Q\delta}} \\ \boldsymbol{J}_{\mathrm{PV}} & \boldsymbol{J}_{\mathrm{P\delta}} \end{bmatrix} \begin{bmatrix} \Delta \boldsymbol{V} \\ \Delta \boldsymbol{\delta} \end{bmatrix} \quad \text{or} \quad \begin{bmatrix} \Delta \mathbf{Q} \\ \Delta \mathbf{P} \end{bmatrix} = [\boldsymbol{J}] \begin{bmatrix} \Delta \mathbf{V} \\ \Delta \boldsymbol{\delta} \end{bmatrix} \tag{8.69}$$

where $\Delta \boldsymbol{P}$ and $\Delta \boldsymbol{Q}$ are vectors of increments of real and reactive powers of all nodes, $\Delta \boldsymbol{V}$ and $\Delta \boldsymbol{\delta}$ are vectors of increments of nodal voltages and their arguments, and $\boldsymbol{J}$ is a Jacobi matrix.

Using partial inversion described in Section A.2, Eq. (8.69) can be transformed into a form

$$\begin{bmatrix} \Delta \boldsymbol{Q} \\ \Delta \boldsymbol{\delta} \end{bmatrix} = \left[\begin{array}{c|c} \boldsymbol{J}_{\mathrm{QV}} - \boldsymbol{J}_{\mathrm{Q\delta}} \boldsymbol{J}_{\mathrm{P\delta}}^{-1} \boldsymbol{J}_{\mathrm{PV}} & \boldsymbol{J}_{\mathrm{Q\delta}} \boldsymbol{J}_{\mathrm{P\delta}}^{-1} \\ \hline -\boldsymbol{J}_{\mathrm{P\delta}}^{-1} \boldsymbol{J}_{\mathrm{PV}} & \boldsymbol{J}_{\mathrm{P\delta}}^{-1} \end{array}\right] \begin{bmatrix} \Delta \boldsymbol{V} \\ \Delta \boldsymbol{P} \end{bmatrix} \tag{8.70}$$

Assuming that sensitivity is tested only for changes in reactive power, one can substitute $\Delta \boldsymbol{P} = \boldsymbol{0}$ and then get

$$\Delta \boldsymbol{Q} = \boldsymbol{J}_{\mathrm{Q}}\, \Delta \boldsymbol{V} \quad \text{and} \quad \Delta \boldsymbol{V} = \boldsymbol{s}\, \Delta \boldsymbol{Q} \tag{8.71}$$

where $\boldsymbol{J}_{\mathrm{Q}} = \left(\boldsymbol{J}_{\mathrm{QV}} - \boldsymbol{J}_{\mathrm{Q\delta}} \boldsymbol{J}_{\mathrm{P\delta}}^{-1} \boldsymbol{J}_{\mathrm{PV}}\right)$ and $\boldsymbol{s} = \boldsymbol{J}_{\mathrm{Q}}^{-1} = \left(\boldsymbol{J}_{\mathrm{QV}} - \boldsymbol{J}_{\mathrm{Q\delta}} \boldsymbol{J}_{\mathrm{P\delta}}^{-1} \boldsymbol{J}_{\mathrm{PV}}\right)^{-1}$. For reactive power changes in individual nodes $\Delta V_i = s_{ii}\, \Delta Q_i$ and that is

$$\frac{\Delta V_i}{\Delta Q_i} = s_{ii} \tag{8.72}$$

where s_{ii} is a diagonal element of the matrix $\boldsymbol{s}$.

It is worth noting that matrix calculation can be simplified considerably using Eq. (A.59) derived in Section A.2 in the appendix. From this equation, it follows that the matrix inverse to the partial inversion, i.e. $\left(\boldsymbol{J}_{\mathrm{QV}} - \boldsymbol{J}_{\mathrm{Q\delta}} \boldsymbol{J}_{\mathrm{P\delta}}^{-1} \boldsymbol{J}_{\mathrm{PV}}\right)^{-1}$, can be calculated as the inverse of a submatrix being part of matrix $\boldsymbol{J}^{-1}$. The matrix $\boldsymbol{J}^{-1}$ is calculated (in an implicit way in the form of factors) at each step of an iterative solving of network equations using the Newton method. So, taking the matrix $\boldsymbol{J}^{-1}$ from the last step of the Newton method, one can calculate the matrix $\boldsymbol{s}$ and determine the sensitivity s_{ii} for the selected nodes.

The sensitivity of the load nodes to changes in reactive power can be used to select weakly supplied nodes, which is important for expansion planning of EPS or identifying threats during the operation of EPS, for example in weakening networks due to maintenance outages or post-fault trippings of network elements.

Testing EPS sensitivity on changes in power demand can be also done using eigenvalues and eigenvectors of reduced Jacobi matrix. Such analysis in the literature is called (not very aptly) *modal analysis* by analogy to the modal analysis of differential equations. This method uses the matrix $\boldsymbol{J}_{\mathrm{Q}} = \left(\boldsymbol{J}_{\mathrm{QV}} - \boldsymbol{J}_{\mathrm{Q\delta}} \boldsymbol{J}_{\mathrm{P\delta}}^{-1} \boldsymbol{J}_{\mathrm{PV}}\right)$ and the first of the equations in (8.71). It is assumed that $\boldsymbol{W}$ and the matrix $\boldsymbol{M}$ are respectively matrices consisting of right and left eigenvectors of matrix $\boldsymbol{J}_{\mathrm{Q}}$. As is known $\boldsymbol{M} = \boldsymbol{W}^{-1}$. New fictitious variables are introduced, such that

$$\Delta \boldsymbol{Q} = \boldsymbol{W} \Delta \boldsymbol{q}, \text{ so } \Delta \boldsymbol{q} = \boldsymbol{M} \Delta \boldsymbol{Q}, \quad \text{and} \quad \Delta \boldsymbol{V} = \boldsymbol{W} \Delta \boldsymbol{v}, \text{ so } \Delta \boldsymbol{v} = \boldsymbol{M} \Delta \boldsymbol{V} \tag{8.73}$$

After substituting dependence (8.73) into the first of the equations in (8.71), it is obtained that

$$\boldsymbol{W} \Delta \boldsymbol{q} = \boldsymbol{J}_{\mathrm{Q}}\, \boldsymbol{W} \Delta \boldsymbol{v}, \text{ so } \Delta \boldsymbol{q} = \boldsymbol{M} \boldsymbol{J}_{\mathrm{Q}}\, \boldsymbol{W} \Delta \boldsymbol{v} \text{ or } \Delta \boldsymbol{q} = \boldsymbol{\Lambda} \Delta \mathbf{v} \tag{8.74}$$

where $\Lambda = \text{diag}\,\lambda_i$ is a diagonal matrix consisting of eigenvalues of the matrix $\boldsymbol{J}_Q$. The last of Eq. (8.74) allows $\Delta\boldsymbol{v} = \Lambda^{-1}\Delta\boldsymbol{q}$ to be obtained. Hence, $\Delta v_i = \Delta q_i/\lambda_i$, so the sensitivity of fictitious voltage to changes in fictitious reactive power is expressed by the formula

$$\frac{\Delta v_i}{\Delta q_i} = \frac{1}{\lambda_i} \tag{8.75}$$

From the above also follows the conclusion that if some eigenvalue tends to zero ($\lambda_i \to 0$), the sensitivity of fictitious voltage tends to infinity and EPS becomes unstable. However, this method has three major disadvantages:

- Owing to the need to calculate eigenvalues and eigenvectors, it is time-consuming.
- Engineers practicing EPS analysis are reluctant to use fictitious variables.
- The method silently assumes that the eigenvalues λ_i of the matrix $\boldsymbol{J}_Q$ are real. This assumption is valid only when the matrix is symmetrical. It can be shown that this condition is theoretically met only if the resistance of network elements is omitted. In practice, this condition is also met for HV networks, for which the resistances are much smaller than the reactances. In the case of MV network, in which resistances play a significant role, the $\boldsymbol{J}_Q$ matrix eigenvalues λ_i can be complex and then formula (8.75) loses meaning.

Owing to these drawbacks, the modal method of testing voltage stability cannot be recommended as a universal method.

8.4.2.3 Other Methods

Singular analysis is a generalization of modal analysis on a rectangular matrix $\boldsymbol{A}$ with dimensions $m \times n$ assuming that $m \geq n$. If λ_i is the eigenvalue of the matrix $\boldsymbol{A}^T\boldsymbol{A}$ and the $\boldsymbol{u}_i$ is the corresponding eigenvector then $\mu_i = \sqrt{\lambda_i}$ is referred to as the *singular value* of the matrix $\boldsymbol{A}$, and eigenvector $\boldsymbol{w}_i$ of matrix $\boldsymbol{A}^T\boldsymbol{A}$ is the *right singular vector* of the matrix $\boldsymbol{A}$. The matrix $\boldsymbol{A}\boldsymbol{A}^T$ has the same eigenvalues as the matrix $\boldsymbol{A}^T\boldsymbol{A}$. The eigenvector $\boldsymbol{u}_i$ of matrix $\boldsymbol{A}\boldsymbol{A}^T$ is the *left singular vector* of the matrix $\boldsymbol{A}$. Generally, the matrix $\boldsymbol{A}$ ($m \times n$) has m right singular vectors $\boldsymbol{w}_i$ and n left singular vectors $\boldsymbol{u}_i$. By creating matrices $\boldsymbol{W}$ and $\boldsymbol{U}$, respectively, from these vectors, one can transform matrix $\boldsymbol{A}$ into the form $\boldsymbol{A} = \boldsymbol{W}\breve{\boldsymbol{A}}\boldsymbol{U}$ where $\breve{\boldsymbol{A}}$ consists of a square diagonal matrix containing singular values μ_i and a rectangular sub-matrix with zero elements. This form of the matrix $\boldsymbol{A}$ is referred to as *singular-value decomposition*. The singular values of Jacobi matrix of equations describing the EPS network can be used to assess voltage stability.

Another method is based on bifurcation. In mathematics, *bifurcation* refers to changes in the properties of a dynamic system resulting from small continuous changes in its parameters. An example of bifurcation is the change in the character of solutions of differential equations describing a generator and infinite bus system in the vicinity of critical power angle $\delta = \delta_{cr}$ (Figure 5.17 in Section 5.4.6). For power angle $\delta < \delta_{cr}$, the dynamic system response is aperiodic, as in point C. However, with a small angle change, such that $\delta > \delta_{cr}$ the system responses with aperiodic increase of power angle. The same is true when changing the system parameter, such as the turbine power. At power $P_m < P_{E_q cr}$ the system oscillates back to the equilibrium point (Figure 5.5 in Section 5.4.1). With $P_m = P_{E_q cr}$ the system moves away from the equilibrium point. At $P_m > P_{E_q cr}$ the system has no equilibrium point and loses synchronism in an aperiodic manner. Similar changes can be observed when the voltage stability limit is exceeded following changes in EPS parameters such as power of loads or changes in network parameters (e.g. resulting from the network element tripping). The application of the bifurcation theory allows for the mathematical evaluation of the voltage stability of EPS.

In practice, in power engineering, evaluation of voltage stability based on bifurcation theory or singular values is not willingly used, owing to the mathematical complexity of these methods.

8.4.3 Voltage Stability Indices

The discussed coefficient $\xi_{cr} = P_{n\ MAX}/P_0$ may be treated as a measure of voltage stability margin from the point of view of demand increase. The voltage stability criteria discussed in Section 8.2 may also be used to determine voltage stability indices also referred to as *voltage proximity indices.*

A voltage stability index based on the classical dQ/dV criterion can be constructed by observing that, as the load demand gets closer to the critical value, both the equilibrium points shown in Figure 8.11 move toward each other until they become one unstable point. As shown in Figure 8.8b, there is always a point between the equilibrium points, of voltage V_X, such that

$$\left.\frac{d(Q_S - Q_L)}{dV}\right|_{V = V_X} = 0 \tag{8.76}$$

As the power demand of the composite load increases, the voltage V_X tends toward the critical voltage V_{cr}. A voltage proximity index can be therefore defined as (Venikov 1978)

$$k_V = \frac{V_S - V_X}{V_S} \tag{8.77}$$

where V_X must satisfy Eq. (8.76). In practice, calculating the proximity index defined by Eq. (8.77) is quite cumbersome as it requires a fragment of the generation characteristic $Q_S(V)$ to be determined by means of a load flow program. An alternative proximity index that is easier to determine is the value of the derivative

$$k_{\Delta Q} = \frac{d(Q_S - Q_L)}{dV}\frac{V_s}{Q_s} \tag{8.78}$$

calculated near a given equilibrium point. As the load demand tends toward the critical value, index (8.78) tends toward zero.

Yet another voltage proximity index can be derived directly from the dQ_G/dQ_L criterion as

$$k_Q = \frac{dQ_G}{dQ_L} \tag{8.79}$$

When the network is lightly loaded, the reactive power absorbed by the network is small so that the increment in the generation caused by the increment in the load demand is almost equal to the increment in the load demand itself and the index defined in Eq. (8.79) is near unity. When the system approaches the critical state, the index tends to infinity (Figure 8.10). The value of this index can be calculated for a multi-node system using a load flow program.

In practice, the nose curves discussed in Section 8.1.1 and 8.4.2 and critical power demands corresponding to the tops of the nose curves are used to calculate voltage stability indices. After determining the nose curves for a given load growth scenario, the following coefficients can be calculated for all the receiving nodes

$$k_{Qi} = \frac{Q(V_{Ci}) - Q_{0i}}{Q_{0i}} \quad \text{or} \quad k_{Pi} = \frac{P(V_{Ci}) - P_{0i}}{P_{0i}} \tag{8.80}$$

where P_{Ci} and Q_{Ci} are real and reactive power for a critical power demand (top of the nose curve) and P_{0i} and Q_{0i} are real and reactive power of load at a given EPS operation point. Both definitions (8.80) are equivalent, because when determining the nose curve a constant phase angle of the load is assumed, i.e. $Q = \tan\varphi \cdot P$.

Currently, the problem of voltage stability is a rapidly developing research area with many new papers appearing either proposing new proximity indices and improved methods of calculation or suggesting how proximity indices can be identified online using local measurements.

8.5 Dynamic Analysis

The voltage stability analysis presented so far in this chapter has assumed that the loads can be represented by their static voltage characteristics. Although this is useful to help understand the principles of voltage stability, and its relation to the network feasibility problem, the static load voltage characteristics can only approximate the real behavior of the composite load for slow voltage variations. In practice the actual behavior of both the composite load and the system is a tightly coupled dynamic process that is influenced by the load dynamics, especially induction motor dynamics, the automatic voltage and frequency control equipment, and by the operation of the protection systems. Any, or all, of these factors may speed up, slow down, or even prevent voltage collapse.

8.5.1 The Dynamics of Voltage Collapse

To complement the static considerations of the previous sections a few examples of some typical voltage collapse scenarios are now briefly discussed (Taylor 1994; Bourgin et al. 1993).

8.5.1.1 Scenario 1: Load Build-up

Section 8.3 explains how voltage collapse can result from a very large build-up of load, particularly during periods of heavy demand, so that the power demand exceeds the critical value as determined by the network parameters. In this scenario the main factors contributing to voltage collapse are:

- The stiffness of the load characteristics continuing to demand high values of active and reactive power despite voltage dips in the load area. As Chapter 3 explains, induction motors are mainly responsible for producing a stiff load characteristic.
- The control of tap-changing transformers in distribution and subtransmission networks maintaining constant voltage, and therefore high active and reactive power demand, when the supply voltage dips. High demand is undesirable in the emergency state.
- The limited ability for reactive power control by the generators. Owing to field and armature current limits, a high reactive power demand by the system loads may cause the generators to lose their ability to act as a constant voltage source. The generator then behaves like a voltage source behind the synchronous reactance and its terminal voltage reduces.

A voltage collapse due to load build-up may be caused by some, or all, of the above factors. The dynamics of the various voltage control devices (generators, compensators, transformers) may interact in such a way that the actual voltage collapse is different to that predicted by static considerations.

Voltage changes during voltage collapse are shown in Figure 8.12. The dynamics contain a long-term drift of voltage caused by a slow increase in the system load.

8.5.1.2 Scenario 2: Network Outages

As seen in previous sections of this chapter the network parameters play a crucial role in determining the maximum power that can be delivered to the load areas. Tripping one of the lines in the power grid increases the equivalent reactance between the equivalent voltage source and the load thus increasing voltage drops in lines and therefore depressing network voltages. Reduced voltages and increased equivalent reactance reduce the critical power and increase the probability of voltage collapse. Generator tripping has a similar effect in that it not only increases the equivalent system reactance but also reduces the system's capability to generate real and reactive power. The dynamics of the various voltage control devices in the system can again influence the actual scenario.

Typical voltage changes during voltage collapse are shown in Figure 8.21. The dynamics contain about 10 seconds of transient oscillations caused by the outage and a drift of voltage caused by a slow increase in reactive power deficit in the system while taking into account the actions of various control devices and their limiters.

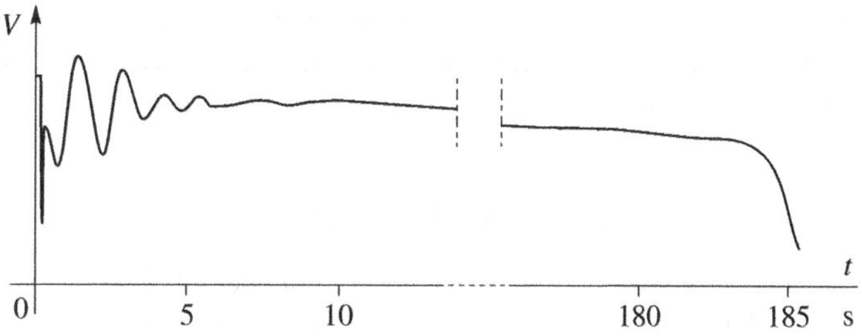

Figure 8.21 An example of voltage collapse initialized by network outage.

8.5.1.3 Scenario 3: Voltage Collapse and Asynchronous Operation

Voltage collapse at one, or a few, of the network nodes may cause the voltage to dip at neighboring nodes leading to voltage collapse at these nodes. The voltage then dips at other nodes so propagating throughout the network and affecting the synchronous generators. If the affected generators are weakly connected with the system they may lose synchronism.

A voltage dip, accompanied by a reduction in the active power demand and an increase in the reactive power demand, has a similar effect on the synchronous generator as a short circuit in the network. An example of just such behavior is shown in Figure 8.22 where the tripping of line L2 results in the generator and the load operating through one, quite long, line. The load is in the critical state and a small increase in the load demand results in voltage collapse and the generator losing synchronism with the rest of the system. The load voltage then undergoes periodic voltage variations characteristic of asynchronous operation. The asynchronously operating generator must now be tripped from the system, which further deteriorates the situation at the load node leading to an eventual total voltage collapse.

8.5.1.4 Scenario 4: Phenomena Inside the Composite Load

As already mentioned, the stiffness of the load characteristics is one of the dominant factors contributing to voltage collapse. However, the dynamic response of the composite load may result in the dynamic and static load characteristics being different. This difference is mainly attributed to induction motors and may result in a reduction in the system stability ultimately leading to voltage collapse. For example, a rapid, severe voltage dip, such as that which occurs during a slowly cleared short circuit, can cause a reduction in the motor torque and consequent motor stalling. As shown in Section 3.4, a stalled induction motor demands reactive power further reducing the voltage stability conditions. This may lead to other nearby motors stalling. In this scenario the voltage continues to fall until the protective equipment, or the electromechanically held contactors, trip the motors from

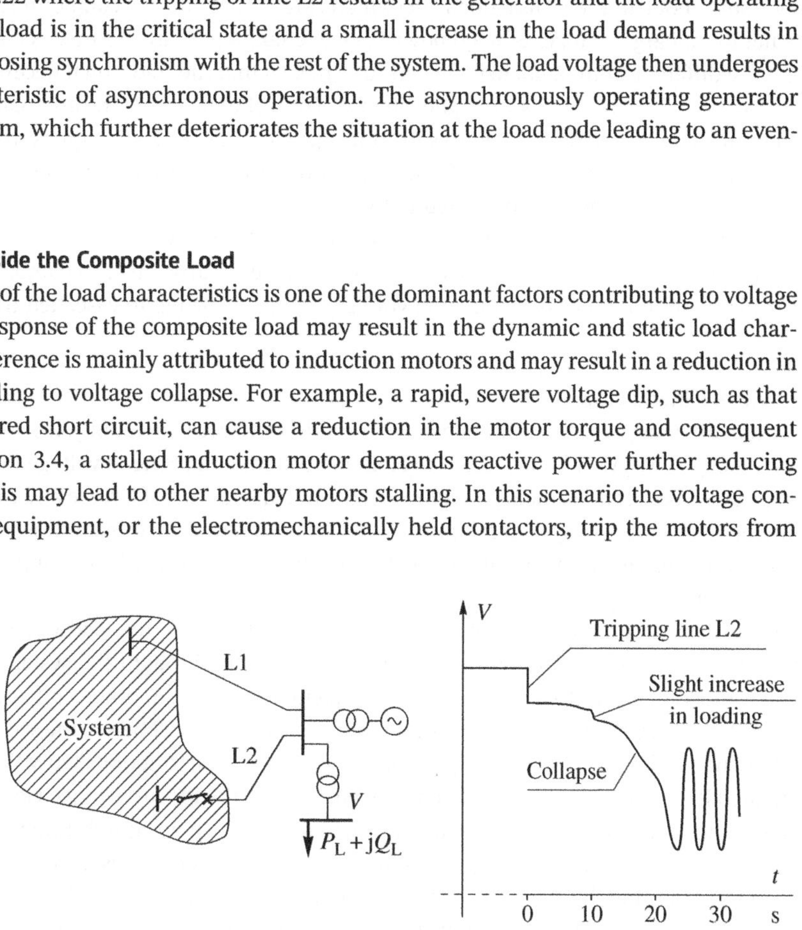

Figure 8.22 Loss of synchronism caused by the voltage collapse: (a) schematic diagram of the system; (b) voltage variations. *Source:* Based on Venikov (1978).

the system thereby reducing the reactive power demand. The voltage will then start to recover but an uncontrolled restoration of the composite load by, for example, heavy induction motor self-starts can again reduce the voltage and lead to a total voltage collapse.

8.5.2 Examples of Power System Blackouts

There have been a number of well-publicized blackouts in Europe and North America in the early years of this millennium. Although in each case the blackout was caused by a specific technical problem, the unprecedented concentration of blackouts has caused many observers to argue that there are underlying systemic reasons for such a large number of disturbances occurring at more or less the same time (Bialek 2007). Increased liberalization of electricity supply industry in the 1990s has resulted in a significant increase in inter-area (or cross-border) trades in interconnected networks of North America and Europe. This means that the interconnected systems are used for purposes they were not designed for. Interconnections grew by connecting self-sustained areas so that tie-lines tend to be relatively weak and require a careful monitoring of tie-line flows (this is further discussed in Chapter 9). It should be emphasized that any transaction in a meshed network may affect all the network flows, sometimes quite far away from the direct contract path linking the source and the sink of a transaction. This is referred to as the *loop flow effect* (Section 3.6). Loop flows may be a problem because inter-area transactions are often not properly accounted for when assessing system security by a TSO (transmission system operator) that does not know about all the transactions affecting its area. That effect is compounded by increased penetration of wind generation. Large changes of wind power due to changing weather patterns mean that actual network flows may be quite different from predicted ones. All this means that the traditional decentralized way of operating systems by TSOs, with each TSO looking after its own control area and with little real-time information exchange, resulted in inadequate and slow responses to contingencies. Bialek (2007) argues that a new mode of coordinated operation with real-time exchange of information for real-time security assessment and control is needed in order to maintain the security of interconnected networks.

The remainder of this section is devoted to a description of blackouts related to voltage problems. The first blackout discussed in this section was a textbook case of a voltage collapse due to a combination of load growth and loss of power plants. In the other blackouts the voltage collapsed, owing to cascaded tripping of transmission lines.

8.5.2.1 Athens Blackout in 2004

The blackout affecting over five million people in southern Greece, including Athens, has been described by Vournas et al. (2006). It was well known that the Hellenic system was prone to voltage instability, owing to long transmission distance between the main generation in the north and west of Greece and the main load center in Athens. Consequently, a number of system reinforcements, such as new transmission lines, autotransformers, and capacitor banks, were ordered in the run-up to the 2004 Olympic Games. The blackout happened when those reinforcements were not yet fully commissioned.

The disturbance started just after midday on a hot July day when the load was on the increase because of air-conditioning. A generating unit near Athens was lost and that brought the system to an emergency state. Load shedding was initiated but it had not been fully implemented when another unit at the same plant tripped. Curve 1 in Figure 8.23 shows the PV nose curve, simulated post-mortem, just before the second unit tripped. The load was 9320 MW which is slightly less than the peak of the PV curve equal to about 9390 MW. Curve 2 in Figure 8.23 shows the simulated PV curve after the second unit tripped. Clearly, the PV curve moved left with the critical load of about 9230 MW, which was about 90 MW less than the actual load. With no equilibrium point, voltage collapse was inevitable. When voltages started to collapse, the undervoltage element of distance relays in the north–south 400 kV lines opened, separating the southern part of the Greek system from the northern part. The remaining generation in the southern part of Greece was tripped by undervoltage protection leading to a blackout.

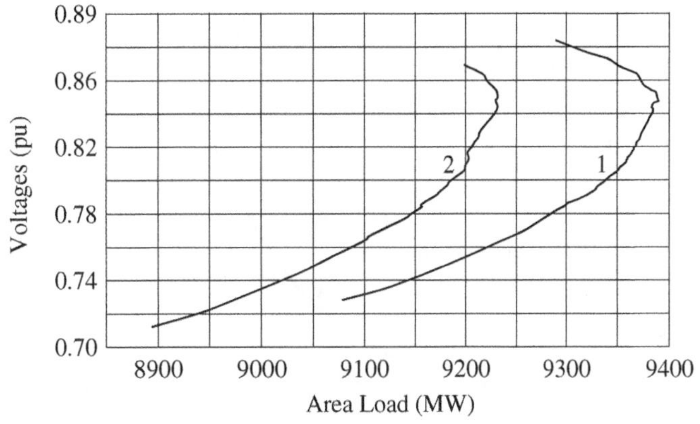

Figure 8.23 PV curves: (1) before Athens blackout; (2) after the second unit tripped (Vournas et al. 2006). *Source:* Reproduced by permission of IEEE.

8.5.2.2 US/Canada Blackout in 2003

That widespread blackout affecting about 50 million customers in the United States/Canada has been described in the report by the US-Canada Power System Outage Task Force (2004). The direct cause of the disturbance was due to undetected cascaded tripping of transmission lines. Each line trip caused increased loading on the remaining lines and the resulting lowering of voltages (Figure 8.24). Subsequent analysis has shown that the blackout was inevitable when a crucial Sammis-Star transmission line was unnecessarily tripped by zone 3 of the distance protection (Section 2.6). The locus of a constant impedance determining a tripping threshold is a circle (Figure 8.25). The normal point of operation is shown by a cross outside the circle. However, high currents and depressed voltages caused by earlier outages of other transmission lines caused a lowering of the measured value of the apparent impedance measured by the relay as shown by a cross inside the circle. The relay tripped the line, causing rapid

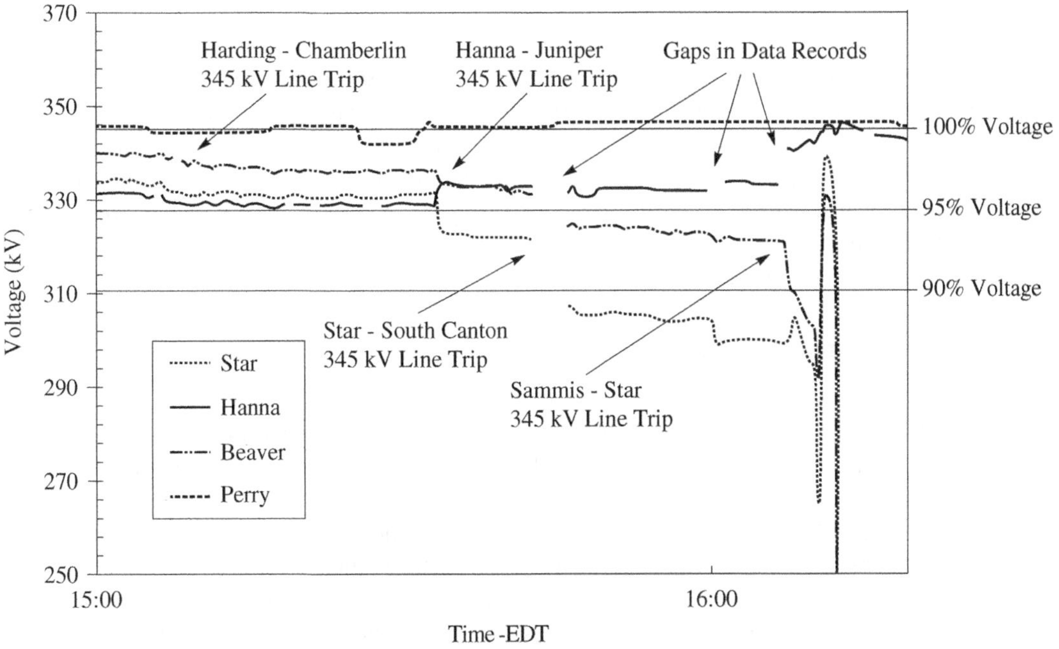

Figure 8.24 Cascaded line trips depressing voltages during US/Canada blackout (US-Canada Power System Outage Task Force 2004).

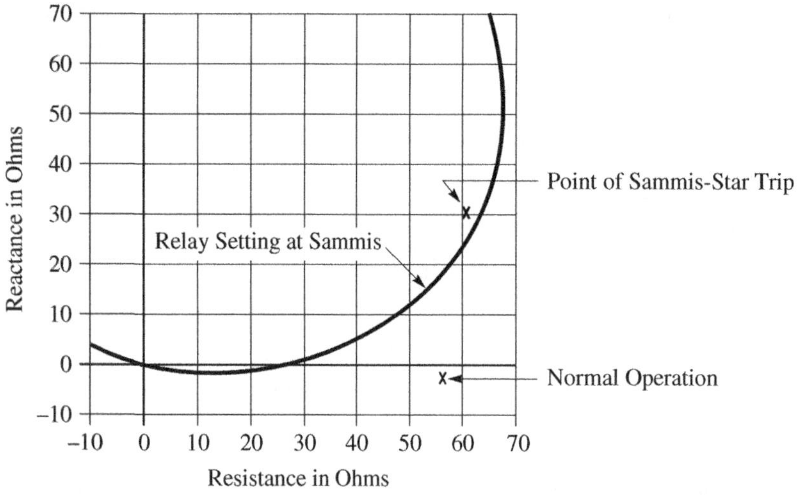

Figure 8.25 Normal operating point and the tripping point of Sammis-Star line during 2001 US/Canada blackout (US-Canada Power System Outage Task Force 2004).

cascaded tripping of other lines and generators and resulting in the blacking out of a large part of the United States and Canada.

8.5.2.3 Scandinavian Blackout in 2003

The blackout affected 2.4 million customers in eastern Denmark and southern Sweden and is described in the report by Elkraft Systems (2003). First a 1200 MW unit at Oskarshamn Power Station tripped following problems with a valve in the feedwater circuit, and 15 minutes later a double-busbar fault occurred at a substation in southern Sweden, which caused four 400 kV lines and two units (1800 MW) at Ringhals nuclear power station to trip (Figure 8.26). Such a heavy loss of generation and transmission caused heavy flows on the remaining transmission lines supplying southern Sweden. Consequently, voltages started to fall heavily in southern Sweden, where there was no generation left, but less so in eastern Denmark (Zealand), where local power plants were able to keep up voltages (Figure 8.27). Overloading caused further tripping of a number of 130 and 220 kV transmission lines in southern Sweden and a consequent halving of voltage there. The combination of heavy flows and low voltages caused, as during the US/Canada blackout, tripping of zone 3 distance relays on a number of 400 kV lines in central and eastern Sweden so that these areas were no longer electrically connected. Finally, 90 seconds after the substation fault, the voltage fell to zero and a full blackout ensued.

8.5.3 Computer Simulation of Voltage Collapse

As mentioned above, voltage collapse is a dynamic process. The creation of a mathematical model of a large-scale power system while taking into account detailed models of structure and dynamics of composite loads is practically impossible. It is, however, possible to create various task-oriented dynamic models devoted to the investigation of some scenarios of voltage collapse.

Voltage collapse caused by load build-up (scenario 1) is a long-term process. The slow drift of voltage may take many minutes or even hours (Figure 8.12). Using dynamic simulation to investigate such a process is not required. It is much easier to investigate a number of snapshots as the disturbance progresses, using the static methods described in Section 8.4. The influence of all reactive power compensation devices, voltage regulators, and limiters of relevant regulators should be taken into account.

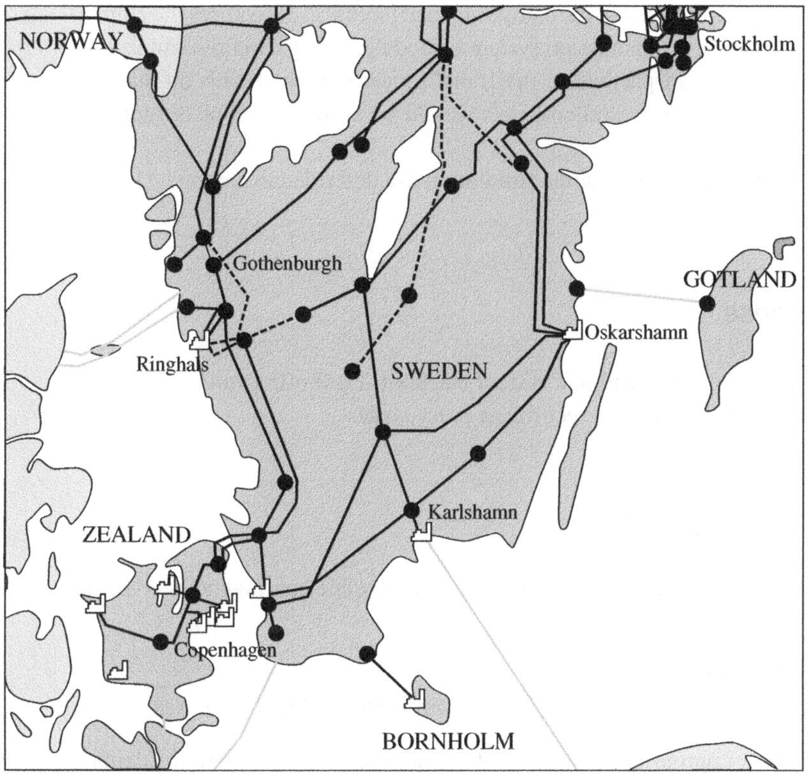

Figure 8.26 The grid 10 seconds after the substation fault. Tripped transmission lines are shown by dashed lines. Karlshamn power station was out of operation (Elkraft Systems 2003).

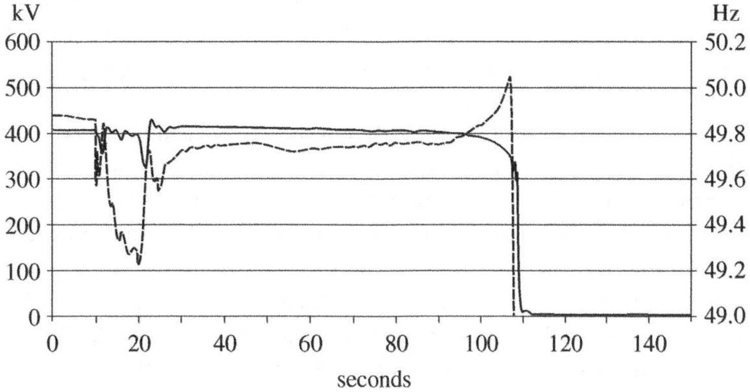

Figure 8.27 Voltage (solid line) and frequency (dashed line) measured at 400 kV connection between Sweden and Denmark (Elkraft Systems 2003).

Voltage collapse caused by network outages (scenario 2) and voltage collapse combined with asynchronous operation (scenario 3) can be simulated using computer programs devoted to short- and mid-term dynamic simulation. Such programs should include models of excitation and voltage control systems of synchronous generators, with their limiters, as well as models of all shunt and series flexible AC transmission systems (FACTS) devices and specialized reactive power compensation devices such as a static compensator (STATCOM). Composite loads should be modeled using their dynamic equivalent models.

The simulation of voltage collapse due to phenomena inside the composite load (scenario 3) requires the employment of specialized computer programs in which the transmission system is modeled using dynamic equivalents, but the distribution network is modeled in detail. In particular, the transformers equipped with on-load tap-changers and their regulators should be modeled and dynamic models should be used for groups of induction motors.

Examples of such models used for short- and mid-term dynamic simulation are described in Chapter 11. Computer algorithms are described in Chapter 13.

8.6 Prevention of Voltage Collapse

Voltage collapse is a dangerous disturbance which may lead to a system blackout. Preventing voltage instability requires action from personnel responsible for power system security at every stage:

- at the network planning stage;
- at the system operational planning stage;
- at the system operation monitoring and control stage.

During the network planning stage, reliability criteria must be satisfied for all possible contingencies of at least N-1 type. It should be ensured that for each contingency:

- maximum allowable voltage drops are not exceeded;
- stability margins for real and reactive power are large enough for each composite load and load area.

Satisfying these criteria can be ensured by an appropriate expansion of the network in response to the demand growth and by installing appropriate reactive power compensation and voltage regulation devices in the system.

During operational planning and real-time monitoring and control, a desired voltage profile should be continuously maintained and appropriate reactive power compensation devices activated. An adequate reserve of real and reactive power should be maintained at the generators. Reserve is the amount of power by which generators in operation can be additionally loaded without exceeding the reactive power capability curve shown in Figure 3.20 and described in Section 3.3.4. For the prevention of voltage collapse, the reserve of reactive power is more important. Such a reserve provided by synchronous generators is necessary to quickly cover a reactive power deficit when voltage collapse symptoms are detected.

The power capability curve (Figure 3.20) and the voltage characteristic (Figure 3.25) of a generating unit clearly show that a unit can support the system with additional reactive power if, at a given operating point, the reactive power output is within the area determined by the power capability curve (curve GFE in Figure 3.20) and the unit operates within the voltage regulation range (curve AB in Figure 3.25) and with an adequate margin (point B in Figure 3.25).

Apart from satisfying the reliability criteria during the operational planning and real-time monitoring and control stages, each system should be equipped with additional defense facilities in order to prevent a voltage collapse following extreme disturbances. The most commonly used methods include:

- Using emergency back-up reactive power reserve not used during normal operation.
- Automatic fast start-up of back-up generation (hydro and gas turbines) when a growing power imbalance appears.
- Emergency increase of reactive power produced by generators in the areas of depressed voltages at the expense of reduction of active power outputs, if there is a possibility of increased imports to make up the resulting active power imbalance. This may be explained as follows. Covering reactive power imbalance in a given area is most easily done locally. If local generators operate at the limit of their power capability curves then any additional

reactive power can be obtained only by reducing their active power output (Figure 3.22). This results in an active power imbalance which can be covered by increased imports (if possible).

- Reduction of power demand in a given area by reducing voltages at load buses using transformers equipped with on-load tap-changers. When reactive power imbalance appears in a given area and it is not possible to increase reactive power generation, active power demand may be reduced by reducing voltages at load buses. This makes use of characteristics of a typical composite load (Figure 3.35), which show that the reactive power demand reduces with reduced voltages. That method can be executed using an auxiliary control loop in the controller of the on-load tap-changer that controls the primary (supply) and secondary (load) side voltage. Normally, if the primary voltage is higher than a threshold value then, when the secondary voltage is dropping, the regulator changes the taps in order to increase the load voltage. However, if there is a deficit of reactive power in the transmission network and the primary voltage of the transformer supplying a distribution network is smaller than a threshold value, when the secondary voltage drops the regulator should act in the opposite direction. It should reduce voltage at the load bus and therefore reduce power demand. This could be referred to as the *reverse action* of the on-load tap-changer regulator. Reverse action of the regulator is justified only if a reduction in the voltage is accompanied by a reduction in reactive power demand. For some consumers operating at reduced voltages (Figure 3.35a) a further reduction of voltages may cause an increase in reactive power demand. If that happens, the reverse action of the regulator must be blocked so that the situation does not deteriorate further.
- With deep voltage sags, when the described prevention methods are not sufficient, customers can be disconnected when a threshold voltage value is reached. This is referred to as *undervoltage load shedding*. Threshold values should be chosen such that the shedding is activated only in extreme situations caused, for example by N-2 or N-3 contingencies.

Choosing the threshold values for the reverse action of the on-load tap-changer regulator or undervoltage load shedding is a delicate matter. Those values can be determined based on off-line voltage stability analyses. In practice, when actual operating conditions are different from the assumed ones, the calculated threshold values can be too optimistic, with a resulting voltage collapse, or too pessimistic, with resulting unnecessary voltage reductions or load shedding. These disadvantages of simple algorithms based on threshold values require further research on new methods of voltage control in the vicinity of voltage collapse.

Two different approaches can be identified here. The first one is based on improving decision methods based on local signals. One example is the application of a very sophisticated method based on Hopf bifurcation theory and the Lyapunov exponents and entropy of the system (i.e. the a-eberle CPR-D Collapse Prediction Relay). The second approach uses measurements from a large area, using wide area measurement system (WAMS) or wide area measurement, protection, and control (WAMPAC) systems (described in Chapter 2 and CIGRE 2007) in order to take a decision.

8.7 Self-excitation of a Generator Operating on a Capacitive Load

Voltage instability may also arise because of *self-excitation*, that is a spontaneous rise in the voltage. The phenomenon is connected with the parametric resonance of *RLC* circuits and may appear when a generator supplies a capacitive load.

8.7.1 Parametric Resonance in RLC Circuits

Figure 8.28 shows an RLC series circuit that has been connected to a voltage source for sufficient time to fully charge the capacitor. The power source is then disconnected instantaneously allowing the capacitor to discharge through resistance R and inductance L. The resulting current is periodic. Its amplitude depends on whether the

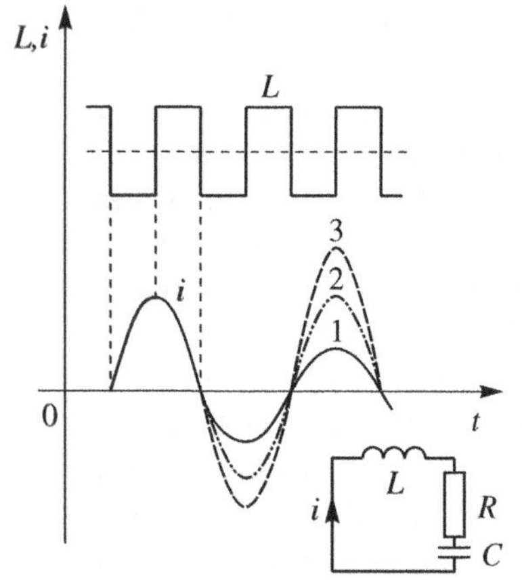

Figure 8.28 Current oscillations in *RLC* circuit depending on the changes in the inductance: (1) constant inductance; (2) small variations in the inductance; (3) large variations in the inductance.

inductance is constant or periodic. When inductance L is constant, the resulting current will be periodic (curve 1) and decays with time as the energy is transferred back and forth from the capacitor to the coil with some of the energy being dissipated in the resistance.

Now assume that during this process an external action is applied to the coil so as to produce a periodic change in its inductance. This could be achieved by changing the length of the air gap in the core. It will also be assumed that the inductance increases when the current reaches a maximum and decreases when the current crosses zero. As the magnetic energy stored in the inductance is $Li^2/2$, the effect of the external force is to increase the stored magnetic energy by an amount that is proportional to the increase in the inductance. This energy is not transferred back to the external force when the inductance is decreased, because the decrease in inductance takes place at zero current when all the energy has already been transferred to the capacitor. This increase in the stored energy results in an increase in the current when the energy is transferred back to the coil in the next period (curve 2). If the changes in L are large, the increase of the energy in the circuit may be greater than the energy lost in the resistance and the current increases every period (curve 3). This increase in the current produces a cyclic increase in the voltage across the capacitor. This phenomenon, known as *parametric resonance*, may also occur when the inductance of the coil is changed sinusoidally as in the self-excitation of the synchronous generator described below.

8.7.2 Self-excitation of a Generator with Open-circuited Field Winding

Consider a salient-pole generator, shown in Figure 8.29, with open-circuited field winding and loaded by an *RC* series circuit. The arrangement is rather unusual for a synchronous generator but it will help to understand self-excitation.

For the salient-pole generator the air gap seen by the armature winding changes periodically (Figure 8.29a) so that the reactance of generator fluctuates between X_d and X_q. The phasor diagram is shown in Figure 8.29b. If the generator has an open-circuited field winding then its synchronous electromotive force is very small, $E_q \cong 0$, and is

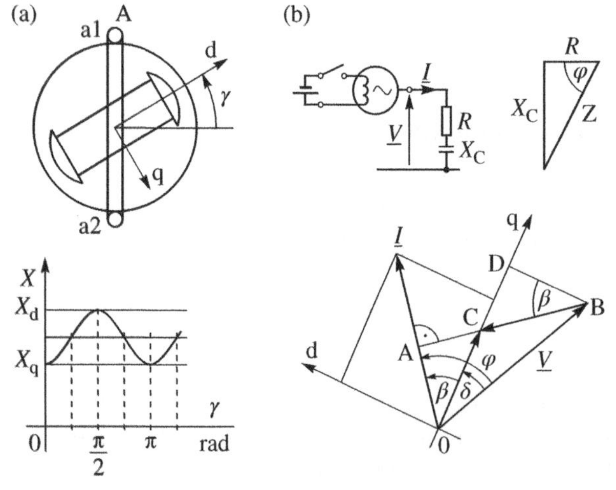

Figure 8.29 Salient-pole generator supplying a capacitive load: (a) periodic changes of the synchronous reactance; (b) block diagram and phasor diagram. For a generator with open-circuited field winding.

due to the remnant magnetism. Hence the generator current is initially very small because it is driven by a small emf induced by remnant magnetism. Generator voltage V is delayed with respect to the current by angle φ and with respect to the q-axis by angle $\delta = \varphi - \beta$. Hence the active power can be expressed by

$$\begin{aligned} P &= VI\cos\varphi = VI\cos(\delta+\beta) \\ &= VI\cos\delta\cos\beta - VI\sin\delta\sin\beta = VI_q\cos\delta - VI_d\sin\delta \end{aligned} \tag{8.81}$$

Inspection of the phasor diagram shown in Figure 8.29b yields

$$I_q X_q = V\sin\delta, \quad E_q + I_d X_d = V\cos\delta \tag{8.82}$$

Now Eq. (8.82) can be transformed to

$$P = \frac{E_q V}{X_d}\sin\delta + \frac{V^2}{2}\frac{X_d - X_q}{X_d X_q}\sin 2\delta \tag{8.83}$$

This equation is analogous to Eq. (3.157) derived in Section 3.3.6. Active power consists of two components, namely synchronous power and reluctance power. The former is due to the interaction between the stator and rotor magnetic fields, while the latter is due to magnetic asymmetry of the rotor and may be developed without any field excitation. In the considered case of the generator with open-circuited field winding, that is for $E_q \cong 0$ the synchronous power can be neglected and (8.83) simplifies to

$$P \cong \frac{V^2}{2}\frac{X_d - X_q}{X_d X_q}\sin 2\delta \tag{8.84}$$

which corresponds to the reluctance power only.

$$\overline{\mathrm{OC}} = E_q \cong 0$$

When the generator is loaded as in Figure 8.29, it may happen that active power I^2R consumed by the resistor R is smaller than the generated reluctance power so that a surplus of energy is created which causes the generator current to increase. This happens when

$$\frac{V^2}{2}\frac{X_d - X_q}{X_d X_q}\sin 2\delta > I^2 R \tag{8.85}$$

or

$$\frac{V^2}{I^2}\frac{X_d - X_q}{X_d X_q}\sin\delta\cos\delta > R \tag{8.86}$$

In the case considered $E_q \cong 0$ so that applying (8.82) gives: $\sin\delta = I_q X_q / V$ and $\cos\delta \cong I_d X_d / V$. Substituting these equations into (8.86) gives

$$(X_d - X_q)\frac{I_d}{I}\frac{I_q}{I} > R \tag{8.87}$$

where $I_d = I\sin\beta$ and $I_q = I\cos\beta$. Consequently, one gets

$$(X_d - X_q)\sin\beta\cos\beta > R \tag{8.88}$$

Angle β of the current phase shift can be determined using the phasor diagram

$$\tan\beta = \frac{\overline{\mathrm{AC}}}{\overline{\mathrm{OA}}} = \frac{\overline{\mathrm{AB}} - \overline{\mathrm{BC}}}{\overline{\mathrm{OA}}} = \frac{V\sin\varphi - IX_q}{V\cos\varphi} = \frac{\frac{V}{I}\sin\varphi - X_q}{\frac{V}{I}\cos\varphi} = \frac{Z\sin\varphi - X_q}{Z\cos\varphi} \tag{8.89}$$

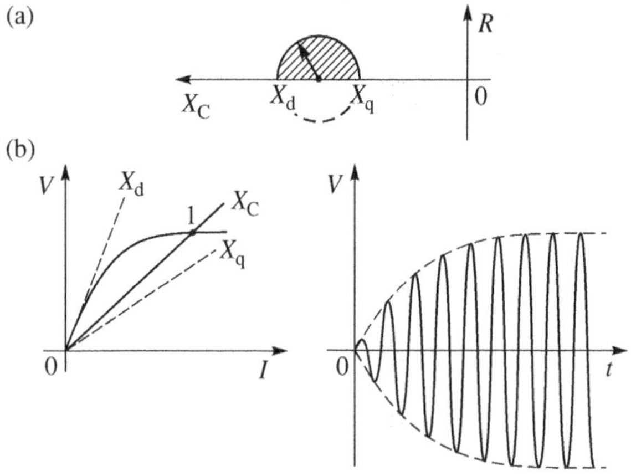

Figure 8.30 Synchronous self-excitation: (a) condition for self-excitation; (b) magnetization characteristic and voltage changes.

The impedance triangle in Figure 8.29 gives $Z \sin \varphi = X_C$ and $Z \cos \varphi = R$. Now Eq. (8.89) yields

$$\tan \beta = \frac{X_C - X_q}{R} \tag{8.90}$$

Noting that $\sin \beta \cos \beta = \tan \beta/(\tan^2\beta + 1)$, and substituting from the above equation for $\tan \beta$ into Eq. (8.88), gives the condition for self-excitation

$$(X_C - X_q)(X_C - X_d) + R^2 < 0 \tag{8.91}$$

This inequality is satisfied for all the points inside the circle in the impedance plane shown in Figure 8.30. Self-excitation arises for all the parameters inside that circle.

Physically self-excitation can be explained by recognizing that the self-inductance of each phase of the stator winding changes sinusoidally, owing to the variable air-gap width, as the rotor rotates. These periodic changes produce parametric resonance, and the resulting increase in the current. If R and X_C are constant, self-excitation also results in an increase in the generator terminal voltage. Outside the circle shown in Figure 8.30, the energy supplied by the generator is too small to overcome the resistance loss and self-excitation cannot happen.

If the generator had a linear magnetization characteristic then self-excitation would lead to an unlimited increase in the current and the terminal voltage of the generator. In reality both reactances X_d and X_q decrease as the current increases. Figure 8.30b shows four characteristics. The bold curve shows the generator voltage–current characteristic with saturation included. The dashed lines X_d and X_q show the idealized characteristics when the generator opposes the current with unsaturated reactances X_d and X_q, respectively. Line X_C corresponds to the capacitor voltage–current characteristic. As X_C is between X_d and X_q, condition (8.91) is satisfied and self-excitation occurs. However, as the voltage and the current increase, the armature saturates and both X_d and X_q start to decrease. The voltage rise stops at point 1 where the saturated reactance becomes equal to X_C. Beyond point 1 the reactance X_C is greater than the saturated reactance X_d and condition (8.91) is not satisfied. Any momentary increase of the voltage above this point results in the resistance loss exceeding the reluctance power, then the current drops and the system returns to point 1. As a result, the generator voltage settles down to a level corresponding to point 1.

As the angle β between the current and q-axis (Figure 8.29) remains constant, self-excitation is referred to the as *synchronous self-excitation*. In practice, synchronous self-excitation is rare but may arise when a salient-pole generator, operating with a small field current, is connected to a lightly loaded long transmission line. In such a situation the terminal voltage will increase but at a slower rate than that shown in Figure 8.30.

8.7.3 Self-excitation of a Generator with Closed Field Winding

Self-excitation can also occur when the excitation circuit is closed, but will be different in character to the synchronous self-excitation described above. If the resistance of the field winding is neglected, the field winding may be treated as a perfect magnetic screen and any change in the armature current will induce an additional field current which completely compensates the change in the armature flux linking with the rotor. This corresponds to the transient state of operation in which the generator opposes the armature current with transient reactance X'_{d} (Chapter 4). There is no q-axis in the field winding so that $X'_{\mathrm{q}} \neq X'_{\mathrm{d}}$. This is referred to as the *transient saliency* (Table 4.2 and 4.3). Active power generated per phase may be expressed by an equation similar to (8.84) as

$$P' = \frac{V^2}{2}\frac{X'_{\mathrm{q}} - X'_{\mathrm{d}}}{X'_{\mathrm{d}}X'_{\mathrm{q}}}\sin 2\delta \tag{8.92}$$

and is referred to as the *dynamic reluctance power*. As before, parametric resonance will only occur if the dynamic reluctance power is greater than the active power I^2R consumed by the resistor R. Similarly, as in Eqs. (8.85) and (8.91), the condition for dynamic self-excitation is satisfied when

$$\left(X_{\mathrm{C}} - X'_{\mathrm{q}}\right)\left(X_{\mathrm{C}} - X'_{\mathrm{d}}\right) + R^2 < 0 \tag{8.93}$$

Condition (8.93) is satisfied inside a circle based on X'_{d} and X'_{q}, as shown in Figure 8.31. If magnetic saturation is neglected, both the stator current and the field current, which opposes the changes in the stator field, will increase to infinity, as energy is continuously supplied in every cycle by the dynamic reluctance torque. Magnetic saturation limits the current increase but in a different way than when the excitation circuit was open.

Initially, the relative position of the phasors is as shown in Figure 8.31a and the dynamic reluctance power causes a rapid increase in the armature current. This, according to the law of constant flux linkage, is accompanied by an increase in the field current, segment OA in Figure 8.31b. As the rotor iron saturates, the emf induced in the closed field winding decreases until the field current experiences no further increase. The resistance of the closed field circuit now plays a deciding role as the magnetic energy stored in the field winding starts to dissipate. This causes the field current to decay and allows the armature flux to enter the rotor iron resulting in an increase in the d-axis

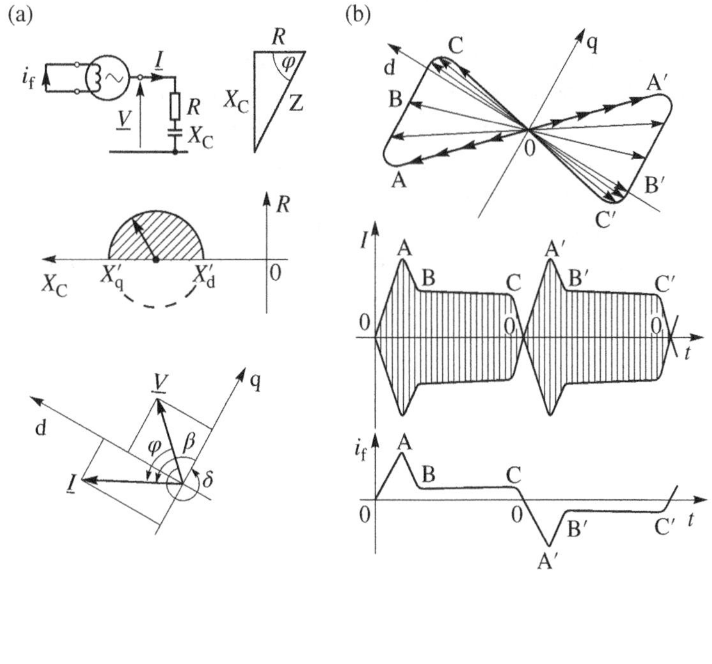

Figure 8.31 Repulsive self-excitation: (a) circuit diagram, area of self-excitation, phasor diagram; (b) changes in the phasor of the armature current, the envelope of the armature current, the field current.

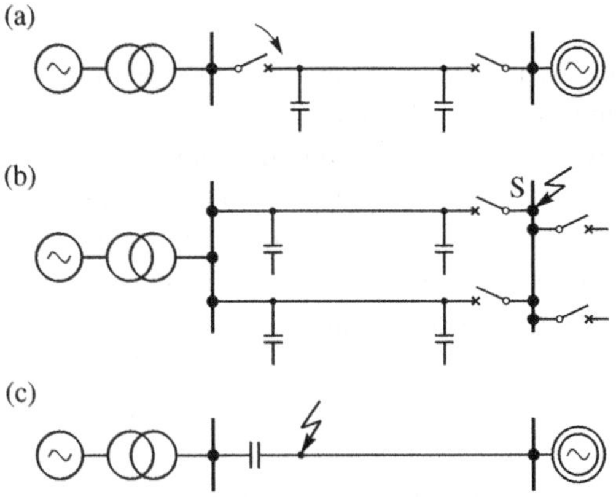

Figure 8.32 Possible cases of self-excitation: (a) connecting a generator to a no-load line; (b) a fault at a neighboring substation; (c) a close fault in a line with series capacitive compensation.

reactance. As a result the dynamic reluctance power decreases, producing a corresponding decrease in the armature current (segment AB in Figure 8.31b). At the same time the stator field begins to slip behind the rotor field as the angle β changes, with the result that the machine operates as an induction generator with its armature current sustained by the asynchronous induction power. This continues until the resulting generator flux reaches a small shift with respect to the d-axis. The conditions sustaining the asynchronous operation now cease to exist and the armature current quickly vanishes (segment CO in Figure 8.31b). The whole cycle then repeats itself as denoted by letters O′, A′, B′, C′, and O in Figure 8.31b. The described resonance is referred to as the *repulsive self-excitation* or *asynchronous self-excitation*.

8.7.4 Practical Possibility of Self-excitation

From a technical point of view, the possibility of a generator meeting the conditions for either synchronous or asynchronous self-excitation is limited. Three possible situations are shown in Figure 8.32.

Figure 8.32a depicts a case when a disconnected long transmission line is first connected to the generating plant. This situation is similar to that shown in Figure 8.31 and may lead to synchronous self-excitation. To prevent this, a transmission line should be connected first to the rest of the network before connecting to the generating plant.

Figure 8.32b shows a situation which may occur after a short circuit in a neighboring substation. The fault on the substation busbars is cleared by opening the circuit-breakers so that the generator operates onto open-circuited lines. The sudden drop in active power load on the generator results in an increase in the angular speed and frequency of the generator. This increase in frequency produces a corresponding increase in the inductive reactances and a drop in the capacitive reactances. With such reactance changes the likelihood of the condition for self-excitation being met is increased.

Figure 8.32c shows the case of a nearby short circuit in a line with series capacitive compensation. A closed circuit is formed which has a high capacitance and a low reactance. This situation may also lead to *subsynchronous resonance*, as described in Section 4.5 and in Anderson et al. (1990).

9

Frequency Stability and Control

Previous chapters describe how the generators in an electric power system (EPS) respond when there is a momentary disturbance in the power balance between the electrical power consumed in the system and the mechanical power delivered by the turbines. Such disturbances are caused by short circuits in the transmission network and are normally cleared without the need to reduce the generated or consumed power. However, if a large load is suddenly connected (or disconnected) to the system, or if a generating unit is suddenly disconnected by the protection equipment, there will be a long-term distortion in the power balance between that delivered by the turbines and that consumed by the loads. This imbalance is initially covered from the kinetic energy of rotating rotors of turbines, generators, and motors and, as a result, the frequency in the system will change. This frequency change can be conveniently divided into a number of stages allowing the dynamics associated with each of these stages to be described separately. This helps illustrate how the different dynamics develop in the system. However, it is first necessary to describe the operation of the automatic generation control (AGC) as this is fundamental in determining the way in which the frequency will change in response to a change in load.

The general framework of frequency control described in this chapter was originally developed under the framework of traditional vertically integrated utilities which used to control generation, transmission, and often distribution in their own service areas and where power interchanges between control areas were scheduled in advance and strictly adhered to. Chapter 2 describes the liberalization of the industry taking place in many countries in the world since the 1990s whereby utilities are no longer vertically integrated generation companies that compete against each other and the coordination necessary for reliable system operation, including frequency control, is undertaken by transmission system operators (TSOs). The framework of frequency control had to be adapted to the market environment in the sense that TSOs have to procure, and pay for, frequency support from individual power plants. The way those services are procured differs from country to country, so the commercial arrangements are not be discussed in this book. However, the overall hierarchical control framework has been largely retained, with only some changes in the tertiary control level, as discussed later in this chapter.

The discussion concentrates on frequency control in an interconnected power system, using the European Union for the Coordination of Transmission of Electricity (UCTE) as the example. It is worth adding that the framework of frequency control in an islanded system, such as in the United Kingdom, may be different. Also, the meaning of some of the terms may differ from country to country.

9.1 Automatic Generation Control

Section 2.1 explains how an EPS consists of many generating units and many loads, while its total power demand varies continuously throughout the day in a more or less anticipated manner. The large, slow changes in demand are met centrally by deciding at regular intervals which generating units will be operating, shut down, or in an

Power System Dynamics: Stability and Control, Third Edition. Jan Machowski, Zbigniew Lubosny, Janusz W. Bialek and James R. Bumby.

intermediate hot reserve state. This process of *unit commitment* may be conducted once per day to give the daily operating schedule, while at shorter intervals, typically every 30 minutes, *economic dispatch* determines the actual power output required from each of the committed generators. Smaller, but faster, load changes are dealt with by AGC so as to:

- maintain frequency at the scheduled value (frequency control);
- maintain the net power interchanges with neighboring control areas at their scheduled values (tie-line control);
- maintain power allocation among the units in accordance with area dispatching needs (energy market, security, or emergency).

In some systems the role of AGC may be restricted to one or two of the above objectives. For example, tie-line power control is only used where a number of separate power systems are interconnected and operate under mutually beneficial contractual agreements.

9.1.1 Generation Characteristic

Section 2.3.3 discusses the operation of turbines and their governing systems. In the steady state the idealized power-speed characteristic of a single generating unit is given by Eq. (2.3). As the rotational speed is proportional to frequency, Eq. (2.3) may be rewritten for the i-th generating unit as

$$\frac{\Delta f}{f_{\mathrm{n}}} = -\rho_i \frac{\Delta P_{\mathrm{m}i}}{P_{\mathrm{n}i}}, \quad \frac{\Delta P_{\mathrm{m}i}}{P_{\mathrm{n}i}} = -K_i \frac{\Delta f}{f_{\mathrm{n}}} \tag{9.1}$$

In the steady state all the generating units operate synchronously at the same frequency when the overall change in the total power ΔP_{T} generated in the system can be calculated as the sum of changes at all generators

$$\Delta P_{\mathrm{T}} = \sum_{i=1}^{N_{\mathrm{G}}} \Delta P_{\mathrm{m}i} = -\frac{\Delta f}{f_{\mathrm{n}}} \sum_{i=1}^{N_{\mathrm{G}}} K_i P_{\mathrm{n}i} = -\Delta f \sum_{i=1}^{N_{\mathrm{G}}} \frac{K_i P_{\mathrm{n}i}}{f_{\mathrm{n}}} \tag{9.2}$$

where N_{G} is the number of generating units in the system. The subscript "T" indicates that ΔP_{T} is the change in generated power as supplied by the turbines. Figure 9.1 illustrates how the characteristics of individual generating units can be added according to Eq. (9.2) to obtain the equivalent *generation characteristic*. This characteristic defines the ability of the system to compensate for a power imbalance at the cost of a deviation in frequency. For an EPS with a large number of generating units, the generation characteristic is almost horizontal such that even a relatively large power change only results in a very small frequency deviation. This is one of the benefits accruing from combining generating units into one large system.

To obtain the equivalent generation characteristic of Figure 9.1 it has been assumed that the speed-droop characteristics of the individual turbine generator units are linear over the full range of power and frequency variations. In practice the output power of each turbine is limited by its technical parameters. For example, coal-burn steam turbines have a lower power limit, because of the need to maintain operational stability of the burners and an

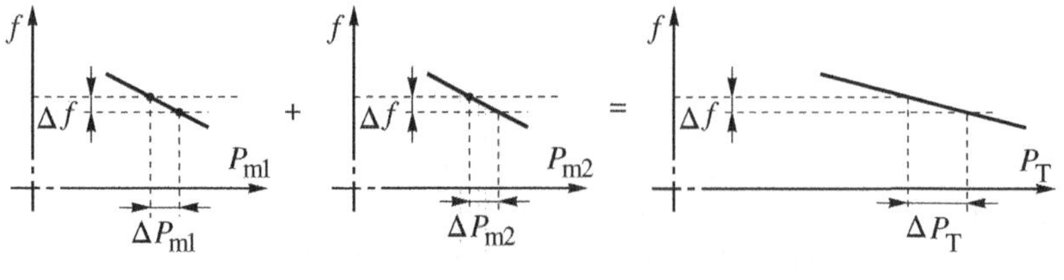

Figure 9.1 Generation characteristic as the sum of the speed-droop characteristics of all the generation units.

upper power limit that is set by thermal and mechanical considerations. In the remainder of this section only the upper limit is considered, so the turbine characteristic is as shown in Figure 9.2.

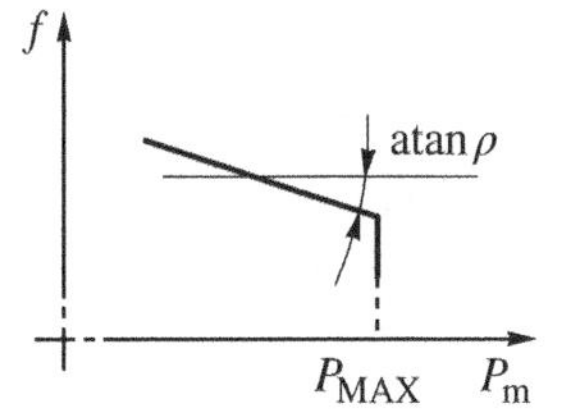

Figure 9.2 Speed-droop characteristic of a turbine with an upper limit.

If a turbine is operating at its upper power limit then a decrease in the system frequency will not produce a corresponding increase in its power output. At the limit $\rho = \infty$ or $K = 0$ and the turbine does not contribute to the equivalent system characteristic. Consequently, the generation characteristic of the system will be dependent on the number of units operating away from their limit at part load, that is it will depend on the *spinning reserve*, where the spinning reserve is the difference between the sum of the power ratings of all the operating units and their actual load. The allocation of spinning reserve is an important factor in power system operation as it determines the shape of the generation characteristic. This is demonstrated in Figure 9.3, which shows a simple case of two generating units. In Figure 9.3a the spinning reserve is allocated proportionally to both units so that they both reach their upper power limits at the same frequency f_1. In this case the equivalent characteristic, obtained from Eq. (9.2), is linear until the upper power limit for the whole system is reached. In Figure 9.3b the same amount of spinning reserve is available but is allocated solely to the second generator with the first unit operating at full load. The generation characteristic is now nonlinear and consists of two sections of different slope.

Similarly, the generation characteristic of an actual power system is nonlinear and consists of many short sections of increasing slope as more and more generating units reach their operating limits as the total load increases until, at maximum system load, there is no more spinning reserve available. The generation characteristic then becomes a vertical line. For small power and frequency disturbances it is convenient to approximate this nonlinear generation characteristic in the vicinity of the operation point by a linear characteristic with a local droop value. The total system generation is equal to the total system load P_L (including transmission losses)

$$\sum_{i=1}^{N_G} P_{mi} = P_L \tag{9.3}$$

where N_G is the number of generating units. Dividing Eq. (9.2) by P_L gives

$$\frac{\Delta P_T}{P_L} = -K_T \frac{\Delta f}{f_n} \quad \text{or} \quad \frac{\Delta f}{f_n} = -\rho_T \frac{\Delta P_T}{P_L} \tag{9.4}$$

where

$$K_T = \frac{\sum_{i=1}^{N_G} K_i P_{ni}}{P_L}; \quad \rho_T = \frac{1}{K_T} \tag{9.5}$$

Equation (9.4) describes the linear approximation of the generation characteristic calculated for a given total system demand. Consequently, the coefficients in Eq. (9.5) are calculated with respect to the total demand, not the sum of the power ratings, so that ρ_T is the local speed droop of the generation characteristic and depends on the spinning reserve and its allocation in the system, as shown in Figure 9.3.

In the first case, shown in Figure 9.3a, it was assumed that the spinning reserve is allocated uniformly between both generators, that is both generators are underloaded by the same amount at the operating point (frequency f_0) and the maximum power of both generators is reached at the same point (frequency f_1). The sum of both characteristics is then a straight line up to the maximum power $P_{MAX} = P_{MAX\,1} + P_{MAX\,2}$. In the second case, shown in Figure 9.3b, the total system reserve is the same but it is allocated to the second generator only. That generator is loaded up to its maximum at the operating point (frequency f_2). The resulting total generation characteristic is

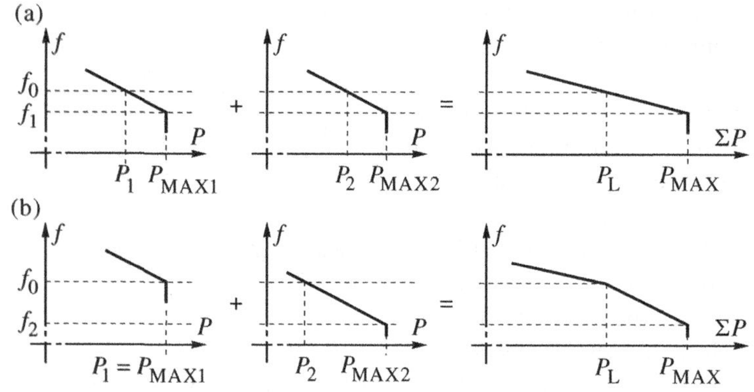

Figure 9.3 Influence of the turbine upper power limit and the spinning reserve allocation on the generation characteristic.

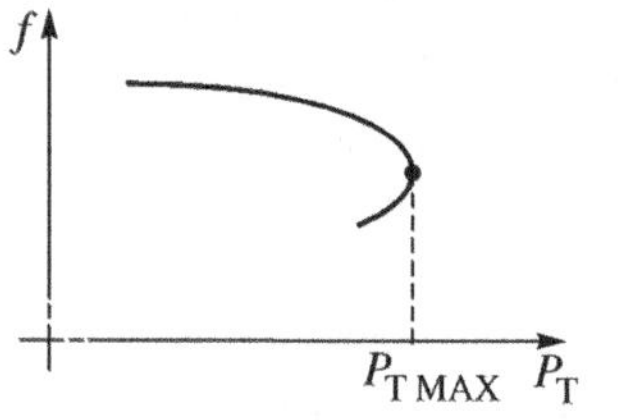

Figure 9.4 Static system generation characteristic.

nonlinear and consists of two lines of different slope. The first line is formed by adding both inverse droops, $K_{T1} \neq 0$ and $K_{T2} \neq 0$, in Eq. (9.5). The second line is formed noting that the first generator operates at maximum load and $K_{T1} = 0$ so that only $K_{T2} \neq 0$ appears in the sum in Eq. (9.5). Hence the slope of that characteristic is higher.

The number of units operating in a real system is large. Some of them are loaded to the maximum but others are partly loaded, generally in a nonuniform way, to maintain a spinning reserve. Adding up all the individual characteristics would give a nonlinear resulting characteristic consisting of short segments with increasingly steeper slopes. That characteristic can be approximated by a curve, shown in Figure 9.4. The higher the system load, the higher the droop until it becomes infinite $\rho_T = \infty$, and its inverse $K_T = 0$, when the maximum power P_{MAX} is reached. If the dependence of a power station's auxiliary requirements on frequency were neglected, that part of the characteristic would be vertical (shown as a dashed line in Figure 9.4). However, power stations tends to have a curled-back characteristic – see curve 4 in Figure 2.15. Similarly curled is the system characteristic shown in Figure 9.4.

For further considerations, the nonlinear generation characteristic shown in Figure 9.4 will be linearized at a given operating point, that is the linear approximation Eq. (9.4) will be assumed with the droop ρ_T given by (9.5).

9.1.2 Primary Control

When the total generation is equal to the total system demand (including losses) then the frequency is constant, the system is in equilibrium, and the generation characteristic is approximated by Eq. (9.4). However, as discussed in Section 3.4, system loads are also frequency dependent and an expression similar to Eq. (9.4) can be used to obtain a linear approximation of the frequency response characteristic of the total system load as

$$\frac{\Delta P_L}{P_L} = K_L \frac{\Delta f}{f_n} \tag{9.6}$$

where K_L is the *frequency sensitivity coefficient of the power demand*. A similar coefficient k_{Pf} is defined in Eq. (3.158) and referred to as the frequency sensitivity of a single composite load and should not be confused with K_L, which relates to the total system demand. Tests conducted on actual systems indicate that the generation response characteristic is much more frequency dependent than the demand response characteristic. Typically, K_L is between 0.5 and 3 (Table 3.3), while $K_T \cong 20$ ($\rho = 0.05$). In Eqs. (9.4) and (9.6) the coefficients K_T and K_L have opposite sign so that an increase in frequency corresponds to a drop in generation and an increase in electrical load.

In the (P, f) plane the intersection of the generation and the load characteristic, Eqs. (9.4) and (9.6), defines the system equilibrium point. A change in the total power demand ΔP_L corresponds to a shift of the load characteristic in the way shown in Figure 9.5 so that the equilibrium point is moved from point 1 to point 2. The increase in the system load is compensated in two ways: first of all, by the turbines increasing the generation by ΔP_T and, secondly, by the system loads reducing the demand by ΔP_L from that required at point 3 to that required at point 2. Figure 9.5 shows that taking both increments into account gives

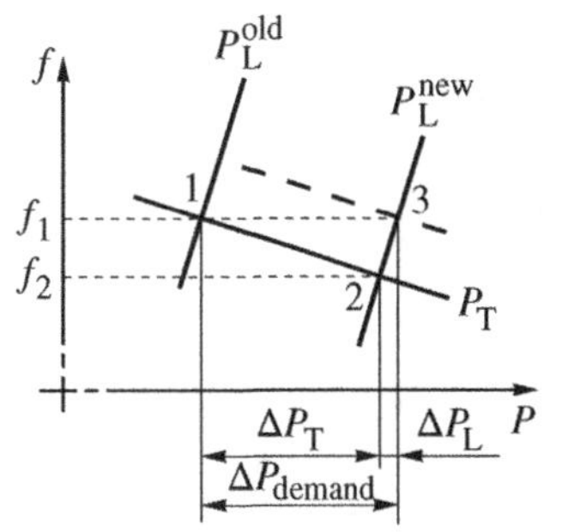

Figure 9.5 Equilibrium points for an increase in the power demand.

$$\Delta P_{demand} = \Delta P_T - \Delta P_L = -(K_T + K_L)P_L\frac{\Delta f}{f_n} = -K_f P_L\frac{\Delta f}{f_n} \tag{9.7}$$

The new operating point of the system corresponds to a new demand and a new frequency. The new value of the system frequency f_2 is lower than the old one f_1. Equation (9.7) represents the frequency response of the system while the coefficient $K_f = K_T + K_L$ is referred to as the *stiffness*. It should be emphasized that $\Delta P_T \gg \Delta P_L$.

A reduction of the demand by ΔP_L is due to the frequency sensitivity of demand. An increase of generation by ΔP_T is due to turbine governors. The action of turbine governors due to frequency changes when reference values of regulators are kept constant is referred to as *primary frequency control.*

When the system demand increases, primary control is activated, obviously only if there are any units which are operating but are not fully loaded. Figure 9.3b shows that if any of the units operate at maximum output, a reduction in frequency cannot increase a unit's output. Only those units that are partly loaded and carry a spinning reserve can be loaded more.

In order to secure safe system operation and a possibility of activating the primary control, the system operator must have an adequate spinning reserve at its disposal. The spinning reserve to be utilized by the primary control should be uniformly distributed around the system, that is at power stations evenly located around the system. Then the reserve will come from a variety of locations and the risk of overloading some transmission corridors will be minimized. Locating the spinning reserve in one region may be dangerous from the point of view of security of the transmission network. If one or more power stations suffer outages, the missing power would come from just one region, some transmission corridors might get overloaded, and the disturbance might spread.

An interconnected system requires coordination so the requirements regarding the primary control are normally the subject of agreements between partners cooperating in a given interconnected network. For the European UCTE system, the requirements are defined in UCTE (2007b).

For the purposes of primary frequency control, each subsystem in the UCTE system has to ensure a large enough spinning reserve proportional to a given subsystem's share in the overall UCTE energy production. This is referred to as the *solidarity principle*. It is required that the spinning reserve is uniformly located within each subsystem and the operating points of individual units providing the reserve are such that the whole reserve in the system is activated when the frequency deviation is not more than 200 mHz. The required time of activation of the reserve should not be longer than (15–23) s. To satisfy this condition, the units participating in the primary reserve should be able to regulate power quickly within ±5% of their rated power. Turbine governors of those units have typically speed-droop characteristics with dead zones, as in Figure 9.6a. The first dead zone has the width $\pm\varepsilon_f = \pm 10$ mHz, which makes the unit operate with constant set power P_{ref} in the presence of small frequency error $|\Delta f| < 10$ mHz. When a frequency error $|\Delta f| > 10$ mHz appears, the unit operates in the primary control with a fast regulation range $\pm\varepsilon_P = \pm 5\%$. The whole range of the reserve is released when the frequency error is about ±200 mHz. These requirements mean the whole primary reserve is released in the system when the frequency error is not greater than 200 mHz. The droop can be calculated in the fast regulation range from Eq. (9.1). Substituting $\Delta f = (200 - 10 = 190)$ mHz = 190 mHz = 0.190 Hz, that is $\Delta f/f_n = 0.190/50 = 0.0038$ and $\Delta P_m/P_n = -0.05$ results

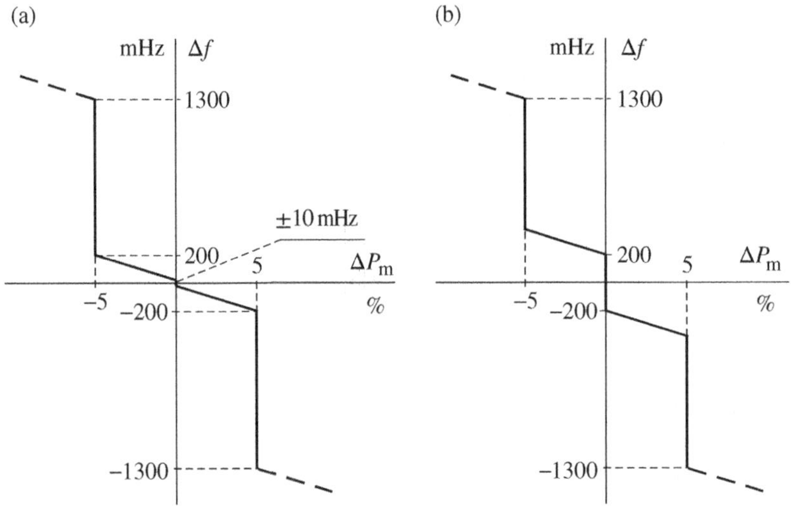

Figure 9.6 Example of speed-droop characteristics with fast regulation range $\pm\varepsilon_P = \pm 5\%$ and (a) small dead zone $\pm\varepsilon_f = \pm 10$ mHz; (b) large dead zone $\pm\varepsilon_f = \pm 200$ mHz.

in $\rho = 0.0038/0.05 = 0.076 = 7.6\%$. This is a typical value as, according to Section 2.2.3, the droop ρ is typically assumed to be between 4 and 9%. For the range beyond fast regulation $\pm 5\%$ the turbine governor maintains constant power until the frequency error reaches $|\Delta f| > 1300$ mHz $= 1.3$ Hz when the governor is switched from the power control regime to speed control.

Governors of units that do not participate in the primary frequency control have the first dead zone of the speed-droop characteristic (Figure 9.6b) set at $\pm\varepsilon_f = \pm 200$ mHz. They form an additional primary reserve that activates only for large disturbances. This is necessary to defend the system against blackouts (details in Section 9.1.6).

9.1.3 Secondary Control

If the turbine-generators are equipped with governing systems, such as those described in Section 2.3.3, then, following a change in the total power demand, the system will not be able to return to the initial frequency on its own, without any additional action. According to Figure 9.5, in order to return to the initial frequency the generation characteristic must be shifted to the position shown by the dashed line. Such a shift can be enforced by changing the P_{ref} setting in the turbine governing system (the load reference set point in Figure 2.14). As shown in Figure 9.7, changes in the settings $P_{ref(1)}$, $P_{ref(2)}$, and $P_{ref(3)}$ enforce a corresponding shift of the characteristic to the positions $P_{m(1)}$, $P_{m(2)}$, and $P_{m(3)}$. To simplify, the first dead zone around P_{ref}, which is shown in Figure 9.6, is neglected in Figure 9.7. Obviously, no change of settings can force a turbine to exceed its maximum power rating P_{MAX}. Changing more settings P_{ref} of individual governors will move upwards the overall generation characteristic of the system. Eventually, this will lead to the restoration of the rated frequency but now at the required increased value of power demand. Such control action on the governing systems of individual turbines is referred to as *secondary control.*

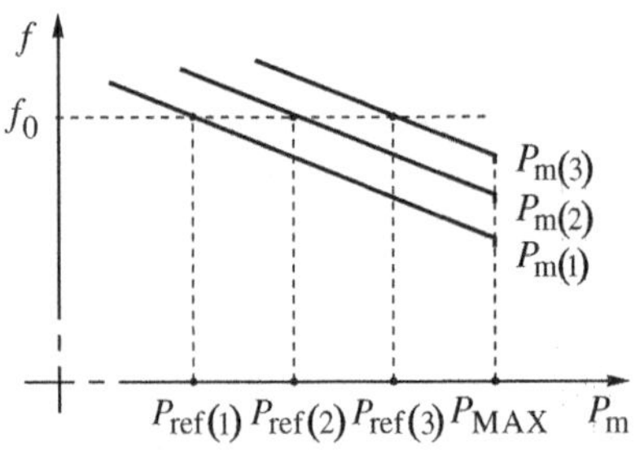

Figure 9.7 Turbine speed-droop characteristics for various settings of P_{ref}.

In an isolated power system automatic secondary control may be implemented as a decentralized control function by adding a supplementary control loop to the turbine-governor system. This modifies the block diagram of the turbine governor (Figure 2.16) to that shown in Figure 9.8 where P_{ref} and P_m are expressed as a fraction of the rated power P_n. The supplementary control loop, shown by the dashed line, consists of an integrating element which adds a control signal ΔP_ω that is proportional to the integral of the speed (or frequency) error to the load reference point. This signal modifies the value of the setting in the P_{ref} circuit thereby shifting the speed-droop characteristic in the way shown in Figure 9.7.

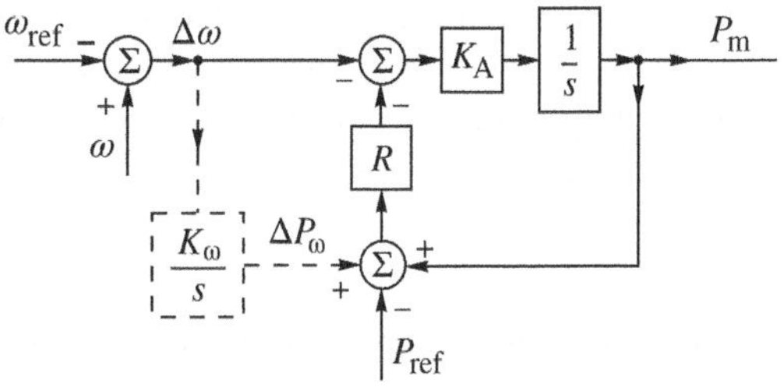

Figure 9.8 Supplementary control added to the turbine governing system.

Not all the generating units in a system that implements decentralized control need be equipped with supplementary loops and participate in secondary control. Usually, medium-sized units are used for frequency regulation, while large base load units are independent and set to operate at a prescribed generation level. In combined cycle gas and steam turbine power plants the supplementary control may affect only the gas turbine or both the steam and the gas turbines.

In an interconnected power system consisting of a number of different control areas, secondary control cannot be decentralized, because the supplementary control loops have no information as to where the power imbalance occurs so that a change in the power demand in one area would result in regulator action in all the other areas. Such decentralized control action would cause undesirable changes in the power flows in the tie-lines linking the systems and the consequent violation of the contracts between the cooperating systems. To avoid this, centralized secondary control is used.

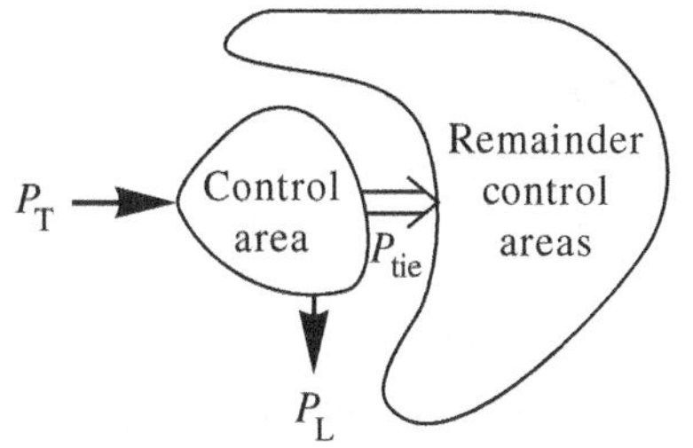

Figure 9.9 Power balance of a control area.

In interconnected power systems, AGC is implemented in such a way that each area, or subsystem, has its own central regulator. As shown in Figure 9.9, the EPS is in equilibrium if, for each area, the total power generation P_T, the total power demand P_L, and the net tie-line interchange power P_{tie} satisfy the condition

$$P_T - (P_L + P_{tie}) = 0 \tag{9.8}$$

The objective of each area regulator is to maintain frequency at the scheduled level (frequency control) and to maintain net tie-line interchanges from the given area at the scheduled values (tie-line control). If there is a large power balance disturbance in one subsystem (caused, for example, by the tripping of a generating unit) then regulators in each area should try to restore the frequency and net tie-line interchanges. This is achieved when the regulator in the area where the imbalance originated enforces an increase in generation equal to the power deficit. In other words, each area regulator should enforce an increased generation covering its own area power imbalance and maintain planned net tie-line interchanges. This is referred to as the *nonintervention rule.*

The regulation is executed by changing the power output of turbines in the area through varying P_{ref} in their governing systems. Figure 9.10 shows a functional diagram of the central regulator. Frequency is measured in the local low-voltage network and compared with the reference frequency to produce a signal that is proportional

Figure 9.10 Functional diagram of a central regulator.

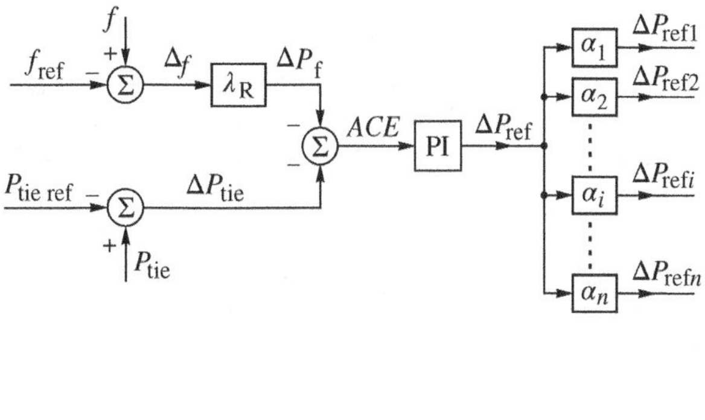

to the frequency deviation Δf. The information on power flows in the tie-lines is sent via telecommunication lines to the central regulator which compares it with the reference value in order to produce a signal proportional to the tie-line interchange error ΔP_{tie}. Before adding these two signals together the frequency deviation is amplified by the factor λ_R, called the *frequency bias factor*, to obtain

$$\Delta P_f = \lambda_R \Delta f \tag{9.9}$$

This represents a change in the generation power which must be forced upon the controlled area in order to compensate for the frequency deviation resulting from the power imbalance in this area.

The choice of the bias factor λ_R plays an important role in the nonintervention rule. According to this rule, each subsystem should cover its own power imbalance and try to maintain planned power interchanges. If the frequency drops following a generation deficit then applying Eq. (9.7) gives $\Delta P_{\text{demand}} = -(K_f P_L / f_n)\Delta f$. The central regulator should enforce an increased generation $\Delta P_f = \lambda_R \Delta f$ covering the deficit, that is $\Delta P_f = -\Delta P_{\text{demand}}$. Hence $\lambda_R \Delta f = (K_f P_L / f_n)\Delta f$ or

$$\lambda_R = \frac{K_f P_L}{f_n} = K_{f\ \text{MW/Hz}} \tag{9.10}$$

That equation indicates that the value of the bias λ_R can be assessed providing that the stiffness K_f of a given area and its total demand are known. $K_{f\ \text{MW/Hz}}$ denotes the frequency stiffness of the EPS expressed in MW/Hz.

Clearly the nonintervention rule requires that the value of the bias set in the central regulator is equal to the area stiffness expressed in MW/Hz. In practice it is difficult to evaluate the exact value of the stiffness so that imprecise setting of the bias λ_R in the central regulator may have some undesirable effects on the regulation process. This is discussed later in this chapter.

The signal ΔP_f is added to the net tie-line interchange error ΔP_{tie} so that the *area control error* (*ACE*) is created

$$ACE = -\Delta P_{\text{tie}} - \lambda_R \Delta f \tag{9.11}$$

Similarly, as in the decentralized regulator shown in Figure 9.8, the central regulator must have an integrating element in order to remove any error and this may be supplemented by a proportional element. For such a PI regulator the output signal is

$$\Delta P_{\text{ref}} = \beta_R \cdot ACE + \frac{1}{T_R}\int_0^t ACE\,\text{d}t \tag{9.12}$$

where β_R and T_R are the regulator parameters. Usually, a regulator with a small, or even zero, participation of the proportional element is used, that is an integral element.

ACE corresponds to the power by which the total area power generation must be changed in order to maintain both the frequency and the tie-line flows at their scheduled values. The regulator output signal ΔP_{ref} is then multiplied by the *participation factors* $\alpha_1, \alpha_2, \ldots, \alpha_n$ which define the contribution of the individual generating units to the total generation control (Figure 9.10). The control signals $\Delta P_{\text{ref1}}, \Delta P_{\text{ref2}}, \ldots, \Delta P_{\text{ref}n}$ obtained in this way are then transmitted to the power plants and delivered to the reference set points of the turbine governing systems (Figure 2.11). As with decentralized control, not all generating units need participate in generation control.

It is worth noting that regulation based on *ACE* defined by Eq. (9.11) does not always finish with the removal of both errors Δf and ΔP_{tie}. According to Eq. (9.11), zeroing of *ACE* can be generally achieved in two ways:

- Zeroing of both errors, that is achieving $\Delta P_{\text{tie}} = 0$ and $\Delta f = 0$. This is a more desirable outcome. When the available regulation power in a given subsystem is large enough to cover its own power deficit, the nonintervention rule enforces a return of the interchange power to a reference value and both frequency and power interchange errors are removed.

- Achieving a compromise between the errors $\Delta P_{\text{tie}} + \lambda_R \Delta f = 0$, or $\Delta P_{\text{tie}} = -\lambda_R \Delta f$. When all the available regulation power in a given subsystem is not large enough to cover its own power imbalance, the regulating units in that subsystem exhaust their capability before the errors are removed and the regulation ends with the violation of a reference value of power interchange. The missing power must be delivered from neighboring networks. In that case the nonintervention rule will be violated and the following errors arise: $\Delta P_{\text{tie}\infty} = -\lambda_R \Delta f_{\infty}$, where $\Delta P_{\text{tie}\infty}$ is power which the subsystem must additionally import to cover the power deficit.

Both cases are be illustrated in Section 9.2 using the results of computer simulation.

The exact determination of the actual stiffness $K_{\text{f MW/Hz}}$ in real time is a difficult task as the stiffness is continuously varying, because of changes in the demand, its structure, and the composition of power stations. Determination of $K_{\text{f MW/Hz}}$ in real time is the subject of ongoing research. Generally, it would require a sophisticated dynamic identification methodology. In practice, the frequency bias factors λ_R are set in central regulators in the European UCTE system using a simplified methodology. In each year, the share of a given control area in the total energy production is determined. Then the value of $K_{\text{f MW/Hz}}$ is estimated for the whole interconnected system. This estimate of $K_{\text{f MW/Hz}}$ is divided between the control areas in proportion to their annual energy share and that value is set to be the frequency bias factor for each area.

For example, let $K_{\text{f MW/Hz}} = 20\,000$ MW/Hz be the estimate of stiffness of the whole interconnected system. Let the shares of the k-th and j-th control area in the annual energy production be, respectively, $\alpha_k = 0.05$ and $\alpha_j = 0.20$. According to the approximate method, the frequency bias factors set in the central regulators are

$$\lambda_{R\,k} = \alpha_k K_{\text{f MW/Hz}} = 0,05 \cdot 20\,000 = 1\,000\,\text{MW/Hz}$$

$$\lambda_{R\,j} = \alpha_j K_{\text{f MW/Hz}} = 0,02 \cdot 20\,000 = 4\,000\,\text{MW/Hz}$$

The described method is simple and results in an approximate fulfillment of the nonintervention rule.

Secondary frequency control is much slower than the primary one. As an example, given below is a description of requirements agreed by members of the European UCTE system.

Tie-line flow measurements have to be sent to the central regulator either cyclically or whenever the flow exceeds a certain value, with a delay not exceeding (1–5) s. Instructions to change the reference values are sent from the central regulator (Figure 9.10) to area regulators approximately every 10 minutes. For the maximum speed of secondary control, activation of the whole range of secondary reserve must be done within 15 minutes.

Power stations contributing to the secondary control, that is those controlled by the central regulator, must operate with a wide range of power regulation. This range depends on the type of the power station. For thermal plants, the typical range of power regulation operated within primary control (Figure 2.13) is from 40% (technical minimum) to 100% (maximum). Usually, secondary control uses about ±5% of the maximum power from within that range. The speed of regulation depends on the type of unit. It is required that the speed of regulation is not less than:

- for gas or oil units: 8% of rated power per minute;
- for coal and lignite units: (2–4)% of rated power per minute;
- for nuclear units: (1–5)% of rated power per minute;
- for hydro units: 30% of rated power per minute.

The sum of regulation ranges, up and down, of all the generating units active in secondary control is referred to as the *bandwidth of secondary control*. The positive value of the bandwidth, that is from the maximum to the actual operating point, forms the *reserve of secondary control*. In the European UCTE system, the required value of the secondary reserve for each control area is in the range of 1% of the power generated in the area. It is additionally required that the secondary reserve is equal at least to the size of the largest unit operating in the area. This requirement is due to the nonintervention rule if the largest unit is suddenly lost. If that happens, secondary control in the area must quickly, in no longer than 15 minutes, increase the power generated in the area by the value of the lost power.

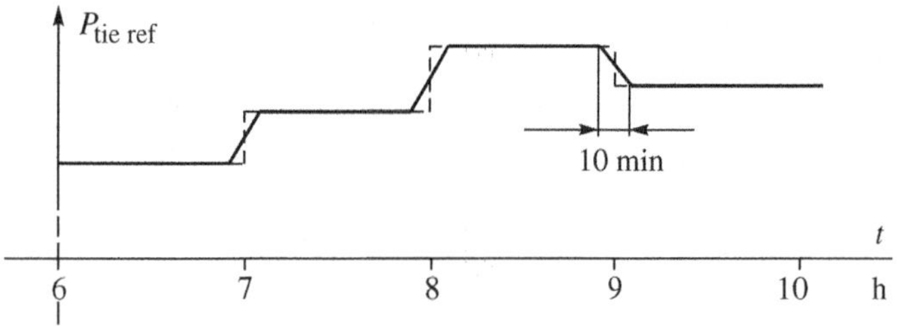

Figure 9.11 A schedule of changes in the required value of power interchange.

The schedule of required values of power interchange $P_{tie\ ref}$ is sent to the central regulator based on a schedule for the whole interconnection. In order to prevent power swings between control areas due to rapid changes in the reference values changes in the values of $P_{tie\ ref}$ are phased in, as shown in Figure 9.11. The ramp of phasing in starts 5 minutes before, and finishes 5 minutes after, the set time of the change.

9.1.4 Tertiary Control

Tertiary control is additional to, and slower than, primary and secondary frequency control. The task of tertiary control depends on the organizational structure of a given power system and the role that power plants play in this structure.

Under the vertically integrated industry structure (Chapter 2), the system operator sets the operating points of individual power plants based on the economic dispatch, or more generally optimal power flow (OPF), which minimizes the overall cost of operating plants subject to network constraints. Hence tertiary control sets the reference values of power in individual generating units to the values calculated by optimal dispatch in such a way that the overall demand is satisfied together with the schedule of power interchanges.

In many parts of the world, electricity supply systems have been liberalized, and privately owned power plants are not directly controlled by the system operator. Economic dispatch is then executed via an energy market. Depending on the actual market structure in place, power plants either bid their prices to a centralized pool or arrange bilateral contracts directly with companies supplying power to individual customers. The main task of the system operator is then to adjust the supplied bids or contracts to make sure that the network constraints are satisfied and to procure the required amount of primary and secondary reserve from individual power plants. In such a market structure the task of tertiary control is to adjust, manually or automatically, the set points of individual turbine governors in order to ensure the following:

- Adequate spinning reserve in the units participating in primary control.
- Optimal dispatch of units participating in secondary control.
- Restoration of the bandwidth of secondary control in a given cycle.

Tertiary control is supervisory with respect to the secondary control that corrects the loading of individual units within an area. Tertiary control is executed via the following:

- Automatic change of the reference value of the generated power in individual units.
- Automatic or manual connection or disconnection of units that are on the reserve of the tertiary control.

The reserve of the tertiary control is made up of those units that can be manually or automatically connected within 15 minutes of a request being made. The reserve should be utilized in such a way that the bandwidth of the secondary control is restored.

9.1.5 AGC as a Multi-level Control

AGC is an excellent example of a multi-level control system, whose overall structure is shown in Figure 9.12.

The turbine governing system with its load reference point is at the lowest primary control level and all the commands from the upper levels are executed at this level. Primary control is decentralized because it is installed in power plants situated at different geographical sites.

Frequency control and tie-line control constitute secondary control and force the primary control to eliminate the frequency and net tie-line interchange deviations. In isolated systems secondary control is limited to frequency control and could be implemented locally without the need for coordination by the central regulator. In interconnected systems secondary control of frequency and tie-line flows is implemented using a central computer. Secondary control should be slower than primary control.

The task of tertiary control is to ensure an appropriate bandwidth of secondary control. Obviously, tertiary control must be slower than both primary and secondary control. Hence it may be neglected when considering the dynamics of coordination between primary and secondary control.

Modern solutions execute the tasks of secondary and tertiary control using a function LFC (load and frequency control) and control algorithms of SCADA-EMS (supervisory control and data acquisition energy management system) that supports the system operator. Apart from controlling frequency and power interchanges, SCADA-EMS contains a number of other optimization and security management functions. A detailed description is beyond the scope of this book but interested readers are referred to Wood and Wollenberg (1996).

The hierarchical structure shown in Figure 9.12 also contains a block UTC (coordinated universal time) corresponding to control of the synchronous time, that is time measured by synchronous clocks based on the system frequency. Power system frequency varies continuously, so synchronous clocks tend to have an error proportional to the integral of the system frequency. This error is eliminated by occasionally changing the reference value of the frequency.

In the European UCTE system the nominal frequency is $f_n = 50$ Hz. The synchronous time deviation is calculated at the control center in Laufenberg (Switzerland) using the UTC time standard. On certain days of each month, the center broadcasts a value of frequency correction to be inserted in secondary control systems in order to eliminate the

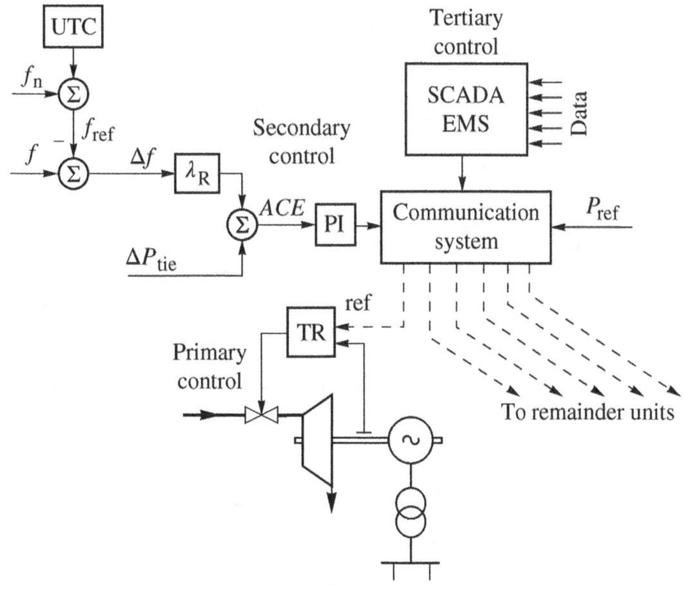

Figure 9.12 Levels of automatic generation control.

synchronous time deviation. If synchronous clocks are delayed then the frequency correction is set to 0.01 Hz, that is the set frequency is $f_{\text{ref}} = 50.01$ Hz. If synchronous clocks are ahead then the frequency correction is set to −0.01 Hz, that is the set frequency is $f_{\text{ref}} = 49.99$ Hz. Usually, the system operates for a few tens of days with a correction of −0.01 Hz, for a few days with +0.01 Hz and then without any correction for the remainder of the year.

Usually, control areas are grouped in large interconnected systems with the central regulator of one area (usually the largest) regulating power interchanges in the given area with respect to other areas. In such a structure the central controller of each area regulates its own power interchanges, while the central controller of the main area additionally regulates power interchanges of the whole group.

9.1.6 Defense Plan Against Frequency Instability

The system of frequency and power control discussed above is adequate for typical disturbances of an active power balance consisting of the unplanned outage of a large generating unit. The largest disturbance which can be handled by the system is referred to as the *reference incident*. In the European UCTE system, the reference incident consists of losing units with a total power of 3000 MW. Larger incidences have to be handled by each system operator using its own defense plan together with appropriate facilities that defend the EPS from the disturbance spreading out. Examples of such a defense plan are shown in Tables 9.1 and 9.2.

Table 9.1 Example of a defense plan against a frequency drop.

f [Hz]	Δf [Hz]	Type of defense action
50.000	<0.200	Normal operation with primary control with small dead zones (Figure 9.6a) and secondary control of frequency and the tie-line power: $ACE = -\Delta P_{\text{tie}} - \lambda_R \Delta f$
49.800	0.200	Central secondary controllers of subsystems are blocked. Generating units operate only with primary control and manual setting of reference power
		Primary control hidden in units operating with large dead zones ±200 mHz (Figure 9.6b) is automatically activated
		Switching of pumped storage plants from the pump mode to the generation mode
		Starting of fast-start units (diesel and open-cycle gas units)
49.000 48.700	1.000 1.300	Underfrequency load shedding (first two stages)
48.700	1.300	Turbine governors (primary control) switch from power regulation according to droop characteristic to speed control (Figure 9.6)
48.500 48.300 48.100	1.500 1.700 1.900	Underfrequency load shedding (next three stages)
47.500	2.500	Generating units are allowed to trip by turbine protections. The units supply their own ancillary services and demand of their islands (if they survived). System operators start system restoration by reconnecting the islands and lost generating units

Table 9.2 Example of a defense plan against frequency rise.

f [Hz]	Δf [Hz]	Type of defense action
51.500	1.500	Generating units are allowed to trip by turbine protections. The units supply their own ancillary services
51.300	1.300	Turbine governors (primary control) switch from power regulation according to droop characteristic to speed control (Figure 9.6)
50.200	0.200	Switching of pumped storage plants from the generation mode to the pump mode
		Stopping of fast-start units (diesel and open-cycle gas units)
		Primary control hidden in units operating with large dead zones ±200 mHz (Figure 9.6b) is automatically activated
		Central secondary controllers of subsystems are blocked. Generating units operate only with primary control and manual setting of reference power
50.000	<0.200	Normal operation with primary control with small dead zones (Figure 9.6a) and secondary control of frequency and the tie-line power: $ACE = -\Delta P_{\text{tie}} - \lambda_{\text{R}}\Delta f$

The defense plan described in these tables should be treated as an example. It includes updates resulting from experience gained from the wide area disturbance in UCTE interconnected power system on 4 November 2006. The interested reader is advised to check the UCTE Master Plan document at www.ucte.org.

9.1.7 Quality Assessment of Frequency Control

The quality assessment of frequency control can be divided into two types:

- Quality assessment of the control during normal system operation.
- Quality assessment during large disturbances such as unplanned outages of a generating unit.

Quality assessment during normal system operation is executed using the standard deviation of the frequency error

$$\sigma = \sqrt{\frac{1}{n}\sum_{i=1}^{n}(f_i - f_{\text{ref}})^2} \tag{9.13}$$

where n is the number of measurements. The measurements are taken every 15 minutes in a month. Additionally, a percentage share of frequency deviations higher than 50 mHz is calculated together with the times of their appearance.

Quality assessment of frequency control following a large disturbance is shown in Figure 9.13. The bold line shows an example of frequency changes following an unplanned outage of a large generating unit. There was a frequency error of Δf_0 just before the disturbance. The disturbance happened during a correction to the synchronous time when $f_{\text{ref}} = 50.01$ Hz. Following the disturbance, the frequency dropped by a maximum of Δf_2. Frequency control restored the frequency to the reference value within the allowed error. The whole range of frequency variations was limited to within an area shown by the dashed lines and referred to as the *trumpet characteristic*. The trumpet characteristic is made up of two exponential curves defined by

$$\begin{aligned} H(t) &= f_{\text{ref}} \pm A\,\mathrm{e}^{-t/T} \quad \text{for} \quad t \le 900\,\mathrm{s} \\ H(t) &= \pm 20\,\mathrm{mHz} \qquad \text{for} \quad t \ge 900\,\mathrm{s} \end{aligned} \tag{9.14}$$

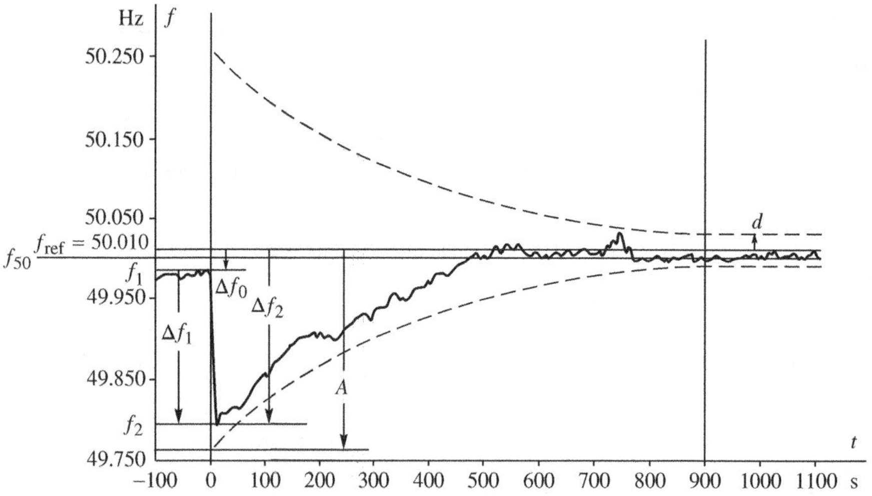

Figure 9.13 Illustration to the quality assessment of frequency control using a trumpet characteristic. Based on UCTE (2007b). *Source:* Reproduced by permission of UCTE.

where the signs ± correspond to the upper and lower parts of the characteristic and A is the initial width. The exponential characteristic finishes when $t = 900$ s that is $t = 15$ min. Then the characteristic consists of two horizontal lines limiting the frequency error to ±20 mHz corresponding to the required accuracy of frequency control during normal operation. The exponential curves must descend smoothly to the horizontal lines. For that to happen, the time constant of the exponential curves must be equal to

$$T = \frac{900}{\ln\left(\frac{A}{d}\right)} \tag{9.15}$$

The initial width A depends on the disturbance size ΔP_0 according to

$$A = 1.2\left(\frac{|\Delta P_0|}{\lambda_R} + 0,030\right) \tag{9.16}$$

where λ_R is the frequency stiffness of the EPS. As an example, if a unit of several hundred megawatts is tripped, the width of the trump characteristic, and therefore also the varying frequency error, is a few hundred millihertz.

If variations of frequency following a disturbance ΔP_0 are inside the trump characteristic then the frequency control is deemed to be satisfactory.

9.2 Stage I – Rotor Swings in the Generators

Having described the AGC, it is now possible to analyze the response of an EPS to a power imbalance caused, for example by the tripping of a generating unit. This response can be divided into four stages depending on the duration of the dynamics involved:

Stage I Rotor swings in the generators (first few seconds).
Stage II Frequency drop (a few seconds to several seconds).
Stage III Primary control by the turbine governing systems (several seconds).
Stage IV Secondary control by the central regulators (several seconds to a minute).

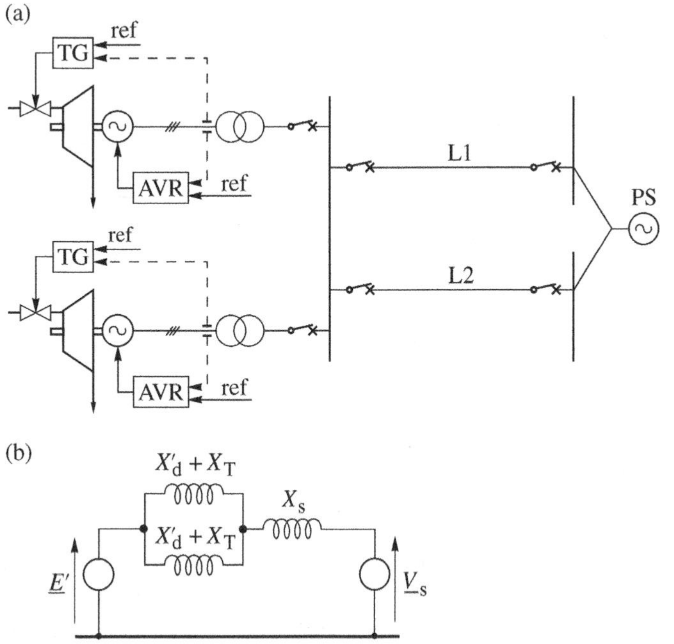

Figure 9.14 Parallel generating units operating onto an infinite busbar: (a) schematic diagram; (b) equivalent circuit.

The dynamics associated with each of these four stages will be described separately in order to illustrate how they develop in the system. To begin the discussion the EPS shown schematically in Figure 9.14a will be considered where a power station transmits its power to the system via two parallel transmission lines. The power station itself is assumed to consist of two identical generating units operating onto the same busbar. The disturbance considered will be the disconnection of one of the generating units. In Stage I of the disturbance the way in which the remaining generating unit contributes to the production of the lost power will be given special attention.

The sudden disconnection of one of the generators will initially produce large rotor swings in the remaining generating unit and much smaller rotor swings in the other generators within the system. For simplicity these small oscillations will be neglected to allow the rest of the system to be replaced by an infinite busbar. The timescale of the rotor swings is such that the generator transient model applies and the mechanical power supplied by the turbine remains constant. To simplify considerations, the classical model representation of the generator will be used (Eqs. [5.1] and [5.26]). Figure 9.14b shows the equivalent predisturbance circuit diagram for the system. As both generators are identical they can be represented by the same transient electromotive force (emf) $\underline{E}'$ behind an equivalent reactance that combines the generator transient reactance X'_d, the transformer reactance X_T, and the system reactance X_s.

Figure 9.15 shows how the equal area criterion can be applied to this problem. In this diagram $P_-(\delta')$ and $P_+(\delta')$ are the transient power-angle characteristics, and P_{m-} and P_{m+} the mechanical powers, before and after the disturbance occurs. Initially, the plant operates at point 1 and the equivalent power angle with respect to the infinite busbar is δ'_0. Disconnection of one of the generators has two effects. First of all, the equivalent reactance of the system increases so that the amplitude of the power-angle characteristic decreases. Consequently, the pre- and post-disturbance power-angle characteristics are

$$P_-\left(\delta'_0\right) = \frac{E'V_s}{\dfrac{X'_d + X_T}{2} + X_s}\sin\delta'_0; \quad P_+\left(\delta'_0\right) = \frac{E'V_s}{X'_d + X_T + X_s}\sin\delta'_0 \tag{9.17}$$

Secondly, the mechanical power delivered to the system drops by an amount equal to the power of the lost unit, that is $P_{m+} = 0.5P_{m-}$.

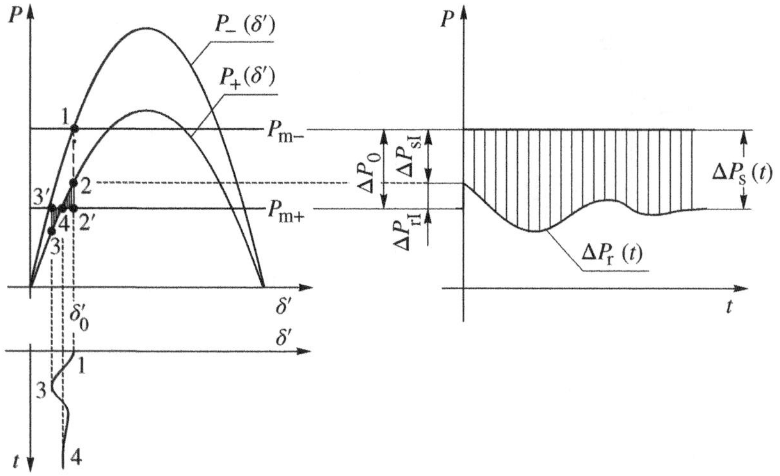

Figure 9.15 Application of the equal area criterion to determine the first stage of the dynamics. Duration of the phenomena: first few seconds.

As the rotor angle of the remaining generator cannot change immediately after the disturbance occurs, the electrical power of the generator is greater than the mechanical power delivered by its prime mover, point 2. The rotor is decelerated and loses kinetic energy corresponding to the area 2-2′-4. Owing to its momentum, the rotor continues to decrease its angle past point 4 until it stops at point 3 when the area 4-3-3′ equals the area 2-2′-4. The damping torques then damp out subsequent oscillations and the rotor tends toward its equilibrium point 4.

The amplitude of the rotor oscillations of any given generator will depend on the amount of lost generation it picks up immediately after the disturbance occurs. Using the notation of Figure 9.15 gives

$$\begin{aligned} \Delta P_0 &= P_-\left(\delta'_0\right) - P_{\mathrm{m}+} = P_{\mathrm{m}-} - P_{\mathrm{m}+} \\ \Delta P_{\mathrm{rI}} &= P_+\left(\delta'_0\right) - P_{\mathrm{m}+} \\ \Delta P_{\mathrm{sI}} &= \Delta P_0 - \Delta P_{\mathrm{rI}} \end{aligned} \tag{9.18}$$

where ΔP_0 is the lost generating power and ΔP_{rI} and ΔP_{sI} are the contributions of the generating units remaining in operation and of the system, respectively, in meeting the power imbalance ΔP_0 at the very beginning of the disturbance. The subscript I indicates that these equations apply to Stage I of the disturbance. Using the first of the equations in (9.18), a formula for ΔP_{rI} can be rewritten as

$$\Delta P_{\mathrm{rI}} = P_+\left(\delta'_0\right) - P_{\mathrm{m}+} = \left[P_+\left(\delta'_0\right) - P_{\mathrm{m}+}\right] \frac{1}{P_-\left(\delta'_0\right) - P_{\mathrm{m}+}} \Delta P_0 \tag{9.19}$$

Substituting Eqs. (9.17) into (9.19) and noting that $P_{\mathrm{m}+} = 0.5 P_{\mathrm{m}-}$ gives

$$\Delta P_{\mathrm{rI}} = \frac{1}{1+\beta} \Delta P_0 \tag{9.20}$$

where $\beta = \left(X'_{\mathrm{d}} + X_{\mathrm{T}}\right)/X_{\mathrm{s}}$. The amount that the system contributes in order to meet the lost generation can now be calculated as

$$\Delta P_{\mathrm{sI}} = \Delta P_0 - \Delta P_{\mathrm{rI}} = \frac{\beta}{\beta+1} \Delta P_0 \tag{9.21}$$

with the ratio of the contributions (9.20) and (9.21) being

$$\frac{\Delta P_{\mathrm{rI}}}{\Delta P_{\mathrm{sI}}} = \frac{1}{\beta} = \frac{X_{\mathrm{s}}}{X'_{\mathrm{d}} + X_{\mathrm{T}}} \tag{9.22}$$

This equation shows that ΔP_r, the contribution of the unit remaining in operation in meeting the lost power, is proportional to the system equivalent reactance X_s. Both contributions ΔP_r and ΔP_s are depicted in Figure 9.15. Owing to the fact that the inertia of the unit is much smaller than that of the EPS, the generator quickly decelerates, loses kinetic energy, and both its rotor angle and generated power decrease (Figure 9.15). As a consequence the power imbalance starts to increase and is met by the system which starts to increase its contribution ΔP_s. This power delivered by the system has been shaded in Figure 9.15 and, when added to the generator's share $P_r(t)$, must equal the load existing before the disturbance. As can be seen from this figure, the proportion of lost generation picked up by the system changes with time so that the system contributes more and more as time progresses.

Although the formula in Eq. (9.22) has been derived for two parallel generating units operating in the system, a similar expression can be derived for the general multi-machine case. This expression would lead to a conclusion that is similar to the two-machine case in that at the beginning of Stage I of the dynamics the share of any given generator in meeting the lost load will depend on its electrical distance from the disturbance (coefficient β). In the case of Figure 9.14 the imbalance appeared at the power plant busbar. Thus X_s is a measure of the electrical distance of the system from the disturbance and $\left(X'_d + X_T\right)$ is a measure of the electrical distance of the unit remaining in operation.

9.3 Stage II – Frequency Drop

The situation shown in Figure 9.15 can only last for a few seconds before the power imbalance causes all the generators in the system to slow down and the system frequency to drop. Thus begins Stage II of the dynamics. During this stage, the share of any one generator in meeting the power imbalance depends solely on its inertia and not on its electrical distance from the disturbance. Assuming that all the generators remain in synchronism, they will slow down at approximately the same rate after a few rotor swings in Stage I of the dynamics. This may be written as

$$\frac{d\Delta\omega_1}{dt} \cong \frac{d\Delta\omega_2}{dt} \cong \ldots \cong \frac{d\Delta\omega_{N_G}}{dt} \cong \varepsilon \tag{9.23}$$

where $\Delta\omega_i$ is the speed deviation of i-th generator, ε is the average acceleration, and N_G is the number of generators.

According to swing equation (5.1), the derivative of the speed deviation can be replaced by the ratio of the accelerating power ΔP_i of the i-th unit over the inertia coefficient M_i. This modifies Eq. (9.23) to

$$\frac{\Delta P_1}{M_1} \cong \frac{\Delta P_2}{M_2} \cong \ldots \cong \frac{\Delta P_n}{M_n} \cong \varepsilon \tag{9.24}$$

If the change in the system load due to the frequency change is neglected, the sum of the extra load taken by each generator must be equal to the power lost ΔP_0

$$\Delta P_0 = \sum_{k=1}^{N_G} \Delta P_k \tag{9.25}$$

Substituting the increment ΔP_i in this equation by $\Delta P_i = M_i\varepsilon$ gives

$$\Delta P_0 = \varepsilon \sum_{k=1}^{N_G} M_k \quad \text{or} \quad \varepsilon = \frac{d\Delta\omega}{dt} = \frac{\Delta P_0}{\sum_{k=1}^{N_G} M_k} \quad \text{so that} \quad \Delta P_i = M_i\varepsilon = \frac{M_i}{\sum_{k=1}^{N_G} M_k}\Delta P_0 \tag{9.26}$$

Taking into account Eq. (4.9) it is obtained

$$M_\Sigma = \sum_{k=1}^{N_G} M_k = \sum_{k=1}^{N_G} (2H_k S_{nk})/\omega_s = \sum_{k=1}^{N_G} J_k\omega_s \tag{9.27}$$

where M_Σ is the angular momentum of all masses rotating in the EPS, also referred to as the *inertia coefficient of the electric power system* (inertia coefficient of the EPS).

Equation (9.26) determines the contribution of the i-th generator in meeting the lost power in Stage II of the dynamics when each generator contributes an amount of power proportional to its inertia. In practice the mechanical starting time constants T_m (Section 4.1) are similar for all the generators, so that substituting for $M_k = T_{mk}S_{nk}/\omega_s$ from Eq. (4.10), then Eq. (9.26) can be written as

$$\Delta P_i \cong \frac{S_{ni}}{\sum_{k=1}^{N_G} S_{nk}} \Delta P_0 \tag{9.28}$$

During Stage II, the contribution of the generator remaining in operation and the rest of the EPS in meeting the lost power can be expressed, using Eq. (9.26), as

$$\Delta P_{rII} = \frac{M_r}{M_r + M_s}\Delta P_0 \quad \text{and} \quad \Delta P_{sII} = \frac{M_s}{M_r + M_s}\Delta P_0 \tag{9.29}$$

Subscript II has been added here to emphasize that these equations are valid during Stage II. In a similar way as in Eq. (9.22), the ratio of the contributions is

$$\frac{\Delta P_{rII}}{\Delta P_{sII}} = \frac{M_r}{M_s} \cong \frac{S_{nr}}{S_{ns}} \tag{9.30}$$

and is equal to the ratio of the inertia coefficients or, approximately, to the ratio of the rated powers. As $S_{ns} \gg S_{nr}$ (infinite busbar), the contribution by the generator remaining in operation in covering the lost power is very small. The assumption made when preparing Figure 9.15 was that $\Delta P_{rII} \cong \Delta P_0 S_{nr}/S_{ns} \cong 0$.

Figure 9.16a illustrates the case when the system has a finite equivalent inertia and the rotor angle of both the remaining generator and the system equivalent generator decreases as the frequency drops. The initial rotor angle oscillations are the same as those shown for the first stage of the dynamics in Figure 9.15. The angles then decrease together as the generators operate synchronously.

Figure 9.16b shows the case of a three-generator system where the load at node 4 suddenly increases. Generators 1 and 2 are close to the disturbance, so they participate more strongly in the first stage of the oscillations than does generator 3. During the second stage, the power angles synchronously decrease and the system frequency drops.

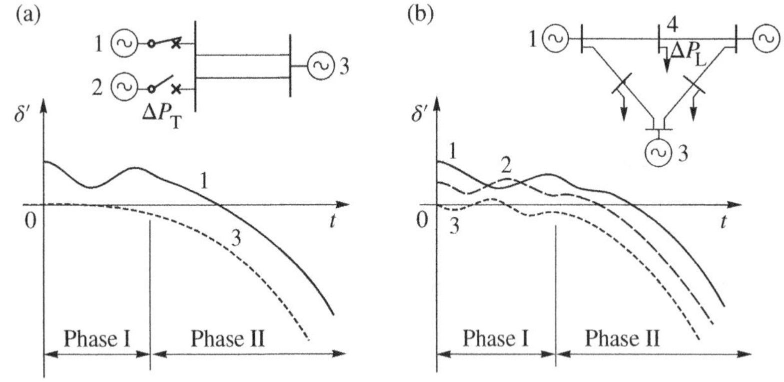

Figure 9.16 Examples of changes in the rotor angles in the case of an active power disturbance: (a) disconnection of generator 2; (b) sudden increase of the load at node 4. The duration of Stage II is a few to several seconds.

9.4 Stage III – Primary Control

Stage III of the dynamics depends on how the generating units and the loads react to the drop in frequency. Section 2.2.3 explains how, as frequency (speed) drops, the turbine governor opens the main control valves to increase the flow of working fluid through the turbine and so increase the turbine mechanical power output. In the steady state, and during very slow changes of frequency, the increase in mechanical power for each generating unit is inversely proportional to the droop of the static turbine characteristic. Equations (9.4) and (9.6) can be rewritten as

$$\begin{aligned} P_{\mathrm{T}} &= P_{\mathrm{T0}} + \Delta P_{\mathrm{T}} = P_{\mathrm{T0}} - K_{\mathrm{T}} \Delta f \frac{P_{\mathrm{T0}}}{f_{\mathrm{n}}} \\ P_{\mathrm{L}} &= P_{\mathrm{L0}} + \Delta P_{\mathrm{L}} = P_{\mathrm{L0}} + K_{\mathrm{L}} \Delta f \frac{P_{\mathrm{L0}}}{f_{\mathrm{n}}} \end{aligned} \tag{9.31}$$

The operating frequency of the system is determined by the point of intersection of these two characteristics.

Typical generation and load characteristics are shown in Figure 9.17, where the generation characteristic before the disturbance is denoted $P_{\mathrm{T-}}$ and after by $P_{\mathrm{T+}}$. If a generating unit is lost, the system generation characteristic P_{T} moves to the left by the value of the lost power ΔP_0 and, according to Eq. (9.4), the slope of the characteristic increases slightly as the value of K_{T} decreases.

Before the disturbance occurs, the system operates at point 1, corresponding to the intersection of the P_{L} and $P_{\mathrm{T-}}$ characteristics. After the generator is disconnected the frequency initially remains unchanged and the generation operating point is shifted to point 2. Then the generation tends to move toward point III, which corresponds to the intersection of the P_{L} and $P_{\mathrm{T+}}$ characteristics but, because of the delay introduced by the time constants of the turbines and their governors, this point cannot be reached immediately. Initially, the difference between the generated power (point 2) and the load power (point 1) is large and the frequency starts to drop, as described in Stage I and Stage II of the dynamics. In Stage III the turbine reacts to the drop in frequency by increasing its power output but, because of the aforementioned time delays in the turbine regulator system, the trajectory of the turbine power f (P_{T}) lies below the static generation characteristic $P_{\mathrm{T+}}$. As the frequency drops, the generated power increases while the power taken by the load decreases. The difference between the load and the generation is zero at point 3. According to the swing equation, a zero value of deceleration power means

$$\frac{\mathrm{d}\Delta\omega}{dt} = 2\pi \frac{\mathrm{d}\Delta f}{dt} = 0 \tag{9.32}$$

and the frequency $f(t)$ reaches a local minimum (Figure 9.17b).

Because of the inherent inertia of the turbine regulation process, the mechanical power continues to increases after point 3 so that the generated power exceeds the load power and the frequency starts to rise. At point 4 the balance of power is again zero and corresponds to the local maximum in the frequency shown in Figure 9.17b. The oscillations continue until the steady-state value of the frequency f_{III} is reached at point III corresponding to the intersection of the P_{L} and $P_{\mathrm{T+}}$ static characteristics. The value of f_{III} can be determined from Figure 9.17a

$$\Delta P_0 = \Delta P_{\mathrm{TIII}} - \Delta P_{\mathrm{LIII}} = -K_{\mathrm{T}} \frac{P_{\mathrm{L}}}{f_{\mathrm{n}}} \Delta f_{\mathrm{III}} - K_{\mathrm{L}} \frac{P_{\mathrm{L}}}{f_{\mathrm{n}}} \Delta f_{\mathrm{III}} = -P_{\mathrm{L}} (K_{\mathrm{T}} + K_{\mathrm{L}}) \frac{\Delta f_{\mathrm{III}}}{f_{\mathrm{n}}} \tag{9.33}$$

with the frequency error f_{III} being

$$\frac{\Delta f_{\mathrm{III}}}{f_{\mathrm{n}}} = \frac{-1}{K_{\mathrm{T}} + K_{\mathrm{L}}} \frac{\Delta P_0}{P_{\mathrm{L}}} \quad \text{or} \quad \frac{\Delta f_{\mathrm{III}}}{f_{\mathrm{n}}} = \frac{-1}{K_{\mathrm{f}}} \frac{\Delta P_0}{P_{\mathrm{L}}} \tag{9.34}$$

where the coefficient $K_{\mathrm{f}} = K_{\mathrm{T}} + K_{\mathrm{L}}$ is the system stiffness introduced in Eq. (9.7). Figure 9.17c shows how the generated power and the load power change during this period. The difference between the two is shaded and corresponds to the power that decelerates ($P_{\mathrm{T}} < P_{\mathrm{L}}$) or accelerates ($P_{\mathrm{T}} > P_{\mathrm{L}}$) the rotor.

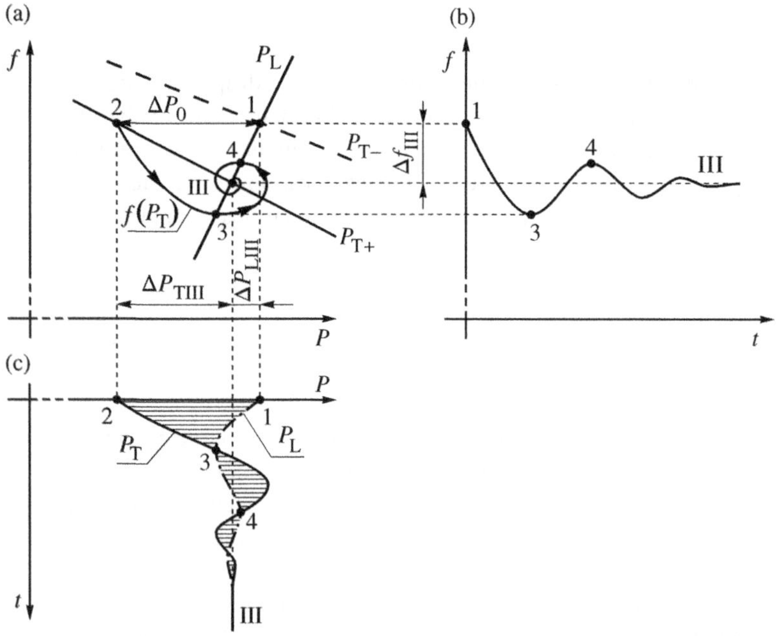

Figure 9.17 Stage III of the dynamics caused by an imbalance in real power: (a) the generation characteristic and the frequency response characteristic; (b) frequency changes; (c) power changes. The duration of Stage III is several seconds.

For the system shown in Figure 9.14 the contributions of the generator remaining in operation and that of the system in covering the lost generation in Stage III of the dynamics can be obtained from Eq. (9.4) as

$$\Delta P_{rIII} = -K_{Tr}\frac{\Delta f_{III}}{f_n}P_{nr};\quad \Delta P_{sIII} = -K_{Ts}\frac{\Delta f_{III}}{f_n}P_{ns} \tag{9.35}$$

where the subscript III is added to emphasize that these equations are valid during Stage III of the dynamics. The ratio of contributions resulting from these equations is

$$\frac{\Delta P_{rIII}}{\Delta P_{sIII}} = \frac{K_{Tr}P_{nr}}{K_{Ts}P_{ns}} \cong \frac{P_{nr}}{P_{ns}} \tag{9.36}$$

The approximate equality in this equation is valid when the droop of all the turbine characteristics is approximately the same. In practice the ratios (9.36) and (9.30) are very similar. This physically corresponds to the fact that, during the transition period between the second and third stages, the generator is in synchronism with the system and there are no mutual oscillations between them.

9.4.1 The Importance of the Spinning Reserve

The discussion so far has assumed that, within the range of frequency variations, each of the generating units has a linear turbine characteristic and that the overall generation characteristic is also linear. In practice each generating unit must operate within the limits that are placed on its thermal and mechanical performance. To ensure that these limits are adhered to its governing system is equipped with the necessary facilities to make certain that the unit does not exceed its maximum power limit or the limit placed on the speed at which it can take up power. These limits can have a substantial impact on how a unit behaves when the system frequency changes.

Figure 9.3 shows how the spinning reserve, and its allocation in the system, influences the shape of the generation characteristic. In order to help quantify the influence of spinning reserve, the following coefficients are defined

$$r = \frac{\sum_{i=1}^{N_G} P_{ni} - P_L}{P_L}; \qquad p = \frac{\sum_{i=1}^{R} P_{ni}}{\sum_{i=1}^{N_G} P_{ni}} \tag{9.37}$$

where $\sum_{i=1}^{N_G} P_{ni}$ is the sum of the power ratings of all the generating units connected to the system and $\sum_{i=1}^{R} P_{ni}$ is the sum of the power ratings of all the units operating on the linear part of their characteristics, that is loaded below their power limit. The coefficient r is the *spinning reserve coefficient* and defines the relative difference between the maximum power capacity of the system and the actual load.

A simple expression for the local droop of the generation characteristic can be obtained by assuming that the droop of all the units which are not fully loaded are approximately identical, that is $\rho_i = \rho$ and $K_i = K = 1/\rho$. For the units operating at their limits, $\rho_i = \infty$ and $K_i = 0$. Under these conditions

$$\begin{aligned} \Delta P_T &= -\sum_{i=1}^{N_G} K_i P_{ni} \frac{\Delta f}{f_n} = -\sum_{i=1}^{R} K_i P_{ni} \frac{\Delta f}{f_n} \cong -K \sum_{i=1}^{R} P_{ni} \frac{\Delta f}{f_n} \\ &= -Kp \sum_{i=1}^{N_G} P_{ni} \frac{\Delta f}{f_n} = -Kp(r+1) P_L \frac{\Delta f}{f_n} \end{aligned} \tag{9.38}$$

Dividing by P_L gives

$$\frac{\Delta P_T}{P_L} = -K_T \frac{\Delta f}{f_n} \tag{9.39}$$

where

$$K_T = p(r+1)K \quad \text{and} \quad \rho_T = \frac{\rho}{p(r+1)} \tag{9.40}$$

Equation (9.39) is similar to Eq. (9.4) and describes the linear approximation of the nonlinear generation characteristic for a given load. The local droop ρ_T increases as the spinning reserve decreases. At the limit, when the load P_L is equal to the system generating capacity, both the coefficients r and p are zero and $\rho_T = \infty$. This corresponds to all the generating units being fully loaded.

The influence of the spinning reserve on the frequency drop in Stage III of the dynamics can now be determined. For the linear approximation to the generator characteristic given in Eq. (9.39) the drop in frequency can be determined from Eq. (9.34) as

$$\frac{\Delta f_{III}}{f_n} = \frac{-1}{p(r+1)K + K_L} \frac{\Delta P_0}{P_L} \tag{9.41}$$

indicating that the smaller the spinning reserve coefficient r, the bigger the drop in frequency due to the loss of power ΔP_0. With a large spinning reserve the static characteristic $P_{T(1)}$ shown in Figure 9.18 has a shallow slope and the drop in frequency is small. On the other hand, when the spinning reserve is small, the $P_{T(2)}$ characteristic is steep, and the frequency drop is increased for the same power disturbance ΔP_0. In the extreme case when no spinning reserve is available, the generating units would be unable to increase their generation and the whole power imbalance ΔP_0 could only be covered by the frequency effect of the loads. As the frequency sensitivity K_L of the loads is generally small, the frequency drop would be very high.

Equation (9.38) defines the way in which each of the generating units contributes to the power imbalance at the end of Stage III, with Eq. (9.39) quantifying the net effect of the primary turbine control in meeting the lost load.

Example 9.1 Consider a 50 Hz system with a total load P_L= 10 000 MW in which $p = 60\%$ of the units give $r =$ 15% of the spinning reserve. The remaining 40% of the units are fully loaded. The average droop of the units with spinning reserve is $\rho = 7\%$ and the frequency sensitivity coefficients of the loads is $K_L = 1$. If the system suddenly loses a large generating unit of $\Delta P_0 = 500$ MW, calculate the frequency drop and the amount of power contributed by the primary control.

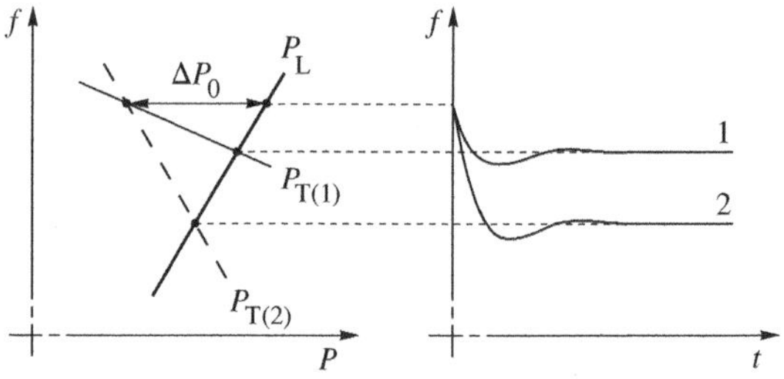

Figure 9.18 The influence of spinning reserve on the frequency drop due to a power imbalance: 1 – large spinning reserve; 2 – small spinning reserve.

From Eq. (9.40) $K_T = 0.6(1 + 0.15)1/0.07 = 9.687$ and, according to Eq. (9.34), the primary control will give the frequency drop in Stage III as

$$\Delta f_{III} = \frac{-1}{9.867 + 1} \times \left[\frac{500}{10\,000}\right] \times 50 \cong 0.23 \text{ Hz}$$

where the turbine governor primary control contributes

$$\Delta P_{T\,III} = 9.867 \times \frac{0.23}{50} \times 10\,000 = 454 \text{ MW}$$

with the remaining deficit of

$$\Delta P_{L\,III} = 1 \times \frac{0.23}{50} \times 10\,000 = 46 \text{ MW}$$

being covered by the frequency effect of the loads. If there were no spinning reserve, the deficit would be covered entirely by the frequency effect of the loads giving a frequency drop of

$$\Delta f_{III} = \frac{-1}{0 + 1} \times \left[\frac{500}{10\,000}\right] \times 50 = 2.5 \text{ Hz}$$

which is about 10 times greater than in the case with (only) 15% spinning reserve.

******

9.4.2 Frequency Collapse

Spinning reserve is much more important than is suggested by the approximate formula in Eq. (9.41). In practice the power output of the turbine is frequency dependent so that if the frequency is much lower than the nominal frequency the turbine power will be less than that given by Eq. (9.39) and the frequency may drop further and further until the system suffers a *frequency collapse.*

The hidden assumption made in the generation characteristics of Figure 9.18 is that when a unit is fully loaded the mechanical driving power delivered by the turbine does not depend on the frequency deviation. This assumption is only true for small frequency deviations, while a larger frequency drop reduces mechanical power due to deterioration in performance of boiler feed pumps (Section 2.3.3). In this case the static generation characteristic takes the form shown in Figure 2.15. Adding the power-frequency characteristics of individual generating units now gives the system generation characteristic P_T shown in Figure 9.19.

On the upper part of the generation characteristic the local droop at each point is positive and depends on the system load according to the formulae given in (9.40). The lower part of the characteristic corresponds to the decrease in the power output due to the deterioration of the performance of the boiler feed pumps. For a given load frequency characteristic P_L the inflection in the generation characteristic produces two equilibrium points. At the upper point s any small disturbance that produces an increase in the frequency $\Delta f > 0$ will result in the load power exceeding the generation power; the generators are decelerated, and the system returns to the equilibrium point s. Similarly, any small disturbance that causes a decrease in frequency will result in the generation power exceeding the load power, the rotors accelerate, the frequency increases, and the system again returns to the equilibrium point. Point s is therefore *locally stable* because for any small disturbance within the vicinity of this point the system returns to point s. The region in which this condition holds is referred to as the *area of attraction*.

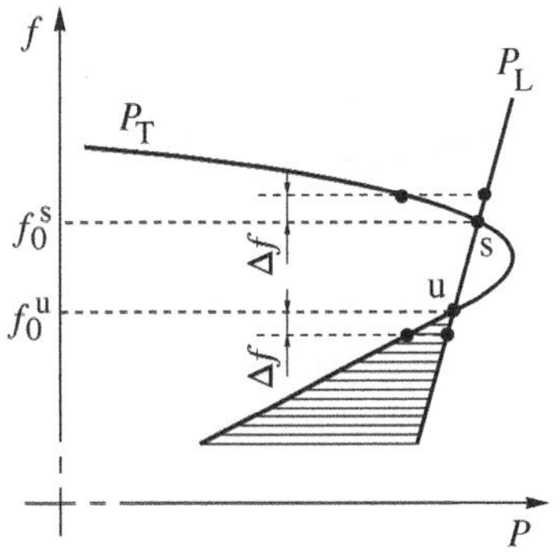

Figure 9.19 The static generation and load characteristics of the EPS and the equilibrium points: s – locally stable equilibrium point; u – locally unstable equilibrium point.

In contrast the lower point u is *locally unstable* because any disturbance within the vicinity of this point will result in the system moving away from the equilibrium point. The region in which this happens is referred to as the *area of repulsion*. For example, if a disturbance reduces the frequency $\Delta f < 0$ then the system moves into the shaded area below point u, the generated power is reduced so that the system load is greater than the system generation, the rotors decelerate, and the frequency drops further.

With these qualifications of turbine performance in mind, assume that the system operates with a low spinning reserve at point 1 on Figure 9.20. A loss ΔP_0 in generation now occurs and the operating point moves from point 1 to point 2. The excess load over generation is large and produces an initial rapid drop in frequency. As the difference between the load and generation reduces, the rate at which the frequency drops slows down and the system generation trajectory $f(P_T)$ approaches the equilibrium point u. When the trajectory $f(P_T)$ enters the area of repulsion of point u, it is forced away and the system suffers a frequency collapse.

The frequency collapse illustrated in Figure 9.20 occurred following the *sudden* appearance of the power imbalance ΔP_0. Had the system generation characteristic changed slowly from P_{T+} to P_{T-} by gradually reducing the generation then it would have met the locally stable point s and stayed there. Obviously, if the power imbalance ΔP_0 is greater than the spinning reserve then the post-disturbance demand characteristic P_L lies to the right of the nose of the generation characteristic P_{T+} and the frequency would collapse regardless of how sudden the power change was.

9.4.3 Underfrequency Load Shedding

Many systems can be protected from frequency collapse by importing large blocks of power from neighboring systems to make up for the lost generation. However, in an islanded system, or in an interconnected system with a shortage of tie-line capacity, this will not be possible and the only way to prevent a frequency collapse following a large disturbance is to employ *automatic load shedding*.

Automatic load shedding is implemented using underfrequency relays. These relays detect the onset of decay in the system frequency and shed appropriate amounts of system load until the generation and load are once again in balance and the EPS can return to its normal operating frequency. Load shedding relays are normally installed in distribution and subtransmission substations as it is from here that the feeder loads can be controlled.

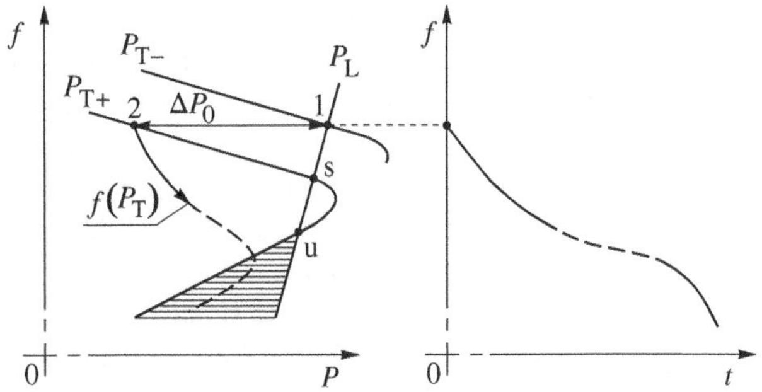

Figure 9.20 An example of frequency collapse.

As load shedding is a somewhat drastic control measure, it is usually implemented in stages with

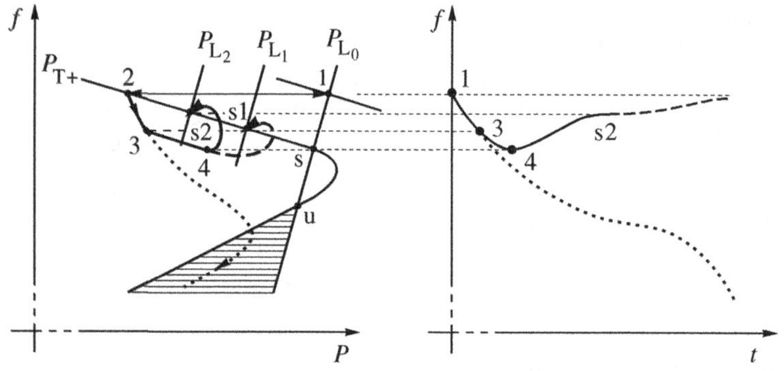

Figure 9.21 Two-stage load shedding to protect against frequency collapse.

each stage triggered at a different frequency level to allow the least important loads to be shed first. Figure 9.21 shows the effect of load shedding when a power imbalance ΔP_0 appears on a system with a low spinning reserve. Without load shedding the system would suffer a frequency collapse, as shown by the dotted line in Figure 9.21. The first stage of load shedding is activated at point 3 and limits the load to a value corresponding to characteristic P_{L_1}. The rate of frequency drop is now much slower as the difference between the load and generation is smaller. At point 4 the second stage of load shedding is triggered, further reducing the load to a value corresponding to characteristic P_{L_2}. Generation is now higher than the load, the frequency increases, and the system trajectory tends toward point s2 where the P_{L_2} and P_{T+} characteristics intersect.

Besides protecting against frequency collapse, load shedding may also be used to prevent deep drops in system frequency.

9.5 Stage IV – Secondary Control

In Stage IV of the dynamics, the drop in system frequency and the deviation in the tie-line power flows activate the central AGC, the basic operation of which is described in Section 9.1.

9.5.1 Islanded System

In an islanded system there are no tie-line connections to neighboring systems, and so the central regulator controls only the system frequency. As the frequency drops, the central regulator transmits control signals to the participating generating units to force them to increase their power output. This lifts the system generation characteristic on the (f, P) plane.

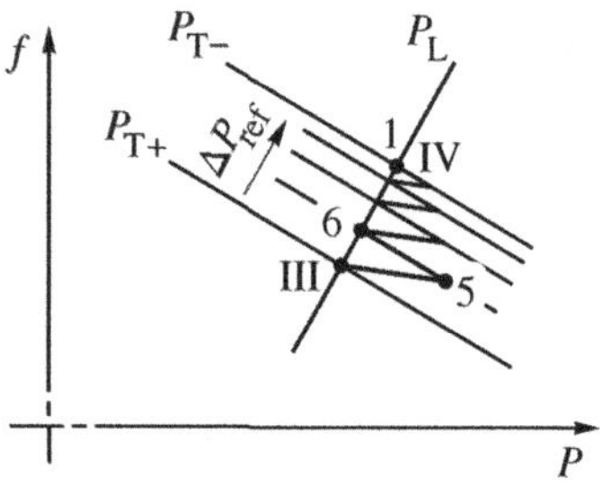

Figure 9.22 Stepped shifting of the generation characteristic by a slow-acting central regulator.

Figure 9.22 illustrates the action of a very slow-acting central regulator which transmits its first control signal at the end of Stage III corresponding to point III. This first control signal shifts the generation characteristic upwards a small amount so that at a given frequency, point 5, there is an excess of generation overload. The generators start to accelerate and the frequency increases until point 6 is reached. The central regulator now sends a further signal to increase power output and the generation characteristic is shifted further up. After a few such steps point IV is reached at which the frequency returns to its reference value and the central regulator ceases operating.

Although the zigzag line in Figure 9.22 is only a rough approximation to the actual trajectory, it illustrates the interaction between the secondary control action of the central regulator, which shifts the generation characteristic upwards, and the primary control of the turbine governing systems, which moves the operating point along the static generation characteristic. In an active power system the inertia within the power regulation process ensures smooth changes in the power around the zigzag line to produce the type of response shown in Figure 9.23. Stages I, II, and III of the frequency change are as described in the previous sections and, at the end of Stage III, the trajectory tends to wrap itself around the temporary equilibrium point III (Figure 9.23a) but does not settle at this point. The AGC of Stage IV now comes into operation and the trajectory tends toward the new equilibrium point IV. Although the difference between the frequency at points III and IV is small, the central regulator acts slowly so that correction of the frequency drop during Stage IV of the dynamics may take a long time, as shown by the broken curves in Figure 9.23b and c.

If the central frequency control acts faster than as shown in Figure 9.23, it will come into operation before the end of Stage III. The trajectory $f(P_T)$ does not now wrap around point III and the frequency starts to increase earlier as shown by curve 2 in Figure 9.24.

The way in which the frequency is returned to its nominal value depends on the dynamics of the central regulator shown in Figure 9.10 and defined by Eq. (9.12). The regulator dynamics consist of proportional and integral action, both of which increase the regulator output signal ΔP_{ref} as the frequency drops. The amount of integral

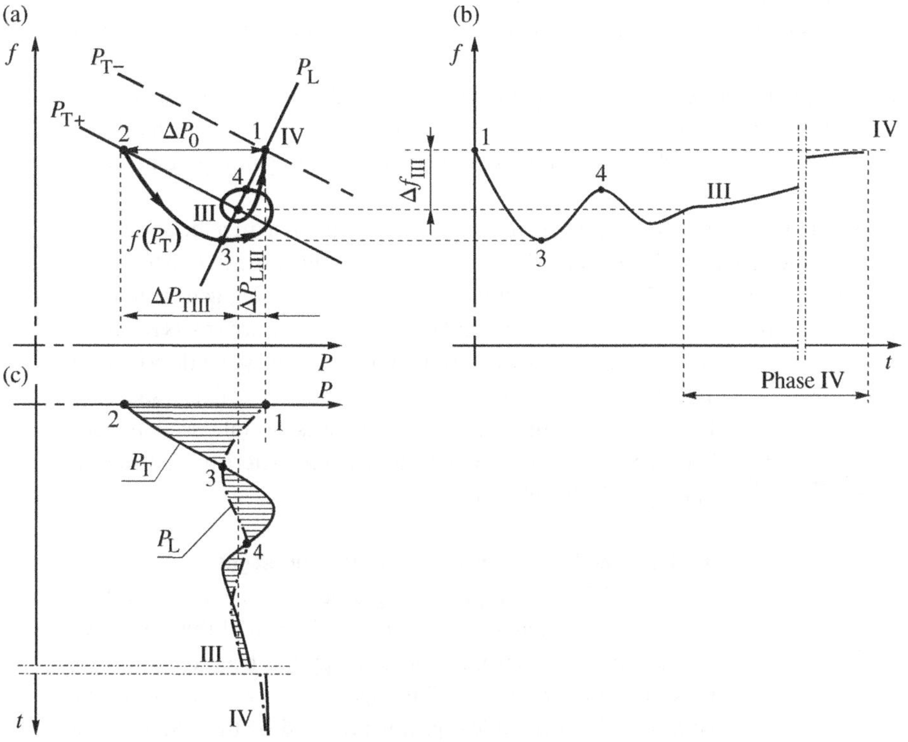

Figure 9.23 Stage III and IV of the frequency variations produced by a disturbance in the balance of real power: (a) generation and load characteristics and the system trajectory; (b) changes in frequency; (c) changes in power. The time duration of Stage IV is several seconds to a minute.

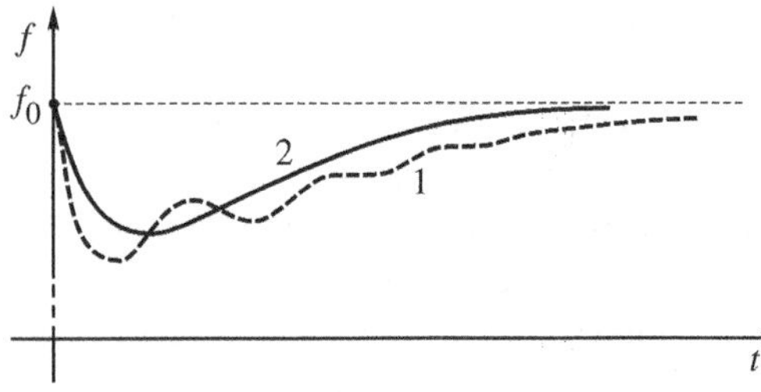

Figure 9.24 Examples of changes in frequency: 1 – with a slow central regulator; 2 – with a fast central regulator.

action is determined by the integral time constant T_R, while the proportional action is dependent on the coefficient β_R. Careful selection of these two coefficients ensures that the frequency returns smoothly to its reference value, as shown in Figure 9.24. Particularly problematic is a small integral time constant, because the smaller is this time constant, the faster the regulation signal increases. This can be compensated to some extent by the coefficient β_R, which will start to decrease the signal ΔP_{ref} as soon as the frequency starts to rise. In the extreme case of small values of β_R and T_R the response is underdamped and the reference frequency value will be reached in an oscillatory way.

Section 9.1 explains how not all the generating units necessarily participate in AGC. This means that only a part of the spinning reserve can be activated by the central regulator during secondary control. That part of the spinning reserve that belongs to the generators participating in the secondary control is referred to as the *available regulation power*.

If the available regulation power is less than the lost power, Stage IV of the dynamics will terminate when all the available regulation power has been used. This corresponds to the system trajectory settling at an equilibrium point somewhere between points III and IV on Figure 9.23 at a frequency that is lower than the reference value. The system operators may now verbally instruct other generating stations not on central control to increase their power output to help remove the frequency offset. In the case of a large power deficit, further action would consist of connecting new generating units from the cold reserve which, when connected, will release the capacity controlled by the central regulator. The frequency changes associated with this process are very slow and are not considered here.

Equation (9.36) defines the contribution of each of the generating units in covering the power imbalance at the end of Stage III. In Stage IV the increase in generation is enforced by the central regulator and the contribution of each unit, and the ratio $\Delta P_{rIV}/\Delta P_{sIV}$, will depend on whether or not the particular generating unit participates in central control.

Provided that the spinning reserve and the available regulation power are large enough, then in many cases they can prevent the system suffering a frequency collapse. Figure 9.25 shows an example of how the frequency variations depend on the value of the spinning reserve coefficient. The disturbance consists of losing generation ΔP_0 equivalent to 10% of the total load power P_L. In the first two cases $r \geq 14\%$ and the frequency returns to its reference value thanks to the operation of primary and secondary control. The third case corresponds to the frequency collapse shown in Figure 9.20. In the fourth case there is no intersection point between the generation and the load characteristic and the frequency quickly collapses.

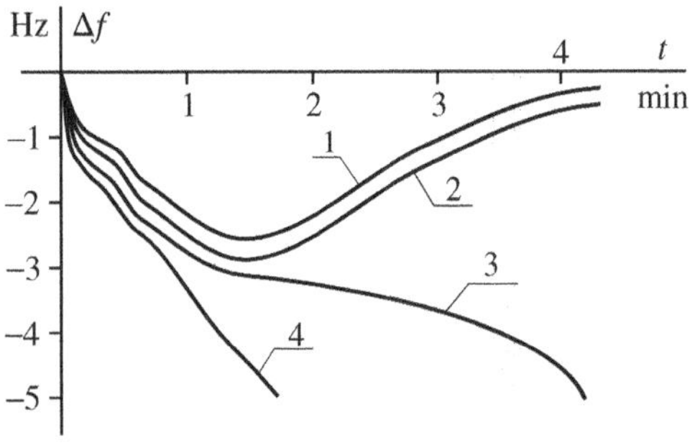

Figure 9.25 An example of the frequency variations when $\Delta P_0 = 10\ \%\ P_L$. Curve 1 corresponds to $r = 16\%$; 2 corresponds to $r = 14\%$; 3 corresponds to $r = 12\%$; and 4 corresponds to $r = 8\%$.

9.5.1.1 The Energy Balance over the Four Stages

When an EPS loses a generating unit it loses a source of both electrical and mechanical energy. By the end of the power system dynamic response, this lost energy has been recovered, as illustrated in Figure 9.26. The upper bold curve shows the variations of the mechanical power provided by the system while the lower bold curve shows the variations of electrical power of the loads due to frequency variations. All these variations are similar to those shown in Figure 9.23c. Initially the energy

Figure 9.26 Share of the individual components in covering the power imbalance: 1 – rotating masses of the generating units; 2 – rotating masses of the loads; 3 – primary control; 4 – secondary control. *Source:* Based on Welfonder (1980). Reproduced by permission of E. Welfonder.

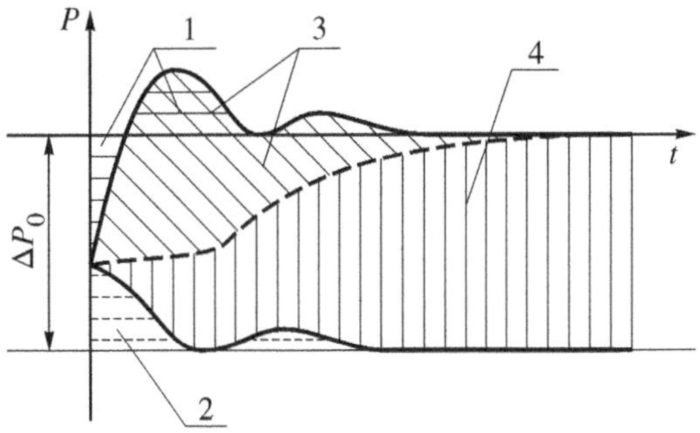

shortfall is produced by converting the kinetic energy of the rotating masses of the generating units and the loads to electrical energy, area 1 and area 2. This reduction in kinetic energy causes a drop in frequency which activates the turbine governor primary control so that the mechanical energy supplied to the system is increased but at a lower frequency, area 3. This energy is used partially to generate the required energy shortfall, and partially to return the kinetic energy borrowed from the rotating masses. Secondary control then further increases the mechanical energy (area 4), which is used to generate the required additional electrical energy and to increase the kinetic energy of the rotating masses so restoring the system frequency.

9.5.2 Interconnected Systems and Tie-line Oscillations

This discussion will be limited to the interconnected system shown in Figure 9.27 consisting of two subsystems of disproportionate size, system A and system B, referred to as the *big system* and the small system respectively. The tie-line interchange power P_{tie} will be assumed to flow from the big system to the small system and an imbalance of power ΔP_0 assumed to arise in the small system. During the first three stages of the dynamics, the influence of the central regulators in both system A and system B may be neglected.

Both systems are replaced by an equivalent generator, as in Figure 9.14b, to obtain the equivalent circuit of Figure 9.28a. In this circuit the reactance X combines the reactances of the tie-line connecting the two systems, the equivalent network reactance of both systems and the transient reactances of all the generators. As one of the subsystems is large compared with the other, the generator-infinite busbar model and the equal area criterion can be used to analyze Stage I of the dynamics.

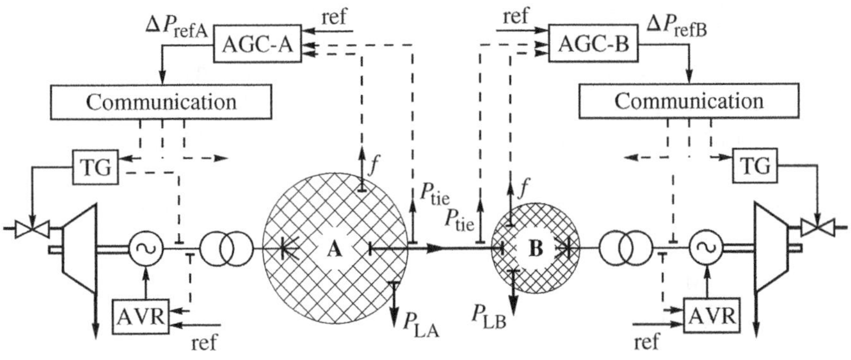

Figure 9.27 The functional diagram of an interconnected system consisting of two subsystems: a big system A and a small system B.

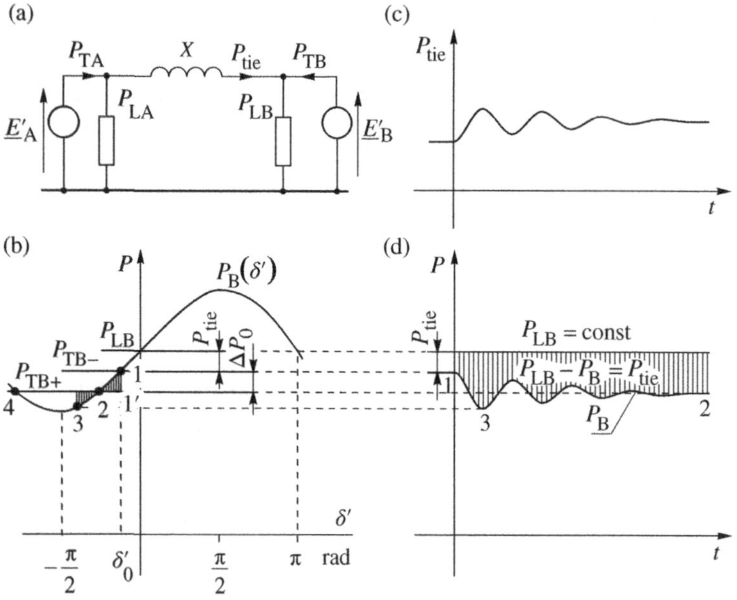

Figure 9.28 Application of the equal area criterion to determine the tie-line power during the first stage of the dynamics: (a) equivalent circuit; (b) the power-angle characteristic taking into account equivalent loads; (c) changes in the tie-line power; (d) changes in the generated power. The time duration of the phenomena is a few seconds.

The power-angle characteristic $P_B(\delta')$ of the small system is a sinusoid, corresponding to a power transfer between the systems, shifted by a constant value corresponding to the power demand in the small system P_{LB}. It is assumed that system B imports power from A so that $P_{TB} < P_{LB}$ and its power angle is negative with respect to system A. Before the disturbance occurs, the system operates at point 1 corresponding to the intersection of the $P_B(\delta')$ characteristic and the horizontal line representing the mechanical power P_{TB-}. The small system now loses generation equal to ΔP_0 and the power generated in this system drops to P_{TB+}. The electrical power exceeds the mechanical power and the rotors of the generators in system B slow down. The equivalent system rotor moves from point 1 to point 2 and then to point 3 in Figure 9.28b. Along this motion the electrical power generated in system B decreases and, as a result, additional power starts to flow from system A (Figure 9.28d). The maximum change in the value of the instantaneous power in the tie-lines increases to almost double the value of the lost power, the difference between points 1 and 3. Oscillations in the power begin during which kinetic energy in both systems is used to cover the lost generation. The angular velocities of the generator rotors drop and the system enters Stage II of the dynamics.

Stage I of the dynamics is dangerous from a stability point of view in that if area 1-1′-2 is greater than area 2-3-4 then the system loses stability and the small system will operate asynchronously with respect to the big system with $\omega_B < \omega_A$. Such a situation may occur when the tie-line capacity is small and the disturbance in power balance large. During asynchronous operation, large oscillations in the tie-line power will occur with the difference between the maximum and the minimum value being twice the amplitude of the sinusoid shown in Figure 9.28b. As the possibility of resynchronizing the systems is small, they would usually be disconnected to avoid damaging equipment in either system.

In the Stage II the power imbalance is covered in proportion to the inertia coefficients of the equivalent systems, as determined by Eq. (9.29), that is

$$\Delta P_{AII} = \frac{M_A}{M_A + M_B}\Delta P_0 \quad \text{and} \quad \Delta P_{BII} = \frac{M_B}{M_A + M_B}\Delta P_0 \tag{9.42}$$

where $M_A = \sum_{i=1}^{N_{GA}} M_i$ and $M_B = \sum_{i=1}^{N_{GB}} M_i$ are the sums of the inertia coefficients of the rotating masses in each systems. Equations (9.42) suggest that the big system A will almost entirely cover the power imbalance as $M_A \gg M_B$, so in Stage II of the frequency variations $\Delta P_{\text{tieII}} \approx \Delta P_0$ and the tie-lines will be additionally loaded with a value of the power lost in system *B*.

At the end of Stage III the frequency drops by a value Δf_{III} that can be calculated from the system stiffness K_{f} defined in Eq. (9.34). $\Delta P_0$is given by

$$\Delta P_0 = (\Delta P_{\text{TAIII}} - \Delta P_{\text{LAIII}}) + (\Delta P_{\text{TBIII}} - \Delta P_{\text{LBIII}}) \tag{9.43}$$

and, when substituting the values for K_{TA}, K_{LA}, K_{TB}, and K_{LB} obtained from Eqs. (9.40) and (9.6), a similar formula to that in Eq. (9.34) for the frequency drop in Stage III is obtained

$$\frac{\Delta f_{\text{III}}}{f_{\text{n}}} = \frac{-1}{K_{\text{fA}} P_{\text{LA}} + K_{\text{fB}} P_{\text{LB}}} \Delta P_0 \tag{9.44}$$

where $K_{\text{fA}} = K_{\text{TA}} + K_{\text{LA}}$ and $K_{\text{fB}} = K_{\text{TB}} + K_{\text{LB}}$ are the stiffness of the big and the small system respectively. The increase in the tie-line interchange can be determined from a power balance on one of the systems, in this case the big system A, when

$$\begin{aligned} \Delta P_{\text{tieIII}} &= \Delta P_{\text{TAIII}} - \Delta P_{\text{LAIII}} = -(K_{\text{TA}} + K_{\text{LA}}) P_{\text{LA}} \frac{\Delta f_{\text{III}}}{f_{\text{n}}} \\ &= -K_{\text{fA}} P_{\text{LA}} \frac{\Delta f_{\text{III}}}{f_{\text{n}}} = \frac{K_{\text{fA}} P_{\text{LA}}}{K_{\text{fA}} P_{\text{LA}} + K_{\text{fB}} P_{\text{LB}}} \Delta P_0 \end{aligned} \tag{9.45}$$

Under the assumption that $P_{\text{LA}} \gg P_{\text{LB}}$ this formula simplifies to $\Delta P_{\text{tieIII}} \approx \Delta P_0$ when, during Stage III, the tie-line power interchange is increased by the value of the lost power, as during Stage II. Such an increase in the power flow may result in system instability when the interconnected system split into two asynchronously operating subsystems. Also, with such a large power transfer the thermal limit on the line may be exceeded when the overcurrent relays trip the line, again leading to asynchronous operation of both systems.

Assuming that the tie-line remains intact, the increase in the tie-line interchange power, combined with the drop of frequency, will force the central regulators in both systems to intervene. The *ACE* at the end of Stage III can be calculated from Eq. (9.11) using the formulae given in Eqs. (9.44) and (9.45) as

$$ACE_{\text{A}} = -\Delta P_{\text{tieIII}} - \lambda_{\text{RA}} \Delta f_{\text{III}} \quad \text{and} \quad ACE_{\text{B}} = +\Delta P_{\text{tieIII}} - \lambda_{\text{RB}} \Delta f_{\text{III}} \tag{9.46}$$

As explained in Section 9.1, Eq. (9.10), the ideal regulator bias settings are

$$\lambda_{\text{RA}} = K_{\text{fA}} \frac{P_{\text{LA}}}{f_{\text{n}}} \quad \text{and} \quad \lambda_{\text{RB}} = K_{\text{fB}} \frac{P_{\text{LB}}}{f_{\text{n}}} \tag{9.47}$$

but this is difficult to achieve in practice as the values of the stiffness K_{fA} and K_{fB} can only be estimated. Consequently, assuming that K_{RA} and K_{RB} are estimates of K_{fA} and K_{fB} respectively, then the bias settings are

$$\lambda_{\text{RA}} = K_{\text{RA}} \frac{P_{\text{LA}}}{f_{\text{n}}} \quad \text{and} \quad \lambda_{\text{RB}} = K_{\text{RB}} \frac{P_{\text{LB}}}{f_{\text{n}}} \tag{9.48}$$

and Eq. (9.46) become

$$ACE_{\text{A}} = -\Delta P_{\text{tieIII}} - \lambda_{\text{RA}} \Delta f_{\text{III}} = \frac{-K_{\text{fA}} P_{\text{LA}} + K_{\text{RA}} P_{\text{LA}}}{K_{\text{fA}} P_{\text{LA}} + K_{\text{fB}} P_{\text{LB}}} \Delta P_0 \tag{9.49}$$

$$ACE_{\text{B}} = +\Delta P_{\text{tieIII}} - \lambda_{\text{RB}} \Delta f_{\text{III}} = \frac{K_{\text{fA}} P_{\text{LA}} + K_{\text{RB}} P_{\text{LB}}}{K_{\text{fA}} P_{\text{LA}} + K_{\text{fB}} P_{\text{LB}}} \Delta P_0 \tag{9.50}$$

9.5.2.1 Ideal Settings of the Regulators

Assume that the stiffness K_{fA} and K_{fB} of both systems are known and that the central regulator bias settings λ_{RA} and λ_{RB} are selected such that

$$K_{\text{RA}} = K_{\text{fA}}; \quad K_{\text{RB}} = K_{\text{fB}} \tag{9.51}$$

In this case, Eqs. (9.49) and (9.50) give

$$\text{ACE}_A = 0 \quad \text{and} \quad \text{ACE}_B = \Delta P_0 \tag{9.52}$$

and the central regulator of the big system (system A) will not demand an increase in the power generation in its area. Only the small system will increase its generation in order to cover its own power imbalance. If the available regulation power in the small system is high enough to cover the lost generation then the big system will not intervene at all and, as the generation in the small system is increased, the tie-line power interchange will drop to its scheduled value.

The value of the system stiffness $K_f = K_T + K_L$ is never constant because it depends on the composition of the system load and generation and on the spinning reserve. Consequently, the condition defined by Eq. (9.51) is almost never satisfied and the central regulator of the bigger system will take part in the secondary control to an amount defined by Eq. (9.49).

Example 9.2 An interconnected system consists of two subsystems of different size. The data of the subsystems are: $f_n = 50$ Hz, $P_{LA} = 37500$ MW, $K_{TA} = 8$ ($\rho_{TA} = 0.125$), $K_{LA} \cong 0$, $K_{RA} = K_{TA}$, $P_{LB} = 4000$ MW, $K_{TB} = 10$ ($\rho_{TB} = 0.1$), $K_{LB} \cong 0$ and $K_{RB} = K_{TB}$.

Two large generating units are suddenly lost in the smaller system producing a power deficit of $\Delta P_0 = 1300$ MW, that is 32.5% of the total generation in this subsystem. The resulting frequency and power variations for both subsystems, assuming no limits on the power generation (linear generation characteristics), are shown in Figure 9.29.

The characteristic P_{TA} has a small slope corresponding to $K_{TA}P_{LA}/f_n = 6000$ MW/Hz. The slope of the characteristic of the smaller system is much higher and corresponds to $K_{TB}P_{LB}/f_n = 800$ MW/Hz. The characteristic P_{TB+} is inverted and shifted by $\Delta P_0 = 1300$ MW, so that the intersection of the characteristics P_{TB+} and P_{TA} determines the frequency deviation Δf_{III} at point III. The central regulator of the small system acts fast enough for the P_{TB} characteristic to move so that point III is not reached. Further action due to the integral term in the regulator shifts the characteristic P_{TB} slowly to the position $P_{TB\infty}$ corresponding to nominal frequency.

******

It is worth noting the small frequency oscillations characteristic of the third stage of the dynamics. Because of the fast-acting central regulator, these oscillations take place above the value Δf_{III} (Figure 9.29b). The tie-line power flow initially increases very quickly in Stage I of the dynamics, and then oscillates around the scheduled value increased by the value of the lost generation. The period of oscillations is around 3 s. Then, as the central regulator in the small system enforces an increase in the generation, the tie-line flow slowly decreases to its scheduled value.

The central PI regulator quite quickly creates the output signal ΔP_{refB}, as shown in Figure 9.29c. The generating units respond to this increased power demand by increasing the generation by ΔP_{TB} (Figure 9.29d). Because of the primary turbine control the mechanical power of the big system P_{TA} initially increases to cover the power imbalance and then quickly drops to the instantaneous equilibrium point corresponding to the end of Stage III. After that P_{TA} slowly decreases as the generation P_{TB}, forced by secondary control, increases. At the same time the frequency increases so that the frequency error of the central regulator of the big system becomes negligibly small and the regulator becomes inactive.

9.5.2.2 Nonideal Settings of the Central Regulator

If $K_{RA} > K_{fA}$ then the signal $\lambda_{RA}\Delta f$ is initially larger than the signal ΔP_{tie} and the regulator tries to increase the generation in system A. Although the increased generation speeds up the rate at which the frequency increases, it slows down the rate at which the tie-line error ΔP_{tie} decreases and in some cases can lead to a temporary increase in this error. Consequently, the signal ΔP_{tie} becomes greater than the signal $\lambda_{RA}\Delta f$ and the central regulator starts to reduce the generation in system A, so essentially withdrawing this subsystem from secondary control.

If $K_{RA} < K_{fA}$ then the signal ΔP_{tie} is initially higher than the signal $\lambda_{RA}\Delta f$ and the regulator of the big system tries to reduce its generation despite the fact that the frequency is smaller than nominal. This drop in generation is not desirable from the frequency regulation point of view but does reduce the tie-line flows. When $\lambda_{RA}\Delta f$ becomes

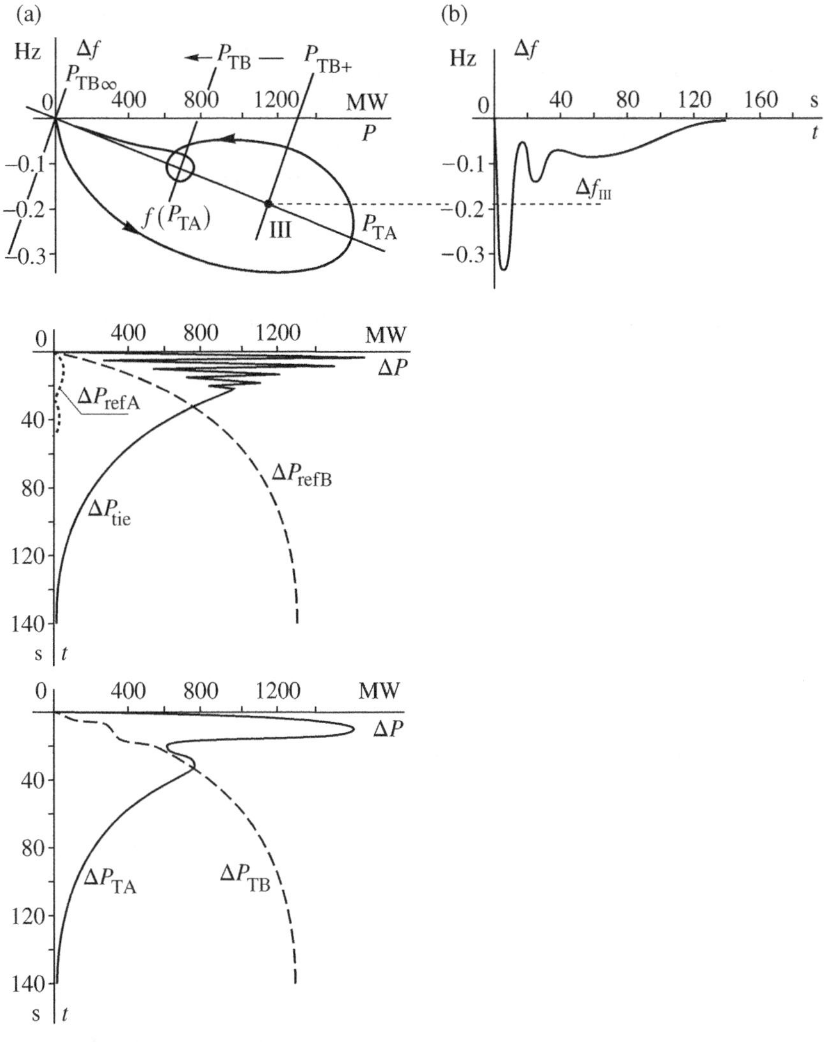

Figure 9.29 Illustration for Example 9.2: (a) the static characteristics of both systems and the trajectory $f(P_{TA})$; (b) frequency variations; (c) tie-line power interchange oscillations and the change of power demanded by the central regulators of both subsystems; (d) change in the mechanical power supplied by the turbines.

larger than ΔP_{tie}, the regulator will start to increase the recently reduced generation so as to re-establish the required power in system A.

In both the above cases the inaccuracy of the frequency bias setting causes unnecessary intervention of the bigger system in covering the generation loss in the smaller system. In the above example a large amount of regulation power was available so that use of the nonideal regulator settings was not dangerous. However, should the amount of regulation power available be insufficient to cover the lost power then the consequence of nonideal regulator settings becomes more significant, as described below.

9.5.2.3 Insufficient Available Regulation Power

When the available regulation power ΔP_{regB} in the small system is less than the generation loss ΔP_0 then system B is unable to cover the power loss on its own and the big system A must intervene to cover part of the lost power. Initially, the dynamics are identical to the case in which the regulation power in the small system was unlimited.

The difference between the two cases appears when the small system has used up all its available regulation power. Further changes can only take place via regulation of the big system. Its central regulator is now subject to two error signals of opposite sign (Figure 9.12). The signal $\lambda_{RA}\Delta f$ generated by the frequency error demands an increase in the generation, while the signal ΔP_{tie} generated by the tie-line power deviation demands a decrease in the generation. The regulation process terminates when the signals balance each other and the total error signal is zero. Denoting the final steady-state values of the error signals as Δf_{∞} and $\Delta P_{tie\infty}$, the regulation equation is

$$\mathrm{ACE_A} = -\Delta P_{\mathrm{tie}\,\infty} - K_{\mathrm{RA}} P_{\mathrm{LA}} \frac{\Delta f_{\infty}}{f_{\mathrm{n}}} = 0 \tag{9.53}$$

On the other hand, the tie-line power interchange must satisfy the overall power balance of the small system

$$\Delta P_0 - \Delta P_{\mathrm{regB}} = \Delta P_{\mathrm{tie}\,\infty} - (K_{\mathrm{TB}} + K_{\mathrm{LB}}) P_{\mathrm{LB}} \frac{\Delta f_{\infty}}{f_{\mathrm{n}}} \tag{9.54}$$

Physically, this means that the power imbalance in the small system B may be covered partly by an increase in the power imported from the big system A, partly by a change in internal generation and partly by a decreased demand resulting from the frequency drop in the whole interconnected system. Solving Eqs. (9.53) and (9.54) gives

$$\Delta P_{\mathrm{tie}\,\infty} = \frac{K_{\mathrm{RA}} P_{\mathrm{LA}}}{K_{\mathrm{RA}} P_{\mathrm{LA}} + K_{\mathrm{fB}} P_{\mathrm{LB}}} \left(\Delta P_0 - \Delta P_{\mathrm{regB}}\right) \tag{9.55}$$

$$\frac{\Delta f_{\infty}}{f_{\mathrm{n}}} = -\frac{1}{K_{\mathrm{RA}} P_{\mathrm{LA}} + K_{\mathrm{fB}} P_{\mathrm{LB}}} \left(\Delta P_0 - \Delta P_{\mathrm{regB}}\right) \tag{9.56}$$

Under the assumption that $P_{LA} \gg P_{LB}$, Eqs. (9.55) and (9.56) may be simplified to

$$\Delta P_{\mathrm{tie}\,\infty} \cong \left(\Delta P_0 - \Delta P_{\mathrm{regB}}\right) \tag{9.57}$$

$$\frac{\Delta f_{\infty}}{f_{\mathrm{n}}} \cong -\frac{1}{K_{\mathrm{RA}} P_{\mathrm{LA}}} \left(\Delta P_0 - \Delta P_{\mathrm{regB}}\right) \tag{9.58}$$

Additional validity is given to this simplification if the small system is fully loaded ($\Delta P_{regB} = 0$ and $K_{TB} = 0$) when the stiffness $K_{fB} = K_{TB} + K_{LB}$ has a small value corresponding to the frequency sensitivity of the loads K_{LB}. In this situation the whole power imbalance is covered by the tie-line interchange.

The frequency deviation Δf_{∞} is inversely proportional to the coefficient $\lambda_{RA} = K_{RA}P_{LA}/f_n$ set at the central regulator. If the available regulation power is not large enough then too low a setting of the central regulator produces a steady-state frequency error. If $K_{RA} = K_{fA}$ then the final value of the frequency will correspond to the frequency level at which the available regulation power of the small system has run out. If $K_{RA} > K_{fA}$ then the regulator of the big system will increase its generation, decreasing the frequency error and allowing the tie-line interchange error to increase. If $K_{RA} < K_{fA}$ then the regulator of the big system will decrease its generation increasing the frequency error, and the tie-line interchange error will not be allowed to increase. This can be illustrated by Example 9.3.

Example 9.3 The available regulating power of the small subsystem considered in Example 9.2 is $\Delta P_{regB} =$ 500 MW. The settings of the central regulators are $K_{RA} = 5.55 < K_{TA}$ and $K_{RB} = 12.5 > K_{TB}$. Neglecting the frequency sensitivity of the load, Eqs. (9.57) and (9.58) give $\Delta P_{tie\infty} = 800$ MW and $\Delta f_{\infty} = -0.16$ Hz. The power and frequency variations are illustrated in Figure 9.30. During the first 20 s of the disturbance, the trajectory is the same as the one shown in Figure 9.29, but now the characteristic P_{TB} eventually settles at $P_{TB\infty}$. Trajectory $f(P_{TA})$ begins to wrap around the instantaneous equilibrium point IV, but the big system regulator decreases its generation shifting the characteristic from position P_{TA+} to position $P_{TA\infty}$. The dynamics end up at point ∞, where the frequency of the interconnected system is lower, by about $\Delta f_{\infty} = -0.16$ Hz, than the required value. A small portion of the trajectory $f(P_{TA})$, between point IV and ∞, corresponds to a slow reduction of frequency over

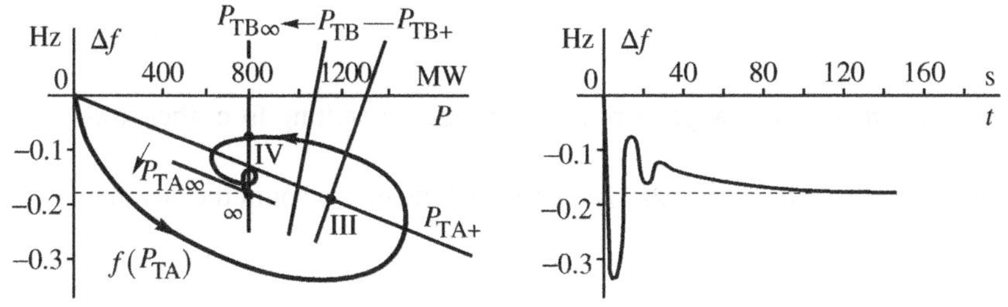

Figure 9.30 Illustration to Example 9.3.

several tens of seconds. The variations in the tie-line power interchange are similar to those shown in Figure 9.29, but settle down at a level corresponding to point ∞, that is $\Delta P_{\text{tie}\infty} = 800$ MW (the generation lost was $\Delta P_0 = 1300$ MW).

******

If the tie-line power interchange determined by Eq. (9.55) is greater than the maximal thermal capacity of the tie-line then the line will be tripped and the systems separated. The imbalance of power in the small system will cause a further drop in frequency which, in the absence of automatic load shedding, may lead to frequency collapse in that subsystem.

9.6 Simplified Simulation Models

Computer simulation of the dynamics associated with AGC (Figure 9.12) can be performed approximately with the help of the simplified model described below, taking into account the swing equation of generating units rotors and simplified models of turbines and their governing systems.

9.6.1 Simplified Model of Islanded System

Rotor dynamics of each generating unit are determined by Eq. (4.15), described in Section 4.1. In a simplified model, one can omit generator rotors swings (Figure 9.16) that appear in the transient state after the occurrence of a power imbalance caused by sudden change in power generation or power demand. With this simplification, it can be assumed that the rotor speed deviations $\Delta\omega_i$ of all generators are the same – Eq. (9.23). After summing up the equations of motion (4.15) and omitting the damping ($D_i = 0$) one gets

$$\sum_{i=1}^{N} M_i \frac{\mathrm{d}\Delta\omega}{\mathrm{d}t} = \sum_{i=1}^{N} P_{\text{m}i} - \sum_{i=1}^{N} P_i \tag{9.59}$$

or

$$M_{\text{EPS}} \frac{\mathrm{d}\Delta\omega}{\mathrm{d}t} = P_{\text{T}} - P_{\text{L}} \tag{9.60}$$

$$P_{\text{L}} = \sum_{i=1}^{N} P_i; \quad P_{\text{T}} = \sum_{i=1}^{N} P_{\text{m}i}; \quad M_{\text{EPS}} = \sum_{i=1}^{N} M_i \tag{9.61}$$

where P_{L} is the sum of active power of all generators or the power demand of all loads increased by losses in the network, P_{T} is the mechanical power in the EPS equal to mechanical power of all turbines, M_{EPS} is, as in Eq. (9.26), the angular momentum of all rotating masses in the EPS also referred to as the *inertia coefficient of the EPS*.

When the power demand changes by ΔP_0, and the frequency differs from the reference value by Δf, then (taking into account Figure 9.5) the powers P_{L} and P_{T} can be expressed using the following formulas

$$P_{\text{T}} = P_{\text{T}}^{\text{old}} + \Delta P_{\text{T}} \tag{9.62}$$

$$P_{\mathrm{L}} = P_{\mathrm{L}}^{\mathrm{old}} + \Delta P_0 + K_{\mathrm{L}} P_{\mathrm{L}}^{\mathrm{old}} \frac{\Delta f}{f_{\mathrm{n}}} \tag{9.63}$$

The last component in formula (9.63) corresponds to the change in power of all loads resulting from their frequency sensitivity – Eq. (9.6).

After substituting Eqs. (9.62) and (9.63) into Eq. (9.59) and $\Delta\omega = 2\pi\Delta f$, and then taking into account the fact that $P_{\mathrm{T}}^{\mathrm{old}} = P_{\mathrm{L}}^{\mathrm{old}}$, the following is obtained

$$2\pi M_{\mathrm{EPS}} \frac{\mathrm{d}\Delta f}{\mathrm{d}t} = \Delta P_{\mathrm{T}} - \Delta P_0 - K_{\mathrm{L}} P_{\mathrm{L}}^{\mathrm{old}} \frac{\Delta f}{f_{\mathrm{n}}} \tag{9.64}$$

so

$$\Delta P_{\mathrm{T}} - \Delta P_0 = 2\pi M_{\mathrm{EPS}} \frac{\mathrm{d}\Delta f}{\mathrm{d}t} + \frac{K_{\mathrm{L}} P_{\mathrm{L}}^{\mathrm{old}}}{f_{\mathrm{n}}} \Delta f \tag{9.65}$$

and using the Laplace operator

$$\Delta P_{\mathrm{T}}(s) - \Delta P_0(s) = 2\pi M_{\mathrm{EPS}}\, s \Delta f(s) + \frac{K_{\mathrm{L}} P_{\mathrm{L}}^{\mathrm{old}}}{f_n} \Delta f(s)$$

hence

$$\Delta P_{\mathrm{T}}(s) - \Delta P_0(s) = \frac{K_{\mathrm{L}} P_{\mathrm{L}}^{\mathrm{old}}}{f_{\mathrm{n}}} \left[1 + s \frac{2\pi f_{\mathrm{n}} M_{\mathrm{EPS}}}{K_{\mathrm{L}} P_{\mathrm{L}}^{\mathrm{old}}}\right] \Delta f(s) \tag{9.66}$$

In this equation, $\Delta f(s)$ can be treated as the system response of the EPS to the change in power demand $\Delta P_0(s)$ and the change in mechanical power $\Delta P_{\mathrm{T}}(s)$ forced by primary and secondary control. Equation (9.66) can be transformed to the following form

$$\Delta f(s) = G_{\mathrm{EPS}}(s) \left[\Delta P_{\mathrm{T}}(s) - \Delta P_0(s)\right] \tag{9.67}$$

where

$$G_{\mathrm{EPS}}(s) = \frac{K_{\mathrm{EPS}}}{1 + s\, T_{\mathrm{EPS}}}; \quad K_{\mathrm{EPS}} = \frac{f_{\mathrm{n}}}{K_{\mathrm{L}} P_{\mathrm{L}}^{\mathrm{old}}}; \quad T_{\mathrm{EPS}} = \frac{2\pi f_{\mathrm{n}} M_{\mathrm{EPS}}}{K_{\mathrm{L}} P_{\mathrm{L}}^{\mathrm{old}}} = 2\pi\, K_{\mathrm{EPS}} M_{\mathrm{EPS}} \tag{9.68}$$

Equation (9.67) is the basis for creating an equivalent block diagram of the EPS with frequency control shown in Figure 9.31. In this figure ΔP and Δf are the input and output signals of the block with transfer function $G_{\mathrm{EPS}}(s)$. The remaining blocks in this figure indicate the appropriate: $G_{\mathrm{T}}(s)$ – equivalent transfer function of turbines, $G_{\mathrm{TG}}(s)$ – equivalent transfer function of turbine governing systems, and $G_{\mathrm{AGC}}(s)$ – transfer function of frequency regulator (secondary control). Coefficient R corresponds to the gain of the feedback loop in the turbine governing system (Figures 2.16 and 9.7) such that $\rho = R/\omega_{\mathrm{n}} = R/2\pi f_{\mathrm{n}}$ is the *speed-droop coefficient* (Section 2.3.3). Symbol λ_{R} corresponds to the frequency bias factor in the central (secondary) regulator (Figure 9.10) defined in Eq. (9.10).

Using the block diagram shown in Figure 9.31, one can use the available simulation tools to examine approximately the frequency control after the occurrence of disturbance in the power balance. Such disturbance is introduced into the model as ΔP_0 and can be a step change or a time function $\Delta P_0 = \Delta P(t)$.

9.6.2 Simplified Model of Two Connected Subsystems

It is assumed that each subsystem of the considered interconnected system has its own central (secondary) regulator operating according to the scheme in Figure 9.12. As in Section 9.6.1, it is also assumed that the speed deviations $\Delta\omega_i$ of all generators in a given subsystem are the same – Eq. (9.23), which means that the transient swings

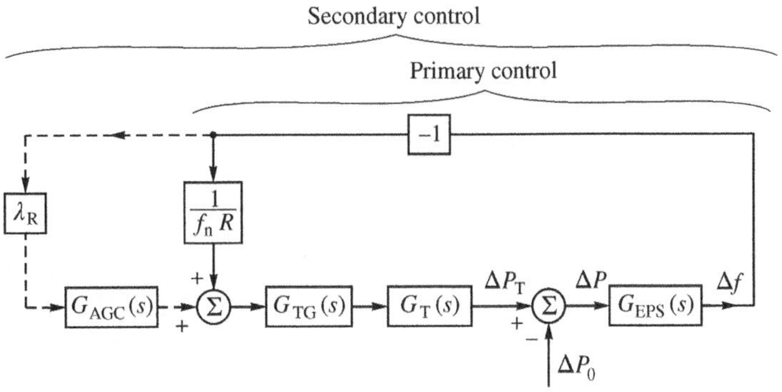

Figure 9.31 Equivalent block diagram of EPS with frequency control.

inside each subsystem are neglected. With such assumptions, each subsystem can be modeled using the transfer function $G_{EPS}(s)$ specified by Eq. (9.68).

In the case of the interconnected EPS consisting of two subsystems (Figure 9.32) in addition to ΔP_T, ΔP_L appears also ΔP_{tie}, where P_{tie} is the tie-line interchange power. From the point of view of power balance the change ΔP_{tie} of the tie-line power can be added to the change ΔP_L of power of loads and Eq. (9.67) takes the following form

$$\Delta f(s) = G_{EPS}(s)\ [\Delta P_T(s) - \Delta P_{tie} - \Delta P_0(s)] \tag{9.69}$$

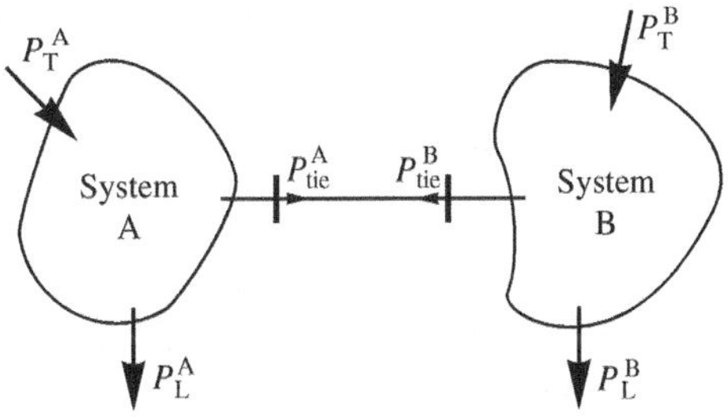

Figure 9.32 Illustration of power balance in an interconnected system consisting of two subsystems.

In the equivalent block diagram (Figure 9.31) signal ΔP_{tie} must be inserted to the summation point producing input signal to the transfer function $G_{EPS}(s)$. However, the fundamental problem in further developing the model discussed here is to find a mathematical relationship between ΔP_{tie} and Δf. There is no doubt that such a link exists, because the changes in active power in the tie-line depend on the arguments of the voltages at the ends of the tie-line and the angular changes are closely related to the frequency changes.

For the sake of simplicity, the interconnected EPS consisting of two systems A and B, as shown on Figure 9.33, will be discussed. It is assumed that in the network model the synchronous generators are modeled by voltage sources with electromotive forces (emf's) $\underline{E}'_i = E'_i e^{j\delta_i}$ and reactances X'_{di}. Arguments δ_i of these emf's correspond to the rotor angles δ_i occurring in the swing equations. The method of creating the equivalent circuit of the two connected subsystems is illustrated on Figure 9.33. In the upper part of this figure (Figure 9.33a) two connected subsystems with generators $\{G_A\}$ and $\{G_B\}$ and load $\{L_A\}$ and $\{L_B\}$ are shown. Each generator is replaced by a classic model consisting of a voltage source behind transient reactance. All load nodes (with the exception of tie-line ends) are eliminated by the method described in Section 18.2.1. In the next step (Figure 9.33b) generation nodes $\{G_A\}$ and $\{G_B\}$ are aggregated using the method described in Section 18.2.3 and groups of generators $\{G_A\}$ and $\{G_B\}$ are replaced by single equivalent generators G_A and G_B, respectively. It is worth noting that the condition in which a group of generators can be replaced by one equivalent generator (aggregation) is the electromechanical coherency of rotors of these generators during the transient state. In the considered case this condition is met because, according to Eq. (9.23), the speed deviations $\Delta\omega_i$ of generators belonging to the subsystem are the same. After aggregation of the generation nodes, the equivalent circuit of the model of both subsystems is as in Figure 9.33c. In this circuit each subsystem is replaced by equivalent π-circuit, in which the shunt branches correspond to the eliminated loads. In the middle of this equivalent circuit there is impedance $\underline{Z}_{AB}$ corresponding to the impedance of the tie-line connecting both subsystems.

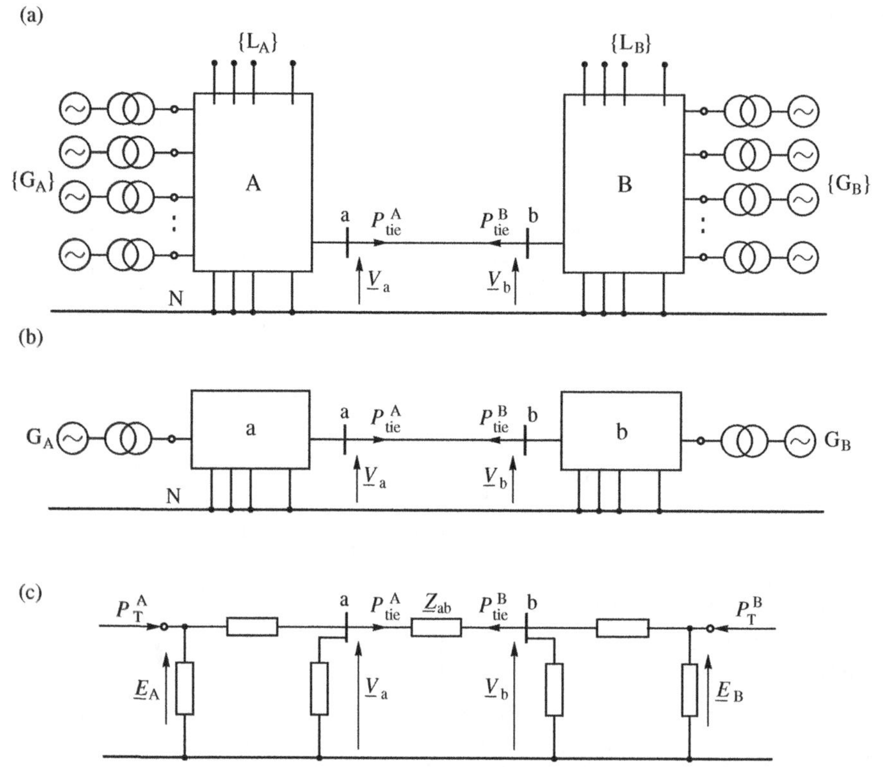

Figure 9.33 Creation of an equivalent circuit of two connected subsystem.

For the equivalent scheme shown in Figure 9.33c, in analogy to Eq. (3.184), while omitting changes in magnitudes of nodal voltages, the following incremental equation can be written

$$\begin{matrix} \mathrm{A} \\ \mathrm{B} \\ \mathrm{a} \\ \mathrm{b} \end{matrix} \quad \begin{bmatrix} \Delta P_{\mathrm{A}} \\ \Delta P_{\mathrm{B}} \\ 0 \\ 0 \end{bmatrix} = \begin{bmatrix} H_{\mathrm{AA}} & 0 & H_{\mathrm{Aa}} & 0 \\ 0 & H_{\mathrm{BB}} & 0 & H_{\mathrm{Bb}} \\ H_{\mathrm{aA}} & 0 & H_{\mathrm{aa}} & H_{\mathrm{ab}} \\ 0 & H_{\mathrm{bB}} & H_{\mathrm{ba}} & H_{\mathrm{bb}} \end{bmatrix} \begin{bmatrix} \Delta \delta_{\mathrm{A}} \\ \Delta \delta_{\mathrm{B}} \\ \Delta \delta_{\mathrm{a}} \\ \Delta \delta_{\mathrm{b}} \end{bmatrix} \tag{9.70}$$

where ΔP_{A} and ΔP_{B} are changes in active power of both equivalent generators, $\Delta \delta_{\mathrm{A}}$ and $\Delta \delta_{\mathrm{B}}$ are changes in rotor angles of these generators, and $\Delta \delta_{\mathrm{a}}$ and $\Delta \delta_{\mathrm{b}}$ are changes in arguments of voltages at the ends of the tie-line connecting two subsystems. For further considerations, the relationships between $\Delta \delta_{\mathrm{a}}$, $\Delta \delta_{\mathrm{b}}$ and $\Delta \delta_{\mathrm{A}}$, $\Delta \delta_{\mathrm{B}}$ are important. From the lower part of Eq. (9.70) it follows that

$$\begin{bmatrix} 0 \\ 0 \end{bmatrix} = \begin{bmatrix} H_{\mathrm{aA}} & 0 \\ 0 & H_{\mathrm{bB}} \end{bmatrix} \begin{bmatrix} \Delta \delta_{\mathrm{A}} \\ \Delta \delta_{\mathrm{B}} \end{bmatrix} + \begin{bmatrix} H_{\mathrm{aa}} & H_{\mathrm{ab}} \\ H_{\mathrm{ba}} & H_{\mathrm{bb}} \end{bmatrix} \begin{bmatrix} \Delta \delta_{\mathrm{a}} \\ \Delta \delta_{\mathrm{b}} \end{bmatrix} \tag{9.71}$$

Hence

$$\begin{bmatrix} \Delta \delta_{\mathrm{a}} \\ \Delta \delta_{\mathrm{b}} \end{bmatrix} = - \begin{bmatrix} H_{\mathrm{aa}} & H_{\mathrm{ab}} \\ H_{\mathrm{ba}} & H_{\mathrm{bb}} \end{bmatrix}^{-1} \begin{bmatrix} H_{\mathrm{aA}} & 0 \\ 0 & H_{\mathrm{bB}} \end{bmatrix} \begin{bmatrix} \Delta \delta_{\mathrm{A}} \\ \Delta \delta_{\mathrm{B}} \end{bmatrix} \tag{9.72}$$

or

$$\begin{bmatrix} \Delta\delta_a \\ \Delta\delta_b \end{bmatrix} = \begin{bmatrix} \kappa_{aa} & \kappa_{ab} \\ \kappa_{ba} & \kappa_{bb} \end{bmatrix} \begin{bmatrix} \Delta\delta_A \\ \Delta\delta_B \end{bmatrix} \tag{9.73}$$

where

$$\begin{aligned} \kappa_{aa} &= +\left(H_{aa}H_{bb} - H_{ba}H_{ab}\right)^{-1}H_{aa}H_{aA} \\ \kappa_{ab} &= -\left(H_{aa}H_{bb} - H_{ba}H_{ab}\right)^{-1}H_{ab}H_{bB} \\ \kappa_{ba} &= -\left(H_{aa}H_{bb} - H_{ba}H_{ab}\right)^{-1}H_{ab}H_{aA} \\ \kappa_{bb} &= +\left(H_{aa}H_{bb} - H_{ba}H_{ab}\right)^{-1}H_{bb}H_{bB} \end{aligned} \tag{9.74}$$

The power in the tie-line connecting both subsystems (Figure 9.33) is a function of the angle $\delta_{ab} = \delta_a - \delta_b$. If power losses in tie-line are omitted, real powers on both ends of the line are identical and given by formula

$$P_{tie} = \frac{V_a V_b}{X_{ab}} \sin\delta_{ab} \tag{9.75}$$

Therefore

$$\Delta P_{tie} \cong \frac{\partial P_{tie}}{\partial \delta_{ab}} \Delta\delta_{ab} = H_{ab}\Delta\delta_{ab} = H_{ab}(\Delta\delta_a - \Delta\delta_b) \tag{9.76}$$

where partial derivative

$$H_{ab} = \frac{\partial P_{tie}}{\partial \delta_{ab}} = \frac{V_a V_b}{X_{ab}} \cos\delta_{ab} \tag{9.77}$$

is the synchronizing power calculated for the tie-line.

Using Eq. (9.73) one can write

$$\begin{aligned} \Delta\delta_a &= \kappa_{aa}\Delta\delta_A + \kappa_{ab}\Delta\delta_B \\ \Delta\delta_b &= \kappa_{ba}\Delta\delta_A + \kappa_{bb}\Delta\delta_B \end{aligned} \tag{9.78}$$

so

$$\Delta\delta_{ab} = \Delta\delta_a - \Delta\delta_b = \left(\kappa_{aa} - \kappa_{ba}\right)\Delta\delta_A + \left(\kappa_{ab} - \kappa_{bb}\right)\Delta\delta_B \tag{9.79}$$

Substituting Eqs. (9.79) into (9.76) one gets

$$\Delta P_{tie} \cong H_{ab}\left[\left(\kappa_{aa} - \kappa_{ba}\right)\Delta\delta_A + \left(\kappa_{ab} - \kappa_{bb}\right)\Delta\delta_B\right] \tag{9.80}$$

This means that the change of the tie-line power can be expressed by changes of the arguments of the equivalent emf's of both subsystems.

Time derivatives of the rotor angles correspond to the speed deviations and the same time to the average frequency changes in subsystems, i.e.

$$\Delta f_A = \frac{1}{2\pi}\frac{d\Delta\delta_A}{dt} \quad \text{and} \quad \Delta f_B = \frac{1}{2\pi}\frac{d\Delta\delta_B}{dt} \tag{9.81}$$

It follows from Eq. (9.81) that changes of rotor angles can be calculated as integrals of frequency changes in subsystems

$$\Delta\delta_A = 2\pi\int_0^t \Delta f_A dt \quad \text{and} \quad \Delta\delta_B = 2\pi\int_0^t \Delta f_B dt \tag{9.82}$$

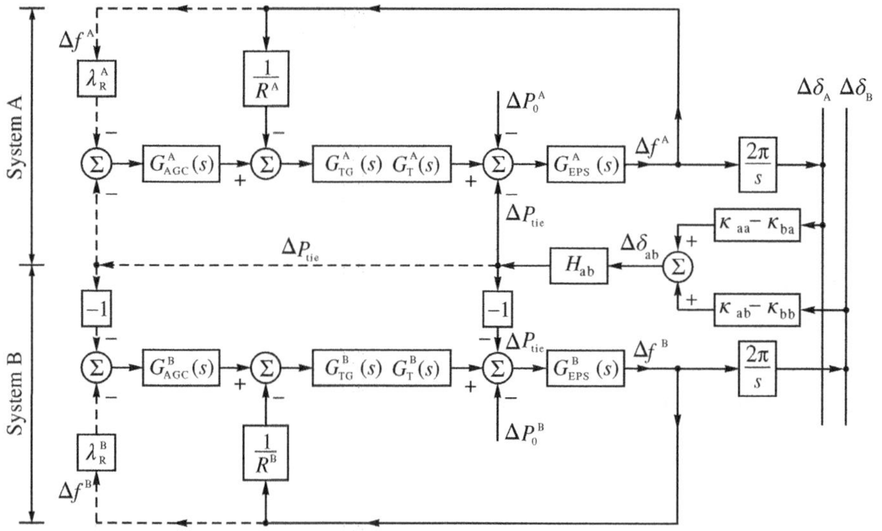

Figure 9.34 Block diagram of the simplified model of two connected subsystems with central AGC.

Hence using the Laplace operator

$$\Delta\delta_{\rm A} = \frac{2\pi}{s}\Delta f_{\rm A}(s) \quad \text{and} \quad \Delta\delta_{\rm B} = \frac{2\pi}{s}\Delta f_{\rm B}(s) \tag{9.83}$$

Equations (9.80) and (9.83) allow block diagrams of both subsystems to be connected in the way shown in Figure 9.34.

The upper and lower parts of Figure 9.34 concern subsystems A and B, respectively. On the right-hand side of the diagram the frequency changes $\Delta f_{\rm A}$, $\Delta f_{\rm B}$ in the subsystems are entered into the integrators to calculate the changes in rotor angles $\Delta\delta_{\rm A}$, $\Delta\delta_{\rm B}$ of equivalent generators. The obtained values multiplied by the corresponding coefficients (with proper signs) are added in the summing point and multiplied by synchronizing power $H_{\rm ab}$ to obtain the change $\Delta P_{\rm tie}$ of the tie-line power. This change is forwarded (dashed line) to the summing points before the transfer function $G_{\rm AGC}(s)$, into which the signals $\lambda_{\rm R}\Delta f$ are also coming and create the signal *ACE* defined by Eq. (9.11). For subsystem B, the change in the tie-line power is multiplied by (−1), which results from the fact that the direction of the given subsystem was taken as the positive direction of this power. Positive flow for subsystem A is negative for subsystem B, and vice versa.

Using the discussed block diagram (Figure 9.34), it is possible to simulate approximately the control of the frequency and the interchange power, taking into account the primary and secondary control in both subsystems. Disturbances $\Delta P^{\rm A}_0$ or $\Delta P^{\rm B}_0$ of power balance are introduced as a step changes or changes defined as a time functions $\Delta P(t)$.

9.7 Series FACTS Devices in Tie-lines

Series FACTS devices, described in Section 2.5.4 may be installed in tie-lines linking control areas in an interconnected power system. Their main function is the execution of steady-state control functions described in Section 3.6.3. During the transient state caused by a sudden disturbance of a power balance in one of the subsystems, series FACTS devices installed in the tie-lines may affect the values of tie-line power interchanges $P_{\rm tie}$ and therefore also the value of the *ACE* given by (9.11) and the dynamics of secondary control executed by the central regulator shown in Figure 9.10. Hence a proper control algorithm and proper parameter selection have to be

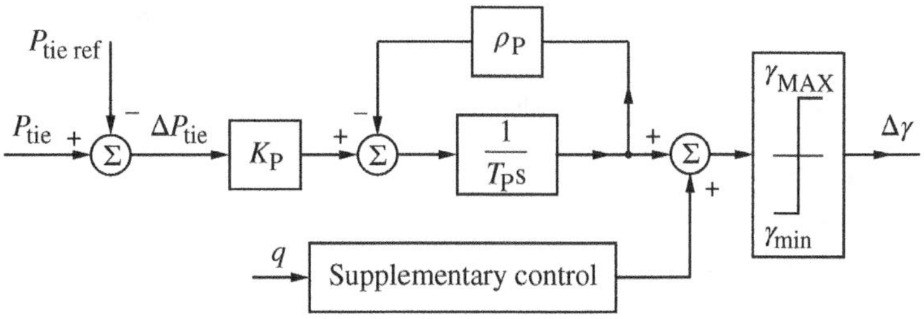

Figure 9.35 Power flow controller installed in a tie-line of an interconnected power system.

implemented at the regulator of the series FACTS device so that the control does not deteriorate the frequency and tie-line power interchange regulation process. This problem is now discussed in detail using as an example a thyristor controlled phase angle regulator (TCPAR) which, from the power system point of view, acts as a fast phase shifting transformer.

A schematic diagram of a TCPAR regulator is shown in Figure 9.35. An integral-type regulator with negative feedback is placed in the main control path. The task of the regulator is to regulate the active power flow in the line in which the FACTS device is installed. The reference value is supplied from the supervisory control system. A supplementary control loop devoted to damping of power swings and improving power stability is shown in the lower part of the diagram.

From the point of view of power system dynamics, an important problem for a series FACTS device installed in a tie-line is the control algorithm executing the supplementary control loop ensuring damping of inter-area power swings in such a way that the frequency control executed by the central LFC regulator is not disturbed. The control algorithm described in this section is based on Nogal (2009).

9.7.1 Incremental Model of a Multi-machine System

Figure 9.36 illustrates the stages of developing a model of a phase shifting transformer installed in a tie-line. A booster voltage, which is in quadrature to the supply voltage, is injected in the transmission line using a booster transformer

$$\Delta V_P = \gamma V_a \tag{9.84}$$

where γ is the controlled variable. The booster transformer reactance has been added to the equivalent line reactance. To simplify considerations, the line and transformer resistances have been neglected.

The following relationships can be derived using the phasor diagram of Figure 9.36

$$\sin\theta = \frac{\Delta V_P}{V_c} = \frac{\gamma V_a}{V_c}; \quad \cos\theta = \frac{V_a}{V_c}; \quad \delta_{cb} = \delta_{ab} - \theta \tag{9.85}$$

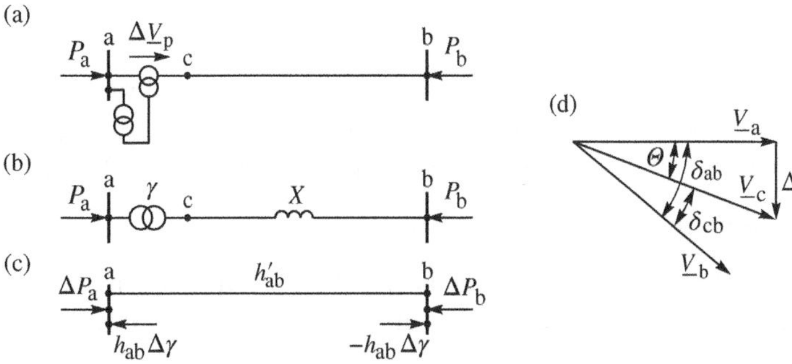

Figure 9.36 Stages of developing an incremental model of a transmission line with a phase shifting transformer: (a) one-line diagram; (b) admittance model with ideal transformation ratio; (c) incremental model; (d) phasor diagram.

According to Eq. (3.15), active power flowing through a transmission line is given by

$$P_a = P_{ab} = P_{cb} = \frac{V_c V_b}{X} \sin \delta_{cb} \tag{9.86}$$

Substituting Eqs. (9.85) gives

$$\begin{aligned} P_a &= \frac{V_c V_b}{X} \sin(\delta_{ab} - \theta) = \frac{V_c V_b}{X} (\sin \delta_{ab} \cos \theta - \cos \delta_{ab} \sin \theta) \\ &= \frac{V_a V_b}{X} \sin \delta_{ab} - \gamma \frac{V_a V_b}{X} \cos \delta_{ab} \end{aligned} \tag{9.87}$$

That equation can also be written as

$$P_a = b_{ab} \sin \delta_{ab} - b_{ab} \cos \delta_{ab} \, \gamma(t) \tag{9.88}$$

where $b_{ab} = V_a V_b / X$ is the amplitude of the power-angle characteristic of the transmission line.

The values of variables at a given operating point are $(\hat{P}_{ab}, \hat{\delta}, \hat{\gamma})$. Using these values, Eq. (9.88) gives

$$\hat{P}_a = b_{ab} \sin \hat{\delta}_{ab} - b_{ab} \cos \hat{\delta}_{ab} \, \hat{\gamma} \tag{9.89}$$

The tie-line flow in Eq. (9.88) depends on both the power angle δ_{ab} and the quadrature transformation ratio $\gamma(t)$. Taking that into account and differentiating Eq. (9.88) in the vicinity of the operating point gives

$$\Delta P_a = \left. \frac{\partial P_a}{\partial \delta_{ab}} \right|_{\delta_{ab} = \hat{\delta}_{ab}} \Delta \delta_{ab} + \left. \frac{\partial P_a}{\partial \gamma} \right|_{\gamma = \hat{\gamma}} \Delta \gamma \tag{9.90}$$

Hence, taking into account Eq. (9.88),

$$\Delta P_a = \left(b_{ab} \cos \hat{\delta}_{ab} + \hat{\gamma} \, b_{ab} \sin \hat{\delta}_{ab} \right) \Delta \delta - \left(b_{ab} \cos \hat{\delta}_{ab} \right) \Delta \gamma \tag{9.91}$$

The coefficients $b_{ab} \cos \hat{\delta}_{ab}$ and $b_{ab} \sin \hat{\delta}_{ab}$ in this equation are the same as those in (9.89). Component $b_{ab} \sin \hat{\delta}_{ab}$ can be eliminated from Eqs. (9.91) using (9.89) in the following way. Equation (9.89) gives

$$b_{ab} \sin \hat{\delta}_{ab} = \hat{P}_a + \hat{\gamma} \, b_{ab} \cos \hat{\delta}_{ab} \tag{9.92}$$

or

$$\hat{\gamma} \, b_{ab} \sin \hat{\delta}_{ab} = \hat{\gamma} \, \hat{P}_a + \hat{\gamma}^2 \, b_{ab} \cos \hat{\delta}_{ab} \tag{9.93}$$

Substituting this equation into Eq. (9.91) gives

$$\Delta P_a = \left[\left(1 + \hat{\gamma}^2\right) \left(b_{ab} \cos \hat{\delta}_{ab} \right) + \hat{\gamma} \, \hat{P}_a \right] \Delta \delta_{ab} - \left(b_{ab} \cos \hat{\delta}_{ab} \right) \Delta \gamma \tag{9.94}$$

The following notation is now introduced

$$h_{ab} = \left. \frac{\partial P_a}{\partial \delta_{ab}} \right|_{\delta_{ab} = \hat{\delta}_{ab}, \hat{\gamma} = 0} = b_{ab} \cos \hat{\delta}_{ab} \tag{9.95}$$

$$h'_{ab} = \left(1 + \hat{\gamma}^2\right) \left(b_{ab} \cos \hat{\delta}_{ab} \right) + \hat{\gamma} \, \hat{P}_a = \left(1 + \hat{\gamma}^2\right) h_{ab} + \hat{\gamma} \, \hat{P}_a \tag{9.96}$$

The variable h_{ab} given by Eq. (9.95) corresponds to the mutual synchronizing power for the line ab calculated neglecting the booster transformer. On the other hand, h'_{ab} given by Eq. (9.96) corresponds to the synchronizing power when the booster transformer has been taken into account. Using that notation, Eq. (9.94) takes the form

$$\Delta P_a + h_{ab} \, \Delta \gamma = h'_{ab} \, \Delta \delta_{ab} = h'_{ab} \, \Delta \delta_a - h'_{ab} \, \Delta \delta_b \tag{9.97}$$

Equation (9.97) describes the incremental model of the transmission line shown in Figure 9.36c. There is an equivalent transmission line between nodes a and b. A change in the flow in that line corresponds to a change in the voltage angles at both nodes. Nodal power injections correspond to flow changes following regulation of the quadrature transformation ratio $\gamma(t)$. The power injections in nodes a and b are $+h_{ab}\,\Delta\gamma$ and $-h_{ab}\,\Delta\gamma$, respectively. To understand this, note that (9.71) holds for node "a," while the same equation for node b is

$$\Delta P_b - h_{ab}\,\Delta\gamma = -h'_{ab}\,\Delta\delta_{ab} = -h'_{ab}\,\Delta\delta_a + h'_{ab}\,\Delta\delta_b \tag{9.98}$$

It will be shown later that the derived incremental model of a branch with a phase shifting transformer is convenient for network analysis, especially for large networks, because it models changes in the quadrature transformation ratio by changes in power injections without changing the parameters of the branches.

Equation (3.182) derived in Section 3.5.1 models the effect of small changes of nodal voltages in a network. In analyzing system frequency regulation, one can assume that changes in voltage magnitudes can be neglected and only changes in voltage angles are considered. Then Eq. (3.182) takes the form

$$\Delta \boldsymbol{P} \cong \boldsymbol{H}\Delta\boldsymbol{\delta} \tag{9.99}$$

where $\Delta\boldsymbol{P}$ and $\Delta\boldsymbol{\delta}$ are the vectors of changes in active power injections and voltage angles, respectively. Matrix $\boldsymbol{H}$ is the Jacobi matrix and consists of the partial derivatives $H_{ij} = \partial P_i / \partial \delta_j$. Equation (9.99) describes the *incremental model* of a network. Including a phase shifting transformer in incremental model of a network is illustrated in Figure 9.37. There are the following node types:

- {G} – generator nodes behind transient generator reactances;
- {L} – load nodes;
- a and b – terminal nodes of a line with a phase shifting transformer (as in Figure 9.36).

The line with the phase shifting transformer (Figure 9.37) is modeled using a transformation ratio and a branch. In the incremental model shown in Figure 9.37 this line is modeled the same way as shown in Figure 9.36. Matrix $\boldsymbol{H}$ describing that network includes branch h'_{ab} from the incremental line model with the phase shifting transformer. There are active power injections in nodes a and b, similar to Figure 9.36c, corresponding to flow changes due to transformation ratio regulation $\gamma(t)$.

Now Eq. (9.99) describing the model shown in Figure 9.37b can be expanded as

$$\begin{matrix}\{G\}\\ a\\ b\\ \{L\}\end{matrix}\quad \begin{bmatrix}\Delta \boldsymbol{P}_G\\ \hline +h_{ab}\Delta\gamma\\ \hline -h_{ab}\Delta\gamma\\ \hline \boldsymbol{0}\end{bmatrix} \cong \begin{bmatrix} & & \\ & \boldsymbol{H} & \\ & & \end{bmatrix} \begin{bmatrix}\Delta\boldsymbol{\delta}_G\\ \hline \Delta\delta_a\\ \hline \Delta\delta_b\\ \hline \Delta\boldsymbol{\delta}_L\end{bmatrix}. \tag{9.100}$$

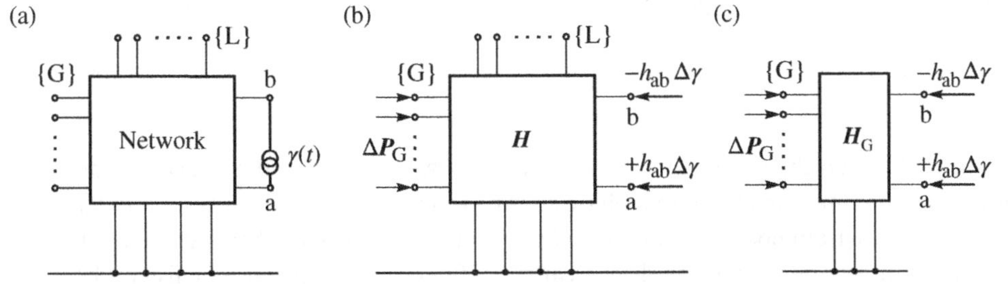

Figure 9.37 Stages of developing the incremental model: (a) admittance model with a phase shifting transformer; (b) incremental model; (c) incremental model after elimination of nodes {L}.

Substitution $\Delta \boldsymbol{P}_{\rm L} = \boldsymbol{0}$ has been made on the left-hand side of Eq. (9.100) because loads at {L} nodes are modeled as constant powers. Eliminating variables related to load nodes {L} in Eq. (9.100) by using the partial inversion method shown in Appendix A.2 makes it possible to transform Eq. (9.100) to the following form

$$\begin{matrix} \{G\} \\ a \\ b \end{matrix} \begin{bmatrix} \Delta \boldsymbol{P}_{\rm G} \\ \hline +h_{\rm ab}\Delta\gamma \\ \hline -h_{\rm ab}\Delta\gamma \end{bmatrix} \cong \left[\begin{array}{c|c|c} \boldsymbol{H}_{\rm GG} & \boldsymbol{H}_{\rm Ga} & \boldsymbol{H}_{\rm Gb} \\ \hline \boldsymbol{H}_{\rm aG} & H_{\rm aa} & H_{\rm ab} \\ \hline H_{\rm bG} & H_{\rm ba} & H_{\rm bb} \end{array}\right] \begin{bmatrix} \Delta \boldsymbol{\delta}_{\rm G} \\ \hline \Delta\delta_{\rm a} \\ \hline \Delta\delta_{\rm b} \end{bmatrix}. \tag{9.101}$$

This equation can be further transformed by partial inversion to the following equations

$$\Delta \boldsymbol{P}_{\rm G} \cong \boldsymbol{H}_{\rm G}\Delta\boldsymbol{\delta}_{\rm G} + [\boldsymbol{K}_{\rm Ga} \mid \boldsymbol{K}_{\rm Gb}] \begin{bmatrix} +h_{\rm ab}\Delta\gamma \\ \hline -h_{\rm ab}\Delta\gamma \end{bmatrix}, \tag{9.102}$$

$$\begin{bmatrix} \Delta\delta_{\rm a} \\ \hline \Delta\delta_{\rm b} \end{bmatrix} \cong -\begin{bmatrix} \boldsymbol{K}_{\rm aG} \\ \hline \boldsymbol{K}_{\rm bG} \end{bmatrix} \Delta\boldsymbol{\delta}_{\rm G} + \left[\begin{array}{c|c} H_{\rm aa} & H_{\rm ab} \\ \hline H_{\rm ba} & H_{\rm bb} \end{array}\right]^{-1} \begin{bmatrix} +h_{\rm ab}\Delta\gamma \\ \hline -h_{\rm ab}\Delta\gamma \end{bmatrix}, \tag{9.103}$$

where

$$\boldsymbol{H}_{\rm G} = \boldsymbol{H}_{\rm GG} - [\boldsymbol{H}_{\rm Ga} \mid \boldsymbol{H}_{\rm Gb}] \left[\begin{array}{c|c} H_{\rm aa} & H_{\rm ab} \\ \hline H_{\rm ba} & H_{\rm bb} \end{array}\right]^{-1} \begin{bmatrix} \boldsymbol{H}_{\rm aG} \\ \hline \boldsymbol{H}_{\rm bG} \end{bmatrix}, \tag{9.104}$$

$$[\boldsymbol{K}_{\rm Ga} \mid \boldsymbol{K}_{\rm Gb}] = [\boldsymbol{H}_{\rm Ga} \mid \boldsymbol{H}_{\rm Gb}] \left[\begin{array}{c|c} H_{\rm aa} & H_{\rm ab} \\ \hline H_{\rm ba} & H_{\rm bb} \end{array}\right]^{-1}, \tag{9.105}$$

$$\begin{bmatrix} \boldsymbol{K}_{\rm aG} \\ \hline \boldsymbol{K}_{\rm bG} \end{bmatrix} = \left[\begin{array}{c|c} H_{\rm aa} & H_{\rm ab} \\ \hline H_{\rm ba} & H_{\rm bb} \end{array}\right]^{-1} \begin{bmatrix} \boldsymbol{H}_{\rm aG} \\ \hline \boldsymbol{H}_{\rm bG} \end{bmatrix}. \tag{9.106}$$

Equations (9.102) and (9.103) describe the incremental model shown in Figure 9.37c. The former describes how a change in the transformation ratio of a phase shifting transformer affects power changes in all power system generators. The latter describes the influence of changes in the transformation ratio on the voltage angle changes in the terminal nodes of the line with the phase shifting transformer.

Equation (9.102) can be transformed to

$$\Delta \boldsymbol{P}_{\rm G} \cong \boldsymbol{H}_{\rm G}\Delta\boldsymbol{\delta}_{\rm G} + \Delta\boldsymbol{K}_{\rm ab}\, h_{\rm ab}\, \Delta\gamma \tag{9.107}$$

where

$$\Delta\boldsymbol{K}_{\rm ab} = \boldsymbol{K}_{\rm Ga} - \boldsymbol{K}_{\rm Gb} \tag{9.108}$$

Hence a power change in the i-th generator can be expressed as

$$\Delta P_i \cong \sum_{j\in\{G\}} H_{ij}\Delta\delta_j + \Delta K_i\, h_{\rm ab}\, \Delta\gamma \tag{9.109}$$

where $\Delta K_i = K_{i\rm a} - K_{i\rm b}$. Thus, if $K_{i\rm a} \cong K_{i\rm b}$ then changes in $\Delta\gamma$ cannot influence power changes in the i-th generator. In other words, that generator cannot be controlled using that phase shifting transformer. Coefficients $K_{i\rm a}$ and $K_{i\rm b}$ can be treated as measures of the distance from nodes a and b to the i-th generator. It means that if nodes a and b are at the same distance from the i-th generator then the device cannot influence that generator. This can be checked using Figure 9.37c, since power injections in nodes a and b have opposite signs. Hence if the distances are the same, the influences on that generator cancel each other out.

The swing equation describing increments of rotor angles, Eq. (5.1) in Section 5.1, is

$$\begin{aligned}\frac{d\Delta\delta_i}{dt} &= \Delta\omega_i \\ M_i\frac{d\Delta\omega_i}{dt} &= -\Delta P_i - D_i\Delta\omega_i\end{aligned} \tag{9.110}$$

for $i \in \{G\}$. As the network equations were derived in matrix form, it is convenient to write the above equation in matrix form, too

$$\begin{aligned}\Delta\dot{\boldsymbol{\delta}}_G &= \Delta\boldsymbol{\omega}_G \\ \boldsymbol{M}\,\Delta\dot{\boldsymbol{\omega}}_G &= -\Delta\boldsymbol{P}_G - \boldsymbol{D}\,\Delta\boldsymbol{\omega}_G\end{aligned} \tag{9.111}$$

where $\boldsymbol{M}$ and $\boldsymbol{D}$ are diagonal matrices of the inertia and damping coefficients, and $\Delta\boldsymbol{\delta}_G$, $\Delta\boldsymbol{\omega}_G$, and $\Delta\boldsymbol{P}_G$ are column matrices of changes in rotor angles, rotor speed deviations, and active power generations, respectively.

Substituting Eq. (9.107) to the second equation of (9.111) gives the following state equation

$$\boldsymbol{M}\,\Delta\dot{\boldsymbol{\omega}}_G = -\boldsymbol{H}_G\Delta\boldsymbol{\delta}_G - \boldsymbol{D}\,\Delta\boldsymbol{\omega}_G - \Delta\boldsymbol{K}_{ab}\,h_{ab}\,\Delta\gamma(t) \tag{9.112}$$

Here $\Delta\gamma(t)$ is the control function corresponding to the transformation ratio change of the phase shifting transformer. Function $\Delta\gamma(t)$ affects rotor motions in proportionally to the coefficients $\Delta K_i h_{ab} = (K_{ia} - K_{ib})h_{ab}$.

The main question is how $\Delta\gamma(t)$ should be changed so that control of a given phase shifting transformer improves damping of oscillations. The control algorithm of $\Delta\gamma(t)$ will be derived using the Lyapunov direct method.

9.7.2 State-variable Control Based on Lyapunov Method

In Section 6.3, the total system energy $V(\delta,\omega) = E_k + E_p$ is used as the Lyapunov function in the nonlinear system model (with line conductances neglected). In the considered linear model Eq. (9.112) the total system energy can be expressed as the sum of rotor speed and angle increments. This corresponds to expanding $V(\delta,\omega) = E_k + E_p$ in a Taylor series in the vicinity of an operating point, as in Eq. (6.18). This equation shows that $V(\boldsymbol{x})$ can be approximated in the vicinity of an operating point using a quadratic form based on the Hesse matrix of function $V(\boldsymbol{x})$.

For the potential energy E_P given by Eq. (6.54), the Hesse matrix corresponds to the gradient of active power generations and therefore also the Jacobi matrix used in the above incremental model

$$\left[\frac{\partial^2 E_p}{\partial\delta_i\partial\delta_j}\right] = \left[\frac{\partial P_i}{\partial\delta_j}\right] = \boldsymbol{H}_G \tag{9.113}$$

Equations (6.18) and (9.111) lead to

$$\Delta E_p = \frac{1}{2}\,\Delta\boldsymbol{\delta}_G^T\,\boldsymbol{H}_G\,\Delta\boldsymbol{\delta}_G \tag{9.114}$$

It is shown in Chapter 12 that if the network conductances are neglected, matrix $\boldsymbol{H}_G$ is positive definite at an operating point (stable equilibrium point). Hence the quadratic form (9.114) is also positive definite.

Using (6.18), the kinetic energy E_k given by (6.53) can be expressed as

$$\Delta E_k = \frac{1}{2}\,\Delta\boldsymbol{\omega}_G^T\,\boldsymbol{M}\,\Delta\boldsymbol{\omega}_G \tag{9.115}$$

This is a quadratic form made up of the vector of speed changes and a diagonal matrix of inertia coefficients. Matrix $\boldsymbol{M}$ is positive definite so the above quadratic form is also positive definite.

The total energy increment $\Delta V(\delta,\omega) = \Delta E_k + \Delta E_p$ is given by

$$\Delta V = \Delta E_k + \Delta E_p = \frac{1}{2}\,\Delta\boldsymbol{\omega}_G^T\,\boldsymbol{M}\,\Delta\boldsymbol{\omega}_G + \frac{1}{2}\,\Delta\boldsymbol{\delta}_G^T\,\boldsymbol{H}_G\,\Delta\boldsymbol{\delta}_G \tag{9.116}$$

This function is positive definite as the sum of positive definite functions and therefore can be used as a Lyapunov function, provided its time derivative at the operating point is negative definite.

Differentiating Eqs. (9.114) and (9.115) gives

$$\Delta \dot{E}_{\mathrm{p}} = \frac{1}{2}\ \Delta\boldsymbol{\omega}_{\mathrm{G}}^{\mathrm{T}}\ \boldsymbol{H}_{\mathrm{G}}\ \Delta\boldsymbol{\delta}_{\mathrm{G}} + \frac{1}{2}\ \Delta\boldsymbol{\delta}_{\mathrm{G}}^{\mathrm{T}}\ \boldsymbol{H}_{\mathrm{G}}\ \Delta\boldsymbol{\omega}_{\mathrm{G}} \tag{9.117}$$

$$\Delta \dot{E}_{\mathrm{k}} = \frac{1}{2}\ \Delta\dot{\boldsymbol{\omega}}_{\mathrm{G}}^{\mathrm{T}}\ \boldsymbol{M}\ \Delta\boldsymbol{\omega}_{\mathrm{G}} + \frac{1}{2}\ \Delta\boldsymbol{\omega}_{\mathrm{G}}^{\mathrm{T}}\ \boldsymbol{M}\ \Delta\dot{\boldsymbol{\omega}}_{\mathrm{G}} \tag{9.118}$$

Now, it is useful to transpose Eq. (9.112)

$$\Delta\dot{\boldsymbol{\omega}}_{\mathrm{G}}^{\mathrm{T}}\ \boldsymbol{M} = -\Delta\boldsymbol{\delta}_{\mathrm{G}}^{\mathrm{T}}\boldsymbol{H}_{\mathrm{G}} - \Delta\boldsymbol{\omega}_{\mathrm{G}}^{\mathrm{T}}\boldsymbol{D} - \Delta\boldsymbol{K}_{\mathrm{ab}}^{\mathrm{T}}\ h_{\mathrm{ab}}\ \Delta\gamma(t) \tag{9.119}$$

Substituting the right-hand side of Eq. (9.119) for $\Delta\dot{\boldsymbol{\omega}}_{\mathrm{G}}^{\mathrm{T}}\ \boldsymbol{M}$ in the first component of Eq. (9.118) gives

$$\begin{aligned}\Delta \dot{E}_{\mathrm{k}} = & -\frac{1}{2}\Delta\boldsymbol{\delta}_{\mathrm{G}}^{\mathrm{T}}\boldsymbol{H}_{\mathrm{G}}\Delta\boldsymbol{\omega}_{\mathrm{G}} - \frac{1}{2}\Delta\boldsymbol{\omega}_{\mathrm{G}}^{\mathrm{T}}\boldsymbol{H}_{\mathrm{G}}\Delta\boldsymbol{\delta}_{\mathrm{G}} - \Delta\boldsymbol{\omega}_{\mathrm{G}}^{\mathrm{T}}\boldsymbol{D}\ \Delta\boldsymbol{\omega}_{\mathrm{G}} \\ & -\frac{1}{2}\left(\Delta\boldsymbol{K}_{\mathrm{ab}}^{\mathrm{T}}\Delta\boldsymbol{\omega}_{\mathrm{G}} + \Delta\boldsymbol{\omega}_{\mathrm{G}}^{\mathrm{T}}\Delta\boldsymbol{K}_{\mathrm{ab}}\right)\ h_{\mathrm{ab}}\Delta\gamma(t)\end{aligned} \tag{9.120}$$

It can be easily checked that both expressions in the last component of Eq. (9.120) are identical scalars as

$$\Delta\boldsymbol{K}_{\mathrm{ab}}^{\mathrm{T}}\Delta\boldsymbol{\omega}_{\mathrm{G}} = \Delta\boldsymbol{\omega}_{\mathrm{G}}^{\mathrm{T}}\Delta\boldsymbol{K}_{\mathrm{ab}} = \sum_{i\in\{\mathrm{G}\}} \Delta K_i \Delta\omega_i \tag{9.121}$$

Hence Eq. (9.120) can be rewritten as

$$\Delta \dot{E}_{\mathrm{k}} = -\frac{1}{2}\Delta\boldsymbol{\delta}_{\mathrm{G}}^{\mathrm{T}}\boldsymbol{H}_{\mathrm{G}}\Delta\boldsymbol{\omega}_{\mathrm{G}} - \frac{1}{2}\Delta\boldsymbol{\omega}_{\mathrm{G}}^{\mathrm{T}}\boldsymbol{H}_{\mathrm{G}}\Delta\boldsymbol{\delta}_{\mathrm{G}} - \Delta\boldsymbol{\omega}_{\mathrm{G}}^{\mathrm{T}}\boldsymbol{D}\ \Delta\boldsymbol{\omega}_{\mathrm{G}} - \Delta\boldsymbol{K}_{\mathrm{ab}}^{\mathrm{T}}\Delta\boldsymbol{\omega}_{\mathrm{G}} h_{\mathrm{ab}}\Delta\gamma(t) \tag{9.122}$$

Adding both sides of Eqs. (9.122) and (9.117) gives

$$\Delta\dot{V} = \Delta\dot{E}_{\mathrm{k}} + \Delta\dot{E}_{\mathrm{p}} = -\Delta\boldsymbol{\omega}_{\mathrm{G}}^{\mathrm{T}}\ \boldsymbol{D}\Delta\boldsymbol{\omega}_{\mathrm{G}} - \Delta\boldsymbol{K}_{\mathrm{ab}}^{\mathrm{T}}\Delta\boldsymbol{\omega}_{\mathrm{G}} h_{\mathrm{ab}}\Delta\gamma(t) \tag{9.123}$$

In a particular case when there is no control, that is when $\Delta\gamma(t) = 0$, the equation in (9.123) gives

$$\Delta\dot{V} = \Delta\dot{E}_{\mathrm{k}} + \Delta\dot{E}_{\mathrm{p}} = -\Delta\boldsymbol{\omega}_{\mathrm{G}}^{\mathrm{T}}\ \boldsymbol{D}\Delta\boldsymbol{\omega}_{\mathrm{G}} \tag{9.124}$$

As matrix $\boldsymbol{D}$ is positive definite, the function above is negative definite. Hence function (9.116) can be treated as the Lyapunov function for the system described by Eq. (9.86).

In order for the considered system to be stable when $\Delta\gamma(t) \neq 0$ changes, the second component in (9.123) should always be positive

$$\Delta\boldsymbol{K}_{\mathrm{ab}}^{\mathrm{T}}\Delta\boldsymbol{\omega}_{\mathrm{G}}\ h_{\mathrm{ab}}\ \Delta\gamma(t) \geq 0 \tag{9.125}$$

This can be ensured using the following control law

$$\Delta\gamma(t) = \kappa\ h_{\mathrm{ab}}\Delta\boldsymbol{K}_{\mathrm{ab}}^{\mathrm{T}}\Delta\boldsymbol{\omega}_{\mathrm{G}} \tag{9.126}$$

With this control law the derivative (9.123) of the Lyapunov function is given by

$$\Delta\dot{V} = -\Delta\boldsymbol{\omega}_{\mathrm{G}}^{\mathrm{T}}\ \boldsymbol{D}\Delta\boldsymbol{\omega}_{\mathrm{G}} - \kappa\ \left(h_{\mathrm{ab}}\Delta\boldsymbol{K}_{\mathrm{ab}}^{\mathrm{T}}\Delta\boldsymbol{\omega}_{\mathrm{G}}\right)^2 \leq 0 \tag{9.127}$$

where κ is the control gain. Taking into account (9.121), the control law (9.126) can be written as

$$\Delta\gamma(t) = \kappa\, h_{\mathrm{ab}} \sum_{i\in\{\mathrm{G}\}} \Delta K_i \Delta\omega_i \tag{9.128}$$

where $\Delta K_i = K_{ia} - K_{ib}$. This control law is valid for any location of the phase shifting transformer. For the particular case when the phase shifting transformer is located in a tie-line, the control law can be simplified, as described below.

The generator set {G} in an interconnected system can be divided into a number of subsets corresponding to subsystems. Let us consider three subsystems as in Figure 9.38, that is $\{G\} = \{G_A\} + \{G_B\} + \{G_C\}$. Now the summation in Eq. (9.128) can be divided into three sums

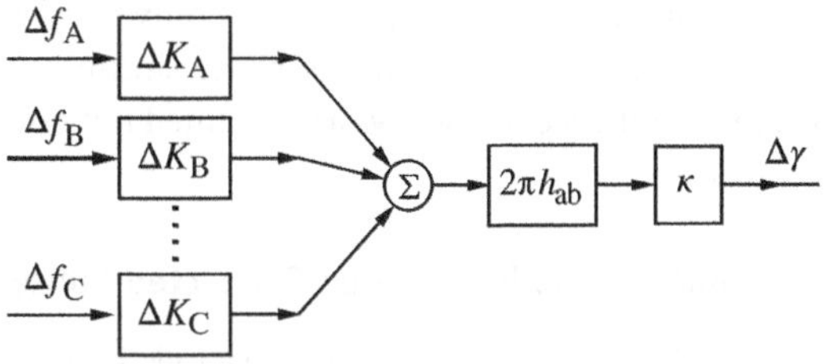

Figure 9.38 Block diagram of the stabilizing control loop of a power flow controller installed in a tie-line of an interconnected power system.

$$\Delta\gamma(t) = \kappa\, h_{ab}\left[\sum_{i\in\{G_A\}} \Delta K_i \Delta\omega_i + \sum_{i\in\{G_B\}} \Delta K_i \Delta\omega_i + \sum_{i\in\{G_C\}} \Delta K_i \Delta\omega_i\right] \quad (9.129)$$

Following a disturbance in one of the subsystems, there are *local swings* of generator rotors inside each subsystem and *inter-area swings* of subsystems with respect to each other. The frequency of local swings is about 1 Hz while the frequency of inter-area swings is much lower, usually about 0.25 Hz. Hence, when investigating the inter-area swings, the local swings can be approximately neglected. Therefore it can be assumed that

$$\begin{aligned}
&\Delta\omega_1 \cong \ldots = \Delta\omega_i \cong \ldots\ldots \cong \Delta\omega_{n_A} \cong 2\pi\,\Delta f_A \quad \text{for} \quad i \in \{G_A\}\\
&\Delta\omega_1 \cong \ldots = \Delta\omega_i \cong \ldots\ldots \cong \Delta\omega_{n_B} \cong 2\pi\,\Delta f_B \quad \text{for} \quad i \in \{G_B\}\\
&\Delta\omega_1 \cong \ldots = \Delta\omega_i \cong \ldots\ldots \cong \Delta\omega_{n_C} \cong 2\pi\,\Delta f_C \quad \text{for} \quad i \in \{G_C\}
\end{aligned} \quad (9.130)$$

Now Eq. (9.129) can be expressed as

$$\Delta\gamma(t) = \kappa 2\pi\, h_{ab}\left[\Delta f_A \sum_{i\in\{G_A\}} \Delta K_i + \Delta f_B \sum_{i\in\{G_B\}} \Delta K_i + \Delta f_C \sum_{i\in\{G_C\}} \Delta K_i\right] \quad (9.131)$$

or, after summing the coefficients,

$$\Delta\gamma(t) = \kappa 2\pi\, h_{ab}(\Delta K_A \Delta f_A + \Delta K_B \Delta f_B + \Delta K_C \Delta f_C) \quad (9.132)$$

where

$$\Delta K_A = \sum_{i\in\{G_A\}} \Delta K_i, \quad \Delta K_B = \sum_{i\in\{G_B\}} \Delta K_i, \quad \Delta K_C = \sum_{i\in\{G_C\}} \Delta K_i \quad (9.133)$$

Equation (9.132) shows that the control of a phase shifting transformer should employ the signals of frequency deviations weighted by coefficients Eq. (9.133).

A block diagram of the supplementary control loop based on Eq. (9.132) is shown in Figure 9.38. The way in which the supplementary control loop is added to the overall regulator is shown in Figure 9.35.

The input signals to the supplementary control are frequency deviations Δf in each subsystem. These signals should be transmitted to the regulator using telecommunication links or wide area measurement system (WAMS) system discussed in Section 2.7. For the frequency of inter-area swings of about 0.25 Hz, the period of oscillation is about 4 s and the speed of signal transmission to the regulator does not have to be high. It is enough if the signals are transmitted every 0.1 s, which is not a tall order for modern telecom systems.

The coefficients h_{ab}, ΔK_A, ΔK_B, and ΔK_C in Eq. (9.132) have to be calculated by an appropriate SCADA/EMS function using current state estimation results and the system configuration. Obviously, those calculations do not have to be repeated frequently. Modifications have to be done only after system configuration changes or after a significant change of power system loading.

When deriving Eq. (9.132), for simplicity only one phase shifting transformer was assumed. Similar considerations can be taken for any number of phase shifting transformers installed in any number of tie-lines. For each transformer, identical control laws are obtained but obviously with different coefficients calculated for the respective tie-lines.

9.7.3 Simulation Model of Three Connected Subsystems

Section 9.6.2 shows that neglecting local swings within the subsystems of an interconnected system allows a simplified system model (Figure 9.34) to be created using the incremental network model. This model takes into account central control of frequency and the tie-line interchange power (secondary control). When the above discussed FACTS devices (such as TCPAR) are installed in tie-lines, the influence of the control variable $\gamma(t)$ on ΔP_{tie} must be taken into account. For this purpose Eq. (9.94) can be used if the TCPAR device is installed (as in Figure 9.36) at the beginning of the tie-line. For further considerations this equation can be rewritten as follows

$$\Delta P_{\text{tie}} = \Delta P_{\text{a}} = \left(1 + \hat{\gamma}^2\right) \hat{H}_{\text{tie}} \Delta\delta_{\text{ab}} + \hat{\gamma}\, \hat{P}_{\text{tie}} \Delta\delta_{\text{ab}} - \hat{H}_{\text{tie}} \Delta\gamma \tag{9.134}$$

where

$$\hat{H}_{\text{tie}} = \hat{H}_{\text{ab}} = b_{\text{ab}} \cos\hat{\delta}_{\text{ab}} \tag{9.135}$$

is the synchronizing power in the pre-fault state. $\hat{P}_{\text{tie}}$ and $\hat{\gamma}$ are respectively the values of the interchange power and control variable in the pre-fault state. Based on Eq. (9.134), one can create the block diagram shown in Figure 9.39. In this diagram the TCPAR controller can be replaced by model shown in Figure 9.35 or in Figure 9.38.

The block diagram shown in Figure 9.39 can be included in the model of two connected subsystems shown in Figure 9.34 in place of the block marked there with H_{ab}. An analogous scheme can also be built for an interconnected power system consisting of any number of subsystems. However, this requires a proper connection of models of individual subsystems. For three systems, a diagram is obtained as in Figure 9.40. In this diagram the abbreviation Dist. means disturbance in power balance in a given subsystem.

9.7.4 Example of Simulation Results

An example of the influence of phase shifting transformer regulation described below is taken from a dissertation published by Rasolomampionona (2007). Figure 9.41 shows a test system with parameters. All three tie-lines contain TCPAR-type devices controlled by the regulators shown in Figure 9.38. The stabilizing controllers use frequency deviations as their input signals.

Figure 9.42 shows the simulation results for a power balance disturbance $\Delta P_0 = 200$ MW consisting of an outage of a generating unit in system A. A thick line shows the responses when a TCPAR device was active and a thin line when the device was not active. Frequency changes in the subsystems are shown in Figure 9.42a. When TCPAR devices are not active, frequency responses are affected by inter-area oscillations (the thin line). Active TCPAR devices quickly damp out the inter-area oscillations and the remaining slow frequency changes are due to the frequency and tie-line flow control (the thick line). The maximum frequency deviation in subsystems B and C is reduced due to the action of TCPAR.

Tie-line flow changes are shown in Figure 9.42b. When TCPAR devices are not acting, changes due to frequency and tie-line flow control are superimposed on inter-area swings (the thin line). Acting TCPAR devices quickly damp out the inter-area swings. The

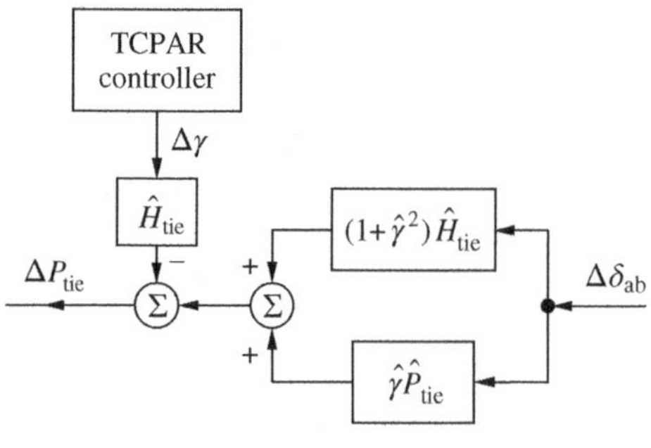

Figure 9.39 Block diagram of the tie-line with TCPAR.

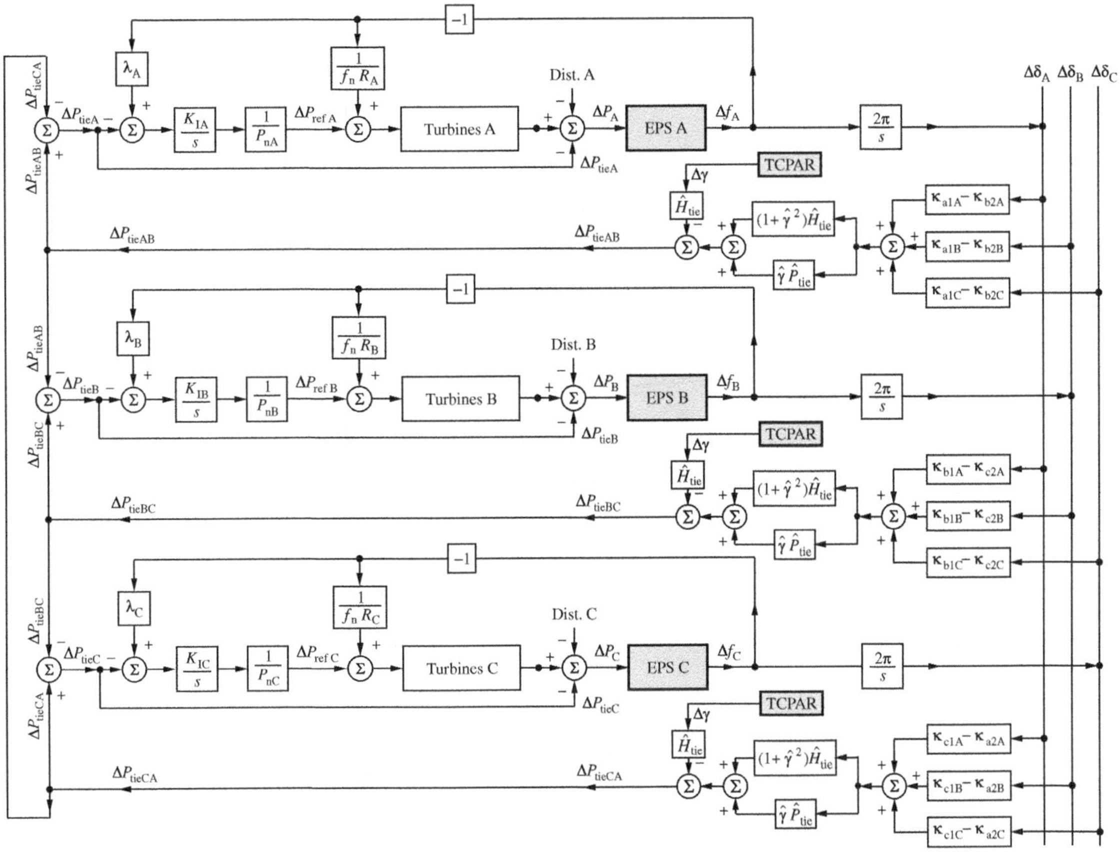

Figure 9.40 Block diagram of the simplified model of three connected subsystems with central AGC and TCPARs in tie-lines.

Figure 9.41 A test system consisting in three subsystems with central AGC.

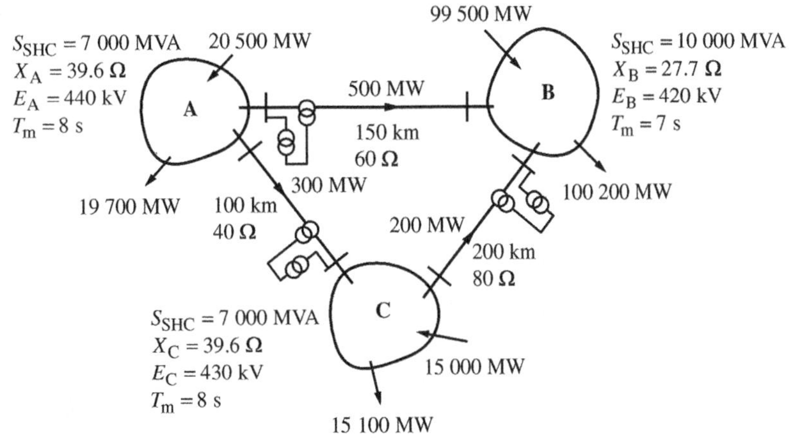

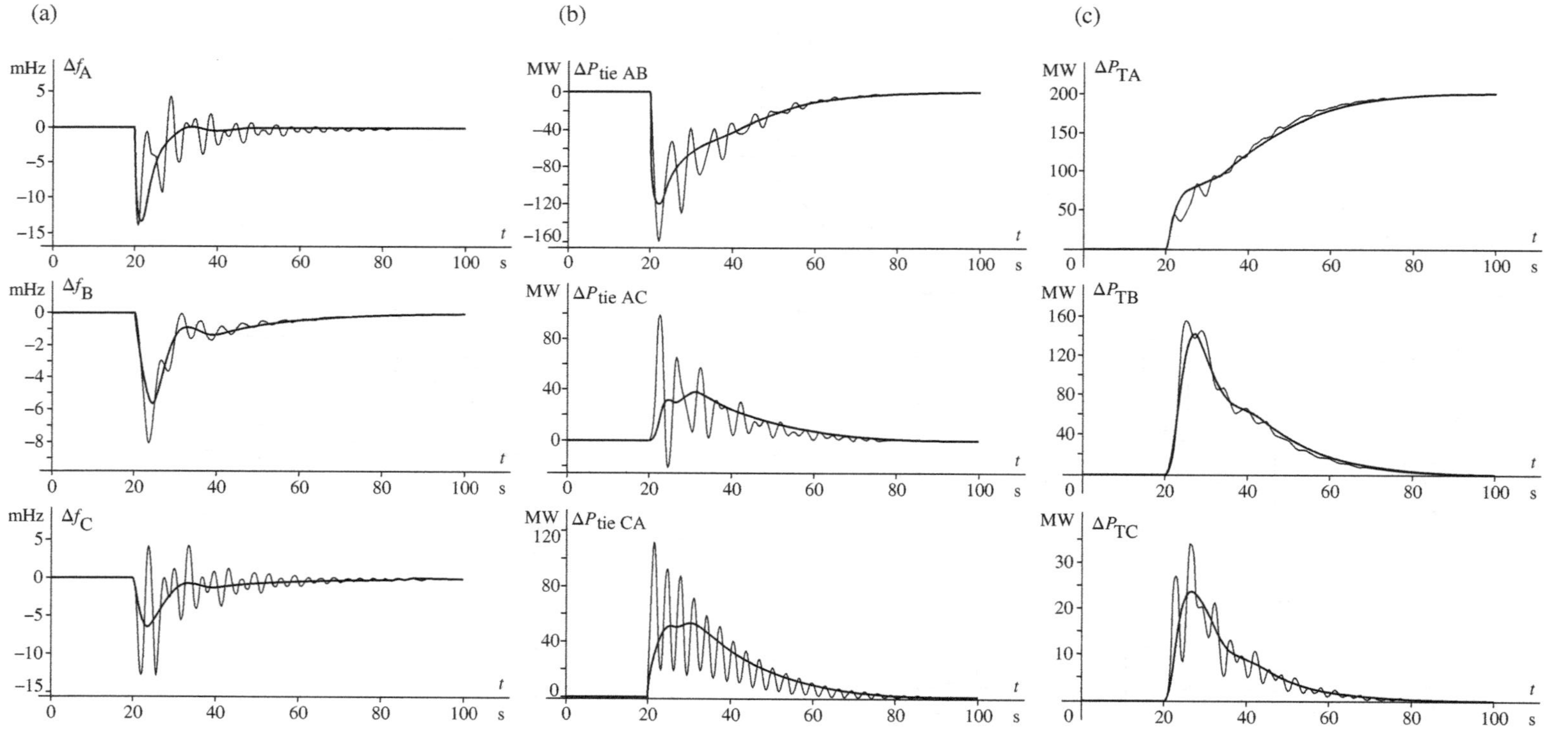

Figure 9.42 Simulation results following a power balance disturbance in subsystem A: (a) local frequency changes; (b) tie-line flow changes; (c) generation changes.

remaining tie-line deviations tend to zero with time which shows that the nonintervention rule is fulfilled. Fulfillment of the rule is also visible in Figure 9.42b showing generation changes. Following the power balance disturbance in subsystem A, subsystems B and C support A for a short time by means of a power injection. As the frequency returns to its reference value and subsystem A increases its generation, generation in B and C returns to its initial value. The diagram also shows inter-area oscillations in generation (especially in subsystem C) when the TCPAR device was not active.

9.7.5 Coordination Between AGC and Series FACTS Devices in Tie-lines

The power flow controller shown in Figure 9.35 can be treated as a multi-level controller consisting of three control paths:

level 1: supplementary control loop with frequency deviations Δf_A, Δf_B, and Δf_C as the input signals (Figure 9.38);
level 2: main control path with active power P_{tie} as the input signal;
level 3: supervisory control at SCADA/EMS level setting $P_{tie\,ref}$.

The actions of these three control loops are superimposed on top of each other and, through changes in $\gamma(t)$, influence tie-line flows and therefore also operation of AGC in individual subsystems of an interconnected power system. In order for both FACTS and AGC controls to be effective and beneficial for the EPS, there must be appropriate coordination. This coordination has to be achieved by adjusting the speed of operation of the three control paths of the FACTS devices to the speed of operation of the three levels of AGC (primary, secondary, tertiary). The three control loops of AGC (Figure 9.12) differ widely in their speed of operation. The three control levels of the FACTS device installed in the tie-line of an interconnected power system must also exhibit a similarly differing speed of operation.

Referring to the description of four stages of power system dynamics due to AGC after a large power imbalance (Sections 9.2–9.5) and the description of operation of a TCPAR-type FACTS device (Section 9.7.4), the following conclusions can be drawn about time coordination of individual control levels.

Supplementary loop control (level 1) should respond quickly, according to the control law (9.132), to frequency changes due to inter-area swings. Hence the speed of reaction of that control level must be the fastest, similar to that of primary control performed by AGC (prime mover control).

Control executed in the main path (level 2) cannot be fast and must be slower than secondary control performed by AGC (frequency and tie-line flow control). This can be explained in the following way. Following an active power imbalance in a given subsystem, a power injection, lasting several tens of seconds, may flow to that subsystem from the other subsystems (Figure 9.42b). This power injection causes P_{tie} to be different from $P_{tie\,ref}$ and a control error appears in that control path. If the controller reacted too quickly, the FACTS device could affect the power injection which would adversely affect the frequency control of the secondary level of AGC. The maximum frequency deviation would increase and the quality of regulation would decrease (Figure 9.13). To prevent this, the discussed control level should act with a long time constant. Figure 9.35 shows that the main control path contains an integrator with a feedback loop. The transfer function of the element is $G(s) = 1/(\rho_P + sT_P)$, which means that the speed of operation of the element is determined by the time constant T_P/ρ_P. If this time constant is several times higher than the duration time of power injection then the discussed control level should not adversely affect the secondary control executed by AGC.

The supervisory control (level 3) setting $P_{tie\,ref}$ executed by SCADA/EMS must be the slowest. Especially important for the dynamic performance is the case shown in Figure 9.42d when insufficient regulation power in the subsystem where the power imbalance occurred must result in a permanent deviation in exchanged power. The FACTS device controlled by the regulator shown in Figure 9.35 will try to regulate P_{tie} to a value $P_{tie\,ref}$. It may turn out that such regulation is not beneficial for the system and result in, for example, overloading of other transmission lines. Regulation at that level must be centrally executed by SCADA/EMS based on the analysis of the whole network.

9.8 Static Analysis by Snapshots of Power Flow

The transients in an EPS following a loss of generation can be simulated in the simplified way by the method described in Section 9.6. A more accurate simulation can be performed by mid-term or long-term simulation programs using the simulation methods described in Chapter 13 and the mathematical models described in Chapter 11. However, when a multiple-case analysis is required (e.g. power system expansion planning, power system security assessment) the use of mid/long-term simulation is too time consuming and is unacceptable from a practical point of view. In such a situation a static approach is preferred in which power system performance is evaluated by snapshots of power flows for characteristic stages of the power system dynamic response described in Sections 9.2–9.5. The following characteristic snapshots of the power flows can be then defined:

- the initial power flow in the pre-fault state;
- the *acceleration power flow* at the first moment of the disturbance, when the power flow in the network depends on the transient emf's of generators and their angles in the pre-fault state;
- the *inertial power flow* at the second stage (Section 9.3), when frequency drops and the power imbalance is distributed among the generating units proportionally to their inertia coefficients;
- the *governor power flow* at the third stage (Section 9.4), when the distribution of power imbalance depends on the frequency characteristics of the turbines governing systems (primary control);
- the *AGC power flow* at the fourth stage (Section 9.5), when at the end of the dynamic response the power imbalance is covered by the generating units controlled by the AGC (secondary control) proportionally to their participation factors (Section 9.1.3).

The acceleration and inertial power flows can be used to assess RMS voltage sags (also referred to as *RMS voltage dips*). Sustained voltage sags (longer then several seconds) and overloads in the network can be evaluated on the basis of the AGC power flow. It is shown below that the inertial, governor, and AGC power flows can be found by power flow programs, which use participation factors to distribute the power mismatch among a number of generation nodes, as described in Section 3.6.1.

In the second stage of the dynamic response (Section 9.3) the power imbalance is covered by all generating units proportionally to their inertia coefficients M_k. On the basis of Eqs. (3.193) and (9.26) it can be written

$$P_k = P_{0k} + \alpha_k\,\Delta P_\Sigma \quad \text{where} \quad \alpha_k = \frac{M_k}{\sum_{i=1}^{N_G} M_i} \tag{9.136}$$

It is worth emphasizing that in the second stage of the dynamic response the electric power of the generating units is not limited by turbine constraints $P_{\min} \geq P_k \leq P_{\max}$ and for this reason the mismatch in Eq. (9.136) is distributed among all generation nodes $k = 1, \ldots, N_G$ and summing appearing in the participation factors α_k defined by Eq. (9.136) is for all $i = 1, \ldots, N_G$.

Another situation is in the third stage described in Section 9.4. As shown in Figure 9.17, the distribution of power imbalance depends then on the static characteristics of all turbines and frequency characteristics of all loads. The drop in frequency is determined by Eq. (9.34). For the k-th generating unit, on the basis of (9.1), it can be written that $P_k = P_{0k} - K_k P_{0k}\,\Delta f_{\mathrm{III}}/f_{\mathrm{n}}$, where $K_k = 1/\rho_k$ and ρ_k is the droop of turbine static characteristic. Substituting in this equation $\Delta f_{\mathrm{III}}/f_{\mathrm{n}}$ by the right-hand side of Eq. (9.34) gives

$$P_k = P_{0i}k + \frac{K_k}{K_{\mathrm{f}}}\frac{P_{0k}}{P_{\mathrm{L}}}\Delta P_\Sigma \tag{9.137}$$

where K_{f} is the system stiffness introduced in Eq. (9.7) and P_{L} is the total system load including network losses. Assuming that $K_{\mathrm{f}} = K_{\mathrm{T}} + K_{\mathrm{L}} \cong K_{\mathrm{T}}$ and taking into account Eq. (9.5), it can be obtained from Eq. (9.137) that

$$P_k = P_{0k} + \frac{K_k P_{0i}}{\sum_{i=1}^{N_G} K_i P_{0i}} \Delta P_\Sigma \tag{9.138}$$

This equation shows that in the governor power flow, the power mismatch must be distributed among generation nodes in the following way

$$P_k = P_{0k} + \alpha_k \, \Delta P_\Sigma \quad \text{where} \quad \alpha_k = \frac{K_k P_{0k}}{\sum_{i=1}^{N_G} K_i P_{0i}} \tag{9.139}$$

Obviously, in the third stage of the power system dynamic response (primary control) the active power generated by each generating unit is limited by turbine constraints $P_{\min} \geq P_k \leq P_{\max}$ and depends on its static characteristic and frequency drop Δf_{III} (Figure 9.17) caused by a given power imbalance. For generating units, which operate with a wide dead zone (Figure 9.6b) and do not participate in the primary frequency control, the droop of static characteristic is equal $\rho_k = \infty$ and $K_k = 1/\rho_k = 0$. In order to select a proper value of K_k for each generating unit, it is necessary to involve an estimation of the frequency drop Δf_{III} in the iterative process of the governor power flow program. This complicates the calculation algorithm because the frequency drop becomes an additional state variable. The estimated frequency drop can be calculated in each step of the iterative process on the basis of Eq. (9.34) assuming (as above) that $K_{\mathrm{f}} \cong K_{\mathrm{T}}$

$$\Delta f_{\mathrm{III}} \cong \frac{-1}{K_{\mathrm{T}}} \frac{\Delta P_0}{P_{\mathrm{L}}} f_{\mathrm{n}} \cong \frac{-1}{\sum_{i=1}^{N_G} K_i} \frac{\Delta P_\Sigma}{\sum_{i=1}^{N_G} P_{0i}} f_{\mathrm{n}} \tag{9.140}$$

A simple expression for the distribution factors α_k can be obtained by assuming, as in Eq. (9.38), that the droop of all the units which participate in the primary control (Figure 9.6a) is approximately identical, that is $\rho_i = \rho_k = \rho$ and $K_i = K_k = K = 1/\rho$. For the units which form an additional primary reserve (Figure 9.6b) and have a wide dead zone, $\rho_i = \infty$ and $K_i = 0$. Under these conditions, Eq. (9.139) shows that the distribution factors are approximately constant and equal to the share of the given generating unit in the active power generation

$$\alpha_i = \frac{P_{0i}}{\sum_{i=1}^{N_R} P_{0i}} \quad \text{and} \quad \sum_{i=1}^{N_R} \alpha_i = 1 \tag{9.141}$$

where N_R is the number of units participating in primary control. This assumption simplifies significantly the algorithm of the governor power flow program.

The algorithm of the AGC power flow program is also simple. In this case the distribution factors α_i in equation

$$P_i = P_{0i} + \alpha_i \, \Delta P_\Sigma \quad \text{and} \quad P_{\min} \geq P_i \leq P_{\max} \tag{9.142}$$

are simply treated as the fixed input data and are equal to the participation factors $\alpha_1, \alpha_2, \ldots, \alpha_n$ of generating units in the secondary control (Figure 9.10 in Section 9.1.3).

For static security assessment and especially evaluation of possible overloads and RMS sustained voltage sags in the EPS equipped with the secondary control the AGC power flow program is a valuable and simple tool. However, for the EPS which operates with only primary control a more complicated governor power flow program must be used.

10

Stability Enhancement

The stability of an electric power system (EPS) is understood as its ability to return to the equilibrium state after being subjected to a physical disturbance. Important variables at power system equilibrium are rotor (power) angles, nodal voltages and frequency. Hence power system stability can be divided into: (i) rotor (power) angle stability, (ii) voltage stability, and (iii) frequency stability. These terms are introduced in Chapter 1 when discussing Figure 1.5. Prevention of voltage instability (voltage collapse) is discussed in Section 8.6. A defense plan against frequency instability is discussed in Section 9.1.6. This chapter deals with the possibilities of counteracting rotor (power) angle instability.

The rotor (power) angle stability of an EPS can be enhanced, and its dynamic response improved, by correct system design and operation. For example, the following features help to improve stability:

- the use of protective equipment and circuit-breakers that ensure the fastest possible fault clearing;
- the use of single-pole circuit-breakers so that during single-phase faults only the faulted phase is cleared and the unfaulted phases remain intact;
- the use of a system configuration that is suitable for the particular operating conditions (e.g. avoiding long, heavily loaded transmission links);
- ensuring an appropriate reserve in transmission capability;
- avoiding operation of the system at low frequency and/or voltage;
- avoiding weakening the network by the simultaneous outage of a large number of lines and transformers.

In practice financial considerations determine the extent to which any of these features can be implemented, and there must always be a compromise between operating a system near to its stability limit with that of operating a system with an excessive reserve of generation and transmission. The risk of losing stability can be reduced by using additional elements inserted into the system to help smooth the system dynamic response. This is commonly referred to as *stability enhancement* and is the subject of this chapter.

Thanks to damping windings, synchronous generators are characterized by quite large electromagnetic torques that suppress their rotor swings in the post-fault states. Under the influence of voltage regulators these swings can undergo quite large changes. As shown in Chapter 5 (Sections 5.4 and 5.5) and Chapter 6 (Section 6.1.5), the voltage regulators can induce additional currents in the rotor circuits (damper and field windings) that oppose the currents induced by the rotor speed deviation. Such additional currents create a negative component of the damping torque and weaken the natural damping provided by the damper and field windings. This results in weakly damped swings, as shown in Figure 6.14. In extreme situations the negative damping forced by the action of the automatic voltage regulator (AVR) can exceed the positive damping forced by the rotor speed deviation and system becomes oscillatorily unstable (Figures 5.13 and 6.12). Stability enhancement can be obtained with additional elements incorporated in the excitation control system.

Power System Dynamics: Stability and Control, Third Edition. Jan Machowski, Zbigniew Lubosny, Janusz W. Bialek and James R. Bumby.

10.1 Excitation Control System

There are two basic methods of improving the dynamic performance of the EPS by using excitation systems of synchronous generators: (a) lag–lead compensation in the control system referred to as TGR (transient gain reduction) and (b) supplementary control loop referred to as the PSS (power system stabilizer).

10.1.1 Transient Gain Reduction

Section 5.5.3 explains that the negative damping increases with the following factors: increased equivalent reactance of the network connecting the generator with remaining part of the EPS, increased generator load resulting in a large value of the power angle, increased gain of the AVR, and finally reduced equivalent time constants of the excitation system.

Figure 10.1 shows an example of rotor angle oscillations for three values of the voltage regulator gain K_A. In this example, with low gain $K_A = 60$ the oscillations decline, owing to resultant damping being positive. With higher gain value $K_A = 120$ and $K_A = 250$ negative damping forced by the voltage regulator exceeds positive damping resulting from rotor speed deviation and oscillations diverge.

Figure 10.2 shows an example of how the maximum voltage regulator gain depends on the power angle, with the equivalent time constant of the excitation system as a parameter. Curves 1, 2, 3, and 4 correspond to different values of this time constant in the ascending order. For a small time constant (curve 1) the system is stable (area highlighted in the figure) at very low gain values. For larger time constants (slower voltage regulation) larger gains limited by curves 2, 3, 4 are permissible.

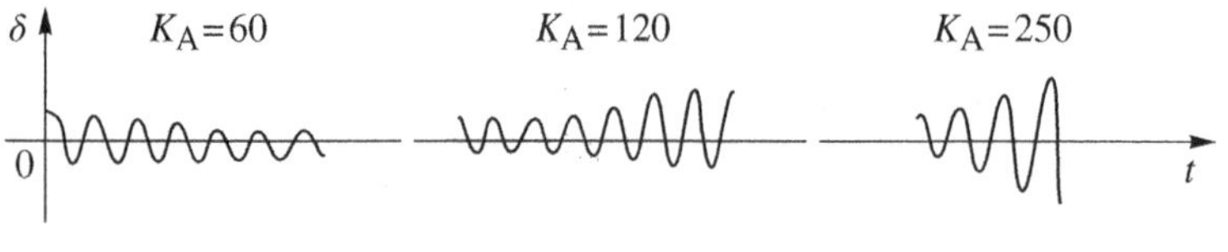

Figure 10.1 Examples of rotor angle oscillations for three values of the voltage regulator gain.

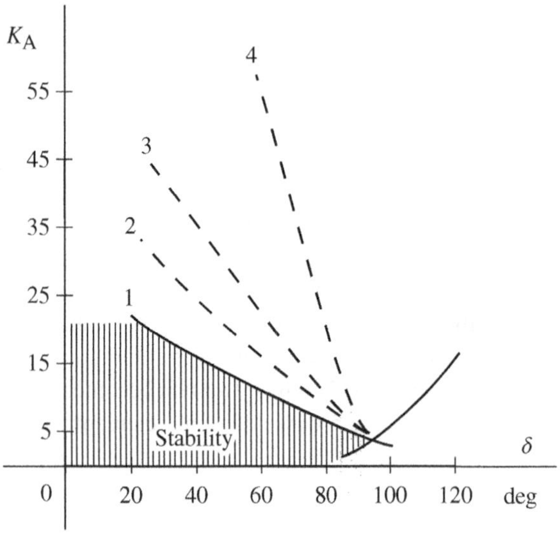

Figure 10.2 Example of the dependence between maximal voltage regulator gain and power angle.

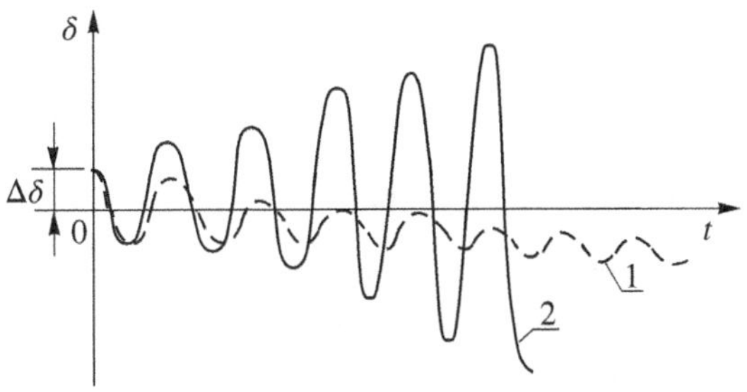

Figure 10.3 Example of rotor angle swings in the case of: 1 – too large voltage regulator gain; 2 – with the manual voltage control.

When voltage regulators of synchronous generators have large gains and the excitation system has small time constants, then weakly damped or diverging swings will appear, as shown with a solid line in Figure 10.3. In such situations a power system operator usually orders to switch off the AVR's of some generators and operates with manual voltage control. Generators working with manual voltage control have only natural positive damping and oscillations decline as shown in Figure 10.3 with a dashed curve.

Therefore increasing the time constant and decreasing the gain of the voltage regulator makes it possible to alleviate the harmful influence of voltage regulation on damping of swings. However, this methods cannot be used in practice because good-quality voltage regulation requires quick regulators with high gain. There is a contradiction between good-quality voltage regulation and strong damping of generator rotor and power swings in the post-fault states. To overcome this contradiction an additional element TGR can be incorporated in the control system of the AVR, as shown in Figure 10.4.

TGR is effectively a cascade lag–lead compensation that attenuates the gain even several times for frequency typical for power swings. An example of such characteristics is shown in Figure 10.5. For this lag–lead element with transfer function $G(s) = (1 + 3s)/(1 + 25s)$ for $\Delta f \geq 0.3$ Hz the magnitude of frequency characteristic is $|G(s)| \cong 0.12$, what means that the gain is reduced $1/0.12 \cong 8.3$ times. For $\Delta f \geq 0.4$ Hz the phase of the transfer function $\arg G(s)$ is very small, what means that for frequency typical for power swings Δf (0.8–1.5) Hz this phase compensator practically does not change the phase of the voltage error.

The TGR can be called a passive method because it only eliminates the possible harmful effects of the AVR on damping of rotor and power swings but does not use excitation control of synchronous generators to enhance the

Figure 10.4 Block diagram of the voltage control with TGR.

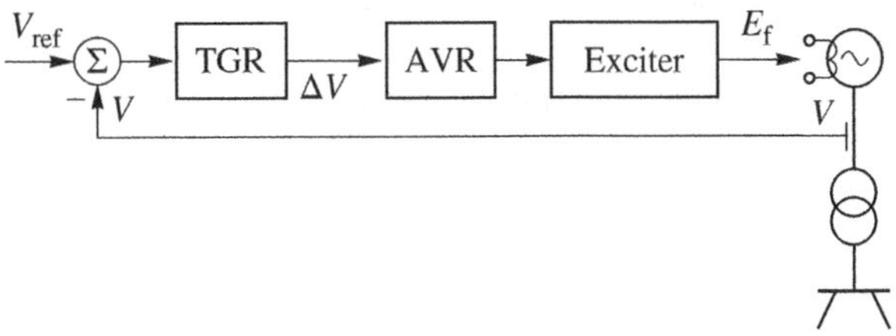

Figure 10.5 Bode plots of the transfer function $(1 + s3)/(1 + s25)$ (a) magnitude, (b) phase.

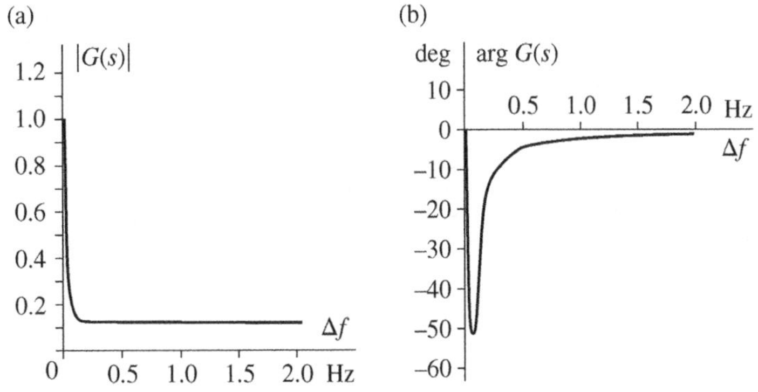

damping in power system. An active method, which assumes the use of the excitation control to improve damping, is described below.

10.1.2 Power System Stabilizers

Adding supplementary control loops to the generator AVR is one of the most common ways of enhancing both small-signal (steady-state) stability and large-signal (transient) stability. Adding such additional control loops must be done with great care; Section 5.5 explains how an AVR (without supplementary control loops) can weaken the damping provided by the damper and field windings. This reduction in the damping torque is primarily due to the voltage regulation effects inducing additional currents in the rotor circuits that oppose the currents induced by the rotor speed deviation $\Delta\omega$. This phase relationship is illustrated in Figure 5.26 for the field winding and in Figure 5.27 for the damper winding. These two figures give an immediate insight into what is required from the PSS.

The main idea of power system stabilization is to recognize that in the steady state, that is when the speed deviation is zero or nearly zero, the voltage controller should be driven by the voltage error ΔV only. However, in the transient state the generator speed is not constant, the rotor swings, and ΔV undergoes oscillations caused by the change in rotor angle. The task of the PSS is to add an additional signal which compensates for the ΔV oscillations and provides a damping component that is in phase with $\Delta\omega$. This is illustrated in Figure 10.6a, where the signal V_{PSS} is added to the main voltage error signal ΔV. In the steady state V_{PSS} must be equal to zero so that it does not distort the voltage regulation process. Figure 10.6b shows the phasor diagram of the signals in the transient state. As in Section 5.5.3, it is assumed that each signal varies sinusoidally with the frequency of rotor swings and may therefore be represented by a phasor. The phasor $\underline{V}_{PSS}$ directly opposes $\Delta\underline{V}$ and is larger than it. The net voltage error phasor $\Delta\underline{V}_{\Sigma} = \Delta\underline{V}_{PSS} - \Delta\underline{V}$ now leads the speed deviation phasor $\Delta\underline{\omega}$ instead of lagging it, as in Figure 5.27c. As explained in Section 5.5.3, the phasor of the incremental excitation electromotive force (emf) $\Delta\underline{E}_{f}$ lags $\Delta\underline{V}_{\Sigma}$ by an angle α introduced by the AVR and the exciter so that the quadrature component (with respect to $\Delta\underline{\delta}$) of the phasor $\Delta\underline{E}'_{q(\Delta E_f)}$ due to the excitation control is now in phase with $\Delta\underline{\omega}$. This, together with $\Delta\underline{E}'_{q(\Delta\delta)}$, introduces a large damping torque into the system. However, if the magnitude of $\underline{V}_{PSS}$ is less than that of $\Delta\underline{V}$ then only partial compensation of the negative damping component introduced by the AVR is achieved.

It is worth remembering that the lagging phase shift α introduced by the AVR and the exciter is in the range of several dozen degrees for rotating exciters (Figure 2.4a–c) and from a few to a dozen of degrees for static exciters (Figure 2.4d–f).

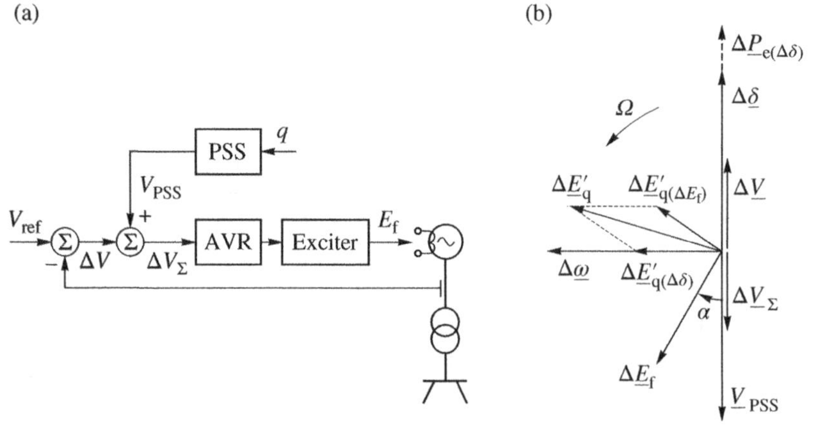

Figure 10.6 Supplementary control loop for the AVR system: (a) block diagram; (b) phasor diagram.

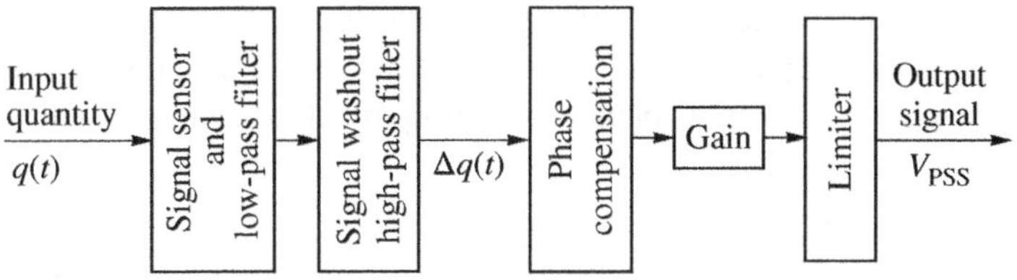

Figure 10.7 The major elements of a PSS.

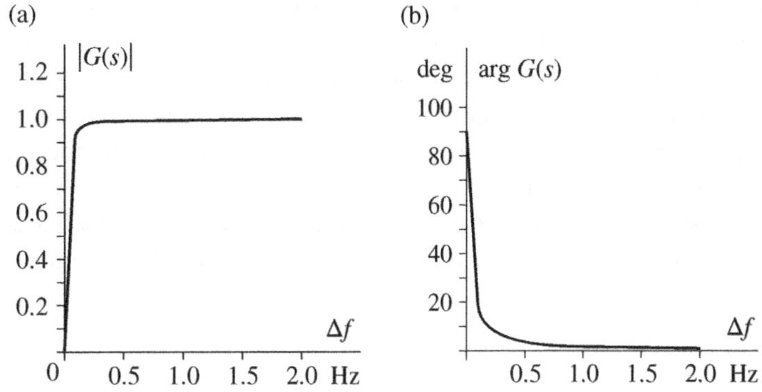

Figure 10.8 Bode plots of the transfer function of the real differentiator with T_w = 5 s (a) magnitude; (b) phase.

The general structure of the PSS is shown in Figure 10.7. The measured quantity $q(t)$ is passed through a low-pass (anti-aliasing) filter cutting out high-frequency components and the noise. Then, the filtered signal is passed through the washout filter eliminating an offset from the waveform signal. Usually, the real differentiator is used with transfer function $G_w(s) = sT_w/(1 + sT_w)$ and $T_w = (2 - 10)$ s as the washout filter. An example of the frequency characteristics of such a function is shown in Figure 10.8. When the frequency of the signal $q(t)$ is $\Delta f \geq 0.3$ Hz, the washout filter passes its alternating part $\Delta q(t)$ almost without a change in the magnitude and phase, because for such frequency $|G(s)| \cong 1.0$ and arg $G(s)$ is very small. The filtered waveform $\Delta q(t)$ is then passed through a lead and/or lag element in order to obtain the required phase shift. The required phase shift depends on the kind of quantity $q(t)$ used as the input signal and time constants of other elements in the circuit. Finally, the signal is amplified and passed to a limiter providing the output signal V_{PSS}.

Dynamic properties of the AVR with the PSS depend very much on what is the quantity of the input signal. In practice, various electric and mechanic quantities are used. Typically, the measured quantities used as input signals to the PSS are the rotor speed deviation, the generator active power, or the frequency of the generator terminal voltage. There are a number of possible ways of constructing a PSS, depending on the signal chosen. They have their advantages and disadvantages. The most commonly used solutions are discussed below.

10.1.2.1 PSS Based on $\Delta\omega$

The oldest type of PSS uses the rotor speed ω measured at the end of the generating unit shaft usually on the side of the generator. Obviously (as in Figure 10.7), this signal must be passed through the low-pass filter (in order to filter out the measurement noise) and through the washout filter (in order to eliminate the synchronous speed and obtain the speed deviation $\Delta\omega$). To get the position of the phasor $\Delta \underline{E}_f$, as in Figure 10.6b, a lead element with quite a big phase shift should be used. This is illustrated in Figure 10.9a. This figure shows phasors with phase shifts for one selected value of the oscillation frequency $\Delta f = \Omega/2\pi$.

. It was assumed that for this frequency the AVR and exciter (Figure 10.6a) introduce the lagging phase shift between $\Delta \underline{V}_\Sigma$ and $\Delta \underline{E}_f$ equal to the angle α. If the PSS did not contain phase compensation, one would obtain $\underline{V}_{PSS}$ in phase with $\Delta\underline{\omega}$ and phasor $\Delta \underline{E}_{f(K\Delta\omega)}$ lying in the second quadrant of the plane, as in Figure 10.9a. In such a case the PSS would weaken the damping or even cause negative damping. In order to obtain phasor $\Delta \underline{E}_f$ lying in the third quarter of the plane as in Figure 10.6b, the phase $\underline{V}_{PSS}$ must be changed by an angle $\gamma = (\alpha + \beta)$. Such a large leading phase shift can be obtained by using two lead elements with transfer function

$$G_c(s) = \frac{1 + sT_3}{1 + sT_4} \frac{1 + sT_5}{1 + sT_6} \tag{10.1}$$

in which $T_3 > T_4$ and $T_5 > T_6$. Exemplary values: $T_3 = T_5 = 0.8$ and $T_4 = T_6 = 0.1$.

This type of PSS performs well in the case of generating units of small and medium power ratings with short shafts. In the case of turbogenerators with very high power ratings and long shafts the problem arises that, during the rotor swings, the rotor torsional oscillations in the drive shaft also occur (Figure 4.39), as described in Section 4.4.3. If the PSS receives a measurement of the rotor speed from only one shaft point (e.g. from the shaft end), it can amplify the torsional oscillations. This can result in a reduction in the shaft fatigue life and possibly shaft failure. In order to overcome this problem in practice, as an input signal to the PSS an average value of rotor speed measured at a number of points along the shaft is used. This, however, requires the use of a complex measuring system. A suitable low-pass filter attenuating signal of a frequency much higher than the rotor swings frequency can also be introduced into the PSS structure. These problems are described by Watson and Coultes (1973) and Kundur et al. (1981).

10.1.2.2 PSS Based on P_e

As shown in Figure 10.6b, also the signal $(-P_e)$ corresponding to the negative value of active power of the generator has a corresponding phase to create the signal V_{PSS}. Obviously (as in Figure 10.7), this signal must be filtered by the low-pass filter and by the washout filter to obtain signal proportional to the change in active power $(-\Delta P_e)$. To obtain the position of the phasor $\Delta \underline{E}_f$, as in Figure 10.6b, it is necessary to use a lag element with a small phase shift. This is illustrated in Figure 10.9b. This figure shows phasors with phase shift for one selected value Δf. It was assumed that for this frequency the AVR and exciter (Figure 10.6a) lag $\Delta \underline{E}_f$ relative to $\Delta \underline{V}_\Sigma$ by angle α. If the PSS did not contain phase compensation, $\underline{V}_{PSS}$ would be obtained in phase with $(-\Delta \underline{P}_e)$ and phasor $\Delta \underline{E}_{f(K\Delta P)}$ lying in the third quadrant of the plane, as in Figure 10.9b. In such a case, the PSS would improve the damping. To obtain even better damping, phasor $\Delta \underline{E}_f$ needs to be lagged by angle $\gamma = \pi/2 - (\alpha + \beta)$ to get phasor $\Delta \underline{E}_{f(PSS)}$ lying in the third quadrant at angle β, as in Figure 10.6b. Such a small phase shift can be obtained by using one lag element with a transfer function

$$G_c(s) = \frac{1 + sT_3}{1 + sT_4} \tag{10.2}$$

wherein $T_3 < T_4$. Exemplary values: $T_3 = 0.85$ s and $T_4 = 4.75$ s.

Transfer function of such PSS corresponds to the multiplication of the transfer function of the measuring element (signal sensor), the washout, and the lag compensator

$$G_{PSS}(s) = G_w(s) \cdot G_c(s) = K \frac{1}{1 + sT_m} \frac{sT_w}{1 + sT_w} \frac{1 + sT_3}{1 + sT_4} \tag{10.3}$$

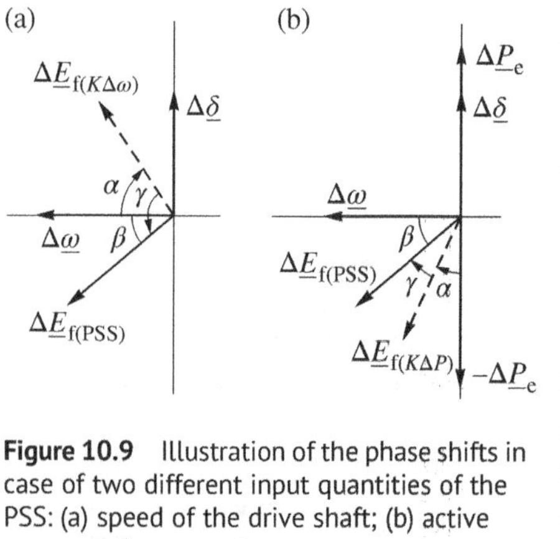

Figure 10.9 Illustration of the phase shifts in case of two different input quantities of the PSS: (a) speed of the drive shaft; (b) active power of the generator.

Figure 10.10 shows the frequency characteristic of the phase of a PSS with active power as input signal and transfer function (10.3) for the following parameters: $K = -1$, $T_m = 0.02$ s, $T_w = 5$, $T_3 = 0.85$ s, and $T_4 = 4.75$ s. As can be seen, for frequency (0.3–2.0) Hz such a PSS introduces the lagging phase shift of a value about −20°. This shift is good for excitation systems with static exciters, when the angle

α is small as in Figure 10.9b. In the case of excitation systems with rotating exciters the lagging phase shift α is larger and phase compensation introduced by the lag element of PSS should be lower and in some cases even unnecessary.

The PSS discussed here, based on the measurement of active power, has two major drawbacks:

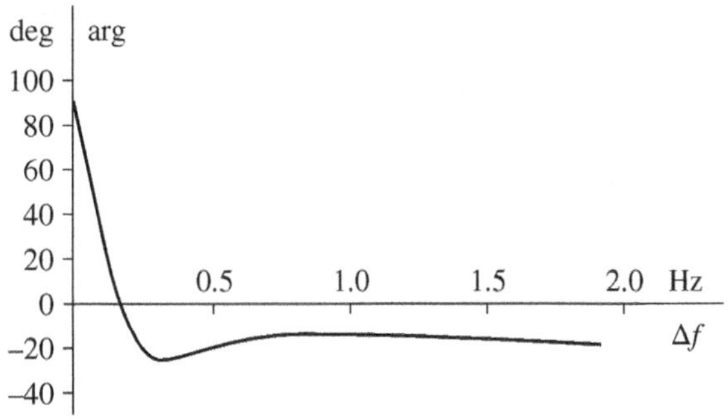

Figure 10.10 Example of frequency characteristic of phase of a PSS with active power as input signal.

- It produces a signal when there is a change in the active power of the generating unit. Such changes occur not only during power swings in the post-fault state. The change in active power may result from a change in turbine power forced by the operation of the automatic generation control (Figure 9.12). In this case, the PSS reacting unnecessarily to ΔP_e generates signal V_{PSS} and acts on the excitation current of the generator and thus causes the oscillations of reactive power.
- As shown in Section 6.1, during a short circuit in the network the active power of the generator drops from its pre-fault value to a small value depending on type of short circuit (Figure 6.3) and distance from the generator (Figure 6.5). When active power P_e decreases then its increment ΔP_e is negative and thus the signal proportional to $(-\Delta P_e)$ is positive and hence $V_{PSS} > 0$. This means that the PSS reduces voltage error ΔV and opposes a rapid increase of the generator excitation current. As shown in Figure 6.10, such an action is detrimental to transient stability and the PSS during fault duration rather worsens instead of improve transient stability. The improvement of transient stability occurs after fault clearing.

Despite the disadvantages discussed here, the PSS based on active power measurement is often used in practice due to its simplicity and low manufacturing costs.

10.1.2.3 PSS Based on P_e and $\Delta\omega$

The need to measure the speed deviation at a number of points along the shaft can be avoided by calculating the average speed deviation from measured electrical quantities. The method calculates the equivalent speed deviation $\Delta\omega_{eq}$ indirectly from the integral of the accelerating power

$$\Delta\omega_{eq} = \frac{1}{M}\int(\Delta P_m - \Delta P_e)dt \tag{10.4}$$

and ΔP_e is calculated from measurements of the generated active power P_e. The integral of the change in the mechanical power ΔP_m can be obtained from

$$\int \Delta P_m dt = M\Delta\omega_{measured} + \int \Delta P_e dt \tag{10.5}$$

where $\omega_{measured}$ is based on the end-of-shaft speed sensing system. Because the mechanical power changes are relatively slow, the derived integral of the mechanical power can be passed through a low-pass filter to remove the torsional frequencies from the speed measurement. The resulting PSS contains two input signals, $\Delta\omega_{measured}$ and ΔP_e, which are used to calculate $\Delta\omega_{eq}$. The final V_{PSS} signal is designed to lead $\Delta\omega_{eq}$. The block diagram of the system is shown in Figure 10.11 (Kundur 1994), where $G(s)$ is the transfer function of the torsional filter. This type of PSS with two input signals allows a large gain to be used so that good damping of power swings is obtained (Lee et al. 1981).

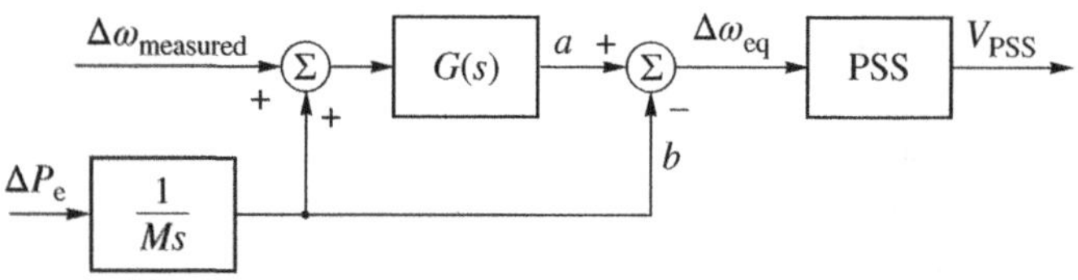

Figure 10.11 Block diagram of a PSS using speed deviation and active power as input signals.

10.1.2.4 PSS Based on f_{V_g} and $f_{E'}$

The measurement of shaft speed can be replaced by a measurement of the generator terminal voltage frequency f_{V_g} (Larsen and Swan 1981). A disadvantage of this solution is that the terminal voltage waveform can contain noise produced by large industrial loads such as arc furnaces. The accuracy of this measured speed signal can be improved by adding the voltage drop across the transient reactance to the generator voltage to obtain the transient emf E' and its frequency $f_{E'}$. The PSS now receives two signals: the generator current and voltage. Similar to the case of the PSS utilizing the measured shaft speed deviation the PSS gain is limited by the effect of shaft torsional oscillations. The advantage of this solution compared with other types of stabilizers is that it improves the damping of inter-area oscillations in interconnected power systems.

10.1.2.5 Influence of PSS on Electromechanical Swings in Transient State

With the right choice of parameters, PSSs improve damping of rotor swings for both small and large disturbances. It is worth coming back to Figure 6.12 and 6.14 in Section 6.1.5 illustrating the influence of the AVR on rotor motion in the post-fault state. There it is mentioned that under certain conditions just after fault clearing in backward rotor swing the AVR can keep the field (excitation) current on the ceiling and can increase the deceleration area and thus increase the second and further rotor swings. The action of the AVR equipped with the PSS (Figure 10.6) is completely different. When a fault is cleared, the PSS generates supplementary signal V_{PSS}, and improves damping and rotor swings and power oscillations decline faster, as illustrated in Figures 10.12 and 10.13.

Figure 10.12 illustrates the rotor swings caused by a short circuit in the network when the EPS is stable. It is clear from the comparison of both waveforms that PSS contributes to the faster disappearance of oscillations. It is worth noting, however, that the influence of the PSS action is visible only from the first backward swing.

Figure 10.13 illustrates oscillations of the active power of the generator and the excitation (field) voltage in the case very close to the critical state. This is a case analogous to the case shown in Figure 6.9 (Section 6.1.3). It can be seen that after a short circuit appearance the voltage control system (AVR + PSS) increases the excitation (field) voltage up to the ceiling. During the backward swing, this voltage is decreased very fast and then increased very fast during the forward swing, etc., up to the disappearance of oscillations.

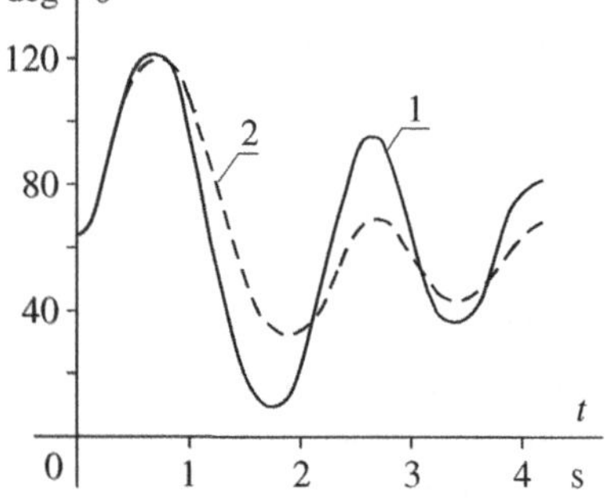

Figure 10.12 Example of post-fault rotor swings in the case of the AVR without PSS (curve 1) and with PSS (curve 2).

10.1.2.6 PSS Design

Designing and applying the PSS is not simple and requires a thorough analysis of the regulator structure and its parameters. A badly designed PSS can become the source of a variety of undesired oscillations. It should be remembered that the phasor diagrams in Figures 5.26, 5.27, and 10.6 are valid only for a simple generator-infinite busbar system with all the resistances and local loads neglected. A more detailed analysis shows that the phase shift between $\Delta \underline{E}_f$ and $\Delta \underline{E}'_{q(\Delta E_f)}$ is not exactly $\pi/2$ and depends on the loading and system parameters (DeMello and Concordia 1969). This requires a more precise matching of the phase compensation to the actual loading and system parameters. Suitable optimization methods are described in a book by Gibbard et al. (2015) and also in Chapter 15 of this book.

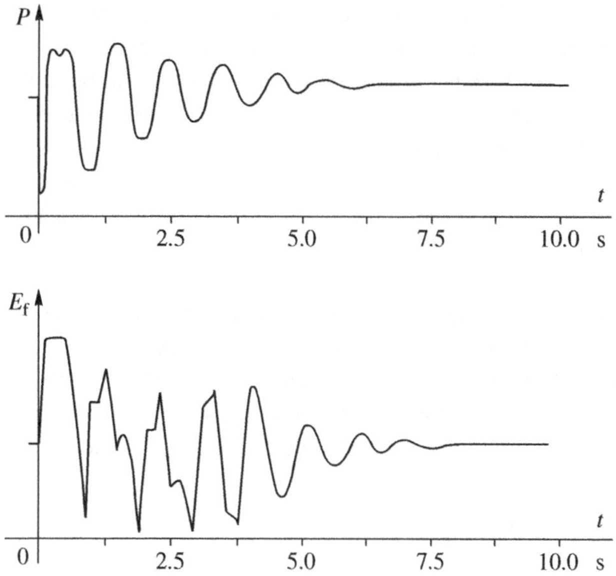

Figure 10.13 Example of post fault oscillations of the active power of the generator and the excitation (field) voltage in the case of the AVR with PSS.

The parameters of a PSS are usually optimized with respect to the damping of small-disturbance power swings. However, a properly designed PSS also improves the damping under large-disturbance conditions. In order to enhance the first-swing transient stability, an additional control loop can be added to the PSS that acts in a similar way as the *forced excitation* in old electromechanical AVR systems. Such forced excitation was executed by short-circuiting resistors in the excitation winding in order to increase E_f to its ceiling value for about 0.5 s. The resistors were then reinserted and E_f decreased. A similar solution is used in so-called *discontinuous excitation control* systems. In the solution described by Kundur (1994), an additional element is switched in by a relay which supplies, in parallel with the PSS, a signal to force the excitation to an increased value. This element is switched off when the sign of the speed deviation changes (i.e. the rotor decelerates).

10.2 Turbine Control System

Stability enhancement can be obtained also by using control of mechanical power. Two methods are described below. The first method is based on the use of a supplementary control loop (PSS) in turbine governor, and the second one is based on the use of an open loop fast control of the main or/and intercept valves of steam turbine referred to as *the fast valving*.

10.2.1 PSS Applied to the Turbine Governor

As all the generators in the EPS are linked by the transmission network, voltage control on one of the generators influences the dynamic response of all the other generators. Consequently, a PSS that improves the damping of one generator does not necessarily improve the damping of the other generators. Therefore a local design may not provide the global optimal solution, and so a coordinated synthesis procedure is desirable. This coordination increases the design computation and is usually valid only for typical network configurations and loading conditions. When a severe fault occurs, the post-fault network configuration, and load, may be significantly different from the pre-fault conditions, and poorly damped swings may result. Because of these factors, interest has focused on utilizing the turbine governor for the damping of local and inter-area oscillations.

Including a PSS signal in the turbine governing systems with the aim of improving damping is not new. Moussa and Yu (1972) describe some solutions regarding hydro turbines. The principle of providing an additional damping torque from the turbine governor is similar to that used when adding a PSS loop to the excitation system. The time constants in the turbine governor introduce a phase shift between the oscillations in the speed deviation $\Delta\omega$ and the turbine mechanical power. As the input signal to the supplementary PSS control loop is equal to $\Delta\omega$, the PSS transfer function must be chosen in such a way that at the frequency of rotor oscillations it compensates the phase shift introduced by the turbine governor. Consequently, the PSS will force changes in the mechanical power ΔP_m that are in phase with $\Delta\omega$ and, according to the swing equation (5.1), provide positive damping.

The main advantage of applying a PSS loop to the turbine governor lies in the fact that the turbine governor dynamics are weakly coupled with those of the rest of the system. Consequently, the parameters of the PSS do not depend on the network parameters. Wang et al. (1993) show interesting simulation results for systems equipped with a PSS applied to the governors of steam turbines. Although this type of PSS is not currently used in practice, such solutions should not be ruled out in the future.

10.2.2 Fast Valving

Chapter 6 explains how a large disturbance near to a generator (e.g. a sudden short circuit) will produce a sudden drop in the generator output power followed by rapid acceleration of the generator rotor. The natural action to counteract this drop in electrical power would be to reduce rapidly the mechanical input power, thereby limiting the acceleration torque. The effect of such action can be explained by considering Figure 10.14, which shows the equal area criterion applied to the system shown in Figure 6.6 when a fault appears on line L2 and is cleared

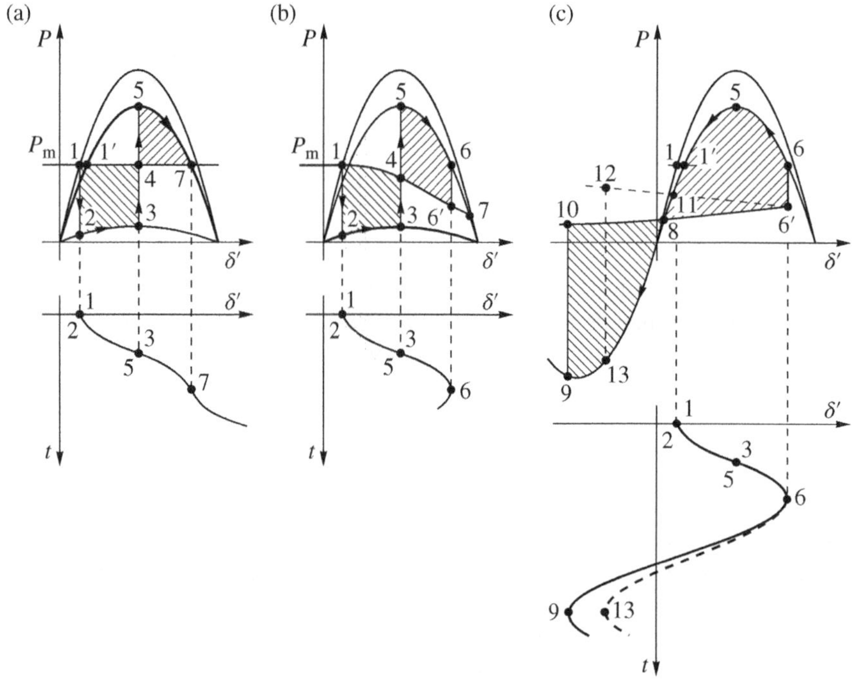

Figure 10.14 Equal area criterion with: (a) constant mechanical power; (b) fast reduction of the mechanical power during the forward swing; (c) influence of the mechanical power on the backward swing.

without auto-reclosure. If it is assumed that the accelerating area 1-2-3-4 in Figure 10.14a is greater than the maximum possible decelerating area 4-5-7, then, when the line is tripped, the rotor will make an asynchronous rotation and the system will lose stability. Now assume that the mechanical power P_m is reduced immediately after the disturbance occurs. The technical possibilities of providing such a fast power reduction is discussed later. Figure 10.14b shows that reduction of P_m during the forward swing reduces the accelerating area 1-2-3-4 and increases the decelerating area to area 4-5-6-6′. The system remains stable with the stability margin being proportional to area 6-7-6′. The maximum improvement in stability is obtained when the reduction in P_m takes place as early, and as fast, as possible.

Figure 10.14c shows the situation to be somewhat different during the backward swing. With the mechanical input power reduced, the system returns toward a rotor angle δ' that is smaller than the initial value and performs deceleration work equal to the integral of the difference between the electrical and mechanical power. In this phase a further decrease in the mechanical power, shown by the solid line 6′-8-10, has a detrimental effect on the system dynamics because the deceleration work 6′-6-5-8 performed during the backward swing can only be balanced by a large accelerating area 8-9-10 giving a large rotor deflection toward negative δ'. The backward swing is increased by the drop in P_m. The dashed line in Figure 10.14c shows how the amplitude of the backward swing could be reduced by increasing P_m during the backward swing. This would result in a smaller decelerating area 6′-6-5-11, which is then balanced by the smaller accelerating area 11-13-12.

After a disturbance, the system shown in Figure 10.14c recovers to the pre-fault value of the mechanical power. In systems operating near to their steady-state stability limit it may happen that the fault clearance increases the system equivalent reactance to a value that would correspond to steady-state instability, that is when the mechanical power is greater than the amplitude of the post-fault characteristic. In order to ensure steady-state stability in such situations it is necessary to reduce the final, post-fault value of the mechanical power. This is illustrated in Figure 10.15a where, on fault clearance, the post-fault power-angle characteristic lies below the dashed line P_{m0} = constant and the system would lose stability even if the fault were cleared instantaneously. To prevent such an instability, the final, post-fault value of mechanical power must be reduced to $P_{m\infty}$, as illustrated in Figure 10.15b.

Fast changes in the mechanical power, as shown in Figures 10.14 and 10.15, require a very fast response from the turbine. The examples shown above indicate that the decrease in the turbine power should take place within the first third of the swing period, that is during the first few tenths of a second following a disturbance. Restoring the power to the required post-fault value should take about half of the swing period, that is less than a second. Such fast control is not possible with hydro turbines, owing to the large change in pressure, and huge torques, necessary to move the control gates. However, a steam turbine can be used for stability enhancement because it can be made to respond almost as quickly as required. The control action required within the turbine to produce this rapid response is referred to as *fast valving*. When the mechanical power is restored to a final value that is equal to the pre-fault value, the fast valving is said to be *momentary fast valving*. When the final value of mechanical

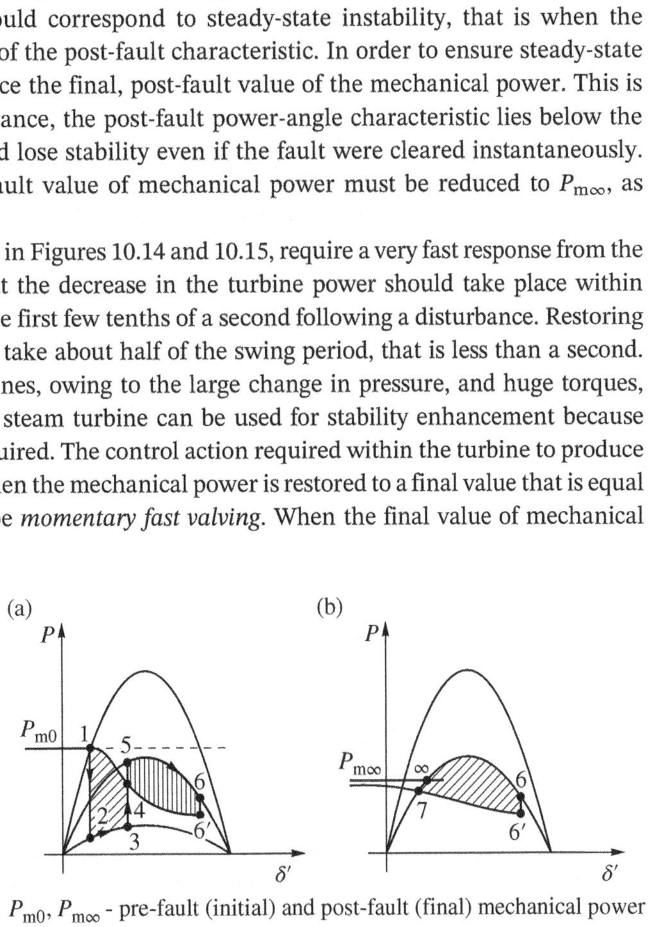

Figure 10.15 Equal area criterion with fast changes of mechanical power in a system operating near its steady-state stability limit: (a) forward swing; (b) backward swing.

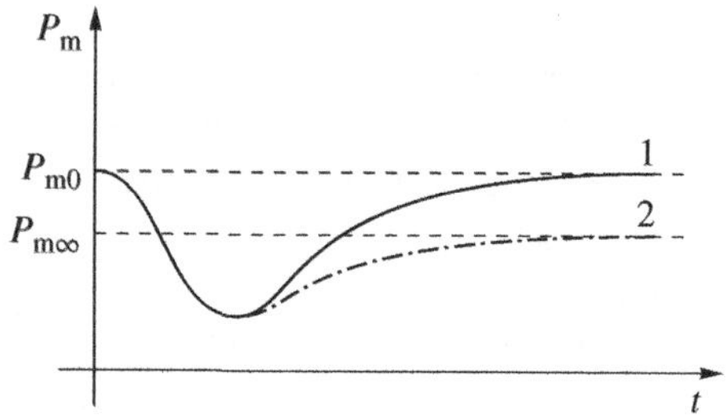

Figure 10.16 Mechanical power variations in the case of: 1 – momentary fast valving; 2 – sustained fast valving.

power is smaller than the pre-fault value, the fast valving is said to be *sustained fast valving*. Power variations corresponding to both these cases are shown in Figure 10.16.

Steam turbine fast valving cannot be achieved using the standard closed-loop turbine governor shown in Figure 2.14 (Section 2.3.3) because a change in the valve setting only occurs after a speed error appears at the controller input. Owing to the large inertia of the turbine-generator rotor the speed can only change slowly so that the response of the closed-loop arrangement, with speed as the control variable, is too slow. As control action is required immediately when the fault is detected, open-loop control systems are normally used.

In a modern steam turbine fast valving is executed using the existing control valves. Consider, for example, the single-reheat unit shown in Figure 2.9 (Section 2.3.3). Here rapid closure of the main governor control valves will not produce a large reduction in the turbine power, because the high-pressure stage produces only about 30% of the power. The reheater stores a large volume of steam and, even with the main governor valve closed, the turbine still supplies about 70% of power through the intermediate- and low-pressure stages. A large, rapid reduction of power can only be obtained by closing the intercept control valves, because it is these which control the steam flow to the intermediate- and low-pressure parts of the turbine.

Momentary fast valving, shown in Figure 10.17a, is achieved by rapid closing of the intercept control valves, holding them closed for a short time and then reopening them in order to restore the power to its initial, pre-fault value. During the short time the intercept valves are shut, the steam flows from the boiler through the high-pressure turbine and is accumulated in the reheater.

Sustained fast valving can be provided by a rapid closing of the intercept valves (Figure 10.17b), followed by slow partial closing of the main governor valve. After a short period the intercept valves are again reopened. This guarantees a rapid initial reduction in the mechanical power followed by its restoration to a value smaller than the initial one.

Fast valving of modern turbines equipped with the electro-hydraulic governing systems shown in Figure 2.14b can be achieved by feeding a fast-valving signal $V_{FV}(t)$ directly to the coil of the electrohydraulic converter. The signal is produced by an additional controller FV operating in open-loop mode. In the case of momentary fast valving the signal $V_{FV}(t)$ consists of two parts, as shown in Figure 10.17a. The first part is a rectangular pulse necessary for rapid closing of the valves. The height of the rectangular signal determines the magnitude of the voltage applied to the converter coil and, if large enough, the servomotor will close the valve in about (0.1–0.4) s. The width of the rectangular pulse t_{FV} determines the duration of the power reduction. The second part of the signal is a pulse that

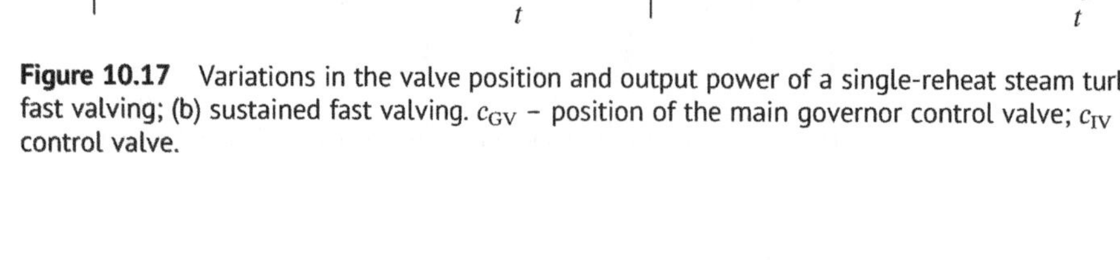

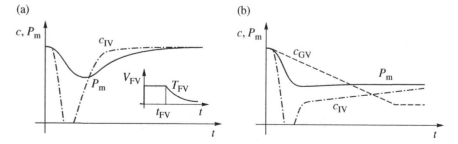

Figure 10.17 Variations in the valve position and output power of a single-reheat steam turbine due to: (a) momentary fast valving; (b) sustained fast valving. c_{GV} – position of the main governor control valve; c_{IV} – position of the intercept control valve.

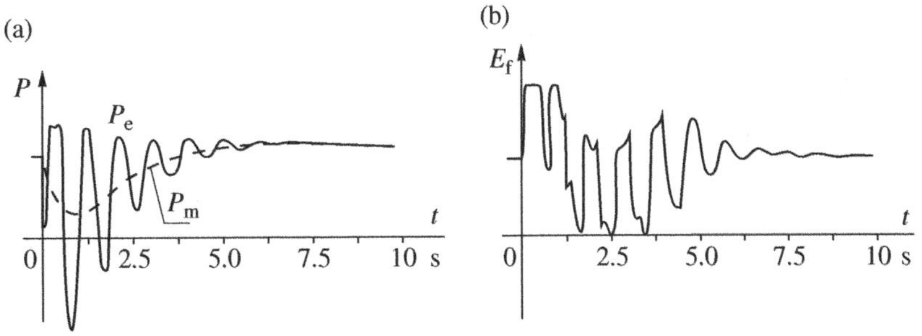

Figure 10.18 Influence of fast valving and PSS on the transient state: (a) mechanical and electrical power variations; (b) excitation voltage variations.

decays with a time constant T_{FV}, where T_{FV} is selected so that the speed at which the valves are reopened does not exceed a preset maximum value. This limit on the speed of valve reopening is mainly set by the strength of the rotor blades so that when reopening the valves a 100% change in the valve position cannot normally be completed in less than about 1 second.

Generally, the fast-valving controller FV operates with a predetermined set of control parameters that are prepared offline, by performing a large number of power system simulations, that take into account:

- the pre-fault network configuration;
- the pre-fault loading condition;
- the location of the fault, that is in which line it occurs;
- the distance and the type of fault measured by, for example, the accelerating power or voltage drop;
- if the fault clearance is with, or without, auto-reclosure.

A control signal is then prepared offline for a large number of possible fault scenarios. When a fault occurs, the controller is fed real-time information on all the above factors so that it can pick a control strategy from its pre-determined set of signals that most closely matches the actual fault condition.

In practice the reduction, and restoration, of power shown in Figure 10.17 tends to be slower than that required from the stability enhancement point of view described in Figure 10.14. If the control valves are reopened too late, the backward swing will increase and may lead to *second-swing instability*. When a generator is equipped with a fast AVR and a PSS, the situation can be improved, as illustrated in Figure 10.18. Fast power reduction takes place within the first few tenths of a second after the fault so that the rotor does not lose stability but swings over the peak of the power-angle characteristic, as indicated by the two humps in the first power swing (compare Figure 6.9a). Restoration of the turbine power takes a few seconds, which is too slow and so causes two deep rotor backward swings that reach the motoring range of operation (negative electrical power). The depth of these backward swings is reduced by rapid control of the excitation voltage enforced by the AVR equipped with PSS. The average value of E_f is decreased and the average value of P_e follows the mechanical power P_m.

Although the cost of implementing fast valving is usually small, the adverse effect on the turbine and boiler may be serious. Generally, fast valving is only used in difficult situations where the AVR and PSS cannot, on their own, prevent instability.

10.3 Braking Resistors

One of the possibilities of affecting the rotor motion following a disturbance is by connecting a *braking resistor* (BR) at the generator or substation terminals. Such an action amounts to *electrical braking* of the accelerating rotor.

The BR may be switched on using a mechanical circuit-breaker, which is discussed in this section, or may be (Figure 2.35) switched on/off or controlled by thyristors, which is discussed in Section 10.5.

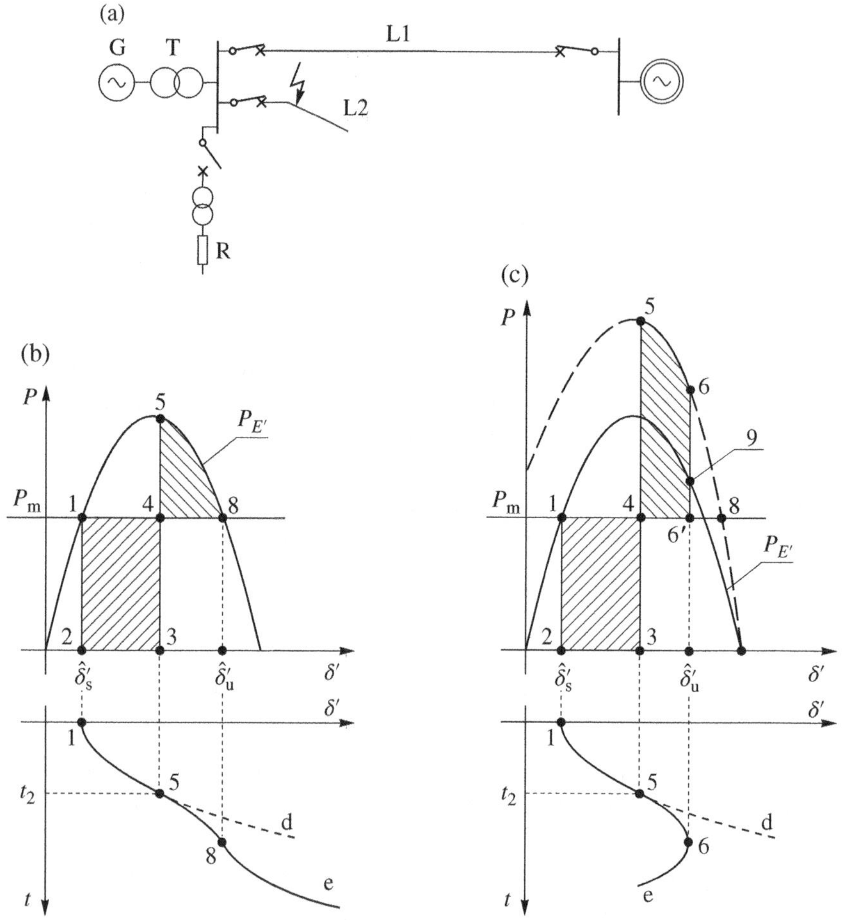

Figure 10.19 Influence of braking resistor: (a) circuit diagram; (b) situation without braking resistor; (c) situation with braking resistor.

The influence of connecting a BR in the generator-infinite busbar system is illustrated in Figure 10.19. To simplify considerations, a short circuit in a radial network was assumed similar to Figure 6.1. The system is unstable without the BR and with the assumed fault duration and pre-fault system loading. This is shown in Figure 6.2b and is repeated here in Figure 10.19b.

Using considerations similar to those accompanying Figure 6.1 but for a shunt resistor, it can be shown that inserting a BR causes an increase in the amplitude of the power-angle characteristic $P(\delta')$ and a shift to the left of the intersection point of that characteristic with the δ'-axis. In Figure 10.19, the power-angle characteristic with the BR inserted is shown using a dashed line.

A signal to insert the BR is obtained from the protection of the transmission line. This protection gives a signal to trip the short-circuited line and to close the circuit-breaker of the BR. In the characteristic shown in Figure 10.19c, the BR is inserted at the instant when the faulted line is tripped. After disconnecting the line and inserting the BR, the electrical power corresponds to point 5. At that point, the generator power is greater than when the BR was absent (Figure 10.19b). The rotor swings to point 6 where the area 4-5-6-6′ is equal to area 1-2-3-4. The system is now stable with a stability margin corresponding to the area 6′-6-8.

Keeping the BR connected during the rotor backward swing is not appropriate as the rotor would do a large amount of deceleration work and would swing deeply in the direction of the motor operation of the generator.

To avoid this, the BR should be disconnected when the rotor speed deviation changes its sign from positive to negative. In the discussed case (Figure 10.19c) this happens at point 6. After the BR has been disconnected, the system moves to point 9 on the characteristic without the BR and then returns toward the equilibrium point doing deceleration work corresponding to the area 9-1-6′.

During the second rotor swing, when the speed deviation becomes positive again, the BR can be inserted again. This type of control is of bang-bang type and consists of inserting the BR when the speed deviation is positive and disconnecting it when the speed deviation is negative. To avoid excessive wear of the circuit-breaker, a maximum of two or three BR insertions are usually made following a fault. When the measurement of rotor speed deviation is not available, a simpler control is used consisting of a single BR insertion for a predetermined period of about (0.3–0.5) s (Kundur 1994).

BRs operate only for a short period of time so they can be made cheaply and volume-efficiently using cast-iron wire strung on towers. They can withstand strong heating up to several hundred degrees centigrade. The mass of the BR is relatively small at about 150 kg per 100 MW of power consumption.

BRs are a relatively cheap and efficient means of preventing loss of synchronism. They are used in hydroelectric power plants where, owing to heavy control gates, fast valving discussed in Section 10.2 cannot be used.

10.4 Generator Tripping

Tripping one or more generators from a group of generators that are operating in parallel on a common busbar is perhaps the simplest, and most effective, means of rapidly changing the torque balance on the generator rotors. Historically, generator tripping was confined to hydropower stations where fast valving could not be used, but now many power companies have extended its use to both fossil fuel and nuclear power generating units in order to try and prevent system instability after severe disturbances.

When a generator is tripped it will normally go through a standard shutdown and start-up cycle which can take several hours. To avoid this long procedure the generator is usually disconnected from the network but still used to supply the station auxiliary demand. This allows the generator to be resynchronized to the system and brought back to full or partial load in several minutes.

The main disadvantage of generator tripping is that it creates a long-term power imbalance characterized by variations in frequency and power interchange between interconnected systems, as shown in Chapter 9. Moreover, generator tripping results in a sudden increase in the electromagnetic torque acting on a generator rotor which can lead to a reduction in the shaft fatigue life. Although unlikely, if two or more units are tripped at exactly the same time, then the shaft loading on the remaining generators can become very severe indeed.

Generator tripping falls into two different categories. *Preventive tripping* is when tripping is coordinated with fault clearing to ensure that the generators remaining in operation maintain synchronism. *Restitutive tripping* is when one, or more, generators are tripped from a group of generators that have already lost synchronism. The objective here is to make resynchronization of the remaining generators easier.

10.4.1 Preventive Tripping

Preventive generator tripping is illustrated in Figure 10.20. In this example both generators are assumed to be identical and, since they are connected in parallel, may be treated as one equivalent generator. This means that the equivalent generator will have a reactance half that of each individual generator but an input power twice that of an individual generator. The pre-fault, fault, and post-fault power-angle characteristics are denoted as I, II, and III, respectively. In the case without tripping (Figure 10.20b), the acceleration area 1-2-3-4 is larger than the available deceleration area 4-5-6. The system is unstable and both generators lose synchronism. In the second case (Figure 10.20c), one of the generators is assumed to be tripped at the same instant as the fault is cleared. The

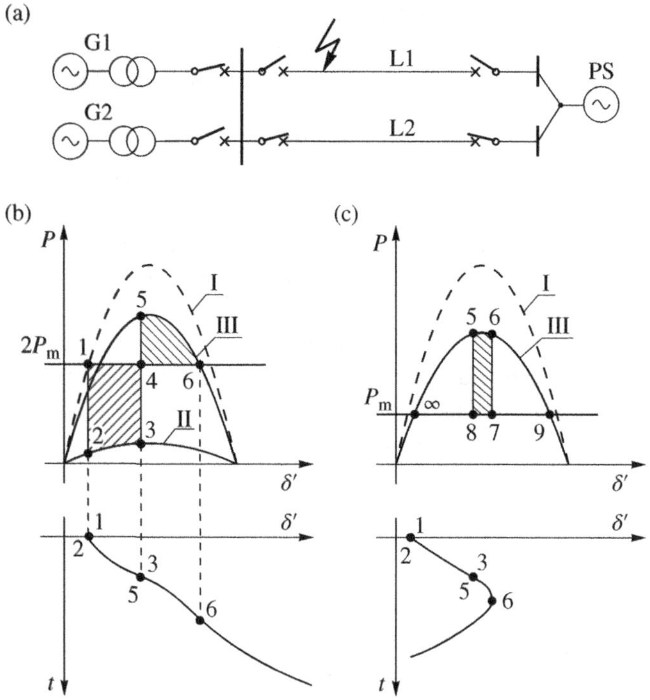

Figure 10.20 Illustration of generator tripping: (a) schematic diagram of the system; (b) acceleration and deceleration area when no generator is tripped; (c) deceleration area when one of the generators is tripped.

post-fault characteristic III now has a smaller amplitude than in Figure 10.20b as the tripping of one of the generators increases the system equivalent reactance. However, the mechanical power of one generator is half that of the equivalent generator so that the acceleration area corresponding to one generator is half the acceleration area 1-2-3-4 shown in Figure 10.20b. The rotor now reaches synchronous speed (speed deviation equal to zero) at point 6 when area 8-5-6-7 is half of area 1-2-3-4. The system remains stable with a stability margin equal to the area 6-9-7. The new stable equilibrium point ∞ corresponds to the intersection between the electrical characteristic III and the mechanical power P_{m}, and will be reached after a number of deep rotor swings. In this example the synchronism of one generator has been saved at the cost of tripping the other.

As the objective of generator tripping is to maintain the stability of a number of generators operating in parallel on the same busbar, tripping one generator may be insufficient and *multi-tripping* may be necessary. The number of generators that must be tripped depends on a number of factors, including the pre-fault loading conditions, fault type and location, and the fault clearing time. The control system that executes the tripping must be able to take into account all of these factors in order to prevent asynchronous operation while minimizing the number of generators tripped.

Broadly, there are two types of control system that can be used to achieve the above objective. The first is similar to that used to select the fast-valving control signal described in the previous section. Central to such a control scheme is the logic predetermined offline on the basis of a detailed stability analysis of the system. Just as in the fast-valving control scheme, the control system obtains real-time information about the fault and from this information selects the number of generators to be tripped. As the predetermined logic cannot always accurately assess the actual fault condition an overly pessimistic assessment may be made resulting in more generators than necessary being tripped. This is the main drawback of preventive generator tripping.

The second type of control system is more sophisticated and is based on real-time simulation by fast microcomputer systems (Kumano et al. 1994). Briefly this system consists of a number of microcomputers connected by a fast telecommunications network. Each microcomputer obtains real-time measurements from the power plant and information about the disturbance. The microcomputers then simulate the dynamic process faster than real time in order to predict instability and compute the minimum number of generators that must be tripped to maintain stability.

10.4.2 Restitutive Tripping

As explained above, preventive tripping systems can be overly pessimistic and trip too many generators, while the more sophisticated control schemes using faster than real-time simulation are very new and expensive. An alternative solution is to use restitutive tripping and utilize signals from the out-of-step relays described in Section 6.6.3. When a group of generators operating in parallel on the same busbars loses synchronism, one of the generators is tripped to make resynchronization of the remaining generators easier. However, if, after a set number of asynchronous rotations, resynchronization is unsuccessful, another generator in the group is tripped and the process repeated until resynchronization is successful.

The main disadvantage of restitutive tripping is in allowing momentary asynchronous operation. The main advantage, when compared with preventive tripping, is that it never disconnects more generators than necessary.

Careful study of Figure 6.34a shows that the optimum instant for restitutive tripping is when the rotor angle passes through the unstable equilibrium point when the large acceleration area (shaded) behind point 6 is reduced by a sudden change in P_m. The unstable equilibrium point is close to $\pi/2$ so that when the system trajectory passes through it the apparent impedance measured by the out-of-step relay crosses the lenticular or offset mho characteristic of the relay shown in Figure 6.42. The signal from the relay can therefore be used to trigger the generator tripping. The circuit-breakers have a certain operating time so that the generator is tripped after a short delay. In order to avoid this delay, and provide for earlier tripping, the relay can be extended with the measurement of the derivative of the apparent resistance dR/dt (Taylor et al. 1983, 1986).

10.5 Shunt FACTS Devices

Power swings can also be damped by changing the parameters in the transmission network that links the generator with the system. Such a parameter change can be accomplished by using additional network elements, such as shunt capacitors and reactors switched on and off at appropriate moments. Optimal system performance can be achieved by correct control of the switching instants and has been the subject of a large number of publications. Rather than describing such control schemes here, the aim in this section is to explain how a proper switching strategy can force damping of power swings and what control signals can be used to execute the switching sequence. The following explanations are based on publications by Machowski and Nelles (1992a, 1992b, 1993, 1994) and Machowski and Bialek (2008).

10.5.1 Power-angle Characteristic

Figure 10.21 shows the situation where a shunt element is connected at a point along a transmission link. The generator is represented by $E_g = E' =$ constant behind the transient reactance and by the swing equation (the classical model). The shunt element is modeled by admittance $\underline{Y}_{sh} = G_{sh} + jB_{sh}$. Depending on the type of shunt element used, the admittance is calculated from the current or power of the shunt element and the actual value of V_{sh}. For example for superconducting magnetic energy storage (SMES) or battery energy storage system (BESS), the admittance is calculated from $\underline{Y}_{sh} = G_{sh} + jB_{sh} = \underline{S}^*_{sh}/V^2_{sh}$. Similarly, as was the case with inequality (2.9), the following constraint has to be satisfied

$$[G_{sh}(t)]^2 + [B_{sh}(t)]^2 \leq |Y_{max}|^2 \quad \text{where} \quad |\underline{Y}_{max}| = \frac{|\underline{S}_{max}|}{|\underline{V}_{sh}|^2} \tag{10.6}$$

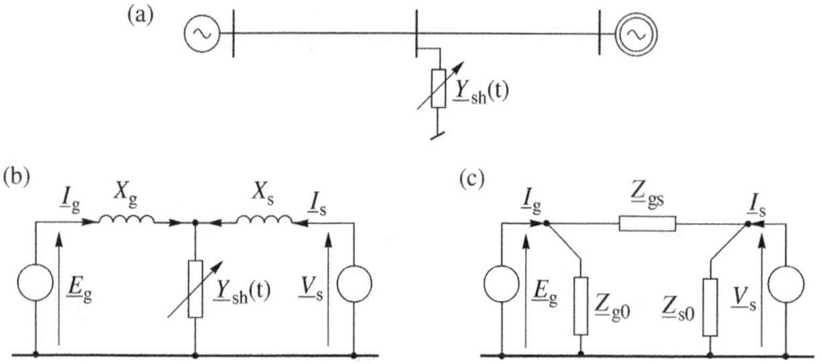

Figure 10.21 Generator-infinite busbar system with the shunt element and its equivalent circuits.

Obviously, for static VAR compensator (SVC) or static compensator (STATCOM) the active power is zero and $G_{sh}(t) = 0$. For the BR the reactive power is zero and $B_{sh}(t) = 0$. To emphasize that the admittance is not constant, the following notation was assumed: $\underline{Y}_{sh}(t) = G_{sh}(t) + jB_{sh}(t)$.

The reactance to the left of the shunt element (including the reactance of the generator) is denoted as X_g. The reactance to the right of the shunt element (including the reactance of the system) is denoted as X_s.

Using the star-delta transformation, a π-equivalent is obtained, which contains a resistance in its equivalent series branch if the shunt element, $\underline{Y}_{sh} = G_{sh} + jB_{sh}$, contains a nonzero conductance. The resulting values of the π-equivalent circuit are

$$\begin{aligned}
\underline{Z}_{gs} &= (X_g + X_s)\left[-X_{SHC}G_{sh} + j(1 - X_{SHC}B_{sh})\right] \\
\tan\mu_{gs} &= \frac{\operatorname{Re}\underline{Z}_{gs}}{\operatorname{Im}\underline{Z}_{gs}} = -\frac{X_{SHC}G_{sh}}{1 - X_{SHC}B_{sh}} \\
\underline{Y}_{g0} &= \frac{1}{\underline{Z}_{g0}} = \frac{1}{X_g}\frac{X_{SHC}G_{sh} + jX_{SHC}B_{sh}}{(1 - X_{SHC}B_{sh}) + jX_{SHC}G_{sh}}
\end{aligned} \tag{10.7}$$

where $X_{SHC} = X_g X_s/(X_g + X_s)$ is the short-circuit reactance of the system as seen from the node where the shunt element is installed. Normally, the short-circuit power $S_{SHC} = V_n^2/X_{SHC}$ tends to be between a few thousand and up to about 20 000 MVA, while the rated power of the shunt element $P_{n\,sh} = G_{sh}V_n^2$ and $Q_{n\,sh} = B_{sh}V_n^2$ tends to be less than 100 MVA. Therefore, as $P_{n\,sh}$ and $Q_{n\,sh}$ are at least 10 times less than S_{SHC}, it can be safely assumed that

$$X_{SHC}G_{sh} \ll 1; \; X_{SHC}B_{sh} \ll 1 \tag{10.8}$$

Now consider the real number α such that for $\alpha \ll 1$ it can be assumed, with good accuracy, that

$$\frac{1}{1-\alpha} \cong 1 + \alpha \quad \text{and} \quad \frac{1}{1+\alpha} \cong 1 - \alpha \tag{10.9}$$

Equations (10.8), and the identity in Eq. (10.9), now allow the equations in (10.7) to be simplified to give

$$\underline{Y}_{gs} = G_{gs} + jB_{gs} = \frac{1}{\underline{Z}_{gs}} \cong \frac{1}{X_g + X_s}\left[-X_{SHC}G_{sh} - j(1 + X_{SHC}B_{sh})\right] \tag{10.10}$$

$$Y_{gs} \cong \frac{1}{X_g + X_s}(1 + X_{SHC}B_{sh}) \tag{10.11}$$

$$\tan\mu_{gs} \cong -X_{SHC}G_{sh}(1 + X_{SHC}B_{sh}) = -X_{SHC}G_{sh} \tag{10.12}$$

$$\underline{Y}_{g0} = G_{g0} + jB_{g0} \cong \frac{1}{X_g}(X_{SHC}G_{sh} + jX_{SHC}B_{sh}) \tag{10.13}$$

$$G_{gg} = G_{g0} + G_{gs} \cong \frac{1}{X_g + X_s}\frac{X_s}{X_g}X_{SHC}G_{sh} \tag{10.14}$$

Diligent readers may find the plus sign in Eq. (10.10) rather surprising but it is correct and arises from the use of the small-value approximations in Eq. (10.9). These derived parameters now allow a formula for the generator active power to be determined using the general formula given in Eq. (3.175)[1] adapted to the π-equivalent circuit of Figure 10.21c to give

$$P(\delta') = G_{g0}E_g^2 + Y_{gs}E_gV_s\sin\left(\delta' - \mu_{gs}\right) \tag{10.15}$$

where $E_g = E' =$ constant and δ' is the transient rotor angle measured with respect to the infinite busbar. The sine function can be expressed as

$$\sin\left(\delta' - \mu_{gs}\right) = \sin\delta'\cos\mu_{gs} - \cos\delta'\sin\mu_{gs} = \cos\mu_{gs}\left(\sin\delta' - \tan\mu_{gs}\cos\delta'\right) \tag{10.16}$$

According to Eqs. (10.8) and (10.12), the angle μ_{gs} is small and $\cos\mu_{gs} \cong 1$. Making this approximation, and substituting Eq. (10.12) into Eq. (10.16), gives

$$\sin\left(\delta' - \mu_{gs}\right) \cong \sin\delta' - X_{SHC}G_{sh}\cos\delta' \tag{10.17}$$

This equation, and the expressions in Eqs. (10.11) and (10.13), can now be substituted into Eq. (10.15) to give

$$P(\delta') \cong b\sin\delta' + b(\xi + \cos\delta')X_{SHC}G_{sh} + (b\sin\delta')X_{SHC}B_{sh} \tag{10.18}$$

where $b = E_gV_s/(X_g + X_s)$ and $\xi = (E_g/V_s)/(X_s/X_g)$. Careful examination of this equation shows that the coefficient b is the amplitude of the power-angle characteristic *without* the shunt element, while the coefficient ξ depends on the location of the shunt element along the transmission link. With the shunt element disconnected, the characteristic defined by Eq. (10.18) is the transient power-angle characteristic defined in Eq. (5.26).

Figure 10.22 shows the influence of both G_{sh} and B_{sh} on the transient power-angle characteristic. Inserting a purely resistive shunt element – $G_{sh} \neq 0$ and $B_{sh} = 0$ – shifts the characteristic to the left, or the right, by the angle

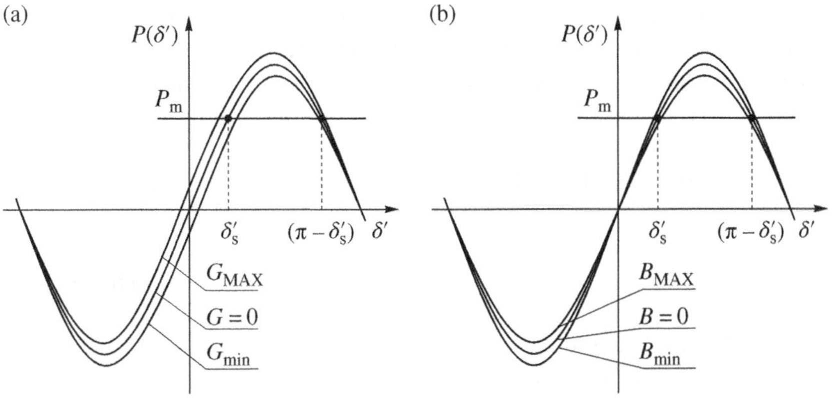

Figure 10.22 Influence of the shunt element on the power-angle characteristic: (a) pure conductance G_{sh}; (b) pure susceptance B_{sh}.

1 Equation (3.175) used the angle, $\theta = \arg(\underline{Y})$, while it is more convenient to use here the angle $\mu = \pi/2 - \theta$.

μ_{GS} depending on the sign of G_{sh} (Figure 10.22a). This shift by the angle μ_{GS}, and controlled by G_{sh}, is clearly demonstrated in Eq. (10.15). Figure 10.22b shows how inserting a pure reactive shunt element – $G_{sh} = 0$ and $B_{sh} \neq 0$ – increases the amplitude of the characteristic by $X_{SHC}B_{sh}$.

Further examination of Figure 10.22 shows that the main influence of G_{sh} is at small values of δ' before the peak of the characteristic is reached. For large values of δ', beyond the peak, the influence of the conductance is negligible. In contrast the influence of B_{sh} is mostly at large values of δ', near the peak of the characteristic.

10.5.2 State-variable Stabilizing Control

Since the shunt element influences the shape of the power-angle characteristic, it can be used to damp rotor swings either by switching it on and off at appropriate moments (bang-bang control) or by suitable continuous control action (provided continuous control of the shunt element is possible). The method used to control the shunt element is called the *control strategy*. Control using the system state variables is referred to as *state-variable control*.

The system state equations can be obtained by substituting the generator active power given by Eq. (10.18) into the swing equation (5.1) to give

$$\begin{aligned} \frac{d\delta'}{dt} &= \Delta\omega \\ M\frac{d\Delta\omega}{dt} &= (P_m - b\sin\delta') - D\frac{d\delta'}{dt} \\ &\quad - b(\xi + \cos\delta')X_{SHC}G_{sh}(t) - (b\sin\delta')X_{SHC}B_{sh}(t) \end{aligned} \tag{10.19}$$

The last two terms in the second equation depend on the conductance and susceptance of the controlled shunt element. Both these parameters are shown as functions of time to emphasize that they are the time-varying control variables.

10.5.2.1 Energy Dissipation

In order to derive the required control strategy, the energy approach described in Section 6.3 will be used. When a disturbance occurs in an EPS, part of the kinetic energy stored in the rotating masses of generators and loads is released and undergoes oscillatory conversions from kinetic to potential energy and then back again during subsequent rotor swings. The oscillations continue until the damping torques dissipate all the released energy and the system trajectory returns to the equilibrium point. The goal is to control $G_{sh}(t)$ and $B_{sh}(t)$ in such a way as to maximize the speed of energy dissipation. This can be accomplished by maximizing the value of the derivative of the total system energy with time along the trajectory of the differential Eq. (10.19).

The total system energy $V = E_k + E_p$ for the generator-infinite busbar system is determined by Eq. (6.25). The speed of energy changes is

$$\frac{dV}{dt} = \frac{\partial V}{\partial \delta'}\frac{d\delta'}{dt} + \frac{\partial V}{\partial \Delta\omega}\frac{d\Delta\omega}{dt} \tag{10.20}$$

The partial derivatives $\partial V/\partial\delta'$ and $\partial V/\partial\Delta\omega$ can be calculated by differentiating the total energy given by Eq. (6.25). The ordinary time derivatives $d\delta'/dt$ and $d\Delta\omega/dt$ can now be substituted for using Eq. (10.19) to give

$$\frac{dV}{dt} = -\left[D\Delta\omega^2 + \Delta\omega\, b(\xi + \cos\delta')X_{SHC}G_{sh}(t) + \Delta\omega\,(b\sin\delta')X_{SHC}B_{sh}(t)\right] \tag{10.21}$$

The first term in this equation corresponds to the energy dissipated by the natural damping torques (coefficient D), while the next two terms are contributed by the shunt element. Proper control of this element can contribute to a faster dissipation of energy. This will happen if $G_{sh}(t)$ and $B_{sh}(t)$ are varied so that the signs of the two last terms in

Eq. (10.21) are always positive. This requires the sign of $G_{sh}(t)$ and $B_{sh}(t)$ to vary depending on the sign of the two state variables $\Delta\omega$ and δ'. This can be realized using either bang-bang control or continuous control.

10.5.2.2 Continuous Control

The aim of the control is, first, to ensure that the last two terms in Eq. (10.21) are of positive sign and, second, to maximize their sum. This can be achieved by enforcing the following values on the control variables

$$\begin{aligned} G_{sh}(t) &= K\ \ \Delta\omega\,[b(\xi+\ \cos\delta')]X_{SHC} \\ B_{sh}(t) &= K\ \ \Delta\omega\,[b\sin\delta']X_{SHC} \end{aligned} \tag{10.22}$$

where K is the gain of the controller. Substituting this equation into Eq. (10.21) gives

$$\frac{dV}{dt} = -D\Delta\omega^2 - D_{sh}\Delta\omega^2 \tag{10.23}$$

where

$$D_{sh} = K\Big\{\,[b(\xi+\ \cos\delta')]^2 + (b\sin\delta')^2\Big\}X_{SHC}^2 \tag{10.24}$$

is the equivalent damping coefficient introduced by the shunt element control. For the control strategy in Eq. (10.22), the swing equation then becomes

$$\begin{aligned} \frac{d\delta'}{dt} &= \Delta\omega \\ M\frac{d\Delta\omega}{dt} &= (P_m - b\sin\delta') - D\frac{d\delta'}{dt} - D_{sh}\frac{d\delta'}{dt} \end{aligned} \tag{10.25}$$

In the first of the two equations in (10.22) the expression $(\xi + \cos\delta')$ is positive over a large range of angle $(-\pi/2 < \delta' < \pi/2)$ so that the sign of $G_{sh}(t)$ will change at the same time as the rotor speed deviation $\Delta\omega$. On the other hand, the sign of $B_{sh}(t)$ depends on both $\Delta\omega$ and $\sin\delta'$ and changes sign whenever the angle crosses through zero.

Figure 10.23 shows the changes in the sign of $G_{sh}(t)$ and $B_{sh}(t)$ that must occur along the system trajectory in order to produce positive damping. Whenever the trajectory crosses the horizontal axis, $\Delta\omega$ changes its sign, and this will be accompanied by a change in sign of both $G_{sh}(t)$ and $B_{sh}(t)$. These points of sign changes have been denoted by small squares on the trajectory. Whenever the trajectory crosses the vertical axis, δ' changes its sign, and this is accompanied by a sign change of $B_{sh}(t)$ only. These points of sign changes have been denoted by the small solid circles on the trajectory.

In Figure 10.23 the characteristic states of the shunt element are illustrated schematically in each quadrant in the phase plane. Negative susceptance is denoted by a coil symbol, while positive susceptance is denoted by a capacitor symbol. Positive conductance is denoted by a resistance symbol, while negative resistance (i.e. a source of real power) is denoted by an arrow. This obviously corresponds to the general case when the power of the shunt element can be adjusted in all four quadrants of the complex power plane.

Figure 10.23 Stability area and the system trajectory.

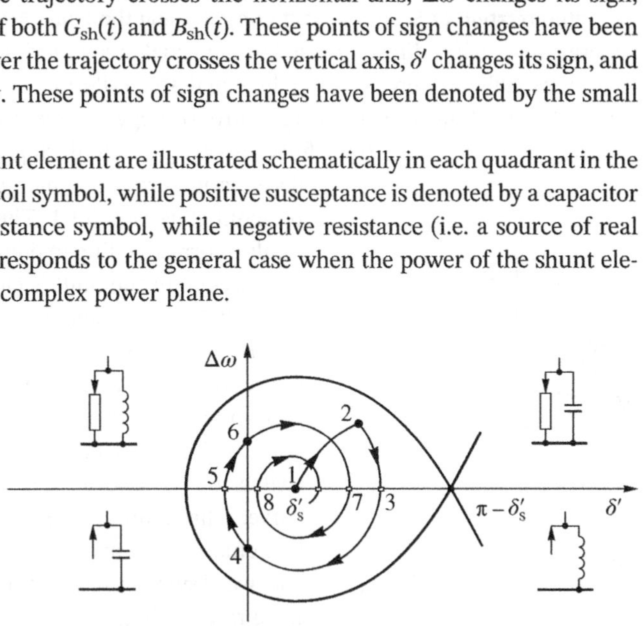

10.5.2.3 Bang-Bang Control

The term *bang-bang control* refers to a control mode in which the element is switched on and off at appropriate moments. This type of control can be used with a shunt element whose parameters cannot be smoothly controlled. The switching then takes place at instants when the system trajectory crosses one of the phase axes shown in Figure 10.23 and passes into the next quadrant.

Equation (10.22) leads to the following bang-bang control strategies

$$G_{sh}(t) = \begin{cases} G_{MAX} & \text{for} \quad [b(\xi + \cos\delta')\Delta\omega] \geq +\varepsilon \\ 0 & \text{for} \quad +\varepsilon > [b(\xi + \cos\delta')\Delta\omega] > -\varepsilon \\ G_{min} & \text{for} \quad [b(\xi + \cos\delta')\Delta\omega] \leq -\varepsilon \end{cases} \tag{10.26}$$

$$B_{sh}(t) = \begin{cases} B_{MAX} & \text{for} \quad [\Delta\omega(b\sin\delta')] \geq +\varepsilon \\ 0 & \text{for} \quad +\varepsilon < [\Delta\omega(b\sin\delta')] > -\varepsilon \\ B_{min} & \text{for} \quad [\Delta\omega(b\sin\delta')] \leq -\varepsilon \end{cases} \tag{10.27}$$

where G_{MAX} and B_{MAX} are the maximum, and G_{min} and B_{min} the minimum, values of the switching element. If the element cannot assume a negative value then the minimum value will be zero. The small positive number determines the dead zone ($\pm\varepsilon$) where the control variables are set to zero. The dead zone is necessary in order to avoid unstable operation of the controller for small signals.

As $(\xi + \cos\delta') > 0$ for a wide range of δ', the strategy for controlling the conductance $G_{sh}(t)$ can be simplified to

$$G_{sh}(t) = \begin{cases} G_{MAX} & \text{for} \quad \Delta\omega \geq +\varepsilon \\ 0 & \text{for} \quad +\varepsilon > \Delta\omega > -\varepsilon \\ G_{min} & \text{for} \quad \Delta\omega \leq -\varepsilon \end{cases} \tag{10.28}$$

and the conductance switching is triggered by a change in sign of the speed deviation.

10.5.2.4 Interpretation of Shunt Element Control Using the Equal Area Criterion

As the energy approach is directly related to the equal area criterion (Section 6.3.4) the control strategies derived above can be usefully interpreted using the equal area criterion.

Figure 10.24 shows how the bang-bang control strategy enlarges the available deceleration area and reduces the acceleration area during every rotor forward swing, while during the backward swing the control reduces the deceleration area and enlarges the available acceleration area. The initial pre-fault state is point 1. The fault reduces the electrical active power to a value corresponding to point 2 and the rotor accelerates. The rotor angle increases until the fault is cleared at point 3, giving the first acceleration area 1-2-3-4. After clearing the fault, $\Delta\omega$ and δ' are positive so that G_{MAX} and B_{MAX} are switched in and the electrical power follows the higher $P(\delta')$ curve. As the maximum available decelerating area is 4-5-6-7, the generator remains stable with a stability margin proportional to area 6-7-8. At point 6 the speed deviation changes sign, the shunt admittance values switch to G_{min} and B_{min}, and the electrical power follows the lower $P(\delta')$ curve. This reduces the deceleration area during the backward swing and enlarges the available acceleration area so that the amplitude of the backward swing is reduced. The cycle then repeats and helps damp consecutive rotor swings.

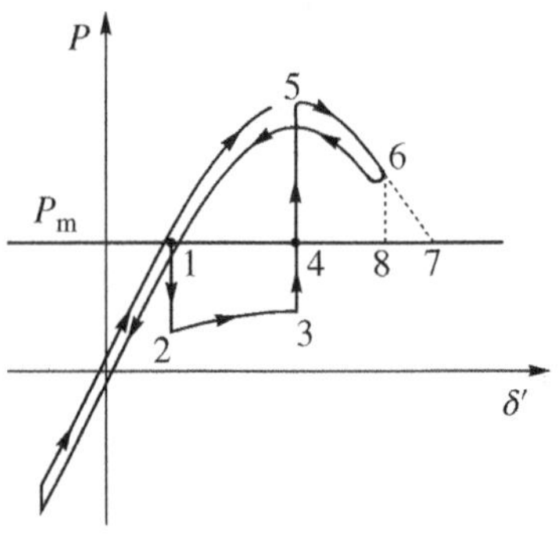

Figure 10.24 Interpretation of the control strategy using the equal area criterion.

10.5.3 Control Based on Local Measurements

The control strategy given by Eq. (10.22) is based on the state variables δ' and $\Delta\omega$. As these quantities are not normally available at the shunt element busbar, the practical implementation of the control must be based on other signals that can be measured locally. How exactly such a local control emulates the state-variable control depends on the choice of the measured quantities and the structure of the controller.

10.5.3.1 Dynamic Properties of Local Measurements

Let q_G and q_B be some quantities used as input signals to the shunt element controller. In the control strategy given by Eq. (10.22) the shunt admittance depends on the rotor speed deviation $\Delta\omega$. If the magnitude of the transient emf is assumed constant (classical model), the derivative with respect to time of any electric quantity q_G can be expressed as

$$\frac{dq_G}{dt} = \frac{\partial q_G}{\partial \delta'}\Delta\omega + \alpha_{GG}\frac{dG_{sh}}{dt} + \alpha_{GB}\frac{dB_{sh}}{dt} \tag{10.29}$$

where the coefficients

$$\alpha_{GG} = \frac{\partial q_G}{\partial G_{sh}}; \quad \alpha_{GB} = \frac{\partial q_G}{\partial B_{sh}} \tag{10.30}$$

determine the sensitivity of q_G to a change in the control variables $G_{sh}(t)$ and $B_{sh}(t)$. Equation (10.29) gives

$$\Delta\omega\frac{\partial q_G}{\partial \delta'} = \frac{dq_G}{dt} - \alpha_{GG}\frac{dG_{sh}}{dt} - \alpha_{GB}\frac{dB_{sh}}{dt} \tag{10.31}$$

If the sensitivity coefficients α_{GG} and α_{GB} are known, the right-hand side of Eq. (10.31) can be computed in real time and used to determine a signal proportional to the rotor speed deviation necessary for the control of $G_{sh}(t)$. Comparing the right-hand side of the first of the equations in (10.22) with the left-hand side of Eq. (10.31) shows that the signal obtained from Eq. (10.31) is the same as the state-variable control signal if

$$\frac{\partial q_G}{\partial \delta'} = [b(\xi + \cos\delta')]X_{SHC} \tag{10.32}$$

Substitution of the right-hand side of the first equation in (10.22) by Eq. (10.31) gives the following control principle

$$G_{sh}(t) = K\left[\frac{dq_G}{dt} - \alpha_{GG}\frac{dG_{sh}}{dt} - \alpha_{GB}\frac{dB_{sh}}{dt}\right] \tag{10.33}$$

This means that if a measured quantity q_G satisfies the condition in Eq. (10.32) then the modulation controller need simply differentiate q_G with respect to time and subtract from the result values proportional to the rate of change of the controlled variables $B_{sh}(t)$ and $G_{sh}(t)$.

The control principle for the shunt susceptance can be obtained in a similar way as

$$B_{sh}(t) = K\left[\frac{dq_B}{dt} - \alpha_{BG}\frac{dG_{sh}}{dt} - \alpha_{BB}\frac{dB_{sh}}{dt}\right] \tag{10.34}$$

where the coefficients

$$\alpha_{BG} = \frac{\partial q_B}{\partial G_{sh}}; \quad \alpha_{BB} = \frac{\partial q_B}{\partial B_{sh}} \tag{10.35}$$

determine the sensitivity of q_B to changes in the control variables $B_{sh}(t)$ and $G_{sh}(t)$. Comparison with the second of the equations in (10.22) shows that the input quantity q_B should satisfy the following condition

$$\frac{\partial q_B}{\partial \delta'} = [b \sin \delta'] X_{SHC} \tag{10.36}$$

It now remains to determine what locally measurable quantities q_B and q_G will satisfy the conditions defined in Eqs. (10.32) and (10.36).

10.5.3.2 Voltage-based Quantities

The current flowing from the network to the shunt element in Figure 10.21 is given by

$$\underline{I}_{sh} = \underline{V}_{sh}\underline{Y}_{sh} = \frac{\underline{E}_g - \underline{V}_{sh}}{jX_g} + \frac{\underline{V}_s - \underline{V}_{sh}}{jX_s} \tag{10.37}$$

where $\underline{Y}_{sh} = G_{sh}(t) + jB_{sh}(t)$ and X_g and X_s are the equivalent reactances denoted in Figure 10.21. Multiplying the current by the short-circuit reactance X_{SHC} and moving $\underline{V}_{sh}$ to the left-hand side gives

$$\underline{V}_{sh}\{[X_{SHC}G_{sh}(t)] + j[1 - X_{SHC}B_{sh}(t)]\} = \frac{\underline{E}_g X_s + \underline{V}_s X_g}{j(X_g + X_s)} \tag{10.38}$$

Substituting for the complex voltages

$$\underline{E}_g = E_g(\cos \delta' + j \sin \delta') \text{ and } \underline{V}_s = V_s \tag{10.39}$$

and multiplying the resulting equation by its conjugate gives, after a little algebra,

$$V_{sh}^2 = \frac{bX_{SHC}\left(\xi + \frac{1}{\xi} + 2\cos\delta'\right)}{[X_{SHC}G_{sh}(t)]^2 + [1 - X_{SHC}B_{sh}(t)]^2} \tag{10.40}$$

where ξ is the coefficient defined in Eq. (10.18). When deriving Eq. (10.40), it is also possible to find the phase angle θ of the shunt element voltage measured with respect to the infinite bus

$$\tan\theta = \frac{\left(\frac{1}{\xi} + \cos\delta'\right)X_{SHC}G_{sh}(t) + \sin\delta'[1 - X_{SHC}B_{sh}(t)]}{\sin\delta' X_{SHC}G_{sh}(t) + \left(\frac{1}{\xi} + \cos\delta'\right)[1 - X_{SHC}B_{sh}(t)]} \tag{10.41}$$

The inequalities in Eq. (10.8) allow Eqs. (10.40) and (10.41) to be simplified to

$$V_{sh}^2 \cong b\left(\xi + \frac{1}{\xi} + 2\cos\delta'\right)X_{SHC}; \; \tan\theta \cong \frac{\sin\delta'}{\frac{1}{\xi} + \cos\delta'} \tag{10.42}$$

Using the first of Eq. (10.42), and after differentiating with respect to δ', gives

$$\frac{\partial V_{sh}^2}{\partial \delta'} = -2[b \sin\delta'] X_{SHC} \tag{10.43}$$

Calculation of the derivative $\partial\theta/\partial\delta'$ is slightly more difficult. The second part of Eq. (10.42) may be written as $f(\theta, \delta') = 0$. Hence θ is an implicit function of δ'. The derivative of that function can be calculated from

$$\frac{\partial\theta}{\partial\delta'} = -\frac{\frac{\partial f}{\partial\delta'}}{\frac{\partial f}{\partial\theta}} \tag{10.44}$$

Using this equation gives

$$\frac{\partial\theta}{\partial\delta'} = \frac{\xi + \cos\delta'}{\xi + \frac{1}{\xi} + 2\cos\delta'} \tag{10.45}$$

The expression in the denominator of Eq. (10.45) is the same as the expression in brackets in the first of the equations in (10.42). Substitution gives

$$V_{\text{sh}}^2\frac{\partial\theta}{\partial\delta'} = [b(\xi + \cos\delta')]X_{\text{SHC}} \tag{10.46}$$

Equations (10.43) and (10.46) show that local measurements of the squared magnitude of the shunt element voltage and its phase angle can give good signals for controlling the shunt element. Comparing Eqs. (10.43) and (10.36) shows that the signal V_{sh}^2 satisfies the condition in Eq. (10.36) for the control strategy of the shunt susceptance. Similarly, comparing Eqs. (10.46) and (10.32) shows that θ satisfies the condition in Eq. (10.32) for the required control of the shunt conductance provided that the derivative is multiplied by V_{sh}^2.

A sensitivity analysis of the effect of changes in V_{sh}^2 and θ on the changes in the controlled variables $B_{\text{sh}}(t)$ and $G_{\text{sh}}(t)$ can be conducted by evaluating the derivatives in Eqs. (10.30) and (10.35) using Eqs. (10.40) and (10.41). This involves a lot of simple, but arduous, algebraic and trigonometric transformations which finally lead to the following simplified formulae

$$\begin{aligned} \alpha_{\text{GG}} &= \frac{\partial q_{\text{G}}}{\partial G_{\text{sh}}} \cong -X_{\text{SHC}}; \qquad & \alpha_{\text{GB}} &= \frac{\partial q_{\text{G}}}{\partial B_{\text{sh}}} \cong 0 \\ \alpha_{\text{BG}} &= \frac{\partial q_{\text{B}}}{\partial G_{\text{sh}}} \cong 0; \qquad & \alpha_{\text{BB}} &= \frac{\partial q_{\text{B}}}{\partial B_{\text{sh}}} \cong -V_{\text{sh}}^2 X_{\text{SHC}} \end{aligned} \tag{10.47}$$

Zero values of α_{GB} and α_{BG} signify that changes in the shunt susceptance/conductance have a negligibly small effect on the given quantity.

10.5.3.3 Control Schemes

Substituting Eq. (10.47) into Eqs. (10.33) and (10.34) and taking the squared voltage magnitude and the voltage phase angle as control signals yields

$$G_{\text{sh}}(t) = KV_{\text{sh}}^2\left[\frac{\mathrm{d}\theta}{\mathrm{d}t} + X_{\text{SHC}}\frac{\mathrm{d}G_{\text{sh}}}{\mathrm{d}t}\right] \tag{10.48a}$$

$$B_{\text{sh}}(t) = K\left[-\frac{1}{2}\frac{\mathrm{d}(V_{\text{sh}}^2)}{\mathrm{d}t} + X_{\text{SHC}}V_{\text{sh}}^2\frac{\mathrm{d}B_{\text{sh}}}{\mathrm{d}t}\right] \tag{10.48b}$$

The second component in formula (10.48a) constitutes a positive feedback of the output signal. The simulation tests show that this feedback hinders the possibility to operate with high regulator gain and it is better to neglect it. Then the conductance is equal to

$$G_{\text{sh}}(t) \cong KV_{\text{sh}}^2\frac{\mathrm{d}\theta}{\mathrm{d}t} \tag{10.49}$$

The rotor angle δ' and the voltage phase angle θ are measured with respect to the infinite busbar voltage or the synchronous reference frame. As the derivative of θ with time is equal to the deviation of the local frequency, that is $\mathrm{d}\theta/\mathrm{d}t = 2\pi\Delta f$, control based on formulas (10.48b) and (10.49) is referred to as *frequency- and voltage-based control*.

Figure 10.25 shows the block diagram of the appropriate control circuits. Differentiation has been replaced by a real differentiating element with a small time constant T. The shunt susceptance controller is nonlinear because

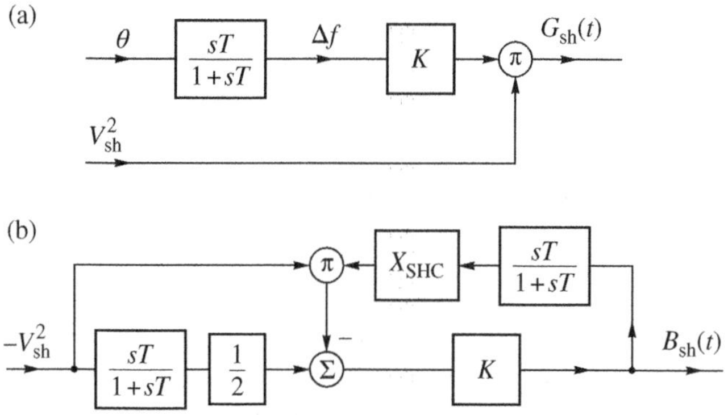

Figure 10.25 Modulation controller employing the frequency-voltage control scheme for: (a) $G_{sh}(t)$; (b) $B_{sh}(t)$.

the output signal in the main feedback loop is multiplied by the main input signal. The shunt conductance controller is linear but its effective gain is modulated by the squared voltage magnitude, which is also the main input signal for the shunt susceptance controller. The short-circuit reactance X_{SHC} plays only a corrective role and its value may be set with a large error. For practical applications its value can be assessed offline and set as a constant parameter.

It is worth noting that the time derivative of the voltage angle $d\theta/dt$ is equal to the deviation of the local frequency Δf. Thus the proposed shunt element controller is a frequency- and voltage-orientated controller. The input signals for the control system may be measured using digital techniques described by Phadke et al. (1983) or Kamwa and Grondin (1992).

10.5.4 Examples of Controllable Shunt Elements

Continuous state-variable control, and its practical implementation based on locally measurable quantities, is possible assuming that both $G_{sh}(t)$ and $B_{sh}(t)$ can be changed smoothly over a range of negative and positive values. Some of the different types of shunt elements that can be thyristor controlled are described in Section 2.5.4, where the ability to change $G_{sh}(t)$ and $B_{sh}(t)$ depends on the particular device in question. When using a particular shunt element, any such limitation must be taken into account in the control structure by inserting appropriate limiters into the control circuits shown in Figure 10.25.

10.5.4.1 Supplementary Control of SVC and STATCOM

SVCs based on conventional thyristors (Figure 2.29) are equipped with a voltage regulator (Figure 2.30) giving the static characteristic (Figure 2.31) with a small droop in the regulation area. In the steady state such regulation is very effective in forcing the steady-state voltage error to zero. In the transient state the regulator is incapable of providing enough damping, because the voltage error does not carry proper information about the system dynamic response. A more robust control can be obtained when the voltage regulator is equipped with the supplementary control loop, as shown in Figure 10.26. This additional supplementary loop can be used to force a control signal for $B_{sh}(t)$ to enhance damping of power swings. This control loop acts much faster than the main voltage controller. The controller in the supplementary control loop will be based on Eq. (10.48) and is shown in Figure 10.25b.

In a similar way the supplementary control loop may be also equipped with the voltage regulator of STATCOM (Figure 2.32).

10.5.4.2 Control of BR

Application of BRs switched by a mechanical circuit-breaker is described in Section 10.3. Another, more expensive, possibility is to switch or control the BR using thyristors, as described in Section 2.4.4.

Thyristor-switched BRs (Figure 2.35) may be equipped with a bang-bang controller to implement the strategy defined in Eq. (10.28) with $G_{min} = 0$. Alternatively, the resistors may be equipped with a continuous controller to implement the strategy defined in Eq. (10.26) with $G_{min} = 0$. If the rotor speed deviation signal $\Delta\omega$ is not available, it may be replaced by the local frequency deviation $2\pi\,\Delta f = d\theta/dt$, as in the shunt controller shown in Figure 10.25a. Obviously, an output limiter and a dead zone would have to be added before the output in the block diagram, as in Figure 10.26. In the steady state the BRs are switched off and the modulation controller (Figure 10.25a) is their only control circuit.

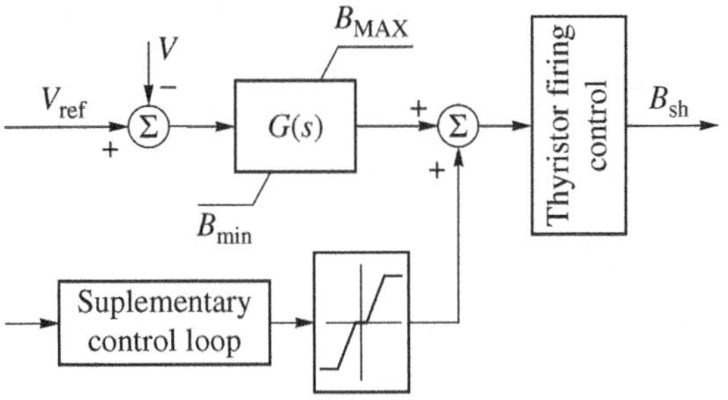

Figure 10.26 Voltage regulator of the SVC with supplementary control loop.

10.5.4.3 STATCOM + BR as a More Effective Device

The influence of an active and reactive shunt element on active power of a generator is shown in Figure 10.22. When power angle δ' is small, the influence of a shunt reactive element is quite small. Consequently, when the generator operates at a small power angle, little damping of the power swings using STATCOM alone can be obtained. This can be improved by adding to STATCOM a thyristor-controlled BR. Such a resistor may be included in the DC circuit of STATCOM in the same way as a battery is included with BESS (Figure 2.34). STATCOM would then act alone during the steady-state operation providing a return on investment from the voltage control and reactive compensation. The additional BR would support STATCOM in the transient state by providing additional damping of power swings.

10.5.4.4 Modulation of Energy Storage SMES or BESS

The schematic diagram of energy storage systems BESS or SMES utilizing voltage source converters is shown in Figure 2.34. The voltage source converter is controlled using a power conditioning system (PCS). The PCS allows BESS or SMES to generate power, for a short time, from any quadrant of the complex power plane assuming that the apparent power (or equivalent admittance) satisfies the limits described by Eq. (2.9). The frequency- and voltage-based controller shown in Figure 10.25 can be used as the modulation controller of an SMES system by forcing the required values of the real and reactive power to be proportional to signals $G_{sh}(t)$ and $B_{sh}(t)$. If the values of these signals are too large then they can be proportionally reduced to satisfy the limits.

10.5.5 Generalization to Multi-machine Systems

The control strategy given by Eqs. (10.22) and (10.48) was derived for the generator-infinite busbar system. These equations can be generalized to a multi-machine system regardless of where the shunt element is located within the system. A detailed proof can be found in Machowski and Bialek (2008). Here only the general framework of the proof and the final equations are shown.

10.5.5.1 Mathematical Model

All the lines and transformers belonging to the modeled network are represented by π-equivalent circuits. Power flowing from the transmission to the distribution network is treated as a load and replaced by a constant admittance included in the network model. All nodes of the network model can be divided into three types:

{G} – internal generator nodes (behind the transient reactances);
k – a chosen node where the considered shunt flexible AC transmission systems (FACTS) device is installed;
{L} – remaining network nodes.

Similar to Figure 10.21, the shunt FACTS device is included in the network model as the varying shunt admittance $\underline{Y}_{sh}(t) = G_{sh}(t) + jB_{sh}(t)$. In the first step of the proof, all the load nodes {L} are eliminated from the network

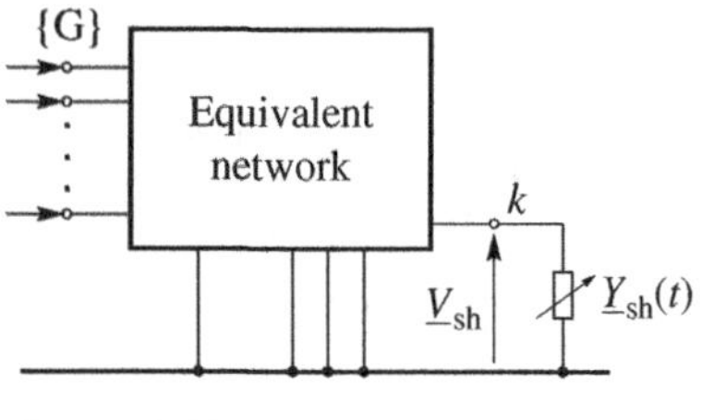

Figure 10.27 Block diagram of equivalent network.

model using the network transformation method described in Section 18.2. Consequently, an equivalent network is obtained that is shown schematically in Figure 10.27. The equivalent network connects all generator nodes {G} with node k in which the considered shunt FACTS device is installed.

As in Figure 10.21, conductances G_{ij} of the equivalent network are neglected and only susceptances B_{ij} are included in the equivalent network model. To retain the balance of power between generation and demand, fictitious loads responsible for active power losses on the conductances G_{ij} of the equivalent network are added in {G} nodes. This is obviously a simplified method of treating conductances and active power losses.

After long and tedious algebraic transformations, the following equation for active power at a generator node is obtained

$$P_i = P^0_{0i} + \sum_{j=1}^{n} b_{ij} \sin \delta'_{ij} + \left[\sum_{j=1}^{n} \beta_{ik}\beta_{kj} \cos \delta'_{ij}\right] G_{sh}(t) + \left[\sum_{j=1}^{n} \beta_{ik}\beta_{kj} \sin \delta'_{ij}\right] B_{sh}(t) \tag{10.50}$$

where P^0_{0i} is active power of the equivalent load at a generator node and $b_{ij} = |\underline{E}'_i||\underline{E}'_j|B_{ij}$ is the magnitude of the power-angle characteristic for an equivalent branch connecting a given pair of internal generator nodes $\{i, j\}$. Coefficients β_{ik} and β_{jk} constitute electric measures of the distance between node k and internal generator nodes $\{i, j\}$, respectively. These coefficients are given by

$$\beta_{ik} = X_{SHC}B_{ik}|\underline{E}'_i|, \beta_{jk} = X_{SHC}B_{jk}|\underline{E}'_j| \tag{10.51}$$

where B_{ik} and B_{jk} are susceptances of the equivalent branches connecting a given pair of internal generator nodes $\{i, j\}$ with node k, and X_{SHC} is the short-circuit reactance of the system seen from node k in which the shunt FACTS device is installed. Equation (10.50) is important in the sense that it shows that a shunt FACTS device introduces two components proportional to G_{sh}(t) and B_{sh}(t), respectively, to the equation expressing active power of a generator. Note similarity of Eqs. (10.18) and (10.50).

10.5.5.2 Control Strategy

Using Eq. (10.50), the swing equations can be formed for all generators, as in Eq. (10.19)

$$\begin{aligned}
\frac{d\delta'_i}{dt} &= \Delta\omega \\
\frac{d\Delta\omega_i}{dt} &= \frac{1}{M_i}\left[P_{mi} - P^0_{0i}\right] - \frac{1}{M_i}\sum_{j=1}^{n} b_{ij} \sin \delta'_{ij} - \frac{D_i}{M_i}\Delta\omega_i \\
&\quad - \frac{1}{M_i}\left[\sum_{j=1}^{n} \beta_{ik}\beta_{kj} \cos \delta'_{ij}\right] G_{sh}(t) - \frac{1}{M_i}\left[\sum_{j=1}^{n} \beta_{ik}\beta_{kj} \sin \delta'_{ij}\right] B_{sh}(t)
\end{aligned} \tag{10.52}$$

where rotor angles δ'_i and speed deviations $\Delta\omega_i$ are the state variables of the system. Equations (10.52) form the nonlinear state-space model describing a dynamic response of the system when changes in the equivalent admittance of the shunt FACTS device are considered.

As a Lyapunov function, the total system energy equal to the sum of the kinetic and potential energy may be used: $V(\delta, \omega) = E_k + E_p$, where E_k and E_p are given by Eqs. (6.53) and (6.54). Using Eqs. (6.62) and (6.63) and the state-space Eqs. (10.52), it can be shown that time derivatives of the kinetic and potential energy are given by

$$\frac{dE_k}{dt} = \sum_{i=1}^{n} \Delta\omega_i\left[P_{mi} - P_{0i}^0\right] - \sum_{i=1}^{n} \Delta\omega_i \sum_{j=1}^{n} b_{ij}\sin\delta'_{ij} - \sum_{i=1}^{n} D_i\Delta\omega_i^2$$
$$- \left[\sum_{i=1}^{n} \Delta\omega_i \sum_{j=1}^{n} \beta_{ik}\beta_{kj}\cos\delta'_{ij}\right] G_{sh}(t) - \left[\sum_{i=1}^{n} \Delta\omega_i \sum_{j=1}^{n} \beta_{ik}\beta_{kj}\sin\delta'_{ij}\right] B_{sh}(t) \tag{10.53}$$

$$\frac{dE_p}{dt} = -\sum_{i=1}^{n} \Delta\omega_i\left(P_{mi} - P_{0i}^0\right) + \sum_{i=1}^{n} \Delta\omega_i \sum_{j\neq i}^{n} b_{ij}\sin\delta'_{ij} \tag{10.54}$$

Note that the first two components of Eq. (10.53) are equal to Eq. (10.54) but with opposite signs. This means that there is a continuous exchange of energy in the transient state between the potential and kinetic energy terms. Moreover, as shown in Eq. (10.53), the shunt FACTS element has a direct influence on the rate of change of the kinetic energy.

Adding Eqs. (10.53) and (10.54) gives

$$\dot{V} = \frac{dV}{dt} = \frac{dE_k}{dt} + \frac{dE_p}{dt} = -\sum_{i=1}^{n} D_i\Delta\omega_i^2 + \dot{V}(\mathrm{sh}) \tag{10.55}$$

where

$$\dot{V}(\mathrm{sh}) = -\left[\sum_{i=1}^{n} \Delta\omega_i\beta_{ik} \sum_{j=1}^{n} \beta_{kj}\cos\delta'_{ij}\right] G_{sh}(t) - \left[\sum_{i=1}^{n} \Delta\omega_i \sum_{j=1}^{n} \beta_{ik}\beta_{kj}\sin\delta'_{ij}\right] B_{sh}(t) \tag{10.56}$$

The first component of the right-hand side of Eq. (10.55) is due to natural damping of generator swings and is always negative for $D_i > 0$. The second component of Eq. (10.55) represents damping introduced by the supplementary control of the shunt FACTS device.

The shunt FACTS device contributes to the system damping if $\dot{V}(\mathrm{sh})$ is negative. Inspection of Eq. (10.56) shows that it is possible to make $\dot{V}(\mathrm{sh})$ always negative by making the values of $G_{sh}(t)$ and $B_{sh}(t)$ to always have the same sign as the relevant values in the square brackets in Eq. (10.56). Hence the stabilizing control based on the measurement of the state variables should follow the following control strategies

$$G_{sh}(t) = K\cdot\sum_{i=1}^{n} \Delta\omega_i\beta_{ik} \sum_{j=1}^{n} \beta_{kj}\cos\delta'_{ij} \tag{10.57}$$

$$B_{sh}(t) = K\cdot\sum_{i=1}^{n} \Delta\omega_i\beta_{ik} \sum_{j=1}^{n} \beta_{kj}\sin\delta'_{ij} \tag{10.58}$$

where K is the control gain.

State-variable control based on the above strategies can be treated as multi-loop control with speed deviations of generators as the input signals and dynamic gains dependent on power angles. This can be shown by expressing formulas (10.57) and (10.58) as

$$G_{sh}(t) = K\cdot\sum_{i=1}^{n} \Delta\omega_i\,\beta_{ik}\,g_i(\boldsymbol{\delta}') \tag{10.59}$$

$$B_{sh}(t) = K\cdot\sum_{i=1}^{n} \Delta\omega_i\,\beta_{ik}\,b_i(\boldsymbol{\delta}') \tag{10.60}$$

where

$$g_i(\boldsymbol{\delta}') = \sum_{j=1}^{n} \beta_{kj}\cos\delta'_{ij} \ \text{ and } \ b_i(\boldsymbol{\delta}') = \sum_{j=1}^{n} \beta_{kj}\sin\delta'_{ij} \tag{10.61}$$

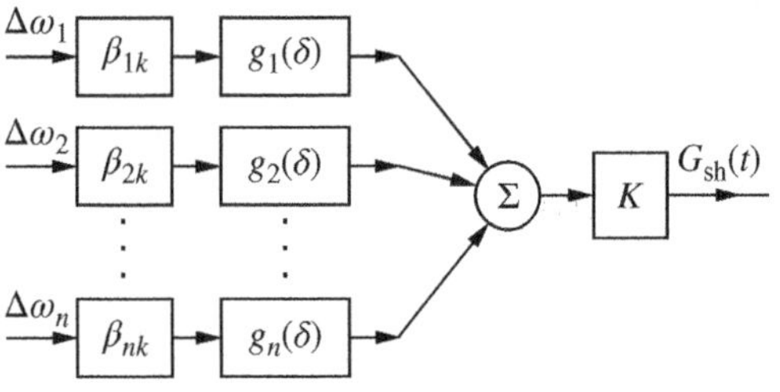

Figure 10.28 Block diagram of the multi-loop control of $G_{sh}(t)$.

are dynamic gains dependent on the current values of the power angles $\boldsymbol{\delta}' = [\delta'_1, \delta'_2, ..., \delta'_n]$ and distance measures β_{ik}, β_{jk} are given by Eq. (10.51). A block diagram of such multi-loop control of $G_{sh}(t)$ is shown in Figure 10.28. For $B_{sh}(t)$ control, the block diagram is identical but $g_i(\boldsymbol{\delta}')$ is replaced by $b_i(\boldsymbol{\delta}')$.

Application of dynamic gains $g_i(\boldsymbol{\delta}')$ and $b_i(\boldsymbol{\delta}')$ significantly influences the dynamic properties of the control process in two ways:

1) Remote generators (small β_{ik}) have little influence on the output signal. The output signal is influenced mostly by generators close to the node k where the shunt device is installed (big β_{ik}). Such dependence of gains on distance measures is justified because shunt FACTS devices have little influence on power produced by remote generators. Hence there would be no point in making control of the device dependent on the state variables of remote generators.
2) Current values of power angles change dynamic gains. In the case of $G_{sh}(t)$, dynamic gains $g_i(\boldsymbol{\delta}')$ decrease as the power angles increase (cosine function). In the case of $B_{sh}(t)$, dynamic gains $b_i(\boldsymbol{\delta}')$ increase as the power angles increase (sine function). Such dynamic changes of gains are justified because the influence of reactive elements $B_{sh}(t)$ on generated power is significant only when the power angle is high (Figure 10.22b). On the other hand, the influence of active elements $G_{sh}(t)$ is reduced when the power angle increases (Figure 10.22a).

The control algorithm satisfying the above properties 1 and 2 is therefore intelligent in the sense that it does not act when the control does not bring the required effects in the system dynamic response. Obviously, the effectiveness of the proposed controller in damping a particular mode of oscillation will depend on its location in the system. This would be revealed by observability and controllability analysis, but such analysis is beyond the scope of this book.

The double summation in Eq. (10.58) corresponds to the sum of elements of a square matrix with elements equal to $\Delta\omega_i\beta_{ik}\beta_{kj}\sin\delta'_{ij}$. The diagonal elements of the matrix are zero because $\sin\delta'_{ii} = \sin 0 = 0$, while the sign of the elements in the upper triangle of the matrix is opposite to that in the lower triangle because $\sin\delta'_{ij} = -\sin\delta'_{ji}$. Hence

$$\sum_{i=1}^{n}\Delta\omega_i\sum_{j=1}^{n}\beta_{ik}\beta_{kj}\sin\delta'_{ij} = \sum_{i=1}^{n}\sum_{j>i}^{n}\Delta\omega_{ij}\beta_{ik}\beta_{kj}\sin\delta'_{ij} \tag{10.62}$$

where $\Delta\omega_{ij} = \Delta\omega_i - \Delta\omega_j$. Equation (10.62) allows the control strategy Eq. (10.60) to be expressed as

$$B_{sh}(t) = K\cdot\sum_{i=1}^{n}\sum_{j>i}^{n}\beta_{ik}\beta_{kj}\,\Delta\omega_{ij}\sin\delta'_{ij} \tag{10.63}$$

which means that $B_{sh}(t)$ depends on the relative speed deviations $\Delta\omega_{ij}$. This property is very important for the response of the supplementary control when disturbances lead to changes in the system frequency. In that case all rotors of all the generators change their speed coherently and the signal produced by formulas (10.63) or (10.60) is equal to zero, $B_{sh}(t) = 0$. This makes sense because reactive shunt elements cannot influence the frequency.

In the case of $G_{sh}(t)$ control, it is not possible to transform the strategy Eq. (10.57) in such a way that a relationship similar to Eq. (10.62) is obtained because $\cos\delta'_{ij} = \cos\delta'_{ji}$. Hence the value of $G_{sh}(t)$ is determined by individual $\Delta\omega_i$ rather than relative values $\Delta\omega_{ij}$. This makes sense because, when there is a surplus of energy in the system, frequency increases and all the loops in Figure 10.28 produce a positive signal. The controlled SMES or BR will then absorb energy, reducing the surplus. On the other hand, when there is an energy deficit in the system, the frequency decreases and all the loops shown in Figure 10.28 produce a negative signal. The SMES will then inject active power into the system thereby reducing the energy deficit.

It is also worth noticing that the coordinates of the post-fault equilibrium point $\hat{\delta}'_i$, which were present in the energy-type Lyapunov function, are not present in the control strategies described by Eqs. (10.57) and (10.58). This means that it is not necessary to calculate the coordinates of $\hat{\delta}'_i$ following a disturbance. Control strategies described by Eqs. (10.57) and (10.58) utilize only the values of angles δ'_i in the transient state.

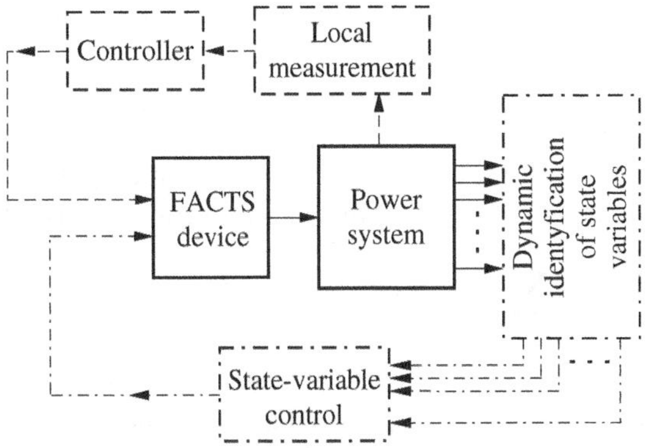

Figure 10.29 Block diagram of the local control and state-variable stabilizing control.

10.5.5.3 Wide Area Control System WAMPAC

Each loop of the supplementary stabilizing control contains a coefficient β_{ik} corresponding to the measures of the distance between a given ith generator and a given node k where the shunt FACTS device is installed. When the distance is long, the distance measure β_{ik} is small, and it may be approximately assumed that $\beta_{ik} \cong 0$ so that the corresponding loop can be neglected. Hence, in practice, the proposed multi-loop controller will contain only a few loops corresponding to generators in a small area surrounding the shunt FACTS device. Hence, from the point of view of the state-variable stabilizing control of shunt FACTS devices, it is not necessary to measure phasors $\underline{E}_i = E_i e^{j\delta'_i}$ in the whole power system. It is sufficient to measure phasors only in a small area around the shunt FACTS device. Such control may be referred to as *area control*.

Control strategies Eqs. (10.59) and (10.60) simplified to area control could be utilized in a WAMPAC-type (Section 2.7) control system making use of phasor measurements. A possible structure of such a system is shown in Figure 10.29. The main steady-state control loop (the upper part of Figure 10.29) is based on measuring a locally observable signal to be controlled by a FACTS device. For example, in the case of STATCOM, it is the voltage at a given node of the system. The supplementary stabilizing loop (the lower part of Figure 10.29) utilizes state variables as input signals and, from the point of view of the whole system, it is a state-variable control. The main problem for such a closed-loop control is the speed of data transmission. Current modern flexible communication platforms (Figure 2.46) cannot transmit data fast enough in order to damp power swings. However, it may be expected that the speed of data transmission will increase in the near future so that practical implementation of a WAMPAC system similar to that shown in Figure 10.29 will be possible.

10.5.5.4 Emulation of State-variable Control by Local Control

As the speed of data transmission is not fast enough to implement the WAMPAC shown in Figure 10.29, it is necessary to look for controllers using local measurements and which are able to emulate the control strategy Eqs. (10.57) and (10.58). How exactly such local control will emulate optimal control depends on the choice of local signals used and the structure of the regulator. A detailed analysis of the choice of signals can be found in EPRI (1999). Here analysis is presented based on Machowski and Nelles (1994). Equations derived in their paper show that the controller of Figure 10.25 implementing Eq. (10.48) is also valid for the multi-machine system regardless of where the shunt FACTS device is located within the system.

For the multi-machine system, an equation similar to Eq. (10.29) can be written

$$\frac{dq_G}{dt} = \sum_{i=1}^{n} \Delta\omega_i \frac{\partial q_G}{\partial \delta'_i} + \alpha_{GG}\frac{dG_{sh}}{dt} + \alpha_{GB}\frac{dB_{sh}}{dt} \tag{10.64}$$

where α_{GG} and α_{GB} are sensitivity factors determined by Eq. (10.30). Equation (10.64) can be transformed to give

$$\sum_{i=1}^{n} \Delta\omega_i \frac{\partial q_G}{\partial \delta_i} = \frac{dq_G}{dt} - \alpha_{GG}\frac{dG_{sh}}{dt} - \alpha_{GB}\frac{dB_{sh}}{dt} \tag{10.65}$$

The left-hand side of Eq. (10.65) depends on $\Delta\omega_i$. It is exactly the same as in the theoretically optimal strategy determined by Eq. (10.57) if the following condition is satisfied

$$\frac{\partial q_G}{\partial \delta_i} = \sum_{j=1}^{n} \beta_{ik}\beta_{kj}\cos\delta'_{ij} \tag{10.66}$$

Substituting the partial derivative on the left-hand side of Eq. (10.65) by the right-hand side of Eq. (10.66) and inserting the result into Eq. (10.57) gives the local control Eq. (10.33).

In the control strategy given by Eq. (10.63) the shunt susceptance depends on the relative speed deviations $\Delta\omega_{ij}$. The time derivative of q_B, expressed in terms of the relative speed deviations, is given by

$$\frac{dq_B}{dt} = \sum_{i=1}^{n}\sum_{j>i}^{n} \Delta\omega_{ij}\frac{\partial q_B}{\partial \delta'_{ij}} + \alpha_{BG}\frac{dG_{sh}}{dt} + \alpha_{BB}\frac{dB_{sh}}{dt} \tag{10.67}$$

where α_{GG} and α_{GB} are sensitivity factors determined by Eq. (10.35). Rearranging Eq. (10.67) gives

$$\sum_{i=1}^{n}\sum_{j>i}^{n} \Delta\omega_{ij}\frac{\partial q_B}{\partial \delta'_{ij}} = \frac{dq_B}{dt} - \alpha_{BG}\frac{dG_{sh}}{dt} - \alpha_{BB}\frac{dB_{sh}}{dt} \tag{10.68}$$

The left-hand side of that equation depends on $\Delta\omega_{ij}$. It is exactly the same as in the theoretically optimal strategy given by Eq. (10.63) if the following condition is satisfied

$$\frac{\partial q_B}{\partial \delta'_{ij}} = \beta_{ik}\beta_{kj}\sin\delta'_{ij} \tag{10.69}$$

Substituting the partial derivative on the left-hand side of Eq. (10.68) by the right-hand side of Eq. (10.69), and inserting the result into Eq. (10.63), gives the local control Eq. (10.34).

The conditions defined by Eqs. (10.66) and (10.69) are the basic conditions under which the local control defined by Eqs. (10.33) and (10.34) can emulate the theoretical optimal control defined by Eqs. (10.57) and (10.63). An additional condition is that for the given input variables the sensitivity factors defined by Eqs. (10.30) and (10.35) must be either known or negligible.

In the considered model shown in Figure 10.27, when the network conductances are neglected it can be shown that the magnitude and angle of voltage $\underline{V}_k$ at node k are given by

$$\underline{V}_k \cong \varphi\,(\mathrm{sh})\,\varphi(\delta') \tag{10.70}$$

where

$$\varphi\,(\mathrm{sh}) = (1 + X_{SHC}B_{sh}) - jX_{SHC}G_{sh} \tag{10.71}$$

$$\varphi(\delta') = -\sum_{j=1}^{n} \beta_{kj}\left[\cos\delta'_j + j\sin\delta'_j\right] \tag{10.72}$$

Both functions φ (sh) and $\varphi(\delta')$ are complex.

Using Eq. (10.8) for a simplified analysis of voltage $\underline{V}_k$, it may be assumed that both components $X_{SHC}B_{sh}$ and $X_{SHC}G_{sh}$ in Eq. (10.71) are negligible. Consequently, the following simplifications are obtained which are important for further considerations

$$\varphi(\mathrm{sh}) \cong 1 \text{ and } \underline{V}_k \cong \varphi(\delta') \tag{10.73}$$

Using the simplifications above, Eqs. (10.70) and (10.72) give

$$\underline{V}_k \cong \varphi(\boldsymbol{\delta}') = -\sum_{j=1}^{n} \beta_{kj}\left[\cos\delta'_j + \mathrm{j}\sin\delta'_j\right] \tag{10.74}$$

Symbol θ denotes the phase angle of voltage $\underline{V}_k$ measured with respect to the reference frame common for all nodes of the network model. Thus Eq. (10.74) gives

$$|\underline{V}_k|\cos\theta = -\sum_{j=1}^{n} \beta_{kj}\cos\delta'_j \text{ and } |\underline{V}_k|\sin\theta = -\sum_{j=1}^{n} \beta_{kj}\sin\delta'_j \tag{10.75}$$

Hence

$$\tan\theta \cong \frac{\sum_{j=1}^{n} \beta_{kj}\sin\delta'_j}{\sum_{j=1}^{n} \beta_{kj}\cos\delta'_j} \tag{10.76}$$

$$|\underline{V}_k|^2 = \left[\sum_{j=1}^{n} \beta_{kj}\sin\delta'_j\right]^2 + \left[\sum_{j=1}^{n} \beta_{kj}\cos\delta'_j\right]^2 \tag{10.77}$$

Similar to the second of Eqs. (10.42) and (10.76) allows θ to be treated as an implicit function of $\boldsymbol{\delta}'$, that is $f(\theta, \boldsymbol{\delta}') = 0$. Using Eqs. (10.44), (10.76), and (10.77) gives

$$\begin{aligned} |\underline{V}_k|^2 \frac{\partial\theta}{\partial\delta_i} &= \beta_{ki}\cos\delta'_i \sum_{j=1}^{n} \beta_{kj}\cos\delta_j + \beta_{ki}\sin\delta'_i \sum_{j=1}^{n} \beta_{kj}\sin\delta'_j \\ &= \sum_{j=1}^{n} \beta_{ki}\beta_{kj}\left[\cos\delta_i\cos\delta'_j + \sin\delta_i\sin\delta'_j\right] = \sum_{j=1}^{n} \beta_{ki}\beta_{kj}\cos\delta'_{ij} \end{aligned} \tag{10.78}$$

Comparison of Eqs. (10.78) and (10.66) leads to

$$|\underline{V}_k|^2 \frac{\partial\theta}{\partial\delta'_i} = \sum_{j=1}^{n} \beta_{ik}\beta_{kj}\cos\delta'_{ij} = \frac{\partial q_{\mathrm{G}}}{\partial\delta'_i} \tag{10.79}$$

The conclusion from this equation is that the time derivative of the voltage phase angle θ multiplied by the voltage squared $|\underline{V}_k|^2$ satisfies the conditions of a good input signal for local control of $G_{\mathrm{sh}}(t)$.

When the sensitivity of θ with respect to $G_{\mathrm{sh}}(t)$ and $B_{\mathrm{sh}}(t)$ is considered, it is necessary to take into account Eq. (10.70) with $\varphi(\mathrm{sh})$ given by Eq. (10.71). Calculating the derivative of the implicit function and further simplifying the result gives

$$|\underline{V}_k|^2 \frac{\partial\theta}{\partial G_{\mathrm{sh}}} = -|\underline{V}_k|^2 X_{\mathrm{SHC}} \text{ and } |\underline{V}_k|^2 \frac{\partial\theta}{\partial B_{\mathrm{sh}}} \cong 0 \tag{10.80}$$

This means that the phase angle θ of the voltage $\underline{V}_k$ is mainly sensitive to the changes of $G_{\mathrm{sh}}(t)$, and its sensitivity factor can be assessed on the basis of the expected value of the short-circuit reactance X_{SHC}.

Substituting the relevant sensitivity factors in Eq. (10.33) by Eqs. (10.80) and (10.79) gives the same control scheme as Eq. (10.48a).

The signal dependent on the magnitude of the squared voltage can be expressed, using Eqs. (10.70)–(10.72), as

$$q_{\mathrm{B}} = -\frac{1}{2}\left|\underline{V}_k\right|^2 = -\frac{1}{2}\underline{V}_k^*\underline{V}_k = -\frac{1}{2}\left|\varphi(\mathrm{sh})\right|^2\left|\varphi(\boldsymbol{\delta}')\right|^2 \tag{10.81}$$

where

$$\left|\varphi(\mathrm{sh})\right|^2 = [1 + X_{\mathrm{SHC}}B_{\mathrm{sh}}(t)]^2 + [X_{\mathrm{SHC}}G_{\mathrm{sh}}(t)]^2 \tag{10.82}$$

$$\left|\varphi(\boldsymbol{\delta}')\right|^2 = \sum_{i=1}^{n}\sum_{j=1}^{n}\beta_{ik}\beta_{kj}\cos\delta'_{ij} = \sum_{i=1}^{n}\beta_{ik}^2 + 2\sum_{i=1}^{n}\sum_{j>1}^{n}\beta_{ik}\beta_{kj}\cos\delta'_{ij} \tag{10.83}$$

In order to calculate the sensitivity of signal Eq. (10.81) with respect to power angles, it may be assumed, as in Eq. (10.73), that $\varphi(\mathrm{sh}) \cong 1$. Then Eqs. (10.81) and (10.83) give

$$\frac{\partial q_{\mathrm{B}}}{\partial \delta'_{ij}} = -\frac{1}{2}\frac{\partial\left|\underline{V}_k\right|^2}{\partial \delta'_{ij}} \cong -\frac{1}{2}\frac{\partial\left|\varphi(\boldsymbol{\delta}')\right|^2}{\partial \delta'_{ij}} = \beta_{ik}\beta_{kj}\sin\delta'_{ij} \tag{10.84}$$

Equations (10.84) and (10.69) are the same, showing that the signal given by Eq. (10.81) satisfies the condition of a good input signal for control of the shunt susceptance $B_{\mathrm{sh}}(t)$.

Sensitivity factors α_{GG} and α_{GB} determined by Eqs. (10.35) can be easily found, assuming that $\underline{V}_k \cong \phi(\delta')$. Under this assumption, differentiation of Eq. (10.81) gives

$$\frac{\partial q_{\mathrm{B}}}{\partial G_{\mathrm{sh}}} \cong -\left|\underline{V}_k\right|^2 X_{\mathrm{SHC}}[X_{\mathrm{SHC}}G_{\mathrm{sh}}(t)] \cong 0 \tag{10.85a}$$

$$\frac{\partial q_{\mathrm{B}}}{\partial B_{\mathrm{sh}}} \cong -\left|\underline{V}_k\right|^2 X_{\mathrm{SHC}}[1 + X_{\mathrm{SHC}}B_{\mathrm{sh}}(t)] \cong -\left|\underline{V}_k\right|^2 X_{\mathrm{SHC}} \tag{10.85b}$$

This means that the magnitude of the squared voltage is sensitive mainly to the changes of $B_{\mathrm{sh}}(t)$. Its sensitivity factor can be assessed on the basis of the expected value of the short-circuit reactance and the measured voltage magnitude.

Substituting the relevant sensitivity factors in Eq. (10.34) by Eqs. (10.85) and (10.84), and the local signal by Eq. (10.81), gives the same control formula as Eq. (10.48b). Resulting controller is the same as in Figure 10.25 (Section 10.5.3). Simulation results can be found in publications by Machowski and Nelles (1992a, 1992b, 1993, 1994) and Machowski et al. (2001).

10.5.6 Example of Simulation Results

Here only the simulations of the state-variable stabilizing control using the WAMPAC-type structure shown in Figure 10.29 are presented. The time delay introduced by the telecommunication system of WAMPAC has been modeled using a first-order block with a time constant of 30 ms. Currently, typical delays recorded in WAMPAC systems are about 100 ms. Such a big delay would significantly worsen the control process.

Simulation have been executed for the CIGRE test system (Figure 10.30). Generator G4 in this system has very high inertia constituting effectively the infinite busbar and providing a reference. The considered test system experiences transient stability problems mainly for generators G7 and G6, especially as a result of a short circuit in line L7 without reclosing. Such a case has been chosen to illustrate the robustness of the proposed control algorithm.

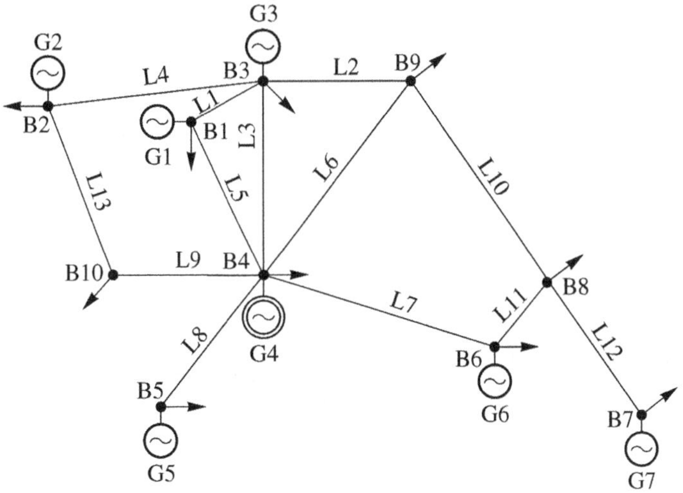

Figure 10.30 Schematic diagram of the CIGRE test system.

Table 10.1 Distance measures β_{ik} for S_{base} = 100 MVA.

—	L7	G1	G2	G3	G4	G5	G6	G7
B6	on	0.024	0.006	0.034	0.406	0.008	0.538	0.134
	off	0.023	0.005	0.038	0.147	0.003	0.762	0.188
B7	on	0.024	0.006	0.037	0.283	0.005	0.279	0.552
	off	0.015	0.003	0.025	0.094	0.002	0.121	0.482
B8	on	0.034	0.008	0.053	0.397	0.008	0.387	0.229
	off	0.034	0.008	0.056	0.214	0.004	0.547	0.268

Table 10.1 shows the values of distance measures β_{ik} for nodes B6, B7, and B8 and all internal generator nodes behind transient reactances of the generator. Symbols "on" and "off" in column L7 in Table 10.1 correspond to appropriate "on" and "off" states of line L7. The results show that only generators G4, G6, and G7 are important for shunt FACTS devices installed at each of the three chosen nodes B6, B7, and B8. For the remaining generators, the distance measures are an order of magnitude smaller, and it may be safely assumed that $\beta_{ik} \cong 0$. Consequently, control of shunt FACTS devices in nodes B6, B7, and B8 can be based only on state variables ω_i and δ'_i for generators G4, G6, and G7. A corresponding multi-loop supplementary controller (Figure 10.28) will then contain only three loops with ω_4, ω_6, ω_7 as input signals.

Figure 10.31 shows the simulation results when one SMES (rated 40 MVA) at bus B8 was controlled using state-variable stabilizing control limited to the closest generators G4, G6, and G7 (local area control). The two lower graphs illustrate the changes in the values of $G_{sh\ B8}$ and $B_{sh\ B8}$ which correspond to real and reactive power of the SMES. The apparent power absorbed from the network or injected into the network is limited, as shown by Eq. (10.6). Consequently, the control algorithm uses more active power (conductance $G_{sh\ B8}$) than reactive power (susceptance $B_{sh\ B8}$). The two upper graphs show the changes in the values of rotor angles of generators G6 and G7. For comparison dashed lines show the case without SMES.

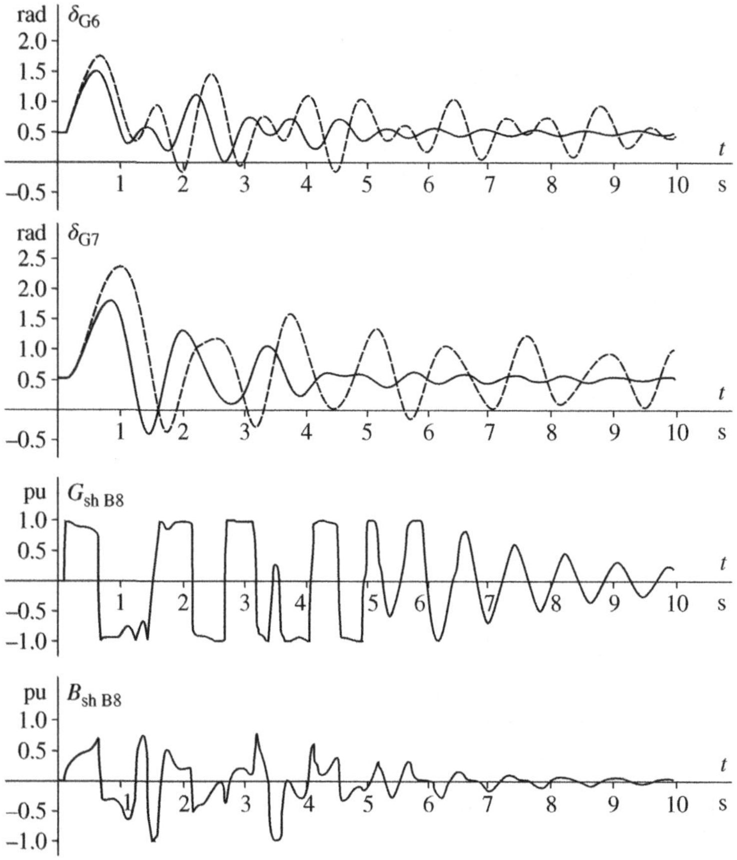

Figure 10.31 Simulation results for one SMES installed at bus B8.

Very good damping can be obtained by using thyristor-controlled BRs connected at generator busbars. Simulation results for such a case are shown in Figure 10.32 when two BRs (each rated 40 MW) are installed at bus B6 and bus B7. In this case rotor swings are damped very quickly.

10.6 Series Compensators

Section 3.1.2 shows how the power transfer capability of a long transmission line depends on its inductive reactance and how this reactance can be offset by inserting a series capacitor. Besides being useful in the steady state, such a reduction in the line reactance is also useful in the transient state as it increases the amplitude of the transient power-angle characteristic thereby increasing the available deceleration area. By proper control of a switched series capacitor this change in amplitude can be used to provide additional damping of power swings. In particular the conventional series capacitor, equipped with a zinc oxide protective scheme (Figure 2.26) and thyristor-switched series capacitor (Figure 2.36a), can be used in a bang-bang control mode as it can be almost instantaneously bypassed and reinserted at appropriate moments. The thyristor-controlled series capacitor (Figure 2.36b) and

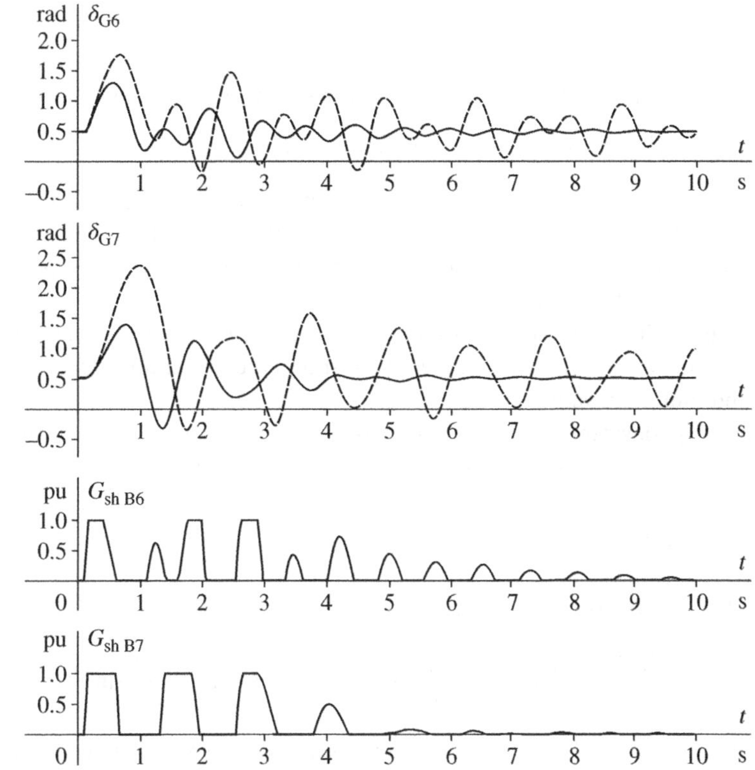

Figure 10.32 Simulation results for two BRs installed at bus B6 and bus B7.

the static synchronous series compensator (Figure 2.37) allow the equivalent reactance to be smoothly controlled so that they can be used in a continuous control mode.

10.6.1 State-variable Control

Consider the simple generator-infinite busbar system shown in Figure 10.33a. Line L2 is assumed to be open-circuited before a fault occurs. The fault results in tripping the line after some clearing time. The generator is represented by the classical model, that is constant transient electromotive force (emf) E' behind the transient reactance X'_d. Neglecting the resistance and shunt capacitance, the generated active power is

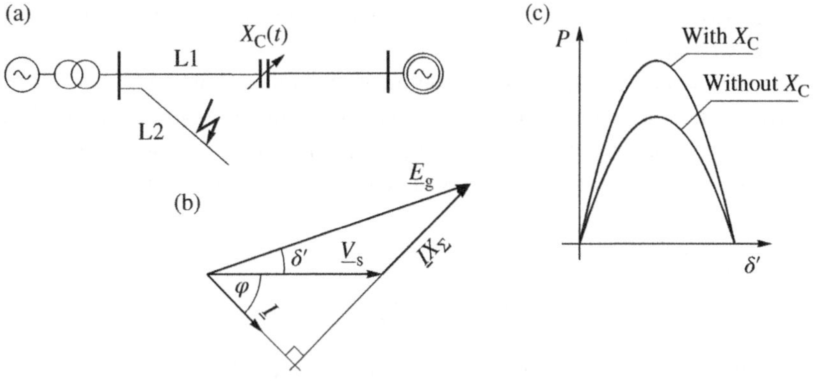

Figure 10.33 System with a shunt capacitor: (a) network diagram; (b) phasor diagram; (c) power-angle characteristic with and without series compensation.

$$P(\delta') = \frac{E' V_s}{X_\Sigma} \sin \delta' \tag{10.86}$$

where V_s is the infinite busbar voltage, δ' is the transient power angle between V_s and E', and

$$X_\Sigma = \left(X'_d + X_T + X_{L1} + X_s\right) - X_C(t) = X - X_C(t) \tag{10.87}$$

is the equivalent reactance of the transmission link, where X_T is the reactance of the transformer, X_{L1} is the reactance of the line, X_s is the equivalent reactance of the infinite busbar, and X is the equivalent reactance of the transmission link without the compensator. A positive value of $X_C(t)$ corresponds to a capacitance, while a negative $X_C(t)$ corresponds to an inductance.

A change in the compensator reactance $X_C(t)$ causes a change in X_Σ and therefore a change in the amplitude of the power-angle characteristic. Figure 10.33c shows the power-angle characteristics corresponding to the maximum and minimum values of X_Σ.

To simplify considerations further, it is worth separating the system reactance into components with and without the series compensator. This can be done in the following way

$$\frac{1}{X_\Sigma} = \frac{1}{X - X_C(t)} = \frac{1}{X} + \frac{1}{X_\Sigma}\frac{X_C}{X} \tag{10.88}$$

so that Eq. (10.86) takes the following form

$$P(\delta') = \frac{E' V_s}{X} \sin \delta' + \frac{E' V_s}{X_\Sigma}\frac{X_C(t)}{X} \sin \delta' = b \sin \delta' + b_\Sigma \frac{X_C(t)}{X} \sin \delta' \tag{10.89}$$

where $b_\Sigma = E'V_s/X_\Sigma$ and $b = E'V_s/X$ are the amplitudes of the power-angle characteristic with and without the series compensator, respectively. The first component in Eq. (10.89) defines power which would flow in the system if the series compensator was not used, that is for $X_C(t) = 0$. The second component is responsible for a change in the power flow due to the variation in the compensator reactance. As series compensation is normally less than 100% of the line reactance, it can be assumed that X_Σ is always positive, which is important for further considerations.

Taking into account Eq. (10.89), the swing equation of the system can now be written as

$$\begin{aligned} \frac{d\delta'}{dt} &= \Delta\omega \\ M\frac{d\Delta\omega}{dt} &= P_m - b \sin \delta' - D\frac{d\delta'}{dt} - (b_\Sigma \sin \delta')\frac{X_C(t)}{X} \end{aligned} \tag{10.90}$$

where $\Delta\omega$ is the speed deviation, M is the inertia coefficient, P_m is the mechanical power input from the prime mover, and D is the damping coefficient. The equilibrium point of this equation has coordinates $(\hat{\delta}', \Delta\hat{\omega} = 0)$. The control variable is $X_C(t)$.

The system described by Eq. (10.90) is nonlinear. A standard approach to derive the optimal state-variable control of such a system would be to linearize the system around its operating point. Here, as before in this book, the optimal state-variable control is derived from the nonlinear model using the Lyapunov direct method.

The Lyapunov function for the system is equal to the sum of the potential and kinetic energy $V = E_k + E_p$, where

$$\begin{aligned} E_k &= \frac{1}{2} M \Delta\omega^2 \\ E_p &= -\left[P_m\left(\delta' - \hat{\delta}'\right) + b\left(\cos \delta' - \cos \hat{\delta}'\right)\right] \end{aligned} \tag{10.91}$$

At the equilibrium point $(\hat{\delta}', \Delta\hat{\omega} = 0)$ the total energy given by Eq. (10.91) is zero. A fault releases some energy, that is it causes an increase in the total energy expressed by Eq. (10.91), which results initially in acceleration of the rotor and an increase in δ' and $\Delta\omega$. The goal of a control strategy is to enforce such changes in the equivalent reactance of the transmission link so that the system is brought back as fast as possible to the equilibrium point $(\hat{\delta}', \Delta\hat{\omega} = 0)$, where $V = 0$. This is equivalent to a fast dissipation of the energy released by the fault and quick damping of rotor swings. The control strategy must therefore maximize the value of the derivative $\dot{V} = \mathrm{d}V/dt$ calculated along the trajectory of the differential Eq. (10.90).

It can be easily proved that, for functions given by Eq. (10.91), the following hold

$$\frac{\mathrm{d}E_\mathrm{k}}{\mathrm{d}t} = \frac{\partial E_\mathrm{k}}{\partial \omega}\frac{\mathrm{d}\Delta\omega}{\mathrm{d}t} = M\frac{\mathrm{d}\Delta\omega}{\mathrm{d}t}\Delta\omega \tag{10.92}$$

$$\frac{\mathrm{d}E_\mathrm{p}}{\mathrm{d}t} = \frac{\partial E_\mathrm{p}}{\partial \delta'}\frac{\mathrm{d}\delta'}{\mathrm{d}t} = \frac{\partial E_\mathrm{p}}{\partial \delta'}\Delta\omega = -[P_\mathrm{m} - b\sin\delta']\Delta\omega \tag{10.93}$$

Substituting the left-hand side of the second of Eq. (10.90) into the right-hand side of Eq. (10.92) gives

$$\frac{\mathrm{d}E_\mathrm{k}}{\mathrm{d}t} = +[P_\mathrm{m} - b\sin\delta']\Delta\omega - D\Delta\omega^2 - (b_\Sigma \sin\delta')\frac{X_\mathrm{C}(t)}{X}\Delta\omega \tag{10.94}$$

Adding Eqs. (10.93) and (10.94) gives a time derivative of the Lyapunov function

$$\dot{V} = \frac{\mathrm{d}V}{\mathrm{d}t} = \frac{\mathrm{d}E_\mathrm{k}}{\mathrm{d}t} + \frac{\mathrm{d}E_\mathrm{p}}{\mathrm{d}t} = -D\Delta\omega^2 - (b_\Sigma \sin\delta')\frac{X_\mathrm{C}(t)}{X}\Delta\omega \tag{10.95}$$

The system is stable if this derivative is negative. Moreover, the speed with which the system returns to the equilibrium point is proportional to $\dot{V}$, that is the greater the negative value of $\dot{V}$, the faster the dissipation of energy released by the fault and the faster the damping of the swings.

The second component of Eq. (10.95) depends on the control variable $X_\mathrm{C}(t)$ and the state variables $(\delta', \Delta\omega)$. This component will be negative if the control strategy of the series compensator is such that

$$X_\mathrm{C}(t) = KX(\sin\delta')\Delta\omega \tag{10.96}$$

where K is the gain of the regulator. Recall that $X =$ constant is the link reactance without the compensator. Substituting Eq. (10.96) into Eq. (10.95) gives

$$\frac{\mathrm{d}V}{\mathrm{d}t} = \frac{\mathrm{d}E_\mathrm{k}}{\mathrm{d}t} + \frac{\mathrm{d}E_\mathrm{p}}{\mathrm{d}t} = -D\Delta\omega^2 - K\,b_\Sigma(\sin\delta')^2\Delta\omega^2 \tag{10.97}$$

which means that assuming the control strategy given by Eq. (10.96), the derivative $\dot{V}$ is always negative. At any moment during the transient state such control will improve the damping. This can be additionally shown by substituting $X_\mathrm{C}(t)$ from Eq. (10.96) into the swing equation (10.90). The swing equation will then take the form

$$\begin{aligned} \frac{\mathrm{d}\delta'}{\mathrm{d}t} &= \Delta\omega \\ M\frac{\mathrm{d}\Delta\omega}{\mathrm{d}t} &= P_\mathrm{m} - b\sin\delta' - D\frac{\mathrm{d}\delta'}{\mathrm{d}t} - D_\mathrm{ser}\frac{\mathrm{d}\delta'}{\mathrm{d}t} \end{aligned} \tag{10.98}$$

where

$$D_\mathrm{ser} = Kb_\Sigma(\sin\delta')^2 \geq 0 \tag{10.99}$$

is a positive damping coefficient due to the control of the series compensator.

10.6.2 Interpretation Using the Equal Area Criterion

The influence of the control given by Eq. (10.96) on the transient stability can be simply explained by assuming that the FACTS device is a thyristor-switched series capacitor (Figure 2.36a). In that case the control given by Eq. (10.96) can be implemented only as bang-bang control

$$X_{\rm C}(t) = \begin{cases} X_{\rm C\,MAX} & \text{for} \quad (\sin\delta')\Delta\omega \geq +\varepsilon \\ 0 & \text{for} \quad (\sin\delta')\Delta\omega < +\varepsilon \end{cases} \tag{10.100}$$

where ε determines the dead zone.

Consider again the simple generator-infinite busbar system of Figure 10.33 with the fault occurring on line L2. Figure 10.34 shows the power-angle curves with $X_{\rm C} = 0$, the lower characteristic, and with $X_{\rm C} = X_{\rm C\,MAX}$, the upper characteristic. The fault causes the generator power to drop so that kinetic energy proportional to area 1-2-3-4 is released. At the instant of fault clearance the expression ($\sin\delta'$) is positive so that the signal $X_{\rm C}(t)$ is set to its maximal value $X_{\rm C} = X_{\rm C\,MAX}$ and the whole capacitor is inserted. The available declaration area is 4-5-6-10. The rotor reaches speed deviation $\Delta\omega = 0$ at point 6 when area 4-5-6-7 becomes equal to area 1-2-3-4 and then starts to swing back. The expression ($\Delta\omega \cdot \sin\delta'$) becomes negative and the control strategy in Eq. (10.96) causes the capacitor to be bypassed $X_{\rm C}(t) = 0$. The rotor follows the lower $P(\delta')$ characteristic. The power jumps from point 6 to point 8 and the rotor swings back along path 8–9 performing deceleration work proportional to area 8-9-7. This backward swing deceleration area is now much smaller than area 6-7-1, the deceleration area available when $X_{\rm C} = X_{\rm C\,MAX}$. When the rotor starts to swing forward, the speed deviation changes sign, $X_{\rm C}(t)$ is increased, and the acceleration area is reduced. The complete switching cycle is then repeated but with rotor swings of reduced amplitude. Thus the control principle is established. During the forward swing, the acceleration area should be minimized and the available deceleration area maximized, while during the backward swing the deceleration area should be minimized and the available acceleration area maximized.

In some circumstances bang-bang control with a small dead zone can lead to large power angle swings and instability. To explain this again consider Figure 10.34 and assume that the rotor reaches point 11 during its forward swing. Sudden bypassing of the capacitor causes the system trajectory to jump to point 12, from which the rotor would be further accelerated, causing a subsequent asynchronous operation. This would not happen with continuous control because both the control signal and $X_{\rm C}(t)$ change smoothly. Such control can be implemented using the thyristor-controlled series capacitor (Figure 2.36b) or the static synchronous series compensator (Figure 2.35). The next section describes a state-variable stabilizing control of the series compensator, based on Eq. (10.96), which can be emulated using local measurements.

10.6.3 Control Strategy Based on the Squared Current

The control strategy Eq. (10.96) utilizes state variables (δ', $\Delta\omega$) which are not readily available at the point of installation of the series compensator. Therefore, it is convenient to emulate this optimal strategy using another strategy based on locally available measurements. Similar considerations are used in Section 10.5.3 but with regard to the shunt compensation.

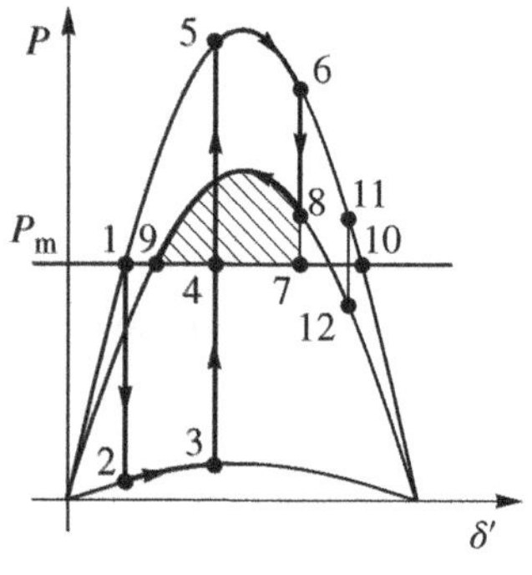

Figure 10.34 Interpretation of series capacitor control using the equal area criterion.

For the case considered, the cosine theorem applied to the voltage triangle from Figure 10.33b gives

$$(IX_\Sigma)^2 = (E')^2 + V_{\rm s}^2 - 2E'V_{\rm s}\cos\delta' \tag{10.101}$$

so that

$$I^2 = \frac{1}{X_\Sigma^2}\left((E')^2 + V_{\rm s}^2 - 2E'V_{\rm s}\cos\delta'\right) \tag{10.102}$$

Assuming constant values of E' and $V_{\rm s}$, the signal given by Eq. (10.102) depends on X_Σ and δ'. Hence the speed of the signal changes can be expressed as

$$\frac{{\rm d}(I^2)}{{\rm d}t} = \frac{\partial(I^2)}{\partial\delta'}\frac{{\rm d}\delta'}{{\rm d}t} + \frac{\partial(I^2)}{\partial X_\Sigma}\frac{{\rm d}X_\Sigma}{{\rm d}t} \tag{10.103}$$

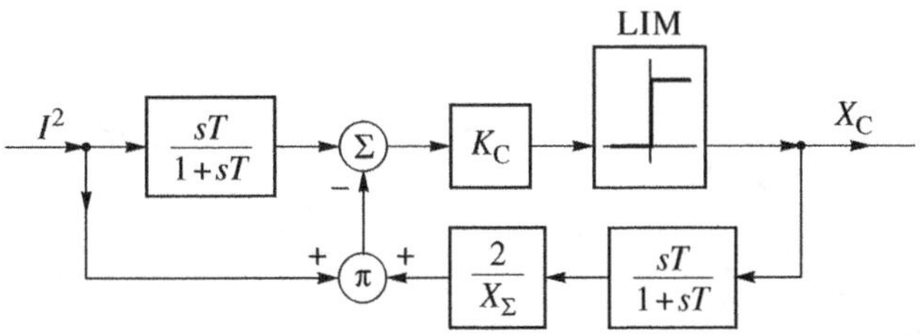

Figure 10.35 Supplementary control of $X_C(t)$.

and the partial derivatives are given by

$$\frac{\partial(I^2)}{\partial\delta'} = \frac{2}{X_\Sigma^2}E'V_s \sin\delta' \quad \text{and} \quad \frac{\partial(I^2)}{\partial X_\Sigma} = -\frac{2}{X_\Sigma}I^2 \tag{10.104}$$

As $X_\Sigma = X - X_C(t)$ and $X = \text{constant}$, it holds that $dX_\Sigma/dt = -dX_C/dt$. Substituting expressions given by Eq. (10.104) into Eq. (10.103) gives

$$\frac{d(I^2)}{dt} = \frac{2}{X_\Sigma^2}E'V_s(\sin\delta')\,\Delta\omega + \frac{2}{X_\Sigma}I^2\frac{dX_C}{dt} \tag{10.105}$$

which, after reordering the terms, gives

$$(\sin\delta')\Delta\omega = \frac{X_\Sigma^2}{2E'V_s}\left[\frac{d(I^2)}{dt} - \frac{2}{X_\Sigma}I^2\frac{dX_C}{dt}\right] \tag{10.106}$$

Substituting Eq. (10.106) into Eq. (10.96) gives

$$X_C(t) = K_C\left[\frac{d(I^2)}{dt} - \frac{2}{X_\Sigma}I^2\frac{dX_C}{dt}\right] \tag{10.107}$$

where K_C is the equivalent gain of the controller.

Figure 10.35 shows the block diagram of a controller executing the control strategy Eq. (10.107). Derivation is executed by a real differentiator with a small time constant T. The limiter at the output of the controller limits the output signal to that applicable for a particular type of series compensator. The controller is nonlinear because it contains a product of the input signal and the derivative of the output signal. Sensitivity of the squared current I^2 to the changes in the control variable $X_C(t)$ is compensated by using a feedback loop with a gain inversely proportional to X_Σ. This feedback plays a secondary role when compared with the main feedback loop. Consequently, the gain in the corrective loop can be determined approximately using an estimate of the equivalent reactance X_Σ.

If the series compensator is equipped with a steady-state power flow controller then the considered controller can be attached as a supplementary control loop for damping of power swings.

10.6.4 Control Based on Other Local Measurements

It is shown above that control based on the squared current allows the optimal control strategy to be executed. Some authors suggest that other locally measured signals such as active power or the current magnitude (but not squared) can be used as input signals for the regulator. A question then arises as to what the differences are between the controllers using these three input signals in the case of large disturbances involving large changes of the power angle.

When the sensitivity of the given signal $q(t)$ to the changes in X_Σ is neglected, then, as in Eq. (10.103), the control signal can be expressed as

$$\frac{dq}{dt} \cong \frac{\partial q}{\partial\delta'}\frac{d\delta'}{dt} = \frac{\partial q}{\partial\delta'}\Delta\omega \tag{10.108}$$

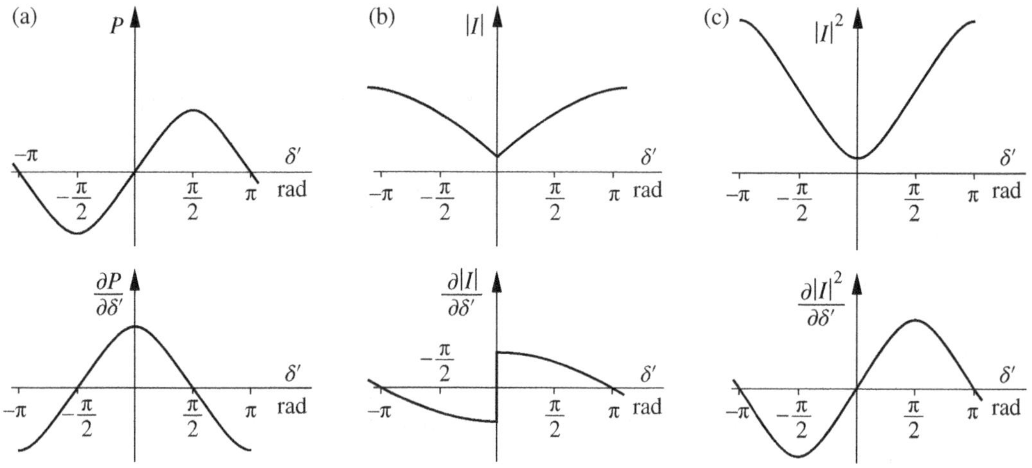

Figure 10.36 Electric quantities and their partial derivatives: (a) real power; (b) squared current magnitude; (c) current magnitude.

For a given speed deviation $\Delta\omega$, the value of the control signal $\mathrm{d}q/\mathrm{d}t$ is determined by the value of the partial derivative $\partial q/\partial \delta'$. A comparison with Eq. (10.96) shows that the partial derivative $\partial q/\partial \delta'$ should ideally be of a sine type, that is it should be positive for $\delta' > 0$ and negative for $\delta' < 0$. Figure 10.36 shows the partial derivative $\partial q/\partial \delta'$ for the active power P, current I and squared current magnitude I^2.

The partial derivative of the real power, $\partial P/\partial \delta'$, is largest at $\delta' = 0$ and then it decreases taking negative values for $\delta' > \pi/2$. The derivative of the current magnitude is large and discontinuous, changing its sign around $\delta' = 0$. The derivative of the squared current magnitude has a sine shape, that is it is zero at $\delta' = 0$ and it reaches a maximum at $\delta' = \pi/2$. This leads to conclusions described in the next section.

10.6.4.1 Controller Based on Active Power

In the vicinity of $\delta' = 0$, when the control is not effective because it does not influence damping in a significant way, the control signal $\mathrm{d}P/\mathrm{d}t$ is unnecessarily large. When the power angle increases and the control action starts to influence damping, the control signal decreases. Around $\delta' = \pi/2$, when the control is the most effective, the control signal is zero. For $\delta' > \pi/2$ the control signal changes sign and becomes negative, causing negative damping which harms the system. Similarly, negative damping occurs for $\delta' < -\pi/2$, that is for a large power angle during backward swing. Obviously, at any operating point when $-\pi/2 < \delta' < \pi/2$ a small disturbance will produce a correct control signal and positive damping. Hence the linearized analysis of the controller in the vicinity of the operating point does not expose the disadvantages of using active power as an input signal for the regulator.

10.6.4.2 Controller Based on Current

The control signal $\mathrm{d}I/\mathrm{d}t$ has the correct sign over the whole range of power angle changes. However, the controller produces an unnecessarily large signal in the vicinity of $\delta' = 0$, when the control is ineffective. Moreover, when crossing the value $\delta' = 0$, the control signal is discontinuous.

10.6.4.3 Controller Cased on Squared Current

The control signal $\mathrm{d}(I^2)/\mathrm{d}t$ has the correct sign and shape over the whole range of power angle changes. In the vicinity of $\delta' = 0$, when the control is ineffective, the signal is small but increases with angle, reaching a maximum at $\delta' = \pi/2$, when the control is the most effective. Also note that the control signal is continuous.

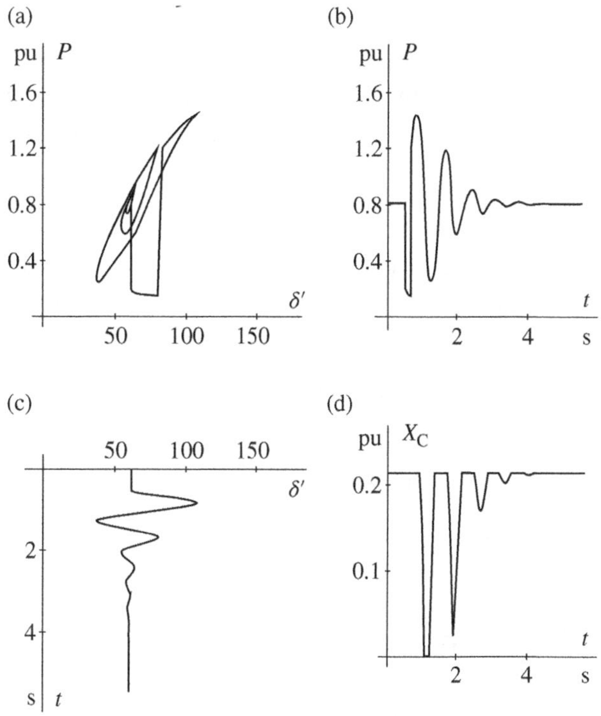

Figure 10.37 Simulation results for the generator-infinite busbar system.

10.6.5 Simulation Results

The proposed controller has been tested using a variety of systems. Owing to the lack of space, the simulation results for the simple generator-infinite busbar system only are presented. A short circuit was assumed at the end of the line, beyond the series compensator. The considered case was stable. Without the series compensator the resulting power swings vanished after about 10 s. The regulation process when the series compensator was used is shown in Figure 10.37.

Figure 10.37a shows the system trajectory in the $P(\delta)$ plane. One can see a sudden change in the value of active power following the fault and then its clearance. The system trajectory during the back swing lies below the trajectory corresponding to the forward swing. Good damping of power swings may be observed as the oscillations vanish after about 3 s (Figure 10.37b and c). After the disturbance, the first two changes in $X_C(t)$ are big (Figure 10.37d) and as the swings disappear the controller enforces smaller changes in $X_C(t)$.

10.7 Unified Power Flow Controller

As discussed in Section 2.5.4, series FACTS devices also include, apart from the controlled series capacitor, also the phase angle regulator (thyristor controlled phase angle regulator [TCPAR]; Figure 2.37) and the unified power flow controller (UPFC; Figure 2.38). The UPFC can control three signals: (i) the quadrature component of the booster voltage, (ii) the direct component of the booster voltage, and (iii) the reactive shunt current. The TCPAR can control only the first signal, that is the quadrature component of the booster voltage. Hence this section discusses the more general case, that is supplementary stabilizing control of the UPFC.

10.7.1 Power-angle Characteristic

To simplify considerations, the generator-infinite busbar system is discussed, as shown in Figure 10.38. The generator is represented by the classical model. The shunt part of the UPFC is modeled by a variable susceptance $B_{sh}(t)$. The series part, inserting the booster voltage, is modeled by a complex transformation ratio defined as

$$\eta = \frac{\underline{V}_a}{\underline{V}_b} = |\eta| e^{j\theta} \text{ and } \frac{\underline{I}_b}{\underline{I}_a} = \eta * = |\eta| e^{-j\theta} \tag{10.109}$$

Booster transformer reactance is added, on the generator side, to the equivalent reactance of the network.

The phasor diagram shown in Figure 10.38c breaks down the booster voltage into its direct ΔV_Q and quadrature ΔV_P components. These components can be expressed as a fraction of the busbar voltage

$$\Delta V_P = \gamma V_b \text{ and } \Delta V_Q = \beta V_b \tag{10.110}$$

where β and γ are the output variables of the supplementary control of the UPFC. To emphasize the time dependency, the variables will be denoted as $\beta(t)$ and $\gamma(t)$. The voltage triangle in Figure 10.38c gives

$$\sin\theta = \frac{\Delta V_P}{V_a} = \frac{\gamma V_b}{V_a} = \frac{\gamma}{|\eta|} \tag{10.111}$$

$$\cos\theta = \frac{V_b + \Delta V_Q}{V_a} = \frac{V_b + \beta V_b}{V_a} = \frac{1+\beta}{|\eta|} \tag{10.112}$$

$$(1+\beta)^2 + \gamma^2 = |\eta|^2 \text{ or } |\eta| = \sqrt{(1+\beta)^2 + \gamma^2} \tag{10.113}$$

Derivation of a formula for the generator power that includes all three control variables $B_{sh}(t) \neq 0$, $\gamma(t) \neq 0$, and $\beta(t) \neq 0$ takes over three pages of algebraic transformations even for the simple generator-infinite busbar system. To illustrate the problem, only a simplified case of neglected shunt compensation, that is $B_{sh}(t) = 0$, is discussed here.

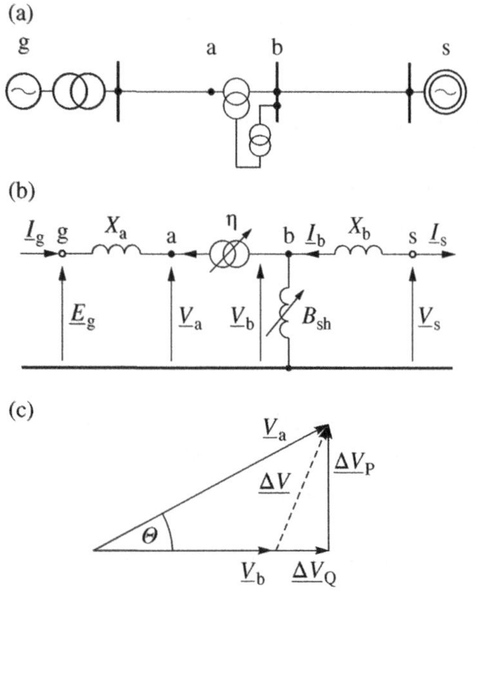

Figure 10.38 Generator-infinite busbar system to investigate UPFC control: (a) block diagram; (b) equivalent network; (c) phasor diagram.

When the shunt susceptance $B_{sh}(t)$ is neglected, the equivalent system reactance as seen by the generator is equal to $X_\Sigma = X_a + |\eta|^2 X_b$, where the second component corresponds to reactance X_b transformed by the transformation ratio to the generator side. The angle between the generator emf and the infinite busbar voltage is δ'. The angle between the voltages on both sides of the booster transformer defined in Eqs. (10.109). is θ. This means that the phase angle of the voltage drop on reactance X_Σ is $(\delta' - \theta)$. The infinite busbar voltage transformed to the generator side is $V_s|\eta|$. Hence, taking into account the general Eq. (1.8), one can write

$$P = \frac{E_g V_s |\eta|}{X_\Sigma} \sin(\delta' - \theta) = \frac{E_g V_s}{X_\Sigma} |\eta| (\sin\delta' \cos\theta - \cos\delta' \sin\theta) \tag{10.114}$$

Substituting Eqs. (10.111) and (10.112) into Eq. (10.114) gives

$$P = b_\Sigma \sin\delta' - b_\Sigma \cos\delta' \gamma(t) + b_\Sigma \sin\delta' \beta(t) \tag{10.115}$$

where $b_\Sigma = E_g V_s / X_\Sigma$ is the amplitude of the power-angle characteristic when the transformation ratio given by Eq. (10.113) is included. When the booster voltage is absent, that is when $\gamma(t) = 0$ and $\beta(t) = 0$, the characteristic corresponds to the first component of Eq. (10.115). The second and third components correspond to the direct and quadrature components of the booster voltage Eq. (10.110), respectively.

The influence of the booster voltages on the power-angle characteristic is illustrated in Figure 10.39. The quadrature component enforces a nonzero value of the angle θ and, according to Eq. (10.114), causes a shift in the power-angle characteristic to the left if $\gamma > 0$, or to the right if $\gamma < 0$. The direct component changes the amplitude of the characteristic, increasing it when $\beta > 0$ and reducing it when $\beta < 0$.

Figure 10.22b shows that the shunt compensation $B_{sh}(t)$ of reactive power also affects the amplitude of the power-angle characteristic. Hence the shunt element $B_{sh}(t)$ neglected in Eq. (10.114) will also have an influence on the amplitude of the characteristic similarly just like the direct component of the booster voltage.

Januszewski (2001) derives a similar equation to Eq. (10.115) but including simultaneous control of all three quantities

$$\begin{aligned} P \cong\ & b_\Sigma \sin\delta' \\ & - b_\Sigma \sin\delta' X_{SHC} B_{sh}(t) \\ & + b_\Sigma \sin\delta' (1 - X_{SHC} B_{sh}) \beta(t) - b_\Sigma \cos\delta' (1 - X_{SHC} B_{sh}) \gamma(t) \end{aligned} \tag{10.116}$$

It is worth remembering that, as in Eq. (10.8), $X_{SHC} B_{sh} \ll 1$ holds. This means that expression $(1 - X_{SHC} B_{sh})$ is positive and $(1 - X_{SHC} B_{sh}) \cong 1$.

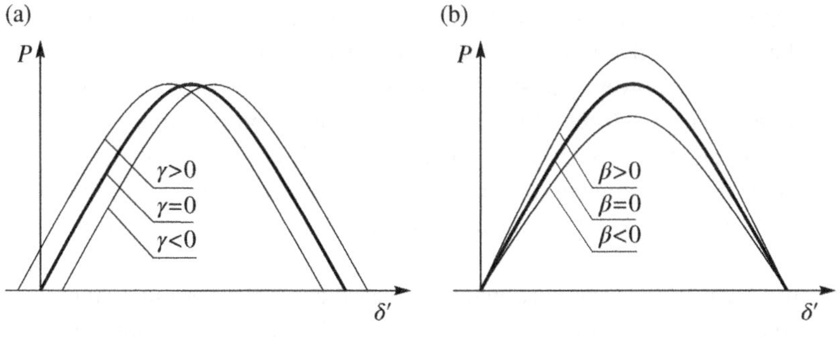

Figure 10.39 Influence of the booster voltages on the power-angle characteristic: (a) influence of the quadrature component; (b) influence of the direct component.

10.7.2 State-variable Control

Taking into account Eq. (10.116), the swing equation of the system can now be written as

$$M\frac{d\Delta\omega}{dt} = P_m - b_\Sigma \sin\delta' - D\frac{d\delta'}{dt} + b_\Sigma X_{SHC}B_{sh}\sin\delta' - (1 - X_{SHC}B_{sh})\left[\beta\, b_\Sigma \sin\delta' - \gamma\, b_\Sigma \cos\delta'\right] \tag{10.117}$$

where $\Delta\omega$ and δ' are the state variables and $X_C(t)$ is the controlled variable. The control law will be derived, similar to the shunt devices, using the Lyapunov direct method.

The total system energy $V = E_k + E_p$ is chosen as the Lyapunov function, as for the shunt devices and the series capacitor

$$V = \frac{1}{2}M\Delta\omega^2 - \left[P_m(\delta' - \hat{\delta}') + b_\Sigma\left(\cos\delta' - \cos\hat{\delta}'\right)\right] \tag{10.118}$$

Calculating the partial derivative along the system trajectory of Eq. (10.117) gives

$$\dot{V} = \frac{dE_k}{dt} + \frac{dE_p}{dt} = -D\,\Delta\omega^2 + \Delta\omega\,(b_\Sigma \sin\delta')\,X_{SHC}B_{sh}(t) + \Delta\omega\,(1 - X_{SHC}B_{sh})(b_\Sigma\cos\delta')\,\gamma(t) - \Delta\omega\,(1 - X_{SHC}B_{sh})\,(b_\Sigma\sin\delta')\,\beta(t) \tag{10.119}$$

Taking into account that $(1 - X_{SHC}B_{sh}) > 0$, it can be concluded that controlling each of the three quantities will introduce a negative term to the derivative (10.119) if the control is according to the following three equations

$$\beta(t) = +K_\beta[b_\Sigma \sin\delta']\,\Delta\omega \tag{10.120}$$

$$\gamma(t) = -K_\gamma[b_\Sigma \cos\delta']\,\Delta\omega \tag{10.121}$$

$$B_{sh}(t) = -K_B[b_\Sigma \sin\delta']\,\Delta\omega \tag{10.122}$$

where K_β, K_γ, and K_B are the control gains. With that control, the system energy will change according to

$$\dot{V} = -D\Delta\omega^2 - \left[K_\beta(\sin\delta')^2 + K_\gamma(\cos\delta')^2\right]\,b_\Sigma^2\Delta\omega^2 + K_B X_{SHC}(\sin\delta')^2 b_\Sigma^2\Delta\omega^2 \tag{10.123}$$

For $K_\beta = K_\gamma = K_\eta$, controlling both booster voltage components gives a constant damping independent of the power angle, because the expression in the square brackets in Eq. (10.123) is equal to K_η, and Eq. (10.123) becomes

$$\dot{V} = -D\Delta\omega^2 - K_\eta\, b_\Sigma^2\Delta\omega^2 + K_B X_{SHC}(\sin\delta')^2 b_\Sigma^2\Delta\omega^2 \tag{10.124}$$

Control using the direct booster voltage component (signal β) and controlling the shunt compensation (signal B_{sh}) is executed the same way as controlling the shunt reactive power compensator, Eq. (10.48). Control action is proportional to the speed deviation $\Delta\omega$ and the sine of the power angle $\sin\delta'$. When the power angle is positive and when $\Delta\omega > 0$ (Figure 10.39b), the control chooses values $\beta > 0$ such that the amplitude of the power-angle characteristic is increased and therefore the available deceleration area is increased. During the backward swing when $\Delta\omega < 0$, the control chooses such values $\beta < 0$ that the amplitude of the characteristic is reduced (Figure 10.39b), causing a reduction in the deceleration work during the backward swing and a reduction of the maximum rotor displacement.

Control of the quadrature voltage component – signal $\gamma(t)$ – is proportional to the speed deviation $\Delta\omega$ and cosine of the power angle $\cos\delta'$. The influence of such control is illustrated in Figure 10.40. During the short circuit, the rotor gains energy corresponding to the acceleration area 1-2-3-4. At the same time, the control system applies control $\gamma > 0$ such that the power-angle characteristic is moved to the left. As a result, electrical power at point

5 is higher than it would otherwise have been (the middle bolder line). Power then changes along line 5–6 on the characteristic, that is shifted to the left. At the peak of the characteristic $\cos\delta'$ tends to zero and signal $\gamma(t)$ tends to zero too. Hence power changes along the middle characteristic corresponding to the lack of control. Then $\cos \delta'$ changes sign, the signal $\gamma(t)$ changes sign too, and power changes according to the characteristic shifted to the right. The available deceleration area is 4-5-6-7-8. In the case considered, this area is bigger than the area 1-2-3-4 and the system is stable. The rotor does not move beyond point 8 because, immediately after moving past point 7, the speed deviation decreases significantly so that the signal $\gamma(t)$ decreases, too. Hence the movement is no longer along line 7–8 but along the middle characteristic. After reaching $\Delta\omega = 0$ the rotor starts the backward swing and the control signal $\gamma(t) > 0$ appears as long as $\cos \delta' < 0$. The backward swing is along the characteristic shifted to the left. Based on that description a trajectory of the rotor motion could be constructed, similar to that shown in Figure 10.24 for the BR.

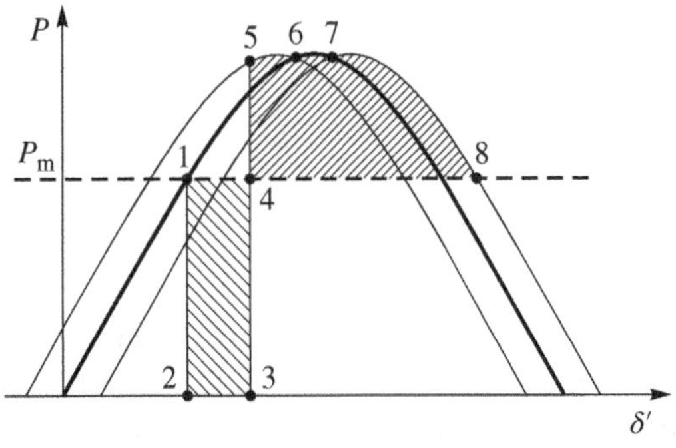

Figure 10.40 Influence of the control strategy in the generator-infinite busbar system.

Control based on Eqs. (10.120)–(10.122) is state-variable control and requires measurements of both the speed deviation and the rotor angle, which are difficult to measure in a multi-machine system. Hence, similar to the case for the shunt FACTS devices and the series capacitor, Eqs. (10.120)–(10.122) have to be replaced by equations making it possible to control the UPFC using local measurements.

10.7.3 Control Based on Local Measurements

In order to obtain a signal proportional to the speed deviation, a locally available signal $q(t)$ should be used. As in Eqs. (10.64) and (10.67) discussed earlier, one can write

$$\dot{q} = \frac{dq}{dt} = \frac{\partial q}{\partial \delta'}\frac{d\delta'}{dt} + \frac{\partial q}{\partial \beta}\frac{d\beta}{dt} + \frac{\partial q}{\partial \gamma}\frac{d\gamma}{dt} + \frac{\partial q}{\partial B_{sh}}\frac{dB_{sh}}{dt} \tag{10.125}$$

In this equation only the first component is proportional to the speed deviation and can be used for state-variable control. The other components should have as small an influence as possible on the value of the derivative. The equation can be rewritten as follows

$$\dot{q} = \frac{dq}{dt} = \frac{\partial q}{\partial \delta'}\Delta\omega + \varepsilon(t) \tag{10.126}$$

where $\Delta\omega = d\delta'/dt$ and $\varepsilon(t)$ are functions of variables β, γ, B_{sh}. If a local measurement $q(t)$ is to emulate well state-variable control, two conditions must be satisfied:

1) For the control of $\beta(t)$ and $B_{sh}(t)$, partial derivatives $\partial q/\partial\delta'$ should be proportional to $\sin\delta'$. For the control of $\gamma(t)$, the partial derivative $\partial q/\partial\delta'$ should be proportional to $\cos\delta'$.
2) The first component, proportional to the speed deviation, should dominate on the right-hand side of Eq. (10.126). The second component should be negligible, that is a given signal should not be sensitive to the changes in β, γ, B_{sh}.

Januszewski (2001) shows that the above conditions are reasonably well satisfied for the reactive power Q and active power P measured in a line where a booster transformer is connected (Figure 10.38). Hence state-variable control defined by Eqs. (10.120)–(10.122) can be approximately replaced by a control that uses local measurements and is defined by

$$\gamma(t) \cong +K_{\gamma}\frac{dP}{dt} \tag{10.127}$$

$$\beta(t) \cong +K_{\beta}\frac{dQ}{dt} \tag{10.128}$$

$$B_{sh}(t) \cong -K_{B}\frac{dQ}{dt} \tag{10.129}$$

where K_{γ},K_{β}, and K_B are appropriately chosen control gains.

According to Eqs. (10.123) and (10.124), booster voltage control through control of $\gamma(t)$ and $\beta(t)$ gives a strong damping independent of the actual value of the power angle. Compared with that, damping introduced by the regulation of shunt compensation $B_{sh}(t)$ is quite weak. Hence the UPFC could be used only to control its nodal voltage without introducing shunt compensation supplementary control.

The series controller may have the structure shown in Figure 10.41. Control of the booster voltage components is achieved by using an integral regulator with a feedback loop and a PSS acting according to Eqs. (10.127) and (10.128). There is an output limiter common for both components of the booster voltage limiting its value ΔV where (see Figure 10.38c)

$$(\Delta V)^2 = (\Delta V_P)^2 + (\Delta V_Q)^2 \tag{10.130}$$

This limit is shown schematically in Figure 10.41 as a circle.

10.7.4 Examples of Simulation Results

Stability enhancement using supplementary UPFC control will now be illustrated using the simple test system shown in Figure 10.42. A UPFC device has been installed to control the flow of power in a transmission link consisting of lines L4 and L5 that is parallel to a link consisting of lines L3 and L2. The UPFC has the structure shown in Figure 10.41. It was assumed that there was a short circuit in line L6 and the fault was cleared by tripping the line. The resulting simulation results are shown in Figure 10.43.

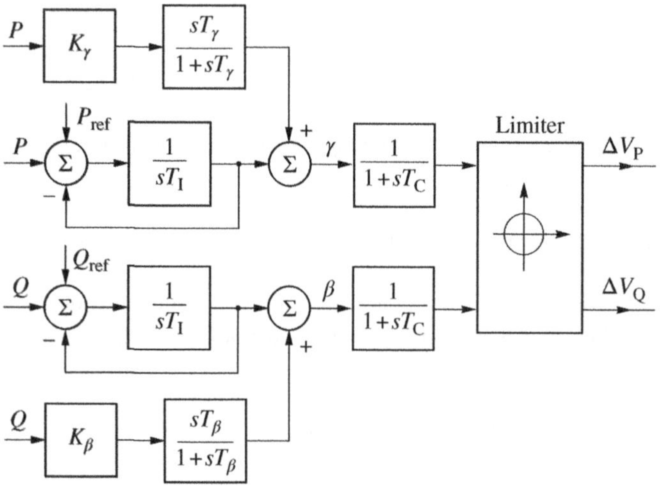

Figure 10.41 Block diagram of the booster voltage controller.

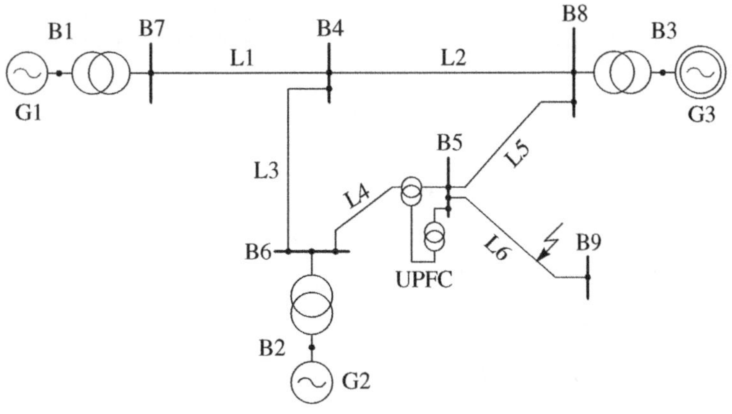

Figure 10.42 Three-machine test system.

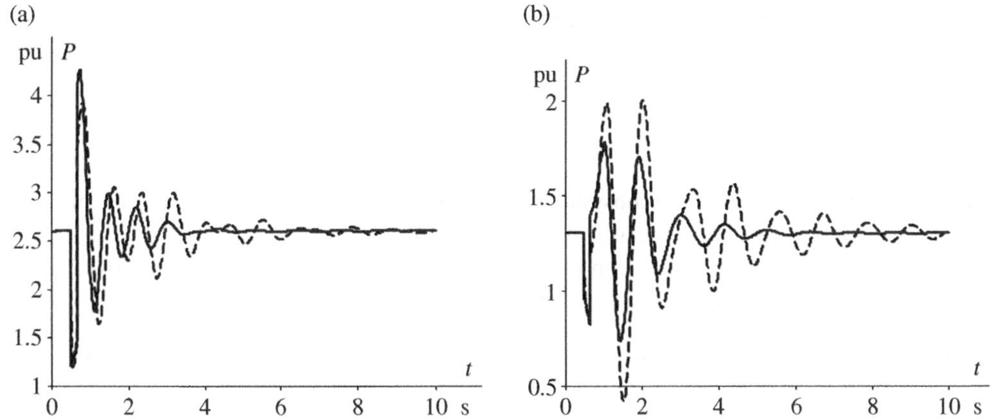

Figure 10.43 Examples of simulation results: (a) active power of generator G1; (b) active power of generator G2.

The fault caused swinging of both generators. Because of their different parameters, the frequencies of rotor swings are also quite different, which makes damping difficult. However, Figure 10.43 shows that the swings are quickly damped and a new steady state is achieved after a few seconds.

10.8 HVDC Links in Transmission Network

The high-voltage direct current (HVDC) transmission lines have various applications within AC power systems such as submarine (underwater) cable transmission, integration of large wind farms, enhancement of the transfer capability in AC transmission network, etc. The control of HVDC transmission lines can also be used to enhance transient stability of AC power systems (Machowski et al. 2013). For that purpose, the HVDC controllers can be equipped with supplementary control loops referred to as PSS, i.e. analogous as supplementary control loops used in excitation control systems of synchronous generators.

In such PSS as input signals, locally measurable quantities or signals obtained from the wide area measurement system (WAMS) described in Section 2.7 can be used. Before discussing these two solutions, it is worth considering what the control algorithm should be that would be optimal from the viewpoint of transient stability of the whole

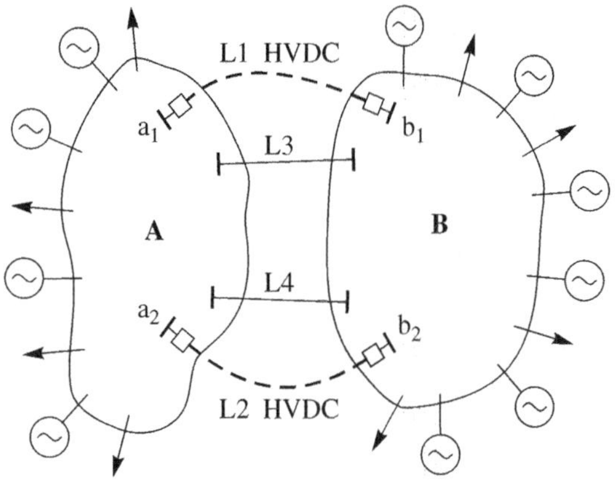

Figure 10.44 Two weakly connected AC systems with two embedded HVDC lines.

AC power system and that damp the power swings in the subsystem where a short-circuit occurred with only slight influence on the other subsystem.

10.8.1 Mathematical Model

In order to enlarge the transfer capability between these subsystems and to enhance transient stability of such an AC power system, additional HVDC lines are embedded, as shown in Figure 10.44.

Control formulas for the supplementary stabilizing loops in the controllers of the HVDC lines will be derived using a linear model $\dot{\boldsymbol{x}} = \boldsymbol{A}\boldsymbol{x}$ of the AC power system with the assumptions described below.

1) It is assumed that subsystem "B" is sending (exporting) the active power via the HVDC lines and subsystem "A" is receiving (importing) this power. Power losses in the HVDC lines are neglected. For the AC network model the sending and receiving powers can be treated as power injections at the terminal nodes $\{a_1, b_1, a_2, b_2\}$

$$P_{a_1} = P_1;\ \ P_{b1} = -P_1;\ \ P_{a_2} = P_2;\ \ P_{b_2} = -P_2 \tag{10.131}$$

The terminal substations of the HVDC lines absorb from the AC network the reactive powers, which taken with the opposite sign can be treated as the nodal reactive power injections, respectively

$$Q_{a_1}, Q_{b_1}, Q_{a_2}, Q_{b_2} \tag{10.132}$$

Control of reactive power at the terminal substations is independent and generally: $Q_{b_1} \neq Q_{a_1}$ and $Q_{b_2} \neq Q_{a_2}$.

2) Each synchronous generator is modeled by a classical model with constant transient emf behind the transient reactance and by the swing equation (5.1).
3) All load nodes are eliminated from the network model by a method described in Chapter 18. The reduced network model contains only fictitious generator nodes {G} behind the transient reactances and the terminal nodes $\{a_1,b_1, a_2,b_2\}$ of the HVDC lines.

It should be emphasized that the model with such assumptions is used only to derive the control formulas.

Taking into account the above assumptions the AC transmission network can be described by the incremental matrix equation similar to Eq. (3.182), described in Section 3.5.1

$$\begin{bmatrix} \Delta \boldsymbol{P} \\ \Delta \boldsymbol{Q} \end{bmatrix} = \begin{bmatrix} \boldsymbol{H} & \boldsymbol{K} \\ \boldsymbol{N} & \boldsymbol{W} \end{bmatrix} \begin{bmatrix} \Delta \boldsymbol{\delta} \\ \Delta \boldsymbol{V} \end{bmatrix} \tag{10.133}$$

where $\Delta \boldsymbol{P}$ and $\Delta \boldsymbol{Q}$ are the column matrices of the real and reactive power increments at nodes {G} and {a_1, b_1, a_2, b_2}, $\Delta \boldsymbol{V}$ and $\Delta \boldsymbol{\delta}$ are column matrices of voltage magnitude and angle increments, and $\boldsymbol{H}$, $\boldsymbol{K}$, $\boldsymbol{N}$, and $\boldsymbol{W}$ are Jacobi submatrices.

The following designations are to be followed: {A} and {B} are sets of generating units in the subsystem A and B, respectively, and {G} = {A} + {B} is the set of generating units in the whole interconnected system.

For the fictitious generator nodes {G} the magnitudes of transient emf's are constant and for those nodes

$$\Delta \boldsymbol{V}_{\mathrm{G}} = \Delta \boldsymbol{E}_{\mathrm{G}} = 0 \tag{10.134}$$

In the considered model at the nodes {a_1, b_1, a_2, b_2}, where the HVDC lines are connected to, owing to Eqs. (10.131a) and (10.132b) there are the following nodal real and reactive power injections

$$\Delta P_{a_1} = \Delta P_1; \quad \Delta P_{b1} = -\Delta P_1; \quad \Delta Q_{a_1}; \Delta Q_{b_1} \tag{10.135}$$

$$\Delta P_{a_2} = \Delta P_2; \quad \Delta P_{b_2} = -\Delta P_2; \quad \Delta Q_{a_2}; \Delta Q_{b_2} \tag{10.136}$$

When Eqs. (10.134), (10.135), and (10.136) are taken into account, it is possible to rewrite Eq. (10.133) in the following way

$$\begin{bmatrix} \Delta \boldsymbol{P}_{\mathrm{G}} \\ \hline \Delta P_{a_1} \\ \Delta P_{b_1} \\ \Delta Q_{a_1} \\ \Delta Q_{b_1} \\ \hdashline \Delta P_{a_2} \\ \Delta P_{b_2} \\ \Delta Q_{a_2} \\ \Delta Q_{b_2} \end{bmatrix} \cong \left[\begin{array}{c|ccc:ccccc} \boldsymbol{H}_{\mathrm{GG}} & \boldsymbol{H}_{\mathrm{Ga_1}} & \boldsymbol{H}_{\mathrm{Gb_1}} & \boldsymbol{K}_{\mathrm{Ga_1}} & \boldsymbol{K}_{\mathrm{Gb_1}} & \boldsymbol{H}_{\mathrm{Ga_2}} & \boldsymbol{H}_{\mathrm{Gb_2}} & \boldsymbol{K}_{\mathrm{Ga_2}} & \boldsymbol{K}_{\mathrm{Gb_2}} \\ \hline \boldsymbol{H}_{a_1\mathrm{G}} & \boldsymbol{H}_{a_1a_1} & \boldsymbol{H}_{a_1b_1} & K_{a_1a_1} & K_{a_1b_1} & H_{a_1a_2} & H_{a_1b_2} & K_{a_1b_2} & K_{a_1b_2} \\ \boldsymbol{H}_{b_1\mathrm{G}} & H_{b_1a_1} & H_{b_1a_1} & K_{b_1a_1} & K_{b_1a_1} & H_{b_1a_2} & H_{b_1b_2} & K_{b_1a_2} & K_{b_1b_2} \\ \boldsymbol{N}_{a_1\mathrm{G}} & N_{a_1a_1} & N_{a_1b_1} & W_{a_1a_1} & W_{a_1b_1} & N_{a_1a_2} & N_{a_1b_2} & W_{a_1b_2} & W_{a_1b_2} \\ \boldsymbol{N}_{b_1\mathrm{G}} & N_{b_1a_1} & N_{b_1b_1} & W_{b_1a_1} & W_{b_1b_1} & N_{b_1a_n} & N_{b_1b_2} & W_{a_1b_2} & W_{b_1b_2} \\ \hdashline \boldsymbol{H}_{a_n\mathrm{G}} & H_{a_12a_1} & H_{a_2b_1} & K_{a_2a_1} & K_{a_2b_1} & H_{a_2a_1} & H_{a_2b_2} & K_{a_2a_2} & K_{a_2b_2} \\ \boldsymbol{H}_{b_2\mathrm{G}} & H_{b_2a_1} & H_{b_2b_1} & K_{b_2a_1} & K_{b_2b_1} & H_{b_2a_2} & H_{a_2b_2} & K_{b_2a_2} & K_{a_2b_2} \\ \boldsymbol{N}_{a_2\mathrm{G}} & \boldsymbol{N}_{a_2a_1} & N_{a_2b_1} & W_{a_2b_1} & W_{a_2b_1} & N_{a_2a_2} & N_{a_2b_2} & W_{a_2a_2} & W_{a_2b_2} \\ \boldsymbol{N}_{b_2\mathrm{G}} & \boldsymbol{N}_{b_2a_1} & N_{b_2b_1} & W_{b_2a_1} & W_{b_2b_1} & N_{b_2a_2} & N_{b_2b_2} & W_{b_2a_2} & W_{b_2b_2} \end{array}\right] \begin{bmatrix} \Delta \delta'_{\mathrm{G}} \\ \hline \Delta \delta_{a_1} \\ \Delta \delta_{b_1} \\ \Delta V_{a_1} \\ \Delta V_{b_1} \\ \hdashline \Delta \delta_{a_2} \\ \Delta \delta_{b_2} \\ \Delta V_{a_2} \\ \Delta V_{b_2} \end{bmatrix} \tag{10.137}$$

Partial inversion (Section A.2 in the appendix) of Eq. (10.137) results in

$$\begin{bmatrix} \Delta \boldsymbol{P}_{\mathrm{G}} \\ \hline \Delta \delta_{a_1} \\ \Delta \delta_{b_1} \\ \Delta V_{a_1} \\ \Delta V_{b_1} \\ \hdashline \Delta \delta_{a_2} \\ \Delta \delta_{b_2} \\ \Delta V_{a_2} \\ \Delta V_{b_2} \end{bmatrix} \cong \left[\begin{array}{c|cccc:cccc} \boldsymbol{h}_{\mathrm{GG}} & \boldsymbol{h}_{\mathrm{Ga_1}} & \boldsymbol{h}_{\mathrm{Gs_1}} & \boldsymbol{k}_{\mathrm{Ga_1}} & \boldsymbol{k}_{\mathrm{Gb_1}} & \boldsymbol{h}_{\mathrm{Ga_2}} & \boldsymbol{h}_{\mathrm{Gb_2}} & \boldsymbol{k}_{\mathrm{Ga_2}} & \boldsymbol{k}_{\mathrm{Gb_2}} \\ \boldsymbol{h}_{a_1\mathrm{G}} & h_{a_1a_1} & h_{a_1b_1} & k_{a_1a_1} & k_{a_1b_1} & h_{a_1a_2} & h_{a_1b_2} & k_{a_1b_2} & k_{a_1b_2} \\ \hline \boldsymbol{n}_{b_1\mathrm{G}} & h_{b_1a_1} & h_{b_1a_1} & k_{b_1a_1} & k_{b_1a_1} & h_{b_1a_2} & h_{b_1b_2} & k_{b_1a_2} & k_{b_1b_2} \\ \boldsymbol{n}_{a_1\mathrm{G}} & n_{a_1a_1} & n_{a_1b_1} & w_{a_1a_1} & w_{a_1b_1} & n_{a_1a_2} & n_{a_1b_2} & w_{a_1b_2} & w_{a_1b_2} \\ \boldsymbol{n}_{b_1\mathrm{G}} & n_{b_1a_1} & n_{b_1b_1} & w_{b_1a_1} & w_{b_1b_1} & n_{b_1a_n} & n_{b_1b_2} & w_{a_1b_2} & w_{b_1b_2} \\ \hdashline \boldsymbol{h}_{a_n\mathrm{G}} & h_{a_12a_1} & h_{a_2b_1} & k_{a_2a_1} & k_{a_2b_1} & h_{a_2a_1} & h_{a_2b_2} & k_{a_2a_2} & k_{a_2b_2} \\ \boldsymbol{h}_{b_2\mathrm{G}} & h_{b_2a_1} & h_{b_2b_1} & k_{b_2a_1} & k_{b_2b_1} & h_{b_2a_2} & h_{a_2b_2} & k_{b_2a_2} & k_{a_2b_2} \\ \boldsymbol{n}_{a_2\mathrm{G}} & n_{a_2a_1} & n_{a_2b_1} & w_{a_2b_1} & w_{a_2b_1} & n_{a_2a_2} & n_{a_2b_2} & w_{a_2a_2} & w_{a_2b_2} \\ \boldsymbol{n}_{b_2\mathrm{G}} & n_{b_2a_1} & n_{b_2b_1} & w_{b_2a_1} & w_{b_2b_1} & n_{b_2a_2} & n_{b_2b_2} & w_{b_2a_2} & w_{b_2b_2} \end{array}\right] \begin{bmatrix} \Delta \delta'_{\mathrm{G}} \\ \hline \Delta P_{a_1} \\ \Delta P_{b_1} \\ \Delta Q_{a_1} \\ \Delta Q_{b_1} \\ \hdashline \Delta P_{a_2} \\ \Delta P_{b_2} \\ \Delta Q_{a_2} \\ \Delta Q_{b_2} \end{bmatrix} \tag{10.138}$$

Equation (10.138) describes the incremental model shown in Figure 10.45. In this model of the AC system the HVDC lines are modeled by power injections. The lower part of Eq. (10.138) determines the influence of the real and reactive powers on voltage arguments and nodal voltages at the HVDC terminal nodes. The upper part of Eq. (10.138) determines the way in which real and reactive power changes influence the changes in power generated by generators in the AC system.

Taking into account Eq. (10.135) and (10.136) and using Eq. (10.138), one can derive the following equation describing changes in the power of the generators in the system

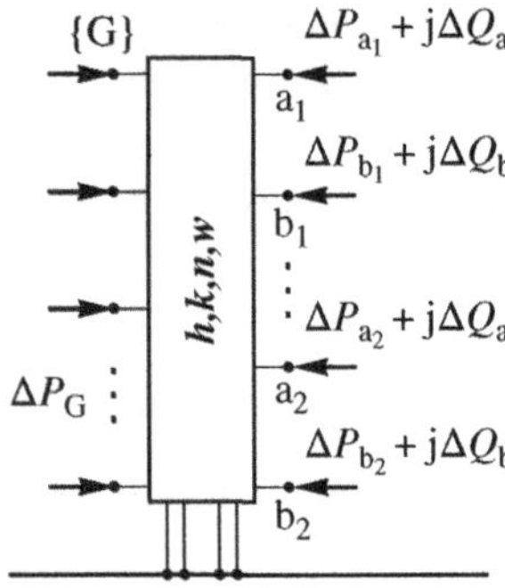

Figure 10.45 Incremental model of the AC system.

$$\begin{aligned} \Delta \boldsymbol{P}_{\mathrm{G}} \cong\ & \boldsymbol{h}_{\mathrm{GG}} \Delta \boldsymbol{\delta}'_{\mathrm{G}} \\ & + (\boldsymbol{h}_{\mathrm{Ga}_1} - \boldsymbol{h}_{\mathrm{Gb}_1})\, \Delta P_1 + \boldsymbol{k}_{\mathrm{Ga}_1} \Delta Q_{\mathrm{a}_1} + \boldsymbol{k}_{\mathrm{Gb}_1} \Delta Q_{\mathrm{b}_1} \\ & + (\boldsymbol{h}_{\mathrm{Ga}_2} - \boldsymbol{h}_{\mathrm{Gb}_2})\, \Delta P_2 + \boldsymbol{k}_{\mathrm{Ga}_2} \Delta Q_{\mathrm{a}_2} + \boldsymbol{k}_{\mathrm{Gb}_2} \Delta Q_{\mathrm{b}_2} \end{aligned} \tag{10.139}$$

where the symbols $\boldsymbol{h}_{\mathrm{Ga}_1}$, $\boldsymbol{h}_{\mathrm{Gb}_1}$, $\boldsymbol{k}_{\mathrm{Ga}_1}$, $\boldsymbol{k}_{\mathrm{Gb}_1}$ and $\boldsymbol{h}_{\mathrm{Ga}_2}$,$\boldsymbol{h}_{\mathrm{Gb}_2}$,$\boldsymbol{k}_{\mathrm{Ga}_2}$, $\boldsymbol{k}_{\mathrm{Gb}_2}$ denote column matrices, while ΔP_1, ΔQ_{a_1}, ΔQ_{b_1} and ΔP_2, ΔQ_{a_2}, ΔQ_{b_2} are scalars. Hence, the change in power of i-th generator can be expressed as

$$\begin{aligned} \Delta P_i \cong\ & \sum_{j \in \{G\}} h_{ij} \Delta \delta'_j \\ & + (h_{i\mathrm{a}_1} - h_{i\mathrm{b}_1})\, \Delta P_1 + k_{i\mathrm{a}_1} \Delta Q_{\mathrm{a}_1} + k_{i\mathrm{b}_1} \Delta Q_{\mathrm{b}_1} \\ & + (h_{i\mathrm{a}_2} - h_{i\mathrm{b}_2})\, \Delta P_2 + k_{i\mathrm{a}_2} \Delta Q_{\mathrm{a}_2} + k_{i\mathrm{b}_2} \Delta Q_{\mathrm{b}_2} \end{aligned} \tag{10.140}$$

Rotor motion is described by the following differential equation in the classical model

$$\boldsymbol{M}\, \Delta \dot{\boldsymbol{\omega}}_{\mathrm{G}} = -\Delta \boldsymbol{P}_{\mathrm{G}} - \boldsymbol{D}\, \Delta \boldsymbol{\omega}_{\mathrm{G}} \tag{10.141}$$

where $\boldsymbol{M}$ is the diagonal matrix of inertia constants, $\boldsymbol{D}$ is the diagonal matrix of damping coefficients, $\Delta \boldsymbol{\omega}_{\mathrm{G}} = \mathrm{d}\Delta \boldsymbol{\delta}'_{\mathrm{G}}/\mathrm{d}t$ is the column matrix of the speed deviations, and $\Delta \boldsymbol{\delta}'_{\mathrm{G}}$ is the column matrix of rotor angles. Substituting Eq.(10.139) into Eq. (10.141) gives

$$\begin{aligned} \boldsymbol{M}\, \Delta \dot{\boldsymbol{\omega}}_{\mathrm{G}} =\ & -\boldsymbol{h}_{\mathrm{GG}} \Delta \boldsymbol{\delta}'_{\mathrm{G}} - \boldsymbol{D}\, \Delta \boldsymbol{\omega}_{\mathrm{G}} \\ & - (\boldsymbol{h}_{\mathrm{Ga}_1} - \boldsymbol{h}_{\mathrm{Gb}_1})\, \Delta P_1 - \boldsymbol{k}_{\mathrm{Ga}_1} \Delta Q_{\mathrm{a}_1} + \boldsymbol{k}_{\mathrm{Gb}_1} \Delta Q_{\mathrm{b}_1} \\ & - (\boldsymbol{h}_{\mathrm{Ga}_2} - \boldsymbol{h}_{\mathrm{Gb}_2})\, \Delta P_2 - \boldsymbol{k}_{\mathrm{Ga}_2} \Delta Q_{\mathrm{a}_2} + \boldsymbol{k}_{\mathrm{Gb}_2} \Delta Q_{\mathrm{b}_2} \end{aligned} \tag{10.142}$$

In Eq. (10.142) quantities ΔP_1,ΔQ_{a_1},ΔQ_{b_1} and ΔP_2, ΔQ_{a_2}, ΔQ_{b_2} are the control variables that correspond to the active power changes in HVDC lines and the reactive power changes at the terminals {a_1, b_1, a_2, b_2} of the HVDC lines.

10.8.2 State-variable Stabilizing Control

The task now is to derive a stabilizing control strategy, that is how the control variables should change so that rotor swings of all the generators are damped. The supplementary control of the HVDC lines is defined as a control that maximizes negative value of the time derivative $\dot{V}(\boldsymbol{x}) = \mathrm{d}V/\mathrm{d}t$ of the energy-type Lyapunov function $V(\boldsymbol{x})$. A similar approach is applied in Section 10.5 for shunt FACTS devices and in Section 10.6 and Section 10.7 for series FACTS devices.

10.8.2.1 Lyapunov Function

As described in Section 6.3.5, for the conservative power system model the total energy $V(\delta', \Delta\omega) = E_{\mathrm{k}} + E_{\mathrm{p}}$ can be used as the Lyapunov function. For the linearized model this function can be expanded into the Taylor series

$$\Delta V(\delta', \Delta\omega) = \Delta E_{\mathrm{p}} + \Delta E_{\mathrm{k}} \tag{10.143}$$

where

$$\Delta E_{\mathrm{p}} = \frac{1}{2}\, \Delta \boldsymbol{\delta}'^{\mathrm{T}}_{\mathrm{G}}\, \boldsymbol{h}_{\mathrm{GG}}\, \Delta \boldsymbol{\delta}'_{\mathrm{G}} \quad \text{and} \quad \Delta E_{\mathrm{k}} = \frac{1}{2}\, \Delta \boldsymbol{\omega}^{\mathrm{T}}_{\mathrm{G}}\, \boldsymbol{M}\, \Delta \boldsymbol{\omega}_{\mathrm{G}} \tag{10.144}$$

are increments of the potential and kinetic energy in the vicinity of the post-fault equilibrium point. Applying to Eqs. (10.143) and (10.144) well-known formula for the derivative of a sum of functions and the derivative of a product of two functions leads to:

$$\Delta\dot{V}(\delta', \Delta\omega) = \Delta\dot{E}_{\rm p} + \Delta\dot{E}_{\rm k} \tag{10.145}$$

$$\Delta\dot{E}_{\rm p} = \frac{1}{2}\,\Delta\omega_{\rm G}^{\rm T}\,\boldsymbol{h}_{\rm GG}\,\Delta\delta'_{\rm G} + \frac{1}{2}\,\Delta\delta'^{\rm T}_{\rm G}\,\boldsymbol{h}_{\rm GG}\,\Delta\omega_{\rm G} \tag{10.146}$$

$$\Delta\dot{E}_{\rm k} = \frac{1}{2}\,\Delta\dot{\omega}_{\rm G}^{\rm T}\,\boldsymbol{M}\,\Delta\omega_{\rm G} + \frac{1}{2}\,\Delta\omega_{\rm G}^{\rm T}\,\boldsymbol{M}\,\Delta\dot{\omega}_{\rm G} \tag{10.147}$$

It.is assumed that the supplementary stabilizing control is optimal when it maximizes the negative value of the time derivative of the Lyapunov function at any moment of the transient state. From the physical point of view it means that the stabilizing control enforces the fastest dissipation of the energy released in the system by a given disturbance.

10.8.2.2 Control Formulas

For the convenience of further considerations it is worthwhile to write Eq. (10.142) in a transposed form

$$\begin{aligned}\Delta\dot{\omega}_{\rm G}^{\rm T}\,\boldsymbol{M} = & -\Delta\delta'^{\rm T}_{\rm G}\boldsymbol{h}_{\rm GG} - \Delta\omega_{\rm G}^{\rm T}\boldsymbol{D} - \\ & -\left(\boldsymbol{h}_{\rm Ga_1}^{\rm T} - \boldsymbol{h}_{\rm Gb_1}^{\rm T}\right)\Delta P_1 - \boldsymbol{k}_{\rm Ga_1}^{\rm T}\Delta Q_{\rm a_1} + \boldsymbol{k}_{\rm Gb_1}^{\rm T}\Delta Q_{\rm b_1} \\ & -\left(\boldsymbol{h}_{\rm Ga_2}^{\rm T} - \boldsymbol{h}_{\rm Gb_2}^{\rm T}\right)\Delta P_2 - \boldsymbol{k}_{\rm Ga_2}^{\rm T}\Delta Q_{\rm a_2} + \boldsymbol{k}_{\rm Gb_2}^{\rm T}\Delta Q_{\rm b_2}\end{aligned} \tag{10.148}$$

Substituting Eqs. (10.148) and (10.142) into (10.147) one gets

$$\begin{aligned}\Delta\dot{E}_{\rm k} = & -\frac{1}{2}\Delta\delta'^{\rm T}_{\rm G}\boldsymbol{h}_{\rm GG}\Delta\omega_{\rm G} - \frac{1}{2}\Delta\omega_{\rm G}^{\rm T}\boldsymbol{h}_{\rm GG}\Delta\delta'_{\rm G} - \Delta\omega_{\rm G}^{\rm T}\boldsymbol{D}\,\Delta\omega_{\rm G} \\ & -\frac{1}{2}\left(\Delta\boldsymbol{h}_{\rm a_1b_1}^{\rm T}\Delta\omega_{\rm G} + \Delta\omega_{\rm G}^{\rm T}\Delta\boldsymbol{h}_{\rm a_1b_1}\right)\Delta P_1(t) \\ & -\frac{1}{2}\left(\boldsymbol{k}_{\rm Ga_1}^{\rm T}\Delta\omega_{\rm G} + \Delta\omega_{\rm G}^{\rm T}\boldsymbol{k}_{\rm Ga_1}\right)\Delta Q_{\rm a_1}(t) - \frac{1}{2}\left(\boldsymbol{k}_{\rm Gb_1}^{\rm T}\Delta\omega_{\rm G} + \Delta\omega_{\rm G}^{\rm T}\boldsymbol{k}_{\rm Gb_1}\right)\Delta Q_{\rm b_1}(t) \\ & -\frac{1}{2}\left(\Delta\boldsymbol{h}_{\rm a_2b_2}^{\rm T}\Delta\omega_{\rm G} + \Delta\omega_{\rm G}^{\rm T}\Delta\boldsymbol{h}_{\rm a_2b_2}\right)\Delta P_2(t) \\ & -\frac{1}{2}\left(\boldsymbol{k}_{\rm Ga_2}^{\rm T}\Delta\omega_{\rm G} + \Delta\omega_{\rm G}^{\rm T}\boldsymbol{k}_{\rm Ga_2}\right)\Delta Q_{\rm a_2}(t) - \frac{1}{2}\left(\boldsymbol{k}_{\rm Gb_2}^{\rm T}\Delta\omega_{\rm G} + \Delta\omega_{\rm G}^{\rm T}\boldsymbol{k}_{\rm Gb_2}\right)\Delta Q_{\rm b_2}(t)\end{aligned} \tag{10.149}$$

It can be easily verified that the components on the right-hand side of (10.149) are scalars and are identical in the following pairs

$$\Delta\boldsymbol{h}_{\rm a_1b_1}^{\rm T}\Delta\omega_{\rm G} = \Delta\omega_{\rm G}^{\rm T}\Delta\boldsymbol{h}_{\rm a_1b_1} = \sum_{i\in\{\rm G\}}(h_{i\rm a_1} - h_{i\rm b_1})\Delta\omega_i \tag{10.150}$$

$$\boldsymbol{k}_{\rm Ga_1}^{\rm T}\Delta\omega_{\rm G} = \Delta\omega_{\rm G}^{\rm T}\boldsymbol{k}_{\rm Ga_1} = \sum_{i\in\{\rm G\}}k_{i\rm a_1}\Delta\omega_i \tag{10.151}$$

$$\boldsymbol{k}_{\rm Gb_1}^{\rm T}\Delta\omega_{\rm G} = \Delta\omega_{\rm G}^{\rm T}\boldsymbol{k}_{\rm Gb_1} = \sum_{i\in\{\rm G\}}k_{i\rm b_1}\Delta\omega_i \tag{10.152}$$

$$\Delta\boldsymbol{h}_{\rm a_2b_2}^{\rm T}\Delta\omega_{\rm G} = \Delta\omega_{\rm G}^{\rm T}\Delta\boldsymbol{h}_{\rm a_2b_2} = \sum_{i\in\{\rm G\}}(h_{i\rm a_2} - h_{i\rm b_2})\Delta\omega_i \tag{10.153}$$

$$\boldsymbol{k}_{\mathrm{Ga}_2}^{\mathrm{T}}\Delta\omega_{\mathrm{G}} = \Delta\omega_{\mathrm{G}}^{\mathrm{T}}\boldsymbol{k}_{\mathrm{Ga}_2} = \sum_{i\in\{\mathrm{G}\}} k_{i\mathrm{a}_2}\Delta\omega_i \tag{10.154}$$

$$\boldsymbol{k}_{\mathrm{Gb}_2}^{\mathrm{T}}\Delta\omega_{\mathrm{G}} = \Delta\omega_{\mathrm{G}}^{\mathrm{T}}\boldsymbol{k}_{\mathrm{Gb}_2} = \sum_{i\in\{\mathrm{G}\}} k_{i\mathrm{b}_2}\Delta\omega_i \tag{10.155}$$

Now Eq. (10.149) can be written as

$$\begin{aligned}\Delta\dot{E}_{\mathrm{k}} = &-\frac{1}{2}\Delta\boldsymbol{\delta}'^{\mathrm{T}}_{\mathrm{G}}\boldsymbol{h}_{\mathrm{GG}}\Delta\boldsymbol{\omega}_{\mathrm{G}} - \frac{1}{2}\Delta\boldsymbol{\omega}_{\mathrm{G}}^{\mathrm{T}}\boldsymbol{h}_{\mathrm{GG}}\Delta\boldsymbol{\delta}'_{\mathrm{G}} - \Delta\boldsymbol{\omega}_{\mathrm{G}}^{\mathrm{T}}\boldsymbol{D}\,\Delta\boldsymbol{\omega}_{\mathrm{G}}\\ &-\Delta\boldsymbol{h}_{\mathrm{a}_1\mathrm{b}_1}^{\mathrm{T}}\Delta\boldsymbol{\omega}_{\mathrm{G}}\Delta P_1(t) - \boldsymbol{k}_{\mathrm{Ga}_1}^{\mathrm{T}}\Delta\boldsymbol{\omega}_{\mathrm{G}}\Delta Q_{\mathrm{a}_1}(t) - \boldsymbol{k}_{\mathrm{Gb}_1}^{\mathrm{T}}\Delta\boldsymbol{\omega}_{\mathrm{G}}\Delta Q_{\mathrm{b}_1}(t)\\ &-\Delta\boldsymbol{h}_{\mathrm{a}_2\mathrm{b}_2}^{\mathrm{T}}\Delta\boldsymbol{\omega}_{\mathrm{G}}\Delta P_2(t) - \boldsymbol{k}_{\mathrm{Ga}_2}^{\mathrm{T}}\Delta\boldsymbol{\omega}_{\mathrm{G}}\Delta Q_{\mathrm{a}_2}(t) - \boldsymbol{k}_{\mathrm{Gb}_2}^{\mathrm{T}}\Delta\boldsymbol{\omega}_{\mathrm{G}}\Delta Q_{\mathrm{b}_2}(t)\end{aligned} \tag{10.156}$$

Adding the sides of Eqs. (10.156) and (10.146) one gets

$$\begin{aligned}\Delta\dot{V} = \Delta\dot{E}_{\mathrm{k}} + \Delta\dot{E}_{\mathrm{p}} = &-\Delta\omega_{\mathrm{G}}^{\mathrm{T}}\boldsymbol{D}\Delta\omega_{\mathrm{G}}\\ &-\Delta\boldsymbol{h}_{\mathrm{a}_1\mathrm{b}_1}^{\mathrm{T}}\Delta\omega_{\mathrm{G}}\Delta P_1(t) - \boldsymbol{k}_{\mathrm{Ga}_1}^{\mathrm{T}}\Delta\omega_{\mathrm{G}}\Delta Q_{\mathrm{a}_1}(t) - \boldsymbol{k}_{\mathrm{Gb}_1}^{\mathrm{T}}\Delta\omega_{\mathrm{G}}\Delta Q_{\mathrm{b}_1}(t)\\ &-\Delta\boldsymbol{h}_{\mathrm{a}_2\mathrm{b}_2}^{\mathrm{T}}\Delta\omega_{\mathrm{G}}\Delta P_2(t) - \boldsymbol{k}_{\mathrm{Ga}_2}^{\mathrm{T}}\Delta\omega_{\mathrm{G}}\Delta Q_{\mathrm{a}_2}(t) - \boldsymbol{k}_{\mathrm{Gb}_2}^{\mathrm{T}}\Delta\omega_{\mathrm{G}}\Delta Q_{\mathrm{b}_2}(t)\end{aligned} \tag{10.157}$$

During the swings the first component in the right-hand side of Eq. (10.157) is always negative. The supplementary controls of the HVDC lines contribute to the system damping if the remaining components in Eq. (10.157) are also negative. Inspection of Eq. (10.157) shows that it is possible to make these components negative if the supplementary controls follow the following control strategies

$$\Delta P_1(t) = \kappa_{\mathrm{P}}\,\Delta\boldsymbol{h}_{\mathrm{a}_1\mathrm{b}_1}^{\mathrm{T}}\Delta\omega_{\mathrm{G}};\quad \Delta Q_{\mathrm{a}_1}(t) = \kappa_{\mathrm{Q}}\,\boldsymbol{k}_{\mathrm{Ga}_1}^{\mathrm{T}}\Delta\omega_{\mathrm{G}};\quad \Delta Q_{\mathrm{b}_1}(t) = \kappa_{\mathrm{Q}}\,\boldsymbol{k}_{\mathrm{Gb}_1}^{\mathrm{T}}\Delta\omega_{\mathrm{G}} \tag{10.158}$$

$$\Delta P_2(t) = \kappa_{\mathrm{P}}\,\Delta\boldsymbol{h}_{\mathrm{a}_2\mathrm{b}_2}^{\mathrm{T}}\Delta\omega_{\mathrm{G}};\quad \Delta Q_{\mathrm{a}_2}(t) = \kappa_{\mathrm{Q}}\,\boldsymbol{k}_{\mathrm{Ga}_2}^{\mathrm{T}}\Delta\omega_{\mathrm{G}};\quad \Delta Q_{\mathrm{b}_2}(t) = \kappa_{\mathrm{Q}}\,\boldsymbol{k}_{\mathrm{Gb}_2}^{\mathrm{T}}\Delta\omega_{\mathrm{G}} \tag{10.159}$$

where κ_{P} and κ_{Q} are gains in control loop. For control formulas (10.158) and (10.159) the time derivative Eq. (10.157) of Lyapunov function is given by the following formula

$$\begin{aligned}\Delta\dot{V} = &-\Delta\omega_{\mathrm{G}}^{\mathrm{T}}\boldsymbol{D}\Delta\omega_{\mathrm{G}}\\ &-\kappa_{\mathrm{P}}\left(\Delta\boldsymbol{h}_{\mathrm{a}_1\mathrm{b}_1}^{\mathrm{T}}\Delta\omega_{\mathrm{G}}\right)^2 - \kappa_{\mathrm{Q}}\left(\boldsymbol{k}_{\mathrm{Ga}_1}^{\mathrm{T}}\Delta\omega_{\mathrm{G}}\right)^2 - \kappa_{\mathrm{Q}}\left(\boldsymbol{k}_{\mathrm{Gb}_1}^{\mathrm{T}}\Delta\omega_{\mathrm{G}}\right)^2\\ &-\kappa_{\mathrm{P}}\left(\Delta\boldsymbol{h}_{\mathrm{a}_2\mathrm{b}_2}^{\mathrm{T}}\Delta\omega_{\mathrm{G}}\right)^2 - \kappa_{\mathrm{Q}}\left(\boldsymbol{k}_{\mathrm{Ga}_2}^{\mathrm{T}}\Delta\omega_{\mathrm{G}}\right)^2 - \kappa_{\mathrm{Q}}\left(\boldsymbol{k}_{\mathrm{Gb}_2}^{\mathrm{T}}\Delta\omega_{\mathrm{G}}\right)^2 \le 0\end{aligned} \tag{10.160}$$

Taking into account Eqs. (10.150) and (10.153), the control strategy expressed by Eqs. (10.158) and (10.159) can be written as

$$\Delta P_1(t) = \kappa_{\mathrm{P}}\sum_{i\in\{\mathrm{G}\}}(h_{i\mathrm{a}_1} - h_{i\mathrm{b}_1})\Delta\omega_i = \kappa_{\mathrm{P}}\sum_{i\in\{\mathrm{G}\}}\Delta h_{i1}\Delta\omega_i \tag{10.161}$$

$$\Delta Q_{\mathrm{a}_1}(t) = \kappa_{\mathrm{Q}}\sum_{i\in\{\mathrm{G}\}}k_{i\mathrm{a}_1}\Delta\omega_i;\Delta Q_{\mathrm{b}_1}(t) = \kappa_{\mathrm{Q}}\sum_{i\in\{\mathrm{G}\}}k_{i\mathrm{b}_1}\Delta\omega_i \tag{10.162}$$

$$\Delta P_2(t) = \kappa_{\mathrm{P}}\sum_{i\in\{\mathrm{G}\}}(h_{i\mathrm{a}_2} - h_{i\mathrm{b}_2})\Delta\omega_i = \kappa_{\mathrm{P}}\sum_{i\in\{\mathrm{G}\}}\Delta h_{i2}\Delta\omega_i \tag{10.163}$$

$$\Delta Q_{\mathrm{a}_2}(t) = \kappa_{\mathrm{Q}}\sum_{i\in\{\mathrm{G}\}}k_{i\mathrm{a}_2}\Delta\omega_i;\Delta Q_{\mathrm{b}_2}(t) = \kappa_{\mathrm{Q}}\sum_{i\in\{\mathrm{G}\}}k_{i\mathrm{b}_2}\Delta\omega_i \tag{10.164}$$

Control expressed by Eqs. (10.161)–(10.164) can be realized using a PSS of the structure similar to that shown in Figure 10.28 for the shunt compensator. Block diagrams for each HVDC line are the same but the coefficients are different and should be calculated for each pair of terminal nodes of a given HVDC line. Such control is based on the state variables and could be utilized in a WAMPAC-type control system making use of phasor measurements. A possible structure of such a system is shown in Figure 10.29. The main problem for such a closed-loop control is the speed of data transmission. Current modern flexible communication platforms (Figure 2.46) cannot transmit data fast enough in order to damp power swings. However, it may be expected that the speed of data transmission will increase in the future so that practical implementation of a WAMPAC system will be possible.

10.8.2.3 Controllability

Coefficients h_{ia_1},h_{ib_1}, h_{ia_2}, and h_{ib_2} appearing in Eq. (10.140) can be treated as the measures of closeness between the nodes $\{a_1, b_1, a_2, b_2\}$ and the i-th generator. A high value of the coefficient means that the nodes are close to a given generator. A small value of the coefficient means that the nodes are distant from the given generator. Nodes {a, b} are equally distant to i-th generator if $h_{ia} = h_{ib}$. Equation (10.140) shows that if $h_{ia_1} \cong h_{ib_1}$ and $h_{ia_1} \cong h_{ib_1}$ then the changes ΔP_1 and ΔP_2 in HVDC lines do not affect power changes of i-th generator. In other words, the generator power is not controllable by using the control of HVDC line as the control actions coming from the nodes $\{a_1, b_1\}$ and $\{a_2, b_2\}$ compensate each other. In other words, if the nodes $\{a_1, b_1\}$ and $\{a_2, b_2\}$ are equally distant to i-th generator then HVDC control cannot influence the power of that generator because nodal injections at $\{a_1, b_1\}$ and $\{a_2, b_2\}$ have opposite signs (Figure 10.45) and the influence of changes in the nodal power of both nodes on the generator compensate each other out.

If the changes in the reactive power ΔQ_{a_1}, ΔQ_{b_1} and ΔQ_{a_2}, ΔQ_{b_2} are independent then, according to Eq. (10.140), each of them can independently influence the power of the i-th generator proportionally to coefficients k_{ia_1}, k_{ib_1} and k_{ia_1}, k_{ib_1}, respectively.

10.8.2.4 Additive Damping

Equation (10.160) proves a very important property of the proposed stabilizing control given by Eqs. (10.161)–(10.164). From Eq. (10.160) it is obvious that each controlled signal Eq. (10.161)–(10.164) introduces to $\Delta \dot{V}$ its own negative component. All these components are independent. Therefore, the conclusion is that the proposed control produces an additive damping. Control of two signals gives better damping than control of one signal. Control of three signals gives better damping than control of two signal, and so on. The more controlled elements, the better resultant damping in AC power system.

10.8.2.5 Simulation Results

The above control formulas (10.161)–(10.164) have been derived with the use of the linearized model. Below described simulations have been performed for a detailed power system model. Simulation tests for various disturbances have been performed by computer simulations with the use of a nonlinear model of the modified New England test system. In the modified version (Figure 10.46) there are three weakly connected subsystems: A, B, and C. Each subsystem has its own automatic generation control (Figure 9.12) taken into account in the simulation model. The northern part is connected with the southern part via only two AC lines L5 and L36. In order to enhance the transfer ability between the northern and southern parts two HVDC lines L2 and L21 have been embedded. Example simulation results are shown in Figure 10.47 and 10.48. In these cases the considered disturbance appears in area B and consists of an outage of a 250 MW generating unit (tripped by a generator protection).

Figure 10.47 shows the variation of the active power in the AC tie-lines in the case without action of the stabilizing control. The assumed disturbance in subsystem B causes the sudden increase in active power exported from subsystem A to subsystem B and large power swings in the AC tie-lines. In several dozen of seconds the automatic generation control (AGC) reduces this exchange of power to the initial (reference) value.

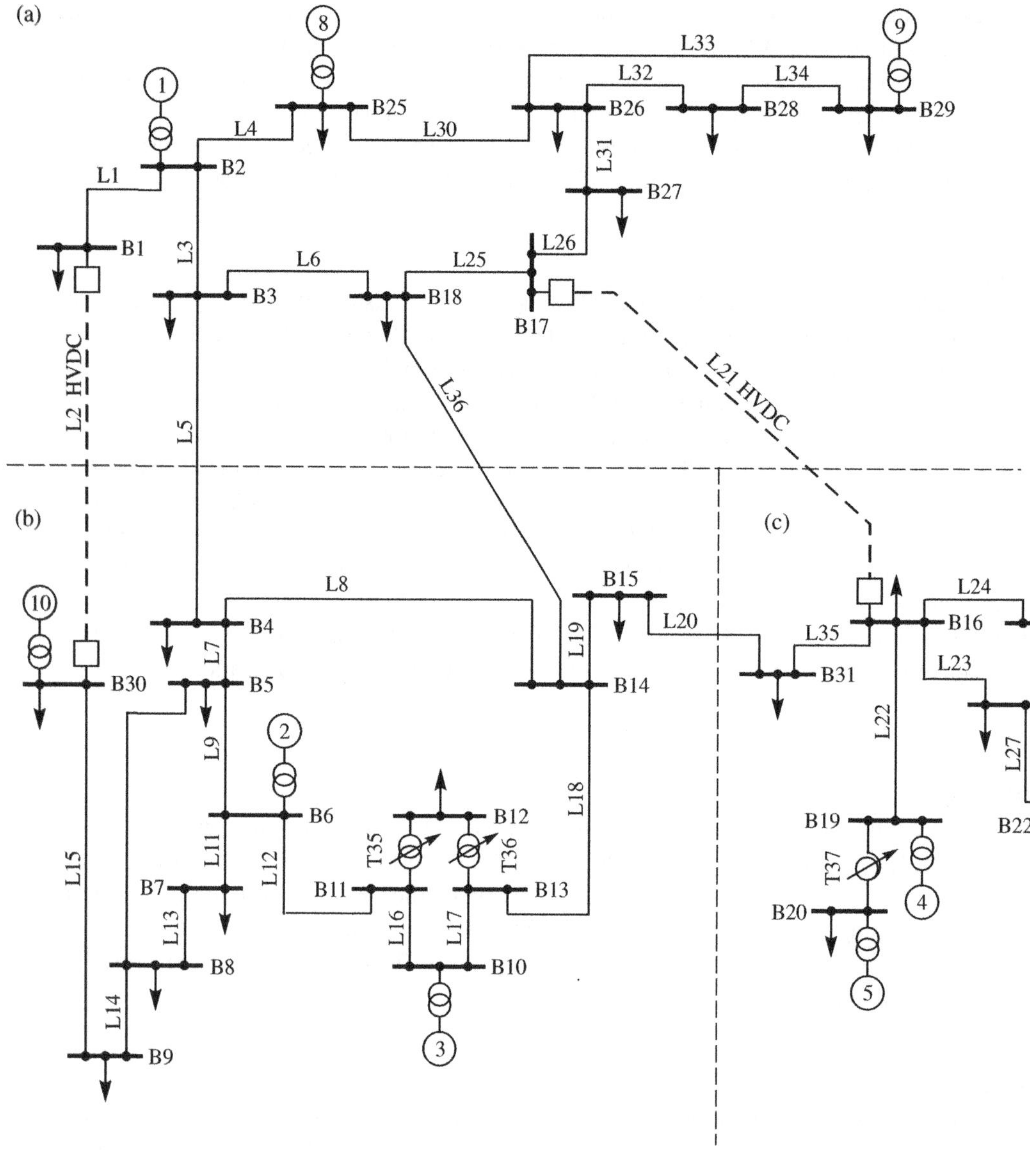

Figure 10.46 Modified New England test system.

Figure 10.48a shows simulation results for the same disturbance but with above described stabilizing control of line L2 HVDC. Figure 10.48b shows simulation results for the same disturbance but with the proposed stabilizing control acting for line L2 HVDC and line L21 HVDC. On both these figures the proposed control quickly damps out the inter-area oscillations. In the case when two HVDC lines are controlled the dynamic response is almost aperiodic. Comparison of Figure 10.47 and Figure 10.48 shows that damping introduced by the stabilizing control is additive (the more controlled devices, the better damping).

Results shown in Figure 10.48 are without reactive power control when only active power control is used. Control of reactive power improves damping only slightly, but has a significant influence on oscillations of voltages at the terminal nodes of the HVDC lines. The proposed control smooths those oscillations.

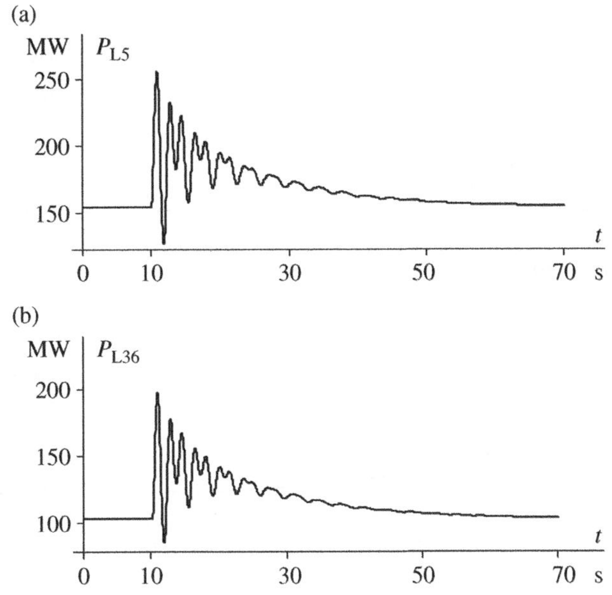

Figure 10.47 Time response of active power in the AC tie-lines for considered disturbance without stabilizing control of HVDC lines.

10.8.2.6 Robustness

Network configuration and operating conditions may change during power system operation. Hence, a power system model and state is always subject to some uncertainty. Therefore, the control strategy should be robust, that is insensitive (or weakly sensitive) to the uncertainty. The simulation results (not presented here) have shown (Machowski et al. 2013) that the control based on formulas (10.161)–(10.164) is very robust and practically insensitive to changes in the network configuration and loading conditions. Even in the case when a short-circuit is cleared by switching off the AC transmission lines close to the HVDC terminals, a considerable enhancement of the power swing damping has been obtained without updating the values of coefficients Δh_{i1} and Δh_{i2}. It means that soon after the occurrence of any fault in the AC power system it is not necessary to recalculate the coefficients. This is a very valuable practical feature of the considered control strategy.

10.8.3 Control Based on Local Measurements

Above, control formulas (10.161)–(10.164) use state variables of the system as the input signals. Such control is currently difficult and expensive to implement. However, these formulas can serve as a source of information on how to implement a stabilizing control based on local measurements that would emulate state-variable control. This will be discussed for the system as in Figure 10.44 and the HVDC line L1 with the additional assumption that the AC tie-lines connecting both subsystems constitute a very weak connection with comparison to the very strong connections inside each subsystem.

As stated in Section 10.8.2, the coefficients h_{ia_1} and h_{ib_1} found in Eq. (10.161) can be regarded as the closeness measures of the nodes $\{a_1, b_1\}$ and the i-th generator. With a poor connection of subsystems A and B it can be assumed that the measures $h_{ia_1} \cong 0$ for $i \in \{B\}$ and $h_{ib_1} \cong 0$ for $i \in \{A\}$. In such a case, Eq. (10.161) can be written as

$$\Delta P_1(t) = \kappa_P \sum_{i\in\{A\}} h_{ia_1} \Delta\omega_i - \kappa_P \sum_{i\in\{B\}} h_{ib_1} \Delta\omega_i \tag{10.165}$$

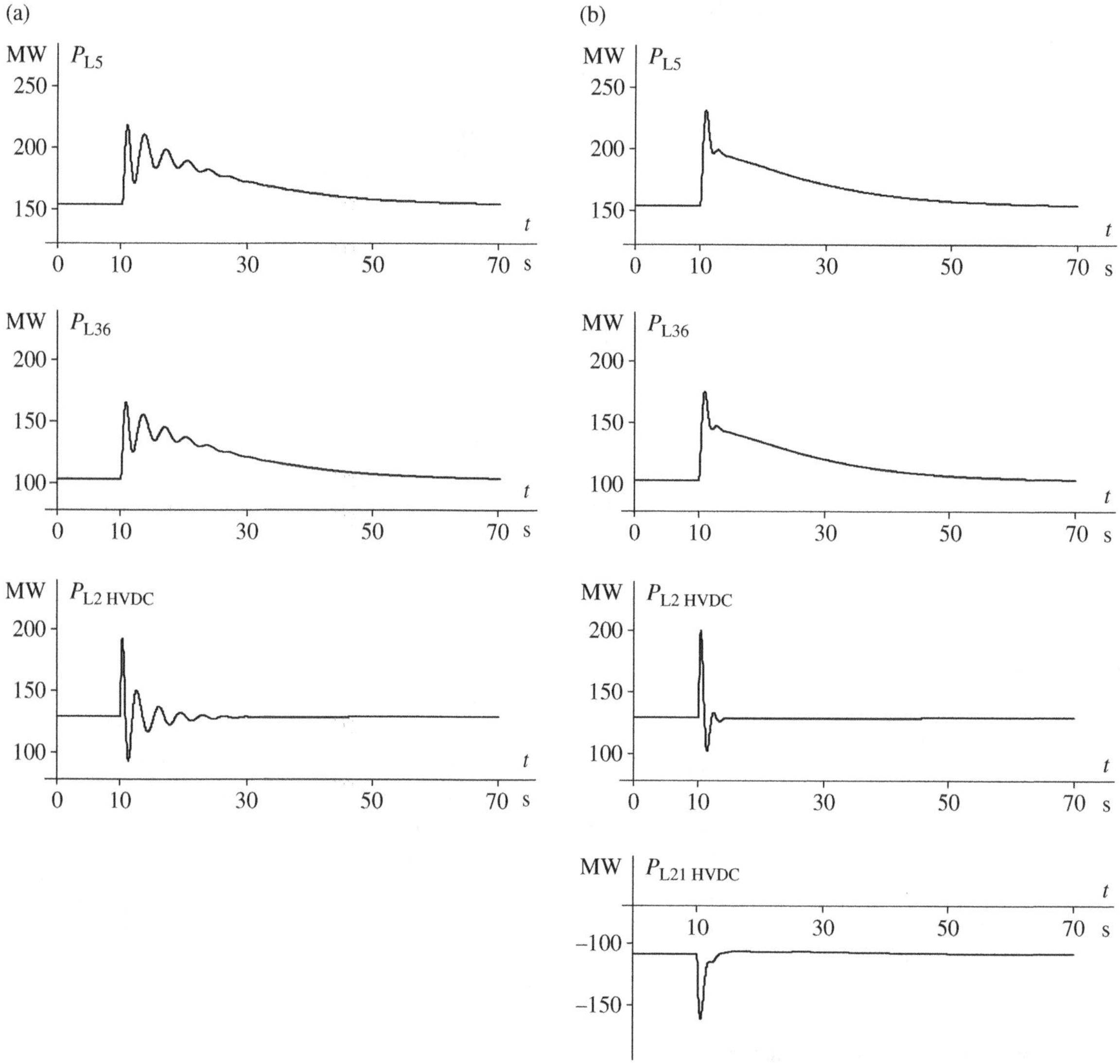

Figure 10.48 Time response for considered disturbance and the stabilizing control acting for: (a) only one line L2 HVDC; (b) two lines L2 HVDC and L21 HVDC.

In Eq. (10.165) the first component results from the rotor swings in subsystem A, and the second one from rotor swings in subsystem B. In the case of weakly connected subsystems each of which is strongly connected inside, one can assume that after the disturbance in power balance the rotor swings inside subsystems are damped out quickly (Figure 9.16 in Section 9.3) and tie-line power oscillations begin (Section 9.5.2). For such oscillations and swings between weakly connected subsystems it can be assumed that in each of the subsystems there are speed and frequency deviations such that

$$\begin{aligned} \Delta\omega_i &\cong \Delta\omega_{\mathrm{A}} = 2\pi\,\Delta f_{\mathrm{A}} \quad \text{for} \quad i \in \{\mathrm{A}\} \\ \Delta\omega_i &\cong \Delta\omega_{\mathrm{B}} = 2\pi\,\Delta f_{\mathrm{B}} \quad \text{for} \quad i \in \{\mathrm{B}\} \end{aligned} \tag{10.166}$$

Substituting Eq. (10.166) into Eq. (10.165) one obtains

$$\Delta P_1(t) = \kappa_P\, 2\pi \left(h_{a_1a_1}\Delta f_A - h_{b_1b_1}\Delta f_B\right) \tag{10.167}$$

where $h_{a_1a_1} = \sum_{i\in\{A\}} h_{ia_1}$ is the synchronizing power of the terminal $\{a_1\}$ of considered HVDC line (line L1 HVDC in Figure 10.44) calculated for the equivalent model of subsystem A. Similarly, $h_{b_1b_1} = \sum_{i\in\{B\}} h_{ib_1}$ is the synchronizing power of the terminal $\{b_1\}$ calculated for the equivalent model of subsystem B. On the basis of Eq. (10.163) an analogous equation can be written for $\Delta P_2(t)$. In general, the control formula for the changes in active power transmitted by the HVDC line required for damping of tie-line oscillations between the AC subsystems has the following form

$$\Delta P(t) = \kappa_P \left(\alpha\,\Delta f_A - \beta\,\Delta f_B\right) \tag{10.168}$$

where α and β are the weighting coefficients depending on the parameters of the connected AC subsystems. In a special case where $h_{a_1a_1} \cong h_{b_1b_1}$ one gets $\beta = \alpha$ and $\Delta P(t) = \kappa_P\,\alpha\,\Delta f_{AB}$ where $\Delta f_{AB} = \Delta f_A - \Delta f_B$. It should be emphasized that in practice the coefficients may vary up to several times and in Eq. (10.168) one must assume $\beta \neq \alpha$.

Frequency deviations Δf_A and Δf_B can be calculated on the basis of local measurements of voltage frequencies at terminal nodes {a, b} whereto the given HVDC line is connected to the AC subsystems. Obviously, in order to obtain the increments the measured frequencies f_a and f_b must be passed through a washout filter with the Bode plots similar to that shown in Figure 10.8.

Part III

Advanced Topics in Power System Dynamics

11

Advanced Power System Modeling

In Chapter 4 the dynamic interactions taking place inside a generator following a disturbance are explained by considering the changes in the armature and rotor magnetomotive force (mmf) and flux linkage. Although this type of approach allows the mechanisms by which the currents and torques are produced to be explained, it is difficult to quantify the behavior of the generator under all operating conditions. In this chapter a more general mathematical approach is adopted that can be used to quantify the changes in the currents and the torque but is slightly more removed from the physics. To produce this *mathematical model* the generator will be represented by a number of electrical circuits, each with its own inductance and resistance and with mutual coupling between the circuits. By making some judicious assumptions with regard to the dominant changes taking place inside the generator during a particular type of disturbance, this detailed mathematical model can be simplified to produce a series of generator models. These simplified models can then be used in the appropriate situation.

To utilize fully these mathematical models of the generator it is also necessary to produce mathematical models of the turbine and its governor, as well as the automatic voltage regulator (AVR). These aspects are covered in the second part of the chapter. The chapter ends by considering suitable models of power system loads and flexible AC transmission systems (FACTS) devices.

11.1 Synchronous Generator

In order to examine what happens inside a synchronous machine when it is subjected to an abrupt change in operating conditions, Section 4.2 describes the behavior of the generator following a sudden short circuit on the machine terminals. The effect of the fault is to cause the current in, and the flux linking, the different windings to change in such a way that three characteristic states can be identified. These characteristic states are termed the *subtransient state*, the *transient state*, and the *steady state*. In each of these three characteristic states the generator may be represented by an electromotive force (emf) behind a reactance, the value of which is linked to the reluctance of the armature reaction flux path, as explained in Section 4.2.3. In reality the transition from one state to another takes place smoothly so that the values of the fictitious internal emf's also change smoothly with time. In previous chapters these smooth changes are neglected and the emf's assumed to be constant in each of the characteristic states. In this chapter the flux changes occurring within the synchronous machine are analyzed more rigorously with the resulting algebraic and differential equations constituting an advanced, dynamic, mathematical model of the synchronous generator.

Power System Dynamics: Stability and Control, Third Edition. Jan Machowski, Zbigniew Lubosny, Janusz W. Bialek and James R. Bumby.

11.1.1 Assumptions

A schematic cross-section of a generator is shown in Figure 11.1. The generator is assumed to have a three-phase stator armature winding (A, B, C), a rotor field winding (f) and two rotor damper windings – one in the d-axis (D) and one in the q-axis (Q). Figure 11.1 also shows the relative position of the windings, and their axes, with the center of phase A taken as the reference. The notation is the same as that used in Figure 4.3 and follows the normal IEEE convention (IEEE Committee Report 1969). In developing the mathematical model the following assumptions are made.

- The three-phase stator winding is symmetrical.
- The capacitance of all the windings can be neglected.
- Each of the distributed windings may be represented by a concentrated winding.
- The change in the inductance of the stator windings due to rotor position is sinusoidal and does not contain higher harmonics.
- Hysteresis loss is negligible but the influence of eddy currents can be included in the model of the damper windings.
- In the transient and subtransient states the rotor speed is near synchronous speed ($\omega \cong \omega_s$).
- The magnetic circuits are linear (not saturated) and the inductance values do not depend on the current.

11.1.2 The Flux Linkage Equations in the Stator Reference Frame

All the generator windings are magnetically coupled so that the flux in each winding depends on the currents in all the other windings. This is represented by the following matrix equation

$$\begin{bmatrix} \Psi_A \\ \Psi_B \\ \Psi_C \\ \hline \Psi_f \\ \Psi_D \\ \Psi_Q \end{bmatrix} = \left[\begin{array}{ccc|ccc} L_{AA} & L_{AB} & L_{AC} & L_{Af} & L_{AD} & L_{AQ} \\ L_{BA} & L_{BB} & L_{BC} & L_{Bf} & L_{BD} & L_{BQ} \\ L_{CA} & L_{CB} & L_{CC} & L_{Cf} & L_{CD} & L_{CQ} \\ \hline L_{fA} & L_{fB} & L_{fC} & L_{ff} & L_{fD} & L_{fQ} \\ L_{DA} & L_{DB} & L_{DC} & L_{Df} & L_{DD} & L_{DQ} \\ L_{QA} & L_{QB} & L_{QC} & L_{Qf} & L_{QD} & L_{QQ} \end{array}\right] \begin{bmatrix} i_A \\ i_B \\ i_C \\ \hline i_f \\ i_D \\ i_Q \end{bmatrix} \quad \text{or} \quad \begin{bmatrix} \Psi_{ABC} \\ \hline \Psi_{fDQ} \end{bmatrix} = \left[\begin{array}{c|c} \boldsymbol{L}_S & \boldsymbol{L}_{SR} \\ \hline \boldsymbol{L}_{SR}^T & \boldsymbol{L}_R \end{array}\right] \begin{bmatrix} \boldsymbol{i}_{ABC} \\ \hline \boldsymbol{i}_{fDQ} \end{bmatrix} \quad (11.1)$$

where $\boldsymbol{L}_S$ is a submatrix of the stator self- and mutual inductances, $\boldsymbol{L}_R$ is a submatrix of the rotor self-and mutual inductances, and $\boldsymbol{L}_{SR}$ is a submatrix of the rotor to stator mutual inductances. Most of these inductances are subject to periodic changes, owing to both the saliency and rotation of the rotor. Consistent with the assumptions outlined above, the higher harmonics of these inductance changes will be neglected and the inductances represented by a constant component and a single periodic component.

In the case of the two-pole machine shown in Figure 11.1 the self-inductance of each stator phase winding will reach a maximum value whenever the rotor d-axis aligns with the axis of the phase winding because, with the rotor in this position, the reluctance of the flux path is a minimum. This minimum reluctance condition occurs twice during each rotation of the rotor so that the stator self-inductances are of the form

$$L_{AA} = L_S + \Delta L_S \cos 2\gamma, \quad L_{BB} = L_S + \Delta L_S \cos\left(2\gamma - \frac{2}{3}\pi\right), \quad L_{CC} = L_S + \Delta L_S \cos\left(2\gamma + \frac{2}{3}\pi\right) \quad (11.2)$$

Both L_S and ΔL_S are constant and $L_S > \Delta L_S$.

As each of the stator windings is shifted in space relative to the others by 120° the mutual inductance between each of the stator windings is negative. The magnitude of the inductance is a maximum when the rotor d-axis is midway between the axes of two of the windings. Referring to Figure 11.1 gives

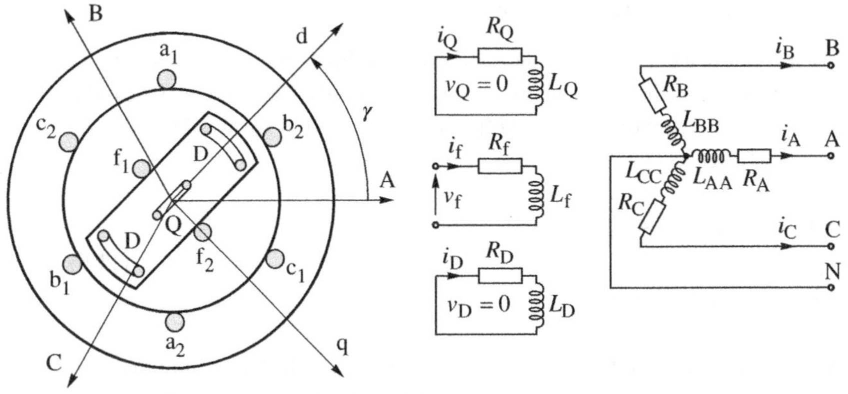

Figure 11.1 The windings in the synchronous generator and their axes.

$$
\begin{aligned}
L_{AB} &= L_{BA} = -M_S - \Delta L_S \cos 2\left(\gamma + \frac{1}{6}\pi\right) \\
L_{BC} &= L_{CB} = -M_S - \Delta L_S \cos 2\left(\gamma - \frac{1}{2}\pi\right) \\
L_{CA} &= L_{AC} = -M_S - \Delta L_S \cos 2\left(\gamma + \frac{5}{6}\pi\right)
\end{aligned}
\tag{11.3}
$$

where $M_s > \Delta L_s$.

The mutual inductances between the stator and rotor windings change with rotor position and have a positive maximum value when the axes of a stator winding and the rotor winding align and have the same positive flux direction. When the flux directions are in opposition the value of the inductance is a negative minimum and when the axes are perpendicular the inductance is zero. Referring again to Figure 11.1 gives

$$
\begin{aligned}
&L_{Af} = L_{fA} = M_f \cos\gamma \\
&L_{AD} = L_{DA} = M_D \cos\gamma \\
&L_{AQ} = L_{QA} = M_Q \sin\gamma \\
&L_{Bf} = L_{fB} = M_f \cos\left(\gamma - \frac{2}{3}\pi\right); \quad L_{Cf} = L_{fC} = M_f \cos\left(\gamma + \frac{2}{3}\pi\right) \\
&L_{BD} = L_{DB} = M_D \cos\left(\gamma - \frac{2}{3}\pi\right); \quad L_{CD} = L_{DC} = M_D \cos\left(\gamma + \frac{2}{3}\pi\right) \\
&L_{BQ} = L_{QB} = M_Q \sin\left(\gamma - \frac{2}{3}\pi\right); \quad L_{CQ} = L_{QC} = M_Q \sin\left(\gamma + \frac{2}{3}\pi\right)
\end{aligned}
\tag{11.4}
$$

The self-and mutual inductances of the rotor windings are constant and do not depend on rotor position. As the d-and q-axis windings are perpendicular to each other, their mutual inductances are zero

$$
L_{fQ} = L_{Qf} = 0 \quad \text{and} \quad L_{DQ} = L_{QD} = 0 \tag{11.5}
$$

Most of the elements forming the inductance matrix $\boldsymbol{L}$ in the flux linkage Eq. (11.1) depend on rotor position and are therefore functions of time.

11.1.3 The Flux Linkage Equations in the Rotor Reference Frame

At any instant in time the position of the rotor relative to the stator reference axis is defined by the angle γ shown in Figure 11.1. Each phasor, whether voltage, current, or flux linkage, in the stator reference frame (A, B, C) can be transformed into the (d, q) reference frame by projecting one reference frame onto the other using trigonometric functions of the angle γ. Using the notation of Figure 11.1, the current vectors are

$$\begin{aligned} i_{\mathrm{d}} &= \beta_{\mathrm{d}}\left[i_{\mathrm{A}}\cos\gamma + i_{\mathrm{B}}\cos\left(\gamma - \frac{2}{3}\pi\right) + i_{\mathrm{C}}\cos\left(\gamma + \frac{2}{3}\pi\right)\right] \\ i_{\mathrm{q}} &= \beta_{\mathrm{q}}\left[i_{\mathrm{A}}\sin\gamma + i_{\mathrm{B}}\sin\left(\gamma - \frac{2}{3}\pi\right) + i_{\mathrm{C}}\sin\left(\gamma + \frac{2}{3}\pi\right)\right] \end{aligned} \tag{11.6}$$

where β_{d} and β_{q} are arbitrary nonzero coefficients introduced due to the change of reference frame. Equation (11.6) describes a unique transformation from the stator (A, B, C) to the rotor (d, q) reference axis. The reverse transformation from (d, q) to (A, B, C) is not unique as the two equations in Eq. (11.6) have three unknowns: i_{A}, i_{B}, and i_{C}. A unique transformation may be achieved by supplementing the (d, q) coordinates by an additional coordinate. It is convenient to assume this additional coordinate to be the zero-sequence coordinate defined in the same way as in the method of symmetrical components

$$i_0 = \beta_0(i_{\mathrm{A}} + i_{\mathrm{B}} + i_{\mathrm{C}}) \tag{11.7}$$

where β_0 is again an arbitrary coefficient introduced, owing to the change in reference frame. Combining Eqs. (11.6) and (11.7) gives the following matrix equation

$$\begin{bmatrix} i_0 \\ i_{\mathrm{d}} \\ i_{\mathrm{q}} \end{bmatrix} = \begin{bmatrix} \beta_0 & \beta_0 & \beta_0 \\ \beta_{\mathrm{d}}\cos\gamma & \beta_{\mathrm{d}}\cos\left(\gamma - \frac{2}{3}\pi\right) & \beta_{\mathrm{d}}\cos\left(\gamma + \frac{2}{3}\pi\right) \\ \beta_{\mathrm{q}}\sin\gamma & \beta_{\mathrm{q}}\sin\left(\gamma - \frac{2}{3}\pi\right) & \beta_{\mathrm{q}}\sin\left(\gamma + \frac{2}{3}\pi\right) \end{bmatrix} \begin{bmatrix} i_{\mathrm{A}} \\ i_{\mathrm{B}} \\ i_{\mathrm{C}} \end{bmatrix} \quad \text{or} \quad \boldsymbol{i}_{0\mathrm{dq}} = \boldsymbol{W}\boldsymbol{i}_{\mathrm{ABC}} \tag{11.8}$$

The coefficients β_0, β_{d}, and β_{q} are nonzero. Matrix $\boldsymbol{W}$ is nonsingular and the inverse transformation is uniquely determined by

$$\boldsymbol{i}_{\mathrm{ABC}} = \boldsymbol{W}^{-1}\boldsymbol{i}_{0\mathrm{dq}} \tag{11.9}$$

A similar transformation can be defined for the phasors of stator voltage and flux linkage.

The rotor currents, voltages, and flux linkages are already in the (d, q) reference frame and no transformation is necessary, allowing the transformation of all the winding currents to be written as

$$\begin{bmatrix} \boldsymbol{i}_{0\mathrm{dq}} \\ \boldsymbol{i}_{\mathrm{fDQ}} \end{bmatrix} = \begin{bmatrix} \boldsymbol{W} & \boldsymbol{0} \\ \boldsymbol{0} & \boldsymbol{1} \end{bmatrix} \begin{bmatrix} \boldsymbol{i}_{\mathrm{ABC}} \\ \boldsymbol{i}_{\mathrm{fDQ}} \end{bmatrix} \tag{11.10}$$

In this equation $\boldsymbol{i}_{\mathrm{fDQ}}$ is a column vector of the currents i_{f}, i_{D}, and i_{Q}, and $\boldsymbol{1}$ is a diagonal unit matrix. A similar transformation can be defined for the rotor voltages and flux linkages. The inverse transformation of Eq. (11.10) is

$$\begin{bmatrix} \boldsymbol{i}_{\mathrm{ABC}} \\ \boldsymbol{i}_{\mathrm{fDQ}} \end{bmatrix} = \begin{bmatrix} \boldsymbol{W}^{-1} & \boldsymbol{0} \\ \boldsymbol{0} & \boldsymbol{1} \end{bmatrix} \begin{bmatrix} \boldsymbol{i}_{0\mathrm{dq}} \\ \boldsymbol{i}_{\mathrm{fDQ}} \end{bmatrix} \tag{11.11}$$

Substituting this inverse transformation, along with a similar transformation for the fluxes, into the flux linkage Eq. (11.1) gives

$$\begin{bmatrix} \boldsymbol{\Psi}_{0\mathrm{dq}} \\ \boldsymbol{\Psi}_{\mathrm{fDQ}} \end{bmatrix} = \begin{bmatrix} \boldsymbol{W} & \boldsymbol{0} \\ \boldsymbol{0} & \boldsymbol{1} \end{bmatrix} \begin{bmatrix} \boldsymbol{L}_{\mathrm{S}} & \boldsymbol{L}_{\mathrm{SR}} \\ \boldsymbol{L}_{\mathrm{SR}}^{\mathrm{T}} & \boldsymbol{L}_{\mathrm{W}} \end{bmatrix} \begin{bmatrix} \boldsymbol{W}^{-1} & \boldsymbol{0} \\ \boldsymbol{0} & \boldsymbol{1} \end{bmatrix} \begin{bmatrix} \boldsymbol{i}_{0\mathrm{dq}} \\ \boldsymbol{i}_{\mathrm{fDQ}} \end{bmatrix} \tag{11.12}$$

which, after multiplying the three square matrices by each other, yields

$$\begin{bmatrix} \boldsymbol{\Psi}_{\mathrm{0dq}} \\ \hline \boldsymbol{\Psi}_{\mathrm{fDQ}} \end{bmatrix} = \left[\begin{array}{c|c} \boldsymbol{W}\boldsymbol{L}_{\mathrm{S}}\boldsymbol{W}^{-1} & \boldsymbol{W}\boldsymbol{L}_{\mathrm{SR}} \\ \hline \boldsymbol{L}_{\mathrm{SR}}^{\mathrm{T}}\boldsymbol{W}^{-1} & \boldsymbol{L}_{\mathrm{W}} \end{array}\right] \begin{bmatrix} \boldsymbol{i}_{\mathrm{0dq}} \\ \hline \boldsymbol{i}_{\mathrm{fDQ}} \end{bmatrix} \tag{11.13}$$

The coefficients introduced following the change in the reference frame are now chosen as $\beta_0 = 1/\sqrt{3}$ and $\beta_\mathrm{d} = \beta_\mathrm{q} = \sqrt{2/3}$ to give the following transformation matrix

$$\boldsymbol{W} = \sqrt{\frac{2}{3}} \left[\begin{array}{c|c|c} \frac{1}{\sqrt{2}} & \frac{1}{\sqrt{2}} & \frac{1}{\sqrt{2}} \\ \hline \cos\gamma & \cos\left(\gamma - \frac{2}{3}\pi\right) & \cos\left(\gamma + \frac{2}{3}\pi\right) \\ \hline \sin\gamma & \sin\left(\gamma - \frac{2}{3}\pi\right) & \sin\left(\gamma + \frac{2}{3}\pi\right) \end{array}\right] \tag{11.14}$$

With this choice of transformation coefficients $\boldsymbol{W}^{-1} = \boldsymbol{W}^{\mathrm{T}}$, where $\boldsymbol{W}^{-1}$ and $\boldsymbol{W}^{\mathrm{T}}$ are, respectively, the inverse and the transpose of $\boldsymbol{W}$. For a matrix that satisfies the condition $\boldsymbol{W}^{-1} = \boldsymbol{W}^{\mathrm{T}}$, then $\boldsymbol{W}\boldsymbol{W}^{\mathrm{T}} = \mathbf{1}$ and the matrix is said to be *orthogonal*. As will be seen later, such an orthogonal transformation is necessary to ensure that the power calculated in both the (A, B, C) and (d, q) reference frames is identical and the transformation is said to be *power invariant*. The transformation matrix $\boldsymbol{W}$ transforms the submatrix of self- and mutual inductances of the stator windings $\boldsymbol{L}_\mathrm{S}$ into

$$\boldsymbol{W}\boldsymbol{L}_{\mathrm{S}}\boldsymbol{W}^{-1} = \boldsymbol{W} \begin{bmatrix} L_{\mathrm{AA}} & L_{\mathrm{AB}} & L_{\mathrm{AC}} \\ L_{\mathrm{BA}} & L_{\mathrm{BB}} & L_{\mathrm{BC}} \\ L_{\mathrm{CA}} & L_{\mathrm{CB}} & L_{\mathrm{CC}} \end{bmatrix} \boldsymbol{W}^{-1} = \begin{bmatrix} L_0 & & \\ & L_\mathrm{d} & \\ & & L_\mathrm{q} \end{bmatrix} \tag{11.15}$$

This is a diagonal matrix in which $L_0 = L_\mathrm{S} - 2M_\mathrm{S}$, $L_\mathrm{d} = L_\mathrm{S} + M_\mathrm{S} + 3\Delta L_\mathrm{S}/2$, and $L_\mathrm{q} = L_\mathrm{S} + M_\mathrm{S} - 3\Delta L_\mathrm{S}/2$. Similarly, the submatrix of the mutual inductances between the stator and the rotor windings is transformed into

$$\boldsymbol{W}\boldsymbol{L}_{\mathrm{SR}} = \boldsymbol{W} \begin{bmatrix} L_{\mathrm{Af}} & L_{\mathrm{AB}} & L_{\mathrm{AQ}} \\ L_{\mathrm{Bf}} & L_{\mathrm{BD}} & L_{\mathrm{BQ}} \\ L_{\mathrm{Cf}} & L_{\mathrm{CD}} & L_{\mathrm{CQ}} \end{bmatrix} = \begin{bmatrix} & & \\ kM_\mathrm{f} & kM_\mathrm{D} & \\ & & kM_\mathrm{Q} \end{bmatrix}$$

where $k = \sqrt{3/2}$. The submatrix of the mutual inductances between the rotor and the stator windings is transformed into the same form, because $\boldsymbol{W}^{-1} = \boldsymbol{W}^{\mathrm{T}}$ and

$$\boldsymbol{L}_{\mathrm{SR}}^{\mathrm{T}}\boldsymbol{W}^{-1} = \boldsymbol{L}_{\mathrm{SR}}^{\mathrm{T}}\boldsymbol{W}^{\mathrm{T}} = (\boldsymbol{W}\boldsymbol{L}_{\mathrm{SR}})^{\mathrm{T}}.$$

The matrix of the self- and mutual inductances of the rotor windings is not changed. As a result of these transformations Eq. (11.13) becomes

$$\begin{bmatrix} \Psi_0 \\ \hline \Psi_\mathrm{d} \\ \hline \Psi_\mathrm{q} \\ \hline \Psi_\mathrm{f} \\ \hline \Psi_\mathrm{D} \\ \hline \Psi_\mathrm{Q} \end{bmatrix} = \left[\begin{array}{c|c|c|c|c|c} L_0 & & & & & \\ \hline & L_\mathrm{d} & & kM_\mathrm{f} & kM_\mathrm{D} & \\ \hline & & L_\mathrm{q} & & & kM_\mathrm{Q} \\ \hline & kM_\mathrm{f} & & L_\mathrm{f} & L_\mathrm{fD} & \\ \hline & kM_\mathrm{D} & & L_\mathrm{fD} & L_\mathrm{D} & \\ \hline & & kM_\mathrm{Q} & & & L_\mathrm{Q} \end{array}\right] \begin{bmatrix} i_0 \\ \hline i_\mathrm{d} \\ \hline i_\mathrm{q} \\ \hline i_\mathrm{f} \\ \hline i_\mathrm{D} \\ \hline i_\mathrm{Q} \end{bmatrix} \tag{11.16}$$

An important feature of this equation is that the matrix of inductances is symmetrical. This is due to the correct choice of the transformation coefficients β_0, β_d, and β_q ensuring orthogonality of the transformation matrix, $\boldsymbol{W}$.

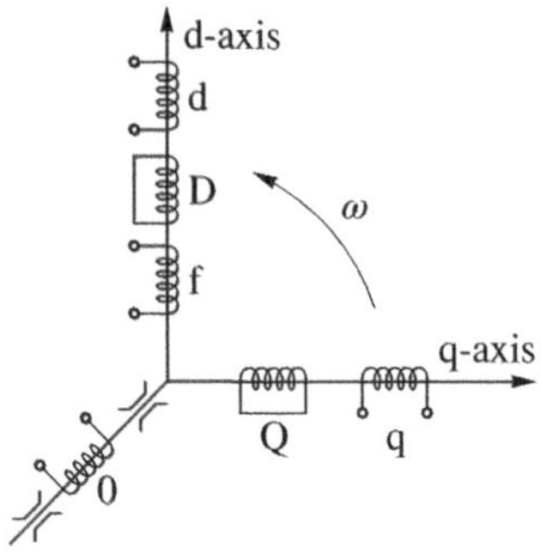

Figure 11.2 Three sets of fictitious perpendicular windings representing the synchronous generator.

The transformation of all the generator windings into the rotor reference frame is referred to as the *0dq transformation*, or *Park's transformation*. The original transformation matrix proposed by Park (Concordia 1951) was not orthogonal, and consequently the resulting matrix of equivalent inductances was not symmetrical. Concordia corrected this, but customarily the transformation is still referred to as Park's, or the modified Park's transformation. All the elements of the inductance matrix in Eq. (11.16) are constant and independent of time. This is the main advantage of Park's transformation.

After reordering the variables Eq. (11.16) can be written as three independent sets of equations

$$\Psi_0 = L_0 i_0 \tag{11.17}$$

$$\begin{bmatrix} \Psi_d \\ \Psi_f \\ \Psi_D \end{bmatrix} = \begin{bmatrix} L_d & kM_f & kM_D \\ kM_f & L_f & L_{fD} \\ kM_D & L_{fD} & L_D \end{bmatrix} \begin{bmatrix} i_d \\ i_f \\ i_D \end{bmatrix} \tag{11.18}$$

$$\begin{bmatrix} \Psi_q \\ \Psi_Q \end{bmatrix} = \begin{bmatrix} L_q & kM_Q \\ kM_Q & L_Q \end{bmatrix} \begin{bmatrix} i_q \\ i_Q \end{bmatrix} \tag{11.19}$$

These equations describe three sets of magnetically coupled windings, as shown in Figure 11.2. Each set of windings is independent of the others in that there is no magnetic coupling between the different winding sets. Figure 11.2 reflects this by showing the three winding sets perpendicular to each other. The first set of windings, represented by Eq. (11.18), consists of three windings in the d-axis. Two of these, f and D, correspond to the real field and damper windings of the rotor. The third winding, denoted as d, is fictitious and represents the effect of the three-phase stator winding in the d-axis of the rotor. This fictitious d-axis winding rotates with the rotor.

The second set of windings, represented by Eq. (11.19), consists of two windings. The first one, denoted by Q, corresponds to the real damper winding in the rotor q-axis, while the second, denoted by q, is a fictitious winding representing the effect of the three-phase stator winding in the q-axis. Obviously, both windings rotate with the rotor.

Equation (11.17) represents the third winding set which consists of a single winding which is magnetically separate from both of the other two sets. This winding is shown in Figure 11.2 to be perpendicular to both the d-axis and the q-axis and to be along the axis of rotation of the equivalent rotor. This winding can be omitted if the three-phase stator winding is connected in star with the neutral point isolated, that is not earthed. With this winding connection the sum of the stator phase currents must be zero and, as $i_0 = (i_A + i_B + i_C)/\sqrt{3} = 0$, the current flowing in this third winding is also zero.

A physical interpretation of the d- and q-axis coils can be obtained by considering Eq. (11.16). This equation defines the flux linkages within the generator but with the actual three-phase stator armature winding replaced by one winding in the d-axis and another in the q-axis. As shown in Chapter 3, currents in the three-phase stator armature winding produce a rotating armature reaction flux which enters the rotor at an angle that depends on the armature loading condition. In the rotor (d, q) reference frame this rotating flux is simply represented by two DC flux components, one acting along the d-axis and the other along the q-axis. These (d, q) component fluxes are produced by the currents flowing in the two fictitious (d, q) armature windings. In light of this, selecting the transformation coefficients $\beta_d = \beta_q = \sqrt{2/3}$ has important implications with regard to the number of turns on the fictitious d and q armature windings. For balanced three-phase currents in the armature, Eq. (3.42) showed the value of the armature mmf rotating with the rotor was equal to $3/2N_a I_m$ where N_a is the effective number of turns in series per phase and $I_m = \sqrt{2}I_g$. However, as will be shown in Eq. (11.82), the same balanced three-phase current gives the maximum values of i_d and i_q as $\sqrt{3/2}I_m$. Consequently, if i_d and i_q are to produce the same mmf as the actual three-phase armature winding the d- and q-axis armature windings must each have $\sqrt{3/2}$ more turns than an actual armature phase winding. This effect is reflected by the factor $k = \sqrt{3/2}$ which appears in the mutual inductance between the d-axis armature winding and both d-axis rotor windings. The same factor is present in the mutual inductance between the q-axis armature winding and the q-axis damper winding.

Other values of transformation coefficient can be used to produce an orthogonal transformation. In particular, a transformation coefficient of 2/3 is favored by a number of authors, for example Adkins (1957). The reason for this is discussed by Harris et al. (1970), who compare different transformation systems. In their discussion these authors argue that a transformation coefficient of 2/3 is more closely related to flux conditions in the generator than a transformation coefficient of $\sqrt{2/3}$. With a current transformation coefficient of 2/3 both the d- and q-axis armature coils have the same number of turns as an individual phase winding. However, in order to maintain a power-invariant transformation, the transformation coefficient used for voltages is different from that used for currents unless a per-unit (pu) system is used. In such a system it is necessary to have a different value of base current in the phase windings to the d, q windings; the (d, q) base current differs by a factor of 3/2 to the (A, B, C) base current.

The three-phase power output of the generator is equal to the scalar product of the stator voltages and currents

$$p_{\mathrm{g}} = v_{\mathrm{A}} i_{\mathrm{A}} + v_{\mathrm{B}} i_{\mathrm{B}} + v_{\mathrm{C}} i_{\mathrm{C}} = \boldsymbol{v}_{\mathrm{ABC}}^{\mathrm{T}} \boldsymbol{i}_{\mathrm{ABC}} \tag{11.20}$$

The orthogonal transformation from the (A, B, C) to the (0, d, q) reference frame ensures that the transformation is power invariant so that the power is also given by

$$p_{\mathrm{g}} = v_0 i_0 + v_{\mathrm{d}} i_{\mathrm{d}} + v_{\mathrm{q}} i_{\mathrm{q}} = \boldsymbol{v}_{0\mathrm{dq}}^{\mathrm{T}} \boldsymbol{i}_{0\mathrm{dq}} \tag{11.21}$$

Equation (11.21) can be verified by substituting the voltage and current transformations $\boldsymbol{v}_{\mathrm{ABC}} = \boldsymbol{W}^{-1}\boldsymbol{v}_{0\mathrm{dq}}$ and $\boldsymbol{i}_{\mathrm{ABC}} = \boldsymbol{W}^{-1}\boldsymbol{i}_{0\mathrm{dq}}$ into Eq. (11.20) to give

$$\begin{aligned} p_{\mathrm{g}} &= \boldsymbol{v}_{\mathrm{ABC}}^{\mathrm{T}} \boldsymbol{i}_{\mathrm{ABC}} = \left(\boldsymbol{W}^{-1}\boldsymbol{v}_{0\mathrm{dq}}\right)^{\mathrm{T}} \boldsymbol{W}^{-1}\boldsymbol{i}_{0\mathrm{dq}} = \boldsymbol{v}_{0\mathrm{dq}}^{\mathrm{T}} \left(\boldsymbol{W}^{-1}\right)^{\mathrm{T}} \boldsymbol{W}^{-1}\boldsymbol{i}_{0\mathrm{dq}} \\ &= \boldsymbol{v}_{0\mathrm{dq}}^{\mathrm{T}} \left(\boldsymbol{W}^{\mathrm{T}}\right)^{\mathrm{T}} \boldsymbol{W}^{-1}\boldsymbol{i}_{0\mathrm{dq}} = \boldsymbol{v}_{0\mathrm{dq}}^{\mathrm{T}} \boldsymbol{W}\boldsymbol{W}^{-1}\boldsymbol{i}_{0\mathrm{dq}} = \boldsymbol{v}_{0\mathrm{dq}}^{\mathrm{T}} \boldsymbol{i}_{0\mathrm{dq}} \end{aligned}$$

and noting that, as the matrix $\boldsymbol{W}$ is orthogonal, $\boldsymbol{W}^{-1} = \boldsymbol{W}^{\mathrm{T}}$.

11.1.4 Voltage Equations

The winding circuits shown in Figure 11.1 can be divided into two characteristic types. The first type, consisting of the stator windings (A, B, C) and the damper windings (D, Q), are circuits in which the emf induced in the winding drives the current in the winding. The application of Kirchhoff's voltage law to such a circuit is shown in Figure 11.3a. The second type of circuit is represented by the rotor field winding f in which the current is supplied by an external voltage source. In this case an emf is induced in the winding which opposes the current. The equivalent circuit for this circuit is shown in Figure 11.3b. The convention for the direction of the voltages is the same as that used in Figure 11.1.

Using this convention, the voltage equation in the (A, B, C) reference frame follows as

$$\begin{bmatrix} v_{\mathrm{A}} \\ v_{\mathrm{B}} \\ v_{\mathrm{C}} \\ -v_{\mathrm{f}} \\ 0 \\ 0 \end{bmatrix} = -\begin{bmatrix} R_{\mathrm{A}} & & & & & \\ & R_{\mathrm{B}} & & & & \\ & & R_{\mathrm{C}} & & & \\ & & & R_{\mathrm{f}} & & \\ & & & & R_{\mathrm{D}} & \\ & & & & & R_{\mathrm{Q}} \end{bmatrix} \begin{bmatrix} i_{\mathrm{A}} \\ i_{\mathrm{B}} \\ i_{\mathrm{C}} \\ i_{\mathrm{f}} \\ i_{\mathrm{D}} \\ i_{\mathrm{Q}} \end{bmatrix} - \frac{\mathrm{d}}{\mathrm{d}t} \begin{bmatrix} \Psi_{\mathrm{A}} \\ \Psi_{\mathrm{B}} \\ \Psi_{\mathrm{C}} \\ \Psi_{\mathrm{f}} \\ \Psi_{\mathrm{D}} \\ \Psi_{\mathrm{Q}} \end{bmatrix} \tag{11.22}$$

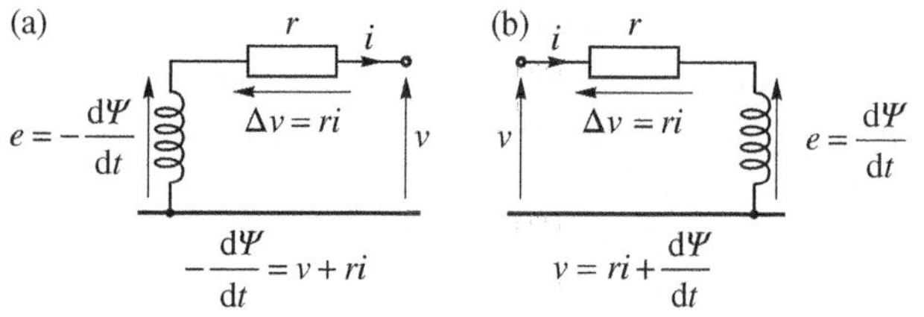

Figure 11.3 Kirchhoff's voltage law applied to the two types of circuits: (a) generator circuit; (b) motor circuit.

or, when written in compact matrix form,

$$\begin{bmatrix} \boldsymbol{v}_{\mathrm{ABC}} \\ \boldsymbol{v}_{\mathrm{fDQ}} \end{bmatrix} = -\begin{bmatrix} \boldsymbol{R}_{\mathrm{ABC}} & \\ & \boldsymbol{R}_{\mathrm{fDQ}} \end{bmatrix} \begin{bmatrix} \boldsymbol{i}_{\mathrm{ABC}} \\ \boldsymbol{i}_{\mathrm{fDQ}} \end{bmatrix} - \frac{\mathrm{d}}{\mathrm{d}t} \begin{bmatrix} \boldsymbol{\Psi}_{\mathrm{ABC}} \\ \boldsymbol{\Psi}_{\mathrm{fDQ}} \end{bmatrix} \tag{11.23}$$

where $\boldsymbol{R}_{\mathrm{ABC}}$ and $\boldsymbol{R}_{\mathrm{fDQ}}$ are diagonal resistance matrices. These equations can be transformed into the rotating reference frame using the transformation Eq. (11.11) for currents, voltages, and flux linkages. After some algebra this gives

$$\begin{bmatrix} \boldsymbol{W}^{-1} & \\ & \boldsymbol{1} \end{bmatrix} \begin{bmatrix} \boldsymbol{v}_{0\mathrm{dq}} \\ \boldsymbol{v}_{\mathrm{fDQ}} \end{bmatrix} = -\begin{bmatrix} \boldsymbol{R}_{\mathrm{ABC}} & \\ & \boldsymbol{R}_{\mathrm{fDQ}} \end{bmatrix} \begin{bmatrix} \boldsymbol{W}^{-1} & \\ & \boldsymbol{1} \end{bmatrix} \begin{bmatrix} \boldsymbol{i}_{0\mathrm{dq}} \\ \boldsymbol{i}_{\mathrm{fDQ}} \end{bmatrix} - \frac{\mathrm{d}}{\mathrm{d}t} \begin{bmatrix} \boldsymbol{W}^{-1} & \\ & \boldsymbol{1} \end{bmatrix} \begin{bmatrix} \boldsymbol{\Psi}_{0\mathrm{dq}} \\ \boldsymbol{\Psi}_{\mathrm{fDQ}} \end{bmatrix}$$

which, when left-multiplied by the transformation matrix $\boldsymbol{W}$, becomes

$$\begin{aligned} \begin{bmatrix} \boldsymbol{v}_{0\mathrm{dq}} \\ \boldsymbol{v}_{\mathrm{fDQ}} \end{bmatrix} = &-\begin{bmatrix} \boldsymbol{W} & \\ & \boldsymbol{1} \end{bmatrix} \begin{bmatrix} \boldsymbol{R}_{\mathrm{ABC}} & \\ & \boldsymbol{R}_{\mathrm{fDQ}} \end{bmatrix} \begin{bmatrix} \boldsymbol{W}^{-1} & \\ & \boldsymbol{1} \end{bmatrix} \begin{bmatrix} \boldsymbol{i}_{0\mathrm{dq}} \\ \boldsymbol{i}_{\mathrm{fDQ}} \end{bmatrix} + \\ &-\begin{bmatrix} \boldsymbol{W} & \\ & \boldsymbol{1} \end{bmatrix} \frac{\mathrm{d}}{\mathrm{d}t} \left\{ \begin{bmatrix} \boldsymbol{W}^{-1} & \\ & \boldsymbol{1} \end{bmatrix} \begin{bmatrix} \boldsymbol{\Psi}_{0\mathrm{dq}} \\ \boldsymbol{\Psi}_{\mathrm{fDQ}} \end{bmatrix} \right\} \end{aligned} \tag{11.24}$$

If the resistance of each of the stator phases is identical $R_{\mathrm{A}} = R_{\mathrm{B}} = R_{\mathrm{C}} = R$, and the product of the first three matrices on the right is a diagonal matrix, then

$$\boldsymbol{W}\boldsymbol{R}_{\mathrm{ABC}}\boldsymbol{W}^{-1} = \boldsymbol{R}_{\mathrm{ABC}} \tag{11.25}$$

According to Eq. (11.14), the transformation matrix $\boldsymbol{W}$ is a function of time and the derivative of the last term on the right in Eq. (11.24) must be calculated as the derivative of a product of two functions

$$\frac{\mathrm{d}}{\mathrm{d}t}\left(\boldsymbol{W}^{-1}\boldsymbol{\Psi}_{0\mathrm{dq}}\right) = \dot{\boldsymbol{W}}^{-1}\boldsymbol{\Psi}_{0\mathrm{dq}} + \boldsymbol{W}^{-1}\dot{\boldsymbol{\Psi}}_{0\mathrm{dq}}$$

where the dot on the top of a symbol denotes a derivative with respect to time. Multiplication by the transformation matrix gives

$$\boldsymbol{W}\frac{\mathrm{d}}{\mathrm{d}t}\left(\boldsymbol{W}^{-1}\boldsymbol{\Psi}_{0\mathrm{dq}}\right) = \left(\boldsymbol{W}\dot{\boldsymbol{W}}^{-1}\right)\boldsymbol{\Psi}_{0\mathrm{dq}} + \dot{\boldsymbol{\Psi}}_{0\mathrm{dq}} = -\left(\dot{\boldsymbol{W}}\boldsymbol{W}^{-1}\right)\boldsymbol{\Psi}_{0\mathrm{dq}} + \dot{\boldsymbol{\Psi}}_{0\mathrm{dq}} \tag{11.26}$$

as the derivative of the product $\boldsymbol{W}\boldsymbol{W}^{-1} = \boldsymbol{1}$ is $\dot{\boldsymbol{W}}\boldsymbol{W}^{-1} + \boldsymbol{W}\dot{\boldsymbol{W}}^{-1} = \boldsymbol{0}$ and $\dot{\boldsymbol{W}}\boldsymbol{W}^{-1} = -\boldsymbol{W}\dot{\boldsymbol{W}}^{-1}$. Calculating $\dot{\boldsymbol{W}}$ as $\dot{\boldsymbol{W}} = \frac{\mathrm{d}}{\mathrm{d}t}\boldsymbol{W}$ and multiplying by $\boldsymbol{W}^{-1} = \boldsymbol{W}^{\mathrm{T}}$ gives $\dot{\boldsymbol{W}}\boldsymbol{W}^{-1}$ as

$$\boldsymbol{\Omega} = \dot{\boldsymbol{W}}\boldsymbol{W}^{-1} = \omega \begin{bmatrix} 0 & 0 & 0 \\ 0 & 0 & -1 \\ 0 & 1 & 0 \end{bmatrix} \tag{11.27}$$

This matrix is referred to as the *rotation matrix* as it introduces terms into the voltage equations which are dependent on the speed of rotation.

The voltage equations in the (d, q) reference frame can be obtained after substituting the formulae Eqs. (11.25)–(11.27) into Eq. (11.24) to give

$$\begin{bmatrix} \boldsymbol{v}_{0dq} \\ \boldsymbol{v}_{fDQ} \end{bmatrix} = -\begin{bmatrix} \boldsymbol{R}_{ABC} & \\ & \boldsymbol{R}_{fDQ} \end{bmatrix} \begin{bmatrix} \boldsymbol{i}_{0dq} \\ \boldsymbol{i}_{fDQ} \end{bmatrix} - \begin{bmatrix} \dot{\boldsymbol{\Psi}}_{0dq} \\ \dot{\boldsymbol{\Psi}}_{fDQ} \end{bmatrix} + \begin{bmatrix} \boldsymbol{\Omega} & \\ & \boldsymbol{0} \end{bmatrix} \begin{bmatrix} \boldsymbol{\Psi}_{0dq} \\ \boldsymbol{\Psi}_{fDQ} \end{bmatrix} \tag{11.28}$$

This equation, without the term $\boldsymbol{\Omega\Psi}_{0dq}$, describes Kirchhoff's voltage law for the equivalent generator circuits shown in Figure 11.2. The rotational term represents the emf's induced in the stator windings, owing to the rotation of the magnetic field. These *rotational emf's* are represented as

$$\boldsymbol{\Omega\Psi}_{0dq} = \omega \begin{bmatrix} 0 & 0 & 0 \\ 0 & 0 & -1 \\ 0 & 1 & 0 \end{bmatrix} \begin{bmatrix} \Psi_0 \\ \Psi_d \\ \Psi_q \end{bmatrix} = \begin{bmatrix} 0 \\ -\omega\Psi_q \\ +\omega\Psi_d \end{bmatrix} \tag{11.29}$$

Importantly, this equation shows that the d-axis rotational emf is induced by the q-axis flux, while the q-axis rotational emf is induced by the d-axis flux. The plus and minus signs are a result of the assumed direction, and rotation, of the rotor axes and the fact that an induced emf must lag the flux which produces it by 90°.

The armature emf's proportional to the rate of change of the flux, that is the $\dot{\Psi}$ terms, are referred to as the *transformation emf's* and are due to changing currents in coils on the same axis as the one being considered. They would be present even if the machine was stationary.

Equation (11.28) may be expanded to give

$$\left.\begin{aligned} v_0 &= -Ri_0 - \dot{\Psi}_0 \\ v_d &= -Ri_d - \dot{\Psi}_d - \omega\Psi_q \\ v_q &= -Ri_q - \dot{\Psi}_q + \omega\Psi_d \end{aligned}\right\} \text{stator} \tag{11.30}$$

$$\left.\begin{aligned} v_f &= R_f i_f + \dot{\Psi}_f \\ 0 &= R_D i_D + \dot{\Psi}_D \\ 0 &= R_Q i_Q + \dot{\Psi}_Q \end{aligned}\right\} \text{rotor} \tag{11.31}$$

If balanced operation only is considered then there are no zero-sequence currents and the first of the stator equations, corresponding to the zero sequence, can be omitted.[1] Generally, changes in the generator speed are small ($\omega \approx \omega_s$), while the transformation emf's ($\dot{\Psi}_d$ and $\dot{\Psi}_q$) are also small when compared with the rotation emf's ($-\omega\Psi_q$ and $+\omega\Psi_d$), whose values are close to the corresponding components of the generator voltage. Neglecting the transformation emf's allows the differential Eqs. (11.30) describing the stator voltage to be replaced by the following two algebraic equations

$$\begin{bmatrix} v_d \\ v_q \end{bmatrix} \approx -\begin{bmatrix} R & 0 \\ 0 & R \end{bmatrix} \begin{bmatrix} i_d \\ i_q \end{bmatrix} + \omega \begin{bmatrix} -\Psi_q \\ +\Psi_d \end{bmatrix} \tag{11.32}$$

The differential voltage equations of the rotor windings Eq. (11.31) remain unchanged and can be rewritten as

$$\begin{bmatrix} \dot{\Psi}_f \\ \dot{\Psi}_D \\ \dot{\Psi}_Q \end{bmatrix} = -\begin{bmatrix} R_f & 0 & 0 \\ 0 & R_D & 0 \\ 0 & 0 & R_Q \end{bmatrix} \begin{bmatrix} i_f \\ i_D \\ i_Q \end{bmatrix} + \begin{bmatrix} v_f \\ 0 \\ 0 \end{bmatrix} \tag{11.33}$$

1 As the zero-axis voltage and flux equations are decoupled from the other two axes, these equations can be solved separately for the unbalanced operation.

The differential Eqs. (11.32) and (11.33) together with the algebraic flux linkage Eqs. (11.18) and (11.19) constitute the full model of the synchronous generator with the transformation emf's neglected.

To be used for power system studies the generator equations must be interfaced to the equations describing the power system transmission network. If the armature transformation emf's are included in the model then the implication of having two differential equations to describe the armature voltage is that the power system transmission network equations themselves must be differential equations. For all but the simplest systems this introduces significant complexity into the system equations, requires a large amount of computation time, implies a parameter accuracy that is often unrealistic, and is usually not necessary when studying electromechanical dynamics, other than shaft torques. By neglecting the transformation emf's the armature differential equations are replaced by two algebraic equations, which permits the electric power system (EPS) to be modeled by a set of algebraic equations as described in Chapter 3, Eq. (3.146). This significantly simplifies the generator–power system interface.

For many power system studies it is possible, and highly desirable, to rephrase and simplify the full set of generator equations – Eqs. (11.32), (11.33), (11.18), and (11.19) – so that they are in a more acceptable form and easier to interface to the power system network equations. Before examining how these changes can be made, it is necessary to relate these circuit equations to the flux conditions inside the generator when it is in the steady state, transient state, or the subtransient state. These flux conditions and characteristic states are extensively described and discussed in Chapter 4.

11.1.5 Generator Reactances in Terms of Circuit Quantities

The d-axis consists of three RL coupled circuits, one each for the d-axis armature coil, the field coil, and the d-axis damper, as shown in Figure 11.4a. Only two coils are on the q-axis, one each for the q-axis armature winding and the q-axis damper, as shown in Figure 11.4b. When viewed from the terminals of the armature, the effective impedance of the armature coil to any current change will depend on the parameters of the different circuits, their mutual coupling, and whether or not the circuits are open or closed.

11.1.5.1 Steady State

When in the steady state the armature flux has penetrated through all the rotor circuits, the field and damper winding currents are constant and the armature current simply sees the synchronous inductance L_d in the direct axis and L_q in the quadrature axis.

11.1.5.2 Transient State

In the transient state the armature flux has penetrated the damper circuits and the field winding screens the rotor body from the armature flux. The damper circuits are no longer effective and can be removed from the model, while the screening behavior of the field winding is modeled by short-circuiting the field winding and setting its resistance to zero (Figure 11.5a). This effectively represents the current changes that would occur in the field

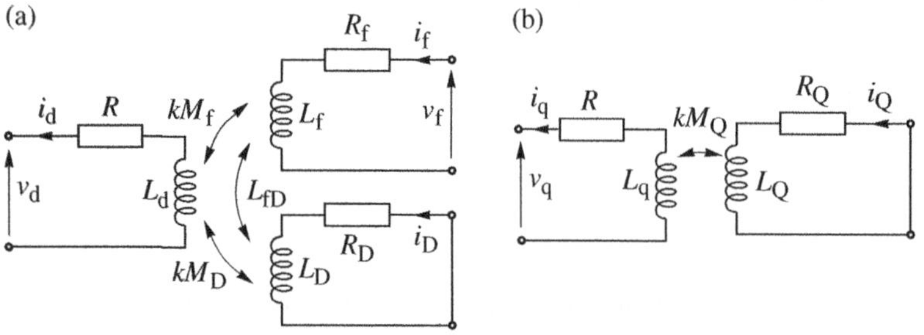

Figure 11.4 The d- and q-axis coupled circuits: (a) d-axis; (b) q-axis.

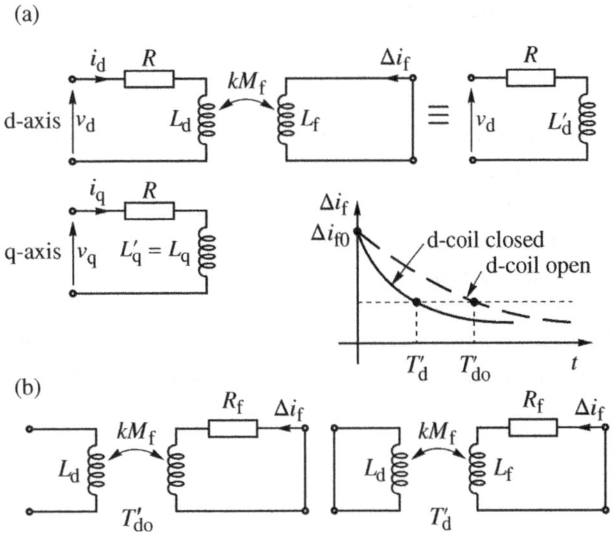

Figure 11.5 The d- and q-axis coupled circuits in the transient state: (a) for determining transient inductance; (b) for determining the field winding time constants.

winding in order to maintain the flux linkage of this winding constant, the definition of the transient state. The circuit equations for the d-axis can be written as

$$\begin{aligned} v_d &= -Ri_d + L_d\frac{di_d}{dt} + kM_f\frac{d\Delta i_f}{dt} \\ \Delta v_f = 0 &= L_f\frac{d\Delta i_f}{dt} + kM_f\frac{di_d}{dt} \end{aligned} \tag{11.34}$$

At this point it is convenient to use Laplace transform techniques to simplify these simultaneous differential equations. As the initial conditions are all zero, d/dt can be replaced by the Laplace operator and the differential equations written in matrix form as

$$\begin{bmatrix} v_d \\ 0 \end{bmatrix} = \begin{bmatrix} -R + sL_d & skM_f \\ skM_f & sL_f \end{bmatrix} \begin{bmatrix} i_d \\ \Delta i_f \end{bmatrix} \tag{11.35}$$

This equation[2] can be solved for v_d by eliminating Δi_f and writing

$$v_d = \left(-R + sL'_d\right) i_d \tag{11.36}$$

where the d-axis transient inductance

$$L'_d = L_d - \frac{k^2M_f^2}{L_f}; \quad X'_d = \omega L'_d \tag{11.37}$$

2 The general matrix equation

$$\begin{bmatrix} \boldsymbol{v}_1 \\ \boldsymbol{0} \end{bmatrix} = \begin{bmatrix} \boldsymbol{z}_{11} & \boldsymbol{z}_{12} \\ \boldsymbol{z}_{21} & \boldsymbol{z}_{22} \end{bmatrix} \begin{bmatrix} \boldsymbol{i}_1 \\ \boldsymbol{i}_2 \end{bmatrix}$$

has the solution $\boldsymbol{v}_1 = \boldsymbol{z}_{eq}\boldsymbol{i}_1$ where $\boldsymbol{z}_{eq} = \left[\boldsymbol{z}_{11} - \boldsymbol{z}_{12}\boldsymbol{z}_{22}^{-1}\boldsymbol{z}_{21}\right]$.

As there is no field winding in the quadrature axis

$$L'_q = L_q, \quad X'_q = \omega L'_q = X_q \tag{11.38}$$

However, in many cases it is convenient to represent the rotor body of a turbogenerator by an additional q-axis rotor coil when an equation for L'_q, similar to that for L'_d with appropriate parameter changes, will result. This important point will arise again later when generator models are considered.

It is also useful to define the decay time constant of the induced field current. This time constant will depend on whether the d-axis armature coil is open circuit or short circuit (Figure 11.5b). The circuit situation is very similar to that used to establish the relationship for the transient inductance, but now viewed from the field winding, so that the same equation can be used for the effective field winding inductance, with appropriate symbol changes. When the direct-axis armature circuit is open circuit the *d-axis transient open-circuit time constant* T'_{do} is obtained as

$$T'_{do} = \frac{L_f}{R_f} \tag{11.39}$$

and, when the armature circuit is short circuit, the time constant becomes the *d-axis transient short-circuit time constant* T'_d

$$T'_d = \left(L_f - \frac{k^2 M_f^2}{L_d}\right)\frac{1}{R_f} = T'_{do}\frac{L'_d}{L_d} \tag{11.40}$$

As there is no quadrature-axis field circuit there are no q-axis transient time constants.

11.1.5.3 Subtransient State

In the subtransient state the armature flux is deflected around the damper winding screening the field winding from the armature flux. The circuit configurations that reflect this flux condition are shown in Figure 11.6a. All the rotor circuits are now represented by short-circuited coils with zero resistance.

In the d-axis the matrix equation for the coupled circuits becomes

$$\begin{bmatrix} v_d \\ 0 \\ 0 \end{bmatrix} = \begin{bmatrix} R + sL_d & skM_f & skM_D \\ skM_f & sL_f & sL_{fD} \\ skM_D & sL_{fD} & L_D \end{bmatrix} \begin{bmatrix} i_d \\ \Delta i_f \\ i_D \end{bmatrix} \tag{11.41}$$

with

$$v_d = \left(-R + sL''_d\right) i_d \tag{11.42}$$

Eliminating the second two rows and columns using the same matrix procedure as before gives

$$L''_d = L_d - \left[\frac{k^2 M_f^2 L_D + k^2 M_D^2 L_f - 2kM_f kM_D L_{fD}}{L_D L_f - L_{fD}^2}\right] \quad \text{and} \quad X''_d = \omega L''_d \tag{11.43}$$

while in the quadrature axis an equation similar to Eq. (11.37) for the d-axis transient reactance gives

$$L''_q = L_q - \frac{k^2 M_Q^2}{L_Q}, \qquad X''_q = \omega L''_q \tag{11.44}$$

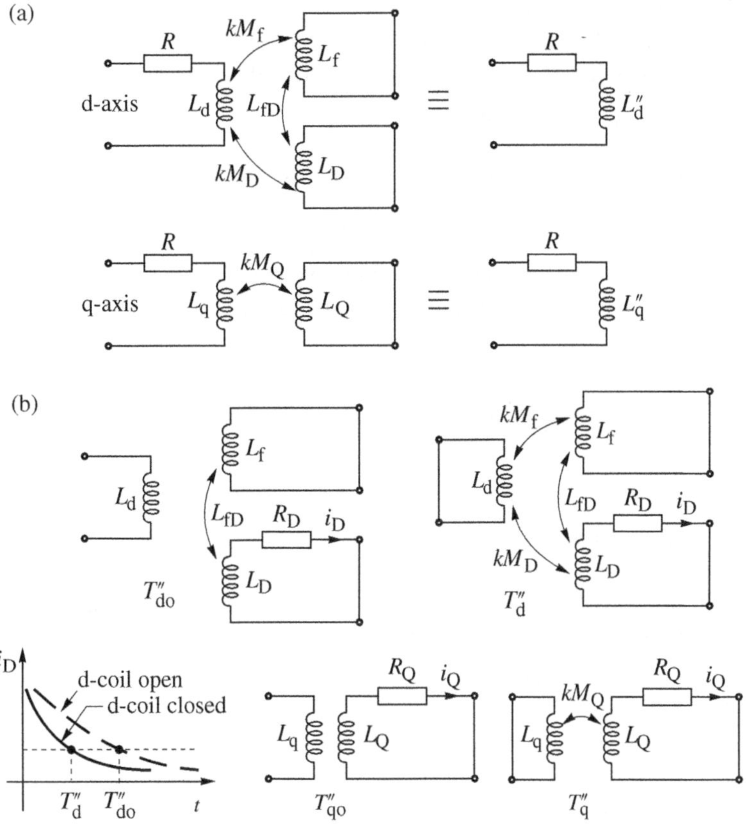

Figure 11.6 The d- and q-axis coupled circuits in the subtransient state: (a) for determining subtransient inductance; (b) for determining the damper winding time constants.

Similarly as for the transient state, direct-axis time constants can be established for the decay of current in the damper winding using the equivalent circuits of Figure 11.6b as

$$T''_{do} = \left(L_D - \frac{L_{fD}^2}{L_f}\right)\frac{1}{R_D}$$
$$T''_d = \left[L_D - \left(\frac{L_{fD}^2 L_d + k^2 M_D^2 L_f - 2L_{fD} kM_D kM_f}{L_d L_f - k^2 M_f^2}\right)\right]\frac{1}{R_D} = T''_{do}\frac{L''_d}{L'_d} \tag{11.45}$$

where T''_{do} is the d-axis subtransient open-circuit time constant and T''_d the d-axis subtransient short-circuit time constant.

If there are no rotor body screening effects in the q-axis then the equivalent quadrature-axis time constants are

$$T''_{qo} = \frac{L_Q}{R_Q}; \quad T''_q = \left(L_Q - \frac{k^2 M_Q^2}{L_q}\right)\frac{1}{R_Q} = T''_{qo}\frac{L''_q}{L'_q} \tag{11.46}$$

Recall that similar relations between open- and short-circuit time constants are derived in a less precise way in Chapter 4, Eq. (4.16).

11.1.6 Synchronous Generator Equations

Having established how the parameters of the coupled circuits are related to the generator reactances and time constants, the set of equations that constitute the full generator model, with the armature transformation emf's neglected, can now be examined more closely with a view to establishing more meaningful expressions.

11.1.6.1 Steady-state Operation

In the steady state the field winding current is constant and the damper winding currents $i_D = i_Q = 0$ so that the armature flux linkages Ψ_d and Ψ_q in Eqs. (11.18) and (11.19) become

$$\Psi_d = L_d i_d + kM_f i_f, \qquad \Psi_q = L_q i_q \tag{11.47}$$

Substituting for these flux linkages into the armature voltage Eqs. (11.32) gives

$$v_d = -Ri_d - X_q i_q, \qquad v_q = -Ri_q + X_d i_d + e_q \tag{11.48}$$

where $e_q = \omega kM_f i_f$ is the open-circuit armature voltage induced by the field current i_f. When on open circuit the armature current is zero and the field current can be related to the self-flux linkage $\Psi_{f(i_d=0)}$ from the flux Eq. (11.18) to give

$$e_q = \omega kM_f i_f = \omega \frac{kM_f}{L_f} \Psi_{f(i_d=0)} \tag{11.49}$$

11.1.6.2 Transient Operation

When the generator is in the transient state, the armature flux has penetrated the damper coils and the damper currents have decayed to a relatively small value. This allows the circuits representing the damper windings to be removed from the equation set so that the flux equations become

$$\begin{bmatrix} \Psi_d \\ \Psi_f \end{bmatrix} = \begin{bmatrix} L_d & kM_f \\ kM_f & L_f \end{bmatrix} \begin{bmatrix} i_d \\ i_f \end{bmatrix}; \qquad \Psi_q = L_q i_q \tag{11.50}$$

while

$$\dot{\Psi}_f = v_f - R_f i_f \tag{11.51}$$

$$v_d = -Ri_d - \omega \Psi_q$$
$$v_q = -Ri_q + \omega \Psi_d \tag{11.52}$$

These equations can be considered in two parts. First of all, the way in which the armature voltage Eqs. (11.52) are influenced by the presence of the field winding and, secondly, how the differential Eq. (11.51) determines the way in which the armature flux penetrates the field winding. Consider first the armature voltage equations and in particular the voltage on the quadrature axis.

The flux linkage equation allows Ψ_d to be written in terms of i_d and Ψ_f, which, when substituted into the quadrature-axis voltage equation, gives

$$v_q = -Ri_q + \omega \left[i_d \left(L_d - \frac{k^2 M_f^2}{L_f} \right) + \frac{kM_f}{L_f} \Psi_f \right] \tag{11.53}$$

This equation is readily simplified by noting that the coefficient of the first term in the square brackets is L'_d while the second term represents a voltage proportional to the field flux linkage Ψ_f. This voltage is given the symbol e'_q and is called the *quadrature-axis transient emf* where

$$e'_q = \omega \left(\frac{kM_f}{L_f} \right) \Psi_f \tag{11.54}$$

This emf can be usefully compared with the q-axis steady-state emf

$$e_q = \omega \frac{kM_f}{L_f} \Psi_{f(i_d=0)}$$

Here $\Psi_{f(i_d = 0)}$ is the field winding self-flux linkage in the steady state and e_q the emf that the corresponding field current would induce in the armature. This emf is equal to the armature open-circuit voltage. In contrast, in the transient state Ψ_f is the field flux linkage that includes the effect of armature reaction. The voltage e'_q is the equivalent armature emf that would be induced by a field current proportional to this flux linkage. As this field flux linkage must remain constant in the short period after the disturbance, Ψ_f only changes its value slowly. Making the substitutions for the transient inductance and transient emf, and assuming that $\omega \cong \omega_s$, gives

$$v_q = -Ri_q + X'_d i_d + e'_q \tag{11.55}$$

As there is no field winding on the quadrature axis $X'_q = X_q$ and

$$v_d = -Ri_d - X'_q i_q \tag{11.56}$$

Although Eq. (11.56) is correct for the rotor model assumed, many generators, and in particular turbogenerators, have a solid-steel rotor body which acts as a screen in the q-axis. It is convenient to represent this by an additional q-axis, short-circuited coil, given the symbol g, when the q-axis flux equation becomes

$$\begin{bmatrix} \Psi_q \\ \Psi_g \end{bmatrix} = \begin{bmatrix} L_q & kM_g \\ kM_g & L_g \end{bmatrix} \begin{bmatrix} i_q \\ i_g \end{bmatrix} \tag{11.57}$$

with the change of flux linking this coil defined by the additional differential equation

$$\dot{\Psi}_g = v_g - R_g i_g = -R_g i_g \qquad (v_g = 0) \tag{11.58}$$

The similarity with the d-axis rotor coils is immediately apparent and the voltage in the d-axis armature coil becomes

$$v_d = -Ri_d - X'_q i_q + e'_d \tag{11.59}$$

where $X'_q \neq X_q$ and

$$e'_d = -\omega \left(\frac{kM_g}{L_g} \right) \Psi_g \tag{11.60}$$

The flux linking the field winding Ψ_f does not remain constant during the entire period but changes slowly as the armature flux penetrates through the winding. This change in the field flux linkage is determined from the differential Eq. (11.51). Although this equation, along with Eq. (11.54) for e'_q, can be used directly to evaluate how e'_q changes with time, it is usually more convenient to rephrase the flux linkage differential equation so that it can be more easily related to the armature. This modification can be accomplished by substituting into the differential Eq. (11.51) for i_f obtained from the flux linkage Eq. (11.50) to give

$$v_f = \dot{\Psi}_f + \frac{R_f}{L_f}\Psi_f - R_f \frac{kM_f}{L_f} i_d \tag{11.61}$$

Differentiating Eq. (11.54) gives

$$\dot{e}'_q = \omega \frac{kM_f}{L_f} \dot{\Psi}_f \tag{11.62}$$

which, when rearranged and substituted into Eq. (11.61), gives after some simplification

$$e_f = \dot{e}'_q T'_{do} + e'_q - (X_d - X'_d) i_d \tag{11.63}$$

where e_f is the field voltage v_f referred to the armature given by

$$e_f = \omega k M_f v_f / R_f \tag{11.64}$$

and e_f is also the output voltage of the exciter referred to the armature. Rearranging Eq. (11.63) gives

$$\dot{e}'_q = \frac{e_f - e'_q + i_d(X_d - X'_d)}{T'_{do}} \tag{11.65}$$

This analysis can be repeated for the quadrature axis when, assuming an additional rotor coil to represent the rotor body

$$\dot{e}'_d = \frac{-i_q\left(X_q - X'_q\right) - e'_d}{T'_{qo}}; \qquad X'_q \neq X_q \tag{11.66}$$

If no additional coil is present $X'_q = X_q$ and $e'_d = 0$.

11.1.6.3 Subtransient Operation

During the subtransient period, the rotor damper coils screen both the field winding and the rotor body from changes in the armature flux. The field flux linkages Ψ_f remain constant during this period while the damper winding flux linkages are constant immediately after the fault or disturbance but then decay with time as the generator moves toward the transient state. These changes can be quantified using a similar procedure as for the transient period. Now the full equation set for the flux linkages, Eqs. (11.18) and (11.19), and the flux decay Eq. (11.33) apply.

The armature voltage Eqs. (11.32) are now modified because of their coupling with the rotor circuits in both the d- and the q-axes. The d-axis flux linkage equations allow the armature flux Ψ_d to be written in terms of i_d, Ψ_D, and Ψ_f as

$$\Psi_d = L''_d i_d + (k_1\Psi_f + k_2\Psi_D) \tag{11.67}$$

where

$$k_1 = \frac{kM_f L_D - kM_D L_{fD}}{L_f L_D - L_{fD}^2}; \quad k_2 = \frac{kM_D L_f - kM_f L_{fD}}{L_f L_D - L_{fD}^2} \tag{11.68}$$

which, when substituted into the armature voltage Eq. (11.32), gives

$$v_q = -Ri_q + X''_d i_d + e''_q \tag{11.69}$$

where

$$e''_q = \omega(k_1\Psi_f + k_2\Psi_D) \tag{11.70}$$

and represents an armature voltage proportional to the d-axis rotor flux linkages. These flux linkages remain constant immediately after the fault and only change as Ψ_D changes.

A similar analysis for the quadrature-axis armature voltage yields

$$v_d = -Ri_d - X''_q i_q + e''_d \tag{11.71}$$

The way in which the subtransient voltage decays can be found using a similar approach as for the transient period. The differential equation governing the decay of the flux through the d-axis damper is given in Eq. (11.33) as

$$\dot{\Psi}_D = -R_D i_D \tag{11.72}$$

From the d-axis flux linkage Eqs. (11.18) i_D can be written in terms of i_d, Ψ_D, and Ψ_f when

$$\dot{\Psi}_D = k_2 i_d + \frac{1}{T''_{do}}\frac{L_{fD}}{L_f}\Psi_f - \frac{1}{T''_{do}}\Psi_D \tag{11.73}$$

Differentiating e''_q, Eq. (11.70) gives

$$\dot{e}''_q \cong \omega k_2 \dot{\Psi}_D \tag{11.74}$$

as the field flux linkages Ψ_f are constant during the subtransient period. The relationships for e''_q and $\dot{e}''_q$ can now be substituted into Eq. (11.73) to give, after some simplification,

$$\dot{e}''_q = \frac{e'_q + (X'_d - X''_d)i_d - e''_q}{T''_{do}} \tag{11.75}$$

A similar analysis for the q-axis armature coil gives

$$\dot{e}''_d = \frac{e'_d - \left(X'_q - X''_q\right)i_q - e''_d}{T''_{qo}} \tag{11.76}$$

11.1.6.4 The Generator (d, q) Reference Frame and the System (a, b) Reference Frame

All the generator equations so far developed have been expressed in the (d, q) reference frame, whereas the network equations developed in Chapter 3 are expressed in phase quantities in the system (a, b) reference frame. It is now necessary to examine how the two reference frames can be linked together. This is most conveniently done by considering the steady-state operation of the generator when the instantaneous phase voltages and currents constitute a balanced set given by

$$\begin{aligned} v_A &= \sqrt{2}V_g \sin(\omega t); & i_A &= \sqrt{2}I_g \sin(\omega t + \phi) \\ v_B &= \sqrt{2}V_g \sin(\omega t - 2\pi/3); & i_B &= \sqrt{2}I_g \sin(\omega t - 2\pi/3 + \phi) \\ v_C &= \sqrt{2}V_g \sin(\omega t - 4\pi/3); & i_C &= \sqrt{2}I_g \sin(\omega t - 4\pi/3 + \phi) \end{aligned} \tag{11.77}$$

At time $t = 0$ the terminal voltage of phase A is zero and the angle between the axis of phase A and the rotor d-axis is the rotor angle δ_g. As the rotational speed of the generator is ω, the position of the rotor, relative to the axis of phase A, at any instant in time is given by $\gamma = \omega t + \delta_g$. Applying the transformation of Eq. (11.8) to the phase voltages in Eq. (11.77) then gives

$$v_d = -\sqrt{3}V_g \sin\delta_g; \quad v_q = \sqrt{3}V_g \cos\delta_g \tag{11.78}$$

The phasor diagram in Figure 11.7 is similar to that in Figure 3.30 and shows how the root-mean-square (RMS) terminal phase voltage V_g can be resolved into two orthogonal components, V_d and V_q, along the d- and q-axes where

$$V_q = V_g \cos\delta_g; \quad V_d = -V_g \sin\delta_g; \quad \text{and} \quad \underline{V}_g = V_q + jV_d \tag{11.79}$$

Substituting Eq. (11.79) into Eq. (11.78) allows the instantaneous generator voltages v_d and v_q to be related to the orthogonal RMS terminal voltage components V_d and V_q by the relationship

$$v_d = \sqrt{3}V_d; \quad v_q = \sqrt{3}V_q \tag{11.80}$$

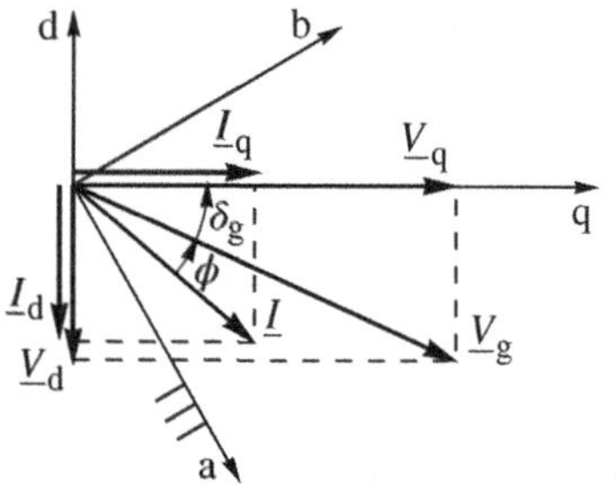

Figure 11.7 Phasor diagram in the (d, q) reference frame showing how the terminal voltage and current can be resolved into two components in quadrature. Also shown is the relative position of the system (a, b) reference frame.

This same process can also be applied to the currents to give

$$I_q = I_g \cos(\delta_g + \phi); \qquad I_d = -I_g \sin(\delta_g + \phi) \tag{11.81}$$

with the instantaneous currents being

$$i_d = -\sqrt{3} I_g \sin(\delta_g + \phi) = \sqrt{3} I_d; \quad i_q = \sqrt{3} I_g \cos(\delta_g + \phi) = \sqrt{3} I_q \tag{11.82}$$

The identities in Eqs. (11.80) and (11.82) show that the instantaneous (d, q) currents and voltages are DC variables that are proportional to the orthogonal components of the RMS phase currents and voltages. Consequently, the instantaneous voltages and currents in the steady-state armature voltage Eqs. (11.48) can be replaced by the orthogonal components of the phase currents and voltages to give

$$\begin{aligned} V_d &= -RI_d - X_q I_q \\ V_q &= -RI_q + X_d I_d + E_q \end{aligned} \tag{11.83}$$

which can be expressed in matrix form as

$$\begin{bmatrix} V_d \\ V_q \end{bmatrix} = \begin{bmatrix} 0 \\ E_q \end{bmatrix} - \begin{bmatrix} R & X_q \\ -X_d & R \end{bmatrix} \begin{bmatrix} I_d \\ I_q \end{bmatrix} \tag{11.84}$$

and

$$E_q = \omega M_f i_f / \sqrt{2} \tag{11.85}$$

The meaning of E_q is now much clearer as it is the RMS voltage that would be induced in each of the armature phases by the field current i_f when the generator is on open circuit. Equation (11.83) is identical to Eq. (3.67) derived in Chapter 3 and the equivalent circuit and phasor diagram corresponding to this equation are shown in Figures 3.17 and 3.18.

This concept can now be extended to the generator in the transient and subtransient states by recognizing that, by neglecting the transformation emf terms in the armature voltage equations, the effect of asymmetric armature currents (DC offset) has been omitted so that the armature currents are always AC quantities of varying magnitude. The effect of the transformation of Eq. (11.8) is therefore to produce (d, q) currents that are solely DC quantities; they have no AC component. The DC (d, q) currents and voltages, (i_d,i_q,v_d,v_q), are related to the armature RMS values (I_g,V_g) by the identities in Eqs. (11.80) and (11.82). This allows the generator to be represented in the (d, q) reference frame by slowly changing DC quantities in all three characteristic states. All the emf equations and armature voltage equations developed in the previous section maintain the same form except that the instantaneous values i_d, i_q, e_d, e_q, and so on are replaced by orthogonal phase values I_d, I_q, E_d, E_q, and so on. This then allows equivalent circuits and phasor diagrams similar to those developed in Section 4.2 to be used to model the generator in the different operating states.

As the network equations are all expressed in the system (a, b) reference frame, the current and voltage equations for each generator must be transformed from its own (d, q) reference frame to the system reference. The two reference frames are linked through the terminal voltage which appears explicitly in the generator equations, (d, q) reference frame, and the system equations, (a, b) reference frame. Although Figure 11.7 correctly shows the relationship between the generator (d, q) and the system (a, b) reference frames, the necessary transformation is more clearly seen by reference to Figure 3.39. A phasor $\underline{E}$ can be defined in either the (d, q) or (a, b) reference frame with the two related by the $\boldsymbol{T}$ transformation defined by Eqs. (3.191) and (3.192).

11.1.6.5 Power, Torque, and the Swing Equation

To complete the set of equations necessary to describe the generator, expressions are required for the three-phase terminal electrical power and the air-gap power. With transformation emf's neglected, the terminal power can be obtained by substituting for v_d, v_q, i_d, and i_q from Eqs. (11.80) and (11.82) into the instantaneous terminal power Eq. (11.21) to obtain

$$P_g = 3(V_d I_d + V_q I_q) \quad \text{W} \tag{11.86}$$

This equation is consistent with the normal three-phase power output expression $V_d I_d + V_q I_q = VI\cos\phi$ and the generator terminal power is, as would be expected, three times the phase power.

The air-gap power is obtained from the terminal power, Eq. (11.86), by adding the resistance loss

$$P_e = 3\left(V_d I_d + V_q I_q + \left(I_d^2 + I_q^2\right)R\right) \quad \text{W} \tag{11.87}$$

As $P = \omega\tau$ the air-gap torque is given by

$$\tau = \frac{3}{\omega}\left(V_d I_d + V_q I_q + \left(I_d^2 + I_q^2\right)R\right) \quad \text{Nm} \tag{11.88}$$

The final equation required to complete the equation set is the swing equation derived in Chapter 5, Eq. (5.15), as

$$\frac{d\Delta\omega}{dt} = \frac{1}{M}(P_m - P_e - D\Delta\omega); \quad \Delta\omega = \omega - \omega_s = \frac{d\delta}{dt} \tag{11.89}$$

where P_e is the air-gap electrical power, P_m is the mechanical turbine power, D is the damping power coefficient, ω is the generator rotational speed, ω_s is the synchronous speed, and $\Delta\omega$ is the speed deviation.

11.1.6.6 Per-unit Notation

It is standard practice in power system analysis to normalize all quantities to a common MVA base and, in the pu system used here (see Appendix A.1), this is taken as the generator rated MVA/phase $S_{1\phi}$. A base voltage V_b is also defined equal to the rated generator phase voltage. This allows the base current and the base impedance to be defined as

$$I_b = \frac{S_{1\phi}}{V_b}; \quad Z_b = \frac{V_b}{I_b} \tag{11.90}$$

Every parameter and equation can now be normalized by dividing by the appropriate base value as described in the appendix. Of particular importance is the impact that normalization has on power and torque as these are normalized to the total three-phase rated generator output $S_{3\phi}$. As shown in the appendix, when currents, voltages, and flux linkages are all expressed in pu notation the three-phase power expressions (11.86) and (11.87) normalized to the three-phase MVA base become

$$\begin{aligned} P_g &= \left(V_d I_d + V_q I_q\right) \quad \text{pu} \\ P_e &= \left[V_d I_d + V_q I_q + \left(I_d^2 + I_q^2\right)R\right] \quad \text{pu} \end{aligned} \tag{11.91}$$

while the pu notation adopted requires $\tau_{pu} = P_{pu}$ so that Eq. (11.88) becomes

$$\tau = \frac{\omega_s}{\omega}\left[V_d I_d + V_q I_q + \left(I_d^2 + I_q^2\right)R\right] \quad \text{pu} \tag{11.92}$$

A full explanation of the pu system is given in the appendix. In general all voltage, current, and flux linkage equations retain the same form whether or not they are expressed in SI or in pu notation. It is only the power and torque equations that change their form with the introduction of a 1/3 factor into the power expressions and a $\omega_s/3$ factor into the torque expressions. The introduction of the 1/3 factor into the total power expression carries the implication that the total generator power output, normalized to the $S_{3\phi}$ base, is the same as the power output per phase, normalized to $S_{1\phi}$, where $S_{1\phi} = S_{3\phi}/3$.

One of the advantages of referring all the rotor quantities to the armature winding is that only base quantities for the armature circuit need to be considered. If the rotor equations had not been referred to the armature then additional base quantities for the rotor circuits would be necessary. Such a pu system based on the concept of *equal mutual flux linkages* (Anderson and Fouad 1977) is described in the appendix and ensures that all the equations developed earlier in this chapter retain exactly the same form whether in pu or SI. This pu system also has the effect that the pu value of all the mutual inductances on one axis are equal and, in pu notation, are given the symbol L_{ad} on the d-axis and L_{aq} on the q-axis, that is

$$kM_f = kM_D = L_{fD} \equiv L_{ad}; \quad kM_Q \equiv L_{aq} \tag{11.93}$$

The self-inductances are also expressed as being equal to the sum of a magnetizing inductance and a leakage component so that

$$\begin{aligned} L_d &= L_{md} + l_l, \quad L_f = L_{mf} + l_f, \quad L_D = L_{mD} + l_D \\ L_q &= L_{mq} + l_l, \quad L_Q = L_{mQ} + l_Q \end{aligned} \tag{11.94}$$

and, in pu notation,

$$L_{md} = L_{mf} = L_{mD} \equiv L_{ad}, \quad L_{mq} = L_{mQ} \equiv L_{aq} \tag{11.95}$$

It is common to use these substitutions when expressing all the reactance and time constant expressions and so on derived in Section 11.1.4 in pu notation.

11.1.7 Synchronous Generator Models

The equations derived in the previous subsection can now be used to model the behavior of a synchronous generator. A number of different models will be developed where the generator is modeled by either its subtransient or transient emf's acting behind appropriate reactances. The way that the armature flux gradually penetrates into the rotor during a fault and affects these emf's is quantified by the differential Eqs. (11.75), (11.76), (11.65), and (11.66). These differential equations are gathered here and expressed in orthogonal phase quantities as

$$T''_{do}\dot{E}''_q = E'_q - E''_q + I_d(X'_d - X''_d) \tag{11.96}$$

$$T''_{qo}\dot{E}''_d = E'_d - E''_d - I_q(X'_q - X''_q) \tag{11.97}$$

$$T'_{do}\dot{E}'_q = E_f - E'_q + I_d(X_d - X'_d) \tag{11.98}$$

$$T'_{qo}\dot{E}'_d = -E'_d - I_q(X_q - X'_q) \tag{11.99}$$

It is worth noting the similar structure of these equations. On the left-hand side is the time derivative of the emf multiplied by the relevant time constant and the right-hand side relates to the equivalent d- or q-axis armature circuit, with resistance neglected, shown in Figure 11.8. These armature circuits were initially introduced in Chapter 4, Figure 4.15. The first component on the right-hand side in the equations constitutes a driving voltage, while the final component constitutes a voltage drop in the relevant reactance.

The right-hand side of Eq. (11.96) constitutes Kirchhoff's voltage law for the middle part of the circuit shown in Figure 11.8a, that is for the driving voltage E'_q, the voltage drop $I_d(X'_d - X''_d)$, and the emf E''_q. Similarly, the right-hand side of Eq. (11.97) constitutes Kirchhoff's voltage law for the middle part of Figure 11.8b, while

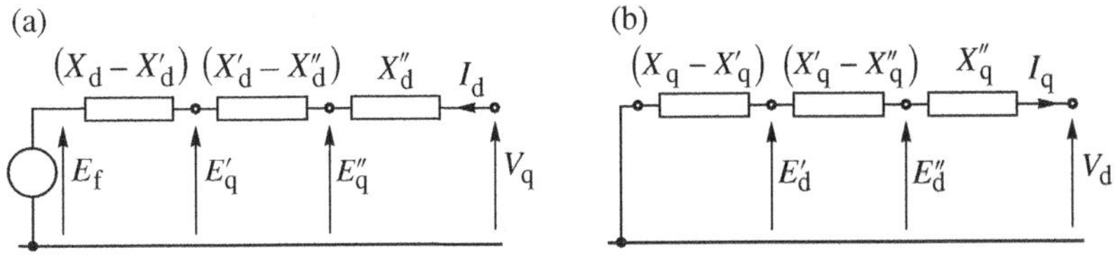

Figure 11.8 Generator equivalent circuits with resistances neglected: (a) d-axis; (b) q-axis.

Eq. (11.98) corresponds to the left part of Figure 11.8a with the driving voltage E_f, the voltage drop $I_d(X_d - X'_d)$, and the emf E'_q. In Eq. (11.99), corresponding to the left-hand part of Figure 11.8b, there is no driving voltage because there is no excitation in the q-axis.

Equations (11.96)–(11.99) allow five different generator models of decreasing complexity and accuracy to be developed. Each model is given a model number that indicates the number of differential equations required in the model. The larger the number, the greater the model complexity and the greater the time required to solve the differential equations. The model number is then followed by a number of terms enclosed in brackets which define the differential equations used by the model. In developing the generator models it is assumed that all quantities are expressed in pu notation.

11.1.7.1 Sixth-order Model – $(\dot{\delta}, \dot{\omega}, \dot{E}''_d, \dot{E}''_q, \dot{E}'_d, \dot{E}'_q)$

In this model the generator is represented by the subtransient emf's E''_q and E''_d behind the subtransient reactances X''_d and X''_q, as defined by the modified armature voltage Eqs. (11.69) and (11.71)

$$\begin{bmatrix} V_d \\ V_q \end{bmatrix} = \begin{bmatrix} E''_d \\ E''_q \end{bmatrix} - \begin{bmatrix} R & X''_q \\ -X''_d & R \end{bmatrix} \begin{bmatrix} I_d \\ I_q \end{bmatrix} \tag{11.100}$$

This equation corresponds to Eq. (4.34) derived in Section 4.2.4. The corresponding equivalent circuit and phasor diagram are shown in Figure 4.15.

The differential Eqs. (11.96)–(11.99) describe the change in these emf's as the flux linking the rotor circuits decays. To these must be added Eqs. (11.89) in order to include the speed deviation and angle change of the rotor. As the differential Eqs. (11.96) and (11.97) include the influence of the damper windings, the damping coefficient in the swing equation need only quantify the mechanical damping due to windage and friction and, as this is usually small, it may be neglected ($D \cong 0$).

Under these assumptions, the full set of six differential equations describing the generator is

$$\begin{aligned} M\Delta\dot{\omega} &= P_m - P_e \\ \dot{\delta} &= \Delta\omega \\ T'_{do}\dot{E}'_q &= E_f - E'_q + I_d(X_d - X'_d) \\ T'_{qo}\dot{E}'_d &= -E'_d - I_q(X_q - X'_q) \\ T''_{do}\dot{E}''_q &= E'_q - E''_q + I_d(X'_d - X''_d) \\ T''_{qo}\dot{E}''_d &= E'_d - E''_d - I_q(X'_q - X''_q) \end{aligned} \tag{11.101}$$

Changes in the mechanical power P_m in the first equation should be calculated from differential equations describing the models of turbines and their governing systems discussed in Section 11.3. Changes in emf E_f in the third equation should be calculated from the differential equations describing the models of excitation systems discussed in Section 11.2.

The air-gap power of the generator can be calculated using Eq. (11.91) which, after substituting for the armature voltages from Eq. (11.100), gives

$$P_e = \left(E''_d I_d + E''_q I_q\right) + \left(X''_d - X''_q\right) I_d I_q \tag{11.102}$$

The second term in this power equation defines the subtransient saliency power discussed in Chapter 4.

11.1.7.2 Fifth-order Model – $(\dot{\delta}, \dot{\omega}, \dot{E}''_d, \dot{E}''_q, \dot{E}'_q)$

In this model the screening effect of the rotor body eddy currents in the q-axis is neglected so that $X'_q = X_q$ and $E'_d = 0$. This model reverts to the classical fifth-order model with armature transformation emf's neglected. Equation (11.99) from the equation set of the sixth-order model is eliminated to give a set of five differential equations

$$\begin{aligned}
M\Delta\dot{\omega} &= P_m - P_e \\
\dot{\delta} &= \Delta\omega \\
T'_{do}\dot{E}'_q &= E_f - E'_q + I_d\left(X_d - X'_d\right) \\
T''_{do}\dot{E}''_q &= E'_q - E''_q + I_d\left(X'_d - X''_d\right) \\
T''_{qo}\dot{E}''_d &= E'_d - E''_d - I_q\left(X'_q - X''_q\right)
\end{aligned} \tag{11.103}$$

Changes in mechanical power P_m and emf E_f should be calculated, as in the sixth-order model. This model can be used for salient pole generators (hydrogenerators).

11.1.7.3 Fourth-order Model – $(\dot{\delta}, \dot{\omega}, \dot{E}'_d, \dot{E}'_q)$

In this model the effect of the damper windings in the sixth-order model are neglected and Eqs. (11.96) and (11.97) are removed from the equation set. The generator is now represented by the transient emf's E'_q and E'_d behind the transient reactances X'_d and X'_q as defined by the equation

$$\begin{bmatrix} V_d \\ V_q \end{bmatrix} = \begin{bmatrix} E'_d \\ E'_q \end{bmatrix} - \begin{bmatrix} R & X'_q \\ -X'_d & R \end{bmatrix} \begin{bmatrix} I_d \\ I_q \end{bmatrix} \tag{11.104}$$

This equation corresponds to Eq. (4.34) derived in Section 4.2.4. The corresponding equivalent circuit and phasor diagram are shown in Figure 4.16.

The changes in the emf's E'_q and E'_d are determined from differential Eqs. (11.98) and (11.99) while the electrical air-gap power is

$$P_e = E'_q I_q + E'_d I_d + \left(X'_d - X'_q\right) I_d I_q \tag{11.105}$$

with the second part of the equation defining the transient saliency power.

As the damper windings are ignored, the air-gap power calculated by this equation neglects the asynchronous torque produced by the damper windings. Consequently, the damping coefficient in the swing equation should be increased by an amount corresponding to the average asynchronous torque, or power, calculated using the simplified formulae in Eq. (5.25).

Under these assumptions, the model is described by the following four differential equations

$$\begin{aligned}
M\Delta\dot{\omega} &= P_m - P_e - D\Delta\omega \\
\dot{\delta} &= \Delta\omega \\
T'_{do}\dot{E}'_q &= E_f - E'_q + I_d\left(X_d - X'_d\right) \\
T'_{qo}\dot{E}'_d &= -E'_d - I_q\left(X_q - X'_q\right)
\end{aligned} \tag{11.106}$$

Changes in mechanical power P_m and emf E_f should be calculated as in the sixth-order model.

This simplified model of the synchronous generator is widely considered to be sufficiently accurate to analyze electromechanical dynamics (Stott 1979). Experience has shown that including the second differential equation to account for rotor body effects in the q-axis improves the accuracy of the model. The main disadvantage of this model is that the equivalent damping coefficient, appearing in the swing equation, can only be calculated approximately.

11.1.7.4 Third-order Model – $(\dot{\delta}, \dot{\omega}, \dot{E}'_q)$

This model is similar to the fourth-order model except that the d-axis transient emf E'_d is assumed to remain constant allowing Eq. (11.97) to be removed from the equation set. The generator is described only by Eqs. (11.98) and (11.89) with active power again given by Eq. (11.105). Besides neglecting the effect of the damper windings by assuming E'_d to be constant, this model also neglects the damping produced by the rotor body eddy currents, even if an additional coil is used to represent the rotor body. If there is no winding in the quadrature axis to represent the rotor body then $E'_d = 0$, $X'_q = X_q$, and Eq. (11.105) reduces to

$$P = E'_q I_q + \left(X'_d - X_q\right) I_d I_q \tag{11.107}$$

As in the fourth-order model, damper winding effects are neglected and so their effect must be included by increasing the value of the damping coefficient in the swing equation. The model is described by the following three differential equations

$$\begin{aligned} M\Delta\dot{\omega} &= P_m - P_e - D\Delta\omega \\ \dot{\delta} &= \Delta\omega \\ T'_{do}\dot{E}'_q &= E_f - E'_q + I_d\left(X_d - X'_d\right) \end{aligned} \tag{11.108}$$

Changes in mechanical power P_m and emf E_f should be calculated as in the sixth-order model.

11.1.7.5 Second-order Model – The Classical Model $(\dot{\delta}, \dot{\omega})$

The classical synchronous generator model, widely used in all the previous chapters for a simplified analysis of power system dynamics, assumes that neither the d-axis armature current I_d nor the internal emf E_f representing the excitation voltage changes very much during the transient state. In this model the generator is represented by a constant emf E' behind the transient reactance X'_d and the swing equations

$$\begin{aligned} M\Delta\dot{\omega} &= P_m - P_e - D\Delta\omega \\ \dot{\delta} &= \Delta\omega \end{aligned} \tag{11.109}$$

The justification of the classical model is that the time constant T'_{do}, appearing in Eq. (11.98), is relatively long, the order of a few seconds, so that E'_q does not change much, providing that changes in E_f and I_d are small. This means that $E'_q \cong$ constant and, because it has already been assumed that $E'_d \cong$ constant, both the magnitude of the transient emf E' and its position α with respect to the rotor may be assumed to be constant. If rotor transient saliency is neglected, $X'_q = X'_d$ and the two equivalent circuits in Figure 4.16a may be replaced by the one equivalent circuit shown in Figure 5.7 for a generator connected to an infinite busbar. As

$$\underline{I} = \underline{I}_q + j\underline{I}_d; \quad \underline{V} = \underline{V}_q + j\underline{V}_d; \quad \underline{E}' = \underline{E}'_q + j\underline{E}'_d \tag{11.110}$$

the two algebraic equations describing the armature voltage in Eqs. (11.104) can now be replaced by one equation

$$\underline{V} = \left(\underline{E}'_q + j\underline{E}'_d\right) - jX'_d\left(\underline{I}_q + j\underline{I}_d\right) = \underline{E}' - jX'_d\underline{I} \tag{11.111}$$

The assumption of small changes in the direct component of the generator current, and in the internal emf, means that only generators located a long way from the point of disturbance should be represented by the classical model.

11.1.7.6 Summary

The number of reactances and time constants representing the generator depends on the number of equivalent windings used in a particular model. The five-winding model has two equivalent rotor windings and time constants (T''_{do}, T'_{do}) on the d-axis and three armature reactances (X''_d, X'_d, and X_d). In the q-axis there is one equivalent rotor winding, with a time constant (T''_{qo}) and two armature reactances (X''_q, X_q). Although generators with rotors constructed from ferromagnetic, electrically insulated sheets of steel are well characterized by these parameters, this is not the case for generators whose rotors are made from solid steel. In such generators the rotor eddy currents play a significant role in the q-axis damping. This damping may be modeled approximately by introducing an additional q-axis winding as in the sixth-order model. This expands the model by an additional reactance $X'_q \neq X'_d$ and a time constant T'_{qo} to represent the flux decay through this circuit. If these parameters are not specified by the manufacturer it is typical to assume that $X'_q = 2X''_q$ and $T'_q = 10\, T''_q$. Owing to the screening effect of the field winding, the influence of the eddy currents in the d-axis is small and there is no need to introduce an additional winding to account for this. For slower changing disturbances, and to enable faster solution in complex systems, the generator model can be simplified by omitting the damper windings from the electrical equations, the fourth- and third-order models, and representing their damping effect by an increased damping coefficient in the swing equation. The simplest second-order model is the model traditionally used in quantitative power system analysis. It is simple, but very approximate, and only really suitable for representing generators well away from the fault. It can also be useful for evaluating generator behavior during the first rotor swing.

In developing all the generator models it has been assumed that $\omega \cong \omega_s$ and that the transformation emf's in the armature voltage equations can be neglected. Speed changes can be accounted for in all the models by introducing a factor ω/ω_s in front of every reactance. Accounting for the armature transformation emf's is significantly more difficult and is only justified if detailed modeling of the electromagnetic transients immediately after the fault is required, for example in computing short-circuit currents.

11.1.8 Saturation Effects

The generator equations derived in the previous subsection ignore the effect of magnetic saturation in the stator and rotor iron and the effect this would have on the generator parameters and the operating conditions. Saturation effects are highly nonlinear and depend on the generator loading conditions so that trying to account for them accurately in the generator model is nigh on impossible. What is required is a relatively simple saturation model that produces acceptable results, is linked to the physical process, and uses easily obtainable data.

At this stage it is interesting to consider the flux paths in Figure 4.8 that are associated with the different generator reactances. As it is only the iron paths that saturate, one might expect those generator reactances where the flux path is mainly in air, that is X'_d, X'_q, X''_d, and X''_q, to be less susceptible to saturation effects than X_d and X_q where a substantial part of the flux path is through iron (at least for the round-rotor generator). This concept of parameter sensitivity will be investigated later in this section, but first it is necessary to investigate the general effects of saturation itself. This is more easily accomplished and understood if all generator parameters are expressed in pu notation.

11.1.8.1 Saturation Characteristic

For any magnetic circuit that comprises an iron path and an air path the magnetomotive force (mmf) and flux will be related by the general curve shown in Figure 11.9a. When the iron path is unsaturated the relationship between the mmf and the flux is linear and is represented by the *air-gap line* 0A. In this situation the reluctance of the magnetic circuit is dominated by the reluctance of the air gap. As the mmf is increased, the iron saturates and the mmf/flux relationship ceases to be linear and follows the *saturation curve* 0B. Consequently, for some flux linkage Ψ_T the mmf required to produce this flux can be considered to have two components, the air-gap mmf F_a and the iron mmf F_i, so that the total mmf is given by

$$F_T = F_a + F_i \tag{11.112}$$

A *saturation factor* S can now be defined as

$$S = \frac{\text{air gap mmf}}{\text{air gap mmf + iron mmf}} = \frac{F_a}{F_a + F_i} = \frac{F_a}{F_T} \tag{11.113}$$

and, by similar triangles,

$$S = \frac{F_a}{F_a + F_i} = \frac{\Psi_T}{\Psi_T + \Psi_s} \tag{11.114}$$

Having defined a saturation factor, it is now necessary to provide a simple method of finding its value for any generator loading condition and then be able to take this into account when computing the generator parameters. To achieve these objectives it is usually assumed that:

- The open-circuit generator saturation curve can be used under load conditions. On no load the mmf in the d-axis is due to the field current, i_f, while on load it is due to $(i_f + i_D + i_d)$. As Kundur (1994) points out, this is usually the only saturation datum available.
- As the leakage flux path is mainly in air, the leakage inductances are assumed to be independent of saturation. This implies that it is only the mutual inductances L_{ad} and L_{aq} that are affected by saturation, and, of course, the self-inductance $L_d = L_{ad} + l_l$, $L_f = L_{ad} + l_f$, etc., as explained in Eq. (11.94) and Section A.4 in the appendix.
- Saturation on the d-axis and saturation on the q-axis are independent of each other.

11.1.8.2 Calculation of the Saturation Factor

When the generator is on no load, Eq. (11.85) shows that the voltage induced in one of the armature phases V_{g0} is

$$V_{g0} = E = E_q = \frac{1}{\sqrt{2}}\omega_s M_f i_{f0} = \frac{1}{\sqrt{3}}\omega_s k M_f i_{f0} = \left[\frac{1}{\sqrt{3}}\omega_s\right] L_{ad} i_{f0} = \left[\frac{1}{\sqrt{3}}\omega_s\right] \Psi_{ad} \tag{11.115}$$

because $\Psi_{ad} = L_{ad} i_{f0}$ and is the mutual flux linkage defined in Eq. (A.34) in the appendix. This equation can now be related to the open-circuit characteristic in Figure 11.9b where the slope of the air-gap line is proportional to L_{ad}. It should also be recognized that as the flux linkage and the voltage are related by Eq. (11.115), the vertical axis can be interpreted in *either* voltage *or* flux linkage.

If at some open-circuit voltage E the required flux linkage is Ψ_{ad} then if there is no saturation of the iron paths the field current required to set up this flux linkage would be i_{f0}; however, if saturation is present then the required field current would be i_{fsat}. From the definition of saturation factor in Eq. (11.114)

$$i_{fsat} = \frac{i_{f0}}{S} \tag{11.116}$$

while

$$L_{ad} = \frac{\Psi_{ad}}{i_{f0}} \quad \text{and} \quad L_{adsat} = \frac{\Psi_{ad}}{i_{fsat}} \tag{11.117}$$

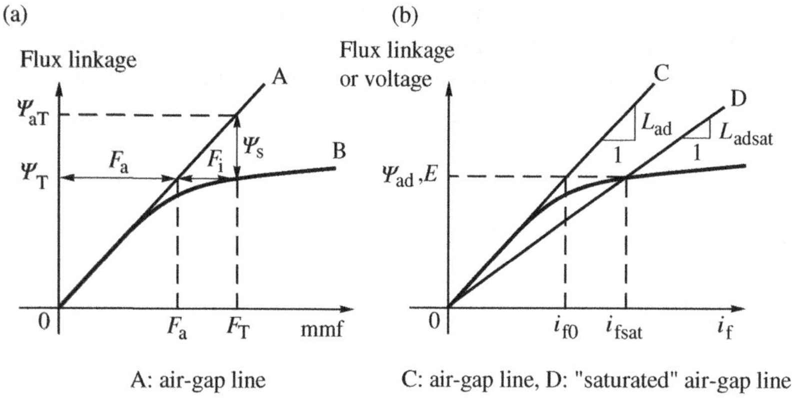

Figure 11.9 Saturation: (a) typical saturation curve for a magnetic circuit whose flux path is in iron and air; (b) generator open-circuit saturation characteristic.

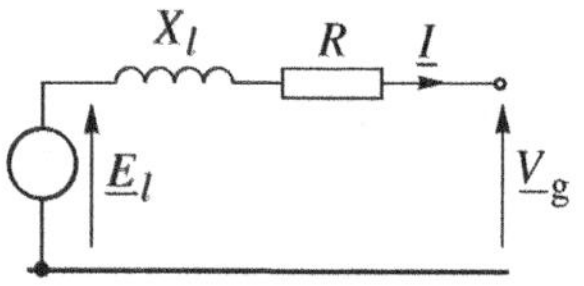

Figure 11.10 Equivalent circuit to calculate the armature voltage behind leakage reactance.

Substituting Eq. (11.116) into (11.117) gives

$$L_{adsat} = SL_{ad} \tag{11.118}$$

where L_{adsat} is the slope of the "saturated" air-gap line in Figure 11.9b. Consequently, if Ψ_{ad} is known then the saturation factor S can be calculated from Eq. (11.114) by recognizing that Ψ_{ad} and Ψ_T are equivalent. The saturated value of L_{ad} is then found from Eq. (11.118) and used to account for saturation in all the generator parameter equations developed in Section 11.1.4.

The flux linkage, Ψ_{ad}, can be calculated, provided that the armature voltage proportional to these flux linkages is known. As shown in Figure 11.10, this voltage is the voltage behind the armature leakage reactance X_l and, since X_l is not affected by saturation,

$$\underline{E}_l = \underline{V}_g + (R + jX_l)\underline{I} \tag{11.119}$$

so that provided $\underline{V}_g$ and $\underline{I}$ are known, E_l and Ψ_{ad} can be computed for any loading condition. In some instances the voltage E_p behind the Potier reactance X_p is used instead of E_l (Arrillaga and Arnold 1990).

The final piece of the jigsaw is to have some method of conveniently storing the saturation curve in computer memory so as to facilitate the calculation of the saturation factor. A number of methods are available with perhaps the simplest being to fit the saturation curve by the function

$$I_f = V + C_n V^n \tag{11.120}$$

where n is 7 or 9 and C_n is a constant for the particular curve (Hammons and Winning 1971; Arrillaga and Arnold 1990). Alternatively, a polynomial can be used.

A slightly different approach is to use a two-stage exponential process (Anderson and Fouad 1977; Anderson et al. 1990; Kundur 1994) where

$$\begin{aligned} &\Psi_{ad} < 0.8; \quad \Psi_s = 0 \\ &\Psi_{ad} \geq 0.8; \quad \Psi_s = A_{sat} e^{B_{sat}(\Psi_{ad} - 0.8)} \end{aligned} \tag{11.121}$$

where A_{sat} and B_{sat} are constants that are easily calculated by taking two points from the known saturation curve. When using this method there is usually a slight discontinuity at $\Psi_{ad} = 0.8$ but is usually of no consequence (Anderson and Fouad 1977).

The procedure for accounting for saturation in the generator equations can now be summarized as:

Step 1: Knowing $\underline{V}_g$ and $\underline{I}$ use Eq. (11.119) to calculate E_l.
Step 2: Divide E_l by $\frac{\omega_s}{\sqrt{3}}$ to obtain Ψ_{ad}.
Step 3: Use Eq. (11.121) to compute Ψ_s.
Step 4: Use Eq. (11.114) to calculate the saturation factor S.
Step 5: Modify L_{ad} to account for saturation using Eq. (11.118).
Step 6: Modify all the generator parameters that are a function of L_{ad}.

The procedure described above defines saturation in the d-axis, and, although a similar procedure can be used in the q-axis, it is usual to assume:

- for round-rotor generators the saturation factor is the same in both axes when $S_d = S_q$;
- for salient-pole generators, since the q-axis reluctance is dominated by air-paths, L_{aq} does not vary as much with saturation as L_{ad} and $S_q = 1$ is assumed under all loading conditions.

Although the saturation factor can be varied during a dynamic simulation, it is quite common to calculate it at the beginning of the simulation and assume that it remains constant during the simulation period. This ensures that the initial conditions are calculated correctly. If the saturation factor is varied during a dynamic simulation then it will be necessary to iterate steps 1 to 6 at each integration step in order to find a solution for L_{adsat}. This is because $\underline{V}_g$ and $\underline{I}$ depend on the degree of saturation while the saturation factor itself is dependent on $\underline{V}_g$ and $\underline{I}$. Other variations to this method of accounting for saturation can be found in Arrillaga and Arnold (1990) and Pavella and Murthy (1994).

11.1.8.3 Sensitivity of Parameters to Saturation

It is explained at the beginning of this section how generator reactances associated with flux paths mainly in air are less susceptible to saturation effects than are those associated with flux paths in iron. Such parameter sensitivity can be quantified by assessing the sensitivity of the particular parameter to variations in L_{ad} (or L_{aq}). For example, assume that the general parameter X is a function of L_{ad} so that for a small change in X

$$X = X_0 + \Delta X \tag{11.122}$$

where ΔX, the change in X, is given by

$$\Delta X = \left.\frac{\partial X}{\partial L_{ad}}\right|_0 \Delta L_{ad} \tag{11.123}$$

and the suffix 0 indicates an initial value.

In the context of the present discussion, Table 11.1 summarizes the pu d-axis steady-state and transient parameters defined in Section 11.1.4 with the sensitivity parameter determined using Eq. (11.122). For example, the synchronous inductance $L_d = L_{ad} + l_l$. Therefore $\Delta L_d = \left.\frac{\partial L_d}{\partial L_{ad}}\right|_0 \Delta L_{ad}$ but as

$$\left.\frac{\partial L_d}{\partial L_{ad}}\right|_0 = 1$$

then $\Delta L_d = \Delta L_{ad}$. Similarly, for the d-axis transient inductance

$$L'_d = L_d - \frac{L_{ad}^2}{L_f} = L_{ad} + l_l - \frac{L_{ad}^2}{L_f} \tag{11.124}$$

Table 11.1 Sensitivity of d-axis parameters to saturation.

Parameter	Equation	Equation numbers	Δ Parameter	Typical sensitivity to a 10% change in mutual flux linkage
L_d	$L_{ad} + l_l$	(11.94)	ΔL_{ad}	21%
L'_d	$L_d - \dfrac{L_{ad}^2}{L_f}$	(11.37)	$\left[\dfrac{l_f}{L_{f0}}\right]^2 \Delta L_{ad}$	1.5%
T'_{do}	$\dfrac{L_f}{R_f}$	(11.39)	$\dfrac{1}{R_f}\Delta L_{ad}$	20%
E_f	$\omega_s L_{ad}\dfrac{v_f}{R_f}$	(11.64)	$\left[\dfrac{\omega_s v_f}{R_f}\right]\Delta L_{ad}$	21%

Source: Based on Anderson et al. 1990.

and differentiating gives

$$\Delta L'_d = \Delta L_{ad} - \left[\frac{2L_{ad0}\Delta L_{ad}}{L_{f0}} - \frac{L_{ad0}^2 \Delta L_f}{L_{f0}^2}\right] \tag{11.125}$$

but $L_f = L_{ad} + l_l$ so that $\Delta L_f = \Delta L_{ad}$ and

$$\Delta L'_d = \Delta L_{ad} - \left[\frac{2L_{ad0}\Delta L_{ad}}{L_{f0}} - \frac{L_{ad0}^2 \Delta L_{ad}}{L_{f0}^2}\right] \tag{11.126}$$

Rearranging and writing $L_{f0} = L_{ad0} + l_f$ gives

$$\Delta L'_d = \left[\frac{l_f}{L_{f0}}\right]^2 \Delta L_{ad} \tag{11.127}$$

Anderson et al. (1990) calculates typical values for these sensitivity parameters assuming a 10% change in the mutual flux linkage Ψ_{ad}, the values of which are shown in Table 11.1. As expected, the synchronous inductance is particularly sensitive to saturation but the transient inductance, associated with flux paths mainly in air, shows little change. Indeed, if the analysis is extended to include subtransient inductance, the change in this parameter is negligible.

11.2 Excitation Systems

In Chapter 2 the different types of excitation system and AVR used to supply, and control, the field current to the generator are described and the complete exciter and AVR subsystem shown in block diagram form in Figure 2.5. Generally, the exciter is equipped with both an AVR and a manual regulator. In some cases a back-up AVR will also be provided. It is the AVR that is of particular interest here. As the power system industry uses a wide range of exciters and AVRs, with the details varying from one manufacturer to another, the aim of this section is to develop generic models of some of the most common types of excitation system. With careful parameter selection these models can then be used to represent different exciter systems from different manufacturers. A comprehensive study of excitation system models can be found in the IEEE Committee Reports on excitation systems (IEEE Committee Report, 1968, 1973a–c, 1981, 1992), and the interested reader should consult these excellent sources.

Ultimately, the exciter and the AVR model must be interfaced with the generator model developed in Section 11.1.6. This interface is through the variable E_f, which represents the generator main field voltage

v_f referred to the generator armature winding; v_f is also the exciter output voltage v_{ex}. Initially, exciter subsystem models will be developed in terms of the currents and voltages at the field winding but will finally be referred to the armature and the values converted to pu notation. This is equivalent to expressing the exciter variables in a pu system where one pu exciter output voltage corresponds to the field voltage required to produce one pu armature voltage on the air-gap line, the pu system used in the armature. This exciter pu system is commonly used by manufacturers.

11.2.1 Transducer and Comparator Model

Figure 11.11 shows the block diagram of the transducer and comparator together with the load compensation element. The first block in this model corresponds to the load compensation element, shown previously in Figure 2.6, which corrects the generator terminal voltage according to the value of the compensation impedance. The second block represents the delay introduced by the signal transducers. Normally, the equivalent time constant of the transducers is small so that this block is often neglected. The third element represents the comparator at which the corrected generator voltage is added algebraically to the reference voltage V_{ref}.

11.2.2 Exciters and Regulators

Section 2.3.2 discusses different types of exciters. Modern excitation systems usually employ either a brushless AC exciter or a static exciter with either an analogue or digital AVR, although digital AVRs are becoming more common. However, many older generators are fitted with DC excitation systems and consequently the models developed in this section are typical of the excitation systems in use today.

11.2.2.1 DC Exciters

Two different DC exciters are shown in Figure 11.12, the first being separately excited and the second self-excited. In order to develop a mathematical model of these two exciters, consider first the separately excited case in Figure 11.12a. A change in the exciter field current i_{exf} can be described by the following equation

$$v_R = R_{exf} i_{exf} + L_{exf} \frac{di_{exf}}{dt} \tag{11.128}$$

where L_{exf} depends on saturation and is an incremental inductance. The relationship between the exciter field current i_{exf} and the emf e_{ex} induced in the exciter armature is nonlinear because of magnetic saturation in the exciter core, while the exciter output voltage v_{ex} depends on both the saturation characteristic and the armature loading. Both of these effects can be included in the modeling process by using the constant resistance load saturation curve shown in Figure 11.13.

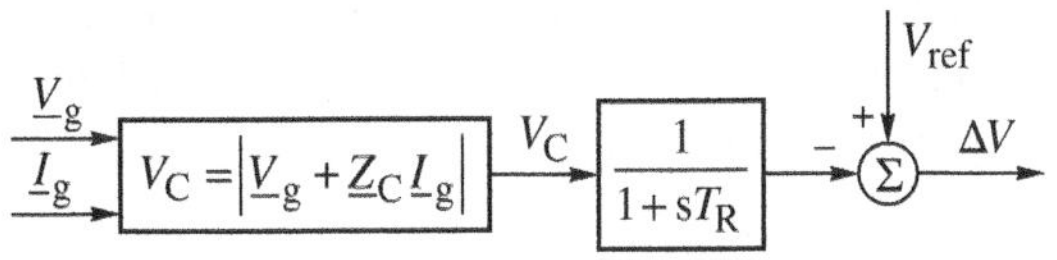

Figure 11.11 Block diagram of the transducer and comparator.

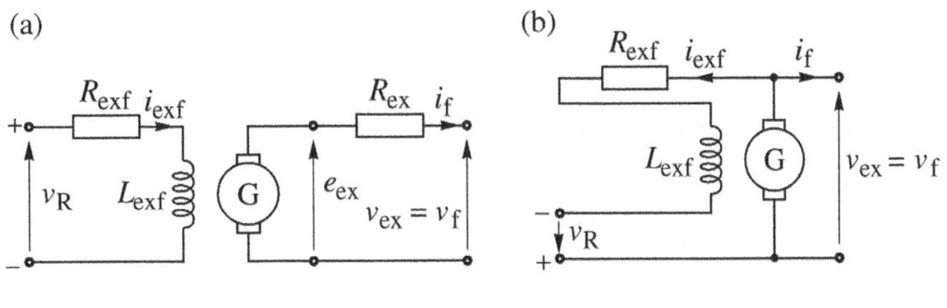

Figure 11.12 Equivalent circuit diagrams of DC exciters: (a) separately excited; (b) self-excited.

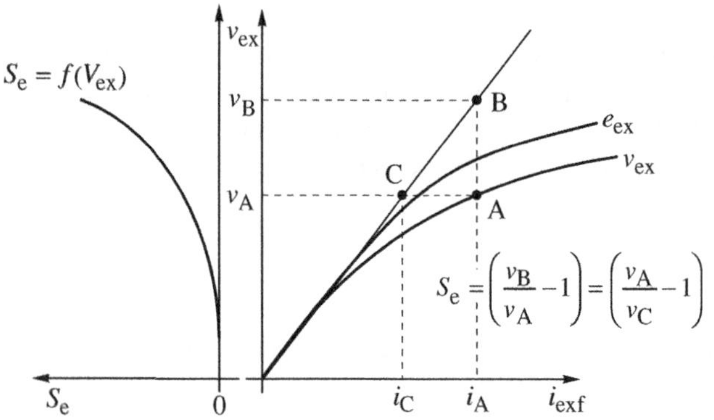

Figure 11.13 Illustration to define the saturation coefficient: e_{ex} is the no-load saturation curve; v_{ex} the constant resistance load saturation curve. The saturation characteristic is shown only around its saturation knee and has been exaggerated for clarity.

The first step in the modeling process is to observe that the slope of the air-gap line in Figure 11.13 is tangential to the linear part of the no-load saturation curve and can be represented by a resistance with the value $R = v_A/i_C$. The exciter field current that corresponds to the nonlinear part of the constant resistance load characteristic can then be expressed using similar triangles as

$$i_A = i_C \frac{v_B}{v_A} = i_C(1 + S_e) = \frac{v_A}{R}(1 + S_e) \tag{11.129}$$

where S_e is a saturation coefficient defined in Figure 11.13 with reference to the constant resistance load saturation curve. This means that any point on the constant resistance load saturation characteristic is defined by

$$i_{exf} = \frac{v_{ex}}{R}(1 + S_e) \tag{11.130}$$

Substituting Eq. (11.130) into Eq. (11.128) and writing

$$L_{exf}\frac{di_{exf}}{dt} = L_{exf}\left[\frac{di_{exf}}{dv_{ex}}\right]_{v_{ex0}}\frac{dv_{ex}}{dt}$$

gives

$$v_R = \frac{R_{exf}}{R}(1 + S_e)v_{ex} + L_{exf}\left[\frac{di_{exf}}{dv_{ex}}\right]_{v_{ex0}}\frac{dv_{ex}}{dt} \tag{11.131}$$

where the derivative term in square brackets is simply the slope of the saturation characteristic at the initial operating point.

To enable the exciter to be easily interfaced with the generator model, the exciter base quantities are defined with the base exciter voltage E_{exb} being that voltage which gives rated open-circuit generator voltage $V_{go/c}$ on the generator air-gap line. This means that E_{exb} and the generator voltage are related by

$$V_{go/c} = \frac{1}{\sqrt{2}}\frac{\omega M_f}{R_f}E_{exb}$$

The base exciter resistance is defined as $R_b = R$ with base current $I_{exfb} = E_{exb}/R_b$. Equation (11.131) can now be expressed in pu notation by dividing through by E_{exb}. It is also convenient to normalize the saturation curve to the base quantities E_{exb} and I_{exfb} when

$$L_{exf}\left[\frac{di_{exf}}{dv_{ex}}\right]_{v_{ex0}} = \frac{L_{exf}}{R}\left[\frac{dI_{exf}}{dV_{ex}}\right]_{V_{ex0}}$$

and Eq. (11.131) becomes

$$V_R = \frac{R_{exf}}{R}(1 + S_e)V_{ex} + \frac{L_{exf}}{R}\left[\frac{dI_{exf}}{dV_{ex}}\right]_{V_{ex0}} \frac{dV_{ex}}{dt} \tag{11.132}$$

Noting that $E_f = V_{ex}$ and writing

$$\frac{L_{exf}}{R}\left[\frac{dI_{exf}}{dV_{ex}}\right]_{V_{ex0}} = T_E$$

gives Eq. (11.132) in normalized form as

$$V_R = (K_E + S_E)E_f + T_E \frac{dE_f}{dt} \tag{11.133}$$

where $K_E = R_{exf}/R$ and $S_E = (R_{exf}/R)S_e$. This equation is shown in block diagram form in Figure 11.14 and comprises an integrating element, with integration time T_E, and two negative feedback loops with gains K_E and S_E, respectively.[3] The negative feedback loop with gain S_E models the saturation in the exciter iron. As the saturation increases so does the value S_E, the magnitude of the negative feedback signal increases, and the influence of the regulator on the exciter voltage, E_f, is reduced. A DC exciter that is separately excited usually operates with $R_f < R$ so that $K_E = (0.8 - 0.95)$. Often, approximated values of $K_E = 1$ and $S_E = S_e$ are assumed in the exciter model. The constant T_E is under 1 s and is often taken to be $T_E \cong 0.5$ s.

If the exciter is self-excited as shown in Figure 11.12b then the voltage v_R is the difference between the exciter internal emf and the excitation voltage v_{ex}. Including this in Eq. (11.128) would give a differential equation that is identical to (11.133) except that $K_E = (R_{exf}/R - 1)$. Consequently, the exciter block diagram in Figure 11.14 is also valid for self-excited exciters. Typically, R_{exf} is slightly less than R so that K_E assumes a small negative value in the range -0.05 to -0.2.

The block diagram of the main part of the excitation system can now be formulated by combining the block diagram of the exciter with that of the regulator and the stabilizing feedback signal as shown in Figure 11.15. The regulator is represented by a first-order transfer function with a time constant T_A and gain K_A. Typical values of these parameters are $T_A = (0.05 - 0.2)$ s and $K_A = (20 - 400)$. The high regulator gain is necessary to ensure small voltage regulation of the order of 0.5%. Unfortunately, although this high gain ensures low steady-state error, when coupled with the length of the time constants the transient performance of the exciter is unsatisfactory. To achieve acceptable transient performance the system must be stabilized in some way that reduces the transient (high-frequency) gain. This is achieved by a feedback stabilization signal represented by the first-order differentiating element with gain K_F and time constant T_F. Typical values of the parameters in this element are $T_F = (0.35 - 1)$ s and $K_F = (0.01 - 0.1)$.

Although the saturation function $S_E = f(E_f)$ can be approximated by any nonlinear function, an exponential function of the form $S_E = A_{ex}e^{B_{ex}E_f}$ is commonly used. As this function must model the saturation characteristic

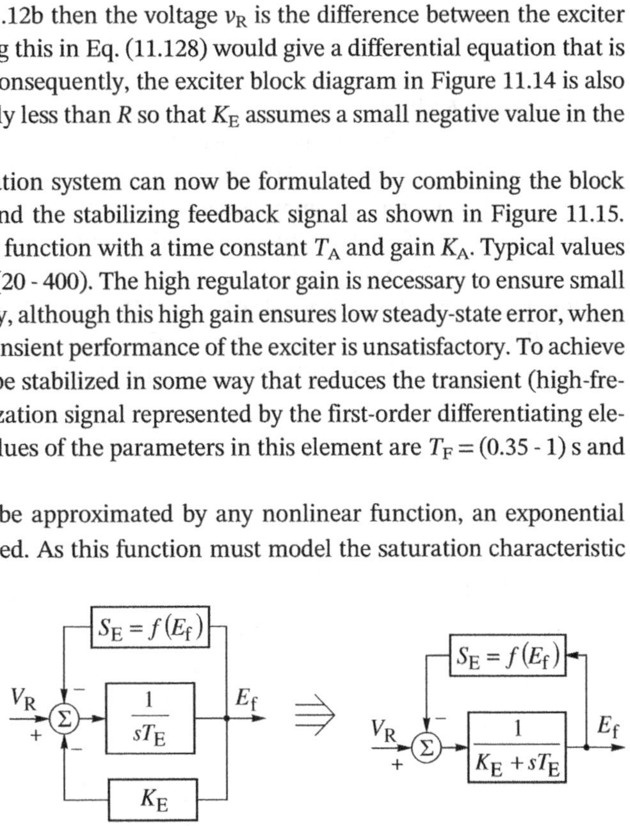

Figure 11.14 Block diagram of the regulated DC exciter.

3 The constant T_E is sometimes called the *exciter time constant*. It is not a time constant. As shown in Figure 11.14 the time constant would depend on K_E and S_E. If saturation is neglected then the effective exciter time constant would be T_E/K_E.

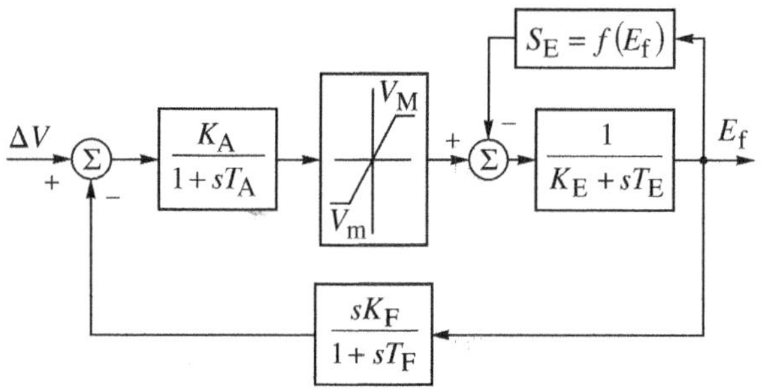

Figure 11.15 Block diagram of the excitation system with a DC exciter. *Source:* Based on IEEE Committee Report (1968).

over a wide range of exciter operating conditions, the parameters A_{ex} and B_{ex} of the exponential function are determined by considering the heavily saturated region of the characteristic corresponding to high excitation voltage and high exciter field current. For example, Anderson et al. (1990) recommend that points on the characteristic corresponding to 100 and 75% of the maximum excitation voltage be used.

It is important to note that the limits on E_f are linked to the regulator limit and the saturation function such that the maximum value E_{fM} is obtained from

$$V_M - (K_E E_{fM} + S_{EM}) = 0. \tag{11.134}$$

11.2.2.2 AC Rotating Exciters

These exciters usually use a three-phase bridge rectifier consisting of six diode modules, as shown in Figure 11.16a. The rectifier is fed from a three-phase AC voltage source of emf V_E and reactance X_E. As with any rectifier system, the output voltage depends on the commutation characteristics of the rectifier as determined by the degree of commutation overlap. As the effect of commutation overlap depends on the current through the rectifiers and the commutating reactance X_E, three main operating states can be identified as a function of field current, as shown in Figure 11.16b. These three states, denoted as I, II, and III on the characteristic, depend on the commutation behavior of the rectifier and allow the voltage at the rectifier terminals to be determined for a given field current I_f and voltage source emf V_E (Witzke et al. 1953; IEEE Committee Report 1981).

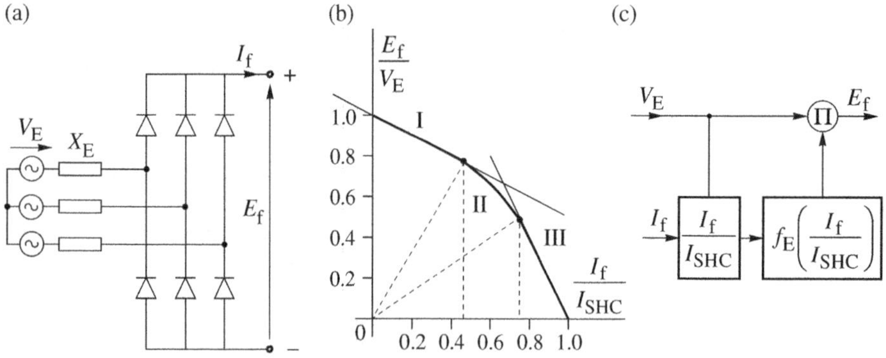

Figure 11.16 Three-phase uncontrolled bridge rectifier: (a) equivalent circuit; (b) voltage–current characteristic; (c) block diagram. V_E and X_E are the emf and the reactance of the voltage source; E_f and I_f are the internal emf and the field current; I_{SHC} is the rectifier short-circuit current. *Source:* Based on IEEE Committee Report (1981).

State I refers to the case where commutation in one branch of the bridge finishes before commutation in another branch starts. During this state, the relationship between the rectifier terminal voltage and the field current is linear and is described by

$$\frac{E_f}{V_E} = 1 - \frac{1}{\sqrt{3}}\frac{I_f}{I_{SHC}} \tag{11.135}$$

where $I_{SHC} = V_E/X_E$ is the rectifier short-circuit current conveniently chosen here to be one pu. This relationship is valid for field currents $I_f < (\sqrt{3}/4)I_{SHC}$.

As the field current increases, commutation overlap also increases and the rectifier reaches state II when each diode can only conduct current when the counter-connected diode of the same phase has finished its conduction. In this state the relationship between the rectifier voltage and the field current is nonlinear and corresponds to a circle with radius $\sqrt{3}/2$ as described by the equation

$$\left(\frac{E_f}{V_E}\right)^2 + \left(\frac{I_f}{I_{SHC}}\right)^2 = \left(\frac{\sqrt{3}}{2}\right)^2 \tag{11.136}$$

This relationship is valid for current

$$\frac{\sqrt{3}}{4}I_{SHC} \leq I_f \leq \frac{3}{4}I_{SHC}$$

As the field current increases further, the rectifier reaches state III where commutation overlap is such that four diodes all conduct at the same time, two in each of the upper and lower arms. In this state the relationship between the rectifier voltage and the field current is linear and is described by the equation

$$\frac{E_f}{V_E} = \sqrt{3}\left(1 - \frac{I_f}{I_{SHC}}\right) \tag{11.137}$$

This relationship is valid for currents $(3/4)I_{SHC} \leq I_f \leq I_{SHC}$.

The block diagram that models the rectifier is shown in Figure 11.16c with all values in pu form. In this diagram the first block on the left calculates the value of the field current relative to the short-circuit current I_{SHC}. The second block calculates the value of the function $f_E(I_f/I_{SHC})$ before passing it to the multiplying element to find the voltage at the terminals of the rectifier.

Figures 2.3b and c show how the rectifier can be fed from either an inductor generator or an inside-out AC generator with the latter being most common as it eliminates the need for slip rings. Although the alternator can be modeled using one of the generator models described in Section 11.1, it is usually sufficiently accurate to simplify the alternator model to be similar to that used to represent the DC exciter. This simplification allows the complete excitation system to be modeled, as shown in Figure 11.17.

As for the DC exciter, the alternator is modeled by an integrating element with three feedback loops. The feedback loops with gain K_E and S_E play the same role as in the DC excitation system. However, the effect of the alternator resistance is now included in the voltage-current characteristics of the rectifier when S_E is determined from the no-load saturation characteristic rather than the load saturation line as used for the DC exciter. As the armature current in the AC exciter is proportional to the current in the main generator field winding, the third feedback loop, with gain K_D, uses this current to model the demagnetizing effect of the armature reaction in the AC exciter. As the output voltage of a diode rectifier cannot drop below zero, this is modeled by the negative feedback loop containing the signal limiter. If the excitation voltage drops below zero a large negative signal is fed back to the summing point that prevents a further drop in the voltage, thereby keeping it at zero.

The system is stabilized by the feedback loop with transfer function $K_G(s) = K_F s/(1 + T_F s)$. In this case the stabilizing loop is supplied by a signal that is proportional to the exciter field current. Alternatively, the system could

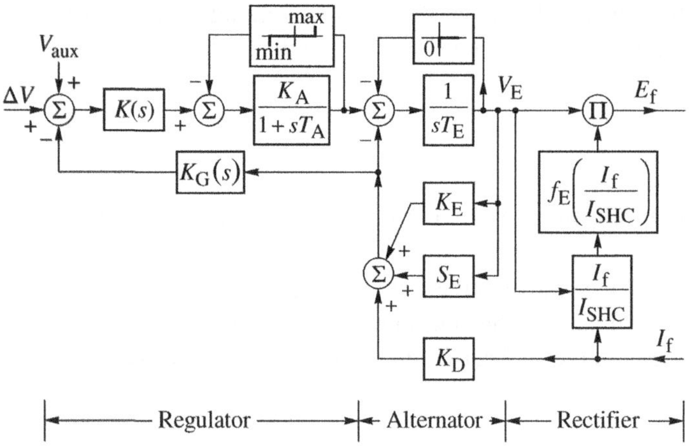

Figure 11.17 Block diagram of the excitation system with AC alternator and the uncontrolled rectifier.

be stabilized by supplying this block directly from the output of the voltage regulator or from the excitation voltage E_f. In this diagram the feedback stabilization is supplemented by an additional block with the transfer function $K(s)$ in the forward path preceding the regulator block. Both $K_G(s)$ and $K(s)$ depend on the specific excitation system and can be implemented by either analogue or digital techniques. Normally, $K(s)$ will have a PI or PID type of structure and is often represented by the transfer function $(1 + sT_C)/(1 + sT_B)$. A major simplification to the model can be made by neglecting the variable effect of the field current on the rectifier voltage. In this case the model reduces to one that is very similar to that shown in Figure 11.15 for the DC exciter.

11.2.2.3 Static Exciters

In static excitation systems the source of the direct current is a controlled three-phase bridge rectifier consisting of six thyristor modules, as shown in Figure 11.18. The output characteristic of the rectifier depends on both the thyristor firing angle α and the system commutation characteristic. In the limiting case $\alpha = 0$ and the output characteristic is similar to that of the uncontrolled rectifier shown in Figure 11.16b. One very important characteristic of the controlled rectifier is the ability to provide a negative exciter output voltage so providing the exciter with *exciter buck* capability. Although the exciter output voltage may go negative, the current cannot and must always flow in the same direction. As the firing angle increases, the output voltage of the controlled rectifier decreases proportionally to cos α (Lander 1987) to give the family of characteristics shown in Figure 11.18.

The firing angle is set by the voltage regulator. The cosinusoidal dependency of the firing angle and the rectifier output voltage can be negated by introducing an inverted cosine function at the regulator output so as to produce a linear relationship. As the slight delays caused by discretization of the firing sequence in each phase are much smaller than the power system time constants, the rectifier can be regarded as a current source with no time delay.

The complete excitation system can then be modeled by the block diagram shown in Figure 11.19. In this figure the regulator and the stabilization element are shown in the upper part of the diagram while the static characteristic of the rectifier is shown in the bottom part of the diagram. The rectifier supply voltage V_E is proportional to both the generator armature voltage and armature current as determined by the constants K_v and K_i. The values of these constants depend on how the rectifier is fed (Figure 2.3d–f). When $K_i = 0$ and $K_v = 1$ the system has no load compensation and corresponds to the rectifier being supplied directly from the terminals of the generator.

The way in which the main generator field current affects the rectifier output voltage is modeled in the same way as the uncontrolled rectifier. The regulator, together with the firing circuits, is modeled by a first-order transfer function with gain K_A and time constant T_A. If the system does not contain cosinusoidal compensation of the voltage-firing angle dependency, the gain K_A will not be constant and should be modeled as a cosine function of the

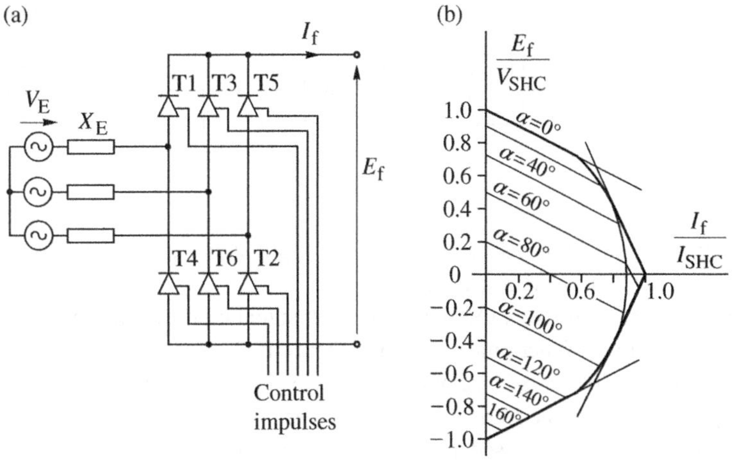

Figure 11.18 Three-phase controlled bridge rectifier: (a) circuit diagram; (b) voltage–current characteristic. α is the firing angle of the thyristors.

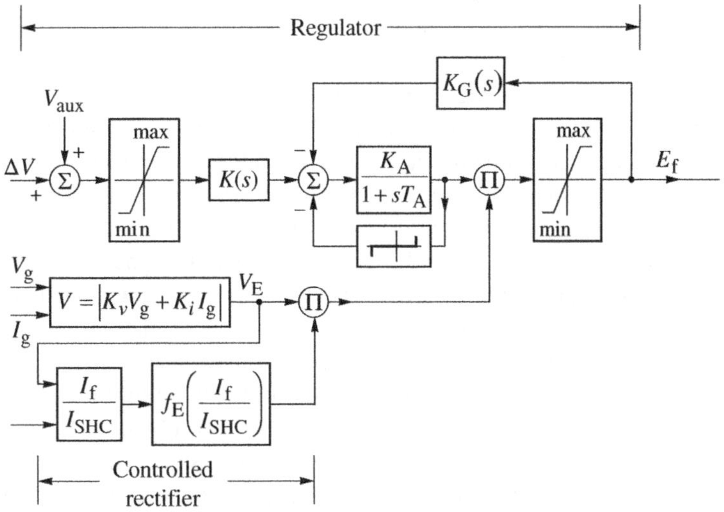

Figure 11.19 Block diagram of the excitation system with a static exciter. *Source:* Based on IEEE Committee Report (1968).

regulator signal. The system is stabilized by the transfer function $K(s)$ in the forward path and by feedback of the exciter output voltage through the block $K_G(s)$. For example, the transfer function $K(s) = K_1(1 + sT_C)/(1 + sT_B)$ and a constant gain $K_G(s) = K_G$ may be used. Although $K(s)$ and $K_G(s)$ can be implemented using digital or analogue methods, digital AVRs are becoming common as they allow more sophisticated functions to be built into the AVR while only software changes are needed between different generators. The exciter output voltage E_f is given by the product of the supply voltage and the regulator output signal, which represents the firing angle. If the influence of the generator field current on the rectifier output voltage is neglected then the exciter block diagram may be simplified to that consisting of the transfer function of the regulator and its stabilization element.

11.2.3 Power System Stabilizer (PSS)

Section 10.1.2 explains how a PSS can help to damp generator rotor oscillations by providing to the excitation system an additional signal that produces a torque component that is in phase with the rotor speed deviation. Dynamic properties of the excitation systems equipped with a PSS are very much dependent on what quantity

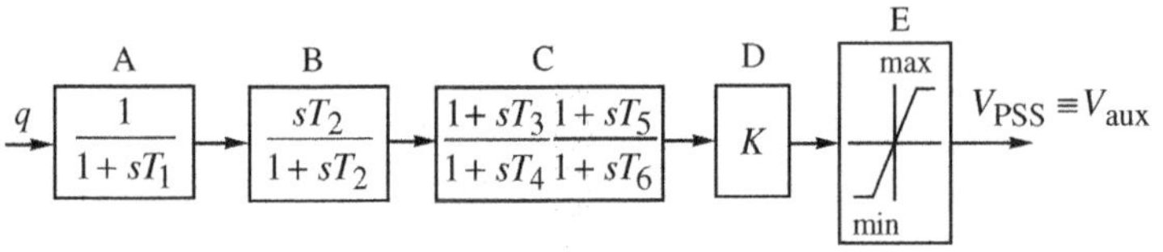

Figure 11.20 Block diagram of the power system stabilizer (PSS): A – signal sensor; B – high-pass filter; C – lead–lag compensation element; D – amplifier; E – limiter.

is the input signal. In practice, PSSs with one input signal or two input signals are used. The most important of them are described in Section 10.1.2.

The general structure of a PSS with one input signal is shown in Figure 10.7. A block diagram of the PSS model corresponding to the functional diagram of Figure 10.7 is shown in Figure 11.20, where the main elements are: the signal sensor A, a low-pass (washout) filter B, a lead–lag compensation elements C, an amplifier D, and a limiter E. The input signal q can be rotor speed, real power, frequency, or some other signal, as described in Section 10.1.2. The output signal V_{PSS} is passed to the AVR as the auxiliary signal V_{aux}. The parameters within the PSS are carefully selected for each PSS, depending on its input signals and location in the system. Typically, $T_1 = 0.02$ s and $T_2 = (2 - 10)$ s. If the input signal is the rotor speed then (as explained in Section 10.1.2) two lead elements are used in which $T_3 > T_4$ and $T_5 > T_6$. Sample data are: $T_3 = T_5 = 0.8$ sand $T_4 = T_6 = 0.1$ s. If the input signal is the active power of the generator then the signal sign changes and one lag element is used, in which $T_3 < T_4$ s. Sample data are: $T_3 = 0.85$ s, $T_4 = 4.75$ s and $T_5 = T_6 = 0$ s.

PSSs with one input signal have significant disadvantages, as discussed in Section 10.1.2. An improvement can be achieved by using more than one input signal. Dual-input stabilizers found wide applications. A block diagram of one of those PSS is shown in Figure 10.11, discussed in Section 10.1.2. A block diagram of another dual-input PSS is shown in Figure 11.21. In this solution the input signal in the first control loop is the frequency of transient electromotive force (emf). In the second control loop the input signal is the real power. Both input signals are calculated on the basis of generator and current voltage. Transient emf is calculated by adding to the generator voltage a voltage drop across the generator transient reactance. The delay caused by the measurement algorithm is represented by a single time constant element. Typical time constant values are $T_{p1} = T_{p2} = 0.02$ s. Increments of the input signals Δf and ΔP are obtained by the use of the washout filters with the frequency characteristics like in Figure 10.8. Typical time constants are: $T_1 = (2 - 5)$ s, $T_2 = (2 - 5)$ s. The increments obtained in this way are subtracted, since according to the description of Figure 10.9 signal ΔP must be taken with a minus sign. Referring to Figure 10.9, it can be said that the phasor of the signal produced by this PSS lies in the third quadrant of the plane, bearing in mind that the angle relative to $\Delta \underline{\omega}$ depends on the proportion between signal Δf and signalΔP, i.e. depends on gains K_1 and K_2.

The specific values of time constants T_1 and T_2 and gains K_1, K_2, and K_3 are selected in the optimization process of the parameters of the control systems.

In practice, other dual-input stabilizers are also used. In order to include in one model the widest possible number of stabilizers types, in computer programs for power system stability analyses, one can use a generic model, as in Figure 11.22. By zeroing some values in this model, one can create various other models.

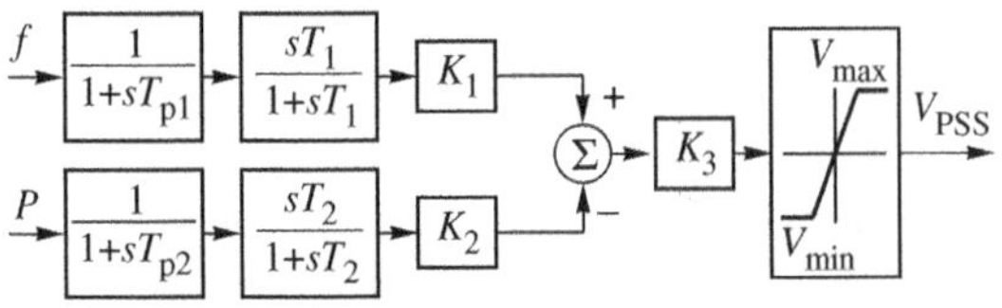

Figure 11.21 Block diagram of a dual-input power system stabilizer (PSS). *Source:* Based on IEEE Committee Report (1996).

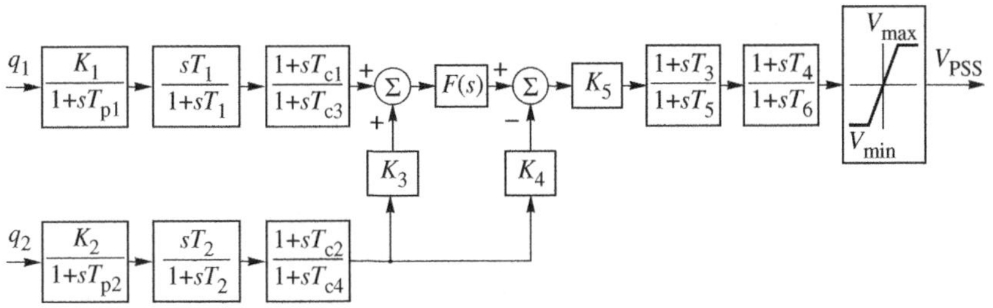

Figure 11.22 Generic block diagram of the dual-input stabilizer. *Source:* Based on IEEE Committee Report (1992).

In the middle part of the block diagram shown in Figure 11.22 there is an element $F(s)$ performing the same role as $G(s)$ in block diagram in Figure 10.11. It is a low-pass filter, which task is to remove frequencies resulting from torsional shaft oscillations. It is assumed that this filter has the following transfer function

$$F(s) = \left[\frac{1 + sT_7}{(1 + sT_8)^m}\right]^n \tag{11.138}$$

Sample values are: $T_7 = 0.1$ s, $T_8 = 0.5$ s, $m = 5$, and $n = 1$.

It is worth noting here that all the above numerical data are examples. For a specific real or benchmark model of EPS the gains and time constants of the AVR and PSS must be selected in the parameter optimization process. One can find extensive information on optimizing PSS parameters included in the book by Gibbard et al. (2015) and also Chapter 15 of this book.

11.3 Turbines and Turbine Governors

Chapter 2 explains how turbines, of one form or another, are used almost exclusively to provide the mechanical input power to the generator. It also explains how the basic governing systems on these turbines work. In this section models of the turbine and its governor are developed so that the input power into the generator model can be regulated in the same way as in the actual system. The turbine models developed are simplified models designed solely to be used for power system analysis rather than for detailed mechanical modeling of the turbine itself. Models of both the mechanical-hydraulic and the electro-hydraulic governor will be developed. As with the AVR, modern generators tend to be fitted with sophisticated digital governors that allow a high degree of functionality. As the structure of these vary between manufacturers, the models developed here are typical of the turbines and governors currently in use. Although the turbine control system allows for both start-up of the turbine and for its control when in operation, it is the latter that is of interest here. A comprehensive discussion on the modeling of both steam and water turbines and their governors can be found in IEEE Committee Reports (1973a–c, 1991, 1992, 1994).

11.3.1 Steam Turbines

Steam turbines are used extensively throughout the world to provide mechanical power to the generator. The mathematical model of the turbine will be derived using a simple model of steam flow through a steam vessel shown in Figure 11.23.

The vessel introduces a time delay into the system as changes in the steam flow at the input take a finite time to appear at the output. This time delay can be quantified by considering a steam vessel of volume V, as shown in Figure 11.23a. In this diagram m is the mass of steam in the vessel, p the steam pressure, and $\dot{m}_1$ and $\dot{m}_2$ the steam

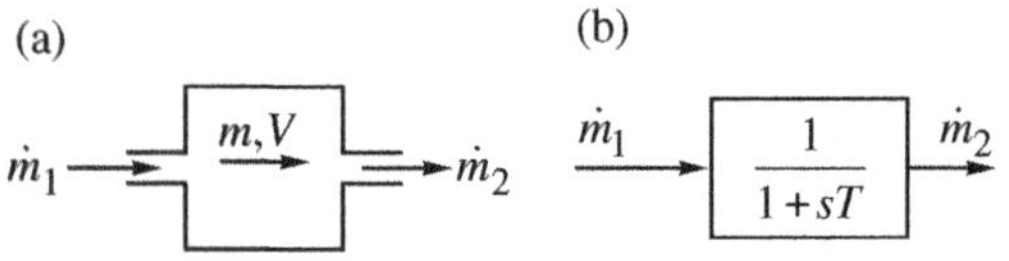

Figure 11.23 Model of the steam flow through a steam vessel: (a) vessel of capacity V; (b) block diagram.

mass flow rates at the input and output. The mass of the steam in the vessel is constant when $\dot{m}_1 = \dot{m}_2$. When the steam input flow rate changes, because of a change in the valve position, the mass of steam in the vessel will change at a rate proportional to the difference between the input flow rate and the output flow rate, that is $dm/dt = (\dot{m}_1 - \dot{m}_2)$. If the steam temperature is constant then the change of mass in the vessel must result in a pressure change when this equation can be written as

$$\dot{m}_1 - \dot{m}_2 = \frac{dm}{dt} = \frac{\partial m}{\partial p}\frac{dp}{dt} = V\frac{\partial}{\partial p}\left(\frac{1}{v}\right)\frac{dp}{dt} \tag{11.139a}$$

where v is the steam specific volume at a given pressure (volume divided by mass). Assuming that the outflow of steam is proportional to the pressure in the vessel

$$\dot{m}_2 = \dot{m}_0\frac{p}{p_0} \quad \text{or} \quad \frac{dp}{dt} = \frac{p_0}{\dot{m}_0}\frac{d\dot{m}_2}{dt} \tag{11.139b}$$

where $\dot{m}_0 = \dot{m}_1(t=0) = \dot{m}_2(t=0)$ and $p_0 = p(t=0)$. Substituting (11.139b) into (11.139a) gives

$$\dot{m}_1 - \dot{m}_2 = T\frac{d\dot{m}_2}{dt} \tag{11.140}$$

where $T = V\frac{p_0}{\dot{m}_0}\frac{\partial}{\partial p}\left(\frac{1}{v}\right)$ is a time constant corresponding to the inertia of the mass of steam in the vessel. Applying Laplace transforms and writing in transfer function form gives

$$\frac{\dot{m}_2(s)}{\dot{m}_1(s)} = \frac{1}{(1+Ts)} \tag{11.141}$$

which corresponds to the inertia block in Figure 11.23b.

Figure 11.24 shows how the above equation can be used to model the tandem compound single reheat turbine described in Section 2.3.3. A schematic diagram of this arrangement, Figure 11.24a, shows how the steam passes through the governor control valves and the inlet piping to the HP steam chest (A). On leaving the steam chest it passes through the HP turbine before entering the reheat stage between the HP and IP steam turbines. After being reheated the steam passes through the intercept valves to the IP turbine and then through the crossover piping to the LP turbine. The mathematical model of such a system is shown in Figure 11.24b and can be divided into two parts. First of all, as the power extracted by a turbine is proportional to the steam mass flow rate $\dot{m}_s$, each turbine stage can be modeled by a constant α, β, γ which corresponds to the portion of the total turbine power developed in the different turbine stage with $\alpha + \beta + \gamma = 1$. As turbine power is proportional to the steam mass flow rate, the input flow rate is changed in this diagram to input power to simplify the diagram further. If the generator MVA base is used then these values must be modified in the ratio of the two bases.

The second part of the turbine model concerns the storage of the steam chests and associated piping and corresponds to Figure 11.23b. Typical values of the parameters for the single-reheat turbine shown in Figure 11.24 are: $T_A = (0.1 - 0.4)$ s, $T_B = (4 - 11)$ s, $T_C = (0.3 - 0.5)$ s, $T_D = (0.3 - 0.5)$ s, $\alpha = 0.3$, $\beta = 0.4$, and $\gamma = 0.3$.

Assuming that there is no control of the intercept valves, the block diagram of Figure 11.24b can be simplified to that shown in Figure 11.24c by combining the four inertia blocks to give an equivalent block with the combined parameters

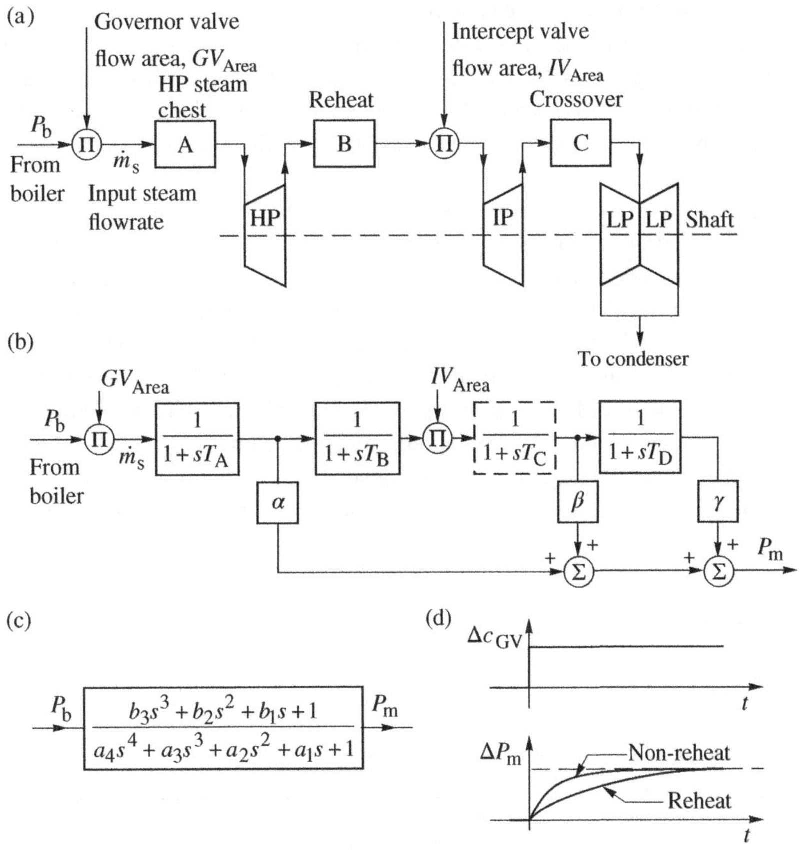

Figure 11.24 Tandem compound single-reheat turbine: (a) schematic diagram; (b) block diagram; (c) transformed block diagram; (d) response of the linear turbine model to a step change in valve position. Symbols P_i and P_m are the power of the inlet steam and of the turbine.

$$b_3 = \alpha T_B T_C T_D;\quad b_2 = \alpha\,(T_B T_C + T_B T_D + T_C T_D);\quad b_1 = \alpha(T_B + T_C + T_D) + \beta T_D;$$
$$a_4 = T_A T_B T_C T_D;\quad a_3 = T_A T_B T_C + T_A T_B T_D + T_A T_C T_D + T_B T_C T_D;$$
$$a_2 = T_A T_B + T_A T_C + T_A T_D + T_B T_C + T_B T_D;\quad a_1 = T_A + T_B + T_C + T_D$$

The reheat time constant T_B is several times higher than the time constant T_C of the intermediate-pressure stage ($T_C << T_B$), and therefore the inertia block drawn with dashed line is often neglected. Moreover, taking into account that also $T_A << T_B$ and $T_D << T_B$, the turbine model can be further simplified by assuming $T_A \cong T_C \cong T_D \cong 0$, reducing the turbine model to a first-order block with time constant T_B.

A steam turbine without reheat can be modeled by a first-order block with time constant $T_A = (0.2 - 0.5)$ s and with $\alpha = 1$, $\beta = \gamma = 0$, and $T_B = T_C = T_D = 0$.

A comparison of the time response of a turbine with and without reheat with an incremental step increase in the opening of the governor valve Δc_{GV} is shown in Figure 11.24d. When the valve opening increases so does the valve flow area GV_{Area} so that for a constant boiler pressure the steam flow rate also increases. However, steam storage effects introduce a delay before the increased steam flow rate can reach the turbine blades and increase the output power. This effect is particular noticeable for a turbine with reheat when the rise time may be as long as 10 s as shown in Figure 11.24d. As this delay between activating the valve and the power changing can be excessively long in reheat turbines, coordinated control of both the governor and the interceptor valves is necessary, as explained in the next section.

11.3.1.1 Governing System for Steam Turbines

Chapter 2 explains that the reheat turbine shown in Figure 11.24a is fitted with two sets of control valves and two sets of emergency stop valves. Normally, turbine control is accomplished by regulating the position of the governor valves and the intercept valves, while the emergency stop valves are kept fully open and only used in an emergency to stop the turbine. As both sets of stop valves are normally kept fully open, they can be neglected for modeling purposes.

Although the way in which control of the governor valves and the intercept valves is coordinated depends on the purpose of the control action, the type of governor and the manufacturer concerned, the generic features of the governor can be included in a general model capable of representing both the mechanical-hydraulic and the electrohydraulic governor. This model allows for overspeed control and load/frequency control, as described in Chapter 9. The fast-valving features described in Chapter 10 can be readily added if necessary.

In order to develop this governor model the functional diagram of the mechanical-hydraulic governor shown in Figure 11.25a will be used. This diagram shows the main elements of a mechanical-hydraulic governor as the speed transducer, regulator, speed relay, servomotor, and the steam control valves. Compared with Figure 2.12 an additional element, the speed relay, is shown that develops an output proportional to the load reference signal less any contribution due to speed deviation. This additional element is required because on some larger generators the

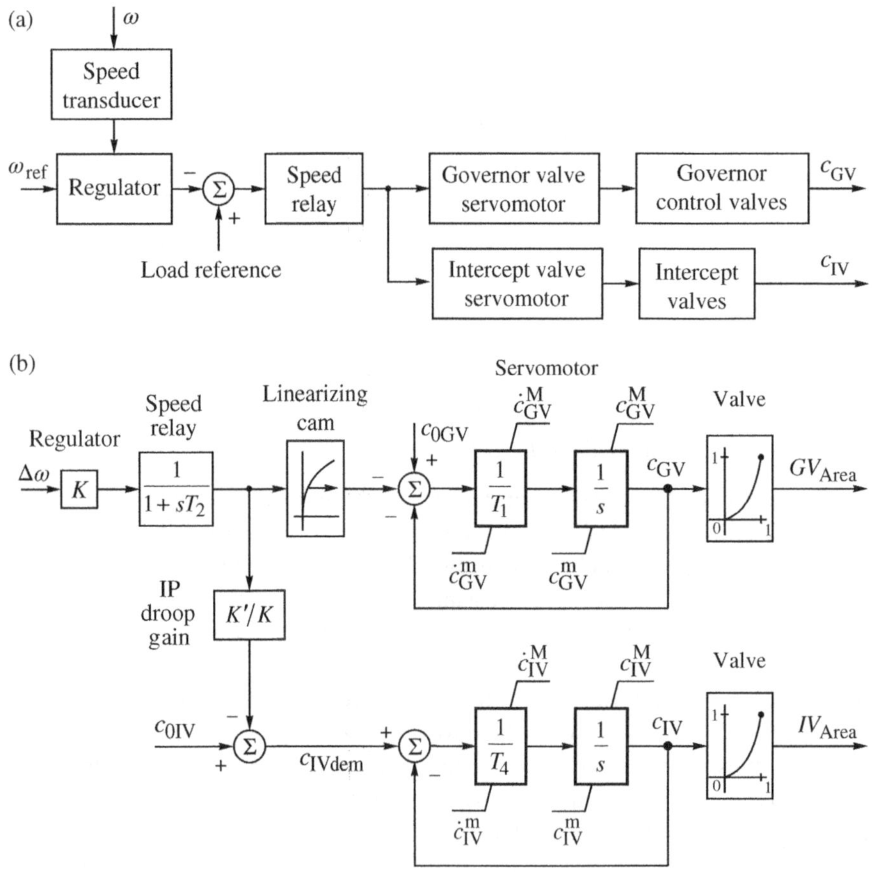

Figure 11.25 Mechanical-hydraulic steam turbine governor of the steam turbine: (a) functional diagram; (b) block diagram; (c) simplified block diagram. Symbols: c – main steam valve position; GV –governor valves; IV – intercept valves.

force required to adjust the position of the main steam valves is very large and an additional spring-loaded servomotor, the speed relay, is required between the speeder linkage and the main servomotor. On some smaller machines the spring-loaded servomotor of the speed relay is used to adjust the position of the control valves rather than the servomotor arrangement shown in Figure 2.12.

The elements in this functional diagram are modeled as shown in the block diagram of Figure 11.25b. The regulator is modeled by a gain $K = 1/\rho$, where ρ is the droop of the regulator static characteristic, while the speed relay is modeled by a first-order lag with time constant T_2. The effect of the load reference has been modeled as the required opening of the main governor control valves c_{OGV}. The main servomotor, which alters the position of the control valves, is modeled by an integrator with direct feedback of valve position along with two limiters. The first limiter is necessary to protect the turbine from rapid opening or closing of the steam valves. The second limiter limits the valve position between fully open and fully closed, or to some set position if a load limiter is present. The final, nonlinear block models the valve characteristics, effectively converting the valve position to normalized flow area. The nonlinearity of the valve characteristic may be compensated in the governor by a linearizing cam located between the speed relay and the servomotor. These two mutually compensating nonlinearities are often neglected in the model. The output of the model is the normalized valve flow area and appears as the main input to the turbine model in Figure 11.24b.

One way in which the intercept valves can be controlled to limit overspeed is also shown in Figure 11.25b. In this example the position of the intercept valves is controlled to follow a demand position signal c_{IVdem} generated according to the equation

$$c_{IVdem} = c_{OIV} - K'\Delta\omega \tag{11.142}$$

where K' is the IP gain and is the inverse of the IP droop $K' > K$.

To examine in further detail how this control loop works, assume that $K = 25$ (main droop 4%) and that $K' = 50$ (IP droop 2%) and that it is necessary to keep the intercept valves fully open until the overspeed is 4%. At this speed the main governor valves will just have fully closed. The reference signal c_{OIV} is set to 3. With $0.04 > \Delta\omega > 0$ then the demand $c_{IVdem} \geq 1$ and the intercept valves are kept fully open. When $\Delta\omega = 0.04$, the intercept valves start to close and are fully closed when $\Delta\omega = 0.06$.

In some cases an auxiliary governor is used that acts in parallel with the normal regulator. This auxiliary governor only comes into operation when large speed deviations are registered and has the effect of increasing the regulator gain K. In this way the regulator output is increased to close quickly both the interceptor valves and the governor control valves.

The governor model can also be used to model the electrohydraulic governor because the main power components are similar, except that now the mechanical components have been replaced by electronic circuits. Such changes provide a greater degree of flexibility and functionality than can be achieved in the traditional mechanical-hydraulic governor. A further advance is the digital-hydraulic governor where the control functions are implemented in software, so providing even greater flexibility. From a modeling point of view this increased flexibility means that the actual governor control used is manufacturer dependent, particularly with regards to the more advanced features, such as fast valving. Nevertheless, the governor model developed in Figure 11.25b with the speed relay replaced by the general transfer function $(1 + sT_3)/(1 + sT_2)$ to provide any phase compensation forms a good basis for modeling an electrohydraulic governor. If a particular control logic is to be implemented then it is relatively straightforward to modify Figure 11.25b accordingly.

When analyzing electromechanical dynamics in a time interval of around 10 s the boiler pressure can be assumed to be constant and can be equated to the initial power P_0 so that $P_b = P_m(t = 0) = P_0 = \text{const}$. With this assumption the governor block diagram can be simplified to that shown in Figure 11.26a by multiplying the limits, and the initial conditions, in the servomotor model by P_0. In this model the regulator and the speed relay transfer functions have been replaced by a general control block $K(1 + sT_3)/(1 + sT_2)$ to enable this model to be used for

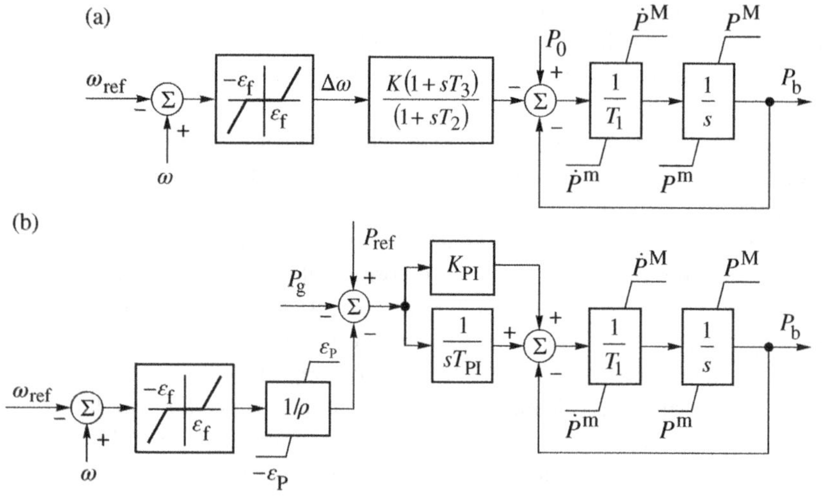

Figure 11.26 Simplified governor models.

both mechanical-hydraulic and electromechanical governors. The additional time constant now allows both types of governor to be modeled by correct choice of values. Intercept valve control is not shown on this diagram as it assumes that they are kept fully open but could be added if required.

Some types of steam turbine governors use PI or PID type regulators. In such a case the simplified governor model takes a form as presented Figure 11.26b. The control error results here both from active power error and from speed (frequency) error. The model is valid with the assumption that boiler steam pressure P_0 is constant. The governor model consists of the PI type regulator, the nonlinear block forming speed-droop characteristic, and the valves servomotor. Parameter ε_f defines the regulator dead zone for frequency input, while ε_P defines limits of the maximum and minimum power deviation. For inputs to the regulator expressed in pus, parameter ρ defines the droop of the regulator static characteristic. These three parameters allow the generating unit speed-droop characteristic to be shaped, as presented in Figure 9.6.

Typical values of the governor parameters for structures presented in Figure 11.26b are: $K = (0.25 - 5)$, $T_1 = (0.1 - 1)$ s, $T_2 = (0 - 1)$ s, $T_3 = (0.2 - 0.25)$ s, $\dot{P}^M = (0.05 - 0.3)$ 1/s, $\dot{P}^m = -(0.05 - 0.3)$ 1/s, $P^M = (1.0 - 1.1)$ pu, $P^m = 0$ pu, $K_{PI} = 1$, and$T_{PI} = (1 - 5)$ s. The governor's power measurement transducer, not indicated in the figure, is modeled by first-order lag block with time constant $T_P \leq 1$ s. The high value of this time constant prevents an undesirable reaction of the governor to active power variation, e.g. during short-circuit.

Other forms of electrohydraulic and mechanical-hydraulic governors are described in IEEE Committee Reports (1973b, 1991) and Kundur (1994).

11.3.1.2 Boiler Control

In the modeling process the boiler output is linked to the turbine via the boiler pressure parameter P_b. It is not the purpose here to describe boiler models, but a brief discussion of boiler/turbine interaction is necessary because throughout this book it is assumed that P_b effectively remains constant. Traditionally, conventional turbine control operated in a "boiler follows turbine" mode where all load changes are carried out by control of the turbine valves with the boiler controls operating to maintain constant steam conditions, that is pressure and flow. Although this control mode allows rapid turbine response and good frequency control by utilizing the stored energy in the boiler, the changes in the boiler pressure and other variables can be quite large. To prevent this an alternative "turbine follows boiler" control mode can be used when all load changes are made via the boiler controls. The turbine valves are controlled to regulate the boiler pressure. As the speed of response of the turbine valves is very fast, almost perfect pressure control can be achieved. However, load changes are now very slow with time constants of

1–2 min, depending on the type of boiler and the fuel, since now no use is made of the stored energy in the boiler. As a compromise, "integrated boiler and turbine" control may be used to achieve both a quick turbine response while limiting the changes in the boiler variables.

11.3.2 Hydraulic Turbines

If water flows from a high level to a low level through a hydraulic turbine, the potential energy of the water stored in a high reservoir is converted into mechanical work on the turbine shaft. The turbine may be either an impulse turbine or a reaction turbine and, although the way in which they operate hydraulically differs, the work done by both types is entirely due to the conversion of kinetic energy. In impulse turbines, such as the Pelton wheel, all the available energy in the water is converted into kinetic energy as the water passes through the nozzle. The water forms a free jet as it leaves the nozzle and strikes the runner where the kinetic energy is converted into mechanical work. In the reaction turbine, such as the Francis turbine, only a part of the energy in the water is converted into kinetic energy as the water passes through the adjustable wicket gates with the remaining conversion taking place inside the runner itself. All passages are filled with water, including the draft tube. In both turbines the power is controlled by regulating the flow into the turbine by wicket gates on the reaction turbine and by a needle, or spear, on the impulse turbine. What is required is a mathematical description of how the turbine power changes as the position of the regulating device is changed.

Figure 11.27a shows a schematic diagram of a turbine installation where water flows down the penstock and through the turbine before exiting into the tailwater. The penstock is modeled by assuming that the flow is incompressible when the rate of change of flow in the penstock is obtained by equating the rate of change of momentum of the water in the penstock to the net force on the water in the penstock when

$$\rho L \frac{\mathrm{d}Q}{\mathrm{d}t} = F_{\text{net}} \tag{11.143}$$

where Q is the volumetric flow rate, L the penstock length, and ρ the mass density of water.

The net force on the water can be obtained by considering the pressure head at the conduit. On entry to the penstock the force on the water is simply proportional to the static head H_s, while at the wicket gate it is proportional to the head H across the turbine. Owing to friction effects in the conduit, there is also a friction force on the water represented by the head loss H_l so that the net force on the water in the penstock is

$$F_{\text{net}} = (H_s - H_l - H)A\rho g \tag{11.144}$$

where A is the penstock cross-sectional area and g the acceleration due to gravity. Substituting the net force into Eq. (11.143) gives

$$\rho L \frac{\mathrm{d}Q}{\mathrm{d}t} = (H_s - H_l - H)A\rho g \tag{11.145}$$

It is usual to normalize this equation to a convenient base. Although this base system is arbitrary, the base head h_{base} is taken as the static head *above* the turbine, in this case H_s, while the base flow rate q_{base} is taken as the flow rate through the turbine with the gates fully

(a)

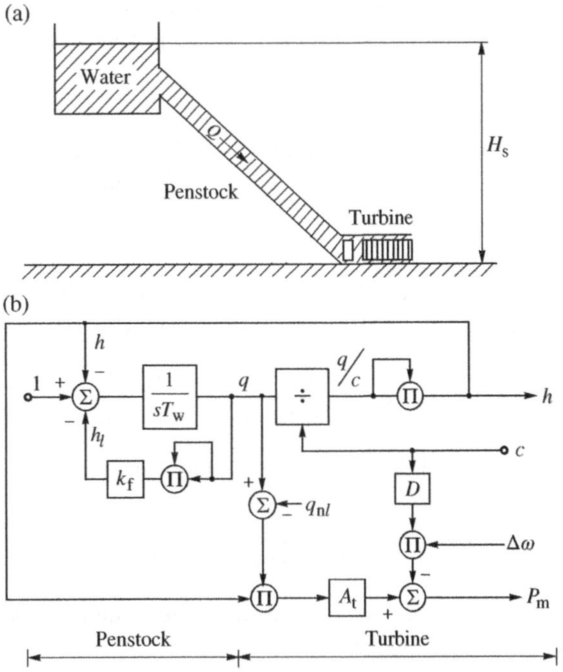

Figure 11.27 Hydraulic turbine: (a) schematic diagram; (b) nonlinear model. *Source:* Based on IEEE committee Report (1968).

open and the head at the turbine equal to h_{base} (IEEE Committee Report 1992). Dividing both sides of Eq. (11.145) by $h_{base}q_{base}$ gives

$$\frac{dq}{dt} = \frac{1}{T_w}(1 - h_l - h) \tag{11.146}$$

where $q = Q/q_{base}$ and $h = H/h_{base}$ are the normalized flow rate s and pressure heads, respectively, and $T_w = \frac{Lq_{base}}{Agh_{base}}$ is the *water starting time*. Theoretically, T_w is defined as the time taken for the flow rate in the penstock to change by a value equal to q_{base} when the head term in brackets changes by a value equal to h_{base}. The head loss h_l is proportional to the flow rate squared and depends on the conduit dimensions and friction factor. It suffices here to assume that $h_l = k_f q^2$ and can often be neglected. This equation defines the penstock model and is shown in block diagram form in the left-hand part of Figure 11.27b.

In modeling the turbine itself both its hydraulic characteristics and mechanical power output must be modeled. First of all, the pressure head across the turbine is related to the flow rate by assuming that the turbine can be represented by the valve characteristic

$$Q = kc\sqrt{H} \tag{11.147}$$

where c is the gate position between 0 and 1 and k is a constant. With the gate fully open $c = 1$ and this equation can be normalized by dividing both sides by $q_{base} = k\sqrt{h_{base}}$ to give, in pu form,

$$q = c\sqrt{h} \tag{11.148}$$

Secondly, the power developed by the turbine is proportional to the product of the flow rate and the head and depends on the efficiency. To account for the turbine not being 100% efficient the no-load flow q_{nl} is subtracted from the actual flow to give, in normalized parameters,

$$P_m = h(q - q_{nl}) \tag{11.149}$$

Unfortunately, this expression is in a different pu system to that used for the generator whose parameters are normalized to the generator MVA base so that Eq. (11.149) is written as

$$P_m = A_t h(q - q_{nl}) \tag{11.150}$$

where the factor A_t is introduced to account for the difference in the bases. The value of the factor A_t can be obtained by considering the operation of the turbine at rated load when

$$P_m = A_t h_r(q_r - q_{nl}) = \frac{\text{turbine power (MW)}}{\text{generator MVA rating}} \tag{11.151}$$

and the suffix r indicates the value of the parameters at rated load. Rearranging Eq. (11.151) gives

$$A_t = \frac{\text{turbine power (MW)}}{\text{generator MVA rating}} \frac{1}{h_r(q_r - q_{nl})} \tag{11.152}$$

A damping effect is also present that is dependent on gate opening so that at any load condition the turbine power can be expressed by

$$P_m = A_t h(q - q_{nl}) - Dc\,\Delta\omega \tag{11.153}$$

where D is the damping coefficient. Equations (11.148) and (11.153) constitute the turbine nonlinear model shown on the right of Figure 11.27b where the wicket gate position is the control variable.

11.3.2.1 Linear Turbine Model

The classical model of the water turbine (IEEE Committee Report 1973a–c) uses a linearized version of the non-linear model. Such a model is valid for small changes of mechanical power and can be obtained by linearizing Eqs. (11.146), (11.148), and (11.150) about an initial operating point to give

$$\frac{d\Delta q}{dt} = -\frac{\Delta h}{T_W}; \quad \Delta q = \frac{\partial q}{\partial c}\Delta c + \frac{\partial q}{\partial h}\Delta h; \quad \Delta P_m = \frac{\partial P_m}{\partial h}\Delta h + \frac{\partial P_m}{\partial q}\Delta q \tag{11.154}$$

Introducing the Laplace operator s and eliminating Δh and Δq from the equations gives

$$\frac{\Delta P_m}{\Delta c} = \frac{\left[\frac{\partial q}{\partial c}\frac{\partial P_m}{\partial q} - sT_W\frac{\partial P_m}{\partial h}\frac{\partial q}{\partial c}\right]}{1 + sT_W\frac{\partial q}{\partial h}} \tag{11.155}$$

where the partial derivatives are

$$\begin{aligned} \frac{\partial q}{\partial h} &= \frac{1}{2}\frac{c_0}{\sqrt{h_0}}; & \frac{\partial q}{\partial c} &= \sqrt{h_0} \\ \frac{\partial P_m}{\partial q} &= A_t h_0; & \frac{\partial P_m}{\partial h} &= A_t(q_0 - q_{nl}) \cong A_t(q_0) \end{aligned} \tag{11.156}$$

and the suffix 0 indicates an initial value. Substituting into Eq. (11.155) and noting that $q_0 = c_0\sqrt{h_0}$ gives

$$\frac{\Delta P_m}{\Delta c} = A_t h_0^{3/2}\frac{1 - sT'_W}{1 + s\frac{T'_W}{2}} \tag{11.157}$$

where

$$T'_W = T_W\frac{q_0}{h_0} = \frac{L}{Ag}\frac{Q_0}{H_0}$$

Typically, T'_W is between 0.5 and 5 s.

This is the classic definition of water starting time but is dependent on the values of the head and flow rate at the linearization point. It therefore varies with load. If required, the constant A_t can be absorbed into the gate position when it effectively converts the gate opening to pu turbine power on the generator MVA base. The block diagram of this linear turbine model is shown in Figure 11.28a.

Equation (11.157) describes an interesting and important characteristic of water turbines. For example, suppose that the position of the gate is suddenly closed slightly so as to reduce the turbine power output. The flow rate in the penstock cannot change instantaneously so the velocity of the flow through the turbine will initially increase. This increase in water velocity will produce an initial increase in the turbine power until, after a short delay, the flow rate in the penstock has time to reduce when the power will also reduce. This effect is reflected in Eq. (11.157) by

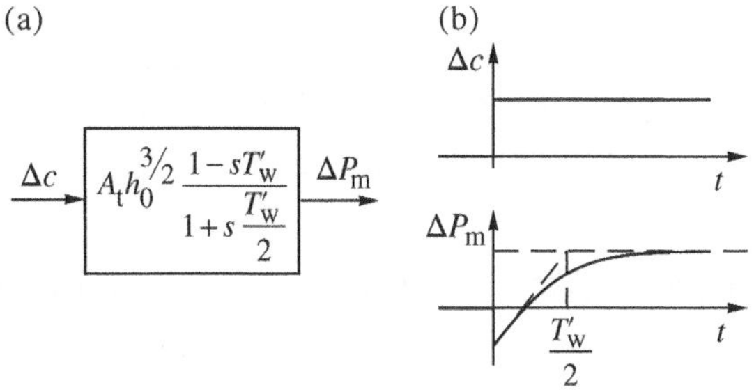

Figure 11.28 Hydraulic turbine: (a) linear model; (b) response of the linear turbine model to a step change in gate position.

the minus sign in the numerator. This characteristic is shown in Figure 11.28b, where a step increase in the gate position Δc initially produces a rapid drop in power output. As the flow rate in the penstock increases the power output increases.

Although the linearized model (11.157) has been successfully used in both steady-state and transient stability studies, IEEE Committee Report (1992) recommends the use of the nonlinear turbine model in power system studies because its implementation using computers is no more difficult than the approximate linear transfer function. Other, more detailed models are discussed in IEEE Committee Reports (1973a–c, 1992).

11.3.2.2 Governing System for Hydraulic Turbines

Governing systems for hydraulic turbines differ from those used in steam turbines in two main ways. First of all, a very high force is required to move the control gate as it must overcome both high water pressure and high friction forces and, secondly, the peculiar response of the hydraulic turbine to changes in valve position must be adequately compensated. To provide the necessary force to move the gate two servomotors are used as shown in the functional diagram in Figure 11.29a. In a similar way as in the steam turbine, the speed regulator acts through a system of levers on the *pilot valve* which controls the flow of hydraulic fluid into the *pilot servomotor*. The pilot servomotor

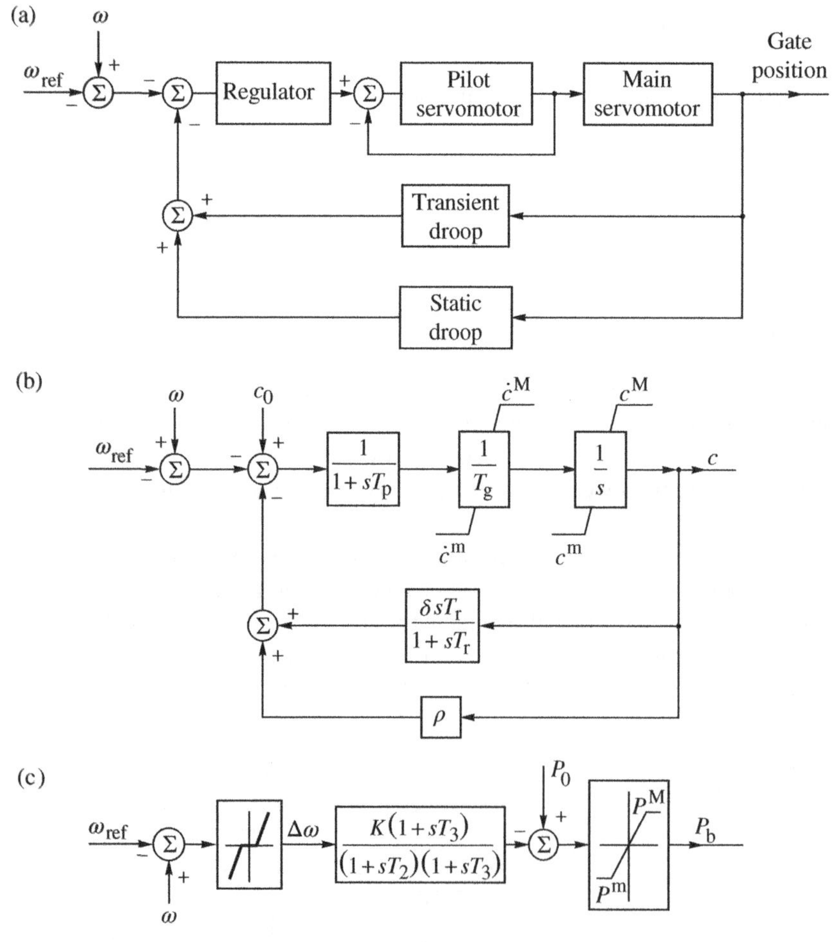

Figure 11.29 Block diagram of hydraulic turbine governing system: (a) functional diagram; (b) full diagram; (c) simplified diagram.

then acts on the *relay valve* of the very high-power *main servomotor*, which controls the gate position. Just like the servomotor on the steam turbine, negative feedback of the position of both the servomotors is necessary to achieve the required movement. To compensate for the peculiar response of the water turbine to changes in the gate position it is necessary to slow down the initial gate movement to allow the water flow in the penstock to catch up. This is achieved by the *transient droop* element which reduces the gain of the governor for fast changes in valve position and, in mechanical-hydraulic governor systems, is achieved by feeding back the position of the gate via a system of levers that includes a dashpot system. Similar to the steam turbine governor shown in Figure 2.11, direct feedback of the gate position through a series of levers controls the static droop.

The block diagram model of the governor, including the transient droop, is shown in Figure 11.29b. The main servomotor is modeled by an integrating element, with integration time T_g, and two limiters. The first limiter limits the gate position between fully open and fully closed, while the second, the rate limiter, limits the rate at which the gate can be moved. This is necessary because, if the gate is closed too rapidly, the resulting high pressure could damage the penstock. The pilot servomotor with its position feedback is modeled by a first-order lag with a time constant T_p. The system has two main feedback loops. The proportional feedback loop provides the static droop characteristic equal to ρ, while the feedback loop with the differentiating element corrects the transient droop to a value δ. Typical values of the parameters recommended by Ramey and Skooglund (1970) are: $T_p = 0.04$ s, $T_g = 0.2$ s, $T_r = 5T'_w$, $\delta = 2.5\,T'_w/T_m$, and $\rho = 0.03$ to 0.06, where T'_w is the water starting time and T_m is the mechanical time constant of the unit.

If the nonlinearities introduced by the limiters are, for the moment, neglected, the system may be described by the third-order transfer function

$$\frac{\Delta c}{\Delta \omega} = \frac{\dfrac{(1 + T_r s)}{\rho}}{\dfrac{T_p T_r T_g}{\rho} s^3 + \dfrac{(T_p + T_r) T_g}{\rho} s^2 + \dfrac{T_g + T_r(\rho + \delta)}{\rho} s + 1} \tag{11.158}$$

As the time constant T_p is several times smaller than the time constants T_g and T_r it may be neglected to give the second-order transfer function

$$\frac{\Delta c}{\Delta \omega} = \frac{(1 + T_3 s)K}{(1 + T_2 s)(1 + T_4 s)} \tag{11.159}$$

where $K = 1/\rho$, $T_2 \cong T_r T_g/[T_g + T_r(\rho + \delta)]$, $T_3 = T_r$, and $T_4 = [T_g + T_r(\rho + \delta)]/\rho$. Usually, $T_4 \gg T_2$ when $T_4 + T_2 \cong T_4$.

If the gate limiters are now added to the transfer function, the simplified governor block diagram shown in Figure 11.29c is obtained and is similar to the simplified system used to represent the steam turbine governor.

Typical parameter values for use in the simplified block diagram of the steam and hydraulic turbine governors are tabulated in Table 11.2.

Table 11.2 Typical values of parameters of turbine governing systems.

	Parameters				
Type of turbine	ρ	T_1	T_2	T_3	T_4
		s	s	s	s
Steam	0.02–0.07	0.1	0.2–0.3	0	—
Hydraulic	0.02–0.04	—	0.5	5	50

11.3.3 Gas-steam Combined-cycle Power Plants

Example of the functional diagram of a gas-steam combined-cycle turbine is shown in Figure 2.11 and described in Section 2.3.3. Owing to their advantages in overall plant efficiency and lower emissions (as compared to conventional coal fired power plants) combined-cycle power plants (CCPP) have gained in popularity and account for a significant portion of the generation in many power systems around the world.

The controls of a CCPP are quite complex. Load control and frequency response of a CCPP are handled by the plant load control system. It receives a load set-point signal from the automatic generation control (AGC) system (Figure 9.12) and determines how the gas turbine should be loaded. The steam turbine, being essentially in sliding pressure mode, follows the gas turbine output with a delay of several minutes. This is illustrated in Figure 11.30 by an example of the dynamic response to a step increase in the load/speed reference set-point of a typical CCPP connected to a large power grid. The gas turbine output increases until it is limited by the temperature control, transiently over-shooting its steady-state maximum power limit. The steam turbine load adjusts automatically with a delay dependent on the response of the heat-recovery boiler (HRB) also referred to as the heat-recovery steam generator (HRSG).

Approximately two-thirds of the generation capacity in CCPPs is produced by gas turbines. Gas turbines and their controls are significantly different from fossil-fuel steam-turbine power plants. In particular, the maximum power output of a gas turbine is highly dependent on the environmental ambient conditions, because the gas turbine thermal cycle is an open cycle using atmospheric air as its working fluid. The maximum power output of the turbine is also dependent on the deviation of its operating frequency from its rated value. The overall block diagram of gas turbine control is shown in Figure 11.31.

The start-up and shutdown controls are key to ramping the unit up during start-up and down during shutdown. These controls are irrelevant to analyses of power system dynamics.

The acceleration control is used primarily during gas turbine start-up to limit the rate of rotor acceleration prior to reaching governor speed, thus ameliorating the thermal stresses. This control serves a secondary function during normal operation, in that it acts to reduce fuel flow and limit the tendency to overspeed in the event that the generating unit separates from the system. The acceleration control can typically be ignored for power system studies for large interconnected systems. However, for islanding studies, smaller power systems with large frequency variations, the acceleration control loop may need to be considered.

Of particular relevance to power system analyses is the speed/load control, which is the main control loop associated with the speed governor. When the generating unit is connected to a power grid the speed/load control operates as the load controller with the speed-droop characteristic similar to that shown in Figure 9.6 and described in Section 9.1. The load error is formed by the difference between the set-point (reference) power and the actual generator load. This error is corrected by the speed error multiplied by a gain equal to the reciprocal of the droop.

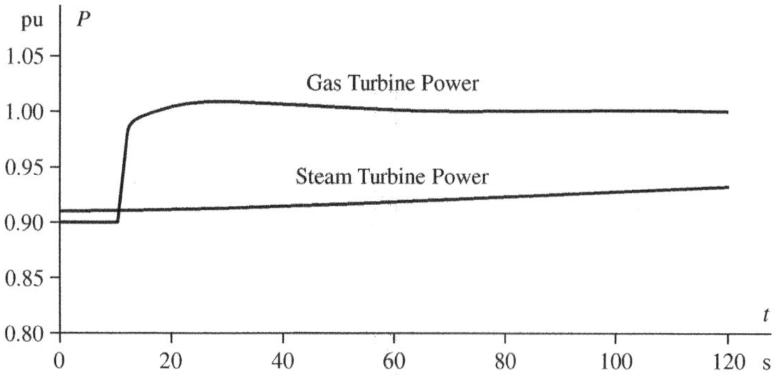

Figure 11.30 Dynamic response to a step increase of a gas turbine load. *Source:* Reproduced with permission from CIGRE Technical Brochure No. 238 (2003).

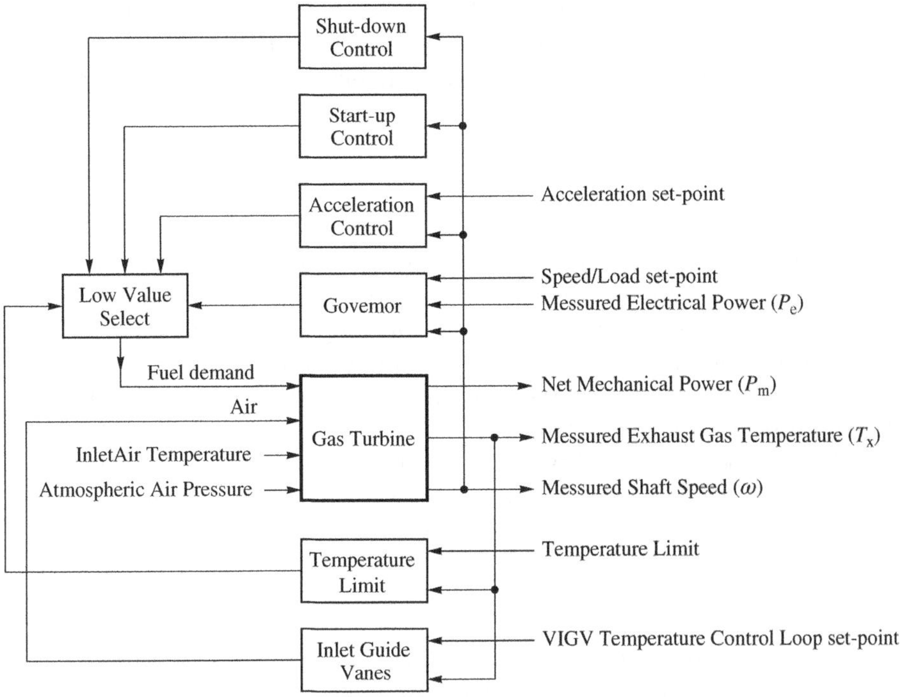

Figure 11.31 Overall block diagram of gas turbine control. *Source:* Reproduced with permission of CIGRE from CIGRE Report 238 (2003).

The speed/load control determines the demand fuel flow. The width of the dead zone depends on the role of a given CCPP in the frequency control of the whole power system.

Temperature control is the normal means of limiting the gas turbine output at a predetermined firing temperature, independent of variation in ambient temperature or fuel characteristics. The exhaust temperature is measured by a number of thermocouples incorporating radiation shields. The design of the temperature controller is intended to compensate for the transients resulting from the time constants of the exhaust temperature measurement system.

The variable inlet guide vanes (VIGV) are located in front of the first stage of the compressor of a gas turbine. The air flow is a function of the VIGV angle, ambient temperature at compressor inlet, atmospheric pressure, and the shaft speed. By changing the angular position of the inlet guide vanes the airflow can be adjusted to the ambient temperature, atmospheric pressure, and the shaft speed. In gas-steam CCPPs the VIGV control is used in the range of low turbine loading only. When the gas turbine is loaded close to the base load, the VIGVs are wide open and for this reason are often omitted in models used for analyses of power system dynamics. For low turbine loading by the VIGV control and reducing the airflow, the exhaust temperature is kept at high levels to maintain the desired level of heat transfer into the HRSG and to maintain overall higher plant efficiency.

As shown in Figure 11.31 the output signals from the above described control loops (speed/load control, acceleration control, temperature control) enter in a low value select block aiming to set the least amount of fuel.

There are several models of gas-steam CCPPs described in publications and used in computer programs for stability studies of power system dynamics.

The first model of a gas turbine was proposed by Rowen (1983). This original model has the three major control loops (speed/load, acceleration, temperature) associated with the dynamic response of the gas turbine during disturbances. Outputs of these control functions are all inputs into a minimum value selector determining the least fuel request. The speed control loop corresponds directly to the governor. The inputs of the speed governor are load demand and speed deviation. In Rowen's model the gas turbine dynamics is essentially made of the delays associated with the combustion process and transport of the exhaust gas as well as by the blocks with the two following functions

$$F_1 = T_X - a_1 \cdot (1 - W_F) - b_1 \cdot \Delta\omega \tag{11.160}$$

$$F_2 = a_2 + b_2 \cdot W_F + c_2 \cdot \Delta\omega \tag{11.161}$$

where a_1, a_2, b_1, b_2, and c_2 stand for coefficients and T_x, W_F, and $\Delta\omega$ refer to the rated exhaust temperature, fuel flow, and speed deviation, respectively. Function F_1 calculates the exhaust temperature. Function F_2 calculates the turbine torque output of the gas turbine. In a later publication by Rowen (1992), this original model is extended to include the VIGV control and its effect on the gas turbine dynamics, especially the exhaust temperature. Coefficients of Rowen's model can be calculated on the design characteristics of different gas turbines (Yee et al. 2008). Many later published models have been based on Rowen's original models.

Among many other gas turbine models are models developed by the IEEE Working Group (1994), WECC Modeling & Validation Work Group (2002), and CIGRE Task Force (2003). An overview and a comparative analysis of these models can be found in Yee et al. (2008). The model described below is the generic CIGRE model.

11.3.3.1 Gas Turbine Model

The generic model of a gas-steam CCPP developed by CIGRE Task Force consists of two parts: (i) the gas turbine model and (ii) the heat-recovery generator and steam turbine model. In the case of a single-shaft CCPP model both parts are connected to provide a common drive power. Both parts of this model were developed under the following assumptions:

- The major control loops associated with the dynamic response must be taken into account, including the dependence of maximum output power on frequency variations up to ±5%.
- Electrical power of the CCPP is controlled by means of the gas turbine only.
- The steam turbine always follows the gas turbine by generating power whatever steam is available.
- The HRSG must be represented in the model in the appropriate way.

A block diagram of the gas turbine model proposed by the CIGRE Task Force is shown in Figure 11.32. Similar to Rowen's model, there are three major control loops (speed/load, acceleration, temperature) feeding into a low value select. However, both models differ with respect to the gas turbine model and calculation of the exhaust temperature.

The speed/load control loop is in the upper left part of the diagram shown in Figure 11.32. The input data to this control are: rotor speed ω in pu, load reference P_{ref} or P_{AGC}, and measured electrical power P_e in pu (on turbine nameplate rating). The resulting signal is passed through a dead-band characteristic and through a transfer function representing the PID controller. dbd and err are the intentional width of dead band and the intentional limit of the error signal. R_v is the governor feedback droop. R_p is the electrical-power feedback droop. It must be emphasized that the CIGRE model is a generic model and some parameters must be set to zero in order to represent desired mode of governor action. Sample data for governor acting as a standard-droop governor with speed feedback alone are: $R_v = 0$, $K_{dg} = 0$, $K_{ig} = 0$, $R_p = 0$, and $K_{pg} = 10$. Sample data for governor acting on speed and electrical-power feedback are: $R_v = 0$, $K_{dg} = 0$, $R_p = 0.05$, $T_p = 5$, and $K_{pg} = 10$. In both cases $P_{ref} \neq 0$ is defined by the user or simulation program during initialization of the model. The speed/load control loop is also augmented with additional outer loop with the input signal P_{AGC} coming from the power plant control. If this outer loop controller is in-service, P_{AGC} must be set equal to the initial steady-state value of turbine output during initialization of the model.

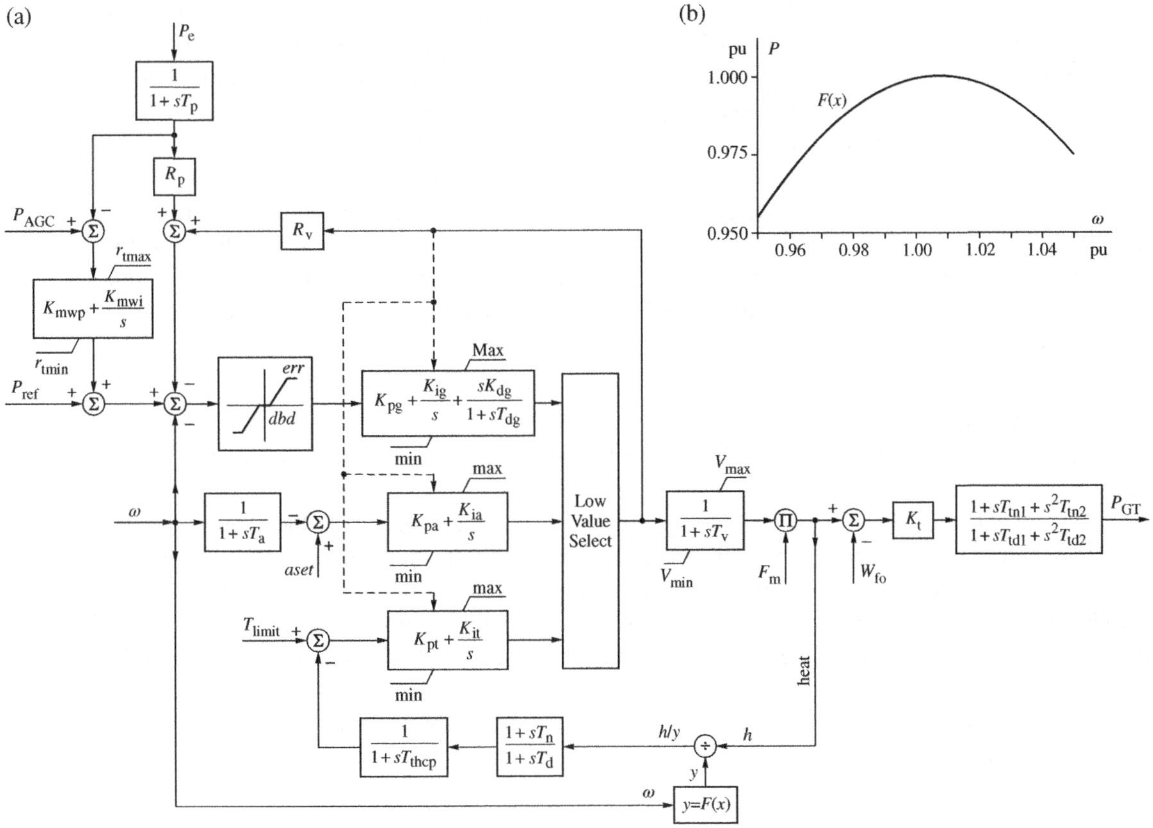

Figure 11.32 Block diagram of the generic gas turbine model (a) and (b) example of function $F(x)$. *Source:* Reproduced with permission from CIGRE Technical Brochure No. 238 (2003).

The acceleration control loop is in the middle part of the diagram shown in Figure 11.32. As mentioned above (description of Figure 11.31), for studies of large power systems, this control can be ignored. It is important for islanding studies and smaller power systems with large frequency variations. If the generating unit begins to accelerate at a rate over *aset* (pu/s^2) then this control loop acts to limit fuel flow. Where such details are important to simulation studies, the user should consult the manufacturer for detailed implementation information, because logic of this control varies among manufacturers. This is depicted in Figure 11.32 by the dashed lines bounding the control loops affected by this logic and feedback from the low value gate output to each control loop. Sample data are: *aset* = 0.01, differentiator time constant $T_a = 0.1$, proportional gain $K_{pa} = 0$, and integral gain $K_{ia} = 10$.

The temperature control loop is in the lower part of the diagram shown in Figure 11.32. The exhaust temperature in CIGRE model is not calculated by using function (11.160) as in Rowen's model but is provided via a signal calculated in block $F(x)$. The temperature limit in pu corresponds to the fuel flow required for 1 pu turbine power. Sample data are: temperature limit $T_{limit} = 0.9167$, proportional gain $K_{pt} = 1$, integral gain $K_{ia} = 0.2$, maximum fuel flow command max = 1.0, minimum fuel flow command min = 0.15, thermocouple time constant $T_{thcp} = 2.5$, heat transfer lead time constant $T_n = 10$, and heat transfer lag time constant $T_d = 15$.

Function $F(x)$ depends on the manufacturer data on the dependency of turbine peak load on system frequency and assumed ambient temperature. An example of such a function is shown in Figure 11.32. In simulation program function $F(x)$ may be implemented as either a look-up table or a piecewise linear function.

The output signals from the above-described control loops (speed/load, acceleration, temperature) enter in a low value select block aiming to set the least amount of fuel. The next block (single time constant element) represents the time constant of the fuel system. Sample value is $T_v = 0.5$. V_{max} and V_{min} represent the maximum and minimum fuel valve opening and thus by definition are typically set to $V_{max} = 1.0$ and $V_{min} = 0.0$. F_m is the fuel flow multiplayer. Typically, $F_m = 1$ is set or in some cases it is equal to the speed ω in pu. W_{fo} represents the amount of fuel flow at full-speed no-load. This is the fuel/power required to run the compensator. Typically, the value is set $W_{fo} = 0.25$. Parameter K_t represents the turbine gain and is a scaling factor between the net fuel consumption (fuel flow minus W_{fo}) to the net mechanical power (gross turbine power minus power consumed by compressor). Sample value is $K_t = 1.5$. In the CIGRE generic model the turbine dynamics is represented be the second-order transfer function. However, usually turbine dynamics is modeled by a single time constant element and then $T_{tn1} = T_{tn2} = T_{td2} = 0$ and $T_{td1} = 0.5$.

11.3.3.2 Steam Turbine Model

A block diagram of the HRSG and steam turbine model proposed by the CIGRE Task Force is shown in Figure 11.33.

In CCPPs (Figure 2.11) the exhaust gases from the gas turbine are directed into the HRSG (boiler) to raise the steam level. The controls of such a boiler are simpler than a conventional steam plant in that the fuel is not controlled to maintain steam conditions but results directly from the gas turbine output. In the CIGRE model the heat Q_g provided by the gas turbine is described by a look-up table or a piecewise function $F(P_{GT})$, where the gas turbine power P_{GT} is the output signal from the gas turbine model (Figure 11.32). The HRSG may be equipped with supplemental firing. Then the production of steam is proportional to the sum $(Q_g + Q_s)$, where Q_g and Q_s are the heat provided by the gas turbine and supplemental firing, respectively. Q_s must be defined by the user and/or program upon initialization of the model.

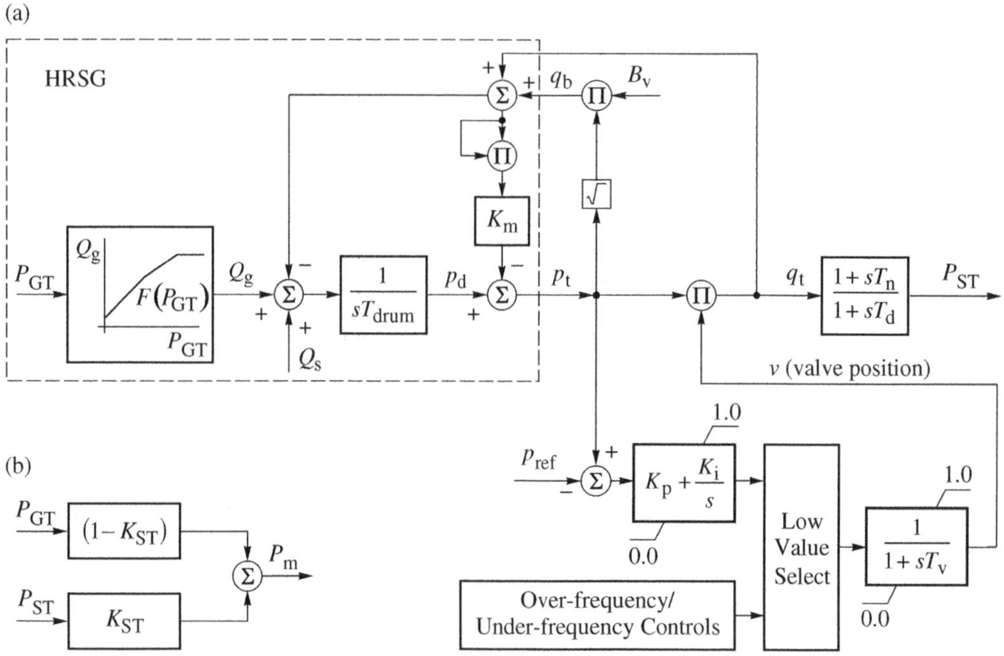

Figure 11.33 Block diagram of the HRSG and steam turbine model (a) and (b) connection of gas and steam turbine models. *Source:* Reproduced with permission from CIGRE Technical Brochure No. 238 (2003).

The HRSG may generate steam at multiple pressures and may have multiple steam drums and steam admission at corresponding multiple points in the turbine. For purposes of grid simulation this complex steam system can be treated as a single drum and single turbine admission valve represented by a simplified model. In the considered CIGRE model (Figure 11.33) the variation of drum pressure p_d (as steam production and consumption are varied) is characterized by the single time constant T_{drum}. The loss of pressure due to flow friction in the boiler tubes and throttling losses upstream of the turbine inlet valves are characterized by the coefficient K_m. The resulting signal p_t represents steam pressure at the turbine admission valve. Diversion of steam for process use or directly to the condenser is represented by opening B_v of a single bypass valve. In the considered model B_v must be defined by the user to simulate a fixed amount of steam extraction. Symbol q_b represents the steam flow through the bypass valve.

The considered CIGRE model includes a simple proportional-integral controller to regulate inlet steam pressure. Symbols K_p and K_i represent the governor proportional and integral gains, respectively. T_v is the actuator time constant for main steam. The controller can be made inactive by specifying a low value for the pressure reference p_{ref}. This allows to represent operation of the steam turbine in either controlled inlet pressure or sliding pressure mode. Dynamics of the steam turbine is modeled by transfer function with the lead T_n and lag T_d time constants.

Sample data are: $T_{drum} = 300$, $K_m = 0.15$, $T_v = 0.5$, $K_p = 10$, $K_i = 2$, $T_n = 3$, $T_d = 10$, $p_{ref} = 0.5$, $F(P_{GT})$ straight line $Q_g = 0.9 \cdot P_{GT} + 0.1$ for $P_{GT} \le 1.0$, and line $Q_g = 1.0$ for $P_{GT} > 1.0$.

Additional over-frequency/under-frequency controls may be added to the model, as indicated in the lower part of the diagram shown in Figure 11.33 in the case of studies involving severe system frequency disturbances. Where studies require such detail, a more detailed model may need to be used and the advice of the manufacturer should be sought.

For a single-shaft CCPP both described models of the gas and steam turbine may be connected, as shown in Figure 11.33b. The constant K_{ST} represents the fraction, in pu, of the total nameplate mechanical power developed by the steam turbine. Then P_m is the resultant total mechanical power on the rotating shaft connecting both turbines and the generator expressed in pu on the nameplate rating of the whole generating unit. In power system studies this power is used in the swing equation of the generating unit.

11.3.3.3 Simplified Models

In commercial computer programs for studies of power system dynamics various simplified models of gas turbine and CCPPs may be found. These type of models usually neglect some control loops presented in Figure 11.31.

The start-up and shutdown controls must be taken into account only for simulation of ramping the unit up during start-up and down during shutdown. For analyses of power system dynamics these controls are irrelevant. Also, the acceleration control and control of the VIGV may be omitted in models used for analyses of power system dynamics. Such a simplified model may be easily obtained from the model shown in Figure 11.32 by canceling relevant control loops.

In some simplified models used for power system studies only the speed/load control and the temperature control are represented. In extremely simplified models, such as the simplified Rowen's model shown in Figure 11.34, only the speed/load control loop associated with the speed governor is represented.

The speed/load controller is depicted on the left part of Figure 11.34. It operates as the load regulator with the speed-droop characteristic with a dead-band. Droop $\rho = 1/K_R$ of this characteristic is determined by gain K_R. The regulator is represented by a single time constant element with time delay T_R. Power is limited by a max/min limiter. The multiplication by rotor speed in pu (on synchronous speed) represents dependency of the air flow from the compressor on the rotor speed. Symbol W_{fo} represents the idle demand, i.e. amount of fuel flow at full-speed no-load required to run the compensator. Factor $(1 - W_{fo})$ determines controlled portion of the power. Further single time constant elements represent time delays introduced respectively by: valve positioner T_R, fuel system T_F, combustor T_S. Symbol W_f represents fuel flow injected to the combustor. The output signal is multiplied by the function F1 to give the turbine power

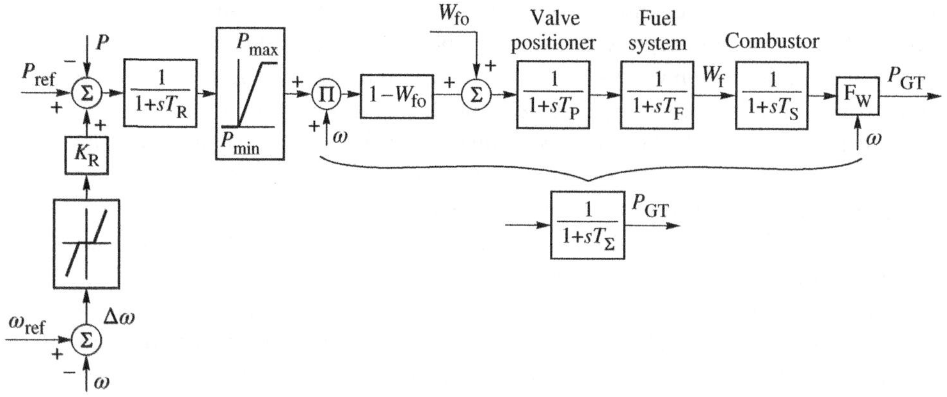

Figure 11.34 Simplified gas turbine model for very small changes in rotor speed. *Source:* Based on Rowen (1983).

$$\mathrm{F_W}: \quad P_{\mathrm{GT}} = \alpha \cdot (W_{\mathrm{F}} - W_{\mathrm{Fo}}) + 0.5 \cdot \Delta\omega \tag{11.162}$$

where $\alpha = 1/(1 - W_{\mathrm{Fo}})$ and $\Delta\omega = (1 - \omega)$. Sample data: $W_{\mathrm{Fo}} = 0.23$, $(1 - W_{\mathrm{Fo}}) = 0.77$, $\alpha = 1/(1 - W_{\mathrm{Fo}}) = 1.3$, $K_{\mathrm{R}} = 1/0.05 = 20$, $T_{\mathrm{R}} = 0.05$, $P_{\mathrm{max}} = 1$, $P_{\mathrm{min}} = 0$, $T_{\mathrm{P}} = 0.05$, $T_{\mathrm{F}} = 0.4$, and $T_{\mathrm{S}} = 0.5$.

Assuming that during simulated transients in an EPS the rotor speed of the considered generating unit is very close to the synchronous speed ($\omega \cong 1$), the second component in Eq. (11.162) can be neglected and then the part of the diagram (Figure 11.34) over the curly bracket may be replaced by one equivalent single time constant element with an equivalent time delay T_Σ. Such an extremely simplified model consists only of the speed/load control loop and equivalent transfer function representing the steam system. This model may be used to represent only those generating units which are very remote from the disturbance in the grid.

It is worth to emphasize here that simplified models may be used only for studies of large power systems in which frequency variations are negligible. For a small power system with large frequency variations or islanding studies more advanced models (for example the above-described CIGRE model) must be used.

11.4 Wind Turbine Generator Systems and Wind Farms

Modeling the behavior of wind turbine generator systems (WTGSs) follows a similar approach to that for conventional steam or water generating units except that now more detailed modeling of the energy resource is often required to take into account its variable nature.

11.4.1 Wind Energy Systems

Figures 7.5–7.11 in Section 7.2 describe the general structure of WTGSs. Although the general structure of all these system options is similar, they differ in the type of electrical generator that is used and the way that it is controlled. However, all the systems can be broken down into a number of subsystems each of which can be modeled individually. Slootweg et al. (2003) suggests a convenient set of subsystems:

1) A wind speed model representative of the site that takes account of turbulence, wind gusts, and so on.
2) A wind turbine model that converts the power in the wind into mechanical power at the turbine low-speed drive shaft.
3) A transmission model that accounts for the effect of the gearbox, if present.

4) A model of the generator and, if necessary, its associated power electronic converter.
5) A power or speed controller to control the power output of the generator, particularly at wind speeds below rated speed, if required.
6) A voltage or reactive power controller, if required.
7) A pitch controller to control the power output of the turbine, particularly at a wind speed above rated, if required.
8) A protection system for limiting converter current, isolating the turbine if voltage or frequency exceeds specified values. This system may also be required to shut the turbine down.

Not all of these subsystems are required in all WTGSs. However, items 1, 2, 4, and 8 will always be required, while the other subsystems will depend on the type of generator and speed control used. For example, the fixed-speed induction generator with stall control (Figure 7.6) will not require items 5 or 7, while the variable speed turbines utilizing pitch control will require all the subsystems to a greater or lesser extent.

Wind farms (also named wind parks) have power ratings from a dozen to several hundred MW. The number of WTGSs to constitute a wind farm varies, and can be as many as several dozen. The rated power of individual WTGSs in a wind farm are different. These powers are currently up to 10 MW.

The wind farm, in addition to a single WTGS, includes: WTGSs step-up transformers, an MV ([10–30] kV) cable grid, MV switchgear, auxiliary services transformer, which is also a grounding transformer, and a plant's step-up transformer, connecting the MV plant's grid with the HV grid. The wind farm element may also be a reactive power compensator in the form of capacitor banks or a reactor. Reactive power compensators are used in the wind farm when the WTGSs are not able to provide the required power factor at the point of the farm connection to the grid (point of common coupling, PCC).

For the purposes of the EPS stability analysis, the wind farm can be modeled in a detailed or simplified way. In the detail model individual WTGSs are modeled as separate dynamic objects. Then, the farm grid together with a WTGS step-up transformers and a farm step-up transformer are modeled using static models, i.e. like other elements of the grid. In turn, the simplified model of the farm (in its simplest form) consists of a dynamic model of one WTGS, scaled to the farm's rated power, and equivalent impedance comprising the WTGS step-up transformers and the farm MV grid. The farm step-up transformer is modeled separately here.

The way wind farms are modeled depends on the size of the EPS and the purpose of the analyzes. Detailed modeling, in the case of a large EPS with high amount of wind farms and in it high number of WTGSs, leads to a model accurately reproducing the given system, including its dynamics. At the same time, however, such a model becomes excessively complex, as the number of WTGS models (that are low-rated power energy sources), step-up transformers and the MV lines (elements of farms grids) can reach thousands. However, in the case of simplified modeling, the number of modeled elements is significantly reduced, but at the same time the accuracy of the model, in the sense of dynamic properties, decreases. It results directly from the fact that the dynamic equivalent of a wind farm in the form of a single WTGS model does not reproduce the dynamic properties of a wind farm well. Nevertheless, in models of large power systems, such as continental Europe or even national systems, simplified wind power farms models are often used.

WTGSs, compared to classic energy sources, are relatively new elements of the EPS. For this reason, information on the construction details and control algorithms used is quite scarce. A good source of information in addition to the literature, for example Hansen et al. (2007) and Clark et al. (2010), are computer programs designed to analyze the stability of power systems, for example GE Energy Consulting (2009) and DigSILENT (2018).

Western Electricity Coordinating Council (WECC) distinguishes four basic types of WTGS:

Type 1 WTGS with a squirrel-cage-rotor induction machine (Figure 7.6). It is the simplest type of a WTGS, with no turbine or generator regulators. The WTGS power as function of wind speed (output power) characteristic (Figure 7.12) is due to the shape of the turbine blades (stall control). For wind speeds greater than the rated speed the power extracted by the wind turbine from the air stream is limited. It results from the detachment of air streams flowing around the blade profile. Reactive power consumed by an induction generator is usually compensated to the reactive power at no-load condition.

Type 2 A WTGS with a wound-rotor induction machine and controlled resistance in the rotor circuit (Figure 7.7). In WTGSs of this type, in contrast to type 1, a generator regulator is used, which controls additional resistance in the rotor circuit. This additional resistance increases the slope of the speed-torque characteristic and thus enables the range of operating speeds of the WTGS to be increased. This, in turn, leads to a better fit to the changing wind conditions and thus to a more efficient takeover of energy from the wind stream. The wind turbine is not controlled here, i.e. for high wind speeds the power extracted by the turbine from the wind stream is limited, as in the case of the WTGS type 1 (stall control).

Type 3 A WTGS with a wound-rotor induction machine and a power electronic converter connected between a rotor winding and a grid. This type of WTGS is called a double-fed induction generator (DFIG) (Figure 7.9). The WTGS is equipped with turbine and generator regulators. The turbine regulator, controlling the blades' pitch angle, is used to limit the power extracted from the wind stream at high wind speeds. The generator regulator controls the rotor current of the generator. It consists of two connected transistor-based power electronics converters with a capacitor as intermediate element, operating as a voltage source converter (VSC). The machine-side inverter (connected to the rotor winding) allows for shaping the WTGS desired power out characteristic (Figure 7.12), and the grid-side inverter maintains the preset voltage value on the intermediate capacitor. The VSC allows the power flow in both directions, i.e. from and to the rotor circuit, what makes possible the WTGS operation with over- and subsynchronous speed. Because the VSC enables control of two signals, it is also possible to control the torque and reactive power, voltage or power factor on the power farm busbars.

Type 4 A WTGS with a synchronous machine connected to the grid through a fully rated converter (FRC). In a WTGS of this type, externally excited synchronous machines (Figure 7.11) or permanent magnet synchronous machine is used (Figure 7.10). Synchronous machine is connected to the grid via a power electronic converter consisted on controlled rectifier (machine-side inverter), intermediate capacitor, and inverter (grid-side inverter). This WTGS, like type 3 (DFIG), is equipped with turbine and generator regulators. The turbine regulator performs the same functions as in the case of the WTGS type 3, i.e. it limits the power extracted from the wind stream to the rated value or lower, reduced owing to operational reasons. The generator regulator controls the machine-side and grid-side inverters. The machine-side inverter regulator MSIR shapes the WTGS power out characteristic (Figure 7.12). The grid-side inverter regulator GSIR controls the voltage on the intermediate capacitor and the voltage, reactive power, or power factor on the farm busbars. In the case of an externally excited synchronous machine, the generator controller also controls the field current, and more precisely the ratio of the machine terminal voltage and the frequency.

The following sections present mathematical models of WTGSs with asynchronous and synchronous generators. The components common to various types of WTGSs are presented only once.

11.4.2 Wind Speed

The wind is not steady. It is time variable and subject to variable amounts of turbulence depending on location. The wind speed can be modeled either by using a suitable spectral density method to predict turbulence or by directly recorded wind speed data. In either case the wind speed produced will be a *point wind speed*. Spectral models use a suitable transfer function to produce a time-domain turbulence variation of wind speed that can be added to a steady wind speed to obtain the net point wind speed (Leithead et al. 1991; Leithead 1992; Stannard and Bumby 2007). In deriving this wind speed, the nature of the terrain and the height above ground level must be taken into account. This wind speed may be valid over a period of tens of seconds but for longer periods the model may need to be augmented by wind gusts and a steady increase in the mean wind speed (Slootweg et al. 2003).

Because of the size of the wind turbine, the point wind speed will vary across the diameter of the turbine and the effective wind speed that the turbine reacts to will be different from the point wind speed. Leithead (1992) identifies four major effects of rotational sampling, wind shear, tower shadow, and disc averaging. *Rotational sampling* accounts for the averaging of the torque over a rotational period, while *tower shadow* takes account of the effect of blades passing in front of the tower (or behind if a downwind turbine) losing lift and hence torque. For a three-bladed wind turbine this tends to produce torque pulsations at three times the rotational speed. *Disc averaging* accounts for the fact that the turbulence will not be constant over the turbine swept area so impacting on the local point wind speed. Similarly, *wind shear* recognizes that the diameter of the turbine is such that the effect of height above ground level cannot be assumed constant in the wind speed calculation. These effects should be included in computing the effective wind speed used in the turbine rotor model. Wind speed modeling is a sophisticated subject and interested readers are referred to the references above and also Wasynczuk et al. (1981) and Anderson and Bose (1983).

11.4.3 Wind Turbine Generator System with Induction Generator

This section presents mathematical models of components of a WTGS with a DFIG (type 3), since these types of farms are the most complex objects. While modeling WTGSs of type 1 (Figure 7.6) or type 2 (Figure 7.7), some of the components described below should be omitted.

11.4.3.1 Wind Turbine

The power from the turbine rotor can be calculated using Eq. (7.1), but here it is modified slightly to make clear that the coefficient of performance depends both on the tip speed ratio and the pitch angle of the blades

$$P_{\mathrm{T}} = c_{\mathrm{P}}(\lambda, \vartheta)P_{\mathrm{wind}} = c_{\mathrm{P}}(\lambda, \vartheta)\ \frac{1}{2}\rho A\nu^3 \tag{11.163}$$

where P_{wind} is the power of the wind stream, ρ is the air density, and A is the turbine swept area ($A = \pi r^2$, where r is the turbine rotor radius approximately equal to the length of the turbine blades), ν is the wind speed and c_{P} is the performance coefficient determining the efficiency of energy conversion of wind flow into mechanical energy. The performance coefficient c_{P} is a function of the tip speed ratio λ and the blades pitch angle ϑ. Generally, as the pitch angle is decreased the performance coefficient c_{p} reduces, as shown in Figure 11.35.

The tip speed ratio λ depends on the turbine rotational speed and the wind speed and is given by Eq. (7.2). To compute the tip speed ratio both the wind speed and the rotational speed of the turbine must be known. Knowing the tip speed ratio, the current value of c_{p} can be computed either from a look-up table (DigSILENT 2018) that models the curves, as presented in Figure 11.35, or by using an appropriate curve fitting to these curves. For example, Slootweg et al. (2003) use the approximation

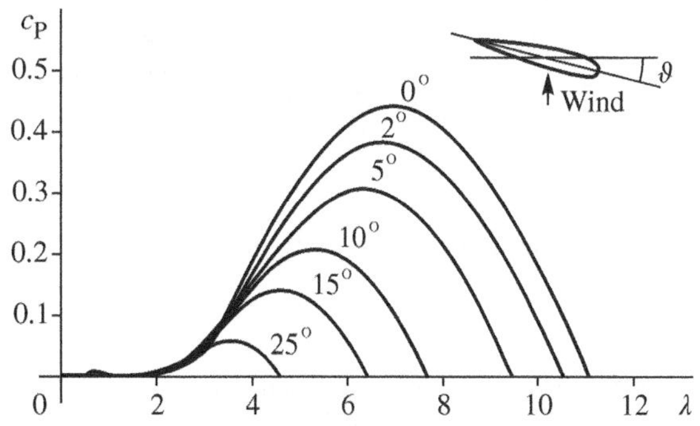

Figure 11.35 Turbine performance coefficient versus pitch angle.

$$c_P = c_1\left(c_2\frac{1}{\lambda_i} - c_3\vartheta - c_4\vartheta^x - c_5\right)e^{-c_6\frac{1}{\lambda_i}}$$
$$\frac{1}{\lambda_i} = \frac{1}{\lambda + 0,02\vartheta} - \frac{0,003}{1+\vartheta^3} \tag{11.164}$$

where exemplary values of coefficients for a type of turbine with a rated power of 2 MW are equal to: $c_1 = 0.46$, $c_2 = 151$, $c_3 = 0.58$, $c_4 = 0.002$, $c_5 = 13.2$, $c_6 = 18.4$, and $c_1 = 2.14$.

Three-bladed WTGSs with rated powers of (1–3) MW obtain the maximum efficiency of energy conversion from the wind stream for the tip speed ratio equal to $\lambda_{\text{opt}} = (7 - 8)$.

The angular velocity of the wind turbine is relatively small. For example, for a turbine with a blade length of $r = 50$ m, for an optimal value of the tip speed ratio $\lambda_{\text{opt}} = 7$ and wind speed $v = 15$ m/s, the angular speed is equal to $\omega_T = \lambda v/r = 7 \cdot 15/50 = 2.1$ rad/s. However, the rated synchronous angular speed of the rotor of an asynchronous generator is equal to 157 rad/s (1500 rpm) or 105 rad/s (1000 rpm).[4] Therefore, the connection of a wind turbine and an asynchronous generator requires the use of a mechanical transmission with a gearbox. Additionally, in order to limit the mechanical stresses, a relatively flexible shaft connecting the turbine and the gearbox is used. A schematic representation of the turbine shaft and gear system is shown in Figure 11.36 and assumes that the gear ratio is υ, so that the gearbox steps up the low speed ω_T at the turbine shaft to the high speed ω_G at the generator rotor

$$\omega_G = \upsilon\omega_T \tag{11.165}$$

For an ideal gearbox (with no losses) the power input and output are equal, and the torque on the high-speed shaft is related to the torque on the low-speed shaft by

$$\tau'_{\text{tr}} = \frac{\tau_{\text{tr}}}{\upsilon} \tag{11.166}$$

where υ is the gear ratio and τ'_{tr} and τ_{tr} are torques, as in Figure 11.36. Two equations of motion (one for the low-speed shaft and one for the high-speed shaft) describe the torque transmission

$$J_T\frac{d\omega_T}{dt} = \tau_T - \tau_{\text{tr}} \quad \text{and} \quad J_G\frac{d\omega_G}{dt} = \tau'_{\text{tr}} - \tau_G \tag{11.167}$$

Combining Eq. (11.167) and referring all torques and speeds to the high-speed shaft gives the combined torque equation of motion referred to the high-speed shaft in SI units as

$$J\frac{d\omega_G}{dt} = \frac{\tau_T}{\upsilon} - \tau_G; \qquad J = J_G + \frac{J_T}{\upsilon^2} \tag{11.168}$$

where J is the combined moment of inertia of the turbine and generator referred to the high-speed shaft.

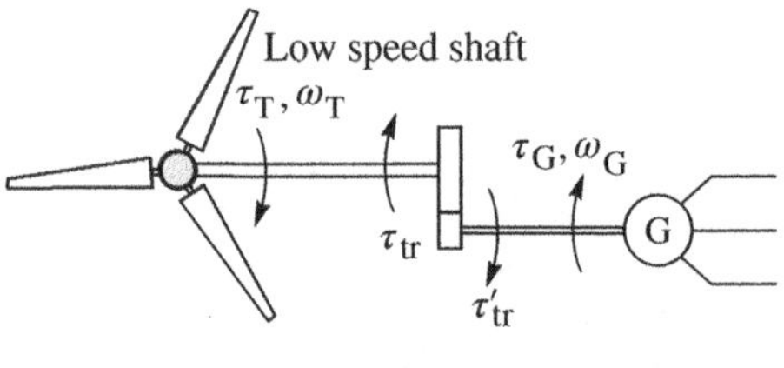

Figure 11.36 Turbine shaft and gear system.

If necessary a finite shaft stiffness and shaft damping can be taken into account (Lubosny 2006; Estanqueiro 2007). Taking into account the shaft elasticity, its dynamic can be described by three equations of motion, one for the low-speed shaft, one for the high-speed shaft, and one describing the shaft elasticity

$$\begin{aligned} \Delta\dot{\delta}_{TG} &= \omega_T - \omega_G/\upsilon \\ J_T\dot{\omega}_T &= \tau_T - K_S\Delta\varphi_{TG} - D_S(\omega_T - \omega_G/\upsilon) \\ J_G\dot{\omega}_G &= \tau_G + \frac{K_S\Delta\varphi_{TG} + D_S(\omega_T - \omega_G/\upsilon)}{\upsilon} \end{aligned} \tag{11.169}$$

4 In some WTGSs with induction generators the number of pole pairs is changed by changing the stator windings connection. The current windings configuration, i.e. number of the pole pairs, depends on the generator's current load.

where $\tau_T = P_T / \omega_T$ and $\tau_G = P_G/\omega_G$ are the torques of the turbine and generator, P_T and P_G are the turbine and generator powers, J_T and J_G are the moments of inertia of the turbine and rotor of the generator, δ_{TG} is the angle of the shaft twist, υ is the gearbox ratio equal to the generator speed and turbine speed quotient, K_S and D_S are the shaft coupling stiffness and coupling damping. Exemplary values of the shaft coupling stiffness and coupling damping of 2 MW WTGS (Abad et al. 2011) are equal to: $K_S = 12500$ Nm/rad, $D_S = 130$ Nms/rad, and exemplary values of moment of inertia of the turbine and generator are respectively equal to: $J_T = 800$ kgm^2 and $J_G = 90$ kgm^2. For another type of 2 MW WTGS, according to DigSILENT (2018) the constants of inertia are equal to $J_T = 1.5 \cdot 10^6$ kgm^2 and $J_G = 75$ kgm^2.

11.4.3.2 Asynchronous Machine

Various types of induction generators used in WTGSs are described in Section 7.1 with the fixed-speed induction generator and double-fed induction generator being most common. The fixed-speed induction generator can be modeled by the induction motor model described in detail in Sections 7.2–7.4. This model is equally valid for the machine operating in generating mode.

In order to model the DFIG, models of the generator, the power electronic converter, and its control system must be developed. Detailed models of these components that include the effect of the rotation emf's (see Section 7.5) are described in detail by Ekanayake et al. (2003b), Holdsworth et al. (2003), and Abad et al. (2011).

A basic type of induction machine model is presented below (Lubosny 2006). A schematic cross-section of an induction machine is shown in Figure 11.37. The machine is assumed to have a three-phase stator armature winding (A_S, B_S, C_S) and a three-phase rotor armature winding (A_R, B_R, C_R). Figure 11.37 also shows the relative position of the windings, and their axes, with the center of phase A taken as the reference. In developing the mathematical model the following assumptions are made:

1) The stator and rotor windings are placed symmetrical along the air-gap as far as the mutual effect with the rotor is concerned.
2) The stator slots cause no appreciable variations of the rotor inductances with rotor position.
3) The rotor slots cause no appreciable variations of the stator inductances with rotor position.
4) Magnetic hysteresis and saturation effects are negligible.

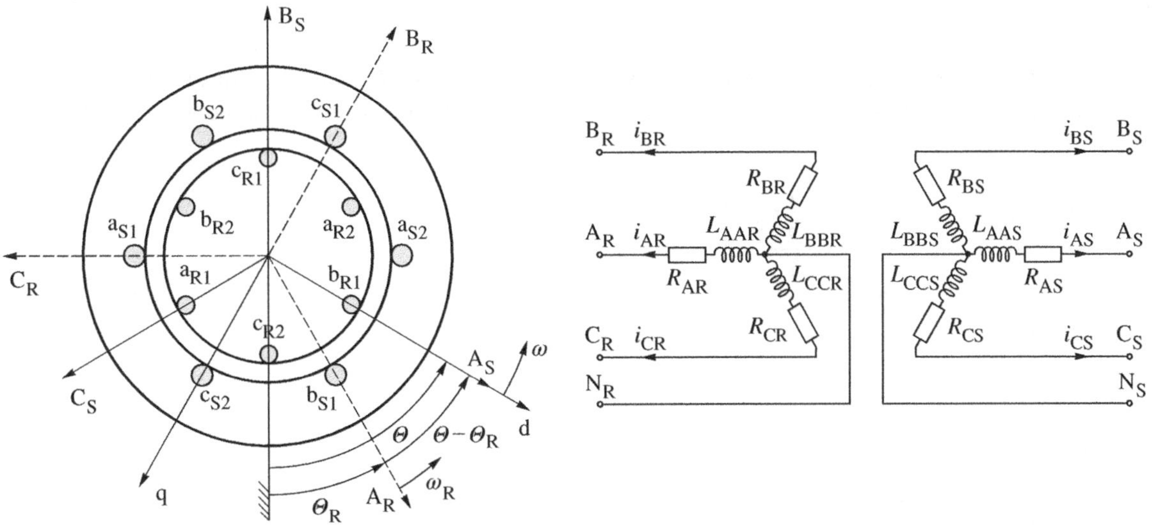

Figure 11.37 The windings in the induction generator and their axes.

5) The stator and rotor windings are symmetrical.
6) The capacitance of all the windings can be neglected.

More detailed modeling of asynchronous machine usually encounters difficulties in getting appropriate data. Additionally, for machine modeling, such a type of model is adequately precise. In the following model of induction machine, the generator convention was adopted, i.e. the currents are marked as flowing from the machine, both on the stator side and on the rotor side. According to this convention, the active and reactive powers expressed by a positive number mean the power flow from the machine to the grid. In the case of a rotor, this means the flow of power from the rotor winding to the machine-side inverter.

For natural axes (A_S, B_S, C_S) and (A_R, B_R, C_R), the model can be formulated by the flux equation, the voltage equation and the mechanical system equation. The flux equation describing the relationship between the flux and current in each winding has the following form

$$\boldsymbol{\Psi}_N = \boldsymbol{L}_N \boldsymbol{i}_N \tag{11.170}$$

where the flux vector is equal to

$$\boldsymbol{\Psi}_N = \begin{bmatrix} \boldsymbol{\Psi}_{ABCS} \\ \boldsymbol{\Psi}_{ABCR} \end{bmatrix} = [\Psi_{AS} \quad \Psi_{BS} \quad \Psi_{CS} \mid \Psi_{AR} \quad \Psi_{BR} \quad \Psi_{CR}]^T \tag{11.171}$$

and where $\boldsymbol{\Psi}_{ABCS}$ is the vector of stator fluxes and $\boldsymbol{\Psi}_{ABCR}$ is the vector of rotor fluxes. The current vector is equal to

$$\boldsymbol{i}_N = \begin{bmatrix} \boldsymbol{i}_{ABCS} \\ \boldsymbol{i}_{ABCR} \end{bmatrix} = [i_{AS} \quad i_{BS} \quad i_{CS} \mid i_{AR} \quad i_{BR} \quad i_{CR}]^T \tag{11.172}$$

where $\boldsymbol{i}_{ABCS}$ is the vector of stator currents and $\boldsymbol{i}_{ABCR}$ is the vector of rotor currents. The stator currents frequency is equal to the grid frequency $f = \omega/2\pi$, where ω is the angular electrical synchronous speed in radians per second. The rotor currents frequency is equal to $f_R = (\omega - \omega_R)/2\pi$, where ω_R is a rotor angular speed in radians per second.

The inductance matrix is equal to

$$\boldsymbol{L}_N = \begin{bmatrix} \boldsymbol{L}_S & \boldsymbol{L}_{SR} \\ \boldsymbol{L}_{SR}^T & \boldsymbol{L}_R \end{bmatrix} \tag{11.173}$$

where $\boldsymbol{L}_S$ is a submatrix of the self and mutual inductances of the stator windings, $\boldsymbol{L}_R$ is a submatrix of the self and mutual inductances of the rotor windings, and $\boldsymbol{L}_{SR}$ is a submatrix of the rotor-to-stator mutual inductances. The inductances are given by

$$\boldsymbol{L}_S = \left[\begin{array}{c:c:c} L_{lS} + L_{mS} & -\frac{1}{2}L_{mS} & -\frac{1}{2}L_{mS} \\ \hdashline -\frac{1}{2}L_{mS} & L_{lS} + L_{mS} & -\frac{1}{2}L_{mS} \\ \hdashline -\frac{1}{2}L_{mS} & -\frac{1}{2}L_{mS} & L_{lS} + L_{mS} \end{array}\right] \tag{11.174}$$

$$\boldsymbol{L}_R = \left[\begin{array}{c:c:c} L_{lR} + L_{mR} & -\frac{1}{2}L_{mR} & -\frac{1}{2}L_{mR} \\ \hdashline -\frac{1}{2}L_{mR} & L_{lR} + L_{mR} & -\frac{1}{2}L_{mR} \\ \hdashline -\frac{1}{2}L_{mR} & -\frac{1}{2}L_{mR} & L_{lR} + L_{mR} \end{array}\right] \tag{11.175}$$

$$
\boldsymbol{L}_{\mathrm{SR}} = \left[\begin{array}{c:c:c} L_{\mathrm{SR}}\cos\Theta_{\mathrm{R}} & L_{\mathrm{SR}}\cos\left(\Theta_{\mathrm{R}}+\frac{2}{3}\pi\right) & L_{\mathrm{SR}}\cos\left(\Theta_{\mathrm{R}}-\frac{2}{3}\pi\right) \\ \hdashline L_{\mathrm{SR}}\cos\left(\Theta_{\mathrm{R}}-\frac{2}{3}\pi\right) & L_{\mathrm{SR}}\cos\Theta_{\mathrm{R}} & L_{\mathrm{SR}}\cos\left(\Theta_{\mathrm{R}}+\frac{2}{3}\pi\right) \\ \hdashline L_{\mathrm{SR}}\cos\left(\Theta_{\mathrm{R}}+\frac{2}{3}\pi\right) & L_{\mathrm{SR}}\cos\left(\Theta_{\mathrm{R}}-\frac{2}{3}\pi\right) & L_{\mathrm{SR}}\cos\Theta_{\mathrm{R}} \end{array}\right] \tag{11.176}
$$

where L_{lS} and L_{lR} are the stator and rotor leakage inductances, L_{mS} and L_{mR} are the magnetizing inductances of the stator and rotor windings, L_{SR} is the amplitude of the mutual inductance, and Θ_{R} is angular displacement of rotor in relation to axis fixed to stator (Figure 11.37).

The voltage equation describing the relationship between the voltage, flux, and current in each winding (according to Figure 11.37) has the following form

$$
\boldsymbol{v}_{\mathrm{N}} = -\boldsymbol{R}_{\mathrm{N}}\boldsymbol{i}_{\mathrm{N}} - \dot{\boldsymbol{\Psi}}_{\mathrm{N}} \tag{11.177}
$$

where the voltage vector is equal to

$$
\boldsymbol{v}_{\mathrm{N}} = \begin{bmatrix} v_{\mathrm{ABCS}} \\ v_{\mathrm{ABCR}} \end{bmatrix} = \begin{bmatrix} v_{\mathrm{AS}} & v_{\mathrm{BS}} & v_{\mathrm{CS}} \mid v_{\mathrm{AR}} & v_{\mathrm{BR}} & v_{\mathrm{CR}} \end{bmatrix}^{\mathrm{T}} \tag{11.178}
$$

and $\boldsymbol{v}_{\mathrm{ABCS}}$ is the vector of stator voltages and $\boldsymbol{v}_{\mathrm{ABCR}}$ is the vector of rotor voltages. The resistance matrix is equal to

$$
\boldsymbol{R}_{N} = \mathrm{diag}\begin{bmatrix} R_{\mathrm{AS}} & R_{\mathrm{BS}} & R_{\mathrm{CS}} & R_{\mathrm{AR}} & R_{\mathrm{BR}} & R_{\mathrm{CR}} \end{bmatrix} \tag{11.179}
$$

The mechanical system can be described by the following equation of motion

$$
J\frac{1}{p}\frac{\mathrm{d}\omega_{\mathrm{R}}}{\mathrm{d}t} = \frac{\tau_{\mathrm{T}}}{\upsilon} - \tau_{\mathrm{G}} - D\frac{1}{p}\omega_{\mathrm{R}} \tag{11.180}
$$

where p is the number of machine pole pairs, τ_{T} is the turbine torque, D is the damping coefficient, and J is the moment of inertia defined by Eq. (11.168). The electromagnetic torque τ_{G} and the angular displacements Θ and Θ_{R} are given by

$$
\tau_{\mathrm{G}} = p\frac{\partial E(\boldsymbol{i}_{\mathrm{ABCS}}, \boldsymbol{i}_{\mathrm{ABCR}}, \Theta)}{\partial \Theta} = \frac{p}{2}\boldsymbol{i}^{T}_{\mathrm{ABCS}}\frac{\partial}{\partial \Theta}\left[\boldsymbol{L}^{T}_{\mathrm{SR}}(\Theta)\right]\boldsymbol{i}_{\mathrm{ABCR}} \tag{11.181}
$$

$$
\Theta = \int_{0}^{t}\omega(\xi)\mathrm{d}\xi + \Theta_{0} \tag{11.182}
$$

$$
\Theta_{\mathrm{R}} = \int_{0}^{t}\omega_{\mathrm{R}}(\xi)\mathrm{d}\xi + \Theta_{\mathrm{R0}} \tag{11.183}
$$

where E is the energy stored in the machine magnetic field, and Θ_{0} and Θ_{R0} are the initial values of the angular displacements.

The stator and rotor terminal powers are, respectively, given by

$$
\begin{aligned} P_{\mathrm{S}} &= \boldsymbol{v}^{T}_{\mathrm{ABCS}}\boldsymbol{i}_{\mathrm{ABCS}} = v_{\mathrm{AS}}i_{\mathrm{AS}} + v_{\mathrm{BS}}i_{\mathrm{BS}} + v_{\mathrm{CS}}i_{\mathrm{CS}} \\ P_{\mathrm{R}} &= \boldsymbol{v}^{T}_{\mathrm{ABCR}}\boldsymbol{i}_{\mathrm{ABCR}} = v_{\mathrm{AR}}i_{\mathrm{AR}} + v_{\mathrm{BR}}i_{\mathrm{BR}} + v_{\mathrm{CR}}i_{\mathrm{CR}} \end{aligned} \tag{11.184}
$$

Analogously like in the case of the synchronous generator (Section 11.1) the asynchronous machine (generator) model for dynamic phenomena in the power system analysis is not usually utilized in the form presented above, i.e. in natural axes (A, B, C). The main disadvantage of the model is the dependence of the inductance matrix $\boldsymbol{L}_{\mathrm{N}}$ on the angular displacement Θ_{R}. Additionally, the number of quantities, which change with time (currents, voltages, and fluxes), is relatively large. Therefore, the set of above equations is converted into equations in the 0dq-reference frame. Axes d, q are related to the rotating stator flux or voltage, as shown in Figure 11.37. In this case, the transformation of the model defined in natural axes is realized by using the transformation matrix $\boldsymbol{W}$ in the following form

$$\boldsymbol{W} = \begin{bmatrix} \boldsymbol{W}_{\mathrm{S}} & \boldsymbol{0} \\ \boldsymbol{0} & \boldsymbol{W}_{\mathrm{R}} \end{bmatrix} \tag{11.185}$$

where the submatrices $\boldsymbol{W}_{\mathrm{S}}$ and $\boldsymbol{W}_{\mathrm{R}}$ are equal to

$$\boldsymbol{W}_{\mathrm{S}} = \frac{2}{3}\begin{bmatrix} \frac{1}{2} & \frac{1}{2} & \frac{1}{2} \\ \cos\Theta & \cos\left(\Theta - \frac{2}{3}\pi\right) & \cos\left(\Theta + \frac{2}{3}\pi\right) \\ \sin\Theta & \sin\left(\Theta - \frac{2}{3}\pi\right) & \sin\left(\Theta + \frac{2}{3}\pi\right) \end{bmatrix} \tag{11.186}$$

$$\boldsymbol{W}_{\mathrm{R}} = \frac{2}{3}\begin{bmatrix} \frac{1}{2} & \frac{1}{2} & \frac{1}{2} \\ \cos(\Theta - \Theta_{\mathrm{R}}) & \cos\left(\Theta - \Theta_{\mathrm{R}} - \frac{2}{3}\pi\right) & \cos\left(\Theta - \Theta_{\mathrm{R}} + \frac{2}{3}\pi\right) \\ \sin(\Theta - \Theta_{\mathrm{R}}) & \sin\left(\Theta - \Theta_{\mathrm{R}} - \frac{2}{3}\pi\right) & \sin\left(\Theta - \Theta_{\mathrm{R}} + \frac{2}{3}\pi\right) \end{bmatrix} \tag{11.187}$$

The way to transform flux and voltage Eqs. (11.170)– (11.177) is similar to the calculations used for a synchronous generator (Sections 11.1.3 and 11.1.4). After transformation, the mathematical model of a wound-rotor induction machine, neglecting the saturation of a magnetic circuit, in 0dq-reference rotating frame form Eqs. (11.188)–(11.194).

The flux equation of the model in the 0dq-reference frame takes the form

$$\boldsymbol{\Psi} = \boldsymbol{L}\boldsymbol{I} \tag{11.188}$$

where the inductance matrix, the flux vector, and the current vector are, respectively, given by

$$\boldsymbol{L} = \begin{bmatrix} L_{0\mathrm{S}} & 0 & 0 & 0 & 0 & 0 \\ 0 & L_{l\mathrm{S}} + \frac{3}{2}L_{\mathrm{mS}} & 0 & 0 & \frac{3}{2}L_{\mathrm{mR}} & 0 \\ 0 & 0 & L_{l\mathrm{S}} + \frac{3}{2}L_{\mathrm{mS}} & 0 & 0 & \frac{3}{2}L_{\mathrm{mR}} \\ 0 & 0 & 0 & L_{0\mathrm{R}} & 0 & 0 \\ 0 & \frac{3}{2}L_{\mathrm{mS}} & 0 & 0 & L_{l\mathrm{R}} + \frac{3}{2}L_{\mathrm{mR}} & 0 \\ 0 & 0 & \frac{3}{2}L_{\mathrm{mS}} & 0 & 0 & L_{l\mathrm{R}} + \frac{3}{2}L_{\mathrm{mR}} \end{bmatrix}$$

$$\boldsymbol{\Psi} = \begin{bmatrix} \Psi_{0\mathrm{S}} & \Psi_{\mathrm{dS}} & \Psi_{\mathrm{qS}} & \Psi_{0\mathrm{R}} & \Psi_{\mathrm{dR}} & \Psi_{\mathrm{qR}} \end{bmatrix}^{\mathrm{T}}$$

$$\boldsymbol{I} = \begin{bmatrix} I_{0\mathrm{S}} & I_{\mathrm{dS}} & I_{\mathrm{qS}} & I_{0\mathrm{R}} & I_{\mathrm{dR}} & I_{\mathrm{qR}} \end{bmatrix}^{\mathrm{T}} \tag{11.189}$$

where subscripts S and R relate respectively to the stator and rotor circuit, L_{0S} and L_{0R} are the zero-sequence component inductances of the stator and rotor, L_{lS} and L_{lR} are the leakage inductances of the stator and rotor, and $L_{mS} = L_{mR} = L_m$ are the inductances of magnetization.

The voltage equation in the 0dq-reference frame takes the following form

$$\dot{\boldsymbol{\Psi}} = -\boldsymbol{RI} - \boldsymbol{V} - \boldsymbol{\Omega\Psi} \tag{11.190}$$

where the voltage vector, the resistance matrix, and the rotation matrix are, respectively, given by

$$\begin{aligned}
&\boldsymbol{V} = [V_{0S}\ \ V_{dS}\ \ V_{qS}\ \ V_{0R}\ \ V_{dR}\ \ V_{qR}]^{\mathrm{T}}\\
&\boldsymbol{R} = \mathrm{diag}[R_{0S}\ \ R_S\ \ R_S\ \ R_{0R}\ \ R_R\ \ R_R]^{\mathrm{T}}\\
&\boldsymbol{\Omega} = \begin{bmatrix} 0 & 0 & 0 & 0 & 0 & 0\\ 0 & 0 & \omega & 0 & 0 & 0\\ 0 & -\omega & 0 & 0 & 0 & 0\\ 0 & 0 & 0 & 0 & 0 & 0\\ 0 & 0 & 0 & 0 & 0 & (\omega-\omega_R)\\ 0 & 0 & 0 & 0 & -(\omega-\omega_R) & 0 \end{bmatrix}
\end{aligned} \tag{11.191}$$

where ω and ω_R, respectively, are pulsation resulting from the grid frequency and rotor angular speed expressed in electrical radians per second, i.e. taking into account the number of pole pairs p. The relationship between the actual rotor speed ω_G and speed ω_R is as follows

$$\omega_R = \frac{p}{2}\omega_G \tag{11.192}$$

The electromagnetic torque of the machine is in turn equal to

$$\tau_G = p\frac{3}{2}\left(\Psi_{dS}I_{qS} - \Psi_{qS}I_{dS}\right) \tag{11.193}$$

As the motion equation can be used Eq. (11.168) or Eq. (11.169). Equation (11.169) should be used, if the wind turbine shaft elasticity is taken into account. If the elasticity of the shaft is neglected, the equation of motion, referred to the rotor of the generator, takes the form of Eq. (11.168).

The active and reactive powers of the stator and rotor are equal to

$$\begin{aligned}
P_S &= \frac{3}{2}\left(2V_{0S}I_{0S} + V_{dS}I_{dS} + V_{qS}I_{qS}\right)\\
P_R &= \frac{3}{2}\left(2V_{0R}I_{0R} + V_{dR}I_{dR} + V_{qR}I_{qR}\right) \cong -P_S\frac{\omega-\omega_R}{\omega}\\
Q_S &= \frac{3}{2}\left(V_{qS}I_{dS} - V_{dS}I_{qS}\right)\\
Q_R &= \frac{3}{2}\left(V_{qR}I_{dR} - V_{dR}I_{qR}\right)
\end{aligned} \tag{11.194}$$

When modeling an induction machine for the WTGS, the equation for the zero-sequence component can be neglected, because the grounding of the star neutral point of the machine windings is not used. In addition, while modeling the WTGS with a squirrel-cage rotor induction machine (type 1), the rotor voltage V_R should be set equal to zero. While modeling the WTGS type 2, i.e. with induction machine with additional resistance connected to the rotor winding, the rotor voltage $\underline{V}_R = 0$ should also be set, but the controllable resistance should be added to the rotor resistance R_R. In case of modeling the WTGS with a DFIG, i.e. type 3, the rotor voltage $\underline{V}_R$ should be set equal to the output voltage of the machine-side inverter $\underline{V}_I$.

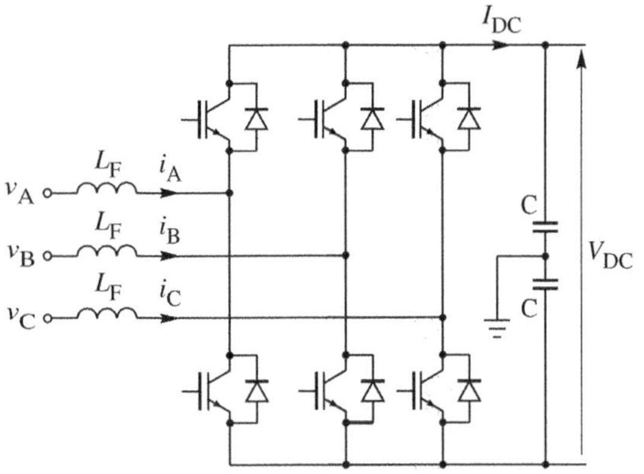

Figure 11.38 Basic structure of the VSC.

During operation of the machine at a speed higher than resulted from the grid frequency ($\omega_R > \omega$), active power flows from the rotor to the machine-side inverter ($P_R > 0$), and the intermediate capacitor is recharged. When the machine operates with subsynchronous speed, the power flows from inverter to the rotor.

Sample data of the DFIG (Abad et al. 2011) with rated power $P_n = 2$ MW are as follows: $\tau_n = 12732$ Nm, $V_{nS} = 690$ V, $\omega_n = 1500$ rpm, speed range $n_n = (900 - 2000)$ rpm, $p = 2$, $L_m = 2.5$ mH, $L_{OS} = L_{OR} = 0.087$ mH, $L_S = L_R = 2.587$ mH, $R_S = 0.026\ \Omega$, and $R_R = 0.029\ \Omega$. The rotor parameters are recalculated to the stator side.

11.4.3.3 Power Electronic Converter

A DFIG used in the WTGS type 3 is a machine powered from the rotor side by voltage with controllable magnitude and frequency. In modern WTGSs, a power electronic converter based on a transistor bridge is used to feed the machine rotor circuit. The second converter connects the DC circuit of the first converter to the AC network (Figure 7.9). The intermediate element, included in the DC circuit, is a capacitor and a DC filter and possibly a chopper. The basic structure of the transistor bridge with capacitor C and AC filter L_F, forming a VSC, is shown in Figure 11.38.

The relationship between the fundamental harmonic of the AC side voltage and the DC voltage of the three-phase transistor based ideal (lossless) bridge is determined by the equation

$$\frac{V}{V_{DC}} = k_o m \tag{11.195}$$

where V is the RMS line-to-line phase voltage, V_{DC} is the DC voltage, m is the amplitude modulation index, and k_o is the coefficient. The coefficient k_o value depends on the type of the pulse-width modulation (PWM) used. In the case of rectangular modulation $k_o = \sqrt{6}/\pi$, and in the case of a sinusoidal modulation, $k_o = \sqrt{3}/(2\sqrt{2})$.

For a converter with sinusoidal modulation, Eq. (11.195) is valid for a modulation index m of no more than one. For larger values of the modulation index the converter saturates, which results in decreasing the slope of the characteristic (Figure 11.39) and decreasing the coefficient k_o.

The Eq. (11.195) in the models presented below is used in form

$$\begin{aligned} V_d &= k_o m_d V_{DC} \\ V_q &= k_o m_q V_{DC} \end{aligned} \tag{11.196}$$

where V_d and V_q are the components of AC voltage, i.e. $\underline{V} = V_d + jV_q$ and m_d and m_q are the components of the modulation index, i.e. $\underline{m} = m_d + jm_q$.

The value of the VSC current on the AC side results from the configuration and parameters of the AC circuit and voltages, including the AC voltage (Eq. [11.195]) at the inverter terminals. The relationship between active and reactive power on the AC side and the power on the DC side of the converter (in case of lossless bridge) is

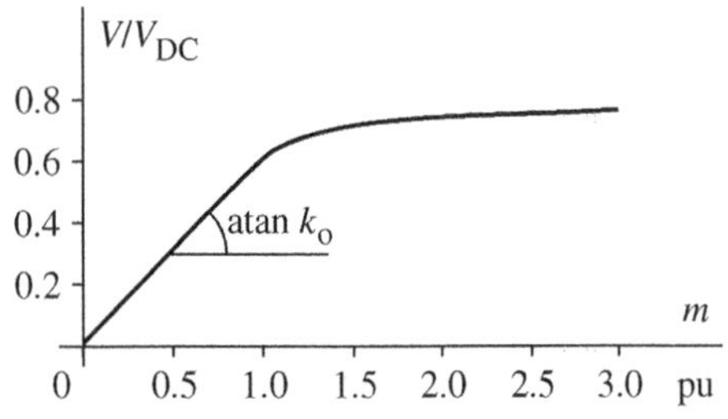

Figure 11.39 PWM converter with sinusoidal modulation characteristic.

$$\begin{aligned} P &= \mathrm{Re}\left(\sqrt{3}\underline{V}\,\underline{I}^*\right) = V_{DC} I_{DC} = P_{DC} \\ Q &= \mathrm{Im}\left(\sqrt{3}\underline{V}\,\underline{I}^*\right) \end{aligned} \tag{11.197}$$

where $\underline{V}$ and $\underline{I}$ are the AC voltage and current phasors (RMS values), and V_{DC} and I_{DC} are the DC voltage and current, respectively. The current on the DC side of the converter results from Eq. (11.197).

For the transistor converter, the power losses in the DC part are equal to

$$\Delta P_{DC} = G_c V_{DC}^2 + R_c I_{DC}^2 + \Delta V_c I_{DC} \tag{11.198}$$

where G_c is the conductance modeling power losses associated with the unloaded converter transistors switching, R_c is the series resistance modeling power losses independent of the transistors switching, and ΔU_c is the voltage drop associated with the transistors switching, defined by function

$$\Delta V_c = \text{sign}(I_{DC})\Delta v_c\left(1 - e^{-200 I_{DC}}\right) \tag{11.199}$$

where Δv_c is the measure of the power losses expressed in kW/A related to the transistors switching, the voltage drop ΔV_c is expressed in kV, and the current I_{DC} is expressed in A.

In the mathematical models of WTGS converters, the VSC is often idealized, i.e. power and energy losses are neglected.

The model shown above applies to both the machine-side inverter and the grid-side inverter. When modeling a system composed of two inverters of this type, the intermediate element (in this case the capacitor) should be taken into account. The equation associated with the intermediate element has the form

$$V_{DC} = \frac{1}{C}\int (I_{DCI} - I_{DCR})\, dt \tag{11.200}$$

where C is the capacitance of intermediate element, I_{DCI} is the direct current of machine-side inverter, I_{DCR} is the direct current of grid-side inverter, and V_{DC} is voltage on the capacitor.

11.4.3.4 Control System

The speed and torque (power) control of a variable speed turbine is a sophisticated control process designed to maximize the power output of the turbine and prevent turbine overspeed. In its simplest form (at wind speeds below rated) controlling the generator electromagnetic torque as an optimized function of rotor speed maximizes energy extracted from the wind, and it controls the turbine speed. However, at high wind speed when the generator is operating at rated output power the torque cannot be increased further without overloading the generator and the speed and power output of the turbine must be controlled by pitch control of the turbine blades.

The idea of a WTGS with a DFIG control in form of characteristics wind speed versus power and rotor angular speed versus power is shown in Figure 11.40 and 11.41. Switching on the WTGS occurs at point A when the wind speed reaches the cut-in (start) speed value υ_s. The rotor angular speed of the generator is then equal to a certain minimum value ω_{min}. The increase in wind speed increases the power generated by the WTGS, but the rotational speed of the generator is kept at the level corresponding to the minimum value ω_{min}. The operating point of the WTGS moves from point A to point B corresponding to the maximum value of the performance coefficient c_P. After reaching this point, in the wind speed range between points B and C, the regulator of generator controls the WTGS (by the machine-side inverter), maximizing the power extracted from the wind stream. The angular reference speed ω_{ref} results directly from formula (7.2). For the known, optimal value of the tip speed ratio λ_{opt}, this speed depends on the wind speed υ, as shown in Figure 11.41b, i.e. there is linear dependence between ω and υ.

Point C corresponds to the maximum rotational speed of the machine ω_{max} allowed in the steady-state operation. This means that the generator regulator, despite the further increase in wind speed, maintains a constant speed equal to its maximum value. The maximum rotational speed of the turbine in the steady state ω_{max} depends on the design of the WTGS, including: mechanical strength of the turbine, electrical parameters of the asynchronous machine, rated power of the power electronic converter, etc. The maximum rotational speed that the turbine can achieve in transient states is $\omega_{max\ dyn} > \omega_{max}$.

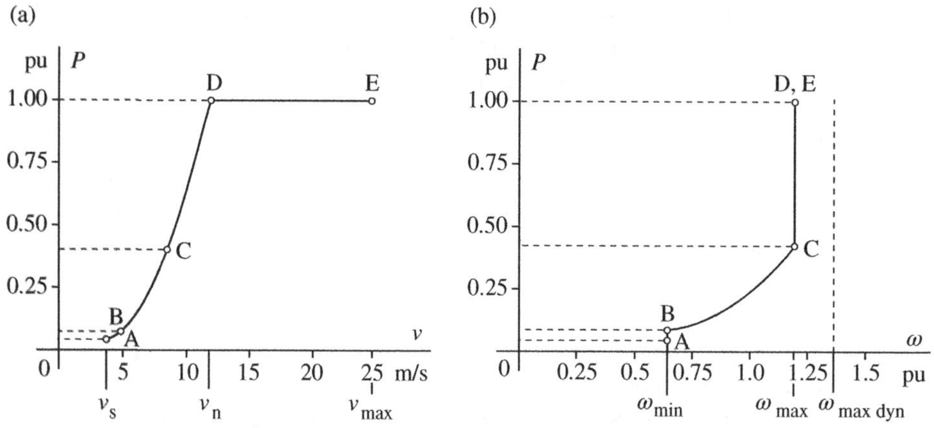

Figure 11.40 Idea of a WTGS with DFIG control: (a) output power characteristic $P(\upsilon)$; (b) induction machine control characteristic $P(\omega)$. *Source:* Based on Hansen et al. (2007). Reproduced by permission of Riso National Laboratory.

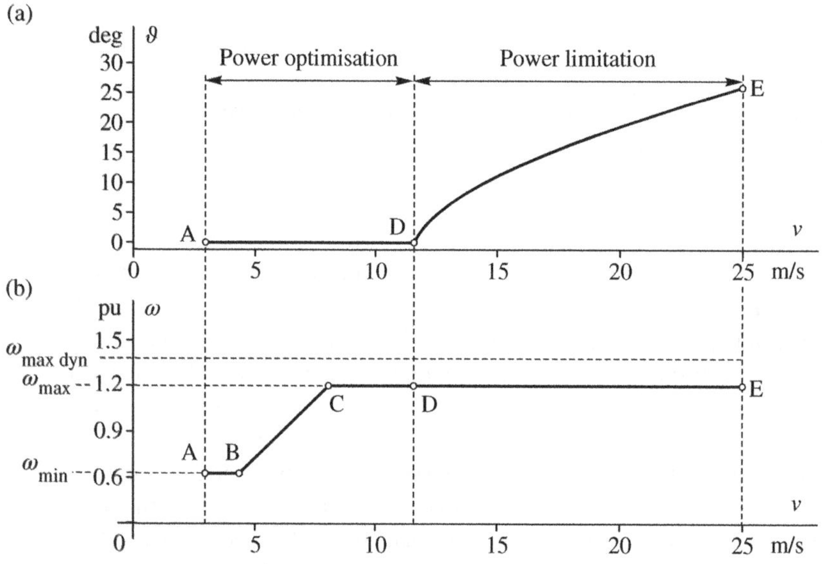

Figure 11.41 Idea of a WTGS control: (a) turbine blades pitch angle and (b) angular velocity of wind turbine as function of wind speed. *Source:* Based on Hansen et al. (2007). Reproduced by permission of Riso National Laboratory.

A further increase in wind speed beyond point C results in a further increase in the power generated by the WTGS, but the energy conversion of the wind stream is no longer optimal. The wind reaching the so-called rated speed υ_n (corresponding to the nominal power, and point D in Figure 11.40 and 11.41) leads to the activation of the turbine regulator. The turbine regulator (which in the steady state, for wind speed lower than the nominal one keeps the minimum blade angle ϑ) when the wind speed increases further, increases the blades pitch angle, limiting the power extracted by the turbine from the wind stream. The turbine regulator keeps power at a level equal to the rated power (Figure 11.41b). At this zone, i.e. between points D and E, the generator controller still maintains the generator speed at the level corresponding to the maximum speed ω_{max} (Figure 11.40b).

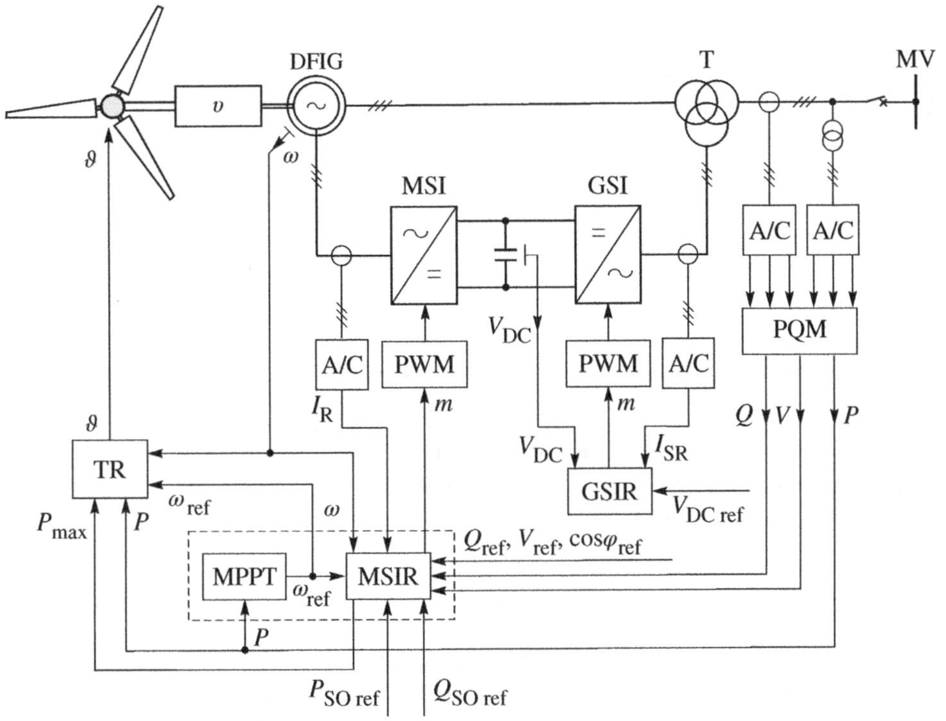

Figure 11.42 Control structure of the WTGS with a DFIG (type 3) υ – gearbox ratio; DFIG – asynchronous machine; T – step-up transformer; GSI – grid-side inverter; MSI – machine-side inverter; PQM – active and reactive power measurement; A/C – voltage and current analogue-to-digital converter; PWM – pulse-width modulation; TR – turbine regulator; MSIR – machine-side inverter regulator; GSIR – grid-side inverter regulator; MPPT – maximum power point tracking system.

One can see that in the presented example the operating range of the WTGS (wind variability range), corresponding to the maximization of energy extracted from the wind stream, i.e. BC interval in relation to the BD interval, is not large. The ability of various WTGSs to maximize the power obtained from the wind stream in the range of wind speed from the cut-in speed to the rated speed (actually in the BD range) is different. Undoubtedly, one of the design goals of this type of WTGS is the minimization of the CD section and maximization of the BC section.

Simplified functional diagram of the control system of the WTGS with a DFIG is shown in Figure 7.29 (Section 7.6.3). More detailed DFIG control structure is shown here in Figure 11.42. It consists of a turbine regulator and a generator regulator, which consists of maximum power point tracking module MPPT, machine-side inverter regulator MSIR, and grid-side inverter regulator GSIR.

The basic task of the turbine regulator is to limit the active power of the WTGS to maximum power P_{max}, not higher than the rated power. The limitation of active power to a value lower than rated may also result from the needs of the EPS, for example as occurs in the case of excessive frequency increase or as the result of control by the system operator (signal $P_{SO\ ref}$ in Figure 11.42).

An example of a structure of a turbine regulator is shown in Figure 11.43 (GE Energy Consulting 2009). The task of limiting the value of active power to value P_{max} is realized by the active power regulator RP. At the same time, in the entire wind speed range, the angular speed regulator RO is active, limiting the momentary increases in the angular velocity of the turbine, caused by increases in the active power as a result of the increase in wind speed. The output signal of the RO regulator in the steady state is equal to the minimum value (e.g. equal to zero), to which the parameter $\Delta = \text{const}$ on the input of this controller corresponds.

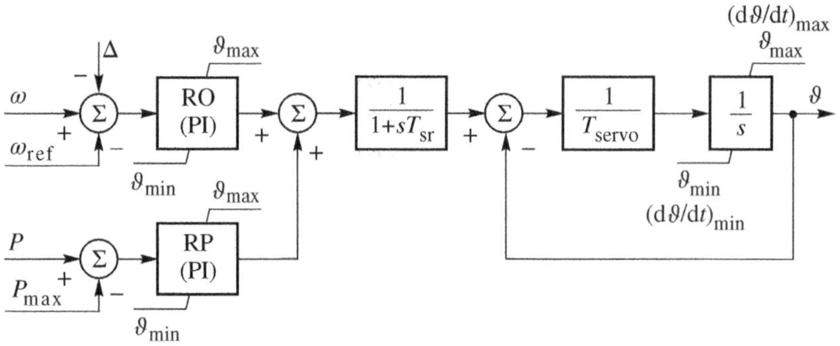

Figure 11.43 Block diagram of the wind turbine regulator. RO – angular speed controller; RP – power controller.

The model of the turbine regulator shown in Figure 11.43 also includes the turbine blade servomotor model. Exemplary values of turbine regulator model parameters and turbine blade servomotor model parameters (GE Energy Consulting 2009, DigSILENT 2018) are equal to

$$\begin{aligned}&\text{RO}: K = 150, T = 0.04\ \text{s};\ \text{RP}: -K = 3, T = 0.3\ \text{s};\ (\text{d}\vartheta/\text{d}t)_{\max} = (5\text{-}10)^{\circ}/\text{s},\\&(\text{d}\vartheta/\text{d}t)_{\min} = -(5\text{-}10)^{\circ}/\text{s},\ \vartheta_{\max} = (27\text{-}70)^{\circ},\ \vartheta_{\min} = 0, T_{\text{sr}} = 0.3\ \text{s}, T_{\text{servo}} = (1\text{-}5)\ \text{s}\end{aligned}$$

Other elements of the WTGS control system are: the power electronic converter consisting of a machine-side inverter MSI and a grid-side inverter GSI. The transistor-based VSC enables simultaneous control of two quantities, which means that in the case of a DFIG it is possible to control the mechanical torque and reactive power on the stator side or active and reactive power on the stator side or the voltage and frequency on the stator side. In the WTGS type 3 the machine-side inverter MSI is used to control:

- active power (and the same time the mechanical torque), shaping the farm power output characteristics (Figure 7.12);
- reactive power or voltage or power factor on the WTGS busbar.

The grid-side inverter GSI is primarily used to control the DC voltage V_{DC} on the intermediate capacitor.

The generator regulator also performs various additional functions, while the basic ones from the EPS stability point of view include:

- Forcing the q-axis (reactive) component of the WTGS current in the event of a significant voltage drop, e.g. during a short-circuit (capability similar to field forcing in synchronous generators). The reactive current component forcing is realized with simultaneous reduction of the d-axis (active) component of the current (and thus the active power fed into the network reduction).
- Rotor winding short-circuiting (crowbar) directly or through an additional impedance at the time of an excessive increase in rotor voltage due to the flow of fault current in the stator winding. This operation is intended to protect the machine-side inverter.
- Fault ride through (FRT) function, which is designed to prevent the WTGS from being shut down during a short-circuit in the power grid and to quickly return the generated power to the value resulting from the wind conditions.
- Over-frequency power limiter function, which aims to reduce active power generated by the WTGS in the event of frequency increase in the grid.

An exemplary generator regulator structure is shown in Figure 11.44. Block maximum power point tracking module MPPT in Figure 11.44 is used to shape the WTGS power-out characteristic. The characteristic shown

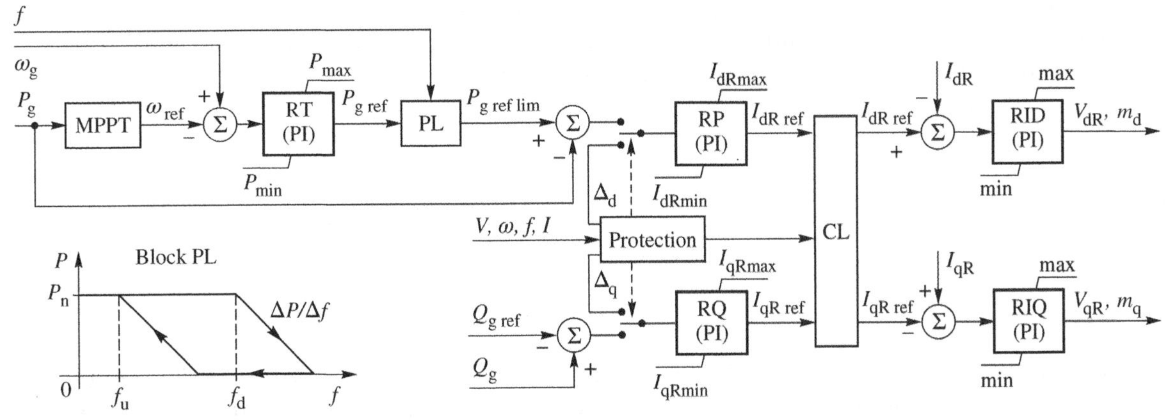

Figure 11.44 Block diagram of the machine-side inverter regulator MSIR of the WTGS with a DFIG. *Source:* Based on DigSILENT (2018).

in Figure 11.40b is used here, where the reference value ω_{ref} of the angular speed is calculated for a given value of the active power P_g. The speed controller RM on its output creates a signal proportional to the reference value $P_{g\ ref}$ of the active power. This signal may be modified in the over-frequency PL or be modified from outside, for example by the system operator. An exemplary characteristic of the PL is shown in the lower left part of Figure 11.44

In the normal operating state, the active power regulator RP on its output creates a signal proportional to the reference value $I_{dR\ ref}$ of the d-axis (active) component of the rotor current. A signal proportional to the reference value $I_{qR\ ref}$ of the q-axis (reactive) component of the rotor current is created in the regulator RQ. This signal is based on the control error of the controlled variable, which in the presented example is the WTGS reactive power, but in general it may also be the WTGS terminal voltage or the WTGS power factor. The rotor current reference signals are then processed in the current limiter. In the normal operating state, when the rotor current exceeds the maximum permissible current I_{Rmax}, the reference value of the q-axis (reactive) component $I_{qR\ ref}$ is limited to the value fulfilling the condition $I_{qR\ ref} \le \sqrt{I_{R\max}^2 - I_{dR\ ref}^2}$. Limiting the current means that the induction machine reaches the operating point corresponding to the permissible operating states. This means change of the operating mode from the terminal voltage or reactive power or power factor control into the mode of maximum current control. However, during a significant decrease of a WTGS terminal voltage, for example caused by a short-circuit in the grid, the q-axis (reactive) component $I_{qR\ ref}$ of the rotor current is forced. At the same time, the d-axis (active) component of rotor current is limited to $I_{dR\ ref} = \sqrt{I_{R\max}^2 - I_{qR\ ref}^2}$.

The last element of the control path, owing to the fact that the generator regulator is the rotor current regulator controlling the rotor voltage, are fast regulators RID and RIQ of the current components. In the case of a WTGS modeling for the purpose of analyzing the EPS stability, modeling of the power electronic converter can be limited to the form given in Eqs. (11.196) and (11.197). Then, as outputs of RID and RIQ regulators, modulation index components P_{md} and P_{mq} (like in Figure 11.44) or rotor voltage components V_{dR} and V_{qR} can be adopted.

The machine-side inverter regulator also performs various protection and logic functions related to voltages, currents, and speeds, including the necessity of rotor winding short-circuiting during the short-circuit in the grid. In the "Protection" block, signals Δ_d and Δ_q are created, which are used next by the active power RP and reactive power RQ regulators during the short-circuit. Details can be found in DigSILENT (2018).

Examples of parameter values of regulators from Figure 11.44 (taken from DigSILENT 2018) are as follows: RT: $K = 1$, $T = 10$ s; RP, RQ: $K = 1$, $T = 0.1$ s; RID, RIQ: $K = 1$, $T = 0.002$ s. The signals of currents and voltages in the regulators are limited to the rated values, and the power reference to about 110% of the rated power.

As described above, the machine-side inverter regulator performs the task of controlling the active and reactive power by independent control of the rotor current components I_{dR} and I_{qR}. In this case, $\Psi_{qS} = \Psi_S$ and $\Psi_{dS} = 0$, and the second and third equation from the matrix Eq. (11.188) take the form[5]

$$\begin{aligned} I_{dS}L_S + I_{dR}L_m &= \Psi_{dS} = 0 \\ I_{qS}L_S + I_{qR}L_m &= \Psi_{qS} = \Psi_S \end{aligned} \tag{11.201}$$

Taking the above into consideration, after transformation, the d, q components of the stator currents are

$$\begin{aligned} I_{dS} &= -\frac{L_m}{L_S} I_{dR} \\ I_{qS} &= \frac{\Psi_S}{L_S} - \frac{L_m}{L_S} I_{qR} \end{aligned} \tag{11.202}$$

Assuming that the resistances of the stator windings are neglected, the algebraic equations describing the steady-state can be obtained by zeroing the flux derivatives in the second and third equation from the matrix Eq. (11.190). Then

$$V_{dS} \cong V_S = -\omega\Psi_{qS} \quad \text{and} \quad V_{qS} = \omega\Psi_{dS} = 0 \tag{11.203}$$

After substitution of Eq. (11.203) to Eq. (11.194), the active and reactive power of the stator are equal to

$$P_S = \frac{3}{2} V_{dS} I_{dS} \quad \text{and} \quad Q_S = -\frac{3}{2} V_{dS} I_{qS} \tag{11.204}$$

Next, substituting the currents from Eq. (11.202) to Eq. (11.204) one obtains equations binding the stator's active power with the d-axis (active) component of the rotor current and the stator's reactive power with the q-axis component

$$\begin{aligned} P_S &= -\frac{3}{2} V_S \frac{L_m}{L_S} I_{dR} \\ Q_S &= -\frac{3}{2} V_S^2 \frac{1}{\omega L_S} + \frac{3}{2} V_S \frac{L_m}{L_S} I_{qR} \end{aligned} \tag{11.205}$$

Equation (11.205) leads to the block diagram shown in Figure 11.45a. The structure enables the separation of active and reactive power control by using the rotor currents I_{dR} and I_{qR}. The output signals are here the reference values of the rotor currents. But, as the reader may notice, this structure is not used in the block diagram in Figure 11.44. The blocks with proportional relations between the current components and powers are replaced by PI controllers RP and RQ, which compensates for machine parameter uncertainties. The separate control of active and reactive power by control of rotor currents components is then realized in a way presented below.

Since the control of the rotor currents in d- and q axes with the VSC use is realized with the rotor voltage components V_{dR} and V_{qR}, the voltage reference values for the inverter should be determined. To achieve this, the fifth and sixth equation from the matrix equations in (11.188) can be used

$$\begin{aligned} I_{dR}L_R + I_{dS}L_m &= \Psi_{dR} \\ I_{qR}L_R + I_{qS}L_m &= \Psi_{qR} \end{aligned} \tag{11.206}$$

In these equations the components of the stator currents can be substituted by I_{dS} and I_{qS} from Eq. (11.202), and by temporarily neglecting the assumption $\Psi_{dS} = 0$, the following is obtained

$$\begin{aligned} \Psi_{dR} &= \left(L_R - \frac{L_m^2}{L_S}\right) I_{dR} + \frac{L_m}{L_S}\Psi_{dS} \\ \Psi_{qR} &= \left(L_R - \frac{L_m^2}{L_S}\right) I_{qR} + \frac{L_m}{L_S}\Psi_{qS} \end{aligned} \tag{11.207}$$

5 In general the d- or q-axis can be aligned with stator voltage or stator flux, which modifies control.

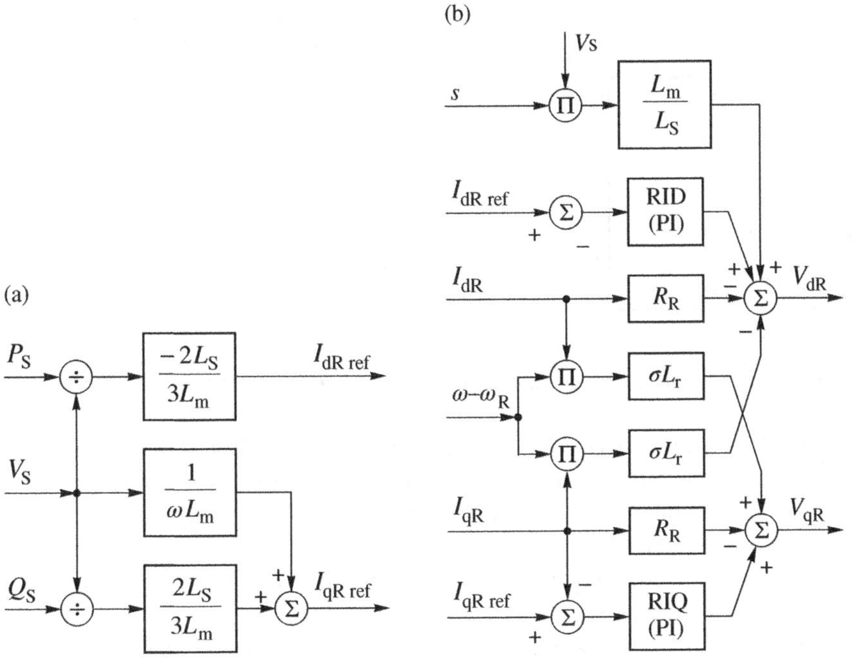

Figure 11.45 Block diagram of the system separating control of: (a) powers; (b) rotor current components, in the d- and q-axes.

The fifth and sixth equation from the matrix Eq. (11.190) can be written in the following way

$$\begin{aligned} V_{dR} &= -R_R I_{dR} - (\omega - \omega_R)\Psi_{qR} - \dot{\Psi}_{dR} \\ V_{qR} &= -R_R I_{qR} + (\omega - \omega_R)\Psi_{dR} - \dot{\Psi}_{qR} \end{aligned} \tag{11.208}$$

Substituting into these equations Eq. (11.207) the following equations binding rotor voltages in d- and q-axes with rotor currents and stator and rotor fluxes are obtained

$$\begin{aligned} V_{dR} &= -R_R I_{dR} - (\omega - \omega_R)\left(L_R - \frac{L_m^2}{L_S}\right)I_{qR} + (\omega - \omega_R)\frac{L_m}{L_S}\Psi_{qS} - \dot{\Psi}_{dR} \\ V_{qR} &= -R_R I_{qR} + (\omega - \omega_R)\left(L_R - \frac{L_m^2}{L_S}\right)I_{dR} + (\omega - \omega_R)\frac{L_m}{L_S}\Psi_{dS} - \dot{\Psi}_{qR} \end{aligned} \tag{11.209}$$

Equation (11.209) after the substitution of the flux derivatives calculated from Eq. (11.207) and taking into account the assumption $\Psi_{dS} = 0$, $\Psi_{qS} \cong -V_{dS}/\omega = -V_S/\omega$ takes for a steady state the form

$$\begin{aligned} V_{dR} &= -R_R I_{dR} - (\omega - \omega_R)\left(L_R - \frac{L_m^2}{L_S}\right)I_{qR} + \frac{\omega - \omega_R}{\omega}\frac{L_m}{L_S}V_S \\ V_{qR} &= -R_R I_{qR} + (\omega - \omega_R)\left(L_R - \frac{L_m^2}{L_S}\right)I_{dR} \end{aligned} \tag{11.210}$$

Equation (11.210) corresponds to the structure illustrated in Figure 11.45b, where $s = (\omega - \omega_R)/\omega$ is the rotor slip and σL_R is the transient inductance of the rotor equal to

$$\sigma L_R = L_R - \frac{L_m^2}{L_S} = L_R\left(1 - \frac{L_m^2}{L_S L_R}\right) \tag{11.211}$$

The above consideration leads to a typical structure of the rotor current regulator, which block diagram is shown in Figure 11.45b. In this block diagram the proportional relations occurring in Eq. (11.210) in the regulators RID

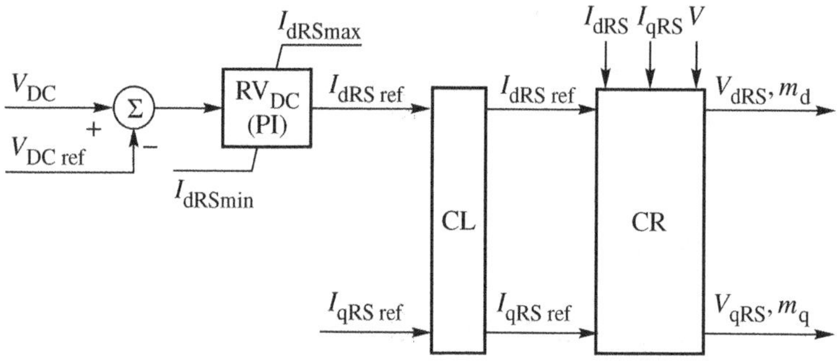

Figure 11.46 Block diagram of the grid-side inverter regulator GSIR of the WTGS with a DFIG (type 3). *Source:* Based on DigSILENT (2018).

and RIQ are replaced by the PI blocks. Their task is to compensate the induction machine parameters uncertainties.

In the block diagram shown in Figure 11.45b there are two cross-couplings between control circuits of V_{dR} and V_{qR}. These cross-couplings provide fast compensation, but in some cases it is neglected. The rotor resistance R_R is also often neglected. In such a case, the current controller of machine-side inverter regulator consists of RID and RIQ regulators only, i.e. like it is presented in Figure 11.44. This structure introduces limitations in the control separation in the d- and q-axes. Therefore, the structure from Figure 11.45b on the machine-side inverter regulator output is recommended.

As mentioned above, the task of the grid-side inverter regulator of the WTGS with a DFIG is maintaining the voltage reference value on the intermediate element, which is the capacitor. An exemplary structure of this regulator is shown in Figure 11.46. The DC voltage reference value is usually determined in the range $V_{DC\ ref} = (1000 - 1200)$ V. The reference value of the q-axis (reactive) component $I_{qRS\ ref}$ is determined as a function of control goal. Its value equal to zero means that the grid-side inverter regulator keeps the zero reactive power value on the inverter AC terminals. In practice, this value may be nonzero, for example to compensate for the reactive power losses on the converter filter or in the transformer feeding the converter. The CL block performs the current limiting task, as described above.

The CR block is a current regulator that simultaneously performs the task of independent control of two quantities using the current components in the d- and q-axes. This type of control, used for the grid-side inverters, is referred to as the *grid voltage oriented vector control*. The control idea is as follows. The VSC is connected to the grid through a filter (or transformer) with impedance $\underline{Z}_F = R_F + j\omega L_F$. The voltage $\underline{V}_I$ at the inverter AC terminals, i.e. generated by the VSC, in the steady state is equal to

$$\underline{V}_I = \underline{V} + \underline{I} \cdot \underline{Z}_F \tag{11.212}$$

where $\underline{V} = V_d + jV_q$ is known (measured) grid voltage in the PCC and $\underline{I}$ is the inverter AC current. Equation (11.212) in the (d, q) reference frame has the form

$$\begin{aligned} V_{dI} &= V_d + R_F I_d - \omega L_F I_q \\ V_{qI} &= V_q + R_F I_q + \omega L_F I_d \end{aligned} \tag{11.213}$$

It is assumed that the (d, q) reference frame rotates at the angular speed resulting from the grid frequency (pulsation $\omega = 2\pi f$). After neglecting the filter resistance R_F and assuming $\omega = \omega_n$, Eqs. (11.213) can be presented in the form

$$\begin{aligned} V_{dI} &= V_d - X_F I_q \\ V_{qI} &= V_q + X_F I_d \end{aligned} \tag{11.214}$$

These equations determine the given values of the VSC voltage components, which lead to independent control of the current components in the d- and q-axes.

In practice, it is assumed that one of the components is equal to the measured voltage $\underline{V}$ module, and the other is equal to zero, e.g. $V_d = V$, $V_q = 0$ or $V_q = V$, $V_d = 0$. In Figure 11.47 the structure of the VSC control (CR block in Figure 11.46) is shown for a variant in which the phasor $\underline{V}$ at the PCC is aligned to axis d. In this structure there are also RID and RIQ of type PI regulators whose task is to compensate for the error of current components in the steady state and during the transient states.

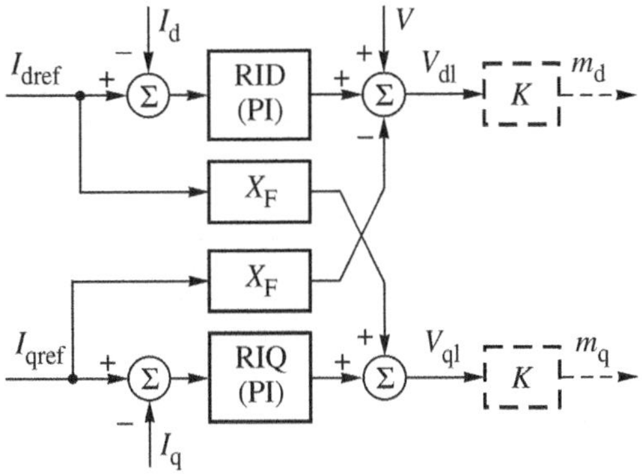

Figure 11.47 Block diagram of the current regulator (block CR in Figure 11.46) in the grid-side inverter regulator GSIR.

Modeling the grid-side inverter regulator as the output signals of the current regulator, one can use the voltages V_{dI} and V_{qI} or possibly the modulation index components m_d and m_q. In the latter case, modulation index components m_d and m_q should be calculated from voltages V_{dI} and V_{qI} (blocks designated as K in Figure 11.47) using Eq. (11.196). For simplification, i.e. to make the coefficient K constant, its value can be calculated assuming the value of the coefficient k_o for the linear range of the PWM converter characteristic (Figure 11.39) and assuming a constant, e.g. a given value of voltage V_{DC}.

Exemplary values of the controller parameters shown in Figure 11.46 and Figure 11.47 (DigSILENT 2018), are as follows: RV_{DC}: $K = 1$, $T = 0.1$ s; RID, RIQ: $K = 1$, $T = 0.015$ s. The current signals in the regulators are limited to the rated values.

11.4.4 WTGS with Synchronous Generator

The structure of the control system of the WTGS (type 4) with synchronous generator connected to a grid through the FRC does not deviate from the structure of the control system of a WTGS with the DFIG. The basic elements of the WTGS with a synchronous generator, from the point of view of modeling for the purpose of analyzing the EPS stability, are: a wind turbine, a synchronous machine, and power electronics converters. The model of the wind turbine and the power electronics converter is shown in Section 11.4.3.1, while the synchronous generator model is discussed in detail in Section 11.1.

11.4.4.1 Control System

Simplified functional diagram of the control system of the WTGS with an FRC is shown in Figure 7.30 (Section 7.7). A more detailed FRC control structure is shown in Figure 11.48. It consists of the turbine regulator and generator regulator. The generator regulator consists of the maximum power point tracking block MPPT, the machine-side inverter regulator, the grid-side inverter regulator, and the excitation regulator AVR.

The turbine regulator has a structure as shown in Figure 11.43 in Section 11.4.3.4. In this case, however, the regulator does not have an angular speed controller RO, which results from the method of shaping the characteristic of the output power of the WTGS in the maximum power point tracking module MPPT block. The input quantity here is (unlike the WTGS with a DFIG) the angular velocity of the turbine and the reference value P_{ref} of the output active power. This reference value can be limited in the PL (Figure 11.49) to the value $P_{max} \leq P_n$. Then this value becomes the reference value of the active power for the turbine regulator RP (Figure 11.43).

The generator excitation regulator AVR is present in a WTGSs with wound rotor synchronous generators. The regulator's task is to maintain a constant value of the excitation flux or to limit it to a maximum permissible value and thus protect the machine magnetic circuit against over-excitation. This is realized by keeping the armature voltage and frequency quotient $V_g/\omega \leq (V/\omega)_{lim}$ at a limited value. The generator excitation regulator AVR is a P-type (proportional) or PI-type (proportional plus integral) regulator controlling a transistor or thyristor exciter supplied from the converter DC circuit or from the machine terminals.

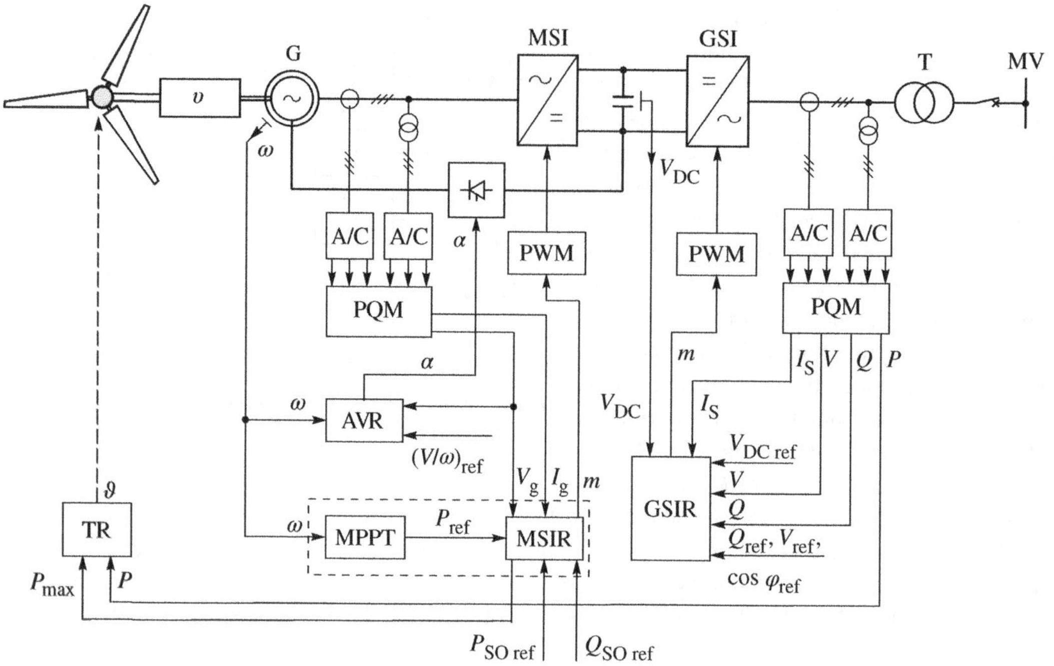

Figure 11.48 Control structure of the WTGS with an FRC (type 4). υ – gearbox ratio; G – synchronous machine; T – step-up transformer; GSI – grid-side inverter; MSI -machine-side inverter; PQM – active and reactive power measurement; A/C – voltage and current analogue-to-digital converter; PWM – pulse-width modulation; TR – turbine regulator; MSIR – machine-side inverter regulator; GSIR – grid-side inverter regulator; AVR – excitation regulator; MPPT – maximum power point tracking system.

In WTGSs with synchronous machines with permanent magnets, the AVR regulator does not exist. In the case of modeling a WTGS of this type, the mathematical model of a synchronous machine with permanent magnets should be used. However, which is an approximation only, it is also possible to use an externally excited synchronous machine model and apply a P-type excitation current regulator with a very high gain, maintaining a constant excitation current value.

The machine-side inverter regulator (Figure 11.49) operates in the control mode of the WTGS active power and the synchronous generator terminals voltage. The WTGS characteristic of the output power is shaped in the

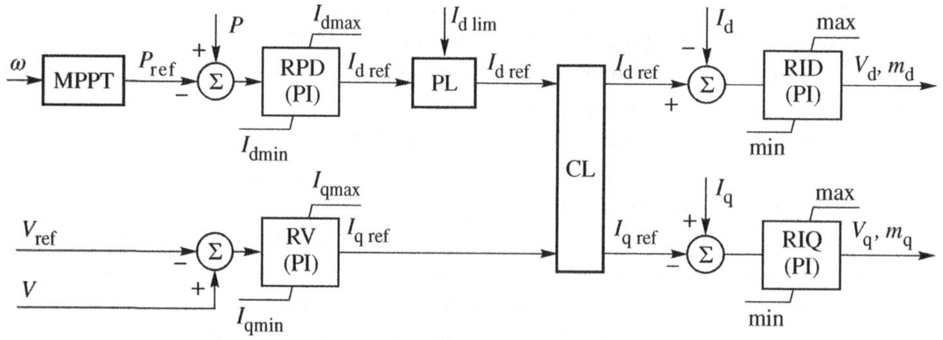

Figure 11.49 Block diagram of the machine-side inverter regulator MSIR with maximum power point tracking module MPPT of the WTGS with an FRC (type 4). *Source:* Based on DigSILENT (2018).

maximum power point tracking module MPPT block. The reference value P_{ref} of the active power results from the defined characteristic $P_{ref} = f(\omega)$. In general, this characteristic does not have to be rectilinear. It may have a shape similar to the one shown in Figure 11.50. Then, the active power deviation is introduced into the power regulator RPD, which calculates the reference value of the d-axis (active) component $I_{d\ ref}$ of the generator current. This current (and at the same time the WTGS active power) can be limited in the PL by an external signal $I_{d\ lim}$, shaped by the WTGS's protection or as the result of an external control, e.g. by the system operator.

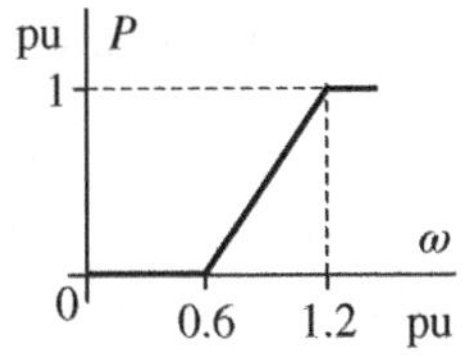

Figure 11.50 An example of a maximum power point tracking module MPPT characteristic.

The control of voltage at the synchronous generator terminals is implemented by control of the q-axis (reactive) component of the generator current using the regulator RV. The reference value of this voltage, usually equal to the rated one, can only be maintained in the rotor speed range of the machine in which the condition $(V_g/\omega) \leq (V_g/\omega)_{lim}$ is met. Controlling the generator voltage in this way is optional. It is used in selected types of WTGSs.

The rest of the control system performs tasks as described for the preceding control system structures, i.e. the reference values of the d- and q-axis components of the generator current are coordinated in a current limiter CL (discussed in Section 11.4.3.4), becoming reference values of fast current regulators RID and RIQ in the current regulator CR (Figure 11.47).

The exemplary parameters of the regulator presented in Figure 11.49 (DigSILENT 2018) are as follows: RPD: $K = 0.5$, $T = 0.005$ s; RV_{DC}: $K = 5$, $T = 0.01$ s; RID: RIQ in CR: $K = 1$, $T = 0.015$ s. Current signals are limited to rated values.

The grid-side inverter regulator operates in the control mode of the DC voltage on the intermediate capacitor and the reactive power (as shown in Figure 11.51) or in the control mode of the voltage or the power factor on the WTGS terminals. The WTGS terminals can be understood here as grid-side inverter AC terminals or step-up transformer busbars on the MV side. Regulator RV_{DC} controls the d-axis (active) component of the inverter current I_d on the basis of the DC voltage deviation. Regulator RQ, based on the reactive power deviation, controls the q-axis (reactive) component of the inverter current I_q. The signal proportional to the reference value of the q-axis current component $I_{q\ ref}$ can be modified by means of protection, for example during a significant decrease of the WTGS terminal voltage. In the structure shown in Figure 11.51 this task is performed by the Protection block, changing the signal input to the current limiter CL (discussed in Section 11.4.3.4). The current regulator CR, shown in Figure 11.47, performs independent control of the inverter current components in the d- and q-axes.

The sample values of the regulators parameters presented in Figure 11.51 (DigSILENT 2018) are as follows: RV_{DC}: $K = 5$, $T = 0.01$ s; RQ: $K = 1$, $T = 0.05$ s; RID, RIQ: $K = 1$, $T = 0.015$ s. Current signals are limited to rated values.

Figure 11.51 Block diagram of the grid-side inverter regulator GSIR of the WTGS with an FRC (type 4). Block CR as in Figure 11.47. *Source:* Based on DigSILENT (2018).

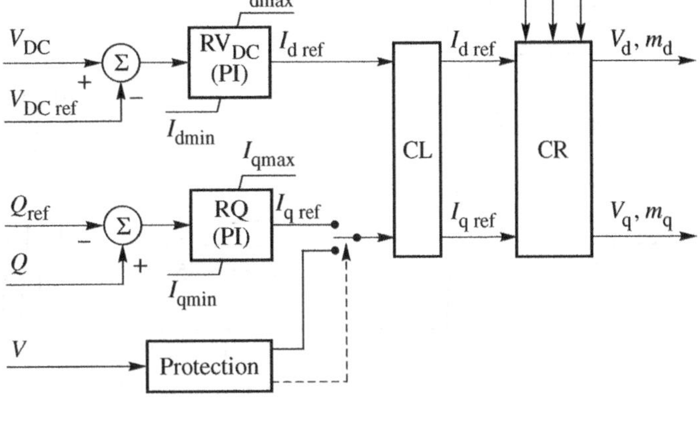

11.5 Photovoltaic Power Plants

Photovoltaic (PV) power plants are sources of electricity, which, as one of the few, do not contain electromechanical converters such as synchronous or asynchronous electric machines. A PV power plant consists of solar panels (PV modules), a power electronic converter (connecting panels with the power grid), and a regulator. Modern PV panels used in PV power plants are converters of electromagnetic radiation energy, mainly in the visible range, of (17–20)% efficiency. The efficiency of radiation energy conversion depends on temperature and time (as a result of aging processes). The typical value of the power temperature coefficient, which defines the efficiency drop with temperature, is equal from −0.38%/K to −0.45%/K.

In turn, the change in efficiency over time is usually given in the form of a drop in efficiency over 25 years. A typical decrease in efficiency over this period is equal to 10%.

A PV panel consists of modules connected in parallel, and those from cells connected in series (60–90 cells with a voltage of 0.6 V). PV cells are divided into: made of monocrystalline or polycrystalline silicon (first generation), thin-film amorphous silicon containing cadmium telluride or a mixture of indium, selenium, gallium, and copper (second generation) and organic and polymeric modules (third generation). Despite the lower efficiency than the second generation cells, owing to durability, lower efficiency variability over time and less dependence of efficiency on temperature, panels with first-generation cells are still the most commonly used commercially.

Regardless of the cell type, the current–voltage characteristics of the PV panels are similar, and typical characteristics are shown in Figure 11.52. The shape of the characteristics depend on the irradiation (sunlight intensity) E, cell temperature T, and (in long-term) from time. As standard values (STC – standard test condition), the intensity of radiation $E_{STC} = 1000\ \mathrm{W/m^2}$, cell temperature $T_{STC} = 25^{\circ}\mathrm{C}$, and an air mass $AM_{STC} = 1.5$ are used to assess panels. The characteristic parameters of PV panels are: peak (maximum) power P_{max}, open circuit voltage V_{oc}, and short-circuit current I_{sc}. The rated power is equal to the maximum power that a panel can generate under standard conditions of radiation and temperature. This power is obtained for maximum power point: voltage V_{mpp} and current I_{mpp}, which are also given in the panel data chart. On the current–voltage characteristics (I–V characteristics), for different values of radiation intensity, shown in Figure 11.52, the operating points corresponding to the maximum power are marked with circles. Typical values of PV panel parameters, corresponding to the standard conditions, are: open circuit voltage of a single panel equal to $V_{oc} = (30 - 65)$ V, short-circuit current equal to $I_{sc} = (6 - 9)$ A, and peak power $P_{max} = (250 - 350)$ W.

Since the rated power of a single PV panel is small, the panels are connected in series in branches. The number of panels in the branch usually does not exceed 20 and results from the maximum and minimum DC voltage of the

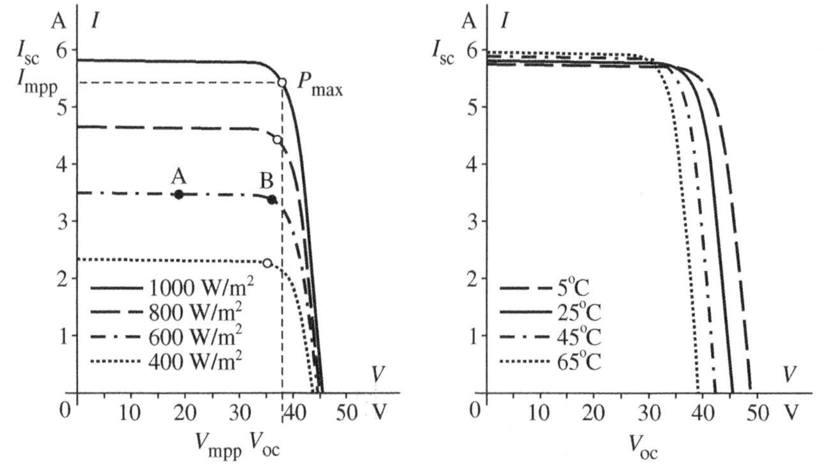

Figure 11.52 Example of current–voltage characteristics of the PV panel.

power electronic converter. The sum of open circuit voltages, resulting from the number of panels connected in series, cannot exceed the maximum input voltage (on the DC side) of the converter. At the same time, the number of panels should be large enough that their total output voltage (which depends on the operating point) for the majority of the operating time is greater than the minimum DC voltage of the converter. The typical minimum value of DC voltage, for which the converter is able to operate, is equal to (25–50)% of the maximum voltage.

The rated power of a PV plant, in addition to the number of panels in a branch, determines the number of branches and it is equal to

$$P = n_b n_p P_{max} \tag{11.215}$$

where n_p and n_b are the number of panels in a branch and the number of branches, respectively.

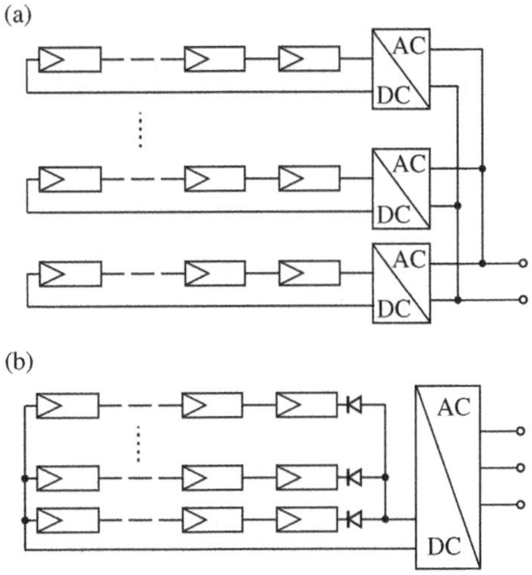

Figure 11.53 Photovoltaic power plants: (a) multi-converter system; (b) single converter system.

Depending on the rated power, small, medium-, or high-power converters are used in PV power plants. The solution used results from the cost and efficiency of the converters. These converters are low- or medium-voltage devices. Transistor-based converters operating as VSC are commonly used here. Examples of PV plant structures are presented in Figure 11.53.

The mathematical model of the VSC is described by Eqs. (11.196)–(11.200), and the structure is shown in Figure 11.38. As mentioned earlier, the transistor converter allows for control of two quantities. In this case, one is a constant voltage V_{DC}. The second controlled variable can be: the AC voltage V in the PCC, reactive power Q, or power factor cosφ. In addition, it is possible, and in the case of parallel operation of inverters it is necessary, to introduce droop in the regulated variable path, i.e. make dependent on the value of regulated voltage V from reactive power Q or reactive current component I_Q, which, for example, leads to dependence

$$V = V_{ref} + \kappa(Q_{ref} - Q) \tag{11.216}$$

where κ is the set value of droop. An example of the characteristics incorporating the regulator droop is shown in Figure 11.54. The inverter regulator also allows to shape the characteristics $Q(P)$, $Q(V)$, and cos$\varphi(P)$, i.e. make the value of regulated reactive power dependent on active power or voltage or make the regulated value of the power factor dependent on active power.

It is worth noting that the regulated quantity is not the active power directly. This is due to the fact that currently there is no restriction on the value of the active power supplied into the grid by PV power plants. PV power plants can inject into the grid the active power resulting from radiation intensity at a given moment. In PV plant case, the task of the plant regulator is to maximize the value of the active power. In general, however, the variable regulated by the VSC can also be reactive power.

From the point of view of modeling phenomena related to electromechanical oscillations in the EPS, the PV panel is a non-inertia object. Since the power converter is also treated as a non-inertia object, the regulator decides about the dynamics of the PV plant. Changes in the power generated by the panels depend on the change in the solar insolation and on the dynamics (structure and parameters) of the regulator. An exemplary structure of the PV plant control system is shown in Figure 11.55. The PV block represents PV panels, C represents the VSC capacitor, and R represents the converter with regulator.

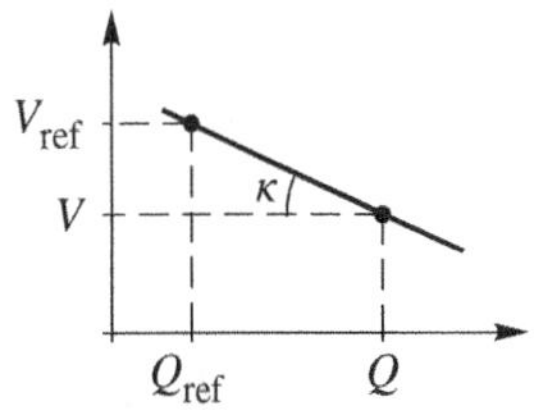

Figure 11.54 Example characteristic of the converter regulator droop.

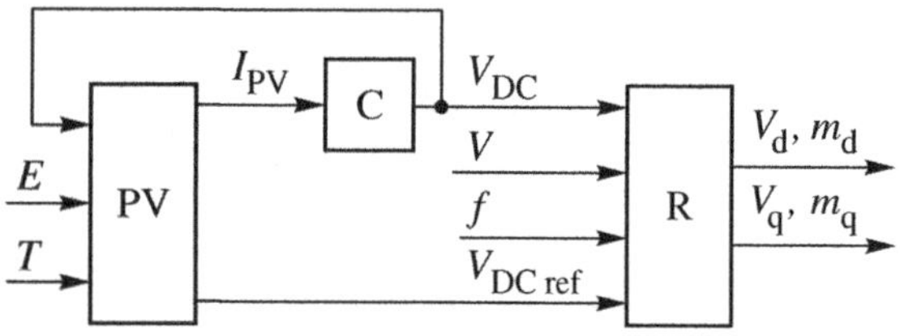

Figure 11.55 Block diagram of photovoltaic plant control system. PV – photovoltaic panels; C – converter's capacitor; R – regulator.

The value of radiation intensity E and panel temperature T at a given time determine the current voltage–current characteristic of PV panels. The operating point on this characteristic, e.g. point A in Figure 11.52, defined by the current value of the panel current $I_{PV} = I_A$, corresponds to the current DC voltage $V_{DC} = V_A$ on capacitor C. This operating point does not have to correspond at a given time to the maximum power that can be obtained from the PV panels under the given weather conditions, i.e. point B. This optimal operating point should be found (calculated) by the regulator. The maximum power corresponding to the maximum power operation point is calculated from the current current–voltage characteristics (here with points A and B). The voltage $V_{mpp} = V_B$ becomes the current value of the voltage reference for the regulator R, i.e. $V_{DC\,ref} = V_{mpp} = V_B$. The regulator, controlling the current component I_d, through the reference value $I_{d\,ref}$ of the current component (Figure 11.56) affects the DC voltage V_{DC} and thus the current I_{PV}, maximizing the power obtained from the PV panels at a given moment.

The method of calculating the reference value $V_{DC\,ref}$ of the DC voltage can be implemented in various ways, for example using a table containing the current–voltage characteristics of PV panels or using functions describing these characteristics or using another system that looks for an operating point corresponding to maximum power. Below, as an example, the voltage $V_{DC\,ref}$ calculation algorithm presented in DigSILENT (2018) is described.

The parameters describing the variability of the current–voltage characteristics of the PV panels are the open-circuit voltage coefficient $\alpha_V < 0$ and the short-circuit current coefficient $\alpha_I > 0$, which allow the effect of the temperature change to be modeled. The sample values of these coefficients are: $\alpha_V = -0.4\,\%/K$ and $\alpha_I = 0.05\,\%/K$. The coefficients are used to calculate the temperature correction factors: voltage κ_V and current κ_I, for temperature T equal to

$$\kappa_V = 1 + \alpha_V(T - 25), \qquad \kappa_I = 1 + \alpha_I(T - 25) \tag{11.217}$$

The input values of the algorithm are: constant voltage V_{DC}, temperature T, radiation intensity E, as well as parameters V_{oc}, I_{sc}, V_{mpp}, and I_{mpp}, describing the basic current–voltage characteristics of the PV panels, i.e. corresponding to the standard radiation intensity $E_{STC} = 1000\ \mathrm{W/m^2}$ and temperature $T_{STC} = 25^\circ\mathrm{C}$.

The characteristic values of the current–voltage characteristic for the temperature T and intensity of radiation E in the i - th step (or a given operating point) are equal to

$$I_{sci} = \kappa_I I_{sc} \frac{E}{E_{STC}}; \qquad V_{oci} = \kappa_V V_{oc} \frac{\ln E}{\ln E_{STC}} \tag{11.218}$$

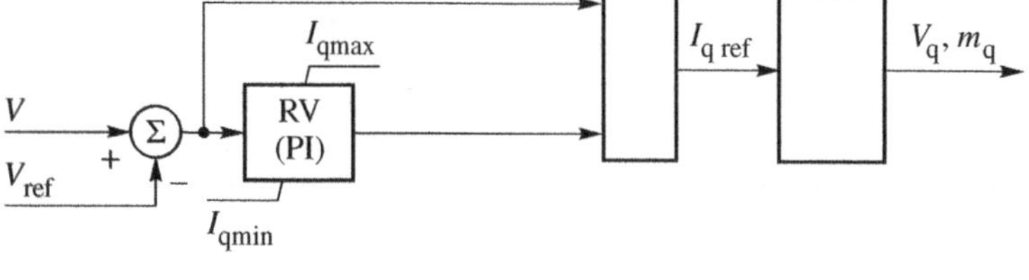

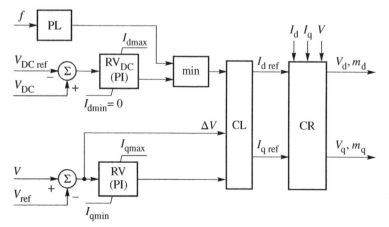

Figure 11.56 Block diagram of the PV plant converter regulator. Block CR as in Figure 11.47. *Source:* Based on DigSILENT (2018).

$$I_{\mathrm{mppi}} = \kappa_{\mathrm{I}} I_{\mathrm{mpp}} \frac{E}{E_{\mathrm{STC}}}; \qquad V_{\mathrm{mppi}} = \kappa_{\mathrm{V}} V_{\mathrm{mpp}} \frac{\ln E}{\ln E_{\mathrm{STC}}} \tag{11.219}$$

The maximum active power achievable under the given conditions is equal to

$$P_{\mathrm{mppi}} = V_{\mathrm{mppi}} I_{\mathrm{mppi}} \tag{11.220}$$

Total panel current I_{PV}, i.e. current resulting from the current current–voltage characteristics, for voltage V_{DC} is determined by the equation

$$I_{\mathrm{PV}} = I_{\mathrm{sc}}\left(1 - \mathrm{e}^{\beta}\right) \quad \text{where } \beta = \frac{\ln\left(1 - I_{\mathrm{mppi}}/I_{\mathrm{sci}}\right)\left(V_{\mathrm{DC}} - V_{\mathrm{oci}}\right)}{V_{\mathrm{mppi}} - V_{\mathrm{oci}}} \tag{11.221}$$

The I_{PV} current is input to block C, which is the VSC capacitor (Figure 11.55). The capacitor state is described by the equation

$$V_{\mathrm{DC}} = \frac{1}{C}\int (I_{\mathrm{PV}} - I_{\mathrm{I}})\,\mathrm{d}t \tag{11.222}$$

where I_{I} is the DC current of the converter. Change in the PV panels' current I_{PV} or converter current I_{I} causes a change in voltage V_{DC}.

The last of the model elements, i.e. the regulator R, may have a structure as shown in Figure 11.56. In the presented structure, the regulated variables are: voltage V_{DC}, whose reference value $V_{\mathrm{DC\ ref}}$ is determined by the system maximizing power provided by the PV panels, and the AC voltage V in the PCC.

The DC voltage control, as mentioned above, is carried out by controlling the d-axis (active) component of the inverter current (regulator RV$_{\mathrm{DC}}$ in Figure 11.56). The current reference value can be additionally modified by over-frequency PL, an example of which characteristic is shown in Figure 11.57a. The necessity to use an active PL with such a characteristic results from the requirements of power grid operators. The characteristic of this type of limiter is defined by one or two frequencies and the slope of the characteristics. The frequency above which the value of the active power generated begins to be decreased is typically equal to $f_{\mathrm{d}} = 50.2$ Hz, and the slope of the characteristics is equal to $\Delta P/\Delta f = 40\,\%$ /Hz. The return frequency of the power generated to the value resulting from sun radiation (solar insolation) is assumed at the level of $f_{\mathrm{u}} = (50.02 - 50.05)$ Hz.

The AC voltage control in the PCC is implemented by inverter current control in the q-axis (controller RV in Figure 11.56). In normal operating conditions, i.e. when the terminal's voltage V of the PV plant does not deviate from the rated value more than the defined value ΔV_{z} (Figure 11.57b), the regulator keeps the voltage reference value V_{ref}. However, if the inverter current exceeds the maximum permissible value I_{Imax}, the reactive current is limited by the current limiter CL to the level $I_{\mathrm{q}} = \sqrt{I_{\mathrm{Imax}}^2 - I_{\mathrm{d}}^2}$, resulting from the maximum permissible current I_{Imax} and (not subject to limitation) the d-axis (active) current component I_{d}. In this case, the regulator switches from the mode of the AC terminals voltage control into the mode of the q-axis (reactive) current component control.

However, in the case of a significant decrease of the AC voltage in the grid, for example as a result of a short-circuit, the q-axis (reactive) current component is forced and the d-axis (active) current component is limited. The value of the q-axis (reactive) current component results then from the characteristics of the current forcing system (an example of the characteristics is presented in Figure 11.57b). The d-axis (active) current component is then

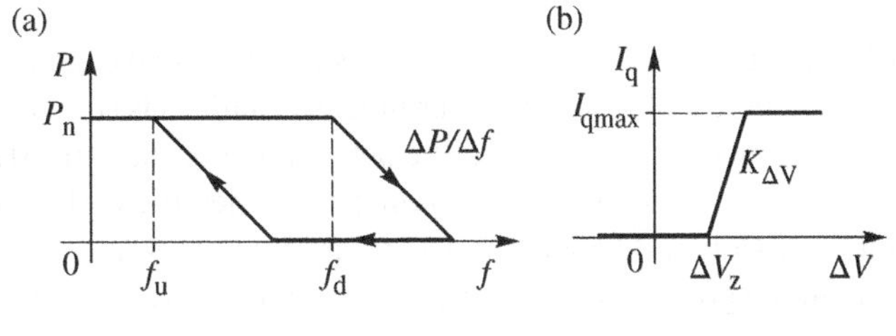

Figure 11.57 Characteristics of: (a) power-frequency limiter; (b) reactive current forcing function.

limited to the level in which the inverter current is equal to the maximum permissible current, i.e. $I_d = \sqrt{I_{I\max}^2 - I_q^2}$, where $I_q \leq I_{q\max}$ is the value of the forced q-axis (reactive) current component.

The output of the current limiter CL are the set values of the d-axis (active) and q-axis (reactive) components $I_{d\,ref}$ and $I_{q\,ref}$ of the inverter current. These components are controlled by the fast RID and RIQ regulators of the PI-type in a current regulator CR with a structure as shown in Figure 11.47. The output values of the current regulator CR are voltage components V_d and V_q or (depending on the way of the VSC modeling) the modulation index components m_d and m_q. The components of the modulation index can be calculated from Eq. (11.196), taking into account the value of voltage V_{DC}.

Example values of the parameters of the regulator shown in Figure 11.56 (DigSILENT 2018) are as follows: RV_{DC}: $K = 0.005$, $T = 0.03$ s; RV: $K = 0.005$, $T = 0.03$ s; RID, RIQ: $K = 1$, $T = 0.002$ s. The voltage and current signals are limited to the maximum permissible values, e.g. rated values.

The structure of the control system presented above not only reflects the structure of PV power plants. It also corresponds to the control systems of other EPS elements connected to the power grid directly by a power electronic converter, enabling the control of the active or reactive power, including: energy storage, energy sources with fuel cells, static reactive power compensators, and WTGSs with DC machines.

11.6 HVDC Links

The HVDC (high voltage direct current) links, also known as DC transmission systems, in EPSs are used to:

- Transmit electricity over long distances, because they eliminate the negative effect of the impedance of long AC transmission lines on the EPS stability.
- Connect power systems that for various reasons, e.g. technical, economic, or political, should not or cannot cooperate synchronously. For example, EPSs that have different nominal frequencies (Japan 50 Hz and 60 Hz) or potentially could create an AC system that is too large in terms of stability (UCTE and Russia IPS/UPS). Example of such HVDC links are also systems with zero length DC lines, referred to as the *back-to-back connection*.
- Connect power systems separated by the sea, e.g. located on islands or at sea, e.g. off-shore wind farms. HVDC links are cheaper than AC links when the distances between these systems or the system and the object exceed a certain distance.

11.6.1 HVDC Link Structure

The HVDC links have different structures. The basic ones are:

- Monopolar (Figure 11.58a). The DC transmission line can consist here of a single conductor, since the return path is usually earth (ground or water). Using the earth as a return path causes that metal constructions embedded in the ground (e.g. pipelines) and located on the route of the line become an element of the circuit. For this reason, and in the case of high ground impedances, metal conductors, e.g. cables or overhead lines, are also used as the return paths. The change of the energy flow direction in this system does not require a mechanical switching of both converters. This is accomplished by controlling the firing angle of the thyristors appropriately.
- Bipolar (Figure 11.58b). The DC transmission line is two conductors with potentials $+V$ and $-V$ with respect to the ground potential. Each of the terminal substations is equipped with two converters. Changing the direction of energy flow requires changing the polarization of the converters. In the event of damage to one of the converters, this link can operate as a monopolar. Then the earth becomes the return path.
- Homopolar (Figure 11.58c). The DC transmission line has two conductors with potentials $-V$ and $-V$ with respect to the ground potential. The choice of negative polarity results from a smaller negative environmental

Figure 11.58 Types of HVDC links: (a) monopolar; (b) bipolar; (c) homopolar. *Source:* Based on Kundur (1994).

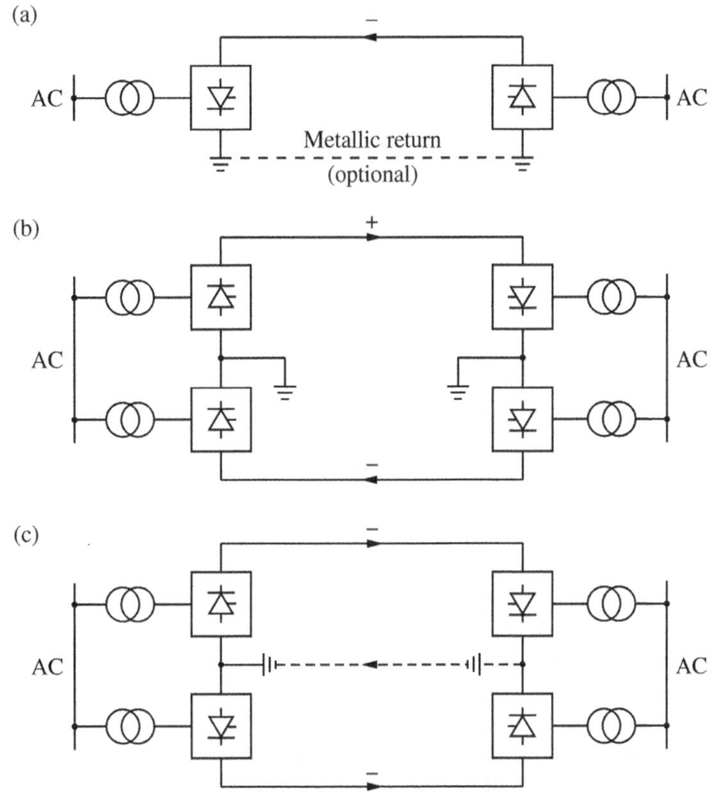

effect (radio interference due to corona effect). The return path is in this case the earth, as in the monopolar link. Switching off one conductor in this case does not eliminate the link from operation.

The DC link consists of two converter substations, a DC transmission line and a control system. The basic elements of the converter substations are power electronic (thyristors or transistors) converters, transformers, harmonic filters, and reactive power sources (capacitor banks). In the case of a thyristor-based converter in order to reduce the current and voltage pulsation in the DC line, the two connected converters are supplied from two transformers with a Y–Y and Y–Δ connection or from a three-winding transformer with a Y–Y–Δ configuration, which leads to a 12-pulse converter. An example of a bipolar HVDC link structure with thyristor converters is shown in Figure 11.59.

The power transmitted via the DC link is equal to

$$P_{\mathrm{DC}} = V_{\mathrm{DC}} I_{\mathrm{DC}} \tag{11.223}$$

where V_{DC} is DC voltage and I_{DC} is DC current.

11.6.2 Power Electronic Converter Models

In the case of a power electronic converter with a thyristor bridge, the relationship between the DC and AC sides of an ideal (lossless) three-phase, not controlled bridge is determined by

$$V_{\mathrm{DC0}} = \frac{s_0 q}{\pi} \sin\left(\frac{\pi}{q}\right) \frac{\sqrt{2}}{\sqrt{3}} V \tag{11.224}$$

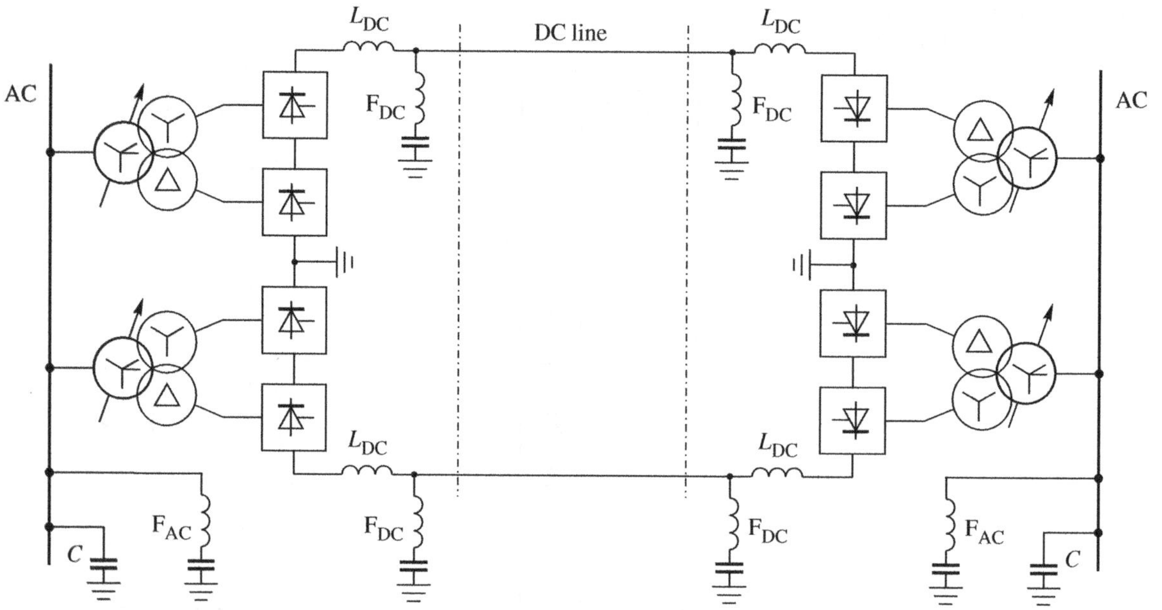

Figure 11.59 An exemplary structure of a bipolar HVDC) link. C – reactive power compensator; F_{AC} – AC filter (harmonic currents filter); F_{DC} – DC filter; L_{DC} – smoothing reactor.

where s_o is number of commutating groups, q is number of rectifier branches, V_{DCO} is DC voltage of the uncontrolled bridge, and V is the RMS line-to-line AC voltage. For a three-phase six-pulse system, where $s_o = 2$ and $q = 3$, the voltage of the ideal not controlled thyristor bridge is equal to

$$V_{DC0} = \frac{3\sqrt{2}}{\pi} V \tag{11.225}$$

For a controlled thyristor bridge, the DC voltage V_{DCO} depends on the firing angle (ignition delay angle) α and commutating angle (overlap angle) μ. The firing angle is defined as the angle between the moment the sine voltage on the thyristor (commutating voltage), for example v_{BA} in Figure 11.60, crosses zero and the moment the thyristor fires. The commutating angle is a measure of the switching time between the thyristors. An example of a commutation process between thyristors T_1 and T_3 is illustrated in Figure 11.60.

Taking into account these two parameters leads to the equation of a six-pulse controlled bridge

$$V_{DC} = V_{DC0} \frac{\cos\alpha + \cos(\alpha + \mu)}{2} = \frac{3\sqrt{2}}{\pi} V \cos\alpha - \frac{3}{\pi} X_c I_{DC} \tag{11.226}$$

where X_c is the commutating reactance, depending on the commutation angle and the reactance of the feeding circuit. Since the reactance of the transformer supplying the rectifier is the largest component of the commutation reactance, the commutation reactance is assumed to be equal to the transformer impedance. Expression $3X_c/\pi = R_c$ is named the equivalent commutating resistance.

Equation (11.226) is proper for the bridge first operation interval (Figure 11.18), i.e. the interval in which, apart from the commutation periods, only two thyristors are simultaneously conducting. During the normal operation of the bridge, but also during short-circuits, the equation correctly describes the behavior of the thyristor converter.

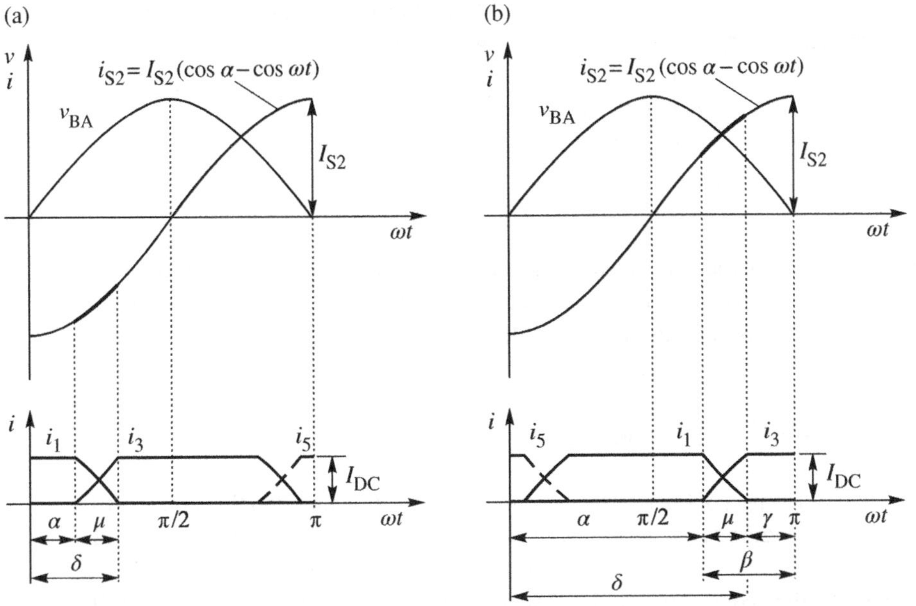

Figure 11.60 Definition of angles in the converter system on the example of commutation between thyristors T1 and T3 shown in Fig. 11.18 (a) rectifier; (b) inverter. v_{BA} and i_{S2} – voltage and current of commutation; i_1, i_3, and i_5 – currents of thyristors T1 and T3; $I_{S2} = \sqrt{2}V/(2X_c)$ – amplitude of commutation current of thyristor T3. *Source:* Based on Kundur (1994).

The relationship between the DC and AC of the six-pulse thyristor bridge, and precisely the first harmonic of this current, determines the equation

$$I = k\frac{\sqrt{6}}{\pi}I_{DC} \tag{11.227}$$

where coefficient k is close to one. For an ideal rectifier, i.e. for which the commutation angle μ is equal to zero, this coefficient equals one, i.e. $k = 1$. However, for nonzero values of the commutation angle, this coefficient can be slightly higher than one.

The power factor, which allows the value of reactive power consumed in the commutation process to be determined, is equal to

$$\cos\varphi = \frac{V_{DC}}{V_{DC0}} \tag{11.228}$$

Hence the active and reactive power on the AC side of the thyristor bridge, for a given value of power on the DC side, is equal to

$$\begin{aligned} P &= P_{DC} \\ Q &= P \cdot \mathrm{tg}\varphi \end{aligned} \tag{11.229}$$

In the case of a system consisting of several serially connected thyristor bridges, such as shown in Figure 11.59, the right side of the Eq. (11.226) should be multiplied by the number of rectifiers. In addition, by binding the rectifier model to the grid model, one should take into account the transformation ratio of the transformer feeding the converter.

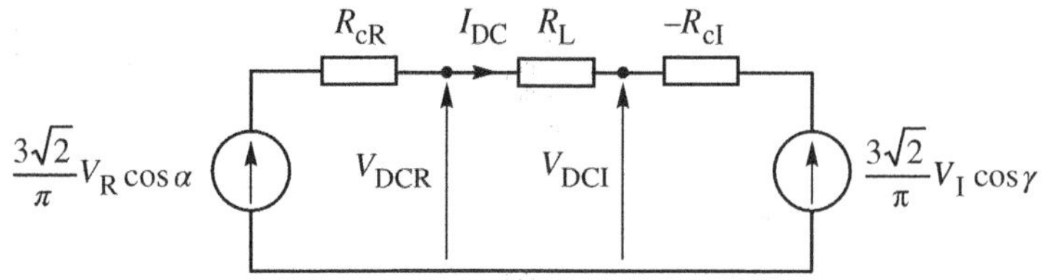

Figure 11.61 Mathematical model of the DC part of the HVDC link with thyristor-based converters. R_L – DC line resistance; R_{cR}, R_{cI} – commutation resistance of the rectifier and inverter.

The above-presented equations are valid for the controlled bridge both rectifier and inverter operations. The bridge inverter operation takes place for ignition delay angles α greater than 90°. However, since in the description of the inverter the extinction advance angle γ is often used (Figure 11.60), Eq. (11.226) for the inverter is usually presented in the form

$$V_{DC} = V_{DC0}\frac{\cos\gamma + \cos(\gamma+\mu)}{2} = \frac{3\sqrt{2}}{\pi}V\cos(\gamma+\mu) + \frac{3}{\pi}X_c I_{DC} \tag{11.230}$$

Based on Eqs. (11.226) and (11.230) the DC part of the HVDC link model, i.e. DC transmission line and thyristor power converters, can be represented as in Figure 11.61. The subscript R refers to the rectifier and the subscript I to the inverter. This model is a static model, because the time constants in this system associated with the switching processes are negligible in relation to the time constants associated with electromechanical oscillations in an EPS. In addition, this model does not include filters on the DC side, which results from the assumption of the ideal conversion of energy from the AC side to the DC side. This assumption is valid for system models based on RMS values of currents and voltages (RMS simulations), but not for models based on instantaneous values (the EMT simulations).

The model of the transistor-based converter of VSC type can be formed by Eqs. (11.196)–(11.199). On the basis of Eqs. the HVDC link model with transistor-based converters can be presented in the form as in Figure 11.62. This model neglects the inductance of DC transmission line and filters on the DC side.

Transformers feeding VSC (rectifier and inverter), unlike the thyristor converter, are not modeled as integral (internal) converter components. They are modeled like other AC grid components, e.g. transmission lines and other transformers. The HVDC links with transistor-based VSC are named VSC-HVDC links.

11.6.3 Control of HVDC Links

Owing to the variety of configurations, applied technologies (transistors, thyristors), role in power systems (EPSs interconnections, links to offshore wind farms, elements of DC grid) detailed structures of the HVDC link control systems can be different. Moreover the settings of controller parameters are usually not available too.

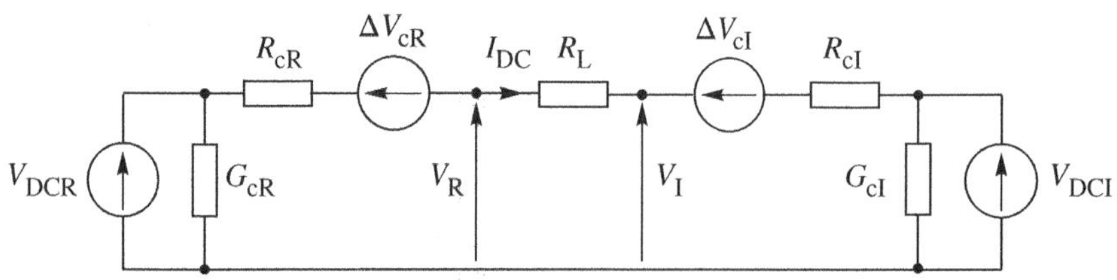

Figure 11.62 Mathematical model of the DC part of the HVDC link with transistor-based converters. R_L – line resistance; G_{cR} and G_{cI} – conductances of rectifier and inverter, modeling losses related to unloaded inverter switching; R_{cR} and R_{cI} – rectifier and inverter resistances modeling losses independent of the transistors switching; ΔV_{cR} and ΔV_{cI} – serial voltage drops related to switching transistor of rectifier and inverter.

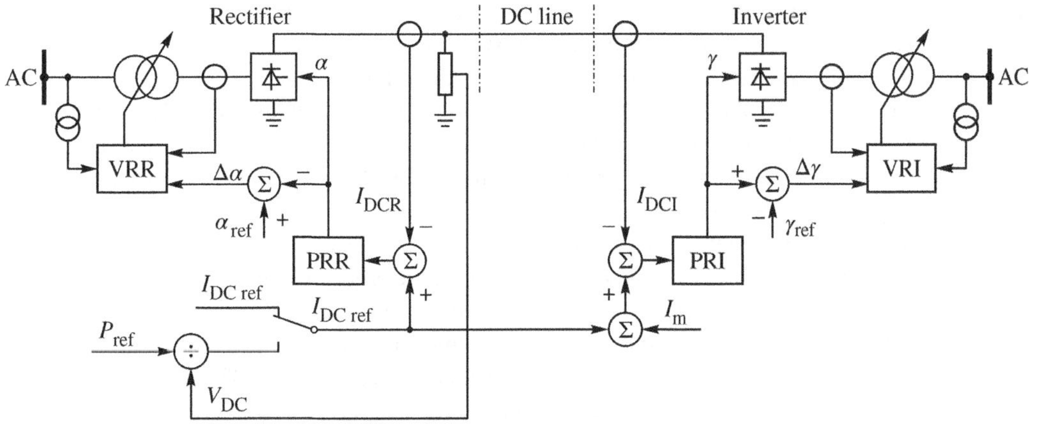

Figure 11.63 Structure of the HVDC link control system.

Below is discussed an example of the structure of the HVDC link control with thyristor-based converters and the HVDC link control system with transistor-based converters.

11.6.3.1 Control of Thyristor-based Converter HVDC Link

The control algorithm of the thyristor-based converter determines the structure of the control system and thus the structure of the link model. The HVDC link with thyristor-based converters can perform various control functions, where the control of current or active power are the basic ones. The general control algorithm has the following form:

- Current regulator of rectifier PRR (Figure 11.63) maintains the set value of direct current $I_{DC} = I_{DC\,ref}$. This corresponds to the characteristics of AB in Figure 11.64. In the case of a PI-type controller use, this characteristics is a vertical line, and in the case of a proportional controller, including with corrective elements, the AB characteristics become oblique. This regulator controls the thyristors firing delay angle α. The current regulator is a fast regulator with time constants in the order of a hundredth of a second.
- Current regulator of inverter PRI operates with a constant value of the extinction advance angle $\gamma = \text{const}$, which corresponds to the GD characteristics. Operating point E is at the intersection of the AB and GD characteristics. The regulator can also operate in constant voltage control mode. In this case, the GD characteristics become horizontal. The inverter regulator is also a fast regulator.
- As it results from Eq. (11.226), the value of direct current I_{DC} depends on the firing delay angle α and the AC voltage V_R supplying the rectifier and thus depends on the DC no load voltage $V_{DC0R} = \frac{3\sqrt{2}}{\pi} V_R \cos\alpha$ (Figure 11.61). Voltage regulator of rectifier VRR, superior to the current regulator of rectifier PRR, controls transformation ratio of the rectifier transformer and can control the reactive power compensator. The regulator's goal is to keep an optimum

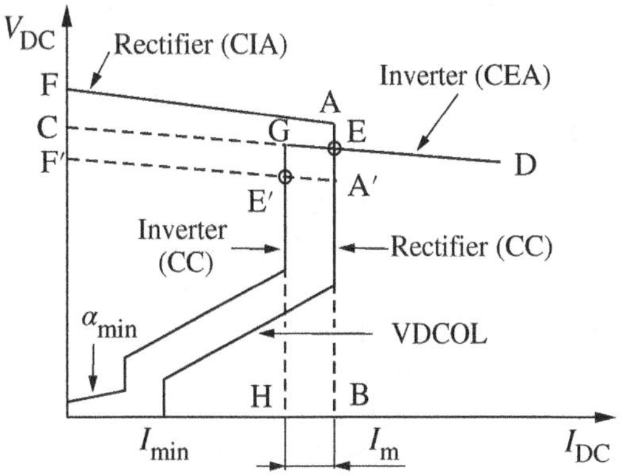

Figure 11.64 Characteristics of the HVDC link control system taking into account the protection, start-up and stop functions. CIA – constant ignition angle control mode α = const; CEA – constant extinction angle control mode γ = const; CC – constant current control mode. *Source:* Based on Kundur (1994).

(from the point of view of the rectifier operation) value of the firing delay angle, equal to $\alpha_{ref} = (10 - 20)^o$. The firing delay angle reference is set usually as close to the minimum, which is equal to $\alpha_{min} = (5 - 10)^o$. The regulator is a slow regulator, with a time constant of a few or several seconds or longer.
- At the same time, the voltage regulator of inverter VRI, superior to the current regulator of inverter PRI, controls the transformation ratio of the inverter transformer and can control the reactive power compensator (Figure 11.59). The regulator goal is to keep an optimum (from the point of view of the inverter) value of extinction advance angle γ, equal to $\gamma_{ref} = (15 - 18)^o$. This regulator is also a slow regulator.
- In the current (power) regulator of the inverter PRI, the current reference value is also set, which corresponds to the GH characteristics. The PRI regulator current reference is less than the reference value $I_{DC\ ref}$ of the PRR rectifier current by current I_m equal to (10 - 15)% of the HVDC link rated current. This is to ensure the correct operation of the link in the event of a voltage drop on the rectifier side, which may result from a sudden drop of the AC voltage. In this case (line F′–A′ in Figure 11.63) a new operating point E′ is established. The inverter goes into operation mode with current control, with current reference equal to $(I_{DC\ ref} - I_m)$, and the rectifier goes into operation mode with a constant value of the ignition delay angle α.

Operation of the HVDC link with thyristor converters in the power control mode P_{DC} is performed by controlling the link current as described above. The reference value $I_{DC\ ref}$ of the DC link current is determined based on the power reference $P_{DC\ ref}$ and voltage V_{DC} from the use of Eq. (11.223). In order to avoid power swings, this control is slow, i.e. change of the DC link current reference $I_{DC\ ref}$, in order to obtain a given power value, is appropriately slowed down at the input of the control system.

The control system of the real HVDC link also performs protection functions, i.e. ensuring correct operation of the link in different states, and functions related to starting and stopping the link operation. These functions have the form of delimiters, modifying the characteristics depicted in Figure 11.63. Among them, one can distinguish: the limiter of the maximum and minimum link current and the voltage-dependent current-order limiter VDCOL. An exemplary characteristics of a link control system including limiters is shown in Figure 11.63.

11.6.3.2 Control of Transistor-based Converter HVDC Link

As mentioned earlier, the transistor-based VSC in contrast to the thyristor-based converter, enables (Eq. [11.196]) independent control of the d-axis (active) and q-axis (reactive) components of the AC voltage. In practice, this means that it is possible to control a pair of values from: DC link voltage and DC link current as well as voltage, current, power, and power factor of the AC side. For typical operating modes of transistor-based converters (VSC-HVDC), depending on the location in the system, the controlled object type and the needs, typical control modes are[6]:

- active power and AC voltage (P, V) – applied to grid-side converters of VSC-HVDC links, and machine-side inverters of WTGSs with synchronous and induction machines;
- active and reactive power (P, Q) – used as (P, V) control;
- DC voltage and reactive power (V_{DC}, Q) – used for STATCOM (static compensator), unified power flow controller (UPFC), PV plants, and grid-side inverters of WTGSs with synchronous and induction machines;
- DC voltage and AC voltage (V_{DC}, V) – used as (V_{DC}, Q) control;
- DC voltage and power factor (V_{DC}, $\cos\varphi$) – used for VSC-HVDC systems and converters of renewable energy sources;
- AC voltage and frequency (V, f) – used for inverters of offshore wind farms.

6 The VSC, as shown in Sections 11.4 and 11.5, are used in a wide class of objects, e.g. power plants, compensators, and HVDC links. Therefore, the information presented here summarizes application of VSC regulators for the discussed objects, including a WTGS, PV power plant, UPFC, and STATCOM.

The above indicates the VSC wide ability to control variables in HVDC link. For example, in the case of an HVDC link connecting two AC nodes in the EPS the VSC regulator in sending node can operate in active power and AC voltage control mode while regulator in receiving node can operate in DC voltage and AC voltage control mode.

Below, as an example of another possible application of an HVDC link, structures of regulators are presented applied to VSC control in the case of offshore wind farms with an onshore substation connection. In this case the offshore VSC regulator controls AC voltage and frequency, while the onshore VSC regulator controls DC voltage and reactive power.

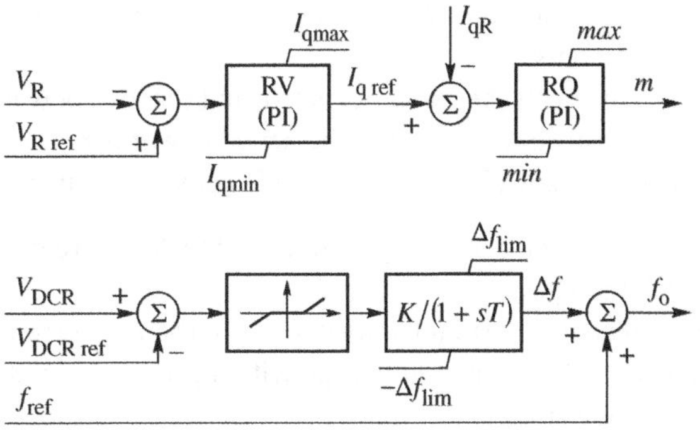

Figure 11.65 Block diagram of the VSC-HVDC wind farm inverter regulator. *Source:* Based on DigSILENT (2018).

An exemplary structure of the regulator of the offshore converter is presented in Figure 11.65. The regulator maintains the set value of voltage V_R (regulator RV) and frequency f in the wind farm AC grid.

The wind farm (offshore) converter AC voltage components V_d and V_q can be calculated by using Eq. (11.196) or the AC voltage can be calculated as

$$\underline{V} = k_o m V_{DCo}(\cos\varphi + \mathrm{j}\sin\varphi) \tag{11.231}$$

where the modulation index m determines the magnitude of the wind farm AC terminal voltage, while the functions cos φ and sin φ define d- and q-axis components of this voltage and simultaneously they define active and reactive power on the converter AC side.

When the output signals of the regulator are components m_d and m_q of the modulation index, (like in previously presented regulators), the modulation index m and functions cos φ and sin φ are calculated as

$$m = \sqrt{m_d^2 + m_q^2}; \qquad \cos\varphi = m_d/m; \qquad \sin\varphi = m_q/m \tag{11.232}$$

Typically, the wind farm frequency is set by the converter regulator as constant, e.g. equal to nominal. But it is also possible to make the wind farm supporting the AC power system in case of disturbances, e.g. in case of frequency deviation in EPS. To achieve the effect of inertial response of a wind farm connected to grid through a HVDC link, both regulators have to be equipped with dedicated control functions. The grid-side converter regulator has to be equipped with a control function which makes the power injected to the grid dependent on frequency while the offshore converter regulator has to be equipped with a control function which makes the wind farm frequency dependent from DC voltage. An example of such a function is presented in the lower block diagram in Figure 11.65. The regulator output, i.e. frequency f_o, is initially set as the reference value for the converter. In normal operating conditions, it is equal to f_{ref}. However, when the DC voltage V_{DC} deviates excessively up or down from the reference value $V_{DC\,ref}$, the regulator modifies the frequency and, through Eq. (11.231), modifies the wind farm AC voltage components. The functions cos φ and sin φ in Eq. (11.231) can be calculated as

$$\cos\varphi = \cos(\varphi_0 + \Delta\varphi); \qquad \sin\varphi = \sin(\varphi_0 + \Delta\varphi) \tag{11.233}$$

where φ_0 is the angle calculated in the previous simulation step and $\Delta\varphi$ is the angle change defined as

$$\frac{\mathrm{d}\Delta\varphi}{\mathrm{d}t} = 2\pi(f_0 - f_{ref}) \tag{11.234}$$

where f_{ref} and f_0, respectively, are the wind farm frequency reference and the regulator output frequency. The initial value of the angle φ_0 defines the initial value of the active and reactive power on the converter AC side. The value of this angle equal to zero means that the converter (which is a slack node for the wind farm grid) transfers active power only.

The above affects the converter active and reactive power and next influences on power generated by WTGSs in the wind farm. And this makes the active power generated by the wind farm dependent on frequency in EPS.

In turn, the controller of the grid-side converter (onshore controller) acts on:

- d-axis (active) current component I_{dI} of the AC current injected to the grid (regulator RV_{DC}), by controlling the DC voltage V_{DCI};
- q-axis (reactive) current component I_{qI} of the AC current injected to the grid (regulator RR) by controlling: the grid voltage V_I or the grid voltage with a predetermined droop (depending on active power) $V_I + \Delta V(P)$ or the reactive power Q_I.

During normal operation of the HVDC link, the reactive component of the grid-side inverter current I_{qI} is usually minimized. The inverter then operates as a source of active power. However, in a short time (e.g. up to 500 ms after the disturbance) resulting in a significant reduction of the grid voltage (AC voltage supplying the inverter), the active component of the current is limited, and the control is carried out by forcing the reactive component of the current I_{qI}. Simultaneously, the grid-side converter regulator can participate in frequency primary control or perform tasks related to the damping of electromechanical oscillations.

The grid-side converter regulator structure is presented in Figure 11.66. The regulator components not listed in Figure 11.66 (i.e. the current limiter CL and the current regulator CR) are discussed in Section 11.4.3.4 and shown in Figure 11.47. Exemplary values of controller parameters from Figure 11.65 and Figure 11.66 (DigSILENT 2018) are as follows: RV: $K = 2$, $T = 0.2$ s; RQ: $K = 2$, $T = 0.02$ s; RV_{DC}: $K = 10$, $T = 0.1$ s; RR: $K = 12$, $T = 0.01$ s; RID, RIQ: $K = 1$, $T = 0.002$ s.

The regulators presented in Figure 11.65 and Figure 11.66 together with the HVDC link model (Eqs. (11.196)–(11.199) and Figure 11.62) form a mathematical model of a DC link of this type. Other models of exemplary HVDC link control systems can be found in Ainsworth (1985), Arrillaga et al. (2007), CIGRE Technical Brochure No. 604 (2014), Clark et al. (2010), Fisher et al. (2012), Guan and Xu (2012), and Machowski et al. (2013) and in libraries and software manuals for analysis of power system operation, such as PowerFactory (DigSILENT 2018), PSLF (GE Energy Consulting 2009), and PSSE, PSCAD/EMTDC.

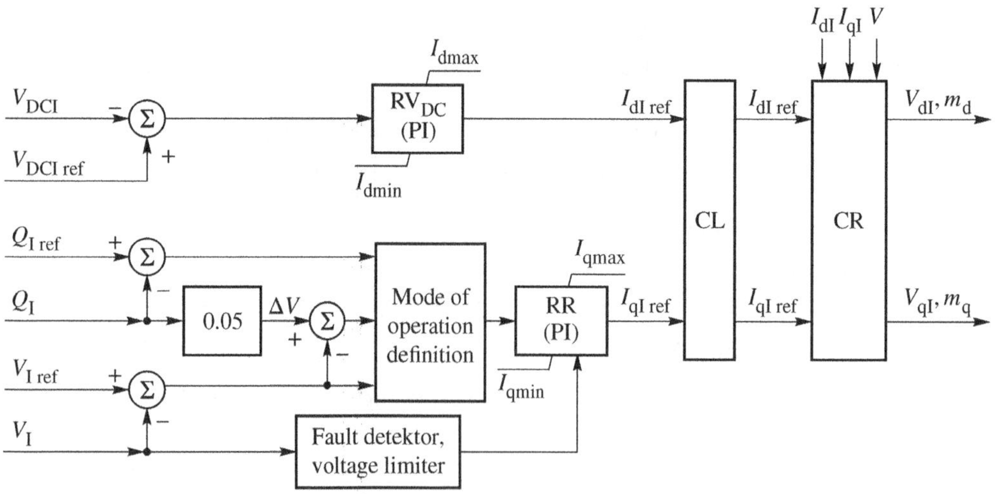

Figure 11.66 Block diagram of the VSC-HVDC grid-side inverter regulator. Block CR as in Figure 11.47. *Source:* Based on DigSILENT (2018).

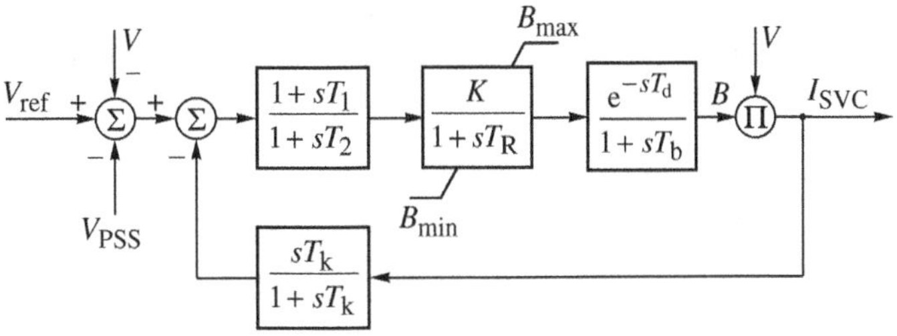

Figure 11.67 Simplified dynamic model of the SVC.

11.7 Facts Devices

In power system stability analysis it is necessary to include dynamic models of those elements whose control is fast enough to influence electromechanical dynamics. FACTS elements belong to that category and are described in Section 2.5. CIGRE (Technical Brochure No. 145) published a detailed report on modeling FACTS devices. Here only some simple models are discussed.

11.7.1 Shunt FACTS Devices

Figure 11.67 shows a dynamic model of the SVC based on conventional thyristors. The model is developed from the simplified model shown in Figure 2.28. The first and second blocks of the dynamic model represent the regulator. Time constants T_1 and T_2 of the correction block are chosen based on stability analysis of the EPS. That choice is heavily influenced by the type of optional PSS whose task is to stabilize control of the compensator (see Section 10.5). The gain of the regulator is $K \cong (10 - 100)$. Such a gain corresponds to the droop of the regulator static characteristic (Figure 2.29) of about (1 - 10)%. The time constant of the regulator is $T_R \cong (20 - 150)$ ms. The third block represents the thyristor firing circuits. The time constants are in the range $T_d \cong 1$ ms and $T_b \cong (10 - 50)$ ms. At the bottom of the diagram there is a differentiating block used in the SVC to stabilize the circuit using the compensator current I_{SVC}.

In the discussed SVC model the most important role is played by the models of the voltage regulator and the PSS. From the point of view of electromechanical dynamics, thyristors and their firing circuits are proportional elements.

The dynamic model of STATCOM is shown in Figure 11.68 and it includes the regulator's transfer function and the feedback loop ρ enforcing a required droop of the compensator static characteristic (Figure 2.31). The converter is modeled by a first-order block with time constant $T_C = (10 - 30)$ ms. The output signal is the compensator current. In the steady state when $t \to \infty$ or $s \to 0$ the model results in the static characteristic shown in Figure 2.31 with droop ρ.

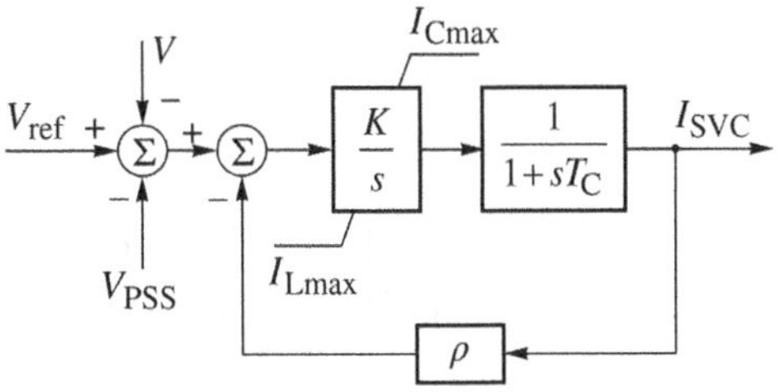

Figure 11.68 Simplified dynamic model of STATCOM.

11.7.2 Series FACTS Devices

The most general series FACTS device is the UPFC described in Section 2.5.4. Hence, here only the dynamic model of the UPFC is discussed. Descriptions of models of other series FACTS devices can be found in CIGRE Technical Brochure No. 145 (1999).

Figure 2.40 shows that the UPFC has two voltage-sourced converters and each of them is equipped with its own PWM controller with two control parameters: m_1 and ψ_1 for CONV 1 and m_2 and ψ_2 for CONV 2. These four parameters are selected by the UPFC regulator controlling the following three important quantities:

- Direct component Re $(\Delta\underline{V})$ of the booster voltage.
- Quadrature component Im $(\Delta\underline{V})$ of the booster voltage.
- Reactive component of the shunt current Im$(\underline{I}_{\text{shunt}})$.

From the point of view of electromechanical dynamics, thyristors and their firing circuits are proportional elements and may be neglected. Hence the dynamic model of the UPFC consists of its regulator model including technical constraints and is shown in Figure 11.69. The input variables in the series part of the regulator (Figure 11.69a) are the transmitted active and reactive power. The output variables are the direct and quadrature components of the booster voltage. The reference values at the input of the regulator are divided by the actual voltage values in order to obtain the required current components. The actual values of the components are subtracted from the reference ones in order to create control errors. The regulator is of integral type with negative feedback. Time constants of the integral block $T_P = T_Q$ are chosen depending on the speed with which power flow has to be regulated and on the type of PSS used. Converters are modeled by first-order blocks with time constants in the range of T_C= (10–30) ms. There is a limiter on the output of the regulator that proportionally reduces the components of the booster voltage when its magnitude exceeds the allowed value.

The input signal of the regulator's shunt part (Figure 11.69b) is the busbar voltage, while the output signal is the reactive component of the shunt current. The regulator is of integral type with constraints. The converter is modeled by a first-order block. The output of the block produces the converter voltage, that is the voltage on the lower side of the excitation (supply) transformer. From that value the voltage on the upper side of the excitation transformer is subtracted in order to obtain the reactive part of the UPFC shunt current. Similar to STATCOM, there is also a negative feedback loop responsible for a required droop of the static characteristic. There is also an additional signal PSS entering the summation point of regulation loops. That signal is due to an additional optional block responsible for the damping of power swings in the system – see Section 10.7.

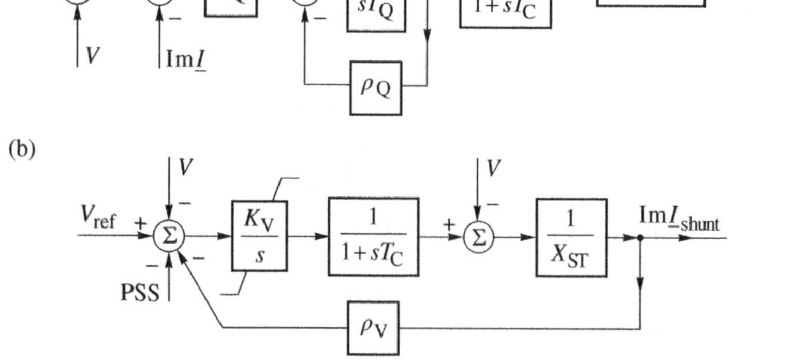

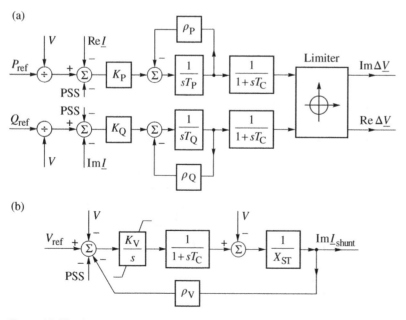

Figure 11.69 Simplified dynamic model of the UPFC with regulators: (a) series part; (b) shunt part.

11.8 Dynamic Load Models

Section 3.4 explains how power system loads, at the transmission and the subtransmission levels, can be represented by a static load characteristic which describes how the active and reactive power at a busbar change with both voltage and frequency. Although this is an adequate method of representing modest changes of voltage and/or frequency for a composite load, there are cases when it is necessary to account for the dynamics of the load components themselves.

Typically, motors consume (60–70)% of the total power system energy, and their dynamics are important for studies of inter-area oscillations, voltage stability, and long-term stability. This section presents the dynamic model of an induction motor. The dynamic model of the synchronous motor is not discussed as it is identical to that of the synchronous generator presented in Section 11.1, except for a sign change on the armature current necessary to reflect the motoring action.

The induction motor can be modeled in the same way as the synchronous generator but with three important differences. First of all, as there is no field winding in the induction motor, the cage rotor is modeled by two coils in quadrature (as for the damper windings in the synchronous machine) while the armature is modeled in the normal way by a d- and a q-axis armature coil. Secondly, as the induction motor does not rotate at synchronous speed, both the stator armature coils and the two rotor coils are transformed into a reference frame that rotates at synchronous speed. This means that the two rotor equations now include a rotational emf term proportional to the rotor slip speed $s\omega_s$ and are of the form

$$\begin{aligned} v_{dR} &= 0 = R_R i_{dR} + \dot{\Psi}_{dR} - s\omega_s \Psi_{qR} \\ v_{qR} &= 0 = R_R i_{qR} + \dot{\Psi}_{qR} + s\omega_s \Psi_{dR} \end{aligned} \tag{11.235}$$

where the suffix R signifies rotor quantities and the slip $s = (\omega_S - \omega)/\omega_S$.

The two rotational voltages are proportional to the rotor slip speed and to the flux linkage of the other rotor coil. During a disturbance, the flux linkages Ψ_{qR} and Ψ_{dR} cannot change instantaneously and, just as in Eqs. (11.54) and (11.60), can be equated to d- and q-axis armature emf's E'_d and E'_q. If Eq. (11.235) is further compared with Eq. (11.51) the rotational voltages $s\omega_s\Psi_{qR}$ and $s\omega_s\Psi_{dR}$ can be seen to play a similar role as the excitation voltage E_f and will therefore appear in the corresponding flux decay equations.

The third important difference is that positive current is now defined for motoring action, requiring a sign change in the appropriate equations.

With these points in mind Figure 11.70 shows how the induction motor can be modeled by a transient emf E' behind a transient impedance X' in the same way as in the fourth-order model of the synchronous generator (Section 11.1.6). However, as the reactance is unaffected by rotor position, and the model is in the synchronously rotating reference frame, the necessary equations are more conveniently expressed in the network (a, b) coordinates (Arrillaga and Arnold 1990) as

$$\begin{bmatrix} V_b \\ V_a \end{bmatrix} = \begin{bmatrix} E'_b \\ E'_a \end{bmatrix} + \begin{bmatrix} R_S & X' \\ -X' & R_S \end{bmatrix} \begin{bmatrix} I_b \\ I_a \end{bmatrix} \tag{11.236}$$

with the change in the emf's E'_b and E'_a given by

$$\begin{aligned} \dot{E}'_b &= -s\omega_s E'_a - \frac{E'_b - (X - X')I_a}{T'_0} \\ \dot{E}'_a &= s\omega_s E'_b - \frac{E'_a + (X - X')I_b}{T'_0} \end{aligned} \tag{11.237}$$

where $s = (\omega_s - \omega)/\omega_s$ is the rotor slip, $X' = X_S + X_\mu X_R/(X_\mu + X_R)$ is the transient reactance and is equal to the blocked-rotor (short-circuit) reactance, $X = X_S + X_\mu$ is the motor no-load (open-circuit) reactance, and

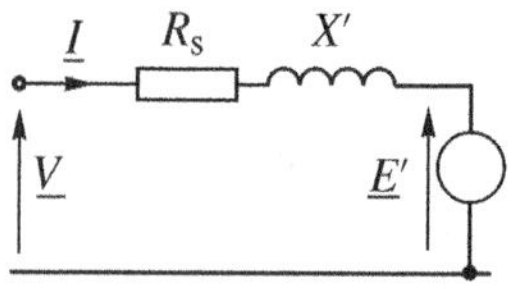

Figure 11.70 Transient state representation of induction motor.

$T_0' = (X_R + X_\mu)/(\omega_s R_R)$ is the transient open-circuit time constant. The other reactances take the meaning explained in the induction motor's steady-state equivalent circuit in Figure 3.28. The stator current can be calculated from Eq. (11.236). Alternatively, Eqs. (11.236) and (11.237) can be written directly in phasor notation as

$$\underline{V} = \underline{E}' + \mathrm{j}X'\underline{I} + \underline{I}R_s$$
$$\dot{\underline{E}}' = -s\omega_s\mathrm{j}\underline{E}' - \frac{[\underline{E}' - \mathrm{j}\underline{I}(X - X')]}{T_0'} \tag{11.238}$$

To evaluate the slip $s = (\omega_s - \omega)/\omega_s$ it is necessary to calculate the rotor speed ω from the equation of motion

$$J\frac{d\omega}{dt} = \tau_e - \tau_m \tag{11.239}$$

where J is the inertia of the motor and load, τ_m is the mechanical load torque, and τ_e is the electromagnetic torque that can be converted into useful work.

Taking into account that the slip is $s = (\omega_s - \omega)/\omega_s$ and using the notation shown in the equivalent circuit of Figure 3.31, the induction motor air-gap power can be expressed as

$$P_{ag} = I^2\frac{R}{s} = I^2R\frac{\omega_s}{\omega_s - \omega} \tag{11.240}$$

while the electromagnetic power P_e is given by

$$P_e = I^2R\frac{(1-s)}{s} = P_{ag}(1-s) = P_{ag}\frac{\omega}{\omega_s} \tag{11.241}$$

The electromagnetic torque may now be expressed using the last two equations as

$$\tau_e = \frac{P_e}{\omega} = \frac{P_{ag}}{\omega_s} \tag{11.242}$$

where for the considered model (Figure 11.28)

$$P_{ag} = \mathrm{Re}\,(\underline{E}'\underline{I}^*) = E_b'I_b + E_a'I_a \tag{11.243}$$

Normally, τ_m will vary with speed and is commonly represented by a quadratic equation of the form

$$\tau_m = \tau_{m0}\left(A\omega^2 + B\omega + C\right) \tag{11.244}$$

where τ_{m0} is the rated load torque, $A\omega_0^2 + B\omega_0 + C = 1$, and ω_0 is the rated speed. For example, a simple pump load, the torque of which is proportional to speed squared, can be represented by setting $B = C = 0$.

The above analysis is derived for single-cage induction motors but can be readily extended to double-cage and deep-bar rotors by using slip-dependent rotor parameters $X_R(s)$ and $R_R(s)$ in the calculation of X' and T_0', as explained by Arrillaga and Arnold (1990). However, in many cases it is sufficient to model the induction motor by its steady-state equivalent circuit when the electrical torque is obtained from the static torque-speed characteristic and the slip calculated via Eq. (11.239). The format of the induction motor dynamic model is similar to that of the synchronous generator, so its inclusion in a simulation program is relatively straightforward.

With the increased penetration of renewable generation often connected at the distribution level, it may become necessary to include dynamic models of distribution networks when assessing the system stability. As that would increase the size of the problem quite considerably, dynamic equivalents of the full distribution network model could be used – see Chapter 14. Often, the actual configuration and models of generators connected may not be known and then the equivalent would be derived from measurements of certain electrical quantities taken inside the distribution network and/or at the border nodes, for example Feng et al. (2007).

12

Steady-state Stability of Multi-machine Systems

The *steady-state*, or *small-signal*, stability of an electric power system (EPS) is the ability of the system to maintain synchronism when subjected to a small disturbance. In Chapter 5 the steady-state stability of the generator-infinite busbar system is discussed and the system is seen to be steady-state stable about an equilibrium point if, following a small disturbance, the system remains within a small region surrounding the equilibrium point. In such cases the system is said to be *locally stable*. Furthermore, if, as time progresses, the system returns to the equilibrium point, it is also said to be *asymptotically stable*. These concepts, and those introduced in Chapter 5, are expanded in this chapter to assess the steady-state stability of a multi-machine power system in which the generators are described by the mathematical models introduced in Chapter 11.

12.1 Mathematical Background

Section 5.4.6 analyzes the classical model of a synchronous generator connected to the infinite busbar and showed that rotor swings around the synchronous speed can be described by a second-order differential equation (the swing equation). Depending on the values of the roots of the characteristic Eq. (5.46), the swings can be aperiodic or oscillatory. A power system consists of many generators and each of the generators is more accurately described by a higher-order model (discussed in Section 11.1.6). Consequently, an active power system will be described by a high number of nonlinear differential equations. This chapter analyzes the steady-state stability of such a large dynamic system using *eigenvalue analysis*. The main aim of eigenvalue analysis is to simplify analysis of a large dynamic system by representing the system response to a disturbance as a linear combination of uncoupled aperiodic and oscillatory responses, similar to those analyzed in Section 5.4.6 and referred to as the *modes*.

12.1.1 Eigenvalues and Eigenvectors

A number λ is referred to as an *eigenvalue* of matrix $\boldsymbol{A}$ if there is a nonzero column vector $\boldsymbol{w}$ satisfying

$$\boldsymbol{A}\boldsymbol{w} = \boldsymbol{w}\lambda \tag{12.1}$$

Each such vector $\boldsymbol{w}$ is referred to as the *right eigenvector* associated with eigenvalue λ. Equation (12.1) shows that eigenvectors are not unique as they can be rescaled by multiplying or dividing their elements by a nonzero number. If $\boldsymbol{w}_i$ is an eigenvector then any other vector $c\boldsymbol{w}_i$ is also an eigenvector, where $c \neq 0$ is a nonzero number. This property makes it possible to multiply or divide eigenvectors by any number. In practice eigenvalues are normalized by dividing their values by the vector length

$$\|\boldsymbol{w}\| = \sqrt{\boldsymbol{w}^{\mathrm{T}*}\boldsymbol{w}} = \sqrt{|w_1|^2 + \ldots + |w_n|^2} \tag{12.2}$$

Power System Dynamics: Stability and Control, Third Edition. Jan Machowski, Zbigniew Lubosny, Janusz W. Bialek and James R. Bumby.

Eq. (12.1) can be rewritten as

$$(\boldsymbol{A} - \lambda \mathbf{1})\, \boldsymbol{w} = \mathbf{0} \tag{12.3}$$

where $\mathbf{1}$ is a diagonal identity matrix while $\mathbf{0}$ is a column vector of zeros. Equation (12.3) has a nontrivial solution $\boldsymbol{w} \neq \mathbf{0}$ if and only if

$$\det(\boldsymbol{A} - \lambda\mathbf{1}) = 0 \tag{12.4}$$

Equation (12.4) is called the *characteristic equation*. It can be written in a polynomial form

$$\det(\boldsymbol{A} - \lambda\mathbf{1}) = \varphi(\lambda) = (-1)^n \left(\lambda^n + c_{n-1}\lambda^{n-1} + \ldots + c_1\lambda + c_0\right) \tag{12.5}$$

where n is the rank of $\boldsymbol{A}$, c_{n-1} is the sum of the main minors of the first degree, c_{n-2} is the sum of the main minors of the second degree, and so on, and finally c_0 is the main minor of the highest degree, that is the determinant of matrix $\boldsymbol{A}$. Polynomial (12.5) is of the n-th order and therefore has n roots $\lambda_1, \lambda_2, \ldots, \lambda_n$ which are at the same time the eigenvalues of matrix $\boldsymbol{A}$.

Numerical methods for determining the eigenvalues and eigenvectors of a matrix can be found in Press et al. (1992) or other textbooks on numerical methods and so they are not dealt with in this book. Instead, we concentrate on the application of eigenvalues and eigenvectors to the stability analysis of power systems.

Knowing eigenvalue λ_i of matrix $\boldsymbol{A}$, it is easy to find its associated eigenvector $\boldsymbol{w}_i$ using the following equation

$$\boldsymbol{w}_i = \text{col}\,(\boldsymbol{A} - \lambda_i\mathbf{1})^{\text{D}} \tag{12.6}$$

where col denotes selection of any nonzero column from a square matrix and upper index D denotes an adjacent matrix. The correctness of Eq. (12.6) can be proved using the definition of the adjacent matrix. For any matrix $\boldsymbol{B}$ it holds that $\boldsymbol{B}\boldsymbol{B}^{\text{D}} = \det \boldsymbol{B} \cdot \mathbf{1}$. This property is true also for matrix $\boldsymbol{B} = (\boldsymbol{A} - \lambda_i\mathbf{1})$, which can be written as $(\boldsymbol{A} - \lambda_i\mathbf{1})(\boldsymbol{A} - \lambda_i\mathbf{1})^{\text{D}} = \det(\boldsymbol{A} - \lambda_i\mathbf{1}) \cdot \mathbf{1}$. Hence, taking into account (12.4), one gets $(\boldsymbol{A} - \lambda_i\mathbf{1})(\boldsymbol{A} - \lambda_i\mathbf{1})^{\text{D}} = 0 \cdot \mathbf{1}$. This equation shows that multiplying matrix $(\boldsymbol{A} - \lambda_i\mathbf{1})$ by any column of matrix $(\boldsymbol{A} - \lambda_i\mathbf{1})^{\text{D}}$ gives $0 \cdot \mathbf{1} = \mathbf{0}$, where $\mathbf{0}$ is a column vector of zeros. This shows that if $\mathbf{a} \neq \mathbf{0}$ is a nonzero column of $(\boldsymbol{A} - \lambda_i\mathbf{1})^{\text{D}}$, then $(\boldsymbol{A} - \lambda_i\mathbf{1})\mathbf{a} = \mathbf{0}$ or $\boldsymbol{A}\boldsymbol{a} = \lambda_i\boldsymbol{a}$. This shows, according to Eq. (12.1), that $\boldsymbol{a} = \boldsymbol{w}_i$ is a right eigenvector associated with λ_i. In other words, any nonzero column of $(\boldsymbol{A} - \lambda_i\mathbf{1})^{\text{D}}$ is an eigenvector of $\boldsymbol{A}$ associated with λ_i.

This methodology can be used to determine eigenvectors for all eigenvalues if the eigenvalues are distinct, that is $\lambda_1 \neq \lambda_2 \neq \ldots \neq \lambda_n$. However, if λ_i is a multiple eigenvalue repeated k times then the described methodology can be used to determine only one eigenvector associated with such a multiple eigenvalue. The other $(k-1)$ linearly independent eigenvectors associated with λ_i have to be determined in a different way. This is not discussed here; the reader can find details in many textbooks, for example Ogata (1967).

A square complex matrix $\boldsymbol{A}$ is Hermitian if $\boldsymbol{A}^{\text{T}*} = \boldsymbol{A}$, that is the transposed conjugate matrix $\boldsymbol{A}^{\text{T}*}$ is equal to $\boldsymbol{A}$. Obviously, a real symmetrical matrix $\boldsymbol{A} = \boldsymbol{A}^{\text{T}}$ satisfies the definition of a Hermitian matrix as for a real matrix $\boldsymbol{A}^* = \boldsymbol{A}$.

Now it will be shown that eigenvalues of a Hermitian matrix are always real. The proof will be indirect by assuming that there is a complex eigenvalue $\lambda = \alpha + \text{j}\Omega$. Then Eq. (12.1) gives

$$\boldsymbol{A}\boldsymbol{w} = \boldsymbol{w}(\alpha + \text{j}\Omega) \tag{12.7}$$

Left-multiplying this equation by $\boldsymbol{w}^{\text{T}*}$ gives

$$\boldsymbol{w}^{\text{T}*}\boldsymbol{A}\boldsymbol{w} = (\alpha + \text{j}\Omega)\boldsymbol{w}^{\text{T}*}\boldsymbol{w} \tag{12.8}$$

Transposing and conjugating Eq. (12.8) gives $\boldsymbol{w}^{\text{T}*}\boldsymbol{A}^{\text{T}*} = (\alpha - \text{j}\Omega)\boldsymbol{w}^{\text{T}*}$. As $\boldsymbol{A}^{\text{T}*} = \boldsymbol{A}$, one can write

$$\boldsymbol{w}^{\text{T}*}\boldsymbol{A} = (\alpha - \text{j}\Omega)\boldsymbol{w}^{\text{T}*} \tag{12.9}$$

Right-multiplying this by $\boldsymbol{w}$ gives

$$\boldsymbol{w}^{\mathrm{T}*}\boldsymbol{A}\boldsymbol{w} = (\alpha - \mathrm{j}\Omega)\boldsymbol{w}^{\mathrm{T}*}\boldsymbol{w} \tag{12.10}$$

Comparing Eqs. (12.8) and (12.10) results in

$$(\alpha + \mathrm{j}\Omega)\boldsymbol{w}^{\mathrm{T}*}\boldsymbol{w} = (\alpha - \mathrm{j}\Omega)\boldsymbol{w}^{\mathrm{T}*}\boldsymbol{w} \tag{12.11}$$

Obviously, $\boldsymbol{w}^{\mathrm{T}*}\boldsymbol{w} \neq 0$ is real and positive as the sum of products of complex conjugate numbers. Hence Eq. (12.11) results in $(\alpha + \mathrm{j}\Omega) = (\alpha - \mathrm{j}\Omega)$, or $\alpha + \mathrm{j}\Omega - \alpha + \mathrm{j}\Omega = 0$. Obviously, the imaginary part must be zero, $\mathrm{j}2\Omega = 0$, which proves that the eigenvalues of a Hermitian matrix are real.

Example 12.1 Using Eq. (12.6), find right eigenvectors of the following symmetrical matrix

$$\boldsymbol{A} = \begin{bmatrix} 5 & 0 & 0 \\ 0 & 2 & -\sqrt{2} \\ 0 & -\sqrt{2} & 3 \end{bmatrix} \tag{12.12}$$

The characteristic equation is

$$\det(\boldsymbol{A} - \lambda\mathbf{1}) = \begin{bmatrix} 5-\lambda & 0 & 0 \\ 0 & 2-\lambda & -\sqrt{2} \\ 0 & -\sqrt{2} & 3-\lambda \end{bmatrix} = (5-\lambda)[(2-\lambda)(3-\lambda)-2] = (5-\lambda)\left(\lambda^2 - 5\lambda + 4\right)] = 0$$

Hence $(5-\lambda)(\lambda-4)(\lambda-1) = 0$, which means that: $\lambda_1 = 5$; $\lambda_2 = 4$; $\lambda_3 = 1$.
For $\lambda_1 = 5$ one gets

$$(\boldsymbol{A}-\lambda_1\mathbf{1}) = \begin{bmatrix} 0 & 0 & 0 \\ 0 & -3 & -\sqrt{2} \\ 0 & -\sqrt{2} & -2 \end{bmatrix}, (\boldsymbol{A}-\lambda_1\mathbf{1})^{\mathrm{T}} = \begin{bmatrix} 0 & 0 & 0 \\ 0 & -3 & -\sqrt{2} \\ 0 & -\sqrt{2} & -2 \end{bmatrix}, (\boldsymbol{A}-\lambda_1\mathbf{1})^{\mathrm{D}} = \begin{bmatrix} 4 & 0 & 0 \\ 0 & 0 & 0 \\ 0 & 0 & 0 \end{bmatrix}$$

$$(\boldsymbol{A}-\lambda_2\mathbf{1}) = \begin{bmatrix} 1 & 0 & 0 \\ 0 & -2 & -\sqrt{2} \\ 0 & -\sqrt{2} & -1 \end{bmatrix}, (\boldsymbol{A}-\lambda_2\mathbf{1})^{\mathrm{T}} = \begin{bmatrix} 1 & 0 & 0 \\ 0 & -2 & -\sqrt{2} \\ 0 & -\sqrt{2} & -1 \end{bmatrix}, (\boldsymbol{A}-\lambda_2\mathbf{1})^{\mathrm{D}} = \begin{bmatrix} 0 & 0 & 0 \\ 0 & -1 & \sqrt{2} \\ 0 & \sqrt{2} & -2 \end{bmatrix}$$

$$(\boldsymbol{A}-\lambda_3\mathbf{1}) = \begin{bmatrix} 4 & 0 & 0 \\ 0 & 1 & -\sqrt{2} \\ 0 & -\sqrt{2} & 2 \end{bmatrix}, (\boldsymbol{A}-\lambda_3\mathbf{1})^{\mathrm{T}} = \begin{bmatrix} 4 & 0 & 0 \\ 0 & 1 & -\sqrt{2} \\ 0 & -\sqrt{2} & 2 \end{bmatrix}, (\boldsymbol{A}-\lambda_3\mathbf{1})^{\mathrm{D}} = \begin{bmatrix} 0 & 0 & 0 \\ 0 & 8 & 4\sqrt{2} \\ 0 & 4\sqrt{2} & 4 \end{bmatrix}$$

For λ_1, only one column is different from zero. This column can therefore be taken as an eigenvector, divided by, for example, four. For λ_2, the second and third columns are nonzero. As an eigenvector, for example, the third column can be chosen divided by four. Similarly, an eigenvector can be chosen for λ_3. Consequently, the following eigenvectors are obtained

$$\boldsymbol{w}_1 = \begin{bmatrix} 1 \\ 0 \\ 0 \end{bmatrix}, \quad \boldsymbol{w}_2 = \begin{bmatrix} 0 \\ -1 \\ \sqrt{2} \end{bmatrix}, \quad \boldsymbol{w}_3 = \begin{bmatrix} 0 \\ \sqrt{2} \\ 1 \end{bmatrix}; \quad \text{hence} \quad \boldsymbol{W} = \begin{bmatrix} 1 & 0 & 0 \\ 0 & -1 & \sqrt{2} \\ 0 & \sqrt{2} & 1 \end{bmatrix} \tag{12.13}$$

******

The eigenvectors in all examples in this chapter are not normalized in order to make manual calculations easy by maintaining round numbers.

A real asymmetric matrix $\boldsymbol{A} \neq \boldsymbol{A}^{\mathrm{T}}$ may have real eigenvalues (Example 12.1), complex eigenvalues, or a mix of real and complex eigenvalues. Regarding complex eigenvalues, the following property holds:

> If matrix $\boldsymbol{A} \neq \boldsymbol{A}^{\mathrm{T}}$ has a complex eigenvalue λ_i then the conjugate complex number λ_i^* is also an eigenvalue of that matrix. Moreover, the eigenvector associated with λ_i^* is equal to the conjugate eigenvector associated with λ_i.

In other words, complex eigenvalue and eigenvectors appear in complex conjugate pairs

$$\lambda_i, \boldsymbol{w}_i \quad \text{and} \quad \lambda_i^*, \boldsymbol{w}_i^* \tag{12.14}$$

Proof of this important property is simple and results from Eq. (12.1) while taking into account that for a real matrix $\boldsymbol{A}^* = \boldsymbol{A}$. Conjugating $\boldsymbol{A}\boldsymbol{w}_i = \boldsymbol{w}_i\lambda_i$ (without transposing) gives $\boldsymbol{A}\boldsymbol{w}_i^* = \boldsymbol{w}_i^*\lambda_i^*$. Hence λ_i^* and $\boldsymbol{w}_i^*$ satisfy the definition of an eigenvalue and eigenvector of matrix $\boldsymbol{A}$.

Obviously, a pair of complex conjugate eigenvalues constitutes two distinct eigenvalues $\lambda_i^* \neq \lambda_i$ and therefore their associated eigenvectors are linearly independent, that is $\boldsymbol{w}_i^* \neq c\boldsymbol{w}_i$ for any $c \neq 0$.

Example 12.2 Calculate the eigenvalues and eigenvectors of the matrix

$$\boldsymbol{A} = \begin{bmatrix} -6 & 0 & 0 \\ 0 & -1 & 5 \\ 0 & -5 & -1 \end{bmatrix}, \quad \det(\boldsymbol{A} - \lambda\boldsymbol{1}) = \begin{bmatrix} -6-\lambda & 0 & 0 \\ 0 & -1-\lambda & 5 \\ 0 & -5 & -1-\lambda \end{bmatrix} \tag{12.15}$$

As there are two zero elements in the first row, expansion of the determinant is easy

$$\det(\boldsymbol{A} - \lambda\boldsymbol{1}) = (-6-\lambda)\,[(1+\lambda)(1+\lambda) + 5\cdot 5] = -(6+\lambda)\left[\lambda^2 + 2\lambda + 26\right] = 0$$

Consider the second-degree polynomial in the square brackets. The determinant of that polynomial is negative giving a pair of complex conjugate roots

$$\det(\boldsymbol{A} - \lambda\boldsymbol{1}) = -(6+\lambda)\,[\lambda - (-1-\mathrm{j}5)]\,[\lambda - (-1+\mathrm{j}5)] = 0 \tag{12.16}$$

Hence matrix $\boldsymbol{A}$ will have the following eigenvalues

$$\lambda_1 = -6, \quad \lambda_2 = (-1-\mathrm{j}5), \quad \lambda_3 = (-1+\mathrm{j}5) = \lambda_2^* \tag{12.17}$$

Eigenvectors can be calculated from Eq. (12.6), using the adjacent matrix similarly, as in Example 12.1

$$(\boldsymbol{A} - \lambda_1\boldsymbol{1}) = \begin{bmatrix} 0 & 0 & 0 \\ 0 & 5 & 5 \\ 0 & -5 & 5 \end{bmatrix}, \quad (\boldsymbol{A} - \lambda_1\boldsymbol{1})^{\mathrm{T}} = \begin{bmatrix} 0 & 0 & 0 \\ 0 & 5 & -5 \\ 0 & 5 & 5 \end{bmatrix}, \quad (\boldsymbol{A} - \lambda_1\boldsymbol{1})^{\mathrm{D}} = \begin{bmatrix} 50 & 0 & 0 \\ 0 & 0 & 0 \\ 0 & 0 & 0 \end{bmatrix}$$

$$(\boldsymbol{A} - \lambda_2\boldsymbol{1}) = \begin{bmatrix} -5+\mathrm{j}5 & 0 & 0 \\ 0 & \mathrm{j}5 & 5 \\ 0 & -5 & \mathrm{j}5 \end{bmatrix}, (\boldsymbol{A} - \lambda_2\boldsymbol{1})^{\mathrm{T}} = \begin{bmatrix} -5+\mathrm{j}5 & 0 & 0 \\ 0 & \mathrm{j}5 & -5 \\ 0 & 5 & \mathrm{j}5 \end{bmatrix}, (\boldsymbol{A} - \lambda_2\boldsymbol{1})^{\mathrm{D}} = \begin{bmatrix} 0 & 0 & 0 \\ 0 & -25-\mathrm{j}25 & 25-\mathrm{j}25 \\ 0 & -25+\mathrm{j}25 & -25-\mathrm{j}25 \end{bmatrix}$$

$$(\boldsymbol{A} - \lambda_3\boldsymbol{1}) = \begin{bmatrix} -5-\mathrm{j}5 & 0 & 0 \\ 0 & -\mathrm{j}5 & 5 \\ 0 & -5 & -\mathrm{j}5 \end{bmatrix}, \quad (\boldsymbol{A} - \lambda_3\boldsymbol{1})^{\mathrm{T}} = \begin{bmatrix} -5-\mathrm{j}5 & 0 & 0 \\ 0 & -\mathrm{j}5 & -5 \\ 0 & 5 & -\mathrm{j}5 \end{bmatrix}, \quad (\boldsymbol{A} - \lambda_3\boldsymbol{1})^{\mathrm{D}} = \begin{bmatrix} 0 & 0 & 0 \\ 0 & -25+\mathrm{j}25 & 25+\mathrm{j}25 \\ 0 & -25-\mathrm{j}25 & -25+\mathrm{j}25 \end{bmatrix}$$

For λ_1, only the first column is nonzero and dividing it by 50 gives eigenvector $\boldsymbol{w}_1$. For λ_2, the second and third columns are nonzero. The second column, divided by 25, may be assumed to be eigenvector $\boldsymbol{w}_2$. Similarly for λ_3, the second and third columns are nonzero. To be consistent, the second column, divided by 25, may be assumed to be eigenvector $\boldsymbol{w}_3$, giving

$$\boldsymbol{w}_1 = \begin{bmatrix} 1 \\ \hline 0 \\ \hline 0 \end{bmatrix}, \quad \boldsymbol{w}_2 = \begin{bmatrix} 0 \\ \hline -1-\mathrm{j}1 \\ \hline -1+\mathrm{j}1 \end{bmatrix}, \quad \boldsymbol{w}_3 = \begin{bmatrix} 0 \\ \hline -1+\mathrm{j}1 \\ \hline -1-\mathrm{j}1 \end{bmatrix} = \boldsymbol{w}_2^* \tag{12.18}$$

Hence $\lambda_3 = \lambda_2^*$ resulted in $\boldsymbol{w}_3 = \boldsymbol{w}_2^*$.

******

Example 12.2 confirmed that complex eigenvalues and eigenvectors form complex conjugate pairs, as in Eq. (12.14). This property is important for further considerations.

12.1.2 Diagonalization of a Square Real Matrix

Let λ_i and $\boldsymbol{w}_i$ be an eigenvalue and a right eigenvector of matrix $\boldsymbol{A}$. Then, for every pair of eigenvalues and eigenvectors, $\boldsymbol{A}\boldsymbol{w}_i = \boldsymbol{w}_i\lambda_i$ holds and

$$\boldsymbol{A}\left[\boldsymbol{w}_1, \boldsymbol{w}_2, ..., \boldsymbol{w}_n\right] = \left[\boldsymbol{w}_1, \boldsymbol{w}_2, ..., \boldsymbol{w}_n\right] \begin{bmatrix} \lambda_1 & 0 & \cdots & 0 \\ 0 & \lambda_2 & \cdots & 0 \\ \vdots & \vdots & \ddots & \vdots \\ 0 & 0 & \cdots & \lambda_n \end{bmatrix} \quad \text{or} \quad \boldsymbol{AW} = \boldsymbol{W\Lambda} \tag{12.19}$$

where $\boldsymbol{W} = [\boldsymbol{w}_1, \boldsymbol{w}_2, ..., \boldsymbol{w}_n]$ is a square matrix whose columns are the right eigenvectors of matrix $\boldsymbol{A}$ and $\boldsymbol{\Lambda} = \text{diag}\ \lambda_i$ is a diagonal matrix of the corresponding eigenvalues.

If all the eigenvalues λ_i are distinct, $\lambda_1 \neq \lambda_2 \neq ... \neq \lambda_n$, then the corresponding eigenvectors are linearly independent. The proof of this property is conducted indirectly by assuming that the eigenvectors are linearly dependent and showing that the false assumption leads to a contradiction. Details can be found in many textbooks, for example Ogata (1967).

If vectors $\boldsymbol{w}_1, \boldsymbol{w}_2, ..., \boldsymbol{w}_n$ are linearly independent then matrix $\boldsymbol{W}$ made up from those vectors is nonsingular and the inverse $\boldsymbol{U} = \boldsymbol{W}^{-1}$ exists. The following notation will be used

$$\boldsymbol{U} = \boldsymbol{W}^{-1} = \left[\boldsymbol{w}_1, \boldsymbol{w}_2, ..., \boldsymbol{w}_n\right]^{-1} = \begin{bmatrix} \boldsymbol{u}_1 \\ \boldsymbol{u}_2 \\ \vdots \\ \boldsymbol{u}_n \end{bmatrix} \tag{12.20}$$

where $\boldsymbol{u}_i$ are the rows of matrix $\boldsymbol{U} = \boldsymbol{W}^{-1}$. Pre-multiplying both sides of Eq. (12.19) by $\boldsymbol{W}^{-1}$ gives

$$\boldsymbol{\Lambda} = \text{diag}\ \lambda_i = \boldsymbol{W}^{-1}\boldsymbol{AW} = \boldsymbol{UAW} \tag{12.21}$$

Right-multiplying this equation by $\boldsymbol{W}^{-1} = \boldsymbol{U}$ gives $\boldsymbol{UA} = \boldsymbol{\Lambda U}$, that is

$$\begin{bmatrix} \boldsymbol{u}_1 \\ \boldsymbol{u}_2 \\ \vdots \\ \boldsymbol{u}_n \end{bmatrix} \boldsymbol{A} = \begin{bmatrix} \lambda_1 & 0 & \cdots & 0 \\ 0 & \lambda_2 & \cdots & 0 \\ \vdots & \vdots & \ddots & \vdots \\ 0 & 0 & \cdots & \lambda_n \end{bmatrix} \begin{bmatrix} \boldsymbol{u}_1 \\ \boldsymbol{u}_2 \\ \vdots \\ \boldsymbol{u}_n \end{bmatrix} \tag{12.22}$$

Hence for each eigenvalue λ_i, $\boldsymbol{u}_i\boldsymbol{A} = \lambda_i\boldsymbol{u}_i$ holds. Neglecting the indices gives

$$\boldsymbol{u}\boldsymbol{A} = \boldsymbol{u}\lambda \tag{12.23}$$

This equation is similar to Eq. (12.1) but now with the row vector $\boldsymbol{u}$ on the left-hand side of matrix $\boldsymbol{A}$. Hence the row vector $\boldsymbol{u}$ is referred to as the *left eigenvector* of matrix $\boldsymbol{A}$ associated with eigenvalue λ.

It should be noted that transposing matrices in Eq. (12.23) gives

$$\boldsymbol{A}^{\mathrm{T}}\boldsymbol{u}^{\mathrm{T}} = \lambda\,\boldsymbol{u}^{\mathrm{T}} \tag{12.24}$$

This equation shows that the column vector $\boldsymbol{u}^{\mathrm{T}}$ is the right eigenvector of matrix $\boldsymbol{A}^{\mathrm{T}}$. This means that the left eigenvector of matrix $\boldsymbol{A}$ has the same values as the right eigenvector of matrix $\boldsymbol{A}^{\mathrm{T}}$. Hence the left eigenvector of matrix $\boldsymbol{A}$ is defined by some authors as the right eigenvector of matrix $\boldsymbol{A}^{\mathrm{T}}$.

Example 12.3 Consider the matrix $\boldsymbol{A}$ given below. Its eigenvalues are $\lambda_1 = 3$, $\lambda_2 = 2$, and $\lambda_3 = 1$. Application of Eq. (12.6) results in the following eigenvectors

$$\boldsymbol{A} = \begin{bmatrix} 2 & -1 & 2 \\ 0 & -1 & 4 \\ 0 & -2 & 5 \end{bmatrix}, \quad \boldsymbol{\Lambda} = \begin{bmatrix} \lambda_1 & & \\ & \lambda_2 & \\ & & \lambda_3 \end{bmatrix} = \begin{bmatrix} 3 & & \\ & 2 & \\ & & 1 \end{bmatrix}, \quad \boldsymbol{W} = \begin{bmatrix} 1 & 1 & 0 \\ 1 & 0 & 2 \\ 1 & 0 & 1 \end{bmatrix} \tag{12.25}$$

Checking the definition of right eigenvectors $\boldsymbol{AW} = \boldsymbol{W\Lambda}$ corresponding to Eq. (12.1)

$$\boldsymbol{AW} = \begin{bmatrix} 2 & -1 & 2 \\ 0 & -1 & 4 \\ 0 & -2 & 5 \end{bmatrix} \cdot \begin{bmatrix} 1 & 1 & 0 \\ 1 & 0 & 2 \\ 1 & 0 & 1 \end{bmatrix} = \begin{bmatrix} 3 & 2 & 0 \\ 3 & 0 & 2 \\ 3 & 0 & 1 \end{bmatrix} = \begin{bmatrix} 1 & 1 & 0 \\ 1 & 0 & 2 \\ 1 & 0 & 1 \end{bmatrix} \cdot \begin{bmatrix} 3 & & \\ & 2 & \\ & & 1 \end{bmatrix} = \boldsymbol{W\Lambda} \tag{12.26}$$

Inverting matrix $\boldsymbol{W}$ gives

$$\boldsymbol{U} = \boldsymbol{W}^{-1} = \begin{bmatrix} 0 & -1 & 2 \\ 1 & 1 & -2 \\ 0 & 1 & -1 \end{bmatrix} \quad \text{or} \quad \begin{aligned} \boldsymbol{u}_1 &= [\,0 \;\; -1 \;\; 2\,] \\ \boldsymbol{u}_2 &= [\,1 \;\; 1 \;\; -2\,] \\ \boldsymbol{u}_3 &= [\,0 \;\; 1 \;\; -1\,] \end{aligned} \tag{12.27}$$

Checking the definition of left eigenvectors $\boldsymbol{UA} = \boldsymbol{\Lambda U}$ in Eq. (12.22)

$$\boldsymbol{UA} = \begin{bmatrix} 0 & -1 & 2 \\ 1 & 1 & -2 \\ 0 & 1 & -1 \end{bmatrix} \cdot \begin{bmatrix} 2 & -1 & 2 \\ 0 & -1 & 4 \\ 0 & -2 & 5 \end{bmatrix} = \begin{bmatrix} 0 & -3 & 6 \\ 2 & 2 & -4 \\ 0 & 1 & -1 \end{bmatrix} = \begin{bmatrix} 3 & & \\ & 2 & \\ & & 1 \end{bmatrix} \cdot \begin{bmatrix} 0 & -1 & 2 \\ 1 & 1 & -2 \\ 0 & 1 & -1 \end{bmatrix} = \boldsymbol{\Lambda U} \tag{12.28}$$

Transposing matrix $\boldsymbol{A}$ gives

$$\boldsymbol{A}^{\mathrm{T}} = \begin{bmatrix} 2 & 0 & 0 \\ -1 & -1 & -2 \\ 2 & 4 & 5 \end{bmatrix} \tag{12.29}$$

Checking that vectors $\boldsymbol{u}_1^{\mathrm{T}}$, $\boldsymbol{u}_2^{\mathrm{T}}$, and $\boldsymbol{u}_3^{\mathrm{T}}$ are indeed right eigenvectors of $\boldsymbol{A}^{\mathrm{T}}$, that is that equation $\boldsymbol{A}^{\mathrm{T}}\boldsymbol{U}^{\mathrm{T}} = \boldsymbol{U}^{\mathrm{T}}\boldsymbol{\Lambda}$ is satisfied, corresponding to Eq. (12.24),

$$\boldsymbol{A}^{\mathrm{T}}\boldsymbol{U}^{\mathrm{T}} = \begin{bmatrix} 2 & 0 & 0 \\ -1 & -1 & -2 \\ 2 & 4 & 5 \end{bmatrix} \cdot \begin{bmatrix} 0 & 1 & 0 \\ -1 & 1 & 1 \\ 2 & -2 & -1 \end{bmatrix} = \begin{bmatrix} 0 & 2 & 0 \\ -3 & 2 & 1 \\ 6 & -4 & -1 \end{bmatrix} = \begin{bmatrix} 0 & 1 & 0 \\ -1 & 1 & 1 \\ 2 & -2 & -1 \end{bmatrix} \cdot \begin{bmatrix} 3 & & \\ & 2 & \\ & & 1 \end{bmatrix} = \boldsymbol{U}^{\mathrm{T}}\boldsymbol{\Lambda} \tag{12.30}$$

******

The following practical note should be borne in mind when calculating eigenvectors. Some professional programs, such as MATLAB, calculate eigenvalues and the corresponding right eigenvectors. In order to calculate left eigenvectors, manuals recommend that similar calculations are performed for the transposed matrix, that is the left eigenvectors of $\boldsymbol{A}$ should be calculated as right eigenvectors of $\boldsymbol{A}^{\mathrm{T}}$. This recommendation may be confusing because the eigenvalues of $\boldsymbol{A}^{\mathrm{T}}$ and the corresponding eigenvectors are usually ordered differently than those obtained for matrix $\boldsymbol{A}$. Consequently, the corresponding pairs of right and left eigenvectors ($\boldsymbol{w}_i$; $\boldsymbol{u}_i$) have to be selected manually based on the identification of identical eigenvalues λ_i. Therefore, it is simpler to calculate left eigenvectors by inverting the matrix $\boldsymbol{W}^{-1} = \boldsymbol{U}$. Then the pairs ($\boldsymbol{w}_i$; $\boldsymbol{u}_i$) can be selected as the columns of $\boldsymbol{W}$ and rows of $\boldsymbol{U}$, respectively.

According to Eq. (12.20), the square matrix $\boldsymbol{U}$ made up of left eigenvectors corresponds to the inverse matrix of $\boldsymbol{W}$ made up of right eigenvectors. Obviously, the product of both matrixes is the identity matrix $\boldsymbol{UW} = \mathbf{1}$, that is

$$\boldsymbol{UW} = \begin{bmatrix} \boldsymbol{u}_1 \\ \boldsymbol{u}_2 \\ \vdots \\ \boldsymbol{u}_n \end{bmatrix} [\boldsymbol{w}_1, \boldsymbol{w}_2, \ldots, \boldsymbol{w}_n] = \begin{bmatrix} \boldsymbol{u}_1\boldsymbol{w}_1 & 0 & \cdots & 0 \\ 0 & \boldsymbol{u}_2\boldsymbol{w}_2 & \cdots & 0 \\ \vdots & \vdots & \ddots & \vdots \\ 0 & 0 & \cdots & \boldsymbol{u}_n\boldsymbol{w}_n \end{bmatrix} = \begin{bmatrix} 1 & & & \\ & 1 & & \\ & & \ddots & \\ & & & 1 \end{bmatrix} = \mathbf{1} \tag{12.31}$$

Hence the following equations are true for left and right eigenvectors

$$\boldsymbol{u}_i\boldsymbol{w}_i = 1 \quad \text{and} \quad \boldsymbol{u}_i\boldsymbol{w}_j = 0 \quad \text{for} \quad j \neq i \tag{12.32}$$

Note that if λ_i is complex then this equation also holds, that is $\boldsymbol{u}_i^*\boldsymbol{w}_i^* = 1$. In that case matrices $\boldsymbol{U}$ and $\boldsymbol{W}$ have the following structure

$$\boldsymbol{U} = \begin{bmatrix} \vdots \\ \hline \boldsymbol{u} \\ \hline \boldsymbol{u}^* \\ \hline \vdots \end{bmatrix}, \quad \boldsymbol{W} = [\ldots \mid \boldsymbol{w} \mid \boldsymbol{w}^* \mid \ldots] \tag{12.33}$$

Equations (12.32) and (12.33) are important when considering the solution of differential equations.

Example 12.4 Calculate the matrix of left eigenvectors $\boldsymbol{U} = \boldsymbol{W}^{-1}$ of matrix $\boldsymbol{A}$ from Example 12.2 and use Eq. (12.21) to diagonalize the matrix $\boldsymbol{\Lambda} = \boldsymbol{UAW}$. Matrix $\boldsymbol{W}$ was calculated in Example 12.2

$$\boldsymbol{A} = \left[\begin{array}{c:c:c} -6 & 0 & 0 \\ \hdashline 0 & 1 & 5 \\ \hdashline 0 & -5 & 1 \end{array}\right], \quad \boldsymbol{W} = \left[\begin{array}{c:c:c} 1 & 0 & 0 \\ \hdashline 0 & -1-\mathrm{j}1 & -1+\mathrm{j}1 \\ \hdashline 0 & -1+\mathrm{j}1 & -1-\mathrm{j}1 \end{array}\right] \tag{12.34}$$

Simple calculations lead to

$$\boldsymbol{U} = \boldsymbol{W}^{-1} = \frac{1}{4}\left[\begin{array}{c:c:c} 4 & 0 & 0 \\ \hline 0 & -1+\mathrm{j}1 & -1-\mathrm{j}1 \\ \hline 0 & -1-\mathrm{j}1 & -1+\mathrm{j}1 \end{array}\right] = \begin{bmatrix} \boldsymbol{u}_1 \\ \boldsymbol{u}_2 \\ \boldsymbol{u}_3 \end{bmatrix} = \begin{bmatrix} \boldsymbol{u}_1 \\ \boldsymbol{u}_2 \\ \boldsymbol{u}_2^* \end{bmatrix} \tag{12.35}$$

$$\boldsymbol{u}_1 = \frac{1}{4}[4 \mid 0 \mid 0], \quad \boldsymbol{u}_2 = \frac{1}{4}[0 \mid -1+\mathrm{j}1 \mid -1-\mathrm{j}1], \quad \boldsymbol{u}_3 = \frac{1}{4}[0 \mid -1-\mathrm{j}1 \mid -1+\mathrm{j}1] = \boldsymbol{u}_2^*$$

It is easy to check that $\boldsymbol{u}_3 = \boldsymbol{u}_2^*$ is a pair of complex conjugate row vectors, which confirms the validity of Eq. (12.33). Multiplying matrices (12.34) gives

$$\boldsymbol{AW} = \begin{bmatrix} -6 & 0 & 0 \\ 0 & 1 & 5 \\ 0 & -5 & 1 \end{bmatrix} \begin{bmatrix} 1 & 0 & 0 \\ 0 & -1-\text{j}1 & -1+\text{j}1 \\ 0 & -1+\text{j}1 & -1-\text{j}1 \end{bmatrix} = \begin{bmatrix} -6 & 0 & 0 \\ 0 & -4+\text{j}6 & -4-\text{j}6 \\ 0 & +6+\text{j}4 & +6-\text{j}4 \end{bmatrix} \tag{12.36}$$

$$\boldsymbol{UAW} = \frac{1}{4}\begin{bmatrix} 4 & 0 & 0 \\ 0 & -1+\text{j}1 & -1-\text{j}1 \\ 0 & -1-\text{j}1 & -1+\text{j}1 \end{bmatrix} \begin{bmatrix} -6 & 0 & 0 \\ 0 & -4+\text{j}6 & -4-\text{j}6 \\ 0 & +6+\text{j}4 & +6-\text{j}4 \end{bmatrix} = \begin{bmatrix} -6 & 0 & 0 \\ 0 & -1-\text{j}5 & 0 \\ 0 & 0 & -1+\text{j}5 \end{bmatrix} \tag{12.37}$$

The obtained diagonal matrix contains previously calculated eigenvalues of matrix $\boldsymbol{A}$ (Example 12.2, Eq. [12.17]).

******

Diagonalization of a square matrix $\boldsymbol{A}$ using matrices $\boldsymbol{W}$ and $\boldsymbol{U}$ made up of right and left eigenvectors is important for the next section that considers the solution of matrix differential equations.

12.1.3 Solution of Matrix Differential Equation

The solution of scalar differential homogeneous equations is discussed in Appendix A3. It is shown that the fundamental solution of ordinary linear differential equations consists of exponential functions $\text{e}^{\lambda t}$, where numbers λ must be chosen so that the Wronskian matrix of solutions is different from zero. In the case of a first-order differential homogeneous equation $\dot{x} - ax = 0$ or $\dot{x} = ax$ the fundamental system of solutions consists of only one exponential function e^{at}. The particular solution is of the form $x(t) = \text{e}^{at}x_0$, where $x_0 = x(t_0)$ is the initial condition. In this section the matrix form of a linear differential homogeneous equation is considered

$$\dot{\boldsymbol{x}} = \boldsymbol{Ax} \tag{12.38}$$

where $\boldsymbol{A}$ is a square real matrix referred to as the *state matrix*. Equation (12.38) is referred to as the *state equation* and vector $\boldsymbol{x}$ is the vector of the *state variables* or in short the *state vector*.

Matrix Eq. (12.38) has a solution in the same form as the scalar equation

$$\boldsymbol{x}(t) = \text{e}^{\boldsymbol{A}t}\boldsymbol{x}_0 \tag{12.39}$$

where $\boldsymbol{x}(t)$ and $\boldsymbol{x}_0$ are column matrices, while $\text{e}^{\boldsymbol{A}t}$ is a square matrix which can be proved by expanding $\text{e}^{\boldsymbol{A}t}$ as a Taylor series

$$\text{e}^{\boldsymbol{A}t} = \boldsymbol{1} + \boldsymbol{A}t + \frac{(\boldsymbol{A}t)^2}{2!} + \frac{(\boldsymbol{A}t)^3}{3!} + \cdots \tag{12.40}$$

The Taylor expansion also proves that for a real matrix $\boldsymbol{A}$ matrix $\text{e}^{\boldsymbol{A}t}$ is also real and the solution given by Eq. (12.39) is also real.

A number of different methods of calculation of $\text{e}^{\boldsymbol{A}t}$ are given in textbooks; see, for example, Ogata (1967) or Strang (1976). In power system analysis practice the calculation of $\text{e}^{\boldsymbol{A}t}$ is replaced by the diagonalization of $\boldsymbol{A}$ and calculation of $\text{e}^{\Lambda t}$, where $\Lambda = \text{diag}\ \ \lambda_i$ is a diagonal matrix (12.21).

In order to utilize the matrix diagonalization for the solution of the state Eq. (12.38), the state vector $\boldsymbol{x}$ can be transformed into a new state vector $\boldsymbol{z}$ using the linear transformation

$$\boldsymbol{x} = \boldsymbol{Wz} \tag{12.41}$$

where $\boldsymbol{W}$ is a square matrix consisting of right eigenvectors of matrix $\boldsymbol{A}$. Note that vector $\boldsymbol{z}$ is generally complex. Using the inverse matrix $\boldsymbol{U} = \boldsymbol{W}^{-1}$ the following inverse transformation can be defined

$$\boldsymbol{z} = \boldsymbol{W}^{-1}\boldsymbol{x} = \boldsymbol{Ux} \tag{12.42}$$

Substituting Eq. (12.41) into Eq. (12.38) gives $\boldsymbol{W}\dot{\boldsymbol{z}} = \boldsymbol{A}\boldsymbol{W}\boldsymbol{z}$, or $\dot{\boldsymbol{z}} = \boldsymbol{W}^{-1}\boldsymbol{A}\boldsymbol{W}\boldsymbol{z}$, which after taking into account Eq. (12.21) gives

$$\dot{\boldsymbol{z}} = \boldsymbol{\Lambda}\boldsymbol{z} \tag{12.43}$$

Equation (12.43) is the *modal form* of the state Eq. (12.38). Matrix $\boldsymbol{\Lambda}$ given by Eq. (12.21) is the modal form of the state matrix, matrix $\boldsymbol{W}$ is the *modal matrix*, and variables $\boldsymbol{z}(t)$ are the *modal variables.*[1]

Because matrix $\boldsymbol{\Lambda}$ is diagonal the matrix Eq. (12.43) describes a set of uncoupled scalar differential equations

$$\dot{z}_i = \lambda_i z_i \quad \text{for} \quad i = 1, 2, \dots, n \tag{12.44}$$

Each of the equations is of first order and its solution is of the form

$$z_i(t) = e^{\lambda_i t} z_{i0} \quad \text{for} \quad i = 1, 2, \dots, n \tag{12.45}$$

where z_{i0} is the initial condition of the modal variable. The set of these scalar solutions can be expressed as the following column vector

$$\boldsymbol{z}(t) = e^{\boldsymbol{\Lambda} t} \boldsymbol{z}_0 \tag{12.46}$$

where

$$e^{\boldsymbol{\Lambda} t} = \begin{bmatrix} e^{\lambda_1 t} & 0 & \cdots & 0 \\ 0 & e^{\lambda_2 t} & \cdots & 0 \\ \vdots & \vdots & \ddots & \vdots \\ 0 & 0 & \cdots & e^{\lambda_n t} \end{bmatrix} = \text{diag}\left[e^{\lambda_i t}\right] \tag{12.47}$$

Equations (12.41) and (12.46) give

$$\boldsymbol{x} = \boldsymbol{W} e^{\boldsymbol{\Lambda} t} \boldsymbol{z}_0 \tag{12.48}$$

where $\boldsymbol{z}_0 = \boldsymbol{z}(t_0)$ is a column of initial conditions for the modal variables $\boldsymbol{z}(t)$. These initial conditions can be found using Eq. (12.42) as

$$\boldsymbol{z}_0 = \boldsymbol{U}\boldsymbol{x}_0 \tag{12.49}$$

Substituting Eq. (12.49) into (12.48) gives

$$\boldsymbol{x} = \boldsymbol{W} e^{\boldsymbol{\Lambda} t} \boldsymbol{U} \boldsymbol{x}_0 \tag{12.50}$$

Obviously the solution given by Eq. (12.50) is equivalent to the solution given by Eq. (12.39) as

$$e^{\boldsymbol{A} t} = \boldsymbol{W} e^{\boldsymbol{\Lambda} t} \boldsymbol{U} \tag{12.51}$$

The correctness of Eq. (12.51) can be proved by using the Taylor series expansion (12.40) and Eq. (12.19). The proof can be found in Strang (1976).

It should be noted that, generally, matrix $\boldsymbol{A}$ may be asymmetric, giving complex eigenvalues λ_i (see Example 12.2). In that case the modal variables $z_i(t)$ given by Eq. (12.45) are complex. The solution of complex differential Eqs. (12.44) in the complex domain is discussed in Section A.3.6 of the appendix. It is shown that for complex λ_i the trajectories of solutions $z_i(t)$ form logarithmic spirals on the complex plane. The spirals are converging for

---

1 Note that $\boldsymbol{z}(t)$ are complex variables while $\boldsymbol{x}(t)$ are real. Many authors refer to $\boldsymbol{z}(t)$ as simply the *modes*. In this book $\boldsymbol{z}(t)$ are referred to as the *modal variables* analogically as $\boldsymbol{x}(t)$ are the *state variables*. It is shown later in this section that a state variable can be expressed as a linear combination of uncorrelated real variables of the form. $e^{\alpha_i t} \cdot \cos(\Omega_i t + \phi_{ki}) e^{\alpha_i t}$ which will be referred to as the *modes* in this book.

$\alpha_i = \text{Re}\,\lambda_i < 0$ and diverging for $\alpha_i = \text{Re}\,\lambda_i > 0$. For $\alpha_i = \text{Re}\,\lambda_i = 0$ the solution $z_i(t)$ represents a circle in the complex plane. The spiral rotates anticlockwise when $\Omega_i = \text{Im}\,\lambda_i > 0$ and clockwise when $\Omega_i = \text{Im}\,\lambda_i < 0$.

In the case of matrix equations (Eq. (12.14) there is always a pair of complex conjugate eigenvalues $\lambda_j = \lambda_i^*$. That pair results in two solutions $z_i(t)$ and $z_j(t)$ forming counter-rotating spirals in the complex plane (Appendix A.3). Obviously, the imaginary parts of the spirals cancel each other out (because they have opposite signs) so that the real solution will be equal to the double real part, that is

$$z_i(t) + z_j(t) = z_i(t) + z_i^*(t) = 2\,\text{Re}\,z_i(t)$$

That canceling of the imaginary parts of the solution will now be proved formally for the discussed matrix state equation (12.38) and its modal form (12.43).

When the eigenvalues are complex, matrix $e^{\mathbf{A}t}$ is complex too. From the definition of eigenvectors it can be concluded that complex eigenvalues correspond to complex eigenvectors (Example 12.4). Hence matrices $\boldsymbol{W}$ and $\boldsymbol{U}$ may be generally complex. In Eq. (12.50) there is a product of three complex matrices $\boldsymbol{W}e^{\mathbf{\Lambda}t}\boldsymbol{U}$ and one real matrix $\boldsymbol{x}_0$. On the other hand, it is clear from Eq. (12.39) that the solution $\boldsymbol{x}(t)$, and therefore also the product $\boldsymbol{W}e^{\mathbf{\Lambda}t}\boldsymbol{U}\boldsymbol{x}_0$, must be real. Hence there is a question about how the product of complex matrices produces a real result. The answer comes from a previous observation that complex eigenvalues and eigenvectors must always form conjugate pairs λ, λ^* and the associated matrices of left and right eigenvectors have the structure shown in Eq. (12.33).

Let the i-th and j-th eigenvalues be a complex conjugate pair

$$\lambda_j = \lambda_i^*, \quad \boldsymbol{w}_j = \boldsymbol{w}_i^*, \quad \boldsymbol{u}_j = \boldsymbol{u}_i^* \tag{12.52}$$

Then the column of initial conditions of modal variables $\boldsymbol{z}(t)$ has the following structure

$$\boldsymbol{z}_0 = \begin{bmatrix} \vdots \\ \hline z_{i0} \\ \hline z_{j0} \\ \hline \vdots \end{bmatrix} = \boldsymbol{U}\boldsymbol{x}_0 = \begin{bmatrix} \vdots \\ \hline \boldsymbol{u}_i \\ \hline \boldsymbol{u}_i^* \\ \hline \vdots \end{bmatrix} \cdot \boldsymbol{x}_0 = \begin{bmatrix} \vdots \\ \hline \boldsymbol{u}_i\boldsymbol{x}_0 \\ \hline \boldsymbol{u}_i^*\boldsymbol{x}_0 \\ \hline \vdots \end{bmatrix} = \begin{bmatrix} \vdots \\ \hline z_{i0} \\ \hline z_{i0}^* \\ \hline \vdots \end{bmatrix} \tag{12.53}$$

that is, for the considered pair, $z_{j0} = z_{i0}^*$ and the two elements of the column $\boldsymbol{z}_0$ are a complex conjugate pair. The product $\boldsymbol{W}e^{\mathbf{\Lambda}t}$ is a square matrix that in this case has the following structure

$$\boldsymbol{W}e^{\mathbf{\Lambda}t} = [\ldots \mid \boldsymbol{w}_i \mid \boldsymbol{w}_i^* \mid \ldots] \cdot \begin{bmatrix} \ddots & \vdots & \vdots & \cdots \\ \cdots & e^{\lambda_i t} & 0 & \cdots \\ \cdots & 0 & e^{\lambda_i^* t} & \cdots \\ \cdots & \vdots & \vdots & \ddots \end{bmatrix} = [\ldots \mid \boldsymbol{w}_i e^{\lambda_i t} \mid \boldsymbol{w}_i^* e^{\lambda_i^* t} \mid \ldots] \tag{12.54}$$

that is the matrix has two columns which are complex conjugate. Taking into account the matrix structure expressed by Eqs. (12.53) and (12.54), Eq. (12.48) gives

$$\boldsymbol{x}(t) = \begin{bmatrix} x_1(t) \\ \hline \vdots \\ \hline x_k(t) \\ \hline \vdots \\ \hline x_n(t) \end{bmatrix} = \boldsymbol{W}e^{\mathbf{\Lambda}t}\boldsymbol{z}_0 = \begin{bmatrix} \cdots & w_{1i}e^{\lambda_i t} & w_{1i}^* e^{\lambda_i^* t} & \cdots \\ \hline & \vdots & \vdots & \\ \hline \cdots & w_{ki}e^{\lambda_i t} & w_{ki}^* e^{\lambda_i^* t} & \cdots \\ \hline & \vdots & \vdots & \\ \hline \cdots & w_{ni}e^{\lambda_i t} & w_{ni}^* e^{\lambda_i^* t} & \cdots \end{bmatrix} \cdot \begin{bmatrix} \vdots \\ \hline z_{i0} \\ \hline z_{i0}^* \\ \hline \vdots \end{bmatrix} \tag{12.55}$$

This structure shows that the solution for the state variable for any k is

$$x_k(t) = \dots + w_{ki}z_{i0}e^{\lambda_i t} + w_{ki}^* z_{i0}^* e^{\lambda_i^* t} + \dots \tag{12.56}$$

or

$$x_k(t) = \dots + c_{ki}e^{\lambda_i t} + c_{ki}^* e^{\lambda_i^* t} + \dots \tag{12.57}$$

where $c_{ki} = w_{ki}z_{i0}$ is a complex number depending on the eigenvector and the initial condition. Denoting $\lambda_i = \alpha_i + j\Omega_i$ gives

$$c_{ki}e^{\lambda_i t} + c_{ki}^* e^{\lambda_i^* t} = c_{ki}e^{\alpha_i t}(\cos\Omega_i t + j\sin\Omega_i t) + c_{ki}^* e^{\alpha_i t}(\cos\Omega_i t - j\sin\Omega_i t)$$

Rearranging the right-hand side gives

$$c_{ki}e^{\lambda_i t} + c_{ki}^* e^{\lambda_i^* t} = e^{\alpha_i t}\left[\left(c_{ki} + c_{ki}^*\right)\cos\Omega_i t + j\left(c_{ki} - c_{ki}^*\right)\sin\Omega_i t\right] \tag{12.58}$$

Note that

$$a_i = \left(c_i + c_i^*\right) = 2\,\mathrm{Re}\,c_i \quad \text{and} \quad b_i = j\left(c_i - c_i^*\right) = -2\cdot\mathrm{Im}\;c_i \tag{12.59}$$

are real numbers equal to the real part and double imaginary part of the integration constant $c_{ki} = w_{ki}z_{i0}$. Hence Eq. (12.58) takes the form

$$c_{ki}e^{\lambda_i t} + c_{ki}^* e^{\lambda_i^* t} = e^{\alpha_i t}[a_{ki}\cdot\cos\Omega_i t - b_{ki}\cdot\sin\Omega_i t] \tag{12.60}$$

Note that the difference between the cosine and sine functions in the square brackets can be replaced by a cosine function (also Figure A.2 in the Appendix)

$$c_{ki}e^{\lambda_i t} + c_{ki}^* e^{\lambda_i^* t} = e^{\alpha_i t}\cdot\sqrt{a_{ki}^2 + b_{ki}^2}\left[\frac{a_{ki}}{\sqrt{a_{ki}^2 + b_{ki}^2}}\cdot\cos\Omega_i t - \frac{b_{ki}}{\sqrt{a_{ki}^2 + b_{ki}^2}}\cdot\sin\Omega_i t\right] \tag{12.61}$$

In the same way as in Appendix A.3 it can be assumed that

$$\cos\varphi_{ki} = \frac{a_{ki}}{\sqrt{a_{ki}^2 + b_{ki}^2}} = \frac{\mathrm{Re}\,c_i}{\sqrt{(\mathrm{Re}\,c_i)^2 + (\mathrm{Im}\,c_i)^2}} = \frac{\mathrm{Re}\,c_i}{|c_i|} \tag{12.62}$$

$$\sin\varphi_{ki} = \frac{b_{ki}}{\sqrt{a_{ki}^2 + b_{ki}^2}} = \frac{\mathrm{Im}\,c_i}{\sqrt{(\mathrm{Re}\,c_i)^2 + (\mathrm{Im}\,c_i)^2}} = \frac{\mathrm{Im}\,c_i}{|c_i|} \tag{12.63}$$

$$\phi_{ki} = \arcsin\left(\mathrm{Im}\,c_i/|c_i|\right) \tag{12.64}$$

$$|c_{ki}| = \sqrt{(\mathrm{Re}\,c_i)^2 + (\mathrm{Im}\,c_i)^2} \quad \text{and} \quad \sqrt{a_{ki}^2 + b_{ki}^2} = 2\cdot\sqrt{(\mathrm{Re}\,c_i)^2 + (\mathrm{Im}\,c_i)^2} = 2\cdot|c_i| \tag{12.65}$$

With this notation Eq. (12.61) gives

$$c_{ki}e^{\lambda_i t} + c_{ki}^* e^{\lambda_i^* t} = 2\cdot|c_{ki}|e^{\alpha_i t}\cdot\cos\left(\Omega_i t + \phi_{ki}\right) \tag{12.66}$$

Finally, substituting Eq. (12.66) into Eq. (12.57) gives

$$x_k(t) = \dots + 2\cdot|c_{ki}|\,e^{\alpha_i t}\cdot\cos\left(\Omega_i t + \varphi_{ki}\right) + \dots \tag{12.67}$$

This means that in the solution for the state variable $x_k(t)$ a pair of complex eigenvalues λ_i, λ_i^* corresponds to an oscillatory term $e^{\alpha_i t}\cdot\cos\left(\Omega_i t + \phi_{ki}\right)$, where $\alpha_i = \mathrm{Re}\,\lambda_i$ and $\Omega_i = \mathrm{Im}\,\lambda_i$ are the real and imaginary parts of the eigenvalue. The component relating with this term is referred to as the *oscillatory mode* and it corresponds to a solution of a second-order underdamped differential equation analyzed in Example A.3.4 in Appendix A.3. Factor $A_i = 2 \cdot |c_{ki}|$ is *the amplitude* of the oscillatory mode, Ω_i is *the frequency* of the oscillatory mode, and $(-1/\alpha_i)$ is *the time constant* of exponential decay of the oscillatory mode.

The above considerations were concerned with complex eigenvalues. The case of real eigenvalues can be obtained from the derived equations substituting $\Omega_i = \text{Im}\,\lambda_i = 0$ and $\phi_{ki} = \arcsin(\text{Im}\,c_i/|c_i|) = 0$. The only difference is that the resulting term will not be multiplied by 2 as real eigenvalues are considered individually, not in pairs. This gives for a real eigenvalue $\lambda_i = \alpha_i$

$$x_k(t) = \ldots + c_{ki}\mathrm{e}^{\alpha_i t} + \ldots \tag{12.68}$$

The component relating with this term $\mathrm{e}^{\alpha_i t}$ is referred to as the *aperiodic mode* and it corresponds to a solution of a first-order differential equation analyzed in Appendix A.3. Factor $A_i = |c_{ki}|$ is *the amplitude* of the aperiodic mode and $(-1/\alpha_i)$ is *the time constant* of exponential decay of the aperiodic mode.

Taking into account in Eq. (12.55), both the real and imaginary parts of eigenvalues after multiplying both matrices for each variable, the following solution is obtained

$$x_k(t) = \sum_{\lambda_i \in \text{Real}} c_{ki} \cdot \mathrm{e}^{\alpha_i t} + \sum_{\lambda_i \in \text{Complex}} 2|c_{ki}| \cdot \mathrm{e}^{\alpha_i t} \cdot \cos(\Omega_i t + \phi_{ki}) \tag{12.69}$$

Analysis of Eq. (12.69) leads to the following conclusions which are important for the analysis of power system dynamics:

- Real eigenvalues $\lambda_i = \alpha_i$ introduce to the response of $x_k(t)$ aperiodic modes that are proportional to $\mathrm{e}^{\alpha_i t}$. If $\alpha_i < 0$ then the corresponding aperiodic mode is stable and $(-1/\alpha_i)$ is the time constant of exponential decay of the mode. If $\alpha_i > 0$ then the corresponding aperiodic mode is unstable and exponentially increasing.
- Each conjugate pair of complex eigenvalues $\lambda_i = \alpha_i \pm \mathrm{j}\Omega_i$ introduces to the response of $x_k(t)$ oscillatory modes proportional to $\mathrm{e}^{\alpha_i t} \cdot \cos(\Omega_i t + \phi_{ki})$. If $\alpha_i < 0$ then the corresponding oscillatory mode is stable. If $\alpha_i > 0$ then the corresponding oscillatory mode is unstable. The term Ω_i is the frequency of oscillation (in rad/s) of the oscillatory mode. The angle ϕ_{ki} is the phase angle of the oscillatory mode and its value depends on the initial conditions.
- The solution $x_k(t)$ of a differential equation is a linear combination of the modes and the coefficients of proportionality in that combination depend on the initial conditions. As an oscillatory mode corresponds to a response of a second-order underdamped system while an aperiodic mode corresponds to a response of a first-order system, effectively a small-signal response of a dynamic system of high order is represented as a linear combination of responses of decoupled second- and first-order systems.
- A dynamic system described by Eq. (12.38) is unstable if any of the modes are unstable.

The definition and the types of modes are given in Table 12.1.

For an oscillatory mode, analogously to Eq. (5.51), the following definition of the *damping ratio* may be introduced

$$\zeta_i = \frac{-\alpha_i}{\sqrt{\alpha_i^2 + \Omega_i^2}} \tag{12.70}$$

In case of power systems, as discussed in Section 5.4.6, damping is considered to be satisfactory if the damping ratio $\zeta \geq 0.05$. The frequency of oscillations (Table 5.1 in Section 5.4.6) varies depending on loading conditions and location and type of fault. This result forms the fact that for given fault usually only a part of oscillatory modes are excited. For large interconnected power systems, the *range of modal frequencies* Ω_i (referred to as range of modal frequencies, or RoMF) is very wide. For example (1.2 - 15.7) rad/s, i.e. (0.2 - 2.5) Hz for an EPS with nominal frequency 50 Hz.

Table 12.1 Illustration of the definition and types of modes.

	Eigenvalue λ_i	
	real	**complex pair**
Notation	$\lambda_i = \alpha_i$	$\lambda_i = \alpha_i + \mathrm{j}\Omega_i,\ \lambda_i^* = \alpha_i - \mathrm{j}\Omega_i$
Mode definition	$\mathrm{e}^{\alpha_i t}$	$\mathrm{e}^{\alpha_i t} \cdot \cos\omega_i t$
Mode type	aperiodic	oscillatory
Corresponding to	response of a first-order system	response of a second-order underdamped system
$\alpha_i < 0$	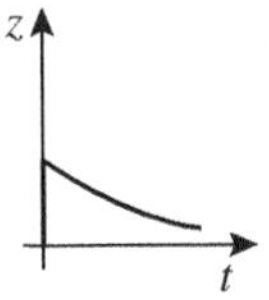	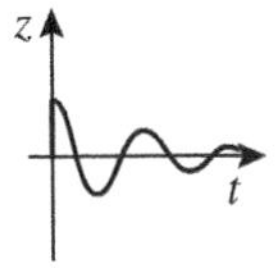
$\alpha_i > 0$	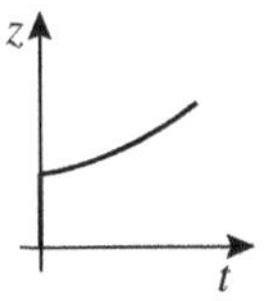	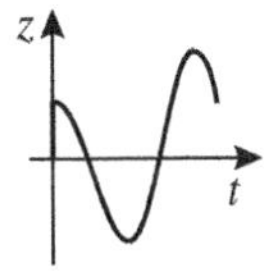

Note that $c_{ki} = w_{ki} z_{i0}$ in Eq. (12.69) depends on the initial conditions z_{i0} of a given modal variable $z_i(t)$. If this modal variable has zero initial conditions then obviously $c_{ki} = 0$ and the mode has no influence on the value of $x_k(t)$. A mode or a modal variable $z_i(t)$ is said to be *excited* if $c_{ki} \neq 0$. Equation (12.69) shows that the trajectory of $x_k(t)$ is influenced only by the excited modes or excited modal variables. Those modes or modal variables that have the largest values of c_{ki} are said to be *dominant modes* or *dominant modal variables*.

The analysis in Section 5.4.6 of the second-order classical model of a synchronous generator connected to an infinite busbar showed that rotor swings around the synchronous speed can be aperiodic or oscillatory, depending on the roots of the characteristic equation. These roots are equal to the eigenvalues of the state matrix – Eqs. (5.56) and (5.57) – and the swings are characterized by the frequency Ω and the damping ratio ζ. This section shows that the response of a multi-machine power system, or of a generator described by a higher-order differential equation, can be expressed as a linear combination of uncoupled aperiodic and oscillatory responses, depending on whether the eigenvalues are real or complex. In other words, rotor swings in a multi-machine power system can be expressed as a linear combination of uncoupled swings of different frequencies Ω_i and damping ratios ζ_i, similar to those analyzed in Section 5.4.6. This finding simplifies significantly the analysis of multi-machine power systems.

Example 12.5 Find the solution of the differential equation $\dot{\boldsymbol{x}} = \boldsymbol{A}\boldsymbol{x}$ for matrix $\boldsymbol{A}$ from Example 12.2. The eigenvalues $\lambda_1 = -6$ and $\lambda_2 = (-1 - \mathrm{j}5)$ and the corresponding matrices $\boldsymbol{W}$ and $\boldsymbol{U}$ were calculated in Examples 12.2 and 12.4, respectively

$$\boldsymbol{A} = \begin{bmatrix} -6 & 0 & 0 \\ 0 & 1 & 5 \\ 0 & -5 & 1 \end{bmatrix}, \quad \boldsymbol{W} = \begin{bmatrix} 1 & 0 & 0 \\ 0 & -1-\mathrm{j}1 & -1+\mathrm{j}1 \\ 0 & -1+\mathrm{j}1 & -1-\mathrm{j}1 \end{bmatrix}, \quad \boldsymbol{U} = \frac{1}{4}\begin{bmatrix} 4 & 0 & 0 \\ 0 & -1+\mathrm{j}1 & -1-\mathrm{j}1 \\ 0 & -1-\mathrm{j}1 & -1+\mathrm{j}1 \end{bmatrix} \tag{12.71}$$

The initial conditions are $x_{10} = x_1(t_0) \neq 0$ and $x_{20} = x_2(t_0) \neq 0$. To simplify complex number manipulations, multiplication of matrices in Eq. (12.50) will be executed in such a way that first the product ($\boldsymbol{W}e^{\boldsymbol{\Lambda}t}$) will be calculated, then $(\boldsymbol{W}e^{\boldsymbol{\Lambda}t})\boldsymbol{U}$, and finally the solution $\boldsymbol{x}(t) = (\boldsymbol{W}e^{\boldsymbol{\Lambda}t}\boldsymbol{U})\boldsymbol{x}_0$

$$\boldsymbol{W}e^{\boldsymbol{\Lambda}t} = \begin{bmatrix} 1 & 0 & 0 \\ 0 & -1-j1 & -1+j1 \\ 0 & -1+j1 & -1-j1 \end{bmatrix} \cdot \begin{bmatrix} e^{\lambda_1 t} & & \\ & e^{\lambda_2 t} & \\ & & e^{\lambda_2^* t} \end{bmatrix} = \begin{bmatrix} e^{\lambda_1 t} & 0 & 0 \\ 0 & (-1-j)\cdot e^{\lambda_2 t} & (-1+j)\cdot e^{\lambda_2^* t} \\ 0 & (-1+j)\cdot e^{\lambda_2 t} & (-1-j)\cdot e^{\lambda_2^* t} \end{bmatrix} \tag{12.72}$$

$$\boldsymbol{W}e^{\boldsymbol{\Lambda}t}\boldsymbol{U} = \frac{1}{4}\cdot \begin{bmatrix} e^{\lambda_1 t} & 0 & 0 \\ 0 & (-1-j)\cdot e^{\lambda_2 t} & (-1+j)\cdot e^{\lambda_2^* t} \\ 0 & (-1+j)\cdot e^{\lambda_2 t} & (-1-j)\cdot e^{\lambda_2^* t} \end{bmatrix} \cdot \begin{bmatrix} 4 & 0 & 0 \\ 0 & -1+j1 & -1-j1 \\ 0 & -1-j1 & -1+j1 \end{bmatrix}$$

Patiently multiplying the matrices and ordering the terms gives

$$\boldsymbol{W}e^{\boldsymbol{\Lambda}t}\boldsymbol{U} = \frac{1}{4}\cdot \begin{bmatrix} 4\cdot e^{\lambda_1 t} & 0 & 0 \\ 0 & 2\cdot\left(e^{\lambda_2 t} + e^{\lambda_2^* t}\right) & 2j\cdot\left(e^{\lambda_2 t} - e^{\lambda_2^* t}\right) \\ 0 & 2j\cdot\left(-e^{\lambda_2 t} + e^{\lambda_2^* t}\right) & 2\cdot\left(e^{\lambda_2 t} + e^{\lambda_2^* t}\right) \end{bmatrix} \tag{12.73}$$

Substituting into this equation $\lambda_1 = \alpha_1$, $\lambda_2 = (\alpha_2 + j\Omega_2)$, and $\lambda_3 = \lambda_2^* = (\alpha_2 - j\Omega_2)$ gives the following matrix

$$\boldsymbol{W}e^{\boldsymbol{\Lambda}t}\boldsymbol{U} = \begin{bmatrix} e^{\alpha_1 t} & 0 & 0 \\ 0 & e^{\alpha_2 t}\cos\Omega_2 t & -e^{\alpha_2 t}\sin\Omega_2 t \\ 0 & +e^{\alpha_2 t}\sin\Omega_2 t & e^{\alpha_2 t}\cos\Omega_2 t \end{bmatrix} \tag{12.74}$$

Substituting this matrix into Eq. (12.50) gives

$$\boldsymbol{x}(t) = \boldsymbol{W}e^{\boldsymbol{\Lambda}t}\boldsymbol{U}\boldsymbol{x}_0 = \begin{bmatrix} x_1(t) \\ x_2(t) \\ x_3(t) \end{bmatrix} = \begin{bmatrix} e^{\alpha_1 t} & 0 & 0 \\ 0 & e^{\alpha_2 t}\cos\Omega_2 t & -e^{\alpha_2 t}\sin\Omega_2 t \\ 0 & +e^{\alpha_2 t}\sin\Omega_2 t & e^{\alpha_2 t}\cos\Omega_2 t \end{bmatrix} \cdot \begin{bmatrix} x_{10} \\ x_{20} \\ x_{30} \end{bmatrix} \tag{12.75}$$

or

$$\begin{aligned} x_1(t) &= x_{10}\cdot e^{\alpha_1 t} \\ x_2(t) &= e^{\alpha_2 t}\cdot[x_{20}\cdot\cos\Omega_2 t - x_{30}\cdot\sin\Omega_2 t] \\ x_3(t) &= e^{\alpha_2 t}\cdot[x_{30}\cdot\cos\Omega_2 t + x_{20}\cdot\sin\Omega_2 t] \end{aligned} \tag{12.76}$$

Obviously, the solutions $x_2(t)$ and $x_3(t)$ corresponding to complex eigenvalues can be expressed in a form containing an oscillatory mode $e^{\alpha_2 t}\cdot\cos\Omega_2 t$ with a phase angle ϕ_2. To do this the following notation is introduced (Figure A2 in the appendix)

$$\sin\phi_2 = \frac{x_{30}}{\sqrt{x_{20}^2 + x_{30}^2}}, \quad \cos\phi_2 = \frac{x_{20}}{\sqrt{x_{20}^2 + x_{30}^2}}, \quad \phi_2 = \arcsin\left(x_{30}/\sqrt{x_{20}^2 + x_{30}^2}\right) \tag{12.77}$$

Now the solutions can be expressed as

$$x_2(t) = e^{\alpha_2 t}\cdot\sqrt{x_{20}^2 + x_{30}^2}\cdot(\cos\phi_2\cdot\cos\Omega_2 t - \sin\phi_2\sin\Omega_2 t)$$

$$x_3(t) = e^{\alpha_2 t}\cdot\sqrt{x_{20}^2 + x_{30}^2}\cdot(\sin\phi_2\cdot\cos\Omega_2 t + \cos\phi_2\cdot\sin\Omega_2 t)$$

and finally

$$\begin{aligned} x_1(t) &= x_{10} \cdot e^{\alpha_1 t} \\ x_2(t) &= \sqrt{x_{20}^2 + x_{30}^2} \cdot e^{\alpha_2 t} \cdot \cos(\Omega_2 t + \phi_2), \quad x_3(t) = \sqrt{x_{20}^2 + x_{30}^2} \cdot e^{\alpha_2 t} \cdot \sin(\Omega_2 t + \phi_2) \end{aligned} \tag{12.78}$$

Substituting the numbers $\alpha_1 = -6$, $\alpha_2 = -1$, and $\Omega_2 = 5$ gives

$$\begin{aligned} x_1(t) &= x_{10} \cdot e^{-6t} \\ x_2(t) &= \sqrt{x_{20}^2 + x_{30}^2} \cdot e^{-t} \cdot \cos(5t + \phi_2), \quad x_3(t) = \sqrt{x_{20}^2 + x_{30}^2} \cdot e^{-t} \cdot \sin(5t + \phi_2) \end{aligned} \tag{12.79}$$

where ϕ_2 is given by Eq. (12.77) and depends on the initial conditions. As oscillatory responses $x_3(t)$ and $x_2(t)$ are proportional to cosine and sine functions, they are shifted in phase by $\pi/2$.

Initial conditions $x_{10} = x_{20} = x_{30} = 1$ give the phase angle of the mode equal to $\phi_2 = 45^\circ = \pi/4$ and the following time responses

$$x_1(t) = e^{-6t}, \quad x_2(t) = \sqrt{2} \cdot e^{-t} \cdot \cos\left(5t + \frac{\pi}{4}\right), \quad x_3(t) = \sqrt{2} \cdot e^{-t} \cdot \sin\left(5t + \frac{\pi}{4}\right)$$

shown in Figure 12.1.

The example shows that the overall system response can be represented as a linear combination of aperiodic and oscillatory modes. The real eigenvalue $\lambda_1 = -6$ produces the aperiodic mode e^{-6t} decaying with a time constant of $1/6 = 0.17$ seconds. The complex conjugate pair of eigenvalues $\lambda_{2,3} = (-1 \pm j5)$ produces the oscillatory mode $e^{-t} \cdot \cos(5t + \phi_2)$ oscillating at frequency 5 rad/s and exponentially decaying with a time constant of 1 second. The system is stable because the real parts of all the eigenvalues are negative, that is the exponential functions are decaying. In this particular case the aperiodic mode shows itself only in the response of $x_1(t)$, while the oscillatory mode shows itself only in the responses of $x_2(t)$ and $x_3(t)$ because the first row of matrix $\boldsymbol{A}$ (corresponding to $x_1(t)$) contains only one diagonal element. In other words, the first row of $\boldsymbol{A}$ is decoupled from the other rows.

******

The considerations so far have assumed that all the eigenvalues of matrix $\boldsymbol{A}$ are distinct, $\lambda_1 \neq \lambda_2 \neq \ldots \neq \lambda_n$. If matrix $\boldsymbol{A}$ has multiple eigenvalues then the situation is more complicated. Nevertheless, it can be shown (Willems 1970) that

> The linear equation $\dot{\boldsymbol{x}} = \boldsymbol{A}\boldsymbol{x}$ is stable if, and only if, all the eigenvalues of matrix $\boldsymbol{A}$ have nonpositive real parts, $\mathrm{Re}(\lambda_i) \leq 0$. The system is asymptotically stable if, and only if, all the eigenvalues of matrix $\boldsymbol{A}$ have negative real parts, $\mathrm{Re}(\lambda_i) < 0$.

The stability of a linear equation does not depend on the initial condition but only on the eigenvalues of the state matrix $\boldsymbol{A}$.

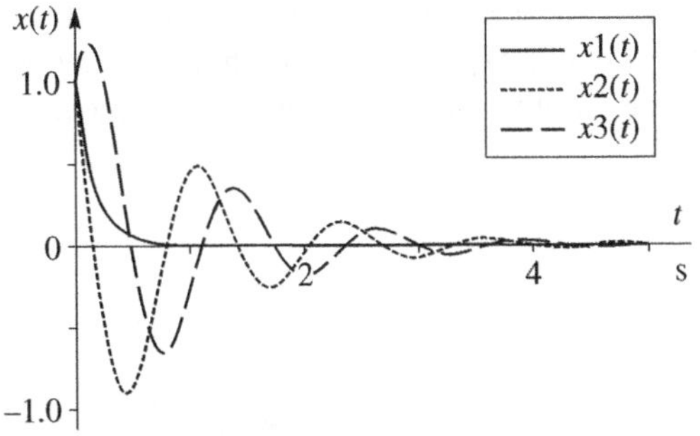

Figure 12.1 Illustration for the solution of Example 12.5.

12.1.4 Modal and Sensitivity Analysis

Equation (12.42) shows that each modal variable $z_i(t)$ can be expressed as a linear combination of the state variables, that is

$$z_i(t) = \sum_{j=1}^{n} u_{ij} x_j(t) \tag{12.80}$$

where u_{ij} is the (i, j) element of matrix $\boldsymbol{U}$ consisting of left eigenvectors. Expanding the sum in Eq. (12.80) gives

$$z_i(t) = u_{i1}x_1(t) + u_{i2}x_2(t) + \ldots + u_{ij}x_j(t) + \ldots + u_{in}x_n(t) \tag{12.81}$$

This equation shows that the left eigenvectors carry information about the controllability of individual modal variable by individual state variables. If the eigenvectors are normalized then u_{ij} determines the magnitude and phase of the share of a given variable $x_j(t)$ in the activity of a given mode $z_i(t)$. Controlling $x_j(t)$ influences a given modal variable $z_i(t)$ only if element u_{ij} is large. If u_{ij} is small then controlling $x_j(t)$ cannot influence modal variable $z_i(t)$.

Equation (12.41) shows that each state variable can be expressed as a linear combination of modal variables

$$x_k(t) = \sum_{i=1}^{n} w_{ki}z_i(t) \tag{12.82}$$

where w_{ki} is the (k, i) element of matrix $\boldsymbol{W}$ consisting of right eigenvectors. Expanding the sum in Eq. (12.82) gives

$$x_k(t) = w_{k1}z_1(t) + w_{k2}z_2(t) + \ldots + w_{kj}z_j(t) + \ldots + w_{kn}z_n(t) \tag{12.83}$$

This equation shows that the right eigenvectors carry information about the observability of individual modal variables in individual state variables. If the eigenvectors are normalized then w_{kj} determines the magnitude and phase of the share of modal variable $z_j(t)$ in the activity of state variable $x_k(t)$. This is referred to as the *mode shape*. Note that the mode shape represents an inherent feature of a linear dynamic system and does not depend on where and how a disturbance is applied. The mode shape plays an important role in power system stability analysis, especially for determining the influence of individual oscillatory modes on swings of rotors of individual generators.

Example 12.6 For a certain large interconnected system, it has been calculated that the damping ratio of one of the oscillatory modes is unsatisfactory, $\zeta < 0.05$. This mode corresponds to a complex conjugate pair of eigenvalues $\lambda_i = -0.451 + \text{j}2.198$ and $\lambda_j = \lambda_i^* = -0.451 - \text{j}2.198$. The frequency of the mode is $2.198/2\pi \cong 0.35$ Hz while the damping ratio is $\zeta = 0.045/\sqrt{0.045^2 + 2.198^2} \cong 0.02$. The important elements of the right eigenvector associated with λ_i and the corresponding mode shapes are shown in Figure 12.2.

The state variables corresponding to the considered mode shapes are the rotor angles of three generators: $x_1(t) = \Delta\delta_1(t)$, $x_k(t) = \Delta\delta_k(t)$, and $x_l(t) = \Delta\delta_l(t)$. For these variables one gets, according to Eq. (12.83)

$$\begin{aligned}
\Delta\delta_1(t) &= \cdots + w_{1i}z_i(t) + \cdots \\
\Delta\delta_k(t) &= \cdots + w_{ki}z_i(t) + \cdots \\
\Delta\delta_l(t) &= \cdots + w_{li}z_i(t) + \cdots
\end{aligned} \tag{12.84}$$

Figure 12.2 shows the mode shapes in the complex plane. Note that the mode shapes $\underline{w}_{ki}$ and $\underline{w}_{li}$ are almost directly in the opposite direction with respect to the mode shape $\underline{w}_{1i}$. The interpretation of this is that if a disturbance excites a mode corresponding to the pair of complex eigenvalues $\lambda_i = \lambda_j^*$ then the rotor of generator 1 will swing at a frequency of 0.35 Hz against the rotors of generators k and l which are coherent with respect to each other at that particular frequency. Generator l swings almost directly against generator 1 (the phase difference is 186°), while the 0.35 Hz swings of generator k are shifted by 159° with respect to generator 1. Obviously, these conclusions may not be true for other modes, that is other frequencies of modal oscillations make up the overall rotor swings

* * * * * *

(a)

$$\boldsymbol{w}_i = \begin{bmatrix} \underline{w}_{1i} \\ \vdots \\ \underline{w}_{ki} \\ \underline{w}_{li} \\ \vdots \end{bmatrix} = \begin{bmatrix} +4.5 + \text{j}0.8 \\ \vdots \\ -3.5 + \text{j}0.7 \\ -5.2 - \text{j}0.5 \\ \vdots \end{bmatrix}$$

(b)

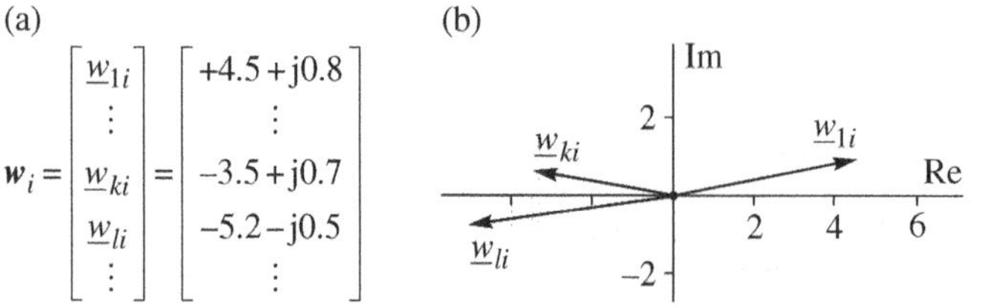

Figure 12.2 Example of mode shapes: (a) right eigenvector associated with the considered eigenvalue; (b) mode shapes in the complex plane.

Interesting examples of application of mode shapes to analyze power swings in Union for the Coordination of Transmission of Electricity (UCTE) interconnected power systems can be found in Breulmann et al. (2000). The article describes how a number of poorly damped modes have been discovered and how they were grouped, using mode shapes, into coherent groups of generators swinging against each other.

Knowledge of matrices $\boldsymbol{W}$ and $\boldsymbol{U}$ allows the sensitivity of a particular modal variable $z_i(t)$ to the changes in a particular system parameter to be determined. This is especially important when choosing the parameter values of any control device installed in the system. Such a *sensitivity analysis* is accomplished in the following way.

Let λ_i be an eigenvalue of matrix $\boldsymbol{A}$ and $\boldsymbol{w}_i$ and $\boldsymbol{u}_i$ be the right and left eigenvectors associated with this eigenvalue. Equation (12.23) shows that $\boldsymbol{u}_i\boldsymbol{A} = \lambda_i\boldsymbol{u}_i$. Right-multiplying by $\boldsymbol{w}_i$ gives $\boldsymbol{u}_i\boldsymbol{A}\boldsymbol{w}_i = \lambda_i\boldsymbol{u}_i\boldsymbol{w}_i$. Substituting $\boldsymbol{u}_i\boldsymbol{w}_i = 1$ to the right-hand side of that equation – see Eq. (12.32) – gives

$$\lambda_i = \boldsymbol{u}_i\boldsymbol{A}\boldsymbol{w}_i \tag{12.85}$$

Now let β be a system parameter. Equation (12.85) shows that

$$\frac{\partial \lambda_i}{\partial \beta} = \boldsymbol{u}_i \frac{\partial \boldsymbol{A}}{\partial \beta} \boldsymbol{w}_i \tag{12.86}$$

If the derivative $\partial\boldsymbol{A}/\partial\beta$ in Eq. (12.86) is known then it is possible to determine whether or not a given parameter improves the system stability by observing if the eigenvalues acquire a larger real part, that is move to the left in the complex plane, when the value of a parameter changes.

A particular case of sensitivity analysis is the investigation of the influence of the diagonal elements of a state matrix on the eigenvalues. Assuming $\beta = A_{kk}$ gives

$$\boldsymbol{A} = \begin{bmatrix} A_{11} & \cdots & A_{1k} & \cdots \\ \vdots & \ddots & \vdots & \\ A_{k1} & \cdots & A_{kk} & \\ \vdots & & \vdots & \ddots \end{bmatrix} \quad \text{and} \quad \frac{\partial \boldsymbol{A}}{\partial \beta} = \frac{\partial \boldsymbol{A}}{\partial A_{kk}} = \begin{bmatrix} 0 & \cdots & 0 & \cdots \\ \vdots & \ddots & \vdots & \\ 0 & \cdots & 1 & \\ \vdots & & \vdots & \ddots \end{bmatrix} \tag{12.87}$$

$$\frac{\partial \lambda_i}{\partial A_{kk}} = \boldsymbol{u}_i \frac{\partial \boldsymbol{A}}{\partial A_{kk}} \boldsymbol{w}_i = [u_{i1} \cdots u_{ik} \cdots] \begin{bmatrix} 0 & \cdots & 0 & \cdots \\ \vdots & \ddots & \vdots & \\ 0 & \cdots & 1 & \\ \vdots & & \vdots & \ddots \end{bmatrix} \begin{bmatrix} w_{1i} \\ \vdots \\ w_{ki} \\ \vdots \end{bmatrix} \tag{12.88}$$

Multiplying the matrices on the right-hand side of Eq. (12.88) gives

$$\frac{\partial \lambda_i}{\partial A_{kk}} = u_{ik} w_{ki} = p_{ki} \tag{12.89}$$

Coefficients $p_{ki} = u_{ik}w_{ki}$ are referred to as the *participation factors*. Each participation factor is a product of the k-th element of the i-th left and right eigenvectors. It quantifies the sensitivity of the i-th eigenvalue to the k-th diagonal element of the state matrix. Element w_{ki} contains information about the observability of the i-th modal variable in the k-th state variable, while u_{ik} contains information about the controllability of the i-th modal variable using the k-th state variable. Hence the product $p_{ki} = u_{ik}w_{ki}$ contains information about the observability and controllability. Consequently, the participation factor $p_{ki} = u_{ik}w_{ki}$ is a good measure of correlation between the i-th modal variable and the k-th state variable. Participation factors can be used to determine the sitting of devices, enhancing system stability. Generally, a damping controller or a stabilizer is preferably sited where the modal variables associated with a given eigenvalue are both well observable and well controllable.

The method of calculation of participation factors for the i-th eigenvalue and all diagonal elements $A_{11}, \cdots, A_{kk}, \cdots, A_{nn}$ of the state matrix $\boldsymbol{A}$ is illustrated below in Eq. (12.90) in which for convenience transposition of the left eigenvalue was used

$$\boldsymbol{u}_i^{\mathrm{T}} = \begin{bmatrix} u_{i1} \\ \vdots \\ u_{ik} \\ \vdots \\ u_{in} \end{bmatrix}, \quad \boldsymbol{w}_i = \begin{bmatrix} w_{1i} \\ \vdots \\ w_{ki} \\ \vdots \\ w_{ni} \end{bmatrix}; \text{ hence } \begin{bmatrix} u_{i1} \\ \vdots \\ u_{ik} \\ \vdots \\ u_{in} \end{bmatrix} \begin{matrix} \rightarrow \cdot \\ \\ \rightarrow \cdot \\ \\ \rightarrow \cdot \end{matrix} \begin{bmatrix} w_{1i} \\ \vdots \\ w_{ki} \\ \vdots \\ w_{ni} \end{bmatrix} \begin{matrix} \Rightarrow \\ \\ \Rightarrow \\ \\ \Rightarrow \end{matrix} \begin{bmatrix} u_{i1}w_{1i} \\ \vdots \\ u_{ik}w_{ki} \\ \vdots \\ u_{in}w_{ni} \end{bmatrix} = \begin{bmatrix} p_{1i} \\ \vdots \\ p_{ki} \\ \vdots \\ p_{ni} \end{bmatrix} = \boldsymbol{up}_i \tag{12.90}$$

Elements of column vector $\boldsymbol{p}_i$ contain participation factors quantifying to what extent individual diagonal elements $A_{11}, \cdots, A_{kk}, \cdots, A_{nn}$ of the state matrix may influence eigenvalue λ_i. If, for example, p_{ki} is large, it means that a diagonal element A_{kk} of the state matrix has large influence on λ_i.

Example 12.7 Consider again the system used previously in Examples 12.2, 12.4, and 12.5 characterized by

$$\boldsymbol{A} = \left[\begin{array}{c|c|c} -6 & 0 & 0 \\ \hline 0 & -1 & 5 \\ \hline 0 & -5 & -1 \end{array}\right]; \quad \lambda_1 = -6, \quad \lambda_2 = (-1-\mathrm{j}5), \quad \lambda_3 = (-1+\mathrm{j}5) = \lambda_2^* \tag{12.91}$$

The concluding remark in Example 12.5 stated that, owing to the block diagonal structure of matrix $\boldsymbol{A}$, the aperiodic mode corresponding to λ_1 was linked only with the first row of $\boldsymbol{A}$, while the oscillatory mode corresponding to $\lambda_2 = \lambda_3^*$ was linked only to the second and third rows of $\boldsymbol{A}$. That conclusion will now be confirmed formally by calculating participation factors.

Right eigenvalues $\boldsymbol{w}_1,\boldsymbol{w}_2$, and $\boldsymbol{w}_3$ are calculated in Example 12.2. Left eigenvalues $\boldsymbol{u}_1,\boldsymbol{u}_2$, and $\boldsymbol{u}_3$ are calculated in Example 12.4. For convenience, left-transposed eigenvectors are used: $\boldsymbol{u}_1^{\mathrm{T}},\boldsymbol{u}_2^{\mathrm{T}}$, and $\boldsymbol{u}_3^{\mathrm{T}}$. For the first pair, that is for λ_1, one gets

$$\boldsymbol{u}_1^{\mathrm{T}} = \frac{1}{4}\begin{bmatrix} 4 \\ 0 \\ 0 \end{bmatrix}, \quad \boldsymbol{w}_1 = \left[\begin{array}{c} 1 \\ \hline 0 \\ \hline 0 \end{array}\right], \quad \text{or } \frac{1}{4}\begin{bmatrix} 4 \\ 0 \\ 0 \end{bmatrix} \begin{matrix} \rightarrow \cdot \\ \rightarrow \cdot \\ \rightarrow \cdot \end{matrix} \left[\begin{array}{c} 1 \\ \hline 0 \\ \hline 0 \end{array}\right] \begin{matrix} \Rightarrow \\ \Rightarrow \\ \Rightarrow \end{matrix} \left[\begin{array}{c} 1 \\ \hline 0 \\ \hline 0 \end{array}\right] = \boldsymbol{p}_1 \tag{12.92}$$

This confirms that the first eigenvalue may be influenced only by changing element A_{11} of matrix $\boldsymbol{A}$. Elements A_{22} and A_{33} have no influence on λ_1 and the corresponding aperiodic mode.
For the second pair, that is for λ_2, one gets

$$\boldsymbol{u}_2^{\mathrm{T}} = \frac{1}{4}\begin{bmatrix} 0 \\ -1+\mathrm{j} \\ -1-\mathrm{j} \end{bmatrix}, \quad \boldsymbol{w}_2 = \left[\begin{array}{c} 0 \\ \hline -1-\mathrm{j} \\ \hline -1+\mathrm{j} \end{array}\right] \text{ or } \frac{1}{4}\begin{bmatrix} 0 \\ -1+\mathrm{j} \\ -1-\mathrm{j} \end{bmatrix} \begin{matrix} \rightarrow \cdot \\ \rightarrow \cdot \\ \rightarrow \cdot \end{matrix} \left[\begin{array}{c} 0 \\ \hline -1-\mathrm{j} \\ \hline -1+\mathrm{j} \end{array}\right] \begin{matrix} \Rightarrow \\ \Rightarrow \\ \Rightarrow \end{matrix} \left[\begin{array}{c} 0 \\ \hline 2 \\ \hline 2 \end{array}\right] \frac{1}{4} = \boldsymbol{p}_2 \tag{12.93}$$

For the third pair, that is for $\lambda_3 = \lambda_2^*$, one gets

$$\boldsymbol{u}_3^{\mathrm{T}} = \frac{1}{4}\begin{bmatrix} 0 \\ -1-\mathrm{j} \\ -1+\mathrm{j} \end{bmatrix}, \quad \boldsymbol{w}_3 = \left[\begin{array}{c} 0 \\ \hline -1+\mathrm{j} \\ \hline -1-\mathrm{j} \end{array}\right] \text{ or } \frac{1}{4}\begin{bmatrix} 0 \\ -1-\mathrm{j} \\ -1+\mathrm{j} \end{bmatrix} \begin{matrix} \rightarrow \cdot \\ \rightarrow \cdot \\ \rightarrow \cdot \end{matrix} \left[\begin{array}{c} 0 \\ \hline -1+\mathrm{j} \\ \hline -1-\mathrm{j} \end{array}\right] \begin{matrix} \Rightarrow \\ \Rightarrow \\ \Rightarrow \end{matrix} \left[\begin{array}{c} 0 \\ \hline 2 \\ \hline 2 \end{array}\right] \frac{1}{4} = \boldsymbol{p}_3 \tag{12.94}$$

This confirms that eigenvalues λ_2 and $\lambda_3 = \lambda_2^*$ may be equally influenced by changing elements A_{22} and A_{33} of $\boldsymbol{A}$. Element A_{11} has no influence on λ_2 and $\lambda_3 = \lambda_2^*$.

******

12.1.5 Modal Form of the State Equation with Inputs

The state Eq. (12.38) is homogeneous, that is it is of the form $\dot{\boldsymbol{x}} - \boldsymbol{A}\boldsymbol{x} = \boldsymbol{0}$. Sometimes it is necessary to consider a nonhomogeneous equation of the form $\dot{\boldsymbol{x}} - \boldsymbol{A}\boldsymbol{x} = \boldsymbol{B}\boldsymbol{u}$. That equation is usually written as

$$\dot{\boldsymbol{x}} = \boldsymbol{A}\boldsymbol{x} + \boldsymbol{B}\boldsymbol{u}, \tag{12.95}$$

where $\boldsymbol{B}$ is a rectangular matrix and $\boldsymbol{u}$ is a column vector containing system inputs.

Equation (12.95) can also be analyzed using modal analysis. Substituting Eqs. (12.41) into Eq. (12.95) gives $\boldsymbol{W}\dot{\boldsymbol{z}} = \boldsymbol{A}\boldsymbol{W}\boldsymbol{z} + \boldsymbol{B}\boldsymbol{u}$ or $\dot{\boldsymbol{z}} = \boldsymbol{W}^{-1}\boldsymbol{A}\boldsymbol{W}\boldsymbol{z} + \boldsymbol{W}^{-1}\boldsymbol{B}\boldsymbol{u}$. After taking into account Eqs. (12.20) and (12.21), this gives

$$\dot{\boldsymbol{z}} = \boldsymbol{\Lambda}\boldsymbol{z} + \boldsymbol{b}\boldsymbol{u} \tag{12.96}$$

where $\boldsymbol{b} = \boldsymbol{W}^{-1}\boldsymbol{B} = \boldsymbol{U}\boldsymbol{B}$. Equation (12.96) represents the modal form of the state Eq. (12.95). This equation may be used to study the influence of inputs $\boldsymbol{u}$ on the excitation of modal variables $\boldsymbol{z}(t)$. Equation (12.96) may be written as

$$\begin{bmatrix} \dot{z}_1 \\ \vdots \\ \dot{z}_i \\ \vdots \\ \dot{z}_n \end{bmatrix} = \begin{bmatrix} \lambda_1 & & & & \\ & \ddots & & & \\ & & \lambda_i & & \\ & & & \ddots & \\ & & & & \lambda_n \end{bmatrix} \begin{bmatrix} z_1 \\ \vdots \\ z_i \\ \vdots \\ z_n \end{bmatrix} + \begin{bmatrix} \boldsymbol{b}_1 \\ \vdots \\ \boldsymbol{b}_i \\ \vdots \\ \boldsymbol{b}_n \end{bmatrix} \boldsymbol{u} \tag{12.97}$$

or $\dot{z}_i = \lambda_i z_i + \boldsymbol{b}_i\boldsymbol{u}$. It shows that the excitation of a given modal variable by input $\boldsymbol{u}$ is decided by row vector $\boldsymbol{b}_i$. It may happen that a certain structure of matrix $\boldsymbol{B}$ results in some modal variables not being excited. An example of this is given in Section 18.6.3.

12.1.6 Nonlinear System

Generally, a nonlinear dynamic system can be described by the differential matrix equation

$$\dot{\boldsymbol{x}} = \boldsymbol{F}(\boldsymbol{x}) \tag{12.98}$$

where $\boldsymbol{x}$ is the vector of the n state variables. The equilibrium points $\hat{\boldsymbol{x}}$ are those points where the system is at rest, that is where all the state variables are constant and their values do not change with time, so that $\boldsymbol{F}(\hat{\boldsymbol{x}}) = \boldsymbol{0}$. Expanding function $\boldsymbol{F}(\boldsymbol{x})$ in the vicinity of $\hat{\boldsymbol{x}}$ in a Taylor series, and neglecting the nonlinear part of the expansion, gives the *linear approximation* of the nonlinear Eq. (12.98) as

$$\Delta\dot{\boldsymbol{x}} = \boldsymbol{A}\Delta\boldsymbol{x} \tag{12.99}$$

where $\Delta\boldsymbol{x} = \boldsymbol{x} - \hat{\boldsymbol{x}}$ and $\boldsymbol{A} = \partial\boldsymbol{F}/\partial\boldsymbol{x}$ is the Jacobi matrix calculated at the point $\hat{\boldsymbol{x}}$. Equation (12.99) is known as the *state equation.*

Lyapunov's first method now defines the stability of the nonlinear system based on its linear approximation as follows:

> A nonlinear system is steady-state stable in the vicinity of the equilibrium point $\hat{\boldsymbol{x}}$ if its linear approximation is asymptotically stable. If the linear approximation is unstable, the nonlinear system is also unstable. If the linear approximation is stable, but not asymptotically stable, then it is not possible to assess the system stability based on its linear approximation.

This theorem, together with Lyapunov's first theorem, leads to the conclusion that if the nonlinear equation $\dot{\boldsymbol{x}} = \boldsymbol{F}(\boldsymbol{x})$ can be approximated by the linear equation $\Delta\dot{\boldsymbol{x}} = \boldsymbol{A}\Delta\boldsymbol{x}$ then the nonlinear system is asymptotically stable if all the eigenvalues of the state matrix $\boldsymbol{A}$ are negative ($\mathrm{Re}(\lambda_i) < 0$). If any of the eigenvalues have a positive real part then the system is unstable. If any of the eigenvalues are zero then no conclusion can be reached with regard to the system stability and some other method, such as Lyapunov's second method, must be used.

12.2 Steady-state Stability of Unregulated System

The *inherent steady-state stability* of an EPS is concerned with analyzing the response of generators when the system is subjected to a small disturbance and when the effect of the voltage regulators is neglected. This form of stability was first discussed in Section 5.4 for the generator-infinite busbar system where the steady-state stability limit (critical power) is determined (Figure 5.5) by the steady-state model of the generator; that is constant electromotive forces (emf's) $E_f = E_q$ and $E_d = 0$ behind the synchronous reactances X_d and X_q in conjunction with the swing equation (5.1). The same assumption is valid in this chapter when each i-th generator is replaced by emf $\underline{E}_i$ behind reactance X_i. In the considered case of an unregulated system is further assumed that

$$\underline{E}_i = \underline{E}_{q\,i} \quad \text{and} \quad X_i = X_{d\,i} = X_{q\,i} \tag{12.100}$$

Nodes behind the generator reactances form the generator nodes set {G}. The system model is shown schematically in Figure 12.3a. {L} is the set of load nodes that are eliminated, which leads to an equivalent transfer network, shown in Figure 12.3b. In the equivalent network, the generator nodes {G} are directly connected with each other.

Assuming small power angle variations, the swing equation (5.1) can be written in the following way for each i-th generator

$$\begin{aligned} \frac{d\Delta\delta_i}{dt} &= \Delta\omega_i \\ \frac{d\Delta\omega_i}{dt} &= -\frac{\Delta P_i}{M_i} - \frac{D_i}{M_i}\Delta\omega_i \quad \text{for} \quad i = 1, 2, \ldots, n \end{aligned} \tag{12.101}$$

where ΔP_i is the change in the generator active power determined using the incremental network Eq. (3.182). Equations. (3.182) and (12.101) constitute the basic linearized system model suitable for assessing small-signal stability. However, the final form of the state equation will depend on additional assumptions regarding the value of the damping coefficients and the load model used.

12.2.1 State-space Equation

Computations are considerably simplified if all the system loads are modeled as constant impedance loads. Further simplifications can be introduced by neglecting steady-state saliency so that for all the generators $X_d = X_q$. The next step is to eliminate all the load nodes, including the generator terminal nodes, in the network model of Figure 12.3

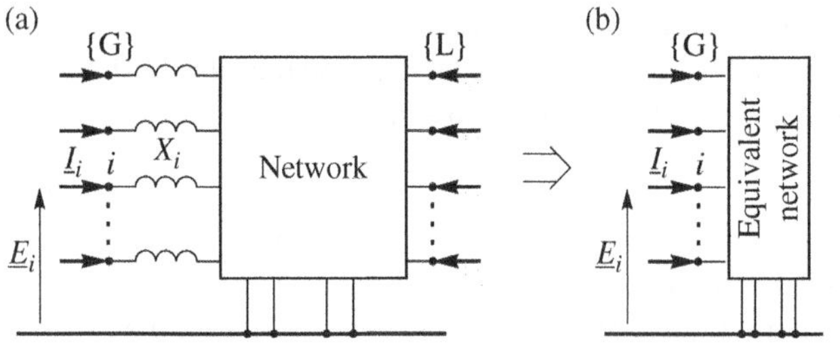

Figure 12.3 Network model for the steady-state stability analysis: (a) before elimination of load nodes; (b) after elimination. {G} is the set of generator nodes behind the synchronous reactances, {L} is the set of load nodes including generator terminals.

using the method described in Section 18.2.1. The only retained nodes are the fictitious generator nodes behind the synchronous reactances. The equivalent transfer network directly links all the generator nodes. For this network, Eqs. (3.179) and (3.182) can be used with the voltages V_i replaced by the synchronous emf's E_i. As $\Delta E_i = 0$, Eq. (3.182) can be simplified to

$$\Delta \boldsymbol{P} = \boldsymbol{H}\Delta\boldsymbol{\delta} \tag{12.102}$$

where, according to Eq. (3.185), the elements of the Jacobi matrix are

$$\begin{aligned} H_{ij} &= \frac{\partial P_i}{\partial \delta_j} = E_i E_j\left(-B_{ij}\cos\delta_{ij} + G_{ij}\sin\delta_{ij}\right) \\ H_{ii} &= \frac{\partial P_i}{\partial \delta_i} = \sum_{j=1}^{n} E_i E_j\left(B_{ij}\cos\delta_{ij} - G_{ij}\sin\delta_{ij}\right) \end{aligned} \tag{12.103}$$

The matrix $\boldsymbol{H}$ is singular because the sum of the elements in each of its rows is zero

$$\sum_{j=1}^{n} H_{ij} = H_{ii} + \sum_{j\neq i}^{n} H_{ij} = 0 \tag{12.104}$$

At this stage it is tempting to define the increments in the power angles as the state variables, but, as a loss of synchronism does not correspond to a simultaneous increase in all the power angles, this is not a valid choice. Rather, as a loss of synchronism is determined by the relative angles $\Delta\delta_{in}$ calculated with respect to a *reference generator*, it is the increment in these relative angles that must be used as the state variables. This is illustrated in Figure 12.4.

The power change in all the generators must now be related to the relative angles $\Delta\delta_{in}$. Assuming that the last generator, numbered n, acts as the reference, Eq. (12.104) gives

$$H_{in} = -\sum_{j\neq n}^{n} H_{ij} \tag{12.105}$$

The power change in any of the generators, Eq. (12.102), can be expressed as

$$\begin{aligned} \Delta P_i &= \sum_{j=1}^{n} H_{ij}\Delta\delta_i = \sum_{j\neq n} H_{ij}\Delta\delta_j + H_{in}\Delta\delta_n = \sum_{j\neq n} H_{ij}\Delta\delta_j - \sum_{j\neq n} H_{ij}\Delta\delta_n \\ &= \sum_{j\neq n} H_{ij}\left(\Delta\delta_j - \Delta\delta_n\right) = \sum_{j\neq n} H_{ij}\Delta\delta_{jn} \end{aligned} \tag{12.106}$$

Equation (12.102) can now be expressed as a function of the relative angles by removing the last column of $\boldsymbol{H}$, that is

$$\begin{bmatrix} \Delta P_1 \\ \Delta P_2 \\ \vdots \\ \Delta P_n \end{bmatrix} = \begin{bmatrix} H_{11} & H_{12} & \cdots & H_{1,n-1} \\ H_{21} & H_{22} & \cdots & H_{2,n-1} \\ \vdots & \vdots & \ddots & \vdots \\ H_{21} & H_{n2} & \cdots & H_{n,n-1} \end{bmatrix} \begin{bmatrix} \Delta\delta_{1n} \\ \Delta\delta_{2n} \\ \vdots \\ \Delta\delta_{n-1,n} \end{bmatrix} \quad \text{or} \quad \Delta\boldsymbol{P} = \boldsymbol{H}_n\Delta\boldsymbol{\delta}_n \tag{12.107}$$

where the matrix $\boldsymbol{H}_n$ is rectangular and is of dimension $n \times (n-1)$, while the vector $\Delta\boldsymbol{\delta}_n$ contains $(n-1)$ relative rotor angles calculated with respect to the n-th reference generator.

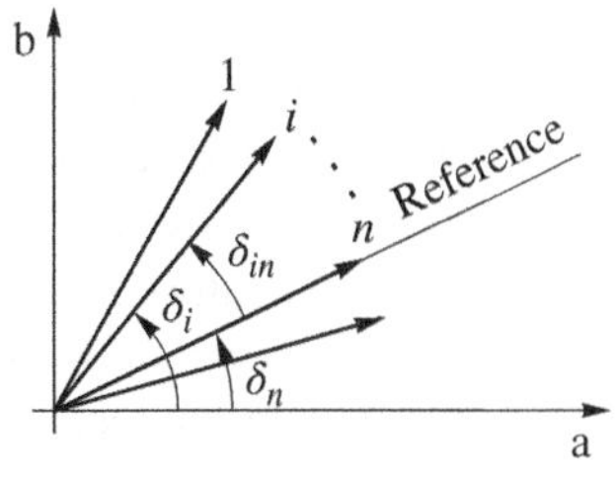

Figure 12.4 The emf's in the complex plane. Finding the relative angles $\delta_{in} = \delta_i - \delta_n$ where n is the reference generator and (a, b) are the rectangular coordinates of network equations.

Substituting Eq. (12.107) into the differential Eq. (12.101) gives

$$
\begin{bmatrix} \Delta\dot{\delta}_{1n} \\ \Delta\dot{\delta}_{2n} \\ \vdots \\ \Delta\dot{\delta}_{n-1,n} \\ \hdashline \Delta\dot{\omega}_1 \\ \Delta\dot{\omega}_2 \\ \vdots \\ \Delta\dot{\omega}_{n-1} \\ \hdashline \Delta\dot{\omega}_n \end{bmatrix}
=
\left[\begin{array}{cccc:cccc:c}
0 & 0 & \cdots & 0 & 1 & 0 & \cdots & 0 & -1 \\
0 & 0 & \cdots & 0 & 0 & 1 & \cdots & 0 & -1 \\
\vdots & \vdots & \ddots & \vdots & \vdots & \vdots & \ddots & \vdots & \vdots \\
0 & 0 & \cdots & 0 & 0 & 0 & \cdots & 1 & -1 \\
\hdashline
-\dfrac{H_{11}}{M_1} & -\dfrac{H_{12}}{M_1} & \cdots & -\dfrac{H_{1,n-1}}{M_1} & -\dfrac{D_1}{M_1} & 0 & \cdots & 0 & 0 \\
-\dfrac{H_{21}}{M_2} & -\dfrac{H_{22}}{M_2} & \cdots & -\dfrac{H_{2,n-1}}{M_2} & 0 & -\dfrac{D_2}{M_2} & \cdots & 0 & 0 \\
\vdots & \vdots & \ddots & \vdots & \vdots & \vdots & \ddots & \vdots & \vdots \\
-\dfrac{H_{n-1,1}}{M_{n-1}} & -\dfrac{H_{n-1,n-1}}{M_{n-1}} & \cdots & -\dfrac{H_{n-1,n-1}}{M_{n-1}} & 0 & 0 & \cdots & -\dfrac{D_{n-1}}{M_{n-1}} & 0 \\
\hdashline
-\dfrac{H_{n1}}{M_n} & -\dfrac{H_{n2}}{M_n} & \cdots & -\dfrac{H_{n,n-1}}{M_n} & 0 & 0 & \cdots & 0 & -\dfrac{D_n}{M_n}
\end{array}\right]
\begin{bmatrix} \Delta\delta_{1n} \\ \Delta\delta_{2n} \\ \vdots \\ \Delta\delta_{n-1,n} \\ \hdashline \Delta\omega_1 \\ \Delta\omega_2 \\ \vdots \\ \Delta\omega_{n-1} \\ \hdashline \Delta\omega_n \end{bmatrix}
\tag{12.108}
$$

Equation (12.108) can be expressed more compactly as

$$
\begin{bmatrix} \Delta\dot{\boldsymbol{\delta}}_n \\ \hdashline \Delta\dot{\boldsymbol{\omega}} \end{bmatrix}
=
\left[\begin{array}{c:c} \mathbf{0} & \mathbf{1}_{-1} \\ \hdashline -\boldsymbol{M}^{-1}\boldsymbol{H}_n & -\boldsymbol{M}^{-1}\boldsymbol{D} \end{array}\right]
\begin{bmatrix} \Delta\boldsymbol{\delta}_n \\ \hdashline \Delta\boldsymbol{\omega} \end{bmatrix}
\quad \text{or} \quad \Delta\dot{\boldsymbol{x}} = \boldsymbol{A}\Delta\boldsymbol{x}
\tag{12.109}
$$

where $\mathbf{1}_{-1}$ denotes a diagonal unit matrix extended by a column whose elements are equal to (-1). In this equation the state vector has $(2n-1)$ elements consisting of $(n-1)$ angle changes and n speed deviations. The matrix $\boldsymbol{A}$ has a rank equal to $(2n-1)$.

For the generator-infinite busbar system Eq. (12.108) takes the form

$$
\begin{bmatrix} \Delta\dot{\delta} \\ \hdashline \Delta\dot{\omega} \end{bmatrix}
=
\left[\begin{array}{c:c} 0 & 1 \\ \hdashline -\dfrac{H}{M} & -\dfrac{D}{M} \end{array}\right]
\begin{bmatrix} \Delta\delta \\ \hdashline \Delta\omega \end{bmatrix}.
\tag{12.110}
$$

The characteristic equation of this matrix is

$$
\lambda^2 + \frac{D}{M}\lambda + \frac{H}{M} = 0
\tag{12.111}
$$

and results in two eigenvalues. This equation is identical to the characteristic Eq. (5.59) but with $H = K_{E_q}$ used instead of the transient synchronizing power coefficient $K_{E'}$. This is a consequence of using Eqs. (12.108) or (12.110) to assess the steady-state stability, not the frequency of rotor oscillations. If the classical transient model was used with the generator represented by the transient reactance X'_d then Eqs. (12.108) or (12.110) could be used to determine the frequency of oscillations but not for assessing the steady-state stability. As shown in Section 5.4 (Figure 5.5), the steady-state stability limit is determined by the steady-state characteristic obtained when the generator is represented by the emf E_q behind the synchronous reactance X_d.

Example 12.8 Using Eq. (12.110) for the generator-infinite busbar system one must calculate the participation factors. In order to simplify the considerations, this equation can be written as

$$
\begin{bmatrix} \Delta\dot{\delta} \\ \hdashline \Delta\dot{\omega} \end{bmatrix}
=
\left[\begin{array}{c:c} 0 & 1 \\ \hdashline -h & -d \end{array}\right]
\begin{bmatrix} \Delta\delta \\ \hdashline \Delta\omega \end{bmatrix}
\quad \text{or} \quad \dot{\boldsymbol{x}} = \boldsymbol{A}\boldsymbol{x}
\tag{12.112}
$$

where $h = H/M$ and $d = D/M$. From Eq. (12.112) it follows that in this case the eigenvalues of the matrix $\boldsymbol{A}$ are

$$\lambda_{1,2} = \frac{-d \pm \sqrt{d^2 - 4h}}{2} \tag{12.113}$$

Right eigenvectors can be calculated based on Eq. (12.6). To this end, an adjacent matrix is calculated

$$[\boldsymbol{A} - \lambda \mathbf{1}] = \left[\begin{array}{c|c} -\lambda & 1 \\ \hline -h & -d-\lambda \end{array}\right]; \quad [\boldsymbol{A} - \lambda \mathbf{1}]^{\mathrm{T}} = \left[\begin{array}{c|c} -\lambda & -h \\ \hline -1 & -d-\lambda \end{array}\right]; \quad [\boldsymbol{A} - \lambda \mathbf{1}]^{\mathrm{D}} = \left[\begin{array}{c|c} -d-\lambda & -1 \\ \hline h & -\lambda \end{array}\right]$$

As a right eigenvector one can assume any column of the above adjacent matrix multiplied by any number, e.g. the second column multiplied by (−1). Substituting for λ, respectively, λ_1 and λ_2 the following eigenvectors are obtained

$$\boldsymbol{w}_1 = \left[\begin{array}{c} 1 \\ \hline \lambda_1 \end{array}\right]; \quad \boldsymbol{w}_2 = \left[\begin{array}{c} 1 \\ \hline \lambda_2 \end{array}\right].$$

The left eigenvectors matrix is obtained from Eq. (12.20)

$$\boldsymbol{W} = [\boldsymbol{w}_1 | \boldsymbol{w}_2] = \left[\begin{array}{c|c} 1 & 1 \\ \hline \lambda_1 & \lambda_2 \end{array}\right]; \quad \det \boldsymbol{W} = \lambda_2 - \lambda_1; \ \boldsymbol{U} = \boldsymbol{W}^{-1} = \frac{1}{\lambda_2 - \lambda_1} \left[\begin{array}{c|c} \lambda_2 & -1 \\ \hline -\lambda_1 & 1 \end{array}\right]. \tag{12.114}$$

therefore, the following left eigenvectors

$$\boldsymbol{u}_1 = \frac{1}{\lambda_2 - \lambda_1} [\lambda_2 \,|\, -1]; \qquad \boldsymbol{u}_2 = \frac{1}{\lambda_2 - \lambda_1} [-\lambda \,|\, 1]. \tag{12.115}$$

Multiplying according to Eq. (12.90) appropriate elements of left and right eigenvectors, the following matrix containing all participation factors is obtained

$$\boldsymbol{x} = \left[\begin{array}{c} \Delta\delta \\ \hline \Delta\omega \end{array}\right]; \quad \mathrm{diag}\boldsymbol{A} = \left[\begin{array}{c} A_{11} \\ \hline A_{22} \end{array}\right]; \quad \boldsymbol{p} = \frac{1}{\lambda_2 - \lambda_1} \begin{array}{c} \begin{array}{cc} \lambda_1 & \lambda_2 \end{array} \\ \left[\begin{array}{c|c} \lambda_2 & -\lambda_1 \\ \hline -\lambda_1 & \lambda_2 \end{array}\right] \end{array} \tag{12.116}$$

In the particular case, when damping is omitted, i.e. for $d = 0$ one gets: $\lambda_1 = \mathrm{j}\sqrt{h}$, $\lambda_2 = -\mathrm{j}\sqrt{h}$, and $\lambda_2 - \lambda_1 = -\mathrm{j}2\sqrt{h}$, which after substituting into Eq. (12.116) gives

$$\boldsymbol{x} = \left[\begin{array}{c} \Delta\delta \\ \hline \Delta\omega \end{array}\right]; \quad \mathrm{diag}\boldsymbol{A} = \left[\begin{array}{c} A_{11} \\ \hline A_{22} \end{array}\right]; \quad \boldsymbol{p} = \begin{array}{c} \begin{array}{cc} \lambda_1 & \lambda_2 \end{array} \\ \left[\begin{array}{c|c} 0.5 & 0.5 \\ \hline 0.5 & 0.5 \end{array}\right] \end{array} \tag{12.117}$$

This means that all participation factors are the same and equal to 0.5. This can be interpreted in such a way that modes corresponding to their own values λ_1 and λ_2 have the same influence on variables $\Delta\delta$ and $\Delta\omega$, which is a trivial conclusion.

******

12.2.2 Simplified Steady-state Stability Conditions

Equation (12.108) is valid when the network conductances are included and for any value of the damping coefficients. Checking the steady-state stability condition requires the calculation of eigenvalues of matrix $\boldsymbol{A}$, which is time consuming. This can be significantly simplified if network conductances are neglected and uniform weak damping of swings of all generators is assumed.

Damping of rotor swing in a system is *uniform* if

$$\frac{D_1}{M_1} = \frac{D_2}{M_2} = \dots = \frac{D_n}{M_n} = d \tag{12.118}$$

With uniform damping the swing equation for any machine, and for the reference machine, can be written as

$$\frac{\mathrm{d}\Delta\omega_i}{\mathrm{d}t} = -\frac{\Delta P_i}{M_i} - d\Delta\omega_i; \qquad \frac{\mathrm{d}\Delta\omega_n}{\mathrm{d}t} = -\frac{\Delta P_n}{M_n} - d\Delta\omega_n \tag{12.119}$$

Subtracting the last equation from the first gives

$$\frac{\mathrm{d}\Delta\omega_{in}}{\mathrm{d}t} = -\left(\frac{\Delta P_i}{M_i} - \frac{\Delta P_n}{M_n}\right) - d\Delta\omega_{in} \quad \text{for} \quad i = 1, 2, \dots, (n-1) \tag{12.120}$$

where $\Delta\omega_{in} = \Delta\omega_i - \Delta\omega_n = \mathrm{d}\,\delta_{in}/\mathrm{d}t$ and is the rotor speed deviation of the i-th generator relative to the reference machine. Substituting the power changes calculated from Eq. (12.107) into Eq. (12.120) gives

$$\begin{bmatrix} \Delta\dot{\boldsymbol{\delta}}_{n-1} \\ \Delta\dot{\boldsymbol{\omega}}_{n-1} \end{bmatrix} = \begin{bmatrix} \mathbf{0} & \mathbf{1} \\ \boldsymbol{a} & -\boldsymbol{d} \end{bmatrix} \begin{bmatrix} \Delta\boldsymbol{\delta}_{n-1} \\ \Delta\boldsymbol{\omega}_{n-1} \end{bmatrix} \quad \text{or} \quad \Delta\dot{\boldsymbol{x}} = \boldsymbol{A}\Delta\boldsymbol{x}, \tag{12.121}$$

where $\mathbf{1}$ is the unit matrix, $\Delta\boldsymbol{\delta}_{n-1}$ and $\Delta\boldsymbol{\omega}_{n-1}$ are the vectors of the $(n-1)$ relative changes in the power angles and speed deviations, and $\boldsymbol{d}$ is a diagonal matrix in which all the diagonal elements are identical and equal to d given by Eq. (12.118). The matrix $\boldsymbol{a}$ is equal to

$$\boldsymbol{a} = -\begin{bmatrix} \frac{H_{11}}{M_1} - \frac{H_{n1}}{M_n} & \frac{H_{12}}{M_1} - \frac{H_{n2}}{M_n} & \cdots & \frac{H_{1,n-1}}{M_1} - \frac{H_{n,n-1}}{M_n} \\ \frac{H_{21}}{M_2} - \frac{H_{n1}}{M_n} & \frac{H_{22}}{M_2} - \frac{H_{n2}}{M_n} & \cdots & \frac{H_{2,n-1}}{M_2} - \frac{H_{n,n-1}}{M_n} \\ \vdots & \vdots & \ddots & \vdots \\ \frac{H_{n-1,1}}{M_{n-1}} - \frac{H_{n1}}{M_n} & \frac{H_{n-1,2}}{M_{n-1}} - \frac{H_{n2}}{M_n} & \cdots & \frac{H_{n-1,n-1}}{M_{n-1}} - \frac{H_{n,n-1}}{M_n} \end{bmatrix} \tag{12.122}$$

Comparing the matrix $\boldsymbol{a}$ with the square state matrix in Eq. (12.108) shows that the matrix $\boldsymbol{a}$ is created by subtracting the last row in Eq. (12.108) from all the upper rows corresponding to the elements H_{ij}. The state vector in Eq. (12.121) has $(2n-2)$ elements, including $(n-1)$ relative angles $\Delta\delta_{in}$ and $(n-1)$ relative speed deviations $\Delta\omega_{in}$. The problem of finding the eigenvalues of the state matrix in Eq. (12.121) can now be simplified to that of determining the eigenvalues of the matrix $\boldsymbol{a}$ which, with rank $(n-1)$, is half the size. This can be explained as follows.

Let λ_i and w_i be the eigenvalues and eigenvectors of the matrix $\boldsymbol{A}$ in Eq. (12.121). The definition of the eigenvalues and eigenvectors gives

$$\begin{bmatrix} \mathbf{0} & \mathbf{1} \\ \boldsymbol{a} & -\boldsymbol{d} \end{bmatrix} \begin{bmatrix} \boldsymbol{w}_i' \\ \boldsymbol{w}_i'' \end{bmatrix} = \lambda_i \begin{bmatrix} \boldsymbol{w}_i' \\ \boldsymbol{w}_i'' \end{bmatrix} \tag{12.123}$$

or, after multiplying, $\boldsymbol{w}_i'' = \lambda_i \boldsymbol{w}_i'$ and $\left(\boldsymbol{a}\boldsymbol{w}_i' - \boldsymbol{d}\boldsymbol{w}_i''\right) = \lambda_i \boldsymbol{w}_i''$. Substituting $\boldsymbol{w}_i''$ calculated from the first equation into the second gives $\boldsymbol{a}\boldsymbol{w}_i' = \left(\lambda_i^2 + d\lambda_i\right)\boldsymbol{w}_i'$ or $\boldsymbol{a}\boldsymbol{w}_i' = \mu_i \boldsymbol{w}_i'$, where $\mu_i = \lambda_i^2 + d\lambda_i$. Hence

$$\lambda_i^2 + d\lambda_i - \mu_i = 0 \tag{12.124}$$

Obviously, μ_i and $\boldsymbol{w}_i'$ also satisfy the definition of eigenvalues and eigenvectors of matrix $\boldsymbol{a}$. The eigenvalue λ_i can be found by solving Eq. (12.124) to give

$$\lambda_i = -\frac{\mathrm{d}}{2} \pm \sqrt{\frac{\mathrm{d}^2}{4} + \mu_i} \tag{12.125}$$

Figure 12.5 Eigenvalues μ and λ when: (a) μ is complex; (b) μ is real and negative; (c) μ is real and negative and there is positive damping.

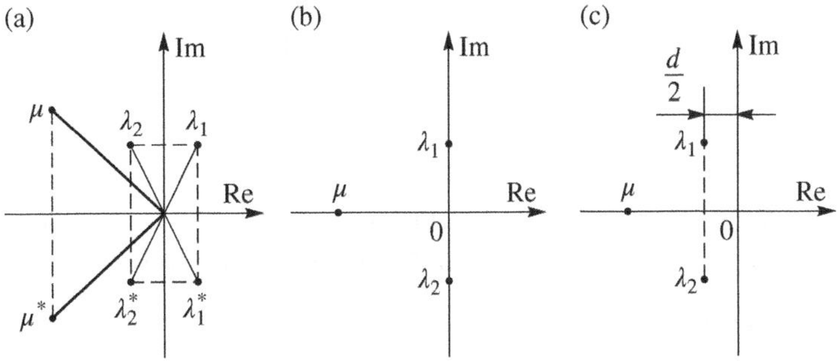

Thus, knowing the eigenvalues μ_i of matrix $\boldsymbol{a}$, it is possible to determine the eigenvalues λ_i. With this information it is possible to evaluate the system stability knowing the eigenvalues μ_i and the damping coefficient d. Obviously, determining the eigenvalues μ_i of matrix $\boldsymbol{a}$ of rank $(n-1)$ is less computationally intensive than determining the eigenvalues λ_i of matrix $\boldsymbol{A}$, which has a rank $(2n-2)$. Thus the assumption of uniform damping, if valid, considerably simplifies stability calculations.

Equation (12.124) shows that if the damping coefficient is zero, $d \cong 0$, then $\lambda_i = \pm\sqrt{\mu_i}$, that is the eigenvalues of the state matrix $\boldsymbol{A}$ are the square root values of the eigenvalues of matrix $\boldsymbol{a}$. Hence if an eigenvalue μ_i is complex (Figure 12.5a) then one of the eigenvalues λ_i lies on the right-hand side of the complex plane and the system is unstable. On the other hand, if an eigenvalue μ_i is real and negative (Figure 12.5b), the corresponding eigenvalues λ_i lie on the imaginary axis. In this case any small positive damping will move the eigenvalues to the left-hand plane and the system will be stable (Figure 12.5c).

When damping is weak, the steady-state stability condition is that eigenvalues μ_i of matrix $\boldsymbol{a}$ should be real and negative

$$\mu_i \in \text{Real} \quad \text{and} \quad \mu_i < 0 \quad \text{for} \quad i = 1, 2, \ldots, n-1 \tag{12.126}$$

Further simplification can be obtained by neglecting network conductances. Equation (12.103) shows that if $G_{ij} \cong 0$ then $H_{ij} \cong H_{ji}$, that is the matrix of synchronizing powers is symmetric $\boldsymbol{H}^{\mathrm{T}} = \boldsymbol{H}$.

When matrix $\boldsymbol{H}$ is symmetric, Eq. (12.104) is valid for rows and also valid for columns of the matrix. Similar to Eq. (12.105) one gets

$$H_{nj} = -\sum_{i \neq n}^{n} H_{ij} \tag{12.127}$$

Under this assumption matrix (12.122) can be written in a product form

$$\boldsymbol{a} = -\boldsymbol{m}\,\boldsymbol{H}_{n-1} \tag{12.128}$$

where

$$\boldsymbol{m} = \begin{bmatrix} \frac{1}{M_1} + \frac{1}{M_n} & \frac{1}{M_n} & \cdots & \frac{1}{M_n} \\ \frac{1}{M_n} & \frac{1}{M_2} + \frac{1}{M_n} & \cdots & \frac{1}{M_n} \\ \vdots & \vdots & \ddots & \vdots \\ \frac{1}{M_n} & \frac{1}{M_n} & \cdots & \frac{1}{M_{n-1}} + \frac{1}{M_n} \end{bmatrix} \tag{12.129}$$

$$\boldsymbol{H}_{n-1}=\begin{bmatrix} H_{11} & H_{12} & \cdots & H_{1,n-1} \\ H_{21} & H_{22} & \cdots & H_{2,n-1} \\ \vdots & \vdots & \ddots & \vdots \\ H_{n-1,1} & H_{n-1,2} & \cdots & H_{n-1,n-1} \end{bmatrix} \tag{12.130}$$

According to the definition of the right eigenvector and eigenvalue, $\boldsymbol{a}\boldsymbol{w}_i = \mu_i \boldsymbol{w}_i$ holds. Substituting for $\boldsymbol{a}$ on the right-hand side of Eq. (12.128) gives $\boldsymbol{m}\boldsymbol{H}_{n-1}\boldsymbol{w}_i = -\mu_i\boldsymbol{w}_i$, that is

$$\boldsymbol{H}_{n-1}\boldsymbol{w}_i = -\mu_i \boldsymbol{m}^{-1}\boldsymbol{w}_i \tag{12.131}$$

and

$$\boldsymbol{w}_i^{*\mathrm{T}}\boldsymbol{H}_{n-1}^{*\mathrm{T}} = -\mu_i^{*}\boldsymbol{w}_i^{*\mathrm{T}}\left(\boldsymbol{m}^{-1}\right)^{*\mathrm{T}} \tag{12.132}$$

Now Eq. (12.131) should be left-multiplied by $\boldsymbol{w}_i^{*\mathrm{T}}$ and Eq. (12.132) right-multiplied by $\boldsymbol{w}_i$. This gives

$$\boldsymbol{w}_i^{*\mathrm{T}}\boldsymbol{H}_{n-1}\boldsymbol{w}_i = -\mu_i \boldsymbol{w}_i^{*\mathrm{T}}\boldsymbol{m}^{-1}\boldsymbol{w}_i \tag{12.133}$$

$$\boldsymbol{w}_i^{*\mathrm{T}}\boldsymbol{H}_{n-1}^{*\mathrm{T}}\boldsymbol{w}_i = -\mu_i^{*}\boldsymbol{w}_i^{*\mathrm{T}}\left(\boldsymbol{m}^{-1}\right)^{*\mathrm{T}}\boldsymbol{w}_i \tag{12.134}$$

Under the discussed assumptions matrix $\boldsymbol{H}_{n-1}$ is real and symmetric and therefore $\boldsymbol{H}_{n-1} = \boldsymbol{H}_{n-1}^{*\mathrm{T}}$ holds. Similarly for the matrix in Eq. (12.129), $\boldsymbol{m}^{-1} = (\boldsymbol{m}^{-1})^{*\mathrm{T}}$ holds. Now comparing Eqs. (12.133) and (12.134) gives $\mu_i^{*} = \mu_i$. This means that under the considered assumptions (neglecting conductances) the eigenvalues μ_i of matrix $\boldsymbol{a}$ are real $\mu_i \in$ Real. Hence the first of conditions (12.126) is satisfied and there is no need to check it. Checking the second of conditions (12.126), that is $\mu_i < 0$, can be simplified by the following observations.

It is easy to check that matrix $\boldsymbol{m}$ is positive definite, hence $\boldsymbol{w}_i^{*\mathrm{T}}\boldsymbol{m}^{-1}\boldsymbol{w}_i > 0$. Equation (12.133) shows that if $\mu_i < 0$ then $\boldsymbol{w}_i^{*\mathrm{T}}\boldsymbol{H}_{n-1}\boldsymbol{w}_i > 0$. This means that, instead of checking if eigenvalues μ_i are negative, it is enough to check whether matrix $\boldsymbol{H}_{n-1}$ is positive definite. If conductances are neglected, the steady-state stability condition is that the matrix of synchronizing powers $\boldsymbol{H}_{n-1}$ is positive definite. According to Sylvester's theorem, this condition is satisfied if the main minors are positive, that is

$$H_{11} > 0, \quad \begin{vmatrix} H_{11} & H_{12} \\ H_{21} & H_{22} \end{vmatrix} > 0, \quad \begin{vmatrix} H_{11} & H_{12} & H_{13} \\ H_{21} & H_{22} & H_{23} \\ H_{31} & H_{32} & H_{33} \end{vmatrix} > 0, \quad \text{etc.} \tag{12.135}$$

Condition (12.135) is a generalization of condition (5.19) in Section 5.4.1 obtained for the generator-infinite busbar system when condition (12.135) simplifies to $H_{11} > 0$ and then $H = (\partial P/\partial \delta) > 0$.

It can be shown that, when the conductances are neglected, the sufficient condition for matrix $\boldsymbol{H}_{n-1}$ to be positive definite is

$$\left|\delta_i - \delta_j\right| < \frac{\pi}{2} \quad \text{for} \quad i,j = 1,2,\ldots,n \tag{12.136}$$

This means that the angle between the emf's of any pair of generators must be less than $\pi/2$. Proof of this will be undertaken in the following way. Matrix $\boldsymbol{H}_{n-1}$ can be expressed as

$$\boldsymbol{H}_{n-1} = \boldsymbol{C}^{\mathrm{T}}\left(\operatorname{diag} H_{ij}\right)\boldsymbol{C} \tag{12.137}$$

where $\boldsymbol{C}$ is the incidence matrix of generator nodes and the branches linking them. Each such a branch corresponds to a parameter

$$H_{ij} = \partial P_i/\partial \delta_j = -E_iE_jB_{ij}\cos\delta_{ij} \tag{12.138}$$

corresponding to the synchronizing power. According to the Cauchy–Binet theorem (Seshu and Reed 1961), the determinant of matrix (12.137) can be expressed as

$$\det \boldsymbol{H}_{n-1} = \sum_{\text{trees}} \left(\prod_{\text{branches}} (-H_{ij}) \right) \tag{12.139}$$

that is as the sum of negative values of synchronizing powers over all the possible branches of the graph. The sufficient condition for the sum to be positive is that each component is positive. The sufficient condition for each component to be positive is $(-H_{ij}) = E_i E_j B_{ij} \cos \delta_{ij} > 0$, or $\cos \delta_{ij} > 0$, as for the inductive branches $B_{ij} > 0$. Condition $\cos \delta_{ij} > 0$ is satisfied if condition (12.136) is satisfied. Similar considerations can be applied for matrices $\boldsymbol{H}_{n-2}$, $\boldsymbol{H}_{n-3}$, and so on, that is the main minors of matrix $\boldsymbol{H}_{n-1}$.

The stability conditions (12.126), (12.135), and (12.136) are all based on a variety of assumptions. Condition (12.126) is the most accurate because it includes network conductances which are neglected in condition (12.135). Condition (12.136) holds when the network conductances are neglected. Compared with (12.126), conditions (12.135) and (12.136) give an approximate assessment of steady-state stability. This will be illustrated using an example.

Example 12.9 Consider a three-machine system in which elimination of the load nodes has resulted in the following admittances (in per-unit notation) of the branches linking the generator nodes: $y_{12} = (0.3 + \text{j}1.0)$, $y_{13} = (2.5 + \text{j}7.0)$, and $y_{23} = (1.5 + \text{j}4.0)$. The inertia coefficients of the generators are $M_1 = 10$, $M_2 = 2$, and $M_3 = 1$. The third generator was assumed to be the reference machine, hence the angles δ_{13} and δ_{23} are the state variables. The steady-state stability area is shown in Figure 12.6 in the $(\delta_{23}, \delta_{13})$ plane for each of the stability conditions (12.126), (12.135), and (12.136).

The solid line in Figure 12.6a corresponds to the case when the network conductances have been included and condition (12.126) used. When the line is exceeded, the system loses stability in the aperiodic way, that is real positive eigenvalues λ will appear. Oscillatory instability corresponds to the small dashed areas when complex eigenvalues λ have positive real parts. When the network conductances are neglected, condition (12.135) corresponds to the area restricted by the dashed line in Figure 12.6a. The area corresponding to condition (12.135) is repeated in Figure 12.6b and compared with the area corresponding to the necessary condition (12.136). Clearly, both areas are quite close to each other.

******

12.2.3 Including the Voltage Characteristics of the Loads

The linear equation in (12.108) was derived in two stages. First, the load nodes in the network admittance model were eliminated (Figure 12.3a) and then the power equations of the equivalent transfer matrix were linearized. By reversing the order of these steps the voltage characteristics of the loads can be included in the model.

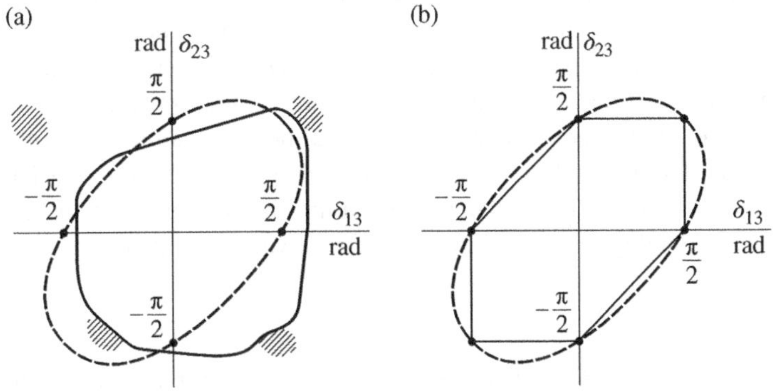

Figure 12.6 Examples of steady-state stability areas in the three-machine system: (a) the area with conductances included (solid line) and conductances neglected (dashed line); (b) the area with conductances neglected (dashed line) and the area due to the necessary condition (dashed line).

To accomplish this it is first necessary to linearize the original network model and replace the power increments of the loads by the increments resulting from the voltage characteristics before eliminating the load nodes in the linearized network equation.

As the voltage magnitude at all the generator nodes {G} is constant (steady-state representation), all the increments $\Delta E_i = 0$ for $i \in \{G\}$. In this case the incremental network Eq. (3.182) has the following form for the original network model (Figure 12.3a)

$$\begin{bmatrix} \Delta \boldsymbol{P}_{\mathrm{G}} \\ \hline \Delta \boldsymbol{P}_{\mathrm{L}} \\ \Delta \boldsymbol{Q}_{\mathrm{L}} \end{bmatrix} = \left[\begin{array}{c|cc} \boldsymbol{H}_{\mathrm{GG}} & \boldsymbol{H}_{\mathrm{GL}} & \boldsymbol{M}_{\mathrm{GL}} \\ \hline \boldsymbol{H}_{\mathrm{LG}} & \boldsymbol{H}_{\mathrm{LL}} & \boldsymbol{M}_{\mathrm{LL}} \\ \boldsymbol{N}_{\mathrm{LG}} & \boldsymbol{N}_{\mathrm{LL}} & \boldsymbol{K}_{\mathrm{LL}} \end{array}\right] \begin{bmatrix} \Delta \boldsymbol{\delta}_{\mathrm{G}} \\ \hline \Delta \boldsymbol{\delta}_{\mathrm{L}} \\ \Delta \boldsymbol{V}_{\mathrm{L}} \end{bmatrix} \tag{12.140}$$

where the submatrices, forming the square matrix, are the partial derivatives of the power with respect to the nodal voltage angles and magnitudes. The indices show which set of nodes, {G} or {L}, a particular submatrix refers to.

In Section 3.4, the voltage sensitivities of the loads k_{PV} and k_{QV} were introduced so that for a small change in the voltage at each of the load nodes $j \in \{L\}$, both $\Delta P_j = k_{\mathrm{PV}j} \Delta V_j$ and $\Delta Q_j = k_{\mathrm{QV}j} \Delta V_j$ hold. These can be written for the whole set {L} as

$$\Delta \boldsymbol{P}_{\mathrm{L}} = -\boldsymbol{k}_{\mathrm{P}} \Delta \boldsymbol{V}_{\mathrm{L}}; \quad \Delta \boldsymbol{Q}_{\mathrm{L}} = -\boldsymbol{k}_{\mathrm{Q}} \Delta \boldsymbol{V}_{\mathrm{L}} \tag{12.141}$$

where $\boldsymbol{k}_{\mathrm{P}} = \mathrm{diag}(k_{\mathrm{PV}i})$ and $\boldsymbol{k}_{\mathrm{Q}} = \mathrm{diag}(k_{\mathrm{QV}i})$ are diagonal matrices of the voltage sensitivities. The minus sign is present because loads in the network equations represent negative injections, that is outflow from the node.

Substituting Eq. (12.141) into Eq. (12.140) gives, after some algebra,

$$\begin{bmatrix} \Delta \boldsymbol{P}_{\mathrm{G}} \\ \hline \boldsymbol{0} \\ \hline \boldsymbol{0} \end{bmatrix} = \left[\begin{array}{c|c|c} \boldsymbol{H}_{\mathrm{GG}} & \boldsymbol{H}_{\mathrm{GL}} & \boldsymbol{M}_{\mathrm{GL}} \\ \hline \boldsymbol{H}_{\mathrm{LG}} & \boldsymbol{H}_{\mathrm{LL}} & \boldsymbol{M}_{\mathrm{LL}} + \boldsymbol{k}_{\mathrm{P}} \\ \hline \boldsymbol{N}_{\mathrm{LG}} & \boldsymbol{N}_{\mathrm{LL}} & \boldsymbol{K}_{\mathrm{LL}} + \boldsymbol{k}_{\mathrm{Q}} \end{array}\right] \begin{bmatrix} \Delta \boldsymbol{\delta}_{\mathrm{G}} \\ \hline \Delta \boldsymbol{\delta}_{\mathrm{L}} \\ \hline \Delta \boldsymbol{V}_{\mathrm{L}} \end{bmatrix}. \tag{12.142}$$

Partial inversion of this equation allows the submatrices $\Delta \boldsymbol{\delta}_{\mathrm{L}}$, $\Delta \boldsymbol{V}_{\mathrm{L}}$ to be eliminated to obtain an equation of the same form as Eq. (12.102), where

$$\boldsymbol{H} = \boldsymbol{H}_{\mathrm{GG}} - [\boldsymbol{H}_{\mathrm{GL}} \mid \boldsymbol{M}_{\mathrm{GL}}] \left[\begin{array}{c|c} \boldsymbol{H}_{\mathrm{LL}} & \boldsymbol{M}_{\mathrm{LL}} + \boldsymbol{k}_{\mathrm{P}} \\ \hline \boldsymbol{N}_{\mathrm{LL}} & \boldsymbol{K}_{\mathrm{LL}} + \boldsymbol{k}_{\mathrm{Q}} \end{array}\right]^{-1} \begin{bmatrix} \boldsymbol{H}_{\mathrm{LG}} \\ \hline \boldsymbol{N}_{\mathrm{LG}} \end{bmatrix}. \tag{12.143}$$

The remaining calculations follow the same pattern as described in the previous section with Eqs. (12.108) and (12.123) being equally valid but now with the elements H_{ij} calculated from the matrix defined in Eq. (12.143).

12.2.4 Transfer Capability of the Network

System operators often want to know the *transfer capability*, that is the maximum power that can be exported from a given power plant to the system or from one subsystem to another. Calculation of the transfer capability can be done using a modified load flow program in which impedances of the generating units are also taken into account while the fictitious nodes behind those nodes are treated as the generator (PV) nodes. The calculations are done step by step. The power demand for a given set of nodes is increased in each step, and load flow together with steady-state stability conditions are determined. The power P_{MAX} at which the steady-stability condition is reached is considered to represent the transfer capability. When P_0 is a relevant transfer power at the operating point, the steady-state stability can be assessed by the following coefficient

$$c_{\mathrm{P}} = \frac{P_{\mathrm{MAX}} - P_0}{P_{\mathrm{MAX}}} \tag{12.144}$$

similar to coefficient (5.20) described in Section 5.4.1 and referred to as the *steady-state stability margin*.

The previous section shows that system stability can be checked using the simple necessary condition (12.136) or (12.135). When checking condition (12.135), one can calculate only the determinant of the Jacobi matrix. This is because, when the operating point is moved toward the stability limit, the first to reach zero is the main minor, that is the determinant of the Jacobi matrix. Some professional load flow programs based on the Newton method calculate the value of the determinant of the Jacobi matrix and hence may be used to determine the stability limit.

12.3 Steady-state Stability of the Regulated System

In Section 5.5 the steady-state stability of the regulated generator-infinite busbar system is discussed. In the present section this analysis is extended to multi-machine systems.

In order to assess the steady-state stability of a regulated power system, detailed models of the generators, the exciters, the turbine governors, and the power system stabilizers must be used. This leads to a large number of state variables being required to describe each generating unit. As a complete description of the full linearized system model would therefore be complicated, and take up a lot of space, the approach adopted here is to describe the overall methodology of creating the system model using relatively simple component models. The creation of more extensive system models would then follow similar lines.

12.3.1 Generator and Network

To simplify considerations, the fifth-order generator model described in Section 11.1.6 will be used. For this model it is convenient to attach the generator subtransient reactance to the network model. The procedure is similar to that shown in Figure 12.3 with the only difference being that for the purpose of assessing the inherent steady-state stability the generator was represented by the synchronous reactances and synchronous emf E_q = constant. As variations in the subtransient emf's now have to be considered, the generator reactances attached to the network are the subtransient (not steady-state) reactances and hence, similar to Eq. (12.100), can be written

$$\underline{E}_i = \underline{E}'' \quad \text{and} \quad X_i = X''_{di} = X''_{qi} \tag{12.145}$$

After all the load nodes, including the generator terminal nodes, have been eliminated, the generator currents can be expressed in terms of the subtransient emf's. Working in rectangular coordinates, the currents can be expressed in the same way as in Eq. (3.178) to give

$$I_{ai} = \sum_{j=1}^{n} \left(G_{ij} E''_{aj} - B_{ij} E''_{bj} \right); \quad I_{bi} = \sum_{j=1}^{n} \left(G_{ij} E''_{bj} + B_{ij} E''_{aj} \right) \tag{12.146}$$

The emf's E''_d and E''_q in the individual generator (d, q) coordinate system can be transformed to the (a, b) coordinate system using Eq. (3.191) to give

$$E''_{aj} = -E''_{dj} \sin\delta_j + E''_{qj} \cos\delta_j; \quad E''_{bj} = E''_{dj} \cos\delta_j + E''_{qj} \sin\delta_j \tag{12.147}$$

while the reverse transformation in Eq. (3.192) can be used for the generator currents

$$I_{di} = -I_{ai} \sin\delta_i + I_{bi} \cos\delta_i, \quad I_{qi} = I_{ai} \cos\delta_i + I_{bi} \sin\delta_i \tag{12.148}$$

It should be noted that the rotor angle δ is the angle of the rotor q-axis measured with respect to the reference frame and not the phase angle of the subtransient emf $\underline{E}''$.

Substituting Eqs. (12.146) and (12.147) into Eq. (12.148) gives, after some simple but arduous algebra,

$$
\begin{aligned}
I_{\mathrm{d}i} &= \sum_{j=1}^{n}\left\{\left(B_{ij}\cos\delta_{ij} - G_{ij}\sin\delta_{ij}\right)E''_{\mathrm{q}j} + \left(B_{ij}\sin\delta_{ij} + G_{ij}\cos\delta_{ij}\right)E''_{\mathrm{d}j}\right\} \\
I_{\mathrm{q}i} &= \sum_{j=1}^{n}\left\{\left(B_{ij}\sin\delta_{ij} + G_{ij}\cos\delta_{ij}\right)E''_{\mathrm{q}j} - \left(B_{ij}\cos\delta_{ij} - G_{ij}\sin\delta_{ij}\right)E''_{\mathrm{d}j}\right\}
\end{aligned}
\tag{12.149}
$$

These are the transfer network equations in the (d, q) coordinates of the individual generators.

Similar to the power increment in Eqs. (12.106), (12.149) allows the current increments to be expressed in terms of the increments in the relative angles $\Delta\delta_{jn}$ and the increments of the component emf's as

$$
\begin{bmatrix} \Delta \boldsymbol{I}_{\mathrm{q}} \\ \hline \Delta \boldsymbol{I}_{\mathrm{d}} \end{bmatrix} = \left[\begin{array}{c|c|c} \dfrac{\partial \boldsymbol{I}_{\mathrm{q}}}{\partial \boldsymbol{\delta}_{n-1}} & \dfrac{\partial \boldsymbol{I}_{\mathrm{q}}}{\partial \boldsymbol{E}''_{\mathrm{q}}} & \dfrac{\partial \boldsymbol{I}_{\mathrm{q}}}{\partial \boldsymbol{E}''_{\mathrm{d}}} \\ \hline \dfrac{\partial \boldsymbol{I}_{\mathrm{d}}}{\partial \boldsymbol{\delta}_{n-1}} & \dfrac{\partial \boldsymbol{I}_{\mathrm{d}}}{\partial \boldsymbol{E}''_{\mathrm{q}}} & \dfrac{\partial \boldsymbol{I}_{\mathrm{d}}}{\partial \boldsymbol{E}''_{\mathrm{d}}} \end{array}\right] \begin{bmatrix} \Delta \boldsymbol{\delta}_{n-1} \\ \hline \Delta \boldsymbol{E}''_{\mathrm{q}} \\ \hline \Delta \boldsymbol{E}''_{\mathrm{d}} \end{bmatrix}, \tag{12.150}
$$

where $\Delta \boldsymbol{I}_{\mathrm{q}}$, $\Delta \boldsymbol{I}_{\mathrm{d}}$, $\Delta \boldsymbol{\delta}_{n-1}$, $\Delta \boldsymbol{E}''_{\mathrm{q}}$, and $\Delta \boldsymbol{E}''_{\mathrm{d}}$ are the appropriate generator increment vectors. The elements of the Jacobi matrix can be found by differentiating the right-hand side of the equations in (12.149) and will not be considered further here. However, it should be remembered that vector $\Delta \boldsymbol{\delta}_{n-1}$ is of size $(n-1)$ so that the submatrices $\partial \boldsymbol{I}_{\mathrm{q}}/\partial \boldsymbol{\delta}_{n-1}$ and $\partial \boldsymbol{I}_{\mathrm{d}}/\partial \boldsymbol{\delta}_{n-1}$ are rectangular with dimensions $n \times (n-1)$.

If transient saliency is neglected $X''_{\mathrm{q}} \cong X''_{\mathrm{d}}$ then Eq. (11.102) simplifies to $P_i = E''_{\mathrm{q}i}I_{\mathrm{q}i} + E''_{\mathrm{d}i}I_{\mathrm{d}i}$, which, after substituting for $I_{\mathrm{d}i}$ and $I_{\mathrm{q}i}$ from Eq. (12.149), yields

$$
\begin{aligned}
P_i = {} & E''_{\mathrm{d}i}\sum_{j=1}^{n}\left\{\left(B_{ij}\cos\delta_{ij} - G_{ij}\sin\delta_{ij}\right)E''_{\mathrm{q}j} + \left(B_{ij}\sin\delta_{ij} + G_{ij}\cos\delta_{ij}\right)E''_{\mathrm{d}j}\right\} \\
& + E''_{\mathrm{q}i}\sum_{j=1}^{n}\left\{\left(B_{ij}\sin\delta_{ij} + G_{ij}\cos\delta_{ij}\right)E''_{\mathrm{q}j} - \left(B_{ij}\cos\delta_{ij} - G_{ij}\sin\delta_{ij}\right)E''_{\mathrm{d}j}\right\}
\end{aligned}
\tag{12.151}
$$

Linearizing and expressing this equation in matrix form gives

$$
[\Delta \boldsymbol{P}] = \left[\begin{array}{c|c|c} \dfrac{\partial \boldsymbol{P}}{\partial \boldsymbol{\delta}_{n-1}} & \dfrac{\partial \boldsymbol{P}}{\partial \Delta \boldsymbol{E}''_{\mathrm{q}}} & \dfrac{\partial \boldsymbol{P}}{\partial \Delta \boldsymbol{E}''_{\mathrm{d}}} \end{array}\right] \begin{bmatrix} \Delta \boldsymbol{\delta}_{n-1} \\ \hline \Delta \boldsymbol{E}''_{\mathrm{q}} \\ \hline \Delta \boldsymbol{E}''_{\mathrm{d}} \end{bmatrix}, \tag{12.152}
$$

where $\Delta \boldsymbol{P}$ is the vector of the power increments in all the generators and $\partial \boldsymbol{P}/\partial \boldsymbol{\delta}_{n-1}$ is a rectangular Jacobi matrix of dimension $n \times (n-1)$. The Jacobi submatrices are calculated by differentiating the right-hand side of the equations in (12.151).

The terminal voltage Eq. (11.100) can be treated in a similar way. Substituting Eq. (12.149) into Eq. (11.100) gives

$$\begin{aligned} V_{qi} &= E''_{qi} + X''_{di} \sum_{j=1}^{n} \left\{ \left(B_{ij}\cos\delta_{ij} - G_{ij}\sin\delta_{ij}\right)E''_{qj} + \left(B_{ij}\sin\delta_{ij} + G_{ij}\cos\delta_{ij}\right)E''_{dj} \right\} \\ V_{di} &= E''_{di} - X''_{qi} \sum_{j=1}^{n} \left\{ \left(B_{ij}\sin\delta_{ij} + G_{ij}\cos\delta_{ij}\right)E''_{qj} - \left(B_{ij}\cos\delta_{ij} - G_{ij}\sin\delta_{ij}\right)E''_{dj} \right\} \end{aligned} \tag{12.153}$$

The quantity of interest with regard to voltage regulator action is the voltage magnitude $V_i = \sqrt{V_{qi}^2 + V_{di}^2}$ so that

$$[\Delta \boldsymbol{V}] = \begin{bmatrix} \dfrac{\partial \boldsymbol{V}}{\partial \boldsymbol{\delta}_{n-1}} & \dfrac{\partial \boldsymbol{V}}{\partial \Delta \boldsymbol{E}''_{q}} & \dfrac{\partial \boldsymbol{V}}{\partial \Delta \boldsymbol{E}''_{d}} \end{bmatrix} \begin{bmatrix} \Delta \boldsymbol{\delta}_{n-1} \\ \Delta \boldsymbol{E}''_{q} \\ \Delta \boldsymbol{E}''_{d} \end{bmatrix}, \tag{12.154}$$

where $\Delta \boldsymbol{V}$ is the vector of the magnitude increments in the terminal voltages of all the generators. The Jacobi submatrices can be calculated by taking the partial derivative of the voltage magnitude with respect to a variable α as

$$\frac{\partial V_i}{\partial \alpha} = \frac{1}{V_i}\left(V_{qi}\frac{\partial V_{qi}}{\partial \alpha} + V_{di}\frac{\partial V_{di}}{\partial \alpha} \right) \tag{12.155}$$

where α signifies E''_{qi}, E''_{qj}, E''_{di}, E''_{dj}, δ_I, or δ_j. The derivatives $\partial V_{qi}/\partial \alpha$ and $\partial V_{di}/\partial \alpha$ are obtained by differentiating the right-hand sides of the equations in (12.153).

The linearized generator differential equations can now be obtained from Eqs. (11.89) and (11.96)–(11.99) as

$$\begin{aligned} &\Delta \dot{\boldsymbol{\delta}}_{n-1} = \boldsymbol{1}_{-1}\Delta \boldsymbol{\omega} \\ &\boldsymbol{M}\Delta \dot{\boldsymbol{\omega}} = -\Delta \boldsymbol{P} - \boldsymbol{D}\Delta \boldsymbol{\omega} \\ &\boldsymbol{T}'_{d0}\Delta \dot{\boldsymbol{E}}'_{q} = -\Delta \boldsymbol{E}'_{q} + \Delta \boldsymbol{X}'_{d}\Delta \boldsymbol{I}_{d} + \Delta \boldsymbol{E}_{f} \\ &\boldsymbol{T}''_{d0}\Delta \dot{\boldsymbol{E}}''_{q} = -\Delta \boldsymbol{E}'_{q} - \Delta \boldsymbol{E}''_{q} + \Delta \boldsymbol{X}''_{d}\Delta \boldsymbol{I}_{d} \\ &\boldsymbol{T}''_{q0}\Delta \dot{\boldsymbol{E}}''_{d} = -\Delta \boldsymbol{E}''_{d} - \Delta \boldsymbol{X}''_{q}\Delta \boldsymbol{I}_{q} \end{aligned} \tag{12.156}$$

where $\Delta \boldsymbol{E}'_{q}$, $\Delta \boldsymbol{E}''_{q}$, $\Delta \boldsymbol{E}''_{d}$, $\Delta \boldsymbol{I}_{q}$, $\Delta \boldsymbol{I}_{d}$, $\Delta \boldsymbol{P}$, and $\Delta \boldsymbol{\omega}$ are the generator increment vectors of size n, while $\Delta \boldsymbol{\delta}_{n-1}$ is the increment vector of the relative angles of size $(n-1)$. The matrices $\boldsymbol{T}'_{d0}$, $\boldsymbol{T}''_{d0}$, $\boldsymbol{T}''_{q0}$, $\boldsymbol{M}$, and $\boldsymbol{D}$ are the respective diagonal matrices of the generator time constants, inertia coefficients, and damping coefficients. Matrices $\Delta \boldsymbol{X}$ are diagonal and contain the following elements

$$\Delta \boldsymbol{X}'_{d} = \operatorname{diag}\left(X_{di} - X'_{d}\right); \quad \Delta \boldsymbol{X}''_{d} = \operatorname{diag}\left(X'_{d} - X''_{d}\right); \quad \Delta \boldsymbol{X}''_{q} = \operatorname{diag}\left(X'_{q} - X''_{q}\right) \tag{12.157}$$

Substituting Eqs. (12.150) and (12.152) into Eq. (12.156) gives the state equation (12.49). The four submatrices in the upper left-hand corner of the state matrix have the same structure as the state matrix used in Eq. (12.110) to assess the steady-state inherent stability. The changes in E_f, appearing in the last component of Eq. (12.158), are due to the action of the generator exciters and automatic voltage regulator (AVR) systems.

$$
\begin{bmatrix} \Delta\dot{\boldsymbol{\delta}}_{n-1} \\ \Delta\dot{\boldsymbol{\omega}} \\ \Delta\dot{\boldsymbol{E}}'_{\mathrm{q}} \\ \Delta\dot{\boldsymbol{E}}''_{\mathrm{q}} \\ \Delta\dot{\boldsymbol{E}}''_{\mathrm{d}} \end{bmatrix}
=
\left[\begin{array}{cc|ccc}
\mathbf{0} & \mathbf{1}_{-1} & \mathbf{0} & \mathbf{0} & \mathbf{0} \\
-\boldsymbol{M}^{-1}\dfrac{\partial \boldsymbol{P}}{\partial \boldsymbol{\delta}_{n-1}} & -\boldsymbol{M}^{-1}\boldsymbol{D} & \mathbf{0} & -\boldsymbol{M}^{-1}\dfrac{\partial \boldsymbol{P}}{\partial \boldsymbol{E}''_{\mathrm{q}}} & -\boldsymbol{M}^{-1}\dfrac{\partial \boldsymbol{P}}{\partial \boldsymbol{E}''_{\mathrm{d}}} \\
\hline
(\boldsymbol{T}'_{\mathrm{do}})^{-1}\Delta \boldsymbol{X}'_{\mathrm{d}}\dfrac{\partial \boldsymbol{I}_{\mathrm{d}}}{\partial \boldsymbol{\delta}_{n-1}} & \mathbf{0} & -(\boldsymbol{T}'_{\mathrm{do}})^{-1} & (\boldsymbol{T}'_{\mathrm{do}})^{-1}\Delta \boldsymbol{X}'_{\mathrm{d}}\dfrac{\partial \boldsymbol{I}_{\mathrm{d}}}{\partial \boldsymbol{E}''_{\mathrm{q}}} & (\boldsymbol{T}'_{\mathrm{do}})^{-1}\Delta \boldsymbol{X}'_{\mathrm{d}}\dfrac{\partial \boldsymbol{I}_{\mathrm{d}}}{\partial \boldsymbol{E}''_{\mathrm{d}}} \\
(\boldsymbol{T}''_{\mathrm{do}})^{-1}\Delta \boldsymbol{X}''_{\mathrm{d}}\dfrac{\partial \boldsymbol{I}_{\mathrm{d}}}{\partial \boldsymbol{\delta}_{n-1}} & \mathbf{0} & -(\boldsymbol{T}''_{\mathrm{do}})^{-1} & (\boldsymbol{T}''_{\mathrm{do}})^{-1}\left(\Delta \boldsymbol{X}''_{\mathrm{d}}\dfrac{\partial \boldsymbol{I}_{\mathrm{d}}}{\partial \boldsymbol{E}''_{\mathrm{q}}} - 1\right) & (\boldsymbol{T}''_{\mathrm{do}})^{-1}\Delta \boldsymbol{X}''_{\mathrm{d}}\dfrac{\partial \boldsymbol{I}_{\mathrm{d}}}{\partial \boldsymbol{E}''_{\mathrm{d}}} \\
-(\boldsymbol{T}''_{\mathrm{qo}})^{-1}\Delta \boldsymbol{X}''_{\mathrm{q}}\dfrac{\partial \boldsymbol{I}_{\mathrm{q}}}{\partial \boldsymbol{\delta}_{n-1}} & \mathbf{0} & \mathbf{0} & -(\boldsymbol{T}''_{\mathrm{qo}})^{-1}\Delta \boldsymbol{X}''_{\mathrm{q}}\dfrac{\partial \boldsymbol{I}_{\mathrm{q}}}{\partial \boldsymbol{E}''_{\mathrm{q}}} & -(\boldsymbol{T}''_{\mathrm{qo}})^{-1}\left(\Delta \boldsymbol{X}''_{\mathrm{q}}\dfrac{\partial \boldsymbol{I}_{\mathrm{q}}}{\partial \boldsymbol{E}''_{\mathrm{d}}} + 1\right)
\end{array}\right]
\begin{bmatrix} \Delta\boldsymbol{\delta}_{n-1} \\ \Delta\boldsymbol{\omega} \\ \Delta\boldsymbol{E}'_{\mathrm{q}} \\ \Delta\boldsymbol{E}''_{\mathrm{q}} \\ \Delta\boldsymbol{E}''_{\mathrm{d}} \end{bmatrix}
+
\begin{bmatrix} \mathbf{0} \\ \mathbf{0} \\ (\boldsymbol{T}'_{\mathrm{do}})^{-1}\Delta\boldsymbol{E}_{\mathrm{f}} \\ \mathbf{0} \\ \mathbf{0} \end{bmatrix}. \qquad (12.158)
$$

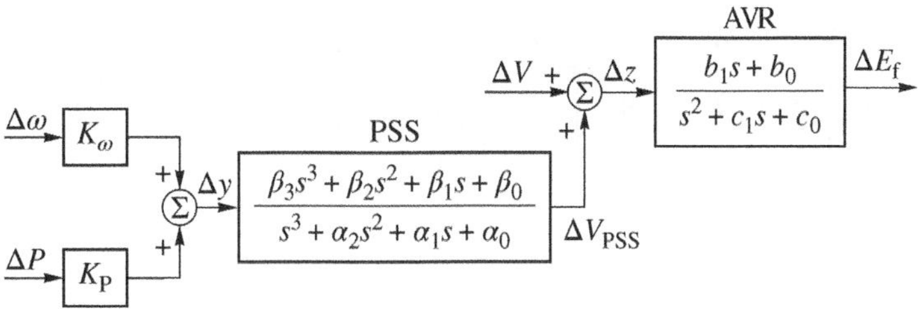

Figure 12.7 Simplified block diagram of the exciter with AVR and PSS.

12.3.2 Including Excitation System Model and Voltage Control

Models are derived in Section 11.2 for the different types of excitation systems. For steady-state stability purposes nonlinearity effects do not play a major role and may be neglected, allowing the excitation systems to be represented by a high-order transfer function. To show how the linearized equations of the excitation system can be included in the complete linearized system model, a second-order transfer function will be used to represent the exciter and the AVR while a third-order transfer function will be used to represent the power system stabilizer (PSS). The block diagram corresponding to such an excitation system is shown in Figure 12.7. In this system the PSS will react to a signal proportional to rotor speed deviation, when $K_P = 0$, or to a signal proportional to the generator real power, when $K_\omega = 0$.

The differential equation describing the excitation system can be obtained from the transfer function in Figure 12.7 by introducing two auxiliary variables, x_1 and x_2, to give

$$\begin{aligned} \dot{x}_{2i} &= \Delta z_i - c_{1i} x_{2i} - c_{0i} x_{1i} \\ \dot{x}_{1i} &= x_{2i} \end{aligned} \tag{12.159}$$

$$\Delta E_{fi} = b_{1i} x_{2i} + b_{0i} x_{1i} \tag{12.160}$$

where i is the generator number. Expressing this equation in matrix form, and substituting Eq. (12.154), gives

$$\begin{bmatrix} \dot{\boldsymbol{x}}_1 \\ \dot{\boldsymbol{x}}_2 \end{bmatrix} = \begin{bmatrix} \mathbf{0} & \mathbf{1} \\ -\boldsymbol{c}_0 & -\boldsymbol{c}_1 \end{bmatrix} \begin{bmatrix} \boldsymbol{x}_1 \\ \boldsymbol{x}_2 \end{bmatrix} + \begin{bmatrix} \mathbf{0} & \mathbf{0} & \mathbf{0} \\ -\dfrac{\partial \boldsymbol{V}}{\partial \boldsymbol{\delta}_{n-1}} & -\dfrac{\partial \boldsymbol{V}}{\partial \boldsymbol{E}''_q} & -\dfrac{\partial \boldsymbol{V}}{\partial \boldsymbol{E}''_d} \end{bmatrix} \begin{bmatrix} \Delta \boldsymbol{\delta}_{n-1} \\ \Delta \boldsymbol{E}''_q \\ \Delta \boldsymbol{E}''_d \end{bmatrix} + \begin{bmatrix} \mathbf{0} \\ -\Delta \boldsymbol{V}_{PSS} \end{bmatrix}$$

$$[\Delta \boldsymbol{E}_f] = [\boldsymbol{b}_0 \; \boldsymbol{b}_1] \begin{bmatrix} \boldsymbol{x}_1 \\ \boldsymbol{x}_2 \end{bmatrix}, \tag{12.161}$$

where $\boldsymbol{c}_0$, $\boldsymbol{c}_1$, $\boldsymbol{b}_0$, and $\boldsymbol{b}_1$ are diagonal submatrices of the coefficients in the excitation system transfer functions.

The vector $\Delta \boldsymbol{V}_{PSS}$ can be obtained from the stabilizer equations, which, because the stabilizer transfer function is third order, can be constructed using three auxiliary variables x_3, x_4, and x_5. This results in

$$\begin{aligned} \dot{x}_{5i} &= \Delta y_i - \alpha_{2i} x_{5i} - \alpha_{1i} x_{4i} - \alpha_{0o} x_{3i} \\ \dot{x}_{4i} &= x_{5i} \\ \Delta V_{PSS} &= \beta_{3i} \Delta y_i - (\alpha_{2i}\beta_{3i} - \beta_{2i}) x_{5i} - (\alpha_{1i}\beta_{3i} - \beta_{1i}) x_{4i} - \beta_{3i} x_{3i} \end{aligned} \tag{12.162}$$

where $\Delta y_i = K_{\omega i} \Delta\omega_i + K_{Pi} \Delta P_i$ is the input signal and i is the generator number. Representing the equations in the matrix form and taking into account Eq. (12.152) gives

$$\begin{bmatrix} \dot{x}_3 \\ \dot{x}_4 \\ \dot{x}_5 \end{bmatrix} = \begin{bmatrix} 0 & 1 & 0 \\ 0 & 0 & 1 \\ -\alpha_0 & -\alpha_1 & -\alpha_2 \end{bmatrix} \begin{bmatrix} x_3 \\ x_4 \\ x_5 \end{bmatrix} + \begin{bmatrix} 0 & 0 & 0 & 0 \\ 0 & 0 & 0 & 0 \\ -K_{\mathrm{P}} \dfrac{\partial P}{\partial \delta_{n-1}} & -K_{\omega} & -K_{\mathrm{P}} \dfrac{\partial P}{\partial E''_{\mathrm{q}}} & -K_{\mathrm{P}} \dfrac{\partial P}{\partial E''_{\mathrm{d}}} \end{bmatrix} \begin{bmatrix} \Delta\delta_{n-1} \\ \Delta\omega \\ \Delta E''_{\mathrm{q}} \\ \Delta E''_{\mathrm{d}} \end{bmatrix}$$

$$[\Delta V_{\mathrm{pss}}] \begin{bmatrix} K_{\mathrm{p}}\beta_3 \dfrac{\partial P}{\partial \delta_{n-1}} & K_{\omega}\beta_3 & K_{\mathrm{p}}\beta_3 \dfrac{\partial P}{\partial E''_{\mathrm{q}}} & K_{\mathrm{p}}\beta_3 \dfrac{\partial P}{\partial E''_{\mathrm{d}}} & -\beta_3 & \beta_1 - \alpha_1\beta_3 & \beta_2 - \alpha_2\beta_3 \end{bmatrix} \begin{bmatrix} \Delta\delta_{n-1} \\ \Delta\omega \\ \Delta E''_{\mathrm{q}} \\ \Delta E''_{\mathrm{d}} \\ x_3 \\ x_4 \\ x_5 \end{bmatrix}, \tag{12.163}$$

where α_0, α_1, α_2, β_0, β_1, β_2, β_3, K_{P}, and K_{ω} are diagonal matrices of the stabilizer coefficients.

12.3.3 Linear State Equation of the System

As the linear equations in (12.158), (12.161), and (12.163) have common state variables they can be combined to form one large state equation that includes the effect of all the generators (fifth-order model), exciters, AVRs, and PSSs. This results in the matrix Eq. (12.164). The solid line in the upper left-hand corner separates the submatrix corresponding to the swing equation. The dashed line separates the submatrix corresponding to the generator equations without voltage control. The lower side of the state matrix below the dashed line corresponds to the voltage control and PSS.

So far the turbine power has been assumed to be constant. If this is not the case then additional equations that describe the turbine and its governor must be added to the state equation, further increasing its size. The size of the state equation will be increased yet further if the effect of control devices, such as static VAR compensator (SVCs) or other flexible AC transmission systems (FACTS) devices, is included. Provided a suitable model is known then including the effect of such elements is straightforward and follows similar lines as for the exciter and AVR and for this reason is not considered further in this book.

As the state matrix in Eq. (12.164) is sparse, it is advantageous to use sparse matrix techniques when calculating the eigenvalues and eigenvectors even if a small system is under investigation. In the case of a large interconnected system, when all the control devices influencing the steady-state stability and damping are included, the dimension of the state matrix may be very large indeed and well outside the range of conventional eigenvalue analysis methods. In such cases special solution techniques, which evaluate a selected subset of the eigenvalues associated with the complete system response, are necessary (Kundur 1994).

$$
\begin{bmatrix}
\Delta\dot{\delta}_{n-1} \\
\Delta\dot{\omega} \\
\hline
\Delta\dot{E}'_{\mathrm{q}} \\
\Delta\dot{E}''_{\mathrm{q}} \\
\Delta\dot{E}''_{\mathrm{d}} \\
\dot{x}_1 \\
\dot{x}_2 \\
\dot{x}_3 \\
\dot{x}_4 \\
\dot{x}_5
\end{bmatrix}
=
\begin{bmatrix}
0 & 1_{-1} & 0 & 0 & 0 & 0 & 0 & 0 & 0 & 0 \\
-M^{-1}\dfrac{\partial P}{\partial \delta_{n-1}} & -M^{-1}D & 0 & -M^{-1}\dfrac{\partial P}{\partial E''_{\mathrm{q}}} & -M^{-1}\dfrac{\partial P}{\partial E''_{\mathrm{d}}} & 0 & 0 & 0 & 0 & \\
\hline
(T'_{\mathrm{do}})^{-1}\Delta X'_{\mathrm{d}}\dfrac{\partial I_{\mathrm{d}}}{\partial \delta_{n-1}} & 0 & -(T'_{\mathrm{do}})^{-1} & (T'_{\mathrm{do}})^{-1}\Delta X'_{\mathrm{d}}\dfrac{\partial I_{\mathrm{d}}}{\partial E''_{\mathrm{q}}} & (T'_{\mathrm{do}})^{-1}\Delta X'_{\mathrm{d}}\dfrac{\partial I_{\mathrm{d}}}{\partial E''_{\mathrm{d}}} & (T'_{\mathrm{do}})^{-1}b_0 & (T'_{\mathrm{do}})^{-1}b_1 & 0 & 0 & 0 \\
(T''_{\mathrm{do}})^{-1}\Delta X''_{\mathrm{d}}\dfrac{\partial I_{\mathrm{d}}}{\partial \delta_{n-1}} & 0 & -(T''_{\mathrm{do}})^{-1} & (T''_{\mathrm{do}})^{-1}\left(\Delta X''_{\mathrm{d}}\dfrac{\partial I_{\mathrm{d}}}{\partial E''_{\mathrm{q}}} - 1\right) & (T''_{\mathrm{do}})^{-1}\Delta X''_{\mathrm{d}}\dfrac{\partial I_{\mathrm{d}}}{\partial E''_{\mathrm{d}}} & 0 & 0 & 0 & 0 & 0 \\
-(T''_{\mathrm{qo}})^{-1}\Delta X''_{\mathrm{q}}\dfrac{\partial I_{\mathrm{q}}}{\partial \delta_{n-1}} & 0 & 0 & -(T''_{\mathrm{qo}})^{-1}\Delta X''_{\mathrm{q}}\dfrac{\partial I_{\mathrm{q}}}{\partial E''_{\mathrm{q}}} & -(T''_{\mathrm{qo}})^{-1}\left(\Delta X''_{\mathrm{q}}\dfrac{\partial I_{\mathrm{q}}}{\partial E''_{\mathrm{d}}} + 1\right) & 0 & 0 & 0 & 0 & 0 \\
\hdashline
0 & 0 & 0 & 0 & 0 & 0 & 0 & 0 & 0 & 0 \\
K_{\mathrm{P}}\beta_3\dfrac{\partial P}{\partial \delta_{n-1}} - \dfrac{\partial V}{\partial \delta_{n-1}} & K_{\omega}\beta_3 & 0 & K_{\mathrm{P}}\beta_3\dfrac{\partial P}{\partial E''_{\mathrm{q}}} - \dfrac{\partial V}{\partial E''_{\mathrm{q}}} & K_{\mathrm{P}}\beta_3\dfrac{\partial P}{\partial E''_{\mathrm{d}}} - \dfrac{\partial V}{\partial E''_{\mathrm{d}}} & -c_0 & -c_1 & -\beta_3 & \beta_1 - a_1\beta_3 & \beta_2 - a_2\beta_3 \\
0 & 0 & 0 & 0 & 0 & 0 & 0 & 0 & 0 & 0 \\
0 & 0 & 0 & 0 & 0 & 0 & 0 & 0 & 0 & 0 \\
K_{\mathrm{P}}\dfrac{\partial P}{\partial \delta_{n-1}} & K_{\omega} & 0 & K_{\mathrm{P}}\dfrac{\partial P}{\partial E''_{\mathrm{q}}} & K_{\mathrm{P}}\dfrac{\partial P}{\partial E''_{\mathrm{d}}} & 0 & 0 & -a_0 & -a_1 & -a_2
\end{bmatrix}
\begin{bmatrix}
\Delta\delta_{n-1} \\
\Delta\omega \\
\hline
\Delta E'_{\mathrm{q}} \\
\Delta E''_{\mathrm{q}} \\
\Delta E''_{\mathrm{d}} \\
x_1 \\
x_2 \\
x_3 \\
x_4 \\
x_5
\end{bmatrix}
. \quad (12.164)
$$

12.3.4 Examples

A number of interesting examples of the application of modal analysis to large interconnected power systems can be found in Wang (1997) and Breulmann et al. (2000). Below are two examples based on data from a paper by Rasolomampionona (2000).

Example 12.10 Consider the generator-infinite busbar system shown in Figure 6.13 when line L1 has an outage due to maintenance. Since this will weaken the connection of the generator with the system, it is necessary to check the steady-state stability. The lines remaining in operation are of the length 250 km (line L2) and 80 km (line L3). The lines operate at the rated voltage of 220 kV and their reactance is $x = 0.4\Omega/\text{km}$. There is a load of $(350 + \text{j}150)$ MVA at node B3. The generator operates with a step-up transformer of reactance $X_T = 0.14$. The generator is salient-pole one with the following parameters: $S_n = 426\text{MVA}$, $X''_d \cong X''_q = 0.160$, $X'_d = 0.21$, $X_d = 1.57$, $X'_q \cong X_q = 0.85$, $T'_{do} = 6.63$, $T''_{do} = 0.051$, $T''_{q0} = 1.2$, and $T_m = 10$. The transfer function of the voltage control and excitation system is

$$\frac{\Delta E_f(s)}{\Delta V(s)} = K_A \frac{2s + 1}{0.3s^2 + 10s + 1} = K_A \frac{6.66s + 3.33}{s^2 + 33.3s + 3.33} = \frac{b_1 s + b_0}{s^2 + c_1 s + c_0} \tag{12.165}$$

where (according to Figure 12.7) $b_0 = 3.33 \cdot K_A$, $b_1 = 6.66 \cdot K_A$, and $c_0 = 3.33$, $c_1 = 33.3$.

It is necessary to compute eigenvalues and the associated eigenvectors and, on the basis of the participation factors, determine which diagonal elements of the state matrix and which state variables are strongly related to which eigenvalues. Assume a small regulator gain $K_A = 30$.

Substituting the data into Eq. (12.164), the following state equation of the fifth-order generator model together with voltage control and excitation system is obtained

$$\begin{bmatrix} \Delta\dot{\delta} \\ \Delta\dot{\omega} \\ \Delta\dot{E}'_q \\ \Delta\dot{E}''_q \\ \Delta\dot{E}''_d \\ \dot{x}_1 \\ \dot{x}_2 \end{bmatrix} = \begin{bmatrix} 0 & 1 & 0 & 0 & 0 & 0 & 0 \\ -20.316 & 0 & 0 & -25.048 & -1.411 & 0 & 0 \\ -0.061 & 0 & -0.773 & -0.083 & 0.018 & 15.06 & 30.12 \\ -0.213 & 0 & 0 & -5.026 & 0.063 & 0 & 0 \\ -2.654 & 0 & 7.050 & -1.463 & -12.958 & 0 & 0 \\ 0 & 0 & 0 & 0 & 0 & 1 & 1 \\ -0.008 & 0 & 0 & -0.565 & 0.971 & -3.33 & -33.33 \end{bmatrix} \begin{bmatrix} \Delta\delta \\ \Delta\omega \\ \Delta E'_q \\ \Delta E''_q \\ \Delta E''_d \\ x_1 \\ x_2 \end{bmatrix}. \tag{12.166}$$

The 2×2 submatrix in the upper left-hand corner corresponds to the swing equation. It relates to the state matrix in Eq. (12.121) corresponding to the inherent stability and the second-order generator model. The matrix has two imaginary eigenvalues $\lambda_{1,2} = \pm\text{j}\,4.507$ which correspond to the frequency of oscillations of about 0.72 Hz. The upper left 5×5 submatrix corresponds to the swing equation together with the equations describing the excitation system and the damping circuits in both axes. It relates to the state Eq. (12.158). The state matrix in Eq. (12.166) has the following eigenvalues

$$\lambda_{1,2} = -0.177 \pm \text{j}4.535,\quad \lambda_3 = -1.239,\quad \lambda_4 = -3.681,\quad \lambda_5 = -13.036,\quad \lambda_6 = -0.342,\quad \lambda_7 = -33.334$$

Eq. (12.90) is used to determine the complex values of participation factors for all eigenvalues. The matrix of participation factors $\boldsymbol{p}$ below contains the magnitudes $|p_{ki}|$

Table 12.2 Physical meaning and values of modes in Example 12.10.

Equation	Physical meaning of the mode	Notation	Eigenvalue
$\Delta\dot{\delta}_{n-1}$, $\Delta\dot{\omega}$	Rotor swings	$\lambda_{1,\,2}$	$-0.177 \pm \text{j}\,4.535$
$\Delta\dot{E}'_{\text{q}}$	Excitation circuit	λ_3	-0.342
$\Delta\dot{E}''_{\text{q}}$	Damping circuit in the d-axis and excitation circuit	λ_4	-3.681
$\Delta\dot{E}''_{\text{d}}$	Damping circuit in the q-axis	λ_5	-13.036
$\dot{x}_1$	Voltage controller and excitation circuit	λ_6	-3.681
$\dot{x}_2$	Voltage controller	λ_7	-33.334

$$
x = \begin{bmatrix} \Delta\dot{\delta} \\ \Delta\dot{\omega} \\ \Delta\dot{E}'_{\text{q}} \\ \Delta\dot{E}''_{\text{q}} \\ \Delta E''_{\text{d}} \\ \dot{x}_1 \\ \dot{x}_2 \end{bmatrix}, \quad \text{diag}\boldsymbol{A} = \begin{bmatrix} A_{11} \\ A_{22} \\ A_{33} \\ A_{44} \\ A_{55} \\ A_{66} \\ A_{77} \end{bmatrix}, \quad \boldsymbol{p} = \begin{array}{c} \begin{array}{ccccccc} \lambda_1 & \lambda_2 & \lambda_3 & \lambda_4 & \lambda_5 & \lambda_6 & \lambda_7 \end{array} \\ \left[\begin{array}{c|c|c|c|c|c|c} \mathbf{0.5} & \mathbf{0.5} & 0.05 & 0.06 & & & \\ \hline \mathbf{0.5} & \mathbf{0.5} & 0.05 & 0.06 & & & \\ \hline 0.04 & 0.04 & \mathbf{1.80} & \mathbf{0.54} & & \mathbf{0.33} & \\ \hline 0.03 & 0.03 & \mathbf{0.29} & \mathbf{1.33} & 0.01 & 0.03 & \\ \hline & & 0.03 & 0.02 & \mathbf{1.0} & & \\ \hline & & \mathbf{0.41} & 0.05 & & \mathbf{1.36} & \\ \hline & & 0.03 & 0.05 & & & \mathbf{1.0} \end{array}\right] \end{array}. \tag{12.167}
$$

The dominant values of $|p_{ki}|$ are shown in bold. Values less than 0.01 are neglected. The majority of eigenvalues are connected with a few diagonal elements of the state matrix and therefore with a few state variables. However, by considering only the dominant values of $|p_{ki}|$, it is possible to link individual eigenvalues approximately with some state variables. The simplest relationship is for λ_7 which has only one dominant participation factor associated with A_{77} and x_2, that is with the voltage controller. Eigenvalue λ_5 is associated with A_{55} and electromotive force (emf) E''_{d}, while E''_{d} corresponds to the damping circuit in the q-axis. Eigenvalue λ_4 is mainly associated with A_{44} and emf E''_{q}, while E''_{q} corresponds to the damping circuit in the d-axis. The same λ_4 is also, but less strongly, associated with A_{33} and emf E'_{q}, while E'_{q} corresponds to the excitation circuit. Eigenvalue λ_6 is mainly associated with A_{66} and variable x_1, while x_1, corresponds to the voltage controller, and to a lesser extent with A_{33} and emf E'_{q}. The complex conjugate pair of eigenvalues $\lambda_{1,\,2}$ is equally associated with A_{11} and A_{22}, that is with variables $\Delta\delta$ and $\Delta\omega$, which correspond to the swing equation. A summary of those associations and the physical meaning of all the eigenvalues is shown in Table 12.2.

******

Example 12.11 Consider the generator-infinite busbar system described in Example 12.10. It is necessary to check the steady-state stability conditions and determine the influence of the regulator gain K_{A} on the position of eigenvalues.

Taking into account the voltage control and excitation system introduces two additional eigenvalues, which are shown in Table 12.3. All the eigenvalues lie in the left half-plane so the system is stable over a wide range of the regulator gain values K_{A}. For small values of the gain, the eigenvalues λ_3 and λ_4 are still real. As the gain increases, the values come close to each other until they become complex values $\lambda_{3,\,4}$. Increased damping causes an increase in the imaginary part. This means that, as the gain increases, electromagnetic phenomena in the excitation

Table 12.3 Physical meaning and values of modes in Example 12.11.

Regulator gain	Rotor swings	Excitation circuit	d-axis damping	q-axis damping	Voltage control and excitation system	
K_A	$\lambda_{1,2}$	λ_3	λ_4	λ_5	λ_6	λ_7
0	$-0.111 \pm j\,4.432$	-0.784	-4.828	-12.931	0	0
50	$-0.193 \pm j\,4.555$	-1.602	-2.585	-12.983	-0.352	-33.468
60	$-0.215 \pm j\,4.584$	$-2.403 \pm j\,0.1531$		-19.056	-0.376	-33.498
80	$-0.257 \pm j\,4.637$	$-2.301 \pm j\,1.1519$		-18.976	-0.403	-33.548
150	$-0.459 \pm j\,4.866$	$-1.918 \pm j\,2.379$		-18.968	-0.448	-33.732
200	$-0.644 \pm j\,5.090$	$-1.619 \pm j\,2.837$		-11.871	-0.462	-33.861
250	$-0.828 \pm j\,5.395$	$-1.331 \pm j\,3.097$		-19.419	-0.469	-33.988
325	$-1.083 \pm j\,5.976$	$-1.082 \pm j\,3.264$		-19.563	-0.477	-34.185

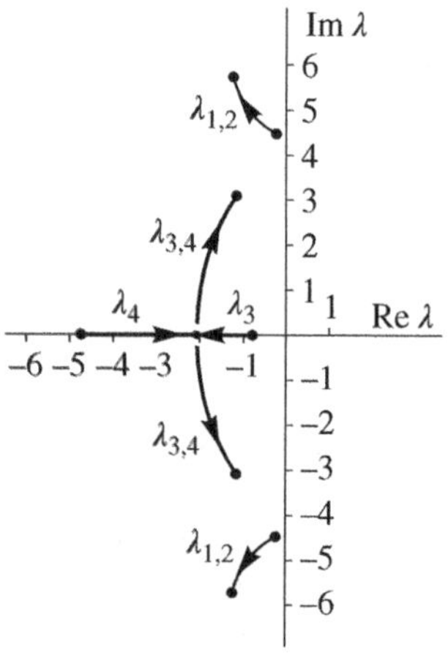

Figure 12.8 Eigenvalue loci when the regulator gain K_A increases.

winding start to be oscillatory. The higher the gain, the higher frequency of oscillations. As the gain increases, the eigenvalues $\lambda_{1,2}$ move left. Figure 12.8 shows the loci of eigenvalues $\lambda_{1,2}$ and $\lambda_{3,4}$ as the gain increases.

Increasing the regulator gain improves the damping of swing modes $\lambda_{1,2}$ but deteriorates the damping of modes $\lambda_{3,4}$ connected with oscillations in the rotor circuits. When $K_A = 325$ the real parts of $\lambda_{1,2}$ and $\lambda_{3,4}$ are the same (Table 12.3). It is reasonable to assume that rotor circuit oscillations should not be less well damped than the rotor oscillations. Assuming that the rotor circuit oscillations should be better damped than the rotor oscillation, the gain is $K_A = 250$. With this gain $\lambda_{1,2} = -0.828 \pm j\ 5.395$ and the damping ratio is $\zeta = 0.828/\sqrt{0.828^2 + 5.395^2} \cong 0.15 = 15\%$. This value is quite high because usually satisfactory damping corresponds to $\zeta \geq 5\%$.

******

Example 12.12 Consider the same generator-infinite busbar system as before by assuming that the transfer function of the voltage control and excitation system is

$$\frac{\Delta E_f(s)}{\Delta V(s)} = \frac{K_A}{s}\frac{(2s+1)}{(0.2s+1)} \tag{12.168}$$

This is an integrating regulator with a corrective circuit of the lead type. It should be emphasized that such a transfer function is not recommended for a voltage regulator. It was selected just to demonstrate how eigenvalue analysis can expose the bad influence of a particular regulator on the damping of rotor swings.

Using the same model as before, eigenvalues have been calculated for a range of values of the regulator gain K_A

$$K_A = 0: \ \lambda_{1,2} = -0.111 \pm j\,4.432, \ \lambda_3 = -0.784, \ \lambda_4 = -4.821$$

$$K_A = 10: \ \lambda_{1,2} = -0.085 \pm j\,4.638, \ \lambda_{3,4} = -1.442 \pm j\,1.588$$

$$K_A = 20: \ \lambda_{1,2} = -0.018 \pm j\,4.941, \ \lambda_{3,4} = -0.977 \pm j\,2.426$$

For $K_A = 0$ the eigenvalues correspond to the regulator being deactivated and therefore inherent damping of swings. Increased gain reduces the negative real parts of $\lambda_{1,2}$ and therefore deteriorates damping. For $K_A = 20$

the eigenvalues $\lambda_{1,\,2}$ lie close to the imaginary axis and damping is weak. When $K_A > 20$ the eigenvalues $\lambda_{1,\,2}$ move to the right half-plane and the system becomes oscillatorily unstable.

It should, however, be remembered that in order to achieve good voltage regulation (Chapter 2) the regulator gain should be high, usually higher than $K_A = 20$. Such a high gain is unacceptable from the point of view of damping. Hence the discussed regulator must be equipped with a supplementary control loop such as a PSS. For example, a PSS with an active power input signal could be used.

Assume that the transfer function of the PSS is

$$\frac{\Delta V_{PSS}(s)}{\Delta P(s)} = K_P \frac{0.05s}{1 + 0.05s} \frac{1 + 0.15s}{1 + 0.05s} \frac{1 + 0.15s}{1 + 3s} \tag{12.169}$$

Now it is necessary to check if the PSS improves damping and select gain K_P assuming that voltage regulation requires the regulator gain $K_A = 20$.

The eigenvalue loci of the influence of the PSS gain K_P are shown in Figure 12.9. For $K_P = 0$ the eigenvalues $\lambda_{1,\,2}$ are close to the imaginary axis. Increasing K_P causes a shift to the left, that is improved damping. At the same time the eigenvalues $\lambda_{3,\,4}$ corresponding to the rotor circuits move to the right, therefore deteriorating damping. For $K_P = 3$ one gets $\lambda_{1,\,2} = -0.441 \pm j\,7.214$ and $\lambda_{3,\,4} = -0.441 \pm j\,1.730$, that is the same real parts. Further increase of K_P causes too high a deterioration of oscillations in the rotor circuits.

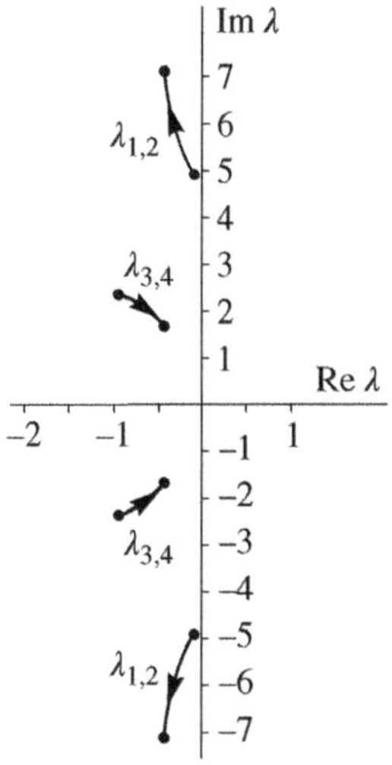

Figure 12.9 Eigenvalue loci when the gain K_P of PSS increases.

For $K_P = 3$ the eigenvalues are $\lambda_{1,2} = -0.441 \pm j7.214$ and the damping ratio is $\zeta = 0.441/\sqrt{0.441^2 + 7.214^2} \cong 0.06 = 6\%$. Such damping is satisfactory, as it is higher than 5%, but much weaker than with the previously discussed regulator with transfer function (12.166). Both examples show the importance of choosing the right value for the transfer function of the voltage regulator.

******

13

Power System Dynamic Simulation

Simulation of power system behavior, which is the subject of this chapter, is a highly useful tool for planning, for the analysis of stability, and for operator training. The individual models of the generator, automatic voltage regulator (AVR), turbine-governor, and the system loads are given by the differential and algebraic equations developed in Chapter 11 and the network is modeled by the algebraic equations developed in Chapter 3. Together these equations form a complete mathematical model of the system, which can be solved numerically to simulate the system behavior.

To develop a power system dynamic simulation the equations used to model the different elements are collected together to form a set of differential equations

$$\dot{\boldsymbol{x}} = \boldsymbol{f}(\boldsymbol{x}, \boldsymbol{y}) \tag{13.1}$$

that describe the system dynamics, primarily contributed by the generating units and the dynamic loads, and a set of algebraic equations

$$\boldsymbol{0} = \boldsymbol{g}(\boldsymbol{x}, \boldsymbol{y}) \tag{13.2}$$

that describe the network, the static loads, and the algebraic equations of the generator. The solution of these two sets of equations defines the electromechanical state of the electric power system (EPS) at any instant in time. A disturbance in the network usually requires a change to both the network configuration and the boundary conditions. These are modeled by changing the coefficients in the functions appearing on the right-hand side of Eqs. (13.1) and (13.2). The computer program for the dynamic power system simulation must then solve the differential and algebraic equations over a period of time for a given sequence of network disturbances.

Equations (13.1) and (13.2) can be solved using either a partitioned solution or a simultaneous solution. The partitioned solution is sometimes referred to as the *alternating solution* and the simultaneous solution as the *combined solution*. In the partitioned solution the differential equations are solved using a standard explicit numerical integration method with the algebraic Eq. (13.2) being solved separately at each time step. The simultaneous solution uses implicit integration methods to convert the differential Eq. (13.1) into a set of algebraic equations which are then combined with the algebraic network Eq. (13.2) to be solved as one set of simultaneous algebraic equations. The effectiveness of these two solutions depends both on the generator model used and on the method of numerical integration.

In order to select the most appropriate integration method it is necessary to understand the timescale of the dynamics included in the model of the generating unit. As explained in Chapter 12, the solution of any set of linear differential equations is in the form of a linear combination of exponential functions each of which describes the individual system modes. These modes are themselves defined by the system eigenvalues which are linked to the timescale of the different dynamics in the model. When the eigenvalues have a range of values that are widely distributed in the complex plane, the solution will consist of the sum of fast-changing dynamics, corresponding

Power System Dynamics: Stability and Control, Third Edition. Jan Machowski, Zbigniew Lubosny, Janusz W. Bialek and James R. Bumby.

to large eigenvalues, and slow-changing dynamics, corresponding to small eigenvalues. In this instance the system of differential equations is referred to as a *stiff system*. A nonlinear system is referred to as stiff if its linear approximation is stiff. Among the power system electromechanical models developed in Chapter 11, all the models that include both the subtransient equations, with their very small time constants, and the relatively slow rotor dynamics constitute stiff models. The model stiffness is further aggravated if the AVR equations, with their small time constants, and the turbine equations, with their long time constants, are included in the model. Consequently, if the power system model includes AVR systems and high-order generator models then the solution method should use integration methods well suited to stiff systems. In contrast, the classical power system model does not constitute a stiff set of differential equations because it only includes slow rotor dynamics. Transient stability programs using this model can use simpler integration formulas.

13.1 Numerical Integration Methods

The analytical solution of nonlinear differential equations is not generally possible and a numerical solution consisting of a series of values $(x_1, x_2, \ldots, x_k, \ldots)$ that satisfies the equation $\dot{x} = f(x, t)$ at the time instants $(t_1, t_2, \ldots, t_k, \ldots)$ must be found. This requires the use of *numerical integration formulas* that calculate the value x_{k+1} knowing all the previous values $\ldots, x_{k-2}, x_{k-1}, x_k)$. These formulas fall into two general categories: the single-step *Runge–Kutta methods* and the multi-step *predictor–corrector methods*. Both are used in power system simulation programs, and the interested reader is referred to Chua and Lin (1975) or Press et al. (1992), where these methods, and in particular the Runge–Kutta methods, are discussed in detail along with examples of the necessary computer code. In this section discussion is mainly concentrated on implicit integration methods because they can be effectively used in both the partitioned solution and the simultaneous solution of stiff differential equations. Such equations are generally used to describe power system behavior. The standard Runge–Kutta methods are restricted to the partitioned solution of nonstiff systems.

When numerically solving differential equations, each calculated value of the solution will differ from the accurate solution by an amount called the *local error*. This error comprises a *round-off error* that depends on the computational accuracy of the particular computer used, and the *method error* that depends on the type, order, and step length of the integration method used. As the local error propagates to subsequent steps, the total error at any one step consists of the local error made at that step plus the local errors transmitted from previous steps. The way in which the error propagates to subsequent steps determines the practical usefulness of a method. If the error does not increase from step to step then the formula is *numerically stable*; otherwise, the formula is *numerically unstable*. In the latter case the cumulative effect of the errors may cause the solution x_k to be drastically different from the accurate solution.

For the differential equation $\dot{x} = f(x, t)$ the value of x_{k+1} can be found by either integrating the function $f(x, t)$ along its time path between t_k and t_{k+1} or integrating $x(t)$ along its path from x_k to x_{k+1}. Each method will result in a different set of formulas known as the *Adams formulas* and the *Gear formulas*, respectively.

The Adams formulas used in predictor–corrector schemes are obtained by approximating the function $f(x, t)$ by a power series $w(t)$ in the time interval over which the function $f(x, t)$ must be integrated. This gives

$$x_{k+1} = x_k + \int_{t_k}^{t_{k+1}} f(x, t)\mathrm{d}t \approx x_k + \int_{t_k}^{t_{k+1}} w(t)\mathrm{d}t \tag{13.3}$$

where the power series $w(t)$ is based on r values of $f(x, t)$. The coefficients in this power series depend on the values of $f(x, t)$ at the individual points so that the integration formula becomes

$$x_{k+1} = x_k + h\left(\sum_{j=1}^{r} b_j f_{k+1-j} + b_0 f_{k+1}\right) \tag{13.4}$$

Table 13.1 Examples of Adams–Bashforth and Adams–Moulton formulas.

Type	Order	Formula	ε_0
Adams–Bashforth (explicit) formulas	1	$x_{k+1} = x_k + hf_k$	1/2
	2	$x_{k+1} = x_k + \frac{h}{2}(3f_k - f_{k-1})$	5/12
	3	$x_{k+1} = x_k + \frac{h}{12}(23f_k - 16f_{k-1} + 5f_{k-2})$	9/24
Adams–Moulton (implicit) formulas	1	$x_{k+1} = x_k + hf_{k+1}$	−1/2
	2	$x_{k+1} = x_k + \frac{h}{2}(f_{k+1} + f_k)$	−1/12
	3	$x_{k+1} = x_k + \frac{h}{12}(5f_{k+1} + 8f_k - f_{k-1})$	−1/24

where the function $f_i = f(x(t_i))$ is the value of the function at a given point t_i in time and h is the integration step length. The number of points r used in the power series is referred to as the *order* of the formula.

If $b_0 = 0$ then the resulting formulas are referred to as the *explicit* or *Adams–Bashforth formulas*. In these formulas the approximating polynomial $w(t)$ is calculated using the known values $(\ldots, f_{k-2}, f_{k-1}, f_k)$ and used to extrapolate the function $f(x, t)$ in the new interval from t_k to t_{k+1}. If $b_0 \neq 0$, the resulting formulas are referred to as the *implicit* or *Adams–Moulton formulas*. In these formulas the approximating polynomial $w(t)$ is calculated using the known values $(\ldots, f_{k-2}, f_{k-1}, f_k)$ and the unknown value of f_{k+1} in order to interpolate function $f(x, t)$ in the interval from t_k to t_{k+1}. Table 13.1 contains Adams–Bashforth and Adams–Moulton formulas up to third order.

The first-order formulas in the Adams family are the *Euler formulas* and they can have either an explicit or an implicit form. The second-order interpolation formula is the *trapezoidal rule*, when the polynomial $w(t)$ corresponds to the area of a trapezium below a line linking the points f_k and f_{k+1}.

The error in the Adams formulas depends on the order and is the integral of the error between the function $f(x, t)$ and the polynomial approximation $w(t)$. This error may be expressed as

$$\varepsilon_{k+1} = \varepsilon_0 h^{r+1} x_k^{r+1}(\tau) \tag{13.5}$$

where $x_k^{(r+1)}(\tau)$ is the $(r+1)$th derivative at a point τ lying in the interval from t_{k-r} to t_{k+1}, while ε_0 is a constant. Table 13.1 shows that for $r > 1$ the size of the error introduced by the implicit formulas is much smaller than that introduced by the explicit formulas.

An additional advantage of the implicit formulas is their better numerical stability. Figure 13.1 shows the area of numerical stability in the complex λh plane, where h is the integration step length and λ is the largest system eigenvalue. The integration process is numerically stable if the integration step length is small enough so that for all the system eigenvalues λ_i the product $h\lambda_i$ lies inside the stability area. The figure shows that the stability area of the explicit formulas is much smaller than that of their implicit counterparts. The first- and second-order implicit formulas are stable over the whole left half-plane and are therefore numerically *absolutely stable*. This stability issue is particularly important for stiff systems where large eigenvalues force the use of a small integration step h. The low-order implicit formulas are numerically stable over a large part of the complex plane and potentially allow the use of a longer integration step. When using these formulas the length of the integration step is limited only by the accuracy required in the calculation and the error as determined by Eq. (13.5).

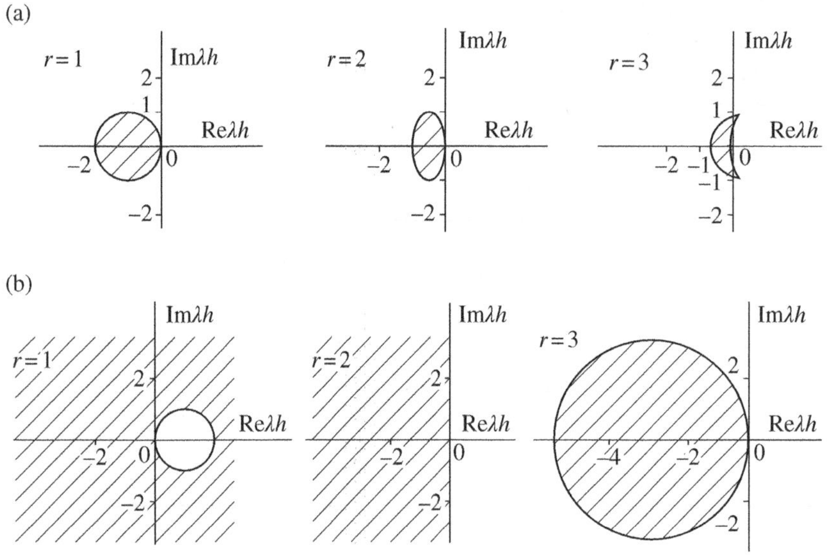

Figure 13.1 Areas of numerical stability for the Adams formulas: (a) explicit; (b) implicit.

A disadvantage of the implicit formulas is that the value of x_{k+1} cannot be calculated directly. When $b \neq 0$ Eq. (13.4) can be expressed as

$$x_{k+1} = \beta_k + hb_0 f(x_{k+1}) \tag{13.6}$$

where $\beta_k = x_k + \sum_{j=1}^{r} b_j f_{k+1-j}$. The unknown value of x_{k+1} now lies on both sides of the equation, which means that, if the function $f(x)$ is nonlinear, it must be found iteratively.

The simplest method of solving Eq. (13.6), referred to as *functional iteration*, consists of a series of substitutions according to the following iterative formula

$$x_{k+1}^{(l+1)} = \beta_k + hb_0 f\left(x_{k+1}^{(l)}\right) \tag{13.7}$$

where the upper indices in brackets denote the iteration number. If the first value $x_{k+1}^{(0)}$ used in the iteration is calculated using an explicit Adams formula, the whole procedure is known as the *prediction–correction method.* The explicit formula is used as a predictor, while the implicit formula is the corrector. The iterative process Eq. (13.7) converges if

$$hb_0 L < 1 \tag{13.8}$$

where L is the *Lipschitz constant*, $L = \sqrt{\alpha_M}$, and α_M is the maximum eigenvalue of the matrix product ($\boldsymbol{A}^T\boldsymbol{A}$), where $\boldsymbol{A} = [\partial f/\partial x]$ is the Jacobi matrix calculated at point $\hat{x}_{k+1}$ being a solution of Eq. (13.6). The smaller the value of the product (hb_0L), the faster the convergence.

Because of the presence of large eigenvalues in stiff systems, the convergence condition of Eq. (13.8) may force a limitation on the integration step length that is more restrictive than that required either by the required accuracy or by numerical stability. In such a case Eq. (13.6) may be solved using *Newton's method* instead of by functional iteration. For any equation $F(x) = 0$ Newton's formula is

$$x^{(l+1)} = x^{(l)} - \left[\frac{\partial F}{\partial x}\right]_{(l)}^{-1} F\left(x^{(l)}\right) \tag{13.9}$$

where the upper index in brackets denotes the iteration number. Applying Eq. (13.9) to the implicit formula of Eq. (13.6) gives

$$x_{k+1}^{(l+1)} = x_{k+1}^{(l)} - \left[1 - hb_0 A_{k+1}^{(l)}\right]^{-1} \left[x_{k+1}^{(l)} - \beta_k - hb_0 f\left(x_{k+1}^{(l)}\right)\right] \tag{13.10}$$

where $A_{k+1}^{(l)}$ is the Jacobi matrix calculated for the lth iteration. Newton's method allows the integration step length to be increased to a much higher value than that defined by Eq. (13.8) but, because the matrix is inverted at each step, the complexity of the method is much greater than functional iteration. However, if the integration step length is sufficiently large then the added complexity of Newton's method may be justified.

In developing the Adams formulas the function $f(x, t)$ was approximated by the power series $w(t)$ in order to find x_{k+1}. An alternative approach is to approximate $x(t)$, rather than the function $f(x, t)$, by the power series $x(t)$. In this case $w(t) \approx x(t)$ and the coefficients in the approximating polynomial $w(t)$ are functions of consecutive values of $(..., x_{k-2}, x_{k-1}, ...)$. Taking the time derivative gives $\dot{x} = \dot{w}(t)$, or $\dot{w}(t) = f(x, t)$. When $w(t)$ is used as the extrapolation formula, this leads to the following explicit integration formula

$$x_{k+1} = \sum_{j=0}^{r} a_j x_{k-j} + b_0 h f_k \tag{13.11}$$

If $w(t)$ is used as the interpolation formula, the following implicit integration formula is obtained

$$x_{k+1} = \sum_{j=0}^{r} a_j x_{k-j} + b_0 h f_{k+1} \tag{13.12}$$

These formulas, known as the Gear formulas, are shown in Table 13.2 (Variant I) up to the third order. The first-order formula is the Euler formula, while the second-order formula is referred to as the *intermediate point formula.*

The solution of Eq. (13.12) may be obtained using either functional iteration or Newton's method, just as for the Adams formulas. If Eq. (13.12) is solved by functional iteration then Eq. (13.11) is used as the predictor and Eq. (13.12) as the corrector.

Table 13.2 Examples of Gear formulas.

Type	Order	Formula	ε_0
Explicit Variant I	1	$x_{k+1} = x_k + hf_k$	
	2	$x_{k+1} = x_{k-1} + 2hf_k$	
	3	$x_{k+1} = -\frac{3}{2}x_k + 3x_{k-1} - \frac{1}{2}x_{k-2} + 3f_k h$	
Explicit Variant II	1	$x_{k+1} = 2x_k - x_{k-1}$	
	2	$x_{k+1} = 3x_k - 3x_{k-1} + x_{k-2}$	
	3	$x_{k+1} = 4x_k - 6x_{k-1} + 4x_{k-2} - x_{k-3}$	
Implicit	1	$x_{k+1} = x_k + hf_{k+1}$	−1/2
	2	$x_{k+1} = \frac{4}{3}x_k - \frac{1}{3}x_{k-1} + \frac{2}{3}hf_{k+1}$	−2/9
	3	$x_{k+1} = \frac{18}{11}x_k - \frac{9}{11}x+_{k-1} + \frac{2}{11}x_{k-2} + \frac{6}{12}hf_{k+1}$	−3/22

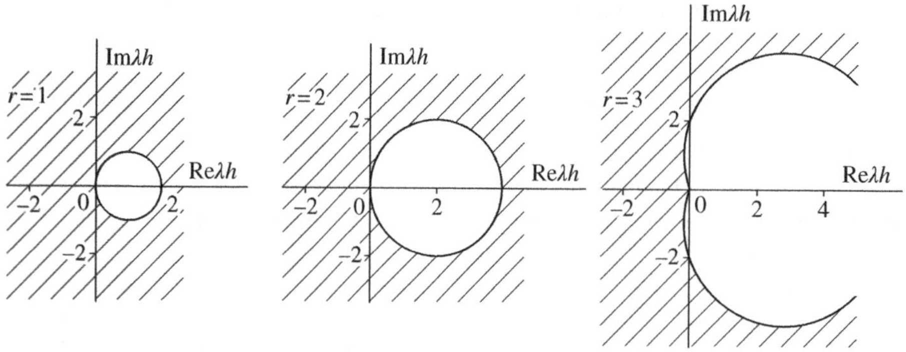

Figure 13.2 Areas of numerical stability of the Gear formulas.

The main advantage of the Gear formulas over the Adams formulas is that they have a larger area of numerical stability, as shown in Figure 13.2. They are therefore better suited to stiff systems. If a large integration step length is used on a stiff systems the predictor in Eq. (13.11) may not give a good approximation and instead $x(t)$ may be extrapolated directly using the values $(\ldots, x_{k-2}, x_{k-1}, x_k)$. The Lagrange extrapolation equations then give

$$x_{k+1} = a_0 x_k + \sum_{j=1}^{r} a_j x_{k-1} \tag{13.13}$$

Values of the coefficients for these formulas up to the third order are given in Table 13.2 (Variant II).

When a set of differential equations is nonlinear, the Jacobi matrix and its eigenvalues are not constant and the criteria that limit the integration step length constantly change with time. For example, the numerical stability depends on the eigenvalues and the step length, the convergence properties on the Lipschitz constant and the step length, while the local error is determined by the derivative $x^{(r+1)}(\tau)$ and the step length so that the correct choice of integration step length is of fundamental importance.

To combat these problems there are two extreme ways in which the difference formulas can be used. The first is to use a low-order formula, which is absolutely stable numerically, together with a constant integration step length that is limited to a value that will guarantee good convergence of the iterative corrector and limited local errors for the whole of the simulation period. Alternatively, the order of the formulas can be automatically varied during the simulation process so that the integration step length can be maximized without compromising either the accuracy or the convergence properties.

Because higher-order formulas have a low area of numerical stability, the highest-order formula that is normally used in practice is the sixth. Simple programs use second- or third-order formulas in order to avoid problems with numerical stability. It is also possible to use a constant order, variable step procedure. Automatic changes in either the integration step length or the order of the formula used require additional calculation. In addition, a change in the integration step length is not easy because if the step length is changed then the coefficients in most of the formulas also change since they depend on the distance between the interpolation nodes. Problems with the use of variable order, constant step formulas are alleviated if the equations are written in *canonical form*, details of which can be found in Chua and Lin (1975).

13.2 The Partitioned Solution

In each step of the numerical integration procedure the partitioned solution alternates between the solutions of the differential Eqs. (13.1) and the algebraic Eqs. (13.2). In order to match the values of the variables $\boldsymbol{y}(t)$ to the values of the variables $\boldsymbol{x}(t)$ it is necessary to solve the algebraic equations before numerically integrating the differential

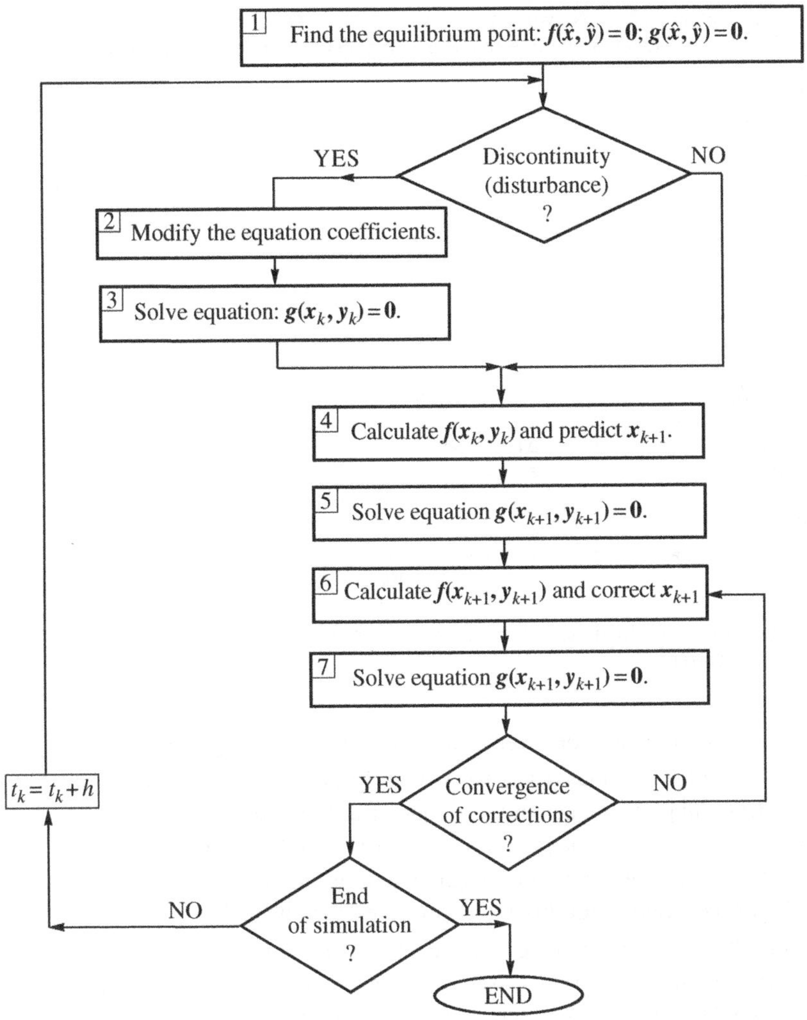

Figure 13.3 Simplified algorithm of the partitioned solution using the predictor–corrector method.

equations. A general solution algorithm for the partitioned solution, using a predictor–corrector numerical integration method, is shown in Figure 13.3. In this algorithm the algebraic equations are solved at three stages. The first solution is at stage 3, and occurs whenever there is a change in the network configuration. This change in the network configuration alters the coefficients in the algebraic equations so that for a given set of state variables $\mathbf{x}_k$ the value of the dependent variables $\mathbf{y}_k$ change. The second solution of the algebraic equations takes place at stage 5, after predicting the new values of the variables $\mathbf{x}_{k+1}$. The third, and final solution, is at stage 7 after correcting $\mathbf{x}_{k+1}$ in stage 6. The solution at stage 7 is repeated each integration step as many times as there are corrections of $\mathbf{x}_{k+1}$.

In the algorithm shown in Figure 13.3 the solution of the algebraic equations takes a significant proportion of the total computation time. It is therefore important to consider the methods available for effectively realizing the solution. In the following discussion the network equations will be solved assuming the network to be represented as shown in Figure 13.4.

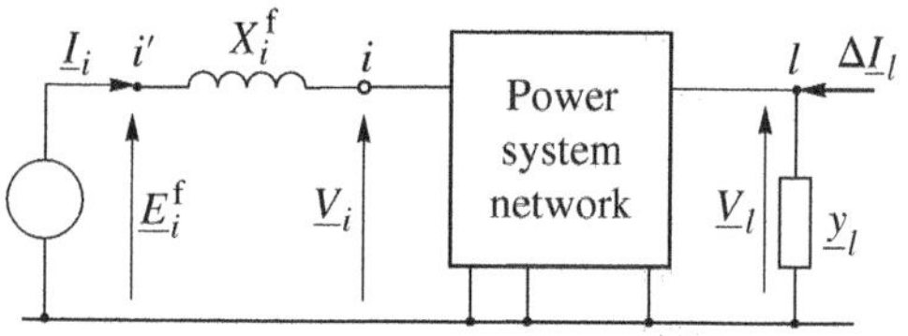

Figure 13.4 Network model with each generator replaced by a Thévenin voltage *source:* i' – a generator node; l – a load node.

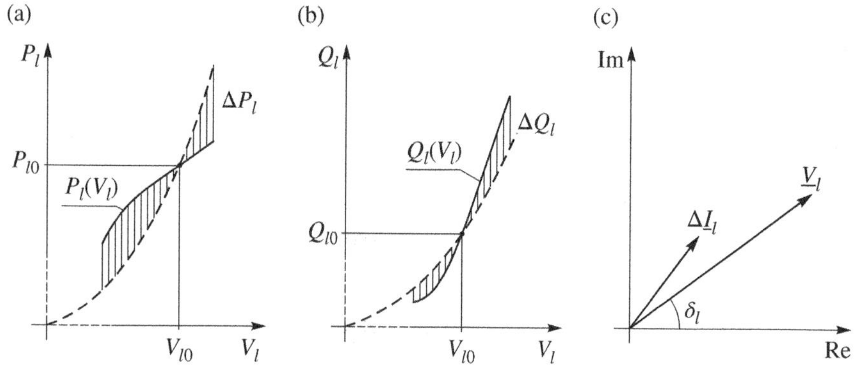

Figure 13.5 Modeling of nonlinear voltage characteristic of a load: (a) correction real power; (b) correction reactive power; (c) phasors of the nodal voltage and nodal correction current.

In the considered network model (Figure 13.4) each generator is represented by an additional generator node i' and a fictitious electromotive force (emf) $\underline{E}_i^{\rm f}$ behind a fictitious reactance $X_i^{\rm f}$ with $\underline{I}_i$ being the injected generator current. If rotor saliency is neglected then, depending on the generator model used, the fictitious emf and the fictitious reactance will correspond to the transient or subtransient values (Section 11.1.6). If, on the other hand, rotor saliency is included then both the reactance and the emf will have some fictitious value.

Each load is represented (Figure 13.4) by an equivalent admittance $\underline{y}_l$ and a correction nodal current $\Delta\underline{I}_l$. Figure 13.5 illustrates the way $\Delta\underline{I}_l$ is calculated. The loads are modeled (see Section 3.4.4) using nonlinear voltage characteristics $P_l(V_l)$ and $Q_l(V_l)$ denoted by solid lines in the diagram. Dashed lines denote parabolas corresponding to, respectively, real and reactive power consumed by admittance $\underline{y}_l = g_l + jb_l$ inserted in the network model. Differences between the required load characteristic and the admittance characteristic are made up by correction powers ΔP_l and ΔQ_l, respectively, which are denoted by vertical lines in Figure 13.5. When the load voltage is $\underline{V}_l$, correction power $\Delta\underline{S}_l = \Delta P_l + j\Delta Q_l$ corresponds to a correction current $\Delta\underline{I}_l = \Delta\underline{S}_l^*/\underline{V}_l$. Note that the voltage is a complex number as nodal voltage is a phasor in the common network reference frame (Figure 13.4).

Under the above assumptions the network is described by the following nodal equation using complex numbers

$$\begin{bmatrix} \underline{\boldsymbol{I}}_{\rm G} \\ \Delta\underline{\boldsymbol{I}}_{\rm L} \end{bmatrix} = \begin{bmatrix} \underline{\boldsymbol{Y}}_{\rm GG} & \underline{\boldsymbol{Y}}_{\rm GL} \\ \underline{\boldsymbol{Y}}_{\rm LG} & \underline{\boldsymbol{Y}}_{\rm LL} \end{bmatrix} \begin{bmatrix} \underline{\boldsymbol{E}}_{\rm G} \\ \underline{\boldsymbol{V}}_{\rm L} \end{bmatrix} \tag{13.14}$$

where {G} is the set of fictitious generator nodes, {L} is the set of all the other nodes (called the load nodes) that includes the generator terminal nodes, $\underline{\boldsymbol{I}}_{\rm G}$ is a vector of the generator currents, $\Delta\underline{\boldsymbol{I}}_{\rm L}$ a vector of the load corrective currents, $\underline{\boldsymbol{E}}_{\rm G}$ a vector of the fictitious emf's, and $\underline{\boldsymbol{V}}_{\rm L}$ a vector of the voltages at the load nodes. The matrix $\underline{\boldsymbol{Y}}_{\rm GG}$ is a diagonal matrix of generator admittance $\underline{y}_i = 1/jX_i^{\rm f}$. $\underline{\boldsymbol{Y}}_{\rm GL}$ is a rectangular matrix comprising a diagonal submatrix with elements equal to $(-\underline{y}_i)$ and all other elements equal to zero. $\underline{\boldsymbol{Y}}_{\rm LG}$ is the transpose of $\underline{\boldsymbol{Y}}_{\rm GL}$. The matrix $\underline{\boldsymbol{Y}}_{\rm LL}$ is a modified version of the nodal admittance matrix, introduced in Chapter 3, whose diagonal terms now include the load and generator admittances at the rows corresponding to the relevant load nodes and generator terminal nodes, respectively. If each load is represented by a constant admittance then the correction currents $\Delta\underline{\boldsymbol{I}}_{\rm L} = \underline{\boldsymbol{0}}$.

13.2.1 Partial Matrix Inversion

Appendix A.2 contains the derivation of partial matrix inversion which may be applied to Eq. (13.14). This equation can be expanded as

$$\underline{I}_G = \underline{Y}_{GG}\underline{E}_G + \underline{Y}_{GL}\underline{V}_L \tag{13.15}$$

$$\Delta\underline{I}_L = \underline{Y}_{LG}\underline{E}_G + \underline{Y}_{LL}\underline{V}_L \tag{13.16}$$

Rearranging Eq. (13.16)

$$\underline{V}_L = -\underline{Y}_{LL}^{-1}\underline{Y}_{LG}\underline{E}_G + \underline{Y}_{LL}^{-1}\Delta\underline{I}_L \tag{13.17}$$

and substituting into Eq. (13.15) gives

$$\underline{I}_G = (\underline{Y}_{GG} - \underline{Y}_{GL}\underline{Y}_{LL}^{-1}\underline{Y}_{LG})\underline{E}_G + \underline{Y}_{GL}\underline{Y}_{LL}^{-1}\Delta\underline{I}_L \tag{13.18}$$

These last two Eqs. (13.17) and (13.18) can be rewritten in matrix form as

$$\begin{bmatrix} \underline{I}_G \\ \underline{V}_L \end{bmatrix} = \begin{bmatrix} \underline{Y}_G & \underline{K}_I \\ \underline{K}_V & \underline{Z}_{LL} \end{bmatrix} \begin{bmatrix} \underline{E}_G \\ \Delta\underline{I}_L \end{bmatrix} \tag{13.19}$$

where $\underline{Y}_G = \underline{Y}_{GG} - \underline{Y}_{GL}\underline{Y}_{LL}^{-1}\underline{Y}_{LG}$, $\underline{K}_I = \underline{Y}_{GL}\underline{Y}_{LL}^{-1}$, $\underline{K}_V = -\underline{Y}_{LL}^{-1}\underline{Y}_{LG}$, and $\underline{Z}_{LL} = \underline{Y}_{LL}^{-1}$. The square matrix in Eq. (13.19) is called the *partial inversion matrix* and refers to the fact that only the submatrix $\underline{Y}_{LL}$ is explicitly inverted to obtain $\underline{Z}_{LL} = \underline{Y}_{LL}^{-1}$.

If rotor saliency is neglected then the fictitious emf's $\underline{E}_i^f$ in the network model are equal to the generator emf's and are calculated during the numerical integration of the differential equations. If, in addition, each load is represented by a constant admittance then the correction currents $\Delta\underline{I}_L$ are equal to zero and the generator currents and the load voltages can be calculated directly from

$$\underline{I}_G = \underline{Y}_G\underline{E}_G; \quad \underline{V}_L = \underline{K}_V\underline{E}_G \tag{13.20}$$

without the need for an iterative solution. However, if the loads are nonlinear, the correction currents are nonzero and depend on the load voltage according to the function $\Delta\underline{I}_L(\underline{V}_L)$. As the unknown load voltages now appear on both sides of Eq. (13.19), they must be calculated iteratively. The lower equation in the Eq. (13.19) can be used to formulate the iterative formula

$$\underline{V}_L^{(l+1)} = \underline{K}_V\underline{E}_G + \underline{Z}_{LL}\Delta\underline{I}_L\left(\underline{V}_L^{(l)}\right) \tag{13.21}$$

where the upper index in brackets denotes the iteration number. When the iteration process is complete, the voltage $\underline{V}_L = \underline{V}_L^{(l+1)}$ and the generator currents may be calculated from the upper equation in Eq. (13.19) as

$$\underline{I}_G = \underline{Y}_G\underline{E}_G + \underline{K}_I\Delta\underline{I}_L \tag{13.22}$$

where $\Delta\underline{I}_L$ are the correction currents corresponding to the calculated values of the voltages.

If rotor saliency is included then the fictitious emf's representing the generators in the network equations must also be calculated iteratively together with the generator currents. To explain how this is done the generator fourth-order transient model $(\dot{E}_d', \dot{E}_q', \dot{\delta}, \dot{\omega})$ will be used, but the technique can equally well be applied to the sixth-order subtransient model $(\dot{E}_d'', \dot{E}_q'', \dot{E}_d', \dot{E}_q', \dot{\delta}, \dot{\omega})$ or the fifth-order subtransient model $(\dot{E}_d'', \dot{E}_q'', \dot{E}_q', \dot{\delta}, \dot{\omega})$ defined in Section 11.1.6.

Figure 13.6 shows three circuit diagrams. The first two diagrams correspond to the generator in the transient state, while the third diagram corresponds to the fictitious Thévenin source used in the network model to replace the generator. The emf of the fictitious generator voltage source $\underline{E}^f$ must satisfy the equation $\underline{E}^f = \underline{V} + jX^f\underline{I}$, or $(E_a^f + jE_b^f) = (V_a + jV_b) + jX^f(I_a + jI_b)$, which expressed in matrix form is

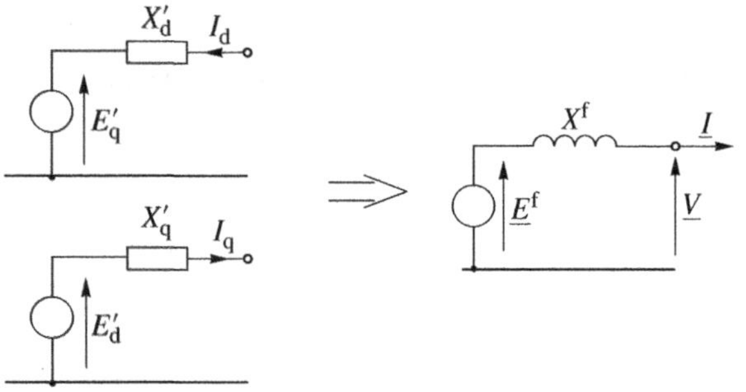

Figure 13.6 Replacing the equivalent d- and q-axis circuit of the generator by one equivalent circuit with fictitious reactance X^f and fictitious emf $\underline{E}^f$.

$$\begin{bmatrix} E_a^f \\ E_b^f \end{bmatrix} = \begin{bmatrix} V_a \\ V_b \end{bmatrix} - \begin{bmatrix} 0 & -X^f \\ X^f & 0 \end{bmatrix} \begin{bmatrix} I_a \\ I_b \end{bmatrix} \quad \text{or} \quad \boldsymbol{E}_{ab}^f = \boldsymbol{V}_{ab} - \boldsymbol{Z}_{ab}^f \boldsymbol{I}_{ab} \tag{13.23}$$

This equation can be transformed from the system (a, b) reference frame to the individual generator (d, q) reference frame using the $\boldsymbol{T}$ transformation defined in Eq. (3.192) to give

$$\begin{bmatrix} E_d^f \\ E_q^f \end{bmatrix} = \begin{bmatrix} V_d \\ V_q \end{bmatrix} - \begin{bmatrix} 0 & -X^f \\ X^f & 0 \end{bmatrix} \begin{bmatrix} I_d \\ I_q \end{bmatrix} \quad \text{or} \quad \boldsymbol{E}_{dq}^f = \boldsymbol{V}_{dq} - \boldsymbol{Z}_{dq}^f \boldsymbol{I}_{dq} \tag{13.24}$$

where $\boldsymbol{Z}_{dq}^f = \boldsymbol{T}^{-1}\boldsymbol{Z}_{ab}^f\boldsymbol{T} = \boldsymbol{Z}_{ab}^f$ and $\boldsymbol{T}^{-1} = \boldsymbol{T}$. On the other hand, the armature voltage equation for the fourth-order model of the generator ($\dot{E}'_d, \dot{E}'_q, \dot{\delta}, \dot{\omega}$) is given by Eq. (11.104) as

$$\begin{bmatrix} V_d \\ V_q \end{bmatrix} = \begin{bmatrix} E'_d \\ E'_q \end{bmatrix} - \begin{bmatrix} 0 & X'_q \\ -X'_d & 0 \end{bmatrix} \begin{bmatrix} I_d \\ I_q \end{bmatrix} \quad \text{or} \quad \boldsymbol{V}_{dq} = \boldsymbol{E}'_{dq} - \boldsymbol{Z}_{dq} \boldsymbol{I}_{dq} \tag{13.25}$$

which, when substituted into Eq. (13.24), gives

$$\begin{bmatrix} E_d^f \\ E_q^f \end{bmatrix} = \begin{bmatrix} E'_d \\ E'_q \end{bmatrix} - \begin{bmatrix} 0 & -\Delta X_q \\ \Delta X_d & 0 \end{bmatrix} \begin{bmatrix} I_d \\ I_q \end{bmatrix} \quad \text{or} \quad \boldsymbol{E}_{dq}^f = \boldsymbol{E}'_{dq} - \Delta \boldsymbol{Z} \boldsymbol{I}_{dq} \tag{13.26}$$

where $\Delta X_q = X'_q - X^f$ and $\Delta X_d = X'_d - X^f$. This equation determines the emf of the fictitious voltage source in terms of the generator current.

Equation (13.26) can be solved iteratively together with the network Eq. (13.19). To show how this is done Eq. (13.26) is rewritten as

$$\begin{bmatrix} E_d^{f(l+1)} \\ E_q^{f(l+1)} \end{bmatrix} = \begin{bmatrix} E'_d \\ E'_q \end{bmatrix} - \begin{bmatrix} 0 & -\Delta X_q \\ \Delta X_d & 0 \end{bmatrix} \begin{bmatrix} I_d^{(l)} \\ I_q^{(l)} \end{bmatrix} \quad \text{or} \quad \boldsymbol{E}_{dq}^{f(l+1)} = \boldsymbol{E}'_{dq} - \Delta \boldsymbol{Z} \boldsymbol{I}_{dq}^{(l)} \tag{13.27}$$

where l is the iteration number. The iterative solution algorithm is then:

1) Estimate the values of E_d^f and E_q^f for every generator and transform them to the system reference frame (a, b) in order to obtain $\underline{E}^f = E_a^f + jE_b^f$.
2) Solve the network Eqs. (13.19). Calculate the current $\underline{I} = I_a + jI_b$ for every generator and transform I_a and I_b to the generator (d, q) reference frame in order to obtain I_d and I_q.
3) Use Eq. (13.27) to correct the values of E_d^f and E_q^f.

4) Compare the result with the pervious iteration; if they differ, transform the new values of $E_{\mathrm{d}}^{\mathrm{f}}$ and $E_{\mathrm{q}}^{\mathrm{f}}$ to the system reference frame (a, b) and repeat step 2 until the voltages converge.

The number of iterations necessary to solve the generator equations and the network equations depends on the value of the reactance chosen for the fictitious Thévenin source. Let $\hat{E}_{\mathrm{d}}^{\mathrm{f}}$, $\hat{E}_{\mathrm{q}}^{\mathrm{f}}$ and $\hat{I}_{\mathrm{d}}$, $\hat{I}_{\mathrm{q}}$ be the solutions to the equations. According to Eq. (13.27), the solution must satisfy

$$\begin{bmatrix} \hat{E}_{\mathrm{d}}^{\mathrm{f}} \\ \hat{E}_{\mathrm{q}}^{\mathrm{f}} \end{bmatrix} = \begin{bmatrix} E'_{\mathrm{d}} \\ E'_{\mathrm{q}} \end{bmatrix} - \begin{bmatrix} 0 & -\Delta X_{\mathrm{q}} \\ \Delta X_{\mathrm{d}} & 0 \end{bmatrix} \begin{bmatrix} \hat{I}_{\mathrm{d}} \\ \hat{I}_{\mathrm{q}} \end{bmatrix} \quad \text{or} \quad \hat{\boldsymbol{E}}_{\mathrm{dq}}^{\mathrm{f}} = \boldsymbol{E}'_{\mathrm{dq}} - \Delta \boldsymbol{Z} \hat{\boldsymbol{I}}_{\mathrm{dq}} \tag{13.28}$$

Subtracting Eq. (13.27) from Eq. (13.28) gives

$$\left(\boldsymbol{E}_{\mathrm{dq}}^{\mathrm{f}(l+1)} - \hat{\boldsymbol{E}}_{\mathrm{dq}}^{\mathrm{f}}\right) = \Delta \boldsymbol{Z}\left(\boldsymbol{I}_{\mathrm{dq}}^{(l)} - \hat{\boldsymbol{I}}_{\mathrm{dq}}\right) \tag{13.29}$$

This equation is important because it means that, for a given error $\Delta \boldsymbol{I}_{\mathrm{dq}}^{(l)} = \left(\boldsymbol{I}_{\mathrm{dq}}^{(l)} - \hat{\boldsymbol{I}}_{\mathrm{dq}}\right)$ in the current estimation, the smaller the elements $\Delta X_{\mathrm{q}} = X'_{\mathrm{q}} - X^{\mathrm{f}}$ and $\Delta X_{\mathrm{d}} = X'_{\mathrm{d}} - X^{\mathrm{f}}$ in the matrix $\Delta \boldsymbol{Z}$, the closer the next estimation of the fictitious emf will be to the final solution. Consequently, the iterative process will converge quickly if the equivalent generator reactance is chosen to have a value equal to an average value of X'_{d} and X'_{q} when the fictitious emf $\underline{E}^{\mathrm{f}}$ then has a value close to the generator transient emf $\underline{E}'$. Generally, one of the following "average" reactance values is used for the fictitious reactance

$$X^{\mathrm{f}} = \frac{1}{2}\left(X'_{\mathrm{d}} + X'_{\mathrm{q}}\right), \quad X^{\mathrm{f}} = 2\frac{X'_{\mathrm{d}} X'_{\mathrm{q}}}{X'_{\mathrm{d}} + X'_{\mathrm{q}}}, \quad X^{\mathrm{f}} = \sqrt{X'_{\mathrm{d}} X'_{\mathrm{q}}} \tag{13.30}$$

The use of one of these fictitious reactance values guaranties a small number of iterations in the solution of Eq. (13.27). If X'_{d} is close to X'_{q} then one or two iterations will suffice. If transient saliency is neglected then $X'_{\mathrm{d}} = X'_{\mathrm{q}} = X^{\mathrm{f}}$ and $\hat{\boldsymbol{E}}_{\mathrm{dq}}^{\mathrm{f}} = \boldsymbol{E}'_{\mathrm{dq}}$ and the solution is obtained without any iterations. When $X'_{\mathrm{d}} \neq X'_{\mathrm{q}}$, the number of iterations depends on the initial choice of the fictitious emf's. If the value of X^{f} is chosen according to one of the formulas in (13.30) then the magnitude and the angle of the fictitious emf are close to the magnitude and the angle of the generator transient emf. In order to achieve a good estimate for the final value, the fictitious emf's should be changed in each integration step in proportion to the generator emf's.

When rotor saliency and load nonlinearity are included, the algorithm will involve both of the iterative processes described above. A simplified algorithm is shown in Figure 13.7.

A disadvantage of the partial matrix inversion technique used in Eq. (13.19) to solve the network equation is that all the submatrices $\underline{\boldsymbol{Y}}_{\mathrm{G}}$, $\underline{\boldsymbol{K}}_{\mathrm{I}}$, $\underline{\boldsymbol{K}}_{\mathrm{V}}$, and $\underline{\boldsymbol{Z}}_{\mathrm{LL}}$ are dense and therefore have a large computer memory requirement while also requiring a large number of arithmetic operations to calculate the currents and the voltages. If the number of loads for which the voltage must be computed is small, in comparison with the total number of nodes, then the partial matrix inversion technique is worth using because then only part of the submatrices $\underline{\boldsymbol{K}}_{\mathrm{I}}$, $\underline{\boldsymbol{K}}_{\mathrm{V}}$, and $\underline{\boldsymbol{Z}}_{\mathrm{LL}}$ need be stored and the number of arithmetic operations is also reduced. If the voltage must be computed for the majority of the load nodes then the partial matrix inversion method is not recommended since the matrix factorization technique described in the next section is more efficient.

13.2.2 Matrix Factorization

The lower equation in (13.14) gives

$$\underline{\boldsymbol{Y}}_{\mathrm{LL}} \underline{\boldsymbol{V}}_{\mathrm{L}} = \left(\Delta \underline{\boldsymbol{I}}_{\mathrm{L}} + \underline{\boldsymbol{I}}_{\mathrm{N}}\right) \tag{13.31}$$

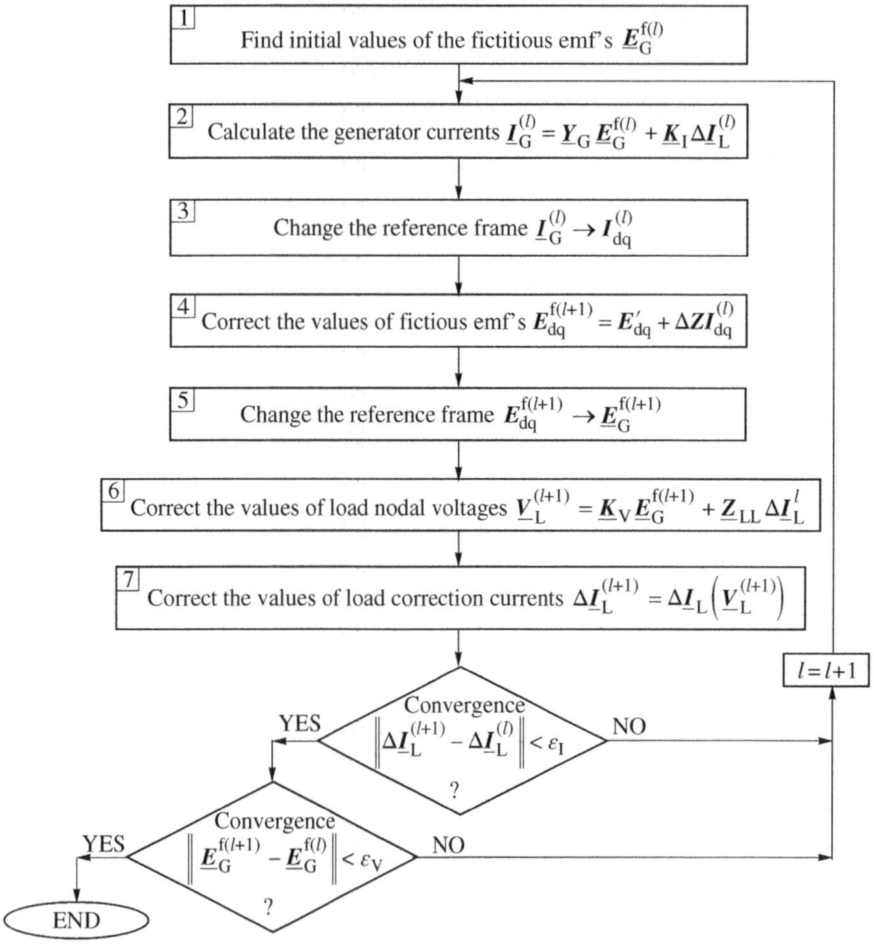

Figure 13.7 A simplified flowchart for the solution of the network equations using partial matrix inversion with both load nonlinearity and rotor saliency included.

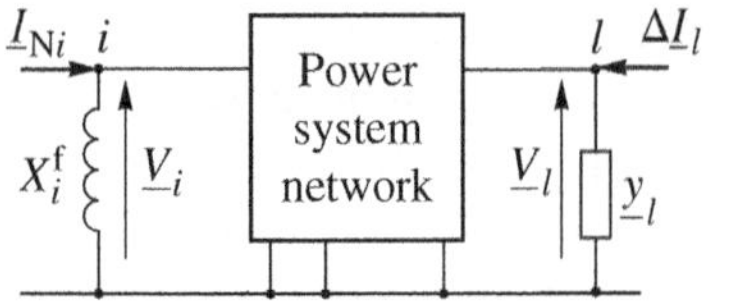

Figure 13.8 Network model with each generator replaced by a Norton source.

where $\underline{I}_N = -\underline{Y}_{LG}\underline{E}_G$ is a vector of the currents having nonzero elements only at the nodes where the generators are connected to the system. Equation (13.31) corresponds to the network shown in Figure 13.8.

Using *triangular factorization* (Press et al. 1992) the square matrix $\underline{Y}_{LL}$ can be replaced by the product of an upper, or right, triangular matrix $\underline{R}_{LL}$ and a lower, or left, triangular matrix $\underline{L}_{LL}$ to give

$$\underline{Y}_{LL} = \underline{L}_{LL}\underline{R}_{LL} \tag{13.32}$$

Equation (13.31) can then be rewritten as two equations

$$\underline{L}_{LL}\underline{b}_L = (\Delta\underline{I}_L + \underline{I}_N), \quad \underline{R}_{LL}\underline{V}_L = \underline{b}_L \tag{13.33}$$

where $\underline{b}_L$ is an unknown vector.

Assuming that the loads are represented by constant admittances, then, for a given set of generator emf's $\underline{E}_G$, Eqs. (13.33) can be solved noniteratively. As the vector $\underline{I}_N = -\underline{Y}_{LG}\underline{E}_G$ and the lower triangular matrix $\underline{L}_{LL}$ are both known and $\Delta\underline{I}_L = \mathbf{0}$, the unknown vector $\underline{b}_L$ can be found using a series of *forward substitutions* starting from the upper left-hand corner of $\underline{L}_{LL}$. After calculating $\underline{b}_L$ the values of $\underline{V}_L$ can be found by a series of *back substitutions* starting from the lower right-hand corner of the upper triangular matrix $\underline{R}_{LL}$. The advantage of this method is that if the matrix $\underline{Y}_{LL}$ is sparse then both the factor submatrices $\underline{R}_{LL}$ and $\underline{L}_{LL}$ are also sparse, thus allowing sparse matrix techniques to be used to save on computer memory and the number of arithmetic operations needed to effect a solution (Tewerson 1973; Brameller et al. 1976; Pissanetzky 1984; Duff et al. 1986).

If nonlinearity of the loads is included, Eq. (13.33) must be solved iteratively since the correction currents $\Delta\underline{I}_L$ depend on the voltages. The iteration formulas are

$$\underline{L}_{LL}\underline{b}_L^{(l+1)} = \left(\Delta\underline{I}_L^{(l)} + \underline{I}_N\right), \quad \underline{R}_{LL}\underline{V}_L^{(l+1)} = \underline{b}_L^{(l+1)} \tag{13.34}$$

where $\Delta\underline{I}_L^{(l)} = \Delta\underline{I}_L\left(\underline{V}_L^{(l)}\right)$ is a vector of the load correction currents. The upper index l denotes the iteration number. If rotor saliency is included then the generator Norton current $\underline{I}_N$ must also be calculated iteratively, in a similar way to the equivalent emf's in Eq. (13.27). These iterations can be executed together with the iterations necessary to calculate the voltages and the correction currents at the load nodes. A simplified algorithm of the method is shown in Figure 13.9.

13.2.3 Newton's Method

Newton's method was introduced in Chapter 3 as a way to solve the network power–voltage equations. In system simulation Newton's method is also used but must now solve a set of current–voltage equations so that the solution algorithm is different from that used in the steady-state load flow. A particularly attractive feature of Newton's method is that if the network nodal current–voltage equations are solved in rectangular coordinates, in the system (a, b) reference frame, then rotor saliency can be conveniently included in the solution. To account for saliency the generator equations are added to the network nodal admittance equation expressed in real numbers as in Eq. (3.179).

Assuming that the generators are represented by the fourth-order transient model $(\dot{E}'_d, \dot{E}'_q, \dot{\delta}, \dot{\omega})$, the generator armature voltage Eqs. (11.104) can be written as

$$\begin{bmatrix} \dot{E}'_d - V_d \\ \dot{E}'_q - V_q \end{bmatrix} = \begin{bmatrix} 0 & -X'_q \\ X'_d & 0 \end{bmatrix} \begin{bmatrix} I_d \\ I_q \end{bmatrix} \tag{13.35}$$

which, when inverted, give

$$\begin{bmatrix} I_d \\ I_q \end{bmatrix} = \frac{1}{X'_d X'_q} \begin{bmatrix} 0 & X'_q \\ -X'_d & 0 \end{bmatrix} \begin{bmatrix} \dot{E}'_d - V_d \\ \dot{E}'_q - V_q \end{bmatrix} \quad \text{or} \quad \boldsymbol{I}_{dq} = \boldsymbol{Y}_{dq}\left(\boldsymbol{E}'_{dq} - \boldsymbol{V}_{dq}\right) \tag{13.36}$$

These voltages and currents are in the generator (d, q) reference frame and are transformed into the system (a, b) reference frame using the transformation matrix $\boldsymbol{T}$ in Eq. (3.192) to give

$$\begin{bmatrix} I_a \\ I_b \end{bmatrix} = \left(\boldsymbol{T}^{-1}\boldsymbol{Y}_{dq}\boldsymbol{T}\right) \begin{bmatrix} \dot{E}'_d - V_d \\ \dot{E}'_q - V_q \end{bmatrix} \quad \text{or} \quad \boldsymbol{I}_{ab} = \boldsymbol{Y}_{ab}\left(\boldsymbol{E}'_{ab} - \boldsymbol{V}_{ab}\right) \tag{13.37}$$

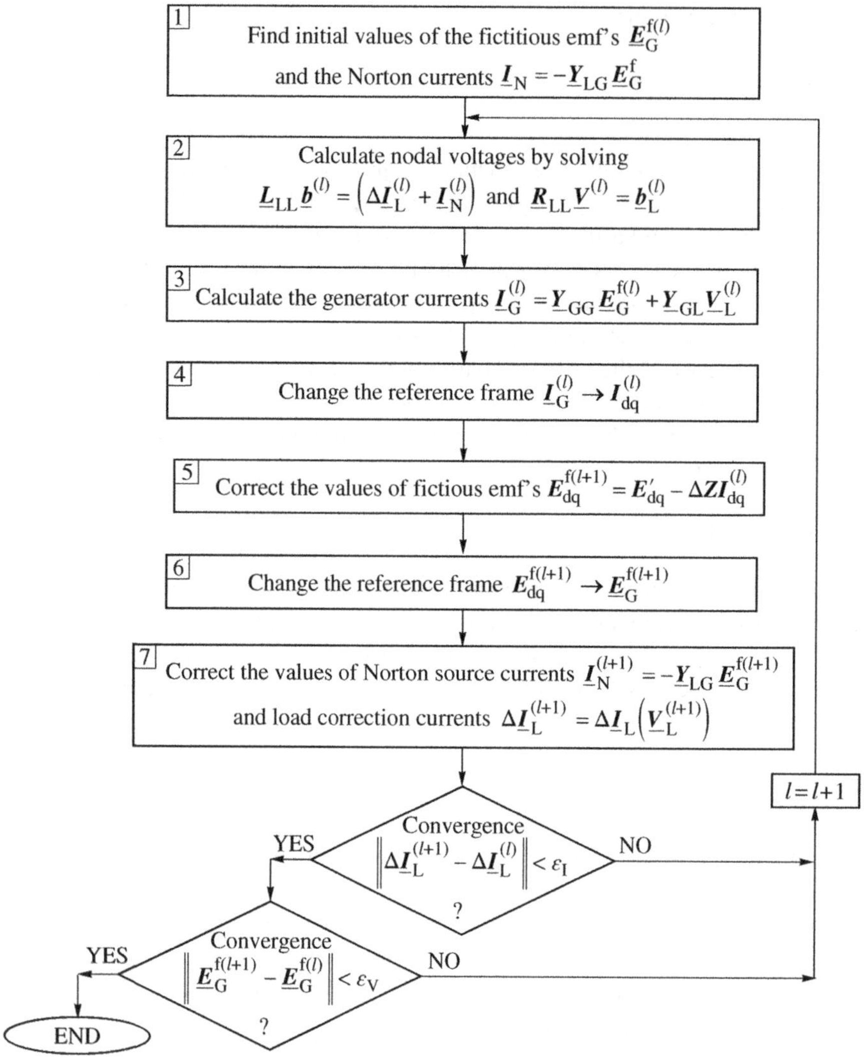

Figure 13.9 Simplified network solution flowchart using triangular factorization with load nonlinearity and rotor saliency included.

where

$$\boldsymbol{Y}_{ab} = \boldsymbol{T}^{-1}\boldsymbol{Y}_{dq}\boldsymbol{T} = \frac{1}{X'_d X'_q}\begin{bmatrix} -\sin\delta & \cos\delta \\ \cos\delta & \sin\delta \end{bmatrix}\begin{bmatrix} 0 & X'_q \\ -X'_d & 0 \end{bmatrix}\begin{bmatrix} -\sin\delta & \cos\delta \\ \cos\delta & \sin\delta \end{bmatrix} =$$
$$= \frac{1}{X'_d X'_q}\begin{bmatrix} \frac{1}{2}\left(X'_d - X'_q\right)\sin 2\delta & -X'_q\sin^2\delta - X'_d\cos^2\delta \\ X'_q\cos^2\delta + X'_d\sin^2\delta & -\frac{1}{2}\left(X'_d - X'_q\right)\sin 2\delta \end{bmatrix} = \begin{bmatrix} g_a & -b_{ab} \\ b_{ba} & g_b \end{bmatrix} \tag{13.38}$$

is a submatrix similar in form to $\boldsymbol{Y}_{ij}$ in Eq. (3.154). In the nodal admittance technique the submatrix $\boldsymbol{Y}_{ab}$ describes a generator with transient saliency. Generally, because $g_a \neq g_b$ and $b_{ab} \neq b_{ba}$, there is no one equivalent branch with admittance $\underline{Y}_{ab}$ so that an equivalent circuit for the salient-pole machine cannot be drawn. In such a case Eq. (13.37) cannot be written in complex form as $\underline{I} = \underline{Y}_{ab}(\underline{E}' - \underline{V})$ since the equivalent admittance $\underline{Y}_{ab}$ does not exist. If transient saliency is neglected, $X'_d = X'_q$, then the submatrix $\boldsymbol{Y}_{ab}$ is skew symmetric as $g_a = g_b = 0$ and

$b_{ab} = b_{ba} = 1/X'_d$ and an equivalent branch with admittance $\underline{Y}_{ab} = 0 + jb_{ab} = j\left(1/X'_d\right)$ now exists to represent the generator, as shown in Figure 13.4 for the Thévenin source or in Figure 13.8 for the Norton source with, in both cases, $X_i^f = X'_d$.

Because of saliency the generator current–voltage equation can only be written using real numbers in the (a, b) coordinate system. Complex notation cannot be used. In order to include Eq. (13.37) with the network Eq. (13.14) or (13.31), the latter must also be written using real numbers in the same way as Eq. (3.155). Then each generator will be represented by a submatrix as in Eq. (13.38), which is added to the respective elements of the real submatrices $\boldsymbol{Y}_{GG}$, $\boldsymbol{Y}_{GL}$, $\boldsymbol{Y}_{LG}$, and $\boldsymbol{Y}_{LL}$. In the case of the Norton source, Eq. (13.31) can be rewritten as

$$\boldsymbol{Y}_{LL}(\delta)\,\boldsymbol{V}_L = \Delta\boldsymbol{I}_L(\boldsymbol{V}_L) + \boldsymbol{I}_N \tag{13.39}$$

where the real matrix $\boldsymbol{Y}_{LL}(\delta)$ is shown as a function of δ to emphasize that the diagonal elements of this matrix that refer to the generator depend on the power angle in the way defined by the submatrix (13.38). The elements of these submatrices change with time as the generator rotor angles change. The Norton source currents $\boldsymbol{I}_N$ are given by

$$\boldsymbol{I}_N = -\boldsymbol{Y}_{LG}(\delta)\boldsymbol{E}_G \tag{13.40}$$

where matrix $\boldsymbol{Y}_{LG}(\delta)$ consists of the submatrices (13.38) and $\boldsymbol{E}_G$ is a vector comprising the individual generator emf's $\boldsymbol{E}'_{ab}$. All variables are in the (a, b) system coordinates.

To solve Eq. (13.39) it can be rewritten in the standard Newton form as

$$\boldsymbol{F}(\boldsymbol{V}_L) = [\boldsymbol{Y}_{LG}(\delta)\boldsymbol{E}_G - \Delta\boldsymbol{I}_L(\boldsymbol{V}_L)] + \boldsymbol{Y}_{LL}(\delta)\boldsymbol{V}_L = \boldsymbol{0} \tag{13.41}$$

Using Newton's iterative formula gives

$$\boldsymbol{V}_L^{(l+1)} = \boldsymbol{V}_L^{(l)} - \left[\frac{\partial \boldsymbol{F}}{\partial \boldsymbol{V}_L}\right]_l^{-1} \boldsymbol{F}\left(\boldsymbol{V}_L^l\right) \tag{13.42}$$

where the Jacobi matrix

$$\left[\frac{\partial \boldsymbol{F}}{\partial \boldsymbol{V}_L}\right] = \boldsymbol{Y}_{LL}(\delta) - \left[\frac{\partial \Delta\boldsymbol{I}_L}{\partial \boldsymbol{V}_L}\right] \tag{13.43}$$

is equal to the nodal admittance matrix minus the matrix of the derivatives of the correction currents with respect to the voltages. This correction matrix is diagonal and its elements are of the form

$$\left[\frac{\partial \Delta I_i}{\partial V_i}\right] = \begin{bmatrix} \dfrac{\partial \Delta I_{ai}}{\partial V_{ai}} & \dfrac{\partial \Delta I_{ai}}{\partial V_{bi}} \\ \dfrac{\partial \Delta I_{bi}}{\partial V_{ai}} & \dfrac{\partial \Delta I_{bi}}{\partial V_{bi}} \end{bmatrix} \tag{13.44}$$

The way in which the derivatives in Eq. (13.44) are calculated needs some explanation. In rectangular coordinates the relationship between the correction powers and the correction currents can be expressed as

$$\begin{bmatrix} \Delta P_i \\ \Delta Q_i \end{bmatrix} = \begin{bmatrix} V_{ai} & V_{bi} \\ V_{bi} & -V_{ai} \end{bmatrix}\begin{bmatrix} \Delta I_{ai} \\ \Delta I_{bi} \end{bmatrix} \quad \text{or} \quad \begin{bmatrix} \Delta I_{ai} \\ \Delta I_{bi} \end{bmatrix} = \frac{1}{|V_i|^2}\begin{bmatrix} V_{ai} & V_{bi} \\ V_{bi} & -V_{ai} \end{bmatrix}\begin{bmatrix} \Delta P_i \\ \Delta Q_i \end{bmatrix} \tag{13.45}$$

Making the substitution

$$\Delta p_i = \frac{\Delta P_i}{|V_i|^2}, \quad \Delta q_i = \frac{\Delta Q_i}{|V_i|^2} \tag{13.46}$$

allows Eq. (13.45) to be rewritten as

$$\begin{bmatrix} \Delta I_{\mathrm{ai}} \\ \Delta I_{\mathrm{bi}} \end{bmatrix} = \begin{bmatrix} V_{\mathrm{ai}} & V_{\mathrm{bi}} \\ V_{\mathrm{bi}} & -V_{\mathrm{ai}} \end{bmatrix} \begin{bmatrix} \Delta p_i \\ \Delta q_i \end{bmatrix} \tag{13.47}$$

Differentiating Eq. (13.47) gives

$$\begin{aligned}
\frac{\partial \Delta I_{\mathrm{ai}}}{\partial V_{\mathrm{ai}}} &= V_{\mathrm{ai}} \frac{\partial \Delta p_i}{\partial V_{\mathrm{ai}}} + \Delta p_i + V_{\mathrm{bi}} \frac{\partial \Delta q_i}{\partial V_{\mathrm{ai}}} \\
\frac{\partial \Delta I_{\mathrm{ai}}}{\partial V_{\mathrm{bi}}} &= V_{\mathrm{bi}} \frac{\partial \Delta q_i}{\partial V_{\mathrm{bi}}} + \Delta q_i + V_{\mathrm{ai}} \frac{\partial \Delta p_i}{\partial V_{\mathrm{bi}}} \\
\frac{\partial \Delta I_{\mathrm{bi}}}{\partial V_{\mathrm{ai}}} &= -V_{\mathrm{ai}} \frac{\partial \Delta q_i}{\partial V_{\mathrm{ai}}} - \Delta q_i + V_{\mathrm{bi}} \frac{\partial \Delta p_i}{\partial V_{\mathrm{ai}}} \\
\frac{\partial \Delta I_{\mathrm{bi}}}{\partial V_{\mathrm{bi}}} &= V_{\mathrm{bi}} \frac{\partial \Delta p_i}{\partial V_{\mathrm{ai}}} + \Delta p_i - V_{\mathrm{ai}} \frac{\partial \Delta q_i}{\partial V_{\mathrm{bi}}}
\end{aligned}$$

which, when substituted into Eq. (13.44) and the matrix rearranged, leads to

$$\begin{aligned}
\left[\frac{\partial \Delta I_i}{\partial V_i}\right] &= \begin{bmatrix} V_{\mathrm{ai}} & V_{\mathrm{bi}} \\ V_{\mathrm{bi}} & -V_{\mathrm{ai}} \end{bmatrix} \begin{bmatrix} \dfrac{\partial \Delta p_i}{\partial V_{\mathrm{ai}}} & \dfrac{\partial \Delta p_i}{\partial V_{\mathrm{bi}}} \\ \dfrac{\partial \Delta q_i}{\partial V_{\mathrm{ai}}} & \dfrac{\partial \Delta q_i}{\partial V_{\mathrm{bi}}} \end{bmatrix} + \begin{bmatrix} \Delta p_i & \Delta q_i \\ -\Delta q_i & \Delta p_i \end{bmatrix} \\
&= \frac{1}{|V_i|} \begin{bmatrix} V_{\mathrm{ai}} & V_{\mathrm{bi}} \\ -V_{\mathrm{bi}} & -V_{\mathrm{ai}} \end{bmatrix} \begin{bmatrix} \dfrac{\partial \Delta p_i}{\partial |V_i|} & \dfrac{\partial \Delta p_i}{\partial |V_i|} \\ \dfrac{\partial \Delta q_i}{\partial |V_i|} & \dfrac{\partial \Delta q_i}{\partial |V_i|} \end{bmatrix} \begin{bmatrix} V_{\mathrm{ai}} & 0 \\ 0 & V_{\mathrm{bi}} \end{bmatrix} + \begin{bmatrix} \Delta p_i & \Delta q_i \\ -\Delta q_i & \Delta p_i \end{bmatrix}
\end{aligned} \tag{13.48}$$

If the derivatives in the second matrix are expressed in the form

$$\frac{\partial \Delta p_i}{\partial V_{\mathrm{ai}}} = \frac{\partial \Delta p_i}{\partial |V_i|} \frac{\partial |V_i|}{\partial V_{\mathrm{ai}}} = \frac{\partial \Delta p_i}{\partial |V_i|} \frac{\partial}{\partial V_{\mathrm{ai}}} \sqrt{V_{\mathrm{ai}}^2 + V_{\mathrm{bi}}^2} = \frac{1}{|V_i|} \frac{\partial \Delta p_i}{\partial |V_i|} V_{\mathrm{ai}}$$

then the partial derivatives $\partial \Delta p_i / \partial |V_i|$ and $\partial \Delta q_i / \partial |V_i|$ can be computed from the static load characteristics $P(V)$ and $Q(V)$.

The solution algorithm consists of the iterative formula (13.42), the Jacobi matrix (13.43), and the correction currents given by Eq. (13.48). The algorithm may be simplified by using the *dishonest Newton method*, where the Jacobi matrix is calculated only once at the beginning of each integration step using the initial values of the correction currents. This simplification should not be used during those integration steps when a network disturbance is being simulated, because the change in the correction currents may be large.

If rotor saliency and the load correction currents are neglected, Eq. (13.42) of the Newton method is identical to the second equation of (13.20) in the partial matrix inversion method. This can be shown by substituting $\Delta \boldsymbol{I}_{\mathrm{L}} = 0$, $\boldsymbol{Y}_{\mathrm{LL}}(\delta) = \boldsymbol{Y}_{\mathrm{LL}} =$ constant, and $\boldsymbol{V}_{\mathrm{L}}^{(l+1)} = \boldsymbol{V}_{\mathrm{L}}^{l}$ into Eqs. (13.42) and (13.41).

13.2.4 Ways of Avoiding Iterations and Multiple Network Solutions

The basic algorithm of the partitioned solution shown in Figure 13.1 attempts to match the values of the variables $\boldsymbol{y}(t)$, for given values of $\boldsymbol{x}(t)$, by solving the linear algebraic Eq. (13.2). This solution is repeated after each prediction and each correction of $\boldsymbol{x}(t)$. One way to speed up the algorithm is to replace the solution of the algebraic equation by an extrapolation of the value of $\boldsymbol{y}(t)$ at some appropriate stage in the solution. As the extrapolated values of $\boldsymbol{y}(t)$ are only approximate, an error is introduced into the right-hand side of Eq. (13.1), called the *interface error*, which influences the accuracy of $\boldsymbol{x}(t)$.

There are three ways of introducing extrapolation into the algorithm:

- The algebraic equations are solved after each prediction, with extrapolated values being used after each correction.
- The algebraic equations are solved after each prediction and after the last correction.
- The algebraic equations are solved after each correction with the prediction of $\boldsymbol{y}(t)$ and $\boldsymbol{x}(t)$ being done together, by extrapolation.

The first method is not recommended as it may introduce large interface errors which force the integration step length to shorten. The second and third methods are much better, with the third method being preferred because, with this method, interface errors generated at the prediction stage are eliminated during correction. In this case extrapolation eliminates one solution of the algebraic equations in each integration step. Obviously, this method is beneficial only if the number of corrections required is small and the corrector converges quickly.

In most cases the variables in $\boldsymbol{y}(t)$ are extrapolated individually and independently of each other. Simple extrapolation formulas are normally employed that use the past values of the variable obtained from the previous two or three steps. Typical of these formulas are those listed as Variant II in Table 13.2 as

$$x_{k+1} = 2x_k - x_{k-1} \quad \text{or} \quad x_{k+1} = 3x_k - 3x_{k-1} + x_{k-2}$$

though Adibi et al. (1974) suggest updating the complex load voltages using

$$\underline{V}_{k+1} = \frac{\underline{V}_k^2}{\underline{V}_{k-1}} \quad \text{or} \quad \underline{V}_{k+1} = \frac{\underline{V}_k^3 \underline{V}_{k-2}}{\underline{V}_{k-1}^2} \tag{13.49}$$

After extrapolating the voltages at all the load nodes using Eq. (13.49) the generator currents are computed from the nodal Eq. (13.14) using the generator emf's obtained from the numerical integration. Extrapolation procedures can also be used to obtain the values of other variables, such as generator real power, voltage error, and so on (Stott 1979). Generally high-order extrapolation formulas are not used to update $\boldsymbol{y}(t)$ as the improvement in the accuracy is small compared with the simple formulas. In addition, complications occur following network disturbances because, at the instant of the discontinuity, all the previous old values of the variable are invalid and extrapolation must start at the step where the disturbance occurs. Sometimes, no previous values are used in the extrapolation process but a linearized equation is formed that links the increments in $\boldsymbol{y}(t)$ with the increments in $\boldsymbol{x}(t)$ (Stott 1979).

Besides reducing the number of times that the algebraic equations need to be solved at each integration step, it is also possible to avoid iterations in the network solution. These iterations result from nonlinear load characteristics and from saliency in the generator rotors. The number of iterations necessary to account for generator saliency can be reduced by using either the sixth- or the fifth-order subtransient models rather than the fourth- or third-order transient models. Table 4.3 (Section 4.2.3) shows that transient saliency, $X'_\mathrm{q} \neq X'_\mathrm{d}$, is usually much larger than subtransient saliency, $X''_\mathrm{q} \neq X''_\mathrm{d}$, so that the iteration process for the subtransient model converges faster, and with fewer iterations, than that for the transient model. Unfortunately, the reduction in computing time due to the faster convergence is partially offset by the shorter integration time step required by the subtransient model to account for the smaller time constants. As subtransient saliency effects are normally quite small, Dandeno and Kundur (1973) suggest that in order to produce a fast noniterative solution the subtransient model with subtransient saliency effects neglected should be used. By adopting this approach a solution is obtained that is more accurate than that produced by the transient model where the damper windings are neglected. In this noniterative algorithm the iterations required to account for load nonlinearity are performed only at discontinuities, that is only at the time of the disturbance. Except at disturbances, the change of the voltage at the load nodes is smooth and slow so that the correction currents can be approximately calculated at each integration step by basing them on the voltages in the previous step, that is

$$\Delta \boldsymbol{I}_{\mathrm{L}(k+1)} = \Delta \boldsymbol{I}_{\mathrm{L}}\left(\boldsymbol{V}_{\mathrm{L}(k)}\right) \tag{13.50}$$

With these assumptions the network equations can be solved noniteratively, apart from the instants of discontinuity, using either partial matrix inversion or matrix factorization. At discontinuities, changes in the voltages may be large and it is necessary to execute a few iterations in order to calculate accurately the correction currents. Adibi et al. (1974) note that small errors resulting from Eq. (13.50) can be partially eliminated by extrapolating the voltages at the load nodes thereby improving the estimated correction currents.

Based on these assumptions, the solution algorithm in Figure 13.3 can be modified. When solving the algebraic equations at the discontinuity, stage 3, Eqs. (13.21) or (13.34) can be used to introduce the iterations necessary to model the nonlinear loads. Solution of the algebraic equations at stage 5, after the prediction, can be replaced by extrapolation of the algebraic variables $\boldsymbol{y}(t)$, while the solution at stage 7, after correction, can be executed using Eqs. (13.20) or (13.33) to solve the network equations noniteratively.

13.3 The Simultaneous Solution Methods

The concept of the simultaneous solution methods is to use implicit integration formulas to change the differential Eq. (13.1) into algebraic form and then to solve these algebraic equations simultaneously with the algebraic network equations in (13.2).

Any implicit integration formula can be written in the general form

$$x_{k+1} = \beta_k + hb_0 f(x_{k+1}) \tag{13.51}$$

where h is the integration step length, b_0 is a coefficient that depends on the actual integration method, $f(x_{k+1})$ is the right-hand side of the differential Eq. (13.1) calculated at the value x_{k+1}, and

$$\beta_k = x_k + \sum_j b_j f_{k+1-j} \tag{13.52}$$

is a coefficient depending on all the previous steps. Using the formula in Eq. (13.51), Eqs. (13.1) and (13.2) can be rewritten as

$$\begin{aligned} \boldsymbol{F}_1(\boldsymbol{x}_{k+1}, \boldsymbol{y}_{k+1}) &= \boldsymbol{f}(\boldsymbol{x}_{k+1}, \boldsymbol{y}_{k+1}) - \frac{1}{hb_0}\boldsymbol{x}_{k+1} - \boldsymbol{\beta}_k = \boldsymbol{0} \\ \boldsymbol{F}_2(\boldsymbol{x}_{k+1}, \boldsymbol{y}_{k+1}) &= \boldsymbol{g}(\boldsymbol{x}_{k+1}, \boldsymbol{y}_{k+1}) = \boldsymbol{0} \end{aligned} \tag{13.53}$$

where $\boldsymbol{\beta}_k$ is a column vector containing the values β_k.

Newton's method gives the iteration formula as

$$\begin{bmatrix} \boldsymbol{x}_{k+1}^{(l+1)} \\ \boldsymbol{y}_{k+1}^{(l+1)} \end{bmatrix} = \begin{bmatrix} \boldsymbol{x}_{k+1}^{(l)} \\ \boldsymbol{y}_{k+1}^{(l)} \end{bmatrix} - \left[\begin{array}{c|c} \boldsymbol{f}_x - \frac{1}{hb_0}\boldsymbol{1} & \boldsymbol{f}_y \\ \hline \boldsymbol{g}_x, & \boldsymbol{g}_y \end{array}\right]^{-1} \begin{bmatrix} \boldsymbol{F}_1\left(\boldsymbol{x}_{k+1}^{(l)}, \boldsymbol{y}_{k+1}^{(l)}\right) \\ \boldsymbol{F}_2\left(\boldsymbol{x}_{k+1}^{(l)}, \boldsymbol{y}_{k+1}^{(l)}\right) \end{bmatrix} \tag{13.54}$$

where $\boldsymbol{1}$ is the unit diagonal matrix and $\boldsymbol{f}_x = \partial \boldsymbol{f}/\partial \boldsymbol{x}$, $\boldsymbol{f}_y = \partial \boldsymbol{f}/\partial \boldsymbol{y}$, $\boldsymbol{g}_x = \partial \boldsymbol{g}/\partial \boldsymbol{x}$, and $\boldsymbol{g}_y = \partial \boldsymbol{g}/\partial \boldsymbol{y}$ are the Jacobi submatrices. The Jacobi matrix in Eq. (13.54) is sparse, so computer programs that simulate large systems do not generally explicitly invert this matrix. Instead Eq. (13.54) is solved using triangular factorization and forward and back substitution. The network equations are expressed in rectangular form in the system (a, b) reference frame so that rotor saliency can be included without any difficulty. The nonlinear load correction currents are modified during each iteration. The effectiveness of the method depends on the choice of the variables to be iterated in the Newton method and on skillful use of sparse matrix techniques. A significant role is also played by appropriate grouping of the variables allowing block matrices to be used.

Vorley (1974) presents one of the variants of this method where Eqs. (13.1) and (13.2) are arranged in such a way that the differential and algebraic equations of the generator are grouped together to produce an equation of the form

$$\left[\begin{array}{ccc|ccc} 1 & \cdots & 0 & & & \\ \vdots & \ddots & \vdots & & & \\ 0 & \cdots & 1 & & & \\ \hline & & & 0 & \cdots & 0 \\ & & & \vdots & \ddots & \vdots \\ & & & 0 & \cdots & 0 \end{array}\right] \left[\begin{array}{c} \dot{x}_1 \\ \vdots \\ \dot{x}_r \\ \hline \dot{x}_{r+1} \\ \vdots \\ \dot{x}_m \end{array}\right] = \left[\begin{array}{c} f_1 \\ \vdots \\ f_r \\ \hline f_{r+1} \\ \vdots \\ f_m \end{array}\right] \quad \text{or} \quad c_i \dot{\boldsymbol{x}}_i = \boldsymbol{f}_i(\boldsymbol{x}_i, \boldsymbol{V}) \tag{13.55}$$

where $(x_1, \ldots, x_r)$ are the variables of the differential equations describing the ith generating unit and $(x_{r+1}, \ldots, x_m)$ are the variables of the algebraic equations describing this unit. As matrix $\boldsymbol{c}_i$ is singular this equation is singular. The whole system is described by

$$\begin{aligned} \boldsymbol{C}\dot{\boldsymbol{x}} &= \boldsymbol{F}(\boldsymbol{x}, \boldsymbol{V}) \\ \boldsymbol{0} &= \boldsymbol{G}(\boldsymbol{x}, \boldsymbol{V}) \end{aligned} \tag{13.56}$$

where the first of the equations consists of Eqs. (13.55) corresponding to individual generating units and the second, describing the network, is the equation of nodal voltages. Using the implicit integration formulas and Newton's equation, as in Eq. (13.54), gives

$$\begin{bmatrix} \boldsymbol{x}_{k+1}^{(l+1)} \\ \boldsymbol{V}_{k+1}^{(l+1)} \end{bmatrix} = \begin{bmatrix} \boldsymbol{x}_{k+1}^{(l)} \\ \boldsymbol{V}_{k+1}^{(l)} \end{bmatrix} - \begin{bmatrix} \boldsymbol{F}_x - \dfrac{1}{hb_0}\boldsymbol{C} & \boldsymbol{F}_v \\ \boldsymbol{G}_x & \boldsymbol{G}_v \end{bmatrix}^{-1} \begin{bmatrix} \boldsymbol{F}\left(\boldsymbol{x}_{k+1}^{(l)}, \boldsymbol{V}_{k+1}^{(l)}\right) - \dfrac{1}{hb_0}\boldsymbol{C}\left(\boldsymbol{x}_{k+1}^{(l)} - \boldsymbol{\beta}_k\right) \\ \boldsymbol{G}\left(\boldsymbol{x}_{k+1}^{(l)}, \boldsymbol{V}_{k+1}^{(l)}\right) \end{bmatrix} \tag{13.57}$$

where $\boldsymbol{F}_x = \partial\boldsymbol{F}/\partial\boldsymbol{x}$, $\boldsymbol{F}_v = \partial\boldsymbol{F}/\partial\boldsymbol{V}$, $\boldsymbol{G}_x = \partial\boldsymbol{G}/\partial\boldsymbol{x}$, and $\boldsymbol{G}_v = \partial\boldsymbol{G}/\partial\boldsymbol{V}$ are the submatrices of the Jacobi matrix.

The Jacobi matrices of the individual generating units have a block structure which simplifies the factorization. To speed up calculations the dishonest Newton method is used, where the iterations at each integration step are executed for a constant matrix calculated from predicted values. It is also possible to simplify the method further by modifying the Jacobi matrix only after network disturbances and after a certain number of integration steps. The number of iterations necessary for convergence can be used as an indicator for when to modify the Jacobi matrix, with this matrix being updated if the number of iterations exceeds a preset value, for example three.

Variable integration step length and variable order interpolation formulas are also used. As the differential and algebraic equations are solved together, there is no interfacing problem, and the use of Newton's method ensures no convergence problems, even when a long integration step length is used with a stiff systems. At the start of each integration step, extrapolated initial values are used in the iterations.

Description of other examples of simultaneous solution method can be found in (Adibi et al. 1974; Harkopf 1978; Stott 1979; Rafian et al. 1987).

13.4 Comparison Between the Methods

The simultaneous solution methods allow rotor saliency and nonlinear loads to be readily included and are especially attractive for simulations that cover a long time period. Newton's method, together with implicit integration formulas, allow the integration step length to be increased when the changes in the variables are not very steep.

The dishonest Newton method can be used to speed up the calculations. Interfacing problems between the algebraic and differential equations do not exist.

In contrast, the partitioned solution methods are attractive for simulations that cover a shorter time interval. They are more flexible, easier to organize and allow a number of simplifications to be introduced that speed up the solution. However, unless care is taken, these simplifications may cause large interfacing errors. The majority of dynamic simulation programs described in the literature are based on partitioned solution methods.

The main characteristics of the partitioned solution methods relate to the way in which the network equations are solved. Partial matrix inversion is only attractive for simplified systems because the submatrices of the partially inverted nodal matrix are dense. If the nodal matrix is large these submatrices take up a lot of computer memory. Additionally, because of the large number of nonzero elements in these submatrices, the number of arithmetic operations needed to solve the network equations is also large. The speed of solution can be improved by assuming that the loads are linear (constant admittances) and by calculating the voltages at only a small number of load nodes thereby limiting the size of the relevant inverted submatrices. This method becomes particularly attractive when model reduction is employed based on the aggregation of coherent generators, as discussed in Chapter 18. In this case, when the algorithm is reorganized, the transfer matrix that is used to predict groups of coherent generators (after certain transformations corresponding to aggregation) can also be used to solve the equations of the reduced network.

If nonlinear loads are included, or the voltage change at a certain number of loads is required, then triangular factorization is superior to partial inversion because the factor matrices remain sparse after factorization. For a typical power network the factor matrices only contain about 50% more elements than the original admittance matrix and the number of arithmetic operations required to solve the network is not very high. If additional modifications that limit the number of iterations due to rotor saliency and nonlinear loads are included, triangular factorization becomes by far the fastest solution method.

The properties of the computer algorithms that use Newton's method are similar to those for the simultaneous solution method. Compared with triangular factorization, Newton's method requires a larger computer memory and more arithmetic calculations per integration step. However, owing to good convergence, Newton's method can use a longer integration step than the factorization method, which partially compensates for the greater number of computations per step. The use of the dishonest Newton method speeds up the calculations quite considerably. Moreover, rotor saliency and nonlinear loads can be included more easily than is the case with triangular factorization.

It is worth adding that fairly recently, with the ever-increasing power of computers, there has been a tendency to develop *real-time simulators* to train operators for dispatch and security monitoring and which can also be used as the core of an online dynamic security assessment system. To make these simulators operate in real time, it is often required to split the program into independent tasks to be executed in parallel (Chai and Bose 1993; Bialek 1996).

13.5 Modeling of Unbalanced Faults

In some computer programs used to simulate power system dynamic response, the transmission network is modeled only for the positive-sequence component. In such programs the unbalanced faults must be modeled using an additional shunt element connected in the positive sequence network model to the faulted node (Figure 13.10).

As explained in Section 6.1.1 (Table 6.1) the shunt impedance $\Delta \underline{z}_F$ depends on the type of fault in the following way

$$\Delta \underline{z}_F = \underline{Z}_2 \underline{Z}_0 / (\underline{Z}_2 + \underline{Z}_0) \quad \text{for} \quad \text{K2E}; \quad \Delta \underline{z}_F = \underline{Z}_2 \quad \text{for} \quad \text{K2}; \quad \Delta \underline{z}_F = (\underline{Z}_2 + \underline{Z}_0) \quad \text{for} \quad \text{K1} \tag{13.58}$$

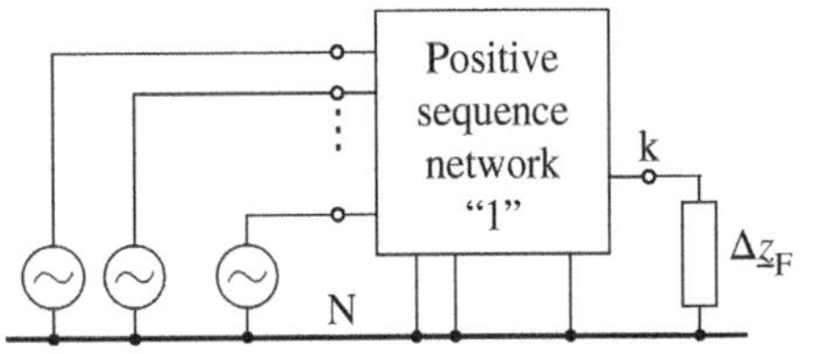

Figure 13.10 Positive-sequence network with shunt impedance modeling an unbalanced short circuit.

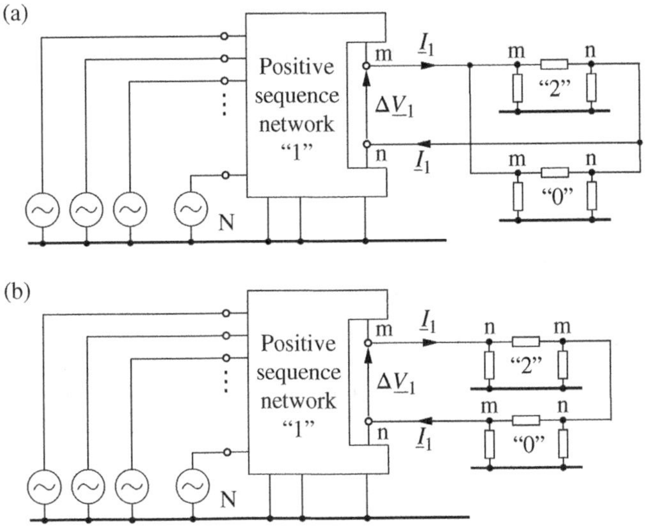

Figure 13.11 Positive-sequence network with π-circuit modeling (a) single-phase interruption; (b) double-phase interruption.

where: K2E – double-phase-to-ground fault, K2 – phase-to-phase fault, K1 – single phase fault and $\underline{Z}_2$ and $\underline{Z}_0$ are the Thévenin's impedance seen from the faulted node in negative-sequence network and zero-sequence network, respectively.

Similarly modeled are the single- or double-phase interruptions occurring during the clearing of unbalanced faults. In such a case, the component of the equivalent π-circuits representing the negative-sequence network and zero-sequence network are connected to the positive sequence network model. This is shown in Figure 13.11. It is assumed that the phase interruption occurred in the network element connected to the nodes m and n. The negative-sequence network and zero-sequence network are replaced by equivalent π-circuits seen from these nodes.

Parameters of an equivalent π-circuit replacing a given network can be calculated by reducing the model of this network to nodes m and n. For this purpose, one can use node elimination described in Section 18.2.1 or simply by using the impedance nodal matrix used by the short-circuit program.

The network model for a given component intended for short-circuit calculations can be described using the nodal admittance and impedance matrices as follows

$$\underline{\mathbf{Y}} = \begin{bmatrix} \underline{\mathbf{Y}}_{RR} & \underline{\mathbf{Y}}_{RE} \\ \underline{\mathbf{Y}}_{ER} & \underline{\mathbf{Y}}_{EE} \end{bmatrix}; \quad \underline{\mathbf{Z}} = \underline{\mathbf{Y}}^{-1} = \begin{bmatrix} \underline{\mathbf{Z}}_{RR} & \underline{\mathbf{Z}}_{RE} \\ \underline{\mathbf{Z}}_{ER} & \underline{\mathbf{Z}}_{EE} \end{bmatrix} \tag{13.59}$$

where {E} and {R} are the groups of the nodes, which must be eliminated and retained, respectively.

On the basis of the partial inversion equations described in Section A.2, one can write

$$\begin{bmatrix} (\underline{Y}_{RR} - \underline{Y}_{RE}\underline{Y}_{EE}^{-1}\underline{Y}_{ER}) & \underline{Y}_{RE}\underline{Y}_{EE}^{-1} \\ -\underline{Y}_{EE}^{-1}\underline{Y}_{ER} & \underline{Y}_{EE}^{-1} \end{bmatrix} = \begin{bmatrix} \underline{Z}_{RR}^{-1} & -\underline{Z}_{RR}^{-1}\underline{Z}_{RE} \\ \underline{Z}_{ER}\underline{Z}_{RR}^{-1} & (\underline{Z}_{EE} - \underline{Z}_{ER}\underline{Z}_{RR}^{-1}\underline{Z}_{RE}) \end{bmatrix} \tag{13.60}$$

Section 18.2.1 shows that the upper left submatrix in the matrix on the left side of Eq. (13.60) describes the equivalent network obtained after eliminating nodes {E}. Equation (13.60) shows also that this matrix is equal to the left upper matrix on the right side of Eq. (13.60). Therefore

$$\underline{Y}_{R} = (\underline{Y}_{RR} - \underline{Y}_{RE}\underline{Y}_{EE}^{-1}\underline{Y}_{ER}) = \underline{Z}_{RR}^{-1} \tag{13.61}$$

If {R} = {m, n} a group of retained nodes consisting of two nodes m and n, then on the basis of Eq. (13.61) it is possible to write

$$\underline{Y}_{R} = \begin{bmatrix} \underline{Y}_{mm} & \underline{Y}_{mn} \\ \underline{Y}_{nm} & \underline{Y}_{nn} \end{bmatrix} = \begin{bmatrix} \underline{Z}_{mm} & \underline{Z}_{mn} \\ \underline{Z}_{nm} & \underline{Z}_{nn} \end{bmatrix}^{-1} \tag{13.62}$$

where $\underline{Y}_{mm}$, $\underline{Y}_{mn}$, $\underline{Y}_{nm}$, and $\underline{Y}_{nn}$ are elements of an admittance matrix describing equivalent π-circuit and $\underline{Z}_{mm}$, $\underline{Z}_{mn}$, $\underline{Z}_{nm}$, and $\underline{Z}_{nn}$ are elements of the nodal impedance matrix used for short-circuit calculations. Based on the elements of the admittance matrix in Eq. (13.62), one can calculate the admittances of π-circuit branches

$$\underline{y}_{mN} = \underline{Y}_{mm} + \underline{Y}_{mn}; \quad \underline{y}_{mn} = -\underline{Y}_{mn}; \quad \underline{y}_{nm} = -\underline{Y}_{nm}; \quad \underline{y}_{nN} = \underline{Y}_{nn} + \underline{Y}_{nm} \tag{13.63}$$

If the matrix describing the network is symmetrical, it happens that $\underline{y}_{mn} = \underline{y}_{nm}$. Hence the impedances of the branches

$$\underline{z}_{mN} = \underline{y}_{mN}^{-1}; \quad \underline{z}_{mn} = \underline{z}_{nm} = \underline{y}_{mn}^{-1}; \quad \underline{z}_{nN} = \underline{y}_{nN}^{-1} \tag{13.64}$$

In Figure 13.11 two series connected shunt branches of π-circuit are parallel connected with the series π-circuit branch. Hence, the resulting impedance, which replaces the entire π-circuit, is given by the formula

$$\Delta\underline{z}_{mn} = \frac{\underline{z}_{mn}(\underline{z}_{mN} + \underline{z}_{nN})}{\underline{z}_{mn} + \underline{z}_{mN} + \underline{z}_{nN}} \tag{13.65}$$

Using Eq. (13.65), the equivalent impedances should be calculated from the nodes m and n respectively in the negative-sequence network $\Delta\underline{z}_{2mn}$ and in the zero-sequence network $\Delta\underline{z}_{0mn}$.

As shown Figure 13.11 in the case of a single-phase interruption, the impedances $\Delta\underline{z}_{2mn}$ and $\Delta\underline{z}_{0mn}$ should be connected in parallel and then connected to the nodes mand n, in the positive sequence network. In the case of the double-phase interruption, impedances $\Delta\underline{z}_{2mn}$ and $\Delta\underline{z}_{0mn}$ must be connected in series.

In some programs used for the simulation of power system dynamic response the three-phase network models with mutual couplings of the phase circuits are used. In this case, the modeling of unbalanced faults is different from the one described above. It is required to carefully read the manual of a given computer program.

13.6 Evaluation of Power System Dynamic Response

The following criteria can be used to evaluate the dynamic response of the EPS:

a) critical clearing time (CCT) and transient stability margin (Eq. 6.8 in Section 6.1.1);
b) transient voltage dip and post-fault voltage variation (Figure 2.8 in Section 2.3.2);
c) damping of post-fault rotor and power swings.

Evaluation of damping of rotor and power swings can be done using one of the following factors.

Damping ratio is defined by the formula (5.51) discussed in Section 5.4.6 or by the formula (12.70) discussed in Section 12.1.3. This coefficient can be calculated when the modal analysis described in Section 12.1.4 and 12.3.4 is used for the local angular stability study. In the modal analysis, a dynamic system is described by matrix differential equation $\dot{\boldsymbol{x}} = \boldsymbol{A}\,\boldsymbol{x}$, in which $\boldsymbol{x}$ is the column vector of state variables, $\boldsymbol{A}$ is the square state matrix dependent on system parameters. If $\lambda_i = \alpha_i + j\Omega_i$ is the i-th eigenvalue of matrix $\boldsymbol{A}$ then the damping ratio of the oscillatory mode $e^{\alpha_i t} \cdot \cos(\Omega_i t + \phi_{ki})$ corresponding to this eigenvalue is given by formula $\xi_i = -\alpha_i/\sqrt{\alpha_i^2 + \Omega_i^2}$. In the case of the EPS it is considered that the damping is satisfactory if the damping ratio for the least damped mode is $\xi \geq 0, 05$ for normal states and $\xi \geq 0, 03$ for post-contingency states (Section 14.2.3).

When stability studies are performed by simulation of power system dynamic response, the following factors can be used to evaluate damping of power swings.

Damping decrement is the quotient of two consecutive peak values with the same sign: x_{k+2}/x_k, where k is the number of the peak value of $x(t)$ waveform being evaluated. In the case of a power system, simulation of waveforms is done for time ≥10 s and it is considered that the damping is good if damping decrement is ≤(0.70 - 0.85). In the case of multi-machine power systems, because of the interaction between generating units, the power swings do not have the character of a damped sine wave and damping decrement may not be constant over time. For this reason, to assess power swings in power systems it is recommended to use the below-described settling time or halving time.

Settling time is the time elapsed from the disturbance in the EPS to the time at which a power swing enters and remains within a specified error band. Error band is specified in percentage of the highest swing magnitude. The analyses of power system dynamics uses the settling time $t_{15\%}$ corresponding to the error band equal to 15%. The damping is considered sufficient when $t_{15\%} < 20$ s, i.e. when for any generating unit after 20 s from the fault occurrence the magnitude of power swings does not exceed 15% of the highest swing magnitude. When the simulation of power system dynamic response is carried out for time < 20 s, e.g. for (5–10) s, the halving time can be used to evaluate the damping.

Halving time is the settling time for error band 50%. So it is the time after which the swing magnitude decreases two times. This definition is illustrated in Figure 13.12. In the case of power system, damping of power swings is considered satisfactory if $t_{50\%} < 10$ s, i.e. when swing magnitude decreases two times in time ≤10 s.

The above values of individual coefficients are not equivalent in the terms of disappearance of power swings. This depends on the frequency of swings. Figure 13.13 shows damped sine waves for two frequencies. For the frequency 1 Hz (typical for local power swings) the strongest damped is the sine wave for the damping ratio $\xi = 0.05$ (Figure 13.13a) and the least damped is the sine wave for settling time $t_{15\%} = 20$ s. For the frequency 0.3 Hz (typical for inter-area power swings) the strongest damped is the sine wave for halving time $t_{50\%} = 20$ s and the least damped for the damping decrement 0.85 (Figure 13.13b). For the frequency 0.3 Hz sine waves for the damping ratio $\xi = 0, 05$ and for settling time $t_{15\%} = 20$ s overlap. From a practical point of view, the easiest way is to use the halving time.

Figure 13.12 Illustration of halving time definition.

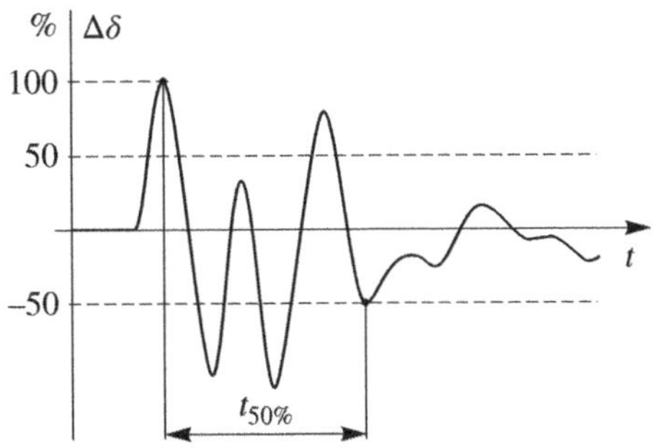

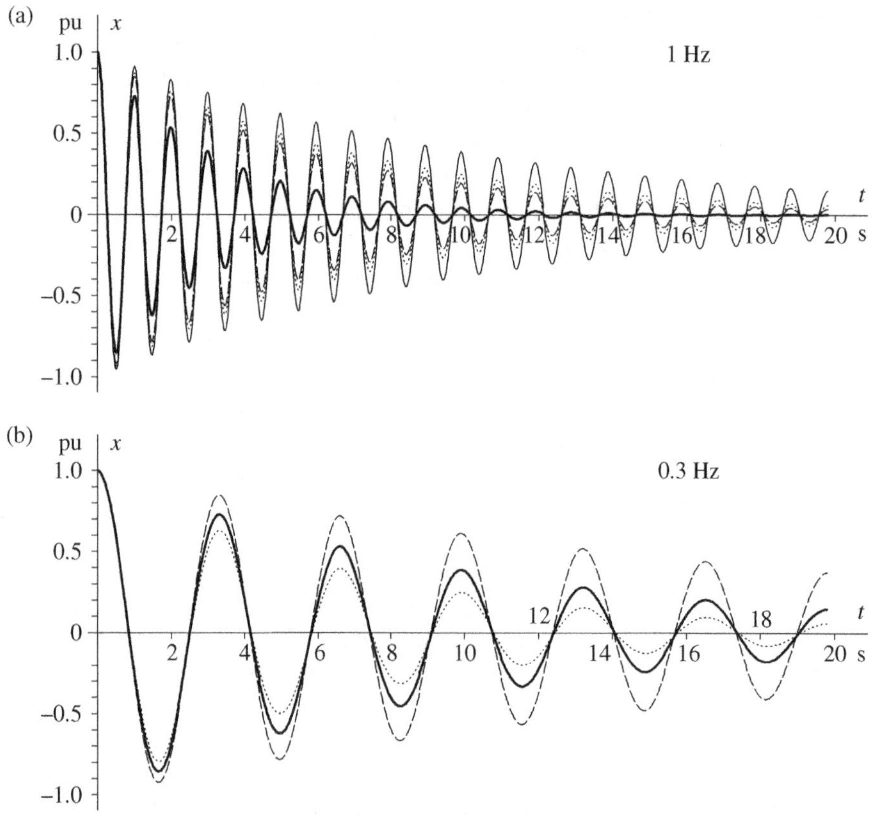

Figure 13.13 Damped sine waves for frequency: (a) 1 Hz; (b) 0.3 Hz. Damping ratio $\xi = 0.05$ – bold continuous line; damping decrement 0.85 – dashed line; settling time $t_{15\%} = 20$ s – thin continuous line; halving time $t_{50\%} = 5$ s – dotted line.

To assess the power swings in power systems, one can also use the Prony analysis discussed in the appendix (Section A.4). Prony analysis has the advantage of being able to estimate damping factors regardless of the frequency, phase, and magnitude of the signal.

14

Stability Studies in Power System Planning

Power system planning may be subdivided into operational planning (day-to-day operation) and expansion planning. In expansion planning three time horizons are distinguished: long term (≥15 years), medium term (10 years), and short term (≤5 years). Transmission system operators (TSOs) are obliged to outline and publish a master plan presenting the ability of the electric power system (EPS) to serve future demands. Contents of such a plan, the scope of studies, and the time horizon are usually defined and approved by the entity responsible for planning or regulatory authority. Most TSOs outline a master plan with a time horizon in (5–10) years ahead and update such a plan every year or every (2–5) years. The creation of the master plan begins with the development of a number of planning scenarios by TSO and all stakeholders taking into account the growth in the power demand and various future environmental and policy objectives. Scenarios include relevant generation increase, possible locations of new power plants, alternatives of network development, and expected wind power penetration according to regional wind power plans and resources. For each scenario detailed network analyses are carried out to check the reliability of the EPS and its ability withstand to various disturbances. Such analyses are briefly described in this chapter.

14.1 Purposes and Kinds of Analyses

Long-term planning concentrates on the power system's capacity expansion and options to cover the growth in the power demand. The network studies are carried out in a very limited way without studies of power system stability or with simplified stability assessment based on the criteria described in Section 6.1.6 and 8.4.1. Network and stability studies in a wide range are performed in short- and medium-term planning. The scope and level of details of the performed analyses depend on the planning time horizon. The shorter the time horizon, the greater the emphasis on the scope and details of the network and stability analyses.

Generally, planning studies may be subdivided into static and dynamic analyses. The scope of these analyses depends on investigated aspects. Figure 14.1 shows that, in the majority of cases, dynamic studies are carried out to check the impact on the system from the connection of new power plants (90%) or the construction of new lines, either internal (70%) or cross-border (55%). The examination of the dynamic impact of new wind farms connected not only directly on the transmission level (38%) but also on the underlying distribution level (23%) turns out to be of growing importance. Dynamic analyses are also required for other reasons (28%) related, for example, to the design of special protection systems (SPSs), installation of static VAR compensator (SVC), series compensations, high voltage direct-current (HVDC) links, and flexible AC transmission systems (FACTS) devices.

Concerning connection of new power plants the common criterion adopted to decide whether dynamic analyses have to be executed or not is based on the size of the new plant. Simplified angle stability criteria based on the short-circuit power (Section 6.1.6) may also be used to limit the multiple-case analysis of steady-state and transient

Power System Dynamics: Stability and Control, Third Edition. Jan Machowski, Zbigniew Lubosny, Janusz W. Bialek and James R. Bumby.

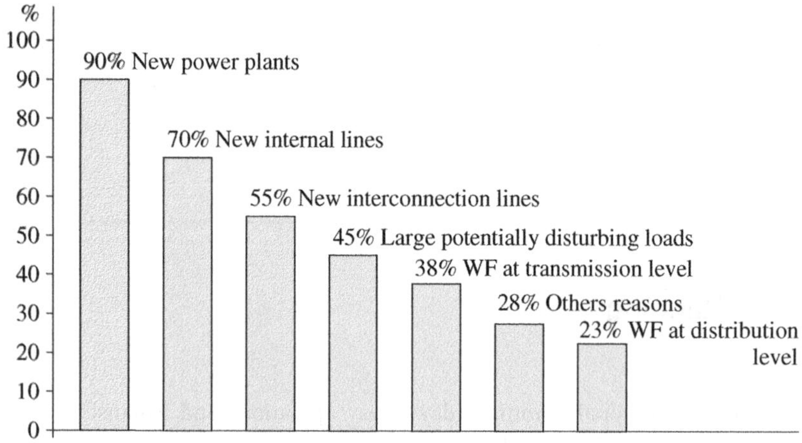

Figure 14.1 Main aspects investigated through dynamic analyses. *Source:* Reprinted with permission from CIGRE Technical Brochure No. 312.

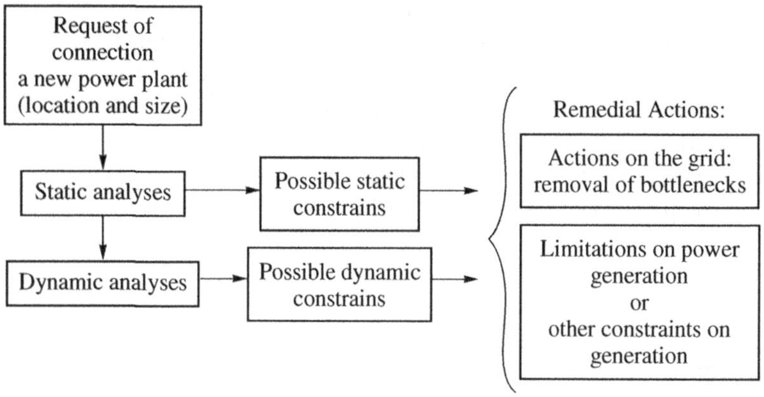

Figure 14.2 Flow to be followed for examination of the impact of a new power plant on the grid. *Source:* Reprinted with permission from CIGRE Technical Brochure No. 312.

stability. Figure 14.2 shows in a highly synthetic way the planning process followed by the identification of a possible impact of a new power plant on the grid. To begin this multi-step process the planners and stakeholders have to determine a number of possible locations for the new power plant and its size. Then, for each agreed variant, the static and dynamic analyses are carried out and possible constraints or bottlenecks must be identified and remedial actions must be proposed and verified. A comparative analysis of all variants must be presented in a report to support the final decision. Similar multi-step planning procedure is performed in the case of other kinds of aspects depicted in Figure 14.1.

14.1.1 Static Analyses

Static analyses concern the ability of high voltage (HV) networks to transmit and distribute electrical energy in a secure way without violation of voltage limits and without overloads of lines and transformers. Static analyses include:

- power flows
- reactive power compensation
- short circuits
- reliability assessment.

Power flow analyses are carried out to check voltages and currents in network elements and to compare them with prescribed limits. If limits are violated the possible bottlenecks on power flows must be identified and countermeasures must be proposed. The representation of power system elements should include models of transmission lines, transformers, generators, reactive sources, and any other equipment which can affect power flow or voltage, such as the automatic voltage regulator (AVR) of generators and reactive power capability curves (Section 3.3.4), tap-changing control of transformers, and off-nominal transformation ratio (Section 3.2). Power flow analysis must be carried out for peak and off-peak loads and other binding conditions justified by conditions of the power system operation. Power factors for delivery points must be established by the active and reactive load forecast.

Analyses of reactive power compensation complement the load flow analyses and define the preliminary location and size of the shunt VAR compensation. Reactive compensation designed to operate on the transmission system should be modeled on the low voltage side of the supply transformers. Reactive compensation designed to operate on the feeder circuits may be netted with the load.

Short-circuit studies are performed to determine the short-circuit current duties of circuit-breakers and other equipment. Results of these studies can also be used to determine appropriate shunt impedances for modeling unbalanced faults in transient stability studies (Section 6.1.1 and 13.5). For such calculations all generators and all transmission system facilities are assumed to be in service and operating as designed. Special consideration may be necessary when assessing generating unit breakers.

Reliability analysis carried out for power system expansion planning assesses deterministic reliability by defining the network topology, outage conditions, load, and generation conditions that the transmission system must be capable of withstanding safely. Reliability analyses must provide an assessment of reliability indexes and identification of possible network bottlenecks and related costs. The transmission system is designed to meet these deterministic criteria to ensure the reliability and efficiency of electric service on the bulk power system and also with the intention of providing an acceptable level of reliability to the customer.

14.1.2 Dynamic Analyses

Dynamic analyses concern the ability of an EPS to withstand various types of disturbances including the influence of network configuration, loading conditions, and action of control systems. Dynamic analyses may consist of the following studies:

- transient angle stability
- steady-state angle stability
- voltage stability
- frequency stability
- subsynchronous resonance (SSR).

Figure 14.3 shows the outcome of the survey carried out by CIGRE among power system planners. As can be seen, a wide spectrum of different dynamic phenomena are taken into account, while in the past only transient stability was examined. Nowadays, most studies are performed on angle transient and steady-state stability and on voltage stability. The smallest interest of planners is for SSR. In the planning process such analyses are carried out only occasionally to identify possible countermeasures to decouple identified resonance between network oscillation frequencies and mechanical oscillation frequencies of the generating units.

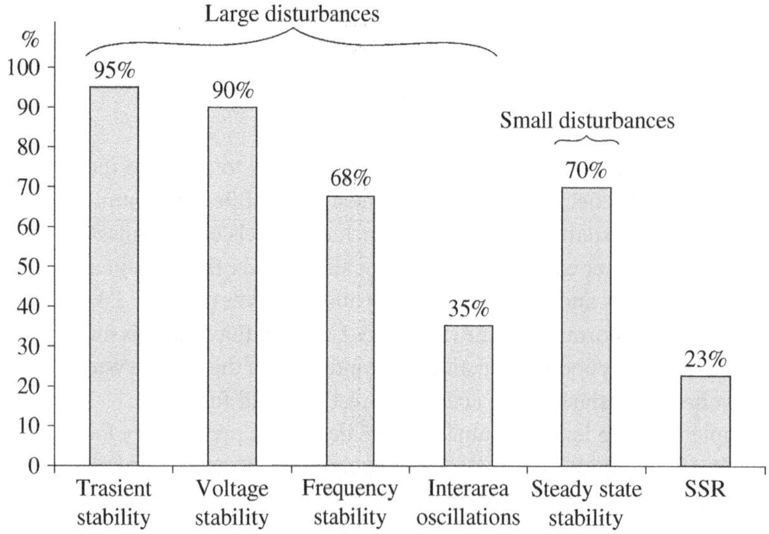

Figure 14.3 Dynamic phenomena examined in planning studies. *Source:* Reprinted with permission from CIGRE Technical Brochure No. 312.

Stability is one of the necessary conditions of power system safety. Transient angle stability is checked and assessed on the basis of the simulation results of power system dynamic response (Chapter 13). Usually, a multiple-case simulation is performed to find the critical fault clearing time and calculation of the transient stability margin (Section 6.1.1). Simulation results can be also used to assess damping of power swings using one of the following parameters described in Section 13.6: damping decrement, settling time, halving time.

Detailed study of the steady-state stability may be carried out with the use of modal analysis (Section 12.1) to enhance damping of power swings through appropriate location and tuning of PSS and AVR of synchronous generators and other control loops associated to HVDC and/or FACTS devices. Such analyses are performed rather for short-term planning (when detailed parameters of controlled devices are known) than for medium-term planning (when there is an uncertainty of system parameters and its operation conditions).

Simplified steady-state stability conditions for unregulated system models (Section 12.2) are used by TSOs to assess various conditions of active power dispatch and determine the transfer capability of the network (Section 12.2.4). Such studies are very suitable for medium-term planning when decisions must be undertaken on transmission network expansion. System operators often want to know the maximum power that can be exported from a given power plant to the system (assessment of steady-state stability limits) or from one subsystem to another (assessment of inter-area power transfer). Studies can be carried with the use of modified load flow program and the stepwise method (continuation power flow) in which in each step the active power is increased in one area and decreased in another area. Simplified steady-state stability condition is checked by calculation of the determinant of the Jacobi matrix (Section 12.2.4).

In the past, mainly the angle stability was examined by power system planners. However, increased load demand and market economics pushed some transmission networks closer to their voltage stability limit. This fact and several blackouts with voltage problems, which happened in the world (Section 8.5.2), are the reasons for which nowadays also various voltage stability studies are carried out. Most of power system planners prefer to assess the voltage stability on the basis of the nose curves (Section 8.4.2), which can be drawn stepwise by a continuation power flow program.

Nowadays, also long-term dynamics and frequency stability are investigated in the planning process to analyze more complex interactions of control loops that could originate instability phenomena in several tens of seconds after the occurrence of a fault. Simulation of frequency control (Section 9.6) enables frequency stability after sudden disturbance in the power balance (caused, e.g. by tripping of large generating units) and the compliance of a load shedding scheme with frequency deviations to be checked. However, not all TSOs undertake the frequency stability studies in the planning process.

14.2 Planning Criteria

In the majority of countries, the planning criteria for power system expansion and day-to-day operation are defined and approved by the entity responsible for planning. In some countries the planning criteria are approved by the ministry of energy or by the regulatory authority. Moreover, in the case of interconnected power systems the planning criteria approved at national level are coordinated with the criteria adopted by the pool to which the country belongs.

The power system should be capable of meeting consumer demand while surviving reasonably anticipated contingency conditions without violating specified thermal, voltage, and stability limits or experiencing uncontrolled cascade tripping. Therefore, formulation of planning criteria must be based on a compromise between low-cost expansion and safe and reliable power system operation. Planning criteria consist of rules and descriptions relating to the following issues:

- planning contingency events;
- permissible constraints in power system operation;
- performance standard.

A contingency event is defined as the tripping of one single or several network elements that cannot be predicted in advance. A scheduled outage is not a contingency event. Events for which the EPS must meet all requirements of performance standards are referred to as the *planning contingency events.*

For some severe events, the planning criteria may permit some constraints in power system operation such as curtailment of transmission service and/or load loss.

Some TSOs use separate planning criteria for operational planning and short- and medium-term expansion planning. However, most TSOs use the same criteria.

"Performance standard" relates to the acceptable power system response to contingency events and voltage constraints at buses, allowed frequency changes, thermal loading of network elements, and power system stability.

14.2.1 Contingency Events and Initial Conditions

Contingency events are events which cannot be predicted in advance, e.g. a short circuit in the network and its clearance or a sudden loss of generating unit. Dynamic performance of an EPS depends on the severity of a given contingency (Chapter 6). Expansion planning of transmission network is based on a compromise between investment costs and power system security. Therefore, the categorization of contingency events according to their severity is a key issue. However, the lists of planning contingencies in many guidelines published in the world differ in both their classification as well as their naming.

An example of classification shown in Table 14.1 is based on the probability of the event's occurrence. However, it is worth emphasizing here that the probability of some contingencies is not the same in all regions of the globe. For example an event such as the loss of many adjacent transmission circuits on common structure is more likely to occur in hard storm or tornado areas. Therefore, the list of planning contingencies may differ among planning guidelines published in the world. Event categorized in one guideline as less credible may be categorized in another guideline as noncredible, etc.

Table 14.1 Contingency events classification.

Type of contingency event		Performance standard	Constraints	Planning analyses
Planning	Credible	Met unconditionally	Not permitted	Obligatory in a full scope
	Less credible	Met with help of transmission system adjustments	Only for most severe contingencies	
Extreme	Noncredible	Not required but desirable	Permitted	Recommended in a limited scope

Typical often and quite often occurring events are referred to as *credible* and *less credible* contingency events, respectively. Usually, the credible contingency event is defined as the loss of a single element such as a single line, a single generating unit, a single transformer or two transformers connected to the same bay, respectively, a phase shifting transformer, a large voltage compensation installation, or a DC link. For credible events all requirements of the performance standard must be met unconditionally.

The less credible contingency event is defined as the uncommon loss of the particular elements such as a double line (which refers to two lines on the same tower over a long distance), a single busbar, the common cause failure with the loss of more than one generating unit, including large wind production, and common cause failure of DC links. For less credible events all requirements of the performance standard must be met, but in some cases with the help of transmission system adjustments.

Both (credible and less credible events) are named the *planning contingency events*. For planning events the analyses must be carried out in full scope.

Events other than planning contingencies are referred to as *noncredible contingency events* or *extreme events*. These are generally events rare in occurrence, such as the combination of a number of credible contingencies, e.g. the independent and simultaneous loss of at least two lines, the loss of a total substation with more than one busbar, the loss of a total power plant with more than two generating units, the loss of a tower with more than two lines, etc. The extreme (noncredible) contingencies are not taken into account, owing to exceeding dimensioning efforts in the single TSO's network. For such events the violation of some performance standards is accepted and the planning analyses do not need to be carried out in full scope. However, it is useful to do some studies in a limited scope in order to assess the risk of cascade tripping and/or blackouts. Cascade tripping is defined as the uncontrolled loss of a sequence of network elements caused by an initial contingency event.

In some planning guidelines other names are used to categorize the contingency events. For example instead of credible, less credible, and noncredible events the names *normal*, *exceptional*, or *out-of-range events* may be used, respectively.

When creating the planning criteria a key issue associated with contingency events is the type of short circuit which causes the loss of power system elements. As explained in Section 6.1.1, the short-circuit type and its distance to the power plant are decisive for transient stability. The hardest disturbances for the EPS are three-phase short circuits in a short distance to the power plant and the easiest are single-phase short circuits. For this reason, the planning criteria must precisely define for what type of short circuit a given contingency event must be analyzed.

Transient stability must be assessed on the basis of the realistic duration of a short circuit (clearing time) resulting from used protection systems and circuit-breakers.

The clearing time depends on the operation time of the protection device and the opening time of the circuit-breaker. It is said that the fault is cleared with a *normal clearing time*, when it is isolated by operation of the main protection. A fault isolated by the operation of the back-up protection is referred to as a fault cleared with a *delayed clearing time*.

The typical operation time of main protections used in HV networks is about (1–2) cycles. Opening time of circuit-breakers is about (2–3) cycles. Both together give the normal clearing time of about (3–5) cycles ([60–100] ms for 50 Hz). Many planning guidelines mention about (80–120) ms as the typical normal clearing time.

The delayed clearing time depends on the type of back-up protection and is about (10–20) cycles ([200–400] ms for 50 Hz). Delay caused by breaker failure protection (BFP) depends on the substation layout and BFP logic (Kasztenny and Thompson 2011). Example values of the absolute maximum fault clearing time are: 300 ms for double bus single breaker scheme and 200 ms for the breaker-and-a-half scheme.

Depending on the number of power system elements switched off (as a result of a fault isolation by protection system), a given contingency event may be classified as a *single event* or *multiple event*.

In the initial condition (prior to the considered contingency event) an EPS may operate in normal state (*N*-0) or in (*N*-*M*) state, where *M* is the number of the scheduled outages (e.g. maintenance). The (*N*-*M*-*K*) state includes *M* scheduled outages and *K* power system elements switched off as a result of the considered contingency event. For example, (*N*-1-2) means the overlapping of a double contingency event (*N*-2) and a single scheduled outage in the initial state (*N*-1).

In the classification of contingency events a fault clearance by back-up protection or BFP may be treated in two ways. In some planning guidelines, the tripping of one network element caused by back-up protection is treated as a single event. In other guidelines (e.g. NERC 2011) both primary devices (generators, lines, transformers, compensators, and circuit-breakers) and secondary devices (protections and telecommunication equipment) are considered power system elements. With such a definition, the tripping of a network element as a result of the operation of a back-up protection or BFP is considered a double contingency (*N*-2).

For the (*N*-*M*-*K*) state and some contingency events it may be assumed that the power system condition resulting from a scheduled outage has been improved prior to a given contingency by manual or automatic actions of *transmission system adjustments*. Typical transmission system adjustments include:

- automatic taps changing on regulating transformers;
- switching on/off VAR compensators;
- automatic or manual re-dispatch of power generation;
- network reconfiguration by busbar sectionalizing and network switching.

Generally, such corrective actions by transmission system adjustments should not result in any load loss and/or curtailment of transmission service.

Each individual TSO is obliged to define (on the basis of the probability assessment and years of experience) the list of credible (normal), less credible (exceptional), and noncredible (out-of-range) contingency events and types of short circuits, which must be analyzed to check all requirements of the performance standard. Such a list must be also augmented with allowed constraints in power system operation (such as interruption or curtailment of transmission service or load loss) and with description of the initial condition for which a given event must be analyzed.

14.2.2 Allowed Constraints in System Operation

Transmission service is the reservation of energy transmission across the grid from the point(s) of delivery to the point(s) of receipt. *Firm transmission service* is the highest-quality (priority) service offered to customers under a *filed rate schedule* that anticipates no planned interruption. By buying firm transmission service a utility (e.g. a power plant that runs almost every hour of a year) is more likely (but not guaranteed) to be able to serve all of its customers. Nonfirm transmission service is provided on an as-available basis and is subject to interruption or curtailment before firm transmission service. If either firm transmission service or nonfirm transmission service could be curtailed to solve a problem on the transmission grid, nonfirm transmission service would be curtailed first. Planning criteria must contain information for which severe contingency events the interruption or curtailment of the firm transmission service is allowed.

In planning criteria it is necessary to distinguish between consequential (load or generation) loss and nonconsequential (load or generation) loss. *Consequential load loss* is defined as a load that is no longer served by the EPS as a result of transmission facilities being removed from the service by a protection system to isolate the fault. *Nonconsequential load loss* is defined as a non-interruptible load loss other than consequential load loss and the load disconnected from the system by end-user equipment. Similar definitions may be formulated for *the consequential generation loss* and *the nonconsequential generation loss.*

Consequential (load or generation) loss is acceptable as a consequence of any contingency event. Nonconsequential (load or generation) loss is acceptable for extreme (out-of-range) events and is not allowed for the planning contingency.

Entities responsible for planning or regulatory authorities accept the nonconsequential load and/or generation loss only for extreme contingency events leading to power system emergency states or blackouts. Load curtailment as a countermeasure against deep frequency drop or frequency collapse (Section 9.4) can be done automatically by the *under-frequency load shedding*. Similarly, the *under-voltage load shedding* may prevent against voltage collapse (Section 8.3). For such events the nonconsequential load and/or generation loss may be forced also by SPSs.

An SPS is defined as an automatic protection system designed to detect abnormal system conditions and take corrective actions to maintain power system stability, acceptable voltage, or power flows. Such action may include changes in power demand, and changes in real and/or reactive power generation or network re-configuration. An SPS does not include under-frequency or under-voltage load shedding or fault and out-of-step relaying. An example of an SPS designed to prevent cascade tripping caused by severe overloads of transmission lines connecting a large power plant with the rest of the system is described by Robak et al. (IEEE 2018). Such an SPS alleviates the overloads by generation curtailment by partial load rejection or tripping generating unit to auxiliaries. Another SPS may use tripping to auxiliaries a part of generating units (Section 10.4) in order to prevent a loss of synchronism after extreme (e.g. multiple) contingency events.

Most of planning guidelines accept temporary use of SPS in transition periods when construction of planned transmission infrastructure is delayed and reliability risk is temporary or during unscheduled multiple outages that have occurred for reasons independent of a TSO (e.g. caused by natural disasters). Permanent use of an SPS is commonly accepted for extreme contingency events to improve power system reliability and when construction of new transmission infrastructure is not possible, because of physical constraints, or obtaining permits is not feasible or the expense associated with permanent transmission upgrades is not justified. Some (but very few) TSOs also use special protections to improve power system reliability for less credible (exceptional) contingencies, such as the loss of a bus section.

14.2.3 Performance Standard

Performance standard concerns static and dynamic response of an EPS to contingency events which may occur at normal initial state (N-1) or at (N-M) states with M scheduled outages. Generally, most of performance standard relates to the following issues:

- thermal constraints on lines and transformers (normal and emergency ratings);
- voltage constraints at buses;
- transient voltage dips (percentage deviation and duration);
- short-circuit power;
- voltage stability (margin to voltage collapse);
- steady-state stability (margin to steady-state instability, damping of small swings);
- transient stability (critical clearing times [CCTs], margin to transient instability, damping of large swings);
- post-contingency cascade tripping.

Thermal ratings of network elements may be divided into *normal* (long-term) *ratings* I_∞ and *emergency* (short-term) *ratings* I_e for a limited period t_e. Both the normal and emergency ratings of the network elements depend on many factors, such as ambient temperature, wind speed, and solar exposure. Because of this, the actual network loadability changes dynamically with changing environmental conditions. For planning purpose it is usually assumed that the emergency ratings are $\alpha > 1$ times larger than the normal ratings. For example, $I_e = 1.2 \cdot I_\infty$ and $t_e = 20$ min for overhead lines and $I_e = 1.5 \cdot I_\infty$ and $t_e = 30$ min for transformers.

All planning guidelines do not permit loading within emergency ratings for normal states (*N*-0). For single outages or contingency events (*N*-1) some planning guidelines also do not accept loading within emergency ratings, but some allow it conditionally with the assumption that loading with normal ratings can be reset in time $<t_e$. Providing the same assumption, most planning guidelines permit loading within emergency ratings for overlapping single outage and single contingency events (*N*-1-1) or double contingency events (*N*-2), such as the loss of a two-circuit transmission line. For multiple contingency events and a loss of bus section the loading within emergency ratings is permitted.

Almost all planning guidelines determine voltage constraints at buses. For example, for normal states and conditions following all credible contingency events, the voltage at each load node must remain within ±5% of voltage ratings V_n and for generation node within (0–5)%. Under conditions resulting from less credible contingency events voltage must remain within ±10% of V_n for each load node and ±5% for generation node, respectively. As regards transient voltage dips the guidelines give values for percentage deviation and allowed duration. For generation nodes supplying the auxiliary service of generating units, some guidelines determine voltage versus time characteristics HVRT and LVRT, as shown in Figure 2.8 (Section 2.3.2).

Short-circuit currents are calculated mainly for the purpose of checking the ability of breakers to interrupt the circuits and isolate the fault. In planning studies the short-circuit power S''_K may also be used to select load nodes for detailed voltage stability analysis (Section 8.4.1) or power plants for detailed transient stability analysis (Section 6.1.6). A part of the nodal impedance matrix (for negative and zero-sequence components) may be used to find parameters of equivalent circuits to model unbalanced faults and phase interruptions, as described in Section 13.5.

Planning guidelines recommend to perform voltage stability studies only for nodes or load areas for which in the post-contingency states a voltage problem occurs. Such nodes or load areas can be identified on the basis of the following criteria:

- For normal states (*N*-0), when the voltage is $\geq V_n$ for each generation node and is $\geq 0.95\ V_n$ for load nodes, the voltage stability analyses do not need to be undertaken.
- For post-contingency states, when the voltage at load nodes is $\leq 0.94\ V_n$ voltage stability analysis must be performed. Moreover, when the voltage at load nodes is $\leq 0.9\ V_n$, a remedial action plan should be prepared.
- For post-contingency states when the short-circuit power at transformer terminals supplying distribution network is $S''_K < 5 \cdot S_{rT}$ (where S_{rT} is the rated power of the transformer) it is recommended to perform voltage stability analysis and compare the critical power demand (top of the nose curve) with emergency loadability of the transformer.

For such nodes or load areas the voltage stability can be assessed with the help of nose curves (Section 8.4). However, only very few planning guidelines refer to values of required voltage stability margin based on the critical power demand (Eq. [8.80]). Values 10% for normal states and 5% for post-contingency state recommended in some guidelines seem to be too small from the point of view of power system security.

Modal analysis (Section 12.1.4) is rarely used in planning studies for the purpose of checking the steady-state stability condition. Most planning guidelines assume that for planning contingency events the post-contingency states are stable for small disturbances and are required to evaluate the transient stability only. Modal analysis, if described in such guidelines, is recommended as the method to calculate damping ratios (Section 12.1.3).

The damping of power swings is supposed to be satisfactory (Section 5.4.6) if for each electromechanical mode the damping ratio $\zeta \geq 0.05$ for the normal state and $\zeta \geq 0.03$ for the post-contingency state.

Simplified steady-state stability condition (Section 12.2.2) based on the determinant of the Jacobi matrix is used in planning studies to determine the capability of the network to transfer electrical energy from one part of the system to another, for example in an interconnected power system from one subsystem to another (inter-area power transfer) or from a given power plant to the system. For such studies the steady-state stability margin (Section 12.2.4) may be used. Very few planning guidelines refer to the required values of such a margin. Values 20% for normal states and 10% for post-contingency state recommended in some guidelines seem to be too small from the point of view of power system security.

In regards to transient stability, in planning guidelines there is a very common requirement that a power system must be stable for all planning contingency events. The transient stability margin (Section 6.1.1) must be ≥20% for single contingency events and ≥10% for overlapping events (*N*-1-1) or double (*N*-2) contingency events. This means that the CCT must be, respectively, at least 20 and 10% larger than the real duration of the fault. This applies to planning events for which the fault is isolated with both a normal and a delayed clearing time.

Simulations of power system dynamic response performed to check transient stability are often used also to evaluate the above-mentioned transient voltage dips and damping of large power swings. Simulated voltage changes are compared with the LVRT and HVRT characteristics (Section 2.3.2) to check the fault ride through of synchronous generating units. To evaluate the damping of large power swings in a transient state, various coefficients may be used such as damping decrement, settling time, or halving time (Section 13.6). Particular guidelines recommend slightly different values of these coefficients. For example, in some guidelines damping is considered good if damping decrement is ≤0.85 and in others if ≤0.70. When settling time is used for evaluation it is assumed that damping is satisfactory if in time 20 s the magnitude of swings becomes less than 15% of its maximal post-fault value. In other guidelines damping is considered good if the halving time is ≤10 s. Some (but very few) guidelines distinguish between the damping of local power swings with frequency of about (0.8 - 1.2) Hz and inter-area power swings with frequency of about (0.25 - 0.40) Hz and impose stricter requirements, e.g. halving time ≤ 5 s for local swings and ≤ 7 s for inter-area swings.

Cascading failure in an EPS could result in a widespread interruption of transmission service and area or system blackout. In all planning guidelines cascade tripping is absolutely forbidden as a result of all planning contingency events. Cascade tripping can be caused by some extreme contingency events such as a loss of three or more network elements, a loss of entire substation, or entire plant with three or more generating units, etc. Most planning guidelines recommend that some studies be carried out for extreme contingencies in order to assess the risk of cascade tripping and/or blackouts. In order to diminish the risk of cascading failures, some TSOs use above-mentioned SPSs (Section 14.2.2) alleviating severe overloads and preventing the loss of synchronism after extreme contingency events.

14.2.4 Examples

A planning guide must give recommendation for the evaluation of all the above issues for all categories of planning contingency events and allowed constraints in EPS operation, such as the curtailment of firm transmission service or nonconsequential load/generation loss. For this reason, it is difficult to create a consistent informative text.

Approaches used in published guidelines differ in terms of form. Usually, the text is divided into parts relating to particular issues. Recommendations are summarized in tables presenting static and dynamic requirements. A fundamental difference between guidelines is that some of them are based on the classification of power system performance and some on the categorization of contingency events. For example, the approach presented by CERTS (2004) is based on the classification of power system performance and the approach presented by NERC (2011) is based on the categorization of contingency events.

In CERTS (2004) the power system performance is specified as five discrete levels: A, B, C, D, and E in descending order of frequency and increasing order of severity:

A) Performance should produce no significant adverse effects outside of the system in which the disturbance occurs. Post-contingency loss of a single element, such as a generator, one circuit, one transformer, DC monopole. Loading with emergency ratings is acceptable. Loss of firm load is not acceptable.
B) Performance allows for some adverse effects that may occur outside of the system in which the disturbance occurs. Post-contingency loss of bus section. Loading with emergency ratings is acceptable. Loss of firm load is not acceptable.
C) Performance allows substantial adverse effects outside of the system in which the disturbance occurs. Post-contingency loss of two elements, such as two generators, two circuits, DC bipole. Loading with emergency ratings is acceptable. Loss of firm load is acceptable.
D) Performance seeks only to prevent cascading and the subsequent blackout of islanded areas. Some additional adverse effects may occur, including load shedding. Post-contingency loss of three or more circuits on common right of way (ROW). Cascading failure is not permitted.
E) Substantial loss of customer demand and generation in a widespread area or areas. Portions or all of the interconnected systems may or may not achieve a new, stable operating point. Post-contingency loss of multiple elements such as: three or more extra high voltage (EHV) or HV circuits, entire substation, entire plant with three or more generating units, etc. Studies must evaluate the risks and consequences.

For such levels performance must be checked for the following contingency events: no fault, three-phase short circuit with normal clearing time, single-phase short circuit with delayed clearing time, DC disturbance. In cases where a prior outage exists on a system, system adjustments will be made to allow the system to meet the required performance specified for the next disturbance. When multiple elements are specified, they are assumed to be lost simultaneously.

CERTS (2004) describes allowable effects on a system, such as post-transient voltage deviation, transient voltage dips, minimum transient frequency, etc.

NERC (2011) specifies the planning contingencies as eight categories and for each of them an initial condition, event, and fault type is described. The following categories have been distinguished:

P0. No contingency and no event. Interruption of firm transmission service is not allowed. Nonconsequential load loss is not allowed.
P1. Single contingency. Normal initial state. Three-phase fault with normal clearing time and post-contingency loss of one system element, such as generator, transmission circuit, transformer, and shunt device. Interruption of firm transmission service is not allowed. Nonconsequential load loss is not allowed.
P2. Single contingency. Normal initial state. Single-phase fault with normal clearing time at bus section or internal breaker fault. Interruption of firm transmission service in an EHV network is not allowed but it is allowed in an HV network. Nonconsequential load loss in an EHV network is not allowed but it is allowed in an HV network.
P3. Multiple contingency. In initial state, a loss of generating unit followed by system adjustment. Three-phase fault with normal clearing time and post-contingency loss of one system element, such as generator, transmission circuit, and transformer. Post-contingency (single-pole fault) loss of single pole of DC line. Interruption of firm transmission service is not allowed. Nonconsequential load loss is not allowed.
P4. Multiple contingency (fault plus stuck breaker). Normal initial state. Single-phase fault with normal clearing time and post-contingency loss of multiple elements caused by stuck breaker attempting to clear a fault on one of the following elements: generator, transmission circuit, transformer, shunt device, or bus section. Interruption of firm transmission service in an EHV network is not allowed but allowed in an HV network. Nonconsequential load loss in an EHV network is not allowed but allowed in an HV network.

P5. Multiple contingency (fault plus relay failure to operate). Normal initial state. Delayed clearing of a single-phase fault. Post-contingency loss of multiple elements caused by stuck breaker attempting to clear a fault on one of the following elements: generator, transmission circuit, transformer, shunt device, bus section. Interruption of firm transmission service in an EHV network is not allowed but allowed in an HV network. Nonconsequential load loss in an EHV network is not allowed but allowed in an HV network.

P6. Multiple contingency (two overlapping single contingencies). In initial state a loss of one system element (transmission circuit, transformer, shunt device, single pole of a DC line) followed by system adjustment. Three-phase fault with normal clearing time and post-contingency loss of one system element, such as transmission circuit, transformer, orshunt device. Post-contingency (single-pole fault) loss of single pole of DC line. Interruption of firm transmission service is allowed. Nonconsequential load loss is allowed.

P7. Multiple contingency (common structure). Normal initial state. Single-phase fault with normal clearing time and post-contingency loss of any two adjacent circuits on common structure or loss a bipolar DC line. Interruption of firm transmission service is allowed. Nonconsequential load loss is allowed.

NERC (2011) has several footnotes that contain explanations and comments that are important for power system planners. Extreme contingencies are not categorized. For extreme contingency events studies must evaluate the risks and potential consequences.

Guidelines based on publication by Robak et al. (2017) are briefly described below. Table 14.2 and 14.3 relate to static analyses and Table 14.4 and 14.5 to dynamic analyses, respectively. Description of planning criteria is based on the categorization of contingency events.

For static analyses four planning contingency levels are specified in Table 14.2 and three levels in Table 14.3 for extreme events. The performance standard for static analyses is similar to that described by CERTS (2004) or NERC (2011), except that for events S2 and S3 it allows a limited loss of generation as a method to alleviate severe

Table 14.2 Planning contingency events and performance standard for static analyses.

Level	Description of planning contingency event	Loading with ratings	Post-contingency voltage deviation[a]	Steady-state stability margin		Steady-state stability margin	
				Voltage[b]	Angle[c]	System adjustment	Loss of generation
S0	Normal state	normal	≅0% for PV ≤ 5% for PQ	≥ 10%	≥ 20%	No	No
S1	Loss of single element (single circuit line, one circuit of multi-circuit line, transformer, generator, shunt device)	normal	≤ 5% for PV ≤ 10% for PQ	≥5%	≥10%	Yes	
S2	Loss of two elements in one area (double circuit line, two single circuit lines, two transformers, line, and transformer)	emergency		≥ 2.5%			≤ 200 MW[d]
S3	Loss of bus section	emergency			≥ 5%		≤ 750 MW[d]

[a] PV – generation nodes; PQ – load nodes.
[b] Voltage stability studies must be performed only for nodes for which a voltage problem occurs.
[c] Relates to capability of the network to transfer electrical energy from one part of the system to another.
[d] Relates to subsystem (generation about 30 GW) operating in a large interconnected system (generation about 200 GW).

Table 14.3 Extreme contingency events and allowed performance for static analyses.

		Allowed emergency actions			Allowed consequences			
						Instability		
Level	**Description of extreme contingency event**	**Transmission system adjustment**	**Load shedding**	**System islanding**	**Violation of voltage or loading limits**	**Voltage**	**Angle**	**Cascade tripping**
S4	Loss of three or more elements (one or multi-circuit transmission lines, transformers) in the same area of transmission network	Yes	Yes	No	Yes	No	No	No
S5	Loss of an entire plant with three or more generating units	Yes	Yes	Yes	Studies must evaluate the risks and potential consequences			
S6	Loss of an entire substation							

overloads. This is justified by the fact that the relevant power system operates as a subsystem in a large interconnected power system and it can be assumed that the generation lost in one subsystem can be replaced by generation in the other subsystems. Voltage stability studies must be performed (as described in Section 14.2.3) only in the case when for some load nodes a voltage problem occurs. The voltage stability margin is based on the critical power demand (top of the nose curve). The nose curve must be drawn while taking into account all reactive power limits – and especially the reactive power capability curves of all generating units supplying the considered subsystem. Steady-state angle stability analysis is performed when the maximum transfer capability of the network must be found (Section 14.2.3).

For dynamic analyses six planning contingency levels are specified in Table 14.4 and four levels in Table 14.5 for extreme events.

Level D0 concerns short circuits in single network elements and D1 is concerned with overlapping single outage and short circuit in a single network element. In order to evaluate properly the power system transient stability the short circuit should be placed consecutively close to the busbars in all network elements connecting power plants with the system. As in guidelines published by NERC (2011), for single contingency event it is assumed that the fault is of the three-phase type and is cleared with normal time (main protection) without or with automatic reclosing (if used).

Level D2 concerns a three-phase short circuit on the bus section of the power plant substation. For substations with a double bus single breaker scheme it is assumed that the fault is cleared by opening the circuit-breakers of all associated network elements and the bus coupler (Figure 14.4a). For a substation with the breaker and a half configuration, the fault is cleared by opening all circuit-breakers associated with faulted bus but the network elements stay in operation through the tie-breakers (Figure 14.4b).

Level D3 concerns a common three-phase short circuit in double circuit line connecting a power plant with the system.

Levels D4 and D5 concern the delayed clearing of the three-phase short circuits resulting from the main protection failure or operation of the BFP. For level D4 it is assumed that a three-phase fault is cleared with the time of the second zone of the distance protection. For level D5 it is assumed that one pole of the breaker has stuck in on position when attempting to clear a three-phase fault. As a result the fault is cleared by the operation of the local circuit-breaker back-up. In the case of a substation with a double bus single breaker scheme the circuit-breakers of all associated network elements and the bus coupler must be opened (similar to the case of the bus section fault). In the case of a substation with a breaker and a half scheme such contingency events must be considered both for failure of a circuit-breaker and for failure of a tie-breaker.

Table 14.4 Planning contingency events and performance standard for dynamic analyses.

		Description of planning contingency events				Transient angle stability		Acceptable remedial actions	
Level	Initial condition	Faulted network element	Fault type[a]	Clearing time	Transient voltage dip	Margin[b]	Damping[c]	SPS[d]	Loss of generation
D0	Normal	Short circuit in a single network element (single circuit line, one circuit of double circuit line, transformer)	K3	Normal	≤HVRT ≥LVRT	≥ 20%	≤ 5 s ≤ 7 s	No	No
D1	Single outage followed by system adjustments	Short circuit in a single network element (single circuit line, one circuit of double circuit line, transformer)				≥10%		Yes	≤ 200 MW[e]
D2	Normal	Short circuit on bus section							≤750 MW[e]
D3	Normal	Common short circuit in two circuits of double circuit line							≤500 MW[e]
D4	Normal	Failure of the main protection and short circuit in a single network element (single circuit line, one circuit of double circuit line, transformer)		Delayed				No	No
D5	Normal	Short circuit in a single network element and stuck breaker (one pole) in on position	K3→K1					Yes	≤750 MW[e]

[a] K3 – three-phase short circuit, K1 – single-phase short circuit.
[b] Margin based on critical clearing time (CCT).
[c] Halving time. **S**maller values for local power swings and larger values for inter-area power swings.
[d] Special protection system (SPS) (e.g. tripping of a part of generating units).
[e] Relates to subsystem (generation about 30 GW) operating in a large interconnected system (generation about 200 GW).

Table 14.5 Extreme contingency events and allowed performance for dynamic analyses.

		Description of extreme contingency event			Transient angle stability		Acceptable remedial actions	
Level	**Initial condition**	**Faulted network element**	**Fault type**	**Clearing time**	**Margin**	**Damping**	**SPS**	**Loss of load or generation**
D6	Single outage followed by system adjustments	Short circuit in a single network element and stuck breaker (one pole) in on position	K3→K1	Delayed	> 0%	≤ 15 s	Yes	Yes
D7		Common short circuit in two circuits of double circuit line	K3	Normal				
D8	Double outage in one area	Short circuit in a single network element (single circuit line, one circuit of double circuit line, transformer)						
D9	Loss of an entire plant with three or more generating units or loss of an entire substation				Studies may be performed with objective to check whether SPS can prevent against instability and/or power system islanding			

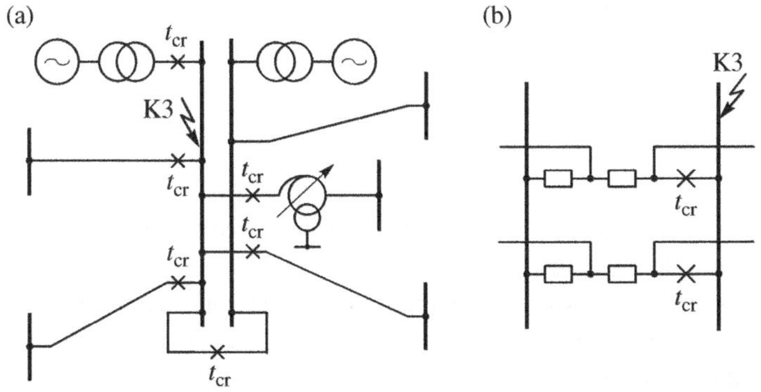

Figure 14.4 Three-phase short circuit on the bus section (a) double bus single breaker scheme; (b) breaker-and-a-half scheme.

The case of the circuit-breaker failure is illustrated in Figure 14.5a. The left part of the figure illustrates the initial situation when a three-phase fault (K3) occurs. The middle part of the figure illustrates the situation when two poles of the circuit-breaker and the tie-breaker are opened. As a result the three phase short circuit (K3) is transformed into a single phase short circuit to ground (K1) supplied through the stuck pole. The right part of the figure illustrates the final situation when after the BFP operation all other circuit-breakers are opened and the faulted line is disconnected.

The case of tie-breaker failure is illustrated in Figure 14.5b. The left part of the figure illustrates the initial situation when a three-phase fault (K3) occurs. In the middle part of the figure the circuit-breaker and two poles of the tie-breaker are opened. As a result the three-phase fault (K3) is transformed into a single-phase fault (K1).

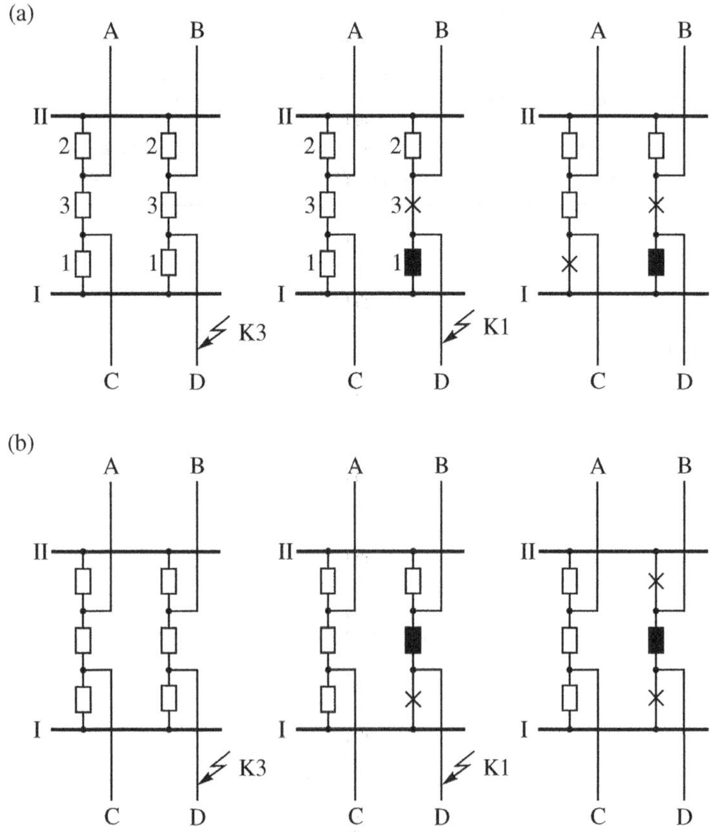

Figure 14.5 Three-phase short circuit and breaker stuck in on position (a) circuit-breaker; (b) tie-breaker.

The right part of the figure illustrates the final situation when after the BFP operation the second circuit-breaker is opened and the faulted line as well as the second line are disconnected.

It must be emphasized that for contingency events such as a stuck breaker the loss of a network elements depends on the layout of the substation. Therefore, such events must be carefully modeled for the simulation of a power system dynamic response and for the calculation of CCT.

Compared to the standard by NERC (2011), contingencies in levels D2, D3, and D5 listed in Table 14.4 are more demanding for an EPS in terms of stability than contingencies in corresponding levels P2, P7, and P4. The NERC standard requires transient stability for single-phase short circuits (K1) while standard in Table 14.4 for an EPS in terms of stability than contingencies in corresponding levels P2, P7, and P4. The NERC standard requires transient stability for single-phase short circuits (K1) while standard in Table 14.4 extends this requirements to three-phase short circuits (K3). From the other side, the NERC standard does not allow the use of a SPS for any planning contingency event while standard in Table 14.5 allows to ensure the transient stability for the three-phase faults by the use of the SPS tripping a part of generating units.

Levels D6, D7, and D8 concern extreme multiple (three or more) overlapping events. For such events in Table 14.5 it is recommended to carry out the simulations to check whether the SPS can prevent instability and/or power system islanding.

14.3 Automation of Analyses and Reporting

The planning analyses discussed above must be performed using a software package that enables advanced power system analyses such as multiple-case load flow studies, drawing nose curves, short-circuit calculations, and simulation of power system dynamic response. Analyses done with the use of these programs must be performed for all planning contingency events as well as for extreme contingency events for a large number of power system elements, such as transmission lines, transformers, generating units, shunt compensators, and busbars. Of course, such analyses are carried out for each scenario of power demand increase and locations of new electrical energy sources. As of consequence, to perform planning analyses, individual computer programs must be run very many times (hundreds of times for a midsize power system and thousands of times for a large power system).

Today, it is difficult to imagine that the planner could run individual specialized programs manually that many times. Therefore, the analysis must be automated. For this purpose, *scripting languages* (also sometimes referred to as *very high-level programming languages* or *control languages*) are used. Programs (scripts) written in such languages allow for an automated execution of user-defined tasks. For example a program created in a scripting language may execute repetitive calculations of load flow for all possible single or multiple outages of transmission network elements and/or may execute repetitive simulations of power system dynamic response to find CCTs for all planning contingency events. Of course, with a huge number of results, such programs that control the execution of the analysis must also perform post-processing of results, for example overloaded network element summaries and CCTs or stability margins values summaries, etc.

Currently, professional software packages designed for power system analyses are integrated with generally available scripting languages or have their own internal scripting languages. For example, package PSLF uses internal Engineering Process Control Language (EPCL), PowerFactory uses internal DigSILENT Programming Language (DPL), PSS/E uses scripting language called Python. Gonzalez-Longatt and Rueda (2014) describe several applications of scripting language DPL for various power system analyses.

The results of the transmission system planning study must be elaborated in detail in a concise report, the general draft of which is defined by the planning or regulatory authority. This report should demonstrate the ability of the EPS to serve future demands and should describe the assumptions, procedures, problems, alternatives, comparison of scenarios, conclusions, and recommendations resulting from the study. The recommendation should include the resolution of any potential violation of the design criteria. The recommended corrective (remedial) action plan should be based on a composite consideration of factors, such as safety, the forecasted performance and reliability, environmental impacts, economics, technical preference, schedule, availability of land and materials, acceptable facility designs, and complexity and lead time to license and permit.

15

Optimization of Control System Parameters

15.1 Grid Code Requirements

In the era of development of competitive energy markets, the funds for the development of power systems and their maintenance has become limited. Additionally, for modern electric power systems (EPSs) the increase of the use of RES in power systems has prompted changes within the operation of the EPS, related to technical and economic issues. This means that modern EPS power systems with RES and classical units (i.e. grid, loads and control systems) operating together, greatly influence the power system's dynamic properties. Because of this, EPSs are operating closer and closer to their technical limitations, including their stability limit. The ability to maintain or improve an EPS's dynamic properties depends on its features, including the structure, dynamic properties of the objects (generators, etc.), their regulators and on the operating point.

Despite the structural changes in EPSs related to the development of RES, regulators of high-rated power synchronous generators may still be perceived as elements determining the dynamic properties of EPSs. The other types of power plants, i.e. connected to a network through power electronic converters, and other types of objects, e.g. FACTS, DC links, SVC, and others, can also be used for stability enhancement. Their presence in EPSs is relatively small and, importantly, they are not very commonly used for this purpose. Thus, the design and use of appropriate, effective, and optimal regulators of synchronous generators is still a very important task for modern EPSs.

The quality of the control process is usually assessed by quality indicators. Integral indices are often used, such as the integral from the control error, the integral from the absolute value of the control error, and the integral from the square of the control error. The latter indicator of the quality of control, owing to its relationship with physical phenomena, is used most often. This is because it can be treated as a measure of energy of the signal, as the root-mean-square (RMS) of the signal (if the control process is associated with sinusoidal signals) or as a measure of variance in the case of stochastic signals.

The assessment of the quality of the control processes is also based on the evaluation of the control system's[1] time domain response to the change in the operating point. Changing the operating point is usually associated with a step change in the reference of the controlled variable or a change in the structure of the system. The change in the simplest case is realized by switching off or on an element of this structure.

In this case, as a measure of the control process quality assessment, parameters (indices) characterizing the time response are used. For the generating unit, these indices (Figure 2.7) are: the settling time t_ε, overshoot ε_p, and rise time t_r, or speed of voltage increase $\Delta V/\Delta t$. Assessing the quality of the control based on time domain response is widely used on the real objects. This results both from the simplicity of the test's execution as well as the ease of

1 The control system is defined as an object with its regulator. Thus, the control system of a synchronous generator is understood as a synchronous generator with exciter and regulator (e.g. AVR).

Power System Dynamics: Stability and Control, Third Edition. Jan Machowski, Zbigniew Lubosny, Janusz W. Bialek and James R. Bumby.

obtaining the parameters from the time domain response. This approach is commonly used to assess the quality of the control process of the real synchronous generator.

The requirements for dynamic properties of the synchronous generator control system operating in EPS are clearly quantified in the Grid Codes issued by transmission system operators (TSOs), distribution system operators (DSOs), or other organizations, e.g. ENTSO-E (European Network of Transmission System Operators for Electricity). Examples of these requirements are presented in Section 2.3.2. The Grid Code requirements are related to the voltage response of the synchronous generator in tests performed on a real object. These tests include the reactive load rejection and a step change in the voltage reference value of the generator operating at idle.

It should be noted that the Grid Codes do not introduce requirements related to the voltage response of the synchronous generator operating synchronously in the power system. It is assumed here that the fulfillment of the requirements for tests for a generating unit operating at idle will ensure the appropriate dynamic properties of the generating unit operating in the EPS. Usually, the response of the generating unit operating in the EPS is characterized by higher damping than the generating unit operating at idle (Section 5.4.6). In general, this assumption does not have to be true. Therefore, in practice, there are also tests performed related to a step change in the voltage reference value during the operation of the generator in the EPS. In this case, step changes in the voltage reference value are introduced at a level usually not exceeding $\pm 3\%$ of the rated voltage. This limited change aim is to limit the change and value of the reactive power in a new steady state.

As described in Chapter 14, the requirements concerning the electromechanical oscillations damping occur in the Grid Code only for a limited number of TSOs. Most of them require that the damping ratios of dominant local or inter-area modes of electromechanical oscillations should exceed a specific value, e.g. $\zeta \geq 0.05$ for normal state and $\zeta \geq 0.03$ for post-contingency state (Section 14.2.3).

Requirements, scope of tests, and methods of evaluating the quality of the synchronous generator control are also discussed in other documents, e.g. in IEEE Std 421.2-2014 and IEEE Std 421.3-2016. These requirements are given in the form of "generally accepted values of indices characterizing good feedback control system performance." For example IEEE Std 421.2-2014, among others, indicates the following requirements: gain margin[2] $GM \geq 6$ dB, phase margin $PM \geq 40^{\circ}$, maximum overshoot $\varepsilon_{\mathrm{p}} = (0 - 15)\%$, peak value of the closed-loop frequency characteristic $M_{\mathrm{T}} = (1.1 - 1.6)$, damping ratio $\xi \geq 0.6$, settling time $t_{\varepsilon} = (0.2 - 10)$ s, accuracy of control, i.e. maximum error of the controlled signal at steady-state $\varepsilon = (0 - 1)\%$, and rise time $t_{\mathrm{r}} = (0.025 - 2.5)$ s. As one can see, the standard presented are ranges of values of accepted indices and not their limit values.

15.2 Optimization Methods

The synchronous generator regulator, in its basic form, consists of a voltage regulator (AVR), limiters of a power capability area (Section 3.3.4), e.g. underexcitation limiter (UEL), overexcitation limiter (OEL), etc., and, in the case of a high rated power generating unit, a power system stabilizer (PSS). As a subject of the regulator commissioning, the PSS is treated as a separate element, although in practice it is one of many functions which can be performed by the regulator. In digital version it is just a fragment of code of the regulator algorithm.

The AVR and PSS allows the dynamic properties of the generating unit to be shaped to a large extent. This applies to the case of a synchronous generator operating in its power capability area. But in the case of a generator operation at a point corresponding to the power capability area border, the dynamic properties of the generating unit determine a corresponding limiter, e.g. UEL, OEL, etc.

2 Terms explained in Section 15.3.1.

From the point of view of the control theory, the task of the AVR and PSS (if this function is on) is to ensure the appropriate dynamic properties of a generating unit operating in an EPS, such as:

- Tracking performance of generator terminal voltage reference V_{ref}. This is a particularly important feature when the generating unit is controlled by the supervisory voltage and reactive power control system. The supervisory control can force frequent changes in the voltage reference.
- Rejection performance of mechanical power (shaft torque) disturbance. From the point of view of the AVR, the mechanical power change is a disturbance. To some extent it can also be regarded as the synchronous generator with AVRs (or additionally with PSS) ability to track the active power change enforced by the governor. Because of that this feature can also be named as the tracking performance of the active power reference P_{ref}. This feature is particularly important when the generating unit participates in automatic generation control (AGC), which is (or can be) associated with frequent changes in the active power reference. This applies to the majority of high-rated power generating units operating in EPSs.
- Rejection performance of generator terminal voltage V_g disturbance. This is related to disturbances originating from the network and leading to changes in generator terminal voltage. Changes in generator voltage are a result of various events in the EPS, such as disturbances in the active or reactive power balance, short-circuits, switching operations, changes in voltage or active power references of other generating units, operation of regulators of other system objects, etc. This feature is particularly important from the power system stability point of view.

Designing effective regulators of synchronous generators is not an easy task, which is due to the features of power system, including: large variation of operating conditions, large variety of disturbances which can occur in the EPS, variation of EPS parameters as a result of network configuration change, complexity and nonlinearity of EPS, etc. Typically, a fast AVR is characterized by a poor ability to damp the oscillations of the active power. An increase of damping of the active power oscillations leads to a deterioration of the quality of the generator voltage control process (Section 10.1.2).

Today, the AVRs used in EPSs are linear, time-invariant regulators with structures that can be called classic. For years classic methods of their design (parameters tuning) have been used. Owing to the problems occurring in the classical way of designing such regulators, including the need to take into account many factors affecting the quality of the control process, limited efficiency and its optimality for a single operating point only, and also because of the development of synthesis and optimization theory, new methods and AVR design techniques are sought after.

The following broad categories of types of regulators (and control methods) can be distinguished:

- *classic regulators*, i.e. linear, time-invariant – commonly used in power systems; regulators of this type are effective, but unfortunately they are not able to perform optimally over the full range of operating conditions and disturbances (IEEE Std 421.5-2016; Larsen and Swan 1981; Lee and Kundur 1986; Kundur 1994; Machowski et al. 2008);
- nonadaptive linear controllers type linear-quadratic regulator (LQR), linear-quadratic-Gaussian (LQG), H_2, H_∞, and based on μ-synthesis, called *optimal regulators* (Skogestad and Postlethwaite 1996; Swarcewicz and Lubosny 2001; Soos and Malik 2002);
- *nonlinear nonadaptive* regulators based on artificial neural networks (ANNs) and fuzzy logic, using heuristic, i.e. designer's experience (knowledge), in the process of synthesis of the regulator (Lubosny 1999; Cirstea et al. 2002; Saccomanno 2003; Zeng and Jun 2010);
- *linear adaptive* regulators based on classical control theory (Gupta 1986; Niederlinski et al. 1995; Soos and Malik 2002);
- *nonlinear adaptive* regulators based on ANNs and the theory of fuzzy logic (Lubosny 1999; Cirstea et al. 2002; Saccomanno 2003; Zeng and Jun 2010).

The AVR design is usually carried out using a mathematical model of a single machine infinite bus (SMIB) system or a reduced model of a multi-machine system. A model of the full or partially reduced multi-machine system is used more often to verify the efficiency of the earlier designed regulator. In general, a model of the full multi-machine system can also be used for the regulator optimization. Usually, the design (synthesis) or parametric optimization of the regulator is performed for a given, selected operating point or for a certain number of operating points.

Verification of the regulator efficiency based on a linear model is related to the analysis of eigenvalues, frequency characteristics, or its time domain response to a disturbance. However, in the case of a nonlinear model (including a multi-machine system), only the time domain simulations (in the form of the EPS dynamic response to a disturbance) are performed. This allows for the optimization of the regulator parameters or for the verification of its dynamic properties. The methods of designing (synthesis) and analyzing synchronous generator regulators in relation to various methods of synchronous generator control systems (types of regulators) are presented in Table 15.1. The indications in Table 15.1 are to be understood as follows:

O – *Parametric optimization*, understood as designing a regulator with a specific structure using heuristic as well as trial-and-error methods or using an optimization algorithm that seeks the optimum of a defined-quality control indicator which is an explicit or implicit function of the regulator parameters.
S – *Synthesis*, understood as a simultaneous search for the structure and parameters of the regulator.
L – *Synthesis through learning*, understood as a search for the parameters of a regulator with a defined structure, in which a "student" is a regulator under design, and a "teacher" is another regulator characterized by appropriate (optimal) dynamic properties. The learning process is carried out by comparing the outputs of the student-regulator and the teacher-regulator for the same inputs and minimizing the function of the difference in their response. The function is minimized by modifying the student-regulator parameters.
T – *Verification of the dynamic properties of a nonlinear regulator or control system*, carried out by the time domain simulations.
G – *Verification of the dynamic properties of the linear regulator or control system*, carried out by classical methods, for example by calculating eigenvalues, calculating and evaluating the response of the system to unit step (Heaviside step function) or to delta pulse (Dirac delta function), using the Laplace transform, calculating and analyzing the magnitude and phase-frequency characteristics of the system transfer function.

Owing to limitations or a lack of analytical methods of verification of dynamic properties of the control systems equipped with regulators based on fuzzy logic, regulators based on ANN, and adaptive regulators, their use in real power systems is smaller than marginal ones. For control systems with regulators of this type, the time domain

Table 15.1 Methods of synthesis and analysis of the synchronous generator regulator.

		Single-machine		Multi-machine	
Type of regulator	**EPS model**	**Linear**	**Nonlinear**	**Linear**	**Nonlinear**
Nonadaptive	Classical	O, G, T	O, T	O, T	O, T
	Optimal	S, G, T	T	T	T
	Fuzzy	O, T	O, T	O, T	O, T
	Neural network	L, T	L, T	L, T	L, T
Adaptive	Classical	S, T	T	T	T
	Fuzzy	O, T	O, T	O, T	O, T
	Neural network	L, T	L, T	L, T	L, T

O – parametric optimization; S – synthesis; L – synthesis through learning; T – verification through simulations; G – verification by use of classical methods, i.e. eigenvalues or frequency characteristics.

simulation is (as mentioned above) the only way to verify their dynamic properties. Probably because of this, TSOs are skeptical about the use of this type of regulator in their systems.

In turn, the so-called optimal regulators are time-invariant linear and nonadaptive regulators, and thus their dynamic properties can be assessed by means of classical analytical methods, such as eigenvalue analysis, frequency characteristics, etc. Therefore, theoretically the probability of using this type of regulators in the future in real power systems seems to be considerable.

In the case of adaptive regulators in applications for controlling synchronous generators, during severe disturbances in the system there are some problems related to the correct parameter's identification of the regulator model. This can lead to control ability degradation and thus significantly limits possibilities of their application in its basic form. Additionally, dedicated algorithms are required to ensure this type of regulator is insensitive to severe disturbances in EPS.

Summarizing the above, it should be noted that, despite the large number of publications regarding the use of various types of regulators to a synchronous generator control, in real EPSs practically only classic, i.e. linear, time-invariant regulators, are used (the only nonlinearities involved in such regulators are limiters). Structures of this type of regulators are published in IEEE Std 421.5-2016, and are also presented in Chapter 11.

15.3 Linear Regulators

15.3.1 Linear Regulator Design

Below the overall structure of the object-regulator system (also known as the control system) illustrated in Figure 15.1 is considered. Individual elements of the control system from Figure 15.1 are: $G(s)$ – object (plant) model, $K(s)$ – regulator, $G_d(s)$ – disturbance model, r – reference inputs (setpoints), y – object (plant) outputs, d – disturbances (process noise), u – control signals, n – measurement noise, y_m – measured output. Regulator $K(s)$ with two degrees-of-freedom (Figure 15.1b) has a structure $K(s) = [K_p(s)\ K_r(s)]^T$, where $K_p(s)$ is a

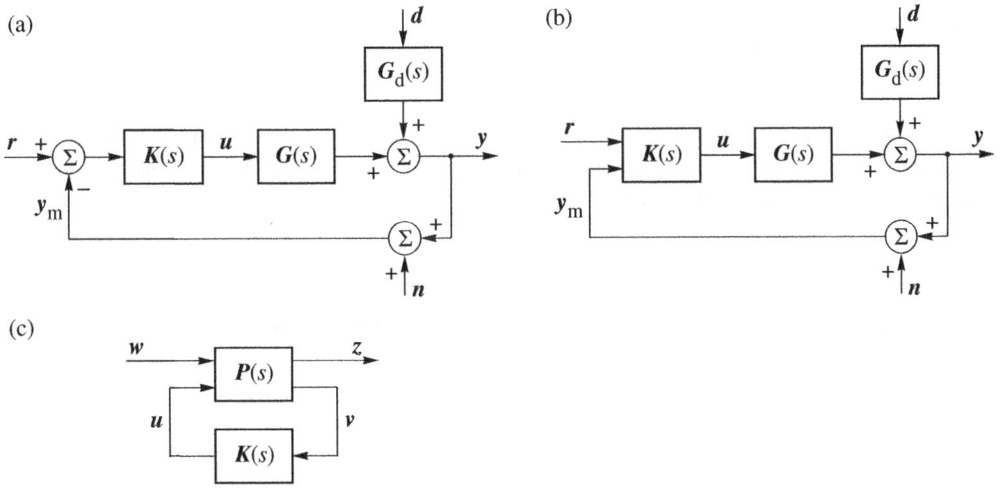

Figure 15.1 Structure of the control systems: (a) control system with one degree-of-freedom; (b) control system with two degrees-of-freedom; (c) general structure of the control system. *Source:* Based on Skogestad and Postlethwaite (1996).

pre-compensator and $\boldsymbol{K}_r(s)$ is a regulator in the feedback.[3] In the case of the general structure of the control system (Figure 15.1c) its constituent elements should be understood as:

- general model of the object $\boldsymbol{P}(s)$, containing models of object $\boldsymbol{G}(s)$, and disturbances $\boldsymbol{G}_d(s)$;
- connections between the object and the regulator;
- signals: $\boldsymbol{w}$ – exogenous inputs (reference inputs, disturbances, and process noise); $\boldsymbol{z}$ – exogenous outputs (error signals to be minimized, e.g. for a system with one degree-of-freedom $\boldsymbol{e} = \boldsymbol{y} - \boldsymbol{r}$); $\boldsymbol{u}$ – control signals; $\boldsymbol{v}$ – inputs to the regulator (for a system with one degree-of-freedom $\boldsymbol{v} = \boldsymbol{r} - \boldsymbol{y}$).

The main task of the control system (in general) is to control the signals $\boldsymbol{u}$, so that the object outputs $\boldsymbol{y}$ do not differ from the reference inputs $\boldsymbol{r}$. An additional task of the signals' $\boldsymbol{u}$ control is to counteract the influence of disturbances $\boldsymbol{d}$ on object outputs $\boldsymbol{y}$. In order to achieve both effects, it is necessary to minimize the control errors $\boldsymbol{e} = \boldsymbol{y} - \boldsymbol{r}$, which is the task of the regulator $\boldsymbol{K}(s)$. In order to design the appropriate regulator $\boldsymbol{K}(s)$ it is necessary to have a priori information about the expected disturbances and the models of the object $\boldsymbol{G}(s)$ and disturbances $\boldsymbol{G}_d(s)$.

EPSs are nonlinear objects. As mentioned above, in order to design (synthesize) the regulator or verify their dynamic properties, linear or nonlinear models of the system are used. In both cases, the EPS component models are models with unchanged time parameters, i.e. they are time-invariant models. Changes in the network configuration or changes in loads are treated as a change in the object's $\boldsymbol{G}(s)$ parameters. The changes can also be transferred to the model of disturbances $\boldsymbol{G}_d(s)$.

In the considerations presented here, it is assumed that the object model $\boldsymbol{G}(s)$ and disturbances model $\boldsymbol{G}_d(s)$ are linear and time-invariant models. Thus, the regulator $\boldsymbol{K}(s)$ is designed as a linear and time-invariant object. For such an object, transfer function between the reference inputs $\boldsymbol{r}$, disturbances, and measurement noise $\boldsymbol{d}$ and $\boldsymbol{n}$, and the object outputs $\boldsymbol{y}$ can be determined. The structure of the control system shown in Figure 15.1a takes the form

$$\boldsymbol{y} = [\boldsymbol{I} + \boldsymbol{G}(s)\boldsymbol{K}(s)]^{-1}\boldsymbol{G}(s)\boldsymbol{K}(s)\boldsymbol{r} + [\boldsymbol{I} + \boldsymbol{G}(s)\boldsymbol{K}(s)]^{-1}\boldsymbol{G}_d(s)\boldsymbol{d} - [\boldsymbol{I} + \boldsymbol{G}(s)\boldsymbol{K}(s)]^{-1}\boldsymbol{G}(s)\boldsymbol{K}(s)\boldsymbol{n} \tag{15.1}$$

The components of the transfer functions defined by Eq. (15.1) are called as follows:

- *Open-loop system transfer function*

$$\boldsymbol{L}(s) = \boldsymbol{G}(s)\boldsymbol{K}(s) \tag{15.2}$$

- *Sensitivity transfer function*

$$\boldsymbol{S}(s) = [\boldsymbol{I} + \boldsymbol{G}(s)\boldsymbol{K}(s)]^{-1} \tag{15.3}$$

  This transfer function defines the ability of the considered control system to attenuate disturbances and, in other words, to increase immunity of the control system to disturbances (process and measurement noise). This transfer function is also referred to as *disturbance transfer function*.
- *Complementary sensitivity transfer function*

$$\boldsymbol{T}(s) = (\boldsymbol{I} + \boldsymbol{G}(s)\boldsymbol{K}(s))^{-1}\boldsymbol{G}(s)\boldsymbol{K}(s) \tag{15.4}$$

  This transfer function defines the tracking properties of the control system and is therefore also called *tracking transfer function*.

3 The blocks of the control system structures shown in Figure 15.1 are defined herein by transfer functions with names corresponding to the names of individual blocks. The above components of the control system structures may be single- or multi-dimensional, i.e. they may be in the form of matrices, vectors, or individual signals.

For such defined transfer functions, the control error $\boldsymbol{e}$ is equal

$$\boldsymbol{e} = \boldsymbol{y} - \boldsymbol{r} = -\boldsymbol{S}(s)\boldsymbol{r} + \boldsymbol{S}(s)\boldsymbol{G}_{\mathrm{d}}(s)\boldsymbol{d} - \boldsymbol{T}(s)\boldsymbol{n} \tag{15.5}$$

and the control signal of the regulator defines the function

$$\boldsymbol{u} = \boldsymbol{K}(s)\boldsymbol{S}(s)\boldsymbol{r} - \boldsymbol{K}(s)\boldsymbol{S}(s)\boldsymbol{G}_{\mathrm{d}}(s)\boldsymbol{d} - \boldsymbol{K}(s)\boldsymbol{S}(s)\boldsymbol{n} \tag{15.6}$$

Equations (15.5) and (15.6) allow relatively easy determination of the requirements for the properties of the sensitivity transfer function $\boldsymbol{S}(s)$ and the complementary sensitivity transfer function $\boldsymbol{T}(s)$. These requirements can be expressed in relation to magnitude or phase of the transfer functions.

The design of a linear, time-invariant regulator can be implemented using various methods, which can be divided into the following groups:

- Methods consisting of shaping the transfer function of the control system. In these methods the regulator $\boldsymbol{K}(s)$ should ensure the appropriate shape of the open-loop frequency characteristic $\boldsymbol{L}(s)$ and the closed-loop frequency characteristics $\boldsymbol{S}(s)$, $\boldsymbol{T}(s)$, and $\boldsymbol{K}(s)\boldsymbol{S}(s)$. The design process is often carried out using heuristic technique, i.e. using the designer experience.
- Signal-based methods in which a regulator $\boldsymbol{K}(s)$ is sought that ensures the minimization of the control quality indicator (norm) defined in the time domain. A typical example of an indicator of the quality of control can be the integral from the square of the error of the controlled variable (H_2 norm) or H_∞ norm. The effect of design are so-called optimal regulators LQR, LQG, LQI, H_2, and H_∞.
- Methods based on online optimization performed by a regulator. Adaptive and MPC (model predictive control) regulators are examples of the result of design. These methods are also associated with multi-criteria optimization of complex control systems.

The design of the regulator shaping the control system transfer function (first group of the enlisted above methods) is based on spectral transfer functions $\boldsymbol{L}(\mathrm{j}\omega)$, $\boldsymbol{S}(\mathrm{j}\omega)$, and $\boldsymbol{T}(\mathrm{j}\omega)$ in the form of magnitude and phase-frequency characteristics, and not Laplace operators $\boldsymbol{L}(s)$, $\boldsymbol{S}(s)$, or $\boldsymbol{T}(s)$.[4] Designing a regulator based on frequency characteristics in comparison to designing based on a step change of operating point (e.g. in the case of an EPS a step change in voltage reference or nodal voltage or load or network configuration, etc.) has the advantage of considering the signal class here with high cardinality, i.e. sinusoidal signals at any frequencies.

In case of a single-input single-output system (SISO) in the regulator $K(s)$ design (based on the frequency characteristics of the open-loop system transfer function $L(\mathrm{j}\omega) = K(\mathrm{j}\omega)G(\mathrm{j}\omega)$) the margin of the closed-loop system stability is determined by a pair of numbers: gain margin *GM* and phase margin *PM* (Figure 15.2). They are defined as follows:

- *Gain margin* is equal to

$$GM = \frac{1}{|L(\mathrm{j}\omega_{180})|} \tag{15.7}$$

where ω_{180} is the pulsation for which the phase-frequency characteristic of the open-loop system is delayed by 180°, i.e. $\arg(L(\mathrm{j}\omega_{180})) = -180°$. The gain margin indicates if the gain of the regulator $K(s)$ can be increased so

4 In order to obtain the spectral transfer function from a Laplace operator based transfer function a substitution $s = \mathrm{j}\omega$ is applied.

that the closed-loop system reaches the stability limit. Typically, condition $GM > 2$ is required, which is equivalent to $GM > 6$ dB.[5]

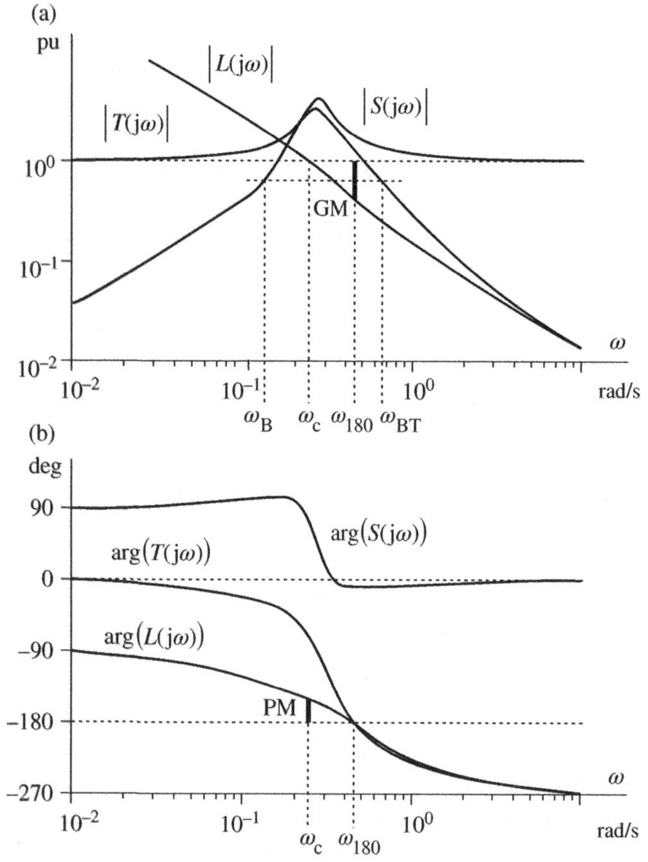

Figure 15.2 Example of (a) magnitude-frequency and (b) phase-frequency characteristics of a control system.

- *Phase margin* is equal to

$$PM = \arg(L(j\omega_c)) - 180^\circ \tag{15.8}$$

where the gain crossover frequency ω_c is a pulsation for which the magnitude-frequency characteristic of the open-loop system is equal to one, i.e. $|L(j\omega_c)| = 1$. The phase margin indicates how big a time delay can be additionally entered in the regulator $K(s)$ for pulsation ω_c, so that the closed-loop system reaches the stability limit. The phase margin should be large enough to compensate time delays in the control system that are not modeled. Typically, the condition $PM \geq 40^\circ$ is required.

The closed system becomes unstable when one enters a time delay of no less than

$$T_{max} = \frac{PM}{\omega_c} \tag{15.9}$$

5 The relationship between the magnitude of the transfer function expressed in decibels (G dB) and per unit (pu) (M in pu) is as follows $G = 20\log_{10}(M)$. The gain margin expressed in dB is equal to $GM = 0 - G$, and expressed in pu is equal to $GM = 1/M$. Graphically, the gain margin corresponds to the section indicated in Figure 15.2 as GM.

As it results from Eq. (15.9), by reducing the gain crossover pulsation ω_c one increases the tolerance range of the control system to the time delays not taken into account in the model, but at the same time the response of the control system is slowed down.

The phase shift of the open-loop frequency characteristic equal to $-180°$ is important because in the control system with negative feedback the control signal with a given pulsation shifted by 180° is added to the reference input, creating a positive feedback loop. If the open-loop system gain for a given pulsation is greater than one then this signal, in the feedback loop, is amplified, theoretically to infinity, which in turn means that the closed-loop system becomes unstable. Therefore, the gain of an open-loop system for pulsation (frequency), for which the phase shift is equal to $-180°$, must be less than one. Formally, according to the Nyquist criterion, and precisely one of its forms, the closed-loop system is stable if and only if the magnitude characteristic of the open-loop system (when it is monotonically decreasing with the increase of pulsation) fulfills the condition

$$|L(j\omega_{180})| < 1 \tag{15.10}$$

This also means that the condition $\omega_c < \omega_{180}$ must be met.

Since the frequency response of the open-loop system does not give full information about the dynamic properties of the closed-loop system, the design process of the regulator uses also the characteristics of the closed-loop system. In this the following quantities (illustrated by Figure 15.2) are defined:

- Maximum value of the magnitude-frequency characteristic of the sensitivity transfer function (disturbance rejection transfer function)

$$M_S = \max_{\omega} |S(j\omega)| \tag{15.11}$$

- Maximum value of the magnitude-frequency characteristic of the complementary sensitivity transfer function (tracking transfer function)

$$M_T = \max_{\omega} |T(j\omega)| \tag{15.12}$$

  In control systems, it is required that $M_S < 2$ and $M_T < 1.25$. For stable objects, the condition $M_S > M_T$ is usually met. However, the closer M_S to the value of unity, the closer to the robust system is the control system. At the same time, it is required that for the pulsations going to zero the transfer function magnitude $S(j\omega)$ also tends to zero.
- Frequency response (bandpass) specified for the sensitivity transfer function $S(j\omega)$, which corresponds to the pulsation range from ω_B to infinity, for which the transfer function magnitude meets the condition $|S(j\omega)| \geq 1/\sqrt{2}$.
- Frequency response (bandpass) specified for a complementary sensitivity transfer function $T(j\omega)$, which corresponds to the pulsation range from zero to ω_{BT}, for which the transfer function magnitude meets the condition $|T(j\omega)| \geq 1/\sqrt{2}$.

Usually, the relationship between the above pulsations is as follows: $\omega_B < \omega_c < \omega_{BT}$. A wider bandpass width of the sensitivity transfer function, i.e. lower value of the cutoff pulsation ω_B, is associated with a faster response of the control system. In turn, it indicates the control system, which is more sensitive to disturbances and changes in the object's parameters. A narrower bandpass width of the transfer function $|S(j\omega)|$ means a system characterized by a slower response to a change in the reference, but leads to a more robust control system.

An important parameter characterizing the control system is also the value of the slope of the magnitude-frequency characteristic in the vicinity of the crossover pulsation ω_c. In general, in the case of the reference input or disturbances of a step type, the required slope of the characteristic equal to -20 dB/decade[6] is sufficient.

---

6 The slope of the amplitude characteristic equal to -20 dB/decade means a 10-fold reduction in the amplitude for a 10-fold increase in pulsation. The slope of the characteristic (but also in general the gain), expressed in decibels, is defined by formula $K = 20\log_{10}(|L(\omega)| / |L(10\omega)|)$.

However, in the case when the change in reference input or disturbance signal is linear over the time, the required slope of the characteristic is equal to −40 dB/decade.

As it results from Eq. (15.5), the control error e depends on the transfer function $S(\mathrm{j}\omega)$ and $T(\mathrm{j}\omega)$, whereby the ability of the control system to minimize the control error as a reference input r and disturbance d requires that $S(\mathrm{j}\omega) \approx 0$ (and therefore $T(\mathrm{j}\omega) \approx 1$), while as a function of measurement disturbances n it requires that $T(\mathrm{j}\omega) \approx 0$ (and thus $S(\mathrm{j}\omega) \approx 1$, what is attainable when $L(\mathrm{j}\omega) \approx 0$). These requirements are contradictory and thus the design of the regulator in a feedback system requires the adoption of a certain compromise between the ability to suppress measurement noise (requiring a small gain in the feedback loop) and appropriate tracking properties and damping disturbances (requiring a high gain).

According to Skogestad and Postlethwaite (1996), the following rules for designing a regulator to SISO control system with feedback, based on open-loop system frequency characteristics, are used:

1) To obtain high attenuation of disturbances, a high gain of $L(\mathrm{j}\omega)$ is required and thus a high gain of regulator $K(\mathrm{j}\omega)$.
2) To achieve good tracking properties, a high gain of $L(\mathrm{j}\omega)$ is required.
3) To attenuate the measurement noise, a small gain of $L(\mathrm{j}\omega)$ is required.
4) To obtain a regulator control u of a small amplitude (small control energy), a small gain of the regulator $K(\mathrm{j}\omega)$ is required and thus a small gain of $L(\mathrm{j}\omega)$.
5) To achieve a stable control system in the case of an unstable object, a high gain of $L(\mathrm{j}\omega)$ is required.
6) To obtain a real strictly proper regulator[7] it is required that the regulator and the open-loop system transfer functions for high frequencies tends to zero, i.e. $K(\mathrm{j}\omega) \to 0$ and $L(\mathrm{j}\omega) \to 0$.
7) To achieve nominal stability[8] (for a stable object), a small gain of $L(\mathrm{j}\omega)$ is required.
8) To achieve robust stability[9] (for a stable object), a small gain of $L(\mathrm{j}\omega)$ is required.

Fortunately, the above-mentioned, conflicting requirements refer to different frequency ranges, and for a large group of objects the use of a large gain of open-loop system transfer function $|L(\mathrm{j}\omega)| > 1$ for pulsation below the gain crossover pulsation ω_c and a small gain $|L(\mathrm{j}\omega)| < 1$ for pulsation above ω_c leads to a stable control system. This is illustrated in Figure 15.2, which also shows the typical frequency characteristics of the closed-loop system: complementary sensitivity (tracking) $T(\mathrm{j}\omega)$ and sensitivity (disturbance) $S(\mathrm{j}\omega)$. The magnitude-frequency characteristic of the tracking transfer function $T(\mathrm{j}\omega)$ in the figure for small pulsations is equal to one, which is appropriate because it means that the reference input is "transferred" to the output invariantly. For large pulsations, the transfer function is characterized by low gain (low sensitivity to reference input), which can be treated as a certain ability to attenuate disturbances in the reference input and attenuate the measuring noise. In turn from the shape of the magnitude-frequency characteristic of the sensitivity transfer function $S(\mathrm{j}\omega)$ it appears that the disturbances with small pulsations are well damped, and those with large pulsations are not amplified (the gain is equal to one), which should be considered appropriate.

In the case of multiple inputs and multiple outputs (MIMO) systems, the regulator design is also based on the frequency characteristics but in this case the magnitude-frequency characteristics are defined as

---

7 A strictly proper transfer function is a transfer function where the degree of the numerator is less than the degree of the denominator. A proper transfer function is a transfer function in which the degree of the numerator does not exceed the degree of the denominator.

8 *Nominal stability* means the stability of a system with a given regulator for a nominal model (object). In the theory of robust regulators, the nominal model is a certain base model.

9 *Robust stability* means stability of the control system with a given regulator for all models from a defined set of object models, including the nominal model. The set of object models is created by supplementing the nominal model with transfer functions (called *uncertainties*) representing the expected deviations of the transfer function of the object from the transfer function of the nominal model. See footnote 11.

$$\sigma(\omega) = \frac{\|\boldsymbol{y}(\omega)\|_2}{\|\boldsymbol{d}(\omega)\|_2} = \frac{\|\boldsymbol{G}(j\omega)\cdot\boldsymbol{d}(\omega)\|_2}{\|\boldsymbol{d}(\omega)\|_2} \tag{15.13}$$

where $\boldsymbol{y}(\omega)$ is the vector of outputs, $\boldsymbol{d}(\omega)$ is the vector of inputs, and $\boldsymbol{G}(j\omega)$ is the matrix of the object's transfer functions. The characteristic, for a given pulsation ω, defines the system gain. The gain is called the *singular value* σ of the dynamic control system. The maximum and minimum values of characteristic (15.13) are named upper $\overline{\sigma}$ and lower $\underline{\sigma}$ singular values. They are defined as

$$\overline{\sigma}(\boldsymbol{G}) = \max_{d\neq 0}\frac{\|\boldsymbol{G}\cdot\boldsymbol{d}\|_2}{\|\boldsymbol{d}\|_2} = \max_{\|\boldsymbol{d}\|_2=1}\|\boldsymbol{G}\cdot\boldsymbol{d}\|_2 \tag{15.14}$$

$$\underline{\sigma}(\boldsymbol{G}) = \min_{d\neq 0}\frac{\|\boldsymbol{G}\cdot\boldsymbol{d}\|_2}{\|\boldsymbol{d}\|_2} = \min_{\|d\|_2=1}\|\boldsymbol{G}\cdot\boldsymbol{d}\|_2 \tag{15.15}$$

Note that such function defined by Eq. (15.13) for SISO systems is equal to the magnitude-frequency characteristic of the control system transfer function. Thus, the upper $\overline{\sigma}$ and lower $\underline{\sigma}$ singular values are equal to the maximum and minimum values of the characteristic.

However, in MIMO systems, the functions $\boldsymbol{\sigma}(\omega)$ differ from the transfer functions between the given input and output. This results from the fact that the norm $\|\boldsymbol{G}\cdot\boldsymbol{d}\|_2$ in Eqs. (15.14) and (15.15) is a result of simultaneous excitation of each input of the transfer function $\boldsymbol{G}$. Thus, the upper and lower singular values related to a given output differ from the maximum and minimum values of the transfer functions magnitude between each pair of input/output.

Example 15.1 Consider a SMIB system consisting of a synchronous generator with rated data: $S_n = 426$ MVA, $V_n = 22$ kV, $\cos\varphi_n = 0.85$, $f_n = 50$ Hz, $R_s = 0.0016$ per unit (pu), $X_d = 2.6$ pu, $X_q = 2.48$ pu, $X'_d = 0.33$ pu, $X'_q = 0.53$ pu, $X''_d = 0.235$ pu, $X''_q = 0.235$ pu, $X_l = 0.199$ pu, $T'_{do} = 9.2$ s, $T'_{qo} = 1.095$ s, $T''_{do} = 0.042$ s, $T''_{qo} = 0.065$ s, $H = 3.225$ s, and a step-up transformer with rated data: $S_n = 426$ MVA, $V_{n1} = 22$ kV, $V_{n2} = 242$ kV, $Z_T = 0.145$ pu, $R_T = 0.0029$ pu, and an external impedance equal to $\underline{Z}_S = 0.007 + j0.08$ pu. For this object with regulator $K_3(s)$ (discussed in Section 15.3.2), the magnitude-frequency characteristic of transfer function $|V_g/V_{ref}|$ is presented in Figure 15.3 and marked by a dashed line.[10] If one considers this control system as an SISO one, i.e. with voltage reference as input V_{ref} and generator voltage as output V_g, the singular values function $\sigma(\omega)$ corresponds to the magnitude-frequency transfer function, and the upper and lower singular values are respectively equal to: $\overline{\sigma} = 2.97$, $\underline{\sigma} = 0$ (dotted line in Figure 15.3).

If in the same model one additionally specifies other inputs or outputs, the shape of the function $\sigma(\omega)$ will differ from the transfer function $|V_g/V_{ref}|$. The example shows the function marked in Figure 15.3 as $\sigma(\omega)$, calculated for the model as above, in which, in addition to the voltage reference V_{ref}, the mechanical torque on the generator shaft τ_m and infinity busbars voltage V_S are used as inputs. The voltage reference is the reference input, while the torque and the infinite busbars voltage are the disturbances. In this case, the upper and lower singular values are, respectively, equal: $\overline{\sigma} = 3.53$ and $\underline{\sigma} = 0.47$. This increase in upper and lower singular values in comparison to the SISO system results from the possibility of simultaneous influence of signals from various inputs on the output. However, the summation of signals on the output (V_g here) is of a phasor character, which results from a different phase shift of signals of the component transfer function: $V_g/V_{ref}(\omega)$, $V_g/\tau_m(\omega)$, and $V_g/V_S(\omega)$. It also means that the

10 In MATLAB, the *sigma* function is used to plot singular values as a function of pulsation for dynamic system defined by transfer function $\boldsymbol{G}(s)$ and the *svd* function is used to calculate singular values of the matrix $\boldsymbol{G}(s) = \boldsymbol{C}(s\boldsymbol{I} - \boldsymbol{A})^{-1}\boldsymbol{B}$ defined for a given pulsation ω. The singular values of the matrix $\boldsymbol{G}_{m\times n}$ of row q are marked σ_i and are equal to squares of non-negative eigenvalues of the matrix $\boldsymbol{G}^*\boldsymbol{G}$ (where $\boldsymbol{G}^*$ is a conjugate transpose of $\boldsymbol{G}$) ordered from the highest to the lowest $\sigma_1 \geq \sigma_2 \geq \ldots \geq \sigma_p \geq 0$, where $p \leq \min\{m, n\}$. If $q < p$, then there are $p - q$ singular values equal to zero $\sigma_{r+1} = \sigma_{r+2} = \ldots = \sigma_p = 0$. The largest singular value is marked $\overline{\sigma}(\boldsymbol{G}) = \sigma_1$. If $\boldsymbol{G}_{n\times n}$ is a square matrix, the last (smallest) singular value is denoted as $\underline{\sigma}(\boldsymbol{G}) = \sigma_n$. These values are called, respectively, the upper and lower singular values.

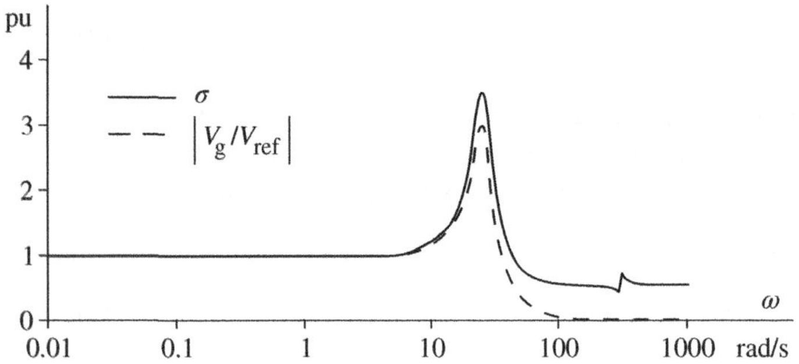

Figure 15.3 An example of the dependence of singular values σ on frequency for an SMIB system.

function $\sigma(\omega)$ is not greater than the sum of the algebraic magnitude-frequency characteristics between the output and each input. The component transfer functions' frequency characteristics, for the considered example, are shown in Figure 15.12. As can be seen, the algebraic sum of the magnitude-frequency characteristics $|V_g/V_{ref}|$, $|V_g/\tau_m|$, and $|V_g/V_S|$ is greater than the characteristic $\sigma(\omega)$ from Figure 15.3.

******

The classic approach to regulator $\boldsymbol{K}(\mathrm{j}\omega)$ design for MIMO control system, based on open-loop system transfer function $\boldsymbol{L}(\mathrm{j}\omega) = \boldsymbol{G}(\mathrm{j}\omega)\boldsymbol{K}(\mathrm{j}\omega)$ and singular values leads to the following requirements (Skogestad and Postlethwaite 1996):

1) To obtain disturbance attenuation, singular values $\overline{\sigma}(\boldsymbol{L}(\mathrm{j}\omega))$ should be large. This is met for the frequencies for which $\overline{\sigma}(\boldsymbol{L}(\mathrm{j}\omega)) \gg 1$.
2) To obtain attenuation of measurement noise, singular values $\overline{\sigma}(\boldsymbol{L}(\mathrm{j}\omega))$ should be small. This is met for the frequencies for which $\overline{\sigma}(\boldsymbol{L}(\mathrm{j}\omega)) \ll 1$.
3) To obtain good tracking properties, singular values $\underline{\sigma}(\boldsymbol{L}(\mathrm{j}\omega))$ should be large. This is met for the frequencies for which $\underline{\sigma}(\boldsymbol{L}(\mathrm{j}\omega)) \gg 1$.
4) To achieve a reduction of control energy, singular values $\overline{\sigma}(\boldsymbol{L}(\mathrm{j}\omega))$ should be small. This is met for the frequencies for which $\overline{\sigma}(\boldsymbol{L}(\mathrm{j}\omega)) \ll 1$.
5) To achieve robust stability in the case of additive uncertainties, singular values $\overline{\sigma}(\boldsymbol{L}(\mathrm{j}\omega))$ should be low. This is met for the frequencies for which $\overline{\sigma}(\boldsymbol{L}(\mathrm{j}\omega)) \ll 1$.
6) To achieve robust stability in the case of multiplicative uncertainties, singular values $\overline{\sigma}(\boldsymbol{L}(\mathrm{j}\omega))$ should be small. This is met for the frequencies for which $\overline{\sigma}(\boldsymbol{L}(\mathrm{j}\omega)) \ll 1$.

The above can also be formulated in form of requirements for a regulator in relation to the closed-loop system (Skogestad and Postlethwaite 1996):

1) To obtain disturbances attenuation, singular values $\overline{\sigma}(\boldsymbol{S}(\mathrm{j}\omega))$ should be small.
2) To obtain attenuation of measurement noise, singular values $\overline{\sigma}(\boldsymbol{T}(\mathrm{j}\omega))$ should be small.
3) To obtain good tracking properties, singular values should fulfill the condition $\overline{\sigma}(\boldsymbol{T}(\mathrm{j}\omega)) \approx \underline{\sigma}(\boldsymbol{T}(\mathrm{j}\omega)) \approx 1$.
4) To achieve a reduction of control energy, singular values $\overline{\sigma}(\boldsymbol{K}(\mathrm{j}\omega)\boldsymbol{S}(\mathrm{j}\omega))$ should be small.
5) To get robust stability in case of additive uncertainties,[11] i.e. $\boldsymbol{G}_p(\mathrm{j}\omega) = \boldsymbol{G}(\mathrm{j}\omega) + \boldsymbol{\Delta}(\mathrm{j}\omega)$, singular values $\overline{\sigma}(\boldsymbol{K}(\mathrm{j}\omega)\boldsymbol{S}(\mathrm{j}\omega))$ should be small.

11 In the process of synthesis and analysis of robust regulators, object $\boldsymbol{G}_p(\mathrm{j}\omega)$ is defined by the nominal object $\boldsymbol{G}(\mathrm{j}\omega)$ and uncertainties $\boldsymbol{\Delta}(\mathrm{j}\omega)$, representing the expected deviations of the object's transfer function from the nominal one. There are distinguished additive and multiplicative uncertainties. In the case of additive uncertainties the object is defined as $\boldsymbol{G}_p(\mathrm{j}\omega) = \boldsymbol{G}(\mathrm{j}\omega) + \boldsymbol{\Delta}(\mathrm{j}\omega)$, i.e. the uncertainties are added to the nominal object. In case of multiplicative uncertainties the object is defined as $\boldsymbol{G}_p(\mathrm{j}\omega) = (\boldsymbol{I} + \boldsymbol{\Delta}(\mathrm{j}\omega))\boldsymbol{G}(\mathrm{j}\omega)$, i.e. the uncertainties and nominal plant transfer functions are multiplied. Taking into account the uncertainty leads to the object model in the form of a set of transfer functions $\boldsymbol{G}_p(\mathrm{j}\omega)$.

6) To obtain robust stability in the case of multiplicative uncertainties, i.e. $\boldsymbol{G}_p(j\omega) = (\boldsymbol{I} + \boldsymbol{\Delta}(j\omega))\boldsymbol{G}(j\omega)$, singular values $\overline{\sigma}(\boldsymbol{T}(j\omega))$ should be small.

As in the case of an SISO control system, in this case it is usually found that it is not possible to meet all the above conditions. This means that there must be a certain compromise between, for example, the ability to attenuate disturbances (referring to low frequencies) and the ability to attenuate measurement noise (relating to high frequencies) or the tracking ability (referring to low frequencies). Fortunately and typically, requirements 1 and 3 relate to low frequencies and other requirements to high frequencies, as shown in Figure 15.4. It shows that, for a given control system, the bandwidth determined by the frequency characteristics of the upper $\overline{\sigma}(\omega)$ and lower $\underline{\sigma}(\omega)$ singular values should not intersect with "forbidden" areas (A and B in Figure 15.4), owing to sensitivity to disturbances, measurement noise, and required tracking properties.

It can therefore be concluded that shaping the characteristics of the open-loop system (its specific values) by selecting the parameters of the regulator $\boldsymbol{K}$ is relatively simple. This process, guaranteeing the stability of the closed-loop system, unfortunately does not always guarantee obtaining the desired dynamic properties of the closed-loop system. In order to obtain the desired properties of a closed-loop system, it is necessary to perform additional verification tests in the regulator design, such as time domain simulations, eigenvalues analysis, etc. The regulator design is usually an iterative process in which in each step the shape of the frequency characteristic is modified (in it the singular values can be considered) and next the dynamic properties of the control system are verified by the time domain simulations.

15.3.2 Voltage Regulators

Design of the regulator based on the frequency characteristic is related to characteristic shaping. The regulator-object frequency characteristic shaping is realized by correctors, which are part of the regulator. There are three distinguished types of correctors: serial correctors, correctors in feedback loop, and correctors in form of a PID regulator. The serial correctors typically have a form of serially (to the regulator amplifier) connected lead–lag blocks. The correctors in feedback typically have a form of serially connected real differentiator and lead–lag blocks.

In the case of AVRs, all of the corrector types mentioned are used. A significant number of AVRs operating in power systems are regulators consisting of serial correctors.

An example of AVR with a serial corrector and corrector in feedback is a regulator with a structure called, in IEEE Std 421.5-2005, an ST1A (Figure 15.5). This regulator is dedicated to static excitation systems. The mathematical model of the ST1A regulator contains two serial correctors with time constants T_B, T_C, T_{B1}, and T_{C1} and a

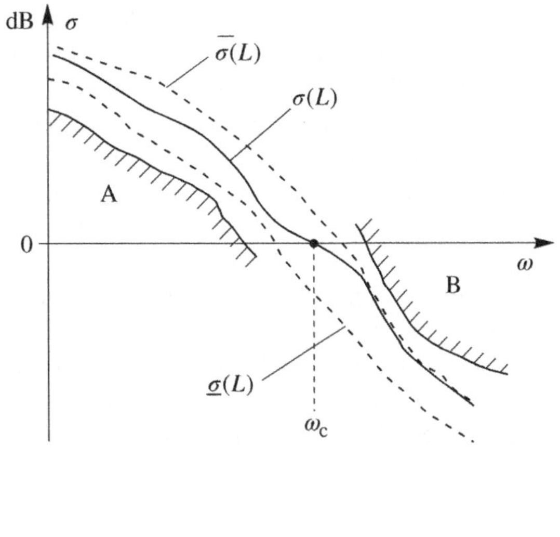

Figure 15.4 Design requirements for the MIMO system regulator based on an open-loop system transfer function: A – area resulting from the requirements in terms of control quality, including tracking properties; B – area resulting from stability of the control system, ability to attenuate disturbances and control energy limitation; $\sigma(L)$ – singular values of the desired characteristics of an open-loop system. *Source:* Based on MathWorks (2019).

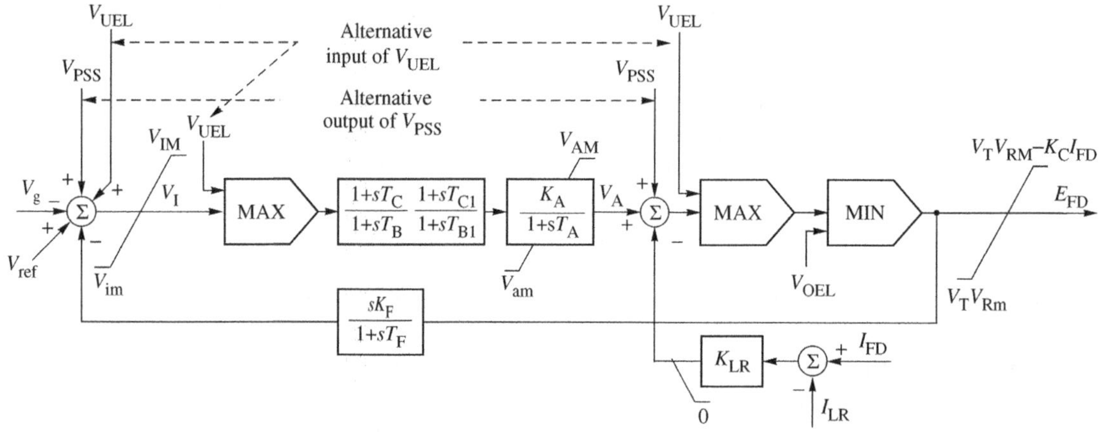

Figure 15.5 Mathematical model of the IEEE ST1A AVR. Based on IEEE Std 421.5-2005. *Source:* Reproduced by permission of IEEE.

corrector in the feedback loop in the form of a real differentiator with parameters K_F and T_F. This regulator can therefore be used with a serial corrector or corrector in the feedback loop. Parameters of the regulator K_A and T_A define the amplifier, and the remaining elements model the static excitation system, i.e. a thyristor rectifier with an excitation transformer. Typically, in the case of real AVRs with a static excitation system, the serial correctors are active while the differentiating element in the feedback is inactive. The regulator with active feedback and an inactive serial corrector is less frequently used.

Regulators dedicated to machine excitation systems practically always operate with active feedback. An example here may be a regulator with the structure of AC5B (IEEE Std 421.5-2005). In this case, the regulator model also includes a machine exciter model.

The PID regulators are also used in synchronous generator control systems. They are used both in the generating units with machine as well as static excitation systems. An example of such regulator is model AC8B (IEEE Std 421.5-2005). The regulator is dedicated to machine excitation systems.

The meaning of limiters appearing in various types of blocks is explained in the appendix (Section A.5).

The basic task of an AVR is to ensure stability and a required dynamic properties in the case of the generating unit operating synchronized with the network and operating at idle. This is accomplished by varying the AVR gain for various frequencies of oscillations in EPS (terminal voltage oscillations). In steady states, the AVR should represent high gain to ensure an appropriately high response to voltage changes and to minimize the voltage control error at steady-state. Typically (owing to the Grid Codes), it is required that control error at steady-state is not greater than $\pm 0.5\%$. In transient states the AVR gain (called the *transient gain*) must be limited to avoid amplification of voltage oscillations resulting from the rotor (power) swings. For large interconnected power systems the frequency range, for which the regulator should represent a lower gain, is very wide, e.g. from (1–1.5) to (10–15) rad/s. This results from the wide range of modal frequencies (RoMF) typical for interconnected power systems (Section 12.1.3), in which all types (internal, local, inter-area) of oscillations can occur. When concern is related to internal and local oscillations only, this reduced gain can refer to the narrower range of frequency, e.g. (6–12) rad/s.

The principles of shaping frequency characteristics and the selection of parameters of the AVR with three mentioned types of correctors – i.e. serial, in the feedback loop, and the PID regulator – are discussed below.

The serial correction of the AVR frequency characteristic is carried out by lead–lag blocks with transfer function $(1 + T_C)/(1 + T_B)$. The blocks are also called transient gain reduction (TGR) blocks (Figure 10.4 in Section 10.1.1). The number of these elements serially connected can be one or more. An AVR example with serial correction, with

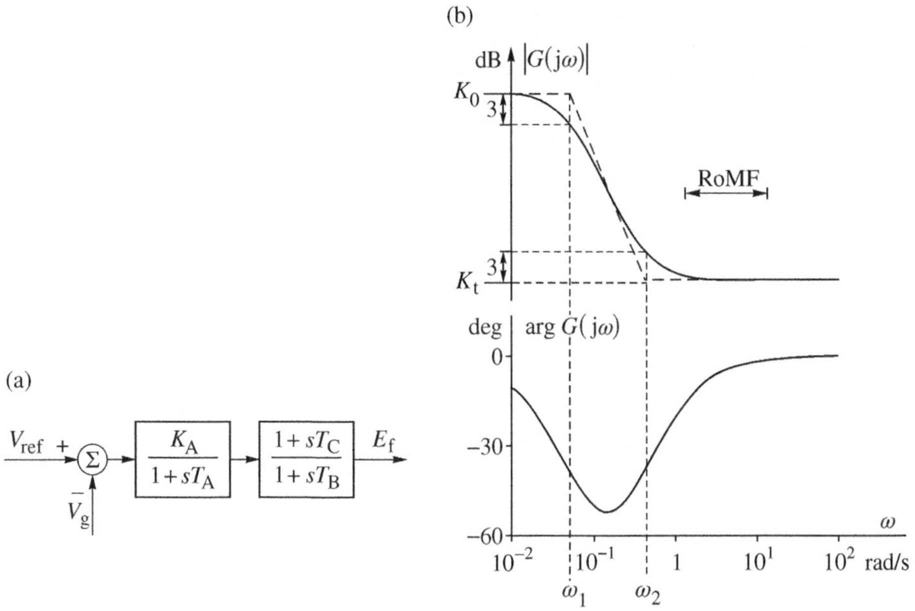

Figure 15.6 AVR with serial corrector: (a) block diagram; (b) transfer function.

one TGR block, is shown in Figure 15.6a. A first-order inertial element with a gain K_A and a time constant T_A models the AVR amplifier. The time constant T_A models delays in the control path, and its value is usually small, i.e. (0.02–0.1) s.

An example of a desired characteristics of an AVR with serial corrector using a single TGR block is shown in Figure 15.6b. This characteristic is, in fact, a characteristic of the TGR block, i.e. it is created assuming $T_A = 0$. The real magnitude-frequency characteristic (marked by continuous line) is imposed on its straight-line approximation (marked by a dashed line) with characteristic points with coordinates (ω_1, K_0) and (ω_2, K_t). Pulsations ω_1 and ω_2 are called the cutoff (corner) pulsations, and the difference between the value of the magnitude of the real characteristic and the approximation for these pulsations is equal to 3 dB. The time constant of the TGR block numerator is equal to $T_C = 1/\omega_2$, while the time constant of the TGR block denominator is equal to $T_B = 1/\omega_1$.

The assumed or calculated parameters in the AVR design are the cutoff pulsations ω_1 and ω_2, and gains K_0 and K_t. The gain of the regulator in steady states is equal K_0, but in transient states it is equal K_t. It should be emphasized that these gains do not correspond to the actual total amplification of the generator control system, because the amplification of the excitation system and the object are neglected here.

The regulator parameters are selected in the following way. The pulsation ω_2 is assumed to be less than the lower limit of the RoMF, i.e. $\omega_2 < (1.0 - 1.5)$ rad/s. The necessity of choosing a value of ω_2 lower than the lower limit of the RoMF results from the fact that for pulsation ω_2 the regulator gain is 3 dB higher than desired K_t, while the gain should be kept as close as possible to K_t in the entire RoMF (Figure 15.6b). The values of the gain K_t according to Demello and Concordia (1969) should fulfill the condition

$$K_t \leq \frac{T'_{do}}{2T_E} \tag{15.16}$$

where T'_{do} is the open-circuit d-axis transient time constant of generator and T_E is the exciter time constant. For example, for a generator characterized by $T'_{do} = 9.2$ s and an exciter characterized by $T_E = 0.1$ s the transient gain should fulfill condition $K_t \leq 46$, while for $T_E = 0.025$ s, e.g. for the static exciter, the condition takes the form $K_t \leq 184$. Gibbard et al. (2015) suggest taking the value of this coefficient from the range $K_t = 25$ - 50 pu.

Next, the gain K_0 should be assumed while pulsation ω_1 (and T_B) should be calculated. Or conversely, the pulsation ω_1 should be assumed while the gain K_0 should be calculated. Correlation among the gains and pulsations results from the transfer function of the considered system for pulsations tends to infinity. Assuming $T_A = 0$, for $\omega \to \infty$ or $s = j\omega \to \infty$, the AVR transient gain becomes equal to

$$K_t = K_A \frac{T_C}{T_B} = K_A \frac{\omega_1}{\omega_2} \tag{15.17}$$

The gain K_0 at steady state (i.e. for $t \to 0$ and hence for $\omega \to 0$ or $s = j\omega \to 0$) becomes equal to $K_0 = K_A$ and should ensure the generator voltage error minimization. For admissible (maximum) voltage error equal to $\pm 0.5\%$ the minimum gain is equal to $K_0 = 1/0.005 = 200$.

As an example, for a generating unit with static exciter considered further in this chapter, the AVR gain is high and equal to $K_A = 1170$ and the TGR time constants are equal to $T_B = 20.4$ s, $T_C = 2.4$ s. This means that the cutoff pulsations are equal to $\omega_1 = 1/T_B = 0.05$ rad/s and $\omega_2 = 1/T_C = 0.42$ rad/s. The pulsation ω_2 fulfills requirement $\omega_2 < (1.0 - 1.5)$ rad/s with adequate margin. The transient gain according to Eq. (15.15) is then equal to $K_t = 137.6$, what fulfills requirement (15.16) for the fast exciter.

It is also worth noting that the phase shift for RoMF is not big (Figure 15.6b), which can be advantageous for PSS design.

Another way of the AVR frequency characteristics correction is realized by the application of the corrector in the feedback loop. The regulator block diagram is presented in Figure 15.7a. The feedback transfer function is equal to $sK_F/(1 + sT_F)$. Output E_f (which is the input to the feedback block), depending on the type of the excitation system G_{ex}, is proportional to the field current of the generator or to the exciter output voltage or to the output signal of the AVR main control path.

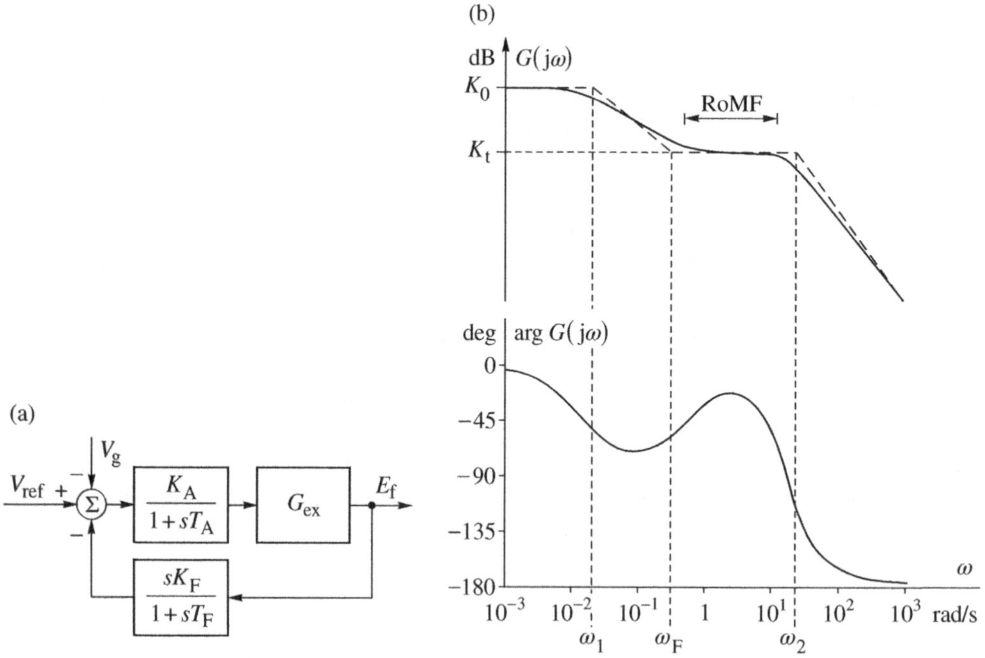

Figure 15.7 AVR with corrector in a feedback loop: (a) block diagram; (b) transfer function.

An example of a desired magnitude-frequency characteristic of the AVR with feedback is shown in Figure 15.7b. As above, the real characteristic (marked by a continuous line) is imposed with the straight-line approximation (marked by a dashed line) with characteristic points with coordinates (ω_1, K_0), (ω_F, K_t), and (ω_2, K_t). The new cutoff pulsation, in comparison to characteristics presented in Figure 15.6b, is pulsation ω_F related to the corrector in feedback time constant $T_F = 1/\omega_F$.

For the analysis simplification it is assumed that the main amplifier and the exciter transfer functions are equal to

$$G(s) = \frac{K_A}{1 + sT_A} G_{ex}(s) = \frac{K_0}{1 + sT} \tag{15.18}$$

where the exciter transfer function is equal to $G_{ex}(s) = K_E/(1 + sT_E)$. Assuming $K_E = 1$, $T_E = 0.1$ s, and $T_A = 0$ one can get $K_0 = K_A$ and $T = T_E$. Then the transfer function of the system (with feedback) shown in Figure 15.7a takes the form

$$\frac{E_f}{V_{ref}} = \frac{K_0(1 + sT_F)}{1 + s(T + T_F + K_0K_F) + s^2TT_F} \tag{15.19}$$

which can be considered as a product of two transfer functions, i.e.

$$\frac{E_f}{V_{ref}} = K_0\frac{(1 + sT_F)}{1 + s/\omega_1} \cdot \frac{1}{1 + s/\omega_2} \tag{15.20}$$

Comparing denominators of Eqs. (15.19) and (15.20) one gets relation between pulsations ω_1, ω_2, and the AVR parameters

$$\omega_1\omega_2 = \frac{1}{TT_F} \tag{15.21}$$

$$\omega_1 + \omega_2 = \frac{T + T_F + K_0K_F}{TT_F} \tag{15.22}$$

It is also required that the condition $\omega_2 \geq 10\omega_F$ be met.

For pulsations tending to infinity, i.e. $\omega \to \infty$, the gain of the left transfer function in Eq. (15.20), including K_0, tends to $K_t = K_0T_F\omega_1$, which means pulsation can be calculated

$$\omega_1 = \frac{K_t}{K_0T_F} \tag{15.23}$$

Then by substituting Eq. (15.23) into Eq. (15.21) pulsation ω_2 is defined as

$$\omega_2 = \frac{K_0}{K_tT} \tag{15.24}$$

Next, by substituting Eqs. (15.23) and (15.24) to Eq. (15.22), the relationship between the gain K_F and the time constant $T_F = 1/\omega_F$ of the corrector becomes equal to

$$K_F = T_F\left(\frac{1}{K_t} - \frac{1}{K_0}\right) + \frac{T}{K_0}\left(\frac{K_t}{K_0} - 1\right) \tag{15.25}$$

According to Eq. (15.25), for a big difference between the gains, i.e. $K_0 \gg K_t$ and the time constants $T_F \gg T$ the relationship between the gain and the time constant of the corrector is equal to $K_F \approx T_F/K_t$.

As an example, it is assumed that the AVR gain is equal to $K_0 = K_A = 1170$, the transient gain is equal to $K_t = 137.6$, and the time constant in feedback block is equal to $T_F = 2.4$ s. The pulsation $\omega_F = 1/T_F = 0.42$ rad/s is located

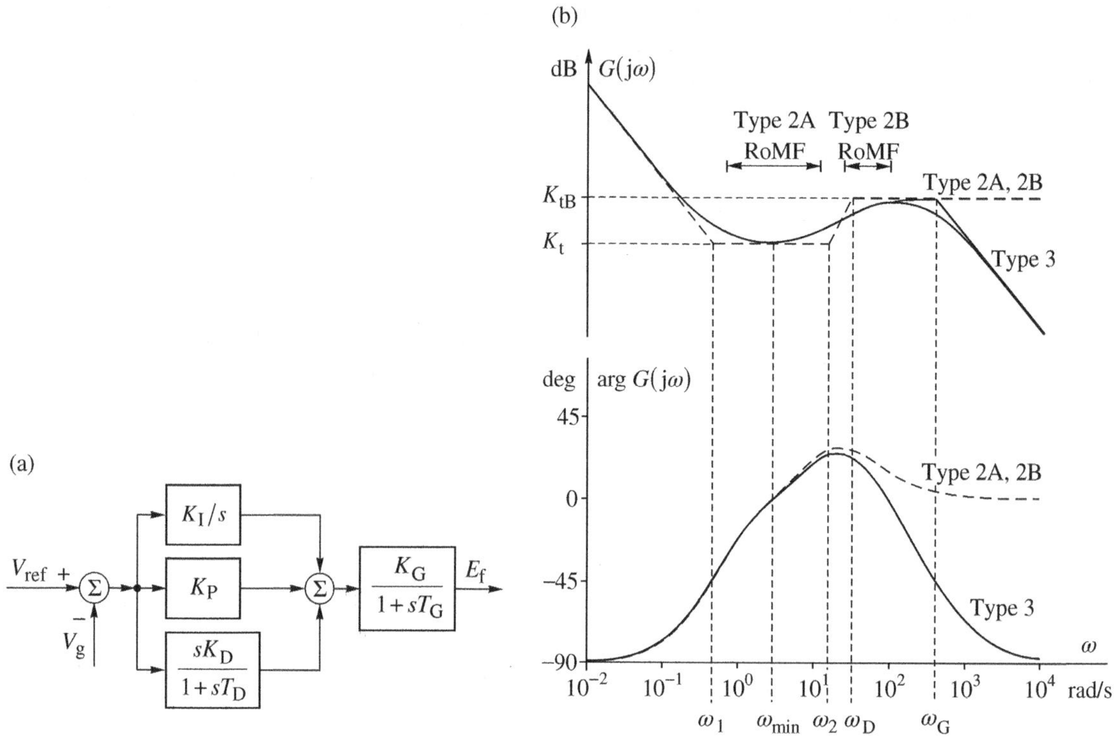

Figure 15.8 AVR with PID regulator: (a) block diagram; (b) transfer function.

below the lower limit of the RoMF with adequate margin. According to Eqs. (15.23) and (15.24) the cutoff pulsations are equal to $\omega_1 = 0.05$ rad/s and $\omega_2 = 85$ rad/s. The RoMF is then located between ω_F and ω_2. According to Eq. (15.25), the feedback gain is equal to $K_F = 0.0153$.

For this type of correction, the AVR phase shift (Figure 15.7b) is more sensitive to the frequency, in comparison to serial correction (Figure 15.6).

The third way of the frequency characteristic correction discussed here provides the PID regulator. The PID regulator block diagram is presented in Figure 15.8a, while its transfer function is equal to

$$G(s) = \left(\frac{K_I}{s} + K_P + \frac{sK_D}{1 + sT_D}\right)\frac{K_G}{1 + sT_G} \tag{15.26}$$

Theoretically, there are three types of PID regulator considered:

- Type 1, which uses an ideal differentiator, i.e. $T_D = 0$. The output inertia block is neglected, i.e. $K_G = 1$ and $T_G = 0$. The presence of the ideal differentiator leads to an infinite increase of the regulator gain for frequencies tending to infinity, which can be a drawback of this type of regulator.
- Type 2, which uses a real differentiator, i.e. $T_D > 0$. The output inertia block is also neglected here, i.e. $T_G = 0$, while its gain can be any positive number, i.e. $K_G > 0$. There are two subtypes of that type of regulator, named Type 2A and Type 2B. For a Type 2A regulator the RoMF is located between the cutoff pulsations ω_1 and ω_2, while for Type 2B regulator the RoMF is located above the cutoff pulsation ω_2 (see Figure 15.8b).
- Type 3, which uses all components of Eq. (15.26). The use of an inertia block leads to the regulator gain reduction for high frequencies, i.e. higher than $\omega_G = 1/T_G$, which can be positive and desirable in some cases.

For the Type 2 PID regulator, with $K_G = 1$, a relation between the cutoff pulsations and the regulator parameters takes the form

$$\omega_{1,2} = \frac{K_P + K_I T_D \pm \sqrt{(K_P + K_I T_D)^2 - 4K_I(K_D + K_P T_D)}}{2(K_D + K_P T_D)} \tag{15.27}$$

where, to keep the cutoff pulsations real, the following condition must be fulfilled

$$K_P \geq K_I T_D \pm 2\sqrt{K_I K_D} \tag{15.28}$$

The parameters of the PID controller can be specified as follows. When the pulsations ω_1 and ω_2 are respectively distant, i.e. $\omega_2 > 10\omega_1$, the ideal PID (Type 1) regulator can be decomposed to PI (proportional-integral) regulator and PD (proportional-derivative) regulator. One can assume here (which is a simplification) that for the cutoff pulsation ω_1 the influence of the PID regulator derivative component on the magnitude of the frequency characteristics can be neglected. Analogously, for the cutoff pulsation ω_2, the influence of the PID regulator integral component on the magnitude of the frequency characteristics can also be neglected.

Then, the PI regulator gain (Figure 15.8b) for pulsation ω_1 and gain $K_G = 1$ is equal to

$$\left| \frac{K_I}{j\omega_1} + K_P \right| = K_{tB} \tag{15.29}$$

which leads to

$$K_I = \omega_1 \sqrt{K_{tB}^2 - K_P^2} \tag{15.30}$$

where K_{tB} is a transient gain for frequency ω_1 (for a PI regulator, which is not indicated in Figure 15.8b), K_I is the integral block gain, K_P is the proportional block gain, equal to the desired transient gain K_t, which is simultaneously equal to the minimum magnitude of the PID regulator transfer function between frequencies ω_1 and ω_2, i.e. $K_t = K_P = K_{min}$.

Analogously, the PD regulator gain (Figure 15.8b) for pulsation ω_2 and gain $K_G = 1$ is equal to

$$|j\omega_2 K_D + K_P| = K_{tB} \tag{15.31}$$

which leads to

$$K_D = \frac{\sqrt{K_{tB}^2 - K_P^2}}{\omega_2} \tag{15.32}$$

where K_D is the differentiator gain. Assuming that the difference between the magnitude of the real frequency characteristic (marked by continuous line) and its straight-line approximation (marked by dashed line) is equal to 3 dB (i.e. $K_{tB} = \sqrt{2}K_P$), from Eqs. (15.30) and (15.32) simple relations among the PID regulator parameters take the form of

$$K_I = \omega_1 K_P \tag{15.33}$$

$$K_D = \frac{K_P}{\omega_2} \tag{15.34}$$

Next, for an ideal PID regulator (Type 1), for pulsation ω_{min}, the regulator gain K_{min} is equal to

$$\left| \frac{K_I}{j\omega_{min}} + K_P + j\omega_{min} K_D \right| = K_{min} \tag{15.35}$$

which leads to

$$K_{\min}^2 = K_P^2 + \frac{\left(K_I - \omega_{\min}^2 K_D\right)^2}{\omega_{\min}^2} \tag{15.36}$$

Therefore, because $K_{\min} = K_P$, from Eq. (15.36), one finally gets

$$\omega_{\min} = \sqrt{\frac{K_I}{K_D}} = \sqrt{\omega_1 \omega_2} \tag{15.37}$$

As an example, the Type 2A PID regulator with transient gain assumed equal to $K_t = 137.6$ and the corner pulsations chosen as equal to $\omega_1 = 0.5$ rad/s and $\omega_2 = 15$ rad/s is taken into consideration. The proportional gain, for $K_G = 1$, is then equal to $K_P = K_t = 137.6$. The integral and differentiator gains according to Eqs. (15.33) and (15.34) are then equal to $K_I = 0.5 \cdot 137.6 = 68.8$ and $K_D = 137.6/15 = 9.2$. In the next step, one of the two parameters, i.e. time constant $T_D = 1/\omega_D$, where $\omega_D > \omega_2$ or the regulator transient gain $K_{tB} > K_t$, has to be assumed, while the second parameter can be calculated from

$$K_{tB} = K_G \left(K_P + \frac{K_D}{T_D} \right) \tag{15.38}$$

Equation (15.38) defines the PID regulator gain derived from Eq. (15.26), for $s = j\omega \rightarrow \infty$. For example, assuming $\omega_D = 20$ rad/s the transient gain becomes equal to $K_{tB} = 321.6$. For these parameters condition (15.28) is fulfilled.

In the case of the Type 3 PID regulator, the value of the inertia block time constant T_G has to be defined as lower than T_D, i.e. condition $\omega_G > \omega_D$ must be fulfilled.

It can be concluded from the above description that designing the AVR by shaping frequency characteristic is relatively simple. Unfortunately, some input information about some parameters, e.g. transient gain, is required. Values of these parameters are determined based on the designer's experience, including knowledge about dynamics of the generating unit operating at idle or loaded. The AVR design is a multi-step process in which various evaluation tools, methods, and indicators for assessing the quality of control, such as frequency characteristics, eigenvalues, time response parameters, etc., are used.

The design process of the AVR based on the frequency characteristics is illustrated below for the IEEE ST1A model shown in Figure 15.5. A regulator with single serial corrector (i.e. with $K_F = 0$) is considered. In general, the AVR can be designed using a single- or multi-machine power system model. For simplicity, without compromising the essence of the design, a mathematical model of an SMIB system with a seventh-order model – Eqs. (11.30) and (11.31) – of a synchronous generator of rated power equal to 360 MW is considered. The model data are presented in Example 15.1. The rated operating point of the generator is assumed. The external impedance (i.e. from the busbars of the step-up transformer to voltage of the infinite busbar) equal to $\underline{Z}_S = (0.007 + j0.08)$ pu is added to the generator model. The nonlinear model of the considered system was linearized at the operating point corresponding to the rated load of the generator. The AVR influence on frequency characteristic is presented in four steps.

Step 0

Consider the open-loop system transfer function $L_0(j\omega) = K_0(j\omega)G(j\omega)$ with a regulator with a gain equal to one $K_0(j\omega) = 1$. This somewhat corresponds to a system without a regulator. Transfer function $G(j\omega)$ is the transfer function of the object, i.e. from the field voltage E_f to the generator terminal voltage V_g. The magnitude and phase-frequency transfer functions of such a defined open-loop system are presented in Figure 15.9, and are denoted as L_0.

The magnitude of this transfer function for any pulsation is less than one, and the phase does not decrease below −180°. It is in the range from 0° to −110°. An object of this type is deemed as structurally stable. A resonant peak with pulsation 8.85 rad/s on the characteristic is also visible. The above means that the closed loop system (i.e. generating unit with P type regulator with gain equal to unity) is stable.

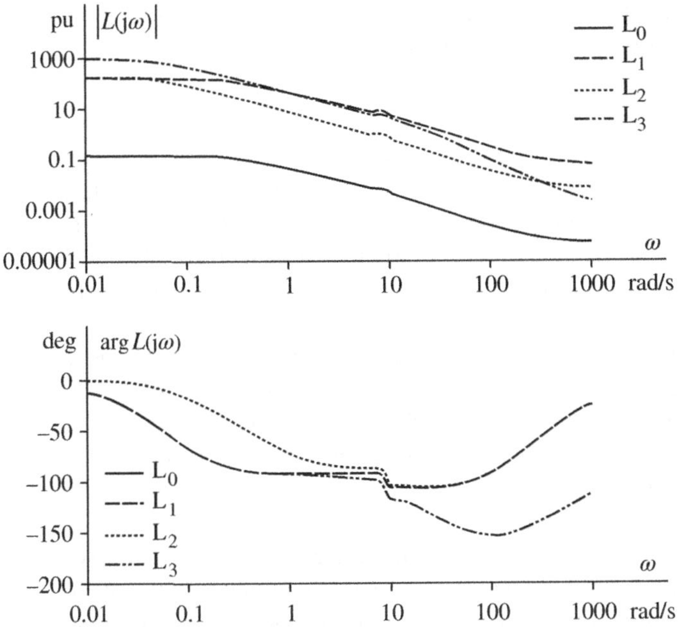

Figure 15.9 Open-loop system frequency characteristic $L(j\omega) = K(j\omega)G(j\omega)$: ($L_0$) $K_0(s) = 1$, (L_1) $K_1(s) = 1170$, (L_2) $K_2(s) = 1170\ (1 + s2.4)/(1 + s20.4)$, ($L_3$) $K_2(s) = 1170\ (1 + s2.4)/(1 + s20.4)$ with static exciter.

Step 1

In the next step, again proportional regulator is used, but with a gain equal to 1170, i.e. the regulator transfer function has the form $K_1(j\omega) = 1170$. The gain corresponds to the default value of the gain K_A in the ST1A model considered before. It is a relatively large gain but used sometimes for synchronous generator with a static excitation system. The open-loop system characteristics $L_1(j\omega) = K_1(j\omega)G(j\omega)$ in Figure 15.9 are denoted as L_1. The regulator $K_1(j\omega)$ shifts the magnitude-frequency characteristic upwards in comparison to the characteristic $L_0(j\omega)$. The position of the phase-frequency characteristic does not change here, i.e. the characteristic arg($L_1(j\omega)$) coincides with the phase-frequency characteristic of the system with regulator $K_0(j\omega)$, i.e. arg($L_0(j\omega)$). The phase margin, for the gain crossover pulsation equal to $\omega_c = 35.3$ rad/s, is equal to $PM = 74°$. The gain margin GM remains undefined because the phase characteristic for all pulsations is located above −180°. This means that the closed-loop system remains stable, and the large phase margin PM indicates the admissibility of a relatively large, unmodeled time delay in the real object.

Step 2

Consider inclusion a single serial corrector with time constants $T_B = 20.4$ s and $T_C = 2.4$ s, which leads to a regulator defined by transfer function $K_2(s) = 1170 \cdot (1 + s2.4)/(1 + s20.4)$, i.e. discussed above. The transfer function of the open-loop system is equal to $L_2(j\omega) = K_2(j\omega)G(j\omega)$, and in Figure 15.9 it is denoted as L_2. The use of a regulator with this structure, in comparison to L_1, causes a shift down of the magnitude-frequency characteristic for pulsations greater than approximately 0.03 rad/s and a shift down of the phase-frequency characteristic for pulsations less than approximately 9 rad/s. This results directly from the phase-frequency characteristic of the corrector. For pulsations larger than 9 rad/s, the phase-frequency characteristic of the open-loop system L_2 does not change as compared to those previously discussed (L_0 and L_1). This is because the phase shift introduced by the corrector is negligibly small for these pulsations. The gain crossover pulsation ω_c shifts from a value of approximately 35.3 rad/s (for L_1) to 5.9 rad/s, and the phase margin increases from 74° (L_1) to 87°. Thus, the use of a corrector reduces the sensitivity of the control system to disturbances (gain in the open-loop is less than one), but at the same time slows down the tracking performance of the control system (lower value of the crossover pulsation).

Step 3

The last of the characteristics shown in Figure 15.9 (denoted as L_3) refer to the system with the regulator as in Step 2, but in which the model of the static exciter, i.e. the excitation transformer with excitation rectifier is added. Simultaneously, time constant of the AVR amplifier was assumed as nonzero and set equal to $T_A = 0.02$ s. The inclusion of the exciter in this form brings the model closer to reality, and also causes the introduction of additional feedback loop from the generator voltage (via the excitation transformer) and from field current (voltage drops in the rectifier).

In this case the magnitude-frequency characteristic $L_3(j\omega) = K_3(j\omega)G(j\omega)$ is located above the characteristic $L_2(j\omega)$ practically in the entire pulsation range. The gain crossover pulsation ω_c increases to a value lower than in case of L_1 and is equal to 25.2 rad/s, while the phase margin decreases to 46°. The phase-frequency characteristic decreases significantly for pulsations greater than 9 rad/s, as a result of the occurrence of an inertial element with a time constant T_A. The closed-loop system is still stable, but its dynamic properties deteriorate in comparison to the system with regulator $K_2(s)$, which does not take into account the exciter.

In general, it can be concluded that decreasing the value of the AVR gain leads to a decrease in the crossover pulsation and thus a slower response of the system. At the same time it increases the immunity of the control system to high-frequency disturbances (noise). The addition of a second corrector would also lead to a decrease in the magnitude-frequency characteristic, which would result in a decrease of the crossover pulsation, a slower tracking response, and a further reduction in the sensitivity to high-frequency disturbances. At the same time, depending on the frequency range in which this element would effectively delay the phase, the phase margin could be reduced.

Since the dynamic properties of the control system depend strongly on the generator operating point, Figures 15.10 and 15.11 show the dependence of the phase margin PM from the generator load (P_g, Q_g) and the external impedance $\underline{Z}_S$. It can be seen here (Figure 15.10) that the influence of the active power on the phase

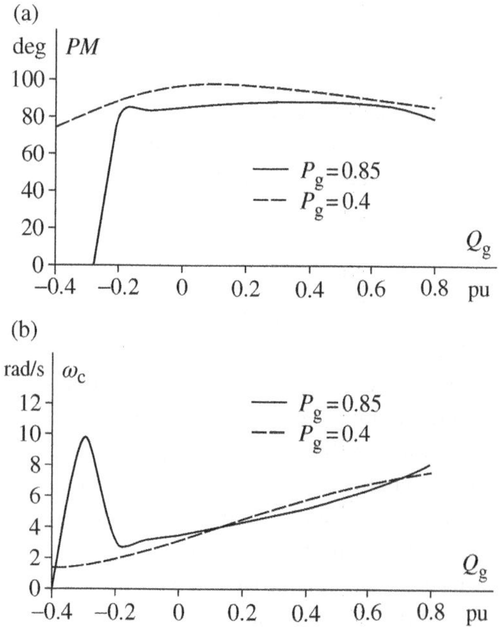

Figure 15.10 Influence of generating unit operating point P_g, Q_g on (a) phase margin PM and (b) gain crossover pulsation ω_c. SMIB model with regulator K_3, $V_g = V_{gn}$.

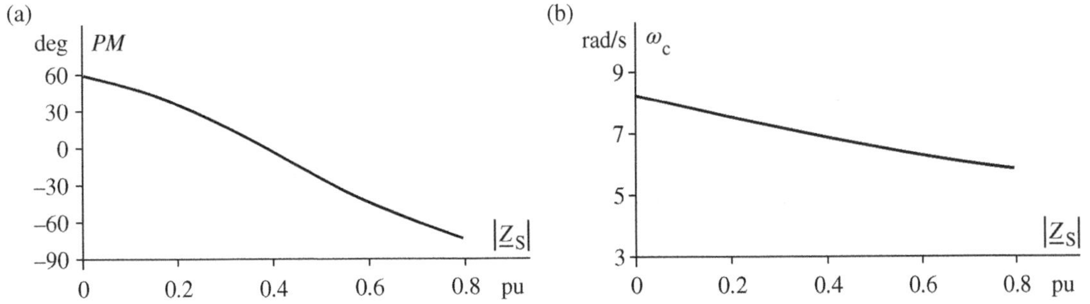

Figure 15.11 Influence of external impedance $\underline{Z}_S$ in SMIB model on (a) phase margin PM and (b) gain crossover pulsation ω_c. Model with regulator K_3, $V_g = V_{gn}$, $P_g = P_{gn}$, and $Q_g = Q_{gn}$.

margin is small. Both presented curves are located close to each other. Also, the influence of the reactive power on the phase margin is small with exception to the generator load near the angle stability limit, i.e. $Q_g < -0.3$ pu and $P_g = P_{gn}$. At the same time, however, a significant effect of the reactive power load on the gain crossover pulsation ω_c is visible, which means a significant impact on the dynamic properties of a generating unit. It decreases as the reactive power load decreases to the value corresponding to the stability limit. An analogous effect (i.e. the decrease of the gain crossover pulsation ω_c and the phase margin PM as it approaches the stability limit) is shown in Figure 15.11 as a result of the external impedance $\underline{Z}_S$ increase. In the presented example, for the impedance bigger than 0.4 pu the phase margin decreases to zero.

The frequency characteristic of the open-loop system $L(s) = K(s)G(s)$ primarily provides information about the stability of the closed-loop system and about margins of phase and gain, which in a general sense is rather an indicator of the quality of control in a closed-loop system. As follows from Eq. (15.5), the quality of the control process is defined by the transfer functions of the closed-loop system, i.e. the sensitivity transfer function $S(s)$ and the complementary sensitivity transfer function (tracking transfer function) $T(s)$.

To evaluate quality of control, Figure 15.12 presents the frequency characteristics of the closed-loop transfer functions: tracking transfer function $T(s) = V_g/V_{ref}$ and transfer functions $V_g/V_S = S(s)G_{dV}(s)$ and $V_g/\tau_m = S(s)G_{d\tau}(s)$. The latter transfer functions are a combination of a sensitivity transfer function $S(s) = [1 + K(s)G(s)]^{-1}$ and disturbance models (Figure 15.1): $G_{dV}(s)$ from disturbances in the system voltage V_S and $G_{d\tau}(s)$ from disturbances at the mechanical torque τ_m. As shown in Figure 15.12a, the magnitude-frequency characteristic of tracking transfer function $T(j\omega)$ is equal to one in a wide pulsation range, i.e. from 0 to about 9 rad/s, which is a desirable effect. However, for pulsation 25.2 rad/s, a resonance peak with a large (larger than suggested in IEEE Std. 421.2-2014) magnitude $|T| = 2.97$ is visible. However, since the generator voltage reference does not change frequently – this is determined by the very slow supervisory regulator – the occurrence of this peak, as well as its magnitude, are not important from the point of view of the quality of the tracking control properties. In the case of frequent changes in the voltage reference, a pre-compensator should be used.

In turn, the magnitude-frequency disturbance characteristic from the system (infinite busbars) voltage (Figure 15.12b) is characterized by very low values for low pulsations, which is desirable. Simultaneously, it is characterized by mean magnitudes at the level $|SG_{dV}| = 0.5$ for high pulsations, which, however, means a certain level of disturbance attenuation. The resonant peak magnitude is here equal to $|SG_{dV}| = 1.95$, which is not beneficial, because it means amplification of disturbances with pulsations close to 26.4 rad/s, i.e. close to RoMF, whereas the magnitude-frequency disturbance characteristic (i.e. from the mechanical torque) is characterized (Figure 15.12c) by very small magnitudes for low and high pulsations, which is desirable. There are visible two resonant peaks, but also of small magnitudes (pulsations 9.4 and 25.2 rad/s). In general, all characteristics of the closed-loop system are characterized by the resonant pulsation, equal to 25.2 rad/s, which does not occur

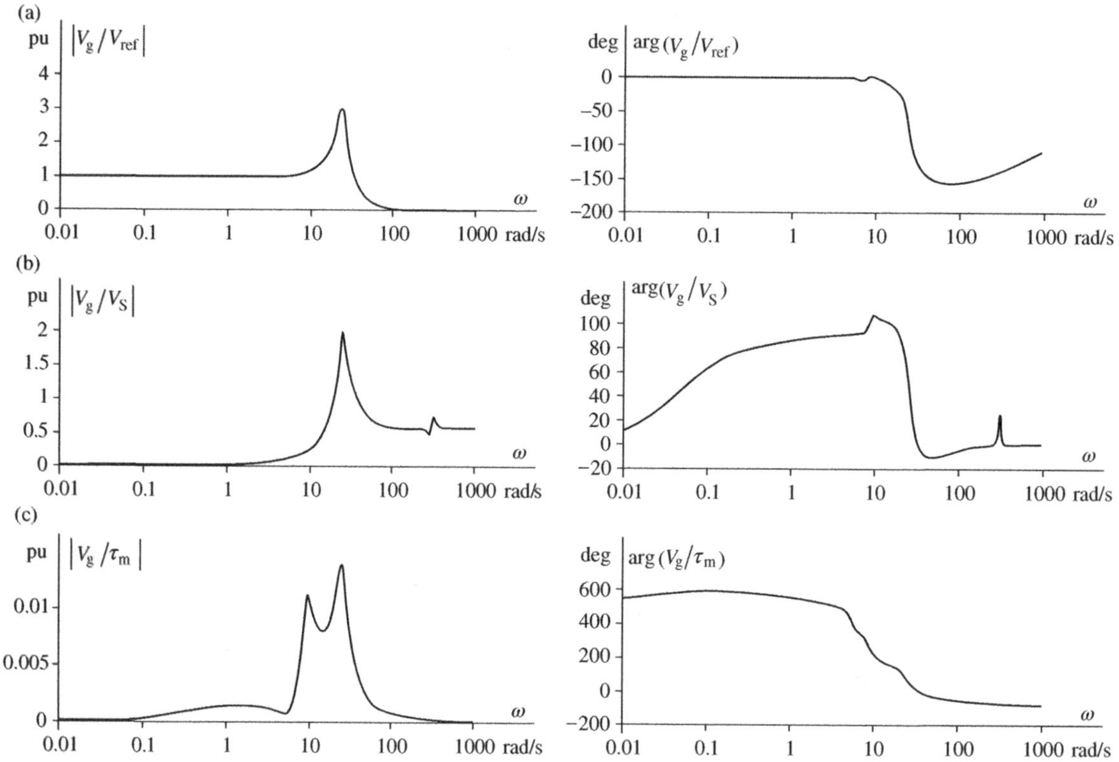

Figure 15.12 Frequency characteristics of a closed-loop system with regulator K_3: (a) $T(s) = V_g/V_{ref}$; (b) $S(s)G_{dV}(s) = V_g/V_S$; (c) $S(s)G_{d\tau}(s) = V_g/\tau_m$.

in the frequency characteristics of the open-loop system. The occurrence of other resonant pulsations in the closed-loop system than in the open-loop system results from the fact that the denominator of a closed-loop system transfer function is equal to the sum of the polynomials of the numerator and the denominator of the open-loop system. This means that the zeros of the open-loop system modify the poles of the closed-loop system. At the same time, however, the zeros of the open-loop system remain zeros of the closed-loop system.

The frequency characteristics determine the dynamic properties of the control system in the qualitative sense, but in a limited way (in the quantitative sense), they define the response of the control system. The quality of control in the quantitative sense, for the linear system, is indicated by eigenvalues of the closed-loop system and by time response of the system to selected disturbances. For the system under consideration with regulator K_3 and the generator rated load, eigenvalues $\underline{\lambda}_i$ are presented in Table 15.2. These values indicate three oscillatory modes with pulsations equal to 313.88, 25.58, and 8.85 rad/s. The first of these is related to the network frequency ($f = 50$ Hz here) and occurs in mathematical models of a synchronous generator (Eqs. [11.30] and [11.31]) in which the transformation electromotive forces are not neglected. The third mode (8.85 rad/s), with frequency $f = 1.41$ Hz, is associated with electromechanical oscillations and is visible in the frequency characteristic of the open-loop system $L_3(j\omega) = K_3(j\omega)G(j\omega)$ and in the disturbance characteristic of the closed-loop system $V_g/\tau_m = S(s)G_{d\tau}(s)$. Whereas the second mode (25.58 rad/s), with frequency $f = 4.07$ Hz, is visible on all presented frequency characteristics of the closed-loop system, and its source is the AVR. This mode is referred to as an *excitation mode* or *control mode*.

As a supplement to considerations related to the eigenvalues, Figure 15.13 shows the eigenvalues' loci associated with the two low-frequency modes as a function of the AVR gain K_A. The AVR gain is changing here from zero to 1200. It can be seen here that the gain increase causes a diminishing of the damping of the excitation mode $\underline{\lambda}_R$ with

Table 15.2 Eigenvalues of the SMIB system.

No.	$\underline{\lambda}_i$ for K_A = 1170 rad/s	$\underline{\lambda}_i$ for K_A = 585 rad/s
1	$-8.25 \pm j313.88$	$-8.25 \pm j313.88$
2	-59.59	$-55.25 \pm j3.45$
3	-56.45	-18.83
4	$-4.73 \pm j25.58$	-7.57 ± 17.51
5	-18.94	-1.32 ± 8.95
6	$-1.34 \pm j8.85i$	-0.42
7	-0.42	–

the increase of its pulsation and a slightly increasing damping of the electromechanical mode $\underline{\lambda}_M$. For the regulator gain $K_A = 1200$, the damping ratio $\xi = \cos\theta$ of the electromechanical and excitation modes becomes the same. This does not mean, however, that the value of the gain is optimal. It may turn out that a given value of a mode pulsation (for example an excitation mode) is, for various reasons, unacceptable for a specific EPS.

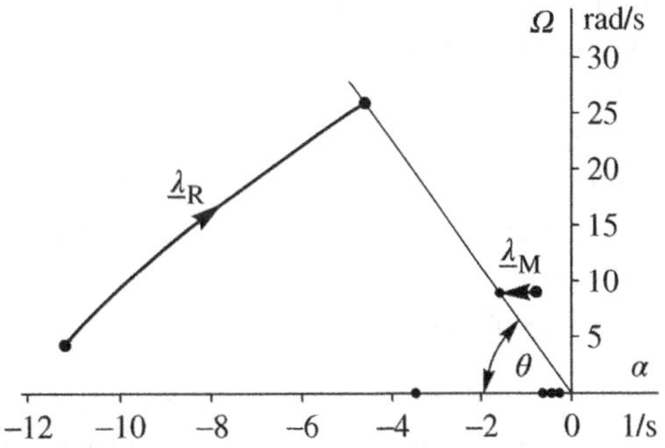

Figure 15.13 Eigenvalues' loci of the SMIB control system with regulator K_3 as a function of the AVR gain K_A.

Figure 15.14 presents responses of the generating unit to the step change in voltage reference V_{ref} and the step change of the infinity busbars voltage V_S equal to 0.01 pu for two values of the gain K_A. Both responses are characterized by damped oscillations. For $K_A = 1170$ the settling time (assuming accuracy of control $\varepsilon = 0.001$ due to the small value of the voltage step change) is equal $t_\varepsilon = 0.51$ s, while the overshoot is equal to 0.45%. The voltage overshoot limitation can be achieved as a result of decreasing the regulator gain. For $K_A = 585$ almost double reduction of the overshoot, shortening of the voltage settling time to $t_\varepsilon = 0.35$ s and reduction of the frequency of the periodic oscillations is achieved. This change in the frequency of the oscillation is also visible in eigenvalues (Table 15.2). There are two oscillatory modes for the gain under consideration, with pulsation 17.51 rad/s (2.79 Hz) and 3.45 rad/s (0.55 Hz). The first of these is clearly visible in the responses shown in Figure 15.14. The second (not shown in Figure 15.13) is a highly damped mode with a relatively low frequency.

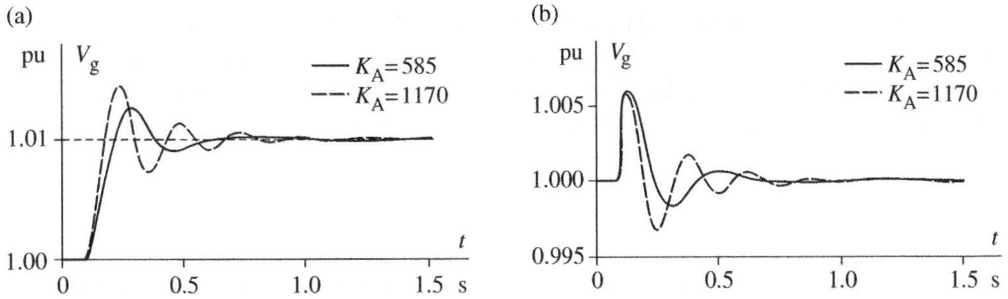

Figure 15.14 Responses of the generating unit to step change in: (a) voltage reference V_{ref} and (b) infinite busbars' voltage V_S, for various values of the AVR gain K_A.

One can see also that decreasing the regulator gain slows down the response of the control system, as shown in Figure 15.14. This effect is not always associated with an elongation of settling time, which is exactly the case here.

As mentioned earlier, the process of the AVR design is a multi-step process and is often performed by trial and error. The above is an analysis presented for the defined values of the regulator parameters. This example should be treated only as a single step in the AVR design process.

15.3.3 Power System Stabilizers

Damping of electromechanical oscillations in power systems is not a new problem. Already in the 1950s, when power systems began to grow and operate with larger loads, regulators were sought to improve oscillation damping. Currently, the most commonly used element to improve power system's stability are PSSs. PSSs introduce an additional, stabilizing signal in the synchronous generator voltage regulator path (Figure 10.6). Typical, practically applicable, PSS structures are described in Section 10.1, and wider in IEEE Std 421.5-2016. Quite commonly single-input PSSs with fixed parameters (time-invariant) are used, although in recent years there has been a visible increase in the popularity of dual-input stabilizers.

The design of a PSS (understood here as the definition of parameters for a defined structure) is usually based on the compensation of the frequency characteristic of the object, the optimization of a specific performance (control quality) index, or on the pole placement (shifting) of the control system transfer function. Other methods of the PSS's synthesis are also used, including methods based on so-called artificial intelligence (neural networks, fuzzy logic), using techniques of synthesis of so-called optimal regulators (LQG/LQR, H_2, H_∞, synthesis μ), and using adaptive algorithms.

The methods of designing PSSs that are in practical use include:

- Method proposed by Demello et al. (1986) based on transfer function P-Vr from the voltage reference input to the electromagnetic torque or active power output. The P-Vr transfer function is calculated from the mathematical model of a single- or multi-machine power system with the shaft dynamics neglected, i.e. with an assumption of the generating unit (or all units in case of multi-machine model) inertia constant equal to infinity ($H = \infty$).
- GEP (generator-exciter-power) method (Larsen and Swann 1981) based on the transfer function from the voltage reference input to the generator terminal voltage output. This is a practical procedure based on measurement taken in the field. According to Gibbard et al. (2015) "the use of the field measured frequency response[12] for PSS design relies on assumption that, because the generator is connected to a large power system, its speed remains more-or-less constant during frequency response measurement. This is equivalent to assuming that the inertia constant of the unit is very large or, alternatively, the speed and angle perturbations are negligible."[13]
- Method proposed by Pagola et al. (1989) based on residues of the transfer function from the voltage reference input to the shaft speed output.

The foundations for the PSS synthesis were presented in Demello and Concordia (1969). Practical methods of synthesis were presented in Larsen and Swann (1981), and their development in the work of Demello et al. (1986). Procedures for the synthesis of stabilizers in multi-machine systems are proposed in the publication of Athay et al. (1979) and then developed in Gibbard (1991). Methods for the PSS synthesis based on GEP and P-Vr transfer functions are compared and discussed in the work of Gibbard and Vowles (2004). The interested reader can refer to the

---

12 Phase-frequency characteristic $\arg(V_g/V_{ref})$.

13 Notice that in case of the Heffron–Phillips model (Figure 15.15) with the shaft dynamics disabled ($H = \infty$) both the GEP and P-Vr transfer functions differ by scalar ratio only, i.e. $GEP(j\omega) = (K_2/K_6)(\text{P - Vr})(j\omega)$, which means that the phase-frequency characteristics of both transfer functions are identical. But when the shaft dynamics is not disabled the GEP phase-frequency characteristics differs from the P-Vr characteristics, especially near the unit's resonant frequencies. This can complicate the PSS design by using the GEP method.

work of Kundur (1994) and Machowski et al. (2008), which contain general considerations regarding system stabilizers and their design, as well as reports and standards (Bérubé and Hajagos 2005; CIGRE Technical Brochure No. 166 2000; IEEE Std. 421.2-2014; IEEE Std. 421.3-2016) assessing the applied synthesis methods. Valuable practical considerations can also be found in the publications of Lee and Kundur (1986), Murdoch et al. (1999), and Gibbard et al. (2015).

To illustrate the idea of the GEP and P-Vr methods, the Heffron–Phillips model (Heffron and Phillips 1952) of single-machine system is usually used. In this model, shown in Figure 15.15, the synchronous generator is equipped with an AVR and a PSS utilizing the rotational angular velocity of the rotor $\Delta\omega$ as an input signal. This signal is obtained from the ideal filter F, while input signal q is not defined at this stage. In consideration below this filter is neglected and it is assumed that $\Delta\omega$ is the input signal of the PSS considered.

Assuming $H = \infty$, the phasor of the electromagnetic torque $\Delta\tau_e$ can be defined as the function of the phasors of the power angle $\Delta\delta$, the generator voltage reference ΔV_{ref}, and the angular velocity of the generator rotor $\Delta\omega$

$$\Delta\tau_e = T_\delta(s)\Delta\delta + T_V(s)\Delta V_{ref} + T_{PSS}(s)\Delta\omega \tag{15.39}$$

where $T_\delta(s)$, $T_V(s)$, and $T_{PSS}(s)$ are certain transfer functions depending on the parameters $K_1, \ldots, K_6$, T_3 of the model (Heffron and Phillips 1952; Demello and Concordia 1969) and from the AVR transfer function $G_{ex}(s)$ and the PSS transfer function $G_{PSS}(s)$.

The first component of Eq. (15.39), containing the transfer function $T_\delta(s)$, can be expressed in the following way

$$T_\delta(s)\Delta\delta = \left(K_1 - \frac{K_2K_3(K_4(1+sT_R) + K_5G_{ex}(s))}{(1+sT_3)(1+sT_R) + K_3K_6G_{ex}(s)}\right)\Delta\delta \tag{15.40}$$

This component, by using substitutions $\Delta\delta = (\omega_0/s)\Delta\omega$ (according to Figure 15.15) and $s = j\omega$, takes a form

$$\begin{aligned} T_\delta(j\omega)\Delta\delta =& \ \mathrm{Re}\{T_\delta(j\omega)\}\Delta\delta + \mathrm{Im}\{T_\delta(j\omega)\}\Delta\delta = \\ =& \ \mathrm{Re}\{T_\delta(j\omega)\}\Delta\delta + \mathrm{Im}\{T_\delta(j\omega)\}\frac{\omega_0}{j\omega}\Delta\omega = \\ =& \ \mathrm{Re}\{T_\delta(j\omega)\}\Delta\delta + \mathrm{Re}\left\{\frac{\omega_0}{j\omega}T_\delta(j\omega)\right\}\Delta\omega \end{aligned} \tag{15.41}$$

The real component of transfer function $T_\delta(j\omega)$, being in phase with the power angle is proportional to the synchronizing torque, and the imaginary component of the transfer function $T_\delta(j\omega)$, multiplied by ω_0/ω, being in phase with the angular velocity, is proportional to the damping torque. On the basis of Eq. (15.41), defining the frequency characteristics, the synchronizing τ_S, and damping τ_D torques can be presented in the form

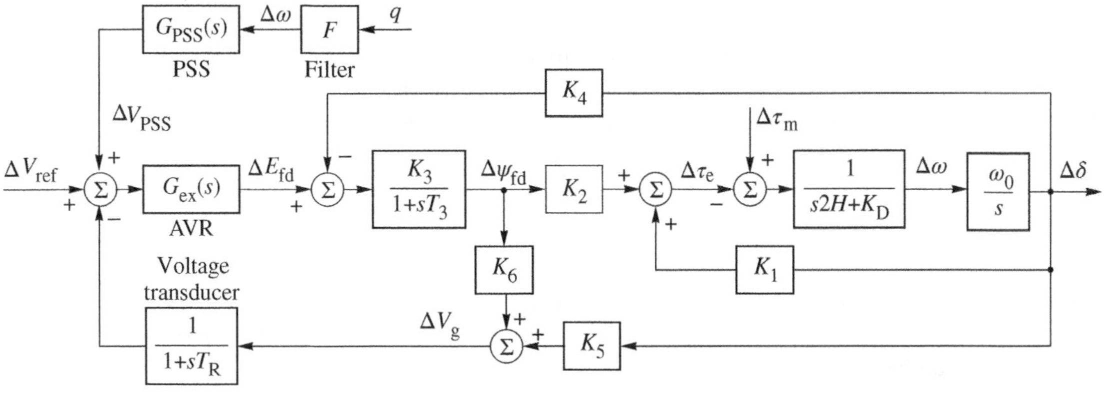

Figure 15.15 Linear model of an SMIB with voltage regulator (AVR) and power system stabilizer (PSS).

$$\tau_S = \text{Re}\{T_\delta(j\omega)\}; \quad \tau_D = \text{Re}\left\{\frac{\omega_0}{j\omega}T_\delta(j\omega)\right\} \tag{15.42}$$

The third component of Eq. (15.39), containing the transfer function $T_{PSS}(s)$, and in it the PSS transfer function $G_{PSS}(s)$, is equal to

$$\Delta\tau_{PSS} = T_{PSS}(s)\Delta\omega = \frac{K_2K_3G_{ex}(s)G_{PSS}(s)(1+sT_R)}{(1+sT_R)(1+sT_3)+K_3K_6G_{ex}(s)}\Delta\omega \tag{15.43}$$

This component is equal to the electromagnetic torque created by the PSS. The real component of transfer function $T_{PSS}(s)$ is proportional to the damping torque and the imaginary component is proportional to the synchronizing torque.

In turn, the second component of Eq. (15.39), proportional to the electromagnetic torque associated with the change of the voltage reference V_{ref}, is equal to

$$\Delta\tau_V = T_V(s)\Delta V_{ref} = \frac{K_2K_3G_{ex}(s)(1+sT_R)}{(1+sT_R)(1+sT_3)+K_3K_6G_{ex}(s)}\Delta V_{ref} \tag{15.44}$$

The transfer function $T_V(s)$ is equal to P-Vr transfer function, and by comparing Eq. (15.43) to Eq. (15.44) it can be seen that

$$T_{PSS}(s) = T_V(s)G_{PSS}(s) \tag{15.45}$$

If one assumes that the ideal PSS should provide only the damping torque, the transfer function $T_{PSS}(s)$ should be real. In addition, if one assumes that the value of the damping torque provided by the PSS should not depend on the frequency, the transfer function $T_{PSS}(s)$ should be a real number, i.e.

$$T_{PSS}(j\omega) = T_V(j\omega)G_{PSS}(j\omega) = K_{PSS} \tag{15.46}$$

This in turn means that the transfer function of ideal PSS should be equal to

$$G_{PSS}(s) = K_{PSS}/T_V(s) \tag{15.47}$$

For the PSS transfer function defined in such a way, the K_{PSS} is equal to damping torque provided by PSS, and sums up with the damping torque τ_D, resulted from the synchronous machine parameters, the AVR parameters, and the operating point of the generating unit.

Transfer function $G_{PSS}(s)$ determined on the basis of Eq. (15.47) only provides compensation for phase shifts resulting from the frequency characteristic $T_V(s)$, i.e. P-Vr transfer function. The other elements necessary for the proper operation of the PSS are the low-pass filter used to attenuate high-frequency oscillations (noise) and the high-pass filter used to eliminate the constant component of the input signal in steady states. Typically, the high-pass filter with the transfer function $sT_w/(1+sT_w)$, the so-called wash-out functions, is used.

The inclusion of these two filters in the PSS structure leads to a structure as shown Figure 10.7 (Section 10.1.2) and to the following form of the PSS transfer function

$$G_{PSS}(s) = G_f(s)G_w(s)G_c(s)K_{s1} \tag{15.48}$$

where the individual components of the PSS transfer function are: $G_f(s)$ – low-pass filter, $G_w(s)$ – high-pass filter, $G_c(s)$ – phase compensation block (equivalent of $G_{PSS}(s)$ transfer function in Eq. (15.47), and K_{s1} – PSS gain. This way of the PSS compensation block design has two drawbacks:

- The transfer function $T_V(s)$ is defined for a given model, which means that the PSS is optimal for a model while not necessarily for the real object.

- The transfer function $T_V(s)$ is calculated for a single operating point and for a single structure of the EPS, i.e. the parameters of the generating unit and rest of the power system. This again means that the PSS in a real system performs sub-optimally.

To minimize these effects, the P-Vr transfer function $T_V(s)$ used in the PSS design process is calculated (or measured) for the operating point and for the EPS parameters ensuring best performance of the PSS over a wide range of operating points.

In real power systems various types of PSSs are used. Their structures are presented and discussed in Section 10.1.2: PSS based on speed (further named PSS-ω), PSS based on active power (further named PSS-P), PSS based on speed and active power (further named PSS-P-ω), and PSS based on frequency. Other structures of PSSs are also considered. Usually, they are modifications of the above structures. An example of such a PSS is a stabilizer consisting of several parallel compensators tuned to various modal frequencies of oscillations (PSS4B, IEEE Std 421.5-2016).

Figure 15.16 t presents two basic (out of a dozen defined by IEEE Std 421.5-2016) structures of PSSs. These are: the single input PSS with active power as input (IEEE type PSS1A) and the dual-input PSS with active power and rotor speed (frequency) as inputs (IEEE type PSS2B). The process of the PSSs parameters' selection (PSS design) is described below. The process can be divided into the following stages:

- Selection of parameters of low-pass and high-pass filter: parameters K_{s2}, K_{s3}, and T_6, ..., T_9 for the PSS2B stabilizer or A_1 and A_2 for the PSS1A stabilizer (filter with transfer function $1/(1 + A_1 s + A_2 s^2)$) and time constants of the high-pass filters T_{w1}, ..., T_{w4}.
- Selection of the time constants of the compensators T_1, ..., T_4, T_{10}, and T_{11}.
- Selection of the PSS gain K_{s1}.
- Selection of limits for the output signal V_{PSS} and active power threshold $P_{blocking}$ causing the PSS blocking.

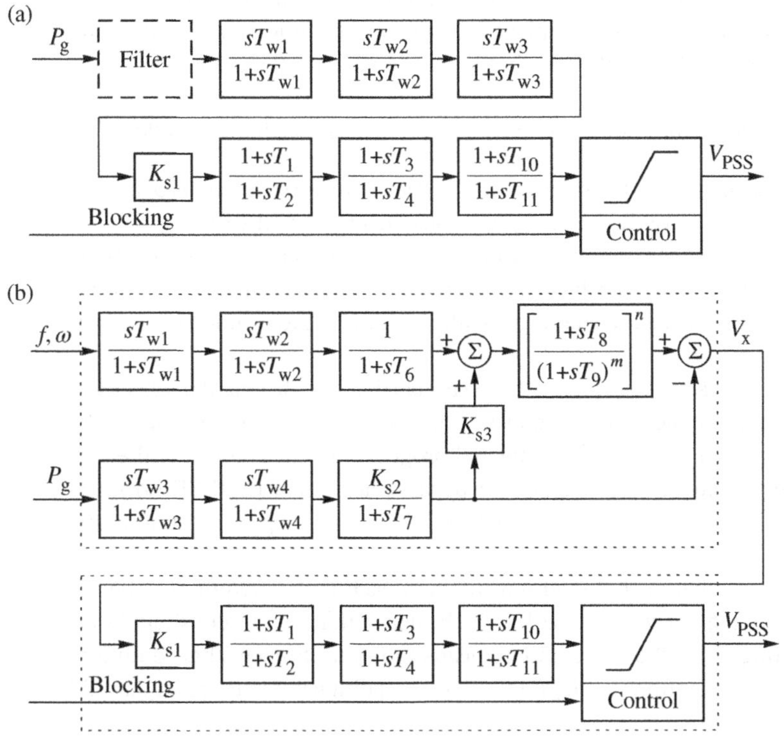

Figure 15.16 Block diagrams of PSSs: (a) single-input type PSS1A (PSS-P) and (b) dual-input type PSS2B (PSS-P-ω). *Source:* Based on IEEE Std 421.5-2016.

15.3.3.1 Selection of Low-pass $G_f(s)$ and High-pass $G_w(s)$ Filters Parameters

The selection of a low-pass filter in the case of a single-input PSS is carried out in a classical way, i.e. depending on the requirements, including the signals occurring in the input signal with undesired frequencies, a low-pass filter is used. As the cutoff frequency of the low-pass filter, i.e. the frequency above which the amplification (magnitude of frequency characteristic), drops to the level of 0.707 (−3 dB), the frequency is usually assumed a decade above the highest frequency of the rotor oscillations – Table 5.1 in Section 5.4.6), i.e. from the range of (8–15) Hz, (50–94) rad/s. If it is necessary to eliminate torsional oscillations with low frequency, e.g. (5–8) Hz, filters of second or higher order are used instead of the first order filters.

For example, for a filter with a transfer function $1/(1 + A_1 s + A_2 s^2)$, for a cutoff frequency of $\omega_c = 50$ rad/s (8 Hz) the filter time constant is equal to $T = 1/\omega_c = 0.02$ s, leading to $A_1 = 0.02$ and $A_2 = 0$ for the first-order filter and $A_1 = 0.04$ and $A_2 = 0.0004$ for the second-order filter.

In turn, in the case of a high-pass (wash-out) filter the cutoff frequency should be about a decade below the rotor modes of oscillations, usually a decade below the lower frequency of inter-area oscillations (Table 5.1 in Section 5.4.6). For example, for the minimum frequency of inter-area mode equal to 0.3 Hz (1.9 rad/s), the filter cutoff pulsation can be assumed to be $\omega_c = 0.19$ rad/s and the time constant of the filter is equal to $T_w = 1/\omega_c = 5.3$ s.

The number of the wash-out blocks in series can be more than one. Usually, the single wash-out filter is used, despite it does not eliminate ramp signal like the two wash-out filters in series do.

Filter selection in the case of a dual-input stabilizer is more complex. In the case of a stabilizer of this type, it is possible to obtain significant insensitivity to the linear during time change of the active power resulting, for example from a change in the governor power reference (power ramping shown in Figure 9.11 in Section 9.1.3). This is particularly valuable in the case of the participation of a given generating unit in frequency control.

This effect is not achievable in the case of a single-input PSS-P with active power as the input. Such stabilizers react to a linear in time change in the active power, causing a significant change in generator voltage in the transient state (Bérubé and Hajagos 2005). An example of the difference in the reaction of a single-input PSS-P and a dual-input PSS-P-ω for a linear change in the active power is shown in Figure 15.17.

The rules for the selection filter parameters of the dual-input stabilizer PSS-P-ω type PSS2B are as follows:

- The time constants $T_{w1} = T_{w2} = T_{w3} = T_7$ are taken from the range of 2.5 to 10 s. There are suggestions of accepting these values equal to $2H$. The adoption of small values of these time constants leads to the filtering of electromechanical oscillations signals with low frequencies and thus the elimination of the ability to suppress these oscillations by the PSS. However, if the purpose of a given PSS is to attenuate only the local-area oscillations (typically with frequencies between 1.0 and 2.0 Hz), then taking smaller values of these time constants is justified.
- For a filter on which output one wants to obtain signal V_x (Figure 15.16) proportional to the angular velocity of the rotor and hence proportional to the integral of accelerating power, time constant $T_{w4} = 0$. This means the elimination (or bypass) of a given block. In real PSSs and in commercial programs for the analysis of the EPS dynamics, setting the value of this time constant equal to zero results in a bypass of a given block. IEEE Std 421.5-2016 introduces the PSS2C stabilizer model, which formalizes this state, i.e. it includes bypass function of the block with time constant T_{w4}.

  In the case of dual-input stabilizers, the number of the wash-out filters in each path is required to be the same. Two wash-out filters are used in the angular velocity path, which means that two filters should also be used in the active power path. Accepting $T_{w4} = 0$ only seemingly leads to failure to meet this condition. The output of the first-order lag block with transfer function $K_{s2}/(1 + sT_7)$ (also point b in Figure 10.11) should

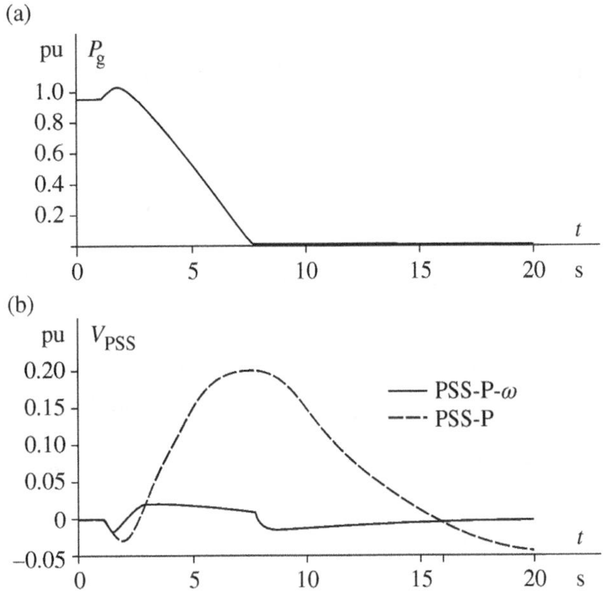

Figure 15.17 Comparison of single-input PSS1A and dual-input PSS2B stabilizers' reactions to active power ramping: (a) change in active power; (b) PSS response (output signal).

be equal to the integral of the generator's active power deviation (Section 10.1.2). The active power path with the two wash-out filters and pure integration in series can be replaced by the single wash-out filter and the inertia function

$$\frac{sT_{w3}}{1+sT_{w3}}\frac{sT_{w4}}{1+sT_{w4}}\frac{1}{s2H}=\frac{sT_{w3}}{1+sT_{w3}}\frac{T_{w4}/2H}{1+sT_{w4}}=\frac{sT_{w3}}{1+sT_{w3}}\frac{K_{s2}}{1+sT_7} \tag{15.49}$$

which apparently eliminates the second wash-out filter and leads to $T_{w4}=0$. The values of the inertial block's gain in this case should be equal to $K_{s2}=T_{w2}/(2H)$ and the time constant to $T_7=T_{w2}$, because T_{w2} is the second wash-out filter time constant in the speed (frequency) path.

- The time constant T_6 is assumed to be equal to $T_6=(0-0.02)$ s. The time constant higher than zero usually worsens the dynamic properties of the whole filter.
- Gain K_{s3}, usually set as $K_{s3}=1$, is used to match the signal's level in the speed and power paths of the filter. This can result from various ways of the input signal's measurements in the real generating unit.
- The block containing time constants T_8 and T_9 and the exponents m and n is a *ramping filter* whose task is to eliminate the PSS sensitivity to linear changes in the generator power, as shown in Figure 15.17. In order to achieve this effect, the dependency $T_8=mT_9$ should be met. The sample values of such a filter are: $T_9=0.1$ s, $T_8=0.5$ s, $m=5$, and $n=1$. They ensure damping of the 7 Hz frequency signal at the level of −40 dB.

For such defined filter parameters, its output signal V_x (Figure 15.16) in the transient states coincides (is in phase) with the course of the frequency deviation (angular speed of the rotor) of the generator. This in essence leads to a state in which the parameters of the compensating blocks T_1 - T_4, T_{10}, and T_{11} are selected as for the stabilizer with angular velocity at the input, i.e. PSS-ω.

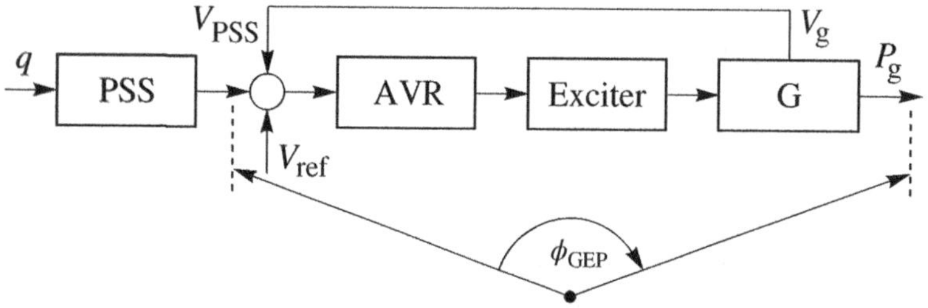

Figure 15.18 The idea of angle ϕ_{GEP} measurement based on a mathematical model.

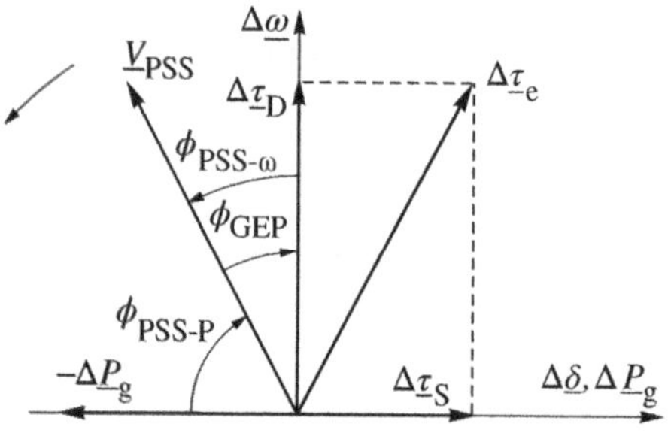

Figure 15.19 The idea of angle ϕ_{GEP} compensation by the power system stabilizer.

15.3.3.2 Selection of Time Constants of Compensating Elements $G_c(s)$

The selection of the time constants of the compensating blocks of the PSS (phase compensation block) according to the P-Vr and GEP method is based on the phase-frequency characteristics, i.e. on angles between the input of the AVR (PSS output V_{PSS}) and the electromagnetic torque or the generator's active power P_g (P-Vr method) or generator terminal voltage V_g (GEP method). When the shaft dynamics is neglected ($H = \infty$) both characteristics become identical.

The basic way to define the PSS phase compensating block parameters for stabilizers of PSS-ω and PSS-P-ω type (i.e. with speed as input to $G_c(s)$) is to calculate the P-Vr transfer function, e.g. denoted $T_V(s)$. Next, according to Eq. (15.47), the compensating block transfer function $G_c(s)$ is calculated as a reciprocal of transfer function $T_V(s)$. The P-Vr transfer function is usually in the form of a quotient of polynomials, where the degree of the polynomial of the numerator is smaller than the degree of the denominator polynomial. This means that the reciprocal of function $T_V(s)$ (i.e. the compensation block transfer function $G_c(s)$ being sought) is not realizable. This is because the degree of its numerator is larger than the degree of its denominator. Increasing the degree of the numerator of the transfer function $G_c(s)$ is naturally implemented by a low-pass filter. The assembly of a low-pass filter and a phase compensating block leads to transfer function $G_f(s)G_c(s)$, whose degree of the polynomial of the numerator is no larger than the degree of the polynomial of denominator.

Below, as a generalization of considerations, the idea of selecting parameters for the basic types of stabilizers, i.e. PSS-P, PSS-ω, and PSS-P-ω is presented. For this purpose, angle ϕ_{GEP} (Figure 15.18) is defined which is an angle of the P-Vr phase-frequency characteristic for a given frequency. This angle reflects the phase shift brought by the AVR, the excitation system, the synchronous generator, and the rest of the EPS.

As it results from the considerations presented in Section 10.1.2, the task of the PSS is to suppress electromechanical oscillations, which is accomplished by creating an additional phasor $\Delta\underline{\tau}_e$ of electromagnetic torque. By projecting the torque phasor $\Delta\underline{\tau}_e$ on the axes of the coordinate system ($\Delta\delta$, $\Delta\omega$) one can distinguish the torque component that is in phase with the power angle phasor $\Delta\underline{\delta}$ and simultaneously the active power phasor $\Delta\underline{P}_g$ as well as the component being in phase with the rotor speed phasor $\Delta\underline{\omega}$ (Figure 15.19). The first component corresponds to the phasor $\Delta\underline{\tau}_S$ of the synchronizing torque, and the second to the phasor $\Delta\underline{\tau}_D$ of the damping torque. The phasors shown in the figure correspond to the state in which the PSS generates an additional damping torque ($\Delta\tau_D > 0$) and an additional synchronizing torque ($\Delta\tau_S > 0$). In general, a condition is possible in which the PSS produces an additional damping torque ($\Delta\tau_D > 0$) and a negative synchronizing torque ($\Delta\tau_S < 0$). In the case of incorrect selection of the PSS parameters, it is also possible to have a negative impact of the PSS on both the damping and synchronizing torque, i.e. reducing the torque generated by the synchronous generator and the AVR.

As mentioned above, theoretically, the ideal PSS should only produce a damping torque. However, opinions can also be found that the PSS should also be a source of some additional synchronizing torque, i.e. be characterized by a certain degree of overcompensation[14] for PSS-P or undercompensation for PSS-ω and PSS-P-ω.

First, it is assumed that a stabilizer producing only the damping torque is considered. In Figure 15.19 phasor $\Delta\underline{V}_{PSS}$ is shifted by the angle ϕ_{GEP} resulting from parameters of the object (GEP) in such a way that the angle between this phasor and the axis corresponding to the angular velocity phasor $\Delta\underline{\omega}$ and the damping torque $\Delta\underline{\tau}_D$ is equal to ϕ_{GEP}. Phasor $\Delta\underline{V}_{PSS}$ leads phasor $\Delta\underline{\tau}_D$ because the object (i.e. the AVR, the exciter, and the generator with the rest of the EPS) is a lagging object. For such an object the PSS with output signal ΔV_{PSS} produces the damping torque $\Delta\tau_D$ only in the following cases:

- When the angular velocity ω (or frequency) is used as the input signal to the PSS and the transfer function of the PSS introduce the phase shift $\phi_{PSS-\omega} = -\phi_{GEP}$ compensating for the phase shift introduced by the object ($\phi_{GEP} < 0$).
- When the active power with the opposite sign ($-P_g$) is used as the input signal to the PSS and the transfer function of the PSS introduces the phase shift $\phi_{PSS-P} = -90^\circ - \phi_{GEP}$. Using an active power with the opposite sign (i.e. shifted by 180°) leads to a lagging stabilizer of less than 90°, which is easily realizable. In addition, the stabilizer is then a low-pass filter, which is advantageous from the point of view of damping signals (disturbances) at higher frequencies. The minus sign can be introduced in the summing point (Figure 15.16b) or using the negative value of the PSS gain $K_{s1} < 0$ (Figure 15.16a).

If the task of the PSS is to generate a certain additional synchronizing torque in addition to generating an additional damping torque, the stabilizer should introduce a phase shift:

- smaller than $|\phi_{GEP}|$ for stabilizers with an angular speed (frequency) at the input;
- higher than $|-90^\circ - \phi_{GEP}|$ in case of stabilizers with active power at the input

which, in both cases, corresponds to the introduction of the phasor $\Delta\underline{V}_{PSS}$ into the first quadrant of the coordinate system ($\Delta\delta$, $\Delta\omega$) (Figure 15.19).

The choice of the time constants of the compensating blocks of the PSS therefore consists in such a determination that the total transfer function of the PSS and the object introduces a corresponding phase shift, as presented above. Contrary to its appearance, this is not a trivial task, because the phase characteristic of the object is not constant, i.e. the angle ϕ_{GEP} is different for different frequencies. That means that the ideal PSS should introduce a phase shift adjusted to the characteristic of the object, i.e. different for different frequencies. Unfortunately, when using a PSS consisting of several serially connected lead–lag blocks, this is not possible in the whole RoMF, i.e. (0.2–2.5) Hz.

The idea of how to choose the time constants of the PSS compensating blocks is illustrated in Figure 15.20. The dash-dotted curves ϕ_{GEP} represent an exemplary phase characteristics of the object. In the case of a PSS with active power at the input (PSS-P), the aim is to shift the phase characteristic of the object with the stabilizer to the angle equal to −90° (Figure 15.20a). However, this is not possible for the whole considered frequency range. Therefore, the pulsation ω_c of the compensator is chosen to be similar but smaller than that for which full compensation of the phase shift introduced by the object has to be achieved. If the pulsation ω_c of the correcting element is equal to the pulsation of the electromechanical oscillations that has to be suppressed, a PSS introducing a certain negative synchronizing torque will be obtained.

In the case under consideration (Figure 15.20a) due to the phase shift introduced by the object for the selected pulsation ω_c smaller than −90° a lagging compensator should be applied, i.e. an element with time constants satisfying the condition $T_1 < T_2$. As a result, the phase-frequency characteristic of the object with the PSS is obtained, as a solid curve $\phi_{GEP} + \phi_{PSS}$ in the figure. The use of a single lead–lag block here should be treated as an example.

14 An overcompensated stabilizer introduces a phase shift that is greater than causing full compensation. Overcompensated PSS-P and undercompensated PSS-ω and PSS-P-ω, in addition to the damping torque, generate an additional synchronizing torque.

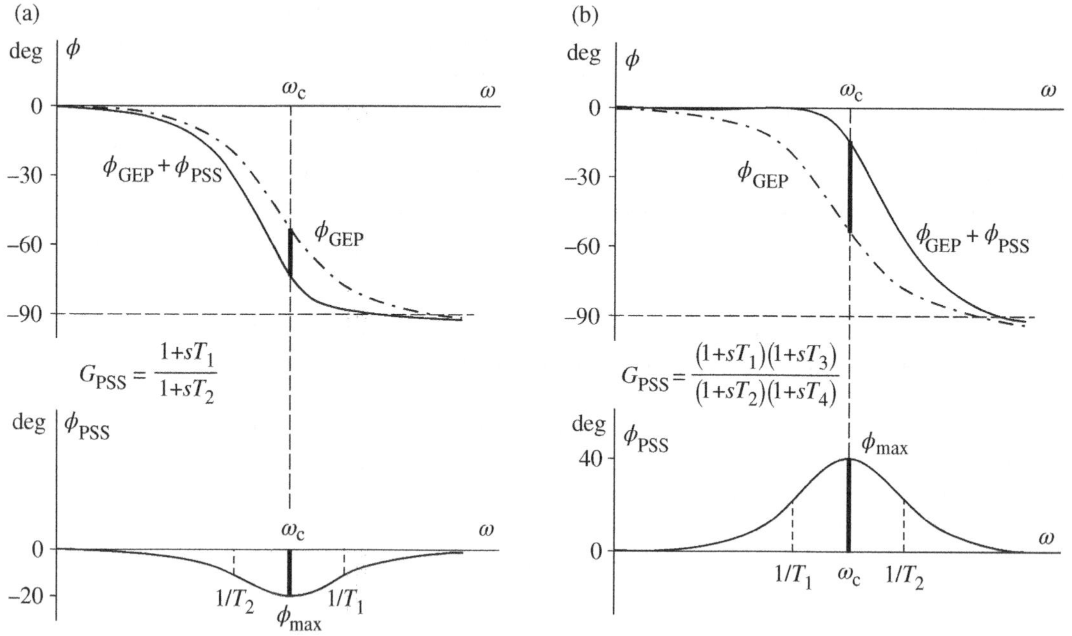

Figure 15.20 The idea of angle ϕ_{GEP} compensation by the power system stabilizer: (a) single-input (PSS-P); (b) single-input (PSS-ω) and dual-input (PSS-P-ω).

In turn, in the case of a PSS with angular velocity (frequency) at the input (PSS-ω but also PSS-P-ω) with filter parameters as defined above, the aim of the compensator is to shift the phase characteristic of the object with the stabilizer to the angle equal to 0° (Figure 15.20b). In this case, the compensator should lead the phase, i.e. a lead–lag block should be used in which the numerator time constants are greater than the denominator time constants, i.e. $T_1 > T_2$, and optionally $T_3 > T_4$. Figure 15.20b shows how to shape a characteristic of an object with a PSS, i.e. $\phi_{GEP} + \phi_{PSS}$, when using two compensators with identical time constants of the numerator and denominator, i.e. $T_1 = T_3$ and $T_2 = T_4$.

For a single compensator (Figure 15.20a), the pulsation corresponding to the maximum value of the phase shift introduced by the lead–lag block is equal to $\omega_c = 1/\sqrt{T_1 T_2}$, and the maximum phase shift depends on the quotient of the time constants. For example, for the lagging compensator ($T_1 < T_2$), for which the quotient of the time constants is equal to $T_1/T_2 = 0.5 = x$, the maximum phase shift is equal to $\phi_{max} = \text{arctg}\,[(x-1)/(2\sqrt{x})] = -19.5°$. The gain of the element for pulsation equal to zero is equal to one, and for pulsation reaching infinity it decreases to zero. On the other hand, for a differentiating compensator ($T_1 > T_2$), for $T_1/T_2 = 2$, the maximum phase shift is equal to $\phi_{max} = 19.5°$, and the gain for pulsation equal to zero is equal to zero and for pulsation going to infinity is equal to $G(\infty) = T_1/T_2 = 2$.

In turn, for two series differentiating compensators (Figure 15.20), when $T_1 = T_3$, $T_2 = T_4$, and $T_1/T_2 = 2$, the pulsation corresponding to the maximum phase shift is equal to $\omega_c = 1/\sqrt{T_1 T_2}$, the maximum phase shift is equal to $\phi_{max} = 41°$, and the gain for pulsation reaching infinity is equal to $G(\infty) = T_1 T_3/(T_2 T_4) = 4$. The PSS gain, greater than one for high frequencies, is not a beneficial feature. In practice, this gain can be compensated by the transfer function of the AVR with the excitation system and the synchronous generator itself.

In real multi-machine power systems the phase-frequency characteristics $\delta_{GEP}(\omega)$ may be nonmonotonic in the frequency range of electromechanical oscillations or have a large change in angular displacement in the vicinity of

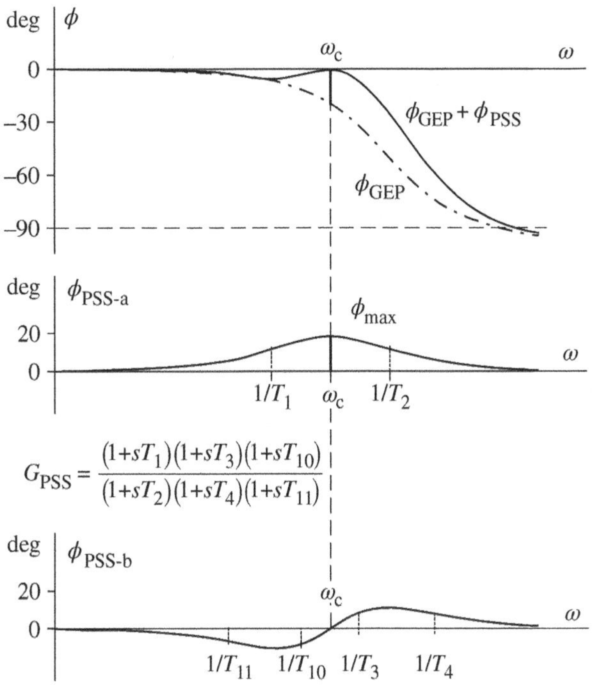

Figure 15.21 The idea of angle ϕ_{GEP} compensation by a dual-input power system stabilizer (PSS-P-ω) using three correction elements.

the oscillations' frequency, which should be suppressed. In this case, it is necessary to use three or more compensators. An example is shown in Figure 15.21. In this case, the task of the compensator with time constants T_1 and T_2 is to shift the characteristic of the object with the PSS by an angle ϕ_{max}, while the task of the compensator with time constants T_3, T_4, T_{10}, and T_{11} is to shape the characteristic in the vicinity of the point corresponding to pulsation ω_c. The leading compensator ($T_3 > T_4$) shifts the $\phi_{GEP} + \phi_{PSS}$ characteristic up (decreases the phase lagging) for pulsations greater than ω_c, and the lagging compensator ($T_{10} < T_{11}$) shifts the characteristic down (increases the phase lag).

The calculation of the PSS time constants can be implemented using various methods. This can be done by a method of trial and error or by using any optimization method. In the latter case, for a given pulsation interval, a certain function (e.g. mean square error) of the phase deviation of the desired phase-frequency characteristic P-Vr (e.g. measured in real EPS or calculated on mathematical model) and characteristic of the PSS being tuned is minimized.

This apparently simple way of shaping the phase-frequency characteristic of the object by defining the values of the time constants of the PSS, however, encounters the problem of the dependence of the amplification of the series connected compensators from pulsation. The magnitude-frequency characteristic of the PSS is determined by time constants and, unfortunately, it cannot be shaped independently of the phase-frequency characteristic. In some cases, this limits the possibility of designing an effective PSS.

The considerations carried out above refer to a certain frequency characteristic P-Vr obtained for the undefined above operating point of the generator as well as the remaining part of the EPS. As is known, the dynamic properties of the generating unit depend on the generator operating point and the impedance of the generator connection with the power system, which in the multi-machine power system is understood as the structure and parameters of the network. The damping of electromechanical oscillations is the smaller the closer to the angle stability limit is the generator operating point and the lower is the apparent short-circuit power at the point of

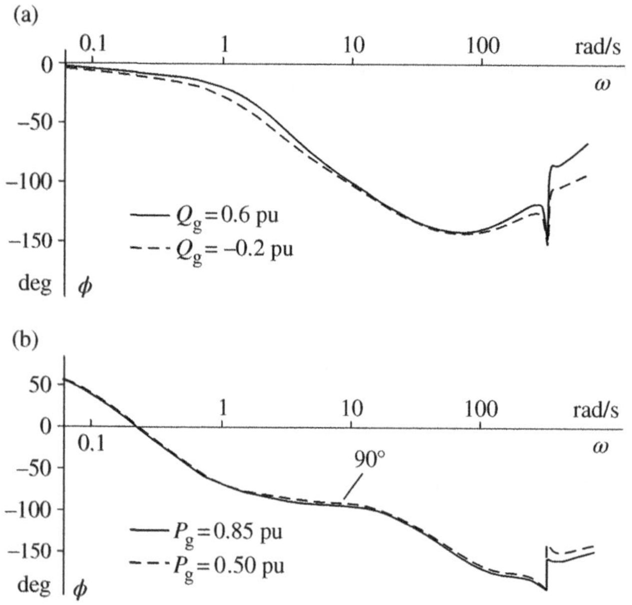

Figure 15.22 Phase characteristics of the open system: (a) without PSS (ϕ_{GEP}) and (b) with PSS ($\phi_{GEP} + \phi_{PSS}$) for different operating points of the synchronous generator ($H = \infty$).

connection of the generator to the EPS. The question then arises as to whether the operating point has an influence on the shape of the frequency characteristic, and if so how great an influence this is. Figure 15.22a presents examples of frequency characteristics of the object (SMIB model, constant inertia $H = \infty$) for different values of reactive power load Q_g and rated active power equal to $P_g = 0.85$ pu. It can be seen here that in the presented case the influence of the reactive power generated on the phase-frequency characteristic is small, at least in the reactive power generation range to the value equal to $Q_g = -0.2$ pu (capacitive load). On the other hand, Figure 15.22b presents phase-frequency characteristics for different active powers, for the case of the stabilizer of PSS-P type switched on. Also in this case, there is visible small dependence of the phase-frequency characteristic from active power P_g. This limited influence of the synchronous generator operating point on the P-Vr characteristic is very important and a positive feature, from the point of view of the possibility of choosing the PSS parameters. This is a feature that ensures the PSS robustness. While designing the PSS, frequency characteristics should be calculated for several extreme operating points of the object and a single one of them should be used for the PSS parameters' tuning. Because damping of electromechanical oscillations is the smallest for capacitive loads, which are as extreme conditions, it is suggested (reasonable) to assume capacitive load and rated active power.

In addition, it is worth noting the effect of the wash-out filter shown in Figure 15.22b on the phase characteristics. This filter for very low frequencies ($\omega < 0.22$ rad/s) introduces a positive phase shift, which in general should be treated as an undesirable effect. It is also worth paying attention (Figure 15.2b) to phase characteristics for frequencies in the (1–15) rad/s range. The phase shift equal to −90° shows that the PSS with active power at the input (PSS-P) in this frequency range provides practically only damping torque, which is a desirable feature.

15.3.3.3 Selection of the Gain

The PSS gain is selected based on the mathematical model and then verified in the real power system. It can also be selected directly in the real power system (Section 15.8).

In the case of using a linear mathematical model of EPS, the "optimal" value of the gain is determined on the basis of eigenvalues, by plotting the eigenvalues' loci as a function of the PSS gain. Then a gain is chosen that provides the strongest damping of modes of electromechanical oscillations.

An example of the shape of the eigenvalues' loci associated with modes of electromechanical oscillations calculated for a SMIB system (considered in Example 15.1) with single-input stabilizer of PSS-P type is shown in Figure 15.23. The eigenvalues' loci are calculated for the generating unit operated at rated load. In this case, there are only two oscillatory modes with frequencies in the range of (0.1–3) Hz. The trajectories shown in the figure correspond to the PSS gain increasing from $K_{s1} = 0$ (starting points of the trajectory) to $K_{s1} = 20$ (end points). The intersection point with the straight line corresponds to the gain $K_{s1} = 10$. As can be seen from the figure, an increase in the PSS gain leads to an increase in the damping of thr mode of local oscillations while the frequency of the mode decreases. The measure of damping of a mode is angle θ, i.e. the relative damping factor is equal $\xi = \cos\theta$. The direction of the eigenvalues' loci of the second mode (with a pulsation equal to about 1.57 rad/s) is opposite to the first one, i.e. an increase in the PSS gain leads to a decrease in the mode damping. The selection of the gain is therefore related to a compromise between the damping of various modes. When choosing the PSS gain, the influence of change of the generator operating point and external impedance change is also taken into account.

Figure 15.24 presents the magnitude-frequency characteristics $G(\omega) = |P_g/V_S|$ for a system with PSS switched off and for a system with PSS switched on with a gain equal to $K_{s1} = 10$.[15] The characteristic of the system with PSS switched off (curve denoted PSS off) shows two oscillatory peaks, which are typical for single-machine systems[16] with a seventh-order synchronous generator model. First peak, with pulsation equal to 8.2 rad/s, corresponds to the frequency of local oscillations, i.e. is related to the starting point of the trajectory from Figure 15.23. The second peak corresponds to pulsation 314.2 rad/s (50 Hz). The use of a PSS with a given gain almost eliminates the oscillatory peak associated with the local oscillations (curve denoted PSS on). The mode of local oscillations is more than five times damped here than in a system with PSS switched off. In order to verify the effectiveness of a particular regulator (AVR + PSS), Figure 15.25 presents the generating unit responses with the PSS switched on and off to step change in the infinite busbars'

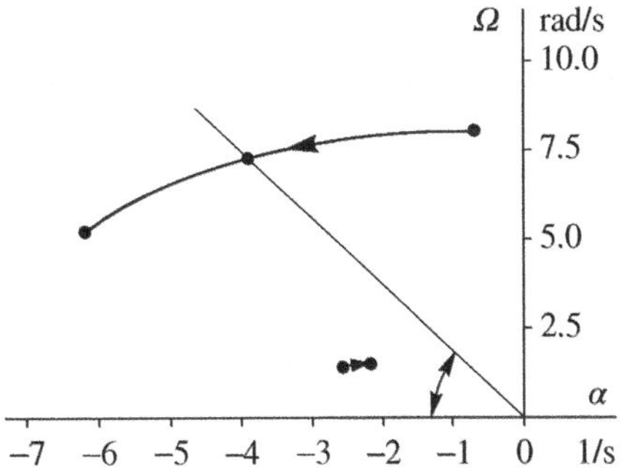

Figure 15.23 Eigenvalues loci of SMIB with a PSS-P as a function of the PSS gain K_{s1}.

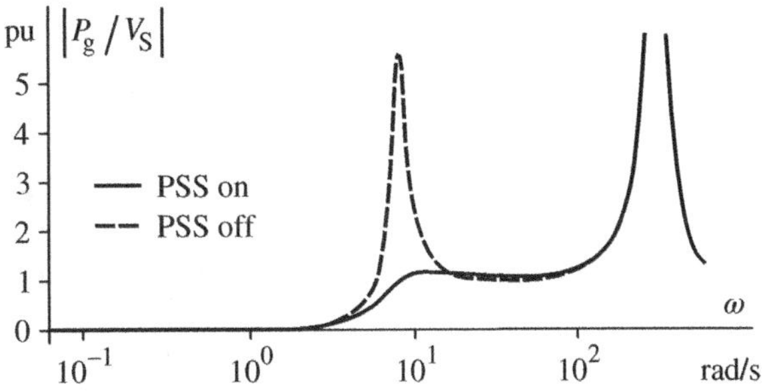

Figure 15.24 Magnitude-frequency characteristic $|P_g/V_S|$ of an SMIB system ($K_{s1} = 10$, rated operating point).

15 The characteristic is calculated for system with inertia constant $H = 3.225$ s. This characteristic should not be confused with the P-Vr characteristic, which is calculated assuming $H = \infty$.

16 In multi-machine models and in real EPSs such frequency characteristic contains more oscillatory peaks.

voltage. As can be seen, the use of PSS highly changes the active power response. Active power oscillations are significantly damped. On the other hand, the initial part of the voltage response of the system with the PSS switched on is deteriorated in relation to the system with PSS off. This is a typical effect of using a PSS, consisting in some deterioration of the voltage control process. However, in the further part of the voltage response, the improvement of the control process is also visible, which results directly from the damping of the periodic component of power response.

Figures 15.26–15.28 refer to the same example, as Figures 15.23–15.25, but present results for a dual-input stabilizer PSS-P-ω. The eigenvalues' loci shown in Figure 15.26 correspond to the PSS gain increasing from $K_{s1} = 0$

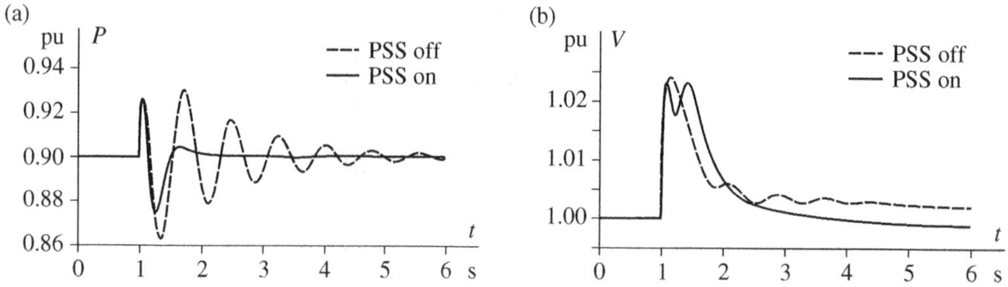

Figure 15.25 Generating unit response to step change in infinity busbars' voltage $\Delta V_S = 0.05$ pu ($K_{s1} = 10$, rated operating point).

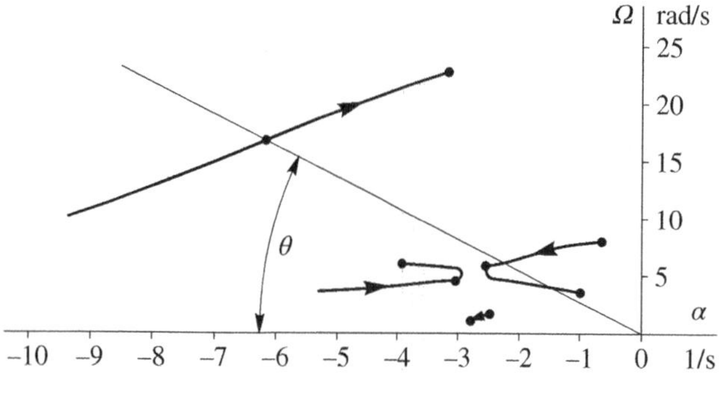

Figure 15.26 Eigenvalues loci of an SMIB with a PSS-P-ω as a function of the PSS gain K_{s1}.

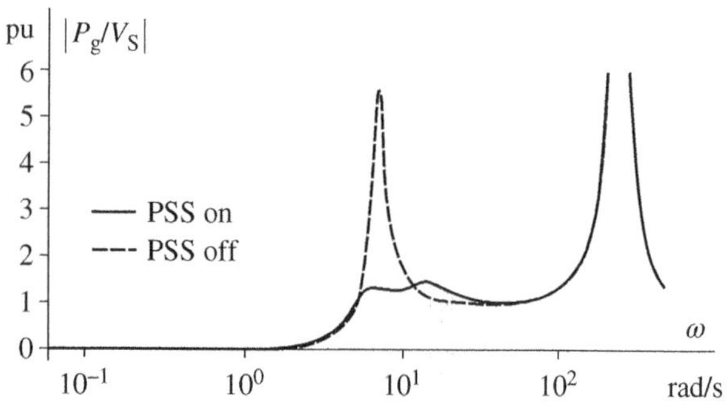

Figure 15.27 Magnitude-frequency characteristic $|P_g/V_S|$ of an SMIB system ($K_{s1} = 1.4$, rated operating point).

(starting points of the trajectory) to $K_{s1} = 4$ (end points). Points denoted with a dot on the trajectory (between ends) correspond to gain equal to $K_{s1} = 1.4$. As above, the generating unit rated load is assumed. In this case, the shape of the eigenvalues associated with oscillatory modes is more complex. The increase in the PSS gain leads to a worsening of the damping of some modes. Further increase of this gain (not shown in Figure 15.26) would lead to system instability. In this case, an optimal gain can be assumed as equal to $K_{s1} = 1.4$. The PSS thus determined also effectively attenuates the oscillatory peak with pulsation 8.2 rad/s, as shown in Figure 15.27. A new mode (regulatory mode) with pulsation 17 rad/s is visible, which can also be seen in Figure 15.26. The generating unit response shown in Figure 15.28 confirms the high efficiency of the dual-input PSS. The terminal voltage response in this case similar to the one presented in Figure 15.25b.

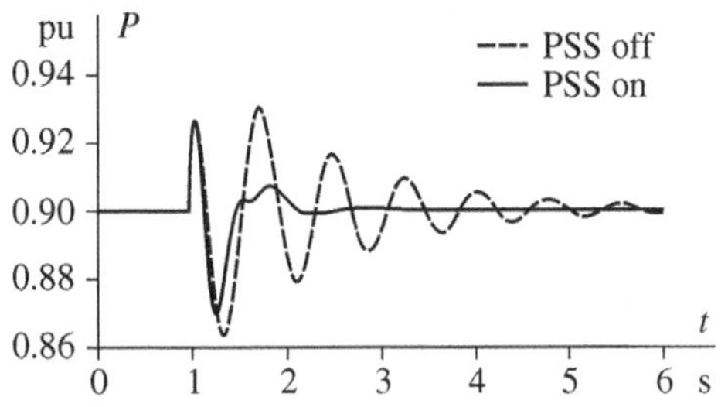

Figure 15.28 Generating unit response to step change in infinity busbars' voltage $\Delta V_S = 0.05$ pu with PSS on and off ($K_{s1} = 1.4$, rated operating point).

15.3.3.4 Selection of Limits for the Output Signal V_{PSS} and the PSS Blocking

The limits of the PSS output signal are typically set in the range of $V_{lim} = \pm(0.05 - 0.1)$ pu. These settings usually are asymmetrical, and their values depend on the PSS influence on generator voltage in transient states. The influence depends on the PSS type and is different for PSS-P, PSS-ω, and PSS-P-ω. Typically, in transient states, a smaller increase and a greater drop of generator voltage caused by the stabilizer is allowed.

The necessity of the PSS blocking (Figure 15.16) results, among other things, from the small impact of PSSs in case of a low active power load. In the case of a power plant with steam turbines, this value is usually set at the level of the boiler lower power limit, i.e. (40 - 50)% the rated active power, $P_{blocking} = (0.4 - 0.5)$ pu. However, for hydro plants, this value is set at the (20 - 30)% of rated power, i.e. $P_{blocking} = (0.2 - 0.3)$ pu.

The parameter adjusted for some of the PSSs is also the time after which the stabilizer blocking is activated or deactivated. This time is set at a level of approximately 1 s.

15.4 Optimal Regulators LQG, LQR, and LQI

The regulators presented in the previous sections are regulators with a defined structure, which means that the process of their designing consists in choosing the parameters for this structure. This parameter selection is made by shaping the frequency characteristics, by analysis of eigenvalues and evaluating the control system time domain responses.

Another group of regulators are the so-called optimal regulators. Their design process is related to the optimization of a certain control objective (performance index). The structure of the regulator is not imposed here, but results from the model of the object. Example of these types of regulator, discussed below, are the *LQR*, *LQG* regulator, and *linear-quadratic-Gaussian regulator with tracking loop based on integral element* (LQI). The regulators can be applied to time-invariant or time-variant objects.

In the LQG regulator design (synthesis) the linear object is driven by additive white Gaussian noise. Then the linear continuous time-invariant model of an object can be described by the system of state equations in form

$$\begin{aligned} \dot{\boldsymbol{x}} &= \boldsymbol{A}\boldsymbol{x} + \boldsymbol{B}\boldsymbol{u} + \boldsymbol{w} \\ \boldsymbol{y} &= \boldsymbol{C}\boldsymbol{x} + \boldsymbol{v} \end{aligned} \tag{15.50}$$

where $\boldsymbol{w}$ and $\boldsymbol{v}$ represent noncorrelated disturbances (process noise) and measurement noise of a Gaussian process characterized with a zero mean value and a constant spectral density equal to: $\boldsymbol{W}$ and $\boldsymbol{V}$, i.e. $\boldsymbol{w}$ and $\boldsymbol{v}$ represent white noise with covariances equal to

$$\begin{aligned} E\left\{\boldsymbol{w}(t)\boldsymbol{w}(\tau)^{\mathrm{T}}\right\} &= \boldsymbol{W}\delta(t-\tau) \\ E\left\{\boldsymbol{v}(t)\boldsymbol{v}(\tau)^{\mathrm{T}}\right\} &= \boldsymbol{V}\delta(t-\tau) \\ E\left\{\boldsymbol{w}(t)\boldsymbol{v}(\tau)^{\mathrm{T}}\right\} &= 0 \\ E\left\{\boldsymbol{v}(t)\boldsymbol{w}(\tau)^{\mathrm{T}}\right\} &= 0 \end{aligned} \tag{15.51}$$

where E denotes expected value and $\delta(t-\tau)$ is the Dirac delta. Matrices $\boldsymbol{W}$ and $\boldsymbol{V}$ are also called *intensity matrices*.

The LQG regulator $\boldsymbol{K}$ optimizes the control $\boldsymbol{u}(t)$ minimizing the following objective function

$$J = E\left\{\lim_{T\to\infty}\frac{1}{T}\int_0^T\left[\boldsymbol{x}^{\mathrm{T}}\boldsymbol{Q}\boldsymbol{x} + \boldsymbol{u}^{\mathrm{T}}\boldsymbol{R}\boldsymbol{u}\right]\mathrm{d}t\right\} \tag{15.52}$$

where weighting matrices $\boldsymbol{Q}$ and $\boldsymbol{R}$ (fulfilling the conditions: $\boldsymbol{Q}=\boldsymbol{Q}^{\mathrm{T}}\geq\boldsymbol{0}$ and $\boldsymbol{R}=\boldsymbol{R}^{\mathrm{T}}\geq\boldsymbol{0}$) are matrices defining requirements in relation to the control system. It can be concluded that these matrices are critical to the effectiveness of the designed regulator, and they are defined by the designer. The structure of the weighting matrices $\boldsymbol{Q}$ and $\boldsymbol{R}$ and values of their elements can be selected on the basis of the designer's experience or can be determined by the trial and error method. Matrices $\boldsymbol{Q}$ and $\boldsymbol{R}$ are usually accepted as diagonal.

The LQG regulator consists formally of two elements, i.e. the LQR and the linear-quadratic-estimator (LQE). The state estimator (LQE) is also named as Kalman filter. The basic structure of a control system with LQG regulator is presented in Figure 15.29 by the part drawn with solid lines. The synthesis of the LQG regulator is performed in two independent stages consisting of: synthesis of the optimal regulator LQR and synthesis of the state estimator LQE.

Synthesis of Linear-quadratic Regulator

Synthesis of the regulator LQR is carried out for the object described by the state Eq. (15.50) with neglected noise $\boldsymbol{w}$ and $\boldsymbol{v}$, and taking the reference input equal to zero, $\boldsymbol{r}(t)=0$. The structure of the regulator is assumed in the form

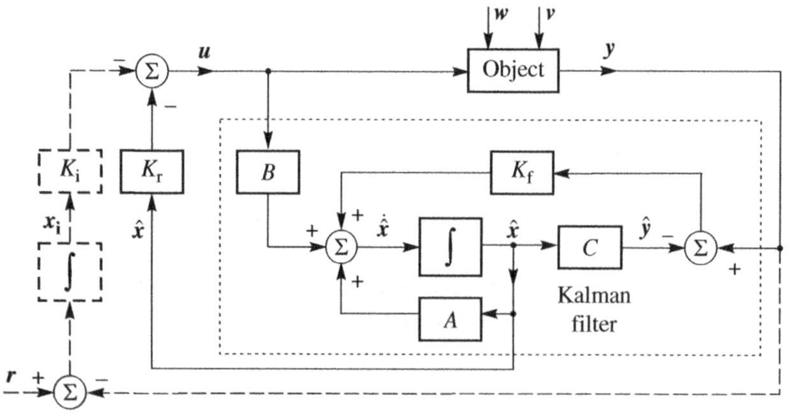

Figure 15.29 Control system with LQG regulator.

$$u(t) = -K_r x(t) \tag{15.53}$$

where K_r is the vector of regulator gains. One can see that such a regulator is a proportional one, i.e. a regulator whose output is proportional to the state variables x. Regulator K_r minimizing function is sought

$$J_r = \int_0^{\infty} \left(x^T Q x + u^T R u \right) dt \tag{15.54}$$

i.e. a regulator which brings the state of the object $x(t)$, after any initial deviation of any state variable, to the initial state $x = 0$ (equilibrium point), in an optimal way. The solution is regulator K_r with coefficients determined by the equation

$$K_r = R^{-1} B^T X \tag{15.55}$$

where matrix X, $(X = X^T \geq 0)$ is solution of the algebraic Riccati equation

$$A^T X + XA + XBR^{-1} B^T X + Q = 0 \tag{15.56}$$

It should be noted that the resulting regulator K_r depends on Q and R matrices.

However, state variables of the mathematical model used for the regulator design are not always available in the real object. This happens when the state variables in the mathematical model do not refer to the actual physical quantities (signals) of the object or when they are difficult to measure. This means that the inaccessible state variables should be estimated for the LQR regulator. A Kalman filter is used for this purpose.

Synthesis of Linear-quadratic Estimator (Kalman Filter)

The Kalman filter has the form of a system of state equations corresponding to the structure shown in Figure 15.29 in a rectangle with a dotted line

$$\dot{\hat{x}} = A\hat{x} + Bu + K_f (y - C\hat{x}) \tag{15.57}$$

The element sought here is the matrix of coefficients K_f of the estimator that minimizes the asymptotic covariance

$$P = \lim_{t \to \infty} E\left\{ [x - \hat{x}]^T [x - \hat{x}] \right\} \tag{15.58}$$

The optimal matrix K_f is defined by equation

$$K_f = YC^T V^{-1} \tag{15.59}$$

where matrix Y, $(Y = Y^T \geq 0)$ is the solution of the algebraic Riccati equation

$$YA^T + AY - YC^T V^{-1} CY + W = 0 \tag{15.60}$$

It should be noted that the resulting gain K_f of the estimator depends on matrices W and V.

The control realized by regulator LQR is implemented here using estimates of the state variables and thus has the form

$$u(t) = -K_r \hat{x}(t) \tag{15.61}$$

The regulator LQG[17] is a combination of regulator LQR and estimator LQE (Kalman filter), i.e. it is a regulator with two exogenous inputs, whose state equations have the following form

17 For the synthesis of the LQR, LQG, and LQI regulator in the MATLAB program there are dedicated functions: *lqgreg*, *lqr*, *lqi*, *lqry*, *lqgtrack*. The algebraic Riccati equation is solved by function *care*.

$$\begin{aligned} \dot{\hat{\boldsymbol{x}}} &= \boldsymbol{A}_{\mathrm{LQG}}\hat{\boldsymbol{x}} + \boldsymbol{B}_{\mathrm{LQG}}\boldsymbol{y} \\ \boldsymbol{u} &= \boldsymbol{C}_{\mathrm{LQG}}\hat{\boldsymbol{x}} \end{aligned} \tag{15.62}$$

where the regulator state matrices are defined as follows

$$\boldsymbol{A}_{\mathrm{LQG}} = \boldsymbol{A} - \boldsymbol{B}\boldsymbol{K}_{\mathrm{r}} - \boldsymbol{K}_{\mathrm{f}}\boldsymbol{C}, \boldsymbol{B}_{\mathrm{LQG}} = \boldsymbol{K}_{\mathrm{f}}, \boldsymbol{C}_{\mathrm{LQG}} = -\boldsymbol{K}_{\mathrm{r}} \tag{15.63}$$

As it results from the above, this type of regulator is a regulator with the number of poles equal to the number of poles of the object model. This directly results from the fact that the LQR regulator order is equal to zero, and the estimator's order is the same as the object model order. In the case of designing an AVR or PSS, this implies the need to minimize the order (number of state variables) of the EPS model, and often also to limit the order of the synchronous generator model for which the regulator is designed.

Example 15.2 Consider the SMIB system with data as presented in Example 15.1, operating at rated terminal voltage, rated active power, and reactive power equal to zero, i.e. $V_{\mathrm{g}} = 1$ pu, $P_{\mathrm{g}} = 0.85$ pu, and $Q_{\mathrm{g}} = 0$. In further consideration the Heffron–Phillips model (Figure 15.15) with $K_1 = 1.6957$, $K_2 = 1.7636$, $K_3 = 0.1918$, $K_4 = 4.1471$, $K_5 = 0.1039$, $K_6 = 0.1275$, and $T_3 = 1.6734$ s is used and the AVR model (Figure 15.6a) with $K_{\mathrm{A}} = 1170$, $T_{\mathrm{A}} = 0$, $T_{\mathrm{B}} = 20.4$ s, and $T_{\mathrm{C}} = 2.4$ s. The voltage transducer and the damping coefficient are neglected, i.e. $T_{\mathrm{R}} = 0$ and $K_{\mathrm{D}} = 0$.

For the LQG PSS of PSS-ω type design, the SISO object with single input $u = \Delta V_{\mathrm{PSS}}$ and single output $y = \Delta\omega$ is considered. The state-space model matrices of the object are then equal to

$$\begin{bmatrix} \dot{x}_1 \\ \dot{x}_2 \\ \dot{x}_3 \\ \dot{x}_4 \end{bmatrix} = \begin{bmatrix} 0 & -0.26 & -0.03 & 0 \\ 314.2 & 0 & 0 & 0 \\ 0 & -18.44 & -2.61 & 0.04 \\ 0 & -121.52 & -17.09 & -0.05 \end{bmatrix} \cdot \begin{bmatrix} x_1 \\ x_2 \\ x_3 \\ x_4 \end{bmatrix} + \begin{bmatrix} 0 \\ 0 \\ 137.6 \\ 1170 \end{bmatrix} \cdot u \tag{15.64}$$

$$y = \begin{bmatrix} 1 & 0 & 0 & 0 \end{bmatrix} \cdot \begin{bmatrix} x_1 \\ x_2 \\ x_3 \\ x_4 \end{bmatrix} + [0] \cdot u$$

The state variables in the model are related to: $x_1 = \Delta\omega$, $x_2 = \Delta\delta$, $x_3 = \Delta\psi_{\mathrm{fd}}$, and x_4 is the AVR compensator (TGR block) state variable.

For the LQG PSS-ω design purpose the matrices $\boldsymbol{Q}$, $\boldsymbol{R}$, $\boldsymbol{W}$, and $\boldsymbol{V}$ are assumed as equal to: $\boldsymbol{Q} = \mathrm{diag}\begin{bmatrix} 0 & 100 & 0 & 0 \end{bmatrix}$, $\boldsymbol{R} = [1]$, $\boldsymbol{W} = \mathrm{diag}\begin{bmatrix} 100 & 1 & 1 & 1 \end{bmatrix}$, and $\boldsymbol{V} = [1]$. The $\boldsymbol{Q}$ matrix has only one nonzero coefficient $Q_{2,2} \neq 0$, i.e. other coefficients are equal to zero. This means that the power angle $x_2 = \Delta\delta$ is the only state variable here influencing the plant dynamic properties through PSS. Because of the power angle and the rotor speed are directly correlated the same control effect can be achieved here when setting $Q_{2,2} = 0$ and $Q_{1,1} = 100\omega_0$, where $\omega_0 = 314.2$ rad/s. Next, in the case of matrix $\boldsymbol{W}$ the highest influence on the overall plant dynamics has coefficient $W_{1,\,1}$. This can be interpreted that the high value of the coefficient "forces" the estimator sensitivity to process noise related to speed $x_1 = \Delta\omega$.

For such defined matrices $\boldsymbol{Q}$ and $\boldsymbol{R}$, according to Eqs. (15.55) and (15.56), the optimal PSS gain is equal to $\boldsymbol{K}_{\mathrm{r}} = \begin{bmatrix} -238.8 & -7.36 & 0.31 & 0.0003 \end{bmatrix}$. One can see that the dominating influence on the PSS output have the first and second state variables, i.e. speed and power angle, which is a known feature. According to Eqs. (15.59) and (15.60), for matrices $\boldsymbol{W}$ and $\boldsymbol{V}$ defined as above, the Kalman estimator gain is equal to $\boldsymbol{K}_{\mathrm{f}} = \begin{bmatrix} 8.36 & 0.002 & 4826 & 2638 \end{bmatrix}$. Then the complete LQG regulator (PSS-ω), according to Eq. (15.63) is defined by the state space system matrices

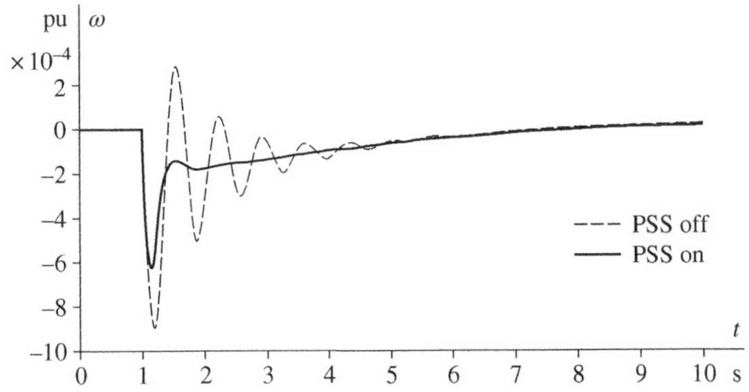

Figure 15.30 Generating unit response to mechanical torque step change $\Delta\tau_m = -0.05$ pu.

$$A_{LQG} = \begin{bmatrix} -8.35 & -0.26 & -0.03 & 0 \\ 314.2 & 0 & 0 & 0 \\ 3.24e4 & 995.2 & -45.45 & 0.0002 \\ 2.77e5 & 8494 & -381.3 & -0.42 \end{bmatrix} B_{LQG} = \begin{bmatrix} 8.36 \\ -0.0002 \\ 482.6 \\ 2638 \end{bmatrix} \tag{15.65}$$

$$C_{LQG} = [238.9 \;\; 7.36 \;\; -0.31 \;\; -0.0003] \quad D_{LQG} = [0]$$

Finally, Figure 15.30 presents a comparison of the considered object (SMIB) response to mechanical torque step change equal to $\Delta\tau_m = -0.05$ pu. Comparing the object response for the PSS switched on and switched off one can see that the LQG PSS effectively damps electromechanical oscillations.

******

The above-described LQG regulator design method does not guarantee achieving the desired tracking properties of the control system. This is not required in the case of a PSS. However, it is necessary for the AVR. Suitable tracking properties can be obtained, for example, by supplementing the regulator with a control path with an integrating element as illustrated in Figure 15.29 by a dashed line (forming LQI regulator) or by using a more complex compensator.

Furthermore, it is worth noting that such a regulator is optimal for a given operating point of the object only, while the optimal gain $\boldsymbol{K}_r$ of the regulator and the estimator gain $\boldsymbol{K}_f$ depends on the operating point. To achieve satisfactory dynamic properties of the control system, i.e. object with regulator, it is necessary for the regulator synthesis purpose to choose the generator load and the external EPS model (including operating point) properly.

15.5 Robust Regulators H_2, h_∞

Another type of regulator, which is different in relation to previously discussed by the design method, is the robust regulator. Robust regulators are optimal regulators in the sense of minimizing the appropriate norm in a system, including uncertainties (Glover 1984; Skogestad and Postlethwaite 1996). The basic norms that are subject to minimization in this case (defining the type of regulator) are norms H_2 and H_∞. For a stable time-invariant linear object, norms H_2 and H_∞ are calculated from: the time response $y(t)$ of the control system, object transfer function $\boldsymbol{G}(j\omega)$, or from singular values σ. The norms are defined as follows

- norm H_2 for MIMO system

$$\|\boldsymbol{G}\|_2 \equiv \sqrt{\frac{1}{2\pi}\int_{-\infty}^{\infty} \text{Trace}\,(\boldsymbol{G}(j\omega)^* \boldsymbol{G}(j\omega))d\omega} = \left[\frac{1}{2\pi}\int_{-\infty}^{\infty} \sum_{i=1}^{p} (\sigma_i(\boldsymbol{G}(j\omega)))^2 d\omega\right] \tag{15.66}$$

- norm H_2 for SISO system

$$\|G\|_2 \equiv \sqrt{\frac{1}{2\pi}\int_{-\infty}^{\infty} |G(j\omega)|^2 d\omega} \tag{15.67}$$

- norm H_∞ for MIMO system

$$\|\boldsymbol{G}\|_\infty \equiv \sup_{\omega} \bar{\sigma}(\boldsymbol{G}(j\omega)) \tag{15.68}$$

- norm H_∞ for SISO system

$$\|G\|_\infty \equiv \sup_{\omega} | G(j\omega) | \tag{15.69}$$

Norm H_2 calculated from the time response in discrete form, i.e. from the vector $\boldsymbol{y}$, is equal to the square root of the sum of the squares of the vector components, while norm H_∞ is equal to the maximum value of the vector $\boldsymbol{y}$.

The design of H_2 and H_∞ regulators is carried out on the basis of the general structure of the control system (Figure 15.1c). In the standard control problem $\boldsymbol{w}$, $\boldsymbol{u}$, $\boldsymbol{z}$, and $\boldsymbol{v}$ are: exogenous signals (reference inputs and disturbances), controlled outputs of the regulator, minimized control errors, and measurements. The control system of Figure 15.1c is formalized as follows

$$\begin{bmatrix} \boldsymbol{z} \\ \boldsymbol{v} \end{bmatrix} = \begin{bmatrix} \boldsymbol{P}_{11}(s) & \boldsymbol{P}_{12}(s) \\ \boldsymbol{P}_{21}(s) & \boldsymbol{P}_{22}(s) \end{bmatrix} \cdot \begin{bmatrix} \boldsymbol{w} \\ \boldsymbol{u} \end{bmatrix} = \boldsymbol{P}(s) \cdot \begin{bmatrix} \boldsymbol{w} \\ \boldsymbol{u} \end{bmatrix} \tag{15.70}$$

$$\boldsymbol{u} = \boldsymbol{K}(s)\boldsymbol{v}$$

where the object $\boldsymbol{P}(s)$ in the form of state equations and algebraic equations has the following structure

$$\begin{bmatrix} \dot{\boldsymbol{x}} \\ \boldsymbol{z} \\ \boldsymbol{v} \end{bmatrix} = \begin{bmatrix} \boldsymbol{A} & \boldsymbol{B}_1 & \boldsymbol{B}_2 \\ \boldsymbol{C}_1 & \boldsymbol{D}_{11} & \boldsymbol{D}_{12} \\ \boldsymbol{C}_2 & \boldsymbol{D}_{21} & \boldsymbol{D}_{22} \end{bmatrix} \cdot \begin{bmatrix} \boldsymbol{x} \\ \boldsymbol{w} \\ \boldsymbol{u} \end{bmatrix} \tag{15.71}$$

The transfer function of the closed-loop system between $\boldsymbol{w}$ and $\boldsymbol{z}$, called the *lower linear fractional transformation* (LLFT), is equal to

$$\boldsymbol{F}_l(\boldsymbol{P}, \boldsymbol{K}) = \boldsymbol{P}_{11} + \boldsymbol{P}_{12}\boldsymbol{K}(\boldsymbol{I} - \boldsymbol{P}_{22}\boldsymbol{K})^{-1}\boldsymbol{P}_{21} \tag{15.72}$$

The synthesis of the H_∞ regulator consists in finding a regulator $\boldsymbol{K}$, for which the transfer function $\boldsymbol{F}_l(\boldsymbol{P}, \boldsymbol{K})$ is minimized in accordance with the norm H_∞, i.e.

$$\|\boldsymbol{F}_l(\boldsymbol{P}, \boldsymbol{K})\|_\infty = \max_{0<\omega<\infty} \bar{\sigma}[\boldsymbol{F}_l(\boldsymbol{P}, \boldsymbol{K})(j\omega)] \tag{15.73}$$

where $\bar{\sigma}$ is an upper singular value of matrix $\boldsymbol{F}_l$. In the case of the H_2 regulator, the transfer function $\boldsymbol{F}_l(\boldsymbol{P}, \boldsymbol{K})$ is minimized by the use of Eq. (15.66) or (15.67).

In the H_2 and H_∞ control the object $\boldsymbol{P}(s)$ is required to meet the following requirements:

- The matrix $(\boldsymbol{A}, \boldsymbol{B}_2, \boldsymbol{C}_2)$ is stabilizable and detectable, i.e. the object $\boldsymbol{P}(s)$ is stabilizable (guarantee of the existence of a regulator) and observable (guarantee of the existence of an observer).

- The matrix $\boldsymbol{D}_{12}$ has full column rank, which means that there must be a direct connection between the input $\boldsymbol{u}$ and the output $\boldsymbol{v}$. The matrix $\boldsymbol{D}_{21}$ has full row rank, which means that there must be a direct connection between the input $\boldsymbol{w}$ and the output $\boldsymbol{z}$. This guarantees obtaining the proper regulator, i.e. on the degree of the transfer function numerator polynomial, not higher than the degree of the denominator polynomial, which means the regulator's realizability.
- The matrices

$$\begin{bmatrix} \boldsymbol{A}-\mathrm{j}\omega\boldsymbol{I} & \boldsymbol{B}_2 \\ \boldsymbol{C}_1 & \boldsymbol{D}_{12} \end{bmatrix} \quad \text{and} \quad \begin{bmatrix} \boldsymbol{A}-\mathrm{j}\omega\boldsymbol{I} & \boldsymbol{B}_1 \\ \boldsymbol{C}_2 & \boldsymbol{D}_{21} \end{bmatrix}$$

have full column rank and full row rank, respectively, for all ω. This means that the regulator does not cancel the poles and zeros of the object's transfer function, which could lead to system instability.
- Matrix $\boldsymbol{D}_{11}=\boldsymbol{0}$, $\boldsymbol{D}_{22}=\boldsymbol{0}$ is required for regulator H_2, which must be a strictly proper regulator. In the case of the regulator H_∞ this condition is not required, but its fulfillment significantly simplifies the controller's design algorithm.

The above problem can be solved iteratively (Doyle 1982) after fulfilling additional assumptions:

- $\boldsymbol{D}_{12}^{\mathrm{T}}(\boldsymbol{C}_1 \;\; \boldsymbol{D}_{12})=[\boldsymbol{0} \;\; \boldsymbol{I}]$
- $\begin{pmatrix} \boldsymbol{B}_1 \\ \boldsymbol{D}_{21} \end{pmatrix}\boldsymbol{D}_{21}^{\mathrm{T}} = \begin{bmatrix} \boldsymbol{0} \\ \boldsymbol{I} \end{bmatrix}$
- $(\boldsymbol{A}, \boldsymbol{B}_1)$ is stabilizable and $(\boldsymbol{A}, \boldsymbol{C}_1)$ is detectable.

If the system $(\boldsymbol{P}, \boldsymbol{K})$ meets the above requirements, the regulator $\boldsymbol{K}$, ensuring the stability of the closed loop system and for which the condition $\|\boldsymbol{F}_l(\boldsymbol{P}, \boldsymbol{K})\|_\infty < \gamma$ is met, exists only if the following assumptions are met:

- eigenvalues of the matrices $\boldsymbol{I}_\infty$ and $\boldsymbol{J}_\infty$

$$\boldsymbol{I}_\infty = \begin{pmatrix} \boldsymbol{A} & \gamma^{-2}\boldsymbol{B}_1\boldsymbol{B}_1^{\mathrm{T}} - \boldsymbol{B}_2\boldsymbol{B}_2^{\mathrm{T}} \\ -\boldsymbol{C}_1^{\mathrm{T}}\boldsymbol{C}_1 & -\boldsymbol{A}^{\mathrm{T}} \end{pmatrix} \tag{15.74}$$

$$\boldsymbol{J}_\infty = \begin{pmatrix} \boldsymbol{A}^{\mathrm{T}} & \gamma^{-2}\boldsymbol{C}_1^{\mathrm{T}}\boldsymbol{C}_1 - \boldsymbol{C}_2^{\mathrm{T}}\boldsymbol{C}_2 \\ -\boldsymbol{B}_1\boldsymbol{B}_1^{\mathrm{T}} & -\boldsymbol{A} \end{pmatrix} \tag{15.75}$$

do not have only imaginary values
- matrices $\boldsymbol{X}_\infty$ and $\boldsymbol{Y}_\infty$ are non-negative solutions of algebraic Riccati equations

$$\boldsymbol{A}^{\mathrm{T}}\boldsymbol{X}_\infty + \boldsymbol{X}_\infty\boldsymbol{A} + \boldsymbol{X}_\infty\left(\gamma^{-2}\boldsymbol{B}_1\boldsymbol{B}_1^{\mathrm{T}} - \boldsymbol{B}_2\boldsymbol{B}_2^{\mathrm{T}}\right)\boldsymbol{X}_\infty + \boldsymbol{C}_1^{\mathrm{T}}\boldsymbol{C}_1 = \boldsymbol{0} \tag{15.76}$$

$$\boldsymbol{A}\boldsymbol{Y}_\infty + \boldsymbol{Y}_\infty\boldsymbol{A}^{\mathrm{T}} + \boldsymbol{Y}_\infty\left(\gamma^{-2}\boldsymbol{C}_1^{\mathrm{T}}\boldsymbol{C}_1 - \boldsymbol{C}_2^{\mathrm{T}}\boldsymbol{C}_2\right)\boldsymbol{Y}_\infty + \boldsymbol{B}_1\boldsymbol{B}_1^{\mathrm{T}} = \boldsymbol{0} \tag{15.77}$$

- $\rho(\boldsymbol{X}_\infty\boldsymbol{Y}_\infty) < \gamma^2$, where ρ is the spectral radius and γ is the defined value of the norm $\|\boldsymbol{F}_l(\boldsymbol{P}, \boldsymbol{K})\|_\infty$.

Then the sought regulator $\boldsymbol{K}$ is described by a system of equations

$$\begin{aligned} \dot{\boldsymbol{p}} &= \boldsymbol{A}_\infty\boldsymbol{p} - \boldsymbol{Z}_\infty\boldsymbol{L}_\infty\boldsymbol{v} \\ \boldsymbol{u} &= \boldsymbol{F}_\infty\boldsymbol{p} \end{aligned} \tag{15.78}$$

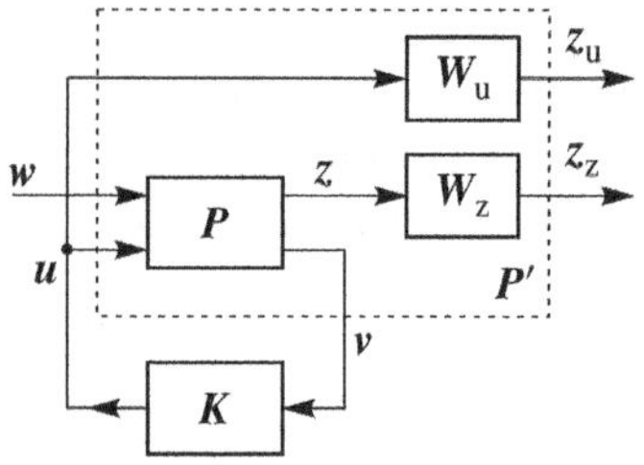

Figure 15.31 The structure of the control system used in the process of H_∞ regulator design: P – object; K – regulator; W_z and W_u – weighting functions; P' – modified object.

where p is the vector of the state variables of the regulator K and the matrices A_∞, F_∞, L_∞, and Z_∞ are equal to:

$$A_\infty = A + \gamma^{-2} B_1 B_1^T X_\infty + B_2 F_\infty + Z_\infty L_\infty C_2 \tag{15.79}$$

$$F_\infty = -B_2^T X_\infty, \quad L_\infty = -Y_\infty C_2^T, \quad Z_\infty = \left(I - \gamma^{-2} Y_\infty X_\infty\right)^{-1} \tag{15.80}$$

The regulator H_2 design process is a one-step process that involves solving the Riccati equations one time, while the H_∞ regulator design process is of an iterative nature, where in the each next step a regulator for decreasing values γ is sought after. The solution is therefore a set of regulators, including a regulator that minimizes the value of the coefficient γ, i.e. is characterized by the smallest value of this coefficient.

Shaping the desired dynamic properties of the control system is accomplished here by introducing requirements for signals z and u. These requirements are introduced in the form of weighting functions, having the form of transfer function between w and z (function W_z) and between w and u (function W_u). For such a modified object P', shown in Figure 15.31, regulator K is being sought that minimizes the norm

$$\left\| \begin{matrix} W_Z \cdot F_l(P, K) \\ W_U \cdot T(P, K) \end{matrix} \right\|_\infty \tag{15.81}$$

where transfer function $F_l(P, K)$ is defined by Eq. (15.72), and transfer function $T(P, K)$ is equal

$$T(P, K) = (I - P_{22} K)^{-1} K P_{21} \tag{15.82}$$

For a robust regulator, the value of the norm (15.81) should not exceed one.

Example 15.3 For example, consider the process of designing a PSS that minimizes the norm H_∞ for an object $P'(s)$ (Figure 15.31), which is an SMIB model with a structure presented in Figure 15.32. The model consists of a seventh-order model of synchronous generator with parameters, as in Example 15.1; IEEE ST1A type AVR with parameters as in Example 15.2, single wash-out filter $sT_w/(1 + sT_w)$, output weighting functions W_{zu} and W_{zp}, and input weighting function W_u. To assess the effectiveness of the H_∞ PSS being synthesized, stabilizer IEEE PSS1A type, with parameters: $K_{S1} = -1.64$, $T_1 = 0$ s, $T_2 = 0.02$ s, $T_3 = 0.55$ s, $T_4 = 6.8$ s, $T_w = 10$ s is used (named further PSS-P). Two configurations of the synchronous generator control system with PSS minimizing the norm are presented. In the first (rare used configuration) labeled as $H_{\infty 1}$, the PSS output signal V_{PSS} is summed with AVR output (dotted line in Figure 15.32). In the second, typical configuration in the real generating units, labeled as $H_{\infty 2}$, the PSS output signal V_{PSS} is injected to the AVR input.

For the object $P'(s)$ as shown in Figure 15.32, the exogenous signals, control inputs, control errors, and measurements are defined as

$$\begin{aligned} w &= [V_S \;\; V_{ref} \;\; \tau_m]^T, \quad u = [V_{PSS}] \\ z_z &= [z_{zu} \;\; z_{zp} \;\; z_u]^T, \quad v = [V_x] \end{aligned} \tag{15.83}$$

where V_S – infinite busbars voltage, V_{ref} – generator voltage reference, V_{PSS} – PSS output, τ_m – mechanical torque, V_x – output signal of the PSS wash-out filter, z_{zu}, z_{zp}, and z_u – outputs of weighting functions W_{zu}, W_{zp}, and W_u. The structure of all weighing functions is adopted uniformly and evenly

$$W(s) = K_W \frac{1 + sT_{Wa}}{1 + sT_{Wb}} \tag{15.84}$$

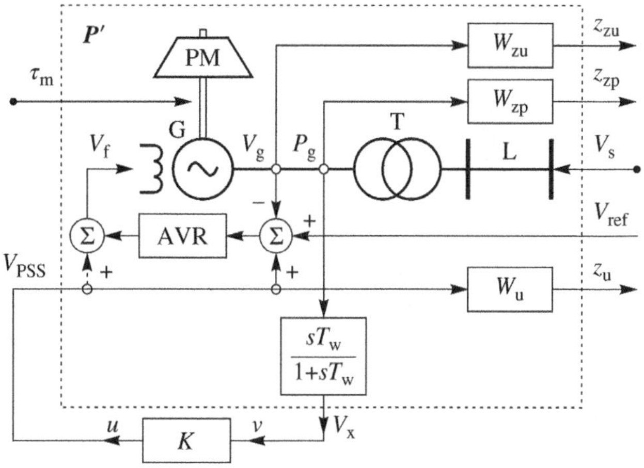

Figure 15.32 Single-machine infinite bus (SMIB) system for power system stabilizer (PSS) minimizing H_∞ norm: W_{zu} and W_{zp} – output weighting functions; W_u – input weighting function; K – synthesized (designed) H_∞ PSS.

For example, it is assumed that the goal of synthesis of the stabilizer minimizing the norm H_∞ is to obtain a control system whose dynamic properties are at least comparable with the dynamic properties of the generating unit equipped with classic PSS, i.e. the PSS-P mentioned above. To achieve this, the weighting functions W_{zu} and W_{zp} are selected so that the magnitude-frequency characteristics of the inverse of these weighting functions are as close as possible to the characteristics of the object with PSS-P. The following values of weighing functions defined by Eq. 15.84 are adopted:

- W_{zu} to ensure fast voltage control and good tracking properties: $K_W = 58.8$, $T_{Wa} = 0.76$ s, $T_{Wb} = 40$ s
- W_{zp} to ensure effective damping of electromechanical oscillations: $K_W = 100$, $T_{Wa} = 0.0003$ s, $T_{Wb} = 12.5$ s
- W_u to limit the PSS output signal V_{PSS} to a realistic level: $K_W = 0.02$, $T_{Wa} = 0.0067$ s, $T_{Wb} = 0.4$ s.

Figure 15.33 presents the magnitude-frequency characteristics of an object with a PSS-P and inverse weighting characteristics, i.e. $1/W_{zu}$ and $1/W_{zp}$, which are used in the $H_{\infty 2}$ PSS design process. This figure also shows the magnitude-frequency characteristics of the control system with the $H_{\infty 2}$ PSS, which is the result of the presented design process.[18]

The magnitude-frequency characteristic $|P_g/V_S|$ of an object equipped with an $H_{\infty 2}$ PSS (Figure 15.33a) is located below the inverse of the weight function $1/W_{zp}$, which is a desirable effect. At the same time, this characteristic is located below the characteristic obtained for an object with a PSS-P in the entire range of the analyzed frequencies. The significant issue is that for frequencies close to the local-area oscillations (around 1 Hz) the characteristic takes on considerably smaller values. This means better damping of electromechanical oscillations by the designed PSS in comparison to the conventional stabilizer. The inverse of the weighing function $1/W_{zp}$ deliberately bypasses the resonant peak with pulsation 314.1 rad/s (50 Hz). The peak results from taking into account the transformation voltage (derivative of stator field) in the synchronous generator model. Limiting this peak is practically impossible.

In turn, the magnitude-frequency characteristic $|V_g/V_S|$ of the control system equipped with a $H_{\infty 2}$ PSS (Figure 15.33b), outside the frequency range corresponding to RoMF (1–15 rad/s), coincides with the characteristic

18 The *hinfsyn* function of MATLAB can be used to synthesize the stabilizer. In the synthesis of the stabilizer minimizing the norm H_2, the *h2syn* function can be used.

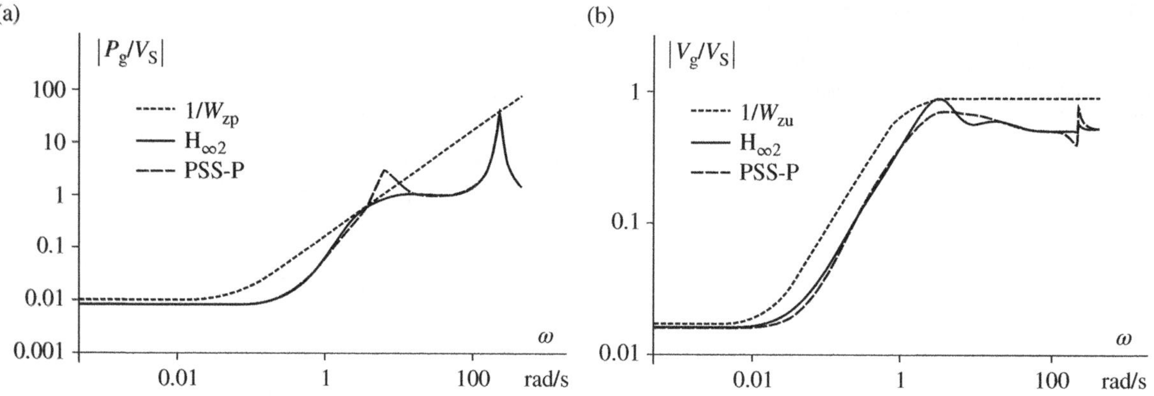

Figure 15.33 Magnitude-frequency characteristics of the synchronous generator control system: (a) $|P_g/V_S|$; (b) $|V_g/V_S|$.

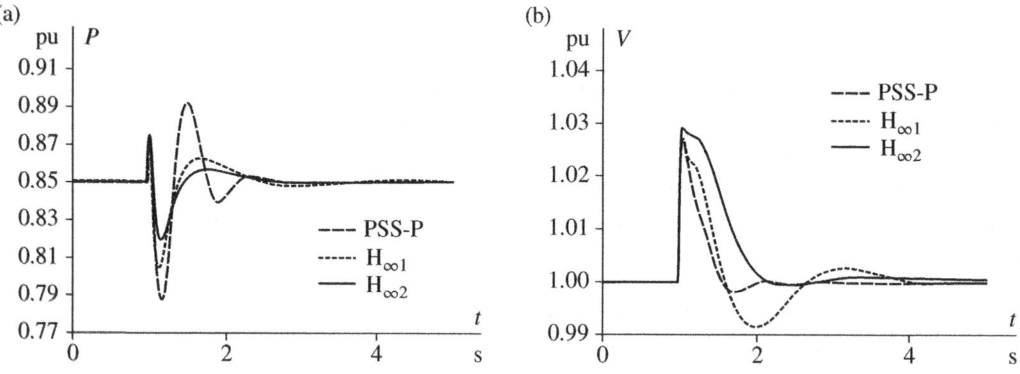

Figure 15.34 Generating unit response to the step change in infinite busbars' voltage $\Delta V_S = 0.05$ pu.

of the control system with PSS-P. This difference means that the control system with the designed PSS differs only to a certain extent, in the sense of the quality of the voltage control, from the system equipped with the PSS-P.

As a confirmation of the H_∞ PSS efficiency, resulting from frequency characteristics, Figures 15.34–15.36 show the responses of the object (with a nonlinear generator model) equipped with $H_{\infty 1}$ and $H_{\infty 2}$ PSSs and equipped with a conventional stabilizer (PSS-P).

The system responses shown in Figure 15.34 show the effective damping of active power oscillations by the designed stabilizers. Electromechanical oscillations are most effectively damped by $H_{\infty 2}$ PSS, but the voltage control in this case is slower than for the generating units equipped with other PSSs. Deterioration of the voltage control is a typical effect, the price that is paid for the strong damping of electromechanical oscillations. A PSS marked as $H_{\infty 1}$, despite a very good damping of the active power oscillations, causes unacceptable oscillations of the generator's voltage.

In case of the system response to a change in the voltage or active power reference, there are no fundamental differences between the three compared systems. However, Figure 15.36 shows a long-lasting increase in the generator voltage, which is typical for single-input conventional PSSs that use the active power as an input signal. Both designed PSSs that minimize the H_∞ norm show an advantage over the classic system. The smallest overshoot of the active power is obtained for a system equipped with $H_{\infty 2}$ PSS. Also, the quality of voltage control in the system with this PSS is acceptable.

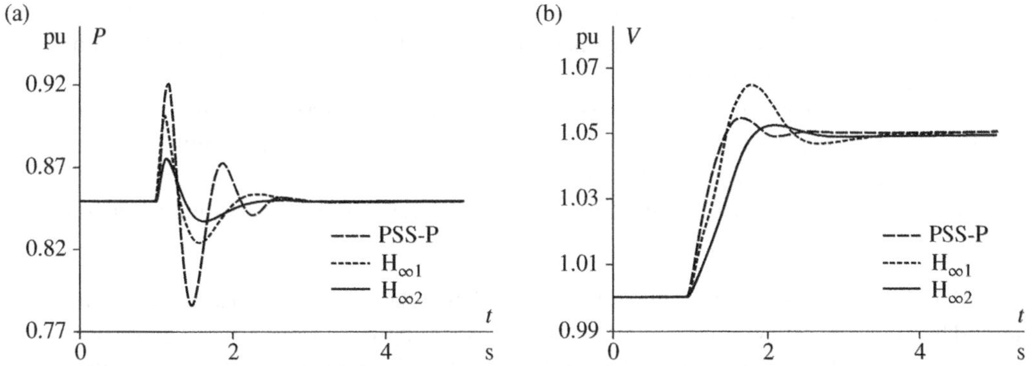

Figure 15.35 Generating unit response to a step change in the voltage reference $\Delta V_{\text{ref}} = 0.05$ pu.

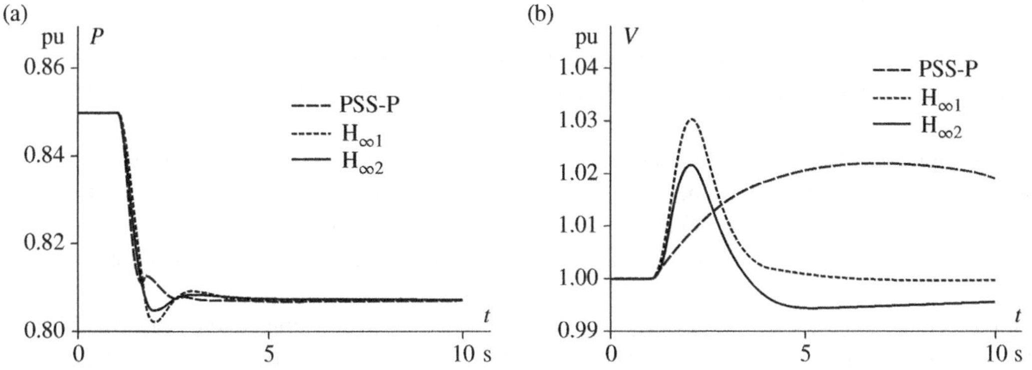

Figure 15.36 Generating unit response to a step change in the power reference $\Delta P_{\text{ref}} = -0.041$ pu.

Figure 15.37 shows the influence of the considered PSSs on the angle stability margin. It can be seen here that both designed PSSs significantly increase the margin of stability.

The PSS synthesis in the presented example is made for the operating point corresponding to the rated active and reactive power of the generator and for the external impedance magnitude equal to $|\underline{Z}_{\text{S}}| = 0.23$ pu. It turns out that the effectiveness of these stabilizers at another operating point, for example P_{gn}, $Q_{\text{g}} = 0$, is similar to that shown in Figure 15.34.

******

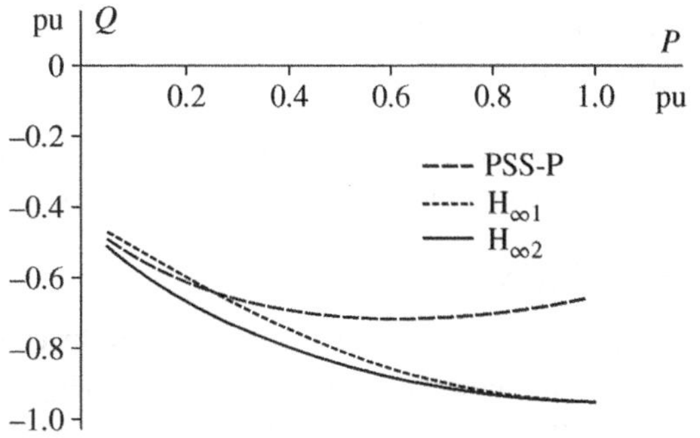

Figure 15.37 The stability limit of the considered single-machine infinite bus (SMIB) system.

The advantages of the above-described method of the regulator design include the possibility of shaping the dynamic properties of the control system by defining the weighting functions. Formally, it is relatively simple, but requires knowledge about the relationships between magnitude-frequency characteristics and the dynamic properties of the object. However, the basic limitations of the method in applications in power systems are:

- Very high sensitivity of the method to the structure of the object, its parameters, and operating point. It turns out that meeting the requirements to the matrices of the components making up the object $\boldsymbol{P}(s)$, Eq. (15.71), is not always possible for a given type of the EPS model.
- Usually, a large size of the synthesized regulator equal to the size of the object $\boldsymbol{P}'(s)$. The order of the regulator is equal to the number of object state variables including weighting functions. That means that the multi-machine model of the EPS cannot be directly used in the regulator design process. This model must be reduced to a low-order model, for example to SMIB model but also with a preferably low-order model of synchronous generator model.

The AVRs and PSSs used in real power systems are regulators with standard, relatively simple structures. Despite this, they are characterized by a high efficiency of operation. One of the ways of applying an H_2 or H_∞ regulator in a real commercial regulator is its transformation to the form corresponding to the standard structure. This may require reducing the regulator's size.

In general, reduction of the high-order linear, time-invariant stable model $G(s)$ of a regulator or object to the lower-order model $G_a(s)$ is based on the minimization of the norm $\|G(s) - G_a(s)\|_\infty$. For the model in the form of a state-space equation, reduction is done by truncation or by residualization.

The truncation is based on the state variables related to the high-frequency dynamics (fastest modes) elimination by simple skipping the differential equations related to the eliminated state variables. The resulting reduced order model $G_a(s)$ properly approximates model $G(s)$ for high frequencies.

In the case of residualization the derivatives of eliminated (unwanted) state variables are assumed equal to zero, i.e. differential equations related to these variables become algebraic equations. The algebraic equations are used next to remove the variables from the remaining differential equations of the model. The resulting reduced order model $G_a(s)$ properly approximates model $G(s)$ for low frequencies and preserves the steady-state gain of the system, i.e. for $\omega = 0$.

Most effective methods of the model truncation are based on a balanced realization of the system. Such realization is an asymptotically stable minimal realization of system $G(s)$ in the form of state matrices ($\boldsymbol{A}, \boldsymbol{B}, \boldsymbol{C}, \boldsymbol{D}$) of N-th order, in which the controllability Gramian $\boldsymbol{P}$ and the observability Gramian $\boldsymbol{Q}$ are equal and diagonal, i.e. $\boldsymbol{P} = \boldsymbol{Q} = diag(\sigma_1, \sigma_2, \ldots, \sigma_N)$, where $\sigma_1 \geq \sigma_2 \geq \ldots \geq \sigma_N$. The controllability and observability Gramians are solutions of the Lyapunov equations

$$\boldsymbol{AP} + \boldsymbol{PA}^{\mathrm{T}} = -\boldsymbol{BB}^{\mathrm{T}} \tag{15.85}$$

$$\boldsymbol{A}^{T}\boldsymbol{Q} + \boldsymbol{QA} = -\boldsymbol{C}^{\mathrm{T}}\boldsymbol{C} \tag{15.86}$$

while the singular values σ_i are named Hankel singular values, and are equal to the square roots of the eigenvalues λ_i of the matrix being the product of the Gramians

$$\sigma_i = \sqrt{\lambda_i(\boldsymbol{PQ})} \tag{15.87}$$

The singular Hankel values should not be confused with singular values defined, for example, by Eq. (15.13). Hankel singular values are a measure of the "energy" associated with each state variable of a dynamic system. Maintaining state variables characterized by higher energy, the majority of dynamic properties of the dynamic system, in the sense of stability, frequency characteristics, and time responses retain. In general, the large controllability Gramians relate to states, which do not require high energy of input signal, and large observability Gramians refer to states that generate high energy of the output signal.

The basic methods of model reduction are: balanced truncation approximation (BTA) (Moore 1981), balanced residualization, and optimal Hankel norm approximation (Glover 1984).[19]

19 The MATLAB program to reduce the size of the system model provides the functions *reduce*, *ncfmr*, *balancmr*, *schurmr*, and *hankelmr*.

In practice, reducing the order of the AVR or PSS model to the transfer function of order corresponding to the standard structure can be difficult or even impossible. To a large extent it depends on the size of the object model used for the synthesis of the regulator and its dynamic properties.

15.6 Nonlinear Regulators

The dynamic response of each power system depends on its operating condition (loading) and the structure. Structure is understood here as configuration and parameters of all system elements such as network, loads, generating units, control, and protection systems. The majority of power system elements are nonlinear and the power system as the whole is nonlinear too. Therefore, a dynamic response and stability of power system depend also on the type of disturbance and the place of its occurrence.

The design and optimization of the AVR, PSS, or regulator of any other object should be implemented on the basis of a mathematical model that best reflects the dynamic properties of EPS in the area being considered. Theoretically, this means that, in the case of designing regulators of synchronous generators as well as regulators of other EPS elements, a nonlinear multi-machine mathematical model must be used. In practice, as shown in the previous sections, the regulator design based on the linear model also allows an effective regulator to be obtained. However, it should be remembered that the use of linear models in the optimization process does not allow for the evaluation of the influence of limiters, hysteresis, switching elements, etc., on the EPS dynamic properties. They can have a great influence on system dynamics, especially in the case of severe disturbances.

The basic method of designing a regulator, e.g. AVR or PSS, as a nonlinear object is usually reduced to the process of the parametric optimization of a regulator of a defined structure.[20] This structure may correspond to the real regulator or can be determined by the designer. The regulator parameters, which may be subject to optimization, are gains and time constants, but also threshold values of limiters, parameters of nonlinear and switching elements, etc. The optimization method used is not critical, although its effectiveness is usually related to the type of optimization problem and to the number of optimized variables.

An important and critical aspect of optimization becomes the indicator of the quality of control. In general, the performance index (function) should be a function of:

1) Power system operating point, including:
   - Operating points of the power system and/or generator, whose regulator is subject to optimization. The set of operating points included in the optimization process, should contain a critical or close to critical points, e.g. be located in a power capability area close to the stability limit. An example can be operating points with capacitive loads of generators, for which damping of electromechanical oscillations is lower than for inductive loads.
   - Parameter of the model of the object that may be subject to change during the EPS operation or whose value is known with limited accuracy. They may be parameters of a synchronous generator or network, the network structure of which may be subject to seasonal changes, e.g. winter and summer network configuration, network configuration in emergency states.
   - Impacts of a specific nature (stochastic signals or signals with defined frequency or signals of defined type) introduced into the network by objects of a certain type, e.g. wind power plants, photovoltaic sources, loads.

20 A separate category of nonlinear regulators are regulators based on artificial intelligence, i.e. fuzzy logic theory and artificial neural networks. They are not discussed here, and examples of applications in synchronous generator control systems can be found in the literature, e.g. Lubosny (1999).

2) Disturbances, including:
   - Type of disturbance (balanced or unbalanced faults, switching on/off network elements, generating units, loads, etc.)
   - Disturbance location in the network or object.
   - Duration time of the disturbance.
   - The extent of a change in the reference input of the controlled variable of a given object. For example, owing to the EPS objects' nonlinearity, their response depends on the value of the reference input change.
   - Disturbance direction. For example, owing to the nonlinearity of the synchronous generator (magnetization curve) the generating unit responses for voltage reference change "up" and "down" differ.

As a performance index, being subject to optimization, integral functions are often used. They are typical for many control problems, not only in power systems. An example of such integral function, previously discussed, is the H_2 norm. The norm is the root square of sum of the squares of the deviation of a certain signal, e.g. the object output signal. The influence of the structure of the performance index being optimized on the efficiency of the AVR, PSS, or other regulator is discussed in literature, e.g. in Kundur (1994), Machowski et al. (2008), Paszek and Nocon (2014), etc. For example, Paszek and Nocon (2014) show that the desired control effects, in the process of the PSSs' gains optimization in a multi-machine system, ensure a performance index

$$J(\boldsymbol{K}) = \sum_{r=1}^{S} \sum_{j=1}^{M} \int_{t_0}^{t_k} \sum_{l=1}^{N} \left(\left| w_\omega \Delta\omega_{ljr} \right| + \left| w_P \Delta P_{ljr} \right| + \left| w_V \Delta V_{ljr} \right| \right) \mathrm{d}t \tag{15.88}$$

where $\Delta\omega_{ljr}$, ΔP_{ljr}, and ΔV_{ljr} are deviations of rotational speed of generators, active power, and voltage calculated for N generating units and for M various critical disturbances and for S various operating points of EPS; w_ω, w_P, and w_V are weighting coefficients; $[t_0, t_k]$ is the interval of time domain simulation used for performance index calculation; and $\boldsymbol{K}$ is a vector of the sought gains of PSSs. The time constants of the PSSs are selected earlier using, for example, the method described in Section 15.3.3.

The simultaneous inclusion of the generating unit voltage and active power time response in the performance index is justified by the fact that there is a strong (usually opposite) relationship between the voltage control process and damping of electromechanical oscillations. As a result of minimizing performance index being only a function of the active power or angular velocity of the rotor, AVR or PSS is obtained ensuring a high attenuation of electromechanical oscillations but not a necessarily satisfactory quality of the voltage control. Conversely, as a result of minimizing performance index being only a function of the generator voltage, one would obtain a regulator ensuring fast voltage control but weak damping of electromechanical oscillations.

When defining the performance index for regulator optimization, it is also necessary to pay attention to the regulator ability to minimize the impact of: disturbances in the network and other generating units and to ensure the desired tracking properties of the regulator.

15.7 Adaptive Regulators

The aim of the regulator design is to ensure a stability of control and appropriate tracking ability and insensitivity to disturbances. In the case of regulators presented in the previous sections, this goal is obtained by optimizing the regulator for a single, properly defined, operating point and single set of the object parameters. The optimization process can also be realized for a number of operating points (Section 15.6). Such regulator ability to ensure appropriate dynamic properties of the control system for other operating points and disturbances is not verified during the optimization process. It requires additional tests. The regulator remains optimal only for the operating point (or points) and the object parameters being used in the optimization process. This is treated as a drawback of this type of regulator.

Adaptive regulators, in contrast to the previously presented ones, adjust their parameters to the changing operating point and, hence, to the object parameters change. Adaptive regulators are divided into two basic groups:

- *Adaptive regulators with programmed changes of regulator parameters.* In this case the selected parameters (or parameter) of the regulator are defined as dependable from selected signals (or signal) measured in the controlled object. The dependence has a form of array or explicit function. These arrays or functions are defined on the basis of analysis on mathematical models and are verified on real objects. The optimality of this type of regulator is limited by the possibility of creating an array or function involving different operating points and disturbances, or the inability to measure a quantity that would be usable as modifying the parameters of the regulator.
- *Adaptive regulators with model identification.* In this case, the parameters of model of the object and/or parameters of the regulator (fulfilling a certain performance criterion) are modified during the control system operation. The identification of the parameters of the object model is carried out on the basis of online measurements of the inputs and outputs of the object. Such regulators, owing to the ability to fulfill the performance criterion (e.g. optimize any objective function) in any operating point and for a wide spectrum of disturbances, outperform control systems with programmed changes of regulators parameters. Their certain shortcoming is the rather high computational complexity in relation to regulators with programmed parameter changes. Adaptive regulators with model identification are divided into:

  - *Adaptive control systems with indirect identification* of object (Figure 15.38a). The parameters of the mathematical model of the object are subject to estimation here. The regulator parameters for the defined performance criterion are calculated based on the estimated parameters of the object model.
  - *Adaptive control systems with direct identification* of object (Figure 15.38b). The regulator model parameters that meet the specific control objective (performance criterion) are directly calculated here. The model parameters are not estimated in explicit form.

The adaptive regulator design can be divided into the following issues discussed below: model of the object definition, estimation of the model parameters, and calculation parameters of the regulator fulfilling the performance criterion.

Model of the Object

In the case of adaptive regulators as mathematical models of objects, whose parameters are subject to estimation, the ARX (autoregressive with exogenous input) and ARMAX (autoregressive moving average with exogenous input) regression models are most often used.

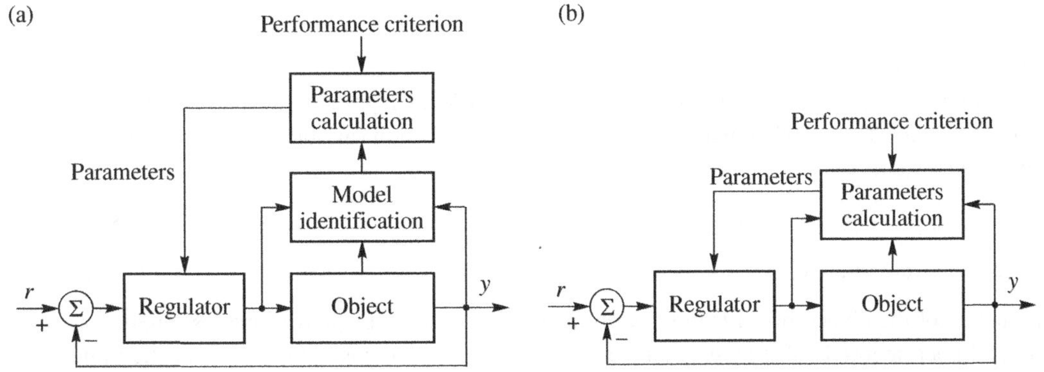

Figure 15.38 Block diagram of an adaptive control system with (a) indirect identification and (b) direct identification.

The ARX model of the object is expressed by

$$y_k = z^{-q}\frac{B}{A}u_k + \frac{1}{A}w_k \tag{15.89}$$

where y_k, u_k, and w_k respectively, are k - th sample of the output, input, and discrete white noise of the variance W, and q is a discrete time delay equal to $\Delta t = qT_p$, where T_p is the sampling period. In general, in addition to disturbances in the form of noise (w_k), this model may also contain a model of deterministic disturbances. The model is defined by polynomials

$$A = A(z^{-1}) = 1 + a_1 z^{-1} + a_2 z^{-2} + \ldots + a_{nA} z^{-nA} \tag{15.90}$$

$$B = B(z^{-1}) = b_0 + b_1 z^{-1} + b_2 z^{-2} + \ldots + b_{nB} z^{-nB} \tag{15.91}$$

where the degree of the denominator polynomial A and the numerator B polynomial is equal to nA and nB, respectively.

In the model identification (model parameters estimation) process, the ARX regression model is used, in the form

$$y_k - \varphi_{k-1}^{\mathrm{T}} \boldsymbol{\Theta} = w_k \tag{15.92}$$

where vector

$$\varphi_{k-1}^{\mathrm{T}} = \left[u_{k-q}, \ldots, u_{k-q-nB}, -y_{k-1}, \ldots, -y_{k-nA}\right] \tag{15.93}$$

is a vector of known regression variables, which are delayed samples of input and output quantities, and vector

$$\boldsymbol{\Theta} = \left[b_0, b_1, \ldots, b_{nB}, a_1, \ldots, a_{nA}\right]^{\mathrm{T}} \tag{15.94}$$

is a vector of unknown (being identified) model parameters.

The ARMAX model of the object is expressed by

$$y_k = z^{-q}\frac{B}{A}u_k + \frac{C}{A}w_k \tag{15.95}$$

where the polynomials A and B are defined by Eqs. (15.90) and (15.91). The numerator polynomial

$$C = C(z^{-1}) = 1 + c_1 z^{-1} + c_2 z^{-2} + \ldots + c_{nC} z^{-nC} \tag{15.96}$$

of a degree nC describes the disturbance model.

Estimation of the Model Parameters

In the model identification (model parameters estimation) process, the ARMAX regression model is used, in the form of the ARX regression model, Eq. (15.95), where the vector of known regression variables and vector of model parameters are, respectively, equal

$$\varphi_{k-1}^{\mathrm{T}} = \left[u_{k-q}, \ldots, u_{k-q-nB}, -y_{k-1}, \ldots, -y_{k-nA}, e_{k-1}, \ldots, e_{k-nC}\right] \tag{15.97}$$

$$\boldsymbol{\Theta} = \left[b_0, b_1, \ldots, b_{nB}, a_1, \ldots, a_{nA}, c_1, \ldots, c_{nC}\right]^{\mathrm{T}} \tag{15.98}$$

Estimation of parameters of the object model is most often made using the least-squares, smallest prediction errors, or maximum likelihood methods. The least-squares method is among the most popular and at the same time the most effective one. This methods consists of searching for estimates $\hat{\boldsymbol{\Theta}}_k$ that minimize the function

$$V(\boldsymbol{\Theta}_k) = \sum_{j=1}^{k} \alpha^{k-j} \left[y_j - \varphi_{j-1}^{\mathrm{T}} \hat{\boldsymbol{\Theta}}_k \right]^2 \tag{15.99}$$

where $0 < \alpha \leq 1$ is the forgetting factor. The factor in the basic form is defined as $\alpha = e^{-T_p/T_z}$, where T_z is the time constant of forgetting. A time constant of forgetting of less than one ensures a faster rate of identifying the model parameters. The number of measurement samples having practical significance in a considered process is equal to $3/(1 - \alpha)$.

The first derivative of the function (15.99) is equal to

$$\frac{\partial V(\boldsymbol{\Theta}_k)}{\partial \boldsymbol{\Theta}_k} = \sum_{j=1}^{k} \alpha^{k-j} \varphi_{j-1} y_j - \sum_{j=1}^{k} \alpha^{k-j} \varphi_{j-1} \varphi_{j-1}^{\mathrm{T}} \hat{\boldsymbol{\Theta}}_k = 0 \tag{15.100}$$

If one defines the matrix $\boldsymbol{P}_k^{-1}$ in the form

$$\boldsymbol{P}_k^{-1} = \sum_{j=1}^{k} \alpha_{j-1}^{k-j} \varphi_{j-1} \varphi_{j-1}^{\mathrm{T}} \tag{15.101}$$

then after the substitution of Eq. (15.101) to Eq. (15.100) the sought estimate of parameters of the object model is equal to

$$\hat{\boldsymbol{\Theta}}_k = \boldsymbol{P}_k \sum_{j=1}^{k} \alpha^{k-j} \varphi_{j-1} y_j \tag{15.102}$$

In adaptive control systems, recursive algorithms are used. The most popular are recursive algorithms of the above-mentioned methods, i.e. recursive least-squares (RLS), recursive extended least-squares (RELS), recursive instrumental variables (RIVs), and recursive maximum likelihood (RML) (Gupta 1986; Niederlinski et al. 1995).

For example, the recursive algorithm of the weighted least-squares (WLS) method has the following form

$$\hat{\boldsymbol{\Theta}}_k = \hat{\boldsymbol{\Theta}}_{k-1} + \boldsymbol{K}_k \left[y_k - \varphi_{k-1}^{\mathrm{T}} \hat{\boldsymbol{\Theta}}_{k-1} \right] \tag{15.103}$$

where the correction vector $\boldsymbol{K}_k$ is calculated using the relationship

$$\boldsymbol{K}_k = \boldsymbol{P}_k \varphi_{k-1} \tag{15.104}$$

or

$$\boldsymbol{K}_k = \frac{\boldsymbol{P}_{k-1} \varphi_{k-1}}{\alpha + \varphi_{k-1}^{\mathrm{T}} \boldsymbol{P}_{k-1} \varphi_{k-1}} \tag{15.105}$$

The latter form of the correction vector (Eq. 15.105) is more advantageous for computational reasons in practical applications than Eq. (15.104). The matrix $\boldsymbol{P}_k^{-1}$ should be nonsingular, and in each step it is calculated as follows

$$\boldsymbol{P}_k = \frac{1}{\alpha} \left[\boldsymbol{P}_{k-1} - \frac{\boldsymbol{P}_{k-1} \varphi_{k-1} \varphi_{k-1}^{\mathrm{T}} \boldsymbol{P}_{k-1}}{\alpha + \varphi_{k-1}^{\mathrm{T}} \boldsymbol{P}_{k-1} \varphi_{k-1}} \right] \tag{15.106}$$

Recursion starts with the estimate of parameters equal to $\hat{\boldsymbol{\Theta}}_0 = \boldsymbol{0}$ and matrix equal to $\boldsymbol{P}_0 = a\boldsymbol{I}$, where $a > > 0$. It is emphasized in literature that, in the estimation process, one should avoid:

- overparameterization of polynomials A and B of the object model. The polynomials' degree should be lower than in the controlled real object;
- too short a sampling period T_p.

In the above considerations the inputs and outputs to the parameters' estimation algorithm are the signals deviations from a steady state. Then while having measurements from the object an important operation is the elimination of the constant component of the inputs and outputs. In the simplest case, into the vector of regression variables – Eq. (15.93) or (15.97) – the difference between signals from the current and previous step is entered into the estimator, i.e. $\Delta u_k = u_k - u_{k-1}$ and $\Delta y_k = y_k - y_{k-1}$.

Calculation Parameters of the Regulator

Assume that the regulator is defined by the equation

$$u_k = -\frac{S}{R}y_k + \frac{T}{R}r_k \tag{15.107}$$

where r_k is the reference input, and polynomials describing the regulator have a form

$$R = R\left(z^{-1}\right) = 1 + r_1 z^{-1} + r_2 z^{-2} + \ldots + r_{nR} z^{-nR} \tag{15.108}$$

$$S = S\left(z^{-1}\right) = s_0 + s_1 z^{-1} + s_2 z^{-2} + \ldots + s_{nS} z^{-nS} \tag{15.109}$$

$$T = T\left(z^{-1}\right) = 1 + t_1 z^{-1} + t_2 z^{-2} + \ldots + t_{nT} z^{-nT} \tag{15.110}$$

where nR, nS and nT are degrees of the polynomials.

The design of the regulator consists in solving the Diophantine equation (e.g. Eq. 15.116), whose constituent elements result from the purpose of control, i.e. from control performance criterion. The basic control performance criterions that are in practice determined for adaptive regulators are: minimal-variance control, poles and zeros placement, and PID control.

The result of solving the Diophantine equation are coefficients of the polynomials of the regulator (15.107), i.e. coefficients of polynomials R, S, and T. The coefficients are a function of object model coefficients, for example defined by Eq. (15.89) when using the ARX model or by Eq. (15.95) for the ARMAX model.

As an example, one should consider a control system consisting of the regulator defined by Eq. (15.107) and an ARX model of object defined by Eq. (15.95), assuming also that the control performance criterion is the location of the poles of the control system in desired points while leaving the zeros unchanged. Then the control system structure takes the form presented in Figure 15.39.

The tracking transfer function of such defined control system (Figure 15.39) is equal to

$$\frac{y(k)}{r(k)} = z^{-q}\frac{TB}{AR + z^{-q}BS} \tag{15.111}$$

There is a sought regulator defined by Eq. (15.107), which will place the poles of the control system defined by Eq. (15.111) in locations defined by the designer. To achieve that assume that the tracking transfer function of the control system with the designed regulator has the form

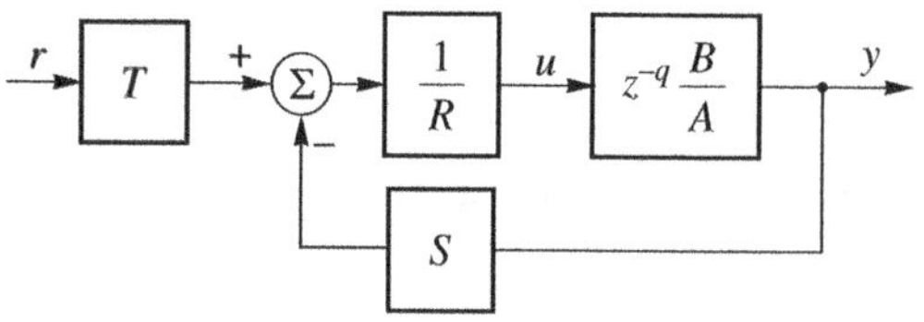

Figure 15.39 Structure of control system with adaptive regulator.

$$\frac{y_k}{r_k} = z^{-q}\frac{K_m B A_o}{A_m A_o} \quad \text{where } K_m = \left(1 + \sum_{i=1}^{nA_m} a_{mi}\right) \Big/ \sum_{i=0}^{nB} b_i \tag{15.112}$$

K_m is a coefficient providing unitary gain in the steady-state, and A_m and A_o are known stable polynomials with given coefficients or roots defined by the designer

$$A_m = A_m(z^{-1}) = 1 + a_{m1}z^{-1} + a_{m2}z^{-2} + \ldots + a_{mnA_m}z^{-nA_m} \tag{15.113}$$

$$A_o = A_o(z^{-1}) = 1 + a_{o1}z^{-1} + a_{o2}z^{-2} + \ldots + a_{onA_o}z^{-nA_o} \tag{15.114}$$

Product $A_mA_o = 0$ is a characteristic equation of the control system, Eq. (15.112), defining the desired location of the poles. The polynomial A_o does not affect the tracking transfer function, which is also shown by Eq. (15.112). On the other hand, the product has an impact on the control system's robustness, i.e. the ability to maintain stability in the event of omitting certain components in the object model. The robustness of control increases with the increase of the polynomial magnitude $|A_o|$. This polynomial, in the extreme case, can also be equal to unity, i.e. $A_o = 1$. In this case, the desired location of the poles is defined by coefficients of polynomial A_m only.

As a result of comparing the numerators and denominators of transfer function defined by Eqs. (15.111) and (15.112) the following is obtained

$$T = A_oK_m \tag{15.115}$$

$$A_mA_o = AR + z^{-q}BS \tag{15.116}$$

From Eqs. (15.115) and (15.116) result the requirements concerning the order of polynomials R and S of the regulator

$$nR = q - 1 + nB \tag{15.117}$$

$$nS = \max(nA - 1, nA_o + nA_m - q - nB) \tag{15.118}$$

Solving Eqs. (15.115) and (15.116) in each step of the adaptive regulator operation the coefficients of polynomials R, S, and T of the regulator fulfilling the control performance criterion are calculated.

Example 15.4 As an example, model of object of type ARX is assumed, whose polynomial of the numerator is of the first degree ($nB = 1$), and the polynomial of the denominator of the second degree ($nA = 2$)

$$A = 1 + a_1z^{-1} + a_2z^{-2} \tag{15.119}$$

$$B = b_0 + b_1z^{-1} \tag{15.120}$$

Coefficients of polynomials A and B are estimated in every step of the regulator's operation according to Eq. (15.103). It is assumed that a regulator is sought for that will locate poles of transfer function of the control system in specific places. This is defined by choosing coefficients of polynomial A_m, respectively. It is assumed that this polynomial will also be of the second degree ($nA_m = 2$). The polynomial A_o is also defined arbitrarily. The order of this polynomial is assumed equal to unity ($nA_o = 1$). Polynomials A_m and A_o then have the form

$$A_m = 1 + a_{m1}z^{-1} + a_{m2}z^{-2} \tag{15.121}$$

$$A_o = 1 + a_{o1}z^{-1} \tag{15.122}$$

For such defined polynomials A and B of the object model, polynomials A_m and A_o, defining the desired dynamic properties of the control system and delays $q = 1$, according to Eqs. (15.117) and (15.118), degree of the regulator polynomials R and S is equal to $nR = 1$ and $nS = 1$. Then, there are sought unknown coefficients r_1, s_0, and s_1 of the polynomials

$$R = 1 + r_1z^{-1} \tag{15.123}$$

$$S = s_0 + s_1z^{-1} \tag{15.124}$$

The substitution of polynomials A, B, A_m, A_o, R, and S to the Diophantine Eq. (15.116) leads to a system of equations

$$\begin{bmatrix} a_{m1} + a_{o1} - a_1 \\ a_{m2} + a_{o1}a_{m1} - a_2 \\ a_{o1}a_{m2} \end{bmatrix} = \begin{bmatrix} 1 & b_0 & 0 \\ a_1 & b_1 & b_0 \\ a_2 & 0 & b_1 \end{bmatrix} \begin{bmatrix} r_1 \\ s_0 \\ s_1 \end{bmatrix} \tag{15.125}$$

whose solution are the coefficients r_1, s_0, and s_1 of the polynomials of the regulator.

The regulator gain is equal to

$$K_m = \frac{1 + a_{m1} + a_{m2}}{b_0 + b_1} \tag{15.126}$$

In turn, the polynomial T of the regulator, according to Eq. (15.115), is equal to

$$T = K_m\left(1 + a_{o1}z^{-1}\right) \tag{15.127}$$

The regulator defined in this way, in each step of the operation, based on the output signal of the object (actual and previous) as well as the reference input, identifies the object and creates the control signal that is input to the real object. Coefficients of the regulator's polynomials change with the change of the identified polynomial coefficients of the object model.

******

15.8 Real Regulators and Field Tests

Regulators of synchronous generators enable the generator operation in one of the following operating modes:

- generator voltage control
- excitation (field) current control
- power factor control
- reactive power control.

The change between operating modes must be carried out bumpless, i.e. without interruption of the generator operation and without a step change in its operating state. Contemporary digital regulators are structurally and functionally complex objects. The basic functions performed by generator regulators include:

- generator start up at idle operation (generator excitation);
- control of a reference electric quantity (quantity depending on the operating mode);
- damping of electromechanical oscillations with the use of a PSS;
- communication with external systems of power plant, e.g. operator workstations, supervisory control and data acquisition (SCADA), and distributed control systems (DCS), supervising a technological or production process in the power plant;
- recording of analog, binary, and alert or alarm data;
- testing (the regulator enables selection and introduction of the generating unit test function and recording of selected quantities from the test);
- self-diagnostics.

As mentioned in Section 2.3.2, the regulators of synchronous generators are also equipped with limiters whose aim is to keep the generating unit operating point in the power capability area. Mathematical derivation of equations describing the curves surrounding the power capability area of synchronous generator is presented in

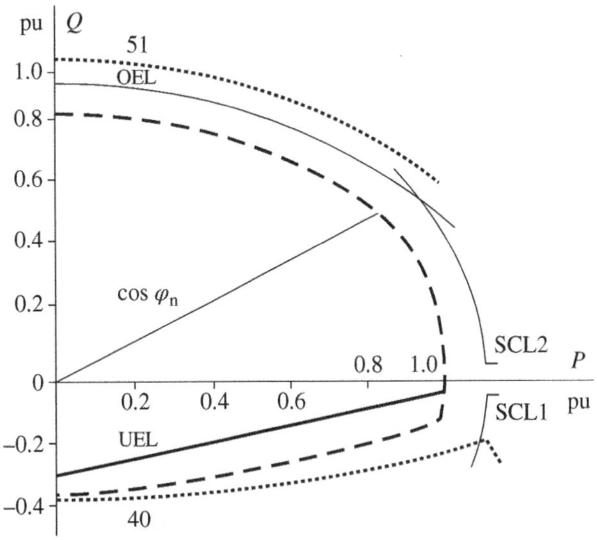

Figure 15.40 Power capability area of the synchronous generator (dashed line) with the characteristics of the limiters: overexcitation limiter (OEL), stator current limiter (SCL1), SCL2, UEL (solid lines), and protections (dotted lines): 40 – loss of field protection; 51 – field overcurrent protection. *Source:* Based on NERC Std. PRC-019-2 (2015). (The numbers in the figure are protection relays codes according to ANSI/IEEE Std. C37.2-2013, Standard for Electrical Power System Device Function Numbers, Acronyms, and Contact Designations.)

Section 3.3.4. For the purpose of this section Figure 15.40 presents a version of the power capability area of the synchronous generator based on NERC Std. PRC-019-2 (2015). The most commonly used limiters are:

- Stator current limiter (SCL) – this limiter consists of one of two elements (e.g. SCL1, SCL2 in Figure 15.40, with a characteristic defined independently for the inductive and capacitive loads of the generator);
- Overexcitation limiter (OEL);
- Minimum field current limiter (MFCL);
- Underexcitation limiter (UEL) – this limiter acts to some extent as a protection against loss of angle stability;
- Limiter of induction, implemented as a V/f (volt per hertz) limiter.

The task of each limiter is to prevent the generator operating point to move outside the power capability area, i.e. outside the limiter's characteristic. For example, a big decrease in network voltage can move the generator operating point beyond the OEL characteristic, i.e. above of the OEL curve in Figure 15.40. This is related to an excessive increase of the field current. Then the task of limiter is to decrease the field current to a value that corresponds to the point on the OEL characteristic. This requires a change of the regulator operating mode from the generator voltage control to the field current control. The change of the operating mode, in the case of the regulator with the structure shown in Figure 15.5, is implemented by the MIN/MAX elements. They select the control signal of the excitation rectifier (input to exciter) from the output signals of the limiters and the voltage regulator. In general, however, structures are also possible in which the output signal of the limiter is added to the output signal of the voltage regulator.

When a generating unit is operating on the power capability area, its control is taken over by the limiters. In such a case the given limiter is the one that affects the dynamic properties of the generating unit, replacing the voltage regulator. Thus, the values of their parameters must be defined so that the operation of the generator on the limit is stable, i.e. does not involve the occurrence of voltage and power oscillations. The limiters are in fact regulators like AVR, i.e. consisted of correctors or they are PI type regulators. The design method of the limiters' parameters is similar to that discussed in Section 15.3.1. An exemplary structure of the UEL is presented in Figure 15.41. This limiter is a reactive power regulator with a power reference being a function of the active power load. In other variants of the limiter, it also takes into account the generator terminal's voltage, i.e. the controlled signal reference is a function of active power and the terminal's voltage.

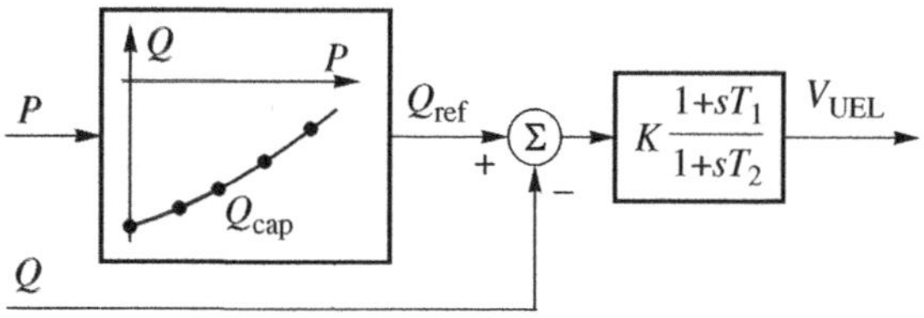

Figure 15.41 Block diagram of an exemplary underexcitation limiter (UEL).

In addition, synchronous generator regulators may be equipped with internal protection functions, such as:

- field overcurrent;
- field overvoltage;
- generator undervoltage and overvoltage;
- loss of generator voltage measurement – loss of voltage measurement is detected on the basis of generator voltage or by comparison of voltage measurements from two measuring transducers;
- underfrequency (e.g. in the Bassler DECS-400 regulator the frequency reference is set at 10 Hz and in the P100C-SX regulator the frequency reference is set at 90% of the nominal speed (frequency);
- loss of field;
- loss of measurement of field voltage or field current – loss of measurement is detected on the basis of value of field current or field voltage;
- excessive temperature of the field winding – used in generating units with static exciters. The temperature is calculated on the basis of the field winding thermal model. Usually, the magnitudes of the field current and field voltage are measured here. In the simplest thermal model, both measurements are used for the calculation of the field winding resistance, which is assumed to be linearly dependent on the temperature;
- damage to the excitation rectifier diodes – used in generating units with machine exciters with rotating diodes.

Equipping regulators with the above protection functions depends on the manufacturer and the type of regulator. These functions do not replace the power system protections of the synchronous generator, which must be installed independently of the protection functions applied in the regulator. Whether the internal protection functions of the regulator are activated or not depends on the TSO's requirements. Examples of usually activated protection functions are: generator rotor overload, rotor earth faults, underfrequency (whose task is to prevent the stopped generator from being excited), excessive temperature of the field winding, loss of field voltage, and loss of generator voltage measurement.

Verification of the synchronous generator voltage regulator settings (AVR commissioning and testing) on the real object is carried out by performing the following tests.

1) Tests performed with the generator operating at idle (not synchronized with the network and not supplying auxiliary services)
   - Synchronous generator excitation test. The test consists in the excitation of the generator from the state of no excitation to the excitation to rated terminal voltage. An example of a generator excitation process is shown in Figure 15.42. This process is implemented in time in accordance with the program stored in the regulator's memory.
   - Generator voltage control quality test. For a synchronous generator operating at idle with a rated voltage and a regulator operating in the voltage control mode, a step change in the generator voltage reference equal to ±10% is introduced. Usually, this test, as well as the subsequent one, is carried out in a single measurement sequence. In such a case the first step change in the voltage reference equal to −10% is introduced, and after the transient process is passing out, a step change in voltage reference +10% is introduced (Figure 15.43). Owing to magnetic saturation effects, the generator voltage after the end of the test does not return to the

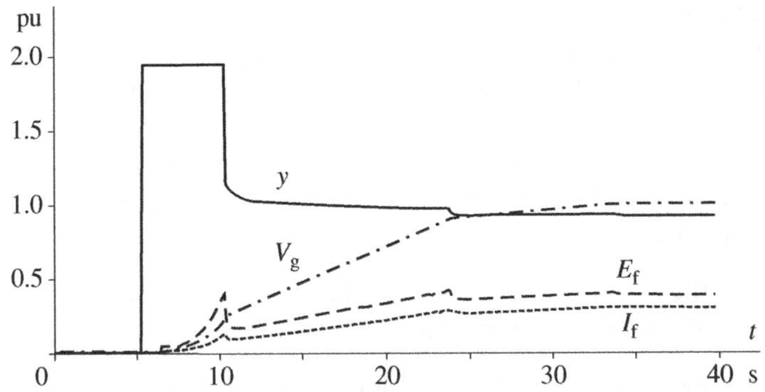

Figure 15.42 Example of the synchronous generator excitation process test: y – regulator output signal; V_g – generator voltage; E_f – field voltage; I_f – field current. *Source:* Reproduced by permission of Institute of Power Engineering, Gdansk Division.

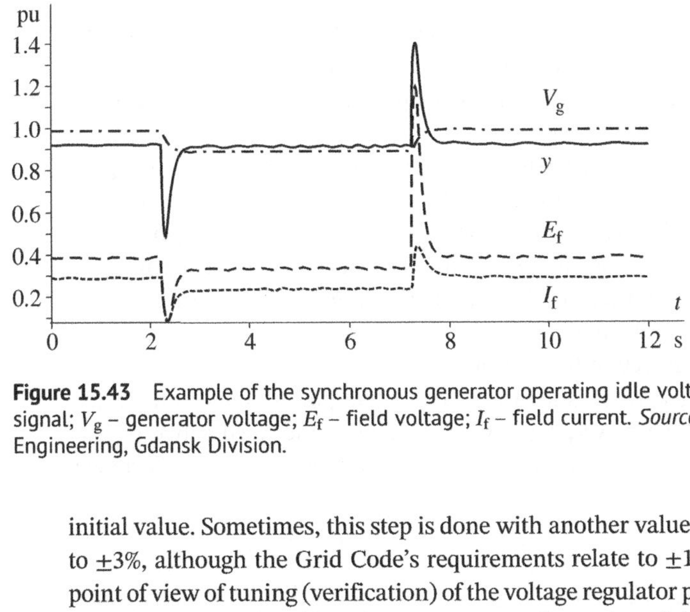

Figure 15.43 Example of the synchronous generator operating idle voltage reference step change test: y – regulator output signal; V_g – generator voltage; E_f – field voltage; I_f – field current. *Source:* Reproduced by permission of Institute of Power Engineering, Gdansk Division.

initial value. Sometimes, this step is done with another value of a step change in voltage reference, e.g. equal to ±3%, although the Grid Code's requirements relate to ±10% voltage step. This test is the basic from the point of view of tuning (verification) of the voltage regulator parameters. If as a result of the test, the regulator fails to comply with Grid Code requirements, the tuning of the regulator parameters can be carried out using the trial-and-error method with following field tests.

- Field current control quality test. For a synchronous generator operating at idle with a rated voltage and a regulator operating in the field current control mode, the field current reference is changed by ±10%. This test is equivalent to the test described above but is performed for another mode of operation of the regulator.
- Limiter of induction test. The test is carried out by a step change (increase) of the generator voltage reference. For example, there is performed change from the generator rated voltage to the voltage causing exceeding the limiter threshold, e.g. (105–107)% of threshold value. The limiter threshold value depends on the synchronous generator protection setting. It should be set lower than the threshold value of the power system protection.
- Loss of generator voltage measurement test. The test is carried out by disconnecting one phase of the voltage transformer or voltage transducer in the generator voltage measurement path or by switching off another element leading to loss of the measurement voltage. The effect of the voltage loss is the change of the regulator's operating mode from voltage control to field current control. Change of the operating mode should be bumpless.

- Regulator switching between channels test. The test is performed in the case of a regulator with channel redundancy by manual switching between channels. The channel switching should be bumpless.
- Synchronous generator de-excitation test. The test is usually performed by manually applying the de-excitation procedure in the regulator.
- Overexcitation limiter (OEL) test. The test is carried out by increasing the generator voltage reference until the limiter threshold value is reached or by a step change in the generator voltage reference leading to the field current exceeding the threshold value of the limiter.

2) Tests performed with the generator synchronized with the network. The tests are carried on with active power defined by the TSO. It is in range from a few percent of rated power to the rated active power.

- Generator voltage control quality test. For a synchronous generator with a regulator operating in the voltage control mode, the generator voltage reference is changed by $\pm(1.5–3)\%$. The change in the voltage reference should be small enough so that the limiters of the power capability area are not activated. The initial value of reactive power is defined by TSO. Some TSOs need to perform tests for two various initial values of reactive power. In this case the powers have to differ by (40–50)% of the rated power. The same test is carried out for the reactive power control mode.
- Stator current limiter (SCL1, SCL2) test. The test is performed by changing the voltage reference of the generator loaded with a correspondingly large power to force stator current that exceeds 110% of the rated current or a temporarily reduced threshold of the limiter. This test is performed independently for the SCL1 and SCL2 limiter, i.e. for capacitive and inductive loads of the generator.
- Underexcitation limiter (UEL) test. The test is carried out by step reduction (and then increase) of the generator voltage reference in order to potentially shift the operating point beyond the limiter characteristic. An example of a test result is shown in Figure 15.44. The overlapping signals y and y_{UEL} indicate the operation of the limiter. In time interval from 1.5 to 7 s, the output signal of the limiter becomes the output signal of the regulator. This means that the limiter takes over control. In this test, the generating unit operating point is shifted toward the stability boarder and therefore, if the limiter does not act or is set incorrectly, it can lead to loss of stability. In such a case the generator's protection has to trip it off.
- Reactive load rejection test. The synchronous generator is loaded with reactive power equal to the rated power and active power equal to minimum active power, e.g. several percent of the rated active power. The test is initiated by the opening of the generator circuit-breaker. An example of a test result is shown in Figure 15.45.

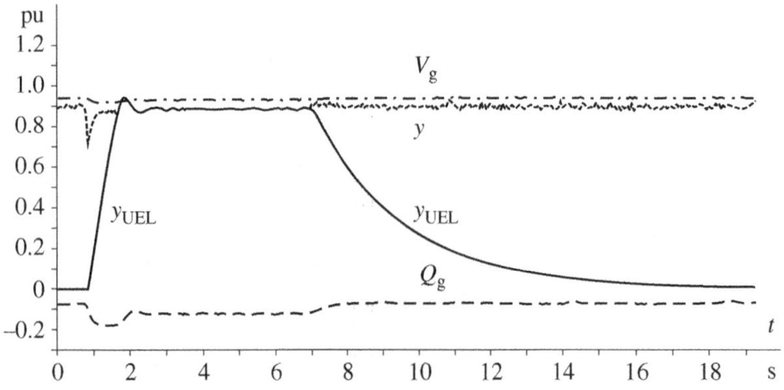

Figure 15.44 Example of a underexcitation limiter (UEL) test: y – regulator output signal; y_{UEL} – UEL limiter output signal; V_{g} – generator voltage; Q_{g} – reactive power. *Source:* Reproduced by permission of Institute of Power Engineering, Gdansk Division.

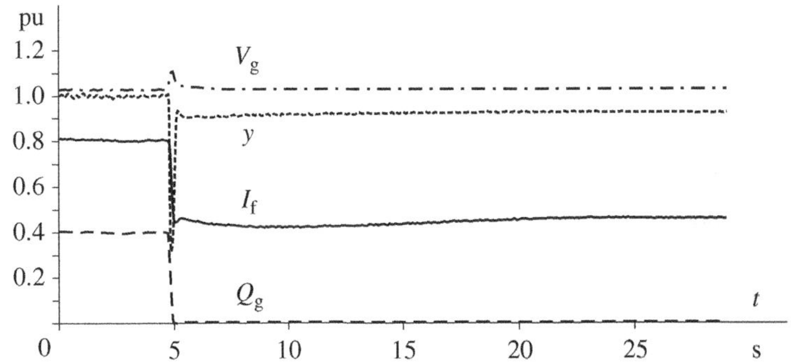

Figure 15.45 Example of the synchronous generator rated reactive load rejection test: y – regulator output signal; V_g – generator voltage; Q_g – reactive power; I_f – field current. *Source:* Reproduced by permission of Institute of Power Engineering, Gdansk Division.

Because in most of the above tests, a change of the generator operating point is made by changing the voltage reference, testing a given limiter often requires blocking another one. It is also possible to test the limiters by changing the network operating point. In this case, the change of the generator operating point is made by changing the voltage on other generators, including ones in nearby power plants. This leads to a change in the voltage in the tested unit or to a change of its reactive power. However, this way of performing the tests is very difficult because it leads to the nodal voltages and the power flow change. Therefore, they are carried out very rarely (usually only at the request of the transmission system operator).

In the case of a PSS, the requirements of any quantitative evaluation of the effectiveness of electromechanical oscillations damping are not defined. The evaluation is made by comparing the result obtained with the PSS on and off.

The selection of the stabilizer gain K_{s1} (even if calculated earlier) is made by gradually increasing its value up to the limit value K_{s1lim}. In this process the gain increase, for the properly defined compensator $G_c(s)$ (Eq. 15.48), initially leads to an increase of electromechanical oscillations damping, but further causes the damping decrease and finally, for some gain K_{s1lim}, undamped oscillations occur. Knowing the gain limit value, according to the principles of control theory, to ensure an adequate stability margin, the PSS gain is defined as equal to K_{s1lim} decreased by approximately (10–12) dB, i.e. the gain is set on value from $K_{s1} = K_{s1lim}/4$ to $K_{s1lim}/3$.

The PSS settings are verified by performing the following tests:

- Step change in the generator voltage reference. The voltage reference step, usually not higher than $\pm 3\%$, is applied to generating unit with the PSS switched on and switched off. Example of the test result is presented in Figure 15.46.
- Change of the network configuration. The aim of network configuration change is to introduce a change of the generating unit operating point. Usually, this is realized by a transmission line switching off or on. The PSS effectiveness is evaluated based on time domain response, similar to the one presented in Figure 15.46.
- Measuring the magnitude-frequency characteristic between the voltage reference V_{ref} or the PSS output V_{PSS} (input to AVR) and active power P_g (output). For comparison purposes, there are measured characteristics of the control system with PSS switched on and switched off. An example of the characteristics is presented in Figure 15.47.

In order to obtain a single point of the characteristic as in Figure 15.47, a sinusoidal signal with a given (appropriately small) magnitude and chosen frequency is injected to the AVR input. The signal is injected for enough time to pass the transient process, i.e. when the magnitude of active power oscillations (output signal) of a given frequency are constant. The magnitude of the characteristic for a given frequency is equal to the ratio of the magnitudes of the output

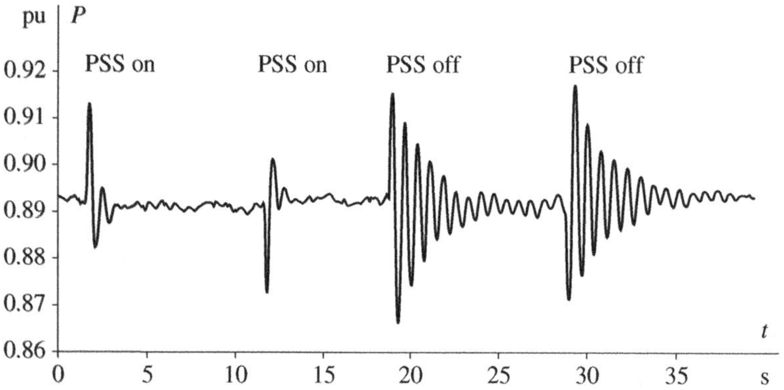

Figure 15.46 Response of the synchronous generator with the power system stabilizer (PSS) on and off to a step change in voltage reference by ±2%, $P_g = 381$ MW, $Q_g = -21$ Mvar, and $V_g = 21.3$ kV. *Source:* Reproduced by permission of Institute of Power Engineering, Gdansk Division.

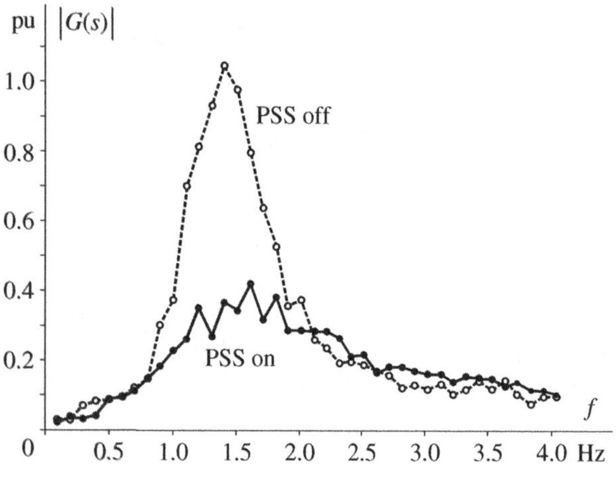

Figure 15.47 Frequency characteristic of generating unit with the power system stabilizer (PSS) on and off. *Source:* Reproduced by permission of Institute of Power Engineering, Gdansk Division.

and input signals, and the phase is equal to the angle of the input and output sine wave. In practice, only the magnitude characteristic is used to evaluate the PSS's efficiency. In order to obtain the desired characteristic, i.e. typically in the frequency range of (0.1–4.0) Hz, a series of the above tests, for various frequencies, is performed. Some regulators (for example voltage generator controller – P100C-SX) allow this characteristic to be obtained automatically. To perform the test automatically it is necessary to define the frequency range, i.e. upper and lower frequencies and the number of points of characteristic.

For a properly designed PSS, the characteristic with PSS switched on, in the frequency range corresponding to the inter-area oscillations, i.e. (0.15–0.6) Hz, should indicate damping no less than with the PSS switched off (see Figure 15.47). At the same time, it should indicate an increase in damping of modes in the range of local-area oscillations, i.e. (1.0–2.0) Hz. This means that the characteristics of a system with the PSS switched on in these frequency ranges should not lie above the characteristics of a system without the PSS switched off, i.e. as in Figure 15.47. For frequencies above 2 Hz, higher values of the characteristic with the PSS switched on are allowed than in the case of the PSS being switched off, but the differences between these characteristics should not be large.

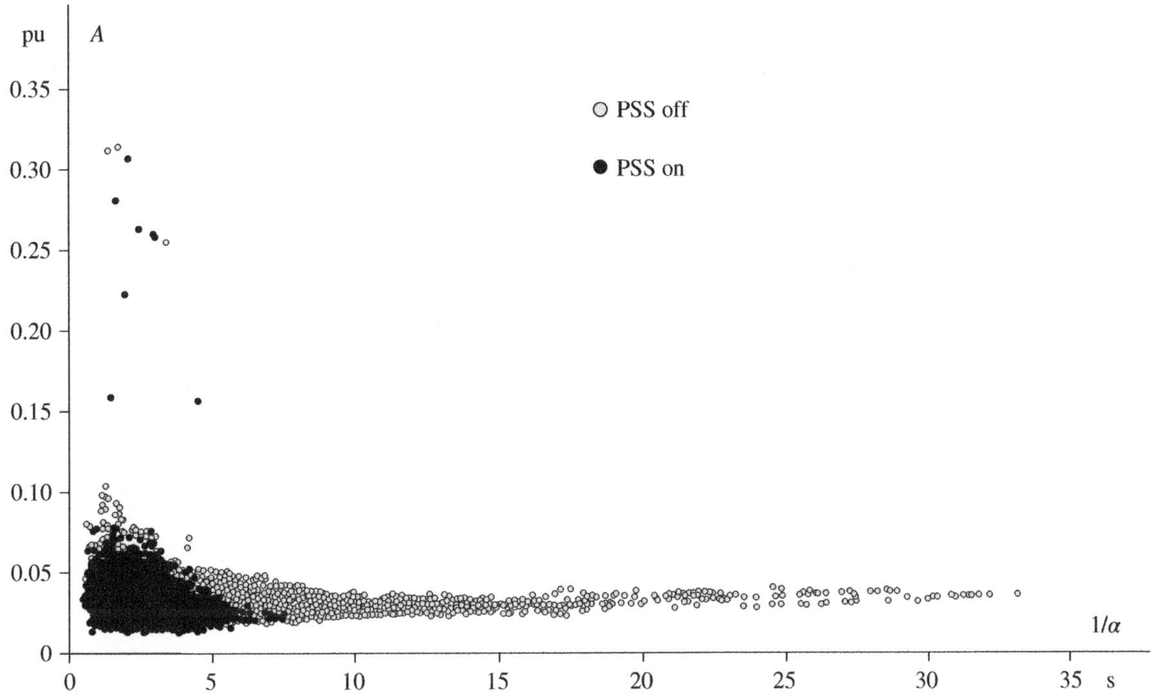

Figure 15.48 Example of the power system stabilizer (PSS) influence on modes with frequencies in the range (1.0–1.8) Hz. *Source:* Reproduced by permission of General Electric Company and/or its Affiliates.

The characteristic presented in Figure 15.47 indicates a PSS which is tuned to damp local-area oscillations. When the PSS is tuned to damp inter-area oscillations, its characteristic has to be shifted to stronger damping of the low frequency oscillations.

The PSS efficiency can also be verified by performing long-term (e.g. months) measurements in the power system. For this purpose WAMS (wide area measurement system) or typical power system recorders can be used. In this case, the active power or voltages and currents are measured at a specific point in the EPS, e.g. at the generator terminals. Then the modes occurring in the measured signals are calculated. These modes, for example those grouped in frequency intervals, can be presented in a graph in a coordinate system, e.g. time constant of amplitude decay $1/\alpha$ versus the mode amplitude A, where both parameters A and $1/\alpha$ are defined by Eq. (12.67) in Section 12.1.3 and Eq. (A.118) in the appendix. In the example shown in Figure 15.48 the gray points shown in the figure correspond to 60 500 modes with frequencies between 1.0 and 1.8 Hz, identified in the signal measured with the PSS switched off, e.g. before the PSS was installed and in periods when it was switched off. The black points correspond to almost 65 000 modes identified during the period of the generating unit operating with the PSS switched on. It can be seen here that a given PSS significantly reduces the time constant of amplitude decay, i.e. it increases the electromechanical oscillations damping at frequencies from a given frequency range.

16

Wide-Area Monitoring and Control

Energy management systems (EMS) and supervisory control and data acquisition (SCADA) systems developed more than a dozen years ago are based on nonsynchronized measurements of the root-mean-square (RMS) values of quantities characterizing the state of the electric power system such as nodal voltages, branch currents, real power, and frequency. Typically, such quantities are measured in power system elements by measuring devices and sent without timestamps to the SCADA/EMS receivers once per second or less.

In the last decade, intensive progress in digital data processing and development of fast telecommunication platforms allowed for the implementation of modern wide area measurement, protection, and control (WAMPAC) systems, which (as described in Section 2.7) integrate WAMS (wide area measurement system) and functions of WAM (wide area monitoring), WAP (wide area protection), and WAC (wide area control). In this chapter some examples of the WAMPAC applications relevant to the power system stability are described.

16.1 Wide Area Measurement Systems

As explained in Section 2.7, the WAMS uses phasor measurement units (PMUs), which on the basis of the momentary values (samples) calculate orthogonal components of the phasors of the nodal voltages and branch currents. Such measurements, together with timestamps, are sent via data concentrators (Figure 2.46) to the local or central SCADA/EMS and various schemes performing the WAM, WAP, and WAC functions. These phasors with timestamps are referred to as *synchrophasors*.

16.1.1 Phasors

For a sinusoidal signal (voltage or current) of the form

$$x(t) = X_{\mathrm{m}} \cos\left[2\pi(f_0 + \Delta f)t + \delta\right] = X_{\mathrm{m}} \cos\left(2\pi f_0 t + 2\pi\Delta f t + \delta\right) \tag{16.1}$$

the phasor is equal to

$$\underline{X}(t) = \frac{X_{\mathrm{m}}}{\sqrt{2}} \mathrm{e}^{\mathrm{j}(2\pi\Delta f t + \delta)} = \frac{X_{\mathrm{m}}}{\sqrt{2}} \left(\cos\left(2\pi\Delta f t + \delta\right) + \mathrm{j}\sin\left(2\pi\Delta f t + \delta\right)\right) \tag{16.2}$$

where X_{m} is the signal amplitude, f_0 is the network nominal frequency, Δf is the frequency deviation from its nominal value, and δ is the phase shift in relation to the selected reference axis.

The real and imaginary components of phasor $\underline{X}(t)$ refer to a reference signal associated with a reference axis rotating at a speed corresponding to a nominal frequency of 50 or 60 Hz. The reference signal is a cosine signal synchronized with UTC (Universal Time Coordinated) time. The relationship between the reference signal and

Power System Dynamics: Stability and Control, Third Edition. Jan Machowski, Zbigniew Lubosny, Janusz W. Bialek and James R. Bumby.

the measured signal and between the reference and measured phasor is illustrated in Figure 16.1. The beginning of the UTC time corresponds to the moment $t = 0$, and the beginning of each second is marked by the 1 PPS (pulse per second) signal obtained from the GPS (global positioning system).

The measured phasor is calculated from N_p discrete samples $x(kT_p)$ obtained by sampling a continuous signal $x(t)$ with the frequency $f_p = 1/T_p$, where T_p is the sampling period. The number of samples N_p in the measurement window is usually the multiplicity of the number of samples N in one period of the signal measured at the nominal frequency. The phasors are calculated using various methods based on discrete Fourier transform or wavelet transform, Kalman filter, genetic algorithms, least squares algorithm, etc. (CIGRE Technical Brochure No. 664, 2016 and No. 702, 2017).

Phasor components of three-phase signals are calculated using the following equations

$$
\begin{aligned}
\underline{X}_{L1} &= \frac{\sqrt{2}}{N_p}\sum_{k=1}^{N_p} x_{L1(n+k)}\left[\cos\left(2\pi\frac{n+k}{N}\right) - \mathrm{j}\sin\left(2\pi\frac{n+k}{N}\right)\right] \\
\underline{X}_{L2} &= \frac{\sqrt{2}}{N_p}\sum_{k=1}^{N_p} x_{L2(n+k)}\left[\cos\left(2\pi\frac{n+k}{N}\right) - \mathrm{j}\sin\left(2\pi\frac{n+k}{N}\right)\right] \\
\underline{X}_{L3} &= \frac{\sqrt{2}}{N_p}\sum_{k=1}^{N_p} x_{L3(n+k)}\left[\cos\left(2\pi\frac{n+k}{N}\right) - \mathrm{j}\sin\left(2\pi\frac{n+k}{N}\right)\right]
\end{aligned} \tag{16.3}
$$

where x_{L1}, x_{L2}, and x_{L3} are samples of continuous signals measured in phases L1, L2, and L3 in moments $t = (n + k)T_p$, where $k = 1, 2, \ldots, N_p$ is the number of the consecutive samples in the measurement window, and $n = 0, 1, 2, \ldots$ determines the beginning of the measuring window that shifts over time. The sampling frequency of modern PMU ranges from 4 to 16 kHz.

In the basic phasor calculation algorithm (Eq. 16.3) the measured signal $x(t)$ is demodulated by multiplying the signal samples x_k by the corresponding values of a reference signal. The reference signal is defined as the cosine function, as in Figure 16.1. If the measured signal is a cosine wave (as a result of the multiplication of the samples of both signals), the real component of the phasor reaches the highest possible value, and the imaginary component is equal to zero. This means that the phase shift of the measured signal relative to the reference is equal to zero. If the measured signal is a sine wave, the effect is reversed, i.e. the real component of the phasor is equal to zero, and the imaginary component achieves a positive or negative maximum value, which depends on the phase shift of the measured signal in relation to the reference signal. The phase shift is then equal to 90° or −90°. In other cases, the real and imaginary phasor components are nonzero, and their quotient is a measure of phase shift.

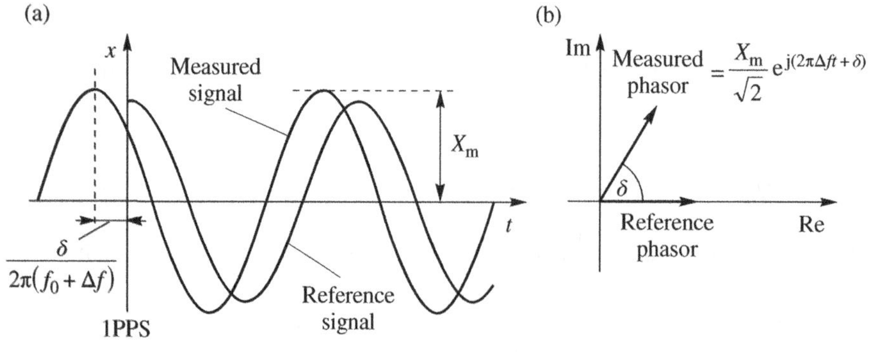

Figure 16.1 The relation between the phasor and the reference signal: (a) the course of the measured and reference signal over time; (b) the phasors of the measured signal and the reference signal. Based on CIGRE report No. 702. *Source:* Reproduced by permission of CIGRE.

If the frequency of the measured signal is equal to the nominal frequency, i.e. $NT_p = 20$ ms, in a 50 Hz system, the phasors $\underline{X}_{L1}$, $\underline{X}_{L2}$, and $\underline{X}_{L3}$ contain only a constant component. However, if the frequency differs from the nominal value and changes over time, these phasors contain an additional constant component and a component with a doubled frequency. This component is eliminated by calculating the desired phasor value from the equation

$$\underline{X} = \underline{X}_1 = \mathrm{Re}\underline{X} + \mathrm{jIm}\underline{X} = \frac{1}{3}\left(\underline{X}_{L1} + \mathrm{a}\underline{X}_{L2} + \mathrm{a}^2\underline{X}_{L3}\right) \tag{16.4}$$

where $\mathrm{a} = \mathrm{e}^{\mathrm{j}120^\circ}$ is the operator shifting the angle by 120°. The calculated phasor $\underline{X}$ is equal to the positive sequence component $\underline{X}_1$ of the measured phasor. Equation (16.4) eliminates the symmetric components: negative sequence and zero sequence if such are present in the measured signal.

The components of phasors are usually calculated at every sampling step T_p and the calculated values are assigned to a time corresponding to the center of the measurement window. This means that theoretically the phasor available at a given moment on the PMU output is delayed by the time resulting from half of the measurement window and by the time consumed by the phasor calculation algorithm. The reporting period of the phasors estimated by PMU is typically equal to (1–5) periods of the measured signals, what for the nominal frequency 50 Hz is equivalent to reporting time interval (20–100) ms and reporting rate (50–10) Hz.

The PMUs in the phasor calculation algorithm use low-pass filters to eliminate high-frequency noise. These filters (*backend filters*) are used at the output, i.e. after calculating the phasor components or are an integral part of the phasor calculation algorithm. In the latter case the computational algorithm and the low-pass filter are given by the following equations

$$\begin{aligned}
\underline{X}_{L1} &= \frac{\sqrt{2}}{K_{DC}} \sum_{k=1}^{N_P} \left[x_{L1(n+k)}\right] h_k \mathrm{e}^{-\mathrm{j}(n+k)T_p 2\pi f_0} \\
\underline{X}_{L2} &= \frac{\sqrt{2}}{K_{DC}} \sum_{k=1}^{N_P} \left[x_{L2(n+k)}\right] h_k \mathrm{e}^{-\mathrm{j}(n+k)T_p 2\pi f_0} \\
\underline{X}_{L3} &= \frac{\sqrt{2}}{K_{DC}} \sum_{k=1}^{N_P} \left[x_{L3(n+k)}\right] h_k \mathrm{e}^{-\mathrm{j}(n+k)T_p 2\pi f_0}
\end{aligned} \tag{16.5}$$

where h_k are the coefficients of a low-pass filter and K_{DC} is the gain of this filter for the constant component, equal to

$$K_{DC} = \sum_{k=1}^{N_P} h_k \tag{16.6}$$

The value of the positive sequence component of the phasor is calculated using Eq. (16.4) based on the result obtained from Eq. (16.5). Equation (16.5), despite the difference in the record, is equivalent to Eq. (16.3), for a filter with a rectangular window, with coefficients equal to $h_k = 1$ for $k = 1, 2, \ldots, N_p$. Then, the filter gain for the constant input signal is equal $K_{DC} = N_p$.

The type of filter used in the algorithms of measurement units depends mainly on the functions performed by the PMU. In the case of measurement units intended for protection purposes (P-class), which calculate the values of phasors based on measurement windows, typically two periods (40 ms) wide, filters with a triangular window characteristic (Figure 16.2, filter P), are often used, whose coefficients define the function

$$h_k = 1 - \frac{|2k - N_p - 1|}{N_p + 2} \tag{16.7}$$

where $k = 1, 2, \ldots, N_p$. This type of filter is intended for measuring windows containing an odd number of samples N_p.

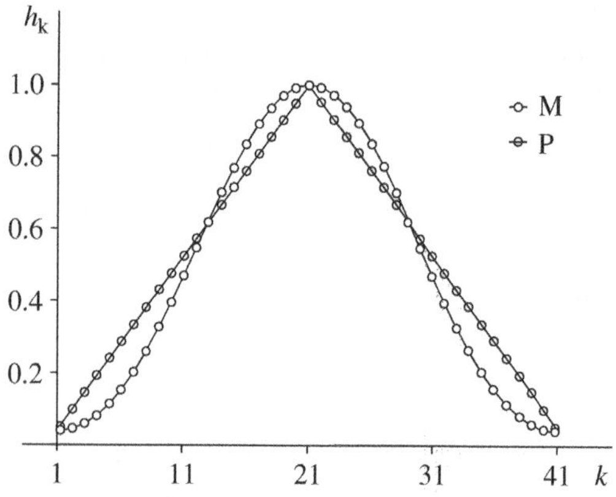

Figure 16.2 Example of filter characteristics of P-class and M-class.

However, in the case of measurement units intended for other than protection purposes (M-class), it is required that the filter be able to suppress the signals above the Nyquist frequency for the selected reporting rate. The width of the measurement window is typically equal to five periods of the measured signal. An example here can be a filter whose coefficients are defined by a function

$$h_k = \frac{\sin\left(2\pi \frac{f_{\rm Fr}}{f_{\rm p}}(2k - N_{\rm p} - 1)\right)}{2\pi \frac{f_{\rm Fr}}{f_{\rm p}}(2k - N_{\rm p} - 1)} W_k \tag{16.8}$$

where $k = 1, 2, \ldots, N_{\rm p}$, $f_{\rm p}$ is the sampling frequency, $f_{\rm Fr}$ is the reference frequency of the filter adequate for the selected reporting rate $f_{\rm r}$, and W_k are the coefficients of the Hamming window, equal to

$$W_k = 0.53836 - 0.46164 \cos\left(\frac{2\pi(k-1)}{N_{\rm p} - 1}\right) \tag{16.9}$$

where $k = 1, 2, \ldots, N_{\rm p}$. The value of the coefficient of this filter for the middle sample, i.e. $h_{(N_{\rm p}+1)/2}$, should be equal to one. As filter reference frequency, values close to those resulting from dependence $f_{\rm Fr} < f_{\rm r}/5$ and not less than 5 Hz are assumed. The recommended values of the reference frequency are given in standard C37.118.2011 (2011). An example of a window characteristic for filters defined by Eqs. (16.7) and (16.8) for the filter window width $N_{\rm p} = 2N + 1$ is shown in Figure 16.2 (M-class). The reference frequency is assumed to be equal to $f_{\rm Fr} = 7.5$ Hz. For simplicity, in Figure 16.2 for both filter types the window width 40 ms is assumed (i.e. required for P-class but less than required for M-class).

The type of filter used affects the dynamic properties of the PMU. As an example, Figure 16.3 shows the dynamic responses of a PMU to a step change in the amplitude of the measured signal by −10% in the cases where the measurement algorithm is based on: Eq. (16.3) – curve A, Eq. (16.5) with filter as in Eq. (16.7) – curve P, Eq. (16.8) – curve M. It is assumed in Figure 16.3 that the change in the measured signal $x_{\rm L1}$, $x_{\rm L2}$, and $x_{\rm L3}$ occurs at the moment $t = 0.02$ s. At this moment the phasor value for $t = 0$, i.e. the center of the measurement window with a width of 40 ms is calculated with the use of samples from the range (−0.02 - 0.02) s. The phasor magnitude is then equal to $X = 1$, because for time $t < 0.02$ s, the measured sinusoidal signal has an amplitude equal to $A_{\rm m} = \sqrt{2}$. The first time after the assumed step change the phasor magnitude equal to $X = 0.9$ is calculated for moment $t = 0.06$ s, which means that the algorithm introduces a time delay equal to the width of the measuring window (Figure 16.3). However, the transient response is different for different filters. Since reporting of the calculation results is performed at a rate much lower than the sampling frequency (and also the phasor calculation), in the PMU output the phasor values from transient process may or may not appear. For example, for a reporting rate of 10 Hz (reporting period 100 ms) for the case presented in Figure 16.3, the phasor values from transient process

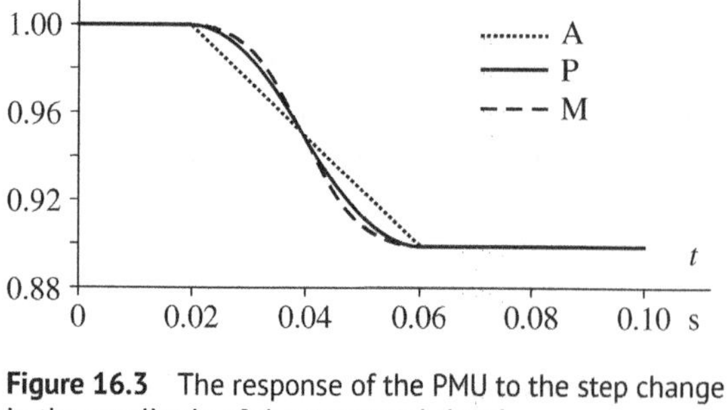

Figure 16.3 The response of the PMU to the step change in the amplitude of the measured signal.

can locate outside reporting moments. Then, only the initial and final steady values of phasor will be reported, i.e. equal to 1.0 or 0.9. In this case, the type of filter used will not affect the reported phasor value. On the other hand, in the case of a phasor reporting with a rate of 50 Hz, i.e. every 20 ms, one or two phasor values from the time interval corresponding to the transient process will appear on the PMU output.

The frequency of measured signal or frequency deviation and rate of change of frequency (RoCoF) are usually calculated less frequently than the phasor components, e.g. with a frequency equal to the reporting rate f_r. The frequency deviation from the nominal value is calculated using the voltage phasor angle

$$\delta_n = \mathrm{arctg}\frac{\mathrm{Im}\underline{X}_n}{\mathrm{Re}\,\underline{X}_n} \tag{16.10}$$

where $\underline{X} = \underline{V}$ is the voltage phasor that results from Eq. (16.4). The frequency deviation is also calculated for the moment corresponding to the center of the measurement window (sample n), based on the phase angle of the voltage δ shifted one sample back and forth with respect to the sample n

$$\Delta f_n = \frac{\delta_{n+1} - \delta_{n-1}}{4\pi T_r} \tag{16.11}$$

where time T_r is equal to $T_r = 1/f_r$, in the case of performing calculations with the reporting rate f_r.

The frequency time derivative, referred to as the *RoCoF*, is equal to

$$RoCoF_n = \frac{\mathrm{d}f_n}{\mathrm{d}t} \cong \frac{\Delta f_n - \Delta f_{n-1}}{T_r} \tag{16.12}$$

After substitution of Eq. (16.11) to Eq. (16.12) a frequency time derivative calculated for the samples from the interval $3T_r$ is obtained. The frequency time derivative is also calculated on the basis of two neighboring samples, i.e. for the interval $2T_r$. Then its value is determined by the equation

$$RoCoF_n = \frac{\mathrm{d}f_n}{\mathrm{d}t} \cong \frac{\delta_{n+1} - 2\delta_n + \delta_{n-1}}{2\pi T_r^2} \tag{16.13}$$

In the next step, the correction factor of the phasor magnitude is calculated, equal to

$$c_{pn} = \frac{1}{\sin\left(\pi\dfrac{f_0 + 1.625\Delta f_n}{2f_0}\right)} \tag{16.14}$$

The corrected value of the phasor magnitude is equal to the product of the correction factor and the phasor magnitude calculated according to Eq. (16.4)

$$X_n = c_{pn}\sqrt{(\mathrm{Re}\,\underline{X}_n)^2 + (\mathrm{Im}\,\underline{X}_n)^2} \tag{16.15}$$

16.1.2 Structure of the WAMS

The basic elements of the WAMS system are: the source or sources of the timestamp, PMUs, phasor data concentrators (PDCs), data transmission networks, and information systems performing superior functions related to the processing of measurement data and control purposes, security, state estimation, etc. An exemplary structure of the WAMS is shown in Figure 2.46 (Section 2.7). In this structure, the lowest layer creates the PMUs. Information from PMUs is sent to PDCs dedicated to the area, object, or function. Data from concentrators are transferred to higher-level concentrators or to systems performing specific protection and control functions (P&C). Schemes implementing protection or control functions, based on information obtained from the PMUs create signals controlling the appropriate objects in the power system, i.e. switches, generators, turbines, etc. These signals are sent to the power system objects through appropriate concentrators or other dedicated links. Data exchange between concentrators is carried out, depending on the needs, one- or two-way. The highest layer of the WAMS is the EMS/SCADA system, which obtains information from all PMUs. This information is used by the system operator for the current control of the electric power station (EPS) and is archived for the needs of various types of analyses.

The source of the timestamp for measuring units in WAMS can be:

- Satellite: American GPS, European Galileo, Russian GLONASS, or currently built Chinese Beidou (Big Bear). The WAMS currently operating in electric power systems use mainly the GPS system. The GPS system is the result of many years of research undertaken by US civil and military institutions aiming to develop a very accurate navigation system. The system has been made available for civil users around the world. The system consists of a space segment, a ground segment, and users.

  The space segment comprises a constellation of 24 satellites located in six orbits, that is with four satellites per orbit. The location of the orbits and the satellites is such that at any time (4–10) satellites can be seen from any point on the earth. Access to a number of satellites is necessary to determine the location of any receiver using three coordinates (longitude, latitude, and altitude) and the reference time. Each satellite transmits two signals (L1 with frequency 1575.42 MHz and L2 with frequency 1227.6 MHz) to earth which are coded messages about the time of transmission and actual coordinates of the satellite with respect to earth. These messages also contain a pulse of 1 PPS informing about the beginning of each second of the universal time. The time transmitted is in SOC (second-of-century) format, i.e. in seconds counted from 1 January 1970. That signal is very important for WAMS as it is used for synchronizing WAMS devices.

  The ground segment of GPS consists of six radio stations located near the equator. One of those stations, in Colorado Springs in the United States, is the master station, while the remaining ones are the monitoring stations. The latter monitor the correctness of operation of the satellites and send information to the main station. The areas of observation of neighboring monitoring stations overlap, which makes it possible to observe the same satellite from two stations. The main station communicates with all the monitoring stations and with the satellites. The main station sends to the satellites corrections to their orbits and a correction to the time of satellite clocks. As the reference time, UTC is used and it is transmitted from a space observatory of the US Navy.

  In the GPS system, the users are only receivers of satellite messages and do not send any information to the system. Consequently, the number of users is not restricted. Calculation of the coordinates of a given receiver and the time is executed in the receiver based on messages received from a few satellites. This means that the receiver is equipped with an algorithm solving an equation in which the message data (the location of a satellite and the reference time) are treated as known data, while the receiver coordinates and the reference time are treated as unknown data. The reference time is implicit in the solved equations because the equations are set up in such a way as to eliminate the transmission time of messages from the satellites to the receiver.

  The accuracy of the GPS reference time of about 0.2 μs and in the Galileo system about 0.008 μs is good enough to measure AC phasors with a frequency of 50 or 60 Hz. For a 50 Hz system the period time corresponding to a full rotation corresponding to 360° is 20 ms $= 20 \cdot 10^3$μs. The time error of 1 μs corresponds to the angle error of $360^\circ/20 \cdot 10^3 = 0.018^\circ$, that is 0.005%. Such an error is small enough from the point of view of phasor measurement.
- Land: based on radio transmission (currently used LORAN-C system), microwave, or fiber optic. The protocols used in computer networks for timing the measurement units of PMU phasors are: PTP (Precision Time Protocol) ensuring the accuracy of time synchronization of (0.1 - 1) μs, and NTP (*Network Time Protocol*) and SNTP (*Simple Network Time Protocol*) ensuring the accuracy of (1 - 100) ms.

The synchronization of PMU clocks with UTC is realized by means of a 1 PPS signal in the case of satellite systems and using the IRIG-B protocol in land systems. The IRIG-B protocol ensures the accuracy of time synchronization such as the PTP protocol, but is suitable for local applications, for example in power stations.

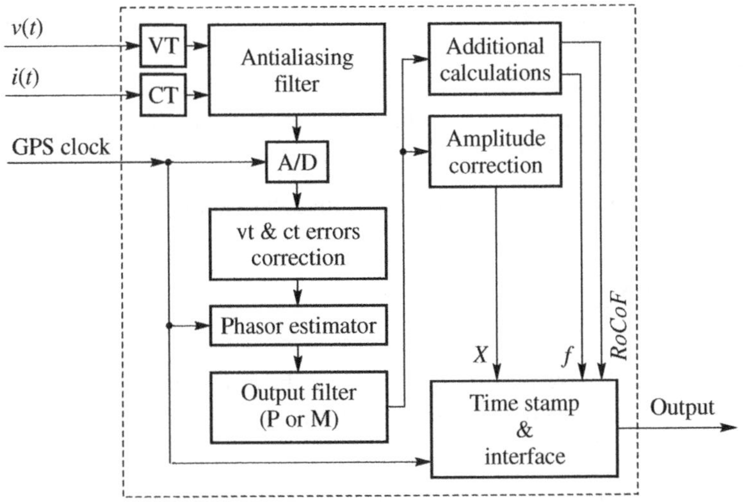

Figure 16.4 Functional diagram of a PMU. Based on CIGRE Report No. 702. *Source:* Reproduced by permission of CIGRE.

The PMU is the basic computational element of the WAMS. The functional diagram of the PMU is shown in Figure 16.4. The input voltage and current signals are passed through internal voltage transformers (VTs) and current transformers (CTs), which step-down signals to a level appropriate for the analog to digital converter (A/D). The stepped-down signals are passed through the anti-aliasing filter. In the case of using a satellite system, the A/C has a frequency sampling synchronized to a GPS clock. The PMU determines its location, and on this basis it calculates the delay of sending the timestamp from the satellite to the PMU. This delay is taken into account in the measurement sample time stamping.

Analog signals are subject to analog filtration, processing into a digital signal and correction of error introduced by CTs and VTs. In the next basic step, the phasors, the frequency deviation, and the frequency derivative are calculated. If there is a frequency deviation from the nominal value, the previously calculated phasors are corrected. In order to achieve a high level of accuracy, the errors of the VTs and CTs are corrected using proprietary algorithms, which are tuned through a calibration exercise.

The output signals of PMU (i.e. the results of the phasor estimation) for power systems with a nominal frequency of 50 Hz can be reported with a rate $f_r = 10,\ 25$, and 50 Hz, and for systems with a nominal frequency of 60 Hz, with a rate $f_r = 10,\ 12,\ 15,\ 20,\ 30$, and 60 Hz. The sampling frequency of the measured signal, as mentioned above, varies and depends on the PMU manufacturer and the calculation algorithms used.

The accuracy of phasor measurement by PMU is evaluated in accordance with the IEEE C37.118.2011 (2011) standard, based on the following defined error types:

- Total vector error *TVE*

$$TVE = \sqrt{\frac{(\mathrm{Re}\,\underline{X}_e - \mathrm{Re}\,\underline{X})^2 + (\mathrm{Im}\,\underline{X}_e - \mathrm{Im}\,\underline{X})^2}{(\mathrm{Re}\,\underline{X})^2 + (\mathrm{Im}\,\underline{X})^2}} = \frac{|\underline{X}_e - \underline{X}|}{X} \tag{16.16}$$

where $\underline{X}_e$ is the estimated phasor (calculated and reported) and $\underline{X}$ is a measured (real) phasor. The *TVE* error in the graphic form on the complex plane is shown in Figure 16.5. The maximum permissible measurement error of the phasor magnitude, when the phase measurement error (angle) is equal to zero, is equal to $TVE = \pm 1\%$. However, if the magnitude measurement error is zero then the maximum permissible phasor measurement error is equal to $TVE = \pm 0.573^{\circ}$. Since the phase measurement is related to the timestamp, the reference time error directly affects the phasor measurement error. For example, a reference time error equal to ±31.8 μs in the frequency system of 50 Hz can lead to an error of $TVE = \pm 1\%$.

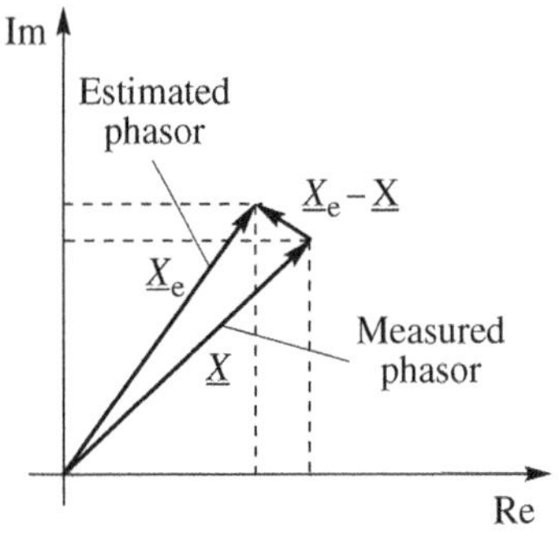

Figure 16.5 Definition of the phasor total vector error *TVE*. Based on CIGRE Report No. 702. *Source:* Reproduced by permission of CIGRE.

- Frequency measurement error *FE*

$$FE = |f - f_e| \tag{16.17}$$

where f and f_e are the measured and estimated frequency, respectively. The permissible value of this error is 0.005 Hz for P-class and M-class.

- Measurement error of a frequency time derivative *RFE* (*RoCoF measurement error*)

$$RFE = \left| \left(\frac{df}{dt}\right) - \left(\frac{df}{dt}\right)_e \right| \tag{16.18}$$

where df/dt and $(df/dt)_e$ are the time derivatives of the measured and estimated frequency, respectively. The permissible value of this error is 0.01 Hz/s for P-class and M-class.

The dynamic properties of the PMUs and delays discussed above play an important role in the assessment of power system dynamics. The method of measurement and reporting of phasors, i.e. digital signal processing, introduces the delay of information reported with reference to information being the basis for calculating the phasor. There are two kinds of delays here: *reporting latency* ΔT_{LRP} and *latency in measurement reporting* ΔT_{LMR}. The first delay is equal to the maximum time between the moment for which the reported change in the value of the phasor occurs for the first time (moment t in Figure 16.6), and the moment when this information is available at the PMU output, i.e. the reporting moment (point C). The second delay is defined as the time between moment when the event occurs at the PMU input (G point) and the moment of reporting the new steady state (point H) calculated for the moment $t + 1/f_r$, i.e. for the measurement window entirely covering this new state.

In the C37.118.2011 standard, two classes of PMUs have been defined (already mentioned above): P-class, which is dedicated to protection systems and M-class provided for other needs related to measurements in the power system. The P-class is characterized by a shorter reaction time to changes in the measured signal and thus a shorter delay of the PMU response, but at the same time its accuracy is limited. In turn, the M-class is characterized by smaller errors than the P-class, but the consequence of this is the slower response to a change in the measured

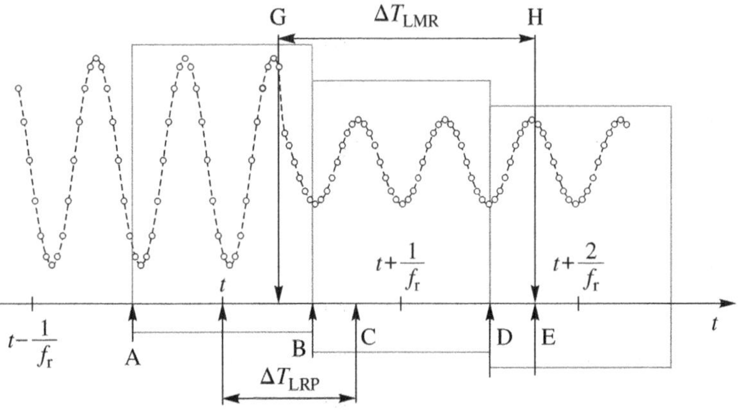

Figure 16.6 Time delays in the PMU. f_r – reporting frequency; AB, BD – measurement windows of the phasor calculated in moments t and $t + 1/f_r$; C – moment of phasor reporting calculated for time t; E – moment of phasor reporting calculated for time $t + 1/f_r$; G – initial moment of the amplitude change of the measured signal; H – the moment of the first new state reporting (measurement at the moment $t + 1/f_r$). Based on CIGRE Report No. 702. *Source:* Reproduced by permission of CIGRE.

signal. For example, in a system with a nominal frequency of 50 Hz, for the assumed reporting rate equal to $f_r = 50$ Hz, for the P-class the maximum required reporting latency is equal to $\Delta T_{LMR} = 2/f_r = 0.04$ s, and for M-class it is equal to $\Delta T_{LMR} = 5/f_r = 0.1$ s. An example of delay formation, shown in Figure 16.6, refers to the case of a sudden change in the amplitude of the signal measured by the PMU with a measurement window equal to two periods of the measured signal, and reporting every two periods ($f_r = 50$ Hz in a 50 Hz system).

PDCs are devices that process data acquired from PMUs and communicate with other data concentrators or schemes performing dedicated functions in the power system (Figure 2.46 in Section 2.7). Functions realized by data concentrators are divided into functions related to data processing, related to communication, and related to the operation of the EPS control.

In the framework of data processing the PDCs perform the following tasks:

- aggregation – grouping of data obtained from various PMUs; in a typical work cycle the PDC waits for data from several PMUs and creates data sets having the same timestamp;
- redirection – routing of processed data sets to different outputs;
- validation – verification of the quality of measurement data, including the fact of obtaining phasors with a specific timestamp from a given PMU; lack of data or lack of adequate quality data is reported (signaled);
- changes in reporting frequency – both down (frequency reduction) and up;
- changing the data format, e.g. from the form of rectangular to polar coordinates and vice versa;
- processing of redundant and duplicated data – processing data from two or more PMUs containing the same measurement data;
- data processing for protection and control purposes, including detection of exceedances and alarms (with different levels), detection of oscillations of selected signals, e.g. frequency, power or angles, trend detection, etc.

In the framework of communication the PDCs perform the following tasks:

- communication with PMUs and other PDCs;
- support for communication protocols, for example: IEC 60870-5-101/104, DNP3, IEC 61850-90-5, IEC 61970, etc.;
- conversion of communication protocols;
- buffering output data.

In the control schemes the PDCs also perform the following tasks:

- management of the PDC configuration to ensure data availability for schemes with which they cooperate, i.e. with other PDCs, control devices, security or EMS/SCADA functions;
- calculation of data transmission delays, for statistical purposes and reports;
- monitoring the quality of the processed data and the quality of communication with PMUs and other PDCs;
- request for retransmission of data in relation to PMUs and other PDCs in order to obtain missing measurement data;
- cybersecurity.

The measurements performed by WAMS are used in WAMPAC systems (Section 2.7) for various purposes, such as:

- WAM of the operation of large-scale power systems, in which WAMS performs the following tasks: analysis of historical events, monitoring of the voltage stability margin, monitoring of local stability, detection of inter-area oscillations, monitoring (calculation) of the EPS inertia, estimation of the parameters of the power system elements (e.g. transmission lines), power system state estimation, islanding detection, detection of subsynchronous resonance, etc.

- Implementation of the WAP covering large areas of the EPS or replacement of selected traditional protection schemes based on local measurements. In such applications, there are schemes that mainly perform tasks related to the prediction of emergency states in the power system or its elements and an appropriate response to these states, e.g. overload emergency control (Cong et al. 2016) or out-of-step protection system (Section 6.6.1) or underfrequency load shedding. WAP schemes do not substitute directly for local measurement based protections, e.g. differential-current protection, distance protection, or overcurrent protection whose reaction to the disturbance must be extremely fast.
- Implementation of the WAC of large-scale power systems based on measurements at remote network points. In such applications, there are schemes that perform the following tasks: damping of inter-area electromechanical oscillations, defense against blackouts, automatic control or system operator support after power system islanding, or after the network splitting (Section 6.6.6). These schemes are implemented as operating in a closed-loop control (e.g. power system stabilizers [PSS]) or with a human operator as the closing element of the control loop. In the latter case, the human operator of the system takes appropriate control/regulation actions based on information obtained from the WAMS.

Summing up the above, it should be emphasized that in applications of WAMS to the WAM and analyzing the power system operation the time delay associated with the measurement and transmission of phasors is not a critical factor. However, in WAP and WAC schemes it is extremely important. Often, this delay is a critical factor that determines the possibility of using phasors to perform specific control tasks and especially protection tasks. Examples of time delays introduced by various elements of the WAMS are as follows: filters: (8–100) ms, calculation algorithms: (20–100) ms, data processing by PMU: (0.005–30) ms, data processing by PDC: (2–2000) ms, and data transfer with various types of telecommunication media: (0.05–8000) ms. Delays related to measurement and data transmission of (100–150) ms are considered acceptable for most possible WAP and WAC applications.

16.2 Examples of WAMS Applications

While the development of WAMSs is a fact, and the number of installed PMUs is increasing, the use of synchronously measured quantities (synchrophasors) for power systems control or as components of protection systems are still in the theoretical or developmental phase. The number of operating schemes of this type, i.e. WAP and WAC, also occurring under the common name of WAMPAC, is relatively small in the world.

In this section, examples of WAMS applications related to power system stability are discussed. The examples presented do not cover all possible applications. The interested reader is referred to the following reports: CIGRE Technical Brochures No. 664, 2016 and No. 702, 2017 and to Phadke and Thorp (2008) and Savulescu (2009).

16.2.1 Evaluation of Power System Operation

The evaluation of the power system operation in real time using WAMS includes: detection of contingency events, evaluation of frequency stability, monitoring of power system inertia, identification of modes of the electromechanical oscillations, monitoring of electromechanical oscillations and identification of modes, evaluation of the voltage stability margin, assessment of small disturbance angle stability, prediction of transient instability, detection of subsynchronous oscillations, estimation of power system state, validation of the mathematical models of power system components, etc. Selected functions of WAMS related to the power system stability are described below.

16.2.1.1 Identification of Modes of Electromechanical Oscillations

The identification of modes of electromechanical oscillations and, possibly, the detection of the sources of these oscillations is carried out on the basis of frequency characteristics (spectra and spectrograms), i.e. functions showing the dependence of the amplitude or power (energy) of the signal on the frequency. The results of calculations are usually presented in the form of spectrograms of the following type: *power spectrum* (PS) or *power spectral density* (PSD), and less frequently in the form of spectral characteristics: *amplitude spectrum* (AS) or *amplitude spectral density* (ASD).[1] Spectrograms are calculated from measurement data obtained at different time intervals, i.e. from minutes to days and weeks. It is reasonable to perform calculations for time intervals in which the configuration of the network and the power generated are not changed. Measurements of frequency or real power in selected transmission lines are often used here. An example of a characteristic being the power spectrum of the measured signal, which is the frequency, is shown in Figure 16.7. The peaks on the characteristic indicate the frequencies of electromechanical oscillations associated with their high amplitude (AS, ASD) or energy (PS, PSD). The spectrogram in Figure 16.7 shows the result of 12-hour measurements performed on a certain node in a Finnish transmission network. Clearly, there is a visible inter-area oscillation peak with a frequency of around 0.5 Hz, characteristic of the Scandinavian power system, and peaks of local electromechanical oscillations equal to 1 Hz and more. The fuzzy peak of the spectral characteristic with frequencies around 0.1 Hz can be interpreted as related to the operation of the turbine control systems. In turn, also the fuzzy peak of the spectral characteristic with frequencies equal to about 0.4 Hz can be associated with other inter-area oscillatory modes in this system, the frequency of which strongly depends on the network configuration and the load flow.

Spectrograms are calculated using various algorithms belonging to the following groups:

- Methods using linear models dedicated to identifying damped modes, named also *linear ringdown analysis methods*. These methods are based on short-term measurements lasting (5–20) s, i.e. on the measurement of transient processes after a disturbance. Examples of such methods are: the Prony method (discussed in Appendix A.4), the *eigensystem realization algorithm* (ERA) (Juang and Pappa 1985), and the *matrix pencil method* (Hua and Sarkar 1990).
- Methods using linear models, dedicated to identifying undamped modes, named also *mode-meter analysis methods*. These methods are based on long-term measurements (minutes, hours, days, weeks, etc.). Examples of methods of this type are: the *fast Fourier transform* (FFT), the Yule–Walker algorithm (Kay 1988), the *least squares method* and the *least mean squares method* (Kariya and Kurata 2004).

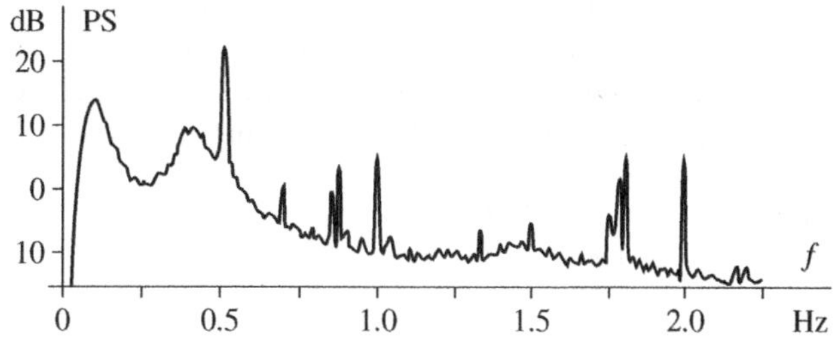

Figure 16.7 An example spectrogram obtained from 12-hour frequency measurements. Based on CIGRE Report No. 664. *Source:* Reproduced by permission of CIGRE.

1 The relationships between the points of these frequency characteristics are as follows: $AS = \sqrt{PS}$ and $ASD = \sqrt{PSD}$. The units of measure of individual characteristics are in turn equal to: PS [x^2], PS [x^2/Hz], AS [x], and ASD $\left[x/\sqrt{\text{Hz}}\right]$, where x is the unit of the measured signal, e.g. voltage, frequency, or power. The characteristics of this type are also presented using logarithmic scale, as in Figure 16.8.

- Methods using nonlinear and nonstationary models based on long-term measurements (*nonlinear analysis methods*). These methods allow for more precise analysis of nonlinear and nonstationary systems than the methods mentioned above, i.e. based on linear models. Examples of methods of this type are: a method based on the *Hilbert–Huang transform analysis* (Huang et al. 1998) and methods based on *energy tracking operators* (Maragos et al. 1993).

An example of an algorithm used in WAMS (in addition to the most popular FFT algorithm) used to identify modes of electromechanical oscillations is the Yule–Walker algorithm. This algorithm for identifying harmonics of a zero-mean time series $x(t)$ uses N samples x_n measured in moments $t = nT_p$ and $n = 1, 2, ..., N$, where: T_p is the continuous signal $x(t)$ sampling step and k is the step number. A discrete linear predictive autoregressive (AR) model of the order K, where $K \le N$ is assumed

$$\sum_{k=0}^{K} a_k x_{n-k} = w_n \tag{16.19}$$

where a_k in addition to $a_0 = 1$ are the desired model coefficients, and w_n is white noise. After multiplying both sides of Eq. (16.19) by x_{n-l}, where $l = 0, 1, 2, ..., N$, one gets

$$\sum_{k=0}^{K} a_k x_{n-k} x_{n-l} = w_n x_{n-l} \tag{16.20}$$

Equation (16.20) of all samples from the measurement window, i.e. for $n = 1, 2, ..., N$ creates N equations which, after adding up, lead to the equation

$$\sum_{k=0}^{K} a_k E\{x_{n-k} x_{n-l}\} = E\{w_n x_{n-l}\} \tag{16.21}$$

where E is the covariance. Values of the covariance of the left side of Eq. (16.21), after dividing by the sum of squares of samples x_n

$$E[x_n] = \sum_{n=1}^{N} x_n^2 \tag{16.22}$$

are equal to the coefficients of autocorrelation of the measured waveform offset from each other by $l - k$ samples and are hereinafter referred to as r_{l-k}. In turn, the covariance values of the right side of Eq. (16.21) for different values l are equal to: the standard deviation of the signal measured σ^2 for $l = 0$ and zero for the remaining values l. Inclusion of Eq. (16.22) transforms Eq. (16.21) into the form

$$\sum_{k=0}^{K} a_k r_{l-k} = \begin{cases} 0 & \text{dla } l > 0 \\ \sigma^2 & \text{dla } l = 0 \end{cases} \tag{16.23}$$

Equation (16.23) for $K = N$ and $l = 1, 2, .., N$ creates a system of linear equations, in which the square matrix of autocorrelation coefficients r_{l-k} is a Hermitian and Toeplitz matrix, which guarantees stable poles of the AR model

$$\begin{bmatrix} r_0 & r_{-1} & \cdots & r_{1-N} \\ r_1 & r_0 & \cdots & r_{2-N} \\ \vdots & \vdots & \ddots & \vdots \\ r_{N-1} & r_{N-2} & \cdots & r_0 \end{bmatrix} \begin{bmatrix} a_1 \\ a_2 \\ \vdots \\ a_N \end{bmatrix} = - \begin{bmatrix} r_1 \\ r_2 \\ \vdots \\ r_N \end{bmatrix} \tag{16.24}$$

The solution of matrix Eq. (16.24) provides all coefficients a_k of the AR model. The standard deviation, which is a measure of the error between model response and measurement, results from Eq. (16.23) for $l = 0$ and is equal to

$$\sigma^2 = \sum_{k=0}^{K} a_k r_{-k} \tag{16.25}$$

The value of the standard deviation σ^2 depends on the assumed order of the model $K \leq N$. The PSD of the continuous signal $x(t)$ sampled with the step T_p, depending on the frequency f, defines the function

$$PSD(f) = \frac{\sigma^2 T_p}{\left| \sum_{k=0}^{K} a_k e^{-j2\pi f k T_p} \right|^2} \tag{16.26}$$

In Eq. (16.24) $K = N$ equations are always used. But when calculating the PSD based on Eqs. (16.25) and (16.26) a parameter K lower than N can be assumed. Neglecting $N - K$ coefficients in the AR model does not have to significantly reduce the accuracy of the calculations. This accuracy depends on the distribution of model coefficients a_k. Small values of these coefficients for $k = K + 1, \ldots, N$ have a small effect on the result, i.e. the PSD characteristics shape.

16.2.1.2 Monitoring of Electromechanical Oscillations

Monitoring of the dynamic properties of the power system is carried out in real time in order to detect weakly damped electromechanical oscillations or high-amplitude electromechanical oscillations and to warn of the alert or emergency states (Figure 1.6). Obtained information can also be used in the offline analyses for seeking sources of weakly damped oscillations, verification of settings, and tuning of control systems (e.g. PSS), etc.

An example of a WAMS-based system that implements the above tasks is a system developed by General Electric (Psymetrix). This system calculates and presents online values of frequency, RoCoF, and parameters of electromechanical modes over time. The parameters of modes are calculated using both a small (3 min) and a large (20 min) measuring window in order to see transient versus persistent modes on the power system. Parameters of modes are presented in graphic form for frequency bands defined by the user (transmission system operator) on the basis of their knowledge of the frequencies of electromechanical oscillations occurring in a given power system. The user may also select the type of electrical quantity, which must be analyzed. Usually, it is frequency deviation[2] at a given measuring point or real power. The monitoring system (developed by Psymetrix) is a real-time system and displays information about the modes identified at a given moment. The modes may be presented in various coordinates, for example (ξ, A), (T, A), (t, f), (t, T), and (t, A), where ξ is the damping ratio, $T = 1/\alpha$ is the mode damping time constant, A is the mode amplitude, t is the time, and f is the frequency. Two examples of information made available to the power system operator are shown in Figure 16.8.

Figure 16.8a shows the first example when the user selected the frequency as the measured signal. This figure shows a locus plot of 0.25 Hz oscillation in measured frequency observed in the band (0.1–0.3) Hz defined by the user. The selected mode is presented in the coordinates (ξ, A), i.e. damping ratio (in %) versus amplitude (in mHz). The trace (broken line in lower part of figure) shows where the mode has moved over time as the state of the EPS changes. The black arrow in this trace shows exactly where the mode is at the present on the root locus and the direction of where it came from. Oscillatory modes with damping ratio greater than 5% ($\xi > 0.05$) or amplitude less than 5 mHz ($A < 5$ mHz) are considered acceptable, corresponding to the normal state of the electric power system operation (Sections 5.4.6, 12.1.3, and 14.2.3). The amplitude of the mode greater than 5 mHz is signaled as an alert state. The occurrence of modes with a damping ratio less than 2% ($\xi < 0.02$) and an amplitude greater than 5 mHz

2 During swings in the EPS the locally measured frequency deviations are equal to $\Delta f(t) = [d\theta/dt]/2\pi$, where θ is the angle (phase) of the voltage measured at a given node.

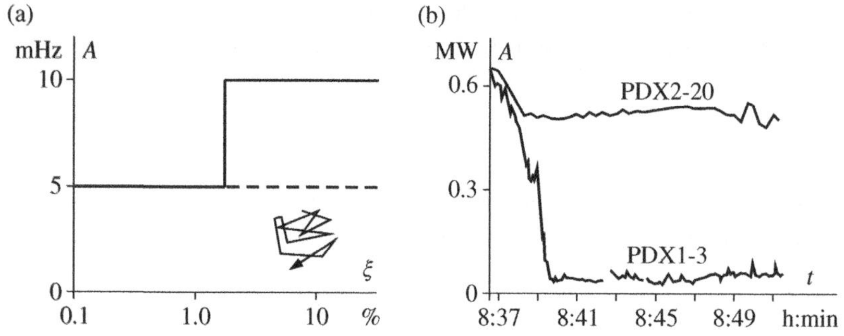

Figure 16.8 Example of mode visualization in the PhasorPoint*. *Source:* Reproduced with permission of General Electric Company and/or its Affiliates. *Trademark of General Electric Company.

or modes with an amplitude greater than 10 mHz independently from damping ratio (i.e. modes with parameters beyond solid line in Figure 16.8a) is signaled as an emergency state.

If the system operator sees that the damping ratio is moving toward the prescribed boundaries (solid or dashed line in Figure 16.8a) then an investigation can be done to determine the countermeasures.

Figure 16.8b shows a second example when the user selected the real power as the measured signal. This figure shows the plots of the mode at a given frequency identified in the measured real power calculated in the small PDX1-3 and large PDX2-20 measuring window. The plot of PDX1-3 indicates the disappearance of a process characterized by oscillations of a given frequency and initial amplitude exceeding 0.6 MW. At the same time the plot of PDX2-20 shows the same mode with not decreasing amplitude. The difference seen in the plot PDX1-3 and the plot PDX2-20 results from the difference in the length of the measurement windows. In the PDX2-20 measurement, samples are used for a period of up to 20 min (from 20 min back to a given moment in time). In the case of PDX2-20 measurement the information about the mode change is delayed because of the difference in the measurement windows length. Therefore, in the following moments one should expect a reduction of the PDX2-20 value, too. Theoretically (which is not the case in the example presented above, e.g. in Figure 16.8), a measurement performed in a small measurement window can be implemented using an algorithm for detecting damped modes, and the measurement performed in a large window using algorithms for detecting undamped modes.

16.2.1.3 Identification of Subsynchronous Oscillations

Subsynchronous oscillations may occur in power systems with series compensation of transmission line impedance. As explained in Section 4.5, these oscillations appear as the result of the resonance between the series compensated lines and synchronous generators. Their effect is usually a fatigue or damage to the shafts of the generating units (Ahlgren et al. 1978).

In a traditional approach, the power system is protected against sub-synchronous resonance by relays, which detect oscillations of predetermined frequencies and switch off the series compensator from the circuit by closing the circuit breaker, thereby bypassing some or all capacitors. In the case of the detection of subsynchronous oscillations in the generator current, the generator is tripped off or the alarm is triggered. Unfortunately, such protection based on local measurements cannot indicate the source of oscillation in the network.

The use of WAMS to detect this type of oscillation consists in measuring and comparing oscillation frequency measured in the electric current flowing through the transmission line (or the series capacitor) and the generators. The detection of subsynchronous oscillations of the same frequency in a given line and a given generator indicates the interaction (resonance) between these elements. Detection of such a state should result in switching off the element (usually capacitor) that causes the oscillations. However, depending on the individual features of the power system, the algorithm of the WAMS-based protection may be different.

A certain problem related to the use of WAMS for the detection of subsynchronous oscillations is the reporting rate of currently used PMUs. The maximum oscillation frequency that can be detected by the PMU reporting at frequency f_r is equal, according to the Kotelnikov–Shannon theorem, to $f_r/2$. Thus, for the maximum reporting rate of today's PMUs equal to 50 Hz, the maximum measured frequency is equal to 25 Hz. This eliminates the possibility of measuring subsynchronous oscillations from the range of 25 to 50 Hz, which may appear in the power system (Table 4.5 in Section 4.5). The solution to the problem is to increase the PMU reporting frequency above 100 Hz. Another solution, indicated in the example of the Swedish system in CIGRE Technical Brochure No. 664 (2016), is to disable the anti-aliasing filter in the PMU. This filter eliminates the side lobes with frequencies equal to $f_r \pm f_{oss}$ from the frequency response, where f_{oss} is in this case the frequency of subsynchronous oscillations. Omitting the anti-aliasing filter allows frequency $f_r - f_{oss}$ and frequency f_{oss} to be identified. For example, one of the subsynchronous oscillation modes identified in the Swedish system is equal to $f_{oss} = 31.25$ Hz. Frequency characteristics calculated on the basis of PMU signals with the anti-aliasing filter turned off show the mode on the frequency $50 - 31.25 = 18.75$ Hz, which is, in fact, the side lobe of subsynchronous oscillation mode with frequency 31.25 Hz.

16.2.2 Detection of Power System Islanding

Generally, power system islanding is a condition in which the power system is divided into large or small isolated subsystems. Such subsystems are referred to as *islands*. Examples of islanding are:

- Division of the EPS into subsystems made intentionally by the transmission system operator (TSO) during heavy conditions threatened with blackout.
- Automatic splitting of the transmission network with the specific configuration (Figure 6.51) made by the out-of-step tripping (OST)relays when asynchronous operation appears.
- Random division of the EPS into subsystems as a consequence of extreme events (Section 14.2) and multiple or cascade tripping.
- An operation of a feeder or a part of a distribution network with sources that continue to feed the circuit, even after the power from the electric utility grid has been cut off.

Controlled islanding made by the TSO or the OST relays usually leads to isolated islands with predetermined boundaries and a real power balance equal or close to zero. The uncontrolled islanding following extreme contingencies and cascade tripping usually leads to unbalanced islands with random boundaries. In such islanded subsystems (islands) the frequency control must be activated (Chapter 9) and (when necessary) also load shedding.

At present, power system operators require switching off the renewable energy sources (RES) from islanded unbalanced subsystems and especially when a given subsystem does not contain sources other than RES. The necessity to switch off the RES in such a case is mainly due to their inability to control the frequency. This takes place when the RES are connected to the network by thyristor-based inverters, which cannot operate in a network where no other sinusoidal voltage sources exist. This, in the case of the growing amount of energy sources of this type, may lead to an increase in the duration of power supply interruptions. In order to eliminate this effect, it is necessary to equip energy sources (especially RES) with control systems. Such frequency control can be activated permanently or in the moment when islanded operation is detected.

Islanding detection using classical protection, based only on local measurements (usually frequency), is practically impossible when loads in the islanded subsystem are supplied by the RES and the real power balance in the island is equal or close to zero. In this case, the frequency in the islanded subsystem practically does not change. In such a case the frequency deviation criterion, as a criterion for islanding detecting, fails.

Detecting of the uncontrolled islanding of the EPS in post-contingency states is an even more difficult task. In this case, the number of transmission lines that can be switched off may be large, and the boundary of the islanded subsystems may be random. Identification of the system splitting would allow for the activation of proper control dedicated to the needs of given subsystems (islands).

The detection of a power system's islanding requires information about the switches' status or frequencies obtained from various places in the system. In the case of uncontrolled and random islanding the number of frequency measurement points and the number of switches whose state must be identified can be very large. The use of WAMS reduces the number of places of measurement necessary to detect the power system islanding or islanded subsystem operation. In a simple network structure, two points of phasor measurement are sufficient. In power systems where the number of possible network splitting can be large, the number of measurement points necessary to detect the power system islanding should be correspondingly greater.

Transmission of information from the measurement points to the monitoring system detecting the system islanding is therefore indispensable both in the case of traditional protection schemes and in the case of using WAMS. At present, the islanded operation detection schemes based on traditional protection are dedicated mainly to distribution networks, while WAMS-based schemes are dedicated to transmission networks.

It should be added that WAMPAC systems are being built that, in addition to islanding detection, carry out power balancing tasks in the potential island. This task requires knowledge of the network structure, the type of loads, their importance in a islanded subsystem, their possibility of shutting down, and the ability to control the switches of the feeders supplying these loads. Such WAMPAC systems are used in large industrial plants. In these systems, power balancing (mainly by switching the loads off, in case of the plant's capacity is insufficient to cover the electricity demand) must take into account the requirements of the technology used in the plant. Power for critical technologies must be maintained, and less important loads are turned off.

The basic structure of the network for the explanation of the islanded operation detection algorithm is shown in Figure 16.9. It is assumed that in islanded subsystem G both the number of energy sources and loads can be greater than one. The total power generated in subsystem G is equal P_g, and the total power consumed, along with the power losses in the network of the islanded subsystem, is equal to P_L. The subsystem marked as Q represents the rest of the power system. The connection of both subsystems (G and Q) in general consists of one or more transmission lines.

Measurement of voltage phasors $\underline{V}_1$ and $\underline{V}_2$ for the islanded operation detection can be implemented anywhere in the subsystem G and the subsystem Q, including at the ends of the transmission line connecting the two subsystems, i.e. as shown in Figure 16.9. The measured voltage phasors in step k are equal to

$$\underline{V}_{1k} = V_{1k}e^{j\delta_{1k}}; \quad \underline{V}_{2k} = V_{2k}e^{j\delta_{2k}} \tag{16.27}$$

and the islanded operation detection algorithms use the differences of angles of these phasors

$$\Delta\delta_k = \delta_{1k} - \delta_{2k} \tag{16.28}$$

The differences of voltage angles, as mentioned above, are perceived as the most important indicators of the dynamic evaluation of the power system state. Islanding detection with the use of WAMS is carried out using various algorithms. The basic ones are:

1) The angle difference algorithm, in which it is assumed that the power system islanding is detected when for a period longer than ΔT_{ref} the difference of the voltage phasor angles $\Delta\delta_k$ is greater than the pre-set threshold $\Delta\delta_{\text{ref}}$, i.e. when the condition is met

$$\forall_{t\in\langle T_m, T_n\rangle}\ |\Delta\delta_k| \geq \Delta\delta_{\text{ref}} \quad \text{where } T_n - T_m > \Delta T_{\text{ref}} \tag{16.29}$$

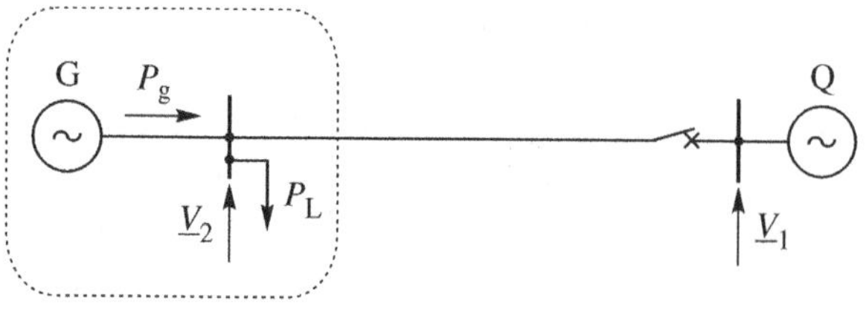

Figure 16.9 Illustration to the explanation of the islanded operation detection.

2) The frequency difference algorithm, in which it is assumed that the power system islanding is detected when for a period longer than ΔT_{ref} the difference of frequency Δf_k is greater than the pre-set threshold Δf_{ref}, i.e. when the condition is met

$$\forall_{t\in\langle T_{\text{m}},T_{\text{n}}\rangle}\ |\Delta f_k| \geq \Delta f_{\text{ref}}\ \text{ where } T_{\text{n}} - T_{\text{m}} > \Delta T_{\text{ref}} \tag{16.30}$$

3) The slip-acceleration algorithm, in which the power system islanding is detected with the help of two quantities. The first quantity is the difference of the frequency in two slipping subsystems, referred to as the *slip frequency* and defined by equation

$$s_k = \frac{\Delta\delta_k - \Delta\delta_{k-1}}{360^\circ} f_{\text{r}} \tag{16.31}$$

where $\Delta\delta_k$ is defined by Eq. (16.28). The second quantity is the rate-of-change of the slip frequency, referred to as the *acceleration* and defined by equation

$$a_k = (s_k - s_{k-1}) f_{\text{r}} \tag{16.32}$$

In both above equations f_{r} is the PMU reporting rate. Islanding detection is carried out based on the location of the values of the above quantities on the plane (s_k,a_k), as illustrated in Figure 16.10. The boundaries of the islanded operation area are defined by linear functions, as shown in Figure 16.10, defined by values s_{lim1}, a_{lim1}, s_{lim2}, and a_{lim2} or by some nonlinear functions, where A is the normal operating region and B are supposed to be islanding regions.

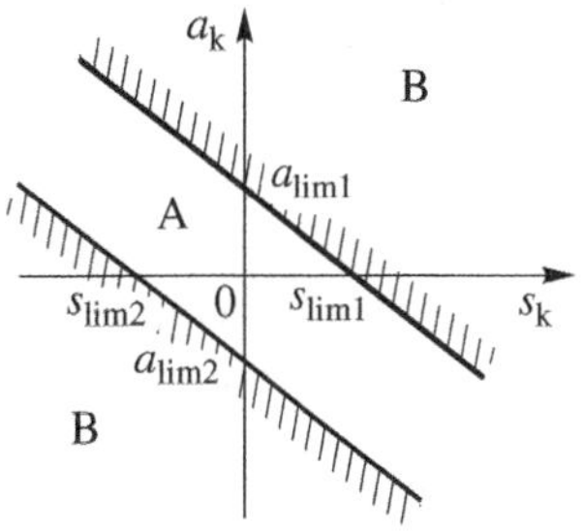

Figure 16.10 Islanding detection characteristic using wide-area measurements. A – normal operating region; B – islanding region. Based on CIGRE Report No. 664. *Source:* Reproduced by permission of CIGRE.

Differences in angles of the nodal voltages in the network depend on the network structure, network element parameters, and the power flow. Therefore, the limit values in the above algorithms are selected individually. Examples of typical (often encountered) pre-set threshold values are: $\Delta\delta_{\text{ref}} = 20$ deg, $\Delta f_{\text{ref}} = 1$ Hz, $(\text{d}f/\text{d}t)_{\text{ref}} = 0.2$ Hz/s, and $\Delta T_{\text{ref}} = 0.2$ s for algorithms 1 and 2. Algorithm 2 and 3 are also used in traditional protections based on the nonsynchronized measurements.

An important difference between traditional islanding detection algorithms and the WAMS-based algorithms using synchrophasors is the detection time, especially if the real power balance in islanded subsystems is equal or close to zero. The advantage of the WAMS-based detection is shown in Figure 16.11. The WAMS-based scheme detects islanding for all conditions, while the traditional protection is unable to detect in a timely manner the islanded operation of the balanced subsystem ($P_{\text{L}}/P_{\text{g}} \cong 1$). Moreover, for all conditions, the detection time ΔT for the WAMS-based schemes is shorter than for the traditional protection.

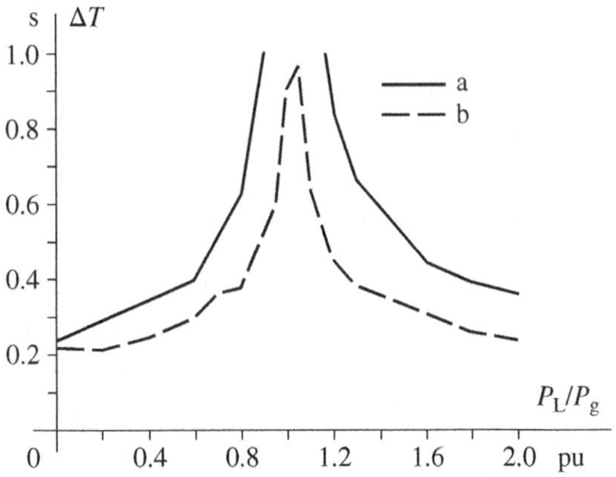

Figure 16.11 Comparison of islanding detection time ΔT for (a) traditional protection based on nonsynchronized measurements and (b) algorithm based on synchrophasors. Based on CIGRE Report No. 664. *Source:* Reproduced by permission of CIGRE.

From the point of view of the dynamics of the processes taking place in the islanded subsystems, the islanding detection time may be critical for the ability of the islanded subsystem to sustain in operation or from the point of view of exposing the EPS elements, for example to excessively high voltages. Shortening the islanding detection time is important for the power system.

16.2.3 Stability Monitoring and Instability Prediction

For several years there is a growing trend for using online stability assessment to quickly inform how far a given operating state is from instability. The state-of-art in the application of the traditional SCADA/EMS is presented in Savulescu (2009). Synchronized phasor measurements and high-speed communication platforms have allowed more-sophisticated WAMPAC systems to be developed. They create a comprehensive solution for wide-area stability monitoring, instability detection, and stability enhancing control. The potential financial benefits of WAMPAC deployment for stability monitoring and instability detection are related to more accurate computations of available stability margins and reducing the risk of blackouts and extensive societal and economic costs associated with such events.

16.2.3.1 Steady-state Inherent Stability

For economic reasons, many power systems currently operate close to the thermal limits of the transmission network and with a small steady-state stability limit. In such a situation a contingency event with the loss of a heavy loaded transmission line may cause steady-state instability in a power system and/or dangerous overloads and cascade tripping. To avoid this situation, some power system operators use automatic alleviation of the overloads by generation curtailment forced by various types of automatic tripping or generation shedding schemes also referred to as overload emergency state control (OLEC). An example of such scheme based on the traditional measurements is described by Robak et al. (2018). An example of such tripping scheme based on WAMS is described by Cong et al. (2016).

Another example of a tripping scheme based on WAMS implemented for a power system with longitudinal structure is described in CIGRE Technical Brochure No. 664 (2016). The power system (Mexican utility) consists of three areas: area 1 with heavy load concentration, area 2 with generation concentration, and area 3 with light load concentration. Generation area 2 is connected with the heavy load area 1 with two transmission links and a sub-transmission network. The light load area 3 is connected to generation area 2. The power system is designed to withstand only a single contingency event. The double contingencies, when two parallel lines are lost between area 2 and area 1, may lead to instability and overloads in the sub-transmission network. The automatic generation shedding schemes (AGSSs) monitor the voltage angle difference between buses and the real power transfer capability between area 2 and area 1. The remedial action scheme to avoid instability and/or uncontrolled power flow over the sub-transmission network is to trip a part of generators in generation area 2. The generators that remain in service feed the load in area 2 and area 3.

The above-described AGSS uses the angle difference as a criterion to trigger emergency control. The use of synchrophasors in such schemes also allows monitoring of steady-state inherent stability margin defined in Section 5.4.1 by Eq. (5.20). Measurement of voltage phasors must be carried out in the power plant node and in the network remote node or nodes away from the power plant. The algorithm of this type can be easily applied to systems with separated areas of generation and consumption of electricity. Synchrophasors provided by WAMS to a control center enable the power system state to be estimated. When the topology of a network and its parameters are known then, by using methods described in Chapter 18, the power system model can be reduced and an impedance between the equivalent source and the equivalent load can be determined. Knowledge of this impedance and voltage phasors allows the power-angle characteristic and the pull-out power P_{cr} and stability margin $c = (P_{cr} - P_m)/P_{cr}$ for given real power P_m in the generation area to be determined.

It should be noted that the identification of the power-angle characteristic may be done periodically during normal operation of the EPS and as soon as possible after occurrence of a contingency event (e.g. tripping of transmission line). Owing to the speed of transient processes resulting from a change in network configuration, the time available to identify the current power-angle characteristics and to trigger the emergency control can be extremely limited. For example, the time needed to identify the transmission lines' connection status reaches up to 100 ms (CIGRE Technical Brochure No. 664, 2016). This is a sufficient time to take actions to ensure required angle stability margin in situations where the change of network configuration, for example by switching the transmission line off, is not preceded by a short-circuit. For contingency events with short-circuits and potential loss of the transient stability, this time is too long. In such cases, also with the use of WAMS, other algorithms are utilized, which are presented below.

16.2.3.2 Transient Instability

Power systems are designed to withstand planning (credible and less credible) contingency events (Chapter 14). Planning performance standards (Section 14.2.3) allow for the potential transient instability for extreme (noncredible) contingency events. However, in order to prevent a loss of synchronism and widespread outages after such events, some transmission system operators use special protection systems (SPSs) tripping to auxiliaries a part of generating units (Section 10.4). In such tripping schemes the WAMS can be used with the aim of predicting impending instabilities through measurements after a disturbance, and trigger emergency control.

In literature, a number of methods have been proposed for predicting impending transient instability. These include: simple indicators, such as time integral of generator rotor angle error; methods based on energy function; machine learning techniques, such as artificial neural networks (ANNs), decision tables, decision trees (DTs), and fuzzy logic; and support vector machines (SVMs). Two examples of WAMS-based tripping schemes are described below.

In the South Korean power system the wide-area measurement, protection, and control system (WAMPAC) has been implemented, which can provide both monitoring and control functionality in real time (CIGRE Technical Brochure No. 664, 2016). One of the control functions is dedicated to the generator tripping as the remedial action against transient instability. The algorithm of this function consists of two parts.

The first part is based on the decision table, which is updated online by the transient stability assessment (TSA) package. For a defined time interval (e.g. equal to 30 min) from the state estimator (also using the WAMS), an updated static model and the operating point of the power system is obtained. On the basis of the dynamic system model with the updated operating state, various contingency events are simulated by the TSA, including those related to short-circuits at various network locations. For contingency events leading to the transient instability the number of generators that should be tripped is determined. Contingency events together with relevant necessary actions are recorded in the decision table. The decision table is updated every 30 min. If a contingency event occurs, the tripping scheme sends a signal to switch off the generator or generators according to the decision table. Reaction of the tripping scheme based on the decision table to the events is very fast ($T < 100$ ms). The tripping command can be generated even before the short-circuit is cleared. This feature is very important. As explained in Section 10.4, the delay in the generator tripping decreases its effectiveness and increases the number of generators which must be switched off.

The second part of the tripping scheme monitors (using also PMU data) the real-time response of the EPS and as the remedial action switches off other generators or a priori defined load. This actions is carried out gradually, which in turn corresponds functionally to classic underfrequency load shedding, but significantly speeds up the defense of the EPS.

Another algorithm for predicting the transient instability is proposed by Gurusinghe and Rajapakse (2016) and earlier by Rajapakse et al. (2010). This algorithm uses voltage phasors obtained from the PMUs installed at the generating stations and monitors the trajectory of the rate of change of voltage magnitude $RoCoV = \mathrm{d}V/\mathrm{d}t$ versus deviation of voltage magnitude ΔV. The trajectory is plotted during the post-disturbance transient state for each three-phase short-circuit in the transmission line adjacent to the power plant. An example of such a trajectory is shown in Figure 16.12. Points A and B in Figure 16.12 correspond to the maximum values of $RoCoV$ and ΔV obtained as a result of short-circuit simulation for the duration of the short-circuit from zero to critical clearing time. Points A and B define the stability limits in the form of straight lines: above point B and to the right of point A. The curve between points A and B is determined as part of an ellipse with vertices at points A and B. The area in Figure 16.12 to the right and above the line marked as the stability limit corresponds to the stable trajectories of the

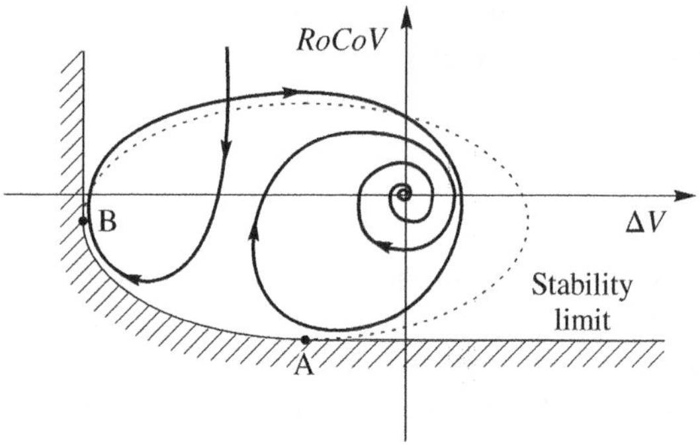

Figure 16.12 The idea of the prediction algorithm of transient stability loss based on the measurement of voltage phasors. Based on CIGRE Report No. 702. *Source:* Reproduced by permission of CIGRE.

generator voltage (or voltage of the upper side of the step-up transformer). The origin of the plane with the coordinates (ΔV, $RoCoV$) is the starting point of the voltage trajectory. In the case presented in Figure 16.12 the post-disturbance equilibrium point is the same as in the pre-disturbance state and the trajectory settles down in the origin of the plane. The initial part of the trajectory is not shown here. The voltage trajectory shown in the figure corresponds to a stable trajectory. An encroachment of the trajectory behind the stability limit means that the given generator impends will lose synchronism in the finite time. This algorithm is intended for online applications and makes it possible to predict the transient instability in advance. Depending on the configuration and operating point of the system, it allows the loss of synchronism up to 600 ms before the loss of stability occurs to be predicted.

16.2.4 Damping of Electromechanical Oscillations

FACTS (flexible AC transmission systems) devices are typically used for voltage or power flow control. They can also be used for damping of electromechanical oscillations in the EPS (improvement of small signal stability). Their regulators can be equipped (as the automatic voltage regulator [AVR] of synchronous generators) with the PSS, also referred to as the *power oscillation damper* (POD). The POD uses locally measured input signal, e.g. voltage, power, or frequency or signals measured in distant points in the EPS. Use of the signals measured in distant points allows for more effective damping of electromechanical oscillations, especially the inter-area oscillations.

An example of using the WAMS for electromechanical oscillations damping is the *wide-area power oscillation damper* (WAPOD) built in the Scandinavian power system (NORDIC Power System). Owing to the geographic shape of this system and the location of energy sources and loads, there are weakly damped modes of inter-area oscillations. The most dominating weakly damped oscillatory modes are between southern Norway and Finland (approximately 0.3 Hz) and between southern Norway and Sweden/Denmark (approximately 0.5 Hz).

In order to increase the damping of these modes, three new static VAR compensators (SVC) were installed in the EPS. The SVC utilized for the WAPOD was installed at the Norwegian side of the Norwegian–Swedish transmission corridor. The regulator of this compensator enables voltage control on the connection of both systems and is equipped with two PSS: the traditional POD and the WAPOD based on phasor measurements. A traditional stabilizer as the input signal uses the active power in the selected transmission line connecting these systems. In contrast, the WAPOD stabilizer uses, as input signal, the difference in voltage phasor angles between two selected nodes in the system – one in Sweden and the other in Norway. The structure of the system is illustrated in Figure 16.13. The SVC control scheme has a switch-over logic that allows for the use of either manual control, local POD, or WAPOD. The SVC switches automatically from the WAPOD to POD in the case of loss of WAPOD communication with PMUs.

The WAPOD was tuned for damping the oscillations at a frequency of 0.5 Hz, but it also improves the damping of inter-area oscillations with other frequencies. End-to-end latency, because of communication and processing of the synchrophasors data used as input signals to the WAPOD, is typically about 30 ms, which is acceptable from the point of view of the task being performed.

In order to damp inter-area oscillations, the WAPOD with the following transfer function is used

$$K(s) = k_s G_f(s) G_w(s) G_c(s) \tag{16.33}$$

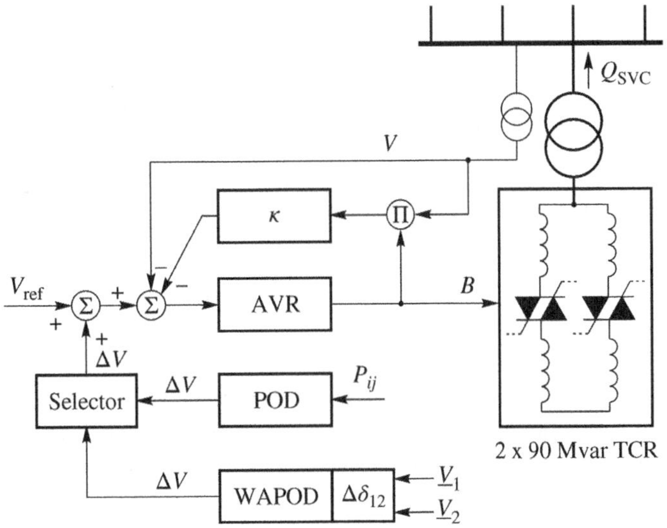

Figure 16.13 Structure of SVC with power oscillations damping function (POD) with the use of phasor measurements; AVR – voltage regulator; κ – voltage droop. Based on CIGRE Report No. 702. *Source:* reproduced by permission of CIGRE.

where its transfer functions respectively are:

- low-pass filter

$$G_f(s) = \frac{\omega_B^2}{s^2 + 2\xi\omega_B s + \omega_B^2} \tag{16.34}$$

- washout filter

$$G_w(s) = \left(\frac{sT_w}{1 + sT_w}\right)^n \tag{16.35}$$

- and phase compensator

$$G_c(s) = \left(\frac{1 + sT_1}{1 + sT_2}\right)^m \tag{16.36}$$

where ω_B is cut-off pulsation of the low-pass filter; ξ is the filter damping factor; k_s is the POD gain; and T_w, T_1, and T_2 are time constants of the washout and the phase compensator.

For the low pass-filter (CIGRE Technical Brochure No. 664, 2016) the parameter values are equal to: $\xi = 0.75$ and $\omega_B = 25$ rad/s, which corresponds to a cut-off frequency of approximately 4 Hz. The time constant of the washout filter, which is a high-pass filter, is assumed to be sufficiently large so as not to damp the inter-area oscillation signal, characterized by low frequency values of (0.2–1.0) Hz. In the presented case, the maximum time constant equal to $T_w = 10$ s used for system stabilizers was adopted. The exponent of the washout filter for the stabilizer using angular velocity or frequency as the input signal often is assumed to be equal $n = 1$, and for the stabilizer using the difference of voltage angles is assumed to be equal $n = 2$. In the discussed case ($n = 2$) a single washout filter can be treated as an imperfect differentiator on which output one gets a signal proportional to the frequency deviation. The difference in voltage angles is indicated as the most effective input signal for stabilizers of inter-area electromechanical oscillations.

The time constants of the phase compensator are calculated using various methods, for example those presented in Chapter 15, based on the P–Vr or GEP characteristics. Below the residue method is presented, which often used to select parameters of PSS (POD function) of power electronics devices, such as SVCs or static compensators (STATCOMs).

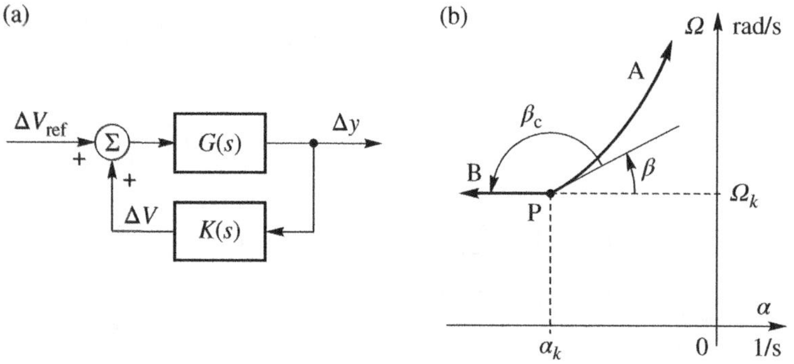

Figure 16.14 Illustration to the design of the POD's parameters: (a) block diagram of the model used; (b) the eigenvalue loci.

In the considered method, the parameters of the phase compensator are sought, enabling to the direction of the eigenvalue loci of the chosen oscillatory mode toward increasing values of the damping coefficient to be changed. This is illustrated in Figure 16.14. The power system (including SVC) is represented (Figure 16.14a) by the transfer function $G(s)$ and the POD is represented by the transfer function $K(s)$ defined by Eq. (16.33). The input signal of the POD is $\Delta y = \Delta\delta$ but generally it could be another signal appropriate for the damping of electromechanical oscillations such as the frequency and real power. The eigenvalue loci of this system are shown in Figure 16.14b. Eigenvalue $\lambda_k = \alpha_k + \mathrm{j}\Omega_k$ at point P is calculated with an assumption that the POD gain is equal to $k_s = 0$, which means that it is the eigenvalue of the system with SVC, but without the POD. The eigenvalue loci A are calculated for the closed-loop system for increasing values of the POD gain k_s starting from $k_s = 0$. In these calculations the phase compensator, low-pass filter, and washout filter are neglected, i.e. $G_f(s)G_w(s)G_c(s) = 1$ and $K(s) = k_s$. The slope angle of the tangent to the eigenvalue loci in point P is equal to β. This tangent defines the direction of the eigenvalue loci in point P. The compensator's aim is to change the given eigenvalue loci from A to B. This means that the compensator has to introduce a phase shift equal to β_c. This angle is named the *compensation angle*.

In the case when the POD is active the transfer function for the closed-loop control is expressed in the following way

$$\frac{\Delta y}{\Delta V_{ref}} = \frac{G(s)}{1 - K(s)G(s)} = \frac{\sum_i \frac{r_i}{s - p_i}}{1 - K(s)\sum_i \frac{r_i}{s - p_i}} \tag{16.37}$$

where p_i are the poles and r_i are the residues. The residues are a measure of the mode's observability and controllability. The higher value of the residue means the higher ability to influence a given mode.

The characteristic equation of the system defined by transfer function (16.37) is then equal to

$$1 - K(s)\sum_i \frac{r_i}{s - p_i} = 0 \tag{16.38}$$

For further consideration it is assumed that an oscillatory mode corresponding to a given eigenvalue $\lambda_k = p_k$ is excited. Then the associated characteristic equation takes the form

$$1 - K(s)\frac{r_k}{s - \lambda_k} = 0 \tag{16.39}$$

Equation (16.38) is a simplification of Eq. (16.37) where the components of $G(s)$, not related to k-th eigenvalue, are neglected.

Next, a shift of the k-th eigenvalue by $\Delta\lambda_k$ is assumed. By setting $s = \lambda_k + \Delta\lambda_k$ into Eq. (16.39) it takes the form

$$1 - K(\lambda_k + \Delta\lambda_k)\frac{r_k}{(\lambda_k + \Delta\lambda_k) - \lambda_k} = 0 \tag{16.40}$$

At point $s = \lambda_k$ the transfer function of the compensator $K(s)$ can be represented by the first-order Taylor series expansion

$$K(\lambda_k + \Delta\lambda_k) = K(\lambda_k) + \Delta\lambda_k\left(\frac{\partial K(s)}{\partial s}\right)_{s=\lambda_k} \tag{16.41}$$

Then, by substituting Eq. (16.41) into Eq. (16.40) and rearranging Eq. (16.40) one gets the relation between the k-th eigenvalue shift and the residue

$$\Delta\lambda_k = \frac{r_k K(\lambda_k)}{1 - r_k\left(\frac{\partial K(s)}{\partial s}\right)_{s=\lambda_k}} \tag{16.42}$$

When the following condition is fulfilled

$$r_k\left(\frac{\partial K(s)}{\partial s}\right)_{s=\lambda_k} \ll 1 \tag{16.43}$$

e.g. the stabilizer gain is small, Eq. (16.42) simplifies to

$$\Delta\lambda_k \cong r_k K(\lambda_k) \tag{16.44}$$

Equation (16.44) is used for the design of the POD's compensator. The compensator design aim is to increase a given mode damping by the eigenvalue shift to the left in the complex plain. This means that the angle of vector defined by $\Delta\lambda_k$ should be equal to $\pm\pi$ rad, which leads to the following condition

$$\arg\{r_k K(\lambda_k)\} = \arg\{K(\lambda_k)\} + \arg\{r_k\} = \pm\pi \tag{16.45}$$

For the compensator parameter's calculation, the transfer function $K(s)$ in Eq. (16.45) is usually reduced to $K(s) = k_s G_c(s)$, i.e. the low-pass filter and the washout filter are neglected. Their parameters are usually defined separately.

Therefore, having k-th residue $\underline{r}_k = r_k e^{j\theta_k}$, Eq. (16.45) can be written as

$$\arg\{G_c(s)\} = \pm\pi - \theta_k = \theta_{kc} \tag{16.46}$$

The basis for the residue method is the observation that the residue argument is equal to the slope angle of the tangent to the eigenvalue loci, i.e. $\theta_k = \beta$ and thus the compensation angle θ_{kc} is equal to β_c (Figure 16.14b). Then, the compensator parameters can be calculated as follows.

When the required compensation angle θ_{kc} is larger than $\pi/3$ rad, usually more than one compensation block is used. For m compensation blocks, the time constants T_1 and T_2 can be calculated as follows

$$\theta_{km} = \frac{\theta_{kc}}{m} \tag{16.47}$$

$$\alpha_0 = \frac{1 - \sin(\theta_{km})}{1 + \sin(\theta_{km})} \tag{16.48}$$

$$T_1 = \frac{1}{\Omega_k\sqrt{\alpha_0}} \tag{16.49}$$

$$T_2 = \alpha_0 T_1 \tag{16.50}$$

Equations (16.47)–(16.50) are derived for a compensator whose maximum phase shift, equal to θ_{km}, is achieved for $s = \lambda_k = \alpha_k + j\Omega_k$ with assumed $\alpha_k = 0$. Gibbard et al. (2015) propose an iterative algorithm for the compensator time constants calculation for eigenvalue with nonzero damping ratio $\alpha_k \neq 0$.

The POD gain k_s is calculated based on the desired value of the damping coefficient or selected based on the eigenvalue loci analysis or, defined in the real system, in field tests. The POD gain can be computed as

$$k_s = \left| \frac{\lambda_{kd} - \lambda_k}{r_k G_c(\lambda_k)} \right| \tag{16.51}$$

where λ_{kd} is the eigenvalue of k -th mode defining a desired damping of the mode. Because of simplification done in Eqs. (16.39) and (16.44), the eigenvalue shift to the desired location is not always possible.

Often, the desired direction of the trajectory of a given mode is obtained only for a small gain of the POD and for a given system's operating point. The direction of the eigenvalue loci for larger gains usually deviates from the one predetermined a priori. In addition, the POD also modifies the eigenvalue loci of some other modes. Thus, the effect of introducing a compensator with the parameters specified above does not have to be positive from the point of view of other modes of electromechanical oscillations. The POD design is complex and requires many factors to be considered.

Analogically to the P-Vr and GEP methods, the residue taken for the compensator design has to be chosen from a set of residues calculated for various operating conditions of the EPS.

In the case of the discussed WAPOD, the optimal value of the POD gain depends on the amplitude of the given mode of inter-area oscillations in the input signal and on the typical values of the difference in voltage angles (which is WAPOD input signal). For small measured values of the difference in voltage angles the bigger values of the POD gain are required to obtain good damping. However, too big a gain can lead to an undesirable influence of the SVC on the EPS operation or even instability. Therefore, the location of the PMU should be carefully determined by simulations to find places where the given oscillatory mode can be observed in the best way.

Example 16.1 Consider a system with single oscillatory mode described by transfer function

$$G(s) = \frac{1}{(s - q_1)(s - q_2)} \tag{16.52}$$

where the transfer function poles are equal to $q_{1,2} = \alpha \pm j\Omega$. The closed-loop transfer function of the system and POD – Figure 16.15 and Eq. (16.37) – in the case of the initial assumption $K(s) = k_s$ is equal to

$$\frac{\Delta\delta}{\Delta V} = \frac{1}{(s - q_1)(s - q_2) + k_s} = \frac{1}{(s - p_1)(s - p_2)} = \frac{r_1}{s - p_1} + \frac{r_2}{s - p_2} \tag{16.53}$$

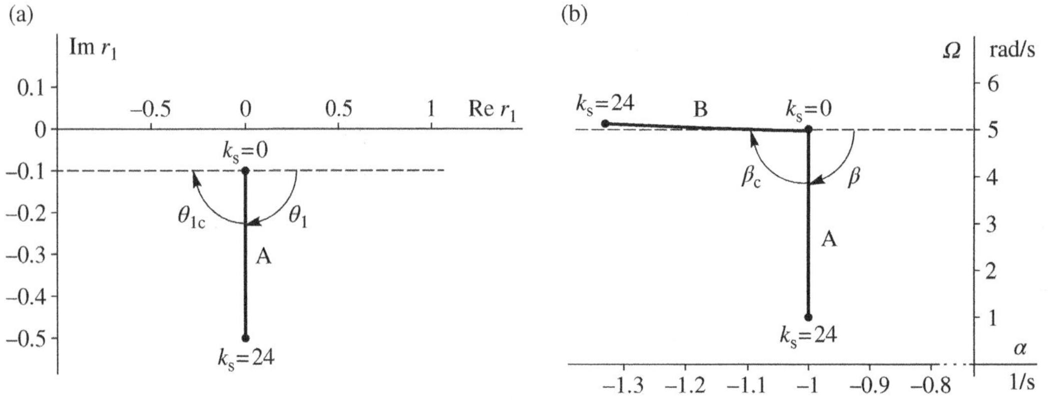

Figure 16.15 Trajectory of (a) residue r_1 and (b) pole p_1 of considered system.

where p_1 and p_2 are poles of the closed-loop system. The poles and residues of the transfer function (16.53) are equal to

$$p_{1,2} = \alpha \pm \sqrt{k_s - \Omega^2}; \quad r_{1,2} = \pm \frac{1}{2\sqrt{k_s - \Omega^2}} \tag{16.54}$$

To illustrate the method, the following values of the mode damping and pulsation are assumed: $\alpha = -1\ \mathrm{s}^{-1}$ and $\Omega = 5$ rad/s. The residue $\underline{r}_1 = re^{j\theta}$ and pole $\underline{p}_1$ for the POD gain range $k_s = \{0...24\}$ are presented in Figure 16.15 and marked by line A. In the considered range of gain k_s the residue r_1 magnitudes are changing from −0.1 to −0.5, while the angle is constant and equal to $\theta_1 = -\pi/2$. The pole p_1 moves down with constant damping α. It should be noted that the residue angle in open loop, i.e. for $k_s = 0$ (but not only for this gain), is equal to the slope angle of the tangent of pole loci, i.e. $\theta_1 = \beta$.

The aim of the POD compensator is to change the direction of the trajectory of the pole loci to the left, i.e. to rotate it by angle $\theta_{1c} = \beta_c = -\pi/2$. Assuming two compensation blocks, i.e. $m = 2$, the block time constants, from Eqs. (16.47)–(16.50), can be calculated as follows

$$\theta_{1m} = \frac{\theta_{1c}}{m} = \frac{-\pi/2}{2} = -\pi/4 \tag{16.55}$$

$$\alpha_0 = \frac{1 - \sin(\theta_{1m})}{1 + \sin(\theta_{1m})} = \frac{1 - \sin(-\pi/4)}{1 + \sin(-\pi/4)} = 5.828 \tag{16.56}$$

$$T_1 = \frac{1}{\Omega\sqrt{\alpha_0}} = \frac{1}{5\sqrt{5.828}} = 0.083\ \mathrm{s} \tag{16.57}$$

$$T_2 = \alpha_0 T_1 = 5.828 \cdot 0.083 = 0.483\ \mathrm{s} \tag{16.58}$$

Then, the compensator transfer function is equal to

$$G_c(s) = \left(\frac{1 + sT_1}{1 + sT_2}\right)^m = \left(\frac{1 + s0.083}{1 + s0.483}\right)^2 \tag{16.59}$$

Such a defined compensator changes the direction of the trajectory of the pole loci of the considered system $G(s)$ to the desired one. In Figure 16.15b the pole loci for gain $k_s = \{0...24\}$ and compensator $G_c(s)$ is shown by line marked as B.

By adding a washout filter and a low-pass filter, e.g. with $T_w = 10$ s, $\omega_B = 25$ rad/s, and $\xi = 0.75$ the system trajectory is subject only to a slight modification.

In general in the compensator design process the phase shift of the low-pass filter and the washout filter introduced for the given eigenvalue pulsation can also be added to the residue angle θ_1.

******

17

Impact of Renewable Energy Sources on Power System Dynamics

17.1 Renewable Energy Sources

In order to compare the dynamic properties of wind and photovoltaic (PV) power plants with conventional energy sources, i.e. generating units with synchronous generators connected to the electric power system (EPS) through step-up transformers, one can consider a single machine infinite bus (SMIB) model, such as the one discussed in Section 6.1.1 (Figure 6.1). In such a model the general Eq. (1.8) can be used to series connected reactances to obtain formula for the active power $P_{\mathrm{E}'}$ injected to the grid during transient state. For example

$$P_{\mathrm{E}'} = \frac{E'V_{\mathrm{g}}}{X'_{\mathrm{d}}}\sin\delta'_{\mathrm{g}} = \frac{E'V_{\mathrm{b}}}{X'_{\mathrm{d}} + X_{\mathrm{T}}}\sin\delta'_{\mathrm{gT}} = \frac{E'V_{\mathrm{s}}}{x'_{\mathrm{d}}}\sin\delta' \tag{17.1}$$

where E', V_{g}, V_{b}, and V_{s} are magnitudes of voltage phasors $\underline{E}'$, $\underline{V}_{\mathrm{g}}$, $\underline{V}_{\mathrm{b}}$, and $\underline{V}_{\mathrm{s}}$, and δ', δ'_{g}, and δ'_{gT} are angles between respective pairs of voltage phasors, $x'_{\mathrm{d}} = X'_{\mathrm{d}} + X_{\mathrm{T}} + X_{\mathrm{L}} + X_{\mathrm{S}}$.

In the steady state, the angular speed of all voltage phasors $\underline{E}'$, $\underline{V}_{\mathrm{g}}$, $\underline{V}_{\mathrm{b}}$, and $\underline{V}_{\mathrm{s}}$ is the same. However, during electromechanical oscillations the angular speeds of voltage phasors $\underline{E}'$, $\underline{V}_{\mathrm{g}}$, $\underline{V}_{\mathrm{b}}$, and $\underline{V}_{\mathrm{s}}$ change over time and are different and not equal to the angular speed of $\underline{V}_{\mathrm{s}}$.

Generally, any thermal or hydro power plant can be treated as a set of series connected energy converters from the primary energy source (coal, crude oil, gas, water, etc.) to electric energy. The synchronous generator is the electromechanical converter of the prime mover mechanical energy to electrical energy. Then, disturbances can be divided into *electrical disturbances* (occurring in electric circuits behind this energy converter) and *nonelectrical disturbances* (occurring before this energy converter i.e. on its nonelectrical side).

Electrical disturbances behind the synchronous generator (listed below) occurring in the network lead to the following effects:

- Change in the magnitude of the generator voltage $\underline{V}_{\mathrm{g}}$ and busbar voltage $\underline{V}_{\mathrm{b}}$. This leads to a change in the active power $P_{\mathrm{E}'}$ expressed by Eq. (17.1) and, according to swing Eq. (5.1), to a change in the rotor speed and the same time a change in the angular speed of phasor $\underline{E}'$ with respect to the angular speed of the voltage phasor $\underline{V}_{\mathrm{s}}$. The change in the magnitude of the generator voltage $\underline{V}_{\mathrm{g}}$ causes a reaction of the automatic voltage regulator (AVR), which acts to restore this voltage by changing the excitation current and indirectly also the magnitude of phasor $\underline{E}'$. Simultaneously, the changes in the active power generated and the changes in the rotor speed lead to the reaction of the turbine regulator, which acts to change its mechanical power.
- Change in the frequency. The change in power system frequency means the rotating voltage phasor $\underline{V}_{\mathrm{s}}$ to slow down or accelerate compared to the phasor $\underline{E}'$. This means also a change in angle δ' between phasors $\underline{E}'$ and $\underline{V}_{\mathrm{s}}$ and, according to Eq. (17.1), a change in active power $P_{\mathrm{E}'}$. The active power imbalance leads to a change in the

Power System Dynamics: Stability and Control, Third Edition. Jan Machowski, Zbigniew Lubosny, Janusz W. Bialek and James R. Bumby.

rotor speed, which means a change in the angular speed of phasor $\underline{E}'$. The changes in power generated and the rotor speed cause a reaction of the turbine regulator, which modifies mechanical power.

Also, nonelectrical disturbances (occurring in the turbine, boiler, etc.) resulting in a change of mechanical power P_m and mechanical torque τ_m lead to a change in the rotor speed. The rotor speed change means a change in the angular speed of the phasor $\underline{E}'$ in relation to the angular speed of the phasor $\underline{V}_s$. A consequence of which is a change in the power angle δ'. The generating unit controller reacting here is also the turbine regulator.

The above shows that a change of the operating point of a synchronous generator is related to a change of relative position of the voltage phasors $\underline{E}'$, $\underline{V}_g$, $\underline{V}_b$, and $\underline{V}_s$. Also, a reaction of the generating unit to disturbance (irrespective of its place of occurrence) is related to a change in the phasor's relative position. The appropriate (in terms of quality and quantity) action of the AVR and the turbine regulator allows the EPS dynamic properties to be shaped.

The rate of change of the angular velocity of the synchronous generator rotor depends on the value of the inertia constant (Section 4.1.1) of the generating unit H, the speed and quality of the turbine control (P_m control), and the speed and quality of the generator control by the AVR (influencing the active power $P_{E'}$). As is well known, the control systems of steam and water turbines are relatively slow, because of the rotor inertia and time constants of turbines and their governors. In turn, the control systems of synchronous generators are much faster, and the rate of their reaction is limited by the exciter time constants and the transient time constant T'_d of the generator. Since the inertia of the classical generating units and the transient time constant are large, despite the fast generator regulators, all in all relatively slow control systems are obtained compared to power electronics.

In contrast to synchronous generating units connected directly to the network, in the case of energy sources connected to the grid by power electronic converters, it is possible to directly and quickly shape the voltage phasor at the point of the converter connection to the network (Sections 11.4 and 11.5). When defining the voltage phasor components one can influence magnitude and angle which are equivalents of electromotive force (emf) E' magnitude and the power angle δ'. Thus, quite large, much larger than in the case of conventional generating units, possibilities of influencing the dynamic properties of energy sources of this type are obtained. This ability of the renewable energy sources (RES) is discussed below in the context of EPS inertia.

17.1.1 Wind Turbine Generator Systems

As explained in Section 11.4, the wind turbine generator systems (WTGS) most often present in EPSs use either double-fed induction generators (DFIGs) or synchronous machines connected to an EPS by fully rated converters (FRCs). Owing to a certain dissimilarity of operation, their impact on power system dynamics is presented below separately.

17.1.1.1 WTGS with a DFIG

The mathematical model of a WTGS with a DFIG is presented in Section 11.4.3. This model can be simplified by eliminating the control system elements irrelevant to the further considerations. After neglecting the resistances and simplifying the induction machine model the simplified model shown in Figure 17.1 is obtained.

In the active power control path there is an angular speed control system with turbine regulator RT and power regulator RP. Whereas the second control path with the RQ regulator is responsible for the control of reactive power or voltage on the stator side of the machine. The output variables of the regulators are reference values of the (d, q) components of rotor currents. A simplified mathematical model of the induction machine has the form of a controlled voltage source $\underline{E} = E_d + jE_q$ and series connected inductance $L = L_F + L_m + L_S$, where L_S and L_m are the machine stator leakage and magnetization inductances (Section 7.3). Inductance L_F corresponds to the external inductance between the stator of the machine and a point in the grid with voltage $\underline{V}$. The external inductance, for example, can be a filter or a step-up transformer inductance. The current $\underline{I}_S$ is the stator current, while $\underline{I}_R$ is the AC current of the DFIG grid-side inverter. The grid-side inverter is connected to a node with voltage $\underline{V}$. For

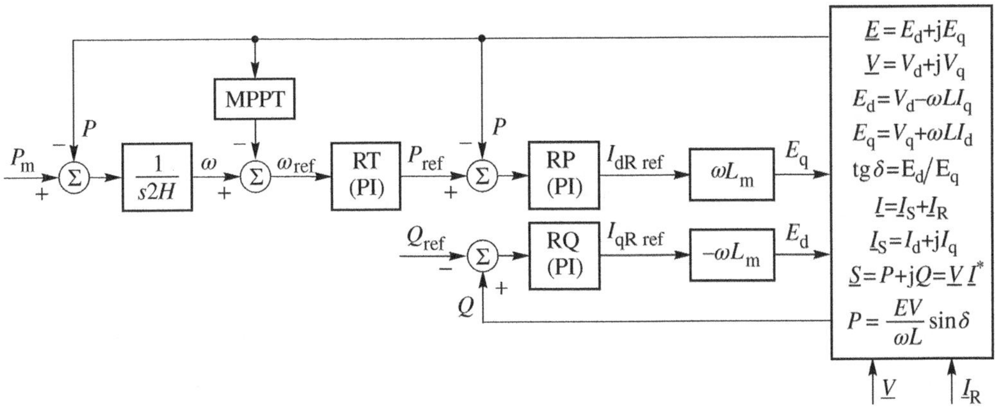

Figure 17.1 Block diagram of simplified control system of a WTGS with a DFIG.

the sake of simplicity, the current of the grid-side converter $\underline{I}_R$ can be neglected. The active power injected into the network by the WTGS, according to the power-angle characteristic, depends on the magnitudes of phasors $\underline{E}$ and $\underline{V}$, the angle between them, and the reactance ωL. The power can be calculated using equations presented in Figure 17.1.

The difference between a system with a synchronous generator (Figure 5.7 in Section 5.4.2) and the one under consideration relies on the power angle δ creation mechanism. In the case of a synchronous generator, the angle between the phasors emf $\underline{E}'$ and voltage $\underline{V}$ is due to the physical shift between the field winding flux (associated with the rotor) and the rotating flux produced by the three-phase stator currents. In the case of a DFIG the components E_d and E_q of phasor $\underline{E}$, as shown in Figure 17.1, do not require changing the relative position of the asynchronous machine rotor flux phasor. They can be created to a certain extent freely and independently in each axis and, more importantly, the change in their value can be implemented very quickly.

This means that the effects of the disturbances in the network, related to the change of voltage, active power, reactive power, or frequency, may not be transferred into the WTGS rotor (mechanical system). Their influence on the WTGS can be (and in practice is) effectively eliminated by fast regulators RP and RQ. For example, disturbance leading to a change in the active power P initially and directly affects the WTGS mechanical system (Figure 17.1), which excites the rotor swings. But quick compensation of the power imbalance by the power regulator RP causes the rotor oscillations to be significantly reduced in terms of amplitude and duration time. Simultaneously, the power response can be nonrelated or only related in a limited way to the rotor oscillations. Then, one can say that the considered disturbance, in the form of power injected to the network by the WTGS, only in a limited way is transferred to the EPS.

The frequency change in power system (EPS) shifts the speed-torque characteristics of the DFIG, which leads to electromagnetic torque shift and at steady-state would lead to the rotor speed change. The rate of frequency change (Figure 9.16 in Section 9.3) in EPSs is much slower than reaction of the WTGS fast regulators. Therefore, the regulators counteract the rotor speed change keeping its value as defined by the power output characteristics. This means that they significantly separate the inertia of the WTGS rotor from the EPS.

On the other hand, disturbances related to the change of mechanical power P_m, as a result of changes in wind speed, can be transferred to the EPS directly. In this case, the inertia of the WTGS rotor can only limit the amplitude of short-term fluctuations in mechanical power, acting as a low-pass filter.

As an example, Figure 17.2 shows the comparison of the dynamic response of a synchronous generating unit and a wind farm with DFIGs to a sudden imbalance of the active power. These simulation results are for modified version of the 39 bus New England Test System in which a wind farm has been connected to node B28. The wind farm has the rated power 300 MW and consists of 60 DFIGs with a rated power of 5 MW each. The model

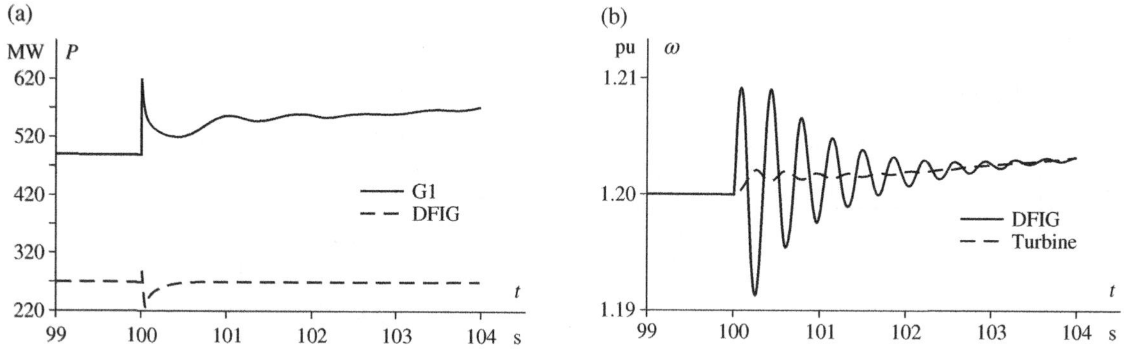

Figure 17.2 Comparison of the dynamic response of a synchronous generating unit and a wind farm with DFIGs to a sudden loss (tripping) of a generating unit: (a) active power; (b) angular velocity of wind turbine and a DFIG rotor.

described in Section 11.4.3 has been used to model this farm. It has been assumed that at moment $t = 100$ s the generating unit G3 is suddenly lost (tripped) causing the active power imbalance $\Delta P_0 = 650$ MW and a long-term dynamic response, as described in Chapter 9. In the first stage (described in Section 9.2) the power imbalance ΔP_0 is distributed among other synchronous generating units and the wind farm too. Figure 17.2a shows the changes in the active power feed to the grid by the generation unit G1 and the wind farm. Increase in active power of generating units causes a decrease in their terminal voltages and reaction of the AVRs.

The further dynamic behavior of the synchronous generating units is as described in Section 9.2. An increase in the active power causes that all synchronous generating units start to slow down with initial deceleration resulting from Eqs. (9.24) and (9.25). In the considered example the initial deceleration is $\varepsilon = \mathrm{d}\omega/\mathrm{d}t \cong -0.28 \cdot 2\pi$ rad/s, which means that the system frequency starts to drop with the initial rate of change equal to $\mathrm{d}f/\mathrm{d}t = -0.28$ Hz/s. During small power swings the average value of the active power slowly increases (curve G1 in Figure 17.2a).

The further dynamic behavior of the wind farm is significantly different. As shown in Figure 17.2a, the transient process related to the WTGS's active power control (DFIG curve), realized by the power regulator RP, is negligibly short in comparison to the duration of the active power oscillations of the synchronous generating unit (curve G1). The reaction of the WTGS's control system is largely due to the change in active power and voltage at the point of connection of the wind farm and the grid. The transient process related to the control of the rotor's angular velocity is, to some extent, separated from the process of the power plant active power control. The rotor's angular velocity is characterized by quickly damped oscillations of the DFIG and turbine rotor. These oscillations, shown in Figure 17.2b, result from the direct connection of the stator of the induction machine to the network and thus the direct influence of changes in active power and the electromagnetic torque on the motion of the rotor[1] with a flexible shaft. The visible further small increase of the rotor angular velocity, results from the wind farm power output characteristic (block MPPT, Figure 11.40b) and the operation of the angular speed regulator RT.

Based on the above, it can be concluded that the inertia of a WTGS with a DFIG is to a certain extent separated from the control processes in the EPS. The dynamic properties of a WTGS of this type are largely determined by the applied control algorithms.

---

1 The oscillations between the generator rotor and the wind turbine shown in Figure 17.2b result from the use of the WTGS with the DFIG model with a dual-mass rotor and a flexible shaft. The swing equation in the model shown in Figure 17.1 refers to the one-mass model. Flexible shaft is used to reduce mechanical stress on wind turbine rotor and blades. Figure 17.2b shows that magnitude of oscillations of the wind turbine angular speed (dashed line) and, at the same time, mechanical stress which is much smaller than the one of the induction machine rotor (continuous line).

17.1.1.2 WTGS with Synchronous Generator and an FRC

A mathematical model of a WTGS with a synchronous generator and an FRC is shown in Section 11.4.4. This model, after introducing simplifications, takes the form as in Figure 17.3.

The active power control path is made up of the power regulator RPD and the DC voltage regulator RV_{DC}. The second path, with the regulator RQ, as in the case of a WTGS with the DFIG, is responsible for controlling the reactive power or AC voltage on the grid-side converter. The output current $I_{d\ ref}$ of the active power regulator RPD is the reference value of the d-axis component of the grid-side converter current. The current in this model, can be treated simultaneously as proportional to the DC current of the machine-side converter and at the same time as proportional to the active power of the synchronous generator P_g.[2] The current $I_{d\ ref}$ is compared here with the current of the grid-side converter I_d, proportional to the active power fed into the network. The difference of these currents, and thus the difference of active power generated by the synchronous generator and power injected into the grid by the inverter, affects the voltage V_{DC} on the capacitor with capacitance C.

The output values of regulators RV_{DC} and RQ are the AC voltage components of the grid-side converter, forming the phasor $\underline{E} = E_d + jE_q$. The power injected to the grid and the currents can be calculated from the set of equations presented in Figure 17.3. Reactance ωL corresponds here to the reactance between the phasor $\underline{E}$ and the voltage $\underline{V}$.

In the considered case, the wind turbine angular velocity control is practically fully separated from the EPS by the DC link created by the power electronics converter (rectifier with the inverter and the intermediate capacitor). The DC voltage regulator RV_{DC} is able to quickly (although temporarily) compensate for the change in active power P injected to the grid, using energy from the capacitor. The fast DC voltage control, which minimizes the change in power P, simultaneously minimizes change in the synchronous generator active power P_g and thus minimizes the change in the rotor's angular velocity ω.

In contrast, the disturbances related to the change of mechanical power P_m, as a result of changes in the wind speed, can be transferred to the EPS directly. As in the case of a WTGS with a DFIG, inertia of the rotor acts here as a low-pass filter, limiting the amplitude of short-term fluctuations in active power.

An example of the quality of active power control performed by the WTGS of this type is shown in Figure 17.4. This figure shows the response of a wind farm with WTGSs with a synchronous generator and an FRC. The wind farm of rated power equal to 300 MW consists of 200 WTGSs with a rated power of 1.5 MW each, connected to node B4 in the New England Test System (Figure 18.17). As in the previous case shown in Figure 17.2, it has been

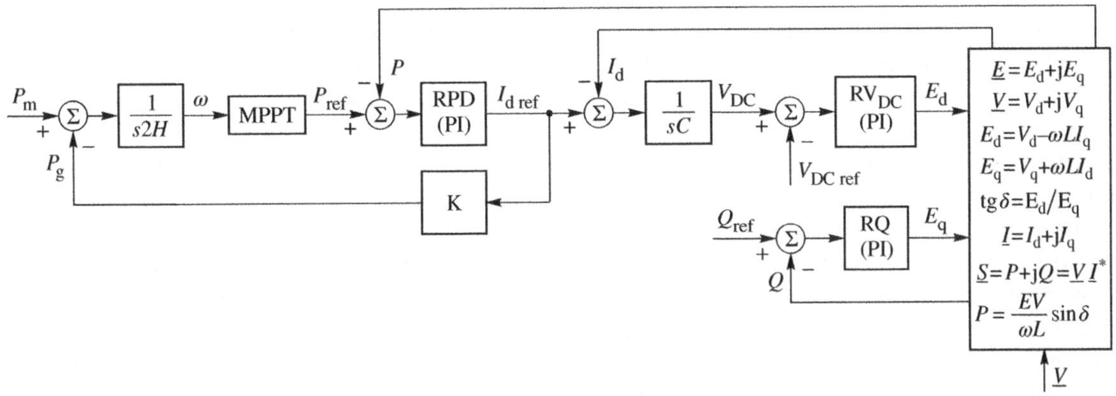

Figure 17.3 Block diagram of simplified control system of the WTGS with an FRC.

---

2 Block K represents a function that links the synchronous generator active power with current in the d-axis. This function has the form of a gain.

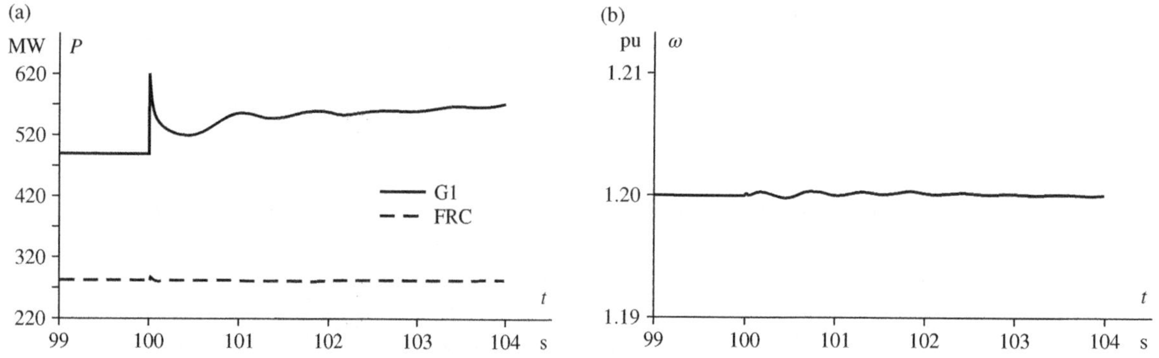

Figure 17.4 Response of the WTGS with the synchronous machine and an FRC to a sudden loss (tripping) of a generating unit: (a) active power; (b) angular velocity of the WTGS rotor.

assumed for this case that at the moment $t = 100$ s the generating unit G3 is suddenly lost (tripped) causing the active power imbalance $\Delta P_0 = 650$ MW.

Figure 17.4a shows that the deviation of the active power of the considered WTGS with an FRC is extremely small, in comparison to the amplitude of the active power oscillation of the generator G1, and also in comparison to the change in active power of the WTGS with a DFIG (Figure 17.2a). The WTGS's rotors' angular speed presented in Figure 17.4b shows, in turn, that this disturbance is practically not transferred to the wind turbine, as evidenced by a very small amplitude of oscillations.

17.1.1.3 Summary Remarks

The above describes dynamic response of a WTGS with a DFIG (Figure 17.2) and the dynamic response of a WTGS with a synchronous machine connected to the EPS by an FRC (Figure 17.4), show that:

- Almost only the control algorithms determine the dynamic properties of the WTGSs. This means that by a proper design of their regulators it is possible to shape the desired dynamic properties of WTGSs. Modern WTGSs are equipped (which is a requirement of system operators) with various control functions supporting the EPS in alert or emergency states. Examples of it include functions dedicated to over-frequency limiter, low-voltage ride through (LVRT), virtual inertia, reactive current forcing during short-circuits, etc. But, thus far, the WTGSs are not equipped with functions supporting the local angular stability, i.e. functions increasing damping of electromechanical oscillations.
- The WTGS control system response to electrical disturbances in the EPS related with the change of voltage, power, or frequency is very fast. It is much faster than the response of control systems of a generating unit with a synchronous generator connected to the EPS directly. Regulators of a WTGS with an FRC practically cut off the mechanical system from the EPS, while regulators of the WTGS with a DFIG, as a result of direct connection of the induction machine stator to the network, transfers the disturbances in the EPS to the mechanical system, but only to a small extent. This means that the regulators of both types of WTGSs do not provide the inertia of their rotors (i.e. energy stored in them) to the power system. This means that the WTGSs do not support EPSs in the event of frequency deviation. In addition, if these plants replace classical generating units, the inertia of the system is reduced, with all negative consequences for the process of frequency control. Incorporation of the WTGS in the process of frequency control requires use of dedicated control functions.
- Nonelectrical disturbances, e.g. those related to the change of power extracted from the wind stream, are transferred to the EPS almost directly. But to some extent, they can be compensated by the WTGS's control system, thanks to the use of a rotor as a temporary energy storage.

17.1.2 Photovoltaic Power Plants

PV power plants, as well as some other sources of electricity, including energy storage, are connected to the EPS by means of a power electronics converter (usually a voltage source converter [VSC]). The description below should therefore be considered as appropriate for a wider class of facilities than just PV power plants.

The mathematical model of a PV power plant control system is presented in Section 11.5. This model, after introducing minor simplifications, has the form as in Figure 17.5. The active power control path is created by the regulator PV with the MPPT system maximizing the acquired active power, and the DC voltage regulator RV_{DC} keeping the set value of voltage on the inverter's capacitor. The second path, with the regulator RQ, as in the case of a WTGS with a synchronous machine and an FRC, is responsible for the control of reactive power or voltage on the inverter's AC side. This model is functionally very similar to the one presented and discussed in Section 17.1.1, for the WTGS with a synchronous generator and an FRC. In this case, however, the energy converter, i.e. the PV panels, is a static system. There is no element to store energy,[3] which in the case of WTGSs is or can be a rotor of wind turbine and generator. The voltage regulator RV_{DC} and the reactive power or AC voltage regulator RQ are fast regulators. Therefore, the reaction of the PV plant control system to the disturbances in the EPS should be similar to the one presented in Section 17.1.1.

An example of the process of active power control by the PV power plant regulator is shown in Figure 17.6. This figure shows the response of a 300 MW power plant consisting of 300 PV plants with a rated power of 1 MW each. The disturbance simulated is the sudden loss (tripping) of the synchronous generator G3 (initially operating with active power 650 MW). The PV plant is connected to node B17 in the New England Test System (Figure 18.17). The PV plant active power response presented in Figure 17.6 is similar to the responses of WTGSs presented in Figures 17.2a and 17.4a. As shown in the considered case, from the point of view of EPS dynamics, the PV plant is a source that introduces practically constant active power to the system, i.e. it is a source that does not participate in the frequency control and thus does not participate in electromechanical oscillations in the EPS. A source of energy of this type (similarly to a DFIG-based or an FRC-based WTGS) does not provide inertia to the system (which is somewhat obvious), and replacing classical sources with synchronous generators leads to a reduction in power system inertia, which is undesirable.

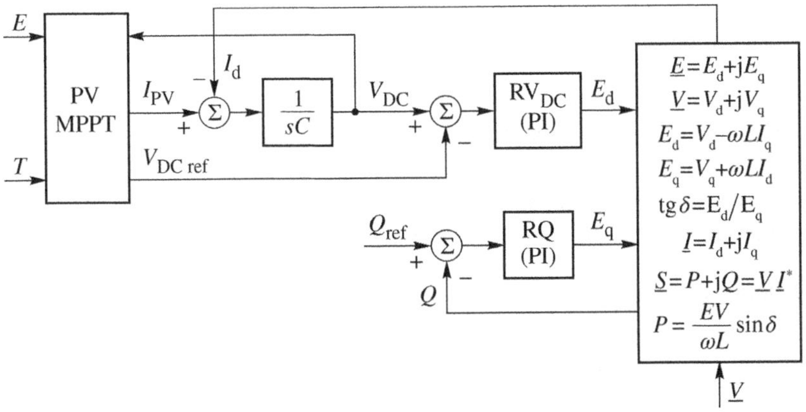

Figure 17.5 Block diagram of simplified control system of the PV power plant. E – radiation intensity; T – air temperature.

---

3 In general, PV plants can be equipped with various types of energy storage, e.g. battery storage. The inverter's capacitor can be treated as a small capacity, temporary energy storage.

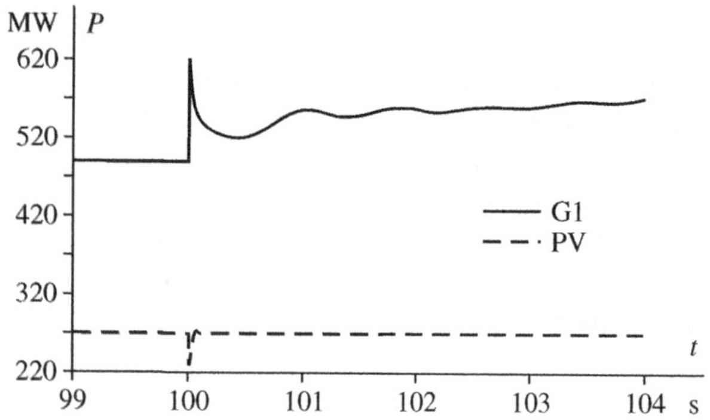

Figure 17.6 Response of a PV power plant to a disturbance in the active power balance in a power system.

In general, however, a source of energy of this type, operating with a power reserve and equipped with appropriate control functions, can support the power system in alert or emergency states. Currently, PV power plants are not equipped with control functions supporting the damping of electromechanical oscillations in EPS.

17.2 Inertia in the Electric Power System

The frequency and active power control, discussed in Chapter 9, is realized using mainly high-rated power generating units with synchronous generators driven by steam, water, and gas turbines. It can be also realized with the use of generating units driven by other engines (e.g. internal combustion or gas) but also with use of other types of power plants, e.g. PV power plants or wind power plants. However, the latter are currently very rarely used for frequency control.

The quality of the frequency and active power control is generally influenced by many parameters of the EPS elements and their control systems. As described in Section 9.3, one of these parameters, which have a particularly large impact on the initial stage of the frequency control process, are the moments of inertia of the rotors of generating units.

In Section 4.1.1 it is explained that the inertia of the rotating masses of a synchronous generation unit can be characterized by the following parameters used in power engineering:

J – moment of inertia;
$H = J\omega_s^2/2S_n$ – inertia constant defined as the kinetic energy stored in rotating masses at synchronous speed ω_s divided by the machine rating S_n;
$M = 2HS_n/\omega_s = J\omega_s$ – angular momentum (also referred to as the *inertia coefficient*);
$T_m = 2H$ – mechanical starting time (also referred to as the *mechanical time constant*).

In this chapter inertia constant H is used, because (as the energy normalized by power rating) it has the physical meaning important for this chapter.

17.2.1 Variability of the Power System Inertia

In the AC power systems with synchronous generating units the frequency in the power system and the angular speed of the rotors of all synchronous machines form a pair of connected variables $\omega = 2\pi f$. Therefore, a change in the frequency means also a change in the angular speed, and vice versa. A change in frequency and/or angular synchronous speed can be caused by a disturbance in the balance of the active power and/or by the frequency and generation control described in Section 9.1.5. With simplifications described in Section 9.6.1 the relation between the change in frequency and the active power imbalance is described by Eq. (9.67), where $G_{\mathrm{EPS}}(s) = K_{\mathrm{EPS}}/(1 + s\,T_{\mathrm{EPS}})$ is the transfer function modeling the EPS as a first-order inertial element with the time constant T_{EPS}. As results from Eq. (9.68) this time constant

$$T_{\mathrm{EPS}} = 2\pi K_{\mathrm{EPS}} M_{\mathrm{EPS}} = 2\pi K_{\mathrm{EPS}}\,\omega_{\mathrm{s}} \sum_{k=1}^{N_{\mathrm{G}}} J_k \tag{17.2}$$

depends on the sum of the moments of inertia of all masses rotating in the EPS. Such a sum is referred to as the *inertia of the power system* (inertia of the EPS).

Inertia of the EPS plays an important role in the dynamic response of the EPS, because it delays the frequency change caused by an active power imbalance. From a physical point of view this delay in the frequency change results from the kinetic energy stored in the rotating masses. The energy of the rotating masses is converted by generating units into electric energy and is provided to the grid when the frequency decreases or is taken from the grid by generating units when the frequency increases. The inertia of rotating masses reduces the effect of active power imbalance on the EPS dynamic response by limiting the rate of change of frequency and magnitude of oscillations. This positive effect is stronger if the EPS inertia and the same time constant T_{EPS} determined by Eq. (17.2) are larger.

The EPS inertia is created by masses of the rotors of the synchronous generating units connected to the grid directly and (to some extent) sources and loads with induction machines connected to the network also directly. On the other hand, as it is shown in Section 17.1, the inertia of the system is not affected by energy sources connected to the grid by power electronic converters, i.e. WTGSs with synchronous and induction generators and PV power plants, as well as drives with motors fed by power electronic converters.

17.2.1.1 Definition of *RoCoF*

As mentioned above, the EPS inertia has a significant influence on the frequency and magnitude of oscillations caused by a sudden change in the active power balance. *RoCoF* is the symbol used for the value of the *rate of change of frequency* of such oscillations in the initial moment t_{0+} just after the occurrence of the disturbance. The formula for *RoCoF* can be derived directly from Eqs. (9.26) and (9.27) derived in Section 9.3. However, for better understanding it is derived here starting from the swing equation of generator rotor described in Section 4.1. For the convenience of the readers this equation, after neglecting the damping coefficient D, can be rewritten here in the form

$$J_i \frac{\mathrm{d}\omega_i}{\mathrm{d}t} = \tau_{\mathrm{m}i} - \tau_{\mathrm{e}i}, \quad \text{or} \quad J_i \omega_i \frac{\mathrm{d}\omega_i}{\mathrm{d}t} = \omega_i(\tau_{\mathrm{m}i} - \tau_{\mathrm{e}i}) \tag{17.3}$$

where J_i is the moment of inertia of i-th generating unit, ω_i is the angular speed, and $\tau_{\mathrm{m}i}$ and $\tau_{\mathrm{e}i}$ are the mechanical shaft torque and electromagnetic torque, respectively. Since the inertia constant H is more often used in the consideration of frequency control than moment of inertia J, in the latter part, the inertia constant H defined in Eq. (4.9) is used. For i-th generating unit it is expressed in the following way

$$H_i = J_i \frac{\omega_{\mathrm{n}i}^2}{2S_{\mathrm{n}i}}, \quad \text{and} \quad J_i = \frac{2H_i S_{\mathrm{n}i}}{\omega_{\mathrm{n}i}^2} \tag{17.4}$$

where $\omega_{\mathrm{n}i}$ is the rated angular speed of i-th machine and $S_{\mathrm{n}i}$ is the rated apparent power of the synchronous generator.

Taking into account that the product of angular speed and torque is equal to the power $\Delta P_i = P_{mi} - P_{ei} = \omega_i(\tau_{mi} - \tau_{ei})$ and substituting in Eq. (17.3) the moment of inertia J_i by the right side of Eq. (17.4) it is obtained

$$\Delta P_i = P_{mi} - P_{ei} = 2S_{ni}H_i\overline{\omega}_i\frac{d\overline{\omega}_i}{dt} \tag{17.5}$$

where $\overline{\omega}_i = \omega_i/\omega_{ni}$ is the angular speed expressed in per units (pu), and P_{mi} and P_{ei} are respectively mechanical and electrical power, both expressed in watts.

The change in active power balance ΔP_0 (defined in Section 9.6.1 as a loss of the power generation or an increase of power demand) is covered by the sum of changes in active power of individual generating units that remain connected to the grid (Section 9.6.1)

$$\sum_i \Delta P_i = \sum_i 2S_{ni}H_i\overline{\omega}_i\frac{d\overline{\omega}_i}{dt} = -\Delta P_0 \tag{17.6}$$

During the frequency drop (Section 9.3) all generating units operating in the EPS can be treated as a single equivalent generating unit with inertia constant H_{EPS} and rotor rotating with the average angular speed of all generating units, i.e. with the speed of the center of inertia (COI) of the EPS, defined in Section 6.2 by Eq. (6.15). With this assumption the following equation, analogous to Eq. (17.6), can be written

$$2S_{nEPS}H_{EPS}\overline{\omega}_{COI}\frac{d\overline{\omega}_{COI}}{dt} = -\Delta P_0 \tag{17.7}$$

where ω_{COI} is the angular speed of the rotor of this equivalent generating unit and S_{nQ} is equal to the sum of the rated apparent powers of the generating units operating at a given moment in the EPS

$$S_{nEPS} = \sum_i S_{ni} \tag{17.8}$$

Comparison of Eq. (17.6) with Eq. (17.7) gives

$$S_{nEPS}H_{EPS}\overline{\omega}_{COI}\frac{d\overline{\omega}_{COI}}{dt} = \sum_i S_{ni}H_i\overline{\omega}_i\frac{d\overline{\omega}_i}{dt} \tag{17.9}$$

Moreover, it can be assumed, as in Eq. (9.23), that during the frequency drop all the generators in a given EPS remain in synchronism and therefore

$$\overline{\omega}_{COI} \cong \overline{\omega}_i \quad \text{and} \quad \frac{d\overline{\omega}_{COI}}{dt} \cong \frac{d\overline{\omega}_i}{dt} \tag{17.10}$$

Taking into account Eqs. (17.10) it is obtained from (17.9) that

$$S_{nEPS}H_{EPS} = \sum_i S_{ni}H_i \tag{17.11}$$

which, after taking into account Eq. (17.8), leads to the equation defining the EPS inertia constant H_{EPS}

$$H_{EPS} = \frac{\sum_i S_{ni}H_i}{\sum_i S_{ni}} \tag{17.12}$$

The other measure of the EPS response to the disturbance of the active power balance is also the energy accumulated in the rotors of the generating units operating in the system. The energy of the rotating masses of the generating units (or of the motors) is defined as follows

$$E_i = J_i\frac{\omega_i^2}{2} \tag{17.13}$$

After substituting Eq. (17.4) into Eq. (17.13) the following is obtained

$$E_i = H_i S_{ni} \frac{\omega_i^2}{\omega_{ni}^2} \tag{17.14}$$

Theoretically, in steady state, the angular velocity of rotors of synchronous machines ω_i is equal to their rated speed ω_{ni}.[4] With this assumption, the Eq. (17.14) takes the form

$$E_{ni} = H_i S_{ni} \tag{17.15}$$

and the total energy accumulated in the rotors of all generating units operating at a given moment in the EPS at rated frequency is equal to the sum of the energy of rotating masses of these generating units

$$E_{\mathrm{nEPS}} = \sum_i H_i S_{ni} \tag{17.16}$$

Again, assuming that the change in frequency at any place in the EPS is the same, based on Eq. (17.7) the following can be written

$$\frac{\mathrm{d}\overline{\omega}_{\mathrm{COI}}}{\mathrm{d}t} = \frac{\Delta P_0}{2S_{\mathrm{nEPS}} H_{\mathrm{EPS}} \overline{\omega}_{\mathrm{COI}}} \tag{17.17}$$

Taking into account that $\omega = 2\pi f$ (where f is in Hz) and assuming that at the first moment after disturbance the frequency does not significantly differ from the nominal one, i.e. $\overline{\omega}_{\mathrm{COI}} \cong 1$, the following is obtained

$$\frac{\mathrm{d}f}{\mathrm{d}t} = \frac{-\Delta P_0 f_{\mathrm{n}}}{2S_{\mathrm{nEPS}} H_{\mathrm{EPS}}} \tag{17.18}$$

where the change in power $\Delta P_0 = (P_{0-} - P_{0+}) > 0$ corresponds to the loss of power generated in the power system, where P_{0-} and P_{0+} are equal to power generated just before and just after the generating unit tripping. Taking into account Eq. (17.11) the following derivative $\mathrm{d}f/\mathrm{d}t$ is obtained

$$RoCoF = \left.\frac{\mathrm{d}f}{\mathrm{d}t}\right|_{t=0+} = \frac{-\Delta P_0 f_{\mathrm{n}}}{2\sum_i S_{ni} H_i} \tag{17.19}$$

which is (expressed in Hertz per second) the rate of change of frequency (*RoCoF*) in the initial moment t_{0+} just after the occurrence of the disturbance.

17.2.1.2 Assessment of the EPS Inertia Constant and *RoCoF*

Calculation of the inertia constant of the EPS and the *RoCoF* after disturbance of the active power balance on the basis of the equations presented above seems to be a simple task. In practice, however, it is not. The difficulty in calculating these parameters results from having limited access to the required data. The transmission system operator (TSO) does not have full information about all generating units operating in a given moment in the EPS and about their parameters. The data of the high rated power generating units controlled by the TSO are available (known), whereas information about small and very small energy sources, and even more about high rated power machine driven loads, is not available to the operator. This leads to a situation in which the system inertia constant and the *RoCoF* are known to the TSO merely approximately.

---

4 In the EPS practice, the operating state of the system is a quasi-steady state in which the frequency and thus the angular velocity of the rotors is constantly fluctuating around a certain average value. This average value for most of the time, i.e. outside the periods of the synchronous time correction, is the speed corresponding to the nominal frequency of 50 or 60 Hz (depending on the nominal frequency of the EPS). The amplitude of frequency fluctuations for a significant part of the EPS operation time does not exceed 10 mHz.

Below, on the example of Continental Europe Synchronous Area (CESA), an assessment method of the inertia constant and *RoCoF* for a given value of the active power balance disturbance is presented. One of the publicly available data sources is the Transparency Platform created by the European Network of Transmission System Operators for Electricity (ENTSO-E). This is a website (www.entsoe.eu) containing information on active power generated by generating units in the systems creating CESA for time intervals of 15 or 60 min. The *Transparency Platform* also provides hourly information about:

- *actual generation per generation unit*, i.e. active power generated within a given technology, broken down into individual generating units;
- *actual generation per production type*, i.e. total hourly active power generation by generating units for a given technology;
- *installed capacity per production unit*, i.e. the total active power achievable in a given technology in a given power plant, but not individual generating units;
- *installed capacity per production* type, i.e. power installed in a given technology.

ENTSO-E distinguishes energy generation technologies and for each technology and provides exemplary parameters for individual technologies of energy production in the CESA, as outlined in Table 17.1.

Table 17.1 Exemplary parameters for individual technologies of energy production in CESA.

Technology	Inertia constant H_k [s]	Load factor α_k	Rated power factor cos φ_{nk}
Biomass	2.00	0.70	0.85
Fossil brown coal/lignite	3.70	0.85	0.85
Fossil coal-derived gas	4.00	0.75	0.85
Fossil gas	3.50	0.70	0.85
Fossil hard coal	4.25	0.75	0.85
Fossil oil	3.50	0.70	0.85
Fossil oil shale	3.50	0.70	0.85
Fossil peat	3.50	0.70	0.85
Geothermal	3.50	0.70	0.85
Hydro pumped storage	6.35	0.80	0.85
Hydro run-of-river and poundage	3.00	0.60	0.85
Hydro WATER REservoir	3.50	0.80	0.85
Marine	0	—	—
Nuclear	7.00	0.90	0.85
Other	2.00	0.75	0.85
Other renewable	2.00	0.70	0.85
Solar	0	—	—
Waste	2.00	0.70	0.85
Wind offshore	0	—	—
Wind onshore	0	—	—

Information on rated apparent power S_{ni} of individual generating units is not available in data published on Transparency Platform. It can be, however, assessed in the following way

$$S_{ni} = \frac{P_{ni}}{\cos\varphi_{ni}} = \frac{P_i}{\alpha_i \cos\varphi_{ni}} \tag{17.20}$$

where $\cos\varphi_{ni}$ is the rated power factor and $\alpha_i = P_i/P_{ni}$ is the load factor (a measure of the utilization rate of given generating unit). Assuming such rated apparent power, Eqs. (17.12) and (17.19) take the following forms

$$H_{\mathrm{aEPS}} = \frac{\sum_i \frac{P_i H_i}{\alpha_i \cos\varphi_{ni}}}{\sum_i \frac{P_i}{\alpha_i \cos\varphi_{ni}}} \quad \text{and} \quad RoCoF_{\mathrm{a}} = \frac{-\Delta P_0 f_n}{2\sum_i \frac{P_i H_i}{\alpha_i \cos\varphi_{ni}}} \tag{17.21}$$

where in these symbols subscript "a" is used to emphasize that they relate to the assessed values.

Equation (17.21) can be modified by distinguishing groups of generating units characterized by common features from the point of view of energy production technology, operating states or data availability (measurement or rated). Dependencies allowing the value of the EPS inertia constant and *RoCoF* to be assessed take the form

$$H_{\mathrm{aEPS}} = \frac{\sum_{k=1}^{N_k}\left(\frac{1}{\alpha_k}\sum_i \frac{P_{i,k}H_{i,k}}{\cos\varphi_{ni,k}}\right)}{\sum_{k=1}^{N_k}\left(\frac{1}{\alpha_k}\sum_i \frac{P_{i,k}}{\cos\varphi_{ni,k}}\right)} \quad \text{and} \quad RoCoF_{\mathrm{a}} = \frac{-\Delta P_0 f_n}{2\sum_{k=1}^{N_k}\left(\frac{1}{\alpha_k}\sum_i \frac{P_{i,k}H_{i,k}}{\cos\varphi_{ni,k}}\right)} \tag{17.22}$$

where N_k is the number of generating units in k-th energy production technology operating in a given moment in the EPS (Wasilewski and Lubosny 2017).

As mentioned above, in the EPS practice the availability of information necessary to calculate the inertia constant H_{EPS} and *RoCoF* is limited. In addition, in case of the interconnected power systems it is different for a subsystem of a given operator and external subsystems. Within the given subsystem it is different for transmission and distribution subsystems. The TSOs have full information about centrally controlled generating units operating at a given moment in the system. However, as mentioned above, they often have only limited or do not have information about medium and especially small units currently operating in the system. In this case, the estimation of the system inertia constant H_{aEPS} and $RoCoF_{\mathrm{a}}$ can be implemented as follows

$$H_{\mathrm{aEPS}} = \frac{\sum_{k=1}^{N_k}\left(\frac{1}{\alpha_k}\sum_i \frac{P_{i,k}H_{i,k}}{\cos\varphi_{ni,k}} + \sum_j S_{nj,k}H_{j,k}\right)}{\sum_{k=1}^{N_k}\left(\frac{1}{\alpha_k}\sum_i \frac{P_{i,k}}{\cos\varphi_{ni,k}} + \sum_j S_{nj,k}\right)} \tag{17.23}$$

$$RoCoF_{\mathrm{a}} = \frac{-\Delta P_0 f_n}{2\sum_{k=1}^{N_k}\left(\frac{1}{\alpha_k}\sum_i \frac{P_{i,k}H_{i,k}}{\cos\varphi_{ni,k}} + \sum_j S_{nj,k}H_{i,k}\right)} \tag{17.24}$$

where $S_{nj,k}$ are known rated apparent powers of generating units and $P_{i,k}$ are active powers generated at a given moment by generating units with limited availability or lack of data.

Another way of the system inertia constant H_{EPS} and *RoCoF* estimation is provided by the online methods based on wide area measurement system (WAMS). They use frequencies and powers measured by phasor measurement units (PMUs) in various points of the EPS. The measured values are used for the identification of coefficients of the EPS assumed models, e.g. autoregressive moving average with exogenous input model (ARMAX) type. The

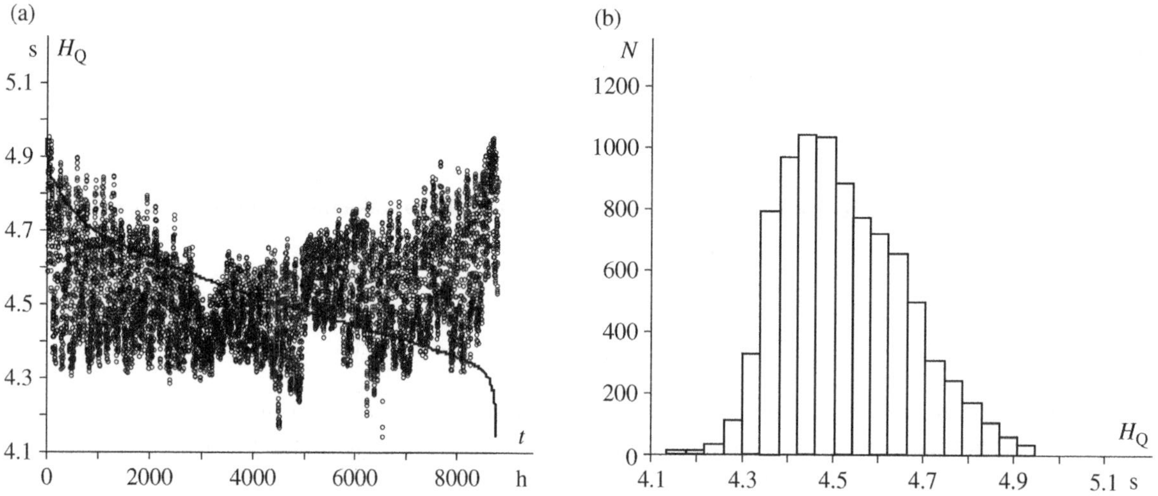

Figure 17.7 Variability of inertia constant of the CESA in 2015: (a) time chart; (b) histogram.

identified models allows the system responses on a given input to be calculated next provide information about EPS inertia. An example of such a method is presented by Tuttelberg et al. (2018).

Below, as an example, the results of calculations of the inertia constant H_{EPS} and *RoCoF* for CESA in year 2015 are presented based on Wasilewski and Lubosny (2017). Figure 17.7 shows the annual variability of the inertia constant H_{EPS} and its histogram. Figure 17.8 shows the annual variation of the *RoCoF* and its histogram for the case of change in active power balance by the reference value $\Delta P = 3000$ MW. Parameter N determines the number of occurrences of inertia constant within the defined range of values.

The smallest value of the inertia constant equal to $H_{EPS} = 4.14$ s was observed in the autumn, on 29th September 2015 at 06:00–07:00 (UTC, i.e. coordinated universal time). This fact was caused by the high share of the generation in Table 17.1 referred to as "Other" ($H_k = 2.0$ s) and the lower share of nuclear plants ($H_k = 7.0$ s). A similar value was observed in summer on 6th July 2015, at 16:00–17:00 (UTC). It resulted from the low share of nuclear units and hydro-pumped storage units ($H_k = 6.35$ s). The average value of the inertia constant was 4.52 s. In turn, the maximum values of the inertia constant were observed in winter in 2nd January

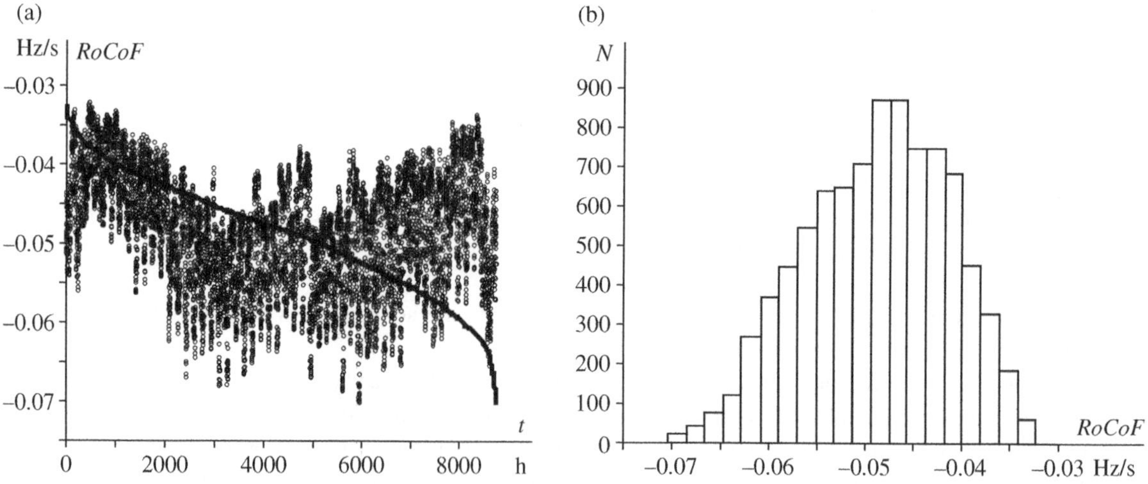

Figure 17.8 Variability of *RoCoF* in the CESA in 2015 for $\Delta P = 3000$ MW: (a) time chart; (b) histogram.

2015 and 28th December 2015, which resulted from the high share of nuclear generation and the operation of pumped storage power plants. The difference between the extreme values of the inertia constant as well as the 0.95 and 0.05 quantiles is small (Figure 17.7a). In the first case it is $H_{\text{EPSmax}} - H_{\text{EPSmin}} = 4.95 - 4.14 = 0.81$ s, and in the second case it is $H_{\text{EPS0.95}} - H_{\text{EPS0.05}} = 4.73 - 4.3 = 0.43$ s. Larger values of the inertia constant are observed in the winter period rather than in the remaining periods of the year. This small difference, however, has a significant influence on the dynamic properties of the power system.

In turn, the smallest value $RoCoF = -0.032$ Hz/s was observed on 20th January 2015, at 17:00–18:00, with the maximum generation from conventional sources equal $P = 360$ GW. At that time, in addition to coal and nuclear units, fossil gas and hydro-pumped storage power plants had a large share in the generation. The maximum value of $RoCoF = -0.070$ Hz/s occurred on 6th September 2015 (Sunday) at 03:00–04:00. This was the case for a minimum conventional generation equal $P = 156$ GW, with a relatively small water generation. The difference between the largest and the smallest value is about two times. These values, from the point of view of the power system dynamics, are very small. The variability of *RoCoF* is determined mainly by the sum of the rated power of the generating units being the source of the inertia. It also depends on the CESA power exchange with the areas operating through HVDC links. The highest values of *RoCoF* are observed in the autumn, compared to the rest of the year.

17.2.1.3 Calculation of *RoCoF* for Individual Generating Units

Using the above equations the value of *RoCoF* is calculated with the assumption that the power angles of all synchronous generators decrease synchronously. This assumption is valid during the frequency change (Figure 9.16 in Section 9.3). At the initial moment t_{0+} (just after the disturbance of the active power balance) the rotors of all individual generating units may slow down (Section 9.2) with different deceleration resulting from the swing equation. The deceleration powers $\Delta P_i = (P_{mi} - P_{ei})$ in Eq. (17.5) can be found by solving the network equations with assumption that the disturbance does not change the transient electromotive forces of all synchronous generators (neither magnitude nor phase)

$$\underline{E}'_i(t = 0_+) = \underline{E}'_i(t = 0_-) = \left|\underline{E}'_i\right| = E'_i\, e^{j\delta'_i} \tag{17.25}$$

Magnitude of the transient emf is not changed suddenly because of the large value of the field winding time constant and the transient time constant T'_{d} (Table 4.3 in Section 4.2.3). Phase of the transient emf is not changed suddenly because of the inertia of the generating unit rotor and the mechanical time constant $T_{\text{m}i} = 2H_i$ resulting from this inertia.

The power flow at the initial moment t_{0+} is referred to as the *acceleration power flow* and is one of the snapshots of the power flows defined in Section 9.8. Such a power flow can be found iteratively by a special type of the power flow computer program, which allows the generating units to be modeled as the voltage sources and allows many slack nodes with predetermined voltage magnitudes and phases (V, δ') to be used. Typical power flow programs usually do not have such abilities.

When all loads are replaced by constant admittances, the acceleration power flow can be found by solving non-iteratively the nodal admittance equations. Three steps of this method are illustrated by Figure 17.9.

The method starts with the calculation of power flow for original network model with load nodes {L} and generation nodes {B}, which are the busbars of the power plants, and $k \in \{B\}$ is the generation node in which in the next step a generating unit must be switched off to cause the assumed disturbance in active power balance. By modeling loads as constant admittances the nodal currents in the load nodes {L} become equal to zero $\underline{\mathbf{I}}_{\text{L}} = \mathbf{0}$. Then the original network model (Figure 17.9a) can be described by the following nodal admittance equation (Section 3.5)

$$\begin{bmatrix} \underline{\boldsymbol{I}}_{\text{B}} \\ \mathbf{0} \end{bmatrix} = \begin{bmatrix} \underline{\boldsymbol{Y}}_{\text{BB}} & \underline{\boldsymbol{Y}}_{\text{BL}} \\ \underline{\boldsymbol{Y}}_{\text{LB}} & \underline{\boldsymbol{Y}}_{\text{LL}} \end{bmatrix} \begin{bmatrix} \underline{\boldsymbol{V}}_{\text{B}} \\ \underline{\boldsymbol{V}}_{\text{L}} \end{bmatrix} \tag{17.26}$$

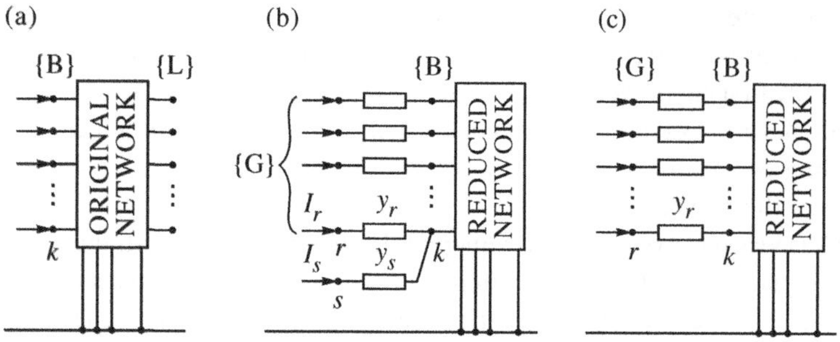

Figure 17.9 Block diagrams of the network models: (a) original network in the initial steady-state; (b) reduced network extended with voltage sources replacing all generating units; (c) reduced network without voltage source replacing the generating unit switched off.

where $\underline{\mathbf{V}}_{\mathrm{B}}$ and $\underline{\mathbf{V}}_{\mathrm{L}}$ are vectors of nodal voltages at generation and load nodes, and $\underline{\mathbf{I}}_{\mathrm{B}}$ is vector of nodal currents at generation nodes. Applying to this equation the partial inversion described in Appendix A.2

$$\underline{\mathbf{I}}_{\mathrm{B}} = \underline{\mathbf{Y}}_{\mathrm{B}}\underline{\mathbf{V}}_{\mathrm{B}} \tag{17.27}$$

is obtained, where matrix $\underline{\mathbf{Y}}_{\mathrm{B}} = \underline{\mathbf{Y}}_{\mathrm{BB}} - \underline{\mathbf{Y}}_{\mathrm{BL}}\underline{\mathbf{Y}}_{\mathrm{LL}}^{-1}\underline{\mathbf{Y}}_{\mathrm{LB}}$ is the nodal admittance matrix of the reduced network obtained by elimination of load nodes (details in Section 18.2.1).

In the second step (Figure 17.9b) the voltage sources $(\underline{E}'_i, \underline{y}_i)$ representing the generating units are connected to the generation nodes {B}, where $\underline{E}'_i = E'_i e^{j\delta'_i}$ are the transient electromotive forces and $\underline{y}_i = 1/(\underline{Z}_{\mathrm{T}i} + jX'_{\mathrm{d}i})$ are admittances of the generating units taking into account the transient reactances $X'_{\mathrm{d}i}$ of the generators and impedances $\underline{Z}_{\mathrm{T}i}$ of the step-up transformers. Two sources are connected to node k: source $(\underline{E}'_s, \underline{y}_s)$ representing a generating unit which will be switched off and source $(\underline{E}'_r, \underline{y}_r)$ representing remaining generating units operating on the same busbar (Figure 17.9b). Currents in these two branches must satisfy the Kirchhoff's current law $\underline{I}_r + \underline{I}_s = \underline{I}_k$, where $\underline{I}_k$ is the nodal current $\underline{I}_k$ in the original network model and $\underline{I}_s$ is the current of the generating unit which will be switched. As shown in Figure 17.9 set {G} includes all fictitious nodes with the exception of node s belonging to the source which will be switched off. Transient emf's at all voltage sources are calculated from Kirchhoff's voltage law

$$\underline{E}'_i = \underline{V}_i + \underline{I}_i/\underline{y}_i; \quad \underline{E}'_r = \underline{V}_k + \underline{I}_r/\underline{y}_r; \quad \underline{E}'_s = \underline{V}_k + \underline{I}_s/\underline{y}_s \tag{17.28}$$

where $\underline{V}_i$ and $\underline{V}_k$ are the nodal voltages taken from the vector $\underline{\mathbf{V}}_{\mathrm{B}}$. In special cases when all generating units connected to the node k will be switched off it is necessary to substitute $\underline{y}_r = 0$ and $\underline{E}'_r = 0$.

In the third step, the source $(\underline{E}'_s, \underline{y}_s)$ is disconnected from the reduced network model (Figure 17.9c), which can be described by the following nodal admittance equation

$$\begin{bmatrix} \underline{\mathbf{I}}_{\mathrm{G}}^{+} \\ \mathbf{0} \end{bmatrix} = \begin{bmatrix} \underline{\mathbf{Y}}_{\mathrm{GG}} & -\underline{\mathbf{Y}}_{\mathrm{GG}} \\ -\underline{\mathbf{Y}}_{\mathrm{GG}} & (\underline{\mathbf{Y}}_{\mathrm{B}} + \underline{\mathbf{Y}}_{\mathrm{GG}}) \end{bmatrix} \begin{bmatrix} \underline{\mathbf{E}}_{\mathrm{G}} \\ \underline{\mathbf{V}}_{\mathrm{B}}^{+} \end{bmatrix}; \quad \underline{\mathbf{Y}}_{\mathrm{GG}} = \begin{bmatrix} \underline{y}_1 & & & & \\ & \ddots & & & \\ & & \underline{y}_i & & \\ & & & \ddots & \\ & & & & \underline{y}_r \end{bmatrix} \tag{17.29}$$

where $\underline{\mathbf{E}}_{\mathrm{G}}$ is the vector of the emf's calculated from Eq. (17.28), $\underline{\mathbf{I}}_{\mathrm{G}}^{+}$ and $\underline{\mathbf{V}}_{\mathrm{B}}^{+}$ are nodal currents of generating units and nodal voltages at the initial moment t_{0+} (just after the disturbance of the active power balance), and $\underline{\mathbf{Y}}_{\mathrm{GG}}$ is the matrix consisting of the admittances of the voltage sources connected to the network model. This matrix is diagonal because individual nodes {G} and {B} are connected directly one-to-one.

Applying to Eq. (17.29) the partial inversion described in Appendix A.2

$$\underline{I}_G^+ = \underline{Y}_G \underline{E}_G \quad \text{and} \quad \underline{V}_B^+ = \underline{K}_V \underline{E}_G \tag{17.30}$$

is obtained, where

$$\underline{Y}_G = \underline{Y}_{GG} - \underline{Y}_{GG}(\underline{Y}_B + \underline{Y}_{GG})^{-1}\underline{Y}_{GG} \quad \text{and} \quad \underline{K}_V = (\underline{Y}_B + \underline{Y}_{GG})^{-1}\underline{Y}_{GG} \tag{17.31}$$

When currents $\underline{I}_i^+$ and voltages $\underline{V}_i^+$ are calculated from Eq. (17.30) for all generation nodes the powers injected by sources to the network and their changes can be calculated from equations

$$\underline{S}_i^+ = 3\underline{V}_i^+ (\underline{I}_i^+)^* \quad \text{and} \quad \Delta\underline{S}_i = \underline{S}_i^+ - \underline{S}_i \quad \text{and} \quad \Delta P_i = \text{Re}\,\Delta\underline{S}_i \tag{17.32}$$

Finally, according to Eq. (17.19), the value of the $RoCoF_i$ for individual generating units can be calculated

$$RoCoF_i = \frac{-\Delta P_i f_n}{2 S_{ni} H_i} \tag{17.33}$$

Example 17.1 Exemplary EPS with total rated power of generating units equal to 30.3 GW, with 4398 nodes and 744 generating units is considered. Generator tripping is made at a 400 kV generating node with generator with a rated power of 885 MW, located in the middle of the network. Figure 17.10 shows the dependence of the $RoCoF_i$ of i-th generating unit on the generator tripping. The presented values of $RoCoF_i$ for chosen generators with basic data are: G1: $S_{nG1} = 459$ MVA, $H_{G1} = 2.9$ s; G2: $S_{nG2} = 447$ MVA, $H_{G2} = 3.225$ s; G3: $S_{nG3} = 435$ MVA, $H_{G3} = 3.225$ s; G4: $S_{nG4} = 274$ MVA, $H_{G4} = 5$ s; G5: $S_{nG5} = 297$ MVA, $H_{G5} = 3.5$ s; G6: $S_{nG6} = 426$ MVA, $H_{G6} = 3.225$ s. The calculations were made using Eq. (17.33) for chosen state of the EPS operation. Generators G1, G2 and G3, G6 operate in two power plants in the central part of the EPS. The generators G1, G3, and G4 are connected to the 400 kV network, generators G2 and G5 to the 220 kV network, and generator G6 to the 110 kV network. Generators G4 and G5 are located in power plants in the western part of the system (far from generators G1, G2, G3, and G6).

The largest (as far as absolute) values of the frequency derivative, reaching −0.71 Hz/s for $\Delta P = 885$ MW, occur for generator G1, the nearest electrically with the generator being tripped. Smaller values of $RoCoF_i$ refer to generators G2 and G3. They are located in various plants. Generator G2 is connected to 220 kV, while G3 is connected to 400 kV network. Electrical distance of these generators to the generator being tripped is various but their various inertia constants compensate it and their $RoCoF_i$ is similar. The smallest $RoCoF_i$, presented, refer to the generator G4, which is located most far from the tripped generator and additionally is characterized by high inertia constant.

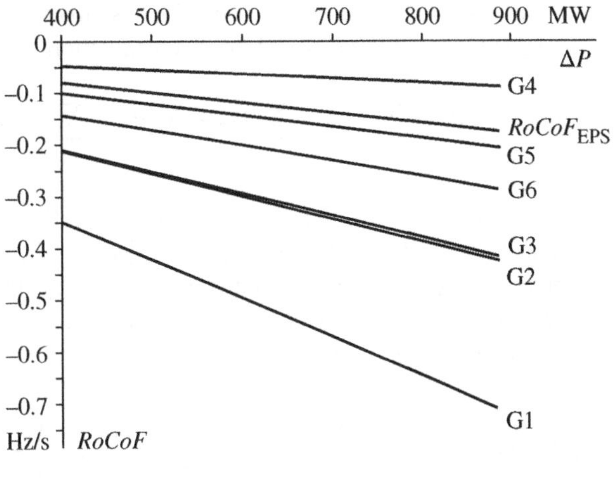

Figure 17.10 Change in the $RoCoF_i$ of selected synchronous generators operating in exemplary EPS as a function of power ΔP switched off in a 400 kV generating node.

It is worth noting that the largest value of $RoCoF_i$ (as an absolute value), characterized generator G1, is about four times greater than the "average" one, defined by Eq. (17.19), calculated for the whole system, equal to $RoCoF_{EPS} = -0.175$ Hz/s for $\Delta P = 885$ MW. It is also worth to underline that for most of the generators the $RoCoF_i$ is close to the one calculated from Eq. (17.19).

******

The significance of the *RoCoF* for EPSs results from the fact that the system operators, e.g. constituting ENTSO-E, are obliged to specify in their grid code the maximum permissible values of the *RoCoF*. On one hand, this results from the desire to reduce excessively fast frequency changes in the system, and on the other hand to protect energy sources from the effects of these changes. European system operators define different limit values of *RoCoF* from the range of −0.5 to −2.5 Hz/s. These values are usually referred to the measurement carried out in the 0.5 s measurement window. In general the period over which *RoCoF* is measured should correspond to the phenomenon of interest, e.g. overall change in the system frequency, not local or inter-area oscillations. This relatively large width of the measurement window is aimed at detuning from the instantaneous values associated with the duration time of non-coherent swings of rotors of synchronous generators.

17.2.2 Impact of Inertia Constant on Power System Dynamics

It is assumed that in a given power system N conventional generating units with synchronous generators participate in the process of primary frequency control and M conventional generating units with synchronous generators do not participate in this control. Both groups of generating units hereinafter are referred to as *inertial*. For the first group, the small dead zones (Figure 9.6a) in the speed-droop characteristics are neglected and the linear speed-droop characteristics are assumed. The generating units from the second group operate with active power independent on frequency, which can result from large dead zones (Figure 9.6b) set in the speed-droop characteristics or the operation of the generating units in the active power control mode or in the mode of steam pressure control.

The swing equations of generating units from these two groups then have the form

$$2H_iS_{ni}\overline{\omega}_i\frac{d\overline{\omega}_i}{dt} = P_{Ti} - P_{Li} \quad \text{for} \quad i = 1, ..., N \tag{17.34}$$

$$2H_jS_{nj}\overline{\omega}_j\frac{d\overline{\omega}_j}{dt} = P_{Tj} - P_{Lj} \quad \text{for} \quad j = 1, ..., M \tag{17.35}$$

where the subscript i refers to the generating units of the first group and the subscript j to the second. Powers P_{Ti} and P_{Tj} are the net shaft power input to the generator; P_{Li} and P_{Lj} are electrical air-gap power.

It is assumed that in the considered system the active power is also generated by non-inertial sources of energy that do not participate in frequency control. These sources may be PV power plants or wind power plants. The balance of active power in the system under consideration has the form

$$\sum_i P_{Ti} + \sum_j P_{Tj} + \sum_k P_{PVk} = \sum_r P_{Lr} + \sum_s P_{Ls} + \sum_t P_{Lt} = P_L \tag{17.36}$$

where i and j relate to inertial and k to non-inertial generating units, and P_L is the total power demand in the EPS including transmission losses. The powers $\sum_r P_{Lr}$, $\sum_s P_{Ls}$, and $\sum_t P_{Lt}$ are parts of the total power demand fed respectively by the inertial generating units of two groups and by the non-inertial generating units.

Assuming coherent oscillations of the rotors of all inertial generating units, i.e. $\overline{\omega}_i \cong \overline{\omega}_j \cong \overline{\omega}$, and adding up Eqs. (17.34) and (17.35), one gets

$$2\left(\sum_i H_iS_{ni} + \sum_j H_jS_{nj}\right)\overline{\omega}\frac{d\overline{\omega}}{dt} = \left(\sum_i P_{Ti} - \sum_r P_{Lr}\right) + \left(\sum_j P_{Tj} - \sum_s P_{Ls}\right) \tag{17.37}$$

Taking into account the kinetic energy, Eq. (17.14), stored in rotors of inertial generating units

$$E = E_i + E_j = \sum_i H_i S_{ni} \overline{\omega}_i^2 + \sum_j H_j S_{nj} \overline{\omega}_j^2 = \overline{\omega}^2 \left(\sum_i H_i S_{ni} + \sum_j H_j S_{nj} \right) \tag{17.38}$$

Equation (17.14) can be presented in the form

$$\frac{2E}{\overline{\omega}} \frac{d\overline{\omega}}{dt} = \sum_i P_{Ti} + \sum_j P_{Tj} - \sum_i P_{Li} - \sum_j P_{Lj} = \sum_i P_{Ti} + \sum_j P_{Tj} - \left(P_L - \sum_k P_{Lk} \right) \tag{17.39}$$

In this equation E defines energy stored in the rotors before disturbance with an assumption that the frequency in the power system (and the rotors speed) can differ from the nominal one, i.e. $E \neq E_n$ – Eqs. (17.14) and (17.15).

It is assumed that ΔP_{Ti} is a change in turbine power caused by the primary frequency control and $\Delta P_T = \sum_i \Delta P_{Ti}$ is the sum of these changes. Then, new power generated by inertial energy sources participating in the primary frequency control is equal to

$$\sum_i P_{Ti} = \sum_i P_{Ti}^{old} + \Delta P_T \tag{17.40}$$

where superscript "old" refers to power before disturbance. The power generated by inertial sources that do not participate in frequency control and by non-inertial sources does not change.

At the same time, considering (as in Section 9.6.1) the characteristics of the power demand as linearly dependent on the frequency, the total power demand in the EPS can be defined as

$$P_L = \sum_i P_{Li}^{old} + \sum_j P_{Lj}^{old} + \sum_k P_{Lk}^{old} = P_L^{old} + K_L P_L^{old} \Delta\overline{\omega} + \Delta P_0 \tag{17.41}$$

where K_L is frequency sensitivity coefficient of power demand – Eq. (3.158) in Section 3.4 – and ΔP_0 is a power imbalance caused by a sudden increase of power demand or a sudden loss of generation. Then, by substituting Eqs. (17.40) and (17.41) into Eq. (17.39) and taking into account Eq. (17.36), one obtains a differential equation of the EPS model in the form

$$2E\,\overline{\omega}^{-1} \frac{d\Delta\overline{\omega}}{dt} + K_L P_L^{old} \Delta\overline{\omega} = \Delta P_T - \Delta P_0 \tag{17.42}$$

Equation (17.42) is a nonlinear differential equation of the Bernoulli type. In application to EPS it is usually simplified by assumption $\overline{\omega} \cong 1$ and $E \cong E_n$, which leads to the form

$$\frac{2E_n}{f_0} \frac{d\Delta f}{dt} + \frac{K_L P_L^{old}}{f_0} \Delta f = \Delta P_T - \Delta P_0 \tag{17.43}$$

where $\Delta f = f - f_0$ is frequency deviation and f_0 is frequency in the EPS before disturbance. The frequency f_0 most of the time during the EPS operation is equal to the nominal frequency f_n. Solutions of both equations applied to the power system model give very small differences, which justifies simplification.

Equation (17.43) defines the mathematical model of the EPS whose transfer function $G_{EPS}(s)$ is defined by first-order inertial element with gain K_{EPS} and time constant T_{EPS} equal to

$$K_{EPS} = \frac{f_0}{K_L P_L^{old}} \quad \text{and} \quad T_{EPS} = \frac{2E_n}{K_L P_L^{old}} = \frac{2K_{EPS}E_n}{f_0} \tag{17.44}$$

This corresponds formally to the model defined by Eq. (9.68) for $f_0 = f_n$ and to a simplified model of frequency control in the EPS, as shown in Figure 9.31.

It is worth noting here that the total system demand P_L influences both the power system gain K_{EPS} and time constant T_{EPS}, while the kinetic energy E_n influences the time constant T_{EPS} only. The energy is a function of inertia constants of the inertial generating units only, which means that only the inertial generating unit tripping changes the system time constant.

Additionally, the power change ΔP_T in the Eq. (17.43) and model from Figure 9.31 refers to change of power injected only by the inertial generating units, participating in the process of the frequency primary control. This means that parameters in the considered model, i.e. in transfer function $G_{TG}(s)G_T(s)$ (Figure 9.31), should be related to this group of generating units. The limitation on the range of the output power related to the primary control (Figure 9.6a) should also be taken into account.

Below, based on the EPS model defined by Eq. (17.43), three examples are presented showing the influence of selected factors on the process of frequency control in EPS.

Example 17.2 Consider EPS operating with inertial generating units only, with a structure as in Figure 9.31, and described by Eq. (17.43). There are no non-inertial generating units in the considered system. The rated power of generating units with synchronous generators is equal to $S_{nQ} = 20$ GVA. It is assumed that 70% of generating units (in terms of rated power) participate in primary control of frequency, and active power generated by other generating units does not depend on frequency. Assuming the active power of generating units equal to 90% of their rated active power, and the rated power factor equal to $\cos\varphi_n = 0.85$, the total active power generated by the above generating units is equal to $P_g = 20 \cdot 0.85 \cdot 0.9 = 15.3$ GW. The power generated by inertial units involved in primary control is equal to $P_{g1} = 10.71$ GW, while power generated by inertial units not involved in frequency control is equal to $P_{g2} = 4.59$ GW. The load demand is equal to $P_L = 15.3$ GW. The system inertia constant is equal to $H = 5$ s, and the frequency sensitivity coefficient of power demand is equal to $K_L = 1.1$.

As the turbine regulator, a PI type regulator (Figure 11.26b) is adopted, with a gain $K_{PI} = 1$ and a time constant $T_{PI} = 3$ s. The droop coefficient is equal to $\rho = 0.05$. As the turbine model, the reheat-turbine model with three stages (Figure 11.24) was used with: $T_A = 0.3$ s, $T_B = 8$ s, $T_C = 0.3$ s, $T_D = 0.45$ s, $\alpha = 0.277$, $\beta = 0.3$, and $\gamma = 0.423$.

The EPS gain, Eq. (17.44), is equal to $K_{EPS} = 0.003$ and does not change when the inertia constant changes. The system time constant T_{SEE} decreases as the inertia constant decreases, which is shown in Figure 17.11a. Also, the kinetic energy of rotating masses E, calculated for $f_0 = f_n$ according to Eq. (17.16) decreases as the inertia constant of the system decreases.

The system considered is characterized by two oscillatory modes and two non-oscillatory modes. For an inertia constant of the system equal to $H = 5$ s the eigenvalues of oscillatory modes are equal to $\lambda_1 = -0.15 \pm j0.4$ rad/s (point A in Figure 17.12) and $\lambda_2 = -3.69 \pm j0.09$ rad/s, while the damping coefficients of non-oscillatory modes are equal to α_i={−2.1, −0.26} 1/s. The decrease in the value of the inertia constant to $H = 3$ s leads to a small change in the value of damping ratios. The eigenvalues of oscillatory modes are equal to $\lambda_1 = -0.23 \pm j0.47$ rad/s (point B in

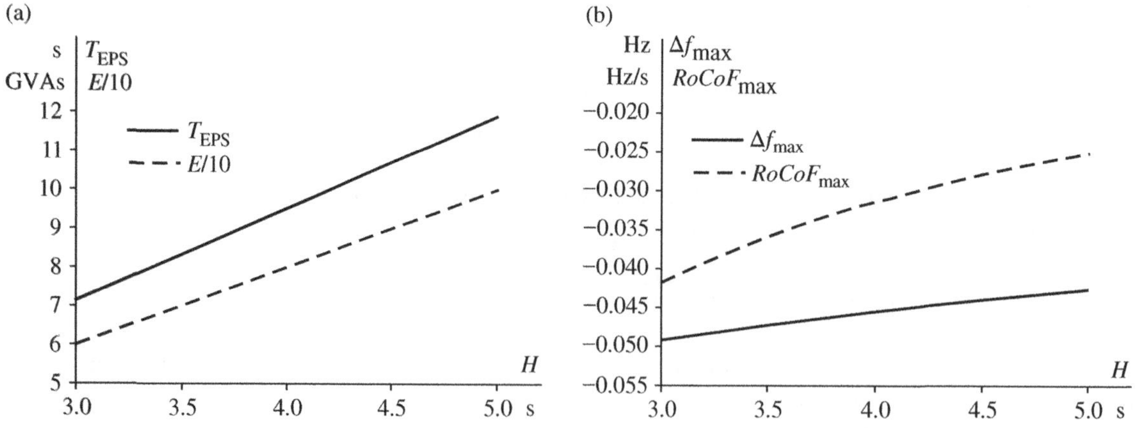

Figure 17.11 Influence of inertia constant in EPS with inertial generating units on: (a) system time constant and kinetic energy stored in rotors; (b) maximum frequency deviation and $RoCoF$ for power imbalance equal to $\Delta P_0 = 100$ MW.

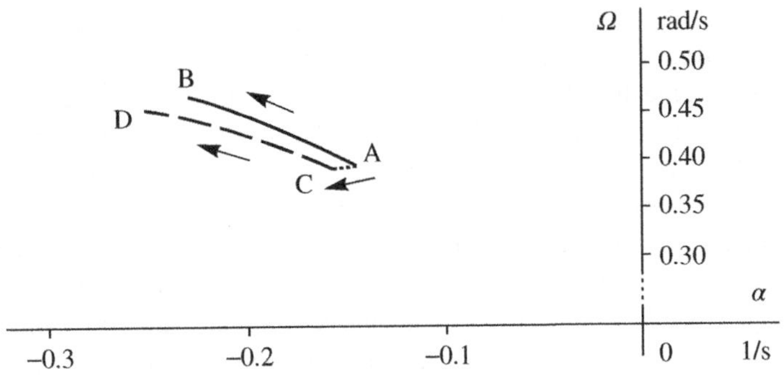

Figure 17.12 Loci of eigenvalue associated with oscillatory mode λ_1 with the weakest damping: AB – decrease in the value of the inertia constant of the system without non-inertial generating units (Example 17.2); AC – increase in load demand covered by non-inertial generating units (Example 17.3); CD – decrease in the value of the inertia constant of the system with inertial and non-inertial generating units (Example 17.4).

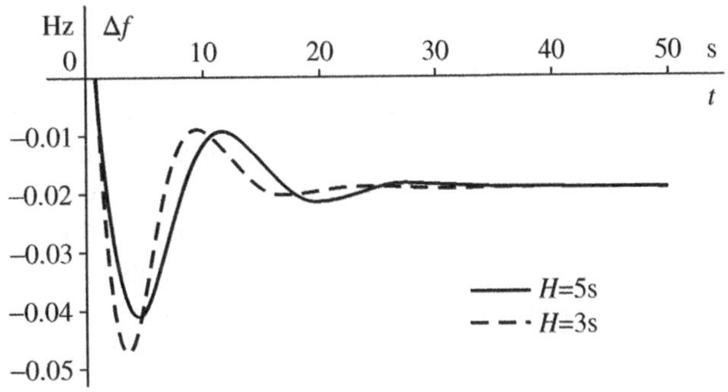

Figure 17.13 Frequency response of the EPS ($P_L = 20.3$ GW) with inertial ($P_g = 15.3$ GW) and non-inertial ($P_{PV} = 5$ GW) generating units for two values of the system inertia constant.

Figure 17.12) and $\lambda_2 = -3.56 \pm j0.19$ rad/s, while the damping coefficients of the non-oscillatory modes are equal to $\alpha_i = \{-2.0, -0.28\}$ 1/s. The loci of eigenvalue λ_1 associated with oscillatory mode with the weakest damping is presented in Figure 17.12 by the continuous curve AB. The beginning of the trajectory corresponds to an inertia constant of 5 s (point A), the end (point B) equal to 3 s. It can be seen here that the decrease in the inertia constant leads to an increase in the damping of frequency oscillations. However, the decrease of the inertia constant leads to the increase of the maximum frequency deviation Δf_{max} and the absolute value of the rate of change of frequency $RoCoF_{max}$. This is due to the reduction of kinetic energy E stored in the rotors of the generating units and thus the decrease of the time constant T_{EPS}. Influence of decreasing inertia constant on the parameters of the system frequency response, after $\Delta P_0 = 100$ MW power change, is shown in Figure 17.11b, while an example of the system time domain response to the same power balance disturbance is shown in Figure 17.13.

******

Example 17.3 Consider EPS with inertial and non-inertial generating units. The inertial generating units are like those considered in Example 17.2. Their total rated apparent power is equal to $S_{nQ} = 20$ GVA, the inertia constant is equal to $H = 5$ s, and the active power generated by these units is equal $P_g = 15.3$ GW. In it the power generated by inertial units involved in primary control is equal to $P_{g1} = 10.71$ GW, while power generated by inertial units not involved in frequency control is equal to $P_{g2} = 4.59$ GW. It is assumed here that the demand for electricity increases and this demand is covered by new, but only non-inertial, generating units. The active power generated by these units is therefore equal to $P_{PV} = (P_L - 15.3)$ GW, where the load demand fulfills inequality $P_L > P_g = 15.3$ GW. The increase of the load demand, according to the Eq. (17.44), leads to a decrease of the system time constant T_{EPS}, while the energy accumulated in the masses of rotating inertial sources does not change. The increase in the load demand from 15.3 GW to 20.3 GW leads to a change in the system time constant T_{EPS} from 11.9 s to 9 s. Along with the increase in the power generated by the non-inertial sources, the pulsation of the oscillatory mode is reduced (line AC in

Figure 17.12). The damping ratio of this mode increases slightly, i.e. from $\xi = 0.34$ to $\xi = 0.37$, with the increase of the load demand in the system and at the same time with an increase of the power generated by non-inertial units.

In this case, the disturbance of the active power balance equal to $\Delta P_0 = 100$ MW results in a small decrease in the maximum frequency deviation (from −0.043 Hz to −0.041 Hz), with no effect on the maximum absolute value of *RoCoF*. The decreasing value of maximum deviation of frequency results from decreasing the EPS gain K_{EPS} as a result of increasing value of the load demand P_L – Eq. (17.19). The lack of change of *RoCoF* results from the fact that it depends on the energy stored in the rotating masses, which in the considered case does not change.

The above shows that the increase in the load demand, covered by non-inertial generating units, with the unchanged structure and power generated by inertial generating units, does not have negative consequences for the power system in terms of the frequency response of the system. It is worth remembering that the conclusion is valid for the considered simple model.

******

Example 17.4 Consider the EPS as in Example 17.3, i.e. with inertial and non-inertial generating units. Assume the load demand equal to $P_L = 20.3$ GW. The power generated by the inertial units is still equal to $P_g = 15.3$ GW, and the power generated by the non-inertial units is equal to $P_{PV} = 5$ GW. Other system parameters remain unchanged. Changing the inertia constant in this system causes effects similar to those shown in Example 17.2. The time constant of the system T_{EPS} decreases from 9 s to 5.4 s with the inertia constant decreasing from 5 s to 3 s. The energy accumulated in the rotors of generating units and values of *RoCoF* change in the same way as in Example 17.2. Time constants of the system and maximum deviations of frequency are smaller than in the system considered in Example 17.2. This results from the higher load demand P_L, which leads to a lower value of the EPS gain and the smaller relative disturbance of the power balance.

The loci of eigenvalue associated with the oscillatory mode with the weakest damping is presented in Figure 17.12 as dashed line CD. A decrease in the inertia constant leads to an increase in the damping ratio of this mode in a way similar to the one considered in Example 17.2 (curve AB). The change in the damping coefficient of other modes is similar to that presented for the system with inertial generating units only.

Figure 17.13 presents the system frequency response to the power balance disturbance equal to $\Delta P_L = 100$ MW for the extreme values of the inertia constant considered.

Figure 17.14 presents the power balance during the system dynamic response to the step change in the load demand ΔP_0. Immediately after the occurrence of the disturbance leading to the frequency decrease, the rotors of the generating units become the source of energy injected into the system. The rotors speed decrease causes changes in the energy of the rotating masses and the energy surplus is delivered to the system. This means that the rotors inject into the EPS a power P_ω. Simultaneously, the control systems of turbines respond to the frequency change, introducing into the system additional power ΔP_T. This response is with a certain delay resulting from time

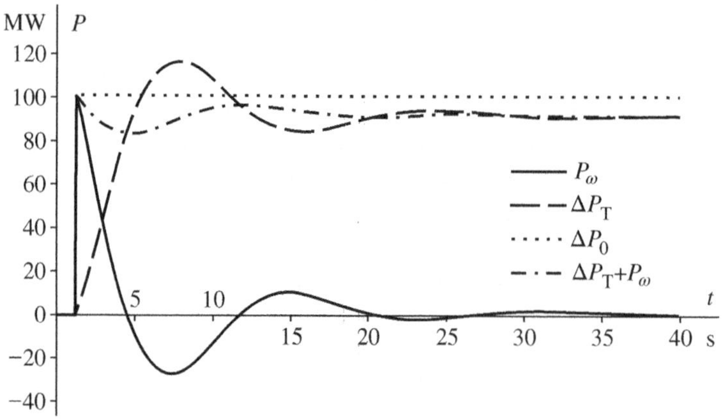

Figure 17.14 Dynamic response of the EPS with inertial and non-inertial generating units to power balance disturbance equal to $\Delta P_0 =$ 100 MW; P_ω – power resulting from the change of the rotational speed of the rotors; ΔP_T – change in net shaft power of turbines involved in the frequency control.

constants of turbines and their governors. The difference in the change in loads and the sum of power provided by generating units and rotating masses, i.e. $\Delta P_L - (\Delta P_T + P_\omega)$, is equal to the change in active power consumed by loads as a result of a change in frequency, i.e. it is related to the frequency sensitivity coefficient of power demand K_L.

******

Example 17.5 Example 17.2 considers a situation in which the load demand is increasing and this additional load is covered by new non-inertial generating units while the inertial generating units (rated power, etc.) do not change. Here a more realistic situation is considered in which the load demand does not change but non-inertial units replace some inertial units. The load demand is equal to $P_L = 15.3$ GW, power generated by inertial units involved in frequency primary control is equal to $P_g = 6.12$ GW, and the power generated by the non-inertial units is equal to $P_{PV} = 4.59$ GW. This means that the inertial generating units not participating in the frequency control are switched off (are replaced by the non-inertial units). The other system parameters remain unchanged. Changing the inertia constant in this system from $H = 5$ s to 3 s causes effects similar to those shown in previous examples. The time constant of the system decreases from $T_{EPS} = 9$ s to 8.3 s, the EPS gain K_{EPS} does not change, the maximum deviation of frequency changes from $\Delta f_{max} = -0.043$ Hz to -0.047 Hz, the $RoCoF_{max}$ changes from -0.25 Hz/s to -0.36 Hz/s, and the eigenvalue of the locus of the weakly damped oscillatory mode shifts from point A to point B (Figure 17.12). This again indicates an increase of the mode damping with worsening other parameters of the frequency response, i.e. maximum frequency deviation and $RoCoF_{max}$.

******

The above considerations, based on a very simple mathematical model of EPS (in which the control of synchronous generators and other system elements are neglected), show that the decrease in inertia of the system negatively affects the frequency response parameters, i.e. Δf_{max} and $RoCoF_{max}$, although it leads to the increase in the damping of oscillatory modes.

The real EPS's operation and analyses carried out on mathematical models, taking into account the elements of control systems neglected above, point to the negative impact of the growing share of non-inertial (e.g. PV, DFIG, and FRC) generating units on the EPS's dynamics. Among others, this was shown in Ulbig et al. (2014), in which the model of the CESA system was used. The model was verified on the basis of a real emergency shutdown of a 1200 MW generator in the Spanish system. The response of the modeled system on the generator tripping is shown in Figure 17.15. The presented response corresponds to two variants of wind generation. In both variants, the system load is 400 GW. In the so-called northern variant 20% (78.5 GW) and in the southern variant 14% (57.5 GW) of the load demand is covered by the wind power plants. In the northern variant, the wind generation in the north of the CESA is equal to 60.5 GW, and in the south 18 GW. In the southern variant, the wind generation in the north is

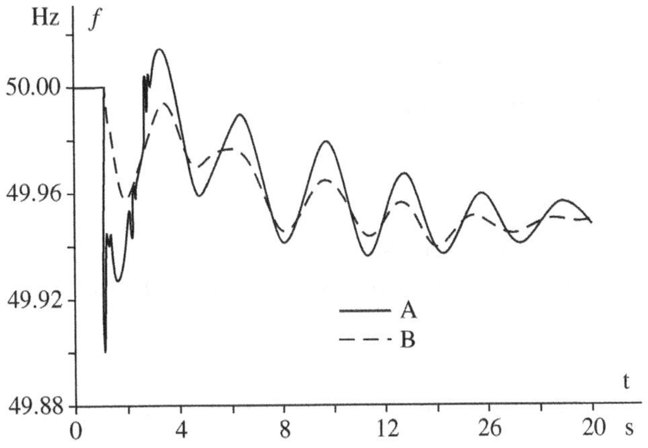

Figure 17.15 CESA's frequency response to the 1200 MW generator tripping in: A – the northern variant; B – the southern variant of RES power generation in EPS.

Table 17.2 Eigenvalues associated with weakly damped oscillatory modes of the sample system.

	39 Bus New England System		39 Bus New England System with RES	
Mode no.	$\underline{\lambda}$ **[rad/s]**	ξ **[-]**	$\underline{\lambda}$ **[rad/s]**	ξ **[-]**
1	$-0.30 \pm j7.06$	0.042	$-0.29 \pm j7.15$	0.040
2	$-0.41 \pm j6.66$	0.061	$-0.49 \pm j11.28$	0.043
3	$-0.47 \pm j7.56$	0.062	$-0.38 \pm j6.76$	0.056
4	$-0.41 \pm j6.22$	0.066	$-0.46 \pm j7.56$	0.061
5	$-0.39 \pm j5.58$	0.070	$-0.40 \pm j6.02$	0.064
6	$-0.58 \pm j7.48$	0.077	$-1.17 \pm j17.67$	0.066
7	$-0.70 \pm j8.93$	0.078	$-0.43 \pm j5.59$	0.078
8	$-0.75 \pm j8.87$	0.084	$-0.71 \pm j8.88$	0.079
9	$-0.82 \pm j9.11$	0.089	$-0.77 \pm j8.88$	0.086
10	$-0.21 \pm j0.81$	0.249	$-4.10 \pm j33.58$	0.121

equal to 19.5 GW, and in the south 38 GW. The CESA's frequency response to the 1200 MW emergency tripping of generating unit shows that the increase of power generated by wind sources of energy (but also by other non-inertial sources of energy) leads to worsening in the damping of frequency oscillations in the system.

A similar effect (worsening in the damping of oscillatory modes as a result of the increase of power generated by an RES and at the same time a reduction in the number of conventional generating units) is presented in Table 17.2. The table presents 10 eigenvalues λ that are characterized by the smallest damping ratio ξ for each model of EPS under consideration. The 39 Bus New England System and the modified one called the *39 Bus New England System with RES* were used here. This modified system was created by replacing the generators G4 ($P_g = 632$ MW) and G5 ($P_g = 254$ MW) and by three RES plants with a capacity of 300 MW each (described in Section 17.1). The wind farm with a DFIG is connected to node B28, the wind farm with synchronous generators (FRC) is connected to node B4, and the PV power plant is connected to node B17.

As presented in Table 17.2 the damping of oscillatory modes of the system with an RES is worse than for the system without an RES. This results from the replacement of inertial sources by non-inertial sources and from the different dynamic properties of control systems of the PV, DFIG, FRC power plants and generating units with synchronous generators. The explanation of this effect is as follows. In real power systems the non-inertial generating units usually replace the conventional units. This leads to a reduction in the number of various controllers, including power system stabilizers. Therefore, the damping of electromechanical oscillations in such power systems is usually worsened.

17.3 Virtual Inertia

The decrease in the value of the power system's inertia leads to higher-frequency fluctuations, faster frequency changes after the disturbance, and also to the worsening of electromechanical oscillations damping. It is therefore natural to strive for power system operators to limit these negative effects. Their reduction or elimination can be obtained by introducing into the systems additional inertia, which may take the form of natural or virtual inertia.[5]

5 The following expressions are also used in literature: *virtual*, *artificial*, *emulated*, *simulated*, and *synthetic inertia*.

Natural inertia is associated with the rotating masses of generating units or drives connected to the EPS directly. This corresponds to conventional generating units or synchronous compensators with massive rotors or sometimes with additional rotating masses. Synchronous compensators also introduce the ability to voltage control or reactive power compensation. Additionally, the synchronous compensators become a source of short-circuit current which is necessary for the proper operation of many types of power system protection, e.g. over-current, distance, differential-current, etc. Today sources of natural inertia are built as new objects or synchronous generators of decommissioned plants are used.

Virtual inertia has the form of a quick injection of the active power to the EPS forced by an appropriate control scheme. The source of such injection can be some types of generating units, controlled loads, energy storage or flexible AC transmission systems (FACTS) devices. The virtual inertia control algorithm allows the power injected to the network dependent on the system frequency to be made. The virtual inertia seems to be a cheaper solution comparing to facilities delivering natural inertia. This is because the virtual inertia is only one of many functions performed by the microprocessor-based regulators of the generating units. Unfortunately, the virtual inertia in comparison to natural inertia is characterized by a number of drawbacks.

Table 17.3 presents a comparison of selected properties of natural and virtual inertia. The basic limitation of virtual inertia in comparison to natural inertia is the lack of symmetry of the reaction to frequency change and the delay in response to this change.

17.3.1 The Algorithm of a Virtual Inertia

It is assumed that a change (decrease) in the inertia of the power system is caused by the tripping of a number of inertial generating units. A control compensating for this decrease is sought, i.e. a control leading to a frequency response identical to that of a system with full inertia. Below an algorithm of virtual inertia is derived using the swing equation of the rotors of the generating units.

Table 17.3 Comparison of natural and virtual inertia.

Property	Natural inertia	Virtual inertia
What is it	Physical quantity	Physical quantity preceded by the signal correcting the power reference in the converter control system
Time of inertia reaction	Immediate	Delay introduced by measuring devices, low-pass filters, dynamic elements of the power regulator, equal to several ms to several hundred ms
Frequency dead-band for inertia operation	None	Dead-band set up to (15–50) mHz
Direction of inertia action	Symmetrical reaction to both increase and decrease in frequency	Asymmetric reaction when the energy source operates with maximum power resulting from wind speed or insolation Symmetrical reaction in the case of generating unit operation with power reserve
Steadiness of accumulated energy	Relatively constant. Change of energy stored in rotating masses result from the power system frequency	Variable
The range of energy available	Proportional to square of frequency deviation	Energy that can be injected to the EPS in a given period of time after disturbance depends on the generating unit's operating point, mode of operation, and the control system settings

A disturbance ΔP_0 in active power balance in EPS with N inertial generating units causes changes in the angular velocity of generating units described by the following swing equation

$$\Delta P_0 = \sum_{i=1}^{N} 2S_{\mathrm{n}i}H_i\overline{\omega}_i\frac{\mathrm{d}\overline{\omega}_i}{\mathrm{d}t} = \sum_{i=1}^{N}(P_{\mathrm{T}i} - P_{\mathrm{L}i}) \tag{17.45}$$

For simplicity, as in Eq. (17.7), the swings between generating units are neglected and coherency of their rotors is assumed. Then Eq. (17.45) can be written in the form

$$\Delta P_0 = 2\overline{\omega}_{\mathrm{COI(N)}}\frac{\mathrm{d}\overline{\omega}_{\mathrm{COI(N)}}}{\mathrm{d}t}\sum_{i=1}^{N}H_iS_{\mathrm{n}i} = \sum_{i=1}^{N}(P_{\mathrm{T}i} - P_{\mathrm{L}i}) \tag{17.46}$$

where $\overline{\omega}_{\mathrm{COI(N)}}$ is the average angular speed (expressed in pu) equal to the angular speed of COI in the system with N coherently oscillating generating units.

If in the considered EPS there is a change in the number of operating generating units from N to M then the disturbance of the active power balance by ΔP_0 is related to the swing equations of M generating units as follows

$$\Delta P_0 = \sum_{j=1}^{M} 2S_{\mathrm{n}j}H_j\overline{\omega}_j\frac{\mathrm{d}\overline{\omega}_j}{\mathrm{d}t} = \sum_{j=1}^{M}(P_{\mathrm{T}j} - P_{\mathrm{L}j}) \tag{17.47}$$

Again, if it is assumed that the oscillations of rotors in a given system are still coherent, Eq. (17.47) takes the form

$$\Delta P_0 = 2\overline{\omega}_{\mathrm{COI(M)}}\frac{\mathrm{d}\overline{\omega}_{\mathrm{COI(M)}}}{\mathrm{d}t}\sum_{j=1}^{M}H_jS_{\mathrm{n}j} = \sum_{j=1}^{M}(P_{\mathrm{T}j} - P_{\mathrm{L}j}) \tag{17.48}$$

where $\overline{\omega}_{\mathrm{COI(M)}}$ is the average angular speed (COI) in the EPS with M generating units.

If then one assumes that the change of the angular velocity of rotors (and at the same time the frequency) in both systems (i.e. before and after the change of the inertia constant) is to be equal, i.e. $\overline{\omega}_{\mathrm{COI(N)}} = \overline{\omega}_{\mathrm{COI(M)}} = \overline{\omega}$, then, for $M < N$, equality of the right sides of Eqs. (17.46) and (17.48) can be achieved by introducing in the system of M generating units the additional active power ΔP_{VI}

$$2\overline{\omega}\frac{\mathrm{d}\overline{\omega}}{\mathrm{d}t}\sum_{i=1}^{N}H_iS_{\mathrm{n}i} = 2\overline{\omega}\frac{\mathrm{d}\overline{\omega}}{\mathrm{d}t}\sum_{j=1}^{M}H_jS_{\mathrm{n}j} + \Delta P_{\mathrm{VI}} \tag{17.49}$$

where subscript VI is the abbreviation of the term virtual inertia. Equation (17.49) after transformation, defines the virtual inertia control rule, in the form of

$$\Delta P_{\mathrm{VI}} = 2\overline{\omega}\frac{\mathrm{d}\overline{\omega}}{\mathrm{d}t}\left(\sum_{i=1}^{N}H_iS_{\mathrm{n}i} - \sum_{j=1}^{M}H_jS_{\mathrm{n}j}\right) = 2\overline{\omega}\frac{\mathrm{d}\overline{\omega}}{\mathrm{d}t}\Delta E_{\mathrm{n}} \tag{17.50}$$

The virtual inertia control rule from Eq. (17.50) with assumption $\overline{\omega} \cong 1$ and use frequency deviation (in Hz/s) as input signal takes the following form

$$\Delta P_{\mathrm{VI}} = 2\frac{\mathrm{d}\Delta f}{\mathrm{d}t}\left(\sum_{i=1}^{N}H_iS_{\mathrm{n}i} - \sum_{j=1}^{M}H_jS_{\mathrm{n}j}\right) = 2\frac{\mathrm{d}\Delta f}{\mathrm{d}t}\frac{\Delta E_{\mathrm{n}}}{f_{\mathrm{n}}} \tag{17.51}$$

As it results from Eqs. (17.50) and (17.51), the factor ΔE_{n} is equal to the difference in the sum of kinetic energies stored in rotors of generating units before and just after the change of the EPS operating state consisting of the generating unit or generating units tripping. In the case of a single generating unit tripping, the parameter ΔE_n is equal to the energy stored in rotor of this unit at nominal frequency.

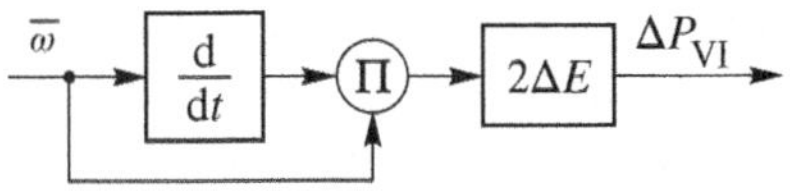

Figure 17.16 Block diagram of the virtual inertia scheme.

Figure 17.16 shows the block diagram of the virtual inertia scheme resulting from Eq. (17.50). In this diagram, the input signal (the angular velocity $\overline{\omega}$) is expressed in pu and the output signal (power ΔP_{VI}) is expressed in watts. The virtual inertia gain is equal to $K_{VI} = 2\Delta E_n$. In the case of the use of the virtual inertia defined by Eq. (17.51), the gain is equal to $K_{VI} = 2\Delta E_n/f_n$.

If the purpose, or one of the objectives, of using virtual inertia is to compensate for the loss of a generating unit, for example with the highest rated power in the system, e.g. equal to S_{ni} and inertia constant H_i, then $K_{VI} = 2\Delta E_n = 2S_{ni}H_i$ for the virtual inertia defined by Eq. (17.50) and $K_{VI} = 2\Delta E_n/f_n = 2S_{ni}H_i/f_n$ for the virtual inertia defined by Eq. (17.51).

One can also imagine that the purpose of using virtual inertia is to introduce some additional inertia to the system in order to achieve a specific control effect. In this case, it does not have to be connected with the rated power and inertia of specific or defined generating units. Then the gain K_{VI} can take any desired, technically achievable, value.

Example 17.6 Consider EPS in which operates $N = 10$ identical generating units, with a rated apparent power $S_{ni} = 1000$ MVA and inertia constant $H_i = 5$ s. Nominal frequency is equal to $f_n = 50$ Hz. The equivalent inertia constant of such a system, according to Eq. (17.12), is equal to

$$H_{EPS(N)} = \frac{\sum_{i=1}^{N} S_{ni}H_i}{\sum_{i=1}^{N} S_{ni}} = \frac{10 \cdot 1000 \cdot 5}{10 \cdot 1000} = 5\text{ s} \tag{17.52}$$

If one generating unit is tripped off in this system, there remain $M = 9$ units, but the inertia constant remains unchanged and is equal to

$$H_{EPS(M)} = \frac{\sum_{j=1}^{M} S_{nj}H_j}{\sum_{j=1}^{M} S_{nj}} = \frac{9 \cdot 1000 \cdot 5}{9 \cdot 1000} = 5\text{ s} \tag{17.53}$$

Despite the lack of a change in the inertia constant, defined by Eq. (17.12), the energy of the rotating masses is reduced, with all the consequences for the frequency control process, such as an increase in the maximum frequency deviation and an increase in the rate of change of frequency. The energy of the rotating masses change, according to the Eq. (17.16), is here equal to

$$\Delta E_n = E_{nEPS(N)} - E_{nEPS(M)} = \sum_{i=1}^{N} S_{ni}H_i - \sum_{j=1}^{M} S_{nj}H_j = 10 \cdot 1000 \cdot 5 - 9 \cdot 1000 \cdot 5 = 5000\text{ MVAs}. \tag{17.54}$$

Therefore, the gain of the virtual inertia (Figure 17.16) to compensate the inertia loss should be equal to $K_{VI} = 2S_{EPS}\Delta H = 2\Delta E_n = 10000$ MVAs for the virtual inertia defined by Eq. (17.50) and equal to $K_{VI} = 2S_{EPS}\Delta H/f_n = 2\Delta E_n/f_n = 200$ MVAs^2 for the virtual inertia defined by Eq. (17.51).

******

Example 17.7 Consider EPS as above, i.e. consisting of $N = 10$ identical generating units with rated apparent power $S_{ni} = 1000$ MVA and inertia constant $H_i = 5$ s. Nominal frequency is equal to $f_n = 50$ Hz. Then the equivalent inertia constant of this system is equal to $H_{EPS(N)} = 5$ s, and the energy of the rotating masses is equal to $E_{nEPS(N)} = 10 \cdot 5 \cdot 1000 = 50000$ MVAs. If one of these generating units is replaced by a generating unit of rated power $S_n = 1000$ MVA (i.e. rated apparent power does not change) and smaller inertia constant $H = 2$ s, the equivalent inertia constant becomes equal to

$$H_{EPS(M)} = \frac{\sum_{j=1}^{M} S_{nj} H_j}{\sum_{j=1}^{M} S_{nj}} = \frac{9 \cdot 1000 \cdot 5 + 1000 \cdot 2}{9 \cdot 1000} = 4.7 \text{ s} \tag{17.55}$$

In this case, the change in inertia constant corresponds to the change in energy of rotating masses equal to

$$\begin{aligned} \Delta E_n &= E_{nQN} - E_{nQM} = \sum_{i=1}^{N} S_{ni} H_i - \sum_{j=1}^{M} S_{nj} H_j = \\ &= 10 \cdot 1000 \cdot 5 - (9 \cdot 1000 \cdot 5 + 1000 \cdot 2) = 3000 \text{ MVAs} \end{aligned} \tag{17.56}$$

and the virtual inertia gain which allows the inertia decrease in the considered EPS should be equal to $K_{VI} = 2\Delta E = 6000$ MVAs for virtual inertia defined by Eq. (17.50) and equal to $K_{VI} = 2\Delta E_n / f_n = 120$ MVAs^2 for virtual inertia defined by Eq. (17.51) to be compensated.

******

It should be noted that the power ΔP_{VI} resulting from Eq. (17.50) or Eq. (17.51) is the total power that should be delivered to the EPS by all generating units, which must compensate for the inertia decrease. This additional power injected to the EPS should be properly distributed among individual sources. This means that the gains in the virtual inertia of these sources should be appropriately selected. In practice, the additional power which can be injected depends on the power reserve available in generating units at a given moment, and the structure and settings in the virtual inertia control system. At present, not all wind power plants are equipped with control systems to achieve the effect of virtual inertia. Examples of power plant manufacturers who implement such systems are GE, Enercon, and Senvion (Repower). The schemes of the additional active power control implemented and proposed for use in wind power plants are shown in Figure 17.17. The scheme presented in Figure 17.17a (GE WindINERTIA™) is the closest to the scheme of the virtual inertia scheme shown in Figure 17.16. In this scheme, the additional element is a low-pass filter. The scheme in Figure 17.17b performs control proportional to the frequency

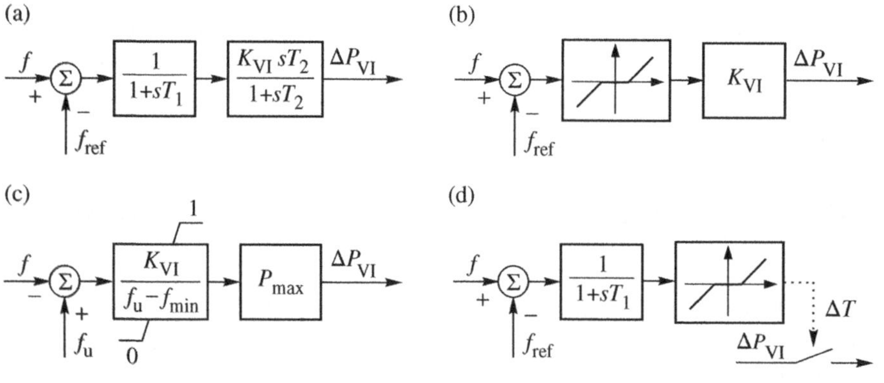

Figure 17.17 Block diagrams of virtual inertia schemes. Based on Shang et al. 2017.

deviation, which functionally corresponds to a conventional generating unit operating in power and frequency control mode. In the scheme in Figure 17.17c (Enercon Inertia Emulation™) the control is proportional to the frequency deviation for frequencies less than f_u. The coefficient P_{max} determines the maximum power provided under the additional control, and the frequency f_{min}, together with the parameters f_u and K, determines the virtual inertia gain in the frequency range $f_u - f_{min}$. However, in the scheme in Figure 17.17d, dedicated to generating units with synchronous machines, the active power activated by time ΔT as part of the additional power control is constant and equal to ΔP_{VI}.

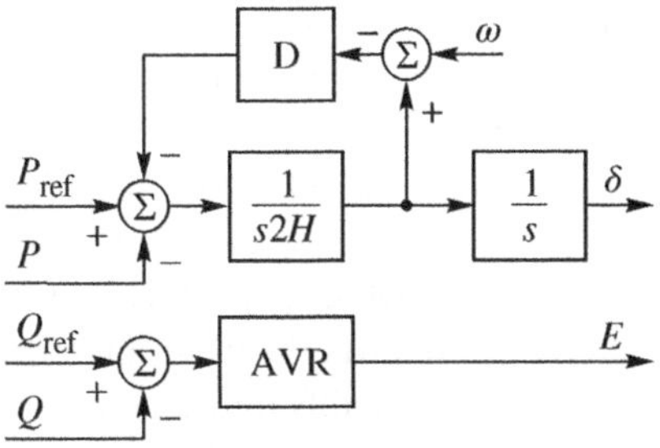

Figure 17.18 Block diagram of the inverter regulator that emulates the dynamics of the synchronous generator.

Another way to compensate for the decrease of inertia in the EPS is to control the RES converter emulating the dynamics of synchronous generator connected to the network directly. The basics of this type of control are presented in Section 7.6.3, and an example of a block diagram implementing such control is shown in Figure 17.18. The inertia constant H and damping coefficient D are adjustable parameters. The structure and parameters of the reactive power (or voltage) regulator (AVR) are assumed to reflect the synchronous generator control system, i.e. the controller along with the dynamics of the exciter. The output signals are: the power angle δ in the active power control path and the voltage magnitude E in the voltage control path. This type of system is dedicated to control the grid-side converter of the WTGS with a fully rated converter (FRC) and the PV plant as well as the machine-side converter of the WTGS with a DFIG.

17.3.2 The Impact of Virtual Inertia on Power System Dynamics

To evaluate the influence of virtual inertia on EPS dynamics, the frequency control system, as shown in Figure 17.19, is considered. It corresponds to the block diagram shown in Figure 9.31, supplemented by a virtual inertia of transfer function $G_{VI}(s)$. The components of this model are discussed in Section 9.6.1. The EPS is modeled as a first-order inertia block with a gain K_{EPS} and a time constant T_{EPS}, defined by Eq. (9.68). The transfer function $G_{AGC}(s) = \lambda(1 + 1/sT_{AGC})$ defines an automatic generation control (AGC) system that performs secondary frequency control, where λ is frequency bias factor. Transfer function $G_{RD}(s) = 1/(Rf_n)$ is a part of the turbine regulator that shapes its characteristics $P(f)$, where $R = 1/\rho$ is the inverse of the regulator droop ρ. Transfer functions $G_T(s)$ and $G_{TG}(s)$ model the turbine and its regulator.

For such a structure, the transfer function defining the relationship between the disturbance of the active power balance ΔP_0 and the frequency change Δf is described by

$$\frac{\Delta f}{\Delta P_0} = \frac{-G_{EPS}(s)}{1 + G_{EPS}(s)[G_T(s)G_{TG}(s)(G_{RD}(s) + G_{AGC}(s)) + G_{VI}(s)]} \tag{17.57}$$

Figure 17.19 EPS model with frequency control and virtual inertia.

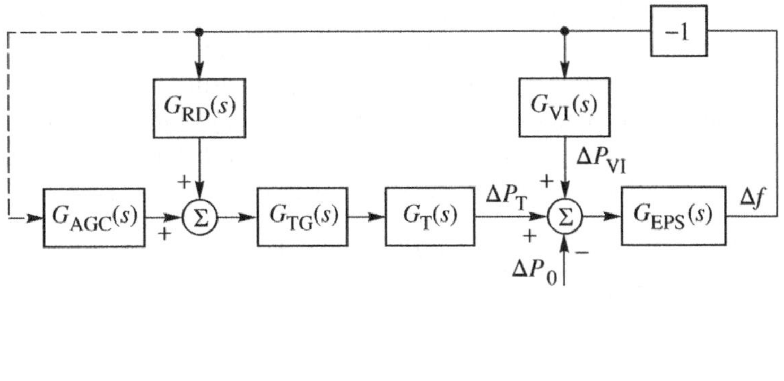

The frequency response of such a defined model is determined by parameters of all transfer functions defined above. In the case of the real EPSs, their dynamic response to the disturbance in active power balance is usually oscillatory in nature.

The change in inertia in the EPS, resulting from a decrease in the number of inertial generating units with synchronous generators, affects the change (decrease) of the time constant T_{EPS} – Eqs. (9.67) and (17.2) – and thus modifies transfer function $G_{EPS}(s)$. The change in the number of generating units also modifies transfer functions $G_T(s)$ and $G_{TG}(s)$. In turn, the inclusion of virtual inertia into the EPS involves in Eq. (17.57) the transfer function $G_{VI}(s)$. These factors modify the numerator and denominator (zeros and poles) of transfer function (17.57) and thus affect the frequency response (dynamic properties) of the system model under consideration.

In order to determine the impact of the virtual inertia on the EPS dynamics in the first step the following (extreme) simplifications relating to the EPS model and the transfer functions in Eq. (17.57) are assumed: the secondary frequency control is neglected and therefore $G_{AGC}(s) = 0$ and the turbine with the regulator is treated as the proportional block $G_T(s)G_{TG}(s)G_{RD}(s) = K_{RTT}$. Moreover, the transfer function of virtual inertia is equal to $G_{VI}(s) = sK_{VI}$, which results from the Eq. (17.51) and the assumption that the grid frequency differs only slightly from the nominal one. The adoption of such a simple transfer function of the virtual inertia is tantamount to neglecting the dynamics of energy sources in which the virtual inertia is applied. In the case under consideration, this results from the need for extreme simplification of the model, and on the other hand it is justified by the very small time constants of the control systems of these energy sources.

After substituting the specified transfer functions to Eq. (17.57) one gets

$$\frac{\Delta f}{\Delta P_0} = \frac{-K_{EPS}(s)}{1 + K_{EPS}K_{RTT} + s(T_{EPS} + K_{EPS}K_{VI})} \tag{17.58}$$

According to Eq. (9.68), the EPS time constant is equal to $T_{EPS} = 2\pi K_{EPS}M = 2K_{EPS}E_n/f_n$, where M is the inertia coefficient of the EPS and E_n is the kinetic energy stored in the rotating masses at rated speed. Transfer function (17.58), substituting the time constant T_{EPS} in the form as above, is then equal to

$$\frac{\Delta f}{\Delta P} = \frac{-K_{SVI}}{1 + sT_{SVI}} \tag{17.59}$$

where the transfer function parameters are equal to

$$K_{SVI} = \frac{K_{EPS}}{1 + K_{EPS}K_{RTT}}; T_{SVI} = \frac{K_{EPS}(2E_n/f_n + K_{VI})}{1 + K_{EPS}K_{RTT}} \tag{17.60}$$

Transfer function (17.59) has one pole (one eigenvalue) equal to

$$\underline{\lambda} = \alpha + \mathrm{j}\Omega = -\frac{1}{T_{SVI}} = -\frac{1 + K_{EPS}K_{RTT}}{K_{EPS}(2E_n/f_n + K_{VI})} \tag{17.61}$$

From Eq. (17.61) it follows that in a system without a virtual inertia (i.e. when $K_{VI} = 0$) a decrease in kinetic energy E_n causes the transfer function pole to move on the s-plane of the variable (α, Ω) to the left, i.e. toward greater damping. This decrease can be compensated by virtual inertia, i.e. in the case when $K_{VI} \neq 0$. For example, if before the inertial generating unit tripping the energy stored in rotors is equal $E_{n(-)}$, and after tripping it decreases to $E_{n(+)}$, then system transfer functions (17.59) before and after the disturbance of the power balance should be equal, which in the considered case is met when $T_{SVI(-)} = T_{SVI(+)}$ or $\underline{\lambda}_{(-)} = \underline{\lambda}_{(+)}$, that is

$$\frac{2K_{EPS(-)}E_{n(-)}/f_n}{1 + K_{EPS(-)}K_{RTT(-)}} = \frac{K_{EPS(+)}(2E_{n(+)}/f_n + K_{VI})}{1 + K_{EPS(+)}K_{RTT(+)}} \tag{17.62}$$

This leads to the desired value of the gain of the artificial inertia system, equal to

$$K_{\mathrm{VI}} = \frac{K_{\mathrm{EPS}(-)}}{K_{\mathrm{EPS}(+)}}\frac{1+K_{\mathrm{EPS}(+)}K_{\mathrm{RTT}(+)}}{1+K_{\mathrm{EPS}(-)}K_{\mathrm{RTT}(-)}}\frac{2E_{\mathrm{n}(-)}}{f_{\mathrm{n}}} - \frac{2E_{\mathrm{n}(+)}}{f_{\mathrm{n}}} \tag{17.63}$$

Assuming that the EPS gains and the turbine regulator gain do not change, that is $K_{\mathrm{EPS}(-)} = K_{\mathrm{EPS}(+)}$ and $K_{\mathrm{RTT}(-)} = K_{\mathrm{RTT}(+)}$, Eq. (17.63) simplifies to the form

$$K_{\mathrm{VI}} = \frac{2\left(E_{\mathrm{n}(-)} - E_{\mathrm{n}(+)}\right)}{f_{\mathrm{n}}} = \frac{2\left(S_{\mathrm{n}(-)}H_{(-)} - S_{\mathrm{n}(+)}H_{(+)}\right)}{f_{\mathrm{n}}} = \frac{2\Delta E}{f_{\mathrm{n}}} \tag{17.64}$$

So the determined gain of the virtual inertia is equal to the gain resulting from Eq. (17.50).

If, on the other hand, a slightly more complex transfer function of the turbine control system is adopted, for example equal to $G_{\mathrm{T}}(s)G_{\mathrm{TG}}(s)G_{\mathrm{RD}}(s) = K_{\mathrm{RTT}}/(1 + sT_{\mathrm{RTT}})$, while keeping the virtual inertia transfer function equal to $G_{\mathrm{VI}}(s) = sK_{\mathrm{VI}}$ then it is obtained from Eq. (17.57) that

$$\frac{\Delta f}{\Delta P} = \frac{-K_{\mathrm{EPS}}(1+sT_{\mathrm{RTT}})}{1+K_{\mathrm{EPS}}K_{\mathrm{RTT}} + s(T_{\mathrm{RTT}}+T_{\mathrm{SI}}) + s^2T_{\mathrm{RTT}}T_{\mathrm{SI}}} \tag{17.65}$$

where $T_{\mathrm{SI}} = 2K_{\mathrm{EPS}}E_{\mathrm{n}}/f_{\mathrm{n}} + K_{\mathrm{VI}}$ is time constant of the EPS with the virtual inertia. In this case, the frequency response of the system may be aperiodic or oscillatory, which depends on the value of the system parameters. The transfer function poles are equal here to

$$\underline{\lambda} = -\frac{1}{2}\left[\frac{1}{T_{\mathrm{RTT}}} + \frac{1}{T_{\mathrm{SI}}} \pm \sqrt{\left(\frac{1}{T_{\mathrm{SI}}} - \frac{1}{T_{\mathrm{RTT}}}\right)^2 - \frac{4K_{\mathrm{EPS}}K_{\mathrm{RTT}}}{T_{\mathrm{RTT}}T_{\mathrm{SI}}}}\right] \tag{17.66}$$

The kinetic energy E_{n} stored in rotating masses of the EPS and the gain of the virtual inertia K_{VI} is a component of only the time constant T_{SI}. This means that changes of the EPS dynamic properties (resulting from the changes in inertia constant) can be compensated by a virtual inertia with K_{VI} determined by Eq. (17.64).

Example 17.8 It is assumed that in a power system with the parameters $R = 0.05$, $S_{\mathrm{nQ}} = 20$ GVA, $f_{\mathrm{n}} = 50$ Hz, $T_{\mathrm{RTT}} = 0.2$ s, and $K_{\mathrm{EPS}} = 0.003$, the inertia constant decreases from $H = 5$ s to 4 s, while the total rated power S_{nQ} does not change. In this case, the kinetic energy of rotating masses decreases from $E_{\mathrm{n}} = 5 \cdot 20 = 100$ GVAs to 80 GVAs, and the system time constant decreases from $T_{\mathrm{EPS}} = 12$ s to 9.6 s. One pole of transfer function (17.65) is constant and equal $\lambda_1 = -5\ \mathrm{s}^{-1}$, and the other is changing from $\lambda_2 = -0.0834\ \mathrm{s}^{-1}$ to $-0.1043\ \mathrm{s}^{-1}$, which means an improvement in damping. In order to preserve the dynamic properties of the system as for an inertia constant equal to $H = 5$ s, when it decreases to $H = 4$ s, the gain of the virtual inertia according to Eq. (17.64) should be equal $K_{\mathrm{VI}} = 2(100 - 80)/50 = 0.8\ \mathrm{GVAs}^2$. The same effect will be obtained for cases of more complex structures (higher-order transfer function) of the turbine control system, while maintaining the transfer function of the virtual inertia system as considered, i.e. $G_{\mathrm{VI}}(s) = sK_{\mathrm{VI}}$.

******

In general, using the virtual inertia, one can also adjust the dynamic properties of the system to the desired ones, and not only compensate for the inertia decrease. The general form of the desired transfer function of the virtual inertia, compensating for inertia changes, resulting from Eq. (17.57), describes the equation

$$G_{\mathrm{VI}}(s) = \frac{G_{\mathrm{EPS}(+)}(s) - G_{\mathrm{EPS}(-)}(s)}{G_{\mathrm{EPS}(+)}(s)G_{\mathrm{EPS}(-)}(s)} + G_{\mathrm{T}(-)}(s)G_{\mathrm{TG}(-)}(s)G_{\mathrm{RD}(-)}(s) - G_{\mathrm{T}(+)}(s)G_{\mathrm{TG}(+)}(s)G_{\mathrm{RD}(+)}(s) \tag{17.67}$$

where subscript (+) refers to a system with virtual inertia and subscript (−) to a system without virtual inertia. In the case of unchanged structure of turbine control systems the desired transfer function of the virtual inertia system results from the system transfer function in both states. For a system defined by transfer function $G_{EPS}(s) = K_{EPS}/(1 + sT_{EPS})$, when $K_{EPS(-)} = K_{EPS(+)} = K_{EPS}$, the required transfer function of virtual inertia is equal to

$$G_{VI}(s) = sK_{VI} = s\frac{2\left(E_{n(-)} - E_{n(+)}\right)}{f_n} \tag{17.68}$$

where the gain is equal to that resulting from Eq. (17.64).

The virtual inertia can also be treated as a technical facility enabling the change of the dynamic properties of the power system. In this case, the right side of the Eq. (17.57) should be compared with a certain transfer function $G_P(s)$ defining the desired dynamic properties of the system

$$G_P(s) = \frac{G_{EPS}(s)}{1 + G_{EPS}(s)[G_T(s)G_{TG}(s)(G_{RD}(s) + G_{AGC}(s)) + G_{VI}(s)]} \tag{17.69}$$

The sought for transfer function of the virtual inertia, or rather the transfer function of the additional power control system of selected generating units, is then equal to

$$G_{VI}(s) = \frac{G_{EPS}(s) - G_P(s)}{G_{EPS}(s)G_P(s)} - G_T(s)G_{TG}(s)(G_{RD}(s) + G_{AGC}(s)) \tag{17.70}$$

A separate issue in the case of transfer functions calculated as in Eqs. (17.67) or (17.70) is the transfer function feasibility. Only the proper and strictly proper transfer functions are realizable transfer functions.[6]

In fact, it is not possible to build an ideal derivative element. Such an element is approximated by real differentiator $s/(1 + sT)$. The lower the time constant T, the closer the real and the ideal differentiator responses are. Considerations of use of the virtual inertia defined by a real differentiator are presented below.

The adoption of the virtual inertia of transfer function $G_{VI}(s) = K_{VI}s/(1 + sT_{VI})$ increases the order of numerator and denominator polynomials of transfer function (17.57) and leads to the appearance of an additional eigenvalue. At the same time, it means that it is not possible to compensate precisely, as by using transfer function as in Eq. (17.68), for the inertia change.

For example, for an extremely simplified EPS model, as illustrated in Example 17.8, the virtual inertia defined by transfer function $G_{VI}(s) = sK_{VI}$ with gain equal to $K_{VI} = 800$ MVAs^2 fully compensates for the inertia change. The eigenvalues in such a system are equal to those shown in Example 17.8. In contrast, the use of a virtual inertia with transfer function $G_{VI}(s) = K_{VI}s/(1 + sT_{VI})$, while keeping the gain equal to $K_{VI} = 800$ MVAs^2, leads to the eigenvalues equal to: $\lambda_1 = -5$ s^{-1}, $\lambda_2 = -0.0833$ s^{-1}, and $\lambda_3 = -12.52$ s^{-1} for $T_{VI} = 0.1$ s, and $\lambda_1 = -5$ s^{-1}, $\lambda_2 = -0.0827$ s^{-1}, and $\lambda_3 = -2.52$ s^{-1} for $T_{VI} = 0.5$ s. This means that the increase in the value of the real differentiator time constant T_{VI} the virtual inertia shifts the poles of the transfer function (17.57) on the s-plane (α, Ω) to the right, i.e. it causes a decrease in the damping coefficient α.

Table 17.4 presents eigenvalues calculated for EPS from Example 17.2. The model contains the PI-type turbine regulator model and the model of a three-stage steam turbine with inter-stage reheating. The eigenvalues are calculated for EPS with inertia constant $H = 5$ s (named as basic), for EPS with inertia constant reduced to $H = 4$ s and for EPS with inertia constant $H = 4$ s and with a virtual inertia included.

There are two oscillatory and two non-oscillatory modes in the system without virtual inertia. As results from the presented data, a decrease in the inertia of the system leads to an increase in the damping coefficient of the weakest damped mode λ_3, as already discussed in the previous section. On the other hand, the inclusion of the virtual inertia with the gain resulting from the Eq. (17.64) causes the oscillatory mode to shift in the direction of the mode obtained in the system with inertia constant $H = 5$ s. This gives almost complete restoration of the EPS dynamic properties. The value of the damping ratio of the oscillatory mode, defined by eigenvalue λ_3 in the system with

6 A proper transfer function is a transfer function in which the degree of the numerator does not exceed the degree of the denominator. A strictly proper transfer function is a transfer function where the degree of the numerator is less than the degree of the denominator.

Table 17.4 The influence of virtual inertia on eigenvalues λ_i in EPS from Example 17.2.

Mode no.	λ_i in EPS with $H = 5$ s	λ_i in EPS with $H = 4$ s	λ_i in EPS with $H = 4$ s and virtual inertia with $T_{VI} = 0.1$ s, $K_{VI} = 800$ MVAs^2
1	−2.099	−2.133	−2.092
2	$-3.688 \pm j0.086$	$-3.644 \pm j0.146$	$-3.071 \pm j0.028$
3	$-0.146 \pm j0.397$	$-0.176 \pm j0.429$	$-0.146 \pm j0.396$
4	−0.257	−0.268	−0.256
5	—	—	−12.49

virtual inertia and in the basic one differs by 0.2% only. The biggest difference is related to the second oscillatory mode, defined by eigenvalue λ_2. The mode damping ratio is higher in EPS with virtual inertia, while the damping coefficient of both modes is close to 1.

Further decreasing the time constant of the virtual inertia moves the eigenvalues of EPS with virtual inertia to the values of the basic system. For $T_{VI} = 0.1$ s the eigenvalues in EPS with reduced inertia constant to $H = 4$ s becomes practically equal to the one as in EPS with $H = 5$ s.

The influence of virtual inertia on the frequency oscillations, presented above on simple system models, qualitatively and quantitatively corresponds to phenomena occurring in real EPSs. This can be also confirmed by the analyzes carried out on the basis of more advanced mathematical models (Chapter 11) of generating units and their controllers and detailed model of the transmission network.

However, real virtual inertia schemes are not able to fully compensate for changes in EPS dynamic properties resulting from the change of this inertia in the EPS. This results, in addition to the nonideal structure of the virtual inertia scheme, largely from the fact that the reduction in inertia in the EPS is associated with elimination, i.e. the decreasing number of inertial generator units with its regulators. The virtual inertia allows only to effectively influence the process of frequency control after disturbance of the active power balance in the system. Efficiency of the virtual inertia schemes applied to an RES (here understood as wind and PV power plants) also depends strongly on the operating point and mode of operation and the type of disturbance, and the direction of frequency change. In the case of decreasing the frequency and the RES operation without a power reserve (i.e. with the load resulting from wind speed in the case of wind power plants or insolation in the case of PV power plants) the ability to increase the generated power is extremely limited or impossible. In order to obtain the ability to affect the frequency, in the case of its decreasing, it is required that these sources should operate with a power reserve. Then it is possible to quickly inject this additional active power into the network. In the case of disturbances in the system leading to an increase in frequency, these sources are able to limit the generated active power very quickly. This asymmetry of the virtual inertia operation leads to its smaller (than potentially possible) effect on power system frequency stability (Table 17.2).

As an illustration of the above, Figure 17.20 shows the response of the 39 Bus New England System model to the disturbance of active power balance caused by load rejection. In this system, all modeled generating units (WTGSs with a DFIG, FRC, and PV power plant) were equipped with a virtual inertia with transfer function $G_{VI}(s) = K_{VI}s/(1 + sT_{VI})$. The gains and time constants for the three RES were the same.

The three RES have been operating close to their rated power. Therefore, the potential emergency tripping off of the synchronous generator, leading to a drop in frequency in the EPS, does not lead to effective support of the power system by the RES. This is due to the fact that they operate without a reserve of active power. A different situation occurs in the case of disturbance of active power balance leading to an increase in frequency. Then, the virtual inertia of RES operate toward reducing the generated active power. Figure 17.20 shows the RES

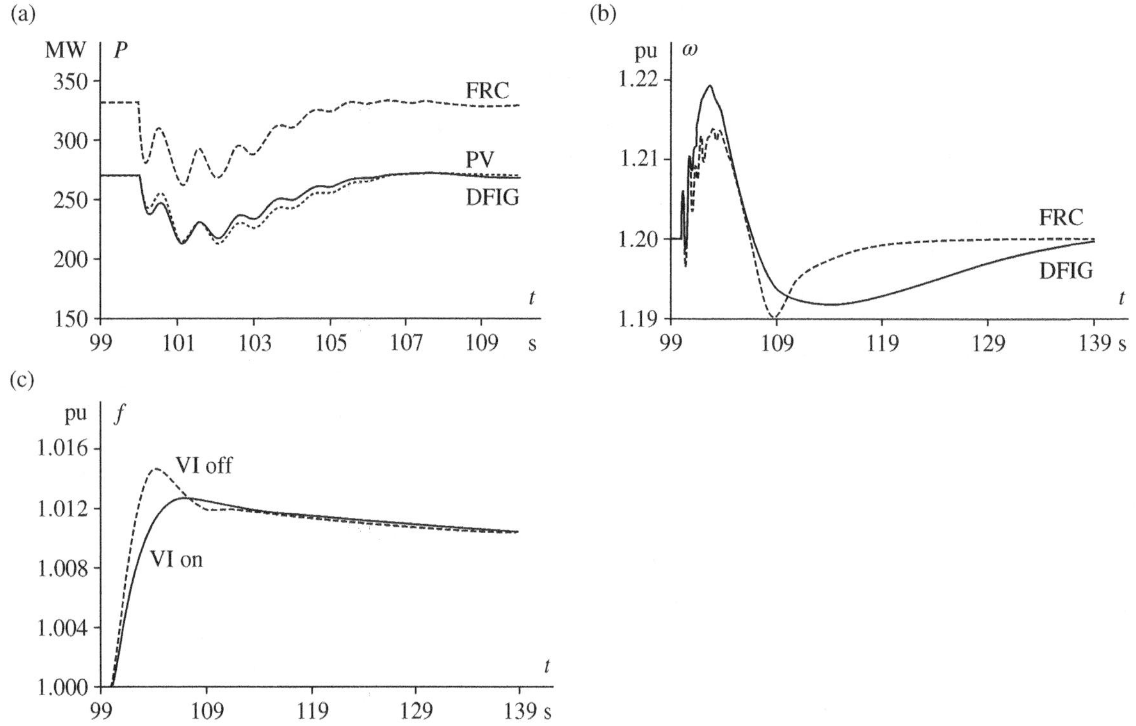

Figure 17.20 Response of RES with a virtual inertia to a 515 MW load rejection at node B8: (a) active power; (b) angular velocity of a DFIG and FRC rotors; (c) frequency in EPS (average angular speed of synchronous generator rotors).

response to the rejection of large load 515 MW. The change in the active power generated by these three RES (Figure 17.20a) in response to a change in the frequency is qualitatively and quantitatively similar. It results from the same structure and parameters of the virtual inertia and a similar control schemes. The control algorithms of a GSIR of a WTGS with an FRC and PV plant are almost identical (Sections 11.4.4.1, 11.5). This also applies to control algorithm of a GSIR of a WTGS with a DFIG, despite the control is carried out on the rotor side of the induction machine. The delay brought by the induction machine is negligible in relation to the RES connected to the EPS by the power electronic converter.

Figure 17.20b shows the speed of wind turbines (WTGSs rotors). In the first stage of the transient state the reduction of the active power injected to the grid by a DFIG and an FRC generating units leads to an increase in the speed of their rotors, which accumulate part of excess energy. In the second stage of the transient state, this excessive energy stored in the rotors is returned to the system. This process occurs with an undershooting of the rotors speed below the steady-state speed. Thus, in the presented example, the process of supporting the EPS by the RES takes about 5 s, and then the generating units in time about 30 s rebuild their steady-state, injecting to the system a small (relative to the steady state) surplus power. This second part of the transient state is against the frequency control need, i.e. the need to limit the active power being injected into the system when the frequency is higher than the nominal one.

In turn, Figure 17.20c shows the frequency response, and more precisely the average angular velocity of the rotors of synchronous generators. The response marked as "VI off" refers to the model in which the RES operate without virtual inertia, and the response marked as "VI on" refers to the system with RES operating with virtual inertia. Since the load rejection was simulated here, the disturbance did not change the inertia of the EPS. This causes the frequency response of the system with virtual inertia to be different from without it. Virtual inertia systems cause a RES response that limits the active power generated and thus reduces the rate of change of frequency (*RoCoF*) and limits maximum deviation of frequency.

17.3.3 Virtual Inertia and Dynamics of Interconnected Systems

The simplified model of two interconnected subsystems A and B with virtual inertia is shown in Figure 17.21. Recall that a similar model, but without virtual inertia, is discussed in Section 9.6.2. In the considered case the virtual inertia is modeled by transfer functions $G_{VI}^{A}(s)$ and $G_{VI}^{B}(s)$. The connection of both subsystems is modeled by transfer functions

$$\alpha(s) = \frac{2\pi}{s}H_{ab}(\kappa_{aa} - \kappa_{ba}) \quad \text{and} \quad \beta(s) = \frac{2\pi}{s}H_{ab}(\kappa_{ab} - \kappa_{bb}) \tag{17.71}$$

where H_{ab} is the synchronizing power and κ_{aa}, κ_{ba}, κ_{ab}, and κ_{bb} are the connection coefficients described in Section 9.6.2. The synchronizing power (Eq. [9.77]) is a function of the tie line impedance $\underline{Z}_{ab}$, the voltage at its ends $\underline{V}_a$ and $\underline{V}_b$, and the angle between phasors of the voltages.

To simplify expressions presented below, it is assumed that

$$\begin{aligned} A(s) &= 1 + G_{RV}^{A}(s)G_{EPS}^{A}(s) + \alpha(s)G_{EPS}^{A}(s) \\ B(s) &= 1 + G_{RV}^{B}(s)G_{EPS}^{B}(s) + \beta(s)G_{EPS}^{B}(s) \end{aligned} \tag{17.72}$$

where $G_{RV}^{A}(s)$ and $G_{RV}^{B}(s)$ are sums of transfer function of turbine control systems (turbine with governor) and transfer functions modeling the virtual inertia

$$\begin{aligned} G_{RV}^{A}(s) &= G_{T}^{A}(s)G_{RG}^{A}(s)G_{RD}^{A}(s) + sK_{VI}^{A} \\ G_{RV}^{B}(s) &= G_{T}^{B}(s)G_{RG}^{B}(s)G_{RD}^{B}(s) + sK_{VI}^{B} \end{aligned} \tag{17.73}$$

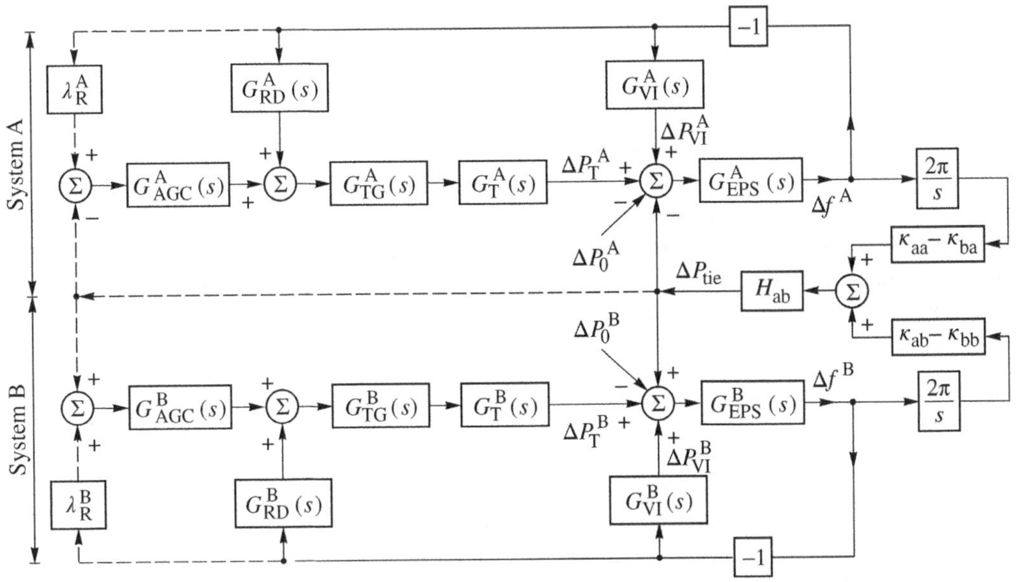

Figure 17.21 Block diagram of the simplified model of two connected subsystems with frequency control and a virtual inertia.

For such defined systems, the relations for both subsystems, in the form of transfer function with power as input and frequency as output have the following form

$$\begin{aligned} \Delta f_{\mathrm{A}} &= -\frac{G_{\mathrm{EPS}}^{\mathrm{A}}\, B}{AB + \alpha\beta\, G_{\mathrm{EPS}}^{\mathrm{A}} G_{\mathrm{EPS}}^{\mathrm{B}}} \Delta P_0^{\mathrm{A}} - \frac{\beta\, G_{\mathrm{EPS}}^{\mathrm{B}}}{AB + \alpha\beta\, G_{\mathrm{EPS}}^{\mathrm{A}} G_{\mathrm{EPS}}^{\mathrm{B}}} \Delta P_0^{\mathrm{B}} \\ \Delta f_{\mathrm{B}} &= -\frac{\alpha\, G_{\mathrm{EPS}}^{\mathrm{A}}}{AB + \alpha\beta\, G_{\mathrm{EPS}}^{\mathrm{A}} G_{\mathrm{EPS}}^{\mathrm{B}}} \Delta P_0^{\mathrm{A}} - \frac{A\, G_{\mathrm{EPS}}^{\mathrm{B}}}{AB + \alpha\beta\, G_{\mathrm{EPS}}^{\mathrm{A}} G_{\mathrm{EPS}}^{\mathrm{B}}} \Delta P_0^{\mathrm{B}} \end{aligned} \tag{17.74}$$

For clarity, the Laplace operator s in symbols of transfer functions is neglected here. The change in the tie line power (i.e. change in the power flow between subsystems) is equal to

$$\Delta P_{\mathrm{tie}} = \alpha\, \Delta f_{\mathrm{A}} + \beta\, \Delta f_{\mathrm{B}} \tag{17.75}$$

After substitution of Eq. (17.74) to Eq. (17.75), the tie line power deviation as a function of disturbance of active power balance in subsystems is defined by a transfer function in form

$$\Delta P_{\mathrm{tie}} = -\frac{\alpha\,(B+\beta) G_{\mathrm{EPS}}^{\mathrm{A}}}{AB + \alpha\beta\, G_{\mathrm{EPS}}^{\mathrm{A}} G_{\mathrm{EPS}}^{\mathrm{B}}} \Delta P_0^{\mathrm{A}} - \frac{\beta\,(\alpha + A) G_{\mathrm{EPS}}^{\mathrm{B}}}{AB + \alpha\beta\, G_{\mathrm{EPS}}^{\mathrm{A}} G_{\mathrm{EPS}}^{\mathrm{B}}} \Delta P_0^{\mathrm{B}} \tag{17.76}$$

Equations (17.74) and (17.76) show that the virtual inertia modifies the components of transfer functions $A(s)$ and $B(s)$ given by Eq. (17.72). The change in active power (i.e. the disturbance in the active power balance) in one subsystem is transferred to the second one. The quantitative effect for a given subsystem depends on parameters of the transfer function $\alpha(s)$ and $\beta(s)$, i.e. depends on synchronizing power H_{ab} and total rated power of each subsystem.

Transfer functions modeling virtual inertias are elements of transfer functions $G_{\mathrm{RV}}^{\mathrm{A}}(s)$ and $G_{\mathrm{RV}}^{\mathrm{B}}(s)$ and occur in the numerator and denominator of transfer functions (17.74) and (17.76). Thus, they modify the dynamic properties of both connected subsystems. The following conclusions can be formulated:

1) In the case of reduction in inertia in each of the subsystems, the restoration of its value should be performed independently in each of them. In this case (i.e. when virtual inertias are present in both subsystems) conditions should be fulfilled relating to transfer functions defined by Eq. (17.72), in form $A_{(-)}(s) = A_{(+)}(s)$ and $B_{(-)}(s) = B_{(+)}(s)$, that is

$$\begin{aligned} G_{\mathrm{RV}(-)}^{\mathrm{A}}(s) G_{\mathrm{EPS}(-)}^{\mathrm{A}}(s) + \alpha(s) G_{\mathrm{EPS}(-)}^{\mathrm{A}}(s) &= G_{\mathrm{RV}(+)}^{\mathrm{A}}(s) G_{\mathrm{EPS}(+)}^{\mathrm{A}}(s) + \alpha(s) G_{\mathrm{EPS}(+)}^{\mathrm{A}}(s) \\ G_{\mathrm{RV}(-)}^{\mathrm{B}}(s) G_{\mathrm{EPS}(-)}^{\mathrm{B}}(s) + \beta(s) G_{\mathrm{EPS}(-)}^{\mathrm{B}}(s) &= G_{\mathrm{RV}(+)}^{\mathrm{B}}(s) G_{\mathrm{EPS}(+)}^{\mathrm{B}}(s) + \beta(s) G_{\mathrm{EPS}(+)}^{\mathrm{B}}(s) \end{aligned} \tag{17.77}$$

where subscript (−), like in Section 17.3.2, refers to both subsystems before the change of inertia and subscript (+) to both subsystems just after the change of inertia and the inclusion of virtual inertias. Equation (17.77), assuming the invariance of subsystems structures and parameters and invariance of turbines and governors, lead to a known condition, i.e. Eq. (17.64), defining the requirements for the value of the virtual inertia gain in both subsystems

$$K_{\mathrm{VI}}^{\mathrm{A}} = \frac{2\left(E_{\mathrm{n}(+)}^{\mathrm{A}} - E_{\mathrm{n}(-)}^{\mathrm{A}}\right)}{f_{\mathrm{n}}} \quad \text{and} \quad K_{\mathrm{VI}}^{\mathrm{B}} = \frac{2\left(E_{\mathrm{n}(+)}^{\mathrm{B}} - E_{\mathrm{n}(-)}^{\mathrm{B}}\right)}{f_{\mathrm{n}}} \tag{17.78}$$

2) The inclusion of virtual inertia in one subsystem, in order to compensate for the change of inertia in this subsystem (with the structure of the other subsystem unchanged), affects the dynamic properties of both systems. The required value of the virtual inertia gain results from Eq. (17.77) for a given subsystem and is equal to the value given by Eq. (17.64) or (17.78). The effect obtained here depends on the difference in size between the two subsystems. The virtual inertia in general makes shift of the eigenvalues of the selected oscillatory modes on the

s-plane (α, Ω) to the right (Example 17.9 below). This means that overuse of the virtual inertia (in sense of power injected) can lead to a worsening of the EPS dynamic response.

3) The inclusion of virtual inertia in one subsystem, to the extent different from compensating for the inertia change in the subsystem, modifies the dynamic properties of both subsystems operating synchronously. The quantitative effect depends here on the power provided by the virtual inertia, i.e. the virtual inertia gain, as well as on the size of both subsystems, the impedance of the connection between them and the operating point (including the tie lines power). The weaker the subsystems are connected to one another, the smaller is the effect of introducing additional active power by virtual inertia in a given subsystem into the second one (or for other subsystems if more than two of them are interconnected).

Transfer functions (17.74) and (17.76) even in extremely simple forms of the transfer function of constituent elements have a complex form, in which practically all polynomial coefficients of the numerator and the denominator (and thus the characteristic equation) are functions of gains of virtual inertia K_{VI}^{A} and K_{VI}^{B}, energies stored in rotating masses E_{n}^{A} and E_{n}^{B}, synchronizing power H_{ab}, etc. Therefore, the influence of a given parameter on frequency or small signal stability cannot be represented as a simple function. Therefore, the selected calculation results obtained for exemplary, synchronously operating subsystems are presented below.

Example 17.9 The model of two subsystems, as shown in Figure 17.21, is considered. Subsystem A has parameters: $\rho^{A} = 0.05$, $S_{n}^{A} = 20$ GVA, $P_{g}^{A} = 15.3$ GW, $P_{L}^{A} = 20.3$ GW, $P_{PV}^{A} = 5$ GW, $K_{L}^{A} = 1.1$, $H^{A} = 5$ s, and $f_{n} = 50$ Hz. Inertia constant $H^{A} = 5$ s is an initial value of inertia of system A only and further is subject to change. Subsystem B is 10 times bigger than subsystem A and has parameters: $\rho^{B} = 0.05$, $S_{n}^{B} = 200$ GVA, $P_{g}^{B} = 153$ GW, $P_{L}^{B} = 203$ GW, $P_{PV}^{B} = 50$ GW, $K_{L}^{B} = 1.1$, and $H^{B} = 5$ s. The synchronizing power is equal to $H_{ab} = -0.76$ GW/deg. For the above data, the transfer functions parameters $G_{EPS}^{A}(s)$ and $G_{EPS}^{B}(s)$ of subsystems A and B are equal to $K_{EPS}^{A} = 0.0022$, $T_{EPS}^{A} = 8.96$ s, $K_{EPS}^{B} = 0.00022$, and $T_{EPS}^{B} = 8.96$ s. Virtual inertias in both subsystems are modeled as $G_{VI}(s) = K_{VI}s/(1 + sT_{VI})$, with gains and time constants equal to $K_{VI}^{A} = 800$ MVAs^2, $K_{VI}^{B} = 8000$ MVAs^2, and $T_{VI}^{A} = T_{VI}^{B} = 0.1$ s. The gains are calculated to compensate the inertia change equal to $\Delta H^{A} = \Delta H^{B} = 5 - 4 = 1$ s in each of the subsystems.

Figure 17.22 shows the influence of the inertia H^{A} of subsystem A on the parameters characterizing the frequency control in subsystem A, i.e. maximum frequency deviation Δf_{max}^{A} and rate of change of frequency $RoCoF_{max}^{A}$, and the tie line power maximum deviation $\Delta P_{tie\ max} = \Delta P_{ab\ max}$ after active power balance disturbance in subsystem A, equal to $\Delta P_{0}^{A} = 0.5$ GW. Inertia of subsystem B is constant and equal to $H^{B} = 5$ s. These values are presented for the subsystem A with (VI on) and without (VI off) virtual inertia only in subsystem A. Virtual inertia in subsystem B is not included. As can be seen from the comparison of the curves in the figures, the virtual inertia reduces the maximum frequency deviation Δf_{max}^{A}, decreases the maximum initial value of $RoCoF_{max}^{A}$, and decreases the maximum deviation of tie line power $\Delta P_{tie\ max}$. The last means that the virtual inertia applied in subsystem A decreases the support of subsystem A by subsystem B in comparison to the variant without the virtual inertia. Owing to the parameters of the connection between the subsystems and the difference in their size, the maximum frequency deviation Δf_{max}^{B} in subsystem B, for the considered disturbance $\Delta P_{0}^{A} = 0.5$ GW in subsystem A, did not exceed -0.003 Hz, and the maximum value of $RoCoF_{max}^{B}$ has not exceeded -0.002 Hz/s.

The considered parameters of frequency response shows also the ability of virtual inertia to restore dynamic properties of the interconnected subsystems. The deviation of frequencies $\Delta f_{max\ VI}^{A}$ in subsystem A with a constant inertia $H^{A} = 4$ s and the virtual inertia on and gain equal to $K_{VI}^{A} = 800$ MVAs^2 is practically the same as the frequency deviation Δf_{max}^{A} in subsystem A, with inertia equal to $H^{A} = 5$ s (Figure 17.22a). The same effect is visible for the initial values of $RoCoF_{max}^{A}$ and $RoCoF_{max\ VI}^{A}$ and also for the maximum values of tie line power $\Delta P_{tie\ max}$ and $\Delta P_{tie\ max\ VI}$. The influence of the use of the real differentiator in the virtual inertia scheme is here very small.

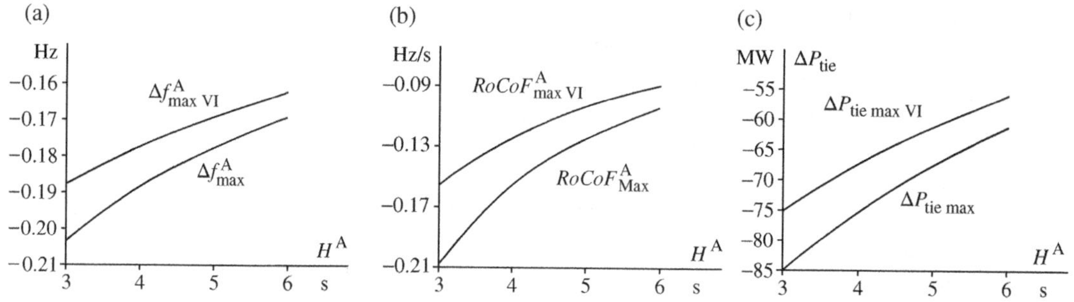

Figure 17.22 Influence of virtual inertia H^{A} of subsystem A on (a) and (b) parameters of frequency response of subsystem A and (c) maximum value of tie line power deviation after disturbance of active power balance in subsystem A equal to $\Delta P_0^{A} = 0.5$ GW.

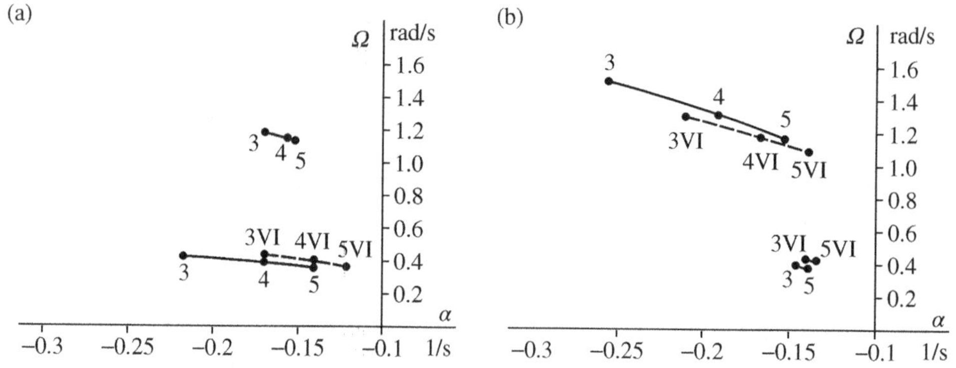

Figure 17.23 Loci of eigenvalues corresponding to the worst damped oscillatory modes in the case of change in inertia: (a) in small subsystem A with large subsystem inertia constant $H^{B} = 5$ s; (b) in large subsystem B with small subsystem inertia constant $H^{A} = 5$ s

Figure 17.23 shows the loci of eigenvalues corresponding to two worst damped oscillatory modes. Figure 17.23a shows the effect of decreasing inertia in the small subsystem A, with constant and equal to $H^{B} = 5$ s inertia constant in subsystem B. The numbers below the dots correspond to the inertia constant values in subsystem A without virtual inertia. One can see here that the decrease of the inertia constant H^{A} leads to a shift of the eigenvalues to the left, which means improvement of the modes' damping. This effect is shown in Section 17.2.2. The inertia change in subsystem A mainly influences the mode characterized by pulsation 0.4 rad/s. Influence of the inertia constant H^{A} on the second mode (with pulsation 1.2 rad/s) is much smaller.

Application of virtual inertia in subsystem A shifts the eigenvalues' trajectories to the right. This is shown by dashed line for mode with pulsation 0.4 rad/s. Points related to a given inertia constant H^{A} in the system with virtual inertia are marked by letters VI. Number before the letters VI define value of the subsystem A inertia constant.

The same effect, i.e. the trajectory shifts to the right, is related to the mode with pulsation 1.2 rad/s. In this case trajectory for inertia H^{A} from 3 to 5 s is located between the middle and right-hand dot.

One can see here that points marked as 5 (means $H^{A} = 5$ s) and 4VI (means $H^{A} = 4$ s and virtual inertia on) are located almost on the same place. This means that the virtual inertia with a given gain, applied in subsystem A compensates the inertia change in subsystem A for the lower frequency mode. Practically, the same effect takes place for the higher frequency mode.

In contrast, the curves presented in Figure 17.23b show the effect of inertia change in large subsystem B. The inertia constant in smaller subsystem A is kept constant and equal to $H^{A} = 5$ s. The virtual inertia in subsystem A is not included. As shown in the figure, the inertia of subsystem B has a greater impact on the higher pulsation mode and small on the lower pulsation mode. This is opposite to that presented in Figure 17.23a for inertia change in subsystem A. The virtual inertia application in subsystem B (dashed line with points with description VI) shifts the eigenvalues with higher pulsation to the right slightly decreasing its pulsation. One can see here that points marked as 5 and 4VI are not located at the same point. This means that the virtual inertia included in subsystem B with gain $K_{VI}^{B} = 8000$ MVAs^2 does not fully compensate change of inertia constant in subsystem B. This can suggest that the assumed gain of the virtual inertia in subsystem B is too small and that the full compensation can be achieved by calculating the gain using data of both interconnected subsystems. This would shift point marked as 4VI to the right, i.e. closer to point marked as 5, which means better compensation of the inertia decrease.

******

18

Power System Model Reduction – Equivalents

Because contemporary power systems are so large, power system analysis programs do not usually model the complete system in detail. This problem of modeling a large system arises for a number of reasons, including:

- Practical limitations on the size of computer memory.
- The excessive computing time required by large power systems, particularly when running dynamic simulation and stability programs.
- Parts of the system far away from a disturbance have little effect on the system dynamics and it is therefore unnecessary to model them with great accuracy.
- Often parts of large interconnected systems belong to different utilities, each having its own control center which treats the other parts of the system as external subsystems.
- In some countries private utilities compete with each other and are reluctant to disclose detailed information about their business. This means that vital data regarding the whole system may not be available.
- Even assuming that full system data are available, maintaining the relevant data bases would be very difficult and expensive.

To avoid all these problems, only a part of the system, called the *internal subsystem*, is modeled in detail. The remainder of the system, called the *external subsystem*, is represented by simple models referred to as the *equivalent system* or simply as the *equivalent*.

18.1 Types of Equivalents

The methods by which the equivalent of an external subsystem can be produced can be broadly divided into two groups, depending on whether they require any knowledge of the configuration and parameters of the external subsystem itself. Methods that do not require any knowledge of the external subsystem are used for online security assessment and will not be considered further here, but details of these methods can be found in Dopazo et al. (1977), Contaxis and Debs (1978), and Feng et al. (2007). Typically, these methods use the measurement of certain electrical quantities taken inside the internal subsystem and at the border nodes and tie-lines to form the equivalent. Methods that do require knowledge of the subsystem are called *model reduction methods*. These methods are used for offline system analysis and are the subject of this chapter.

Model reduction methods can be further divided into the following three groups:

- *Physical reduction*, which consists of choosing appropriate models for the system elements (generators, loads, etc.) depending on how influential an individual element is in determining the system response to a particular disturbance. Generally, elements electrically close to the disturbance are modeled more accurately, while elements further away are modeled more simply.

- *Topological reduction* which consists of eliminating and/or aggregating selected nodes in order to reduce the size of the equivalent network and the number of generating units modeled.
- *Modal reduction* techniques which use linearized models of the external subsystem that eliminate, or neglect, the unexcited modes.

The equivalent model obtained using modal reduction is in the form of a reduced set of linear differential equations (Undrill and Turner 1971). This requires extending the standard power system software to take into account the special form of the equivalent. As standardization of the software is difficult to achieve with modal reduction, this type of equivalent is rarely used in practice.

Topological reduction, used together with physical reduction, gives an equivalent model that comprises standard system elements, such as equivalent generating units, equivalent lines, equivalent nodes, and so on. Consequently, topological equivalents are easy to attach to the internal subsystem model and allow the whole system to be analyzed using standard software.

If the topological reduction is performed using one of the methods described in this chapter then the reduced model obtained will generally be a good representation of both the system static performance and the system dynamic performance for the first few seconds following a disturbance. The reduced model can therefore be used for load flow analysis and transient stability analysis when disturbances occur in the internal subsystem.

The division of the whole system into external and internal components is illustrated in Figure 18.1. A reduced model of the external subsystem is created assuming that the disturbance occurs only inside the internal subsystem. The *border nodes* between the internal and external subsystems are sometimes referred to as the *boundary nodes* or *tearing nodes*. Topological reduction consists of transforming a large external network that consists of load nodes and/or generation nodes into a smaller network by eliminating and/or aggregating the nodes. Eliminated nodes are removed completely from the network while every group of aggregated nodes is replaced by one equivalent node.

18.2 Network Transformation

Topological reduction is achieved by transforming a large network into a smaller equivalent network by either elimination or aggregation of nodes.

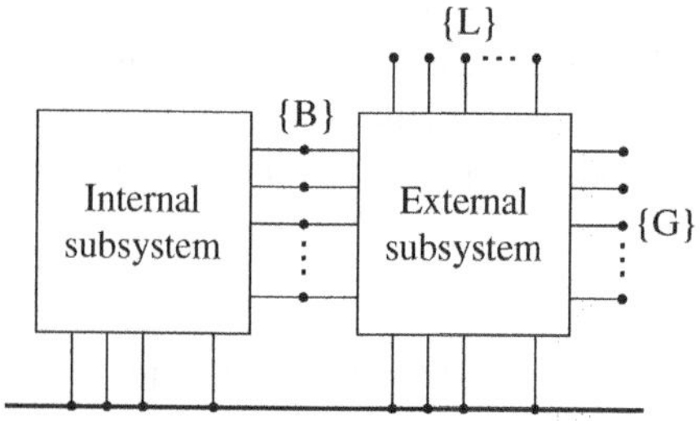

Figure 18.1 Internal and external subsystems: {B}; boundary nodes; {L}; load nodes; {G}; generator nodes of the external subsystem.

18.2.1 Elimination of Nodes

Figure 18.2 illustrates that when nodes are eliminated from the network model, set {E}, they must be removed in such a way that the currents and nodal voltages at the retained nodes, set {R}, are unchanged.

Before any nodes are eliminated the network is described by the following nodal equation (see Section 3.5)

$$\begin{bmatrix} \underline{I}_{\mathrm{R}} \\ \underline{I}_{\mathrm{E}} \end{bmatrix} = \begin{bmatrix} \underline{Y}_{\mathrm{RR}} & \underline{Y}_{\mathrm{RE}} \\ \underline{Y}_{\mathrm{ER}} & \underline{Y}_{\mathrm{EE}} \end{bmatrix} \begin{bmatrix} \underline{V}_{\mathrm{R}} \\ \underline{V}_{\mathrm{E}} \end{bmatrix} \tag{18.1}$$

where the subscripts refer to the eliminated {E} and retained {R} sets of nodes. The eliminated voltages and currents can be swapped using simple matrix algebra to give

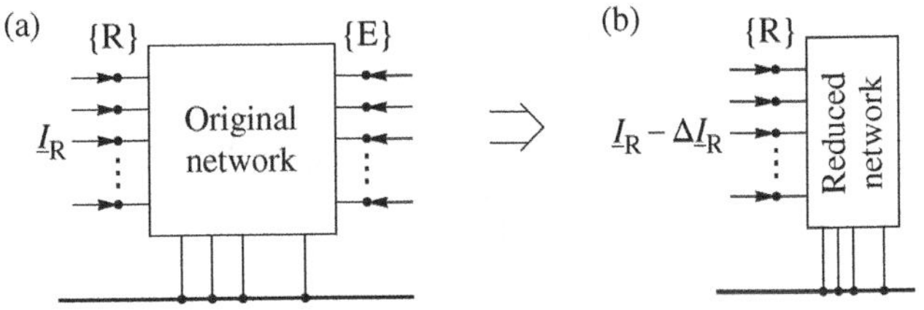

Figure 18.2 Elimination of nodes: (a) network before elimination; (b) network after elimination; {E} – set of eliminated nodes; {R} – set of retained nodes.

$$\begin{bmatrix} \underline{\boldsymbol{I}}_R \\ \underline{\boldsymbol{V}}_E \end{bmatrix} = \begin{bmatrix} \underline{\boldsymbol{Y}}_R & \underline{\boldsymbol{K}}_I \\ \underline{\boldsymbol{K}}_V & \underline{\boldsymbol{Y}}_{EE}^{-1} \end{bmatrix} \begin{bmatrix} \underline{\boldsymbol{V}}_R \\ \underline{\boldsymbol{I}}_E \end{bmatrix} \tag{18.2}$$

where

$$\underline{\boldsymbol{Y}}_R = \underline{\boldsymbol{Y}}_{RR} - \underline{\boldsymbol{Y}}_{RE}\underline{\boldsymbol{Y}}_{EE}^{-1}\underline{\boldsymbol{Y}}_{ER}, \quad \underline{\boldsymbol{K}}_I = \underline{\boldsymbol{Y}}_{RE}\underline{\boldsymbol{Y}}_{EE}^{-1}, \quad \underline{\boldsymbol{K}}_V = -\underline{\boldsymbol{Y}}_{EE}^{-1}\underline{\boldsymbol{Y}}_{ER} \tag{18.3}$$

The square matrix in Eq. (18.2) is the *partial inversion of the admittance matrix* and is described in detail in Appendix A.2. The nodal currents in the set {R} are

$$\underline{\boldsymbol{I}}_R = \underline{\boldsymbol{Y}}_R\underline{\boldsymbol{V}}_R + \Delta\underline{\boldsymbol{I}}_R \tag{18.4}$$

where

$$\Delta\underline{\boldsymbol{I}}_R = \underline{\boldsymbol{K}}_I\underline{\boldsymbol{I}}_E$$

Equation (18.4) describes the relationship between the currents and voltages of the retained nodes in the reduced network. As any electrical network is uniquely described by its admittance matrix, the matrix $\underline{\boldsymbol{Y}}_R$ corresponds to a reduced equivalent network that consists of the retained nodes and equivalent branches linking them. This network is often referred to as the *transfer network* and the matrix describing it as the *transfer admittance matrix*. Matrix $\underline{\boldsymbol{K}}_I$ passes the nodal currents from the eliminated nodes to the retained nodes and is referred to as the *distribution matrix*. Each equivalent current is a combination of the eliminated currents.

Another form of Eq. (18.4) can be obtained by replacing the nodal power injection at each eliminated node by a constant shunt admittance $\underline{Y}_{Ei} = \underline{S}_i^*/V_i^2$ added, with an appropriate sign, to the diagonal elements of the submatrix $\underline{\boldsymbol{Y}}_{EE}$ (and to the network diagram as a shunt connection). The nodal injections at the eliminated nodes then become zero ($\underline{\boldsymbol{I}}_E = \boldsymbol{0}$) and the reduced model does not contain any equivalent currents ($\Delta\underline{\boldsymbol{I}}_R = \boldsymbol{0}$). This is quite convenient but has a drawback. The equivalent shunt branches have large conductance values, corresponding to the active power injections, which become part of the equivalent branches in the reduced model. Consequently, the branches of the equivalent network may have a poor X/R ratio causing convergence problems for some load flow computer programs.

Different authors give different names to the elimination of the network nodes using Eqs. (18.2) and (18.4). Edelmann (1963) refers to it as *Gauss–Rutishauser elimination*, while Brown (1975) and Grainger and Stevenson (1994) call the reduced circuit a *Ward equivalent*.

18.2.1.1 Sparse Matrix Techniques

Equation (18.4) formally describes the elimination algorithm. In practice sparse matrix techniques are used and the nodes are processed one at a time in order to minimize the complexity and memory requirements of the elimination process (Tewerson 1973; Brameller et al. 1976). This is equivalent to Gaussian elimination of a corresponding row and column from the admittance matrix.

Consider one elimination step, namely that of eliminating node k of set {E}. Matrix $\underline{\boldsymbol{Y}}_{EE} = \underline{Y}_{kk}$ is a scalar, $\underline{\boldsymbol{Y}}_{RE}$ is a column, and $\underline{\boldsymbol{Y}}_{ER}$ is a row. The second component of matrix $\underline{\boldsymbol{Y}}_R$ becomes

$$\underline{\mathbf{Y}}_{\text{RE}}\underline{\mathbf{Y}}_{\text{EE}}^{-1}\underline{\mathbf{Y}}_{\text{ER}} = \frac{1}{\underline{Y}_{kk}}\begin{bmatrix}\underline{Y}_{1k}\\ \vdots \\ \underline{Y}_{ik} \\ \vdots \\ \underline{Y}_{nk}\end{bmatrix}\begin{bmatrix}\underline{Y}_{k1} & \cdots & \underline{Y}_{kj} & \cdots & \underline{Y}_{kn}\end{bmatrix} = \frac{1}{\underline{Y}_{kk}}\begin{bmatrix} & \vdots & \\ \cdots & \underline{Y}_{ik}\underline{Y}_{kj} & \cdots \\ & \vdots & \end{bmatrix} \begin{matrix} \\ i \\ \\ \end{matrix} \tag{18.5}$$
$$\qquad\qquad\qquad\qquad\qquad\qquad\qquad\qquad\qquad j$$

where n is the number of nodes in set {R}. Assume now that $\underline{Y}_{ij}^{\text{old}}$ is an element of matrix $\underline{\mathbf{Y}}_{\text{RR}}$ while $\underline{Y}_{ij}^{\text{new}}$ is an element of matrix $\underline{\mathbf{Y}}_{\text{R}}$. Equations (18.3) and (18.5) show that elimination of node k modifies each element of the "new" matrix $\underline{\mathbf{Y}}_{\text{R}}$ to

$$\underline{Y}_{ij}^{\text{new}} = \underline{Y}_{ij}^{\text{old}} - \frac{\underline{Y}_{ik}\underline{Y}_{kj}}{\underline{Y}_{kk}} \quad \text{for} \quad i \neq k,\ j \neq k \tag{18.6}$$

If node i is directly connected to the eliminated node k then it is called its *neighbor* and the corresponding mutual admittance is $\underline{Y}_{ik} \neq 0$. The mutual admittance's $\underline{Y}_{ik}$ of a node i which is not a neighbor of node k are all zero. Equation (18.6) shows that:

- If nodes i and j are not neighbors of node k then elimination of k does not modify the admittance $\underline{Y}_{ij}$.
- Elimination of node k modifies the admittance between all its neighbors, which creates additional connections between the neighbors replacing the original connections of node k.
- Self-admittances of all the neighbors of node k are also modified according to Eq. (18.6) when $i = j$.

Both situations are illustrated in Figure 18.3. Nodes {1, 2, 3} are the neighbors of node k so that its elimination creates additional connections between the nodes. Nodes {4, 5} are not neighbors of node k and their connections do not change.

When using sparse matrix techniques, the order in which the rows/columns of a matrix (or nodes of the network) are processed is important from the point of view of preserving the sparsity of the resultant matrix and minimizing the number of algebraic manipulations required. Although it is not possible to devise a general optimal elimination ordering strategy, simple heuristic schemes usually work well (Tinney and Walker 1967; Brameller et al. 1976). Typically, these node elimination schemes adopt one of the following procedures at each elimination step:

- eliminate the node which has the smallest number of neighbors; or
- eliminate the node which introduces the smallest number of new connections.

It is worth noting that elimination defined by Eq. (18.6) is a generalization of the star-delta transformation. In that particular case three branches are connected to the eliminated node (Figure 18.4). Note that an off-diagonal element of the admittance matrix is equal to the branch admittance with a reversed sign, while the diagonal element is equal to the sum of branches connected to the node. Taking into account that admittance is the reciprocal of impedance, Eq. (18.6) for the delta connection (Figure 18.4b) gives

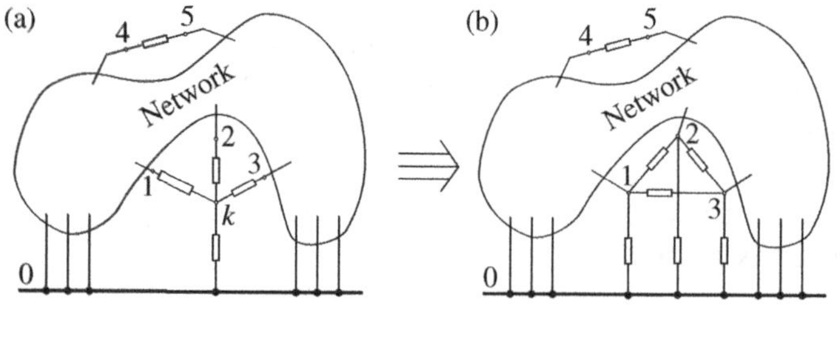

Figure 18.3 Elimination of a single node: (a) situation before elimination; (b) situation after elimination.

$$\underline{y}_{AB} = \frac{\underline{y}_A \underline{y}_B}{\underline{y}_A + \underline{y}_B + \underline{y}_C} \tag{18.7}$$

Changing to impedances

$$\underline{Z}_{AB} = \frac{\underline{y}_A + \underline{y}_B + \underline{y}_C}{\underline{y}_A \underline{y}_B} = \underline{Z}_A \underline{Z}_B \left(\frac{1}{\underline{Z}_A} + \frac{1}{\underline{Z}_B} + \frac{1}{\underline{Z}_C} \right) \tag{18.8}$$

And finally

$$\underline{Z}_{AB} = \underline{Z}_A + \underline{Z}_B + \frac{\underline{Z}_A \underline{Z}_B}{\underline{Z}_C} \tag{18.9}$$

A similar procedure can be applied to the remaining branches $\underline{Z}_{AC}$ and $\underline{Z}_{BC}$. Equation (18.9) is the well-known formula for the star-delta transformation.

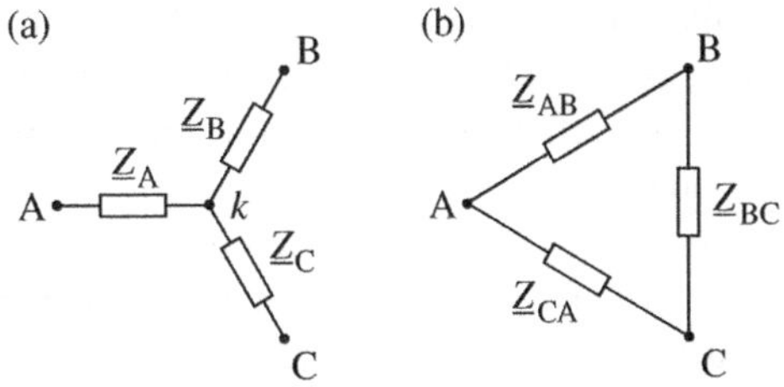

Figure 18.4 Replacing the star connection by the delta connection: (a) the star connection; (b) the equivalent delta connection.

18.2.2 Aggregation of Nodes Using Dimo's Method

This method is illustrated in Figure 18.5 and consists of replacing a group of nodes {A} by an equivalent node a. As before, {R} is the set of retained nodes.

In the first step of the transformation (Dimo 1971), some fictitious branches are added to the aggregated nodes, set {A}. Each branch admittance is chosen in such a way as to make the terminal voltage of all the added branches equal. The terminal equipotential nodes can then be connected together to form a fictitious auxiliary node f. The admittance of each of the fictitious branches can be chosen freely, provided they all give the same terminal voltage. Usually, these admittances are made to correspond to the nodal injections (at a given voltage) in the aggregated nodes

$$\underline{Y}_{fi} = \frac{\underline{S}_i^*}{V_i^2} \quad \text{for} \quad i \in \{A\} \tag{18.10}$$

and then the voltage at the fictitious node f is zero. As it is inconvenient to have an equivalent node operating at zero voltage, an extra fictitious branch with negative admittance is usually added to node f. This branch raises the voltage at its terminal node a to the value close to the rated network voltage. A typical choice of the negative admittance is

$$\underline{Y}_{fa} = -\frac{\underline{S}_a^*}{V_a^2} \quad \text{where} \quad \underline{S}_a = \sum_{i \in \{A\}} \underline{S}_i \tag{18.11}$$

This makes the voltage $\underline{V}_a$ at the equivalent node equal to the weighted average of the voltages at the aggregated nodes

$$\underline{V}_a = \frac{\underline{S}_a}{\underline{I}_a^*} = \frac{\sum_{i \in \{A\}} \underline{S}_i}{\sum_{i \in \{A\}} \left(\frac{\underline{S}_i}{\underline{V}_i} \right)^*} \tag{18.12}$$

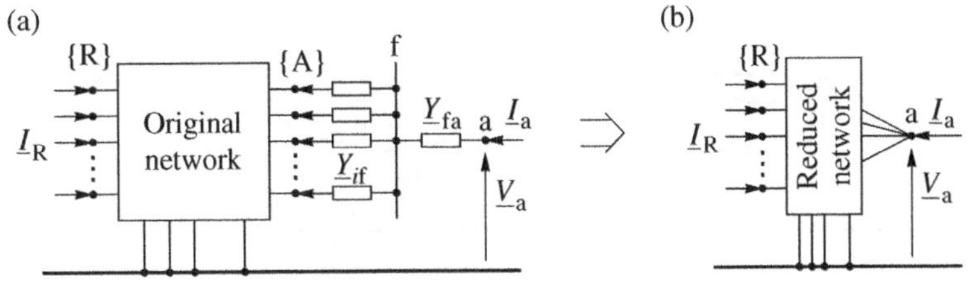

Figure 18.5 Node aggregation using Dimo's method: (a) network with fictitious branches; (b) network after elimination of the nodes and fictitious branches.

The auxiliary node f is eliminated together with the nodes belonging to set {A} giving an equivalent network, referred to as the *radial equivalent independent (REI)* circuit, connecting the equivalent node a with the retained nodes {R}. As well as the REI circuit, the elimination process also creates additional connections between the retained nodes.

If the operating conditions are different from the ones for which the reduction was performed then the obtained equivalent will only imitate the external network accurately if the admittances of the fictitious branches, Eq. (18.10), can be assumed to remain constant. For load nodes this is equivalent to assuming that the loads can be modeled as constant admittances and is only valid for loads with a power–voltage characteristic of the form $\underline{S}_i = V_i^2 \underline{Y}_{fi}^*$ and $\underline{Y}_{fi} = \text{constant}$. The generation nodes operate at a constant voltage and the condition $\underline{Y}_{fi} = \underline{S}_i^*/V_i^2 = \text{constant}$ is only satisfied by those generators where the real and reactive power output can be assumed to be constant.

Dimo's method produces a large number of fictitious branches, because of the elimination of node f and nodes {A}. As aggregation introduces a branch with negative admittance, Eq. (18.11), the branches in the final network model may have negative admittances. Moreover, large nodal injections in the aggregated nodes produce large resistance values in the equivalent branches, Eq. (18.10). Negative branch admittances combined with large resistances may cause convergence problems for some load flow programs.

18.2.3 Aggregation of Nodes Using Zhukov's Method

This method of aggregation was first formulated by Zhukov (1964). The matrix formulation described below was developed by Bernas (Machowski and Bernas 1989), but in view of Zhukov's early publication it is referred to here as *Zhukov's aggregation*.

Aggregation consists of replacing a set of nodes {A} by a single equivalent node a, as shown in Figure 18.6. {R} denotes the set of retained nodes. Aggregation must satisfy the following conditions:

1) It does not change the currents and voltages, $\underline{\boldsymbol{I}}_R$ and $\underline{\boldsymbol{V}}_R$, at the retained nodes.
2) The real and reactive power injection at the equivalent node must be equal to the sum of injections at the aggregated nodes, $\underline{S}_a = \sum_{i \in \{A\}} \underline{S}_i$.

The transformation of the network can then be described by

$$\begin{bmatrix} \underline{\boldsymbol{I}}_R \\ \underline{\boldsymbol{I}}_A \end{bmatrix} = \begin{bmatrix} \underline{\boldsymbol{Y}}_{RR} & \underline{\boldsymbol{Y}}_{RA} \\ \underline{\boldsymbol{Y}}_{AR} & \underline{\boldsymbol{Y}}_{AA} \end{bmatrix} \begin{bmatrix} \underline{\boldsymbol{V}}_R \\ \underline{\boldsymbol{V}}_A \end{bmatrix} \Rightarrow \begin{bmatrix} \underline{\boldsymbol{I}}_R \\ \underline{I}_a \end{bmatrix} = \begin{bmatrix} \underline{\boldsymbol{Y}}_{RR} & \underline{\boldsymbol{Y}}_{Ra} \\ \underline{\boldsymbol{Y}}_{aR} & \underline{Y}_{aa} \end{bmatrix} \begin{bmatrix} \underline{\boldsymbol{V}}_R \\ \underline{V}_a \end{bmatrix} \tag{18.13}$$

where the subscripts refer to the appropriate sets. As a is a single node, $\underline{\boldsymbol{Y}}_{Ra}$ is a column, $\underline{\boldsymbol{Y}}_{aR}$ is a row, and $\underline{Y}_{aa}$ is a scalar.

The first condition is satisfied when

$$\underline{\boldsymbol{Y}}_{RR}\underline{\boldsymbol{V}}_R + \underline{\boldsymbol{Y}}_{RA}\underline{\boldsymbol{V}}_A = \underline{\boldsymbol{Y}}_{RR}\underline{\boldsymbol{V}}_R + \underline{\boldsymbol{Y}}_{Ra}\underline{V}_a \quad \text{or} \quad \underline{\boldsymbol{Y}}_{RA}\underline{\boldsymbol{V}}_A = \underline{\boldsymbol{Y}}_{Ra}\underline{V}_a \tag{18.14}$$

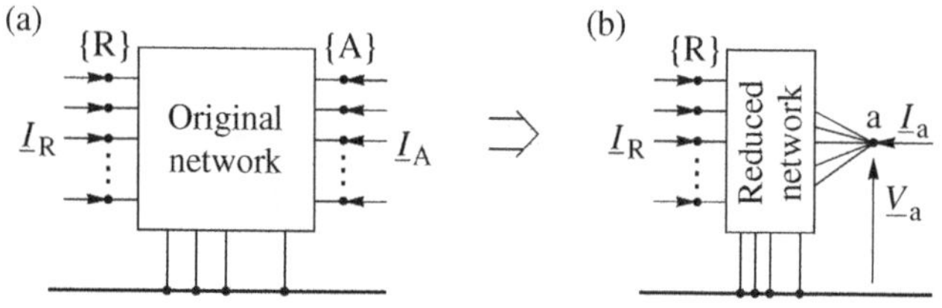

Figure 18.6 Node aggregation using Zhukov's method: (a) network before aggregation; (b) network after aggregation.

If this condition is to be satisfied for any vector $\underline{\boldsymbol{V}}_{\rm A}$, the following must hold

$$\underline{\boldsymbol{Y}}_{\rm Ra} = \underline{\boldsymbol{Y}}_{\rm RA}\underline{\boldsymbol{\vartheta}} \tag{18.15}$$

where

$$\underline{\boldsymbol{\vartheta}} = \underline{V}_{\rm a}^{-1}\underline{\boldsymbol{V}}_{\rm A} = \begin{bmatrix} \underline{\vartheta}_1 \\ \underline{\vartheta}_2 \\ \vdots \end{bmatrix} \tag{18.16}$$

is the vector of voltage transformation ratios between the aggregated nodes and the equivalent node.

The second assumption is satisfied when

$$\underline{V}_{\rm a}\underline{I}_{\rm a}^* = \underline{\boldsymbol{V}}_{\rm A}^{\rm T}\underline{\boldsymbol{I}}_{\rm A}^* \tag{18.17}$$

where the left-hand side expresses the injection at the equivalent node and the right-hand side expresses the sum of all the aggregated injections. Substituting into Eq. (18.17) for $\underline{I}_{\rm a}$ and $\underline{\boldsymbol{I}}_{\rm A}$ calculated from Eqs. (18.13) gives

$$\underline{V}_{\rm a}\underline{\boldsymbol{Y}}_{\rm aR}^*\underline{\boldsymbol{V}}_{\rm R}^* + \underline{V}_{\rm a}\underline{Y}_{\rm aa}^*\underline{V}_{\rm a}^* = \underline{\boldsymbol{V}}_{\rm A}^{\rm T}\underline{\boldsymbol{Y}}_{\rm AR}^*\underline{\boldsymbol{V}}_{\rm R}^* + \underline{\boldsymbol{V}}_{\rm A}^{\rm T}\underline{\boldsymbol{Y}}_{\rm AA}^*\underline{\boldsymbol{V}}_{\rm A}^* \tag{18.18}$$

If this equation is to be satisfied for any vector of $\underline{\boldsymbol{V}}_{\rm A}$, the following two conditions must hold

$$\underline{\boldsymbol{Y}}_{\rm aR} = \underline{\boldsymbol{\vartheta}}^{*\rm T}\underline{\boldsymbol{Y}}_{\rm AR} \tag{18.19}$$

$$\underline{\boldsymbol{Y}}_{\rm aa} = \underline{\boldsymbol{\vartheta}}^{*\rm T}\underline{\boldsymbol{Y}}_{\rm AA}\underline{\boldsymbol{\vartheta}} \tag{18.20}$$

Substituting Eqs. (18.15), (18.19), and (18.20) into the right-hand side of the second of Eqs. (18.13) finally gives

$$\begin{bmatrix} \underline{\boldsymbol{I}}_{\rm R} \\ \underline{I}_{\rm a} \end{bmatrix} = \left[\begin{array}{c|c} \underline{\boldsymbol{Y}}_{\rm RR} & \underline{\boldsymbol{Y}}_{\rm RA}\underline{\boldsymbol{\vartheta}} \\ \hline \underline{\boldsymbol{\vartheta}}^{*\rm T}\underline{\boldsymbol{Y}}_{\rm AR} & \underline{\boldsymbol{\vartheta}}^{*\rm T}\underline{\boldsymbol{Y}}_{\rm AA}\underline{\boldsymbol{\vartheta}} \end{array}\right] \begin{bmatrix} \underline{\boldsymbol{V}}_{\rm R} \\ \underline{V}_{\rm a} \end{bmatrix} \tag{18.21}$$

Equations (18.15), (18.19), and (18.20) describe the admittances of the equivalent network. The admittances of the equivalent branches linking the equivalent node with the retained nodes depend on the vector of transformation ratios $\underline{\boldsymbol{\vartheta}}$, and hence on the voltage angle at the equivalent node. As it is convenient to have equivalent branches of low resistances, the voltage angle $\delta_{\rm a}$ at the equivalent node is assumed to be equal to the weighted average of voltage angles at the aggregated nodes

$$\delta_{\rm a} = \frac{\sum\limits_{i\in\{\rm A\}} S_i\delta_i}{\sum\limits_{i\in\{\rm A\}} S_i} \quad \text{or} \quad \delta'_{\rm a} = \frac{\sum\limits_{i\in\{\rm A\}} M_i\delta'_i}{\sum\limits_{i\in\{\rm A\}} M_i} \tag{18.22}$$

where S_i is the apparent power injection at the aggregated node i and $M_i = T_{{\rm m}i}S_{{\rm n}i}/\omega_{\rm s}$ is the inertia coefficient of the unit installed at the i-th aggregated node. The first formula can be used for forming the equivalent to be used for steady-state analysis and the second formula can be applied for aggregation of a group of generators represented by the classical transient stability model (constant transient electromotive forces [emf's] E'_i). In this case angle $\delta'_{\rm a}$ given by Eq. (18.22) of the equivalent transient emf is equal to the angle of the center of inertia (COI) of the given group {A} of the generating units.

When compared with Dimo's method, the advantage of Zhukov's method is that it does not introduce fictitious branches between the retained nodes {R}. This is because submatrix $\underline{\boldsymbol{Y}}_{\rm RR}$ is unchanged by aggregation. It does, however, introduce fictitious shunt branches at the retained nodes. To understand this, examine the i-th diagonal element of $\underline{\boldsymbol{Y}}_{\rm RR}$, which is equal to the sum of admittances of all the series and shunt branches connected to i

$$\underline{Y}_{ii} = \underline{y}_{i0} + \sum_{j\in\{\rm R\}} \underline{y}_{ij} + \sum_{k\in\{\rm A\}} \underline{y}_{ik} \quad \text{for} \quad i \in \{\rm R\} \tag{18.23}$$

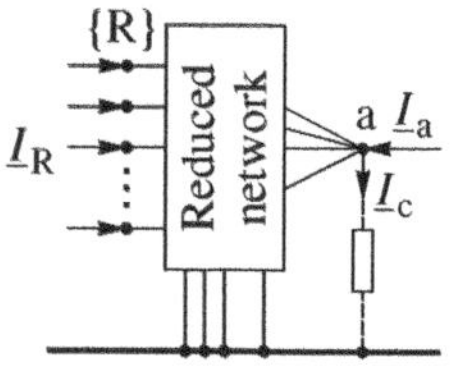

Figure 18.7 Symmetry of the equivalent network.

where $\underline{y}_{i0}$ is the sum of admittances of all shunt branches connected to i and $\underline{y}_{ij}$ is the admittance of a branch linking nodes i and j. During aggregation, all the branches of admittance $\underline{y}_{ik}$ that link node $i \in \{R\}$ with the aggregated nodes $k \in \{A\}$ are replaced by a single branch with admittance $\underline{y}_{ia}$ generally not equal to $\sum_{k\in\{A\}} \underline{y}_{ik}$. As $\underline{Y}_{ii}$ and $\sum_{j\in\{R\}} \underline{y}_{ij}$ must remain unchanged, replacing $\sum_{k\in\{A\}} \underline{y}_{ik}$ by $\underline{y}_{ia}$ must be compensated by a change in the value $\underline{y}_{i0}$. The interpretation of this in network terms is that Zhukov's aggregation introduces some equivalent shunt admittances at the retained nodes {R}.

If the vector $\underline{\boldsymbol{\vartheta}}$ is complex then Zhukov's equivalent admittance matrix is not generally symmetric ($\underline{\mathbf{Y}}_{aR} \neq \underline{\mathbf{Y}}_{Ra}^{T}$). This means that if $\underline{Y}_{ia} \neq \underline{Y}_{ai}$ for $i \in \{R\}$ then the value of the admittances in the equivalent branches obtained after aggregation are direction dependent. From a computational point of view, asymmetry of the admittance matrix is inconvenient and Figure 18.7 shows how this asymmetry can be removed by inserting a correction current $\underline{I}_c$ at the equivalent node a. The nodal equation of the system then takes the form

$$\begin{bmatrix} \underline{\mathbf{I}}_R \\ \underline{I}_a \end{bmatrix} = \begin{bmatrix} \underline{\mathbf{Y}}_{RR} & \underline{\mathbf{Y}}_{Ra} \\ \underline{\mathbf{Y}}_{Ra}^{T} & \underline{Y}_{aa} + \dfrac{\underline{I}_c}{\underline{V}_a} \end{bmatrix} \begin{bmatrix} \underline{\mathbf{V}}_R \\ \underline{V}_a \end{bmatrix} \tag{18.24}$$

where $\underline{I}_c = \left[(\underline{\boldsymbol{\vartheta}}^{*} - \underline{\boldsymbol{\vartheta}})^{T} \underline{\mathbf{Y}}_{AR} \right] \underline{\mathbf{V}}_R$ is the correction current. This current is not constant, because it depends on the voltages in set {R}. The correction current is small (negligible when compared with $\underline{I}_a$) when the difference $(\underline{\boldsymbol{\vartheta}}^{*} - \underline{\boldsymbol{\vartheta}})$ is small, that is when the imaginary parts of the transformation ratios are small. This condition is usually satisfied because the angle of the equivalent voltage, Eq. (18.22), is averaged over the aggregated nodes. Consequently, variations of the correction current can be neglected and a constant current replaced by a constant admittance $(\underline{I}_c/\underline{V}_a)$ added to the self-admittance of the equivalent node a. This is shown using the dotted line in Figure 18.7.

18.2.4 Coherency

The admittances in the equivalent Zhukov network depend on the transformation ratios $\underline{\vartheta}_i = \underline{V}_i/\underline{V}_a$ between the aggregated nodes $i \in \{A\}$ and the equivalent node a. This means that an equivalent network obtained for an initial (pre-fault) state is only valid for other states (transient or steady state) if the transformation ratios (see Figure 18.15) can be assumed to remain constant for all nodes $i \in \{A\}$ in a given group

$$\frac{\underline{V}_i(t)}{\underline{V}_a(t)} = \frac{\hat{\underline{V}}_i}{\hat{\underline{V}}_a} = \underline{\vartheta}_i = \text{constant} \qquad \text{for} \quad i \in \{A\} \tag{18.25}$$

where the circumflex denotes the initial state (stable equilibrium point) for which the reduced model has been constructed. For any two nodes $i, j \in \{A\}$ this condition is equivalent to

$$\frac{\underline{V}_i(t)}{\underline{V}_j(t)} = \frac{V_i(t)}{V_j(t)} e^{j[\delta_i(t) - \delta_j(t)]} = \frac{\hat{V}_i}{\hat{V}_j} e^{j\,[\hat{\delta}_i - \hat{\delta}_j]} = \text{constant} \qquad \text{for} \quad i,j \in \{A\} \tag{18.26}$$

Nodes satisfying this condition are referred to as *electrically coherent nodes* or simply *coherent nodes*. If the voltage magnitude of the aggregated node can be assumed to be constant (as in Section 3.6.1 for PV nodes in the steady-state power flow problem), the above coherency condition (18.26) simplifies to

$$\delta_i(t) - \delta_j(t) = \hat{\delta}_{ij} \qquad \text{for} \quad i,j \in \{A\} \tag{18.27}$$

where $\hat{\delta}_{ij} = \hat{\delta}_i - \hat{\delta}_j$ are the initial values.

Practical experience with power system simulation shows that load nodes are almost never electrically coherent. Only the load nodes very far away from a disturbance maintain constant voltage magnitude and angle. On the other hand, it is usually possible to find groups of coherent generation nodes because some groups of generators in the system have a natural tendency to swing together. This means that Zhukov's method is well suited for the aggregation of groups of electrically coherent generation nodes.

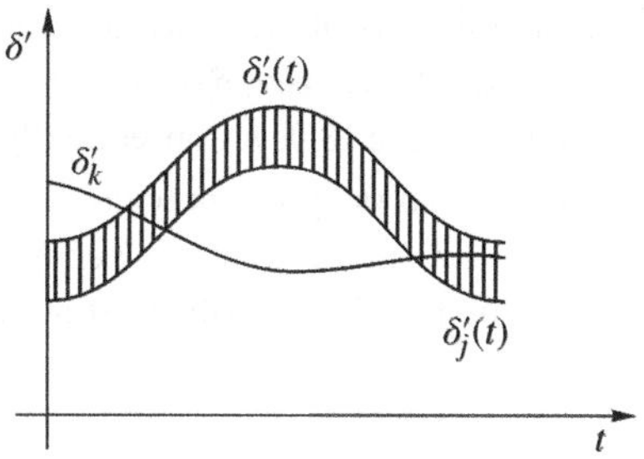

Figure 18.8 Example of variation of rotor angles for three generators.

For the generators modeled by the classical generator model (Figure 5.7) the nodal voltage at the generator nodes is equal to the transient emf $\underline{E}_i'$, the magnitude of which is assumed to be constant $E_i' = \text{constant}$, and the angle corresponds to the rotor angle $\delta_i'(t)$. For these generator nodes the coherency condition (18.26) simplifies to

$$\delta_i'(t) - \delta_j'(t) = \overset{\circ}{\delta}{}_{ij}' \qquad \text{for} \quad i,j \in \{A\} \tag{18.28}$$

where $\overset{\circ}{\delta}{}_{ij}' = \overset{\circ}{\delta}{}_i' - \overset{\circ}{\delta}{}_j'$ are the initial values. The coherency defined by Eq. (18.28) is valid also for generator rotors and is therefore referred to as the *electromechanical coherency*.

An example of rotor swings for three generators is shown in Figure 18.8. Generators i and j are electromechanically coherent because the difference between their rotor angles is almost constant despite both angles undergoing quite deep oscillations. Generator k is not coherent with the other two because its rotor angle variations are different.

Condition (18.28) may also be written as $\left[\delta_i'(t) - \overset{\circ}{\delta}{}_i'\right] - \left[\delta_j'(t) - \overset{\circ}{\delta}{}_j'\right] = 0$ or $\left[\Delta\delta_i'(t) - \Delta\delta_j'(t)\right] = 0$. For practical considerations it may be assumed that coherency is only approximate with accuracy $\varepsilon_{\Delta\delta}$, which corresponds to the condition

$$\left|\Delta\delta_i'(t) - \Delta\delta_j'(t)\right| < \varepsilon_{\Delta\delta} \quad \text{for} \quad i,j \in \{A\} \quad \text{and} \quad t \le t_c \tag{18.29}$$

where $\varepsilon_{\Delta\delta}$ is a small positive number, t_c is the duration time of coherency.

The generators are said to be *exactly coherent generators* if

$$\varepsilon_{\Delta\delta} = 0 \quad \text{and} \quad t_c = \infty \tag{18.30}$$

Exactly coherent generators rarely occur in practice, but the definition is useful for theoretical considerations.

It should be noted that swings of coherent generators can be treated as a constrained motion – as illustrated in Figure 18.9, where generators i and j are electromechanically coherent. In the plane with coordinates $\delta_i'(t)$

Figure 18.9 Illustration of exact coherency: (a) rotor swings; (b) trajectory in the rotor angle space.

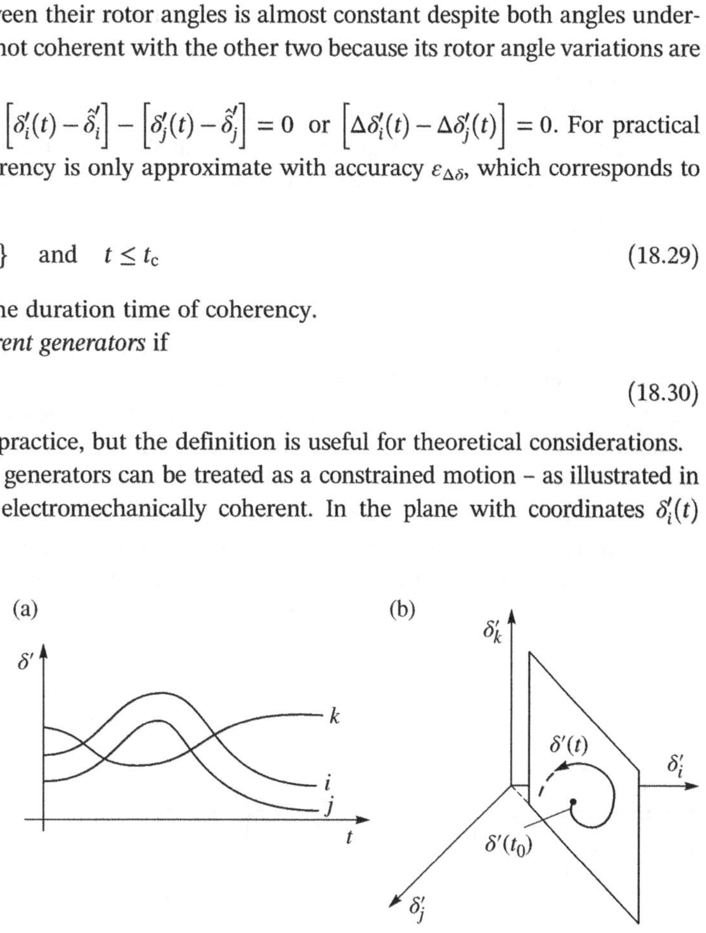

and $\delta'_j(t)$, the trajectory of these two coherent generators is given by $\delta'_i(t) = \delta'_j(t) + \hat{\delta}'_{ij}$ resulting from the coherency condition (18.28). Obviously, $\delta'_i(t)$ as the function of $\delta'_j(t)$ is a straight line. Generator k is not coherent with generators i and j (Figure 18.9a). Consequently, the trajectory $\boldsymbol{\delta}'(t)$ in the space with coordinates $\delta'_i(t)$, $\delta'_j(t)$, and $\delta'_k(t)$ lies on the plane crossing the previously mentioned straight line (Figure 18.9b).

When there are more coherent generators, the trajectory lies on the intersection line of the planes. The intersection line may be described using the following equation

$$\varphi(\boldsymbol{\delta}') = \mathbf{0} \tag{18.31}$$

where $\varphi(\boldsymbol{\delta}')$ is a vector function consisting of the following functions

$$\varphi_j(\boldsymbol{\delta}) = \delta'_1(t) - \delta'_j(t) - \hat{\delta}'_{1j} = 0 \qquad \text{for} \quad j \in \{\mathrm{A}\} \text{ and } j > 1 \tag{18.32}$$

where $\delta'_{1j0} = \delta'_{10} - \delta'_{j0}$. For every generator belonging to a coherent group {A}, the above equation can be treated as a constraint for the rotor motion.

18.3 Aggregation of Generating Units

The elimination and aggregation of nodes considered so far will produce a reduced network model for use in steady-state analysis. If the reduced model is to be used for dynamic analysis then equivalent generating units must be added to the equivalent nodes.

From a mechanical point of view, the rotors of electromechanically coherent generators can be treated as if they rotated on one common rigid shaft (Figure 18.10). A group {A} containing n such generators can be replaced by one equivalent generator with inertia coefficient M_a and mechanical power input P_ma given by

$$M_\mathrm{a} = \sum_{i \in \{\mathrm{A}\}} M_i; \qquad P_\mathrm{ma} = \sum_{i \in \{\mathrm{A}\}} P_{\mathrm{m}i} \tag{18.33}$$

where $M_i = T_{\mathrm{m}i} S_{\mathrm{n}i} / \omega_\mathrm{s}$ is the inertia coefficient and $P_{\mathrm{m}i}$ is the mechanical power input of the i-th aggregated generator. This is consistent with Zhukov's aggregation, which sets the power injection at the equivalent node equal to the sum of power injections to all the aggregated nodes, Eq. (18.17).

The equivalent model of a group of electromechanically coherent generation units is therefore created by Zhukov's aggregation of the generation nodes and by replacing the aggregated generators by one equivalent generator with inertia coefficient and mechanical power given by Eq. (18.33). The equivalent generator is represented by the classical model with constant equivalent transient emf and by the swing equation.

If more detailed models are used then parameters of the equivalent unit can be found by matching the frequency response characteristics of the equivalent unit to the characteristics of the aggregated units. Details can be found in Garmond and Podmore (1978) and Cai and Wu (1986).

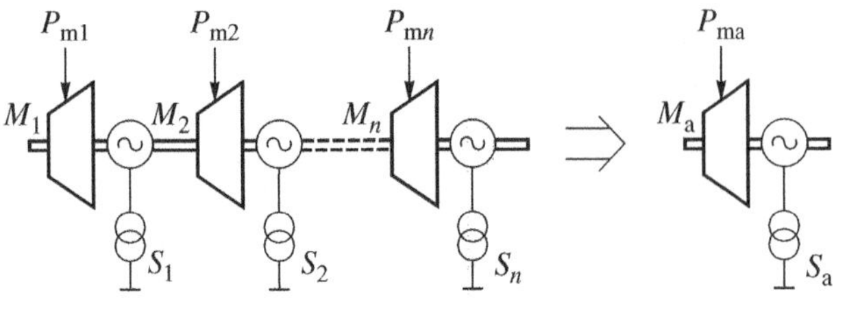

Figure 18.10 Mechanical aggregation of coherent rotors.

18.4 Equivalent Model of External Subsystem

The described method of creating a dynamic equivalent model is based on the following assumptions:

1) The system is divided as in Figure 18.1 into internal and external part.
2) In the internal part, detailed generator and load models are used as described in Chapter 11.
3) In the external part, the loads are replaced by constant admittances while the generators are modeled using the classical model (the rotor swing equation and a constant transient emf behind a transient reactance).

Under these assumptions the creation of the dynamic equivalent model is significantly simplified and consists of three steps:

1) Elimination of the load nodes in the external subsystem.
2) Identification of coherent groups of generators in the external subsystem.
3) Aggregation of the coherent groups.

All three steps are briefly described in the following subsections.

All the load nodes in the external subsystem can be eliminated using the method described in Section 18.2.1. The resulting external equivalent network is referred to as the *PV equivalent network* because, with the exception of the border nodes, it contains only generation nodes (the terminology used with respect to load flow calculations in Section 3.6 referred to such nodes as PV nodes). The power demand of the external system is then distributed among the border nodes and generation nodes. All the generation nodes, and the border nodes, are connected by the equivalent network, which is much denser than the original.

For some power system analysis problems, it may be more convenient not to eliminate the load nodes altogether, but to replace a few of them by equivalent load nodes using Dimo's aggregation method. These equivalent nodes can then be used to change the power demand of the external subsystem if a change in tie-line flows is required.

Coherency recognition is the most difficult step in creating a dynamic equivalent model of the external subsystem. Coherency criteria and coherency recognition algorithms are described in the next section.

When all the groups of coherent generators in the external subsystems have been recognized, the next step is to use Zhukov's method to aggregate the nodes in these groups. Equivalent generating units with parameters calculated, as described in Section 18.3, are connected to the equivalent nodes obtained by Zhukov's method.

Figure 18.11 illustrates the whole process of forming an equivalent model of the external subsystem. The original model of the subsystem contains a large number of load nodes and a large number of generation nodes $\{G\} = \{G_1\} + \{G_2\} + \ldots + \{G_g\}$. The load nodes are either completely eliminated or aggregated into a few equivalent nodes using Dimo's method. The generator nodes are divided into groups of approximately coherent nodes $\{G_1\}, \{G_2\}, \ldots, \{G_g\}$ and each of the group is replaced by one equivalent node with an equivalent generating unit.

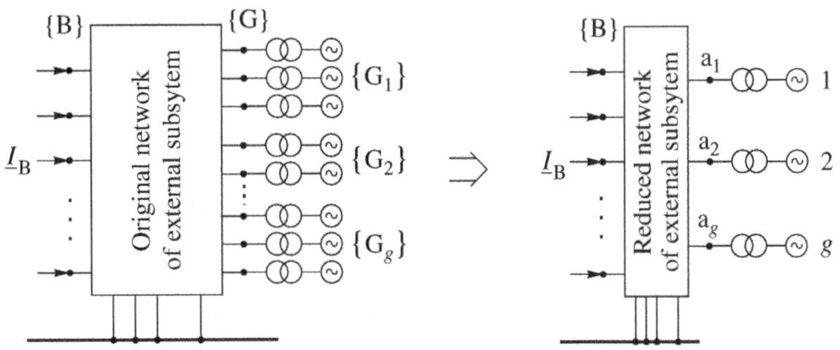

Figure 18.11 Model reduction of the external system.

18.5 Coherency Recognition

The topological network equivalent obtained by the aggregation of generation nodes will only give valid results if, following a disturbance inside the internal subsystem, the generators within each aggregated group are coherent. The problem is therefore how to assess generator coherency without first completing a detailed dynamic simulation of the complete system for a particular disturbance. Fortunately, methods are available that will assess generator coherency without the need for such detailed simulation and this is referred to as *coherency recognition*. The simplest solution to this problem is to assume that all the generators installed at the aggregated nodes can be modeled by the classical generator model and electromechanical coherency must be recognized.

Several methods of coherency recognition have been reported in the literature. The approach described here is based on coherency criteria derived by Machowski et al. (1988).

Coherency criteria for nonlinear dynamic system models are mathematically derived by Machowski (1985) using the theory of motion with prescribed trajectory due to Olas (1975). The methodology is based on an observation that electromechanical coherency is a case of the constrained motion (Figure 18.9) with constraints given by Eq. (18.32). As the proof is rather complicated, the idea of coherency recognition is described here using a single disturbance when the voltage angle at one of the nodes $k \in \{B\}$ is changed.

Providing that all the load nodes of the external subsystem have been eliminated, any disturbance inside the internal subsystem influences the generators in the external subsystem through equivalent branches of the transfer network. According to Eq. (3.181), if the mutual conductances G_{ij} in the transfer admittance matrix are neglected, the active power produced by generator $i \in \{G\}$ in the external subsystem (Figure 18.11) can be expressed as

$$P_i = \left(E_i'\right)^2 G_{ii} + \sum_{k \in \{B\}} E_i' V_k B_{ik} \sin \delta_{ik}' + \sum_{l \in \{G\}} E_i' E_l' B_{il} \sin \delta_{il}' \tag{18.34}$$

where E_i' for $i \in \{G\}$ is the generator transient emf, V_k for $k \in \{B\}$ is the voltage at the border node, $\delta_{ik}' = \delta_i' - \delta_k$, $\delta_{il}' = \delta_i' - \delta_l'$, and G_{ii}, B_{ik}, and B_{il} are the appropriate elements of the transfer admittance matrix.

Assuming that the disturbance is caused by a change in the voltage angle of the border node k from the initial value $\hat{\delta}_k$ to a value $\delta_k = \hat{\delta}_k + \Delta\delta_k$ and assuming that the voltages at the other nodes are constant, this change in angle will cause a change in the power generation at node $i \in \{G\}$ equal to

$$\Delta P_i(\Delta\delta_k) = b_{ik}\left[\sin\left(\hat{\delta}_{ik}' + \Delta\delta_k\right) - \sin\hat{\delta}_{ik}'\right] \tag{18.35}$$

where $b_{ik} = E_i' V_k B_{ik}$ is the maximum power transfer in the equivalent branch linking a generator node $i \in \{G\}$ with a border node $k \in \{B\}$. As $\Delta\delta_k$ is small, it holds that $\cos\Delta\delta_k \cong 1$ and $\sin\Delta\delta \cong \Delta\delta_k$. Expanding the sine term in Eq. (18.35) gives

$$\Delta P_i(\Delta\delta_k) \cong H_{ik}\Delta\delta_k \tag{18.36}$$

where $H_{ik} = b_{ik}\cos\delta_{ik0}$ is the synchronizing power between a given generator $i \in \{G\}$ and a given border node $k \in \{B\}$. The considered disturbance causes the rotor acceleration

$$\varepsilon_i = \frac{\Delta P_i(\Delta\delta_k)}{M_i} = \frac{H_{ik}}{M_i}\Delta\delta_k \qquad \text{for} \quad k \in \{B\} \tag{18.37}$$

where M_i is the inertia coefficient. A similar expression for acceleration can be written for another generator in the external subsystem as

$$\varepsilon_j = \frac{\Delta P_j(\Delta\delta_k)}{M_j} = \frac{H_{jk}}{M_j}\Delta\delta_k \qquad \text{for} \quad k \in \{B\} \tag{18.38}$$

Generators $i, j \in \{G\}$ are ***electromechanically exactly coherent generators*** – see Eq. (18.30) – if their rotor accelerations ε_i and ε_j caused by the disturbance are the same, that is when

$$\frac{H_{ik}}{M_i} = \frac{H_{jk}}{M_j} \qquad \text{for } i,j \in \{A\};\ k \in \{B\} \tag{18.39}$$

Equation (18.39) constitutes the exact coherency condition during the post-fault state and means that the synchronizing powers divided by the inertia constants must be identical.

Section 18.6 shows how exact coherency has an elegant and interesting modal interpretation and that the exact coherency condition (18.39) may be derived using modal analysis.

Equation (18.39) is the condition for exact coherency. In active power systems (apart from the trivial case of identical generating units operating in parallel on the same busbar) the exact coherency practically does not appear. This is not a significant problem as simulation of the internal system gives results of satisfactory accuracy if the external subsystem can be replaced by an approximate equivalent created by the aggregation of generators that are only approximately coherent. For practical purposes the equality (18.39) can be replaced by the following inequality

$$\frac{\max\limits_{i\in\{G\}} \dfrac{H_{ik}}{M_i} - \min\limits_{j\in\{G\}} \dfrac{H_{jk}}{M_j}}{d_{\{G\}}} \le \rho_{\rm h} \qquad \text{for } i,j \in \{G\},\ k \in \{B\} \tag{18.40}$$

where ρ_h is a small number determining the admissible error and $d_{\{G\}}$ is a density measure of the considered group {A}. The density measure for a pair of generators i, j is defined using the following parameter

$$d_{ij} = \min_{i,j\in\{G\}} \left(\frac{H_{ij}}{M_i}; \frac{H_{ji}}{M_j} \right) \tag{18.41}$$

which relates to the direct connection of this pair and appropriate inertia coefficients. For the group of generators {G} this parameter can be used to define the following density measure

$$d_{\{G\}} = \min_{i,j\in\{T\}} d_{ij} \tag{18.42}$$

where {T} is a tree made up of the equivalent branches with the highest values of $d_{\{i,j\}}$. Such a definition is justified by the fact that, inside the group, the nodes with weak direct connections can be strongly connected via other nodes. This is illustrated in Figure 18.12. For example, nodes 5 and 2 are directly very weakly connected by a branch of parameters $d_{\{2,5\}} = 0.03$. However, these nodes are strongly connected via nodes 3 and 4. In Figure 18.12 the tree {T} is denoted using bold lines. The weakest branch of this tree is the branch connecting nodes 1 and 4. Hence in the discussed example the density measure of the group $\{G\} = \{1, 2, 3, 4, 5\}$ is equal to $d_{\{G\}} = d_{14} = 0.33$.

Justification of the density measure (18.40) results from the following observations of simulations of power system dynamic response. Each strongly connected group of generators has a natural tendency to maintain synchronism. The synchronism may be disturbed only by disturbances close to that group. For remote disturbances the further away the disturbance, the less the synchronism is disturbed and the more the group motion is close to exact coherency.

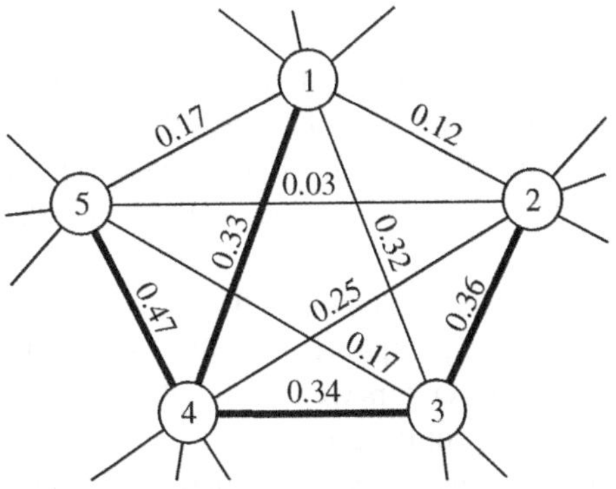

Figure 18.12 Illustration of the definition of the density measure.

Inequality (18.40) was derived based on condition (18.39) for exact coherency when the above observations were taken into account. Hence inequality (18.40) is referred to as the *coherency criterion*, while Eq. (18.39) constitutes the coherency condition.

Another important observation in the simulation of power system dynamic response concerns the influence of the coherency error of a group of generators aggregated in the external subsystem on the simulation accuracy of the internal subsystem. The more the group of aggregated generators is remote from the internal subsystem, the smaller is the influence of the coherency error on the simulation accuracy of dynamic response in internal subsystems. This observation makes it possible to make ρ_h in criterion (18.40) dependent on the distance of group {A} from the border nodes

$$\rho_{\text{h}} = \rho_{\text{h0}} + \Delta\rho_{\text{h}} \frac{d_{\{G\}}}{\max\limits_{k\in\{B\};\ i\in\{G\}} d_{ik}} \tag{18.43}$$

where ρ_{h0} and $\Delta\rho_{\text{h}}$ are small positive numbers ($\rho_{\text{h0}} = 0.2 - 0.5$ and $\Delta\rho_{\text{h}} = 0.1 - 0.3$). For the border nodes {B} the coefficient of inertia is zero and according Eq. (18.41), it holds $d_{ik} = h_{ik}/M_i$ for $k \in \{B\}$ and $i \in \{G\}$.

The coherency recognition algorithm based on criterion (18.40) works as follows:

1) Determine the transfer admittance matrix for the border nodes {B} and all the generation nodes in the external subsystem.
2) Mark all the generators of the external subsystem as eligible generators for grouping.
3) Order all the equivalent branches in ascending order according to the values of the distance measure d_{ij}. Create an ordered list of those branches containing, for each branch, the value of d_{ij} and numbers of the terminal nodes i and j.
4) Take from the list created in step 3 the data of the next equivalent branch. Memorize its terminal nodes i and j and density measure d_{ij}. If there are no branches left, stop the algorithm.
5) If generator i or j is not eligible for grouping then return to step 4.
6) If criterion (18.40) is not satisfied for pair $\{i,j\}$, then return to step 4. Otherwise, create group {G} consisting of two generators $\{i,j\}$.
7) Search all the eligible generators for a new generator x which satisfies criterion (18.40) for the extended group substituting $\{G, x\}$ and gives a minimum value for the left-hand side of (18.40). If such a generator cannot be found, store group {G} as a new group and return to step 4. Otherwise, go to step 8.
8) Mark generator x as not eligible and add it to group {G}. Return to step 7.

The algorithm is very fast and gives good results in practice. Test results of the above coherency recognition algorithm for test systems and real, large interconnected power systems can be found in Machowski (1985), and Machowski et al. (1986a and b, 1988). Owing to a lack of space here, only one example is presented.

Example 18.1 The diagram of a 25-machine test system is shown in Figure 18.13. In order to show in such a small system the influence of the disturbance distance on the grouping of generators, it was assumed that the internal subsystem is very small, it is on the verge of the test system, and it contains two power plants with generators 7 and 18 (bottom right-hand corner of the diagram). The remaining part of the test system was treated as the external subsystem.

For the parameter values $\rho_{\text{h0}} = 0.3$ and $\Delta\rho_{\text{h}} = 0$ a number of groups were obtained that are shown by solid lines in Figure 18.13. The groups are: {4, 5, 6}, {10, 12}, {9, 11, 13, 25}, {14, 15, 16}, {1, 2, 3, 8}, {17, 20, 21, 22}, and {23, 24}. Altogether 22 generators were replaced by 7 equivalent generators. Generator 19 close to the internal subsystem did not enter any of the groups.

After introducing a dependence of the parameter ρ_{h} on the distance to the disturbance and assuming $\Delta\rho_{\text{h}} = 0.2$, three groups were obtained close to the internal system. The first three groups were identical to the previous case:

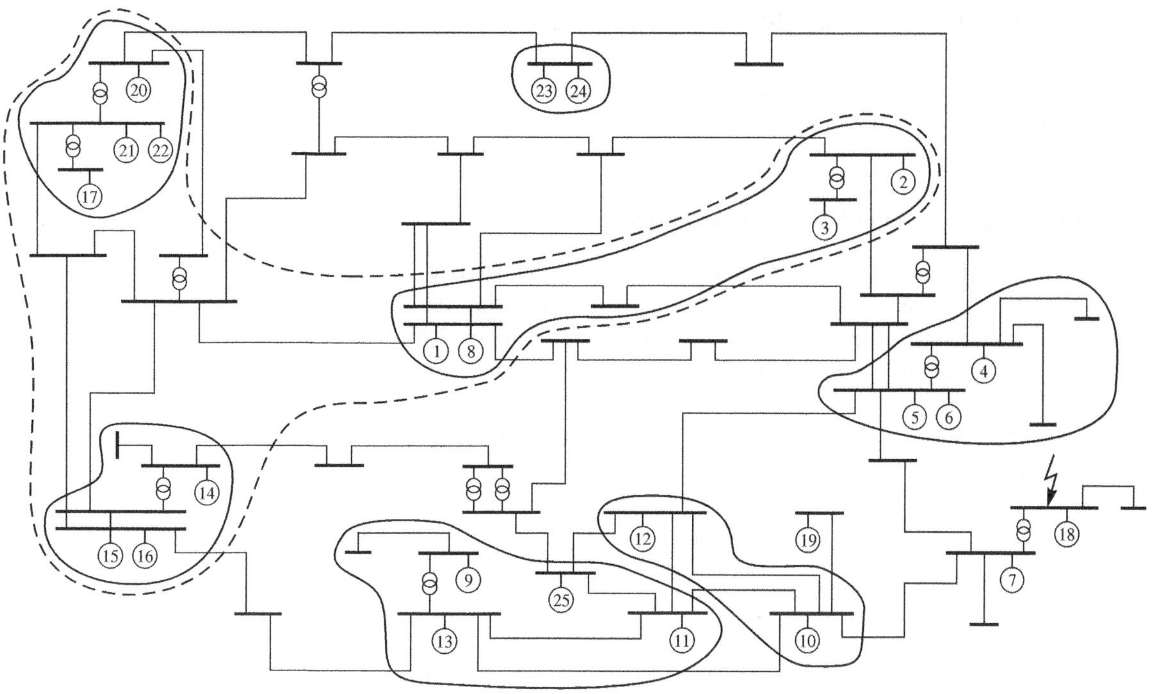

Figure 18.13 Test system and recognized coherent groups.

Figure 18.14 Simulation results for original and reduced system model.

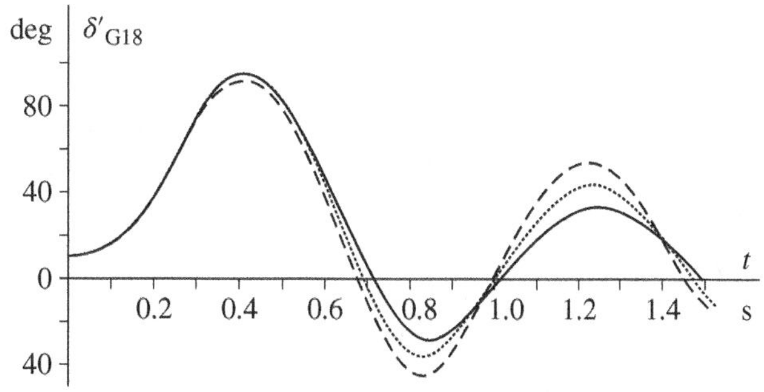

{4, 5, 6}, {10, 12}, and {9, 11, 13, 25}. Three further groups were joined together to form one large group: {14, 15, 16, 1, 2, 3, 8, 17, 20, 21, 22} encircled by a dashed line in Figure 18.13. Group {23, 24} was not included in any other group. Altogether 22 generators were replaced by 5 equivalent generators.

Figure 18.14 shows the simulation results for the original system consisting of 25 generators and the reduced models. The assumed disturbance was an intermittent short circuit on the busbars of power plant number 18. The solid line corresponds to the rotor swing of generator 18 for the original (unreduced) system model. The dotted line corresponds to the group obtained using $\Delta\rho_{\mathrm{h}} = 0$. The dashed line corresponds to the group obtained using $\Delta\rho_{\mathrm{h}} = 0.2$, that is when larger coherency tolerance was assumed for remote generators. In the transient state of about 1.5 s the rotor swings obtained for the reduced model were quite close to the swings for the original (unreduced) model.

******

18.6 Properties of Coherency Based Equivalents

Coherency-based equivalents using Zhukov's aggregation exhibit many interesting static and dynamic properties, which are discussed in this section.

18.6.1 Electrical Interpretation of Zhukov's Aggregation

Demello et al. (1975) propose an aggregation method (Figure 18.15). whereby all the nodes to be aggregated are connected together through ideal transformers with transformation ratios that give a common secondary voltage $\underline{V}_a$. The method of coherency recognition for such aggregation is proposed by Podmore (1978). Now it will be shown that such an aggregation method is, from the mathematical point of view, equivalent to Zhukov's aggregation.

Let $\underline{\tau}$ be a diagonal matrix containing transformation ratios of all the ideal transformers used to aggregate the nodes by using Demello's method

$$\underline{\tau} = \begin{bmatrix} \underline{\vartheta}_1 & & \\ & \underline{\vartheta}_2 & \\ & & \ddots \end{bmatrix} \tag{18.44}$$

Note that there is the following relationship between the above diagonal matrix and vector (18.16) defined in the Zhukov's aggregation

$$\underline{\boldsymbol{\vartheta}} = \begin{bmatrix} \underline{\vartheta}_1 \\ \underline{\vartheta}_2 \\ \vdots \end{bmatrix} = \begin{bmatrix} \underline{\vartheta}_1 & & \\ & \underline{\vartheta}_2 & \\ & & \ddots \end{bmatrix} \begin{bmatrix} 1 \\ 1 \\ \vdots \end{bmatrix} = \underline{\tau}\, \mathbf{1}_\mathrm{A} \tag{18.45}$$

where $\mathbf{1}_\mathrm{A}$ is a column vector with all elements equal to one.

The network shown in Figure 18.15, but without the ideal transformers, is defined by the following nodal admittance equation

$$\begin{bmatrix} \underline{\boldsymbol{I}}_\mathrm{R} \\ \underline{\boldsymbol{I}}'_\mathrm{A} \end{bmatrix} = \begin{bmatrix} \underline{\boldsymbol{Y}}_\mathrm{RR} & \underline{\boldsymbol{Y}}_\mathrm{RA} \\ \underline{\boldsymbol{Y}}_\mathrm{AR} & \underline{\boldsymbol{Y}}_\mathrm{AA} \end{bmatrix} \begin{bmatrix} \underline{\boldsymbol{V}}_\mathrm{R} \\ \underline{\boldsymbol{V}}'_\mathrm{A} \end{bmatrix} \tag{18.46}$$

where the prime denotes variables on the primary side of the ideal transformers. For the ideal transformers

$$\underline{\boldsymbol{I}}''_\mathrm{A} = \underline{\tau}^* \underline{\boldsymbol{I}}'_\mathrm{A}, \qquad \underline{\boldsymbol{V}}'_\mathrm{A} = \underline{\tau} \underline{\boldsymbol{V}}_\mathrm{a}, \qquad \underline{\boldsymbol{V}}_\mathrm{a} = \mathbf{1}_\mathrm{A} \underline{V}_\mathrm{a} \tag{18.47}$$

where $\underline{\boldsymbol{V}}_\mathrm{a}$ is a vector column with all the elements identical and equal to $\underline{V}_\mathrm{a}$. For the equivalent node

$$\underline{I}_\mathrm{a} = \mathbf{1}_\mathrm{A}^\mathrm{T} \underline{\boldsymbol{I}}''_\mathrm{A} \tag{18.48}$$

which is the mathematical expression of the fact that the sum of secondary currents in the ideal transformers is equal to the nodal current at the equivalent node (Figure 18.15).

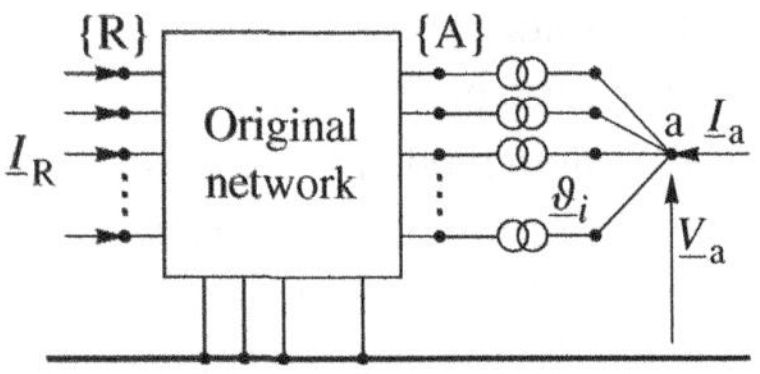

Figure 18.15 Electrical interpretation of Zhukov's aggregation.

Equation (18.46) can be written as

$$\underline{\boldsymbol{I}}_\mathrm{R} = \underline{\boldsymbol{Y}}_\mathrm{RR} \underline{\boldsymbol{V}}_\mathrm{R} + \underline{\boldsymbol{Y}}_\mathrm{RA} \underline{\boldsymbol{V}}'_\mathrm{A} \tag{18.49a}$$

$$\underline{\boldsymbol{I}}'_\mathrm{A} = \underline{\boldsymbol{Y}}_\mathrm{AR} \underline{\boldsymbol{V}}_\mathrm{R} + \underline{\boldsymbol{Y}}_\mathrm{AA} \underline{\boldsymbol{V}}'_\mathrm{A} \tag{18.49b}$$

Vector $\underline{\boldsymbol{V}}'_{\mathrm{A}}$ in these equations can be, according to Eq. (18.47), replaced by $\underline{\boldsymbol{\tau}}\,\underline{\boldsymbol{V}}_{\mathrm{a}}$. Taking this into account and left-multiplying Eq. (18.49b) by $\underline{\boldsymbol{\tau}}^*$ gives

$$\underline{\boldsymbol{I}}_{\mathrm{R}} = \underline{\boldsymbol{Y}}_{\mathrm{RR}}\underline{\boldsymbol{V}}_{\mathrm{R}} + \underline{\boldsymbol{Y}}_{\mathrm{RA}}\underline{\boldsymbol{\tau}}\,\underline{\boldsymbol{V}}_{\mathrm{a}} \tag{18.50a}$$

$$\underline{\boldsymbol{\tau}}^*\underline{\boldsymbol{I}}'_{\mathrm{A}} = \underline{\boldsymbol{\tau}}^*\underline{\boldsymbol{Y}}_{\mathrm{AR}}\underline{\boldsymbol{V}}_{\mathrm{R}} + \underline{\boldsymbol{\tau}}^*\underline{\boldsymbol{Y}}_{\mathrm{AA}}\underline{\boldsymbol{\tau}}\,\underline{\boldsymbol{V}}_{\mathrm{a}} \tag{18.50b}$$

Now, according to Eq. (18.47), $\underline{\boldsymbol{\tau}}^*\underline{\boldsymbol{I}}'_{\mathrm{A}}$ may be replaced by $\underline{\boldsymbol{I}}''_{\mathrm{A}}$ and $\underline{\boldsymbol{V}}_{\mathrm{a}}$ by $\mathbf{1}_{\mathrm{A}}\underline{V}_{\mathrm{a}}$. Equations (18.50) will then take the form

$$\underline{\boldsymbol{I}}_{\mathrm{R}} = \underline{\boldsymbol{Y}}_{\mathrm{RR}}\underline{\boldsymbol{V}}_{\mathrm{R}} + \underline{\boldsymbol{Y}}_{\mathrm{RA}}\underline{\boldsymbol{\tau}}\,\mathbf{1}_{\mathrm{A}}\underline{V}_{\mathrm{a}} \tag{18.51a}$$

$$\underline{\boldsymbol{I}}''_{\mathrm{A}} = \underline{\boldsymbol{\tau}}^*\underline{\boldsymbol{Y}}_{\mathrm{AR}}\underline{\boldsymbol{V}}_{\mathrm{R}} + \underline{\boldsymbol{\tau}}^*\underline{\boldsymbol{Y}}_{\mathrm{AA}}\underline{\boldsymbol{\tau}}\;\mathbf{1}_{\mathrm{A}}\underline{V}_{\mathrm{a}} \tag{18.51b}$$

Equation (18.49b) is left-multiplied by $\mathbf{1}_{\mathrm{A}}^{\mathrm{T}}$ which, after taking into account (18.46), gives

$$\underline{I}_{\mathrm{a}} = \mathbf{1}_{\mathrm{A}}^{\mathrm{T}}\underline{\boldsymbol{\tau}}^*\underline{\boldsymbol{Y}}_{\mathrm{AR}}\underline{\boldsymbol{V}}_{\mathrm{R}} + \mathbf{1}_{\mathrm{A}}^{\mathrm{T}}\underline{\boldsymbol{\tau}}^*\underline{\boldsymbol{Y}}_{\mathrm{AA}}\underline{\boldsymbol{\tau}}\,\mathbf{1}_{\mathrm{A}}\underline{V}_{\mathrm{a}} \tag{18.51c}$$

Taking into account (18.45), it is possible to write Eqs. (18.51a) and (18.51c) as

$$\underline{\boldsymbol{I}}_{\mathrm{R}} = \underline{\boldsymbol{Y}}_{\mathrm{RR}}\underline{\boldsymbol{V}}_{\mathrm{R}} + \underline{\boldsymbol{Y}}_{\mathrm{RA}}\underline{\boldsymbol{\vartheta}}\,\underline{V}_{\mathrm{a}} \tag{18.52a}$$

$$\underline{I}_{\mathrm{a}} = \underline{\boldsymbol{\vartheta}}^{*\mathrm{T}}\underline{\boldsymbol{Y}}_{\mathrm{AR}}\underline{\boldsymbol{V}}_{\mathrm{R}} + \underline{\boldsymbol{\vartheta}}^{*\mathrm{T}}\underline{\boldsymbol{Y}}_{\mathrm{AA}}\underline{\boldsymbol{\vartheta}}\,\underline{V}_{\mathrm{a}} \tag{18.52b}$$

or in matrix form

$$\begin{bmatrix} \underline{\boldsymbol{I}}_{\mathrm{R}} \\ \underline{I}_{\mathrm{a}} \end{bmatrix} = \left[\begin{array}{c:c} \underline{\boldsymbol{Y}}_{\mathrm{RR}} & \underline{\boldsymbol{Y}}_{\mathrm{RA}}\underline{\boldsymbol{\vartheta}} \\ \hdashline \underline{\boldsymbol{\vartheta}}^{*\mathrm{T}}\underline{\boldsymbol{Y}}_{\mathrm{AR}} & \underline{\boldsymbol{\vartheta}}^{*\mathrm{T}}\underline{\boldsymbol{Y}}_{\mathrm{AA}}\underline{\boldsymbol{\vartheta}} \end{array}\right] \begin{bmatrix} \underline{\boldsymbol{V}}_{\mathrm{R}} \\ \underline{V}_{\mathrm{a}} \end{bmatrix} \tag{18.53}$$

Note that Eq. (18.53) is identical with Eq. (18.21) obtained for Zhukov's aggregation. This means that the methods proposed by Demello et al. (1975) and Zhukov (1964) are equivalent. The advantage of Demello's approach is that it gives an electrical interpretation of mathematical transformations.

18.6.2 Incremental Equivalent Model

For the system shown in Figure 18.6 in which nodes {R} and {A} have been emphasized, the incremental Eq. (3.182) takes the following form

$$\begin{bmatrix} \Delta\boldsymbol{P}_{\mathrm{R}} \\ \Delta\boldsymbol{P}_{\mathrm{A}} \end{bmatrix} = \begin{bmatrix} \boldsymbol{H}_{\mathrm{RR}} & \boldsymbol{H}_{\mathrm{RA}} \\ \boldsymbol{H}_{\mathrm{AR}} & \boldsymbol{H}_{\mathrm{AA}} \end{bmatrix} \begin{bmatrix} \Delta\boldsymbol{\delta}'_{\mathrm{R}} \\ \Delta\boldsymbol{\delta}'_{\mathrm{A}} \end{bmatrix} \tag{18.54}$$

where elements in matrix $\boldsymbol{H} = [\partial\boldsymbol{P}/\partial\boldsymbol{\delta}']$ are the synchronizing powers.

In the case of exact coherency (Section 18.2.4) the increments of rotor angles are identical and can be written as

$$\Delta\delta'_i(t) = \Delta\delta'_j(t) = \Delta\delta'_{\mathrm{a}}(t) \quad \text{for} \quad i,j \in \{\mathrm{A}\} \tag{18.55}$$

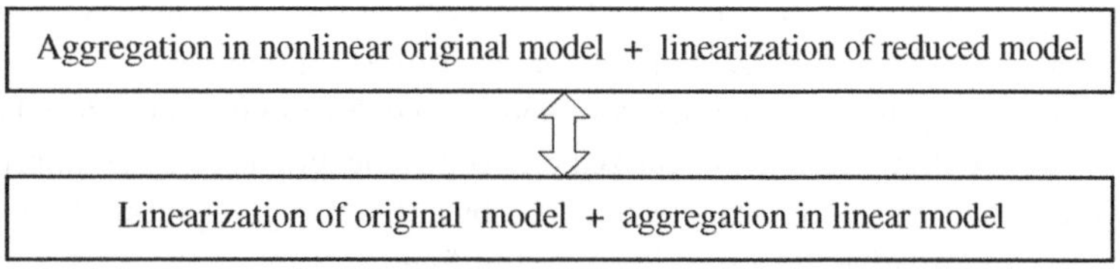

Figure 18.16 Illustration of the fact that aggregation and linearization are commutative.

where $\Delta\delta'_{\mathrm{a}}(t)$ is a common change of angles for the node group {A}. Equation (18.55) can be expressed in matrix form as

$$\Delta\boldsymbol{\delta}'_{\mathrm{A}} = \mathbf{1}_{\mathrm{A}} \cdot \Delta\delta'_{\mathrm{a}}(t) \tag{18.56}$$

where $\mathbf{1}_{\mathrm{A}}$ is a vector of unity elements of the size of group {A}.

When a group of generators is replaced by one equivalent generating node (Figure 18.6), it is assumed that $\underline{S}_{\mathrm{a}} = \sum_{i\in\{\mathrm{A}\}} \underline{S}_i$, which also means $P_{\mathrm{a}} = \sum_{i\in\{\mathrm{A}\}} P_i$. In the case of the incremental model

$$\Delta P_{\mathrm{a}} = \sum_{i\in\{\mathrm{A}\}} \Delta P_i \tag{18.57}$$

which means that a change of power in the equivalent node is equal to the sum of changes of power in the replaced nodes.

Equation (18.57) can be written in matrix form as

$$\Delta P_{\mathrm{a}} = \mathbf{1}_{\mathrm{A}}^{\mathrm{T}} \cdot \Delta\boldsymbol{P}_{\mathrm{A}} \tag{18.58}$$

Equation (18.54) gives

$$\Delta\boldsymbol{P}_{\mathrm{R}} = \boldsymbol{H}_{\mathrm{RR}}\Delta\boldsymbol{\delta}_{\mathrm{R}} + \boldsymbol{H}_{\mathrm{RA}}\Delta\boldsymbol{\delta}'_{\mathrm{A}} \tag{18.59a}$$

$$\Delta\boldsymbol{P}_{\mathrm{A}} = \boldsymbol{H}_{\mathrm{AR}}\Delta\boldsymbol{\delta}_{\mathrm{R}} + \boldsymbol{H}_{\mathrm{AA}}\Delta\boldsymbol{\delta}'_{\mathrm{A}} \tag{18.59b}$$

Substituting into Eqs. (18.59) the value from Eq. (18.56) gives

$$\Delta\boldsymbol{P}_{\mathrm{R}} = \boldsymbol{H}_{\mathrm{RR}}\Delta\boldsymbol{\delta}_{\mathrm{R}} + \boldsymbol{H}_{\mathrm{RA}}\mathbf{1}_{\mathrm{A}} \cdot \Delta\delta'_{\mathrm{a}} \tag{18.60a}$$

$$\Delta\boldsymbol{P}_{\mathrm{A}} = \boldsymbol{H}_{\mathrm{AR}}\Delta\boldsymbol{\delta}_{\mathrm{R}} + \boldsymbol{H}_{\mathrm{AA}}\mathbf{1}_{\mathrm{A}} \cdot \Delta\delta'_{\mathrm{a}} \tag{18.60b}$$

Left-multiplying (18.60b) by $\mathbf{1}_{\mathrm{A}}^{\mathrm{T}}$ and taking into account (18.58) gives

$$\Delta P_{\mathrm{a}} = \mathbf{1}_{\mathrm{A}}^{\mathrm{T}}\boldsymbol{H}_{\mathrm{AR}}\Delta\boldsymbol{\delta}_{\mathrm{R}} + \mathbf{1}_{\mathrm{A}}^{\mathrm{T}}\boldsymbol{H}_{\mathrm{AA}}\mathbf{1}_{\mathrm{A}} \cdot \Delta\delta'_{\mathrm{a}} \tag{18.61}$$

Equations (18.60a) and (18.61) may be merged in matrix form as

$$\begin{bmatrix} \Delta\boldsymbol{P}_{\mathrm{R}} \\ \Delta P_{\mathrm{a}} \end{bmatrix} = \begin{bmatrix} \boldsymbol{H}_{\mathrm{RR}} & \boldsymbol{H}_{\mathrm{RA}}\mathbf{1}_{\mathrm{A}} \\ \mathbf{1}_{\mathrm{A}}^{\mathrm{T}}\boldsymbol{H}_{\mathrm{AR}} & \mathbf{1}_{\mathrm{A}}^{\mathrm{T}}\boldsymbol{H}_{\mathrm{AA}}\mathbf{1}_{\mathrm{A}} \end{bmatrix} \begin{bmatrix} \Delta\boldsymbol{\delta}_{\mathrm{R}} \\ \Delta\delta'_{\mathrm{a}} \end{bmatrix} \tag{18.62}$$

or

$$\begin{bmatrix} \Delta\boldsymbol{P}_{\mathrm{R}} \\ \Delta P_{\mathrm{a}} \end{bmatrix} = \begin{bmatrix} \boldsymbol{H}_{\mathrm{RR}} & \boldsymbol{H}_{\mathrm{Ra}} \\ \boldsymbol{H}_{\mathrm{aR}} & H_{\mathrm{aa}} \end{bmatrix} \begin{bmatrix} \Delta\boldsymbol{\delta}_{\mathrm{R}} \\ \Delta\delta'_{\mathrm{a}} \end{bmatrix} \tag{18.63}$$

where

$$\boldsymbol{H}_{\mathrm{aR}} = \mathbf{1}_{\mathrm{A}}^{\mathrm{T}}\boldsymbol{H}_{\mathrm{AR}}, \quad \boldsymbol{H}_{\mathrm{Ra}} = \boldsymbol{H}_{\mathrm{RA}}\mathbf{1}_{\mathrm{A}}, \quad H_{\mathrm{aa}} = \mathbf{1}_{\mathrm{A}}^{\mathrm{T}}\boldsymbol{H}_{\mathrm{AA}}\mathbf{1}_{\mathrm{A}} \tag{18.64}$$

and $\boldsymbol{H}_{\mathrm{aR}}$ is a row vector, $\boldsymbol{H}_{\mathrm{Ra}}$ is a column vector, while H_{aa} is a scalar.

Note that $\boldsymbol{H}_{\mathrm{aR}}$ is created from $\boldsymbol{H}_{\mathrm{AR}}$ by adding up its rows and $\boldsymbol{H}_{\mathrm{Ra}}$ is created from $\boldsymbol{H}_{\mathrm{RA}}$ by adding up its columns. Element H_{aa} is created by adding up all the elements of $\boldsymbol{H}_{\mathrm{AA}}$. This means that the aggregation of generators in the incremental model in effect adds up all the synchronizing powers.

The described aggregation method corresponding to Eq. (18.62) is proposed by Di Caprio and Marconato (1975).

Now it will be shown that the linearized reduced model proposed by Di Caprio and Marconato corresponds to the linearized form of the reduced model obtained by Zhukov's aggregation. The operations of linearization and aggregation are commutative, which is illustrated in Figure 18.16.

A simple proof of this can be conducted in the complex-number domain by calculating the derivatives $\underline{J}_{ij} = \partial \underline{S}_i / \partial \delta_j$ directly from the apparent power $\underline{S}_i$ instead of active power $P_i = \mathrm{Re}\, \underline{S}_i$. Obviously $\underline{S}_i = P_i + \mathrm{j}Q_i$ and hence $H_{ij} = \partial P_i / \partial \delta'_j = \mathrm{Re}\, \underline{J}_{ij}$. Thus the proof conducted for $\underline{\boldsymbol{J}} = [\partial \underline{\boldsymbol{S}} / \partial \boldsymbol{\delta}']$ is at the same time valid also for $\boldsymbol{H} = [\partial \boldsymbol{P} / \partial \boldsymbol{\delta}']$. Calculation of the derivative in the complex-number domain makes it possible to avoid complicated transformations of trigonometric functions which appear in the equations for real power.

18.6.2.1 Aggregation in the Linear Model

In the original model before aggregation (Figure 18.6a) the apparent power for node $i \in \{\mathrm{R}\}$ can be expressed as

$$\underline{S}_i = \underline{V}_i \sum_{j \in \{\mathrm{R}\}} \underline{Y}^*_{ij} \underline{V}^*_j + \underline{V}_i \sum_{k \in \{\mathrm{A}\}} \underline{Y}^*_{ik} \underline{E}^*_k \tag{18.65}$$

where

$$\underline{V}_i = V_i \mathrm{e}^{\mathrm{j}\delta_i}, \quad \underline{V}^*_j = V_j \mathrm{e}^{-\mathrm{j}\delta_j}, \quad \underline{E}^*_k = E_k \mathrm{e}^{-\mathrm{j}\delta'_k} \tag{18.66}$$

Differentiating gives

$$\underline{J}_{ij} = \frac{\partial \underline{S}_i}{\partial \delta_j} = -\mathrm{j}\, \underline{V}_i \underline{Y}^*_{ij} \underline{V}^*_j \quad \text{and} \quad \underline{J}_{ik} = \frac{\partial \underline{S}_i}{\partial \delta'_k} = -\mathrm{j}\, \underline{V}_i \underline{Y}^*_{ik} \underline{E}^*_k \tag{18.67}$$

Similarly, one gets for $l \in \{\mathrm{A}\}$

$$\underline{S}_l = \underline{E}_l \sum_{j \in \{\mathrm{R}\}} \underline{Y}^*_{lj} \underline{V}^*_j + \underline{E}_l \sum_{k \in \{\mathrm{A}\}} \underline{Y}^*_{lk} \underline{E}^*_k \tag{18.68}$$

where

$$\underline{E}_l = E_l \mathrm{e}^{\mathrm{j}\delta'_l}, \quad \underline{V}^*_j = V_j \mathrm{e}^{-\mathrm{j}\delta_j}, \quad \underline{E}^*_k = E_k \mathrm{e}^{-\mathrm{j}\delta'_k} \tag{18.69}$$

After calculating the derivatives, one gets

$$\underline{J}_{lj} = \frac{\partial \underline{S}_l}{\partial \delta_j} = -\mathrm{j}\, \underline{E}_l \underline{Y}^*_{lj} \underline{V}^*_j \quad \text{and} \quad \underline{J}_{lk} = \frac{\partial \underline{S}_l}{\partial \delta'_k} = -\mathrm{j}\, \underline{E}_l \underline{Y}^*_{lk} \underline{E}^*_k \tag{18.70}$$

After aggregation of group {A} using the method of Di Caprio and Marconato, that is by adding the synchronizing powers, one gets

$$\underline{J}_{i\mathrm{a}} = \sum_{k \in \{\mathrm{A}\}} \underline{J}_{ik} = -\mathrm{j}\, \underline{V}_i \sum_{k \in \{\mathrm{A}\}} \underline{Y}^*_{ik} \underline{E}^*_k \tag{18.71a}$$

$$\underline{J}_{\mathrm{a}j} = \sum_{l \in \{\mathrm{A}\}} \underline{J}_{lj} = -\mathrm{j}\, \underline{V}^*_j \sum_{l \in \{\mathrm{A}\}} \underline{Y}^*_{lj} \underline{E}_l \tag{18.71b}$$

Elements $\underline{J}_{ij}$ for $i, j \in \{\mathrm{R}\}$ do not change during aggregation of group {A}.

18.6.2.2 Linearization of the Reduced Nonlinear Model

In the reduced model obtained after aggregation by Zhukov's method (Figure 18.6b), the apparent power for node $i \in \{\mathrm{R}\}$ can be expressed as

$$\underline{S}_i = \underline{V}_i \sum_{j \in \{\mathrm{R}\}} \underline{Y}^*_{ij} \underline{V}^*_j + \underline{V}_i \underline{Y}^*_{i\mathrm{a}} \underline{E}^*_\mathrm{a} \tag{18.72}$$

where

$$\underline{V}_i = V_i e^{j\delta_i}, \quad \underline{V}_j^* = V_j e^{-j\delta_j}, \quad \underline{E}_a^* = E_a e^{-j\delta'_a} \tag{18.73}$$

Differentiation gives

$$\underline{J}_{ij} = \frac{\partial \underline{S}_i}{\partial \delta_j} = -j\,\underline{V}_i \underline{Y}_{ij}^* \underline{V}_j^*, \quad \underline{J}_{ia} = \frac{\partial \underline{S}_i}{\partial \delta'_a} = -j\,\underline{V}_i \underline{Y}_{ia}^* \underline{E}_a^* \tag{18.74}$$

Utilizing Eq. (18.15) one can write

$$\underline{Y}_{ia} = \sum_{k\in\{A\}} \underline{Y}_{ik} \frac{\underline{E}_k}{\underline{E}_a}$$

and after substituting in the second of Eq. (18.74) one finally gets

$$\underline{J}_{ia} = \frac{\partial \underline{S}_i}{\partial \delta'_a} = -j\,\underline{V}_i \sum_{k\in\{A\}} \underline{Y}_{ik}^* \frac{\underline{E}_k^*}{\underline{E}_a^*} \underline{E}_a^* = -j\,\underline{V}_i \sum_{k\in\{A\}} \underline{Y}_{ik}^* \underline{E}_k^* \tag{18.75}$$

The apparent power in the equivalent node (Figure 18.6) is given by

$$\underline{S}_a = \underline{E}_a \sum_{j\in\{R\}} \underline{Y}_{aj}^* \underline{V}_j^* + \underline{E}_a \underline{Y}_{aa}^* \underline{E}_a^* \tag{18.76}$$

Differentiating gives

$$\underline{J}_{aj} = \frac{\partial \underline{S}_a}{\partial \delta_j} = -j\,\underline{E}_a \underline{Y}_{aj}^* \underline{V}_j^* \tag{18.77}$$

Utilizing Eq. (18.19) one can write

$$\underline{Y}_{aj} = \sum_{l\in\{A\}} \frac{\underline{E}_l^*}{\underline{E}_a^*} \underline{Y}_{lj}$$

and after substituting into Eq. (18.77) one finally gets

$$\underline{J}_{aj} = \frac{\partial \underline{S}_a}{\partial \delta_j} = -j\,\underline{V}_j^* \underline{E}_a \sum_{l\in\{A\}} \frac{\underline{E}_l}{\underline{E}_a} \underline{Y}_{lj}^* = -j\,\underline{V}_j^* \sum_{l\in\{A\}} \underline{Y}_{lj}^* \underline{E}_l \tag{18.78}$$

Comparing (18.71a) with (18.75) and (18.71b) with (18.78) clearly shows that the values obtained through aggregation in the linear model, and by linearization in the reduced model, are the same. Equivalence of the synchronizing power $\underline{J}_{aa} = \partial \underline{S}_a / \partial \delta'_a$ in both cases is due to the self-synchronizing power being equal to the sum of mutual synchronizing powers taken with the opposite sign – Eqs. (3.189) and (3.190) in Section 3.5. This concludes the proof that aggregation and linearization are commutative.

18.6.3 Modal Interpretation of Exact Coherency

In Chapter 12 power swings in the linearized power system model are analyzed using modal analysis. Each mode (corresponding to an eigenvalue of the state matrix) has a frequency of oscillations and a damping ratio. Now it will be shown here that exact coherency can also be analyzed using modal analysis. Also, a proof of the exact coherency condition given by (18.39) will be conducted.

Partial inversion (Appendix A.2) of Eq. (18.54) gives

$$\Delta \boldsymbol{P}_{\rm A} = \boldsymbol{H}_{\rm A} \Delta \boldsymbol{\delta}'_{\rm A} + \boldsymbol{R}_{\rm A} \Delta \boldsymbol{P}_{\rm R} \tag{18.79}$$

where

$$\boldsymbol{H}_{\rm A} = \boldsymbol{H}_{\rm AA} - \boldsymbol{H}_{\rm AR} \boldsymbol{H}_{\rm RR}^{-1} \boldsymbol{H}_{\rm RA} \tag{18.80}$$

$$\boldsymbol{R}_{\rm A} = \boldsymbol{H}_{\rm AR} \boldsymbol{H}_{\rm RR}^{-1} \tag{18.81}$$

The matrix equation for the motion of rotors in group {A} can be written in a similar way to Eq. (12.109), while for further considerations it is more convenient to express the equations as

$$\boldsymbol{M}_{\rm A} \Delta \ddot{\boldsymbol{\delta}}'_{\rm A} = -\boldsymbol{H}_{\rm A} \Delta \boldsymbol{\delta}'_{\rm A} - \boldsymbol{R}_{\rm A} \Delta \boldsymbol{P}_{\rm R} - \boldsymbol{D}_{\rm A} \Delta \dot{\boldsymbol{\delta}}'_{\rm A} \tag{18.82}$$

where $\boldsymbol{M}_{\rm A}$ and $\boldsymbol{D}_{\rm A}$ are diagonal matrices containing inertia and damping coefficients. Neglecting damping, the equation can be written as

$$\Delta \ddot{\boldsymbol{\delta}}'_{\rm A} = -\boldsymbol{M}_{\rm A}^{-1} \boldsymbol{H}_{\rm A} \Delta \boldsymbol{\delta}'_{\rm A} - \boldsymbol{M}_{\rm A}^{-1} \boldsymbol{R}_{\rm A} \Delta \boldsymbol{P}_{\rm R} \tag{18.83}$$

This is the state equation of group {A} in which changes of power $\Delta \boldsymbol{P}_{\rm R}$ in nodes {R} are treated as inputs. This is a second-order equation, which can be replaced by a first-order matrix equation (similar to Eq. [12.110] in Section 12.2.1)

$$\begin{bmatrix} \Delta \dot{\boldsymbol{\delta}}'_{\rm A} \\ \Delta \dot{\boldsymbol{\omega}}_{\rm A} \end{bmatrix} = \left[\begin{array}{c:c} \boldsymbol{0} & \boldsymbol{1} \\ \hdashline -\boldsymbol{M}_{\rm A}^{-1} \boldsymbol{H}_{\rm A} & \boldsymbol{0} \end{array}\right] \left[\begin{array}{c} \Delta \boldsymbol{\delta}'_{\rm A} \\ \hdashline \Delta \boldsymbol{\omega}_{\rm A} \end{array}\right] - \left[\begin{array}{c} \boldsymbol{0}_{\rm A} \\ \hdashline \boldsymbol{M}_{\rm A}^{-1} \boldsymbol{R}_{\rm A} \Delta \boldsymbol{P}_{\rm R} \end{array}\right] \tag{18.84}$$

where $\boldsymbol{0}_{\rm A}$ is a zero column vector and $\Delta \boldsymbol{\omega}_{\rm A} = \Delta \dot{\boldsymbol{\delta}}'_{\rm A}$. Equation (18.84) has the form of Eq. (12.95) described in Section 12.1.5. Now changes of $\Delta \boldsymbol{\delta}'_{\rm A}$ enforced by changes of $\Delta \boldsymbol{P}_{\rm R}$ will be considered using modal analysis.

Let μ_i be an eigenvalue of the state matrix $\boldsymbol{a} = -\boldsymbol{M}_{\rm A}^{-1} \boldsymbol{H}_{\rm A}$ from Eq. (18.83) and let $\boldsymbol{w}_i$ and $\boldsymbol{u}_i$ be, respectively, the right and left eigenvectors of this matrix. It is shown in Section 12.2.2 that a system is stable when all the eigenvalues μ_i are real and negative. The eigenvalues λ_i of the state matrix in Eq. (18.84) are equal to $\lambda_i = \sqrt{\mu_i}$ and, as $\mu_i < 0$, they are imaginary numbers (see Figure 12.5). To simplify considerations, eigenvalues μ_i of the state matrix in Eq. (18.83) will be analyzed rather than λ_i.

From Section 12.1.1

$$\boldsymbol{W} = [\boldsymbol{w}_1 \quad \boldsymbol{w}_2 \quad \cdots \quad \boldsymbol{w}_N] \quad \text{and} \quad \boldsymbol{U} = \boldsymbol{W}^{-1} = \begin{bmatrix} \boldsymbol{u}_1 \\ \boldsymbol{u}_2 \\ \vdots \\ \boldsymbol{u}_N \end{bmatrix} \tag{18.85}$$

where $\boldsymbol{W}$ and $\boldsymbol{U}$ are square matrices consisting of right and left eigenvectors.

Equation (3.189) proved in Section 3.5 shows that the sum of elements in each row of matrix $\boldsymbol{H} = [\partial \boldsymbol{P}/\partial \boldsymbol{\delta}']$ is equal to zero. Let $\boldsymbol{1}_{\rm A}$, $\boldsymbol{1}_{\rm R}$ and $\boldsymbol{0}_{\rm A}$, $\boldsymbol{0}_{\rm R}$ be column vectors with all elements equal to zero or one, respectively. Then Eq. (18.54) can be transformed, using (3.189), to

$$\begin{bmatrix} \boldsymbol{0}_{\rm R} \\ \boldsymbol{0}_{\rm A} \end{bmatrix} = \begin{bmatrix} \boldsymbol{H}_{\rm RR} & \boldsymbol{H}_{\rm RA} \\ \boldsymbol{H}_{\rm AR} & \boldsymbol{H}_{\rm AA} \end{bmatrix} \begin{bmatrix} \boldsymbol{1}_{\rm R} \\ \boldsymbol{1}_{\rm A} \end{bmatrix} \tag{18.86}$$

That is for $\Delta \boldsymbol{\delta}'_{\rm A} = \boldsymbol{1}_{\rm A}$ and $\Delta \boldsymbol{\delta}_{\rm R} = \boldsymbol{1}_{\rm R}$ one gets $\Delta \boldsymbol{P}_{\rm A} = \boldsymbol{0}_{\rm A}$ and $\Delta \boldsymbol{P}_{\rm R} = \boldsymbol{0}_{\rm R}$. Substituting these equations into (18.79) gives $\boldsymbol{H}_{\rm A} \boldsymbol{1}_{\rm A} = \boldsymbol{0}_{\rm A}$. This means that partial inversion given by (18.80) maintains property (3.189) that the sum of elements in each row is equal to zero. Left-multiplying the last equation by $\boldsymbol{M}_{\rm A}^{-1}$ gives

$$\boldsymbol{a} \cdot \boldsymbol{1}_{\rm A} = -\boldsymbol{M}_{\rm A}^{-1} \boldsymbol{H}_{\rm A} \cdot \boldsymbol{1}_{\rm A} = \boldsymbol{0}_{\rm A} = 0 \cdot \boldsymbol{1}_{\rm A} \tag{18.87}$$

that is $\boldsymbol{a} \cdot \mathbf{1}_{\mathrm{A}} = 0 \cdot \mathbf{1}_{\mathrm{A}}$. This equation is the same as Eq. (12.1) defining the eigenvalue and the right eigenvector. Hence finally

$$\mu_1 = 0 \quad \text{and} \quad \boldsymbol{w}_1 = \mathbf{1}_{\mathrm{A}} \tag{18.88}$$

This leads to an important conclusion that one of the eigenvalues of the state matrix $\boldsymbol{a} = -\boldsymbol{M}_{\mathrm{A}}^{-1}\boldsymbol{H}_{\mathrm{A}}$ in Eq. (18.83) is equal to zero and the corresponding right eigenvector consists of ones.

As $\boldsymbol{UW} = \mathbf{1}$ is a diagonal identity matrix, the following hold for the right and left eigenvectors defined by (18.85): $\boldsymbol{u}_1\boldsymbol{w}_1 = 1,\ \boldsymbol{u}_2\boldsymbol{w}_1 = 0,\ \ldots;\ \boldsymbol{u}_n\boldsymbol{w}_1 = 0$. Substituting $\boldsymbol{w}_1 = \mathbf{1}_{\mathrm{A}}$ gives

$$\begin{aligned} \boldsymbol{u}_1\mathbf{1}_{\mathrm{A}} &= 1 \\ \boldsymbol{u}_2\mathbf{1}_{\mathrm{A}} &= 0 \\ &\vdots \\ \boldsymbol{u}_n\mathbf{1}_{\mathrm{A}} &= 0 \end{aligned} \tag{18.89}$$

These relationships are crucial for further considerations.

As in Eq. (12.41), new variables $\boldsymbol{z}$ are introduced, referred to as the *modal variables*, which are related to the state variables $\Delta\delta'_{\mathrm{A}}$ by

$$\Delta\delta'_{\mathrm{A}} = \boldsymbol{Wz} \quad \text{and} \quad \boldsymbol{z} = \boldsymbol{U}\Delta\delta'_{\mathrm{A}} \tag{18.90}$$

Expanding the second equation gives

$$\begin{bmatrix} z_1 \\ z_2 \\ \vdots \\ z_n \end{bmatrix} = \begin{bmatrix} \boldsymbol{u}_1 \\ \boldsymbol{u}_2 \\ \vdots \\ \boldsymbol{u}_n \end{bmatrix} \cdot \Delta\delta'_{\mathrm{A}} = \begin{bmatrix} \boldsymbol{u}_1\Delta\delta'_{\mathrm{A}} \\ \boldsymbol{u}_2\Delta\delta'_{\mathrm{A}} \\ \vdots \\ \boldsymbol{u}_n\Delta\delta'_{\mathrm{A}} \end{bmatrix} \tag{18.91}$$

The generators from group (A) are assumed to be exactly coherent, that is they satisfy (18.28) and (18.56). Substituting $\Delta\delta'_{\mathrm{A}}$ for the right-hand side of Eq. (18.56) and taking into account (18.89) leads to

$$\begin{bmatrix} z_1 \\ z_2 \\ \vdots \\ z_n \end{bmatrix} = \begin{bmatrix} \boldsymbol{u}_1\mathbf{1}_{\mathrm{A}} \\ \boldsymbol{u}_2\mathbf{1}_{\mathrm{A}} \\ \vdots \\ \boldsymbol{u}_n\mathbf{1}_{\mathrm{A}} \end{bmatrix} \cdot \Delta\delta'_{\mathrm{a}} = \begin{bmatrix} 1 \\ 0 \\ \vdots \\ 0 \end{bmatrix} \cdot \Delta\delta'_{\mathrm{a}} = \begin{bmatrix} \Delta\delta'_{\mathrm{a}} \\ 0 \\ \vdots \\ 0 \end{bmatrix} \tag{18.92}$$

This means that if group {A} is exactly coherent then among its n modal variables there is only one modal variable $z_1(t)$ excited. This modal variable is responsible for the swinging of the whole group against the rest of the system. The remaining modal variables $z_2(t), \ldots, z_n(t)$ corresponding to the swings inside the group are not excited, that is $z_2(t) = , \ldots, = z_n(t) = 0$. These considerations lead to the following conclusion:

> In modal analysis, the exact electromechanical coherency of generators belonging to the external subsystem corresponds to a situation where modal variables representing the swinging of generator rotors inside the coherent group are not excited by disturbances in the internal subsystem. Disturbances in the internal subsystem excite only that modal variable that represents the swinging of the whole coherent group with respect to the rest of the system.

Now it will be investigated what the structure of matrix $\boldsymbol{M}_{\mathrm{A}}^{-1}\boldsymbol{R}_{\mathrm{A}}$ in Eq. (18.84) must be so that disturbances in the internal subsystem represented by $\Delta\boldsymbol{P}_{\mathrm{R}}$ cannot excite modal variables $z_2(t), \ldots, z_n(t)$. Section 12.1.5 outlines the general conditions for a particular modal variable not to be excited and this theory will be applied now. Substituting (18.90) into (18.83) gives

$$\ddot{\boldsymbol{z}} = \boldsymbol{\Lambda}\boldsymbol{z} - \boldsymbol{r}\cdot\Delta\boldsymbol{P}_{\mathrm{R}} \tag{18.93}$$

where

$$\boldsymbol{r} = \boldsymbol{W}^{-1}\boldsymbol{M}_{\mathrm{A}}^{-1}\boldsymbol{R}_{\mathrm{A}} \tag{18.94}$$

Substituting (18.81) for $\boldsymbol{R}_{\mathrm{A}}$ gives

$$\boldsymbol{r} = \boldsymbol{W}^{-1}\boldsymbol{M}_{\mathrm{A}}^{-1}\boldsymbol{H}_{\mathrm{AR}}\boldsymbol{H}_{\mathrm{RR}}^{-1} \tag{18.95}$$

and

$$\boldsymbol{W}\boldsymbol{r}\boldsymbol{H}_{\mathrm{RR}} = \boldsymbol{M}_{\mathrm{A}}^{-1}\boldsymbol{H}_{\mathrm{AR}} \tag{18.96}$$

Equation (18.93) shows that excitation of modal variables enforced by $\Delta\boldsymbol{P}_{\mathrm{R}}$ is decided by matrix $\boldsymbol{r}$ given by (18.94). Hence investigation of the structure of matrix $\boldsymbol{r}$ should lead to the derivation of a condition for only one modal variable to be excited, that is the condition for exact coherency. To simplify considerations further, Eq. (18.93) can be rewritten as

$$\begin{bmatrix} \ddot{z}_1 \\ \ddot{z}_2 \\ \vdots \\ \ddot{z}_n \end{bmatrix} = \begin{bmatrix} \lambda_1 & & & \\ & \lambda_2 & & \\ & & \ddots & \\ & & & \lambda_n \end{bmatrix} \begin{bmatrix} z_1 \\ z_2 \\ \vdots \\ z_n \end{bmatrix} - \begin{bmatrix} \boldsymbol{r}_1 \\ \boldsymbol{r}_2 \\ \vdots \\ \boldsymbol{r}_n \end{bmatrix} \cdot \Delta\boldsymbol{P}_{\mathrm{R}} \tag{18.97}$$

Equation (18.97) shows that any input $\Delta\boldsymbol{P}_{\mathrm{R}}$ will excite only one modal variable $z_1(t)$ if matrix $\boldsymbol{r}$ has the following structure

$$\boldsymbol{r} = \begin{bmatrix} \boldsymbol{r}_1 \\ \boldsymbol{r}_2 \\ \vdots \\ \boldsymbol{r}_n \end{bmatrix} = \begin{bmatrix} \boldsymbol{r}_1 \\ \boldsymbol{0} \\ \vdots \\ \boldsymbol{0} \end{bmatrix} \tag{18.98}$$

that is it will have only one row (the first) that is nonzero. The first column of $\boldsymbol{W}$ consists of ones – see (18.88). Now taking into account (18.98) leads to

$$\boldsymbol{W}\boldsymbol{r} = [\boldsymbol{1}_{\mathrm{A}} \quad \boldsymbol{w}_2 \quad \cdots \quad \boldsymbol{w}_n] \cdot \begin{bmatrix} \boldsymbol{r}_1 \\ \boldsymbol{0} \\ \vdots \\ \boldsymbol{0} \end{bmatrix} = \begin{bmatrix} \boldsymbol{r}_1 \\ \boldsymbol{r}_1 \\ \vdots \\ \boldsymbol{r}_1 \end{bmatrix} \tag{18.99}$$

which means that the rows of a matrix equal to the product $\boldsymbol{W}\boldsymbol{r}$ are all identical. Then the left-hand side of Eq. (18.96) will be

$$\boldsymbol{W}\boldsymbol{r}\boldsymbol{H}_{\mathrm{RR}} = \begin{bmatrix} \boldsymbol{r}_1 \\ \boldsymbol{r}_1 \\ \vdots \\ \boldsymbol{r}_1 \end{bmatrix} \cdot \boldsymbol{H}_{\mathrm{RR}} = \begin{bmatrix} \boldsymbol{h}_1 \\ \boldsymbol{h}_1 \\ \vdots \\ \boldsymbol{h}_1 \end{bmatrix} \tag{18.100}$$

that is it will also be a matrix of identical rows. Substituting (18.100) into (18.96) leads to

$$\boldsymbol{M}_{\mathrm{A}}^{-1}\boldsymbol{H}_{\mathrm{AR}} = \begin{bmatrix} \boldsymbol{h}_1 \\ \boldsymbol{h}_1 \\ \vdots \\ \boldsymbol{h}_1 \end{bmatrix} \tag{18.101}$$

which means that the rows of a matrix equal to the product $\boldsymbol{M}_{\mathrm{A}}^{-1}\boldsymbol{H}_{\mathrm{AR}}$ are all identical. Hence all the elements in its column k are identical, which may be written as

$$\frac{H_{ik}}{M_i} = \frac{H_{jk}}{M_j} \quad \text{for } i,j \in \{\mathrm{A}\},\ \ k \in \{\mathrm{R}\} \tag{18.102}$$

This means that the necessary and sufficient condition for exciting only one modal variable $z_1(t)$ by any disturbance $\Delta\boldsymbol{P}_{\mathrm{R}}$ in Eq. (18.97), and therefore for group {A} to be exactly coherent, is that condition (18.102) be satisfied. Remember that the excited modal variable $z_1(t)$ corresponds to swings of group {A} with respect to the rest of the system and the modes $z_2(t) = , ..., = z_n(t) = 0$ correspond to the internal swinging modes inside group {A} that are not excited.

Clearly Eqs. (18.102) and (18.39) are identical. The conclusion is that modal analysis confirms the considerations in Section 18.5.

18.6.4 Eigenvalues and Eigenvectors of the Equivalent Model

The analysis in the previous subsection was undertaken under an assumption that, in the state Eq. (18.83), changes of power in the remaining part of the system constitute a disturbance. Such a model was used to investigate internal group swings and external swings between the group and the rest of the system. The model could not be used to assess the influence of aggregation of nodes in group {A} on the modes corresponding to oscillations in the rest of the system. That task will require the creation of the incremental model of the whole system and an investigation of how aggregation of group {A} influences eigenvalues and eigenvectors in the whole system. This difficult task will be simplified by reducing the system model using aggregation which will be shown as a projection of the state space on a subspace.

Let $\boldsymbol{x}$ be the state vector of a dynamic system described by the state equation

$$\dot{\boldsymbol{x}} = \boldsymbol{A}\boldsymbol{x} \tag{18.103}$$

System reduction will be undertaken by projecting vector $\boldsymbol{x}$ onto a smaller vector

$$\boldsymbol{x}_{\mathrm{e}} = \boldsymbol{C}\boldsymbol{x} \tag{18.104}$$

where $\boldsymbol{C}$ is a rectangular matrix defining this projection and further referred to as the *projection matrix*. The lower index comes from the word "equivalent." The reduced model is described by

$$\dot{\boldsymbol{x}}_{\mathrm{e}} = \boldsymbol{a}\,\boldsymbol{x}_{\mathrm{e}} \tag{18.105}$$

where $\boldsymbol{a}$ is a square matrix that will be now expressed using matrices $\boldsymbol{A}$ and $\boldsymbol{C}$.

Equation (18.105) describes a reduced dynamic system obtained by the reduction of the state vector using transformation (18.104).

Differentiating both sides of Eq. (18.104) gives $\dot{\boldsymbol{x}}_{\mathrm{e}} = \boldsymbol{C}\dot{\boldsymbol{x}}$. Substitution of $\dot{\boldsymbol{x}}_{\mathrm{e}}$ by the right-hand side of Eq. (18.105) leads to $\boldsymbol{a}\,\boldsymbol{x}_{\mathrm{e}} = \boldsymbol{C}\boldsymbol{A}\boldsymbol{x}$. Substitution of $\boldsymbol{x}_{\mathrm{e}}$ by the right-hand side of (18.104) gives $\boldsymbol{a}\,\boldsymbol{C}\boldsymbol{x} = \boldsymbol{C}\boldsymbol{A}\boldsymbol{x}$, which finally leads to

$$\boldsymbol{a}\,\boldsymbol{C} = \boldsymbol{C}\boldsymbol{A} \tag{18.106}$$

Right-multiplying by $\boldsymbol{C}^{\mathrm{T}}$ gives $\boldsymbol{a}\boldsymbol{C}\boldsymbol{C}^{\mathrm{T}} = \boldsymbol{C}\boldsymbol{A}\boldsymbol{C}^{\mathrm{T}}$ leading to

$$\boldsymbol{a} = \boldsymbol{C}\boldsymbol{A}\boldsymbol{C}^{\mathrm{T}}\left(\boldsymbol{C}\boldsymbol{C}^{\mathrm{T}}\right)^{-1} \tag{18.107}$$

where matrix $\boldsymbol{C}\boldsymbol{C}^{\mathrm{T}}$ is a square matrix with rank equal to the number of state variables in the reduced model.

The relationship given by (18.106) is very important because it will make it possible to show that the reduced model (18.105) obtained from reducing the state vector using transformation (18.104) partially retains eigenvalues and eigenvectors of the original (unreduced) system (18.103).

Let λ_i be an eigenvalue of the state matrix $\boldsymbol{A}$ in Eq. (18.103) and let $\boldsymbol{w}_i$ be a right eigenvector of that matrix. Then, according to the definition of eigenvectors, $\boldsymbol{A}\boldsymbol{w}_i = \lambda_i\boldsymbol{w}_i$. Left-multiplying by $\boldsymbol{C}$ gives $\boldsymbol{C}\boldsymbol{A}\boldsymbol{w}_i = \lambda_i\boldsymbol{C}\boldsymbol{w}_i$. Substituting of $\boldsymbol{C}\boldsymbol{A}$ by the left-hand side of (18.106) results in $\boldsymbol{a}\ \boldsymbol{C}\boldsymbol{w}_i = \lambda_i\boldsymbol{C}\boldsymbol{w}_i$ or

$$\boldsymbol{a}\,\boldsymbol{w}_{\mathrm{e}i} = \lambda_i\boldsymbol{w}_{\mathrm{e}i} \tag{18.108}$$

where

$$\boldsymbol{w}_{\mathrm{e}i} = \boldsymbol{C}\boldsymbol{w}_i \tag{18.109}$$

Equation (18.108) shows that for each $\boldsymbol{w}_{\mathrm{e}i} \neq \boldsymbol{0}$ the number λ_i is an eigenvalue of matrix $\boldsymbol{a}$ and $\boldsymbol{w}_{\mathrm{e}i}$ is the corresponding right eigenvector. Obviously, λ_i is also an eigenvalue of $\boldsymbol{A}$. Equation (18.109) shows that vector $\boldsymbol{w}_{\mathrm{e}i}$ is created by the reduction of vector $\boldsymbol{w}_i$. This means that by satisfying the condition

$$\boldsymbol{w}_{\mathrm{e}i} = \boldsymbol{C}\boldsymbol{w}_i \neq \boldsymbol{0} \tag{18.110}$$

the reduced dynamic system (18.105) obtained by reducing the state vector using (18.104) partially retains eigenvalues and eigenvectors of the original (unreduced) system (18.103). Note that the relationship between eigenvector $\boldsymbol{w}_{\mathrm{e}i}$ of the reduced model and eigenvector $\boldsymbol{w}_i$ of the original (unreduced) model is the same as that between the state vector $\boldsymbol{x}_{\mathrm{e}}$ and the state vector $\boldsymbol{x}$. This means that $\boldsymbol{w}_{\mathrm{e}i}$ corresponds to the projection of $\boldsymbol{w}_i$ obtained using the projection matrix $\boldsymbol{C}$.

Obviously, condition (18.110) is satisfied not for every matrix $\boldsymbol{C}$ and the reduced model does not maintain all eigenvalues and eigenvectors of the original (unreduced) model.

In the case of the incremental model of an EPS, Machowski (1985) shows that the projection matrix should be of the following form

$$\boldsymbol{C} = \left[\begin{array}{ccc|cccc} 1 & & & & & & \\ & \ddots & & & \boldsymbol{0} & & \\ & & 1 & & & & \\ \hline & \boldsymbol{0} & & \frac{1}{n} & \frac{1}{n} & \cdots & \frac{1}{n} \end{array}\right] = \left[\begin{array}{c|c} \boldsymbol{1} & \boldsymbol{0} \\ \hline \boldsymbol{0} & \frac{1}{n}\boldsymbol{1}_{\mathrm{A}}^{\mathrm{T}} \end{array}\right] \tag{18.111}$$

where $\boldsymbol{1}_{\mathrm{A}}$ is a vector of ones and n is the number of generators in group {A}.

The discussed reduction using the projection matrix may be applied to Eq. (18.84) or (18.83). This will be shown for the latter since: (i) the state matrix in (18.83) is simpler than in (18.84) and (ii) there is an exact relationship $\lambda_i = \sqrt{\mu_i}$ between eigenvalues of both matrices.

When applying the projection matrix (18.111), vector $\Delta\boldsymbol{\delta}'$ in Eq. (18.62) is transformed in the following way

$$\boldsymbol{C}\begin{bmatrix} \Delta\delta'_{\mathrm{R}} \\ \Delta\delta'_{\mathrm{A}} \end{bmatrix} = \begin{bmatrix} \Delta\delta'_{\mathrm{R}} \\ \Delta\delta'_{\mathrm{a}} \end{bmatrix} \tag{18.112}$$

where

$$\Delta\delta'_{\mathrm{a}} = \frac{1}{n}\sum_{j\in\{\mathrm{A}\}}\Delta\delta'_j \tag{18.113}$$

Equation (18.113) shows that, when using the discussed reduction method, the rotor of the equivalent generator moves on average with respect to all the rotors of aggregated generators. Obviously, for exactly coherent generators this movement is the same for all the generators in group {A} and its average value is equal simply to the value for each generator in the group.

For matrix $\boldsymbol{C}$ in the structure (18.111) it can be shown that

$$\boldsymbol{C}^{\mathrm{T}}\left(\boldsymbol{C}\boldsymbol{C}^{\mathrm{T}}\right)^{-1} = \left[\begin{array}{ccc|c} 1 & & & \\ & \ddots & & \mathbf{0}_{\mathrm{A}} \\ & & 1 & \\ \hline & & & 1 \\ & & & 1 \\ & \mathbf{0}_{\mathrm{RA}} & & \vdots \\ & & & 1 \end{array}\right] = \left[\begin{array}{c|c} \mathbf{1} & \mathbf{0}_{\mathrm{A}} \\ \hline \mathbf{0}_{\mathrm{RA}} & \mathbf{1}_{\mathrm{A}} \end{array}\right] = \boldsymbol{B}^{\mathrm{T}} \tag{18.114}$$

Thus the state matrix of the reduced model given by Eq. (18.107) takes the simple form

$$\boldsymbol{a} = \boldsymbol{C}\boldsymbol{A}\boldsymbol{B}^{\mathrm{T}} \tag{18.115}$$

The matrix equation of motion, with damping neglected, for the original model (Figure 18.6a) containing generators in groups {A} and {R} can be written in the following way

$$\begin{bmatrix} \Delta\ddot{\boldsymbol{\delta}}'_{\mathrm{R}} \\ \Delta\ddot{\boldsymbol{\delta}}'_{\mathrm{A}} \end{bmatrix} = -\left[\begin{array}{c|c} \boldsymbol{M}_R^{-1}\boldsymbol{H}_{\mathrm{RR}} & \boldsymbol{M}_R^{-1}\boldsymbol{H}_{\mathrm{RA}} \\ \hline \boldsymbol{M}_{\mathrm{A}}^{-1}\boldsymbol{H}_{\mathrm{AR}} & \boldsymbol{M}_{\mathrm{A}}^{-1}\boldsymbol{H}_{\mathrm{AA}} \end{array}\right] \begin{bmatrix} \Delta\boldsymbol{\delta}'_{\mathrm{R}} \\ \Delta\boldsymbol{\delta}'_{\mathrm{A}} \end{bmatrix} \tag{18.116}$$

After applying reduction using matrix $\boldsymbol{C}$ in the form (18.111), the state vector is reduced to the form (18.112) while Eq. (18.116) reduces to

$$\left[\begin{array}{c} \Delta\ddot{\boldsymbol{\delta}}'_{\mathrm{A}} \\ \hline \Delta\ddot{\delta}'_{\mathrm{a}} \end{array}\right] = -\left[\begin{array}{c|c} \boldsymbol{M}_R^{-1}\boldsymbol{H}_{\mathrm{RR}} & \boldsymbol{M}_R^{-1}\boldsymbol{H}_{\mathrm{RA}}\mathbf{1}_{\mathrm{A}} \\ \hline \frac{1}{n}\mathbf{1}_{\mathrm{A}}^{\mathrm{T}}\boldsymbol{M}_{\mathrm{A}}^{-1}\boldsymbol{H}_{\mathrm{AR}} & \frac{1}{n}\mathbf{1}_{\mathrm{A}}^{\mathrm{T}}\boldsymbol{M}_{\mathrm{A}}^{-1}\boldsymbol{H}_{\mathrm{AA}}\mathbf{1}_{\mathrm{A}} \end{array}\right] \left[\begin{array}{c} \Delta\boldsymbol{\delta}'_{\mathrm{R}} \\ \hline \Delta\delta'_{\mathrm{a}} \end{array}\right] \tag{18.117}$$

where the state matrix has been calculated according to (18.115). As shown previously, the reduced model given by (18.117) partially retains eigenvalues and eigenvectors of the original (unreduced) model of (18.116).

It is easy to see some similarity between the described reduced model (18.117) and the reduced model obtained using the Di Caprio and Marconato (1975) aggregation described in Section 18.6.2. In both cases there is a summation of matrix elements corresponding to multiplication by $\mathbf{1}_{\mathrm{A}}$ and $\mathbf{1}_{\mathrm{A}}^{\mathrm{T}}$. Using Eq. (18.62) obtained from the Di Caprio and Marconato aggregation it is possible, as in (18.117), to write the following state equation

$$\left[\begin{array}{c} \Delta\ddot{\boldsymbol{\delta}}'_{\mathrm{A}} \\ \hline \Delta\ddot{\delta}'_{\mathrm{a}} \end{array}\right] = -\left[\begin{array}{c|c} \boldsymbol{M}_R^{-1}\boldsymbol{H}_{\mathrm{RR}} & \boldsymbol{M}_R^{-1}\boldsymbol{H}_{\mathrm{RA}}\mathbf{1}_{\mathrm{A}} \\ \hline M_{\mathrm{a}}^{-1}\mathbf{1}_{\mathrm{A}}^{\mathrm{T}}\boldsymbol{H}_{\mathrm{AR}} & M_{\mathrm{a}}^{-1}\mathbf{1}_{\mathrm{A}}^{\mathrm{T}}\boldsymbol{H}_{\mathrm{AA}}\mathbf{1}_{\mathrm{A}} \end{array}\right] \left[\begin{array}{c} \Delta\boldsymbol{\delta}'_{\mathrm{R}} \\ \hline \Delta\delta'_{\mathrm{a}} \end{array}\right] \tag{18.118}$$

where, according to (18.33), the inertia coefficient of the equivalent machine are $M_{\mathrm{a}} = \sum_{i\in\{A\}} M_i$.

It is also easy to see, when comparing Eqs. (18.117) and (18.118), that they differ in the bottom row corresponding to the equivalent generator. The difference lies in the different order of factors, which is important for the result as the multiplication of matrices is generally not commutative. A detailed analysis leads to the conclusion that the elements in the bottom row of Eq. (18.117) are given by

$$a_{\mathrm{a}k} = -\frac{1}{n}\sum_{i\in\{A\}} \frac{H_{ik}}{M_i} \tag{18.119}$$

and those in Eq. (18.118) are given by

$$a_{ak} = -\frac{\sum_{i\in\{A\}} H_{ik}}{\sum_{i\in\{A\}} M_i} \tag{18.120}$$

This is obvious because generally both elements given by Eqs. (18.119) and (18.120) are not the same. In the particular case when Eq. (18.39) is satisfied, that is when the group is exactly coherent, the following holds

$$\frac{H_{ik}}{M_i} = h_k \quad \text{for} \quad i,j \in \{A\}, \quad k \in \{B\} \tag{18.121}$$

Hence $H_{ik} = h_k M_i$. Substituting this into (18.120) gives

$$a_{ak} = -\frac{\sum_{i\in\{A\}} h_k M_i}{\sum_{i\in\{A\}} M_i} = -\frac{h_k \sum_{i\in\{A\}} M_i}{\sum_{i\in\{A\}} M_i} = -h_k \tag{18.122}$$

The same value of $a_{ak} = -h_k$ can be obtained by substituting (18.121) into (18.119). This shows that when the exact coherency condition (18.39) is satisfied the matrices in (18.117) and (18.118) are the same.

Example 18.2 To illustrate how the reduced model partially retains eigenvalues and eigenvectors, a simple three-machine system will be studied in which two generators satisfy the exact coherency condition given by (18.121). The state matrix given by Eq. (18.116) is

$$\left[\begin{array}{c|cc} -6 & 3 & 3 \\ \hline 2 & -4 & 2 \\ 2 & 3 & -5 \end{array}\right]$$

The eigenvalues and eigenvectors are

$$\mu_1 = 0 \quad \text{and} \quad \boldsymbol{w}_1 = [1 \quad 1 \quad 1]^{\mathrm{T}}$$

$$\mu_2 = -8 \quad \text{and} \quad \boldsymbol{w}_2 = [-3 \quad 1 \quad 1]^{\mathrm{T}}$$

$$\mu_3 = -7 \quad \text{and} \quad \boldsymbol{w}_3 = [3 \quad -4 \quad 3]^{\mathrm{T}}$$

The state matrix reduces using Eq. (18.117) to

$$\left[\begin{array}{c|c} -6 & 6 \\ \hline 2 & -2 \end{array}\right]$$

The eigenvalues and eigenvectors of this state matrix are

$$\mu_1 = 0 \quad \text{and} \quad \boldsymbol{w}_{e1} = [1 \quad 1]^{\mathrm{T}}$$

$$\mu_2 = -8 \quad \text{and} \quad \boldsymbol{w}_{e2} = [-3 \quad 1]^{\mathrm{T}}$$

The reduced system also retained, apart from the zero eigenvalue, the eigenvalue $\mu_2 = -8$ and the associated right eigenvector $\boldsymbol{w}_{e2} = [-3 \quad 1]^{\mathrm{T}}$, which is a part of the original eigenvector $\boldsymbol{w}_2 = [-3 \quad 1 \quad 1]^{\mathrm{T}}$. Equation (18.110) is satisfied as

$$\boldsymbol{C}\boldsymbol{w}_2 = \left[\begin{array}{c|c|c} 1 & 0 & 0 \\ \hline 0 & \frac{1}{2} & \frac{1}{2} \end{array}\right] \begin{bmatrix} -3 \\ 1 \\ 1 \end{bmatrix} = \begin{bmatrix} -3 \\ 1 \end{bmatrix} = \boldsymbol{w}_{e2}$$

This illustrates that the reduced model partially retains eigenvalues and eigenvectors of the original (unreduced) model.

******

To summarize the observations contained in this chapter:

- The operations of aggregation and linearization are commutative (proof in Section 18.6.2).
- The reduced linear model (18.62) obtained using the method of Di Caprio and Marconato corresponds to the linearized form of the reduced model obtained by Zhukov's aggregation (proof in Section 18.6.2).
- When the exact coherency condition given by Eq. (18.39) is satisfied, the reduced linear model (18.118) is equivalent to the reduced model (18.117) obtained using transformation (18.111) and the projection matrix (18.111).
- The reduced model (18.117) partially retains the eigenvalues of the original (unreduced) model.

These observations clearly show that, when the exact coherency condition (18.39) is satisfied, the reduced model obtained by Zhukov's aggregation (Section 18.2.3) also partially retains the eigenvalues of the original (unreduced) model. This is a very important property of the coherency-based dynamic equivalent model obtained by Zhukov's aggregation.

In practice, exact coherency rarely occurs in active power systems apart from identical generators operating on the same busbar. Reduced dynamic models are created by aggregation of generators for which the coherency definition is satisfied within accuracy $\varepsilon_{\Delta\delta}$ as in condition (18.29). Obviously, any inaccuracy of coherency means that all the dynamic properties of the original (unreduced) model will be maintained only to some degree by the equivalent (reduced) model. Hence it may be expected that also eigenvalues and eigenvectors of the equivalent (reduced) model will be only approximately equal to eigenvalues and eigenvectors of the original (unreduced) model. It is important here that the equivalent (reduced) model maintains as precisely as possible those modal variables that are strongly excited by disturbances in the internal subsystem and which therefore have the strongest influence on power swings in the internal subsystem. These modal variables will be referred to as *dominant modal variables*. Modal analysis (Section 12.1) shows that matrices $\boldsymbol{U}$ and $\boldsymbol{W}$ built from right and left eigenvectors decide which modal variables are most strongly excited and influence power swings. The example below will show that a coherency-based equivalent model quite accurately retains the dominant modes.

Example 18.3 Figure 18.17 shows a 15-machine test system. Plant 7 was assumed to constitute the internal system. For that internal system, the algorithm described in Section 18.5 was used to identify coherent groups which are encircled in Figure 18.17 using solid lines.

The dominant modes have been identified assuming that the initial disturbance is a rotor angle change of generator 7, that is $\Delta\boldsymbol{\delta}' = \left[0 \ \cdots \ 0 \,\vdots\, \Delta\delta'_7 \,\vdots\, 0 \ \cdots \ 0\right]^{\mathrm{T}}$. With this disturbance, the equation $\boldsymbol{z} = \boldsymbol{U} \cdot \Delta\boldsymbol{\delta}'$ results in $\boldsymbol{z} = \boldsymbol{u}_{\circ 7} \cdot \Delta\delta'_7$ where $\boldsymbol{u}_{\circ 7}$ denotes the seventh column of matrix $\boldsymbol{U}$. For the assumed data (Machowski et al. 1986a,b), the following results were obtained:

- for the original (unreduced) model

$$\boldsymbol{u}_{\circ 7} = 10^{-3} \cdot [-\mathbf{184} \,\vdots\, 0 \,\vdots\, \underline{\mathbf{914}} \,\vdots\, -11 \,\vdots\, \mathbf{160} \,\vdots\, -3 \,\vdots\, -56 \,\vdots\, -9 \,\vdots\, -71 \,\vdots\, -18 \,\vdots\, -64 \,\vdots\, -1 \,\vdots\, -90 \,\vdots\, 0 \,\vdots\, 20]^{\mathrm{T}}$$

- for the equivalent (reduced) model

$$\boldsymbol{u}_{\circ 7} = 10^{-3} \cdot [-\mathbf{154} \,\vdots\, 29 \,\vdots\, \underline{\mathbf{915}} \,\vdots\, -47 \,\vdots\, \mathbf{124} \,\vdots\, -2 \,\vdots\, 0]^{\mathrm{T}}.$$

The largest values correspond to the third modal variable z_3 and are shown in bold and underlined. Note that they are almost the same for the original and the equivalent model, which means that excitation of the third modal variable in both system is the same. Also strongly exited are the first modal variable z_1 and the fifth z_5. That excitation is, nevertheless, several times weaker than excitation of the third modal variable z_3. The remaining values

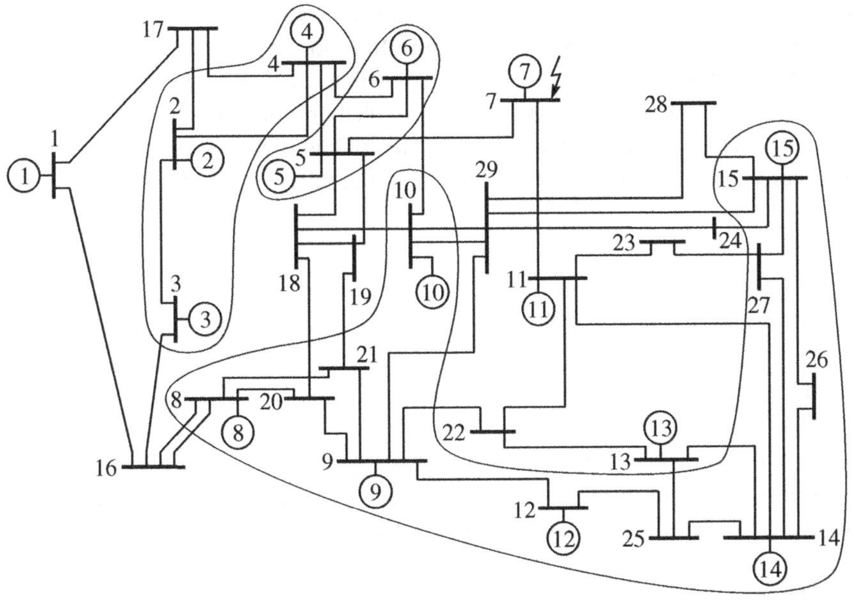

Figure 18.17 New England test system and recognized coherent groups.

are much smaller, so it may be assumed that the remaining modes are either weakly excited or not excited at all. The excited modal variables are associated with the following eigenvalues

original model:	equivalent model:
$\mu_1 = -11.977$	$\mu_1 = -13.817$
$\mu_3 = -42.743$	$\mu_3 = -44.170$
$\mu_5 = -72.499$	$\mu_5 = -119.390$

Clearly, the third eigenvalue corresponding to the most excited modal variable is almost the same for both the equivalent and the original model. The first eigenvalue has similar values for both models, while the fifth is quite different. However, it should be remembered that the first and fifth modal variables are weakly excited and do not have to be accurately modeled.

Matrix $\boldsymbol{W}$ decides how individual modal variables influence power swings in the internal subsystem. The equation $\boldsymbol{z}\Delta\delta' = \boldsymbol{W}\boldsymbol{z}$ results in $\Delta\delta'_7 = \boldsymbol{w}_{7^\circ}\boldsymbol{z}$, where $\boldsymbol{w}_{7^\circ}$ denotes the seventh row of $\boldsymbol{W}$. For the assumed data the following results were obtained:

- for the original (unreduced) model

$$\boldsymbol{w}_{7^\circ} = 10^{-1} \cdot [-\mathbf{34} \,\vdots\, -26 \,\vdots\, \underline{\mathbf{98}} \,\vdots\, -10 \,\vdots\, \mathbf{21} \,\vdots\, 0 \,\vdots\, -50 \,\vdots\, -10 \,\vdots\, -80 \,\vdots\, -20 \,\vdots\, -40 \,\vdots\, 0 \,\vdots\, 0 \,\vdots\, 0 \,\vdots\, -20]$$

- for the equivalent (reduced) model

$$\boldsymbol{w}_{7^\circ} = 10^{-1} \cdot [-\mathbf{52} \,\vdots\, 6 \,\vdots\, \underline{\mathbf{99}} \,\vdots\, -12 \,\vdots\, \mathbf{60} \,\vdots\, 0 \,\vdots\, 7]$$

The largest values again correspond to the third modal variable z_3 and they are almost the same for both models. This means that the influence of the third modal variable on power swings in the internal system is the same in both models. The values for the first modal variable z_1 and the fifth z_5 are quite different, but those modal variables are weakly excited. Nevertheless, model reduction by aggregation causes some differences between power swings simulated in both models – see the simulation results shown previously in Figure 18.14 for a different test system.

By making use of $\boldsymbol{u}_{\circ 7}$ (seventh column of $\boldsymbol{U}$) and $\boldsymbol{w}_{7\circ}$ (seventh row of $\boldsymbol{W}$) it is possible to calculate participation factors defined in Section 11.1. According to Eq. (12.90), it is necessary to multiply elements of matrix column $\boldsymbol{u}_{\circ 7}$ by elements of matrix row $\boldsymbol{w}_{7\circ}$. For example, the first participation factor for the original (unreduced) model is: $10^{-4} \cdot 184 \cdot 34 \cong 63 \cdot 10^{-2}$. The calculated participation factors can be expressed in the following way:

- for the original (unreduced) model

$$10^{-2} \cdot [\mathbf{63} \;\vdots\; 0 \;\vdots\; \underline{\mathbf{896}} \;\vdots\; 1 \;\vdots\; \mathbf{34} \;\vdots\; 0 \;\vdots\; 28 \;\vdots\; 1 \;\vdots\; 57 \;\vdots\; 4 \;\vdots\; 26 \;\vdots\; 0 \;\vdots\; 0 \;\vdots\; 0 \;\vdots\; -4]$$

- for the equivalent (reduced) model

$$10^{-2} \cdot [\mathbf{80} \;\vdots\; 2 \;\vdots\; \underline{\mathbf{906}} \;\vdots\; 6 \;\vdots\; \mathbf{74} \;\vdots\; 0 \;\vdots\; 0]$$

Based on the values of participation factors, it can be concluded that there is a strong relationship between the investigated variable $\Delta\delta_7'$ in the internal subsystem and the third modal variable z_3. The relationships between $\Delta\delta_7'$ and the first modal variable z_1 and the fifth z_5 are an order of magnitude weaker.

******

When analyzing Example 18.3, it should be remembered that the calculated eigenvalues μ_i are the eigenvalues of a matrix in the second-order equation, respectively (18.116) and (18.117). These values are real and negative. The corresponding eigenvalues λ_i of first-order equations of the type (18.84) are complex numbers $\lambda_i = \sqrt{\mu_i}$.

18.6.5 Equilibrium Points of the Equivalent Model

The coherency-based equivalent model obtained by Zhukov's aggregation is constructed for a stable equilibrium point which is at the same time the steady-state operating point of the system. Consequently, the equivalent model must partially retain the coordinates of the stable equilibrium. This can be illustrated in the following way when denoting the nodes, as in Figure 18.6. Let r be the number of generators in group {R} and N be the total number of system generators, that is in both groups {R} and {A}. Then the coordinates of the stable equilibrium point of the original (unreduced) model and the equivalent (reduced) model can be written as

$$\hat{\delta}' = \left[\hat{\delta}_1' \;\cdots\; \hat{\delta}_r' \;\vdots\; \hat{\delta}_{r+1}' \;\cdots\; \hat{\delta}_N'\right]^{\mathrm{T}} \tag{18.123}$$

$$\hat{\delta}_{\mathrm{e}}' = \left[\hat{\delta}_1' \;\cdots\; \hat{\delta}_r' \;\vdots\; \hat{\delta}_{\mathrm{a}}'\right]^{\mathrm{T}} \tag{18.124}$$

where $\hat{\delta}_{\mathrm{a}}'$ is the power angle of the equivalent generator given by Eq. (18.22). Now the question arises whether, and which, unstable equilibrium points are retained by the reduced (equivalent) model. This question is especially important from the point of view of the Lyapunov direct method. It is shown in Section 6.3.5 (Figure 6.28) that, when transient stability is lost, each unstable equilibrium point corresponds to the system splitting in a certain way into groups of asynchronously operating generators. From that point of view the reduced (equivalent) model is a good model if it partially retains those unstable equilibrium points which are important for disturbances in the internal subsystem (Figure 18.1).

The coordinates of an unstable equilibrium point of the original (unreduced) model and the equivalent (reduced) model will be denoted as follows

$$\tilde{\delta}' = \left[\tilde{\delta}_1' \;\cdots\; \tilde{\delta}_r' \;\vdots\; \tilde{\delta}_{r+1}' \;\cdots\; \tilde{\delta}_N'\right]^{\mathrm{T}} \tag{18.125}$$

$$\tilde{\delta}_{\mathrm{e}}' = \left[\tilde{\delta}_{\mathrm{e}1}' \;\cdots\; \tilde{\delta}_{\mathrm{e}r}' \;\vdots\; \tilde{\delta}_{\mathrm{a}}'\right]^{\mathrm{T}} \tag{18.126}$$

The equivalent model will be said to partially retain the unstable equilibrium point of the original model if

$$\tilde{\delta}'_{\mathrm{e}k} = \tilde{\delta}'_k \quad \text{for} \quad k \in \{\mathrm{R}\} \tag{18.127}$$

The electrical interpretation of Zhukov's aggregation shown in Figure 18.15 will reveal which particular unstable equilibrium points satisfy Eq. (18.127). Aggregation will not distort the coordinates of an unstable equilibrium point if at that point the ratio of the voltages will be equal to the transformation ratio used for aggregation, that is the ratio of voltages at the stable equilibrium point. As in Eq. (18.16), the condition may be written as

$$\underline{\tilde{V}}_{\mathrm{a}}^{-1}\underline{\tilde{\boldsymbol{V}}}_{\mathrm{A}} = \underline{\boldsymbol{\vartheta}} = \underline{\hat{V}}_{\mathrm{a}}^{-1}\underline{\hat{\boldsymbol{V}}}_{\mathrm{A}} \tag{18.128}$$

For the classical generator model (constant magnitudes of electromotive forces) the condition simplifies to

$$\tilde{\delta}'_i - \tilde{\delta}'_{\mathrm{a}} = \hat{\delta}'_i - \hat{\delta}'_{\mathrm{a}} \quad \text{for} \quad i \in \{\mathrm{A}\} \tag{18.129}$$

or $\tilde{\delta}'_i - \hat{\delta}'_i = \tilde{\delta}'_{\mathrm{a}} - \hat{\delta}'_{\mathrm{a}}$. This equation must be satisfied for each $i \in \{\mathrm{A}\}$ and therefore for each $i, j \in \{\mathrm{A}\}$. Hence $\tilde{\delta}'_i - \hat{\delta}'_i = \tilde{\delta}'_j - \hat{\delta}'_j = \tilde{\delta}'_{\mathrm{a}} - \hat{\delta}'_{\mathrm{a}}$ must be satisfied, or

$$\tilde{\delta}'_i - \hat{\delta}'_i = \tilde{\delta}'_j - \hat{\delta}'_j \quad i,j \in \{\mathrm{A}\} \tag{18.130}$$

This means that for each generator belonging to a given group $i, j \in \{\mathrm{A}\}$ the distance between an unstable equilibrium point and the stable equilibrium point must be the same. Such unstable equilibrium points can be called *partially equidistant points* with respect to a given group of variables belonging to group {A}.

The equivalent model obtained using Zhukov's aggregation partially retains each unstable equilibrium point equidistant with respect to a given group of variables belonging to group {A}. Aggregation destroys only those unstable equilibrium points that are not partially equidistant. This property is shown in Example 18.4, using an illustration that is intuitively simple to understand.

Example 18.4 Figure 18.18 shows an example of two parallel generators, 1 and 2, operating on an infinite busbar represented by a generator of large capacity, 3. For each external short circuit in the transmission line 4–3, the two parallel generators are exactly coherent. Oscillations between the generators may appear only in the case of an internal short circuit inside the power plants at nodes 5 or 6. The lower part of Figure 18.18a shows the equivalent diagram after elimination of load nodes. The parameters have symbols following the notation in Eq. (6.42). Figure 18.18b shows equiscalar lines of potential energy similar to Figure 6.28.

There are three unstable equilibrium points: u1, u2, and u3. The saddle point u1 corresponds to the loss of synchronism of generator 1 with respect to generators 2 and 3. This may happen when a short circuit appears at node 5. The saddle point u2 corresponds to a loss of synchronism of generator 2 with respect to generators 1 and 3. This may happen when there is a short circuit at node 6. Point u3 is of the maximum type. It corresponds to a loss of synchronism of generators 1 and 2 with respect to generator 3. This may happen when there is a short circuit in line 4–3 at, for example, point 7. For point u3, condition (18.130) is satisfied as $\tilde{\delta}'_{13} - \hat{\delta}'_{13} = \tilde{\delta}'_{23} - \hat{\delta}'_{23}$. Point u3 is at the same time partially equidistant. Note that, when the exact coherency condition is satisfied, trajectory $\boldsymbol{\delta}'(t)$ lies on the straight line AB crossing the origin, point s and point u3. The line is defined by

$$\delta'_{13}(t) - \delta'_{23}(t) = \hat{\delta}'_{13} - \hat{\delta}'_{23} = \hat{\delta}'_{12} = \text{constant}$$

similar to Eq. (18.31). Aggregation of generators 1 and 2 reduces the three-machine system to a two-machine system and destroys the unstable equilibrium points u1 and u2. After aggregation the unstable equilibrium point u3 is retained. A plot of potential energy for the reduced model (two-machine model) corresponds to a cross-section of the diagram in Figure 18.18 along line AB. This plot has the same shape as shown previously in Figure 6.25b.

******

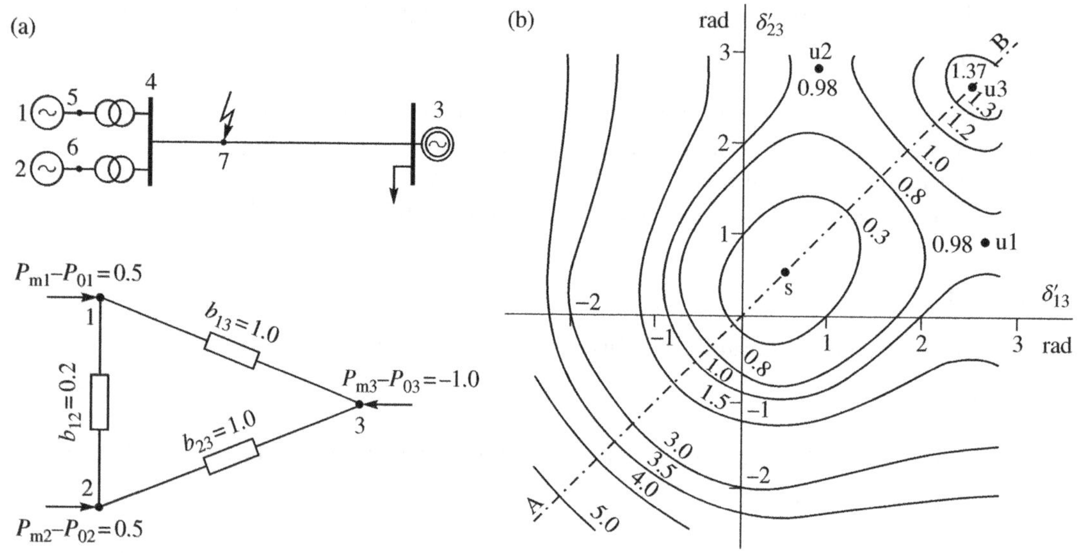

Figure 18.18 Illustration to the definition of the partially equidistant equilibrium point: (a) network diagrams; (b) equiscalar lines of potential energy.

The next important issue for direct the Lyapunov direct method is the question whether the dynamic equivalent (reduced) model retains the values of the Lyapunov function during the transient state and at unstable equilibrium points of the original (unreduced) model. For the Lyapunov function $V(\delta', \Delta\omega) = E_k + E_p$ given by Eq. (6.59) the answer to this question is positive, which will now be proved.

For kinetic energy E_k the proof is trivial. It is enough to separate Eq. (6.53) into two sums

$$E_k = \frac{1}{2}\sum_{i=1}^{N} M_i\Delta\omega_i^2 = \frac{1}{2}\sum_{i\in\{R\}} M_i\Delta\omega_i^2 + \frac{1}{2}\sum_{i\in\{A\}} M_i\Delta\omega_i^2 = \frac{1}{2}\sum_{i\in\{R\}} M_i\Delta\omega_i^2 + \frac{1}{2}M_a\Delta\omega_a^2$$

where for $i \in \{A\}$ the definition of exact coherency gives $\Delta\omega_1 = \ldots = \Delta\omega_n = \Delta\omega_a$ and, according to Eq. (18.33), $M_a = \sum_{i\in\{A\}} M_i$. This concludes the proof.

For potential energy given by Eq. (6.58) the proof is also simple but long. Here only an outline will be given:

1) The sum of components $(P_{mi} - P_{0i})\left(\delta'_i - \hat{\delta}'_i\right)$ should be broken down (similarly as for kinetic energy) into two sums: one for $i \in \{R\}$ and one for $i \in \{A\}$. Then it should be noted that when the exact coherency condition is satisfied for $i \in \{A\}$ then $\left(\delta'_i - \hat{\delta}'_i\right) = \left(\delta'_a - \hat{\delta}'_a\right)$, while, according to the principles of aggregation, $\sum_{i\in(A)} (P_{mi} - P_{0i}) = (P_{ma} - P_{0a})$.
2) The double sum of components $b_{ij}\left(\cos\delta'_{ij} - \cos\hat{\delta}'_{ij}\right)$ in Eq. (6.59) should be broken down into three sums: (i) for $i, j \in \{R\}$, (ii) for $i \in \{R\}\, j \in \{A\}$, and (iii) for $i, j \in \{A\}$. Then it should be noted that components $b_{ij}\cos\delta'_{ij}$ and $b_{ij}\cos\hat{\delta}'_{ij}$ correspond to synchronizing powers. It was shown in Section 18.6.2 that, for the equivalent (reduced) model, synchronizing powers are equal to the sum of synchronizing powers of aggregated generators. Hence the corresponding sums of components give the same values as for the equivalent (reduced) model.

Conclusions from the above points 1 and 2 conclude the proof for potential energy. This is illustrated in Example 18.5 using the results of calculations conducted for a test system.

Example 18.5 Consider again the test system shown in Figure 18.17. In this example the internal subsystem is assumed to consist of power plant 11 located in the middle of the test network. Treating the test system as the original (unreduced) model, the gradient method was used to calculate the coordinates of the stable equilibrium point and the unstable equilibrium point corresponding to the loss of synchronism of generator 11. The coordinates of those points are shown in Table 18.1 in columns under the heading "Original." For the assumed internal subsystem, the coherency recognition algorithm has identified two groups: {2, 3, 4} and {5, 6, 8, 9, 10, 12, 13, 14, 15}. The groups have been aggregated using Zhukov's method. For the equivalent (reduced) model, the stable equilibrium point and unstable equilibrium point corresponding to the loss of synchronism of generator 11 have been calculated. The coordinates of these points are shown in Table 18.1 in columns under the heading "Reduced." The results show that for generators {1, 7, 11}, the coordinates for both the stable and unstable equilibrium points have been well retained. The lower rows of Table 18.1 show the values of the Lyapunov function calculated for the unstable equilibrium point of the original (unreduced) and equivalent (reduced) model. Clearly, the values are quite close, similar to the values of the critical clearing time for a short circuit in busbar 11.

******

Similar results have been obtained for the same and other test systems when choosing different internal subsystems. More examples can be found in publications by Machowski (1985) and Machowski et al. (1986a and b, 1988).

Table 18.1 Results for a fault at bus 11.

		Coordinates of equilibrium points			
		Stable		Unstable	
Generator No	Group No	Original	Reduced	Original	Reduced
1	—	0.00	0.00	0.00	0.00
7	—	23.36	23.40	50.76	49.65
11	—	14.22	14.30	**183.80**	**181.81**
2	1	20.54	19.65	26.42	24.50
3		19.84		25.10	
4		10.56		19.02	
5	2	13.25	18.24	28.22	34.68
6		12.48		27.02	
8		15.39		26.58	
9		12.73		28.28	
10		11.15		26.59	
12		14.23		33.02	
13		14.14		34.44	
14		31.08		52.63	
15		25.55		44.67	
		Value of Lyapunov function		11.05	10.95
		Critical clearing time		0.322	0.325

Appendix

A.1 Per-unit System

Perhaps the one area in power system analysis that causes more confusion than any other is that of per-unit (pu) systems. This confusion is further compounded when a synchronous machine is included in the system. However, the pu system is well established and has a number of attractions. For example, by normalizing the generator equations derived in Chapter 11, the parameters of generators of the same type, but different ratings, will fall within the same range thereby providing the engineer with an intuitive understanding of the generator's performance. Such a normalized set of parameters can also lead to computational efficiencies.

In the following subsections the pu system used in this book is described. First of all, the base system used in the stator armature is described followed by a brief discussion on power invariance in both SI and pu. The pu system is then examined in more detail in order to derive base values for the different rotor circuits before finally explaining how the generator and the network pu systems fit together.

A.1.1 Stator Base Quantities

The principal armature base values used are:

Base voltage, V_b = generator line to neutral RMS terminal voltage, $V_{L\text{-}N}$ (this will normally be the rated voltage)
Base power, S_b = the generator MVA rating/phase, $S_{1\phi}$
Base time, $t_b = 1$ s

These principal base values lead to the following derived base values:

$$\text{Base current}, I_b = \frac{S_b}{V_b} = \frac{S_{1\phi}}{V_b} \quad \text{A}$$

$$\text{Base impedance}, Z_b = \frac{V_b}{I_b} \quad \Omega$$

$$\text{Base inductance}, L_b = \frac{V_b t_b}{I_b} \quad \text{H}$$

Base flux linkage, $\Psi_b = L_b I_b = V_b t_b \equiv V_b$ Vs
Base electrical angle $\theta_b = 1$ electrical radian
Base electrical speed $\omega_b = 1$ electrical rad/s
Base mechanical angle $\theta_{mb} = 1$ mechanical radian

Base mechanical speed $\omega_{mb} = 1$ mechanical rad/s
Base machine power $S_{3\phi} = 3S_{1\phi}$ VA
Base torque, $\tau_b = \frac{S_{3\phi}}{\omega_{sm}}$ Nm

To use these base values any particular current, voltage etc. in SI is simply divided by the corresponding base value to obtain the pu value (or vice versa)

$$\text{per unit value} = \frac{\text{actual value}}{\text{base value}} \tag{A.1}$$

It is important to note that with the voltage and current ABC/dq transformation equations introduced in Chapter 11, the same base values are used for the armature coils in both the A, B, C and the d, q reference frames. This is not the case with other values of transformation coefficient (Harris et al. 1970). As explained in Chapter 11 the ABC/dq transformation is power invariant such that

$$v_a i_a + v_b i_b + v_c i_c = v_d i_d + v_q i_q \tag{A.2}$$

With the base values defined above the transformations are power invariant in both SI and pu notation.
The following points should be noted:

1) With a base time of 1 s all time constants are expressed in seconds.
2) A pu reactance is related to a pu inductance by $X_{pu} = \omega L_{pu}$ so that the normal relationship between inductance and reactance is maintained. The pu inductance is *not equal* to the pu reactance.
3) The definition of base torque is such that at synchronous speed pu torque is equal to pu power, e.g. a turbine torque of 0.8 pu corresponds to a turbine power of 0.8 pu. In general

$$P = \tau\,\omega_m \text{ (SI)} \tag{A.3}$$

dividing by $S_{3\phi}$

$$\frac{P}{S_{3\phi}} = \frac{\tau\omega_m}{S_{3\phi}} = \frac{\tau\omega_m}{\tau_b\omega_{sm}}; \qquad P_{pu} = \tau_{pu}\frac{\omega_m}{\omega_{sm}} \tag{A.4}$$

but as $\omega_m = \omega/p$ and $\omega_{sm} = \omega_s/p$

$$P_{pu} = \tau_{pu}\frac{\omega_m}{\omega_{sm}} = \tau_{pu}\frac{\omega}{\omega_s} \tag{A.5}$$

at synchronous speed $\omega = \omega_s$ and

$$P_{pu} = \tau_{pu} \tag{A.6}$$

4) Under balanced operation the power output of a single phase, normalized to $S_{1\phi}$, is numerically the same in pu as the generator power output normalized to $S_{3\phi}$. Under balanced operation

$$P_{1\phi} = V_{rms}I_{rms}\cos\phi; \quad P_{3\phi} = 3V_{rms}I_{rms}\cos\phi \tag{A.7}$$

dividing by $S_{1\phi}$ and $S_{3\phi}$, respectively, gives

$$P_{pu} = V_{pu}I_{pu}\cos\phi \tag{A.8}$$

This is an extremely useful identity particularly when balanced operation is being studied by means of a phasor diagram.
5) Because of the pu notation adopted, most of the equations developed in this book are the same whether the quantities are expressed in SI or in pu. The two important exceptions to this are generator power and torque, both of which must be normalized to the generator MVA base rather than the phase MVA base. Consequently

$$P_{\text{pu}} = \frac{P_{\text{SI}}}{S_{3\phi}} = \frac{1}{3}\left[\frac{P_{\text{SI}}}{V_{\text{b}}I_{\text{b}}}\right] \tag{A.9}$$

while

$$\tau_{\text{pu}} = \frac{\tau_{\text{SI}}}{\tau_{\text{b}}} = \tau_{\text{SI}}\frac{\omega_{\text{s}}}{S_{3\phi}} = \frac{\omega_{\text{s}}}{3}\left[\frac{\tau_{\text{SI}}}{V_{\text{b}}I_{\text{b}}}\right] \tag{A.10}$$

The implication of these two equations is that generator power or torque equations derived in SI can be simply converted to pu by multiplying by 1/3 and $\omega_s/3$, respectively. See, for example, the torque expressions in Section 4.2.7 and Section 4.3.5.

6) Full load power (and torque) should not be confused with 1 pu power (and torque). They are not the same. In general

$$\text{Full load power} = S_{3\phi}\cos\phi_{\text{rated}}$$

$$\text{Full load torque} = \tau_{\text{b}}\cos\phi_{\text{rated}}$$

Mechanical engineers like to refer to a shaft rated, for example, as four times full load torque. This does not mean four times τ_b – they are different by $\cos\phi_{\text{rated}}$.

7) Because of the base values used the relationship between v_d, v_q and V_d, V_q and i_d, i_q and I_d, I_q derived in Chapter 11 are valid in both SI and pu that is

$$v_{\text{dpu}} = \sqrt{3}\,V_{\text{dpu}} \qquad i_{\text{dpu}} = \sqrt{3}\,I_{\text{dpu}} \tag{A.11}$$

$$v_{\text{qpu}} = \sqrt{3}\,V_{\text{qpu}} \qquad i_{\text{qpu}} = \sqrt{3}\,I_{\text{qpu}} \tag{A.12}$$

A.1.2 Power Invariance

A check on power invariance in SI is useful. Under balanced conditions, and using the current and voltage identities of Eqs. (11.80) and (11.82)

$$\begin{aligned} P_{3\phi} &= v_d i_d + v_q i_q = 3\left(V_d I_d + V_q I_q\right) \\ &= 3V_g I_g(\sin\delta_0 \sin(\delta_0+\phi) + \cos\delta_0\cos(\delta_0+\phi)) \\ &= 3V_g I_g\cos\phi \qquad \text{W} \end{aligned} \tag{A.13}$$

showing that power invariance is maintained.

As $P_{3\phi} = 3V_g I_g\cos\phi$ dividing both sides by $S_{3\phi}$ gives the generator power in pu

$$\frac{P_{3\phi}}{S_{3\phi}} = \frac{v_d i_d + v_q i_q}{3V_b I_b} = \frac{3\left(V_d I_d + V_q I_q\right)}{3V_b I_b} = \frac{3V_g I_g\cos\phi}{3V_b I_b} = V_{\text{g pu}} I_{\text{g pu}}\cos\phi \tag{A.14}$$

$$P_{\text{pu}} = \frac{1}{3}\left(v_{\text{dpu}}i_{\text{dpu}} + v_{\text{qpu}}i_{qpu}\right) = \left(V_{\text{dpu}}I_{\text{dpu}} + V_{\text{qpu}}I_{\text{qpu}}\right) = V_{\text{g pu}}I_{\text{g pu}}\cos\phi \tag{A.15}$$

and power invariance is also maintained in the pu system.

A.1.3 Rotor Base Quantities

Although a number of pu systems are possible (Harris et al. 1970), the system considered here is that of *equal mutual flux linkages*, as expounded by Anderson and Fouad (1977) and also explained in depth by Pavella and Murthy (1994). In this system the base field current, or base d-axis damper current, is defined so that each will produce the same fundamental air-gap flux wave as that produced by the base armature current acting in the fictitious d-axis armature coil. As will be seen as a consequence of this choice of pu system all the pu mutual inductances on a particular axis are equal.

It is convenient at this stage to separate each individual winding self-inductance into a magnetizing inductance and a leakage inductance so that

$$\begin{aligned} L_d &= L_{md} + l_l & L_q &= L_{mq} + l_l \\ L_D &= L_{mD} + l_D & L_Q &= L_{mQ} + l_Q \\ L_f &= L_{mf} + l_f \end{aligned} \tag{A.16}$$

where l represents the winding leakage inductance. The pu system requires the mutual flux linkage in each winding to be equal, that is

$$\begin{aligned} &\text{d}-\textit{coil}: && L_{md}I_b = kM_f I_{fb} = kM_D I_{Db} \\ &\text{f}-\textit{coil}: && kM_f I_b = L_{mf} I_{fb} = L_{fD} I_{Db} \\ &\text{D}-\textit{coil}: && kM_D I_b = L_{fD} I_{fb} = L_{mD} I_{Db} \\ &\text{q}-\textit{coil}: && L_{mq} I_b = kM_Q I_{Qb} \\ &\text{Q}-\textit{coil}: && kM_Q I_b = L_{mQ} I_{Qb} \end{aligned} \tag{A.17}$$

Multiplying each of these winding mutual flux linkages by the coil base current gives the fundamental constraint between the base currents as

$$\begin{aligned} &L_{md} I_b^2 = L_{mf} I_{fb}^2 = L_{mD} I_{Db}^2 = kM_f I_{fb} I_b = kM_D I_{Db} I_b = L_{fD} I_{fb} I_{Db} \\ &L_{mq} I_b^2 = L_{mQ} I_{Qb}^2 = kM_Q I_b I_{Qb} \end{aligned} \tag{A.18}$$

As the MVA base for each winding must be the same and equal to $S_b = V_b I_b$ this gives

$$\begin{aligned} &\frac{V_{fb}}{V_b} = \frac{I_b}{I_{fb}} = \sqrt{\frac{L_{mf}}{L_{md}}} = \frac{kM_f}{L_{md}} = \frac{L_{mf}}{kM_f} = \frac{L_{fD}}{kM_D} \equiv k_f \\ &\frac{V_{Db}}{V_b} = \frac{I_b}{I_{Db}} = \sqrt{\frac{L_{mD}}{L_{md}}} = \frac{kM_D}{L_{md}} = \frac{L_{mD}}{kM_D} = \frac{L_{fD}}{kM_f} \equiv k_D \\ &\frac{V_{Qb}}{V_b} = \frac{I_b}{I_{Qb}} = \sqrt{\frac{L_{mQ}}{L_{mq}}} = \frac{kM_Q}{L_{mq}} = \frac{L_{mQ}}{kM_Q} \equiv k_Q \end{aligned} \tag{A.19}$$

As Eq. (A.19) defines the base currents and voltages in all the windings as a function of the stator base quantities V_b and I_b

$$Z_{fb} = \frac{V_{fb}}{I_{fb}} = k_f^2 Z_b \ \ \Omega; \quad Z_{Db} = \frac{V_{Db}}{I_{Db}} = k_D^2 Z_b \ \ \Omega; \quad Z_{Qb} = \frac{V_{Qb}}{I_{Qb}} = k_Q^2 Z_b \ \ \Omega \tag{A.20}$$

and

$$L_{fb} = \frac{V_{fb} t_b}{I_{fb}} = k_f^2 L_b \ \ \text{H}; \quad L_{Db} = \frac{V_{Db} t_b}{I_{Db}} = k_D^2 L_b \ \ \text{H}; \quad L_{Qb} = k_Q^2 L_b \ \ \text{H} \tag{A.21}$$

while the base mutual inductances are

$$\begin{aligned} &M_{fb} = \frac{V_{fb} t_b}{I_b} = \frac{V_b t_b}{I_{fb}} = k_f L_b \ \ \text{H}; \qquad M_{Db} = k_D L_b \ \ \text{H} \\ &M_{Qb} = k_Q L_b \ \ \text{H}; \qquad L_{fDb} = k_f k_D L_b \ \ \text{H} \end{aligned} \tag{A.22}$$

With the base values now defined Eqs. (11.18), (11.19), (10.30), and (10.31) can now be normalized and expressed in pu. As an example consider the normalization of the field flux linkage Ψ_f in Eq. (11.18) where

$$\Psi_f = kM_f i_d + L_f i_f + L_{fD} i_D \tag{A.23}$$

divide by $\Psi_{fb} = L_{fb} I_{fb}$ to give

$$\Psi_{fpu} = \frac{kM_f}{L_{fb}}\frac{i_d}{I_{fb}} + \frac{L_f}{L_{fb}}\frac{i_f}{I_{fb}} + \frac{L_{fD}}{L_{fb}}\frac{i_D}{I_{fb}} \tag{A.24}$$

when substituting for I_{fb} and L_{fb} from Eqs. (A.19) and (A.20) gives

$$\Psi_{fpu} = \left[\frac{kM_f}{k_f L_b}\right]\left[\frac{i_d}{I_b}\right] + \left[\frac{L_f}{L_{fb}}\right]\left[\frac{i_f}{I_{fb}}\right] + \left[\frac{L_{fD}}{k_f k_D L_b}\right]\left[\frac{i_D}{I_{Db}}\right] \tag{A.25}$$

and

$$\Psi_{fpu} = kM_{fpu} i_{dpu} + L_{fpu} i_{fpu} + L_{fDpu} i_{Dpu} \tag{A.26}$$

This normalized equation is of exactly the same form as the equation in SI and this is true for all other equations in Sections 11.1.2–11.1.5. In other words all the voltage, current and flux equations in all these sections are of the same form whether in pu or SI.

One further interesting feature of the pu system is that the pu values of all the mutual inductances on one axis are equal, that is L_{md}, L_{mf}, L_{mD}, kM_f, kM_D, and L_{fD} are all equal. For example

$$\begin{aligned} kM_{fpu} &= \frac{kM_f}{M_{fb}} = \frac{kM_f}{k_f L_b} = \frac{k_f L_{md}}{k_f L_b} = L_{mdpu} \\ L_{fDpu} &= \frac{L_{fD}}{L_{fDb}} = \frac{L_{fD}}{k_f k_D L_b} = \frac{L_{fD}}{\frac{kM_f}{L_{md}}\frac{L_{fD}}{kM_f} L_b} = \frac{L_{md}}{L_b} = L_{mdpu} \end{aligned} \tag{A.27}$$

It is common practice to replace all these pu mutual values by a pu mutual inductance L_{ad} so that

$$L_{ad} \equiv L_{md} = L_{mf} = L_{mD} = kM_f = kM_D = L_{fD} \tag{A.28}$$

and, in the q-axis,

$$L_{aq} \equiv L_{mq} = L_{mQ} = kM_Q \tag{A.29}$$

All the equations in Sections 11.1.2–11.1.5 can now be written in pu in terms of the mutual inductances L_{ad} and L_{aq}. For example, Eq. (11.43) discussed in Section 11.1.5 for the d-axis subtransient inductance would become, in pu

$$L_d'' = L_d - \left[\frac{L_{ad}^2 L_D + L_{ad}^2 L_f - 2L_{ad}^3}{L_D L_f - L_{ad}^2}\right] \tag{A.30}$$

where $L_d = L_{ad} + l_l, L_D = L_{ad} + l_D$ and $L_f = L_{ad} + l_f$

With this knowledge it is constructive to examine the pu flux linking each winding. Using Eq. (11.18), and dropping the pu symbol for simplicity, the flux linkage of the d-axis coil is

$$\Psi_d = L_d i_d + kM_f i_f + kM_D i_D \tag{A.31}$$

when substituting for L_{ad} and introducing the winding leakage inductance gives

$$\Psi_d = L_{ad}(i_d + i_f + i_D) + l_l i_d \tag{A.32}$$

and similarly for the field and d-axis damper coil

$$\begin{aligned} \Psi_f &= L_{ad}(i_d + i_f + i_D) + l_f i_f \\ \Psi_D &= L_{ad}(i_d + i_f + i_D) + l_D i_D \end{aligned} \tag{A.33}$$

Thus, if the pu leakage flux linkage of a particular winding is subtracted from the total flux linkage, the remaining mutual flux linkage in all the windings on each axis is equal. This mutual flux linkage is often given the symbol Ψ_{ad} and, for the q-axis, Ψ_{aq}, where

$$\begin{aligned} \Psi_{ad} &= L_{ad}(i_d + i_f + i_D) \\ \Psi_{aq} &= L_{aq}(i_q + i_Q) \end{aligned} \tag{A.34}$$

A.1.4 Power System Base Quantities

It is customary in three-phase power system analysis to use rated line-to-line voltage as the base voltage and an arbitrary three-phase volt-amp base, typically 10 MVA, 100 MVA, etc. Such a base system would, at first sight, seem to be totally inconsistent with the generator armature base defined in Section A.1.1. In fact, the two are entirely consistent.

For the power system

$$\begin{aligned} V_{L-L,b} &= V_{L-L} = \sqrt{3}V_{L-N} && \text{V} \\ S_b &= S_{3\phi} && \text{VA} \\ I_b &= \frac{S_{3\phi}}{\sqrt{3}\ V_{L-L,b}} && \text{A} \\ Z_b &= \frac{V_{L-L,b}}{\sqrt{3}\ I_b} = \frac{V_{L-L,b}^2}{S_{3\phi}} && \Omega \end{aligned} \tag{A.35}$$

Ignoring any transformer effects and assuming that the system MVA base $S_{3\varphi}$ is equal to the generator MVA rating then $V_{L-N} = V_b$, where V_b is the generator base voltage and

$$\begin{aligned} Z_b &= \frac{V_{L-L,b}^2}{S_{3\phi}} = \frac{3V_{L-N}^2}{S_{3\phi}} = \frac{V_b^2}{S_{1\phi}} \\ I_b &= \frac{S_{3\phi}}{\sqrt{3}\ V_{L-L,b}} = \frac{S_{1\phi}}{V_b} \end{aligned} \tag{A.36}$$

showing that the power system base and the generator base are totally consistent.

However, there is one complication and that is $S_{3\phi}$ for the system is chosen arbitrarily, while $S_{1\phi}$ for the generator is the rated MVA per phase. In fact, for all the equipment making up the power system their pu impedance values will be defined with respect to their individual MVA ratings. Consequently, in the system analysis it will be necessary to either:

- convert all the generator parameters to be on the system base; or
- have a base conversion between the individual generator equations and the system equations in the computer software. This is easily achieved and has the advantage that not only do the generator pu values retain their familiarity but they are as exactly as provided by the equipment manufacture.

Both methods are used and converting pu values from one system (base 1) to another (base 2) is readily achieved via Eq. (A.1) to obtain

$$\text{per unit value (base 2)} = \text{per unit value (base 1)}\frac{\text{Base 1 value}}{\text{Base 2 value}} \tag{A.37}$$

A.1.5 Transformers

Section 3.2 shows how a transformer could be represented by either the primary or the secondary equivalent circuit shown in Figure A.1a and b. In these equivalent circuits the primary equivalent impedance Z_1 and the secondary equivalent impedance Z_2 are related by

$$Z_1 = n^2 Z_2 \tag{A.38}$$

where n is the nominal turns ratio. However, the primary and secondary base values are defined as

$$\begin{aligned} V_{pb} &= V_{1,L-N} & V_{sb} &= V_{2,L-N} \\ I_{pb} &= \frac{S_{3\phi}}{3V_{pb}} & I_{sb} &= \frac{S_{3\phi}}{3V_{sb}} \\ Z_{pb} &= \frac{V_{pb}}{I_{pb}} & Z_{sb} &= \frac{V_{sb}}{I_{sb}} \end{aligned} \tag{A.39}$$

where

$$V_{pb} = nV_{sb} \tag{A.40}$$

and implies that

$$I_{pb} = \frac{I_{sb}}{n} \qquad Z_{pb} = n^2 Z_{sb} \tag{A.41}$$

Consequently, as 1 pu voltage on the high voltage side of the transformer must be 1 pu on the low voltage side and

$$Z_{pu} = \frac{Z_1}{Z_{pb}} = \frac{Z_2 n^2}{n^2 Z_{sb}} = \frac{Z_2}{Z_{sb}} = Z_{pu} \tag{A.42}$$

the pu value of the primary and secondary equivalent impedance is the same so that the transformer at nominal taps can be represented by the pu equivalent shown in Figure A.1c. If the tap setting changes from the nominal the equivalent circuit is modified to that shown in Figure 3.8.

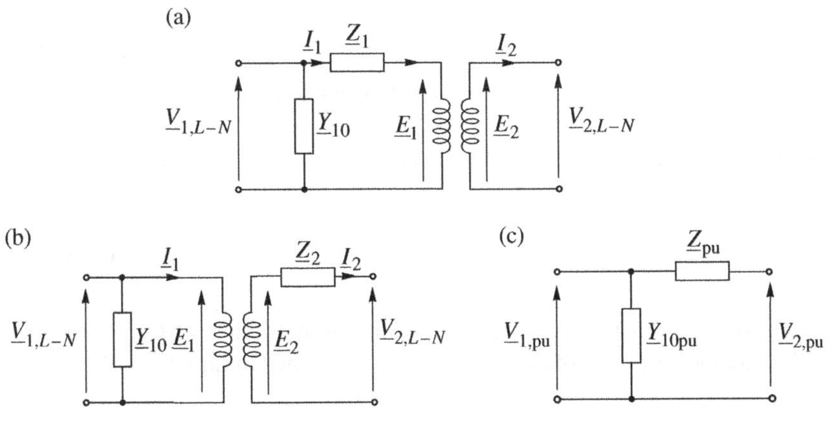

Figure A.1 Transformer equivalent circuit: (a) primary; (b) secondary; (c) pu at nominal tap setting.

A.2 Partial Inversion

Consider the linear equation

$$\boldsymbol{x} = \boldsymbol{A}\,\boldsymbol{y} \tag{A.43}$$

where $\boldsymbol{A}$ is square matrix and $\boldsymbol{x}$ and $\boldsymbol{y}$ are column vectors. When the variables in these vectors are grouped into two groups {R} and {E}, then Eq. (A.43) can be rewritten in the following way

$$\begin{bmatrix} \boldsymbol{x}_{\mathrm{R}} \\ \boldsymbol{x}_{\mathrm{E}} \end{bmatrix} = \left[\begin{array}{c|c} \boldsymbol{A}_{\mathrm{RR}} & \boldsymbol{A}_{\mathrm{RE}} \\ \hline \boldsymbol{A}_{\mathrm{ER}} & \boldsymbol{A}_{\mathrm{EE}} \end{array}\right] \begin{bmatrix} \boldsymbol{y}_{\mathrm{R}} \\ \boldsymbol{y}_{\mathrm{E}} \end{bmatrix}. \tag{A.44}$$

Expanding the equation gives

$$\boldsymbol{x}_{\mathrm{R}} = \boldsymbol{A}_{\mathrm{RR}}\boldsymbol{y}_{\mathrm{R}} + \boldsymbol{A}_{\mathrm{RE}}\boldsymbol{y}_{\mathrm{E}} \tag{A.45}$$

$$\boldsymbol{x}_{\mathrm{E}} = \boldsymbol{A}_{\mathrm{ER}}\boldsymbol{y}_{\mathrm{R}} + \boldsymbol{A}_{\mathrm{EE}}\boldsymbol{y}_{\mathrm{E}} \tag{A.46}$$

Simple manipulations result in

$$\boldsymbol{y}_{\mathrm{E}} = -\boldsymbol{A}_{\mathrm{EE}}^{-1}\boldsymbol{A}_{\mathrm{ER}}\,\boldsymbol{y}_{\mathrm{R}} + \boldsymbol{A}_{\mathrm{EE}}^{-1}\boldsymbol{x}_{\mathrm{E}} \tag{A.47}$$

Substituting (A.47) into (A.45) gives

$$\boldsymbol{x}_{\mathrm{R}} = \left(\boldsymbol{A}_{\mathrm{RR}} - \boldsymbol{A}_{\mathrm{RE}}\boldsymbol{A}_{\mathrm{EE}}^{-1}\boldsymbol{A}_{\mathrm{ER}}\right)\boldsymbol{y}_{\mathrm{R}} + \boldsymbol{A}_{\mathrm{RE}}\boldsymbol{A}_{\mathrm{EE}}^{-1}\boldsymbol{x}_{\mathrm{E}} \tag{A.48}$$

Equations (A.47) and (A.48) can be written as

$$\begin{bmatrix} \boldsymbol{x}_{\mathrm{R}} \\ \boldsymbol{y}_{\mathrm{E}} \end{bmatrix} = \left[\begin{array}{c|c} \boldsymbol{A}_{\mathrm{RR}} - \boldsymbol{A}_{\mathrm{RE}}\boldsymbol{A}_{\mathrm{EE}}^{-1}\boldsymbol{A}_{\mathrm{ER}} & \boldsymbol{A}_{\mathrm{RE}}\boldsymbol{A}_{\mathrm{EE}}^{-1} \\ \hline -\boldsymbol{A}_{\mathrm{EE}}^{-1}\boldsymbol{A}_{\mathrm{ER}} & \boldsymbol{A}_{\mathrm{EE}}^{-1} \end{array}\right] \begin{bmatrix} \boldsymbol{y}_{\mathrm{R}} \\ \boldsymbol{x}_{\mathrm{E}} \end{bmatrix}. \tag{A.49}$$

Comparing to (A.44), $\boldsymbol{y}_{\mathrm{E}}$ has been moved to the left-hand side of the equation, while $\boldsymbol{x}_{\mathrm{E}}$ to the right. This is referred to as the *partial inversion* of a matrix. Equation (A.49) can be written as

$$\begin{bmatrix} \boldsymbol{x}_{\mathrm{R}} \\ \boldsymbol{y}_{\mathrm{E}} \end{bmatrix} = \left[\begin{array}{c|c} \boldsymbol{A}_{\mathrm{R}} & \boldsymbol{B}_{\mathrm{RE}} \\ \hline -\boldsymbol{B}_{\mathrm{ER}} & \boldsymbol{C}_{\mathrm{EE}} \end{array}\right] \begin{bmatrix} \boldsymbol{y}_{\mathrm{R}} \\ \boldsymbol{x}_{\mathrm{E}} \end{bmatrix} \tag{A.50}$$

where

$$\boldsymbol{A}_{\mathrm{R}} = \boldsymbol{A}_{\mathrm{RR}} - \boldsymbol{A}_{\mathrm{RE}}\boldsymbol{A}_{\mathrm{EE}}^{-1}\boldsymbol{A}_{\mathrm{ER}}; \quad \boldsymbol{B}_{\mathrm{RE}} = \boldsymbol{A}_{\mathrm{RE}}\boldsymbol{A}_{\mathrm{EE}}^{-1}; \quad \boldsymbol{B}_{\mathrm{ER}} = \boldsymbol{A}_{\mathrm{EE}}^{-1}\underline{\mathbf{Y}}_{\mathrm{ER}}; \quad \boldsymbol{C}_{\mathrm{EE}} = \boldsymbol{A}_{\mathrm{EE}}^{-1} \tag{A.51}$$

In the particular case when $\boldsymbol{x}_{\mathrm{E}} = \mathbf{0}$ Eq. (A.50) gives

$$\boldsymbol{x}_{\mathrm{R}} = \boldsymbol{A}_{\mathrm{R}}\,\boldsymbol{y}_{\mathrm{R}} \tag{A.52}$$

The derived equations are useful when dealing with transformation of the admittance and incremental network models.

Let $\boldsymbol{V} = \boldsymbol{A}^{-1}$ be the inverse matrix of the matrix $\boldsymbol{A}$ appearing in Eq. (A.43). Using the same groups of variables {R} and {E} as in (A.44) the following equation can be written

$$\begin{bmatrix} \boldsymbol{y}_{\mathrm{R}} \\ \boldsymbol{y}_{\mathrm{E}} \end{bmatrix} = \left[\begin{array}{c|c} \boldsymbol{V}_{\mathrm{RR}} & \boldsymbol{V}_{\mathrm{RE}} \\ \hline \boldsymbol{V}_{\mathrm{ER}} & \boldsymbol{V}_{\mathrm{EE}} \end{array}\right] \begin{bmatrix} \boldsymbol{x}_{\mathrm{R}} \\ \boldsymbol{x}_{\mathrm{E}} \end{bmatrix} \tag{A.53}$$

The essence of the following considerations is to determine the relationship between sub-matrices in matrix $\boldsymbol{V}$ and matrix $\boldsymbol{A}$. For this purpose, a partial inversion analogous to that of Eq. (A.49) will be performed in Eq. (A.53). With Eq. (A.53) it appears that

$$\boldsymbol{y}_{\rm R} = \boldsymbol{V}_{\rm RR}\boldsymbol{x}_{\rm R} + \boldsymbol{V}_{\rm RE}\boldsymbol{x}_{\rm E} \tag{A.54}$$

$$\boldsymbol{y}_{\rm E} = \boldsymbol{V}_{\rm ER}\boldsymbol{x}_{\rm R} + \boldsymbol{V}_{\rm EE}\boldsymbol{x}_{\rm E} \tag{A.55}$$

By converting Eq. (A.54) one obtains

$$\boldsymbol{x}_{\rm R} = \boldsymbol{V}_{\rm RR}^{-1}\boldsymbol{y}_{\rm R} - \boldsymbol{V}_{\rm RR}^{-1}\boldsymbol{V}_{\rm RE}\boldsymbol{x}_{\rm E} \tag{A.56}$$

which after substituting into (A.55) gives

$$\boldsymbol{y}_{\rm E} = \boldsymbol{V}_{\rm ER}\boldsymbol{V}_{\rm RR}^{-1}\boldsymbol{y}_{\rm R} + \left(\boldsymbol{V}_{\rm EE} - \boldsymbol{V}_{\rm ER}\boldsymbol{V}_{\rm RR}^{-1}\boldsymbol{V}_{\rm RE}\right)\boldsymbol{x}_{\rm E} \tag{A.57}$$

Equations (A.56) and (A.57) can be written together in the form of the following matrix equation

$$\begin{bmatrix} \boldsymbol{x}_{\rm R} \\ \boldsymbol{y}_{\rm E} \end{bmatrix} = \begin{bmatrix} \boldsymbol{V}_{\rm RR}^{-1} & -\boldsymbol{V}_{\rm RR}^{-1}\boldsymbol{V}_{\rm RE} \\ \boldsymbol{V}_{\rm ER}\boldsymbol{V}_{\rm RR}^{-1} & \left(\boldsymbol{V}_{\rm EE} - \boldsymbol{V}_{\rm ER}\boldsymbol{V}_{\rm RR}^{-1}\boldsymbol{V}_{\rm RE}\right) \end{bmatrix} \begin{bmatrix} \boldsymbol{y}_{\rm R} \\ \boldsymbol{x}_{\rm E} \end{bmatrix} \tag{A.58}$$

From the comparison of Eqs. (A.49) and (A.58) result the following equalities

$$\boldsymbol{A}_{\rm RR} - \boldsymbol{A}_{\rm RE}\boldsymbol{A}_{\rm EE}^{-1}\boldsymbol{A}_{\rm ER} = \boldsymbol{V}_{\rm RR}^{-1} \tag{A.59}$$

$$\boldsymbol{A}_{\rm RE}\boldsymbol{A}_{\rm EE}^{-1} = -\boldsymbol{V}_{\rm RR}^{-1}\boldsymbol{V}_{\rm RE} \tag{A.60}$$

$$-\boldsymbol{A}_{\rm EE}^{-1}\boldsymbol{A}_{\rm ER} = \boldsymbol{V}_{\rm ER}\boldsymbol{V}_{\rm RR}^{-1} \tag{A.61}$$

$$\boldsymbol{A}_{\rm EE}^{-1} = \boldsymbol{V}_{\rm EE} - \boldsymbol{V}_{\rm ER}\boldsymbol{V}_{\rm RR}^{-1}\boldsymbol{V}_{\rm RE} \tag{A.62}$$

Particularly important are the equalities (A.59) and (A.62).

A.3 Linear Ordinary Differential Equations

There are many good mathematical textbooks dealing with solving ordinary differential equations. A well-written textbook aimed at engineers is Arnold (1992). This appendix contains the essential information regarding scalar linear differential equations necessary for the understanding of this textbook.

A.3.1 Fundamental System of Solutions

For real variables x, $t \in$ Real the linear scalar homogeneous differential equation is of the form

$$\frac{{\rm d}^n x}{{\rm d}t^n} + a_1\frac{{\rm d}^{n-1}x}{{\rm d}t^{n-1}} + \ldots + a_{n-2}\frac{{\rm d}^2 x}{{\rm d}t^2} + a_{n-1}\frac{{\rm d}x}{{\rm d}t} + a_n x = 0 \tag{A.63}$$

where a_1, a_2, ..., a_n are constant coefficients.

Each function $x(t)$ that satisfies Eq. (A.63) is its solution. Without specifying some initial conditions, the solution of Eq. (A.63) is not unique and there may be an infinite number of solutions. For example, if function $x_1(t)$ is a solution then a solution is also any function $cx_1(t)$ where $c \neq 0$ is a constant. Additionally, if functions $x_1(t)$, $x_2(t)$, $x_3(t)$, ... are solutions then any linear combination of the functions $c_1x_1(t) + c_2x_2(t) + c_3x_3(t) + \ldots$ is also a solution, as substituting that combination into Eq. (A.63) gives a sum of zeros, i.e. zero.

Solutions $x_1(t)$, $x_2(t)$, $x_3(t)$, ... are linearly independent if neither can be expressed as a linear combination of the remaining solutions. For example, if $x_i(t)$, $x_j(t)$, $x_k(t)$, ... are linearly independent then there exist no constants c_j, c_k, ... for which it would hold $x_i(t) = c_jx_j(t) + c_kx_k(t) + \ldots$

The largest set of linearly independent solutions $x_1(t)$, $x_2(t)$, $x_3(t)$, ..., $x_n(t)$ of Eq. (A.63) is referred to as the *fundamental system of solutions*. Whether a given set of solutions is fundamental (i.e. the solutions are linearly independent) can be checked by investigating the determinant of the matrix below, the columns of which contain individual solutions and their derivatives

$$\det \boldsymbol{W} = \det \begin{bmatrix} x_1 & x_2 & x_3 & \cdots & x_n \\ \dot{x}_1 & \dot{x}_2 & \dot{x}_3 & \cdots & \dot{x}_n \\ \ddot{x}_1 & \ddot{x}_2 & \ddot{x}_3 & \cdots & \ddot{x}_n \\ \dddot{x}_1 & \dddot{x}_2 & \dddot{x}_3 & \cdots & \dddot{x}_n \\ \vdots & \vdots & \vdots & \ddots & \vdots \end{bmatrix} \neq 0 \tag{A.64}$$

where $\dot{x} = \mathrm{d}x/\mathrm{d}t$, $\ddot{x} = \mathrm{d}^2x/\mathrm{d}t^2$, $\dddot{x} = \mathrm{d}^3x/\mathrm{d}t^3$, etc. denote time derivatives. That determinant is referred to as the *Wronskian*, after a mathematician Józef Maria Hoene-Wroński. It may be shown that solutions $x_1(t)$, $x_2(t)$, $x_3(t)$, ..., $x_n(t)$ are linearly independent and form the fundamental system of solutions if, and only if, $\det \boldsymbol{W} \neq 0$.

It follows then that a linear combination of the fundamental system of solutions of the form

$$x(t) = A_1x_1(t) + A_2x_2(t) + A_3x_3(t) + \ldots + A_nx_n(t) \tag{A.65}$$

is also a solution of Eq. (A.63). Such a solution is referred to as the *general solution*. It is general because it contains all the fundamental solutions. Coefficients $A_1, A_2, A_3, \ldots, A_n$ are referred to as the *integration constants*.

For a linear equation the fundamental solutions are of the exponential form

$$x(t) = \mathrm{e}^{\lambda t}; \quad \frac{\mathrm{d}x}{\mathrm{d}t} = \lambda \mathrm{e}^{\lambda t}; \quad \frac{\mathrm{d}^2x}{\mathrm{d}t^2} = \lambda^2 \mathrm{e}^{\lambda t}; \quad \frac{\mathrm{d}^3x}{\mathrm{d}t^3} = \lambda^3 \mathrm{e}^{\lambda t}; \text{etc.} \tag{A.66}$$

Substituting Eq. (A.66) into Eq. (A.63) gives

$$\lambda^n \mathrm{e}^{\lambda t} + a_1\lambda^{n-1}\mathrm{e}^{\lambda t} + \ldots + a_{n-2}\lambda^2\mathrm{e}^{\lambda t} + a_{n-1}\lambda \mathrm{e}^{\lambda t} + a_n\mathrm{e}^{\lambda t} = 0 \tag{A.67}$$

For each t it holds $\mathrm{e}^{\lambda t} \neq 0$ and Eq. (A.67) may be simplified to the following form

$$\lambda^n + a_1\lambda^{n-1} + \ldots + a_{n-2}\lambda^2 + a_{n-1}\lambda + a_n = 0 \tag{A.68}$$

This equation is referred to as the *characteristic equation*. It determines the values of λ for which function $x(t) = \mathrm{e}^{\lambda t}$ is a solution of (A.63). The characteristic equation is an algebraic equation of n-th order and it generally has n roots $\lambda_1, \lambda_2, \lambda_3, \ldots, \lambda_n$. The roots of the characteristic equation form n solutions of the form

$$x_1(t) = \mathrm{e}^{\lambda_1 t}, \quad x_2(t) = \mathrm{e}^{\lambda_2 t}, \quad x_3(t) = \mathrm{e}^{\lambda_3 t}, \ldots, x_n(t) = \mathrm{e}^{\lambda_n t} \tag{A.69}$$

The Wronskian of the solutions is

$$\det \boldsymbol{W} = \det \begin{bmatrix} \mathrm{e}^{\lambda_1 t} & \mathrm{e}^{\lambda_2 t} & \mathrm{e}^{\lambda_3 t} & \cdots & \mathrm{e}^{\lambda_n t} \\ \lambda_1\mathrm{e}^{\lambda_1 t} & \lambda_2\mathrm{e}^{\lambda_2 t} & \lambda_3\mathrm{e}^{\lambda_3 t} & \cdots & \lambda_n\mathrm{e}^{\lambda_n t} \\ \lambda_1^2\mathrm{e}^{\lambda_1 t} & \lambda_2^2\mathrm{e}^{\lambda_2 t} & \lambda_3^2\mathrm{e}^{\lambda_3 t} & \cdots & \lambda_n^2\mathrm{e}^{\lambda_n t} \\ \lambda_1^3\mathrm{e}^{\lambda_1 t} & \lambda_2^3\mathrm{e}^{\lambda_3 t} & \lambda_3^3\mathrm{e}^{\lambda_3 t} & \cdots & \lambda_n^3\mathrm{e}^{\lambda_n t} \\ \vdots & \vdots & \vdots & \ddots & \vdots \end{bmatrix} \tag{A.70}$$

Multiplying a matrix column by a number corresponds to multiplying the determinant of the matrix by that number. Hence the terms $\mathrm{e}^{\lambda_1 t}$, $\mathrm{e}^{\lambda_2 t}$, and $\mathrm{e}^{\lambda_3 t}$, etc., can be extracted in front of the Wronskian Eq. (A.70). As

$$\mathrm{e}^{\lambda_1 t} \cdot \mathrm{e}^{\lambda_2 t} \cdot \mathrm{e}^{\lambda_3 t} \cdot \ldots \cdot \mathrm{e}^{\lambda_n t} = \mathrm{e}^{(\lambda_1 + \lambda_2 + \lambda_3 + \ldots + \lambda_n)t} \tag{A.71}$$

Equation (A.70) can be expressed as

$$\det \boldsymbol{W} = e^{(\lambda_1 + \lambda_2 + \lambda_3 + \ldots + \lambda_n)t} \cdot \det \begin{bmatrix} 1 & 1 & 1 & \cdots & 1 \\ \lambda_1 & \lambda_2 & \lambda_3 & \cdots & \lambda_n \\ \lambda_1^2 & \lambda_2^2 & \lambda_3^2 & \cdots & \lambda_n^2 \\ \lambda_1^3 & \lambda_2^3 & \lambda_3^3 & \cdots & \lambda_n^3 \\ \vdots & \vdots & \vdots & \ddots & \vdots \end{bmatrix} \tag{A.72}$$

This determinant is made up of successive powers of the roots and is referred to as the *Vandermonde's determinant.* It can be shown using mathematical induction that Vandermonde's determinant is equal to the sum of products of differences between the pairs of roots

$$\det \begin{bmatrix} 1 & 1 & 1 & \cdots & 1 \\ \lambda_1 & \lambda_2 & \lambda_3 & \cdots & \lambda_n \\ \lambda_1^2 & \lambda_2^2 & \lambda_3^2 & \cdots & \lambda_n^2 \\ \lambda_1^3 & \lambda_2^3 & \lambda_3^3 & \cdots & \lambda_n^3 \\ \vdots & \vdots & \vdots & \ddots & \vdots \end{bmatrix} = \prod_{1 \le i \le j \le n} (\lambda_j - \lambda_i) \tag{A.73}$$

where

$$\prod_{1 \le i \le j \le n} (\lambda_j - \lambda_i) = (\lambda_n - \lambda_{n-1})(\lambda_n - \lambda_{n-2})(\lambda_n - \lambda_{n-3}), \ldots, (\lambda_3 - \lambda_2)(\lambda_3 - \lambda_1)(\lambda_2 - \lambda_1) \tag{A.74}$$

Equation (A.73) is useful for a fast determination of Vandermonde's determinant. The proof can be found, for example in Ogata (1967).

A.3.2 Real and Distinct Roots

The sufficient condition for the Vandermonde's determinant given by Eq. (A.73), and therefore also the Wronskian given by Eq. (A.72), to be different from zero is that the roots of the characteristic equation are distinct

$$\lambda_1 \neq \lambda_2 \neq \lambda_3 \neq, \ldots, \neq \lambda_n \tag{A.75}$$

If that conditions is satisfied, the functions given by Eq. (A.69) form the fundamental system of solutions of Eq. (A.63). Hence the general solution Eq. (A.65) is

$$x(t) = A_1 e^{\lambda_1 t} + A_2 e^{\lambda_2 t} + A_3 e^{\lambda_3 t} + \ldots + A_n e^{\lambda_n t} \tag{A.76}$$

When the integration constants A_1, A_2, A_3, ..., A_n are not specified, the general solution gives an infinite number of solutions. The Cauchy's problem consists of finding such a *particular solution* that satisfies the initial conditions for the solution and its derivatives: $x(t_0), \dot{x}(t_0), \ddot{x}(t_0), \dddot{x}(t_0), \ldots$ In order to solve the Cauchy's problem, it is necessary to find such integration constants $A_1, A_2, A_3, \ldots, A_n$ for the general solution that the initial conditions are satisfied.

Often the initial conditions are assumed to be a nonzero value of the solution and zero values of its derivatives

$$x(t_0) = \Delta x \neq 0, \quad \dot{x}(t_0) = 0, \quad \ddot{x}(t_0) = 0 \quad \dddot{x}(t_0) = 0, \text{etc.} \tag{A.77}$$

Substituting function Eq. (A.76) and its derivatives calculated at time instant t_0 into Eq. (A.77) results in an algebraic equation

$$\begin{bmatrix} 1 & 1 & 1 & \cdots & 1 \\ \lambda_1 & \lambda_2 & \lambda_3 & \cdots & \lambda_n \\ \lambda_1^2 & \lambda_2^2 & \lambda_3^2 & \cdots & \lambda_n^2 \\ \lambda_1^3 & \lambda_2^3 & \lambda_3^3 & \cdots & \lambda_n^3 \\ \vdots & \vdots & \vdots & \ddots & \vdots \end{bmatrix} \begin{bmatrix} A_1 \\ A_2 \\ A_3 \\ A_4 \\ \vdots \end{bmatrix} = \begin{bmatrix} \Delta x \\ 0 \\ 0 \\ 0 \\ \vdots \end{bmatrix} \tag{A.78}$$

The matrix on the left-hand side is the Vandermonde's matrix. Equation (A.73) shows that under the assumption of distinct roots of the characteristic equation, the determinant of Vandermonde's matrix is different from zero, which means that the matrix is not singular and there is only one solution for the integration constants $A_1, A_2, A_3, \ldots, A_n$.

Example A.3.1 Solve a third-order equation $\dddot{x} + 6\ddot{x} + 11\dot{x} + 6x = 0$ under the initial solutions given by Eq. (A.77).

The characteristic equation is $\lambda^3 + 6\lambda^2 + 11\lambda + 6 = 0$ with the distinct roots: $\lambda_1 = -3$, $\lambda_2 = -2$, and $\lambda_3 = -1$. The general solution Eq. (A.76) is of the form: $x(t) = A_1 e^{-3t} + A_2 e^{-2t} + A_3 e^{-t}$. Equation (A.78) is

$$\begin{bmatrix} 1 & 1 & 1 \\ -3 & -2 & -1 \\ 9 & 4 & 1 \end{bmatrix} \begin{bmatrix} A_1 \\ A_2 \\ A_3 \end{bmatrix} = \begin{bmatrix} \Delta x \\ 0 \\ 0 \end{bmatrix} \quad \text{or} \quad \begin{bmatrix} A_1 \\ A_2 \\ A_3 \end{bmatrix} = \frac{1}{2} \begin{bmatrix} 2 & 3 & 1 \\ -6 & -8 & -2 \\ 6 & 5 & 1 \end{bmatrix} \begin{bmatrix} \Delta x \\ 0 \\ 0 \end{bmatrix} \tag{A.79}$$

Hence: $A_1 = \Delta x$, $A_2 = -3 \cdot \Delta x$, $A_3 = 3 \cdot \Delta x$. Finally: $x(t) = \Delta x \cdot (e^{-3t} - 3e^{-2t} + 3e^{-t})$.

******

For the dynamics considered in this book of a particular attention is a second-order scalar equation corresponding to the equation of motion of the synchronous generator (Section 5.1). Hence now a solution to the second-order equation is discussed here when the roots of the characteristic equation are initially assumed to be real.

Example A.3.2 Solve the second-order equation $\ddot{x} - (\alpha_1 + \alpha_2)\dot{x} + \alpha_1\alpha_2 x = 0$ with the initial conditions given by Eq. (A.77).

The characteristic equation is $\lambda^2 - (\alpha_1 + \alpha_2)\lambda + \alpha_1\alpha_2 = 0$ with the distinct roots $\lambda_1 = \alpha_1$, $\lambda_2 = \alpha_2$, and $\alpha_2 \neq \alpha_1$. The general solution Eq. (A.76) is: $x(t) = A_1 e^{\alpha_1 t} + A_2 e^{\alpha_2 t}$. Equation (A.78) takes the form

$$\begin{bmatrix} 1 & 1 \\ \alpha_1 & \alpha_2 \end{bmatrix} \begin{bmatrix} A_1 \\ A_2 \end{bmatrix} = \begin{bmatrix} \Delta x \\ 0 \end{bmatrix} \quad \text{or} \quad \begin{bmatrix} A_1 \\ A_2 \end{bmatrix} = \frac{1}{\alpha_2 - \alpha_1} \begin{bmatrix} \alpha_2 & -1 \\ -\alpha_1 & 1 \end{bmatrix} \begin{bmatrix} \Delta x \\ 0 \end{bmatrix} \tag{A.80}$$

Hence

$$A_1 = \Delta x \cdot \alpha_2/(\alpha_2 - \alpha_1) \quad \text{and} \quad A_2 = -\Delta x \cdot \alpha_1/(\alpha_2 - \alpha_1)$$

Finally

$$x(t) = \Delta x \cdot [\alpha_2 e^{\alpha_1 t} - \alpha_1 e^{\alpha_2 t}]/(\alpha_2 - \alpha_1)$$

******

A.3.3 Repeated Real Roots

If condition (A.75) is not satisfied, and there are repeated real roots of the characteristic equation, then the fundamental system of equations can be built from those solutions that are linearly independent and correspond to distinct roots. Obviously then, there will be fewer than n solutions corresponding to those roots, i.e. too few to solve the Cauchy's problem of finding a particular solution for given initial conditions. In order to obtain a unique solution, one has to supplement the fundamental system of solutions by additional linearly independent solutions such that there is overall n solutions, where n is the order of the differential equation.

Let λ_i be a root of the characteristic equation repeated k times. Then one of the solutions belonging to the fundamental system of solutions corresponding to that root is of the form $x_{i1}(t) = e^{\lambda_i t}$. There are still missing $(k-1)$ linearly independent solutions which have to supplement the fundamental system of solutions. For a root λ_i repeated k times, a solution is formed in the following way

$$x_{i_2}(t) = A_{i_2}(t)\cdot e^{\lambda_i t}, \quad x_{i_3}(t) = A_{i_3}(t)\cdot e^{\lambda_i t}, ..., x_{i_k}(t) = A_{i_k}(t)\cdot e^{\lambda_i t} \tag{A.81}$$

where $A_{i_2}(t), A_{i_3}(t),..., A_{i_k}(t)$ are the required functions chosen in such a way that the solutions are linearly independent. It can be shown (Arnold 1992) that the required functions are orthogonal polynomials $t, t^2, t^3, ..., t^{k-1}$. The complete set of additional solutions corresponding to a root λ_i repeated k times is

$$x_{i1}(t) = e^{\lambda_i t}, \quad x_{i_2}(t) = t\cdot e^{\lambda_i t}, \quad x_{i_3}(t) = t^2\cdot e^{\lambda_i t}, ..., x_{i_k}(t) = t^{k-1}\cdot e^{\lambda_i t} \tag{A.82}$$

Obviously, the complete set of fundamental solutions contains also the solutions corresponding to other roots.

Example A.3.3 Solve a second-order equation $\ddot{x} - 2\alpha\dot{x} + \alpha^2 x = 0$ with the initial conditions given by Eq. (A.77).

The characteristic equation is $\lambda^2 - 2\alpha\,\lambda + \alpha^2 = 0$. It has two repeated roots $\lambda_1 = \lambda_2 = \alpha$. The fundamental system of solutions consists of the following functions: $e^{\alpha t}$ and $t\cdot e^{\alpha t}$. The corresponding general solution is: $x(t) = A_1 e^{\alpha t} + A_2 t e^{\alpha t}$.

Hence: $\dot{x}(t) = \alpha A_1 e^{\alpha t} + A_2[1+\alpha t]\cdot e^{\alpha t}$. Substituting the initial conditions $x(t_0) = \Delta x$ and $\dot{x}(t) = 0$ gives $A_1 = \Delta x$ and $\alpha A_1 + A_2(1+\alpha t) = 0$, hence $A_2 = -\Delta x\cdot\alpha$. Finally $x(t) = \Delta x\cdot e^{\alpha t}(1-\alpha t)$.

******

A.3.4 Complex and Distinct Roots

It is known from the theory of polynomials that if polynomial Eq. (A.68) with real coefficients $a_1, ..., a_{n-2}, a_{n-1}, a_n$ has complex roots then the roots form complex conjugate pairs λ_i, λ_i^*, etc.

Assume the following notation

$$\lambda_i = \alpha_i + j\Omega_i \quad \text{and} \quad \lambda_i^* = \alpha_i - j\Omega_i \tag{A.83}$$

Obviously, the condition of distinct roots Eq. (A.75) is satisfied for that pair as $\lambda_i \neq \lambda_i^*$. The Vandermonde's determinant can be expressed using Eq. (A.73) as

$$\prod_{1\le i\le j\le n}(\lambda_j - \lambda_i) = (\lambda_n - \lambda_{n-1})(\lambda_n - \lambda_{n-2}), ..., (\lambda_i - \lambda_i^*), ..., (\lambda_3 - \lambda_2)(\lambda_3 - \lambda_1)(\lambda_2 - \lambda_1) \neq 0 \tag{A.84}$$

and it is different from zero as $(\lambda_i - \lambda_i^*) = j2\Omega_i \neq 0$. This makes it possible to assume the following fundamental system of solutions

$$e^{\lambda_1 t}, ..., e^{\lambda_i t}, e^{\lambda_i^* t}, ..., e^{\lambda_n t} \tag{A.85}$$

which contains exponential functions of λ_i and λ_i^*.

Using Eq. (A.78) for given integration constants $A_1, ..., A_i, ..., A_n$ makes it possible to find the particular solution. As the Vandermonde's matrix in Eq. (A.78) and its determinant are complex, it may be expected that the integration constants in the fundamental set of solutions will also be complex, i.e.

$$x(t) = ... + A_i e^{\lambda_i t} + B_i e^{\lambda_i^* t} + ... \tag{A.86}$$

where variables x, $t \in$ Real and the integration constants A_i, $B_i \in$ Complex. Differentiation of Eq. (A.86) gives

$$\dot{x}(t) = \ldots + \lambda_i A_i e^{\lambda_i t} + \lambda_i^* B_i e^{\lambda_i^* t} + \ldots \tag{A.87}$$

Integration constants A_i, B_i can be calculated from the initial conditions assuming

$$\begin{aligned} x(t = 0) &= \ldots + \Delta x_i + \ldots = \Delta x \\ \dot{x}(t = 0) &= \ldots + 0 + \ldots = 0 \end{aligned} \tag{A.88}$$

Substituting those initial conditions into Eqs. (A.86) and (A.87) gives the following two simple equations: $A_i + B_i = \Delta x_i$ and $\lambda_i A_i + \lambda_i^* B_i = 0$. Solving those equations requires care as both A_i, B_i and λ_i, λ_i^* are complex numbers. Expressing the equation in the matrix form gives

$$\begin{bmatrix} 1 & 1 \\ \lambda_i & \lambda_i^* \end{bmatrix} \begin{bmatrix} A_i \\ B_i \end{bmatrix} = \begin{bmatrix} \Delta x_i \\ 0 \end{bmatrix} \quad \text{or} \quad \begin{bmatrix} A_i \\ B_i \end{bmatrix} = \frac{1}{-\mathrm{j}2\Omega_i} \begin{bmatrix} \lambda_i^* & -1 \\ -\lambda_i & 1 \end{bmatrix} \begin{bmatrix} \Delta x_i \\ 0 \end{bmatrix} \tag{A.89}$$

where according to Eq. (A.83) Ω_i is the imaginary part of λ_i. Now one gets

$$\begin{aligned} A_i &= \Delta x \frac{1}{-\mathrm{j}2\Omega_i} \lambda_i^* = \Delta x \frac{\Omega_i + \mathrm{j}\alpha_i}{2\Omega_i} \\ B_i &= \Delta x \frac{1}{-\mathrm{j}2\Omega_i} (-\lambda_i) = \Delta x \frac{\Omega_i - \mathrm{j}\alpha_i}{2\Omega_i} = A_i^* \end{aligned} \tag{A.90}$$

This shows that $B_i = A_i^*$. The general important conclusion is that for each pair of solutions $e^{\lambda_i t}$ and $e^{\lambda_i^* t}$ the integration constants resulting from initial conditions form a complex conjugate pair A_i, A_i^*. Hence the solutions of Eq. (A.86) is

$$x(t) = \ldots + A_i e^{\lambda_i t} + A_i^* e^{\lambda_i^* t} + \ldots \tag{A.91}$$

where

$$\begin{aligned} A_i e^{\lambda_i t} + A_i^* e^{\lambda_i^* t} &= A_i e^{\alpha_i t}(\cos\Omega_i t + \mathrm{j}\sin\Omega_i t) + A_i^* e^{\alpha_i t}(\cos\Omega_i t - \mathrm{j}\sin\Omega_i t) \\ &= e^{\alpha_i t}\left[(A_i + A_i^*)\cos\Omega_i t + \mathrm{j}(A_i - A_i^*)\sin\Omega_i t\right] \end{aligned} \tag{A.92}$$

Obviously, $(A_i + A_i^*) = 2\,\mathrm{Re}\,A_i$ and $\mathrm{j}(A_i - A_i^*) = -2 \cdot \mathrm{Im}A_i$ are real numbers equal to the real and imaginary parts of the integration constant A_i, respectively. Hence Eq. (A.92) is now

$$A_i e^{\lambda_i t} + A_i^* e^{\lambda_i^* t} = e^{\alpha_i t}[2 \cdot \mathrm{Re}A_i \cdot \cos\Omega_i t - 2 \cdot \mathrm{Im}\,A_i \cdot \sin\Omega_i t] \tag{A.93}$$

Note that the left-hand side of the equation contains operations on real numbers, while the right-hand side contains operation on imaginary numbers. This means that appropriate operations on complex numbers A_i, A_i^*, $e^{\lambda_i t}$, and $e^{\lambda_i^* t}$ must result in the imaginary part of the term $A_i e^{\lambda_i t} + A_i^* e^{\lambda_i^* t}$ to be equal to zero so that the overall result is a real number. This is an important observation leading to a conclusion that for the discussed case of complex conjugate pairs of roots the particular solution is of the form

$$x(t) = \ldots + 2 \cdot \mathrm{Re}A_i \cdot e^{\alpha_i t}\cos\Omega_i t - 2 \cdot \mathrm{Im}\;A_i \cdot e^{\alpha_i t}\sin\Omega_i t + \ldots \tag{A.94}$$

Hence it can be concluded that operations on complex numbers connected with looking for the particular solution are unnecessary, as, instead of the fundamental system of solution given by Eq. (A.85), one can consider a fundamental system of solutions of the form

$$e^{\lambda_1 t}, \ldots, e^{\alpha_i t}\cos\Omega_i t, e^{\alpha_i t}\sin\Omega_i t, \ldots, e^{\lambda_n t} \tag{A.95}$$

consisting of real functions. As sine and cosine functions are orthogonal the solutions $e^{\alpha_i t}\cos\Omega_i t$ and $e^{\alpha_i t}\sin\Omega_i t$ are linearly independent. This can be checked by calculating the Wronskian of the fundamental system of solutions (A.95) and the corresponding Vandermonde's determinant. The latter will contain terms proportional to $(\cos\Omega_i t - \sin\Omega_i t) \neq 0$.

Those considerations lead to an important conclusion

> Each complex conjugate pair of the roots λ_i, λ_i^* in the solution $x(t)$ of the differential Eq. (A.63) corresponds to real exponential functions $e^{\alpha_i t}\cos\Omega_i t$ and $e^{\alpha_i t}\sin\Omega_i t$ as the imaginary parts of the solutions corresponding to the pairs λ_i, λ_i^* cancel each other out.

There is another proof of the above statement using a theorem that if a complex function is a fundamental solution of a linear ordinary differential equation then both real and imaginary parts of that function also form the general solution. Proof of that can be found in a number of textbooks including Arnold (1992).

Examining Eq. (A.95) shows that the real roots λ_i of the characteristic equation will produce exponential terms of the form $e^{\lambda_i t}$ so that the roots are the reciprocals of time constants of the exponential terms. The complex conjugate root pairs $\lambda_i = \lambda_i^* = \alpha_i + j\Omega_i$ of the characteristic equation will produce oscillatory terms $e^{\alpha_i t}\cos\Omega_i t$ and $e^{\alpha_i t}\sin\Omega_i t$. The imaginary parts of the roots are therefore equal to the frequencies of oscillation of each term, while the real parts of the roots are the reciprocals of time constants of the exponential envelope of the oscillatory terms. The overall solution is stable if real parts of all the roots are negative.

For the dynamics considered in this book of particular attention is a second-order scalar equation corresponding to the equation of motion of the synchronous generator (Section 5.1). Hence now a solution to the second-order equation will be discussed when the roots of the characteristic equation are complex.

Example A.3.4 Solve a second-order equation $\ddot{x} - 2\alpha\dot{x} + (\alpha^2 + \Omega^2)x = 0$ with the initial conditions given by Eq. (A.77).

The characteristic equations is $\lambda^2 - 2\alpha\,\lambda + (\alpha^2 + \Omega^2) = 0$. The roots are $\lambda_1 = \alpha + j\Omega$ and $\lambda_2 = \lambda_1^* = \alpha - j\Omega$. The fundamental system of solutions $e^{\lambda_1 t}$ and $e^{\lambda_1^* t}$ results in the following Vandermonde's determinant

$$\det\begin{bmatrix} 1 & 1 \\ \lambda_1 & \lambda_1^* \end{bmatrix} = \lambda_1^* - \lambda_1 = -j2\Omega \neq 0 \tag{A.96}$$

which shows that the fundamental system of solutions was well chosen and the general solutions is of the form

$$x(t) = A_1 e^{\lambda_1 t} + B_1 e^{\lambda_1^* t} \tag{A.97}$$

Equation (A.89) takes the form

$$\begin{bmatrix} 1 & 1 \\ \lambda_1 & \lambda_1^* \end{bmatrix}\begin{bmatrix} A_1 \\ B_1 \end{bmatrix} = \begin{bmatrix} \Delta x \\ 0 \end{bmatrix} \quad \text{or} \quad \begin{bmatrix} A_1 \\ B_1 \end{bmatrix} = \frac{1}{-j2\Omega}\begin{bmatrix} \lambda_1^* & -1 \\ -\lambda_1 & 1 \end{bmatrix}\begin{bmatrix} \Delta x \\ 0 \end{bmatrix} \tag{A.98}$$

Hence

$$A_1 = \Delta x \cdot \frac{\Omega + j\alpha}{2\Omega} \quad \text{and} \quad B_1 = \Delta x \cdot \frac{\Omega - j\alpha}{2\Omega} = A_1^* \tag{A.99}$$

After substituting Eqs. (A.99) into (A.97) simple algebra gives the following particular solution

$$x(t) = \frac{\Delta x}{\Omega} e^{\alpha t}[\omega\cos\Omega t - \alpha\sin\Omega t] \tag{A.100}$$

Obviously, the solution can be obtained in a simpler way by assuming straight away the fundamental system of solutions given by Eq. (A.95): $e^{\alpha t}\cos\Omega t$ and $e^{\alpha t}\sin\Omega t$ and the general solution

$$x(t) = C_1 e^{\alpha t}\Omega\cos\omega t + C_2 e^{\alpha t}\alpha\sin\Omega t \tag{A.101}$$

Substituting the initial condition $x(t_0) = \Delta x$ leads to $C_1 = \Delta x/\Omega$. Differentiating Eq. (A.101) and substituting $\dot{x}(t_0) = 0$ gives $C_2 = -C_1$. Substituting the calculated constants $C_1 = -C_2 = \Delta x/\Omega$ to Eq. (A.101) gives the solution given by Eq. (A.100).

The solution (A.100) contains an expression $[\Omega\cos\Omega t - \alpha\sin\Omega t]$. It corresponds to the cosine of angle differences: $\cos(\Omega t + \phi) = [\cos\Omega t\cos\phi - \sin\Omega t\sin\phi]$. In order to obtain exactly that form it is necessary to transform Eq. (A.100) in the following way

$$x(t) = \frac{\Delta x}{\Omega}e^{\alpha t}\sqrt{\Omega^2+\alpha^2}\left[\frac{\Omega}{\sqrt{\Omega^2+\alpha^2}}\cos\Omega t - \frac{\alpha}{\sqrt{\Omega^2+\alpha^2}}\sin\Omega t\right] \tag{A.102}$$

where the expression in front of the brackets was multiplied by $\sqrt{\Omega^2+\alpha^2}$ while the components in the brackets were divided by the same term. Assume the notation

$$\sin\varphi = \frac{\alpha}{\sqrt{\Omega^2+\alpha^2}} \quad \text{and} \quad \cos\phi = \frac{\Omega}{\sqrt{\Omega^2+\alpha^2}} \tag{A.103}$$

It is easy to check that $\sin^2\phi + \cos^2\phi = 1$. With that definition of angle ϕ Eq. (A.102) becomes

$$x(t) = \frac{\Delta x}{\cos\phi}e^{\alpha t}\cos(\Omega t + \phi) \tag{A.104}$$

That form of the second-order equation is convenient as Eq. (A.104) clearly shows that the solution is in the form of a cosine function with exponentially decaying amplitude for $\alpha < 1$ and exponentially increasing amplitude for $\alpha > 1$ or a constant amplitude for $\alpha = 0$. Inspection of Eq. (A.104) shows that the solution satisfies the initial condition $x(t = 0) = \Delta x$.

******

Second-order equations represent many physical problems. It is convenient to express a second-order equation in the *standard form* investigated in Example A.3.5.

Example A.3.5 Consider the standard form of a second-order equation $\ddot{x} + 2\zeta\Omega_{nat}\dot{x} + \Omega_{nat}^2 x = 0$, where Ω_{nat} is the *natural frequency of oscillations* and ζ is the *damping ratio*. The initial conditions are given by Eq. (A.77). The characteristic equation is $\lambda^2 + 2\zeta\Omega_{nat}\lambda + \Omega_{nat}^2 = 0$. When $\Delta = -4\Omega_{nat}^2(1-\zeta^2) \geq 0$, i.e. the damping ratio $\zeta \geq 1$, the roots are real and the solution will contain exponential terms discussed in Examples A.3.2 and A.3.3. In this example the case of the *underdamped second-order system* is discussed when $0 \leq \zeta < 1$. The characteristic equation will have then two roots forming a complex conjugate pair

$$\lambda_{1,2} = -\zeta\Omega_{nat} \pm j\Omega_{nat}\sqrt{1-\zeta^2} \quad \text{or} \quad \lambda_{1,2} = -\zeta\Omega_{nat} \pm j\Omega_d \tag{A.105}$$

where $\Omega_d = \Omega_{nat}\sqrt{1-\zeta^2}$ is the *damped frequency of oscillation* (in rad/s) as Ω_{nat} is the natural frequency of oscillations (in rad/s) when damping is neglected, i.e. when $\zeta = 0$ and $\lambda_{1,2} = \pm j\Omega_{nat}$. The solution $x(t)$ can be obtained similarly as in the previous example or by using the solution (A.104) and substituting $\Omega = \Omega_d$ and $\alpha = -\zeta\Omega_{nat}$. Hence

$$x(t) = \frac{\Delta x}{\cos\phi}e^{-\zeta\Omega_{nat}t}\cos(\Omega_d t + \phi) \tag{A.106}$$

where $\phi = -\arcsin\zeta$.

******

A.3.5 Repeated Complex Roots

As previously shown, each complex conjugate pair of roots λ_i and λ_i^* corresponds to a solution (A.89) containing the terms $\mathrm{e}^{\alpha_i t}\cos\Omega_i t$ and $\mathrm{e}^{\alpha_i t}\sin\Omega_i t$. When the roots λ_i and λ_i^* are repeated k times then, similarly as in Eq. (A.96), the general solution has to be complemented by the same terms multiplied by orthogonal polynomials: $t, t^2, t^3, \ldots, t^{k-1}$. For a pair of complex roots repeated k times the following solutions are obtained

$$\begin{array}{llll} \mathrm{e}^{\alpha_i t}\cos\Omega_i t, & t\cdot\mathrm{e}^{\alpha_i t}\cos\Omega_i t, & t^2\cdot\mathrm{e}^{\alpha_i t}\cos\Omega_i t,\ldots, & t^{k-1}\cdot\mathrm{e}^{\alpha_i t}\cos\Omega_i t \\ \mathrm{e}^{\alpha_i t}\sin\Omega_i t, & t\cdot\mathrm{e}^{\alpha_i t}\sin\Omega_i t, & t^2\cdot\mathrm{e}^{\alpha_i t}\sin\Omega_i t,\ldots, & t^{k-1}\cdot\mathrm{e}^{\alpha_i t}\sin\Omega_i t \end{array} \tag{A.107}$$

Obviously, the complete set of fundamental solutions contains also the solutions corresponding to other roots.

A.3.6 First-order Complex Differential Eq.

A particular case of a linear first-order differential equation is a homogeneous equation of the form $\dot{z}-\lambda z=0$, where λ is a complex number. The equation can be rewritten as

$$\dot{z}=\lambda z \tag{A.108}$$

According to the theory developed earlier, the solution will be of the form

$$z(t)=\mathrm{e}^{\lambda t}z_0 \tag{A.109}$$

where $z_0=z(t_0)$ is an initial condition (a complex number). Assume the following notation

$$z(t)=x(t)+\mathrm{j}\,y(t);\quad z_0=x_0+\mathrm{j}\,y_0;\quad \lambda=\alpha+\mathrm{j}\Omega \tag{A.110}$$

The solution will be interpreted on the complex plane of coordinates $x=\mathrm{Re}\,z$ and $y=\mathrm{Im}\,z$. Substituting Eqs. (A.110) into (A.109) gives

$$x(t)+\mathrm{j}\,y(t)=\mathrm{e}^{(\alpha+\mathrm{j}\Omega)t}(x_0+\mathrm{j}\,y_0)$$

or

$$x(t)+\mathrm{j}\,y(t)=\mathrm{e}^{\alpha t}(x_0+\mathrm{j}\,y_0)(\cos\Omega t+\mathrm{j}\sin\Omega t)$$

Multiplying and ordering the terms gives

$$x(t)=\mathrm{e}^{\alpha t}(x_0\cos\Omega t-y_0\sin\Omega t) \tag{A.111a}$$

$$y(t)=\mathrm{e}^{\alpha t}(y_0\cos\Omega t+x_0\sin\Omega t) \tag{A.111b}$$

Figure A.2 shows that the initial condition $z_0=x_0+\mathrm{j}\,y_0$ is a point on the complex plane where

$$x_0=r_0\cos\phi_0;\quad y_0=r_0\sin\phi_0;\quad r_0=\sqrt{x_0^2+y_0^2} \tag{A.112}$$

Substituting Eqs. (A.112) into (A.111a,b) gives

$$x(t)=r_0\,\mathrm{e}^{\alpha t}(\cos\phi_0\cos\Omega t-\sin\phi_0\sin\Omega t) \tag{A.113a}$$

$$y(t)=r_0\,\mathrm{e}^{\alpha t}(\sin\phi_0\cos\Omega t+\cos\phi_0\sin\Omega t) \tag{A.113b}$$

Equations (A.113a,b) can be expressed as

$$x(t)=r_0\,\mathrm{e}^{\alpha t}\cos(\Omega t+\phi_0) \tag{A.114a}$$

$$y(t)=r_0\,\mathrm{e}^{\alpha t}\sin(\Omega t+\phi_0) \tag{A.114b}$$

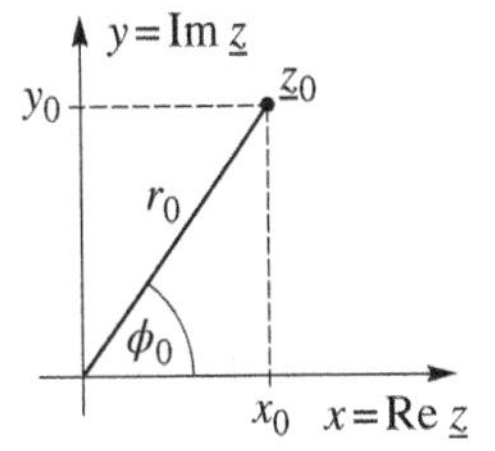

Figure A.2 Initial condition on the complex plane.

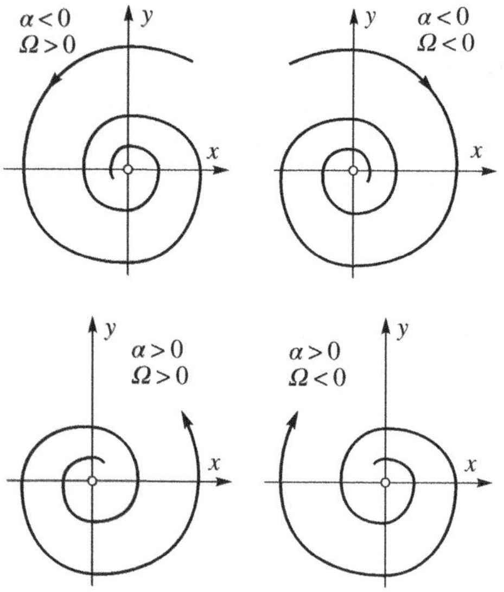

Figure A.3 Logarithmic spirals.

Obviously, the solutions $x(t)$ and $y(t)$ given by (A.114a and b) are proportional to sine and cosine and are therefore shifted in time by $\pi/2$. Squaring and adding both sides of (A.114a and b) gives

$$r(t) = r_0\,e^{\alpha t} \quad \text{where} \quad r(t) = \sqrt{[x(t)]^2 + [y(t)]^2} \tag{A.115}$$

Creating again a complex number $z(t) = x(t) + \text{j}\,y(t)$ from the solutions (A.114a,b) gives

$$\begin{aligned} z(t) &= r_0\,e^{\alpha t}[\cos(\Omega t + \phi_0) + \text{j}\sin(\Omega t + \phi_0)] \\ &= r_0\,e^{\alpha t}e^{\text{j}(\Omega t + \phi_0)} = r(t)\cdot e^{\text{j}(\Omega t + \phi_0)} \end{aligned} \tag{A.116}$$

Figure A.3 shows that function Eq. (A.116) describes the *logarithmic spiral* on the complex plane starting at a point (x_0, y_0) corresponding to the initial condition. The spiral rotates anticlockwise if $\Omega = \text{Im}\,\lambda > 0$ and clockwise if $\Omega = \text{Im}\,\lambda < 0$. For $\alpha = \text{Re}\,\lambda < 0$ the spiral is converging toward the origin of coordinates, while for $\alpha = \text{Re}\,\lambda > 0$ the spiral is diverging. For $\alpha = \text{Re}\,\lambda = 0$ the solution $z(t)$ corresponds to a circle on the complex plane.

Obviously, for a conjugate value $\lambda^* = \alpha - \text{j}\Omega = \alpha + \text{j}(-\Omega)$ the imaginary part of λ^* has the opposite sign than for λ. This means that the spiral corresponding to λ^* rotates in the opposite direction to the spiral corresponding to λ. Hence if a function is the sum of solutions for complex conjugate pairs λ, λ^* then the imaginary parts of the solutions will cancel each other out and the only remaining part will be the double real part of the spiral, i.e.

$$z_i(t) + z_j(t) = z_i(t) + z_i^*(t) = 2\,\text{Re}\,z_i(t) = 2x(t) = 2r_0\,e^{\alpha t}\cos(\Omega t + \phi_0) \tag{A.117}$$

This observation is important for the solution of matrix differential equations discussed in Section 12.1.

A.4 Prony Analysis

It is assumed that the signal $x(t)$ containing the oscillatory damped components is the response of any dynamic system or its model. The Prony method allows a signal to be approximated with a function

$$\hat{x}(t) = \sum_{p=1}^{N/2} A_p e^{-\lambda_p t} = \sum_{p=1}^{N/2} A_p e^{\alpha_p t}\cos(\Omega_p t + \Theta_p) \tag{A.118}$$

being the sum of $N/2$ oscillatory exponential components with the following parameters: A_p – amplitude, α_p – damping coefficient, Ω_p – pulsation, and Θ_p – phase shift of p-th component. The number of exponential components identified $N/2$ is the input parameter of the method, i.e. user-defined, although there are also algorithms for searching for this number. The ability to identify exponential components in the time series $x(t)$ is the basic element that differs the Prony method from the Fourier transform, which is only suitable for periodic signals, i.e. not containing damped components.

In the Prony method, to identify the exponential components of the time series $x(t)$ M measurement samples are used that make up the vector

$$\boldsymbol{X}_\text{M} = [x(0) \mid x(\Delta t) \mid \cdots \mid x(k\Delta t) \mid \cdots \mid x((M-1)\Delta t)]^\text{T}, \tag{A.119}$$

where $M > N$ and Δt is the sampling step of $x(t)$, k is the next sampling moment number, and $k\Delta t$ is the next sampling moment.

Since vector $\boldsymbol{X}_{\mathrm{M}}$ represents a discrete signal $x(k\Delta t)$, further considerations refer to discrete signals. Function (A.118) can be represented in the following discrete form

$$\hat{x}_k = \sum_{p=1}^{N} B_p \mathrm{e}^{\left[\left(\alpha_p + \mathrm{j}\Omega_p\right)k\Delta t + \mathrm{j}\Theta_p\right]} \tag{A.120}$$

Function (A.120) is the sum of N components, because N eigenvalues of the complex conjugate ones define $N/2$ oscillating components (and thus the modes) determined by function (A.118).

The Prony method used for oscillatory signals uses the following form of the function (A.120)

$$\hat{x}_k = \sum_{p=1}^{N} \underline{B}_p \underline{z}_p^k \tag{A.121}$$

where quantities $\underline{B}_p$ given by formula

$$B_p = \frac{A_p}{2}\mathrm{e}^{\mathrm{j}\Theta_p} \tag{A.122}$$

determine complex amplitudes and $\underline{z}_p$ given by formula

$$\underline{z}_p = \mathrm{e}^{\underline{\lambda}_p \Delta t} = \mathrm{e}^{\left(\alpha_p + \mathrm{j}\Omega_p\right)\Delta t} \tag{A.123}$$

are discrete equivalents of modal variables (discussed in Section 12.1.4) calculated for a single sampling step Δt.

The Prony method is a three-step method in which the components of the Eqs. (A.122) and (A.123) are calculated in subsequent steps, i.e. the amplitude, phase shift, attenuation coefficient, and frequency of each mode.

Step I: Calculation of system model coefficients

Looking for the approximation of the $x(t)$ time series, it is assumed that the output signal $x_k = x(k\Delta t)$ of a certain dynamic system model, in subsequent sampling moments, will be consistent (with a correspondingly small error) with the corresponding element of the measurement vector $\boldsymbol{X}_M$. Below (as in the original Prony method) a discrete linear predictive AR model (AutoRegressive model) is assumed

$$x_k = -\left(a_1 x_{k-1} + a_2 x_{k-2} + \ldots + a_N x_{k-N}\right) \tag{A.124}$$

where a_i are the sought coefficients forming the vector

$$\boldsymbol{A} = \left[a_1 \,\vdots\, a_2 \,\vdots\, \cdots \,\vdots\, a_N\right]^{\mathrm{T}} \tag{A.125}$$

This model essentially defines the response of some linear time-invariant discrete system to be stimulated by a certain input signal, for example a unitary step. For the considered measurement window, i.e. the number of vector elements $\boldsymbol{X}_{\mathrm{M}}$ equal to M and the assumed number N of searched function components (A.121), which is also the order of the system under consideration (A.124), a system of equations is obtained

$$\begin{bmatrix} x_N \\ x_{N+1} \\ \vdots \\ x_{M-1} \end{bmatrix} = -\begin{bmatrix} x_{N-1} & x_{N-2} & \cdots & x_0 \\ x_N & x_{N-1} & \cdots & x_1 \\ \vdots & \vdots & \ddots & \vdots \\ x_{M-2} & x_{N-3} & \cdots & x_{M-N-1} \end{bmatrix} \begin{bmatrix} a_1 \\ a_2 \\ \vdots \\ a_N \end{bmatrix} \quad \text{or} \quad \boldsymbol{X} = -\boldsymbol{D}\boldsymbol{A} \tag{A.126}$$

where $\boldsymbol{X}$ corresponds to the part of the vector $\boldsymbol{X}_{\mathrm{M}}$ defined by Eq. (A.119) containing the measuring samples of the time series $x(t)$. The matrix $\boldsymbol{D}$ also containing the measured values is a rectangular matrix. Therefore, the

Eq. (A.126) defines an overdetermined system which has no unique (exact) solution. The vector $\boldsymbol{A}$ of the coefficients sought is calculated in the least-squares sense, minimizing the mean square error of the difference of the model response $\hat{x}_k$ and the measured values x_k

$$\varepsilon = \sum_{k=0}^{M-1} |x_k - \hat{x}_k|^2 \tag{A.127}$$

Various algorithms are used to solve the above problem (Burg 1978; Marple 1980). The *singular value decomposition* (SVD) algorithm is considered the most effective.

Step II: Calculation of eigenvalues

It is assumed that function Eq. (A.121) is a solution of a homogeneous differential equation with constant coefficients (being a model of the dynamical system under examination) for which a characteristic polynomial can be presented in the form

$$\boldsymbol{W}(\underline{z}) = \prod_{p=1}^{N} \left(\underline{z} - \underline{z}_p\right) \tag{A.128}$$

analogous to the polynomial Eq. (A.68) discussed in Section A.3.1. In this case, however, $\underline{z}_p$ are the poles of the transfer function of this discrete system.

The polynomial Eq. (A.128) can, analogically to Eq. (A.68), be developed into a power form, and a characteristic equation of the model of the studied discrete dynamic system can be derived from it

$$\boldsymbol{W}(\underline{z}) = \sum_{p=0}^{N} a_p \underline{z}^{N-p} = a_0 \underline{z}^N + a_1 \underline{z}^{N-1} + \ldots + a_{N-2} \underline{z}^2 + a_{N-1} \underline{z} + a_N = 0 \tag{A.129}$$

where $a_0 = 1$, and other coefficients a_i are elements of vector $\boldsymbol{A}$, calculated in Step I. Equation (A.129) is solved using the algorithm shown, for example, in Anderson et al. (1999). The roots of this equation can be treated as poles of transfer function of a discrete system. Then, the function is used to convert from a discrete time system to a continuous time system

$$\underline{\lambda}_p = \frac{1}{\Delta t} \ln\left(\underline{z}_p\right) \tag{A.130}$$

In this way, the roots of the characteristic equation for a continuous time system are obtained, which are the eigenvalues $\underline{\lambda}_p$ of this system. This allows the damping coefficients α_p and the frequency Ω_p of the individual exponential components of the identified waveform to be calculated, using Function (A.123).

The essence of the Prony method is the statement that the coefficients of the characteristic Eq. (A.129) are equal to the coefficients of the assumed AR model defined by Eq. (A.124). This is justified as follows. Vector $\boldsymbol{K}$ is created, with size M equal to the number of measurement samples, containing model coefficients on subsequent positions Eq. (A.124), i.e. $\kappa_{m+1} = a_N, \kappa_{m+2} = a_{N-1}, \ldots, \kappa_{m+1+N} = a_0 = 1$, where m can be any number in the range $m = 0, \ldots, M - N - 1$. The remaining positions of the vector $\boldsymbol{K}$ are assumed to be equal to zero. If then the Eq. (A.133) is multiplied by the vector $\boldsymbol{K}$ and the expressions are grouped accordingly, the resulting equation takes the form

$$\sum_{p=1}^{N+1} a_{p-1} x_{N-p+1+m} = \sum_{p=1}^{N} \underline{B}_p \underline{z}_p^m \left(\sum_{n=1}^{N+1} a_{n-1} \underline{z}_p^{N-n+1+m} \right) = 0 \tag{A.131}$$

For example, for $m = 1$ it takes the form

$$0 = a_0 x_{N+1} + a_1 x_N + \dots + a_N x_1 = \begin{array}{l} \underline{B}_1 \underline{z}_1 \left(a_0 \underline{z}_1^N + a_1 \underline{z}_1^{N-1} + \dots + a_{N-1} \underline{z}_1 + a_N\right) + \\ + \underline{B}_2 \underline{z}_2 \left(a_0 \underline{z}_2^N + a_1 \underline{z}_2^{N-1} + \dots + a_{N-1} \underline{z}_2 + a_N\right) + \\ \vdots \\ + \underline{B}_N \underline{z}_N \left(a_0 \underline{z}_N^N + a_1 \underline{z}_N^{N-1} + \dots + a_{N-1} \underline{z}_N + a_N\right) \end{array} \tag{A.132}$$

The expression on the left side of the Eq. (A.131), after transferring the factors to one side and supplementing them with the factor $a_0 = 1$ at x_k, corresponds to the model (A.124), while the expression on the right side of the equation in brackets corresponds to relationship Eq. (A.129). The left side of the Eq. (A.131) is equal to zero from the definition, while the expressions in the right side in brackets are also equal to zero for all the values $\underline{z}_p$ that solve the Eq. (A.129), which completes the proof.

It is worth noting that the polynomial $\boldsymbol{W}(\underline{z})$ found in Eq. (A.129) can be treated as a polynomial occurring in parentheses in Eq. (A.131). The structure of this polynomial corresponds to the structure of the characteristic equation. The above proof shows that the coefficients of this polynomial are equal to the coefficients of the model (A.124). This allows the unknown values of variables $\underline{z}_p$ to be calculated.

Step III: Calculation of complex amplitudes

Using the eigenvalues $\underline{\lambda}_p$ calculated in the previous step and the samples (measurement data) $\boldsymbol{X}_\mathrm{M}$ and assuming $\hat{x}_k = x_k$, Eq. (A.121) can be written in the matrix form

$$\begin{bmatrix} x_0 \\ x_1 \\ \vdots \\ x_{M-1} \end{bmatrix} = \begin{bmatrix} 1 & 1 & \cdots & 1 \\ \underline{z}_1 & \underline{z}_2 & \cdots & \underline{z}_N \\ \vdots & \vdots & \ddots & \vdots \\ \underline{z}_1^{M-1} & \underline{z}_2^{M-1} & \cdots & \underline{z}_N^{M-1} \end{bmatrix} \begin{bmatrix} \underline{B}_1 \\ \underline{B}_2 \\ \vdots \\ \underline{B}_N \end{bmatrix} \quad \text{or} \quad \boldsymbol{X}_\mathrm{M} = \underline{\boldsymbol{Z}}\underline{\boldsymbol{B}} \tag{A.133}$$

where vector $\underline{\boldsymbol{B}}$ containing complex amplitudes $\underline{B}_p$ is unknown. Matrix $\underline{\boldsymbol{Z}}$ is similar to the Vandermonde determinant in formulas (A.72) and (A.73) discussed in Section A.3.1. In this case, however, the matrix is rectangular. Equation (A.133) has no unique solution and is resolved in the least-squares sense, like Eq. (A.126). Then, using Eq. (A.122), the amplitude modules B_p and phase shifts Θ_p of individual modes are calculated.

The Prony method, despite its efficiency, is not without restrictions. The literature of the subject indicates examples of functions for which there are problems with the mode's identification. The following limitations or effects are indicated:

1) The result of the calculation may depend on the number of modes identified $N/2$.
2) The result of the calculation may depend on the type of disturbance, which results from the fact that after various disturbances, e.g. grid element disconnection, load change, etc., the electric power system (EPS) structure becomes different.
3) The result of the calculation may depend on the fault features (e.g. the fault duration), which is a consequence of the EPS's nonlinearity.
4) The result of the calculation may depend on the fragment of the object response, chosen for analysis, i.e. the initial moment of time response and the length of the vector $\boldsymbol{X}_\mathrm{M}$ (the number of samples M).
5) Using too long vectors $\boldsymbol{X}_\mathrm{M}$ containing object response may result in losing some information about modes.
6) The result of calculations may depend on oscillatory mode differences. If some eigenvalues $\underline{\lambda}_p$ are located close together on the complex plane then there are difficulties in computing the modal components associated with

them. Such a situation often occurs in the calculation of modal components related to an electromechanical phenomenon in the EPS.

With regard to the EPS, the Prony method is used to identify modes in the time series obtained from measurements in real systems as well as being the responses to some disturbances on the mathematical models of large systems. In order to increase the accuracy of the mode's identification, in complex and nonlinear EPSs, it is recommended (Hauer et al. 1990) to apply the following rules:

1) The sampling frequency of the measured signal $f = 1/\Delta t$ should not be greater than 2–3 times the frequency of the signal $x(t)$ component with the highest frequency, i.e. the frequency of the fastest mode. An increase in the sampling rate leads to a reduction in the accuracy of the calculation.
2) The length of the measurement window M should not be shorter than 1.5 times the component period of the measured signal with the lowest frequency, i.e. the frequency of the slowest mode.
3) The assumed model row N should not be larger than one-third of the number of measuring window samples M, i.e. the condition should be met $N \leq M/3$.
4) The analysis must be related to the response of the object to a single disturbance, i.e. to the change of the value of the input signal (e.g. set point of the controller), to the change of the structure of the object (e.g. disconnection the system element), or in the case of a mathematical model, to change the value of the state variable. For example, when considering a short circuit, one should select a fragment of the response after the fault clearing, not a fragment containing a response during short circuit and a period after a short circuit clearing.

Example A.4.1 As an example, the signal $x(t)$ being a combination of two sinusoidal damped waveforms, expressed by a formula, is considered

$$x(t) = 0.3 \cdot \mathrm{e}^{-0.3t} \cos(3.7699 \cdot t + \pi/3) + 1 \cdot \mathrm{e}^{-0.5t} \cos(10.6814 \cdot t) \tag{A.134}$$

This signal is characterized by the following parameters: $A_1 = 0.3$, $\alpha_1 = -0.3\ \mathrm{s}^{-1}$, $\Omega_1 = 3.7699$ rad/s, $\Theta_1 = \pi/3$ rad, $A_2 = 1$, $\alpha_2 = -0.5\ \mathrm{s}^{-1}$, $\Omega_2 = 10.6814$ rad/s, and $\Theta_2 = 0$. For simplicity, the noise is omitted here. The length of the measurement window (number of samples) equal to $M = 12$ and sampling period $\Delta t = 0.2$ s is assumed. Assuming that the first sample corresponds to the moment $t = 0$, the measurement vector $\boldsymbol{X}_{\mathrm{M}}$ takes the form

$$\boldsymbol{X}_{\mathrm{M}} = [1.1500 \,\vdots\, -0.5494 \,\vdots\, -0.5702 \;\; 0.4879 \,\vdots\, -0.5700 \,\vdots\, -0.1642 \,\vdots\, 0.6900 \,\vdots\, -0.1651 \,\vdots\, 0.0457 \,\vdots\, 0.3817 \,\vdots\, -0.4078 \,\vdots\, -0.1754]^{\mathrm{T}}$$

The original signal $x(t)$ and the samples of the measured signal x_k, i.e. elements of the vector $\boldsymbol{X}_{\mathrm{M}}$, are shown in Figure A.4.

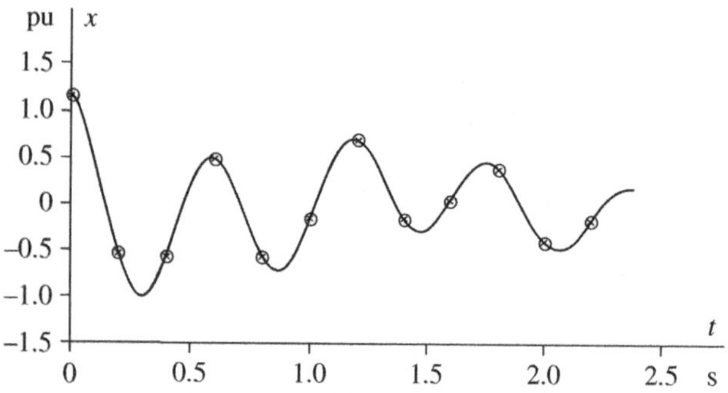

Figure A.4 Tested time series: solid line – signal $x(t)$; o – x_k measured samples; × – $\hat{x}$ samples of approximated signal.

The approximation $\hat{x}(t)$ of the signal $x(t)$ is searched for with the function (A.121) with four exponential components, i.e. $N = 4$ (number of mods is equal to $N/2 = 2$). Despite a long sampling step Δt, a small number of samples M, with quite a large number N of assumed exponential components relative to M, the above requirements related to the efficiency of the Prony method are met. For such a defined set of data, matrix $\boldsymbol{D}$ and vector $\boldsymbol{X}$ take the form

$$\boldsymbol{D} = \begin{bmatrix} 1.1500 & -0.5494 & -0.5702 & 0.4879 \\ -0.5494 & -0.5702 & 0.4879 & -0.5700 \\ -0.5702 & 0.4879 & -0.5700 & -0.1642 \\ 0.4879 & -0.5700 & -0.1642 & 0.6900 \\ -0.5700 & -0.1642 & 0.6900 & -0.1651 \\ -0.1642 & 0.6900 & -0.1651 & 0.0457 \\ 0.6900 & -0.1651 & 0.0457 & 0.3817 \\ -0.1651 & 0.0457 & 0.3817 & 0.4078 \end{bmatrix}$$

$$\boldsymbol{X} = [-0.5700 \;\vdots\; -0.1642 \;\vdots\; 0.6900 \;\vdots\; -0.1651 \;\vdots\; 0.0457 \;\vdots\; 0.3817 \;\vdots\; -0.4078 \;\vdots\; -0.1754]^{\mathrm{T}}$$

The solution of Eq. (A.126) is vector $\boldsymbol{A}$ of the model coefficients Eq. (A.124).

$$\boldsymbol{A} = [0.7261 \;\vdots\; -0.2641 \;\vdots\; 0.3743 \;\vdots\; -0.4034]^{\mathrm{T}} \tag{A.135}$$

The Matlab program *linsolve* function was used to solve Eq. (A.126). Model (A.124) with vector $\boldsymbol{A}$ coefficients, for $k > 4$ with the first four samples entered $x_1, \ldots, x_4$ accurately reproduces the values of the measurement vector $\boldsymbol{X}_{\mathrm{M}}$, i.e. in Figure A.4 the crosses corresponding to values $\hat{x}_k$ are exactly in circles corresponding to the values x_k.

Vector $\boldsymbol{A}$ coefficients in the form of a vector $[1 \;\vdots\; -0.4034 \;\vdots\; 0.3743 \;\vdots\; -0.2641 \;\vdots\; 0.7261]^{\mathrm{T}}$, i.e. supplemented by $a_0 = 1$, making coefficients of the discrete characteristic Eq. (A.129), make it possible to calculate the values $\underline{z}_p$ of the discrete system (MATLAB's *roots* function was used). In the form of a vector $\underline{\Lambda}_z = [\underline{z}_1 \;\; \underline{z}_2 \;\; \underline{z}_3 \;\; \underline{z}_4]^{\mathrm{T}}$ these are equal

$$\underline{\Lambda}_z = [0.6865 + \mathrm{j}0.6447 \;\vdots\; 0.6865 - \mathrm{j}0.6447 \;\vdots\; -0.4848 + \mathrm{j}0.7640 \;\vdots\; -0.4848 - \mathrm{j}0.7640]^{\mathrm{T}}$$

These values, related to a discrete system, using Eq. (A.130), are then converted to the form of a continuous time system. The eigenvalues of the continuous system $\underline{\lambda}_p$ in the form of vector $\underline{\Lambda} = [\underline{\lambda}_1 \;\; \underline{\lambda}_2 \;\; \underline{\lambda}_3 \;\; \underline{\lambda}_4]^{\mathrm{T}}$ are equal to

$$\underline{\Lambda} = [-0.3000 + \mathrm{j}3.7699 \;\vdots\; -0.3000 - \mathrm{j}3.7699 \;\vdots\; -0.5000 + \mathrm{j}10.6814 \;\vdots\; -0.5000 - \mathrm{j}10.6814]^{\mathrm{T}}$$

The result of the calculations are two pairs of mutually coupled eigenvalues corresponding to two damped oscillatory modes. The eigenvalues $\underline{\lambda}_p$ allow the pulsation Ω_p of searched modes and their damping coefficients α_p to be calculated

$$\boldsymbol{\Omega} = [3.7699 \;\vdots\; -3.7699 \;\vdots\; 10.6814 \;\vdots\; -10.6814]^{\mathrm{T}}\ \mathrm{rad/s}$$

$$\boldsymbol{\alpha} = [-0.3 \;\vdots\; -0.3 \;\vdots\; -0.5 \;\vdots\; -0.5]^{\mathrm{T}}\ \mathrm{s}^{-1}$$

The calculated values of damping and pulsation coefficients of both oscillatory modes are equal to the values defining the signal $x(t)$, i.e. Ω_1, α_1, Ω_2, and α_2.

In the next step, in order to calculate complex amplitudes $\underline{B}_p$, the Eq. (A.133) is solved using the vector $\boldsymbol{X}_{\mathrm{M}}$ and the matrix $\underline{\boldsymbol{Z}}$ of the form

$$\underline{\boldsymbol{Z}} = \begin{bmatrix} 1.0000 & 1.0000 & 1.0000 & 1.0000 \\ 0.6865 + \mathrm{j}0.6447 & 0.6865 - \mathrm{j}0.6447 & -0.4848 + \mathrm{j}0.7640 & -0.4848 - \mathrm{j}0.7640 \\ 0.0557 + \mathrm{j}0.8852 & 0.0557 - \mathrm{j}0.8852 & -0.3486 - \mathrm{j}0.7408 & -0.3486 + \mathrm{j}0.7408 \\ -0.5324 + \mathrm{j}0.6436 & -0.5324 - \mathrm{j}0.6436 & 0.7350 + \mathrm{j}0.0928 & 0.7350 - \mathrm{j}0.0928 \\ -0.7804 + \mathrm{j}0.0986 & -0.7804 - \mathrm{j}0.0986 & -0.4273 + \mathrm{j}0.5165 & -0.4273 - \mathrm{j}0.5165 \\ -0.5993 - \mathrm{j}0.4354 & -0.5993 + \mathrm{j}0.4354 & -0.1874 - \mathrm{j}0.5768 & -0.1874 + \mathrm{j}0.5768 \\ -0.1307 - \mathrm{j}0.6853 & -0.1307 + \mathrm{j}0.6853 & 0.5316 + \mathrm{j}0.1365 & 0.5316 - \mathrm{j}0.1365 \\ 0.3521 - \mathrm{j}0.5548 & 0.3521 + \mathrm{j}0.5548 & -0.3620 + \mathrm{j}0.3399 & -0.3620 - \mathrm{j}0.3399 \\ 0.5993 - \mathrm{j}0.1539 & 0.5993 + \mathrm{j}0.1539 & -0.0842 - \mathrm{j}0.4414 & -0.0842 + \mathrm{j}0.4414 \\ 0.5107 + \mathrm{j}0.2807 & 0.5107 - \mathrm{j}0.2807 & 0.3780 + \mathrm{j}0.1497 & 0.3780 - \mathrm{j}0.1497 \\ 0.1696 + \mathrm{j}0.5220 & 0.1696 - \mathrm{j}0.5220 & -0.2976 + \mathrm{j}0.2162 & -0.2976 - \mathrm{j}0.2162 \\ -0.2201 + \mathrm{j}0.4677 & -0.2201 - \mathrm{j}0.4677 & -0.0209 - \mathrm{j}0.3322 & -0.0209 + \mathrm{j}0.3322 \end{bmatrix}$$

The solution of Eq. (A.133) is vector $\underline{\boldsymbol{B}}$ of relative amplitudes

$$\underline{\boldsymbol{B}} = [0.0750 + \mathrm{j}0.1299 \;\vdots\; 0.0750 - \mathrm{j}0.1299 \;\vdots\; 0.5000 + \mathrm{j}0.0000 \;\vdots\; 0.5000 - \mathrm{j}0.0000]^{\mathrm{T}}$$

which coefficients allow amplitudes B_p and relative angular displacement Θ_p of individual modes to be calculated

$$\boldsymbol{B} = [0.1500 \;\vdots\; 0.1500 \;\vdots\; 0.5000 \;\vdots\; 0.5000]^{\mathrm{T}}$$

$$\boldsymbol{\Theta} = [1.0472 \;\vdots\; -1.0472 \;\vdots\; 0.0000 \;\vdots\; -0.0000]^{\mathrm{T}}\ \mathrm{rad}$$

After substitution of the above calculated values B_p, Θ_p, α_p, and Ω_p to formula (A.120) or (A.121), the values $\hat{x}_k$ are obtained corresponding to the vector coefficients $\boldsymbol{X}$, and after substituting to formula (A.118) a time series corresponding to $x(t)$ is obtained. The identity of the approximated $\hat{x}(t)$ and the original $x(t)$ signals results from the fact of considering a simple signal containing two known modes and no containing noise.

******

The Prony method is one of many methods that allow analyzing waveforms in dynamic systems based on measurements of selected signals. Discussion of other methods can be found in Paszek and Nocon (2014). Prony analysis is among other tools offered as part of the Matlab package.

A.5 Limiters and Symbols in Block Diagrams

Dynamic models of control systems used in power system stability studies are represented by block diagrams. In order to avoid misunderstandings, the meaning of some blocks is described below.

A.5.1 Addition, Multiplication, and Division

In block diagrams presented in this book the basic algebraic operations (addition, subtraction, multiplication, or division) are realized as shown in Figure A.5). The summing point has only one output and two or more input signals. Product and division points have also one output signal and usually only two input signals.

A.5.2 Simple Integrator

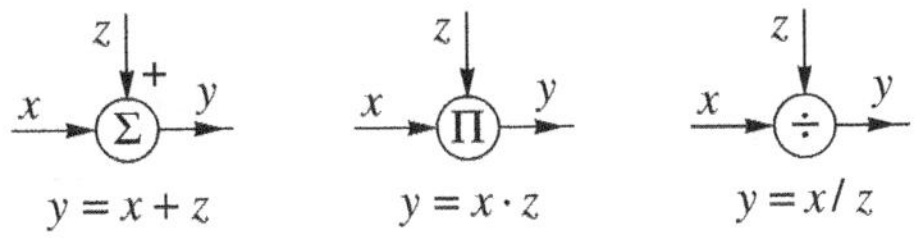

Figure A.5 Symbols of basic algebraic operations in block diagrams: (a) summing point; (b) product point; (c) divide point.

The output signal y of the integrator is equal to the time-integral of the input signal u plus the initial value $y(t = 0)$ of the output. The integrator block may operate with two distinct types of limiters, as shown in Figure A.6.

When the windup limiter (Figure A.6a) is used, the output signal y of the integrator is equal to the integral of the input signal u. As the integration is the reverse of the differentiation it is valid that $\mathrm{d}y/\mathrm{d}t = u$. On the basis of the value of y the limiter determine the output signal in the following way

$$\begin{array}{lll} \text{if} \quad y > A & \text{then} \quad x = A \\ \text{if} \quad B \leq y \leq A & \text{then} \quad x = y \\ \text{if} \quad y < B & \text{then} \quad x = B \end{array} \tag{A.136}$$

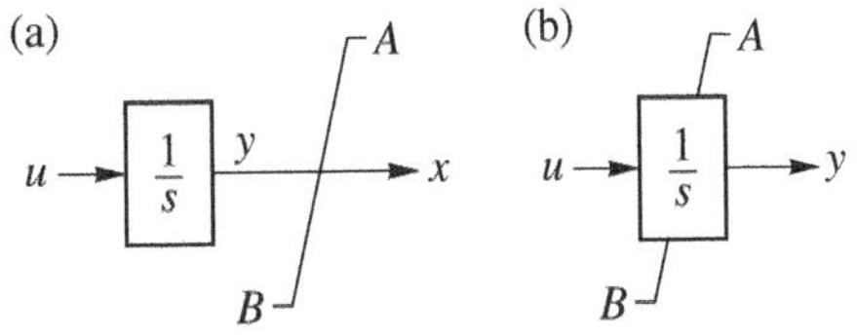

Figure A.6 Simple integrator with: (a) windup limiter; (b) non-windup limiter.

In this case, the limiter has no influence on the integration and the input signal u is integrated regardless of whether the signal y exceeds the limit A or B. The integration process is not stopped and signal y winds up. When the system output x reaches the commanded value, the sign of the error e.g. u reverses, causing the integrator to begin winding down. The output of the integrator y is far beyond the operating range and the integration takes a significant amount of time to recover within the operating range and so causes a lag in response.

There are many techniques to prevent windup known as *anti-windup control mechanisms*. One of them is to use the non-windup limiter (Figure A.6b) operating in the following way

$$\begin{array}{llll} \text{if} \quad y > A & \text{then} \quad x = A & \text{and} & \mathrm{d}y/\mathrm{d}t = 0 \\ \text{if} \quad B \leq y \leq A & \text{then} \quad x = y & \text{and} & \mathrm{d}y/\mathrm{d}t = u \\ \text{if} \quad y < B & \text{then} \quad x = B & \text{and} & \mathrm{d}y/\mathrm{d}t = 0 \end{array} \tag{A.137}$$

In this case, signal y is stopped by zeroing the time derivative $\mathrm{d}y/\mathrm{d}t$ at the moment when the value of y exceeds limit A or B. In physical implementation this is realized by a switch in the circuit of the integrator. Such a switch is opened when the signal exceeds the limits and is closed when the signal returns to a value between the limits.

Windup is not limited to integrators. Windup can actually occur in any control element that contains memory, for example a first-order lag element or any filter can create a windup. If the designer permits a windup to occur, the control system can exhibit excessive overshoot, sustained oscillations, and/or lengthy settling times.

A.5.3 Simple Time Constant

The simple time constant element is equivalent to the integrator with a negative feedback loop with proportional gain equal to one. This results from a well-known formula for the transfer function of a closed-loop system: $G(s) = G_A(s)/[1 + G_A(s) \cdot G_F(s)]$, where $G_A(s)$ and $G_F(s)$ are the transfer functions of the main circuit and the feedback, respectively. In the considered case when $G_A(s) = 1/sT$ and $G_F(s) = 1$, $G(s) = 1/[1 + sT]$ is obtained. Such a relation between a single time constant element and integrator allows to use the same limiters as for integrators (Figure A.7) for the single time constant element.

The input and output signals of the block shown in Figure A.7a are linked with the following relationship $\mathrm{d}y/\mathrm{d}t = (u - y)/T$. In this case the limiter is referred to as the *windup limiter* and its operation is determined by rules (A.136).

A single time constant element with a non-windup limiter is shown in Figure A.7b. The lower part of this figure illustrates how such an element is physically implemented using an integrator and switch. This switch is open when signal excides limit A or B and is closed when the signal returns to the allowed value

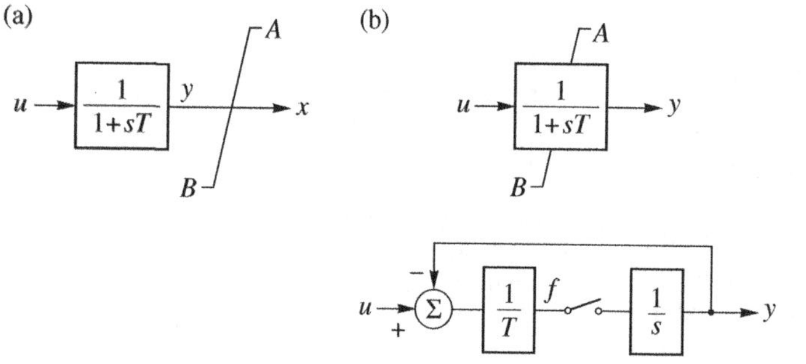

Figure A.7 Simple time constant block with (a) windup limiter; (b) non-windup limiter.

$$\begin{array}{llll} \text{if} & y = A \text{ and } f > 0 & \text{then} & dy/dt = 0 \\ \text{if} & B \le y \le A & \text{then} & dy/dt = f \\ \text{if} & y = B \text{ and } f < 0 & \text{then} & dy/dt = 0 \end{array} \tag{A.138}$$

In models of control systems described in this book the single time constant element is used to represent delays introduced by measuring devices, regulators and other power system elements.

A.5.4 Lead–lag Block

The phase compensators, such as lead–lag blocks, may be also supplemented by limiters as shown in Figure A.8. It is assumed that $T_1 \neq 0$, $T_2 \neq 0$, and $T_2 \neq T_1$.

The operation of the windup limiter in the lead–lag block shown in Figure A.8a is determined by rules (A.136). In the case of the non-windup limiter shown in Figure A.8b operation of the limiter is also determined by rules (A.136), but the meaning of signal y is different. This results from the diagram shown in Figure A.8c presenting the physical implementation of the non-windup limiter in the lead–lag block.

Verification of the equivalence of blocks shown in Figure A.8c and Figure A.8b is simple. It is necessary to use for the diagram shown in Figure A.8c two times the formula for the equivalent transfer function of the closed-loop system. From the lower part of this diagram a simple time constant element is obtained. This element together with the proportional block $(T_2/T_1 - 1)$ constitute a positive feedback loop for the main circuit with proportional element T_2/T_1. After replacing such a closed loop system by one equivalent transfer function the lead–lag block $(1 + sT_1)/(1 + sT_2)$ is obtained, as shown in Figure A.8b.

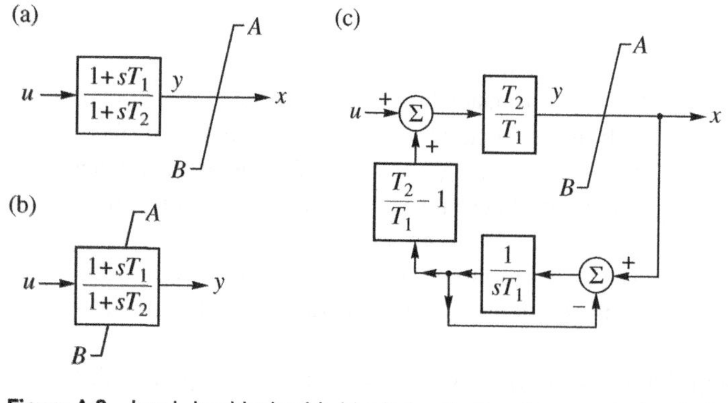

Figure A.8 Lead–lag block with (a) windup limiter; (b) non-windup limiter; (c) implementation of non-windup limiter.

References

Abad, G., Lopez, J., Rodriguez, M.A. et al. (2011). *Doubly Fed Induction Machine*. Hoboken, NJ: IEEE Press.

ABB (1991). *An Introduction to ABB Series Capacitors*. ABB Information Publication.

Abe, S. and Isono, A. (1983). Determination of power system voltage stability. *Electrical Engineering in Japan* 103 (3): 349–356.

Ackermann, T. (2005). *Wind Power in Power Systems*. Chichester: Wiley.

Adibi, M.M., Hirsch, P.M., and Jordan, J.A. (1974). Solution methods for transient and dynamic stability. *Proceedings of the IEEE* 62 (7): 951–958.

Adkins, B. (1957). *The General Theory of Electrical Machines*. Chapman & Hall.

A-eberle (2019). CPR-D collapse prediction relay. https://www.a-eberle.de/sites/default/files/media/ba_cpr_d_d.pdf (accessed 11 October 2019).

Ahlgren, L., Johansson, K.E., and Gadhammar, A. (1978). Estimated life expenditure of turbine-generator shafts at network faults and risk for subsynchronous resonance in the Swedish 400 kV system. *IEEE Transactions on Power Apparatus and Systems* PAS-97 (6): 2005–2018.

Ainsworth, J.D. (1985). Proposed benchmark model for study of HVDC controls by simulator or digital computer. Proc. CIGRE SC-14 Colloq. HVDC with Weak AC Systems, Maidstone. https://e-cigre.org (accessed 11 October 2019).

Ajjarapu, V. and Christy, C. (1992). The continuation power flow: a tool for steady-state voltage stability analysis. *IEEE Transactions on Power Systems* PWRS-7 (1): 416–423.

Akagi, H., Watanabe, E.H., and Aredes, M. (2007). *Instantaneous Power Theory and Application to Power Conditioning*. Wiley.

Anderson, P.M., Agrawal, B.L., and Van Ness, J.E. (1990). *Subsynchronous Resonance in Power Systems*. IEEE Press.

Anderson, E., Bai, Z., Bischof, C. et al. (1999). *LAPACK User's Guide*, 3e. Philadelphia: SIAM.

Anderson, P.M. and Bose, A. (1983). Stability simulation of wind turbine systems. *IEEE Transactions on Power Applications and Systems* 102: 3791–3795.

Anderson, P.M. and Fouad, A.A. (1977). *Power System Control and Stability*. The Iowa State University Press.

Arnold, W.I. (1992). *Ordinary Differential Equations*. Springer-Verlag.

Arrillaga, J. and Arnold, C.P. (1990). *Computer Analysis of Power Systems*. Wiley.

Arrillaga, J., Liu, Y.H., and Watson, N.R. (2007). *Flexible Power Transmission: The HVDC Options*. Chichester: Wiley.

Ashok Kumar, B.S.A., Parthasawathy, K., Prabhakara, F.S., and Khincha, H.P. (1970). Effectiveness of series capacitors in long distance transmission lines. *IEEE Transactions on Power Apparatus and Systems* PAS-89 (4): 941–950.

Athay, T., Podmore, R., and Virmani, S. (1979). A practical method for the direct analysis of transient stability. *IEEE Transactions on Power Apparatus and Systems* 98 (2): 573–584.

Bartylak, A. (2011). The ARC settings requirements on the ESCOM transmission system. CIGRE Study Committee B5 Colloquium (12–17 September 2011). https://e-cigre.org (accessed 11 October 2019).

Bellman, R. (1970). *Introduction to Matrix Analysis*, 2e. McGraw-Hill.

Power System Dynamics: Stability and Control, Third Edition. Jan Machowski, Zbigniew Lubosny, Janusz W. Bialek and James R. Bumby.

Bérubé, G.R. and Hajagos, L.M. (2005). Accelerating-power based power system stabilizers. http://kestrelpower.com/Docs/PSS_Tutorial_Chapter_Accelerating_Power_R2.pdf (accessed 26 September 2019).

Bialek, J. (1996). Tracing the flow of electricity. *IEE Proceedings: Generation, Transmission and Distribution* 143 (4): 313–320.

Bialek, J. (2007). Why has it happened again? Comparison between the 2006 UCTE blackout and the blackouts of 2003. IEEE PowerTech 2007, Lausanne, Switzerland (1–5 July 2007). https://ieeexplore.ieee.org/Xplore/home.jsp (accessed 11 October 2019).

Bölder, P., Kulig, T., and Lambrecht, D. (1975). Beurteilung der Torsionsbeanspruchung in den Wellen von Turbosätzen bei wiederholt auftrenden Störungen im Laufe der Betriebszeit. *Elektrotechnische Zeitschrift Ausgabe A* 96 (4): 172–177.

Bourgin, F., Testud, G., Heilbronn, B., and Verseille, J. (1993). Present practices and trends on the French power system to prevent voltage collapse. *IEEE Transactions on Power Systems* PWRS-8 (3): 778–788.

Brameller, A., Allan, R.N., and Hamam, Y.M. (1976). *Sparsity: Its Practical Application to System Analysis*. London: Pitman.

Breulmann, H., Grebe, E., Lösing, M. et al. (2000). Analysis and damping of inter-area oscillations in the UCTE/CENTREL power system. CIGRE Paper No. 38–113. https://e-cigre.org (accessed 11 October 2019).

Brown, H.E. (1975). *Solution of Large Networks by Matrix Methods*. Wiley.

Brown, P.G., Demello, F.P., Lenfest, E.H., and Mills, R.J. (1970). Effects of excitation, turbine energy control and transmission on transient stability. *IEEE Transactions on Power Apparatus and Systems* PAS-89 (6).

Bumby, J.R. (1982). Torsional natural frequencies in the superconducting turbogenerator. *IEE Proceedings* 129 (Pt C, 4): 141–151.

Bumby, J.R. and Wilson, J.M. (1983). Structural modes and undamped torsional natural frequencies of a superconducting turbogenerator rotor. *Journal of Sound and Vibration* 87 (4): 589–602.

Burg, J.P. (1978). A new analysis technique for time series data. In: *Modern Spectrum Analysis* (ed. D.G. Childers), 42–48. New York: IEEE Press.

Cai, Y.Q. and Wu, C.S. (1986). A novel algorithm for aggregating coherent generating units. IFAC Symposium on Power Systems and Power Plant Control, Beijing, China (12–15 August 1986). https://www.sciencedirect.com/journal/ifac-papersonline (accessed 11 October 2019.

Carpentier, J., Girard, R., and Scano, E. (1984). Voltage collapse proximity indicators computed from an optimal power flow. Proceedings of the 8th Power System Computing Conference, Helsinki. https://www.elsevier.com/books (accessed 11 October 2019).

Castro Luis, M., Claudio, R., Fuerte-Esquivel, J., and Tovar-Hernández, H. (2012). *Solution of power flow with automatic load-frequency control devices including wind farms. IEEE Transactions on Power Systems* 27 (4): 2186–2195.

CERTS Consortium for Electric Reliability Technology Solutions (2004). Integrated security analysis. http://certs.lbl.gov/pdf/certs-isa-final.pdf (accessed 11 October 2019).

Chai, J.S. and Bose, A. (1993). Bottlenecks in parallel algorithms for power system stability analysis. *IEEE Transactions on Power Systems* PWRS-8 (1): 9–15.

Ching, Y.K. and Adkins, B.A. (1954). Transient theory of synchronous generators under unbalanced conditions. *IEE Proceedings* 101 (Pt 4): 166–182.

Christiansen, P. (2003). A sea of turbines. *IEE Power Engineer* 17 (1): 22–24.

Chua, L.O. and Lin, P.-M. (1975). *Computer-Aided Analysis of Electronic Circuits*. Prentice Hall.

CIGRE Paper No. 37/38-01. Working Group 38.04 (1994). Ultra high voltage technology. https://e-cigre.org (accessed 11 October 2019).

CIGRE Task Force C4.02.25 (2003). Technical brochure No 238: Modeling of gas turbines and steam turbines in combined cycle power plants. https://e-cigre.org (accessed 11 October 2019).

CIGRE Technical Brochure No 312. Working Group C1.04 (2007). Application and required developments of dynamic models to support practical planning. https://e-cigre.org (accessed 11 October 2019).

CIGRE Technical Brochure No 604 (2014). Guide for the development of models for HVDC converters in a HVDC grid. https://e-cigre.org (accessed 11 October 2019).

CIGRE Technical Brochure No 664 (2016). Wide area protection and control technologies. https://e-cigre.org (accessed 11 October 2019).

CIGRE Technical Brochure No 702 (2017). Application of phasor measurement units for monitoring power system dynamic performance. https://e-cigre.org (accessed 11 October 2019).

CIGRE Technical Brochure No. 145 (1999). Modeling of power electronics equipment (FACTS) in load flow and stability programs. https://e-cigre.org (accessed 11 October 2019).

CIGRE Technical Brochure No. 166 (2000). *Impact of Interactions among Power Systems.*

CIGRE Technical Brochure No. 316 (2007). Defense plan against extreme contingencies. https://e-cigre.org (accessed 11 October 2019).

CIGRE Technical Brochure No. 325 (2007). Review of on-line dynamic security assessment tools and techniques. https://e-cigre.org (accessed 11 October 2019).

CIGRE Working Group (2006). A3.11: Report No 304: Guide for application of IEC 62271–100 and IEC 62271–1. https://e-cigre.org (accessed 11 October 2019).

Cirstea, M.N., Dinu, A., Khor, J.G., and McCormick, M. (2002). *Neural and Fuzzy Logic Control of Drives and Power Systems.* Elsevier Science.

Clark, K., Miller, N.W., and Sanchez-Gasca, J.J. (2010). *Modeling of GE Wind Turbine-Generators for Grid Studies.* Schenectady, NY: General Electric International Inc.

Concordia, C. (1951). *Synchronous Machines. Theory and Performance.* New York: Wiley.

Concordia, C. and Ihara, S. (1982). Load representation in power system stability studies. *IEEE Transactions on Power Apparatus and Systems* PAS-101 (4): 969–977.

Cong, Y., Regulski, P., Wall, P. et al. (2016). On the use of dynamic thermal-line ratings for improving operational tripping schemes. *IEEE Transactions on Power Delivery* 31 (4): 1891–1900.

Contaxis, G. and Debs, A.S. (1978). Identification of external equivalents for steady-state security assessment. *IEEE Power Apparatus and Systems* PAS-87 (2): 409–414.

Cook, V. (1985). *Analysis of Distance Protection.* Research Studies Press, Wiley.

Dahl, O.G.C. (1938). *Electric Power Circuits: Theory and Applications.* New York: McGraw-Hill.

Dandeno, P. and Kundur, P. (1973). Non-iterative transient stability program including the effects of variable load voltage characteristics. *IEEE Transactions on Power Apparatus and Systems* PAS-92 (5): 1478–1484.

Demello, F.P. and Concordia, C. (1969). Concepts of synchronous machine stability as affected by excitation control. *IEEE Transactions on Power Apparatus and Systems* 88 (4): 316–329.

Demello, F.P., Czuba, J.S., Rushe, P.A., and Willis, J.R. (1986). Developments in applications of stabilising measures through excitation control. https://e-cigre.org (accessed 11 October 2019).

De Mello, R.W., Podmore, R., and Stanton, K.N. (1975). *Coherency based dynamic equivalents: applications in transient stability studies.* PICA Conference.

Di Caprio, U. and Marconato, R. (1975). A novel criterion for the development of multi-area simplified models oriented to on-line evaluation of power system dynamic security. Proceedings of the Fifth Power System Computation Conference, Cambridge (1–5 September 1975). https://www.tib.eu/en/search/id/TIBKAT%3A668028629/PSCC-proceedings-Fifth-Power-Systems-Computation (accessed 11 October 2019).

DigSILENT (2018). PowerFactory manual. https://www.digsilent.de/en/downloads.html (accessed 11 October 2019).

Dimo, P. (1971). *L'analyse Nodale des Reseaux D'energie.* Paris: Eyrolles.

Dopazo, J.F., Dwarakanath, M.H., Li, J.J., and Sasson, A.M. (1977). An external system equivalent model using real-time measurements for system security evaluation. *IEEE Transactions on Power Apparatus and Systems* PAS-96: 431–446.

Doyle, J.C. (1982). Analysis of feedback systems with structured uncertainties. *IEE Proceedings, Part D* 129 (6): 242–250.

Duff, I.S., Erisman, A.M., and Reid, J.K. (1986). *Direct Methods for Sparse Matrices.* Oxford University Press.

Dy Liacco, T.E. (1968). Control of power systems via multi-level concept. Report SRC-68-19. Case Western Reserve University, Cleveland, OH.

Edelmann, H. (1963). *Berechung elektrischer Verbundnetze*. Berlin: Springer-Verlag.

Ekanayake, J., Holdsworth, L., and Jenkins, K. (2003a). Control of DFIG wind turbines. *IEE Power Engineer* 17 (1): 28–32.

Ekanayake, J., Holdsworth, L., Wu, X.G., and Jenkins, K. (2003b). Dynamic modelling of doubly fed induction generator wind turbines. *IEEE Transactions on Power Systems* 18 (2): 803–809.

Elgerd, O. (1982). *Electric Energy Systems Theory: An Introduction*, 2e. New York: McGraw-Hill.

Elkraft Systems (2003). Power failure in Eastern Denmark and Southern Sweden on 23 September 2003. https://link.springer.com/article/10.1007%2FBF03055480 (accessed 11 October 2019).

ENTSO-E (2011). Draft requirements for grid connection applicable to all generators. https://www.entsoe.eu/fileadmin/user_upload/_library/resources/RfG/130308_Final_Version_NC_RfG.pdf (accessed 11 October 2019).

EPRI (1999). Decentralized damping of power swings: feasibility study: final report TR-112417. Agreement WO 8555-01. https://pes-spdc.org/system/files/filedepot/145/epri_downloads.pdf (accessed 11 October 2019).

Estanqueiro, A. (2007). A dynamic wind generation model for power system studies. *IEEE Transactions on Power Systems* 22 (3): 920–928.

Feng, X., Lubosny, Z., and Bialek, J.W. (2007). Dynamic equivalent of a network with high penetration of distributed generation. IEEE PowerTech 2007, Lausanne, Switzerland (1–5 July 2007). https://ieeexplore.ieee.org/Xplore/home.jsp (accessed 11 October 2019).

Fisher, P., Angquist, L., and Nee, H.P. (2012). A new control scheme for an HVDC transmission link with capacitor-commutated converters having the inverter operating with constant alternating voltage. The 44th CIGRE Session, Palais des Congrès, Paris (26–31 August 2012), Paper No. B4_106_2012. https://e-cigre.org (accessed 11 October 2019).

Fouad, A.A. and Vittal, V. (1992). *Power System Transient Stability Analysis Using the Transient Energy Function Method*. Englewood Cliffs, NJ: Prentice Hall.

Garmond, A.J. and Podmore, R. (1978). Dynamic aggregation of generating unit models. *IEEE Transactions on Power Apparatus and Systems* PAS-97 (4): 1366–1372.

GE Energy Consulting (2009). PSLF Manual, 17. https://www.geenergyconsulting.com/practice-area/software-products/pslf-re-envisioned (accessed 10 October 2019).

Gibbard, M.J. (1991). Robust design of fixed-parameter power system stabilisers over a wide range of operating conditions. *IEEE Transactions on Power Systems* 6 (2): 794–800.

Gibbard, M.J., Pourbeik, P., and Vowles, D.J. (2015). *Small-Signal Stability, Control and Dynamic Performance of Power Systems*. Adelaide: University of Adelaide Press.

Gibbard, M.J. and Vowles, D.J. (2004). Reconciliation of methods of compensation for PSSs in multimachine system. *IEEE Transactions on Power Systems* 19 (1).

Giles, R.L. (1970). *Layout of E.H.V. Substations*. Cambridge: Cambridge University Press.

Gless, G.E. (1966). Direct method of Liapunov applied to transient power system stability. *IEEE Transactions on Power Apparatus and Systems* PAS-85 (2).

Glover, K. (1984). All optimal Hankel-norm approximations of linear multivariable systems and their $L\infty$ error bounds. *International Journal of Control* 39 (6): 1115–1193.

Gonzalez-Longatt, F.M. and Rueda, J.L. (2014). *PowerFactory Applications for Power System Analysis*. Springer.

Grainger, J.J. and Stevenson, W.D. (1994). *Power System Analysis*. McGraw-Hill.

Gross, C.A. (1986). *Power System Analysis*, 2e. New York: Wiley.

Guan, M. and Xu, Z. (2012). Modeling and control of a modular multilevel converter-based HVDC system under unbalanced grid conditions. *IEEE Transactions on Power Electronics* 27 (12): 4858–4867.

Gupta, M.M. (1986). *Adaptive Methods for Control System Design*. New York: IEEE Press.

Gurusinghe, D.R. and Rajapakse, A.D. (2016). Post-disturbance transient stability status prediction using synchrophasor measurements. *IEEE Transactions on Power Systems* 31 (5): 3656–3664.

Hammons, T.J. and Winning, D.J. (1971). Comparisons of synchronous machine models in the study of the transient behaviour of electrical power systems. *Proceedings of the IEE* 118 (10): 1442–1458.

Hansen, A.D., Iov, F., Sorensen, P.E. et al. (2007). *Dynamic wind turbine models in power system simulation tool DIgSILENT.* Roskilde, Denmark: Technical University of Denmark, Riso National Laboratory https://orbit.dtu.dk/files/7703047/ris_r_1400_ed2.pdf (accessed 10 October 2019).

Harkopf, T. (1978). Simulation of power system dynamics using trapezoidal rule and Newton's method. Proceedings of the Sixth PSCC Conference, Darmstadt, Germany (21–25 August 1978).

Harris, M.R., Lawrenson, P.J., and Stephenson, J.M. (1970). *Per Unit Systems with Special Reference to Electrical Machines.* Cambridge: Cambridge University Press.

Haubrich, H.J. and Fritz, W. (1999). *Study on Cross-Border Electricity Transmission Tariffs by Order of the European Commission, DG XVII/C1.* Aachen, Germany: Aachen University of Technology Institute of Power Systems and Power Economics.

Hauer, J.F., Demeure, C.J., and Scharf, L.L. (1990). Initial results in Prony analysis of power system response signals. *IEEE Transactions on Power Systems* 5 (1): 80–89.

Hazarika, D. and Sinha, A. (1998). Standing phase-angle reduction for power system restoration. *IEE Proceedings-Generation, Transmission and Distribution* 145 (1): 82–88.

Heffron, W.G. and Phillips, R.A. (1952). Effect of modern amplidyne voltage regulators on underexcited operation of large turbine generators. *American Institution of Electrical Engineers* 71: 692–697.

Hicklin, J. and Grace, A. (1992). *Simulink.* MathWorks Inc.

Hingorani, N.G. and Gyugyi, L. (2000). *Understanding FACTS: Concepts and Technology of Flexible AC Transmission Systems.* IEEE Press.

Holdsworth, L., Jenkins, N., and Strbac, G. (2001). Electrical stability of large offshore wind farms. IEE Seventh International Conference on AC–DC Power Transmission (28–30 November 2001). https://ieeexplore.ieee.org/document/988402 (accessed 10 October 2019).

Holdsworth, L., Wu, X.G., Ekanayake, J., and Jenkins, K. (2003). Comparison of fixed speed and doubly-fed induction wind turbines during power system disturbances. *IEE Proceedings: Generation, Transmission and Distribution* 150 (3): 343–352.

Hua, Y. and Sarkar, T.K. (1990). Matrix pencil method for estimating parameters of exponentially damped/undamped sinusoid noise. *IEEE Transactions on Acoustics and Signal Processing* 38 (5): 814–824.

Huang, N.E., Shen, Z., Long, S.R. et al. (1998). The empirical mode decomposition and the Hilbert spectrum for nonlinear and non-stationary time series analysis. *Proceedings of the Royal Society of London, Series A: Mathematical, Physical and Engineering Sciences* 454: 903–998.

Hughes, F.M., Anaya-Lara, O., Jenkins, N., and Strbac, G. (2006). A power system stabilizer for DFIG-based wind generation. *IEEE Transactions on Power Systems* 21 (2): 763–772.

IEEE A report prepared by the IEEE Working Group on the Effects of Switching on Turbine-Generators (1980). IEEE screening guide for planned steady-state switching operations to minimize harmful effects on steam turbine-generators. *IEEE Transactions on Power Apparatus and Systems* PAS-99 (4): 1519–1521.

IEEE Committee Report (1968). Computer representation of excitation systems. *IEEE Transactions on Power Apparatus and Systems* PAS-87 (6): 1460–1464.

IEEE Committee Report (1969). Recommended phasor diagrams for synchronous machines. *IEEE Transactions on Power Apparatus and Systems* PAS-88 (11): 1593–1610.

IEEE Committee Report (1973a). Excitation system dynamic characteristic. *IEEE Transactions on Power Apparatus and Systems* PAS-92 (1).

IEEE Committee Report (1973b). Dynamic models for steam and hydroturbines in power system studies. *IEEE Transactions on Power Apparatus and Systems* PAS-92 (6): 1904–1915.

IEEE Committee Report (1973c). System load dynamics simulation effects and determination of load constants. *IEEE Transactions on Power Apparatus and Systems* PAS-92 (2): 600–609.

IEEE Committee Report (1981). Excitation system models for power system stability studies. *IEEE Transactions on Power Apparatus and Systems* PAS-100 (2): 494–509.

IEEE Committee Report (1991). Dynamic models for fossil fuelled steam units in power system studies. *IEEE Transactions on Power Systems* PWRS-6 (2): 753–761.

IEEE Committee Report (1992). Hydraulic turbine and turbine control models for system dynamic studies. *IEEE Transactions on Power Systems* PWRS-7 (1): 167–179.

IEEE Committee Report (1994). Static VAR compensator models for power flow and dynamic performance simulation. *IEEE Transactions on Power Systems* PWRS-9 (1): 229–240.

IEEE Digital Excitation Task Force of the Equipment Working Group, and jointly sponsored by the Performance and Modeling Working Group, of the Excitation System Subcommittee (1996). Computer models for representation of digital-based excitation systems. *IEEE Transactions on Energy Conversion* 11 (3): 607–615.

IEEE Power System Relaying Committee Report (1977). Out-of-step relaying for generators. *IEEE Transactions on Power Apparatus and Systems* PAS-96 (5): 1556–1564.

IEEE Standard C37.118.1-2011. (2011). IEEE standard for synchrophasor measurements for power systems. https://ieeexplore.ieee.org/Xplore/home.jsp. (accessed 11 October 2019).

IEEE Std 421.5-2005 (2005). *IEEE recommended practice for excitation system models for powers system stability studies. IEEE Power Engineering Society.*

IEEE Std 421.5-2016 (2016). IEEE recommended practice for excitation system models for powers system stability studies. *IEEE Power Engineering Society* https://ieeexplore.ieee.org/Xplore/home.jsp (accessed 11 October 2019).

IEEE Std. 421.2-2014 (2014). IEEE guide for identification, testing, and evaluation of the dynamic performance of excitation control systems. *IEEE Power and Energy Society* https://ieeexplore.ieee.org/Xplore/home.jsp (accessed 11 October 2019).

IEEE Std. 421.3-2016 (2016). IEEE standard for high-potential test requirements for excitation systems for synchronous machines. *IEEE Power and Energy Society* https://ieeexplore.ieee.org/Xplore/home.jsp (accessed 11 October 2019).

IEEE Synchronous Machinery Subcommittee, Rotating Machinery Committee (1982). Effects of switching network disturbances on turbine-generator shaft systems. *IEEE Working Group Interim Report, IEEE Transactions on Power Apparatus and Systems* PER-2 (9): 32–33.

IEEE Task Force on Load Representation for Dynamic Performance (1993). *Load representation for dynamic performance analysis. IEEE Transactions on Power Systems* 8 (2): 472–482, 1003.

IEEE Task Force on Load Representation for Dynamic Performance (1995). Standard load models for power flow and dynamic performance simulation. *IEEE Transactions on Power Systems* 10 (3): 1302–1312.

IEEE Working Group on Prime Mover and Energy Supply Models for System Dynamic Performance Studies (1994). Dynamic models for combined cycle power plants in power system studies. *IEEE Transactions on Power Systems* PWRS-9 (3): 1698–1708.

Iliceto, F. and Cinieri, E. (1977). Comparative analysis of series and shunt compensation schemes for AC transmission systems. *IEEE Transactions on Power Apparatus and Systems* PAS-96 (1): 167–179.

Januszewski, M. (2001) Transient stability enhancement by using FACTS devices. PhD thesis. Warsaw University of Technology, Poland (in Polish).

Jones, C.V. (1967). *The Unified Theory of Electrical Machines*. Butterworth.

Joyce, J.S., Kulig, T., and Lambrecht, D. (1978). Torsional fatigue of turbine-generator shafts caused by different electrical system faults and switching operations. *IEEE Transactions* PAS-97: 1965–1973.

Juang, J.-N. and Pappa, R.S. (1985). An Eigensystem realization algorithm for modal parameter identification and model reduction. *Journal of Guidance, Control, and Dynamics* 8 (5): 620–627.

Kamwa, I. and Grondin, R. (1992). Fast adaptive scheme for tracking voltage phasor and local frequency in power transmission and distribution systems. *IEEE Transactions on Power Delivery* PWRD-7 (2): 789–795.

Kariya, T. and Kurata, H. (2004). *Generalized Least Squares*. Wiley.

Kasztenny, B. and Thompson, M.J. (2011). Breaker failure protection: standalone or integrated with zone protection relays? Schweitzer Engineering Laboratories, Inc. Presented at the 64 Texas A&M Conference for Protective Relay Engineers, Texas A&M University (11 April 2011). https://ieeexplore.ieee.org/xpl/tocresult.jsp?reload=true&isnumber=6035494 (accessed 11 October 2019).

Kay, S.M. (1988). *Modern Spectral Estimation: Theory and Application*. Englewood Cliffs, NJ: Prentice Hall.

Kessel, P. and Glavitsch, H. (1986). Estimating the voltage stability of a power system. *IEEE Transactions on Power Delivery, PWRD* 1 (3): 346–354.

Kimbark, E.W. (1995). *Power System Stability*, Vols I, II, III. New York: Wiley 1948, 1950, 1956, reprinted by IEEE in 1995.

Kirby, N.M., Xu, L., Luckett, M., and Siepmann, W. (2002). HVDC transmission for large offshore wind farms. *IEE Power Engineering Journal* 16 (3): 135–141.

Kulicke, B. and Webs, A. (1975). Elektromechanisches Verhalten von Turbosetzen bei Kurzschlüssen in Kraftwerksnähe. *Elektrotechnische Zeitschrift Ausgabe A* 96 (4): 194–201.

Kumano, S., Miwa, Y., Kokai, Y. et al. (1994). Evaluation of transient stability controller system model. CIGRE Session 38–303. https://e-cigre.org./publication/38-303_1994-evaluation-of-transients-stability-controller-system-model (accessed 11 October 2019).

Kundur, P. (1994). *Power System Stability and Control*. New York: McGraw-Hill.

Kundur, P., Lee, D.C., and Zein El-Din, H.M. (1981). Power system stabilizers for thermal units: analytical techniques and on-site validation. *IEEE Transactions on Power Apparatus and Systems* PAS-100: 81–95.

Läge, K. and Lambrecht, D. (1974). Die Auswirkung dreipoliger Netzkurzschlüsse mit Kurzschlussfortschaltung auf die mechanische Beansprachung von Turbosätzen. *Elektrotechnische Zeitschrift Ausgabe A* 95 (10): 508–514.

Lander, C.W. (1987). *Power Electronics*, 2e. McGraw-Hill.

Larsen, E.V. and Swan, D.A. (1981). Applying power system stabilizers, parts I, II, and III. *IEEE Transactions on Power Apparatus and Systems* PAS-100: 3017–3046.

Lee, D.C., Beaulieu, R.E., and Service, J.R.R. (1981). A power system stabilizer using speed and electrical power inputs: Design and field experience. *IEEE Transactions on Power Apparatus and Systems* PAS-100: 4151–4167.

Lee, D.C. and Kundur, P. (1986). Advanced excitation controls for power system stability enhancement. Paper Ref. No. 38–01, CIGRE Session (27 August–4 September 1986). https://e-cigre.org/publication/38-01_1986-advanced-excitation-controls-for-power-system-stability-enhancement (accessed 11 October 2019).

Leithead, W.E. (1992) Effective wind speed models for simple wind turbines simulations. Proceedings of the 14th Annual British Wind Energy Association Conference, Nottingham, England (25–27 March 1992). London: Mechanical Engineering Publications.

Leithead, W.E., Delasalle, S., and Reardon, D. (1991). Role and objectives of control for wind turbines. *IEE Proceedings C: Generation, Transmission and Distribution* 138 (2): 135–148.

Löf, P.A., Smed, T., Andersson, G., and Hill, D.J. (1992). Fast calculation of a voltage stability index. *IEEE Transactions on Power Systems* PWRS-7 (1): 54–64.

Lotfalian, M., Schlueter, R., Idizior, D. et al. (1985). Inertial, governor, and AGC/economic dispatch load flow simulations of loss of generation contingencies. *IEEE Transactions on Power Apparatus and Systems* PAS-104 (11): 3020–3028.

Lubosny, Z. (1999). *Self-Organising Controllers of Generating Unit in Electric Power System*. Gdansk University of Technology Press Monograph No. 4.

Lubosny, Z. (2006). *Wind Turbine Operation in Electric Power Systems*. Berlin, Heidelberg, New York: Springer-Verlag.

Lubosny, Z. and Bialek, J.W. (2007). Supervisory control of a wind farm. *IEEE Transactions on Power Systems* 22 (3): 985–994.

Lüders, G.A. (1971). Transient stability of multimachine power system via the direct method of Lyapunov. *IEEE Transactions on Power Apparatus and Systems* PAS-90 (1): 23–36.

MAAC (Mid Atlantic Area Council) (2003). Protective relaying philosophy and design standards. PJM Interconnection. https://www.pjm.com/-/media/planning/design-engineering/RS_stnd.ashx?la=en (accessed 29 September 2019).

Machowski, J. (1985). Dynamic equivalents for transient stability studies of electrical power systems. *International Journal of Electrical Power & Energy Systems* 7 (4): 215–223.

Machowski, J. (2012). *Selectivity of power system protections at power swings in power system. Acta Energetica, Power Engineering Quarterly*, ISSN 2080-7570 (4/12): 96–123.

Machowski, J. and Bernas, S. (1989). *Stany nieustalone i stabilnosc systemu elektroenergetycznego*. Warsaw: Wydawnictwa Naukowo-Techniczne (in Polish).

Machowski, J. and Białek, J. (2008). State-variable control of shunt FACTS devices using phasor measurements. *Electric Power Systems Research* 78 (1): 39–48.

Machowski, J., Bialek, J.W., and Bumby, J.R. (2008). *Power System Dynamics: Stability and Control*, 2e. Chichester: Wiley.

Machowski, J., Cichy, A., Gubina, F., and Omahen, P. (1986a). Modified algorithm for coherency recognition in large electrical power systems. IFAC Symposium on Power Systems and Power Plant Control, Beijing, China (12–15 August 1986). https://www.sciencedirect.com/journal/ifac-proceedings-volumes/vol/20/issue/6 (accessed 11 October 2019).

Machowski, J., Cichy, A., Gubina, F., and Omahen, P. (1988). External subsystem equivalent model for steady-state and dynamic security assessment. *IEEE Transactions on Power Systems* PWRS-3 (4): 1456–1463.

Machowski, J., Gubina, F., and Omahen, P. (1986b). Power system transient stability studies by Lyapunov method using coherency based aggregation. IFAC Symposium on Power Systems and Power Plant Control, Beijing, China 12–15 August 1986. https://www.sciencedirect.com/journal/ifac-proceedings-volumes/vol/20/issue/6 (accessed 11 October 2019).

Machowski, J. and Kacejko, P. (2016). Influence of automatic control of a tap changing step-up transformer on power capability area of generating unit. *Electric Power Systems Research* 140: 46–53.

Machowski, J., Kacejko, P., and Miller, P. (2014). Impedance method used to calculate initial switching currents in transmission networks and generator real power. *GSTF Journal of Engineering Technology (JET)* 3 (1): 42–52.

Machowski, J., Kacejko, P., Nogal, L., and Wancerz, M. (2013). Power system stability enhancement by WAMS - based supplementary control of multi-terminal HVDC networks. *Control Engineering Practice* 21 (5): 583–592.

Machowski, J., Kacejko, P., Robak, S. et al. (2015). Simplified angle and voltage stability criteria for power system planning based on the short-circuit power. *International Transactions on Electrical Energy Systems* 25 (11): 3096–3108.

Machowski, J. and Nelles, D. (1992a). Optimal control of superconducting magnetic energy storage unit. *Electric Machines and Power Systems* 20 (6): 623–640.

Machowski, J. and Nelles, D. (1992b). Power system transient stability enhancement by optimal control of static VAR compensators. *International Journal of Electrical Power & Energy Systems* 14 (5): 411–421.

Machowski, J. and Nelles, D. (1993). Simple robust adaptive control of static VAR compensator. *European Transactions on Electric Power Engineering* 3 (6): 429–435.

Machowski, J. and Nelles, D. (1994). Optimal modulation controller for superconducting magnetic energy storage. *International Journal of Electrical Power & Energy Systems* 16 (5): 291–300.

Machowski, J., Smolarczyk, A., and Bialek, J. (2001). Damping of power swings by control of braking resistors. *International Journal of Electrical Power & Energy Systems* 23: 539–548.

Maragos, P., Kaiser, J.F., and Quatieri, T.F. (1993). On amplitude and frequency demodulation using energy operators. *IEEE Transactions on Signal Processing* 41 (4): 1532–1550.

Marple, L. (1980). A new autoregressive spectrum analysis algorithm. *IEEE Transactions on Acoustics, Speech, and Signal Processing* 28 (4): 441–454.

MathWorks (2019). Loop shaping of HIMAT pitch axis controller. https://ch.mathworks.com/help/robust/examples/loop-shaping-of-himat-pitch-axis-controller.html (accessed 29 September 2019).

McDonald, J.D. (2003). *Electric Power Substations Engineering*. CRC Press.

McPherson, G. and Laramore, R.D. (1990). *Introduction to Electric Machines and Transformers*, 2e. New York: Wiley.

Moore, B. (1981). Principal component analysis in linear systems: Controllability, observability, and model reduction. *IEEE Transactions on Automatic Control* AC-26: 17–31.

Moussa, H.A.M. and Yu, Y.N. (1972). Improving power system damping through supplementary governor control. *PES Summer Meeting, Paper C* 72: 470–473.

Muller, S., Deicke, M., and De Donker, R.W. (2002). Doubly fed induction generator systems. *IEEE Industry Applications Magazine*: 26–33.

Murdoch, A., Venkataraman, S., Lawson, R.A., and Pearson, W.R. (1999). Integral of accelerating power type PSS: Part 1: Theory, design, and tuning methodology. Part 2: Field testing and performance verification. *IEEE Transactions on Energy Conversion* 14 (4): 1658–1663.

Nagao, T. (1975). Voltage collapse at load ends of power systems. *Electrical Engineering in Japan* 95 (4): 1975.

NEMA Publications (2016). *Motor and Generator Standards*. NEMA Publications.

NERC North American Electric Reliability Corporation (2011). *Reliability Standards for the Bulk Electric Systems in North America*. NERC.

NERC Std. PRC-019-2 (2015). The North American Electric Reliability Corporation: *Coordination of Generating Unit or Plant Capabilities, Voltage Regulating Controls, and Protection.*

Niederlinski, A., Moscinski, J., and Ogonowski, Z. (1995). *Adaptive Control: (Regulacja Adaptacyjna).* Warsaw: Wydawnictwo Naukowe PWN (in Polish).

Nogal, L. (2009). Application of wide area measurements to stability enhancing control of FACTS devices installed in tie-lines. PhD thesis. Warsaw University of Technology (in Polish).

Ogata, K. (1967). *State Space Analysis of Control Systems*. Prentice Hall.

O'Kelly, D. (1991). *Performance and Control of Electrical Machines*. McGraw-Hill.

Olas, A. (1975). Synthesis of systems with prescribed trajectories. *Proceedings of Non-linear Vibrations* 16: 277–300.

Omahen, P. (1994). Unified approach to power system analysis in its multi-time scale dynamic response. PhD thesis. Warsaw University of Ethnology (in Polish).

Oprea, L., Popescu, V., and Sattinger, W. (2007) Coordinated synchronism check settings for optimal use of critical transmission network corridors. IEEE PowerTech 2007. Lausanne, Switzerland (1–5 July 2007). https://ieeexplore.ieee.org/Xplore/home.jsp (accessed 10 October 2019).

Pagola, F.L., Perez_Arrillaga, I.J., and Verghese, G.C. (1989). On sensitivities, residues and participations: applications to oscillatory stability analysis and control. *IEEE Transactions on Power Systems* 4: 278–285.

Pai, M.A. (1981). *Power System Stability: Analysis by the Direct Method of Lyapunov*. Amsterdam: North-Holland.

Pai, M.A. (1989). *Energy Function Analysis for Power System Stability*. Kluwer Academic.

Paszek, S. and Nocon, A. (2014). *Optimisation and Polyoptimisation of Power System Stabiliser Parameters*. Saarbrucken: LAP LAMBERT Academic Publishing, OmniScriptum.

Pavella, M., Ernst, D., and Ruiz-Vega, D. (2000). *Transient Stability of Power Systems: A Unified Approach to Assessment and Control*. Kluwer Academic.

Pavella, M. and Murthy, P.G. (1994). *Transient Stability of Power Systems. Theory and Practice*. Wiley.

Phadke, A.G., Thorap, J.S., and Adamiak, M.G. (1983). A new measurement technique for tracking voltage phasors, local system frequency and rate of change of frequency. *IEEE Transactions on Power Apparatus and Systems* PAS-102 (5): 1025–1038.

Phadke, A.G. and Thorap, J.S. (1988). *Computer Relaying for Power Systems*. Wiley.

Phadke, A.G. and Thorap, J.S. (2008). *Synchronized Phasor Measurements and their Applications*. Springer Science + Business Media LLC.

Pissanetzky, S. (1984). *Sparse Matrix Technology*. Academic Press.

Podmore, R. (1978). Identification of coherent generators for dynamic equivalents. *IEEE Transactions on Power Apparatus and Systems* PAS-97: 1344–1354.

Press, W.H., Teukolsky, S.A., Vetterling, W.T., and Flannery, B.P. (1992). *Numerical Recipes in C: The Art of Scientific Computing*, 2e. Cambridge University Press.

Rafian, M., Sterling, M.J.H., and Irving, M.R. (1987). Real time power system simulation. *IEE Proceedings* 134 (Pt C, 3): 206–223.

Rajapakse, A.D., Gomez, F.R., Nanayakkara, K. et al. (2010). Rotor angle instability prediction using post-disturbance voltage trajectories. *IEEE Transactions on Power Apparatus and Systems* 25 (2): 947–956.

Ramey, D.G. and Skooglund, J.W. (1970). Detailed hydrogovernor representation for system stability studies. *IEEE Transactions on Power Apparatus and Systems* PAS-89 (1): 106–112.

Rasolomampionona, D.D. (2000). Analysis of the power system steady-state stability: influence of the load characteristics. *Archives of Electrical Engineering* XLIX: 191–191.

Rasolomampionona, D.D. (2007) Optimisation of parameters of TCPAR installed in tie lines with regard to their interaction with LFC. Prace Naukowe Elektryka, z. 134 Oficyna Wydawnicza WPW. Warsaw University of Technology (in Polish).

Reimert, D. (2006). *Protective Relaying for Power Generation Systems.* CRC Press.

Robak, S., Machowski, J., and Gryszpanowicz, K. (2017). Contingency selection for power system stability analysis. 18th IEEE International Scientific Conference on Electric Power Engineering (EPE), Czech Republic (17–19 May 2017).

Robak, S., Machowski, J., and Gryszpanowicz, K. (2018). Automatic alleviation of overloads in transmission network by generation curtailment. *IEEE Transactions on Power Systems* 33 (4): 4424–4432.

Rowen, W.I. (1983). Simplified mathematical representations of heavy-duty gas turbines. *Journal of Engineering for Power* 105: 865–869.

Rowen, W.I. (1992). Simplified mathematical representations of single shaft gas turbine models in mechanical drive service. The International Gas Turbine and Aeroengine Congress and Expo, Cologne, Germany (1–4 June 1992). https://asmedigitalcollection.asme.org/GT/proceedings/GT1992/78972/Cologne,%20Germany/235860 (accessed 11 October 2019).

Saccomanno, F. (2003). *Electric Power System Control: Analysis and Control.* IEEE Press.

Savulescu, S.C. (ed.) (2009). *Real Time Stability Assessment in Modern Power System Centers.* IEEE Press.

Seshu, S. and Reed, M.B. (1961). *Linear Graphs and Electrical Networks.* Addison-Wesley.

Shang, L., Hu, J., Yuan, X., and Chi, Y. (2017). Understanding inertial response of variable-speed wind turbines by defined internal potential vector. *Energies* 10 (1): 22. https://doi.org/10.3390/en10010022.

Skogestad, S. and Postlethwaite, I. (1996). *Multivariable Feedback Control: Analysis and Design.* Chichester: Wiley.

Slootweg, J.G., de Hann, S.W.H., Polinder, H., and Kling, W.L. (2003). General model for representing variable speed wind turbines in power system dynamic simulations. *IEEE Transactions on Power Systems* 18 (1): 144–151.

Slootweg, J.G., Polinder, H., and Kling, W.L. (2001) Dynamic modelling of a wind turbine with doubly fed induction generator. Power Engineering Society Summer Meeting, Vancouver, BC, Canada (15–19 July 2001). https://ieeexplore.ieee.org/document/969969 (accessed 11 October 2019).

Soos, A. and Malik, O.P. (2002). An H2 optimal power system stabilizer. *IEEE Power Engineering Review* 22 (2): 59.

Stannard, N. and Bumby, J.R. (2007). Performance aspects of mains connected small scale wind turbines. *Proceedings of the IET – Generation, Transmission and Distribution* 1 (2): 348–356.

Stott, B. (1974). Review of load flow calculation methods. *Proceedings of the IEEE* 62: 916–929.

Stott, B. (1979). Power system dynamic response calculations. *Proceedings of the IEEE* 67: 219–241.

Strang, G. (1976). *Linear Algebra and Its Applications.* New York: Academic Press.

Swarcewicz, A. and Lubosny, Z. (2001). The μ-synthesis in power system stabilizer design. *Archives of Electrical Engineering* 1 (3): 235–248.

Taylor, C.W. (1994). *Power System Voltage Stability.* McGraw-Hill.

Taylor, C.W., Haner, J.M., Hill, L.A. et al. (1983). A new out-of-step relay with rate of change of apparent resistance augmentation. *IEEE Transactions on Power Apparatus and Systems* PAS-102 (3): 631–639.

Taylor, C.W., Haner, J.M., and Laughlin, T.D. (1986). Experience with the R-Rdot out-of-step relay. *IEEE Transactions on Power Delivery* PWRD-1 (2): 35–39.

Tewerson, R.P. (1973). *Sparse Matrices.* New York: Academic Press.

Tinney, W.F. and Walker, J.W. (1967). Direct solutions of sparse network equations by optimally ordered triangular factorization. *Proceedings of the IEEE* 55 (11): 1801–1809.

Tiranuchit, A. and Thomas, R.J. (1988). A posturing strategy action against voltage instabilities in electric power systems. *IEEE Transactions on Power Systems* PWRS-3 (1): 87–93.

Troskie, H.J. and de Villiers, L.N.F. (2004). Impact of Long Duration Faults on Out-Of-Step Protection. Eight IEE International Conference on Developments in Power System Protection (5–8 April 2004). Amsterdam, The Netherlands: RAI Centre. https://ieeexplore.ieee.org/Xplore/home.jsp (accessed 11 October 2019).

Tuttelberg, K., Kilter, J., Wilson, D., and Uhlen, K. (2018). Estimation of power system inertia from ambient wide area measurements. *IEEE Transactions on Power Systems* 33 (6): 7249–7257.

UCTE (2003). Final report of the investigation committee on the 28 September 2003 blackout in Italy. https://www.ucte.org/resources/publications/otherreports (accessed 11 October 2019).

UCTE (2007a). Final report: system disturbance on 4 November 2006. https://www.ucte.org/resources/publications/otherreports (accessed 10 October 2019).

UCTE (2007b). Operation handbook. Policy 1: load-frequency control and performance. https://www.entsoe.eu/fileadmin/user_upload/_library/publications/ce/report_2007_2.pdf (accessed 10 October 2019).

Ulbig, A., Borsche, T.S., and Andersson, G. (2014). Impact of low rotational inertia on power system stability and operation. *Proceedings of the 19th IFAC World Congress* https://doi.org/10.3182/20140824-6-ZA-1003.02615.

Undrill, J.M. and Turner, A.E. (1971). Construction of power system electromechanical equivalents by modal analysis. *IEEE Transactions on Power Apparatus and Systems* PAS-90 (5): 2049–92059.

Ungrad, H., Winkler, W., and Wiszniewski, A. (1995). *Protection Techniques in Electrical Energy Systems*. Marcel Dekker.

US–Canada Power System Outage Task Force (2004). Final report on the August 14, 2003 blackout in the United States and Canada. https://www3.epa.gov/region1/npdes/merrimackstation/pdfs/ar/AR-1165.pdf (accessed 10 October 2019).

Vaahedi, E., El-Kady, M.A., Libaque-Esaine, J.A., and Carvalho, V.F. (1987). Load models for large-scale stability studies from end-user consumption. *IEEE Transactions on Power Systems* PWRS-7 (4): 864–862.

Van Cutsem, T. (1991). A method to compute reactive power margins with respect to voltage collapse. *IEEE Transactions on Power Systems, PWRS* PWRS-6: 145–156.

Van Cutsem, T. and Vournas, C. (1998). *Voltage Stability of Electric Power Systems*. Springer-Verlag.

Van Der Hoven, I. (1957). Power spectrum of horizontal wind speed in the frequency range from 0.0007 to 900 cycles per hour. American. *Journal of Meteorology* 14: 160–164.

Venikov [Vênikov], V.A (1978). *Transient Processes in Electrical Power Systems*. Moscow: Mir.

Vorley, D.H. (1974). Numerical techniques for analysing the stability of large power systems. PhD thesis. University of Manchester.

Vournas, C.D., Nikolaidis, V.C., and Tassoulis, A.A. (2006). Postmortem analysis and data validation in the wake of the 2004 Athens Blackout. *IEEE Transactions on Power Systems* 21 (3): 1331–1339.

Wang, X. (1997). *Modal Analysis of Large Interconnected Power System*, Reihe 6: Energietechnik, Nr 380. Düsseldorf: VDI-Verlag.

Wang, H.F., Hao, Y.S., Hogg, B.W., and Yang, Y.H. (1993). Stabilization of power systems by governor-turbine control. *Electrical Power and Energy Systems* 15 (6): 351–361.

Wasilewski, J. and Lubosny, Z. (2017). Analiza rocznej zmienności inercji mas wirujących w elektroenergetycznym systemie synchronicznym kontynentalnej Europy w kontekście rozwoju i pracy OZE. *Wiadomości elektrotechniczne* 2 (85): 8–14. (in Polish).

Wasynczuk, O., Man, D.T., and Sullivan, J.P. (1981). Dynamic behaviour of a class of wind turbine generators during random wind fluctuations. *IEEE Transactions on Power Applications and Systems* PAS-100: 2837–2845.

Watson, W. and Coultes, M.E. (1973). Static exciter stabilizing signals on large generators – mechanical problems. *IEEE Transactions on Power Apparatus and Systems* PAS-92: 204–211.

WECC Modeling & Validation Work Group (2002). Guidelines for thermal governor model data selection, validation, and submittal to WECC. https://www.wecc.org/Reliability/WECC%20Guidelines%20for%20Thermal%20Governor%20Modeling.pdf (accessed 22 September 2019).

Weedy, B.M. (1987). *Electric Power Systems*, 3e. Chichester: Wiley.

Welfonder, E. (1980). Regeldynamisches Zusammenvirken von Kraftwerken und Verbrauchern im Netzverbund-betrieb. *Elektrizitätswirtschaft* 79 (20): 730–741.

Westlake, A.J., Bumby, J.R., and Spooner, E. (1996). Damping the power angle oscillations of a permanent magnet synchronous generator with particular reference to wind turbine applications. *IEE Proceedings: Electric Power Applications* 143 (3): 269–280.

Willems, J.L. (1970). *Stability Theory of Dynamical Systems*. London: Nelson.

Wilson, D., Bialek, J.W., and Lubosny, Z. (2006). Banishing blackouts. *IEE Power Engineering Journal* 20 (2): 38–41.

Witzke, R.L., Kresser, J.V., and Dillard, J.K. (1953). Influence of AC reactance on voltage regulation of 6-phase rectifiers. *AIEE Transactions* 72: 244–253.

Wood, A.J. and Wollenberg, B.F. (1996). *Power Generation Operation and Control*, 2e. Wiley.

Wright, A. and Christopoulos, C. (1993). *Electrical Power System Protection*. London: Chapman & Hall.

Wunderlich, S., Adibi, M., Fischl, R., and Nwankpa, C. (1994). An approach to standing phase angle reduction. *IEEE Transactions on Power Systems* 93 (1): 470–478.

Xiang, D., Ran, L., Tavner, P.J., and Yang, S. (2006). Control of a doubly fed induction generator in a wind turbine during grid fault ride-through. *IEEE Transactions on Energy Conversion* 21 (2): 652–662.

Yee, S.K., Milanowić, J.V., and Hughes, M.F. (2008). Overview and comparative analysis of gas turbine model for system stability studies. *IEEE Transactions on Power Systems* 23 (1): 108–118.

Yu, Y.N. (1983). *Electric Power System Dynamics*. New York: Academic Press.

Zeng, Z. and Jun, W. (eds.) (2010). *Advances in Neural Network Research and Applications*. Berlin, Heidelberg: Springer-Verlag.

Zhukov [Žukov], L.A. (1964). Simplified transformation of circuit diagrams of complex electric power systems. *Izvestia Akademii Nauk SSSR, Energetika i Transport* 2 (in Russian).

Index

Power System Dynamics: Stability and Control, Third Edition. Jan Machowski, Zbigniew Lubosny, Janusz W. Bialek and James R. Bumby.

Printed and bound by CPI Group (UK) Ltd, Croydon, CR0 4YY
16/04/2025
14658836-0001